More Than Just [...] k

Internet Resources

StudentWorks™ *Plus* Online This interactive **eBook** includes the complete Student Edition with audio, Math in Motion, Personal Tutor, Self-Check Quizzes, and much more – all at point of use!

Step 1 **Connect to** NY Math Online glencoe.com

Step 2 **Connect to resources by using simple and convenient QuickPass codes.**

NY5006c1

"NY" for "New York"

This edition, ISBN 978-0-07-888500-6

Enter the appropriate chapter number. c1 = Chapter 1

For Students

Connect to the Student Edition **eBook** that contains all of the following online resources. You don't need to take your textbook home every night.

- Personal Tutor
- Self-Check Quizzes
- Chapter Readiness Quizzes
- Math in Motion: Animation
- Math in Motion: BrainPOP®
- Math in Motion: Interactive Lab
- Extra Examples
- Chapter Test Practice
- Standardized Test Practice
- Study to Go
- Vocabulary Review Games
- Graphing Calculator Keystrokes
- Multilingual eGlossary
- Scavenger Hunts
- Workbooks
- Hotmath Math Homework Help Homework Help

For Teachers

- Teaching Today
- **Advance Tracker**
  - Diagnostic, formative, and summative assessment
  - Progress reports
  - Differentiated instruction
- State Resources
- Professional Development at www.mhpd.com
  - Video Clips
  - Online Credit Courses
- Research
  - White Papers
  - Efficacy Studies

For Parents

Connect to www.glencoe.com to access **StudentWorks *Plus* Online** and all of the resources for students and teachers listed above.

7

Glencoe McGraw-Hill

New York Algebra 2 and Trigonometry

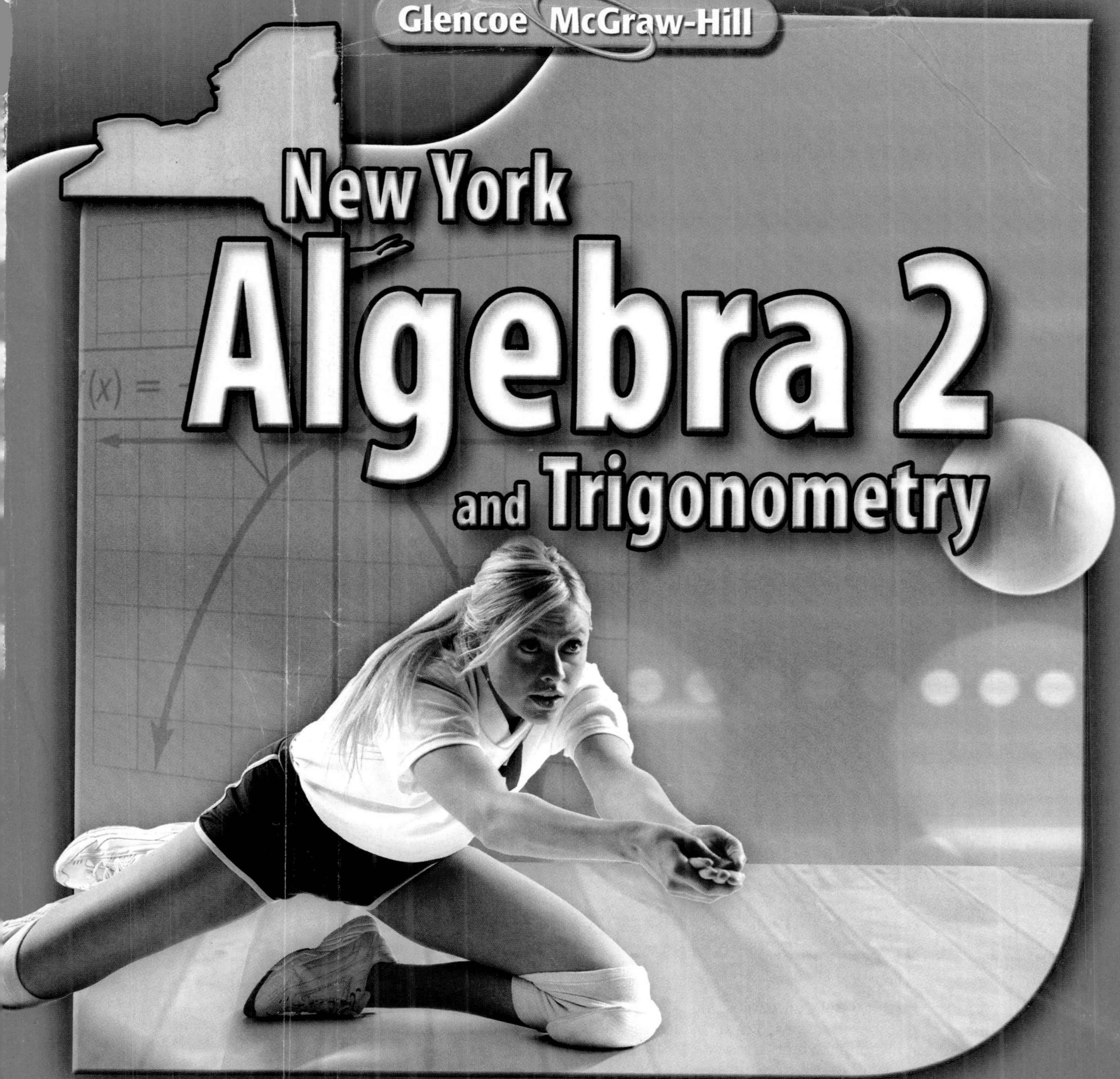

Authors

Carter · Cuevas · Day · Malloy · Casey · Holliday

McGraw Hill Glencoe

About the Cover

After a volleyball is hit upward, it travels along a path shaped like a parabola. This parabola can be modeled mathematically using a quadratic function. With a quadratic function you can find out how long the volleyball will be in the air, how high it will go, and where it will land. You will study quadratic functions in Chapter 5.

The McGraw-Hill Companies

Glencoe

Send all inquiries to:
Glencoe/McGraw-Hill
8787 Orion Place
Columbus, OH 43240-4027

ISBN: 978-0-07-888500-6
MHID: 0-07-888500-0

Printed in the United States of America.

1 2 3 4 5 6 7 8 9 10 079/043 17 16 15 14 13 12 11 10 09 08

CONTENTS IN BRIEF

About the Authors

Macmillan/McGraw-Hill and Glencoe/McGraw-Hill K–12 Mathematics Lead Authors

Our lead authors ensure that Macmillan/McGraw-Hill and Glencoe/McGraw-Hill mathematics programs are truly vertically aligned by beginning with the end in mind—success in Algebra 1 and beyond. By "backmapping" the content from the high school programs, all of our mathematics programs are well articulated in their scope and sequence, ensuring that the content in each program provides a solid foundation for moving forward. These authors also worked closely with the entire K–12 author team to ensure vertical alignment of the instructional approach and visual design.

Dr. John A. Carter, Ph.D.
Assistant Principal for Teaching and Learning
Adlai E. Stevenson High School
Lincolnshire, Illinois
Areas of Expertise: Using technology and manipulatives to visualize concepts, mathematics achievement of English-language learners

Dr. Gilbert J. Cuevas, Ph.D.
Professor of Mathematics Education
Texas State University–San Marcos
San Marcos, Texas
Areas of Expertise: Applying concepts and skills in mathematically rich contexts, mathematical representations

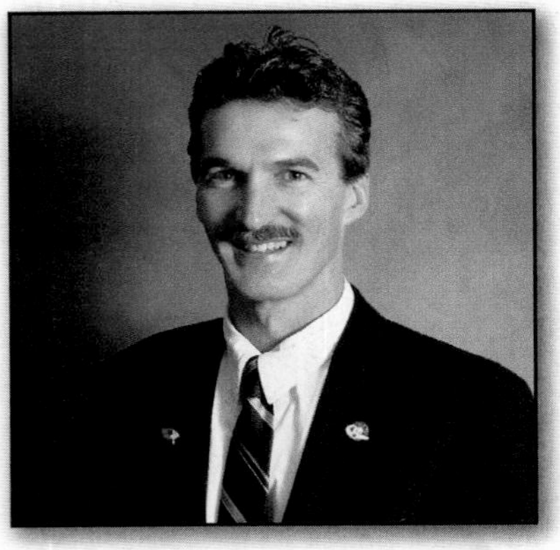

Dr. Roger Day, Ph.D.
Mathematics Department Chairperson
Pontiac Township High School
Pontiac, Illinois
Areas of Expertise: Understanding and applying probability and statistics, mathematics teacher education

Dr. Carol Malloy, Ph.D.
Associate Professor
University of North Carolina at Chapel Hill
Chapel Hill, NC
Areas of Expertise: Representations and critical thinking, student success in Algebra 1

Additional Algebra 2 Authors

The entire Algebra 2 author team strives to create a program that can be used by all types of Algebra 2 teachers with all types of Algebra 2 students. Each author brings their special expertise to making a program that will contribute to the success of every student who uses this instructional resource.

Dr. Berchie Holliday, Ed.D.
National Mathematics Consultant
Silver Spring, Maryland
Areas of Expertise: Using mathematics to model and understand real-world data, the effect of graphics on mathematical understanding

Ruth Casey
Mathematics Consultant,
Regional Teacher Partner
University of Kentucky
Lexington, KY
Areas of Expertise: Graphing technology and mathematics

Contributing Author

This program is the beneficiary of the imagination of Dinah Zike through the contribution of the Foldables Study Organizers.

Dinah Zike
Educational Consultant
Dinah-Might Activities, Inc.
San Antonio, TX

NY Math Online Meet the authors at glencoe.com.

WELCOME TO

New York Algebra 2

Master the New York State Core Curriculum in 3 Easy Steps

1 Practice the Standards Daily

NYS Core Curriculum

A2.A.1 Solve absolute value equations and inequalities **involving linear expressions in**

Each lesson addresses the New York State Performance Indicators covered in that lesson.

NYSRE Practice Questions aligned to the standards in a format like those on the New York State Regents Examination provide you with ongoing opportunities to sharpen your test-taking skills.

Visions of America, LLC/Alamy Images

2 Practice the Standards throughout the Chapter

NYSRE Example Every chapter contains a completely worked-out New York State Mathematics Test Example to help you solve problems that are similar to those you might find on that test.

NYSRE Practice Every chapter contains two full pages of NYSRE Practice with Test-Taking Tips.

3 Practice with a Review Book

To further help you prepare for the New York State Regents Examination in Algebra 2 and Trigonometry, you can also use *Glencoe's New York Review Series: Algebra 2 and Trigonometry*. The extra practice gained from the solved examples and practice problems can help your performance both in your class and on the exam.

New York Reviewers

Each New York Reviewer reviewed at least two chapters of the Student Edition, giving feedback and suggestions for improving the effectiveness of the mathematics instruction.

Dawn Brown
Mathematics Department Chair
Kenmore West High School
Buffalo, New York

Franco DiPasqua
Director of K–12 Mathematics
West Seneca Central Schools
West Seneca, New York

Debra Harley
Director of Math & Science
East Meadow School District
Westbury, New York

Katherine Lohrman
Teacher, Math Specialist, New Teacher Mentor
John Marshall High School
Rochester, New York

Consultants and Reviewers

Glencoe/McGraw-Hill wishes to thank the following professionals for their feedback. They were instrumental in providing valuable input toward the development of this program in these specific areas.

Mathematical Content

Viken Hovsepian
Professor of Mathematics
Rio Hondo College
Whittier, California

Grant A. Fraser, Ph.D.
Professor of Mathematics
California State University, Los Angeles
Los Angeles, California

Arthur K. Wayman, Ph.D.
Professor of Mathematics Emeritus
California State University, Long Beach
Long Beach, California

Gifted and Talented

Shelbi K. Cole
Research Assistant
University of Connecticut
Storrs, Connecticut

College Readiness

Robert Lee Kimball, Jr.
Department Head, Math and Physics
Wake Technical Community College
Raleigh, North Carolina

Differentiation for English-Language Learners

Susana Davidenko
State University of New York
Cortland, New York

Alfredo Gómez
Mathematics/ESL Teacher
George W. Fowler High School
Syracuse, New York

Graphing Calculator

Jerry Cummins
Former President
National Council of Supervisors of Mathematics
Western Springs, Illinois

Mathematical Fluency and STEM

Robert M. Capraro
Associate Professor
Texas A&M University
College Station, Texas

Pre-AP

Dixie Ross
Lead Teacher for Advanced Placement Mathematics
Pflugerville High School
Pflugerville, Texas

Reading and Writing

ReLeah Cossett Lent
Author and Educational Consultant
Morganton, Georgia

Lynn T. Havens
Director of Project CRISS
Kalispell, Montana

Consultants and Reviewers

Each Reviewer reviewed at least two chapters of the Student Edition, giving feedback and suggestions for improving the effectiveness of the mathematics instruction.

Corey Andreasen
Mathematics Teacher
North High School
Sheboygan, WI

Mark B. Baetz
Mathematics Coordinating Teacher
Salem City Schools
Salem, Virginia

Kathryn Ballin
Mathematics Supervisor
Newark Public Schools
Newark, New Jersey

Kevin C. Barhorst
Mathematics Department Chair
Independence High School
Columbus, Ohio

Brenda S. Berg
Mathematics Teacher
Carbondale Community High School
Carbondale, Illinois

Dawn Brown
Mathematics Department Chair
Kenmore West High School
Buffalo, New York

Sheryl Pernell Clayton
Mathematics Teacher
Hume Fogg Magnet School
Nashville, Tennessee

Bob Coleman
Mathematics Teacher
Cobb Middle School
Tallahassee, Florida

Jane E. Cotts
Mathematics Teacher
O'Fallon Township High School
O'Fallon, Illinois

Michael D. Cuddy
Mathematics Instructor
Zypherhills High School
Zypherhills, Florida

Melissa M. Dalton, NBCT
Mathematics Instructor
Rural Retreat High School
Rural Retreat, Virginia

Trina Louise Davis
Teacher
Fort Mill High School
Fort Mill, South Carolina

Tina S. Dohm
Mathematics Teacher
Naperville Central High School
Naperville, Illinois

Laurie L.E. Ferrari
Teacher
L'Anse Creuse High School–North
Macomb, Michigan

Patricia R. Frazier
Mathematics Department Chair/Instructor
Celina High School
Celina, Ohio

Steve Freshour
Mathematics
Parkersburg South High School
Parkersburg, West Virginia

Shirley D. Glover
Mathematics Teacher
TC Roberson High School
Asheville, North Carolina

Caroline W. Greenough
Mathematics Teacher
Cape Fear Academy
Wilmington, North Carolina

Michelle Hanneman
Mathematics Teacher
Moore High School
Moore, Oklahoma

Theresalynn Haynes
Mathematics Teacher
Glenbard East High School
Lombard, Illinois

Sandra Hester
Mathematics Teacher/AIG Specialist
North Henderson High School
Hendersonville, North Carolina

Jacob K. Holloway
Mathematics Teacher
Capitol Heights Junior High School
Montgomery, Alabama

Robert Hopp
Mathematics Teacher
Harrison High School
Harrison, Michigan

Eileen Howanitz
Mathematics Teacher/Department Chairperson
Valley View High School
Archbald, Pennsylvania

Charles R. Howard, NBCT
Mathematics Teacher
Tuscola High School
Waynesville, North Carolina

Sue Hvizdos
Mathematics Department Chairperson
Wheeling Park High School
Wheeling, West Virginia

Elaine Keller
Mathematics Teacher
Mathematics Curriculum Director K-12
Northwest Local Schools
Canal Fulton, Ohio

Sheila A. Kotter
Mathematics Educator
River Ridge High School
New Port Richey, Florida

Frank Lear
Mathematics Department Chair
Cleveland High School
Cleveland, Tennessee

Jennifer Lewis
Mathematics Teacher
Triad High School
Troy, Illinois

Catherine McCarthy
Mathematics Teacher
Glen Ridge High School
Glen Ridge, New Jersey

Jacqueline Palmquist
Mathematics Department Chair
Waubonsie Valley High School
Aurora, Illinois

Thom Schacher
Mathematics Teacher
Otsego High School
Otsego, Michigan

Laurie Shappee
Teacher/Mathematics Coordinator
Larson Middle School
Troy, Michigan

Jennifer J. Southers
Mathematics Teacher
Hillcrest High School
Simpsonville, South Carolina

Sue Steinbeck
Mathematics Department Chair
Parkersburg High School
Parkersburg, West Virginia

Kathleen D. Van Sise
Mathematics Teacher
Mandarin High School
Jacksonville, Florida

Karen Wiedman
Mathematics Teacher
Taylorville High School
Taylorville, Illinois

New York Performance Indicators, Algebra 2 and Trigonometry, Correlated to *New York Algebra 2 and Trigonometry*

Lessons in which the standard is the primary focus are indicated in **bold.**

Number	Performance Indicators	Student Edition Lesson(s)	Student Edition Page(s)
Problem Solving Strand			
A2.PS.1	Use a variety of problem solving strategies to understand new mathematical content	0-4, 1-4, 2-5, 3-5, 4-4, 6-1, 7-3, 8-8, 9-1, 10-4, 11-4, 12-5, 13-5, 14-3, 14-5, Problem-Solving Handbook	P9–P11, 27–32, 92–98, 167–173, 209–217, 333–339, 424–430, 533–539, 553–561, 639–646, 705–711, 773–778, 841–846, 904–909, 919–925, XX
A2.PS.2	Recognize and understand equivalent representations of a problem situation or a mathematical concept	Throughout the text; for example, 1-3, 6-1, 9-6, 11-7, 13-2	18–25, 333–339, 594–602, 727–731, 817–823
A2.PS.3	Observe and explain patterns to formulate generalizations and conjectures	Throughout the text; for example, 1-5, Extend 5-7, 8-1, 11-2, Explore 13-1	33–39, 311, 475–482, 688–695, 807
A2.PS.4	Use multiple representations to represent and explain problem situations (e.g., verbally, numerically, algebraically, graphically)	1-1, 1-2, 1-3, 1-4, 1-5, 1-6, 2-1, 2-3, 2-7, 2-8, 3-1, 4-1, 4-6, 5-1, 5-2, 5-3, 5-4, 5-5, 5-6, 5-8, 6-2, 6-3, 6-8, 7-1, 7-3, 7-4, 7-5, 7-6, 8-1, 8-2, 8-3, 8-4, 8-6, 9-1, 9-5, 10-2, 10-3, 10-4, 10-5, 11-1, 11-2, 11-4, 11-5, 11-7, 12-1, 12-4, 12-5, 13-2, 13-4, 13-6, 13-9, 14-1, 14-4, 14-5	5–10, 11–17, 18–25, 27–32, 33–39, 41–48, 61–68, 76–82, 109–116, 117–121, 135–141, 185–191, 229–235, 249–257, 259–266, 268–275, 276–282, 284–290, 292–300, 312–318, 341–347, 348–355, 391–396, 409–416, 424–430, 431–436, 439–445, 446–452, 475–482, 485–491, 492–499, 502–507, 516–522, 553–561, 586–593, 623–629, 631–637, 639–646, 648–655, 681–687, 688–695, 705–711, 714–719, 727–731, 745–750, 764–771, 773–778, 817–823, 832–839, 848–854, 871–876, 891–897, 911–917, 919–925
A2.PS.5	Choose an effective approach to solve a problem from a variety of strategies (numeric, graphic, algebraic)	Throughout the text; for example, 2-7, 3-2, 4-6, 5-3, 11-6	109–115, 143–150, 229–235, 268–275, 721–725
A2.PS.6	Use a variety of strategies to extend solution methods to other problems	Throughout the text; for example, 1-3, 6-5, 9-3, Extend 11-5, 14-5	18–25, 368–375, 569–575, 720, 919–925
A2.PS.7	Work in collaboration with others to propose, critique, evaluate, and value alternative approaches to problem solving	1-1, 1-2, 1-3, 1-4, 1-5, 1-6, 2-1, 2-3, 2-7, 2-8, 3-1, 4-1, 4-6, 5-1, 5-2, 5-3, 5-4, 5-5, 5-6, 5-8, 6-2, 6-3, 6-8, 7-1, 7-3, 7-4, 7-5, 7-6, 8-1, 8-2, 8-3, 8-4, 8-6, 9-1, 9-5, 10-2, 10-3, 10-4, 10-5, 11-1, 11-2, 11-4, 11-5, 11-7, 12-1, 12-4, 12-5, 13-2, 13-4, 13-6, 13-9, 14-1, 14-4, 14-5	5–10, 11–17, 18–25, 27–32, 33–39, 41–48, 61–68, 76–82, 109–116, 117–121, 135–141, 185–191, 229–235, 249–257, 259–266, 268–275, 276–282, 284–290, 292–300, 312–318, 341–347, 348–355, 391–396, 409–416, 424–430, 431–436, 439–445, 446–452, 475–482, 485–491, 492–499, 502–507, 516–522, 553–561, 586–593, 623–629, 631–637, 639–646, 648–655, 681–687, 688–695, 705–711, 714–719, 727–731, 745–750, 764–771, 773–778, 817–823, 832–839, 848–854, 871–876, 891–897, 911–917, 919–925
A2.PS.8	Determine information required to solve the problem, choose methods for obtaining the information, and define parameters for acceptable solutions	1-4, 1-5, 2-6, 3-2, 3-5, 4-3, 5-2, 6-7, 7-1, 8-2, 8-5, 8-8, 9-6, 10-4, 10-5, 11-1, 11-3, 13-5, 14-3	27–32, 33–39, 101–107, 143–150, 167–173, 200–207, 259–266, 383–390, 409–416, 485–491, 509–515, 533–539, 594–602, 639–646, 648–655, 681–687, 696–702, 841–846, 904–909
A2.PS.9	Interpret solutions within the given constraints of a problem	Throughout the text; for example, 0-6, Extend 2-5, 6-5, 9-6, 13-5	P15–P16, 90, 368–375, 594–602, 841–846
A2.PS.10	Evaluate the relative efficiency of different representations and solution methods of a problem	Throughout the text; for example, 1-6, Extend 6-5, 9-6, Extend 11-3, 13-9	41–48, 376, 594–602, 703–704, 871–876

Number	Performance Indicators	Student Edition Lesson(s)	Student Edition Page(s)
Reasoning and Proof Strand			
A2.RP.1	Support mathematical ideas using a variety of strategies	Throughout the text; for example, 3-2, 5-4, 7-6, 9-2	143–150, 276–282, 446–452, 562–568
A2.RP.2	Investigate and evaluate conjectures in mathematical terms, using mathematical strategies to reach a conclusion	Throughout the text; for example, 4-2, 4-4, 8-6, Extend 8-8, 11-7	193–199, 209–217, 516–522, 540, 727–731
A2.RP.3	Evaluate conjectures and recognize when an estimate or approximation is more appropriate than an exact answer	3-2, Extend 5-1, 5-2, Extend 11-3	143–150, 258, 259–266, 703–704
A2.RP.4	Recognize when an approximation is more appropriate than an exact answer	3-2, Extend 5-1, 5-2, Extend 11-3	143–150, 258, 259–266, 703–704
A2.RP.5	Develop, verify, and explain an argument, using appropriate mathematical ideas and language	Throughout the text; for example, 2-2, 4-3, 7-2, 8-8, 11-2	69–74, 200–207, 417–422, 533–539, 688–695
A2.RP.6	Construct logical arguments that verify claims or counterexamples that refute claims	Throughout the text; for example, 1-6, 4-3, 6-7, 11-7, 14-3	41–48, 200–207, 383–390, 727–731, 904–909
A2.RP.7	Present correct mathematical arguments in a variety of forms	Throughout the text; for example, 1-2, 6-1, 11-3, 13-5	11–17, 333–339, 696–702, 841–846
A2.RP.8	Evaluate written arguments for validity	1-1, 1-2, 1-3, 1-4, 1-5, 1-6, 2-1, 2-3, 2-7, 2-8, 3-1, 4-1, 4-6, 5-1, 5-2, 5-3, 5-4, 5-5, 5-6, 5-8, 6-2, 6-3, 6-8, 7-1, 7-3, 7-4, 7-5, 7-6, 8-1, 8-2, 8-3, 8-4, 8-6, 9-1, 9-5, 10-2, 10-3, 10-4, 10-5, 11-1, 11-2, 11-4, 11-5, 11-7, 12-1, 12-4, 12-5, 13-2, 13-4, 13-6, 13-9, 14-1, 14-4, 14-5	5–10, 11–17, 18–25, 27–32, 33–39, 41–48, 61–68, 76–82, 109–116, 117–121, 135–141, 185–191, 229–235, 249–257, 259–266, 268–275, 276–282, 284–290, 292–300, 312–318, 341–347, 348–355, 391–396, 409–416, 424–430, 431–436, 439–445, 446–452, 475–482, 485–491, 492–499, 502–507, 516–522, 553–561, 586–593, 623–629, 631–637, 639–646, 648–655, 681–687, 688–695, 705–711, 714–719, 727–731, 745–750, 764–771, 773–778, 817–823, 832–839, 848–854, 871–876, 891–897, 911–917, 919–925
A2.RP.9	Support an argument by using a systematic approach to test more than one case	1-4, 13-4	27–32, 832–839
A2.RP.10	Devise ways to verify results, using counterexamples and informal indirect proof	1-2, 1-6, 2-6, 3-3, 3-5, 4-2, 4-3, 4-4, 4-6, 6-1, 6-2, 6-5, 6-7, 7-2, 8-2, 8-5, 8-6, 8-7, 11-2, 11-3, 11-4, 13-9, 14-1, 14-2, 14-3, 14-4	11–17, 41–48, 101–107, 151–157, 167–173, 193–199, 200–207, 209–217, 229–235, 333–339, 341–347, 368–375, 383–390, 417–422, 485–491, 509–515, 516–522, 525–531, 688–695, 696–702, 705–711, 871–876, 891–897, 898–903, 904–909, 911–917
A2.RP.11	Extend specific results to more general cases	Throughout the text; for example, 4-4, 5-3, 7-4, 9-6	209–217, 268–275, 431–436, 594–602
A2.RP.12	Apply inductive reasoning in making and supporting mathematical conjectures	**11-7**	**727–731**
Communication Strand			
A2.CM.1	Communicate verbally and in writing a correct, complete, coherent, and clear design (outline) and explanation for the steps used in solving a problem	Throughout the text; for example, 1-4, 4-3, 6-2, 8-1, 12-6	27–32, 200–207, 341–347, 475–482, 780–784
A2.CM.2	Use mathematical representations to communicate with appropriate accuracy, including numerical tables, formulas, functions, equations, charts, graphs, and diagrams	Throughout the text; for example, 3-5, Extend 4-4, 6-1, 7-5, Explore 10-4	167–173, 218–219, 333–339, 439–445, 638

New York Performance Indicators, Algebra 2 and Trigonometry

Number	Performance Indicators	Student Edition Lesson(s)	Student Edition Page(s)
A2.CM.3	Present organized mathematical ideas with the use of appropriate standard notations, including the use of symbols and other representations when sharing an idea in verbal and written form	Throughout the text; for example, 1-4, Explore 1-6, 4-1, Extend 4-4, Extend 11-6	27–32, 40, 185–191, 218–219, 726
A2.CM.4	Explain relationships among different representations of a problem	5-7, 6-2, 8-6	305–310, 341–347, 516–522
A2.CM.5	Communicate logical arguments clearly, showing why a result makes sense and why the reasoning is valid	Throughout the text; for example, 1-4, 3-3, 5-8, 8-5, 12-6	27–32, 151–157, 312–317, 509–515, 780–784
A2.CM.6	Support or reject arguments or questions raised by others about the correctness of mathematical work	1-1, 1-2, 1-3, 1-4, 1-5, 1-6, 2-1, 2-3, 2-7, 2-8, 3-1, 4-1, 4-6, 5-1, 5-2, 5-3, 5-4, 5-5, 5-6, 5-8, 6-2, 6-3, 6-8, 7-1, 7-3, 7-4, 7-5, 7-6, 8-1, 8-2, 8-3, 8-4, 8-6, 9-1, 9-5, 10-2, 10-3, 10-4, 10-5, 11-1, 11-2, 11-4, 11-5, 11-7, 12-1, 12-4, 12-5, 13-2, 13-4, 13-6, 13-9, 14-1, 14-4, 14-5	5–10, 11–17, 18–25, 27–32, 33–39, 41–48, 61–68, 76–82, 109–116, 117–121, 135–141, 185–191, 229–235, 249–257, 259–266, 268–275, 276–282, 284–290, 292–300, 312–318, 341–347, 348–355, 391–396, 409–416, 424–430, 431–436, 439–445, 446–452, 475–482, 485–491, 492–499, 502–507, 516–522, 553–561, 586–593, 623–629, 631–637, 639–646, 648–655, 681–687, 688–695, 705–711, 714–719, 727–731, 745–750, 764–771, 773–778, 817–823, 832–839, 848–854, 871–876, 891–897, 911–917, 919–925
A2.CM.7	Read and listen for logical understanding of mathematical thinking shared by other students		
A2.CM.8	Reflect on strategies of others in relation to one's own strategy		
A2.CM.9	Formulate mathematical questions that elicit, extend, or challenge strategies, solutions, and/or conjectures of others	4-5, 5-3, 14-2	220–228, 268–275, 898–903
A2.CM.10	Use correct mathematical language in developing mathematical questions that elicit, extend, or challenge other students' conjectures	4-5, 5-3, 14-2	220–228, 268–275, 898–903
A2.CM.11	Represent word problems using standard mathematical notation	Throughout the text; for example, **1-3**, 5-2, 7-1, 13-7	**18–25**, 259–266, 409–416, 855–861
A2.CM.12	Understand and use appropriate language, representations, and terminology when describing objects, relationships, mathematical solutions, and rationale	Throughout the text; for example, Extend 2-1, 10-1, 13-2, 14-3	68, 617–622, 817–823, 904–909
A2.CM.13	Draw conclusions about mathematical ideas through decoding, comprehension, and interpretation of mathematical visuals, symbols, and technical writing	Throughout the text; for example, 3-4, 5-5, 6-5, 10-2, Extend 11-3	160–166, 284–290, 368–375, 623–629, 703–704
Connections Strand			
A2.CN.1	Understand and make connections among multiple representations of the same mathematical idea	Throughout the text; for example, 1-4, Extend 2-2, 10-6, Extend 11-6	27–32, 75, 656–661, 726
A2.CN.2	Understand the corresponding procedures for similar problems or mathematical concepts	Throughout the text; for example, 5-8, 8-5, 10-5, Extend 11-5	312–318, 509–515, 648–655, 720
A2.CN.3	Model situations mathematically, using representations to draw conclusions and formulate new situations	Throughout the text; for example, 2-4, 3-4, Extend 6-4, 7-1, Extend 8-8	83–89, 160–166, 365–366, 409–416, 540
A2.CN.4	Understand how concepts, procedures, and mathematical results in one area of mathematics can be used to solve problems in other areas of mathematics	Throughout the text; for example, 1-3, 3-1, 6-5, 9-1, 11-7	18–25, 135–141, 368–375, 553–561, 727–731
A2.CN.5	Understand how quantitative models connect to various physical models and representations	Throughout the text; for example, 1-3, 5-8, 7-4, 9-1, 14-1	18–25, 312–318, 431–436, 553–561, 891–897
A2.CN.6	Recognize and apply mathematics to situations in the outside world	Throughout the text; for example, 2-2, 4-3, 13-7	69–74, 200–207, 855–861

Number	Performance Indicators	Student Edition Lesson(s)	Student Edition Page(s)
A2.CN.7	Recognize and apply mathematical ideas to problem situations that develop outside of mathematics	Throughout the text; for example, 8-8, 10-1, 12-4	533–539, 617–622, 764–771
A2.CN.8	Develop an appreciation for the historical development of mathematics	1-3, 2-6, 3-2, 4-6, 5-6, 7-5, 8-4, 8-6, 9-6, 11-3, 12-4, 13-6, 14-1	18–25, 101–107, 143–150, 229–235, 292–300, 439–445, 502–507, 516–522, 594–602, 696–702, 764–771, 848–854, 891–897
Representation Strand			
A2.R.1	Use physical objects, diagrams, charts, tables, graphs, symbols, equations, or objects created using technology as representations of mathematical concepts	1-1, 1-2, 1-3, 1-5, 2-1, 2-2, 2-6, 3-2, 4-1, 4-4, 5-1, 5-3, 5-4, 5-5, 6-1, 6-2, 6-3, 6-4, 6-6, 8-1, 8-2, 8-4, 8-6, 8-7, 9-1, 9-3, 9-6, 10-2, 10-3, 10-4, 10-5, 10-6, 10-7, 11-2, 11-7, 12-4, 13-2, 13-6, 13-9, 14-1, 14-2, 14-3	5–10, 11–17, 18–25, 33–39, 61–67, 69–74, 101–107, 143–150, 185–191, 209–217, 249–257, 268–275, 276–282, 284–290, 333–339, 341–347, 348–355, 357–364, 377–382, 475–482, 485–491, 502–507, 516–522, 525–531, 553–561, 569–575, 594–602, 623–629, 631–637, 639–646, 648–655, 656–661, 662–667, 688–695, 727–731, 764–771, 817–823, 848–854, 871–876, 891–897, 898–903, 904–909
A2.R.2	Recognize, compare, and use an array of representational forms		
A2.R.3	Use representation as a tool for exploring and understanding mathematical ideas		
A2.R.4	Select appropriate representations to solve problem situations	5-3, 8-2, 8-8, 9-6, 10-3	268–275, 485–491, 533–539, 594–602, 631–637
A2.R.5	Investigate relationships among different representations and their impact on a given problem	5-7, 6-2, 8-6	305–310, 341–347, 516–522
A2.R.6	Use mathematics to show and understand physical phenomena (e.g., investigate sound waves using the sine and cosine functions)	Throughout the text; for example, 5-2, 8-8, 10-4, 11-4, 14-2	259–265, 533–539, 639–646, 705–711, 898–903
A2.R.7	Use mathematics to show and understand social phenomena (e.g., interpret the results of an opinion poll)	Throughout the text; for example, 2-1, 6-3, 8-5, 11-5, 12-1	61–67, 348–355, 509–515, 714–719, 745–750
A2.R.8	Use mathematics to show and understand mathematical phenomena (e.g., use random number generator to simulate a coin toss)	Throughout the text; for example, 2-7, 5-7, 8-5, 12-7	109–116, 305–310, 509–515, 786–793
Number Sense and Operations Strand			
A2.N.1	Evaluate numerical expressions with negative and/or fractional exponents, without the aid of a calculator (when the answers are rational numbers)	**6-1, 7-6**	**333–339, 446–452**
A2.N.2	Perform arithmetic operations (addition, subtraction, multiplication, division) with expressions containing irrational numbers in radical form	7-5	439–445
A2.N.3	Perform arithmetic operations with polynomial expressions containing rational coefficients	**7-6**	**446–452**
A2.N.4	Perform arithmetic operations on irrational expressions	7-5	439–445
A2.N.5	Rationalize a denominator containing a radical expression	**7-5**	**439–445**
A2.N.6	Write square roots of negative numbers in terms of i	**5-4**	**276–282**
A2.N.7	Simplify powers of i	**5-4**	**276–282**
A2.N.8	Determine the conjugate of a complex number	**5-4**	**276–282**
A2.N.9	Perform arithmetic operations on complex numbers and write the answer in the form $a + bi$	**5-4**	**276–282**

Number	Performance Indicators	Student Edition Lesson(s)	Student Edition Page(s)
A2.N.10	Know and apply sigma notation	11-4, 11-5	705–711, 714–719
Algebra Strand			
A2.A.1	Solve absolute value equations and inequalities involving linear expressions in one variable	1-4, 1-6	27–32, 41–48
A2.A.2	Use the discriminant to determine the nature of the roots of a quadratic equation	5-6	292–300
A2.A.3	Solve systems of equations involving one linear equation and one quadratic equation algebraically	10-7	662–667
A2.A.4	Solve quadratic inequalities in one and two variables, algebraically and graphically	5-8	312–318
A2.A.5	Use direct and inverse variation to solve for unknown values	9-5	586–593
A2.A.6	Solve an application which results in an exponential function	8-1, 8-2	475–482, 485–491
A2.A.7	Factor polynomial expressions completely, using any combination of the following techniques: common factor extraction, difference of two perfect squares, quadratic trinomials	5-3	268–275
A2.A.8	Apply the rules of exponents to simplify expressions involving negative and/or fractional exponents	6-1, 7-6	333–339, 446–452
A2.A.9	Rewrite algebraic expressions that contain negative exponents using only positive exponents	6-1	333–339
A2.A.10	Rewrite algebraic expressions with fractional exponents as radical expressions	7-6	446–452
A2.A.11	Rewrite algebraic expressions in radical form as expressions with fractional exponents	7-6	446–452
A2.A.12	Evaluate exponential expressions, including those with base *e*	8-2, 8-7, 8-8	485–491, 525–531, 533–539
A2.A.13	Simplify radical expressions	7-4	431–436
A2.A.14	Perform addition, subtraction, multiplication, and division of radical expressions	7-5	439–445
A2.A.15	Rationalize denominators involving algebraic radical expressions	7-5	439–445
A2.A.16	Perform arithmetic operations with rational expressions and rename to lowest terms	9-1, 9-2	553–561, 562–568
A2.A.17	Simplify complex fractional expressions	9-1, 9-2	553–561, 562–568
A2.A.18	Evaluate logarithmic expressions in any base	8-4, 8-5, 8-6, 8-7	502–507, 509–515, 516–522, 525–531
A2.A.19	Apply the properties of logarithms to rewrite logarithmic expressions in equivalent forms	8-4, 8-5, 8-6, 8-7	502–507, 509–515, 516–522, 525–531
A2.A.20	Determine the sum and product of the roots of a quadratic equation by examining its coefficients	Extend 5-6	301–302
A2.A.21	Determine the quadratic equation, given the sum and product of its roots	Extend 5-6	301–302
A2.A.22	Solve radical equations	7-7, Extend 7-7	453–459, 460–461
A2.A.23	Solve rational equations and inequalities	9-6, Extend 9-6	594–602, 603–604
A2.A.24	Know and apply the technique of completing the square	5-5	284–290

Number	Performance Indicators	Student Edition Lesson(s)	Student Edition Page(s)
A2.A.25	Solve quadratic equations, using the quadratic formula	**5-6**	**292–300**
A2.A.26	Find the solution to polynomial equations of higher degree that can be solved using factoring and/or the quadratic formula	**6-5, 6-6, 6-7, 6-8**	**368–375, 377–382, 383–390, 391–396**
A2.A.27	Solve exponential equations with and without common bases	**Explore 8-2, 8-2, Extend 8-6**	**483–484, 485–491, 523–524**
A2.A.28	Solve a logarithmic equation by rewriting as an exponential equation	**8-4, 8-5, 8-6,** Extend 8-6	**502–507, 509–515, 516–522,** 523–524
A2.A.29	Identify an arithmetic or geometric sequence and find the formula for its *n*th term	**11-2, 11-3**	**688–695, 696–702**
A2.A.30	Determine the common difference in an arithmetic sequence	**11-1**	**681–687**
A2.A.31	Determine the common ratio in a geometric sequence	**11-1**	**681–687**
A2.A.32	Determine a specified term of an arithmetic or geometric sequence	**11-1**	**681–687**
A2.A.33	Specify terms of a sequence, given its recursive definition	**11-5**	**714–719**
A2.A.34	Represent the sum of a series, using sigma notation	**11-4, 11-6**	**705–711, 721–725**
A2.A.35	Determine the sum of the first *n* terms of an arithmetic or geometric series	**11-2, 11-3, 11-4**	**688–695, 696–702, 705–711**
A2.A.36	Apply the binomial theorem to expand a binomial and determine a specific term of a binomial expansion	**11-7**	**727–731**
A2.A.37	Define a relation and function	**0-1**	**P4–P5**
A2.A.38	Determine when a relation is a function	**0-1**	**P4–P5**
A2.A.39	Determine the domain and range of a function from its equation	**0-1**	**P4–P5**
A2.A.40	Write functions in functional notation	**2-1**	**61–67**
A2.A.41	Use functional notation to evaluate functions for given values in the domain	**2-1,** 5-1, 6-3, 7-3, 8-1, 8-3, 9-4	**61–67,** 249–257, 348–355, 424–430, 475–482, 492–499
A2.A.42	Find the composition of functions	**7-1**	**409–416**
A2.A.43	Determine if a function is one-to-one, onto, or both	**2-1**	**61–67**
A2.A.44	Define the inverse of a function	**7-2**	**417–422**
A2.A.45	Determine the inverse of a function and use composition to justify the result	**7-2,** Extend 7-2	**417–422,** 423
A2.A.46	Perform transformations with functions and relations: $f(x+a)$, $f(x)+a$, $f(-x)$, $-f(x)$, $af(x)$	7-3	424–430
A2.A.47	Determine the center-radius form for the equation of a circle in standard form	Explore 10-3, **10-3**	630, **631–637**
A2.A.48	Write the equation of a circle, given its center and a point on the circle	**10-3**	**631–637**
A2.A.49	Write the equation of a circle from its graph	**10-3**	**631–637**
A2.A.50	Approximate the solution to polynomial equations of higher degree by inspecting the graph	**6-4**	**357–364**
A2.A.51	Determine the domain and range of a function from its graph	**2-1**	**61–67**

Number	Performance Indicators	Student Edition Lesson(s)	Student Edition Page(s)
A2.A.52	Identify relations and functions, using graphs	**2-1**	**61–67**
A2.A.53	Graph exponential functions of the form $y = b^x$ for positive values of b, including $b = e$	**8-1,** 8-7	**475–482,** 525–531
A2.A.54	Graph logarithmic functions, using the inverse of the related exponential function	**8-3**	**492–499**
A2.A.55	Express and apply the six trigonometric functions as ratios of the sides of a right triangle	**13-1**	**808–816**
A2.A.56	Know the exact and approximate values of the sine, cosine, and tangent of 0°, 30°, 45°, 60°, 90°, 180°, and 270° angles	**13-3,** 13-6	**825–831,** 848–854
A2.A.57	Sketch and use the reference angle for angles in standard position	**13-3**	**825–831**
A2.A.58	Know and apply the co-function and reciprocal relationships between trigonometric ratios	**13-1, 14-1**	**808–816, 891–897**
A2.A.59	Use the reciprocal and co-function relationships to find the value of the secant, cosecant, and cotangent of 0°, 30°, 45°, 60°, 90°, 180°, and 270° angles	**13-3**	**825–831**
A2.A.60	Sketch the unit circle and represent angles in standard position	**13-6**	**848–854**
A2.A.61	Determine the length of an arc of a circle, given its radius and the measure of its central angle	**13-2**	**817–823**
A2.A.62	Find the value of trigonometric functions, if given a point on the terminal side of angle q	**13-3**	**825–831**
A2.A.63	Restrict the domain of the sine, cosine, and tangent functions to ensure the existence of an inverse function	**13-9**	**871–876**
A2.A.64	Use inverse functions to find the measure of an angle, given its sine, cosine, or tangent	**13-9**	**871–876**
A2.A.65	Sketch the graph of the inverses of the sine, cosine, and tangent functions	**13-9**	**871–876**
A2.A.66	Determine the trigonometric functions of any angle, using technology	13-1, 13-3, 13-4, 13-5, Explore 13-8	808–816, 825–831, 832–839, 841–846, 862
A2.A.67	Justify the Pythagorean identities	**14-2**	**898–903**
A2.A.68	Solve trigonometric equations for all values of the variable from 0° to 360°	**13-9, 14-5**	**871–876**
A2.A.69	Determine amplitude, period, frequency, and phase shift, given the graph or equation of a periodic function	**13-7, 13-8**	**855–861, 863–870**
A2.A.70	Sketch and recognize one cycle of a function of the form $y = A \sin Bx$ or $y = A \cos Bx$	**13-7, 13-8**	**855–861, 863–870**
A2.A.71	Sketch and recognize the graphs of the functions $y = \sec(x)$, $y = \csc(x)$, $y = \tan(x)$, and $y = \cot(x)$	**13-7**	**855–861**
A2.A.72	Write the trigonometric function that is represented by a given periodic graph	**13-7, 13-8**	**855–861, 863–870**
A2.A.73	Solve for an unknown side or angle, using the Law of Sines or the Law of Cosines	**13-4, 13-5**	**832–839, 841–846**
A2.A.74	Determine the area of a triangle or a parallelogram, given the measure of two sides and the included angle	**Extend 13-2, 13-4**	**824, 832–839**

Number	Performance Indicators	Student Edition Lesson(s)	Student Edition Page(s)
A2.A.75	Determine the solution(s) from the SSA situation (ambiguous case)	**13-4**	**832–839**
A2.A.76	Apply the angle sum and difference formulas for trigonometric functions	**14-3**	**904–909**
A2.A.77	Apply the double-angle and half-angle formulas for trigonometric functions	**14-4**	**911–917**
Measurement Strand			
A2.M.1	Define radian measure	**13-2**	**817–823**
A2.M.2	Convert between radian and degree measures	**13-2**	**817–823**
Statistics and Probability Strand			
A2.S.1	Understand the differences among various kinds of studies (e.g., survey, observation, controlled experiment)	**12-1**	**745–750**
A2.S.2	Determine factors which may affect the outcome of a survey	**12-1**	**745–750**
A2.S.3	Calculate measures of central tendency with group frequency distributions	**12-5,** 12-7	**773–778,** 786–793
A2.S.4	Calculate measures of dispersion (range, quartiles, interquartile range, standard deviation, variance) for both samples and populations	**12-2**	**752–758**
A2.S.5	Know and apply the characteristics of the normal distribution	**12-5, Extend 12-5**	**773–778, 779**
A2.S.6	Determine from a scatter plot whether a linear, logarithmic, exponential, or power regression model is most appropriate	**Extend 8-3**	**492–499**
A2.S.7	Determine the function for the regression model, using appropriate technology, and use the regression function to interpolate and extrapolate from the data	Extend 5-1, Extend 6-4, **Extend 8-3,** Extend 13-9	258, 365–366, **500–501,** 877
A2.S.8	Interpret within the linear regression model the value of the correlation coefficient as a measure of the strength of the relationship	**2-5**	**92–98**
A2.S.9	Differentiate between situations requiring permutations and those requiring combinations	**0-5**	**P12–P14**
A2.S.10	Calculate the number of possible permutations ($_nP_r$) of *n* items taken *r* at a time	**0-5**	**P12–P14**
A2.S.11	Calculate the number of possible combinations ($_nC_r$) of *n* items taken *r* at a time	**0-5**	**P12–P14**
A2.S.12	Use permutations, combinations, and the Fundamental Principle of Counting to determine the number of elements in a sample space and a specific subset (event)	**0-5**	**P12–P14**
A2.S.13	Calculate theoretical probabilities, including geometric applications	**12-4**	**764–771**
A2.S.14	Calculate empirical probabilities	**12-4**	**764–771**
A2.S.15	Know and apply the binomial probability formula to events involving the terms *exactly*, *at least*, and *at most*	**12-7**	**786–793**
A2.S.16	Use the normal distribution as an approximation for binomial probabilities	**12-7**	**786–793**

Unit 1
Linear Relations and Functions

CHAPTER 1

Equations and Inequalities

Chapter 1 Support

Helping You Learn

- **New Vocabulary** 5, 11, 18, 27, 33, 41
- **Key Concepts** 5, 11, 12, 19, 27, 33, 34, 41, 42, 43
- **Check Your Progress** 5, 6, 11, 12, 13, 18, 19, 20, 21, 27, 28, 29, 34, 35, 36, 42, 43, 44
- **Check Your Understanding** 7, 14, 22, 30, 36, 45
- **Multiple Representations** 9, 16, 24, 31, 38
- **H.O.T. Problems** 9, 16, 24, 31, 38, 47
- **Skills Check** 10, 17, 25, 32, 39, 48

NY Math Online

- **Math in Motion: Animation** 12, 40, 42
- **Personal Tutor** 5, 6, 11, 12, 13, 18, 19, 20, 21, 27, 28, 29, 33, 34, 35, 36, 41, 42, 43, 44
- **Self-Check Quizzes** 5, 18, 27, 33, 41
- **Extra Examples** 5, 18, 27, 33, 41
- **Homework Help** 5, 18, 27, 33, 41

Preparing for NYSRE

- **Extended Response** 17
- **Multiple Choice** 10, 17, 21, 22, 25, 26, 32, 39, 48, 53
- **Short/Gridded Response** 10, 25, 32, 39
- **Worked Out Example** 21

Unit 1
Linear Relations and Functions

CHAPTER 2

Linear Relations and Functions

Chapter 2 Support

Helping You Learn

NY Math Online

Preparing for NYSRE

Unit 1
Linear Relations and Functions

CHAPTER 3

Systems of Equations and Inequalities

Chapter 3 Support

Helping You Learn

NY Math Online

Preparing for NYSRE

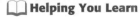

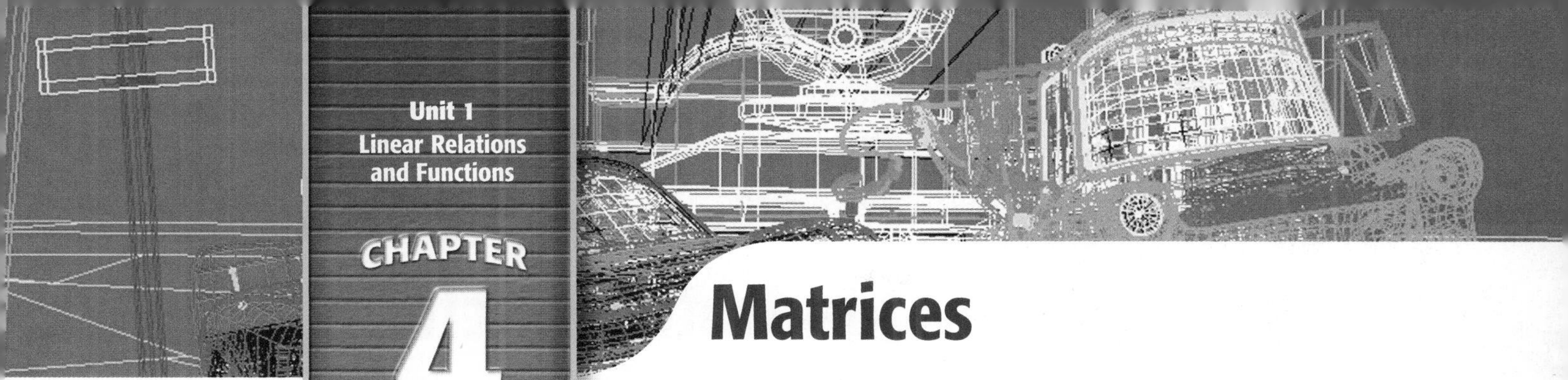

Unit 1
Linear Relations and Functions

CHAPTER 4

Matrices

Chapter 4 Support

Helping You Learn

NY Math Online

Preparing for NYSRE

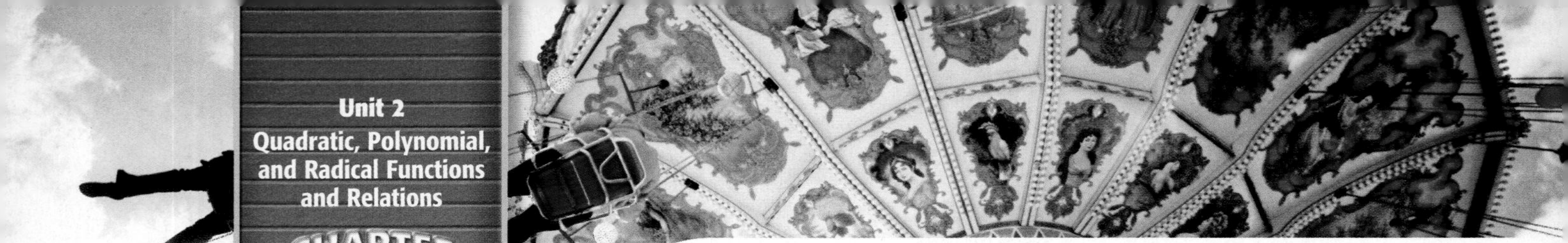

Unit 2
Quadratic, Polynomial, and Radical Functions and Relations

CHAPTER 5

Quadratic Functions and Relations

Chapter 5 Support

Helping You Learn

NY Math Online

Preparing for NYSRE

Unit 2
Quadratic, Polynomial, and Radical Functions and Relations

CHAPTER 6

Polynomials and Polynomial Functions

NYSCC

Chapter 6 Support

Helping You Learn

- **New Vocabulary** 333, 341, 348, 357, 368, 377
- **Key Concepts** 333, 334, 343, 350, 351, 358, 368, 369, 371, 377, 379, 383, 384, 385, 387, 391
- **Check Your Understanding** 337, 345, 352, 361, 372, 380, 388, 393
- **Multiple Representations** 338, 346, 354, 363, 381
- **H.O.T. Problems** 338, 346, 354, 363, 374, 381, 389, 395
- **Skills Check** 339, 347, 355, 364, 375, 382, 390, 396

NY Math Online

- **Math in Motion: Animation** 336, 340, 350, 356, 358, 385
- **Personal Tutor** 333, 334, 335, 336, 341, 342, 343, 344, 348, 349, 351, 357, 358, 359, 360, 368, 369, 370, 371, 377, 378, 379, 383, 384, 385, 386, 387, 391, 392, 393
- **Self-Check Quizzes, Extra Examples, Homework Help** 333, 341, 348, 357, 368, 377, 383, 391

Preparing for NYSRE

- **Extended Response** 355
- **Multiple Choice** 339, 342, 345, 347, 355, 364, 367, 375, 382, 390, 396, 401
- **Short/Gridded Response** 339, 347, 355
- **Worked Out Example** 342

Unit 2
Quadratic, Polynomial, and Radical Functions and Relations

CHAPTER 7

Inverses and Radical Functions and Relations

Chapter 7 Support

Unit 3
Advanced Functions and Relations

CHAPTER 8

Exponential and Logarithmic Functions and Relations

Chapter 8 Support

Helping You Learn

NY Math Online

Preparing for NYSRE

Unit 3
Advanced Functions and Relations

CHAPTER 9

Rational Functions and Relations

Chapter 9 Support

Unit 3
Advanced Functions and Relations

CHAPTER 10

Conic Sections

Chapter 10 Support

Helping You Learn

NY Math Online

Preparing for NYSRE

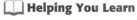

Unit 4
Discrete Mathematics

CHAPTER 11

Sequences and Series

Chapter 11 Support

Helping You Learn

NY Math Online

Preparing for NYSRE

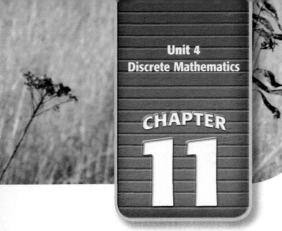

CHAPTER 12

Probability and Statistics

Chapter 12 Support

Helping You Learn

- **New Vocabulary** 745, 752, 759, 764, 773, 780, 786
- **Key Concepts** 752, 753, 754, 759, 764, 773, 774, 780, 781, 786, 787, 789
- **Check Your Progress** 745, 746, 747, 752, 753, 755, 759, 760, 764, 765, 766, 767, 774, 775, 780, 781, 786, 787, 788, 789
- **Check Your Understanding** 748, 755, 761, 767, 775, 782, 790
- **Multiple Representations** 770
- **H.O.T. Problems** 749, 757, 762, 770, 777, 783, 792
- **Skills Check** 750, 758, 763, 771, 778, 784, 793

NY Math Online

- **Math in Motion: Animation** 774, 779, 785
- **Personal Tutor** 745, 746, 747, 752, 753, 755, 759, 760, 764, 765, 766, 767, 773, 774, 780, 781, 786, 787, 788, 789
- **Self-Check Quizzes** 745, 752, 759, 764, 773, 780, 786
- **Extra Examples** 745, 752, 759, 764, 773, 780, 786
- **Homework Help** 745, 752, 759, 764, 773, 780, 786

Preparing for NYSRE

- **Extended Response** 793
- **Multiple Choice** 750, 758, 760, 761, 763, 771, 772, 778, 784, 793, 799
- **Short/Gridded Response** 750, 758, 763, 771, 778, 784
- **Worked Out Example** 760

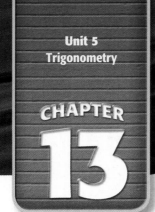

Unit 5
Trigonometry

CHAPTER 13

Trigonometric Functions

Chapter 13 Support

Helping You Learn

- **New Vocabulary** 808, 817, 825, 832, 841, 848, 855, 863, 871
- **Key Concepts** 808, 809, 811, 817, 819, 820, 825, 826, 827, 832, 833, 834, 841, 842, 848, 855, 856, 857, 858, 863, 864, 865, 871
- **Check Your Understanding** 813, 820, 829, 836, 843, 851, 859, 867, 874
- **Multiple Representations** 822, 853, 875
- **H.O.T. Problems** 815, 822, 830, 838, 845, 853, 860, 869, 875
- **Skills Check** 816, 823, 831, 839, 846, 854, 861, 870, 876

NY Math Online

- **Math in Motion: Animation** 817, 826, 855, 863, 864, 871
- **Math in Motion: Interactive Labs** 819
- **Personal Tutor** 808, 809, 810, 811, 812, 817, 818, 819, 820, 825, 826, 827, 828, 832, 833, 834, 835, 841, 842, 843, 848, 849, 850, 855, 856, 857, 863, 864, 865, 866, 871, 872, 873
- **Self-Check Quizzes, Extra Examples, Homework Help** 808, 817, 825, 832, 841, 848, 855, 863, 871

Preparing for NYSRE

- **Extended Response**
- **Multiple Choice** 816, 823, 831, 839, 846, 847, 854, 861, 870, 873, 874, 876, 883
- **Short/Gridded Response** 816, 823, 831, 839, 846, 854, 861, 870, 876
- **Worked Out Example** 873

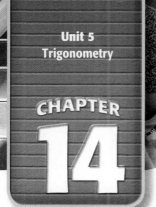

Unit 5
Trigonometry

CHAPTER 14

Trigonometric Identities and Equations

NYSCC

Chapter 14 Support

Helping You Learn

NY Math Online

Preparing for NYSRE

Student Handbook

Contents

CHAPTER 0

Preparing for Advanced Algebra

Chapter 0 contains lessons on topics from previous courses. You can use this chapter in various ways.

- Begin the school year by taking the Pretest. If you need additional review, complete the lessons in this chapter. To verify that you have successfully reviewed the topics, take the Posttest.
- As you work through the text, you may find that there are topics you need to review. When this happens, complete the individual lessons that you need.
- Use this chapter for reference. When you have questions about any of these topics, flip back to this chapter to review definitions or key concepts.

Get Started on Chapter 0

You will review several concepts, skills, and vocabulary terms as you study Chapter 0. To get ready, identify important terms and organize your resources.

FOLDABLES® Study Organizer

Throughout this text, you will be invited to use Foldables to organize your notes.

Why should you use them?

- They help you organize, display, and arrange information.
- They make great study guides, specifically designed for you.
- You can use them as your math journal for recording main ideas, problem-solving strategies, examples, or questions you may have.
- They give you a chance to improve your math vocabulary.

How should you use them?

- Write general information–titles, vocabulary terms, concepts, questions, and main ideas–on the front tabs of your Foldable.
- Write specific information–ideas, your thoughts, answers to questions, steps, notes, and definitions–under the tabs.
- Use the tabs for:
  - math concepts in parts, like types of triangles,
  - steps to follow, or
  - parts of a problem, like *compare and contrast* (2 parts) or *what*, *where*, *when*, *why*, and *how* (5 parts).
- You may want to store your Foldables in a plastic zipper bag that you have three-hole punched to fit in your notebook.

When should you use them?

- Set up your Foldable as you begin a chapter, or when you start learning a new concept.
- Write in your Foldable every day.
- Use your Foldable to review for homework, quizzes, and tests.

New Vocabulary

English		Español
domain	• p. P4 •	dominio
range	• p. P4 •	rango
mapping	• p. P4 •	transformaciones
function	• p. P4 •	función
outcome	• p. P9 •	resultados
sample space	• p. P9 •	espacio muestral
event	• p. P9 •	evento
independent events	• p. P9 •	eventos independientes
dependent events	• p. P9 •	eventos dependientes
factorial	• p. P10 •	factorial
permutation	• p. P12 •	permutación
linear permutation	• p. P12 •	permutación lineal
combination	• p. P12 •	combinacion
congruent	• p. P15 •	congruente
similar	• p. P15 •	semejantes

Multilingual eGlossary glencoe.com

NY Math Online glencoe.com

- Study the chapter online
- Explore **Math in Motion**
- Get extra help from your own **Personal Tutor**
- Use **Extra Examples** for additional help
- Take a **Self-Check Quiz**
- **Review Vocabulary** in fun ways

CHAPTER 0 Pretest

State the domain and range of each relation. Then determine whether each relation is a function. Write *yes* or *no*.

1. $\{(14, 1), (-3, 6), (8, 4)\}$

2.

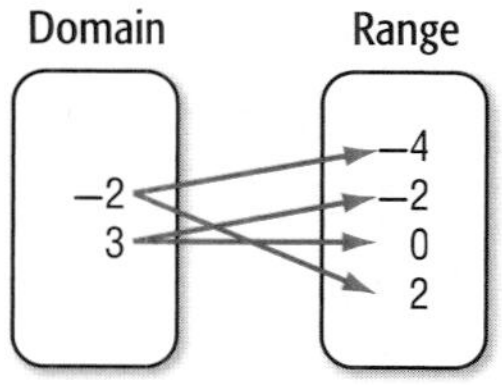

Name the quadrant in which each point is located.

3. $(-6, -2)$

4. $(10, 11)$

5. $(4, -3)$

6. $(-5, -7)$

Find each product.

7. $(x + 1)(x + 4)$

8. $(a - 3)(a + 6)$

9. $(m - 2)(m - 5)$

10. $(d + 7)(d + 7)$

11. $(t - 9)(t + 4)$

12. $(c + 8)(c - 8)$

13. NUMBER THEORY There are two integers. One is 5 more than a number, and the other is 1 less than the same number.

a. Write expressions for the two numbers.

b. Write a polynomial expression for the product of the numbers.

Factor each polynomial.

14. $6a^2 + 2a$

15. $10ab^2 + 5b$

16. $15d - 12cd^2$

17. $x^2 + 5x + 6$

18. $y^2 + 6y - 7$

19. $a^2 - 13a + 36$

State whether the events are *independent* or *dependent*.

20. selecting three playing cards from a standard deck without replacing any of the cards

21. rolling two dice

22. choosing a type of car and selecting a brand of tire

23. selecting two baseballs from a carton of five without replacement

24. BOOKS A bookshelf holds 4 different biographies and 5 different mystery novels. How many ways can one book of each type be selected?

25. ICE CREAM An ice cream shop offers a choice of two types of cones and 15 flavors of ice cream. How many different 1-scoop ice cream cones can a customer order?

Determine whether each situation involves a *permutation* or a *combination*. Then find the number of possibilities.

26. placing an algebra book, a geometry book, a chemistry book, an English book, and a health book on a shelf

27. selecting 3 of 15 flavors of juice at the grocery store

28. Determine whether the triangles are *similar*, *congruent*, or *neither*.

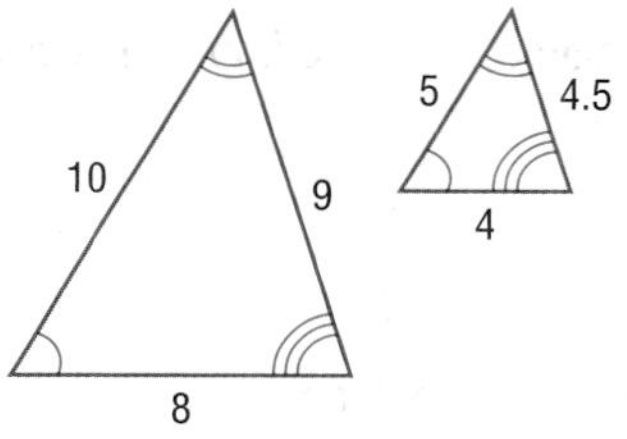

29. PHOTOGRAPHS A photo that is 3 inches wide by 5 inches long is being enlarged so that it is 12 inches long. How wide will the enlarged photo be?

Find each missing measure. Round to the nearest tenth, if necessary.

30.

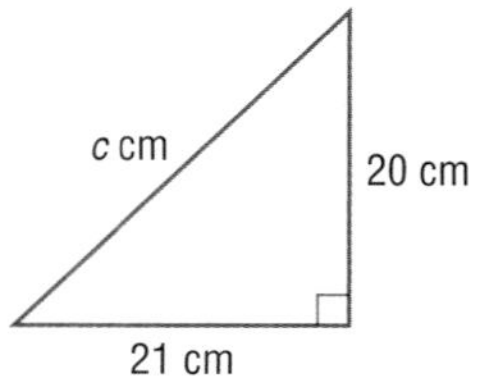

31. $a = 6$ yd, $b = 9$ yd, $c = ?$

The lengths of three sides of a triangle are given. Determine whether each triangle is a right triangle.

32. 12 yd, 14 yd, 16 yd

33. 15 km, 20 km, 25 km

34. 45 mm, 60 mm, 75 mm

0-1

Representing Functions

Objective
- Identify the domain and range of functions.

NYS Core Curriculum

A2.A.38 Determine when a relation is a function
A2.A.39 Determine the domain and range of a function from its equation
Also addresses A2.A.37.

New Vocabulary
domain
range
quadrants
mapping
function

NY Math Online
glencoe.com
- Extra Examples
- Personal Tutor
- Self-Check Quiz
- Homework Help

Recall that a *relation* is a set of ordered pairs. The **domain** of a relation is the set of all first coordinates (x-coordinates) from the ordered pairs, and the **range** is the set of all second coordinates (y-coordinates) from the ordered pairs.

EXAMPLE 1 Domain and Range

State the domain and the range of the relation.
{(−3, 3), (0, −7), (1, −5), (2, 4)}

The domain is the set of x-coordinates.
D = {−3, 0, 1, 2}

The range is the set of y-coordinates.
R = {−7, −5, 3, 4}

A relation can be graphed on a coordinate plane. A coordinate plane is formed by the intersection of the horizontal axis, or x-axis, and the vertical axis, or y-axis. The axes meet at the origin (0, 0) and divide the plane into four **quadrants**. Any ordered pair in the coordinate plane can be written in the form (x, y).

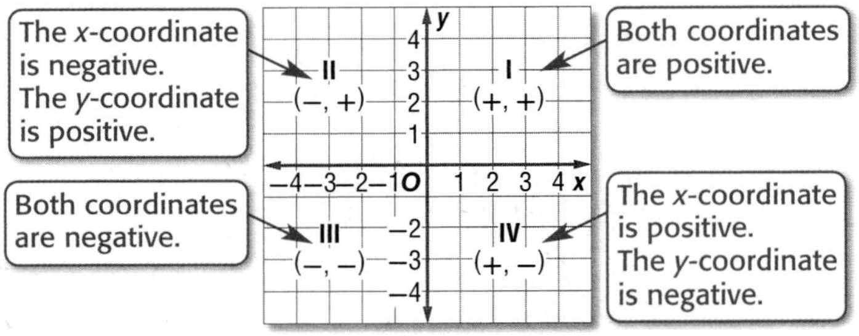

EXAMPLE 2 Locate Coordinates

Name the quadrant in which $T(-8, 5)$ is located.

Point T has a negative x-coordinate and a positive y-coordinate. The point is located in Quadrant II.

A relation can also be represented by a table or a mapping. A **mapping** illustrates how each element of the domain is paired with an element in the range.

Ordered Pairs

(1, 2)
(−2, 3)
(0, −3)

Table

x	y
1	2
−2	3
0	−3

Graph

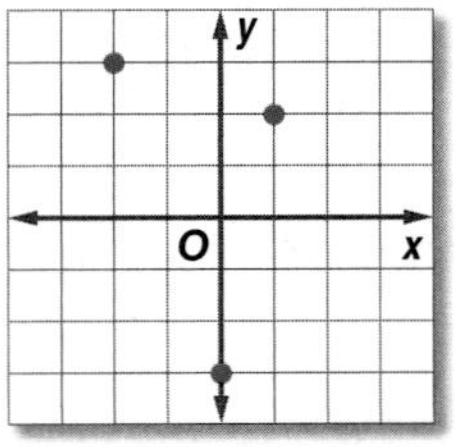

Mapping

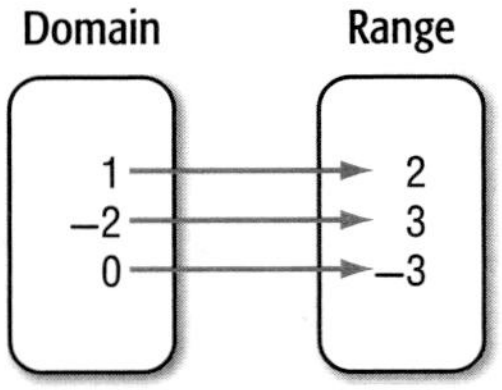

A **function** is a relation in which each element of the domain is paired with *exactly one* element of the range.

EXAMPLE 3 Identify Domain and Range

State the domain and range of each relation. Then determine whether each relation is a function.

a. {(10, 3), (6, −2), (7, 4), (−8, −9)}

D = {−8, 6, 7, 10}
R = {−9, −2, 3, 4}
For each element of the domain, there is only one corresponding element in the range. So, this relation is a function.

b.

x	y
1	3, 4
2	7
3	4

D = {1, 2, 3}
R = {3, 4, 7}

Because 1 is paired with 3 and 4, this is not a function.

c.

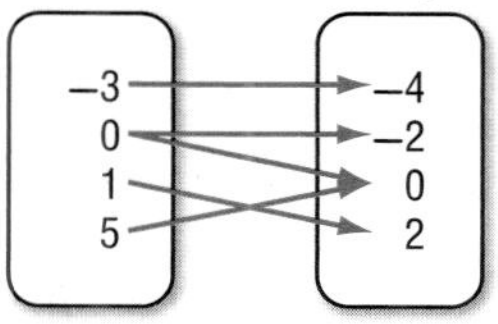

D = {−3, 0, 1, 5}
R = {−4, −2, 0, 2}
Because 0 is paired with −2 and 0, this is not a function.

d.

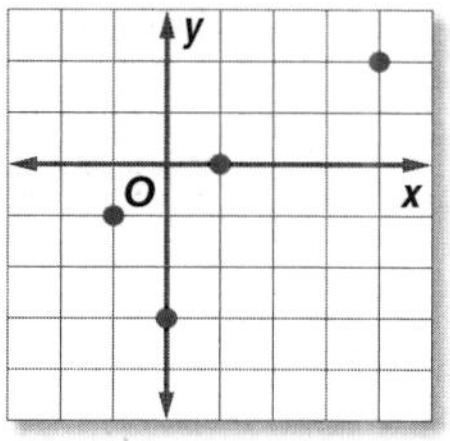

D = {−1, 0, 1, 4}
R = {−3, −1, 0, 2}
This is a function.

Watch Out!

Functions Remember that in a function, an element of the range can be paired with more than one element of the domain. But an element of the domain cannot be paired with more than one element of the range.

Exercises

State the domain and range of each relation. Then determine whether each relation is a function. Write *yes* or *no*.

1. {(2, 7), (3, 10), (1, 6)}

2. {(−6, 0), (5, 5), (9, −2), (−2, −9)}

3.

x	y
1	5
2	7
1	9

4.

x	y
−12	0
−10	1
−8	2
−6	4

5.

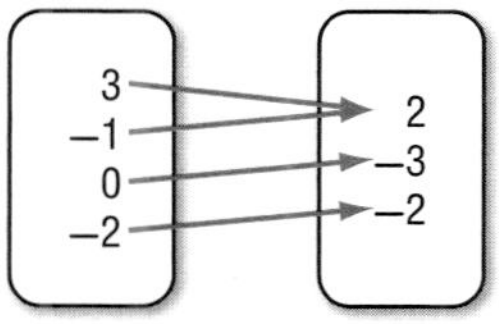

6.

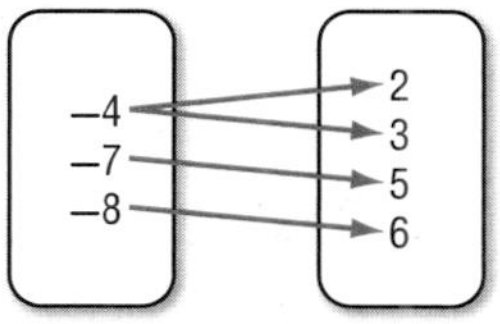

7.

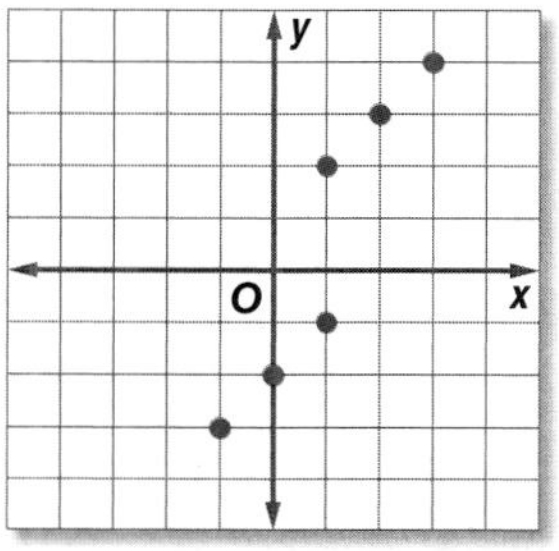

8.

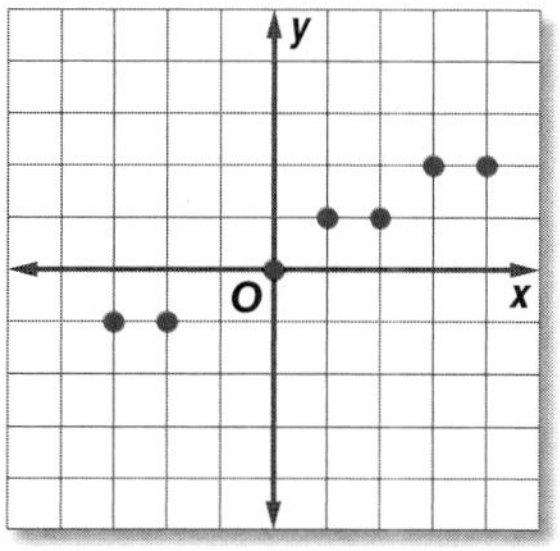

Name the quadrant in which each point is located.

9. (5, 3) **10.** (8, −6) **11.** (2, 0) **12.** (−7, −1)

0-2 FOIL

Objective
- Use the FOIL method to multiply binomials.

NY Math Online
glencoe.com
- Extra Examples
- Personal Tutor
- Self-Check Quiz
- Homework Help

The product of two binomials is the sum of the products of

the *first* terms, the *outer* terms, the *inner* terms, and the *last* terms.

F **O** **I** **L**

EXAMPLE 1 Use the FOIL Method

Find each product.

a. $(x + 3)(x - 5)$

F L First Outer Inner Last

$$(x + 3)(x - 5) = x \cdot x + x \cdot (-5) + 3 \cdot x + 3 \cdot (-5)$$

I

O

$$= x^2 - 5x + 3x - 15$$
$$= x^2 - 2x - 15$$

b. $(3y + 2)(5y + 4)$

$$(3y + 2)(5y + 4) = 3y \cdot 5y + 3y \cdot 4 + 2 \cdot 5y + 2 \cdot 4$$
$$= 15y^2 + 12y + 10y + 8$$
$$= 15y^2 + 22y + 8$$

Exercises

Find each product.

1. $(a + 2)(a + 4)$

2. $(v - 7)(v - 1)$

3. $(h + 4)(h - 4)$

4. $(d - 1)(d + 1)$

5. $(b + 4)(b - 3)$

6. $(t - 9)(t + 11)$

7. $(r + 3)(r - 8)$

8. $(k - 2)(k + 5)$

9. $(p + 8)(p + 8)$

10. $(x - 15)(x - 15)$

11. $(2c + 1)(c - 5)$

12. $(7n - 2)(n + 3)$

13. $(3m + 4)(2m - 5)$

14. $(5g + 1)(6g + 9)$

15. $(2q - 17)(q + 2)$

16. $(4t - 7)(3t - 12)$

17. **NUMBERS** I am thinking of two integers. One is 7 less than a number, and the other is 2 greater than the same number.

a. Write expressions for the two numbers.

b. Write a polynomial expression for the product of the numbers.

18. **OFFICE SPACE** Monica's current office is square. Her office in the company's new building will be 3 feet wider and 5 feet longer.

a. Write expressions for the dimensions of Monica's new office.

b. Write a polynomial expression for the area of Monica's new office.

c. Suppose Monica's current office is 7 feet by 7 feet. How much larger will her new office be?

0-3 Factoring Polynomials

Objective

- Use various techniques to factor polynomials.

NY Math Online

glencoe.com

- Extra Examples
- Personal Tutor
- Self-Check Quiz
- Homework Help

Some polynomials can be factored using the Distributive Property.

EXAMPLE 1 Use the Distributive Property

Factor $4a^2 + 8a$.

Find the GCF of $4a^2$ and $8a$.

$4a^2 = 2 \cdot 2 \cdot a \cdot a \qquad 8a = 2 \cdot 2 \cdot 2 \cdot a \quad \rightarrow \quad \text{GCF: } 2 \cdot 2 \cdot a \text{ or } 4a$

$4a^2 + 8a = 4a(a) + 4a(2)$ **Rewrite each term using the GCF.**

$= 4a(a + 2)$ **Distributive Property**

Thus, the completely factored form of $4a^2 + 8a$ is $4a(a + 2)$.

To factor quadratic trinomials of the form $x^2 + bx + c$, find two integers m and p with a product equal to c and a sum equal to b. Then write $x^2 + bx + c$ using the pattern $(x + m)(x + p)$.

EXAMPLE 2 Use Factors and Sums

Factor each polynomial.

a. $x^2 + 5x + 6$ ← Both b and c are positive.

In this trinomial, b is 5 and c is 6. Find two numbers with a product of 6 and a sum of 5.

Factors of 6	Sum of factors
1, 6	7
2, 3	5

The correct factors are 2 and 3.

$x^2 + 5x + 6 = (x + m)(x + p)$ **Write the pattern.**

$= (x + 2)(x + 3)$ **$m = 2$ and $p = 3$**

b. $x^2 - 8x + 12$ ← b is negative, and c is positive.

In this trinomial, $b = -8$ and $c = 12$. This means that $m + p$ is negative and mp is positive. So m and p must both be negative.

Factors of 12	Sum of factors
−1, −12	−13
−2, −6	−8

The correct factors are −2 and −6.

$x^2 - 8x + 12 = (x + m)(x + p)$ **Write the pattern.**

$= [x + (-2)][x + (-6)]$ **$m = -2$ and $p = -6$**

$= (x - 2)(x - 3)$ **Simplify.**

c. $x^2 + 14x - 15$ ← b is positive, and c is negative.

In this trinomial, $b = 14$ and $c = -15$. This means that $m + p$ is positive and mp is negative. So either m or p must be negative, but not both.

Factors of −15	Sum of factors
1, −15	−15
−1, 15	14

The correct factors are −1 and 15.

$x^2 + 14x - 15 = (x + m)(x + p)$ **Write the pattern.**

$= [x + (-1)](x + 15)$ **$m = -1$ and $p = 15$**

$= (x - 1)(x + 15)$ **Simplify.**

To factor quadratic trinomials of the form $ax^2 + bx + c$, find two integers m and p with a product equal to ac and a sum equal to b. Write $ax^2 + bx + c$ using the pattern $ax^2 + mx + px + c$. Then factor by grouping.

EXAMPLE 3 Use Factors and Sums

Factor $6x^2 + 7x - 3$.

In this trinomial, $a = 6$, $b = 7$, and $c = -3$. This means that $m + p$ is positive and mp is negative. So either m or p must be negative, but not both.

Factors of −18	Sum of factors	
1, −18	−17	
−1, 18	17	
2, −9	−7	
−2, 9	7	**The correct factors are −2 and 9.**

$6x^2 + 7x - 3 = 6x^2 + mx + px - 3$ **Write the pattern.**

$= 6x^2 + (-2)x + 9x - 3$ **$m = -2$ and $p = 9$**

$= (6x^2 - 2x) + (9x - 3)$ **Group terms with common factors.**

$= 2x(3x - 1) + 3(3x - 1)$ **Factor the GCF from each group.**

$= (2x + 3)(3x - 1)$ **Distributive Property**

StudyTip

Checking Solutions You can check to see that you have factored correctly by multiplying the factors and comparing the product to the original polynomial.

Here are some special products.

Perfect Square Trinomials

$(a + b)^2 = (a + b)(a + b) = a^2 + 2ab + b^2$

$(a - b)^2 = (a - b)(a - b) = a^2 - 2ab + b^2$

Difference of Squares

$a^2 - b^2 = (a + b)(a - b)$

EXAMPLE 4 Use Special Products

Factor each polynomial.

a. $4x^2 + 20x + 25$

The first and last terms are perfect squares. The middle term is equal to $2(2x)(5)$. This is a perfect square trinomial of the form $(a + b)^2$.

$4x^2 + 20x + 25 = (2x)^2 + 2(2x)(5) + 5^2$ **Write as $a^2 + 2ab + b^2$.**

$= (2x + 5)^2$ **Factor using the pattern.**

b. $x^2 - 4$

This is a difference of squares.

$x^2 - 4 = x^2 - (2)^2$ **Write in the form $a^2 - b^2$.**

$= (x + 2)(x - 2)$ **Factor the difference of squares.**

Exercises

Factor each polynomial.

1. $12x^2 + 4x$

2. $6x^2y + 2x$

3. $8ab^2 - 12ab$

4. $x^2 + 5x + 4$

5. $y^2 + 12y + 27$

6. $x^2 + 6x + 8$

7. $3y^2 + 13y + 4$

8. $7x^2 + 51x + 14$

9. $3x^2 + 28x + 32$

10. $x^2 - 5x + 6$

11. $y^2 - 5y + 4$

12. $6x^2 - 13x + 5$

13. $6a^2 - 50ab + 16b^2$

14. $11x^2 - 78x + 7$

15. $18x^2 - 31xy + 6y^2$

16. $x^2 + 4xy + 4y^2$

17. $9x^2 - 24x + 16$

18. $4a^2 + 12ab + 9b^2$

19. $x^2 - 144$

20. $4c^2 - 9$

21. $16y^2 - 1$

22. $25x^2 - 4y^2$

23. $36y^2 - 16$

24. $9a^2 - 49b^2$

0-4 The Counting Principle

Objective

- Use the Fundamental Counting Principle to find outcomes involving independent and dependent events.

New Vocabulary

outcome
sample space
event
independent event
dependent event
factorial

NY Math Online

glencoe.com

- Extra Examples
- Personal Tutor
- Self-Check Quiz
- Homework Help

An **outcome** is the result of a single trial. For example, the trial of tossing a coin has two outcomes: head or tail. The set of all possible outcomes is called the **sample space**. An **event** consists of one or more outcomes of a trial. When two or more events occur, the events can be independent or dependent.

Key Concept — For Your FOLDABLE

Independent Events

Words If the outcome of an event does not affect the outcome of another event, the two events are independent.

Example tossing a coin and rolling a die

Dependent Events

Words If the outcome of an event does affect the outcome of another event, the two events are dependent.

Example taking a marble from a bag and then taking another marble from the bag without replacing the first

You can use the **Fundamental Counting Principle** to find the number of possible outcomes when two or more events occur.

Key Concept — **Fundamental Counting Principle** — For Your FOLDABLE

Words If event M can occur in m ways and is followed by event N that can occur in n ways, then the event M followed by event N can occur in $m \cdot n$ ways.

Example If event M can occur in 2 ways and is followed by event N that can occur in 3 ways, then the event M followed by event N can occur in $2 \cdot 3$ or 6 ways.

EXAMPLE 1 Independent Events

CONTESTS Kim won a contest on a radio station. The prize was a restaurant gift certificate and tickets to a sporting event. She can select one of three different restaurants and tickets to a football, baseball, basketball, or hockey game. How many different ways can she select a restaurant followed by a sporting event?

Her choice of a restaurant does not affect her choice of a sporting event, so these events are independent.

There are 3 ways she can choose a restaurant, and there are 4 ways she can choose the sporting event. Use the Fundamental Counting Principle to find the number of ways she can choose her two prizes.

number of restaurants		number of sporting events		number of ways to choose prizes
3	$\cdot$	4	=	12

So, there are 12 different ways that Kim can choose her prizes.

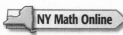

You can also use the Fundamental Counting Principle when there are more than two events.

EXAMPLE 2 More than Two Independent Events

CODES Many answering machines allow owners to call home and get their messages by entering a 3-digit code. How many codes are possible?

The choice of any digit does not affect the other two digits, so the choices of the digits are independent events. Each digit can be any numeral from 0 to 9.

There are 10 possible first digits in the code, 10 possible second digits, and 10 possible third digits. So, there are $10 \cdot 10 \cdot 10$ or 1000 possible different code numbers.

The mathematical notation 5! also means $5 \cdot 4 \cdot 3 \cdot 2 \cdot 1$. The symbol 5! is read five **factorial**. $n!$ means the product of all counting numbers beginning with n and counting backward to 1. By definition, $0! = 1$.

Problem-Solving Tip

Use a Diagram or Model It is sometimes helpful to sketch objects or use models such as counters to represent independent or dependent events. In Example 3, a different color counter could represent each course Charlita could take.

EXAMPLE 3 Dependent Events

SCHOOL Charlita wants to take 6 different classes next year. Assuming that each class is offered each period, how many different schedules could she have?

When Charlita schedules a class for a given period, she cannot schedule that class for any other period. Therefore, the choices of which class to schedule each period are dependent events.

There are 6 classes Charlita can take during first period. That leaves 5 classes she can take second period. After she chooses which classes to take the first two periods, there are 4 remaining choices for third period, and so on.

Period	1	2	3	4	5	6
Number of Choices	6	5	4	3	2	1

There are $6 \cdot 5 \cdot 4 \cdot 3 \cdot 2 \cdot 1$ or 720 schedules that Charlita could have. Note that $6 \cdot 5 \cdot 4 \cdot 3 \cdot 2 \cdot 1 = 6!$.

Exercises

State whether the events are *independent* or *dependent*.

1. selecting a fiction book and a nonfiction book at the library
2. choosing a president, vice president, secretary, and treasurer for Student Council, assuming that a person can hold only one office
3. choosing a style, color, and size of mountain bike
4. selecting two pens from a box of 12 without replacing the first

Real-World Link

"Black tie" is worn to any formal event after 6:00 P.M. The following is appropriate: traditional black tuxedo, white shirt, vest or cummerbund in black or dark color, black tie, and black patent shoes.

Source: Skeffington's FormalWear

Solve.

5. **HOMEWORK** Carlos has homework in math, chemistry, and English. How many ways can Carlos choose the order in which he does homework?

6. **COMPUTERS** A mail-order computer company offers a choice of 4 amounts of memory, 2 sizes of hard drives, and 2 sizes of monitors. How many different systems are available to a customer?

7. **DINING** A cafeteria offers the choices shown in the table. How many different combinations of drink and salad are possible?

Drink	Salad
water	pasta
coffee	fruit
juice	chicken
milk	

8. **DANCES** Dane is renting a tuxedo for prom. Once he has chosen his jacket, he must choose from three types of pants and six colors of vests. How many different ways can he select his attire for prom?

9. **MANUFACTURING** A baseball glove manufacturer makes a glove with the different options shown in the table. How many different gloves are possible?

Option	Number of Choices
sizes	4
types by position	3
materials	2
levels of quality	2

10. **CLOTHES** How many different outfits can be made if Jessica chooses 1 each from 11 skirts, 9 blouses, 3 belts, and 7 pairs of shoes?

11. **PASSWORDS** Abby is registering at a Web site. She must select a password containing six digits from 1 to 9 to be able to use the site. How many passwords are allowed if no digit may be used more than once?

12. **QUIZZES** Each question on a five-question multiple-choice quiz has answer choices labeled A, B, C, and D. How many different ways can a student answer the five questions?

13. **GAMES** The letters A through Z are written on pieces of paper and placed in a jar. Four of them are selected one after the other without replacing any of them. How many ways are there to select the letters?

14. **LICENSE PLATES** How many different license plates are possible with two letters followed by three digits between 0 and 9?

15. **FOOD** How many different combinations of sandwich, side, and beverage are possible for lunch?

Sandwiches	Sides	Beverages
• Hot dog	• Chips	• Bottled water
• Hamburger	• Apple	• Soda
• Veggie burger	• Pasta salad	• Juice
• Bratwurst		• Milk
• Grilled chicken		

16. **AREA CODES** Prior to 1995, area codes were in the following format.

$$(ABC): \quad A = 2, 3, 4, 5, 6, 7, 8, 9$$
$$B = 0, 1$$
$$C = 0, 1, 2, 3, 4, 5, 6, 7, 8, 9$$

a. How many area codes were possible before 1995?

b. In 1995, the restriction on the middle digit was removed, allowing any digit in that position. How many total codes were possible after this change was made?

0-5 Permutations and Combinations

Objective
- Solve problems involving permutations and combinations.

NYS Core Curriculum

A2.S.9 Differentiate between situations requiring permutations and those requiring combinations
A2.S.12 Use permutations, combinations, and the Fundamental Principle of Counting to determine the number of elements in a sample space and a specific subset (event) *Also addresses A2.S.10 and A2.S.11.*

New Vocabulary
permutation
linear permutation
combination

NY Math Online

glencoe.com
- Extra Examples
- Personal Tutor
- Self-Check Quiz
- Homework Help

When a group of objects or people is arranged in a certain order, the arrangement is called a **permutation**. In a permutation, the *order* of the objects is important. The arrangement of objects or people in a line is called a **linear permutation**.

An arrangement or selection of objects in which order is *not* important is called a **combination**.

When solving probability problems, it is helpful to be able to determine whether situations involve permutations or combinations. Often words in a problem give clues as to which type of arrangement is involved.

Type of Arrangement	Description	Clue Words	Examples
permutation	The order of objects or people is important.	• arranging x • an arrangement of first, second, and third	• arranging 4 books on a bookshelf • an arrangement of the letters in *math*
combination	The order of objects or people is not important.	• selecting x of y • choosing x from y	• selecting 3 of 8 flavors • choosing 2 people from a group of 7

EXAMPLE 1 Permutations and Combinations

Determine whether each situation involves a *permutation* or a *combination*.

a. choosing 6 students from a class of 25

Because the order of the students does not matter, this is a combination.

b. an arrangement of the letters in *drive*

Because the order of the letters is important, this is a permutation.

c. selecting 2 of 9 different side dishes

Because the order in which the side dishes are chosen does not matter, this is a combination.

d. choosing 3 classes from a list of 12 to schedule for first, second, and third periods

Because the order of the periods is important, this is a permutation.

You can use the following rule to find the number of permutations of objects.

Key Concept Permutations

For Your FOLDABLE

The number of permutations of n distinct objects taken r at a time is given by

$${}_nP_r = \frac{n!}{(n-r)!}.$$

ReadingMath

Permutations The expression ${}_nP_r$ reads *the number of permutations of n objects taken r at a time.* A permutation can also be written as $P(n, r)$.

EXAMPLE 2 Permutation

There are 10 finalists in a figure skating competition. How many ways can gold, silver, and bronze medals be awarded?

Because each winner will receive a different medal, order is important. You must find the number of permutations of 10 finalists taken 3 at a time.

$${}_nP_r = \frac{n!}{(n-r)!}$$ **Permutation formula**

$${}_{10}P_3 = \frac{10!}{(10-3)!}$$ **$n = 10$ and $r = 3$**

$$= \frac{10!}{7!}$$ **Simplify.**

$$= \frac{10 \cdot 9 \cdot 8 \cdot \overset{1}{\cancel{7}} \cdot \overset{1}{\cancel{6}} \cdot \overset{1}{\cancel{5}} \cdot \overset{1}{\cancel{4}} \cdot \overset{1}{\cancel{3}} \cdot \overset{1}{\cancel{2}} \cdot \overset{1}{\cancel{1}}}{\underset{1}{\cancel{7}} \cdot \underset{1}{\cancel{6}} \cdot \underset{1}{\cancel{5}} \cdot \underset{1}{\cancel{4}} \cdot \underset{1}{\cancel{3}} \cdot \underset{1}{\cancel{2}} \cdot \underset{1}{\cancel{1}}}$$ **Divide by common factors.**

$$= 720$$ **Simplify.**

The gold, silver, and bronze medals can be awarded in 720 ways.

When some letters or objects are alike, use the rule below to find the number of permutations.

Key Concept Permutations with Repetition

For Your FOLDABLE

The number of permutations of n of which p are alike and q are alike is

$$\frac{n!}{p!q!}.$$

EXAMPLE 3 Permutation with Repetition

How many different ways can the letters of the word *MISSISSIPPI* be arranged?

The letter I occurs 4 times, S occurs 4 times, and P occurs 2 times.

You need to find the number of permutations of 11 letters of which 4 of one letter, 4 of another letter, and 2 of another letter are the same.

There are 11 letters. →
There are 4 *I*s, 4 *S*s, and 2 *P*s. →

$$\frac{11!}{4!4!2!} = \frac{11 \cdot 10 \cdot 9 \cdot 8 \cdot 7 \cdot 6 \cdot 5 \cdot 4!}{4!4!2!}$$

$$= 34{,}650$$ **Use a calculator.**

There are 34,650 ways to arrange the letters.

ReadingMath

Combinations The expression ${}_nC_r$ reads *the number of combinations of n objects taken r at a time.* A combination can also be written as $C(n, r)$.

Key Concept Combinations

For Your FOLDABLE

The number of combinations of n distinct objects taken r at a time is given by

$${}_nC_r = \frac{n!}{(n-r)!r!}.$$

EXAMPLE 4 Combination

A group of seven students working on a project needs to choose two students to present the group's report. How many ways can they choose the two students?

Because the order for choosing the students is not important, you must find the number of combinations of 7 students taken 2 at a time.

${}_nC_r = \frac{n!}{(n-r)!r!}$	**Combination formula**
${}_7C_2 = \frac{7!}{(7-2)!2!}$	**$n = 7$ and $r = 2$**
$= \frac{7!}{5!2!}$	**Subtract.**
$= \frac{7 \cdot 6 \cdot \cancel{5}^{1} \cdot \cancel{4}^{1} \cdot \cancel{3}^{1} \cdot \cancel{2}^{1} \cdot \cancel{1}^{1}}{\cancel{5}_{1} \cdot \cancel{4}_{1} \cdot \cancel{3}_{1} \cdot \cancel{2}_{1} \cdot \cancel{1}_{1}}$	**Divide by common factors.**
$= 21$	**Simplify.**

There are 21 possible ways to choose the two students.

Exercises

Find each permutation or combination.

1. ${}_5P_3$ **2.** ${}_6P_3$ **3.** ${}_7P_5$ **4.** ${}_4C_2$

5. ${}_6C_1$ **6.** ${}_{10}C_4$ **7.** ${}_9P_5$ **8.** ${}_{12}C_7$

Determine whether each situation involves a *permutation* or a *combination*. Then find the number of possibilities.

9. 7 shoppers in line at a checkout counter

10. an arrangement of the letters in the word *intercept*

11. selecting 4 of 13 different colored balloons

12. an arrangement of 4 blue tiles, 2 red tiles, and 3 black tiles in a row

13. choosing 2 different pizza toppings from a list of 6

14. the winner and first, second, and third runners-up in a contest with 10 finalists

15. an arrangement of the letters in the word *parallel*

16. choosing 2 CDs to buy from 10 that are on sale

17. SOFTBALL The manager of a softball team has 7 possible players in mind for the top 4 spots in the lineup. How many ways can she choose the top 4 spots?

18. NEWSPAPERS A newspaper has 9 reporters available to cover 4 different stories. How many ways can the reporters be assigned to cover the stories?

19. READING Jack has a reading list of 12 books. How many ways can he select 9 books from the list to check out of the library?

20. BANDS A band is choosing 3 new backup singers from a group of 18 who try out. How many ways can they choose the new singers?

Real-World Link

There are almost half a million words in the English language. But one third of all English writing is made up of only twenty-two words.

Source: *The Reading Solution*

0-6 Congruent and Similar Figures

Objective
- Identify and use congruent and similar figures.

New Vocabulary
congruent
similar

NY Math Online
glencoe.com
- Extra Examples
- Personal Tutor
- Self-Check Quiz
- Homework Help

Congruent figures have the same size and the same shape.

Two polygons are congruent if their corresponding sides are congruent and their corresponding angles are congruent.

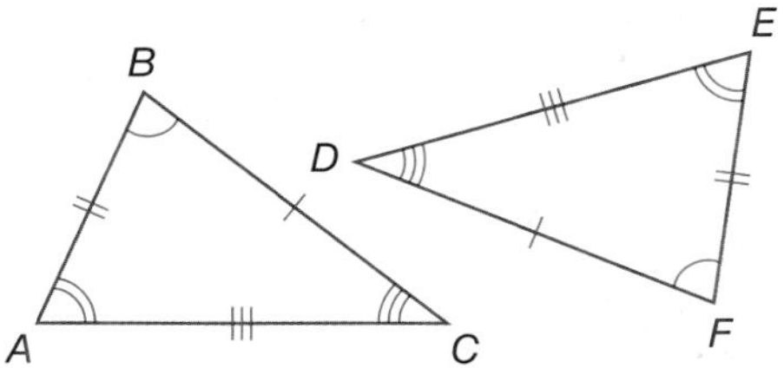

$\triangle ABC \cong \triangle EFD$

The order of the vertices indicates the corresponding parts.

Congruent Angles	Congruent Sides
$\angle A \cong \angle E$	$\overline{AB} \cong \overline{EF}$
$\angle B \cong \angle F$	$\overline{BC} \cong \overline{FD}$
$\angle C \cong \angle D$	$\overline{AC} \cong \overline{ED}$

Read the symbol $\cong$ as *is congruent to*.

EXAMPLE 1 Congruence Statements

The corresponding parts of two congruent triangles are marked on the figure. Write a congruence statement for the two triangles.

List the congruent angles and sides.

$\angle A \cong \angle D$	$\overline{AB} \cong \overline{DE}$
$\angle B \cong \angle E$	$\overline{AC} \cong \overline{DC}$
$\angle ACB \cong \angle DCE$	$\overline{BC} \cong \overline{EC}$

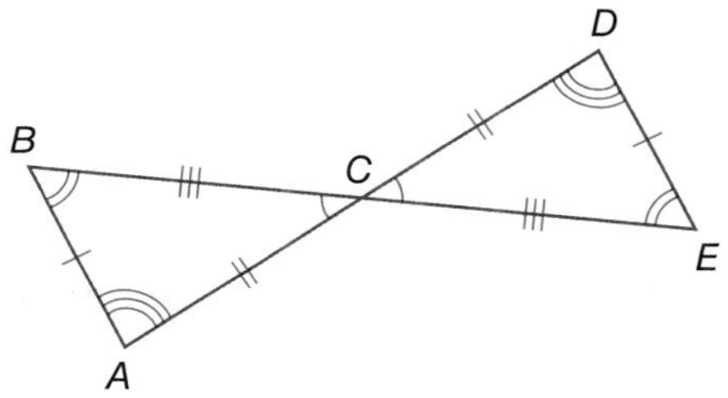

Match the vertices of the congruent angles. Therefore, $\triangle ABC \cong \triangle DEC$.

Similar figures have the same shape, but not necessarily the same size.

In similar figures, corresponding angles are congruent, and the measures of corresponding sides are proportional. (They have equivalent ratios.)

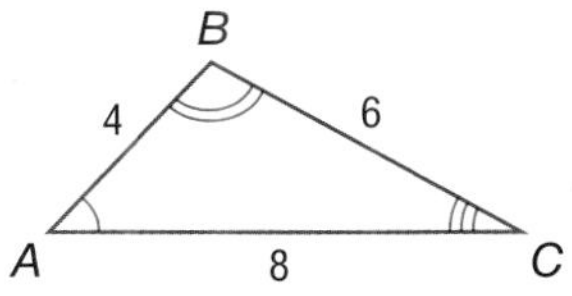

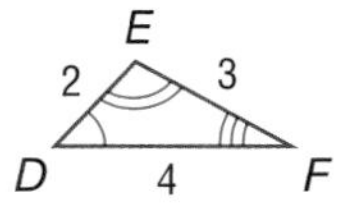

Congruent Angles
$\angle A \cong \angle D, \angle B \cong \angle E, \angle C \cong \angle F$

Proportional Sides
$\frac{AB}{DE} = \frac{BC}{EF} = \frac{AC}{DF}$

$\triangle ABC \sim \triangle DEF$

Read the symbol $\sim$ as *is similar to*.

EXAMPLE 2 Determine Similarity

Determine whether the polygons are similar. Justify your answer.

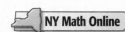

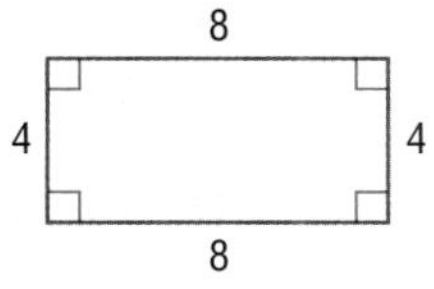

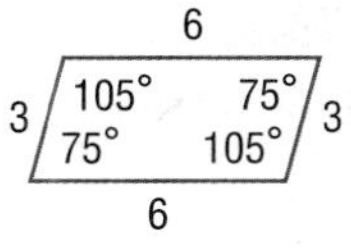

Because $\frac{4}{3} = \frac{8}{6} = \frac{4}{3} = \frac{8}{6}$, the measures of the sides are proportional. However, the corresponding angles are not congruent. The polygons are not similar.

EXAMPLE 3 Solve a Problem Involving Similarity

CIVIL ENGINEERING The city of Mansfield plans to build a bridge across Pine Lake. Use the information in the diagram at the right to find the distance across Pine Lake.

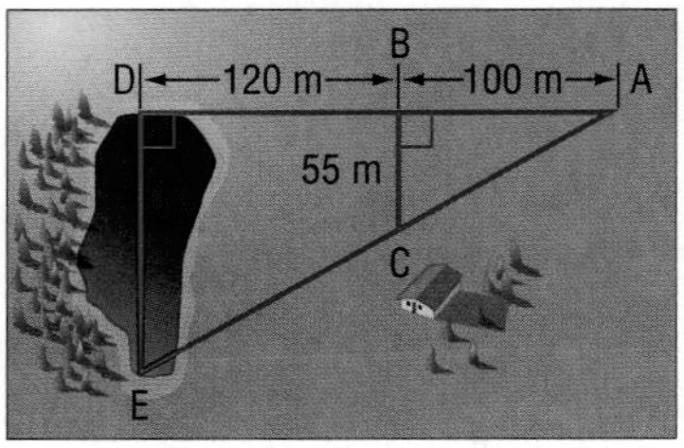

$\triangle ABC \sim \triangle ADE$

$\frac{AB}{AD} = \frac{BC}{DE}$	**Definition of similar polygons**
$\frac{100}{220} = \frac{55}{DE}$	**$AB = 100$, $AD = 100 + 120$ or 220, $BC = 55$**
$100DE = 220(55)$	**Cross products**
$100DE = 12{,}100$	**Simplify.**
$DE = 121$	**Divide each side by 100.**

The distance across the lake is 121 meters.

StudyTip

Reasonableness When solving a problem using a proportion, examine your solution for reasonableness. In Example 3, *DA* is more than twice *BA*, so *DE* should be more than twice *BC*. The solution is reasonable.

Exercises

Determine whether each pair of figures is *similar*, *congruent*, or *neither*.

1.

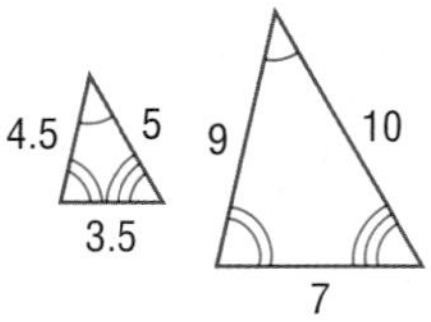

2.

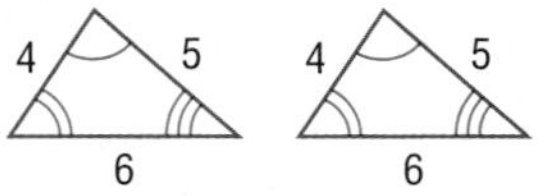

3.

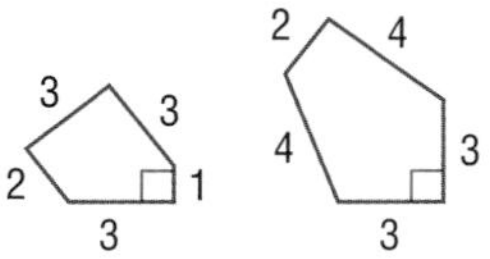

4.

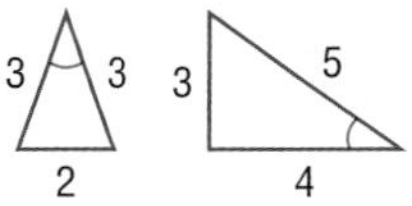

5.

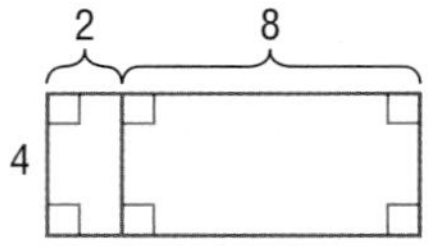

6.

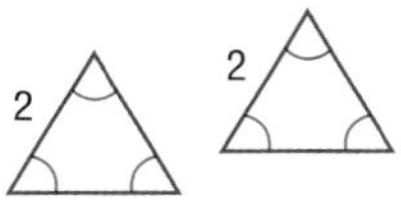

Each pair of polygons is similar. Find the values of *x* and *y*.

7.

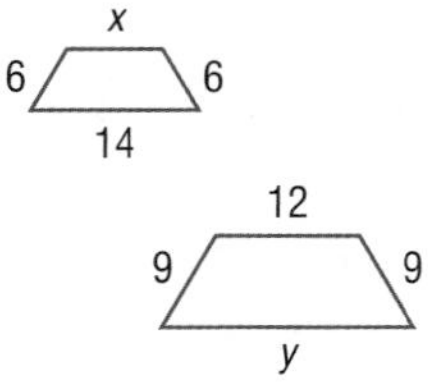

8.

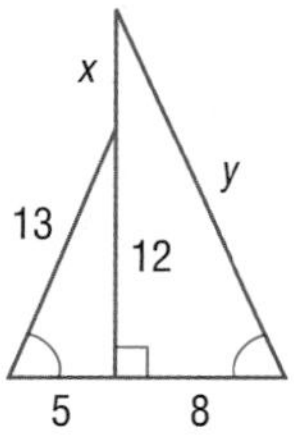

9.

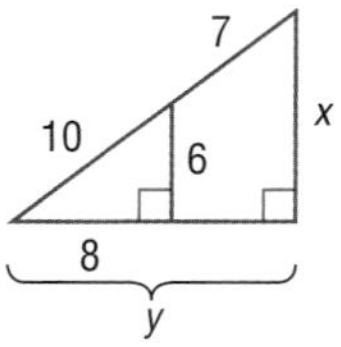

10. **SHADOWS** On a sunny day, Jason measures the length of his shadow and the length of a tree's shadow. Use the figures at the right to find the height of the tree.

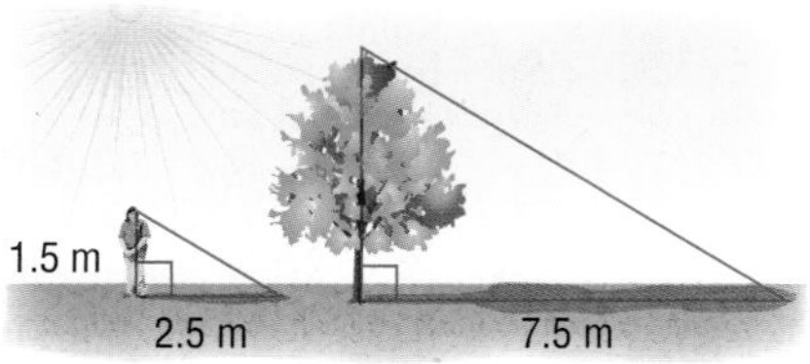

11. **PHOTOGRAPHY** A photo that is 4 inches wide by 6 inches long must be reduced to fit in a space 3 inches wide. How long will the reduced photo be?

12. **SURVEYING** Surveyors use instruments to measure objects that are too large or too far away to measure by hand. They can use the shadows that objects cast to find the height of the objects without measuring them. A surveyor finds that a telephone pole that is 25 feet tall is casting a shadow 20 feet long. A nearby building is casting a shadow 52 feet long. What is the height of the building?

0-7

The Pythagorean Theorem

Objective
- Use the Pythagorean Theorem and its converse.

NY Math Online

glencoe.com
- Extra Examples
- Personal Tutor
- Self-Check Quiz
- Homework Help

The **Pythagorean Theorem** states that in a right triangle, the square of the length of the hypotenuse c is equal to the sum of the squares of the lengths of the legs a and b.

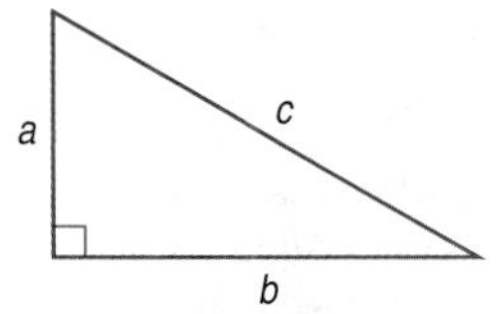

That is, in any right triangle, $c^2 = a^2 + b^2$.

EXAMPLE 1 Find Hypotenuse Measures

Find the length of the hypotenuse of each right triangle.

a.

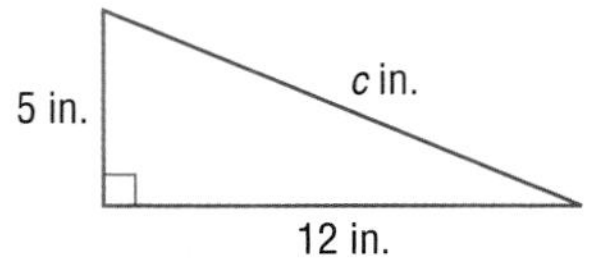

$c^2 = a^2 + b^2$	**Pythagorean Theorem**
$c^2 = 5^2 + 12^2$	**$a = 5$ and $b = 12$**
$c^2 = 25 + 144$	**Simplify.**
$c^2 = 169$	**Add.**
$c = \sqrt{169}$	**Take the square root of each side.**
$c = 13$	**The length of the hypotenuse is 13 inches.**

b.

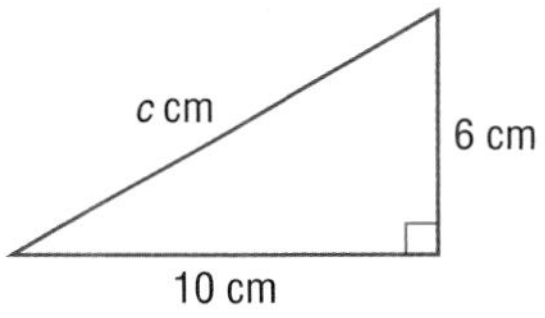

$c^2 = a^2 + b^2$	**Pythagorean Theorem**
$c^2 = 6^2 + 10^2$	**$a = 6$ and $b = 10$**
$c^2 = 36 + 100$	**Simplify.**
$c^2 = 136$	**Add.**
$c^2 = \sqrt{136}$	**Take the square root of each side.**
$c \approx 11.7$	**Use a calculator.**

To the nearest tenth, the length of the hypotenuse is 11.7 centimeters.

You can also find the length of a leg of a right triangle given the lengths of the hypotenuse and the other leg.

EXAMPLE 2 Find Leg Measures

Find the length of the missing leg in each right triangle.

a.

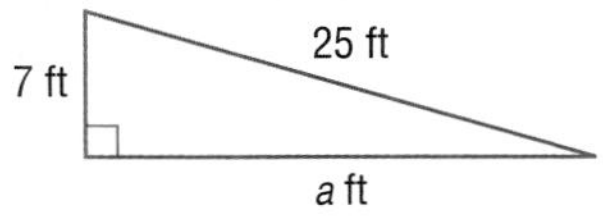

$c^2 = a^2 + b^2$	**Pythagorean Theorem**
$25^2 = a^2 + 7^2$	**$c = 25$ and $b = 7$**
$625 = a^2 + 49$	**Simplify.**
$625 - 49 = a^2 + 49 - 49$	**Subtract 49 from each side.**
$576 = a^2$	**Simplify.**
$\sqrt{576} = a$	**Take the square root of each side.**
$24 = a$	**The length of the leg is 24 feet.**

> **Watch Out!**
>
> **Positive Square Root** When finding the length of a side using the Pythagorean Theorem, use only the positive and not the negative square root, since length cannot be negative.

b.

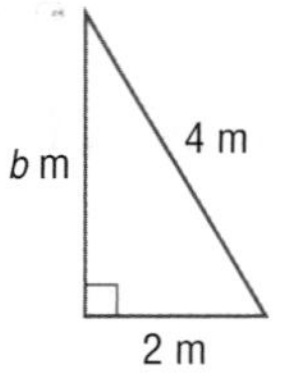

$c^2 = a^2 + b^2$	**Pythagorean Theorem**
$4^2 = 2^2 + b^2$	**$c = 4$ and $a = 2$**
$16 = 4 + b^2$	**Simplify.**
$16 - 4 = 4 - 4 + b^2$	**Subtract 4 from each side.**
$12 = b^2$	**Simplify.**
$\sqrt{12} = b$	**Take the square root of each side.**
$3.5 \approx b$	**Use a calculator.**

To the nearest tenth, the length of the leg is 3.5 meters.

The **converse of the Pythagorean Theorem** states that if the sides of a triangle have lengths a, b, and c, and $c^2 = a^2 + b^2$, then the triangle is a right triangle.

EXAMPLE 3 Identify a Right Triangle

The lengths of the three sides of a triangle are 5, 7, and 9 inches. Determine whether this triangle is a right triangle.

Because the longest side is 9 inches, use 9 as c, the measure of the hypotenuse.

$c^2 = a^2 + b^2$	**Pythagorean Theorem**
$9^2 \stackrel{?}{=} 5^2 + 7^2$	**$c = 9$, $a = 5$, and $b = 7$**
$81 \stackrel{?}{=} 25 + 49$	**Evaluate 9^2, 5^2, and 7^2.**
$81 \neq 74$	**Simplify.**

Because $c^2 \neq a^2 + b^2$, the triangle is *not* a right triangle.

Exercises

Find each missing measure. Round to the nearest tenth, if necessary.

1.

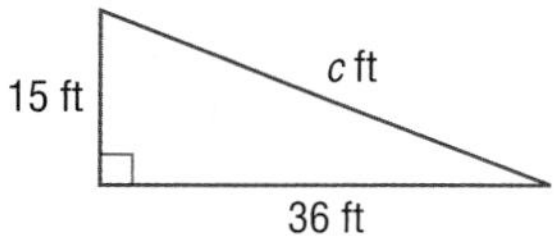

2.

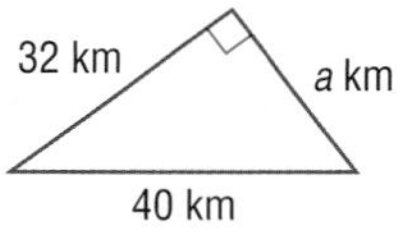

3.

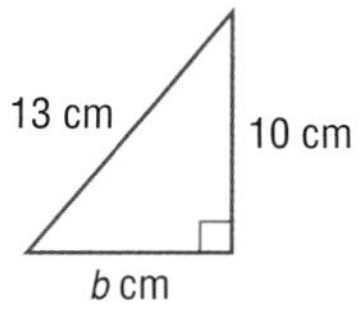

4. $a = 3, b = 4, c = ?$

5. $a = ?, b = 12, c = 13$

6. $a = 14, b = ?, c = 50$

7. $a = 2, b = 9, c = ?$

8. $a = 6, b = ?, c = 13$

9. $a = ?, b = 7, c = 11$

The lengths of three sides of a triangle are given. Determine whether each triangle is a right triangle.

10. 5 in., 7 in., 8 in.

11. 9 m, 12 m, 15 m

12. 6 cm, 7 cm, 12 cm

13. 11 ft, 12 ft, 16 ft

14. 10 yd, 24 yd, 26 yd

15. 11 km, 60 km, 61 km

16. **FLAGPOLES** Mai-Lin wants to find the distance from her feet to the top of the flagpole. If the flagpole is 30 feet tall and Mai-Lin is standing a distance of 15 feet from the flagpole, what is the distance from her feet to the top of the flagpole?

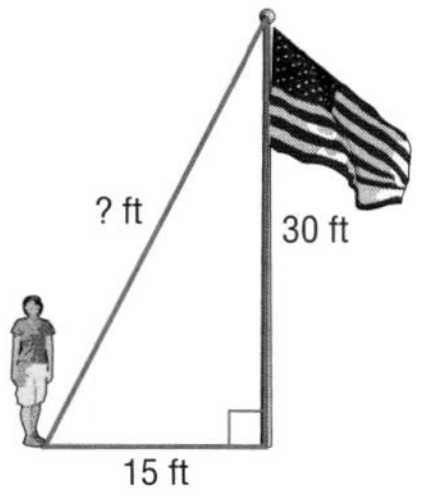

17. **CONSTRUCTION** The walls of a recreation center are being covered with paneling. The doorway into one room is 0.9 meter wide and 2.5 meters high. What is the length of the longest rectangular panel that can be taken through this doorway?

CHAPTER 0 Posttest

State the domain and range of each relation. Then determine whether each relation is a function. Write *yes* or *no*.

1. $\{(4, 5), (5, -1), (0, 12), (0, -2), (7, 9)\}$

2.

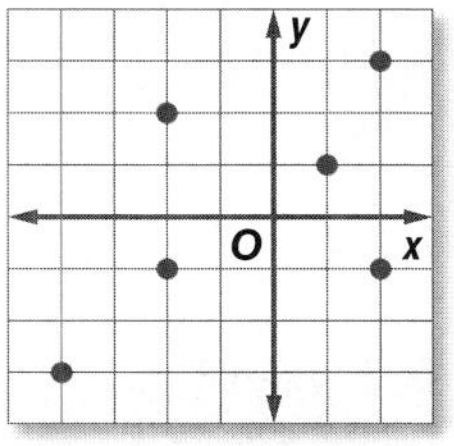

Name the quadrant in which each point is located.

3. $(-3, 7)$

4. $(10, -11)$

5. $(42, 5)$

6. $(-15, 3)$

Find each product.

7. $(2r + 9)(r + 1)$

8. $(4n - 3)(2n + 2)$

9. $(5p - 1)(6p - 10)$

10. $(2r - 5)(r + 5)$

11. $(7x + 4)(7x + 4)$

12. $(3k - 2)(6k + 9)$

13. **GEOMETRY** The height of a rectangle is 3 millimeters less than twice the width.

a. Write an expression for each measure.

b. Write a polynomial expression for the area of the rectangle.

Factor each polynomial.

14. $4x^2 + 4xy + y^2$

15. $25a^2 - 20a + 4$

16. $4a^2 + 16ab + 16b^2$

17. $100n^2 - 1$

18. $81t^2 - 36$

19. $16x^2 - 25y^2$

State whether the events are *independent* or *dependent*.

20. choosing the color and size of a pair of shoes

21. choosing the winner and runner-up at a dog show

22. answering two multiple choice questions

23. answering two matching test questions where each solution is used only once

24. **COLLEGE** For a college application, Macawi must select one of five topics on which to write a short essay. She must also select a different topic from the list for a longer essay. How many ways can she choose the topics of the two essays?

25. **STUDENT COUNCIL** A student council has 6 seniors, 5 juniors, and 1 sophomore as members. In how many ways can a 3-member council committee be formed that includes one member from each class?

Determine whether each situation involves a *permutation* or a *combination*. Then find the number of possibilities.

26. selecting 2 of 8 employees to attend a business seminar

27. forming a team of 12 athletes from a group of 25 who try out

28. Determine whether the rectangles are *similar*, *congruent*, or *neither*.

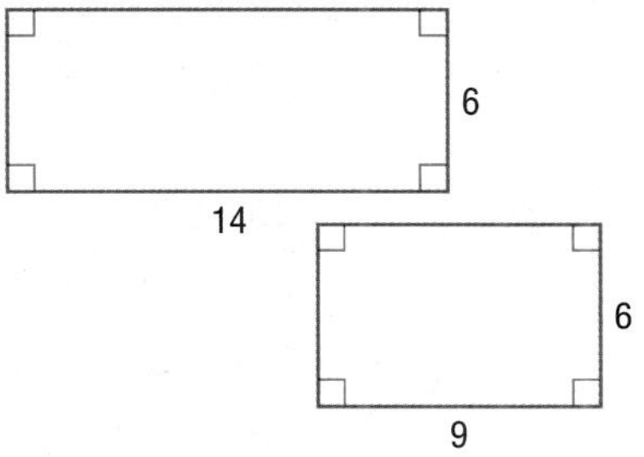

29. **COMPUTERS** A computer image of a painting is 320 pixels wide by 240 pixels high. If the actual painting is 42 inches wide, how high is it?

Find each missing measure. Round to the nearest tenth, if necessary.

30.

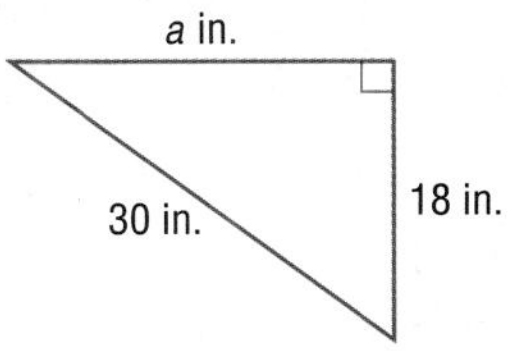

31. $a = 33$ cm, $b = ?$ cm, $c = 45$ cm

The lengths of three sides of a triangle are given. Determine whether each triangle is a right triangle.

32. 6 in., 8 in., 12 in.

33. 30 m, 34 m, 16 m

34. 21 cm, 72 cm, 75 cm

CHAPTER 1

Equations and Inequalities

Then

In Algebra 1, you wrote expressions with variables.

Now

In Chapter 1, you will:

- Simplify and evaluate algebraic expressions.
- Solve linear and absolute value equations.
- Solve and graph inequalities.

NYS Core Curriculum

Reinforcement of
A.N.1 Identify and apply the properties of real numbers
A2.A.1 Solve absolute value equations and inequalities involving linear expressions in one variable

Why?

MONEY Connecting money to mathematics is one of the most practical skills you can learn. As long as you use money, you will be using mathematics. In this chapter, you will explore money topics such as sales tax, income, and budgeting for your first apartment.

Math *in Motion,* Animation glencoe.com

Get Ready for Chapter 1

Diagnose Readiness You have two options for checking Prerequisite Skills.

Text Option

Take the Quick Check below. Refer to the Quick Review for help.

QuickCheck

Simplify.

1. $15.7 + (-3.45)$
2. $-18.54 - (-32.05)$
3. $-9.8 \cdot 6.75$
4. $4 \div (-0.5)$
5. $3\frac{2}{3} + \left(-1\frac{4}{5}\right)$
6. $\frac{54}{7} - \frac{26}{6}$
7. $\left(\frac{6}{5}\right)\left(-\frac{10}{9}\right)$
8. $-3 \div \frac{7}{8}$
9. **CRAFTS** Felisa needs $\frac{7}{8}$ yard of one type of material to make a quilt. How much of this material will she need to make 12 quilts?

Evaluate each power.

10. 6^3
11. $(-4)^3$
12. $-(0.6)^2$
13. $-(-2.5)^3$
14. $\left(\frac{4}{5}\right)^2$
15. $\left(\frac{7}{3}\right)^4$
16. $\left(-\frac{7}{10}\right)^2$
17. $\left(-\frac{15}{2}\right)^3$
18. **FOOD** Nate's Deli offers 3 types of bread, 3 types of meat, and 3 types of cheese. How many different sandwiches can be made with 1 type each of bread, meat, and cheese?

Identify each statement as *true* or *false*.

19. $-6 \geq -7$
20. $8 > -5$
21. $\frac{1}{7} \leq \frac{1}{9}$
22. $\frac{5}{6} \leq \frac{25}{30}$
23. **MEASUREMENT** Christy has a board that is 0.6 yard long. Marissa has a board that is $\frac{2}{3}$ yard long. Marissa states that $\frac{2}{3} > 0.6$. Is she correct?

QuickReview

EXAMPLE 1

Simplify $\left(\frac{3}{16}\right)\left(-\frac{4}{5}\right)$.

$\left(\frac{3}{16}\right)\left(-\frac{4}{5}\right) = -\frac{3(4)}{16(5)}$	**Multiply the numerators and the denominators.**
$= -\frac{12}{80}$	**Simplify.**
$= -\frac{12 \div 4}{80 \div 4}$	**Divide the numerator and denominator by the GCF, 4.**
$= -\frac{3}{20}$	**Simplify.**

EXAMPLE 2

Evaluate $(-1.5)^3$.

$(-1.5)^3 = (-1.5)(-1.5)(-1.5)$	**$(-1.5)^3$ means 1.5 is a factor 3 times.**
$= -3.375$	**Simplify.**

EXAMPLE 3

Identify $\frac{3}{8} > \frac{12}{24}$ as *true* or *false*.

$\frac{3}{8} \overset{?}{>} \frac{12 \div 3}{24 \div 3}$	**Divide 12 and 24 by 3 to get a denominator of 8.**
$\frac{3}{8} \not> \frac{4}{8}$	**Simplify.**

False; $\frac{3}{8} \not> \frac{4}{8}$ because $\frac{3}{8} < \frac{4}{8}$.

Online Option

NY Math Online Take a self-check Chapter Readiness Quiz at glencoe.com.

Get Started on Chapter 1

You will learn several new concepts, skills, and vocabulary terms as you study Chapter 1. To get ready, identify important terms and organize your resources. You may wish to refer to **Chapter 0** to review prerequisite skills.

FOLDABLES® Study Organizer

Equations and Inequalities Make this Foldable to help you organize your Chapter 1 notes about equations and inequalities. Begin with one sheet of 11" × 17" paper.

1 **Fold** 2" tabs on each of the short sides.

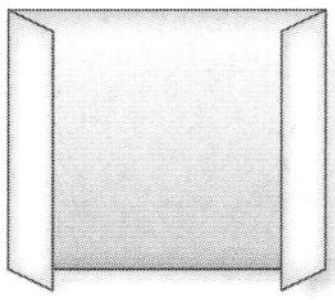

2 **Then** fold in half in both directions. Open and cut as shown.

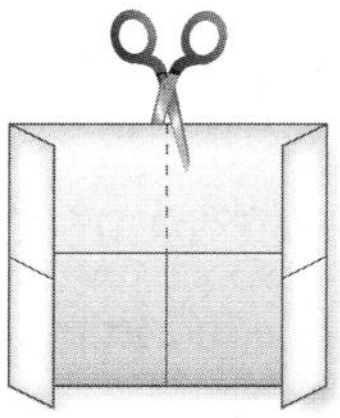

3 **Refold** along the width. Staple each pocket. Label pockets as *Algebraic Expressions, Properties of Real Numbers, Solve Equations,* and *Solve and Graph Inequalities.* Place index cards for notes in each pocket.

New Vocabulary

English		Español
variable	• p. 5 •	variable
algebraic expression	• p. 5 •	expressión algebraica
order of operations	• p. 5 •	orden de las operaciones
formula	• p. 6 •	formula
real numbers	• p. 11 •	números reales
rational numbers	• p. 11 •	números racional
irrational numbers	• p. 11 •	números irracional
integers	• p. 11 •	enteros
whole numbers	• p. 11 •	números enteros
natural numbers	• p. 11 •	números naturales
open sentence	• p. 18 •	enuciado abierto
equation	• p. 18 •	ecuación
solution	• p. 18 •	solución
absolute value	• p. 27 •	valor absolute
empty set	• p. 28 •	conjunto vacío
set-builder notation	• p. 35 •	notación de construcción de conjuntos
compound inequality	• p. 41 •	desigualdad compuesta
intersection	• p. 41 •	intersección
union	• p. 42 •	unión

Review Vocabulary

evaluate • p. 5 • evaluar to find the value of an expression

inequality • p. 33 • desigualdad an open sentence that contains the symbol $<$, \leq, $>$, or \geq

power • p. 6 • potencia an expression of the form x^n, read *x to the nth power*

The *base* is the number that is multiplied.

The *exponent* tells how many times the base is used as a factor.

x^n

The number that can be expressed using an exponent is called a *power*.

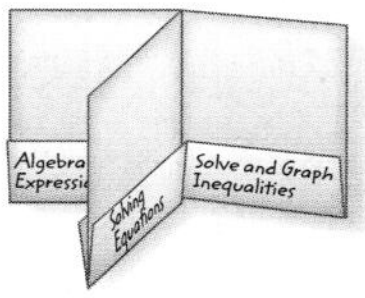

NY Math Online ▷ glencoe.com

- Study the chapter online
- Explore **Math in Motion**
- Get extra help from your own **Personal Tutor**
- Use **Extra Examples** for additional help
- Take a **Self-Check Quiz**
- **Review Vocabulary** in fun ways

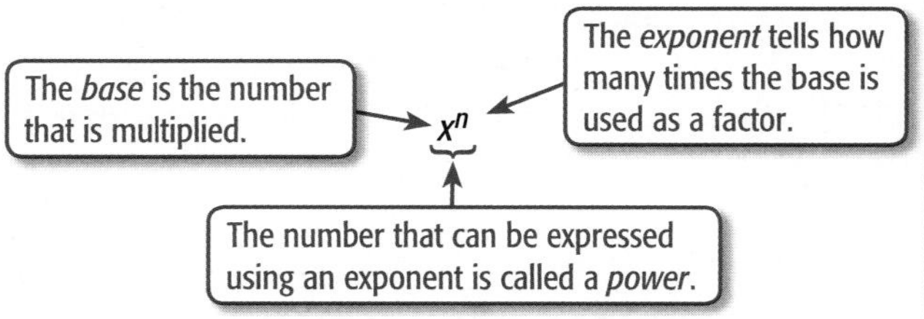

▷ **Multilingual eGlossary glencoe.com**

1-1 Expressions and Formulas

Why?

The following formula can be used to calculate a baseball player's on-base percentage x.

$$x = \frac{h + w + p}{b + w + p + s}$$

- h is the number of hits.
- w is the number of walks.
- p is the number of times the players has been hit by a pitch.
- b is the number of times at bat.
- s is the number of sacrifice flies.

During the first twenty games of a season, Ian has 9 hits, 2 walks, 38 at bats, 5 sacrifice flies, and he is hit by 1 pitch. The expression $\frac{9 + 2 + 1}{38 + 2 + 1 + 5}$ gives Ian's on-base percentage.

Then
You used the rules of exponents. (Lesson 0-3)

Now
- Use the order of operations to evaluate expressions.
- Use formulas.

NYS Core Curriculum
Reinforcement of A.A.1 Translate a quantitative verbal phrase into an algebraic expression
Reinforcement of A.A.2 Write a verbal expression that matches a given mathematical expression

New Vocabulary
variables
algebraic expressions
order of operations
formula

NY Math Online
glencoe.com
- Extra Examples
- Personal Tutor
- Self-Check Quiz
- Homework Help

Order of Operations **Variables** are letters used to represent unknown quantities. Expressions that contain at least one variable are called **algebraic expressions**. You can evaluate an algebraic expression by replacing each variable with a number and then applying the **order of operations**.

Key Concept — Order of Operations (For Your FOLDABLE)

Step 1 Evaluate expressions inside grouping symbols.
Step 2 Evaluate all powers.
Step 3 Multiply and/or divide from left to right.
Step 4 Add and/or subtract from left to right.

EXAMPLE 1 Evaluate Algebraic Expressions

Evaluate $m + (p - 1)^2$ if $m = 3$ and $p = -4$.

$m + (p - 1)^2 = 3 + (-4 - 1)^2$ — Replace m with 3 and p with -4.
$= 3 + (-5)^2$ — Add -4 and -1.
$= 3 + 25$ — Evaluate $(-5)^2$.
$= 28$ — Add 3 and 25.

Check Your Progress

Evaluate each expression if $m = 12$ and $q = -1$.

1A. $m + (3 - q)^2$

1B. $m \div 2q + 4$

Personal Tutor glencoe.com

ReadingMath

Exponents x^2 may be read as *x squared* or *x to the second power*. x^3 may be read as *x cubed* or *x to the third power.*

EXAMPLE 2 Evaluate Algebraic Expressions

a. Evaluate $a + b^2(b - a)$ if $a = 5$ and $b = -3.2$.

$$a + b^2(b - a) = 5 + (-3.2)^2(-3.2 - 5) \quad a = 5 \text{ and } b = -3.2$$
$$= 5 + (-3.2)^2(-8.2) \quad \text{Subtract 5 from } -3.2.$$
$$= 5 + 10.24(-8.2) \quad \text{Evaluate } (-3.2)^2.$$
$$= 5 + (-83.968) \quad \text{Multiply 10.24 and } -8.2.$$
$$= -78.968 \quad \text{Add 5 and } -83.968.$$

b. Evaluate $\frac{x^4 - 3wy}{y^3 + 2w}$ if $w = 4$, $x = -3$, and $y = -5$.

$$\frac{x^4 - 3wy}{y^3 + 2w} = \frac{(-3)^4 - 3(4)(-5)}{(-5)^3 + 2(4)} \quad w = 4, x = -3, \text{ and } y = -5$$
$$= \frac{81 - 3(4)(-5)}{-125 + 2(4)} \quad \text{Evaluate the numerator and denominator separately.}$$
$$= \frac{81 - (-60)}{-125 + 8} \quad \text{Multiply in the numerator and denominator.}$$
$$= \frac{141}{-117} \text{ or } -\frac{47}{39} \quad \text{Simplify the numerator and denominator. Then simplify the fraction.}$$

Check Your Progress

Evaluate each expression if $h = 4$, $j = -1$, and $k = 0.5$.

2A. $h^2k + h(h - k)$ **2B.** $j + (3 - h)^2$ **2C.** $\frac{j^2 - 3h^2k}{j^3 + 2}$

Personal Tutor glencoe.com

StudyTip

Grouping Symbols Remember that in the order of operations, expressions inside of grouping symbols are evaluated first. A fraction bar is one type of grouping symbol. Evaluate the expressions in the numerator and denominator separately before dividing.

Formulas A **formula** is a mathematical sentence that expresses the relationship between certain quantities. If you know the value of every variable in the formula except one, you can find the value of the remaining variable.

Real-World EXAMPLE 3 Use a Formula

TORNADOES The formula for the volume of a cone, $V = \frac{1}{3}\pi r^2 h$, can be used to approximate the volume of a tornado. Find the approximate volume of the tornado at the right.

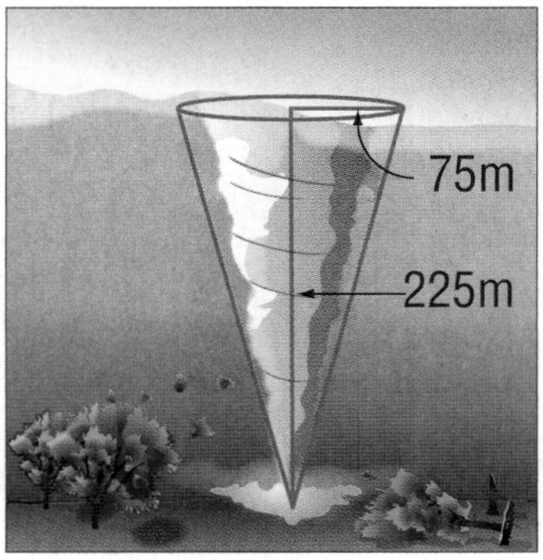

$$V = \frac{1}{3}\pi r^2 h \quad \text{Volume of a cone}$$
$$= \frac{1}{3}\pi(75)^2(225) \quad r = 75 \text{ and } h = 225$$
$$= \frac{1}{3}\pi(5625)(225) \quad \text{Evaluate } 75^2.$$
$$\approx 1{,}325{,}359 \quad \text{Multiply.}$$

The approximate volume of the tornado is about 1,325,359 cubic meters.

Check Your Progress

3. GEOMETRY The formula for the volume V of a rectangular prism is $V = \ell wh$, where ℓ represents the length, w represents the width, and h represents the height. Find the volume of a rectangular prism with a length of 4 feet, a width of 2 feet, and a height of 3.5 feet.

Personal Tutor glencoe.com

Check Your Understanding

Example 1 p. 5

Evaluate each expression if $a = -2$, $b = 3$, and $c = 4.2$.

1. $a - 2b + 3c$

2. $2a + (b + 3)^2$

3. $a + 3[b^2 - (a + c)]$

Example 2 p. 6

4. $5c - 2[(b - a) + c]$

5. $4(2a + 3b) - 2c$

6. $\frac{a^2 + 4c}{3b + 2a}$

7. $\frac{b^3 + ac}{ab + 2bc}$

8. $\frac{3b + 2a}{5 - c}$

9. $\frac{3a - 2c}{4ab}$

Example 3 p. 6

10. **VOLLEYBALL** A player's attack percentage A is calculated using the formula $A = \frac{k - e}{t}$, where k represents the number of kills, e represents the number of attack errors including blocks, and t represents the total attacks attempted. Find the attack percentage given each set of values.

a. $k = 22, e = 11, t = 35$

b. $k = 33, e = 9, t = 50$

Practice and Problem Solving

● = **Step-by-Step Solutions** begin on page R20.
Extra Practice begins on page 947.

Example 1 p. 5

Evaluate each expression if $w = -3$, $x = 4$, $y = 2.6$, and $z = \frac{1}{3}$.

11. $y + x - z$

12. $w - 2x + y \div 2$

13. $4(x - w)$

14. $6(y + x)$

15. $9z - 4y + 2w$

16. $3y - 4z + x$

17. **GAS MILEAGE** The gasoline used by a car is measured in miles per gallon and is related to the distance traveled by the following formula.

miles per gallon × number of gallons = distance traveled

a. During a trip your car used a total of 46.2 gallons of gasoline. If your car gets 33 miles to the gallon, how far did you travel?

b. Your friend has decided to buy a hybrid car that gets 60 miles per gallon. The gasoline tank holds 12 gallons. How far can the car go on one tank of gasoline?

Example 2 p. 6

Evaluate each expression if $a = -4$, $b = -0.8$, $c = 5$, and $d = \frac{1}{5}$.

18. $\frac{a + b}{c - d}$

19. $\frac{a - b}{bd}$

20. $\frac{ac}{d + b}$

21. $\frac{b^2c^2}{ad}$

22. $\frac{b + 6}{4(d + c)}$

23. $\frac{5(d + a)}{2ab^2}$

24. **TEMPERATURE** The formula $C = \frac{5(F - 32)}{9}$ can be used to convert temperatures in degrees Fahrenheit to degrees Celsius.

a. Room temperature commonly ranges from 64°F to 73°F. Determine room temperature range in degrees Celsius.

b. The normal average human body temperature is 98.6°F. A temperature above this indicates a fever. If your temperature is 42°C, do you have a fever? Explain your reasoning.

Example 3 p. 6

25. **GEOMETRY** The formula for the area A of a triangle with height h and base b is $A = \frac{1}{2}bh$. Write an expression to represent the area of the triangle.

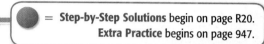

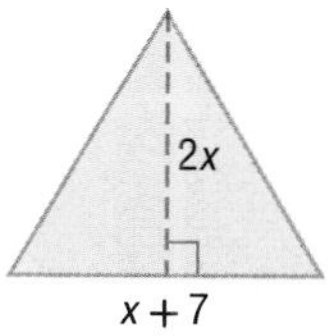

26. **BUSINESS** The profit that a business made during a year is $536,897,000. If the business divides the profit evenly for each share, estimate how much each share made if there are 10,995,000 shares.

27. **EARTH** The radius of Earth's orbit is 93,000,000 miles.

a. Find the circumference of Earth's orbit assuming that the orbit is a circle. The formula for the circumference of a circle is $2\pi r$.

b. Earth travels at a speed of 66,698 miles per hour around the Sun. Use the formula $T = \frac{C}{V}$, where T is time in hours, C is circumference, and V is velocity to find the number of hours it takes Earth to revolve around the sun

c. Did you prove that it takes 1 year for Earth to go around the Sun? Explain.

Real-World Link

Worldwide, over 6000 Web searches are performed every second.

Source: Nielsen Research

28. **ANCIENT PYRAMID** The Great Pyramid in Cairo, Egypt, is approximately 146.7 meters high, and each side of its base is approximately 230 meters.

a. Find the area of the base of the pyramid. Remember $A = \ell w$.

b. The volume of a pyramid is $\frac{1}{3}Bh$, where B is the area of the base and h is the height. What is the volume of the Great Pyramid?

Evaluate each expression if $w = \frac{3}{4}$, $x = 8$, $y = -2$, and $z = 0.4$.

29. $x^3 + 2y^4$

30. $(x - 6z)^2$

31. $2(6w - 2y) - 8z$

32. $\frac{(y + z)^2}{xw}$

33. $\frac{12w - 6y}{z^2}$

34. $\frac{wx + yz}{wx - yz}$

35. **GEOMETRY** The formula for the volume V of a cone with radius r and height h is $V = \frac{1}{3}\pi r^2 h$. Write an expression for the volume of the cone at the right.

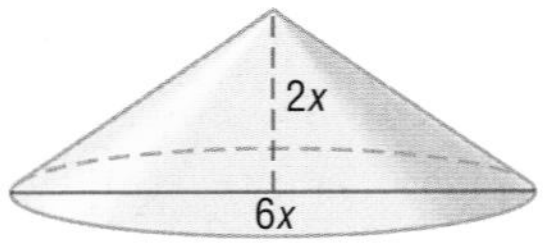

36. **SEARCH ENGINES** Page rank is a numerical value that represents how important a page is on the Web. One formula used to calculate the page rank for a page is $PR = 0.15 + 0.85L$, where L is the page rank of the linking page divided by the number of outbound links on the page. Determine the page rank of a page in which $L = 10$.

37. **WEATHER** In 1898, A.E. Dolbear studied various species of crickets to determine their "chirp rate" based on temperatures. He determined that the formula $t = 50 + \frac{n - 40}{4}$, where n is the number of chirps, could be used to find the temperature t in degrees Fahrenheit. What is the temperature if the number of chirps is 120?

38. **FOOTBALL** The following formula can be used to calculate a quarterback efficiency rating.

$$\left(\frac{\frac{C}{A} - 0.3}{0.2} + \frac{\frac{Y}{A} - 3}{4} + \frac{\frac{T}{A}}{0.05} + \frac{0.095 - \frac{I}{A}}{0.04}\right) \cdot \frac{100}{6}$$

- C is the number of passes completed.
- A is the number of passes attempted.
- Y is passing yardage.
- T is the number of touchdown passes.
- I is the number of interceptions.

Find Byron Leftwich's efficiency rating to the nearest tenth for the season statistics shown.

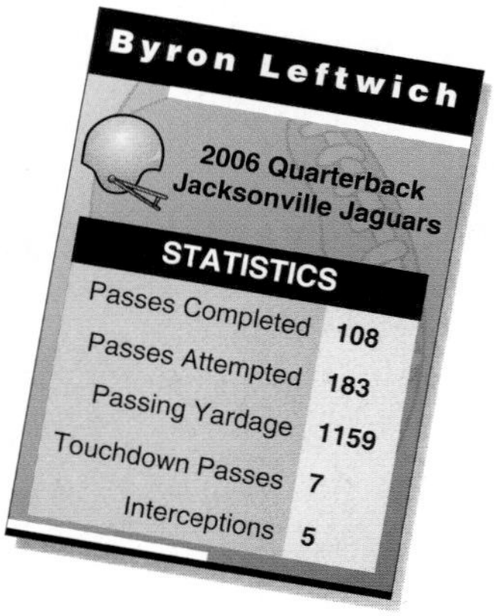

39. **MOVIES** The average price for a movie ticket can be represented by $P = \frac{y^2}{400} + \frac{7y}{100} + 2.96$ where y is the number of years since 1980.

a. Find the average price of a ticket in 1990, 2000, and 2010.

b. Another equation that can be used to represent ticket prices is $P = \frac{y^3}{2500} - \frac{y^2}{100} + \frac{6y}{25} + 2.62$. Find the price of a ticket in 1990, 2000, and 2010. How do these values compare to those you found in part a?

StudyTip

Look Back To review simplifying square root expressions, see Concepts and Skills Bank Lesson 2.

40. **GEOMETRY** The area of a triangle can be found using Heron's Formula, $A = \sqrt{s(s-a)(s-b)(s-c)}$, where a, b, and c are the lengths of the three sides of the triangle, and $s = \frac{a+b+c}{2}$. Find the area of the triangle at the right.

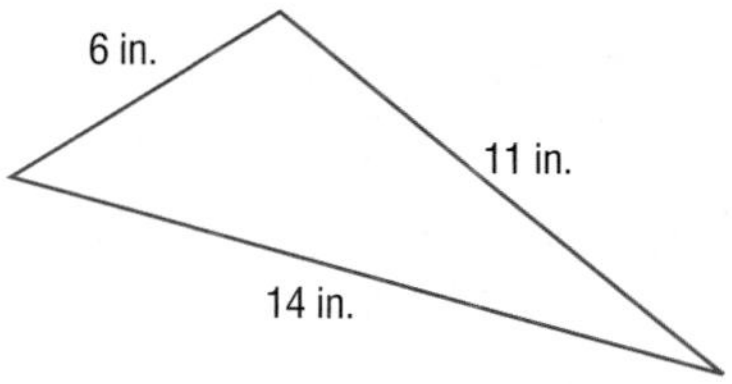

41. Evaluate $y = \sqrt{b^2\left(1 - \frac{x^2}{a^2}\right)}$ if $a = 6$, $b = 8$, and $x = 3$. Round to the nearest tenth.

42. **MULTIPLE REPRESENTATIONS** You will write expressions using the formula for the volume of a cylinder. Recall that the volume of a cylinder can be found using the formula $v = \pi r^2 h$, in which v = volume, r = radius, and h = height.

   a. **GEOMETRIC** Draw two cylinders of different sizes.

   b. **TABULAR** Use a ruler to measure the radius and height of each cylinder. Organize the measures for each cylinder into a table. Include a column in your table to calculate the volume of each cylinder.

   c. **VERBAL** Write a verbal expression for the difference in volume of the two cylinders.

   d. **ALGEBRAIC** Write and solve an algebraic expression for the difference in volume of the two cylinders.

H.O.T. Problems

Use Higher-Order Thinking Skills

43. **FIND THE ERROR** Lauren and Rico are evaluating $\frac{-3d-4c}{2ab}$ for $a = -2$, $b = -3$, $c = 5$, and $d = 4$. Is either of them correct? Explain your reasoning.

Lauren

$\frac{-3d-4c}{2ab} = \frac{-3(4)-4(5)}{2(-2)(-3)}$

$= \frac{-12-20}{12} = \frac{-32}{12} = -\frac{8}{3}$

Rico

$\frac{-3d-4c}{2ab} = \frac{-3(4)-4(5)}{2(-2)(-3)}$

$= \frac{-12-20}{12} = \frac{8}{12} = \frac{2}{3}$

44. **CHALLENGE** For any three distinct numbers a, b, and c, $a\$b\$c$ is defined as $a\$b\$c = \frac{-a-b-c}{c-b-a}$. Find $-2\$(-4)\$5$.

45. **REASONING** Explain the steps involved in finding the value of k such that $4\left(\frac{k+6}{3} - 12\right) + 8 = -48$.

46. **CHALLENGE** Let m, n, p, and q represent nonzero positive integers. Find a number in terms of m, n, p, and q that is halfway between $\frac{m}{n}$ and $\frac{p}{q}$.

47. **OPEN ENDED** Write an algebraic expression using $x = -2$, $y = -3$, and $z = 4$ and all four operations for which the value of the expression is 10.

48. **WRITING IN MATH** Provide an example of a formula used in everyday situations. Explain the usefulness of this formula and what happens if the formula is not used correctly.

49. **WRITING IN MATH** Use the information for on-base percentage given at the beginning of the lesson to explain how formulas are used in baseball to calculate a player's stats. Explain why a formula for on-base percentage is more useful than a table of specific percentages.

NYSRE PRACTICE Reinforcement of A.A.1, A.A.2

50. SAT/ACT If the area of a square with side x is 9, what is the area of a square of side $4x$?

A 36
B 144
C 212
D 324

51. SHORT RESPONSE A coffee shop owner wants to open a second shop when his daily customer average reaches 800 people. He has calculated the daily customer average in the table below for each month since he has opened.

Month	Daily Customer Average
1	225
2	298
3	371
4	444

If the trend continues, during what month can he open a second shop?

52. GEOMETRY In $\triangle DFG$, $\overline{FH}$ and $\overline{HG}$ are angle bisectors and $m\angle D = 84$. How many degrees are in $\angle FHG$?

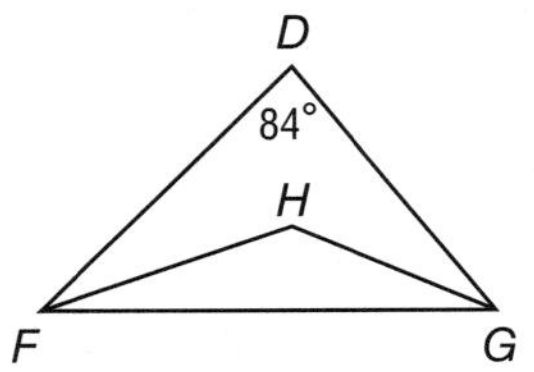

F 96
G 132
H 145
J 192

53. A skydiver free-falls from a height of 3000 m at a rate of 55 meters per second. Which equation can be used to find h, the height of the skydiver after t seconds of free fall?

A $h = -55t - 3000$
B $h = -55t + 3000$
C $h = 3000t - 55$
D $h = 3000t + 55$

Spiral Review

54. The lengths of the three sides of a triangle are 10, 14, and 18 inches. Determine whether this triangle is a right triangle. (Lesson 0-7)

55. The legs of a right triangle measure 6 centimeters and 8 centimeters. Find the length of the hypotenuse. (Lesson 0-7)

56. PHOTOGRAPHY A photo that is 4 inches wide by 6 inches long must be reduced to fit in a space that is 3 inches wide. What will be the length of the reduced photo? (Lesson 0-6)

57. Factor $6x^2 + 12x$. (Lesson 0-3)

58. Find the product of $(a + 2)(a - 4)$. (Lesson 0-2)

59. NUMBER An integer is 2 less than a number, and another integer is 1 greater than double that same number. What are the two integers? (Lesson 0-2)

Skills Review

Evaluate each expression. (Concepts and Skills Bank, Lesson 2)

60. $\sqrt{4}$
61. $\sqrt{25}$
62. $\sqrt{81}$
63. $\sqrt{121}$
64. $-\sqrt{9}$
65. $-\sqrt{16}$
66. $\sqrt{\frac{49}{100}}$
67. $\sqrt{\frac{25}{64}}$

1-2 Properties of Real Numbers

Then

You identified and used the arithmetic properties of real numbers. (Algebra 1)

Now

- Classify real numbers.
- Use the properties of real numbers to evaluate expressions.

NYS Core Curriculum

Reinforcement of A.N.1 Identify and apply the properties of real numbers (closure, commutative, associative, distributive, identity, inverse)
Reinforcement of A.A.1 Translate a quantitative verbal phrase into an algebraic expression

New Vocabulary

real numbers
rational numbers
irrational numbers
integers
whole numbers
natural numbers

NY Math Online

glencoe.com

- Extra Examples
- Personal Tutor
- Self-Check Quiz
- Homework Help
- Math in Motion

Why?

The Central High School Boosters sell snacks and beverages at school functions. The items are priced the same to make determining the total cost easy.

You can use the Distributive Property to calculate the total cost when multiple items are purchased.

Real Numbers **Real numbers** consist of several different kinds of numbers.

- **Rational numbers** can be expressed as a ratio $\frac{a}{b}$, where a and b are integers and b is not zero. The decimal form of a rational number is either a terminating or repeating decimal.
- The decimal form of an **irrational number** neither terminates nor repeats. Square roots of numbers that are not perfect squares are irrational numbers.
- The sets of **integers**, $\{\ldots, -3, -2, -1, 0, 1, 2, 3, \ldots\}$, **whole numbers**, $\{0, 1, 2, 3, 4, \ldots\}$, and **natural numbers**, $\{1, 2, 3, 4, 5, \ldots\}$, are subsets of the rational numbers. These numbers are subsets of the rational numbers because every integer n is equal to $\frac{n}{1}$.

Key Concept Real Numbers (R) — For Your FOLDABLE

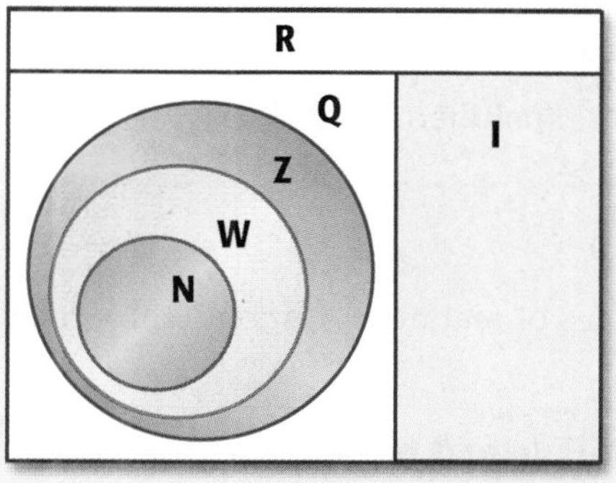

Letter	Set	Examples
Q	rationals	$0.125, -\frac{7}{8}, \frac{2}{3} = 0.66\ldots$
I	irrationals	$\pi = 3.14159\ldots$ $\sqrt{3} = 1.73205\ldots$
Z	integers	$-5, 17, -23, 8$
W	wholes	$2, 96, 0, \sqrt{36}$
N	naturals	$3, 17, 6, 86$

EXAMPLE 1 Classify Numbers

Name the sets of numbers to which each number belongs.

a. -23 integers (Z), rationals (Q), reals (R)

b. $\sqrt{50}$ irrationals (I), reals (R)

c. $-\frac{4}{9}$ rationals (Q), reals (R)

Check Your Progress

1A. -185 **1B.** $-\sqrt{49}$ **1C.** $\sqrt{95}$ **1D.** $-\frac{7}{8}$

Personal Tutor glencoe.com

StudyTip

Real Numbers A number can belong to more than one set of numbers. For example, if a number is natural, it is also whole, an integer, rational, and real.

Properties of Real Numbers Some of the properties of real numbers are summarized below.

Concept Summary **Real Number Properties** For Your FOLDABLE

For any real numbers a, b, and c:

Property	Addition	Multiplication
Commutative	$a + b = b + a$	$a \cdot b = b \cdot a$
Associative	$(a + b) + c = a + (b + c)$	$(a \cdot b) \cdot c = a \cdot (b \cdot c)$
Identity	$a + 0 = a = 0 + a$	$a \cdot 1 = a = 1 \cdot a$
Inverse	$a + (-a) = 0 = (-a) + a$	$a \cdot \frac{1}{a} = 1 = \frac{1}{a} \cdot a, a \neq 0$
Closure	$a + b$ is a real number.	$a \cdot b$ is a real number.
Distributive	$a(b + c) = ab + ac$ and $(b + c)a = ba + ca$	

Math *in Motion,* Animation glencoe.com

EXAMPLE 2 **Name Properties of Real Numbers**

Name the property illustrated by $5 \cdot (4 \cdot 13) = (5 \cdot 4) \cdot 13$.

Associative Property of Multiplication

The Associative Property of Multiplication states that the way in which you group factors does not affect the product.

Check Your Progress

2. Name the property illustrated by $2(x + 3) = 2x + 6$.

Personal Tutor glencoe.com

You can use the properties of real numbers to identify related values.

StudyTip

Additive and Multiplicative Inverses The additive inverse of a number has the opposite sign as the number. The multiplicative inverse of a number has the same sign as the number.

EXAMPLE 3 **Additive and Multiplicative Inverses**

Find the additive inverse and multiplicative inverse for $-\frac{5}{8}$.

Since $-\frac{5}{8} + \frac{5}{8} = 0$, the additive inverse of $-\frac{5}{8}$ is $\frac{5}{8}$.

Since $\left(-\frac{5}{8}\right)\left(-\frac{8}{5}\right) = 1$, the multiplicative inverse of $-\frac{5}{8}$ is $-\frac{8}{5}$.

Check Your Progress

Find the additive and multiplicative inverse for each number.

3A. 1.25 **3B.** $2\frac{1}{2}$

Personal Tutor glencoe.com

Many real-world applications involve working with real numbers.

Real-World EXAMPLE 4 Distributive Property

MONEY The prices of the components of a computer package offered by Computer Depot are shown in the table. If a 6% sales tax is added to the purchase price, how much sales tax is charged for this computer package?

Component	Price ($)
Computer	359.95
Monitor	219.99
Printer	79.00
Digital Camera	149.50
Software Bundle	99.00

There are two ways to determine the total sales tax.

Method 1 Multiply, then add.

Multiply each dollar amount by 6% or 0.06 and then add.

$T = 0.06(359.95) + 0.06(219.99) + 0.06(79.00) + 0.06(149.50) + 0.06(99.00)$

$= 21.60 + 13.20 + 4.74 + 8.97 + 5.94$

$= 54.45$

Method 2 Add, then multiply.

Find the total cost of the computer package, and then multiply the total by 0.06.

$T = 0.06(359.95 + 219.99 + 79.00 + 149.50 + 99.00)$

$= 0.06(907.44)$

$= 54.45$

The sales tax charged is $54.45p. Notice that both methods result in the same answer.

Check Your Progress

4. JOBS Kayla makes $8 per hour working at a grocery store. The number of hours Kayla worked each day in one week are 3, 2.5, 2, 1, and 4. How much money did Kayla earn this week?

Personal Tutor glencoe.com

Real-World Career

Retail Store Manager
Store managers are responsible for the day-to-day operations of a retail store. A store manager may range from being a high school graduate to having a 4-year degree, depending on the business.

The properties of real numbers can be used to simplify algebraic expressions.

EXAMPLE 5 Simplify an Expression

Simplify $3(2q + r) + 5(4q - 7r)$.

$3(2q + r) + 5(4q - 7r)$

$= 3(2q) + 3(r) + 5(4q) - 5(7r)$	**Distributive Property**
$= 6q + 3r + 20q - 35r$	**Multiply.**
$= 6q + 20q + 3r - 35r$	**Commutative Property (+)**
$= (6 + 20)q + (3 - 35)r$	**Distributive Property**
$= 26q - 32r$	**Simplify.**

Check Your Progress

5. Simplify $3(4x - 2y) - 2(3x + y)$.

Personal Tutor glencoe.com

Check Your Understanding

Example 1 p. 11

Name the sets of numbers to which each number belongs.

1. 62 **2.** $\frac{5}{4}$ **3.** $\sqrt{11}$ **4.** -12

Example 2 p. 12

Name the property illustrated by each equation.

5. $(6 \cdot 8) \cdot 5 = 6 \cdot (8 \cdot 5)$

6. $7(9 - 5) = 7 \cdot 9 - 7 \cdot 5$

7. $84 + 16 = 16 + 84$

8. $(12 + 5)6 = 12 \cdot 6 + 5 \cdot 6$

Example 3 p. 12

Find the additive inverse and multiplicative inverse for each number.

9. -7 **10.** $\frac{4}{9}$ **11.** 3.8 **12.** $\sqrt{5}$

Example 4 p. 13

13. MONEY Melba is mowing lawns for \$22 each to earn money for a video game console that costs \$550.

a. Write an expression to represent the total amount of money Melba earned during this week.

b. Evaluate the expression from part a by using the Distributive Property.

c. When do you think Melba will earn enough for the video game console? Is this reasonable? Explain.

Lawns Mowed in One Week

Day	Lawns Mowed
Monday	2
Tuesday	4
Wednesday	3
Thursday	1
Friday	5
Saturday	6
Sunday	7

Example 5 p. 13

Simplify each expression.

14. $5(3x + 6y) + 4(2x - 9y)$

15. $6(6a + 5b) - 3(4a + 7b)$

16. $-4(6c - 3d) - 5(-2c - 4d)$

17. $-5(8x - 2y) - 4(-6x - 3y)$

Practice and Problem Solving

● = **Step-by-Step Solutions** begins on page R20.
Extra Practice begins on page 947.

Example 1 p. 11

Name the sets of numbers to which each number belongs.

18. $-\frac{4}{3}$ **19.** -8.13 **20.** $\sqrt{25}$ **21.** $0.\overline{61}$

22. $\frac{9}{3}$ **23.** $-\sqrt{144}$ **24.** $\frac{21}{7}$ **25.** $\sqrt{17}$

Example 2 p. 12

Name the property illustrated by each equation.

26. $-7y + 7y = 0$

27. $8\sqrt{11} + 5\sqrt{11} = (8 + 5)\sqrt{11}$

28. $(16 + 7) + 23 = 16 + (7 + 23)$

29. $\left(\frac{22}{7}\right)\left(\frac{7}{22}\right) = 1$

Example 3 p. 12

Find the additive inverse and multiplicative inverse for each number.

30. -8 **31.** 12.1 **32.** -0.25

33. $\frac{6}{13}$ **34.** $-\frac{3}{8}$ **35.** $\sqrt{15}$

Example 4 p. 13

36. CONSTRUCTION Jorge needs two different kinds of concrete: quick drying and slow drying. The quick-drying concrete mix calls for $2\frac{1}{2}$ pounds of dry cement, and the slow-drying concrete mix calls for $1\frac{1}{4}$ pounds of dry cement. He needs 5 times more quick-drying concrete and 3 times more slow-drying concrete than the mixes make.

a. How many pounds of dry cement mix will he need?

b. Use the properties of real numbers to show how Jorge could compute this amount mentally. Justify each step.

Example 5 p. 13

Simplify each expression.

37. $8b - 3c + 4b + 9c$

38. $-2a + 9d - 5a - 6d$

39. $4(4x - 9y) + 8(3x + 2y)$

40. $6(9a - 3b) - 8(2a + 4b)$

41. $-2(-5g + 6k) - 9(-2g + 4k)$

42. $-5(10x + 8z) - 6(4x - 7z)$

43. FOOTBALL Illustrate the Distributive Property by writing two expressions for the area of a college football field. Then find the area of the football field.

44. PETS The chart shows the percent of dogs registered with the American Kennel Club that are of the eight most popular breeds.

a. Illustrate the Distributive Property by writing two expressions to represent the number of registered dogs of the top four breeds.

b. Evaluate the expressions you wrote to find the number of registered dogs of the top four breeds.

Top Dogs

Breed	Percent of Registered Dogs
Labrador Retrievers	14.2
Yorkshire Terriers	5.6
German Shepherds	5.0
Golden Retrievers	4.9
Beagles	4.5
Dachshunds	4.1
Boxers	4.1
Poodles	3.4
Total Registered Dogs	**870,192**

Source: American Kennel Club

45 **MONEY** Billie is given \$20 in lunch money by her parents once every two weeks. On some days, she packs her lunch, and on other days, she buys her lunch. A hot lunch from the cafeteria costs \$4.50, and a cold sandwich from the lunch line costs \$2.

a. Billie decides that she wants to buy a hot lunch on Thursday and Friday of the first week and on Wednesday of the second week. Use the Distributive Property to determine how much that will cost.

b. How many cold sandwiches can Billie buy with the amount left over?

c. Assuming that both weeks are Monday through Friday, how many times will Billie have to pack her lunch?

Real-World Link

The National School Lunch Program provides nutritionally balanced lunches to children each school day in more than 101,000 schools and residential childcare institutions.

Source: USDA

Simplify each expression.

46. $\frac{1}{3}(5x + 8y) + \frac{1}{4}(6x - 2y)$

47. $\frac{2}{5}(6c - 8d) + \frac{3}{4}(4c - 9d)$

48. $-6(3a + 5b) - 3(6a - 8c)$

49. $-9(3x + 8y) - 3(5x + 10z)$

50. DECORATING Mary is making curtains out of the same fabric for 5 windows. The two larger windows are the same size, and the three smaller windows are the same size. One larger window requires $3\frac{3}{4}$ yards of fabric, and one smaller window needs $2\frac{1}{3}$ yards of fabric.

a. How many yards of material will Mary need?

b. Use the properties of real numbers to show how Mary could compute this amount mentally.

51. **MULTIPLE REPRESENTATIONS** Consider the following real numbers.

$$-\sqrt{6}, 3, \frac{-15}{3}, 4.1, \pi, 0, \frac{3}{8}, \sqrt{36}$$

a. **TABULAR** Organize the numbers into a table according to the sets of numbers to which each belongs.

b. **ALGEBRAIC** Convert each number to decimal form. Then list the numbers from least to greatest.

c. **GRAPHICAL** Graph the numbers on a number line.

d. **VERBAL** Make a conjecture about using decimal form to list real numbers in order.

52. **CLOTHING** A department store sells shirts for \$12.50 each. Dalila buys 2, Latisha buys 3, and Pilar buys 1.

a. Illustrate the Distributive Property by writing two expressions to represent the cost of these shirts.

b. Use the Distributive Property to find how much money the store received from selling these shirts.

Real-World Link

The average teen in the United States spends \$3400 each year, 42% of which is for clothing.

Source: U.S. Department of Commerce

H.O.T. Problems

Use Higher-Order Thinking Skills

53. **WHICH ONE DOESN'T BELONG?** Identify the number that does not belong with the other three. Explain your reasoning.

$\sqrt{21}$	$\sqrt{35}$	$\sqrt{67}$	$\sqrt{81}$

54. **CHALLENGE** If $12(5r + 6t) = w$, then in terms of w, what is $48(30r + 36t)$?

55. **FIND THE ERROR** Luna and Sophia are simplifying $4(14a - 10b) - 6(b + 4a)$. Is either of them correct? Explain your reasoning.

Luna	Sophia
$4(14a - 10b) - 6(b + 4a)$	$4(14a - 10b) - 6(b + 4a)$
$56a - 40b - 6b + 24a$	$56a - 40b - 6a - 24b$
$80a - 46b$	$50a - 64b$

56. **REASONING** Determine whether the following statement is *sometimes*, *always*, or *never* true. Explain your reasoning.

An irrational number is a real number underneath a radical sign.

57. **OPEN ENDED** Determine whether the Closure Property of Multiplication applies to irrational numbers. If not, provide a counterexample.

OPEN ENDED The set of all real numbers is *dense*, meaning between any two distinct members of the set there lies infinitely many other members of the set. Find an example of (a) a rational number, and (b) an irrational number between the given numbers.

58. 2.45 and 2.5

59. π and $\frac{10}{3}$

60. $1.\overline{9}$ and 2.01

61. **WRITING IN MATH** Explain and provide examples to show why the Commutative Property does not hold true for subtraction or division.

NYSRE PRACTICE Reinforcement of A.N.1, A.A.1

62. EXTENDED RESPONSE Lenora bought several pounds of cashews and several pounds of almonds for a party. The cashews cost \$8 per pound, and the almonds cost \$6 per pound. Lenora bought a total of 7 pounds and paid a total of \$48. Write and solve equations to determine the pounds of cashews and the pounds of almonds that Lenora purchased.

63. SAT/ACT Find the 10th term in the series 2, 4, 7, 11, 16, ...

A 46
B 56
C 67
D 72

64. GEOMETRY What are the coordinates of point A in the parallelogram?

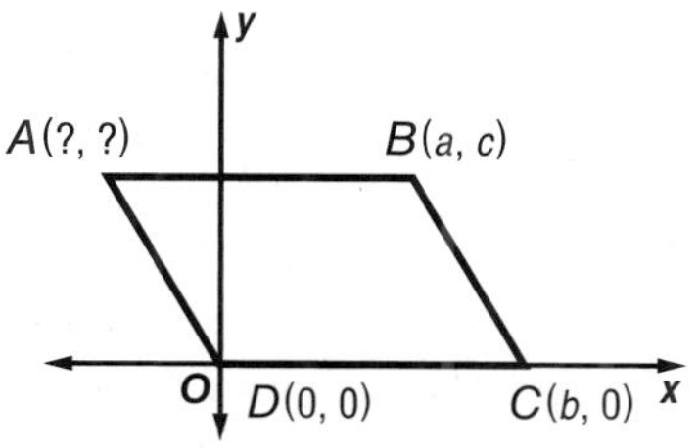

F $(b - a, c)$
G $(a - b, c)$
H (b, c)
J (c, c)

65. What is the domain of the function that contains the points $(-3, 0)$, $(0, 4)$, $(-2, 5)$, and $(6, 4)$?

A $\{-3, 6\}$
B $\{-3, -2, 0, 6\}$
C $\{0, 4, 5, 6\}$
D $\{-3, -2, 0, 4, 5, 6\}$

Spiral Review

66. Evaluate $8(4 - 2)^3$. (Lesson 1-1)

67. Evaluate $a + 3(b + c) - d$, if $a = 5$, $b = 4$, $c = 3$, and $d = 2$. (Lesson 1-1)

68. GEOMETRY The formula for the area A of a circle with diameter d is $A = \pi\left(\frac{d}{2}\right)^2$. Write an expression to represent the area of the circle. (Lesson 1-1)

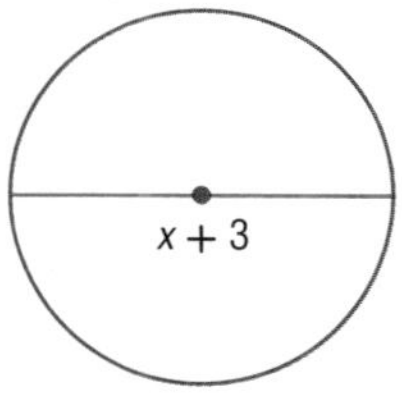

69. CONSTRUCTION The walls of a recreation center are being covered with paneling. The doorway into one room is 0.9 meter wide and 2.5 meters high. What is the width of the widest rectangular panel that can be taken through this doorway? (Lesson 0-7)

Factor each polynomial. (Lesson 0-3)

70. $14x^2 + 10x - 8$
71. $9x^2 - 3x + 18$
72. $8x^2 + 16x + 12$
73. $10x^2 - 20x$
74. $7x^2 - 14x - 21$
75. $12x^2 - 18x - 24$

Find each product. (Lesson 0-2)

76. $(x + 2)(x - 3)$
77. $(y + 2)(y - 1)$
78. $(a - 5)(a + 4)$
79. $(b - 7)(b - 3)$
80. $(n + 6)(n + 8)$
81. $(p - 9)(p + 1)$

Skills Review

Evaluate each expression if $a = 3$, $b = \frac{2}{3}$, and $c = -1.7$. (Lesson 1-1)

82. $6b - 5$
83. $\frac{1}{6}b + 1$
84. $2.3c - 7$
85. $-8(a - 4)$
86. $a + b + c$
87. $\frac{a \cdot b}{c}$
88. $a^2 - c$
89. $\frac{a \cdot c}{a}$

1-3 Solving Equations

Then

You used properties of real numbers to evaluate expressions. (Lesson 1-2)

Now

- Translate verbal expressions into algebraic expressions and equations, and vice versa.
- Solve equations using the properties of equality.

NYS Core Curriculum

A2.CM.11 Represent word problems using standard mathematical notation

New Vocabulary

open sentence
equation
solution

NY Math Online

glencoe.com

- Extra Examples
- Personal Tutor
- Self-Check Quiz
- Homework Help

Why?

The United States is one of the few countries in the world that measures distances in miles. When traveling by car in different countries, it is often useful to convert miles to kilometers. To find the approximate number of kilometers k in miles m, divide the number of miles by 0.62137.

$$m \text{ miles} \times \frac{1 \text{ kilometer}}{0.62137 \text{ mile}} \approx k \text{ kilometers}$$

$$\frac{m}{0.62137} \approx k \text{ kilometers}$$

Verbal Expressions and Algebraic Expressions Verbal expressions can be translated into algebraic expressions by using the language of algebra.

EXAMPLE 1 Verbal to Algebraic Expression

Write an algebraic expression to represent each verbal expression.

a. 2 more than 4 times the cube of a number $4x^3 + 2$

b. the quotient of 5 less than a number and 12 $\frac{n-5}{12}$

Check Your Progress

1A. the cube of a number increased by 4 times the same number

1B. three times the difference of a number and 8

Personal Tutor glencoe.com

A mathematical sentence containing one or more variables is called an **open sentence**. A mathematical sentence stating that two mathematical expressions are equal is called an **equation**.

EXAMPLE 2 Algebraic to Verbal Sentence

Write a verbal sentence to represent each equation.

a. $6x = 72$ The product of 6 and a number is 72.

b. $n + 15 = 91$ The sum of a number and 15 is ninety-one.

Check Your Progress

2A. $g - 5 = -2$

2B. $2c = c^2 - 4$

Personal Tutor glencoe.com

Open sentences are neither true nor false until the variables have been replaced by numbers. Each replacement that results in a true sentence is called a **solution** of the open sentence.

Math History Link

Diophantus of Alexandria (c. 200–284) Diophantus was famous for his work in algebra. His main work was titled *Arithmetica* and introduced symbolism to Greek algebra as well as propositions in number theory and polygonal numbers.

Properties of Equality To solve equations, we can use properties of equality. Some of these properties are listed below.

Key Concept — Properties of Equality

For Your FOLDABLE

Property	Symbols	Examples
Reflexive	For any real number a, $a = a$.	$b + 12 = b + 12$
Symmetric	For all real numbers a and b, if $a = b$, then $b = a$.	If $18 = -2n + 4$, then $-2n + 4 = 18$.
Transitive	For all real numbers a, b, and c, if $a = b$ and $b = c$, then $a = c$.	If $5p + 3 = 48$ and $48 = 7p - 15$, then $5p + 3 = 7p - 15$.
Substitution	If $a = b$, then a may be replaced by b and b may be replaced by a.	If $(6 + 1)x = 21$, then $7x = 21$.

EXAMPLE 3 Identify Properties of Equality

Name the property illustrated by each statement.

a. **If $3a - 4 = b$, and $b = a + 17$, then $3a - 4 = a + 17$.**
Transitive Property of Equality

b. **If $2g - h = 62$, and $h = 24$, then $2g - 24 = 62$.**
Substitution Property of Equality

Check Your Progress

3. If $-11a + 2 = -3a$, then $-3a = -11a + 2$.

Personal Tutor glencoe.com

StudyTip

Checking Answers When solving for a variable, you can use substitution to check your answer by replacing the variable in the original equation with your answer.

To solve most equations, you will need to perform the same operation on each side of the equals sign. The properties of equality allow for the equation to be solved in this way.

Key Concept

For Your FOLDABLE

Addition and Subtraction Properties of Equality

Symbols For any real numbers, a, b, and c, if $a = b$, then $a + c = b + c$ and $a - c = b - c$.

Examples If $x - 6 = 14$, then $x - 6 + 6 = 14 + 6$.
If $n + 5 = -32$, then $n + 5 - 5 = -32 - 5$.

Multiplication and Division Properties of Equality

Symbols For any real numbers, a, b, and c, if $a = b$, then $a \cdot c = b \cdot c$ and, if $c \neq 0$, $\frac{a}{c} = \frac{b}{c}$.

Examples If $\frac{m}{8} = -7$, then $8 \cdot \frac{m}{8} = 8 \cdot (-7)$.
If $-2y = 12$, then $\frac{-2y}{-2} = \frac{12}{-2}$.

EXAMPLE 4 Solve One-Step Equations

Solve each equation. Check your solution.

a. $n - 3.24 = 42.1$

$n - 3.24 = 42.1$	**Original equation**
$n - 3.24 + 3.24 = 42.1 + 3.24$	**Add 3.24 to each side.**
$n = 45.34$	**Simplify.**

The solution is 45.34.

CHECK		
	$n - 3.24 = 42.1$	**Original equation**
	$45.34 - 3.24 \stackrel{?}{=} 42.1$	**Substitute 45.34 for *n*.**
	$42.1 = 42.1$ ✔	**Simplify.**

b. $-\frac{5}{8}x = 20$

$-\frac{5}{8}x = 20$	**Original equation**
$-\frac{8}{5}\left(-\frac{5}{8}\right)x = -\frac{8}{5}(20)$	**Multiply each side by $-\frac{8}{5}$.**
$x = -32$	**Simplify.**

The solution is −32.

CHECK		
	$-\frac{5}{8}x = 20$	**Original equation**
	$-\frac{5}{8}(-32) \stackrel{?}{=} 20$	**Replace *x* with −32.**
	$20 = 20$ ✔	**Simplify.**

StudyTip

Multiplication and Division Properties of Equality Example 4b could also have been solved using the Division Property of Equality. Note that dividing each side of the equation by $-\frac{5}{8}$ is the same as multiplying each side by $-\frac{8}{5}$.

Check Your Progress

4A. $x - 14.29 = 25$

4B. $\frac{2}{3}y = -18$

Personal Tutor glencoe.com

To solve an equation with more than one operation, undo operations by working backward.

EXAMPLE 5 Solve a Multi-Step Equation

Solve $5(x + 3) + 2(1 - x) = 14$.

$5(x + 3) + 2(1 - x) = 14$	**Original equation**
$5x + 15 + 2 - 2x = 14$	**Apply the Distributive Property.**
$3x + 17 = 14$	**Simplify the left side.**
$3x = -3$	**Subtract 17 from each side.**
$x = -1$	**Divide each side by 3.**

Check Your Progress

Solve each equation.

5A. $-10x + 3(4x - 2) = 6$

5B. $2(2x - 1) - 4(3x + 1) = 2$

Personal Tutor glencoe.com

You can use properties to solve an equation for a variable.

EXAMPLE 6 Solve for a Variable

GEOMETRY The formula for the area A of a trapezoid is $A = \frac{1}{2}h(b_1 + b_2)$, where h represents the height, and b_1 and b_2 represent the measures of the bases. Solve the formula for b_2.

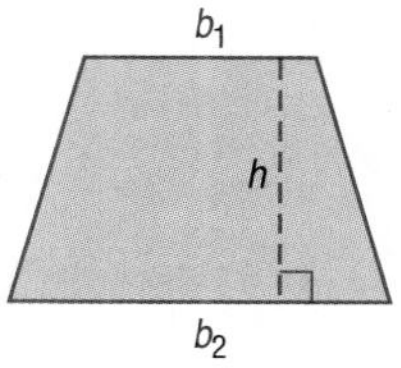

$A = \frac{1}{2}h(b_1 + b_2)$	**Area formula**
$2A = 2\left[\frac{1}{2}h(b_1 + b_2)\right]$	**Multiply each side by 2.**
$2A = h(b_1 + b_2)$	**Simplify.**
$\frac{2A}{h} = \frac{h(b_1 + b_2)}{h}$	**Divide each side by *h*.**
$\frac{2A}{h} = b_1 + b_2$	**Simplify.**
$\frac{2A}{h} - b_1 = b_1 + b_2 - b_1$	**Subtract b_1 from each side.**
$\frac{2A}{h} - b_1 = b_2$	**Simplify.**

Check Your Progress

6. The formula for the surface area S of a cylinder is $S = 2\pi r^2 + 2\pi rh$, where r is the radius of the base and h is the height of the cylinder. Solve the formula for h.

Personal Tutor glencoe.com

NYSRE EXAMPLE 7 Reinforcement of A.A.6, A.A.22

If $6x - 12 = 18$, what is the value of $6x + 5$?

A 5 B 11 C 35 D 41

Read the Test Item

You are asked to find the value of $6x + 5$. Note that you do not have to find the value of x. Instead, you can use the Addition Property of Equality to make the left side of the equation $6x + 5$.

Solve the Test Item

$6x - 12 = 18$	**Original equation**
$6x - 12 + 17 = 18 + 17$	**Add 17 to each side because $-12 + 17 = 5$.**
$6x + 5 = 35$	**Simplify.**

The answer is C.

Test-Taking Tip

Using Properties There are often many ways to solve a problem. Using the properties of equality can help you find a simpler way.

Check Your Progress

7. If $5y + 2 = \frac{8}{3}$, what is the value of $5y - 6$?

F $\frac{-20}{3}$ G $\frac{-16}{3}$ H $\frac{16}{3}$ J $\frac{32}{3}$

Personal Tutor glencoe.com

Check Your Understanding

Example 1 p. 18

Write an algebraic expression to represent each verbal expression.

1. the product of 12 and the sum of a number and negative 3
2. the difference between the product of 4 and a number and the square of the number

Example 2 p. 18

Write a verbal sentence to represent each equation.

3. $5x + 7 = 18$
4. $x^2 - 9 = 27$
5. $5y - y^3 = 12$
6. $\frac{x}{4} + 8 = -16$

Example 3 p. 19

Name the property illustrated by each statement.

7. $(8x - 3) + 12 = (8x - 3) + 12$
8. If $a = -3$ and $-3 = d$, then $a = d$.

Examples 4 and 5 p. 20

Solve each equation. Check your solution.

9. $z - 19 = 34$
10. $x + 13 = 7$
11. $-y = 8$
12. $-6x = 42$
13. $5x - 3 = -33$
14. $-6y - 8 = 16$
15. $3(2a + 3) - 4(3a - 6) = 15$
16. $5(c - 8) - 3(2c + 12) = -84$
17. $-3(-2x + 20) + 8(x + 12) = 92$
18. $-4(3m - 10) - 6(-7m - 6) = -74$

Example 6 p. 21

Solve each equation or formula for the specified variable.

19. $8r - 5q = 3$, for q
20. $Pv = nrt$, for n

Example 7 p. 21

21. **MULTIPLE CHOICE** If $\frac{y}{5} + 8 = 7$, what is the value of $\frac{y}{5} - 2$?

A -10 **B** -3 **C** 1 **D** 5

Practice and Problem Solving

● = **Step-by-Step Solutions** begin on page R20.
Extra Practice begins on page 947.

Example 1 p. 18

Write an algebraic expression to represent each verbal expression.

22. the difference between the product of four and a number and 6
23. the product of the square of a number and 8
24. fifteen less than the cube of a number
25. five more than the quotient of a number and 4

Example 2 p. 18

Write a verbal sentence to represent each equation.

26. $8x - 4 = 16$
27. $\frac{x + 3}{4} = 5$
28. $4y^2 - 3 = 13$

29. **BASEBALL** During a recent season, Miguel Cabrera and Mike Jacobs of the Florida Marlins hit a combined total of 46 home runs. Cabrera hit 6 more home runs than Jacobs. How many home runs did each player hit? Define a variable, write an equation, and solve the problem.

Example 3 p. 19

Name the property illustrated by each statement.

30. If $x + 9 = 2$, then $x + 9 - 9 = 2 - 9$
31. If $y = -3$, then $7y = 7(-3)$
32. If $g = 3h$ and $3h = 16$, then $g = 16$
33. If $-y = 13$, then $-(-y) = -13$

34. **MONEY** Aiko and Kendra arrive at the state fair with $32.50. What is the total number of rides they can go on if they each pay the entrance fee?

Examples 4 and 5 p. 20

Solve each equation. Check your solution.

35. $3y + 4 = 19$

36. $-9x - 8 = 55$

37. $7y - 2y + 4 + 3y = -20$

38. $5g + 18 - 7g + 4g = 8$

39. $5(-2x - 4) - 3(4x + 5) = 97$

40. $-2(3y - 6) + 4(5y - 8) = 92$

41. $\frac{2}{3}(6c - 18) + \frac{3}{4}(8c + 32) = -18$

42. $\frac{3}{5}(15d + 20) - \frac{1}{6}(18d - 12) = 38$

43. **GEOMETRY** The perimeter of a regular pentagon is 100 inches. Find the length of each side.

44. **MEDICINE** For Nina's illness her doctor gives her a prescription for 28 pills. The doctor says that she should take 4 pills the first day and then 2 pills each day until her prescription runs out. For how many days does she take 2 pills?

Example 6 p. 21

Solve each equation or formula for the specified variable.

45. $E = mc^2$, for m

46. $c(a + b) - d = f$, for a

47. $z = \pi q^3 h$, for h

48. $\frac{x + y}{z} - a = b$, for y

49. $y = ax^2 + bx + c$, for a

50. $wx + yz = bc$, for z

51. **GEOMETRY** The formula for the volume of a cylinder with radius r and height h is π times the radius times the radius times the height.

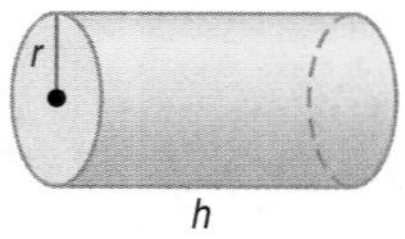

a. Write this as an algebraic expression.

b. Solve the expression in part a for h.

52. **AWARDS BANQUET** A banquet room can seat a maximum of 69 people. The coach, principal, and vice principal have invited the award-winning girls' tennis team to the banquet. If the tennis team consists of 22 girls, how many guests can each student bring?

Solve each equation. Check your solution.

53. $5x - 9 = 11x + 3$

54. $\frac{1}{x} + \frac{1}{4} = \frac{7}{12}$

55. $5.4(3k - 12) + 3.2(2k + 6) = -136$

56. $8.2p - 33.4 = 1.7 - 3.5p$

57. $\frac{4}{9}y + 5 = -\frac{7}{9}y - 8$

58. $\frac{3}{4}z - \frac{1}{3} = \frac{2}{3}z + \frac{1}{5}$

59. **MONEY** Benjamin spent $10,734 on his living expenses last year. Most of these expenses are listed at the right. Benjamin's only other expense last year was rent. If he paid rent 12 times last year, how much is Benjamin's rent each month?

Expense	Annual Cost
Electric	$622
Gas	$428
Water	$240
Renter's Insurance	$144

60. **BRIDGES** The Sunshine Skyway Bridge spans Tampa Bay, Florida. Suppose one crew began building south from St. Petersburg, and another crew began building north from Bradenton. The two crews met 10,560 feet south of St. Petersburg approximately 5 years after construction began.

a. Suppose the St. Petersburg crew built an average of 176 feet per month. Together the two crews built 21,120 feet of bridge. Determine the average number of feet built per month by the Bradenton crew.

b. About how many miles of bridge did each crew build?

c. Is this answer reasonable? Explain.

Real-World Link

The Milliau Viaduct, located in southern France, is the world's tallest vehicular bridge. The bridge stands 1122 feet tall, 1.5 miles long, and 4 lanes wide.

Source: National Geographic Society

61. **MULTIPLE REPRESENTATIONS** The absolute value of a number describes the distance of the number from zero.

a. **GEOMETRIC** Draw a number line. Label the integers from −5 to 5.

b. **TABULAR** Create a table of the integers on the number line and their distance from zero.

c. **GRAPHICAL** Make a graph of each integer x and its distance from zero y using the data points in the table.

d. **VERBAL** Make a conjecture about the integer and its distance from zero. Explain the reason for any changes in sign.

H.O.T. Problems Use Higher-Order Thinking Skills

62. **FIND THE ERROR** Steven and Jade are solving $A = \frac{1}{2}h(b_1 + b_2)$ for b_2. Is either of them correct? Explain your reasoning.

Steven

$$A = \frac{1}{2}h(b_1 + b_2)$$
$$\frac{2A}{h} = (b_1 + b_2)$$
$$\frac{2A - b_1}{h} = b_2$$

Jade

$$A = \frac{1}{2}h(b_1 + b_2)$$
$$\frac{2A}{h} = (b_1 + b_2)$$
$$\frac{2A}{h} - b_1 = b_2$$

63. **CHALLENGE** Solve $d = \sqrt{(x_2 - x_1)^2 + (y_2 - y_1)^2}$ for y_1.

64. **REASONING** Use what you have learned in this lesson to explain why the following number trick works.

- Take any number.
- Multiply it by ten.
- Subtract 30 from the result.
- Divide the new result by 5.
- Add 6 to the result.
- Your new number is twice your original.

65. **OPEN ENDED** Provide one example of an equation involving the Distributive Property that has no solution and another example that has infinitely many solutions.

66. **WRITING IN MATH** Compare and contrast the Substitution Property of Equality and the Transitive Property of Equality.

NYSRE PRACTICE A2.CM.11

67. The graph shows the solution of which inequality?

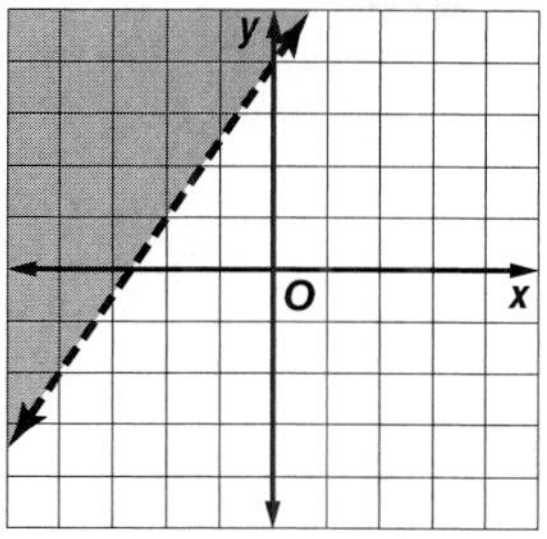

A $y < \frac{2}{3}x + 4$ **C** $y < \frac{3}{2}x + 4$

B $y > \frac{2}{3}x + 4$ **D** $y > \frac{3}{2}x + 4$

68. SAT/ACT What is $1\frac{1}{3}$ subtracted from its reciprocal?

F $-\frac{7}{12}$ **H** $\frac{1}{4}$

G $-\frac{1}{12}$ **J** $\frac{3}{4}$

69. GEOMETRY Which of the following describes the transformation of $\triangle ABC$ to $\triangle A'B'C'$?

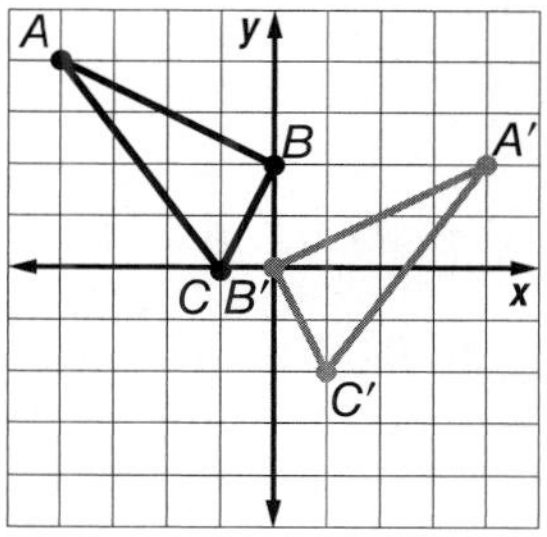

A a reflection across the y-axis and a translation down 2 units

B a reflection across the x-axis and a translation down 2 units

C a rotation 90° to the right and a translation down 2 units

D a rotation 90° to the right and a translation right 2 units

70. SHORT RESPONSE A local theater sold 1200 tickets during the opening weekend of a movie. On the following weekend, 840 tickets were sold. What was the percent decrease of tickets sold?

Spiral Review

71. Simplify $3x + 8y + 5z - 2y - 6x + z$. (Lesson 1-2)

72. BAKING Tamera is making two types of bread. The first type of bread needs $2\frac{1}{2}$ cups of flour, and the second needs $1\frac{3}{4}$ cups of flour. Tamera wants to make 2 loaves of the first recipe and 3 loaves of the second recipe. How many cups of flour does she need? (Lesson 1-2)

73. LANDMARKS Suppose the Space Needle in Seattle, Washington, casts a 220-foot shadow at the same time a nearby tourist casts a 2-foot shadow. If the tourist is $5\frac{1}{2}$ feet tall, how tall is the Space Needle? (Lesson 0-6)

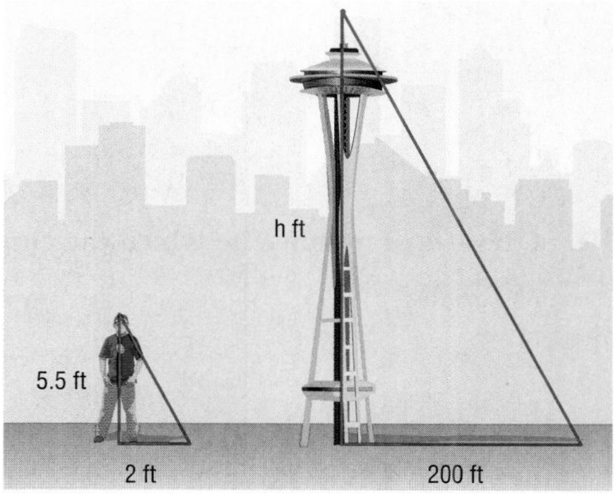

74. Evaluate $a - [c(b - a)]$, if $a = 5$, $b = 7$, and $c = 2$. (Lesson 1-1)

Skills Review

Identify the additive inverse for each number or expression. (Lesson 1-2)

75. $-4\frac{1}{5}$

76. 3.5

77. $-2x$

78. $6 - 7y$

79. $3\frac{2}{3}$

80. -1.25

81. $5x$

82. $4 - 9x$

Mid-Chapter Quiz

Lessons 1-1 through 1-3

NYS Core Curriculum
Reinforcement of A.A.1, A.A.6

1. Evaluate $3c - 4(a + b)$ if $a = -1$, $b = 2$ and $c = \frac{1}{3}$. (Lesson 1-1)

2. **TRAVEL** The distance that Maurice traveled in 2.5 hours riding his bicycle can be found by using the formula $d = rt$, where d is the distance traveled, r is the rate, and t is the time. How far did Maurice travel if he traveled at a rate of 16 miles per hour? (Lesson 1-1)

3. Evaluate $(5 - m)^3 + n(m - n)$ if $m = 6$ and $n = -3$. (Lesson 1-1)

4. **GEOMETRY** The formula for the surface area of the rectangular prism below is given by the formula $S = 2xy + 2yz + 2xz$. What is the surface area of the prism if $x = 2.2$, $y = 3.5$, and $z = 5.1$? (Lesson 1-1)

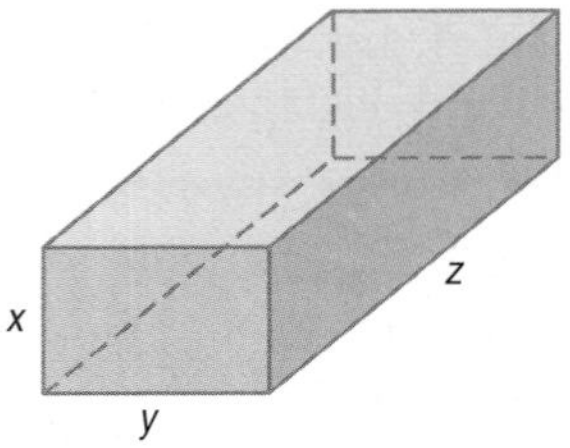

5. **MULTIPLE CHOICE** What is the value of $\frac{q^2 + rt}{qr - 2t}$ if $q = -4$, $r = 3$, and $t = 8$? (Lesson 1-1)

   A $-\frac{17}{6}$

   B $-\frac{1}{6}$

   C $-\frac{10}{7}$

   D $-\frac{2}{7}$

Name the sets of numbers to which each number belongs. (Lesson 1-2)

6. $\frac{25}{11}$
7. $-\frac{128}{32}$
8. $\sqrt{50}$
9. -32.4

10. What is the property illustrated by the equation $(4 + 15)7 = 4 \cdot 7 + 15 \cdot 7$? (Lesson 1-2)

11. Simplify $-3(7a - 4b) + 2(-3a + b)$. (Lesson 1-2)

12. **CLOTHES** Brittany is buying T-shirts and jeans for her new job. T-shirts cost \$10.50, and jeans cost \$26.50. She buys 3 T-shirts and 3 pairs of jeans. Illustrate the Distributive Property by writing two expressions representing how much Brittany spent. (Lesson 1-2)

13. **MULTIPLE CHOICE** Which expression is equivalent to $\frac{2}{3}(4m - 5n) + \frac{1}{5}(2m + n)$? (Lesson 1-2)

    F $\frac{46}{15}m - \frac{47}{15}n$

    G $46m - 47n$

    H $-\frac{mn}{15}$

    J $\frac{5}{4}m - \frac{9}{8}n$

14. Identify the additive inverse and the multiplicative inverse for $\frac{7}{6}$. (Lesson 1-2)

15. Write a verbal sentence to represent the equation $\frac{a}{a - 3} = 1$. (Lesson 1-3)

16. Solve $6x + 4y = -1$ for x. (Lesson 1-3)

17. **MULTIPLE CHOICE** Which algebraic expression represents the verbal expression, *the product of 4 and the difference of a number and 13*? (Lesson 1-3)

    A $4n - 13$

    B $4(n - 13)$

    C $\frac{4}{n - 13}$

    D $\frac{4n}{13}$

18. Solve $-3(6x + 5) + 2(4x) = 20$. (Lesson 1-3)

19. What is the height of the trapezoid below? (Lesson 1-3)

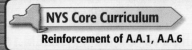

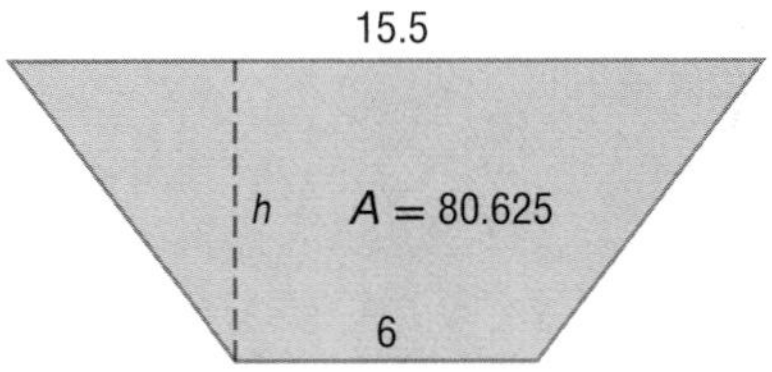

20. **GEOMETRY** The formula for the surface area of a sphere is $SA = 4\pi r^2$, and the formula for the volume of a sphere is $V = \frac{4}{3}\pi r^3$. (Lesson 1-3)

    a. Find the volume and surface area of a sphere with radius 2 inches. Write your answers in terms of π.

    b. Is it possible for a sphere to have the same numerical value for the surface area and volume? If so, find the radius of such a sphere.

1-4 Solving Absolute Value Equations

Then

You solved equations using properties of equality. (Lesson 1-3)

Now

- Evaluate expressions involving absolute values.
- Solve absolute value equations.

NYS Core Curriculum

A2.A.1 Solve absolute value equations and inequalities **involving linear expressions in one variable**

New Vocabulary

absolute value
empty set
extraneous solution

NY Math Online

glencoe.com
- Extra Examples
- Personal Tutor
- Self-Check Quiz
- Homework Help

Why?

Sailors sometimes use a laser range finder to determine distances. Suppose one such range finder is accurate to within ±0.5 yard. This means that if a sailor estimating the distance to shore reads 323.1 yards on the laser range finder, the distance to shore might actually be as close as 322.6 or as far away as 323.6 yards. These extremes can be described by the equation $|E - 323.1| = 0.5$.

Absolute Value Expressions The **absolute value** of a number is its distance from 0 on the number line. Since distance is nonnegative, the absolute value of a number is always nonnegative. The symbol $|x|$ is used to represent the absolute value of a number x.

Key Concept — Absolute Value

For Your FOLDABLE

Words For any real number a, if a is positive or zero, the absolute value of a is a. If a is negative, the absolute value of a is the opposite of a.

Symbols For any real number a, $|a| = a$ if $a \geq 0$, and $|a| = -a$ if $a < 0$.

Model $|-4| = 4$ and $|4| = 4$

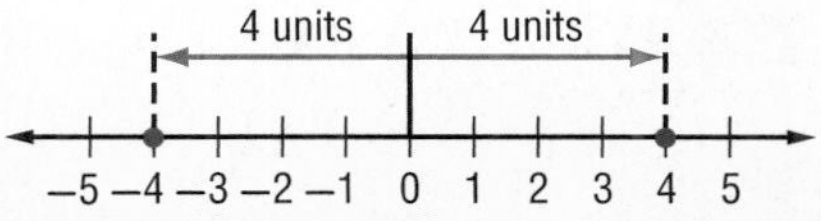

When evaluating expressions, absolute value bars act as a grouping symbol. Perform any operations inside the absolute value bars first.

EXAMPLE 1 Evaluate an Expression with Absolute Value

Evaluate $8.4 - |2n + 5|$ if $n = -7.5$.

$8.4 - |2n + 5| = 8.4 - |2(-7.5) + 5|$	**Replace n with -7.5.**
$= 8.4 - |-15 + 5|$	**Multiply 2 and -7.5.**
$= 8.4 - |-10|$	**Add -15 and 5.**
$= 8.4 - 10$	**$|-10| = 10$**
$= -1.6$	**Subtract 10 from 8.4.**

Check Your Progress

1A. Evaluate $|4x + 3| - 3\frac{1}{2}$ if $x = -2$.

1B. Evaluate $1\frac{1}{3} - |2y + 1|$ if $y = -\frac{2}{3}$.

Personal Tutor glencoe.com

Absolute Value Equations Some equations contain absolute value expressions. The definition of absolute value is used in solving these equations. For any real numbers a and b, where $b \geq 0$, if $|a| = b$, then $a = b$ or $-a = b$. This second case is often written as $a = -b$.

Problem-Solving Tip

Write an Equation Frequently, the best way to solve a problem is to use the given information to write and solve an equation.

EXAMPLE 2 Solve an Absolute Value Equation

TENNIS A standard adult tennis racket has a 100-square-inch head, plus or minus 20 square inches. Write and solve an absolute value equation to determine the least and greatest possible sizes for the head of an adult tennis racket.

Understand We need to determine the greatest and least possible sizes for the head of a tennis racket given the middle size and the range in sizes.

Plan When writing an absolute value equation, the middle or *central* value is always placed inside the absolute value symbols. The *range* is always placed on the other side of the equality symbol.

central value → c; range → r

$$|x - c| = r$$

Solve $|x - c| = r$ **Absolute value equation**

$|x - 100| = 20$ **$c = 100$, and $r = 20$**

Case 1 $a = b$	**Case 2** $a = -b$
$x - 100 = 20$	$x - 100 = -20$
$x - 100 + 100 = 20 + 100$	$x - 100 + 100 = -20 + 100$
$x = 120$	$x = 80$

Check

$\|x - 100\| = 20$	$\|x - 100\| = 20$
$\|120 - 100\| \stackrel{?}{=} 20$	$\|80 - 100\| \stackrel{?}{=} 20$
$\|20\| \stackrel{?}{=} 20$	$\|-20\| \stackrel{?}{=} 20$
$20 = 20$ ✔	$20 = 20$ ✔

On a number line, you can see that both solutions are 20 units away from 100.

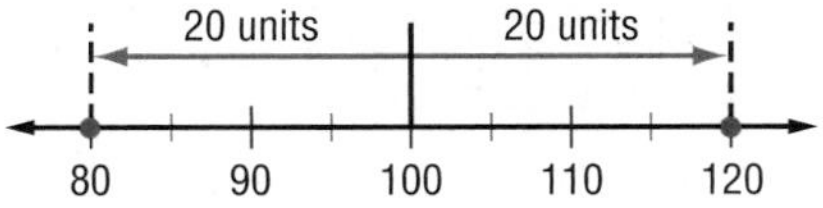

The solutions are 120 and 80. The greatest size is 120 square inches and the least is 80 square inches.

Real-World Link

Originally, players used leather gloves to hit tennis balls. Soon after, the glove was placed at the end of a stick to extend the reach of the "hand."

Source: The Cliff Richard Tennis Foundation

Check Your Progress

Solve each equation. Check your solutions.

2A. $9 = |x + 12|$

2B. $8 = |y + 5|$

Personal Tutor glencoe.com

Because the absolute value of a number is always positive or zero, an equation like $|x| = -4$ is never true. Thus, it has no solution. The solution set for this type of equation is the **empty set**, symbolized by { } or ∅.

EXAMPLE 3 No Solution

Solve $|3x - 2| + 8 = 1$.

$\|3x - 2\| + 8 = 1$	**Original equation**
$\|3x - 2\| + 8 - 8 = 1 - 8$	**Subtract 8 from each side.**
$\|3x - 2\| = -7$	**Simplify.**

This sentence is *never* true. The solution set is $\varnothing$.

✓ Check Your Progress

Solve each equation. Check your solutions.

3A. $-2|3a| = 6$

3B. $|4b + 1| + 8 = 0$

Personal Tutor glencoe.com

It is important to check your answers when solving absolute value equations. Even if the correct procedure for solving the equation is used, the answers may not be actual solutions to the original equation. Such a number is called an **extraneous solution**.

StudyTip

Absolute Value It is possible for an absolute value equation to have only one solution. Remember to set up two cases. Then check your solutions.

EXAMPLE 4 One Solution

Solve $|x + 10| = 4x - 8$. Check your solutions.

Case 1 $a = b$

$$x + 10 = 4x - 8$$
$$10 = 3x - 8$$
$$18 = 3x$$
$$6 = x$$

Case 2 $a = -b$

$$x + 10 = -(4x - 8)$$
$$x + 10 = -4x + 8$$
$$5x + 10 = 8$$
$$5x = -2$$
$$x = -\frac{2}{5}$$

There appear to be two solutions, 6 and $-\frac{2}{5}$.

CHECK Substitute each value in the original equation.

$$|x + 10| = 4x - 8$$
$$|6 + 10| \stackrel{?}{=} 4(6) - 8$$
$$|16| \stackrel{?}{=} 24 - 8$$
$$16 = 16 \checkmark$$

$$|x + 10| = 4x - 8$$
$$\left|-\frac{2}{5} + 10\right| \stackrel{?}{=} 4\left(-\frac{2}{5}\right) - 8$$
$$\left|9\frac{3}{5}\right| \stackrel{?}{=} -1\frac{3}{5} - 8$$
$$9\frac{3}{5} \neq -9\frac{3}{5} \times$$

Because $9\frac{3}{5} \neq -9\frac{3}{5}$, the only solution is 6. The solution set is $\{6\}$.

✓ Check Your Progress

Solve each equation. Check your solutions.

4A. $2|x + 1| - x = 3x - 4$

4B. $3|2x + 2| - 2x = x + 3$

Personal Tutor glencoe.com

Check Your Understanding

Example 1 p. 27

Evaluate each expression if $x = -4$ and $y = -9$.

1. $|x - 8|$
2. $|7y|$
3. $-3|xy|$
4. $-2|3x + 8| - 4$

5. **FISH** Most freshwater tropical fish thrive if the water is within 2°F of 78°F.

   a. Write an equation to determine the least and greatest optimal temperatures.
   b. Solve the equation you wrote in part a.
   c. If your aquarium's thermometer is accurate to within plus or minus 1°F, what should the temperature of the water be to ensure that it reaches the minimum temperature? Explain.

Examples 2–4 pp. 28, 29

Solve each equation. Check your solutions.

6. $|x + 8| = 12$
7. $|y - 4| = 11$
8. $|a - 5| + 4 = 9$
9. $|b - 3| + 8 = 3$
10. $3|2x - 3| - 5 = 4$
11. $-2|5y - 1| = -10$
12. $|a - 4| = 3a - 6$
13. $|b + 5| = 2b + 3$

Practice and Problem Solving

● = **Step-by-Step Solutions** begins on page R20.
Extra Practice begins on page 947.

Example 1 p. 27

Evaluate each expression if $a = -3$, $b = -5$, and $c = 4.2$.

14. $|-3c|$
15. $|5b|$
16. $|a - b|$
17. $|b - c|$
18. $|3b - 4a|$
19. $2|4a - 3c|$
20. $-|3c - a|$
21. $-|abc|$

22. **FOOD** To make cocoa powder, cocoa beans are roasted. The ideal temperature for roasting is 300°F, plus or minus 25°. Write and solve an equation describing the maximum and minimum roasting temperatures for cocoa beans.

Examples 2–4 pp. 28, 29

Solve each equation. Check your solutions.

23. $|z - 13| = 21$
24. $|w + 9| = 17$
25. $9 = |d + 5|$
26. $35 = |x - 6|$
27. $5|q + 6| = 20$
28. $-3|r + 4| = -21$
29. $3|2a - 4| = 0$
30. $8|5w - 1| = 0$
31. $2|3x - 4| + 8 = 6$
32. $4|7y + 2| - 8 = -7$
33. $-3|3t - 2| - 12 = -6$
34. $-5|3z + 8| - 5 = -20$

35. **MONEY** The U.S. Mint produces quarters that weigh about 5.67 grams each. After the quarters are produced, a machine weighs them. If the quarter weighs 0.02 gram more or less than the desired weight, the quarter is rejected. Write and solve an equation to find the heaviest and lightest quarters the machine will approve.

Evaluate each expression if $q = -8$, $r = -6$, and $t = 3$.

36. $12 - t|3r + 2|$
37. $2q + |2rt + q|$
38. $-5t - q|8r - t|$

Real-World Link

During the past century, sea levels along the Mid-Atlantic and Gulf coasts have risen about 6 inches more than the global average.

Source: Environmental Protection Agency

Solve each equation. Check your solutions.

39. $8x = 2|6x - 2|$

40. $-6y + 4 = |4y + 12|$

41. $8z + 20 = -|2z + 4|$

42. $-3y - 2 = |6y + 25|$

43. **SEA LEVEL** Florida is on average 100 feet above sea level. This level varies by as much as 245 feet depending on precipitation and your location. Write and solve an equation describing the maximum and minimum sea levels for Florida. Is this solution reasonable? Explain.

44. **MULTIPLE REPRESENTATIONS** Draw a number line.

a. GEOMETRIC Label any 5 integers on the number line points A, B, C, D, and F.

b. TABULAR Fill in each blank in the table with either > or < using the points from the number line.

A ___ B	$A + C$ ___ $B + C$ $A + D$ ___ $B + D$ $A + F$ ___ $B + F$
B ___ A	$B + C$ ___ $A + C$ $B + D$ ___ $A + D$ $B + F$ ___ $A + F$

A ___ B	$A - C$ ___ $B - C$ $A - D$ ___ $B - D$ $A - F$ ___ $B - F$
B ___ A	$B - C$ ___ $A - C$ $B - D$ ___ $A - D$ $B - F$ ___ $A - F$

c. VERBAL Describe the patterns in the table.

d. ALGEBRAIC Describe the patterns algebraically, using the variable x to replace C, D, and F.

H.O.T. Problems

Use **H**igher-**O**rder **T**hinking Skills

45. **FIND THE ERROR** Ana and Ling are solving $|3x + 14| = -6x$. Is either of them correct? Explain your reasoning.

Ana

$|3x + 14| = -6x$

$3x + 14 = -6x$ or $3x + 14 = 6x$

$9x = -14$ $14 = 3x$

$x = -\frac{14}{9}$ ✓ $x = \frac{14}{3}$ ✓

Ling

$|3x + 14| = -6x$

$3x + 14 = -6x$ or $3x + 14 = 6x$

$9x = -14$ $14 = 3x$

$x = -\frac{14}{9}$ X $x = \frac{14}{3}$ ✓

46. **CHALLENGE** Solve $|2x - 1| + 3 = |5 - x|$. List all cases and resulting equations. (*Hint*: There are four possible cases to examine as potential solutions.)

REASONING If a, x, and y are real numbers, determine whether each statement is *sometimes*, *always*, or *never* true. Explain your reasoning.

47. If $|a| > 7$, then $|a + 3| > 10$.

48. If $|x| < 3$, then $|x| + 3 > 0$.

49. If y is between 1 and 5, then $|y - 3| \leq 2$.

50. **OPEN ENDED** Write an absolute value equation of the form $|ax + b| = cx + d$ that has no solution. Assume that a, b, c, and $d \neq 0$.

51. **WRITING IN MATH** Explain step by step how you solve an absolute value equation of the form $a|x - b| + c = d$ for x.

NYSRE PRACTICE A2.A.1

52. If $4x - y = 3$ and $2x + 3y = 19$, what is the value of y?

A 2

B 3

C 4

D 5

53. **GRIDDED RESPONSE** Two male and 2 female students from each of the 9th, 10th, 11th, and 12th grades comprise the Student Council. If a Student Council representative is chosen at random to attend a board meeting, what is the probability that the student will be either an 11th grader or male?

54. Which equation is equivalent to $4(9 - 3x) = 7 - 2(6 - 5x)$?

F $8x = 41$

G $22x = 41$

H $8x = 24$

J $22x = 24$

55. **SAT/ACT** A square with side length 4 units has one vertex at the point (1, 2). Which one of the following points *cannot* be diagonally opposite that vertex?

A $(-3, -2)$

B $(-3, 6)$

C $(5, -3)$

D $(5, 6)$

Spiral Review

Solve each equation. Check your solution. (Lesson 1-3)

56. $4x + 6 = 30$

57. $5p - 10 = 4(7 + 6p)$

58. $\frac{3}{5}y - 7 = \frac{2}{5}y + 3$

59. **MONEY** Nhu is saving to buy a car. In the first 6 months, his savings were \$80 less than $\frac{3}{4}$ the price of the car. In the second six months, Nhu saved \$50 more than $\frac{1}{5}$ the price of the car. He still needs \$370. (Lesson 1-3)

a. What is the price of the car?

b. What is the average amount of money Nhu saved each month?

c. If Nhu continues to save the average amount each month, in how many months will he be able to afford the car?

Name the property illustrated by each equation. (Lesson 1-2)

60. $(1 + 8) + 11 = 11 + (1 + 8)$

61. $z(9 - 4) = z \cdot 9 - z \cdot 4$

Simplify each expression. (Lesson 1-2)

62. $7a + 3b - 4a - 5b$

63. $3x + 5y + 7x - 3y$

64. $3(15x - 9y) + 5(4y - x)$

65. $2(10m - 7a) + 3(8a - 3m)$

66. $8(r + 7t) - 4(13t + 5r)$

67. $4(14c - 10d) - 6(d + 4c)$

68. **GEOMETRY** The formula for the surface area of a rectangular prism is $SA = 2\ell w + 2\ell h + 2wh$, where ℓ represents the length, w represents the width, and h represents the height. Find the surface area of the rectangular prism at the right. (Lesson 1-1)

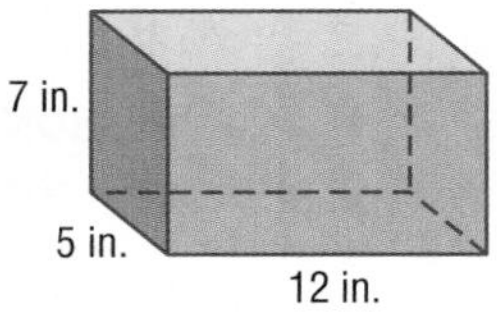

Skills Review

Solve each equation. (Lesson 1-3)

69. $15x + 5 = 35$

70. $2.4y + 4.6 = 20$

71. $8a + 9 = 6a - 7$

72. $3(w - 1) = 2w - 6$

73. $\frac{1}{2}(2b - 4) = 2 + 8b$

74. $\frac{1}{3}(6p - 24) = 18 + 3p$

1-5 Solving Inequalities

Then

You solved equations involving absolute values. (Lesson 1-4)

Now

- Solve one-step inequalities.
- Solve multi-step inequalities.

NYS Core Curriculum

Reinforcement of A.A.6 Analyze and solve verbal problems whose solution requires solving a linear equation in one variable or linear inequality in one variable
Reinforcement of A.A.24 Solve linear inequalities in one variable

New Vocabulary

set-builder notation

NY Math Online

glencoe.com

- Extra Examples
- Personal Tutor
- Self-Check Quiz
- Homework Help

Why?

Josh is trying to decide between two text messaging plans offered by a wireless telephone company.

	Plan 1	Plan 2
Monthly Access Fee	\$25	\$40
Text Messages Included	400	650
Additional Text Messages	\$0.45	\$0.30

To compare these two rate plans, we can use inequalities. The monthly access fee for Plan 1 is less than the fee for Plan 2, \$25 < \$40. However, the additional text messaging fee for Plan 1 is greater than that of Plan 2, \$0.45 > \$0.30.

One-Step Inequalities For any two real numbers, a and b, exactly one of the following statements is true.

$$a < b \qquad a = b \qquad a > b$$

Adding the same number to, or subtracting the same number from, each side of an inequality does not change the truth of the inequality.

Key Concept — For Your FOLDABLE

Addition Property of Inequality

Words For any real numbers, a, b, and c:

Models

If $a > b$, then $a + c > b + c$.

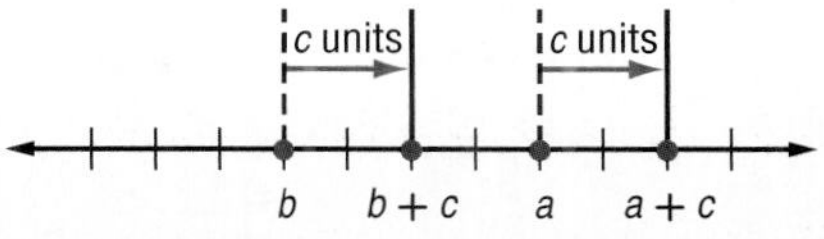

If $a < b$, then $a + c < b + c$.

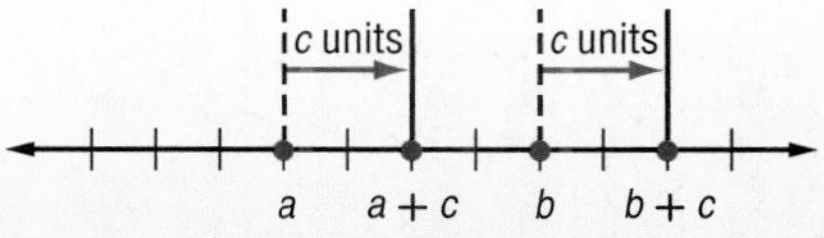

Subtraction Property of Inequality

Words For any real numbers, a, b, and c:

Models

If $a > b$, then $a - c > b - c$.

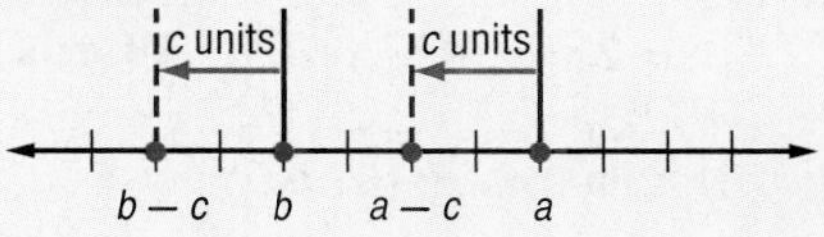

If $a < b$, then $a - c < b - c$.

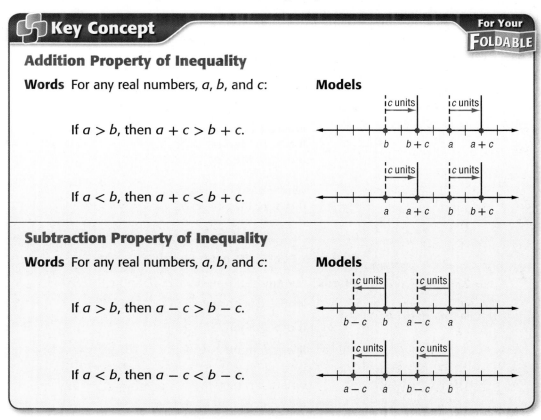

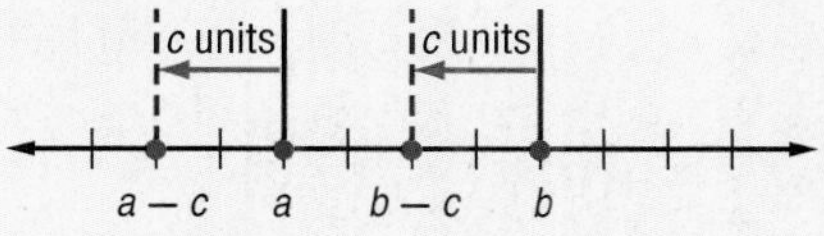

These properties are also true for $\leq$, $\geq$, and $\neq$.

These properties can be used to solve inequalities. The solution sets of inequalities in one variable can then be graphed on number lines.

Review Vocabulary

Inequality Symbols
- $>$ greater than; is more than
- $<$ less than; is fewer than
- $\geq$ greater than or equal to; is at least; is no less than
- $\leq$ less than or equal to; is at most; is no more than

EXAMPLE 1 Solve an Inequality Using Addition or Subtraction

Solve $y - 6 < 3$. Graph the solution set on a number line.

$y - 6 < 3$	**Original inequality**
$y - 6 + 6 < 3 + 6$	**Add 6 to each side.**
$y < 9$	**Simplify.**

Any real number less than 9 is a solution of this inequality. The graph of the solution set is shown at the right.

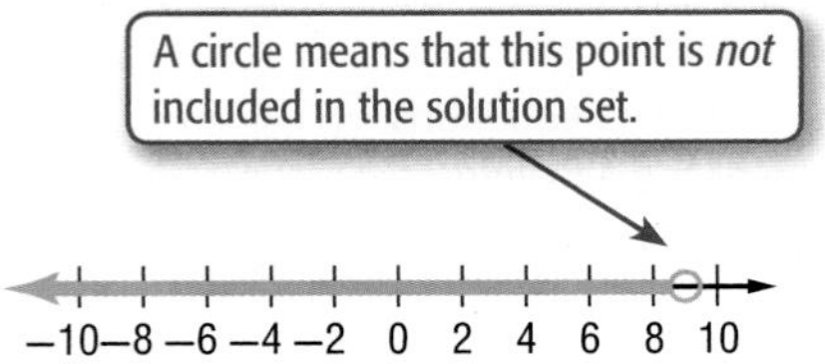
A circle means that this point is *not* included in the solution set.
−10 −8 −6 −4 −2 0 2 4 6 8 10

CHECK Substitute 8 and then 10 for y in $8y - 6 < 7y + 3$. The inequality should be true for $y = 8$ and false for $y = 10$. ✓

Check Your Progress

Solve each inequality. Graph the solution set on a number line.

1A. $5w + 3 > 4w + 9$

1B. $5x - 3 > 4x + 2$

Personal Tutor glencoe.com

StudyTip

Graphing Inequalities A circle is used for $<$ and $>$. A dot is used for $\leq$ and $\geq$.

Multiplying or dividing each side of an inequality by a positive number does not change the truth of the inequality. However, multiplying or dividing each side of an inequality by a *negative* number requires that the order of the inequality be *reversed*. For example, to reverse $\leq$, replace it with $\geq$.

Key Concept

For Your FOLDABLE

Multiplication Property of Inequality

Words	**Examples**
For any real numbers, a, b, and c, where c is positive:	$-5 < -3$
If $a > b$, then $ac > bc$.	$-5(6) < -3(6)$
If $a < b$, then $ac < bc$.	$-30 < -18$
where c is negative:	$12 > -7$
If $a > b$, then $ac < bc$.	$12(-4) < -7(-4)$
If $a < b$, then $ac > bc$.	$-48 < 28$

Division Property of Inequality

Words	**Examples**
For any real numbers, a, b, and c, where c is positive:	$-12 < -8$
If $a > b$, then $\frac{a}{c} > \frac{b}{c}$.	$\frac{-12}{4} < \frac{-8}{4}$
If $a < b$, then $\frac{a}{c} < \frac{b}{c}$.	$-3 < -2$
where c is negative:	$-21 < -14$
If $a > b$, then $\frac{a}{c} < \frac{b}{c}$.	$\frac{-21}{-7} > \frac{-14}{-7}$
If $a < b$, then $\frac{a}{c} > \frac{b}{c}$.	$3 > 2$

These properties are also true for $\leq$, $\geq$, and $\neq$.

ReadingMath

Set-Builder Notation $\{y \mid y < 9\}$ is read *the set of all numbers y such that y is less than 9.*

The solution set of an inequality can be expressed by using **set-builder notation**. For example, the solution set in Example 1 can be expressed as $\{y \mid y < 9\}$.

EXAMPLE 2 Solve an Inequality Using Multiplication or Division

Solve $-4.2x \leq 29.4$. Graph the solution set on a number line.

$-4.2x \leq 29.4$	**Original inequality**
$\frac{-4.2x}{-4.2} \geq \frac{29.4}{-4.2}$	**Divide each side by −4.2, reversing the inequality symbol.**
$x \geq 7$	**Simplify.**

The solution set is $\{x|x \geq 7\}$. The graph of the solution is shown below.

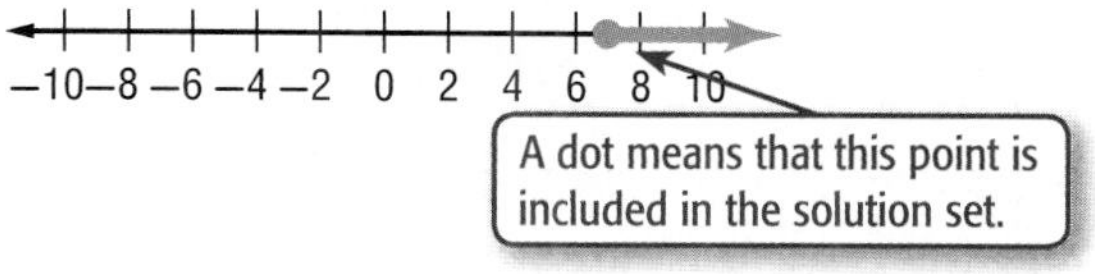

CHECK Substitute 6 and then 8 for x in $-4.2x \leq 29.4$. The inequality should be true for $x = 8$ and false for $x = 6$. ✓

Check Your Progress

Solve each inequality. Graph the solution set on a number line.

2A. $-4x \geq -24$ **2B.** $-9.2y < 23$

Personal Tutor glencoe.com

StudyTip

When multiplying or dividing by a negative number, remember to reverse the inequality symbol.

Multi-Step Inequalities Solving multi-step inequalities is similar to solving multi-step equations.

EXAMPLE 3 Solve Multi-Step Inequalities

Solve $-4c \leq \frac{5c + 58}{6}$. Graph the solution set on a number line.

$-4c \leq \frac{5c + 58}{6}$	**Original inequality**
$-24c \leq 5c + 58$	**Multiply each side by 6.**
$-29c \leq 58$	**Add −5c to each side.**
$c \geq 2$	**Divide each side by −29, reversing the inequality symbol.**

The solution set is $\{c|c \geq -2\}$ and is graphed below.

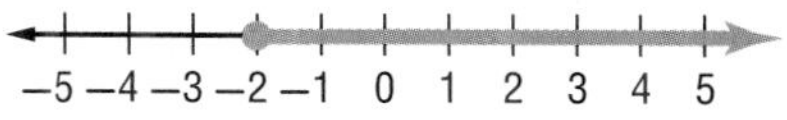

CHECK Substitute −3 and then −1 for x in $-4c \leq \frac{5c + 58}{6}$. The inequality should be true for $x = -1$ and false for $x = -3$. ✓

Check Your Progress

Solve each inequality. Graph the solution set on a number line.

3A. $-3x \leq \frac{-4x + 22}{5}$ **3B.** $8y \geq \frac{-5y + 9}{-4}$

3C. $-6(-4v + 3) \leq 2(10v + 3)$ **3D.** $-5(3d - 7) > 3(2d + 14)$

Personal Tutor glencoe.com

EXAMPLE 4 Write and Solve an Inequality

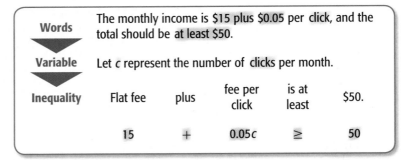

Real-World Link

In 2007, the Netcraft Web Server Survey found over 108,000,000 distinct Web sites.

Source: Netcraft

WEB SITES Enrique's company pays Salim to advertise on Salim's Web site. Salim's Web site earns \$15 per month plus \$0.05 every time a visitor clicks on the advertisement. What is the least number of clicks per month that Salim needs in order to earn \$50 per month or more?

Understand Let c = the number of clicks on the advertisement. Salim earns \$15 per month and \$0.05 per click, and he wants to earn a minimum of \$50 for the advertisement.

Plan Write an inequality.

Words	The monthly income is \$15 plus \$0.05 per click, and the total should be at least \$50.				
Variable	Let c represent the number of clicks per month.				
Inequality	Flat fee	plus	fee per click	is at least	\$50.
	15	+	0.05c	≥	50

Solve

$15 + 0.05c \geq 50$	**Original inequality**
$0.05c \geq 35$	**Subtract 15 from each side.**
$c \geq 700$	**Divide each side by 0.05.**

Check

$15 + 0.05c \geq 50$	**Original inequality**
$5 + 0.05(700) \overset{?}{\geq} 50$	**Replace c with 700.**
$15 + 35 \overset{?}{\geq} 50$	**Multiply.**
$50 \geq 50$ ✓	**Add.**

Visitors to Salim's Web site need to click on Enrique's advertisement at least 700 times per month in order for Salim to earn \$50 or more from Enrique's company.

Check Your Progress

4. Rosa's cell phone plan costs her \$50 per month plus \$0.25 for each minute she goes beyond her free minutes. How many minutes can she go beyond her free minutes and still pay less than a total of \$70?

Personal Tutor glencoe.com

Check Your Understanding

Examples 1–3 pp. 34, 35

Solve each inequality. Then graph the solution set on a number line.

1. $b + 6 < 14$
2. $12 - d > -8$
3. $18 \leq -3x$
4. $-5y \geq -35$
5. $-4w - 13 > -21$
6. $8z - 9 \geq -15$
7. $s \geq \frac{s+6}{5}$
8. $\frac{2x-9}{4} \leq x + 2$

Example 4 p. 36

9. **YARD WORK** Tara is delivering bags of mulch. Each bag weighs 48 pounds, and the push cart weighs 65 pounds. If her flat-bed truck is capable of hauling 2000 pounds, how many bags of mulch can Tara safely take on each trip?

Practice and Problem Solving

● = **Step-by-Step Solutions** begin on page R20.
Extra Practice begins on page 947.

Examples 1–3
pp. 34, 35

Solve each inequality. Then graph the solution set on a number line.

10. $m - 8 > -12$ **11.** $n + 6 \leq 3$ **12.** $6r < -36$

13. $-12t \geq -6$ **14.** $-\frac{w}{4} \leq -7$ **15.** $\frac{k}{3} - 14 < -5$

16. $4x - 15 \leq 21$ **17.** $-6z - 14 > -32$ **18.** $-16 \geq 5(2z - 11)$

19. $12 < -4(3c - 6)$ **20.** $\frac{3y - 4}{0.2} - 8 > 12$ **21.** $\frac{9z + 5}{4} + 18 < 26$

Example 4
p. 36

22. GYMNASTICS In a gymnastics competition, an athlete's final score is calculated by taking 75% of the average technical score and adding 25% of the artistic score. All scores are out of 10, and one gymnast has a 7.6 average technical score. What artistic score does the gymnast need to have a final score of at least 8.0?

Define a variable and write an inequality for each problem. Then solve.

23. Twelve less than the product of three and a number is less than 21.

24. The quotient of three times a number and 4 is at least −16.

25. The difference of 5 times a number and 6 is greater than the number.

26. The quotient of the sum of 3 and a number and 6 is less than −2.

Real-World Link

Stretching 2175 miles from Maine to Georgia, the Appalachian trail is the longest marked footpath in the world.

Source: Appalachian Trail Conservancy

27. HIKING Danielle can hike 3 miles in an hour, but she has to take a one-hour break for lunch and a one-hour break for dinner. If Danielle wants to hike at least 18 miles, solve $3(x - 2) \geq 18$ to determine how many hours the hike should take.

Solve each inequality. Then graph the solution set on a number line.

28. $18 - 3x < 12$ **29.** $-8(4x + 6) < -24$ **30.** $\frac{1}{4}n + 12 \geq \frac{3}{4}n - 4$

31. $0.24y - 0.64 > 3.86$ **32.** $10x - 6 \leq 4x + 42$ **33.** $-6v + 8 > -14v - 28$

34. $n > \frac{-3n - 15}{8}$ **35.** $-2r < \frac{6 - 2r}{5}$ **36.** $\frac{9z - 4}{5} \leq \frac{7z + 2}{4}$

37. MONEY Jin is selling advertising space in *Central City Magazine* to local businesses. Jin earns 3% commission for every advertisement he sells plus a salary of $250 a week. If the average amount of money that a business spends on an advertisement is $500, how many advertisements must he sell each week to make a salary of at least $700 that week?

a. Write an inequality to describe this situation.

b. Solve the inequality and interpret the solution.

Define a variable and write an inequality for each problem. Then solve.

38. One third of the sum of 5 times a number and 3 is less than one fourth the sum of six times that number and 5.

39. The sum of one third a number and 4 is at most the sum of twice that number and 12.

40. GEOMETRY The sides of square *ABCD* are extended to form rectangle *DEFG*. If the perimeter of the rectangle is at least twice the perimeter of the square, what is the maximum length of a side of square *ABCD*?

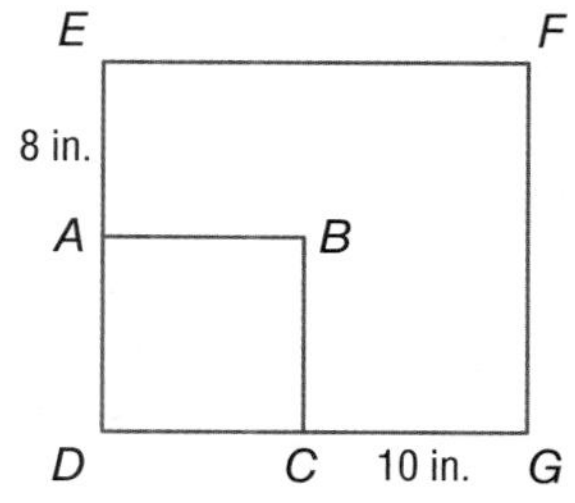

41 **MARATHONS** Jamie wants to be able to run at least the standard marathon distance of 26.2 miles. A good rule for training is that runners generally have enough endurance to finish a race that is up to 3 times his or her average daily distance.

a. If the length of her current daily run is 5 miles, write an inequality to find the amount by which she needs to increase her daily run to have enough endurance to finish a marathon.

b. Solve the inequality and interpret the solution.

Real-World Link

The Boston Marathon was first held in 1897, making it the longest-running marathon in the world.

Source: Boston Athletic Association

42. **MONEY** The costs for renting a car from Ace Car Rental and from Basic Car Rental are shown in the table. For what mileage does Basic have the better deal? Use the inequality $38 + 0.1x > 42 + 0.05x$. Explain why this inequality works.

Rental Car Costs

Company	Cost per Day	Cost per Mile
Ace	\$38	\$0.10
Basic	\$42	\$0.05

43. **MULTIPLE REPRESENTATIONS** In this exercise, you will explore graphing inequalities on a coordinate plane.

a. **TABULAR** Organize the following into a table. Substitute 5 points into the inequality $y \geq -\frac{1}{2}x + 3$. State whether the resulting statement is true or false.

b. **GRAPHICAL** Graph $y = -\frac{1}{2}x + 3$. Also graph the 5 points from the table. Label all points that resulted in a true statement with a T. Label all points that resulted in a false statement with an F.

c. **VERBAL** Describe the pattern produced by the points you have labeled. Make a conjecture about which points on the coordinate plane would result in true and false statements.

H.O.T. Problems

Use Higher-Order Thinking Skills

44. **CHALLENGE** If $-4 < x < 5$ and $0.25 < y < 4$, then $a < \frac{x}{y} < b$. What is $a + b$?

45. **FIND THE ERROR** Madlynn and Emilie were comparing their homework. Is either of them correct? Explain your reasoning.

Madlynn

$$\frac{4x+5}{-2} - 1 > -3$$
$$\frac{4x+5}{-2} < -2$$
$$4x + 5 > 4$$
$$4x > -1$$
$$x > -\frac{1}{4}$$

Emilie

$$\frac{4x+5}{-2} - 1 > -3$$
$$\frac{4x+5}{-2} > -2$$
$$4x + 5 > 4$$
$$4x > -1$$
$$x > -\frac{1}{4}$$

46. **REASONING** Determine whether the following statement is *sometimes*, *always*, or *never* true. Explain your reasoning.

The opposite of the absolute value of a negative number is less than the opposite of that number.

47. **CHALLENGE** Given $\triangle ABC$ with sides $AB = 3x + 4$, $BC = 2x + 5$, and $AC = 4x$, determine the values of x such that $\triangle ABC$ exists.

48. **OPEN ENDED** Write an inequality for which the solution is all real numbers in the form $ax + b > c(x + d)$. Explain how you know this.

49. **WRITING IN MATH** Why does the inequality symbol need to be reversed when multiplying or dividing by a negative number?

NYSRE PRACTICE Reinforcement of A.A.6, A.A.24

50. SHORT RESPONSE Rogelio found a cookie recipe that requires $\frac{3}{4}$ cup of sugar and 2 cups of flour. How many cups of sugar would he need if he used 6 cups of flour?

51. STATISTICS The mean score for Samantha's first six algebra quizzes was 88. If she scored a 96 on her next quiz, what will her mean score be for all 7 quizzes?

A 89
B 90
C 91
D 92

52. SAT/ACT The average of five numbers is 9. The average of 7 other numbers is 8. What is the average of all 12 numbers?

F $8\frac{5}{12}$
G $8\frac{7}{12}$
H $8\frac{3}{4}$
J $8\frac{11}{12}$

53. What is the complete solution of the equation $|8 - 4x| = 40$?

A $x = 8; x = 12$
B $x = 8; x = -12$
C $x = -8; x = -12$
D $x = -8; x = 12$

Spiral Review

Solve each equation. Check your solutions. (Lesson 1-4)

54. $|x - 5| = 12$

55. $7|3y - 4| = 35$

56. $|a + 6| = a$

57. ASTRONOMY Pluto travels in a path that is not circular. Pluto's farthest distance from the Sun is 4539 million miles, and its closest distance is 2756 million miles. Write an equation that can be solved to find the minimum and maximum distances from the Sun to Pluto. (Lesson 1-4)

58. POPULATION In 2005, the population of Bay City was 19,611. For each of the next five years, the population decreased by an average of 715 people per year. (Lesson 1-3)

a. What was the population in 2010?

b. If the population continues to decline at the same rate as from 2005 to 2010, what would you expect the population to be in 2025?

59. GEOMETRY The formula for the surface area of a cylinder is $SA = 2\pi r^2 + 2\pi rh$. (Lesson 1-2)

a. Use the Distributive Property to rewrite the formula by factoring out the greatest common factor of the two terms.

b. Find the surface area for a cylinder with radius 3 centimeters and height 10 centimeters using both formulas. Leave the answer in terms of π.

c. Which formula do you prefer? Explain your reasoning.

60. CONSTRUCTION The Sawyers are adding a family room to their house. The dimensions of the room are 26 feet by 28 feet. Show how to use the Distributive Property to mentally calculate the area of the room. (Lesson 1-2)

Skills Review

Solve each equation. Check your solutions. (Lesson 1-4)

61. $|x| = 9$

62. $|x + 3| = 10$

63. $|4y - 15| = 13$

64. $18 = |3x - 9|$

65. $16 = 4|w + 2|$

66. $|y + 3| + 4 = 20$

EXPLORE **1-6**

Algebra Lab
Interval Notation

NY Math Online glencoe.com
Math *in Motion,* Animation

Objective
Use interval notation to describe sets of numbers.

The solution set of an inequality can be described by using **interval notation**. The **infinity** symbols below are used to indicate that a set is unbounded in the positive or negative direction, respectively.

Read as *positive infinity*. → $+\infty$ $-\infty$ ← Read as *negative infinity*.

To indicate that an endpoint is *not* included in the set, a parenthesis, (or), is used. Parentheses are always used with the symbols $+\infty$ and $-\infty$, because they do not include endpoints.

$x < 2$

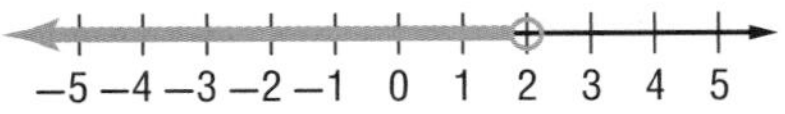

interval notation
$(-\infty, 2)$

A bracket is used to indicate that the endpoint, -2, *is* included in the solution set below.

$x \geq -2$

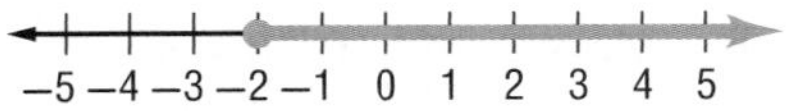

interval notation
$[-2, +\infty)$

In interval notation, the symbol for the union of the two sets is ∪. The compound inequality $y \leq -7$ or $y > -1$ is written as $(-\infty, -7] \cup (-1, +\infty)$.

Exercises

Write each inequality using interval notation.

1. $\{a \mid a \leq -3\}$

2. $\{n \mid n > -8\}$

3. $\{y \mid y < 2 \text{ or } y \geq 14\}$

4. $\{b \mid b \leq -9 \text{ or } b > 1\}$

5. $\{t \mid 1 < t < 3\}$

6. $\{m \mid m \geq 4 \text{ or } m \leq -7\}$

7. $\{x \mid x \geq 0\}$

8. $\{r \mid -3 < r < 4\}$

9.

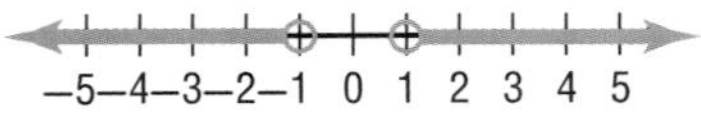

10.

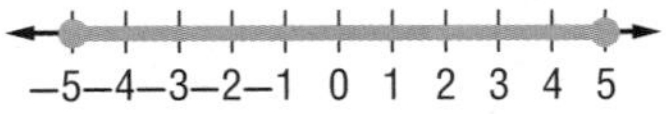

11.

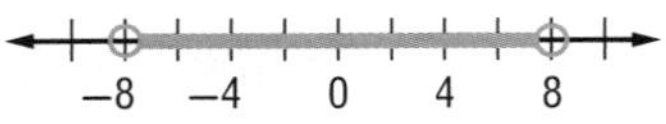

12.

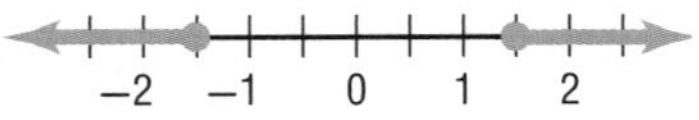

13.

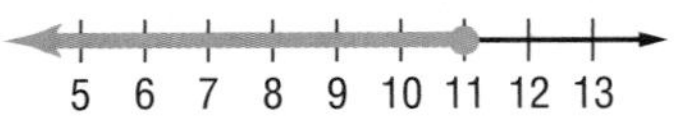

14.

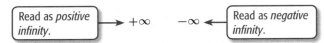

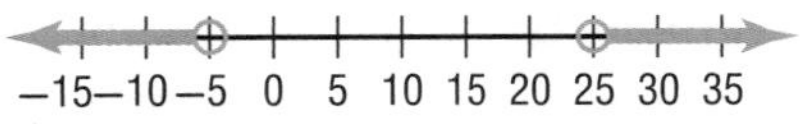

Graph each solution set on a number line.

15. $(-1, \infty)$

16. $(-\infty, 4]$

17. $(-\infty, 5] \cup (7, +\infty)$

18. **WRITING IN MATH** Write in words the meaning of $(-\infty, 3) \cup [10, +\infty)$. Then write the compound inequality that this notation represents.

1-6 Solving Compound and Absolute Value Inequalities

Then
You solved one-step and multi-step inequalities. (Lesson 1-5)

Now
- Solve compound inequalities.
- Solve absolute value inequalities.

NYS Core Curriculum

A2.A.1 Solve absolute value equations and **inequalities involving linear expressions in one variable**

New Vocabulary
compound inequality
intersection
union

NY Math Online

glencoe.com
- Extra Examples
- Personal Tutor
- Self-Check Quiz
- Homework Help
- Math in Motion

Why?

Marine biologists often have to transplant a dolphin from its natural habitat to a pool. Dolphins prefer the temperature of water to be at least 22°C but no more than 29°C. The acceptable temperature of water t for dolphins can be described by the following compound inequality.

$$t \geq 22 \text{ and } t \leq 29$$

Compound Inequalities A **compound inequality** consists of two inequalities joined by the word *and* or the word *or*. To solve a compound inequality, you must solve each part of the inequality. The graph of a compound inequality containing *and* is the **intersection** of the solution sets of the two inequalities.

Key Concept — "And" Compound Inequalities (For Your FOLDABLE)

Words A compound inequality containing the word *and* is true if and only if *both* inequalities are true.

Example

$x \geq -4$

$x < 3$

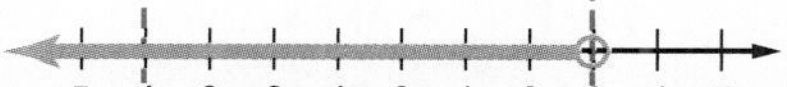

$x \geq -4$ and $x < 3$

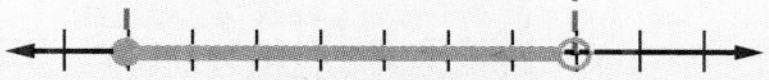

Another way of writing $x \geq -4$ and $x < 3$ is $-4 \leq x < 3$.
Both forms are read *x is greater than or equal to −4 and less than 3.*

EXAMPLE 1 Solve an "And" Compound Inequality

Solve $8 < 3y - 7 \leq 23$. Graph the solution set on a number line.

Method 1 Solve separately.
Write the compound inequality using the word *and*. Then solve each inequality.

$$8 < 3y - 7 \quad \text{and} \quad 3y - 7 \leq 23$$
$$15 < 3y \qquad\qquad 3y \leq 30$$
$$5 < y \qquad\qquad y \leq 10$$
$$5 < y \leq 10$$

Method 2 Solve both together.
Solve both parts at the same time by adding 7 to each part. Then divide by 3.

$$8 < 3y - 7 \leq 23$$
$$15 < 3y \leq 30$$
$$5 < y \leq 10$$
$$5 < y \leq 10$$

(continued on the next page)

Graph the solution set for each inequality and find their intersection.

$5 < y$

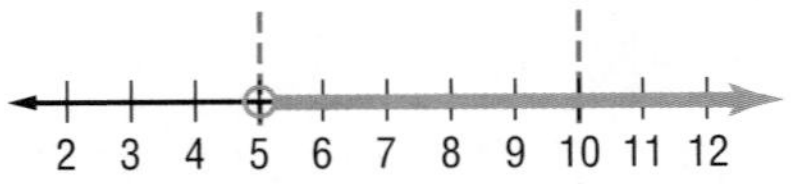

$y \leq 10$

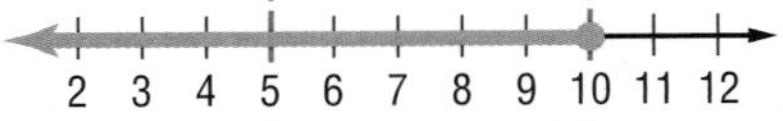

$5 < y \leq 10$

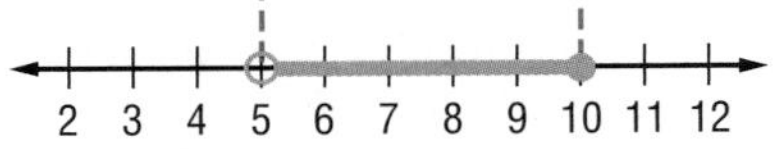

The solution set is $\{y \mid 5 < y \leq 10\}$.

Check Your Progress

Solve each inequality. Graph the solution set on a number line.

1A. $-12 \leq 4x + 8 \leq 32$

1B. $-5 \geq 3z - 2 > -14$

Personal Tutor glencoe.com

The graph of a compound inequality containing *or* is the **union** of the solution sets of the two inequalities.

StudyTip

***Or* Inequality** If there is no value for which either of the inequalities is true, then there is no solution. When every possible value is true for either inequality, then there are infinite solutions.

Key Concept "Or" Compound Inequalities

For Your FOLDABLE

Words A compound inequality containing the word *or* is true if one or more of the inequalities is true.

Example

$x \geq 5$

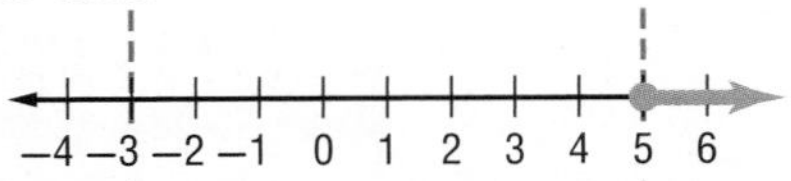

$x < -3$

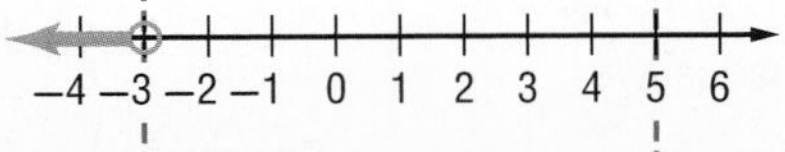

$x \geq 5$ or $x < -3$

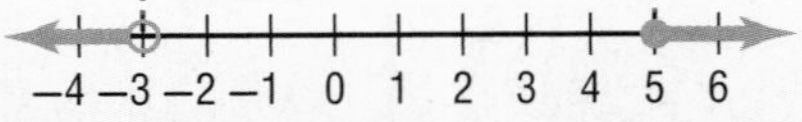

Math *in Motion,* Animation glencoe.com

EXAMPLE 2 Solve an "Or" Compound Inequality

Solve $k + 6 < -4$ or $3k \geq 14$. Graph the solution set.

Solve each inequality separately.

$k + 6 < -4$	or	$3k \geq 14$
$k < -10$		$k \geq \frac{14}{3}$

$k < -10$

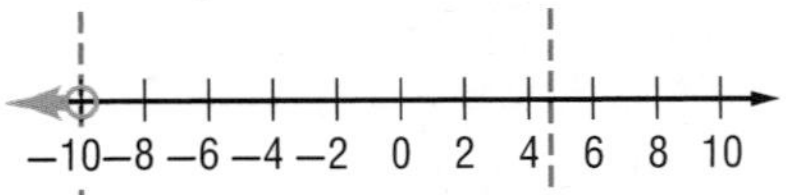

$k \geq \frac{14}{3}$

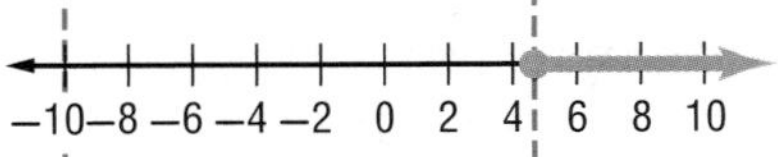

$k < -10$ or $k \geq \frac{14}{3}$

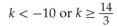

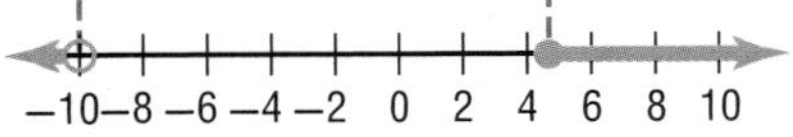

Check Your Progress

Solve each inequality. Graph the solution set on a number line.

2A. $5j \geq 15$ or $-3j \geq 21$

2B. $g - 6 > -11$ or $2g + 4 < -15$

Personal Tutor glencoe.com

Absolute Value Inequalities In Lesson 1-4, you learned that the absolute value of a number is its distance from 0 on the number line. You can use this definition to solve inequalities involving absolute value.

> **ReadingMath**
>
> ***Within* and *Between*** When solving problems involving inequalities, *within* is meant to be inclusive. Use $\leq$ or $\geq$.
>
> *Between* is meant to be exclusive. Use $<$ or $>$.

EXAMPLE 3 Solve Absolute Value Inequalities

Solve each inequality. Graph the solution set on a number line.

a. $|x| < 3$

$|x| < 3$ means that the distance between x and 0 on a number line is less than 3 units. To make $|x| < 3$ true, substitute numbers for x that are fewer than 3 units from 0.

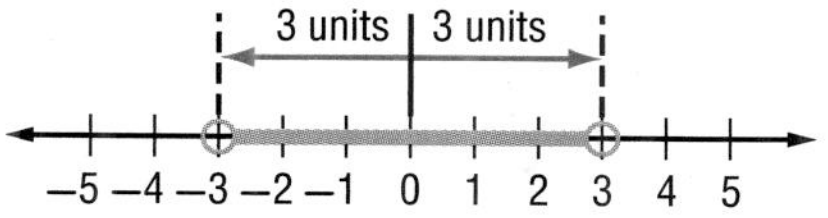

Notice that the graph of $|x| < 3$ is the same as the graph of $x > -3$ and $x < 3$.

All of the numbers between −3 and 3 are less than 3 units from 0. The solution set is $\{x \mid -3 < x < 3\}$.

b. $|x| > 5$

$|x| > 5$ means that the distance between x and 0 on a number line is more than 5 units. To make $|x| > 5$ true, substitute numbers for x that are more than 5 units from 0.

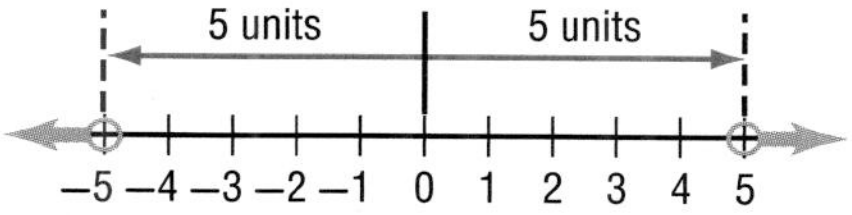

Notice that the graph of $|x| > 5$ is the same as the graph of $x < -5$ or $x > 5$.

All of the numbers between and including −5 and 5 are no more than 5 units from 0. The solution set is $\{x \mid -5 > x \text{ or } x > 5\}$.

Check Your Progress

Solve each inequality. Graph the solution set on a number line.

3A. $|t| < 6$

3B. $|u| < -3$

3C. $|t| > 3$

3D. $|u| > -2$

Personal Tutor glencoe.com

An absolute value inequality can be solved by rewriting it as a compound inequality.

Key Concept Absolute Value Inequalities

For Your FOLDABLE

For all real numbers a, b, c, and x, $c > 0$, the following statements are true.

Absolute Value Inequality	Compound Inequality	Example
$\lvert ax + b \rvert > c$	$ax + b > c$ or $ax + b < c$	If $\lvert 4x + 5 \rvert > 7$, then $4x + 5 > 7$ or $4x + 5 < -7$.
$\lvert ax + b \rvert < c$	$-c < ax + b < c$	If $\lvert 4x + 5 \rvert < 7$, then $-7 < 4x + 5 < 7$.

These statements are also true for $\leq$ and $\geq$, respectively.

StudyTip

Absolute Value Inequalities Because the absolute value of a number is never negative, solutions involving negative numbers are as follows.

$|x| < -5$ is the empty set.

$|x| > -5$ is infinite solutions.

EXAMPLE 4 Solve a Multi-Step Absolute Value Inequality

Solve $|6y - 5| \geq 13$. Graph the solution set on a number line.

$|6y - 5| \geq 13$ is equivalent to $6y - 5 \geq 13$ or $6y - 5 \leq -13$. Solve the inequality.

$6y - 5 \geq 13$	or	$6y - 5 \leq -13$	**Rewrite the inequality.**
$6y \geq 18$		$6y \leq -8$	**Add 5 to each side.**
$y \geq 3$		$y \leq -\frac{8}{6}$ or $\frac{4}{3}$	**Divide each side by 6.**

The solution set is $\left\{y \middle| y \leq -\frac{4}{3} \text{ or } y \geq 3\right\}$.

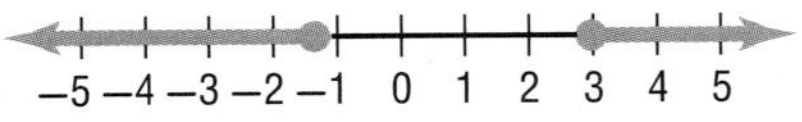

Check Your Progress

Solve each inequality. Graph the solution set on a number line.

4A. $|4x - 7| > 13$

4B. $|5z + 2| \leq 17$

Personal Tutor glencoe.com

Real-World Link

Apartment costs vary greatly depending on location. Of the major U.S. cities, New York has the highest average monthly rent of $2400, while Oklahoma City is lowest at $543.

Source: MSN

Absolute value inequalities can be used to solve real-world problems.

EXAMPLE 5 Write and Solve an Absolute Value Inequality

MONEY Amanda is apartment hunting in a specific area. She discovers that the average monthly rent for a 2-bedroom apartment is $600 a month, but the actual price could differ from the average as much as $225 a month.

a. Write an absolute value inequality to describe this situation.

Let r = average monthly rent.

$|600 - r| \leq 225$

b. Solve the inequality to find the range of monthly rent.

Rewrite the absolute value inequality as a compound inequality. Then solve for r.

$$-225 \leq 600 - r \leq 225$$

$$-225 - 600 \leq 600 - r - 600 \leq 225 - 600$$

$$-825 \leq -r \leq -375$$

$$825 \geq r \geq 375$$

The solution set is $\{r \mid 375 \leq r \leq 825\}$. Thus, monthly rent could fall between $375 and $825, inclusive.

Check Your Progress

5. TUITION Rachel is considering colleges to attend and determines that the average tuition among her choices is $3725 per year, but the tuition at a school could differ by as much as $1650 from the average. Write and solve an absolute value inequality to find the range of tuition.

Personal Tutor glencoe.com

Check Your Understanding

Examples 1–4
pp. 41–44

Solve each inequality. Graph the solution set on a number line.

1. $-4 < g + 8 < 6$

2. $-9 \leq 4y - 3 \leq 13$

3. $z + 6 > 3$ or $2z < -12$

4. $m - 7 \geq -3$ or $-2m + 1 \geq 11$

5. $|c| \geq 8$

6. $|q| \geq -1$

7. $|z| < 6$

8. $|x| \leq -4$

9. $|3v + 5| > 14$

10. $|4t - 3| \leq 7$

Example 5
p. 44

11. MONEY Khalid is considering several types of paint for his bedroom. He estimates that he will need between 2 and 3 gallons. The table at the right shows the price per gallon for each type of paint Khalid is considering. Write a compound inequality and determine how much he could be spending.

Paint Type	Price per Gallon
Flat	\$21.98
Satin	\$23.98
Semi-Gloss	\$24.98
Gloss	\$25.98

Practice and Problem Solving

● = **Step-by-Step Solutions** begin on page R20.
Extra Practice begins on page 947.

Examples 1–4
pp. 41–44

Solve each inequality. Graph the solution set on a number line.

12. $8 < 2v - 4 < 16$

13. $-7 \leq 4d - 3 \leq -1$

14. $4r + 3 < -6$ or $3r - 7 > 2$

15. $6y - 3 < -27$ or $-4y + 2 < -26$

16. $|6h| < 12$

17. $|-4k| > 16$

18. $|3x - 4| > 10$

19 $|8t + 3| \leq 4$

20. $|-9n - 3| < 6$

21. $|-5j - 4| \geq 12$

Example 5
p. 44

22. ANATOMY Forensic scientists use the equation $h = 2.6f + 47.2$ to estimate the height h of a woman given the length in centimeters f of her femur bone.

a. Suppose the equation has a margin of error of ± 3 centimeters. Write an inequality to represent the height of a woman given the length of her femur bone.

b. If the length of a female skeleton's femur is 50 centimeters, write and solve an absolute value inequality that describes the woman's height in centimeters.

Write an absolute value inequality for each graph.

23. 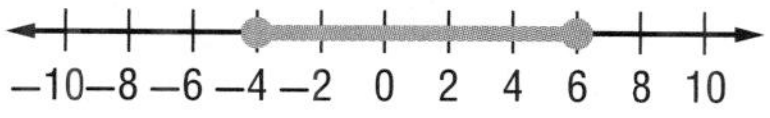

24. 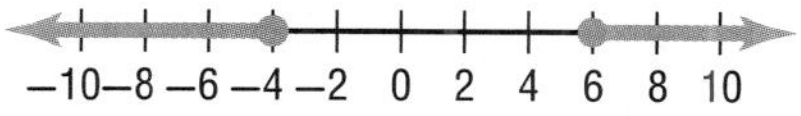

25. 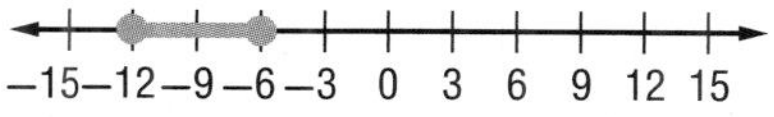

26. 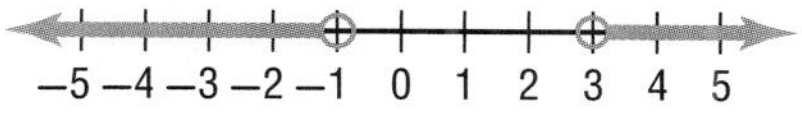

27. 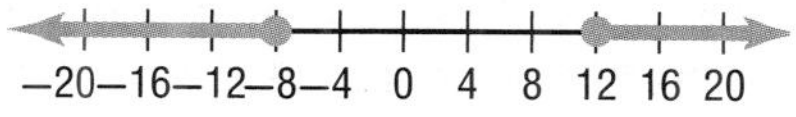

28. 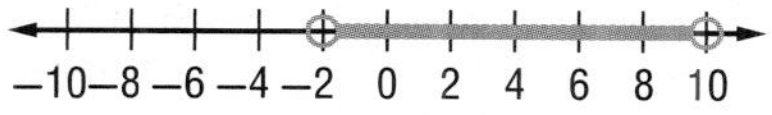

29. 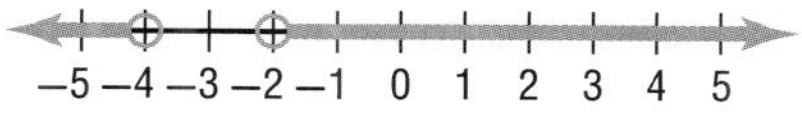

30. 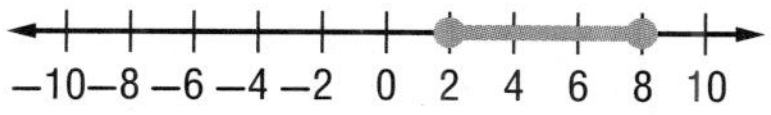

31. **DOGS** The Labrador retriever is one of the most recognized and popular dogs kept as a pet. Using the information given, write a compound inequality to describe the range of healthy weights for a fully grown female Labrador retriever.

Healthy Heights and Weights for Labrador Retrievers

Gender	Height (in.)	Weight
Male	22.5–24.5	65–80
Female	21.5–23.5	55–70

32. **GEOMETRY** The *Exterior Angle Inequality Theorem* states that an exterior angle measure is greater than the measure of either of its corresponding remote interior angles. Write two inequalities to express the relationships among the measures of the angles of $\triangle ABC$.

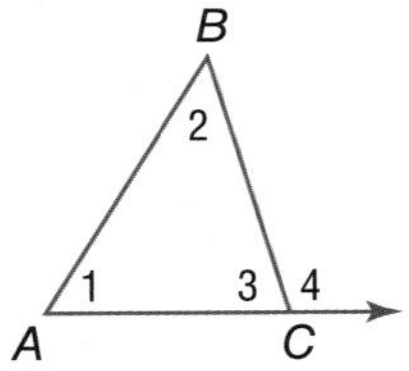

Solve each inequality. Graph the solution set on a number line.

33. $28 > 6k + 4 > 16$

34. $m - 7 > -12$ or $-3m + 2 > 38$

35. $|-6h| > 90$

36. $-|-5k| > 15$

37. $3|2z - 4| - 6 > 12$

38. $6|4p + 2| - 8 < 34$

39. $\dfrac{|5f - 2|}{6} > 4$

40. $\dfrac{|2w + 8|}{5} \geq 3$

Write an algebraic expression to represent each verbal expression.

41. numbers that are at least 4 units from -5

42. numbers that are no more than $\frac{3}{8}$ unit from 1

43. numbers that are at least 6 units but no more than 10 units from 2

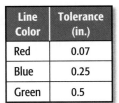

Real-World Link

NASCAR drivers work together. By driving in certain formations, cars create a vacuum behind them that another car can then fill. This reduces drag, which increases speed for all the drivers in the formation.

Source: *Popular Mechanics*

44. **AUTO RACING** NASCAR rules stipulate that a car must conform to a set of 32 templates, each shaped to fit a different contour of the car. When a template is placed on a car, the gap between it and the car cannot exceed the specified tolerance. Each template is marked on its edge with a colored line that indicates the tolerance for the template.

a. Suppose a certain template is 24.42 inches long. Use the information in the table at the right to write an absolute value inequality for templates with each line color.

Line Color	Tolerance (in.)
Red	0.07
Blue	0.25
Green	0.5

b. Find the acceptable lengths for that part of a car if the template has each line color.

c. Graph the solution set for each line color on a number line.

d. The tolerance of which line color includes the tolerances of the other line colors? Explain your reasoning.

Solve each inequality. Graph the solution set on a number line.

45. $n + 6 > 2n + 5 > n - 2$

46. $y + 7 < 2y + 2 < 0$

47. $2x + 6 < 3(x - 1) \leq 2(x + 3)$

48. $a - 16 \leq 2(a - 4) < a + 2$

49. $4g + 8 \geq g + 6$ or $7g - 14 \geq 2g - 4$

50. $5t + 7 > 2t + 4$ and $3t + 3 < 24 - 4t$

51. **HEALTH** Hypoglycemia (low blood sugar) and hyperglycemia (high blood sugar) are potentially dangerous and occur when a person's blood sugar fluctuates by more than 38 mg from the normal blood sugar level of 88 mg. Write and solve an absolute value inequality to describe blood sugar levels that are considered potentially dangerous.

52. **AIR TRAVEL** The airline on which Drew is flying has weight restrictions for checked baggage. Drew is checking one bag.

a. Describe the range of weights that would classify Drew's bag as free, \$25, \$50, and unacceptable.

b. If Drew's bag weighs 68 pounds, how much will he pay to take it on the plane?

Cost for Checked Baggage

Weight	Cost
Up to 50 lb	free
20 lb over	\$25
More than 20 lb over, but less than 50 lb over	\$50
More than 50 lb over	not accepted

H.O.T. Problems

Use Higher-Order Thinking Skills

53. **FIND THE ERROR** David and Sarah are solving $4|-5x - 3| - 6 \geq 34$. Is either of them correct? Explain your reasoning.

David

$4|-5x - 3| - 6 \geq 34$

$|-5x - 3| \geq 10$

$-5x - 3 \geq 10$ or $-5x - 3 \leq -10$

$-5x \geq 13$ $\quad$ $-5x \leq -7$

$x \leq -\frac{13}{5}$ $\quad$ $x \geq \frac{7}{5}$

Sarah

$4|-5x - 3| - 6 \geq 34$

$|-5x - 3| \geq 10$

$-5x - 3 \leq 10$ or $-5x - 3 \geq -10$

$-5x \leq 13$ $\quad$ $-5x \geq -7$

$x \geq -\frac{13}{5}$ $\quad$ $x \leq \frac{7}{5}$

54. **CHALLENGE** Solve $|x - 2| - |x + 2| > x$.

REASONING **Determine whether each statement is *true* or *false*. If false, provide a counterexample.**

55. The graph of a compound inequality involving an *and* statement is bounded on the left and right by two values of x.

56. The graph of a compound inequality involving an *or* statement contains a region of values that are not solutions.

57. The graph of a compound inequality involving an *and* statement includes values that make all parts of the given statement true.

58. **WRITING IN MATH** An alternate definition of absolute value is to define $|a - b|$ as the distance between a and b on the number line. Explain how this definition can be used to solve inequalities of the form $|x - c| < r$.

ReadingMath

Compare to describe *similar* features or characteristics of two or more items
Contrast to describe *different* features or characteristics of two or more items

59. **REASONING** The graphs of the solutions of two different absolute value inequalities are shown. Compare and contrast the absolute value inequalities.

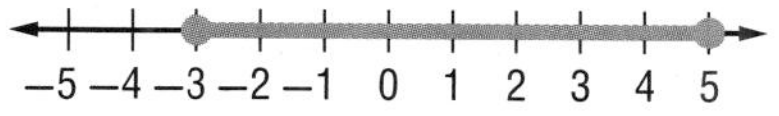

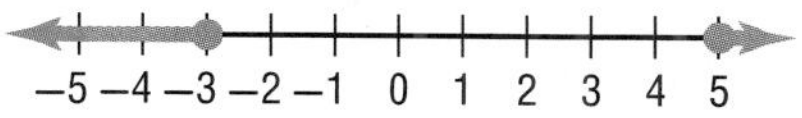

60. **OPEN ENDED** Write an absolute value inequality with a solution of $a \leq x \leq b$.

61. **WHICH ONE DOESN'T BELONG?** Identify the compound inequality that is not the same as the other three. Explain your reasoning.

$-3 < x < 5$	$x > 2$ and $x < 3$	$x > 5$ and $x < 1$	$x > -4$ and $x > -2$

62. **WRITING IN MATH** Summarize the difference between *and* compound inequalities and *or* compound inequalities.

NYSRE PRACTICE A2.A.1

63. Which of the following best describes the graph of the equations below?

$$24y = 8x + 11$$
$$36y = 12x + 11$$

A The lines have the same x-intercept.
B The lines have the same y-intercept.
C The lines are parallel.
D The lines are perpendicular.

64. SAT/ACT Find an expression equivalent to $\left(\frac{3x^3}{y}\right)^3$.

F $\frac{9x^6}{3y}$ H $\frac{27x^6}{3y}$

G $\frac{9x^9}{y^3}$ J $\frac{27x^9}{y^3}$

65. GEOMETRY How many cubes that measure 4 centimeters on each side can be placed completely inside the box below?

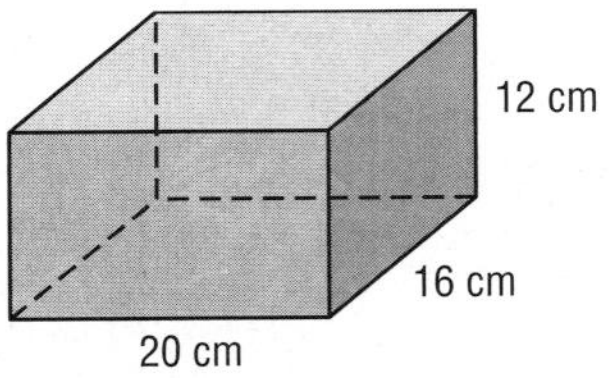

66. Which graph represents the solution set for $|3x - 6| + 8 \geq 17$?

A
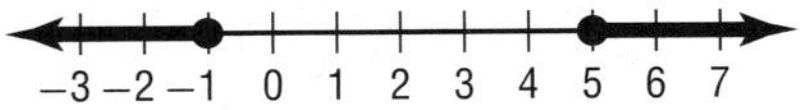

B
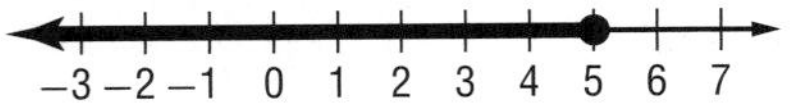

C
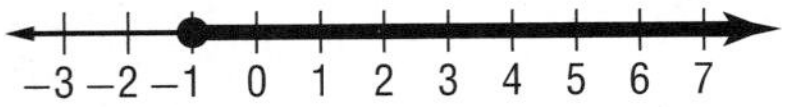

D

Spiral Review

67. HEALTH The National Heart Association recommends that less than 30% of a person's total daily caloric intake come from fat. One gram of fat yields nine Calories. Consider a healthy 21-year-old whose average caloric intake is between 2500 and 3300 Calories. (Lesson 1-5)

a. Write an inequality that represents the suggested fat intake for the person.

b. What is the greatest suggested fat intake for the person?

68. TRAVEL Maggie is planning a 5-day trip to a family reunion. She wants to spend no more than \$1000. Her plane ticket is \$375, and the hotel is \$85 per night. (Lesson 1-5)

a. Let f represent the cost of food for one day. Write an inequality to represent this situation.

b. Solve the inequality and interpret the solution.

Solve each equation. Check your solutions. (Lesson 1-4)

69. $4|x - 5| = 20$ **70.** $|3y + 10| = 25$ **71.** $|7z + 8| = -9$

Skills Review

Name the property illustrated by each statement. (Lesson 1-3)

72. If $5x = 7$, then $5x + 3 = 7 + 3$.

73. If $-3x + 9 = 11$ and $6x + 2 = 11$, then $-3x + 9 = 6x + 2$.

74. If $[x + (-2)] + (-4) = 5$, then $x + [-2 + (-4)] = 5$.

Study Guide and Review

NY Math Online glencoe.com
- STUDY TO GO
- Vocabulary Review

Chapter Summary

Key Concepts

Expressions and Formulas (Lesson 1-1)

- Use the order of operations to solve equations.

Properties of Real Numbers (Lesson 1-2)

- Real numbers can be classified as rational (Q) or irrational (I). Rational numbers can be classified as natural numbers (N), whole numbers (W), integers (Z), and/or quotients of these.

Solving Equations (Lessons 1-3 and 1-4)

- Verbal expressions can be translated into algebraic expressions.
- The absolute value of a number is the number of units it is from 0 on a number line.
- For any real numbers a and b, where $b \geq 0$, if $|a| = b$, then $a = b$ or $-a = b$.

Solving Inequalities (Lessons 1-5 and 1-6)

- Adding or subtracting the same number from each side of an inequality does not change the truth of the inequality.
- When you multiply or divide each side of an inequality by a negative number, the direction of the inequality symbol must be reversed.
- The graph of an *and* compound inequality is the intersection of the solution sets of the two inequalities. The graph of an *or* compound inequality is the union of the solution sets of the two inequalities.
- An *and* compound inequality can be expressed in two different ways. For example, $-2 \leq x \leq 3$ is equivalent to $x \geq -2$ and $x \leq 3$.
- For all real numbers a and b, where $b > 0$, the following statements are true.
  1. If $|a| < b$ then $-b < a < b$.
  2. If $|a| > b$ then $a > b$ or $a < -b$.

FOLDABLES® Study Organizer

Be sure the Key Concepts are noted in your Foldable.

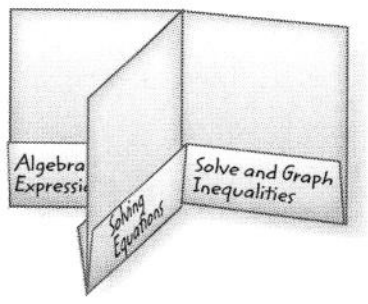

Key Vocabulary

absolute value (p. 27)	**irrational numbers** (p. 11)
algebraic expressions (p. 5)	**natural numbers** (p. 11)
compound inequality (p. 41)	**open sentence** (p. 18)
empty set (p. 28)	**order of operations** (p. 5)
equation (p. 18)	**rational numbers** (p. 11)
extraneous solution (p. 29)	**real numbers** (p. 11)
formula (p. 6)	**set-builder notation** (p. 35)
infinity (p. 40)	**solution** (p. 18)
integers (p. 11)	**union** (p. 42)
intersection (p. 41)	**variables** (p. 5)
interval notation (p. 40)	**whole numbers** (p. 11)

Vocabulary Check

State whether each sentence is *true* or *false*. If *false*, replace the underlined term to make a true sentence.

1. The absolute value of a number is always negative.
2. $\sqrt{12}$ belongs to the set of rational numbers.
3. An equation is a statement that two expressions have the same value.
4. A solution of an equation is a value that makes the equation false.
5. The empty set contains no elements.
6. A mathematical sentence containing one or more variables is called an open sentence.
7. The graph of a compound inequality containing *and* is the union of the solution sets of the two inequalities.
8. Variables are used to represent unknown quantities.
9. The set of rational numbers includes terminating and repeating decimals.
10. Expressions that contain at least one variable are called algebraic expressions.

CHAPTER 1

Study Guide and Review

Lesson-by-Lesson Review

1-1 Expressions and Formulas (pp. 5–10)

A.A.1, A.A.2

Evaluate each expression.

11. $[28 - (16 + 3)] \div 3$

12. $\frac{2}{3}(3^3 + 12)$

13. $\frac{15(9 - 7)}{3}$

Evaluate each expression if $w = 0.2$, $x = 10$, $y = \frac{1}{2}$, and $z = -4$.

14. $4w - 8y$

15. $z^2 + xy$

16. $\frac{5w - xy}{z}$

17. GEOMETRY The formula for the volume of a cylinder is $V = \pi r^2 h$, where V is volume, r is radius, and h is the height. What is the volume of a cylinder that is 6 inches high and has a radius of 3 inches?

EXAMPLE 1

Evaluate $(12 - 15) \div 3^2$.

$(12 - 15) \div 3^2 = -3 \div 3^2$	**Subtract.**
$= -3 \div 9$	**$3^2 = 9$**
$= -\frac{1}{3}$	**Divide.**

EXAMPLE 2

Evaluate $\frac{a^2}{2ac - b}$ if $a = -6$, $b = 5$, and $c = 0.25$.

$\frac{a^2}{2ac - b} = \frac{(-6)^2}{2(-6)(0.25) - 5}$	**$a = -6$, $b = 5$, and $c = 0.25$**
$= \frac{36}{2(-1.5) - 5}$	**Evaluate the numerator and denominator separately.**
$= \frac{36}{-8}$ or $-\frac{9}{2}$	**Simplify.**

1-2 Properties of Real Numbers (pp. 11–17)

A.N.1, A.A.1

Name the sets of numbers to which each value belongs.

18. $1.\overline{3}$ **19.** $\sqrt{4}$ **20.** $-\frac{3}{4}$

Simplify each expression.

21. $4x - 3y + 7x + 5y$

22. $2(a + 3) - 4a + 8b$

23. $4(2m + 5n) - 3(m - 7n)$

24. MONEY At Fun City Amusement Park, hot dogs sell for \$3.50 and sodas sell for \$2.50. Dion bought 3 hot dogs and 3 sodas during one day at the park.

a. Illustrate the Distributive Property by writing two expressions to represent the cost of the hot dogs and the sodas.

b. Use the Distributive Property to find how much money Dion spent on food and drinks.

EXAMPLE 3

Name the sets of numbers to which $\sqrt{50}$ belongs.

$\sqrt{50} = 5\sqrt{2}$ **Irrationals (I), and reals (R)**

EXAMPLE 4

Simplify $-4(a + 3b) + 5b$.

$-4(a + 3b) + 5b$	**Original expression**
$= -4(a) + -4(3b) + 5b$	**Distributive Property**
$= -4a - 12b + 5b$	**Multiply.**
$= -4a - 7b$	**Simplify.**

MIXED PROBLEM SOLVING
For mixed problem-solving practice, see page 979.

1-3 Solving Equations (pp. 18–25)

A2.CM.11

Solve each equation. Check your solution.

25. $8 + 5r = -27$

26. $4w + 10 = 6w - 13$

27. $\frac{x}{6} + \frac{x}{3} = \frac{3}{4}$

28. $6b - 5 = 3(b + 2)$

29. MONEY It cost Lori $14 to go to the movies. She bought popcorn for $3.50 and a soda for $2.50. How much was her ticket?

Solve each equation or formula for the specified variable.

30. $2k - 3m = 16$ for k

31. $\frac{r + 5}{mn} = p$ for m

32. $A = \frac{1}{2}h(a + b)$ for h

33. GEOMETRY Yu-Jun wants to fill the water container at the right. He knows that the radius is 2 inches and the volume is 100.48 cubic inches. What is the height of the water bottle? Use the formula for the volume of a cylinder, $V = \pi r^2 h$, to find the height of the bottle.

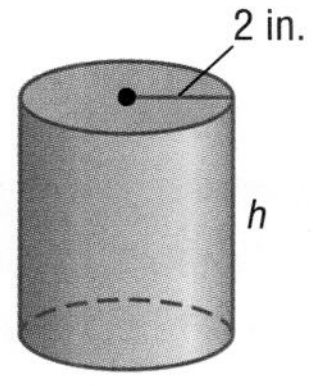

EXAMPLE 5

Solve $-3(a - 3) + 2(3a - 2) = 14$.

$-3(a - 3) + 2(3a - 2) = 14$	Original equation
$-3a + 9 + 6a - 4 = 14$	Distributive Property
$-3a + 6a + 9 - 4 = 14$	Commutative Property
$3a + 5 = 14$	Substitution Property
$3a = 9$	Subtraction Property
$a = 3$	Division Property

EXAMPLE 6

Solve each equation or formula for the specified variable.

a. $y = 2x + 3z$ **for** x

$y = 2x + 3z$	Original equation
$y - 3z = 2x$	Subtract $3z$ from each side.
$\frac{y - 3z}{2} = x$	Divide each side by 2.

b. $V = \frac{\pi r^2 h}{3}$ **for** h

$V = \frac{\pi r^2 h}{3}$	Original equation
$3V = \pi r^2 h$	Multiply each side by 3.
$\frac{3V}{\pi r^2} = h$	Divide each side by πr^2.

1-4 Solving Absolute Value Equations (pp. 27–33)

A2.A.1

Solve each equation. Check your solution.

34. $|r + 5| = 12$

35. $4|a - 6| = 16$

36. $|3x + 7| = -15$

37. $|b + 5| = 2b - 9$

38. MEASUREMENT Marcos is cutting ribbons for a craft project. Each ribbon needs to be $\frac{3}{4}$ yard long. If each piece is always within plus or minus $\frac{1}{16}$ yard, how long are the shortest and longest pieces of ribbon?

EXAMPLE 7

Solve $|3m + 7| = 13$.

Case 1	Case 2
$a = b$	$a = -b$
$3m + 7 = 13$	$3m + 7 = -13$
$3m = 6$	$3m = -20$
$m = 2$	$m = -\frac{20}{3}$

The solutions are 2 and $-\frac{20}{3}$.

CHAPTER 1 Study Guide and Review

1-5 Solving Inequalities (pp. 34–39)

A.A.6, A.A.24

Solve each inequality. Then graph the solution set on a number line.

39. $-4a \leq 24$

40. $\frac{r}{5} - 8 > 3$

41. $4 - 7x \geq 2(x + 3)$

42. $-p - 13 < 3(5 + 4p) - 2$

43. MONEY Ms. Hawkins is taking her science class on a field trip to a museum. She has \$572 to spend on the trip. There are 52 students that will go to the museum. The museum charges \$5 per student, and Ms. Hawkins gets in for free. If the students will have slices of pizza for lunch that cost \$2 each, how many slices can each student have?

EXAMPLE 8

Solve $2m - 7 < -11$. Graph the solution set on a number line.

$2m - 7 < -11$	**Original inequality**
$2m < -4$	**Add 7 to each side.**
$m < -2$	**Divide each side by 2.**

The solution set is $\{m \mid m < -2\}$.

The graph of the solution set is shown below.

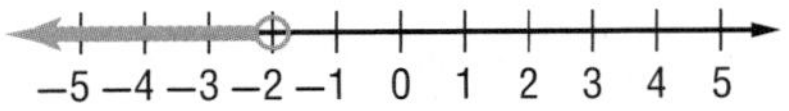

1-6 Solving Compound and Absolute Value Inequalities (pp. 41–48)

A2.A.1

Solve each inequality. Graph the solution set on a number line.

44. $2m + 4 < 7$ or $3m + 5 > 14$

45. $-5 < 4x + 3 < 19$

46. $6y - 1 > 17$ or $8y - 6 \leq -10$

47. $-2 \leq 5(m - 3) < 9$

48. $|a| + 2 < 15$

49. $|p - 14| \leq 19$

50. $|6k - 1| < 15$

51. $|2r + 7| < -1$

52. $\frac{1}{3}|8q + 5| \geq 7$

53. MONEY Cara is making a beaded necklace for a gift. She wants to spend between \$20 and \$30 on the necklace. The bead store charges \$2.50 for large beads and \$1.25 for small beads. If she buys 3 large beads, how many small beads can she buy to stay within her budget? Write and solve a compound inequality to describe the range of possible beads.

EXAMPLE 9

Solve each inequality. Graph the solution set on a number line.

a. $\mathbf{-14 \leq 3x - 8 < 16}$

$-14 \leq 3x - 8 < 16$	**Original inequality**
$-6 \leq 3x < 24$	**Add 8 to each part.**
$-2 \leq x < 8$	**Divide each part by 3.**

The solution set is $\{x \mid -2 \leq x < 8\}$.

b. $\mathbf{|3a - 5| > 13}$

$|3a - 5| > 13$ is equivalent to $3a - 5 > 13$ or $3a - 5 < -13$.

$3a - 5 > 13$	or	$3a - 5 < -13$	
$3a > 18$		$3a < -8$	**Subtract.**
$a > 6$		$a < -\frac{8}{3}$	**Divide.**

The solution set is $\left\{a \mid a > 6 \text{ or } a < -\frac{8}{3}\right\}$.

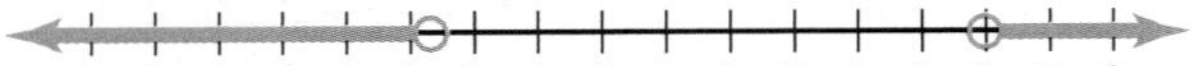

CHAPTER 1 Practice Test

NY Math Online glencoe.com
Chapter Test

1. Evaluate $x + y^2(2 + x)$ if $x = 3$ and $y = -1$.

2. Simplify $-4(3a + b) - 2(a - 5b)$.

3. **MULTIPLE CHOICE** If $3m + 5 = 23$, what is the value of $2m - 3$?

   A 105

   B 9

   C $\frac{47}{3}$

   D 6

4. Solve $r = \frac{1}{2}m^2p$ for p.

Write an algebraic expression to represent each verbal expression.

5. twice the difference of a number and 11

6. the product of the square of a number and 5

7. Evaluate $2|3y - 8| + y$ if $y = 2.5$.

8. Solve $-2b > \frac{18 - b}{5}$. Graph the solution set on a number line.

9. **MONEY** Carson has $35 to spend at the water park. The admission price is $25 and each soda is $2.50. Write an inequality to show how many sodas he can buy.

10. Solve $r - 3 < -5$ or $4r + 1 > 15$. Graph the solution set.

11. Solve $|p - 4| \leq 11$. Graph the solution set on a number line.

12. **MULTIPLE CHOICE** Which graph represents the solution set for $4 < 6t + 1 \leq 43$?

    F
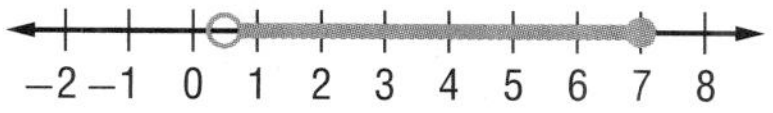

    G
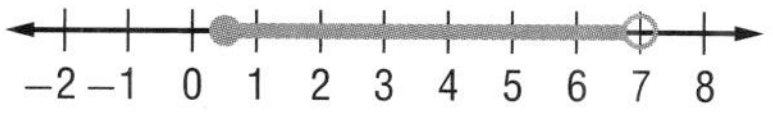

    H
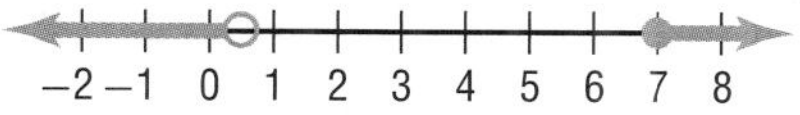

    J
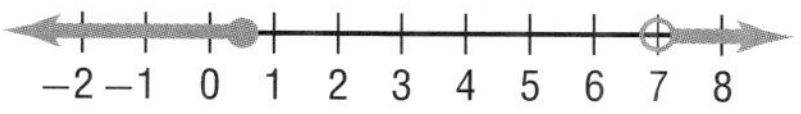

13. **MONEY** Sofia is buying new skis. She finds that the average price of skis is $500 but the actual price could differ from the average by as much as $250. Write and solve an absolute value inequality to describe this situation.

14. **GARDENING** Andy is making 3 trapezoidal garden boxes for his backyard. Each trapezoid will be the size of the trapezoid below. He will place stone blocks around the borders of the boxes. How many feet of stones will Andy need?

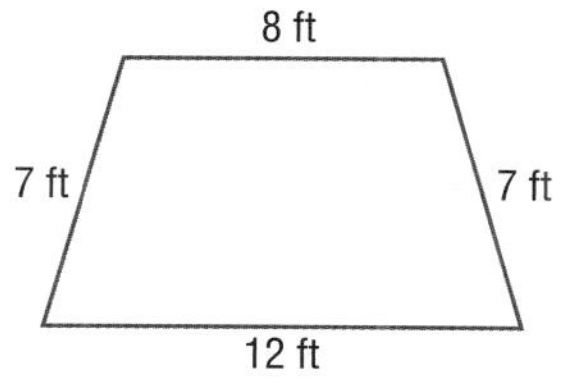

Solve each equation.

15. $|x + 4| = 3$

16. $|3m + 2| = 1$

17. $|3a + 2| = -4$

18. $|2t + 5| - 7 = 4$

19. $|5n - 2| - 6 = -3$

20. $|p + 6| + 9 = 8$

21. **GEOMETRY** The volume of a cylinder is given by the formula $V = \pi r^2 h$. What is the volume of the cylinder below?

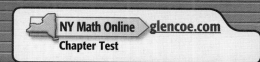

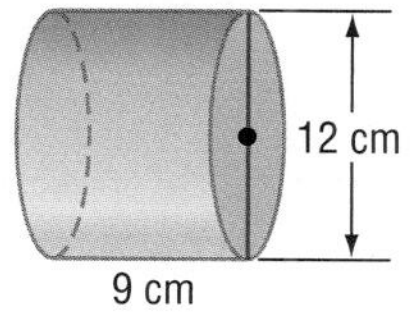

22. Solve $-3b - 5 \geq -6b - 13$. Graph the solution set on a number line.

23. Evaluate $\frac{3(x + y)}{4xy^2}$ if $x = \frac{2}{3}$ and $y = -2$.

24. Name the set(s) of numbers to which $-\frac{1}{3}$ belongs.

25. **MONEY** The costs for making necklaces at two craft stores are shown in the table. For what quantity of beads does The Accessory Store have a better deal? Use the inequality $15 + 3.25b < 20 + 2.50b$.

Shop	Cost per Chain	Cost per Bead
The Accessory Store	$15	$3.25
Finishing Touch	$20	$2.50

Eliminate Unreasonable Answers

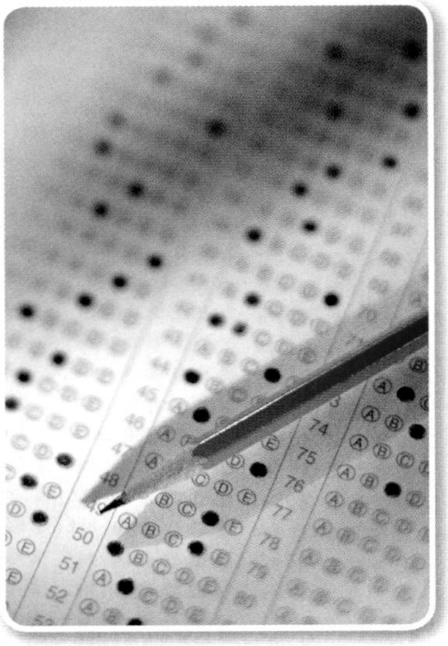

You can eliminate unreasonable answers to help you find the correct answer when solving multiple-choice test items.

Strategies for Eliminating Unreasonable Answers

Step 1

Read the problem statement carefully to determine exactly what you are being asked to find.

Ask yourself:

- What am I being asked to solve?
- What format (that is, fraction, number, decimal, percent, type of graph) will the correct answer be?
- What units (if any) will the correct answer have?

Step 2

Carefully look over each possible answer choice and evaluate for reasonableness.

- Identify any answer choices that are clearly incorrect and eliminate them.
- Eliminate any answer choices that are not in the proper format.
- Eliminate any answer choices that do not have the correct units.

Step 3

Solve the problem and choose the correct answer from those remaining. Check your answer.

EXAMPLE

Read the problem. Identify what you need to know. Then use the information in the problem to solve.

The formula for the area A of a trapezoid with height h and bases b_1 and b_2 is $A = \frac{h}{2}(b_1 + b_2)$. Write an expression to represent the area of the trapezoid at the right.

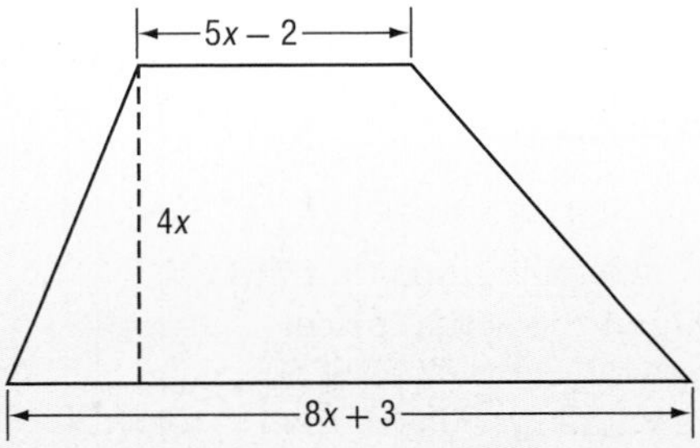

A $26x^2 + 2x$

B $52x^2 + 4x$

C $13x + 1$

D $28x + 10$

To compute the area of the trapezoid, you need to multiply half the height, $2x$, by another linear factor in x. So, the correct answer will contain an x^2 term. Since choices C and D are both linear, they can be eliminated. The correct answer is either A or B. Multiply to find the expression for the area.

$$A = \frac{h}{2}(b_1 + b_2)$$
$$= \frac{4x}{2}(8x + 3 + 5x - 2)$$
$$= 2x(13x + 1)$$
$$= 26x^2 + 2x$$

The correct answer is A.

Exercises

Read each problem. Eliminate any unreasonable answers. Then use the information in the problem to solve.

1. The graph below shows the solution to which inequality?

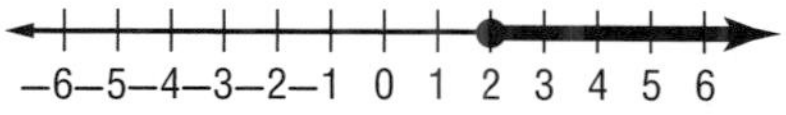

A $8x - 9 \leq 5x - 3$

B $8x - 9 < 5x - 3$

C $8x - 9 \geq 5x - 3$

D $8x - 9 > 5x - 3$

2. Einstein's theory of relativity relates the energy, E, of an object to its mass, m, and the speed of light, c. This relationship can be represented by the formula $E = mc^2$. Solve the formula for m.

F $m = \frac{c^2}{E}$

G $m = \frac{E}{c^2}$

H $m = \frac{c}{E^2}$

J $m = \frac{E^2}{c}$

3. A rectangle has a width of 8 inches and a perimeter of 30 inches. What is the perimeter, in inches, of a similar rectangle with a width of 12 inches?

A 40

B 45

C 48

D 360

4. The rectangular prism below has a volume of 82 cubic inches. What will the volume be if the length, width, and height of the prism are all doubled?

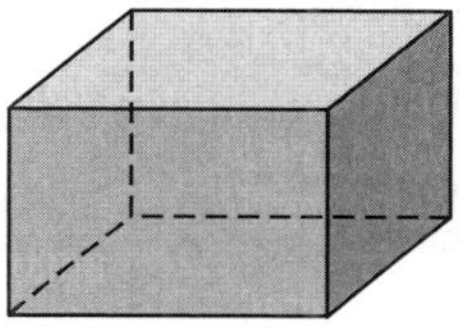

$V = 82$ in^3

F 41 in^3

G 164 in^3

H 482 in^3

J 656 in^3

5. Evaluate $a + (b + 1)^2$ if $a = 3$ and $b = 2$.

A −6

B −1

C 12

D 15

6. At a veterinarian's office, 2 cats and 4 dogs are seen in a random order. What is the probability that the 2 cats are seen in a row?

F $\frac{1}{3}$

G $\frac{2}{3}$

H $\frac{1}{2}$

J $\frac{3}{5}$

CHAPTER 1

NYS Regents Exam Practice

Chapter 1

Multiple Choice

Answer all questions in this part. Select the answer that best completes the statement or answers the question.

1. The state of New York covers about 54,475 square miles, making it the 27th largest state in the country. To which set of numbers does 54,475 *not* belong?

(1) integers
(2) irrational numbers
(3) rational numbers
(4) whole numbers

2. Evaluate $\frac{2a^2 - 4ab}{ab}$ if $a = 6$ and $b = -2$.

(1) −10
(2) −6
(3) 5
(4) 8

Test-TakingTip

Question 2 Substitute 6 for *a* and −2 for *b* in the expression. Then use the order of operations to evaluate the expression.

3. Which property of equality is illustrated by the expression below?

$$d - 1 = 9 \rightarrow 9 = d - 1$$

(1) Reflexive Property
(2) Substitution Property
(3) Symmetric Property
(4) Transitive Property

4. A fountain soft drink machine is supposed to dispense 16 ounces for a medium drink. There are minor variations in the filling process, so in order to be "within spec," the actual fill amounts must satisfy the absolute value inequality $|a - 16| \leq 0.25$. Which of the following fill amounts is not "within spec?"

(1) 15.7 oz
(2) 15.75 oz
(3) 16.2 oz
(4) 16.25 oz

5. Which number line shows the solutions of the absolute value equation $|x - 5| - 2 = 0$?

(1) 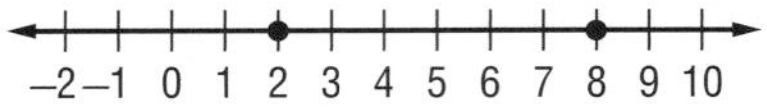

(2) 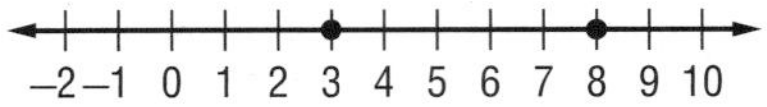

(3) 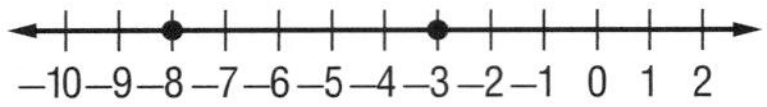

(4) 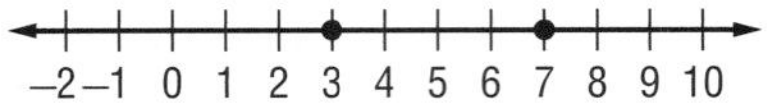

6. Shawn is transporting paver stones and a table saw in his truck for a landscaping project. The saw weighs 120 pounds, and each paver stone weighs 30 pounds. If the maximum weight his truck is capable of carrying is 1500 pounds, what is the greatest number of paver stones Shawn can haul?

(1) 42
(2) 46
(3) 48
(4) 54

7. The volume of a cone with height *h* and radius *r* can be found by multiplying one-third π by the product of the height and the square of the radius. Which equation represents the volume of a cone?

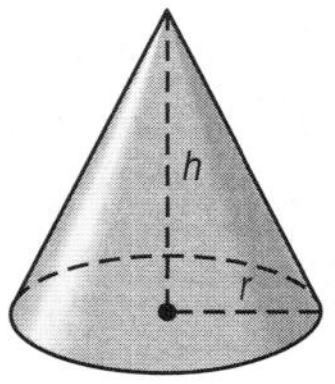

(1) $V = \frac{1}{3}\pi r^2 h$
(2) $V = 3\pi r^2 h$
(3) $V = \frac{1}{3}\pi rh$
(4) $V = \frac{1}{3}\pi rh^2$

8. Which number line shows the solution of the inequality $3n + 6 < 2n + 3$?

(1) 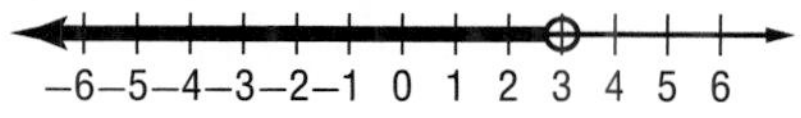

(2) 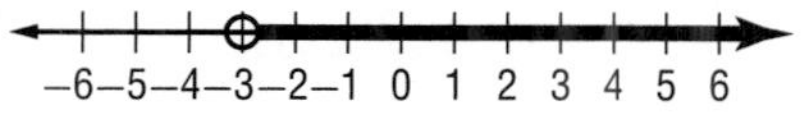

(3) 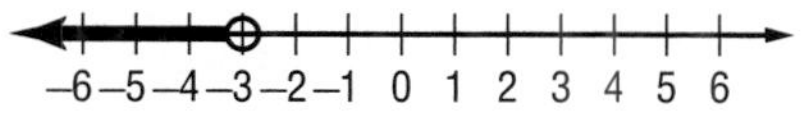

(4) 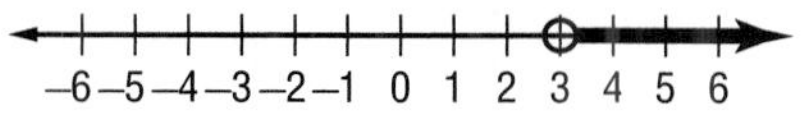

9. Write an algebraic expression to represent the verbal expression below.

five less than the product of 7 and a number

(1) $7n - 5$

(2) $\frac{n}{5} - 7$

(3) $5n - 7$

(4) $\frac{n}{7} - 5$

10. Suppose a thermometer is accurate to within plus or minus 0.15°F. If the thermometer reads 64.5°F, which absolute value inequality represents the actual temperature T?

(1) $|T - 64.5| < 0.15$

(2) $|T - 64.5| \leq 0.15$

(3) $|T - 0.15| < 64.5$

(4) $|T - 0.15| \leq 64.5$

Open-Ended Response

Solve each problem. Clearly indicate the necessary steps, including appropriate formula substitutions, diagrams, graphs, charts, etc.

11. The table below shows Penelope's scores on the first 5 math quizzes this quarter. Each quiz is worth 100 points. There will be 1 more quiz this quarter.

Quiz	Score
1	86
2	79
3	80
4	85
5	77

a. In order to receive a B, Penelope must have a quiz average of 82 or better. Write an inequality that can be solved to find the minimum score she must earn on Quiz 6.

b. Solve the inequality you wrote in part **a**.

c. What does the solution mean?

12. Use the absolute value equation below to answer each question.

$$|x + 8| - 4 = 0$$

a. How many solutions are there of the absolute value equation?

b. Solve the equation.

13. Evaluate the expression $\frac{-5m - 3n}{-2p + r}$ for $m = 1$, $n = -4$, $p = -3$, and $r = -2$.

Need Extra Help?													
If you missed Question...	1	2	3	4	5	6	7	8	9	10	11	12	13
Go to Lesson or Page...	1-1	1-3	1-3	1-6	1-2	1-5	1-3	1-4	1-6	1-2	1-6	1-4	1-1
NYS Core Curriculum	A.CM.11	A.N.1	A.N.1	A2.A.1	A2.A.1	A.A.24	A2.CM.11	A.A.24	A.A.1	A2.A.1	A.A.6	A2.A.1	A.N.1

CHAPTER 2 Linear Relations and Functions

Then

In Chapter 1, you solved equations and inequalities.

Now

In Chapter 2, you will:

- Use equations of relations and functions.
- Determine the slope of a line.
- Use scatter plots and prediction equations.
- Graph linear inequalities.

NYS Core Curriculum

A2.A.41 Use functional notation to evaluate functions for given values in the domain
A2.S.8 Interpret within the linear regression model the value of the correlation coefficient as a measure of the strength of the relationship

Why?

RECREATION Linear functions can be used to model many aspects of recreational activities such as distance ridden on a bicycle, the amount of money a group of people would spend at a state fair, the height of a water slide at various points, or the amount of money you could earn from a hobby.

Math *in Motion*, Animation glencoe.com

Get Ready for Chapter 2

Diagnose Readiness You have two options for checking Prerequisite Skills.

Text Option

Take the Quick Check below. Refer to the Quick Review for help.

QuickCheck

Write the ordered pair for each point. (Lesson 0-1)

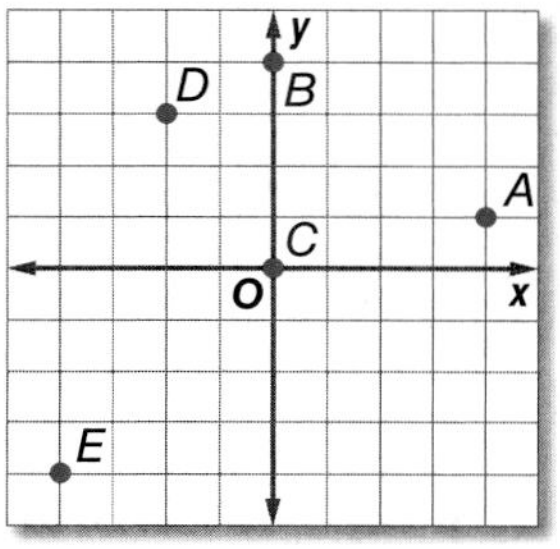

1. A
2. B
3. C
4. D
5. E
6. **BABYSITTING** Aliza earns $6 per hour babysitting. Make a table in which the x-coordinate represents the number of hours Aliza babysits, and the y-coordinate represents the amount of money she earns.

Evaluate each expression if $a = -3$, $b = 4$, and $c = -2$. (Lesson 1-1)

7. $4a - 3$
8. $2b - 5c$
9. $b^2 - 3b + 6$
10. $\frac{2a + 4b}{c}$
11. **PHONE SERVICE** A cell phone company uses the expression $20 + 0.25m$ to determine the monthly charge for m minutes of air time. Find the monthly charge for 80 minutes of air time.

Solve each equation for the given variable. (Lesson 1-3)

12. $4x + 2y = 12$ for y
13. $a = 3b + 9$ for b
14. $15w - 10 = 5v$ for v
15. $3x - 4y = 8$ for x
16. $\frac{d}{6} + \frac{f}{3} = 4$ for d

QuickReview

EXAMPLE 1

Write the ordered pair for point M.

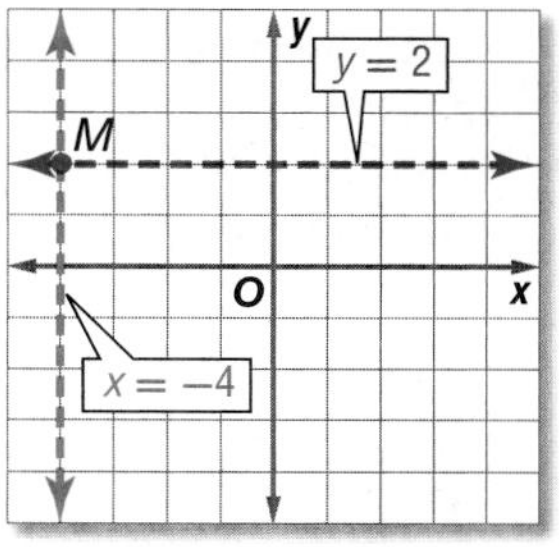

Step 1 Follow a vertical line through the point to find the x-coordinate on the x-axis.

Step 2 Follow a horizontal line through the point to find the y-coordinate on the y-axis.

Step 3 The ordered pair for point M is $(-4, 2)$. It can also be written as $M(-4, 2)$.

EXAMPLE 2

Evaluate $3a^2 - 2ab + b^2$ if $a = 4$ and $b = -3$.

$$\begin{aligned} 3a^2 - 2ab + b^2 &= 3(4^2) - 2(4)(-3) + (-3)^2 \\ &= 3(16) - 2(4)(-3) + 9 \\ &= 48 - (-24) + 9 \\ &= 48 + 24 + 9 \\ &= 81 \end{aligned}$$

EXAMPLE 3

Solve $3x + 6y = 24$ for y.

$3x + 6y = 24$	**Original equation**
$3x + 6y - 3x = 24 - 3x$	**Subtract 3x from each side.**
$6y = 24 - 3x$	**Simplify.**
$\frac{6y}{6} = \frac{24}{6} - \frac{3x}{6}$	**Divide each side by 6.**
$y = 4 - \frac{1}{2}x$	**Simplify.**

Online Option

NY Math Online Take a self-check Chapter Readiness Quiz at glencoe.com.

Get Started on Chapter 2

You will learn several new concepts, skills, and vocabulary terms as you study Chapter 2. To get ready, identify important terms and organize your resources. You may wish to refer to **Chapter 0** to review prerequisite skills.

FOLDABLES® Study Organizer

Linear Relations and Functions Make this Foldable to help you organize your Chapter 2 notes about linear relations and functions. Begin with four sheets of notebook paper.

1. **Fold** each sheet of paper in half from top to bottom.

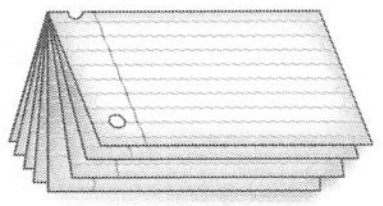

2. **Cut** along the fold. Staple the eight half-sheets together to form a booklet.

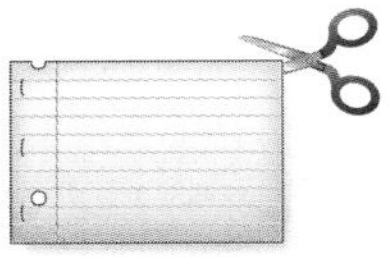

3. **Cut** tabs into the margin. The top tab is 2 lines deep, the next tab is 6 lines deep, and so on.

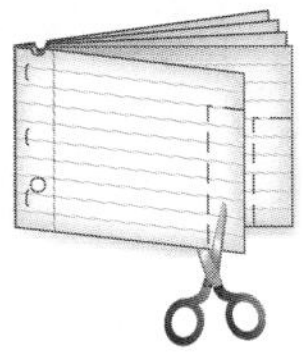

4. **Label** each of the tabs with a lesson number.

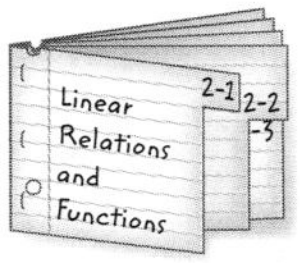

NY Math Online glencoe.com

- Study the chapter online
- Explore **Math in Motion**
- Get extra help from your own **Personal Tutor**
- Use **Extra Examples** for additional help
- Take a **Self-Check Quiz**
- **Review Vocabulary** in fun ways

New Vocabulary

English		Español
one-to-one function	p. 61	función biunívoca
onto function	p. 61	función
discrete relation	p. 62	relación discreta
continuous relation	p. 62	relación continua
vertical line test	p. 62	prudba de la recta vertical
independent variable	p. 64	variable independiente
dependent variable	p. 64	variable dependiente
linear equation	p. 69	ecuación lineal
linear function	p. 69	función lineal
rate of change	p. 76	tasa de cambio
bivariate data	p. 92	datos bivariados
positive correlation	p. 92	correlación positiva
negative correlation	p. 92	correlación negativa
line of fit	p. 92	recta de ajuste
regression line	p. 94	línea de regresión
piecewise-linear function	p. 102	función a intervalos lineal
absolute value function	p. 103	función del valor absoluto
parent function	p. 109	función madre
quadratic function	p. 109	función cuadrática
linear inequality	p. 117	desigualdad lineal

Review Vocabulary

equation • p. 18 • ecuación a mathematical sentence stating that two mathematical expressions are equal

function • p. 7 • función a relation in which each *x*-coordinate is paired with exactly one *y*-coordinate

relation • p. 7 • relación a set of ordered pairs

Relation

Domain Range

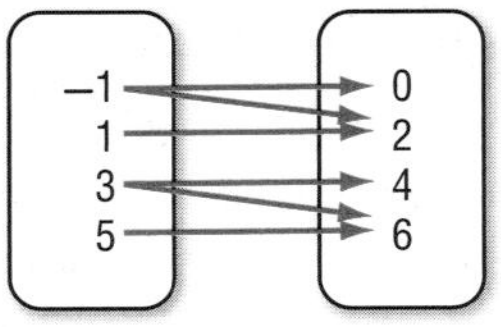

Function

Domain Range

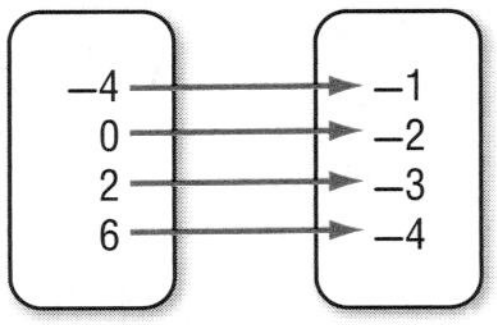

Multilingual eGlossary glencoe.com

2-1 Relations and Functions

Then
You identified domains and ranges for given situations. (Lesson 0-1)

Now
- Analyze relations and functions.
- Use equations of relations and functions.

NYS Core Curriculum

A2.A.41 Use functional notation to evaluate functions for given values in the domain A2.A.43 Determine if a function is one-to-one, onto, or both *Also addresses A2.A.40, A2.A.51, and A2.A.52.*

New Vocabulary
one-to-one function
onto function
discrete relation
continuous relation
vertical line test
independent variable
dependent variable
function notation

NY Math Online
glencoe.com
- Extra Examples
- Personal Tutor
- Self-Check Quiz
- Homework Help

Why?

The table shows the monthly average low and high temperatures for Charlotte, North Carolina. Each month's average temperatures can be represented by the ordered pair (average low, average high). For example, January's average temperatures can be expressed as (32, 51).

Monthly Average Temperature (°F) Charlotte, NC												
Month	Jan	Feb	Mar	Apr	May	Jun	Jul	Aug	Sep	Oct	Nov	Dec
Low	32	34	42	49	58	66	71	69	63	51	42	35
High	51	56	64	73	80	87	90	88	82	73	63	54

Source: The Weather Channel

Relations and Functions Recall that a function is a relation in which each element of the domain is paired with exactly one element in the range. All functions map elements of the domain to elements of the range, but they may differ in the way the elements of the domain and range are paired.

Key Concept Functions

For Your FOLDABLE

one-to-one function
Each element of the domain pairs to exactly one unique element of the range.

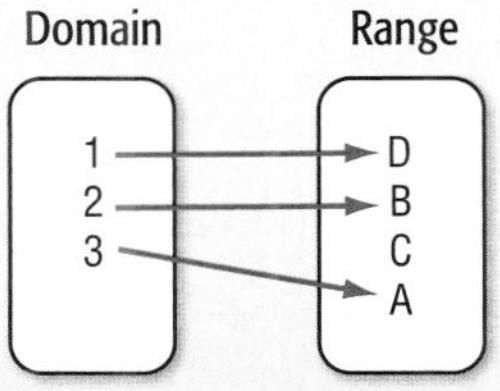

onto function
Each element of the range corresponds to an element of the domain.

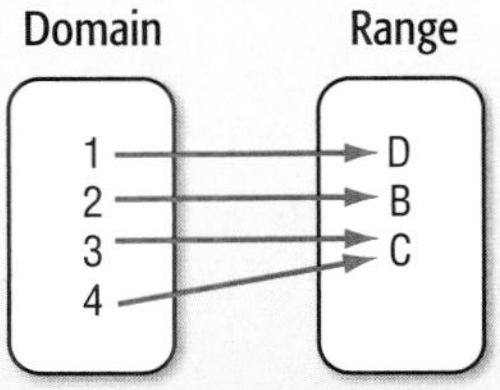

both one-to-one and onto
Each element of the domain is paired to exactly one element of the range, and each element of the range corresponds to a unique element of the domain.

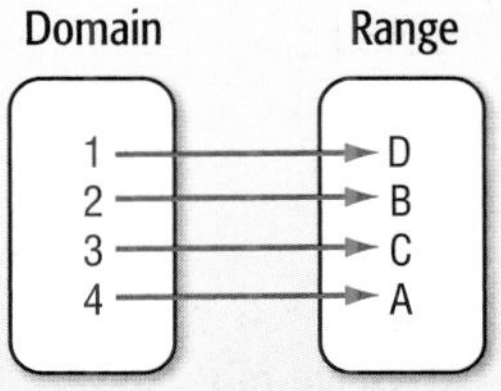

EXAMPLE 1 Domain and Range

State the domain and range of each relation. Then determine whether each relation is a *function*. If it is a function, determine if it is *one-to-one*, *onto*, *both*, or *neither*.

a. {(−6, −1), (−5, −9), (−3, −7), (−1, 7), (6, −9)}

Domain: {−6, −5, −3, −1, 6}   Range: {−9, −7, −1, 7}

function: yes, because each element of the domain is paired with one element of the range

one-to-one: no, because each element of the domain is not paired with a unique element of the range

onto: yes, because each element of the range corresponds to an element of the domain

b.

x	2	−1	−2	−1	2
y	−2	−1	0	1	2

Domain: {−2, −1, 2} Range: {−2, −1, 0, 1, 2}

The relation is not a function because 2 is mapped to both −2 and 2, and −1 is mapped to both −1 and 1.

Check Your Progress

1A.

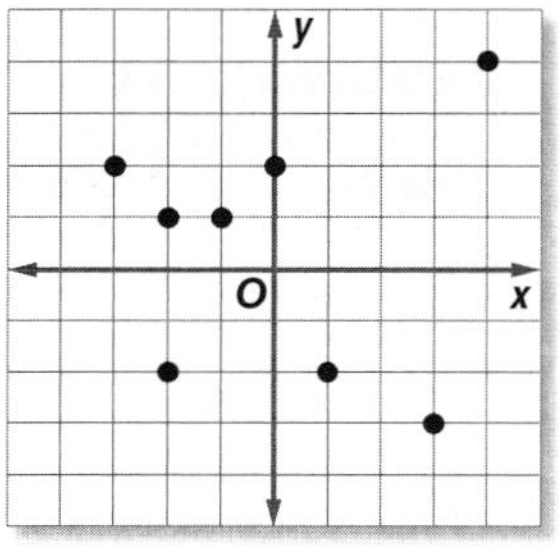

1B.

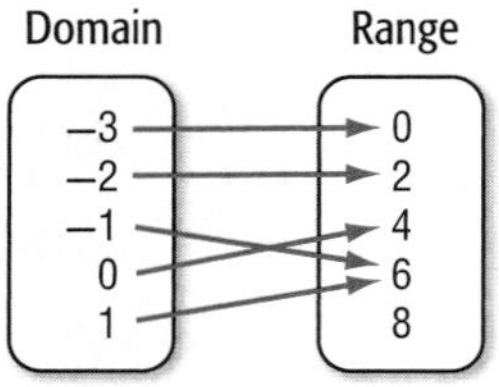

Personal Tutor glencoe.com

A relation in which the domain is a set of individual points, like the relation in Graph A, is said to be a **discrete relation**. Notice that its graph consists of points that are not connected. When the domain of a relation has an infinite number of elements and the relation can be graphed with a line or smooth curve, the relation is a **continuous relation**.

StudyTip

Continuous Relations If you can draw the graph of a relation without lifting your pencil from the paper, this relation is continuous.

Graph A

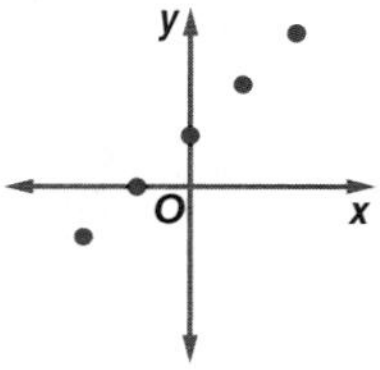

discrete relation

Graph B

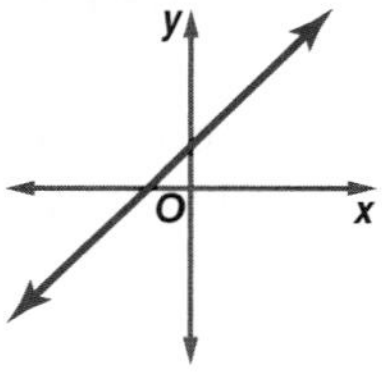

continuous relation

With both discrete and continuous graphs, you can use the **vertical line test** to determine whether the relation is a function.

Key Concept Vertical Line Test

For Your FOLDABLE

Words If no vertical line intersects a graph in more than one point, the graph represents a function.

If a vertical line intersects a graph in two or more points, the graph does not represent a function.

Models

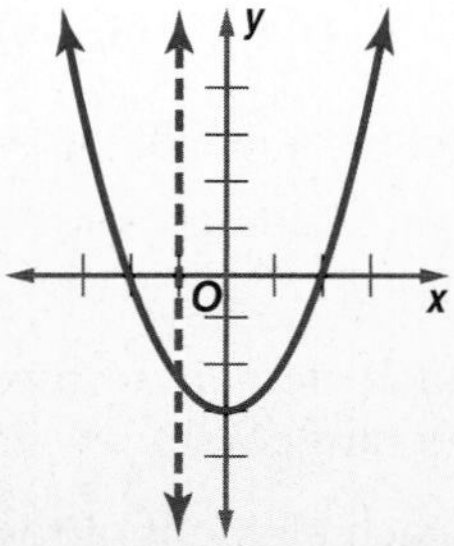

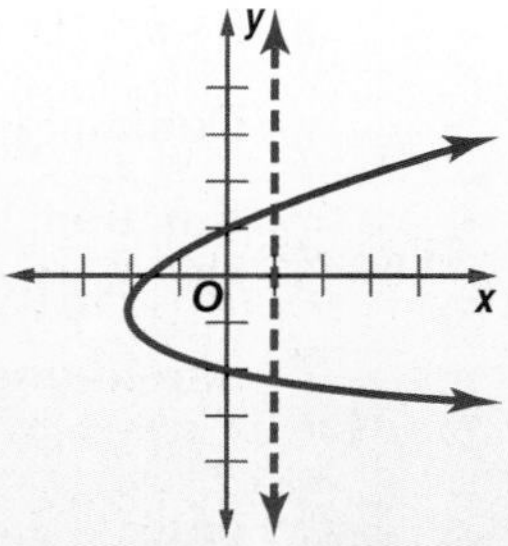

Real-World Link

Lance Armstrong has won the Tour de France more than any other rider, having won 7 consecutive races from 1999 through 2005.

Source: USA Cycling

Real-World EXAMPLE 2

BICYCLING The graph shows the length of the Tour de France in kilometers each year from 1998 through 2006. Is the relation *discrete* or *continuous*? Does the graph represent a function?

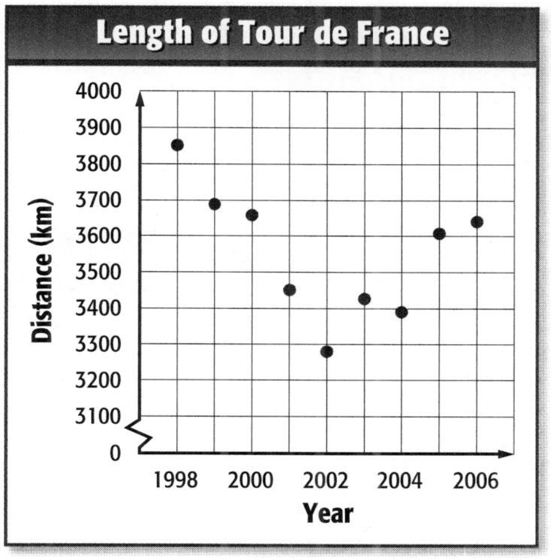

Because the graph consists of distinct points, the function is discrete. Use the vertical line test. No vertical line can be drawn that contains more than one of the data points. Therefore, the relation is a function.

Check Your Progress

2. The number of employees a company had in each year from 2004 to 2009 were 25, 28, 34, 31, 27, and 29. Graph this information and determine whether the relation is *discrete* or *continuous*. Does the graph represent a function?

Personal Tutor glencoe.com

Equations of Relations and Functions Relations and functions can also be represented by equations. The solutions of an equation in x and y are the set of ordered pairs (x, y) that make the equation true. To determine whether an equation represents a function, it is often simplest to look at the graph of the relation.

EXAMPLE 3 Graph a Relation

Graph $y = \frac{1}{2}x - 3$, and determine the domain and range. Then determine whether the equation is a *function*, is *one-to-one*, *onto*, *both*, or *neither*. State whether it is *discrete* or *continuous*.

Make a table of values that satisfy the equation. Then graph the equation.

x	y
−4	−5
−2	−4
0	−3
2	−2
4	−1

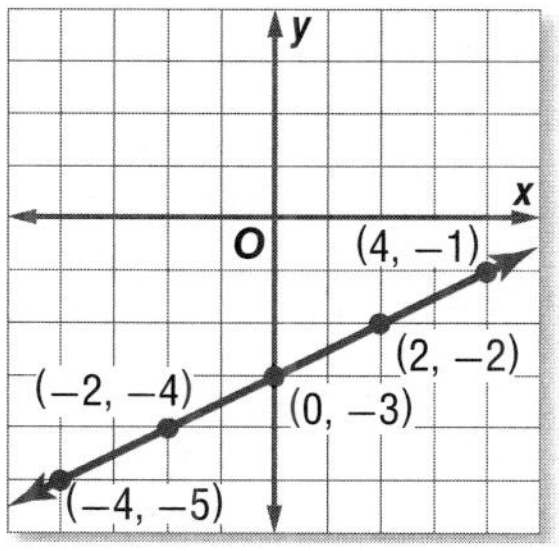

Every real number is the x-coordinate of some point on the line, and every real number is the y-coordinate of some point on the line. So the domain and range are both all real numbers.

The graph passes the vertical line test, so the equation is a function. Every x-value is paired with exactly one unique y-value, and every y-value corresponds to an x-value. Thus, the function is both one-to-one and onto.

Because the graph is a solid line without breaks, the function is continuous.

Check Your Progress

3. Graph $y = x^2 + 1$, and determine the domain and range. Then determine whether the equation is a *function*, is *one-to-one*, *onto*, *both*, or *neither*. State whether it is *discrete* or *continuous*.

Personal Tutor glencoe.com

When an equation represents a function, the variable, often x, with values making up the domain is called the **independent variable**. The other variable, often y, is called the **dependent variable** because its values depend on x.

ReadingMath

Function Notation The symbol $f(x)$ replaces the y and is read "f of x." The f is just the name of the function. It is not a variable that is multiplied by x.

Equations that represent functions are often written in **function notation**. The equation $y = 5x - 1$ can be written as $f(x) = 5x - 1$. Suppose you want to find the value in the range that corresponds to the element -6 in the domain of the function. The value $f(-6)$ is found by substituting -6 for each x in the equation. Therefore, $f(-6) = 5(-6) - 1$ or -31.

EXAMPLE 4 Evaluate a Function

Given $f(x) = 2x^2 - 8$, find each value.

a. $f(6)$

$f(x) = 2x^2 - 8$ **Original function**

$f(6) = 2(6)^2 - 8$ **Substitute.**

$= 2(36) - 8$ **Evaluate 6^2.**

$= 72 - 8$ or 64 **Simplify.**

b. $f(2y)$

$f(x) = 2x^2 - 8$ **Original function**

$f(2y) = 2(2y)^2 - 8$ **Substitute.**

$= 2(4y^2) - 8$ **$(2y)^2 = 2^2y^2$**

$= 8y^2 - 8$ **Simplify.**

Check Your Progress

Given $g(x) = 0.5x^2 - 5x + 3.5$, find each value.

4A. $g(2.8)$

4B. $g(4a)$

Personal Tutor glencoe.com

Check Your Understanding

Example 1 p. 61

State the domain and range of each relation. Then determine whether each relation is a *function*. If it is a function, determine if it is *one-to-one*, *onto*, *both*, or *neither*.

1.

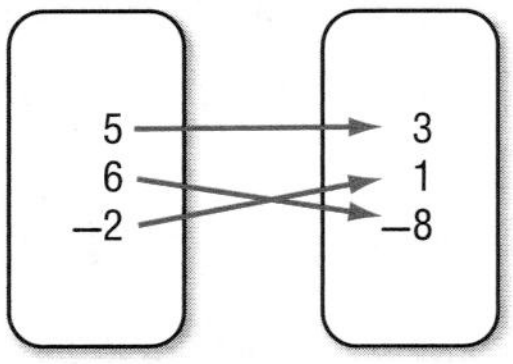

2.

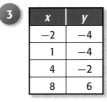

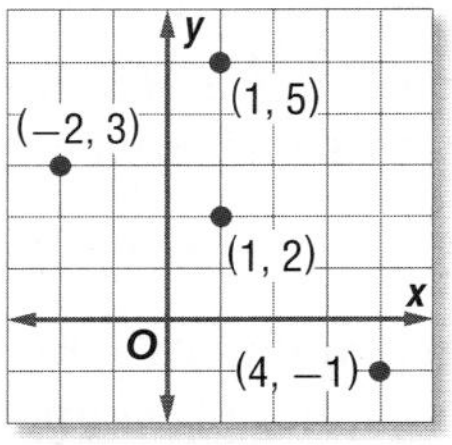

3.

x	y
−2	−4
1	−4
4	−2
8	6

Example 2 p. 63

4. BASKETBALL The table shows the average points per game for Dwayne Wade of the Miami Heat for four years.

SEASON	DWAYNE WADE'S AGE	AVERAGE POINTS PER GAME
2003–2004	22	16.2
2004–2005	23	24.1
2005–2006	24	27.2
2006–2007	25	23.5

a. Assume that the ages are the domain. Identify the domain and range.

b. Write a relation of ordered pairs for the data.

c. State whether the relation is *discrete* or *continuous*.

d. Graph the relation. Is this relation a function?

Example 3 p. 63

Graph each equation, and determine the domain and range. Determine whether the equation is a *function*, is *one-to-one*, *onto*, *both*, or *neither*. Then state whether it is *discrete* or *continuous*.

5. $y = 5x + 4$

6. $y = -4x - 2$

7. $y = 3x^2$

8. $x = 7$

Example 4
p. 64

Evaluate each function.

9. $f(-3)$ if $f(x) = -4x - 8$

10. $g(5)$ if $g(x) = -2x^2 - 4x + 1$

Practice and Problem Solving

= **Step-by-Step Solutions** begin on page R20.
Extra Practice begins on page 947.

Example 1
p. 61

State the domain and range of each relation. Then determine whether each relation is a *function*. If it is a function, determine if it is *one-to-one, onto, both,* or *neither*.

11.

x	y
−0.3	−6
0.4	−3
1.2	−1
1.2	4

12.

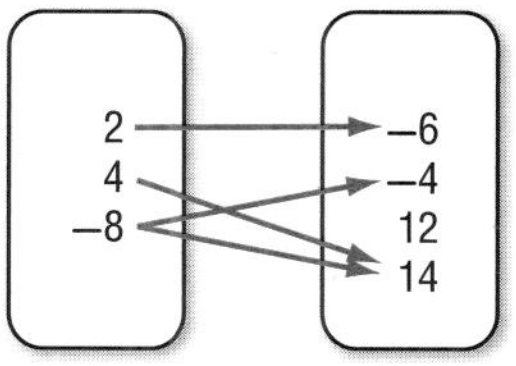

13. $\{(-3, -4), (-1, 0), (3, 0), (5, 3)\}$

Example 2
p. 63

14. POLITICS The table below shows the population of several states and the number of U.S. representatives from those states.

a. Make a graph of the data with population on the horizontal axis and representatives on the vertical axis.

b. Identify the domain and range.

c. Is the relation *discrete* or *continuous*?

d. Does the graph represent a function? Explain your reasoning.

State	Population (millions)	Number of Representatives
California	33.93	53
Florida	16.03	25
Illinois	12.44	19
New York	19.00	29
North Carolina	8.07	13
Texas	20.90	32

Source: U.S. Bureau of the Census

Example 3
p. 63

Graph each equation, and determine the domain and range. Determine whether the equation is a *function*, is *one-to-one, onto, both,* or *neither*. Then state whether it is *discrete* or *continuous*.

15. $y = -3x + 2$

16. $y = 0.5x - 3$

17. $y = 2x^2$

18. $y = -5x^2$

19. $y = 4x^2 - 8$

20. $y = -3x^3 - 1$

Example 4
p. 64

Evaluate each function.

21 $f(-8)$ if $f(x) = 5x^3 + 1$

22. $f(2.5)$ if $f(x) = 16x^2$

23. DIVING The table below shows the pressure on a diver at various depths.

Depth (ft)	0	20	40	60	80	100
Pressure (atm)	1	1.6	2.2	2.8	3.4	4

a. Write a relation to represent the data.

b. Graph the relation.

c. Identify the domain and range. Is the relation *discrete* or *continuous*?

d. Is the relation a function? Explain your reasoning.

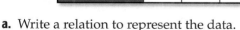

Real-World Link

It takes approximately 32 hours of classroom sessions and practice diving to complete entry-level scuba diving certification.

Source: YMCA Scuba

Find each value if $f(x) = 3x + 2$, $g(x) = -2x^2$, and $h(x) = -4x^2 - 2x + 5$.

24. $f(-5)$

25. $f(9)$

26. $g(-3)$

27. $g(-6)$

28. $h(3)$

29. $h(8)$

30. $f\left(\frac{2}{3}\right)$

31. $g\left(\frac{3}{2}\right)$

32. $h\left(\frac{1}{5}\right)$

33 **PODCASTS** Chaz has a collection of 15 podcasts downloaded on his digital audio player. He decides to download 3 more podcasts each month. The function $P(t) = 15 + 3t$ counts the number of podcasts $P(t)$ he has after t months. How many podcasts will he have after 8 months?

Real-World Link

Introduced in the early 2000s, podcasts allow subscribers to download audio and video programs to their portable media players. Podcasts can include news and information, entertainment, and educational content.

Source: ZDNet

34. **MULTIPLE REPRESENTATIONS** In this problem you will investigate one-to-one and onto functions.

a. **GRAPHICAL** Graph each function on a separate graphing calculator screen.

$f(x) = x^2$ $\qquad g(x) = 2^x$ $\qquad h(x) = x^3 - 3x^2 - 5x + 6$ $\qquad j(x) = x^3$

b. **TABULAR** Use the graphs to create a table showing the number of times a horizontal line could intersect the graph of each function. List all possibilities.

c. **ANALYTICAL** For a function to be one-to-one, a horizontal line on the graph of the function can intersect the function at most once. Which functions meet this condition? Which do not? Explain your reasoning.

d. **ANALYTICAL** For a function to be onto, every possible horizontal line on the graph of the function must intersect the function at least once. Which functions meet this condition? Which do not? Explain your reasoning.

e. **GRAPHICAL** Create a table showing whether each function is one-to-one and/or onto.

H.O.T. Problems

Use Higher-Order Thinking Skills

35. **FIND THE ERROR** Omar and Madison are finding $f(3d)$ for the function $f(x) = -4x^2 - 2x + 1$. Is either of them correct? Explain your reasoning.

Omar	Madison
$f(3d) = -4(3d)^2 - 2(3d) + 1$ $= -4(9d^2) - 6d + 1$ $= -36d^2 - 6d + 1$	$f(3d) = -4(3d)^2 - 2(3d) + 1$ $= 12d^2 - 6d + 1$

36. **CHALLENGE** Consider the functions $f(x)$ and $g(x)$. $f(a) = 19$ and $g(a) = 33$, while $f(b) = 31$ and $g(b) = 51$. If $a = 5$ and $b = 8$, find two possible functions to represent $f(x)$ and $g(x)$.

37. **REASONING** If the graph of a relation crosses the y-axis at more than one point, is the relation *sometimes*, *always*, or *never* a function? Explain your reasoning.

38. **OPEN ENDED** Graph a relation that can be used to represent each of the following.

a. the height of a baseball that is hit into the outfield

b. the speed of a car that travels to the store, stopping at two lights along the way

c. the height of a person from age 5 to age 80

d. the temperature on a typical day from 6 A.M. to 11 P.M.

39. **REASONING** Determine whether the following statement is *true* or *false*. Explain your reasoning.

If a function is onto, then it must be one-to-one as well.

40. **WRITING IN MATH** Explain why the vertical line test can determine if a relation is a function.

NYSRE PRACTICE A2.CM.11

41. Patricia's swimming pool contains 19,500 gallons of water. She drains the pool at a rate of 6 gallons per minute. Which of these equations represents the number of gallons of water g, remaining in the pool after m minutes?

A $g = 19{,}500 - 6m$

B $g = 19{,}500 + 6m$

C $g = \frac{19{,}500}{6m}$

D $g = \frac{6m}{19{,}500}$

42. SHORT RESPONSE Look at the pattern below.

$$-\frac{5}{2}, -2, -\frac{3}{2}, -1, \ldots$$

If the pattern continues, what will the next term be?

43. GEOMETRY Which set of dimensions represents a triangle similar to the triangle shown below?

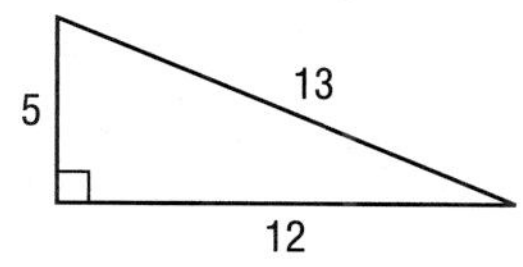

5
13
12

F 1 unit, 2 units, 3 units

G 7 units, 11 units, 12 units

H 10 units, 23 units, 24 units

J 20 units, 48 units, 52 units

44. ACT/SAT If $g(x) = x^2$, which expression is equal to $g(x + 1)$?

A 1

B $x^2 + 1$

C $x^2 + 2x + 1$

D $x^2 - x$

Spiral Review

Solve each inequality. (Lesson 1-6)

45. $48 > 7y + 6 > 20$

46. $z + 12 > 18$ or $-2z + 16 > 12$

47. $2|4x + 2| + 3 > 21$

48. CLUBS Mr. Willis is starting a chess club at his high school. He sent the advertisement at the right to all of the homerooms. Write an absolute value inequality representing the situation. (Lesson 1-6)

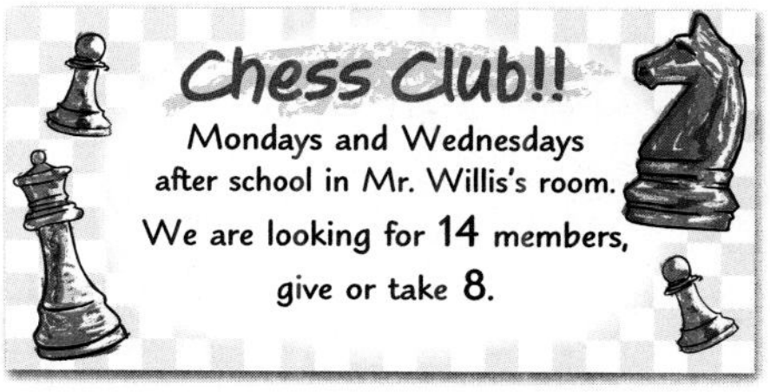

Chess Club!!
Mondays and Wednesdays
after school in Mr. Willis's room.
We are looking for 14 members,
give or take 8.

49. SALES Ling can spend no more than $120 at the summer sale of a department store. She wants to buy shirts on sale for $15 each. Write and solve an inequality to determine the number of shirts she can buy. (Lesson 1-5)

Solve each equation. Check your solutions. (Lesson 1-4)

50. $18 = 2|2a + 6| - 2$

51. $2 = -3|4c - 5| + 8$

52. $-5 = 2|3b + 4| - 9$

Simplify each expression. (Lesson 1-2)

53. $6(3a - 2b) + 3(5a + 4b)$

54. $-4(5x - 3y) + 2(y + 3x)$

55. $-7(2c - 4d) + 8(3c + d)$

Skills Review

Solve each equation. Check your solutions. (Lesson 1-3)

56. $5x + 2 = 32$

57. $6a - 3 = 21$

58. $-2x + 5 = 5x + 19$

59. $6b + 4 = -2b - 28$

60. $2(x + 5) - 3(x - 4) = 19$

61. $4(2y - 3) + 5(3y + 1) = -99$

62. $5c - 8 + 2c = 4c + 10$

63. $8d - 4 + 3d = 2d - 100 - 7d$

64. $10y - 5 - 3y = 4(2y + 3) - 20$

EXTEND 2-1

Algebra Lab
Discrete and Continuous Functions in the Real World

NY Math Online glencoe.com
Math *in Motion,* Animation

Objective
Use discrete and continuous functions to solve real-world problems.

A cup of frozen yogurt costs $2 at the Yogurt Shack. We might describe the cost of x cups of yogurt using the continuous function $y = 2x$, where y is the total cost in dollars. The graph of that function is shown at the right.

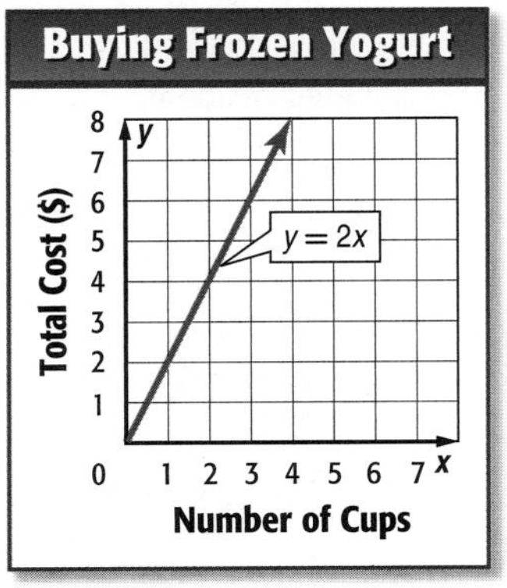

From the graph, you can see that 2 cups of yogurt cost $4, 3 cups cost $6, and so on. The graph also shows that 1.5 cups of yogurt cost 2(1.5) or $3. However, the Yogurt Shack probably will not sell partial cups of yogurt. This function is more accurately modeled with a discrete function.

The graph of the discrete function at the right also models the cost of buying cups of frozen yogurt. The domain in this graph makes sense in this situation.

When choosing a discrete function or a continuous function to model a real-world situation, consider whether all real numbers make sense as part of the domain.

Exercises

Determine whether each function is correctly modeled using a discrete or continuous function. Explain your reasoning.

1. Converting Units

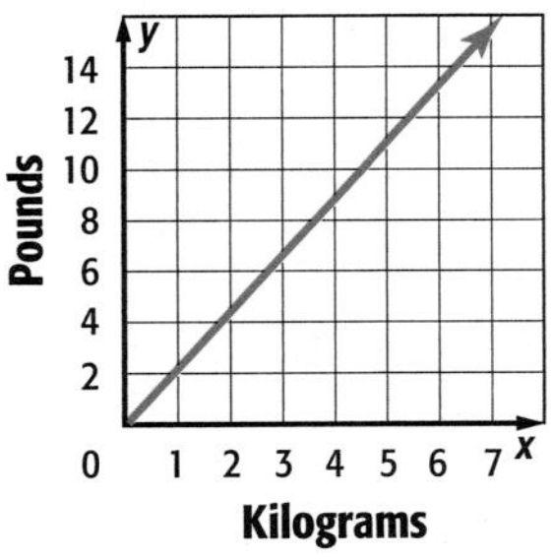

2. E-Mails Received

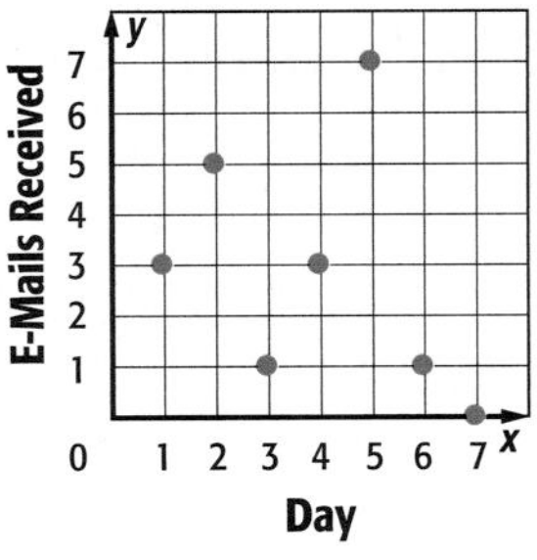

3. y represents the distance a car travels in x hours.

4. y represents the total number of riders who have ridden on a roller coaster after x rides.

5. **WRITING IN MATH** Give an example of a real-world function that is discrete and a real-world function that is continuous. Explain your reasoning.

2-2 Linear Relations and Functions

Then

You analyzed relations and functions. (Lesson 2-1)

Now

- Identify linear relations and functions.
- Write linear equations in standard form.

NYS Core Curriculum

Reinforcement of A.G.4 Identify and graph linear, quadratic (parabolic), absolute value, and exponential functions

New Vocabulary

linear relations
linear equation
linear function
standard form
y-intercept
x-intercept

NY Math Online

glencoe.com

- Extra Examples
- Personal Tutor
- Self-Check Quiz
- Homework Help

Why?

Laura does yard work to earn money during the summer. She either cuts grass x or does general gardening y, and she schedules 5 jobs per day. The equation $x + y = 5$ can be used to relate how many of each task Laura can do in a day.

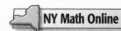

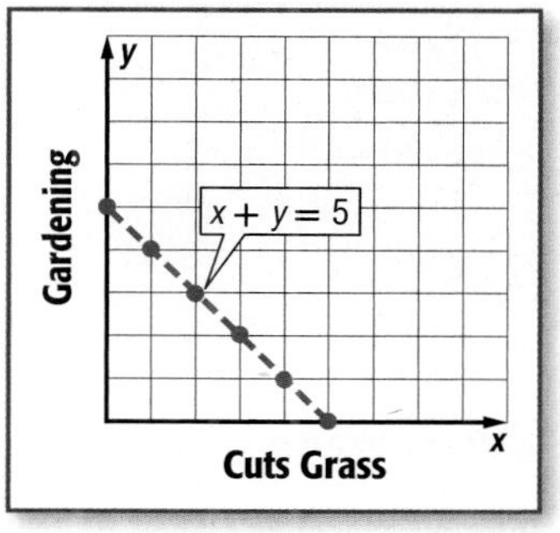

Linear Relations and Functions The points on the graph above lie along a straight line. Relations that have straight line graphs are called **linear relations**.

An equation such as $x + y = 5$ is called a linear equation. A **linear equation** has no operations other than addition, subtraction, and multiplication of a variable by a constant. The variables may not be multiplied together or appear in a denominator. A linear equation does not contain variables with exponents other than 1. The graph of a linear equation is always a line.

Linear equations	Nonlinear equations
$4x - 5y = 16$	$2x + 6y^2 = -25$
$x = 10$	$y = \sqrt{x} + 2$
$y = -\frac{2}{3}x - 1$	$x + xy = -\frac{5}{8}$
$y = \frac{1}{2}x$	$y = \frac{1}{x}$

A **linear function** is a function with ordered pairs that satisfy a linear equation. Any linear function can be written in the form $f(x) = mx + b$, where m and b are real numbers.

EXAMPLE 1 Identify Linear Functions

State whether each function is a linear function. Write *yes* or *no*. Explain.

a. $f(x) = 8 - \frac{3}{4}x$

Yes, because it can be written as $f(x) = -\frac{3}{4}x + 8$.
$m = -\frac{3}{4}, b = 8$

b. $f(x) = \frac{2}{x}$

No, because the expression includes division by the variable.

c. $g(x, y) = 3xy - 4$

No, because the two variables are multiplied together.

Check Your Progress

1A. $f(x) = \frac{5}{x + 6}$

1B. $g(x) = -\frac{3}{2}x + \frac{1}{3}$

Personal Tutor glencoe.com

You can evaluate linear functions by substituting values for x or $f(x)$.

Real-World Link

The largest member of the grass family, bamboo, is capable of growing from 1 to 4 feet per day.

Source: Infoplease

Real-World EXAMPLE 2 Evaluate a Linear Function

PLANTS The growth rate of a sample of Bermuda grass is given by the function $f(x) = 5.9x + 3.25$, where $f(x)$ is the total height in inches x days after an initial measurement.

a. How tall is the sample after 3 days?

$f(x) = 5.9x + 3.25$	**Original function**
$f(3) = 5.9(3) + 3.25$	**Substitute 3 for x.**
$= 20.95$	**Simplify.**

The height of the sample after 3 days is 20.95 inches.

b. The term 3.25 in the function represents the height of the grass when it was initially measured. The sample is how many times as tall after 3 days?

Divide the height after 3 days by the initial height. $\frac{20.95}{3.25} \approx 6.4$

The height after 3 days is about 6.4 times as great as the initial height.

Check Your Progress

2A. If the Bermuda grass is 50.45 inches tall, how many days has it been since it was last cut?

2B. Is it reasonable to think that this rate of growth can be maintained for long periods of time? Explain.

Personal Tutor glencoe.com

Standard Form Any linear equation can be written in **standard form**, $Ax + By = C$, where A, B, and C are integers with a greatest common factor of 1.

Key Concept Standard Form of a Linear Equation

For Your FOLDABLE

Words The standard form of a linear equation is $Ax + By = C$, where A, B, and C are integers with a greatest common factor of 1, $A \geq 0$, and A and B are not both zero.

Examples $3x + 5y = 12$; $A = 3$, $B = 5$, and $C = 12$

EXAMPLE 3 Standard Form

Write $-\frac{3}{10}x = 8y - 15$ in standard form. Identify A, B, and C.

$-\frac{3}{10}x = 8y - 15$	**Original equation**
$-\frac{3}{10}x - 8y = -15$	**Subtract $8y$ from each side.**
$3x + 80y = 150$	**Multiply each side by -10.**

$A = 3$, $B = 80$, and $C = 150$

Check Your Progress

Write each equation in standard form. Identify A, B, and C.

3A. $2y = 4x + 5$

3B. $3x - 6y - 9 = 0$

Personal Tutor glencoe.com

Since two points determine a line, one way to graph a linear function is to find the points at which the graph intersects each axis and connect them with a line. The y-coordinate of the point at which a graph crosses the y-axis is called the **y-intercept**. Likewise, the x-coordinate of the point at which it crosses the x-axis is called the **x-intercept**.

EXAMPLE 4 Use Intercepts to Graph a Line

Find the x-intercept and the y-intercept of the graph of $2x - 3y + 8 = 0$. Then graph the equation.

The x-intercept is the value of x when $y = 0$.

$2x - 3y + 8 = 0$	**Original equation**
$2x - 3(0) + 8 = 0$	**Substitute 0 for y.**
$2x = -8$	**Subtract 8 from each side.**
$x = -4$	**Divide each side by 2.**

The x-intercept is -4.

Likewise, the y-intercept is the value of y when $x = 0$.

$2x - 3y + 8 = 0$	**Original equation**
$2(0) - 3y + 8 = 0$	**Substitute 0 for x.**
$-3y = -8$	**Subtract 8 from each side.**
$y = \frac{8}{3}$	**Divide each side by 3.**

The y-intercept is $\frac{8}{3}$.

Use these ordered pairs to graph the equation.

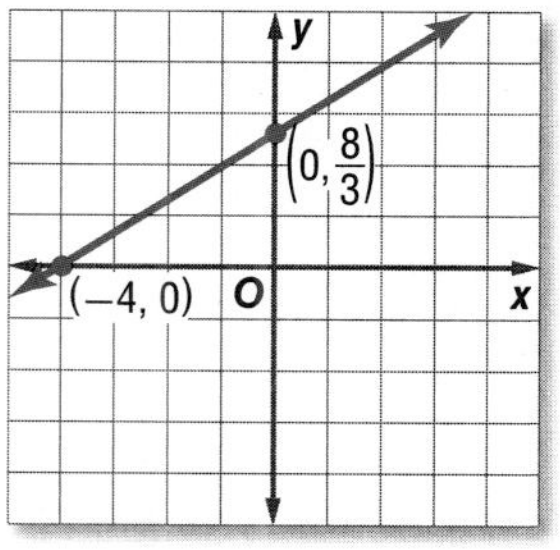

StudyTip

Vertical and Horizontal Lines When C represents a constant, an equation of the form $x = C$ represents a vertical line with only an x-intercept. The equation $y = C$ represents a horizontal line with only a y-intercept.

Check Your Progress

4. Find the x-intercept and the y-intercept of the graph of $2x + 5y - 10 = 0$. Then graph the equation.

Personal Tutor glencoe.com

Check Your Understanding

Example 1 p. 69

State whether each function is a linear function. Write *yes* or *no*. Explain.

1. $f(x) = \frac{x + 12}{5}$ **2.** $g(x) = \frac{7 - x}{x}$ **3.** $p(x) = 3x^2 - 4$ **4.** $q(x) = -8x - 21$

Example 2 p. 70

5 **RECREATION** You want to make sure that you have enough music for a car trip. If each CD is an average of 45 minutes long, the linear function $m(x) = 0.75x$ could be used to find out how many CDs you need to bring.

a. If you have 4 CDs, how many hours of music is that?

b. If the trip you are taking is 6 hours, how many CDs should you bring?

Example 3 p. 70

Write each equation in standard form. Identify A, B, and C.

6. $y = -4x - 7$ **7.** $y = 6x + 5$ **8.** $3x = -2y - 1$

9. $-8x = 9y - 6$ **10.** $12y = 4x + 8$ **11.** $4x - 6y = 24$

Example 4 p. 71

Find the x-intercept and the y-intercept of the graph of each equation. Then graph the equation using the intercepts.

12. $y = 5x + 12$ **13.** $y = 4x - 10$ **14.** $2x + 3y = 12$ **15.** $3x - 4y - 6 = 15$

Practice and Problem Solving

● = **Step-by-Step Solutions** begin on page R20.
Extra Practice begins on page 947.

Example 1 p. 69

State whether each equation or function is a linear function. Write *yes* or *no*. Explain.

16. $3y - 4x = 20$

17. $y = x^2 - 6$

18. $h(x) = 6$

19. $j(x) = 2x^2 + 4x + 1$

20. $g(x) = 5 + \frac{6}{x}$

21. $f(x) = \sqrt{7 - x}$

22. $4x + \sqrt{y} = 12$

23. $\frac{1}{x} + \frac{1}{y} = 1$

24. $f(x) = \frac{4x}{5} + \frac{8}{3}$

Example 2 p. 70

25. ROLLER COASTERS The speed of the Steel Dragon 2000 roller coaster in Mie Prefecture, Japan, can be modeled by $y = 10.4x$, where y is the distance traveled in meters in x seconds.

a. How far does the coaster travel in 25 seconds?

b. The speed of Kingda Ka in Jackson, New Jersey, can be described by $y = 33.9x$. Which coaster travels faster? Explain your reasoning.

Example 3 p. 70

Write each equation in standard form. Identify *A*, *B*, and *C*.

26. $-7x - 5y = 35$

27. $8x + 3y + 6 = 0$

28. $10y - 3x + 6 = 11$

29. $-6x - 3y - 12 = 21$

30. $3y = 9x - 12$

31. $2.4y = -14.4x$

32. $\frac{2}{3}y - \frac{3}{4}x + \frac{1}{6} = 0$

33. $\frac{4}{5}y + \frac{1}{8}x = 4$

34. $-0.08x = 1.24y - 3.12$

Example 4 p. 71

Find the *x*-intercept and the *y*-intercept of the graph of each equation. Then graph the equation using the intercepts.

35. $y = -8x - 4$

36. $5y = 15x - 90$

37. $-4y + 6x = -42$

38. $-9x - 7y = -30$

39. $\frac{1}{3}x - \frac{2}{9}y = 4$

40. $\frac{3}{4}y - \frac{2}{3}x = 12$

41. COMMISSION Latonya earns a commission of \$1.75 for each magazine subscription she sells and \$1.50 for each newspaper subscription she sells. Her goal is to earn a total of \$525 in commissions in the next two weeks.

a. Write an equation that is a model for the different numbers of magazine and newspaper subscriptions that can be sold to meet the goal.

b. Graph the equation. Does this equation represent a function? Explain.

c. If Latonya sells 100 magazine subscriptions and 200 newspaper subscriptions, will she meet her goal? Explain.

42. SNAKES Suppose the body length L in inches of a baby snake is given by $L(m) = 1.5 + 2m$, where m is the age of the snake in months until it becomes 12 months old.

a. Find the length of an 8-month-old snake.

b. Find the snake's age if the length of the snake is 25.5 inches.

43. STATE FAIR The Ohio State Fair charges \$8 for admission and \$5 for parking. After Joey pays for admission and parking, he plans to spend all of his remaining money at the ring game, which costs \$3 per game.

a. Write an equation representing the situation.

b. How much did Joey spend at the fair if he paid \$6 for food and drinks and played the ring game 4 times?

Write each equation in standard form. Identify *A*, *B*, and *C*.

44. $\frac{x+5}{3} = -2y + 4$

45. $\frac{4x-1}{5} = 8y - 12$

46. $\frac{-2x-8}{3} = -12y + 18$

Find the *x*-intercept and the *y*-intercept of the graph of each equation.

47 $\frac{6x+15}{4} = 3y - 12$

48. $\frac{-8x+12}{3} = 16y + 24$

49. $\frac{15x+20}{4} = \frac{3y+6}{5}$

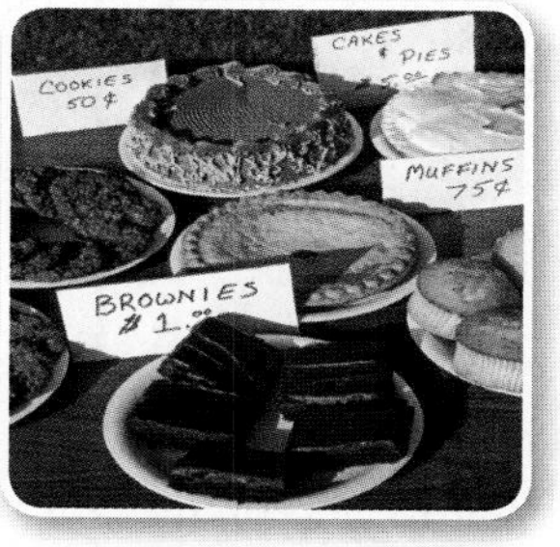

Real-World Link

Bake sales, car washes, and coupon campaigns are among the most popular ideas to make money for schools because it gives students something to offer, making it easier to approach people.

Source: School-Link

50. **FUNDRAISING** The Freshman Class Student Council wanted to raise money by giving car washes. The students spent \$10 on supplies and charged \$2 per car wash.

a. Write an equation to model the situation.

b. Graph the equation.

c. How much money did they earn after 20 car washes?

d. How many car washes are needed for them to earn \$100?

51. **MULTIPLE REPRESENTATIONS** Consider the following linear functions.

$$f(x) = -2x + 4 \qquad g(x) = 6 \qquad h(x) = \frac{1}{3}x + 5$$

a. **GRAPHICAL** Graph the linear functions on separate graphs.

b. **TABULAR** Use the graphs to complete the table.

Function	One-to-One	Onto
$f(x) = -2x + 4$		
$g(x) = 6$		
$h(x) = \frac{1}{3}x + 5$		

c. **VERBAL** Are all linear functions one-to-one and/or onto? Explain your reasoning.

H.O.T. Problems
Use Higher-Order Thinking Skills

52. **CHALLENGE** Write a function with an *x*-intercept of $(a, 0)$ and a *y*-intercept of $(0, b)$.

53. **OPEN ENDED** Write an equation of a line with an *x*-intercept of 3.

54. **REASONING** Determine whether an equation of the form $x = a$, where a is a constant, is *sometimes*, *always*, or *never* a function. Explain your reasoning.

55. **WHICH ONE DOESN'T BELONG?** Of the four equations shown, identify the one that does not belong. Explain your reasoning.

$y = 2x + 3$	$2x + y = 5$	$y = 5$	$y = 2xy$

56. **WRITING IN MATH** Consider the graph of the relationship between hours worked and earnings.

a. Why do you think the graph of this relationship should only be in the first quadrant?

b. Provide another example of a situation in which only the first quadrant is needed. Explain your reasoning.

NYSRE PRACTICE A2.CM.11, A2.A.1

57. Tom bought n DVDs for a total cost of $15n - 2$ dollars. Which expression represents the cost of each DVD?

A $n(15n - 2)$
B $n + (15n - 2)$
C $(15n - 2) \div n;\ n \neq 0$
D $(15n - 2) - n$

58. SHORT RESPONSE What is the complete solution of the equation?

$$|9 - 3x| = 18$$

59. NUMBER THEORY If a, b, c, and d are consecutive odd integers and $a < b < c < d$, how much greater is $c + d$ than $a + b$?

F 2 **H** 6
G 4 **J** 8

60. ACT/SAT Which function is linear?

A $f(x) = x^2$
B $g(x) = 2.7$
C $f(x) = \sqrt{9 - x^2}$
D $g(x) = \sqrt{x - 1}$

Spiral Review

State the domain and range of each relation. Then determine whether each relation is a *function*. If it is a function, determine if it is *one-to-one*, *onto*, *both*, or *neither*. (Lesson 2-1)

61.

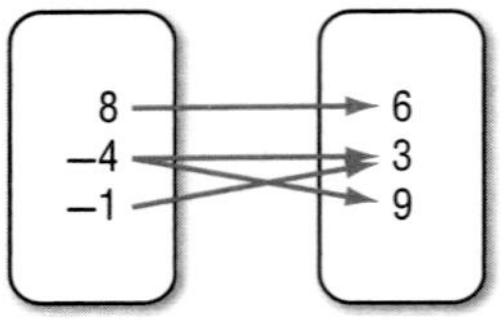

62.

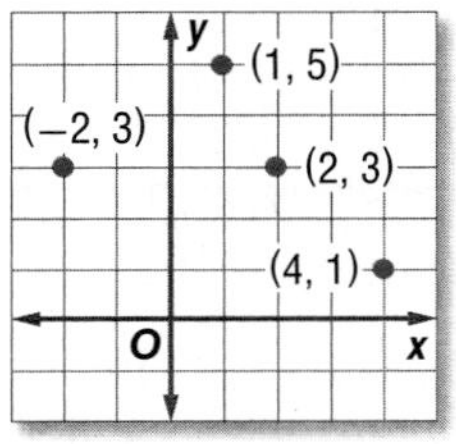

63.

x	y
−4	−2
−3	−1
−3	−1
7	9

64. SHOPPING Claudio is shopping for a new television. The average price of the televisions he likes is \$800, and the actual prices differ from the average by up to \$350. Write and solve an absolute value inequality to determine the price range of the televisions. (Lesson 1-6)

Evaluate each expression if $a = -6$, $b = 5$, and $c = 3.6$. (Lesson 1-1)

65. $\frac{6a - 3c}{2ab}$ **66.** $\frac{a + 7b}{4bc}$ **67.** $\frac{b - c}{a + c}$

68. FOOD Brandi can order a small, medium, or large pizza with pepperoni, mushrooms, or sausage. How many different one-topping pizzas can she order? (Lesson 0-4)

Skills Review

Evaluate each expression. (Lesson 1-1)

69. $\frac{12 - 8}{4 - (-2)}$ **70.** $\frac{5 - 9}{-3 - (-6)}$ **71.** $\frac{-2 - 8}{3 - (-5)}$ **72.** $\frac{-2 - (-6)}{-1 - (-8)}$

73. $\frac{-7 - (-11)}{-3 - 9}$ **74.** $\frac{-1 - 8}{7 - (-3)}$ **75.** $\frac{-12 - (-3)}{-6 - (-5)}$ **76.** $\frac{4 - 3}{2 - 5}$

EXTEND 2-2

Algebra Lab: Roots of Equations and Zeros of Functions

NY Math Online glencoe.com
Math *in Motion*, Animation

The *solution* of an equation is called the *root* of the equation.

EXAMPLE **Determine Roots**

Find the root of $0 = 5x - 10$.

$0 = 5x - 10$	**Original equation**
$10 = 5x$	**Add 10 to each side.**
$2 = x$	**Divide each side by 5.**

The root of the equation is 2.

You can also find the root of an equation by finding the *zero* of its related function. Values of x for which $f(x) = 0$ are called *zeros* of the function f.

Linear Equation	Related Linear Function
$0 = 5x - 10$	$f(x) = 5x - 10$ or $y = 5x - 10$

The zero of a function is the *x-intercept* of its graph. Since the graph of $y = 5x - 10$ intercepts the x-axis at 2, the zero of the function is 2.

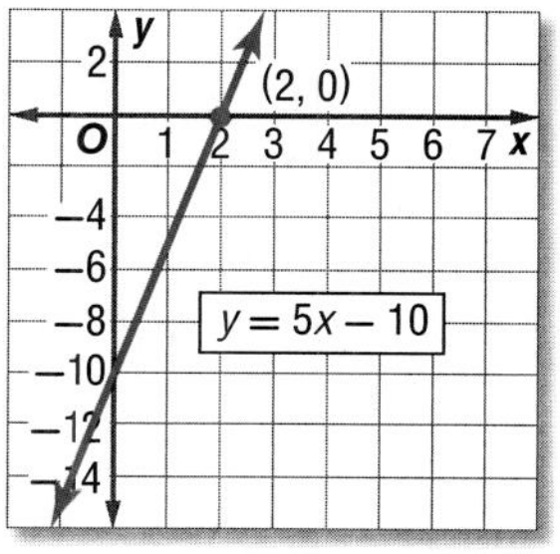

Exercises

1. Use $0 = 4x + 10$ and $f(x) = 4x + 10$ to distinguish among roots, solutions, and zeros.
2. Relate solutions of equations and x-intercepts of graphs.

Determine whether each statement is *true* or *false*. Explain your reasoning.

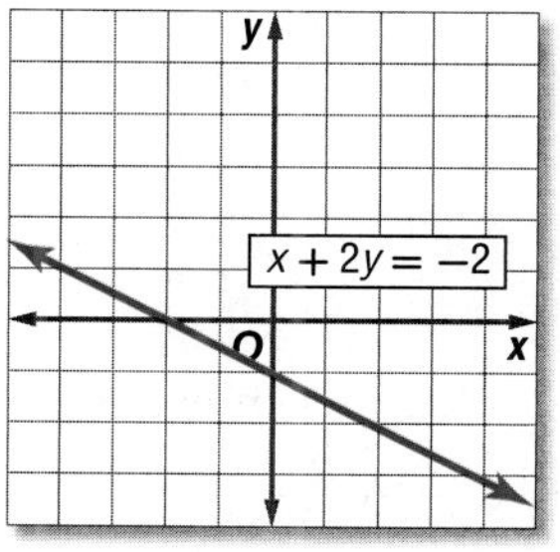

3. The function graphed at the right has two zeros, -2 and -1.
4. The root of $6x + 9 = 0$ is -1.5.
5. $f(0)$ is a zero of the function $f(x) = -\frac{2}{3}x + 12$.
6. **FUNDRAISERS** The function $y = 150 - 2x$ represents the money raised y when the Boosters sell x soft drinks at a basketball game. Find the zero and describe what it means in the context of this situation. Make a connection between the zero of the function and the root of $0 = 150 - 2x$.

2-3 Rate of Change and Slope

Why?

The table at the right shows the total distance a car traveled over various time intervals. You can determine how fast the car was traveling using a formula for distance.

$$rt = d$$
$$r = \frac{d}{t}$$

Time (h)	Distance (mi)
1	68
2.5	170
3	204
4.5	306
5	340

Then
You graphed linear relations. (Lesson 2-2)

Now
- Find rate of change.
- Determine the slope of a line.

NYS Core Curriculum
Reinforcement of A.A.32 Explain slope as a rate of change between dependent and independent variables
Reinforcement of A.A.33 Determine the slope of a line, given the coordinates of two points on the line

New Vocabulary
rate of change
slope

NY Math Online
glencoe.com
- Extra Examples
- Personal Tutor
- Self-Check Quiz
- Homework Help

Rate of Change **Rate of change** is a ratio that compares how much one quantity changes, on average, relative to the change in another quantity. If x is the independent variable and y is the dependent variable, then rate of change $= \frac{\text{change in } y}{\text{change in } x}$. This is sometimes referred to as $\frac{\Delta y}{\Delta x}$.

Real-World EXAMPLE 1 Constant Rate of Change

CHEMISTRY The table shows the temperature of a solution after it has been removed from a heat source. Find the rate of change in temperature for the solution.

Time (min)	Temperature (°C)
0	143.6
2	139.4
5	133.1
8	126.8
12	118.4

Use the ordered pairs (2, 139.4) and (5, 133.1).

$$\text{rate of change} = \frac{\text{change in } y}{\text{change in } x}$$
$$= \frac{\text{change in temperature}}{\text{change in time}}$$
$$= \frac{133.1 - 139.4}{5 - 2}$$
$$= \frac{-6.3}{3} \text{ or } \frac{-2.1}{1}$$

The rate of change is −2.1. This means that the temperature is decreasing by 2.1°C each minute.

Check Your Progress

1. **RECREATION** The graph at the right shows the number of gallons of water in a swimming pool as it is being filled. At what rate is the pool being filled?

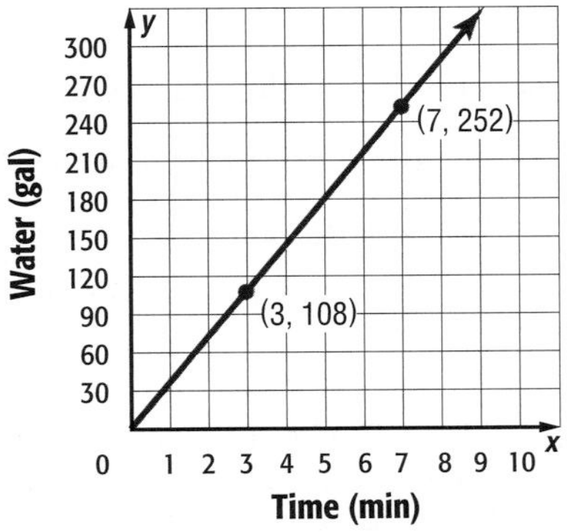

Personal Tutor glencoe.com

Up to this point, you have used rates of change that are constant. Many real-world situations involve rates of change that are not constant. These situations are often described using an average rate of change over a specified interval.

Real-World EXAMPLE 2 Average Rate of Change

MUSIC Refer to the graph at the right. Find the average rate of change of the percent of total music sales for both CDs and downloads from 2001 to 2006. Compare the rates.

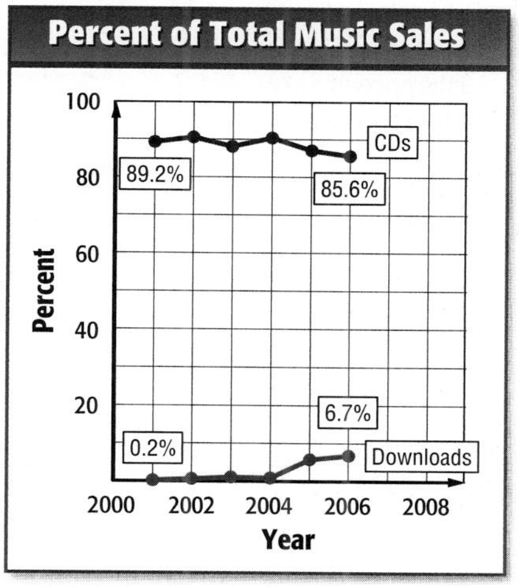

Source: Recording Industry Association of America

CDs:

$$\text{rate of change} = \frac{\text{change in } y}{\text{change in } x}$$
$$= \frac{\text{change in percent}}{\text{change in time}}$$
$$= \frac{85.6 - 89.2}{2006 - 2001}$$
$$= \frac{-3.6}{5} \text{ or } \frac{-0.72}{1}$$

Downloads:

$$\text{rate of change} = \frac{\text{change in } y}{\text{change in } x}$$
$$= \frac{\text{change in percent}}{\text{change in time}}$$
$$= \frac{6.7 - 0.2}{2006 - 2001}$$
$$= \frac{6.5}{5} \text{ or } \frac{1.3}{1}$$

The percent of CD music sales declined at an average rate of 0.72% per year, while the percent of downloaded music sales increased at an average rate of 1.3% per year.

Check Your Progress

2. **EDUCATION** In 2000, 23,142 students applied to State College, and 34,689 students applied to Central University. In 2008, 29,563 students applied to State College, and 36,107 applied to Central University. Determine the average rate of change in applicants for both schools from 2000 to 2008.

Personal Tutor glencoe.com

StudyTip

Independent Quantities Rates of change often include a measure of *time* as the independent variable.

Real-World Link

As of April 2007, 36% of teens preferred to purchase CDs and 64% preferred to download music online.

Source: CNN

Slope The **slope** of a line is the ratio of the change in the y-coordinates to the corresponding change in the x-coordinates. The slope of a line is the same as its rate of change.

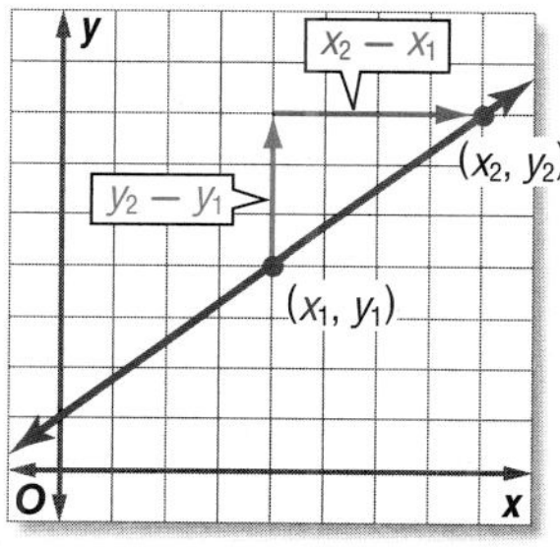

Suppose a line passes through points at (x_1, y_1) and (x_2, y_2).

$$\text{Slope} = \frac{\text{change in } y\text{-coordinates}}{\text{change in } x\text{-coordinates}} = \frac{y_2 - y_1}{x_2 - x_1}$$

StudyTip

Slope The formula for slope is often remembered as *rise over run*, where the rise is the difference in y-coordinates and the run is the difference in x-coordinates.

Key Concept — Slope of a Line

For Your FOLDABLE

Words The slope of a line is the ratio of the change in y-coordinates to the change in x-coordinates.

Example The slope m of a line passing through (x_1, y_1) and (x_2, y_2) is given by $m = \frac{y_2 - y_1}{x_2 - x_1}$, where $x_1 \neq x_2$.

EXAMPLE 3 Find Slope Using Coordinates

Find the slope of the line that passes through (−4, 3) and (2, 5).

$$m = \frac{y_2 - y_1}{x_2 - x_1} \quad \text{Slope Formula}$$

$$= \frac{5 - 3}{2 - (-4)} \quad (x_1, y_1) = (-4, 3), (x_2, y_2) = (2, 5)$$

$$= \frac{2}{6} \text{ or } \frac{1}{3} \quad \text{Simplify.}$$

Check Your Progress

Find the slope of the line that passes through each pair of points.

3A. (1, −3) and (3, 5)

3B. (−8, 11) and (24, −9)

Personal Tutor glencoe.com

You can choose any two points from the graph of a line to find the slope.

StudyTip

Slope is Constant The slope of a line is the same, no matter what two points on the line are used.

EXAMPLE 4 Find Slope Using a Graph

Find the slope of the line shown at the right.

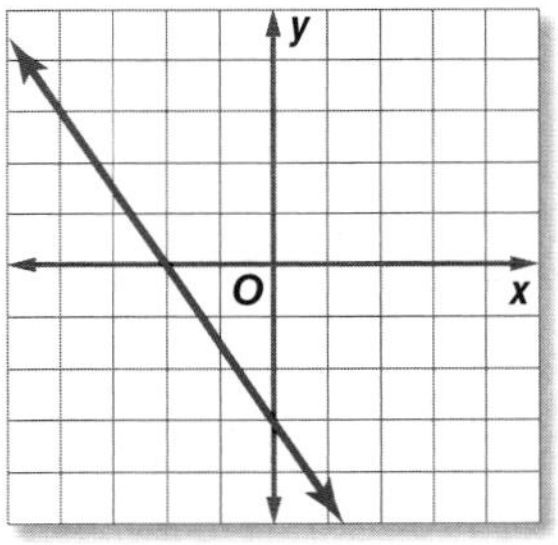

The line passes through (−2, 0) and (0, −3).

$$m = \frac{y_2 - y_1}{x_2 - x_1} \quad \text{Slope Formula}$$

$$= \frac{-3 - 0}{0 - (-2)} \quad (x_1, y_1) = (-2, 0), (x_2, y_2) = (0, -3)$$

$$= \frac{-3}{2} \text{ or } -\frac{3}{2} \quad \text{Simplify.}$$

Check Your Progress

Find the slope of each line.

4A.

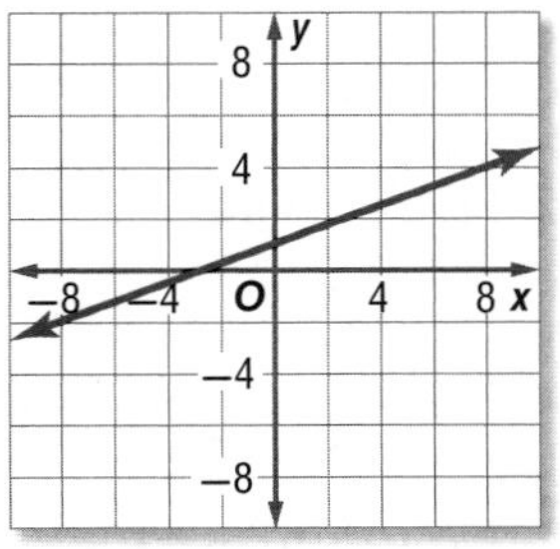

4B.

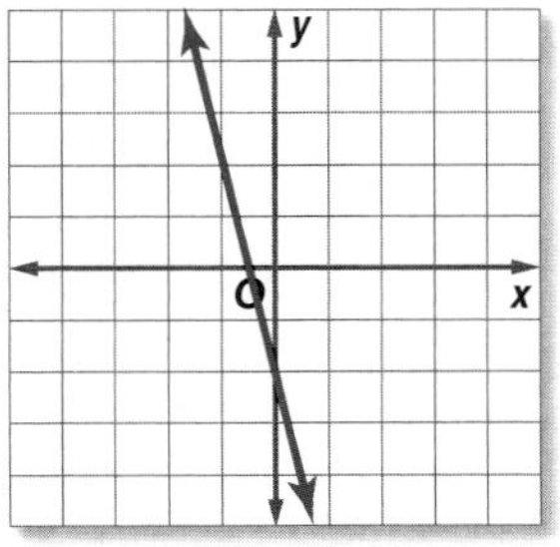

Personal Tutor glencoe.com

Check Your Understanding

Example 1 p. 76

Find the rate of change for each set of data.

1.

Time (min)	2	4	6	8	10
Distance (ft)	12	24	36	48	60

2.

Time (sec)	5	10	15	20	25
Volume (cm^3)	16	32	48	64	80

Example 2 p. 77

3. **CAMERAS** The graph shows the number of digital still cameras and film cameras sold by Yellow Camera Stores in recent years.
   a. Find the average rate of change of the number of digital cameras sold from 2004 to 2009.
   b. Find the average rate of change of the number of film cameras sold from 2004 to 2009.
   c. What do the signs of each rate of change represent?

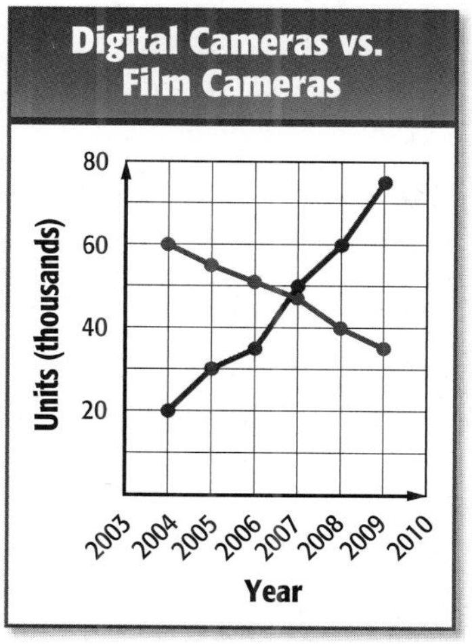

Example 3 p. 78

Find the slope of the line that passes through each pair of points.

4. (3, 2), (8, 12)
5. (−1, 4), (3, −8)
6. (−2, −5), (−7, 10)

Example 4 p. 78

Determine the rate of change of each graph.

7.

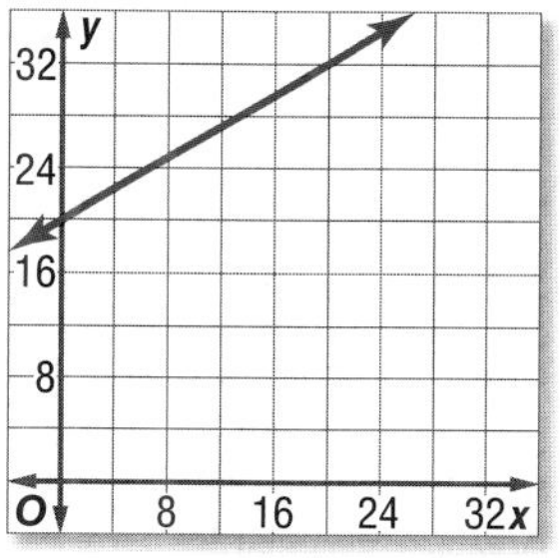

8.

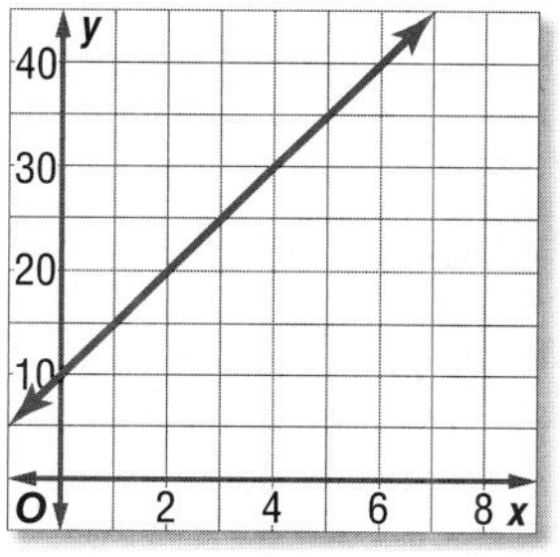

Practice and Problem Solving

● = **Step-by-Step Solutions** begin on page R20.
Extra Practice begins on page 947.

Example 1 p. 76

Find the rate of change for each set of data.

9.

Time (day)	3	6	9	12	15
Height (mm)	20	40	60	80	100

10.

Cost ($)	8	16	24	32	40
Weight (lb)	11	22	33	44	55

Example 2
p. 77

11. HEALTH The table below shows Lisa's temperature during an illness over a 3-day period.

Day	Monday		Tuesday		Wednesday	
Time	8:00 A.M.	8:00 P.M.	8:00 A.M.	8:00 P.M.	8:00 A.M.	8:00 P.M.
Temp (°F)	100.5	102.3	103.1	100.7	99.9	98.6

a. What was the average rate of change in Lisa's temperature from 8:00 A.M. on Monday to 8:00 P.M. on Monday?

b. What was the average rate of change in Lisa's temperature from 8:00 A.M. on Tuesday to 8:00 P.M. on Wednesday? Is your answer reasonable? What does the sign of the rate mean?

c. During which 12-hour period was the average rate of change in Lisa's temperature the greatest?

Example 3
p. 78

Find the slope of the line that passes through each pair of points. Express as a fraction in simplest form.

12. $(-2, 11), (5, 6)$
13. $(-9, -11), (6, 3)$
14. $(-1.5, 3.5), (4.5, 6)$
15. $(-4.5, 9.5), (-1, 2.5)$
16. $(-8, -0.5), (-4, 5)$
17. $(-6, -2), (-1.5, 5.5)$

Example 4
p. 78

Determine the rate of change of each graph.

18.
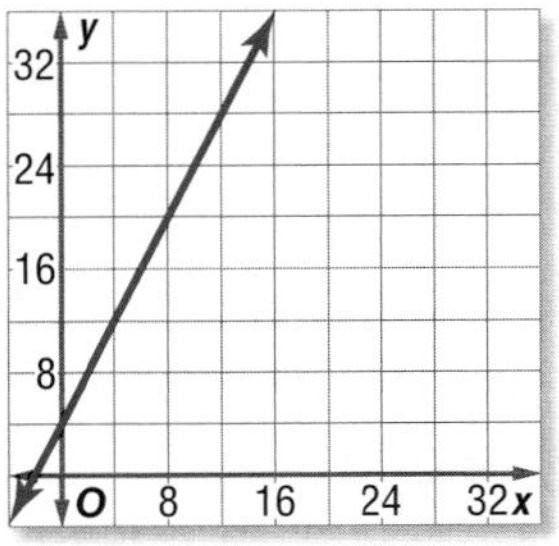

19.
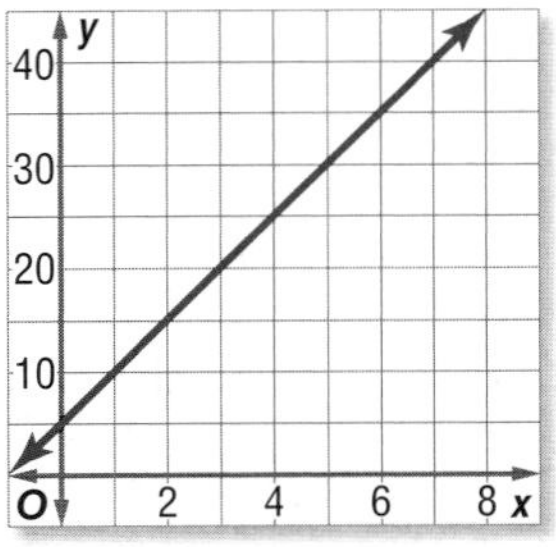

20.
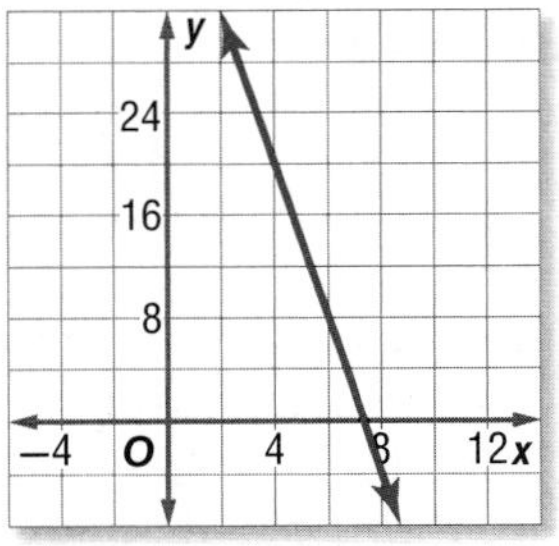

21
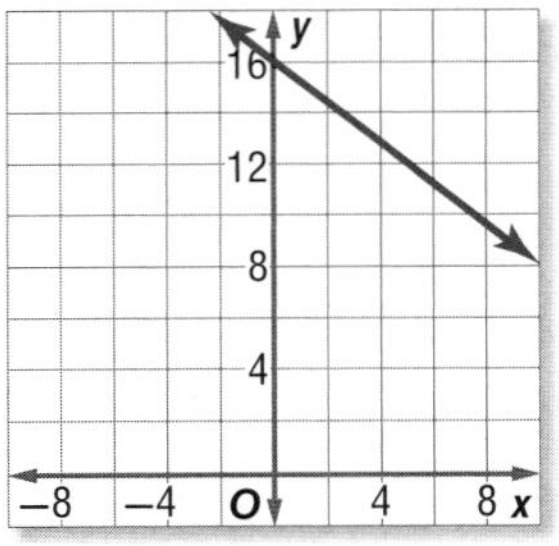

Real-World Link

The Summit Plummet at Walt Disney World in Orlando, Florida, is the tallest water slide in the U.S.

Source: About, Inc.

22. RECREATION The table shows your height on a water slide at various time intervals.

Time (s)	Height (ft)
0	120
1	90
2	60
3	30
4	0
5	0

a. Graph the height versus the time on the water slide.

b. Find the average rate of change of a rider between 1 and 3 seconds.

c. Find the average rate of change of a rider between 0 and 5 seconds.

d. What is another word for *rate of change* in this situation?

Determine the rate of change for each equation.

23. $6y = 8x - 40$
24. $-2y - 16x = 41$
25. $12x - 4y + 5 = 18$
26. $20x + 85y = 120$
27. $\frac{3}{2}x - \frac{5}{4}y = 15$
28. $\frac{1}{6}y + \frac{3}{8}x = 24$

29. **WASHINGTON MONUMENT** The Washington Monument is 555 feet $5\frac{1}{8}$ inches tall and weighs 90,854 tons. The monument is topped by an aluminum square pyramid. The sides of the pyramid's base measure 5.6 inches, and the pyramid is 8.9 inches tall. Estimate the slope that a face of the pyramid makes with its base.

Real-World Link

Construction on the Washington Monument was begun in 1848 and ceased in 1854 due to lack of funds. Construction resumed in 1879 and was completed in 1884.

Source: National Park Service

30. **MARINE LIFE** The illustrations show the growth of a starfish over time.

   a. Find the average rate of change in the measure over time.

   b. Predict the size of the starfish in 2009.

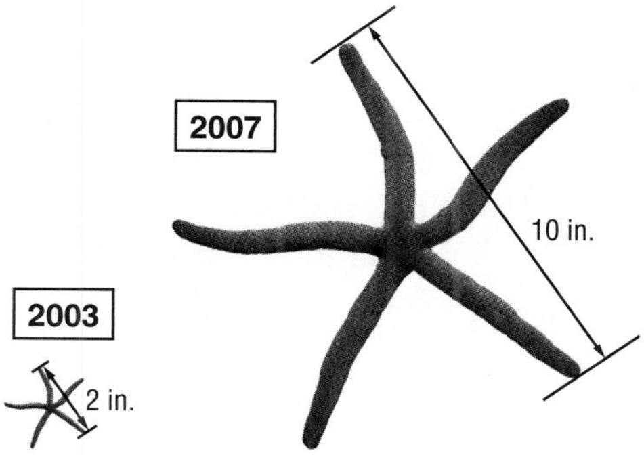

Find the value of r so that the line that passes through each pair of points has the given slope.

31. $(6, r), (3, 3), m = 2$

32. $(8, 1), (5, r), m = \frac{1}{3}$

33. $(10, r), (4, -3), m = \frac{4}{3}$

34. $(8, -2), (r, -6), m = -4$

35. **MULTIPLE REPRESENTATIONS** In this problem, you will explore the rate of change for the function $f(x) = x^2$.

   a. **GRAPHICAL** Graph $f(x) = x^2$.

   b. **TABULAR** Complete the table.

x	−4	−3	−2	−1	0	1	2	3	4
$f(x)$									
slope									

   c. **VERBAL** Describe what happens to the rate of change for $f(x) = x^2$ as x increases.

H.O.T. Problems Use Higher-Order Thinking Skills

36. **FIND THE ERROR** Patty and Tim are asked to find the slope of the line passing through the points (4, 3) and (7, 9). Is either of them correct? Explain.

Patty

$$m = \frac{9-3}{7-4} = \frac{6}{3} \text{ or } 2$$

Tim

$$m = \frac{7-4}{9-3} = \frac{3}{6} \text{ or } \frac{1}{2}$$

37. **CHALLENGE** The graph of a line passes through the points (2, 3) and (5, 8). Explain how you would find the y-coordinate of the point $(11, y)$ on the same line. Then find y.

38. **OPEN ENDED** Write an example of a function with a rate of change four times as large as its x-intercept.

39. **REASONING** Determine whether the statement *A line has a slope that is a real number* is *sometimes*, *always*, or *never* true. Explain your reasoning.

40. **WRITING IN MATH** Describe the process of finding the rate of change for each.

   a. a table of values b. a graph c. an equation

NYSRE PRACTICE Reinforcement of A.A.33

41. GRIDDED RESPONSE What is the slope of the line shown in the graph?

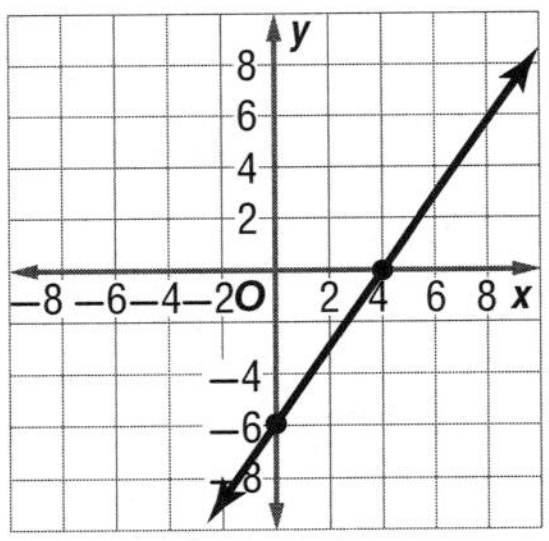

42. ACT/SAT In the figure below, the large square contains two smaller squares. If the areas of the two smaller squares are 4 and 25, what is the sum of the perimeters of the two shaded rectangles?

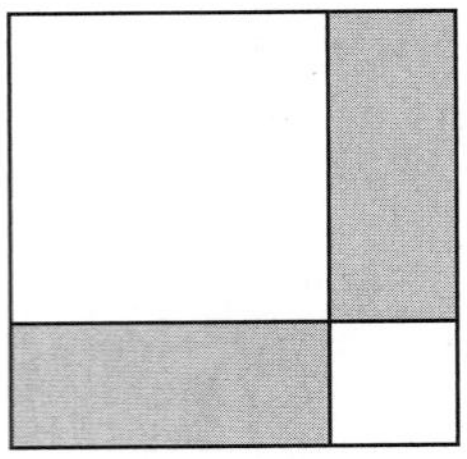

A 14
B 20
C 24
D 28

43. GEOMETRY In $\triangle ABC$ shown, $AC = 16$ and $m\angle DAB = 60$. What is the measure of $\overline{BD}$?

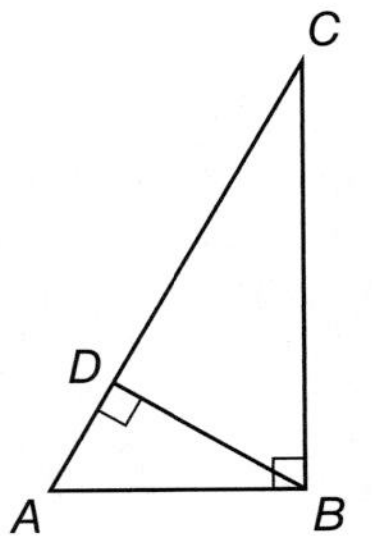

F $9\sqrt{2}$
G $4\sqrt{3}$
H 9
J 4

44. The table shows the cost of bananas depending on the amount purchased. Which conclusion can be made based on information in the table?

Cost of Bananas	
Number of Pounds	Cost ($)
5	1.45
20	4.60
50	10.50
100	19.00

A The cost of 10 pounds of bananas would be more than $4.
B The cost of 200 pounds of bananas would be at most $38.
C The cost of bananas is always more than $0.20 per pound.
D The cost of bananas is always less than $0.28 per pound.

Spiral Review

State whether each equation or function is a linear function. Write *yes* or *no*. Explain. (Lesson 2-2)

45. $6y - 8x = 19$

46. $4x^2 = 2y - 9$

47. $18 = 2xy + 6$

Evaluate each function. (Lesson 2-1)

48. $f(-9)$ if $f(x) = -7x + 8$

49. $g(-4)$ if $g(x) = -3x^2 + 2$

50. $h(12)$ if $h(x) = 4x^2 - 10x$

51. RACING There are 8 contestants in a 400-meter race. In how many different ways can the top three runners finish? (Lesson 0-5)

Determine the quadrant of the coordinate plane where each point is located. (Lesson 0-1)

52. $(-4, -8)$

53. $(-2, 6)$

54. $(3, -1)$

Skills Review

Solve each equation. (Lesson 1-3)

55. $8 = 4m - 6$

56. $-6 = 3(8) + b$

57. $-2 = -3x + 5$

2-4 Writing Linear Equations

Then

You determined slopes of lines. (Lesson 2-3)

Now

- Write an equation of a line given the slope and a point on the line.
- Write an equation of a line parallel or perpendicular to a given line.

NYS Core Curriculum

Reinforcement of A.A.34 Write the equation of a line, given its slope and the coordinates of a point on the line
Reinforcement of A.A.36 Write the equation of a line parallel to the x- or y-axis

New Vocabulary

slope-intercept form
point-slope form
parallel
perpendicular

NY Math Online

glencoe.com

- Extra Examples
- Personal Tutor
- Self-Check Quiz
- Homework Help
- Math in Motion

Why?

Medical insurance companies often require their customers to make a co-payment for every doctor's office visit in addition to an annual insurance premium.

If an insurance company charges \$2280 annually and requires a copayment of \$35 per doctor's office visit, then the linear equation $y = 35x + 2280$ can represent the total annual cost y for x doctor's office visits.

Forms of Equations Consider the line through $A(0, b)$ and $C(x, y)$. Notice that b is the y-intercept. You can use these two points to find the slope of $\overleftrightarrow{AC}$.

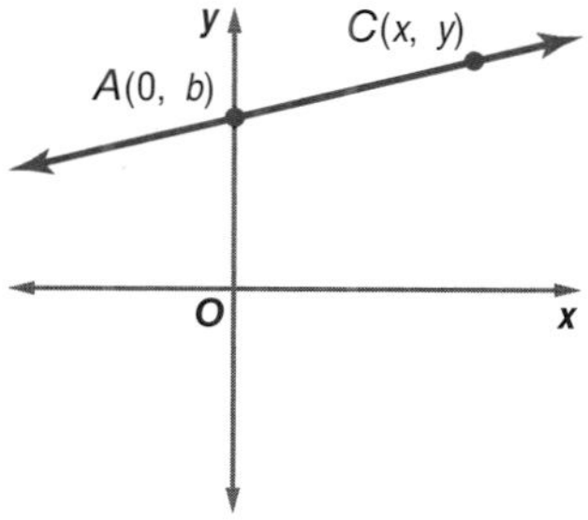

$m = \dfrac{y_2 - y_1}{x_2 - x_1}$ **Slope Formula**

$= \dfrac{y - b}{x - 0}$ **$(x_1, y_1) = (0, b), (x_2, y_2) = (x, y)$**

$= \dfrac{y - b}{x}$ **Simplify.**

Now solve the equation for y.

$mx = y - b$ **Multiply each side by x.**

$mx + b = y$ **Add b to each side.**

$y = mx + b$ **Symmetric Property of Equality**

Equations written in this format are in **slope-intercept form**.

Key Concept: Slope-Intercept Form (For Your FOLDABLE)

Words The slope-intercept form of the equation of a line is $y = mx + b$, where m is the slope and b is the y-intercept.

Symbols $y = mx + b$ (slope: m; y-intercept: b)

Model

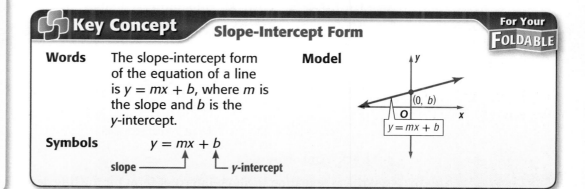

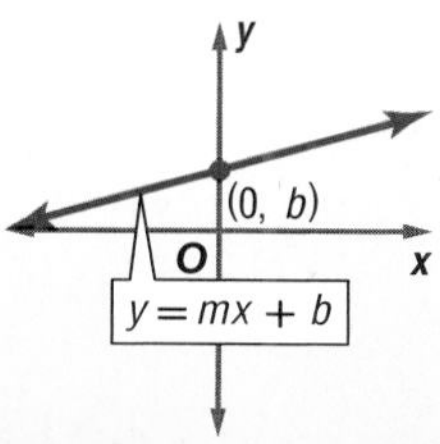

If you are given the slope and y-intercept of a line, you can find an equation of the line by substituting the values of m and b into the slope-intercept form.

Watch Out!

Slope-Intercept Form The equation of a vertical line cannot be written in slope-intercept form because its slope is undefined.

Sometimes it is necessary to calculate the slope before you can write an equation.

EXAMPLE 1 Write an Equation in Slope-Intercept Form

Write an equation in slope-intercept form for the line.

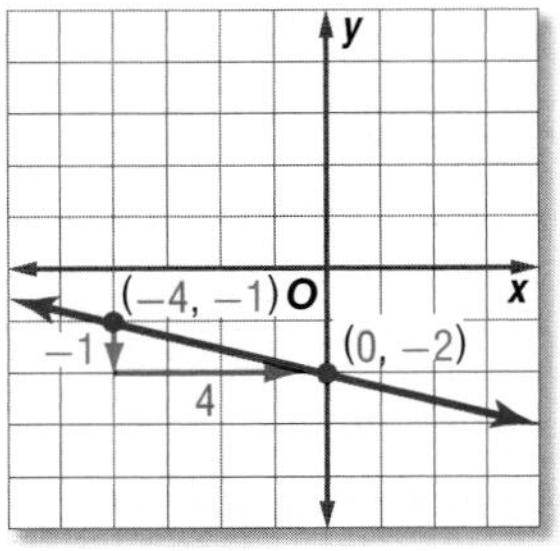

The graph intersects the y-axis at -2. So $b = -2$.

Step 1 Find the slope.

$m = \frac{y_2 - y_1}{x_2 - x_1}$ **Slope Formula**

$= \frac{-2 - (-1)}{0 - (-4)}$ **$(x_1, y_1) = (-4, -1), (x_2, y_2) = (0, -2)$**

$= \frac{-1}{4}$ or $-\frac{1}{4}$ **Simplify.**

Step 2 Substitute the values into the slope-intercept equation.

$y = mx + b$ **Slope-intercept form**

$y = -\frac{1}{4}x - 2$ **$m = -\frac{1}{4}, b = -2$**

StudyTip

Check Your Results You can check that your equation satisfies the conditions by graphing it.

Check Your Progress

Write an equation in slope-intercept form for the line described.

1A. slope $\frac{4}{3}$, passes through (0, 4)

1B. passes through (0, −6) and (−4, 10)

Personal Tutor glencoe.com

If you know the slope of a line and the coordinates of a point on the line, you can use the **point-slope form** to find an equation of the line.

Key Concept Point-Slope Form

For Your FOLDABLE

Words The point-slope form of the equation of a line is $y - y_1 = m(x - x_1)$, where (x_1, y_1) are the coordinates of a point on the line and m is the slope of the line.

Symbols

slope

$y - y_1 = m(x - x_1)$

coordinates of a point on the line

Math *in Motion,* Interactive Lab glencoe.com

EXAMPLE 2 Write an Equation Given Slope and One Point

Write an equation of the line through (6, −2) with a slope of −4.

$y - y_1 = m(x - x_1)$ **Point-slope form**

$y - (-2) = -4(x - 6)$ **$(x_1, y_1) = (6, -2), m = -4$**

$y + 2 = -4x + 24$ **Simplify.**

$y = -4x + 22$ **Subtract 2 from each side.**

Check Your Progress

Write an equation in slope-intercept form for the line described.

2A. passes through (2, 3); $m = \frac{1}{2}$

2B. passes through (−2, −1); $m = -3$

Personal Tutor glencoe.com

You can use any two points on a line to write an equation.

NYSRE EXAMPLE 3 Reinforcement of A.A.34

Which is an equation of the line that passes through (−2, 7) and (3, −3)?

A $y = -\frac{1}{2}x - \frac{3}{2}$

B $y = -2x + 3$

C $y = \frac{1}{2}x + 8$

D $y = 2x + 11$

Read the Test Item

You are given the coordinates of two points on the line.

Solve the Test Item

Step 1 Find the slope of the line.

$m = \frac{y_2 - y_1}{x_2 - x_1}$ **Slope Formula**

$= \frac{-3 - 7}{3 - (-2)}$ **$(x_1, y_1) = (-2, 7)$, $(x_2, y_2) = (3, -3)$**

$= -\frac{10}{5}$ or -2 **Simplify.**

Step 2 Write an equation. Use either ordered pair for (x_1, y_1).

$y - y_1 = m(x - x_1)$ **Point-slope form**

$y - (-3) = -2(x - 3)$ **$(x_1, y_1) = (3, -3)$ and $m = -2$**

$y + 3 = -2x + 6$ **Simplify.**

$y = -2x + 3$ **Subtract 3 from each side.**

The answer is B.

> **Test-Taking Tip**
>
> **Definitions** Be certain to review key vocabulary, such as *y*-intercept, so that you understand what is being asked in a question.

Check Your Progress

3. Which is an equation of the line that passes through (4, −9) and (2, −4)?

F $y = -\frac{5}{2}x + 1$

G $y = -\frac{5}{2}x - 1$

H $y = -\frac{2}{5}x + \frac{37}{5}$

J $y = -\frac{2}{5}x - \frac{37}{5}$

Personal Tutor glencoe.com

Parallel and Perpendicular Lines Slopes can help you determine whether two lines are parallel or perpendicular.

Key Concept Parallel and Perpendicular Lines

For Your FOLDABLE

Parallel Lines	Perpendicular Lines
Two nonvertical lines are **parallel** if and only if they have the same slope. All vertical lines are parallel.	Two nonvertical lines are **perpendicular** if and only if the product of the slopes is −1. Vertical lines and horizontal lines are perpendicular.
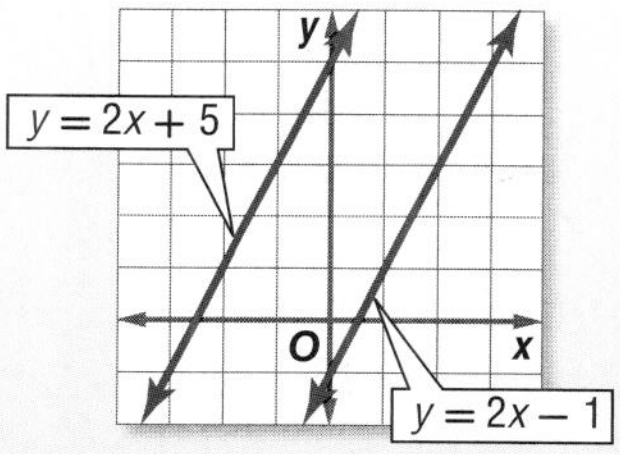	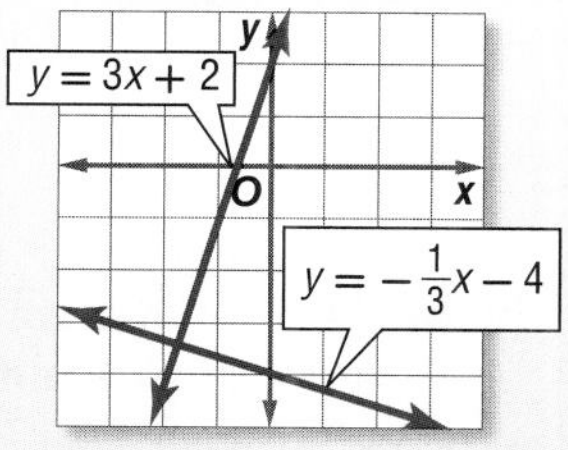
$y = 2x + 5$ and $y = 2x - 1$	$y = 3x + 2$ and $y = -\frac{1}{3}x - 4$

EXAMPLE 4 Write an Equation of a Perpendicular Line

Write an equation in slope-intercept form for the line that passes through (5, −6) and is perpendicular to the line with equation $y = -\frac{3}{2}x + 7$.

The slope of the given line is $-\frac{3}{2}$. Because the slopes of perpendicular lines are opposite reciprocals, the slope of the line perpendicular to the given line is $\frac{2}{3}$.

Use the point-slope form and the ordered pair (5, −6).

$y - y_1 = m(x - x_1)$	**Point-slope form**
$y - (-6) = \frac{2}{3}(x - 5)$	**$(x_1, y_1) = (5, -6)$ and $m = \frac{2}{3}$**
$y + 6 = \frac{2}{3}x - \frac{10}{3}$	**Distributive Property**
$y = \frac{2}{3}x - \frac{28}{3}$	**Subtract 6 from each side and simplify.**

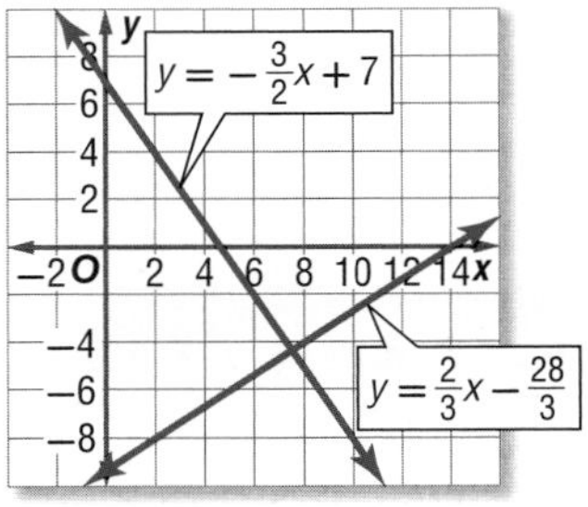

CHECK Graph both equations to verify the solution.

Check Your Progress

4. Write an equation in slope-intercept form for the line that passes through (3, 7) and is parallel to the line with equation $y = \frac{3}{4}x - 5$.

Personal Tutor glencoe.com

Check Your Understanding

Example 1 p. 84

Write an equation in slope-intercept form for the line described.

1. slope 1.5, passes through (0, 5)

2. passes through (−2, 3) and (0, 1)

Example 2 p. 84

3. passes through (3, 5); $m = -2$

4. passes through (−8, −2); $m = \frac{5}{2}$

Example 3 p. 85

5. MULTIPLE CHOICE Which is an equation of the line?

A $y = -\frac{11}{3}x - 22$

B $y = -\frac{2}{3}x - 5$

C $y = \frac{4}{5}x + \frac{29}{25}$

D $y = 6x + 35$

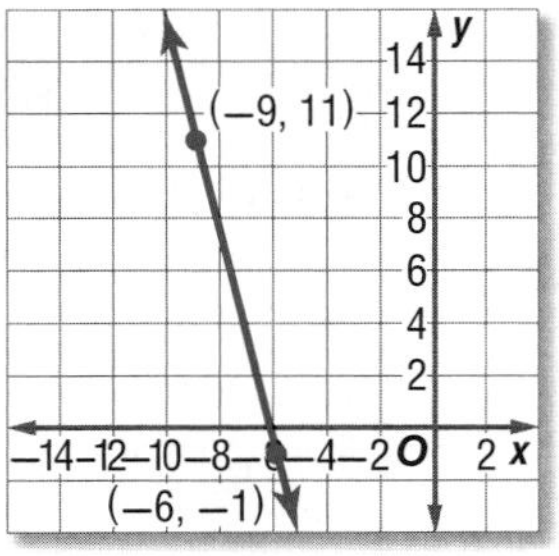

Example 4 p. 86

Write an equation in slope-intercept form for the line that satisfies each set of conditions.

6. passes through (−9, −3), perpendicular to $y = -\frac{5}{3}x - 8$

7 passes through (4, −10), parallel to $y = \frac{7}{8}x - 3$

Practice and Problem Solving

● = **Step-by-Step Solutions** begin on page R20.
Extra Practice begins on page 947.

Example 1 p. 84

Write an equation in slope-intercept form for the line described.

8. slope 3, passes through $(0, -2)$
9. slope $-\frac{1}{2}$, passes through $(0, 5)$
10. slope $-\frac{6}{5}$, passes through $(0, 8)$
11. slope $\frac{9}{2}$, passes through $\left(0, -\frac{13}{2}\right)$

Example 2 p. 84

12. slope -2, passes through $(-3, 14)$
13. slope 4, passes through $(6, 9)$
14. slope $\frac{3}{5}$, passes through $(-6, -8)$
15. slope $-\frac{1}{4}$, passes through $(12, -4)$
16. **PART-TIME JOB** Each week, Carmen earns a base pay of \$15 plus \$0.17 for every pamphlet that she delivers. Write an equation that can be used to find how much Carmen earns each week. How much will she earn the week that she delivers 300 pamphlets?

Example 3 p. 85

Write an equation of the line passing through each pair of points.

17. $(-2, -6), (4, 6)$
18. $(-8, -5), (-3, 10)$
19. $(-4, 12), (-2, -4)$
20. $(4.6, 3.4), (2.2, 2.8)$
21. $(5.5, 0.6), (1.1, 2.8)$
22. $(-25, -16), (-29, 12)$

Example 4 p. 86

Write an equation in slope-intercept form for the line that satisfies each set of conditions.

23. passes through $(4, 2)$, perpendicular to $y = -2x + 3$
24. passes through $(-6, -6)$, parallel to $y = \frac{4}{3}x + 8$
25. passes through $(12, 0)$, parallel to $y = -\frac{1}{2}x - 3$
26. passes through $(10, 2)$, perpendicular to $y = 4x + 6$
27. **CAR EXPENSES** Julio buys a used car for \$5900. Monthly expenses for the car—which include insurance, maintenance, and gas—total \$180 per month. Write an equation that represents the total cost of buying and owning the car for x months.
28. **DELI** The sales of a sandwich store increased from \$52,000 to \$116,000 during the first five years of business. Write an equation that models the sales y after x years. Determine what the sales will be at the end of 12 years if the pattern continues.
29. **WHALES** In 2007, it was estimated that there were 250 mature right whales in existence. The population of right whales is expected to decline by at least 25 whales each generation. Write an equation that represents the number of right whales that will be in existence in x generations.

Real-World Link

The northern right whale has been listed as endangered since 1973. The eastern North Atlantic population is nearly extinct, and the western North Atlantic population numbers around 300 individuals.

Source: NOAA Fisheries Service

Write an equation in slope-intercept form for each graph.

30.

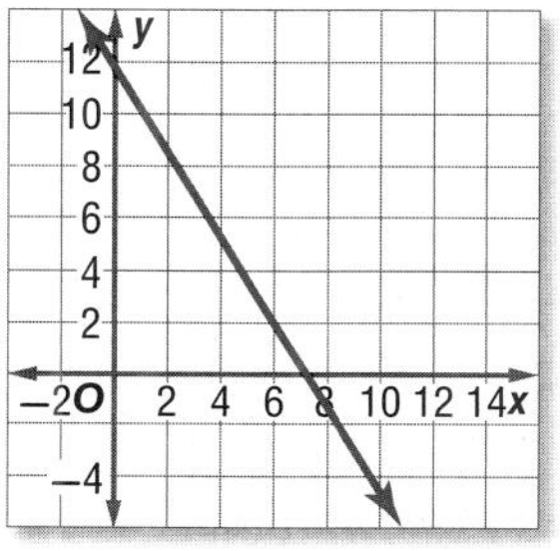

31.

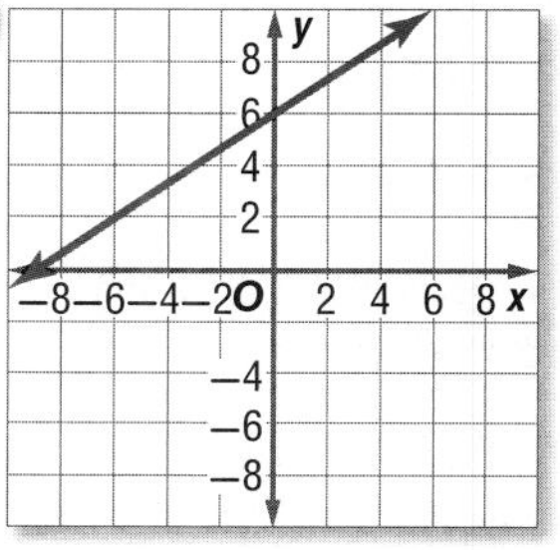

32.

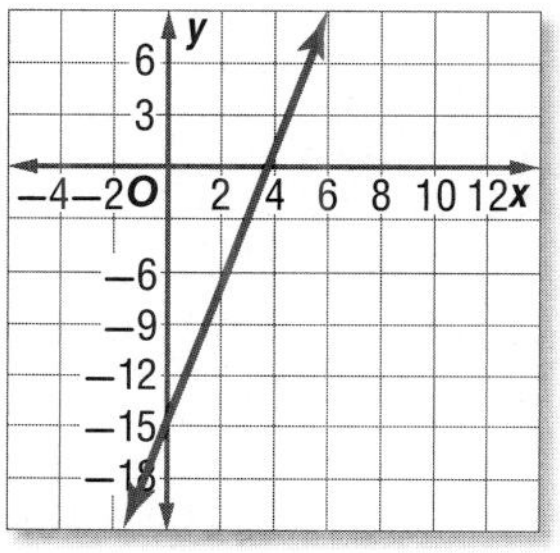

In current years, over 1,200,000,000 roses have been purchased by U.S. flower buyers. This works out to a consumption rate of 4.67 roses per person.

Source: U.S. Department of Agriculture

33. ROSES Brad wants to send his girlfriend Kelli a dozen roses. He visits two stores. For what distance do the two stores charge the same amount to deliver a dozen roses?

34. TYPING The equation $y = 55(23 - x)$ can be used to model the number of words y you have left to type after x minutes.

a. Write this equation in slope-intercept form.

b. Identify the slope and y-intercept.

c. Find the number of words you have left to type after 20 minutes.

35. RECRUITING As an army recruiter, Ms. Cooper is paid a daily salary plus commission. When she recruits 10 people, she earns $100. When she recruits 14 people, she earns $120.

a. Write a linear equation to model this situation.

b. What is Ms. Cooper's daily salary?

c. How much would Ms. Cooper earn in a day if she recruits 20 people?

36. TRAVEL Refer to the table at the right.

Miles	Kilometers
100	161
50	80.5

a. Write and graph the linear equation that gives the distance y in kilometers in terms of the number x in miles.

b. What distance in kilometers corresponds to 20 miles?

c. What number is the same in kilometers and miles? Explain your reasoning.

H.O.T. Problems
Use Higher-Order Thinking Skills

37. REASONING Determine whether the following statement is *always*, *sometimes*, or *never* true. Explain your reasoning.

The quadrilateral formed by any two parallel lines and two lines perpendicular to those lines is a square.

38. CHALLENGE Given ▱$ABCD$ with vertices $A(a, b)$, $B(c - a, d)$, $C(c + a, d)$, and $D(c, b)$, write an equation of a line perpendicular to diagonal $\overline{BD}$ that contains A.

39. REASONING Write $y = ax + b$ in point-slope form.

40. OPEN ENDED Write the equations of two parallel lines with negative slopes.

41. REASONING Write an equation in point-slope form of a line with an x-intercept of c and y-intercept of d.

42. WRITING IN MATH Describe the process of finding the slope-intercept form of the equation of a line containing two given points.

NYSRE PRACTICE Reinforcement of A.A.6, Reinforcement of A.A.33, A2.S.12

43. The total cost c in dollars to go to a water park and ride n water rides is given by the equation

$$c = 15 + 3n.$$

If the total cost was \$33, how many water rides were ridden?

A 6 B 7 C 8 D 9

44. SHORT RESPONSE To raise money, the service club bought 1000 candy bars for \$0.60 each. If the club sells all of the candy bars for \$1 each, what will be their total profit?

45. PROBABILITY A fair six-sided die is tossed. What is the probability that a number less than 3 will show on the face of the die?

F $\frac{1}{2}$ G $\frac{1}{3}$ H $\frac{2}{3}$ J $\frac{1}{6}$

46. ACT/SAT What is an equation of the line through $\left(\frac{1}{2}, -\frac{3}{2}\right)$ and $\left(-\frac{1}{2}, \frac{1}{2}\right)$?

A $y = -2x - \frac{1}{2}$

B $y = -3x$

C $y = 2x - 5$

D $y = \frac{1}{2}x + 1$

Spiral Review

Determine the rate of change of each graph. (Lesson 2-3)

47.

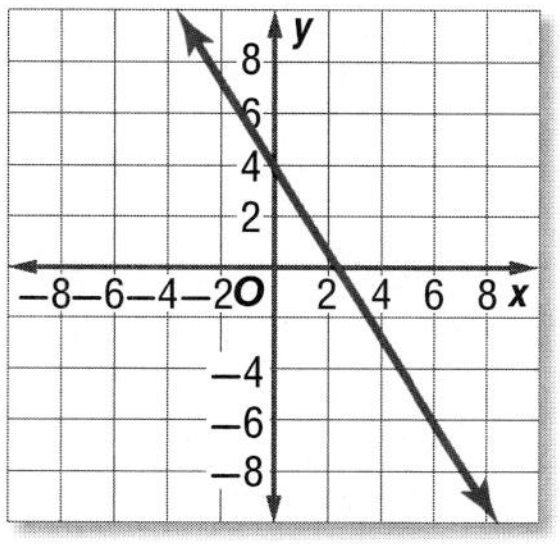

48.

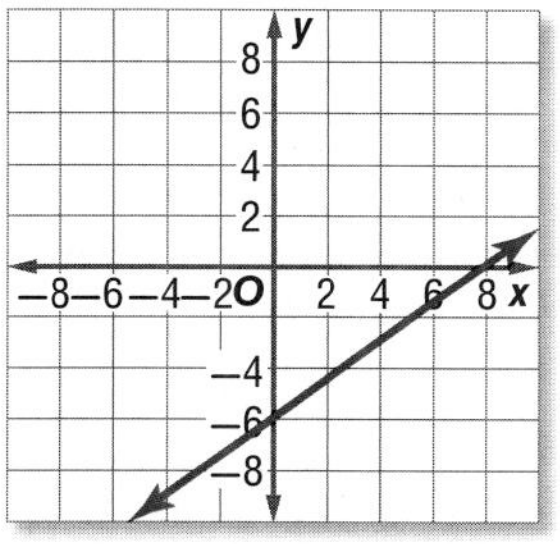

49.

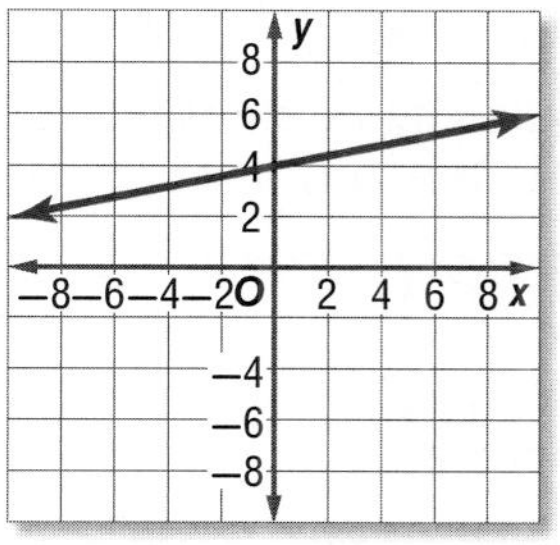

50. RECREATION Scott is currently on page 210 of an epic novel that is 980 pages long. He plans to read 30 pages per day until he finishes the novel. Write and solve a linear relation to determine how many days it will take Scott to complete the novel. (Lesson 2-2)

Solve each inequality. (Lesson 1-5)

51. $-6x - 4 \leq 12 - 2x$

52. $\frac{x+2}{5} > -3x + 1$

53. $\frac{5x+3}{3} \geq \frac{4x-2}{5}$

Determine if the triangles with the following lengths are right triangles. (Lesson 0-7)

54. 5, 12, 13

55. 36, 48, 60

56. 7, 23, 25

Multiply. (Lesson 0-2)

57. $(4c - 6)(2c + 5)$

58. $(-3b + 2)(b + 3)$

59. $(2a - 5)(-3a - 4)$

Skills Review

Find the slope of the line that passes through each pair of points. Express as a fraction in simplest form. (Lesson 2-3)

60. (4, 8), (−2, −6)

61. (−6, 3), (−2, 9)

62. (−4, −1), (−8, −8)

63. (12, 4), (42, 10)

64. (10.5, −3), (18, −8)

65. (3.5, −2.5), (−1, −2)

Graphing Technology Lab
Direct Variation

NY Math Online glencoe.com
Other Calculator Keystrokes

An equation of a direct variation is a special case of a linear equation. A **direct variation** can be expressed in the form $y = kx$. This means that y is a multiple of x. The k in this equation is a constant and is called the **constant of variation**.

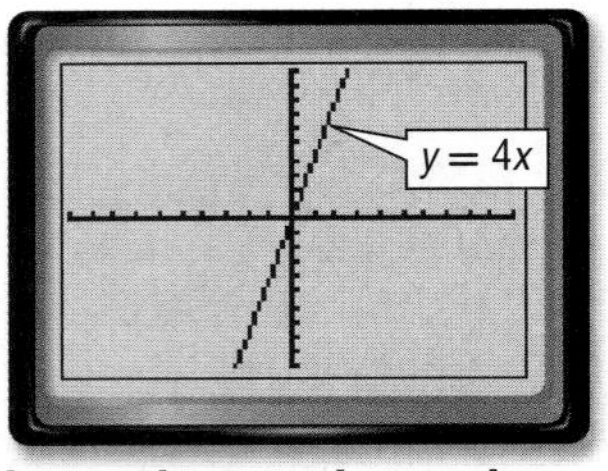

[−10, 10] scl: 1 by [−10, 10] scl: 1

Notice that the graph of $y = 4x$ is a straight line through the origin. An equation of a direct variation is a special case of an equation written in slope-intercept form, $y = mx + b$. When $m = k$ and $b = 0$, $y = mx + b$ becomes $y = kx$. So the slope of a direct variation equation is its constant of variation.

To express a direct variation, we say that y varies directly as x. In other words, as x increases, y increases or decreases at a constant rate.

Key Concept — Direct Variation

For Your FOLDABLE

y varies directly as x if there is some nonzero constant k such that $y = kx$. k is called the *constant of variation*.

ACTIVITY

GOLD The karat rating r of a gold object varies directly as the percentage p of gold in the object. A 14-karat ring is 58.25% gold.

a. Write and graph a direct variation equation relating r and p.

Use the point (0.5825, 14) to find the constant of variation.

$y = kx$	**Direct variation equation**
$14 = k(0.5825)$	**$x = 0.5825$, $y = 14$**
$24.03 \approx k$	**Divide each side by 0.5825.**

The direct variation equation is $r = 24.03p$.

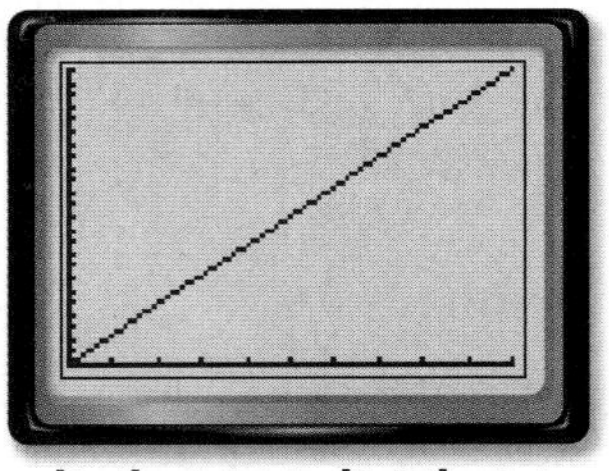

[0, 1] scl: 0.1 by [0, 24] scl: 1

b. Find the karat rating of a ring that is 75% gold.

Use the calculator to find the karat rating.

KEYSTROKES: [2nd] [CALC] 0.75 [ENTER] 18.0225

The karat rating of a ring that is 75% gold is 18 karats.

Exercises

1. **SWIMMING** When you swim underwater, the pressure on your ears varies directly with the depth at which you are swimming. If you are swimming in 8 feet of water, the pressure on your ears is 3.44 pounds per square inch. Write and graph a direct variation equation relating pressure and depth. Then find the pressure at a depth of 65 feet.

2. Graph the direct variation equations $y = -4x$, $y = -2x$, $y = 4x$, and $y = 2x$. Compare and contrast the graphs of the equations.

Mid-Chapter Quiz

Lessons 2-1 through 2-4

NYS Core Curriculum
Reinforcement of A.A.32, A2.A.41, A2.A.43

1. State the domain and range of the relation {(−3, 2), (4, 1), (0, 3), (5, −2), (2, 7)}. Then determine whether the relation is a function.

2. Graph $y = 2x - 3$ and determine whether the equation is a *function*, is *one-to-one*, *onto*, *both*, or *neither*. State whether it is *discrete* or *continuous*.

Given $f(x) = 3x^3 - 2x + 7$, find each value.

3. $f(-2)$
4. $f(2y)$
5. $f(1.4)$

6. State whether $f(x) = 2x^2 - 9$ is a linear function. Explain.

7. **MULTIPLE CHOICE** The daily pricing for renting a mid-sized car is given by the function $f(x) = 0.35x + 49$, where $f(x)$ is the total rental price for a car driven x miles. Find the rental cost for a car driven 250 miles.

   A $84

   B $112.50

   C $136.50

   C $215

Write each equation in standard form. Identify A, B, and C.

8. $y = -6x + 5$
9. $y = 10x$
10. $-\frac{5}{8}x = 2y + 11$
11. $0.5x = 3$

Find the x-intercept and the y-intercept of the graph of each equation. Then graph the equation using the intercepts.

12. $4x - 3y + 12 = 0$
13. $10 - x = 2y$

14. **SPEED** The table shows the distance traveled by a car after each time given in minutes. Find the rate of change in distance for the car.

Time (min)	Distance (mi)
10	12
30	40
45	55
60	75
90	110

Find the slope of the line that passes through each pair of points. Express as a fraction in simplest form.

15. (−2, 6), (1, 15)
16. (3, 5), (7, 15)
17. (4, 8), (4, −3)
18. (−2.5, 4), (1.5, −2)

19. Find the slope of the line shown.

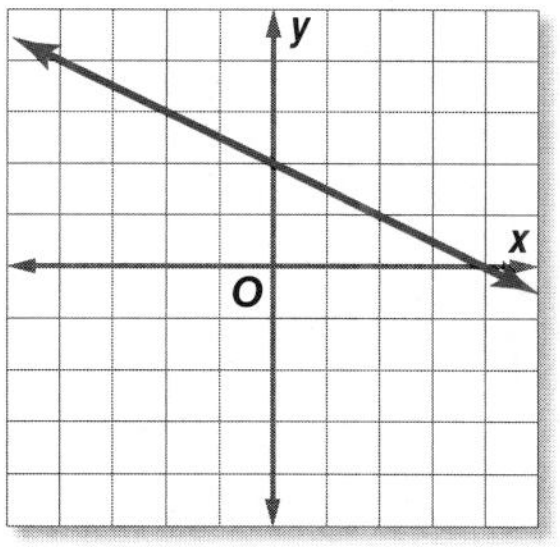

Write an equation of the line through each set of points.

20. slope $\frac{2}{3}$, passes through (3, −4)
21. slope −2.5, passes through (1, 2)

Write an equation of the line through each set of points.

22. (−2, 3), (4, 1)
23. (4.2, 3.6), (1.8, −1.2)

24. **MULTIPLE CHOICE** Each week, Jaya earns $32 plus $0.25 for each newspaper she delivers. Write an equation that can be used to determine how much Jaya earns each week. How much will she earn during a week in which she delivers 240 papers?

   F $75

   G $92

   H $148

   J $212

25. **PART-TIME JOB** Jesse is a pizza delivery driver. Each day his employer gives him $20 plus $0.50 for every pizza that he delivers.

   a. Write an equation that can be used to determine how much Jesse earns each day if he delivers x pizzas.

   b. How much will he earn the day he delivers 20 pizzas?

2-5 Scatter Plots and Lines of Regression

Why?

The scatter plot shows the number of visitors to Isle Royale National Park in Michigan. The linear function $f(x) = -0.43x + 24.6$ can be used to model the data.

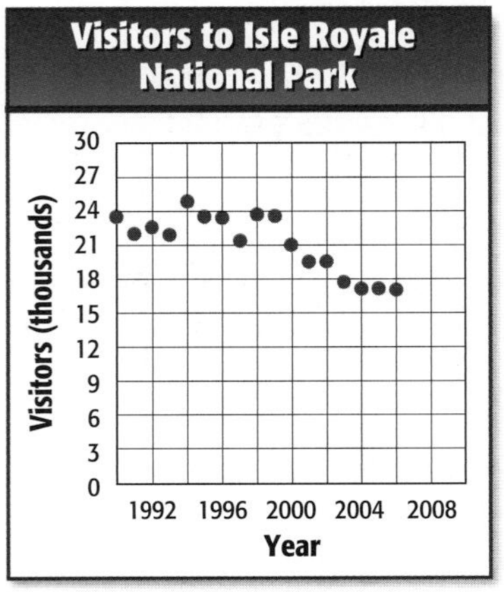

Source: National Park Service

Then
You wrote linear equations. (Lesson 2-4)

Now
- Use scatter plots and prediction equations.
- Model data using lines of regression.

NYS Core Curriculum
A2.S.8 Interpret within the linear regression model the value of the correlation coefficient as a measure of the strength of the relationship

New Vocabulary
bivariate data
scatter plot
dot plot
positive correlation
negative correlation
line of fit
prediction equation
regression line
correlation coefficient

NY Math Online
glencoe.com
- Extra Examples
- Personal Tutor
- Self-Check Quiz
- Homework Help
- Math in Motion

Scatter Plots and Prediction Equations Data with two variables, such as year and number of visitors, are called **bivariate data**. A set of bivariate data graphed as ordered pairs in a coordinate plane is called a **scatter plot** or **dot plot**.

A scatter plot can show whether there is a positive, negative, or no correlation between two variables. Correlations are usually described as *strong* or *weak*. In a strong correlation, the points of the scatterplot are closer to the graph of a line than the points representing a weak correlation.

Key Concept Scatter Plots — For Your FOLDABLE

Positive Correlation

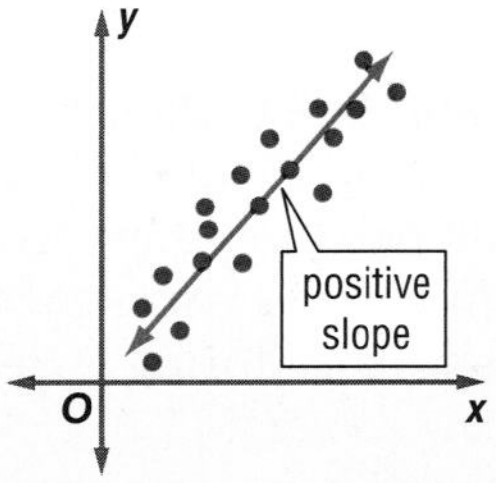

Strong Positive Correlation

The slope of the line is positive and the points are close to the line.

Negative Correlation

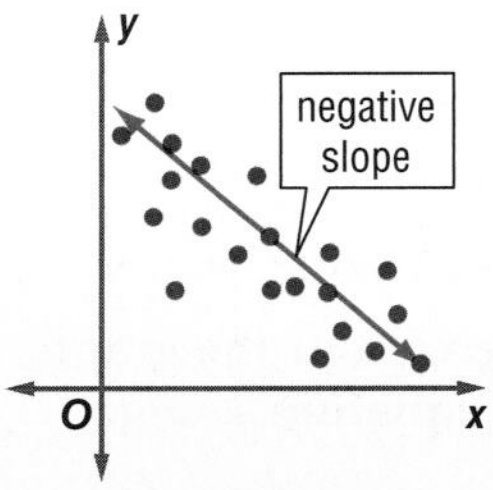

Weak Negative Correlation

The slope of the line is negative and the points are not close to the line.

No Correlation

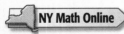

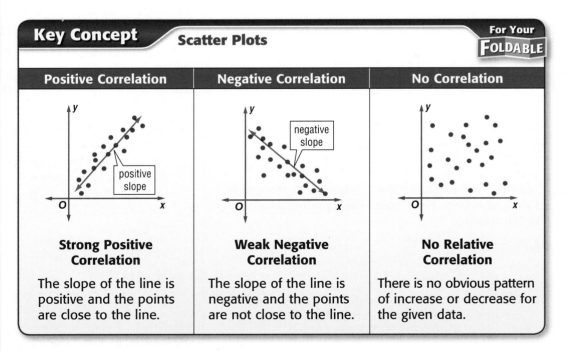

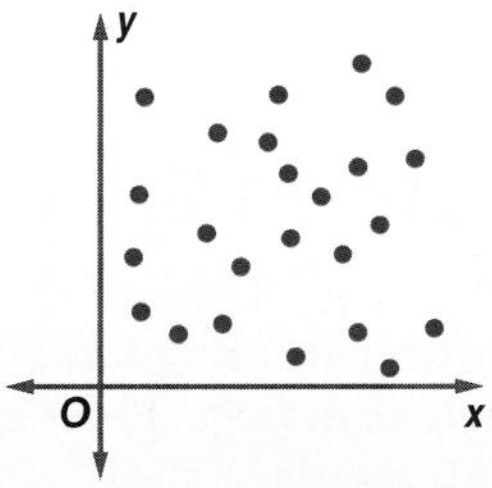

No Relative Correlation

There is no obvious pattern of increase or decrease for the given data.

When you find a line that closely approximates a set of data, you are finding a **line of fit** for the data. An equation of such a line is often called a **prediction equation** because it can be used to predict one of the variables given the other variable.

To find a line of fit and a prediction equation for a set of data, select two points that appear to represent the data well. This is a matter of personal judgment, so your line and prediction equation may be different from someone else's.

Real-World Link

According to the National Technology Scan, about 70% of U.S. households have computers connected to the Internet.

Source: The National Technology Scan

Real-World EXAMPLE 1 Use a Scatter Plot and Prediction Equation

TECHNOLOGY **The table shows the percent of U.S. households with at least one personal computer.**

Year	1984	1989	1993	1997	2001	2003
Percent	8.2	15.0	22.8	36.6	56.3	61.8

Source: U.S. Census Bureau

a. Make a scatter plot and a line of fit, and describe the correlation.

Graph the data as ordered pairs with the number of years since 1984 on the horizontal axis and the percent of households on the vertical axis.

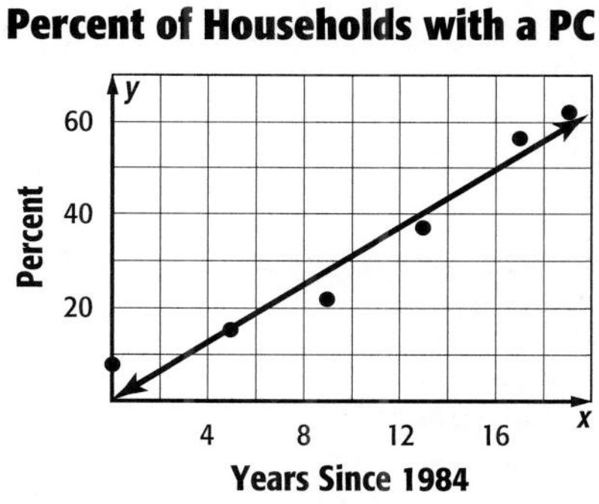

The points (5, 15.0) and (19, 61.8) appear to represent the data well. Draw a line through these two points. The data show a strong positive correlation.

b. Use two ordered pairs to write a prediction equation.

Find an equation of the line through (5, 15.0) and (19, 61.8).

$m = \frac{y_2 - y_1}{x_2 - x_1}$ **Slope Formula**

$= \frac{61.8 - 15.0}{19 - 5}$ **Substitute.**

≈ 3.34 **Simplify.**

$y - y_1 = m(x - x_1)$ **Point-Slope form**

$y - 15 = 3.34(x - 5)$ **Substitute.**

$y - 15 = 3.34x - 16.7$ **Distributive Property**

$y = 3.34x - 1.7$ **Simplify.**

One prediction equation is $y = 3.34x - 1.7$. The slope indicates that the average rate of change is 3.34% per year.

c. Predict the percent of households with at least one personal computer in 2012.

The year 2012 is 28 years after 1984, so find y when $x = 28$.

$y = 3.34x - 1.7$ **Prediction equation**

$= 3.34(28) - 1.7$ **$x = 28$**

$= 91.82$ **Simplify.**

The model predicts that 91.82% of U.S. households will have at least one personal computer in 2012.

d. How accurate does your prediction appear to be?

Except for the outlier at (0, 8.2), the line fits the data well, so the prediction value should be fairly accurate.

Check Your Progress

HOUSING **The table shows the mean selling price of new, privately-owned, single-family homes for some recent years.**

Year	1994	1996	1998	2000	2002	2004
Price ($1000)	154.5	166.4	181.9	207.0	228.7	273.5

Source: U.S. Census Bureau and U.S. Department of Housing and Urban Development

1A. Make a scatter plot and a line of fit, and describe the correlation.

1B. Write a prediction equation.

1C. Predict the selling price of a new, privately-owned, single-family home in 2010.

1D. How accurate does your prediction appear to be?

Review Vocabulary

outlier a data point that does not appear to belong to the rest of the set

Personal Tutor glencoe.com

Lines of Regression Another method for writing a line of fit is to use a line of regression. A **regression line** is determined through complex calculations to ensure that the distance of all data points to the line of fit are at a minimum. Most graphing calculators and spreadsheets can perform these calculations easily.

The **correlation coefficient** r, $-1 \leq r \leq 1$, is a measure that shows how well data are modeled by a linear equation.

- When r is close to -1, the data have a negative correlation.
- When $r = 0$, the data have no correlation.
- When r is close to 1, the data have a positive correlation.

StudyTip

Choosing the Independent Variable Letting x be the number of years since the first year in the data set simplifies the math involved in finding a function to model the data.

Real-World EXAMPLE 2 Regression Line

The table shows the life expectancy for people born in the United States.

Year of Birth	1983	1990	1993	1997	2000	2003
Life Expectancy (yr)	74.6	75.4	75.5	76.5	76.9	77.5

Source: U.S. Department of Health and Human Services

Use a graphing calculator to make a scatter plot of the data. Find an equation for and graph a line of regression. Then use the equation to predict the life expectancy of a person born in 2025.

Step 1 Make a scatter plot.

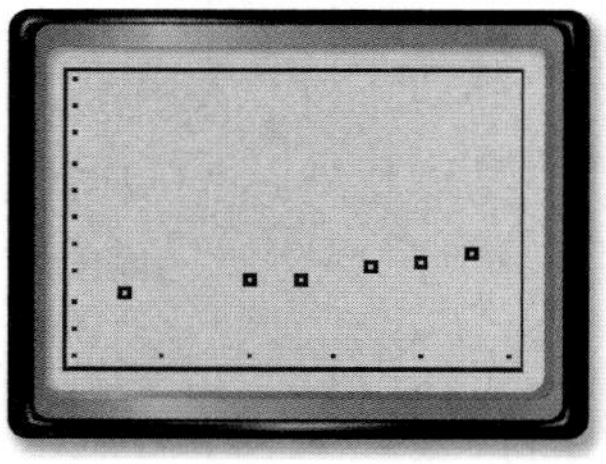

- Enter the years of birth in L1 and the ages in L2.

  KEYSTROKES: [STAT] [ENTER] 1983 [ENTER] 1990 [ENTER] 1993 [ENTER] ...

- Set the viewing window to fit the data.

  KEYSTROKES: [WINDOW] 1980 [ENTER] 2005 [ENTER] 5 [ENTER] 70 [ENTER] 90 [ENTER] 2

- Use STAT PLOT to graph the scatter plot.

  KEYSTROKES: [2nd] [STAT PLOT] [ENTER] [ENTER] [Graph]

Step 2 Find the equation of the line of regression.

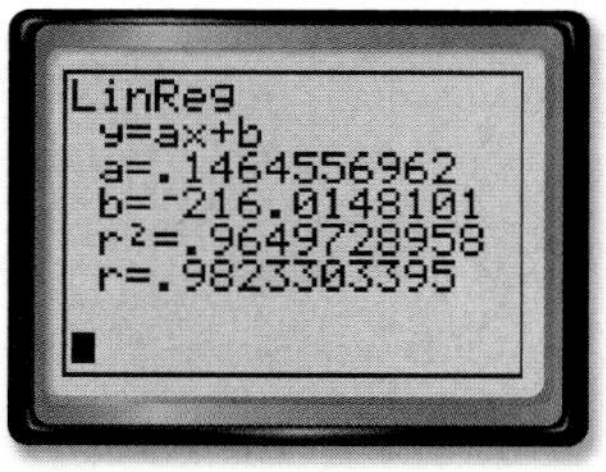

- Find the regression equation by selecting LinReg(ax + b) on the STAT CALC menu.

  KEYSTROKES: [STAT] [▶] 4 [ENTER]

The regression equation is about $y = 0.15x - 216.01$. The slope indicates that the life expectancy increases at a rate of about 0.15 per year. The correlation coefficient r is about 0.98, which is very close to 1. So, the data fit the regression line very well.

Step 3 Graph the regression equation.

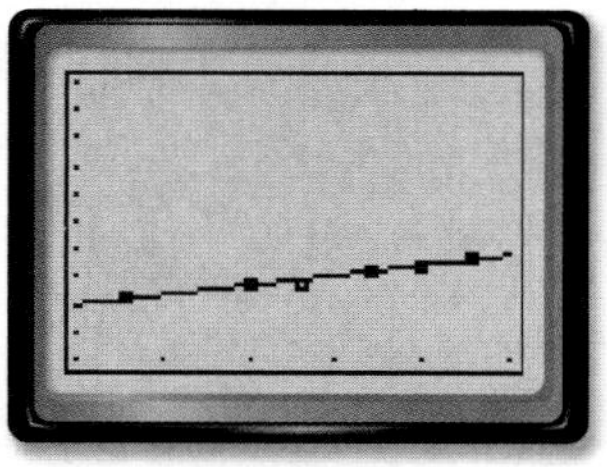

- Copy the equation to the Y= list and graph.

  KEYSTROKES: [Y=] [VARS] 5 [▶] [▶] 1 [Graph]

  Notice that the regression line comes close to all of the data points. As the correlation coefficient indicated, the line fits the data very well.

Math *in Motion,* Interactive Lab
glencoe.com

> **ReadingMath**
>
> **Predictions**
> When you are predicting an x-value greater than or less than any in the data set, the process is known as **extrapolation**. When you are predicting an x-value between the least and greatest in the data set, the process is known as **interpolation**.

Step 4 Predict using the function.

- Find y when $x = 2025$. Use **VALUE** on the **CALC** menu. Reset the window size to accommodate the x-value of 2025.

KEYSTROKES: [2nd] [CALC] 1 2025 [ENTER]

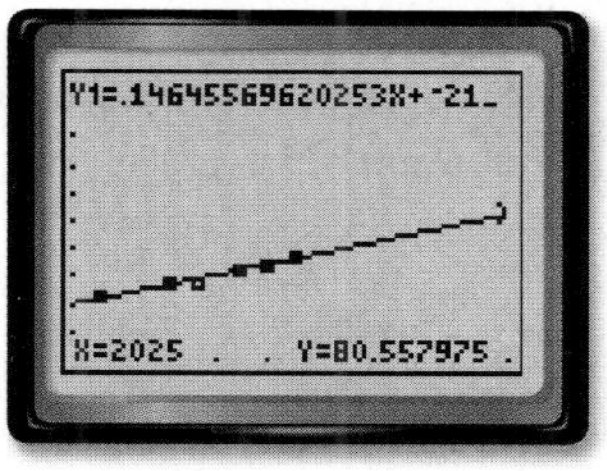

According to the function, the life expectancy of a person born in 2025 will be about 80.6 years.

Check Your Progress

2. MUSIC The table at the right shows the percent of sales that were made in music stores in the United States for the period 1995–2004. Use a graphing calculator to make a scatter plot of the data. Find and graph a line of regression. Then use the function to predict the percent of sales made in a music store in 2015.

Music Store Sales

Year	Sales (percent)
1995	52.0
1996	49.9
1997	51.8
1998	50.8
1999	44.5
2000	42.4
2001	42.5
2002	36.8
2003	33.2
2004	32.5

Source: Recording Industry Association of America

Personal Tutor glencoe.com

Check Your Understanding

Example 1 p. 93

1. OCEANS The table shows the temperature in the ocean at various depths.

Depth (in meters)	0	300	500	1000	2000	2500
Temp (°C)	22	20	13	7	6	?

Source: NOAA

a. Make a scatter plot and a line of fit, and describe the correlation.

b. Use two ordered pairs to write a prediction equation.

c. Use your prediction equation to predict the missing value.

Example 2 p. 94

2. INCOME The table shows the median income of families in North Carolina by family size in a recent year. Use a graphing calculator to make a scatter plot of the data. Find an equation for and graph a line of regression. Then use the equation to predict the median income of a North Carolina family of 9.

Family Size	Income ($)
1	33,265
2	44,625
3	50,528
4	59,481

Source: U.S. Department of Justice

Practice and Problem Solving

● = **Step-by-Step Solutions** begin on page R20.
Extra Practice begins on page 947.

Example 1 p. 93

For Exercises 3–6, complete parts a–c.

a. **Make a scatter plot and a line of fit, and describe the correlation.**

b. **Use two ordered pairs to write a prediction equation.**

c. **Use your prediction equation to predict the missing value.**

3. **COMPACT DISC SALES** The table shows the number of CDs sold in recent years at Jerome's House of Music.

Year	2004	2005	2006	2007	2008	2009
Number of CDs sold	49,300	39,440	31,552	25,242	20,193	?

4. **BASKETBALL** The table shows the number of field goals and assists for some of the members of the Miami Heat in a recent NBA season.

Field Goals	472	353	278	283	238	265	186	162	144
Assists	384	97	81	79	18	130	94	95	?

Source: NBA

Real-World Link

The average number of licks to polish off a single scoop ice cream cone is approximately 50.

Source: Edy's®

5 **ICE CREAM** The table shows the amount of ice cream Sunee's Homemade Ice Creams sold for eight months.

Month	Jan	Feb	Mar	Apr	May	June	July	Aug	Sept
Gallons sold	37	44	72	80	105	110	119	131	?

6. **DRAMA CLUB** The table shows the total revenue of all of Central High School's plays in recent school years.

School Year	2002	2003	2004	2005	2006	2007
Revenue ($)	603	666	643	721	771	?

Example 2 p. 94

7. **SALES** The table shows the sales of Chayton's Computers. Use a graphing calculator to make a scatter plot of the data. Find an equation for and graph a line of regression. Then use the function to predict the sales in 2012.

Year	Sales ($ thousands)
2004	640
2005	715
2006	791
2007	852
2008	910
2009	944

8. **BUSINESS** The table shows the number of employees of a small company. Use a graphing calculator to make a scatter plot of the data. Find an equation for and graph a line of regression. Then use the function to predict the number of employees in 2015.

Year	Number of Employees
2002	4
2003	7
2004	11
2005	14
2006	20
2007	23

Real-World Link

The largest crowd for a single baseball game was 92,706 during the 1959 World Series between the Chicago White Sox and the Los Angeles Dodgers.

Source: Major League Baseball

9. **BASEBALL** The table at the right shows the total attendance for the Florida Marlins in some recent years.

Year	Attendance
2007	1,370,511
2006	1,164,134
2005	1,852,608
2004	1,723,105
2003	1,303,215
2002	813,118

   a. Make a scatter plot of the data.

   b. Find a regression equation for the data.

   c. Predict the attendance in 2020.

   d. How reasonable is your prediction? Explain.

10. **CLASS SIZE** The table at the right shows the relationship between the number of students in a mathematics class and the average grade for each class.

Class Size	Class Average
16	81.2
19	80.6
24	82.5
26	79.9
27	78.6
29	79.3
32	77.7

   a. Make a scatter plot of the data, and find a regression equation for the data. Then sketch a graph of the regression line.

   b. What is the correlation coefficient of the data?

   c. Describe the correlation. How accurate is the regression equation?

11. **SALES** Jocelyn is analyzing the sales of her company. The table at the right shows the total sales for each of six years.

Year	Sales ($ millions)
2003	31.2
2004	34.6
2005	18.9
2006	37.7
2007	41.3
2008	45.1

   a. Find a regression equation and correlation coefficient for the data.

   b. Use the regression equation to predict the sales in 2015.

   c. Remove the outlier from the data set and find a new regression equation and correlation coefficient.

   d. Use the new regression equation to predict the sales in 2015.

   e. Compare the correlation coefficients for the two regression equations. Which function fits the data better? Which prediction should Jocelyn expect to be more accurate?

H.O.T. Problems

Use Higher-Order Thinking Skills

12. **REASONING** What is the relevance of the correlation coefficient of a linear regression line? Explain your reasoning.

13. **CHALLENGE** If statements a and b have a positive correlation, b and c have a negative correlation, and c and d have a positive correlation, what can you determine about the correlation between statements a and d? Explain your reasoning.

14. **OPEN ENDED** Provide real-world quantities that represent each of the following.

   a. positive correlation **b.** negative correlation **c.** no correlation

15. **CHALLENGE** Draw a scatter plot for the following data set.

x	1.0	1.5	2.0	2.8	3.2	4.0	4.8	5.8
y	3.5	4.7	5.1	6.8	7.1	7.5	8.8	10.3

   Which of the following best represents the correlation coefficient r for the data? Justify your answer.

   a. 0.99 **b.** −0.98 **c.** 0.62 **d.** 0.08

16. **WRITING IN MATH** Explain why a linear equation is useful when working with data.

NYSRE PRACTICE Reinforcement of A.A.6, Reinforcement of A.G.4

17. SHORT RESPONSE What is the value of the expression below?

$$17 - 3[-1 + 2(7 - 4)]$$

18. Anna took brownies to a club meeting. She gave half of her brownies to Selena. Selena gave a third of her brownies to Randall. Randall gave a fourth of his brownies to Trina. If Trina has 3 brownies, how many brownies did Anna have in the beginning?

A 12

B 36

C 72

D 144

19. GEOMETRY Which is always true?

F A parallelogram is a square.

G A parallelogram is a rectangle.

H A quadrilateral is a trapezoid.

J A square is a rectangle.

20. ACT/SAT Which line best fits the data in the graph?

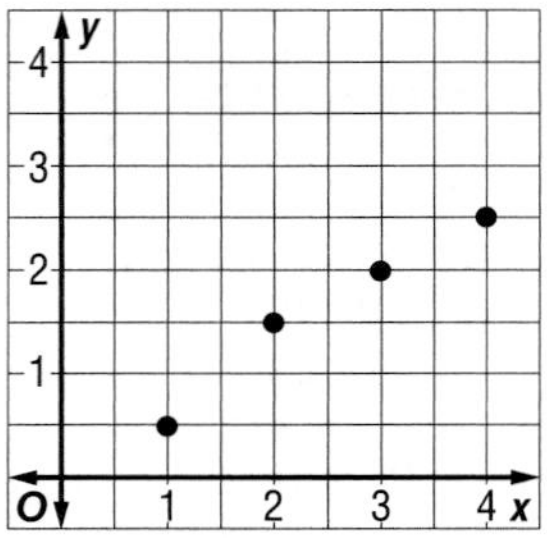

A $y = x$

B $y = -0.5x + 4$

C $y = -0.5x - 4$

D $y = 0.5x + 0.5$

Spiral Review

Write an equation in slope-intercept form for each graph. (Lesson 2-4)

21.

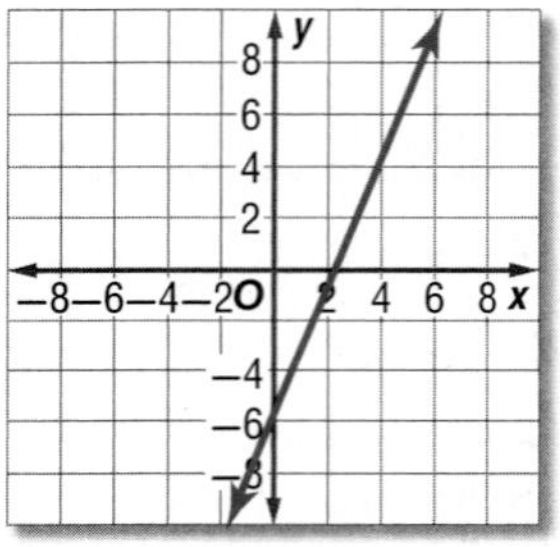

22.

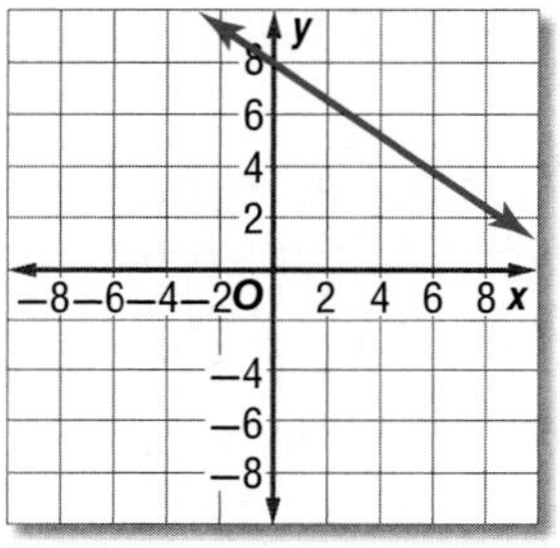

23.

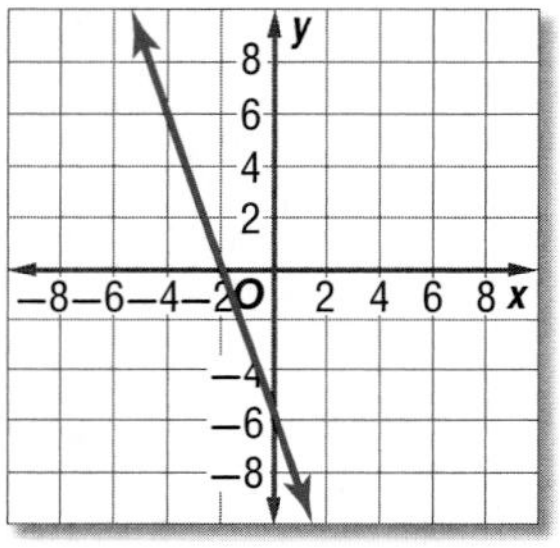

Find the rate of change for each set of data. (Lesson 2-3)

24.

Time (day)	3	6	9	12	15
Height (mm)	12	24	36	48	60

25.

Time (h)	2	4	6	8
Distance (mi)	35	70	105	140

26.

Time (sec)	12	16	20	24	28
Volume (cm^3)	45	60	75	90	105

27.

Force (*N*)	32	40	48	56	64
Work (*J*)	48	60	72	84	96

28. RECREATION Ramona estimates that she will need 50 tennis balls for every player that signs up for the tennis club and at least 150 more just in case. Write an inequality to express the situation. (Lesson 1-5)

29. DODGEBALL Six teams played in a dodgeball tournament. In how many ways can the top three teams finish? (Lesson 0-5)

Skills Review

Solve each equation. (Lesson 1-4)

30. $-4|x - 2| = -12$

31. $|3x + 4| = 21$

32. $2|4x - 1| + 3 = 9$

EXTEND 2-5

Algebra Lab
Median Fit Lines

NY Math Online glencoe.com
Math *in Motion,* Animation

Objective
Use median fit lines to make predictions.

NYS Core Curriculum

A2.S.7, A2.S.8

Many teens would love to own a car. Of course, you know there is more to owning a car than just the cost of the car. Over the years, the cost of gas has been rising. What might the cost per gallon be when you get a car? Graphing data can help you predict the future cost. The table below shows the cost of one gallon of unleaded gas from 1990 through 2005.

Year	1990	1991	1992	1993	1994	1995	1996	1997
Cost	\$1.35	\$1.32	\$1.32	\$1.30	\$1.31	\$1.34	\$1.41	\$1.42
Year	1998	1999	2000	2001	2002	2003	2004	2005
Cost	\$1.25	\$1.36	\$1.69	\$1.66	\$1.56	\$1.78	\$2.07	\$2.49

Source: *The World Almanac and Book of Facts*

ACTIVITY

Follow the steps to find a median-fit line for a data set and make a prediction.

Step 1 As accurately as possible, graph the ordered pairs (year, cost). For example, one ordered pair is (1990, 1.35).

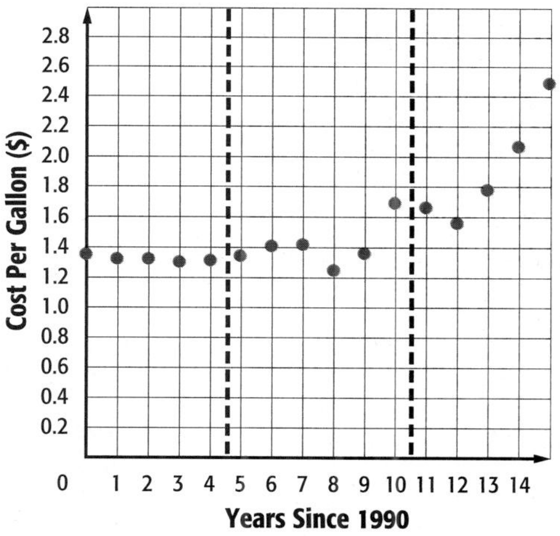

Step 2 Divide the data points into 3 groups as equal in size as possible. If there is one extra point, place it in the middle group. If there are two extra points, place one in each of the two outer groups. Separate the groups with dashed lines.

Step 3 For each group, find the median of the x- and y-values.

Group 1	Group 2	Group 3
(0, 1.35)	(5, 1.34)	(11, 1.66)
(1, 1.32)	(6, 1.41)	(12, 1.56)
(2, 1.32) ← **median**	(7, 1.42) ← **median**	(13, 1.78) ← **median**
(3, 1.30)	(8, 1.25) ← **median**	(14, 2.07)
(4, 1.31)	(9, 1.36)	(15, 2.49)
	(10, 1.69)	
median: (2, 1.32)	median: (7.5, 1.335)	median: (13, 1.78)

(continued on the next page)

Step 4 Plot the median points on the graph. Draw a dashed line through the median points for Groups 1 and 3. Then about one third of the way down toward the median for Group 2, draw a solid line parallel to the dashed line. This is the median-fit line.

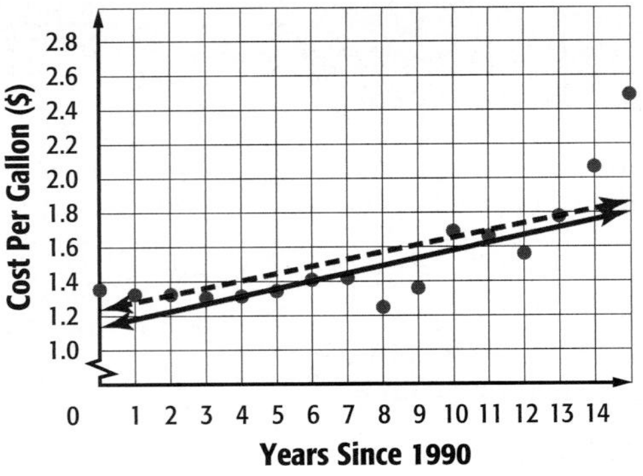

Step 5 When two different people draw a median-fit line, the results may vary slightly due to inaccuracy in measurements and estimation of coordinates. Pick two points on the solid line, for example, (4, 1.32) and (14, 1.76). Then find the slope of the median-fit line and write an equation.

$$m = \frac{y_2 - y_1}{x_2 - x_1}$$
$$= \frac{1.76 - 1.32}{14 - 4}$$
$$= \frac{0.44}{10} \text{ or } 0.044$$

$$(y - y_1) = m(x - x_1)$$
$$(y - 1.32) = 0.044(x - 4)$$
$$y - 1.32 = 0.044x - 0.176$$
$$y = 0.044x + 1.144$$

Exercises

1. Suppose you buy your own car in 2012. Use the equation in **Step 5** to predict the price per gallon of unleaded gasoline for that year. Explain whether you think the prediction is reasonable.

2. Write an equation for another type of best-fit line, and use the equation to predict for the year 2012.

3. The table shows the number of licensed drivers in the U.S., in millions, for various years. Make a scatter plot of the data, draw a median-fit line, and predict the number of licensed drivers in 2012. Then explain whether you think the prediction is reasonable.

Year	1960	1965	1970	1975	1980	1985	1990	1995
Number	87	99	112	130	145	157	167	177
Year	1998	1999	2000	2001	2002	2003	2004	
Number	185	187	191	191	195	196	199	

Source: infoplease

4. The table shows the median sale price for a house in the U.S. for various years. Make a scatter plot of the data, draw a median-fit line, and predict the median house sales price in 2012. Then explain whether you think the prediction is reasonable.

Year	1970	1975	1980	1985	1990	1995	1998
Price	$23,000	$35,300	$62,200	$75,500	$92,000	$110,500	$128,400
Year	1999	2000	2002	2003	2004	2005	
Price	$133,300	$139,000	$158,100	$178,800	$195,400	$219,600	

Source: *The New York Times Almanac*

2-6

Special Functions

Why?

The table shows a recent federal income tax rate schedule. The amount of federal income tax an individual is required to pay is a function of income.

Federal Tax Rate Schedule – Filing Single		
If taxable income is over	But not over	The tax is:
$0	$7,825	10% of the amount over $0
$7,825	$31,850	$782.50 plus 15% of the amount over $7,825
$31,850	$77,100	$4,386.25 plus 25% of the amount over $31,850
$77,100	$160,850	$15,698.75 plus 28% of the amount over $77,100
$160,850	$349,700	$39,148.75 plus 33% of the amount over $160,850
$349,700	no limit	$101,469.25 plus 35% of the amount over $349,700

Source: Internal Revenue Service

Then

You modeled data using lines of regression. (Lesson 2-5)

Now

- Write and graph piecewise-defined functions.
- Write and graph step and absolute value functions.

NYS Core Curriculum

Reinforcement of A.G.4 Identify and graph linear, quadratic (parabolic), absolute value, and exponential functions

New Vocabulary

piecewise-defined function
piecewise-linear function
step function
greatest integer function
absolute value function

NY Math Online

glencoe.com

- Extra Examples
- Personal Tutor
- Self-Check Quiz
- Homework Help
- Math in Motion

Piecewise-Defined Functions The function relating income and tax is not a linear function because each interval, or piece, of the function is defined by a different expression. A function that is written using two or more expressions is called a **piecewise-defined function**. On the graph of a piecewise-defined function, a dot indicates that the point is included in the graph. A circle indicates that the point is not included in the graph.

EXAMPLE 1 Piecewise-Defined Function

Graph $f(x) = \begin{cases} x - 2 \text{ if } x < -1 \\ x + 3 \text{ if } x \geq -1 \end{cases}$. Identify the domain and range.

Step 1 Graph $f(x) = x - 2$ for $x < -1$.

$f(x) = x - 2$
$= (-1) - 2$
$= -3$

Because -1 does not satisfy the inequality, begin with a circle at $(-1, -3)$.

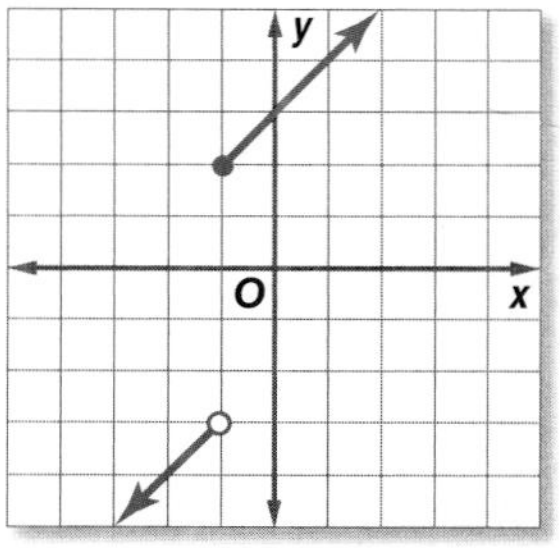

Step 2 Graph $f(x) = x + 3$ for $x \geq -1$.

$f(x) = x + 3$
$= (-1) + 3$
$= 2$

Because -1 satisfies the inequality, begin with a dot at $(-1, 2)$.

The function is defined for all values of x, so the domain is all real numbers.

The y-coordinates of points on the graph are all real numbers less than -3 and all real numbers greater than or equal to 2, so the range is $\{y \mid y < -3 \text{ and } y \geq 2\}$.

Check Your Progress

1. Graph $f(x) = \begin{cases} x + 2 \text{ if } x < 0 \\ x \quad \text{ if } x \geq 0 \end{cases}$. Identify the domain and range.

Personal Tutor glencoe.com

Piecewise-defined functions are often defined by several linear functions.

StudyTip

Graphs of Piecewise-Defined Functions The graphs of different parts of a piecewise-defined function may or may not connect. A graph may also stop at a given x-value and then begin at a different y-value for the same value of x.

EXAMPLE 2 Write a Piecewise-Defined Function

Write the piecewise-defined function shown in the graph.

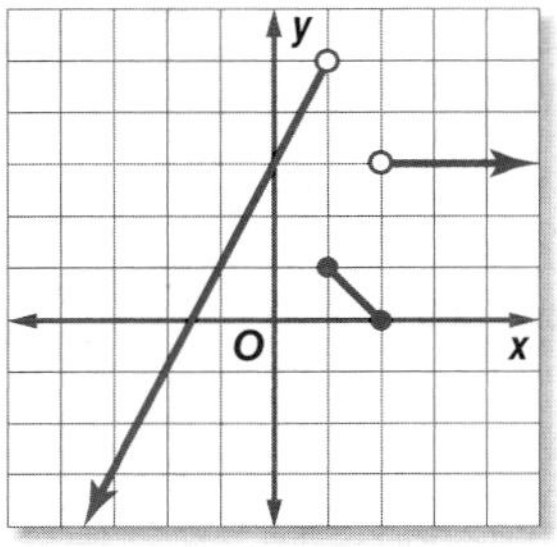

Examine and write a function for each portion of the graph.

The left portion of the graph is the graph of $f(x) = 2x + 3$. There is a circle at (1, 5), so the linear function is defined for $\{x \mid x < 1\}$.

The center portion of the graph is the graph of $f(x) = -x + 2$. There are dots at (1, 1) and (2, 0), so the linear function is defined for $\{x \mid 1 \leq x \leq 2\}$.

The right portion of the graph is the constant function $f(x) = 3$. There is a circle at (2, 3), so the constant function is defined for $\{x \mid x > 2\}$.

Write the piecewise-defined function.

$$f(x) = \begin{cases} 2x + 3 \text{ if } x < 1 \\ -x + 3 \text{ if } 1 \leq x \leq 2 \\ 3 \text{ if } x > 2 \end{cases}$$

CHECK The graph shows a portion of a line with positive slope for $x < 1$. The graph has negative slope for $1 \leq x \leq 2$ and constant slope for $x > 2$. The function is reasonable for the graph.

Check Your Progress

Write the piecewise-defined function shown in each graph.

2A.

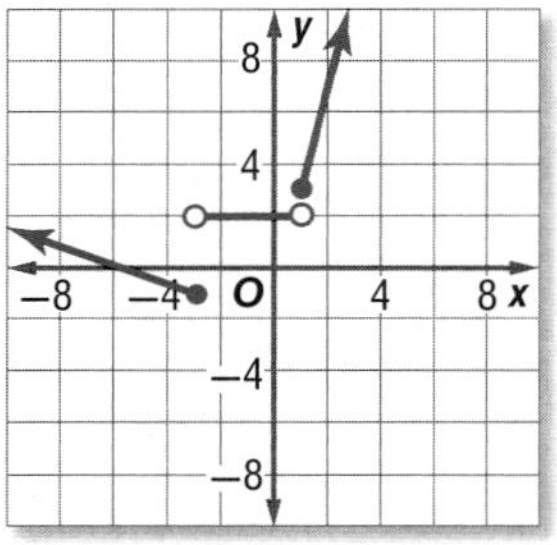

2B.

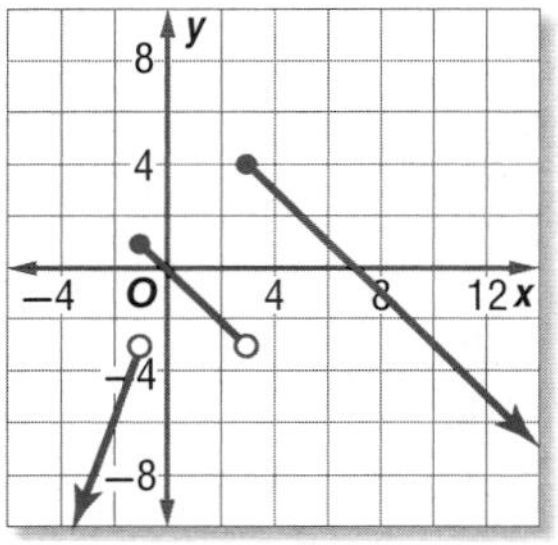

Personal Tutor glencoe.com

StudyTip

Greatest Integer Function Notice that the domain of this step function is all real numbers and the range is all integers.

Step Functions and Absolute Value Functions Unlike a piecewise-defined function, a **piecewise-linear function** contains a single expression. A common piecewise-linear function is the step function. The graph of a **step function** consists of line segments.

The **greatest integer function**, written $f(x) = [\![x]\!]$, is one kind of step function. The symbol $[\![x]\!]$ means *the greatest integer less than or equal to x*. For example, $[\![3.25]\!] = 3$ and $[\![-4.6]\!] = -5$.

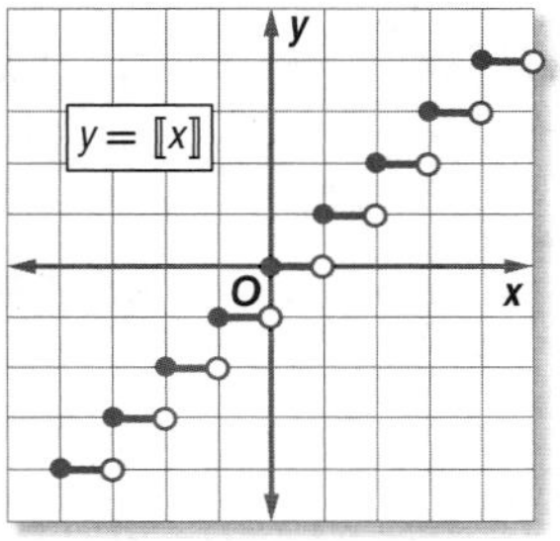

Math *in Motion*, Animation glencoe.com

Real-World EXAMPLE 3 Use a Step Function

BUSINESS An automotive repair center charges $50 for any part of the first hour of labor, and $35 for any part of each additional hour. Draw a graph that represents this situation.

Understand The total labor charge is $50 for the first hour plus $35 for each additional fraction of an hour, so the graph will be a step function.

Plan If the time spent on labor is greater than 0 hours, but less than or equal to 1 hour, then the labor charge is $50. If the time is greater than 1 hour but less than 2 hours, then the labor charge is $85, and so on.

Solve Use the pattern of times and costs to make a table, where x is the number of hours of labor and $T(x)$ is the total labor charge. Then graph.

x	$T(x)$
$0 < x \leq 1$	$50
$1 < x \leq 2$	$85
$2 < x \leq 3$	$120
$3 < x \leq 4$	$155
$4 < x \leq 5$	$190

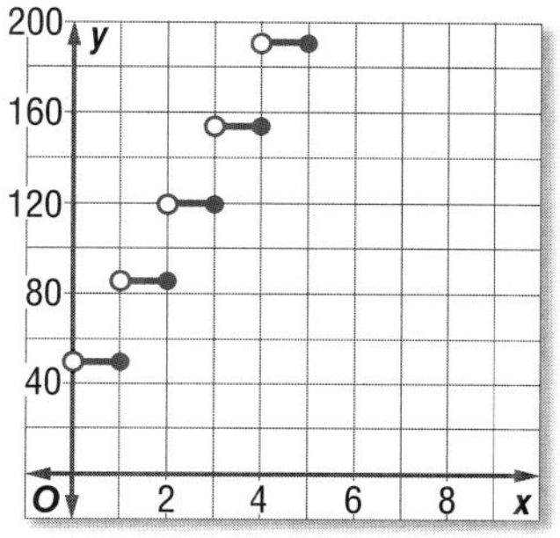

Check Since the repair center rounds any fraction of an hour up to the next whole number, each segment of the graph has a circle at the left endpoint and a dot at the right endpoint.

Check Your Progress

3. **RECYCLING** A recycling company pays $5 for every full box of newspaper. They do not give any money for partial boxes. Draw a graph that shows the amount of money $P(x)$ for the number of boxes x brought to the recycling center.

Personal Tutor glencoe.com

Math History Link

Karl Weierstrass (1815–1897) At the wishes of his father, Weierstrass studied law, economics, and finance at the University of Bonn, but then dropped out to study his true interest, mathematics, at the University of Münster. In an 1841 essay, Weierstrass first used | | to denote absolute value.

Source: Earth 911

Another piecewise-linear function is the absolute value function. An **absolute value function** is a function that contains an algebraic expression within absolute value symbols.

Key Concept

For Your FOLDABLE

Parent Function of Absolute Value Functions

Parent function: $f(x) = |x|$, defined as

$$f(x) = \begin{cases} x \text{ if } x > 0 \\ 0 \text{ if } x = 0 \\ -x \text{ if } x < 0 \end{cases}$$

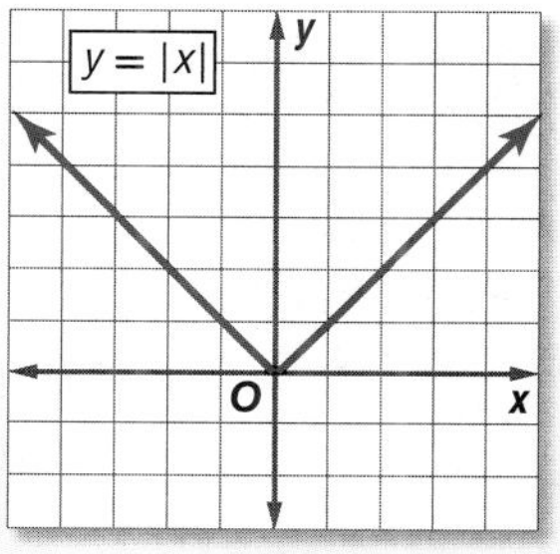

Type of graph: V-shaped

Domain: all real numbers

Range: all nonnegative real numbers

Intercepts: $x = 0, y = 0$

Not defined: $y < 0$

EXAMPLE 4 Absolute Value Functions

Graph $f(x) = |2x| - 4$. Identify the domain and range.

Create a table of values.

x	$\lvert 2x \rvert - 4$
-3	2
-2	0
-1	-2
0	-4
1	-2
2	0
3	2

Graph the points and connect them.

The domain is the set of all real numbers. The range is $\{y \mid y \geq -4\}$.

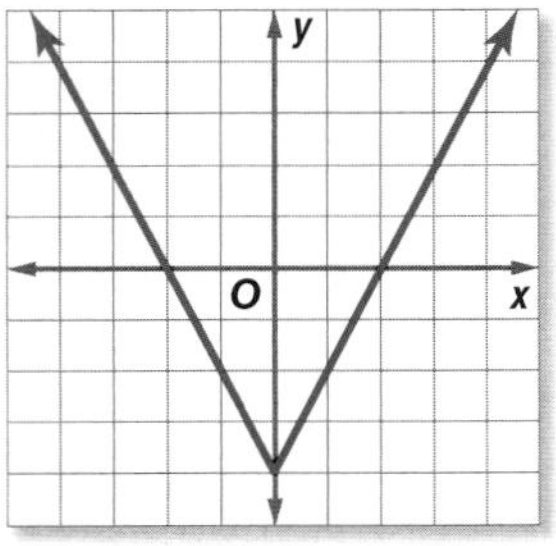

Check Your Progress

Graph each function. Identify the domain and range.

4A. $f(x) = |x - 2|$ **4B.** $f(x) = -|x| + 1$

Personal Tutor glencoe.com

Check Your Understanding

Example 1 p. 101

Graph each function. Identify the domain and range.

1. $g(x) = \begin{cases} -3 \text{ if } x \leq -4 \\ x \text{ if } -4 < x < 2 \\ -x + 6 \text{ if } x \geq 2 \end{cases}$

2. $f(x) = \begin{cases} 8 \text{ if } x \leq -1 \\ 2x \text{ if } -1 < x < 4 \\ -4 - x \text{ if } x \geq 4 \end{cases}$

Example 2 p. 102

Write the piecewise-defined function shown in each graph.

3.

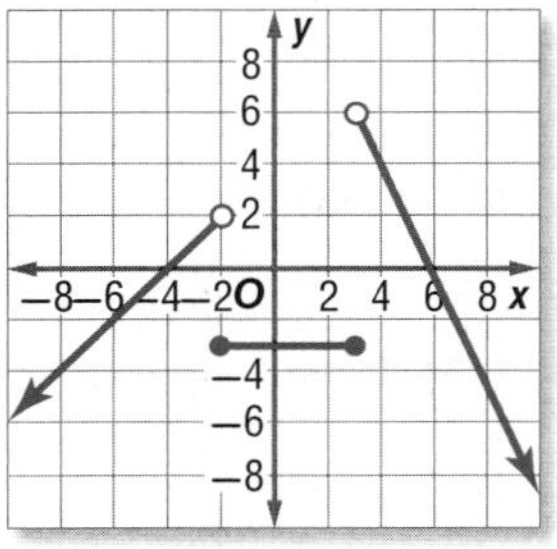

4.

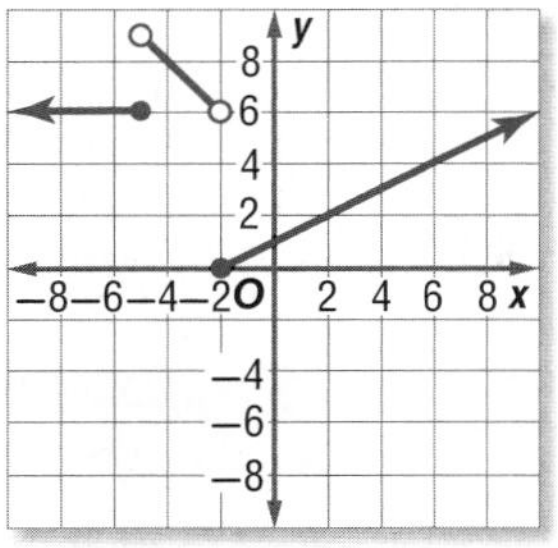

Example 3 p. 103

5. **THEATER** Springfield High School's theater can hold 250 students. The drama club is performing a play in the theater. Draw a graph of a step function that shows the relationship between the number of tickets sold x and the minimum number of performances y that the drama club must do.

Graph each function. Identify the domain and range.

6. $g(x) = -2[\![x]\!]$ **7.** $h(x) = [\![x - 5]\!]$

Example 4 p. 104

Graph each function. Identify the domain and range.

8. $g(x) = |-3x|$ **9.** $f(x) = 2|x|$

10. $h(x) = |x + 4|$ **11.** $s(x) = |-2x| + 6$

Practice and Problem Solving

● = Step-by-Step Solutions begin on page R20.
Extra Practice begins on page 947.

Example 1 p. 101

Graph each function. Identify the domain and range.

12. $f(x) = \begin{cases} -3x \text{ if } x \le -4 \\ x \text{ if } 0 < x \le 3 \\ 8 \text{ if } x > 3 \end{cases}$

13. $f(x) = \begin{cases} 2x \text{ if } x \le -6 \\ 5 \text{ if } -6 < x \le 2 \\ -2x + 1 \text{ if } x > 4 \end{cases}$

14. $g(x) = \begin{cases} 2x + 2 \text{ if } x < -6 \\ x \text{ if } -6 \le x \le 2 \\ -3 \text{ if } x > 2 \end{cases}$

15. $g(x) = \begin{cases} -2 \text{ if } x < -4 \\ x - 3 \text{ if } -1 \le x \le 5 \\ 2x - 15 \text{ if } x > 7 \end{cases}$

Example 2 p. 102

Write the piecewise-defined function shown in each graph.

16.

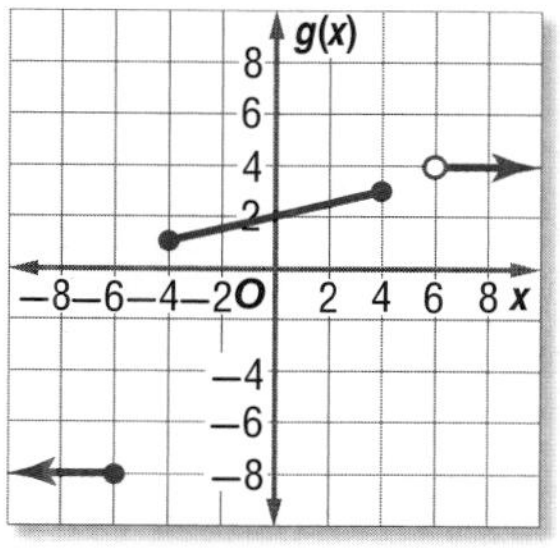

17. ●

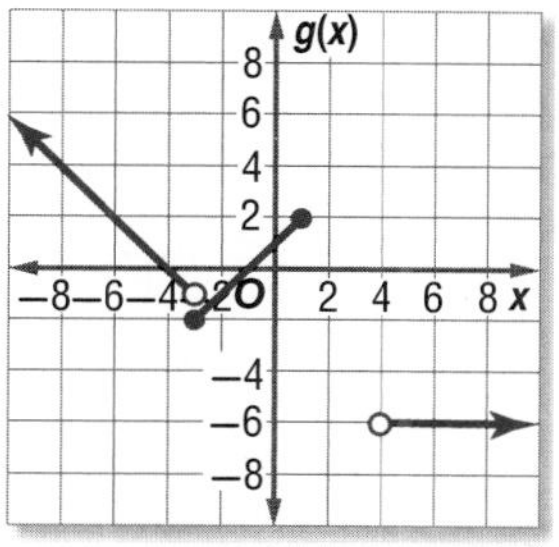

18.

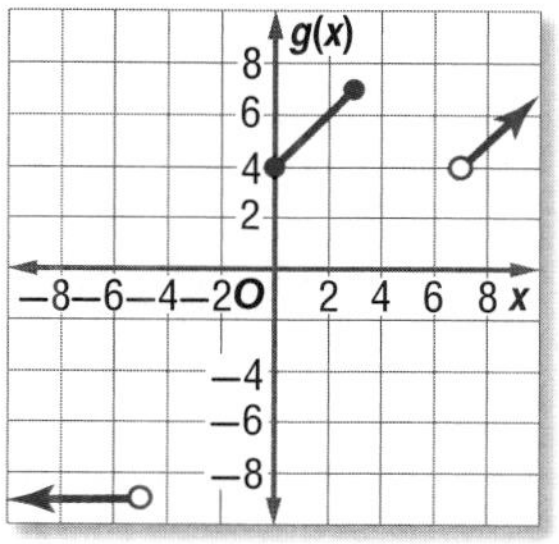

19.

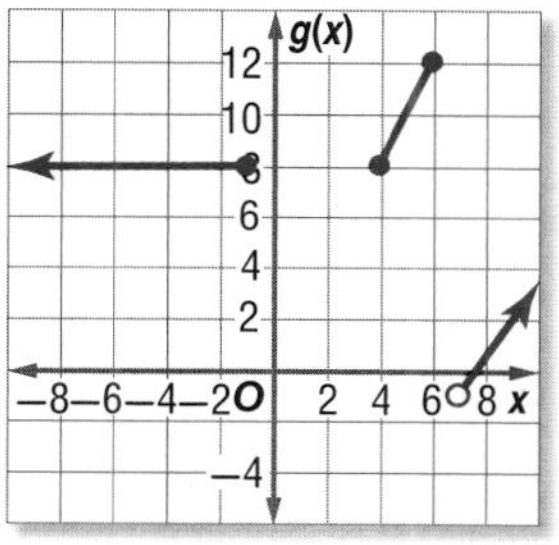

Example 3 p. 103

Graph each function. Identify the domain and range.

20. $f(x) = [\![x]\!] - 6$

21. $h(x) = [\![3x]\!] - 8$

22. $f(x) = [\![3x + 2]\!]$

23. $g(x) = 2[\![0.5x + 4]\!]$

Example 4 p. 104

Graph each function. Identify the domain and range.

24. $f(x) = |x - 5|$

25. $g(x) = |x + 2|$

26. $h(x) = |2x| - 8$

27. $k(x) = |-3x| + 3$

28. $f(x) = 2|x - 4| + 6$

29. $h(x) = -3|0.5x + 1| - 2$

30. **VOLUNTEERING** Patrick is donating and volunteering his time to an organization that restores homes for the needy. He initially donates \$10 and works on one home. He decides to donate \$4 for every additional home on which he works.

a. Identify the type of function that models this situation.

b. Write and graph a function for the situation.

31. **CARS** A car's speedometer reads 60 miles an hour.

a. Write an absolute value function for the difference between the car's actual speed a and the reading on the speedometer.

b. What is an appropriate domain for the function? Explain your reasoning.

c. Use the domain to graph the function.

32. RECREATION The charge for renting a bicycle from a rental shop for different amounts of time is shown at the right.

a. Identify the type of function that models this situation.

b. Write and graph a function for the situation.

Bicycle Rentals

Time	Price
$\frac{1}{2}$ hour	\$6
1 hour	\$10
2 hours	\$16
Daily	\$24

Real-World Link

Cycle tourism has begun to grow rapidly in many parts of the world. Cycle tourism is defined as recreational visits that involve leisure cycling.

Source: Cycle Tourism

Use each graph to write the absolute value function.

33.

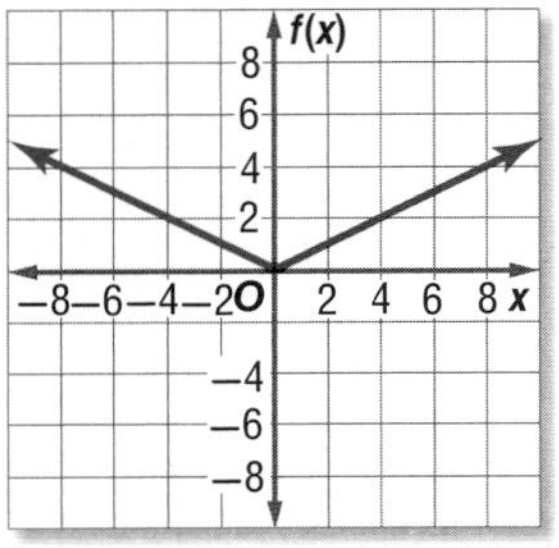

34.

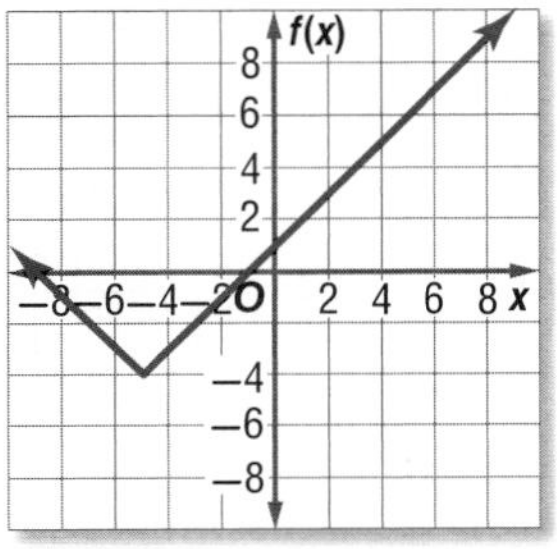

Graph each function. Identify the domain and range.

35. $f(x) = [\![\,|0.5x|\,]\!]$

36. $g(x) = |\,[\![2x]\!]\,|$

37. $g(x) = \begin{cases} [\![x]\!] \text{ if } x < -4 \\ x + 1 \text{ if } -4 \leq x \leq 3 \\ -|x| \text{ if } x > 3 \end{cases}$

38. $h(x) = \begin{cases} -|x| \text{ if } x < -6 \\ |x| \text{ if } -6 \leq x \leq 2 \\ |-x| \text{ if } x > 2 \end{cases}$

39. MULTIPLE REPRESENTATIONS Consider the following absolute value functions.

$$f(x) = |x| - 4 \qquad g(x) = |3x|$$

a. TABULAR Create a table of $f(x)$ and $g(x)$ values for $x = -4$ to $x = 4$.

b. GRAPHICAL Graph the functions on separate graphs.

c. NUMERICAL Determine the slope between each two consecutive points in the table.

d. VERBAL Describe how the slopes of the two sections of an absolute value graph are related.

H.O.T. Problems

Use Higher-Order Thinking Skills

40. OPEN ENDED Write an absolute value function in which the domain is all positive numbers and the range is all real numbers.

41. CHALLENGE Graph $|y| = 2|x + 3| - 5$.

42. REASONING Find a counterexample to the following statement and explain your reasoning.

In order to find the greatest integer function of x when x is not an integer, round x to the nearest integer.

43. OPEN ENDED Write an absolute value function in which $f(5) = -3$.

44. WRITING IN MATH Explain how piecewise functions can be used to accurately represent real-world problems.

NYSRE PRACTICE Reinforcement of A.A.1, A2.A.41

45. SHORT RESPONSE What expression gives the nth term of the linear pattern defined by the table?

2	4	6	8	n
7	13	19	25	?

46. Solve: $5(x + 4) = x + 4$
Step 1: $5x + 20 = x + 4$
Step 2: $4x + 20 = 4$
Step 3: $4x = 24$
Step 4: $x = 6$

Which is the first *incorrect* step in the solution shown above?

A Step 4
B Step 3
C Step 2
D Step 1

47. NUMBER THEORY Twelve consecutive integers are arranged in order from least to greatest. If the sum of the first six integers is 381, what is the sum of the last six integers?

F 345
G 381
H 387
J 417

48. ACT/SAT For which function does

$$f\left(-\frac{1}{2}\right) \neq -1?$$

A $f(x) = 2x$
B $f(x) = |-2x|$
C $f(x) = [\![x]\!]$
D $f(x) = [\![2x]\!]$

Spiral Review

49. FOOTBALL The table shows the relationship between the total number of male students per school and the number of students who tried out for the football team. (Lesson 2-5)

a. Find a regression equation for the data.

b. Determine the correlation coefficient.

c. Predict how many students will try out for football at a school with 800 male students.

Number of Male Students	Number of Tryouts
180	46
212	51
274	62
401	75
513	81
589	90

Write an equation in slope-intercept form for the line described. (Lesson 2-4)

50. passes through $(-3, -6)$, perpendicular to $y = -2x + 1$

51. passes through $(4, 0)$, parallel to $3x + 2y = 6$

52. passes through the origin, perpendicular to $4x - 3y = 12$

Find each value if $f(x) = -4x + 6$, $g(x) = -x^2$, and $h(x) = -2x^2 - 6x + 9$. (Lesson 2-1)

53. $f(2c)$

54. $g(a + 1)$

55. $h(6)$

56. Determine whether the figures below are similar. (Lesson 0-6)

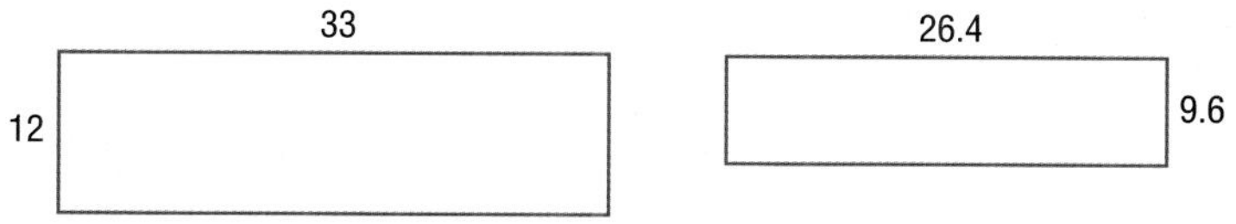

Skills Review

Graph each equation. (Lesson 2-1)

57. $y = -0.25x + 8$

58. $y = \frac{4}{3}x + 2$

59. $8x + 4y = 32$

Graphing Technology Lab
Families of Lines

NY Math Online glencoe.com
Other Calculator Keystrokes

The parent function of the family of linear functions is $f(x) = x$. You can use a graphing calculator to investigate how changing the parameters m and b in $f(x) = mx + b$ affects the graphs as compared to the parent function.

ACTIVITY 1 b in $f(x) = mx + b$

Graph $f(x) = x$, $f(x) = x + 3$, and $f(x) = x - 5$ in the standard viewing window.

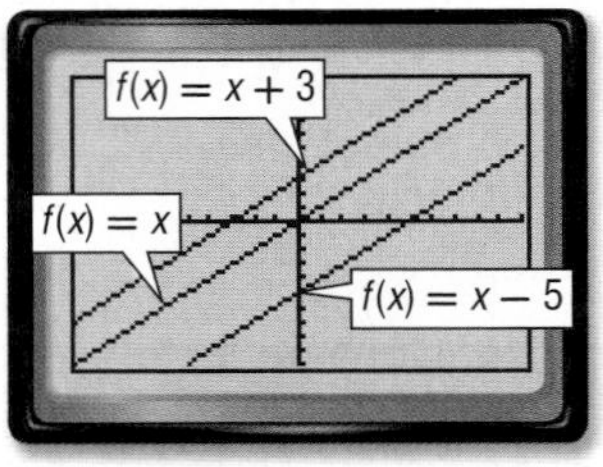

Enter the equations in the Y= list as Y1, Y2, and Y3. Then graph the equations.

KEYSTROKES: [Y=] [X,T,θ,n] [ENTER] [X,T,θ,n] [+] 3 [ENTER] [X,T,θ,n] [−] 5 [ENTER]

1A. Compare and contrast the graphs.

1B. How would you obtain the graphs of $f(x) = x + 3$ and $f(x) = x - 5$ from the graph of $f(x) = x$?

The parameter m in $f(x) = mx + b$ affects the graphs in a different way than b.

ACTIVITY 2 m in $f(x) = mx + b$

Graph $f(x) = x$, $f(x) = 3x$, and $f(x) = \frac{1}{2}x$ in the standard viewing window.

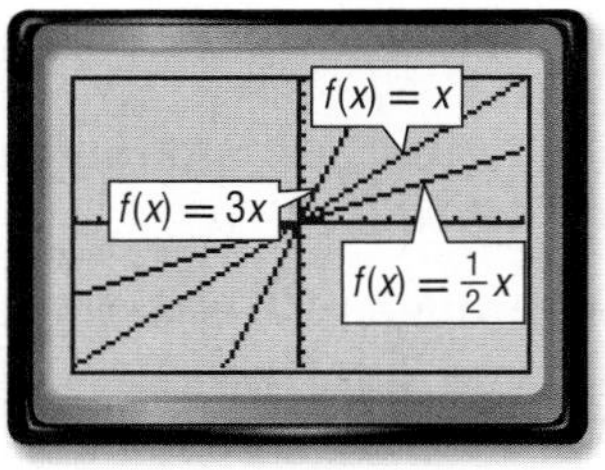

Enter the equations in the Y= list and graph.

2A. How do the graphs compare?

2B. Which graph is steepest? Which graph is the least steep?

2C. Graph $f(x) = -x$, $f(x) = -3x$, and $f(x) = -\frac{1}{2}x$ in the standard viewing window. How do these graphs compare?

Analyze the Results

Graph each set of equations on the same screen. Describe the similarities or differences among the graphs.

1. $f(x) = 3x$
$f(x) = 3x + 1$
$f(x) = 3x - 2$

2. $f(x) = x + 2$
$f(x) = 5x + 2$
$f(x) = \frac{1}{2}x + 2$

3. $f(x) = x - 3$
$f(x) = 2x - 3$
$f(x) = 0.75x - 3$

4. What do the graphs of equations of the form $f(x) = mx + b$ have in common?

5. How do the values of b and m affect the graph of $f(x) = mx + b$ as compared to the parent function $f(x) = x$?

6. Summarize your results. How can knowing about the effects of m and b help you sketch the graph of a function?

2-7 Parent Functions and Transformations

Then

You analyzed and used relations and functions. (Lesson 2-1)

Now

- Identify and use parent functions.
- Describe transformations of functions.

NYS Core Curriculum

Reinforcement of A.G.5 Investigate and generalize how changing the coefficients of a function affects its graph

New Vocabulary

family of graphs
parent graph
parent function
constant function
identity function
quadratic function
translation
reflection
line of reflection
dilation

NY Math Online

glencoe.com

- Extra Examples
- Personal Tutor
- Self-Check Quiz
- Homework Help

Why?

Nick makes $8 an hour working at a pizza shop. The blue line represents his wages. If he also delivers the pizzas, he is paid $2 more per hour. The red line represents Nick's wages when he delivers.

These graphs are examples of transformations.

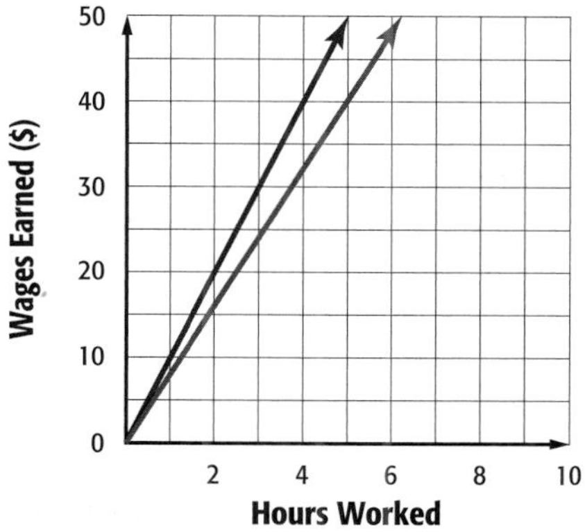

Parent Graphs A **family of graphs** is a group of graphs that display one or more similar characteristics. The **parent graph**, which is the graph of the **parent function**, is the simplest of the graphs in a family. This is the graph that is transformed to create other members in a family of graphs.

Key Concept — Parent Functions (For Your FOLDABLE)

Constant Function

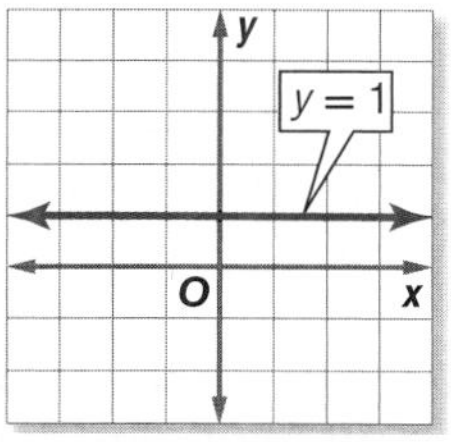

The general equation of a **constant function** is $f(x) = a$, where a is any number. The domain is all real numbers, and the range consists of a single real number a.

Identity Function

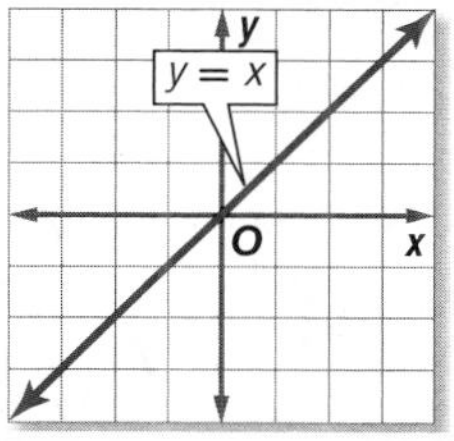

The **identity function** $f(x) = x$ passes through all points with coordinates (a, a). It is the parent function of most linear functions. Its domain and range are all real numbers.

Absolute Value Function

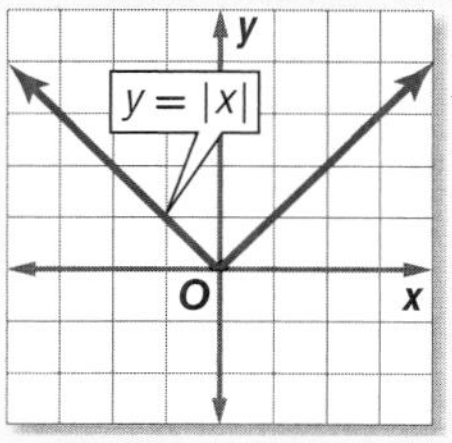

Recall that the parent function of absolute value functions is $f(x) = |x|$. The domain of $f(x) = |x|$ is the set of real numbers, and the range is the set of real numbers greater than or equal to 0.

Quadratic Function

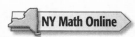

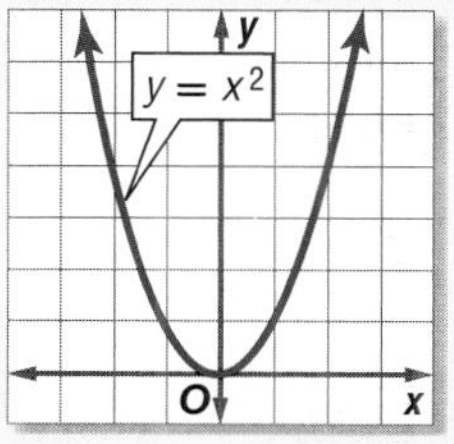

The parent function of **quadratic functions** is $f(x) = x^2$. The domain of $f(x) = x^2$ is the set of real numbers, and the range is the set of real numbers greater than or equal to 0.

EXAMPLE 1 Identify a Function Given the Graph

Identify the type of function represented by each graph.

a.

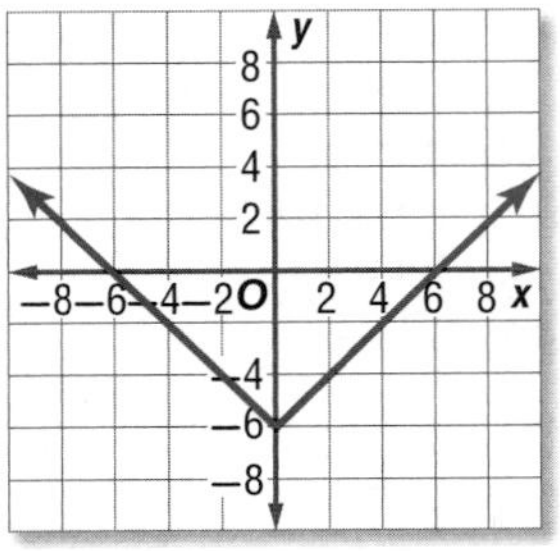

The graph is in the shape of a V. The graph represents an absolute value function.

b.

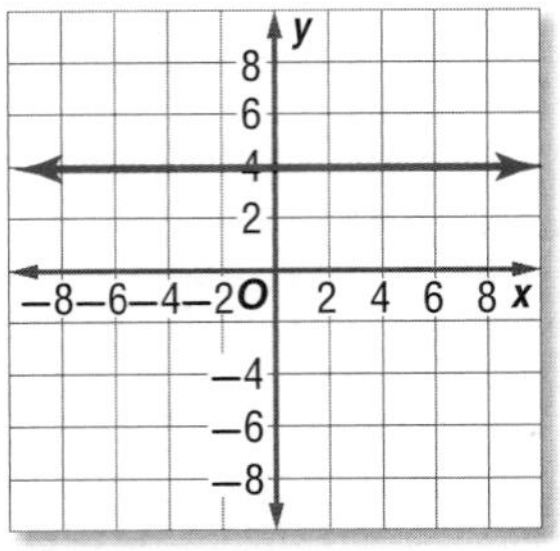

The graph is a horizontal line that crosses the y-axis at 4. The graph represents a constant function.

Check Your Progress

1A.

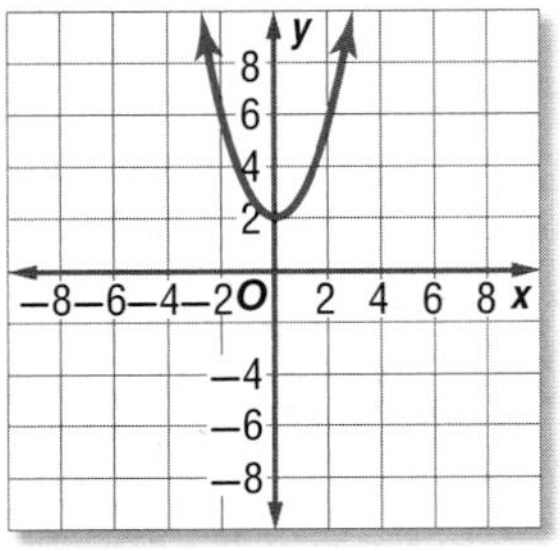

1B.

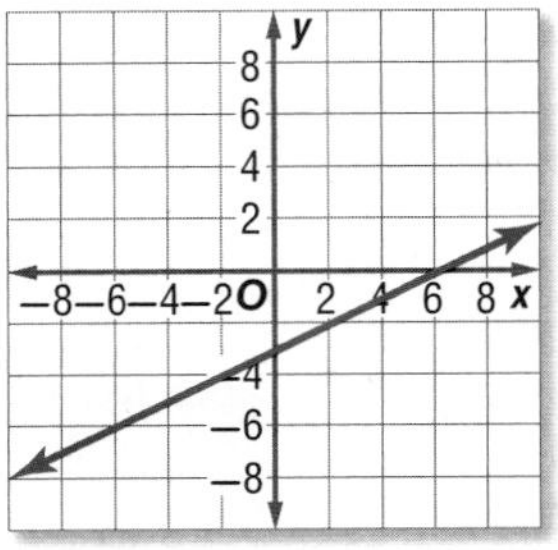

Personal Tutor glencoe.com

Transformations Transformations of a parent graph may appear in a different location, flip over an axis, or appear to have been stretched or compressed. The transformed graph may resemble the parent graph, or it may not.

ReadingMath

translation A translation is also called a *slide*, a *shift*, or a *glide*.

A **translation** moves a figure up, down, left, or right.

- When a constant k is added to or subtracted from a parent function, the result $f(x) \pm k$ is a translation of the graph up or down.
- When a constant h is added to or subtracted from x before evaluating a parent function, the result, $f(x \pm h)$, is a translation left or right.

EXAMPLE 2 Describe and Graph Translations

Describe the translation in $y = |x| + 2$. Then graph the function.

The graph of $y = |x| + 2$ is a translation of the graph of $y = |x|$ up 2 units.

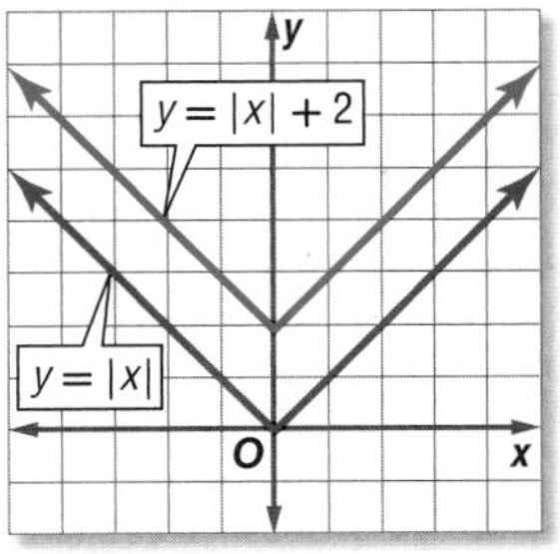

Check Your Progress

Describe the translation in each function. Then graph the function.

2A. $y = |x + 3|$

2B. $y = x^2 - 4$

Personal Tutor glencoe.com

A **reflection** flips a figure over a line called the **line of reflection**.

- When a parent function is multiplied by -1, the result $-f(x)$ is a reflection of the graph in the x-axis.
- When only the variable is multiplied by -1, the result $f(-x)$ is a reflection of the graph in a line of reflection through the vertex.

Review Vocabulary

vertex the maximum or minimum point on the graph of a quadratic or absolute value function (Lesson 2-7)

EXAMPLE 3 Describe and Graph Reflections

Describe the reflection in $y = -x^2$. Then graph the function.

The graph of $y = -x^2$ is a reflection of the graph of $y = x^2$ in the x-axis.

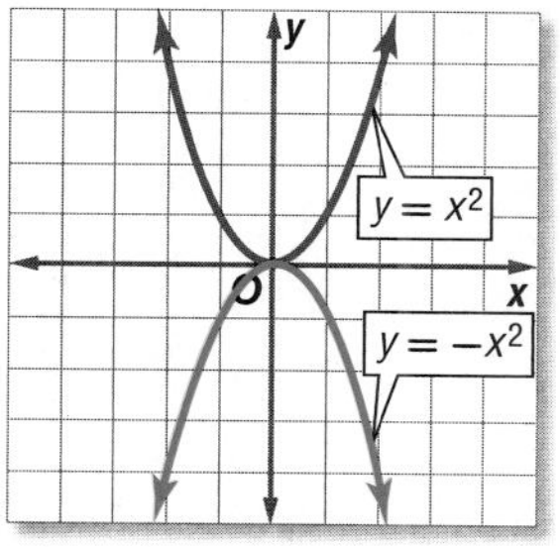

Check Your Progress

Describe the reflection in each function. Then graph the function.

3A. $y = -|x|$

3B. $y = -x$

Personal Tutor glencoe.com

A **dilation** shrinks or enlarges a figure proportionally. When the variable in a linear parent function is multiplied by a nonzero number, the slope of the graph changes.

- When a nonlinear parent function is multiplied by a nonzero number, the function is stretched or compressed vertically.
- Coefficients greater than 1 cause the graph to be stretched vertically, and coefficients between 0 and 1 cause the graph to be compressed vertically.

EXAMPLE 4 Describe and Graph Dilations

Describe the dilation in $y = 4x$. Then graph the function.

The graph of $y = 4x$ is a dilation of the graph of $y = x$. The slope of the graph of $y = 4x$ is steeper than that of the graph of $y = x$.

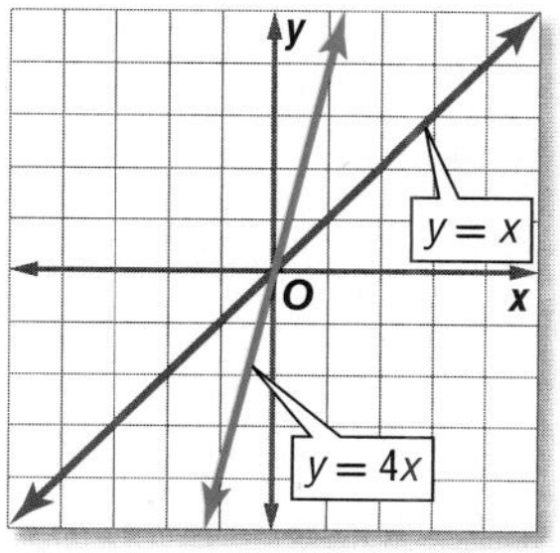

Check Your Progress

Describe the dilation in each function. Then graph the function.

4A. $y = 2x^2$

4B. $y = \left|\frac{1}{3}x\right|$

Personal Tutor glencoe.com

Real-World EXAMPLE 5 Identify Transformations

LANDSCAPING Ethan is going to add a brick walkway around the perimeter of his vegetable garden. The area of the walkway can be represented by the function $f(x) = 4(x + 2.5)^2 - 25$. Describe the transformations in the function. Then graph the function.

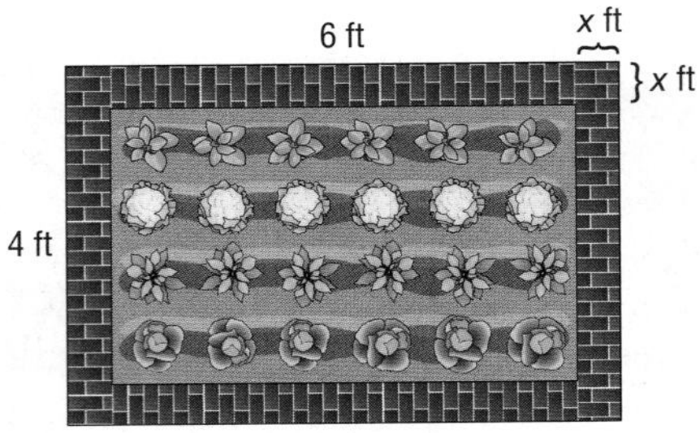

The graph of $f(x) = 4(x + 2.5)^2 - 25$ is a combination of transformations of the parent graph $f(x) = x^2$. Determine how each transformation affects the parent graph.

$f(x) = 4(x + 2.5)^2 - 25$

$+ 2.5$ translates $f(x) = x^2$ left 2.5 units.

$- 25$ translates $f(x) = x^2$ down 25 units.

4 stretches $f(x) = x^2$ vertically.

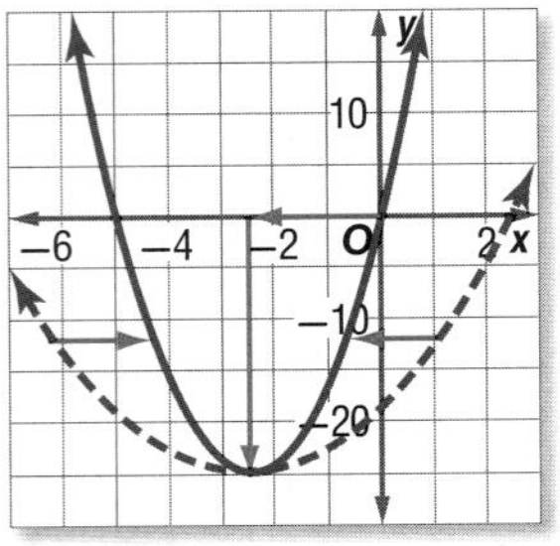

Real-World Link

The average American homeowner spends over $350 each year on indoor and outdoor plants.

Source: American Nursery & Landscape Association

Check Your Progress

5. **SCIENCE** The function $C(x) = \frac{5}{9}(x - 32)$ can be used to determine the temperature in degrees Celsius when given the temperature in degrees Fahrenheit. Describe the transformations in the function. Then graph the function.

Personal Tutor glencoe.com

The table summarizes the changes to the parent graph under different transformations.

Concept Summary — Transformations of Functions

For Your FOLDABLE

Transformation	Change to Parent Graph
Translation	
$f(x + h)$	Translates graph h units left.
$f(x - h)$	Translates graph h units right.
$f(x) + k$	Translates graph k units up.
$f(x) - k$	Translates graph k units down.
Reflection	
$-f(x)$	Reflects graph in the x-axis.
$f(-x)$	Reflects graph in the y-axis.
Dilation	
$a \cdot f(x), a > 1$	Stretches graph vertically.
$a \cdot f(x), 0 < a < 1$	Compresses graph vertically
$f(bx), a > 1$	Compresses graph horizontally.
$f(bx), 0 < a < 1$	Expands graph horizontally.

StudyTip

Describing Transformations Ask yourself these questions to help you identify transformations.

1. What type of function is it?
2. Does the graph open up or down?
3. Does the vertex lie on an axis?

Check Your Understanding

Example 1 p. 110 — **Identify the type of function represented by each graph.**

1. 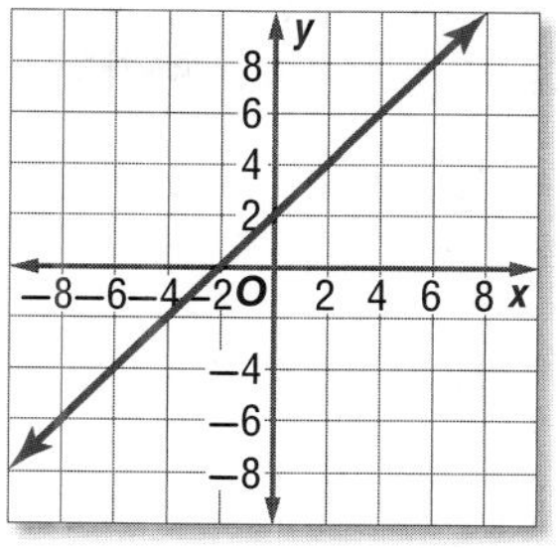

2. 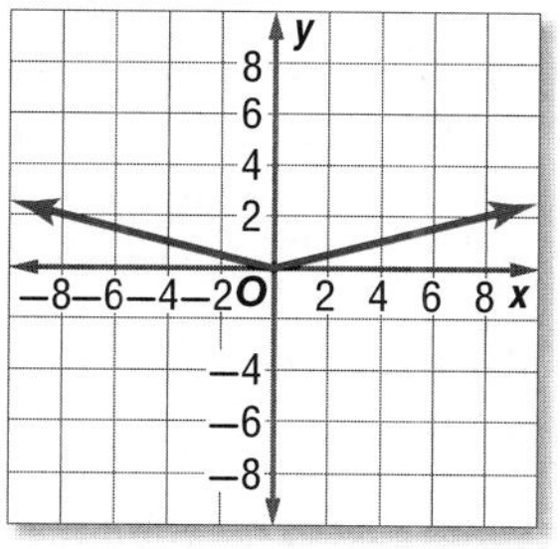

Example 2 p. 110 — **Describe the translation in each function. Then graph the function.**

3. $y = x^2 - 4$

4. $y = |x + 1|$

Example 3 p. 111 — **Describe the reflection in each function. Then graph the function.**

5. $y = -|x|$

6. $y = (-x)^2$

Example 4 p. 111 — **Describe the dilation in each function. Then graph the function.**

7. $y = \frac{3}{5}x$

8. $y = 3x^2$

Example 5 p. 112

9. **FOOD** The manager of a coffee shop is randomly checking cups of coffee drinks prepared by employees to ensure that the correct amount of coffee is in each cup. Each 12-ounce drink should contain half coffee and half steamed milk. The amount of coffee by which each drink varies can be represented by $f(x) = \frac{1}{2}|x - 12|$. Describe the transformations in the function. Then graph the function.

Practice and Problem Solving

● = **Step-by-Step Solutions** begin on page R20.
Extra Practice begins on page 947.

Example 1 p. 110 — **Identify the type of function represented by each graph.**

10. 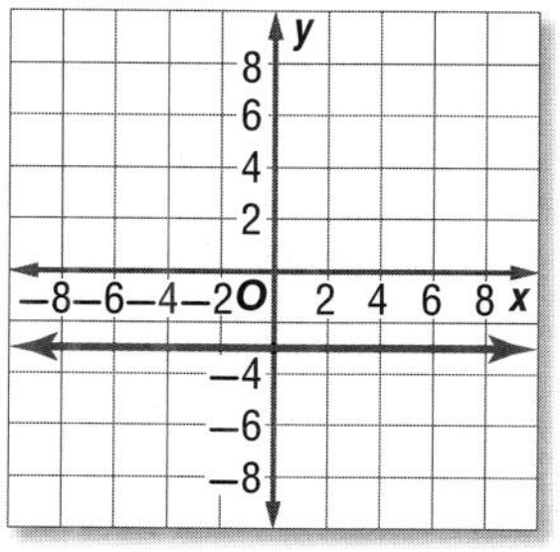

11. 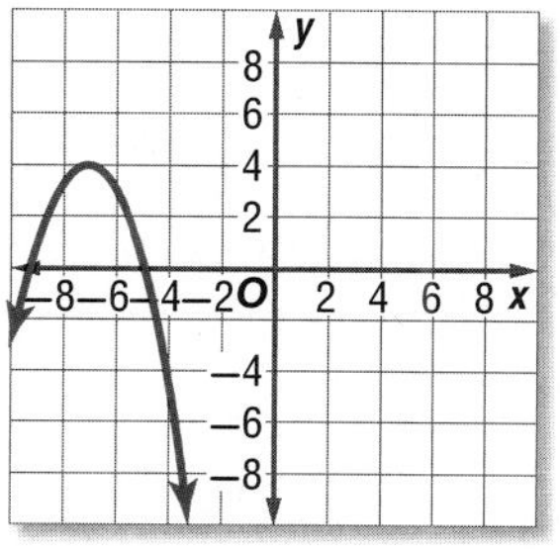

12. 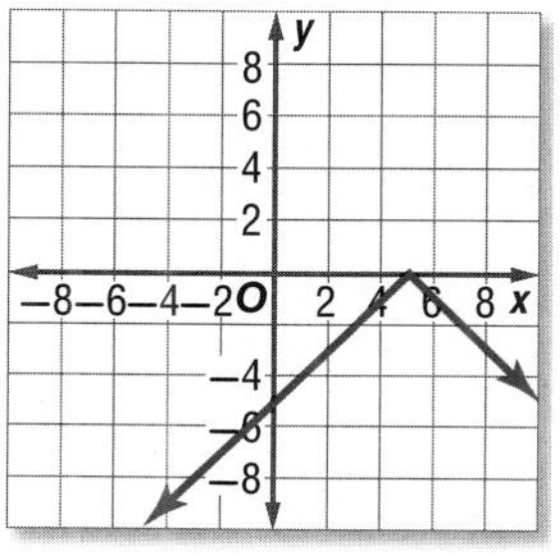

13. 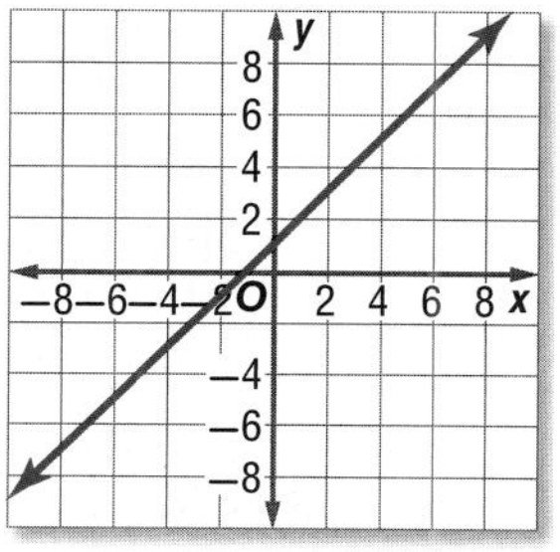

Example 2 p. 110

Describe the translation in each function. Then graph the function.

14. $y = x^2 + 4$ **15.** $y = |x| - 3$ **16.** $y = x - 1$

17. $y = x + 2$ **18.** $y = (x - 5)^2$ **19.** $y = |x + 6|$

Example 3 p. 111

Describe the reflection in each function. Then graph the function.

20. $y = -x$ **21.** $y = -x^2$ **22.** $y = (-x)^2$

23. $y = |-x|$ **24.** $y = -|x|$ **25.** $y = (-x)$

Example 4 p. 111

Describe the dilation in each function. Then graph the function.

26. $y = (3x)^2$ **27.** $y = 6x$ **28.** $y = 4|x|$

29. $y = |2x|$ **30.** $y = \frac{2}{3}x$ **31.** $y = \frac{1}{2}x^2$

Example 5 p. 112

32. HEALTH A non-impact workout can burn up to 7.5 Calories per minute. The equation to represent how many Calories a person burns after m minutes of the workout is $C(m) = 7.5m$. Identify the transformation in the function. Then graph the function.

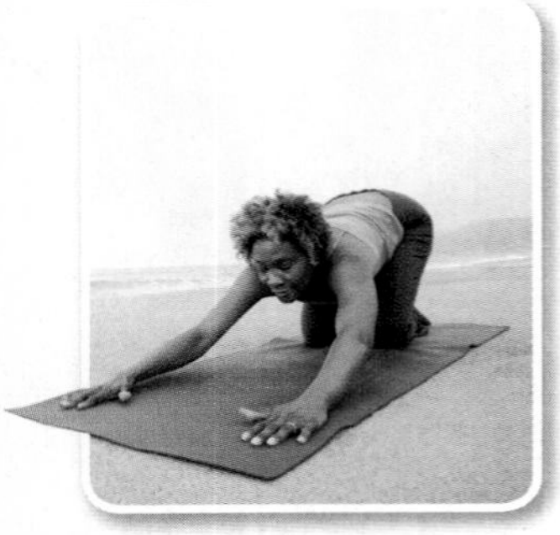

Real-World Link

Non-impact workouts usually include a sequence of exercises that combine strength, flexibility, balance, and inner awareness.

Source: *Medical Encyclopaedia*

Write an equation for each function.

33.

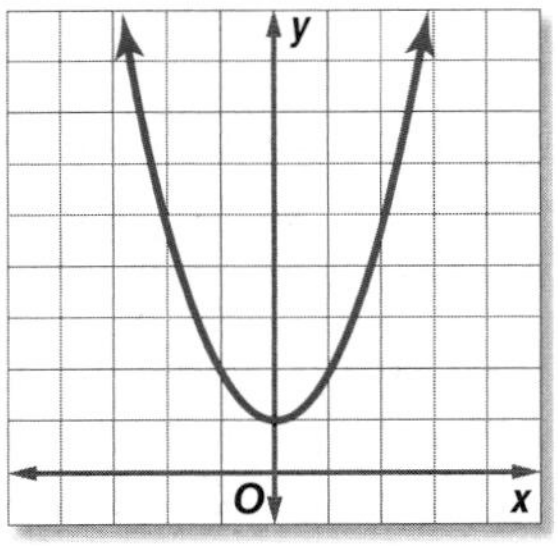

34.

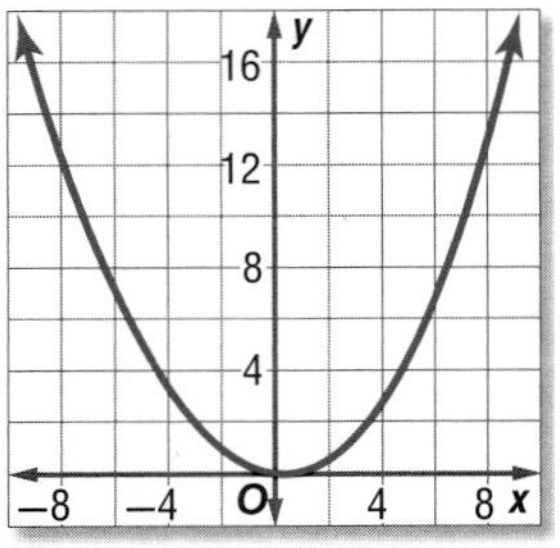

35.

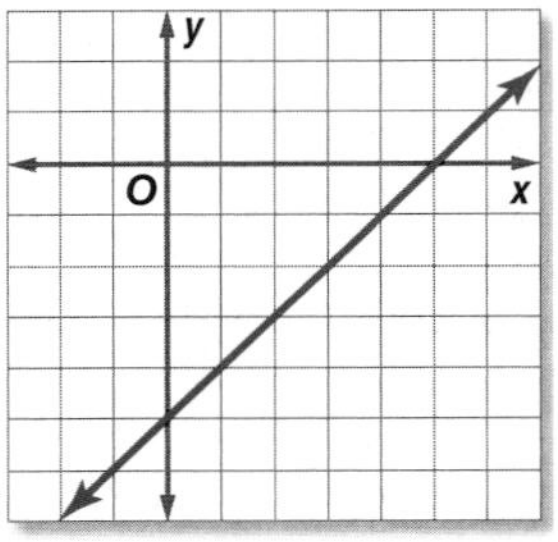

36.

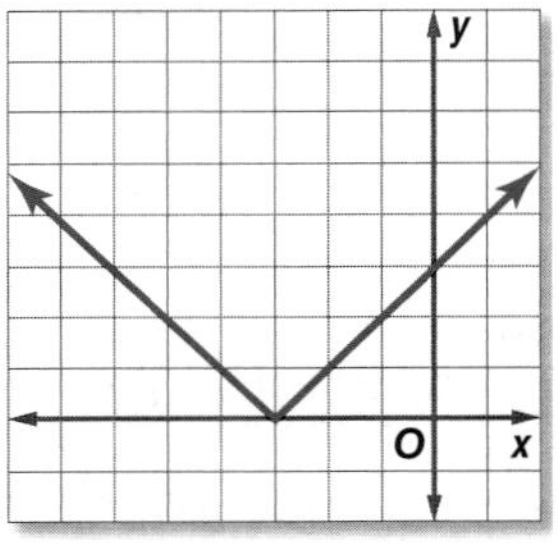

37.

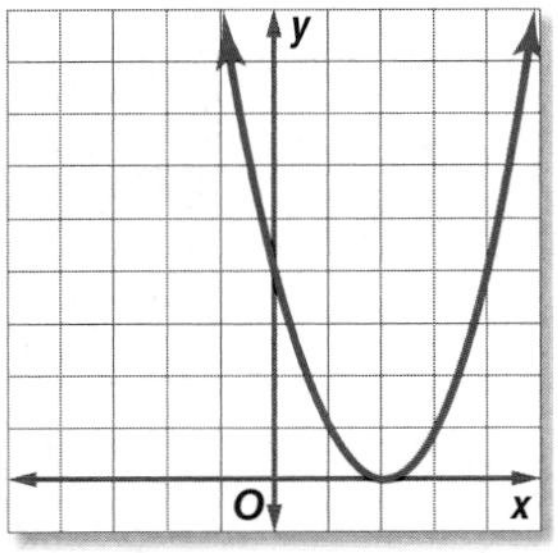

38.

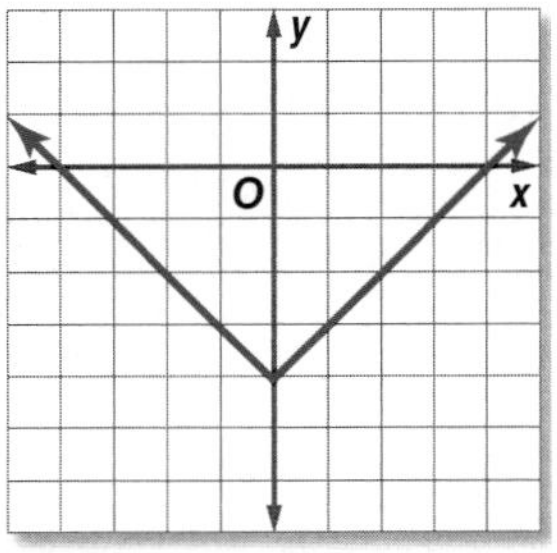

39 **BUSINESS** The graph of the cost of producing x widgets is represented by the blue line in the graph. After hiring a consultant, the cost of producing x widgets is represented by the red line in the graph. Write the equations of both lines and describe the transformation from the blue line to the red line.

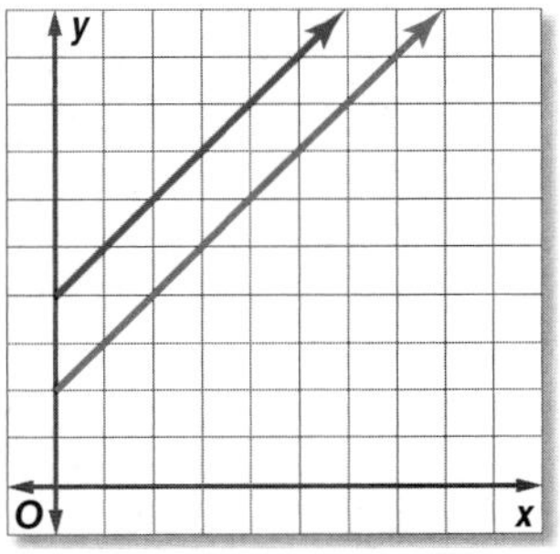

Real-World Link

Model rocketry was developed during the "space-race" era. Most rockets are constructed of cardboard, plastic, and balsa wood, and are fueled by single-use rocket motors.

40. ROCKETRY Kenji launched a toy rocket from ground level. The height $h(t)$ of Kenji's rocket after t seconds is shown in blue. Emily believed that her rocket could fly higher and longer than Kenji's. The flight of Emily's rocket is shown in red.

a. Identify the type of function shown.

b. How much longer than Kenji's rocket did Emily's rocket stay in the air?

c. How much higher than Kenji's rocket did Emily's rocket go?

d. Describe the type of transformation between the two graphs.

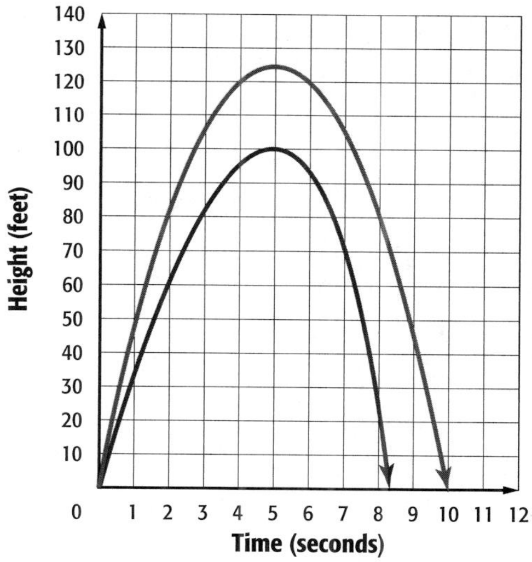

Write an equation for each function.

41

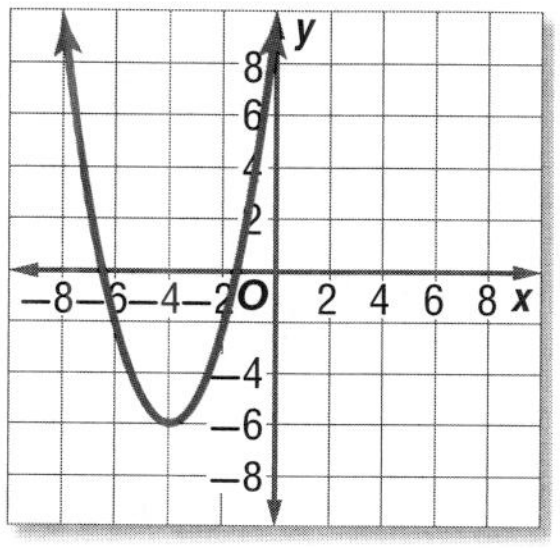

42.

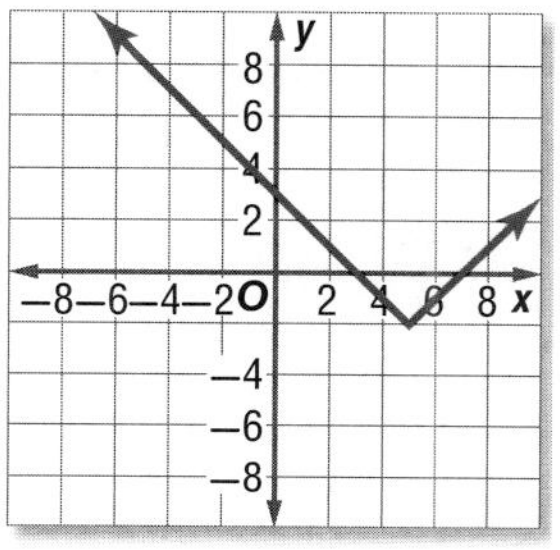

H.O.T. Problems

Use **H**igher-**O**rder **T**hinking Skills

43. CHALLENGE Explain why performing a horizontal translation followed by a vertical translation ends up being the same transformation as performing a vertical translation followed by a horizontal translation.

44. FIND THE ERROR Carla and Kimi are determining if $f(x) = 2x$ is the *identity function*. Is either of them correct? Explain your reasoning.

Carla	Kimi
$f(x) = 2x$ is the identity function because it is linear and goes through the origin.	$f(x) = 2x$ is not the identity function because the values in the domain do not correspond to their duplicates in the range.

45. OPEN ENDED Draw a figure in Quadrant II. Use any of the transformations you learned in this lesson to move your figure to Quadrant IV. Describe your transformation.

46. REASONING Study the parent graphs at the beginning of this lesson. Select a parent graph with positive y-values at its leftmost points and positive y-values at its rightmost points.

47. WRITING IN MATH Explain why the reflection of the graph of $f(x) = x^2$ in the y-axis is the same as the graph of $f(x) = x^2$. Is this true for all reflections of quadratic equations? If not, describe a case when it is false.

NYSRE PRACTICE Reinforcement of A.A.24, Reinforcement of A.G.6

48. What is the solution set of the inequality?

$$6 - |x + 7| \leq -2$$

A $\{x \mid -15 \leq x \leq 1\}$
B $\{x \mid x \leq -1 \text{ or } x \geq 3\}$
C $\{x \mid -1 \leq x \leq 3\}$
D $\{x \mid x \leq -15 \text{ or } x \geq 1\}$

49. GEOMETRY The measures of two angles of a triangle are x and $4x$. Which of these expressions represents the measure of the third angle?

F $180 + x + 4x$
G $180 - x - 4x$
H $180 - x + 4x$
J $180 + x - 4x$

50. GRIDDED RESPONSE Find the value of x that makes $\frac{1}{2} = \frac{x-2}{x+2}$ true.

51. ACT/SAT Which could be the inequality for the graph?

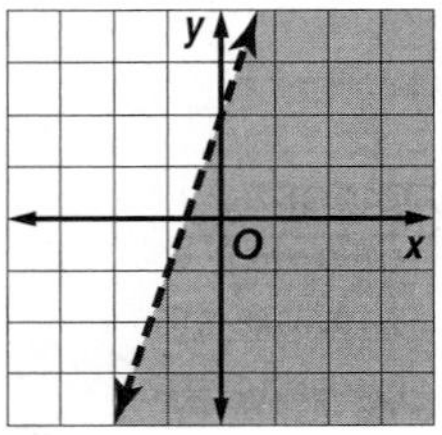

A $y < 3x + 2$
B $y \leq 3x + 2$
C $y > 3x + 2$
D $y \geq 3x + 2$

Spiral Review

Graph each function. Identify the domain and range. (Lesson 2-6)

52. $f(x) = |x - 3|$

53. $h(x) = [\![x]\!] - 5$

54. $f(x) = \begin{cases} -2x \text{ if } x \leq -2 \\ x \text{ if } -2 < x \leq 1 \\ 4 \text{ if } x > 1 \end{cases}$

55. ATTENDANCE The table shows the annual attendance to West High School's Summer Celebration. (Lesson 2-5)

a. Find a regression equation for the data.

b. Determine the correlation coefficient.

c. Predict how many people will attend the Summer Celebration in 2010.

Year	Attendance
2004	61
2005	83
2006	85
2007	92
2008	97
2009	106

Solve each inequality. (Lesson 1-6)

56. $-12 \leq 2x + 4 \leq 8$

57. $-4 < -3y + 2 < 11$

58. $|x - 3| > 7$

59. CARS Loren is buying her first car. She is considering 4 different models and 5 different colors. How many different cars could she buy? (Lesson 0-4)

Determine if each relation is a function. (Lesson 0-1)

60.

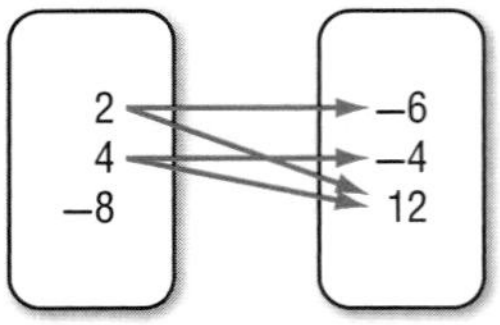

61.

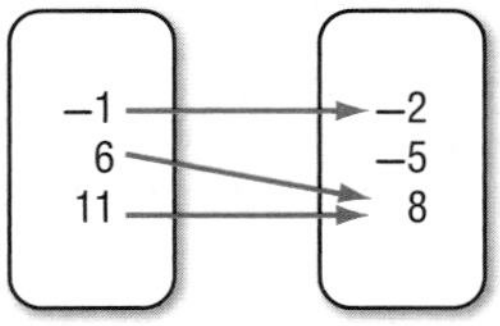

62.

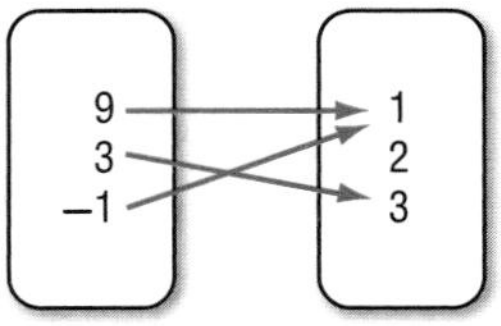

Skills Review

Evaluate each expression if $x = -4$ and $y = 6$. (Lesson 1-1)

63. $4x - 8y + 12$

64. $5y + 3x - 8$

65. $-12x + 10y - 24$

2-8

Graphing Linear and Absolute Value Inequalities

Then

You described transformations on functions. (Lesson 2-7)

Now

- Graph linear inequalities.
- Graph absolute value inequalities.

NYS Core Curriculum

Reinforcement of A.G.6 Graph linear inequalities

New Vocabulary

linear inequality

boundary

NY Math Online

glencoe.com

- Extra Examples
- Personal Tutor
- Self-Check Quiz
- Homework Help

Why?

Randy is planning to treat his lacrosse team to a pizza party after the championship game, but he does not want to spend more than $200.

Randy can use the linear inequality $11p + 2.25d \leq 200$, where p represents the number of pizzas and d represents the number of soft drinks, to check whether certain combinations of pizzas and drinks will fall within his budget.

Graph Linear Inequalities A **linear inequality** resembles a linear equation, but with an inequality symbol instead of an equals symbol. For example, $y > -3x - 2$ is a linear inequality and $y = -3x - 2$ is the related linear equation.

The graph of the inequality $y > -3x - 2$ is shown at the right as a shaded region. Every point in the shaded region satisfies the inequality. The graph of $y = -3x - 2$ is the **boundary** of the region. It is drawn as a dashed line to show that points on the line do not satisfy the inequality. If the symbol were $\leq$ or $\geq$, then points on the boundary would satisfy the inequality, so the boundary would be drawn as a solid line.

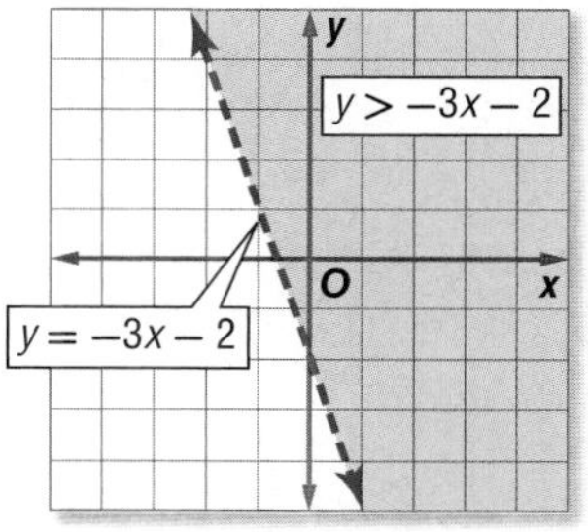

EXAMPLE 1 Dashed Boundary

Graph $x + 4y > 2$.

Step 1 The boundary of the graph is the graph of $x + 4y = 2$. Since the inequality symbol is $>$, the boundary will be dashed.

Step 2 Test the point (0, 0).

$x + 4y > 2$	**Original inequality**
$0 + 4(0) \overset{?}{>} 2$	**$(x, y) = (0, 0)$**
$0 > 2$ ✗	**False**

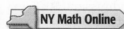

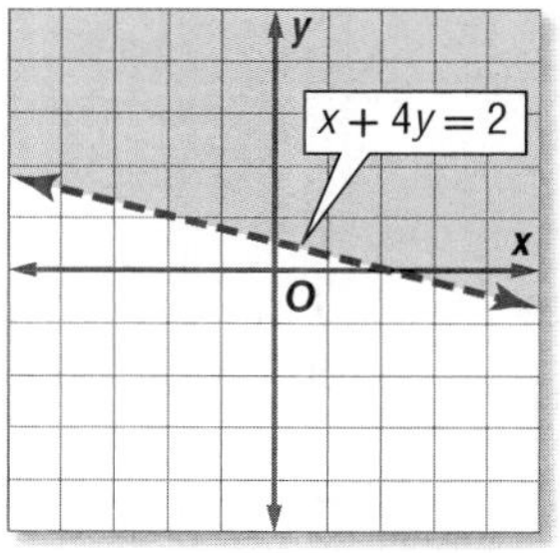

The region that does *not* contain (0, 0) is shaded.

CHECK The graph indicates that (0, 3) is a solution.

$x + 4y > 2$	**Original inequality**
$0 + 4(3) \overset{?}{>} 2$	**$(x, y) = (0, 3)$**
$12 > 2$ ✓	**True**

The solution checks.

Check Your Progress

1A. Graph $3x + \frac{1}{2}y < 2$.

1B. Graph $-x + 2y > 4$.

Personal Tutor glencoe.com

Real-World Career

Recreation Workers
Recreation workers plan, organize, and manage recreational activities to meet the needs of a variety of people. Educational requirements range from a high school diploma to graduate degrees.

Real-World EXAMPLE 2 Solid Boundary

RECREATION A recreation center offers various 30-minute and 60-minute art classes. The recreation director has allotted up to 20 hours per week for art classes.

a. Write an inequality to represent the number of classes that can be offered per week. Graph the inequality.

Let x represent the number of 30-minute or $\frac{1}{2}$-hour art classes, and let y represent the number of 60-minute or 1-hour art classes. Because the sum can equal the maximum, the inequality symbol is $\leq$, and the boundary is solid. The inequality is $\frac{1}{2}x + y \leq 20$.

Step 1 Graph the boundary $\frac{1}{2}x + y = 20$.

Step 2 Test the point (0, 0).

$\frac{1}{2}x + y \leq 20$ **Original inequality**

$\frac{1}{2}(0) + (0) \overset{?}{\leq} 20$ **$(x, y) = (0, 0)$**

$0 \leq 20$ ✓ **True**

The region that contains (0, 0) is shaded.

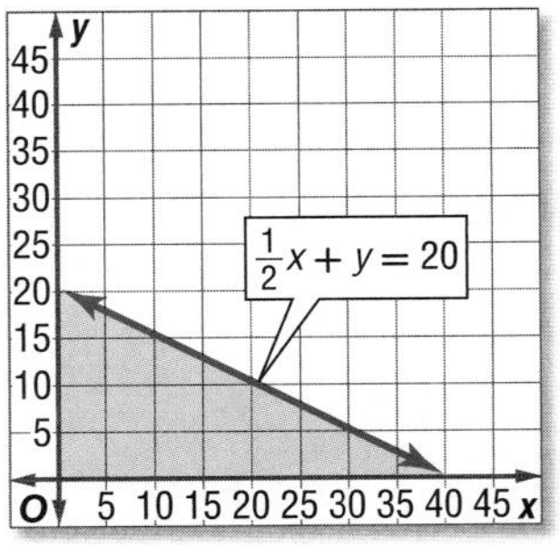

b. Can the recreation director schedule 25 of the 30-minute classes and 15 of the 60-minute classes during a given week? Explain your reasoning.

The point (25, 15) lies outside the shaded region, so it does not satisfy the inequality. Thus, the recreation director cannot schedule 25 30-minute and 15 60-minute classes.

Check Your Progress

2. Manuel has \$15 to spend at the county fair. The fair costs \$5 for admission, \$0.75 for each ride ticket, and \$0.25 for each game ticket. Write an inequality, and draw a graph that represents the number of ride and game tickets that Manuel can buy.

Personal Tutor glencoe.com

Graph Absolute Value Inequalities Graphing absolute value inequalities is similar to graphing linear inequalities. First you graph the absolute value equation. Then you determine whether the boundary is dashed or solid and which region should be shaded.

EXAMPLE 3 Absolute Value Inequality

Graph $y \geq |x| - 4$.

Since the inequality symbol is $\geq$, the boundary is solid. Graph the equation. Then test (0, 0).

$y \geq |x| - 4$ **Original inequality**

$0 \overset{?}{\geq} |0| - 4$ **$(x, y) = (0, 0)$**

$0 \geq -4$ ✓ **True**

The region that includes (0, 0) is shaded.

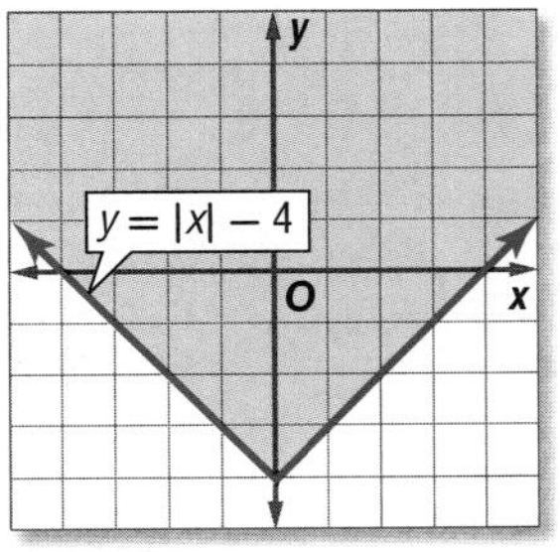

Check Your Progress

3A. Graph $y \leq 2|x| + 3$.

3B. Graph $y \geq 3|x + 1|$.

Personal Tutor glencoe.com

Check Your Understanding

Example 1 p. 117

Graph each inequality.

1. $y < 4$
2. $x \geq -6$
3. $x + 4y \leq 2$
4. $3x + y > -8$

Example 2 p. 118

5. **CAR MAINTENANCE** Gregg needs to buy gas and oil for his car. Gas costs \$3.45 a gallon, and oil costs \$2.41 a quart. He has \$50 to spend.
   a. Write an inequality to represent the situation, where g is the number of gallons of gas he buys and q is the number of quarts of oil.
   b. Graph the inequality.
   c. Can Gregg buy 10 gallons of gasoline and 8 quarts of oil? Explain.

Example 3 p. 118

Graph each inequality.

6. $y \geq |x + 3|$
7. $y - 6 \leq |x|$

Practice and Problem Solving

● = **Step-by-Step Solutions** begin on page R20.
Extra Practice begins on page 947.

Example 1 p. 117

Graph each inequality.

8. $x + 2y > 6$
9. $y \geq -3x - 2$
10. $2y + 3 \leq 11$
11. $4x - 3y > 12$
12. $6x + 4y \leq -24$
13. $y \geq \frac{3}{4}x + 6$

Example 2 p. 118

14. **COLLEGE** April's guidance counselor says that she needs a combined score of at least 1700 on her college entrance exams to be eligible for the college of her choice. The highest possible score is 2400—1200 on the math portion and 1200 on the verbal portion.
   a. The inequality $x + y \geq 1700$ represents this situation, where x is the verbal score and y is the math score. Graph this inequality.
   b. Refer to your graph. If she scores a 680 on the math portion of the test and 910 on the verbal portion of the test, will April be eligible for the college of her choice?

Example 3 p. 118

Graph each inequality.

15. $y > |3x|$
16. $y + 4 \leq |x - 2|$
17. $y - 6 < |-2x|$
18. $y + 8 < 2\left|\frac{2}{3}x + 6\right|$
19. $2y > |4x - 5|$
20. $-y \leq |3x - 4|$

21. **SCHOOL DANCE** Carlos estimates that he will need to earn at least \$700 to take his girlfriend to the prom. Carlos works two jobs as shown in the table.
   a. Write an inequality to represent this situation.
   b. Graph the inequality.
   c. Will he make enough working 50 hours at each job?

Job	Pay
Main St. Deli	\$8 an hour
Babysitting	\$6 an hour

Graph each inequality.

22. $y \geq |-2x - 6|$
23. $y \leq |x - 3| + 4$
24. $y - 3 > -2|x + 4|$
25. $|y| > |x|$
26. $|x - y| > 5$
27. $|x + 3y| \geq -2$

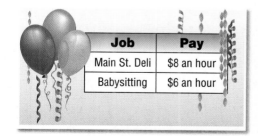

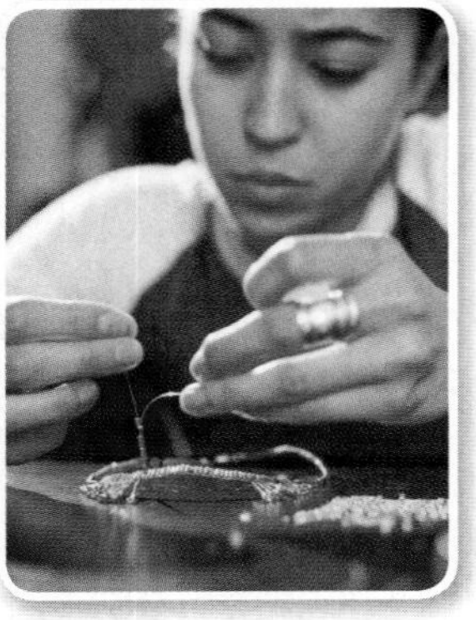

Real-World Link

Making art jewelry is one of the top nine businesses for teen entrepreneurs.

Source: *Entrepreneur Magazine*

28. JEWELRY Mei is making necklaces and bracelets to sell at a craft show. She has enough beads to make 50 pieces. Let x represent the number of bracelets and y represent the number of necklaces.

a. Write an inequality that shows the possible number of necklaces and bracelets Mei can make.

b. Graph the inequality.

c. Give three possible solutions for the number of necklaces and bracelets that can be made.

29. GIFT CARDS Susan received a gift card from an electronics store for $400. She wants to spend the money on DVDs, which cost $20 each, and CDs, which cost $15 each.

a. Let d equal the number of DVDs, and let c equal the number of CDs. Write an inequality that shows the possible combinations of DVDs and CDs that Susan can purchase.

b. Graph the inequality.

c. Give three possible solutions for the number of DVDs and CDs she can buy.

Graph each inequality.

30. $y \geq [\![x]\!]$

31. $y < [\![x + 2]\!]$

32. $y \geq |[\![x]\!]|$

H.O.T. Problems

Use Higher-Order Thinking Skills

33. OPEN ENDED Create an absolute value inequality in which all of the possible solutions fall in the first or fourth quadrant.

34. CHALLENGE Graph the following inequality.

$$g(x) > \begin{cases} |x + 1| \text{ if } x \leq -4 \\ -|x| \text{ if } -4 < x < 2 \\ |x - 4| \text{ if } x \geq 2 \end{cases}$$

35. FIND THE ERROR Paulo and Janette are graphing $x - y \geq 2$. Is either of them correct? Explain your reasoning.

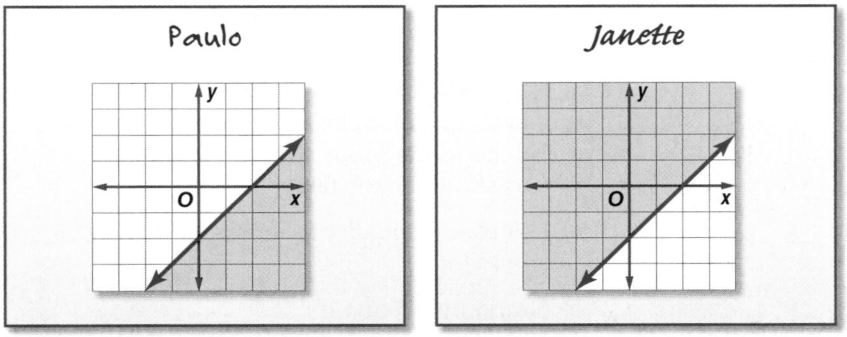

36. REASONING When will it be possible to shade two different areas when graphing a linear absolute value inequality? Explain your reasoning.

37. WRITING IN MATH Describe a situation in which there are no solutions to an absolute value inequality. Explain your reasoning.

NYSRE PRACTICE Reinforcement of A.N.1, A2.A.1

38. EXTENDED RESPONSE Craig scored 85%, 96%, 79%, and 81% on his first four math tests. He hopes to score high enough on the final test to earn a 90% average. If the final test is worth twice as much as one of the other tests, determine if Craig can earn a 90% average. If so, what score does Craig need to get on the final test to accomplish this? Explain how you found your answer.

39. Which of the following sets of numbers represents an infinite set?

A $\{2, 4, 6\}$

B {whole numbers between −50 and 50}

C {integers}

D $\left\{\frac{1}{2}, \frac{3}{4}, \frac{4}{5}, \frac{5}{6}\right\}$

40. SHORT RESPONSE Which theorem of congruence should be used to prove $\triangle ABC \cong \triangle XYZ$?

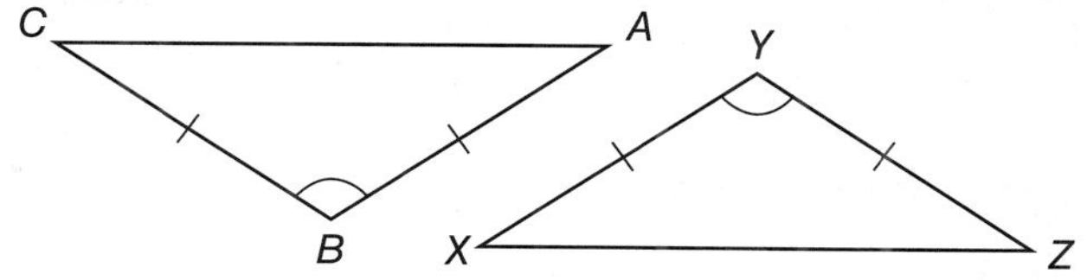

41. ACT/SAT For which function is the range $\{y \mid y \leq 0\}$?

F $f(x) = -x$

G $f(x) = [\![x]\!]$

H $f(x) = |x|$

J $f(x) = -|x|$

Spiral Review

Write an equation for each graph. (Lesson 2-7)

42.

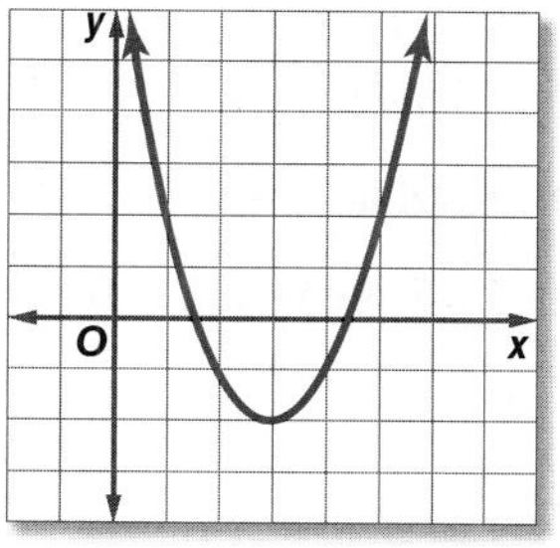

43.

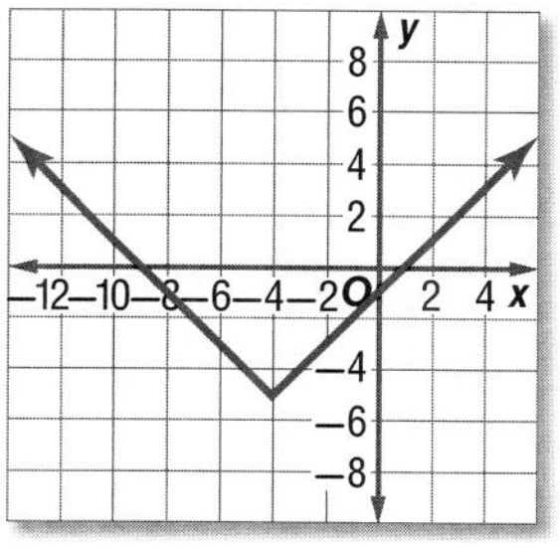

44.

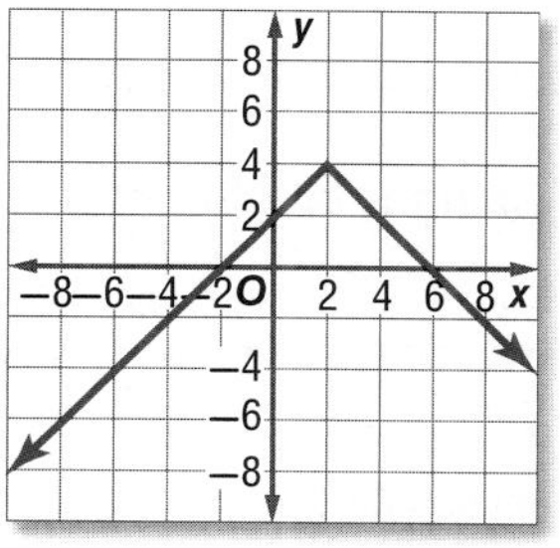

Graph each function. (Lesson 2-6)

45. $f(x) = \begin{cases} x \text{ if } x < 1 \\ 3 \text{ if } 1 \leq x \leq 3 \\ -2x \text{ if } x > 3 \end{cases}$

46. $f(x) = \begin{cases} x + 3 \text{ if } x < -2 \\ 2x \text{ if } -2 \leq x \leq 2 \\ -3x \text{ if } x > 2 \end{cases}$

47. $f(x) = \begin{cases} -2x \text{ if } x \leq -2 \\ x + 1 \text{ if } 0 < x \leq 6 \\ x - 5 \text{ if } x > 6 \end{cases}$

Write each equation in standard form. Identify *A*, *B*, and *C*. (Lesson 2-2)

48. $-6y = 8x - 3$

49. $12y + x = -3y + 5x - 6$

50. $\frac{x + 3}{4} + \frac{y - 1}{2} = 3$

51. TENNIS Sixteen players signed up for tennis lessons. The instructor plans to use 50 tennis balls for every player and have 200 extra. How many tennis balls are needed for the lessons? (Lesson 1-3)

Multiply. (Lesson 0-2)

52. $(3x - 4)(2x + 1)$

53. $(6x + 5)(-x - 3)$

54. $(5x + 2)(-2x + 3)$

Skills Review

Graph each linear equation. (Lesson 2-2)

55. $y = 2x - 8$

56. $y = -\frac{3}{4}x + 2$

57. $3y - 4x = 24$

CHAPTER 2

Study Guide and Review

NY Math Online glencoe.com
- STUDY TO GO
- Vocabulary Review

Chapter Summary

Key Concepts

Relations and Functions (Lesson 2-1)

- A function is a relation where each member of the domain is paired with exactly one member of the range.

Linear Equations and Slope (Lessons 2-2 to 2-4)

- Standard Form: $Ax + By = C$, where A, B, and C are integers whose greatest common factor is 1, $A \geq 0$, and A and B are not both zero.
- Slope-Intercept Form: $y = mx + b$
- Point-Slope Form: $y - y_1 = m(x - x_1)$

Scatter Plots and Lines of Regression (Lesson 2-5)

- A prediction equation can be used to predict the value of one of the variables given the value of the other variable.
- A line of regression can be used to model data.

Special Functions and Parent Functions (Lessons 2-6 and 2-7)

- A piecewise-defined function is made up of two or more expressions.
- Translations, reflections, and dilations to a parent graph form a family of graphs.

Graphing Linear and Absolute Value Inequalities (Lesson 2-8)

- You can graph an inequality by following these steps.

  Step 1 Determine whether the boundary is solid or dashed. Graph the boundary.

  Step 2 Choose a point not on the boundary and test it in the inequality.

  Step 3 If a true inequality results, shade the region containing your test point. If a false inequality results, shade the other region.

FOLDABLES® Study Organizer

Be sure the Key Concepts are noted in your Foldable.

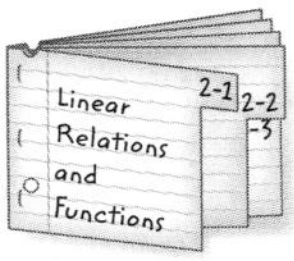

Key Vocabulary

absolute value function (p. 103)
bivariate data (p. 92)
continuous relation (p. 62)
correlation coefficient (p. 94)
dependent variable (p. 64)
dilation (p. 111)
direct variation (p. 90)
discrete relation (p. 62)
family of graphs (p. 109)
greatest integer function (p. 102)
independent variable (p. 64)
linear equation (p. 69)
linear function (p. 69)
linear inequality (p. 117)
line of fit (p. 92)
negative correlation (p. 92)
parent function (p. 109)
piecewise-defined function (p. 101)
point-slope form (p. 84)
positive correlation (p. 92)
prediction equation (p. 92)
quadratic function (p. 109)
rate of change (p. 76)
reflection (p. 111)
regression line (p. 94)
scatter plot (p. 92)
slope (p. 77)
slope-intercept form (p. 83)
standard form (p. 70)
step function (p. 102)
translation (p. 110)
vertical line test (p. 62)

Vocabulary Check

Choose the correct term to complete each sentence.

1. A function is (discrete, one-to-one) if each element of the domain is paired to exactly one unique element of the range.
2. The (domain, range) of a relation is the set of all first coordinates from the ordered pairs which determine the relation.
3. The (constant, identity) function is a linear function described by $f(x) = x$.
4. If you are given the coordinates of two points on a line, you can use the (slope-intercept, point-slope) form to find the equation of the line that passes through them.
5. A set of bivariate data graphed as ordered pairs in a coordinate plane is called a (scatter plot, line of fit).
6. A function that is written using two or more expressions is called a (linear, piecewise) function.

Lesson-by-Lesson Review

2-1 Relations and Functions (pp. 61–67)

A2.A.41, A2.A.43

State the domain and range of each relation. Then determine whether the relation is a function. If it is a function, determine if it is *one-to-one, onto, both,* or *neither*.

7. $\{(1, 2), (3, 4), (5, 6), (7, 8)\}$

8. $\{(-3, 0), (0, 2), (2, 4), (4, 5), (5, 2)\}$

9. $\{(-4, 1), (3, 3), (1, 1), (-2, 5), (3, -4)\}$

10. $\{(7, -4), (5, -2), (3, 0), (1, 2), (-1, 4)\}$

Find each value if $f(x) = -3x + 2$.

11. $f(4)$ **12.** $f(-3)$

13. $f(0)$ **14.** $f(y)$

15. $f(-a)$ **16.** $f(2w)$

17. BOWLING A bowling alley charges \$2.50 for shoe rental and \$3.25 per game bowled. The amount a bowler is charged can be expressed as $y = 2.50 + 3.25x$, when $x \geq 1$. Graph the equation and find the domain and range. Then determine whether the equation is a function. Is the equation discrete or continuous?

EXAMPLE 1

State the domain and range of the relation $\{(-4, 3), (-1, 0), (-2, 4), (3, -1), (2, 6)\}$. Then determine whether the relation is a function. If it is a function, determine if it is *one-to-one, onto, both,* or *neither*.

Domain: $\{-4, -1, -2, 3, 2\}$
Range: $\{3, 0, 4, -1, 6\}$

Each element of the domain is paired with one element of the range, so the relation is a function. The function is both because each element of the domain is paired with a unique element of the range and each element of the range is paired with a unique element of the domain.

EXAMPLE 2

Find $f(-2)$ if $f(x) = 4x - 3$.

$f(-2) = 4(-2) - 3$	**Substitute −3 for *x*.**
$= -8 - 3$	**Multiply.**
$= -11$	**Simplify.**

2-2 Linear Relations and Functions (pp. 69–74)

A.G.4

State whether each function is a linear function. Write *yes* or *no*. Explain.

18. $3x + 4y = 12$ **19.** $x^2 + y^2 = 4$

20. $y = x^3 - 6$ **21.** $y = 6x - 19$

22. $f(x) = -2x + 9$ **23.** $\frac{1}{x} + 3y = -5$

Write each equation in standard form. Identify *A*, *B*, and *C*.

24. $2x + 5y = 10$ **25.** $y = 12x$

26. $-4y = 3x - 24$ **27.** $4x = 8y - 12$

28. TRAVEL The distance the Green family traveled during their family vacation is given by the equation $y = 65x$, where x represents the number of hours spent driving. How far does the Green family travel in 8 hours?

EXAMPLE 3

State whether $f(x) = 3x^2$ is a linear function. Write *yes* or *no*. Explain.

No, because the expression includes a variable raised to the second power.

EXAMPLE 4

Write the equation $y = -5x + 8$ in standard form. Identify *A*, *B*, and *C*.

$y = -5x + 8$	**Original equation**
$5x + y = 8$	**Add 5*x* to each side.**

$A = 5$, $B = 1$, and $C = 8$

Study Guide and Review

2-3 Rate of Change and Slope (pp. 76–82)

A.A.32, A.A.33

29. RETAIL The table shows the number of DVDs sold each week at the Super Movie store. Find the average rate of change of the number of DVDs sold from week 2 to week 5.

Week	1	2	3	4	5
DVDs Sold	76	58	94	83	112

Find the slope of the line that passes through each pair of points.

30. (2, 5), (6, −3)

31. (8, 2), (2, 8)

32. Determine the rate of change of the graph.

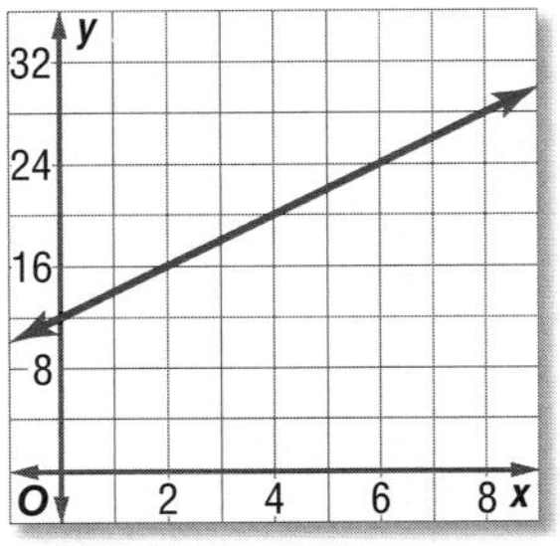

EXAMPLE 5

Find the slope of the line that passes through each pair of points.

a. (−2, 9), (1, 4)

$m = \frac{y_2 - y_1}{x_2 - x_1}$ **Slope Formula**

$= \frac{4 - 9}{1 - (-2)}$ **$(x_1, y_1) = (-2, 9)$, $(x_2, y_2) = (1, 4)$**

$= -\frac{5}{3}$ **Simplify.**

b. (−3, 6), (4, 6)

$m = \frac{y_2 - y_1}{x_2 - x_1}$ **Slope Formula**

$= \frac{6 - 6}{4 - (-3)}$ **$(x_1, y_1) = (-3, 6)$, $(x_2, y_2) = (4, 6)$**

$= \frac{0}{7}$ **Simplify.**

The line has no slope.

2-4 Writing Linear Equations (pp. 83–89)

A.A.34, A.A.36

Write an equation in slope-intercept form for the line that satisfies each set of conditions.

33. slope −2, passes through (−3, −5)

34. slope $\frac{2}{3}$, passes through (4, −1)

35. passes through (−2, 4) and (0, 8)

36. passes through (3, 5) and (−1, 5)

Write an equation of the line passing through each pair of points.

37. (6, 1), (4, 9)

38. (−4, 2), (6, 8)

Write an equation in slope-intercept form for the line that satisfies each set of conditions.

39. through (1, 2), parallel to $y = 4x - 3$

40. through (−3, 5), perpendicular to $y = \frac{2}{3}x - 8$

41. PETS Drew paid a $250 fee when he adopted a puppy. The average monthly cost of feeding and caring for the puppy is $32. Write an equation that represents the total cost of adopting and caring for the puppy for x months.

EXAMPLE 6

Write an equation of the line through (−2, 5) and (0, −9).

Find the slope of the line.

$m = \frac{y_2 - y_1}{x_2 - x_1}$ **Slope Formula**

$= \frac{-9 - 5}{0 - (-2)}$ **$(x_1, y_1) = (-2, 5)$, $(x_2, y_2) = (0, -9)$**

$= \frac{-14}{2}$ or −7 **Simplify.**

Write an equation.

$y - y_1 = m(x - x_1)$ **Point-slope form**

$y - 5 = -7(x - (-2))$ **Substitute.**

$y - 5 = -7(x + 2)$ **Simplify.**

$y - 5 = -7x - 14$ **Distributive Property**

$y = -7x - 9$ **Add 5 to each side.**

The equation is $y = -7x - 9$.

MIXED PROBLEM SOLVING
For mixed problem-solving practice, see page 980.

2-5 Scatter Plots and Lines of Regression (pp. 92–98)

A2.S.8

Make a scatter plot and a line of fit and describe the correlation for each set of data. Then, use two ordered pairs to write a prediction equation.

42. HEATING The table shows the monthly heating cost for a large home.

Month	Sep	Oct	Nov	Dec	Jan	Feb
Bill ($)	72	114	164	198	224	185

43. AMUSEMENT PARK The table shows the annual attendance in thousands at an amusement park during the last 5 years.

Year	1	2	3	4	5
People (× 1000)	44	42	39	31	24

EXAMPLE 7

SCHOOL ENROLLMENT The table shows the number of students each year at a school.

Year	'00	'01	'02	'03	'04	'05
Students	125	116	142	154	146	175

Use (2000, 125) and (2005, 175) to find a prediction equation.

$m = \frac{y_2 - y_1}{x_2 - x_1}$ **Slope Formula**

$= \frac{175 - 125}{2005 - 2000}$ **Substitution**

$= \frac{50}{5}$ or 10 **Simplify.**

$y - y_1 = m(x - x_1)$ **Point-slope form**

$y - 125 = 10(x - 2000)$ **Substitution**

$y - 125 = 10x + 20{,}000$ **Distributive Property**

$y = 10x - 19{,}875$ **Add 125 to each side.**

2-6 Special Functions (pp. 101–107)

A.G.4

Graph each function. Identify the domain and range.

44. $f(x) = \begin{cases} -2x \text{ if } x \leq -1 \\ x + 1 \text{ if } -1 < x < 3 \\ x \text{ if } x \geq 3 \end{cases}$

45. $f(x) = \begin{cases} -3 \text{ if } x < -1 \\ 4x - 3 \text{ if } -1 \leq x \leq 3 \\ x \text{ if } x > 3 \end{cases}$

46. Write the piecewise function shown in the graph.

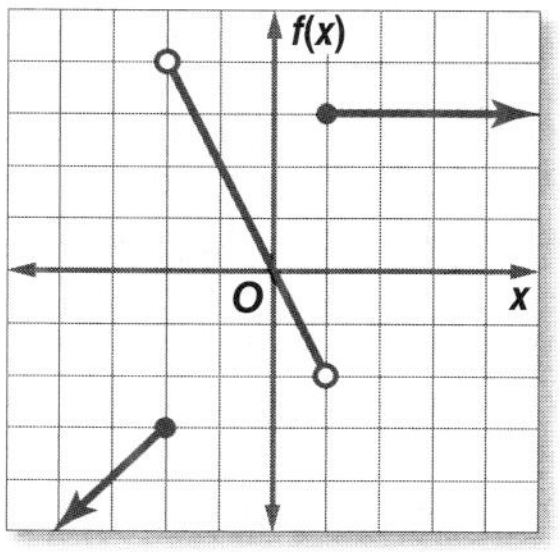

Graph each function. Identify the domain and range.

47. $f(x) = [\![x]\!] + 2$

48. $f(x) = [\![x + 3]\!]$

EXAMPLE 8

Write the piecewise function shown in the graph.

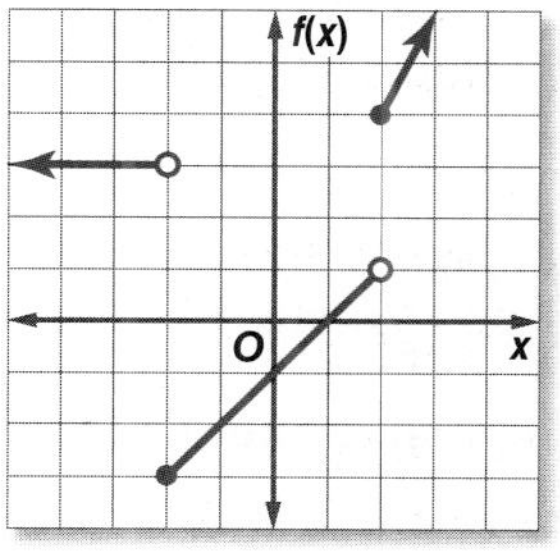

The left portion of the graph is the graph of $f(x) = 3$. There is a circle at $(-2, 3)$, so the linear function is defined for $x < -2$.

The center portion of the graph is the graph of $f(x) = x - 1$. There is a dot at $(-2, -3)$ and a circle at $(2, 1)$, so the linear function is defined for $-2 \leq x < 2$.

The right portion of the graph is the graph of $f(x) = 2x$. There is a dot at $(2, 4)$, so the linear function is defined for $x \geq 2$.

$f(x) = \begin{cases} 3 \text{ if } x < -2 \\ x - 1 \text{ if } -2 \leq x < 2 \\ 2x \text{ if } x \geq 2 \end{cases}$

Study Guide and Review

2-7 Parent Functions and Transformations (pp. 109–116)

A.G.5

Identify the type of function represented by each graph.

49.

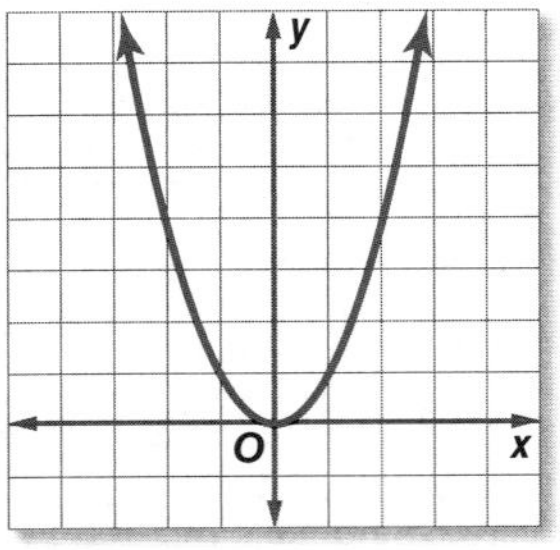

50.

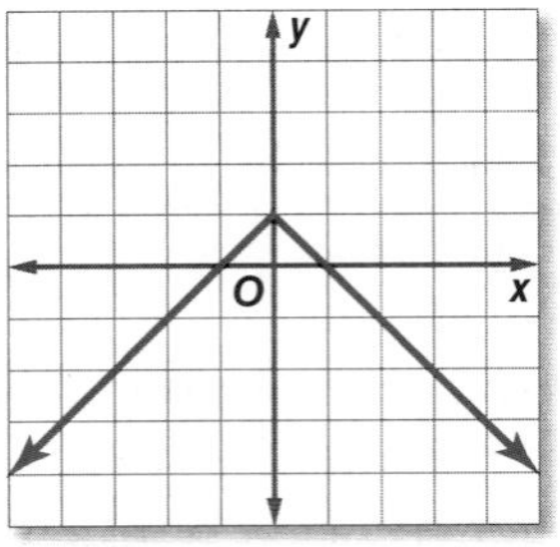

51. Describe the translation in $y = x^2 - 3$.

52. Describe the reflection in $y = -x^2$.

53. **CONSTRUCTION** A large arch is being constructed at the entrance of a new city hall building. The shape of the arch resembles the graph of the function $f(x) = -0.025x^2 + 3.64x - 0.038$. Describe the shape of the arch.

EXAMPLE 9

Identify the type of function represented by the graph.

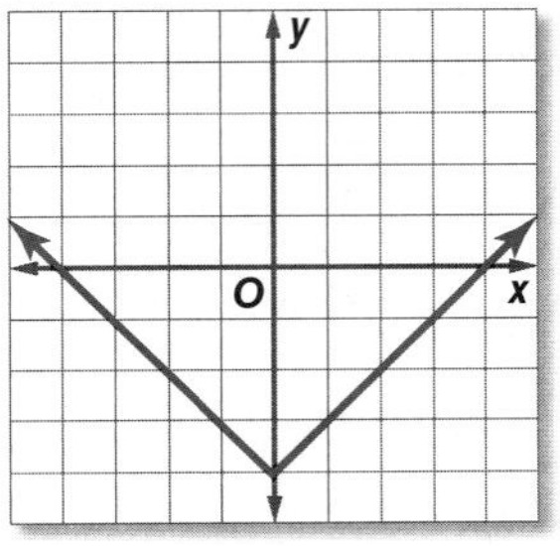

The graph is in the shape of a V. The graph represents an absolute value function.

EXAMPLE 10

Describe the translation in $y = |x + 6|$.

The graph of $y = |x + 6|$ is a translation of the graph of $y = |x|$ 6 units left.

2-8 Graphing Linear and Absolute Value Inequalities (pp. 117–121)

A.G.6

Graph each inequality.

54. $x - 3y < 6$

55. $y \geq 2x + 1$

56. $2x + 4y \leq 12$

57. $y > -3x - 5$

58. $y > |2x|$

59. $y \geq |2x - 2|$

60. $y + 3 < |x + 1|$

61. $2y \leq |x - 3|$

62. **BOOKS** Spencer has saved \$96 for a trip to his favorite bookstore. Each paperback book costs \$8 and each hardback book costs \$12. Write and graph an inequality that shows the number of paperback books and hardback books Spencer can purchase.

EXAMPLE 11

Graph the inequality $x - 2y > 6$.

Since the inequality symbol is >, the graph of the boundary line should be dashed. Graph the equation $x - 2y = 6$.

Test the inequality $x - 2y > 6$ at (0, 0).

$$x - 2y > 6$$
$$0 - 2(0) \stackrel{?}{>} 6$$
$$0 > 6 \text{ ✗}$$

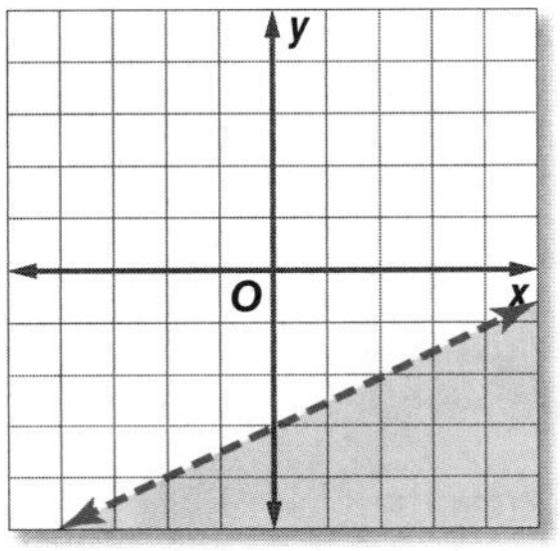

CHAPTER 2 Practice Test

NY Math Online glencoe.com
Chapter Test

1. State the domain and range of the relation shown in the table. Then determine if it is a function. If it is a function, determine if it is *one-to-one*, *onto*, *both*, or *neither*.

x	y
−2	3
4	−1
3	2
6	3

Find each value if $f(x) = -2x + 3$.

2. $f(-4)$
3. $f(3y)$

4. Write $2y = -6x + 4$ in standard form. Identify A, B and C.

5. Find the x-intercept and the y-intercept for $3x - 4y = -24$.

6. **MULTIPLE CHOICE** The cost of producing x pumpkin pies at a small bakery is given by $C(x) = 49 + 1.75x$. Find the cost of producing 25 pies.

   A $74.00

   B $81.50

   C $92.75

   D $108.25

Find the slope of the line that passes through each pair of points.

7. (1, 6), (3, 10)
8. (−2, 7), (3, −1)

9. **MULTIPLE CHOICE** Find the equation of the line that passes through (0, −3) and (4, 1).

   A $y = -x + 3$

   B $y = -x - 3$

   C $y = x - 3$

   D $y = x + 3$

10. Write an equation in slope-intercept form for the line that has slope −2 and passes through the point (3, −4).

11. Write an equation of the line that passes through the points (2, −4) and (1, 6).

12. Write an equation in slope-intercept form for the line that passes through (−3, 5) and is parallel to $y = -6x + 1$.

13. **EMERGENCY ROOM A hospital tracks the number of emergency room visits during the fall and winter months.**

Month	Oct	Nov	Dec	Jan	Feb
Visits	124	163	155	171	192

   a. Make a scatter plot and describe the correlation.

   b. Use two ordered pairs to write a prediction equation.

   c. Use your prediction equation to predict the number of emergency room visits for March.

14. Graph $f(x) = \begin{cases} -x \text{ if } x < -2 \\ x + 2 \text{ if } -2 \leq x \leq 2. \\ 5 \text{ if } x > 2 \end{cases}$

15. Write the piecewise function shown.

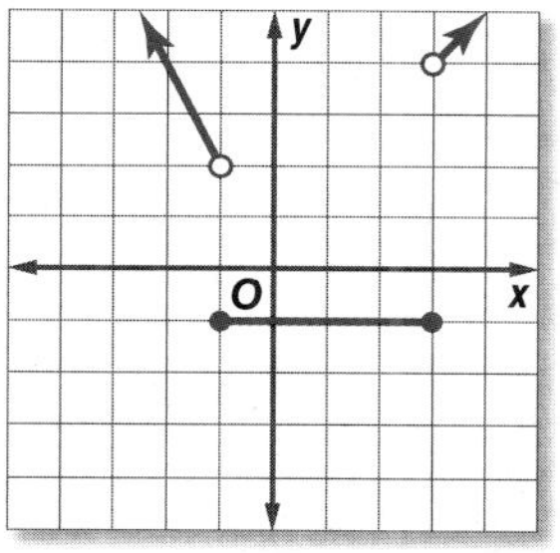

16. Identify the domain and range of $y = [\![x]\!] + 2$.

17. Describe the translation to $y = x^2 + 5$.

18. Describe the reflection in $y = -|x|$.

Graph each inequality.

19. $y \geq 4x - 1$

20. $2x + 6y < -12$

Reading Math Problems

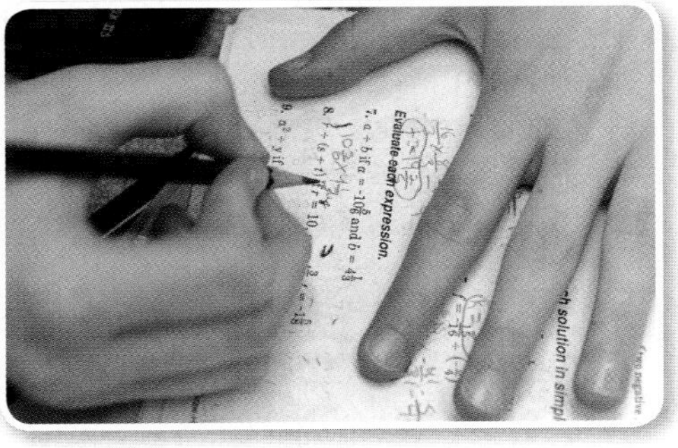

The first step to solving any math problem is to read the problem. When reading a math problem to get the information you need to solve, it is helpful to use special reading strategies.

Strategies for Reading Math Problems

Step 1

Read the problem quickly to gain a general understanding of it.

- **Ask yourself:** "What do I know?" "What do I need to find out?"
- **Think:** "Is there enough information to solve the problem? Is there extra information?"
- **Highlight:** If you are allowed to write in your test booklet, underline or highlight important information. Cross out any information you do not need.

Step 2

Reread the problem to identify relevant facts.

- **Analyze:** Determine how the facts are related.
- **Key Words:** Look for key words to solve the problem.
- **Vocabulary:** Identify mathematical terms. Think about the concepts and how they are related.
- **Plan:** Make a plan to solve the problem.
- **Estimate:** Quickly estimate the answer.

Step 3

Identify any obvious wrong answers.

- **Eliminate:** Eliminate any choices that are very different from your estimate.
- **Units of Measure:** Identify choices that are possible answers based on the units of measure in the question. For example, if the question asks for area, only answers in square units will work.

Step 4

Look back after solving the problem.

Check: Make sure you have answered the question.

EXAMPLE

Read the problem. Identify what you need to know. Then use the information in the problem to solve.

Sandy heated a solution over a burner and then removed it from the heat source. The temperature of the solution decreased linearly as it cooled. The temperatures after 0, 2, 5, and 9 minutes are shown in the table. What is the rate of change in the temperature of the solution as it cools?

Time (min)	Temperature (°C)
0	133.2
2	130.4
5	126.2
9	120.6

A −1.4 degrees per minute

B −0.8 degrees per minute

C 0.8 degrees per minute

D 1.4 degrees per minute

Read the problem carefully. There is extra information in the problem. To determine the slope, you only need information from two points on the linear function. Use two of the points to find the slope.

$$m = \frac{130.4 - 133.2}{2 - 0} = -1.4$$

The correct answer is A.

Exercises

Read each question. Then fill in the correct answer on the answer document provided by your teacher or on a sheet of paper.

1. The graph shows the cost of shipping packages. How much would it cost to ship a package that weighs 2 pounds 8 ounces?

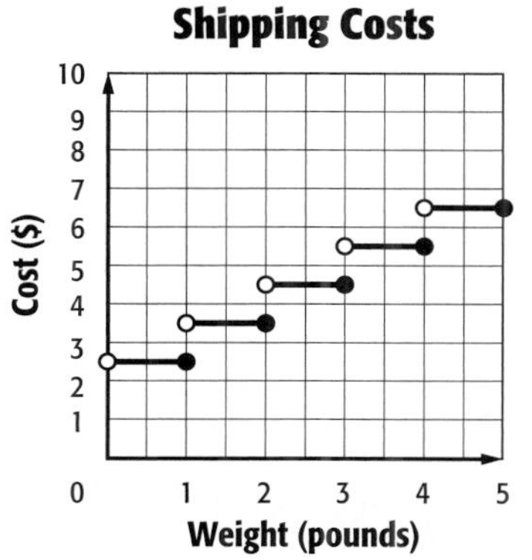

A \$3.50

B \$4.50

C \$5.00

D \$5.50

2. What is the slope of the line shown in the graph?

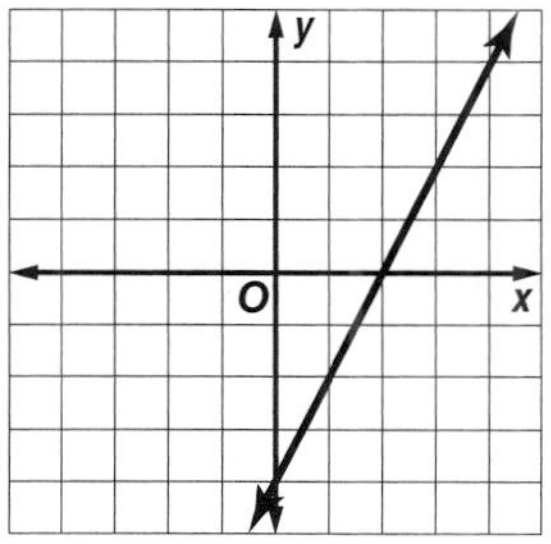

F −2

G $-\frac{1}{2}$

H $\frac{1}{2}$

J 2

NYS Regents Exam Practice

Cumulative, Chapters 1 and 2

Multiple Choice

Answer all questions in this part. Select the answer that best completes the statement or answers the question.

1. What is the range of the relation shown below?

x	y
−4	6
−2	−5
3	1
5	0

(1) $\{-5, 0, 1, 6\}$ **(3)** $\{0, 1, 3, 5, 6\}$

(2) $\{-4, -2, 3, 5\}$ **(4)** $\{-5, 6\}$

2. How do you write the equation shown in the graph in standard form?

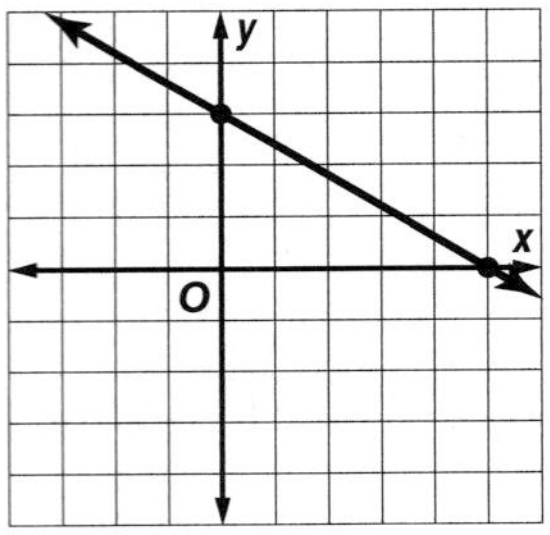

(1) $5x + 3y = 1$

(2) $5x + 3y = 15$

(3) $3x + 5y = 1$

(4) $3x + 5y = 15$

3. According to the U.S. Census Bureau, the median income in New York was \$32,965 per worker in 1990. By 2000, this number had increased to \$43,393 per worker. What was the annual rate of change in the median income per worker during the decade?

(1) \$784.35 per year

(2) \$962.40 per year

(3) \$1042.80 per year

(4) \$1115.75 per year

4. Refer to the information given in Exercise 3. Which of the following best describes the real-world meaning of the rate of change?

(1) The amount the median income in New York decreased each year.

(2) The amount the median income in New York decreased over 10 years.

(3) The amount the median income in New York increased over 10 years.

(4) The amount the median income in New York increased each year.

5. Solve the absolute value equation below.

$$|4x - 8| - 24 = 0$$

(1) −4, 6 **(3)** 4, 8

(2) −4, 8 **(4)** −8, −4

6. What is the range of function values over the domain $-4 \leq x \leq 4$?

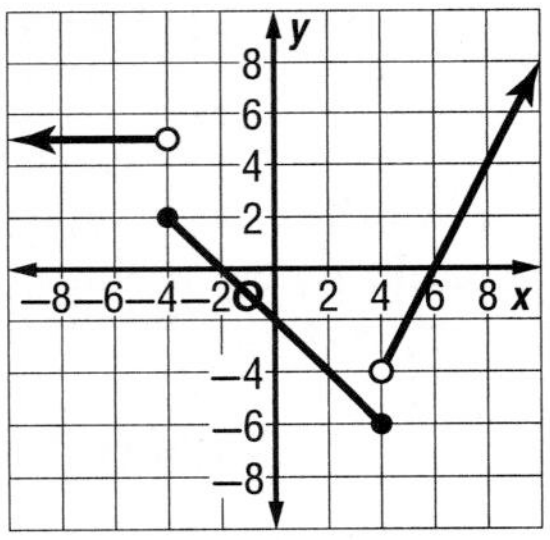

(1) $y = 5$

(2) $y = 2x - 12$

(3) $-6 \leq y \leq 2$

(4) $-2 \leq y \leq 6$

Test-TakingTip

Question 6 What values does y take over the domain $-4 \leq x \leq 4$?

7. Valeria works at a home goods retailer. She earns a weekly salary of \$525 plus a commission of 4% on her weekly sales. Write a linear equation for Valeria's weekly earnings E if she has d dollars in sales.

(1) $E = (525 + 4)d$

(2) $E = (525 + 0.04)d$

(3) $E = 525 + 0.04d$

(4) $E = 525 + 4d$

8. The lowest recorded temperature in the state of New York occurred in Old Forge on February 18, 1979, when the temperature dropped to -52°F. To which set of numbers does -52 *not* belong?

(1) integer
(3) real numbers
(2) rational numbers
(4) whole numbers

9. Suppose a laser range finder is accurate to within plus or minus 0.5 yard. If James measures the distance to the flag on a par 3 golf hole to be 161 yards, which absolute value inequality represents the actual distance d?

(1) $|d - 161| < 0.5$
(3) $|d - 0.5| < 161$
(2) $|d - 161| \leq 0.5$
(4) $|d - 0.5| \leq 161$

10. Which parent function is shown on the coordinate grid below?

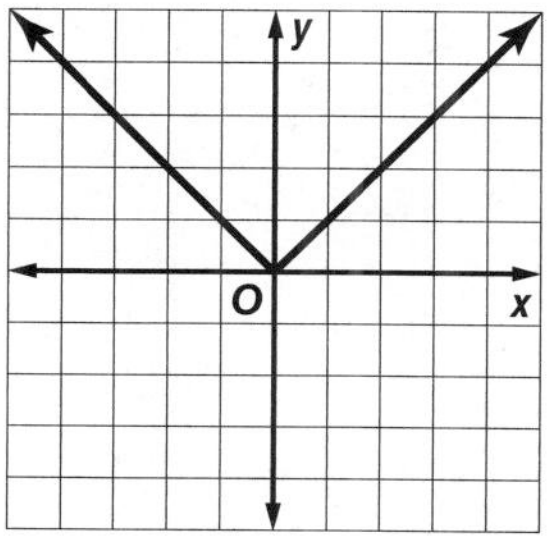

(1) $y = x$
(3) $y = |x|$
(2) $y = x^2$
(4) $y = a$

Open-Ended Response

Solve each problem. Clearly indicate the necessary steps, including appropriate formula substitutions, diagrams, graphs, charts, etc.

11. Use the scatter plot at the right to answer each question.

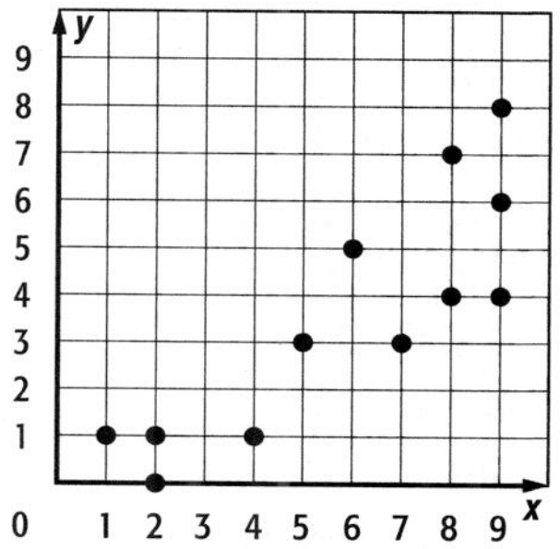

a. What type of correlation is shown by the data in the plot?

b. Determine a regression line for the data.

c. Use your regression line to predict the value of y when $x = 12$.

12. While grilling steaks, Washington likes to keep the grill temperature at 425°, plus or minus 15°.

a. Write an absolute value inequality to model this situation. Let t represent the temperature of the grill.

b. Within what range of temperatures does Washington like the grill to be when he cooks his steaks?

13. Find the slope of segment AB.

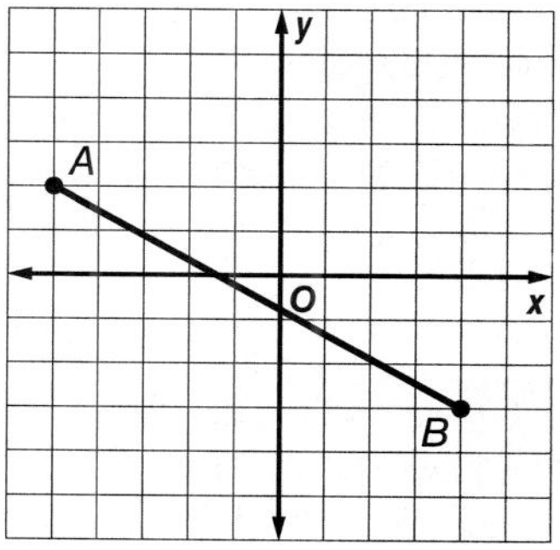

Need Extra Help?

If you missed Question...	1	2	3	4	5	6	7	8	9	10	11	12	13
Go to Lesson or Page...	2-1	2-3	2-3	2-4	1-4	1-3	2-7	2-6	2-6	2-7	2-8	2-5	2-3
NYS Core Curriculum	A2.CM.2	A2.CN.1	A.A.32	A.A.32	A2.A.1	A2.A.39	A2.CN.3	A.CM.11	A2.A.1	A.G.5	A2.S.8	A2.A.1	A.A.36

CHAPTER 3 Systems of Equations and Inequalities

Then

In Chapter 2, you solved and graphed linear equations.

Now

In Chapter 3, you will:

- Solve systems of linear equations graphically and algebraically.
- Solve systems of linear inequalities graphically.
- Solve problems by using linear programming.

NYS Core Curriculum

Reinforcement of A.A.7 Analyze and solve verbal problems whose solution requires solving systems of linear equations in two variables

Reinforcement of A.G.7 Graph and solve systems of linear equations and inequalities with rational coefficients in two variables

Why?

BUSINESS Most of the time, being successful in business means that you have to have good math skills. In this chapter, you will learn how to maximize your profits and minimize your costs. By doing this you will earn the most money possible.

Math *in Motion,* Animation glencoe.com

Get Ready for Chapter 3

Diagnose Readiness You have two options for checking Prerequisite Skills.

Text Option

Take the Quick Check below. Refer to the Quick Review for help.

QuickCheck

Graph each equation. (Lesson 2-2)

1. $x = 4y$

2. $y = \frac{1}{3}x + 5$

3. $x + 2y = 4$

4. $y = -x + 6$

5. $3x + 5y = 15$

6. $3y - 2x = -12$

7. **BUSINESS** A museum charges $8.50 for adult tickets and $5.25 for children's tickets. On Friday they made $650. (Lesson 2-4)

a. Write an equation that can be used to model the ticket sales.

b. Graph the equation.

QuickReview

EXAMPLE 1

Graph $2y + 5x = -10$.

Find the x- and y-intercepts.

$$\begin{aligned} 2(0) + 5x &= -10 \\ 5x &= -10 \\ x &= -2 \end{aligned} \qquad \begin{aligned} 2y + 5(0) &= -10 \\ 2y &= -10 \\ y &= -5 \end{aligned}$$

The graph crosses the x-axis at $(-2, 0)$ and the y-axis at $(0, -5)$. Use these ordered pairs to graph the equation.

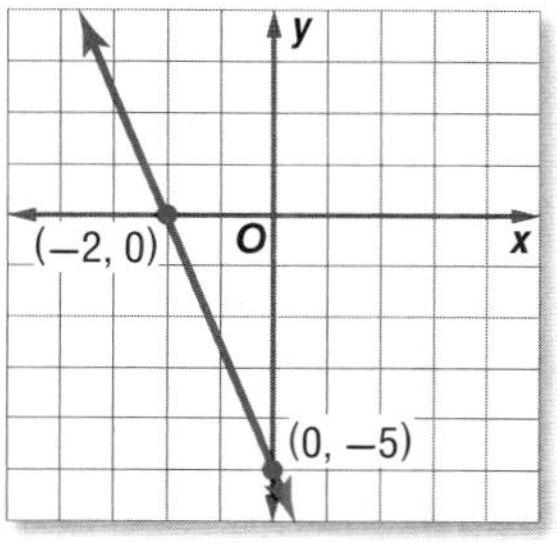

QuickCheck

Graph each inequality. (Lesson 2-8)

8. $y < 3$

9. $x + y \geq 1$

10. $3x - y > 6$

11. $x + 2y \leq 5$

12. $y > 4x - 1$

13. $5x - 4y < 12$

14. **FUNDRAISER** The student council is selling T-shirts for $15 and sweatshirts for $25. They must make $2500 to cover expenses. Write and graph an inequality to show the number of T-shirts and sweatshirts that they must sell.

QuickReview

EXAMPLE 2

Graph $y \geq 3x - 2$.

The boundary is the graph of $y = 3x - 2$. Since the inequality symbol is $\geq$, the boundary will be solid.

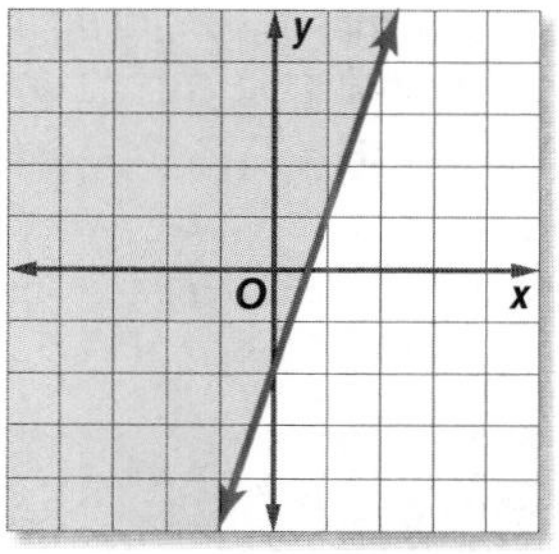

Test the point (0, 0).

$0 \stackrel{?}{\geq} 3(0) - 2$ $\quad$ $(x, y) = (0, 0)$

$0 \geq -2$ ✓

Shade the region that includes (0, 0).

Online Option

NY Math Online Take a self-check Chapter Readiness Quiz at glencoe.com.

Get Started on Chapter 3

You will learn several new concepts, skills, and vocabulary terms as you study Chapter 3. To get ready, identify important terms and organize your resources. You may wish to refer to **Chapter 0** to review prerequisite skills.

FOLDABLES® Study Organizer

Systems of Equations and Inequalities Make this Foldable to help you organize your Chapter 3 notes about systems of equations and inequalities. Begin with a sheet of $8\frac{1}{2}''$ by $11''$ paper.

1. **Fold** in half along the height.

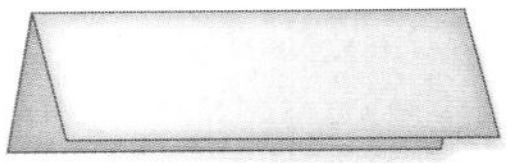

2. **Cut** along the fold.

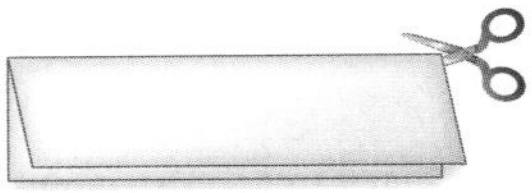

3. **Fold** each sheet along the width into fourths.

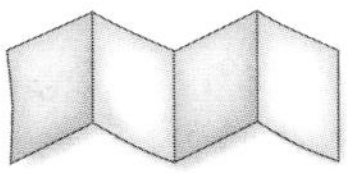

4. **Tape** the ends of two sheets together.

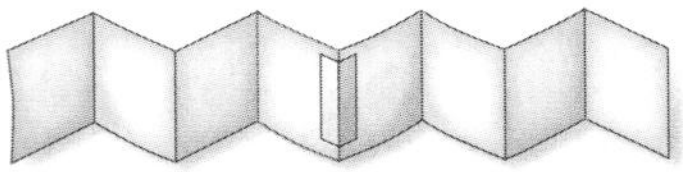

5. **Label** the tabs with *Solve by Graphing, Substitution Method, Elimination Method, No Solution, Infinite Solutions, Graphing Inequalities, Optimization*, and *Systems in Three Variables.*

NY Math Online glencoe.com

- Study the chapter online
- Explore **Math in Motion**
- Get extra help from your own **Personal Tutor**
- Use **Extra Examples** for additional help
- Take a **Self-Check Quiz**
- **Review Vocabulary** in fun ways

New Vocabulary

English		Español
system of equations	• p. 135 •	sistema de ecuasiones
break-even point	• p. 136 •	punto de equilibrio
consistent	• p. 137 •	consistente
inconsistent	• p. 137 •	inconsistente
independent	• p. 137 •	independiente
dependent	• p. 137 •	dependiente
substitution method	• p. 143 •	método de substitución
elimination method	• p. 144 •	método de eliminación
system of inequalities	• p. 151 •	sistema de desigualdades
constraints	• p. 160 •	restricciones
linear programming	• p. 160 •	programación lineal
feasible region	• p. 160 •	región viable
bounded	• p. 160 •	acotada
unbounded	• p. 160 •	no acotado
optimize	• p. 162 •	optimizer
ordered triple	• p. 167 •	triple ordenado

Review Vocabulary

equation • p. 135 • ecuación a mathematical sentence stating that two mathematical expressions are equal

inequality • p. 151 • desigualdad an open sentence that contains the symbol $<, \leq, >$, or $\geq$

linear equation • p. 135 • ecuación lineal an equation that has no operations other than addition, subtraction, and multiplication of a variable by a constant

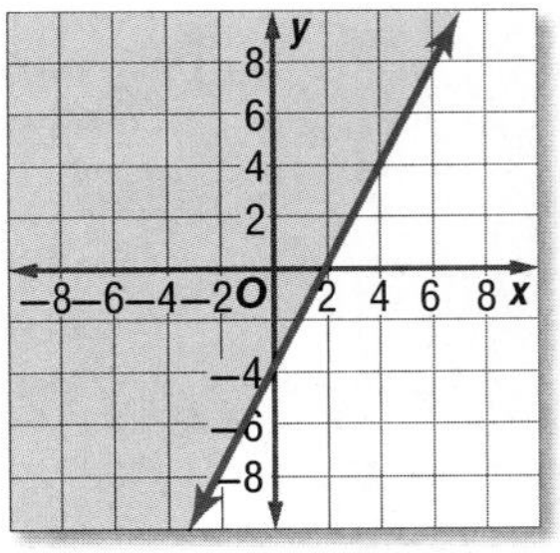

solution • p. 135 • solución a replacement for the variable in an open sentence that results in a true sentence

Multilingual eGlossary glencoe.com

3-1 Solving Systems of Equations by Graphing

Then

You graphed and solved linear equations. (Lesson 2-2)

Now

- Solve systems of linear equations by using tables and graphs.
- Classify systems of linear equations.

NYS Core Curriculum

Reinforcement of A.A.7 Analyze and solve verbal problems whose solution requires solving systems of linear equations in two variables

Reinforcement of A.G.7 Graph and solve systems of linear equations and inequalities with rational coefficients in two variables

New Vocabulary

system of equations
break-even point
consistent
inconsistent
independent
dependent

NY Math Online

glencoe.com

- Extra Examples
- Personal Tutor
- Self-Check Quiz
- Homework Help

Why?

Miguel is moving and needs to rent a moving truck. Rico's Truck Rental charges a flat fee of \$50 plus \$0.29 per mile. Home Movers charges a flat fee of \$30 plus \$0.34 per mile. If Miguel expects to put 500 miles on the rental truck, from which company should he rent the truck? This situation can be modeled by a system of linear equations.

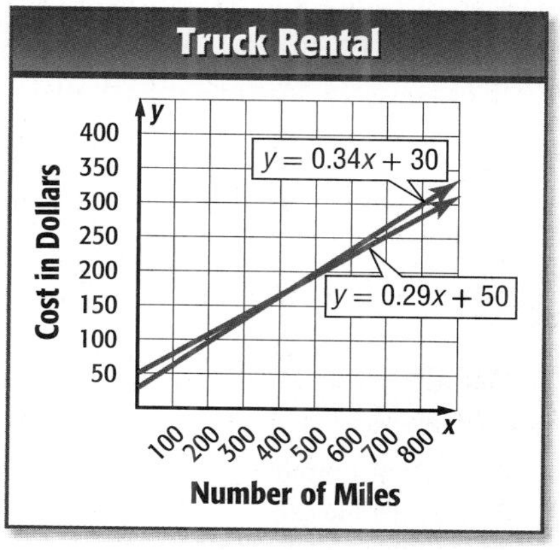

Solve Systems Using Tables and Graphs A **system of equations** is two or more equations with the same variables. To solve a system of equations with two variables, find the ordered pair that satisfies all of the equations.

To solve a system of equations by using a table, first write each equation in slope-intercept form. Then substitute different values for x and solve for the corresponding y-values. For ease of use, choose 0 and 1 as your first x-values.

$y_1 = -2x + 8$
$y_2 = 4x - 7$

x	y_1	y_2	Difference
0	8	−7	15
1	6	−3	9
2	4	1	3
3	2	5	−3

Because the difference between the y-values is closer to 0 for $x = 1$ than $x = 0$, a value greater than 1 should be tried next.

Because the difference between the y-values changed signs from $x = 2$ to $x = 3$, a value between these should be tried next.

The solution is between 2 and 3.

EXAMPLE 1 Solve by Using a Table

Solve the system of equations.

$3x + 2y = -2$
$-4x + 5y = -28$

Write each equation in slope-intercept form.

$3x + 2y = -2 \rightarrow y = -1.5x - 1$
$-4x + 5y = -28 \rightarrow y = 0.8x - 5.6$

Use a table to find the solution that satisfies both equations.

x	y_1	y_2	(x, y_1)	(x, y_2)
0	−1	−5.6	(0, −1)	(0, −5.6)
1	−2.5	−4.8	(1, −2.5)	(1, −4.8)
2	**−4**	**−4**	**(2, −4)**	**(2, −4)**

The solution of the system is (2, −4).

Check Your Progress

1A. $2x - 5y = 11$
$-3x + 4y = -13$

1B. $4x + 3y = -17$
$-7x - 2y = 20$

Personal Tutor glencoe.com

Another method for solving a system of equations is to graph the equations on the same coordinate plane. The point of intersection represents the solution.

EXAMPLE 2 Solve by Graphing

Solve the system of equations by graphing.

$2x - y = -1$
$2y + 5x = -16$

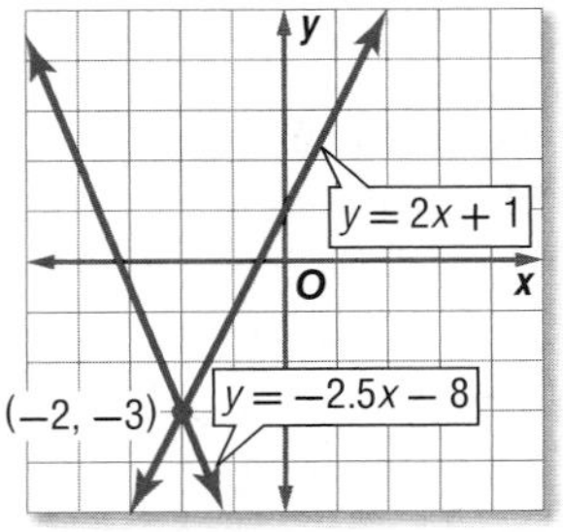

Write each equation in slope-intercept form.

$2x - y = -1 \quad \rightarrow \quad y = 2x + 1$
$2y + 5x = -16 \quad \rightarrow \quad y = -2.5x - 8$

The graphs of the lines appear to intersect at (−2, −3).

CHECK Substitute the coordinates into each original equation.

$2x - y = -1$	$2y + 5x = -16$	**Original equations**
$2(-2) - (-3) \stackrel{?}{=} -1$	$2(-3) + 5(-2) \stackrel{?}{=} -16$	**$x = -2$ and $y = -3$**
$-1 = -1$ ✓	$-16 = -16$ ✓	**Simplify.**

The solution of the system is (−2, −3).

StudyTip

Checking Solutions Always check to see if the values work for *both* of the original equations.

Check Your Progress

2A. $4x + 3y = 12$
$-6x + 4y = -1$

2B. $-3y + 8x = -41$
$6x + y = -21$

Personal Tutor glencoe.com

Systems of equations are used by businesses to determine the break-even point. The **break-even point** is the point at which the income equals the cost.

Real-World EXAMPLE 3 Break-Even Point Analysis

BUSINESS Libby borrowed \$450 to start a lawn-mowing business. She charges \$35 per lawn and incurs \$8 in operating costs per lawn. How many lawns must she mow to make a profit?

Let x = the number of lawns Libby mows, and let y = the number of dollars.

Total Income	**Total Cost**
$y = 35x$	$y = 8x + 450$

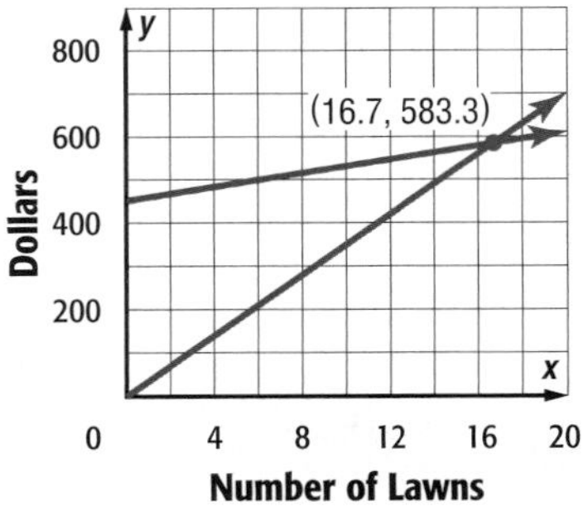

The graphs intersect at (16.7, 583.3). This is the break-even point. She can only mow a whole number of lawns. If she mows 17 lawns, she will make a profit of \$595 − \$586 or \$9. If she mows fewer than 17 lawns, she will lose money.

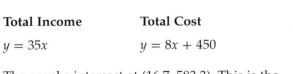

Real-World Link

The Internet provides countless business opportunities for teenage entrepreneurs. Roughly 1.6 million U.S. teens make some money using the Internet.

Source: *Business Week*

Check Your Progress

3. Southeast Publishing paid \$96,000 to an author for the rights to publish her book. If one book costs \$12 to produce and they plan to sell them for \$24 each, how many books must Southeast Publishing sell to break even? Show your results by graphing.

Personal Tutor glencoe.com

Classify Systems of Equations

Systems of equations can be classified by the number of solutions. A system of equations is **consistent** if it has at least one solution and **inconsistent** if it has no solutions. If it has exactly one solution, it is **independent**, and if it has an infinite number of solutions, it is **dependent**.

EXAMPLE 4 Classify Systems

Graph each system of equations and describe them as *consistent and independent, consistent and dependent,* or *inconsistent*.

a. $4x + 3y = 24$
$-3x + 5y = 30$

$y = -\frac{4}{3}x + 8$

$y = \frac{3}{5}x + 6$

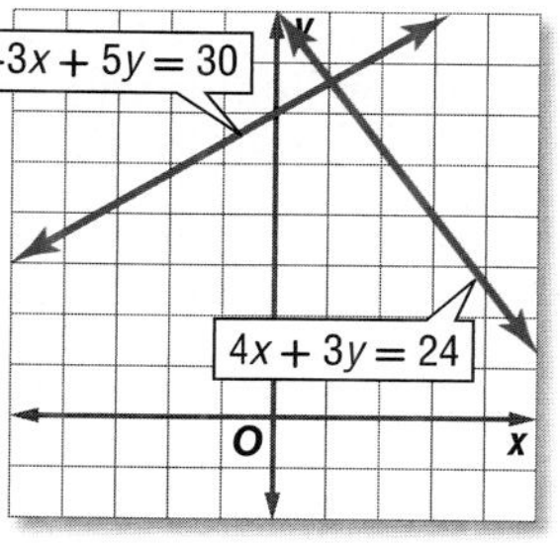

The graphs of the lines intersect at one point, so there is one solution. The system is consistent and independent.

b. $-2x + 5y = 10$
$4x - 10y = -20$

$y = \frac{2}{5}x + 2$

$y = \frac{2}{5}x + 2$

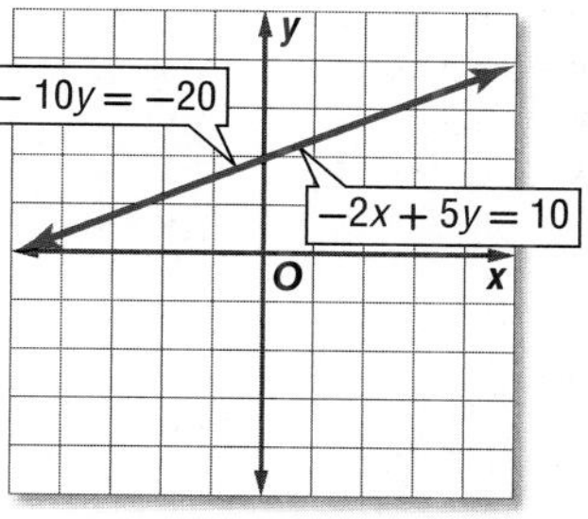

Because the equations are equivalent, their graphs are the same line. The system is consistent and dependent.

c. $-8x + 2y = 20$
$-4x + y = 12$

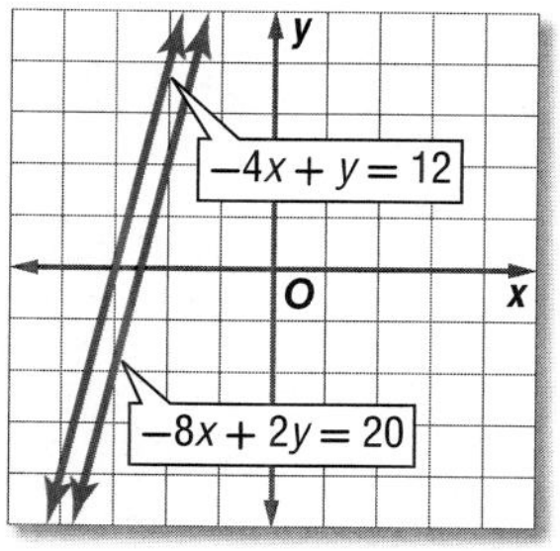

The graphs of the lines do not intersect, so the graphs are parallel and there is no solution. The system is inconsistent.

d. $f(x) = 2x - 8$
$g(x) = 2x - 4$
$h(x) = -3x + 4$

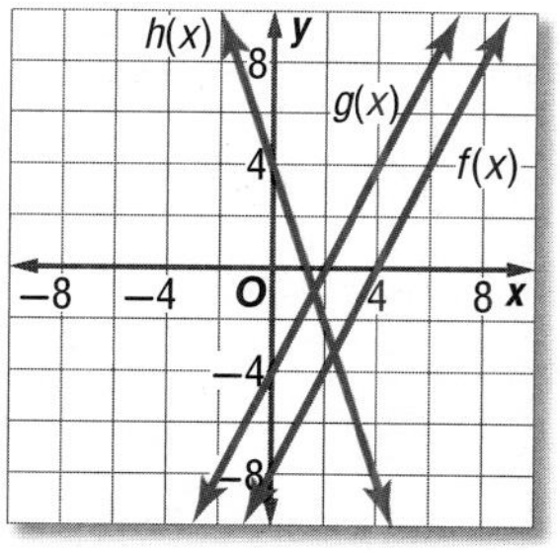

$f(x)$ and $g(x)$ are inconsistent. $f(x)$ and $h(x)$ are consistent and independent. $g(x)$ and $h(x)$ are consistent and independent.

StudyTip

Slope and Classifying Systems If the equations have different slopes, then the system is consistent and independent.

StudyTip

Graphing Calculator Most graphing calculators have a function for determining a point of intersection of two lines.

✓ Check Your Progress

4A. $6x - 4y = 15$
$-6x + 4y = 18$

4B. $-4x + 5y = -17$
$-4x - 2y = 15$

4C. $10x - 12y = 40$
$-5x + 6y = -20$

4D. $6x + y = -13$
$6x - y = 13$

Personal Tutor glencoe.com

The relationship between the graph of a system of equations and the number of solutions is summarized below.

Concept Summary — **Characteristics of Linear Systems** — For Your FOLDABLE

Consistent and Independent	Consistent and Dependent	Inconsistent
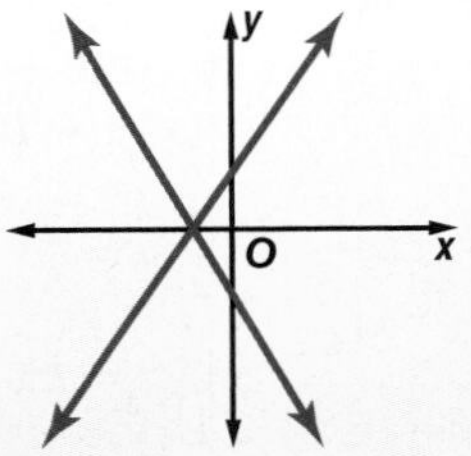	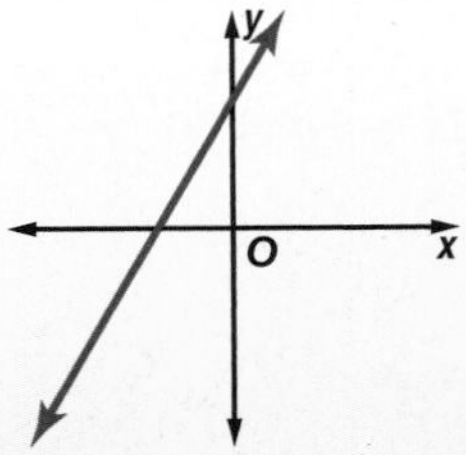	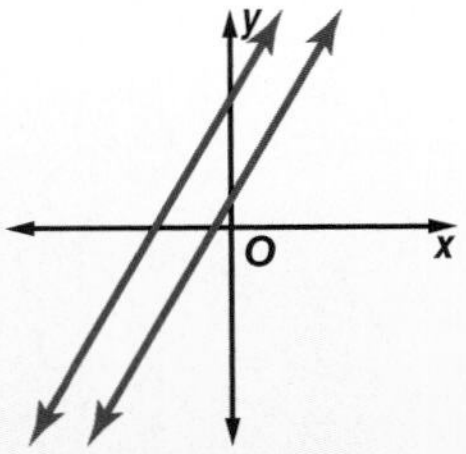
intersecting lines; one solution	same line; infinitely many solutions	parallel lines; no solution

Check Your Understanding

Example 1 p. 135 — **Solve each system of equations by using a table.**

1. $y = 3x - 4$
$y = -2x + 11$

2. $4x - y = 1$
$5x + 2y = 24$

Example 2 p. 136 — **Solve each system of equations by graphing.**

3. $y = -3x + 6$
$2y = 10x - 36$

4. $y = -x - 9$
$3y = 5x + 5$

5. $y = 0.5x + 4$
$3y = 4x - 3$

6. $-3y = 4x + 11$
$2x + 3y = -7$

7. $4x + 5y = -41$
$3y - 5x = 5$

8. $8x - y = 50$
$x + 4y = -2$

Example 3 p. 136

9. **DIGITAL PHOTOS** Refer to the graphic at the right.

a. Write equations that represent the cost of printing digital photos at each lab.

b. Under what conditions is the cost to print digital photos the same at both stores?

c. When is it best to use EZ Online Digital Photos and when is it best to use the local pharmacy?

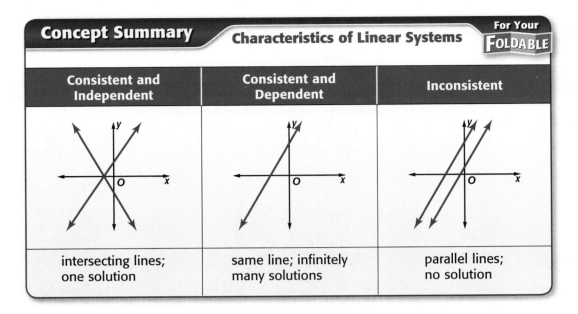

Example 4 p. 137 — **Graph each system of equations and describe it as *consistent and independent, consistent and dependent,* or *inconsistent*.**

10. $y + 4x = 12$
$3y = 8 - 12x$

11. $-2x - 3y = 9$
$4x + 6y = -18$

12. $9x - 2y = 11$
$5x + 4y = 13$

Practice and Problem Solving

● = **Step-by-Step Solutions** begin on page R20.
Extra Practice begins on page 947.

Example 1 p. 135

Solve each system of equations by using a table.

13. $y = 5x + 3$
$y = x - 9$

14. $3x - 4y = 16$
$-6x + 5y = -29$

15. $2x - 5 = y$
$-3x + 4y = 0$

16. $x + y = 6$
$-y + 8x = -15$

Example 2 p. 136

Solve each system of equations by graphing.

17. $4x + 3y = -24$
$8x - 2y = -16$

18. $-3x + 2y = -6$
$-5x + 10y = 30$

19. $-3x - 8y = 12$
$12x + 32y = -48$

20. $6x - 5y = 17$
$6x + 2y = 31$

21. $-10x + 4y = 7$
$2x - 5y = 7$

22. $y - 3x = -29$
$9x - 6y = 102$

Example 3 p. 136

23. SHOPPING Jerilyn has a \$10 coupon and a 15% discount coupon for her favorite store. The store has a policy that only one coupon may be used per purchase. When is it best for Jerilyn to use the \$10 coupon, and when is it best for her to use the 15% discount coupon?

24. HABITATS A zoo is building a new habitat for the wolves. The boundaries for the habitat are $y = 8$, $x = 4$, $y = 2x + 2$, and $y = 0.5x - 1$.

a. Graph the system of equations that models the area of the new wolf enclosure.

b. Find the coordinates of the vertices of the quadrilateral that represents the wolves' new habitat.

c. What is the approximate area of the wolves' new home?

25. WATER SKIING A water ski jump is in the shape of a triangle. The graphs of $y - 2x = 1$, $4x + y = 7$, and $2y - x = -4$ contain the sides of the triangle. Find the coordinates of the vertices of the triangle.

Example 4 p. 137

Graph each system of equations and describe it as *consistent and independent*, *consistent and dependent*, or *inconsistent*.

26. $y = 2x - 1$
$y = 2x + 6$

27. $y = 3x - 4$
$y = 6x - 8$

28. $x - 6y = 12$
$3x + 18y = 14$

29. $2x + 5y = 10$
$-4x - 10y = 20$

30. $8y - 3x = 15$
$-16y + 6x = -30$

31 ● $-5x - 6y = 13$
$12y + 10x = -26$

32. $3.2x - 2.4y = 28$
$0.6y = -0.8x + 7$

33. $4.9x - 7.7y = -14$
$-1.4x + 2.2y = 4$

34. $3.5x - 1.2y = 8.2$
$-5.25x + 1.8y = 12.3$

Solve each system of equations by graphing.

35. $y - \frac{3}{4}x = -\frac{5}{2}$
$\frac{3}{2}x + 6y = 9$

36. $5x + 2y = -\frac{5}{2}$
$\frac{3}{2}y + \frac{1}{2}x = 3$

37. $\frac{5}{6}x + \frac{4}{3}y = -1$
$\frac{1}{4}x + \frac{5}{3}y = \frac{7}{2}$

Use a graphing calculator to solve each system of equations. Round the coordinates of the intersection to the nearest hundredth.

38. $12y = 5x - 15$
$4.2y + 6.1x = 11$

39. $5.8x - 6.3y = 18$
$-4.3x + 8.8y = 32$

40. $-3.8x + 2.9y = 19$
$6.6x - 5.4y = -23$

Real-World Link

Track and Field runners Paavo Nurmi (Finland) and Carl Lewis (U.S.) along with gymnast Larisa Latynina (former Soviet Union) currently hold the record for the most Olympic gold medals with 9 each. Latynina also holds the record for the most Olympic medals with 18.

Source: NBC Sports

41 **OLYMPICS** The table shows the winning times in seconds for the 100-meter dash at the Olympics between 1964 and 2004.

Years Since 1964, x	Men's Gold Medal Time	Women's Gold Medal Time
0	10.0	11.4
4	9.90	11.0
8	10.14	11.07
12	10.06	11.08
16	10.25	11.06
20	9.99	10.97
24	9.92	10.54
28	9.96	10.82
32	9.84	10.94
36	9.87	10.75
40	9.85	10.93

a. Write equations that represent the winning times for men and women since 1964. Assume that both times continue along the same trend.

b. Graph both equations. Estimate when the women's performance will catch up to the men's performance.

c. Do you think that your prediction is reasonable? Explain.

H.O.T. Problems

Use **H**igher-**O**rder **T**hinking Skills

42. **REASONING** If $f(x)$ is consistent and dependent with $g(x)$, $g(x)$ is inconsistent with $h(x)$, and $h(x)$ is consistent and independent with $j(x)$, then $f(x)$ will *sometimes*, *always*, or *never* be consistent and independent with $j(x)$. Explain your reasoning.

43. **FIND THE ERROR** Victor and Alvin are using their calculators to solve the system $y = 3x - 1$ and $x + y = 4$. Is either of them correct? Explain your reasoning.

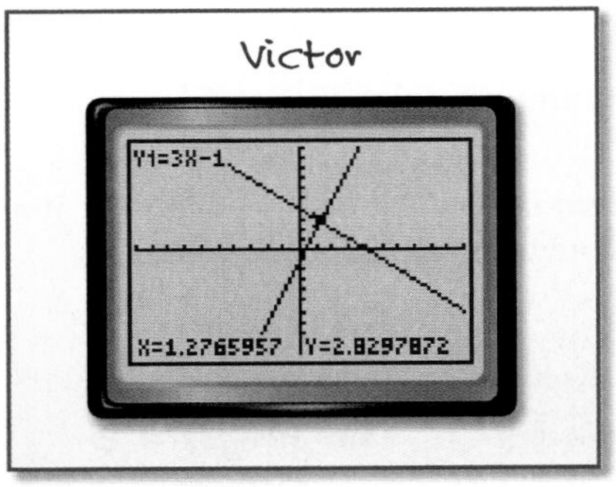

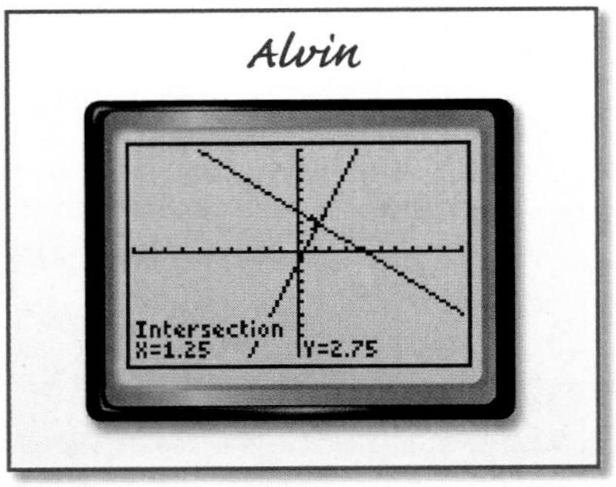

44. **CHALLENGE** Consider the system of equations $kx + 3y = 6$ and $8x + 5y = k$. Find a value of k, if it exists, for each condition.

a. The system is inconsistent.

b. The system is consistent and independent.

c. The system is consistent and dependent.

45. **OPEN ENDED** Write a system to illustrate each of the following.

a. an inconsistent system

b. a consistent and dependent system

c. a consistent and independent system

46. **WRITING IN MATH** Explain how you can determine the consistency and dependence of a system without graphing the system.

NYSRE PRACTICE Reinforcement of A.A.7, A.G.7

47. SHORT RESPONSE Simplify $3y(4x + 6y - 5)$.

48. ACT/SAT Which of the following best describes the graph of the equations?

$$4y = 3x + 8$$
$$-6x = -8y + 24$$

A The lines are parallel.
B The lines are perpendicular.
C The lines have the same x-intercept.
D The lines have the same y-intercept.

49. GEOMETRY Which set of dimensions corresponds to a triangle similar to the one shown at the right?

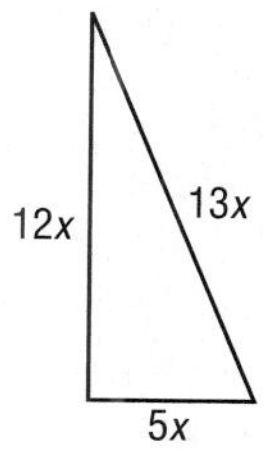

F 1 unit, 2 units, 3 units
G 7 units, 11 units, 12 units
H 10 units, 23 units, 24 units
J 20 units, 48 units, 52 units

50. Move-A-Lot Rentals will rent a moving truck for \$100 plus \$0.10 for every mile it is driven. Which equation can be used to find C, the cost of renting a moving truck and driving it for m miles?

A $C = 0.1(100 + m)$
B $C = 100 + 0.1m$
C $C = 100m + 0.1$
D $C = 100(m + 0.1)$

Spiral Review

51. CRAFTS Priscilla sells stuffed animals at a local craft show. She charges \$10 for the small ones and \$15 for the large. To cover her expenses, she needs to sell at least \$350. (Lesson 2-8)

a. Write an inequality for this situation.
b. Graph the inequality.
c. If she sells 10 small and 15 large animals, will she cover her expenses?

Write an equation for each function. (Lesson 2-7)

52.

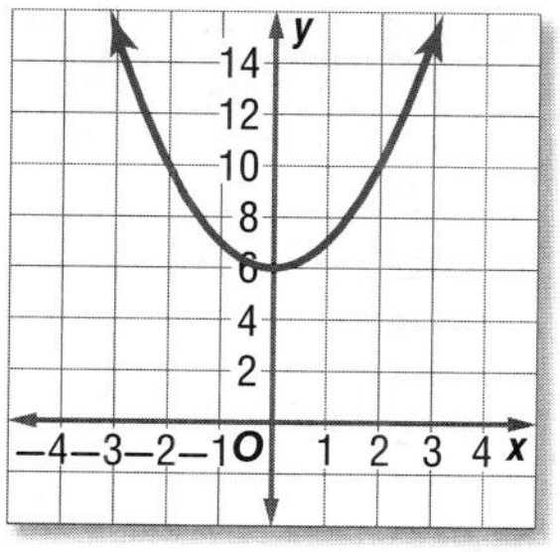

53.

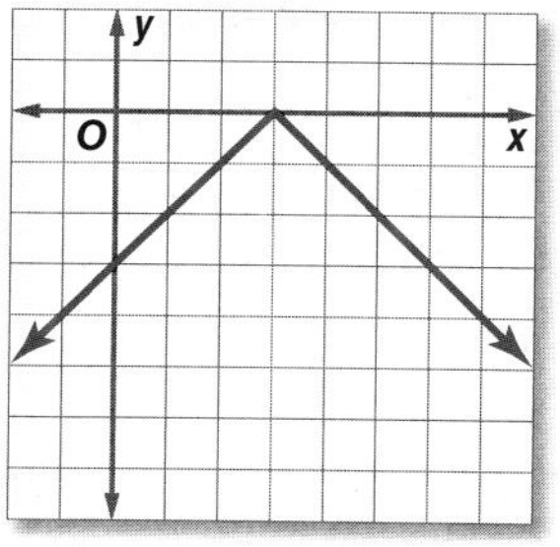

54.

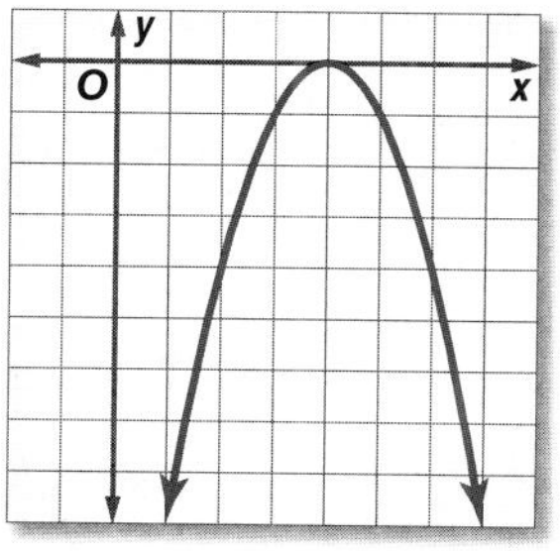

Solve each equation. Check your solution. (Lesson 1-3)

55. $2p = 14$

56. $-14 + n = -6$

57. $7a - 3a + 2a - a = 16$

58. $x + 9x - 6x + 4x = 20$

59. $27 = -9(y + 5) + 6(y + 8)$

60. $-7(p + 7) + 3(p - 4) = -17$

Skills Review

Simplify each expression. (Lesson 1-2)

61. $(2x + 4) - (3x + 2)$

62. $(y - 10) + (5y + 2)$

63. $(4x - 2y) + (-6x + 5y)$

64. $3(6x + 2y - 1)$

65. $4(5x + 2y - 2x + 8)$

66. $6(x + 4y) - 5(x + 9y)$

EXTEND **3-1**

Graphing Technology Lab
Systems of Equations

NY Math Online glencoe.com
Other Calculator Keystrokes

Objective
Use a graphing calculator to solve systems of equations.

You can use a TI-83/84 Plus graphing calculator to solve systems of equations. You can use the Y= menu to graph each equation on the same set of axes.

EXAMPLE Intersection of Two Graphs

Graph the system of equations in the standard viewing window.

$3x + y = 9$

$x - y = -1$

Step 1 Write each equation in the form $y = mx + b$.

$3x + y = 9$ $\qquad$ $x - y = 1$

$y = -3x + 9$ $\qquad$ $-y = -x - 1$

$y = x + 1$

Step 2 Enter $y = -3x + 9$ as **Y1** and $y = x + 1$ as **Y2**. Then graph the lines.

KEYSTROKES: [Y=] [(−)] 3 [X,T,θ,n] [+] 9 [ENTER]
[X,T,θ,n] [+] 1 [ENTER] [ZOOM] 6

Step 3 Find the intersection of the lines.

KEYSTROKES: [2nd] [CALC] 5 [ENTER] [ENTER] [ENTER]

The solution is (2, 3).

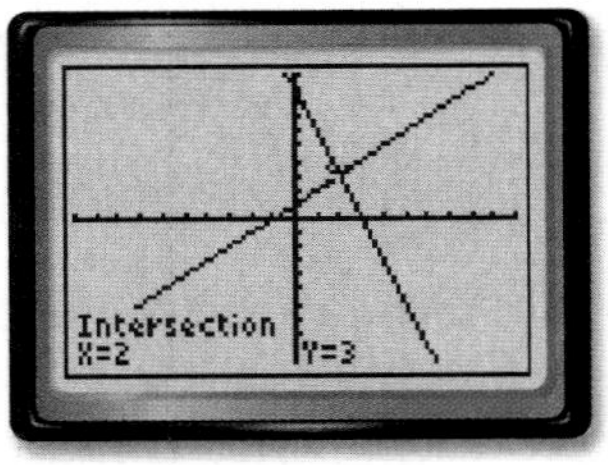

Exercises

Use a graphing calculator to solve each system of equations.

1. $2x + 4y = 36$
   $10y - 5x = 0$

2. $2y - 3x = 7$
   $5x = 4y - 12$

3. $4x - 2y = 16$
   $7x + 3y = 15$

4. $2x + 4y = 4$
   $x + 3y = 13$

5. $5x + y = 13$
   $3x = 15 - 3y$

6. $4y - 5 = 20 - 3x$
   $4x - 7y + 16 = 0$

7. $\frac{1}{4}x + y = \frac{11}{4}$
   $x - \frac{1}{2}y = 2$

8. $3x + 2y = -3$
   $x + \frac{1}{3}y = -4$

9. $3x - 6y = 6$
   $2x - 4y = 4$

10. $6x + 8y = -16$
    $3x + 4y = 12$

3-2 Solving Systems of Equations Algebraically

Then

You solved systems of linear equations by using tables and graphs. (Lesson 3-1)

Now

- Solve systems of linear equations by using substitution.
- Solve systems of linear equations by using elimination.

NYS Core Curriculum

Reinforcement of A.A.7 Analyze and solve verbal problems whose solution requires solving systems of linear equations in two variables

Reinforcement of A.A.10 Solve systems of two linear equations in two variables algebraically

New Vocabulary

substitution method

elimination method

NY Math Online

glencoe.com

- Extra Examples
- Personal Tutor
- Self-Check Quiz
- Homework Help

Why?

Alejandro has a computer support business. He estimates that the cost to run his business can be represented by $y = 65x - 145$, where x is the number of customers. He also estimates that his income can be represented by $y = 48x + 500$. A system of equations can be used to determine the number of customers he will need in order to break even.

Substitution Algebraic methods are used to find exact solutions of systems of equations. One algebraic method is called the **substitution method**.

Key Concept — Substitution Method

For Your FOLDABLE

Step 1 Solve one equation for one of the variables.

Step 2 Substitute the resulting expression into the other equation to replace the variable. Then solve the equation.

Step 3 Substitute to solve for the other variable.

Real-World EXAMPLE 1 — Use the Substitution Method

BUSINESS How many customers will Alejandro need in order to break even? What will his profit be if he has 60 customers?

Understand Alejandro wants to know how many customers he needs for his income to equal his costs. He also wants to know what his profit will be if he has 60 customers.

Plan Solve the system of equations.

Income: $y = 65x - 145$

Cost: $y = 48x + 500$

Solve

$y = 65x - 145$	**Income equation**
$48x + 500 = 65x - 145$	**Substitute $48x + 500$ for y.**
$500 = 17x - 145$	**Subtract $48x$ from each side.**
$645 = 17x$	**Add 145 to each side.**
$37.9 \approx x$	**Divide each side by 17.**

Alejandro needs 38 customers to break even. If he has 60 customers, his income will be 65(60) − 145 or \$3755, and his costs will be 48(60) + 500 or \$3380, so his profit will be 3755 − 3380 or \$375.

Check You can use a graphing calculator to check this solution. The break-even point is near (37.9, 2321.2). Use the CALC function to find the cost and income with 60 customers.

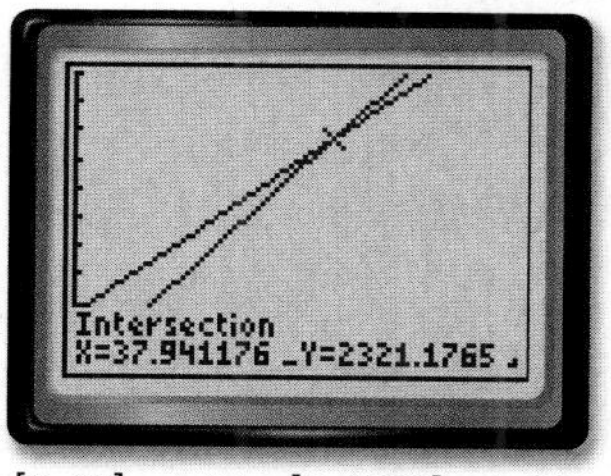

[0, 65] scl: 5 by [0, 3000] scl: 300

Check Your Progress

Use substitution to solve each system of equations.

1A. $5x - 3y = 23$
$2x + y = 7$

1B. $x - 7y = 11$
$5x + 4y = -23$

1C. $-6x - y = 27$
$3x + 8y = 9$

Personal Tutor glencoe.com

Elimination You can use the **elimination method** to solve a system when one of the variables has the same coefficient in both equations.

Key Concept Elimination Method

For Your FOLDABLE

Step 1 Multiply one or both equations by a number to result in two equations that contain opposite terms.

Step 2 Add the equations, eliminating one variable. Then solve the equation.

Step 3 Substitute to solve for the other variable.

Variables can be eliminated by addition or subtraction.

EXAMPLE 2 Solve by Using Elimination

Use the elimination method to solve the system of equations.

$\mathbf{5x + 3y = -19}$
$\mathbf{8x + 3y = -25}$

Notice that solving by substitution would involve fractions.

Step 1 Multiply one equation by -1 so the equations contain $3y$ and $-3y$.

$8x + 3y = -25$ **Multiply by −1.** $-8x - 3y = 25$

Step 2 Add the equations to eliminate one variable.

$5x + 3y = -19$ **Equation 1**
$(+)\ -8x - 3y = 25$ **Equation 2 × (−1)**
$-3x = 6$ **Add the equations.**
$x = -2$ **Divide each side by −3.**

Step 3 Substitute -2 for x into either original equation.

$8x + 3y = -25$ **Equation 2**
$8(-2) + 3y = -25$ **$x = -2$**
$-16 + 3y = -25$ **Multiply.**
$3y = -9$ **Add 16 to each side.**
$y = -3$ **Divide each side by 3.**

The solution is $(-2, -3)$.

StudyTip

Elimination Remember when you add or subtract one equation from another to add or subtract *every* term, including the constant on the other side of the equal sign.

Check Your Progress

2A. $4x - 3y = -22$
$2x + 3y = 16$

2B. $6x - 5y = -8$
$4x - 5y = -12$

2C. $2x - 9y = 34$
$-2x + 6y = -28$

Personal Tutor glencoe.com

Review Vocabulary

Least Common Multiple the least number that is a common multiple of two or more numbers

Sometimes, adding or subtracting equations will not eliminate either variable. You can use multiplication and least common multiples to find a common coefficient.

NYSRE EXAMPLE 3 Reinforcement of A.A.7, A.A.10

Solve the system of equations.

$$5x + 3y = 52$$
$$9x - 4y = 56$$

A (4, 1) B (8, 4) C (8, 0) D (12, 3)

Read the Test Item

You are given a system of two linear equations and are asked to find the solution.

Solve the Test Item

Neither variable has a common coefficient. The coefficients of the y-variables are 3 and 4 and their least common multiple is 12, so multiply each equation by the value that will make the y-coefficient 12.

$5x + 3y = 52$ **Multiply by 4.** $20x + 12y = 208$

$9x - 4y = 56$ **Multiply by 3.** $(+)\ 27x - 12y = 168$

$47x = 376$ **Add the equations.**

$x = 8$ **Divide each side by 47.**

Solve for y by substituting $x = 8$ into either of the original equations.

$5x + 3y = 52$ **Equation 1**

$5(8) + 3y = 52$ **Replace x with 8.**

$40 + 3y = 52$ **Multiply.**

$3y = 12$ **Subtract 40 from each side.**

$y = 4$ **Divide each side by 3.**

The correct answer is B.

Check Your Progress

3. Solve the system of equations. $6a - 5b = -62$, $8a + 7b = 54$

F (4, 6) G (2, −12) H (0, −8) J (−2, 10)

Personal Tutor glencoe.com

> **StudyTip**
> **Adding and Subtracting Equations** If you add or subtract two equations in a system and the result is an equation that is never true, then the system is inconsistent. When you add or subtract two equations in a system and the result is an equation that is always true, then the system is dependent.

> **StudyTip**
> **Subtraction** When a variable has identical coefficients in two equations, it can be eliminated by subtraction.

EXAMPLE 4 No Solution and Infinite Solutions

Use the elimination method to solve each system of equations.

a. $5x + 6y = 45$
$-5x - 6y = 38$

$5x + 6y = 45$
$(+)\ -5x - 6y = 38$
$0 = 83$

Because $0 = 83$ is not true, this system has no solutions.

b. $2x + 3y = 5$
$6x + 9y = 15$

$3(2x + 3y = 5) = 6x + 9y = 15$ **Multiply by 3.**
$6x + 9y = 15$
$(-)\ 6x + 9y = 15$
$0 = 0$

Because the equation $0 = 0$ is always true, there are an infinite number of solutions.

Check Your Progress

4A. $9a - 7b = 14$
$-18a + 14b = -28$

4B. $-6c + 12d = 81$
$-5c + 10d = -61$

Personal Tutor glencoe.com

The following summarizes the various methods for solving systems.

Math History Link

Nina Karlovna Bari (1901–1961) Russian mathematician Nina Karlovna Bari was considered the principal leader of mathematics at Moscow State University. She is best known for her textbooks *Higher Algebra* and *The Theory of Series*.

Concept Summary — Solving Systems of Equations

For Your FOLDABLE

Method	The Best Time to Use
Table	to estimate the solution, since a table may not provide an exact solution
Graphing	to estimate the solution, since graphing usually does not give an exact solution
Substitution	if one of the variables in either equation has a coefficient of 1 or −1
Elimination Using Addition	if one of the variables has opposite coefficients in the two equations
Elimination Using Subtraction	if one of the variables has the same coefficient in the two equations
Elimination Using Multiplication	if none of the coefficients are 1 or −1 and neither of the variables can be eliminated by simply adding or subtracting the equations

Check Your Understanding

Example 1 p. 143

1. **FUNDRAISER** To raise money for new uniforms, the band boosters sell T-shirts and hats. The cost and sale price of each item is shown. The boosters spend a total of \$2000 on T-shirts and hats. They sell all of the merchandise, and make \$3375. How many T-shirts did they sell?

Item	Cost	Sale Price
T-Shirt	\$6	\$10
Hat	\$4	\$7

Solve each system of equations by using substitution.

2. $y = 2x - 10$
   $y = -4x + 8$
3. $x + 5y = 3$
   $3x - 2y = -8$
4. $a - 3b = -22$
   $4a + 2b = -4$
5. $2a + 8b = -8$
   $3a - 5b = 22$
6. $9c - 3d = -33$
   $6c + 5d = -8$
7. $6x - 7y = 23$
   $8x + 4y = 44$

Examples 2–4 pp. 144, 145

Solve each system of equations by using elimination.

8. $4x - 3y = 29$
   $4x + 3y = 35$
9. $-6w - 8z = -44$
   $3w + 6z = 36$
10. $8a - 3b = -11$
    $5a + 2b = -3$
11. $3a + 5b = -27$
    $4a + 10b = -46$
12. $6x - 4y = 30$
    $12x + 5y = -18$
13. $5a + 15b = -24$
    $-2a - 6b = 28$

14. **MULTIPLE CHOICE** What is the solution of the linear system?

$$4x + 3y = 2$$
$$4y - 2x = 12$$

A (8, −10) **B** (2, −2) **C** (−10, 14) **D** no solution

Practice and Problem Solving

● = **Step-by-Step Solutions** begin on page R20.
Extra Practice begins on page 947.

Example 1
p. 143

Solve each system of equations by using substitution.

15. $-4r - 3t = -26$
$5r + t = 27$

16. $6u + 3v = -15$
$8u - 5v = 7$

17. $4x + 12y = 0$
$-3x - 4y = -10$

18. $c - 5d = -16$
$3c + 2d = -14$

19. $-5p - t = 17$
$4p + 6t = 2$

20. $-6y + 5z = -35$
$7y - z = 36$

21. $9y + 3x = 18$
$-3y - x = -6$

22. $5x - 20y = 70$
$6x + 5y = -32$

23. $-4x - 16y = -96$
$7x + 3y = 68$

24. $-4a - 5b = 14$
$9a + 3b = -48$

25. $-9c - 4d = 31$
$6c + 6d = -24$

26. $8f + 3g = 12$
$-32f - 12g = 48$

27. TENNIS At a park, there are 38 people playing tennis. Some are playing doubles, and some are playing singles. There are 13 matches in progress. A doubles match requires 4 players, and a singles match requires 2 players.

a. Write a system of two equations that represents the number of singles and doubles matches going on.

b. How many matches of each kind are in progress?

Examples 2–4
pp. 144, 145

Real-World Link

The Wimbledon Championship is broadcast to over 170 countries with an audience of over 1 billion people.

Source: *Tennis Magazine*

Solve each system of equations by using elimination.

28. $8x + y = 27$
$-3x + 4y = 3$

29. $2a - 5b = -20$
$2a + 5b = 20$

30. $6j + 4k = -46$
$2j + 4k = -26$

31. $3x - 8y = 24$
$-12x + 32y = 96$

32. $5a - 2b = -19$
$8a + 5b = -55$

33. $r - 6t = 44$
$9r + 12t = 0$

34. $6d + 5f = -32$
$5d - 9f = 26$

35. $11u = 5v + 35$
$8v = -6u + 62$

36. $-1.2c + 3.4d = 6$
$6c = -30 + 17d$

37. $6g + 8h = 16$
$-4g - 7h = -19$

38. $15j - 3k = 129$
$-6j + 8k = -72$

39. $-4m - 9n = 30$
$10m + 3n = 42$

40. The sum of four times a number and six times a second number is 36. The difference of five times the second number and three times the first number is 49. Find the numbers.

41. Twice the sum of a number and 3 times a second number is 4. The difference of ten times the second number and five times the first is 90. Find the numbers.

42. Three times the difference of four times a number and three times a second number is 117. Four times the sum of 6 times the second number and 8 times the first number is 72. Find the numbers.

43 **CYCLING** The total distance of the cycling course is 104.8 miles. Julian starts the course at 8:00 A.M. and rides at 12 miles an hour. Peter starts two hours later than Julian but decides to try to catch up with him. Peter rides at a speed of 16 miles an hour.

a. Solve the system of equations to find when Peter will catch up to Julian.

b. Peter wants to reduce the time it takes him to catch up to Julian by 1 hour. Explain how he could do this by changing his starting time. Explain how he could do this by changing his speed. Are your answers reasonable?

Solve each system of equations.

44. $11p + 3q = 6$
$-0.75q - 2.75p = -1.5$

45. $8r - 5t = -60$
$6r + 3t = -18$

46. $10t + 4v = 13$
$-4t - 7v = 11$

47. $6w = 12 - 4x$
$6x = -9w + 18$

48. $\frac{3}{2}y + z = 3$
$-y - \frac{2}{3}z = -2$

49. $\frac{5}{2}a - \frac{3}{4}b = 46$
$-\frac{7}{8}a - 3b = 10$

50. ROWING Allison can row a boat 1 mile upstream (against the current) in 24 minutes. She can row the same distance downstream in 13 minutes. Assume that both the rowing speed and the speed of the current are constant.

a. Find the speed at which Allison is rowing and the speed of the current.

b. If Allison plans to meet her friends 3 miles upstream one hour from now, will she be on time? Explain.

51. EARTH WEEK To reduce waste, The Green Café offers a reduced refill rate on coffee for anyone buying a Green mug. The mug costs \$2.95 and is filled with 16 ounces of coffee. The refill price is \$0.50. A 16-ounce coffee in a disposable cup costs \$0.85.

a. What is the approximate break-even point for buying the mug and refills in comparison to buying coffee in disposable cups? What does this mean?

b. Which offer do you think is best? Explain your reasoning.

c. How would your decision change if the refillable mug offer was extended for a year?

Real-World Link

Earth Day began in San Francisco in 1969 and is now observed by over 500 million people worldwide.

Source: *Earth Week Charlottesville*

52. SKATING PARTY Anita invites 21 friends to the skating rink for a birthday party. The rink rents roller skates for \$3 and inline skates for \$5. The total rental bill for all 22 students is \$96.

a. Write a system of equations that represents the number of students who rented the two types of skates.

b. How many students rented roller skates and how many rented inline skates?

53. GEOMETRY Angles A and B are supplementary and the measure of angle A is 18 degrees greater than the measure of angle B. Find the angle measures.

54. JOBS Levi has a job offer in which he will receive \$800 per month plus a commission of 2% of the total price of cars he sells. At his current job, he receives \$1200 per month plus a commission of 1.5% of his total sales. How much must he sell per month to make the new job a better deal?

55 TRAVEL A youth group went on a trip to an amusement park, travelling in two vans. The number of people in each van and the total cost of admission are shown in the table. Find the adult price and student price of admission.

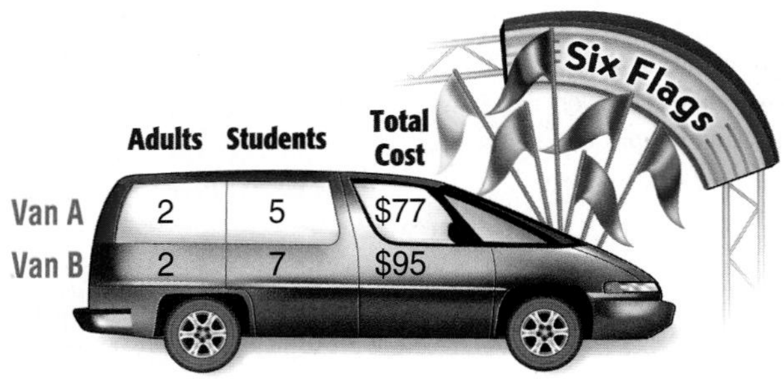

	Adults	Students	Total Cost
Van A	2	5	\$77
Van B	2	7	\$95

Solve each system of equations.

56. $4.1x - 3.4y = 19.97$
$6.3x + 2.2y = 7.67$

57. $3.7x - 4.6y = 15.37$
$-5.1x - 2.8y = 40.97$

58. $3.65x + 6.83y = -34.526$
$-41.3x - 2.68y = 77.336$

59. $-6.79a + 3.29b = -43.792$
$-9.14a - 6.28b = 10.658$

GEOMETRY Find the point at which the diagonals of the quadrilaterals intersect.

60.

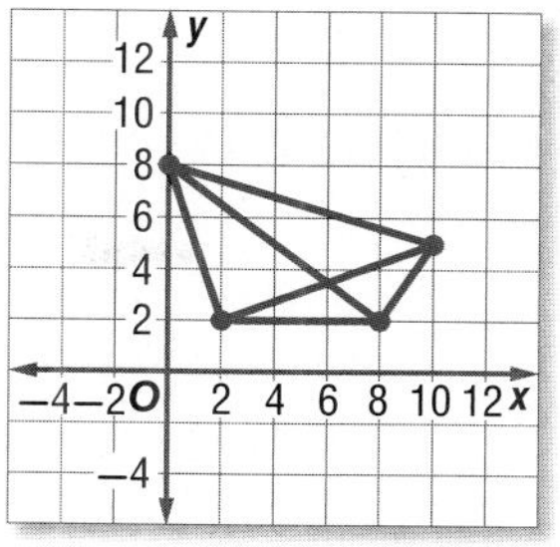

61

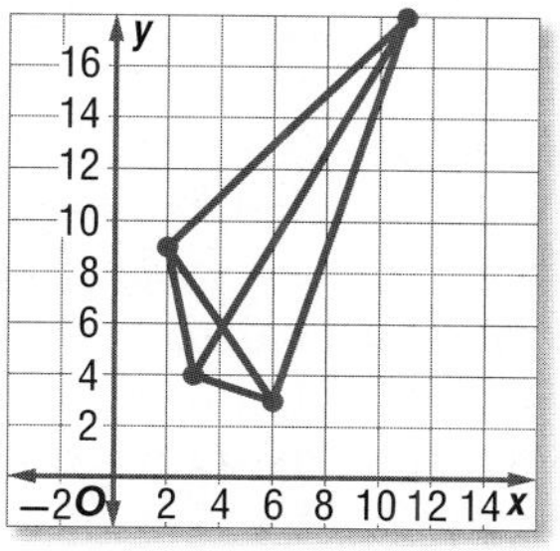

StudyTip

Reasonable Answers For Exercises 60 and 61, you can use the graphs to determine whether your answers are reasonable.

62. **ELECTIONS** In the election for student council, Candidate A received 55% of the total votes, while Candidate B received 1541 votes. If Candidate C received 40% of the votes that Candidate A received, how many total votes were cast?

63. **MULTIPLE REPRESENTATIONS** In this problem, you will explore systems of equations with three lines and two variables.

$$3y + x = 16$$
$$y - 2x = -4$$
$$y + 5x = 10$$

a. **TABULAR** Make a table of x- and y-values for each equation.

b. **ANALYTICAL** Which values from the table indicate intersections? Is there a solution that satisfies all three equations?

c. **GRAPHICAL** Graph the three equations on a single coordinate plane.

d. **VERBAL** What conditions must be met for a system of three equations with two variables to have a solution? What conditions result in no solution?

H.O.T. Problems Use Higher-Order Thinking Skills

64. **FIND THE ERROR** Gloria and Syreeta are solving the system $6x - 4y = 26$ and $-3x + 4y = -17$. Is either of them correct? Explain your reasoning.

Gloria

$6x - 4y = 26$
$\underline{-3x + 4y = -17}$
$3x = 9$
$x = 3$

$6(3) - 4y = 26$
$18 - 4y = 26$
$-4y = 8$
$y = -2$

The solution is $(3, -2)$.

Syreeta

$6x - 4y = 26$
$\underline{-3x + 4y = -17}$
$3x = -9$
$x = -3$

$6(-3) - 4y = 26$
$-18 - 4y = 26$
$-4y = 44$
$y = -11$

The solution is $(3, -11)$.

65. **CHALLENGE** Find values of a and b for which the following system has a solution of $(b - 1, b - 2)$.

$$-8ax + 4ay = -12a$$
$$2bx - by = 9$$

66. **REASONING** Katie says that if the coefficients of each variable are identical, then she does not need to solve the system since it will always have infinite solutions. Is she correct? Explain your reasoning.

67. **OPEN ENDED** Write a system of equations in which one equation needs to be multiplied by 3 and the other needs to be multiplied by 4 in order to solve the system with elimination. Then solve your system.

68. **WRITING IN MATH** Explain why substitution is sometimes more helpful than elimination and why elimination is sometimes more helpful than substitution.

NYSRE PRACTICE Reinforcement of A.A.7, A.A.10

69. GRIDDED RESPONSE A caterer bought several pounds of chicken salad and several pounds of tuna salad. The chicken salad costs \$9 per pound, and the tuna salad costs \$6 per pound. He bought a total of 14 pounds of salad and paid a total of \$111. How many pounds of chicken salad did he buy?

70. A rectangular room is shown below. Which expression represents the width of the door?

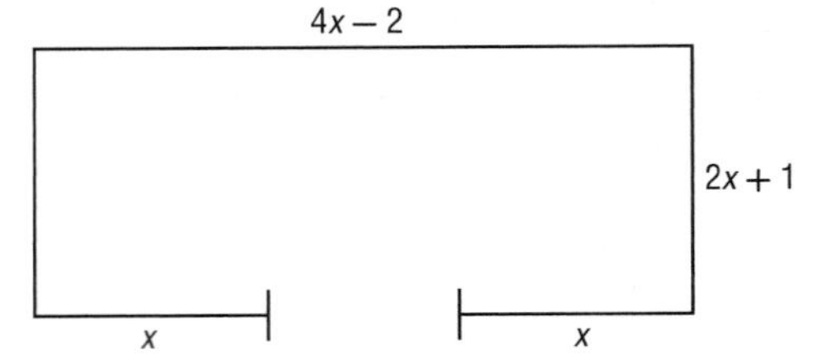

A $(4x - 2) - x - x$
B $(2x + 1) - x - x$
C $(4x - 2) - (2x + 1)$
D $(4x - 2) + x + x$

71. PROBABILITY Which of the following is an example of dependent events?

F rolling a 6-sided die twice and getting different numbers
G choosing two cards from a stack of colored cards, with replacement, and both cards are red
H flipping a coin twice and getting heads both times
J choosing the starting line-up for a football game

72. ACT/SAT Peni bought a basketball and a volleyball that cost a total of \$67. If the price of the basketball b is \$4 more than twice the cost of the volleyball v, which system of linear equations could be used to determine the cost of each ball?

A $b + v = 67$, $b = 2v - 4$
B $b + v = 67$, $b = 2v + 4$
C $b + v = 4$, $b = 2v - 67$
D $b + v = 4$, $b = 2v + 67$

Spiral Review

73. EXERCISE Refer to the graphic. (Lesson 3-1)

a. For each option, write an equation that represents the cost of belonging to the gym.

b. Graph the equations. Estimate the break-even point for the gym memberships.

c. Explain what the break-even point means.

d. If you plan to visit the gym at least once per week during the year, which option should you chose?

Graph each inequality. (Lesson 2-8)

74. $x + y \geq 6$ **75.** $4x - 3y < 10$ **76.** $5x + 7y \geq -20$

Write an equation of the line passing through each pair of points. (Lesson 2-4)

77. $(3, 5), (7, -3)$ **78.** $(8, -2), (4, 8)$ **79.** $(-6, -1), (-9, 11)$ **80.** $(-4, -4), (12, -8)$

Skills Review

Determine whether the given point satisfies each inequality. (Lesson 2-8)

81. $4x + 5y \leq 15; (2, -2)$ **82.** $3x + 5y \geq 8; (1, 1)$ **83.** $6x + 9y < -1; (0, 0)$

3-3 Solving Systems of Inequalities by Graphing

Then

You solved systems of linear equations using tables and graphs. (Lesson 3-1)

Now

- Solve systems of inequalities by graphing.
- Determine the coordinates of the vertices of a region formed by the graph of a system of inequalities.

NYS Core Curriculum

Reinforcement of A.A.40 Determine whether a given point is in the solution set of a system of linear inequalities
Reinforcement of A.G.7 Graph and solve systems of linear equations and inequalities with rational coefficients in two variables

New Vocabulary

system of inequalities

NY Math Online

glencoe.com

- Extra Examples
- Personal Tutor
- Self-Check Quiz
- Homework Help
- Math in Motion

Why?

Many weather conditions need to be met before a space shuttle can launch. The temperature must be greater than 35°F and less than 100°F, and the wind speed cannot exceed 30 knots. These three conditions can be represented by the system of inequalities shown at the right.

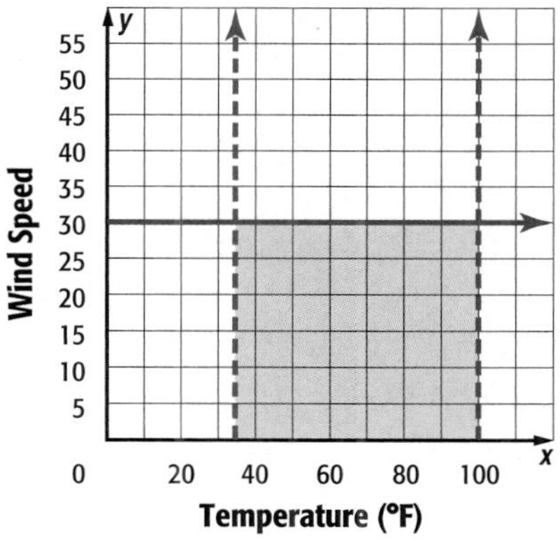

Systems of Inequalities Solving a **system of inequalities** means finding the ordered pairs that satisfy all of the inequalities in the system.

Key Concept — Solving Systems of Inequalities

For Your FOLDABLE

Step 1 Graph each inequality, shading the correct area.

Step 2 Identify the region that is shaded for all of the inequalities. This is the solution of the system.

Math *in Motion*, Animation glencoe.com

EXAMPLE 1 Intersecting Regions

Solve the system of inequalities.

$y > 2x - 4$

$y \le -0.5x + 3$

Solution of $y > 2x - 4 \rightarrow$ Regions 1 and 3

Solution of $y \le -0.5x + 3 \rightarrow$ Regions 2 and 3

Region 3 is part of the solution of both inequalities, so it is the solution of the system.

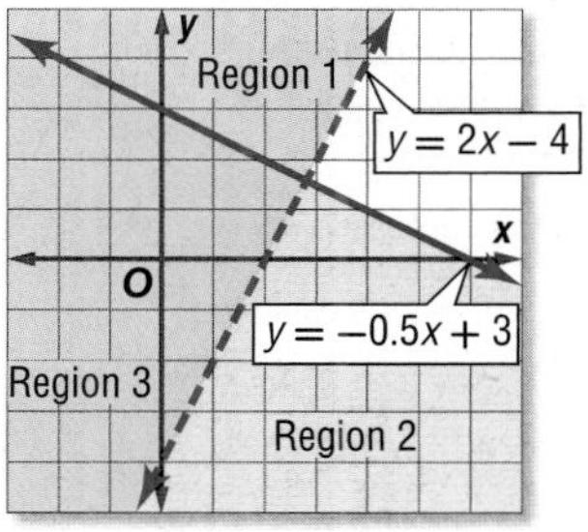

CHECK

Notice that the origin is part of the solution of the system. The origin can be used as a test point. You can test the solution by substituting (0, 0) for x and y in each equation.

$y > 2x - 4$	$y \le -0.5x + 3$
$0 \overset{?}{>} 2(0) - 4$	$0 \overset{?}{\le} -0.5(0) + 3$
$0 \overset{?}{>} 0 - 4$	$0 \overset{?}{\le} 0 + 3$
$0 > -4$ ✓	$0 \le 3$ ✓

Check Your Progress

1A. $y \le -2x + 5$
$y > -\frac{1}{4}x - 6$

1B. $y \ge |x|$
$y < \frac{4}{3}x + 5$

Personal Tutor glencoe.com

ReadingMath

Empty Set The empty set is also called the *null set*. It can be represented by ∅ or {}.

It is possible that the regions do not intersect. When this occurs, there is no solution of the system or the solution set is the *empty set*.

EXAMPLE 2 Separate Regions

Solve the system of inequalities by graphing.
$y \geq x + 5$
$y < x - 4$

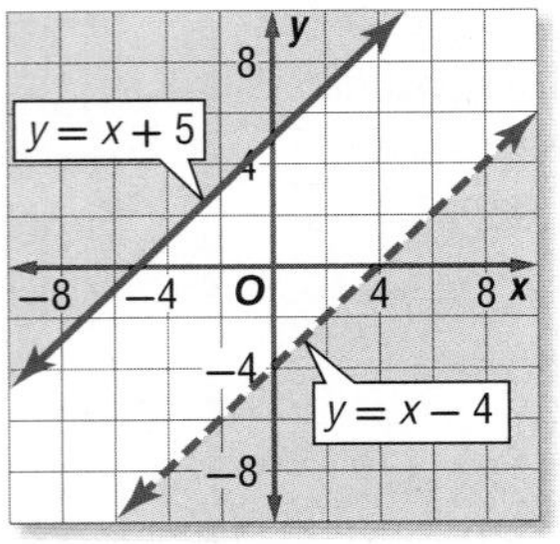

Graph both inequalities.

Since the graphs of the inequalities do not overlap, there are no points in common and there is no solution to the system.

The solution set is the empty set.

Check Your Progress

2A. $y \geq -4x + 8$
$y < -4x + 6$

2B. $y \geq |x|$
$y < 2x - 24$

Personal Tutor glencoe.com

Real-World Link

Typical incoming freshmen will spend more than 3 times as many hours studying in college as in high school.

Source: *National Survey of Student Engagement*

Real-World EXAMPLE 3 Write and Use a System of Inequalities

TIME MANAGEMENT Chelsea has final exams in calculus, physics, and history. She has up to 25 hours to study for the exams. She plans to study history for 2 hours. She needs to spend at least 7 hours studying for calculus, but over 14 is too much. She hopes to spend between 8 and 12 hours on physics. Write and graph a system of inequalities to represent the situation.

Calculus: at least 7 hours, but no more than 14
$7 \leq c \leq 14$

Physics: at least 8 hours, but no more than 12
$8 \leq p \leq 12$

Chelsea has 25 hours available, and 2 of those will be spent on history. She has up to 23 hours left for calculus and physics.
$c + p \leq 23$

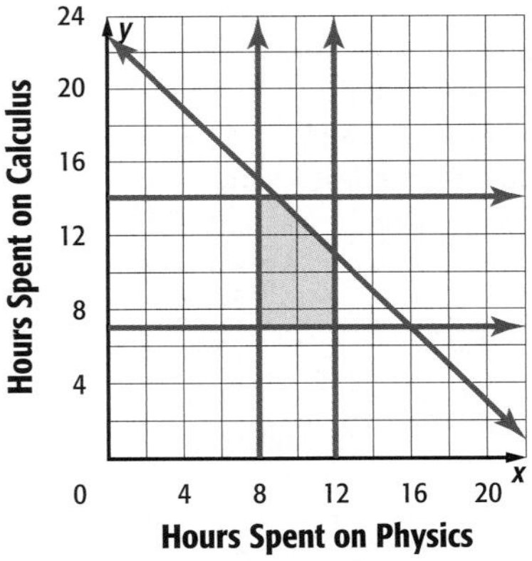

Graph all of the inequalities. Any ordered pair in the intersection is a solution of the system. One solution is 10 hours on physics and 12 hours on calculus.

Check Your Progress

3. TRAVEL Mr. and Mrs. Rodriguez are driving across the country with their two children. They plan on driving a maximum of 10 hours each day. Mr. Rodriguez wants to drive at least 4 hours a day but no more than 8 hours a day. Mrs. Rodriguez can drive in between 2 and 5 hours per day. Write and graph a system of inequalities that represents this information.

Personal Tutor glencoe.com

Boundaries If the inequality that forms the boundary is < or >, then the boundary is not included in the solution, and the line should be dashed.

Find Vertices of an Enclosed Region Sometimes the graph of a system of inequalities produces an enclosed region in the form of a polygon. To find the vertices of the region, determine the coordinates of the points at which the boundaries intersect.

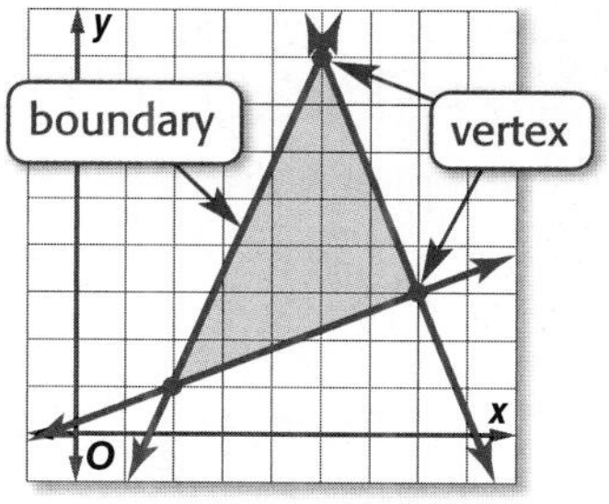

EXAMPLE 4 Find Vertices

Find the coordinates of the vertices of the triangle formed by $y \geq 2x - 8$, $y \leq -\frac{1}{4}x + 6$, and $4y \geq -15x - 32$.

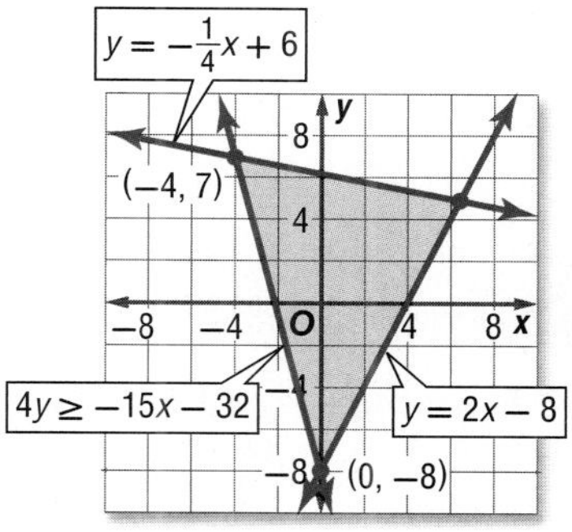

Step 1 Graph each inequality.
The coordinates (−4, 7) and (0, −8) can be determined from the graph. To find the coordinates of the third vertex, solve the system of equations $y = 2x - 8$ and $y = -\frac{1}{4}x + 6$.

Step 2 Substitute for y in the second equation.

$2x - 8 = -\frac{1}{4}x + 6$ **Replace y with $2x - 8$.**

$2x = -\frac{1}{4}x + 14$ **Add 8 to each side.**

$\frac{9}{4}x = 14$ **Add $\frac{1}{4}x$ to each side.**

$x = \frac{56}{9}$ or $6\frac{2}{9}$ **Divide each side by $\frac{9}{4}$.**

Step 3 Find y.

$y = 2\left(6\frac{2}{9}\right) - 8$ **Replace x with $6\frac{2}{9}$.**

$= 12\frac{4}{9} - 8$ **Distributive Property**

$= 4\frac{4}{9}$ **Simplify.**

CHECK Compare the coordinates to the coordinates on the graph. The x-coordinate of the third vertex is between 6 and 7, so $6\frac{2}{9}$ is reasonable. The y-coordinate of the third vertex is between 4 and 5, so $4\frac{4}{9}$ is reasonable.

The vertices of the triangle are at (−4, 7), (0, −8), and $\left(6\frac{2}{9}, 4\frac{4}{9}\right)$.

Check Your Progress

Find the coordinates of the vertices of the triangle formed by each system of inequalities.

4A. $y \geq -3x - 6$
$2y \geq x - 16$
$11y + 7x \leq 12$

4B. $5y \leq 2x + 9$
$y \leq -x + 6$
$9y \geq -2x + 5$

Personal Tutor glencoe.com

Check Your Understanding

Examples 1 and 2 pp. 151, 152

Solve each system of inequalities by graphing.

1. $y \leq 6$
$y > -3 + x$

2. $y \leq -3x + 4$
$y \geq 2x - 1$

3. $y > -2x + 4$
$y \leq 3x - 3$

Example 3 p. 152

4. **COOKOUT** The most Kala can spend on hot dogs and buns for her cookout is \$35. A package of 10 hot dogs costs \$3.50. A package of buns costs \$2.50 and contains 8 buns. She needs to buy at least 40 hot dogs and 40 buns.

a. Graph the region that shows how many packages of each item she can purchase.

b. Give an example of three different purchases she can make.

Example 4 p. 153

Find the coordinates of the vertices of the triangle formed by each system of inequalities.

5. $y \geq 2x + 1$
$y \leq 8$
$4x + 3y \geq 8$

6. $y \geq -2x - 4$
$6y \leq x + 28$
$y \geq 13x - 34$

Practice and Problem Solving

● = **Step-by-Step Solutions** begin on page R20.
Extra Practice begins on page 947.

Examples 1 and 2 pp. 151, 152

Solve each system of inequalities by graphing.

7. $x < 3$
$y \geq -4$

8. $y > 3x - 5$
$y \leq 4$

9. $y < -3x + 4$
$3y + x > -6$

10. $y \geq 0$
$y < x$

11. $6x - 2y \geq 12$
$3x + 4y > 12$

12. $-8x > -2y - 1$
$-4y \geq 2x - 5$

13. $5y < 2x + 10$
$y - 4x > 8$

14. $3y - 2x \leq -24$
$y \geq \frac{2}{3}x - 1$

15. $y > -\frac{2}{5}x + 2$
$5y \leq -2x - 15$

Example 3 p. 152

16. **RECORDING** Jane's band wants to spend no more than \$575 recording their first CD. The studio charges at least \$35 an hour to record. Graph a system of inequalities to represent this situation.

17. **SUMMER TRIP** Rondell has to save at least \$925 to go to Rome with his Latin class in 8 weeks. He earns \$9 an hour working at the Pizza Palace and \$12 an hour working at a car wash. By law, he cannot work more than 25 hours per week. Graph two inequalities that Rondell can use to determine the number of hours he needs to work at each job if he wants to make the trip.

Example 4 p. 153

Find the coordinates of the vertices of the triangle formed by each system of inequalities.

18. $x \geq 0$
$y \geq 0$
$x + 2y < 4$

19. $y \geq 3x - 7$
$y \leq 8$
$-4x + -4y < -4$

20. $x \leq 4$
$y > -3x + 12$
$y \leq 9$

21. $-3x + 4y \leq 15$
$2y + 5x > -12$
$10y + 60 \geq 27x$

22. $8y - 19x < 74$
$38y + 26x \leq 119$
$54y - 12x \geq -198$

23. $6y - 24x \geq -168$
$8y + 7x > 10$
$20y - 2x \leq 64$

24. **BAKING** Rebecca wants to bake cookies and cupcakes for a bake sale. She can bake 15 cookies at a time and 12 cupcakes at a time. She needs to make at least 120 baked goods, but no more than 360, and she wants to have at least three times as many cookies as cupcakes. What combination of batches of each could Rebecca make?

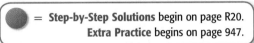

25 CELL PHONES Dale has a maximum of 800 minutes on his cell phone plan that he can use each month. Daytime minutes cost \$0.15, and nighttime minutes cost \$0.10. Dale plans to use at least twice as many daytime minutes as nighttime minutes. If Dale uses at least 200 nighttime minutes and does not go over his limit, what is his maximum bill? his minimum bill?

26. TREES Trees are divided into four categories according to height and trunk circumference. In one forest, the trees are categorized by the heights and circumferences described in the table.

Crown Class	dominant	co-dominant	intermediate	suppressed
Height (in feet)	over 72	56–72	40–55	under 39
Trunk Circumference (in inches)	over 60	48–60	34–48	under 33

Source: USDA Forest Service

a. Write and graph the system of inequalities that represents the range of heights h and circumferences c for a co-dominant tree.

b. Determine the crown class of a basswood that is 48 feet tall. Find the expected trunk circumference.

Real-World Link

The four crown classes are:

- dominant (D): trees well above the general canopy of the forest
- co-dominant (CD): part of the general canopy that receives sunlight from above and the sides
- intermediate (I): trees below the general canopy that only receive sunlight from above
- suppressed (S): trees well below the general canopy

Source: Silva Ecosystem Consultants

27. CAMPING On a camping trip, Jessica needs at least 3 pounds of food and 0.5 gallon of water per day. Marc needs at least 5 pounds of food and 0.5 gallon of water per day. Jessica's equipment weighs 10 pounds, and Marc's equipment weighs 20 pounds. A gallon of water weighs approximately 8 pounds. Each of them carries their own supplies, and Jessica is capable of carrying 35 pounds while Marc can carry 50 pounds.

a. Graph the inequalities that represent how much they can carry.

b. How many days can they camp, assuming that they bring all their supplies in at once?

c. Who will run out of supplies first?

Solve each system of inequalities by graphing.

28. $y \geq |2x + 4| - 2$
$3y + x \leq 15$

29. $y \geq |6 - x|$
$|y| \leq 4$

30. $|y| \geq x$
$y < 2x$

31. $y > -3x + 1$
$4y \leq x - 8$
$3x - 5y < 20$

32. $6y + 2x \leq 9$
$2y + 18 \geq 5x$
$y > -4x - 9$

33. $|x| > y$
$y \leq 6$
$y \geq -2$

34. $2x + 3y \geq 6$
$y \leq |x - 6|$

35. $8x + 4y < 10$
$y > |2x - 1|$

36. $y \geq |x - 2| + 4$
$y \leq [\![x]\!] - 3$

37. MUSIC Steve is trying to decide what to put on his MP3 player. Audio books are 3 hours long and songs are 2.5 minutes long. Steve wants no more than 4 audio books on his MP3 player, but at least ten songs and one audio book. Each book costs \$15.00 and each song costs \$0.95. Steve has \$63 to spend on books and music. Graph the inequalities to show possible combinations of books and songs that Steve can have.

38. JOBS Louie has two jobs and can work no more than 25 total hours per week. He wants to earn at least \$150 per week. Graph the inequalities to show possible combinations of hours worked at each job that will help him reach his goal.

Job	Pay
Busboy	\$6.50
Clerk	\$8.00

Real-World Link

Dodgeball has been played in the United States for over 100 years and is still played in schools today.

Source: Kidzworld

39. **TIME MANAGEMENT** Ramir uses his spare time to write a novel and to exercise. He has budgeted 35 hours per week. He wants to exercise at least 7 hours a week but no more than 15. He also hopes to write between 20 and 25 hours per week. Write and graph a system of inequalities that represents this situation.

Find the coordinates of the vertices of the figure formed by each system of inequalities.

40. $y \geq 2x - 12$
$y \leq -4x + 20$
$4y - x \leq 8$
$y \geq -3x + 2$

41. $y \geq -x - 8$
$2y \geq 3x - 20$
$4y + x \leq 24$
$y \leq 4x + 22$

42. $2y - x \geq -20$
$y \geq -3x - 6$
$y \leq -2x + 2$
$y \leq 2x + 14$

43. **INVESTING** Mr. Hoffman is investing \$10,000 in two funds. One fund will pay 6% interest, and a riskier second fund will pay 10% interest. What is the least amount Mr. Hoffman can invest in the risky fund and still earn at least \$740 after one year?

44. **DODGEBALL** A high school is selecting a dodgeball team to play in a fund-raising exhibition against their rival. There can be between 10 and 15 players on the team and there must be more girls than boys on the team.

a. Write and graph a system of inequalities to represent the situation.

b. List all of the possible combinations of boys and girls for the team.

c. Explain why there is not an infinite number of possibilities.

H.O.T. Problems

Use **H**igher-**O**rder **T**hinking Skills

45. **CHALLENGE** Find the area of the region defined by the following inequalities.

$$y \geq -4x - 16$$
$$4y \leq 26 - x$$
$$3y + 6x \leq 30$$
$$4y - 2x \geq -10$$

46. **OPEN ENDED** Write a system of two inequalities in which the solution:

a. lies only in the third quadrant.

b. does not exist.

c. lies only on two lines.

d. lies on exactly one point.

47. **CHALLENGE** Write a system of equations to represent the solution shown at the right.

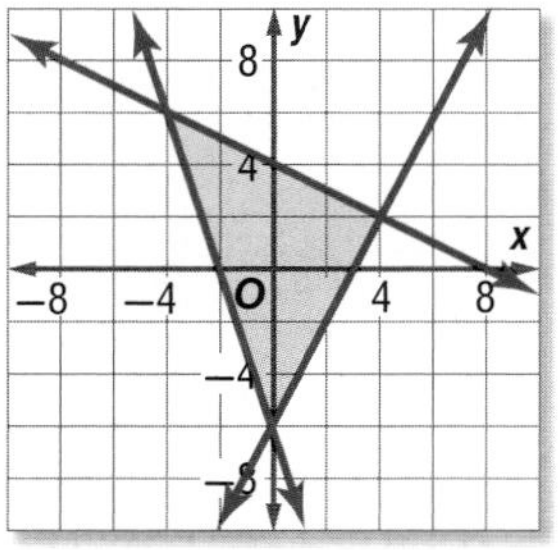

48. **REASONING** Determine whether the following statement is *true* or *false*. If false, give a counterexample.

A system of two linear inequalities has either no points or infinitely many points in its solution.

49. **WRITING IN MATH** Explain how to determine where to shade when graphing a system of inequalities.

50. **WRITING IN MATH** Explain how you would test to see whether (−4, 6) is a solution of a system of inequalities.

NYSRE PRACTICE Reinforcement of A.A.40, A.G.7

51. To be a member of the marching band, a student must have a grade-point average of at least 2.0 and must have attended at least five after-school practices. Choose the system of inequalities that best represents this situation.

A $x \geq 2$, $y \geq 5$

B $x \leq 2$, $y \leq 5$

C $x < 2$, $y < 5$

D $x > 2$, $y > 5$

52. ACT/SAT The table at the right shows a relationship between x and y. Which equation represents this relationship?

x	y
1	5
2	8
3	11
4	14
5	17
6	20

F $3x - 2$

G $3x + 2$

H $4x + 1$

J $4x - 1$

53. SHORT RESPONSE If $3x = 2y$ and $5y = 6z$, what is the value of x in terms of z?

54. GEOMETRY Look at the graph below. Which of these statements describes the relationship between the two lines?

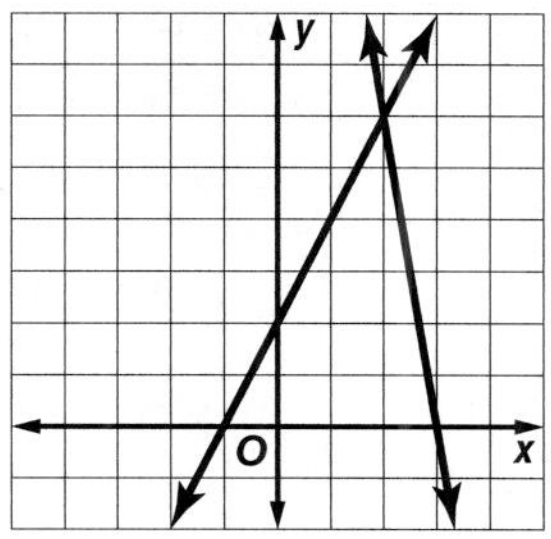

A They intersect at (6, 2).

B They intersect at (0, 2).

C They intersect at (3.5, 0).

D They intersect at (2, 6).

Spiral Review

55. GEOMETRY Find the coordinates of the vertices of the parallelogram whose sides are contained in the lines with equations $y = 3$, $y = 7$, $y = 2x$, and $y = 2x - 13$. (Lesson 3-2)

Solve each system of equations by graphing. (Lesson 3-1)

56. $y = 6 - x$
$y = x + 4$

57. $x + 2y = 2$
$2x + 4y = 8$

58. $x - 2y = 8$
$\frac{1}{2}x - y = 4$

Graph each function. Identify the domain and range. (Lesson 2-7)

59. $g(x) = \begin{cases} -1 \text{ if } x < 0 \\ -x + 2 \text{ if } x \geq 0 \end{cases}$

60. $h(x) = \begin{cases} x + 3 \text{ if } x \leq -1 \\ 2x \text{ if } x > -1 \end{cases}$

61. $h(x) = \begin{cases} -1 \text{ if } x < -2 \\ 1 \text{ if } x > 2 \end{cases}$

62. EXTENDED RESPONSE For each meeting of the Putnam High School book club, $25 is taken from the activities account to buy snacks and materials. After their sixth meeting, there will be $350 left in the activities account. (Lesson 2-4)

a. If no money is put back into the account, what equation can be used to show how much money is left in the activities account after having x number of meetings?

b. How much money was originally in the account?

c. After how many meetings will there be no money left in the activities account?

Skills Review

Find each value if $f(x) = 2x + 5$ and $g(x) = 3x - 4$. (Lesson 2-1)

63. $f(-3)$

64. $g(-2)$

65. $f(-1)$

66. $g(-0.5)$

67. $f(-0.25)$

68. $g(-0.75)$

EXTEND 3-3

Graphing Technology Lab
Systems of Linear Inequalities

NY Math Online glencoe.com
Other Calculator Keystrokes

Objective
Use a graphing calculator to solve systems of inequalities.

You can graph systems of linear inequalities with a graphing calculator by using the Y= menu. You can choose different graphing styles to shade above or below a line.

EXAMPLE Intersection of Two Graphs

Graph the system of inequalities in the standard viewing window.
$y \geq -3x + 4$
$y \leq 2x + 1$

Step 1 Enter $-3x + 4$ as Y1. Because y is greater than $-3x + 4$, shade above the line.

KEYSTROKES: [Y=] [◀] [◀] [ENTER] [ENTER] [▶] [▶] [(−)] 3 [X,T,θ,*n*] [+] 4 [ENTER]

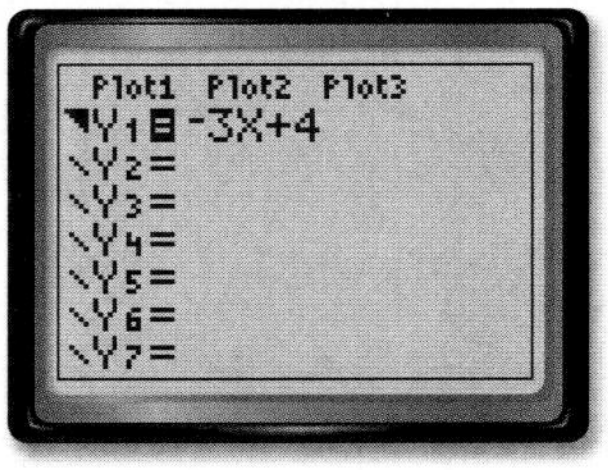

Step 2 Enter $2x - 1$ as Y2. Because y is less than $2x - 1$, shade below the line.

KEYSTROKES: [◀] [◀] [ENTER] [ENTER] [ENTER] [▶] [▶] 2 [X,T,θ,*n*] [−] 1 [ENTER]

Step 3 Display the graphs in the standard viewing window.

KEYSTROKES: [ZOOM] 6

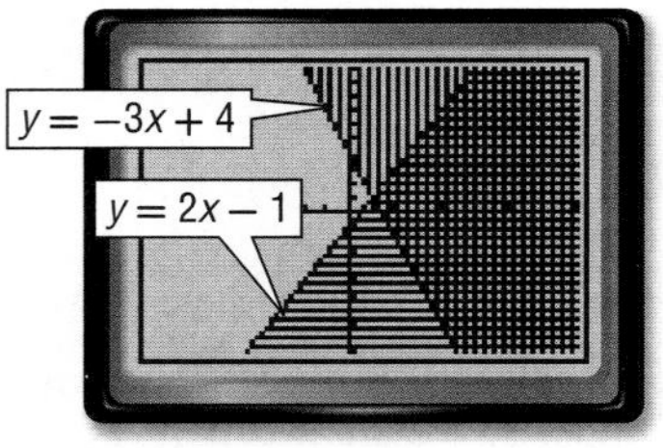

[−10, 10] scl: 1 by [−10, 10] scl: 1

Notice the shading pattern above the line $y = -2x + 3$ and the shading pattern below the line $y = x + 5$. The intersection of the graphs is the region where the patterns overlap. This region includes all the points that satisfy the system $y \geq -2x + 3$ and $y \leq x + 5$.

Exercises

Use a graphing calculator to solve each system of inequalities.

1. $y \geq 3$
$y \leq -x + 1$

2. $y \geq -4x$
$y \leq -5$

3. $y \geq 2 - x$
$y \leq x + 3$

4. $y \geq 2x + 1$
$y \leq -x - 1$

5. $2y \geq 3x - 1$
$3y \leq -x + 7$

6. $y + 5x \geq 12$
$y - 3 \leq 10$

7. $5y + 3x \geq 11$
$3y - x \leq -8$

8. $10y - 7x \geq -19$
$7y - 5x \leq 11$

9. $\frac{1}{6}y - x \geq -3$
$\frac{1}{5}y + x \leq 7$

Mid-Chapter Quiz

Lessons 3-1 through 3-3

NYS Core Curriculum
Reinforcement of A.A.7, A.A.10

Solve each system of equations by graphing. (Lesson 3-1)

1. $y = 2x + 4$
$y = -x - 2$

2. $5x + 2y = 3$
$5x - 4y = 9$

3. $x = 2y - 4$
$x = -3y + 1$

4. $2x - 5y = 14$
$4x + 3y = -24$

5. MULTIPLE CHOICE What type of system is shown? (Lesson 3-1)

$$2x + 4y = 5$$
$$3x + 6y = 11$$

A consistent and dependent

B consistent and independent

C inconsistent

D none of the above

Solve each system of equations by using either substitution or elimination. (Lesson 3-2)

6. $y = x + 4$
$x + y = -12$

7. $3x + 5y = -7$
$6x - 4y = 0$

8. $\frac{1}{3}x - \frac{3}{8}y = 28$
$\frac{1}{7}x + \frac{5}{7}y = -37$

9. $\frac{1}{3}x = y + 2$
$x = 5y - 2$

10. $5x + 2y = 4$
$3y - 4x = -40$

11. $8x - 3y = -13$
$-3x + 5y = 1$

12. $6x - 5y = 92$
$9x + 2y = 100$

13. $4y + 7x = -92$
$5x - 6y = 14$

14. SHOPPING Main St. Media sells all DVDs for one price and all books for another price. Alex bought 4 DVDs and 6 books for \$170, while Matt bought 3 DVDs and 8 books for \$180. What is the cost of a DVD and the cost of a book? (Lesson 3-2)

15. MULTIPLE CHOICE On Thursday, the art museum sold 56 fewer tickets than they sold on Friday. They sold a total of 406 tickets on Thursday and Friday. Which system of equations can be used to find the number of tickets sold on each day? (Lesson 3-2)

F $f - t = 56$
$f + t = 406$

H $f - t = 406$
$f + t = 56$

G $t - 56 = f$
$f + t = 406$

J $f - t = 56$
$f + 406 = t$

16. MULTIPLE CHOICE Which graph shows the solution of the system of inequalities? (Lesson 3-3)

$$y \leq 2x + 3$$
$$y < \frac{1}{3}x + 5$$

A

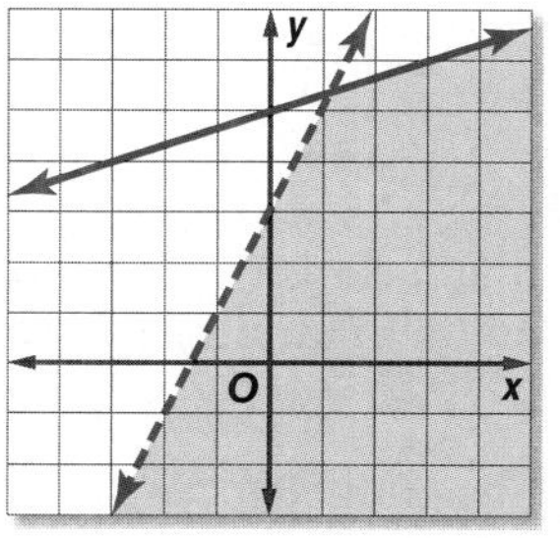

C

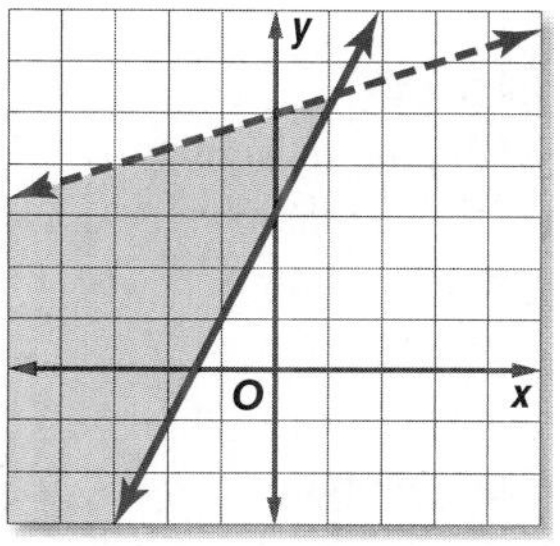

B

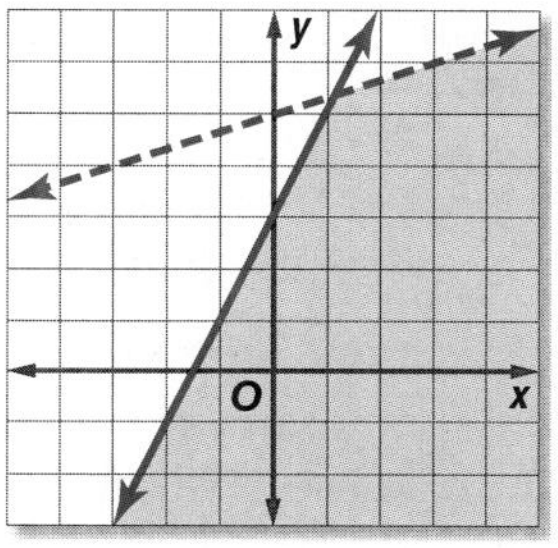

D

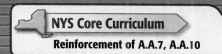

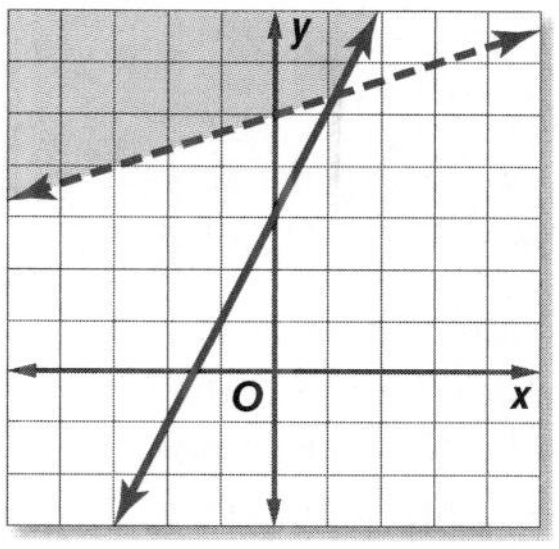

Solve each system of inequalities by graphing. (Lesson 3-3)

17. $x + y > 6$
$x - y < 0$

18. $y \geq 2x - 5$
$y \leq x + 4$

19. $3x + 4y \leq 12$
$6x - 3y \geq 18$

20. $5y + 2x \leq 20$
$4x + 3y > 12$

21. MULTIPLE CHOICE Tia spent \$42 on 2 cans of primer and 1 can of paint for her room. If the price of paint p is 150% of the price of primer r, which system of equations can be used to find the price of paint and primer? (Lesson 3-1)

F $p = r + \frac{1}{2}r$
$p + 2r = 42$

H $r = p + \frac{1}{2}p$
$p + 2p = 42$

G $p = r + 2r$
$p + \frac{1}{2}r = 42$

J $r = p + 2p$
$p + \frac{1}{2} = 42$

22. ART Rai can spend no more than \$225 on the art club's supply of brushes and paint. A box of 3 brushes costs \$7.50. A box of 10 tubes of paint costs \$21.45. She needs at least 20 brushes and 56 tubes of paint. Graph the region that shows how many packages of each item can be purchased. (Lesson 3-3)

3-4 Optimization with Linear Programming

Then

You solved systems of linear inequalities by using graphs. (Lesson 3-3)

Now

- Find the maximum and minimum values of a function over a region.
- Solve real-world optimization problems using linear programming.

NYS Core Curriculum

Reinforcement of A.A.40 Determine whether a given point is in the solution set of a system of linear inequalities

New Vocabulary

constraints
linear programming
feasible region
bounded
unbounded
optimize

NY Math Online

glencoe.com

- Extra Examples
- Personal Tutor
- Self-Check Quiz
- Homework Help
- Math in Motion

Why?

An electronics company produces digital audio players and digital video players. A sign on the company bulletin board is shown.

Keeping Costs Down: We Can Do It!

Our Goal: Production per Shift			
Player	Minimum	Maximum	Cost per Unit
audio	600	1500	$55
video	800	1700	$95

If at least 2000 items must be produced per shift, how many of each type should be made to minimize costs?

The company is experiencing limitations, or **constraints**, on production caused by customer demand, shipping, and the productivity of their factory. A system of inequalities can be used to represent these constraints.

Maximum and Minimum Values Situations often occur in business in which a company hopes to either maximize profits or minimize costs and many constraints need to be considered. These issues can often be addressed by the use of systems of inequalities in linear programming.

Linear programming is a method for finding maximum or minimum values of a function over a given system of inequalities with each inequality representing a constraint. After the system is graphed and the vertices of the solution set, called the **feasible region**, are substituted into the function, you can determine the maximum or minimum value.

Key Concept Feasible Regions — For Your FOLDABLE

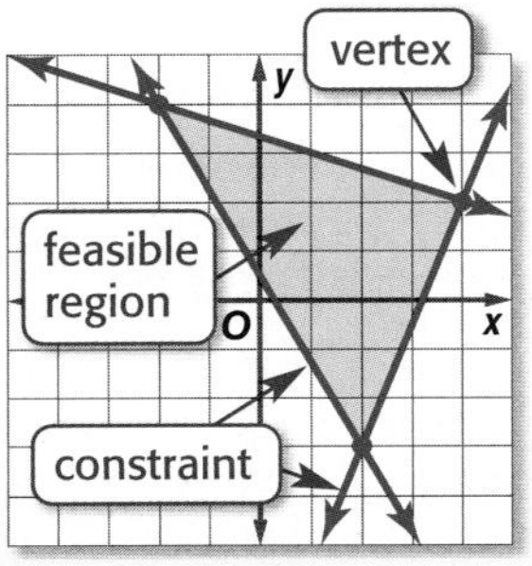

The feasible region is enclosed, or **bounded**, by the constraints. The maximum or minimum value of the related function *always* occurs at a vertex of the feasible region.

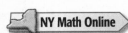

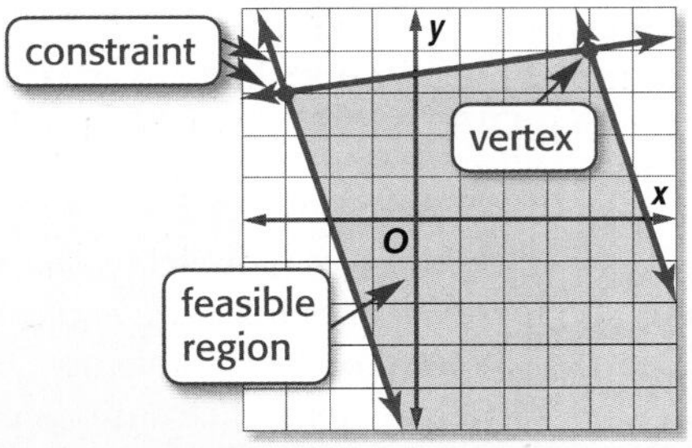

The feasible region is open and can go on forever. It is **unbounded**. Unbounded regions have either a maximum or a minimum.

ReadingMath

Function Notation The notation $f(x, y)$ is used to represent a function with two variables, x and y. It is read *f of x and y*.

EXAMPLE 1 Bounded Region

Graph the system of inequalities. Name the coordinates of the vertices of the feasible region. Find the maximum and minimum values of the function for this region.

$3 \leq y \leq 6$
$y \leq 3x + 12$
$y \leq -2x + 6$
$f(x, y) = 4x - 2y$

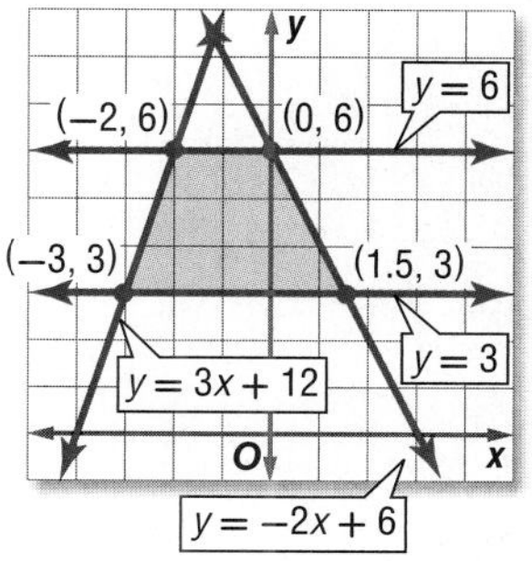

Step 1 Graph the inequalities and locate the vertices.

Step 2 Evaluate the function at each vertex.

(x, y)	$4x - 2y$	$f(x, y)$	
(−3, 3)	4(−3) − 2(3)	−18	
(1.5, 3)	4(1.5) − 2(3)	0	← **maximum**
(0, 6)	4(0) − 2(6)	−12	
(−2, 6)	4(−2) − 2(6)	−20	← **minimum**

The maximum value is 0 at (1.5, 3). The minimum value is −20 at (−2, 6).

Check Your Progress

1A. $-2 \leq x \leq 6$
$1 \leq y \leq 5$
$y \leq x + 3$
$f(x, y) = -5x + 2y$

1B. $-6 \leq y \leq -2$
$y \leq -x + 2$
$y \leq 2x + 2$
$f(x, y) = 6x + 4y$

Personal Tutor glencoe.com

When a system of inequalities does not form a closed region, it is unbounded.

Watch Out!

Maximum Value Do not assume that there is no maximum if the feasible region is unbounded above the vertices. Test points are needed to determine if there is a minimum or maximum.

EXAMPLE 2 Unbounded Region

Graph the system of inequalities. Name the coordinates of the vertices of the feasible region. Find the maximum and minimum values of the function for this region.

$2y + 3x \geq -12$
$y \leq 3x + 12$
$y \geq 3x - 6$
$f(x, y) = 9x - 6y$

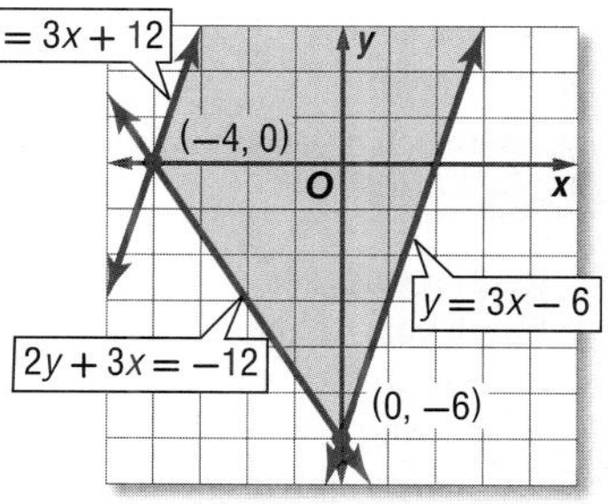

Graph the inequalities and evaluate the function at each vertex.

(x, y)	$9x - 6y$	$f(x, y)$
(−4, 0)	9(−4) − 6(0)	−36
(0, −6)	9(0) − 6(−6)	36

The maximum value is 36 at (0, −6). There is no minimum value. Notice that another point in the feasible region, (0, 8), yields a value of −48, which is less than −36.

Check Your Progress

2A. $y \leq 8$
$y \geq -x + 4$
$y \leq -x + 10$
$f(x, y) = -6x + 8y$

2B. $y \geq x - 9$
$y \leq -4x + 16$
$y \geq -4x - 4$
$f(x, y) = 10x + 7y$

Personal Tutor glencoe.com

Optimization To **optimize** means to seek the best price or amount to minimize costs or maximize profits. This is often obtained with the use of linear programming.

Key Concept Optimization with Linear Programming

For Your FOLDABLE

Step 1 Define the variables.

Step 2 Write a system of inequalities.

Step 3 Graph the system of inequalities.

Step 4 Find the coordinates of the vertices of the feasible region.

Step 5 Write a linear function to be maximized or minimized.

Step 6 Substitute the coordinates of the vertices into the function.

Step 7 Select the greatest or least result. Answer the problem.

Math *in Motion,* Animation glencoe.com

Real-World Career

Operations Manager
Operations management is an area of business that is concerned with the production of goods and services, and involves the responsibility of ensuring that business operations are efficient and effective. A master's degree in business and experience in operations are preferred.

Real-World EXAMPLE 3 Optimization with Linear Programming

BUSINESS Refer to the application at the beginning of the lesson. Use linear programming to determine how many of each type of digital player should be made per shift.

Step 1 Let a = number of audio players produced.
Let v = number of video players produced.

Step 2
$600 \le a \le 1500$
$800 \le v \le 1700$
$a + v \ge 2000$

Steps 3 and 4 The system is graphed at the right. Note the vertices of the feasible region.

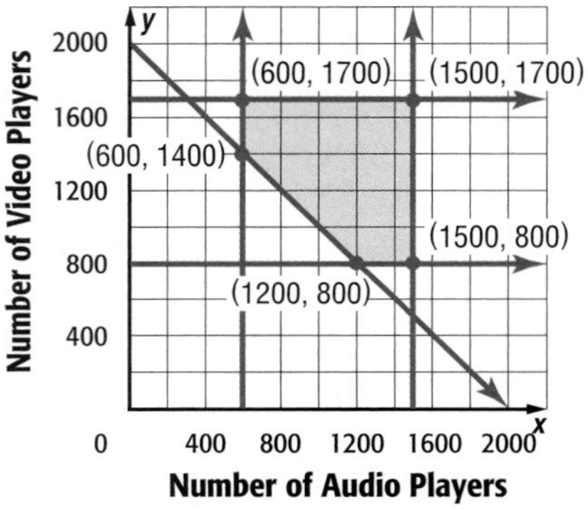

Step 5 The function to be minimized is $f(a, v) = 55a + 95v$.

Step 6

(a, v)	$55a + 95v$	$f(a, v)$
(600, 1700)	55(600) + 95(1700)	194,500
(600, 1400)	55(600) + 95(1400)	166,000
(1500, 1700)	55(1500) + 95(1700)	244,000 ← maximum
(1500, 800)	55(1500) + 95(800)	158,500
(1200, 800)	55(1200) + 95(800)	142,000 ← minimum

Step 7 1200 audio players and 800 video players should be produced to minimize costs.

StudyTip

Reasonableness Check your solutions for reasonableness by thinking of the context of the problem.

Check Your Progress

3. JEWELRY Each week, Mackenzie can make between 10 and 25 necklaces and 15 to 40 pairs of earrings. If she earns profits of \$3 on each pair of earrings and \$5 on each necklace, and she plans to sell at least 30 pieces of jewelry, how many earrings and necklaces should she make to maximize profit?

Personal Tutor glencoe.com

Check Your Understanding

Examples 1 and 2 p. 161

Graph each system of inequalities. Name the coordinates of the vertices of the feasible region. Find the maximum and minimum values of the given function for this region.

1. $y \leq 5$
$x \leq 4$
$y \geq -x$
$f(x, y) = 5x - 2y$

2. $y \leq -3x + 6$
$-y \leq x$
$y \leq 3$
$f(x, y) = 8x + 4y$

3. $y \geq -3x + 2$
$9x + 3y \leq 24$
$y \geq -4$
$f(x, y) = 2x + 14y$

4. $-2 \leq y \leq 6$
$3y \leq 4x + 26$
$y \leq -2x + 2$
$f(x, y) = -3x - 6y$

5. $-3 \leq y \leq 7$
$4y \geq 4x - 8$
$6y + 3x \leq 24$
$f(x, y) = -12x + 9y$

6. $y \leq 2x + 6$
$y \geq 2x - 8$
$y \geq -2x - 18$
$f(x, y) = 5x - 4y$

Example 3 p. 162

7. **BUSINESS** The total number of workers' hours per day available for production in a skateboard factory is 85 hours. There are 40 hours available for finishing decks and quality control each day. The table shows the number of hours needed in each department for two different types of skateboards.

Skateboard Manufacturing Time

Board Type	Production Time	Deck Finishing/Quality Control
Pro Boards	1.5 hours	2 hours
Specialty Boards	1 hour	0.5 hour

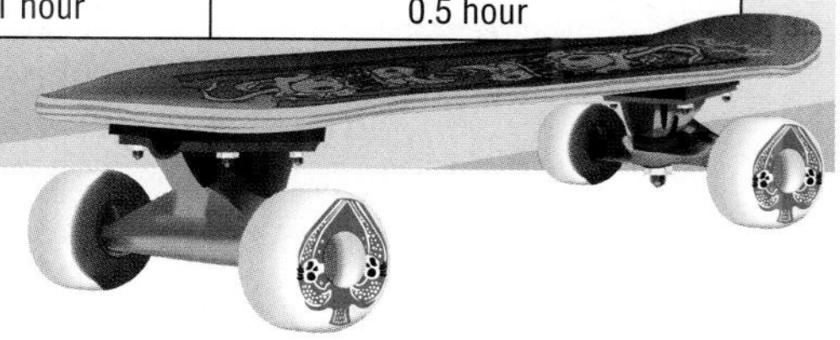

a. Write a system of inequalities to represent the situation.

b. Draw the graph showing the feasible region.

c. List the coordinates of the vertices of the feasible region.

d. If the profit on a pro board is \$50 and the profit on a specialty board is \$65, write a function for the total profit on the skateboards.

e. Determine the number of each type of skateboard that needs to be made to have a maximum profit. What is the maximum profit?

Practice and Problem Solving

● = **Step-by-Step Solutions** begin on page R20.
Extra Practice begins on page 947.

Examples 1 and 2 p. 161

Graph each system of inequalities. Name the coordinates of the vertices of the feasible region. Find the maximum and minimum values of the given function for this region.

8. $1 \leq y \leq 4$
$4y - 6x \geq -32$
$2y \geq -x + 4$
$f(x, y) = -6x + 3y$

9. $2 \geq x \geq -3$
$y \geq -2x - 6$
$4y \leq 2x + 32$
$f(x, y) = -4x - 9y$

10. $-2 \leq x \leq 4$
$5 \leq y \leq 8$
$2x + 3y \leq 26$
$f(x, y) = 8x - 10y$

11. $-8 \leq y \leq -2$
$y \leq x$
$y \leq -3x + 10$
$f(x, y) = 5x + 14y$

12. $x + 4y \geq 2$
$2x + 4y \leq 24$
$2 \leq x \leq 6$
$f(x, y) = 6x + 7y$

13. $3 \leq y \leq 7$
$2y + x \leq 8$
$y - 2x \leq 23$
$f(x, y) = -3x + 5y$

Examples 1 and 2 p. 161

Graph each system of inequalities. Name the coordinates of the vertices of the feasible region. Find the maximum and minimum values of the given function for this region.

14. $-9 \leq x \leq -3$
$-9 \leq y \leq -5$
$3y + 12x \leq -75$
$f(x, y) = 20x + 8y$

15. $x \geq -8$
$3x + 6y \leq 36$
$2y + 12 \geq 3x$
$f(x, y) = 10x - 6y$

16. $y \geq |x - 2|$
$y \leq 8$
$8y + 5x \leq 49$
$f(x, y) = -5x - 15y$

17. $x \geq -6$
$y + x \leq -1$
$2y + x \geq -8$
$f(x, y) = -10x - 12y$

18. $-5 \geq y \geq -17$
$y \leq 3x + 19$
$y \geq -4x + 15$
$f(x, y) = 8x - 3y$

19. $-8 \leq x \leq 16$
$y \geq 2x - 10$
$2y + x \leq 80$
$f(x, y) = 12x + 15y$

20. $y \leq x + 4$
$y \geq x - 4$
$y \leq -x + 10$
$y \geq -x - 10$
$f(x, y) = -10x + 9y$

21. $-4 \leq x \leq 8$
$-8 \leq y \leq 6$
$y \geq x - 6$
$4y + 7x \leq 31$
$f(x, y) = 12x + 8y$

22. $y \geq |x + 1| - 2$
$0 \leq y \leq 6$
$-6 \leq x \leq 2$
$x + 3y \leq 14$
$f(x, y) = 5x + 4y$

Example 3 p. 162

23. COOKING Jenny's Bakery makes two types of birthday cakes: yellow cake, which sells for \$25, and strawberry cake, which sells for \$35. Both cakes are the same size, but the decorating and assembly time required for the yellow cake is 2 hours, while the time is 3 hours for the strawberry cake. There are 450 hours of labor available for production. How many of each type of cake should be made to maximize revenue?

24. BUSINESS The manager of a travel agency is printing brochures and fliers to advertise special discounts on vacation spots during the summer months. Each brochure costs \$0.08 to print, and each flier costs \$0.04 to print. A brochure requires 3 pages, and a flier requires 2 pages. The manager does not want to use more than 600 pages, and she needs at least 50 brochures and 150 fliers. How many of each should she print to minimize the cost?

25. PAINTING Sean has 20 days to paint play houses and sheds. The sheds can be painted at a rate of 2.5 per day, and the play houses can be painted at a rate of 2 per day. He has 45 structures that need to be painted.

a. Write a system of inequalities to represent the possible ways Sean can paint the structures.

b. Draw a graph showing the feasible region and list the coordinates of the vertices of the feasible region.

c. If the profit is \$26 per shed and \$30 per play house, how many of each should he paint?

d. What is the maximum profit?

Real-World Link

43–47% of teens see a movie in theaters at least once a month. The average American sees 5.8 movies per year in theaters.

Source: Chapel Hill

26. MOVIES Employees at a local movie theater work 8-hour shifts from noon to 8 P.M. or from 4 P.M. to midnight. The table below shows the number of employees needed and their corresponding pay. Find the numbers of day-shift workers and night-shift workers that should be scheduled to minimize the cost. What is the minimal cost?

Time	noon to 4 P.M.	4 P.M. to 8 P.M.	8 P.M. to midnight
Number of Employees Needed	at least 5	at least 14	6
Rate per Hour	\$5.50	\$7.50	\$7.50

Real-World Link

Plastic bottles can be recycled into rugs, swimming goggles, dog bowls, back packs, skateboard wheels, winter jackets, and more.

Source: Earth 911

27. **BUSINESS** Each car on a freight train can hold 4200 pounds of cargo and has a capacity of 480 cubic feet. The freight service handles two types of packages: small, which weigh 25 pounds and are 3 cubic feet each, and large, which are 50 pounds and are 5 cubic feet each. The freight service charges \$5 for each small package and \$8 for each large package.

a. Find the number of each type of package that should be placed on a train car to maximize revenue.

b. What is the maximum revenue per train car?

c. In this situation, is maximizing the revenue necessarily the best thing for the company to do? Explain.

28. **RECYCLING** A recycling plant processes used plastic into food or drink containers. The plant processes up to 1200 tons of plastic per week. At least 300 tons must be processed for food containers, while at least 450 tons must be processed for drink containers. The profit is \$17.50 per ton for processing food containers and \$20 per ton for processing drink containers. What is the profit if the plant maximizes processing?

H.O.T. Problems

Use Higher-Order Thinking Skills

29. **OPEN ENDED** Create a set of inequalities that forms a bounded region with an area of 20 units^2 and lies only in the fourth quadrant.

30. **CHALLENGE** Find the area of the bounded region formed by the following constraints: $y \geq |x| - 3$, $y \leq -|x| + 3$, and $x \geq |y|$.

31. **WHICH ONE DOESN'T BELONG?** Identify the system of inequalities that is not the same as the other three. Explain your reasoning.

a.

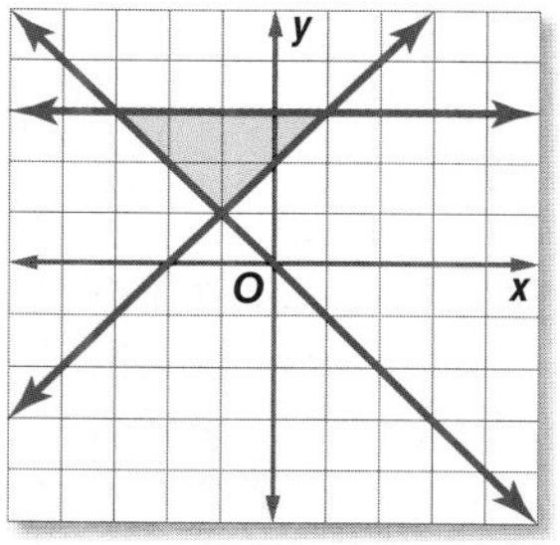

c.

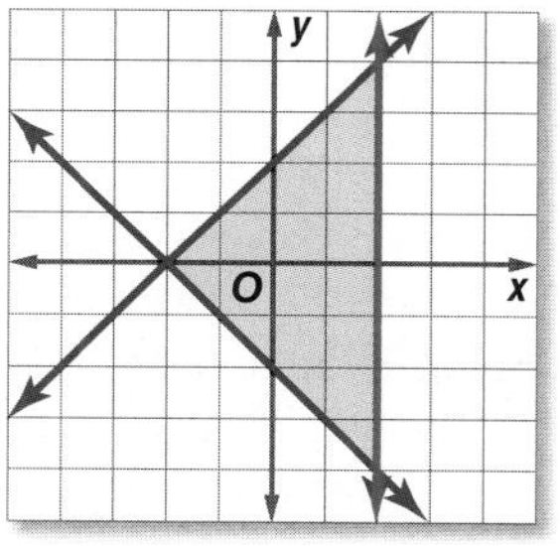

b.

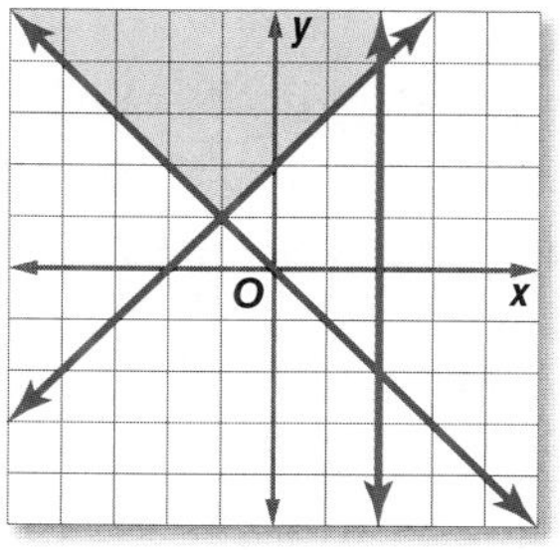

d.

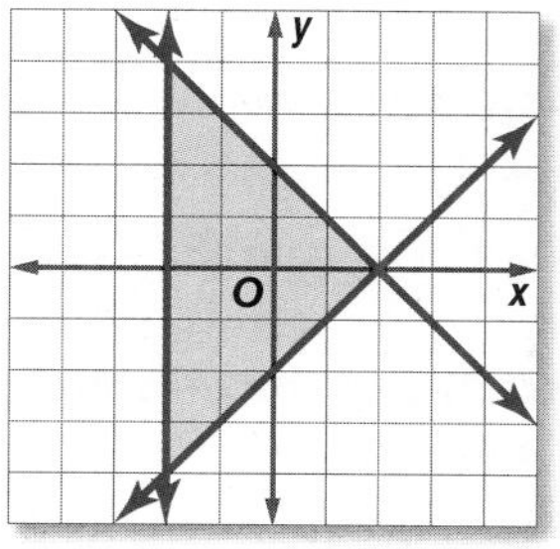

32. **REASONING** Determine whether the following statement is *sometimes*, *always*, or *never* true. Explain your reasoning.

An unbounded region will not have both a maximum and minimum value.

33. **WRITING IN MATH** Upon determining a bounded feasible region, Ayumi noticed that vertices $A(-3, 4)$ and $B(5, 2)$ yielded the same maximum value for $f(x, y) = 16y + 4x$. Kelvin confirmed that her constraints were graphed correctly and her vertices were correct. Then he said that those two points were not the only maximum values in the feasible region. Explain how this could have happened.

NYSRE PRACTICE Reinforcement of A.A.40

34. Kelsey worked 350 hours during the summer and earned \$2978.50. She earned \$6.85 per hour when she worked at a video store and \$11 per hour as an architectural intern. Let x represent the number of hours she worked at the video store and y represent the number of hours that she interned. Which system of equations represents this situation?

A $x + y = 350$
$11x + 6.85y = 2978.50$

B $x + y = 350$
$6.85x + 11y = 2978.50$

C $x + y = 2978.50$
$6.85x + 11y = 350$

D $x + y = 2978.50$
$11x + 6.85y = 350$

35. GRIDDED RESPONSE A family of four went out to dinner. Their bill, including tax, was \$60. They left a 17% tip on the total cost of their bill. What is the total cost of the dinner including tip?

36. ACT/SAT For a game she is playing, Liz must draw a card from a deck of 26 cards, one with each letter of the alphabet on it, and roll a die. What is the probability that Liz will draw a letter in her name and roll an odd number?

F $\frac{2}{3}$ H $\frac{1}{26}$

G $\frac{1}{13}$ J $\frac{3}{52}$

37. GEOMETRY Which of the following best describes the graphs of $y = 3x - 5$ and $4y = 12x + 16$?

A The lines have the same y-intercept.
B The lines have the same x-intercept.
C The lines are perpendicular.
D The lines are parallel.

Spiral Review

Solve each system of inequalities by graphing. (Lesson 3-3)

38. $3x + 2y \geq 6$
$4x - y \geq 2$

39. $4x - 3y < 7$
$2y - x < -6$

40. $3y \leq 2x - 8$
$y \geq \frac{2}{3}x - 1$

41. BUSINESS Last year the chess team paid \$7 per hat and \$15 per shirt for a total purchase of \$330. This year they spent \$360 to buy the same number of shirts and hats because the hats now cost \$8 and the shirts cost \$16. Write and solve a system of two equations that represents the number of hats and shirts bought each year. (Lesson 3-2)

Write an equation in slope-intercept form for the line that satisfies each set of conditions. (Lesson 2-4)

42. passes through (5, 1) and (8, −4)

43. passes through (−3, 5) and (3, 2)

Find the x-intercept and the y-intercept of the graph of each equation. Then graph the equation. (Lesson 2-2)

44. $5x + 3y = 15$

45. $2x - 6y = 12$

46. $3x - 4y - 10 = 0$

47. $2x + 5y - 10 = 0$

48. $y = x$

49. $y = 4x - 2$

Skills Review

Evaluate each expression if $x = -1$, $y = 3$, and $z = 7$. (Lesson 1-1)

50. $x + y + z$

51. $2x - y + 2z$

52. $-x + 4y - 3z$

53. $4x + 2y - z$

54. $5x - y + 4z$

55. $-3x - 3y + 3z$

3-5 Systems of Equations in Three Variables

Then
You solved linear equations with two variables. (Lesson 3-2)

Now
- Solve systems of linear equations in three variables.
- Solve real-world problems using systems of linear equations in three variables.

NYS Core Curriculum

Reinforcement of A.A.7 Analyze and solve verbal problems whose solution requires solving systems of linear equations in two variables
Reinforcement of A.G.7 Graph and solve systems of linear equations and inequalities with rational coefficients in two variables

New Vocabulary
ordered triple

NY Math Online
glencoe.com
- Extra Examples
- Personal Tutor
- Self-Check Quiz
- Homework Help
- Math in Motion

Why?

Seats closest to an amphitheater stage cost \$30. The seats in the next section cost \$25, and lawn seats are \$20. There are twice as many seats in section B as in section A. When all 19,200 seats are sold, the amphitheater makes \$456,000.

A system of equations in three variables can be used to determine the number of seats in each section.

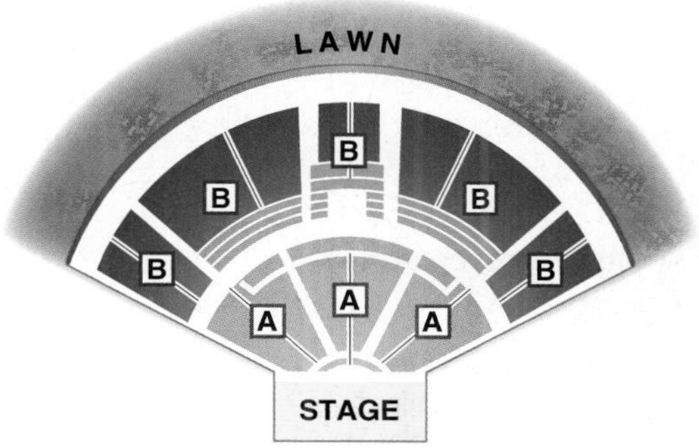

Systems in Three Variables Like systems of equations in two variables, systems in three variables can have one solution, infinite solutions, or no solution. A solution of such a system is an **ordered triple** (x, y, z).

The graph of an equation in three variables is a three-dimensional graph in the shape of a plane. The graphs of a system of equations in three variables form a system of planes.

One Solution

The three individual planes intersect at a specific point.

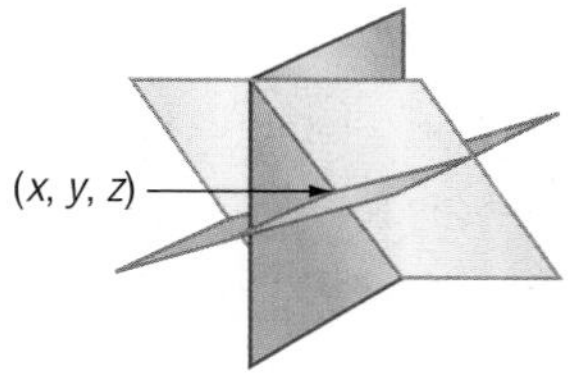

Infinitely Many Solutions

The planes intersect in a line.

Every coordinate on the line represents a solution of the system.

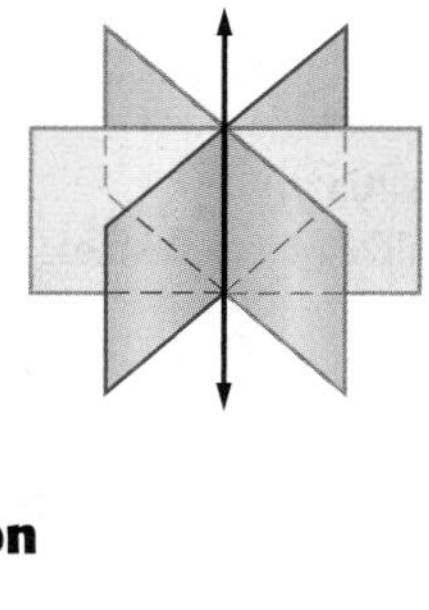

The planes intersect in the same plane.

Every equation is equivalent.
Every coordinate in the plane represents a solution of the system.

No Solution

There are no points in common with all three planes.

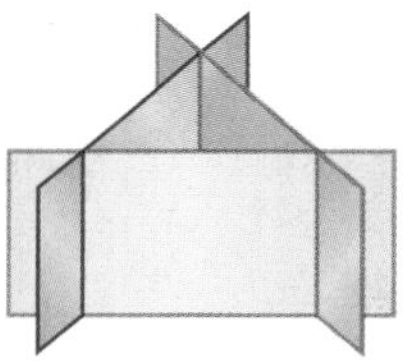

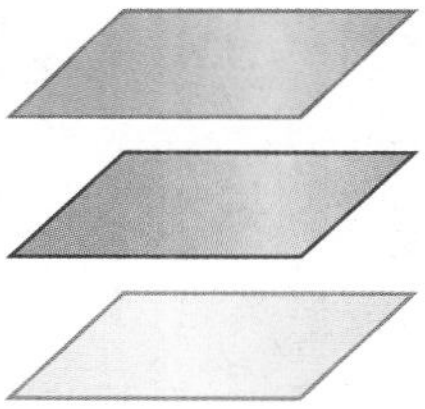

Solving systems of equations in three variables is similar to solving systems of equations in two variables. Use the strategies of substitution and elimination to find the ordered triple that represents the solution of the system.

EXAMPLE 1 A System with One Solution

Solve the system of equations.

$$\begin{aligned} 3x - 2y + 4z &= 35 \\ -4x + y - 5z &= -36 \\ 5x - 3y + 3z &= 31 \end{aligned}$$

The coefficient of 1 in the second equation makes y a good choice for elimination.

Step 1 Eliminate one variable by using two pairs of equations.

$3x - 2y + 4z = 35$
$-4x + y - 5z = -36$ **Multiply by 2.**

$$\begin{aligned} 3x - 2y + 4z &= 35 && \text{Equation 1} \\ (+)\ -8x + 2y - 10z &= -72 && \text{Equation 2} \times 2 \\ \hline -5x \quad - 6z &= -37 \end{aligned}$$

$-4x + y - 5z = -36$ **Multiply by 3.**
$5x - 3y + 3z = 31$

$$\begin{aligned} -12x + 3y - 15z &= -108 && \text{Equation 2} \times 3 \\ (+)\ 5x - 3y + 3z &= 31 && \text{Equation 1} \\ \hline -7x \quad - 12z &= -77 \end{aligned}$$

The y-terms in each equation have been eliminated. We now have a system of two equations and two variables, x and z.

Step 2 Solve the system of two equations.

$-5x - 6z = -37$ **Multiply by −2.**
$-7x - 12z = -77$

$$\begin{aligned} 10x + 12z &= 74 \\ (+)\ -7x - 12z &= -77 \\ \hline 3x \quad &= -3 && \text{Eliminate } z. \\ x &= -1 && \text{Divide by 3.} \end{aligned}$$

Use substitution to solve for z.

$$\begin{aligned} -5x - 6z &= -37 && \text{Equation with two variables} \\ -5(-1) - 6z &= -37 && \text{Substitution} \\ 5 - 6z &= -37 && \text{Multiply.} \\ -6z &= -42 && \text{Subtract 5 from each side.} \\ z &= 7 && \text{Divide each side by } -6. \end{aligned}$$

The result is $x = -1$ and $z = 7$.

Step 3 Substitute the two values into one of the original equations to find y.

$$\begin{aligned} -4x + y - 5z &= -36 && \text{Equation 1} \\ -4(-1) + y - 5(7) &= -36 && \text{Substitution} \\ 4 + y - 35 &= -36 && \text{Multiply.} \\ y &= -5 && \text{Add 31 to each side.} \end{aligned}$$

CHECK

$$\begin{aligned} -4x + y - 5z &= -36 && \text{Equation 2} \\ -4(-1) + (-5) - 5(7) &= -36 && x = -1,\ y = -5,\ z = 7 \\ 4 + (-5) - 35 &= -36 && \text{Simplify.} \\ -36 &= -36\ \checkmark \end{aligned}$$

The solution is $(-1, -5, 7)$.

Check Your Progress

1A. $2x + 4y - 5z = 18$
$-3x + 5y + 2z = -27$
$-5x + 3y - z = -17$

1B. $4x - 3y + 6z = 18$
$-x + 5y + 4z = 48$
$6x - 2y + 5z = 0$

Personal Tutor glencoe.com

StudyTip

Choosing Variables If one of the equations is missing a variable already, use the other two equations and eliminate the same variable.

Math *in Motion,* Interactive Lab glencoe.com

StudyTip

Checking Solutions Always substitute your answer into all of the original equations to confirm your answer.

When solving a system of three linear equations with three variables, it is important to check your answer using all three of the original equations. This is necessary because it is possible for a solution to work for two of the equations but not the third.

EXAMPLE 2 No Solution and Infinite Solutions

Solve each system of equations.

a. $\mathbf{5x + 4y - 5z = -10}$
$\mathbf{-4x - 10y - 8z = -16}$
$\mathbf{6x + 15y + 12z = 24}$

Eliminate x in the second two equations.

$-4x - 10y - 8z = -16$ **Multiply by 3.** $-12x - 30y - 24z = -48$

$6x + 15y + 12z = 24$ **Multiply by 2.** $(+)\ 12x + 30y + 24z = 48$

$0 = 0$

The equation $0 = 0$ is always true. This indicates that the last two equations represent the same plane. Check to see if this plane intersects the first plane.

$5x + 4y - 5z = -10$ **Multiply by 4.** $20x + 16y - 20z = -40$

$-4x - 10y - 8z = -16$ **Multiply by 5.** $(+)\ -20x - 50y - 40z = -80$

$-34y - 60z = -120$

The planes intersect in a line. So, there are an infinite number of solutions.

b. $\mathbf{-6a + 9b - 12c = 21}$
$\mathbf{-2a + 3b - 4c = 7}$
$\mathbf{10a - 15b + 20c = -30}$

Eliminate a in the first two equations.

$-6a + 9b - 12c = 21$ → $-6a + 9b - 12c = 21$
$-2a + 3b - 4c = 7$ **Multiply by −3.** $(+)\ 6a - 9b + 12c = -21$

$0 = 0$

The equation $0 = 0$ is always true. This indicates that the first two equations represent the same plane. Check to see if this plane intersects the last plane.

$-2a + 3b - 4c = 7$ **Multiply by 5.** $-10a + 15b - 20c = 35$
$10a - 15b + 20c = -30$ → $(+)\ 10a - 15b + 20c = -30$

$0 = 5$

The equation $0 = 5$ is never true. So, there is no solution of this system.

StudyTip

Infinite and No Solutions When solving a system of more than two equations, $0 = 5$ will always yield no solution, while $0 = 0$ may not always yield infinite solutions.

Check Your Progress

2A. $-4x - 2y - z = 15$
$12x + 6y + 3z = 45$
$2x + 5y + 7z = -29$

2B. $3x + 5y - 2z = 13$
$-5x - 2y - 4z = 20$
$-14x - 17y + 2z = -19$

Personal Tutor glencoe.com

Real-World Problems When solving problems involving three variables, use the four-step plan to help organize the information. Identify the three variables and what they represent. Then use the information in the problem to form equations using the variables. Once you have three equations and all three variables are represented, you can solve the problem.

Real-World Career

Music Management Music management involves acting as the talent manager to the artists. Other duties include negotiating with record labels, music promoters, and tour promoters. Managers usually earn a percentage of the artist's income. A bachelor's degree is usually required.

Problem-SolvingTip

Look for a Pattern Looking for a pattern can help to identify a function and write an equation.

EXAMPLE 3 Write and Solve a System of Equations

CONCERTS Refer to the application at the beginning of the lesson. Write and solve a system of equations to determine how many seats are in each section of the amphitheater.

Understand Define the variables.

x = seats in section A
y = seats in section B
z = lawn seats

Plan There are 19,200 seats.

$x + y + z = 19{,}200$ **Equation 1**

The total revenue is $456,000.

$30x + 25y + 20z = 456{,}000$ **Equation 2**

There are twice as many seats in section B as in section A.

$y = 2x$ **Equation 3**

Solve Solve the system.

Step 1 Substitute $y = 2x$ in the first two equations.

$x + y + z = 19{,}200$	**Equation 1**
$x + 2x + z = 19{,}200$	**$y = 2x$**
$3x + z = 19{,}200$	**Add.**
$30x + 25y + 20z = 456{,}000$	**Equation 2**
$30x + 25(2x) + 20z = 456{,}000$	**$y = 2x$**
$80x + 20z = 456{,}000$	**Simplify.**

Step 2 Solve the system of two equations in two variables.

$3x + z = 19{,}200$ **Multiply by −20.** $\rightarrow$ $-60x - 20z = -384{,}000$
$80x + 20z = 456{,}000$ $\rightarrow$ $(+)\ 80x + 20z = 456{,}000$

$20x = 72{,}000$
$x = 3600$

Step 3 Substitute to find z.

$3x + z = 19{,}200$	**Remaining equation in two variables**
$3(3600) + z = 19{,}200$	**$x = 3600$**
$10{,}800 + z = 19{,}200$	**Distribute.**
$z = 8400$	**Simplify.**

Step 4 Substitute to find y.

$y = 2x$	**Equation 3**
$y = 2(3600)$ or 7200	**$x = 3600$**

The solution is (3600, 7200, 8400). There are 3600 seats in section A, 7200 in section B, and 8400 lawn seats.

Check Substitute the values into either of the first two equations.

Check Your Progress

3. Ms. Garza invested $50,000 in three different accounts. She invested three times as much money in an account that paid 8% interest than an account that paid 10% interest. The third account earned 12% interest. If she earned a total of $5160 in interest in a year, how much did she invest in each account?

Personal Tutor glencoe.com

Check Your Understanding

Examples 1 and 2 pp. 168, 169

Solve each system of equations.

1. $-3a - 4b + 2c = 28$
$a + 3b - 4c = -31$
$2a + 3c = 11$

2. $3y - 5z = -23$
$4x + 2y + 3z = 7$
$-2x - y - z = -3$

3. $3x + 6y - 2z = -6$
$2x + y + 4z = 19$
$-5x - 2y + 8z = 62$

4. $-4r - s + 3t = -9$
$3r + 2s - t = 3$
$r + 3s - 5t = 29$

5. $3x + 5y - z = 12$
$-2x - 3y + 5z = 14$
$4x + 7y + 3z = 38$

6. $2a - 3b + 5c = 58$
$-5a + b - 4c = -51$
$-6a - 8b + c = 22$

Example 3 p. 170

7. DOWNLOADING Heather downloaded some television shows. A sitcom uses 0.3 gigabyte of memory; a drama, 0.6 gigabyte; and a talk show, 0.6 gigabyte. She downloaded 7 programs totaling 3.6 gigabytes. There were twice as many episodes of the drama as the sitcom.

a. Write a system of equations for the number of episodes of each type of show.

b. How many episodes of each show did she download?

Practice and Problem Solving

● = **Step-by-Step Solutions** begin on page R20.
Extra Practice begins on page 947.

Examples 1 and 2 pp. 168, 169

Solve each system of equations.

8. $-5x + y - 4z = 60$
$2x + 4y + 3z = -12$
$6x - 3y - 2z = -52$

9. $4a + 5b - 6c = 2$
$-3a - 2b + 7c = -15$
$-a + 4b + 2c = -13$

10. $-2x + 5y + 3z = -25$
$-4x - 3y - 8z = -39$
$6x + 8y - 5z = 14$

11. $4r + 6s - t = -18$
$3r + 2s - 4t = -24$
$-5r + 3s + 2t = 15$

12. $-2x + 15y + z = 44$
$4x + 3y + 3z = 18$
$-3x + 6y - z = 8$

13. $4x + 2y + 6z = 13$
$-12x + 3y - 5z = 8$
$-4x + 7y + 7z = 34$

14. $8x + 3y + 6z = 43$
$-3x + 5y + 2z = 32$
$5x - 2y + 5z = 24$

15. $-6x - 5y + 4z = 53$
$5x + 3y + 2z = -11$
$8x - 6y + 5z = 4$

16. $-9a + 3b - 2c = 61$
$8a + 7b + 5c = -138$
$5a - 5b + 8c = -45$

17. $2x - y + z = 1$
$x + 2y - 4z = 3$
$4x + 3y - 7z = -8$

18. $x + 2y = 12$
$3y - 4z = 25$
$x + 6y + z = 20$

19. $r - 3s + t = 4$
$3r - 6s + 9t = 5$
$4r - 9s + 10t = 9$

Example 3 p. 170

20. SWIMMING A friend e-mails you the results of a recent high school swim meet. The e-mail states that 24 individuals placed, earning a combined total of 53 points. First place earned 3 points, second place earned 2 points, and third place earned 1 point. There were as many first-place finishers as second- and third-place finishers combined.

a. Write a system of three equations that represents how many people finished in each place.

b. How many swimmers finished in first place, in second place, and in third place?

c. Suppose the e-mail had said that the athletes scored a combined total of 47 points. Explain why this statement is false and the solution is unreasonable.

21. AMUSEMENT PARKS Nick goes to the amusement park to ride roller coasters, bumper cars, and water slides. The wait for the roller coasters is 1 hour, the wait for the bumper cars is 20 minutes long, and the wait for the water slides is only 15 minutes long. Nick rode 10 total rides during his visit. Because he enjoys roller coasters the most, the number of times he rode the roller coasters was the sum of the times he rode the other two rides. If Nick waited in line for a total of 6 hours and 20 minutes, how many of each ride did he go on?

Real-World Link

Most mechanics would recommend you get an oil change every 6 months and a radiator flush every other year.

Source: *Popular Mechanics*

22. BUSINESS Ramón usually gets one of the routine maintenance options at Annie's Garage. Today however, he needs a different combination of work than what is listed.

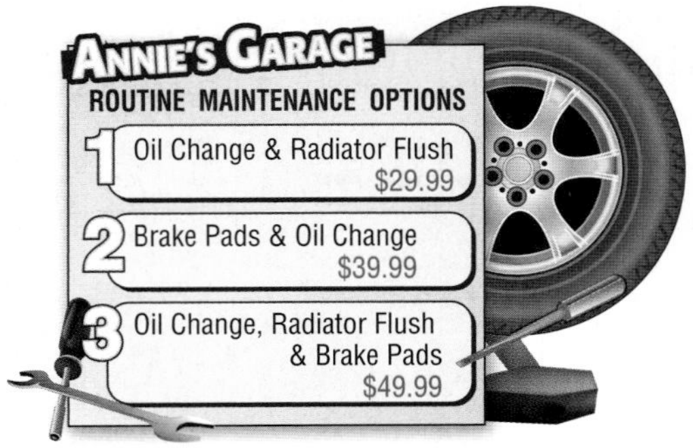

a. Assume that the price of an option is the same price as purchasing each item separately. Find the prices for an oil change, a radiator flush, and a brake pad replacement.

b. If Ramón wants his brake pads replaced and his radiator flushed, how much should he plan to spend?

23. INVESTING Kate invested \$100,000 in three different accounts. If she invested \$30,000 more in account A than account C and is expected to earn \$6300 in interest, how much did she invest in each account?

Account	Expected Interest
A	4%
B	8%
C	10%

H.O.T. Problems
Use Higher-Order Thinking Skills

24. REASONING Write a system of equations to represent the three rows of figures below. Use the system to find the number of red triangles that will balance one green circle.

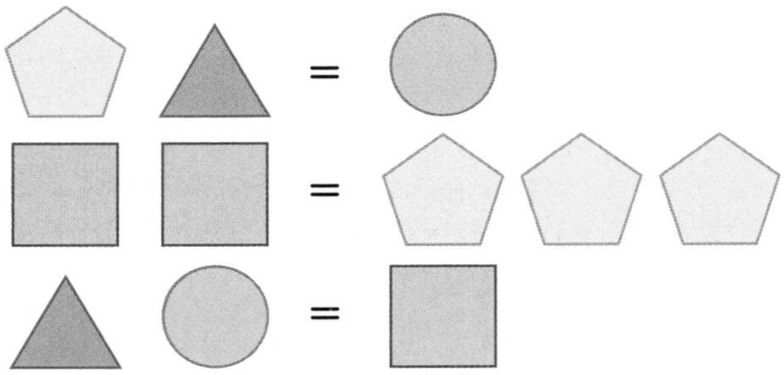

25. CHALLENGE The general form of an equation for a parabola is $y = ax^2 + bx + c$, where (x, y) is a point on the parabola. If three points on a parabola are $(2, -10)$, $(-5, -101)$, and $(6, -90)$, determine the values of a, b, and c and write the general form of the equation.

26. PROOF Consider the following system and prove that if $b = c = -a$, then $ty = a$.

$$rx + ty + vz = a$$
$$rx - ty + vz = b$$
$$rx + ty - vz = c$$

27. OPEN ENDED Write a system of three linear equations that has a solution of $(-5, -2, 6)$. Show that the ordered triple satisfies all three equations.

28. OPEN ENDED Write a system of three equations in three variables for each.

a. no solution

b. infinitely many solutions

29. WRITING IN MATH Use your knowledge of solving a system of three linear equations with three variables to explain how to solve a system of four equations with four variables.

NYSRE PRACTICE Reinforcement of A.A.7, A.G.7

30. What is the solution of the system of equations shown below?

$$\begin{cases} x - y + z = 0 \\ -5x + 3y - 2z = -1 \\ 2x - y + 4z = 11 \end{cases}$$

A (0, 3, 3)
B (2, 5, 3)
C no solution
D infinitely many solutions

31. Which of the following represents a correct procedure for solving each equation?

F
$$\begin{aligned} -3(x - 7) &= -16 \\ -3x - 21 &= -16 \\ -3x &= 5 \\ x &= -\frac{5}{3} \end{aligned}$$

G
$$\begin{aligned} 7 - 4x &= 3x + 27 \\ 7 - 7x &= 27 \\ -7x &= -\frac{20}{7} \\ x &= 20 \end{aligned}$$

H
$$\begin{aligned} 2(x - 4) &= 20 \\ 2x - 8 &= 20 \\ 2x &= 12 \\ x &= 6 \end{aligned}$$

J
$$\begin{aligned} 6(2x + 1) &= 30 \\ 12x + 6 &= 30 \\ 12x &= 24 \\ x &= 2 \end{aligned}$$

32. EXTENDED RESPONSE Use the graph to find the solution of the systems of equations. Describe one way to check the solution.

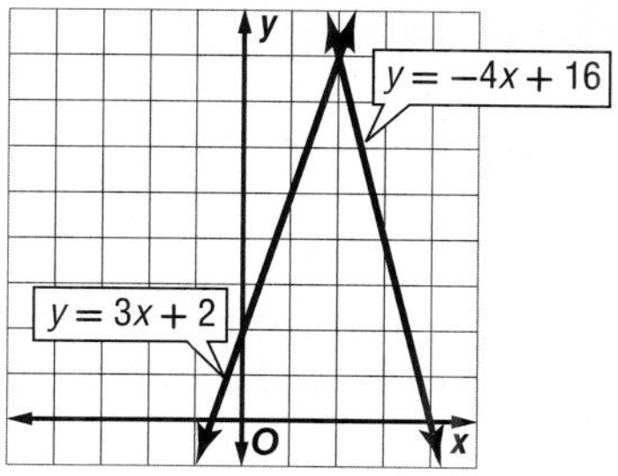

33. ACT/SAT The graph shows which system of equations?

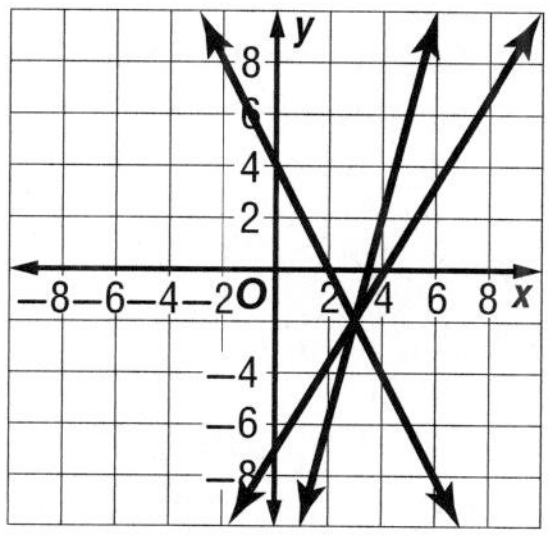

A $y + 14 = 4x$
$y = 4 - 2x$
$-7 = y - \frac{5}{3}x$

B $y - 14 = 4x$
$y = 4 + 2x$
$-7 = y + \frac{5}{3}x$

C $y + 14x = 4$
$-2y = 4 + y$
$-7 = y - \frac{5}{3}x$

D $y - 14x = 4$
$2x = 4 + y$
$7 = y - \frac{5}{3}x$

Spiral Review

A feasible region has vertices at (−3, 2), (1, 3), (6, 1), and (2, −2). Find the maximum and minimum values of each function. (Lesson 3-4)

34. $f(x, y) = 2x - y$

35. $f(x, y) = x + 5y$

36. $f(x, y) = y - 4x$

37. $f(x, y) = -x + 3y$

38. SKI CLUB The ski club's budget for the year is \$4250. They are able to find skis for \$75 per pair and boots for \$40 per pair. They know they should buy more boots than skis because the skis are adjustable to several sizes of boots. (Lesson 3-3)

a. Give an example of three different purchases that the ski club can make.

b. Suppose the ski club wants to spend all of its budget. What combination of skis and boots should they buy? Explain.

Skills Review

Solve each system of equations. (Lesson 3-2)

39. $x = y + 5$
$3x + y = 19$

40. $3x - 2y = 1$
$4x + 2y = 20$

41. $5x + 3y = 25$
$4x + 7y = -3$

42. $y = x - 7$
$2x - 8y = 2$

CHAPTER 3

Study Guide and Review

NY Math Online glencoe.com
- STUDY TO GO
- Vocabulary Review

Chapter Summary

Key Concepts

Systems of Equations (Lessons 3-1 and 3-2)

- The solution of a system of equations can be found by graphing the two equations and determining at what point they intersect.
- In the substitution method, one equation is solved for a variable and substituted to find the value of another variable.
- In the elimination method, one variable is eliminated by adding or subtracting the equations.

Systems of Inequalities (Lesson 3-3)

- The solution of a system of inequalities is found by graphing the inequalities and determining the intersection of the graphs.

Linear Programming (Lesson 3-4)

- Finding maximum or minimum values of a function over a given system of inequalities with each inequality representing a constraint.
- To optimize means to seek the optimal price or amount that is desired to minimize costs or maximize profits.

Systems of Equations in Three Variables (Lesson 3-5)

- A system of equations in three variables can be solved algebraically by using the substitution method or the elimination method.

FOLDABLES® Study Organizer

Be sure the Key Concepts are noted in your Foldable.

Key Vocabulary

bounded (p. 160)
break-even point (p. 136)
consistent (p. 137)
constraints (p. 160)
dependent (p. 137)
elimination method (p. 144)
feasible region (p. 160)
inconsistent (p. 137)
independent (p. 137)
linear programming (p. 160)
optimize (p. 162)
ordered triple (p. 167)
substitution method (p. 143)
system of equations (p. 135)
system of inequalities (p. 151)
unbounded (p. 160)

Vocabulary Check

Choose the term from the list above to complete each sentence.

1. ____________ is a method for finding maximum or minimum values of a function over a given system of inequalities with each inequality representing a constraint.
2. The ____________ is an algebraic method for solving systems of linear equations that eliminates a variable by adding or subtracting the equations.
3. A system of equations is ____________ if it has at least one solution.
4. To ____________ means to seek the best price or profit using linear programming.
5. A feasible region that is open and can go on forever is called ____________.
6. The ____________ is the point at which the income equals the cost.
7. A system of equations is ____________ if it has an infinite number of solutions.
8. A(n) ____________ is two or more equations with the same variables.
9. A system of equations is ____________ if it has no solutions.
10. To solve a system of equations by the ____________, solve one equation for one variable in terms of the other. Then substitute in the other equation.

MIXED PROBLEM SOLVING
For mixed problem-solving practice, see page 981.

Lesson-by-Lesson Review

3-1 Solving Systems of Equations by Graphing (pp. 135–141)

A.A.7, A.G.7

Solve each system of equations by graphing.

11. $3x + 4y = 8$
$x - 3y = -6$

12. $x + \frac{8}{3}y = 12$
$\frac{1}{2}x + \frac{4}{3}y = 6$

13. $y - 3x = 13$
$y = \frac{1}{3}x + 5$

14. $6x - 14y = 5$
$3x - 7y = 5$

15. **LAWN CARE** André and Paul each mow lawns. André charges a \$30 service fee and \$10 per hour. Paul charges a \$10 service fee and \$15 per hour. After how many hours will André and Paul charge the same amount?

EXAMPLE 1

Solve the system of equations by graphing.
$x + y = 4 \qquad x + 2y = 5$

Graph both equations on the coordinate plane.

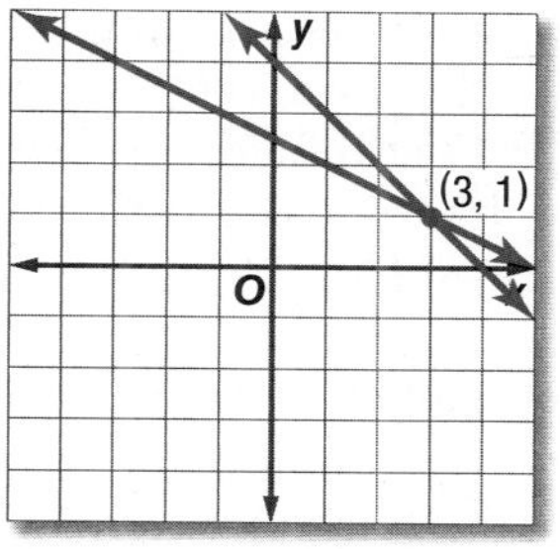

The solution of the system is (3, 1).

3-2 Solving Systems of Equations Algebraically (pp. 143–150)

A.A.7, A.A.10

Solve each system of equations by using either substitution or elimination.

16. $x + y = 6$
$3x - 2y = -2$

17. $5x - 2y = 4$
$-2y + x = 12$

18. $x + y = 3.5$
$x - y = 7$

19. $3y - 5x = 0$
$2y - 4x = -2$

20. **SCHOOL SUPPLIES** At an office supply store, Emilio bought 3 notebooks and 5 pens for \$13.75. If a notebook costs \$1.25 more than a pen, how much does a notebook cost? How much does a pen cost?

EXAMPLE 2

Solve the system of equations by using either substitution or elimination.
$3x + 2y = 1 \qquad y = -x + 1$

Substitute $-x + 1$ for y in the first equation. Then solve for y.

$$3x + 2y = 1$$
$$3x + 2(-x + 1) = 1$$
$$3x - 2x + 2 = 1$$
$$x + 2 = 1$$
$$x = -1$$

$$y = -x + 1$$
$$= -(-1) + 1 \text{ or } 2$$

The solution is (−1, 2).

3-3 Solving Systems of Inequalities by Graphing (pp. 151–157)

A.A.40, A.G.7

Solve each system of inequalities by graphing.

21. $y < 2x - 3$
$y \geq 4$

22. $|y| > 2$
$x > 3$

23. $y \geq x + 3$
$2y \leq x - 5$

24. $y > x + 1$
$x < -2$

25. **JEWELRY** Payton makes jewelry to sell at her mother's clothing store. She spends no more than 3 hours making jewelry on Saturdays. It takes her 15 minutes to set up her supplies and 25 minutes to make each bracelet. Draw a graph that represents this.

EXAMPLE 3

Solve the system of inequalities by graphing.
$y \geq x - 3$
$y < 4 - 2x$

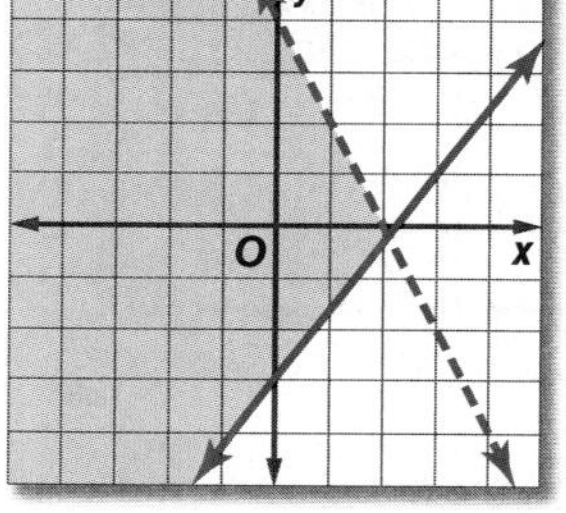

The solution of the system is the region that satisfies both inequalities. The solution of this system is the shaded region.

Study Guide and Review

3-4 Linear Programming (pp. 160–166)

A.A.40

26. FLOWERS A florist can make a grand arrangement in 18 minutes or a simple arrangement in 10 minutes. The florist makes at least twice as many of the simple arrangements as the grand arrangements. The florist can work only 40 hours per week. The profit on the simple arrangements is \$10 and the profit on the grand arrangements is \$25. Find the number and type of arrangements that the florist should produce to maximize profit.

27. MANUFACTURING A shoe manufacturer makes outdoor and indoor soccer shoes. There is a two-step process for both kinds of shoes. Each pair of outdoor shoes requires 2 hours in step one and 1 hour in step two, and produces a profit of \$20. Each pair of indoor shoes requires 1 hour in step one and 3 hours in step 2, and produces a profit of \$15. The company has 40 hours of labor available per day for step one and 60 hours available for step two. What is the manufacturer's maximum profit? What is the combination of shoes for this profit?

EXAMPLE 4

A gardener is planting two types of herbs in a 5184-square-inch garden. Herb A requires 6 square inches of space, and herb B requires 24 square inches of space. The gardener will plant no more than 300 plants. If herb A can be sold for \$8 and herb B can be sold for \$12, how many of each herb should be sold to maximize income?

Let a = the number of herb A and b = the number of herb B.

$a \geq 0$, $b \geq 0$,
$6a + 24b \leq 5184$,
and $a + b \leq 300$

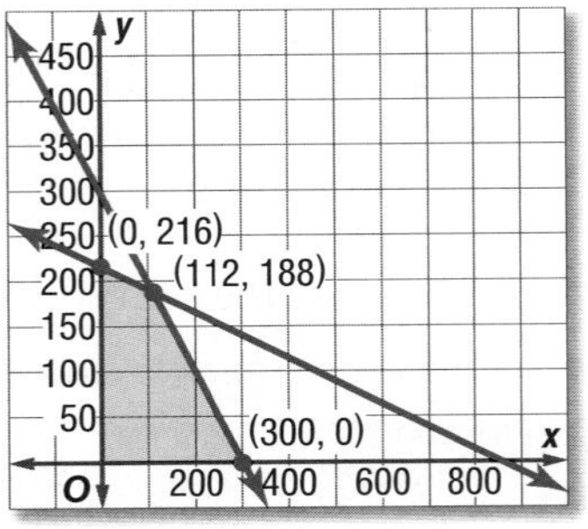

Graph the inequalities. The vertices of the feasible region are (0, 0), (300, 0), (0, 216), and (112, 188).

The profit function is $f(a, b) = 8a + 12b$.

The maximum value of \$3152 occurs at (112, 188). So the gardener should plant 112 of herb A and 188 of herb B.

3-5 Solving Systems of Equations in Three Variables (pp. 167–173)

A.A.7, A.G.7

Solve each system of equations.

28. $a - 4b + c = 3$
$b - 3c = 10$
$3b - 8c = 24$

29. $2x - z = 14$
$3x - y + 5z = 0$
$4x + 2y + 3z = -2$

30. AMUSEMENT PARKS Dustin, Luis, and Marci went to an amusement park. They purchased snacks from the same vendor. Their snacks and how much they paid are listed in the table. How much did each snack cost?

Name	Hot Dogs	Popcorn	Soda	Price
Dustin	1	2	3	\$15.25
Luis	2	0	3	\$14.00
Marci	1	2	1	\$10.25

EXAMPLE 5

Solve the system of equations. $x + y + 2z = 6$
$2x + 5z = 12$
$x + 2y + 3z = 9$

$$\begin{aligned} 2x + 2y + 4z &= 12 \quad \text{Equation 1 } \times 2 \\ (-)\ x + 2y + 3z &= 9 \quad \text{Equation 3} \\ \hline x + z &= 3 \quad \text{Subtract.} \end{aligned}$$

Solve the system of two equations.

$$\begin{aligned} 2x + 5z &= 12 \quad \text{Equation 1} \\ (-)\ 2x + 2z &= 6 \quad 2 \times (x + z = 3) \\ \hline 3z &= 6 \quad \text{Subtract.} \\ z &= 2 \quad \text{Divide each side by 2.} \end{aligned}$$

Substitute 2 for z in one of the equations with two variables, and solve for y. Then, substitute 2 for z and the value you got for y into an equation from the original system to solve for x.

The solution is (1, 1, 2).

Practice Test

NY Math Online glencoe.com
Chapter Test

Solve each system of equations.

1. $2x - 3y = 9$
   $4x + 3y = 9$

2. $x + 2y = 7$
   $y = 5x - 2$

3. $-x + y = 2$
   $4x - 3y = -3$

4. $\frac{1}{2}x + \frac{1}{3}y = 7$
   $\frac{1}{5}x - \frac{2}{3}y = -2$

Solve each system of inequalities by graphing.

5. $x + y \leq 4$
   $y \geq x$

6. $2x + 3y > 12$
   $3x - y < 21$

7. $x - y > 0$
   $4 + y \leq 2x$

8. $2y - 5x \leq 6$
   $4x + y < -4$

9. **MULTIPLE CHOICE** Which statement best describes the graphs of the two equations?

$$x + 4y = 8$$
$$3x + 12y = 2$$

A The lines are parallel.

B The lines are the same.

C The lines intersect in only one point.

D The lines intersect in more than one point, but are not the same.

Solve each system of equations.

10. $x - 2y + 3z = 1$
    $4y - 4z = 12$
    $8y - 14z = 0$

11. $x + y + z = 4$
    $x + 3y + 3z = 10$
    $2x + y - z = 3$

12. $2x - y - 2z = 5$
    $10x + 8z = -4$
    $3x - y = 1$

13. $2x + 3y + z = 0$
    $3x + y = 1$
    $x - 2y + z = 9$

14. **MULTIPLE CHOICE** Seela rented a raft from River Rafter's Inc. She paid $100 to rent the raft and $25 for each hour. Martin rented a raft from Oscar's Outdoor Shop. He paid $50 to rent the raft and $35 per hour. For what number of hours will both rafting companies charge the same amount?

F 0

G 4

H 5

J 10

15. **CARPENTRY** Cal's Carpentry makes tables and chairs. The process involves some carpentry time and some finishing time. The carpentry times and finishing times are listed in the table below.

Product	Carpentry Time (hr)	Finishing Time (hr)
chair	3	0.5
table	2	1

Cal's Carpentry can work for a maximum of 108 carpentry hours and 20 finishing hours per day. The profit is $35 for a table and $25 for a chair. How many tables and chairs should be made each day to maximize profit?

a. Using c for the number of chairs and t for the number of tables, write a system of inequalities to represent this situation.

b. Draw the graph showing the feasible region.

c. Determine the number of tables and chairs that need to be made to maximize profit. What is the maximum profit?

16. **DRAMA** On opening night of the drama club's play, they made $1366. They sold a total of 199 tickets. They charged $8.50 for each adult ticket and $5.00 for each child's ticket. Write a system of equations that can be used to find the number of adult tickets and the number of children's tickets sold.

Graph each system of inequalities. Name the coordinates of the vertices of the feasible region. Find the maximum and the minimum values of the given function.

17. $5 \geq y \geq -3$
    $4x + y \leq 5$
    $-2x + y \leq 5$
    $f(x, y) = 4x - 3y$

18. $x \geq -10$
    $1 \geq y \geq -6$
    $3x + 4y \leq -8$
    $2y \geq x - 10$
    $f(x, y) = 2x + y$

19. **GEOMETRY** An isosceles trapezoid has shorter base of measure a, longer base of measure c, and congruent legs of measure b. The perimeter of the trapezoid is 58 inches. The average of the bases is 19 inches and the longer base is twice the leg plus 7.

a. Find the lengths of the sides of the trapezoid.

b. Find the area of the trapezoid.

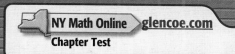

CHAPTER 3 Preparing for Standardized Tests

Short Answer Questions

Short answer questions require you to provide a solution to the problem, along with a method, explanation, and/or justification used to arrive at the solution.

Strategies for Solving Short Answer Questions

Step 1

Short answer questions are typically graded using a **rubric**, or a scoring guide. The following is an example of a short answer question scoring rubric.

Scoring Rubric	
Criteria	**Score**
Full Credit: The answer is correct and a full explanation is provided that shows each step.	2
Partial Credit: • The answer is correct, but the explanation is incomplete. • The answer is incorrect, but the explanation is correct.	1
No Credit: Either an answer is not provided or the answer does not make sense.	0

Step 2

In solving short answer questions, remember to…

- explain your reasoning or state your approach to solving the problem.
- show all of your work or steps.
- check your answer if time permits.

EXAMPLE

Read the problem. Identify what you need to know. Then use the information in the problem to solve.

> Company A charges a monthly fee of $14.50 plus $0.05 per minute for cell phone service. Company B charges $20.00 per month plus $0.04 per minute. For what number of minutes would the total monthly charge be the same with each company?

Read the problem carefully. You are given information about two different cell phone companies and their monthly charges. Since the situation involves a fixed amount and a variable rate, you can set up and solve a system of equations.

Example of a 2-point response:

Set up and solve a system of equations.

flat fee + rate × minutes = total charges
y = total charges, x = minutes used

$y = 14.5 + 0.05x$ (Company A)
$y = 20 + 0.04x$ (Company B)

Solve the system by graphing.

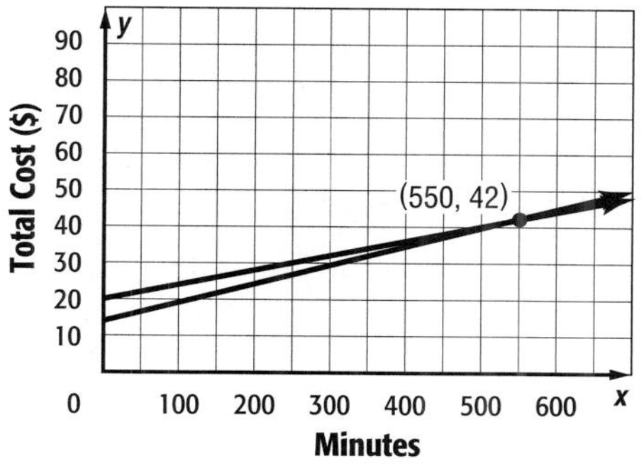

The solution is (550, 42). So, with each company, if the customer uses 550 minutes the total monthly charge is $42.

The steps, calculations, and reasoning are clearly stated. The student also arrives at the correct answer. So, this response is worth the full 2 points.

Exercises

Read each problem. Eliminate any unreasonable answers. Then use the information in the problem to solve.

1. Shawn and Jerome borrowed $1400 to start a lawn mowing business. They charge their customers $45 per lawn, and with each lawn that they mow, they incur $10.50 in operating expenses. How many lawns must they mow in order to start earning a profit?

2. A circle of radius r is circumscribed about a square. What is the exact ratio of the area of the circle to the area of the square?

3. Mr. Williams can spend no more than $50 on art supplies. Packages of paint brushes cost $4.75 each, and boxes of colored pencils cost $6.50 each. He wants to buy at least 2 packages of each supply. Write a system of inequalities and plot the feasible region on a coordinate grid. Give three different solutions to the system.

4. Marla sells engraved necklaces over the Internet. She purchases 50 necklaces for $400, and it costs her an additional $3 for each personalized engraving. If she charges $20 for each necklace, how many will she need to sell in order to make a profit of at least $225?

NYS Regents Exam Practice

Cumulative, Chapters 1 through 3

Multiple Choice

Answer all questions in this part. Select the answer that best completes the statement or answers the question.

1. What is the solution of the system of equations graphed below?

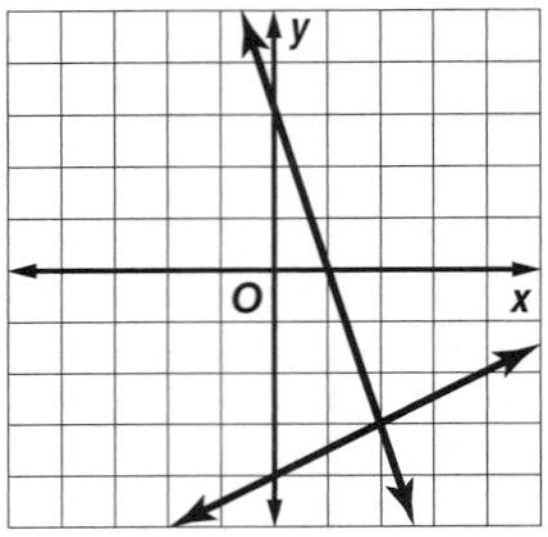

(1) (2, −3) **(3)** (−2, 3)

(2) (−3, 2) **(4)** (3, −2)

2. Atticus bought a new car for $18,750. Six years later the car was worth $8550. What was the annual rate of change in the value of the car?

(1) −$1550 per year **(3)** −$1700 per year

(2) −$1625 per year **(4)** −$1775 per year

3. Suppose Raymond sells apples and tomatoes at a farmer's market. If he sold 211 items one morning and earned $69.85, how many apples did he sell?

Item	Cost
apple	$0.40
tomato	$0.25

(1) 97 **(3)** 123

(2) 114 **(4)** 130

Test-Taking Tip

Question 3 Write one equation using the number of items sold and one using the amount earned. Then solve the system.

4. Which region represents the solution of the system of inequalities below?

$$y \geq \frac{1}{2}x - 2$$

$$y \geq -\frac{2}{3}x - 1$$

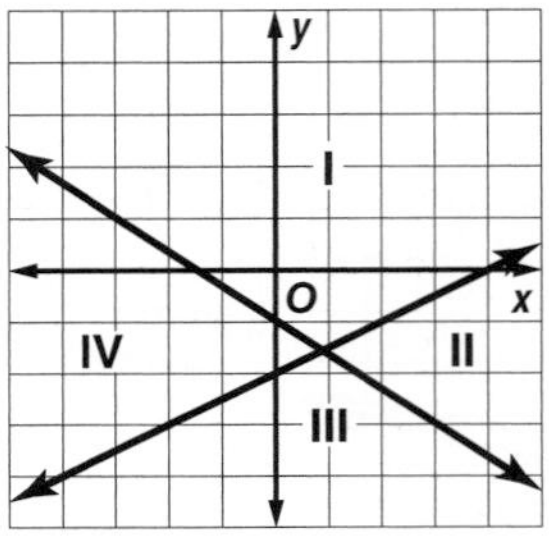

(1) Region I

(2) Region II

(3) Region III

(4) Region IV

5. A manufacturing plant produces engine parts that must be 4 mm thick with a tolerance of plus or minus 0.015 mm. Which absolute value equation can be used to find the minimum and maximum thickness t of the engine parts?

(1) $|t - 4| = 0.015$

(2) $|t + 4| = 0.015$

(3) $|t - 0.015| = 4$

(4) $|t + 0.015| = 4$

6. The point (10, 4) lies on the graph of $f(x) = \sqrt{\frac{x}{5} + 7a}$. Which ordered pair below lies on the graph of $g(x) = \frac{1}{2}\sqrt{\frac{x}{2} + 5a}$?

(1) (5, 2)

(2) (12, 2)

(3) (5, 4)

(4) (10, 4)

7. The New York state legislature is comprised of senators and assembly members. The total number of elected officials in the two bodies of government is 212. There are 26 more assembly members than twice the number of senators. Solve the system of equations below to find the number of state senators s and assembly members m.

$$\begin{cases} s + m = 212 \\ m = 2x + 26 \end{cases}$$

(1) 48 senators, 164 assembly members

(2) 55 senators, 157 assembly members

(3) 62 senators, 150 assembly members

(4) 64 senators, 148 assembly members

8. A bakery sells cookies, doughnuts, and bagels. Cindy bought 3 cookies, 2 doughnuts, and 1 bagel for $4.84. Andrew bought 10 cookies, 12 doughnuts, and 6 bagels for $25.12. Linda bought 12 doughnuts and 10 bagels for $27.38. How much does a doughnut cost?

(1) $0.49 (3) $0.69

(2) $0.59 (4) $0.79

9. A survey at a local shopping center shows that 31.5% of customers use coupons when they shop. Results of surveys of this size can be off by as much as 3.5 percentage points. Which inequality describes the results of the survey compared to the actual percentage of shoppers who use coupons x?

(1) $|x - 0.315| > 0.035$

(2) $|x - 0.315| \leq 0.035$

(3) $x - 0.315 > 0.035$

(4) $x - 0.315 \leq 0.035$

Open-Ended Response

Solve each problem. Clearly indicate the necessary steps, including appropriate formula substitutions, diagrams, graphs, charts, etc.

10. Angelo has some nickels, dimes, and quarters in a jar. There are 43 coins in all, and the total value of the coins is $5.95. The number of nickels and quarters is 11 more than the number of dimes.

 a. Let n represent the number of nickels, d the number of dimes, and q the number of quarters that Angelo has. Write a system of equations to model the situation.

 b. Solve the system of equations. How many of each type of coin does Angelo have?

11. What are the minimum and maximum values for the function $f(x, y) = 4x + 2y$ over the feasible region shown below?

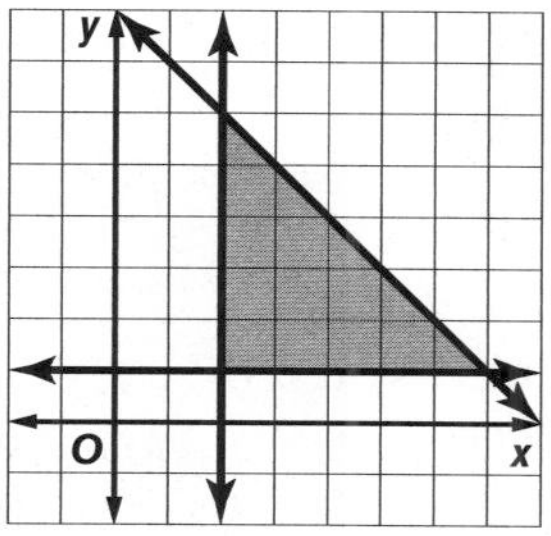

12. For the absolute value equation below, write an expression to describe x when $y < 7$.

$$y = 2|3 - 4x| + 1$$

Need Extra Help?

If you missed Question...	1	2	3	4	5	6	7	8	9	10	11	12
Go to Lesson or Page...	3-2	3-1	3-5	1-3	3-3	3-4	2-4	3-1	2-4	3-2	2-3	3-3
NYS Core Curriculum	A.G.7	A.A.32	A.A.7	A.A.40	A2.A.1	A.G.5	A.A.10	A.A.7	A2.A.1	A.A.7	A.A.40	A.A.40

CHAPTER 4 Matrices

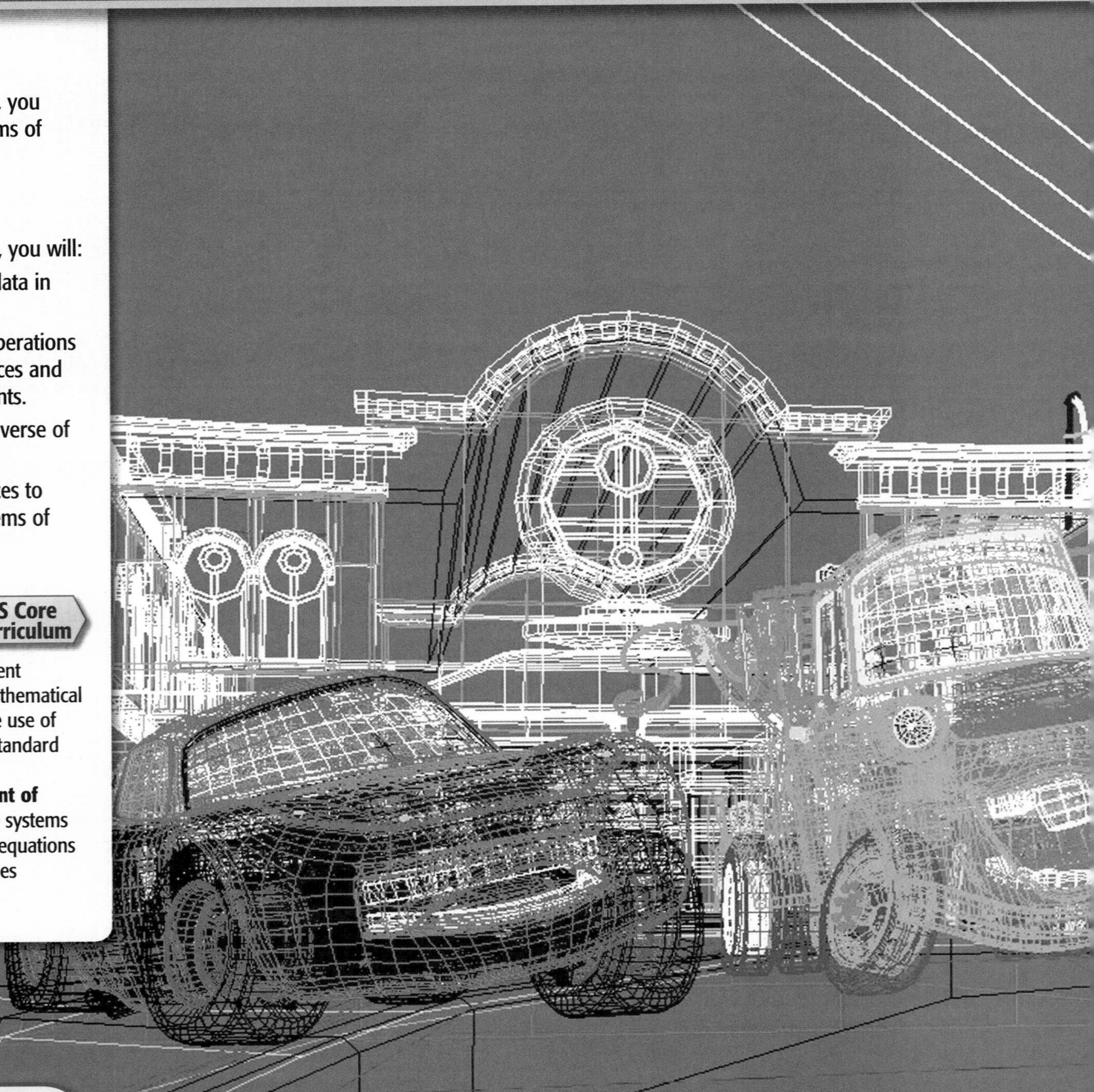

Then

In Chapter 3, you solved systems of equations.

Now

In Chapter 4, you will:

- Organize data in matrices.
- Perform operations with matrices and determinants.
- Find the inverse of a matrix.
- Use matrices to solve systems of equations.

NYS Core Curriculum

A.CM.3 Present organized mathematical ideas with the use of appropriate standard notations

Reinforcement of A.A.10 Solve systems of two linear equations in two variables algebraically

Why?

GRAPHICS Computer graphics and animation use complex models for characters, objects, and scenery. These computer models describe the shapes of objects and the motions of characters. The animation models use matrices to describe the locations of specific points in the images.

Math *in Motion,* Animation glencoe.com

Get Ready for Chapter 4

Diagnose Readiness You have two options for checking Prerequisite Skills.

Text Option

Take the Quick Check below. Refer to the Quick Review for help.

*Quick*Check

Name the additive inverse and the multiplicative inverse for each number. (Lesson 1-2)

1. 4
2. -15
3. 0.2
4. -1.35
5. $-\frac{3}{4}$
6. $2\frac{1}{3}$

Simplify each expression. (Lesson 1-2)

7. $6(x + 2y)$
8. $4(x + 5) - 3$
9. $-4(3x) - (7x - 6)$
10. $5(2x - 5) - \frac{1}{3}(4x + 1)$
11. $6(2x - 1) - 3(y - x) + 0.5(4x - 6)$

Solve each system of equations by using either substitution or elimination. (Lesson 3-2)

12. $y = x + 3$
$2x - y = -1$

13. $2x - 5y = -18$
$3x + 4y = 19$

14. $4y + 6x = -6$
$5y - x = 35$

15. $x = y - 8$
$4x + 2y = 4$

16. MONEY The student council paid \$15 per registration for a conference. They also paid \$10 for T-shirts for a total of \$180. Last year, they spent \$12 per registration and \$9 per T-shirt for a total of \$150 to buy the same number of registrations and T-shirts. Write and solve a system of two equations that represents the number of registrations and T-shirts bought each year.

*Quick*Review

EXAMPLE 1

Name the additive inverse and the multiplicative inverse for −5.

The additive inverse of -5 is a number x such that $-5 + x = 0$. So, $x = 5$.

The multiplicative inverse of -5 is a number x, such that $-5x = 1$. So, $x = -\frac{1}{5}$.

EXAMPLE 2

Simplify $\frac{3}{4}(8x - 4) + 3x$.

$\frac{3}{4}(8x - 4) + 3x$

$= \frac{3}{4}(8x) - \frac{3}{4}(4) + 3x$	**Distributive Property**
$= 6x - 3 + 3x$	**Simplify.**
$= 9x - 3$	**Add.**

EXAMPLE 3

Solve the system of equations algebraically.
$3y = x - 9$
$4x + 5y = 2$

Since x has a coefficient of 1 in the first equation, use the substitution method. First, solve that equation for x.

$3y = x - 9 \quad \rightarrow \quad x = 3y + 9$

$4(3y + 9) + 5y = 2$	**Substitute $3y + 9$ for x.**
$12y + 36 + 5y = 2$	**Distributive Property**
$17y = -34$	**Combine like terms.**
$y = -2$	**Divide each side by 17.**

To find x, use $y = -2$ in the first equation.

$3(-2) = x - 9$	**Substitute −2 for y.**
$-6 = x - 9$	**Multiply.**
$3 = x$	**Add 9 to each side.**

The solution is $(3, -2)$.

Online Option

NY Math Online Take a self-check Chapter Readiness Quiz at glencoe.com.

Get Started on Chapter 4

You will learn several new concepts, skills, and vocabulary terms as you study Chapter 4. To get ready, identify important terms and organize your resources. You may wish to refer to **Chapter 0** to review prerequisite skills.

FOLDABLES® Study Organizer

Matrices Make this Foldable to help you organize your Chapter 4 notes about matrices. Begin with one sheet of 11″ by 17″ paper.

1. **Fold** 2″ tabs on each of the short sides.

2. **Fold** in half in both directions. Open and cut as shown.

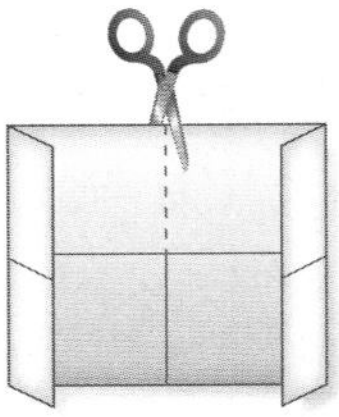

3. **Refold** along the width. Staple each pocket.

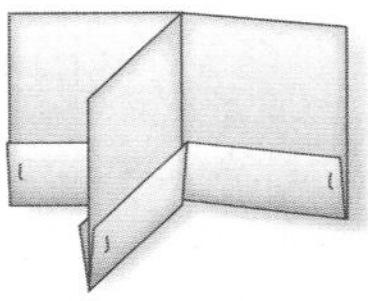

4. **Label** pockets as *Operations, Transformations, Determinants/ Cramer's Rule*, and *Inverses/ Systems*. Place index cards for notes in each pocket.

New Vocabulary

English		Español
element	• p. 185 •	elemento
dimensions	• p. 185 •	tamaño
row matrix	• p. 186 •	matriz fila
column matrix	• p. 186 •	matriz columna
square matrix	• p. 186 •	matriz cuadrada
zero matrix	• p. 186 •	matriz nula
equal matrices	• p. 186 •	matrices iguales
scalar	• p. 194 •	escalar
vertex matrix	• p. 209 •	matriz de vértice
preimage	• p. 209 •	preimagen
image	• p. 209 •	imagen
translation	• p. 209 •	translación
dilation	• p. 211 •	homotecia
reflection	• p. 212 •	reflexión
rotation	• p. 212 •	rotación
determinant	• p. 220 •	determinante
Cramer's Rule	• p. 223 •	regula de Crámer
coefficient matrix	• p. 223 •	matriz coefficiente
identity matrix	• p. 229 •	matriz identidad
inverse	• p. 229 •	inversa
matrix equation	• p. 231 •	ecuación matriz
variable matrix	• p. 231 •	matriz variables
constant matrix	• p. 231 •	matriz constante

Review Vocabulary

coordinate plane • Algebra 1 • plano de coordenadas the plane connecting the *x*- and *y*-axes

system of equations • p. 135 • sistema de ecuaciones a set of equations with the same variables

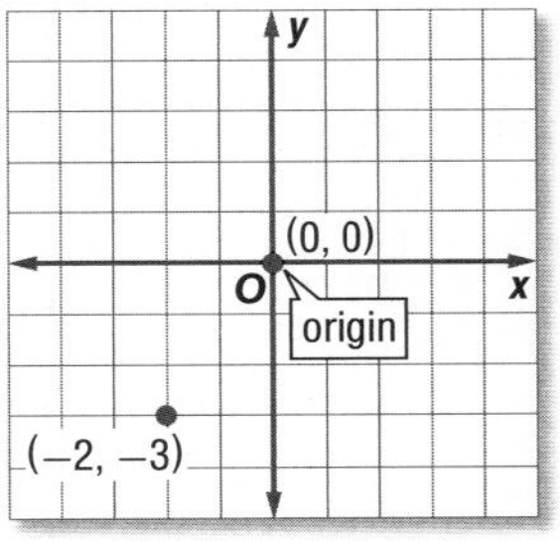

Multilingual eGlossary glencoe.com

NY Math Online glencoe.com

- Study the chapter online
- Explore **Math in Motion**
- Get extra help from your own **Personal Tutor**
- Use **Extra Examples** for additional help
- Take a **Self-Check Quiz**
- **Review Vocabulary** in fun ways

4-1 Introduction to Matrices

Why?

SHOPPING Julie is shopping for a new smartphone and discovers that many different options are available. She wants to be able to easily compare the options, so she decides to organize the data in a matrix.

	Price	Memory	Color	Interface
Choice 1	\$420	512	24	infrared
Choice 2	\$399	512	24	Bluetooth
Choice 3	\$315	256	24	infrared
Choice 4	\$289	128	18	wi-fi

Then
You solved problems by organizing data in tables.

Now
- Organize data in matrices.
- Use matrix row and column operations to analyze data.

NYS Core Curriculum
A.CM.3 Present organized mathematical ideas with the use of appropriate standard notations, including the use of symbols and other representations when sharing an idea in verbal and written form

New Vocabulary
matrix
element
dimensions
row matrix
column matrix
square matrix
zero matrix
equal matrices

NY Math Online
glencoe.com
- Extra Examples
- Personal Tutor
- Self-Check Quiz
- Homework Help

Organize and Analyze Data A **matrix** is a rectangular array of variables or constants in horizontal rows and vertical columns, usually enclosed in brackets. In a matrix, the numbers or data are organized so that each position in the matrix has a purpose. Each value in the matrix is called an **element**. A matrix is usually named using an uppercase letter.

$$A = \begin{bmatrix} 8 & -2 & 5 & 6 \\ -1 & 3 & -3 & 6 \\ 7 & -8 & 1 & 4 \end{bmatrix} \quad \text{4 rows, 4 columns}$$

The element −1 is in Row 2, Column 1, depicted by a_{21}.

The element −8 is in Row 3, Column 2, depicted by a_{32}.

A matrix can be described by its **dimensions**. A matrix with m rows and n columns is an $m \times n$ matrix (read "m by n"). Matrix A above is a 3×4 matrix because it has 3 rows and 4 columns. **a_{12} refers to an element of A, whereas b_{12} refers to an element of B.**

EXAMPLE 1 Dimensions and Elements of a Matrix

Use $A = \begin{bmatrix} -18 & 6 & 38 \\ 9 & -9 & 22 \end{bmatrix}$ to answer the following.

a. State the dimensions of A.

$$\begin{bmatrix} -18 & 6 & 38 \\ 9 & -9 & 22 \end{bmatrix} \quad \text{2 rows, 3 columns}$$

Since A has 2 rows and 3 columns, the dimensions of A are 2×3.

b. Find the value of a_{21}.

$$\begin{bmatrix} -18 & 6 & 38 \\ 9 & -9 & 22 \end{bmatrix}$$

Row 2 → 9 ← Column 1

Since a_{21} is the element in row 2, column 1, the value of a_{21} is 9.

✓ Check Your Progress

Use $B = \begin{bmatrix} 10 & -8 \\ -2 & 19 \\ 6 & -1 \end{bmatrix}$ to answer the following.

1A. State the dimensions of B.

1B. Find the value of b_{32}.

Personal Tutor glencoe.com

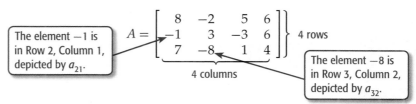

Certain matrices have special names.

row matrix one row: $[8 \quad -5 \quad 2 \quad 4]$

column matrix one column: $\begin{bmatrix} 8 \\ -1 \end{bmatrix}$

square matrix same number of rows and columns: $\begin{bmatrix} -4 & 2 \\ -3 & 9 \end{bmatrix}$

zero matrix Every element is zero.: $\begin{bmatrix} 0 & 0 & 0 \\ 0 & 0 & 0 \end{bmatrix}$

StudyTip

Corresponding Elements *Corresponding* refers to elements that are in the exact same position in each matrix.

Two matrices are considered **equal matrices** if they have the same dimensions and if each element of one matrix is equal to the corresponding element in the other matrix.

$$\begin{bmatrix} -3 & 4 \\ 6 & 3 \\ 5 & 2 \end{bmatrix} \neq \begin{bmatrix} 4 & 3 & 2 \\ -3 & 6 & 5 \end{bmatrix}$$

Matrices with different dimensions *cannot* be equal.

$$\begin{bmatrix} 1 & 4 \\ 6 & -5 \end{bmatrix} \neq \begin{bmatrix} 1 & 6 \\ 4 & -5 \end{bmatrix}$$

The *corresponding* elements are not equal, so the matrices are not equal.

$$\begin{bmatrix} 5 & 6 & 0 \\ 0 & 7 & 2 \\ 3 & 1 & 4 \end{bmatrix} = \begin{bmatrix} 5 & 6 & 0 \\ 0 & 7 & 2 \\ 3 & 1 & 4 \end{bmatrix}$$

The matrices are equal.

Matrices are used to organize and analyze data.

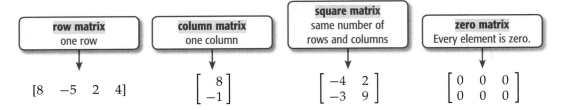

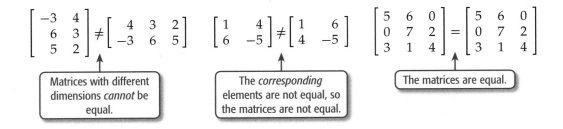

Real-World Link

In 2000, Tyler Ebell of Ventura, California, rushed for a high school record 4494 yards in one season.

Source: National Football League

Real-World EXAMPLE 2 Organize Data into a Matrix

FOOTBALL The West High School football team used five running backs throughout its season. Coach Williams wanted to compare the statistics of each player.

Joey: 11 games, 72 attempts, 439 yards, 6.10 average, 8 TDs

DeShawn: 9 games, 143 attempts, 1024 yards, 7.16 average, 12 TDs

Dario: 11 games, 164 attempts, 885 yards, 5.40 average, 15 TDs

Leo: 11 games, 84 attempts, 542 yards, 6.45 average, 7 TDs

Alex: 10 games, 151 attempts, 966 yards, 6.40 average, 11 TDs

a. Organize the data in a matrix, listing players in the first column, in order from most attempts to least attempts.

b. What are the dimensions of the matrix? What value is a_{34}?

a.

Player	Games	Attempts	Yards	Average	TDs
Dario	11	164	885	5.40	15
Alex	10	151	966	6.40	11
DeShawn	9	143	1024	7.16	12
Leo	11	84	542	6.45	7
Joey	11	72	439	6.10	8

b. There are five rows and five columns, so the dimensions are 5×5. The value a_{34}, which is in the third row and fourth column, is 7.16.

$$\begin{bmatrix} 11 & 164 & 885 & 5.40 & 15 \\ 10 & 151 & 966 & 6.40 & 11 \\ 9 & 143 & 1024 & \boxed{7.16} & 12 \\ 11 & 84 & 542 & 6.45 & 7 \\ 11 & 72 & 439 & 6.10 & 8 \end{bmatrix}$$

Check Your Progress

2. **SUBS** The figure at the right shows the prices of small, medium, and large subs.

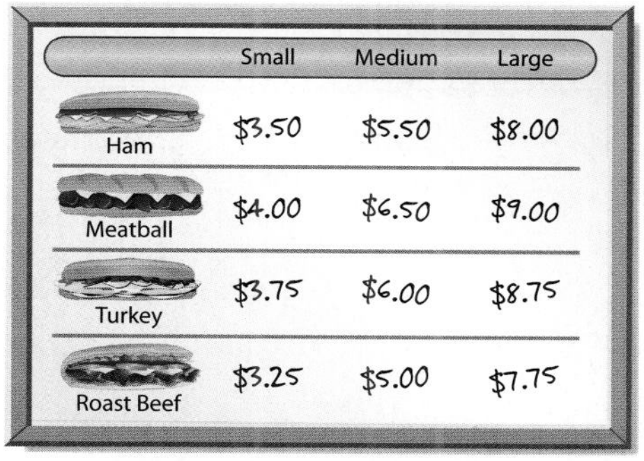

Small Medium Large
Ham $3.50 $5.50 $8.00
Meatball $4.00 $6.50 $9.00
Turkey $3.75 $6.00 $8.75
Roast Beef $3.25 $5.00 $7.75

A. Organize the data in a matrix, listing the subs from least to most expensive.

B. What are the dimensions of the matrix?

C. What is the value of a_{21}?

Personal Tutor glencoe.com

Once data are organized in a matrix, they can be analyzed and interpreted. Sometimes, the sums or averages of rows or columns provide further analysis. Other times, the sums or averages provide data that are meaningless.

StudyTip

Displaying Data Matrices displaying real-world data can often be flipped, with the rows changing places with the columns.

EXAMPLE 3 Analyze Data with Matrices

FOOTBALL Coach Williams would like to use the matrix from Example 2 to further analyze his players' statistics.

$$\begin{array}{c} \begin{array}{ccccc} G & Att & Yd & Avg. & TD \end{array} \\ \begin{bmatrix} 11 & 164 & 885 & 5.40 & 15 \\ 10 & 151 & 966 & 6.40 & 11 \\ 9 & 143 & 1024 & 7.16 & 12 \\ 11 & 84 & 542 & 6.45 & 7 \\ 11 & 72 & 439 & 6.10 & 8 \end{bmatrix} \end{array}$$

a. Add the elements in columns 2 and 3 and interpret the results.

The sum of column 2 is 614.
This is the total number of attempts for the players.

The sum of column 3 is 3856.
This is the total number of yards gained.

b. Coach Williams wants to determine the average yards per attempt for his five running backs combined. He decides to add the elements in column 4 and divide by 5, the number of players. What is this average?

The average is 6.302.

c. Is this an accurate average? Explain.

No. The players did not have the same number of attempts, so finding the average of column 4 would not determine an accurate average. Instead, Coach Williams needs to divide the sum of column 3 by the sum of column 2. The accurate average is about 6.28.

d. Would adding the rows provide any meaningful data for Coach Williams? Explain your reasoning.

No. The sum of a row includes five different forms of data.

Check Your Progress

3. **POPULATION** The table displays some of the U.S. Census data.

Latino Population in the U.S. (millions)

Age	Male	Female
0–19	7.1	6.6
20–39	6.8	5.9
40–59	3.2	2.2
60+	1.1	1.4

A. Organize the data in a matrix.

B. Add the elements in the columns and interpret the results.

C. Add the elements in the rows and interpret the results.

D. Would finding the average of the rows or columns provide any meaningful data?

Personal Tutor glencoe.com

Check Your Understanding

Example 1 p. 185

State the dimensions of each matrix.

1. $\begin{bmatrix} 1 & 4 & -4 & 0 \\ -2 & 3 & 6 & -8 \end{bmatrix}$

2. $\begin{bmatrix} 1 \\ -2 \\ 5 \\ -7 \end{bmatrix}$

3. $\begin{bmatrix} -1 & 4 \\ 2 & 9 \\ 17 & 21 \end{bmatrix}$

Identify each element of matrix $A = \begin{bmatrix} 1 & -6 & x & -4 \\ -2 & 3 & -1 & 9 \\ 5 & -8 & 2 & 12 \end{bmatrix}$.

4. a_{32}

5. a_{11}

6. a_{33}

7. a_{24}

Examples 2 and 3 pp. 186, 187

8. **GAS MILEAGE** Use the table that shows the city and highway gas mileage of five different types of vehicles.

Vehicle	SUV	Mini-van	Sedan	Compact	APV
City	23	21	21	42	61
Highway	25	24	32	49	70

Source: Auto Hoppers

a. Organize the gas mileages in a matrix.

b. Which type of vehicle has the best gas mileage?

c. Add the elements of each row and interpret the results.

d. Add the elements of each column and interpret the results.

Practice and Problem Solving

● = **Step-by-Step Solutions** begin on page R20.
Extra Practice begins on page 947.

Example 1 p. 185

State the dimensions of each matrix.

9. $\begin{bmatrix} -9 & 6 \end{bmatrix}$

10. $\begin{bmatrix} 15 & y \\ 8 & -9 \end{bmatrix}$

11. $\begin{bmatrix} 6 & 11 & -4 & -2 \\ -8 & 5 & -1 & 0 \end{bmatrix}$

12. $\begin{bmatrix} 4 & -3 & -1 \\ x & 3y & 0 \\ 8 & 12 & 11 \end{bmatrix}$

13. $\begin{bmatrix} 2 \\ x \\ -3 \end{bmatrix}$

14. $\begin{bmatrix} 115 \end{bmatrix}$

Identify each element for the following matrices.

$$A = \begin{bmatrix} 6 & y \\ -9 & 31 \\ 11 & 5 \end{bmatrix}, B = \begin{bmatrix} 10 & -8 & 2x \\ -2 & 19 & 4 \end{bmatrix}$$

15. a_{21}

16. b_{22}

17 b_{13}

18. a_{12}

Example 2 p. 186

Organize the information in a matrix.

19.

Name	Game 1	Game 2	Game 3	Series
John	221	201	185	607
Hideo	168	233	159	560
Paulo	187	189	211	587

20.

Name	Cell Phone Minutes	Text Messages	Picture Messages
Chee	95	227	138
Emelia	83	213	189
Lina	101	199	202

Example 3
p. 187

21. SHOES A consumer service company rated several pairs of shoes by cost, level of comfort, look, and longevity using a scale of 1–5, with 1 being low and 5 being high.

Brand	Cost	Comfort	Look	Longevity
A	3	2	2	1
B	4	3	2	3
C	5	5	4	4
D	1	5	5	2

a. Write a 4×4 matrix to organize this information.

b. Which shoe would you buy based on this information, and why?

c. Would finding the sum of the rows or columns provide any useful information? Explain your reasoning.

Identify each element for the following matrices.

$$A = \begin{bmatrix} 23 & 11 \\ x & -5 \\ -12 & 15 \end{bmatrix}, B = \begin{bmatrix} 9 & -3 & 7 \\ 4x & 18 & -6 \end{bmatrix}$$

22. a_{32} **23.** b_{21} **24.** b_{12} **25.** a_{21}

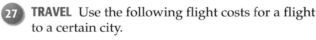

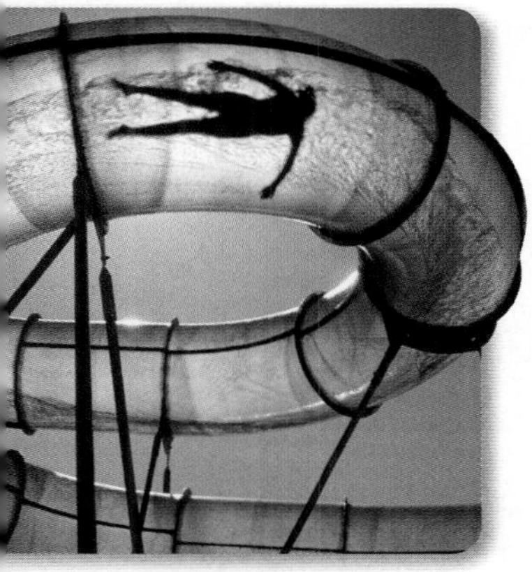

Real-World Link

There are more than 1000 waterparks in the U.S., with about 70 million visitors each year. The first waterpark to open was Wet 'N Wild in Orlando, Florida.

Source: World Waterpark Association

26. WATER PARK Use the sign at the entrance of the park shown at the right.

a. Write a matrix for the prices of admission for adults, children, and students.

b. What are the dimensions of the matrix?

Fun Time Water Park Ticket Information

Before 5 P.M.	After 5 P.M.
Adult $34	Adult $24
Child $19	Child $9
Student . . $27	Student . . $17

27. TRAVEL Use the following flight costs for a flight to a certain city.

Coach: $249 weekday; $259 weekend
Business class: $279 weekday; $289 weekend
First class: $319 weekday; $339 weekend

a. Write a 3×2 matrix that represents the cost of each flight.

b. Write a 2×3 matrix that represents the cost of each flight.

28. INVENTORY Mr. Kelley owns three golf supply stores. Store 1 has 200 white, 100 red, and 150 yellow golf balls. Store 2 has 300 white, 175 red, and 225 yellow golf balls. Store 3 has 275 white, 150 red, and 220 yellow golf balls.

a. Organize this information into a matrix with store numbers as the column heads.

b. Find the sum of the columns. What does the sum represent?

c. Find the sum of the rows. What does the sum represent?

Identify each element for the following matrices.

$$A = \begin{bmatrix} x^2 + 4 & y + 6 \\ x - y & 2 - y \end{bmatrix}, B = \begin{bmatrix} 0 & x & -2y \\ 5x & 3y & -4x \\ -y & 0 & 0 \end{bmatrix}$$

29. a_{11} **30.** a_{22} **31.** b_{31} **32.** b_{23}

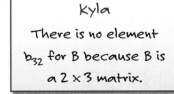

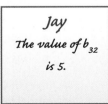

Real-World Link

In 2006, after a redefinition of the word "planet," it was decided that Pluto would no longer be considered a planet. It is now one of more than 40 dwarf planets.

Source: *National Geographic*

33. **PLANETS** Use the table that shows the distance of the other planets from Earth and the Sun.

a. Organize the distances in a matrix.

b. What are the dimensions of the matrix?

c. What is the value of a_{42}?

Planet	Distance from Sun (millions of miles)	Distance from Earth (millions of miles)
Mercury	36.00	57
Venus	67.24	26
Mars	141.71	35
Jupiter	483.88	370
Saturn	887.14	744
Uranus	1783.98	1607
Neptune	2796.46	2680

Source: FactMonster

34. **MULTIPLE REPRESENTATIONS** In this problem, you will explore reversing rows and columns of matrices.

Name	Goals	Assists
Amy	8	3
Tama	6	5
Kristen	1	8
Catalina	4	2

a. **TABULAR** Convert the data into a matrix with the names of the players along the columns.

b. **ALGEBRAIC** Find the sums of the columns.

c. **TABULAR** Switch the data in the matrix, now having the names of the players along the rows.

d. **ALGEBRAIC** Find the sums of the rows.

e. **ANALYTICAL** Make a conjecture about the effect on the data when the rows and columns of a matrix are switched.

H.O.T. Problems

Use Higher-Order Thinking Skills

REASONING Determine whether each statement is *true* or *false*. Explain.

35. C is a square matrix with 4 columns. It contains element c_{53}.

36. D is a square matrix with 3 rows. It contains element c_{22}.

37. **FIND THE ERROR** Kyla and Jay were asked to identify b_{32} for $B = \begin{bmatrix} -6 & 7 \\ 0 & 5 \\ 8 & 2 \end{bmatrix}$. Is either of them correct? Explain your answer.

Kyla	Jay
There is no element b_{32} for B because B is a 2×3 matrix.	The value of b_{32} is 5.

38. **CHALLENGE** Solve the following for x, y, and z.

$$\begin{bmatrix} 2x - y & 3x + 4z \\ 7x - 8z & 5y + 12z \end{bmatrix} = \begin{bmatrix} 9z - 5x + 1 & 5y - 2x \\ 3y - 4z & 12x + 2y \end{bmatrix}$$

39. **OPEN ENDED** Form a matrix using real-world data in which the sum of the columns is relevant and the sum of the rows is irrelevant.

40. **WRITING IN MATH** Explain how a matrix can be helpful when deciding what college you want to attend.

NYSRE PRACTICE Reinforcement of A.A.34, A.CM.3

41. What is the equation of the line that has a slope of 3 and passes through the point $(2, -9)$?

A $y = 3x + 11$
B $y = 3x + 15$
C $y = 3x - 11$
D $y = 3x - 15$

42. GEOMETRY Line q is shown below. Which equation best represents a line parallel to line q?

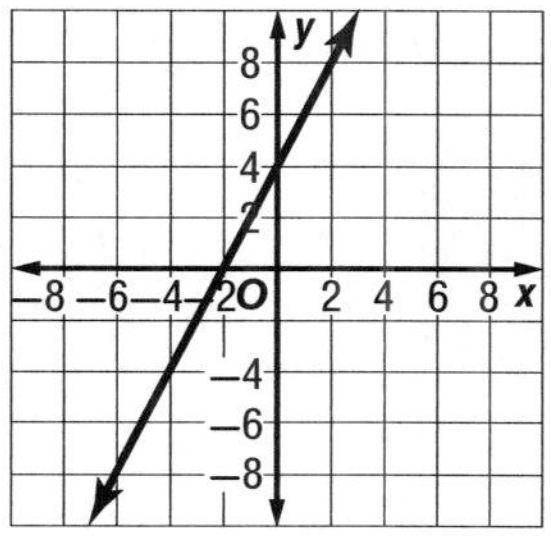

F $y = x + 2$
G $y = 2x - 3$
H $y = 2x + 4$
J $y = -2x + 2$

43. SHORT RESPONSE What is the area of the shaded part of the rectangle below?

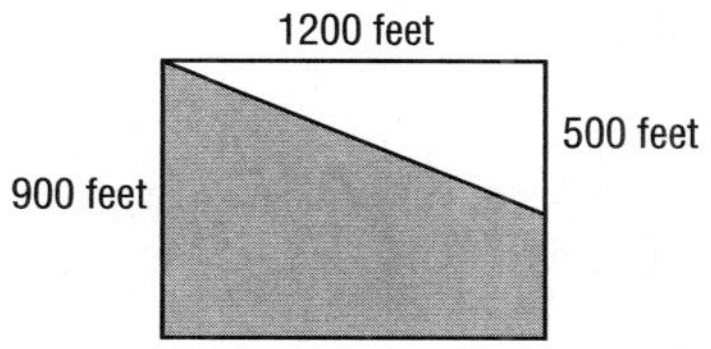

44. SAT/ACT The results of a recent poll are organized in the matrix.

	For	Against
Proposition 1	1553	771
Proposition 2	689	1633
Proposition 3	2088	229

Based on these results, which conclusion is *not* valid?

A There were 771 votes cast against Proposition 1.
B More people voted against Proposition 1 than voted for Proposition 2.
C Proposition 2 has little chance of passing.
D More people voted for Proposition 1 than for Proposition 3.

Spiral Review

45. COLLEGE FOOTBALL In a recent year, Darren McFadden of Arkansas placed second overall in the Heisman Trophy voting. Players are given 3 points for every first-place vote, 2 points for every second-place vote, and 1 point for every third-place vote. McFadden received 490 total votes for first, second, and third place, for a total of 878 points. If he had 4 more than twice as many second-place votes as third-place votes, how many votes did he receive for each place? (Lesson 3-5)

46. PACKAGING The Cookie Factory sells chocolate chip and peanut butter cookies in combination packages that contain between six and twelve cookies. At least three of each type of cookie should be in each package. How many of each type of cookie should be in each package to maximize the profit? (Lesson 3-4)

Cookie	chocolate chip	peanut butter
Cost	\$0.19	\$0.13
Price	\$0.44	\$0.39

Find the slope of the line that passes through each pair of points. (Lesson 2-3)

47. $(-3, -6), (-1, -9)$
48. $(-2, 6), (4, -1)$
49. $(5, -3), (8, 2)$

Multiply. (Lesson 0-2)

50. $(2x + 1)(-3x - 2)$
51. $(y + 6)(y - 8)$
52. $(x + y)(x - 2y)$

Skills Review

Evaluate each expression if $w = 3$, $x = -2$, $y = 4$, and $z = 0.5$. (Lesson 1-2)

53. $4x - 6y + 2z$
54. $5w + 2(x - z) + 2y$
55. $4[3(2z + y) - 2(w + x)]$

EXTEND 4-1

Spreadsheet Lab: Organizing Data

NY Math Online glencoe.com
Other Calculator Keystrokes

NYS Core Curriculum **A.CM.3** Present organized mathematical ideas with the use of appropriate standard notations, including the use of symbols and other representations when sharing an idea in verbal and written form

People in the workforce often use computer **spreadsheets** to organize, display, and analyze data. Similar to a matrix, data in a spreadsheet are entered into rows and columns. Then the data can be used to create graphs or perform calculations.

EXAMPLE

The manager of a gourmet food store has gathered data on the number of pounds of bulk coffees they have sold each week in January. Enter the data into a spreadsheet.

Weekly Sales for January

Coffee	1/5	1/12	1/19	1/26
Hawaiian Kona	17	22	11	23
Mocha Java	31	34	22	29
House Blend	55	61	44	71
Espresso	41	36	60	77
Decaf Espresso	23	29	19	44
Breakfast Blend	8	18	19	31
Decaf Breakfast Blend	22	18	30	32
Organic Italian Roast	26	16	31	39

Use Column A for the types of coffee, Column B for the sales in the week starting 1/5, Column C for sales in the week starting 1/12, and Columns D and E for the sales in the weeks starting 1/19 and 1/26.

	A	B	C	D	E
1	Hawaiian Kona	17	22	11	23
2	Mocha Java	31	34	22	29
3	House Blend	55	61	44	71
4	Espresso	41	36	60	77
5	Decaf Espresso	23	29	19	44
6	Breakfast Blend	8	18	19	31
7	Decaf Breakfast Blend	22	18	30	32
8	Organic Italian Roast	26	16	31	39

Each **row** contains data for the same type of coffee. Row 2 represents Mocha Java.

Each **cell** of the spreadsheet contains one piece of data. Cell 7D contains the value 30, representing the number of pounds of Decaf Breakfast Blend sold the week of 1/19.

Exercises

1. Enter the data on smartphones on page 185 into a spreadsheet.
2. Compare and contrast how data are organized in a spreadsheet and how they are organized in a matrix.
3. A SUM formula allows you to find the sum of the entries in a column or row.
   a. The formula =SUM(B1:B8) finds the sum of column B. Enter formulas in cells B9, C9, D9, and E9 to find the sums of those columns. What do the sums of the columns represent in the situation?
   b. Enter formulas in cells F1 through F8 to find the sums of rows 1 through 8. What do the sums of the rows represent in the situation?
   c. Find the sum of row 9 and the sum of column F. What do you observe? Explain.

4-2 Operations with Matrices

Then

You organized data into matrices.

Now

- Add and subtract matrices.
- Multiply a matrix by a scalar.

NYS Core Curriculum

A2.RP.10 Devise ways to verify results, using counterexamples and informal indirect proof

New Vocabulary

scalar

scalar multiplication

NY Math Online

glencoe.com

- Extra Examples
- Personal Tutor
- Self-Check Quiz
- Homework Help

Why?

Coastal Sales Company has three locations in Florida. The matrices below show the average daily wages and sales of all of the representatives.

	Miami Wages	Miami Sales	Tampa Wages	Tampa Sales	Tallahassee Wages	Tallahassee Sales
Entry	900	145,000	900	122,000	1050	109,500
Assistant	2400	225,000	1800	145,500	1800	135,000
Associate	2700	290,000	1800	160,000	1800	150,500

Add and Subtract Matrices Matrices can be added or subtracted if and only if they have the same dimensions.

Key Concept — Adding and Subtracting Matrices (For Your FOLDABLE)

Words To add or subtract two matrices with the same dimensions, add or subtract their corresponding elements.

Symbols

$$\underset{A}{\begin{bmatrix} a & b \\ c & d \end{bmatrix}} + \underset{B}{\begin{bmatrix} e & f \\ g & h \end{bmatrix}} = \underset{A+B}{\begin{bmatrix} a+e & b+f \\ c+g & d+h \end{bmatrix}}$$

$$\underset{A}{\begin{bmatrix} a & b \\ c & d \end{bmatrix}} - \underset{B}{\begin{bmatrix} e & f \\ g & h \end{bmatrix}} = \underset{A-B}{\begin{bmatrix} a-e & b-f \\ c-g & d-h \end{bmatrix}}$$

Example

$$\begin{bmatrix} 3 & -5 \\ 1 & 7 \end{bmatrix} + \begin{bmatrix} 2 & 0 \\ -9 & 10 \end{bmatrix} = \begin{bmatrix} 3+2 & -5+0 \\ 1+(-9) & 7+10 \end{bmatrix}$$

EXAMPLE 1 Add and Subtract Matrices

Find each of the following for $A = \begin{bmatrix} 16 & 2 \\ -9 & 8 \end{bmatrix}$, $B = \begin{bmatrix} -4 & -1 \\ -3 & -7 \end{bmatrix}$, and $C = \begin{bmatrix} 8 \\ 6 \end{bmatrix}$.

a. $A + B$

$$A + B = \begin{bmatrix} 16 & 2 \\ -9 & 8 \end{bmatrix} + \begin{bmatrix} -4 & -1 \\ -3 & -7 \end{bmatrix} \quad \text{Substitution}$$

$$= \begin{bmatrix} 16+(-4) & 2+(-1) \\ -9+(-3) & 8+(-7) \end{bmatrix} \quad \text{Add corresponding elements.}$$

$$= \begin{bmatrix} 12 & -1 \\ -12 & 1 \end{bmatrix} \quad \text{Simplify.}$$

b. $B - C$

$$B - C = \begin{bmatrix} -4 & -1 \\ -3 & -7 \end{bmatrix} - \begin{bmatrix} 8 \\ 6 \end{bmatrix} \quad \text{Substitution}$$

Since the dimensions of B and C are different, you cannot subtract the matrices.

Check Your Progress

1A. $\begin{bmatrix} -3 & 4 \\ -9 & -5 \end{bmatrix} - \begin{bmatrix} -4 & 12 \\ 8 & -7 \end{bmatrix}$

1B. $\begin{bmatrix} -9 & 8 & 3 \\ -2 & 4 & -7 \end{bmatrix} + \begin{bmatrix} -4 & -3 & 6 \\ -9 & -5 & 18 \end{bmatrix}$

Personal Tutor glencoe.com

Scalar Multiplication You can multiply any matrix by a constant called a **scalar**. When you do this, you multiply each individual element by the value of the scalar. This operation is called **scalar multiplication**.

Key Concept Multiplying by a Scalar

For Your FOLDABLE

Words To multiply a matrix by a scalar k, multiply each element by k.

Symbols
$$k \cdot A = kA$$
$$k\begin{bmatrix} a & b \\ c & d \end{bmatrix} = \begin{bmatrix} ka & kb \\ kc & kd \end{bmatrix}$$

Example
$$-3\begin{bmatrix} 4 & 1 \\ 7 & -2 \end{bmatrix} = \begin{bmatrix} -3(4) & -3(1) \\ -3(7) & -3(-2) \end{bmatrix}$$

ReadingMath

Scalar Think of a scalar as a coefficient for a variable, but instead it is for a matrix.

StudyTip

Scalar Multiplication The bracket of a matrix is treated just like a regular grouping symbol. So when multiplying by a scalar, distribute the same way as with a grouping symbol.

EXAMPLE 2 Multiply a Matrix by a Scalar

If $R = \begin{bmatrix} -12 & 8 & 6 \\ -16 & 4 & 19 \end{bmatrix}$, find $5R$.

$$5R = 5\begin{bmatrix} -12 & 8 & 6 \\ -16 & 4 & 19 \end{bmatrix}$$ Substitution

$$= \begin{bmatrix} 5(-12) & 5(8) & 5(6) \\ 5(-16) & 5(4) & 5(19) \end{bmatrix}$$ Distribute the scalar.

$$= \begin{bmatrix} -60 & 40 & 30 \\ -80 & 20 & 95 \end{bmatrix}$$ Multiply.

Check Your Progress

2. If $S = \begin{bmatrix} 8 & 0 & 3 & -2 \\ -1 & -4 & -2 & 9 \end{bmatrix}$, find $-4S$.

Personal Tutor glencoe.com

Many properties of real numbers also hold true for matrices. A summary of these properties is listed below.

Key Concept Properties of Matrix Operations

For Your FOLDABLE

For any matrices A, B, and C for which the matrix sum and product are defined and any scalar k, the following properties are true.

Commutative Property of Addition	$A + B = B + A$
Associative Property of Addition	$(A + B) + C = A + (B + C)$
Left Scalar Distributive Property	$k(A + B) = kA + kB$
Right Scalar Distributive Property	$(A + B)k = kA + kB$

Multi-step operations can be performed on matrices. The order of these operations is the same as with real numbers.

EXAMPLE 3 Multi-Step Operations

If $A = \begin{bmatrix} -9 & 12 \\ 2 & -6 \end{bmatrix}$ and $B = \begin{bmatrix} -4 & -8 \\ 2 & -3 \end{bmatrix}$, find $-4B - 3A$.

$-4B - 3A = -4\begin{bmatrix} -4 & -8 \\ 2 & -3 \end{bmatrix} - 3\begin{bmatrix} -9 & 12 \\ 2 & -6 \end{bmatrix}$ **Substitution**

$= \begin{bmatrix} -4(-4) & -4(-8) \\ -4(2) & -4(-3) \end{bmatrix} - \begin{bmatrix} 3(-9) & 3(12) \\ 3(2) & 3(-6) \end{bmatrix}$ **Distribute the scalars in each matrix.**

$= \begin{bmatrix} 16 & 32 \\ -8 & 12 \end{bmatrix} - \begin{bmatrix} -27 & 36 \\ 6 & -18 \end{bmatrix}$ **Multiply.**

$= \begin{bmatrix} 16-(-27) & 32-36 \\ -8-6 & 12-(-18) \end{bmatrix}$ **Subtract corresponding elements.**

$= \begin{bmatrix} 43 & -4 \\ -14 & 30 \end{bmatrix}$ **Simplify.**

Check Your Progress

3. If $A = \begin{bmatrix} -5 & 3 \\ 6 & -8 \\ 2 & 9 \end{bmatrix}$ and $B = \begin{bmatrix} 12 & 5 \\ 5 & -4 \\ 4 & -7 \end{bmatrix}$, find $-6B + 7A$.

Personal Tutor glencoe.com

Real-World Career

Financial Planner Financial planners often use matrices to organize and describe the data they use. Financial planners need a bachelor's degree. Usually they have degrees in accounting, finance, economics, business, marketing, or commerce.

Matrices can be used in many business applications.

EXAMPLE 4 Use Multi-Step Operations with Matrices

BUSINESS Refer to the application at the beginning of the lesson. Express the average wages and sales for the entire company for a 5-day week.

To calculate the 5-day sales for the entire company, each matrix needs to be multiplied by 5 and the totals added together.

$5\begin{bmatrix} 900 & 145{,}000 \\ 2400 & 225{,}000 \\ 2700 & 290{,}000 \end{bmatrix} + 5\begin{bmatrix} 900 & 122{,}000 \\ 1800 & 145{,}500 \\ 1800 & 160{,}000 \end{bmatrix} + 5\begin{bmatrix} 1050 & 109{,}500 \\ 1800 & 135{,}000 \\ 1800 & 150{,}500 \end{bmatrix}$ **Write matrices.**

$= \begin{bmatrix} 4500 & 725{,}000 \\ 12{,}000 & 1{,}125{,}000 \\ 13{,}500 & 1{,}450{,}000 \end{bmatrix} + \begin{bmatrix} 4500 & 610{,}000 \\ 9000 & 727{,}500 \\ 9000 & 800{,}000 \end{bmatrix} + \begin{bmatrix} 5250 & 547{,}500 \\ 9000 & 675{,}000 \\ 9000 & 752{,}500 \end{bmatrix}$ **Multiply scalars.**

$= \begin{array}{r} \\ \text{Entry} \\ \text{Assistant} \\ \text{Associate} \end{array} \begin{array}{c} \begin{array}{cc} \text{Wages} & \text{Sales} \end{array} \\ \begin{bmatrix} 14{,}250 & 1{,}882{,}500 \\ 30{,}000 & 2{,}527{,}500 \\ 31{,}500 & 3{,}002{,}500 \end{bmatrix} \end{array}$ **Add matrices.**

The final matrix indicates the average weekly sales and wages for all of the representatives of the company.

StudyTip

Corresponding Elements When representing quantities with multiple matrices, make sure the corresponding elements represent corresponding quantities.

Check Your Progress

4. Use the data above to calculate the average yearly sales and wages for the company, assuming 260 working days.

Personal Tutor glencoe.com

Check Your Understanding

Example 1 p. 193

Perform the indicated operations. If the matrix does not exist, write *impossible*.

1. $\begin{bmatrix} -8 & 2 & 6 \end{bmatrix} + \begin{bmatrix} 11 & -7 & 1 \end{bmatrix}$

2. $\begin{bmatrix} 9 & -8 & 4 \end{bmatrix} + \begin{bmatrix} 12 & 2 \end{bmatrix}$

3. $\begin{bmatrix} 7 & -12 \\ 15 & 4 \end{bmatrix} - \begin{bmatrix} 9 & 6 \\ 4 & -9 \end{bmatrix}$

4. $\begin{bmatrix} 5 & 13 & -6 \\ 3 & -17 & 2 \end{bmatrix} - \begin{bmatrix} -2 & -18 & 8 \\ 2 & -11 & 0 \end{bmatrix}$

Example 2 p. 194

Perform the indicated operations. If the matrix does not exist, write *impossible*.

5. $3\begin{bmatrix} 6 & 4 & 0 \\ -2 & 14 & -8 \\ -4 & -6 & 7 \end{bmatrix}$

6. $-6\begin{bmatrix} 15 & -9 & 2 & 3 \\ 6 & -11 & 14 & -2 \\ 4 & -8 & -10 & 27 \end{bmatrix}$

Example 3 p. 195

Use matrices A, B, C, and D to find the following.

$A = \begin{bmatrix} 6 & -4 \\ 3 & -5 \end{bmatrix}$ $B = \begin{bmatrix} 8 & -1 \\ -2 & 7 \end{bmatrix}$ $C = \begin{bmatrix} -4 & -6 \\ 12 & -7 \end{bmatrix}$ $D = \begin{bmatrix} 9 & 6 & 0 \\ -2 & 8 & 0 \end{bmatrix}$

7. $4B - 2A$

8. $-8C + 3A$

9. $-5B - 2D$

10. $-4C - 5B$

Example 4 p. 195

11. **GRADES** Geraldo, Olivia, and Nikki have had two tests in their math class. The table shows the test grades for each student.

Student	Test 1	Test 2
Geraldo	85	72
Olivia	75	74
Nikki	96	83

a. Write a matrix for the information.

b. Find the sum of the scores from the two tests expressed as a matrix.

c. Express the difference in scores from test 1 to test 2 as a matrix.

Practice and Problem Solving

= **Step-by-Step Solutions** begin on page R20.
Extra Practice begins on page 947.

Example 1 p. 193

Perform the indicated operations. If the matrix does not exist, write *impossible*.

12. $\begin{bmatrix} 12 & -5 \\ -8 & -3 \end{bmatrix} + \begin{bmatrix} -6 & 11 \\ -7 & 2 \end{bmatrix}$

13. $\begin{bmatrix} 9 & 5 \\ -2 & 16 \end{bmatrix} + \begin{bmatrix} -6 & -3 & 7 \\ 12 & 2 & -4 \end{bmatrix}$

Examples 2 and 3 pp. 194, 195

14. **BUSINESS** The drink menu from a fast-food restaurant is shown at the right. The store owner has decided that all of the prices must be increased by 10%.

Drink	Small	Medium	Large
Soda	$0.95	$1.00	$1.05
Iced tea	$0.75	$0.80	$0.85
Lemonade	$0.75	$0.80	$0.85
Coffee	$1.00	$1.10	$1.20

a. Write matrix C to represent the current prices.

b. What scalar can be used to determine a matrix N to represent the new prices?

c. Find N.

d. What is $N - C$? What does this represent in this situation?

Example 4 p. 195

Perform the indicated operations. If the matrix does not exist, write *impossible*.

15. $\begin{bmatrix} -5 \\ 8 \\ 1 \\ 0 \end{bmatrix} - \begin{bmatrix} 19 \\ -2 \\ 4 \\ 7 \end{bmatrix}$

16. $\begin{bmatrix} 4 & -3 & 3 \\ -8 & 12 & 1 \\ 0 & -1 & 5 \\ 7 & -9 & 4 \end{bmatrix} - \begin{bmatrix} -3 & -8 & 12 \\ -11 & -5 & 3 \\ -1 & 22 & -9 \\ -6 & 31 & 9 \end{bmatrix}$

17. $\begin{bmatrix} 62 \\ -37 \\ -4 \end{bmatrix} + \begin{bmatrix} 34 & 76 & -13 \end{bmatrix}$

18. $\begin{bmatrix} 2 & 4 & 11 \\ -6 & 12 & -3 \end{bmatrix} - \begin{bmatrix} 8 & -9 & -3 \\ 5 & 14 & 0 \end{bmatrix}$

19. $\begin{bmatrix} 5 \\ -9 \end{bmatrix} + \begin{bmatrix} -3 \\ -7 \end{bmatrix} - \begin{bmatrix} 9 \\ 16 \end{bmatrix}$

20. $\begin{bmatrix} 5 \\ 3 \end{bmatrix} - \begin{bmatrix} -4 \\ 2 \end{bmatrix} + \begin{bmatrix} -2 & 3 \\ 8 & -3 \end{bmatrix}$

Real-World Link

The Library of Congress in Washington, D.C., is the largest library in the world. The library has more than 32 million books, 2.8 million recordings, 12.5 million photographs, 5.3 million maps, and 59.5 million manuscripts.

Source: The Library of Congress

21. BOOKS Library A has 10,000 novels, 5000 biographies, and 5000 children's books. Library B has 15,000 novels, 10,000 biographies, and 2500 children's books. Library C has 4000 novels, 700 biographies, and 800 children's books.

a. Express each library's number of books as a matrix. Label the matrices *A*, *B*, and *C*.

b. Find the total number of each type of book in all 3 libraries. Express as a matrix.

c. How many more books of each type does Library *A* have than Library *C*?

d. Find $A + B$. Does the matrix have meaning in this situation? Explain.

Perform the indicated operations. If the matrix does not exist, write *impossible*.

22. $-3\begin{bmatrix} 18 & -6 & -8 \\ -5 & -3 & 12 \\ 0 & 3x & -y \end{bmatrix}$

23. $8\begin{bmatrix} -a & 4b & c-b \\ -13 & 10 & -5c \end{bmatrix}$

24. $-4\begin{bmatrix} -7 \\ 4 \\ -3 \end{bmatrix} + 3\begin{bmatrix} -8 \\ 3x \\ -9 \end{bmatrix} - 5\begin{bmatrix} 4 \\ x-6 \\ 12 \end{bmatrix}$

25 $-5\left(\begin{bmatrix} 4 & -8 \\ 8 & -9 \end{bmatrix} + \begin{bmatrix} 4 & -2 \\ -3 & -6 \end{bmatrix}\right)$

26. $-6\left(\begin{bmatrix} 6 & 3y \\ 4x+1 & -2 \\ -9 & xy \end{bmatrix} + \begin{bmatrix} -5 & -6 \\ 8 & -7 \\ x+2 & 2x \end{bmatrix}\right)$

27. $-4\begin{bmatrix} 9 & -5y \\ 11 & -3 \\ -1 & 2-x \end{bmatrix} - 7\begin{bmatrix} 8 & -y & 12 \\ -4 & -8 & 9 \\ 3x & 2+x & -y \end{bmatrix}$

28. WEATHER The table shows snowfall in inches.

City	Normal Snowfall			2007 Snowfall		
	Jan	Feb	Mar	Jan	Feb	Mar
Grand Rapids, MI	21.1	12.2	9.0	15.4	33.6	13.6
Boston, MA	13.3	11.3	8.3	1.0	4.6	10.2
Buffalo, NY	26.1	17.8	12.4	15.5	33.5	5.4
Pittsburgh, PA	12.3	8.5	7.9	11.3	14.0	9.3

Source: National Weather Service

a. Express the normal snowfall data and the 2007 data in two 4×3 matrices.

b. Subtract the matrix of normal data from the matrix of 2007 data. What does the difference represent in the context of the situation?

c. Explain the meaning of positive and negative numbers in the difference matrix. What trends do you see in the data?

Perform the indicated operations. If the matrix does not exist, write *impossible*.

29. $\begin{bmatrix} 12.5 & -16.4 \\ 4.31 & -2.43 \\ -6.8 & -14.1 \end{bmatrix} - 3\begin{bmatrix} -18.7 & -11.8 \\ 8.1 & -6.91 \\ -6.21 & -17.6 \end{bmatrix}$

30. $-2\begin{bmatrix} -9.2 & -8.4 \\ 5.6 & -4.3 \end{bmatrix} - 4\begin{bmatrix} 4.1 & -2.9 \\ 7.2 & -8.2 \end{bmatrix}$

31. $-\frac{3}{4}\begin{bmatrix} 12 & -16 \\ 15 & 8 \end{bmatrix} + \frac{2}{3}\begin{bmatrix} 21 & 18 \\ -4 & -6 \end{bmatrix}$

32. $-4\begin{bmatrix} \frac{4}{5} & 2 & -\frac{3}{4} \\ -1 & -\frac{2}{5} & -6 \end{bmatrix} + 3\begin{bmatrix} \frac{1}{5} & -4 & \frac{1}{8} \\ 4 & \frac{2}{3} & -3 \end{bmatrix}$

Real-World Link

At the age of 40, Dara Torres eclipsed her own American record in winning the women's 50-meter freestyle during the U.S. Swimming Championships in 2007.

33 **SWIMMING** The table shows some of the world, Olympic, and American women's freestyle swimming records.

Distance (m)	World	Olympic	American
50	24.13 s	24.13 s	24.63 s
100	53.52 s	53.52 s	53.99 s
200	1:56.54 min	1:57.65 min	1:57.41 min
800	8:16.22 min	8:19.67 min	8:16.22 min

Source: USA Swimming

a. Find the difference between the American and World records expressed as a column matrix.

b. What is the meaning of each row in the column?

c. In which events were the fastest times set at the Olympics?

34. **MULTIPLE REPRESENTATIONS** In this problem, you will investigate using matrices to represent transformations.

a. ALGEBRAIC The matrix $\begin{bmatrix} -3 & -4 & 1 \\ 8 & 6 & 0 \end{bmatrix}$ represents a triangle with vertices at (−3, 8), (−4, 6), and (1, 0). Write a matrix to represent $\triangle ABC$.

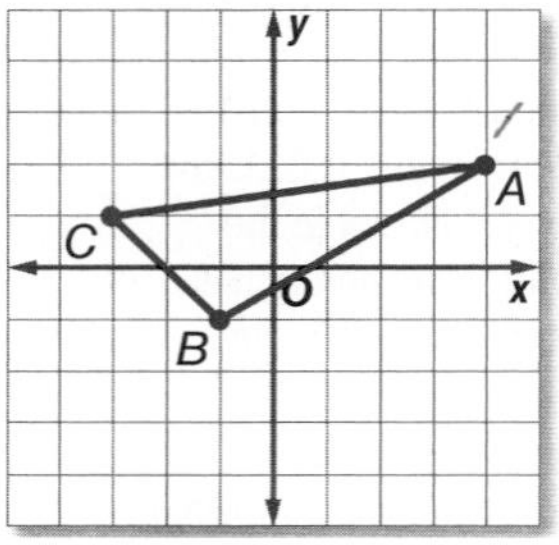

b. GEOMETRIC Multiply the vertex matrix you wrote by 2. Then graph the figure represented by the new matrix.

c. ANALYTICAL How do the figures compare? Make a conjecture about the result of multiplying the matrix by 0.5. Verify your conjecture.

H.O.T. Problems

Use Higher-Order Thinking Skills

35. **PROOF** Prove that matrix addition is commutative for 2×2 matrices.

36. **PROOF** Prove that matrix addition is associative for 2×2 matrices.

37. **CHALLENGE** Find the elements of C if:

$A = \begin{bmatrix} -3 & -4 \\ 8 & 6 \end{bmatrix}$, $B = \begin{bmatrix} 5 & -1 \\ 2 & -4 \end{bmatrix}$, and $3A - 4B + 6C = \begin{bmatrix} 13 & 22 \\ 10 & 4 \end{bmatrix}$.

38. **REASONING** Determine whether each statement is *sometimes*, *always*, or *never* true for matrices A and B. Explain your reasoning.

a. If $A + B$ exists, then $A - B$ exists.

b. If k is a real number, then kA and kB exist.

c. If $A - B$ does not exist, then $B - A$ does not exist.

d. If A and B have the same number of elements, then $A + B$ exists.

e. If kA exists and kB exists, then $kA + kB$ exists.

39. **OPEN ENDED** Give an example of matrices A and B if $4B - 3A = \begin{bmatrix} -6 & 5 \\ -2 & -1 \end{bmatrix}$.

40. **WRITING IN MATH** Explain how to find $4D - 3C$ for two given matrices, C and D with the same dimensions.

NYSRE PRACTICE Reinforcement of A.A.10, A2.S.12

41. What is the solution of the system of equations?

$$0.06p + 4q = 0.88$$
$$p - q = -2.25$$

A $(-0.912, -1.338)$ C $(-2, 0.25)$
B $(0.912, -3.162)$ D $(-2, -4.25)$

42. SHORT RESPONSE Find $A + B$ if $A = \begin{bmatrix} -7 & 3 \\ 2 & 6 \end{bmatrix}$ and $B = \begin{bmatrix} 4 & 2 \\ 0 & 1 \end{bmatrix}$.

43. SAT/ACT Solve for x and y.

$$x + 3y = 16$$
$$7 - x = 12$$

F $x = -5, y = 7$ H $x = 7, y = 5$
G $x = 7, y = 3$ J $x = 5, y = 7$

44. PROBABILITY A local pizzeria offers 5 different meat toppings and 6 different vegetable toppings. You decide to get two vegetable toppings and one meat topping. How many different types of pizzas can you order?

A 60 C 120
B 75 D 150

Spiral Review

Identify each element for the following matrices. (Lesson 4-1)

$$A = \begin{bmatrix} -3 & 6 \\ -5 & x \\ 8 & 4y \end{bmatrix}, B = \begin{bmatrix} 16 & 4 & x \\ -2 & 9 & y \end{bmatrix}, C = \begin{bmatrix} 9 & -5 & 3 & 2 \\ 0 & -6 & 8 & 1 \end{bmatrix}$$

45. a_{32} **46.** c_{13} **47.** b_{32}

Solve each system of equations. (Lesson 3-5)

48. $2x + 3y - z = -1$
$5x + y + 4z = 30$
$-8x - 2y + 5z = -2$

49. $3x - 4y + 6z = 26$
$5x + 3y + 2z = 5$
$-2x + 5y - 3z = -9$

50. $5x + 2y - 4z = 22$
$6x + 3y + 5z = 5$
$-2x - 4y + z = 2$

Solve each system of inequalities by graphing. (Lesson 3-3)

51. $x + 2y > 4$
$y < -2x - 3$

52. $y \geq -4x + 6$
$3y < 2x + 9$

53. $4x + 2y > 8$
$4y - 3x \leq 12$

54. RAKING LEAVES Two students can each earn \$20 plus an extra \$5 for each trash bag they completely fill with leaves. Write and solve an equation to determine how many bags they will need to fill in order to earn \$100 each. (Lesson 2-4)

55. SPORTS There are 15,991 more student athletes in New York than in Illinois. Write and solve an equation to find the number of student athletes in Illinois. (Lesson 1-3)

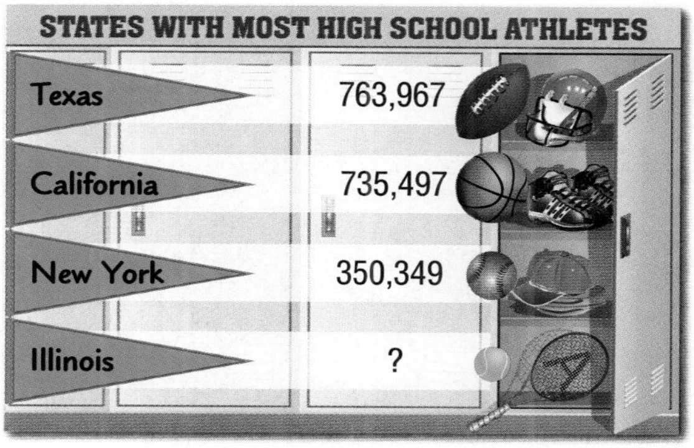

STATES WITH MOST HIGH SCHOOL ATHLETES
Texas
763,967
California
735,497
New York
350,349
Illinois
?

Skills Review

Simplify each expression. (Lesson 1-2)

56. $4(2x - 3y) + 2(5x - 6y)$ **57.** $-3(2a - 5b) - 4(4b + a)$ **58.** $-7(x - y) + 5(y - x)$

4-3 Multiplying Matrices

Then

You multiplied matrices by a scalar. (Lesson 4-2)

Now

- Multiply matrices.
- Use the properties of matrix multiplication.

NYS Core Curriculum

A2.RP.10 Devise ways to verify results, using counterexamples and informal indirect proof

NY Math Online

glencoe.com

- Extra Examples
- Personal Tutor
- Self-Check Quiz
- Homework Help
- Math in Motion

Why?

The table shows the scoring summary for Lisa Leslie, the WNBA's all-time scoring leader, during her highest scoring seasons. Her total points can be summarized in the scoring matrix S. The point values for each type of basket made can be organized in the point value matrix P.

Lisa Leslie Regular Season Scoring

Type	2003	2004	2005	2006
Field Goal	153	217	197	249
3-Point Field Goal	12	6	7	8
Free Throw	82	146	102	158

Source: WNBA

You can use matrix multiplication to find the points scored during each season.

Scoring **Point Values**

$$\begin{bmatrix} 153 & 217 & 197 & 249 \\ 12 & 6 & 7 & 8 \\ 82 & 146 & 102 & 158 \end{bmatrix} \qquad P = \begin{bmatrix} 2 & 3 & 1 \end{bmatrix}$$

Multiply Matrices You can multiply two matrices A and B if and only if the number of columns in A is equal to the number of rows in B. When you multiply two matrices $A_{m \times r}$ and $B_{r \times t}$, the resulting matrix AB is an $m \times t$ matrix.

$$\underset{m \times r}{A} \qquad \underset{r \times t}{B} \qquad = \qquad \underset{m \times t}{AB}$$

same (inner dimensions r); dimensions of AB (outer dimensions m, t)

EXAMPLE 1 Dimensions of Matrix Products

Determine whether each matrix product is defined. If so, state the dimensions of the product.

a. $A_{3 \times 4}$ **and** $B_{4 \times 2}$

$$\underset{3 \times 4}{A} \cdot \underset{4 \times 2}{B} = \underset{3 \times 2}{AB}$$

The inner dimensions are equal, so the product is defined. Its dimensions are 3×2.

b. $A_{5 \times 3}$ **and** $B_{5 \times 4}$

$$\underset{5 \times 3}{A} \cdot \underset{5 \times 4}{B}$$

The inner dimensions are not equal, so the matrix product is not defined.

Check Your Progress

1A. $A_{4 \times 6}$ and $B_{6 \times 2}$

1B. $A_{3 \times 2}$ and $B_{3 \times 2}$

Personal Tutor glencoe.com

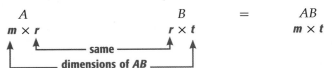

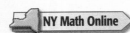

The product of two matrices is found by multiplying columns and rows.

Key Concept **Multiplying Matrices** *For Your FOLDABLE*

Words The element in the mth row and rth column of matrix AB is the sum of the products of the corresponding elements in row m of matrix A and column r of matrix B.

Symbols

$$\underset{A}{\begin{bmatrix} a & b \\ c & d \end{bmatrix}} \cdot \underset{B}{\begin{bmatrix} e & f \\ g & h \end{bmatrix}} = \underset{AB}{\begin{bmatrix} ae+bg & af+bh \\ ce+dg & cf+dh \end{bmatrix}}$$

Math *in Motion*, Animation glencoe.com

Watch Out!

Saving Your Place It is easy to lose your place as you multiply matrices. It may help to cover rows or columns not being multiplied as you find elements of the product matrix.

EXAMPLE 2 Multiply Square Matrices

Find XY if $X = \begin{bmatrix} 6 & -3 \\ -10 & -2 \end{bmatrix}$ and $Y = \begin{bmatrix} -5 & -4 \\ 3 & 3 \end{bmatrix}$.

$$XY = \begin{bmatrix} 6 & -3 \\ -10 & -2 \end{bmatrix} \cdot \begin{bmatrix} -5 & -4 \\ 3 & 3 \end{bmatrix}.$$

Step 1 Multiply the numbers in the first row of X by the numbers in the first column of Y, add the products, and put the result in the first row, first column of XY.

$$\begin{bmatrix} 6 & -3 \\ -10 & -2 \end{bmatrix} \cdot \begin{bmatrix} -5 & -4 \\ 3 & 3 \end{bmatrix} = \begin{bmatrix} 6(-5)+(-3)(3) & \\ & \end{bmatrix}$$

Step 2 Follow the same procedure as in Step 1 using the first row and the second column numbers. Write the result in the first row, second column.

$$\begin{bmatrix} 6 & -3 \\ -10 & -2 \end{bmatrix} \cdot \begin{bmatrix} -5 & -4 \\ 3 & 3 \end{bmatrix} = \begin{bmatrix} 6(-5)+(-3)(3) & 6(-4)+(-3)(3) \\ & \end{bmatrix}$$

Step 3 Follow the same procedure with the second row and the first column numbers. Write the result in the second row, first column.

$$\begin{bmatrix} 6 & -3 \\ -10 & -2 \end{bmatrix} \cdot \begin{bmatrix} -5 & -4 \\ 3 & 3 \end{bmatrix} = \begin{bmatrix} 6(-5)+(-3)(3) & 6(-4)+(-3)(3) \\ -10(-5)+(-2)(3) & \end{bmatrix}$$

Step 4 The procedure is the same for the numbers in the second row, second column.

$$\begin{bmatrix} 6 & -3 \\ -10 & -2 \end{bmatrix} \cdot \begin{bmatrix} -5 & -4 \\ 3 & 3 \end{bmatrix} = \begin{bmatrix} 6(-5)+(-3)(3) & 6(-4)+(-3)(3) \\ -10(-5)+(-2)(3) & -10(-4)+(-2)(3) \end{bmatrix}$$

Step 5 Simplify the product matrix.

$$\begin{bmatrix} 6(-5)+(-3)(3) & 6(-4)+(-3)(3) \\ -10(-5)+(-2)(3) & -10(-4)+(-2)(3) \end{bmatrix} = \begin{bmatrix} -39 & -33 \\ 44 & 34 \end{bmatrix}$$

Check Your Progress

2. Find UV if $U = \begin{bmatrix} 5 & 9 \\ -3 & -2 \end{bmatrix}$ and $V = \begin{bmatrix} 2 & -1 \\ 6 & -5 \end{bmatrix}$.

Personal Tutor glencoe.com

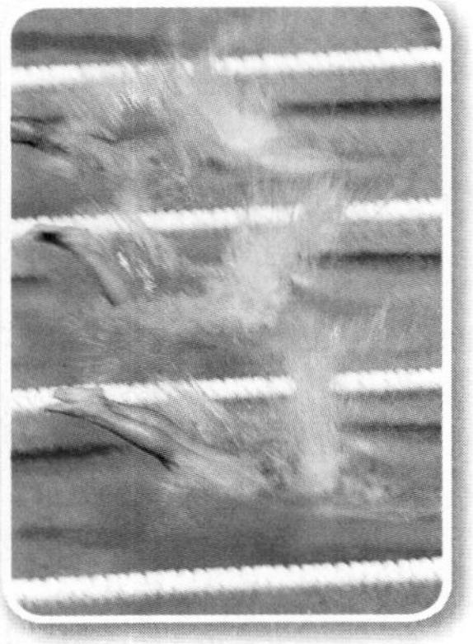

Real-World Link

Swim meets consist of racing and diving competitions. There are more than 241,000 high schools that participate each year.

Source: National Federation of State High School Associations

Matrix multiplication can be used in many real-world situations.

Real-World EXAMPLE 3 Multiply Matrices

SWIM MEET At a particular swim meet, 7 points were awarded for each first-place finish, 4 points for second, and 2 points for third. Find the total number of points for each school. Which school won the meet?

School	First Place	Second Place	Third Place
Central	4	7	3
Franklin	8	9	1
Hayes	10	5	3
Lincoln	3	3	6

Understand The final scores can be found by multiplying the swim results for each school by the points awarded for each first-, second-, and third-place finish.

Plan Write the results of the races and the points awarded in matrix form. Set up the matrices so that the number of rows in the points matrix equals the number of columns in the results matrix.

Results **Points**

$$R = \begin{bmatrix} 4 & 7 & 3 \\ 8 & 9 & 1 \\ 10 & 5 & 3 \\ 3 & 3 & 6 \end{bmatrix} \quad P = \begin{bmatrix} 7 \\ 4 \\ 2 \end{bmatrix}$$

Solve Multiply the matrices.

$$RP = \begin{bmatrix} 4 & 7 & 3 \\ 8 & 9 & 1 \\ 10 & 5 & 3 \\ 3 & 3 & 6 \end{bmatrix} \cdot \begin{bmatrix} 7 \\ 4 \\ 2 \end{bmatrix}$$ **Write an equation.**

$$= \begin{bmatrix} 4(7) + 7(4) + 3(2) \\ 8(7) + 9(4) + 1(2) \\ 10(7) + 5(4) + 3(2) \\ 3(7) + 3(4) + 6(2) \end{bmatrix}$$ **Multiply columns by rows.**

$$= \begin{bmatrix} 62 \\ 94 \\ 96 \\ 45 \end{bmatrix}$$ **Simplify.**

The product matrix shows the scores for Central, Franklin, Hayes, and Lincoln, respectively. Hayes won the swim meet with a total of 96 points.

Check R is a 4×3 matrix and P is a 3×1 matrix, so their product should be a 4×1 matrix.

Check Your Progress

3. BASKETBALL Refer to the beginning of the lesson. Use matrix multiplication to determine in which season Lisa Leslie scored the most points. How many points did she score that season?

Personal Tutor glencoe.com

Multiplicative Properties Recall that the properties of real numbers also held true for matrix addition. However, some of these properties do *not* always hold true for matrix multiplication.

EXAMPLE 4 Test of the Commutative Property

Find each product if $G = \begin{bmatrix} 1 & 3 & -5 \\ 4 & -2 & 0 \end{bmatrix}$ **and** $H = \begin{bmatrix} 2 & 3 \\ -2 & -8 \\ 1 & 7 \end{bmatrix}$.

a. GH

$$GH = \begin{bmatrix} 1 & 3 & -5 \\ 4 & -2 & 0 \end{bmatrix} \cdot \begin{bmatrix} 2 & 3 \\ -2 & -8 \\ 1 & 7 \end{bmatrix} \quad \textbf{Substitution}$$

$$= \begin{bmatrix} 2 - 6 - 5 & 3 - 24 - 35 \\ 8 + 4 + 0 & 12 + 16 + 0 \end{bmatrix} \text{ or } \begin{bmatrix} -9 & -56 \\ 12 & 28 \end{bmatrix}$$

b. HG

$$HG = \begin{bmatrix} 2 & 3 \\ -2 & -8 \\ 1 & 7 \end{bmatrix} \cdot \begin{bmatrix} 1 & 3 & -5 \\ 4 & -2 & 0 \end{bmatrix} \quad \textbf{Substitution}$$

$$= \begin{bmatrix} 2 + 12 & 6 - 6 & -10 + 0 \\ -2 - 32 & -6 + 16 & 10 + 0 \\ 1 + 28 & 3 - 14 & -5 + 0 \end{bmatrix} \text{ or } \begin{bmatrix} 14 & 0 & -10 \\ -34 & 10 & 10 \\ 29 & -11 & -5 \end{bmatrix}$$

Notice that $GH \neq HG$.

Check Your Progress

4. Determine if $AB = BA$ is true for $A = \begin{bmatrix} 4 & -1 \\ 5 & -2 \end{bmatrix}$ and $B = \begin{bmatrix} -3 & 6 \\ -4 & 5 \end{bmatrix}$.

Personal Tutor glencoe.com

StudyTip

Proof and Counterexamples To show that a property is true for all cases, you must show it is true for the general case. To show that a property is *not* always true, you need to find only one counterexample.

Example 4 demonstrates that the Commutative Property of Multiplication does not hold for matrix multiplication. The order in which you multiply matrices is very important.

EXAMPLE 5 Test of the Distributive Property

Find each product if $J = \begin{bmatrix} -2 & 4 \\ -5 & -2 \end{bmatrix}$, $K = \begin{bmatrix} 3 & 2 \\ -1 & 3 \end{bmatrix}$, **and** $L = \begin{bmatrix} -4 & -1 \\ 3 & 0 \end{bmatrix}$.

a. $J(K + L)$

$$J(K + L) = \begin{bmatrix} 2 & 4 \\ -5 & -2 \end{bmatrix} \cdot \left(\begin{bmatrix} 3 & 2 \\ -1 & 3 \end{bmatrix} + \begin{bmatrix} -4 & -1 \\ 3 & 0 \end{bmatrix} \right) \quad \textbf{Substitution}$$

$$= \begin{bmatrix} 2 & 4 \\ -5 & -2 \end{bmatrix} \cdot \begin{bmatrix} -1 & 1 \\ 2 & 3 \end{bmatrix} \quad \textbf{Add.}$$

$$= \begin{bmatrix} -2 + 8 & 2 + 12 \\ 5 - 4 & -5 - 6 \end{bmatrix} \text{ or } \begin{bmatrix} 6 & 14 \\ 1 & -11 \end{bmatrix} \quad \textbf{Multiply.}$$

b. $JK + JL$

$$JK + JL = \begin{bmatrix} 2 & 4 \\ -5 & -2 \end{bmatrix} \cdot \begin{bmatrix} 3 & 2 \\ -1 & 3 \end{bmatrix} + \begin{bmatrix} 2 & 4 \\ -5 & -2 \end{bmatrix} \cdot \begin{bmatrix} -4 & -1 \\ 3 & 0 \end{bmatrix}$$

$$= \begin{bmatrix} 2(3) + 4(-1) & 2(2) + 4(3) \\ -5(3) + (-2)(-1) & -5(2) + (-2)(3) \end{bmatrix} + \begin{bmatrix} 2(-4) + 4(3) & 2(-1) + 4(0) \\ -5(-4) + (-2)(3) & -5(-1) + (-2)(0) \end{bmatrix}$$

$$= \begin{bmatrix} 2 & 16 \\ -13 & -16 \end{bmatrix} + \begin{bmatrix} 4 & -2 \\ 14 & 5 \end{bmatrix} \text{ or } \begin{bmatrix} 6 & 14 \\ 1 & -11 \end{bmatrix}$$

Notice that $J(K + L) = JK + JL$.

Check Your Progress

5. Use the matrices $R = \begin{bmatrix} 2 & -1 \\ 1 & 3 \end{bmatrix}$, $S = \begin{bmatrix} 4 & 6 \\ -2 & 5 \end{bmatrix}$, and $T = \begin{bmatrix} -3 & 7 \\ -4 & 8 \end{bmatrix}$ to determine if $(S + T)R = SR + TR$.

Personal Tutor glencoe.com

The previous example suggests that the Distributive Property is true for matrix multiplication. Some properties of matrix multiplication are shown below.

Key Concept **Properties of Matrix Multiplication** For Your FOLDABLE

For any matrices A, B, and C for which the matrix product is defined and any scalar k, the following properties are true.

Associative Property of Matrix Multiplication	$(AB)C = A(BC)$
Associative Property of Scalar Multiplication	$k(AB) = (kA)B = A(kB)$
Left Distributive Property	$C(A + B) = CA + CB$
Right Distributive Property	$(A + B)C = AC + BC$

Check Your Understanding

Example 1 p. 200

Determine whether each matrix product is defined. If so, state the dimensions of the product.

1. $A_{2 \times 4} \cdot B_{4 \times 3}$

2. $C_{5 \times 4} \cdot D_{5 \times 4}$

3. $E_{8 \times 6} \cdot F_{6 \times 10}$

Examples 2 and 3 pp. 201, 202

Find each product, if possible.

4. $\begin{bmatrix} 2 & 1 \\ 7 & -5 \end{bmatrix} \cdot \begin{bmatrix} -6 & 3 \\ -2 & -4 \end{bmatrix}$

5. $\begin{bmatrix} 10 & -2 \\ -7 & 3 \end{bmatrix} \cdot \begin{bmatrix} 1 & 4 \\ 5 & -2 \end{bmatrix}$

6. $\begin{bmatrix} 9 & -2 \end{bmatrix} \cdot \begin{bmatrix} -2 & 4 \\ 6 & -7 \end{bmatrix}$

7. $\begin{bmatrix} -9 \\ 6 \end{bmatrix} \cdot \begin{bmatrix} -1 & -10 & 1 \end{bmatrix}$

8. $\begin{bmatrix} -8 & 7 & 4 \\ -5 & -3 & 8 \end{bmatrix} \cdot \begin{bmatrix} 10 & 6 \\ 8 & 4 \end{bmatrix}$

9. $\begin{bmatrix} 2 & 8 \\ 3 & -1 \end{bmatrix} \cdot \begin{bmatrix} 6 \\ -7 \end{bmatrix}$

10. $\begin{bmatrix} -4 & 3 & 2 \\ -1 & -5 & 4 \end{bmatrix} \cdot \begin{bmatrix} 2 & 1 & 6 \\ 8 & 4 & -1 \\ 5 & 3 & -2 \end{bmatrix}$

11. $\begin{bmatrix} 2 & 5 & 3 & -1 \\ -3 & 1 & 8 & -3 \end{bmatrix} \cdot \begin{bmatrix} 6 & -3 \\ -7 & 1 \\ 2 & 0 \\ -1 & 0 \end{bmatrix}$

12. FITNESS The table shows the number of people registered for aerobics for the first quarter.

Quinn's Gym charges the following registration fees: class-by-class, \$165; 11-class pass, \$110; unlimited pass, \$239.

Quinn's Gym		
Payment	**Aerobics**	**Step Aerobics**
class-by-class	35	28
11-class pass	32	17
unlimited pass	18	12

a. Write a matrix for the registration fees and a matrix for the number of students.

b. Find the total amount of money the gym received from aerobics and step aerobic registrations.

Examples 4 and 5 p. 203

Use $X = \begin{bmatrix} -10 & -3 \\ 2 & -8 \end{bmatrix}$, $Y = \begin{bmatrix} -5 & 6 \\ -1 & 9 \end{bmatrix}$, **and** $Z = \begin{bmatrix} -5 & -1 \\ -8 & -4 \end{bmatrix}$ **to determine whether the following equations are true for the given matrices.**

13. $XY = YX$

14. $X(YZ) = (XY)Z$

Practice and Problem Solving

● = Step-by-Step Solutions begin on page R20.
Extra Practice begins on page 947.

Example 1 p. 200

Determine whether each matrix product is defined. If so, state the dimensions of the product.

15. $P_{2 \times 3} \cdot Q_{3 \times 4}$
16. $A_{5 \times 5} \cdot B_{5 \times 5}$
17. $M_{3 \times 1} \cdot N_{2 \times 3}$
18. $X_{2 \times 6} \cdot Y_{6 \times 3}$
19. $J_{2 \times 1} \cdot K_{2 \times 1}$
20. $S_{5 \times 2} \cdot T_{2 \times 4}$

Examples 2 and 3 pp. 201, 202

Find each product, if possible.

21. $\begin{bmatrix} 1 & 6 \end{bmatrix} \cdot \begin{bmatrix} -10 \\ 6 \end{bmatrix}$
22. $\begin{bmatrix} 6 \\ -3 \end{bmatrix} \cdot \begin{bmatrix} 2 & -7 \end{bmatrix}$
23. $\begin{bmatrix} -3 & -7 \\ -2 & -1 \end{bmatrix} \cdot \begin{bmatrix} 4 & 4 \\ 9 & -3 \end{bmatrix}$
24. $\begin{bmatrix} -1 & 0 \\ 5 & 2 \end{bmatrix} \cdot \begin{bmatrix} 6 & -3 \\ 7 & -2 \end{bmatrix}$
25. $\begin{bmatrix} -1 & 0 & 6 \\ -4 & -10 & 4 \end{bmatrix} \cdot \begin{bmatrix} 5 & -7 \\ -2 & -9 \end{bmatrix}$
26. $\begin{bmatrix} -6 & 4 & -9 \\ 2 & 8 & 7 \end{bmatrix} \cdot \begin{bmatrix} 7 \\ 2 \\ 4 \end{bmatrix}$
27. $\begin{bmatrix} 2 & 9 & -3 \\ 4 & -1 & 0 \end{bmatrix} \cdot \begin{bmatrix} 4 & 2 \\ -6 & 7 \\ -2 & 1 \end{bmatrix}$
28. $\begin{bmatrix} -4 \\ 8 \end{bmatrix} \cdot \begin{bmatrix} -3 & -1 \end{bmatrix}$

29. **TRAVEL** The Wolf family owns three bed and breakfasts in a vacation spot. A room with a single bed is $220 a night, a room with two beds is $250 a night, and a suite is $360.
    a. Write a matrix for the number of each type of room at each bed and breakfast. Then write a room-cost matrix.
    b. Write a matrix for total daily income, assuming that all the rooms are rented.
    c. What is the total daily income from all three bed and breakfasts, assuming that all the rooms are rented?

Available Rooms at a Wolf Bed and Breakfast

B & B	Single	Double	Suite
1	3	2	2
2	2	3	1
3	4	3	0

Examples 4 and 5 p. 203

Use $P = \begin{bmatrix} 4 & -1 \\ 1 & 2 \end{bmatrix}$, $Q = \begin{bmatrix} 6 & 4 \\ -2 & -5 \end{bmatrix}$, $R = \begin{bmatrix} 4 & 6 \\ -6 & 4 \end{bmatrix}$, **and** $k = 2$ **to determine whether the following equations are true for the given matrices.**

30. $k(PQ) = P(kQ)$
31. $PQR = RQP$
32. $PR + QR = (P + Q)R$
33. $R(P + Q) = PR + QR$

Real-World Link

Retailers report that Mother's Day is the second highest gift-giving holiday in the United States.

Source: Hallmark

34. **FLOWERS** Student Council is selling flowers for Mother's Day. They bought 200 roses, 150 daffodils, and 100 orchids for the purchase prices shown. They sold all of the flowers for the sales prices shown.
    a. Organize the data in two matrices, and use matrix multiplication to find the total amount that was spent on the flowers.
    b. Write two matrices, and use matrix multiplication to find the total amount the student council received for the flower sale.
    c. Use matrix operations to find how much money the student council made on their project.

Flower	Purchase Price	Sales Price
rose	$1.67	$3.00
daffodil	$1.03	$2.25
orchid	$2.59	$4.50

Real-World Link

There are approximately 245 million passenger vehicles registered in the U.S.

Source: Bureau of Transportation Statistics

35. **AUTO SALES** A car lot has four sales associates. At the end of the year, each sales associate gets a bonus of \$1000 for every new car they have sold and \$500 for every used car they have sold.

a. Use a matrix to determine which sales associate earned the most money.

b. What is the total amount of money the car lot spent on bonuses for the sales associates this year?

Cars Sold by Each Associate		
Sales Associate	New Cars	Used Cars
Mason	27	49
Westin	35	36
Gallagher	9	56
Stadler	15	62

Use matrices $X = \begin{bmatrix} 2 & -6 \\ 3y & -4.5 \end{bmatrix}$, $Y = \begin{bmatrix} -5 & -1.5 \\ x+2 & y \\ 13 & 1.2 \end{bmatrix}$, **and** $Z = \begin{bmatrix} -3 \\ x+y \end{bmatrix}$ **to find each of the following. If the matrix does not exist, write** ***undefined.***

36. XY **37.** YX **38.** ZY **39.** YZ

40. $(YX)Z$ **41.** $(XZ)X$ **42.** $X(ZZ)$ **43.** $(XX)Z$

44. CAMERAS Prices of digital cameras depend on features like optical zoom, digital zoom, and megapixels.

Optical Zoom	6 MP	7 MP	10 MP
3 to 4	\$189.99	\$249.99	\$349.99
5 to 6	\$199.99	\$289.99	\$399.99
10 to 12	\$299.99	\$399.99	\$499.99

a. The 10-mp cameras are on sale for 20% off, and the other models are 10% off. Write a new matrix for these changes.

b. Write a new matrix allowing for a 6.25% sales tax on the discounted prices.

c. Describe what the differences in these two matrices represent.

45. BUSINESS The Kangy Studio has packages available for senior portraits.

a. Use matrices to determine the total cost of each package.

b. The studio offers an early bird discount of 15% off any package. Find the early bird price for each package.

Size (price)	Packages			
	A	B	C	D
4 × 5 (\$7)	10	10	8	0
5 × 7 (\$10)	4	4	4	4
8 × 10 (\$14)	2	2	2	2
11 × 14 (\$45)	1	1	0	0
16 × 20 (\$95)	1	0	0	0
Wallets (8 for \$13)	88	56	16	0

H.O.T. Problems

Use Higher-Order Thinking Skills

46. REASONING If the product matrix AB has dimensions 5×8, and A has dimensions 5×6, what are the dimensions of matrix B?

47. PROOF Show that each property of matrices is true for all 2×2 matrices.

a. Scalar Distributive Property

b. Matrix Distributive Property

c. Associative Property of Multiplication

d. Associative Property of Scalar Multiplication

48. OPEN ENDED Write two matrices A and B such that $AB = BA$.

49. CHALLENGE Find the missing values in $\begin{bmatrix} a & b \\ c & d \end{bmatrix} \cdot \begin{bmatrix} 4 & 3 \\ 2 & 5 \end{bmatrix} = \begin{bmatrix} 10 & 11 \\ 20 & 29 \end{bmatrix}$.

50. WRITING IN MATH Use the data on Lisa Leslie found at the beginning of the lesson to explain how matrices can be used in sports statistics. Describe a matrix that represents the total number of points she has scored during her career and an example of a sport in which different point values are used in scoring.

NYSRE PRACTICE Reinforcement of A.A.7

51. GRIDDED RESPONSE The average (arithmetic mean) of r, w, x, and y is 8, and the average of x and y is 4. What is the average of r and w?

52. Carla, Meiko, and Kayla went shopping to get ready for college. Their purchases and total amounts spent are shown in the table below.

Person	Shirts	Pants	Shoes	Total Spent
Carla	3	4	2	$149.79
Meiko	5	3	3	$183.19
Kayla	6	5	1	$181.14

Assume that all of the shirts were the same price, all of the pants were the same price, and all of the shoes were the same price. What was the price of each item?

A shirt, $12.95; pants, $15.99; shoes, $23.49
B shirt, $15.99; pants, $12.95; shoes, $23.49
C shirt, $15.99; pants, $23.49; shoes, $12.95
D shirt, $23.49; pants, $15.99; shoes, $12.95

53. GEOMETRY Rectangle $LMNQ$ has diagonals that intersect at point P.

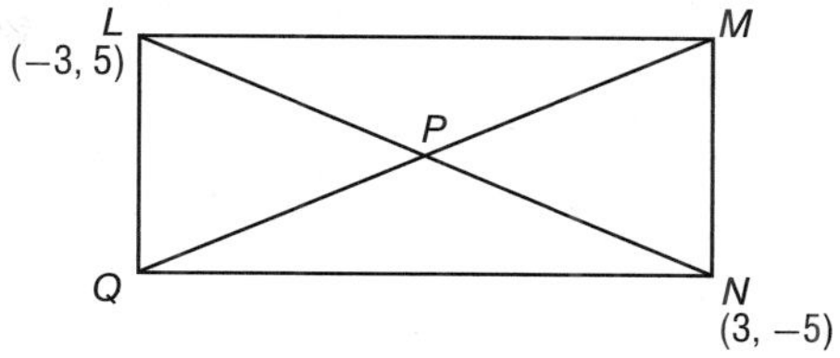

Which of the following represents point P?

F (2, 2) **H** (0, 0)
G (1, 1) **J** (−1, −1)

54. SAT/ACT What are the dimensions of the matrix that results from the multiplication shown?

$$\begin{bmatrix} a & b & c \\ d & e & f \\ g & h & i \\ j & k & l \end{bmatrix} \cdot \begin{bmatrix} 7 \\ 4 \\ 6 \end{bmatrix}$$

A 1×4 **C** 4×1
B 3×3 **D** 4×3

Spiral Review

Perform the indicated operations. If the matrix does not exist, write *impossible*. (Lesson 4-2)

55. $4\begin{bmatrix} 8 & -1 \\ -3 & -4 \end{bmatrix} - 5\begin{bmatrix} -2 & 4 \\ 6 & 3 \end{bmatrix}$

56. $5\left(2\begin{bmatrix} -2 & -5 \\ -1 & 3 \end{bmatrix} - 3\begin{bmatrix} -1 & -2 \\ 6 & 4 \end{bmatrix}\right)$

57. $-4\left(\begin{bmatrix} 8 & 9 \\ -5 & 5 \end{bmatrix} - 2\begin{bmatrix} -6 & -1 \\ 6 & 3 \end{bmatrix}\right)$

State the dimensions of each matrix. (Lesson 4-1)

58. $[\,-2 \quad 1\,]$

59. $\begin{bmatrix} 1 & 6 \\ -8 & -3 \end{bmatrix}$

60. $\begin{bmatrix} 9 & 1 \\ 2 & 3 \\ 5 & -3 \\ -9 & 0 \end{bmatrix}$

61. MEDICINE The graph shows how much Americans spent on doctors' visits in some recent years and a prediction for 2014. (Lesson 2-5)

a. Find a regression equation for the data without the predicted value.

b. Use your equation to predict the expenditures for 2014.

c. Compare your prediction to the one given in the graph.

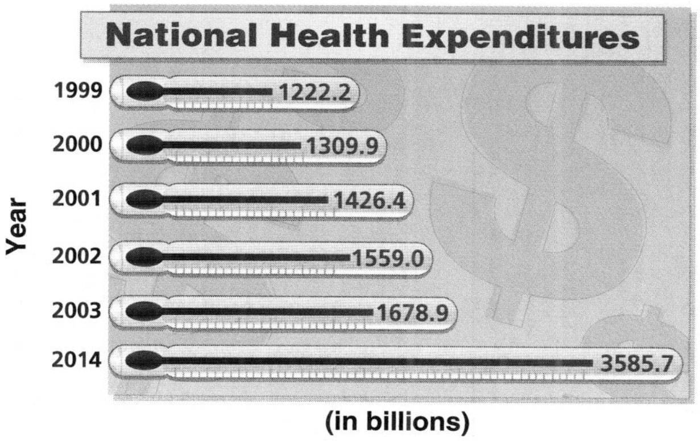

62. How many different ways can the letters of the word *MATHEMATICS* be arranged? (Lesson 0-5)

Skills Review

Describe the transformation in each function. Then graph the function. (Lesson 2-7)

63. $f(x) = |x - 4| + 3$

64. $f(x) = 2|x + 3| - 5$

65. $f(x) = (x + 2)^2 - 6$

Mid-Chapter Quiz

Lessons 4-1 through 4-3

NYS Core Curriculum
A.CM.3, A2.RP.10

State the dimensions of each matrix. (Lesson 4-1)

1. $\begin{bmatrix} 3 & 4 & 5 & 6 & 7 \end{bmatrix}$

2. $\begin{bmatrix} 10 & -6 & 18 & 0 \\ -7 & 5 & 2 & 4 \\ 3 & 11 & 9 & 7 \end{bmatrix}$

Identify each element for the following matrices. (Lesson 4-1)

$$A = \begin{bmatrix} 4 & 3 \\ -5 & 1 \\ -3 & 7 \end{bmatrix}, B = \begin{bmatrix} 1 & -9 & 2 \\ 0 & 10 & 4 \end{bmatrix}$$

3. a_{21}

4. b_{22}

5. FUNDRAISER The ninth and tenth grade classes sold different types of clothing to raise money. Two weeks of sales are shown in the table. (Lesson 4-2)

Grade	Week	Type of Clothing			
		T-shirt	Sweatshirt	Hat	Coat
9	1	25	14	18	5
	2	32	26	15	4
10	1	44	10	13	8
	2	18	38	17	2

a. Write a matrix for each week's sales.

b. Find the sum of the two weeks' sales using matrix addition.

Perform the indicated operations. If the matrix does not exist, write *impossible*. (Lessons 4-2 and 4-3)

6. $\begin{bmatrix} 0 & 15 \\ -6 & -10 \end{bmatrix} - \begin{bmatrix} 8 & 0 \\ -3 & 5 \end{bmatrix}$

7. $-3\begin{bmatrix} 3 & 5 & 12 \\ 0 & -1 & 3 \\ 9 & 6 & -5 \end{bmatrix}$

8. $2\begin{bmatrix} -1 \\ 5 \\ -6 \end{bmatrix} + 4\begin{bmatrix} -3x \\ 2 \\ x \end{bmatrix} - 3\begin{bmatrix} x-2 \\ 3 \\ 1 \end{bmatrix}$

9. MULTIPLE CHOICE Find $2\begin{bmatrix} 3 & 5 \\ -6 & 0 \end{bmatrix} + 4\begin{bmatrix} 9 & -1 \\ 2 & 3 \end{bmatrix}$. (Lesson 4-2)

A $\begin{bmatrix} 42 & 6 \\ -4 & 12 \end{bmatrix}$

B $\begin{bmatrix} 21 & 3 \\ -2 & 6 \end{bmatrix}$

C $\begin{bmatrix} 12 & 4 \\ -4 & 3 \end{bmatrix}$

D $\begin{bmatrix} 27 & -5 \\ -12 & 0 \end{bmatrix}$

Find each product if possible. (Lesson 4-3)

10. $\begin{bmatrix} -2 & 3 \\ 1 & 6 \end{bmatrix} \cdot \begin{bmatrix} 3 & -1 & 4 \\ 0 & 5 & -6 \end{bmatrix}$

11. $\begin{bmatrix} -4 & 0 & -1 \\ 0 & 1 & 8 \end{bmatrix} \cdot \begin{bmatrix} -1 & 0 \\ 0 & 4 \end{bmatrix}$

12. $\begin{bmatrix} 4 & -2 & -7 \\ 6 & 3 & 5 \end{bmatrix} \cdot \begin{bmatrix} -2 \\ 5 \\ 3 \end{bmatrix}$

13. MULTIPLE CHOICE If the product matrix XY has dimensions 3×2 and X has dimensions 3×4, what are the dimensions of matrix Y? (Lesson 4-3)

F 2×3

G 3×2

H 3×4

J 4×2

14. SALES Alex is in charge of stocking shirts for the concession stand at the high school football game. The number of shirts needed for a regular season game is listed in the matrix. Alex plans to double the number of shirts stocked for a playoff game. (Lesson 4-3)

Size	small	medium	large
Child	10	10	15
Adult	25	35	45

a. Write a matrix A to represent the regular season stock.

b. What scalar can be used to determine a matrix M to represent the new numbers? Find M.

c. What is $M - A$? What does this represent in this situation?

15. MULTIPLE CHOICE What is the product of $\begin{bmatrix} 4 & 0 & -2 \end{bmatrix}$ and $\begin{bmatrix} 2 & -1 \\ -3 & 0 \\ 0 & 4 \end{bmatrix}$? (Lesson 4-3)

A $\begin{bmatrix} 8 & -12 \end{bmatrix}$

B $\begin{bmatrix} 8 \\ -12 \end{bmatrix}$

C $\begin{bmatrix} 8 & -4 \\ 0 & 0 \\ 0 & -8 \end{bmatrix}$

D impossible

Determine whether each matrix product is defined. If so, state the dimensions of the product.

16. $A_{2 \times 3} \cdot B_{3 \times 2}$

17. $A_{4 \times 1} \cdot B_{2 \times 1}$

18. $A_{2 \times 5} \cdot B_{5 \times 5}$

19. $A_{1 \times 5} \cdot B_{5 \times 3}$

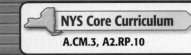

4-4 Transformations with Matrices

Then

You added, subtracted, and multiplied matrices. (Lessons 4-2 and 4-3)

Now

- Use matrices for translations and dilations.
- Use matrices for reflections and rotations.

NYS Core Curriculum

Reinforcement of G.G.61 Investigate, justify, and apply the analytical representations for translations, rotations, reflections, and dilations.

New Vocabulary

vertex matrix
coordinate matrix
transformation
preimage
image
translation
dilation
reflection
rotation

NY Math Online

glencoe.com

- Extra Examples
- Personal Tutor
- Self-Check Quiz
- Homework Help
- Math in Motion

Why?

Video game designers often create detailed settings to add effects to games. One way they do this is to make reflections of objects on a shiny surface, such as a glass table top. To make a reflection, the game designers copy the original object and flip the copy. Matrices are frequently used to define the positions of the objects and to reposition and reorient them.

Translations and Dilations Points on a coordinate plane can be represented by matrices. The ordered pair (x, y) can be represented by the column matrix $\begin{bmatrix} x \\ y \end{bmatrix}$. Likewise, polygons can be represented by placing all of the column matrices of the coordinates of the vertices into one matrix. This is called a **vertex matrix** or **coordinate matrix**.

Triangle ABC with vertices $A(-4, -3)$, $B(-2, 2)$, and $C(3, -1)$ can be represented by the following vertex matrix.

$$\triangle ABC = \begin{array}{c} \quad A \quad\;\; B \quad\;\; C \\ \begin{bmatrix} -4 & -2 & 3 \\ -3 & 2 & -1 \end{bmatrix} \end{array} \begin{array}{l} \leftarrow \textit{x-coordinates} \\ \leftarrow \textit{y-coordinates} \end{array}$$

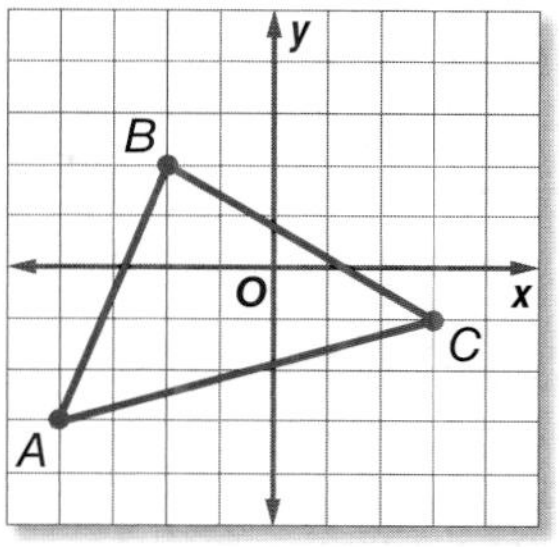

Matrices can be used to perform transformations. **Transformations** are functions that map points of a **preimage** onto its **image**.

One type of transformation is a translation. A **translation** occurs when a figure is moved from one location to another without changing its size, shape, or orientation. You can use matrix addition and a *translation matrix* to determine the coordinates of a translation image. The dimensions of a translation matrix should be the same as the dimensions of the vertex matrix.

Preimage

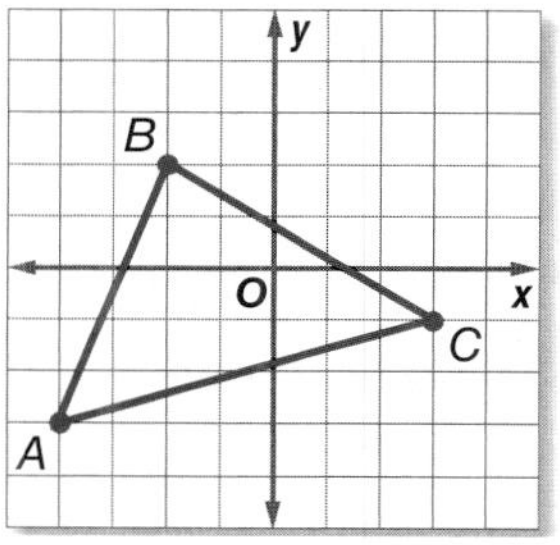

Translation

1 unit down,
2 units right

Image

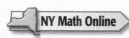

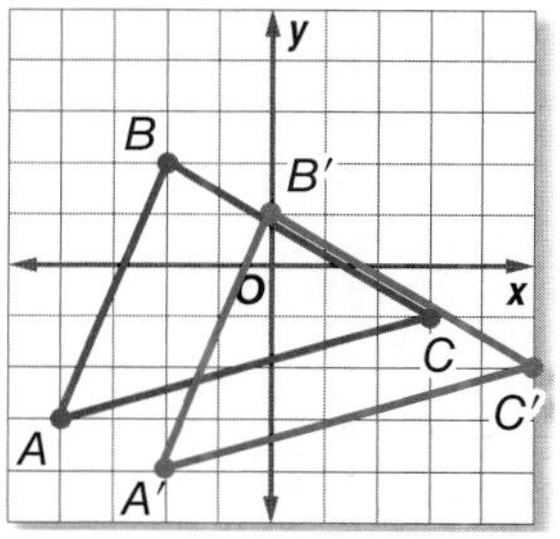

$$\begin{bmatrix} -4 & -2 & 3 \\ -3 & 2 & -1 \end{bmatrix} + \begin{bmatrix} 2 & 2 & 2 \\ -1 & -1 & -1 \end{bmatrix} = \begin{bmatrix} -2 & 0 & 5 \\ -4 & 1 & -2 \end{bmatrix}$$

ReadingMath

Image Coordinates
The notation $D'E'F'G'$ indicates the vertices of the image of $DEFG$ and is read *quadrilateral D prime E prime F prime G prime.*

EXAMPLE 1 Translation

Find the coordinates of the vertices of the image of quadrilateral $DEFG$ with $D(-4, -2)$, $E(0, -1)$, $F(2, -4)$, and $G(-2, -5)$ if it is translated 1 unit left and 4 units up. Then graph $DEFG$ and its image $D'E'F'G'$.

Write the vertex matrix for quadrilateral $DEFG$. $\begin{bmatrix} -4 & 0 & 2 & -2 \\ -2 & -1 & -4 & -5 \end{bmatrix}$

To translate the quadrilateral 1 unit to the left, add -1 to each x-coordinate. To translate the figure 4 units up, add 4 to each y-coordinate. This can be done by adding the translation matrix $\begin{bmatrix} -1 & -1 & -1 & -1 \\ 4 & 4 & 4 & 4 \end{bmatrix}$ to the vertex matrix of $DEFG$.

$$\underset{\text{Vertex Matrix of } DEFG}{\begin{bmatrix} -4 & 0 & 2 & -2 \\ -2 & -1 & -4 & -5 \end{bmatrix}} + \underset{\text{Translation Matrix}}{\begin{bmatrix} -1 & -1 & -1 & -1 \\ 4 & 4 & 4 & 4 \end{bmatrix}} = \underset{\text{Vertex Matrix of } D'E'F'G'}{\begin{bmatrix} -5 & -1 & 1 & -3 \\ 2 & 3 & 0 & -1 \end{bmatrix}}$$

The vertices of $D'E'F'G'$ are $D'(-5, 2)$, $E'(-1, 3)$, $F'(1, 0)$, and $G'(-3, -1)$.

$DEFG$ and $D'E'F'G'$ have the same size, shape, and orientation.

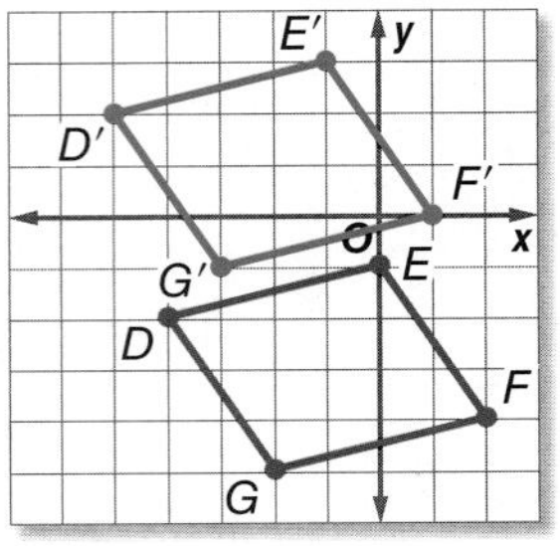

Check Your Progress

1. Find the coordinates of the vertices of the image of triangle RST with $R(-1, 5)$, $S(2, 1)$, and $T(-3, 2)$ if it is moved 3 units to the right and 4 units up. Then graph RST and its image $R'S'T'$.

Personal Tutor glencoe.com

You can work backward to find a translation matrix.

NYSRE EXAMPLE 2 Reinforcement of G.G.61

Rectangle $H'J'K'L'$ is an image of rectangle $HJKL$. A table of the vertices of each rectangle is shown. What are the coordinates of K'?

Rectangle *HJKL*	Rectangle *H'J'K'L'*
$H(4, -5)$	$H'(-1, -8)$
$J(2, -8)$	$J'(-3, -11)$
$K(-4, -4)$	$K'(?, ?)$
$L(-2, -1)$	$L'(-7, -4)$

A $(-5, -3)$ **B** $(-9, -7)$ **C** $(1, -1)$ **D** $(-7, -9)$

Real-World Link

Translation matrices are used for modeling cartoon characters.

Read the Test Item

You are given the coordinates of the preimage and image of points H, J, and L. Use this information to find the translation matrix. Then you can use the translation matrix to find the coordinates of K'.

Test-TakingTip

Two-Step Solutions Sometimes you need to solve for unknown value(s) before you can solve for the value(s) requested in the question.

Solve the Test Item

Step 1 Write a matrix equation. Let (a, b) represent the coordinates of K'.

$$\begin{bmatrix} 4 & 2 & -4 & -2 \\ -5 & -8 & -4 & -1 \end{bmatrix} + \begin{bmatrix} x & x & x & x \\ y & y & y & y \end{bmatrix} = \begin{bmatrix} -1 & -3 & a & -7 \\ -8 & -11 & b & -4 \end{bmatrix}$$

$$\begin{bmatrix} 4+x & 2+x & -4+x & -2+x \\ -5+y & -8+y & -4+y & -1+y \end{bmatrix} = \begin{bmatrix} -1 & -3 & a & -7 \\ -8 & -11 & b & -4 \end{bmatrix}$$

Step 2 The matrices are equal, so corresponding elements are equal. Use the elements from the first row in each matrix.

$4 + x = -1$ **Solve for x.**
$x = -5$

$-5 + y = -8$ **Solve for y.**
$y = -3$

Step 3 Use the values for x and y to find the values for $K'(a, b)$.

$-4 + (-5) = a$
$-9 = a$

$-4 + (-3) = b$
$-7 = b$

So, the coordinates of K' are $(-9, -7)$ and the answer is B. Check by solving different equations for x and y.

Check Your Progress

2. Triangle $X'Y'Z'$ is the image of triangle XYZ. Find the coordinates of Z' using the information shown in the table.

F (3, 2) **H** (7, 0)

G (7, 2) **J** (3, 0)

Triangle XYZ	Triangle $X'Y'Z'$
$X(3, -1)$	$X'(1, 0)$
$Y(-4, 2)$	$Y'(-6, 3)$
$Z(5, 1)$	$Z'(?, ?)$

Personal Tutor glencoe.com

When a geometric figure is enlarged or reduced, the transformation is called a **dilation**. A dilation is performed relative to its center. Unless otherwise specified, the center is always the origin. You can use scalar multiplication to perform dilations.

StudyTip

Dilations In dilation, the image is similar to the preimage. All linear measures of the image change in the same ratio.

EXAMPLE 3 Dilation

Dilate $\triangle MNP$ with $M(-1, 1)$, $N(2, 3)$, and $P(1, -2)$ so that the perimeter of the image is twice that of the preimage. Find the coordinates of the vertices of $\triangle M'N'P'$. Then graph $\triangle MNP$ and $\triangle M'N'P'$.

If the perimeter of the image is twice that of the preimage, then the lengths of the sides of the figure will be twice the measure of the original lengths. Multiply the vertex matrix by the scale factor 2.

$$2\begin{bmatrix} -1 & 2 & 1 \\ 1 & 3 & -2 \end{bmatrix} = \begin{bmatrix} -2 & 4 & 2 \\ 2 & 6 & -4 \end{bmatrix}$$

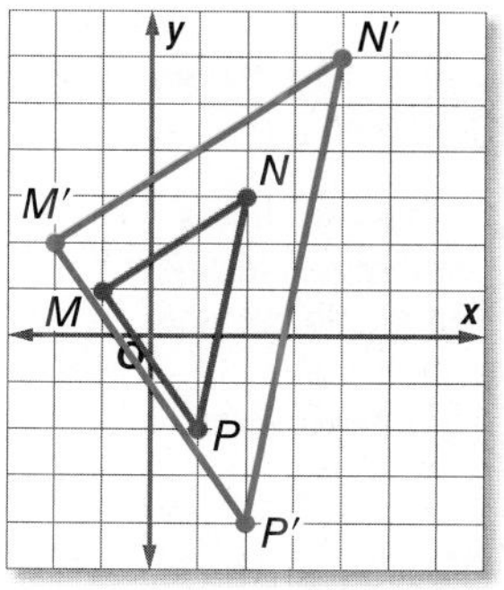

The coordinates of the vertices of $\triangle M'N'P'$ are $M'(-2, 2)$, $N'(4, 6)$, and $P'(2, -4)$.

The preimage and image are similar. Both figures have the same shape.

Check Your Progress

3. Dilate rectangle $WXYZ$ with $W(4, 4)$, $X(4, 12)$, $Y(8, 4)$, and $Z(8, 12)$ so that the perimeter of the image is one fourth that of the preimage. Find the coordinates of the vertices of rectangle $W'X'Y'Z'$.

Personal Tutor glencoe.com

Reflections and Rotations A **reflection** maps every point of a preimage to an image in a line of symmetry using a *reflection matrix*.

Key Concept — Reflection Matrices

For Your FOLDABLE

To reflect in the given line, multiply the vertex matrix by the given matrix.

Line of Reflection	x-axis	y-axis	line $y = x$
Multiply on the left by:	$\begin{bmatrix} 1 & 0 \\ 0 & -1 \end{bmatrix}$	$\begin{bmatrix} -1 & 0 \\ 0 & 1 \end{bmatrix}$	$\begin{bmatrix} 0 & 1 \\ 1 & 0 \end{bmatrix}$
Models	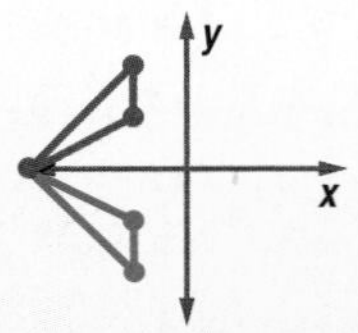	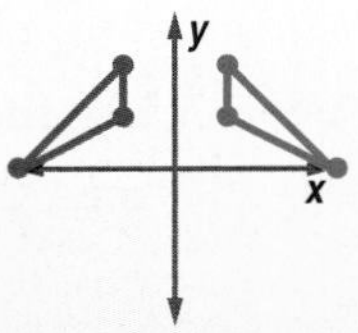	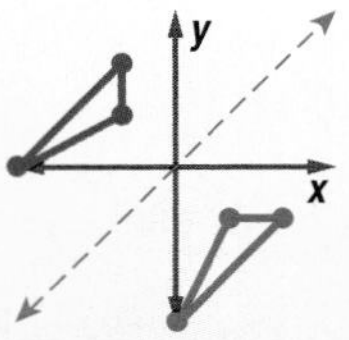

StudyTip

Reflection In a reflection, the preimage and image are congruent.

EXAMPLE 4 Reflection

Find the coordinates of the vertices of the image of *QRST* with *Q*(3, 2), *R*(4, −3), *S*(−3, −4), and *T*(−2, 1) after reflection in $y = x$. Graph the preimage and image.

Write the ordered pairs as a vertex matrix. Then multiply the vertex matrix by the reflection matrix for the line $y = x$.

$$\begin{bmatrix} 0 & 1 \\ 1 & 0 \end{bmatrix} \cdot \begin{bmatrix} 3 & 4 & -3 & -2 \\ 2 & -3 & -4 & 1 \end{bmatrix} = \begin{bmatrix} 2 & -3 & -4 & 1 \\ 3 & 4 & -3 & -2 \end{bmatrix}$$

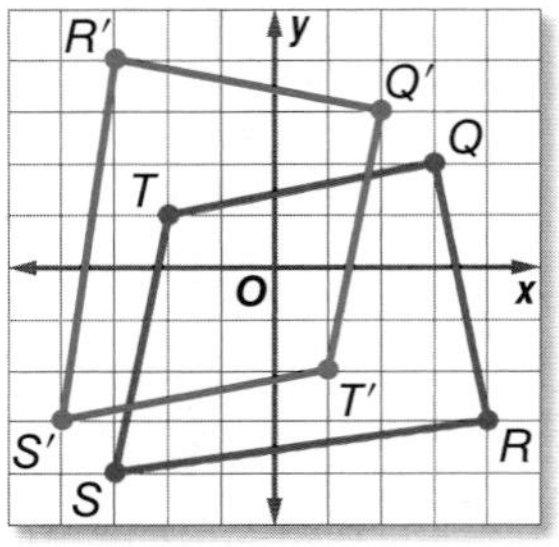

The coordinates of the vertices of quadrilateral $Q'R'S'T'$ are $Q'(2, 3)$, $R'(-3, 4)$, $S'(-4, -3)$, and $T'(1, -2)$.

Check Your Progress

4. Find the coordinates of the vertices of the image of pentagon *ABCDE* with *A*(1, 3), *B*(3, 2), *C*(3, −1), *D*(1, −2), and *E*(−1, 1) reflected across the x-axis.

Personal Tutor glencoe.com

A **rotation** maps every point of a preimage to an image rotated about a center point, usually the origin, using a *rotation matrix*.

StudyTip

Rotation Unless otherwise indicated, all rotations in this text will be counterclockwise.

Key Concept — Rotation Matrices

For Your FOLDABLE

To rotate counterclockwise about the origin, multiply the vertex matrix by the given matrix.

Angle of Rotation	90°	180°	270°
Multiply on the left by:	$\begin{bmatrix} 0 & -1 \\ 1 & 0 \end{bmatrix}$	$\begin{bmatrix} -1 & 0 \\ 0 & -1 \end{bmatrix}$	$\begin{bmatrix} 0 & 1 \\ -1 & 0 \end{bmatrix}$
Models	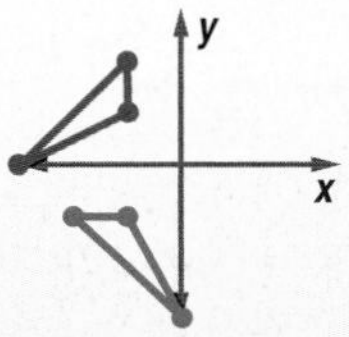	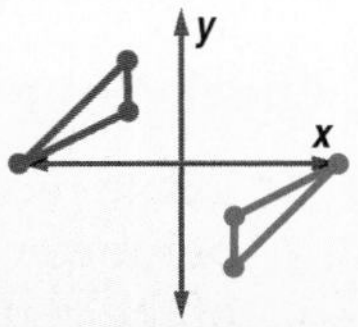	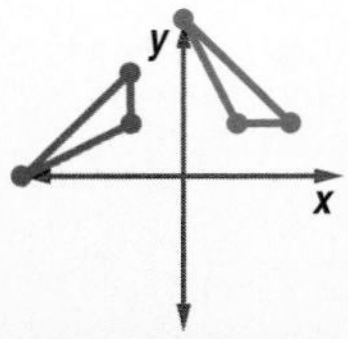

Math *in Motion,* Interactive Lab glencoe.com

StudyTip

Rotation In a rotation, like a reflection, the preimage and image are congruent.

EXAMPLE 5 Rotation

Find the coordinates of the vertices of the image of $\triangle VWX$ with $V(3, 2)$, $W(4, 1)$, and $X(2, 1)$ after it is rotated 270° counterclockwise about the origin.

Write the ordered pairs in a vertex matrix. Then multiply the vertex matrix by the rotation matrix.

$$\begin{bmatrix} 0 & 1 \\ -1 & 0 \end{bmatrix} \cdot \begin{bmatrix} 3 & 4 & 2 \\ 2 & 1 & 1 \end{bmatrix} = \begin{bmatrix} 2 & 1 & 1 \\ -3 & -4 & -2 \end{bmatrix}$$

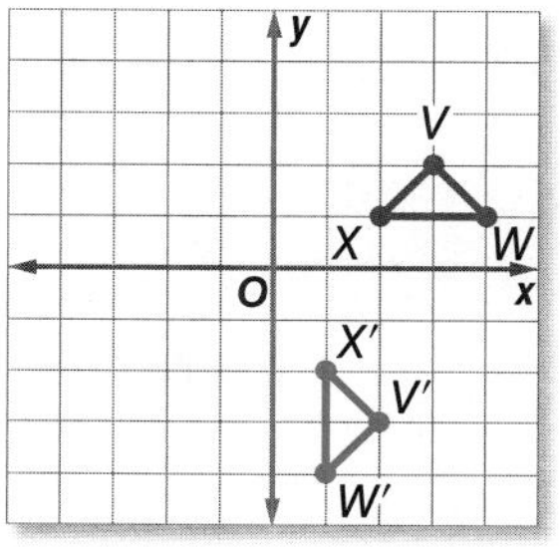

The coordinates of the vertices of $\triangle V'W'X'$ are $V'(2, -3)$, $W'(1, -4)$, and $X'(1, -2)$.

The image is congruent to the preimage. Both figures have the same size and shape.

Check Your Progress

5. Find the coordinates of the vertices of the image of $\triangle XYZ$ with $X(-5, -6)$, $Y(-1, -3)$, and $Z(-2, -4)$ after it is rotated 180° counterclockwise about the origin.

Personal Tutor glencoe.com

Check Your Understanding

Example 1 p. 210

Find the coordinates of the vertices of the image for each figure after the given translation. Then graph the preimage and image.

1. $\triangle ABC$ with vertices $A(-3, -3)$, $B(-4, 2)$, and $C(1, 0)$, translated 5 units right and down 3 units

2. quadrilateral $WXYZ$ with vertices $W(1, -3)$, $X(0, 2)$, $Y(1, 2)$, and $Z(2, 1)$, translated 2 units left and up 4 units

Example 2 p. 210

3. **MULTIPLE CHOICE** Rectangle $RSTU$ with vertices $R(-3, 2)$, $S(1, 2)$, $T(-3, -1)$, and $U(1, -1)$ is translated so that T' is at $(-4, 1)$. What are the coordinates of R' and U'?

A $R'(-4, 4)$, $U'(0, 1)$

B $R'(0, 1)$, $U'(-4, 4)$

C $R'(-2, 0)$, $U'(0, -3)$

D $R'(0, -2)$, $U'(-3, 0)$

Example 3 p. 211

Find the coordinates of the vertices of the image after the given dilation. State the coordinates of the vertices of the image. Then graph the preimage and image.

4. $\triangle DEF$ with vertices $D(-2, -1)$, $E(0, 3)$, and $F(2, -1)$, dilated so that its perimeter is three times the original perimeter

5. square $STUV$ with vertices $S(1, 0)$, $T(4, -3)$, $U(1, -6)$, and $V(-2, -3)$, dilated so that its perimeter is one half the original perimeter

Example 4 p. 212

Find the coordinates of the vertices of the image of each figure after a reflection in the given axis of symmetry. Then graph the preimage and image.

6. rectangle $GHJK$ with vertices $G(1, 5)$, $H(3, 4)$, $J(0, -2)$, and $K(-2, -1)$; x-axis

7. $\triangle PQR$ with vertices $P(-1, 2)$, $Q(4, -4)$, and $R(-1, -4)$; $y = x$

Example 5 p. 213

Find the coordinates of the vertices of the image of each figure after the given rotation. Then graph the preimage and image.

8. $\triangle LMN$ with vertices $L(3, -5)$, $M(1, 2)$, and $N(3, 3)$; 180° about the origin

9. quadrilateral $ABCD$ with $A(-4, 1)$, $B(-3, 4)$, $C(0, 4)$, and $D(1, 1)$; 90° about the origin

Practice and Problem Solving

● = **Step-by-Step Solutions** begin on page R20.
Extra Practice begins on page 947.

Example 1 p. 210

Find the coordinates of the vertices of the image of each figure after the given translation. Then graph the preimage and image.

10. $\triangle MNO$ with vertices $M(-7, 6)$, $N(1, 7)$, and $O(-3, 1)$, translated 2 units right and 6 units down

11. quadrilateral $EFGH$ with vertices $E(-4, 3)$, $F(2, -2)$, $G(-2, -4)$, and $H(-3, -3)$, translated 4 units left and 1 unit up

12. rectangle $PQRS$ with vertices $P(-2, -3)$, $Q(-2, 2)$, $R(1, 2)$, and $S(1, -3)$, translated 2 units left and 5 units down

13. $\triangle JKL$ with vertices $J(1, 4)$, $K(2, 1)$, and $L(-1, -2)$, translated 3 units right and 4 units down

14. $\triangle ABC$ with vertices $A(-1, 2)$, $B(2, 4)$, and $C(3, -2)$, translated 5 units left and 3 units down

15. square $WXYZ$ with vertices $W(3, 1)$, $X(3, 5)$, $Y(7, 5)$, and $Z(7, 1)$, translated 2 units left and 4 units up

Example 2 p. 210

16. Quadrilateral $A'B'C'D'$ is an image after a translation of quadrilateral $ABCD$. A table of the vertices of each quadrilateral is shown.

    a. Find the coordinates of B'.

    b. What are the coordinates of C?

Quadrilateral *ABCD*	Quadrilateral *A'B'C'D'*
$A(4, 8)$	$A'(-1, -2)$
$B(11, -10)$	$B'(?, ?)$
$C(?, ?)$	$C'(-17, -14)$
$D(-4, 6)$	$D'(-9, -4)$

Problem-SolvingTip

Draw a Diagram When you solve problems, it may be helpful to draw a diagram to visualize the situation. In Exercise 17, a drawing can really help.

17. **MAPS** Camila looks at a map of a city she is visiting to find the way to a mall. Currently Camila is standing at an intersection with coordinates (5.5, 7). She figures out that she must go 3 blocks east and 4 blocks north to get to the mall.

    a. Write a translation matrix that Camila can use to find the coordinates of the mall.

    b. Using the translation matrix, what are the coordinates of the mall?

Example 3 p. 211

Find the coordinates of the vertices of the image after the given dilation. State the coordinates of the vertices of the image. Then graph the preimage and image.

18. $\triangle TUV$ with vertices $T(-3, -1)$, $U(2, -1)$, and $V(1, -5)$, dilated so that the perimeter of the image is twice that of the preimage

19. square $DEFG$ with vertices $D(0, 4)$, $E(2, 2)$, $F(0, 0)$, and $G(-2, 2)$, dilated by a scale factor of 4

20. rectangle $KLMN$ with vertices $K(-3, 8)$, $L(-3, 2)$, $M(-6, 2)$, and $N(-6, 8)$, dilated by a scale factor of $\frac{1}{3}$

21. $\triangle QRS$ with vertices $Q(6, 4)$, $R(8, 0)$, and $S(0, 1)$, dilated so that the perimeter of the image is one fourth that of the preimage

Example 4 p. 212

Find the coordinates of the vertices of the image of each figure after a reflection in the given axis of symmetry. Then graph the preimage and image.

22. $\triangle ABC$ with vertices $A(9, -10)$, $B(-5, -6)$, and $C(7, 7)$; y-axis

23. $\triangle DEF$ with vertices $D(7, 4)$, $E(4, 0)$, and $F(-3, 2)$; x-axis

24. quadrilateral $GHJK$ with vertices $G(-5, 6)$, $H(-2, 10)$, $J(0, 8)$, and $K(-4, -4)$; $y = x$

25. square $LMNP$ with vertices $L(1, -2)$, $M(-5, -1)$, $N(-4, 5)$, and $P(2, 4)$; y-axis

26. $\triangle QRS$ with vertices $Q(-2, 2)$, $R(4, 2)$, and $S(2, -6)$; x-axis

27. $\triangle TUV$ with vertices $T(-4, 5)$, $U(1, 3)$, and $V(-2, 0)$; $y = x$

28. **ANIMATION** Yori has plotted all of the locations for her cartoon on a coordinate plane. The vertices of the outline of a character's house have coordinates $(-3, 5)$, $(-3, 7)$, $(5, 5)$, and $(5, 7)$. Yori decides that she wants to move all of the locations so that they are reflected across the y-axis. What will be the new coordinates of the house?

Example 5 p. 213

Find the coordinates of the vertices of the image of each figure after the given rotation. Then graph the preimage and image.

Real-World Link

The Singapore Flyer is 42 stories tall.

29. $\triangle XYZ$ with vertices $X(1, 2)$, $Y(1, 4)$, and $Z(5, 2)$; 180° about the origin

30. quadrilateral $ABCD$ with vertices $A(-1, 1)$, $B(-4, 1)$, $C(-6, 5)$, and $D(-3, 5)$; 270° about the origin

31. rectangle $EFGH$ with vertices $E(3, 2)$, $F(3, -4)$, $G(2, -4)$, and $H(2, 2)$; 270° about the origin

32. square $JKLM$ with vertices $J(-3, 3)$, $K(1, 3)$, $L(1, -1)$, and $M(-3, -1)$; 90° about the origin

33. quadrilateral $NPQR$ with vertices $N(-2, -1)$, $P(-3, -5)$, $Q(-6, -5)$, and $R(-6, -1)$; 90° about the origin

34. $\triangle STU$ with vertices $S(-4, 2)$, $T(1, 5)$, and $U(-2, -1)$; 180° about the origin

35. **RIDES** The world's tallest Ferris wheel, the Singapore Flyer, is 165 meters across. Jeremy and Nicole take a ride on it while on vacation. When they are at the top of the Ferris wheel, they have coordinates $(0, 165)$. Find their coordinates after the wheel has turned 90° counterclockwise.

Find the coordinates of the vertices of the image after the given transformations.

36. reflected in $y = x$, then rotated 270° about the origin

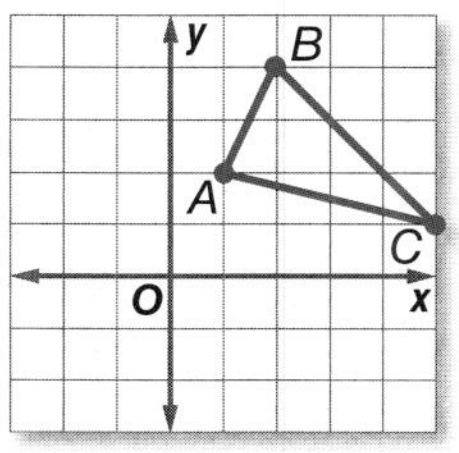

37. translated 4 units left and 2 units up, then reflected in the x-axis

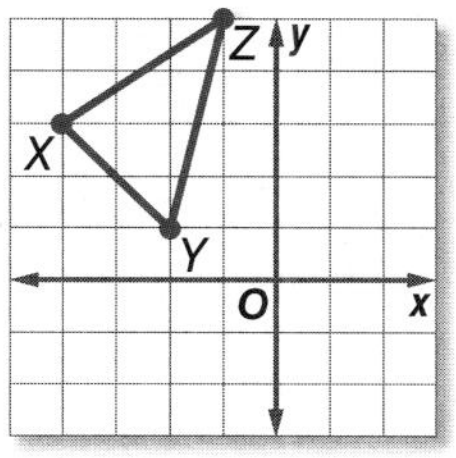

38. rotated 90° about the origin after being reflected in the y-axis

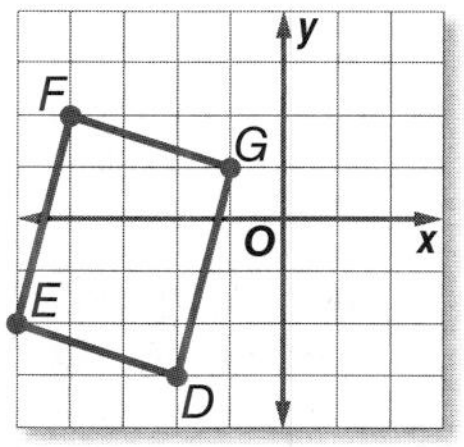

39. rotated 180° about the origin

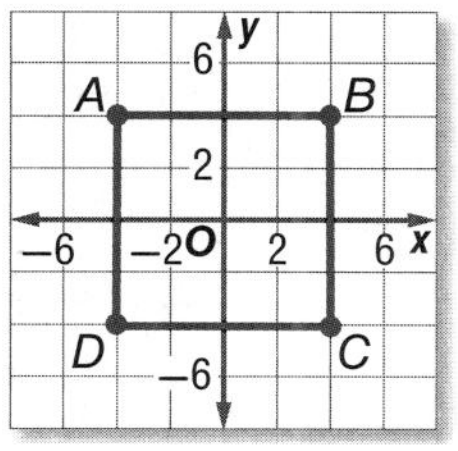

Real-World Link

Blueprint floor plans are typically drawn to a $\frac{1}{4}$ inch per 1 foot scale of the actual size of the home.

Source: The House Designers

40. ARCHITECTURE On the blueprint for a new house, the coordinates of the corners of the garage are (1, 25), (1, 5), (31, 5), and (31, 25). If the scale of the blueprint is $\frac{1}{48}$ that of the actual structure, find the coordinates of the garage when it is built.

41. MULTIPLE REPRESENTATIONS In this problem, you will investigate the results of multiple reflections.

- **a. SYMBOLIC** Write the vertex matrices for $\triangle ABC$ with vertices $A(-2, 3)$, $B(4, 0)$, and $C(2, -5)$ and its image $\triangle A'B'C'$ after reflection in the x-axis.
- **b. GRAPHICAL** Graph $\triangle ABC$ and $\triangle A'B'C'$.
- **c. VERBAL** Make a conjecture about the reflection of $\triangle A'B'C'$ in the x-axis. Perform the matrix multiplication to verify in conjecture.
- **d. ANALYTICAL** Find one matrix that could be used to reflect a triangle in the x-axis twice. How does the matrix relate to in observations?

H.O.T. Problems Use Higher-Order Thinking Skills

42. OPEN ENDED A *composite transformation* is a transformation involving two or more transformations performed in sequence. Write a transformation matrix that could be used to perform a composite transformation involving dilation and rotation of a figure.

43. REASONING Determine whether the following statement is *sometimes*, *always*, or *never* true. Explain your reasoning.

The image of a dilation is congruent to its preimage.

44. CHALLENGE Write a transformation matrix for each of the following.

- **a.** reflection in the line $y = -x$
- **b.** rotation 90° clockwise about the origin

45. OPEN ENDED Consider the point represented by the matrix $\begin{bmatrix} 3 \\ -2 \end{bmatrix}$. Give an example of a translation matrix and a reflection matrix that when applied separately to the point produce the same image point.

46. WHICH ONE DOESN'T BELONG? Determine which of the transformations is not the same as the others. Explain your reasoning.

A

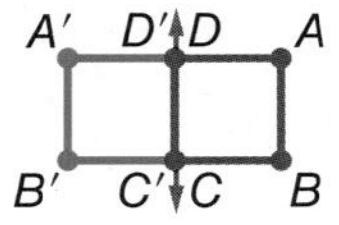

C

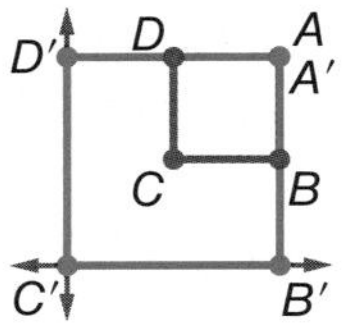

B

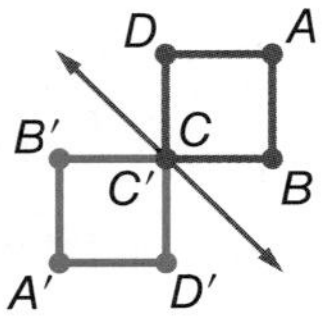

D

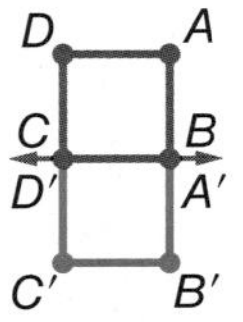

47. PROOF Show that rotating $\triangle ABC$ with vertices $A(x_1, y_1)$, $B(x_2, y_2)$, and $C(x_3, y_3)$ 180° counterclockwise about the origin is the same as reflecting the figure in the x-axis, then in the y-axis.

48. WRITING IN MATH Use the information in this lesson to describe how matrices can be used to transform figures in two-dimensional space. Describe the physical characteristics of a figure after each transformation.

NYSRE PRACTICE A2.S.12, Reinforcement of G.G.61

49. Tonya wanted to find 5 consecutive whole numbers that add up to 95. She wrote the equation $(n - 2) + (n - 1) + n + (n + 1) + (n + 2) = 95$. What does the variable n represent in the equation?

A the least of the 5 whole numbers
B the middle of the 5 whole numbers
C the greatest of the 5 whole numbers
D the difference between the least and the greatest of the 5 whole numbers

50. PROBABILITY There are 12 songs on your MP3 player. The player is set to shuffle, but not repeat—that is, to play the songs in random order without repeating any. What is the probability that any two of the four country songs will be the first two played?

F $P \approx 20\%$
G $P \approx 11.1\%$
H $P \approx 9.09\%$
J $P \approx 0.83\%$

51. EXTENDED RESPONSE The following types of vehicles are rented at The Auto Store. Organize the data into a 4×2 matrix. Then write a new matrix providing the prices after a 15% markup.

	Daily	Weekly
Economy	\$29.99	\$149.99
Mid-Size	\$39.99	\$179.99
Full-Size	\$49.99	\$209.99
SUV	\$69.99	\$349.99

52. SAT/ACT Triangle ABC has vertices $A(-4, 2)$, $B(-4, -3)$, and $C(3, -2)$. After a dilation, triangle $A'B'C'$ has vertices $A'(-12, 6)$, $B'(-12, -9)$, $C'(9, -6)$. How many times as great is the perimeter of $\triangle A'B'C'$ as that of $\triangle ABC$?

A 3
B 6
C 12
D $\frac{1}{3}$

Spiral Review

Find each product, if possible. (Lesson 4-3)

53. $\begin{bmatrix} 4 & 2 \\ -1 & -3 \end{bmatrix} \cdot \begin{bmatrix} 6 & 2 \\ 5 & 1 \end{bmatrix}$

54. $\begin{bmatrix} 8 & -2 \\ -4 & -5 \end{bmatrix} \cdot \begin{bmatrix} -2 \\ 3 \end{bmatrix}$

55. $\begin{bmatrix} -3 \\ -4 \end{bmatrix} \cdot \begin{bmatrix} -6 & -8 \\ -4 & 5 \end{bmatrix}$

56. BUSINESS The table lists the prices at the Sandwich Shoppe. (Lesson 4-2)

a. List the prices in a 4×3 matrix.
b. The manager decides to cut the prices of every item by 20%. List this new set of data in a 4×3 matrix.
c. Subtract the second matrix from the first and determine the savings to the customer for each sandwich.

Sandwich	Small	Medium	Large
ham	\$4.50	\$6.75	\$9.50
salami	\$4.50	\$6.75	\$9.50
veggie	\$4.00	\$6.25	\$8.75
meatball	\$4.75	\$7.50	\$10.25

Graph each inequality. (Lesson 2-8)

57. $|2x + 5| + 3 \geq y$

58. $y \leq 2|x - 4|$

59. $y \geq -2|x + 3| + 1$

State whether the events are *independent* or *dependent*. (Lesson 0-4)

60. choosing a president, vice president, secretary, and treasurer for Key Club, assuming that a person can hold only one office

61. selecting a horror movie and a comedy at the video store

Skills Review

Solve each system of equations. (Lesson 3-2)

62. $y = 3x - 10$
$4x - 3y = 20$

63. $4y + 5x = 21$
$2x + 7y = 3$

64. $-5x - 2y = 27$
$8x + 5y = -54$

Algebra Lab
Vectors and Matrices

NY Math Online glencoe.com
Math in Motion, Animation

A 1×2 or 2×1 matrix can be represented by a vector. A **vector** is a quantity that has both **magnitude**, or length, and **direction**. They are represented as directed segments such as $\overrightarrow{AB}$, read *vector AB*.

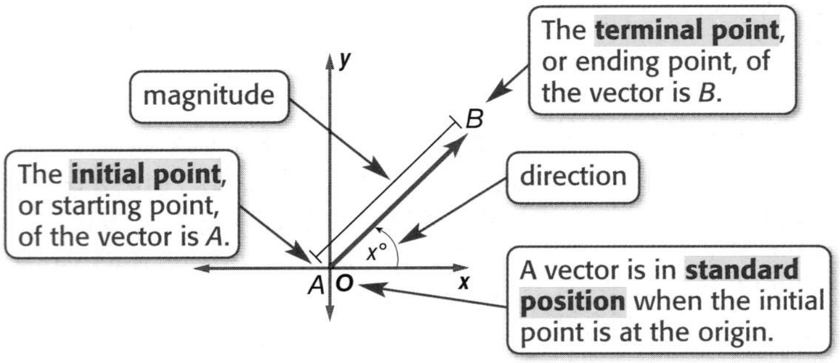
magnitude
The terminal point, or ending point, of the vector is B.
The initial point, or starting point, of the vector is A.
direction
A vector is in standard position when the initial point is at the origin.
y
x
B
A
O
x°

A vector on a coordinate grid can be described in **component form**, or in terms of its horizontal and vertical components. The notation for a vector in component form is ⟨*x* component, *y* component⟩. The vector ⟨2, 3⟩ has initial point (0, 0) and terminal point (2, 3).

Matrix operations can be used to determine the component form of a vector.

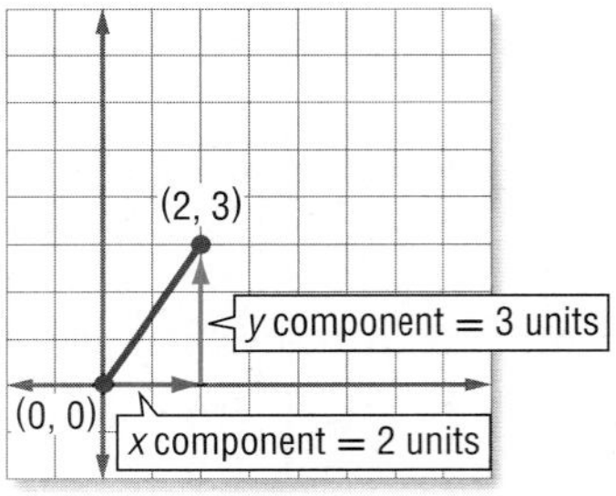
(2, 3)
y component = 3 units
(0, 0)
x component = 2 units

EXAMPLE 1 Determine Component Form

If $A = \begin{bmatrix} -2 \\ 1 \end{bmatrix}$ **and** $B = \begin{bmatrix} 4 \\ 3 \end{bmatrix}$**, express** $\overrightarrow{AB}$ **in component form. Then graph** $\overrightarrow{AB}$**.**

Step 1 Determine the *x*- and *y*-components by finding $B - A$.

$$B - A = \begin{bmatrix} 4 \\ 3 \end{bmatrix} - \begin{bmatrix} -2 \\ 1 \end{bmatrix}$$

$$= \begin{bmatrix} 4 - (-2) \\ 3 - 1 \end{bmatrix} \text{ or } \begin{bmatrix} 6 \\ 2 \end{bmatrix}$$

Step 2 Express $\overrightarrow{AB}$ in component form.

$$\left.\begin{bmatrix} 6 \\ 2 \end{bmatrix} \begin{matrix} \leftarrow x \text{ component} \\ \leftarrow y \text{ component} \end{matrix}\right\} \langle 6, 2 \rangle$$

Step 3 Graph $\overrightarrow{AB}$.

Because $\overrightarrow{AB} = \langle 6, 2 \rangle$, begin the vector at (0, 0) and extend it to (6, 2).

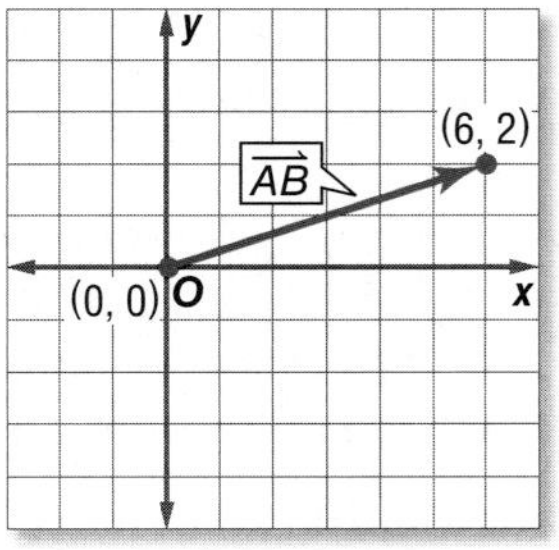
y
(6, 2)
AB
(0, 0)
O
x

EXAMPLE 2 Add Vectors

Find $\overrightarrow{CD} + \overrightarrow{GH}$ for $C(-2, -2)$, $D(1, 3)$, $G(2, 5)$, and $H(4, -1)$.

Step 1 Express the ordered pairs as matrices.

$$C = \begin{bmatrix} -2 \\ -2 \end{bmatrix}, D = \begin{bmatrix} 1 \\ 3 \end{bmatrix}, G = \begin{bmatrix} 2 \\ 5 \end{bmatrix}, H = \begin{bmatrix} 4 \\ -1 \end{bmatrix}$$

Step 2 Use matrix operations to represent each vector.

$$\overrightarrow{CD}: D - C = \begin{bmatrix} 1 - (-2) \\ 3 - (-2) \end{bmatrix} = \begin{bmatrix} 3 \\ 5 \end{bmatrix}$$

$$\overrightarrow{GH}: H - G = \begin{bmatrix} 4 - 2 \\ -1 - 5 \end{bmatrix} = \begin{bmatrix} 2 \\ -6 \end{bmatrix}$$

Step 3 Add the matrices to represent $\overrightarrow{CD} + \overrightarrow{GH}$.

$$\begin{bmatrix} 3 \\ 5 \end{bmatrix} + \begin{bmatrix} 2 \\ -6 \end{bmatrix} = \begin{bmatrix} 3 + 2 \\ 5 + (-6) \end{bmatrix} = \begin{bmatrix} 5 \\ -1 \end{bmatrix}$$

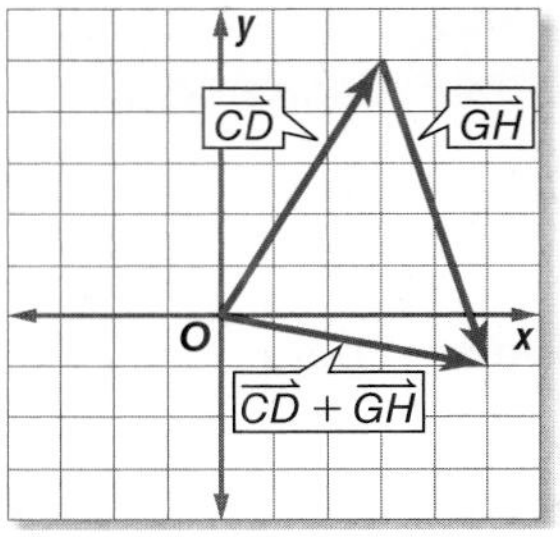

Step 4 Write the matrix as a component vector.

$$\begin{bmatrix} 5 \\ -1 \end{bmatrix} \rightarrow \langle 5, -1 \rangle$$

Thus, $\overrightarrow{CD} + \overrightarrow{GH} = \langle 5, -1 \rangle$.

Exercises

Express $\overrightarrow{XY}$ in component form. Then graph $\overrightarrow{XY}$.

1. $X = \begin{bmatrix} 5 \\ 9 \end{bmatrix}, Y = \begin{bmatrix} -10 \\ -2 \end{bmatrix}$

2. $X = \begin{bmatrix} 8 \\ 0 \end{bmatrix}, Y = \begin{bmatrix} 3 \\ -5 \end{bmatrix}$

3. $X = [\,-8 \quad -9\,], Y = [\,8 \quad 9\,]$

4. $X = [\,-4 \quad -7\,], Y = [\,-4 \quad 6\,]$

Find $\overrightarrow{PQ} + \overrightarrow{RS}$.

5. $P(-1, -3), Q(8, -6), R(-5, -7), S(4, 2)$

6. $P(0, 10), Q(-5, 2), R(-5, 0), S(3, 4)$

7. $P(1, -6), Q(-9, 0), R(6, -6), S(1, 9)$

8. $P(5, -8), Q(2, 5), R(-6, 2), S(-7, 3)$

Vectors can be subtracted by using the same method as addition. Find $\overrightarrow{JK} - \overrightarrow{LM}$.

9. $J(7, -1), K(-3, 9), L(-9, 4), M(6, 0)$

10. $J(6, 7), K(10, -4), L(-2, 8), M(3, 1)$

11. **MAKE A CONJECTURE** Is addition of vectors commutative? Justify your reasoning.

12. **WRITING IN MATH** Explain how you could apply scalar multiplication of matrices to vectors. Describe how a vector is affected by scalar multiplication.

4-5 Determinants and Cramer's Rule

Then

You solved systems of equations algebraically. (Lesson 3-2)

Now

- Evaluate determinants.
- Solve systems of linear equations by using Cramer's Rule.

NYS Core Curriculum

Reinforcement of A.A.7 Analyze and solve verbal problems whose solution requires solving systems of linear equations in two variables
Reinforcement of A.A.10 Solve systems of two linear equations in two variables algebraically

New Vocabulary

determinant
second-order determinant
third-order determinant
diagonal rule
Cramer's Rule
coefficient matrix

NY Math Online

glencoe.com
- Extra Examples
- Personal Tutor
- Self-Check Quiz
- Homework Help

Why?

A zoologist tagged a tiger with a GPS tracker so that she could determine the tiger's territory. After several days, the zoologist determined that the tiger's territory was a triangular region. By using the coordinates of the vertices of this triangle, she could use matrices and determinants to determine the size of the tiger's territory.

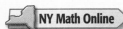

Determinants Every square matrix has a **determinant**. The determinant of a 2×2 matrix is called a **second-order determinant**.

Key Concept — Second-Order Determinant

For Your FOLDABLE

Words The value of a second-order determinant is the difference of the products of the two diagonals.

Symbols $\det \begin{bmatrix} a & b \\ c & d \end{bmatrix} = \begin{vmatrix} a & b \\ c & d \end{vmatrix} = ad - bc$

Example $\begin{vmatrix} 4 & 5 \\ -3 & 6 \end{vmatrix} = 4(6) - (-3)(5) = 39$

EXAMPLE 1 Second-Order Determinant

Evaluate each determinant.

a. $\begin{vmatrix} 5 & -4 \\ 8 & 9 \end{vmatrix}$

$\begin{vmatrix} 5 & -4 \\ 8 & 9 \end{vmatrix} = 5(9) - 8(-4)$ **Definition of determinant**

$= 45 + 32$ **Simplify.**
$= 77$

b. $\begin{vmatrix} 0 & 6 \\ 4 & -11 \end{vmatrix}$

$\begin{vmatrix} 0 & 6 \\ 4 & -11 \end{vmatrix} = 0(-11) - 4(6)$ **Definition of determinant**

$= 0 - 24$ **Simplify.**
$= -24$

Check Your Progress

1A. $\begin{vmatrix} -6 & -7 \\ 10 & 8 \end{vmatrix}$

1B. $\begin{vmatrix} 7 & 5 \\ 9 & -4 \end{vmatrix}$

Personal Tutor glencoe.com

StudyTip

Diagonal Rule The diagonal rule can only be used for 3×3 matrices.

Determinants of 3×3 matrices are called **third-order determinants**. They can be evaluated by using the **diagonal rule**.

Key Concept — Diagonal Rule

For Your FOLDABLE

Step 1 Rewrite the first two columns to the right of the determinant.

Step 2 Draw diagonals, beginning with the upper left-hand element. Multiply the elements in each diagonal. Repeat the process, beginning with the upper right-hand element.

Step 3 Find the sum of the products of the elements in each set of diagonals.

Step 4 Subtract the second sum from the first sum.

EXAMPLE 2 Use Diagonals

Evaluate $\begin{vmatrix} 4 & -8 & 3 \\ -3 & 2 & 6 \\ -4 & 5 & 9 \end{vmatrix}$ **using diagonals.**

Step 1 Rewrite the first two columns to the right of the determinant.

$$\left|\begin{array}{rrr|rr} 4 & -8 & 3 & 4 & -8 \\ -3 & 2 & 6 & -3 & 2 \\ -4 & 5 & 9 & -4 & 5 \end{array}\right.$$

Step 2 Find the products of the elements of the diagonals.

$$\left|\begin{array}{rrr|rr} 4 & -8 & 3 & 4 & -8 \\ -3 & 2 & 6 & -3 & 2 \\ -4 & 5 & 9 & -4 & 5 \end{array}\right. \qquad \left|\begin{array}{rrr|rr} 4 & -8 & 3 & 4 & -8 \\ -3 & 2 & 6 & -3 & 2 \\ -4 & 5 & 9 & -4 & 5 \end{array}\right.$$

$4(2)(9) = 72$	$-4(2)(3) = -24$
$-8(6)(-4) = 192$	$5(6)(4) = 120$
$3(-3)(5) = -45$	$9(-3)(-8) = 216$

Step 3 Find the sum of each group.

$72 + 192 + (-45) = 219$ $\qquad$ $-24 + 120 + 216 = 312$

Step 4 Subtract the sum of the second group from the sum of the first group.

$219 - 312 = -93$

The value of the determinant is -93.

Check Your Progress

Evaluate each determinant.

2A. $\begin{vmatrix} -5 & 9 & 4 \\ -2 & -1 & 5 \\ -4 & 6 & 2 \end{vmatrix}$ $\qquad$ **2B.** $\begin{vmatrix} -8 & -4 & 4 \\ 0 & -5 & -8 \\ 3 & 4 & 1 \end{vmatrix}$

Personal Tutor glencoe.com

Determinants can also be used to find the areas of triangles. If the coordinates of the vertices of the triangle are known, the formula below can be used to calculate the area of the triangle.

Key Concept — Area of a Triangle

For Your FOLDABLE

Words The area of a triangle with vertices (a, b), (c, d), and (e, f) is $|A|$, where

$$A = \frac{1}{2}\begin{vmatrix} a & b & 1 \\ c & d & 1 \\ e & f & 1 \end{vmatrix}.$$

Example $$A = \frac{1}{2}\begin{vmatrix} -4 & 3 & 1 \\ 3 & 1 & 1 \\ -2 & -2 & 1 \end{vmatrix}$$

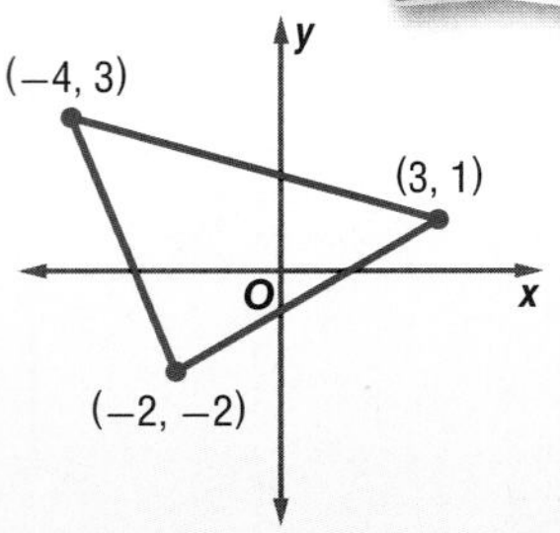

Real-World Link

Tigers are very territorial, solitary animals. Their territories can measure up to 100 square kilometers.

Source: *National Geographic*

Real-World EXAMPLE 3 — Use Determinants

ZOOLOGY Refer to the application at the beginning of the lesson. The coordinates of the vertices of the tiger's territory are shown to the right. Use determinants to find the area of the tiger's territory.

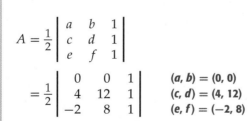

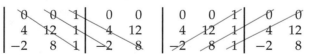

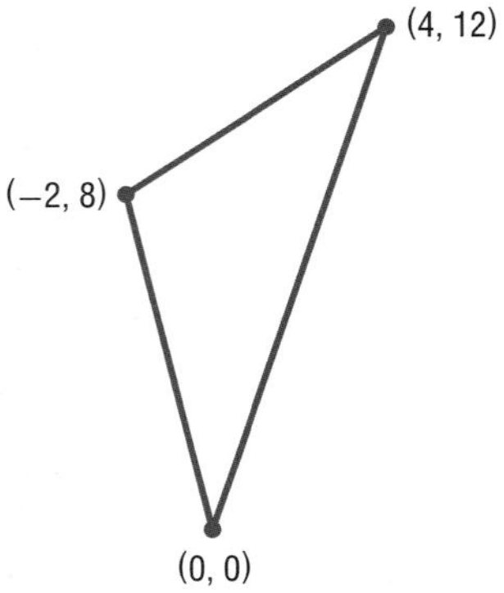

$$A = \frac{1}{2}\begin{vmatrix} a & b & 1 \\ c & d & 1 \\ e & f & 1 \end{vmatrix}$$

$$= \frac{1}{2}\begin{vmatrix} 0 & 0 & 1 \\ 4 & 12 & 1 \\ -2 & 8 & 1 \end{vmatrix}$$ **(a, b) = (0, 0)** **(c, d) = (4, 12)** **(e, f) = (−2, 8)**

$$\left|\begin{matrix} 0 & 0 & 1 \\ 4 & 12 & 1 \\ -2 & 8 & 1 \end{matrix}\right|\begin{matrix} 0 & 0 \\ 4 & 12 \\ -2 & 8 \end{matrix} \qquad \left|\begin{matrix} 0 & 0 & 1 \\ 4 & 12 & 1 \\ -2 & 8 & 1 \end{matrix}\right|\begin{matrix} 0 & 0 \\ 4 & 12 \\ -2 & 8 \end{matrix}$$ **Diagonal Rule**

$0 + 0 + 32 = 32$ $\qquad -24 + 0 + 0 = -24$ **Sum of products of diagonals**

$$A = \frac{1}{2}\begin{vmatrix} 0 & 0 & 1 \\ 4 & 12 & 1 \\ -2 & 8 & 1 \end{vmatrix}$$ **Area of a Triangle**

$$= \left(\frac{1}{2}\right)[32 - (-24)] \text{ or } 28$$ **Simplify.**

The area of the tiger's territory is 28 square kilometers.

Check Your Progress

3. CAR WASH To raise money for their rowing club, Hannah, Christina, and Dario are advertising a car wash at three different street corners in a neighborhood. On a map, the coordinates for the corners are (3, 15), (6, 4), and (11, 9). Each unit represents 0.5 kilometer. What is the area of the neighborhood in which they are advertising?

Personal Tutor glencoe.com

ReadingMath

Determinants The determinant is used to *determine* whether a system has a unique solution.

Cramer's Rule You can use determinants to solve systems of equations. If a determinant is nonzero, then the system has a unique solution. If a determinant is 0, then the system either has no solution or infinite solutions. A method called **Cramer's Rule** uses the coefficient matrix. The **coefficient matrix** is a matrix that contains only the coefficients of the system.

Key Concept — Cramer's Rule

For Your FOLDABLE

Let C be the coefficient matrix of the system $\begin{aligned} ax + by &= m \\ fx + gy &= n \end{aligned} \rightarrow \begin{vmatrix} a & b \\ f & g \end{vmatrix}$

The solution of this system is $x = \dfrac{\begin{vmatrix} m & b \\ n & g \end{vmatrix}}{|C|}$ and $y = \dfrac{\begin{vmatrix} a & m \\ f & n \end{vmatrix}}{|C|}$, if $C \neq 0$.

StudyTip

Cramer's Rule When the determinant of the coefficient matrix C is 0, the system does not have a unique solution.

EXAMPLE 4 Solve a System of Two Equations

Solve the system by using Cramer's Rule.

$5x - 6y = 15$
$3x + 4y = -29$

$x = \dfrac{\begin{vmatrix} m & b \\ n & g \end{vmatrix}}{\lvert C \rvert}$	**Cramer's Rule**	$y = \dfrac{\begin{vmatrix} a & m \\ f & n \end{vmatrix}}{\lvert C \rvert}$
$= \dfrac{\begin{vmatrix} 15 & -6 \\ -29 & 4 \end{vmatrix}}{\begin{vmatrix} 5 & -6 \\ 3 & 4 \end{vmatrix}}$	**Substitute values.**	$= \dfrac{\begin{vmatrix} 5 & 15 \\ 3 & -29 \end{vmatrix}}{\begin{vmatrix} 5 & -6 \\ 3 & 4 \end{vmatrix}}$
$= \dfrac{15(4) - (-29)(-6)}{5(4) - (3)(-6)}$	**Evaluate.**	$= \dfrac{5(-29) - 3(15)}{5(4) - (3)(-6)}$
$= \dfrac{60 - 174}{20 + 18}$	**Multiply.**	$= \dfrac{-145 - 45}{20 + 18}$
$= -\dfrac{114}{38}$	**Add and subtract.**	$= -\dfrac{190}{38}$
$= -3$	**Simplify.**	$= -5$

The solution of the system is $(-3, -5)$.

CHECK

$5(-3) - 6(-5) \stackrel{?}{=} 15$ — $x = -3, y = -5$
$-15 + 30 \stackrel{?}{=} 15$ — Simplify.
$15 = 15$ ✓

$3(-3) + 4(-5) \stackrel{?}{=} -29$ — $x = -3, y = -5$
$-9 - 20 \stackrel{?}{=} -29$ — Simplify.
$-29 = -29$ ✓

Check Your Progress

Solve each system using Cramer's Rule.

4A. $7x + 3y = 37$
$-5x - 7y = -41$

4B. $8x - 5y = 70$
$9x + 7y = 3$

Personal Tutor glencoe.com

Cramer's Rule can also be used for systems of three equations.

Key Concept — For Your FOLDABLE

Cramer's Rule for a System of Three Equations

Let C be the coefficient matrix of the system
$$\begin{aligned} ax + by + cz &= m \\ fx + gy + hz &= n \\ jx + ky + \ell z &= p \end{aligned} \rightarrow \begin{vmatrix} a & b & c \\ f & g & h \\ j & k & \ell \end{vmatrix}.$$

The solution of this system is $x = \dfrac{\begin{vmatrix} m & b & c \\ n & g & h \\ p & k & \ell \end{vmatrix}}{|C|}$, $y = \dfrac{\begin{vmatrix} a & m & c \\ f & n & h \\ j & p & \ell \end{vmatrix}}{|C|}$,

and $z = y = \dfrac{\begin{vmatrix} a & b & m \\ f & g & n \\ j & k & p \end{vmatrix}}{|C|}$, if $C \neq 0$.

EXAMPLE 5 Solve a System of Three Equations

Solve the system by using Cramer's Rule.

$$\begin{aligned} 4x + 5y - 6z &= -14 \\ 3x - 2y + 7z &= 47 \\ 7x - 6y - 8z &= 15 \end{aligned}$$

$$x = \frac{\begin{vmatrix} m & b & c \\ n & g & h \\ p & k & \ell \end{vmatrix}}{|C|} = \frac{\begin{vmatrix} -14 & 5 & -6 \\ 47 & -2 & 7 \\ 15 & -6 & -8 \end{vmatrix}}{\begin{vmatrix} 4 & 5 & -6 \\ 3 & -2 & 7 \\ 7 & -6 & -8 \end{vmatrix}} = \frac{3105}{621} \text{ or } 5$$

$$y = \frac{\begin{vmatrix} a & m & c \\ f & n & h \\ j & p & \ell \end{vmatrix}}{|C|} = \frac{\begin{vmatrix} 4 & -14 & -6 \\ 3 & 47 & 7 \\ 7 & 15 & -8 \end{vmatrix}}{\begin{vmatrix} 4 & 5 & -6 \\ 3 & -2 & 7 \\ 7 & -6 & -8 \end{vmatrix}} = -\frac{1242}{621} \text{ or } -2$$

$$z = \frac{\begin{vmatrix} a & b & m \\ f & g & n \\ j & k & p \end{vmatrix}}{|C|} = \frac{\begin{vmatrix} 4 & 5 & -14 \\ 3 & -2 & 47 \\ 7 & -6 & 15 \end{vmatrix}}{\begin{vmatrix} 4 & 5 & -6 \\ 3 & -2 & 7 \\ 7 & -6 & -8 \end{vmatrix}} = \frac{2484}{621} \text{ or } 4$$

The solution of the system is (5, −2, 4).

CHECK

$$\begin{aligned} 4(5) + 5(-2) - 6(4) &\stackrel{?}{=} -14 \\ 20 - 10 - 24 &\stackrel{?}{=} -14 \\ -14 &= -14 \checkmark \end{aligned}$$

$$\begin{aligned} 3(5) - 2(-2) + 7(4) &\stackrel{?}{=} 47 \\ 15 + 4 + 28 &\stackrel{?}{=} 47 \\ 47 &= 47 \checkmark \end{aligned}$$

$$\begin{aligned} 7(5) - 6(-2) - 8(4) &\stackrel{?}{=} 15 \\ 35 + 12 - 32 &\stackrel{?}{=} 15 \\ 15 &= 15 \checkmark \end{aligned}$$

StudyTip

Check for Accuracy Always substitute your answers into the initial equations to confirm accuracy.

Check Your Progress

Solve each system using Cramer's Rule.

5A. $3x + 5y + 2z = -7$
$-4x + 3y - 5z = -19$
$5x + 4y - 7z = -15$

5B. $6x + 5y + 2z = -1$
$-x + 3y + 7z = 12$
$5x - 7y - 3z = -52$

Personal Tutor glencoe.com

Check Your Understanding

Example 1
p. 220

Evaluate each determinant.

1. $\begin{vmatrix} 8 & 6 \\ 5 & 7 \end{vmatrix}$

2. $\begin{vmatrix} -6 & -6 \\ 8 & 10 \end{vmatrix}$

3. $\begin{vmatrix} -4 & 12 \\ 9 & 5 \end{vmatrix}$

4. $\begin{vmatrix} 16 & -10 \\ -8 & 5 \end{vmatrix}$

Example 2
p. 221

Evaluate each determinant using diagonals.

5. $\begin{vmatrix} 3 & -2 & 2 \\ -4 & 2 & -5 \\ -3 & 1 & 4 \end{vmatrix}$

6. $\begin{vmatrix} 2 & -3 & 5 \\ -4 & 6 & -2 \\ 4 & -1 & -6 \end{vmatrix}$

7. $\begin{vmatrix} 8 & 4 & 0 \\ -2 & -6 & -1 \\ 5 & -3 & 6 \end{vmatrix}$

8. $\begin{vmatrix} -5 & -3 & 4 \\ -2 & -4 & -3 \\ 8 & -2 & 4 \end{vmatrix}$

9. $\begin{vmatrix} 8 & 3 & 4 \\ 2 & 4 & 2 \\ 1 & 6 & 5 \end{vmatrix}$

10. $\begin{vmatrix} -4 & 3 & 0 \\ 1 & 5 & -2 \\ -1 & -8 & -3 \end{vmatrix}$

11. $\begin{vmatrix} 2 & -6 & -3 \\ 7 & 9 & -4 \\ -6 & 4 & 9 \end{vmatrix}$

12. $\begin{vmatrix} -5 & -6 & 7 \\ 4 & 0 & 5 \\ -3 & 8 & 2 \end{vmatrix}$

Example 3
p. 222

Use Cramer's Rule to solve each system of equations.

13. $4x - 5y = 39$
$3x + 8y = -6$

14. $5x + 6y = 20$
$-3x - 7y = -29$

15. $-8a - 5b = -27$
$7a + 6b = 22$

16. $10c - 7d = -59$
$6c + 5d = -63$

Examples 4 and 5
pp. 223, 224

17. **GEOGRAPHY** The "Bermuda Triangle" is an area located off the southeastern Atlantic coast of the United States, and noted for reports of unexplained losses of ships, small boats, and aircraft.

a. Find the area of the triangle on the map.

b. Suppose each grid represents 175 miles. What is the area of the Bermuda Triangle?

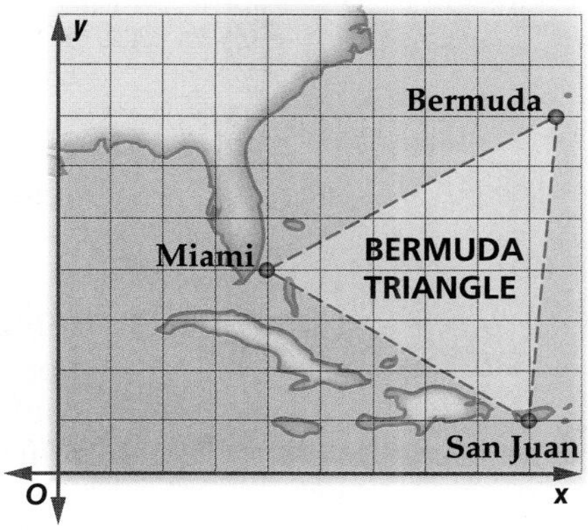

Real-World Link

The Bermuda Islands is a British territory located about 640 miles from Cape Hatteras, North Carolina. Although commonly referred to just as Bermuda, the territory consists of about 140 islands.

Use Cramer's Rule to solve each system of equations.

18. $4x - 2y + 7z = 26$
$5x + 3y - 5z = -50$
$-7x - 8y - 3z = 49$

19. $-3x - 5y + 10z = -4$
$-8x + 2y - 3z = -91$
$6x + 8y - 7z = -35$

20. $6x - 5y + 2z = -49$
$-5x - 3y - 8z = -22$
$-3x + 8y - 5z = 55$

21. $-9x + 5y + 3z = 50$
$7x + 8y - 2z = -60$
$-5x + 7y + 5z = 46$

22. $x + 2y = 12$
$3y - 4z = 25$
$x + 6y + z = 20$

23. $9a + 7b = -30$
$8b + 5c = 11$
$-3a + 10c = 73$

24. $2n + 3p - 4w = 20$
$4n - p + 5w = 13$
$3n + 2p + 4w = 15$

25. $x + y + z = 12$
$6x - 2y - z = 16$
$3x + 4y + 2z = 28$

Practice and Problem Solving

● = **Step-by-Step Solutions** begin on page R20.
Extra Practice begins on page 947.

Examples 1 and 2
pp. 220, 221

Evaluate each determinant.

26. $\begin{vmatrix} -7 & 12 \\ 5 & 6 \end{vmatrix}$

27. $\begin{vmatrix} -8 & -9 \\ 11 & 12 \end{vmatrix}$

28. $\begin{vmatrix} -5 & 8 \\ -6 & -7 \end{vmatrix}$

29. $\begin{vmatrix} 3 & 5 & -2 \\ -1 & -4 & 6 \\ -6 & -2 & 5 \end{vmatrix}$

30. $\begin{vmatrix} 2 & 0 & -6 \\ -3 & -4 & -5 \\ -2 & 5 & 8 \end{vmatrix}$

31. $\begin{vmatrix} -5 & -1 & -2 \\ 1 & 8 & 4 \\ 0 & -6 & 9 \end{vmatrix}$

32. $\begin{vmatrix} 6 & -3 & -5 \\ 0 & -7 & 0 \\ 3 & -6 & -4 \end{vmatrix}$

33. $\begin{vmatrix} -8 & -3 & -9 \\ 0 & 0 & 0 \\ 8 & -2 & -4 \end{vmatrix}$

34. $\begin{vmatrix} 1 & 6 & 7 \\ -2 & -5 & -8 \\ 4 & 4 & 9 \end{vmatrix}$

35. $\begin{vmatrix} 1 & -8 & -9 \\ 6 & 5 & -6 \\ -2 & -8 & 10 \end{vmatrix}$

36. $\begin{vmatrix} 5 & -5 & -5 \\ -8 & -3 & -2 \\ -2 & 4 & 6 \end{vmatrix}$

37. $\begin{vmatrix} -4 & 1 & -2 \\ 10 & 12 & 9 \\ -6 & 0 & 13 \end{vmatrix}$

38. TRAVEL Mr. Smith's art class took a bus trip to an art museum. The bus averaged 65 miles per hour on the highway and 25 miles per hour in the city. The art museum is 375 miles away from the school, and it took the class 7 hours to get there. Use Cramer's Rule to find how many hours the bus was on the highway and how many hours it was driving in the city.

Examples 3 and 5
pp. 222, 224

Use Cramer's Rule to solve each system of equations.

39. $6x - 5y = 73$
$-7x + 3y = -71$

40. $10a - 3b = -34$
$3a + 8b = -28$

41. $-4c - 5d = -39$
$5c + 8d = 54$

42. $-6f - 8g = -22$
$-11f + 5g = -60$

43. $9r + 4s = -55$
$-5r - 3s = 36$

44. $-11u - 7v = 4$
$9u + 4v = -24$

45. $5x - 4y + 6z = 58$
$-4x + 6y + 3z = -13$
$6x + 3y + 7z = 53$

46. $8x - 4y + 7z = 34$
$5x + 6y + 3z = -21$
$3x + 7y - 8z = -85$

Real-World Link

Archaeologists use grids at digs to record the locations of their finds.

Source: College of Staten Island Library

47. DOUGHNUTS Mi-Ling is ordering doughnuts for a class party. The box contains 2 dozen doughnuts, some of which are plain and some of which are jelly-filled. The plain doughnuts each cost \$0.50, and the jelly-filled cost \$0.60. If the total cost is \$12.60, use Cramer's Rule to find how many jelly-filled doughnuts Mi-Ling ordered.

Example 4
p. 223

48. MOVIES The salary for each of the stars of a new movie is \$5 million, and the supporting actors each receive \$1 million. The total amount spent for the salaries of the actors and actresses is \$19 million. If the cast has 7 members, use Cramer's Rule to find the number of stars in the movie.

49. ARCHAEOLOGY Archaeologists found whale bones at coordinates (0, 3), (4, 7), and (5, 9). If the units of the coordinates are meters, find the area of the triangle formed by these finds.

Real-World Link

Names for soft drinks vary widely depending on where you live. *Soda* is popular in the southwest and northeast United States; *pop* is used more in the northwest and midwest; *coke* is common in the south.

Use Cramer's Rule to solve each system of equations.

50. $6a - 7b = -55$
$2a + 4b - 3c = 35$
$-5a - 3b + 7c = -37$

51. $3a - 5b - 9c = 17$
$4a - 3c = 31$
$-5a - 4b - 2c = -42$

52. $4x - 5y = -2$
$7x + 3z = -47$
$8y - 5z = -63$

53. $7x + 8y + 9z = -149$
$-6x + 7y - 5z = 54$
$4x + 5y - 2z = -44$

54. GARDENING Rob wants to build a triangular flower garden. To plan out his garden he uses a coordinate grid where each of the squares represents one square foot. The coordinates for the vertices of his garden are $(-1, 7)$, $(2, 6)$, and $(4, -3)$. Find the area of the garden.

55. BUSINESS A vendor sells small drinks for \$1.15, medium drinks for \$1.75, and large drinks for \$2.25. During a week in which he sold twice as many small drinks as medium drinks, his total sales were \$2,238.75 for 1385 drinks.

a. Use Cramer's Rule to determine how many of each drink were sold.

b. The vendor decided to increase the price for small drinks to \$1.25 the next week. The next week, he sold 140 fewer small drinks, 125 more medium drinks, and 35 more large drinks. Calculate his sales for that week.

c. Was raising the price of the small drink a good business move for the vendor? Explain your reasoning.

H.O.T. Problems

Use Higher-Order Thinking Skills

56. REASONING Some systems of equations cannot be solved by using Cramer's Rule.

a. Find the value of $\begin{vmatrix} a & b \\ f & g \end{vmatrix}$. When is the value 0?

b. Choose values for a, b, f, and g to make the determinant of the coefficient matrix 0. What type of system is formed?

57. REASONING What can you determine about the solution of a system of linear equations if the determinant of the coefficients is 0?

58. FIND THE ERROR James and Amber are finding the value of $\begin{vmatrix} 8 & 3 \\ -5 & 2 \end{vmatrix}$.

James

$$\begin{vmatrix} 8 & 3 \\ -5 & 2 \end{vmatrix} = 16 - (-15) = 31$$

Amber

$$\begin{vmatrix} 8 & 3 \\ -5 & 2 \end{vmatrix} = 16 - 15 = 1$$

Is either of them correct? Explain your reasoning.

59. CHALLENGE Find the determinant of a 3×3 matrix defined by

$$a_{mn} = \begin{cases} 0 & \text{if } m + n \text{ is even} \\ m + n & \text{if } m + n \text{ is odd} \end{cases}.$$

60. OPEN ENDED Write a 2×2 matrix with each of the following characteristics.

a. The determinant equals 0.

b. The determinant equals 25.

c. The elements are all negative numbers and the determinant equals -32.

61. WRITING IN MATH Describe the possible graphical representations of a 2×2 system of linear equations if the determinant of the matrix of coefficients is 0.

NYSRE PRACTICE Reinforcement of A.A.7

62. Tyler paid \$25.25 to play three games of miniature golf and two rides on go-karts. Brent paid \$25.75 for four games of miniature golf and one ride on the go-karts. How much does one game of miniature golf cost?

A \$4.25 **C** \$5.25
B \$4.75 **D** \$5.75

63. Use the table to determine the expression that best represents the number of faces of any prism having a base with n sides.

Base	Sides of Base	Faces of Prisms
triangle	3	5
quadrilateral	4	6
pentagon	5	7
hexagon	6	8
heptagon	7	9
octagon	8	10

F $2(n - 1)$ **H** $n + 2$
G $2(n + 1)$ **J** $2n$

64. **SHORT RESPONSE** A right circular cone has radius 4 inches and height 6 inches.

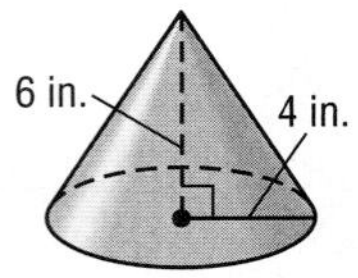

What is the lateral area of the cone? (lateral area of cone $= \pi r \ell$, where $\ell =$ slant height)

65. **SAT/ACT** Find the area of $\triangle ABC$.

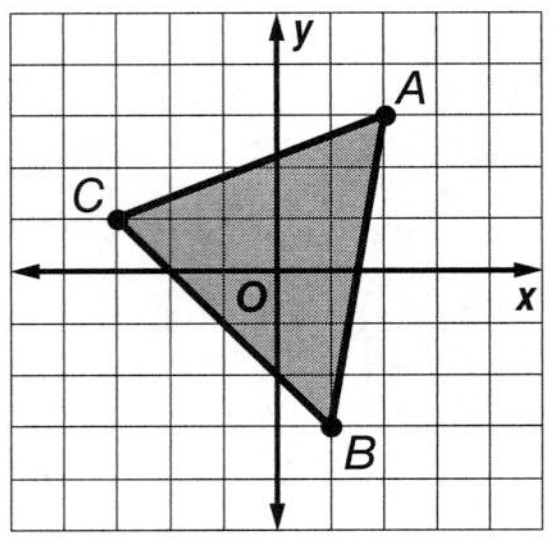

A 10 units^2 **C** 12 units^2
B 14 units^2 **D** 16 units^2

Spiral Review

Determine the coordinates of the vertices of the image of each figure after a reflection in the given axis of symmetry. Then graph the preimage and image. (Lesson 4-4)

66. Rectangle $ABCD$ with vertices $A(-1, 6)$, $B(1, 5)$, $C(-2, -1)$, and $D(-4, 0)$; y-axis

67. $\triangle EFG$ with vertices $E(-2, 5)$, $F(5, -3)$, and $G(-1, -6)$; $y = x$

Determine whether each matrix product is defined. If so, state the dimensions of the product. (Lesson 4-3)

68. $A_{4 \times 2} \cdot B_{2 \times 6}$

69. $C_{5 \times 4} \cdot D_{5 \times 3}$

70. $E_{2 \times 7} \cdot F_{7 \times 1}$

Graph each function. (Lesson 2-6)

71. $f(x) = 2|x - 3| - 4$

72. $f(x) = -3|2x| + 4$

73. $f(x) = |3x - 1| + 2$

Skills Review

Solve each system of equations. (Lesson 3-2)

74. $2x - 5y = -26$
$5x + 3y = -34$

75. $4y + 6x = 10$
$2x - 7y = 22$

76. $-3x - 2y = 17$
$-4x + 5y = -8$

4-6 Inverse Matrices and Systems of Equations

Then

You solved systems of linear equations algebraically. (Lesson 3-2)

Now

- Find the inverse of a 2×2 matrix.
- Write and solve matrix equations for a system of equations.

NYS Core Curriculum

Reinforcement of A.A.7 Analyze and solve verbal problems whose solution requires solving systems of linear equations in two variables

Reinforcement of A.A.10 Solve systems of two linear equations in two variables algebraically

New Vocabulary

identity matrix
inverses
matrix equation
variable matrix
constant matrix

NY Math Online

glencoe.com

- Extra Examples
- Personal Tutor
- Self-Check Quiz
- Homework Help

Why?

Maria's Sandwich Shop offers three lunch options as shown at the right.

To determine how much each individual item costs, you can solve the following matrix equation in which w represents the cost of a sandwich, s the cost of a side, and d the cost of a drink.

$$\begin{bmatrix} 1 & 2 & 0 \\ 2 & 2 & 2 \\ 4 & 3 & 4 \end{bmatrix} \begin{bmatrix} w \\ s \\ d \end{bmatrix} = \begin{bmatrix} 9 \\ 16.50 \\ 30.75 \end{bmatrix}$$

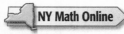

Lunch Options
Basic Brown Bag $9.00
Includes sandwich and 2 sides
Lunch for Two $16.50
Includes 2 sandwiches, 2 sides, and 2 drinks
Family $30.75
Includes 4 sandwiches, 3 sides, and 4 drinks

Identity and Inverse Matrices Recall that in real numbers, two numbers are multiplicative inverses if their product is the identity, 1. Similarly, for matrices, the **identity matrix** is a square matrix that, when multiplied by another matrix, equals that same matrix.

2 × 2 Identity Matrix

$$\begin{bmatrix} 1 & 0 \\ 0 & 1 \end{bmatrix}$$

3 × 3 Identity Matrix

$$\begin{bmatrix} 1 & 0 & 0 \\ 0 & 1 & 0 \\ 0 & 0 & 1 \end{bmatrix}$$

Key Concept — Identity Matrix for Multiplication (For Your FOLDABLE)

Words The identity matrix for multiplication I is a square matrix with 1 for every element of the main diagonal, from upper left to lower right, and 0 in all other positions. For any square matrix A of the same dimension as I, $A \cdot I = I \cdot A = A$.

Symbols If $A = \begin{bmatrix} a & b \\ c & d \end{bmatrix}$, then $I = \begin{bmatrix} 1 & 0 \\ 0 & 1 \end{bmatrix}$ such that

$$\begin{bmatrix} a & b \\ c & d \end{bmatrix} \cdot \begin{bmatrix} 1 & 0 \\ 0 & 1 \end{bmatrix} = \begin{bmatrix} 1 & 0 \\ 0 & 1 \end{bmatrix} \cdot \begin{bmatrix} a & b \\ c & d \end{bmatrix} = \begin{bmatrix} a & b \\ c & d \end{bmatrix}.$$

Two $n \times n$ matrices are **inverses** of each other if their product is the identity matrix. If matrix A has an inverse symbolized by A^{-1}, then $A \cdot A^{-1} = A^{-1} \cdot A = I$.

StudyTip

Verifying Inverses Since multiplication of matrices is not commutative, it is necessary to check the product in both orders.

EXAMPLE 1 Verify Inverse Matrices

Determine whether the matrices in each pair are inverses.

a. $A = \begin{bmatrix} -4 & 2 \\ -2 & 1 \end{bmatrix}$ **and** $B = \begin{bmatrix} \frac{1}{4} & -\frac{1}{2} \\ \frac{1}{2} & -1 \end{bmatrix}$

If A and B are inverses, then $A \cdot B = B \cdot A = I$.

$$A \cdot B = \begin{bmatrix} -4 & 2 \\ -2 & 1 \end{bmatrix} \cdot \begin{bmatrix} \frac{1}{4} & -\frac{1}{2} \\ \frac{1}{2} & -1 \end{bmatrix}$$ **Write an equation.**

$$= \begin{bmatrix} -1+1 & 2-2 \\ -\frac{1}{2}+\frac{1}{2} & 1-1 \end{bmatrix} \text{ or } \begin{bmatrix} 0 & 0 \\ 0 & 0 \end{bmatrix}$$ **Matrix multiplication**

Since $A \cdot B \neq I$, they are *not* inverses.

b. $F = \begin{bmatrix} 3 & -5 \\ -2 & 6 \end{bmatrix}$ **and** $G = \begin{bmatrix} \frac{3}{4} & \frac{5}{8} \\ \frac{1}{4} & \frac{3}{8} \end{bmatrix}$

If F and G are inverses, then $F \cdot G = G \cdot F = I$.

$$F \cdot G = \begin{bmatrix} 3 & -5 \\ -2 & 6 \end{bmatrix} \cdot \begin{bmatrix} \frac{3}{4} & \frac{5}{8} \\ \frac{1}{4} & \frac{3}{8} \end{bmatrix}$$ **Write an equation.**

$$= \begin{bmatrix} \frac{9}{4}-\frac{5}{4} & \frac{15}{8}-\frac{15}{8} \\ -\frac{6}{4}+\frac{6}{4} & -\frac{10}{8}+\frac{18}{8} \end{bmatrix} \text{ or } \begin{bmatrix} 1 & 0 \\ 0 & 1 \end{bmatrix}$$ **Matrix multiplication**

$$G \cdot F = \begin{bmatrix} \frac{3}{4} & \frac{5}{8} \\ \frac{1}{4} & \frac{3}{8} \end{bmatrix} \cdot \begin{bmatrix} 3 & -5 \\ -2 & 6 \end{bmatrix}$$ **Write an equation.**

$$= \begin{bmatrix} \frac{9}{4}-\frac{10}{8} & -\frac{15}{4}+\frac{30}{8} \\ \frac{3}{4}-\frac{6}{8} & -\frac{5}{4}+\frac{18}{8} \end{bmatrix} \text{ or } \begin{bmatrix} 1 & 0 \\ 0 & 1 \end{bmatrix}$$ **Matrix multiplication**

Since $F \cdot G = G \cdot F = I$, F and G are inverses.

Check Your Progress

1. Determine whether $X = \begin{bmatrix} 4 & -1 \\ 2 & -2 \end{bmatrix}$ and $Y = \begin{bmatrix} \frac{1}{3} & -\frac{1}{6} \\ \frac{1}{3} & -\frac{2}{3} \end{bmatrix}$ are inverses of each other.

Personal Tutor glencoe.com

Some matrices do not have inverses. You can determine whether a matrix has an inverse by using the determinant.

Key Concept **Inverse of a 2 × 2 Matrix** *For Your FOLDABLE*

The inverse of matrix $A = \begin{bmatrix} a & b \\ c & d \end{bmatrix}$ is $A^{-1} = \frac{1}{ad - bc}\begin{bmatrix} d & -b \\ -c & a \end{bmatrix}$, where $ad - bc \neq 0$.

Notice that $ad - bc$ is the value of det A. Therefore, if the value of the determinant of a matrix is 0, the matrix cannot have an inverse.

EXAMPLE 2 Find the Inverse of a Matrix

Find the inverse of each matrix, if it exists.

a. $P = \begin{bmatrix} 7 & -5 \\ 2 & -1 \end{bmatrix}$

$\begin{vmatrix} 7 & -5 \\ 2 & -1 \end{vmatrix} = -7 - (-10) \text{ or } 3$ **Find the determinant.**

Since the determinant does not equal 0, P^{-1} exists.

$P^{-1} = \frac{1}{ad - bc}\begin{bmatrix} d & -b \\ -c & a \end{bmatrix}$ **Definition of inverse**

$= \frac{1}{7(-1) - (-5)(2)}\begin{bmatrix} -1 & 5 \\ -2 & 7 \end{bmatrix}$ **$a = 7, b = -5, c = 2, d = -1$**

$= \frac{1}{3}\begin{bmatrix} -1 & 5 \\ -2 & 7 \end{bmatrix} \text{ or } \begin{bmatrix} -\frac{1}{3} & \frac{5}{3} \\ -\frac{2}{3} & \frac{7}{3} \end{bmatrix}$ **Simplify.**

CHECK Find the product of the matrices. If the product is I, then they are inverses.

$$\begin{bmatrix} 7 & -5 \\ 2 & -1 \end{bmatrix} \cdot \begin{bmatrix} -\frac{1}{3} & \frac{5}{3} \\ -\frac{2}{3} & \frac{7}{3} \end{bmatrix} = \begin{bmatrix} -\frac{7}{3} + \frac{10}{3} & \frac{35}{3} - \frac{35}{3} \\ -\frac{2}{3} + \frac{2}{3} & \frac{10}{3} - \frac{7}{3} \end{bmatrix} \text{ or } \begin{bmatrix} 1 & 0 \\ 0 & 1 \end{bmatrix} \checkmark$$

b. $Q = \begin{bmatrix} -8 & -6 \\ 12 & 9 \end{bmatrix}$

$\begin{vmatrix} -8 & -6 \\ 12 & 9 \end{vmatrix} = -72 - (-72) = 0$ **Find the determinant.**

Since the determinant equals 0, Q^{-1} does not exist.

Math History Link

Seki Kowa (1642–1708) Known as The Arithmetical Sage, Seki Kowa was the first to develop the theory of determinants.

Check Your Progress

2A. $\begin{bmatrix} 3 & 7 \\ 1 & -4 \end{bmatrix}$

2B. $\begin{bmatrix} 2 & 1 \\ -4 & 3 \end{bmatrix}$

Personal Tutor glencoe.com

Matrix Equations Matrices can be used to represent and solve systems of equations. Consider the system of equations below. You can write a **matrix equation** to solve this system.

$$\begin{aligned} x + 2y &= 9 \\ 3x - 6y &= 3 \end{aligned} \quad \rightarrow \quad \begin{bmatrix} x + 2y \\ 3x - 6y \end{bmatrix} = \begin{bmatrix} 9 \\ 3 \end{bmatrix}$$

Write the left side of the matrix equation as the product of the coefficient matrix and the variable matrix. Write the right side as a constant matrix.

$$\underset{A}{\begin{bmatrix} 1 & 2 \\ 3 & -6 \end{bmatrix}} \cdot \underset{X}{\begin{bmatrix} x \\ y \end{bmatrix}} = \underset{B}{\begin{bmatrix} 9 \\ 3 \end{bmatrix}}$$

coefficient matrix

variable matrix only the variables of a system

constant matrix only the constants of a system

StudyTip

Inverse You can use this method to solve systems of equations only if A has an inverse. If A does not have an inverse, then the system either has no solution or infinitely many solutions.

Then solve the matrix equation in the same way that you would solve any other equation.

$ax = b$	**Write the equation.**	$AX = B$
$\left(\frac{1}{a}\right)ax = \left(\frac{1}{a}\right)b$	**Multiply each side by the inverse of the coefficient, if it exists.**	$A^{-1}AX = A^{-1}B$
$1x = \frac{b}{a}$	$\left(\frac{1}{a}\right)a = 1$, $A^{-1}A = I$	$IX = A^{-1}B$
$x = \frac{b}{a}$	$1x = x$, $IX = X$	$X = A^{-1}B$

Notice that the solution of the matrix equation is the product of the inverse of the coefficient matrix and the constant matrix.

Real-World EXAMPLE 3 Solve a System of Equations

TRAVEL Helena stopped for gasoline twice during a car trip. The price of gasoline at the first station where she stopped was \$3.75 per gallon. At the second station, the price was \$3.50 per gallon. Helena bought a total of 24.2 gallons of gasoline and spent \$88.05. How much gasoline did Helena buy at each gas station?

A system of equations to represent the situation is as follows.

$x + y = 24.2$
$3.75x + 3.50y = 88.05$

The matrix equation is $\begin{bmatrix} 1 & 1 \\ 3.75 & 3.50 \end{bmatrix} \cdot \begin{bmatrix} x \\ y \end{bmatrix} = \begin{bmatrix} 24.2 \\ 88.05 \end{bmatrix}$.

Step 1 Find the inverse of the coefficient matrix.

$$A^{-1} = \frac{1}{3.50 - 3.75}\begin{bmatrix} 3.50 & -1 \\ -3.75 & 1 \end{bmatrix} \text{ or } -\frac{1}{0.25}\begin{bmatrix} 3.50 & -1 \\ -3.75 & 1 \end{bmatrix}$$

Step 2 Multiply each side of the matrix equation by the inverse matrix.

$$-\frac{1}{0.25}\begin{bmatrix} 3.50 & -1 \\ -3.75 & 1 \end{bmatrix} \cdot \begin{bmatrix} 1 & 1 \\ 3.75 & 3.50 \end{bmatrix} \cdot \begin{bmatrix} x \\ y \end{bmatrix} = -\frac{1}{0.25}\begin{bmatrix} 3.50 & -1 \\ -3.75 & 1 \end{bmatrix} \cdot \begin{bmatrix} 24.2 \\ 88.05 \end{bmatrix}$$

$$\begin{bmatrix} 1 & 0 \\ 0 & 1 \end{bmatrix} \cdot \begin{bmatrix} x \\ y \end{bmatrix} = -\frac{1}{0.25}\begin{bmatrix} -3.35 \\ -2.70 \end{bmatrix}$$

$$\begin{bmatrix} x \\ y \end{bmatrix} = \begin{bmatrix} 13.4 \\ 10.8 \end{bmatrix}$$

The solution is (13.4, 10.8), where x represents the amount of gas Helena purchased at the first gas station, and y represents the amount purchased at the second gas station.

CHECK You can check your answer by using inverses.

Enter $\begin{bmatrix} 1 & 1 \\ 3.75 & 3.50 \end{bmatrix}$ as matrix A.

Enter $\begin{bmatrix} 24.2 \\ 88.05 \end{bmatrix}$ as matrix B.

Multiply the inverse of A by B.

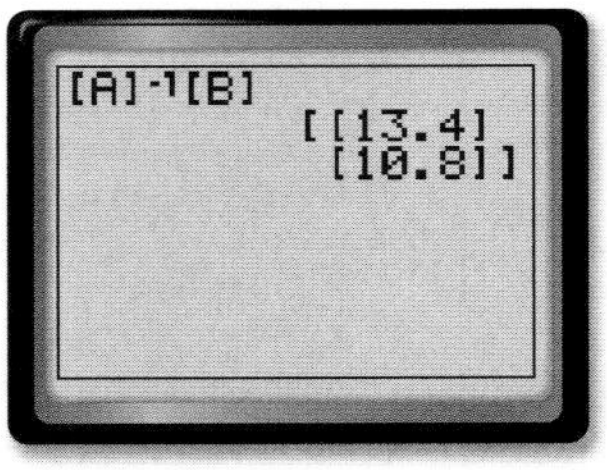

Real-World Link

Average gas prices increased fivefold from \$0.70 per gallon in 1977 to \$3.50 per gallon in 2007.

Source: U.S. Department of Energy

Check Your Progress

3. COMIC BOOKS Dante and Erica just returned from a comic book store that sells new and used comics. Dante spent \$11.25 on 3 new and 4 old books, and Erica spent \$15.75 on 10 used and 3 new ones. If comics of one type are sold at the same price, what is the price in dollars of a new comic book?

Personal Tutor glencoe.com

Check Your Understanding

Example 1 p. 230

Determine whether the matrices in each pair are inverses.

1. $A = \begin{bmatrix} 2 & 1 \\ -1 & 0 \end{bmatrix}, B = \begin{bmatrix} 1 & 2 \\ 2 & 1 \end{bmatrix}$

2. $C = \begin{bmatrix} 2 & 1 \\ 5 & 3 \end{bmatrix}, D = \begin{bmatrix} 2 & 1 \\ 5 & -3 \end{bmatrix}$

3. $F = \begin{bmatrix} -1 & 1 \\ 0 & -1 \end{bmatrix}, G = \begin{bmatrix} -1 & -1 \\ 0 & -1 \end{bmatrix}$

4. $H = \begin{bmatrix} -3 & -1 \\ -4 & -2 \end{bmatrix}, J = \begin{bmatrix} -1 & 2 \\ 3 & -4 \end{bmatrix}$

Example 2 p. 231

Find the inverse of each matrix, if it exists.

5. $\begin{bmatrix} 6 & -3 \\ -1 & 0 \end{bmatrix}$

6. $\begin{bmatrix} 2 & -4 \\ -3 & 0 \end{bmatrix}$

7. $\begin{bmatrix} -3 & 0 \\ 5 & 2 \end{bmatrix}$

8. $\begin{bmatrix} 2 & 4 \\ 1 & 2 \end{bmatrix}$

Example 3 p. 232

Use a matrix equation to solve each system of equations.

9. $-2x + y = 9$
$x + y = 3$

10. $4x - 2y = 22$
$6x + 9y = -3$

11. $-2x + y = -4$
$3x + y = 1$

Example 3 p. 232

12. **MONEY** Kevin had 25 quarters and dimes. The total value of all the coins was $4. How many quarters and dimes did Kevin have?

Practice and Problem Solving

● = **Step-by-Step Solutions** begin on page R20.
Extra Practice begins on page 947.

Example 1 p. 230

Determine whether each pair of matrices are inverses of each other.

13. $K = \begin{bmatrix} 1 & 2 \\ 3 & 0 \end{bmatrix}, L = \begin{bmatrix} 0 & 1 \\ 2 & -1 \end{bmatrix}$

14. $M = \begin{bmatrix} 0 & 2 \\ 4 & 5 \end{bmatrix}, N = \begin{bmatrix} 1 & 1 \\ 0 & 0 \end{bmatrix}$

15. $P = \begin{bmatrix} 4 & 0 \\ 3 & 0 \end{bmatrix}, Q = \begin{bmatrix} -1 & -1 \\ \frac{2}{3} & 5 \end{bmatrix}$

16. $R = \begin{bmatrix} \frac{1}{2} & -\frac{1}{4} \\ \frac{1}{4} & -\frac{1}{2} \end{bmatrix}, S = \begin{bmatrix} 2 & 4 \\ 4 & 2 \end{bmatrix}$

Example 2 p. 231

Find the inverse of each matrix, if it exists.

17. $\begin{bmatrix} 3 & 0 \\ 0 & 2 \end{bmatrix}$

18. $\begin{bmatrix} 2 & 3 \\ 3 & 2 \end{bmatrix}$

19. $\begin{bmatrix} 3 & 0 \\ 5 & 1 \end{bmatrix}$

20. $\begin{bmatrix} 1 & -1 \\ -6 & -1 \end{bmatrix}$

21. $\begin{bmatrix} -5 & -4 \\ 4 & 2 \end{bmatrix}$

22. $\begin{bmatrix} -5 & 9 \\ 4 & -8 \end{bmatrix}$

23. $\begin{bmatrix} 6 & -5 \\ 4 & 9 \end{bmatrix}$

24. $\begin{bmatrix} -4 & -2 \\ 7 & 8 \end{bmatrix}$

25. $\begin{bmatrix} -6 & 8 \\ 8 & -7 \end{bmatrix}$

Example 3 p. 232

26. **BAKING** Peggy is preparing a colored frosting for a cake. For the right shade of purple, she needs 25 milliliters of a 44% concentration food coloring. The store has a 25% red and a 50% blue concentration of food coloring. How many milliliters each of blue food coloring and red food coloring should be mixed to make the necessary amount of purple food coloring?

Use a matrix equation to solve each system of equations.

27. $-x + y = 4$
$-x + y = -4$

28. $-x + y = 3$
$-2x + y = 6$

29. $x + y = 4$
$-4x + y = 9$

30. $3x + y = 3$
$5x + 3y = 6$

31. $y - x = 5$
$2y - 2x = 8$

32. $4x + 2y = 6$
$6x - 3y = 9$

33. $1.6y - 0.2x = 1$
$0.4y - 0.1x = 0.5$

34 $4y - x = -2$
$3y - x = 6$

35. $2y - 4x = 3$
$4x - 3y = -6$

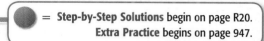

36. **POPULATIONS** The diagram shows the annual percent migration between a city and its suburbs.

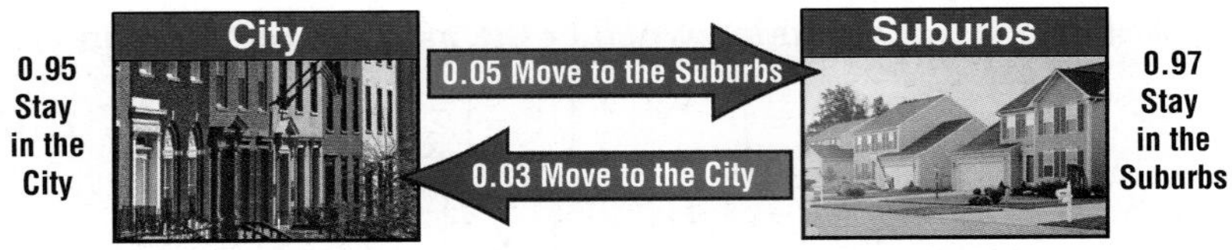

a. Write a matrix to represent the transitions in city population and suburb population.

b. There are currently 16,275 people living in the city and 17,552 people living in the suburbs. Assuming that the trends continue, predict the number of people who will live in the suburbs next year.

c. Use the inverse of the matrix from part **b** to find the number of people who lived in the city last year.

Real-World Link

The first commercially successful digital audio player was released in 1998 and weighed less than three ounces.

Source: CNet

37. **MUSIC** The diagram shows the trends in digital audio player and portable CD player ownership over the past five years for Central City. Every person in Central City has either a digital audio player or a portable CD player. Central City has a stable population of 25,000 people, of whom 17,252 own digital audio players and 7748 own portable CD players.

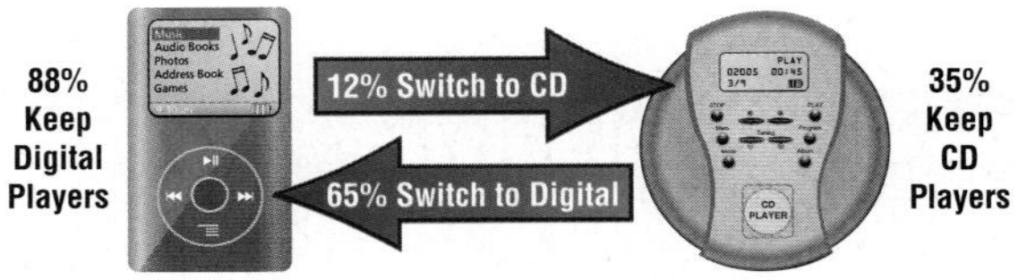

a. Write a matrix to represent the transitions in player ownership.

b. Assume that the trends continue. Predict the number of people who will own digital audio players next year.

c. Use the inverse of the matrix from part **b** to find the number of people who owned digital audio players last year.

H.O.T. Problems

Use **H**igher-**O**rder **T**hinking Skills

38. **FIND THE ERROR** Cody and Megan are setting up matrix equations for the system $5x + 7y = 19$ and $3y + 4x = 10$. Is either of them correct? Explain your reasoning.

Cody

$$\begin{bmatrix} 5 & 7 \\ 3 & 4 \end{bmatrix}\begin{bmatrix} x \\ y \end{bmatrix} = \begin{bmatrix} 19 \\ 10 \end{bmatrix}$$

Megan

$$\begin{bmatrix} 5 & 7 \\ 4 & 3 \end{bmatrix}\begin{bmatrix} x \\ y \end{bmatrix} = \begin{bmatrix} 19 \\ 10 \end{bmatrix}$$

39. **CHALLENGE** Describe what a matrix equation with infinite solutions looks like.

40. **REASONING** Determine whether the following statement is *always, sometimes,* or *never* true. Explain your reasoning.

A square matrix has a multiplicative inverse.

41. **OPEN ENDED** Write a matrix equation that does not have a solution.

42. **WRITING IN MATH** Explain how matrix equations can be used to solve systems of equations.

NYSRE PRACTICE Reinforcement of A.A.7, A2.RP.10

43. The Yogurt Shoppe sells cones in three sizes: small, \$0.89; medium, \$1.19; and large, \$1.39. One day Santos sold 52 cones. He sold seven more medium cones than small cones. If he sold \$58.98 in cones, how many medium cones did he sell?

A 11 **B** 17 **C** 24 **D** 36

44. The chart shows an expression evaluated for different values of x.

x	$x^2 + x + 1$
1	3
2	7
3	13
4	31

A student concludes that for all values of x, $x^2 + x + 1$ produces a prime number. Which value of x serves as a counterexample to prove this conclusion false?

F −4 **G** −3 **H** 2 **J** 4

45. What is the solution of the system of equations $6a + 8b = 5$ and $10a - 12b = 2$?

46. SAT/ACT Each year at Capital High School the students vote to choose the theme of the homecoming dance. The theme "A Night Under the Stars" received 225 votes, and "The Time of My Life" received 480 votes. If 40% of girls voted for "A Night Under the Stars" and 75% of boys voted for "The Time of My Life," how many girls and boys voted?

A 854 boys and 176 girls
B 705 boys and 325 girls
C 395 boys and 310 girls
D 380 boys and 325 girls

Spiral Review

Evaluate each determinant. (Lesson 4-5)

47. $\begin{vmatrix} 8 & -3 \\ 6 & -9 \end{vmatrix}$

48. $\begin{vmatrix} 9 & -7 \\ -5 & -3 \end{vmatrix}$

49. $\begin{vmatrix} 8 & 6 & -1 \\ -4 & 5 & 1 \\ -3 & -2 & 9 \end{vmatrix}$

50. Quadrilateral $A'B'C'D'$ is an image after a translation of quadrilateral $ABCD$. A table of the vertices of each rectangle is shown. (Lesson 4-4)

a. Determine the coordinates of C'.

b. Determine the coordinates of B'.

Quadrilateral *ABCD*	Quadrilateral *A'B'C'D'*
$A(2, 5)$	$A'(-1, 13)$
$B(9, -15)$	$B'(?, ?)$
$C(-14, -9)$	$C'(?, ?)$
$D(-6, 1)$	$D'(-9, 9)$

51. MILK The Yoder Family Dairy produces at most 200 gallons of skim and whole milk each day for delivery to large bakeries and restaurants. Regular customers require at least 15 gallons of skim and 21 gallons of whole milk each day. If the profit on a gallon of skim milk is \$0.82 and the profit on a gallon of whole milk is \$0.75, how many gallons of each type of milk should the dairy produce each day to maximize profits? (Lesson 3-4)

Skills Review

Identify the type of function represented by each graph. (Lesson 2-7)

52.

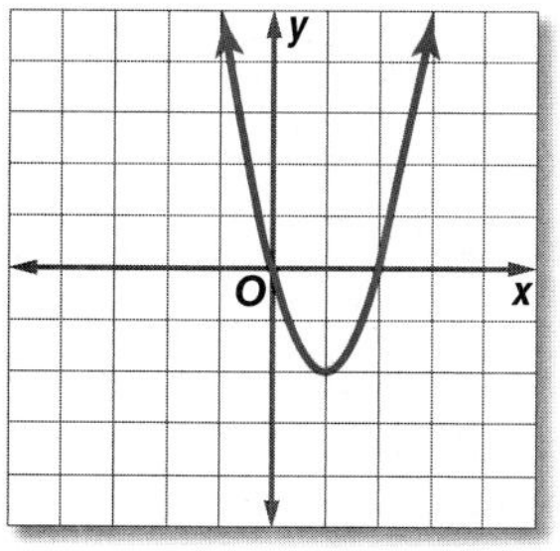

53.

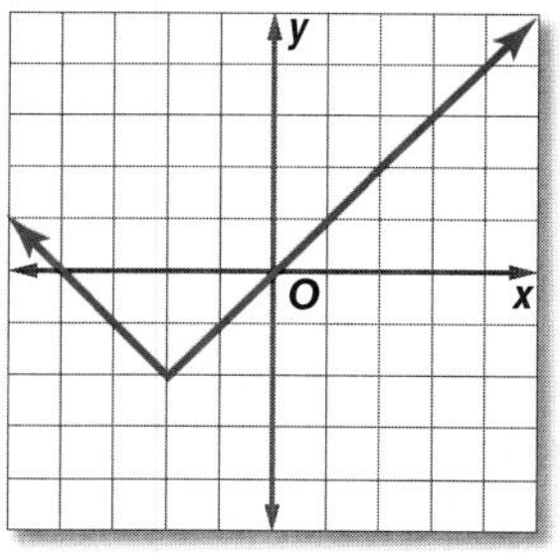

54.

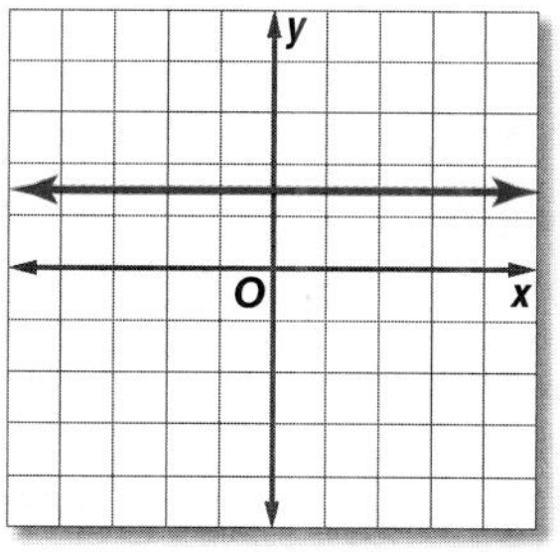

Graphing Technology Lab
Augmented Matrices

NY Math Online glencoe.com
Other Calculator Keystrokes

NYS Core Curriculum **Reinforcement of A.A.7** Analyze and solve verbal problems whose solution requires solving systems of linear equations in two variables
Reinforcement of A.A.10 Solve systems of two linear equations in two variables algebraically

Using a TI-83/84 Plus graphing calculator, you can solve a system of linear equations using the MATRIX function. An **augmented matrix** contains the coefficient matrix with an extra column containing the constant terms. You can use a graphing calculator to reduce the augmented matrix so that the solution of the system of equations can be easily determined.

EXAMPLE

Write an augmented matrix for the following system of equations. Then solve the system by using a graphing calculator.

$2x + y + z = 1$
$3x + 2y + 3z = 12$
$4x + y + 2z = -1$

Step 1 Write the augmented matrix and enter it into a calculator.

The augmented matrix $B = \left[\begin{array}{ccc|c} 2 & 1 & 1 & 1 \\ 3 & 2 & 3 & 12 \\ 4 & 1 & 2 & -1 \end{array}\right]$.

Begin by entering the matrix.

KEYSTROKES: 2nd [MATRIX] ▶ ▶ ENTER 3 ENTER 4 ENTER 2 ENTER 1 ENTER 1 ENTER 1 ENTER 3 ENTER 2 ENTER 3 ENTER 12 ENTER 4 ENTER 1 ENTER 2 ENTER (−) 1 ENTER

Step 2 Find the reduced row echelon form (rref) using the graphing calculator.

KEYSTROKES: 2nd [QUIT] 2nd [MATRIX] ▶ ALPHA [B] 2nd [MATRIX] ENTER) ENTER

Study the reduced echelon matrix. The first three columns are the same as a 3 × 3 identity matrix. The first row represents $x = -4$, the second row represents $y = 3$, and the third row represents $z = 6$. The solution is $(-4, 3, 6)$.

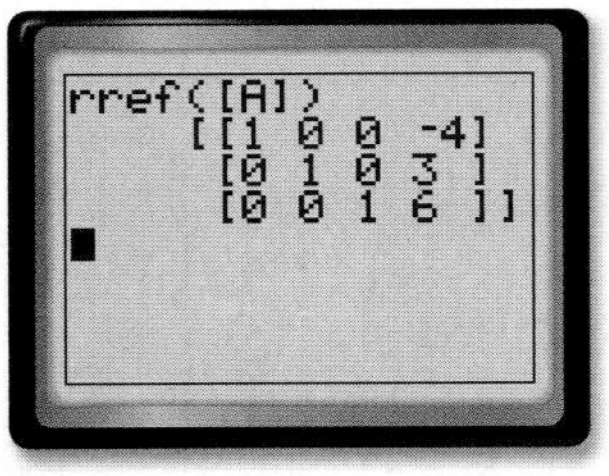

Exercises

Write an augmented matrix for each system of equations. Then solve with a graphing calculator.

1. $3x + 2y = -4$
   $4x + 7y = 13$
2. $2x + y = 6$
   $6x - 2y = 0$
3. $2x + 2y = -4$
   $7x + 3y = 10$
4. $4x + 6y = 0$
   $8x - 2y = 7$
5. $6x - 4y + 2z = -4$
   $2x - 2y + 6z = 10$
   $2x + 2y + 2z = -2$
6. $5x - 5y + 5z = 10$
   $5x - 5z = 5$
   $5y + 10z = 0$

NY Math Online glencoe.com
- STUDY TO GO
- Vocabulary Review

Chapter Summary

Key Concepts

Matrices (Lesson 4-1)

- A matrix is a rectangular array of variables or constants in horizontal rows and vertical columns.
- Equal matrices have the same dimensions and corresponding elements are equal.

Operations (Lessons 4-2 and 4-3)

- Matrices can be added or subtracted if they have the same dimensions. Add or subtract corresponding elements.
- To multiply a matrix by a scalar k, multiply each element in the matrix by k.
- Two matrices can be multiplied if and only if the number of columns in the first matrix is equal to the number of rows in the second matrix.

Transformations (Lesson 4-4)

- Use matrix addition and a translation matrix to find the coordinates of a translated figure.
- Use scalar multiplication to perform dilations.
- To rotate a figure counterclockwise about the origin, multiply the vertex matrix on the left by a rotation matrix.

Identity and Inverse Matrices (Lesson 4-6)

- An identity matrix is a square matrix with ones on the diagonal and zeros in the other positions.
- Two matrices are inverses of each other if their product is the identity matrix.

Matrix Equations (Lesson 4-6)

- To solve a matrix equation, find the inverse of the coefficient matrix. Then multiply each side of the equation by the inverse matrix.

FOLDABLES® Study Organizer

Be sure the Key Concepts are noted in your Foldable.

Key Vocabulary

coefficient matrix (p. 223)
column matrix (p. 186)
constant matrix (p. 231)
Cramer's rule (p. 223)
determinant (p. 220)
diagonal rule (p. 221)
dilation (p. 211)
dimension (p. 185)
element (p. 185)
equal matrices (p. 186)
identity matrix (p. 229)
image (p. 209)
inverse (p. 229)
matrix (p. 185)
matrix equation (p. 231)
preimage (p. 209)
reflection (p. 212)
rotation (p. 212)
row matrix (p. 186)
scalar (p. 194)
scalar multiplication (p. 194)
second-order determinant (p. 220)
square matrix (p. 186)
third-order determinant (p. 221)
transformation (p. 209)
translation (p. 209)
variable matrix (p. 231)
vertex matrix (p. 209)
zero matrix (p. 186)

Vocabulary Check

Choose the correct term from the list above to complete each sentence.

1. A(n) ____________ is a rectangular array of constants or variables.
2. A matrix can be multiplied by a constant called a(n) ____________.
3. A(n) ____________ occurs when a figure is moved from one location to another without changing its size, shape, or orientation.
4. Each value in a matrix is called a(n) ____________.
5. When a geometric figure is enlarged or reduced, the transformation is called a(n) ____________.
6. A(n) ____________ occurs when a figure is moved about a center point, usually the origin.
7. The ____________ matrix is a square matrix that, when multiplied by another matrix, equals that same matrix.
8. In a(n) ____________ matrix, every element is zero.
9. The ____________ of $\begin{bmatrix} -1 & 2 \\ 2 & -3 \end{bmatrix}$ is -1.
10. If the product of two matrices is the identity matrix, they are ____________.

Study Guide and Review

Lesson-by-Lesson Review

4-1 Introduction to Matrices (pp. 185–191)

A.CM.3

11. SALES Three competing retail stores recorded the number and type of customers that purchased items at their stores one day. The following table shows the numbers.

Type of Customer	Store A	Store B	Store C
adult (18 and older)	64	108	31
student (under 18)	42	9	68

a. Write a matrix for the numbers of customers.

b. What are the dimensions of the matrix?

c. What value is a_{23}?

d. What value is a_{11}?

e. Add the elements in columns 1 and 2 and interpret the results.

f. Would finding the sum of the rows provide any meaningful data? Explain.

EXAMPLE 1

A movie house has three theatres; each theatre shows a different movie. The number of people who attended each movie is shown.

Type of Movie	Theatre 1	Theatre 2	Theatre 3
matinee	37	19	26
evening	69	58	75

a. Write a matrix for the number of customers. $\begin{bmatrix} 37 & 19 & 26 \\ 69 & 58 & 75 \end{bmatrix}$

b. What are the dimensions of the matrix? 2×3

c. Add the elements in rows 1 and 2, and interpret the results.
The sum of row 1 is 82, which is the total number of customers at the matinee. The sum of row 2 is 202, which is the total number of customers at the evening show.

4-2 Operations with Matrices (pp. 193–199)

A2.RP.10

Perform the indicated operations. If the matrix does not exist, write *impossible*.

12. $\begin{bmatrix} 2 \\ -6 \end{bmatrix} - \begin{bmatrix} -3 \\ 2 \end{bmatrix} + \begin{bmatrix} 6 \\ 0 \end{bmatrix}$

13. $3\left(\begin{bmatrix} -2 & 0 \\ 6 & 8 \end{bmatrix} + \begin{bmatrix} 1 & 9 \\ -3 & -4 \end{bmatrix}\right)$

14. RETAIL Current Fashions buys shirts, jeans and shoes from a manufacturer, marks them up, and then sells them. The table shows the purchase price and the selling price.

Item	Purchase Price	Selling Price
shirts	$15	$35
jeans	$25	$55
shoes	$30	$85

a. Write a matrix for the purchase price.

b. Write a matrix for the selling price.

c. Use matrix operations to find the profit on 1 shirt, 1 pair of jeans, and 1 pair of shoes.

EXAMPLE 2

Find $2B + 3A$ if $A = \begin{bmatrix} 9 & 1 \\ 1 & 2 \end{bmatrix}$ and $B = \begin{bmatrix} 1 & 4 \\ 3 & 7 \end{bmatrix}$.

$2B = 2\begin{bmatrix} 1 & 4 \\ 3 & 7 \end{bmatrix}$ or $\begin{bmatrix} 2 & 8 \\ 6 & 14 \end{bmatrix}$

$3A = 3\begin{bmatrix} 9 & 1 \\ 1 & 2 \end{bmatrix}$ or $\begin{bmatrix} 27 & 3 \\ 3 & 6 \end{bmatrix}$

$2B + 3A = \begin{bmatrix} 2 & 8 \\ 6 & 14 \end{bmatrix} + \begin{bmatrix} 27 & 3 \\ 3 & 6 \end{bmatrix}$ or $\begin{bmatrix} 29 & 11 \\ 9 & 20 \end{bmatrix}$

MIXED PROBLEM SOLVING
For mixed problem-solving practice, see page 982.

4-3 Multiplying Matrices (pp. 200–207)

A2.RP.10

Find each product, if possible.

15. $\begin{bmatrix} 3 & -6 \end{bmatrix} \cdot \begin{bmatrix} 8 \\ -3 \end{bmatrix}$

16. $\begin{bmatrix} -3 & 0 & 2 \\ 6 & -1 & 5 \end{bmatrix} \cdot \begin{bmatrix} 8 & -1 \\ -4 & 3 \\ 6 & 7 \end{bmatrix}$

17. $\begin{bmatrix} 2 & 11 \\ 0 & -3 \\ -6 & 7 \end{bmatrix} \cdot \begin{bmatrix} 0 & 8 & -5 \\ 12 & 0 & 9 \\ 4 & -6 & 7 \end{bmatrix}$

18. GROCERIES Martin bought 1 gallon of milk, 2 apples, 4 frozen dinners, and 1 box of cereal. The following matrix shows the prices for each item respectively.

$$\begin{bmatrix} \$2.59 & \$0.49 & \$5.25 & \$3.99 \end{bmatrix}$$

Use matrix multiplication to find the total amount of money Martin spent at the grocery store.

EXAMPLE 3

Find XY if $X = \begin{bmatrix} 0 & -6 \\ 3 & 5 \end{bmatrix}$ and $Y = \begin{bmatrix} 8 \\ -1 \end{bmatrix}$.

$XY = \begin{bmatrix} 0 & -6 \\ 3 & 5 \end{bmatrix} \cdot \begin{bmatrix} 8 \\ -1 \end{bmatrix}$ **Write an equation.**

$= \begin{bmatrix} 0(8) + (-6)(-1) \\ 3(8) + 5(-1) \end{bmatrix}$ **Multiply columns by rows.**

$= \begin{bmatrix} 6 \\ 19 \end{bmatrix}$ **Simplify.**

4-4 Transformations with Matrices (pp. 209–217)

G.G.61

Use $\triangle ABC$ to find the coordinates of the image after each transformation.

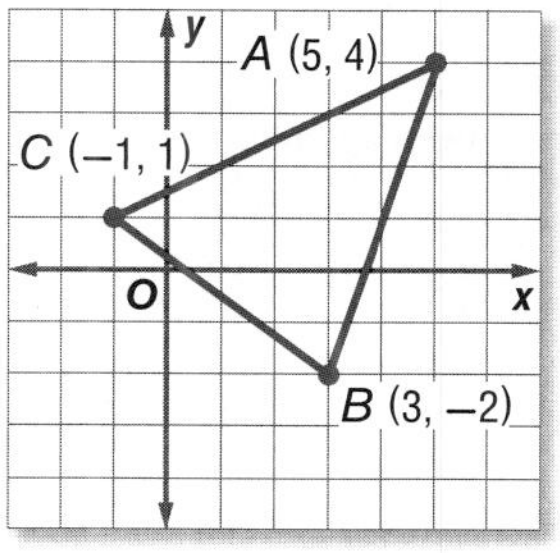

19. translation 2 units left and 3 units up

20. dilation by a scale factor of 3

21. reflection in the x-axis

22. rotation of 180 degrees

23. QUILTS Carol used the pattern of a rectangle shown at the right for a quilt piece.

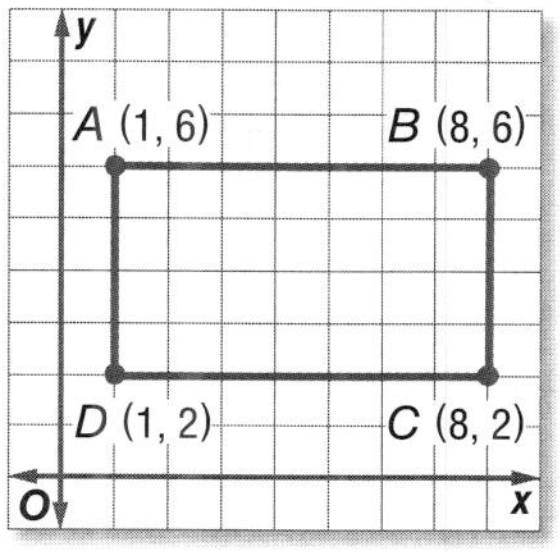

Carol wants to make a rectangle pattern that is twice the size. What will the new coordinates of the rectangle be?

EXAMPLE 4

Find the coordinates of the vertices of the image of $\triangle XYZ$ with $X(-2, 3)$, $Y(6, 6)$, and $Z(8, -3)$ after a rotation of 270° counterclockwise about the origin.

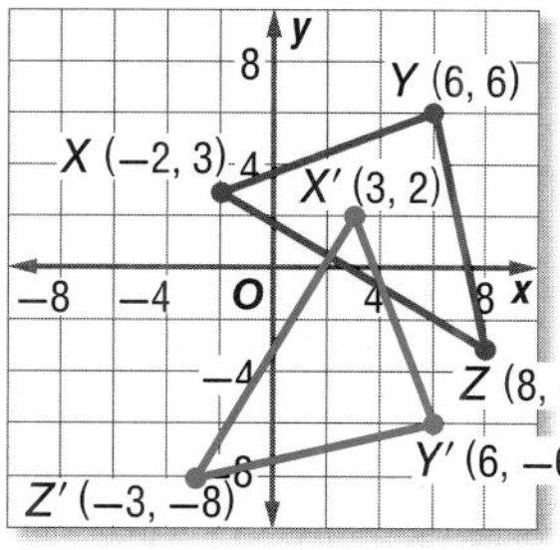

Write the ordered pairs in a vertex matrix. Then multiply by the rotation matrix.

$$\begin{bmatrix} 0 & 1 \\ -1 & 0 \end{bmatrix} \cdot \begin{bmatrix} -2 & 6 & 8 \\ 3 & 6 & -3 \end{bmatrix} = \begin{bmatrix} 3 & 6 & -3 \\ 2 & -6 & -8 \end{bmatrix}$$

The vertices of $\triangle X'Y'Z'$ are $X'(3, 2)$, $Y'(6, -6)$, and $Z'(-3, -8)$.

Study Guide and Review

4-5 Determinants and Cramer's Rule (pp. 220–228)

A.A.7, A.A.10

Evaluate each determinant.

24. $\begin{vmatrix} 2 & 4 \\ 7 & -3 \end{vmatrix}$

25. $\begin{vmatrix} 2 & 3 & -1 \\ 0 & 2 & 4 \\ -2 & 5 & 6 \end{vmatrix}$

Use Cramer's Rule to solve each system of equations.

26. $3x - y = 0$
$5x + 2y = 22$

27. $5x + 2y = 4$
$3x + 4y + 2z = 6$
$7x + 3y + 4z = 29$

28. JEWELRY Alana paid \$98.25 for 3 necklaces and 2 pairs of earrings. Petra paid \$133.50 for 2 necklaces and 4 pairs of earrings. Use Cramer's Rule to find out how much 1 necklace costs and how much 1 pair of earrings costs.

EXAMPLE 5

Evaluate $\begin{vmatrix} 4 & -6 \\ 2 & 5 \end{vmatrix}$.

$\begin{vmatrix} 4 & -6 \\ 2 & 5 \end{vmatrix} = 4(5) - 2(-6)$ **Definition of determinant**

$= 20 + 12 \text{ or } 32$ **Simplify.**

EXAMPLE 6

Use Cramer's Rule to solve $2a + 6b = -1$ and $a + 8b = 2$.

$a = \dfrac{\begin{vmatrix} -1 & 6 \\ 2 & 8 \end{vmatrix}}{\begin{vmatrix} 2 & 6 \\ 1 & 8 \end{vmatrix}}$ **Cramer's Rule** $b = \dfrac{\begin{vmatrix} 2 & -1 \\ 1 & 2 \end{vmatrix}}{\begin{vmatrix} 2 & 6 \\ 1 & 8 \end{vmatrix}}$

$= \dfrac{-8 - 12}{16 - 6}$ **Evaluate each determinant.** $= \dfrac{4 + 1}{16 - 6}$

$= \dfrac{-20}{10} \text{ or } -2$ **Simplify.** $= \dfrac{5}{10} \text{ or } \dfrac{1}{2}$

The solution is $\left(-2, \frac{1}{2}\right)$.

4-6 Inverse Matrices and Systems of Equations (pp. 229–235)

A.A.7, A.A.10

Find the inverse of each matrix, if it exists.

29. $\begin{bmatrix} 7 & 4 \\ 3 & 2 \end{bmatrix}$

30. $\begin{bmatrix} 2 & 5 \\ -5 & -13 \end{bmatrix}$

31. $\begin{bmatrix} 6 & -3 \\ -8 & 4 \end{bmatrix}$

Use a matrix equation to solve each system of equations.

32. $\begin{bmatrix} 5 & 3 \\ 3 & 2 \end{bmatrix} \cdot \begin{bmatrix} x \\ y \end{bmatrix} = \begin{bmatrix} 4 \\ 0 \end{bmatrix}$

33. $\begin{bmatrix} 3 & -1 \\ 1 & 2 \end{bmatrix} \cdot \begin{bmatrix} a \\ b \end{bmatrix} = \begin{bmatrix} 5 \\ 4 \end{bmatrix}$

34. HEALTH FOOD Heath sells nuts and raisins by the pound. Sonia bought 2 pounds of nuts and 2 pounds of raisins for \$23.50. Drew bought 3 pounds of nuts and 1 pound of raisins for \$22.25. What is the cost of 1 pound of nuts and 1 pound of raisins?

EXAMPLE 7

Solve $\begin{bmatrix} 2 & -5 \\ 3 & -6 \end{bmatrix} \cdot \begin{bmatrix} x \\ y \end{bmatrix} = \begin{bmatrix} 15 \\ 36 \end{bmatrix}$.

Step 1 Find the inverse of the coefficient matrix.

$$A^{-1} = \frac{1}{-12 - (-15)}\begin{bmatrix} -6 & 5 \\ -3 & 2 \end{bmatrix} \text{ or } \frac{1}{3}\begin{bmatrix} -6 & 5 \\ -3 & 2 \end{bmatrix}$$

Step 2 Multiply each side by the inverse matrix.

$$\frac{1}{3}\begin{bmatrix} -6 & 5 \\ -3 & 2 \end{bmatrix} \cdot \begin{bmatrix} 2 & -5 \\ 3 & -6 \end{bmatrix} \cdot \begin{bmatrix} x \\ y \end{bmatrix} = \frac{1}{3}\begin{bmatrix} -6 & 5 \\ -3 & 2 \end{bmatrix} \cdot \begin{bmatrix} 15 \\ 36 \end{bmatrix}$$

$$\begin{bmatrix} 1 & 0 \\ 0 & 1 \end{bmatrix} \cdot \begin{bmatrix} x \\ y \end{bmatrix} = \frac{1}{3}\begin{bmatrix} 90 \\ 27 \end{bmatrix}$$

$$\begin{bmatrix} x \\ y \end{bmatrix} = \begin{bmatrix} 30 \\ 9 \end{bmatrix}$$

Practice Test

NY Math Online glencoe.com
Chapter Test

Identify each element of matrix

$$A = \begin{bmatrix} 2 & 2 & 7 \\ 9 & 1 & 1 \\ 8 & 0 & 8 \end{bmatrix}.$$

1. a_{22}

2. a_{31}

Perform the indicated operations. If the matrix does not exist, write *impossible*.

3. $-3\begin{bmatrix} 4a \\ 0 \\ -3 \end{bmatrix} + 4\begin{bmatrix} -2 \\ 3 \\ -1 \end{bmatrix}$

4. $\begin{bmatrix} -3 & 0 \\ 1 & 5 \end{bmatrix} \cdot \begin{bmatrix} 2 & 4 \\ -6 & 0 \end{bmatrix}$

5. $\begin{bmatrix} 2 & 0 \\ -3 & 5 \\ 1 & 4 \end{bmatrix} \cdot \begin{bmatrix} 3 \\ -2 \end{bmatrix}$

6. $\begin{bmatrix} -5 & 7 \\ 6 & 8 \end{bmatrix} - \begin{bmatrix} 4 & 0 & -2 \\ 9 & 0 & 1 \end{bmatrix}$

7. **SALES** Soren sells children's educational books door to door to save for college. The following table shows the cost of a set of books and the selling price. He sold 20 sets of encyclopedias, 32 sets of science books, and 14 sets of literature books.

Type of Book	Cost	Selling Price
Encyclopedia set	\$35	\$105
Science books	\$25	\$85
Literature books	\$40	\$125

a. Organize the data into two matrices. Then use matrix multiplication to find the total amount that Soren paid for the books.

b. Use matrix multiplication to find the total amount that buyers paid for the books.

c. Use matrix operations to find how much money Soren made on his sales.

8. Find $AB - AC$ if $A = \begin{bmatrix} 3 & 8 \\ -3 & -4 \end{bmatrix}$, $B = \begin{bmatrix} -7 & 5 \\ 5 & -4 \end{bmatrix}$, and $C = \begin{bmatrix} -4 & 7 \\ 2 & 0 \end{bmatrix}$.

9. Triangle ABC with $A(0, 2)$, $B(1.5, -1.5)$, and $C(-2.5, 0)$ is dilated so that its perimeter is three times the original perimeter. Write the vertex matrix for $\triangle ABC$. Then find the coordinates of the image after the dilation.

10. **MULTIPLE CHOICE** Triangle $A'B'C'$ is an image of $\triangle ABC$. A table of the vertices of each triangle is shown. What are the coordinates of C'?

$\triangle ABC$	$\triangle A'B'C'$
$A(3, 4)$	$A'(5, 3)$
$B(1, -2)$	$B'(3, -3)$
$C(5, -4)$	$C'(?, ?)$

A $(-5, 7)$
B $(7, -5)$
C $(4, -2)$
D $(7, -3)$

Triangle XYZ has vertices $X(1, 2)$, $Y(3, 6)$, and $Z(-1, 4)$.

11. Find the coordinates of the vertices of a triangle that is a dilation of $\triangle XYZ$ with a perimeter two times that of $\triangle XYZ$.

12. Find the coordinates of the vertices of $\triangle XYZ$ after it is rotated 90 degrees counterclockwise about the origin.

13. Use a determinant to find the area of $\triangle XYZ$.

14. **MULTIPLE CHOICE** What is the value of $\begin{vmatrix} 2 & 3 & -1 \\ 0 & 2 & 4 \\ -2 & 5 & 6 \end{vmatrix}$?

F -44
G 44
H $\frac{1}{44}$
J $-\frac{1}{44}$

Find the inverse of each matrix, if it exists.

15. $\begin{bmatrix} 5 & 0 \\ 0 & 1 \end{bmatrix}$

16. $\begin{bmatrix} 1 & 2 \\ 2 & 1 \end{bmatrix}$

17. $\begin{bmatrix} 6 & 3 \\ 8 & 4 \end{bmatrix}$

18. $\begin{bmatrix} -3 & -2 \\ 6 & 4 \end{bmatrix}$

Use Cramer's Rule to solve the following system of equations.

19. $2x - y = -9$
$x + 2y = 8$

20. $x - y + 2z = 0$
$3x + z = 11$
$-x + 2y = 0$

21. $6x + 2y + 4z = 2$
$3x + 4y - 8z = -3$
$-3x - 6y + 12z = 5$

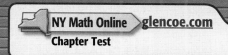

CHAPTER 4 Preparing for Standardized Tests

Gridded Response Questions

In addition to multiple choice, short answer, and extended response questions, you will likely encounter gridded response questions on standardized tests. For gridded response questions, you must print your answer on an answer sheet and mark in the correct circles on the grid to match your answer.

Strategies for Solving Gridded Response Questions

Step 1

Read the problem carefully and solve.

- Be sure your answer makes sense.
- If time permits, check your answer

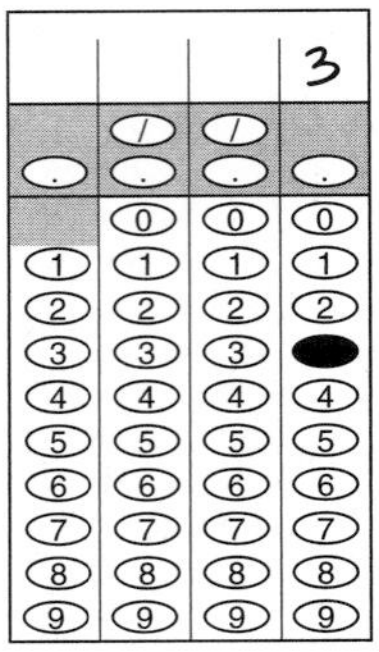

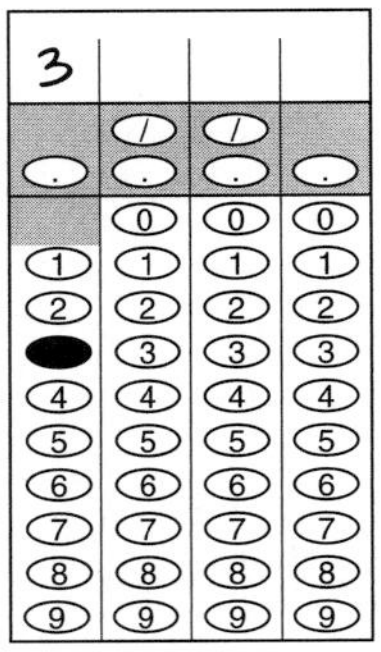

Step 2

Print your answer in the answer boxes.

- Print only one digit or symbol in each answer box.
- Do not write any digits or symbols outside the answer boxes.
- Answers to gridded response questions may be whole numbers, decimals, or fractions.

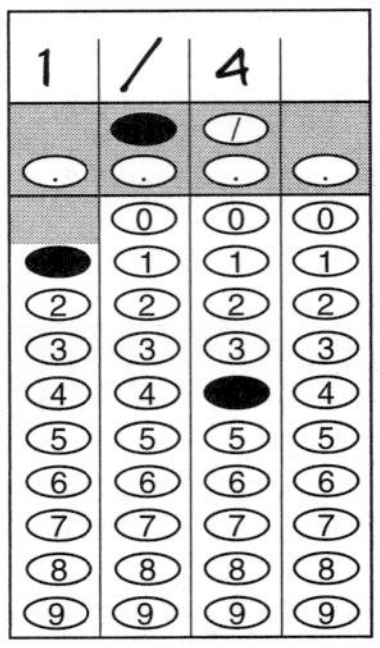

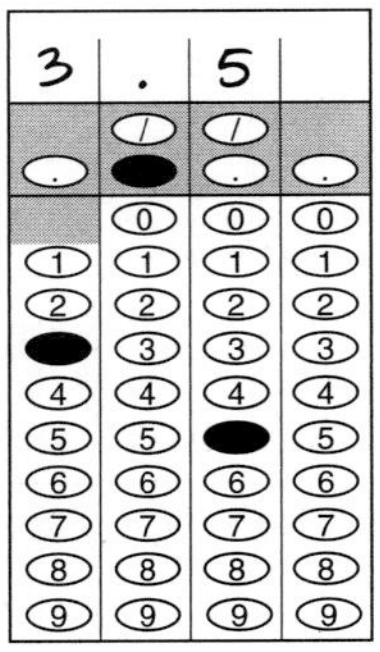

Step 3

Fill in the grid.

- Fill in only one bubble for every answer box that you have written in. Be sure not to fill in a bubble under a blank answer box.
- Fill in each bubble completely and clearly.

EXAMPLE

Read the problem. Identify what you need to know. Then use the information in the problem to solve.

> Christopher stopped for gasoline twice during a road trip. The price of gasoline at the first gas station where he stopped was \$2.96 per gallon. At the second gas station, the price was \$3.15 per gallon. Christopher bought a total of 28.3 gallons of gasoline between the two stops and spent a total of \$86.39. How many gallons did he buy at the first gas station?

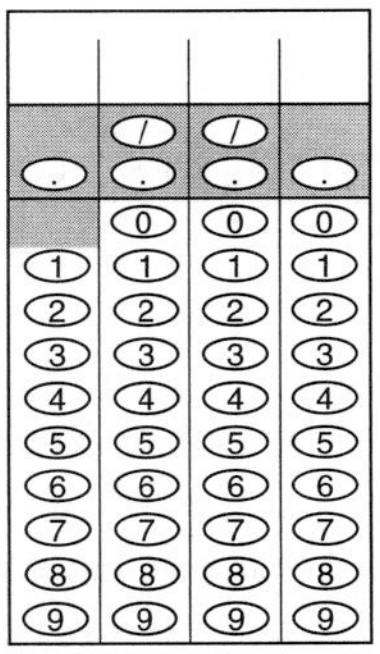

Read the problem carefully. The problem can be solved using a system of equations. Let x represent the number of gallons bought at the first gas station, and let y represent the number of gallons bought at the second station. The following system models the situation.

$$x + y = 28.3$$
$$2.96x + 3.15y = 86.39$$

The system of equations can be solved algebraically. However, if time is a concern, it may be faster and simpler to use matrices and a calculator to solve the system.

Solve the Problem

Enter the coefficient and constant matrices into a graphing calculator and solve using inverses.

$$\begin{bmatrix} 1 & 1 \\ 2.96 & 3.15 \end{bmatrix}\begin{bmatrix} x \\ y \end{bmatrix} = \begin{bmatrix} 28.3 \\ 86.39 \end{bmatrix}$$

$$A = \begin{bmatrix} 1 & 1 \\ 2.96 & 3.15 \end{bmatrix}, B = \begin{bmatrix} 28.3 \\ 86.39 \end{bmatrix}$$

$$\begin{bmatrix} x \\ y \end{bmatrix} = A^{-1}B = \begin{bmatrix} 14.5 \\ 13.8 \end{bmatrix}$$

So, Christopher bought 14.5 gallons of gasoline at the first gas station. Carefully fill in the grid to show 14.5.

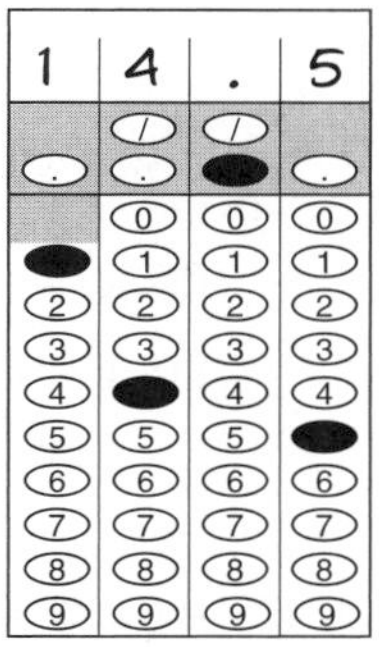

Exercises

Read each problem. Identify what you need to know. Then use the information in the problem to solve. Copy and complete an answer grid on your paper.

1. Heather has 23 nickels, dimes, and quarters. The total value of all the coins is \$3.40. If she has twice as many quarters as dimes, how many nickels does Heather have?

2. Find the value of the determinant $\begin{vmatrix} -1 & 4 \\ -3 & 0 \end{vmatrix}$.

3. Kendra is displaying eight sweaters in a store window. There are four identical red sweaters, three identical brown sweaters, and one white sweater. How may different arrangements of the eight sweaters are possible?

4. Evaluate the determinant of matrix H.

$$H = \begin{bmatrix} -2 & 0 & 3 \\ -5 & -7 & -1 \\ 4 & -8 & 1 \end{bmatrix}$$

5. Polygon $DABC$ is rotated 90° counterclockwise and then reflected over the line $y = x$. What is the x-coordinate of the final image of A?

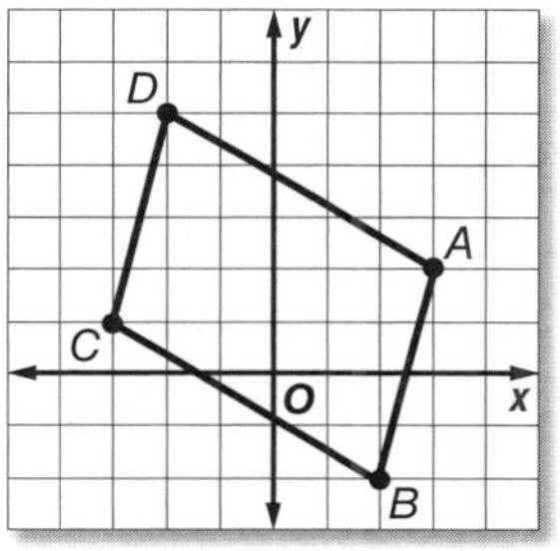

NYS Regents Exam Practice

Cumulative, Chapters 1 through 4

Multiple Choice

Answer all questions in this part. Select the answer that best completes the statement or answers the question.

1. What are the dimensions of matrix *D* below?

$$D = \begin{bmatrix} -1 & 8 & -6 \\ 7 & 4 & 9 \\ 3 & 0 & -2 \\ -7 & 5 & 10 \end{bmatrix}$$

(1) 3 by 4 **(3)** 4 by 3

(2) 4 by 4 **(4)** 12 by 1

2. Which of the following terms best describes a pair of parallel lines in terms of the number of solutions the system has?

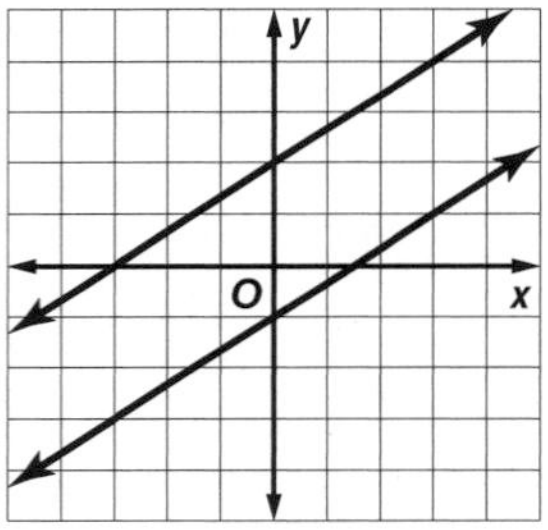

(1) consistent

(2) dependent

(3) inconsistent

(4) independent

3. Evaluate the determinant of matrix *H* below.

$$H = \begin{bmatrix} -1 & 3 & -3 \\ 4 & -2 & -1 \\ 0 & -5 & 2 \end{bmatrix}$$

(1) −12

(2) 19

(3) 33

(4) 45

4. Find the product, if possible.

$$\begin{bmatrix} 5 & -3 \end{bmatrix} \cdot \begin{bmatrix} 4 \\ -2 \end{bmatrix}$$

(1) −13

(2) 26

(3) $\begin{bmatrix} 20 & -12 \\ -10 & 6 \end{bmatrix}$

(4) undefined

Test-Taking Tip

Question 4 The product of a 1 by 2 matrix and a 2 by 1 matrix is a 1 by 1 matrix. So, answer choices 3 and 4 can be eliminated since they are not in the proper format.

5. What is the domain of the relation shown below?

x	*y*
−6	8
−2	−1
0	7
5	−4

(1) {0, 5, 7, 8}

(2) {−6, −4, −2, −1}

(3) {−4, −1, 7, 8}

(4) {−6, −2, 0, 5}

6. Triangle *LMN* has vertices *L*(−4, 3), *M*(6, 2), and *N*(9, −5). Evaluate the determinant below to find the area of the triangle. (Remember to take the absolute value if the determinant is negative.)

$$\frac{1}{2}\begin{vmatrix} -4 & 3 & 1 \\ 6 & 2 & 1 \\ 9 & -5 & 1 \end{vmatrix}$$

(1) 33.5 square units **(3)** 51 square units

(2) 47.5 square units **(4)** 67 square units

7. In the system below, x represents New York's national rank according to state populations, and y represents the population, in millions. Which of the following statements is true?

$$\begin{cases} 10x - 3y = -27 \\ -7m + 2y = 17 \end{cases}$$

(1) New York is the 5th most populous state at 18 million people.

(2) New York is the 3rd most populous state at 19 million people.

(3) New York is the 6th most populous state at 12 million people.

(4) New York is the 4th most populous state at 15 million people.

8. Matrix A shows the number of soft drinks sold at a concessions stand during the first two home games. Matrix B shows the dollar amount per soft drink in terms of revenue generated and cost.

Number Sold

$$A = \begin{array}{r} \textbf{Game 1} \\ \textbf{Game 2} \end{array} \overset{\begin{array}{ccc} \text{Diet} & \text{Grape} & \text{Cola} \end{array}}{\begin{bmatrix} 42 & 51 & 44 \\ 48 & 39 & 47 \end{bmatrix}}$$

Dollar Amount Per Soda

$$B = \begin{array}{r} \textbf{Diet} \\ \textbf{Grape} \\ \textbf{Cola} \end{array} \overset{\begin{array}{cc} \text{Revenues (\$)} & \text{Cost (\$)} \end{array}}{\begin{bmatrix} 0.75 & 0.12 \\ 0.85 & 0.14 \\ 0.75 & 0.12 \end{bmatrix}}$$

If matrix C is the product of A and B, which element in matrix C represents the revenue generated from soda sales in Game 1?

$$C = A \times B = \begin{bmatrix} c_{11} & c_{12} \\ c_{21} & c_{22} \end{bmatrix}$$

(1) c_{11} **(3)** c_{21}

(2) c_{12} **(4)** c_{22}

Open-Ended Response

Solve each problem. Clearly indicate the necessary steps, including appropriate formula substitutions, diagrams, graphs, charts, etc.

9. Craig is checking in a shipment of ink jet printers that cost \$200 each and laser printers that cost \$500 each. There are 40 boxes in the shipment, and the invoice total is \$11,600.

a. Write a system of equations to model the situation. Let x represent the number of ink jet printers, and let y represent the number of laser printers.

b. Write a matrix equation that can be used to solve the system of equations you wrote in part **a**.

c. Find the inverse of the coefficient matrix and solve the matrix equation. How many ink jet printers and laser printers were included in the shipment?

10. Cindy is evaluating the expression $\frac{-5m - 3n}{-2p + r}$ for $m = 2$, $n = -5$, $p = -1$, and $r = -3$. Her work is shown below.

$$\frac{-5m - 3n}{-2p + r} = \frac{-5(2) - 3(-5)}{-2(-1) + (-3)}$$

$$= \frac{-10 - 15}{2 - 3} = \frac{-25}{-1} = 25$$

a. What error did Cindy make in her computation?

b. What is the correct answer?

Need Extra Help?

If you missed Question...	1	2	3	4	5	6	7	8	9	10
Go to Lesson or Page...	4-2	4-3	1-2	4-5	3-2	4-1	2-8	4-6	4-5	4-6
NYS Core Curriculum	A.CM.3	A.G.7	A2.R.2	A2.R.2	A.CM.11	A2.R.2	A.A.7	A2.R.2	A.A.10	A2.N.1

CHAPTER 5 Quadratic Functions and Relations

Then

In Chapter 2, you graphed linear equations and inequalities.

Now

In Chapter 5, you will:

- Graph quadratic functions.
- Solve quadratic equations.
- Perform operations with complex numbers.
- Graph and solve quadratic inequalities.

NYS Core Curriculum

A2.A.7 Factor polynomial expressions completely

A2.A.24, A2.A.25 Solve quadratic equations

Why?

MOTION The path that a soccer ball or a firework takes can be modeled by a quadratic function. Quadratic functions can map an object in motion. In this chapter you will look at a pumpkin catapult, an amusement park ride, and a diver in motion.

Math *in Motion,* Animation glencoe.com

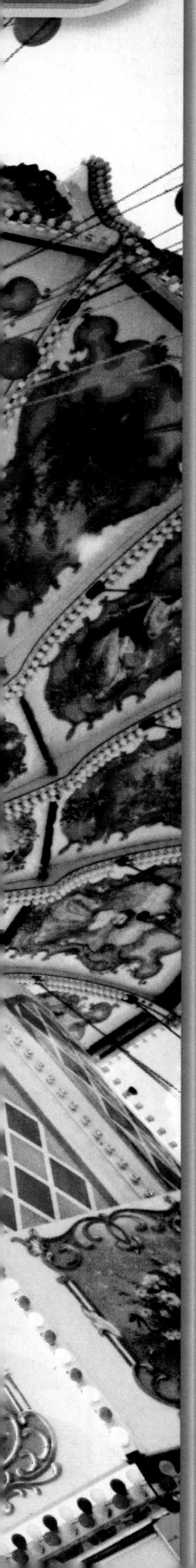

Get Ready for Chapter 5

Diagnose Readiness You have two options for checking Prerequisite Skills.

Text Option

Take the Quick Check below. Refer to the Quick Review for help.

*Quick*Check

Given $f(x) = 2x^2 + 4$ and $g(x) = -x^2 - 2x + 3$, find each value. (Lesson 2-1)

1. $f(-1)$
2. $f(3)$
3. $f(0)$
4. $g(4)$
5. $g(0)$
6. $g(-3)$

7. **FISH** Tuna swim at a steady rate of 9 miles per hour until they die, and they never stop moving. (Lesson 2-1)

a. Write a function that is a model for the situation.

b. Evaluate the function to estimate how far a 2-year old tuna has traveled.

8. **BUDGET** Marla has budgeted \$65 per day on food during a business trip. Write a function that is a model for the situation and evaluate what she would spend on a 2 week business trip. (Lesson 2-4)

Factor completely. If the polynomial is not factorable, write *prime*. (Lesson 0-3)

9. $x^2 + 13x + 40$
10. $x^2 - 10x + 21$
11. $2x^2 + 7x - 4$
12. $2x^2 - 7x - 15$
13. $x^2 - 11x + 15$
14. $x^2 + 12x + 36$

15. **FLOOR PLAN** The rectangular room pictured below has an area of $x^2 + 14x + 48$ square feet. If the width of the room is $(x + 6)$ feet, what is the length? (Lesson 0-3)

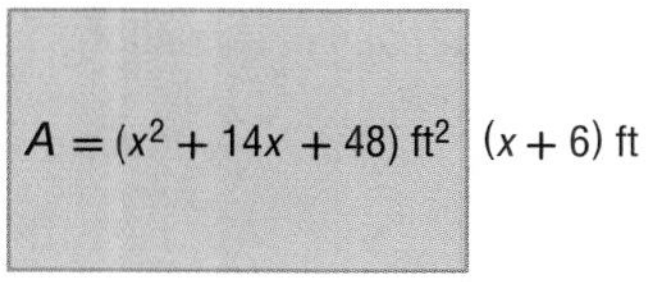

*Quick*Review

EXAMPLE 1

Given $f(x) = -2x^2 + 3x - 1$ and $g(x) = 3x^2 - 5$, find each value.

a. $f(2)$

$f(x) = -2x^2 + 3x - 1$	**Original function**
$f(2) = -2(2)^2 + 3(2) - 1$	**Substitute 2 for *x*.**
$= -8 + 6 - 1$ or -3	**Simplify.**

b. $g(-2)$

$g(x) = 3x^2 - 5$	**Original function**
$g(-2) = 3(-2)^2 - 5$	**Substitute −2 for *x*.**
$= 12 - 5$ or 7	**Simplify.**

EXAMPLE 2

Factor $2x^2 - x - 3$ completely. If the polynomial is not factorable, write *prime*.

To find the coefficients of the x-terms, you must find two numbers whose product is $2(-3)$ or -6, and whose sum is -1. The two coefficients must be 2 and -3 since $2(-3) = -6$ and $2 + (-3) = -1$. Rewrite the expression and factor by grouping.

$2x^2 - x - 3$

$= 2x^2 + 2x - 3x - 3$	**Substitute $2x - 3x$ for $-x$.**
$= (2x^2 + 2x) + (-3x - 3)$	**Associative Property**
$= 2x(x + 1) + -3(x + 1)$	**Factor out the GCF.**
$= (2x - 3)(x + 1)$	**Distributive Property**

Online Option

NY Math Online Take a self-check Chapter Readiness Quiz at glencoe.com.

Get Started on Chapter 5

You will learn several new concepts, skills, and vocabulary terms as you study Chapter 5. To get ready, identify important terms and organize your resources. You may wish to refer to **Chapter 0** to review prerequisite skills.

FOLDABLES® Study Organizer

Quadratic Functions Make this Foldable to help you organize your Chapter 5 notes about quadratic functions and relations. Begin with one sheet of 11″ by 17″ paper.

1. **Fold** in half lengthwise.

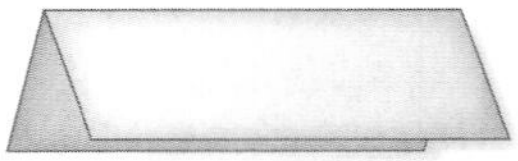

2. **Fold** in fourths crosswise.

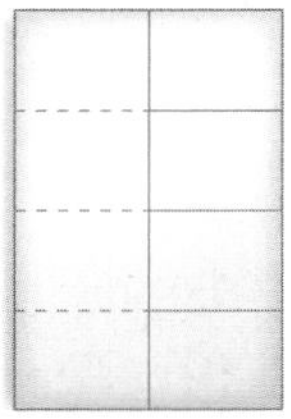

3. **Cut** along the middle fold from the edge to the last crease as shown.

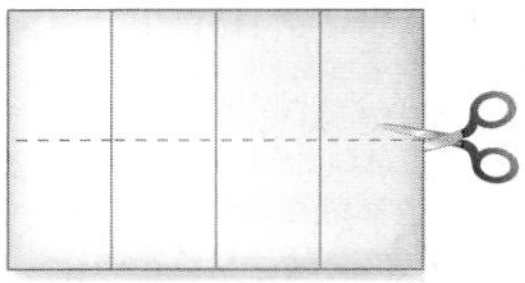

4. **Refold** along the lengthwise fold and tape the uncut section at the top. Label each section with a lesson number and close to form a booklet.

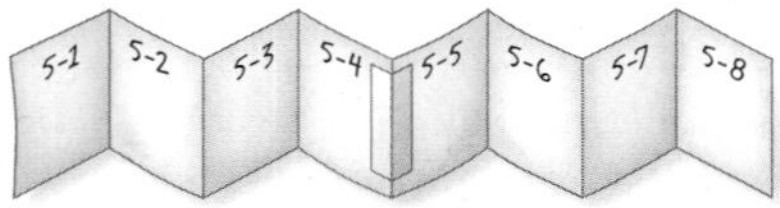

NY Math Online glencoe.com

- Study the chapter online
- Explore **Math in Motion**
- Get extra help from your own **Personal Tutor**
- Use **Extra Examples** for additional help
- Take a **Self-Check Quiz**
- **Review Vocabulary** in fun ways

New Vocabulary

English		Español
quadratic term	• p. 249 •	término cuadrático
linear term	• p. 249 •	término lineal
constant term	• p. 249 •	término constante
vertex	• p. 250 •	vértice
maximum value	• p. 252 •	valor máximo
minimum value	• p. 252 •	valor mínimo
quadratic equation	• p. 259 •	ecuación cuadrática
standard form	• p. 259 •	forma estándar
root	• p. 259 •	raíz
zero	• p. 259 •	cero
imaginary unit	• p. 276 •	unidad imaginaria
pure imaginary number	• p. 276 •	número imaginario puro
complex number	• p. 277 •	número complejo
complex conjugates	• p. 279 •	conjugados complejos
completing the square	• p. 285 •	completar el cuadrado
Quadratic Formula	• p. 292 •	fórmula cuadrática
discriminant	• p. 295 •	discriminante
vertex form	• p. 305 •	forma de vértice
quadratic inequality	• p. 312 •	desigualdad cuadrática

Review Vocabulary

domain • p. P7 • dominio the set of all x-coordinates of the ordered pairs of a relation

function • p. P7 • función a relation in which each x-coordinate is paired with exactly one y-coordinate

range • p. P7 • rango the set of all y-coordinates of the ordered pairs of a relation

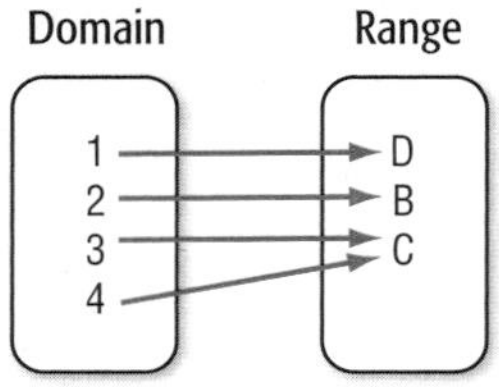

Multilingual eGlossary glencoe.com

5-1 Graphing Quadratic Functions

Then

You identified and manipulated graphs of functions. (Lesson 2-6)

Now

- Graph quadratic functions.
- Find and interpret the maximum and minimum values of a quadratic function.

NYS Core Curriculum

A2.A.41 Use functional notation to evaluate functions for given values in the domain

New Vocabulary

quadratic function
quadratic term
linear term
constant term
parabola
axis of symmetry
vertex
maximum value
minimum value

NY Math Online

glencoe.com

- Extra Examples
- Personal Tutor
- Self-Check Quiz
- Homework Help

Why?

Eddie is organizing a charity tournament. He plans to charge a $20 entry fee for each of the 80 players. He recently decided to raise the entry fee by $5, and 5 fewer players entered with the increase. He used this information to determine how many fee increases will maximize the money raised.

The quadratic function at the right represents this situation. The tournament prize pool increases when he first increases the fee, but eventually the pool starts to decrease as the fee gets even higher.

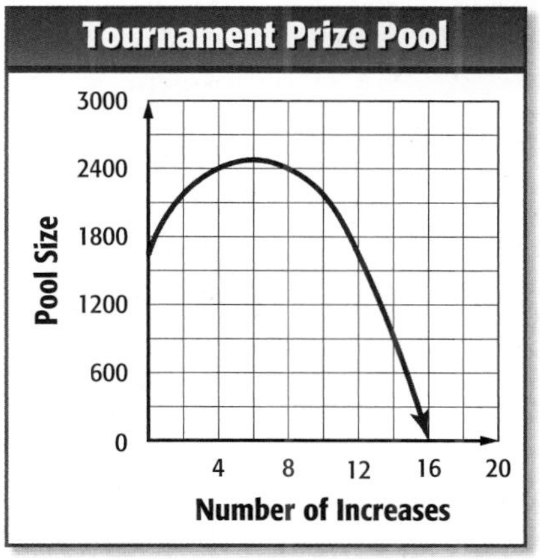

Graph Quadratic Functions In a **quadratic function**, the greatest exponent is 2. These functions can have a **quadratic term**, a **linear term**, and a **constant term**. The general quadratic function is shown below.

$$f(x) = ax^2 + bx + c, \text{ where } a \neq 0$$

quadratic term (ax^2) | linear term (bx) | constant term (c)

The graph of a quadratic function is called a **parabola**. To graph a quadratic function, graph ordered pairs that satisfy the function.

EXAMPLE 1 Graph a Quadratic Function by Using a Table

Graph $f(x) = 3x^2 - 12x + 6$ by making a table of values.

Choose integer values for x, and evaluate the function for each value. Graph the resulting coordinate pairs, and connect the points with a smooth curve.

x	$3x^2 - 12x + 6$	$f(x)$	$(x, f(x))$
0	$3(0)^2 - 12(0) + 6$	6	(0, 6)
1	$3(1)^2 - 12(1) + 6$	−3	(1, −3)
2	$3(2)^2 - 12(2) + 6$	−6	(2, −6)
3	$3(3)^2 - 12(3) + 6$	−3	(3, −3)
4	$3(4)^2 - 12(4) + 6$	6	(4, 6)

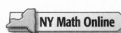

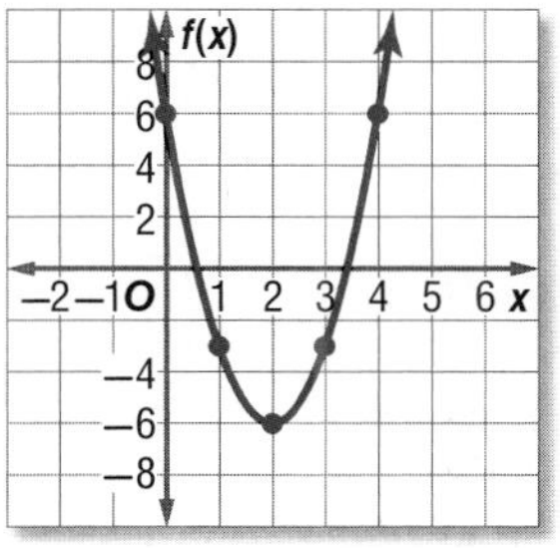

Check Your Progress

Graph each function by making a table of values.

1A. $g(x) = -2x^2 + 8x - 3$

1B. $h(x) = 4x^2 - 8x + 1$

Personal Tutor glencoe.com

Notice in Example 1 that there seemed to be a pattern in the values for $f(x)$. This is due to the axis of symmetry of parabolas. The **axis of symmetry** is a line through the graph of a parabola that divides the graph into two congruent halves. Each side of the parabola is a reflection of the other side.

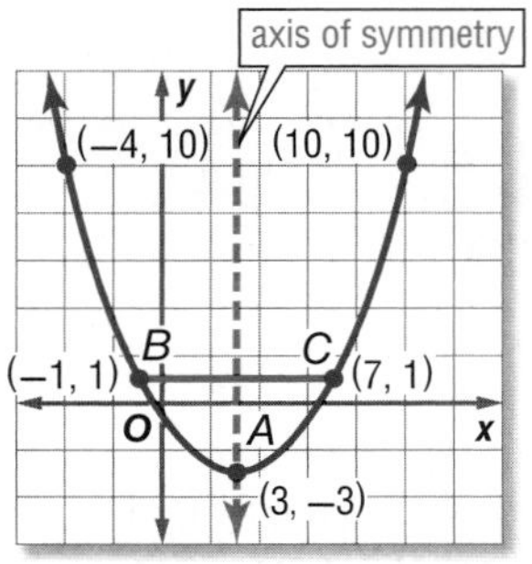

The axis of symmetry will intersect a parabola at only one point, called the **vertex**. The vertex of the graph at the right is $A(3, -3)$.

Review Vocabulary

Symmetry When something is symmetrical, its opposite sides are mirror images of each other.

Notice that the x-coordinates of points B and C are both 4 units away from the x-coordinate of the vertex, and they have the same y-coordinate. This is due to the symmetrical nature of the graph.

Key Concept For Your FOLDABLE

Graph of a Quadratic Function

Words Consider the graph of $y = ax^2 + bx + c$, where $a \neq 0$.

- The y-intercept is $a(0)^2 + b(0) + c$ or c.
- The equation of the axis of symmetry is $x = -\frac{b}{2a}$.
- The x-coordinate of the vertex is $-\frac{b}{2a}$.

Model

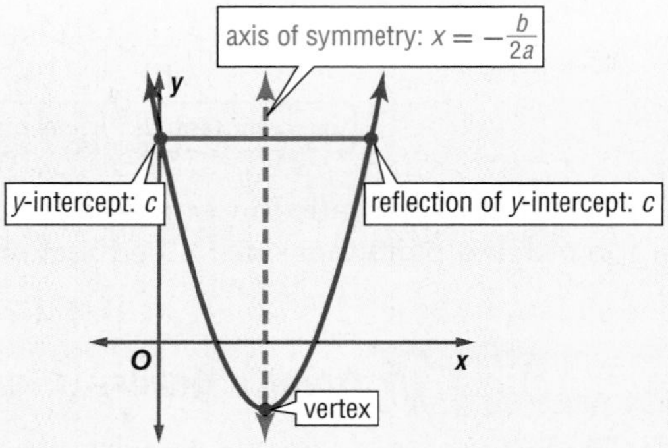

Now you can use the axis of symmetry to help plot points and graph a parabola. For $y = x^2 + 6x - 2$ below, the axis of symmetry is $x = -\frac{b}{2a} = -\frac{6}{2(1)}$ or $x = -3$.

StudyTip

Plotting Reflections The reflection of a point is its mirror image on the other side of the axis of symmetry.

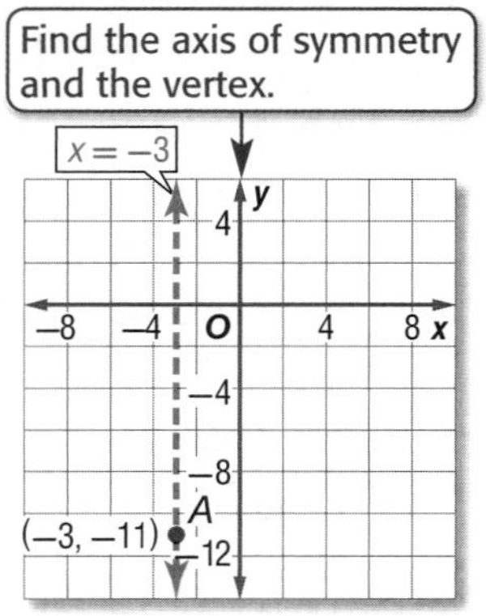

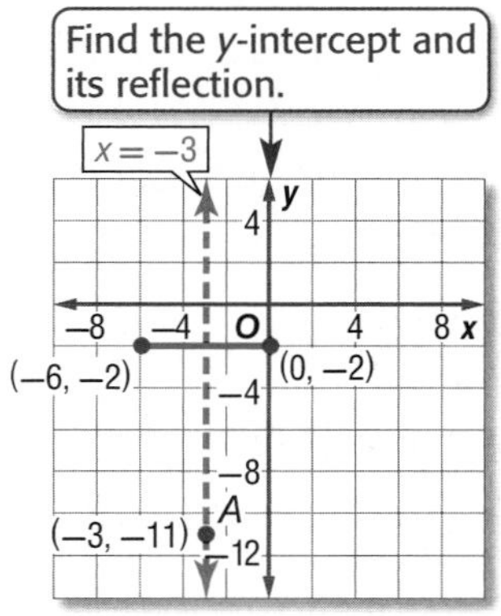

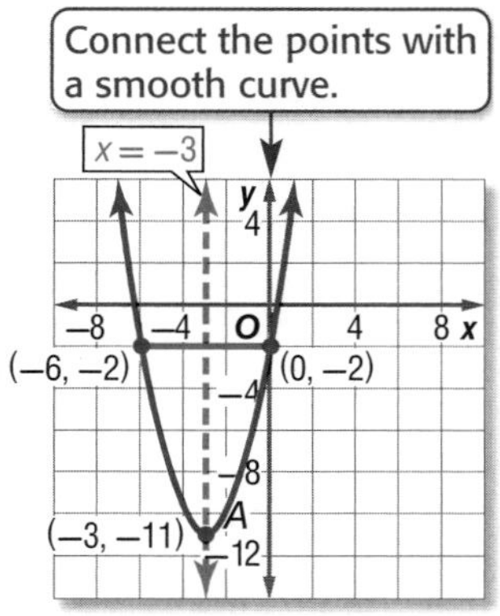

EXAMPLE 2 Axis of Symmetry, y-intercept, and Vertex

Consider $f(x) = x^2 + 4x - 3$.

a. Find the y-intercept, the equation of the axis of symmetry, and the x-coordinate of the vertex.

The function is of the form $f(x)$ is $ax^2 + bx + c$, so we can identify a, b, and c.

$$f(x) = ax^2 + bx + c$$
$$\downarrow \quad \downarrow \quad \downarrow$$
$$f(x) = 1x^2 + 4x - 3 \quad \rightarrow \quad a = 1, b = 4, \text{ and } c = 3$$

Use a and b to find the equation of the axis of symmetry.

$x = -\frac{b}{2a}$ **Equation of the axis of symmetry**

$= -\frac{4}{2(1)}$ **$a = 1$ and $b = 4$**

$= -2$ **Simplify.**

The equation of the axis of symmetry is $x = -2$. Therefore, the x-coordinate of the vertex is -2.

b. Make a table of values that includes the vertex.

The y-intercept is $c = -3$.

Select five specific points, with the vertex in the middle and two points on either side of the vertex, including the y-intercept and its reflection. Use symmetry to determine the y-values of the reflections.

x	$x^2 + 4x - 3$	$f(x)$	$(x, f(x))$
−6	$(-6)^2 + 4(-6) - 3$	9	(−6, 9)
−4	$(-4)^2 + 4(-4) - 3$	−3	(−4, −3)
−2	$(-2)^2 + 4(-2) - 3$	−7	(−2, −7)
0	$(0)^2 + 4(0) - 3$	−3	(0, −3)
2	$(2)^2 + 4(2) - 3$	9	(2, 9)

c. Use this information to graph the function.

Graph the points from the table and the y-intercept, connecting them with a smooth curve.

Draw the axis of symmetry, $x = -2$, as a dashed line. The graph should be symmetrical about this line.

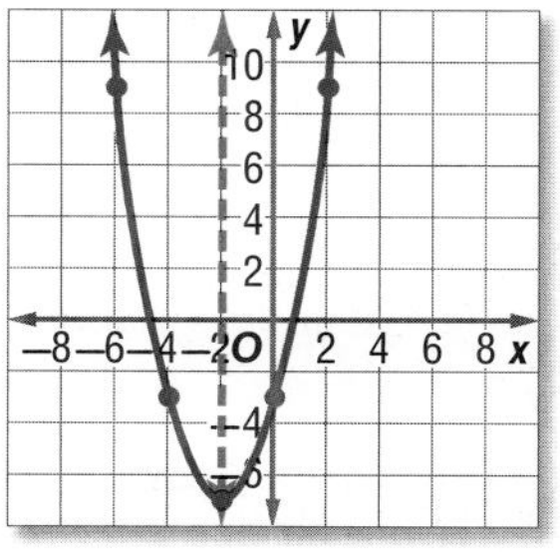

StudyTip

Quadratic Form Make sure the function is in standard quadratic form, $y = ax^2 + bx + c$, before graphing.

StudyTip

Fractions When the x-coordinate of the vertex is a fraction, select the nearest integer for the next point to avoid using fractions and simplify the calculations.

Check Your Progress

Consider $f(x) = -5x^2 - 10x + 6$.

2A. Find the y-intercept, the equation of the axis of symmetry, and the x-coordinate of the vertex.

2B. Make a table of values that includes the vertex.

2C. Use this information to graph the function.

Personal Tutor glencoe.com

> **Watch Out!**
>
> **Maxima and Minima** The terms *minimum point* and *minimum value* are not interchangeable. The minimum point on the graph of a quadratic function is the set of coordinates that describe the location of the vertex. The minimum value of a function is the y-coordinate of the minimum point. It is the smallest value obtained when $f(x)$ is evaluated for all values of x.

Maximum and Minimum Values The y-coordinate of the vertex of a quadratic function is the **maximum value** or the **minimum value** of the function. These values represent the greatest or lowest possible value the function can reach.

Key Concept — Maximum and Minimum Value

For Your FOLDABLE

Words The graph of $f(x) = ax^2 + bx + c$, where $a \neq 0$,

- opens up and has a minimum value when $a > 0$, and
- opens down and has a maximum value when $a < 0$.

Models

a is positive.

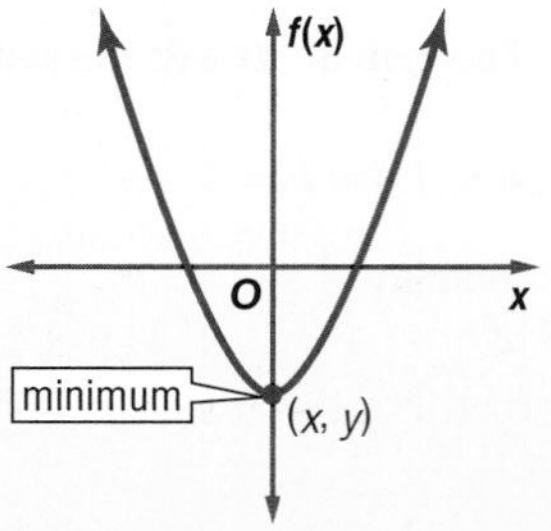

The y-coordinate is the minimum value.

a is negative.

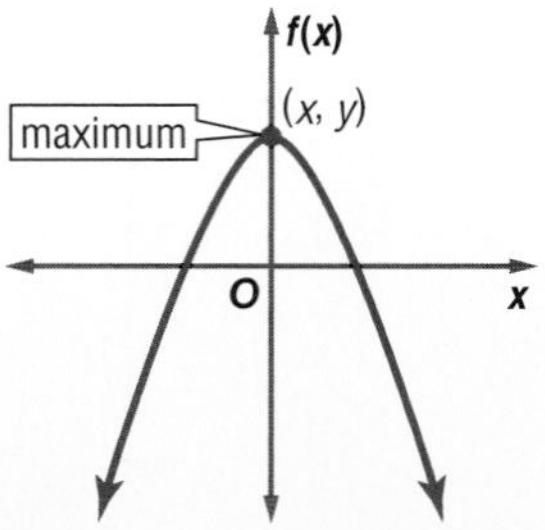

The y-coordinate is the maximum value.

EXAMPLE 3 Maximum or Minimum Values

Consider $f(x) = -4x^2 + 12x + 18$.

a. Determine whether the function has a *maximum* or *minimum* value.

For this function, $a = -4$, so the graph opens down and the function has a maximum value.

b. State the maximum or minimum value of the function.

The maximum value of the function is the y-coordinate of the vertex.

The x-coordinate of the vertex is $-\frac{12}{2(-4)}$ or 1.5.

Find the y-coordinate of the vertex by evaluating the function for $x = 1.5$.

$f(x) = -4x^2 + 12x + 18$	**Original function**
$= -4(1.5)^2 + 12(1.5) + 18$	**$x = 1.5$**
$= -9 + 18 + 18$ or 27	The maximum value of the function is 27.

c. State the domain and range of the function.

The domain is all real numbers. The range is all real numbers less than or equal to the maximum value, or $\{f(x) \mid f(x) \leq 27\}$.

> **StudyTip**
>
> **Domain and Range** The domain of a quadratic function will always be all real numbers. The range will either be all real numbers less than or equal to the maximum or all real numbers greater than or equal to the minimum.

Check Your Progress

Consider $f(x) = 4x^2 - 24x + 11$.

3A. Determine whether the function has a maximum or minimum value.

3B. State the maximum or minimum value of the function.

3C. State the domain and range of the function.

Personal Tutor glencoe.com

Real-World Link

As of 2006, there were approximately 900,000 public charities in the United States.

Source: National Center for Charitable Statistics

Real-World EXAMPLE 4 Quadratic Equations in the Real World

CHARITY **Refer to the beginning of the lesson.**

a. How much should Eddie charge in order to maximize charity income?

Words	Total	equals	fee	times	number of entrants.
Variables	Let x = the number of price increases. Let $P(x)$ = the total pool as a function of x.				
Equation	$P(x)$	$=$	$20 + 5x$	$\cdot$	$(80 - 5x)$

Solve for x in the equation.

$P(x) = (20 + 5x) \cdot (80 - 5x)$

$= 20(80) + 20(-5x) + 5x(80) + 5x(-5x)$ **Distribute.**

$= 1600 - 100x + 400x - 25x^2$ **Multiply.**

$= 1600 + 300x - 25x^2$ **Simplify.**

$= -25x^2 + 300x + 1600$ **$ax^2 + bx + c$ form**

Use the formula for the axis of symmetry, $x = -\frac{b}{2a}$, to find the x-coordinate.

$x = -\frac{300}{2(-25)}$ or 6 **$a = -25$ and $b = 300$**

Eddie needs to have 6 price increases, so he should charge $20 + 6(5)$ or \$50.

b. What will be the maximum value of the pool?

Find the maximum value of the quadratic function $P(x)$ by evaluating $P(6)$.

$P(x) = -25x^2 + 300x + 1600$ **Total pool function**

$P(6) = -25(6)^2 + 300(6) + 1600$ **$x = 6$**

$= -900 + 1800 + 1600$ or 2500 **Simplify.**

Thus, the maximum prize pool is \$2500 after 6 price increases.

CHECK Graph the function on a graphing calculator and use the **CALC:Maximum** function to confirm the solution.

Select a left bound of 0 and a right bound of 10. The calculator will display the coordinates of the maximum at the bottom of the screen.

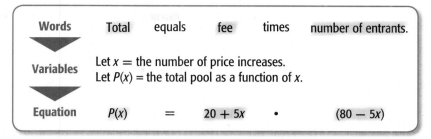

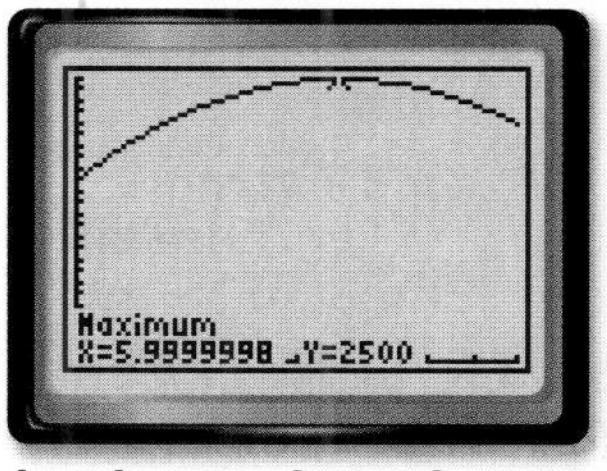

[0, 10] scl: 1 by [0, 2500] scl: 100

The domain is $\{x \mid x \geq 0\}$ because there can be no negative increases in price. The range is $\{y \mid 0 \leq y \leq 2500\}$ because the prize pool cannot have a negative monetary value.

StudyTip

Domain and Range in Real-World Problems Use logic and the information from the problem to determine the domain and range that are reasonable in the situation.

Check Your Progress

4. Suppose a different tournament that Eddie organizes has 120 players and the entry fee is \$40. Each time he increases the fee by \$5, he loses 10 players. Determine what the entry fee should be to maximize the value of the pool.

Personal Tutor glencoe.com

Check Your Understanding

Examples 1 and 2 pp. 249–251

Complete parts a–c for each quadratic function.

a. Find the y-intercept, the equation of the axis of symmetry, and the x-coordinate of the vertex.

b. Make a table of values that includes the vertex.

c. Use this information to graph the function.

1. $f(x) = 3x^2$

2. $f(x) = -6x^2$

3. $f(x) = x^2 - 4x$

4. $f(x) = -x^2 - 3x + 4$

5. $f(x) = 4x^2 - 6x - 3$

6. $f(x) = 2x^2 - 8x + 5$

Example 3 p. 252

Determine whether each function has a *maximum* or *minimum* value, and find that value. Then state the domain and range of the function.

7. $f(x) = -x^2 + 6x - 1$

8. $f(x) = x^2 + 3x - 12$

9. $f(x) = 3x^2 + 8x + 5$

10. $f(x) = -4x^2 + 10x - 6$

Example 4 p. 253

11. **BUSINESS** A store rents 1400 videos per week at \$2.25 per video. The owner estimates that they will rent 100 fewer videos for each \$0.25 increase in price. What price will maximize the income of the store?

Practice and Problem Solving

● = **Step-by-Step Solutions** begin on page R20. **Extra Practice** begins on page 947.

Examples 1 and 2 pp. 249–251

Complete parts a–c for each quadratic function.

a. Find the y-intercept, the equation of the axis of symmetry, and the x-coordinate of the vertex.

b. Make a table of values that includes the vertex.

c. Use this information to graph the function.

12. $f(x) = 4x^2$

13. $f(x) = -2x^2$

14. $f(x) = x^2 - 5$

15. $f(x) = x^2 + 3$

16. $f(x) = 4x^2 - 3$

17. $f(x) = -3x^2 + 5$

18. $f(x) = x^2 - 6x + 8$

19. $f(x) = x^2 - 3x - 10$

20. $f(x) = -x^2 + 4x - 6$

21. $f(x) = -2x^2 + 3x + 9$

Example 3 p. 252

Determine whether each function has a *maximum* or *minimum* value, and find that value. Then state the domain and range of the function.

22. $f(x) = 5x^2$

23. $f(x) = -x^2 - 12$

24. $f(x) = x^2 - 6x + 9$

25. $f(x) = -x^2 - 7x + 1$

26. $f(x) = 8x - 3x^2 + 2$

27. $f(x) = 5 - 4x - 2x^2$

28. $f(x) = 15 - 5x^2$

29. $f(x) = x^2 + 12x + 27$

30. $f(x) = -x^2 + 10x + 30$

31. $f(x) = 2x^2 - 16x - 42$

Example 4 p. 253

32. **PRODUCTION** A financial analyst determined the cost in thousands of dollars of producing bicycle frames is $C = 0.000025f^2 - 0.04f + 40$, where f is the number of frames produced.

a. Find the number of frames that minimizes cost.

b. What is the total cost for that number of frames?

Complete parts a–c for each quadratic function.

a. Find the y-intercept, the equation of the axis of symmetry, and the x-coordinate of the vertex.

b. Make a table of values that includes the vertex.

c. Use this information to graph the function.

33. $f(x) = 2x^2 - 6x - 9$

34. $f(x) = -3x^2 - 9x + 2$

35. $f(x) = -4x^2 + 5x$

36. $f(x) = 2x^2 + 11x$

37. $f(x) = 0.25x^2 + 3x + 4$

38. $f(x) = -0.75x^2 + 4x + 6$

39. $f(x) = \frac{3}{2}x^2 + 4x - \frac{5}{2}$

40. $f(x) = \frac{2}{3}x^2 - \frac{7}{3}x + 9$

41. BABYSITTING A babysitting club sits for 50 different families. They would like to increase their current rate of \$9.50 per hour. After surveying the families, the club finds that the number of families will decrease by about 2 for each \$0.50 increase in the hourly rate.

a. Write a quadratic equation that models this situation.

b. State the domain and range of this function as it applies to the situation.

c. What hourly rate will maximize the club's income? Is this reasonable?

d. What is the maximum income the club can expect to make?

Real-World Link

Participation in a drama club is not limited to acting. Other involvement includes set construction, lighting and sound, promotion, and business management.

42. ACTIVITIES Last year, 300 people attended the Franklin High School Drama Club's winter play. The ticket price was \$8. The advisor estimates that 20 fewer people would attend for each \$1 increase in ticket price.

a. What ticket price would give the greatest income for the Drama Club?

b. If the Drama Club raised its tickets to this price, how much income should it expect to bring in?

GRAPHING CALCULATOR Use a calculator to find the maximum or minimum of each function. Round to the nearest hundredth if necessary.

43. $f(x) = -9x^2 - 12x + 19$

44. $f(x) = 12x^2 - 21x + 8$

45. $f(x) = -8.3x^2 + 14x - 6$

46. $f(x) = 9.7x^2 - 13x - 9$

47. $f(x) = 28x - 15 - 18x^2$

48. $f(x) = -16 - 14x - 12x^2$

Determine whether each function has a *maximum* or *minimum* value, and find that value. Then state the domain and range of the function.

49. $f(x) = -5x^2 + 4x - 8$

50. $f(x) = -4x^2 - 3x + 2$

51. $f(x) = -9 + 3x + 6x^2$

52. $f(x) = 2x - 5 - 4x^2$

53. $f(x) = \frac{2}{3}x^2 + 6x - 10$

54. $f(x) = -\frac{3}{5}x^2 + 4x - 8$

Determine the function represented by each graph.

55.

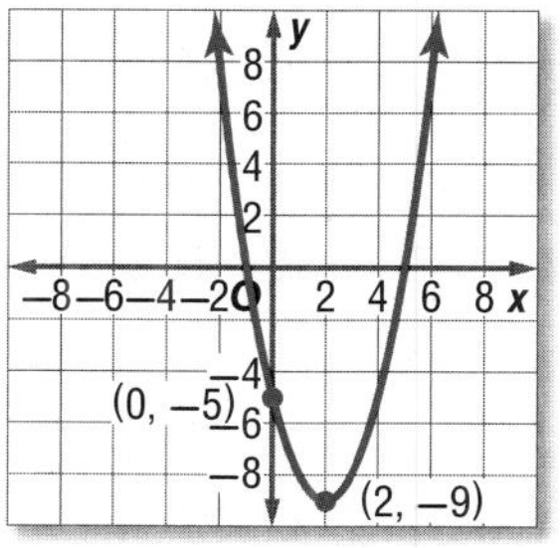

56.

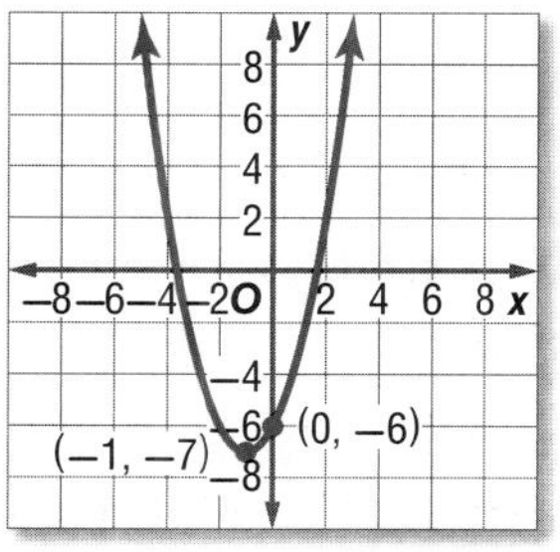

57.

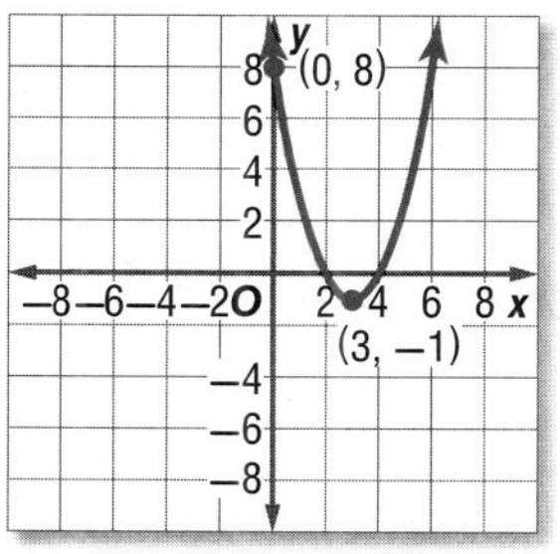

58. **MULTIPLE REPRESENTATIONS** Consider $f(x) = x^2 - 4x + 8$ and $g(x) = 4x^2 - 4x + 8$.

a. **TABULAR** Make a table of values for $f(x)$ and $g(x)$ if $-4 \leq x \leq 4$.

b. **GRAPHICAL** Graph $f(x)$ and $g(x)$.

c. **VERBAL** Explain the difference in the shapes of the graphs of $f(x)$ and $g(x)$. What value was changed to cause this difference?

d. **ANALYTICAL** Predict the appearance of the graph of $h(x) = 0.25x^2 - 4x + 8$. Confirm your prediction by graphing all three functions if $-10 \leq x \leq 10$.

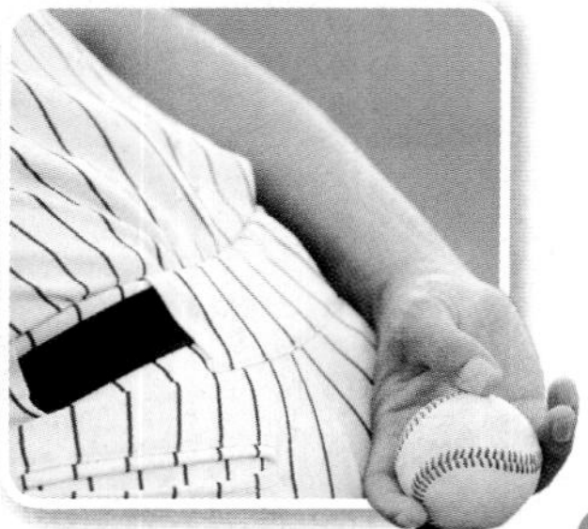

Real-World Link

To get the most distance and velocity when throwing a baseball, experienced baseball players grip a ball with their fingertips across the seams.

Source: QCBaseball

59. **VENDING MACHINES** Omar owns a vending machine in a bowling alley. He currently sells 600 cans of soda per week at \$0.65 per can. He estimates that he will lose 100 customers for every \$0.05 increase in price and gain 100 customers for every \$0.05 decrease in price. (*Hint*: The charge *must* be a multiple of 5.)

a. Write and graph the related quadratic equation for a price increase.

b. If Omar lowers the price, what price should he charge in order to maximize his income?

c. What will be his income per week from the vending machine?

60. **BASEBALL** Lolita throws a baseball into the air and the height h of the ball in feet at a given time t in seconds after she releases the ball is given by the function $h(t) = -16t^2 + 30t + 5$.

a. State the domain and range for this situation.

b. Find the maximum height the ball will reach.

H.O.T. Problems

Use **H**igher-**O**rder **T**hinking Skills

61. **FIND THE ERROR** Trent and Madison are asked to find the maximum of $f(x) = -4x^2 + 8x - 6$. Is either of them correct? Explain your reasoning.

Madison

$x = -\frac{8}{2(-4)} = 1$

$f(1) = -4(1)^2 + 8(1) - 6$ or -2

The maximum is -2.

Trent

$x = -\frac{8}{2(-4)} = -1$

$f(-1) = -4(-1)^2 + 8(-1) - 6$

or -18

The maximum is -18.

62. **REASONING** Determine whether the following is *sometimes*, *always*, or *never* true. Explain your reasoning.

In a quadratic function, if two x-coordinates are equidistant from the axis of symmetry, then they will have the same y-coordinate.

63. **CHALLENGE** The table at the right represents some points on the graph of a quadratic function.

a. Find the values of a, b, c, and d.

b. What is the x-coordinate of the vertex?

c. Does the function have a maximum or a minimum?

x	y
-19	2
c	14
-3	$b - 6$
-2	22
d	a
3	$10d$
6	14
19	d

64. **OPEN ENDED** Give an example of a quadratic function with a

a. maximum of 8. b. minimum of -4. c. vertex of $(-2, 6)$.

65. **WRITING IN MATH** Describe how you determine whether a function is quadratic and if it has a maximum or minimum value.

NYSRE PRACTICE Reinforcement of A.A.40

66. Which expression is equivalent to $\frac{8!}{5!}$?

A $\frac{8}{5}$ C 3!

B $8 \cdot 7 \cdot 6$ D $8 \cdot 7 \cdot 6 \cdot 5$

67. SAT/ACT The price of coffee beans is d dollars for 6 ounces, and each ounce makes c cups of coffee. In terms of c and d, what is the cost of the coffee beans required to make 1 cup of coffee?

F $\frac{cd}{6}$ H $6cd$

G $\frac{6c}{d}$ J $\frac{d}{6c}$

68. SHORT RESPONSE Each side of the square base of a pyramid is 20 feet, and the pyramid's height is 90 feet. What is the volume of the pyramid?

69. Which ordered pair is the solution of the following system of equations?

$$3x - 5y = 11$$
$$3x - 8y = 5$$

A (2, 1) C (7, 2)

B (7, −2) D $\left(\frac{1}{3}, -2\right)$

Spiral Review

Find the inverse of each matrix, if it exists. (Lesson 4-6)

70. $\begin{bmatrix} 3 & -4 \\ 2 & -1 \end{bmatrix}$ **71.** $\begin{bmatrix} -4 & -1 \\ 0 & 6 \end{bmatrix}$ **72.** $\begin{bmatrix} 2 & 8 \\ -3 & -5 \end{bmatrix}$

Evaluate each determinant. (Lesson 4-5)

73. $\begin{vmatrix} 6 & -3 \\ -1 & 8 \end{vmatrix}$ **74.** $\begin{vmatrix} -3 & -5 \\ -1 & -9 \end{vmatrix}$ **75.** $\begin{vmatrix} 8 & 6 \\ 4 & 3 \end{vmatrix}$

76. MANUFACTURING The Community Service Committee is making canvas tote bags and leather tote bags for a fundraiser. They will line both types of bags with canvas and use leather handles on both. For the canvas bags, they need 4 yards of canvas and 1 yard of leather. For the leather bags, they need 3 yards of leather and 2 yards of canvas. The committee leader purchased 56 yards of leather and 104 yards of canvas. (Lesson 3-4)

a. Let c represent the number of canvas bags, and let ℓ represent the number of leather bags. Write a system of inequalities for the number of bags that can be made.

b. Draw the graph showing the feasible region.

c. List the coordinates of the vertices of the feasible region.

d. If the club plans to sell the canvas bags at a profit of \$20 each and the leather bags at a profit of \$35 each, write a function for the total profit on the bags.

e. How can the club make the maximum profit?

f. What is the maximum profit?

State whether each function is a linear function. Write *yes* or *no*. Explain. (Lesson 2-2)

77. $y = 4x^2 - 3x$ **78.** $y = -2x - 4$ **79.** $y = 4$

Skills Review

Evaluate each function for the given value. (Lesson 2-1)

80. $f(x) = 3x^2 - 4x + 6, x = -2$ **81.** $f(x) = -2x^2 + 6x - 5, x = 4$ **82.** $f(x) = 6x^2 + 18, x = -5$

EXTEND 5-1

Graphing Technology Lab
Modeling Real-World Data

NY Math Online glencoe.com
Other Calculator Keystrokes

NYS Core Curriculum **A2.S.7** Determine the function for the regression model, using appropriate technology, and use the regression function to interpolate and extrapolate from the data

You can use a TI-83/84 Plus graphing calculator to model data points for which a curve of best fit is a quadratic function.

WATER A bottle is filled with water. The water is allowed to drain from a hole made near the bottom of the bottle. The table shows the level of the water y measured in centimeters from the bottom of the bottle after x seconds.

Time (s)	0	20	40	60	80	100	120	140	160	180	200	220
Water level (cm)	42.6	40.7	38.9	37.2	35.8	34.3	33.3	32.3	31.5	30.8	30.4	30.1

Find and graph a linear regression equation and a quadratic regression equation. Determine which equation is a better fit for the data.

Step 1 **Find and graph a linear regression equation.**

- Enter the times in L1 and the water levels in L2. Then find a linear regression equation.

  KEYSTROKES: *Refer to Lesson 2-5.*

- Use STAT PLOT to graph a scatter plot. Copy the equation to the Y= list and graph.

  KEYSTROKES: *Review statistical plots and graphing a regression equation in Lesson 2-5.*

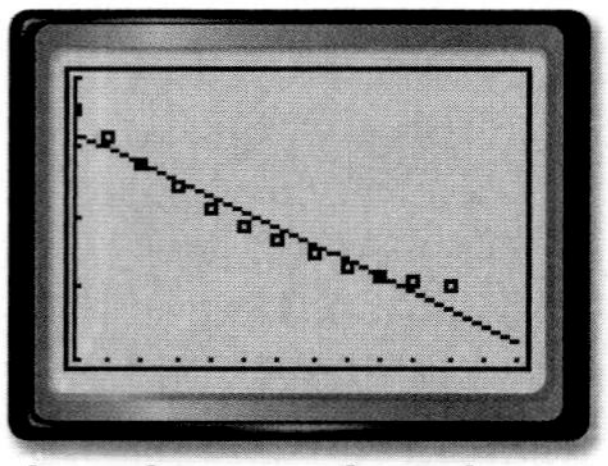

[0, 260] scl: 20 by [25, 45] scl: 5

Step 2 **Find and graph a quadratic regression equation.**

- Find the quadratic regression equation. Then copy the equation to the Y= list and graph.

  KEYSTROKES: STAT ▶ 5 ENTER Y= VARS 5 ▶ ▶ ENTER Graph

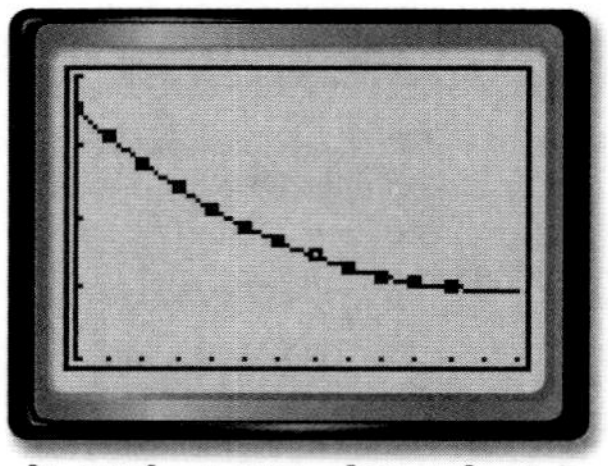

[0, 260] scl: 20 by [25, 45] scl: 5

Notice that the graph of the linear regression equation appears to pass through just two data points. However, the graph of the quadratic regression equation fits the data very well.

Exercises

Refer to the table.

Height of Player Making a Slam Dunk

Time (s)	Height (ft)
0.1	3.04
0.2	5.76
0.3	8.16
0.4	10.24
0.5	12
0.6	13.44
0.7	14.56

1. Find and graph a linear regression equation and a quadratic regression equation for the data. Determine which equation is a better fit for the data.
2. Use the CALC menu with each regression equation to estimate the height of the player after 1 second and 1.5 seconds.
3. Compare and contrast the estimates you found in Exercise 2.
4. How might choosing a regression equation that does not fit the data well affect predictions made by using the equation?

5-2 Solving Quadratic Equations by Graphing

Then

You solved systems of equations by graphing. (Lesson 3-1)

Now

- Solve quadratic equations by graphing.
- Estimate solutions of quadratic equations by graphing.

NYS Core Curriculum

Reinforcement of A.A.8 Analyze and solve verbal problems that involve quadratic equations
Reinforcement of A.G.8 Find the roots of a parabolic function graphically

New Vocabulary

quadratic equation
standard form
root
zero

NY Math Online

glencoe.com

- Extra Examples
- Personal Tutor
- Self-Check Quiz
- Homework Help

Why?

Arielle works in the marketing department of a major retailer. Her job is to set prices for new products sold in the stores. Arielle determined that for a certain product, the function $f(p) = -6p^2 + 192p - 1440$ tells the profit $f(p)$ made at price p.

Arielle can determine the price range by finding the prices for which the profit is equal to \$0. This can be done by finding the solution of the related quadratic equation, which is zero.

The graph of the function indicates that the profit is zero at 12 and 20, so the profitable price range of the item is between \$12 and \$20.

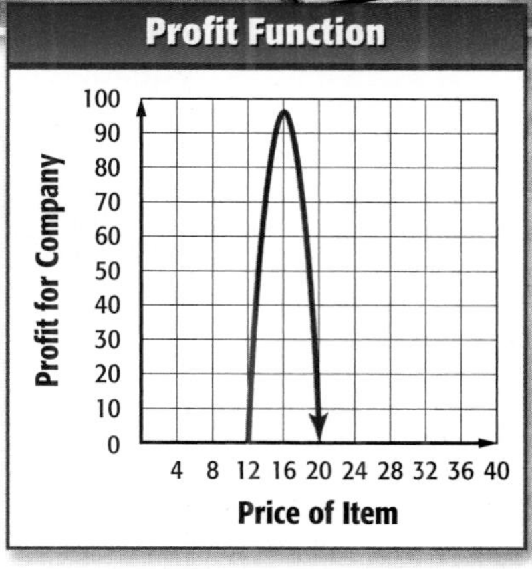

Solve Quadratic Equations **Quadratic equations** are quadratic functions that are set equal to a value. The **standard form** of a quadratic equation is $ax^2 + bx + c = 0$, where $a \neq 0$ and a, b, and c are integers.

The solutions of a quadratic equation are called the **roots** of the equation. One method for finding the roots of a quadratic equation is to find the **zeros** of the related quadratic function.

The zeros of the function are the x-intercepts of its graph.

Quadratic Function

$$f(x) = x^2 - x - 6$$

$$f(-2) = (-2)^2 - (-2) - 6 \text{ or } 0$$
$$f(3) = 3^2 - 3 - 6 \text{ or } 0$$

−2 and 3 are zeros of the function.

Quadratic Equation

$$x^2 - x - 6 = 0$$

$$(-2)^2 - (-2) - 6 \text{ or } 0$$
$$3^2 - 3 - 6 \text{ or } 0$$

−2 and 3 are roots of the equation.

Graph of Function

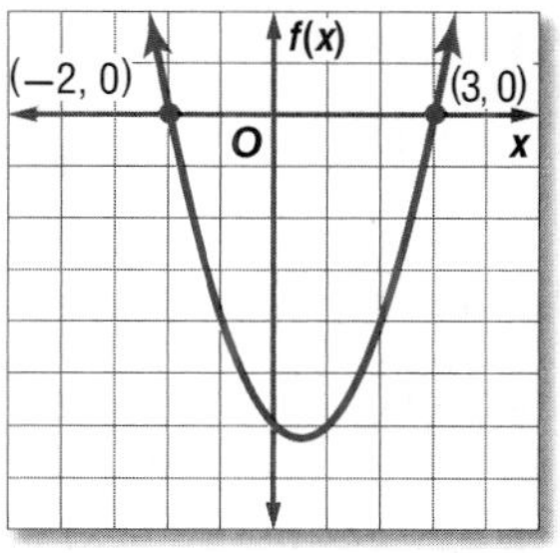

The x-intercepts are −2 and 3.

StudyTip

Graphing Sometimes more than five points will be needed before the zeros are located.

EXAMPLE 1 Two Real Solutions

Solve $x^2 - 3x - 4 = 0$ by graphing.

Graph the related function, $f(x) = x^2 - 3x - 4$. The equation of the axis of symmetry is $x = -\frac{-3}{2(1)}$ or 1.5. Make a table using x-values around 1.5. Then graph each point.

x	−1	0	1	1.5	2	3	4
$f(x)$	0	−4	−6	−6.25	−6	−4	0

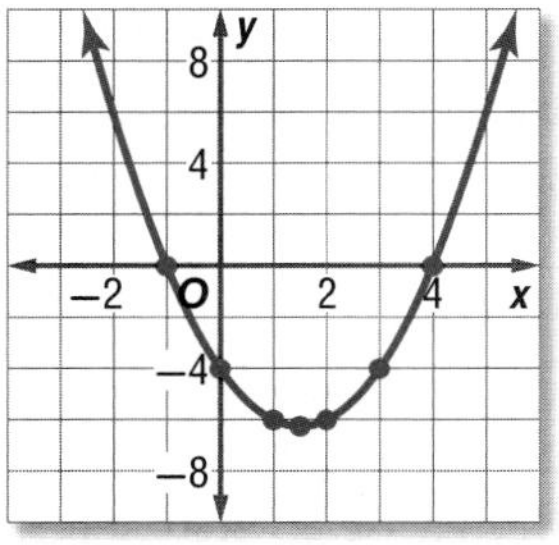

The zeros of the function are −1 and 4.
Therefore, the solutions of the equation are −1 and 4.

Check Your Progress

Solve each equation by graphing.

1A. $x^2 + 2x - 15 = 0$

1B. $x^2 - 8x = -12$

Personal Tutor glencoe.com

The graph of the related function in Example 1 has two zeros; therefore, the quadratic equation has two real solutions. This is one of the three possible outcomes when solving a quadratic equation.

Key Concept Solutions of a Quadratic Equation

For Your FOLDABLE

Words A quadratic equation can have one real solution, two real solutions, or no real solutions.

Models

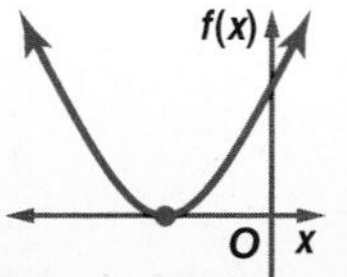

one real solution

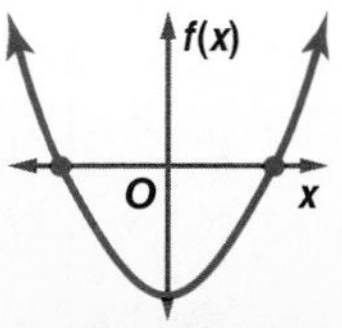

two real solutions

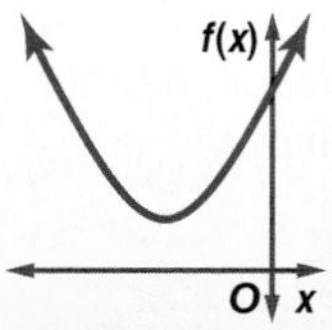

no real solution

EXAMPLE 2 One Real Solution

Solve $14 - x^2 = -6x + 23$ by graphing.

$14 - x^2 = -6x + 23$ **Original equation**

$14 = x^2 - 6x + 23$ **Add x^2 to each side.**

$0 = x^2 - 6x + 9$ **Subtract 14.**

Graph the related function $f(x) = x^2 - 6x + 9$.

x	1	2	3	4	5
$f(x)$	4	1	0	1	4

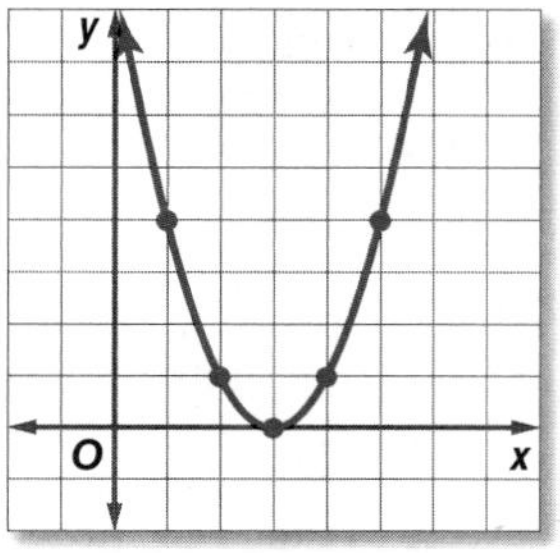

Notice that the function has only one zero, 3. Therefore, the only solution is 3.

StudyTip

Optional Graph $f(x) = -x^2 + 6x - 9$ could also have been graphed for this example. The graph would appear different, but it would have the same solution.

Check Your Progress

Solve each equation by graphing.

2A. $x^2 + 5 = -8x - 11$

2B. $12 - x^2 = 48 - 12x$

Personal Tutor glencoe.com

EXAMPLE 3 No Real Solution

NUMBER THEORY Use a quadratic equation to find two real numbers with a sum of 15 and a product of 63.

Understand Let x represent one of the numbers. Then $15 - x$ is the other number.

Plan

$x(15 - x) = 63$ **The product of the numbers is 63.**

$15x - x^2 = 63$ **Distributive Property**

$-x^2 + 15x - 63 = 0$ **Subtract 63.**

Solve Graph the related function.

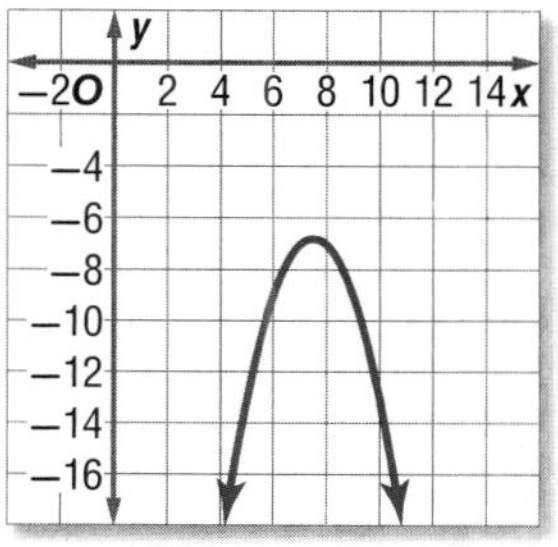

The graph has no x-intercepts. This means the original equation has no real solution. Thus, it is not possible for two numbers to have a sum of 15 and a product of 63.

Check Try finding the product of several pairs of numbers with sums of 15. Is each product less than 63 as the graph suggests?

Check Your Progress

3. Find two real numbers with a sum of 6 and a product of −55, or show that no such numbers exist.

Personal Tutor glencoe.com

Estimate Solutions Often exact roots cannot be found by graphing. You can estimate the solutions by stating the integers between which the roots are located.

When the value of the function is positive for one value and negative for a second value, then there is at least one zero between those two values.

x	−3	−2	−1	0	1	2	3
$f(x)$	12	3	−6	−2	4	8	14

zero (between −2 and −1) zero (between 0 and 1)

Watch Out!

Zeros You will see in later chapters that many zeros can appear within small intervals.

EXAMPLE 4 Estimate Roots

Solve $x^2 - 6x + 4 = 0$ by graphing. If exact roots cannot be found, state the consecutive integers between which the roots are located.

x	0	1	2	3	4	5	6
$f(x)$	4	−1	−4	−5	−4	−1	4

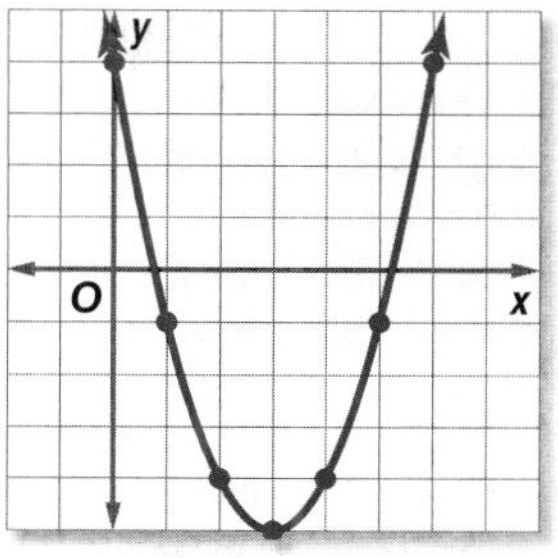

The x-intercepts of the graph indicate that one solution is between 0 and 1, and the other solution is between 5 and 6.

Check Your Progress

4. Solve $x^2 - x - 10 = 0$ by graphing. If exact roots cannot be found, state the consecutive integers between which the roots are located.

Personal Tutor glencoe.com

You can also use tables to solve quadratic equations. After entering the equation in your calculator, scroll through the table to locate the solutions.

EXAMPLE 5 Solve by Using a Table

Solve $x^2 - 6x + 2 = 0$.

Enter $y_1 = x^2 - 6x + 2$ in your graphing calculator. Use the **TABLE** window to find where the sign of **Y1** changes. Change **△Tbl** to 0.1 and look again for the sign change. Repeat the process with 0.01 and 0.001 to get a more accurate location of the zero.

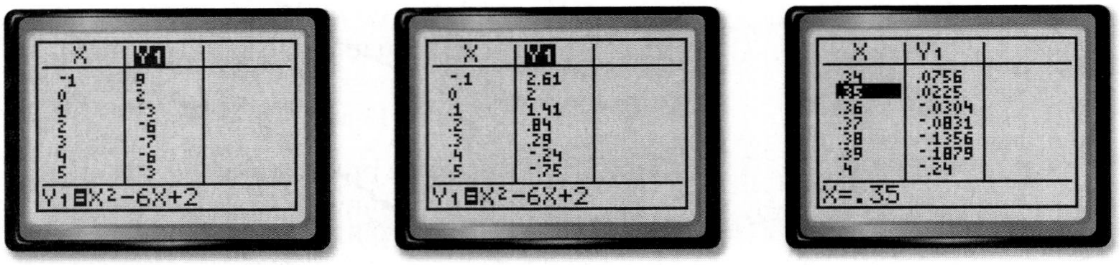

One solution is approximately 0.354.

Check Your Progress

5. Locate the second zero in the function above to the nearest thousandth.

Personal Tutor glencoe.com

Quadratic equations can be solved with a calculator as well. After entering the equation, use the **ZERO** operation in the **CALC** menu.

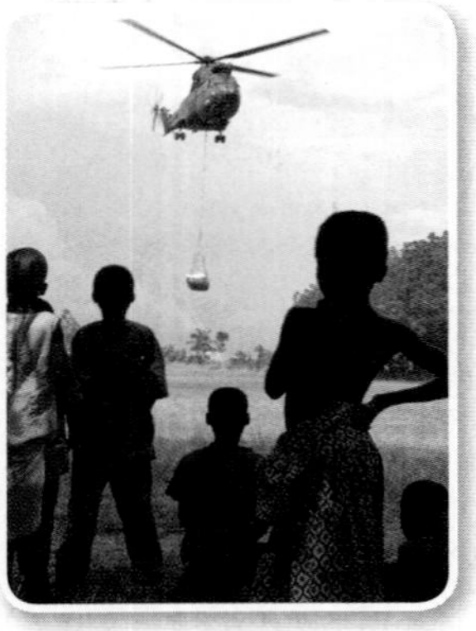

Real-World Link

Each year, the American Red Cross responds to over 70,000 disaster situations ranging from house fires to natural disasters such as hurricanes and earthquakes.

Real-World EXAMPLE 6 Solve by Using a Calculator

RELIEF A package of supplies is dropped from a helicopter at an altitude of 200 feet. The package's height above the ground is modeled by $h(t) = -16t^2 + 28t + 200$, where t is the time in seconds after it is dropped. How long will it take the package to reach the ground?

We need to find t when $h(t)$ is 0. Solve $0 = -16t^2 + 28t + 200$. Then graph the related function $f(t) = -16t^2 + 28t + 200$ on a graphing calculator.

- Use the **ZERO** feature in the **CALC** menu to find the positive zero of the function, since time cannot be negative.
- Use the arrow keys to select a left bound and press ENTER.
- Locate a right bound and press ENTER twice.
- The positive zero of the function is about 4.52. The package would take about 4.52 seconds to reach the ground.

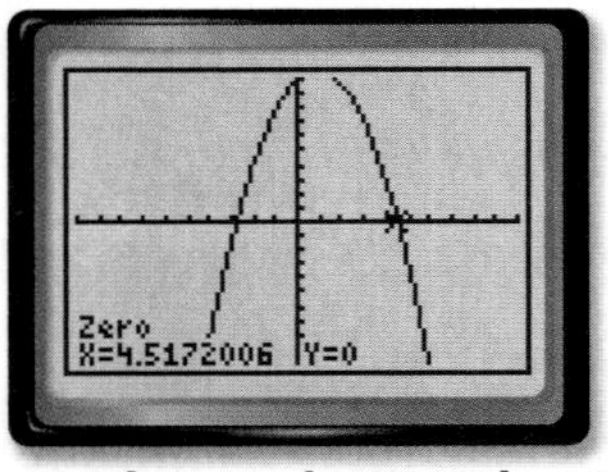

[−10, 10] scl: 1 by [−200, 200] scl: 20

Check Your Progress

6. How long would it take to reach the ground if the height was modeled by $h(t) = -16t^2 + 48t + 400$?

Personal Tutor glencoe.com

Check Your Understanding

Example 1 p. 260

Use the related graph of each equation to determine its solutions.

1. $x^2 + 2x + 3 = 0$

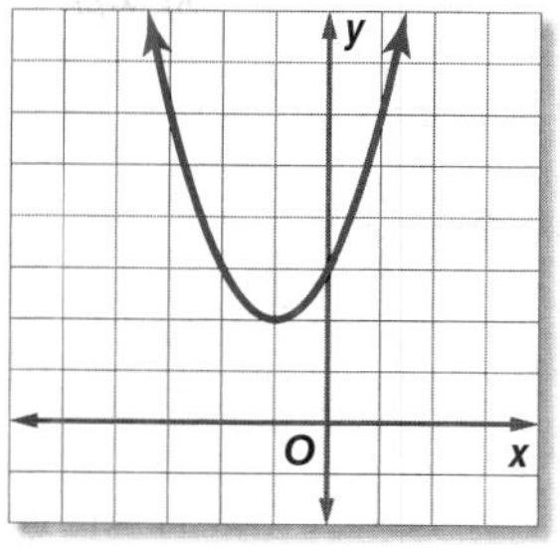

2. $x^2 - 3x - 10 = 0$

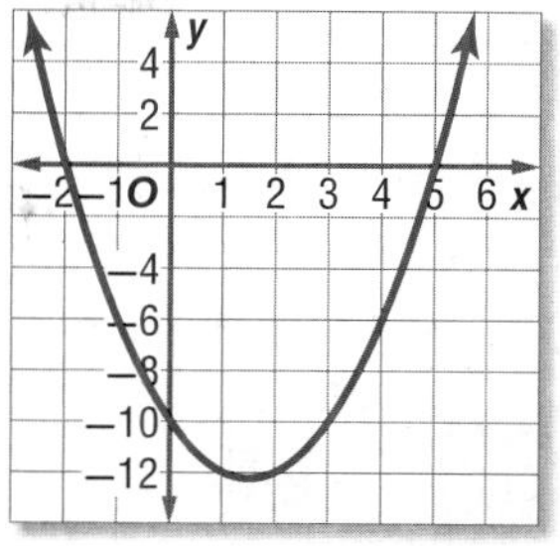

3. $-x^2 - 8x - 16 = 0$

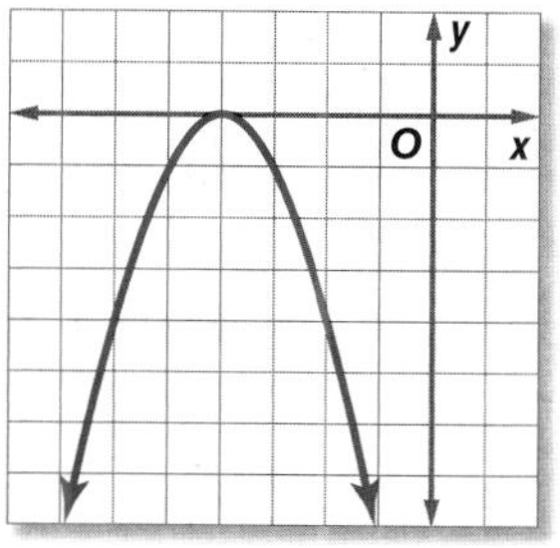

Examples 2–5 pp. 260–262

Solve each equation. If exact roots cannot be found, state the consecutive integers between which the roots are located.

4. $x^2 + 8x = 0$
5. $x^2 - 3x - 18 = 0$
6. $4x - x^2 + 8 = 0$
7. $-12 - 5x + 3x^2 = 0$
8. $x^2 - 6x + 4 = -8$
9. $9 - x^2 = 12$
10. $5x^2 + 10x - 4 = -6$
11. $x^2 - 20 = 2 + x$

12. **NUMBER THEORY** Use a quadratic equation to find two real numbers with a sum of 2 and a product of −24.

Example 6 p. 262

13. **PHYSICS** How long will it take an object to fall from the roof of a building 400 feet above ground? Use the formula $h(x) = -16t^2 + h_0$, where t is the time in seconds and the initial height h_0 is in feet.

Practice and Problem Solving

● = **Step-by-Step Solutions** begin on page R20. **Extra Practice** begins on page 947.

Example 1 p. 260

Use the related graph of each equation to determine its solutions.

14. $x^2 + 4x = 0$

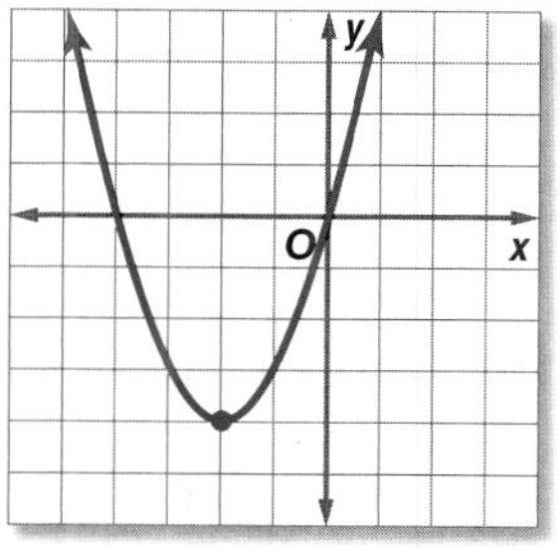

15. $-2x^2 - 4x - 5 = 0$

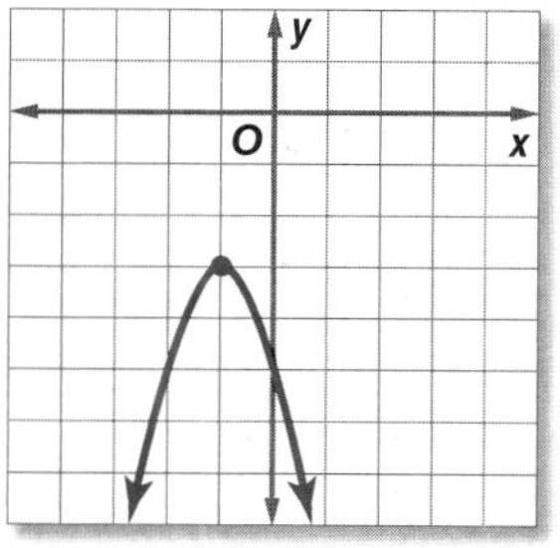

16. $0.5x^2 - 2x + 2 = 0$

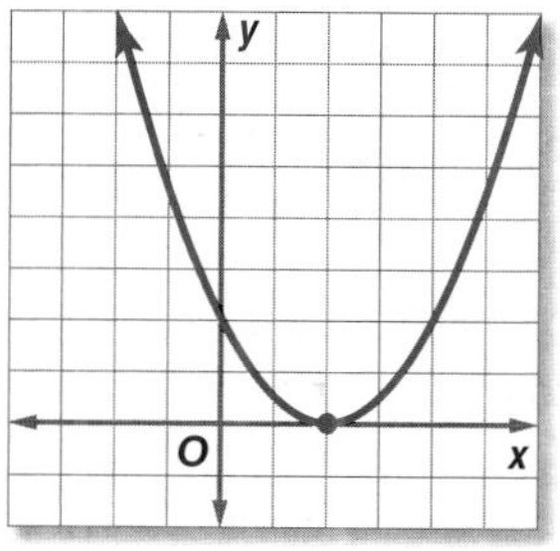

17. $-0.25x^2 - x - 1 = 0$

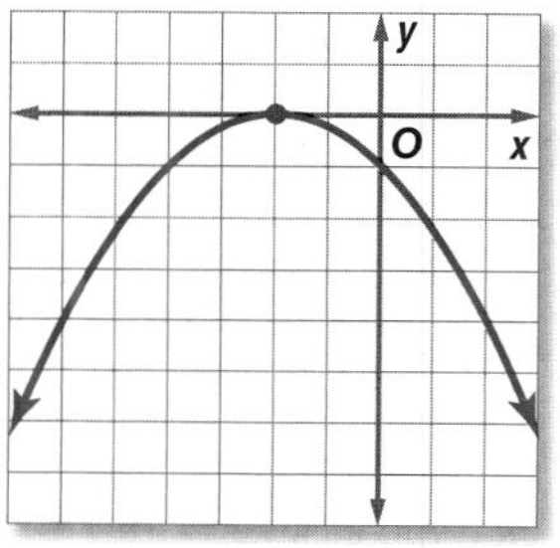

18. $x^2 - 6x + 11 = 0$

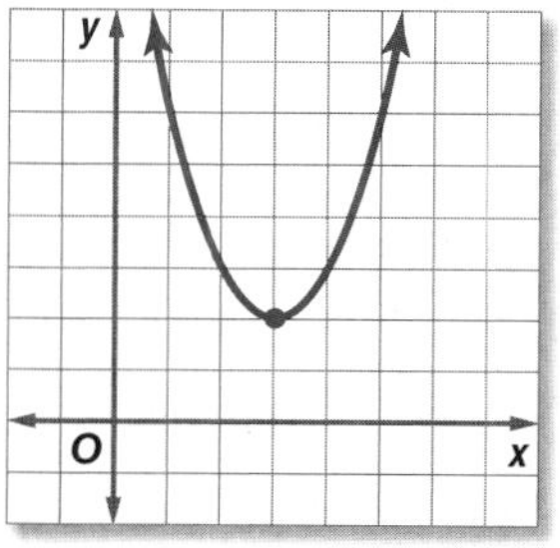

19. $-0.5x^2 + 0.5x + 6 = 0$

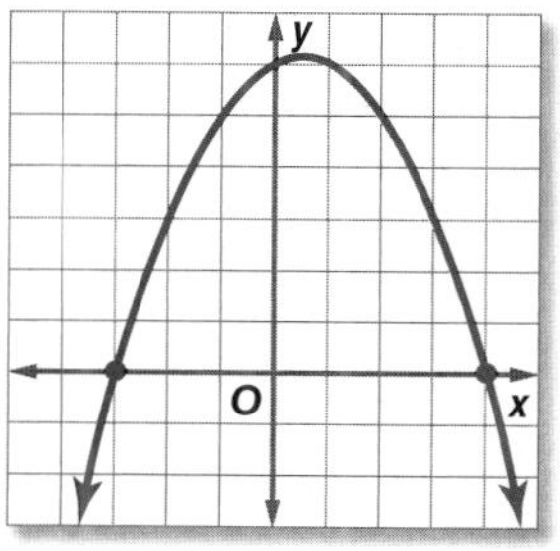

Examples 2–4 pp. 260–261

Solve each equation. If exact roots cannot be found, state the consecutive integers between which the roots are located.

20. $x^2 = 5x$

21. $-2x^2 - 4x = 0$

22. $x^2 - 5x - 14 = 0$

23. $-x^2 + 2x + 24 = 0$

24. $x^2 - 18x = -81$

25. $2x^2 - 8x = -32$

26. $2x^2 - 3x - 15 = 4$

27. $-3x^2 - 7 + 2x = -11$

28. $-0.5x^2 + 3 = -5x - 2$

29. $-2x + 12 = x^2 + 16$

Example 5 p. 262

Use the tables to determine the location of the zeros of each quadratic function.

30.

x	−7	−6	−5	−4	−3	−2	−1	0
$f(x)$	−8	−1	4	4	−1	−8	−22	−48

31.

x	−2	−1	0	1	2	3	4	5
$f(x)$	32	14	2	−3	−3	2	14	32

32.

x	−6	−3	0	3	6	9	12	15
$f(x)$	−6	−1	3	5	3	−1	−6	−14

Example 6 p. 262

NUMBER THEORY **Use a quadratic equation to find two real numbers that satisfy each situation, or show that no such numbers exist.**

33. Their sum is −15, and their product is −54.

34. Their sum is 4, and their product is −117.

35. Their sum is 12, and their product is −84.

36. Their sum is −13, and their product is 42.

37. Their sum is −8 and their product is −209.

For Exercises 38–40, use the formula $h(t) = v_0t - 16t^2$, where $h(t)$ is the height of an object in feet, v_0 is the object's initial velocity in feet per second, and t is the time in seconds.

38. BASEBALL A baseball is hit directly upward with an initial velocity of 80 feet per second. Ignoring the height of the baseball player, how long does it take for the ball to hit the ground?

39. CANNONS A cannonball is shot directly upward with an initial velocity of 55 feet per second. Ignoring the height of the cannon, how long does it take for the cannonball to hit the ground?

40. GOLF A golf ball is hit directly upward with an initial velocity of 100 feet per second. How long will it take for it to hit the ground?

Solve each equation. If exact roots cannot be found, state the consecutive integers between which the roots are located.

41. $2x^2 + x = 15$

42. $-5x - 12 = -2x^2$

43. $4x^2 - 15 = -4x$

44. $-35 = -3x - 2x^2$

45. $-3x^2 + 11x + 9 = 1$

46. $13 - 4x^2 = -3x$

47. $-0.5x^2 + 18 = -6x + 33$

48. $0.5x^2 + 0.75 = 0.25x$

49. WATER BALLOONS Tony wants to drop a water balloon so that it splashes on his brother. Use the formula $h(t) = -16t^2 + h_0$, where t is the time in seconds and the initial height h_0 is in feet, to determine how far his brother should be from the target when Tony lets go of the balloon.

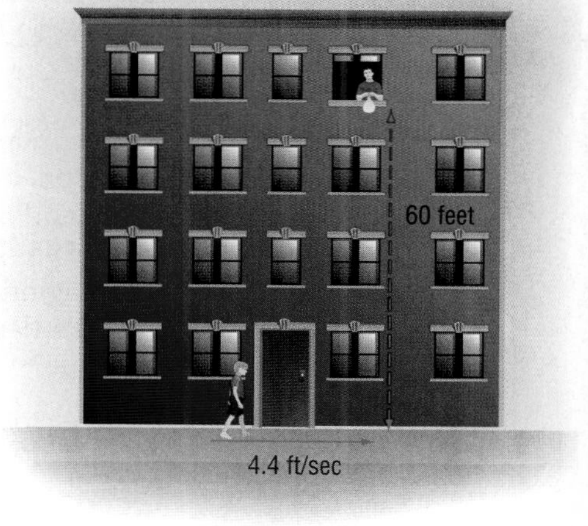

50. WATER HOSES A water hose can spray water at an initial velocity of 40 feet per second. Use the formula $h(t) = v_0t - 16t^2$, where $h(t)$ is the height of the water in feet, v_0 is the initial velocity in feet per second, and t is the time in seconds.

a. How long will it take the water to hit the nozzle on the way down?

b. Assuming the nozzle is 5 feet up, what is the maximum height of the water?

Real-World Link

The longest free fall distance by one person is 24,500 meters.

Source: Fédération Aéronautique Internationale

51. SKYDIVING In 2003, John Fleming and Dan Rossi became the first two blind skydivers to be in free fall together. They jumped from an altitude of 14,000 feet and free fell to an altitude of 4000 feet before their parachutes opened. Ignoring air resistance and using the formula $h(t) = -16t^2 + h_0$, where t is the time in seconds and the initial height h_0 is in feet, determine how long they were in free fall.

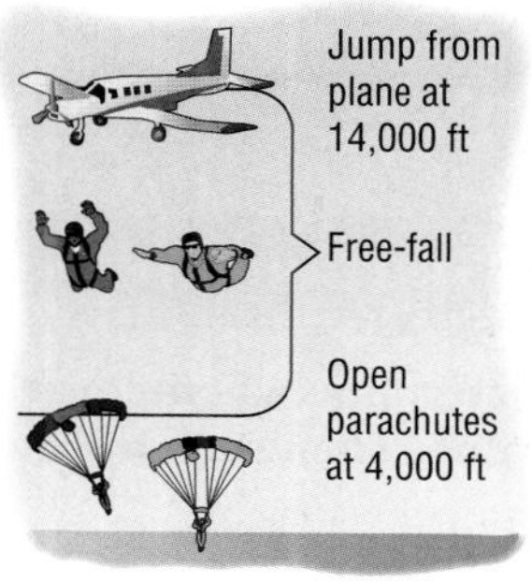

H.O.T. Problems

Use Higher-Order Thinking Skills

52. FIND THE ERROR Hakeem and Tanya were asked to find the location of the roots of the quadratic function represented by the table. Is either of them correct? Explain.

x	−4	−2	0	2	4	6	8	10
$f(x)$	52	26	8	−2	−4	2	16	38

Hakeem

The roots are between 4 and 6 because f(x) stops decreasing and begins to increase between x = 4 and x = 6.

Tanya

The roots are between −2 and 0 because x changes signs at that location.

53. CHALLENGE Find the value of a positive integer k such that $f(x) = x^2 - 2kx + 55$ has roots at $k + 3$ and $k - 3$.

54. REASONING If a quadratic function has a minimum at (−6, −14) and a root at $x = -17$, what is the other root? Explain your reasoning.

55. OPEN ENDED Write a quadratic function with a maximum at (3, 125) and roots at −2 and 8.

56. WRITING IN MATH Explain how to solve a quadratic equation by graphing its related quadratic function.

NYSRE PRACTICE A2.S.9, A2.S.12

57. SHORT RESPONSE A bag contains five different colored marbles. The colors of the marbles are black, silver, red, green, and blue. A student randomly chooses a marble. Then, without replacing it, chooses a second marble. What is the probability that the student chooses the red and then the green marble?

58. Which number would be closest to zero on the number line?

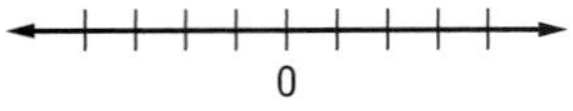

A -0.6 **C** $\frac{\sqrt{2}}{2}$

B $\frac{2}{5}$ **D** 0.5

59. SAT/ACT A salesman's monthly gross pay consists of \$3500 plus 20 percent of the dollar amount of his sales. If his gross pay for one month was \$15,500, what was the dollar amount of his sales for that month?

F \$12,000 **H** \$60,000

G \$16,000 **J** \$70,000

60. Find the next term in the sequence below.

$$\frac{2x}{5}, \frac{3x}{5}, \frac{4x}{5}, \ldots$$

A x **C** $\frac{x}{5}$

B $5x$ **D** $\frac{5x}{4}$

Spiral Review

Determine whether each function has a *maximum* or *minimum* value, and find that value. Then state the domain and range of the function. (Lesson 5-1)

61. $f(x) = -4x^2 + 8x - 16$

62. $f(x) = 3x^2 + 12x - 18$

63. $f(x) = 4x + 13 - 2x^2$

Determine whether each pair of matrices are inverses of each other. (Lesson 4-6)

64. $\begin{bmatrix} 4 & -3 \\ -1 & -6 \end{bmatrix}$ and $\begin{bmatrix} \frac{3}{13} & -\frac{1}{18} \\ -\frac{1}{26} & -\frac{2}{13} \end{bmatrix}$

65. $\begin{bmatrix} 6 & -3 \\ 4 & 8 \end{bmatrix}$ and $\begin{bmatrix} \frac{1}{10} & \frac{1}{20} \\ -\frac{1}{15} & \frac{2}{15} \end{bmatrix}$

66. $\begin{bmatrix} 2 & 4 \\ -3 & -2 \end{bmatrix}$ and $\begin{bmatrix} -\frac{1}{4} & -\frac{1}{2} \\ \frac{3}{8} & \frac{1}{4} \end{bmatrix}$

67. FOOTPRINTS The combination of a reflection and a translation is called a *glide reflection*. An example is a set of footprints. (Lesson 4-4)

a. Describe the reflection and transformation combination shown at the right.

b. Write two matrix operations that can be used to find the coordinates of point C.

c. Does it matter which operation you do first? Explain.

d. What are the coordinates of the next two footprints?

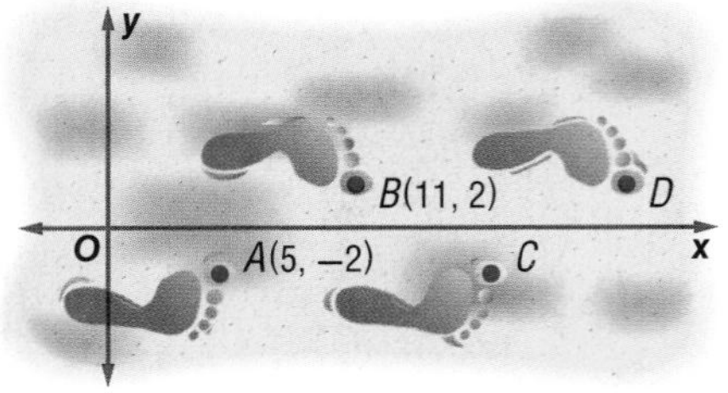

Solve each system of equations. (Lesson 3-2)

68. $4x - 7y = -9$
$5x + 2y = -22$

69. $8y - 2x = 38$
$5x - 3y = -27$

70. $3x + 8y = 24$
$-16y - 6x = 48$

Solve each inequality. (Lesson 1-5)

71. $3x - 6 \leq -14$

72. $6 - 4x \leq 2$

73. $-6x + 3 \geq 3x - 16$

Skills Review

Find the GCF of each set of numbers.

74. 16, 48, 128

75. 15, 21, 49

76. 12, 28, 36

EXTEND
5-2
Graphing Technology Lab
Solving Quadratic Equations by Graphing

NY Math Online glencoe.com
Other Calculator Keystrokes

NYS Core Curriculum **Reinforcement of A.G.8** Find the roots of a parabolic function graphically

You can use a TI-83/84 Plus graphing calculator to solve quadratic equations.

ACTIVITY Solving Quadratic Equations

Solve $x^2 - 8x + 15 = 0$.

Step 1 Rewrite the equation in the form $y = ax^2 + bx + c$.
$y = x^2 - 8x + 15$

Step 2 Graph $y = x^2 - 8x + 15$ in a standard viewing window.

KEYSTROKES: [Y=] [X,T,θ,n] [x²] [(–)] 8 [X,T,θ,n] [+] 15 [ZOOM] 6

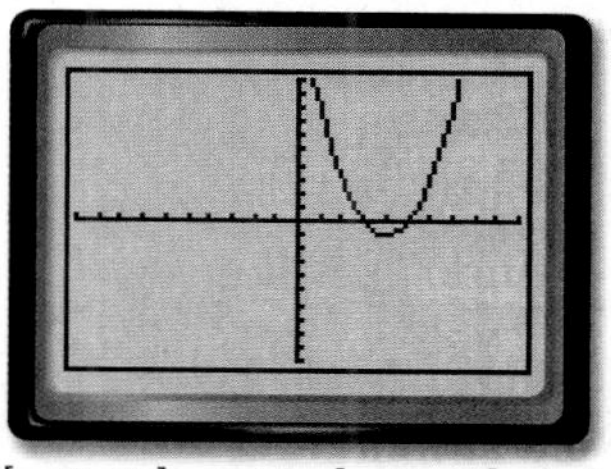

[–10, 10] scl: 1 by [–10, 10] scl: 1

Step 3 Find an x-intercept.

KEYSTROKES: [2nd] [CALC] 2

Use [◀] or [▶] to position the cursor to the left of the first x-intercept. Press [ENTER]. Then use [▶] to position the cursor to the right of the first x-intercept. Press [ENTER] [ENTER] to display the x-intercept.

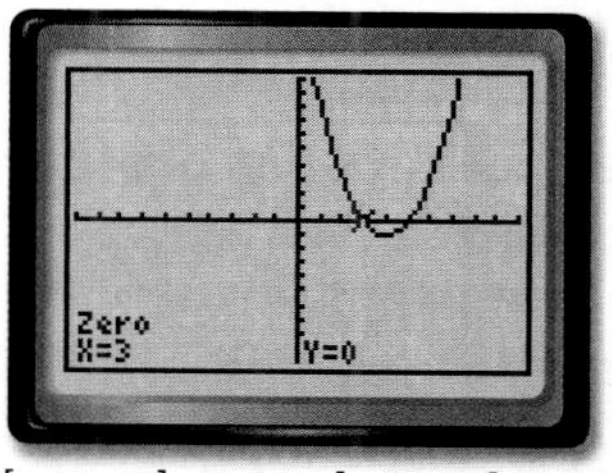

[–10, 10] scl: 1 by [–10, 10] scl: 1

Step 4 Find the second x-intercept.

KEYSTROKES: Use [▶] to position the cursor to the left of the second x-intercept. Press [ENTER]. Then use [▶] to position the cursor to the right of the second x-intercept. Press [ENTER] [ENTER] to display the x-intercept.

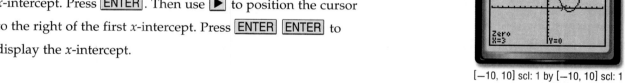

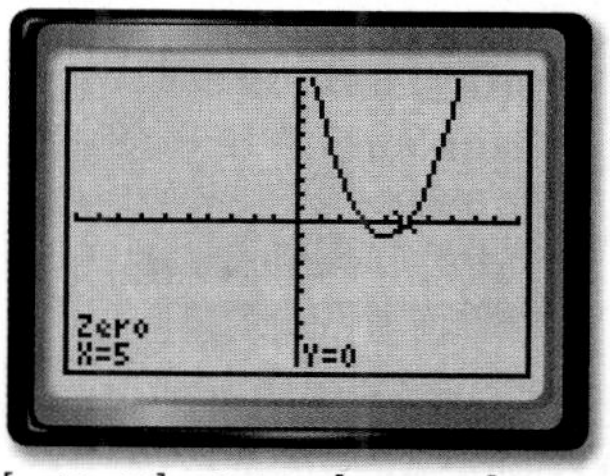

[–10, 10] scl: 1 by [–10, 10] scl: 1

The x-intercepts are 3 and 5, so $x = 3$ and $x = 5$.

Exercises

Solve each equation. Round to the nearest tenth if necessary.

1. $x^2 - 7x + 12 = 0$
2. $x^2 + 5x + 6 = 0$
3. $x^2 - 3 = 2x$
4. $x^2 + 5x + 6 = 12$
5. $x^2 + 5x = 0$
6. $x^2 - 4 = 0$
7. $x^2 + 8x + 16 = 0$
8. $x^2 - 10x = -25$
9. $9x^2 + 48x + 64 = 0$
10. $2x^2 + 3x - 1 = 0$
11. $5x^2 - 7x = -2$
12. $6x^2 + 2x + 1 = 0$

5-3 Solving Quadratic Equations by Factoring

Then

You factored polynomials. (Lesson 0-3)

Now

- Write quadratic equations in intercept form.
- Solve quadratic equations by factoring.

NYS Core Curriculum

A2.A.7 Factor polynomial expressions completely, using any combination of the following techniques: common factor extraction, difference of two perfect squares, quadratic trinomials

New Vocabulary

factored form

FOIL method

NY Math Online

glencoe.com

- Extra Examples
- Personal Tutor
- Self-Check Quiz
- Homework Help

Why?

The **factored form** of a quadratic equation is $0 = a(x - p)(x - q)$. In the equation, p and q represent the x-intercepts of the graph of the equation.

The x-intercepts of the graph at the right are 2 and 6. In this lesson, you will learn how to change a quadratic equation in factored form into standard form and vice versa.

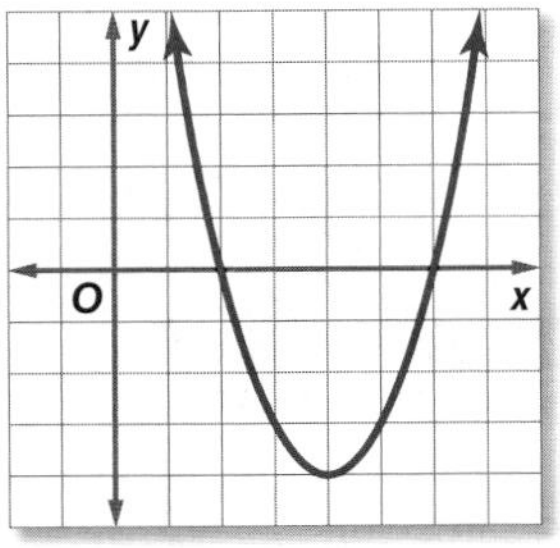

Related Graph
2 and 6 are
x-intercepts.

Standard Form	Factored Form
$0 = x^2 - 8x + 12$	$0 = (x - 6)(x - 2)$

Factors (pointing to $(x - 6)$ and $(x - 2)$)

Factored Form You can use the FOIL method to write a quadratic equation that is in factored form in standard form. The **FOIL method** uses the Distributive Property to multiply binomials.

Key Concept — For Your FOLDABLE

FOIL Method for Multiplying Binomials

Words To multiply two binomials, find the sum of the products of ***F*** the *First* terms, ***O*** the *Outer* terms, ***I*** the *Inner* terms, and ***L*** the *Last* terms.

Examples

	Product of **First** Terms		Product of **Outer** Terms		Product of **Inner** Terms		Product of **Last** Terms
$(x - 6)(x - 2) =$	$(x)(x)$	$+$	$(x)(-2)$	$+$	$(-6)(x)$	$+$	$(-6)(-2)$

F O I L

$= x^2 - 2x - 6x + 12$ or $x^2 - 8x + 12$

EXAMPLE 1 Write an Equation Given Roots

Write a quadratic equation in standard form with $-\frac{1}{3}$ and 6 as its roots.

$(x - p)(x - q) = 0$ — **Write the pattern.**

$\left[x - \left(-\frac{1}{3}\right)\right](x - 6) = 0$ — **Replace p with $-\frac{1}{3}$ and q with 6.**

$\left(x + \frac{1}{3}\right)(x - 6) = 0$ — **Simplify.**

$x^2 - \frac{17}{3}x - 2 = 0$ — **Multiply.**

$3x^2 - 17x - 6 = 0$ — **Multiply each side by 3 so that b and c are integers.**

Check Your Progress

1. Write a quadratic equation in standard form with $\frac{3}{4}$ and -5 as its roots.

Personal Tutor glencoe.com

StudyTip

Writing Equations $(x - p)(x - q) = 0$ produces one equation with roots p and q. In fact, there are an infinite number of equations that have these same roots. This is represented by the a coefficient.

Solve Equations by Factoring You have learned various techniques for factoring polynomials. A summary of them is listed below.

Concept Summary — Factoring Techniques

For Your FOLDABLE

Factoring Technique	General Case
Greatest Common Factor (GCF)	$a^3b^2 - 8ab^2 = ab^2(a^2 - 8)$
General Trinomials	$acx^2 + (ad + bc)x + bd = (ax + b)(cx + d)$
Difference of Two Squares	$a^2 - b^2 = (a + b)(a - b)$
Perfect Square Trinomials	$a^2 \pm 2ab + b^2 = (a \pm b)^2$

EXAMPLE 2 Factor GCF and by Grouping

Factor each polynomial.

a. $\mathbf{16x^2 + 8x}$

$16x^2 + 8x = 8x(2x) + 8x(1)$ — **Factor the GCF.**

$= 8x(2x + 1)$ — **Distributive Property**

b. $\mathbf{7x^2 + 6xy^2 + 14xy + 12y^3}$

$7x^2 + 6xy^2 + 14xy + 12y^3$ — **Original expression**

$= (7x^2 + 14xy) + (6xy^2 + 12y^3)$ — **Group terms with common factors.**

$= 7x(x + 2y) + 6y^2(x + 2y)$ — **Factor the GCF from each group.**

$= (7x + 6y^2)(x + 2y)$ — **Distributive Property**

Check Your Progress

2A. $20x^2y - 15xy^2$ **2B.** $4x^2y - 16xy - y^2$ **2C.** $ab + 3cd + 4a^2b + 12acd$

Personal Tutor glencoe.com

Review Vocabulary

perfect square a number with a positive square root that is a whole number

Trinomials and binomials that are perfect squares have special factoring rules. In order to use these rules, the first and last terms need to be perfect squares and the middle term needs to be twice the product of the square roots of the first and last terms.

EXAMPLE 3 Perfect Squares and Differences of Squares

Factor each polynomial.

a. $\mathbf{x^2 + 16x + 64}$

$x^2 = (x)^2;\ 64 = (8)^2$ — **First and last terms are perfect squares.**

$16x = 2(x)(8)$ — **Middle term equals 2*ab*.**

$x^2 + 16x + 64$ is a perfect square trinomial.

$x^2 + 16x + 64 = (x + 8)^2$ — **Factor using the pattern.**

b. $\mathbf{36a^2 - 64y^4}$

$36a^2 - 64y^4 = 4(9a^2 - 16y^4)$ — **Factor the GCF.**

$= 4[(3a)^2 - (4y^2)^2]$ — **Write in form $a^2 - b^2$.**

$= 4(3a + 4y^2)(3a - 4y^2)$ — **Factor the difference of squares.**

Check Your Progress

3A. $4x^2 - 12x + 9$ **3B.** $81x^2 - y^6$ **3C.** $75x^2y - 27y$

Personal Tutor glencoe.com

StudyTip

Trinomials If values for m and p exist, then the trinomial can always be factored.

A special pattern is used when factoring trinomials of the form $ax^2 + bx + c$. First, multiply the values of a and c. Then, we must find two values, m and p, such that their product equals ac and their sum equals b.

Consider $6x^2 + 13x - 5$: $ac = 6(-5) = -30$.

Factors of −30	Sum	Factors of −30	Sum
1, −30	−29	−1, 30	29
2, −15	−13	**−2, 15**	**13**
3, −10	−7	−3, 10	7
5, −6	−1	−5, 6	1

Now the middle term, $13x$, can be rewritten as $-2x + 15x$.

This polynomial can now be factored by grouping.

$6x^2 + 13x - 5 = 6x^2 + mx + px - 5$ **Write the pattern.**
$= 6x^2 - 2x + 15x - 5$ **$m = -2$ and $p = 15$**
$= (6x^2 - 2x) + (15x - 5)$ **Group terms.**
$= 2x(3x - 1) + 5(3x - 1)$ **Factor the GCF.**
$= (2x + 5)(3x - 1)$ **Distributive Property**

EXAMPLE 4 Factor Trinomials

Factor each polynomial.

a. $x^2 + 9x + 20$

$ac = 20$ **$a = 1, c = 20$**

Factors of 20	Sum	Factors of 20	Sum
1, 20	21	−1, −20	−21
2, 10	12	−2, −10	−12
4, 5	**9**	−4, −5	−9

StudyTip

Trinomials It does not matter if the values of m and p are switched when grouping.

$x^2 + 9x + 20$ **Original expression**
$= x^2 + mx + px + 20$ **Write the pattern.**
$= x^2 + 4x + 5x + 20$ **$m = 4, p = 5$**
$= (x^2 + 4x) + (5x + 20)$ **Group terms with common factors.**
$= x(x + 4) + 5(x + 4)$ **Factor the GCF from each grouping.**
$= (x + 5)(x + 4)$ **Distributive Property**

b. $6y^2 - 23y + 20$

$ac = 120$ **$a = 6, c = 20$**
$m = -8, p = -15$ **$-8(-15) = 120; -8 + (-15) = -23$**

$6y^2 - 23y + 20$ **Original expression**
$= 6y^2 + my + py + 20$ **Write the pattern.**
$= 6y^2 - 8y - 15y + 20$ **$m = -8, p = -15$**
$= (6y^2 - 8y) + (-15y + 20)$ **Group terms with common factors.**
$= 2y(3y - 4) - 5(3y - 4)$ **Factor the GCF from each grouping.**
$= (2y - 5)(3y - 4)$ **Distributive Property**

Check Your Progress

4A. $x^2 - 11x + 30$

4B. $x^2 - 4x - 21$

4C. $15x^2 - 8x + 1$

4D. $-12x^2 + 8x + 15$

Personal Tutor glencoe.com

Solving quadratic equations by factoring is an application of the Zero Product Property.

Key Concept — Zero Product Property

For Your FOLDABLE

Words For any real numbers a and b, if $ab = 0$, then either $a = 0$, $b = 0$, or both a and b equal zero.

Examples If $(x + 3)(x - 5) = 0$, then $x + 3 = 0$ or $x - 5 = 0$.

Real-World EXAMPLE 5 Solve Equations by Factoring

TRACK AND FIELD The height of a javelin in feet is modeled by $h(t) = -16t^2 + 79t + 5$, where t is the time in seconds after the javelin is thrown. How long is it in the air?

To determine how long the javelin is in the air, we need to find when the height equals 0. We can do this by solving $-16t^2 + 79t + 5 = 0$.

$-16t^2 + 79t + 5 = 0$	**Original equation**
$m = 80; p = -1$	**$-16(5) = -80$, $80 \cdot (-1) = -80$, $80 + (-1) = 79$**
$-16t^2 + 80t - t + 5 = 0$	**Write the pattern.**
$(-16t^2 + 80t) + (-t + 5) = 0$	**Group terms with common factors.**
$16t(-t + 5) + 1(-t + 5) = 0$	**Factor GCF from each group.**
$(16t + 1)(-t + 5) = 0$	**Distributive Property**
$16t + 1 = 0$ or $-t + 5 = 0$	**Zero Product Property**
$16t = -1$ $\quad -t = -5$	**Solve both equations.**
$t = -\frac{1}{16}$ $\quad t = 5$	**Solve.**

CHECK We have two solutions.

- The first solution is negative and since time cannot be negative, this solution can be eliminated.
- The second solution of 5 seconds seems reasonable for the time a javelin spends in the air.
- The answer can be confirmed by substituting back into the original equation.

$$-16t^2 + 79t + 5 = 0$$
$$-16(5)^2 + 79(5) + 5 \stackrel{?}{=} 0$$
$$-400 + 395 + 5 \stackrel{?}{=} 0$$
$$0 = 0 \checkmark$$

The javelin is in the air for 5 seconds.

Real-World Link

Cuba's Osleidys Menendez broke the javelin world record in 2002 with a distance of 234 feet and 8 inches.

Source: *New York Times*

Check Your Progress

5. BUNGEE JUMPING Juan recorded his brother bungee jumping from a height of 1200 feet. At the time the cord lifted his brother back up, he was 50 feet above the ground. If Juan started recording as soon as his brother fell, how much time elapsed when the cord snapped back? Use $f(t) = -16t^2 + c$, where c is the height in feet.

Personal Tutor glencoe.com

Check Your Understanding

Example 1 p. 268

Write a quadratic equation in standard form with the given root(s).

1. $-8, 5$
2. $\frac{3}{2}, \frac{1}{4}$
3. $-\frac{2}{3}, \frac{5}{2}$

Examples 2–4 pp. 269–270

Factor each polynomial.

4. $35x^2 - 15x$
5. $18x^2 - 3x + 24x - 4$
6. $x^2 - 12x + 32$
7. $x^2 - 4x - 21$
8. $2x^2 + 7x - 30$
9. $16x^2 - 16x + 3$
10. $x^2 - 36$
11. $12x^2y - 18xy$
12. $12x^2 - 2x - 2$

Example 5 p. 271

Solve each equation by factoring.

13. $x^2 - 9x = 0$
14. $x^2 - 3x - 28 = 0$
15. $2x^2 - 24x = -72$
16. **GARDENING** Tamika wants to double the area of her garden by increasing the length and width by the same amount. What will be the dimensions of her garden then?

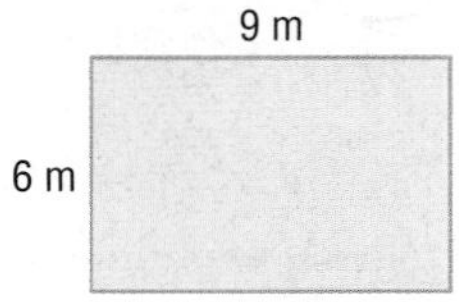

Practice and Problem Solving

= Step-by-Step Solutions begin on page R20.
Extra Practice begins on page 947.

Example 1 p. 268

Write a quadratic equation in standard form with the given root(s).

17. 7
18. $-5, \frac{1}{2}$
19. $\frac{1}{5}, 6$

Examples 2–4 pp. 269–270

Factor each polynomial.

20. $40a^2 - 32a$
21. $51c^3 - 34c$
22. $32xy + 40bx - 12ay - 15ab$
23. $3x^2 - 12$
24. $15y^2 - 240$
25. $48cg + 36cf - 4dg - 3df$
26. $x^2 + 13x + 40$
27. $x^2 - 9x - 22$
28. $3x^2 + 12x - 36$
29. $15x^2 + 7x - 2$
30. $4x^2 + 29x + 30$
31. $18x^2 + 15x - 12$
32. $8x^2z^2 - 4xz^2 - 12z^2$
33. $9x^2 - 25$
34. $18x^2y^2 - 24xy^2 + 36y^2$
35. $15x^2 - 84x - 36$
36. $12x^2 + 13x - 14$
37. $12xy^2 - 108x$

Example 5 p. 271

Solve each equation by factoring.

38. $x^2 + 4x - 45 = 0$
39. $x^2 - 5x - 24 = 0$
40. $x^2 = 121$
41. $x^2 + 13 = 17$
42. $-3x^2 - 10x + 8 = 0$
43. $-8x^2 + 46x - 30 = 0$
44. **GEOMETRY** The hypotenuse of a right triangle is 1 centimeter longer than one side and 4 centimeters longer than three times the other side. Find the dimensions of the triangle.
45. **NUMBER THEORY** Find two consecutive even integers with a product of 624.

GEOMETRY Find x and the dimensions of each rectangle.

46.

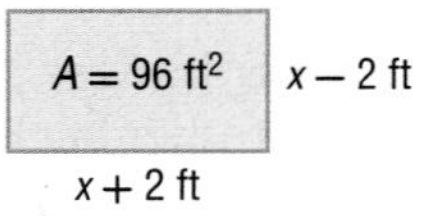

47.

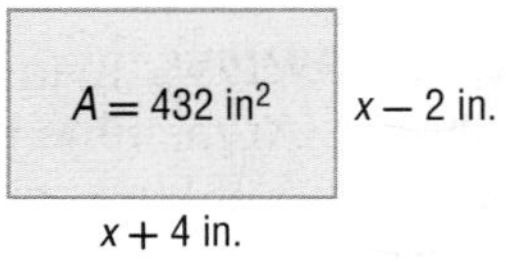

48.

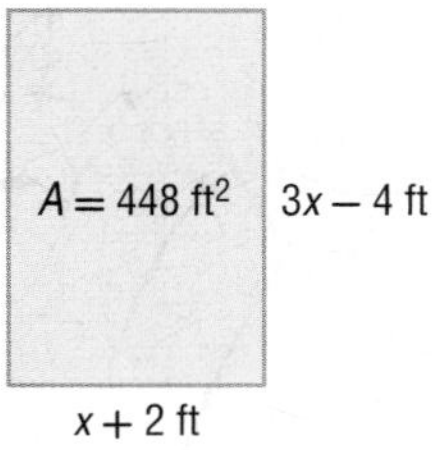

Solve each equation by factoring.

49. $12x^2 - 4x = 5$ **50.** $5x^2 = 15x$ **51.** $16x^2 + 36 = -48x$

52. $75x^2 - 60x = -12$ **53.** $4x^2 - 144 = 0$ **54.** $-7x + 6 = 20x^2$

Real-World Link

The average movie theater costs approximately $1 million per screen to build.

Source: *National Real Estate Investor*

55. **MOVIE THEATER** A company plans to build a large multiplex theater. The financial analyst told her manager that the profit function for their theater was $P(x) = -x^2 + 48x - 512$, where x is the number of movie screens, and $P(x)$ is the profit earned in thousands of dollars. Determine the range of production of movie screens that will guarantee that the company will not lose money.

Write a quadratic equation in standard form with the given root(s).

56. $-\frac{4}{7}, \frac{3}{8}$ **57.** 3.4, 0.6 **58.** $\frac{2}{11}, \frac{5}{9}$

Solve each equation by factoring.

59. $10x^2 + 25x = 15$ **60.** $27x^2 + 5 = 48x$ **61.** $x^2 + 0.25x = 1.25$

62. $48x^2 - 15 = -22x$ **63.** $3x^2 + 2x = 3.75$ **64.** $-32x^2 + 56x = 12$

65. **DESIGN** A square is cut out of the figure at the right. Write an expression for the area of the figure that remains, and then factor the expression.

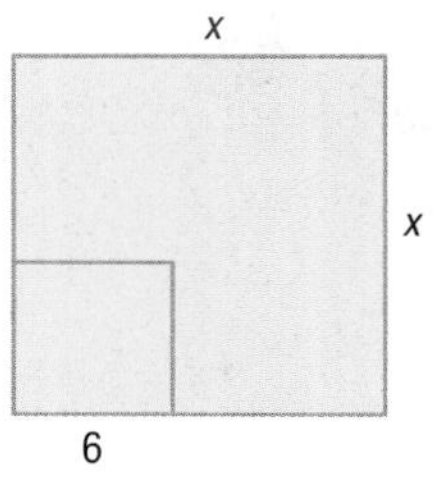

66. **BUSINESS** After analyzing the market, a company that sells Web sites determined the profitability of their product was modeled by $P(x) = -16x^2 + 368x - 2035$, where x is the price of each Web site and $P(x)$ is the company's profit. Determine the price range of the Web sites that will be profitable for the company.

67. **PAINTINGS** Enola wants to add a border to her painting, distributed evenly, that has the same area as the painting itself. What are the dimensions of the painting with the border included?

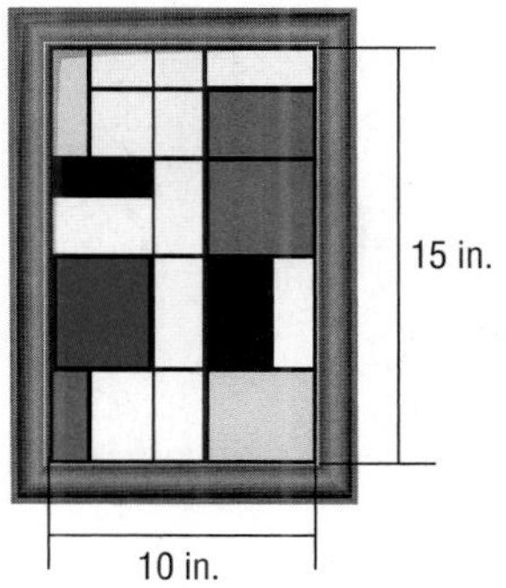

68. **MULTIPLE REPRESENTATIONS** In this problem, you will consider $a(x - p)(x - q) = 0$.

a. **GRAPHICAL** Graph the related function for $a = 1$, $p = 2$, and $q = -3$.

b. **ANALYTICAL** What are the solutions of the equation?

c. **GRAPHICAL** Graph the related functions for $a = 4, -3$, and $\frac{1}{2}$ on the same graph.

d. **VERBAL** What are the similarities and differences between the graphs?

e. **VERBAL** What conclusion can you make about the relationship between the factored form of a quadratic equation and its solutions?

69. **GEOMETRY** The area of the triangle is 26 square centimeters. Find the length of the base.

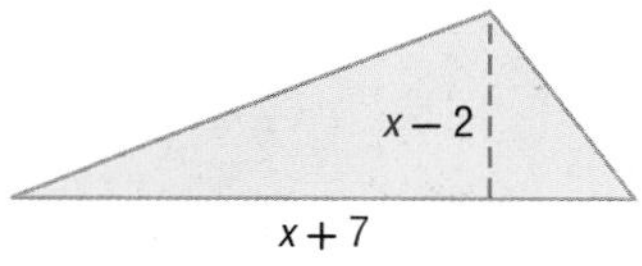

Real-World Link

Kicking accuracy is usually highest when the velocity of the ball is about 80% of the maximum. Professional players can kick a ball between 32 to 35 meters per second.

Source: Coach's Information Service

70. SOCCER When a ball is kicked in the air, its height in meters above the ground can be modeled by $h(t) = -4.9t^2 + 14.7t$ and the distance it travels can be modeled by $d(t) = 16t$, where t is the time in seconds.

a. How long was the ball in the air?

b. How far did it travel before it hit the ground? (*Hint*: Ignore air resistance.)

c. What was the maximum height of the ball?

Factor each polynomial.

71. $18a - 24ay + 48b - 64by$

72. $3x^2 + 2xy + 10y + 15x$

73 $6a^2b^2 - 12ab^2 - 18b^3$

74. $12a^2 - 18ab + 30ab^3$

75. $32ax + 12bx - 48ay - 18by$

76. $30ac + 80bd + 40ad + 60bc$

77. $5ax^2 - 2by^2 - 5ay^2 + 2bx^2$

78. $12c^2x + 4d^2y - 3d^2x - 16c^2y$

H.O.T. Problems

Use **H**igher-**O**rder **T**hinking Skills

79. FIND THE ERROR Gwen and Morgan are solving $-12x^2 + 5x + 2 = 0$. Is either of them correct? Explain your reasoning.

Gwen	Morgan
$-12x^2 + 5x + 2 = 0$	$-12x^2 + 5x + 2 = 0$
$-12x^2 + 8x - 3x + 2 = 0$	$-12x^2 + 8x - 3x + 2 = 0$
$4x(-3x + 2) - (3x + 2) = 0$	$4x(-3x + 2) + (-3x + 2) = 0$
$(4x - 1)(3x + 2) = 0$	$(4x - 1)(-3x + 2) = 0$
$x = \frac{1}{4}$ or $-\frac{2}{3}$	$x = \frac{1}{4}$ or $\frac{2}{3}$

80. CHALLENGE Solve $3x^6 - 39x^4 + 108x^2 = 0$ by factoring.

81. CHALLENGE The rule for factoring a difference of cubes is shown below. Use this rule to factor $40x^5 - 135x^2y^3$.

$$a^3 - b^3 = (a - b)(a^2 + ab + b^2)$$

82. OPEN ENDED Choose two integers. Then write an equation in standard form with those roots. How would the equation change if the signs of the two roots were switched?

83. CHALLENGE For a quadratic equation of the form $(x - p)(x - q) = 0$, show that the axis of symmetry of the related quadratic function is located halfway between the x-intercepts p and q.

84. WRITE A QUESTION A classmate is using the guess and check strategy to factor trinomials of the form $x^2 + bx + c$. Write a question to help him think of a way to use that strategy for $ax^2 + bx + c$.

85. REASONING Determine whether the following statement is *sometimes*, *always*, or *never* true. Explain your reasoning.

In a quadratic equation in standard form where a, b, and c are integers, if b is odd, then the quadratic cannot be a perfect square trinomial.

86. WRITING IN MATH Explain how to factor a trinomial in standard form with $a > 1$.

NYSRE PRACTICE A2.S.9, A2.S.12

87. SHORT RESPONSE If *ABCD* is transformed by $(x, y) \rightarrow (3x, 4y)$, determine the area of $A'B'C'D'$.

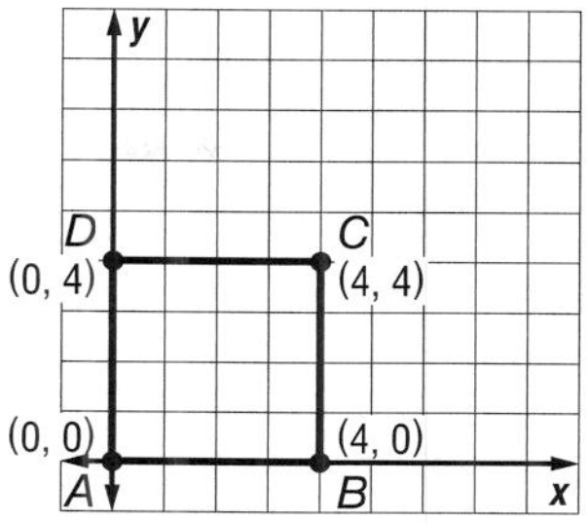

88. For $y = 2|6 - 3x| + 4$, which set describes x when $y < 6$?

A $\left\{x \,\middle|\, \frac{5}{3} < x < \frac{7}{3}\right\}$

B $\left\{x \,\middle|\, x < \frac{5}{3} \text{ or } x > \frac{7}{3}\right\}$

C $\left\{x \,\middle|\, x < \frac{5}{3}\right\}$

D $\left\{x \,\middle|\, x > \frac{7}{3}\right\}$

89. PROBABILITY A 5-character password can contain the numbers 0 through 9 and 26 letters of the alphabet. None of the characters can be repeated. What is the probability that the password begins with a consonant?

F $\frac{21}{26}$

G $\frac{21}{35}$

H $\frac{21}{36}$

J $\frac{5}{36}$

90. SAT/ACT If $c = \frac{8a^3}{b}$, what happens to the value of c when both a and b are doubled?

A c is unchanged.

B c is halved.

C c is doubled.

D c is multiplied by 4.

Spiral Review

Use the related graph of each equation to determine its solutions. (Lesson 5-2)

91. $x^2 - 2x - 8 = 0$

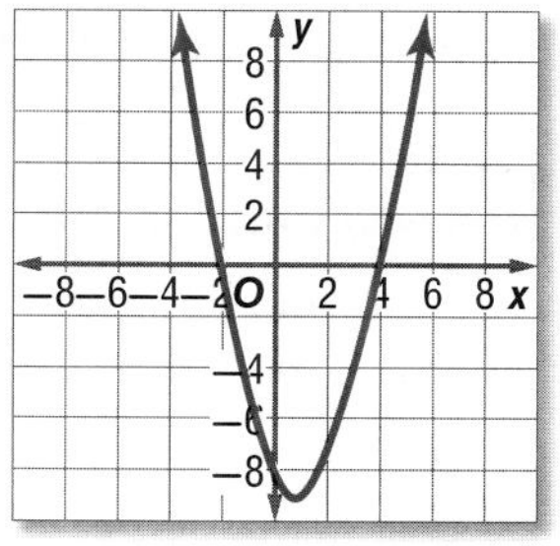

92. $x^2 + 4x = 12$

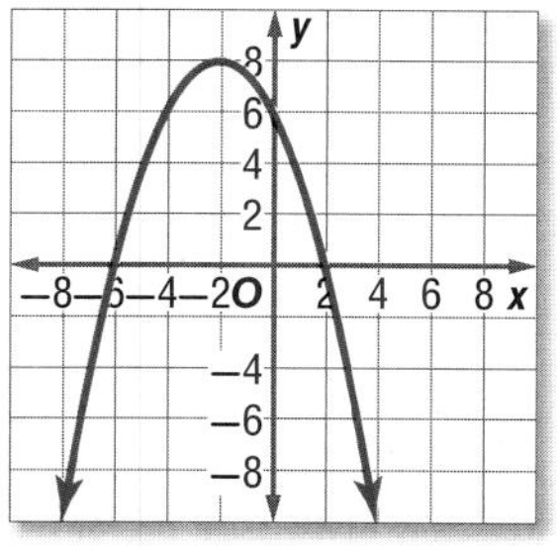

93. $x^2 - 4x + 4 = 0$

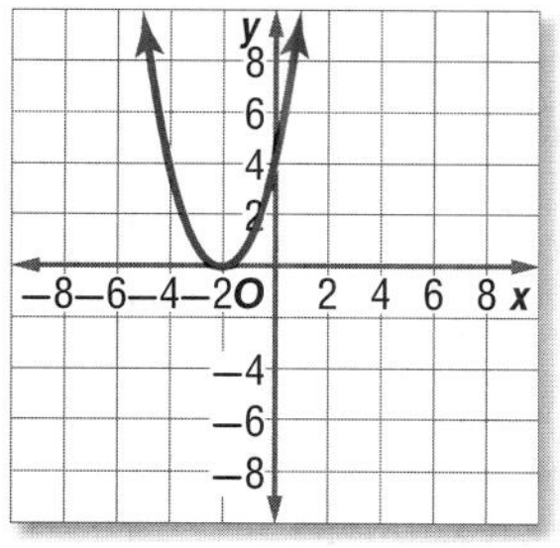

Graph each function. (Lesson 5-1)

94. $f(x) = x^2 - 6x + 2$

95. $f(x) = -2x^2 + 4x + 1$

96. $f(x) = (x - 3)(x + 4)$

97. FUNDRAISING Lawrence High School sold wrapping paper and boxed cards for their fundraising event. The school gets \$1.00 for each roll of wrapping paper sold and \$0.50 for each box of cards sold. (Lesson 4-3)

a. Write a matrix that represents the amounts sold for each class and a matrix that represents the amount of money the school earns for each item sold.

b. Write a matrix that shows how much each class earned.

c. Which class earned the most money?

d. What is the total amount of money the school made from the fundraiser?

Total Amounts for Each Class

Class	Wrapping Paper	Cards
Freshmen	72	49
Sophomores	68	63
Juniors	90	56
Seniors	86	62

Skills Review

Simplify. (Prerequisite Skills 2)

98. $\sqrt{5} \cdot \sqrt{15}$

99. $\sqrt{8} \cdot \sqrt{32}$

100. $2\sqrt{3} \cdot \sqrt{27}$

5-4 Complex Numbers

Then

You simplified square roots. (Prior Course)

Now

- Perform operations with pure imaginary numbers.
- Perform operations with complex numbers.

NYS Core Curriculum

A2.N.6 Write square roots of negative numbers in terms of *i*
A2.N.7 Simplify powers of *i*
A2.N.8 Determine the conjugate of a complex number *Also addresses A2.N.9.*

New Vocabulary

imaginary unit
pure imaginary number
complex number
complex conjugates

NY Math Online

glencoe.com

- Extra Examples
- Personal Tutor
- Self-Check Quiz
- Homework Help
- Math in Motion

Why?

Consider the graph of $y = x^2 + 2x + 4$ at the right. Notice how this graph has no x-intercepts and therefore does not have any roots. Does this mean there are no solutions?

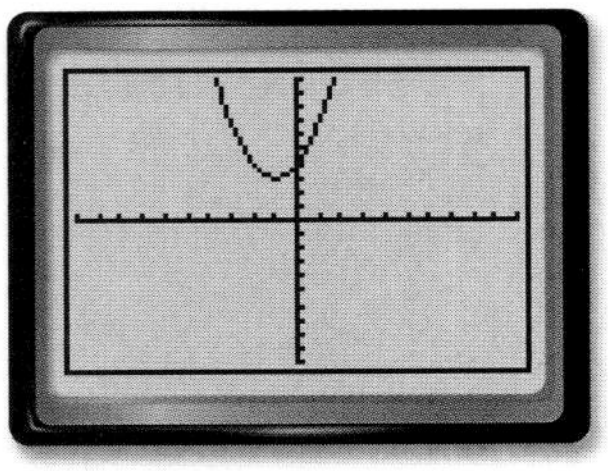

Use the Solver function located in the math menu of a graphing calculator. Enter the equation and select $x = 2$ as your *guess* to a solution.

Press ALPHA ENTER and the calculator will attempt to solve the equation. The calculator indicates there is no solution with the error message. So there are no real solutions. However, there are *imaginary* solutions.

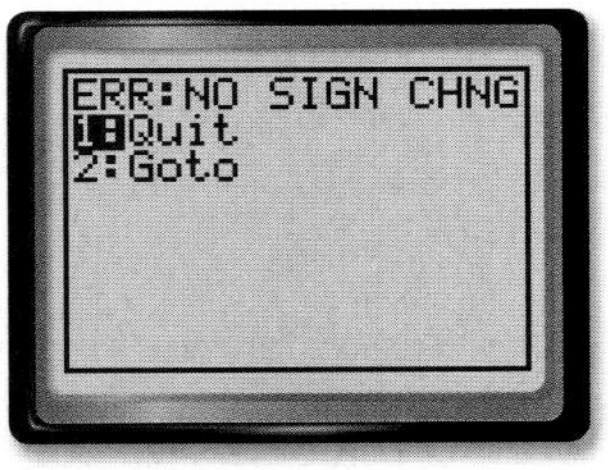

Pure Imaginary Numbers In your math studies so far, you have worked with real numbers. Equations like the one above led mathematicians to define imaginary numbers. The **imaginary unit *i*** is defined to be $i^2 = -1$. The number i is the principal square root of -1; that is, $i = \sqrt{-1}$.

Numbers of the form $6i$, $-2i$, and $i\sqrt{3}$ are called **pure imaginary numbers**. Pure imaginary numbers are square roots of negative real numbers. For any positive real number b, $\sqrt{-b^2} = \sqrt{b^2} \cdot \sqrt{-1}$ or bi.

EXAMPLE 1 Square Roots of Negative Numbers

Simplify.

a. $\sqrt{-27}$

$$\sqrt{-27} = \sqrt{-1 \cdot 3^2 \cdot 3}$$
$$= \sqrt{-1} \cdot \sqrt{3^2} \cdot \sqrt{3}$$
$$= i \cdot 3 \cdot \sqrt{3} \text{ or } 3i\sqrt{3}$$

b. $\sqrt{-216}$

$$\sqrt{-216} = \sqrt{-1 \cdot 6^2 \cdot 6}$$
$$= \sqrt{-1} \cdot \sqrt{6^2 \cdot 6}$$
$$= i \cdot 6 \cdot \sqrt{6} \text{ or } 6i\sqrt{6}$$

Check Your Progress

1A. $\sqrt{-18}$ **1B.** $\sqrt{-125}$

Personal Tutor glencoe.com

The Commutative and Associative Properties of Multiplication hold true for pure imaginary numbers. The first few powers of i are shown below.

$i^1 = i$	$i^2 = -1$	$i^3 = i^2 \cdot i$ or $-i$	$i^4 = (i^2)^2$ or 1
$i^5 = i^4 \cdot i$ or i	$i^6 = i^4 \cdot i^2$ or -1	$i^7 = i^4 \cdot i^3$ or $-i$	$i^8 = (i^2)^4$ or 1

EXAMPLE 2 Products of Pure Imaginary Numbers

Simplify.

a. $-5i \cdot 3i$

$-5i \cdot 3i = -15i^2$ **Multiply.**

$= -15(-1)$ **$i^2 = -1$**

$= 15$ **Simplify.**

b. $\sqrt{-6} \cdot \sqrt{-15}$

$\sqrt{-6} \cdot \sqrt{-15} = i\sqrt{6} \cdot i\sqrt{15}$ **$i = \sqrt{-1}$**

$= i^2\sqrt{90}$ **Multiply.**

$= -1 \cdot \sqrt{9} \cdot \sqrt{10}$ **Simplify.**

$= -3\sqrt{10}$ **Multiply.**

StudyTip

Square Root Property Refer to Concepts and Skills Lesson 3 to review the Square Root Property.

Check Your Progress

2A. $3i \cdot 4i$ **2B.** $\sqrt{-20} \cdot \sqrt{-12}$ **2C.** i^{31}

Personal Tutor glencoe.com

You can solve some quadratic equations by using the **Square Root Property**.

EXAMPLE 3 Equation with Pure Imaginary Solutions

Solve $4x^2 + 256 = 0$.

$4x^2 + 256 = 0$ **Original equation**

$4x^2 = -256$ **Subtract 256 from each side.**

$x^2 = -64$ **Divide each side by 4.**

$x = \pm\sqrt{-64}$ **Square Root Property**

$x = \pm 8i$ **$\sqrt{-64} = \sqrt{64} \cdot \sqrt{-1}$ or $8i$**

Check Your Progress

Solve each equation.

3A. $4x^2 + 100 = 0$ **3B.** $x^2 + 4 = 0$

Personal Tutor glencoe.com

Operations with Complex Numbers Consider $2 + 3i$. Since 2 is a real number and $3i$ is a pure imaginary number, the terms are not like terms and cannot be combined. This type of expression is called a **complex number**.

Key Concept Complex Numbers

For Your FOLDABLE

Words A complex number is any number that can be written in the form $a + bi$, where a and b are real numbers and i is the imaginary unit. a is called the real part, and b is called the imaginary part.

Examples $5 + 2i$ $1 - 3i = 1 + (-3)i$

Math in Motion, Interactive Lab glencoe.com

The Venn diagram shows the set of complex numbers.

- If $b = 0$, the complex number is a real number.
- If $b \neq 0$, the complex number is imaginary.
- If $a = 0$, the complex number is a pure imaginary number.

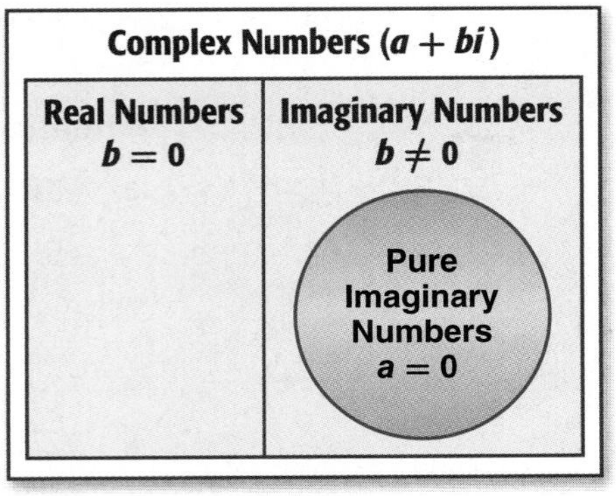

Two complex numbers are equal if and only if their real parts are equal and their imaginary parts are equal. That is, $a + bi = c + di$ if and only if $a = c$ and $b = d$.

EXAMPLE 4 Equate Complex Numbers

Find the values of x and y that make $3x - 5 + (y - 3)i = 7 + 6i$ true.

Set the real parts equal to each other and the imaginary parts equal to each other.

$3x - 5 = 7$	**Real parts**	$y - 3 = 6$	**Imaginary parts**
$3x = 12$	**Add 5 to each side.**	$y = 9$	**Add 3 to each side.**
$x = 4$	**Divide each side by 3.**		

Check Your Progress

4. Find the values of x and y that make $5x + 1 + (3 + 2y)i = 2x - 2 + (y - 6)i$ true.

Personal Tutor glencoe.com

StudyTip

Complex Numbers Whereas all real numbers are also complex, the term *complex number* usually refers to a number that is not real.

The Commutative, Associative, and Distributive Properties of Multiplication and Addition hold true for complex numbers. To add or subtract complex numbers, combine like terms. That is, combine the real parts, and combine the imaginary parts.

EXAMPLE 5 Add and Subtract Complex Numbers

Simplify.

a. $(5 - 7i) + (2 + 4i)$

$(5 - 7i) + (2 + 4i) = (5 + 2) + (-7 + 4)i$ — **Commutative and Associative Properties**

$= 7 - 3i$ — **Simplify.**

b. $(4 - 8i) - (3 - 6i)$

$(4 - 8i) - (3 - 6i) = (4 - 3) + [-8 - (-6)]i$ — **Commutative and Associative Properties**

$= 1 - 2i$ — **Simplify.**

Check Your Progress

5A. $(-2 + 5i) + (1 - 7i)$

5B. $(4 + 6i) - (-1 + 2i)$

Personal Tutor glencoe.com

StudyTip

Reading Math Electrical engineers use ***j*** as the imaginary unit to avoid confusion with the ***i*** for current.

Complex numbers are used with electricity. In these problems, j usually represents the imaginary unit. In a circuit with alternating current, the voltage, current, and impedance, or hindrance to current, can be represented by complex numbers. To multiply these numbers, use the FOIL method.

Real-World Link

An example of a series circuit is a string of holiday lights. The number of bulbs on a circuit affects the strength of the current, which in turn affects the brightness of the lights.

Source: *Popular Science*

Real-World EXAMPLE 6 Multiply Complex Numbers

ELECTRICITY In an AC circuit, the voltage V, current C, and impedance I are related by the formula $V = C \cdot I$. Find the voltage in a circuit with current $2 + 4j$ amps and impedance $9 - 3j$ ohms.

$V = C \cdot I$	**Electricity formula**
$= (2 + 4j) \cdot (9 - 3j)$	**$C = 2 + 4j$ and $I = 9 - 3j$**
$= 2(9) + 2(-3j) + 4j(9) + 4j(-3j)$	**FOIL Method**
$= 18 - 6j + 36j - 12j^2$	**Multiply.**
$= 18 + 30j - 12(-1)$	**$j^2 = -1$**
$= 30 + 30j$	**Add.**

The voltage is $30 + 30j$ volts.

Check Your Progress

6. Find the voltage in a circuit with current $2 - 4j$ amps and impedance $3 - 2j$ ohms.

Personal Tutor glencoe.com

Two complex numbers of the form $a + bi$ and $a - bi$ are called **complex conjugates**. The product of complex conjugates is always a real number. You can use this fact to simplify the quotient of two complex numbers.

EXAMPLE 7 Divide Complex Numbers

Simplify.

a. $\frac{2i}{3 + 6i}$

$\frac{2i}{3 + 6i} = \frac{2i}{3 + 6i} \cdot \frac{3 - 6i}{3 - 6i}$	**$3 + 6i$ and $3 - 6i$ are complex conjugates.**
$= \frac{6i - 12i^2}{9 - 36i^2}$	**Multiply.**
$= \frac{6i - 12(-1)}{9 - 36(-1)}$	**$i^2 = -1$**
$= \frac{6i + 12}{45}$	**Simplify.**
$= \frac{4}{15} + \frac{2}{15}i$	**$a + bi$ form**

b. $\frac{4 + i}{5i}$

$\frac{4 + i}{5i} = \frac{4 + i}{5i} \cdot \frac{i}{i}$	**Multiply by $\frac{i}{i}$.**
$= \frac{4i + i^2}{5i^2}$	**Multiply.**
$= \frac{4i - 1}{-5}$	**$i^2 = -1$**
$= \frac{1}{5} - \frac{4}{5}i$	**$a + bi$ form**

Check Your Progress

7A. $\frac{-2i}{3 + 5i}$

7B. $\frac{2 + i}{1 - i}$

Personal Tutor glencoe.com

Check Your Understanding

Examples 1 and 2 pp. 276–277

Simplify.

1. $\sqrt{-81}$
2. $\sqrt{-32}$
3. $(4i)(-3i)$
4. $3\sqrt{-24} \cdot 2\sqrt{-18}$
5. i^{40}
6. i^{63}

Example 3 p. 277

Solve each equation.

7. $4x^2 + 32 = 0$
8. $2x^2 + 24 = 0$

Example 4 p. 278

Find the values of a and b that make each equation true.

9. $3a + (4b + 2)i = 9 - 6i$
10. $4b - 5 + (-a - 3)i = 7 - 8i$

Examples 5 and 7 pp. 278–279

Simplify.

11. $(-1 + 5i) + (-2 - 3i)$
12. $(7 + 4i) - (1 + 2i)$
13. $(6 - 8i)(9 + 2i)$
14. $(3 + 2i)(-2 + 4i)$
15. $\frac{3 - i}{4 + 2i}$
16. $\frac{2 + i}{5 + 6i}$

Example 6 p. 279

17. **ELECTRICITY** The current in one part of a series circuit is $5 - 3j$ amps. The current in another part of the circuit is $7 + 9j$ amps. Add these complex numbers to find the total current in the circuit.

Practice and Problem Solving

● = **Step-by-Step Solutions** begin on page R20.
Extra Practice begins on page 947.

Examples 1 and 2 pp. 276–277

Simplify.

18. $\sqrt{-121}$
19. $\sqrt{-169}$
20. $\sqrt{-100}$
21. $\sqrt{-81}$
22. $(-3i)(-7i)(2i)$
23. $4i(-6i)^2$
24. i^{11}
25. i^{25}
26. $(10 - 7i) + (6 + 9i)$
27. $(-3 + i) + (-4 - i)$
28. $(12 + 5i) - (9 - 2i)$
29. $(11 - 8i) - (2 - 8i)$
30. $(1 + 2i)(1 - 2i)$
31. $(3 + 5i)(5 - 3i)$
32. $(4 - i)(6 - 6i)$
33. $\frac{2i}{1 + i}$
34. $\frac{5}{2 + 4i}$
35. $\frac{5 + i}{3i}$

Example 3 p. 277

Solve each equation.

36. $4x^2 + 4 = 0$
37. $3x^2 + 48 = 0$
38. $2x^2 + 50 = 0$
39. $2x^2 + 10 = 0$
40. $6x^2 + 108 = 0$
41. $8x^2 + 128 = 0$

Example 4 p. 278

Find the values of x and y that make each equation true.

42. $9 + 12i = 3x + 4yi$
43. $x + 1 + 2yi = 3 - 6i$
44. $2x + 7 + (3 - y)i = -4 + 6i$
45. $5 + y + (3x - 7)i = 9 - 3i$
46. $a + 3b + (3a - b)i = 6 + 6i$
47. $(2a - 4b)i + a + 5b = 15 + 58i$

Examples 5 and 7
pp. 278–279

Simplify.

48. $\sqrt{-10} \cdot \sqrt{-24}$

49. $4i\left(\frac{1}{2}i\right)^2(-2i)^2$

50. i^{41}

51. $(4 - 6i) + (4 + 6i)$

52. $(8 - 5i) - (7 + i)$

53. $(-6 - i)(3 - 3i)$

54. $\frac{(5 + i)^2}{3 - i}$

55. $\frac{6 - i}{2 - 3i}$

56. $(-4 + 6i)(2 - i)(3 + 7i)$

57. $(1 + i)(2 + 3i)(4 - 3i)$

58. $\frac{4 - i\sqrt{2}}{4 + i\sqrt{2}}$

59. $\frac{2 - i\sqrt{3}}{2 + i\sqrt{3}}$

Example 6
p. 279

60. **ELECTRICITY** The impedance in one part of a series circuit is $7 + 8j$ ohms, and the impedance in another part of the circuit is $13 - 4j$ ohms. Add these complex numbers to find the total impedance in the circuit.

ELECTRICITY **Use the formula $V = C \cdot I$.**

61. The current in a circuit is $3 + 6j$ amps, and the impedance is $5 - j$ ohms. What is the voltage?

62. The voltage in a circuit is $20 - 12j$ volts, and the impedance is $6 - 4j$ ohms. What is the current?

63. Find the sum of $ix^2 - (4 + 5i)x + 7$ and $3x^2 + (2 + 6i)x - 8i$.

64. Simplify $[(2 + i)x^2 - ix + 5 + i] - [(-3 + 4i)x^2 + (5 - 5i)x - 6]$.

65. **MULTIPLE REPRESENTATIONS** In this problem, you will explore adding complex numbers in the complex plane. The *complex plane* is a lot like the *real plane*, but it has real numbers along the x-axis and the imaginary numbers along the y-axis.

a. **GRAPHICAL** Graph $3 + 4i$ in the complex plane by drawing a segment from the origin to (3, 4). Label this point A.

b. **GRAPHICAL** Graph $-2 - 5i$ in the complex plane by drawing a segment from the origin to $(-2, -5)$. Label this point B.

c. **GRAPHICAL** Given three vertices of a parallelogram, complete it by adding a point C.

d. **ANALYTICAL** What complex number does C represent? What is the relationship between A, B, and C?

Real-World Link

Electricity is the flow of an electrical charge. The electricity we use is a secondary energy source converted from fossil fuels, nuclear reactors, and natural sources.

Source: Energy Information Administration

H.O.T. Problems

Use **H**igher-**O**rder **T**hinking Skills

66. **FIND THE ERROR** Joe and Sue are simplifying $(2i)(3i)(4i)$. Is either of them correct? Explain your reasoning.

Joe	Sue
$24i^3 = -24$	$24i^3 = -24i$

67. **CHALLENGE** Simplify $(1 + 2i)^3$.

68. **REASONING** Determine whether the following statement is *always*, *sometimes*, or *never* true. Explain your reasoning.

Every complex number has both a real part and an imaginary part.

69. **OPEN ENDED** Write two complex numbers with a product of 20.

70. **WRITING IN MATH** Explain how complex numbers are related to quadratic equations. How do you know when a quadratic equation will have only complex solutions?

NYSRE PRACTICE A2.N.7

71. EXTENDED RESPONSE Refer to the figure to answer the following.

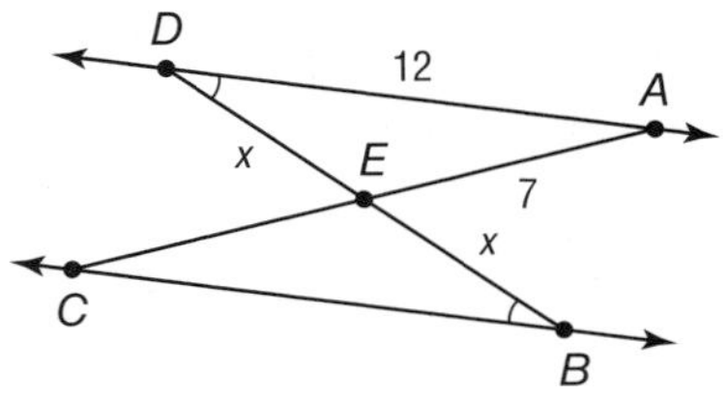

a. Name two congruent triangles with vertices in correct order.

b. Explain why the triangles are congruent.

c. What is the length of $\overline{EC}$? Explain your procedure.

72. $(3 + 6)^2 =$

A $2 \times 3 + 2 \times 6$
B 9^2
C $3^2 + 6^2$
D $3^2 \times 6^2$

73. SAT/ACT A store charges $49 for a pair of pants. This price is 40% more than the amount it costs the store to buy the pants. After a sale, any employee is allowed to purchase any remaining pairs of pants at 30% off the store's cost.

How much would it cost an employee to purchase the pants after the sale?

F $12.50
G $13.72
H $24.50
J $35.00

74. What are the values of x and y when $(5 + 4i) - (x + yi) = (-1 - 3i)$?

A $x = 6, y = 7$
B $x = 4, y = i$
C $x = 6, y = i$
D $x = 4, y = 7$

Spiral Review

Solve each equation by factoring. (Lesson 5-3)

75. $2x^2 + 7x = 15$ **76.** $4x^2 - 12 = 22x$ **77.** $6x^2 = 5x + 4$

NUMBER THEORY Use a quadratic equation to find two real numbers that satisfy each situation, or show that no such numbers exist. (Lesson 5-2)

78. Their sum is −3, and their product is −40.

79. Their sum is 19, and their product is 48.

80. Their sum is−15, and their product is 56.

81. Their sum is −21, and their product is 108.

82. RECREATION Refer to the table. (Lesson 4-2)

a. Write a matrix that represents the cost of admission for residents and a matrix that represents the cost of admission for nonresidents.

b. Write the matrix that represents the additional cost for nonresidents.

c. Write a matrix that represents the difference in cost if a child or adult goes after 6:00 P.M. instead of before 6:00 P.M.

Daily Admission Fees

Residents		
Time of day	**Child**	**Adult**
Before 6:00 P.M.	$3.00	$4.50
After 6:00 P.M.	$2.00	$3.50
Nonresidents		
Time of day	**Child**	**Adult**
Before 6:00 P.M.	$4.50	$6.75
After 6:00 P.M.	$3.00	$5.25

83. PART-TIME JOBS Terrell makes $10 an hour cutting grass and $12 an hour for raking leaves. He cannot work more than 15 hours per week. Graph two inequalities that Terrell can use to determine how many hours he needs to work at each job if he wants to earn at least $120 per week. (Lesson 3-3)

Skills Review

Determine whether each trinomial is a perfect square trinomial. Write *yes* or *no*. (Lesson 5-3)

84. $x^2 + 16x + 64$ **85.** $x^2 - 12x + 36$ **86.** $x^2 + 8x - 16$

87. $x^2 - 14x - 49$ **88.** $x^2 + x + 0.25$ **89.** $x^2 + 5x + 6.25$

CHAPTER 5

Mid-Chapter Quiz

Lessons 5-1 through 5-4

NYS Core Curriculum
A2.A.7, A2.N.7

1. Find the y-intercept, the equation of the axis of symmetry, and the x-coordinate of the vertex for $f(x) = 2x^2 + 8x - 3$. Then graph the function by making a table of values. (Lesson 5-1)

2. **MULTIPLE CHOICE** For which equation is the axis of symmetry $x = 5$? (Lesson 5-1)

A $f(x) = x^2 - 5x + 3$

B $f(x) = x^2 - 10x + 7$

C $f(x) = x^2 + 10x - 3$

D $f(x) = x^2 + 5x + 2$

3. Determine whether $f(x) = 5 - x^2 + 2x$ has a maximum or a minimum value. Then find this maximum or minimum value and state the domain and range of the function. (Lesson 5-1)

4. **PHYSICAL SCIENCE** From 4 feet above the ground, Maya throws a ball upward with a velocity of 18 feet per second. The height $h(t)$ of the ball t seconds after Maya throws the ball is given by $h(t) = -16t^2 + 18t + 4$. Find the maximum height reached by the ball and the time that this height is reached. (Lesson 5-1)

5. Solve $3x^2 - 17x + 5 = 0$ by graphing. If exact roots cannot be found, state the consecutive integers between which the roots are located. (Lesson 5-2)

Use a quadratic equation to find two real numbers that satisfy each situation, or show that no such numbers exist. (Lesson 5-2)

6. Their sum is 15, and their product is 36.

7. Their sum is 7, and their product is 15.

8. **MULTIPLE CHOICE** Using the graph of the function $f(x) = x^2 + 6x - 7$, what are the solutions to the equation $x^2 + 6x - 7 = 0$? (Lesson 5-2)

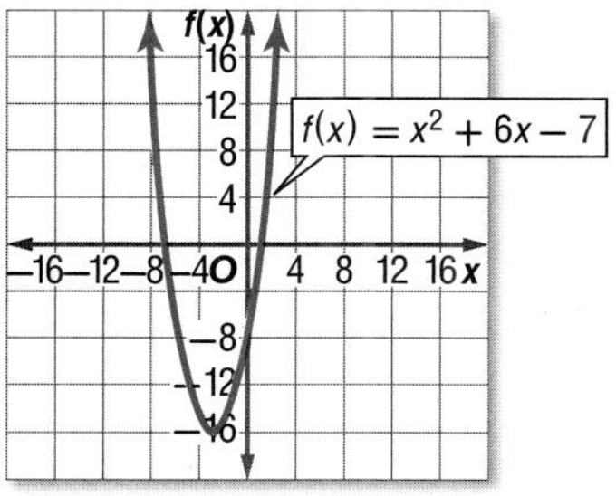

F $-1, 6$ **H** $-1, 7$

G $1, -6$ **J** $1, -7$

9. **BASEBALL** A baseball is hit upward with a velocity of 40 feet per second. Ignoring the height of the baseball player, how long does it take for the ball to fall to the ground? Use the formula $h(t) = v_0t - 16t^2$ where $h(t)$ is the height of an object in feet, v_0 is the object's initial velocity in feet per second, and t is the time in seconds. (Lesson 5-2)

Solve each equation by factoring. (Lesson 5-3)

10. $x^2 - x - 12 = 0$

11. $3x^2 + 7x + 2 = 0$

12. $x^2 - 2x - 15 = 0$

13. $2x^2 + 5x - 3 = 0$

14. Write a quadratic equation in standard form with roots -6 and $\frac{1}{4}$. (Lesson 5-3)

15. **TRIANGLES** Find the dimensions of a triangle if the base is $\frac{2}{3}$ the measure of the height and the area is 12 square centimeters. (Lesson 5-3)

16. **PATIO** Eli is putting a cement slab in his backyard. The original slab was going to have dimensions of 8 feet by 6 feet. He decided to make the slab larger by adding x feet to each side. The area of the new slab is 120 square feet. (Lesson 5-3)

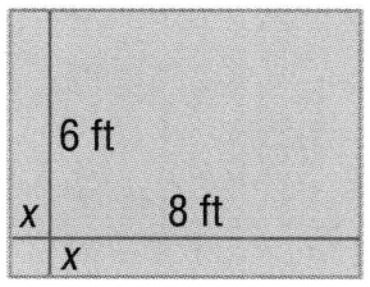

a. Write a quadratic equation that represents the area of the new slab.

b. Find the new dimensions of the slab.

Simplify. (Lesson 5-4)

17. $\sqrt{-81}$

18. $\sqrt{-25x^4y^5}$

19. $(15 - 3i) - (4 - 12i)$

20. i^{37}

21. $(5 - 3i)(5 + 3i)$

22. $\frac{3 - i}{2 + 5i}$

23. The impedance in one part of a series circuit is $3 + 4j$ ohms and the impedance in another part of the circuit is $6 - 7j$ ohms. Add these complex numbers to find the total impedance in the circuit. (Lesson 5-4)

5-5 Completing the Square

Then

You factored perfect square trinomials. (Lesson 5-3)

Now

- Solve quadratic equations by using the Square Root Property.
- Solve quadratic equations by completing the square.

NYS Core Curriculum

A2.A.24 Know and apply the technique of completing the square

New Vocabulary

completing the square

NY Math Online

glencoe.com

- Extra Examples
- Personal Tutor
- Self-Check Quiz
- Homework Help
- Math in Motion

Why?

When going through a school zone, drivers must slow to a speed of 20 miles per hour. Once they are out of the school zone, the drivers can increase their speed.

Suppose Arturo is leaving school to go home for lunch, and he lives 5000 feet from the school zone. If Arturo accelerates at a constant rate of 8 feet per second squared, the equation $t^2 + 2t + 8 = 16$ represents the time t it takes him to reach home.

To solve this equation, you can use the Square Root Property.

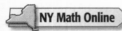

Square Root Property You have solved equations like $x^2 - 25 = 0$ by factoring. You have also used the Square Root Property to solve such an equation. This method can be useful with equations like the one above that describes the car's speed. In this case, the quadratic equation contains a perfect square trinomial set equal to a constant.

EXAMPLE 1 Equation with Rational Roots

Solve $x^2 + 6x + 9 = 36$ by using the Square Root Property.

$x^2 + 6x + 9 = 36$	**Original equation**
$(x + 3)^2 = 36$	**Factor the perfect square trinomial.**
$x + 3 = \pm\sqrt{36}$	**Square Root Property**
$x + 3 = \pm 6$	$\sqrt{36} = 6$
$x = -3 \pm 6$	**Subtract 3 from each side.**
$x = -3 + 6$ or $x = -3 - 6$	**Write as two equations.**
$= 3$ $\quad = -9$	**Simplify.**

The solution set is {3, −9}.

CHECK Substitute both values into the original equation.

$x^2 + 6x + 9 = 36$	**Original equation**	$x^2 + 6x + 9 = 36$
$3^2 + 6(3) + 9 \stackrel{?}{=} 36$	**Substitute 3 and −9.**	$(-9)^2 + 6(-9) + 9 \stackrel{?}{=} 36$
$9 + 18 + 9 \stackrel{?}{=} 36$	**Simplify.**	$81 - 54 + 9 \stackrel{?}{=} 36$
$36 = 36$ ✓	**Both solutions are correct.**	$36 = 36$ ✓

Check Your Progress

Solve each equation by using the Square Root Property.

1A. $x^2 - 12x + 36 = 25$ **1B.** $x^2 - 16x + 64 = 49$

Personal Tutor glencoe.com

Roots that are irrational numbers may be written as exact answers in radical form or as *approximate* answers in decimal form when a calculator is used.

StudyTip

Plus or Minus When using the Square Root Property, remember to put a $\pm$ sign before the radical.

EXAMPLE 2 Equation with Irrational Roots

Solve $x^2 - 10x + 25 = 27$ by using the Square Root Property.

$x^2 - 10x + 25 = 27$	**Original equation**
$(x - 5)^2 = 27$	**Factor the perfect square trinomial.**
$x - 5 = \pm\sqrt{27}$	**Square Root Property**
$x = 5 \pm 3\sqrt{3}$	**Add 5 to each side; $\sqrt{27} = 3\sqrt{3}$.**
$x = 5 + 3\sqrt{3}$ or $x = 5 - 3\sqrt{3}$	**Write as two equations.**
≈ 10.2 $\quad\quad \approx -0.2$	**Use a calculator.**

The exact solutions of this equation are $5 + 3\sqrt{3}$ and $5 - 3\sqrt{3}$. The approximate solutions are -0.2 and 10.2. Check these results by finding and graphing the related quadratic function.

$x^2 - 10x + 25 = 27$	**Original equation**
$x^2 - 10x - 2 = 0$	**Subtract 27 from each side.**
$y = x^2 - 10x - 2$	**Related quadratic function**

CHECK Use the **ZERO** function of a graphing calculator. The approximate zeros of the related function are -0.2 and 10.2.

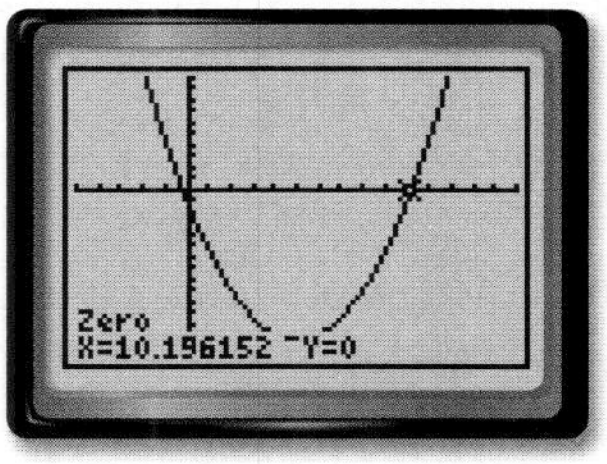

Check Your Progress

Solve each equation by using the Square Root Property.

2A. $x^2 + 8x + 16 = 20$

2B. $x^2 - 6x + 9 = 32$

Personal Tutor glencoe.com

Complete the Square All quadratic equations can be solved using the Square Root Property by manipulating the equation until one side is a perfect square. This method is called **completing the square**.

Consider $x^2 + 16x = 9$. Remember to perform each operation on each side of the equation.

$x^2 + 16x + \blacksquare = 9$	**What value is needed for the perfect square?**
$x^2 + 16x + 64 = 9 + 64$	$\left(\frac{16}{2}\right)^2$ **= 64; add 64 to each side.**
$x^2 + 16x + 64 = 73$	**We can now use the Square Root Property.**

Use this pattern of coefficients to complete the square of a quadratic expression.

Key Concept — Completing the Square

For Your FOLDABLE

Words To complete the square for any quadratic expression of the form $x^2 + bx$, follow the steps below.

Step 1 Find one half of b, the coefficient of x.
Step 2 Square the result in Step 1.
Step 3 Add the result of Step 2 to $x^2 + bx$.

Symbols $x^2 + bx + \left(\frac{b}{2}\right)^2 = \left(x + \frac{b}{2}\right)^2$

Math *in Motion*, Animation glencoe.com

EXAMPLE 3 Complete the Square

Find the value of c that makes $x^2 + 16x + c$ a perfect square. Then write the trinomial as a perfect square.

Step 1 Find one half of 16. $\frac{16}{2} = 8$

Step 2 Square the result in Step 1. $8^2 = 64$

Step 3 Add the result of Step 2 to $x^2 + 16x$. $x^2 + 16x + 64$

The trinomial $x^2 + 16x + 64$ can be written as $(x + 8)^2$.

Check Your Progress

3. Find the value of c that makes $x^2 - 14x + c$ a perfect square. Then write the trinomial as a perfect square.

Personal Tutor glencoe.com

You can solve any quadratic equation by completing the square. Because you are solving an equation, add the value you use to complete the square to each side.

EXAMPLE 4 Solve an Equation by Completing the Square

Solve $x^2 + 10x - 11 = 0$ by completing the square.

$x^2 + 10x - 11 = 0$	**Notice that $x^2 + 10x - 11 = 0$ is not a perfect square.**
$x^2 + 10x = 11$	**Rewrite so the left side is of the form $x^2 + bx$.**
$x^2 + 10x + 25 = 11 + 25$	**Since $\left(\frac{10}{2}\right)^2 = 25$, add 25 to each side.**
$(x + 5)^2 = 36$	**Write the left side as a perfect square.**
$x + 5 = \pm 6$	**Square Root Property**
$x = -5 \pm 6$	**Subtract 5 from each side.**
$x = -5 + 6$ or $x = -5 - 6$	**Write as two equations.**
$= 1$ $= -11$	**Simplify.**

The solution set is $\{-11, 1\}$. Check the result by using factoring.

Watch Out!

***Each* Side** When solving equations by completing the square, don't forget to add $\left(\frac{b}{2}\right)^2$ to *each* side of the equation.

Check Your Progress

Solve each equation by completing the square.

4A. $x^2 - 10x + 24 = 0$

4B. $x^2 + 10x + 9 = 0$

Personal Tutor glencoe.com

When the coefficient of the quadratic term is not 1, you must divide the equation by that coefficient before completing the square.

EXAMPLE 5 Equation with $a \neq 1$

Solve $2x^2 - 7x + 5 = 0$ by completing the square.

$2x^2 - 7x + 5 = 0$	Notice that $2x^2 - 7x + 5 = 0$ is not a perfect square.
$x^2 - \frac{7}{2}x + \frac{5}{2} = 0$	Divide by the coefficient of the quadratic term, 2.
$x^2 - \frac{7}{2}x = -\frac{5}{2}$	Subtract $\frac{5}{2}$ from each side.
$x^2 - \frac{7}{2}x + \frac{49}{16} = -\frac{5}{2} + \frac{49}{16}$	Since $\left(-\frac{7}{2} \div 2\right)^2 = \frac{49}{16}$, add $\frac{49}{16}$ to each side.
$\left(x - \frac{7}{4}\right)^2 = \frac{9}{16}$	Write the left side as a perfect square by factoring. Simplify the right side.
$x - \frac{7}{4} = \pm\frac{3}{4}$	Square Root Property
$x = \frac{7}{4} \pm \frac{3}{4}$	Add $\frac{7}{4}$ to each side.
$x = \frac{7}{4} + \frac{3}{4}$ or $x = \frac{7}{4} - \frac{3}{4}$	Write as two equations.
$= \frac{5}{2}$ $\quad = 1$	

The solution set is $\left\{1, \frac{5}{2}\right\}$.

Check Your Progress

Solve each equation by completing the square.

5A. $3x^2 + 10x - 8 = 0$

5B. $3x^2 - 14x + 16 = 0$

Personal Tutor glencoe.com

Not all solutions of quadratic equations are real numbers. In some cases, the solutions are complex numbers of the form $a + bi$, where $b \neq 0$.

StudyTip

Check by Graphing A graph of the related function shows that the equation has no real solutions since the graph has no x-intercepts. Imaginary solutions must be checked algebraically by substituting them in the original equation.

EXAMPLE 6 Equation with Imaginary Solutions

Solve $x^2 + 8x + 22 = 0$ by completing the square.

$x^2 + 8x + 22 = 0$	Notice that $x^2 + 8x + 22$ is not a perfect square.
$x^2 + 8x = -22$	Rewrite so the left side is of the form $x^2 + bx$.
$x^2 + 8x + 16 = -22 + 16$	Since $\left(\frac{8}{2}\right)^2 = 16$, add 16 to each side.
$(x + 4)^2 = -6$	Write the left side as a perfect square.
$x + 4 = \pm\sqrt{-6}$	Square Root Property
$x + 4 = \pm i\sqrt{6}$	$\sqrt{-1} = i$
$x = -4 \pm i\sqrt{6}$	Subtract 4 from each side.

The solution set is $\{-4 + i\sqrt{6}, -4 - i\sqrt{6}\}$.

Check Your Progress

Solve each equation by completing the square.

6A. $x^2 + 2x + 2 = 0$

6B. $x^2 - 6x + 25 = 0$

Personal Tutor glencoe.com

Check Your Understanding

Examples 1 and 2
pp. 284–285

Solve each equation by using the Square Root Property. Round to the nearest hundredth if necessary.

1. $x^2 + 12x + 36 = 6$

2. $x^2 - 8x + 16 = 13$

3. $x^2 + 18x + 81 = 15$

4. $9x^2 + 30x + 25 = 11$

5. **LASER LIGHT SHOW** The area A in square feet of a projected laser light show is given by $A = 0.16d^2$, where d is the distance from the laser to the screen in feet. At what distance will the projected laser light show have an area of 100 square feet?

Example 3
p. 286

Find the value of c that makes each trinomial a perfect square. Then write the trinomial as a perfect square.

6. $x^2 - 10x + c$

7. $x^2 - 5x + c$

Examples 4–6
pp. 286–287

Solve each equation by completing the square.

8. $x^2 + 2x - 8 = 0$

9. $x^2 - 4x + 9 = 0$

10. $2x^2 - 3x - 3 = 0$

11. $2x^2 + 6x - 12 = 0$

12. $x^2 + 4x + 6 = 0$

13. $x^2 + 8x + 10 = 0$

Practice and Problem Solving

● = **Step-by-Step Solutions** begin on page R20.
Extra Practice begins on page 947.

Examples 1 and 2
pp. 284–285

Solve each equation by using the Square Root Property. Round to the nearest hundredth if necessary.

14. $x^2 + 4x + 4 = 10$

15. $x^2 - 6x + 9 = 20$

16. $x^2 + 8x + 16 = 18$

17. $x^2 + 10x + 25 = 7$

18. $x^2 + 12x + 36 = 5$

19. $x^2 - 2x + 1 = 4$

20. $x^2 - 5x + 6.25 = 4$

21. $x^2 - 15x + 56.25 = 8$

22. $x^2 + 32x + 256 = 1$

23. $x^2 - 3x + \frac{9}{4} = 6$

24. $x^2 + 7x + \frac{49}{4} = 4$

25. $x^2 - 9x + \frac{81}{4} = \frac{1}{4}$

Example 3
p. 286

Find the value of c that makes each trinomial a perfect square. Then write the trinomial as a perfect square.

26. $x^2 + 8x + c$

27. $x^2 + 16x + c$

28. $x^2 - 11x + c$

29. $x^2 + 9x + c$

Examples 4–6
pp. 286–287

Solve each equation by completing the square.

30. $x^2 - 4x + 12 = 0$

31. $x^2 + 2x - 12 = 0$

32. $x^2 + 2x - 8 = 0$

33. $x^2 - 4x + 3 = 0$

34. $2x^2 + x - 3 = 0$

35. $2x^2 - 3x + 5 = 0$

36. $2x^2 + 5x + 7 = 0$

37. $3x^2 - 6x - 9 = 0$

38. $x^2 - 2x + 3 = 0$

39. $x^2 + 4x + 11 = 0$

40. $x^2 - 6x + 18 = 0$

41. $x^2 - 10x + 29 = 0$

42. $3x^2 - 4x = 2$

43. $2x^2 - 7x = -12$

44. $x^2 - 2.4x = 2.2$

45. $x^2 - 5.3x = -8.6$

46. $x^2 - \frac{1}{5}x - \frac{11}{5} = 0$

47. $x^2 - \frac{9}{2}x - \frac{24}{5} = 0$

48. **ARCHITECTURE** An architect's blueprints call for a dining room measuring 13 feet by 13 feet. The customer would like the dining room to be a square, but with an area of 250 square feet. How much will this add to the dimensions of the room?

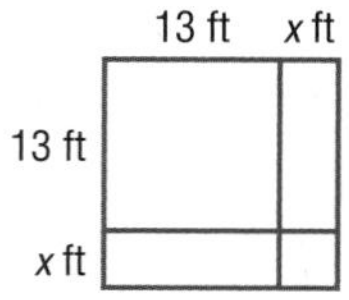

Solve each equation. Round to the nearest hundredth if necessary.

49. $4x^2 - 28x + 49 = 5$

50. $9x^2 + 30x + 25 = 11$

51. $x^2 + x + \frac{1}{3} = \frac{2}{3}$

52. $x^2 + 1.2x + 0.56 = 0.91$

Real-World Link

Each year, 250 to 300 million pounds of fireworks are used in the U.S.

Source: American Pyrotechnics Association

53. **FIREWORKS** A firework's distance d meters from the ground is given by $d = -1.5t^2 + 25t$, where t is the number of seconds after the firework has been lit.

a. How many seconds have passed since the firework was lit when the firework explodes if it explodes at the maximum height of its path?

b. What is the height of the firework when it explodes?

Find the value of c that makes each trinomial a perfect square. Then write the trinomial as a perfect square.

54. $x^2 + 0.7x + c$ **55.** $x^2 - 3.2x + c$ **56.** $x^2 - 1.8x + c$

57. **MULTIPLE REPRESENTATIONS** In this problem, you will use quadratics to investigate golden rectangles and the golden ratio.

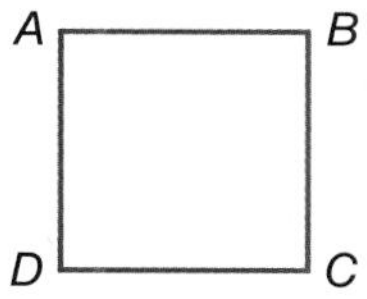

a. GEOMETRIC
- Draw square $ABCD$.
- Locate the midpoint of $\overline{CD}$. Label the midpoint P.
- Draw $\overline{PB}$.
- Construct an arc with a radius of $\overline{PB}$ from B clockwise past the bottom of the square.
- Extend $\overline{CD}$ until it intersects the arc. Label this point Q.
- Construct rectangle $ARQD$.

b. ALGEBRAIC Let $AD = x$ and $CQ = 1$. Use completing the square to solve $\frac{DQ}{AD} = \frac{QR}{CQ}$ for x.

c. TABULAR Make a table of x and values for $CQ = 2, 3,$ and 4.

d. VERBAL What do you notice about the x-values? Write an equation you could use to determine x for $CQ = n$, where n is a nonzero real number.

H.O.T. Problems

Use Higher-Order Thinking Skills

58. **FIND THE ERROR** Alonso and Aida are solving $x^2 + 8x - 20 = 0$ by completing the square. Is either of them correct? Explain your reasoning.

Alonso	Aida
$x^2 + 8x - 20 = 0$	$x^2 + 8x - 20 = 0$
$x^2 + 8x = 20$	$x^2 + 8x = 20$
$x^2 + 8x + 16 = 20 + 16$	$x^2 + 8x + 16 = 20$
$(x + 4)^2 = 36$	$(x + 4)^2 = 20$
$x + 4 = \pm 6$	$x + 4 = \pm\sqrt{20}$
$x = -4 \pm 6$	$x = -4 \pm \sqrt{20}$

59. **CHALLENGE** Solve $x^2 + bx + c = 0$ by completing the square. Your answer will be an expression for x in terms of b and c.

60. **REASONING** Explain why certain quadratic equations such as $x^2 + 22x + 121 = 246$ cannot be solved by factoring.

61. **OPEN ENDED** Write a perfect square trinomial equation in which the linear coefficient is negative and the constant term is a fraction. Then solve the equation.

62. **WRITING IN MATH** Explain what it means to complete the square. Include a description of the steps you would take.

NYSRE PRACTICE A2.N.7

63. SAT/ACT If $x^2 + y^2 = 2xy$, then y must equal

A -1 B 1 C $-x$ D x

64. GEOMETRY Find the area of the shaded region.

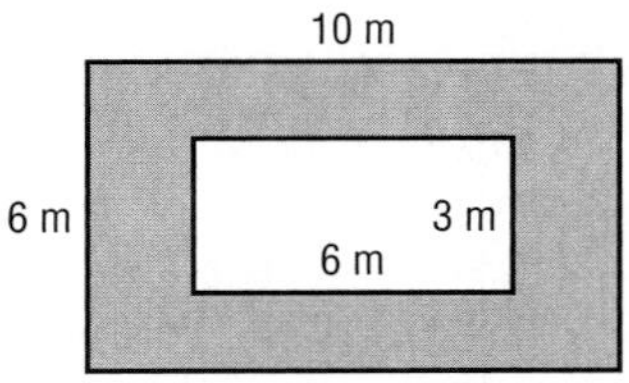

F 14 m^2 G 18 m^2 H 42 m^2 J 60 m^2

65. SHORT RESPONSE Ten consecutive integers are arranged in order from least to greatest. If the sum of the first five integers is 100, what is the sum of the last five integers?

66. If $5 - 3i$ is a solution for $x^2 + ax + b = 0$, where a and b are real numbers, what is the value of b?

A 10 C 34

B 14 D 40

Spiral Review

Simplify. (Lesson 5-4)

67. $(8 + 5i)^2$ **68.** $4(3 - i) + 6(2 - 5i)$ **69.** $\frac{5 - 2i}{6 + 9i}$

Write a quadratic equation in standard form with the given root(s). (Lesson 5-3)

70. $\frac{4}{5}, \frac{3}{4}$ **71.** $-\frac{2}{5}, 6$ **72.** $-\frac{1}{4}, -\frac{6}{7}$

73. MOVIES Refer to the table. (Lesson 4-1)

a. Write a matrix for the prices of movie tickets for adults, children, and seniors.

b. What are the dimensions of the matrix?

74. TRAVEL Yoko is going with the Spanish Club to Costa Rica. She buys 10 traveler's checks in denominations of \$20, \$50, and \$100, totaling \$370. She has twice as many \$20 checks as \$50 checks. How many of each denomination of traveler's checks does she have? (Lesson 3-5)

Graph each inequality. (Lesson 2-8)

75. $y \geq 4x - 3$ **76.** $2x - 3y < 6$ **77.** $5x + 2y + 3 \leq 0$

Write the piecewise function shown in each graph. (Lesson 2-6)

78.

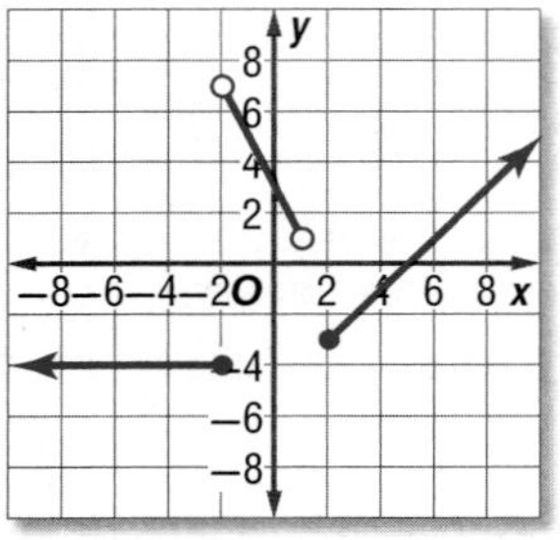

79.

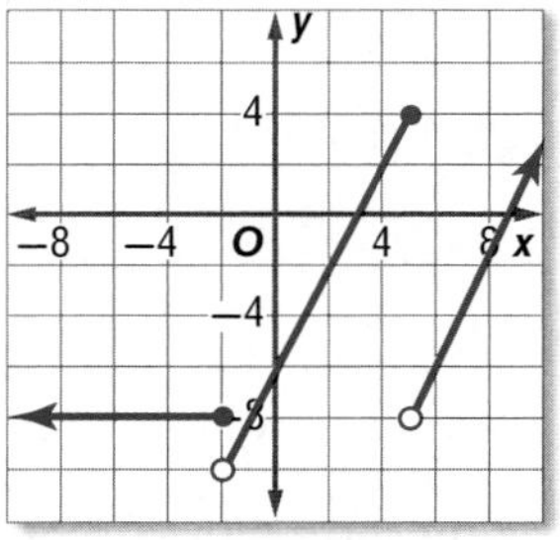

80.

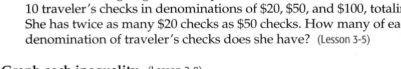

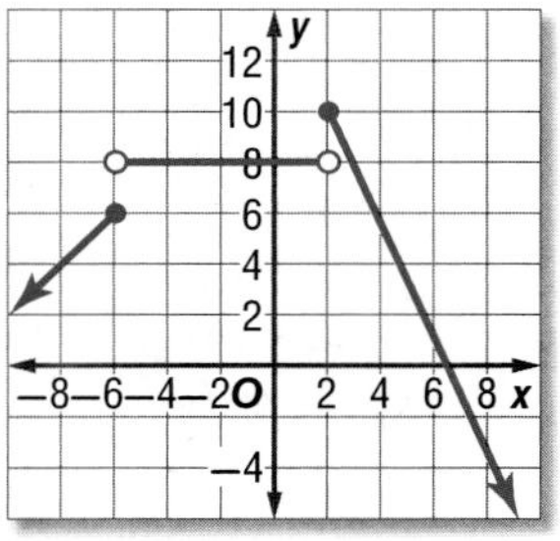

Skills Review

Evaluate $b^2 - 4ac$ for the given values of a, b, and c. (Lesson 1-2)

81. $a = 5, b = 6, c = 2$ **82.** $a = -2, b = -7, c = 3$ **83.** $a = -5, b = -8, c = -10$

Graphing Technology Lab
Solving Quadratic Equations

You can use a TI-*n*spire CAS to solve quadratic equations.

ACTIVITY Finding Roots

Solve each equation.

a. $3x^2 - 4x + 1 = 0$

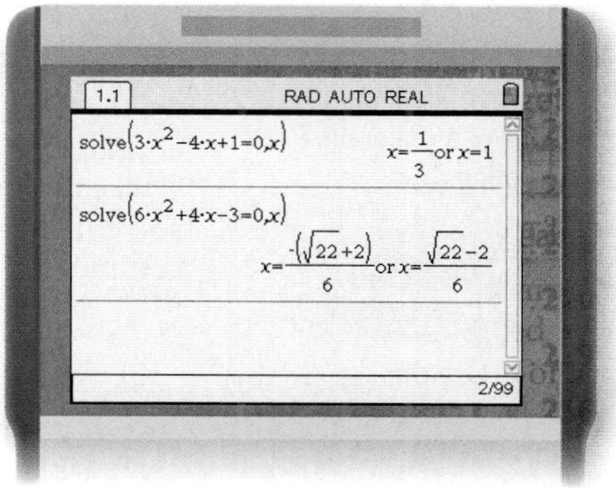

Step 1 From the Home screen, select New Document. Then select Add Calculator.

Step 2 Under menu, select Algebra, then select Solve.

Step 3 Enter the equation.

KEYSTROKES: 3 [X] [x²] [−] 4 [X] [+] 1 [=] 0 [,] [X] [enter]

The solutions are $x = \frac{1}{3}$ or $x = 1$.

b. $6x^2 + 4x - 3 = 0$

Step 1 Under menu, select Algebra, then select Solve.

Step 2 Enter the equation.

KEYSTROKES: 6 [X] [x²] [+] 4 [X] [−] 3 [=] 0 [,] [X] [enter]

The solutions are $x = \frac{-2 \pm \sqrt{22}}{6}$.

c. $x^2 - 6x + 10 = 0.$

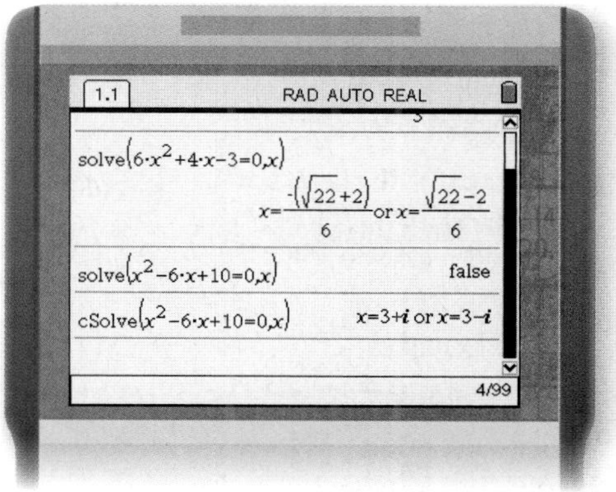

Step 1 Under menu, select Algebra, then select Solve.

Step 2 Enter the equation.

KEYSTROKES: [X] [x²] [−] 6 [X] [+] 10 [=] 0 [,] [X] [enter]

The calculator returns a value of *false*, meaning that there are no real solutions.

Step 3 Under menu, select Algebra, Complex, then Solve. Reenter the equation.

The solutions are $x = 3 \pm i$.

Exercises

Solve each equation.

1. $x^2 - 2x - 24 = 0$

2. $-x^2 + 4x - 1 = 0$

3. $0 = -3x^2 - 6x + 9$

4. $x^2 - 2x + 5 = 0$

5. $0 = 4x^2 - 8$

6. $0 = 2x^2 - 4x + 1$

7. $x^2 + 3x + 8 = 5$

8. $25 + 4x^2 = -20x$

9. $x^2 - x = -6$

5-6 The Quadratic Formula and the Discriminant

Then

You solved equations by completing the square. (Lesson 5-5)

Now

- Solve quadratic equations by using the Quadratic Formula.
- Use the discriminant to determine the number and type of roots of a quadratic equation.

NYS Core Curriculum

A2.A.2 Use the discriminant to determine the nature of the roots of a quadratic equation
A2.A.25 Solve quadratic equations, using the quadratic formula

New Vocabulary

Quadratic Formula
discriminant

NY Math Online

glencoe.com

- Extra Examples
- Personal Tutor
- Self-Check Quiz
- Homework Help

Why?

Pumpkin catapult is an event in which a contestant builds a catapult and launches a pumpkin at a target.

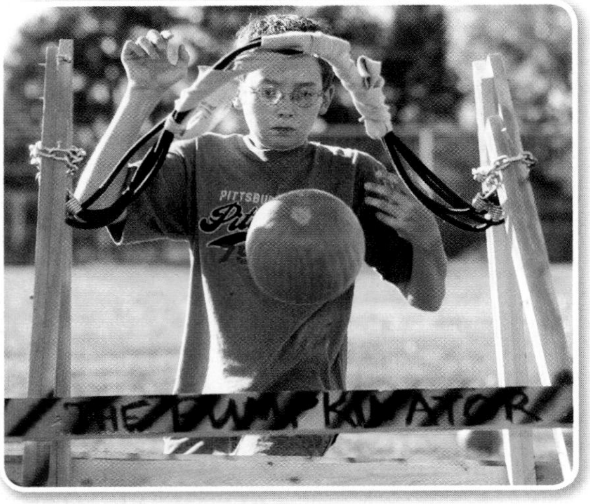

The path of the pumpkin can be modeled by the quadratic function $h = -4.9t^2 + 117t + 42$, where h is the height of the pumpkin and t is the number of seconds.

To predict when the pumpkin will hit the target, you can solve the equation $0 = -4.9t^2 + 117t + 42$. This equation would be difficult to solve using factoring, graphing, or completing the square.

Quadratic Formula You have found solutions of some quadratic equations by graphing, by factoring, and by using the Square Root Property. There is also a formula that can be used to solve any quadratic equation. This formula can be derived by solving the standard form of a quadratic equation.

General Case		Specific Case
$ax^2 + bx + c = 0$	**Standard quadratic equation**	$2x^2 + 8x + 1 = 0$
$x^2 + \frac{b}{a}x + \frac{c}{a} = 0$	**Divide each side by *a*.**	$x^2 + 4x + \frac{1}{2} = 0$
$x^2 + \frac{b}{a}x = -\frac{c}{a}$	**Subtract $\frac{c}{a}$ from each side.**	$x^2 + 4x = -\frac{1}{2}$
$x^2 + \frac{b}{a}x + \frac{b^2}{4a^2} = -\frac{c}{a} + \frac{b^2}{4a^2}$	**Complete the square.**	$x^2 + 4x + \left(\frac{4}{2}\right)^2 = -\frac{1}{2} + \left(\frac{4}{2}\right)^2$
$\left(x + \frac{b}{2a}\right)^2 = -\frac{c}{a} + \frac{b^2}{4a^2}$	**Factor the left side.**	$(x + 2)^2 = -\frac{1}{2} + \left(\frac{4}{2}\right)^2$
$\left(x + \frac{b}{2a}\right)^2 = \frac{b^2 - 4ac}{4a^2}$	**Simplify the right side.**	$(x + 2)^2 = \frac{7}{2}$
$x + \frac{b}{2a} = \pm\frac{\sqrt{b^2 - 4ac}}{2a}$	**Square Root Property**	$x + 2 = \pm\sqrt{\frac{7}{2}}$
$x = -\frac{b}{2a} \pm \frac{\sqrt{b^2 - 4ac}}{2a}$	**Subtract $\frac{b}{2a}$ from each side.**	$x = -2 \pm \sqrt{\frac{7}{2}}$
$x = \frac{-b \pm \sqrt{b^2 - 4ac}}{2a}$	**Simplify.**	$x = \frac{-4 \pm \sqrt{14}}{2}$

The equation $x = \frac{-b \pm \sqrt{b^2 - 4ac}}{2a}$ is known as the **Quadratic Formula**.

StudyTip

Quadratic Formula Although factoring may be an easier method to solve some of the equations, the Quadratic Formula can be used to solve any quadratic equation.

Key Concept — Quadratic Formula

For Your FOLDABLE

Words The solutions of a quadratic equation of the form $ax^2 + bx + c = 0$, where $a \neq 0$, are given by the following formula.

$$x = \frac{-b \pm \sqrt{b^2 - 4ac}}{2a}$$

Example $x^2 + 5x + 6 = 0 \rightarrow x = \dfrac{-5 \pm \sqrt{5^2 - 4(1)(6)}}{2(1)}$

EXAMPLE 1 Two Rational Roots

Solve $x^2 - 10x = 11$ by using the Quadratic Formula.

First, write the equation in the form $ax^2 + bx + c = 0$ and identify a, b, and c.

$$\begin{array}{ccc} & & ax^2 + bx + c = 0 \\ & & \downarrow \quad \downarrow \quad \downarrow \\ x^2 - 10x = 11 & \rightarrow & 1x^2 - 10x - 11 = 0 \end{array}$$

Then, substitute these values into the Quadratic Formula.

$x = \dfrac{-b \pm \sqrt{b^2 - 4ac}}{2a}$ **Quadratic Formula**

$= \dfrac{-(-10) \pm \sqrt{(-10)^2 - 4(1)(-11)}}{2(1)}$ **Replace *a* with 1, *b* with −10, and *c* with −11.**

$= \dfrac{10 \pm \sqrt{100 + 44}}{2}$ **Multiply.**

$= \dfrac{10 \pm \sqrt{144}}{2}$ **Simplify.**

$= \dfrac{10 \pm 12}{2}$ **$\sqrt{144} = 12$**

$x = \dfrac{10 + 12}{2}$ or $x = \dfrac{10 - 12}{2}$ **Write as two equations.**

$= 11$ $= -1$ **Simplify.**

The solutions are −1 and 11.

CHECK Substitute both values into the original equation.

$$\begin{aligned} x^2 - 10x &= 11 \\ (-1)^2 - 10(-1) &\stackrel{?}{=} 11 \\ 1 + 10 &\stackrel{?}{=} 11 \\ 11 &= 11 \checkmark \end{aligned} \qquad \begin{aligned} x^2 - 10x &= 11 \\ (11)^2 - 10(11) &\stackrel{?}{=} 11 \\ 121 - 110 &\stackrel{?}{=} 11 \\ 11 &= 11 \checkmark \end{aligned}$$

Review Vocabulary

radicand the value underneath the radical symbol (Concepts and Skills Bank, Lesson 2)

Check Your Progress

Solve each equation by using the Quadratic Formula.

1A. $x^2 + 6x = 16$

1B. $2x^2 + 25x + 33 = 0$

Personal Tutor glencoe.com

When the value of the radicand in the Quadratic Formula is 0, the quadratic equation has exactly one rational root.

Math History Link

Brahmagupta (598–668) Indian mathematician Brahmagupta offered the first general solution of the quadratic equation $ax^2 + bx = c$, now know as the Quadratic Formula.

EXAMPLE 2 One Rational Root

Solve $x^2 + 8x + 16 = 0$ by using the Quadratic Formula.

Identify a, b, and c. Then, substitute these values into the Quadratic Formula.

$$x = \frac{-b \pm \sqrt{b^2 - 4ac}}{2a}$$ **Quadratic Formula**

$$= \frac{-(8) \pm \sqrt{(8)^2 - 4(1)(16)}}{2(1)}$$ **Replace a with 1, b with 8, and c with 16.**

$$= \frac{-8 \pm \sqrt{0}}{2}$$ **Simplify.**

$$= \frac{-8}{2} \text{ or } -4$$ **$\sqrt{0} = 0$**

The solution is -4.

CHECK A graph of the related function shows that there is one solution at $x = -4$.

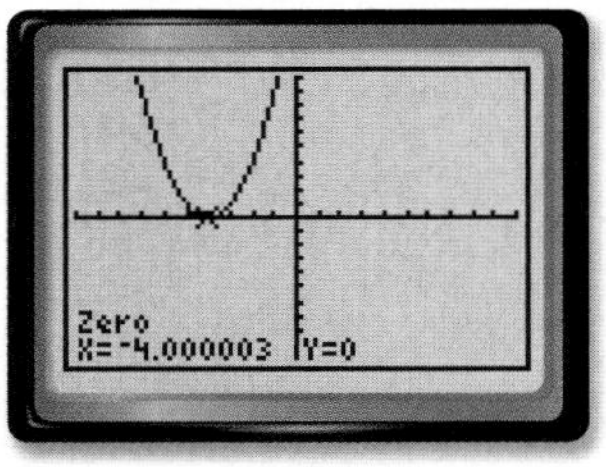

Check Your Progress

Solve each equation by using the Quadratic Formula.

2A. $x^2 - 16x + 64 = 0$

2B. $x^2 + 34x + 289 = 0$

Personal Tutor glencoe.com

You can express irrational roots exactly by writing them in radical form.

EXAMPLE 3 Irrational Roots

Solve $2x^2 + 6x - 7 = 0$ by using the Quadratic Formula.

$$x = \frac{-b \pm \sqrt{b^2 - 4ac}}{2a}$$ **Quadratic Formula**

$$= \frac{-(6) \pm \sqrt{(6)^2 - 4(2)(-7)}}{2(2)}$$ **Replace a with 2, b with 6, and c with -7.**

$$= \frac{-6 \pm \sqrt{92}}{4}$$ **Simplify.**

$$= \frac{-6 \pm 2\sqrt{23}}{4} \text{ or } \frac{-3 \pm \sqrt{23}}{2}$$ **$\sqrt{92} = \sqrt{4 \cdot 23}$ or $2\sqrt{23}$**

The approximate solutions are -3.9 and 0.9.

CHECK Check these results by graphing the related quadratic function, $y = 2x^2 + 6x - 7$. Using the ZERO function of a graphing calculator, the approximate zeros of the related function are -3.9 and 0.9.

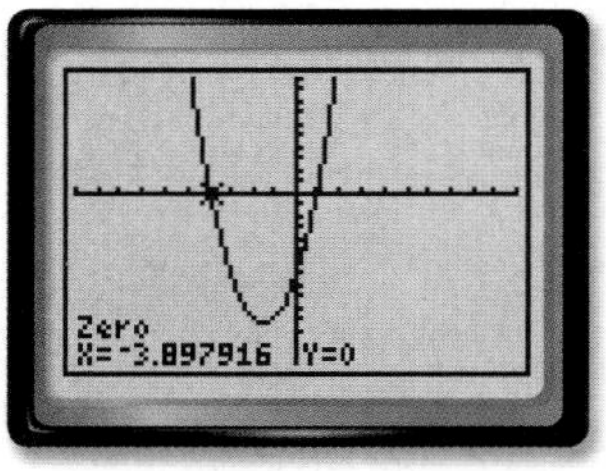

Check Your Progress

Solve each equation by using the Quadratic Formula.

3A. $3x^2 + 5x + 1 = 0$

3B. $x^2 - 8x + 9 = 0$

Personal Tutor glencoe.com

When using the Quadratic Formula, if the value of the radicand is negative, the solutions will be complex. Complex solutions always appear in conjugate pairs.

StudyTip

Complex Numbers Remember to write your solutions in the form $a + bi$, sometimes called the **standard form** of a complex number.

EXAMPLE 4 Complex Roots

Solve $x^2 - 6x = -10$ by using the Quadratic Formula.

$x = \dfrac{-b \pm \sqrt{b^2 - 4ac}}{2a}$ — **Quadratic Formula**

$= \dfrac{-(-6) \pm \sqrt{(-6)^2 - 4(1)(10)}}{2(1)}$ — **Replace a with 1, b with −6, and c with 10.**

$= \dfrac{6 \pm \sqrt{-4}}{2}$ — **Simplify.**

$= \dfrac{6 \pm 2i}{2}$ — **$\sqrt{-4} = \sqrt{4 \cdot (-1)}$ or $2i$**

$= 3 \pm i$ — **Simplify.**

The solutions are the complex numbers $3 + i$ and $3 - i$.

CHECK A graph of the related function shows that the solutions are complex, but it cannot help you find them. To check complex solutions, substitute them into the original equation.

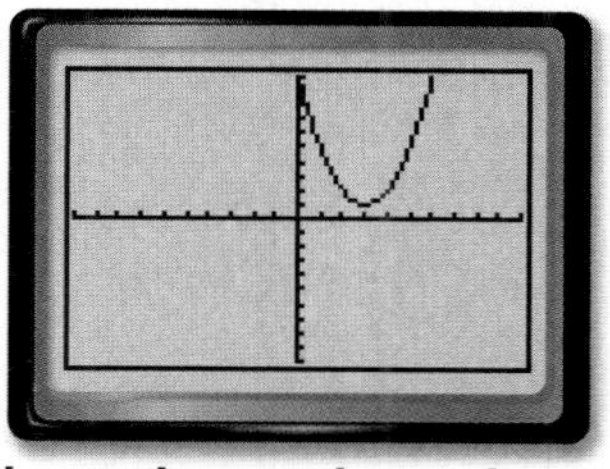

[−10, 10] scl: 1 by [−10, 10] scl: 1

$x^2 - 6x = -10$	**Original equation**
$(3 + i)^2 - 6(3 + i) \stackrel{?}{=} -10$	**$x = 3 + i$**
$9 + 6i + i^2 - 18 - 6i \stackrel{?}{=} -10$	**Square of a sum; Distributive Property**
$-9 + i^2 \stackrel{?}{=} -10$	**Simplify.**
$-9 - 1 = -10$ ✓	**$i^2 = -1$**
$x^2 - 6x = -10$	**Original equation**
$(3 - i)^2 - 6(3 - i) \stackrel{?}{=} -10$	**$x = 3 - i$**
$9 - 6i + i^2 - 18 + 6i \stackrel{?}{=} -10$	**Square of a sum; Distributive Property**
$-9 + i^2 \stackrel{?}{=} -10$	**Simplify.**
$-9 - 1 = -10$ ✓	**$i^2 = -1$**

Check Your Progress

Solve each equation by using the Quadratic Formula.

4A. $3x^2 + 5x + 4 = 0$

4B. $x^2 - 4x = -13$

Personal Tutor glencoe.com

Roots and the Discriminant In the previous examples, observe the relationship between the value of the expression under the radical and the roots of the quadratic equation. The expression $b^2 - 4ac$ is called the **discriminant**.

$$x = \frac{-b \pm \sqrt{b^2 - 4ac}}{2a} \quad \leftarrow \textbf{discriminant}$$

The value of the discriminant can be used to determine the number and type of roots of a quadratic equation. The table on the following page summarizes the possible types of roots.

The discriminant can also be used to confirm the number and type of solutions after you solve the quadratic equation.

StudyTip

Roots Remember that the solutions of an equation are called *roots* or *zeros* and are the value(s) where the graph crosses the x-axis.

Key Concept Discriminant

For Your FOLDABLE

Consider $ax^2 + bx + c = 0$, where a, b, and c are real numbers and $a \neq 0$.

Value of Discriminant	Type and Number of Roots	Example of Graph of Related Function
$b^2 - 4ac > 0$; $b^2 - 4ac$ is a perfect square.	2 real, rational roots	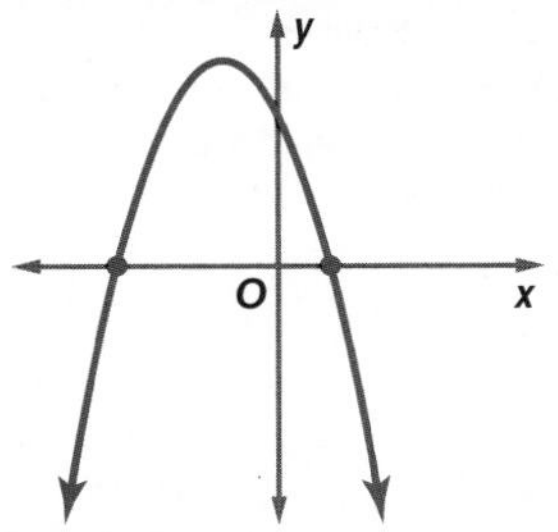
$b^2 - 4ac > 0$; $b^2 - 4ac$ is *not* a perfect square.	2 real, irrational roots	
$b^2 - 4ac = 0$	1 real root	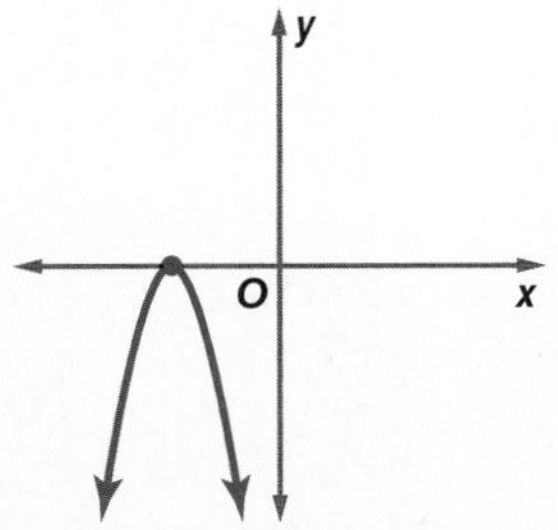
$b^2 - 4ac < 0$	2 complex roots	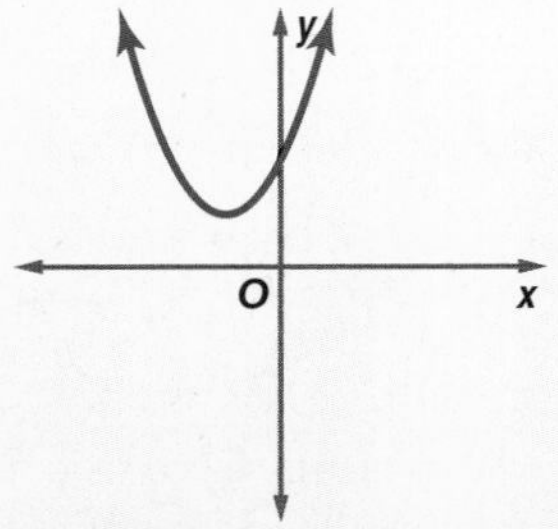

EXAMPLE 5 Describe Roots

Find the value of the discriminant for each quadratic equation. Then describe the number and type of roots for the equation.

a. $7x^2 - 11x + 5 = 0$

$a = 7, b = -11, c = 5$

$$\begin{aligned} b^2 - 4ac &= (-11)^2 - 4(7)(5) \\ &= 121 - 140 \\ &= -19 \end{aligned}$$

The discriminant is negative, so there are two complex roots.

b. $x^2 + 22x + 121 = 0$

$a = 1, b = 22, c = 121$

$$\begin{aligned} b^2 - 4ac &= (22)^2 - 4(1)(121) \\ &= 484 - 484 \\ &= 0 \end{aligned}$$

The discriminant is 0, so there is one rational root.

Check Your Progress

5A. $-5x^2 + 8x - 1 = 0$

5B. $-7x + 15x^2 - 4 = 0$

Personal Tutor glencoe.com

StudyTip

Study Notebook You may wish to copy this list of methods to your math notebook or Foldable to keep as a reference as you study.

You have studied a variety of methods for solving quadratic equations. The table below summarizes these methods.

Concept Summary **Solving Quadratic Equations** For Your FOLDABLE

Method	Can be Used	When to Use
graphing	sometimes	Use only if an exact answer is not required. Best used to check the reasonableness of solutions found algebraically.
factoring	sometimes	Use if the constant term is 0 or if the factors are easily determined. **Example** $x^2 - 7x = 0$
Square Root Property	sometimes	Use for equations in which a perfect square is equal to a constant. **Example** $(x - 5)^2 = 18$
completing the square	always	Useful for equations of the form $x^2 + bx + c = 0$, where b is even. **Example** $x^2 + 6x - 14 = 0$
Quadratic Formula	always	Useful when other methods fail or are too tedious. **Example** $2.3x^2 - 1.8x + 9.7 = 0$

Check Your Understanding

Examples 1–4 pp. 293–295

Solve each equation by using the Quadratic Formula.

1. $x^2 + 12x - 9 = 0$
2. $x^2 + 8x + 5 = 0$
3. $4x^2 - 5x - 2 = 0$
4. $9x^2 + 6x - 4 = 0$
5. $10x^2 - 3 = 13x$
6. $22x = 12x^2 + 6$
7. $-3x^2 + 4x = -8$
8. $x^2 + 3 = -6x + 8$

Examples 3 and 4 pp. 294–295

9. **AMUSEMENT PARK** An amusement park ride takes riders to the top of a tower and drops them at speeds reaching 80 feet per second. A function that models this ride is $h = -16t^2 + 64t - 60$, where h is the height in feet and t is the time in seconds. About how many seconds does it take for riders to drop from 60 feet to 0 feet?

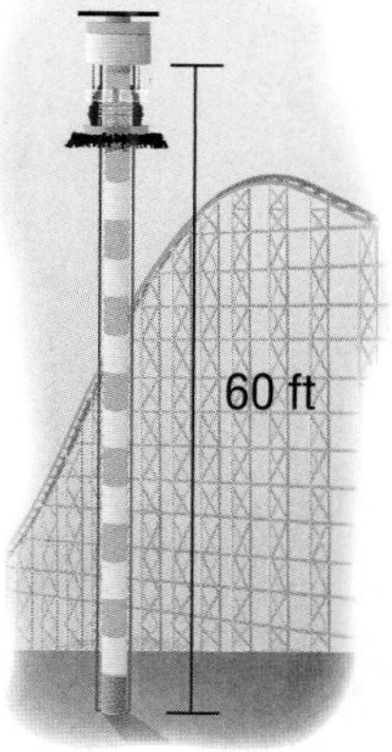

Example 5 p. 296

Complete parts a and b for each quadratic equation.
a. Find the value of the discriminant.
b. Describe the number and type of roots.

10. $3x^2 + 8x + 2 = 0$
11. $2x^2 - 6x + 9 = 0$
12. $-16x^2 + 8x - 1 = 0$
13. $5x^2 + 2x + 4 = 0$

Practice and Problem Solving

● = Step-by-Step Solutions begin on page R20.
Extra Practice begins on page 947.

Examples 1–4
pp. 293–295

Solve each equation by using the Quadratic Formula.

14. $x^2 + 45x = -200$

15. $4x^2 - 6 = -12x$

16. $3x^2 - 4x - 8 = -6$

17. $4x^2 - 9 = -7x - 4$

18. $5x^2 - 9 = 11x$

19. $12x^2 + 9x - 2 = -17$

20. DIVING Competitors in the 10-meter platform diving competition jump upward and outward before diving into the pool below. The height h of a diver in meters above the pool after t seconds can be approximated by the equation $h = -4.9t^2 + 3t + 10$.

a. Determine a domain and range for which this function makes sense.

b. When will the diver hit the water?

Example 5
p. 296

Complete parts a–c for each quadratic equation.
a. Find the value of the discriminant.
b. Describe the number and type of roots.
c. Find the exact solutions by using the Quadratic Formula.

21 $2x^2 + 3x - 3 = 0$

22. $4x^2 - 6x + 2 = 0$

23. $6x^2 + 5x - 1 = 0$

24. $6x^2 - x - 5 = 0$

25. $3x^2 - 3x + 8 = 0$

26. $2x^2 + 4x + 7 = 0$

27. $-5x^2 + 4x + 1 = 0$

28. $x^2 - 6x = -9$

29. $-3x^2 - 7x + 2 = 6$

30. $-8x^2 + 5 = -4x$

31. $x^2 + 2x - 4 = -9$

32. $-6x^2 + 5 = -4x + 8$

33. VIDEO GAMES While Darnell is grounded his friend Jack brings him a video game. Darnell stands at his bedroom window, and Jack stands directly below the window. If Jack tosses a game cartridge to Darnell with an initial velocity of 35 feet per second, an equation for the height h feet of the cartridge after t seconds is $h = -16t^2 + 35t + 5$.

a. If the window is 25 feet above the ground, will Darnell have 0, 1, or 2 chances to catch the video game cartridge?

b. If Darnell is unable to catch the video game cartridge, when will it hit the ground?

34. ENGINEERING Civil engineers are designing a section of road that is going to dip below sea level. The road's curve can be modeled by the equation $y = 0.00005x^2 - 0.06x$, where x is the horizontal distance in feet between the points where the road is at sea level and y is the elevation. The engineers want to put stop signs at the locations where the elevation of the road is equal to sea level. At what horizontal distances will they place the stop signs?

Complete parts a–c for each quadratic equation.
a. Find the value of the discriminant.
b. Describe the number and type of roots.
c. Find the exact solutions by using the Quadratic Formula.

35. $5x^2 + 8x = 0$

36. $8x^2 = -2x + 1$

37. $4x - 3 = -12x^2$

38. $0.8x^2 + 2.6x = -3.2$

39. $0.6x^2 + 1.4x = 4.8$

40. $-4x^2 + 12 = -6x - 8$

41. **SMOKING** A decrease in smoking in the United States has resulted in lower death rates caused by lung cancer. The number of deaths per 100,000 people y can be approximated by $y = -0.26x^2 - 0.55x + 91.81$, where x represents the number of years after 2000.

Year	Deaths per 100,000
2000	91.9
2002	89.4
2004	85.3
2010	?
2015	?

a. Calculate the number of deaths per 100,000 people for 2010 and 2015.

b. Use the Quadratic Formula to solve for x when $y = 50$.

c. According to the quadratic function, when will the death rate be 0 per 100,000? Do you think that this prediction is reasonable? Why or why not?

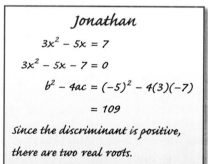

Real-World Link

In 1965, 42% of American adults smoked, compared with fewer than 22% in 2003.

Source American Lung Association

42. **NUMBER THEORY** The sum S of consecutive integers 1, 2, 3, …, n is given by the formula $S = \frac{1}{2}n(n + 1)$. How many consecutive integers, starting with 1, must be added to get a sum of 666?

H.O.T. Problems

Use **H**igher-**O**rder **T**hinking Skills

43. **FIND THE ERROR** Tama and Jonathan are determining the number of solutions of $3x^2 - 5x = 7$. Is either of them correct? Explain your reasoning.

Tama

$3x^2 - 5x = 7$

$b^2 - 4ac = (-5)^2 - 4(3)(7)$

$= -59$

Since the discriminant is negative, there are no real solutions.

Jonathan

$3x^2 - 5x = 7$

$3x^2 - 5x - 7 = 0$

$b^2 - 4ac = (-5)^2 - 4(3)(-7)$

$= 109$

Since the discriminant is positive, there are two real roots.

44. **CHALLENGE** Find the solutions of $4ix^2 - 4ix + 5i = 0$ by using the Quadratic Formula.

45. **REASONING** Determine whether each statement is *sometimes*, *always*, or *never* true. Explain your reasoning.

a. In a quadratic equation in standard form, if a and c are different signs, then the solutions will be real.

b. If the discriminant of a quadratic equation is greater than 1, the two roots are real irrational numbers.

46. **OPEN ENDED** Sketch the corresponding graph and state the number and type of roots for each of the following.

a. $b^2 - 4ac = 0$

b. A quadratic function in which $f(x)$ never equals zero.

c. A quadratic function in which $f(a) = 0$ and $f(b) = 0$; $a \neq b$.

d. The discriminant is less than zero.

e. a and b are both solutions and can be represented as fractions.

47. **CHALLENGE** Find the value(s) of m in the quadratic equation $x^2 + x + m + 1 = 0$ such that it has one solution.

48. **WRITING IN MATH** Describe three different ways to solve $x^2 - 2x - 15 = 0$. Which method do you prefer, and why?

NYSRE PRACTICE A2.A.25

49. A company determined that its monthly profit P is given by $P = -8x^2 + 165x - 100$, where x is the selling price for each unit of product. Which of the following is the best estimate of the maximum price per unit that the company can charge without losing money?

A \$10 **B** \$20 **C** \$30 **D** \$40

50. **SAT/ACT** For which of the following sets of numbers is the mean greater than the median?

F {4, 5, 6, 7, 8}
G {4, 6, 6, 6, 8}
H {4, 5, 6, 7, 9}
J {3, 5, 6, 7, 8}

51. **SHORT RESPONSE** In the figure below, P is the center of the circle with radius 15 inches. What is the area of $\triangle APB$?

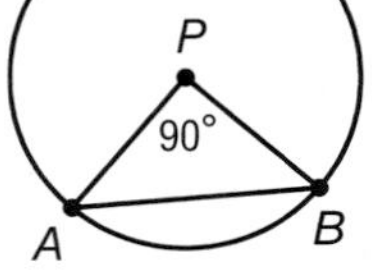

52. 75% of 88 is the same as 60% of what number?

A 100 **B** 101 **C** 108 **D** 110

Spiral Review

Find the value of c that makes each trinomial a perfect square. Then write the trinomial as a perfect square. (Lesson 5-5)

53. $x^2 + 13x + c$

54. $x^2 + 2.4x + c$

55. $x^2 + \frac{4}{5}x + c$

Simplify. (Lesson 5-4)

56. i^{26}

57. $\sqrt{-16}$

58. $4\sqrt{-9} \cdot 2\sqrt{-25}$

59. **PILOT TRAINING** Evita is training for her pilot's license. Flight instruction costs \$105 per hour, and the simulator costs \$45 per hour. She spent 4 more hours in airplane training than in the simulator. If Evita spent \$3870, how much time did she spend training in an airplane and in a simulator? (Lesson 4-6)

60. **BUSINESS** Ms. Larson owns three fruit farms on which she grows apples, peaches, and apricots. She sells apples for \$22 a case, peaches for \$25 a case, and apricots for \$18 a case. (Lesson 4-3)

a. Write an inventory matrix for the number of cases for each type of fruit for each farm and a cost matrix for the price per case for each type of fruit.

b. Find the total income of the three fruit farms expressed as a matrix.

c. What is the total income from all three fruit farms?

Number of Cases in Stock of Each Type of Fruit

Fruit	Farm 1	Farm 2	Farm 3
apples	290	175	110
peaches	165	240	75
apricots	210	190	0

Skills Review

Write an equation for each graph. (Lesson 2-7)

61.

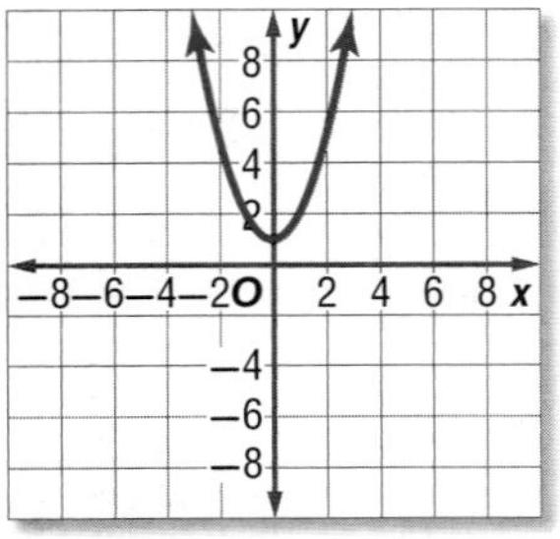

62.

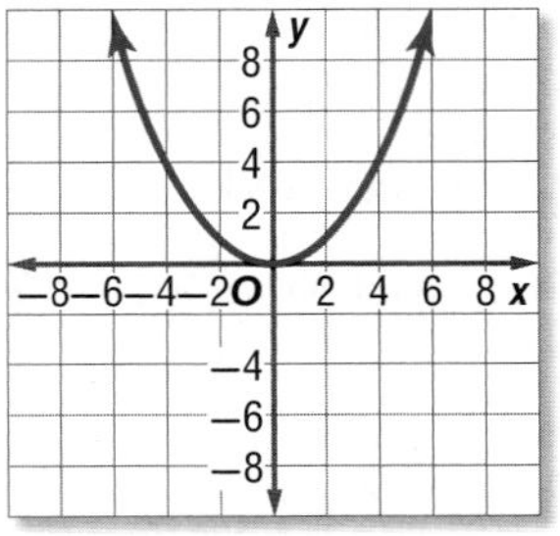

63.

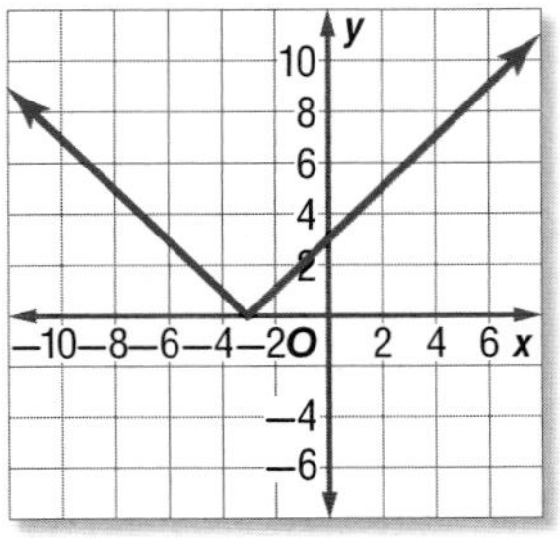

EXTEND 5-6

Algebra Lab
Sums and Products of Roots

NY Math Online glencoe.com
Math *in Motion*, Animation

Objective
Use sums and products of roots to write quadratic equations.

NYS Core Curriculum

A2.A.20 Determine the sum and product of the roots of a quadratic equation by examining its coefficients
A2.A.21 Determine the quadratic equation, given the sum and product of its roots

If you know the roots of a quadratic equation, you can use the sum and product of the roots to determine the equation.

Write a quadratic equation that has roots 3 and −8.

When you factored quadratic equations, you used the Zero Product Property to determine the roots. You can use the roots and work backward to find the equation.

$x = 3$ or $x = -8$	**Start with the solutions.**
$x - 3 = 0$ $\quad x + 8 = 0$	**Rewrite equations equal to 0.**
$(x - 3)(x + 8) = 0$	**Multiplicative Property of Zero**
$x^2 + 5x - 24 = 0$	**Multiply.**

The equation $x^2 + 5x - 24 = 0$ has roots 3 and −8 and is written in standard form. The sum and product of the roots can also be used to determine the equation.

Add the roots. $\quad 3 + (-8) = -5 \quad$ **−5 is the opposite of the coefficient of *x*.**

$$x^2 + 5x - 24 = 0$$

Multiply the roots. $\quad 3(-8) = -24 \quad$ **−24 is the constant term.**

The pattern above can be generalized for any quadratic equation by using the Quadratic Formula. Let r_1 and r_2 represent the roots of a quadratic equation.

$$r_1 = \frac{-b + \sqrt{b^2 - 4ac}}{2a} \qquad r_2 = \frac{-b - \sqrt{b^2 - 4ac}}{2a}$$

Add the roots.

$$r_1 + r_2 = \frac{-b + \sqrt{b^2 - 4ac}}{2a} + \frac{-b - \sqrt{b^2 - 4ac}}{2a} \qquad \textbf{Add the roots.}$$

$$= \frac{-2b + 0}{2a} \text{ or } -\frac{b}{a} \qquad \textbf{Simplify.}$$

The sum of the roots is $-\frac{b}{a}$.

Multiply the roots.

$$r_1 \cdot r_2 = \frac{-b + \sqrt{b^2 - 4ac}}{2a} \cdot \frac{-b - \sqrt{b^2 - 4ac}}{2a} \qquad \textbf{Multiply the roots.}$$

$$= \frac{b^2 - (b^2 - 4ac)}{4a^2} \qquad \textbf{Multiply.}$$

$$= \frac{b^2 - b^2 + 4ac)}{4a^2} \qquad \textbf{Distributive Property}$$

$$= \frac{4ac}{4a^2} \text{ or } \frac{c}{a} \qquad \textbf{Simplify.}$$

The product of the roots is $\frac{c}{a}$.

The following rule can be used to write a quadratic equation when you know the roots.

Key Concept **Sum and Product of Roots** For Your FOLDABLE

If the roots of $ax^2 + bx + c = 0$, with $a \neq 0$, are r_1 and r_2, then

$$r_1 + r_2 = -\frac{b}{a} \text{ and } r_1 \cdot r_2 = \frac{c}{a}.$$

ACTIVITY 1 Use the Sum and Product of Roots

Write a quadratic equation that has roots 2 and −7.

Step 1 Find the sum of the roots.

$$r_1 + r_2 = 2 + (-7)$$
$$= -5$$

Step 2 Find the product of the roots.

$$r_1 \cdot r_2 = 2 \cdot (-7)$$
$$= -14$$

Step 3 Write the equation.

$-5 = \frac{b}{a}$ and $-14 = \frac{c}{a}$. So, $a = 1$, $b = -5$, and $c = -14$. Thus, the equation is $x^2 - 5x - 14 = 0$.

ACTIVITY 2 Use the Sum and Product of Roots

Write a quadratic equation that has roots $\frac{3}{4}$ and $-\frac{12}{5}$.

Step 1 Find the sum of the roots.

$$r_1 + r_2 = \frac{3}{4} + \left(-\frac{12}{5}\right)$$
$$= \frac{15}{20} - \frac{48}{20} \text{ or } -\frac{33}{20}$$

Step 2 Find the product of the roots.

$$r_1 \cdot r_2 = \frac{3}{4} \cdot \left(-\frac{12}{5}\right)$$
$$= -\frac{36}{20}$$

Step 3 Write the equation.

$-\frac{33}{20} = -\frac{b}{a}$ and $-\frac{36}{20} = \frac{c}{a}$

So, $a = 20$, $b = 33$, and $c = -36$. Thus, the equation is $20x^2 + 33x - 36 = 0$.

Exercises

Write a quadratic equation that has the given roots.

1. $-\frac{3}{4}, \frac{5}{8}$

2. $-7, \frac{2}{3}$

3. $\pm\frac{2}{5}$

4. $4 \pm \sqrt{3}$

5. $1 \pm \sqrt{6}$

6. $\frac{-2 \pm 3\sqrt{5}}{7}$

7. $7 \pm 3i$

8. $\sqrt{5} \pm 8i$

Write a quadratic equation with roots that satisfy the following conditions.

9. The sum of the roots is 4. The product of the roots is $\frac{13}{12}$.

10. The sum of the roots is $\frac{1}{6}$. The product of the roots is $\frac{5}{21}$.

Graphing Technology Lab
Families of Parabolas

NY Math Online glencoe.com
Other Calculator Keystrokes

Objective
Use a graphing calculator to investigate changes to parabolas.

NYS Core Curriculum

Reinforcement of A.G.5 Investigate and generalize how changing the coefficients of a function affects its graph

The general form of a quadratic function is $y = a(x - h)^2 + k$. Changing the values of a, h, and k results in a different parabola in the family of quadratic functions. You can use a TI-83/84 Plus graphing calculator to analyze the effects that result from changing each of these parameters.

ACTIVITY 1

Graph each set of functions on the same screen in the standard viewing window. Describe any similarities and differences among the graphs.

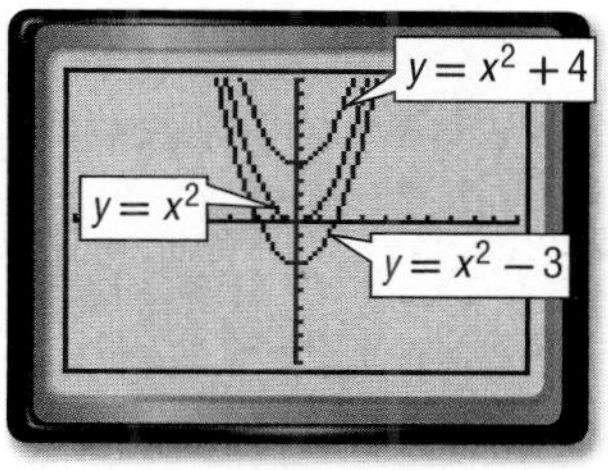

$y = x^2$, $y = x^2 + 4$, $y = x^2 - 3$

The graphs have the same shape, and all open up. The vertex of each graph is on the y-axis. However, the graphs have different vertical positions.

Example 1 shows how changing the value of k in the function $y = a(x - h)^2 + k$ *translates* the parabola along the y-axis. If $k > 0$, the parabola is translated k units up, and if $k < 0$, it is translated k units down.

How do you think changing the value of h will affect the graph of $y = x^2$?

ACTIVITY 2

Graph each set of functions on the same screen in the standard viewing window. Describe any similarities and differences among the graphs.

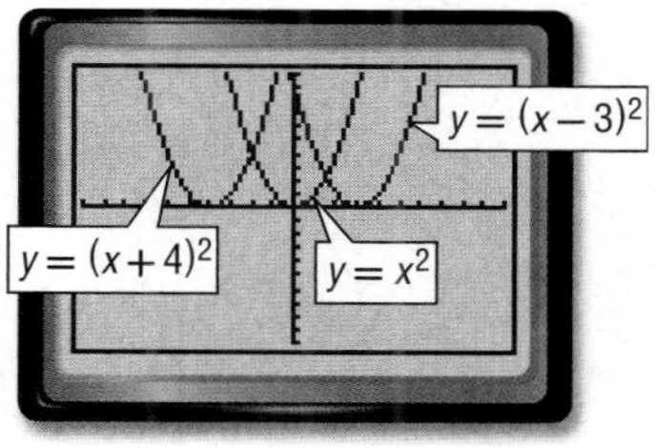

$y = x^2$, $y = (x + 4)^2$, $y = (x - 3)^2$

These three graphs all open up and have the same shape. The vertex of each graph is on the x-axis. However, the graphs have different horizontal positions.

Example 2 shows how changing the value of h in the equation $y = a(x - h)^2 + k$ *translates* the graph horizontally. If $h > 0$, the graph translates to the right h units. If $h < 0$, the graph translates to the left h units.

ACTIVITY 3

Graph each set of functions on the same screen in the standard viewing window. Describe any similarities and differences among the graphs.

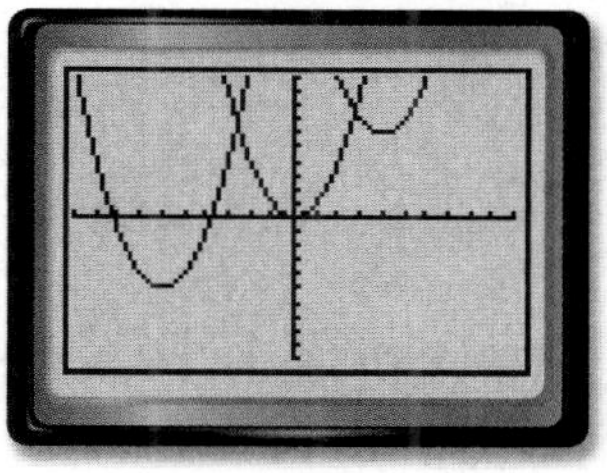

$y = x^2$, $y = (x + 6)^2 - 5$, $y = (x - 4)^2 + 6$

These three graphs all open up and have the same shape. However, the graphs have different horizontal and vertical positions.

How does the value a affect the graph of $y = x^2$?

ACTIVITY 4

Graph each set of functions on the same screen in the standard viewing window. Describe any similarities and differences among the graphs.

a. $y = x^2$, $y = -x^2$

The graphs have the same vertex and the same shape. However, the graph of $y = x^2$ opens up and the graph of $y = -x^2$ opens down.

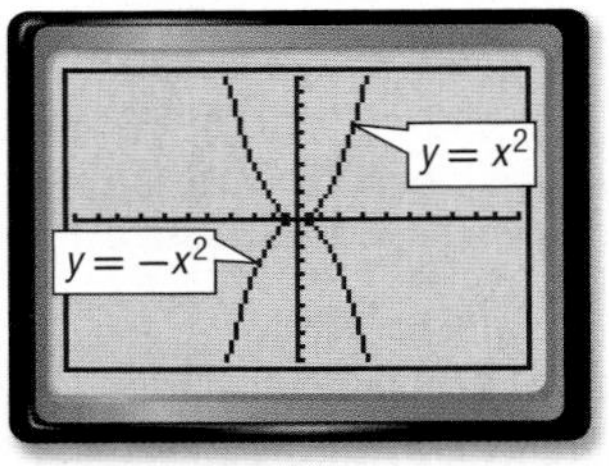

b. $y = x^2$, $y = 5x^2$, $y = \frac{1}{5}x^2$

The graphs have the same vertex, (0, 0), but each has a different shape. The graph of $y = 5x^2$ is narrower than the graph of $y = x^2$. The graph of $y = \frac{1}{5}x^2$ is wider than the graph of $y = x^2$.

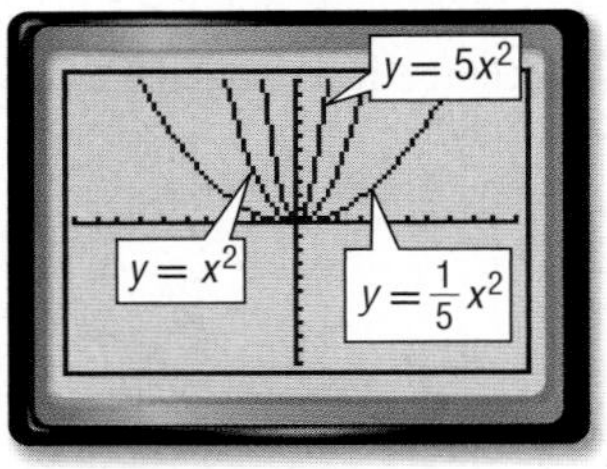

Changing the value of a in the function $y = a(x - h)^2 + k$ can affect the direction of the opening and the shape of the graph. If $a > 0$, the graph opens up, and if $a < 0$, the graph opens down or is *reflected* over the x-axis. If $|a| > 1$, the graph is expanded vertically and is narrower than the graph of $y = x^2$. If $|a| < 1$, the graph is compressed vertically and is wider than the graph of $y = x^2$. Thus, a change in the absolute value of a results in a *dilation* of the graph of $y = x^2$.

Analyze the Results

1. How does changing the value of h in $y = a(x - h)^2 + k$ affect the graph? Give an example.

2. How does changing the value of k in $y = a(x - h)^2 + k$ affect the graph? Give an example.

3. How does using $-a$ instead of a in $y = a(x - h)^2 + k$ affect the graph? Give an example.

Examine each pair of functions and predict the similarities and differences in their graphs. Use a graphing calculator to confirm your predictions. Write a sentence or two comparing the two graphs.

4. $y = x^2$, $y = x^2 + 3.5$

5. $y = -x^2$, $y = x^2 - 7$

6. $y = x^2$, $y = 4x^2$

7. $y = x^2$, $y = -8x^2$

8. $y = x^2$, $y = (x + 2)^2$

9. $y = -\frac{1}{6}x^2$, $y = -\frac{1}{6}x^2 + 2$

10. $y = x^2$, $y = (x - 5)^2$

11. $y = x^2$, $y = 2(x + 3)^2 - 6$

12. $y = x^2$, $y = -\frac{1}{8}x^2 + 1$

13. $y = (x + 5)^2 - 4$, $y = (x + 5)^2 + 7$

14. $y = 2(x + 1)^2 - 4$, $y = 5(x + 3)^2 - 1$

15. $y = 5(x - 2)^2 - 3$, $y = \frac{1}{4}(x - 5)^2 - 6$

5-7 Transformations with Quadratic Functions

Then

You transformed graphs of functions. (Lesson 2-7)

Now

- Write a quadratic function in the form $y = a(x - h)^2 + k$.
- Transform graphs of quadratic functions of the form $y = a(x - h)^2 + k$.

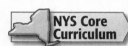

NYS Core Curriculum

Reinforcement of A.G.5 Investigate and generalize how changing the coefficients of a function affects its graph

New Vocabulary

vertex form

NY Math Online

glencoe.com

- Extra Examples
- Personal Tutor
- Self-Check Quiz
- Homework Help
- Math in Motion

Why?

Recall that a family of graphs is a group of graphs that display one or more similar characteristics. The parent graph is the simplest graph in the family. For the family of quadratic functions, $y = x^2$ is the parent graph.

Other graphs in the family of quadratic functions, such as $y = (x - 2)^2$ and $y = x^2 - 4$, can be drawn by transforming the graph of $y = x^2$.

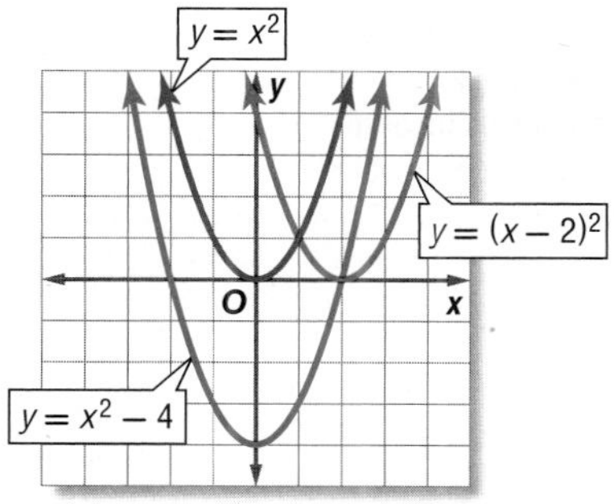

Write Quadratic Functions in Vertex Form Each function above is written in **vertex form**, $y = a(x - h)^2 + k$, where (h, k) is the vertex of the parabola, $x = h$ is the axis of symmetry, and a determines the shape of the parabola and the direction in which it opens.

When a quadratic function is in the form $y = ax^2 + bx + c$, you can complete the square to write the function in vertex form. If the coefficient of the quadratic term is not 1, then factor the coefficient from the quadratic and linear terms *before* completing the square.

EXAMPLE 1 Write Functions in Vertex Form

Write each function in vertex form.

a. $y = x^2 + 6x - 5$

$y = x^2 + 6x - 5$	**$x^2 + 6x - 5$ is not a perfect square. Complete the square by adding $\left(\frac{6}{2}\right)^2$ or 9.**
$y = (x^2 + 6x + 9) - 5 - 9$	**Balance the equation by subtracting 9.**
$y = (x + 3)^2 - 14$	**Write $x^2 + 6x + 9$ as a perfect square.**

b. $y = -2x^2 + 8x - 3$

$y = -2x^2 + 8x - 3$	**Original function**
$y = -2(x^2 - 4x) - 3$	**Group $ax^2 + bx$ and factor, dividing by a.**
$y = -2(x^2 - 4x + 4) - 3 - (-2)(4)$	**Complete the square by adding 4 inside the parentheses. This is an overall addition of −2(4). Balance the equation by subtracting −2(4).**
$y = -2(x - 2)^2 + 5$	**Write $x^2 - 4x + 4$ as a perfect square.**

Check Your Progress

1A. $y = x^2 + 4x + 6$

1B. $y = 2x^2 - 12x + 17$

Personal Tutor glencoe.com

Test-TakingTip

The Meaning of *a* The sign of a in the vertex form does not determine the width of the parabola. The sign indicates whether the parabola opens up or down. The width of a parabola is determined by the absolute value of a.

If the vertex and one additional point on the graph of a parabola are known, you can write the equation of the parabola in vertex form.

NYSRE EXAMPLE 2 Reinforcement of A.G.5

Which is an equation of the function shown in the graph?

A $y = -4(x - 3)^2 + 2$

B $y = -\frac{1}{4}(x - 3)^2 + 2$

C $y = \frac{1}{4}(x + 3)^2 - 2$

D $y = 4(x + 3)^2 - 2$

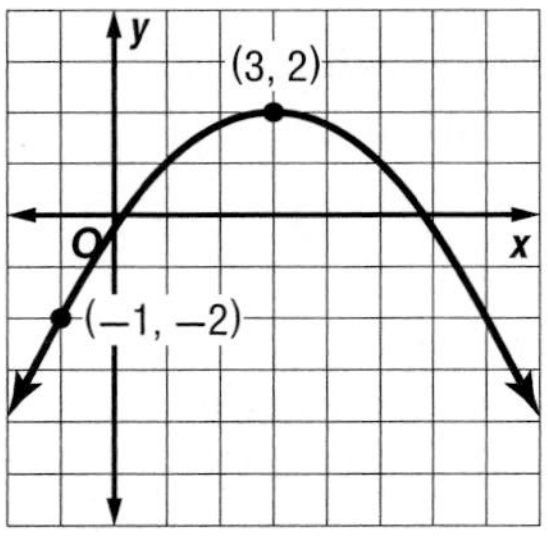

Read the Test Item

You are given a graph of a parabola with the vertex and a point on the graph labeled. You need to find an equation of the parabola.

Solve the Test Item

The vertex of the parabola is at (3, 2), so $h = 3$ and $k = 2$. Since $(-1, -2)$ is a point on the graph, let $x = -1$ and $y = -2$. Substitute these values into the vertex form of the equation and solve for a.

$y = a(x - h)^2 + k$	**Vertex form**
$-2 = a(-1 - 3)^2 + 2$	**Substitute −2 for *y*, −1 for *x*, 3 for *h* and 2 for *k*.**
$-2 = a(16) + 2$	**Simplify.**
$-4 = 16a$	**Subtract 2 from each side.**
$-\frac{1}{4} = a$	**Divide each side by 16.**

The equation of the parabola in vertex form is $y = -\frac{1}{4}(x - 3)^2 + 2$.

The answer is B.

Check Your Progress

2. Which is an equation of the function shown in the graph?

F $y = \frac{9}{25}(x - 1)^2 + 2$

G $y = \frac{3}{5}(x + 1)^2 - 2$

H $y = \frac{5}{3}(x + 1)^2 - 2$

J $y = \frac{25}{9}(x - 1)^2 + 2$

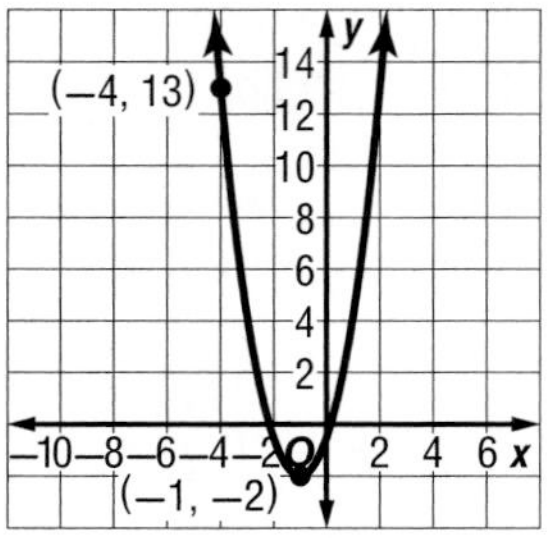

Personal Tutor glencoe.com

Transformations of Quadratic Functions In Lesson 2-7, you learned how different transformations affect the graphs of parent functions. The following summarizes these transformations for quadratic functions.

StudyTip

Absolute Value
$0 < |a| < 1$ means that a is a rational number between 0 and 1, such as $\frac{3}{4}$, or a rational number between -1 and 0, such as -0.3.

Concept Summary — Transformations of Quadratic Functions

For Your **FOLDABLE**

$f(x) = a(x - h)^2 + k$

h, Horizontal Translation
$|h|$ units to the right if h is positive
$|h|$ units to the left if h is negative

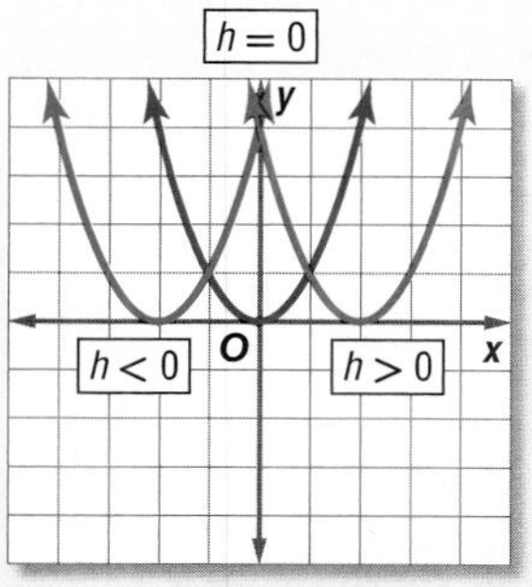

k, Vertical Translation
$|k|$ units up if k is positive
$|k|$ units down if k is negative

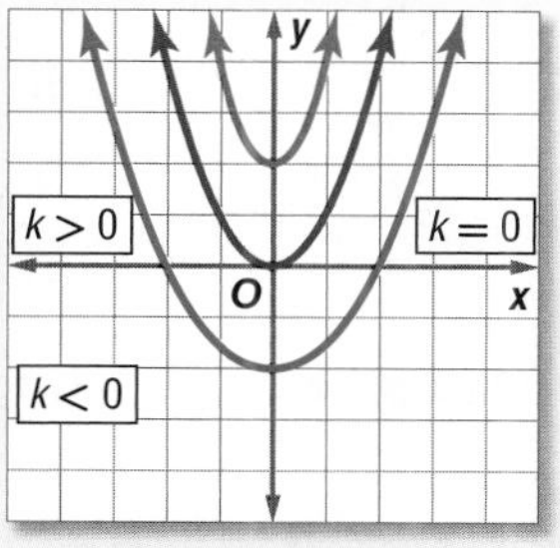

a, Reflection
If $a > 0$, the graph opens up.
If $a < 0$, the graph opens down.

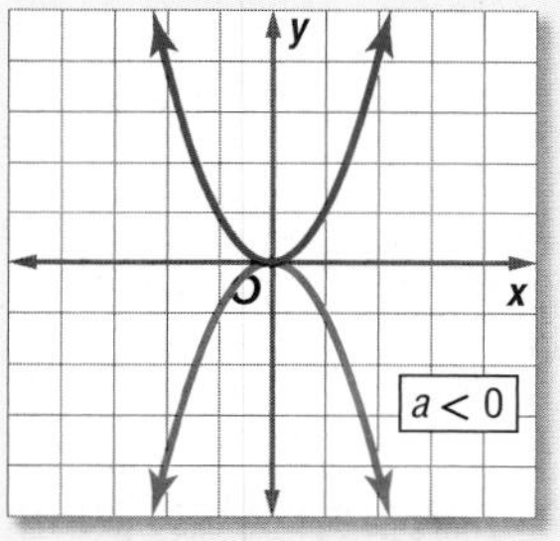

a, Dilation
If $|a| > 1$, the graph is stretched vertically. If $0 < |a| < 1$, the graph is compressed vertically.

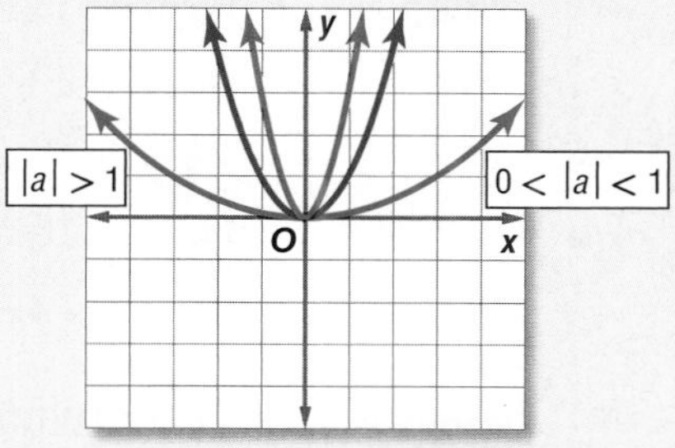

Math *in Motion*, Animation glencoe.com

EXAMPLE 3 Graph Equations in Vertex Form

Graph $y = 4x^2 - 16x - 40$

Step 1 Rewrite the equation in vertex form.

$y = 4x^2 - 16x - 40$	**Original equation**
$y = 4(x^2 - 4x) - 40$	**Distributive Property**
$y = 4(x - 4x + 4) - 40 - 4(4)$	**Complete the square.**
$y = 4(x - 2)^2 - 56$	**Simplify.**

Step 2 The vertex is at $(2, -56)$. The axis of symmetry is $x = 2$. Because $a = 4$, the graph is narrower than the graph of $y = x^2$.

Step 3 Plot additional points to help you complete the graph.

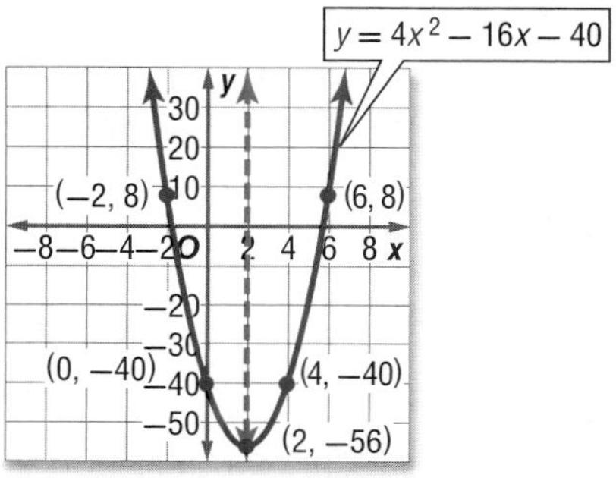

Check Your Progress

3A. $y = (x - 3)^2 - 2$

3B. $y = 0.25(x + 1)^2$

Personal Tutor glencoe.com

Check Your Understanding

Example 1 p. 305

Write each function in vertex form.

1. $y = x^2 + 6x + 2$
2. $y = -2x^2 + 8x - 5$
3. $y = 4x^2 + 24x + 24$

Example 2 p. 306

4. **MULTIPLE CHOICE** Which function is shown in the graph?

   A $y = -(x + 3)^2 + 6$

   B $y = -(x - 3)^2 - 6$

   C $y = -2(x + 3)^2 + 6$

   D $y = -2(x - 3)^2 - 6$

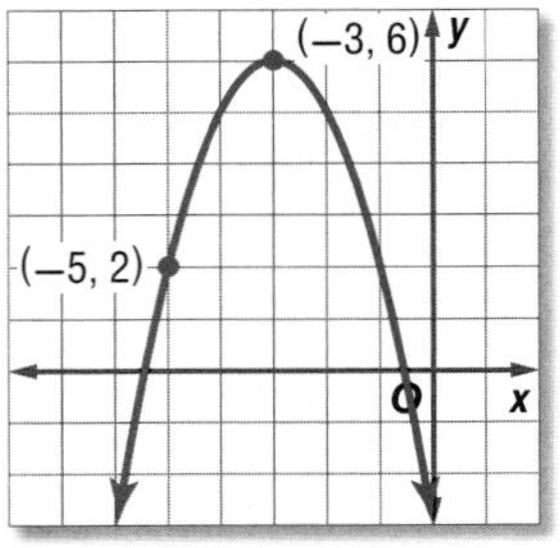

Example 3 p. 307

Graph each function.

5. $y = (x - 3)^2 - 4$
6. $y = -2x^2 + 5$
7. $y = \frac{1}{2}(x + 6)^2 - 8$

Practice and Problem Solving

= **Step-by-Step Solutions** begin on page R20.
Extra Practice begins on page 947.

Example 1 p. 305

Write each function in vertex form.

8. $y = x^2 + 9x + 8$
9. $y = x^2 - 6x + 3$
10. $y = -2x^2 + 5x$
11. $y = x^2 + 2x + 7$
12. $y = -3x^2 + 12x - 10$
13. $y = x^2 + 8x + 16$
14. $y = 2x^2 - 4x - 3$
15. $y = 3x^2 + 10x$
16. $y = x^2 - 4x + 9$
17. $y = -4x^2 - 24x - 15$
18. $y = x^2 - 12x + 36$
19. $y = -x^2 - 4x - 1$

Example 2 p. 306

20. **FIREWORKS** During an Independence Day fireworks show, the height h in meters of a specific rocket after t seconds can be modeled by $h = -4.9(t - 4)^2 + 80$. Graph the function.

21. **BICYCLES** A bicycle rental shop rents an average of 120 bicycles per week and charges \$25 per day. The manager estimates that there will be 15 additional bicycles rented for each \$1 reduction in the rental price. The maximum income the manager can expect can be modeled by $y = -15x^2 + 255x + 3000$. Write this function in vertex form. Then graph.

Example 3 p. 307

Graph each function.

22. $y = (x - 5)^2 + 3$
23. $y = 9x^2 - 8$
24. $y = -2(x - 5)^2$
25. $y = \frac{1}{10}(x + 6)^2 + 6$
26. $y = -3(x - 5)^2 - 2$
27. $y = -\frac{1}{4}x^2 - 5$
28. $y = 2x^2 + 10$
29. $y = -(x + 3)^2$
30. $y = \frac{1}{6}(x - 3)^2 - 10$
31. $y = (x - 9)^2 - 7$
32. $y = -\frac{5}{8}x^2 - 8$
33. $y = -4(x - 10)^2 - 10$

34. **SAILBOARDING** A sailboard manufacturer uses an automated process to manufacture the masts for its sailboards. The function $f(x) = \frac{1}{250}x^2 + \frac{3}{5}x$ is programmed into a computer to make one such mast.

    a. Write the quadratic function in vertex form. Then graph the function.

    b. Describe how the manufacturer can adjust the function to make its masts with a greater or smaller curve.

Write an equation in vertex form for each parabola.

35.

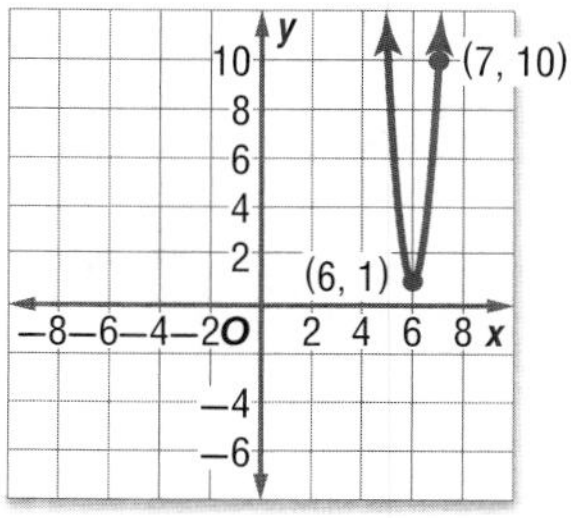

36.

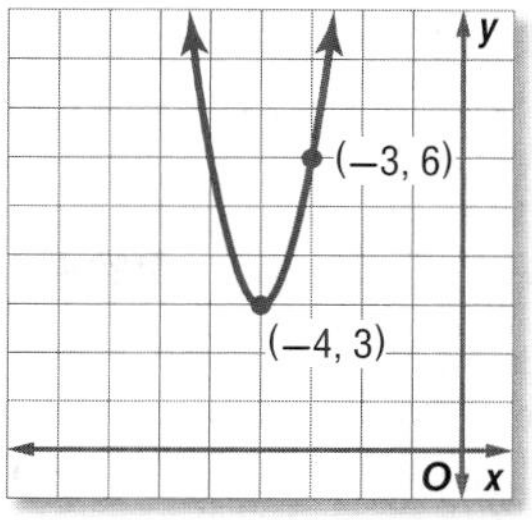

37.

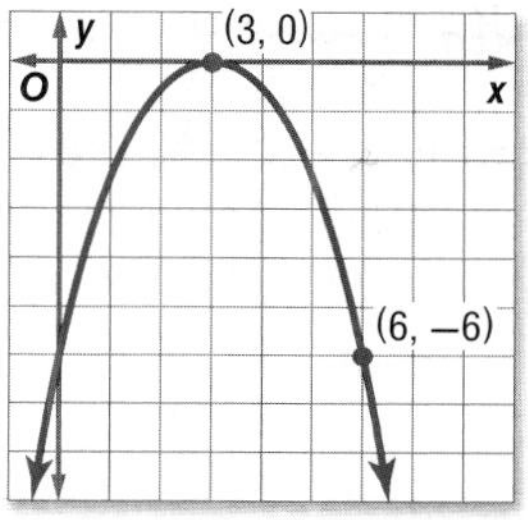

38.

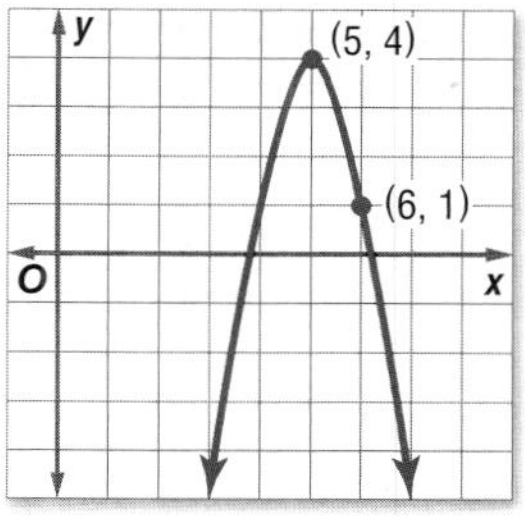

39.

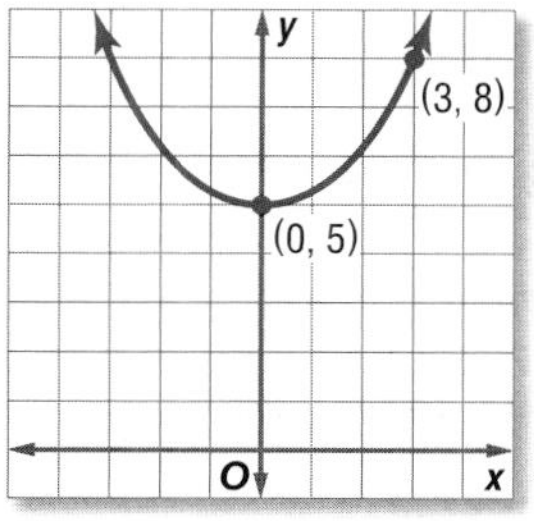

40.

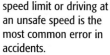

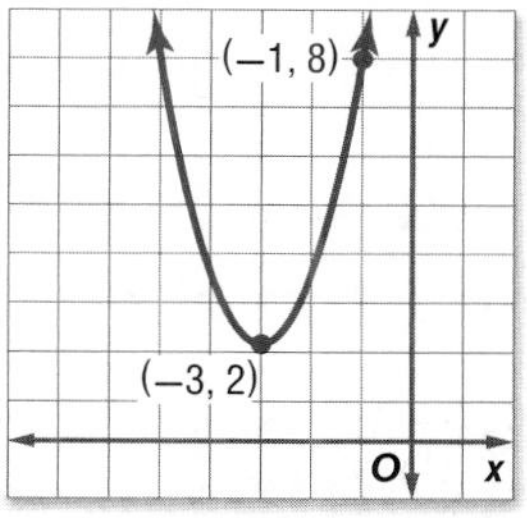

Real-World Link

Exceeding the posted speed limit or driving at an unsafe speed is the most common error in accidents.

Source: National Safety Council

Write each function in vertex form. Then identify the vertex, axis of symmetry, and direction of opening.

41. $3x^2 - 4x = 2 + y$ **42.** $-2x^2 + 7x = y - 12$ **43.** $-x^2 - 4.7x = y - 2.8$

44. $x^2 + 1.4x - 1.2 = y$ **45.** $x^2 - \frac{2}{3}x - \frac{26}{9} = y$ **46.** $x^2 + 7x + \frac{49}{4} = y$

47 **CARS** The formula $S(t) = \frac{1}{2}at^2 + v_0t$ can be used to determine the position $S(t)$ of an object after t seconds at a rate of acceleration a with initial velocity v_0. Valerie's car can accelerate 0.002 miles per second squared.

a. Express $S(t)$ in vertex form as she accelerates from 35 miles per hour to enter highway traffic.

b. How long will it take Valerie to match the average speed of highway traffic of 68 miles per hour?

c. If the entrance ramp is $\frac{1}{8}$-mile long, will Valerie have sufficient time to match the average highway speed? Explain.

H.O.T. Problems

Use Higher-Order Thinking Skills

48. OPEN ENDED Write an equation for a parabola that has been translated, compressed, and reflected in the x-axis.

49. CHALLENGE Explain how you can find an equation of a parabola using the coordinates of three points on the graph.

50. CHALLENGE Write the standard form of a quadratic function $ax^2 + bx + c = y$ in vertex form. Identify the vertex and the axis of symmetry.

51. REASONING Describe the graph of $f(x) = a(x - h)^2 + k$ when $a = 0$. Is the graph the same as that of $g(x) = ax^2 + bx + c$ when $a = 0$? Explain.

52. WRITING IN MATH Explain how the graph of $y = x^2$ can be used to graph any quadratic function. Include a description of the effects produced by changing a, h, and k in the equation $y = a(x - h)^2 + k$, and a comparison of the graph of $y = x^2$ and the graph of $y = a(x - h)^2 + k$ using values you choose for a, h, and k.

NYSRE PRACTICE A2.A.1

53. Flowering bushes need a mixture of 70% soil and 30% vermiculite. About how many buckets of vermiculite should you add to 20 buckets of soil?

A 6.0
B 8.0
C 14.0
D 24.0

54. SAT/ACT The sum of the integers x and y is 495. The units digit of x is 0. If x is divided by 10, the result is equal to y. What is the value of x?

F 40
G 245
H 250
J 450

55. What is the solution set of the inequality $|4x - 1| < 9$?

A $\{x \mid 2.5 < x \text{ or } x < -2\}$
B $\{x \mid x < 2.5\}$
C $\{x \mid x > -2\}$
D $\{x \mid -2 < x < 2.5\}$

56. SHORT RESPONSE At your store, you buy wrenches for $30.00 a dozen and sell them for $3.50 each. What is the percent markup for the wrenches?

Spiral Review

Solve each equation by using the method of your choice. Find exact solutions. (Lesson 5-6)

57. $4x^2 + 15x = 21$

58. $-3x^2 + 19 = 5x$

59. $6x - 5x^2 + 9 = 3$

Find the value of c that makes each trinomial a perfect square. (Lesson 5-5)

60. $x^2 - 12x + c$

61. $x^2 + 0.1x + c$

62. $x^2 - 0.45x + c$

Determine whether each function has a maximum or minimum value, and find that value. (Lesson 5-1)

63. $f(x) = 6x^2 - 8x + 12$

64. $f(x) = -4x^2 + x - 18$

65. $f(x) = 3x^2 - 9 + 6x$

66. ARCHAEOLOGY A coordinate grid is laid over an archeology dig to identify the location of artifacts. Three corners of a building have been partially unearthed at $(-1, 6)$, $(4, 5)$, and $(-1, -2)$. If each square on the grid measures one square foot, estimate the area of the floor of the building. (Lesson 4-5)

67. HOTELS Use the costs for an overnight stay at a hotel provided at the right. (Lesson 4-1)

a. Write a 3×2 matrix that represents the cost of each room.

b. Write a 2×3 matrix that represents the cost of each room.

HOTEL

	Weekday	Weekend
Single Room	$60.00	$79.00
Double Room	$70.00	$89.00
Suite	$75.00	$95.00

Solve each system of equations by graphing. (Lesson 3-1)

68. $y = 3x - 4$
$y = -2x + 16$

69. $2x + 5y = 1$
$6y - 5x = 16$

70. $4x + 3y = -30$
$3x - 2y = 3$

Evaluate each function. (Lesson 2-1)

71. $f(3)$ if $f(x) = x^2 - 4x + 12$

72. $f(-2)$ if $f(x) = -4x^2 + x - 8$

73. $f(4)$ if $f(x) = 3x^2 + x$

Skills Review

Determine whether the given value satisfies the inequality. (Lesson 1-6)

74. $3x^2 - 5 > 6; x = 2$

75. $-2x^2 + x - 1 < 4; x = -2$

76. $4x^2 + x - 3 \leq 36; x = 3$

EXTEND **5-7**

Algebra Lab: Quadratics and Rate of Change

NY Math Online glencoe.com
Math *in Motion,* Animation

NYS Core Curriculum **Reinforcement of A.A.32** Explain slope as a rate of change between dependent and independent variables

You have learned that a linear function has a constant rate of change. You will investigate the rate of change for quadratic functions.

ACTIVITY Determine Rate of Change

Consider $f(x) = 0.1875x^2 - 3x + 12$.

Step 1 Make a table like the one below. Use values from 0 through 16 for x.

x	0	1	2	3	...	16
y	12	9.1875	6.75			
First Order Differences						
Second Order Differences						

Step 2 Find each y-value. For example, when $x = 1$, $y = 0.1875(1)^2 - 3(1) + 12$ or 9.1875.

Step 3 Graph the ordered pairs (x, y). Then connect the points with a smooth curve. Notice that the function *decreases* when $0 < x < 8$ and *increases* when $8 < x < 16$.

Step 4 The rate of change from one point to the next can be found by using the slope formula. From (0, 12) to (1, 9.1875), the slope is $\frac{9.1875 - 12}{1 - 0}$ or -2.8125. This is the first-order difference at $x = 1$. Complete the table for all the first-order differences. Describe any patterns in the differences.

Step 5 The second-order differences can be found by subtracting consecutive first-order differences. For example, the second-order difference at $x = 2$ is found by subtracting the first order difference at $x = 1$ from the first-order difference at $x = 2$. Describe any patterns in the differences.

Exercises

For each function make a table of values for the given x-values. Graph the function. Then determine the first-order and second-order differences.

1. $y = -x^2 + 2x - 1$ for $x = -3, -2, -1, 0, 1, 2, 3$
2. $y = 0.5x^2 + 2x - 2$ for $x = -5, -4, -3, -2, -1, 0, 1$
3. $y = -3x^2 - 18x - 26$ for $x = -6, -5, -4, -3, -2, -1, 0$
4. **MAKE A CONJECTURE** Repeat the activity for a cubic function. At what order difference would you expect $g(x) = x^4$ to be constant? $h(x) = x^n$?

5-8 Quadratic Inequalities

Why?

A water balloon launched from a slingshot can be represented by several different quadratic equations and inequalities.

Suppose the height of a water balloon $h(t)$ in meters above the ground t seconds after being launched is modeled by the quadratic function $h(t) = -4.9t^2 + 32t + 1.2$. You can solve a quadratic inequality to determine how long the balloon will be a certain distance above the ground.

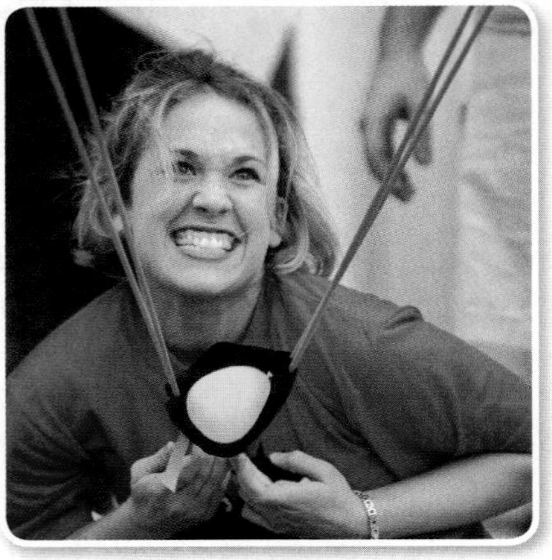

Then

You solved linear inequalities. (Lesson 2-8)

Now

- Graph quadratic inequalities in two variables.
- Solve quadratic inequalities in one variable.

NYS Core Curriculum

A2.A.4 Solve quadratic inequalities in one and two variables, algebraically and graphically

New Vocabulary

quadratic inequality

NY Math Online

glencoe.com

- Extra Examples
- Personal Tutor
- Self-Check Quiz
- Homework Help
- Math in Motion

Graph Quadratic Inequalities You can graph **quadratic inequalities** in two variables by using the same techniques used to graph linear inequalities in two variables.

Step 1 Graph the related function.

Step 2 Test a point not on the parabola.

Step 3 Shade accordingly.

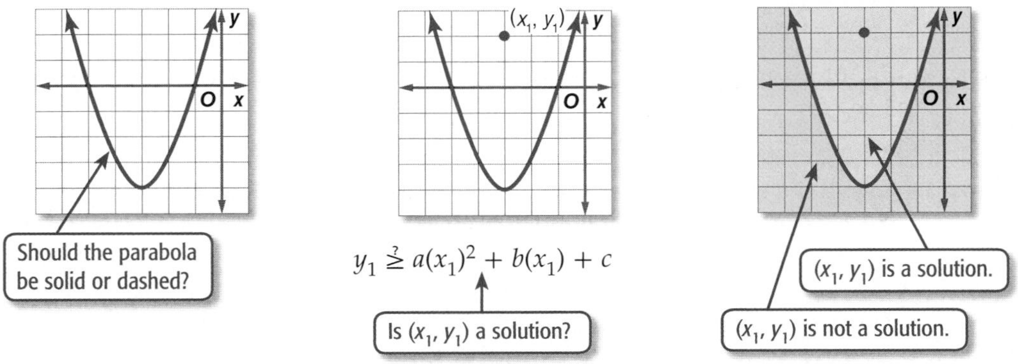

y
O x
Should the parabola be solid or dashed?
(x_1, y_1)
$y_1 \overset{?}{\geq} a(x_1)^2 + b(x_1) + c$
Is (x_1, y_1) a solution?
(x_1, y_1) is a solution.
(x_1, y_1) is not a solution.

EXAMPLE 1 Graph a Quadratic Inequality

Graph $y > x^2 + 2x + 1$.

Step 1 Graph the related function, $y = x^2 + 2x + 1$. The parabola should be dashed.

Step 2 Test a point not on the graph of the parabola.

$$y > x^2 + 2x + 1$$

$$-1 \overset{?}{>} 0^2 + 2(0) + 1$$

$$-1 \not> 1$$

So, $(0, -1)$ is *not* a solution of the inequality.

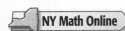

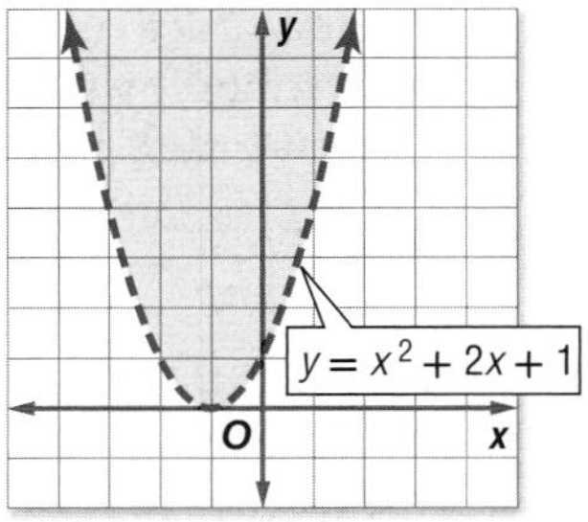

y
$y = x^2 + 2x + 1$
O
x

Step 3 Shade the region that does not contain the point $(0, -1)$.

Check Your Progress

Graph each inequality.

1A. $y \leq x^2 + 2x + 4$

1B. $y < -2x^2 + 3x + 5$

Personal Tutor glencoe.com

Solve Quadratic Inequalities Quadratic inequalities in one variable can be solved using the graphs of the related quadratic functions.

$ax^2 + bx + c < 0$

Graph $y = ax^2 + bx + c$ and identify the x-values for which the graph lies *below* the x-axis.

For $\leq$, include the x-intercepts in the solution.

$a > 0$

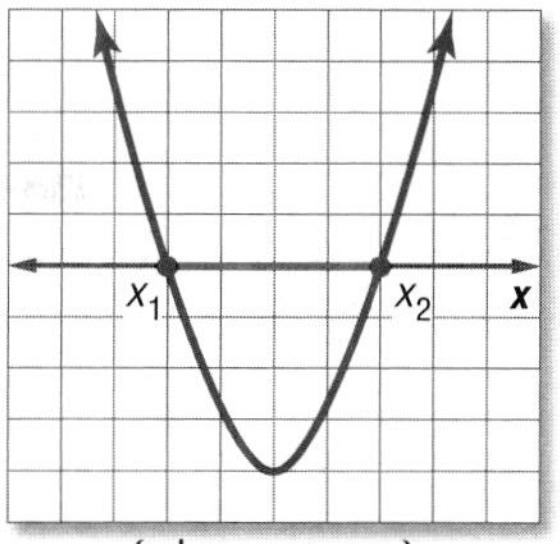

$\{x \mid x_1 < x < x_2\}$

$a < 0$

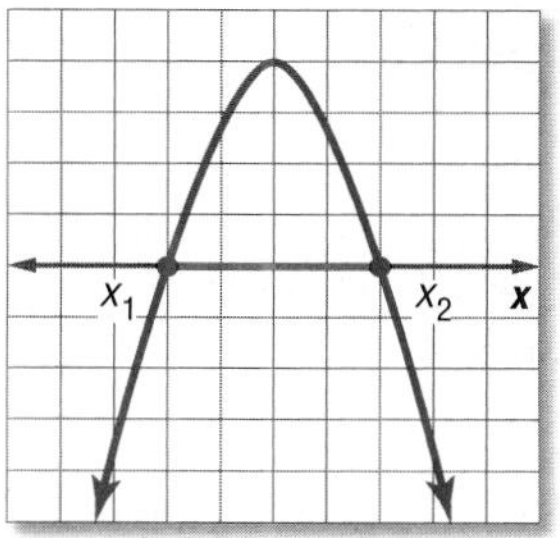

$\{x \mid x < x_1 \text{ or } x > x_2\}$

Math *in Motion,* Interactive Lab glencoe.com

$ax^2 + bx + c > 0$

Graph $y = ax^2 + bx + c$ and identify the x-values for which the graph lies *above* the x-axis.

For $\geq$, include the x-intercepts in the solution.

$a > 0$

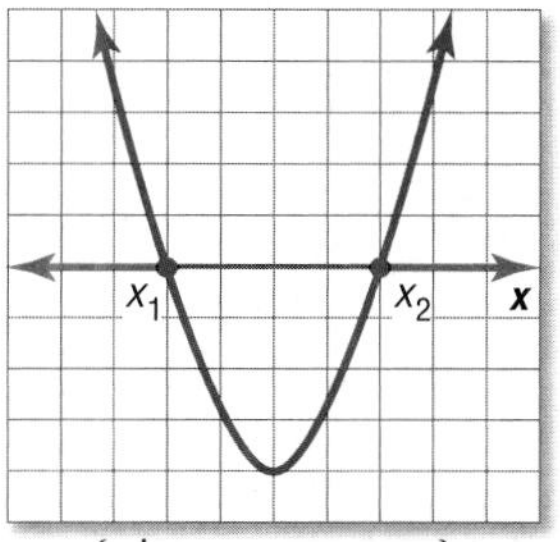

$\{x \mid x < x_1 \text{ or } x > x_2\}$

$a < 0$

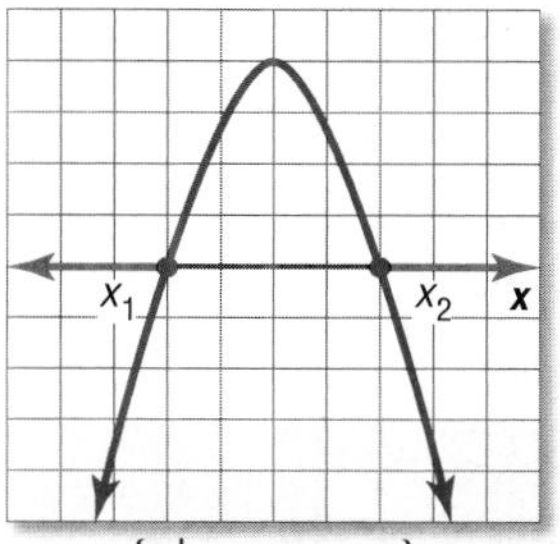

$\{x \mid x_1 < x < x_2\}$

StudyTip

Solving Quadratic Inequalities by Graphing A precise graph of the related quadratic function is not necessary since the zeros of the function were found algebraically.

EXAMPLE 2 Solve $ax^2 + bx + c < 0$ by Graphing

Solve $x^2 + 2x - 8 < 0$ by graphing.

The solution consists of x-values for which the graph of the related function lies *below* the x-axis. Begin by finding the roots of the related function.

$x^2 + 2x - 8 = 0$		**Related equation**
$(x - 2)(x + 4) = 0$		**Factor.**
$x - 2 = 0$	or $x + 4 = 0$	**Zero Product Property**
$x = 2$	$x = -4$	**Solve each equation.**

Sketch the graph of a parabola that has x-intercepts at -4 and 2. The graph should open up because $a > 0$.

The graph lies below the x-axis between $x = -4$ and $x = 2$. Thus, the solution set of the inequality is $\{x \mid -4 < x < 2\}$.

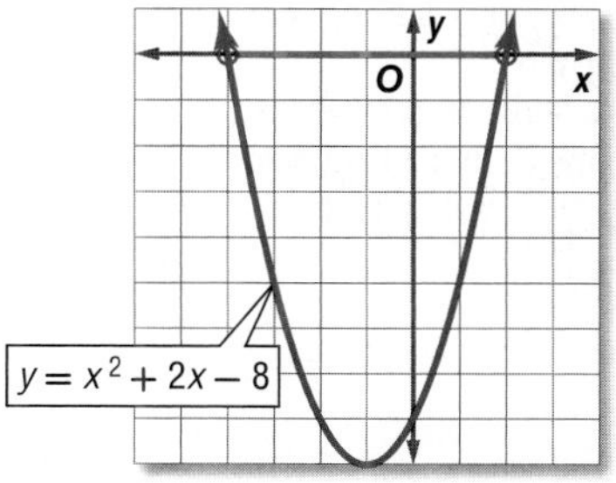

CHECK Test one value of x less than -4, one between -4 and 2, and one greater than 2 in the original inequality.

Test $x = -6$.	**Test $x = 0$.**	**Test $x = 5$.**
$x^2 + 2x - 8 < 0$	$x^2 + 2x - 8 < 0$	$x^2 + 2x - 8 < 0$
$(-6)^2 + 2(-6) - 8 \overset{?}{<} 0$	$0^2 + 2(0) - 8 \overset{?}{<} 0$	$5^2 + 2(5) - 8 \overset{?}{<} 0$
$16 < 0$ ✗	$-8 < 0$ ✓	$27 < 0$ ✗

Check Your Progress

Solve each inequality by graphing.

2A. $0 > x^2 + 5x - 6$ **2B.** $-x^2 + 3x + 10 \leq 0$

Personal Tutor glencoe.com

EXAMPLE 3 Solve $ax^2 + bx + c \geq 0$ by Graphing

Solve $2x^2 + 4x - 5 \geq 0$ by graphing.

The solution consists of x-values for which the graph of the related function lies *on and above* the x-axis. Begin by finding the roots of the related function.

$2x^2 + 4x - 5 = 0$	**Related equation**
$x = \frac{-b \pm \sqrt{b^2 - 4ac}}{2a}$	**Use the Quadratic Formula**
$x = \frac{-4 \pm \sqrt{4^2 - 4(2)(-5)}}{2(2)}$	**Replace *a* with 4, *b* with 2, and *c* with −5.**
$x = \frac{-4 + \sqrt{56}}{4}$ or $x = \frac{-4 - \sqrt{56}}{4}$	**Simplify and write as two equations.**
≈ 0.87 $\quad\approx -2.87$	**Simplify.**

Sketch the graph of a parabola with x-intercepts at -2.87 and 0.87. The graph opens up since $a > 0$. The graph lies on and above the x-axis at about $x \leq -2.87$ and $x \geq 0.87$. Therefore, the solution set of the inequality is approximately $\{x \mid x \leq -2.87 \text{ or } x \geq 0.87\}$.

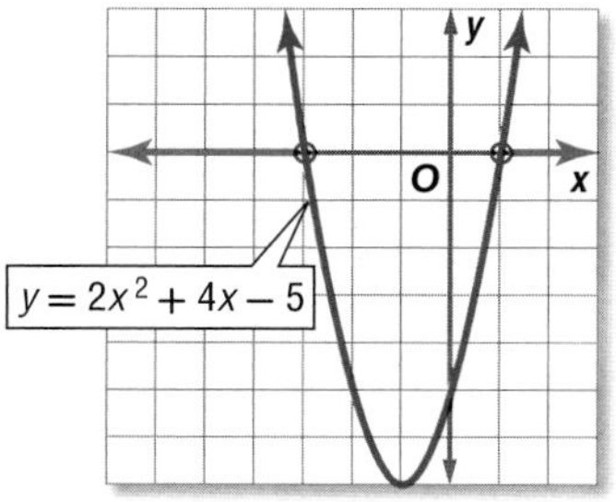

Check Your Progress

Solve each inequality by graphing.

3A. $x^2 - 6x + 2 > 0$

3B. $-4x^2 + 5x + 7 \geq 0$

Personal Tutor glencoe.com

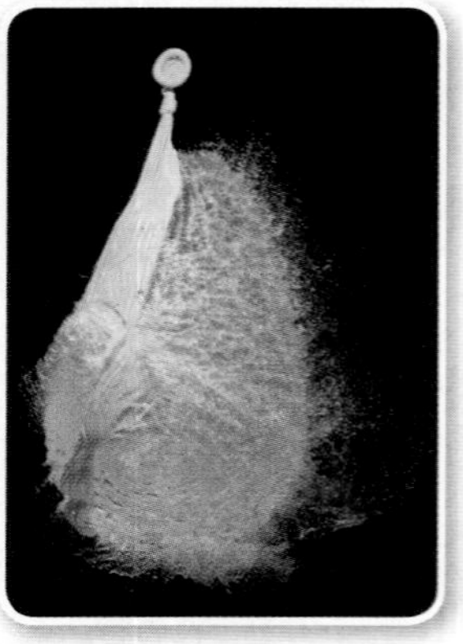

Real-World Link

It takes just milliseconds for a water balloon to break. A high-speed camera can capture the impact on the fluid before gravity makes it fall.

Source: NASA

Real-world problems can be solved by graphing quadratic inequalities.

Real-World EXAMPLE 4 Solve a Quadratic Inequality

WATER BALLOONS Refer to the application at the beginning of the lesson. At what time will a water balloon be within 3 meters of the ground after it has been launched?

The function $h(t) = -4.9t^2 + 32t + 1.2$ describes the height of the water balloon. Therefore, you want to find the values of t for which $h(t) \leq 3$.

$h(t) \leq 3$	**Original inequality**
$-4.9t^2 + 32t + 1.2 \leq 3$	**$h(t) = -4.9t^2 + 32t + 1.2$**
$-4.9t^2 + 32t - 1.8 \leq 0$	**Subtract 3 from each side.**

Graph the related function $y = -4.9t^2 + 32x - 1.8$ using a graphing calculator. The zeros of the function are about 0.06 and 6.47, and the graph lies below the x-axis when $x < 0.06$ and $x > 6.47$.

So, the water balloon is within 3 meters of the ground during the first 0.06 second after being launched and again after about 6.47 seconds until it hits the ground.

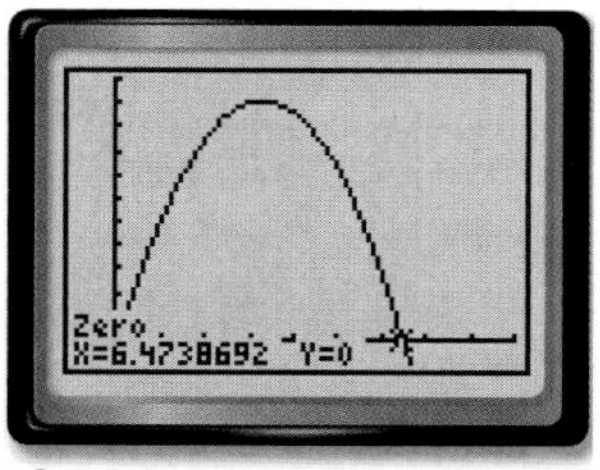

[−1, 9] scl: 1 by [−5, 55] scl: 5

Check Your Progress

4. ROCKETS The height $h(t)$ of a model rocket in feet t seconds after its launch can be represented by the function $h(t) = -16t^2 + 82t + 0.25$. During what interval is the rocket at least 100 feet above the ground?

Personal Tutor glencoe.com

StudyTip

Solving Quadratic Inequalities Algebraically The solution set of a quadratic inequality is all real numbers when all three test points satisfy the inequality. It is the empty set when none of the test points satisfy the inequality.

EXAMPLE 5 Solve a Quadratic Inequality Algebraically

Solve $x^2 - 3x \leq 18$ algebraically.

Step 1 Solve the related quadratic equation $x^2 - 3x = 18$.

$x^2 - 3x = 18$	**Related quadratic equation**
$x^2 - 3x - 18 = 0$	**Subtract 18 from each side.**
$(x + 3)(x - 6) = 0$	**Factor.**
$x + 3 = 0$ or $x - 6 = 0$	**Zero Product Property**
$x = -3$ $\quad$ $x = 6$	**Solve each equation.**

Step 2 Plot −3 and 6 on a number line. Use dots since these values are solutions of the original inequality. Notice that the number line is divided into three intervals.

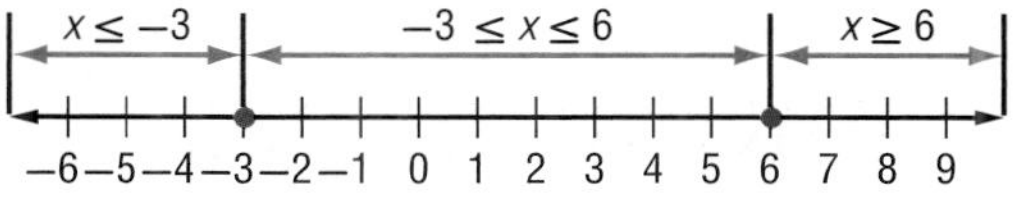

Step 3 Test a value from each interval to see if it satisfies the original inequality.

$x \leq -3$	$-3 \leq x \leq 6$	$x \geq 6$
Test $x = -5$.	Test $x = 0$.	Test $x = 8$.
$x^2 - 3x \leq 18$	$x^2 - 3x \leq 18$	$x^2 - 3x \leq 18$
$(-5)^2 - 3(-5) \overset{?}{\leq} 18$	$(0)^2 - 3(0) \overset{?}{\leq} 18$	$(8)^2 - 3(8) \overset{?}{\leq} 18$
$40 \not\leq 18$	$0 \leq 18$	$40 \not\leq 18$

The solution set is $\{x \mid -3 \leq x \leq 6\}$. This is shown on the number line below.

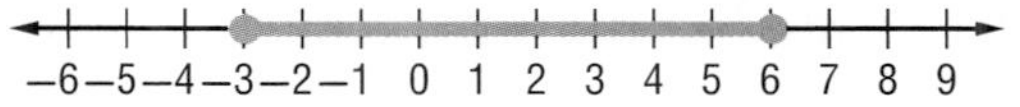

Check Your Progress

Solve each inequality algebraically.

5A. $x^2 + 5x < -6$ $\quad$ **5B.** $x^2 + 11x + 30 \geq 0$

Personal Tutor glencoe.com

Check Your Understanding

Example 1 p. 312

Graph each inequality.

1. $y \leq x^2 - 8x + 2$ $\quad$ **2.** $y > x^2 + 6x - 2$ $\quad$ **3.** $y \geq -x^2 + 4x + 1$

Examples 2 and 3 pp. 313–314

Solve each inequality by graphing.

4. $0 < x^2 - 5x + 4$ $\quad$ **5.** $x^2 + 8x + 15 < 0$

6. $-2x^2 - 2x + 12 \geq 0$ $\quad$ **7.** $0 \geq 2x^2 - 4x + 1$

Example 4 p. 314

8. **SOCCER** A midfielder kicks a ball toward the goal during a match. The height of the ball in feet above the ground $h(t)$ at time t can be represented by $h(t) = -0.1t^2 + 2.4t + 1.5$. If the height of the goal is 8 feet, at what time during the kick will the ball be able to enter the goal?

Example 5 p. 315

Solve each inequality algebraically.

9. $x^2 + 6x - 16 < 0$ $\quad$ **10.** $x^2 - 14x > -49$

11 $-x^2 + 12x \geq 28$ $\quad$ **12.** $x^2 - 4x \leq 21$

Practice and Problem Solving

● = **Step-by-Step Solutions** begin on page R20.
Extra Practice begins on page 947.

Example 1
p. 312

Graph each inequality.

13. $y \geq x^2 + 5x + 6$ **14.** $x^2 - 2x - 8 < y$ **15.** $y \leq -x^2 - 7x + 8$

16. $-x^2 + 12x - 36 > y$ **17.** $y > 2x^2 - 2x - 3$ **18.** $y \geq -4x^2 + 12x - 7$

Examples 2 and 3
pp. 313–314

Solve each inequality by graphing.

19. $x^2 - 9x + 9 < 0$ **20.** $x^2 - 2x - 24 \leq 0$ **21.** $x^2 + 8x + 16 \geq 0$

22. $x^2 + 6x + 3 > 0$ **23.** $0 > -x^2 + 7x + 12$ **24.** $-x^2 + 2x - 15 < 0$

25. $4x^2 + 12x + 10 \leq 0$ **26.** $-3x^2 - 3x + 9 > 0$ **27.** $0 > -2x^2 + 4x + 4$

28. $3x^2 + 12x + 36 \leq 0$ **29.** $0 \leq -4x^2 + 8x + 5$ **30.** $-2x^2 + 3x + 3 \leq 0$

Example 4
p. 314

31 **ARCHITECTURE** An arched entry of a room is shaped like a parabola that can be represented by the equation $f(x) = -x^2 + 6x + 1$. How far from the sides of the arch is its height at least 7 feet?

32. **MANUFACTURING** A box is formed by cutting 4-inch by 4-inch squares from each corner of a square piece of cardboard and then folding the sides. If $V(x) = 4x^2 - 64x + 256$ represents the volume of the box, what should the dimensions of the original piece of cardboard be if the volume of the box cannot exceed 750 cubic inches?

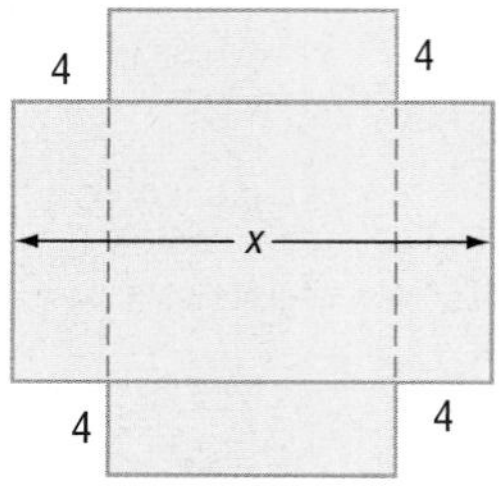

Example 5
p. 315

Solve each inequality algebraically.

33. $x^2 - 9x < -20$ **34.** $x^2 + 7x \geq -10$ **35.** $2 > x^2 - x$

36. $-3 \leq -x^2 - 4x$ **37.** $-x^2 + 2x \leq -10$ **38.** $-6 > x^2 + 4x$

39. $2x^2 + 4 \geq 9$ **40.** $3x^2 + x \geq -3$ **41.** $-4x^2 + 2x < 3$

42. $-11 \geq -2x^2 - 5x$ **43.** $-12 < -5x^2 - 10x$ **44.** $-3x^2 - 10x > -1$

Real-World Link

As a rule of thumb, total poolside area should be at least equal to the surface area of a swimming pool.

Source: About

45. **SWIMMING POOLS** The Sanchez family is adding a deck along two sides of their swimming pool. The deck width will be the same on both sides and the total area of the pool and deck cannot exceed 750 square feet.

a. Graph the quadratic inequality.

b. Determine the maximum width of the deck.

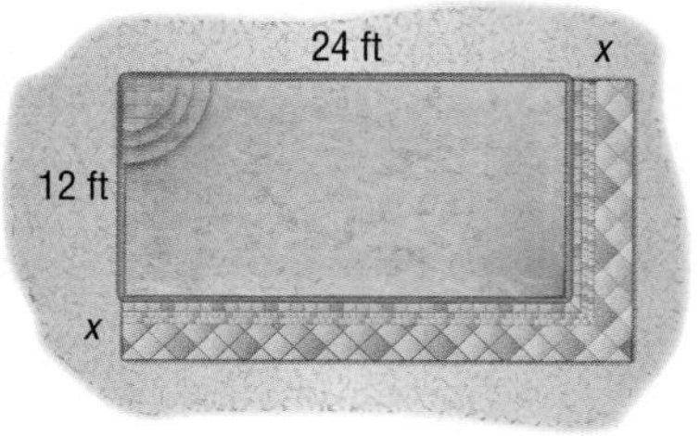

Write a quadratic inequality for each graph.

46.

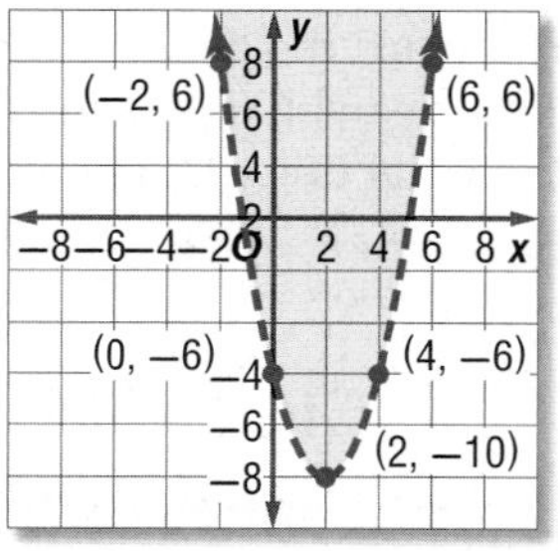

47.

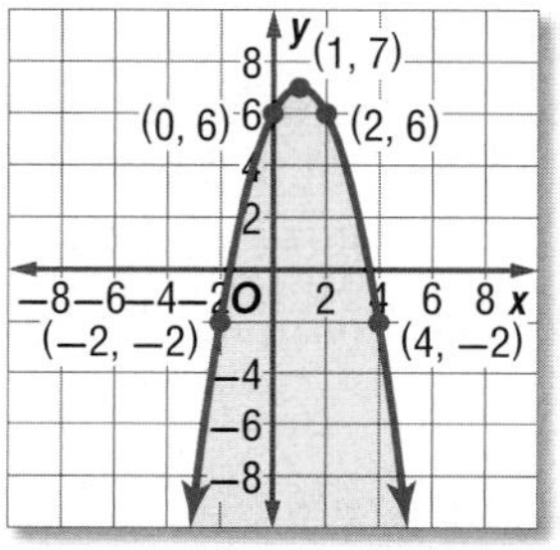

48.

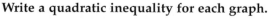

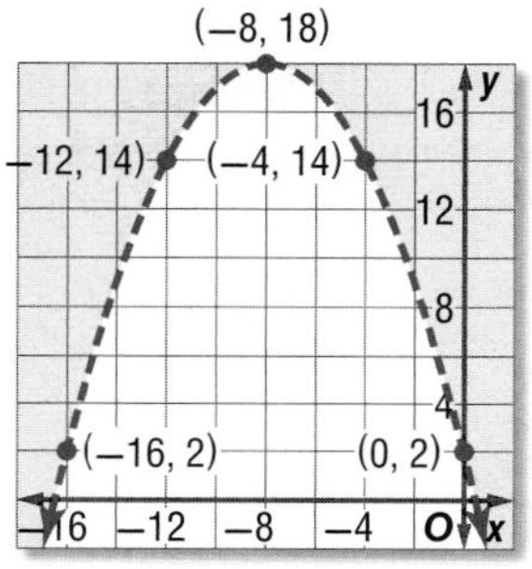

Solve each quadratic inequality by using a graph, a table, or algebraically.

49. $-2x^2 + 12x < -15$ **50.** $5x^2 + x + 3 \geq 0$ **51** $11 \leq 4x^2 + 7x$

52. $x^2 - 4x \leq -7$ **53.** $-3x^2 + 10x < 5$ **54.** $-1 \geq -x^2 - 5x$

55. BUSINESS An electronics manufacturer uses the function $P(x) = x(-27.5x + 3520) + 20{,}000$ to model their monthly profits when selling x thousand digital audio players.

a. Graph the quadratic inequality for a monthly profit of at least \$100,000.

b. How many digital audio players must the manufacturer sell to earn a profit of at least \$100,000 in a month?

c. Suppose the manufacturer has an additional monthly expense of \$25,000. Explain how this affects the graph of the profit function. Then determine how many digital audio players the manufacturer needs to sell to have at least \$100,000 in profits.

56. UTILITIES A contractor is installing drain pipes for a shopping center's parking lot. The outer diameter of the pipe is to be 10 inches. The cross sectional area of the pipe must be at least 35 square inches and should not be more than 42 square inches.

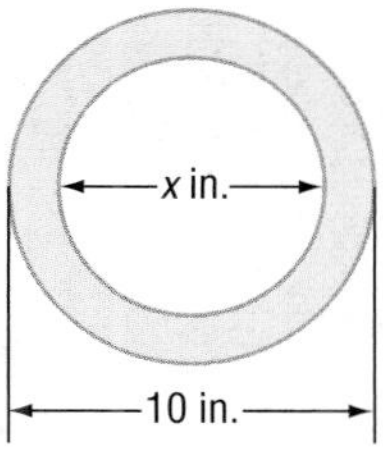

a. Graph the quadratic inequalities.

b. What thickness of drain pipe can the contractor use?

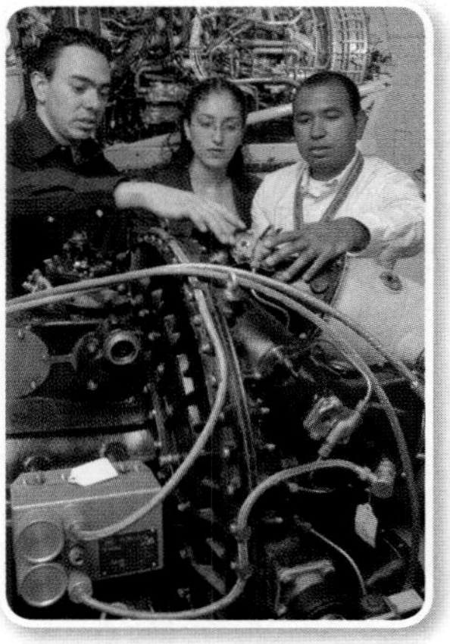

Real-World Career

Electrical Engineer
Electrical engineers design, develop, test, and supervise the making of electrical equipment such as digital music players, electric motors, lighting, and radar and navigation systems. A bachelor's degree in engineering is required for almost all entry-level engineering jobs.

H.O.T. Problems

Use **H**igher-**O**rder **T**hinking Skills

57. OPEN ENDED Write a quadratic inequality for each condition.

a. The solution set is all real numbers.

b. The solution set is the empty set.

58. FIND THE ERROR Don and Diego used a graph to solve the quadratic inequality $x^2 - 2x - 8 > 0$. Is either of them correct? Explain.

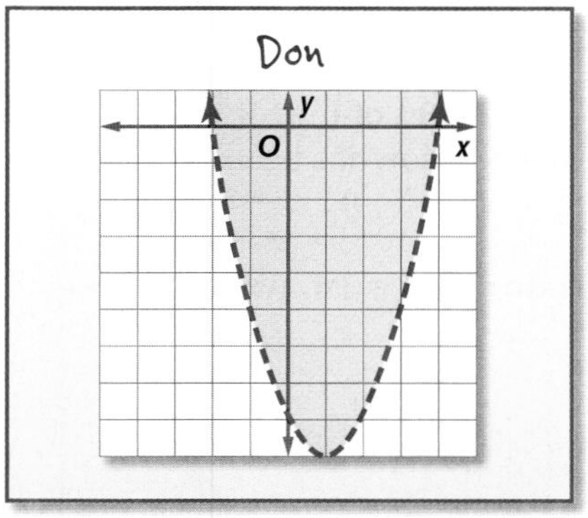

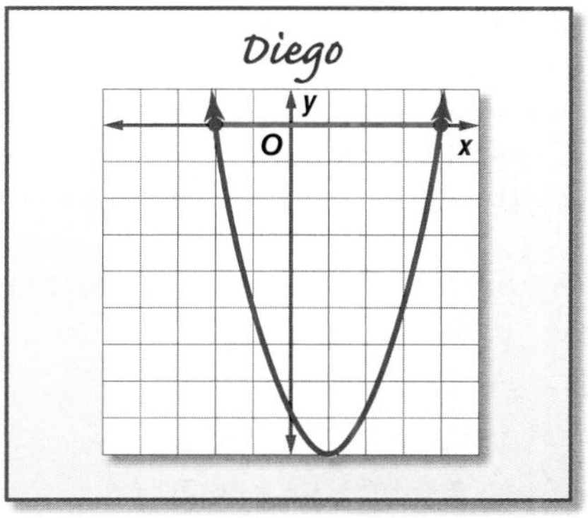

59. REASONING Are the boundaries of the solution set of $x^2 + 4x - 12 \leq 0$ twice the value of the boundaries of $\frac{1}{2}x^2 + 2x - 6 \leq 0$? Explain.

60. REASONING Determine if the following statement is *sometimes*, *always*, or *never* true. Explain your reasoning.

The intersection of $y \leq -ax^2 + c$ *and* $y \geq ax^2 - c$ *is the empty set.*

61. CHALLENGE Graph the intersection of the graphs of $y \leq -x^2 + 4$ and $y \geq x^2 - 4$.

62. WRITING IN MATH Compare and contrast graphing linear and quadratic inequalities.

NYSRE PRACTICE A2.N.8, A2.N.7

63. GRIDDED RESPONSE You need to seed an area that is 80 feet by 40 feet. Each bag of seed can cover 25 square yards of land. How many bags of seed will you need?

64. SAT/ACT The product of two integers is between 107 and 116. Which of the following *cannot* be one of the integers?

A 5   C 12
B 10   D 15

65 PROBABILITY Five students are to be arranged side by side with the tallest student in the center and the two shortest students on the ends. If no two students are the same height, how many different arrangements are possible?

F 2   H 5
G 4   J 6

66. SHORT RESPONSE Simplify $\frac{5+i}{6-3i}$.

Spiral Review

Write an equation in vertex form for each parabola. (Lesson 5-7)

67.
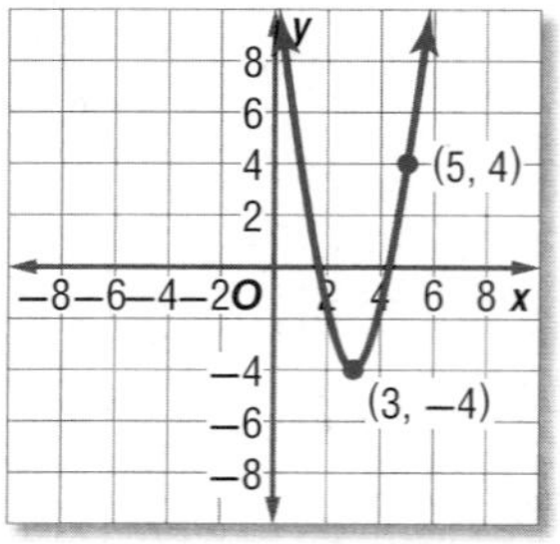

68.
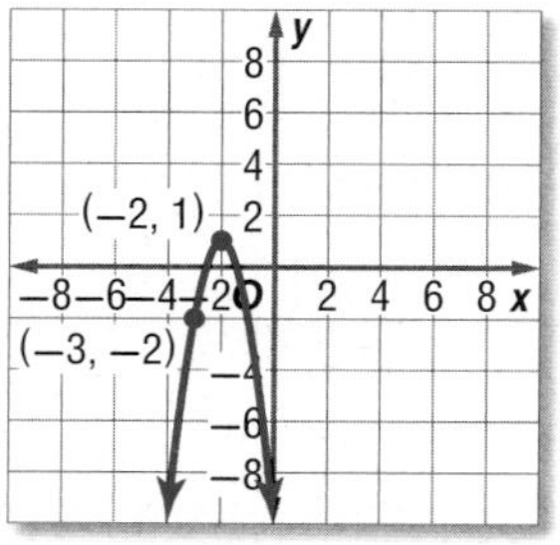

69.
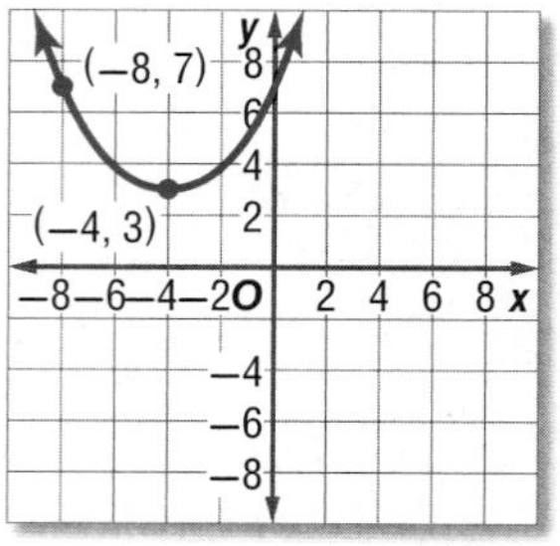

Complete parts a and b for each quadratic equation.
a. Find the value of the discriminant.
b. Describe the number and type of roots. (Lesson 5-6)

70. $4x^2 + 7x - 3 = 0$

71. $-3x^2 + 2x - 4 = 9$

72. $6x^2 + x - 4 = 12$

73. GYMNASTICS Suppose the drawing is placed on a coordinate grid with the hand grips at $H(0, 0)$ and the toe of the figure in the upper right corner at $T(7, 8)$. Find the coordinates of the toes at the other three positions, if each successive position has been rotated 90° counterclockwise about the origin. (Lesson 4-4)

High Bar

A routine with continuous flow to quick changes in body position.

Key move:
Giant swing. As the body swings around the bar the body should be straight with a slight hollow to the chest.

Height: $8\frac{1}{2}$ feet
Length: 8 feet

Perform the indicated operation. If the matrix does not exist, write *impossible*. (Lesson 4-2)

74. $4\begin{bmatrix} 3 & -6 \\ -5 & 2 \end{bmatrix} - 3\begin{bmatrix} 4 & -1 \\ -2 & 8 \end{bmatrix}$

75. $-2\begin{bmatrix} 5 & -9 \\ 5 & 11 \end{bmatrix} - 6\begin{bmatrix} 3 & -7 \\ -5 & 8 \end{bmatrix}$

76. $\begin{bmatrix} 2 & -6 \\ -4 & 6 \end{bmatrix} \cdot \begin{bmatrix} 2 & -1 & 1 \\ -1 & 6 & 4 \end{bmatrix}$

Skills Review

Use the Distributive Property to find each product. (Lesson 1-2)

77. $-6(x - 4)$

78. $8(w + 3x)$

79. $-4(-2y + 3z)$

80. $-1(c - d)$

81. $0.5(5x + 6y)$

82. $-3(-6y - 4z)$

EXTEND 5-8

Graphing Technology Lab
Modeling Motion

NY Math Online glencoe.com
Other Calculator Keystrokes

Set Up the Lab

- Place a board on a stack of books to create a ramp.
- Connect the data collection device to the graphing calculator. Place at the top of the ramp so that the data collection device can read the motion of the car on the ramp.
- Hold the car still about 6 inches up from the bottom of the ramp and zero the collection device.

ACTIVITY

Step 1 One group member should press the button to start collecting data.

Step 2 Another group member places the car at the bottom of the ramp. After data collection begins, gently but quickly push the car so it travels up the ramp toward the motion detector.

Step 3 Stop collecting data when the car returns to the bottom of the ramp. Save the data as Trial 1.

Step 4 Remove one book from the stack. Then repeat the experiment. Save the data as Trial 2. For Trial 3, create a steeper ramp and repeat the experiment.

Analyze the Results

1. What type of function could be used to represent the data? Justify your answer.
2. Use the CALC menu to find the vertex of the graph. Record the coordinates in a table like the one at the right.
3. Use the TRACE feature of the calculator to find the coordinates of another point on the graph. Then use the coordinates of the vertex and the point to find an equation of the graph.
4. Find an equation for each of the graphs of Trials 2 and 3.
5. How do the equations for Trials 1, 2, and 3 compare? Which graph is widest and which is most narrow? Explain what this represents in the context of the situation. How is this represented in the equations?
6. What do the x-intercepts and vertex of each graph represent?
7. Why were the values of h and k different in each trial?

Trial	Vertex (h, k)	Point (x, y)	Equation
1			
2			
3			

Study Guide and Review

NY Math Online glencoe.com
• STUDY TO GO
• Vocabulary Review

Chapter Summary

Key Concepts

Graphing Quadratic Functions (Lesson 5-1)

- The graph of $y = ax^2 + bx + c, a \neq 0$, opens up, and the function has a minimum value when $a > 0$. The graph opens down, and the function has a maximum value when $a < 0$.

Solving Quadratic Equations (Lessons 5-2 and 5-3)

- Roots of a quadratic equation are the zeros of the related quadratic function. You can find the zeros of a quadratic function by finding the x-intercepts of the graph.

Complex Numbers (Lesson 5-4)

- i is the imaginary unit; $i^2 = -1$ and $i = \sqrt{-1}$.

Solving Quadratic Equations (Lessons 5-5 and 5-6)

- Completing the square: **Step 1** Find one half of b, the coefficient of x. **Step 2** Square the result in Step 1. **Step 3** Add the result of Step 2 to $x^2 + bx$.
- Quadratic Formula: $x = \frac{-b \pm \sqrt{b^2 - 4ac}}{2a}$

Transformations with Quadratic Functions (Lesson 5-7)

- The graph of $y = (x - h)^2 + k$ is the graph of $y = x^2$ translated $|h|$ units left if h is negative or $|h|$ units right if h is positive and $|k|$ units up if k is positive or $|k|$ units down if k is negative.
- Consider $y = a(x - h)^2 + k, a \neq 0$. If $a > 0$, the graph opens up; if $a < 0$ the graph opens down. If $|a| > 1$, the graph is narrower than the graph of $y = x^2$. If $|a| < 1$, the graph is wider than the graph of $y = x^2$.

Quadratic Inequalities (Lesson 5-8)

- Graph the related function, test a point on the parabola and determine if it is a solution, and shade accordingly.

FOLDABLES® Study Organizer

Be sure the Key Concepts are noted in your Foldable.

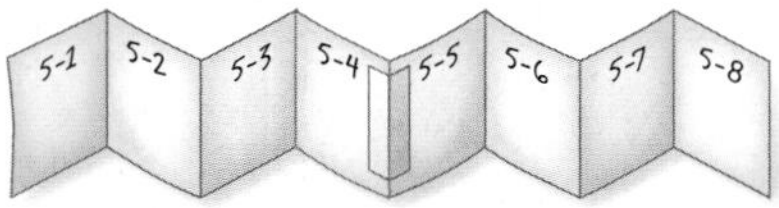

Key Vocabulary

- **axis of symmetry** (p. 250)
- **complex conjugates** (p. 279)
- **complex number** (p. 277)
- **completing the square** (p. 285)
- **constant term** (p. 249)
- **discriminant** (p. 295)
- **factored form** (p. 268)
- **FOIL method** (p. 268)
- **imaginary unit** (p. 276)
- **linear term** (p. 249)
- **maximum value** (p. 252)
- **minimum value** (p. 252)
- **parabola** (p. 249)
- **pure imaginary number** (p. 276)
- **quadratic equation** (p. 259)
- **Quadratic Formula** (p. 292)
- **quadratic function** (p. 249)
- **quadratic inequality** (p. 312)
- **quadratic term** (p. 249)
- **root** (p. 259)
- **Square Root Property** (p. 277)
- **standard form** (p. 259)
- **vertex** (p. 250)
- **vertex form** (p. 305)
- **zero** (p. 259)

Vocabulary Check

State whether each sentence is *true* or *false*. If *false*, replace the underlined term to make a true sentence.

1. The <u>factored form</u> of a quadratic equation is $ax^2 + bx + c = 0$ where $a \neq 0$ and a, b, and c are integers.
2. The graph of a quadratic function is called a <u>parabola</u>.
3. The <u>vertex form</u> of a quadratic function is $y = a(x - p)(x - q)$.
4. The axis of symmetry will intersect a parabola in one point called the <u>vertex</u>.
5. A method called <u>FOIL method</u> is used to make a quadratic expression a perfect square in order to solve the related equation.
6. The equation $x = \frac{-b \pm \sqrt{b^2 - 4ac}}{2a}$ is known as the <u>discriminant</u>.
7. The number $6i$ is called a <u>pure imaginary number</u>.
8. The two numbers $2 + 3i$ and $2 - 3i$ are called <u>complex conjugates</u>.

Lesson-by-Lesson Review

5-1 Graphing Quadratic Functions (pp. 249–257)

A2.A.41

Complete parts a–c for each quadratic function.

a. **Find the y-intercept, the equation of the axis of symmetry, and the x-coordinate of the vertex.**

b. **Make a table of values that includes the vertex.**

c. **Use this information to graph the function.**

9. $f(x) = x^2 + 5x + 12$ **10.** $f(x) = x^2 - 7x + 15$

11. $f(x) = -2x^2 + 9x - 5$ **12.** $f(x) = -3x^2 + 12x - 1$

Determine whether each function has a maximum or minimum value and find the maximum or minimum value. Then state the domain and range of the function.

13. $f(x) = -x^2 + 3x - 1$ **14.** $f(x) = -3x^2 - 4x + 5$

15. **BUSINESS** Sal's Shirt Store sells 100 T-shirts per week at a rate of \$10 per shirt. Sal estimates that he will sell 5 less shirts for each \$1 increase in price. What price will maximize Sal's T-shirt income?

EXAMPLE 1

Consider the quadratic function $f(x) = x^2 - 4x + 11$. Find the y-intercept, the equation for the axis of symmetry, and the x-coordinate of the vertex.

In the function, $a = 1$, $b = -4$, and $c = 11$. The y-intercept is $c = 11$.

Use a and b to find the equation of the axis of symmetry.

$x = -\frac{b}{2a}$	Equation of the axis of symmetry.
$= -\frac{-4}{2(1)}$	$a = 1, b = -4$
$= 2$	Simplify.

The equation of the axis of symmetry is $x = 2$. Therefore, the x-coordinate of the vertex is 2.

5-2 Solving Quadratic Functions by Graphing (pp. 259–266)

A.A.8, A.G.8

Solve each equation by graphing. If exact roots cannot be found, state the consecutive integers between which the roots are located.

16. $x^2 - x - 20 = 0$

17. $2x^2 - x - 3 = 0$

18. $4x^2 - 6x - 15 = 0$

19. **BASEBALL** A baseball is hit upward at 120 feet per second. Use the formula $h(t) = v_0t - 16t^2$, where $h(t)$ is the height of an object in feet, v_0 is the object's initial velocity in feet per second, and t is the time in seconds. Ignoring the height of the ball when it was hit, how long does it take for the ball to hit the ground?

EXAMPLE 2

Solve $2x^2 - 7x + 3 = 0$ by graphing.

The equation of the axis of symmetry is $-\frac{-7}{2(2)}$ or $x = \frac{7}{4}$.

x	0	1	$\frac{7}{4}$	2	3
$f(x)$	3	−2	$-2\frac{5}{8}$	−3	0

The zeros of the related function are $\frac{1}{2}$ and 3. Therefore, the solutions of the equation are $\frac{1}{2}$ and 3.

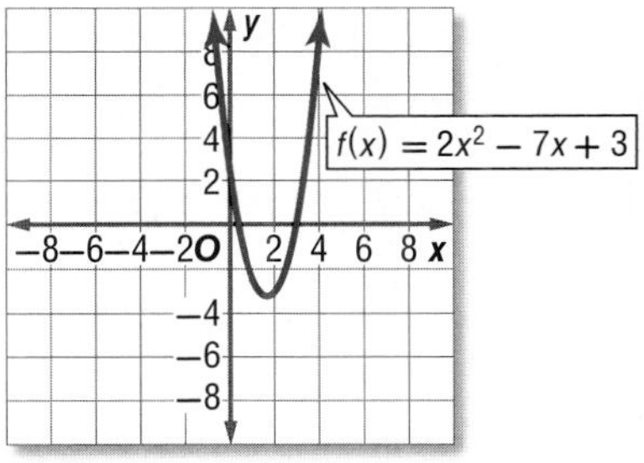

CHAPTER 5

Study Guide and Review

5-3 Solving Quadratic Equations by Factoring (pp. 268–275)

A2.A.7

Write a quadratic equation in standard form with the given roots.

20. 5, 6

21. −3, −7

22. −4, 2

23. $-\frac{2}{3}, 1$

24. $\frac{1}{6}, 5$

25. $-\frac{1}{4}, -1$

Solve each equation by factoring.

26. $2x^2 - 2x - 24 = 0$

27. $2x^2 - 5x - 3 = 0$

28. $3x^2 - 16x + 5 = 0$

29. Find x and the dimensions of the rectangle below.

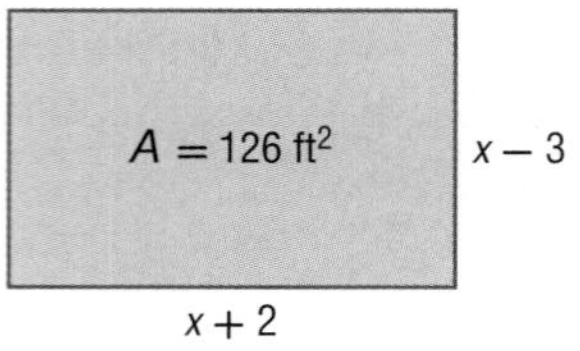

EXAMPLE 3

Write a quadratic equation in standard form with $-\frac{1}{2}$ and 4 as its roots.

$(x - p)(x - q) = 0$ **Write the pattern.**

$\left[x - \left(-\frac{1}{2}\right)\right](x - 4) = 0$ **Replace p with $-\frac{1}{2}$ and q with 4.**

$\left(x + \frac{1}{2}\right)(x - 4) = 0$ **Simplify.**

$x^2 - \frac{7}{2}x - 2 = 0$ **Multiply.**

$2x^2 - 7x - 4 = 0$ **Multiply each side by 2 so that *b* and *c* are integers.**

EXAMPLE 4

Solve $2x^2 - 3x - 5 = 0$ by factoring.

$2x^2 - 3x - 5 = 0$ **Original equation**

$(2x - 5)(x + 1) = 0$ **Factor the trinomial.**

$2x - 5 = 0$ or $x + 1 = 0$ **Zero Product Property**

$x = \frac{5}{2}$ or $x = -1$

The solution set is $\left\{-1, \frac{5}{2}\right\}$.

5-4 Complex Numbers (pp. 276–282)

A2.N.6, A2.N.7, A2.N.8

Simplify.

30. $\sqrt{-8}$

31. $(2 - i) + (13 + 4i)$

32. $(6 + 2i) - (4 - 3i)$

33. $(6 + 5i)(3 - 2i)$

34. **ELECTRICITY** The impedance in one part of a series circuit is $3 + 2j$ ohms, and the impedance in the other part of the circuit is $4 - 3j$ ohms. Add these complex numbers to find the total impedance in the circuit.

Solve each equation.

35. $2x^2 + 50 = 0$

36. $4x^2 + 16 = 0$

37. $3x^2 + 15 = 0$

38. $8x^2 + 16 = 0$

39. $4x^2 + 1 = 0$

EXAMPLE 5

Simplify $(12 + 3i) - (-5 + 2i)$.

$(12 + 3i) - (-5 + 2i)$

$= [12 - (-5)] + (3 - 2)i$ **Group the real and imaginary parts.**

$= 17 + i$ **Add.**

EXAMPLE 6

Solve $3x^2 + 12 = 0$.

$3x^2 + 12 = 0$ **Original equation**

$3x^2 = -12$ **Subtract 12 from each side.**

$x^2 = -4$ **Divide each side by 3.**

$x = \pm\sqrt{-4}$ **Square Root Property**

$x = \pm 2i$ $\sqrt{-4} = \sqrt{4} \cdot \sqrt{-1}$

MIXED PROBLEM SOLVING
For mixed problem-solving practice, see page 983.

5-5 Completing the Square (pp. 284–290)

A2.A.24

Find the value of c that makes each trinomial a perfect square. Then write the trinomial as a perfect square.

40. $x^2 + 18x + c$

41. $x^2 - 4x + c$

42. $x^2 - 7x + c$

43. $x^2 + 2.4x + c$

44. $x^2 - \frac{1}{2}x + c$

45. $x^2 + \frac{6}{5}x + c$

Solve each equation by completing the square.

46. $x^2 - 6x - 7 = 0$

47. $x^2 - 2x + 8 = 0$

48. $2x^2 + 4x - 3 = 0$

49. $2x^2 + 3x - 5 = 0$

50. **FLOOR PLAN** Mario's living room has a length 6 feet wider than the width. The area of the living room is 280 square feet. What are the dimensions of his living room?

EXAMPLE 7

Find the value of c that makes $x^2 + 14x + c$ a perfect square. Then write the trinomial as a perfect square.

Step 1 Find one half of 14.

Step 2 Square the result of Step 1.

Step 3 Add the result of Step 2 to $x^2 + 14x$.

The trinomial $x^2 + 14x + 49$ can be written as $(x + 7)^2$.

EXAMPLE 8

Solve $x^2 + 12x - 13 = 0$ by completing the square.

$$x^2 + 12x - 13 = 0$$
$$x^2 + 12x = 13$$
$$x^2 + 12x + 36 = 13 + 36$$
$$(x + 6)^2 = 49$$
$$x + 6 = \pm 7$$

$x + 6 = 7$ or $x + 6 = -7$

$x = 1$ $\quad$ $x = -13$

The solution set is $\{-13, 1\}$.

5-6 The Quadratic Formula and the Discriminant (pp. 292–300)

A2.A.2, A2.A.25

Complete parts a–c for each quadratic equation.

a. **Find the value of the discriminant.**

b. **Describe the number and type of roots.**

c. **Find the exact solutions by using the Quadratic Formula.**

51. $x^2 - 10x + 25 = 0$

52. $x^2 + 4x - 32 = 0$

53. $2x^2 + 3x - 18 = 0$

54. $2x^2 + 19x - 33 = 0$

55. $x^2 - 2x + 9 = 0$

56. $4x^2 - 4x + 1 = 0$

57. $2x^2 + 5x + 9 = 0$

58. **PHYSICAL SCIENCE** Lauren throws a ball with an initial velocity of 40 feet per second. The equation for the height of the ball is $h = -16t^2 + 40t + 5$, where h represents the height in feet and t represents the time in seconds. When will the ball hit the ground?

EXAMPLE 9

Solve $x^2 - 4x - 45 = 0$ by using the Quadratic Formula.

In $x^2 - 4x - 45 = 0$, $a = 1$, $b = -4$, and $c = -45$.

$$x = \frac{-b \pm \sqrt{b^2 - 4ac}}{2a} \quad \text{Quadratic Formula}$$

$$= \frac{-(-4) \pm \sqrt{(-4)^2 - 4(1)(-45)}}{2(1)}$$

$$= \frac{4 \pm 14}{2}$$

Write as two equations.

$x = \frac{4 + 14}{2}$ or $x = \frac{4 - 14}{2}$

$= 9$ $\quad$ $= -5$

The solution set is $\{-5, 9\}$.

5-7 Transformations with Quadratic Functions (pp. 305–310)

A.G.5

Write each quadratic function in vertex form, if not already in that form. Then identify the vertex, axis of symmetry, and direction of opening. Then graph the function.

59. $y = -3(x - 1)^2 + 5$ **60.** $y = 2x^2 + 12x - 8$

61. $y = -\frac{1}{2}x^2 - 2x + 12$ **62.** $y = 3x^2 + 36x + 25$

63. The graph at the right shows a product of 2 numbers with a sum of 10. Find a function that models this product and use it to determine the two numbers that would give a maximum product.

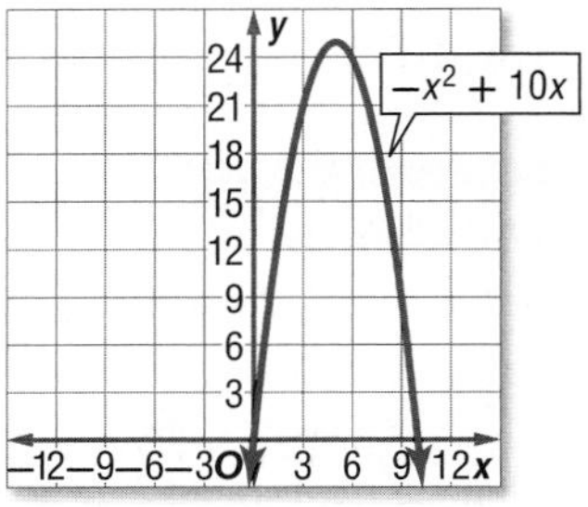

EXAMPLE 10

Write the quadratic function $y = 3x^2 + 24x + 15$ in vertex form. Then identify the vertex, axis of symmetry, and direction of opening.

$y = 3x^2 + 24x + 15$	**Original equation**
$y = 3(x^2 + 8x) + 15$	**Group and factor.**
$y = 3(x^2 + 8x + 16) + 15 - 3(16)$	**Complete the square.**
$y = 3(x + 4)^2 - 33$	**Rewrite $x^2 + 8x + 16$ as a perfect square.**

So, $a = 3$, $h = -4$, and $k = -33$. The vertex is at $(-4, -33)$ and the axis of symmetry is $x = -4$. Since a is positive, the graph opens up.

5-8 Quadratic Inequalities (pp. 312–318)

A2.A.4

Graph each quadratic inequality.

64. $y \geq x^2 + 5x + 4$ **65.** $y < -x^2 + 5x - 6$

66. $y > x^2 - 6x + 8$ **67.** $y \leq x^2 + 10x - 4$

68. Solomon wants to put a deck along two sides of his garden. The deck width will be the same on both sides and the total area of the garden and deck cannot exceed 500 square feet. How wide can the deck be?

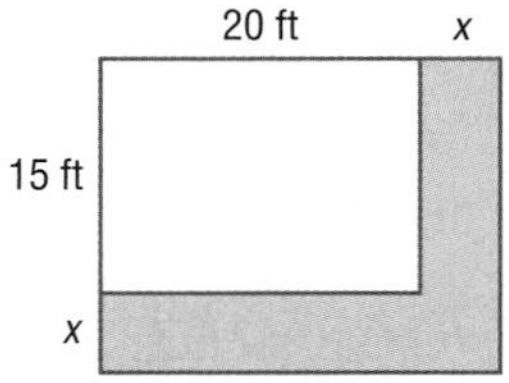

Solve each inequality using a graph or algebraically.

69. $x^2 + 8x + 12 > 0$

70. $6x + x^2 \geq -9$

71. $2x^2 + 3x - 20 > 0$

72. $4x^2 - 3 < -5x$

73. $3x^2 + 4 > 8x$

EXAMPLE 11

Graph $y > x^2 + 3x + 2$.

Step 1 Graph the related function, $y > x^2 + 3x + 2$. Because the inequality symbol > is used, the parabola should be dashed.

Step 2 Test a point not on the graph of the parabola such as (0, 0).

$y > x^2 + 3x + 2$
$(0) \stackrel{?}{>} 0^2 + 3(0) + 2$
$0 \not> 2$

So, (0, 0) is not a solution of the inequality.

Step 3 Shade the region that does not contain the point (0, 0).

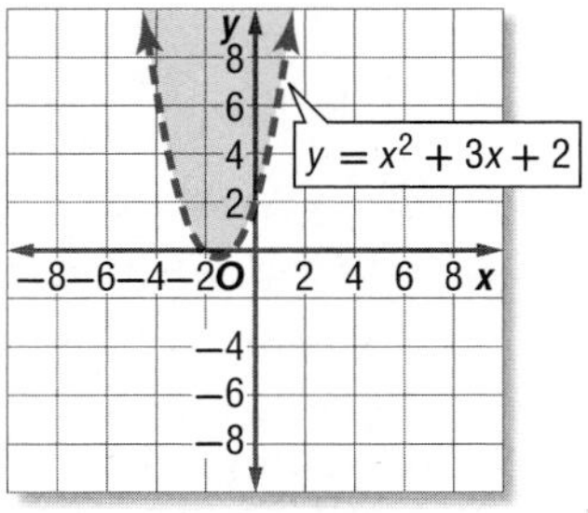

CHAPTER 5 Practice Test

NY Math Online glencoe.com
Chapter Test

Complete parts a–c for each quadratic function.
a. Find the y-intercept, the equation of the axis of symmetry, and the x-coordinate of the vertex.
b. Make a table of values that includes the vertex.
c. Use this information to graph the function.

1. $f(x) = x^2 + 4x - 7$
2. $f(x) = -2x^2 + 5x$
3. $f(x) = -x^2 - 6x - 9$

Determine whether each function has a maximum or minimum value. State the maximum or minimum value of each function.

4. $f(x) = x^2 + 10x + 25$
5. $f(x) = -x^2 + 6x$

Solve each equation using the method of your choice. Find exact solutions.

6. $x^2 - 8x - 9 = 0$
7. $-4.8x^2 + 1.6x + 24 = 0$
8. $12x^2 + 15x - 4 = 0$
9. $x^2 - 7x - \frac{17}{4} = 0$
10. $4x^2 + x = 3$
11. $-9x^2 + 40x + 84 = 0$
12. **PHYSICAL SCIENCE** Parker throws a ball off the top of a building. The building is 350 feet high and the initial velocity of the ball is 30 feet per second. Find out how long it will take the ball to hit the ground by solving the equation $-16t^2 - 30t + 350 = 0$.
13. **MULTIPLE CHOICE** Which equation below has roots at -6 and $\frac{1}{5}$?

    A $0 = 5x^2 - 29x - 6$

    B $0 = 5x^2 + 31x + 6$

    C $0 = 5x^2 + 29x - 6$

    D $0 = 5x^2 - 31x + 6$
14. **PHYSICS** A ball is thrown into the air vertically with a velocity of 112 feet per second. The ball was released 6 feet above the ground. The height above the ground t seconds after release is modeled by $h(t) = -16t^2 + 112t + 6$.

    a. When will the ball reach 130 feet?

    b. Will the ball ever reach 250 feet? Explain.

    c. In how many seconds after its release will the ball hit the ground?
15. The rectangle below has an area of 104 square inches. Find the value of x and the dimensions of the rectangle.

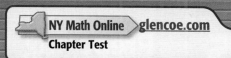

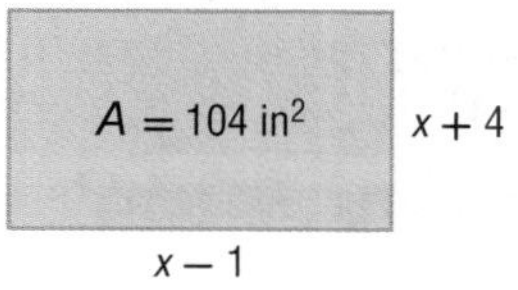

Simplify.

16. $(3 - 4i) - (9 - 5i)$
17. $\frac{4i}{4 - i}$
18. **MULTIPLE CHOICE** Which value of c makes the trinomial $x^2 - 12x + c$ a perfect square trinomial?

    F 6

    G 12

    H 36

    J 144

Complete parts a–c for each quadratic equation.
a. Find the value of the discriminant.
b. Describe the number and type of roots.
c. Find the exact solution by using the Quadratic Formula.

19. $6x^2 + 7x = 0$
20. $5x^2 = -6x + 1$
21. $2x^2 + 5x - 8 = -13$

Write each quadratic function in vertex form. Then identify the vertex, axis of symmetry, and direction of opening.

22. $3x^2 + 6x = 2 + y$
23. $x^2 + 9x + \frac{81}{4} = y$
24. Graph the quadratic inequality $0 < -3x^2 + 4x + 10$.

Solve each inequality by using a graph or algebraically.

25. $x^2 + 6x > -5$
26. $4x^2 - 19x \leq -12$

Preparing for Standardized Tests

Use a Graph

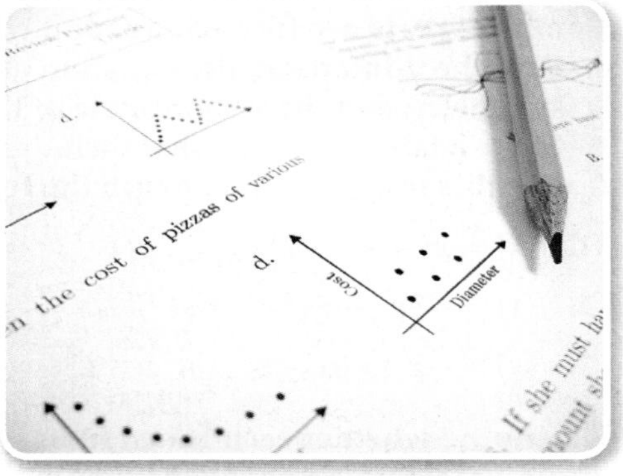

Using a graph can help you solve many different kinds of problems on standardized tests. Graphs can help you solve equations, evaluate functions, and interpret solutions to real-world problems.

Strategies for Using a Graph

Step 1

Read the problem statement carefully.

Ask yourself:

- What am I being asked to solve?
- What information is given in the problem?
- How could a graph help me solve the problem?

Step 2

Create your graph.

- Sketch your graph on scrap paper if appropriate.
- If allowed, you can also use a graphing calculator to create the graph.

Step 3

Solve the problem.

- Use your graph to help you model and solve the problem.
- Check to be sure your answer makes sense.

EXAMPLE

Read the problem. Identify what you need to know. Then use the information in the problem to solve.

The students in Mr. Himebaugh's physics class built a model rocket. The rocket is launched in a large field with an initial upward velocity of 128 feet per second. The function $h(t) = -16t^2 + 128t$ models the height of the rocket above the ground (in feet) t seconds after it is launched. How long will it take for the rocket to reach its maximum height?

A 4 seconds

B 5 seconds

C 6 seconds

D 8 seconds

Graphing the quadratic function will allow you to determine the peak height of the rocket and when it occurs. A graphing calculator can help you quickly graph the function and analyze it.

KEYSTROKES: [Y=] [(−)] 16 [X,T,θ,n] [x^2] [+] 128 [X,T,θ,n] [Graph]

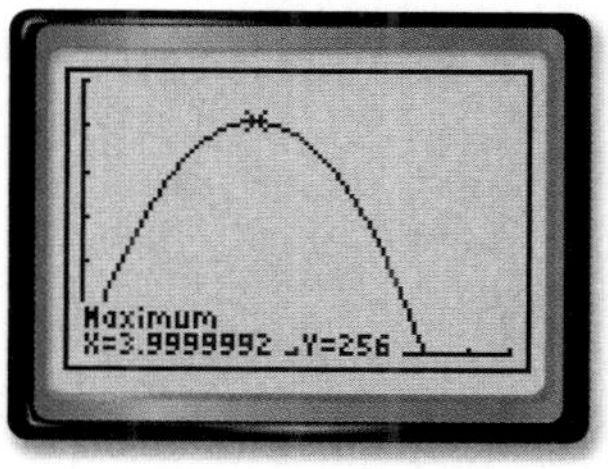

After graphing the equation, use **maximum** under the **CALC** menu.

Press [2nd] [CALC] 4. Then use [◀] to place the cursor to the left of the maximum point and press [ENTER]. Use [▶] to place the cursor to the right of the maximum point and press [ENTER] [ENTER].

The graph shows that the rocket takes 4 seconds to reach its maximum height of 256 feet. The correct answer is A.

Exercises

Read each problem. Identify what you need to know. Then use the information in the problem to solve.

1. What are the roots of $y = 2x^2 + 10x - 48$?

   A $-5, 4$

   B $-6, 1$

   C $-8, 3$

   D $2, 3$

2. How many times does the graph of $f(x) = x^3 + 2x^2 - 3x + 2$ cross the x-axis?

   F 0   H 2

   G 1   J 3

3. Which statement best describes the graphs of the two equations?

$$16x - 2y = 24$$
$$12x = 3y - 36$$

   A The lines are parallel.

   B The lines are the same.

   C The lines intersect in only one point.

   D The lines intersect in more than one point, but are not the same.

4. Adrian is using 120 feet of fencing to enclose a rectangular area for her puppy. One side of the enclosure will be her house.

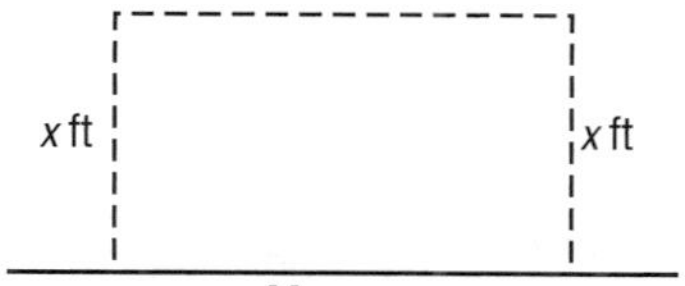

   The function $f(x) = x(120 - 2x)$ represents the area of the enclosure. What is the greatest area Adrian can enclose with the fencing?

   F 1,650 ft^2 H 1,980 ft^2

   G 1,800 ft^2 J 2,140 ft^2

5. For which equation is the x-coordinate of the vertex at 4?

   A $f(x) = x^2 - 8x + 15$ C $f(x) = x^2 + 6x + 8$

   B $f(x) = -x^2 - 4x + 12$ D $f(x) = -x^2 - 2x + 2$

6. For what value of x does $f(x) = x^2 + 5x + 6$ reach its minimum value?

   F -5 H $-\frac{5}{2}$

   G -3 J -2

CHAPTER 5 NYS Regents Exam Practice

Cumulative, Chapters 1 through 5

Multiple Choice

Answer all questions in this part. Select the answer that best completes the statement or answers the question.

1. The quadratic formula shows the height (h, in feet) of a football t seconds after it was kicked up into the air. How long does it take for the ball to reach its maximum height?

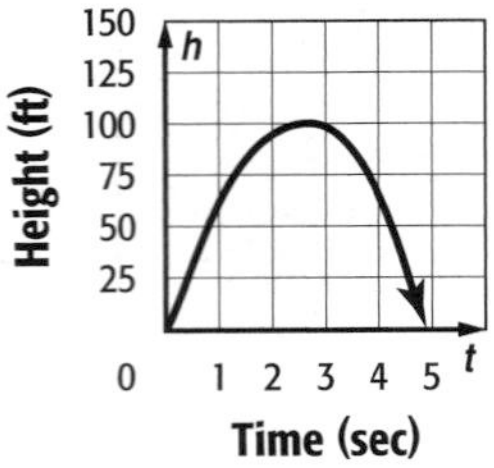

(1) 2 seconds

(2) 2.5 seconds

(3) 3.5 seconds

(4) 5 seconds

2. The New York state quarter was released on January 2, 2001. It features the state outline and the Statue of Liberty along with the slogan "Gateway to Freedom." The diameter of the quarter is 24.26 millimeters. To which set of numbers does the number 24.26 belong?

(1) integers

(2) irrational numbers

(3) rational numbers

(4) whole numbers

3. Solve for x in the matrix equation below.

$$\begin{bmatrix} 42 & 5y+1 \\ 3x-7 & 50 \\ 12 & z+8 \end{bmatrix} = \begin{bmatrix} 42 & 26 \\ 35 & 50 \\ 12 & 17 \end{bmatrix}$$

(1) 12 **(3)** 16

(2) 14 **(4)** 18

4. The soccer team is selling T-shirts and hats to raise money for their new uniforms. The table below shows how much each item costs and how much they will be sold for.

Item	Cost	Price
T-shirt	\$4.50	\$10.00
Hat	\$2.25	\$7.00

If the team spends a total of \$209.25 on T-shirts and hats and sells the merchandise for \$507.00, how many T-shirts will they sell? Set up and solve a system of equations.

(1) 36 T-shirts **(3)** 29 T-shirts

(2) 32 T-shirts **(4)** 21 T-shirts

5. The function $y = (x - h)^2$ is graphed below. What is the value of h?

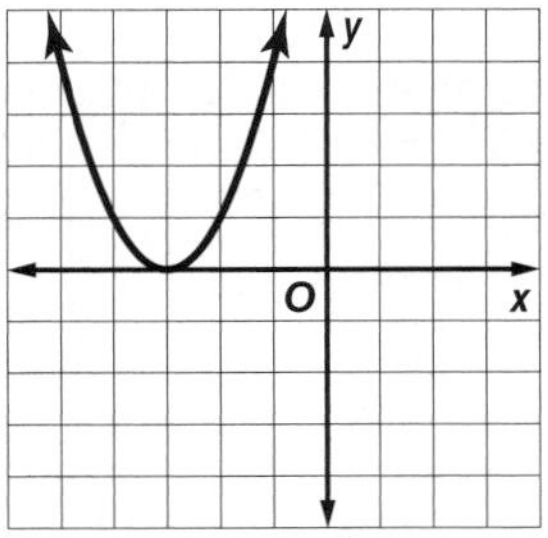

(1) 3 **(3)** $-\frac{1}{3}$

(2) $\frac{1}{3}$ **(4)** -3

Test-Taking Tip

Question 5 Remember, you can quickly check your answer using a graphing calculator.

6. Which of the following best describes the correlation between the grade earned on a final exam and the amount of time spent studying for the exam?

   (1) positive correlation

   (2) negative correlation

   (3) weak correlation

   (4) no correlation

7. Evaluate $ab - b^2 \div c + 2$ if $a = -2$, $b = 6$, and $c = 4$.

   (1) -7 **(3)** -19

   (2) -18 **(4)** -22

8. Solve the quadratic equation below by completing the square.

$$x^2 - 8x = -15$$

   (1) $-1, 4$ **(3)** $3, 7$

   (2) $3, 5$ **(4)** $5, 10$

9. Suppose the graph of $g(x)$ is translated down 12 units to produce the graph of the function $h(x)$. Which of the following could be $h(x)$?

$$g(x) = -\frac{1}{4}x^2 - 8x + 20$$

   (1) $h(x) = -\frac{1}{4}x^2 + 4x + 20$

   (2) $h(x) = -\frac{1}{4}x^2 - 20x + 20$

   (3) $h(x) = -\frac{1}{4}x^2 - 8x + 32$

   (4) $h(x) = -\frac{1}{4}x^2 - 8x + 8$

Open-Ended Response

Solve each problem. Clearly indicate the necessary steps, including appropriate formula substitutions, diagrams, graphs, charts, etc.

10. Use the quadratic function below to answer each question.

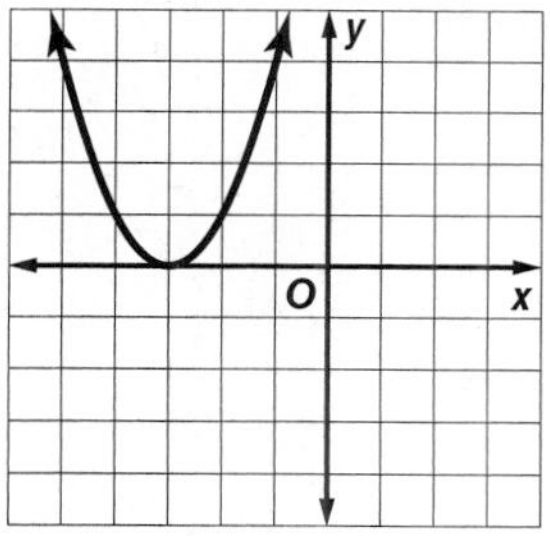

   a. How many real roots does the quadratic function have?

   b. How many complex roots does the function have?

   c. What do you know about the discriminant of the quadratic equation? Explain.

11. The general quadratic equation is $y = ax^2 + bx + c$ for real coefficients a, b, and c.

   a. Describe what the discriminant $b^2 - 4ac$ tells you about the roots of the equation.

   b. Give an example of sets of coefficients that would result in 2 real solutions; in 1 real solution; in 2 complex solutions.

Need Extra Help?											
If you missed Question...	1	2	3	4	5	6	7	8	9	10	11
Go to Lesson or Page...	5-1	5-2	2-4	1-3	5-5	5-7	4-4	5-6	3-2	5-4	5-2
NYS Core Curriculum	A.A.8	A.CM.11	A2.R.2	A.A.7	A.G.5	A2.S.8	A2.N.1	A2.A.24	A.G.5	A2.A.2	A2.A.2

CHAPTER 6 Polynomials and Polynomial Functions

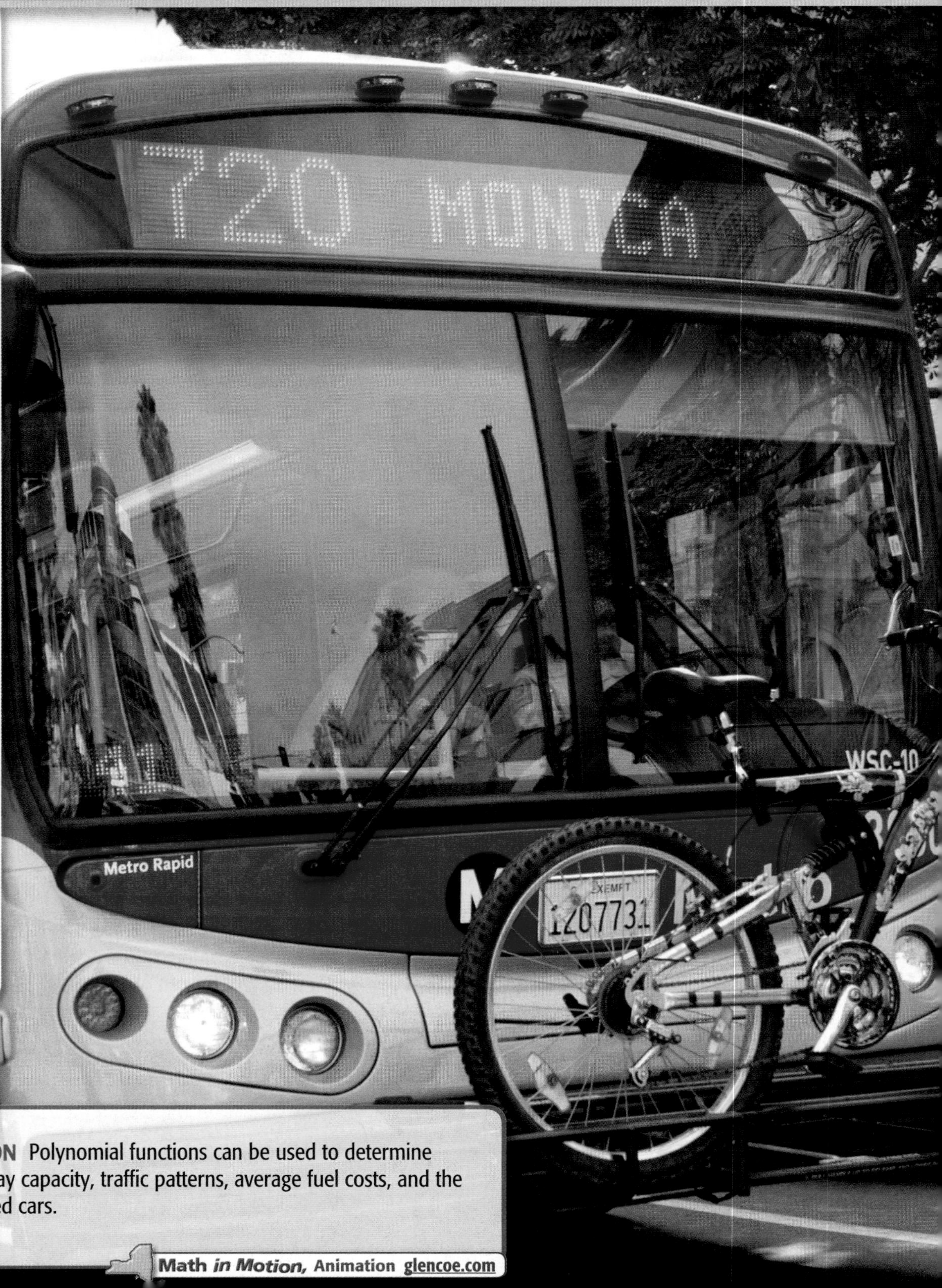

Then

In Chapter 5, you graphed quadratic functions and solved quadratic equations.

Now

In Chapter 6, you will:

- Add, subtract, multiply, divide, and factor polynomials.
- Analyze and graph polynomial functions.
- Evaluate polynomial functions and solve polynomial equations.
- Find factors and zeros of polynomial functions.

NYS Core Curriculum

A2.A.8 Apply the rules of exponents to simplify expressions involving negative and/or fractional exponents

A2.A.26 Find solutions to polynomial equations of higher degree

Why?

TRANSPORTATION Polynomial functions can be used to determine bus schedules, highway capacity, traffic patterns, average fuel costs, and the prices of new and used cars.

Math *in Motion*, Animation glencoe.com

Get Ready for Chapter 6

Diagnose Readiness You have two options for checking Prerequisite Skills.

Text Option

Take the Quick Check below. Refer to the Quick Review for help.

QuickCheck

Rewrite each difference as a sum. (Lesson 1-2)

1. $-5 - 13$
2. $5 - 3y$
3. $5mr - 7mp$
4. $3x^2y - 14xy^2$
5. **PARTIES** Twenty people attended a going away party for Zach. The guests left in groups of 2. By 9:00, x groups had left. Rewrite the number of guests remaining at 9:00 as a sum.

Use the Distributive Property to rewrite each expression without parentheses. (Lesson 1-2)

6. $-4(a + 5)$
7. $-1(3b^2 + 2b - 1)$
8. $-\frac{1}{2}(2m - 5)$
9. $-\frac{3}{4}(3z + 5)$
10. **MONEY** Mr. Chávez is buying pizza and soda for the members of the science club. A slice of pizza costs \$2.25, and a soda costs \$1.25. Write an expression to represent the amount that Mr. Chávez will spend on 15 students. Evaluate the expression by using the Distributive Property.

Solve each equation. (Lesson 5-6)

11. $x^2 + 2x - 8 = 0$
12. $2x^2 + 7x + 3 = 0$
13. $6x^2 + 5x - 4 = 0$
14. $4x^2 - 2x - 1 = 0$
15. **PHYSICS** If an object is dropped from a height of 50 feet above the ground, then its height after t seconds is given by $h = -16t^2 + 50$. Use the equation $0 = -16t^2 + 50$ to find how long it will take until the ball reaches the ground.

QuickReview

EXAMPLE 1

Rewrite $2xy - 3 - z$ as a sum.

$2xy - 3 - z$ **Original expression**

$= 2xy + (-3) + (-z)$ **Rewrite using addition.**

EXAMPLE 2

Use the Distributive Property to rewrite $-3(a + b - c)$.

$-3(a + b - c)$ **Original expression**

$= -3(a) + (-3)(b) + (-3)(-c)$ **Distributive Property**

$= -3a - 3b + 3c$ **Simplify.**

EXAMPLE 2

Solve $2x^2 + 8x + 1 = 0$.

$x = \dfrac{-b \pm \sqrt{b^2 - 4ac}}{2a}$ **Quadratic Formula**

$= \dfrac{-8 \pm \sqrt{8^2 - 4(2)(1)}}{2(2)}$ **$a = 2, b = 8, c = 1$**

$= \dfrac{-8 \pm \sqrt{56}}{4}$ **Simplify.**

$= -2 \pm \dfrac{\sqrt{14}}{2}$ **$\sqrt{56} = \sqrt{4 \cdot 14}$ or $2\sqrt{14}$**

The exact solutions are $-2 + \frac{\sqrt{14}}{2}$ and $-2 - \frac{\sqrt{14}}{2}$.

The approximate solutions are -0.13 and -3.88.

Online Option

NY Math Online Take a self-check Chapter Readiness Quiz at glencoe.com.

Get Started on Chapter 6

You will learn several new concepts, skills, and vocabulary terms as you study Chapter 6. To get ready, identify important terms and organize your resources. You may wish to refer to **Chapter 0** to review prerequisite skills.

FOLDABLES® Study Organizer

Polynomials and Polynomial Functions Make this Foldable to help you organize your Chapter 6 notes about polynomials and polynomial functions. Begin with one sheet of $8\frac{1}{2}''$ by 14″ paper.

1. **Fold** a 2″ tab along the bottom of a long side.

2. **Fold** along the width into thirds.

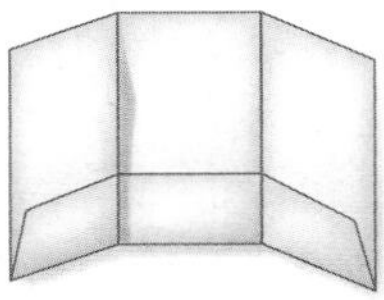

3. **Staple** the outer edges of the tab.

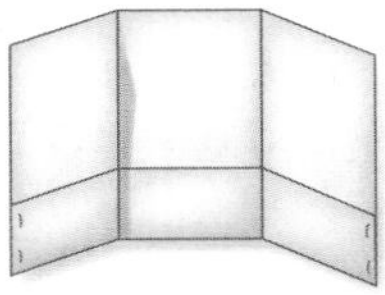

4. **Label** the tabs *Polynomials, Polynomial Functions and Graphs*, and *Solving Polynomial Equations*.

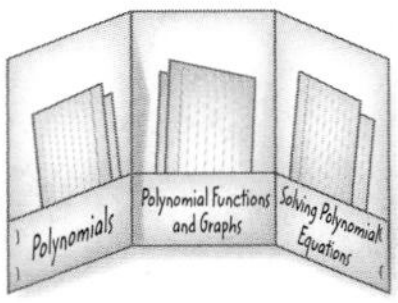

NY Math Online glencoe.com

- Study the chapter online
- Explore **Math in Motion**
- Get extra help from your own **Personal Tutor**
- Use **Extra Examples** for additional help
- Take a **Self-Check Quiz**
- **Review Vocabulary** in fun ways

New Vocabulary

English		Español
simplify	• p. 333 •	reducer
degree of a polynomial	• p. 335 •	grado de un polinomio
synthetic division	• p. 342 •	división sintética
polynomial in one variable	• p. 348 •	polinomio de una variable
leading coefficient	• p. 348 •	coeficiente líder
polynomial function	• p. 349 •	función polinomial
power function	• p. 349 •	función potencia
end behavior	• p. 350 •	comportamiento final
relative maximum	• p. 358 •	máximo relativo
relative minimum	• p. 358 •	mínimo relativo
extrema	• p. 358 •	extrema
turning points	• p. 358 •	momentos cruciales
prime polynomials	• p. 368 •	polinomios primeros
quadratic form	• p. 371 •	forma de ecuación cuadrática
synthetic substitution	• p. 377 •	sustitución sintética
depressed polynomial	• p. 379 •	polinomio reducido

Review Vocabulary

factoring • p. 368 • factorización to express a polynomial as the product of monomials and polynomials

function • p. 348 • función a relation in which each element of the domain is paired with exactly one element in the range

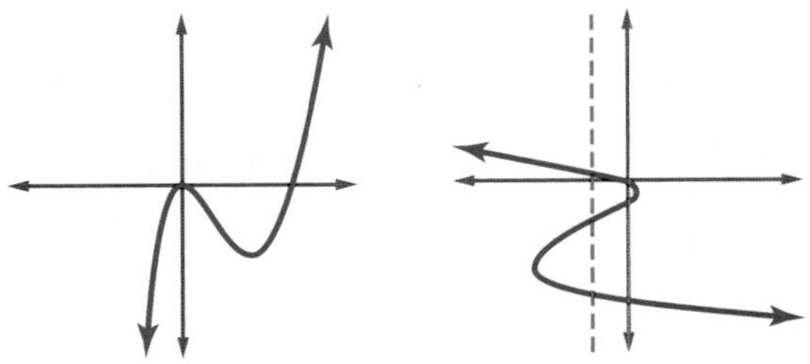

Function   Not a Function

polynomial • p. 335 • polinomio a monomial or sum of monomials

Multilingual eGlossary glencoe.com

6-1 Operations with Polynomials

Why?

The light from the Sun takes approximately 8 minutes to reach Earth. So if you are outside right now you are basking in sunlight that the Sun emitted approximately 8 minutes ago.

Light travels very fast, at a speed of about 3×10^8 meters per second. How long would it take light to get here from the Andromeda galaxy, which is approximately 2.367×10^{21} meters away?

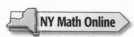

Then

You evaluated powers. (Lesson 1-1)

Now

- Multiply, divide, and simplify monomials and expressions involving powers.
- Add, subtract, and multiply polynomials.

NYS Core Curriculum

A2.A.8 Apply the rules of exponents to simplify expressions involving negative and/or fractional exponents
A2.A.9 Rewrite algebraic expressions that contain negative exponents using only positive exponents
Also addresses A2.N.1.

New Vocabulary

simplify
degree of a polynomial

NY Math Online

glencoe.com

- Extra Examples
- Personal Tutor
- Self-Check Quiz
- Homework Help
- Math in Motion

Multiply and Divide Monomials To **simplify** an expression containing powers means to rewrite the expression without parentheses or negative exponents. Negative exponents are a way of expressing the multiplicative inverse of a number. The following table summarizes the properties of exponents.

Concept Summary — **Properties of Exponents** — For Your FOLDABLE

For any real numbers x and y, integers a and b:

Property	Definition	Examples
Product of Powers	$x^a \cdot x^b = x^{a+b}$	$3^2 \cdot 3^4 = 3^{2+4}$ or 3^6 $p^2 \cdot p^9 = p^{2+9}$ or p^{11}
Quotient of Powers	If $x \neq 0$, $\frac{x^a}{x^b} = x^{a-b}$.	$\frac{9^5}{9^2} = 9^{5-2}$ or 9^3 $\frac{b^6}{b^4} = b^{6-4}$ or b^2
Negative Exponent	$x^{-a} = \frac{1}{x^a}$ and $\frac{1}{x^{-a}} = x^a$	$3^{-5} = \frac{1}{3^5}$ $\frac{1}{b^{-7}} = b^7$
Power of a Power	$(x^a)^b = x^{ab}$	$(3^3)^2 = 3^{3 \cdot 2}$ or 3^6 $(d^2)^4 = d^{2 \cdot 4}$ or d^8
Power of a Product	$(xy)^a = x^a y^a$	$(2k)^4 = 2^4k^4$ or $16k^4$ $(ab)^3 = a^3b^3$
Power of a Quotient	$\left(\frac{x}{y}\right)^a = \frac{x^a}{y^a}$, $y \neq 0$, and $\left(\frac{x}{y}\right)^{-a} = \left(\frac{y}{x}\right)^a$ or $\frac{y^a}{x^a}$, $x \neq 0$, $y \neq 0$	$\left(\frac{x}{y}\right)^2 = \frac{x^2}{y^2}$ $\left(\frac{a}{b}\right)^{-5} = \frac{b^5}{a^5}$
Zero Power	$x^0 = 1, x \neq 0$	$7^0 = 1$

Note that an expression such as x^{-2} is not a monomial. **Why?**

When simplifying a monomial, check to be sure that it has been simplified fully.

Key Concept — Simplifying Monomials

For Your FOLDABLE

A monomial expression is in simplified form when:

- there are no powers of powers,
- each base appears exactly once,
- all fractions are in simplest form, and
- there are no negative exponents.

Problem-Solving Tip

Check You can always check your answer using the definition of exponents.

$$\frac{q^2}{q^7} = \frac{q \cdot q}{q \cdot q \cdot q \cdot q \cdot q \cdot q \cdot q} = \frac{1}{q^5}$$

EXAMPLE 1 Simplify Expressions

Simplify each expression. Assume that no variable equals 0.

a. $(2a^{-2})(3a^3b^2)(c^{-2})$

$(2a^{-2})(3a^3b^2)(c^{-2})$	**Original expression**
$= 2\left(\frac{1}{a^2}\right)(3a^3b^2)\left(\frac{1}{c^2}\right)$	**Definition of negative exponents**
$= \left(\frac{2}{a \cdot a}\right)(3 \cdot a \cdot a \cdot a \cdot b \cdot b)\left(\frac{1}{c \cdot c}\right)$	**Definition of exponents**
$= \left(\frac{2}{\cancel{a} \cdot \cancel{a}}\right)(3 \cdot \cancel{a} \cdot \cancel{a} \cdot a \cdot b \cdot b)\left(\frac{1}{c \cdot c}\right)$	**Divide out common factors.**
$= \frac{6ab^2}{c^2}$	**Simplify.**

b. $\frac{q^2r^4}{q^7r^3}$

$\frac{q^2r^4}{q^7r^3} = q^{2-7} \cdot r^{4-3}$	**Quotient of powers**
$= q^{-5}r$	**Subtract powers.**
$= \frac{r}{q^5}$	**Simplify.**

c. $\left(\frac{-2a^4}{b^2}\right)^3$

$\left(\frac{-2a^4}{b^2}\right)^3 = \frac{(-2a^4)^3}{(b^2)^3}$	**Power of a quotient**
$= \frac{(-2)^3(a^4)^3}{(b^2)^3}$	**Power of a product**
$= \frac{-8a^{12}}{b^6}$	**Power of a power**

Check Your Progress

1A. $(2x^{-3}y^3)(-7x^5y^{-6})$

1B. $\frac{15c^5d^3}{-3c^2d^7}$

1C. $\left(\frac{a}{4}\right)^{-3}$

1D. $(-2x^3y^2)^5$

Personal Tutor glencoe.com

StudyTip

Power of 1 Remember that a variable with no exponent indicated can be written as a power of 1.

Operations With Polynomials The **degree of a polynomial** is the degree of the monomial with the greatest degree. For example, the degree of the polynomial $x^2 + 4x + 58$ is 2.

EXAMPLE 2 Degree of a Polynomial

Determine whether each expression is a polynomial. If it is a polynomial, state the degree of the polynomial.

a. $\frac{1}{4}x^4y^3 - 8x^5$

This expression is a polynomial because each term is a monomial. The degree of the first term is 4 + 3 or 7, and the degree of the second term is 5. The degree of the polynomial is 7.

b. $\sqrt{x} + x + 4$

This expression is not a polynomial because $\sqrt{x}$ is not a monomial.

c. $x^{-3} + 2x^{-2} + 6$

This expression is not a polynomial because x^{-3} and x^{-2} are not monomials: $x^{-3} = \frac{1}{x^3}$ and $x^{-2} = \frac{1}{x^2}$. Monomials cannot contain variables in the denominator.

Check Your Progress

2A. $\frac{x}{y} + 3x^2$ **2B.** $x^5y + 9x^4y^3 - 2xy$

Personal Tutor glencoe.com

You can simplify a polynomial just like you simplify a monomial. Perform the operations indicated, and combine like terms.

StudyTip

Alternative Methods Notice that Example 3a uses a horizontal method, and Example 3b uses a vertical method to simplify. Either method will yield a correct solution.

EXAMPLE 3 Simplify Polynomial Expressions

Simplify each expression.

a. $(4x^2 - 5x + 6) - (2x^2 + 3x - 1)$

Remove parentheses, and group like terms together.

$(4x^2 - 5x + 6) - (2x^2 + 3x - 1)$

$= 4x^2 - 5x + 6 - 2x^2 - 3x + 1$ **Distribute the −1.**

$= (4x^2 - 2x^2) + (-5x - 3x) + (6 + 1)$ **Group like terms.**

$= 2x^2 - 8x + 7$ **Combine like terms.**

b. $(6x^2 - 7x + 8) + (-4x^2 + 9x - 5)$

Align like terms vertically and add.

$$\begin{array}{r} 6x^2 - 7x + 8 \\ (+)\ -4x^2 + 9x - 5 \\ \hline 2x^2 + 2x + 3 \end{array}$$

Check Your Progress

3A. $(-x^2 - 3x + 4) - (x^2 + 2x + 5)$ **3B.** $(3x^2 - 6) + (-x + 1)$

Personal Tutor glencoe.com

You can use the Distributive Property to multiply polynomials.

EXAMPLE 4 Simplify by Using the Distributive Property

Find $3x(2x^2 - 4x + 6)$.

$3x(2x^2 - 4x + 6) = 3x(2x^2) + 3x(-4x) + 3x(6)$ **Distributive Property**

$= 6x^3 - 12x^2 + 18x$ **Multiply the monomials.**

Check Your Progress

Find each product.

4A. $\frac{4}{3}x^2(6x^2 + 9x - 12)$

4B. $-2a(-3a^2 - 11a + 20)$

Personal Tutor glencoe.com

Polynomials can be used to represent real-world situations.

Real-World EXAMPLE 5 Write a Polynomial Expression

DRIVING The U.S. Department of Transportation limits the time a truck driver can work between periods of rest to ten hours. For the first part of his shift, Tom drives at a speed of 60 miles per hour, and for the second part of the shift, he drives at a speed of 70 miles per hour. Write a polynomial to represent the distance driven.

Words	60 mph for	some time, and		70 mph	for the rest
Variable	Let x = the number of hours he drives at 60 miles per hour.				
Expression	60	x	+	70	$(10 - x)$

$60x + 70(10 - x)$ **Original expression**

$= 60x + 700 - 70x$ **Distributive Property**

$= 700 - 10x$ **Combine like terms.**

The polynomial is $700 - 10x$.

Check Your Progress

5. Paul has \$900 to invest in a savings account that has an annual interest rate of 1.8%, and a money market account that pays 4.2% per year. Write a polynomial for the interest he will earn in one year if he invests x dollars in the savings account.

Personal Tutor glencoe.com

Real-World Career

Truck Driver Truck drivers are considered technical professionals because they are required to obtain specialized education and professional licensure. Although state motor vehicle departments administer the Commercial Driver's License program, federal law spells out the requirements to obtain one.

EXAMPLE 6 Multiply Polynomials

Find $(n^2 + 4n - 6)(n + 2)$.

$(n^2 + 4n - 6)(n + 2)$

$= n^2(n + 2) + 4n(n + 2) + (-6)(n + 2)$ **Distributive Property**

$= n^2 \cdot n + n^2 \cdot 2 + 4n \cdot n + 4n \cdot 2 + (-6) \cdot n + (-6) \cdot 2$ **Distributive Property**

$= n^3 + 2n^2 + 4n^2 + 8n - 6n - 12$ **Multiply monomials.**

$= n^3 + 6n^2 + 2n - 12$ **Combine like terms.**

Check Your Progress

Find each product.

6A. $(x^2 + 4x + 16)(x - 4)$

6B. $(2x^2 - 4x + 5)(3x - 1)$

Personal Tutor glencoe.com

Math *in Motion*, Animation glencoe.com

Check Your Understanding

Example 1 p. 334

Simplify. Assume that no variable equals 0.

1. $(2a^3b^{-2})(-4a^2b^4)$
2. $\dfrac{12x^4y^2}{2xy^5}$
3. $\left(\dfrac{2a^2}{3b}\right)^3$
4. $(6g^5h^{-4})^3$

Example 2 p. 335

Determine whether each expression is a polynomial. If it is a polynomial, state the degree of the polynomial.

5. $3x + 4y$
6. $\frac{1}{2}x^2 - 7y$
7. $x^2 + \sqrt{x}$
8. $\dfrac{ab^3 - 1}{az^4 + 3}$

Examples 3, 4, and 6 pp. 335–336

Simplify.

9. $(x^2 - 5x + 2) - (3x^2 + x - 1)$
10. $(3a + 4b) + (6a - 6b)$
11. $2a(4b + 5)$
12. $3x^2(2xy - 3xy^2 + 4x^2y^3)$
13. $(n - 9)(n + 7)$
14. $(a + 4)(a - 6)$

Example 5 p. 336

15. **EXERCISE** Tara exercises 75 minutes a day. She does cardio, which burns an average of 10 Calories a minute, and weight training, which burns an average of 7.5 Calories a minute. Write a polynomial to represent the amount of Calories Tara burns in one day if she does x minutes of weight training.

Practice and Problem Solving

● = **Step-by-Step Solutions** begin on page R20.
Extra Practice begins on page 947.

Example 1 p. 334

Simplify. Assume that no variable equals 0.

16. $(5x^3y^{-5})(4xy^3)$
17. $(-2b^3c)(4b^2c^2)$
18. $\dfrac{a^3n^7}{an^4}$
19. $\dfrac{-y^3z^5}{y^2z^3}$
20. $\dfrac{-7x^5y^5z^4}{21x^7y^5z^2}$
21. $\dfrac{9a^7b^5c^5}{18a^5b^9c^3}$
22. $(n^5)^4$
23. $(z^3)^6$

Example 2 p. 335

Determine whether each expression is a polynomial. If it is a polynomial, state the degree of the polynomial.

24. $2x^2 - 3x + 5$
25. $a^3 - 11$
26. $\dfrac{5np}{n^2} - \dfrac{2g}{h}$
27. $\sqrt{m - 7}$

Examples 3, 4, and 6 pp. 335–336

Simplify.

28. $(6a^2 + 5a + 10) - (4a^2 + 6a + 12)$
29. $(7b^2 + 6b - 7) - (4b^2 - 2)$
30. $3p(np - z)$
31. $4x(2x^2 + y)$
32. $(x - y)(x^2 + 2xy + y^2)$
33. $(a + b)(a^3 - 3ab - b^2)$
34. $4(a^2 + 5a - 6) - 3(2a^3 + 4a - 5)$
35. $5c(2c^2 - 3c + 4) + 2c(7c - 8)$
36. $5xy(2x - y) + 6y^2(x^2 + 6)$
37. $3ab(4a - 5b) + 4b^2(2a^2 + 1)$
38. $(x - y)(x + y)(2x + y)$
39. $(a + b)(2a + 3b)(2x - y)$

Example 5 p. 336

40. **PAINTING** Connor has hired two painters to paint his house. The first painter charges \$12 an hour and the second painter charges \$11 an hour. It will take 15 hours to paint the house.
    a. Write a polynomial to represent the total cost of the job if Connor hires the first painter for x hours.
    b. Write a polynomial to represent the total cost of the job if Connor hires the second painter for y hours.

Simplify. Assume that no variable equals 0.

41. $\left(\dfrac{8x^2y^3}{24x^3y^2}\right)^4$ **42.** $\left(\dfrac{12a^3b^5}{4a^6b^3}\right)^3$ **43.** $\left(\dfrac{4x^{-2}y^3}{xy^{-4}}\right)^{-2}$ **44.** $\left(\dfrac{5a^{-7}b^2}{ab^{-6}}\right)^{-3}$

45. $(a^2b^3)(ab)^{-2}$ **46.** $(-3x^3y)^2(4xy^2)$ **47.** $\dfrac{3c^2d(2c^3d^5)}{15c^4d^2}$

48. $\dfrac{-10g^6h^9(g^2h^3)}{30g^3h^3}$ **49.** $\dfrac{5x^4y^2(2x^5y^6)}{20x^3y^5}$ **50.** $\dfrac{-12n^7p^5(n^2p^4)}{36n^6p^7}$

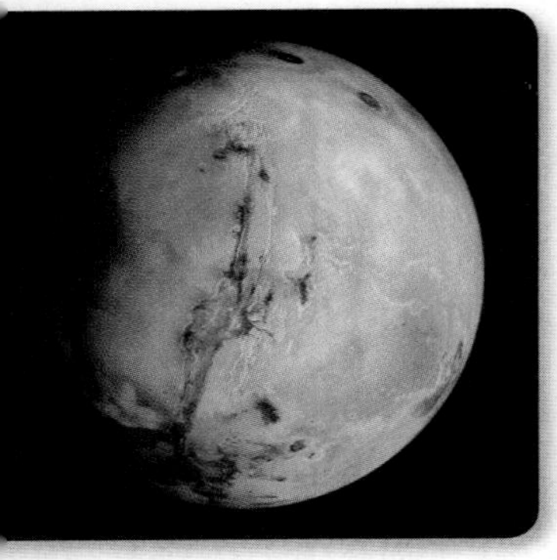

Real-World Link

The environment on Mars is very harsh. Mars has unpredictable weather patterns, as well as the deepest canyon and largest volcano in the solar system.

Source: NASA

51 **ASTRONOMY** Refer to the beginning of the lesson.

a. How long does it take light from Andromeda to reach Earth?

b. The average distance from the Sun to Mars is approximately 2.28×10^{11} meters. How long does it take light from the Sun to reach Mars?

Simplify.

52. $\frac{1}{4}g^2(8g + 12h - 16gh^2)$ **53.** $\frac{1}{3}n^3(6n - 9p + 18np^4)$ **54.** $x^{-2}(x^4 - 3x^3 + x^{-1})$

55. $a^{-3}b^2(ba^3 + b^{-1}a^2 + b^{-2}a)$ **56.** $(g^3 - h)(g^3 + h)$ **57.** $(n^2 - 7)(2n^3 + 4)$

58. $(2x - 2y)^3$ **59.** $(4n - 5)^3$ **60.** $(3z - 2)^3$

61. **EDUCATION** The polynomials $0.108x^2 - 0.876x + 474.1$ and $0.047x^2 + 9.694x + 361.7$ approximate the number of bachelor's degrees, in thousands, earned by males and females, respectively, where x is the number of years after 1971.

a. Find the polynomial that represents the total number of bachelor's degrees (in thousands) earned by both men and women.

b. Find the polynomial that represents the difference between bachelor's degrees earned by men and by women.

62. If $5^{k+7} = 5^{2k-3}$, what is the value of k?

63. What value of k makes $q^{41} = q^{4k} \cdot q^5$ true?

64. **MULTIPLE REPRESENTATIONS** Use the model at the right that represents the product of $x + 3$ and $x + 4$.

a. **GEOMETRIC** The area of the each rectangle is the product of its length and width. Use the model to find the product of $x + 3$ and $x + 4$.

b. **ALGEBRAIC** Use FOIL to find the product of $x + 3$ and $x + 4$.

c. **VERBAL** Explain how each term of the product is represented in the model.

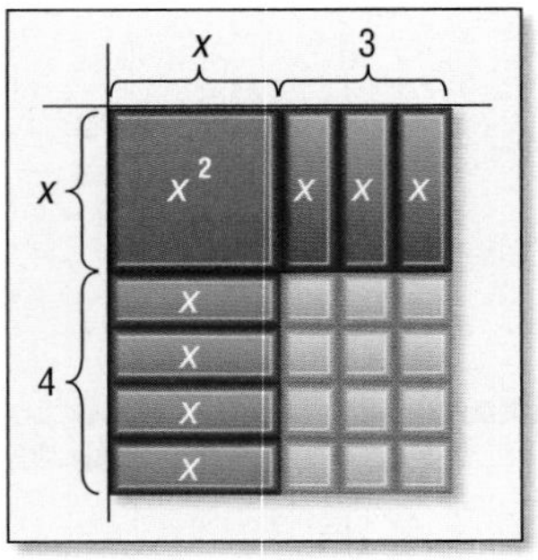

H.O.T. Problems

Use **H**igher-**O**rder **T**hinking Skills

65. **PROOF** Show how the property of negative exponents can be proven using the Quotient of Powers Property and the Zero Power Property.

66. **CHALLENGE** What happens to the quantity of x^{-y} as y increases, for $y > 0$ and $x \geq 1$?

67. **REASONING** Explain why the expression 0^{-2} is undefined.

68. **OPEN ENDED** Write three different expressions that are equivalent to x^{12}.

69. **WRITING IN MATH** Explain why properties of exponents are useful in astronomy. Include an explanation of how to find the amount of time it takes for light from a source to reach a planet.

NYSRE PRACTICE A2.A.8

70. SHORT RESPONSE Simplify $\frac{(2x^2)^3}{12x^4}$.

71. STATISTICS For the numbers a, b, and c, the average (arithmetic mean) is twice the median. If $a = 0$ and $a < b < c$, what is the value of $\frac{c}{b}$?

A 2 | B 3 | C 4 | D 5

72. Which is not a factor of $x^3 - x^2 - 2x$?

F x | G $x + 1$ | H $x - 1$ | J $x - 2$

73. SAT/ACT The expression $(-6 + i)^2$ is equivalent to which of the following expressions?

A $-12i$ | B $-12 + i$ | C $35 - 12i$ | D $37 - 12i$

Spiral Review

Solve each inequality algebraically. (Lesson 5-8)

74. $x^2 - 6x \leq 16$ **75.** $x^2 + 3x > 40$ **76.** $2x^2 - 12 \leq -5x$

Graph each function. (Lesson 5-7)

77. $y = 3(x - 2)^2 - 4$ **78.** $y = -2(x + 4)^2 + 3$ **79.** $f(x) = \frac{1}{3}(x + 1)^2 + 6$

80. BASEBALL A baseball player hits a high pop-up with an initial upward velocity of 30 meters per second, 1.4 meters above the ground. The height $h(t)$ of the ball in meters t seconds after being hit is modeled by $h(t) = -4.9t^2 + 30t + 1.4$. How long does an opposing player have to get under the ball if he catches it 1.7 meters above the ground? Does your answer seem reasonable? Explain. (Lesson 5-3)

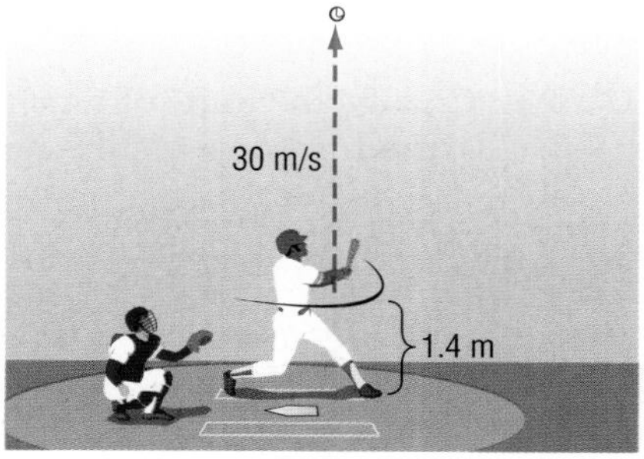

Evaluate each determinant. (Lesson 4-5)

81. $\begin{vmatrix} 3 & 0 & -2 \\ -1 & 4 & 3 \\ 5 & -2 & -1 \end{vmatrix}$ **82.** $\begin{vmatrix} -2 & -4 & -6 \\ 0 & 6 & -5 \\ -1 & 3 & -1 \end{vmatrix}$ **83.** $\begin{vmatrix} -3 & -1 & -2 \\ -2 & 3 & 4 \\ 6 & 1 & 0 \end{vmatrix}$

84. INVESTMENTS A couple is planning to invest $15,000 in certificates of deposit (CDs). For tax purposes, they want their total interest the first year to be $800. They want to put $1000 more in a 2-year CD than in a 1-year CD and then invest the rest in a 3-year CD. How much should they invest in each type of CD? (Lesson 3-5)

Years	1	2	3
Rate	3.4%	5.0%	6.0%

Find the slope of the line that passes through each pair of points. (Lesson 2-3)

85. (6, −2) and (−2, −9) **86.** (−4, −1) and (3, 8) **87.** (3, 0) and (−7, −5)

88. $\left(\frac{1}{2}, \frac{2}{3}\right)$ and $\left(\frac{1}{4}, \frac{1}{3}\right)$ **89.** $\left(\frac{2}{5}, \frac{1}{4}\right)$ and $\left(\frac{1}{10}, \frac{1}{12}\right)$ **90.** (−4.5, 2.5) and (−3, −1)

Skills Review

Factor each polynomial. (Lesson 0-3)

91. $12ax^3 + 20bx^2 + 32cx$ **92.** $x^2 + 2x + 6 + 3x$ **93.** $12y^2 + 9y + 8y + 6$

94. $2my + 7x + 7m + 2xy$ **95.** $8ax - 6x - 12a + 9$ **96.** $10x^2 - 14xy - 15x + 21y$

EXTEND **6-1**

Algebra Lab: Dimensional Analysis

NY Math Online glencoe.com
Math *in Motion*, Animation

Real-world problems often involve units of measure. Performing operations with units is called **dimensional analysis** or **unit analysis**. You can use dimensional analysis to convert units or to perform calculations.

EXAMPLE

A car is traveling at 65 miles per hour. How fast is the car traveling in meters per second?

You want to find the speed in meters per second, so you need to change the unit of distance from miles to meters and the unit of time from hours to seconds. To make the conversion, use fractions that you can multiply.

Step 1 Change the units of length from miles to meters.
Use the relationships of miles to feet and feet to meters.

$$\frac{65 \text{ miles}}{1 \text{ hour}} \cdot \frac{5280 \text{ feet}}{1 \text{ mile}} \cdot \frac{1 \text{ meter}}{3.3 \text{ feet}}$$

Step 2 Change the units of time from hours to seconds.
Write fractions relating hours to minutes and minutes to seconds.

$$\frac{65 \text{ miles}}{1 \text{ hour}} \cdot \frac{5280 \text{ feet}}{1 \text{ mile}} \cdot \frac{1 \text{ meter}}{3.3 \text{ feet}} \cdot \frac{1 \text{ hour}}{60 \text{ minutes}} \cdot \frac{1 \text{ minute}}{60 \text{ seconds}}$$

Step 3 Simplify and check by canceling the units.

$$\frac{65 \cancel{\text{ miles}}}{1 \cancel{\text{ hour}}} \cdot \frac{5280 \cancel{\text{ feet}}}{1 \cancel{\text{ mile}}} \cdot \frac{1 \text{ meter}}{3.3 \cancel{\text{ feet}}} \cdot \frac{1 \cancel{\text{ hour}}}{60 \cancel{\text{ minutes}}} \cdot \frac{1 \cancel{\text{ minute}}}{60 \text{ seconds}}$$

$$= \frac{65 \cdot 5280}{3.3 \cdot 60 \cdot 60} \text{ m/s}$$ **Simplify.**

$$\approx 28.9 \text{ m/s}$$ **Use a calculator.**

So, 65 miles per hour is about 28.9 meters per second. This answer is reasonable because the final units are m/s, not m/hr, ft/s, or mi/hr.

Exercises

Solve each problem by using dimensional analysis. Include the appropriate units with your answer.

1. A zebra can run 40 miles per hour. How far can a zebra run in 3 minutes?
2. A cyclist traveled 43.2 miles at an average speed of 12 miles per hour. How long did the cyclist ride?
3. If you are going 50 miles per hour, how many feet per second are you traveling?
4. The equation $d = \frac{1}{2}(9.8 \text{ m/s}^2)(3.5 \text{ s})^2$ represents the distance d that a ball falls 3.5 seconds after it is dropped from a tower. Find the distance.
5. **WRITING IN MATH** Explain how dimensional analysis can be useful in checking the reasonableness of your answer.

6-2 Dividing Polynomials

Then

You divided monomials. (Lesson 6-1)

Now

- Divide polynomials using long division.
- Divide polynomials using synthetic division.

NYS Core Curriculum

Reinforcement of A.A.14 Divide a polynomial by a monomial or binomial, where the quotient has no remainder

New Vocabulary

synthetic division

NY Math Online

glencoe.com

- Extra Examples
- Personal Tutor
- Self-Check Quiz
- Homework Help

Why?

Arianna needed $140x^2 + 60x$ square inches of paper to make a book jacket $10x$ inches tall. In figuring the area, she allowed for a front and back flap. If the spine is $2x$ inches wide, and the front and back are $6x$ inches wide, how wide are the front and back flaps? You can use a quotient of polynomials to help you find the answer.

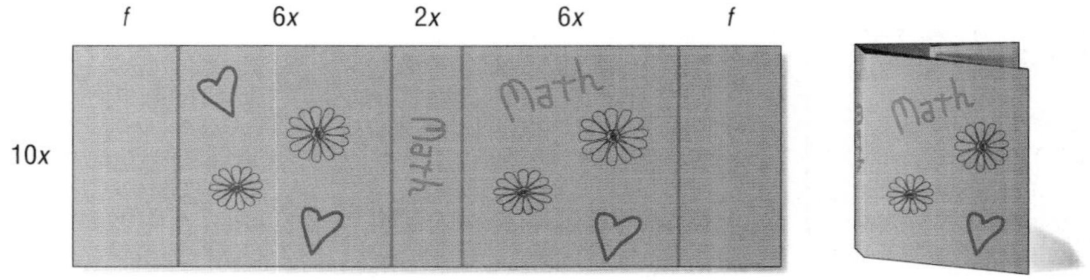

f = flap width

Long Division In Lesson 6-1, you learned to divide monomials. You can divide a polynomial by a monomial by using those same skills.

EXAMPLE 1 Divide a Polynomial by a Monomial

Simplify $\dfrac{6x^4y^3 + 12x^3y^2 - 18x^2y}{3xy}$.

$$\frac{6x^4y^3 + 12x^3y^2 - 18x^2y}{3xy} = \frac{6x^4y^3}{3xy} + \frac{12x^3y^2}{3xy} - \frac{18x^2y}{3xy} \qquad \textbf{Sum of quotients}$$

$$= \frac{6}{3} \cdot x^{4-1}y^{3-1} + \frac{12}{3} \cdot x^{3-1}y^{2-1} - \frac{18}{3} \cdot x^{2-1}y^{1-1} \qquad \textbf{Divide.}$$

$$= 2x^3y^2 + 4x^2y - 6x \qquad y^{1-1} = y^0 \text{ or } 1$$

Check Your Progress Simplify.

1A. $(20c^4d^2f - 16cdf^2 + 4cdf) \div (4cdf)$

1B. $(18x^2y + 27x^3y^2z)(3xy)^{-1}$

Personal Tutor glencoe.com

You can use a process similar to long division to divide a polynomial by a polynomial with more than one term. The process is known as the *division algorithm*.

EXAMPLE 2 Division Algorithm

Use long division to find $(x^2 + 3x - 40) \div (x - 5)$.

$$\begin{array}{r l}
x + 8 & \\
x - 5\overline{)x^2 + 3x - 40} & \\
\underline{(-)\,x^2 - 5x} & \textbf{Multiply divisor by } x \textbf{ since } \frac{x^2}{x} = x. \\
8x - 40 & \textbf{Subtract. Bring down next term.} \\
\underline{(-)\,8x - 40} & \textbf{Multiply divisor by 8 since } \frac{8x}{x} = 8. \\
0 & \textbf{Subtract.}
\end{array}$$

The quotient is $x + 8$. The remainder is 0.

Check Your Progress Use long division to find each quotient.

2A. $(x^2 + 7x - 30) \div (x - 3)$

2B. $(x^2 - 13x + 12) \div (x - 1)$

Personal Tutor glencoe.com

Just as with the division of whole numbers, the division of two polynomials may result in a quotient with a remainder. Remember that $11 \div 3 = 3 + R2$, which is often written as $3\frac{2}{3}$. The result of a division of polynomials with a remainder can be written in a similar manner.

NYSRE EXAMPLE 3 Reinforcement of A.A.14

Which expression is equal to $(a^2 + 7a - 11)(3 - a)^{-1}$?

A $a + 10 - \frac{19}{3 - a}$

B $-a + 10$

C $-a - 10 + \frac{19}{3 - a}$

D $-a - 10 - \frac{19}{3 - a}$

Test-Taking Tip

Multiple Choice You may be able to eliminate some of the answer choices by substituting the same value for *a* in the original expression and the answer choices, and then evaluating.

Read the Test Item

Since the second factor has an exponent of -1, this is a division problem.

$$(a^2 + 7a - 11)(3 - a)^{-1} = \frac{a^2 + 7a - 11}{3 - a}$$

Solve the Test Item

$$\begin{array}{r} -a - 10 \\ -a + 3\overline{)a^2 + 7a - 11} \\ \underline{(-)\ a^2 - 3a} \\ 10a - 11 \\ \underline{(-)\ 10a - 30} \\ 19 \end{array}$$

For ease in dividing, rewrite $3 - a$ as $-a + 3$.
$-a(-a + 3) = a^2 - 3a$
$7a - (-3a) = 10a$
$-10(-a + 3) = 10a - 30$
$-11 - (-30) = 19$

The quotient is $-a - 10$, and the remainder is 19.

Therefore, $(a^2 + 7a - 11)(3 - a)^{-1} = -a - 10 + \frac{19}{3 - a}$. The answer is **C**.

Check Your Progress

3. Which expression is equal to $(r^2 + 5r + 7)(1 - r)^{-1}$?

F $-r - 6 + \frac{13}{1 - r}$

G $r + 6$

H $r - 6 + \frac{13}{1 - r}$

J $r + 6 - \frac{13}{1 - r}$

Personal Tutor glencoe.com

Synthetic Division **Synthetic division** is a simpler process for dividing a polynomial by a binomial. Suppose you want to divide $2x^3 - 13x^2 + 26x - 24$ by $x - 4$ using long division. Compare the coefficients in this division with those in Example 4.

$$\begin{array}{r} 2x^2 - 5x + 6 \\ x - 4\overline{)2x^3 - 13x^2 + 26x - 24} \\ \underline{(-)\ 2x^3 - 8x^2} \\ -5x^2 + 26x \\ \underline{(-)\ -5x^2 + 20x} \\ 6x - 24 \\ \underline{(-)\ 6x - 24} \\ 0 \end{array}$$

When the polynomial in the dividend is missing a term, a zero must be used to represent the missing term. So, with a dividend of $2x^3 - 4x^2 + 6$, a 0 will be used as a placeholder for the *x*-term.

$$\overline{)2x^3 - 4x^2 + 0x + 6}$$

Key Concept — Synthetic Division

For Your FOLDABLE

Step 1 Write the coefficients of the dividend so that the degrees of the terms are in descending order. Write the constant r of the divisor $x - r$ in the box. Bring the first coefficient down.

Step 2 Multiply the first coefficient by r, and write the product under the second coefficient.

Step 3 Add the product and the second coefficient.

Step 4 Repeat Steps 2 and 3 until you reach a sum in the last column. The numbers along the bottom row are the coefficients of the quotient. The power of the first term is one less than the degree of the dividend. The final number is the remainder.

EXAMPLE 4 Synthetic Division

Use synthetic division to find $(2x^3 - 13x^2 + 26x - 24) \div (x - 4)$.

Step 1 Write the coefficients of the dividend. Write the constant r in the box. In this case, $r = 4$. Bring the first coefficient, 2, down.

$$\begin{array}{r|rrrr} 4 & 2 & -13 & 26 & -24 \\ & \downarrow & & & \\ \hline & 2 & & & | \end{array}$$

Step 2 Multiply the first coefficient by r: $2 \cdot 4 = 8$. Write the product under the second coefficient.

$$\begin{array}{r|rrrr} 4 & 2 & -13 & 26 & -24 \\ & & 8 & & \\ \hline & 2 \nearrow & & & | \end{array}$$

Step 3 Add the product and the second coefficient: $-13 + 8 = -5$.

$$\begin{array}{r|rrrr} 4 & 2 & -13 & 26 & -24 \\ & & 8 & & \\ \hline & 2 & -5 & & | \end{array}$$

Step 4 Multiply the sum, -5, by r: $-5 \times 4 = -20$. Write the product under the next coefficient, and add: $26 + (-20) = 6$. Multiply the sum, 6, by r: $6 \cdot 4 = 24$. Write the product under the next coefficient and add: $-24 + 24 = 0$.

$$\begin{array}{r|rrrr} 4 & 2 & -13 & 26 & -24 \\ & & 8 & -20 & 24 \\ \hline & 2 & -5 \nearrow & 6 \nearrow & |\,0 \end{array}$$

CHECK Multiply the quotient by the divisor. The answer should be the dividend.

$$\begin{array}{rrrr} & 2x^2 & - 5x & + 6 \\ (\times) & & x & - 4 \\ \hline & -8x^2 & + 20x & - 24 \\ 2x^3 & - 5x^2 & + 6x & \\ \hline 2x^3 & - 13x^2 & + 26x & - 24 \end{array}$$

The quotient is $2x^2 - 5x + 6$. The remainder is 0.

Check Your Progress

Use synthetic division to find each quotient.

4A. $(2x^3 + 3x^2 - 4x + 15) \div (x + 3)$

4B. $(3x^3 - 8x^2 + 11x - 14) \div (x - 2)$

4C. $(4a^4 + 2a^2 - 4a + 12) \div (a + 2)$

4D. $(6b^4 - 8b^3 + 12b - 14) \div (b - 2)$

Personal Tutor glencoe.com

Watch Out!

Synthetic Division Remember to *add* terms when performing synthetic division.

To use synthetic division, the divisor must be of the form $x - r$. If the coefficient of x in a divisor is not 1, you can rewrite the division expression so that you can use synthetic division.

Watch Out!

Divide Throughout Remember to divide *all* terms in the numerator and denominator.

EXAMPLE 5 Divisor with First Coefficient Other than 1

Use synthetic division to find $(3x^4 - 5x^3 + x^2 + 7x) \div (3x + 1)$.

$$\frac{3x^4 - 5x^3 + x^2 + 7x}{3x + 1} = \frac{(3x^4 - 5x^3 + x^2 + 7x) \div 3}{(3x + 1) \div 3}$$

Rewrite the divisor with a leading coeffient of 1. Then divide the numerator and denominator by 3.

$$= \frac{x^4 - \frac{5}{3}x^3 + \frac{1}{3}x^2 + \frac{7}{3}x}{x + \frac{1}{3}}$$

Simplify the numerator and the denominator.

Since the numerator does not have a constant term, use a coefficient of 0 for the constant term.

$x - r = x + \frac{1}{3}$, so $r = -\frac{1}{3}$. $\rightarrow$

$$\begin{array}{r|rrrrr} -\frac{1}{3} & 1 & -\frac{5}{3} & \frac{1}{3} & \frac{7}{3} & 0 \\ & & -\frac{1}{3} & \frac{2}{3} & -\frac{1}{3} & -\frac{2}{3} \\ \hline & 1 & -2 & 1 & 2 & \big| -\frac{2}{3} \end{array}$$

The result is $x^3 - 2x^2 + x + 2 - \frac{\frac{2}{3}}{x + \frac{1}{3}}$. Now simplify the fraction.

$$\frac{\frac{2}{3}}{x + \frac{1}{3}} = \frac{2}{3} \div \left(x + \frac{1}{3}\right)$$ **Rewrite as a division expression.**

$$= \frac{2}{3} \div \frac{3x + 1}{3}$$ $x + \frac{1}{3} = \frac{3x}{3} + \frac{1}{3} = \frac{3x + 1}{3}$

$$= \frac{2}{3} \cdot \frac{3}{3x + 1}$$ **Multiply by the reciprocal.**

$$= \frac{2}{3x + 1}$$ **Simplify.**

The solution is $x^3 - 2x^2 + x + 2 - \frac{2}{3x + 1}$.

CHECK Divide using long division.

$$\begin{array}{rl} & x^3 - 2x^2 + x + 2 \\ 3x + 1 \big) & 3x^4 - 5x^3 + x^2 + 7x \\ & \underline{(-)\ 3x^4 + x^3} \\ & -6x^3 + x^2 \\ & \underline{(-)\ -6x^3 - 2x^2} \\ & 3x^2 + 7x \\ & \underline{(-)\ 3x^2 + x} \\ & 6x + 0 \\ & \underline{(-)\ 6x + 2} \\ & -2 \end{array}$$

The result is $x^3 - 2x^2 + x + 2 - \frac{2}{3x + 1}$. ✓

Check Your Progress

Use synthetic division to find each quotient.

5A. $(8x^4 - 4x^2 + x + 4) \div (2x + 1)$

5B. $(8y^5 - 2y^4 - 16y^2 + 4) \div (4y - 1)$

5C. $(15b^3 + 8b^2 - 21b + 6) \div (5b - 4)$

5D. $(6c^3 - 17c^2 + 6c + 8) \div (3c - 4)$

Personal Tutor glencoe.com

Check Your Understanding

Examples 1, 2, and 4 pp. 341–343

Simplify.

1. $\frac{4xy^2 - 2xy + 2x^2y}{xy}$
2. $(3a^2b - 6ab + 5ab^2)(ab)^{-1}$
3. $(x^2 - 6x - 20) \div (x + 2)$
4. $(2a^2 - 4a - 8) \div (a + 1)$
5. $(3z^4 - 6z^3 - 9z^2 + 3z - 6) \div (z + 3)$
6. $(y^5 - 3y^2 - 20) \div (y - 2)$

Example 3 p. 342

7. **MULTIPLE CHOICE** Which expression is equal to $(x^2 + 3x - 9)(4 - x)^{-1}$?

A $-x - 7 + \frac{19}{4 - x}$ B $-x - 7$ C $x + 7 - \frac{19}{4 - x}$ D $-x - 7 - \frac{19}{4 - x}$

Example 5 p. 344

Simplify.

8. $(10x^2 + 15x + 20) \div (5x + 5)$
9. $(18a^2 + 6a + 9) \div (3a - 2)$
10. $\frac{12b^2 + 23b + 15}{3b + 8}$
11. $\frac{27y^2 + 27y - 30}{9y - 6}$

Practice and Problem Solving

● = **Step-by-Step Solutions** begin on page R20.
Extra Practice begins on page 947.

Example 1 p. 341

Simplify.

12. $\frac{24a^3b^2 - 16a^2b^3}{8ab}$
13. $\frac{5x^2y - 10xy + 15xy^2}{5xy}$
14. $\frac{7g^3h^2 + 3g^2h - 2gh^3}{gh}$
15. $\frac{4a^3b - 6ab + 2ab^2}{2ab}$
16. $\frac{16c^4d^4 - 24c^2d^2}{4c^2d^2}$
17. $\frac{9n^3p^3 - 18n^2p^2 + 21n^2p^3}{3n^2p^2}$

18. **ENERGY** Compact fluorescent light (CFL) bulbs reduce energy waste. The amount of energy waste that is reduced each day in a certain community can be estimated by $-b^2 + 8b$, where b is the number of bulbs. Divide by b to find the average amount of energy saved per CFL bulb.

19. **BAKING** The number of cookies produced in a factory each day can be estimated by $-w^2 + 16w + 1000$, where w is the number of workers. Divide by w to find the average number of cookies produced per worker.

Examples 2, 4, and 5 pp. 341–344

Simplify.

20. $(a^2 - 8a - 26) \div (a + 2)$
21. $(b^3 - 4b^2 + b - 2) \div (b + 1)$
22. $(z^4 - 3z^3 + 2z^2 - 4z + 4)(z - 1)^{-1}$
23. $(x^5 - 4x^3 + 4x^2) \div (x - 4)$
24. $\frac{y^3 + 11y^2 - 10y + 6}{y + 2}$
25. $(g^4 - 3g^2 - 18) \div (g - 2)$
26. $(6a^2 - 3a + 9) \div (3a - 2)$
27. $\frac{6x^5 + 5x^4 + x^3 - 3x^2 + x}{3x + 1}$
28. $\frac{4g^4 - 6g^3 + 3g^2 - g + 12}{4g - 4}$
29. $(2b^3 - 6b^2 + 8b) \div (2b + 2)$
30. $(6z^6 + 3z^4 - 9z^2)(3z - 6)^{-1}$
31. $(10y^6 + 5y^5 + 10y^3 - 20y - 15)(5y + 5)^{-1}$

32. **GEOMETRY** A rectangular box for a new product is designed in such a way that the three dimensions always have a particular relationship defined by the variable x. The volume of the box can be written as $6x^3 + 31x^2 + 53x + 30$, and the height is always $x + 2$. What are the width and height of the box?

33. **PHYSICS** The voltage V is related to current I and power P by the equation $V = \frac{P}{I}$. The power of a generator is modeled by $P(t) = t^3 + 9t^2 + 26t + 24$. If the current of the generator is $I = t + 4$, write an expression that represents the voltage.

34. **ENTERTAINMENT** A magician gives these instructions to a volunteer.

- Choose a number and multiply it by 4.
- Then add the sum of your number and 15 to the product you found.
- Now divide by the sum of your number and 3.

   a. What number will the volunteer always have at the end?

   b. Explain the process you used to discover the answer.

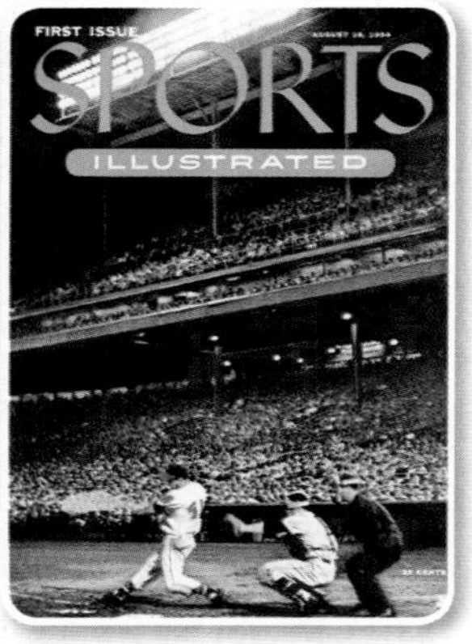

Real-World Link

Created in 1954, *Sports Illustrated* is read by over 23 million adults each week.

Source: *Sports Illustrated*

35. **BUSINESS** The number of magazine subscriptions sold can be estimated by $n = \frac{3500a^2}{a^2 + 100}$, where a is the amount of money the company spent on advertising in hundreds of dollars and n is the number of subscriptions sold.

   a. Perform the division indicated by $\frac{3500a^2}{a^2 + 100}$.

   b. About how many subscriptions will be sold if \$1500 is spent on advertising?

Simplify.

36. $(x^4 - y^4) \div (x - y)$

37. $(28c^3d^2 - 21cd^2) \div (14cd)$

38. $(a^3b^2 - a^2b + 2b)(-ab)^{-1}$

39. $\frac{n^3 + 3n^2 - 5n - 4}{n + 4}$

40. $\frac{p^3 + 2p^2 - 7p - 21}{p + 3}$

41. $\frac{3z^5 + 5z^4 + z + 5}{z + 2}$

42. **MULTIPLE REPRESENTATIONS** Consider a rectangle with area $2x^2 + 7x + 3$ and length $2x + 1$.

   a. CONCRETE Use algebra tiles to represent this situation. Use the model to find the width.

   b. SYMBOLIC Write an expression to represent the model.

   c. NUMERICAL Solve this problem algebraically using synthetic or long division. Does your concrete model check with your algebraic model?

H.O.T. Problems

Use **H**igher-**O**rder **T**hinking Skills

43. **FIND THE ERROR** Sharon and Jamal are dividing $2x^3 - 4x^2 + 3x - 1$ by $x - 3$. Sharon claims that the remainder is 26. Jamal argues that the remainder is -100. Is either of them correct? Explain your reasoning.

44. **CHALLENGE** If a polynomial is divided by a binomial and the remainder is 0, what does this tell you about the relationship between the binomial and the polynomial?

45. **REASONING** Review any of the division problems in this lesson. What is the relationship between the degrees of the dividend, the divisor, and the quotient?

46. **OPEN ENDED** Write a quotient of two polynomials for which the remainder is 3.

47. **WHICH ONE DOESN'T BELONG?** Identify the expression that does not belong with the other three. Explain your reasoning.

$3xy + 6x^2$	$\frac{5}{x^2}$	$x + 5$	$5b + 11c - 9ad^2$

48. **WRITING IN MATH** Use the information at the beginning of the lesson to explain how you can use the division of polynomials to make a paper cover for your textbook.

NYSRE PRACTICE Reinforcement of A.A.14

49. An office employs x women and 3 men. What is the ratio of the total number of employees to the number of women?

A $1 + \frac{3}{x}$

B $\frac{x}{x+3}$

C $\frac{3}{x}$

D $\frac{x}{3}$

50. $(-4x^2 + 2x + 3) - 3(2x^2 - 5x + 1) =$

F $2x^2$

G $-10x^2$

H $-10x^2 + 17x$

J $2x^2 + 17x$

51. GRIDDED RESPONSE In the figure below, $m + n + p = ?$

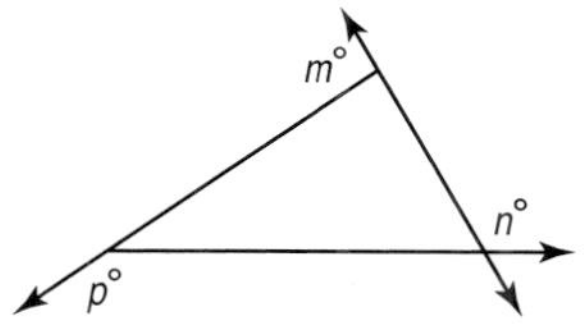

52. SAT/ACT Which polynomial has degree 3?

A $x^3 + x^2 - 2x^4$

B $-2x^2 - 3x + 4$

C $x^2 + x + 12^3$

D $1 + x + x^3$

Spiral Review

Simplify. (Lesson 6-1)

53. $(5x^3 + 2x^2 - 3x + 4) - (2x^3 - 4x)$

54. $(2y^3 - 3y + 8) + (3y^2 - 6y)$

55. $4a(2a - 3) + 3a(5a - 4)$

56. $(c + d)(c - d)(2c - 3d)$

57. $(xy)^2(2xy^2z)^3$

58. $(3ab^2)^{-2}(2a^2b)^2$

59. LANDSCAPING Amado wants to plant a garden and surround it with decorative stones. He has enough stones to enclose a rectangular garden with a perimeter of 68 feet, but he wants the garden to cover no more than 240 square feet. What could the width of his garden be? (Lesson 5-8)

Solve each equation by completing the square. (Lesson 5-5)

60. $x^2 + 6x + 2 = 0$

61. $x^2 - 8x - 3 = 0$

62. $2x^2 + 6x + 5 = 0$

State the consecutive integers between which the zeros of each quadratic function are located. (Lesson 5-2)

63.

x	−7	−6	−5	−4	−3	−2	−1	0
$f(x)$	4	1	−3	−8	−1	2	8	16

64.

x	−2	−1	0	1	2	3	4	5
$f(x)$	−16	−7	−4	3	3	−4	−7	−16

65.

x	−2	−1	0	1	2	3	4	5
$f(x)$	6	1	−3	−5	−3	1	6	14

66. BUSINESS A landscaper can mow a lawn in 30 minutes and perform a small landscape job in 90 minutes. He works at most 10 hours per day, 5 days per week. He earns \$35 per lawn and \$125 per landscape job. He cannot do more than 3 landscape jobs per day. Find the combination of lawns mowed and completed landscape jobs per week that will maximize income. Then find the maximum income. (Lesson 3-4)

Skills Review

Find each value if $f(x) = 4x + 3$, $g(x) = -x^2$, and $h(x) = -2x^2 - 2x + 4$. (Lesson 2-1)

67. $f(-6)$

68. $g(-8)$

69. $h(3)$

70. $f(c)$

71. $g(3d)$

72. $h(2b + 1)$

6-3 Polynomial Functions

Why?

The volume of air in the lungs during a 5-second respiratory cycle can be modeled by $v(t) = -0.037t^3 + 0.152t^2 + 0.173t$, where v is the volume in liters and t is the time in seconds. This model is an example of a polynomial function.

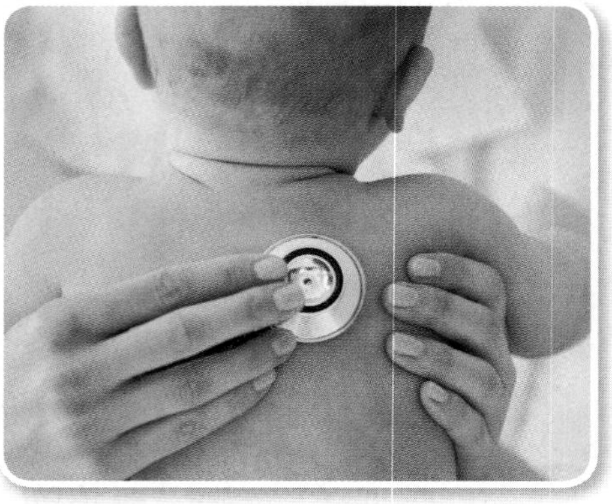

Then

You analyzed graphs of quadratic functions. (Lesson 5-1)

Now

- Evaluate polynomial functions.
- Identify general shapes of graphs of polynomial functions.

NYS Core Curriculum

A2.A.41 Use functional notation to evaluate functions for given values in the domain

New Vocabulary

polynomial in one variable
leading coefficient
polynomial function
power function
end behavior
quartic function
quintic function

NY Math Online

glencoe.com

- Extra Examples
- Personal Tutor
- Self-Check Quiz
- Homework Help
- Math in Motion

Polynomial Functions A **polynomial in one variable** is an expression of the form $a_nx^n + a_{n-1}x^{n-1} + \cdots + a_2x^2 + a_1x + a_0$, where $a_n \neq 0$, a_{n-1}, a_2, a_1, and a_0 are real numbers, and n is a nonnegative integer.

The polynomial is written in standard form when the values of the exponents are in descending order. The degree of the polynomial is the value of the greatest exponent. The coefficient of the first term of a polynomial in standard form is called the **leading coefficient**.

Polynomial	Expression	Degree	Leading Coefficient
Constant	12	0	12
Linear	$4x - 9$	1	4
Quadratic	$5x^2 - 6x - 9$	2	5
Cubic	$8x^3 + 12x^2 - 3x + 1$	3	8
General	$a_nx^n + a_{n-1}x^{n-1} + \cdots + a_1x + a_0$	n	a_n

EXAMPLE 1 Degrees and Leading Coefficients

State the degree and leading coefficient of each polynomial in one variable. If it is not a polynomial in one variable, explain why.

a. $8x^5 - 4x^3 + 2x^2 - x - 3$

This is a polynomial in one variable. The greatest exponent is 5, so the degree is 5 and the leading coefficient is 8.

b. $12x^2 - 3xy + 8x$

This is not a polynomial in one variable. There are two variables, x and y.

c. $3x^4 + 6x^3 - 4x^8 + 2x$

This is a polynomial in one variable. The greatest exponent is 8, so the degree is 8 and the leading coefficient is -4.

Check Your Progress

1A. $5x^3 - 4x^2 - 8x + \frac{4}{x}$ **1B.** $5x^6 - 3x^4 + 12x^3 - 14$ **1C.** $8x^4 - 2x^3 - x^6 + 3$

Personal Tutor glencoe.com

A **polynomial function** is a continuous function that can be described by a polynomial equation in one variable. For example, $f(x) = 3x^3 - 4x + 6$ is a cubic polynomial function. The simplest polynomial functions of the form $f(x) = ax^b$ where a and b are real numbers are called **power functions**.

If you know an element in the domain of any polynomial function, you can find the corresponding value in the range.

Real-World Link

Total lung capacity is approximately 6 liters in a healthy young adult.

Source: *Family Practice Notebook*

Real-World EXAMPLE 2 Evaluate a Polynomial Function

RESPIRATION Refer to the beginning of the lesson. Find the volume of air in the lungs 2 seconds into the respiratory cycle.

By substituting 2 into the function we can find $v(2)$, the volume of air in the lungs 2 seconds into the respiration cycle.

$v(t) = -0.0374t^3 + 0.1522t^2 + 0.1729t$	**Original function**
$v(2) = -0.0374(2)^3 + 0.1522(2)^2 + 0.1729(2)$	**Replace t with 2.**
$= -0.2992 + 0.6088 + 0.3458$	**Simplify.**
$= 0.6554$ L	**Add.**

Check Your Progress

2. Find the volume of air in the lungs 4 seconds into the respiratory cycle.

Personal Tutor glencoe.com

You can also evaluate functions for variables and algebraic expressions.

EXAMPLE 3 Function Values of Variables

Find $f(3c - 4) - 5f(c)$ if $f(x) = x^2 + 2x - 3$.

To evaluate $f(3c - 4)$, replace the x in $f(x)$ with $3c - 4$.

$f(x) = x^2 + 2x - 3$	**Original function**
$f(3c - 4) = (3c - 4)^2 + 2(3c - 4) - 3$	**Replace x with $3c - 4$.**
$= 9c^2 - 24c + 16 + 6c - 8 - 3$	**Multiply.**
$= 9c^2 - 18c + 5$	**Simplify.**

To evaluate $5f(c)$, replace x with c in $f(x)$, then multiply by 5.

$f(x) = x^2 + 2x - 3$	**Original function**
$5f(c) = 5(c^2 + 2c - 3)$	**Replace x with c.**
$= 5c^2 + 10c - 15$	**Distributive Property**

Now evaluate $f(3c - 4) - 5f(c)$.

$f(3c - 4) - 5f(c) = (9c^2 - 18c + 5) - (5c^2 + 10c - 15)$	
$= 9c^2 - 18c + 5 - 5c^2 - 10c + 15$	**Distribute.**
$= 4c^2 - 28c + 20$	**Simplify.**

Check Your Progress

3A. Find $g(5a - 2) + 3g(2a)$ if $g(x) = x^2 - 5x + 8$.

3B. Find $h(-4d + 3) - 0.5h(d)$ if $h(x) = 2x^2 + 5x + 3$.

Personal Tutor glencoe.com

Graphs of Polynomial Functions The general shapes of the graphs of several polynomial functions show the *maximum* number of times the graph of each function may intersect the x-axis. This is the same number as the degree of the polynomial.

Constant function
Degree 0

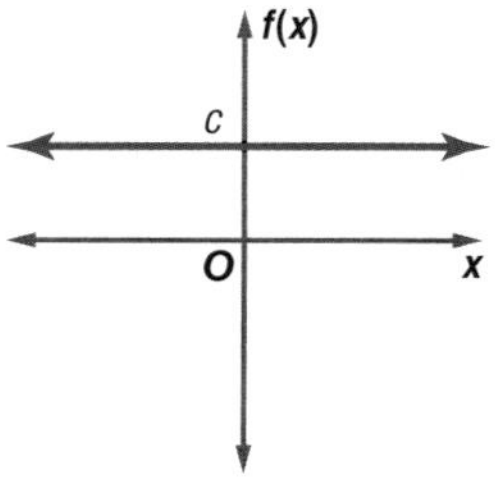

Linear function
Degree 1

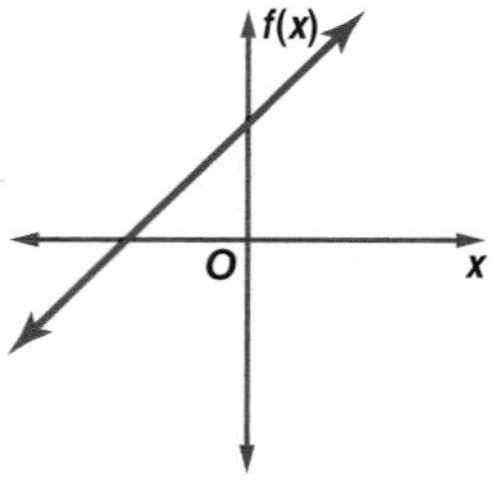

Quadratic function
Degree 2

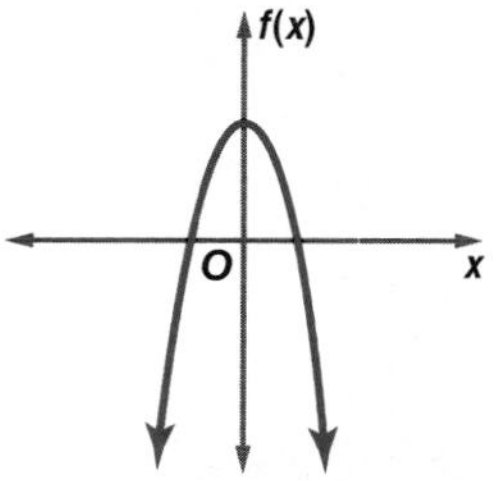

Cubic function
Degree 3

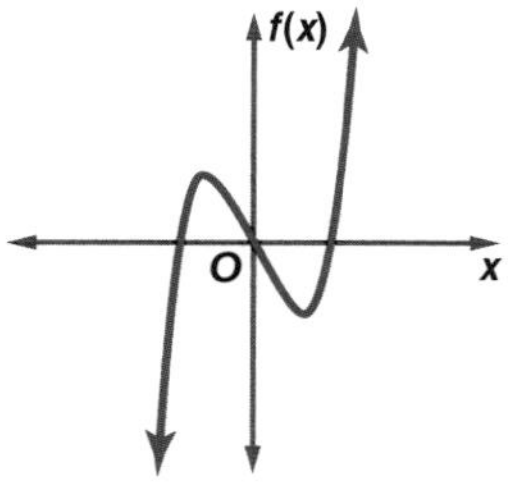

Quartic function
Degree 4

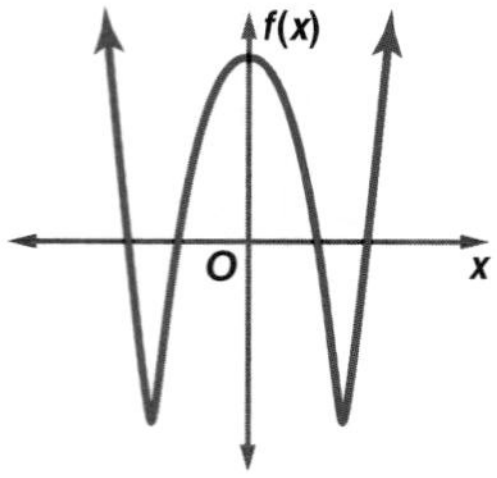

Quintic function
Degree 5

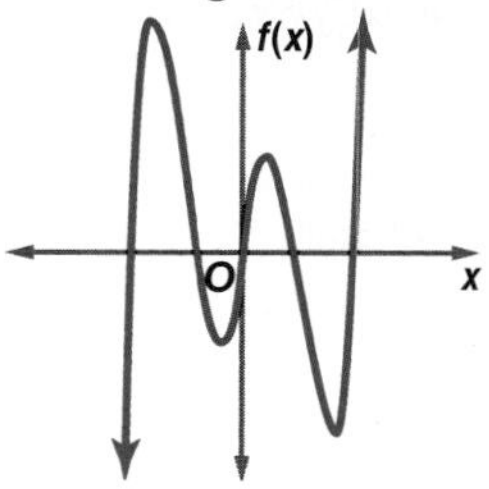

The domain of any polynomial function is all real numbers. The **end behavior** is the behavior of the graph of $f(x)$ as x approaches positive infinity ($x \to +\infty$) or negative infinity ($x \to -\infty$). The degree and leading coefficient of a polynomial function determine the end behavior of the graph and the range of the function.

StudyTip

End behavior The leading coefficient and degree are the sole determining factors for the end behavior of a polynomial. With very large or very small numbers, the rest of the polynomial is insignificant in the appearance of the graph.

Review Vocabulary

infinity endless or boundless (Explore Lesson 1-6)

Key Concept End Behavior of a Polynomial Function

For Your FOLDABLE

Degree: even
Leading Coefficient: positive
End Behavior:

$f(x) \to +\infty$ as $x \to -\infty$

$f(x) \to +\infty$ as $x \to +\infty$

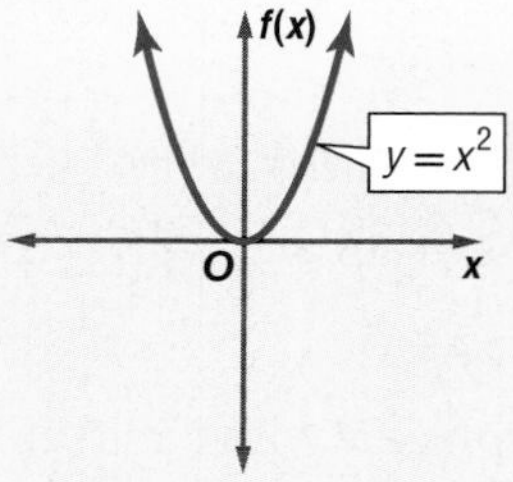

Domain: all reals
Range: all reals $\geq$ minimum

Degree: odd
Leading Coefficient: positive
End Behavior:

$f(x) \to -\infty$ as $x \to -\infty$

$f(x) \to +\infty$ as $x \to +\infty$

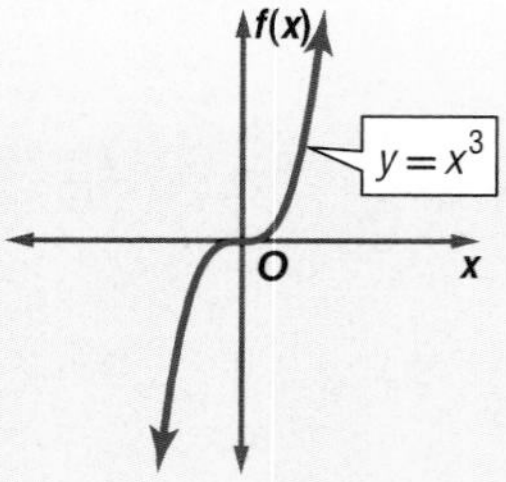

Domain: all reals
Range: all reals

Degree: even
Leading Coefficient: negative
End Behavior:

$f(x) \to -\infty$ as $x \to -\infty$

$f(x) \to -\infty$ as $x \to +\infty$

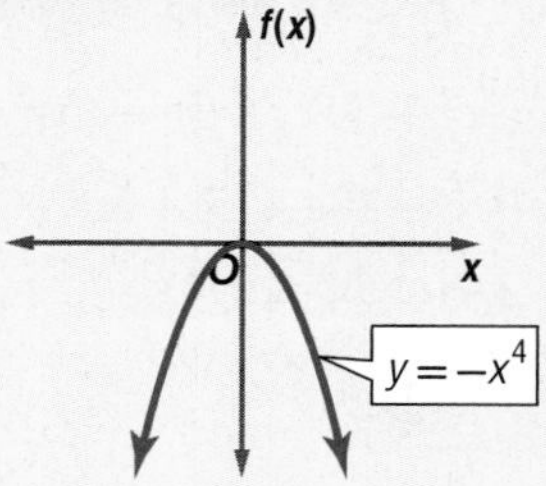

Domain: all reals
Range: all reals $\leq$ maximum

Degree: odd
Leading Coefficient: negative
End Behavior:

$f(x) \to +\infty$ as $x \to -\infty$

$f(x) \to -\infty$ as $x \to +\infty$

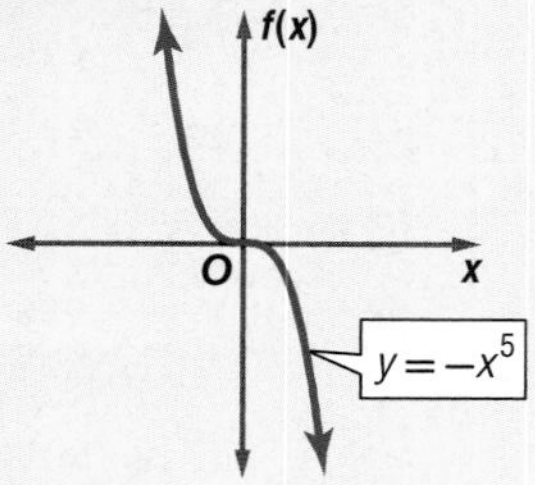

Domain: all reals
Range: all reals

Math *in Motion*, Animation glencoe.com

The number of real zeros of a polynomial equation can be determined from the graph of its related polynomial function. Recall that real zeros occur at x-intercepts, so the number of times a graph crosses the x-axis equals the number of real zeros.

StudyTip

Double roots When a graph is tangent to the x-axis, there is a *double root*, which represents two of the same root.

Key Concept Zeros of Even- and Odd-Degree Functions

For Your FOLDABLE

Odd-degree functions will always have an odd number of real zeros. Even-degree functions will always have an even number of real zeros or no real zeros at all.

Even-Degree Polynomials

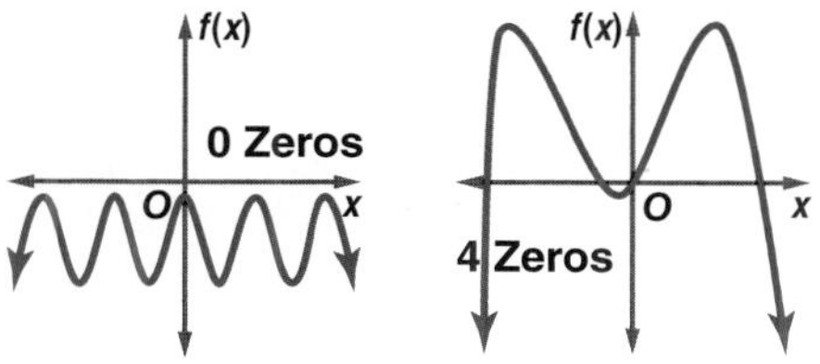

Odd-Degree Polynomials

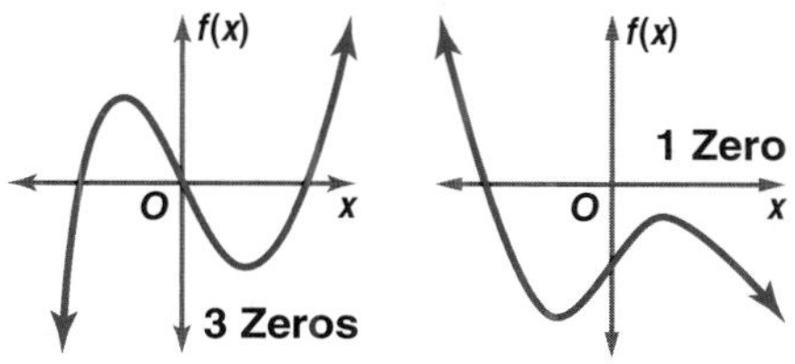

Review Vocabulary

zero the x-coordinate of the point at which a graph intersects the x-axis (Lesson 5-2)

EXAMPLE 4 Graphs of Polynomial Functions

For each graph,

- **describe the end behavior,**
- **determine whether it represents an odd-degree or an even-degree polynomial function, and**
- **state the number of real zeros.**

a.

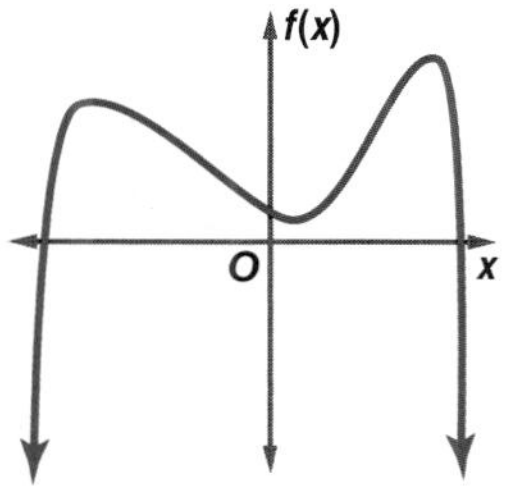

$f(x) \to -\infty$ as $x \to -\infty$.
$f(x) \to -\infty$ as $x \to +\infty$.

Since the end behavior is in the same direction, it is an even-degree function. The graph intersects the x-axis at two points, so there are two real zeros.

b.

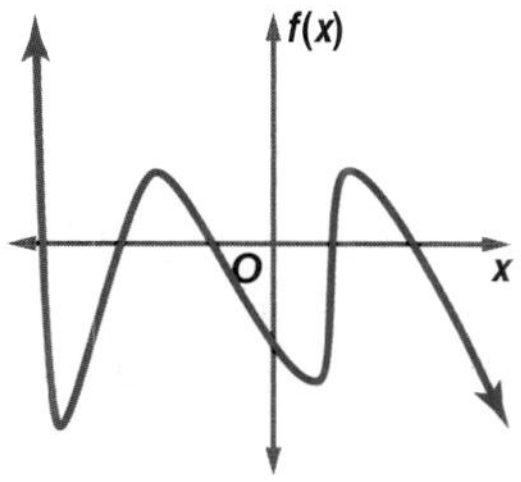

$f(x) \to +\infty$ as $x \to -\infty$.
$f(x) \to -\infty$ as $x \to +\infty$.

Since the end behavior is in opposite directions, it is an odd-degree function. The graph intersects the x-axis at five points, so there are five real zeros.

Check Your Progress

4A.

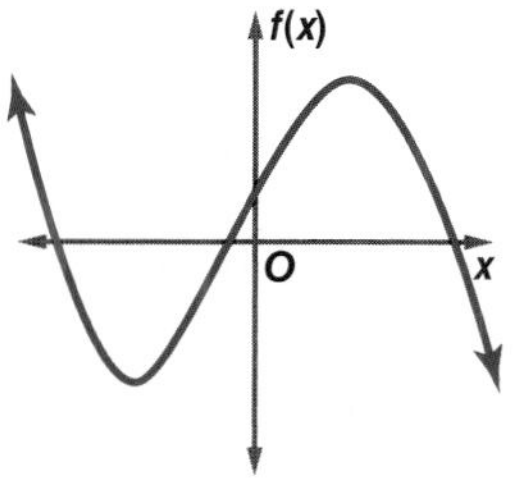

4B.

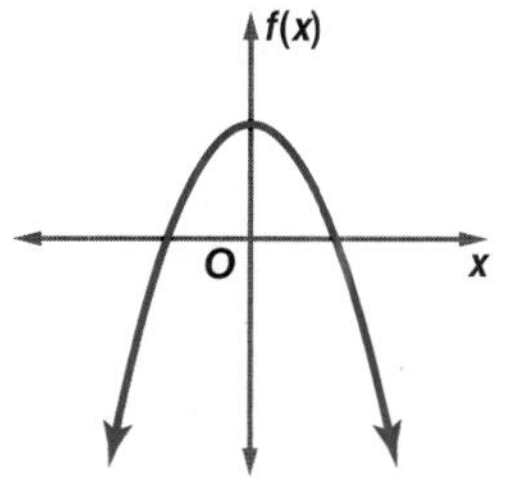

Personal Tutor glencoe.com

Check Your Understanding

Example 1 p. 348

State the degree and leading coefficient of each polynomial in one variable. If it is not a polynomial in one variable, explain why.

1. $11x^6 - 5x^5 + 4x^2$

2. $-10x^7 - 5x^3 + 4x - 22$

3. $14x^4 - 9x^3 + 3x - 4y$

4. $8x^5 - 3x^2 + 4xy - 5$

Example 2 p. 349

Find $w(5)$ and $w(-4)$ for each function.

5. $w(x) = -2x^3 + 3x - 12$

6. $w(x) = 2x^4 - 5x^3 + 3x^2 - 2x + 8$

Example 3 p. 349

If $c(x) = 4x^3 - 5x^2 + 2$ and $d(x) = 3x^2 + 6x - 10$, find each value.

7. $c(y^3)$

8. $-4[d(3z)]$

9. $6c(4a) + 2d(3a - 5)$

10. $-3c(2b) + 6d(4b - 3)$

Example 4 p. 351

For each graph,

a. **describe the end behavior,**

b. **determine whether it represents an odd-degree or an even-degree function, and**

c. **state the number of real zeros.**

11.

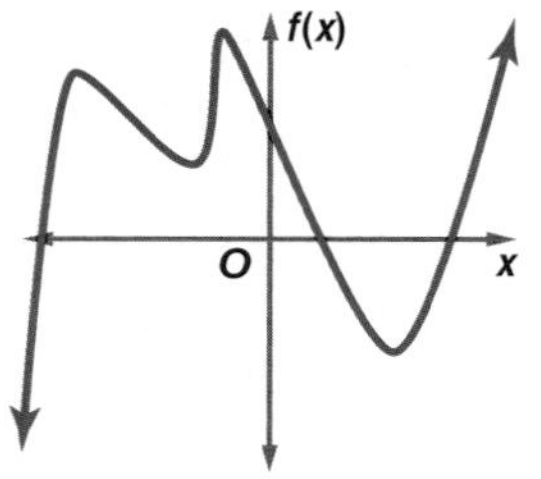

12.

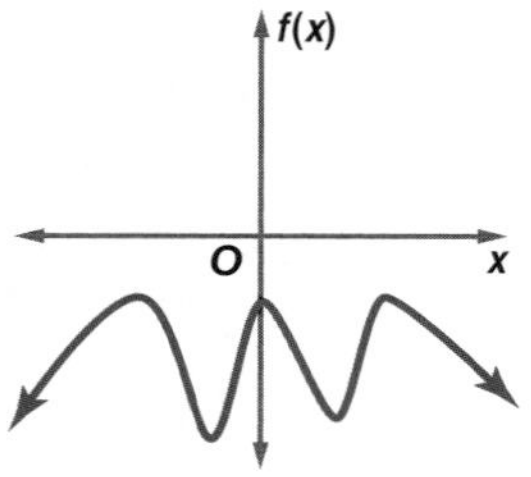

Practice and Problem Solving

● = **Step-by-Step Solutions** begin on page R20.
Extra Practice begins on page 947.

Example 1 p. 348

State the degree and leading coefficient of each polynomial in one variable. If it is not a polynomial in one variable, explain why.

13. $-6x^6 - 4x^5 + 13xy$

14. $3a^7 - 4a^4 + \frac{3}{a}$

15. $8x^5 - 12x^6 + 14x^3 - 9$

16. $-12 - 8x^2 + 5x - 21x^7$

17. $15x - 4x^3 + 3x^2 - 5x^4$

18. $13b^3 - 9b + 3b^5 - 18$

19. $(d + 5)(3d - 4)$

20. $(5 - 2y)(4 + 3y)$

21. $6x^5 - 5x^4 + 2x^9 - 3x^2$

22. $7x^4 + 3x^7 - 2x^8 + 7$

Example 2 p. 349

Find $p(-6)$ and $p(3)$ for each function.

23. $p(x) = x^4 - 2x^2 + 3$

24. $p(x) = -3x^3 - 2x^2 + 4x - 6$

25. $p(x) = 2x^3 + 6x^2 - 10x$

26. $p(x) = x^4 - 4x^3 + 3x^2 - 5x + 24$

27 $p(x) = -x^3 + 3x^2 - 5$

28. $p(x) = 2x^4 + x^3 - 4x^2$

Example 3 p. 349

If $c(x) = 2x^2 - 4x + 3$ and $d(x) = -x^3 + x + 1$, find each value.

29. $c(3a)$

30. $5d(2a)$

31. $c(b^2)$

32. $d(4a^2)$

33. $d(4y - 3)$

34. $c(y^2 - 1)$

Example 4
p. 351

For each graph,

a. **describe the end behavior,**

b. **determine whether it represents an odd-degree or an even-degree function, and**

c. **state the number of real zeros.**

35.

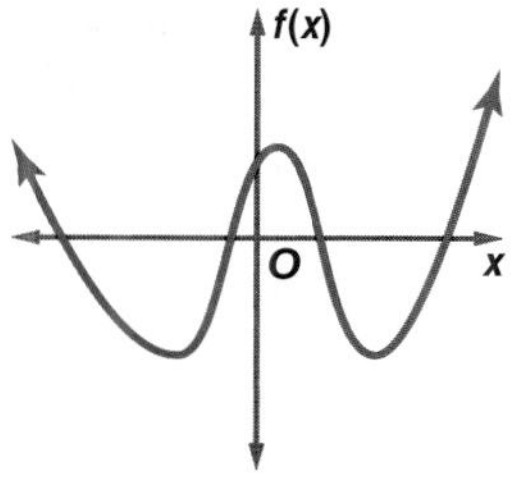

36.

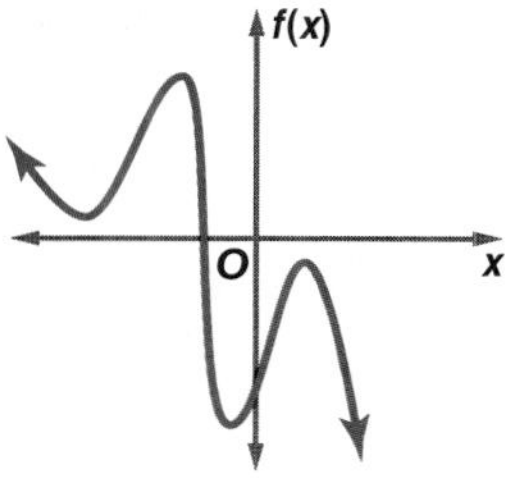

37.

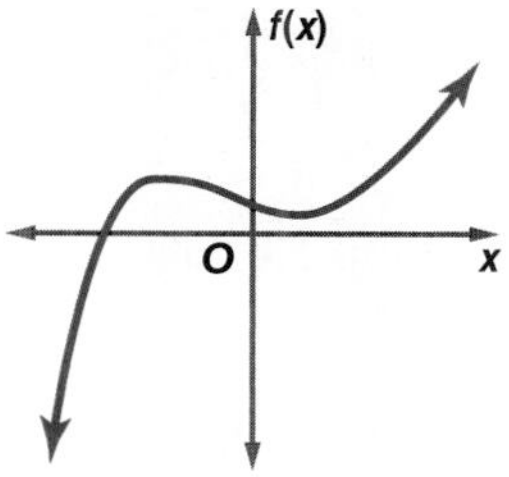

38.

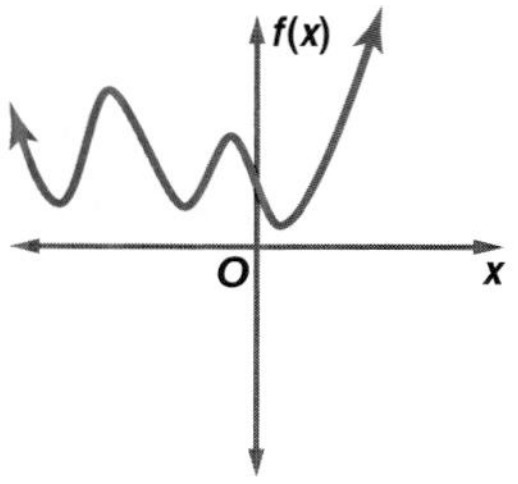

39.

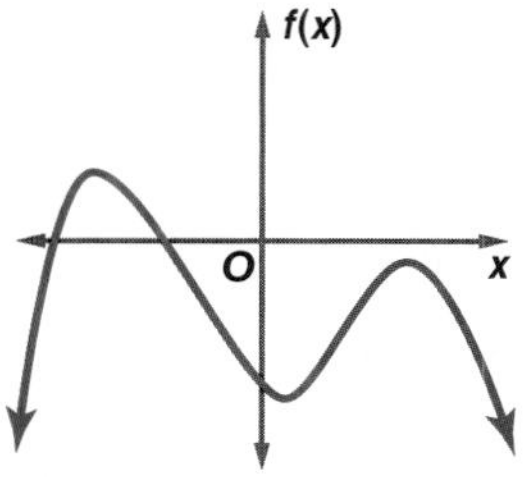

40.

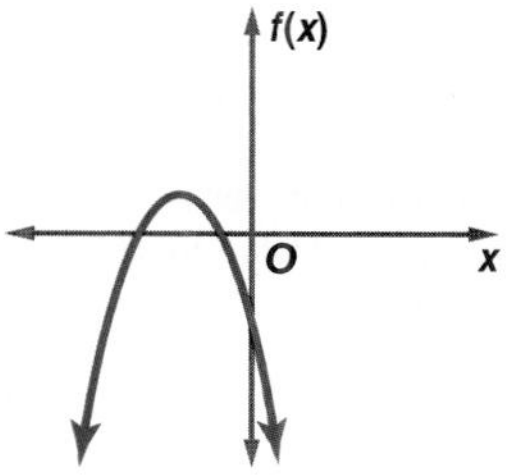

Real-World Link

In 1947, the first commercial microwave oven hit the market. These early units were gigantic and enormously expensive, standing $5\frac{1}{2}$ feet tall, weighing over 750 pounds, and costing about $5000 each.

Source: Microtech

41 **PHYSICS** For a moving object with mass m in kilograms, the kinetic energy KE in joules is given by the function $KE(v) = 0.5mv^2$, where v represents the speed of the object in meters per second. Find the kinetic energy of an all-terrain vehicle with a mass of 171 kilograms moving at a speed of 11 meters/second.

42. **BUSINESS** A microwave manufacturing firm has determined that their profit function is $P(x) = -0.0014x^3 + 0.3x^2 + 6x - 355$, where x is the number of microwaves sold annually.

a. Graph the profit function using your calculator.

b. Determine a reasonable viewing window for the function.

c. Locate all of the zeros of the function using the CALC menu.

d. What must be the range of microwaves sold in order for the firm to have a profit?

Find $p(-2)$ and $p(8)$ for each function.

43. $p(x) = \frac{1}{4}x^4 + \frac{1}{2}x^3 - 4x^2$

44. $p(x) = \frac{1}{8}x^4 - \frac{3}{2}x^3 + 12x - 18$

45. $p(x) = \frac{3}{4}x^4 - \frac{1}{8}x^2 + 6x$

46. $p(x) = \frac{5}{8}x^3 - \frac{1}{2}x^2 + \frac{3}{4}x + 10$

Use the degree and end behavior to match each polynomial to its graph.

A

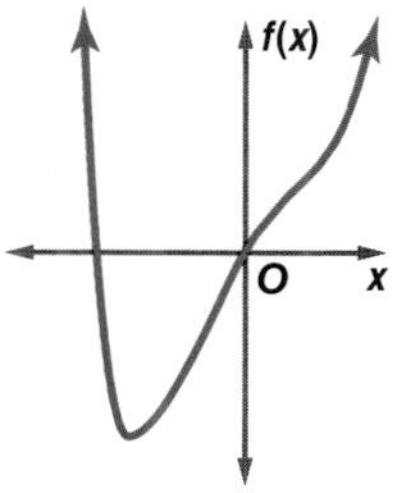

B

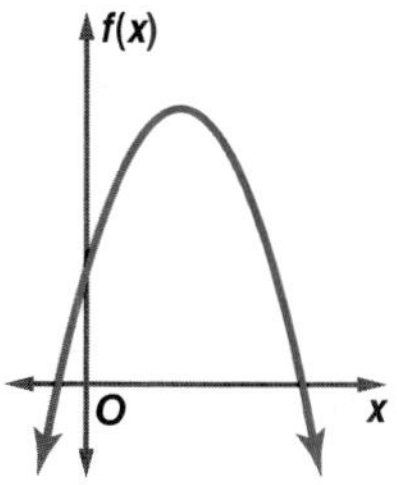

C

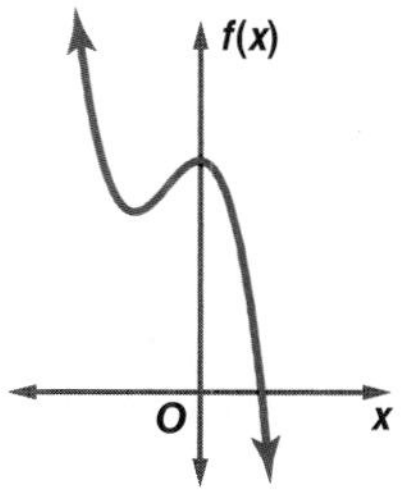

D

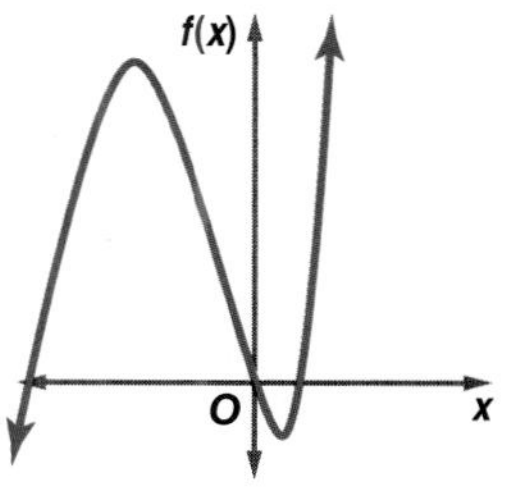

47. $f(x) = x^3 + 3x^2 - 4x$

48. $f(x) = -2x^2 + 8x + 5$

49. $f(x) = x^4 - 3x^2 + 6x$

50. $f(x) = -4x^3 - 4x^2 + 8$

If $c(x) = x^3 - 2x$ and $d(x) = 4x^2 - 6x + 8$, find each value.

51. $3c(a - 4) + 3d(a + 5)$

52. $-2d(2a + 3) - 4c(a^2 + 1)$

53. $5c(a^2) - 8d(6 - 3a)$

54. $-7d(a^3) + 6c(a^4 + 1)$

Real-World Link

Teens are important customers to clothing makers. Clothing tops the lists of both what teens plan to buy and what they actually purchase.

Source: NOPWorld

55. **BUSINESS** A clothing manufacturer's profitability can be modeled by $p(x) = -x^4 + 40x^2 - 144$, where x is the number of items sold in thousands and $p(x)$ is the company's profit in thousands of dollars.
    a. Use a table of values to sketch the function.
    b. Determine the zeros of the function.
    c. Between what two values should the company sell in order to be profitable?
    d. Explain why only two of the zeros are considered in part c.

56. **MULTIPLE REPRESENTATIONS** Consider $g(x) = (x - 2)(x + 1)(x - 3)(x + 4)$.
    a. **ANALYTICAL** Determine the x- and y-intercepts, roots, degree, and end behavior of $g(x)$.
    b. **ALGEBRAIC** Write the function in standard form
    c. **TABULAR** Make a table of values for the function.
    d. **GRAPHICAL** Sketch a graph of the function by plotting points and connecting them with a smooth curve.

Describe the end behavior of the graph of each function.

57. $f(x) = -5x^4 + 3x^2$
58. $g(x) = 2x^5 + 6x^4$
59. $h(x) = -4x^7 + 8x^6 - 4x$
60. $f(x) = 6x - 7x^2$
61. $g(x) = 8x^4 + 5x^5$
62. $h(x) = 9x^6 - 5x^7 + 3x^2$

H.O.T. Problems

Use Higher-Order Thinking Skills

63. **FIND THE ERROR** Shenequa and Virginia are determining the number of zeros of the graph at the right. Is either of them correct? Explain your reasoning.

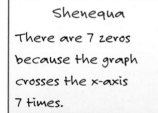

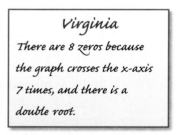

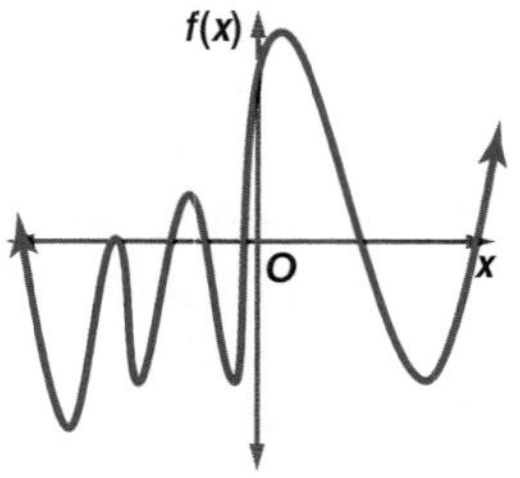

Shenequa

There are 7 zeros because the graph crosses the x-axis 7 times.

Virginia

There are 8 zeros because the graph crosses the x-axis 7 times, and there is a double root.

64. **CHALLENGE** Use the table to determine the minimum number of real roots and the minimum degree of the polynomial function $f(x)$.

x	−24	−18	−12	−6	0	6	12	18	24
$f(x)$	−8	−1	3	−2	4	7	−1	−8	5

65. **CHALLENGE** If $f(x)$ has a degree of 5 and a positive leading coefficient and $g(x)$ has a degree of 3 and a positive leading coefficient, determine the end behavior of $\frac{f(x)}{g(x)}$. Explain your reasoning.

66. **OPEN ENDED** Sketch the graph of an even-degree polynomial with 8 real roots, one of them a double root.

67. **REASONING** Determine whether the following statement is *always*, *sometimes*, or *never* true. Explain.

    A polynomial function that has four real roots is a fourth-degree polynomial.

68. **WRITING IN MATH** Describe what the end behavior of a polynomial function is and how to determine it.

NYSRE PRACTICE A2.A.41

69. SHORT RESPONSE Four students solved the same math problem. Each student's work is shown below. Who is correct?

Student A

$x^2x^{-5} = \frac{x^2}{x^5}$

$= \frac{1}{x^3}, x \neq 0$

Student C

$x^2x^{-5} = \frac{x^2}{x^{-5}}$

$= x^7, x \neq 0$

Student B

$x^2x^{-5} = \frac{x^2}{x^{-5}}$

$= x^{-7}, x \neq 0$

Student D

$x^2x^{-5} = \frac{x^2}{x^5}$

$= x^3, x \neq 0$

70. SAT/ACT What is the remainder when $x^3 - 7x + 5$ is divided by $x + 3$?

A -11
B -1
C 1
D 11

71. EXTENDED RESPONSE A company manufactures tables and chairs. It costs \$40 to make each table and \$25 to make each chair. There is \$1440 available to spend on manufacturing each week. Let t = the number of tables produced and c = the number of chairs produced.

a. The manufacturing equation is $40t + 25c = 1500$. Construct a graph of this equation.
b. The company always produces two chairs with each table. Write and graph an equation to represent this situation on the same graph as the one in part **a**.
c. Determine the number of tables and chairs that the company can produce each week.
d. Explain how to determine this answer using the graph.

72. If $i = \sqrt{-1}$, then $5i(7i) =$

F 70
G 35
H -35
J -70

Spiral Review

Simplify. (Lesson 6-2)

73. $\frac{16x^4y^3 + 32x^6y^5z^2}{8x^2y}$

74. $\frac{18ab^4c^5 - 30a^4b^3c^2 + 12a^5bc^3}{6abc^2}$

75. $\frac{18c^5d^2 - 3c^2d^2 + 12a^5c^3d^4}{3c^2d^2}$

Determine whether each expression is a polynomial. If it is a polynomial, state the degree of the polynomial. (Lesson 6-1)

76. $8x^2 + 5xy^3 - 6x + 4$

77. $9x^4 + 12x^6 - 16$

78. $3x^4 + 2x^2 - x^{-1}$

79. FOUNTAINS The height of a fountain's water stream can be modeled by a quadratic function. Suppose the water from a jet reaches a maximum height of 8 feet at a distance 1 foot away from the jet.

a. If the water lands 3 feet away from the jet, find a quadratic function that models the height $h(d)$ of the water at any given distance d feet from the jet. Then compare the graph of the function to the parent function.

b. Suppose a worker increases the water pressure so that the stream reaches a maximum height of 12.5 feet at a distance of 15 inches from the jet. The water now lands 3.75 feet from the jet. Write a new quadratic function for $h(d)$. How do the changes in h and k affect the shape of the graph? (Lesson 5-7)

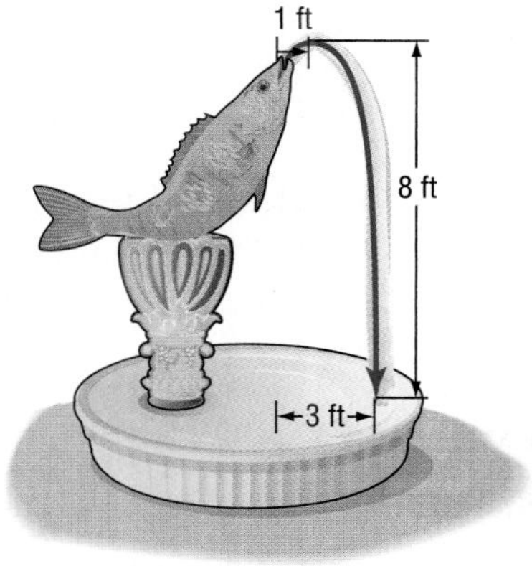

Solve each inequality. (Lesson 1-6)

80. $|2x + 4| \leq 8$

81. $|-3x + 2| \geq 4$

82. $|2x - 8| - 4 \leq -6$

Skills Review

Determine whether each function has a maximum or minimum value, and find that value. (Lesson 5-1)

83. $f(x) = 3x^2 - 8x + 4$

84. $f(x) = -4x^2 + 2x - 10$

85. $f(x) = -0.25x^2 + 4x - 5$

EXTEND 6-3

Algebra Lab: Polynomial Functions and Rate of Change

NY Math Online glencoe.com
Math *in Motion*, Animation

NYS Core Curriculum **Reinforcement of A.A.32** Explain slope as a rate of change between dependent and independent variables

In Chapter 2, you learned that linear functions have a constant first-order difference. Then in Chapter 5, you learned that quadratic functions have a constant second-order difference. Now, you will examine the differences for polynomial functions with degree greater than 2.

ACTIVITY

Consider $f(x) = -2x^3$.

Step 1 Copy the table and complete row 2 for x-values from −4 through 4.

x	−4	−3	−2	−1	...	4
y						
First-order Differences						
Second-order Differences						
Third-order Differences						

Step 2 Graph the ordered pairs (x, y) and connect the points with a smooth curve.

Step 3 Find the first-order differences and complete row 3. Describe any patterns in the differences.

Step 4 Complete rows 4 and 5 by finding the second- and third-order differences. Describe any patterns in the differences. Make a conjecture about the differences for a third-degree polynomial function.

Step 5 Repeat Steps 1 through 4 using a fourth-degree polynomial function. Make a conjecture about the differences for an nth-degree polynomial function.

Exercises

State the degree of each polynomial function described.

1. constant second-order difference of −3

2. constant fifth-order difference of 12

3. constant first-order difference of −1.25

4. CHALLENGE Write an equation for a polynomial function with real-number coefficients for the ordered pairs and differences in the table. Make sure your equation is satisfied for all of the ordered pairs (x, y).

x	−3	−2	−1	0	1	2	3
y	−25	?	?	?	?	?	?
First-order Differences		19	7	1	1	7	19
Second-order Differences			−12	−6	0	6	12

6-4 Analyzing Graphs of Polynomial Functions

Then

You used maxima and minima and graphs of polynomials.
(Lessons 5-1 and 6-3)

Now

- Graph polynomial functions and locate their zeros.
- Find the relative maxima and minima of polynomial functions.

NYS Core Curriculum

A2.A.50 Approximate the solution to polynomial equations of higher degree by inspecting the graph

New Vocabulary

relative maximum
relative minimum
extrema
turning points

NY Math Online

glencoe.com
- Extra Examples
- Personal Tutor
- Self-Check Quiz
- Homework Help
- Math in Motion

Why?

Annual attendance at the movies has fluctuated since the first movie theater, the Nickelodeon, opened in Pittsburgh in 1906. Overall attendance peaked during the 1920s, and it was at its lowest during the 1970s. A graph of the annual attendance to the movies can be represented by a polynomial function.

Graphs of Polynomial Functions To graph a polynomial function, make a table of values to find several points and then connect them to make a smooth continuous curve. Knowing the end behavior of the graph will assist you in completing the graph.

EXAMPLE 1 Graph of a Polynomial Function

Graph $f(x) = -x^4 + x^3 + 3x^2 + 2x$ by making a table of values.

x	$f(x)$	x	$f(x)$
-2.5	≈ -41	0.5	≈ 1.8
-2.0	-16	1.0	5.0
-1.5	≈ -4.7	1.5	≈ 8.1
-1.0	-1.0	2.0	8.0
-0.5	≈ -0.4	2.5	≈ 0.3
0.0	0.0	3.0	-21

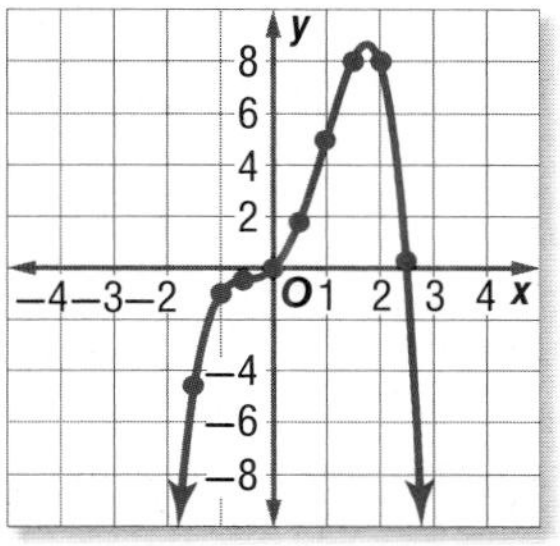

This is an even-degree polynomial with a negative leading coefficient, so $f(x) \to -\infty$ as $x \to -\infty$ and $f(x) \to -\infty$ as $x \to +\infty$. Notice that the graph intersects the x-axis at two points, indicating there are two zeros for this function.

Check Your Progress

1. Graph $f(x) = x^4 - x^3 - 2x^2 + 4x - 6$ by making a table of values.

Personal Tutor glencoe.com

In Example 1, one of the zeros occurred at $x = 0$. Another zero occurred between $x = 2.5$ and $x = 3.0$. Because $f(x)$ is positive for $x = 2.5$ and negative for $x = 3.0$ and all polynomial functions are continuous, we know there is a zero between these two values.

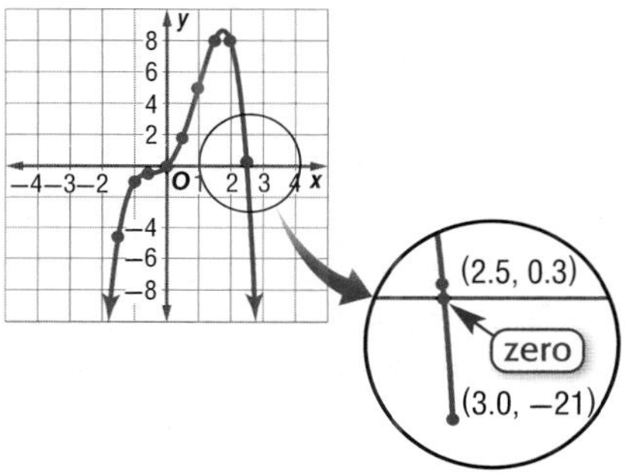

So, if the value of $f(x)$ *changed signs* from one value of x to the next, then there is a zero between those two x-values. This idea is called the **Location Principle**.

StudyTip

Degree Recall that the degree of the function is also the maximum number of zeros the function can have.

Key Concept Location Principle

For Your FOLDABLE

Words Suppose $y = f(x)$ represents a polynomial function and a and b are two real numbers such that $f(a) < 0$ and $f(b) > 0$. Then the function has at least one real zero between a and b.

Model

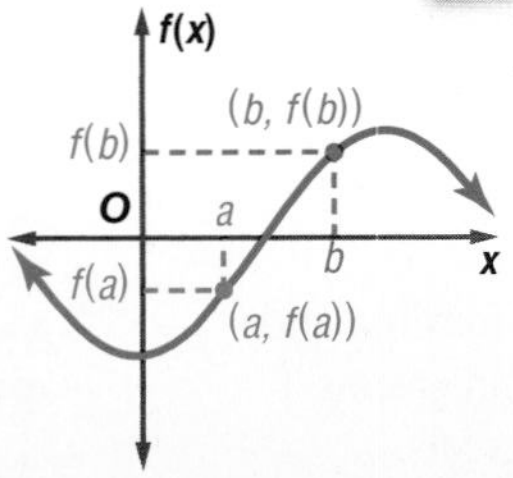

Math *in Motion,* Animation glencoe.com

EXAMPLE 2 Locate Zeros of a Function

Determine consecutive integer values of x between which each real zero of $f(x) = x^3 - 4x^2 + 3x + 1$ is located. Then draw the graph.

Make a table of values. Since $f(x)$ is a third-degree polynomial function, it will have either 3 or 1 real zeros. Look at the values of $f(x)$ to locate the zeros. Then use the points to sketch a graph of the function.

x	$f(x)$	
−2	−29	
−1	−7	
		← change in sign
0	1	
1	1	
		← change in sign
2	−1	
		← change in sign
3	1	
4	13	

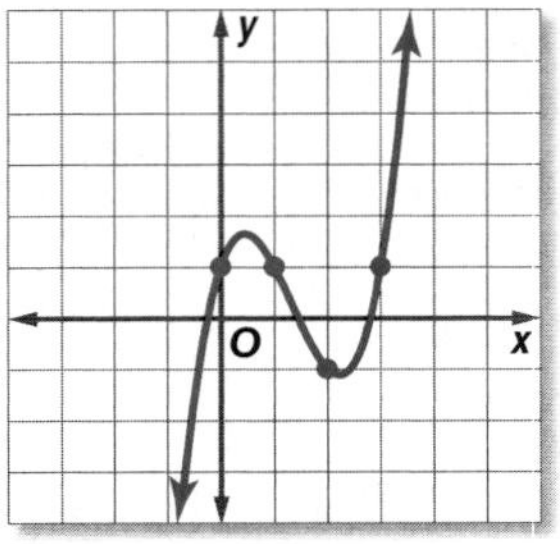

The changes in sign indicate that there are zeros between $x = -1$ and $x = 0$, between $x = 1$ and $x = 2$, and between $x = 2$ and $x = 3$.

StudyTip

Odd Functions Some odd functions, like $f(x) = x^3$, have no turning points.

Check Your Progress

2. Determine consecutive integer values of x between which each real zero of the function $f(x) = x^4 - 3x^3 - 2x^2 + x + 1$ is located. Then draw the graph.

Personal Tutor glencoe.com

Maximum and Minimum Points The graph below shows the general shape of a third-degree polynomial function.

Point A on the graph is a **relative maximum** of the function since no other nearby points have a greater y-coordinate. The graph is increasing as it approaches A and decreasing as it moves from A.

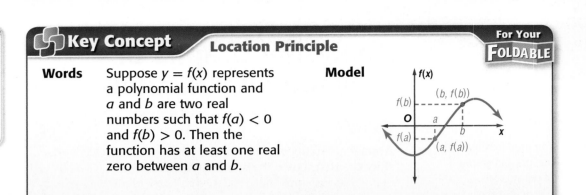

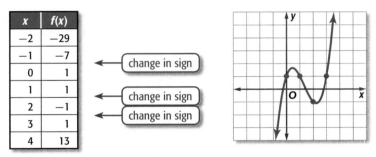

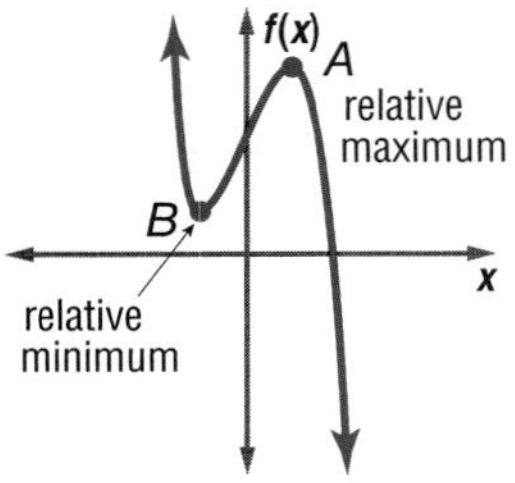

Likewise, point B is a **relative minimum** since no other nearby points have a lesser y-coordinate. The graph is decreasing as it approaches B and increasing as it moves from B. The maximum and minimum values of a function are called the **extrema**.

These points are often referred to as **turning points**. The graph of a polynomial function of degree n has at most $n - 1$ turning points.

StudyTip

Maximum and Minimum A polynomial with a degree greater than 3 may have more than one relative maximum or relative minimum.

StudyTip

Rational Values Zeros and turning points will not always occur at integral values of x.

EXAMPLE 3 Maximum and Minimum Points

Graph $f(x) = x^3 - 4x^2 - 2x + 3$. Estimate the x-coordinates at which the relative maxima and relative minima occur.

Make a table of values and graph the function.

x	$f(x)$	
−2	−17	
−1	0	zero
0	3	indicates a relative maximum
1	−2	
2	−9	
3	−12	indicates a relative minimum
4	−4	zero between 4 and 5
5	18	

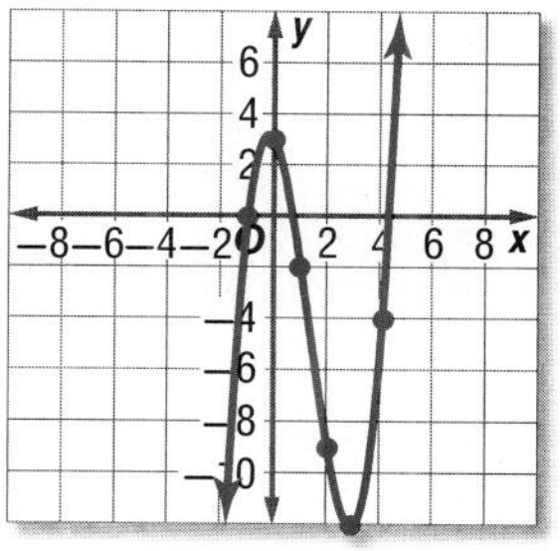

Look at the table of values and the graph.

The value of $f(x)$ changes signs between $x = 4$ and $x = 5$, indicating a zero of the function.

The value of $f(x)$ at $x = 0$ is greater than the surrounding points, so there must be a relative maximum *near* $x = 0$.

The value of $f(x)$ at $x = 3$ is less than the surrounding points, so there must be a relative minimum *near* $x = 3$.

CHECK You can use a graphing calculator to find the relative maximum and relative minimum of a function and confirm your estimates.

Enter $y = x^3 - 4x^2 - 2x + 3$ in the **Y=** list and graph the function.

Use the **CALC** menu to find each maximum and minimum.

When selecting the left bound, move the cursor to the left of the maximum or minimum. When selecting the right bound, move the cursor to the right of the maximum or minimum.

Press ENTER twice.

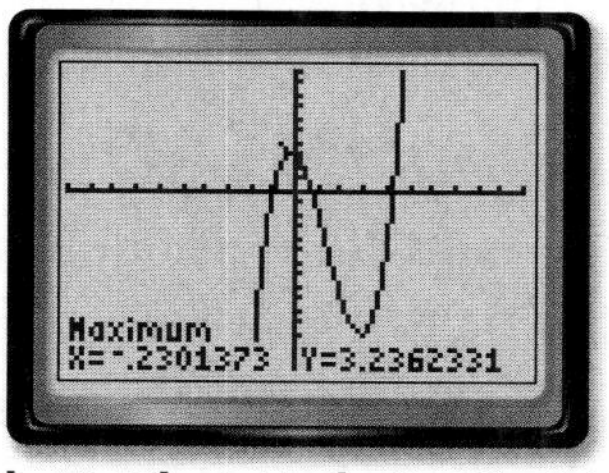

[−10, 10] scl: 1 by [−15, 10] scl: 1

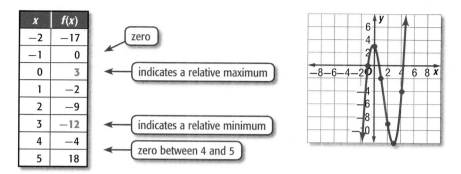

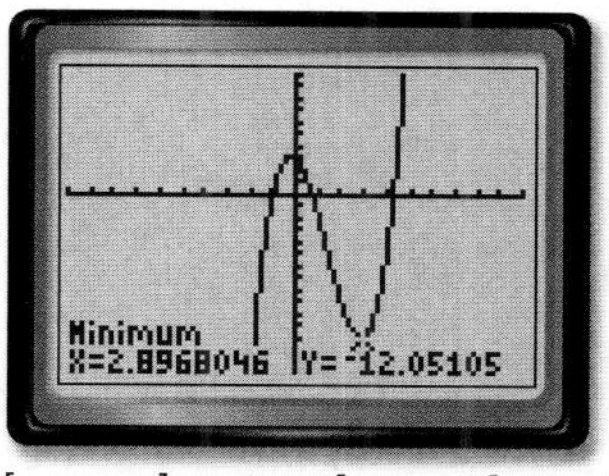

[−10, 10] scl: 1 by [−15, 10] scl: 1

The estimates for a relative maximum near $x = 0$ and a relative minimum near $x = 3$ are accurate.

Check Your Progress

3. Graph $f(x) = 2x^3 + x^2 - 4x - 2$. Estimate the x-coordinates at which the relative maxima and relative minima occur.

Personal Tutor glencoe.com

Real-World Link

Over 1.4 billion movie tickets were sold in the United States in 2006.

Source: CNN

The graph of a polynomial function can reveal trends in real-world data. It is often helpful to note when the graph is increasing or decreasing.

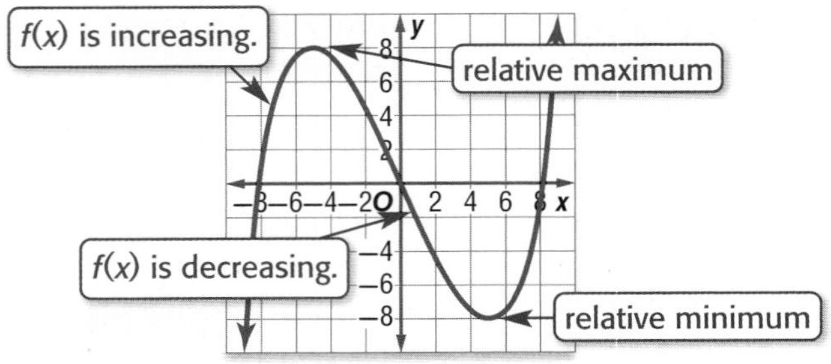

Real-World EXAMPLE 4 Graph a Polynomial Model

MOVIES Refer to the beginning of the lesson. Annual admissions to movies in the United States can be modeled by the function $f(x) = -0.0017x^4 + 0.31x^3 - 17.66x^2 + 277x + 3005$, where x is the number of years since 1926 and $f(x)$ is the annual admissions in millions.

a. Graph the function.

Make a table of values for the years 1926–2006. Plot the points and connect with a smooth curve. Finding and plotting the points for every tenth year gives a good approximation of the graph.

x	$f(x)$
0	3005
10	4302
20	3689
30	2414
40	1317
50	830
60	977
70	1374
80	1229

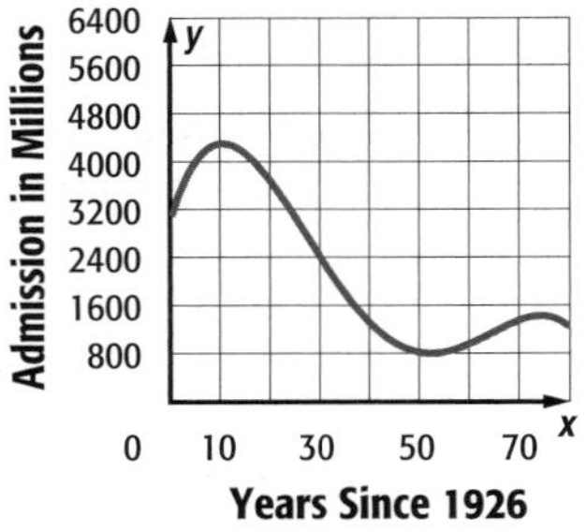

b. Describe the turning points of the graph and its end behavior.

There is a relative maximum near 1936 and a relative minimum between 1976 and 1981. $f(x) \to -\infty$ as $x \to -\infty$ and $f(x) \to -\infty$ as $x \to \infty$.

c. What trends in movie admissions does the graph suggest? Is it reasonable that the trend will continue indefinitely?

Movie attendance peaked around 1936 and declined until about 1978. It then increased until 2000 and began a decline.

d. Is it reasonable that the trend will continue indefinitely?

This trend may continue for a couple of years, but the graph will soon become unreasonable as it predicts negative attendance for the future.

Check Your Progress

4. **FAX MACHINES** The annual sales of fax machines for home use can be modeled by $f(x) = -0.17x^4 + 6.29x^3 - 77.65x^2 + 251x + 1100$, where x is the number of years after 1990 and $f(x)$ is the annual sales in millions of dollars.

   A. Graph the function.

   B. Describe the turning points of the graph and its end behavior.

   C. What trends in fax machine sales does the graph suggest?

   D. Is it reasonable that the trend will continue indefinitely?

Personal Tutor glencoe.com

Check Your Understanding

Example 1 p. 357

Graph each polynomial equation by making a table of values.

1. $f(x) = 2x^4 - 5x^3 + x^2 - 2x + 4$
2. $f(x) = -2x^4 + 4x^3 + 2x^2 + x - 3$
3. $f(x) = 3x^4 - 4x^3 - 2x^2 + x - 4$
4. $f(x) = -4x^4 + 5x^3 + 2x^2 + 3x + 1$

Example 2 p. 358

Determine the consecutive integer values of x between which each real zero of each function is located. Then draw the graph.

5. $f(x) = x^3 - 2x^2 + 5$
6. $f(x) = -x^4 + x^3 + 2x^2 + x + 1$
7. $f(x) = -3x^4 + 5x^3 + 4x^2 + 4x - 8$
8. $f(x) = 2x^4 - x^3 - 3x^2 + 2x - 4$

Example 3 p. 359

Graph each polynomial function. Estimate the x-coordinates at which the relative maxima and relative minima occur. State the domain and range for each function.

9. $f(x) = x^3 + x^2 - 6x - 3$
10. $f(x) = 3x^3 - 6x^2 - 2x + 2$
11. $f(x) = -x^3 + 4x^2 - 2x - 1$
12. $f(x) = -x^3 + 2x^2 - 3x + 4$

Example 4 p. 360

13. **MUSIC SALES** Annual compact disc sales can be modeled by the quartic function $f(x) = 0.48x^4 - 9.6x^3 + 53x^2 - 49x + 599$, where x is the number of years after 1995 and $f(x)$ is annual sales in millions.
    a. Graph the function for $0 \leq x \leq 10$.
    b. Describe the turning points of the graph and its end behavior.
    c. Continue the graph for $x = 11$ and $x = 12$. What trends in compact disc sales does the graph suggest?
    d. Is it reasonable that the trend will continue indefinitely? Explain.

Practice and Problem Solving

● = **Step-by-Step Solutions** begin on page R20.
Extra Practice begins on page 947.

Examples 1–3 pp. 357–359

Complete each of the following.

a. Graph each function by making a table of values.

b. Determine the consecutive integer values of x between which each real zero is located.

c. Estimate the x-coordinates at which the relative maxima and minima occur.

14. $f(x) = x^3 + 3x^2$
15. $f(x) = -x^3 + 2x^2 - 4$
16. $f(x) = x^3 + 4x^2 - 5x$
17. $f(x) = x^3 - 5x^2 + 3x + 1$
18. $f(x) = -2x^3 + 12x^2 - 8x$
19. $f(x) = 2x^3 - 4x^2 - 3x + 4$
20. $f(x) = x^4 + 2x - 1$
21. $f(x) = x^4 + 8x^2 - 12$

Example 4 p. 360

22. **GASOLINE PRICES** The average annual price of gasoline can be modeled by the cubic function $f(x) = 0.0007x^3 - 0.014x^2 + 0.08x + 0.96$, where x is the number of years after 1987 and $f(x)$ is the price in dollars.
    a. Graph the function for $0 \leq x \leq 30$.
    b. Describe the turning points of the graph and its end behavior.
    c. What trends in gasoline prices does the graph suggest?
    d. Is it reasonable that the trend will continue indefinitely? Explain.

Use a graphing calculator to estimate the x-coordinates at which the maxima and minima of each function occur. Round to the nearest hundredth.

23. $f(x) = x^3 + 3x^2 - 6x - 6$
24. $f(x) = -2x^3 + 4x^2 - 5x + 8$
25. $f(x) = -2x^4 + 5x^3 - 4x^2 + 3x - 7$
26. $f(x) = x^5 - 4x^3 + 3x^2 - 8x - 6$

Sketch the graph of polynomial functions with the following characteristics.

27. an odd function with zeros at $-5, -3, 0, 2$ and 4

28. an even function with zeros at $-2, 1, 3,$ and 5

29. a 4-degree function with a zero at -3, maximum at $x = 2$, and minimum at $x = -1$

30. a 5-degree function with zeros at $-4, -1,$ and 3, maximum at $x = -2$

31. an odd function with zeros at $-1, 2,$ and 5 and a negative leading coefficient

32. an even function with a minimum at $x = 3$ and a positive leading coefficient

33 **DIVING** The deflection d of a 10-foot-long diving board can be calculated using the function $d(x) = 0.015x^2 - 0.0005x^3$, where x is the distance between the diver and the stationary end of the board in feet.

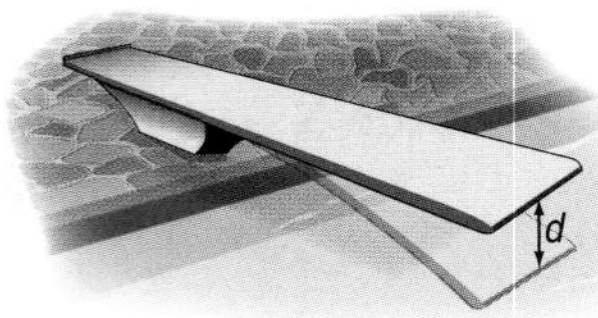

a. Make a table of values of the function for $0 \leq x \leq 10$.

b. Graph the function.

c. What does the end behavior of the graph suggest as x increases?

d. Will this trend continue indefinitely? Explain your reasoning.

Real-World Link

There are about 250,000 swimmers and divers on high school teams nationwide. The sport ranks 8th in popularity for girls and 10th for boys.

Source: National Federation of State High School Associations

Complete each of the following.

a. Estimate the x-coordinate of every turning point and determine if those coordinates are relative maxima or relative minima.

b. Estimate the x-coordinate of every zero.

c. Determine the smallest possible degree of the function.

d. Determine the domain and range of the function.

34.

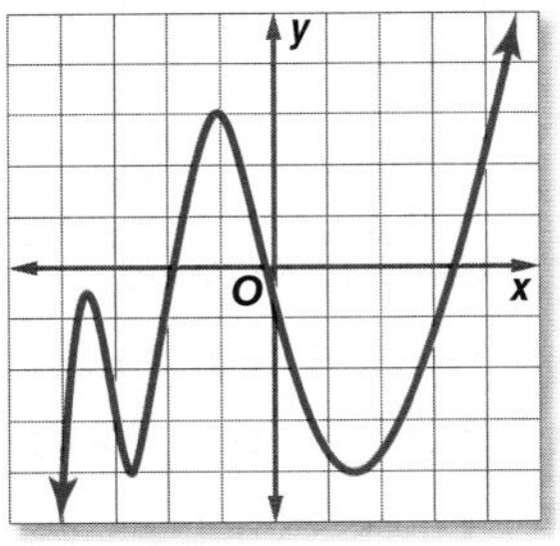

35.

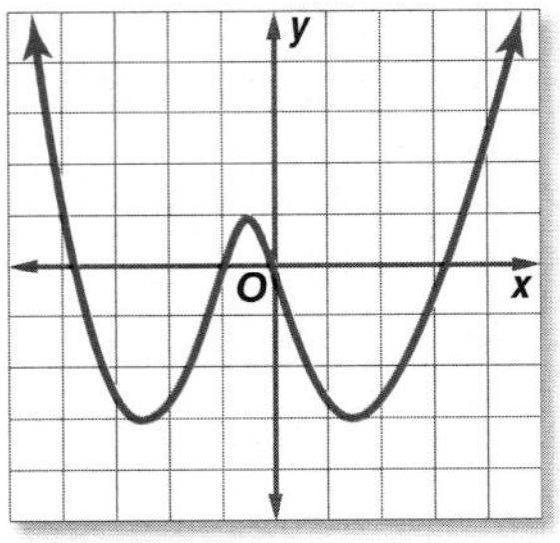

36.

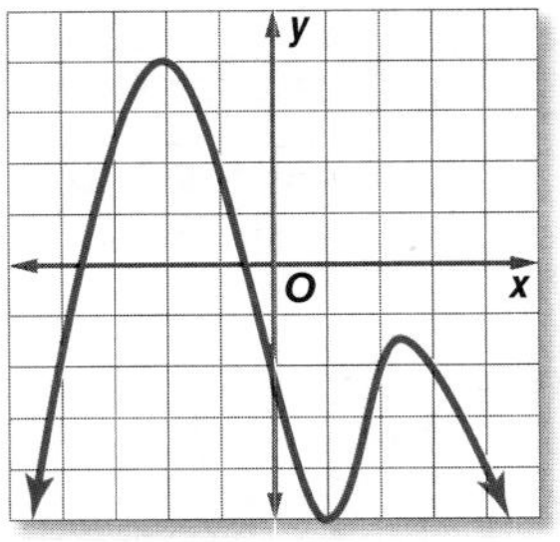

37.

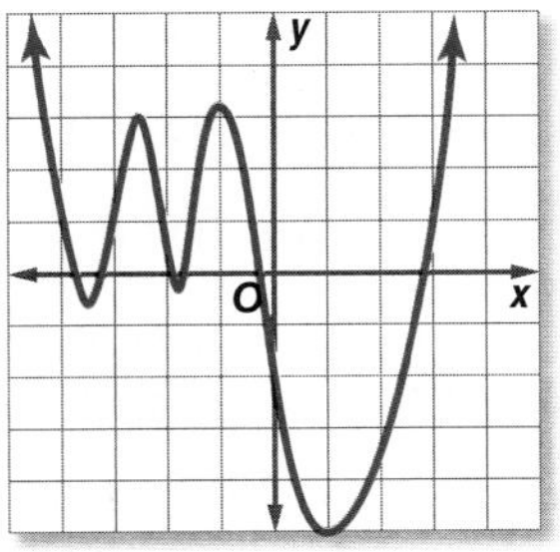

38.

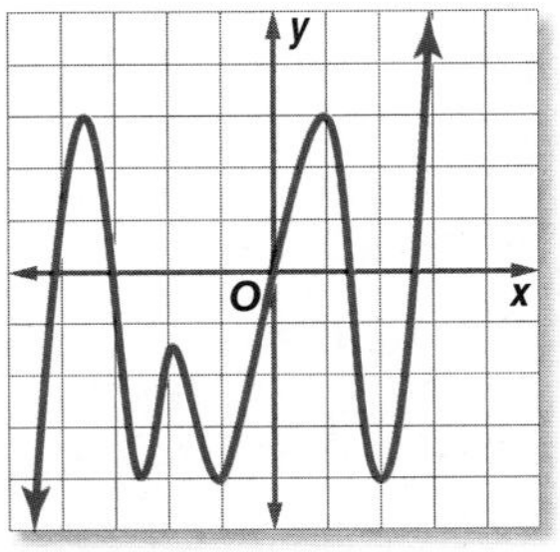

39.

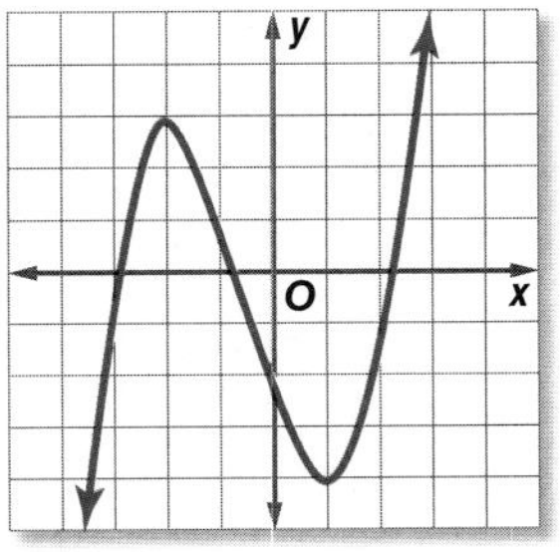

40. **PAGERS** The number of subscribers using pagers in the United States can be modeled by $f(x) = 0.015x^4 - 0.44x^3 + 3.46x^2 - 2.7x + 9.68$, where x is the number of years after 1990 and $f(x)$ is the number of subscribers in millions.

a. Graph the function.

b. Describe the end behavior of the graph.

c. What does the end behavior suggest about the number of pager subscribers?

d. Will this trend continue indefinitely? Explain your reasoning.

Real-World Link

Recent innovations in vending machines include telephone numbers that allow customers to call a machine and have the cost of the snack billed directly to their mobile phones.

Source: Reuters

41. PRICING Jin's vending machines currently sell an average of 3500 beverages per week at a rate of \$0.75 per can. She is considering increasing the price. Her weekly earnings can be represented by $f(x) = -5x^2 + 100x + 2625$, where x is the number of \$0.05 increases. Graph the function and determine the most profitable price for Jin.

For each function,

a. determine the zeros, x- and y-intercepts, and turning points,

b. determine the axis of symmetry, and

c. determine the intervals for which it is increasing, decreasing, or constant.

42. $f(x) = x^4 - 8x^2 + 16$

43 $f(x) = x^5 - 3x^3 + 2x - 4$

44. $f(x) = -2x^4 + 4x^3 - 5x$

45. $f(x) = \begin{cases} x^2 \text{ if } x \le -4 \\ 5 \text{ if } -4 < x \le 0 \\ x^3 \text{ if } x > 0 \end{cases}$

46. MULTIPLE REPRESENTATIONS Consider the following function.

$$f(x) = x^4 - 8.65x^3 + 27.34x^2 - 37.2285x + 18.27$$

a. **ANALYTICAL** What are the degree, leading coefficient, and end behavior?

b. **TABULAR** Make a table of integer values $f(x)$ if $-4 \le x \le 4$. How many zeros does the function appear to have from the table?

c. **GRAPHICAL** Graph the function by using a graphing calculator.

d. **GRAPHICAL** Change the viewing window to [0, 4] scl: 1 by [−0.4, 0.4] scl: 0.2. What conclusions can you make from this new view of the graph?

H.O.T. Problems

Use **H**igher-**O**rder **T**hinking Skills

47. REASONING Explain why the leading coefficient and the degree are the only determining factors in the end behavior of a polynomial function.

48. REASONING The table below shows the values of $g(x)$, a cubic function. Could there be a zero between $x = 2$ and $x = 3$? Explain your reasoning.

x	−2	−1	0	1	2	3
$g(x)$	4	−2	−1	1	−2	−2

49. OPEN ENDED Sketch the graph of an odd polynomial function with 6 turning points and 2 double roots.

50. REASONING Determine whether the following statement is *sometimes*, *always*, or *never* true. Explain your reasoning.

For any continuous polynomial function, the y-coordinate of a turning point is also either a relative maximum or relative minimum.

51. REASONING A function is said to be even if for every x in the domain of f, $f(x) = f(-x)$. Is every even-degree polynomial function also an even function? Explain.

52. REASONING A function is said to be *odd* if for every x in the domain, $-f(x) = f(-x)$. Is every odd-degree polynomial function also an odd function? Explain.

53. WRITING IN MATH Describe the process of sketching the graph of a polynomial function using its degree, leading coefficient, zeros, and turning points.

NYSRE PRACTICE A2.A.50

54. Which of the following is the factorization of $2x - 15 + x^2$?

A $(x - 3)(x - 5)$

B $(x - 3)(x + 5)$

C $(x + 3)(x - 5)$

D $(x + 3)(x - 5)$

55. SHORT RESPONSE In the figure below, if $x = 35$ and $z = 50$, what is the value of y?

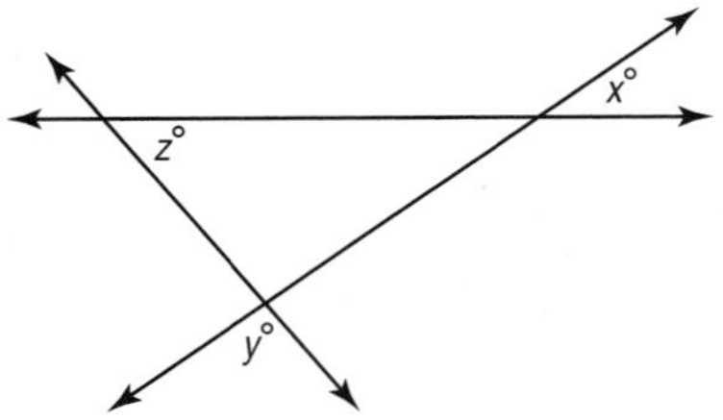

56. Which polynomial represents $(4x^2 + 5x - 3)(2x - 7)$?

F $8x^3 - 18x^2 - 41x - 21$

G $8x^3 + 18x^2 + 29x - 21$

H $8x^3 - 18x^2 - 41x + 21$

J $8x^3 + 18x^2 - 29x + 21$

57. SAT/ACT The figure at the right shows the graph of a polynomial function f(x). Which of the following could be the degree of f(x)?

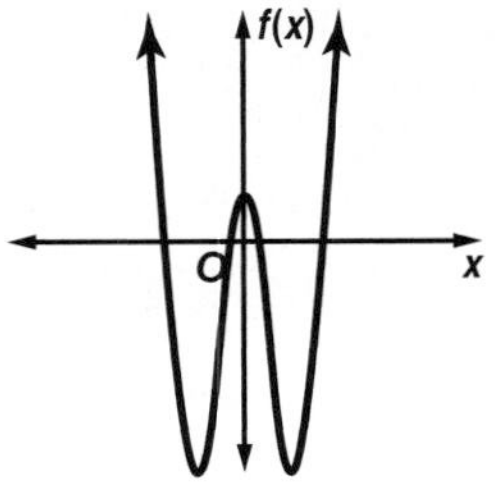

A 2

B 3

C 4

D 5

Spiral Review

For each graph,

a. describe the end behavior,

b. determine whether it represents an odd-degree or an even-degree function, and

c. state the number of real zeros. (Lesson 6-3)

58.

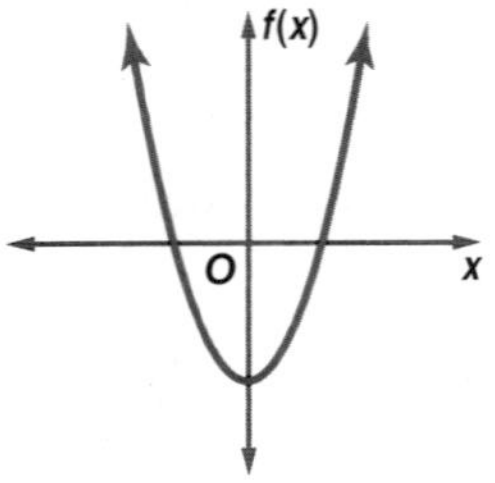

59.

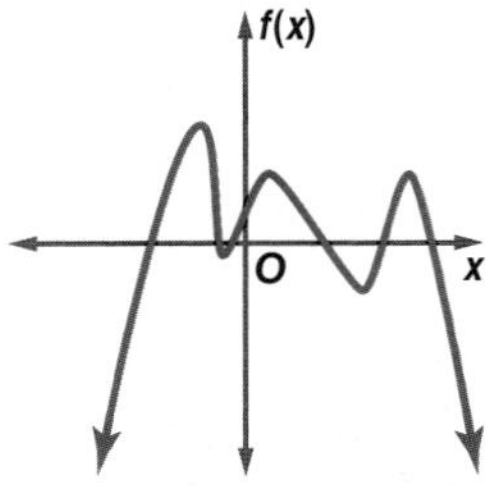

60.

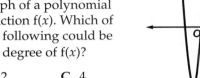

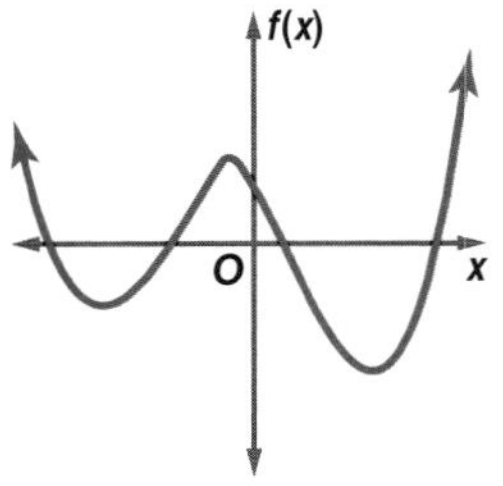

Simplify. (Lesson 6-2)

61. $(x^3 + 2x^2 - 5x - 6) \div (x + 1)$

62. $(4y^3 + 18y^2 + 5y - 12) \div (y + 4)$

63. $(2a^3 - a^2 - 4a) \div (a - 1)$

64. CHEMISTRY Tanisha needs 200 milliliters of a 48% concentration acid solution. She has 60% and 40% concentration solutions in her lab. How many milliliters of 40% acid solution should be mixed with 60% acid solution to make the required amount of 48% acid solution? (Lesson 4-6)

Skills Review

Factor. (Lesson 5-3)

65. $x^2 + 6x + 3x + 18$

66. $y^2 - 5y - 8y + 40$

67. $a^2 + 6a - 16$

68. $b^2 - 4b - 21$

69. $6x^2 - 5x - 4$

70. $4x^2 - 7x - 15$

EXTEND 6-4

Graphing Technology Lab
Modeling Data Using Polynomial Functions

NY Math Online glencoe.com
Other Calculator Keystrokes

NYS Core Curriculum **A2.S.7** Determine the function for the regression model, using appropriate technology, and use the regression function to interpolate and extrapolate from the data

You can use a TI-83/84 Plus graphing calculator to model data points when a curve of best fit is a polynomial function.

EXAMPLE

The table shows the distance a seismic wave produced by an earthquake travels from the epicenter. Draw a scatter plot and a curve of best fit to show how the distance is related to time. Then determine approximately how far away from the epicenter a seismic wave will be felt 8.5 minutes after an earthquake occurs.

Travel Time (min)	1	2	5	7	10	12	13
Distance (km)	400	800	2500	3900	6250	8400	10,000

Source: University of Arizona

Step 1 Enter time in L1 and distance in L2.

KEYSTROKES: [STAT] 1 1 [ENTER] 2 [ENTER] 5 [ENTER] 7 [ENTER] 10 [ENTER] 12 [ENTER] 13 [ENTER] [▶] 400 [ENTER] 800 [ENTER] 2500 [ENTER] 3900 [ENTER] 6250 [ENTER] 8400 [ENTER] 10000 [ENTER]

Step 2 Graph the scatter plot.

KEYSTROKES: [2nd] [STAT PLOT] 1 [ENTER] [▼] [ENTER] [ZOOM] 9

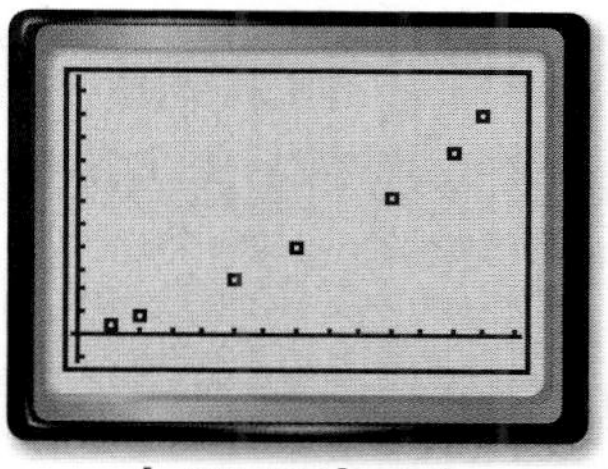

[−0.2, 14.2] scl: 1 by [−1232, 11632] scl: 1000

Step 3 Determine and graph the equation for a curve of best fit. Use a quartic regression for the data.

KEYSTROKES: [STAT] [▶] 7 [ENTER] [Y=] [VARS] 5 [▶] [▶] 1 [GRAPH]

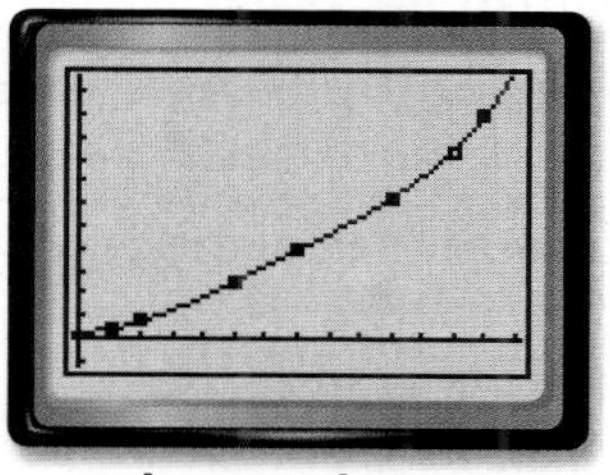

[−0.2, 14.2] scl: 1 by [−1232, 11632] scl: 1000

Step 4 Use the [CALC] feature to find the value of the function for $x = 8.5$.

KEYSTROKES: [2nd] [CALC] 1 8.5 [ENTER] [▼] [ENTER] [ZOOM] 9

After 8.5 minutes, the wave could be expected to be felt approximately 4980 kilometers from the epicenter.

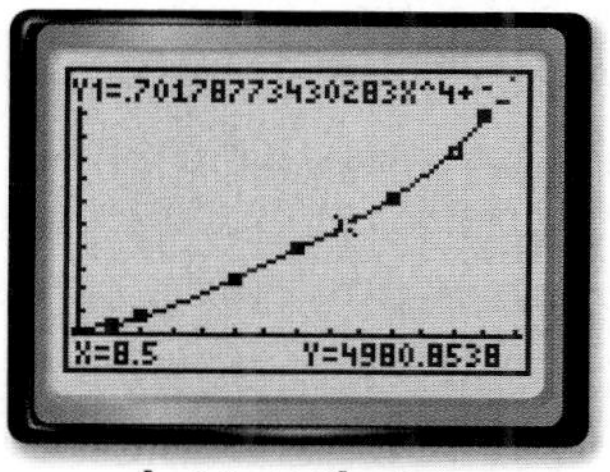

[−0.2, 14.2] scl: 1 by [−1232, 11632] scl: 1000

Exercises

The table shows how many minutes out of each eight-hour work day are used to pay one day's worth of taxes.

Year	Minutes
1930	56
1940	86
1950	119
1960	134
1970	144
1980	147
1990	148
2000	163
2005	151

Source: Tax Foundation

1. Draw a scatter plot of the data. Then graph several curves of best fit that relate the number of minutes to the number of years. Try LinReg, QuadReg, and CubicReg.
2. Write the equation for the curve that best fits the data.
3. Based on this equation, how many minutes should you expect to work each day in the year 2020 to pay one day's taxes?

The table shows the estimated number of alternative-fueled vehicles in use in the United States per year.

Year	Number of Vehicles	Year	Number of Vehicles
1995	246,855	2000	394,664
1996	265,006	2001	425,457
1997	280,205	2002	471,098
1998	295,030	2003	510,805
1999	322,302	2004	547,904

Source: U.S. Department of Energy

4. Draw a scatter plot of the data. Then graph several curves of best fit that relate the distance to the month.
5. Which curve best fits the data? Is that curve best for predicting future values?
6. Use the best-fit equation you think will give the most accurate prediction for how many alternative-fuel vehicles will be in use in the year 2012.

The table shows the average distance from the Sun to Earth during each month of the year.

Month	Distance (astronomical units)
January	0.9840
February	0.09888
March	0.9962
April	1.0050
May	1.0122
June	1.0163
July	1.0161
August	1.0116
September	1.0039
October	0.9954
November	0.9878
December	0.9837

Source: The Astronomy Cafe

7. Draw a scatter plot of the data. Then graph several curves of best fit that relate the distance to the month.
8. Write the equation for the curve that best fits the data.
9. Based on your regression equation, what is the distance from the Sun to Earth halfway through September?
10. Would you use this model to find the distance from the Sun to Earth in subsequent years? Explain your reasoning.

Extension

11. Write a question that could be answered by examining data. For example, you might estimate the number of people living in your town 5 years from now or predict the future cost of a car.
12. Collect and organize the data you need to answer the question you wrote. You may need to research your topic on the Internet or conduct a survey to collect the data you need.
13. Make a scatter plot and find a regression equation for your data. Then use the regression equation to answer the question.

CHAPTER 6

Mid-Chapter Quiz

Lessons 6-1 through 6-4

NYS Core Curriculum
A2.A.41, A2.A.50

Simplify. Assume that no variable equals 0. (Lesson 6-1)

1. $(3x^2y^{-3})(-2x^3y^5)$

2. $4t(3rt - r)$

3. $\frac{3a^4b^3c}{6a^2b^5c^3}$

4. $\left(\frac{p^2r^3}{pr^4}\right)^2$

5. $(4m^2 - 6m + 5) - (6m^2 + 3m - 1)$

6. $(x + y)(x^2 + 2xy - y^2)$

7. **MULTIPLE CHOICE** The volume of the rectangular prism is $6x^3 + 19x^2 + 2x - 3$. Which polynomial expression represents the area of the base? (Lesson 6-1)

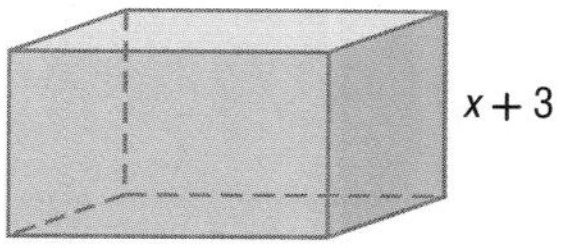

A $6x^4 + 37x^3 + 59x^2 + 3x - 9$

B $6x^2 + x + 1$

C $6x^2 + x - 1$

D $6x + 1$

Simplify. (Lesson 6-2)

8. $(4r^3 - 8r^2 - 13r + 20) \div (2r - 5)$

9. $\frac{3x^3 - 16x^2 + 9x - 24}{x - 5}$

10. Describe the end behavior of the graph. Then determine whether it represents an odd-degree or an even-degree polynomial function and state the number of real zeros. (Lesson 6-3)

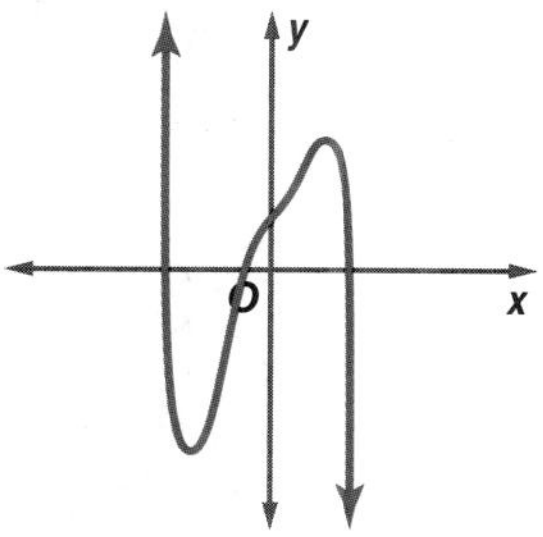

11. **MULTIPLE CHOICE** Find $p(-3)$ if $p(x) = \frac{2}{3}x^3 + \frac{1}{3}x^2 - 5x$. (Lesson 6-3)

F 0

G 11

H 30

J 36

12. **PENDULUMS** The formula $L(t) = \frac{8t^2}{\pi^2}$ can be used to find the length of a pendulum in feet when it swings back and forth in t seconds. Find the length of a pendulum that makes one complete swing in 4 seconds. (Lesson 6-3)

13. **MULTIPLE CHOICE** Find $3f(a - 4) - 2h(a)$ if $f(x) = x^2 + 3x$ and $h(x) = 2x^2 - 3x + 5$. (Lesson 6-3)

A $-a^2 + 15a - 74$

B $-a^2 - 2a - 1$

C $a^2 + 9a - 2$

D $-a^2 - 9a + 2$

14. **ENERGY** The power generated by a windmill is a function of the speed of the wind. The approximate power is given by the function $P(s) = \frac{s^3}{1000}$, where s represents the speed of the wind in kilometers per hour. Find the units of power $P(s)$ generated by a windmill when the wind speed is 18 kilometers per hour. (Lesson 6-3)

Use $f(x) = x^3 - 2x^2 - 3x$ for Exercises 15–17. (Lesson 6-4)

15. Graph the function.

16. Estimate the x-coordinates at which the relative maxima and relative minima occur.

17. State the domain and range of the function.

18. Determine the consecutive integer values of x between which each real zero is located for $f(x) = 3x^2 - 3x - 1$. (Lesson 6-4)

Refer to the graph. (Lesson 6-4)

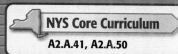

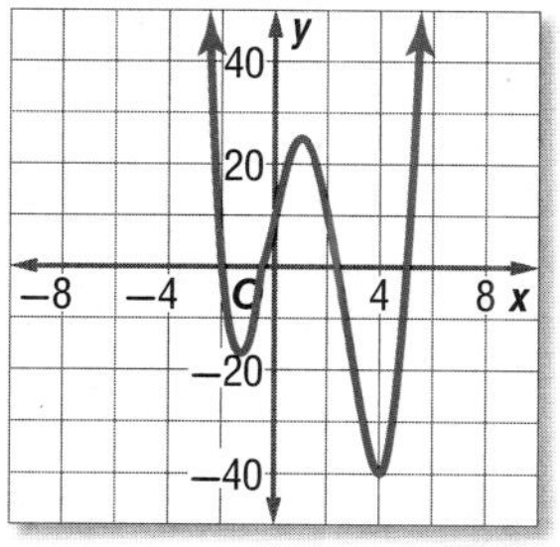

19. Estimate the x-coordinate of every turning point, and determine if those coordinates are relative maxima or relative minima.

20. Estimate the x-coordinate of every zero.

21. What is the least possible degree of the function?

6-5 Solving Polynomial Equations

Then
You solved quadratic functions by factoring. (Lesson 5-3)

Now
- Factor polynomials.
- Solve polynomial equations by factoring.

NYS Core Curriculum

A2.A.26 Find the solution to polynomial equations of higher degree that can be solved using factoring and/or the quadratic formula

New Vocabulary
prime polynomials
quadratic form

NY Math Online
glencoe.com
- Extra Examples
- Personal Tutor
- Self-Check Quiz
- Homework Help

Why?

A small cube is cut out of a larger cube. The volume of the remaining figure is given and the dimensions of each cube need to be determined.

This can be accomplished by factoring the cubic polynomial $x^3 - y^3$.

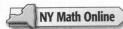

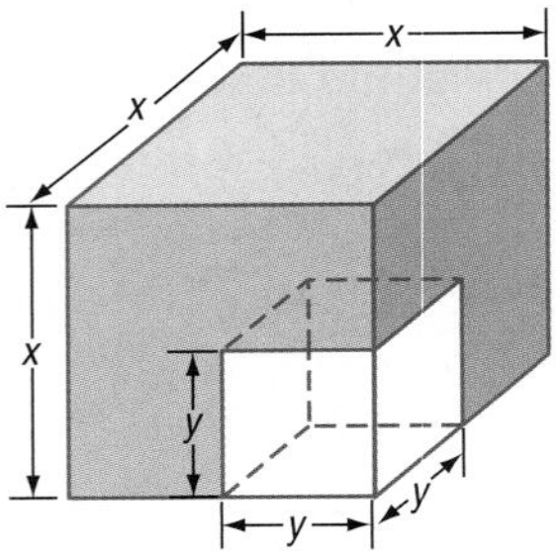

Factor Polynomials In Lesson 5-3, you learned that quadratics can be factored just like whole numbers. Their factors, however, are other polynomials. Like quadratics, some cubic polynomials can also be factored with special rules.

Key Concept Sum and Difference of Cubes — For Your FOLDABLE

Factoring Technique	General Case
Sum of Two Cubes	$a^3 + b^3 = (a + b)(a^2 - ab + b^2)$
Difference of Two Cubes	$a^3 - b^3 = (a - b)(a^2 + ab + b^2)$

Polynomials that cannot be factored are called **prime polynomials**.

EXAMPLE 1 Sum and Difference of Cubes

Factor each polynomial. If the polynomial cannot be factored, write *prime*.

a. $16x^4 + 54xy^3$

$16x^4 + 54xy^3 = 2x(8x^3 + 27y^3)$ — **Factor out the GCF.**

$8x^3$ and $27y^3$ are both perfect cubes, so we can factor the sum of two cubes.

$8x^3 + 27y^3 = (2x)^3 + (3y)^3$ — **$(2x)^3 = 8x^3$; $(3y)^3 = 27y^3$**

$= (2x + 3y)[(2x)^2 - (2x)(3y) + (3y)^2]$ — **Sum of two cubes**

$= (2x + 3y)(4x^2 - 6xy + 9y^2)$ — **Simplify.**

$16x^4 + 54xy^3 = 2x(2x + 3y)(4x^2 - 6xy + 9y^2)$ — **Replace the GCF.**

b. $9y^3 + 5x^3$

The first term is a perfect cube, but the second term is not. So, the polynomial cannot be factored using the sum of two cubes pattern. The polynomial also cannot be factored using quadratic methods or the GCF. Therefore, it is a prime polynomial.

Check Your Progress

1A. $5y^4 - 320yz^3$

1B. $-54w^4 - 250wz^3$

Personal Tutor glencoe.com

Math History Link

Sophie Germain (1776–1831)
Sophie Germain taught herself mathematics with books from her father's library during the French Revolution, when she was confined for safety. Germain discovered the identity $x^4 + 4y^4 = (x^2 + 2y^2 + 2xy)(x^2 + 2y^2 - 2xy)$, which is named for her.

The table below summarizes the most common factoring techniques used with polynomials. Whenever you factor a polynomial, always look for a common factor first. Then determine whether the resulting polynomial factors can be factored again using one or more of the methods below.

Concept Summary — Factoring Techniques

For Your FOLDABLE

Number of Terms	Factoring Technique	General Case
any number	Greatest Common Factor (GCF)	$4a^3b^2 - 8ab = 4ab(a^2b - 2)$
two	Difference of Two Squares Sum of Two Cubes Difference of Two Cubes	$a^2 - b^2 = (a + b)(a - b)$ $a^3 + b^3 = (a + b)(a^2 - ab + b^2)$ $a^3 - b^3 = (a - b)(a^2 + ab + b^2)$
three	Perfect Square Trinomials	$a^2 + 2ab + b^2 = (a + b)^2$ $a^2 - 2ab + b^2 = (a - b)^2$
	General Trinomials	$acx^2 + (ad + bc)x + bd$ $= (ax + b)(cx + d)$
four or more	Grouping	$ax + bx + ay + by$ $= x(a + b) + y(a + b)$ $= (a + b)(x + y)$

EXAMPLE 2 Factoring by Grouping

Factor each polynomial. If the polynomial cannot be factored, write *prime*.

a. $8ax + 4bx + 4cx + 6ay + 3by + 3cy$

$8ax + 4bx + 4cx + 6ay + 3by + 3cy$	Original expression
$= (8ax + 4bx + 4cx) + (6ay + 3by + 3cy)$	Group to find a GCF.
$= 4x(2a + b + c) + 3y(2a + b + c)$	Factor the GCF.
$= (4x + 3y)(2a + b + c)$	Distributive Property

b. $20fy - 16fz + 15gy + 8hz - 10hy - 12gz$

$20fy - 16fz + 15gy + 8hz - 10hy - 12gz$	Original expression
$= (20fy + 15gy - 10hy) + (-16fz - 12gz + 8hz)$	Group to find a GCF.
$= 5y(4f + 3g - 2h) - 4z(4f + 3g - 2h)$	Factor the GCF.
$= (5y - 4z)(4f + 3g - 2h)$	Distributive Property

StudyTip

Answer Checks Multiply the factors to check your answer.

Check Your Progress

2A. $30ax - 24bx + 6cx - 5ay^2 + 4by^2 - cy^2$

2B. $13ax + 18bz - 15by - 14az - 32bx + 9ay$

Personal Tutor glencoe.com

Factoring by grouping is the only method that can be used to factor polynomials with four or more terms. For polynomials with two or three terms, it may be possible to factor according to one of the patterns listed above.

When factoring two terms in which the exponents are 6 or greater, look to factor perfect squares before factoring perfect cubes.

EXAMPLE 3 Combine Cubes and Squares

Factor each polynomial. If the polynomial cannot be factored, write *prime*.

a. $x^6 - y^6$

This polynomial could be considered the difference of two squares or the difference of two cubes. The difference of two squares should always be done before the difference of two cubes for easier factoring.

$x^6 - y^6 = (x^3 + y^3)(x^3 - y^3)$ — **Difference of two squares**

$= (x + y)(x^2 - xy + y^2)(x - y)(x^2 + xy + y^2)$ — **Sum and difference of two cubes**

b. $a^3x^2 - 6a^3x + 9a^3 - b^3x^2 + 6b^3x - 9b^3$

With six terms, factor by grouping first.

$a^3x^2 - 6a^3x + 9a^3 - b^3x^2 + 6b^3x - 9b^3$

$= (a^3x^2 - 6a^3x + 9a^3) + (-b^3x^2 + 6b^3x - 9b^3)$ — **Group to find a GCF.**

$= a^3(x^2 - 6x + 9) - b^3(x^2 - 6x + 9)$ — **Factor the GCF.**

$= (a^3 - b^3)(x^2 - 6x + 9)$ — **Distributive Property**

$= (a - b)(a^2 + ab + b^2)(x^2 - 6x + 9)$ — **Difference of cubes**

$= (a - b)(a^2 + ab + b^2)(x - 3)^2$ — **Perfect squares**

StudyTip

Grouping 6 or more terms Group the terms that have the *most* common values.

Check Your Progress

3A. $a^6 + b^6$

3B. $x^5 + 4x^4 + 4x^3 + x^2y^3 + 4xy^3 + 4y^3$

Personal Tutor glencoe.com

Solve Polynomial Equations In Chapter 5, you learned to solve quadratic equations by factoring and using the Zero Product Property. You can extend these techniques to solve higher-degree polynomial equations.

Real-World EXAMPLE 4 Solve Polynomial Functions by Factoring

GEOMETRY Refer to the beginning of the lesson. If the small cube is half the length of the larger cube and the figure is 7000 cubic centimeters, what should be the dimensions of the cubes?

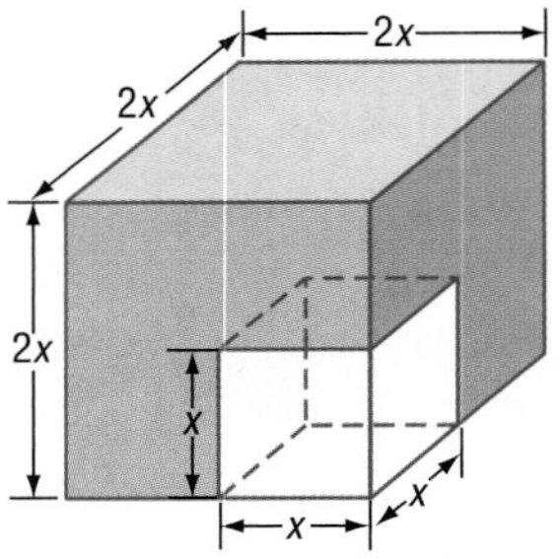

Since the length of the smaller cube is half the length of the larger cube, then their lengths can be represented by x and $2x$, respectively. The volume of the object equals the volume of the larger cube minus the volume of the smaller cube.

$(2x)^3 - x^3 = 7000$	**Volume of object**
$8x^3 - x^3 = 7000$	**$(2x)^3 = 8x^3$**
$7x^3 = 7000$	**Subtract.**
$x^3 = 1000$	**Divide.**
$x^3 - 1000 = 0$	**Subtract 1000 from each side.**
$(x - 10)(x^2 + 10x + 100) = 0$	**Difference of cubes**
$x - 10 = 0$ or $x^2 + 10x + 100 = 0$	**Zero Product Property**
$x = 10$ $x = -5 \pm 5i\sqrt{3}$	

Since 10 is the only real solution, the lengths of the cubes are 10 cm and 20 cm.

Check Your Progress

4. Determine the dimensions of the cubes if the length of the smaller cube is one third of the length of the larger cube, and the volume of the object is 3250 cubic centimeters.

Personal Tutor glencoe.com

In some cases, you can rewrite a polynomial in x in the form $au^2 + bu + c$. For example, by letting $u = x^2$, the expression $x^4 + 12x^2 + 32$ can be written as $(x^2)^2 + 12(x^2) + 32$ or $u^2 + 12u + 32$. This new, but equivalent, expression is said to be in **quadratic form**.

Key Concept Quadratic Form

For Your FOLDABLE

Words An expression that is in quadratic form can be written as $au^2 + bu + c$ for any numbers a, b, and c, $a \neq 0$, where u is some expression in x. The expression $au^2 + bu + c$ is called the quadratic form of the original expression.

Example $12x^6 + 8x^6 + 1 = 3(2x^3)^2 + 2(2x^3)^2 + 1$

StudyTip

Quadratic Form When writing a polynomial in quadratic form, choose the expression equal to u by examining the terms with variables. Pay special attention to the exponents in those terms. Not every polynomial can be written in quadratic form.

EXAMPLE 5 Quadratic Form

Write each expression in quadratic form, if possible.

a. $150n^8 + 40n^4 - 15$

$150n^8 + 40n^4 - 15 = 6(5n^4)^2 + 8(5n^4) - 15$ $\quad$ **$(5n^4)^2 = 25n^8$**

b. $y^8 + 12y^3 + 8$

This cannot be written in quadratic form since $y^8 \neq (y^3)^2$.

Check Your Progress

5A. $x^4 + 5x + 6$

5B. $8x^4 + 12x^2 + 18$

Personal Tutor glencoe.com

You can use quadratic form to solve equations with larger degrees.

EXAMPLE 6 Solve Equations in Quadratic Form

Solve $18x^4 - 21x^2 + 3 = 0$.

$18x^4 - 21x^2 + 3 = 0$		**Original equation**
$2(3x^2)^2 - 7(3x^2) + 3 = 0$		**$2(3x^2)^2 = 18x^4$**
$2u^2 - 7u + 3 = 0$		**Let $u = 3x^2$.**
$(2u - 1)(u - 3) = 0$		**Factor.**
$u = \frac{1}{2}$ or	$u = 3$	**Zero Product Property**
$3x^2 = \frac{1}{2}$	$3x^2 = 3$	**Replace u with $3x^2$.**
$x^2 = \frac{1}{6}$	$x^2 = 1$	**Divide by 3.**
$x = \pm\frac{\sqrt{6}}{6}$	$x = \pm 1$	**Take the square root.**

The solutions of the equation are $1, -1, \frac{\sqrt{6}}{6}$, and $-\frac{\sqrt{6}}{6}$.

Check Your Progress

6A. $4x^4 - 8x^2 + 3 = 0$

6B. $8x^4 + 10x^2 - 12 = 0$

Personal Tutor glencoe.com

Check Your Understanding

Examples 1–3 pp. 368–370

Factor completely. If the polynomial is not factorable, write *prime*.

1. $3ax + 2ay - az + 3bx + 2by - bz$
2. $2kx + 4mx - 2nx - 3ky - 6my + 3ny$
3. $2x^3 + 5y^3$
4. $16g^3 + 2h^3$
5. $12qw^3 - 12q^4$
6. $3w^2 + 5x^2 - 6y^2 + 2z^2 + 7a^2 - 9b^2$
7. $a^6x^2 - b^6x^2$
8. $x^3y^2 - 8x^3y + 16x^3 + y^5 - 8y^4 + 16y^3$
9. $8c^3 - 125d^3$
10. $6bx + 12cx + 18dx - by - 2cy - 3dy$

Example 4 p. 370

Solve each equation.

11. $x^4 - 19x^2 + 48 = 0$
12. $x^3 - 64 = 0$
13. $x^3 + 27 = 0$
14. $x^4 - 33x^2 + 200 = 0$
15. **LANDSCAPING** A boardwalk that is x feet wide is built around a rectangular pond. The pond is 30 feet wide and 40 feet long. The combined area of the pond and the boardwalk is 2000 square feet. What is the width of the boardwalk?

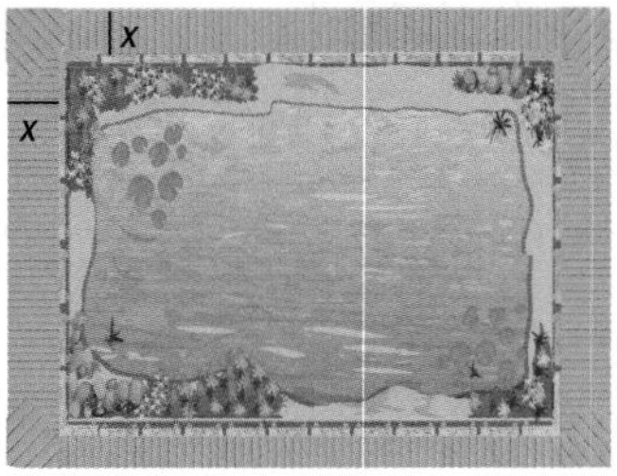

Example 5 p. 371

Write each expression in quadratic form, if possible.

16. $4x^6 - 2x^3 + 8$
17. $25y^6 - 5y^2 + 20$

Example 6 p. 371

Solve each equation.

18. $x^4 - 6x^2 + 8 = 0$
19. $y^4 - 18y^2 + 72 = 0$

Practice and Problem Solving

● = **Step-by-Step Solutions** begin on page R20.
Extra Practice begins on page 947.

Examples 1–3 pp. 368–370

Factor completely. If the polynomial is not factorable, write *prime*.

20. $8c^3 - 27d^3$
21. $64x^4 + xy^3$
22. $a^8 - a^2b^6$
23. $x^6y^3 + y^9$
24. $18x^6 + 5y^6$
25. $w^3 - 2y^3$
26. $gx^2 - 3hx^2 - 6fy^2 - gy^2 + 6fx^2 + 3hy^2$
27. $12ax^2 - 20cy^2 - 18bx^2 - 10ay^2 + 15by^2 + 24cx^2$
28. $a^3x^2 - 16a^3x + 64a^3 - b^3x^2 + 16b^3x - 64b^3$
29. $8x^5 - 25y^3 + 80x^4 - x^2y^3 + 200x^3 - 10xy^3$

Example 4 p. 370

Solve each equation.

30. $x^4 + x^2 - 90 = 0$
31. $x^4 - 16x^2 - 720 = 0$
32. $x^4 - 7x^2 - 44 = 0$
33. $x^4 + 6x^2 - 91 = 0$
34. $x^3 + 216 = 0$
35. $64x^3 + 1 = 0$

Example 5 p. 371

Write each expression in quadratic form, if possible.

36. $x^4 + 12x^2 - 8$
37. $-15x^4 + 18x^2 - 4$
38. $8x^6 + 6x^3 + 7$
39. $5x^6 - 2x^2 + 8$
40. $9x^8 - 21x^4 + 12$
41. $16x^{10} + 2x^5 + 6$

Example 6 p. 371

Solve each equation.

42. $x^4 + 6x^2 + 5 = 0$
43. $x^4 - 3x^2 - 10 = 0$
44. $4x^4 - 14x^2 + 12 = 0$
45. $9x^4 - 27x^2 + 20 = 0$
46. $4x^4 - 5x^2 - 6 = 0$
47. $24x^4 + 14x^2 - 3 = 0$

48. **ZOOLOGY** A species of animal is introduced to a small island. Suppose the population of the species is represented by $P(t) = -t^4 + 9t^2 + 400$, where t is the time in years. Determine when the population becomes zero.

Real-World Link

Brought by Spanish explorers in the 16th century, the modern horse is one of the most important species ever introduced to North America. Closely related horse ancestors had become extinct on the continent some 10,000 years before.

Source: Buffalo Bill Historical Center

Factor completely. If the polynomial is not factorable, write *prime*.

49. $x^4 - 625$

50. $x^6 - 64$

51. $x^5 - 16x$

52. $8x^5y^2 - 27x^2y^5$

53. $15ax - 10bx + 5cx + 12ay - 8by + 4cy + 15az - 10bz + 5cz$

54. $6a^2x^2 - 24b^2x^2 + 18c^2x^2 - 5a^2y^3 + 20b^2y^3 - 15c^2y^3 + 2a^2z^2 - 8b^2z^2 + 6c^2z^2$

55. $6x^5 - 11x^4 - 10x^3 - 54x^3 + 99x^2 + 90x$

56. $20x^6 - 7x^5 - 6x^4 - 500x^4 + 175x^3 + 150x^2$

57. **GEOMETRY** The volume of the figure at the right is 440 cubic centimeters. Find the value of x and the length, height, and width.

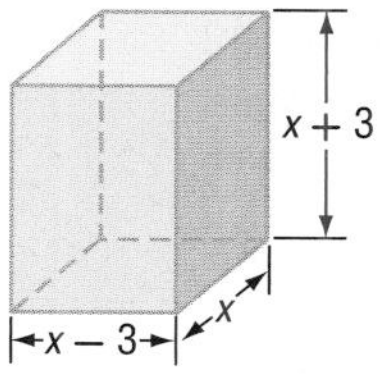

Solve each equation.

58. $8x^4 + 10x^2 - 3 = 0$

59. $6x^4 - 5x^2 - 4 = 0$

60. $20x^4 - 53x^2 + 18 = 0$

61. $18x^4 + 43x^2 - 5 = 0$

62. $8x^4 - 18x^2 + 4 = 0$

63. $3x^4 - 22x^2 - 45 = 0$

64. $x^6 + 7x^3 - 8 = 0$

65. $x^6 - 26x^3 - 27 = 0$

66. $8x^6 + 999x^3 = 125$

67. $4x^4 - 4x^2 - x^2 + 1 = 0$

68. $x^6 - 9x^4 - x^2 + 9 = 0$

69. $x^4 + 8x^2 + 15 = 0$

70. **GEOMETRY** A rectangular prism with dimensions $x - 2$, $x - 4$, and $x - 6$ has a volume equal to $40x$ cubic units.

a. Write out a polynomial equation using the formula for volume.

b. Use factoring to solve for x.

c. Are any values for x unreasonable? Explain.

d. What are the dimensions of the prism?

71. **POOL DESIGN** Andrea wants to build a pool following the diagram at the right. The pool will be surrounded by a sidewalk of a constant width.

a. If the total area of the pool itself is to be 336 ft^2, what is x?

b. If the value of x were doubled, what would be the new area of the pool?

c. If the value of x were halved, what would be the new area of the pool?

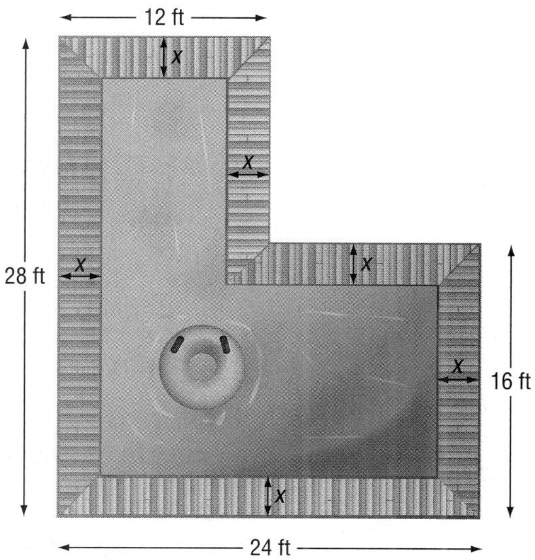

72. BIOLOGY During an experiment, the number of cells of a virus can be modeled by $P(t) = -0.012t^3 - 0.24t^2 + 6.3t + 8000$, where t is the time in hours and P is the number of cells. Jack wants to determine the times at which there are 8000 cells.

a. Solve for t by factoring.

b. What method did you use to factor?

c. Which values for t are reasonable and which are unreasonable? Explain.

d. Graph the function for $0 \le t \le 20$ using your calculator.

Real-World Link

Americans have tripled their spending on home theater systems to nearly $1 billion a year. An estimated one fourth of U.S. households have some kind of home theater.

Source: *USA TODAY*

73. HOME BUILDING Alicia's parents want their basement home theater designed according to the diagram at the right.

a. Write a function in terms of x for the area of the basement.

b. If the basement is to be 1366 square feet, what is x?

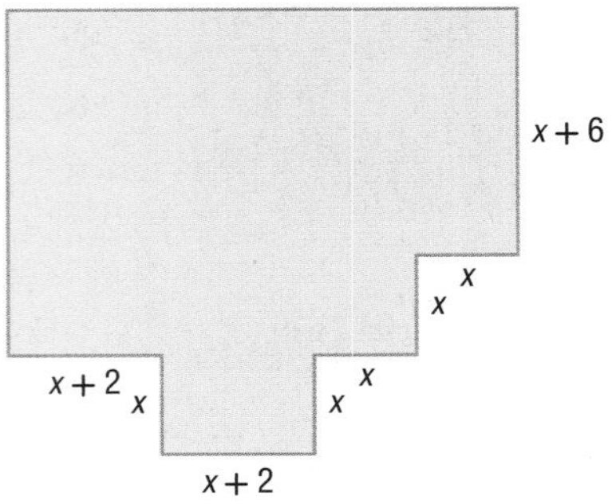

74. BIOLOGY A population of parasites in an experiment can be modeled by $f(t) = t^3 + 5t^2 - 4t - 20$, where t is the time in days.

a. Use factoring by grouping to determine the values of t for which $f(t) = 0$.

b. At what times does the population reach zero?

c. Are any of the values of t unreasonable? Explain.

Factor completely. If the polynomial is not factorable, write *prime*.

75. $x^6 - 4x^4 - 8x^4 + 32x^2 + 16x^2 - 64$

76. $y^9 - y^6 - 2y^6 + 2y^3 + y^3 - 1$

77. $x^6 - 3x^4y^3 + 3x^2y^4 - y^6$

78. CORRALS Fredo's corral, an enclosure for livestock, is currently 32 feet by 40 feet. He wants to enlarge the area to 4.5 times its current area by increasing the length and width by the same amount.

a. Draw a diagram to represent the situation.

b. Write a polynomial equation for the area of the new corral. Then solve the equation by factoring.

c. Graph the function.

d. Which solution is irrelevant? Explain.

H.O.T. Problems

Use Higher-Order Thinking Skills

79. CHALLENGE Factor $36x^{2n} + 12x^n + 1$.

80. CHALLENGE Solve $6x - 11\sqrt{3x} + 12 = 0$.

81. REASONING Find a counterexample to the statement $a^2 + b^2 = (a + b)^2$.

82. OPEN ENDED The cubic form of an equation is $ax^3 + bx^2 + cx + d = 0$. Write an equation with degree 6 that can be written in *cubic* form.

83. WRITING IN MATH Explain how the graph of a polynomial function can help you factor the polynomial.

NYSRE PRACTICE A2.A.26

84. SHORT RESPONSE Tiles numbered from 1 to 6 are placed in a bag and are drawn to determine which of six tasks will be assigned to six people. What is the probability that the tiles numbered 5 and 6 are the last two drawn?

85. STATISTICS Which of the following represents a negative correlation?

A
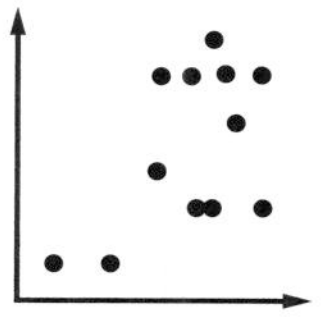

C
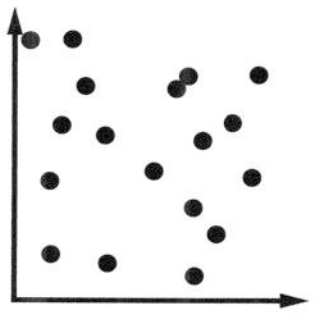

B
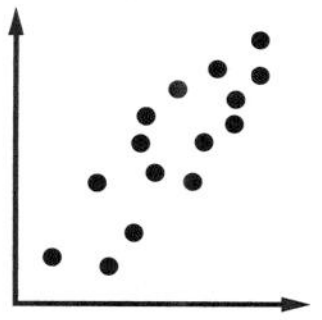

D
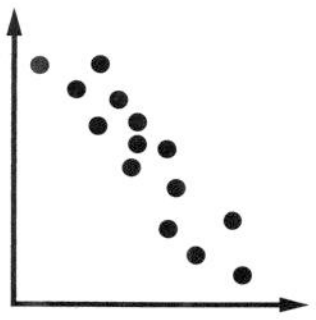

86. Which of the following most accurately describes the translation of the graph $y = (x + 4)^2 - 3$ to the graph of $y = (x - 1)^2 + 3$?

F down 1 and to the right 3

G down 6 and to the left 5

H up 1 and to the left 3

J up 6 and to the right 5

87. SAT/ACT The positive difference between k and $\frac{1}{12}$ is the same as the positive difference between $\frac{1}{3}$ and $\frac{1}{5}$. Which of the following is the value of k?

A $\frac{1}{60}$

B $\frac{1}{20}$

C $\frac{1}{15}$

D $\frac{13}{60}$

Spiral Review

Graph each polynomial function. Estimate the x-coordinates at which the relative maxima and relative minima occur. (Lesson 6-4)

88. $f(x) = 2x^3 - 4x^2 + x + 8$

89. $f(x) = -3x^3 + 6x^2 + 2x - 1$

90. $f(x) = -x^3 + 3x^2 + 4x - 6$

State the degree and leading coefficient of each polynomial in one variable. If it is not a polynomial in one variable, explain why. (Lesson 6-3)

91. $f(x) = 4x^3 - 6x^2 + 5x^4 - 8x$

92. $f(x) = -2x^5 + 5x^4 + 3x^2 + 9$

93. $f(x) = -x^4 - 3x^3 + 2x^6 - x^7$

94. ELECTRICITY The impedance in one part of a series circuit is $3 + 4j$ ohms, and the impedance in another part of the circuit is $2 - 6j$. Add these complex numbers to find the total impedance of the circuit. (Lesson 5-4)

95. SKIING All 28 members of a ski club went on a trip. The club paid a total of \$478 for the equipment. How many skis and snowboards did they rent? (Lesson 3-2)

96. GEOMETRY The sides of an angle are parts of two lines whose equations are $2y + 3x = -7$ and $3y - 2x = 9$. The angle's vertex is the point where the two sides meet. Find the coordinates of the vertex of the angle. (Lesson 3-1)

Skills Review

Divide. (Lesson 6-2)

97. $(x^2 + 6x - 2) \div (x + 4)$

98. $(2x^2 + 8x - 10) \div (2x + 1)$

99. $(8x^3 + 4x^2 + 6) \div (x + 2)$

EXTEND
6-5 Graphing Technology Lab
Solving Polynomial Inequalities

NY Math Online glencoe.com
Other Calculator Keystrokes

You can use a TI-83/84 Plus graphing calculator to solve polynomial inequalities.

ACTIVITY

Solve $x^4 + 2x^3 \leq 7$.

Method 1

Step 1 Graph each side of the related equation separately.

KEYSTROKES: [Y=] [X,T,θ,n] [^] 4 [+] 2 [X,T,θ,n] [^] 3 [ENTER] 7 [ZOOM] 6

Step 2 Find the points of intersection.

KEYSTROKES: [2nd] [CALC] 5

Use [◀] or [▶] to position the cursor on Y1, near the first point of intersection. Press [ENTER] [ENTER] [ENTER].

Then use [▶] to position the cursor near the second intersection point. Press [ENTER] [ENTER] [ENTER].

Step 3 Examine the graphs.

Determine where the graph of $y = x^4 + 2x^3$ intersects and is below $y = 7$.

The solution is approximately $-2.47 \leq x \leq 1.29$.

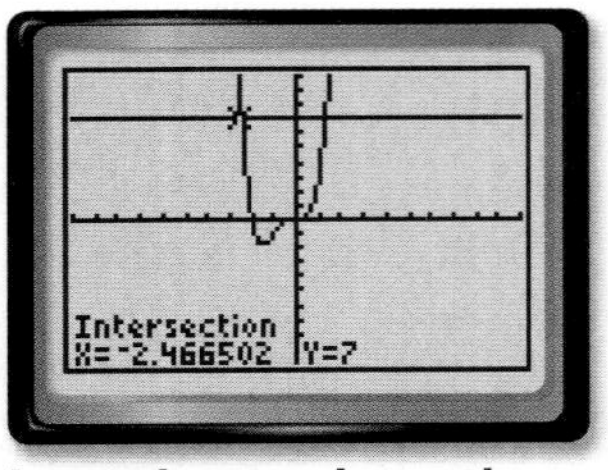

[−10, 10] scl: 1 by [−10, 10] scl: 1

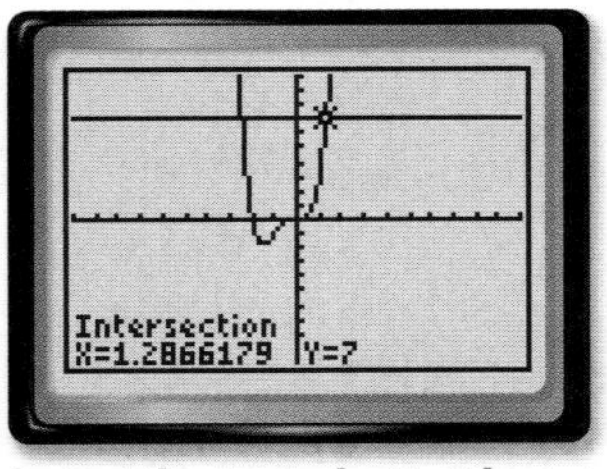

[−10, 10] scl: 1 by [−10, 10] scl: 1

Method 2

Step 1 Graph the related equation equal to 0.

$$x^4 + 2x^3 = 7$$
$$x^4 + 2x^3 - 7 = 0$$

KEYSTROKES: [Y=] [X,T,θ,n] [^] 4 [+] 2 [X,T,θ,n] [^] 3 [−] 7 [ZOOM] 6

Step 2 Find the x-intercepts.

KEYSTROKES: [2nd] [CALC] 2

Use [◀] or [▶] to position the cursor to the left of the first x-intercept. Press [ENTER]. Then use [▶] to position the cursor to the right of the first x-intercept. Press [ENTER] [ENTER] to display the x-intercept. Then, repeat the procedure for any remaining x-intercepts.

Step 3 Examine the graphs.

Determine where the graph of $y = x^4 + 2x^3 - 7$ is below the x-axis.

The solution is approximately $-2.47 \leq x \leq 1.29$.

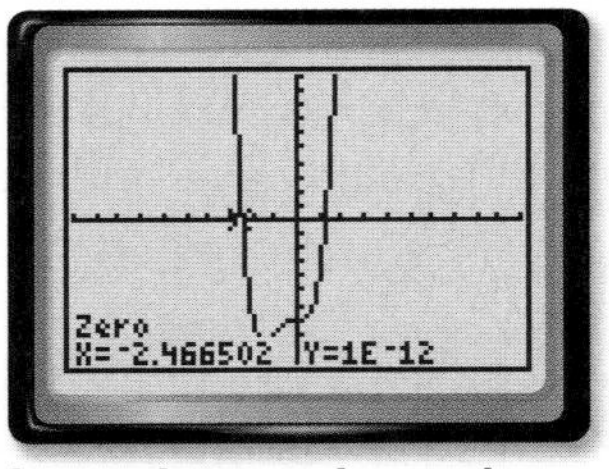

[−10, 10] scl: 1 by [−10, 10] scl: 1

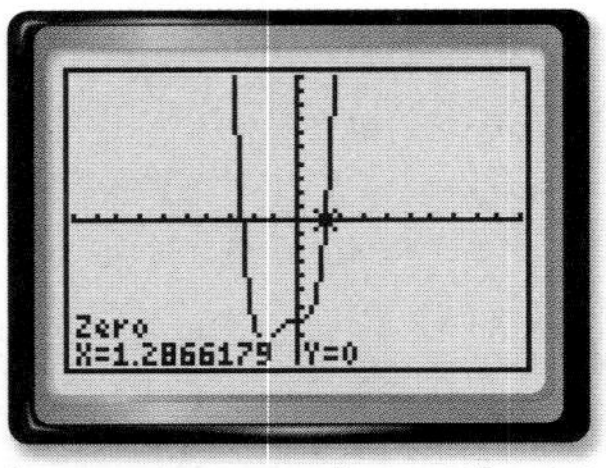

[−10, 10] scl: 1 by [−10, 10] scl: 1

Exercises

Solve each equation. Round to the nearest hundredth.

1. $\frac{2}{3}x^3 + x^2 - 5x \geq -9$
2. $x^3 - 9x^2 + 27x \leq 20$
3. $x^3 + 1 \geq 4x^2$
4. $x^6 - 15 \leq 5x^4 - x^2$
5. $\frac{1}{2}x^5 \geq \frac{1}{5}x^2 - 2$
6. $x^8 < -x^7 + 3$
7. $x^4 - 15x^2 > -24$
8. $x^3 - 6x^2 + 4x < -6$
9. $x^4 - 15x^2 + x + 65 > 0$

6-6 The Remainder and Factor Theorems

Then

You used the Distributive Property and factoring to simplify algebraic expressions. (Lesson 5-3)

Now

- Evaluate functions by using synthetic substitution.
- Determine whether a binomial is a factor of a polynomial by using synthetic substitution.

NYS Core Curriculum

A2.A.26 Find the solution to polynomial equations of higher degree that can be solved using factoring and/or the quadratic formula

New Vocabulary

synthetic substitution

depressed polynomial

NY Math Online

glencoe.com

- Extra Examples
- Personal Tutor
- Self-Check Quiz
- Homework Help

Why?

The number of college students from the United States who study abroad can be modeled by the function $S(x) = 0.02x^4 - 0.52x^3 + 4.03x^2 + 0.09x + 77.54$, where x is the number of years since 1993 and $S(x)$ is the number of students in thousands.

You can use this function to estimate the number of U.S. college students studying abroad in 2013 by evaluating the function for $x = 20$. Another method you can use is *synthetic substitution*.

Synthetic Substitution Synthetic division can be used to find the value of a function. Consider the polynomial function $f(x) = -3x^2 + 5x + 4$. Divide the polynomial by $x - 3$.

Method 1 Long Division

$$\begin{array}{r} -3x - 4 \\ x - 3\overline{)-3x^2 + 5x + 4} \\ \underline{-3x^2 + 9x} \\ -4x + 4 \\ \underline{-4x + 12} \\ -8 \end{array}$$

Method 2 Synthetic Division

$$\begin{array}{r|rrr} 3 & -3 & 5 & 4 \\ & & -9 & -12 \\ \hline & -3 & -4 & -8 \end{array}$$

Compare the remainder of -8 to $f(3)$.

$f(3) = -3(3)^2 + 5(3) + 4$ **Replace *x* with 3.**

$= -27 + 15 + 4$ **Multiply.**

$= -8$ **Simplify.**

Notice that the value of $f(3)$ is the same as the remainder when the polynomial is divided by $x - 3$. This illustrates the **Remainder Theorem**.

Key Concept — Remainder Theorem

For Your FOLDABLE

Words If a polynomial $P(x)$ is divided by $x - r$, the remainder is a constant $P(r)$, and

Dividend	equals	quotient	times	divisor	plus	remainder.
$P(x)$	$=$	$Q(x)$	$\cdot$	$(x - r)$	$+$	$P(r)$,

where $Q(x)$ is a polynomial with degree one less than $P(x)$.

Example $x^2 + 6x + 2 = (x - 4) \cdot (x + 10) + 42$

Applying the Remainder Theorem using synthetic division to evaluate a function is called **synthetic substitution**. It is a convenient way to find the value of a function, especially when the degree of the polynomial is greater than 2.

EXAMPLE 1 Synthetic Substitution

If $f(x) = 3x^4 - 2x^3 + 5x + 2$, find $f(4)$.

Method 1 Synthetic Substitution

By the Remainder Theorem, $f(4)$ should be the remainder when the polynomial is divided by $x - 4$.

4	3	−2	0	5	2
		12	40	160	660
	3	10	40	165	662

Because there is no x^2 term, a zero is placed in this position as a placeholder.

The remainder is 662. Therefore, by using synthetic substitution, $f(4) = 662$.

Method 2 Direct Substitution

Replace x with 4.

$f(x) = 3x^4 - 2x^3 + 5x + 2$ **Original function**

$f(4) = 3(4)^4 - 2(4)^3 + 5(4) + 2$ **Replace x with 4.**

$= 768 - 128 + 20 + 2$ or 662 **Simplify.**

By using direct substitution, $f(4) = 662$. Both methods give the same result.

Check Your Progress

1A. If $f(x) = 3x^3 - 6x^2 + x - 11$, find $f(3)$.

1B. If $g(x) = 4x^5 + 2x^3 + x^2 - 1$, find $f(-1)$.

Personal Tutor glencoe.com

Real-World Link

Some benefits of studying abroad include learning a new language, becoming more independent, and improving communication skills. Studying abroad also gives students a chance to experience different cultures and customs as well as try new foods.

Source: StudyAbroad

Synthetic substitution can be used in situations in which direct substitution would involve cumbersome calculations.

Real-World EXAMPLE 2 Find Function Values

COLLEGE Refer to the beginning of the lesson. How many U.S. college students will study abroad in 2013?

Use synthetic substitution to divide $0.02x^4 - 0.52x^3 + 4.03x^2 + 0.09x + 77.54$ by $x - 20$.

20	0.02	−0.52	4.03	0.09	77.54
		0.4	−2.4	32.6	653.8
	0.02	−0.12	1.63	32.69	731.34

In 2013, there will be about 731,340 U.S. college students studying abroad.

Check Your Progress

2. COLLEGE The function $C(x) = 2.46x^3 - 22.37x^2 + 53.81x + 548.24$ can be used to approximate the number, in thousands, of international college students studying in the United States x years since 2000. How many international college students can be expected to study in the U.S. in 2012?

Personal Tutor glencoe.com

Factors of Polynomials The synthetic division below shows that the quotient of $2x^3 - 3x^2 - 17x + 30$ and $x + 3$ is $2x^2 - 9x + 10$.

$$\begin{array}{r|rrrr} -3 & 2 & -3 & -17 & 30 \\ & & -6 & 27 & -30 \\ \hline & 2 & -9 & 10 & 0 \end{array}$$

When you divide a polynomial by one of its binomial factors, the quotient is called a depressed polynomial. A **depressed polynomial** has a degree that is one less than the original polynomial. From the results of the division, and by using the Remainder Theorem, we can make the following statement.

Dividend	equals	quotient	times	divisor	plus	remainder.
$2x^3 - 3x^2 - 17x + 30$	$=$	$(2x^2 - 9x + 10)$	$\cdot$	$(x + 3)$	$+$	0

Since the remainder is 0, $f(-3) = 0$. This means that $x + 3$ is a factor of $2x^3 - 3x^2 - 17x + 30$. This illustrates the **Factor Theorem**, which is a special case of the Remainder Theorem.

Key Concept **Factor Theorem** For Your FOLDABLE

The binomial $x - r$ is a factor of the polynomial $P(x)$ if and only if $P(r) = 0$.

The Factor Theorem can be used to determine whether a binomial is a factor of a polynomial. It can also be used to determine all of the factors of a polynomial.

EXAMPLE 3 Use the Factor Theorem

Determine whether $x - 5$ is a factor of $x^3 - 7x^2 + 7x + 15$. Then find the remaining factors of the polynomial.

The binomial $x - 5$ is a factor of the polynomial if 5 is a zero of the related polynomial function. Use the Factor Theorem and synthetic division.

$$\begin{array}{r|rrrr} 5 & 1 & -7 & 7 & 15 \\ & & 5 & -10 & -15 \\ \hline & 1 & -2 & -3 & 0 \end{array}$$

Because the remainder is 0, $x - 5$ is a factor of the polynomial. The polynomial $x^3 - 7x^2 + 7x + 15$ can be factored as $(x - 5)(x^2 - 2x - 3)$. The polynomial $x^2 - 2x - 3$ is the depressed polynomial. Check to see if this polynomial can be factored.

$x^2 - 2x - 3 = (x + 1)(x - 3)$ **Factor the trinomial.**

So, $x^3 - 7x^2 + 7x + 15 = (x - 5)(x + 1)(x - 3)$.

You can check your answer by multiplying out the factors and seeing if you come up with the initial polynomial.

StudyTip

Factoring The factors of a polynomial do not have to be binomials. For example, the factors of $x^3 + x^2 - x + 15$ are $x + 3$ and $x^2 - 2x + 5$.

Check Your Progress

3. Show that $x - 2$ is a factor of $x^3 - 7x^2 + 4x + 12$. Then find the remaining factors of the polynomial.

Personal Tutor glencoe.com

Check Your Understanding

Example 1 p. 378

Use synthetic substitution to find $f(4)$ and $f(-2)$ for each function.

1. $f(x) = 2x^3 - 5x^2 - x + 14$
2. $f(x) = x^4 + 8x^3 + x^2 - 4x - 10$

Example 2 p. 378

3. **NATURE** The approximate number of bald eagle nesting pairs in the United States can be modeled by the function $P(x) = -0.16x^3 + 15.83x^2 - 154.15x + 1147.97$, where x is the number of years since 1970. About how many nesting pairs of bald eagles can be expected in 2018?

Example 3 p. 379

Given a polynomial and one of its factors, find the remaining factors of the polynomial.

4. $x^3 - 6x^2 + 11x - 6; x - 1$
5. $x^3 + x^2 - 16x - 16; x + 1$
6. $3x^3 + 10x^2 - x - 12; x - 1$
7. $2x^3 - 5x^2 - 28x + 15; x + 3$

Practice and Problem Solving

● = **Step-by-Step Solutions** begin on page R20. **Extra Practice** begins on page 947.

Example 1 p. 378

Use synthetic substitution to find $f(-5)$ and $f(2)$ for each function.

8. $f(x) = x^3 + 2x^2 - 3x + 1$
9. $f(x) = x^2 - 8x + 6$
10. $f(x) = 3x^4 + x^3 - 2x^2 + x + 12$
11. $f(x) = 2x^3 - 8x^2 - 2x + 5$
12. $f(x) = x^3 - 5x + 2$
13. $f(x) = x^5 + 8x^3 + 2x - 15$
14. $f(x) = x^6 - 4x^4 + 3x^2 - 10$
15. $f(x) = x^4 - 6x - 8$

Example 2 p. 378

16. **FUEL ECONOMY** A specific car's fuel economy in miles per gallon can be approximated by $f(x) = 0.00000056x^4 - 0.000018x^3 - 0.016x^2 + 1.38x - 0.38$, where x represents the car's speed in miles per hour. Determine the fuel economy when the car is traveling 40, 50 and 60 miles per hour.

Example 3 p. 379

Given a polynomial and one of its factors, find the remaining factors of the polynomial.

17. $x^3 - 3x + 2; x + 2$
18. $x^4 + 2x^3 - 8x - 16; x + 2$
19. $x^3 - x^2 - 10x - 8; x + 2$
20. $x^3 - x^2 - 5x - 3; x - 3$
21. $2x^3 + 17x^2 + 23x - 42; x - 1$
22. $2x^3 + 7x^2 - 53x - 28; x - 4$
23. $x^4 + 2x^3 + 2x^2 - 2x - 3; x - 1$
24. $x^3 + 2x^2 - x - 2; x + 2$
25. $6x^3 - 25x^2 + 2x + 8; 2x + 1$
26. $16x^5 - 32x^4 - 81x + 162; 2x - 3$

27. **BOATING** A motor boat traveling against waves accelerates from a resting position. Suppose the speed of the boat in feet per second is given by the function $f(t) = -0.04t^4 + 0.8t^3 + 0.5t^2 - t$, where t is the time in seconds.
    a. Find the speed of the boat at 1, 2, and 3 seconds.
    b. It takes 6 seconds for the boat to travel between two buoys while it is accelerating. Use synthetic substitution to find $f(6)$ and explain what this means.

Real-World Link

The average American spends about $8500 annually on consumer electronics.

Source: IT Facts

28. **SALES** A company's sales, in millions of dollars, of consumer electronics can be modeled by $S(x) = -1.7x^3 + 18x^2 + 26.4x + 678$, where x is the number of years since 2005.
    a. Use synthetic substitution to estimate the sales for 2010 and 2015.
    b. Do you think this model is useful in estimating future sales? Explain.

Use the graphs to find all of the factors for each polynomial function.

29.

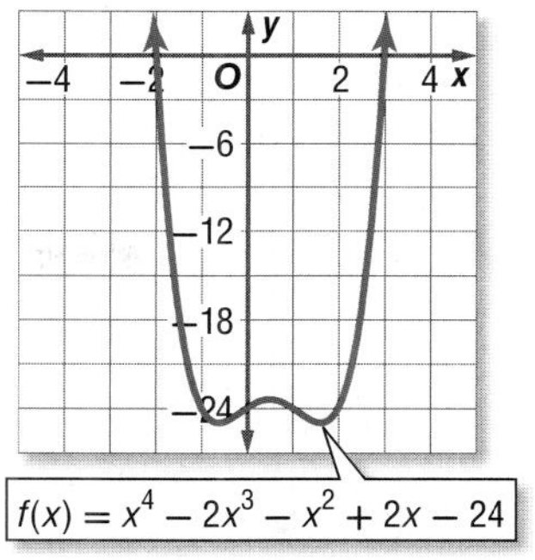

30.

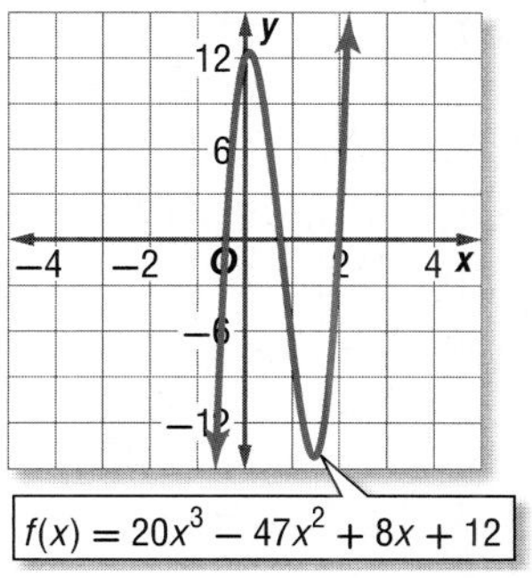

31. **MULTIPLE REPRESENTATIONS** In this problem, you will consider the function $f(x) = -9x^5 + 104x^4 - 249x^3 - 456x^2 + 828x + 432$.

a. **ALGEBRAIC** If $x - 6$ is a factor of the function, find the depressed polynomial.

b. **TABULAR** Make a table of values for $-5 \leq x \leq 5$ for the depressed polynomial.

c. **ANALYTICAL** What conclusions can you make about the locations of the other zeros based on the table? Explain your reasoning.

d. **GRAPHICAL** Graph the original function to confirm your conclusions.

Watch Out!

Synthetic Substitution Remember that synthetic substitution is used to divide a polynomial by $(x - a)$. If the binomial is $(x - a)$, use a. If the binomial is $(x + a)$, use $-a$.

Find values of k so that each remainder is 3.

32. $(x^2 - x + k) \div (x - 1)$

33 $(x^2 + kx - 17) \div (x - 2)$

34. $(x^2 + 5x + 7) \div (x + k)$

35. $(x^3 + 4x^2 + x + k) \div (x + 2)$

H.O.T. Problems

Use Higher-Order Thinking Skills

36. **OPEN ENDED** Write a polynomial function that has a double root of 1 and a double root of −5. Graph the function.

CHALLENGE **Find the solutions of each polynomial function.**

37. $(x^2 - 4)^2 - (x^2 - 4) - 2 = 0$

38. $(x^2 + 3)^2 - 7(x^2 + 3) + 12 = 0$

39. **REASONING** Polynomial $f(x)$ is divided by $x - c$. What can you conclude if:

a. the remainder is 0?

b. the remainder is 1?

c. the quotient is 1, and the remainder is 0?

40. **CHALLENGE** Review the definition for the Factor Theorem. Provide a proof of the theorem.

41. **OPEN ENDED** Write a cubic function that has a remainder of 8 for $f(2)$ and a remainder of −5 for $f(3)$.

42. **CHALLENGE** Show that the quartic function $ax^4 + bx^3 + cx^2 + dx + f = 0$ will always have a rational root when the numbers 1, −2, 3, 4, and −6 are randomly assigned to replace a through f, and all of the numbers are used.

43. **WRITING IN MATH** Explain how the zeros of a function can be located by using the Remainder Theorem and making a table of values for different input values and then comparing the remainders.

NYSRE PRACTICE A2.A.26

44. $27x^3 + y^3 =$

A $(3x + y)(3x + y)(3x + y)$
B $(3x + y)(9x^2 - 3xy + y^2)$
C $(3x - y)(9x^2 + 3xy + y^2)$
D $(3x - y)(9x^2 + 9xy + y^2)$

45. GEOMETRY In the figure, a square with side length $2\sqrt{2}$ is inscribed in a circle. The area of the circle is $k\pi$. What is the exact value of k?

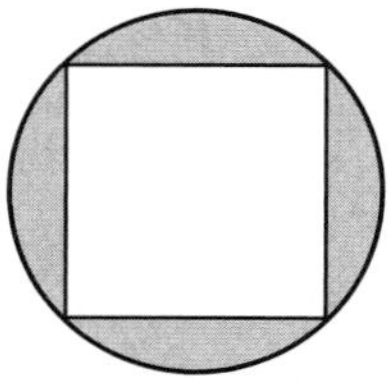

46. What is the product of the complex numbers $(4 + i)(4 - i)$?

F 15
G $16 - i$
H 17
J $17 - 8i$

47. SAT/ACT The measure of the largest angle of a triangle is 14 less than twice the measure of the smallest angle. The third angle measure is 2 more than the measure of the smallest angle. What is the measure of the smallest angle?

A 46
B 48
C 50
D 82

Spiral Review

Solve each equation. (Lesson 6-5)

48. $x^4 - 4x^2 - 21 = 0$

49. $x^4 - 6x^2 = 27$

50. $4x^4 - 8x^2 - 96 = 0$

Complete each of the following. (Lesson 6-4)

a. **Estimate the x-coordinate of every turning point and determine if those coordinates are relative maxima or relative minima.**

b. **Estimate the x-coordinate of every zero.**

c. **Determine the smallest possible degree of the function.**

d. **Determine the domain and range of the function.**

51.

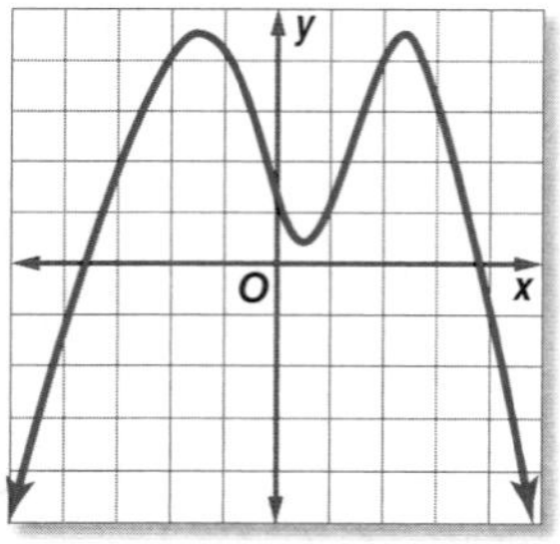

52.

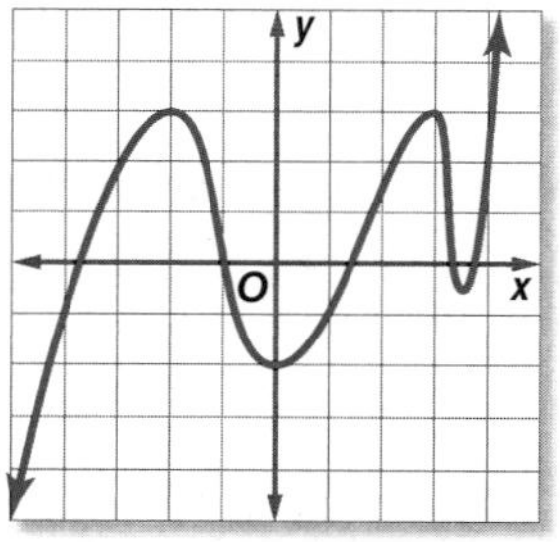

53.

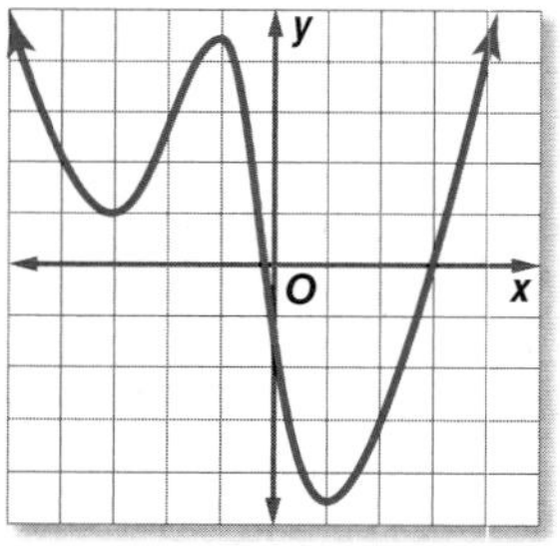

54. HIGHWAY SAFETY Engineers can use the formula $d = 0.05v^2 + 1.1v$ to estimate the minimum stopping distance d in feet for a vehicle traveling v miles per hour. If a car is able to stop after 125 feet, what is the fastest it could have been traveling when the driver first applied the brakes? (Lesson 5-6)

Solve by graphing. (Lesson 3-1)

55. $y = 3x - 1$
$y = -2x + 4$

56. $3x + 2y = 8$
$-4x + 6y = 11$

57. $5x - 2y = 6$
$3y + 2x = 10$

Skills Review

If $c(x) = x^2 - 2x$ and $d(x) = 3x^2 - 6x + 4$, find each value. (Lesson 6-3)

58. $c(a + 2) - d(a - 4)$

59. $c(a - 3) + d(a + 1)$

60. $c(-3a) + d(a + 4)$

61. $3d(3a) - 2c(-a)$

62. $c(a) + 5d(2a)$

63. $-2d(2a + 3) - 4c(a^2 + 1)$

6-7 Roots and Zeros

Why?

The function $g(x) = 1.384x^4 - 0.003x^3 + 0.28x^2 - 0.078x + 1.365$ can be used to model the average price of a gallon of gasoline in a given year if x is the number of years since 1990. To find the average price of gasoline in a specific year, you can use the roots of the related polynomial equation.

Then

You used complex numbers to describe solutions of quadratic equations. (Lesson 5-4)

Now

- Determine the number and type of roots for a polynomial equation.
- Find the zeros of a polynomial function.

NYS Core Curriculum

A2.A.26 Find the solution to polynomial equations of higher degree that can be solved using factoring and/or the quadratic formula

NY Math Online

glencoe.com

- Extra Examples
- Personal Tutor
- Self-Check Quiz
- Homework Help
- Math in Motion

Synthetic Types of Roots Previously, you learned that a zero of a function $f(x)$ is any value c such that $f(c) = 0$. When the function is graphed, the real zeros of the function are the x-intercepts of the graph.

Concept Summary — For Your FOLDABLE

Zeros, Factors, Roots, and Intercepts

Words Let $P(x) = a_nx^n + \cdots + a_1x + a_0$ be a polynomial function. Then the following statements are equivalent.

- c is a zero of $P(x)$.
- c is a root or solution of $P(x) = 0$.
- $x - c$ is a factor of $a_nx^n + \cdots + a_1x + a_0$.
- If c is a real number, then $(c, 0)$ is an x-intercept of the graph of $P(x)$.

Example Consider the polynomial function $P(x) = x^4 + 2x^3 - 7x^2 - 8x + 12$.

The zeros of $P(x) = x^4 + 2x^3 - 7x^2 - 8x + 12$ are -3, -2, 1, and 2.

The roots of $x^4 + 2x^3 - 7x^2 - 8x + 12 = 0$ are -3, -2, 1, and 2.

The factors of $x^4 + 2x^3 - 7x^2 - 8x + 12$ are $(x + 3)$, $(x + 2)$, $(x - 1)$, and $(x - 2)$.

The x-intercepts of the graph of $P(x) = x^4 + 2x^3 - 7x^2 - 8x + 12$ are $(-3, 0)$, $(-2, 0)$, $(1, 0)$, and $(2, 0)$.

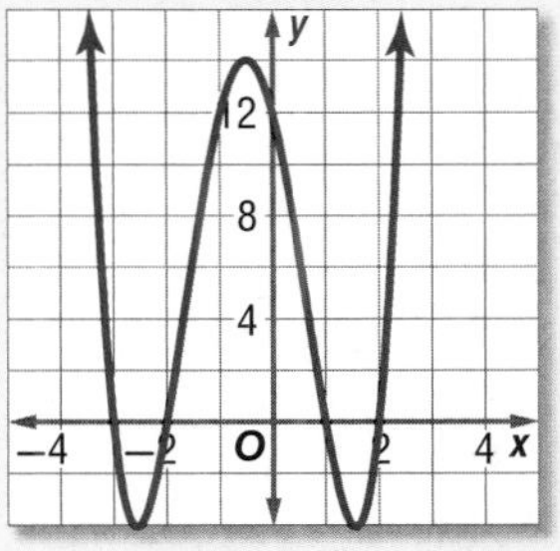

When solving a polynomial equation with degree greater than zero, there may be one or more real roots or no real roots (the roots are imaginary numbers). Since real numbers and imaginary numbers both belong to the set of complex numbers, all polynomial equations with degree greater than zero will have at least one root in the set of complex numbers. This is the **Fundamental Theorem of Algebra**.

Key Concept — **Fundamental Theorem of Algebra** — For Your FOLDABLE

Every polynomial equation with degree greater than zero has at least one root in the set of complex numbers.

EXAMPLE 1 Determine Number and Type of Roots

Solve each equation. State the number and type of roots.

a. $x^2 + 6x + 9 = 0$

$x^2 + 6x + 9 = 0$ — **Original equation**

$(x + 3)^2 = 0$ — **Factor.**

$x + 3 = 0$ — **Take the root of each side.**

$x = -3$ — **Solve for x.**

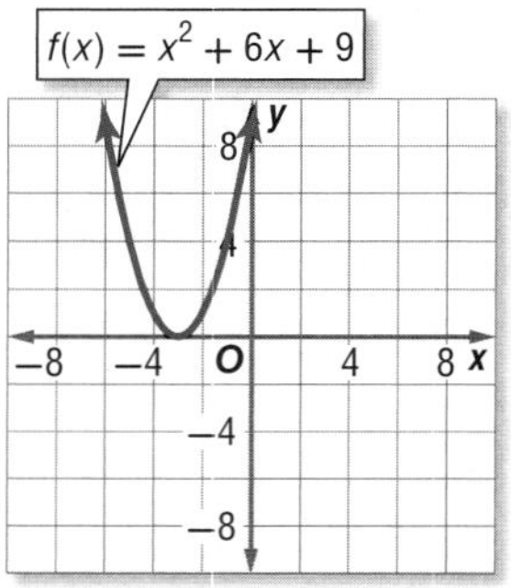

Because $(x + 3)$ is twice a factor of $x^2 + 6x + 9$, -3 is a double root. Thus, the equation has one real repeated root, -3.

CHECK The graph of the equation touches the x-axis at $x = -3$. Since -3 is a double root, the graph does not cross the axis. ✓

b. $x^3 + 25x = 0$

$x^3 + 25x = 0$ — **Original equation**

$x(x^2 + 25) = 0$ — **Factor.**

$x = 0$ or $x^2 + 25 = 0$

$x^2 = -25$

$x = \pm\sqrt{-25}$ or $\pm 5i$

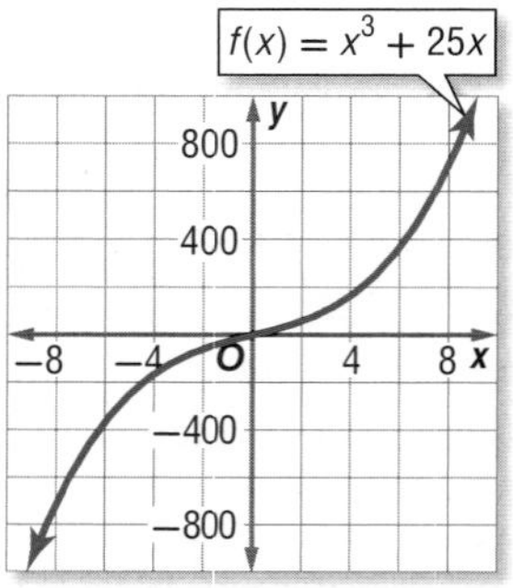

This equation has one real root, 0, and two imaginary roots, $5i$ and $-5i$.

CHECK The graph of this equation crosses the x-axis at only one place, $x = 0$. ✓

Check Your Progress

1A. $x^3 + 2x = 0$

1B. $x^4 - 16 = 0$

1C. $x^3 + 4x^2 - 7x - 10 = 0$

1D. $3x^3 - x^2 + 9x - 3 = 0$

Personal Tutor glencoe.com

> **ReadingMath**
>
> **Repeated Roots** Polynomial equations can have double roots, triple roots, quadruple roots, and so on. In general, these are referred to as *multiple roots*.

Examine the solutions for each equation in Example 1. Notice that the number of solutions for each equation is the same as the degree of each polynomial. The following corollary to the Fundamental Theorem of Algebra describes this relationship between the degree and the number of roots of a polynomial equation.

Key Concept

For Your FOLDABLE

Corollary to the Fundamental Theorem of Algebra

Words A polynomial equation of degree n has exactly n roots in the set of complex numbers, including repeated roots.

Example

$x^3 + 2x^2 + 6$	$4x^4 - 3x^3 + 5x - 6$	$-2x^5 - 3x^2 + 8$
3 roots	4 roots	5 roots

Similarly, an nth degree polynomial function has exactly n zeros.

Additionally, French mathematician René Descartes discovered a relationship between the signs of the coefficients of a polynomial function and the number of positive and negative real zeros.

StudyTip

Zero at the Origin If a zero of a function is at the origin, the sum of the number of positive real zeros, negative real zeros, and imaginary zeros is reduced by how many times 0 is a zero of the function.

Key Concept Descartes' Rule of Signs

For Your FOLDABLE

Let $P(x) = a_n x^n + \cdots + a_1 x + a_0$ be a polynomial function with real coefficients. Then

- the number of positive real zeros of $P(x)$ is the same as the number of changes in sign of the coefficients of the terms, or is less than this by an even number, and
- the number of negative real zeros of $P(x)$ is the same as the number of changes in sign of the coefficients of the terms of $P(-x)$, or is less than this by an even number.

Math *in Motion,* Animation glencoe.com

EXAMPLE 2 Find Numbers of Positive and Negative Zeros

State the possible number of positive real zeros, negative real zeros, and imaginary zeros of $f(x) = x^6 + 3x^5 - 4x^4 - 6x^3 + x^2 - 8x + 5$.

Because $f(x)$ has degree 6, it has six zeros, either real or imaginary. Use Descartes' Rule of Signs to determine the possible number and type of *real* zeros.

Count the number of changes in sign for the coefficients of $f(x)$.

$$f(x) = x^6 \quad + \quad 3x^5 \quad - \quad 4x^4 \quad - \quad 6x^3 \quad + \quad x^2 \quad - \quad 8x \quad + \quad 5$$

no	yes	no	yes	yes	yes
+ to +	+ to −	− to −	− to +	+ to −	− to +

There are 4 sign changes, so there are 4, 2, or 0 positive real zeros.

Count the number of changes in sign for the coefficients of $f(-x)$.

$$f(-x) = (-x)^6 + 3(-x)^5 - 4(-x)^4 - 6(-x)^3 + (-x)^2 - 8(-x) + 5$$
$$= x^6 - 3x^5 - 4x^4 + 6x^3 + x^2 + 8x + 5$$

yes	no	yes	no	no	no
+ to −	− to −	− to +	+ to +	+ to +	+ to +

There are 2 sign changes, so there are 2, or 0 negative real zeros.
Make a chart of the possible combinations of real and imaginary zeros.

Number of Positive Real Zeros	Number of Negative Real Zeros	Number of Imaginary Zeros	Total Number of Zeros
4	2	0	$4 + 2 + 0 = 6$
4	0	2	$4 + 0 + 2 = 6$
2	2	2	$2 + 2 + 2 = 6$
2	0	4	$2 + 0 + 4 = 6$
0	2	4	$0 + 2 + 4 = 6$
0	0	6	$0 + 0 + 6 = 6$

Check Your Progress

2. State the possible number of positive real zeros, negative real zeros, and imaginary zeros of $h(x) = 2x^5 + x^4 + 3x^3 - 4x^2 - x + 9$.

Personal Tutor glencoe.com

Find Zeros You can use the various strategies and theorems you have learned to find all of the zeros of a function.

EXAMPLE 3 Use Synthetic Substitution to Find Zeros

Find all zeros of $f(x) = x^4 - 18x^2 + 12x + 80$.

Step 1 Determine the total number of zeros.
Since $f(x)$ has degree 4, the function has 4 zeros.

Step 2 Determine the type of zeros.
Examine the number of sign changes for $f(x)$ and $f(-x)$.

$f(x) = x^4 - 18x^2 + 12x + 80$ — **yes** **yes** **no**

$f(-x) = x^4 - 18x^2 - 12x + 80$ — **yes** **no** **yes**

Because there are 2 sign changes for the coefficients of $f(x)$, the function has 2 or 0 positive real zeros. Because there are 2 sign changes for the coefficients of $f(-x)$, $f(x)$ has 2 or 0 negative real zeros. Thus, $f(x)$ has 4 real zeros, 2 real zeros and 2 imaginary zeros, or 4 imaginary zeros.

Step 3 Determine the real zeros.
List some possible values, and then use synthetic substitution to evaluate $f(x)$ for real values of x.

x	1	0	−18	12	80
−3	1	−3	−9	39	−37
−2	1	−2	−14	40	0
−1	1	−1	−17	29	51
0	1	0	−18	12	80
1	1	1	−17	−5	75
2	1	2	−14	−2	76

Each row shows the coefficients of the depressed polynomial and the remainder.

From the table, we can see that one zero occurs at $x = -2$. Since there are 2 negative real zeros, use synthetic substitution with the depressed polynomial function $f(x) = x^3 - 2x^2 - 14x + 40$ to find a second negative zero.

A second negative zero is at $x = -4$. Since the depressed polynomial $x^2 - 6x + 10$ is quadratic, use the Quadratic Formula to find the remaining zeros of $f(x) = x^2 - 6x + 10$.

x	1	−2	−14	40
−4	1	−6	10	0
−5	1	−7	21	−65
−6	1	−8	34	−164

$x = \frac{-b \pm \sqrt{b^2 - 4ac}}{2a}$ **Quadratic Formula**

$= \frac{-(-6) \pm \sqrt{(-6)^2 - 4(1)(10)}}{2(1)}$ **Replace *a* with 1, *b* with −6, and *c* with 10.**

$= 3 \pm i$ **Simplify.**

The function has zeros at -4, -2, $3 + i$, and $3 - i$.

CHECK Graph the function on a graphing calculator. The graph crosses the x-axis two times, so there are two real zeros. Use the **zero** function under the **CALC** menu to locate each zero. The two real zeros are -4 and -2.

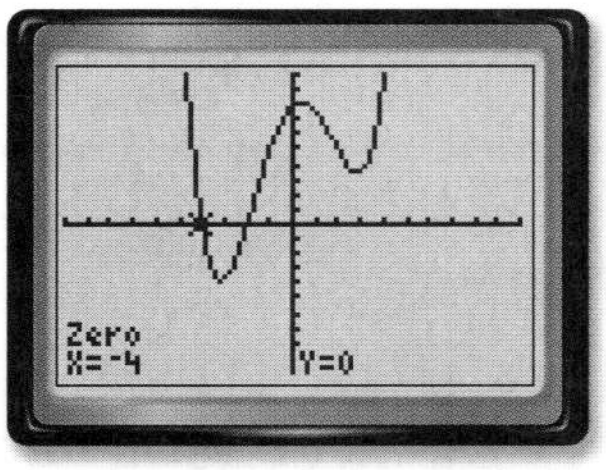

[−10, 10] scl: 1 by [−100, 100] scl: 10

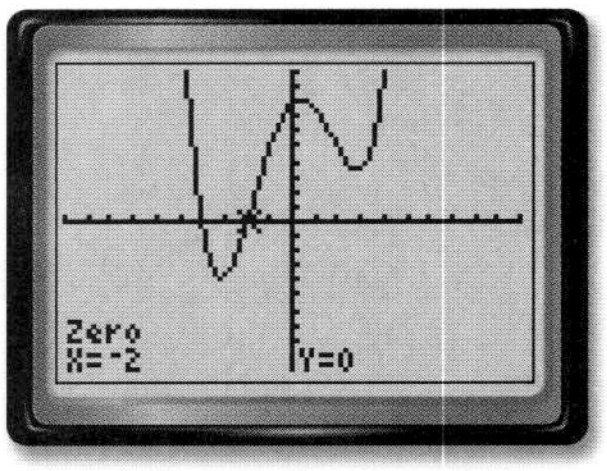

[−10, 10] scl: 1 by [−100, 100] scl: 10

Check Your Progress **3.** Find all of the zeros of $h(x) = x^3 + 2x^2 + 9x + 18$.

Personal Tutor glencoe.com

StudyTip

Testing for Zeros If a value is not a zero for a polynomial, then it will not be a zero for the depressed polynomial either, so it does not need to be checked again.

StudyTip

Locating Zeros Refer to Lesson 5-2 on how to use the CALC menu to locate a zero on your calculator.

Review Vocabulary

complex conjugates two complex numbers of the form $a + bi$ and $a - bi$ (Lesson 5-4)

In Chapter 5, you learned that the product of complex conjugates is always a real number and that complex roots always come in conjugate pairs. For example, if one root of $x^2 - 8x + 52 = 0$ is $4 + 6i$, then the other root is $4 - 6i$.

This applies to the zeros of polynomial functions as well. For any polynomial function with real coefficients, if an imaginary number is a zero of that function, its conjugate is also a zero. This is called the **Complex Conjugates Theorem**.

Key Concept **Complex Conjugates Theorem** For Your FOLDABLE

Words Let a and b be real numbers, and $b \neq 0$. If $a + bi$ is a zero of a polynomial function with real coefficients, then $a - bi$ is also a zero of the function.

Example If $3 + 4i$ is a zero of $f(x) = x^3 - 4x^2 + 13x + 50$, then $3 - 4i$ is also a zero of the function.

When you are given all of the zeros of a polynomial function and are asked to determine the function, convert the zeros to factors and then multiply all of the factors together. The result is the polynomial function.

EXAMPLE 4 Use Zeros to Write a Polynomial Function

Write a polynomial function of least degree with integral coefficients, the zeros of which include −1 and $5 - i$.

Understand If $5 - i$ is a zero, then $5 + i$ is also a zero according to the Complex Conjugates Theorem. So, $x + 1$, $x - (5 - i)$, and $x - (5 + i)$ are factors of the polynomial.

Plan Write the polynomial function as a product of its factors.

$P(x) = (x + 1)[x - (5 - i)][x - (5 + i)]$

Solve Multiply the factors to find the polynomial function.

$P(x) = (x + 1)\,[x - (5 - i)][x - (5 + i)]$	**Write the equation.**
$= (x + 1)\,[(x - 5) + i][(x - 5) - i]$	**Regroup terms.**
$= (x + 1)\,[(x - 5)^2 - i^2]$	**Difference of squares**
$= (x + 1)\,[(x^2 - 10x + 25 - (-1)]$	**Square terms.**
$= (x + 1)\,(x^2 - 10x + 26)$	**Simplify.**
$= x^3 - 10x^2 + 26x + x^2 - 10x + 26$	**Multiply.**
$= x^3 - 9x^2 + 16x + 26$	**Combine like terms.**

Check Because there are 3 zeros, the degree of the polynomial function must be 3, so $P(x) = x^3 - 9x^2 + 16x + 26$ is a polynomial function of least degree with integral coefficients and zeros of -1, $5 - i$, and $5 + i$.

Check Your Progress

4. Write a polynomial function of least degree with integral coefficients having zeros that include -1 and $1 + 2i$.

Personal Tutor glencoe.com

Check Your Understanding

Example 1 p. 384

Solve each equation. State the number and type of roots.

1. $x^2 - 3x - 10 = 0$
2. $x^3 + 12x^2 + 32x = 0$
3. $16x^4 - 81 = 0$
4. $0 = x^3 - 8$

Example 2 p. 385

State the possible number of positive real zeros, negative real zeros, and imaginary zeros of each function.

5. $f(x) = x^3 - 2x^2 + 2x - 6$
6. $f(x) = 6x^4 + 4x^3 - x^2 - 5x - 7$
7. $f(x) = 3x^5 - 8x^3 + 2x - 4$
8. $f(x) = -2x^4 - 3x^3 - 2x - 5$

Example 3 p. 386

Find all zeros of each function.

9. $f(x) = x^3 + 9x^2 + 6x - 16$
10. $f(x) = x^3 + 7x^2 + 4x + 28$
11. $f(x) = x^4 - 2x^3 - 8x^2 - 32x - 384$
12. $f(x) = x^4 - 6x^3 + 9x^2 + 6x - 10$

Example 4 p. 387

Write a polynomial function of least degree with integral coefficients that have the given zeros.

13. $4, -1, 6$
14. $3, -1, 1, 2$
15. $-2, 5, -3i$
16. $-4, 4 + i$

Practice and Problem Solving

● = **Step-by-Step Solutions** begin on page R20.
Extra Practice begins on page 947.

Example 1 p. 384

Solve each equation. State the number and type of roots.

17. $2x^2 + x - 6 = 0$
18. $4x^2 + 1 = 0$
19. $x^3 + 1 = 0$
20. $2x^2 - 5x + 14 = 0$
21. $-3x^2 - 5x + 8 = 0$
22. $8x^3 - 27 = 0$
23. $16x^4 - 625 = 0$
24. $x^3 - 6x^2 + 7x = 0$
25. $x^5 - 8x^3 + 16x = 0$
26. $x^5 + 2x^3 + x = 0$

Example 2 p. 385

State the possible number of positive real zeros, negative real zeros, and imaginary zeros of each function.

27. $f(x) = x^4 - 5x^3 + 2x^2 + 5x + 7$
28. $f(x) = 2x^3 - 7x^2 - 2x + 12$
29. $f(x) = -3x^5 + 5x^4 + 4x^2 - 8$
30. $f(x) = x^4 - 2x^2 - 5x + 19$
31. $f(x) = 4x^6 - 5x^4 - x^2 + 24$
32. $f(x) = -x^5 + 14x^3 + 18x - 36$

Example 3 p. 386

Find all zeros of each function.

33. $f(x) = x^3 + 7x^2 + 4x - 12$
34. $f(x) = x^3 + x^2 - 17x + 15$
35. $f(x) = x^4 - 3x^3 - 3x^2 - 75x - 700$
36. $f(x) = x^4 + 6x^3 + 73x^2 + 384x + 576$
37. $f(x) = x^4 - 8x^3 + 20x^2 - 32x + 64$
38. $f(x) = x^5 - 8x^3 - 9x$
39. $f(x) = x^3 - 5x^2 + 17x - 85$
40. $f(x) = x^3 + 2x$
41. $f(x) = 4x^4 + 15x^2 - 4$
42. $f(x) = 9x^4 + 9x^3 + 4x^2 + 4x$

Example 4 p. 387

Write a polynomial function of least degree with integral coefficients that have the given zeros.

43. $5, -2, -1$
44. $-4, -3, 5$
45. $-1, -1, 2i$
46. $-3, 1, -3i$
47. $0, -5, 3 + i$
48. $-2, -3, 4 - 3i$

49 **BUSINESS** A computer manufacturer determines that for each employee, the profit for producing x computers per day is $P(x) = -0.006x^4 + 0.15x^3 - 0.05x^2 - 1.8x$.

a. How many positive real zeros, negative real zeros, and imaginary zeros exist?

b. What is the meaning of the zeros in this situation?

Match each graph to the given zeros.

a. $-3, 4, i, -i$ **b.** $-4, 3$ **c.** $-4, 3, i, -i$

50.

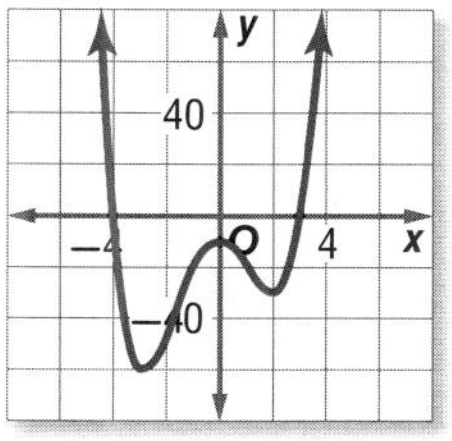

51.

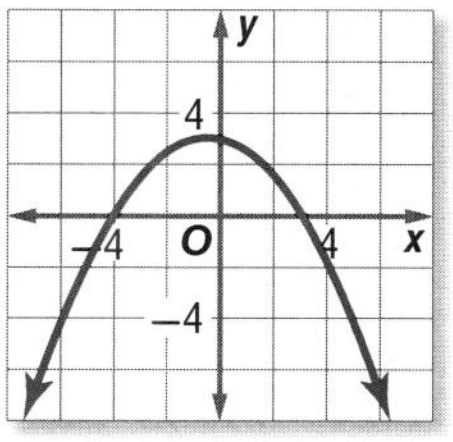

52.

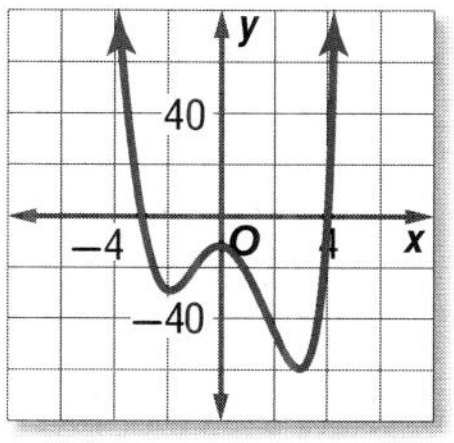

53. **CONCERTS** The amount of money Hoshi's Music Hall took in from 2003 to 2010 can be modeled by $M(x) = -2.03x^3 + 50.1x^2 - 214x + 4020$, where x is the years since 2003.

a. How many positive real zeros, negative real zeros, and imaginary zeros exist?

b. Graph the function using your calculator.

c. Approximate all real zeros to the nearest tenth. What is the significance of each zero in the context of the situation?

Real-World Link

On average, 3500 people attend a concert in the U.S. each night.

Source: *Billboard Magazine*

Determine the number of positive real zeros, negative real zeros, and imaginary zeros for each function. Explain your reasoning.

54.

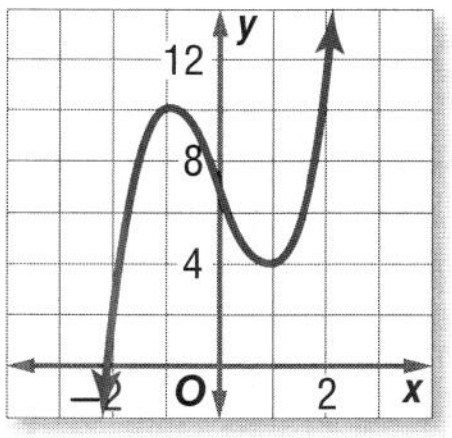

degree: 3

55.

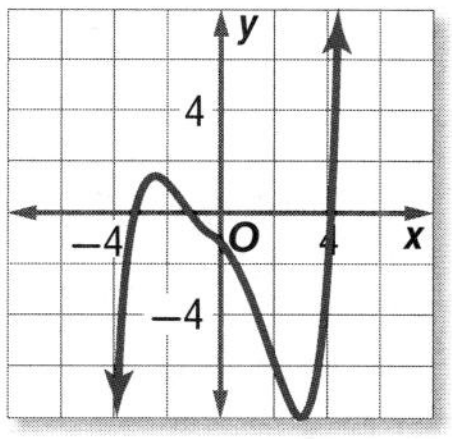

degree: 5

H.O.T. Problems

Use **H**igher-**O**rder **T**hinking Skills

56. **OPEN ENDED** Sketch the graph of a polynomial function with:

a. 3 real, 2 imaginary zeros **b.** 4 real zeros **c.** 2 imaginary zeros

57. **CHALLENGE** Write an equation in factored form of a polynomial function of degree 5 with 2 imaginary zeros, 1 nonintegral zero, and 2 irrational zeros. Explain.

58. **WHICH ONE DOESN'T BELONG** Determine which equation is not like the others. Explain.

$r^4 + 1 = 0$	$r^3 + 1 = 0$	$r^2 - 1 = 0$	$r^3 - 8 = 0$

59. **REASONING** Provide a counterexample for each statement.

a. All polynomial functions of degree greater than 2 have at least 1 negative real root.

b. All polynomial functions of degree greater than 2 have at least 1 positive real root.

60. **WRITING IN MATH** Explain to a friend how you would use Descartes' Rule of Signs to determine the number of possible positive real roots and the number of possible negative roots of the polynomial function $f(x) = x^4 - 2x^3 + 6x^2 + 5x - 12$.

NYSRE PRACTICE A2.A.26

61. SAT/ACT Use the graph of the polynomial function below. Which is not a factor of the polynomial $x^5 + x^4 - 3x^3 - 3x^2 - 4x - 4$?

A $x - 2$
B $x + 2$
C $x - 1$
D $x + 1$

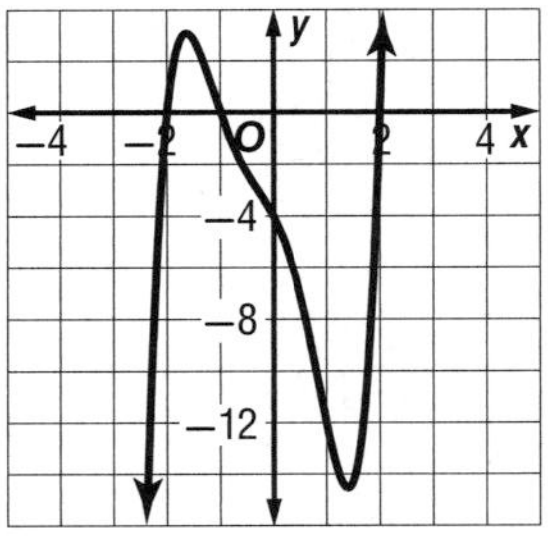

62. GEOMETRY A window is in the shape of an equilateral triangle. Each side of the triangle is 8 feet long. The window is divided in half by a support from one vertex to the midpoint of the side of the triangle opposite the vertex. Approximately how long is the support?

63. GEOMETRY In rectangle $ABCD$, $\overline{AD}$ is 8 units long. What is the length of $\overline{AB}$?

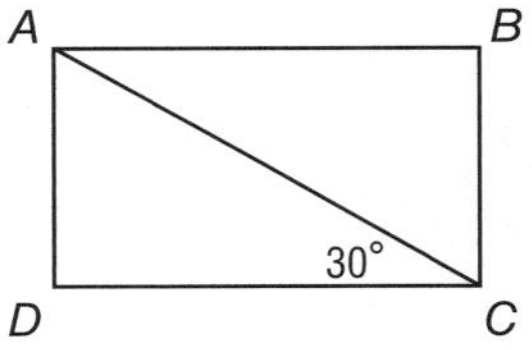

F 4 units
G 8 units
H $8\sqrt{3}$ units
J 16 units

64. The total area of a rectangle is $25a^4 - 16b^2$ square units. Which factors could represent the length and width?

A $(5a^2 + 4b)$ units and $(5a^2 + 4b)$ units
B $(5a^2 + 4b)$ units and $(5a^2 - 4b)$ units
C $(5a - 4b)$ units and $(5a - 4b)$ units
D $(5a + 4b)$ units and $(5a - 4b)$ units

Spiral Review

Use synthetic substitution to find $f(-8)$ and $f(4)$ for each function. (Lesson 6-6)

65. $f(x) = 4x^3 + 6x^2 - 3x + 2$

66. $f(x) = 5x^4 - 2x^3 + 4x^2 - 6x$

67. $f(x) = 2x^5 - 3x^3 + x^2 - 4$

Factor completely. If the polynomial is not factorable, write *prime*. (Lesson 6-5)

68. $x^6 - y^6$

69. $a^6 + b^6$

70. $4x^2y + 8xy + 16y - 3x^2z - 6xz - 12z$

71. $5a^3 - 30a^2 + 40a + 2a^2b - 12ab + 16b$

72. BUSINESS A mall owner has determined that the relationship between monthly rent charged for store space r (in dollars per square foot) and monthly profit $P(r)$ (in thousands of dollars) can be approximated by $P(r) = -8.1r^2 + 46.9r - 38.2$. Solve each quadratic equation or inequality. Explain what each answer tells about the relationship between monthly rent and profit for this mall. (Lesson 5-8)

a. $-8.1r^2 + 46.9r - 38.2 = 0$
b. $-8.1r^2 + 46.9r - 38.2 > 0$
c. $-8.1r^2 + 46.9r - 38.2 > 10$
d. $-8.1r^2 + 46.9r - 38.2 < 10$

73. DIVING To avoid hitting any rocks below, a cliff diver jumps up and out. The equation $h = -16t^2 + 4t + 26$ describes her height h in feet t seconds after jumping. Find the time at which she returns to a height of 26 feet. (Lesson 5-3)

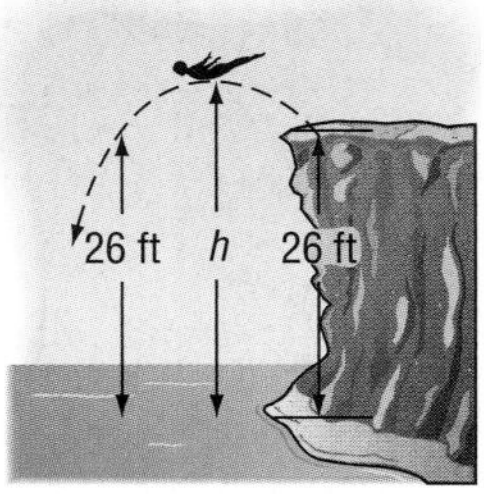

Skills Review

Find all of the possible values of $\pm\frac{b}{a}$ for each replacement set.

74. $a = \{1, 2, 4\}$; $b = \{1, 2, 3, 6\}$

75. $a = \{1, 5\}$; $b = \{1, 2, 4, 8\}$

76. $a = \{1, 2, 3, 6\}$; $b = \{1, 7\}$

6-8 Rational Zero Theorem

Why?

Annual sales of recorded music in the United States can be approximated by $d(t) = 30x^3 - 478x^2 + 1758x + 12{,}392$, where $d(t)$ is the total sales in millions of dollars and t is the number of years since 1997. You can use this function to estimate when music sales will be \$20 billion.

Then

You found zeros of quadratic functions of the form $f(x) = ax^2 + bx + c$. (Lesson 5-3)

Now

- Identify possible rational zeros of a polynomial function.
- Find all of the rational zeros of a polynomial function.

NYS Core Curriculum

A2.A.26 Find the solution to polynomial equations of higher degree that can be solved using factoring and/or the quadratic formula

NY Math Online

glencoe.com

- Extra Examples
- Personal Tutor
- Self-Check Quiz
- Homework Help

Identify Rational Zeros Usually it is not practical to test all possible zeros of a polynomial function using synthetic substitution. The **Rational Zero Theorem** can help you choose some possible zeros to test. **If the leading coefficient is 1, the corollary applies.**

Key Concept — Rational Zero Theorem

For Your FOLDABLE

Words If $P(x)$ is a polynomial function with integral coefficients, then every rational zero of $P(x) = 0$ is of the form $\frac{p}{q}$, a rational number in simplest form, where p is a factor of the constant term and q is a factor of the leading coefficient

Example Let $f(x) = 6x^4 + 22x^3 + 11x^2 - 80x - 40$. If $\frac{4}{3}$ is a zero of $f(x)$, then 4 is a factor of -40, and 3 is a factor of 6.

Corollary to the Rational Zero Theorem

If $P(x)$ is a polynomial function with integral coefficients, a leading coefficient of 1, and a nonzero constant term, then any rational zeros of $P(x)$ must be factors of the constant term.

EXAMPLE 1 Identify Possible Zeros

List all of the possible rational zeros of each function.

a. $f(x) = 4x^5 + x^4 - 2x^3 - 5x^2 + 8x + 16$

If $\frac{p}{q}$ is a rational zero, then p is a factor of 16 and q is a factor of 4.

p: $\pm1, \pm2, \pm4, \pm8, \pm16$ $\quad$ q: $\pm1, \pm2, \pm4$

Write the possible values of $\frac{p}{q}$ in simplest form.

$\frac{p}{q} = \pm1, \pm2, \pm4, \pm8, \pm16, \pm\frac{1}{2}, \pm\frac{1}{4}$

b. $f(x) = x^3 - 2x^2 + 5x + 12$

If $\frac{p}{q}$ is a rational zero, then p is a factor of 12 and q is a factor of 1.

p: $\pm1, \pm2, \pm3, \pm4, \pm6, \pm12$ $\quad$ q: ±1

So, $\frac{p}{q} = \pm1, \pm2, \pm3, \pm4, \pm6$, and ±12

Check Your Progress

1A. $g(x) = 3x^3 - 4x + 10$

1B. $h(x) = x^3 + 11x^2 + 24$

Personal Tutor glencoe.com

Find Rational Zeros Once you have written the possible rational zeros, you can test each number using synthetic substitution and use the other tools you have learned to determine the zeros of a function.

Real-World EXAMPLE 2 Find Rational Zeros

WOODWORKING Adam is building a computer desk with a separate compartment for the computer. The compartment for the computer is a rectangular prism and will be 8019 cubic inches. Find the dimensions of the computer compartment.

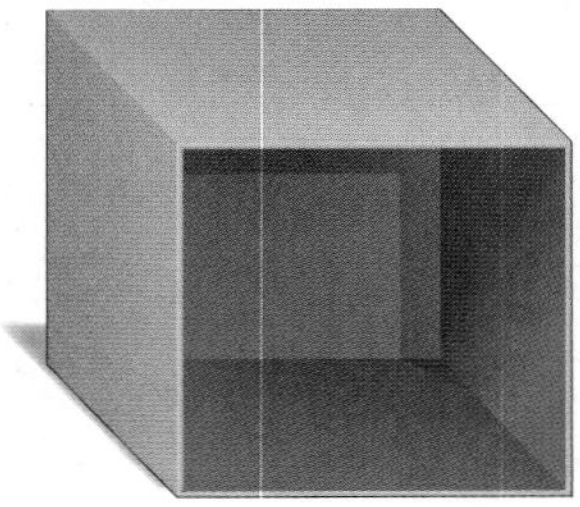

Let x = width, $x + 24$ = length, and $x + 18$ = height.

Write an equation for the volume.

$\ell wh = V$	**Formula for volume**
$(x + 24)(x)(x + 18) = 8019$	**Substitute.**
$x^3 + 42x^2 + 432x = 8019$	**Multiply.**
$x^3 + 42x^2 + 432x - 8019 = 0$	**Subtract 8019 from each side.**

The leading coefficient is 1, so the possible rational zeros are factors of 8019.

$\pm1, \pm3, \pm9, \pm11, \pm27, \pm33, \pm81, \pm99, \pm243, \pm297, \pm729, \pm891, \pm2673$, and ±8019

Since length can only be positive, we only need to check positive values.

There is one change of sign of the coefficients, so by Descartes' Rule of Signs, there is only one positive real zero. Make a table for synthetic division and test possible values.

p	1	42	432	−8019
1	1	43	475	−7544
3	1	45	567	−6318
9	1	51	891	0

One zero is 9. Since there is only one positive real zero, we do not have to test the other numbers. The other dimensions are 9 + 24 or 33 inches, and 9 + 18 or 27 inches.

CHECK Multiply the dimensions and see if they equal the volume of 8019 cubic inches.
$9 \times 33 \times 27 = 8019$ ✓

StudyTip

Descartes' Rule of Signs Examine the signs of the coefficients of the equation. In this case, there is only one change of sign, so there is only one positive real zero.

Check Your Progress

2. GEOMETRY The volume of a rectangular prism is 1056 cubic centimeters. The length is 1 centimeter more than the width, and the height is 3 centimeters less than the width. Find the dimensions of the prism.

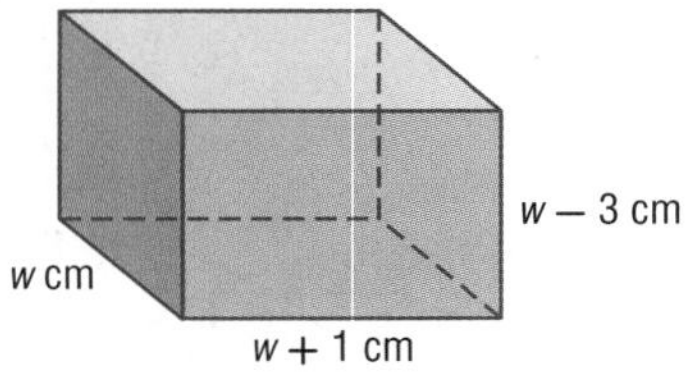

Personal Tutor glencoe.com

You usually do not need to test all of the possible zeros. Once you find a zero, you can try to factor the depressed polynomial to find any other zeros.

EXAMPLE 3 Find All Zeros

Find all of the zeros of $f(x) = 5x^4 - 8x^3 + 41x^2 - 72x - 36$.

From the corollary to the Fundamental Theorem of Algebra, there are exactly 4 complex zeros. According to Descartes' Rule of Signs, there are 3 or 1 positive real zeros and exactly 1 negative real zero. The possible rational zeros are $\pm1, \pm2, \pm3, \pm4, \pm6, \pm9, \pm12, \pm18, \pm36, \pm\frac{1}{5}, \pm\frac{2}{5}, \pm\frac{3}{5}, \pm\frac{4}{5}, \pm\frac{6}{5}, \pm\frac{9}{5}, \pm\frac{12}{5}, \pm\frac{18}{5}$, and $\pm\frac{36}{5}$.

Make a table and test some possible rational zeros.

$\frac{p}{q}$	5	−8	41	−72	−36
−1	5	−13	54	−126	90
1	5	−3	38	−34	−70
2	5	2	45	18	0

Because $f(2) = 0$, there is a zero at $x = 2$. Factor the depressed polynomial $5x^3 + 2x^2 + 45x + 18$.

$5x^3 + 2x^2 + 45x + 18 = 0$ **Write the depressed polynomial.**

$(5x^3 + 2x^2) + (45x + 18) = 0$ **Group terms.**

$x^2(5x + 2) + 9(5x + 2) = 0$ **Factor.**

$(x^2 + 9)(5x + 2) = 0$ **Distributive Property**

$x^2 + 9 = 0$ or $5x + 2 = 0$ **Zero Product Property**

$x^2 = -9$ $\quad 5x = -2$

$x = \pm 3i$ $\quad x = -\frac{2}{5}$

There is another real zero at $x = -\frac{2}{5}$ and two imaginary zeros at $x = 3i$ and $x = -3i$.

The zeros of the function are $-\frac{2}{5}$, 2, $3i$, and $-3i$.

Check Your Progress

Find all of the zeros of each function.

3A. $h(x) = 9x^4 + 5x^2 - 4$

3B. $k(x) = 2x^4 - 5x^3 + 20x^2 - 45x + 18$

Personal Tutor glencoe.com

Check Your Understanding

Example 1 p. 391

List all of the possible rational zeros of each function.

1. $f(x) = x^3 - 6x^2 - 8x + 24$
2. $f(x) = 2x^4 + 3x^2 - x + 15$

Example 2 p. 392

3. **GEOMETRY** The volume of the triangular pyramid is 210 cubic inches. Find the dimensions of the solid.

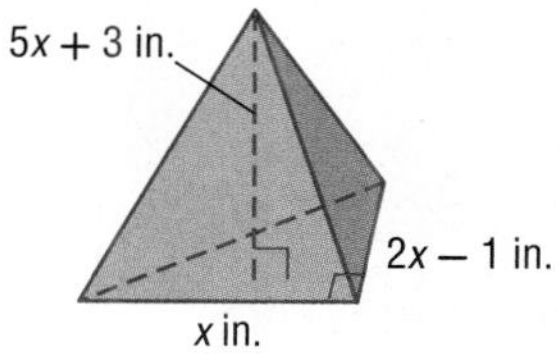

Find all of the rational zeros of each function.

4. $f(x) = x^3 - 6x^2 - 13x + 42$
5. $f(x) = 2x^4 + 11x^3 + 26x^2 + 29x + 12$

Example 3 p. 393

Find all of the zeros of each function.

6. $f(x) = 3x^3 - 2x^2 - 8x + 5$
7. $f(x) = 8x^3 + 14x^2 + 11x + 3$
8. $f(x) = 4x^4 + 13x^3 - 8x^2 + 13x - 12$
9. $f(x) = 4x^4 - 12x^3 + 25x^2 - 14x - 15$

Practice and Problem Solving

● = **Step-by-Step Solutions** begin on page R20.
Extra Practice begins on page 947.

Example 1 p. 391

List all of the possible rational zeros of each function.

10. $f(x) = x^4 + 8x - 32$

11. $f(x) = x^3 + x^2 - x - 56$

12. $f(x) = 2x^3 + 5x^2 - 8x - 10$

13. $f(x) = 3x^6 - 4x^4 - x^2 - 35$

14. $f(x) = 6x^5 - x^4 + 2x^3 - 3x^2 + 2x - 18$

15. $f(x) = 8x^4 - 4x^3 - 4x^2 + x + 42$

16. $f(x) = 15x^3 + 6x^2 + x + 90$

17. $f(x) = 16x^4 - 5x^2 + 128$

Example 2 p. 392

18. MANUFACTURING A box is to be constructed by cutting out equal squares from the corners of a square piece of cardboard and turning up the sides.

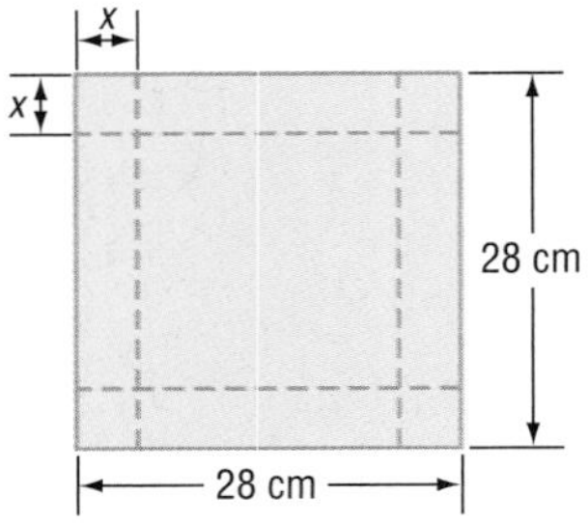

a. Write a function $V(x)$ for the volume of the box.

b. For what value of x will the volume of the box equal 1152 cubic centimeters?

c. What will be the volume of the box if $x = 6$ centimeters?

Find all of the rational zeros of each function.

19. $f(x) = x^3 + 10x^2 + 31x + 30$

20. $f(x) = x^3 - 2x^2 - 56x + 192$

21. $f(x) = 4x^3 - 3x^2 - 100x + 75$

22. $f(x) = 4x^4 + 12x^3 - 5x^2 - 21x + 10$

23. $f(x) = x^4 + x^3 - 8x - 8$

24. $f(x) = 2x^4 - 3x^3 - 24x^2 + 4x + 48$

25. $f(x) = 4x^3 + x^2 + 16x + 4$

26. $f(x) = 81x^4 - 256$

Example 3 p. 393

Find all of the zeros of each function.

27. $f(x) = x^3 + 3x^2 - 25x + 21$

28. $f(x) = 6x^3 + 5x^2 - 9x + 2$

29. $f(x) = x^4 - x^3 - x^2 - x - 2$

30. $f(x) = 10x^3 - 17x^2 - 7x + 2$

31. $f(x) = x^4 - 3x^3 + x^2 - 3x$

32. $f(x) = 6x^3 + 11x^2 - 3x - 2$

33. $f(x) = 6x^4 + 22x^3 + 11x^2 - 38x - 40$

34. $f(x) = 2x^3 - 7x^2 - 8x + 28$

35. $f(x) = 9x^5 - 94x^3 + 27x^2 + 40x - 12$

36. $f(x) = x^5 - 2x^4 - 12x^3 - 12x^2 - 13x - 10$

37. $f(x) = 48x^4 - 52x^3 + 13x - 3$

38. $f(x) = 5x^4 - 29x^3 + 55x^2 - 28x$

39. SWIMMING POOLS A diagram of the swimming pool at the Midtown Community Center is shown below. The pool can hold 9175 cubic feet of water.

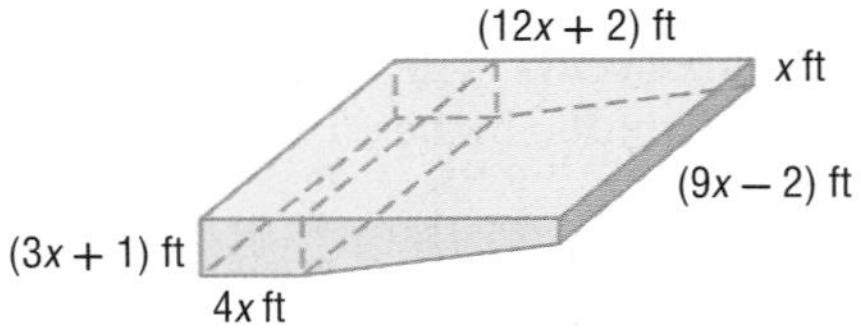

a. Write a polynomial function that represents the volume of the swimming pool.

b. What are the possible values of x? Which of these values are reasonable?

40. ROLLER COASTERS The path of a certain roller coaster can be modeled by $f(t) = t^4 - 31t^3 + 408t^2 - 1700t + 2000$ where t represents the time in seconds and $f(t)$ represents when the height of the roller coaster is at a relative maximum. Use the Rational Zero Theorem to determine the four times in which the roller coaster is at a relative maximum.

Real-World Link

The world's largest swimming pool is along the coastline in San Alfonso del Mar, Chile. It is 1 kilometer long, about 6000 times as large as an average pool.

Source: OhGizmo!

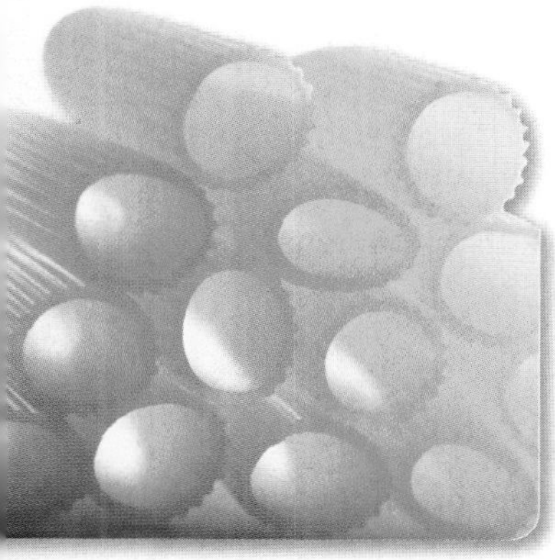

Real-World Link

Thomas Jefferson is credited with introducing macaroni to the United States after serving as the U.S. Ambassador to France.

Source: National Pasta Association

41. **FOOD** A restaurant orders spaghetti sauce in cylindrical metal cans. The volume of each can is about 160π cubic inches, and the height of the can is 6 inches more than the radius.

a. Write a polynomial equation that represents the volume of a can. Use the formula for the volume of a cylinder, $V = \pi r^2 h$.

b. What are the possible values of r? Which of these values are reasonable for this situation?

c. Find the dimensions of the can.

42. **Refer to the graph at the right.**

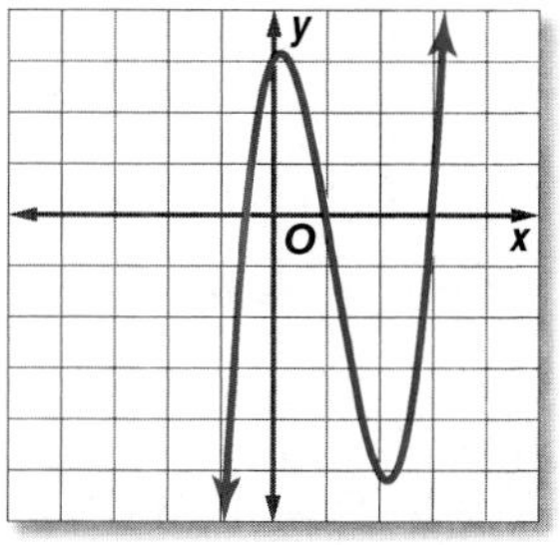

a. Find all of the zeros of $f(x) = 2x^3 + 7x^2 + 2x - 3$ and $g(x) = 2x^3 - 7x^2 + 2x + 3$.

b. Determine which function, f or g, is shown in the graph at the right.

43. **MUSIC SALES** Refer to the beginning of the lesson.

a. Write a polynomial equation that could be used to determine the year in which music sales would be about \$20,000,000,000.

b. List the possible whole number solutions for your equation in part a.

c. Determine the approximate year in which music sales will reach \$20,000,000,000.

d. Does the model represent a realistic estimate for all future music sales? Explain your reasoning.

Find all of the zeros of each function.

44. $f(x) = x^5 + 3x^4 - 19x^3 - 43x^2 + 18x + 40$

45. $f(x) = x^5 - x^4 - 23x^3 + 33x^2 + 126x - 216$

H.O.T. Problems

Use **H**igher-**O**rder **T**hinking Skills

46. **FIND THE ERROR** Doug and Mika are listing all of the possible rational zeros for $f(x) = 4x^4 + 8x^5 + 10x^2 + 3x + 16$. Is either of them correct? Explain your reasoning.

Doug	Mika
$\pm 1, \pm 2, \pm 4, \pm 8, \pm 16, \pm\frac{1}{2}, \pm\frac{1}{4}$	$\pm 1, \pm 2, \pm 4, \pm 8, \pm 16, \pm\frac{1}{2}, \pm\frac{1}{4}, \pm\frac{1}{8}$

47. **CHALLENGE** Give a polynomial function that has zeros at $1 + \sqrt{3}$ and $5 + 2i$.

48. **REASONING** Determine if the following statement is *sometimes, always,* or *never* true. Explain your reasoning.

If all of the possible zeros of a polynomial function are integers, then the leading coefficient of the function is 1.

49. **OPEN ENDED** Write a function that has possible zeros of $\pm 18, \pm 9, \pm 6, \pm 3, \pm 2, \pm 1, \pm\frac{9}{4}, \pm\frac{9}{2}, \pm\frac{3}{2}, \pm\frac{3}{4}, \pm\frac{1}{2}$, and $\pm\frac{1}{4}$.

50. **CHALLENGE** The roots of $x^2 + bx + c = 0$ are M and N. If $|M - N| = 1$, express c in terms of b.

51. **WRITING IN MATH** Explain the process of using the Rational Zero Theorem to determine the number of possible rational zeros of a function.

NYSRE PRACTICE A2.A.26

52. **ALGEBRA** Which of the following is a zero of the function $f(x) = 12x^5 - 5x^3 + 2x - 9$?

A -6

B $-\frac{2}{3}$

C $\frac{3}{8}$

D 1

53. **SAT/ACT** How many negative real zeros does $f(x) = x^5 - 2x^4 - 4x^3 + 4x^2 - 5x + 6$ have?

F 3

G 2

H 1

J 0

54. **ALGEBRA** For all nonnegative numbers n, let $\boxed{n}$ be defined by $\boxed{n} = \frac{\sqrt{n}}{2}$. If $\boxed{n} = 4$, what is the value of n?

A 2

B 4

C 16

D 64

55. **ALGEBRA** What is the y-intercept of a line that contains the point $(-1, 4)$ and has the same x-intercept as $x + 2y = -3$?

Spiral Review

Write a polynomial function of least degree with integral coefficients that has the given zeros. (Lesson 6-7)

56. $6, -3, \sqrt{2}$

57. $5, -1, 4i$

58. $-4, -2, i\sqrt{2}$

Given a polynomial and one of its factors, find the remaining factors of the polynomial. (Lesson 6-6)

59. $x^4 + 5x^3 + 5x^2 - 5x - 6; x + 3$

60. $a^4 - 2a^3 - 17a^2 + 18a + 72; a - 3$

61. $x^4 + x^3 - 11x^2 + x - 12; x + i$

62. **BRIDGES** The supporting cables of the Golden Gate Bridge approximate the shape of a parabola. The parabola can be modeled by the quadratic function $y = 0.00012x^2 + 6$, where x represents the distance from the axis of symmetry and y represents the height of the cables. The related quadratic equation is $0.00012x^2 + 6 = 0$. (Lesson 5-6)

a. Calculate the value of the discriminant.

b. What does the discriminant tell you about the supporting cables of the Golden Gate Bridge?

63. **RIDES** An amusement park ride carries riders to the top of a 225-foot tower. The riders then free-fall in their seats until they reach 30 feet above the ground. (Lesson 5-2)

a. Use the formula $h(t) = -16t^2 + h_0$, where the time t is in seconds and the initial height h_0 is in feet, to find how long the riders are in free-fall.

b. Suppose the designer of the ride wants the riders to experience free-fall for 5 seconds before stopping 30 feet above the ground. What should be the height of the tower?

Skills Review

Simplify. (Lesson 6-1)

64. $(x - 4)(x + 3)$

65. $3x(x^2 + 4)$

66. $x^2(x - 2)(x + 1)$

Find each value if $f(x) = 6x + 2$ and $g(x) = -4x^2$. (Lesson 2-1)

67. $f(5)$

68. $g(-3)$

69. $f(3c)$

NY Math Online glencoe.com
- STUDY TO GO
- Vocabulary Review

Chapter Summary

Key Concepts

Operations with Polynomials (Lessons 6-1 and 6-2)

- To add or subtract: Combine like terms.
- To multiply: Use the Distributive Property.
- To divide: Use long division or synthetic division.

Polynomial Functions and Graphs (Lessons 6-3 and 6-4)

- Turning points of a function are called *relative maxima* and *relative minima*.

Solving Polynomial Equations (Lesson 6-5)

- You can factor polynomials by using the GCF, grouping, or quadratic techniques.

The Remainder and Factor Theorems (Lesson 6-6)

- Factor Theorem: The binomial $x - a$ is a factor of the polynomial $f(x)$ if and only if $f(a) = 0$.

Roots, Zeros, and the Rational Zero Theorem (Lessons 6-7 and 6-8)

- Complex Conjugates Theorem: If $a + bi$ is a zero of a function, then $a - bi$ is also a zero.
- Integral Zero Theorem: If the coefficients of a polynomial function are integers such that $a_0 = 1$ and $a_n = 0$, any rational zeros of the function must be factors of a_n.
- Rational Zero Theorem: If $P(x)$ is a polynomial function with integral coefficients, then every rational zero of $P(x) = 0$ is of the form $\frac{p}{q}$, a rational number in simplest form, where p is a factor of the constant term and q is a factor of the leading coefficient.

FOLDABLES® Study Organizer

Be sure the Key Concepts are noted in your Foldable.

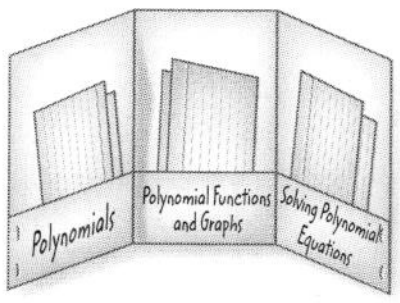

Key Vocabulary

degree of a polynomial (p. 335)
depressed polynomial (p. 379)
end behavior (p. 350)
extrema (p. 358)
leading coefficient (p. 348)
location principle (p. 357)
polynomial function (p. 349)
polynomial in one variable (p. 348)
power function (p. 349)
prime polynomials (p. 368)
quadratic form (p. 371)
relative maximum (p. 358)
relative minimum (p. 358)
simplify (p. 333)
synthetic division (p. 342)
synthetic substitution (p. 377)
turning points (p. 358)

Vocabulary Check

State whether each sentence is *true* or *false*. If *false*, replace the underlined term to make a true sentence.

1. The coefficient of the first term of a polynomial in standard form is called the leading coefficient.
2. Polynomials that cannot be factored are called polynomials in one variable.
3. A prime polynomial has a degree that is one less than the original polynomial.
4. A point on the graph of a function where no other nearby point has a greater y-coordinate is called a relative maximum.
5. A polynomial function is a continuous function that can be described by a polynomial equation in one variable.
6. To simplify an expression containing powers means to rewrite the expression without parentheses or negative exponents.
7. Synthetic division is a shortcut method for dividing a polynomial by a binomial.
8. The relative maximum and relative minimum of a function are often referred to as end behavior.
9. When a polynomial is divided by one of its binomial factors, the quotient is called a depressed polynomial.
10. $(x^3)^2 + 3x^3 - 8 = 0$ is a power function.

Study Guide and Review

Lesson-by-Lesson Review

6-1 Operations with Polynomials (pp. 333–339)

A2.A.8, A2.A.9

Simplify. Assume that no variable equals 0.

11. $\dfrac{14x^4y}{2x^3y^5}$

12. $3t(tn - 5)$

13. $(4r^2 + 3r - 1) - (3r^2 - 5r + 4)$

14. $(x^4)^3$

15. $(m + p)(m^2 - 2mp + p^2)$

16. $3b(2b - 1) + 2b(b + 3)$

EXAMPLE 1

Simplify each expression.

a. $\mathbf{(-4a^3b^5)(5ab^3)}$

$(-4a^3b^5)(5ab^3) = (-4)(5)a^{3+1}b^{5+3}$ **Product of Powers**

$= -20a^4b^8$ **Simplify.**

b. $\mathbf{(2x^2 + 3x - 8) + (3x^2 - 5x - 7)}$

$(2x^2 + 3x - 8) + (3x^2 - 5x - 7)$
$= (2x^2 + 3x^2) + (3x - 5x) + [-8 + (-7)]$
$= 5x^2 - 2x - 15$

6-2 Dividing Polynomials (pp. 341–347)

A.A.14

Simplify.

17. $\dfrac{12x^4y^5 + 8x^3y^7 - 16x^2y^6}{4xy^5}$

18. $(6y^3 + 13y^2 - 10y - 24) \div (y + 2)$

19. $(a^4 + 5a^3 + 2a^2 - 6a + 4)(a + 2)^{-1}$

20. $(4a^6 - 5a^4 + 3a^2 - a) \div (2a + 1)$

21. **GEOMETRY** The volume of the rectangular prism is $3x^3 + 11x^2 - 114x - 80$ cubic units. What is the area of the base?

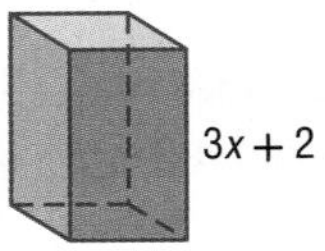

EXAMPLE 2

Simplify $\mathbf{(6x^3 - 31x^2 - 34x + 22) \div (2x - 1)}$.

$$
\begin{array}{r}
3x^2 - 14x - 24 \\
2x - 1\,\overline{)\,6x^3 - 31x^2 - 34x + 22} \\
\underline{(-)\ 6x^3 - 3x^2} \qquad\qquad \\
-28x^2 - 34x \qquad \\
\underline{(-)\ -28x^2 + 14x} \qquad \\
-48x + 22 \\
\underline{(-)\ -48x + 24} \\
-2
\end{array}
$$

The result is $3x^2 - 14x - 24 - \dfrac{2}{2x - 1}$.

6-3 Polynomial Functions (pp. 348–355)

A2.A.41

State the degree and leading coefficient of each polynomial in one variable. If it is not a polynomial in one variable, explain why.

22. $5x^6 - 3x^4 + x^3 - 9x^2 + 1$

23. $6xy^2 - xy + y^2$

24. $12x^3 - 5x^4 + 6x^8 - 3x - 3$

Find $p(-2)$ and $p(x + h)$ for each function.

25. $p(x) = x^2 + 2x - 3$

26. $p(x) = 3x^2 - x$

27. $p(x) = 3 - 5x^2 + x^3$

EXAMPLE 3

What are the degree and leading coefficient of $\mathbf{4x^3 + 3x^2 - 7x^7 + 4x - 1}$?

The greatest exponent is 7, so the degree is 7. The leading coefficient is -7.

EXAMPLE 4

Find $\mathbf{p(a - 2)}$ if $\mathbf{p(x) = 3x + 2x^2 - x^3}$.

$p(a - 2) = 3(a - 2) + 2(a - 2)^2 - (a - 2)^3$
$= 3a - 6 + 2a^2 - 8a + 8 - (a^3 - 6a^2 + 12a - 8)$
$= -a^3 + 8a^2 - 17a + 10$

MIXED PROBLEM SOLVING
For mixed problem-solving practice, see page 984.

6-4 Analyzing Graphs of Polynomial Functions (pp. 357–364)

A2.A.50

Complete each of the following.

a. **Graph each function by making a table of values.**

b. **Determine the consecutive integer values of x between which each real zero is located.**

c. **Estimate the x-coordinates at which the relative maxima and minima occur.**

28. $h(x) = x^3 - 4x^2 - 7x + 10$

29. $g(x) = 4x^4 - 21x^2 + 5$

30. $f(x) = x^3 - 3x^2 - 4x + 12$

31. $h(x) = 4x^3 - 6x^2 + 1$

32. $p(x) = x^5 - x^4 + 1$

33. **BUSINESS** Milo tracked the monthly profits for his sports store business for the first six months of the year. They can be modeled by using the following six points: (1, 675), (2, 950), (3, 550), (4, 250), (5, 600), and (6, 400). How many turning points would the graph of a polynomial function through these points have? Describe them.

EXAMPLE 5

Graph $f(x) = x^3 + 3x^2 - 4$ by making a table of values.

Make a table of values for several values of x.

x	−3	−2	−1	0	1	2
$f(x)$	−4	0	−2	−4	0	16

Plot the points and connect the points with a smooth curve.

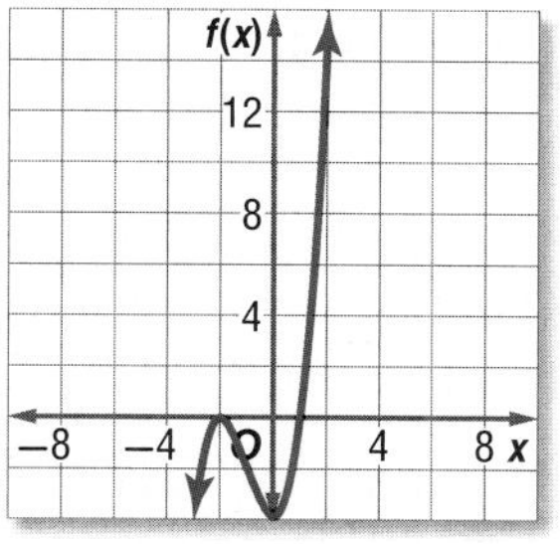

6-5 Solving Polynomial Equations by Factoring (pp. 368–375)

A2.A.26

Factor completely. If the polynomial is not factorable, write *prime*.

34. $a^4 - 16$

35. $4x^3 + 6y^3$

36. $54x^3y - 16y^4$

37. $6ay + 4by - 2cy + 3az + 2bz - cz$

Solve each equation.

38. $x^3 + 2x^2 - 35x = 0$

39. $8x^4 - 10x^2 + 3 = 0$

40. **GEOMETRY** The volume of the prism is 315 cubic inches. Find the value of x and the length, height, and width.

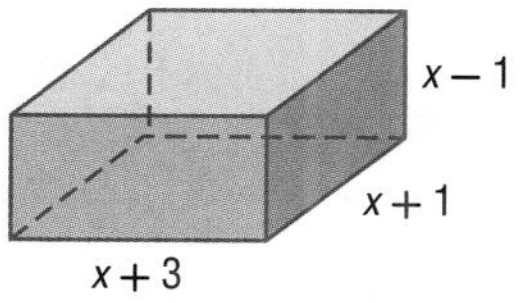

EXAMPLE 6

Factor $r^7 + 64r$.

$r^7 + 64r = r(r^6 + 64)$ — **Factor by GCF.**

$= r[(r^2)^3 + 4^3]$ — **Write as cubes.**

$= r(r^2 + 4)(r^4 - 4r^2 + 16)$

EXAMPLE 7

Solve $4x^4 - 25x^2 + 36 = 0$.

$(x^2 - 4)(4x^2 - 9) = 0$

$x^2 - 4 = 0$ or $4x^2 - 9 = 0$

$x^2 = 4$ $\quad$ $x^2 = \frac{9}{4}$

$x = \pm 2$ $\quad$ $x = \pm\frac{3}{2}$

The solutions are -2, 2, $-\frac{3}{2}$, and $\frac{3}{2}$.

Study Guide and Review

6-6 The Remainder and Factor Theorems (pp. 377–382)

A2.A.26

Use synthetic substitution to find $f(-2)$ and $f(4)$ for each function.

41. $f(x) = x^2 - 3$

42. $f(x) = x^2 - 5x + 4$

43. $f(x) = x^3 + 4x^2 - 3x + 2$

44. $f(x) = 2x^4 - 3x^3 + 1$

Given a polynomial and one of its factors, find the remaining factors of the polynomial.

45. $3x^3 + 20x^2 + 23x - 10;\ x + 5$

46. $2x^3 + 11x^2 + 17x + 5;\ 2x + 5$

47. $x^3 + 2x^2 - 23x - 60;\ x - 5$

EXAMPLE 8

Determine whether $x - 6$ is a factor of $x^3 - 2x^2 - 21x - 18$.

6	1	−2	−21	−18
		6	24	18
	1	4	3	0

$x - 6$ is a factor because $r = 0$.

$x^3 - 2x^2 - 21x - 18 = (x - 6)(x^2 + 4x + 3)$

6-7 Roots and Zeros (pp. 383–390)

A2.A.26

State the possible number of positive real zeros, negative real zeros, and imaginary zeros of each function.

48. $f(x) = -2x^3 + 11x^2 - 3x + 2$

49. $f(x) = -4x^4 - 2x^3 - 12x^2 - x - 23$

50. $f(x) = x^6 - 5x^3 + x^2 + x - 6$

51. $f(x) = -2x^5 + 4x^4 + x^2 - 3$

52. $f(x) = -2x^6 + 4x^4 + x^2 - 3x - 3$

EXAMPLE 9

State the possible number of positive real zeros, negative real zeros, and imaginary zeros of $f(x) = 3x^4 + 2x^3 - 2x^2 - 26x - 48$.

$f(x)$ has one sign change, so there is 1 positive real zero.

$f(-x)$ has 3 sign changes, so there are 3 or 1 negative real zeros.

There are 0 or 2 imaginary zeros.

6-8 Rational Zero Theorem (pp. 391–396)

A2.A.26

Find all of the zeros of each function.

53. $f(x) = x^3 + 4x^2 + 3x - 2$

54. $f(x) = 4x^3 + 4x^2 - x - 1$

55. $f(x) = x^3 + 2x^2 + 4x + 8$

56. STORAGE Melissa is building a storage box that is shaped like a rectangular prism. It will have a volume of 96 cubic feet. Using the diagram below, find the dimensions of the box.

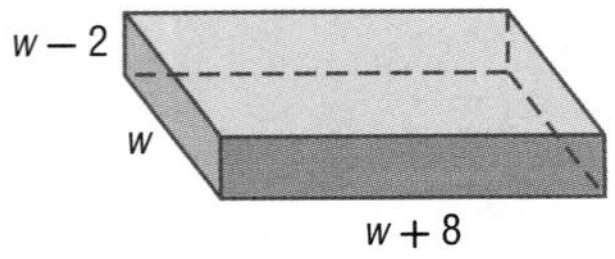

EXAMPLE 10

Find all of the zeros of $f(x) = x^3 + 4x^2 - 11x - 30$.

There are exactly 3 zeros.

There are 1 positive real zero and 2 negative real zeros. The possible rational zeros are $\pm1, \pm2, \pm3, \pm5, \pm6, \pm10, \pm15, \pm30$.

3	1	4	−11	−30
		3	21	30
	1	7	10	0

$$x^3 + 4x^2 - 11x - 30 = (x - 3)(x^2 + 7x + 10) = (x - 3)(x + 2)(x + 5)$$

Thus, the zeros are 3, −2, and −5.

NY Math Online glencoe.com
Chapter Test

Simplify.

1. $(3a)^2(7b)^4$
2. $(7x - 2)(2x + 5)$
3. $(2x^2 + 3x - 4) - (4x^2 - 7x + 1)$
4. $(4x^3 - x^2 + 5x - 4) + (5x - 10)$
5. $(x^4 + 5x^3 + 3x^2 - 8x + 3) \div (x + 3)$
6. $(3x^3 - 5x^2 - 23x + 24) \div (x - 3)$

7. **MULTIPLE CHOICE** How many unique real zeros does the graph have?

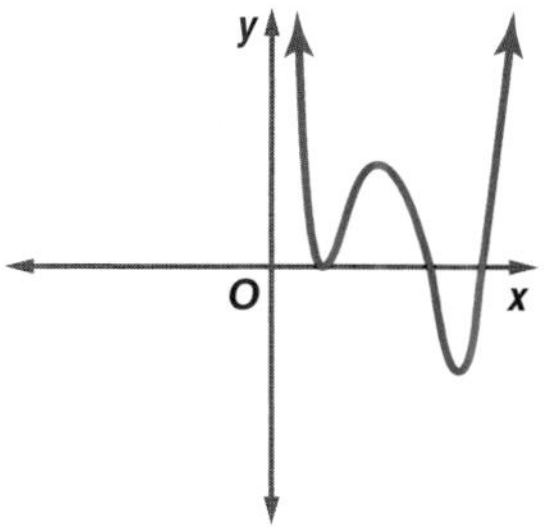

A 0 C 3
B 2 D 5

8. If $c(x) = 3x^3 + 5x^2 - 4$, what is the value of $4c(3b)$?

Complete each of the following.

a. Graph each function by making a table of values.

b. Determine consecutive integer values of x between which each real zero is located.

c. Estimate the x-coordinates at which the relative maxima and relative minima occur.

9. $g(x) = x^3 + 4x^2 - 3x + 1$
10. $h(x) = x^4 - 4x^3 - 3x^2 + 6x + 2$

Factor completely. If the polynomial is not factorable, write *prime*.

11. $8y^4 + x^3y$
12. $2x^2 + 2x + 1$
13. $a^2x + 3ax + 2x - a^2y - 3ay - 2y$

Solve each equation.

14. $8x^3 + 1 = 0$
15. $x^4 - 11x^2 + 28 = 0$

16. **FRAMING** The area of the picture and frame shown below is 168 square inches. What is the width of the frame?

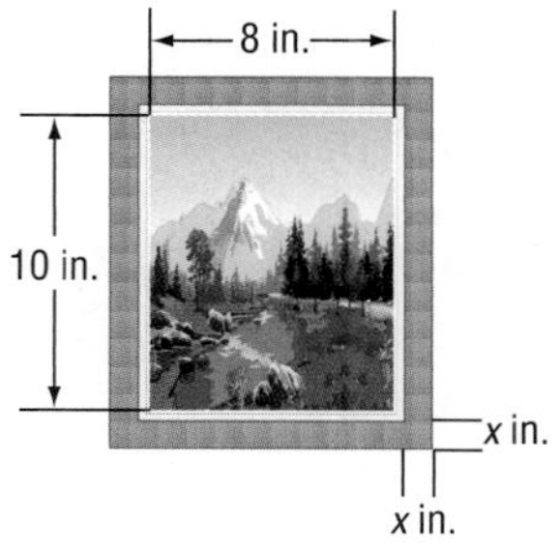

17. **MULTIPLE CHOICE** Let $f(x) = x^4 - 3x^3 + 5x - 3$. Use synthetic substitution to find $f(-2)$.

F 37 H −33
G 27 J −21

Given a polynomial and one of its factors, find the remaining factors of the polynomial.

18. $2x^3 + 15x^2 + 22x - 15; x + 5$
19. $x^3 - 4x^2 + 10x - 12; x - 2$

State the possible number of positive real zeros, negative real zeros, and imaginary zeros of each function.

20. $p(x) = x^3 - x^2 - x - 3$
21. $p(x) = 2x^6 + 5x^4 - x^3 - 5x - 1$

Find all zeros of each function.

22. $p(x) = x^3 - 4x^2 + x + 6$
23. $p(x) = x^3 + 2x^2 + 4x + 8$

24. **GEOMETRY** The volume of the rectangular prism shown is 612 cubic centimeters. Find the dimensions of the prism.

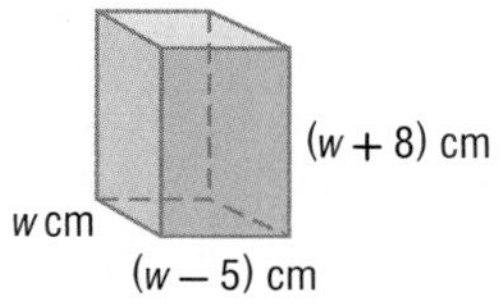

25. List all possible rational zeros of $f(x) = 2x^4 + 3x^2 - 12x + 8$.

CHAPTER 6

Preparing for Standardized Tests

Draw a Picture

Drawing a picture can be a helpful way for you to visualize how to solve a problem. Sketch your picture on scrap paper or in your test booklet (if allowed). Do not make any marks on your answer sheet other than your answers.

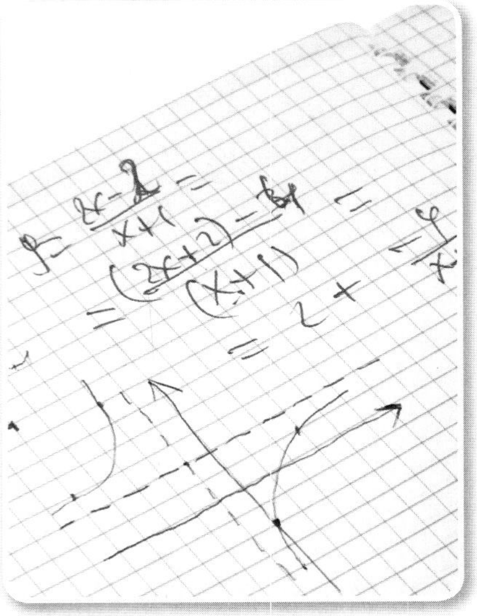

Strategies for Drawing a Picture

Step 1

Read the problem statement carefully.

Ask yourself:

- What am I being asked to solve?
- What information is given in the problem?
- What are the unknowns that I need to model and solve for?

Step 2

Sketch and label your picture.

- Draw your picture as clearly and accurately as possible.
- Label the picture carefully. Be sure to include all of the information given in the problem statement.

Step 3

Solve the problem.

- Use your picture to help you model the problem situation with an equation. Solve the equation.
- Check to be sure your answer makes sense.

EXAMPLE

Read the problem. Identify what you need to know. Then use the information in the problem to solve.

Mr. Nolan has a rectangular swimming pool that measures 25 feet by 14 feet. He wants to have a cement walkway installed around the perimeter of the pool. The combined area of the pool and walkway will be 672 square feet. What will be the width of the walkway?

A 2.75 ft

B 3 ft

C 3.25 ft

D 3.5 ft

Draw a picture to help you visualize the problem situation. Let x represent the unknown width of the cement walkway.

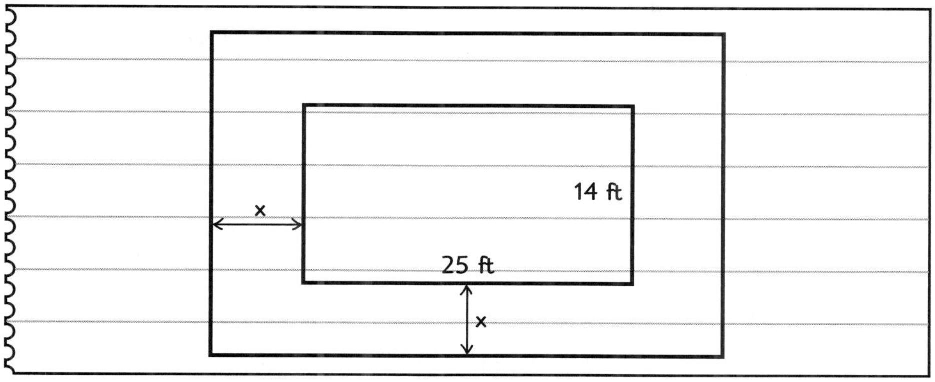

The width of the pool and walkway is $14 + 2x$, and the length is $25 + 2x$. Multiply these polynomial expressions and set the result equal to the combined area, 672 square feet. Then solve for x.

$$(14 + 2x)(25 + 2x) = 672$$
$$350 + 78x + 4x^2 = 672$$
$$4x^2 + 78x - 322 = 0$$
$$x = -23 \text{ or } 3.5$$

Since the width cannot be negative, the width of the walkway will be 3.5 feet. The correct answer is D.

Exercises

Read each problem. Identify what you need to know. Then use the information in the problem to solve.

1. A farmer has 240 feet of fencing that he wants to use to enclose a rectangular area for his chickens. He plans to build the enclosure using the wall of his barn as one of the walls. What is the maximum amount of area he can enclose?

   A 7200 ft^2

   B 4960 ft^2

   C 3600 ft^2

   D 3280 ft^2

2. Metal washers are made by cutting a hole in a circular piece of metal. Suppose a washer is made by removing the center of a piece of metal with a 1.8-inch diameter. What is the radius of the hole if the washer has an area of 0.65π square inches?

   F 0.35 in.

   G 0.38 in.

   H 0.40 in.

   J 0.42 in.

CHAPTER 6

NYS Regents Exam Practice

Cumulative, Chapters 1 through 6

Multiple Choice

Answer all questions in this part. Select the answer that best completes the statement or answers the question.

1. The speed of light is approximately 3×10^8 meters per second. If the distance from the Sun to Earth is about 1.5×10^{11} meters, simplify the expression below to find how long it takes for light to reach Earth from the Sun.

$$\frac{1.5 \times 10^{11}}{3 \times 10^8}$$

(1) 125 seconds **(3)** 500 seconds

(2) 420 seconds **(4)** 1200 seconds

2. What are the solutions of the quadratic equation graphed below?

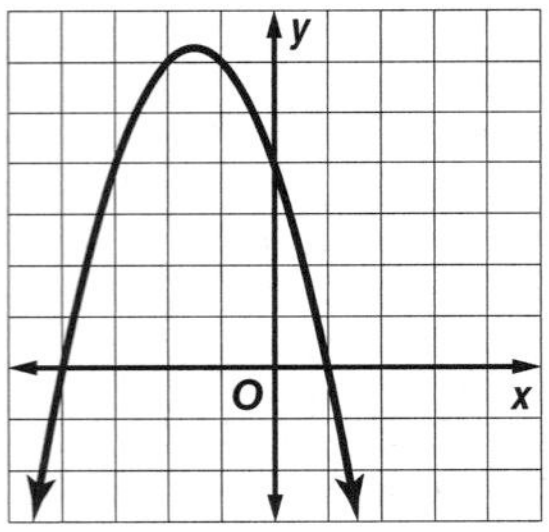

(1) $-4, -1$ **(3)** $-1, 4$

(2) $1, 4$ **(4)** $-4, 1$

3. Which of the following points is *not* a vertex of the feasible region for the system of linear inequalities below?

$$x \geq 0, y \geq 0$$
$$y \leq -3x + 15$$

(1) (0, 0) **(3)** (5, 0)

(2) (0, 5) **(4)** (0, 15)

Test-TakingTip

Question 3 Graph the system of inequalities and find the coordinates of the corners.

4. Simplify the following expression.

$$(3n^2 + 9n - 4) - (n^2 - 1)$$

(1) $3n^2 + 8n - 5$

(2) $3n^2 + 8n - 3$

(3) $2n^2 + 9n - 5$

(4) $2n^2 + 9n - 3$

5. A movie theater charges price x for an adult's ticket and price y for a child's ticket. Two adult tickets and 3 child tickets cost \$35.75. One adult ticket and 2 child tickets cost \$21.00. Which matrix could be multiplied by $\begin{bmatrix} 35.75 \\ 21.00 \end{bmatrix}$ to find x and y?

(1) $\begin{bmatrix} -2 & 3 \\ 1 & -2 \end{bmatrix}$ **(3)** $\begin{bmatrix} 2 & -3 \\ -1 & 2 \end{bmatrix}$

(2) $\begin{bmatrix} \frac{3}{4} & -\frac{3}{4} \\ \frac{1}{2} & -\frac{1}{4} \end{bmatrix}$ **(4)** $\begin{bmatrix} -\frac{1}{4} & \frac{3}{4} \\ \frac{1}{2} & -\frac{1}{2} \end{bmatrix}$

6. How many real zeros does the polynomial function graphed below have?

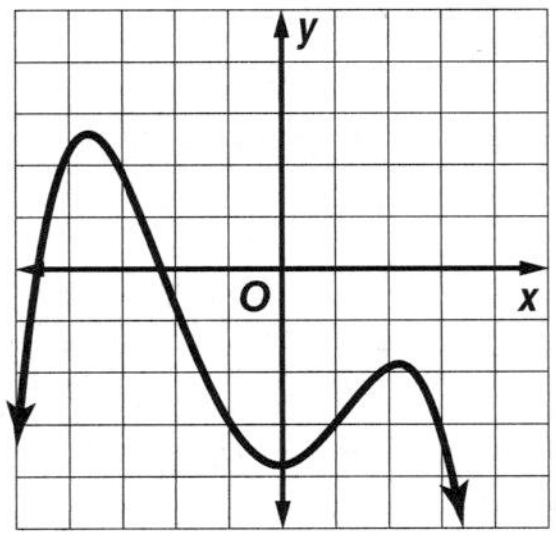

(1) 0

(2) 1

(3) 2

(4) 4

7. The volume of the rectangular prism below is 864 cubic centimeters. Which of the following is ***not*** a measure of one of its sides?

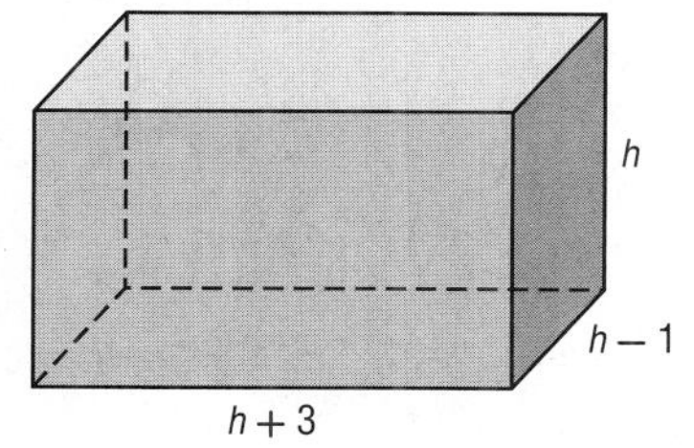

(1) 8 cm

(2) 9 cm

(3) 10 cm

(4) 12 cm

8. Evaluate the piecewise defined function below for $x = -2$.

$$f(x) = \begin{cases} 2x - 5, x \leq -1 \\ 4x + 1, x > -1 \end{cases}$$

(1) -9

(2) -7

(3) -3

(4) -1

9. Which of the following is *not* a solution to the cubic equation below?

$$x^3 - 37x - 84 = 0$$

(1) -4

(2) -3

(3) 6

(4) 7

Open-Ended Response

Solve each problem. Clearly indicate the necessary steps, including appropriate formula substitutions, diagrams, graphs, charts, etc.

10. The population of New York is about 1.9×10^7 people. The state covers a land area of about 4.7×10^4 square miles. Write and simplify an expression for the population density of the state (number of people per square mile). Round your answer to the nearest whole person.

11. Simplify the expression below.

$$-3(5m - 2n) + 2(-m + 5n)$$

12. The height, in feet, of a toy rocket t seconds after it is launched into the air can be modeled by the polynomial function $h(t) = -16t^2 + 96t$. This function is graphed below.

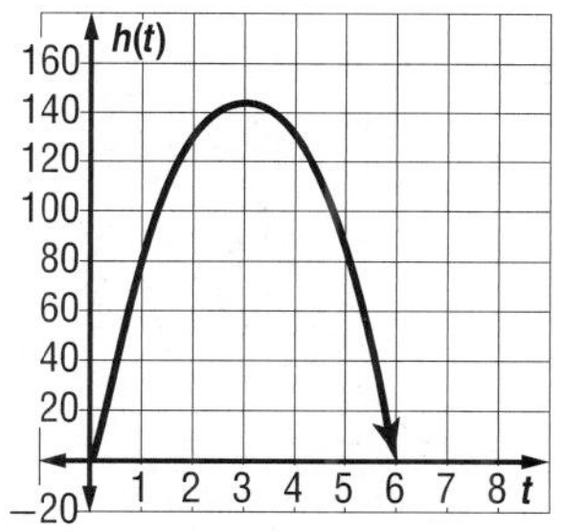

a. What is the vertex of the parabola?

b. What is the real world interpretation of the vertex?

Need Extra Help?

If you missed Question...	1	2	3	4	5	6	7	8	9	10	11	12
Go to Lesson or Page...	6-2	5-7	2-4	6-5	5-5	6-7	3-3	6-3	1-3	6-5	6-5	6-2
NYS Core Curriculum	A2.A.8	A.G.8	A.A.40	A.A.13	A.A.7	A2.A.50	A2.A.26	A2.A.39	A2.A.26	A2.A.8	A.A.13	A2.A.50

CHAPTER 7 Inverses and Radical Functions and Relations

Then

In Chapter 6, you simplified polynomial expressions.

Now

In Chapter 7, you will:

- Find compositions and inverses of functions.
- Graph and analyze square root functions and inequalities.
- Simplify and solve equations involving roots, radicals, and rational exponents.

NYS Core Curriculum

A2.A.45 Determine the inverse of a function and use composition to justify the result
A2.A.13 Simplify radical expressions
A2.A.22 Solve radical equations

Why?

FINANCE Connecting finances to mathematics is a skill that, once mastered, you will use your entire life. Learning to manage your finances entails creating a budget and living within that budget. In this chapter, you will explore financial topics such as saving for college, income, profit, inflation, and converting money when traveling.

Math *in Motion,* Animation glencoe.com

Get Ready for Chapter 7

Diagnose Readiness You have two options for checking Prerequisite Skills.

Text Option

Take the Quick Check below. Refer to the Quick Review for help.

QuickCheck

Use the related graph of each equation to determine its roots. If exact roots cannot be found, state the consecutive integers between which the roots are located. (Lesson 5-2)

1. $x^2 - 4x + 1 = 0$
2. $2x^2 + x - 6 = 0$
3. **PHYSICS** Allie drops a ball from the top of a 30-foot building. How long does it take for the ball to reach the ground, assuming there is no air resistance? Use the formula $h(t) = -16t^2 + h_0$, where t is the time in seconds and the initial height h_0 is in feet.

Simplify each expression by using synthetic division. (Lesson 6-2)

4. $(5x^2 - 22x - 15) \div (x - 5)$
5. $(3x^2 + 14x - 12) \div (x + 4)$
6. $(2x^3 - 7x^2 - 36x + 36) \div (x - 6)$
7. $(3x^4 - 13x^3 + 17x^2 - 18x + 15) \div (x - 3)$
8. **FINANCE** The number of specialty coffee mugs sold at a coffee shop can be estimated by $n = \frac{4000x^2}{x^2 + 50}$, where x is the amount of money spent on advertising in hundreds of dollars and n is the number of mugs sold.
   a. Perform the division indicated by $\frac{4000x^2}{x^2 + 50}$.
   b. About how many mugs will be sold if $1000 is spent on advertising?

QuickReview

EXAMPLE 1

Use the related graph of $0 = 3x^2 - 4x + 1$ to determine its roots. If exact roots cannot be found, state the consecutive integers between which the roots are located.

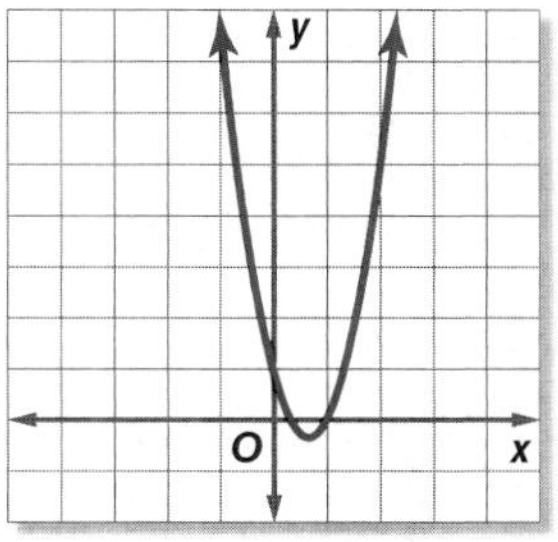

The roots are the x-coordinates where the graph crosses the x-axis.

The graph crosses the x-axis between 0 and 1 and at 1.

EXAMPLE 2

Simplify $(3x^4 + 4x^3 + x^2 + 9x - 6) \div (x + 2)$ by using synthetic division.

$x - r = x + 2$, so $r = -2$.

−2	3	4	1	9	−6
	↓	−6	4	−10	2
	3	−2	5	−1	−4

The result is $3x^3 - 2x^2 + 5x - 1 - \frac{4}{x + 2}$.

Online Option

NY Math Online Take a self-check Chapter Readiness Quiz at glencoe.com.

Get Started on Chapter 7

You will learn several new concepts, skills, and vocabulary terms as you study Chapter 7. To get ready, identify important terms and organize your resources. You may wish to refer to **Chapter 0** to review prerequisite skills.

FOLDABLES® Study Organizer

Inverses and Radical Functions Make this Foldable to help you organize your Chapter 7 notes about radical equations and inequalities. Begin with four sheets of notebook paper.

1. **Stack** three sheets of notebook paper so that each sheet is one inch higher than the sheet in front of it.

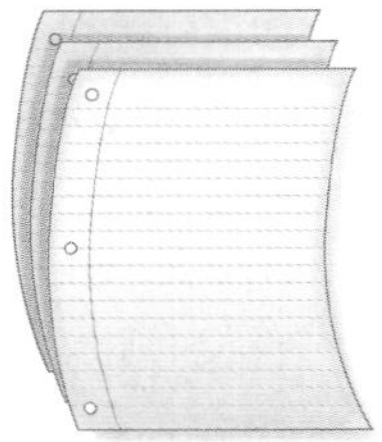

2. **Bring** the bottom of all the sheets upward and align the edges so that all of the layers or tabs are the same distance apart.

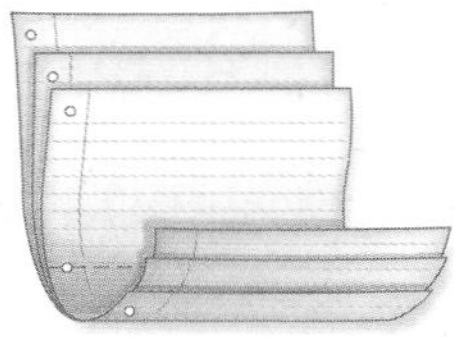

3. **When** all the tabs are an equal distance apart, fold the papers and crease well. Open the papers and staple them together along the valley or inner center fold. Label the pages with lesson titles.

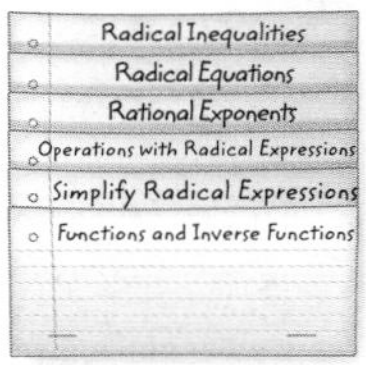

NY Math Online glencoe.com

- Study the chapter online
- Explore **Math in Motion**
- Get extra help from your own **Personal Tutor**
- Use **Extra Examples** for additional help
- Take a **Self-Check Quiz**
- **Review Vocabulary** in fun ways

New Vocabulary

English		Español
composition of functions	p. 411	composición de funciones
inverse relations	p. 418	relaciones inversas
inverse function	p. 418	función inversa
square root function	p. 424	función raíz cuadrada
radical function	p. 424	función radical
square root inequality	p. 426	desigualdad raíz cuadrada
*n*th root	p. 431	raíz enésima
principal root	p. 431	raíz principal
radical sign	p. 431	signo radical
radicand	p. 431	radicando
index	p. 431	índice
rationalizing the denominator	p. 440	racionalizar el denominador
conjugates	p. 442	conjugados
radical equation	p. 453	ecuación radical
extraneous solution	p. 453	solución extraña
radical inequality	p. 455	desigualdad radical

Review Vocabulary

absolute value • p. 432 • valor absoluto a number's distance from zero on the number line, represented by $|x|$

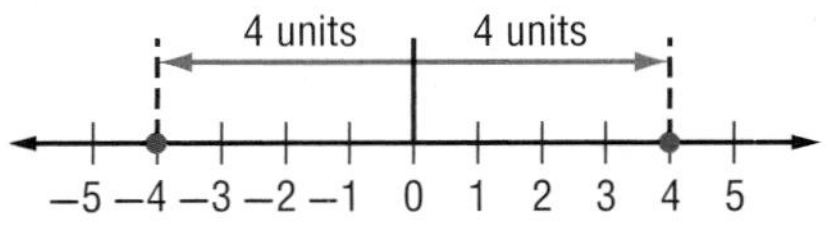

rational number • p. 11 • número racional Any number $\frac{m}{n}$, where m and n are integers and n is not zero; the decimal form is either a terminating or repeating decimal.

relation • p. 418 • relación a set of ordered pairs

Multilingual eGlossary glencoe.com

7-1 Operations on Functions

Then

You performed operations on polynomials. (Lesson 6-1)

Now

- Find the sum, difference, product, and quotient of functions.
- Find the composition of functions.

NYS Core Curriculum

A2.A.42 Find the composition of functions

New Vocabulary

composition of functions

NY Math Online

glencoe.com

- Extra Examples
- Personal Tutor
- Self-Check Quiz
- Homework Help
- Math in Motion

Why?

The graphs model the income for the Brooks family since 2000, where $m(x)$ represents Mr. Brooks' income and $f(x)$ represents Mrs. Brooks' income.

The total household income for the Brooks household can be represented by $f(x) + m(x)$.

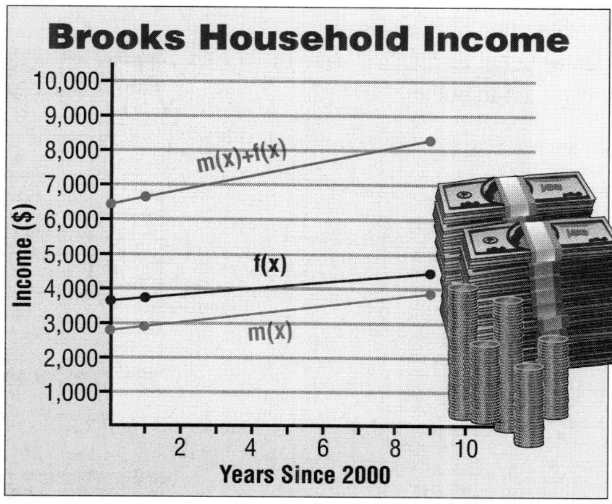

Arithmetic Operations In Chapter 6, you performed arithmetic operations with polynomials. You can also use addition, subtraction, multiplication, and division with functions.

You can perform arithmetic operations according to the following rules.

Key Concept Operations on Functions — For Your FOLDABLE

Operation	Definition	Example Let $f(x) = 2x$ and $g(x) = -x + 5$.
Addition	$(f + g)(x) = f(x) + g(x)$	$2x + (-x + 5) = x + 5$
Subtraction	$(f - g)(x) = f(x) - g(x)$	$2x - (-x + 5) = 3x - 5$
Multiplication	$(f \cdot g)(x) = f(x) \cdot g(x)$	$2x(-x + 5) = -2x^2 + 10x$
Division	$\left(\frac{f}{g}\right)(x) = \frac{f(x)}{g(x)}, g(x) \neq 0$	$\frac{2x}{-x + 5}, x \neq 5$

EXAMPLE 1 Add and Subtract Functions

Given $f(x) = x^2 - 4$ and $g(x) = 2x + 1$, find each function.

a. $(f + g)(x)$

$$\begin{aligned}(f + g)(x) &= f(x) + g(x) && \text{Addition of functions}\\ &= (x^2 - 4) + (2x + 1) && f(x) = x^2 - 4 \text{ and } g(x) = 2x + 1\\ &= x^2 + 2x - 3 && \text{Simplify.}\end{aligned}$$

b. $(f - g)(x)$

$$\begin{aligned}(f - g)(x) &= f(x) - g(x) && \text{Subtraction of functions}\\ &= (x^2 - 4) - (2x + 1) && f(x) = x^2 - 4 \text{ and } g(x) = 2x + 1\\ &= x^2 - 2x - 5 && \text{Simplify.}\end{aligned}$$

Check Your Progress

Given $f(x) = x^2 + 5x - 2$ and $g(x) = 3x - 2$, find each function.

1A. $(f + g)(x)$ **1B.** $(f - g)(x)$

Personal Tutor glencoe.com

Review Vocabulary

intersection the intersection of two sets is the set of elements common to both

You can graph sum and difference functions by graphing each function involved separately, then adding their corresponding functional values. Let $f(x) = x^2$ and $g(x) = x$. Examine the graphs of $f(x)$, $g(x)$, and their sum and difference.

Find $(f + g)(x)$.

x	$f(x) = x^2$	$g(x) = x$	$(f + g)(x) = x^2 + x$
−3	9	−3	$9 + (-3) = 6$
−2	4	−2	$4 + (-2) = 2$
−1	1	−1	$1 + (-1) = 0$
0	0	0	$0 + 0 = 0$
1	1	1	$1 + 1 = 2$
2	4	2	$4 + 2 = 6$
3	9	3	$9 + 3 = 12$

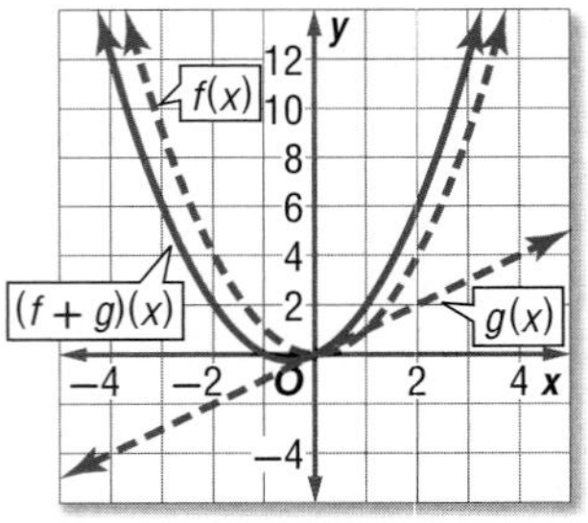

Find $(f - g)(x)$.

x	$f(x) = x^2$	$g(x) = x$	$(f - g)(x) = x^2 + x$
−3	9	−3	$9 - (-3) = 12$
−2	4	−2	$4 - (-2) = 6$
−1	1	−1	$1 - (-1) = 2$
0	0	0	$0 - 0 = 0$
1	1	1	$1 - 1 = 0$
2	4	2	$4 - 2 = 2$
3	9	3	$9 - 3 = 6$

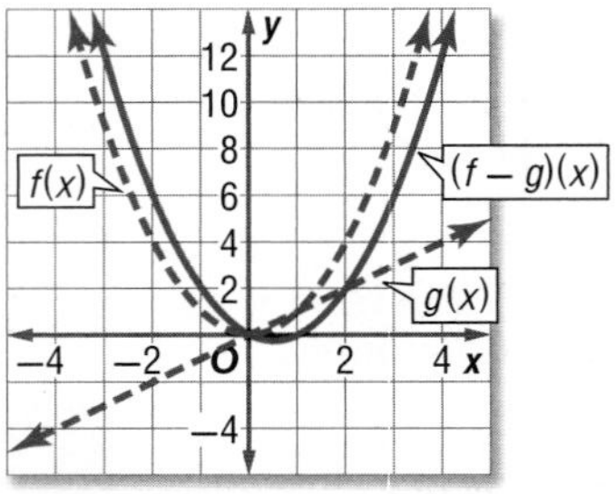

In Example 1, the functions $f(x)$ and $g(x)$ have the same domain of all real numbers. The functions $(f + g)(x)$ and $(f - g)(x)$ also have domains that include all real numbers. For each new function, the domain consists of the intersection of the domains of $f(x)$ and $g(x)$. Under division, the domain of the new function is restricted by excluded values that cause the denominator to equal zero.

EXAMPLE 2 Multiply and Divide Functions

Given $f(x) = x^2 + 7x + 12$ and $g(x) = 3x - 4$, find each function.

a. $(f \cdot g)(x)$

$$\begin{aligned}(f \cdot g)(x) &= f(x) \cdot g(x) && \textbf{Multiplication of functions}\\ &= (x^2 + 7x + 12)(3x - 4) && \textbf{Substitution}\\ &= 3x^3 + 21x^2 + 36x - 4x^2 - 28x - 48 && \textbf{Distributive Property}\\ &= 3x^3 + 17x^2 + 8x - 48 && \textbf{Simplify.}\end{aligned}$$

b. $\left(\frac{f}{g}\right)(x)$

$$\begin{aligned}\left(\frac{f}{g}\right)(x) &= \frac{f(x)}{g(x)} && \textbf{Division of functions}\\ &= \frac{x^2 + 7x + 12}{-3x - 4},\ x \neq -\frac{4}{3} && \textbf{Substitution}\end{aligned}$$

Because $x = -\frac{4}{3}$ makes the denominator $-3x - 4 = 0$, $-\frac{4}{3}$ is excluded from the domain of $\left(\frac{f}{g}\right)(x)$.

Check Your Progress

Given $f(x) = x^2 - 7x + 2$ and $g(x) = x + 4$, find each function.

2A. $(f \cdot g)(x)$

2B. $\left(\frac{f}{g}\right)(x)$

Personal Tutor glencoe.com

Composition of Functions Another method used to combine functions is a composition of functions. In a **composition of functions**, the results of one function are used to evaluate a second function.

ReadingMath

Composite Functions The composition of f and g, denoted by $f \circ g$ or $f[g(x)]$, is read f of g.

Key Concept — Composition of Functions

For Your FOLDABLE

Words Suppose f and g are functions such that the range of g is a subset of the domain of f. Then the composition function $f \circ g$ can be described by

$$[f \circ g](x) = f[g(x)].$$

Model

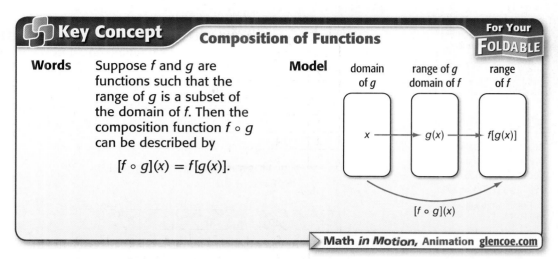

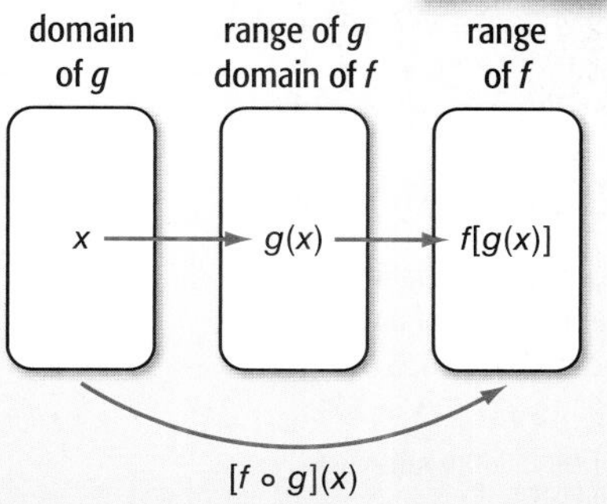

Math *in Motion*, Animation glencoe.com

The composition of two functions may not exist. Given two functions f and g, $[f \circ g](x)$ is defined only if the range of $g(x)$ is a subset of the domain of f. Likewise, $[g \circ f](x)$ is defined only if the range of $f(x)$ is a subset of the domain of g.

EXAMPLE 3 Compose Functions

For each pair of functions, find $[f \circ g](x)$ and $[g \circ f](x)$, if they exist.

a. $f = \{(1, 8), (0, 13), (15, 11), (14, 9)\}$, $g = \{(8, 15), (5, 1), (10, 14), (9, 0)\}$

To find $f \circ g$, evaluate $g(x)$ first. Then use the range as the domain of f and evaluate $f(x)$.

$f[g(8)] = f(15)$ or 11   **$g(8) = 15$**

$f[g(5)] = f(1)$ or 8   **$g(5) = 1$**

$f[g(10)] = f(14)$ or 9   **$g(10) = 14$**

$f[g(9)] = f(0)$ or 13   **$g(9) = 0$**

$f \circ g = \{(8, 11), (5, 8), (10, 9), (9, 13)\}$

To find $g \circ f$, evaluate $f(x)$ first. Then use the range as the domain of g and evaluate $g(x)$.

$g[f(1)] = g(8)$ or 15   **$f(1) = 8$**

$g[f(0)] = g(13)$ **$g(13)$ is undefined.**

$g[f(15)] = g(11)$ **$g(11)$ is undefined.**

$g[f(14)] = g(9)$ or 0   **$f(14) = 0$**

Because 11 and 13 are not in the domain of g, $g \circ f$ is undefined for $x = 11$ and $x = 13$. However, $g[f(1)] = 15$ and $g[f(14)] = 0$, so $g \circ f = \{(1, 15), (14, 0)\}$.

b. $f(x) = 2a - 5$, $g(x) = 4a$

$[f \circ g](x) = f[g(x)]$	**Composition of functions**	$[g \circ f](x) = g[f(x)]$
$= f(4a)$	**Substitute.**	$= g(2a - 5)$
$= 2(4a) - 5$	**Substitute.**	$= 4(2a - 5)$
$= 8a - 5$	**Simplify.**	$= 8a - 20$

StudyTip

Composition Be careful not to confuse a composition $f[g(x)]$ with multiplication of functions $(f \cdot g)(x)$.

Check Your Progress

3A. $f(x) = \{(3, -2), (-1, -5), (4, 7), (10, 8)\}$, $g(x) = \{(4, 3), (2, -1), (9, 4), (3, 10)\}$

3B. $f(x) = x^2 + 2$ and $g(x) = x - 6$

Personal Tutor glencoe.com

Notice that in most cases, $f \circ g \neq g \circ f$. Therefore, the order in which two functions are composed is important.

Real-World Link

Adjusted for inflation, the average price of a new car declined from $23,014 in 1995 to $22,013 in 2005.

Source: U.S. Department of Commerce

Real-World EXAMPLE 4 Use Composition of Functions

SHOPPING A new car dealer is discounting all new cars by 12%. At the same time, the manufacturer is offering a $1500 rebate on all new cars. Mr. Navarro is buying a car that is priced $24,500. Will the final price be lower if the discount is applied before the rebate or if the rebate is applied before the discount?

Understand Let x represent the original price of a new car, $d(x)$ represent the price of a car after the discount, and $r(x)$ the price of the car after the rebate.

Plan Write equations for $d(x)$ and $r(x)$.

The original price is discounted by 12%.
$d(x) = x - 0.12x$

There is a $1500 rebate on all new cars.
$r(x) = x - 1500$

Solve If the discount is applied *before* the rebate, then the final price of Mr. Navarro's new car is represented by $[r \circ d](24{,}500)$.

$[r \circ d](x) = r[d(x)]$

$$\begin{aligned}[r \circ d](24{,}500) &= r[24{,}500 - 0.12(24{,}500)] \\ &= r(24{,}500 - 2940) \\ &= r(21{,}560) \\ &= 21{,}560 - 1500 \\ &= 20{,}060\end{aligned}$$

If the rebate is given *before* the discount is applied, then the final price of Mr. Navarro's car is represented by $[d \circ r](24{,}500)$.

$[d \circ r](x) = d[r(x)]$

$$\begin{aligned}[d \circ r](24{,}500) &= d(24{,}500 - 1500) \\ &= d(23{,}000) \\ &= 23{,}000 - 0.12(23{,}000) \\ &= 23{,}000 - 2760 \\ &= 20{,}240\end{aligned}$$

$[r \circ d](24{,}500) = 20{,}060$ and $[d \circ r](24{,}500) = 20{,}240$. So, the final price of the car is less when the discount is applied before the rebate.

Check The answer seems reasonable because the 12% discount is being applied to a greater amount. Thus, the dollar amount of the discount is greater.

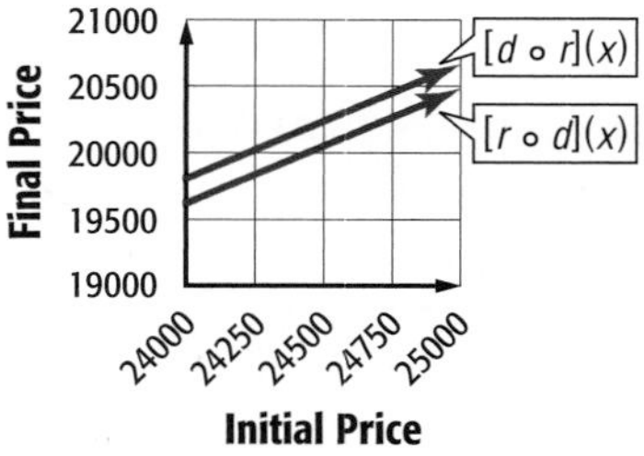

Check Your Progress

4. SHOPPING Sounds-to-Go offers both an in-store $35 rebate and a 15% discount on a digital audio player that normally sells for $300. Which provides the better price: taking the discount before the rebate or taking the discount after the rebate?

Personal Tutor glencoe.com

Check Your Understanding

Examples 1 and 2 pp. 409–410

Find $(f + g)(x)$, $(f - g)(x)$, $(f \cdot g)(x)$, and $\left(\frac{f}{g}\right)(x)$ for each $f(x)$ and $g(x)$.

1. $f(x) = x + 2$
$g(x) = 3x - 1$

2. $f(x) = x^2 - 5$
$g(x) = -x + 8$

Example 3 p. 411

For each pair of functions, find $f \circ g$ and $g \circ f$, if they exist.

3. $f = \{(2, 5), (6, 10), (12, 9), (7, 6)\}$
$g = \{(9, 11), (6, 15), (10, 13), (5, 8)\}$

4. $f = \{(-5, 4), (14, 8), (12, 1), (0, -3)\}$
$g = \{(-2, -4), (-3, 2), (-1, 4), (5, -6)\}$

Find $[f \circ g](x)$ and $[g \circ f](x)$, if they exist.

5. $f(x) = -3x$
$g(x) = 5x - 6$

6. $f(x) = x + 4$
$g(x) = x^2 + 3x - 10$

Example 4 p. 412

7. **FINANCE** Dora has 8% of her earnings deducted from her paycheck for a college savings plan. She can choose to take the deduction either before taxes are withheld, which reduces her taxable income, or after taxes are withheld. Dora's tax rate is 17.5%. If her pay before taxes and deductions is \$950, will she save more money if the deductions are taken before or after taxes are withheld? Explain.

Practice and Problem Solving

● = **Step-by-Step Solutions** begin on page R20.
Extra Practice begins on page 947.

Examples 1 and 2 pp. 409–410

Find $(f + g)(x)$, $(f - g)(x)$, $(f \cdot g)(x)$, and $\left(\frac{f}{g}\right)(x)$ for each $f(x)$ and $g(x)$.

8. $f(x) = 2x$
$g(x) = -4x + 5$

9. $f(x) = x - 1$
$g(x) = 5x - 2$

10. $f(x) = x^2$
$g(x) = -x + 1$

11 $f(x) = 3x$
$g(x) = -2x + 6$

12. $f(x) = x - 2$
$g(x) = 2x - 7$

13. $f(x) = x^2$
$g(x) = x - 5$

14. $f(x) = -x^2 + 6$
$g(x) = 2x^2 + 3x - 5$

15. $f(x) = 3x^2 - 4$
$g(x) = x^2 - 8x + 4$

16. **WALKING** Isaac is walking on a moving walkway. His speed is given by the function $I(x) = 3x - 4$, and the speed of the walkway is $W(x) = 4x + 7$, where x is time in seconds.

a. What is his total speed as he walks along the moving walkway?

b. Isaac turned around because he left his cell phone at a restaurant. What was his speed as he walked against the moving walkway?

Example 3 p. 411

For each pair of functions, find $f \circ g$ and $g \circ f$, if they exist.

17. $f = \{(-8, -4), (0, 4), (2, 6), (-6, -2)\}$
$g = \{(4, -4), (-2, -1), (-4, 0), (6, -5)\}$

18. $f = \{(-7, 0), (4, 5), (8, 12), (-3, 6)\}$
$g = \{(6, 8), (-12, -5), (0, 5), (5, 1)\}$

19. $f = \{(5, 13), (-4, -2), (-8, -11), (3, 1)\}$
$g = \{(-8, 2), (-4, 1), (3, -3), (5, 7)\}$

20. $f = \{(-4, -14), (0, -6), (-6, -18), (2, -2)\}$
$g = \{(-6, 1), (-18, 13), (-14, 9), (-2, -3)\}$

For each pair of functions, find $f \circ g$ and $g \circ f$, if they exist.

21. $f = \{(-15, -5), (-4, 12), (1, 7), (3, 9)\}$
$g = \{(3, -9), (7, 2), (8, -6), (12, 0)\}$

22. $f = \{(-1, 11), (2, -2), (5, -7), (4, -4)\}$
$g = \{(5, -4), (4, -3), (-1, 2), (2, 3)\}$

23. $f = \{(7, -3), (-10, -3), (-7, -8), (-3, 6)\}$
$g = \{(4, -3), (3, -7), (9, 8), (-4, -4)\}$

24. $f = \{(1, -1), (2, -2), (3, -3), (4, -4)\}$
$g = \{(1, -4), (2, -3), (3, -2), (4, -1)\}$

25. $f = \{(-4, -1), (-2, 6), (-1, 10), (4, 11)\}$
$g = \{(-1, 5), (3, -4), (6, 4), (10, 8)\}$

26. $f = \{(12, -3), (9, -2), (8, -1), (6, 3)\}$
$g = \{(-1, 5), (-2, 6), (-3, -1), (-4, 8)\}$

Find $[f \circ g](x)$ and $[g \circ f](x)$, if they exist.

27 $f(x) = 2x$
$g(x) = x + 5$

28. $f(x) = -3x$
$g(x) = -x + 8$

29. $f(x) = x + 5$
$g(x) = 3x - 7$

30. $f(x) = x - 4$
$g(x) = x^2 - 10$

31. $f(x) = x^2 + 6x - 2$
$g(x) = x - 6$

32. $f(x) = 2x^2 - x + 1$
$g(x) = 4x + 3$

33. $f(x) = 4x - 1$
$g(x) = x^3 + 2$

34. $f(x) = x^2 + 3x + 1$
$g(x) = x^2$

35. $f(x) = 2x^2$
$g(x) = 8x^2 + 3x$

36. **FINANCE** A ceramics store manufactures and sells coffee mugs. The revenue $r(x)$ from the sale of x coffee mugs is given by $r(x) = 6.5x$. Suppose the function for the cost of manufacturing x coffee mugs is $c(x) = 0.75x + 1850$.

a. Write the profit function.

b. Find the profit on 500, 1000, and 5000 coffee mugs.

Real-World Link

The main distinguishing characteristic of HDTV is the exclusive use of the 16:9 aspect ratio to deliver a widescreen viewing experience that more closely resembles the human field of vision.

Source: TV Authority

37. **SHOPPING** Ms. Smith wants to buy an HDTV, which is on sale for 35% off the original price of \$2299. The sales tax is 6.25%.

a. Write two functions representing the price after the discount $p(x)$ and the price after sales tax $t(x)$.

b. Which composition of functions represents the price of the HDTV, $[p \circ t](x)$ or $[t \circ p](x)$? Explain your reasoning.

c. How much will Ms. Smith pay for the HDTV?

Perform each operation if $f(x) = x^2 + x - 12$ and $g(x) = x - 3$. State the domain of the resulting function.

38. $(f - g)(x)$

39. $2(g \cdot f)(x)$

40. $\left(\frac{f}{g}\right)(x)$

If $f(x) = 5x$, $g(x) = -2x + 1$, and $h(x) = x^2 + 6x + 8$, find each value.

41. $f[g(-2)]$

42. $g[h(3)]$

43. $h[f(-5)]$

44. $h[g(2)]$

45. $f[h(-3)]$

46. $h[f(9)]$

47. $f[g(3a)]$

48. $f[h(a + 4)]$

49. $g[f(a^2 - a)]$

50. **MULTIPLE REPRESENTATIONS** Let $f(x) = x^2$ and $g(x) = x$.

a. **TABULAR** Make a table showing values for $f(x)$, $g(x)$, $(f + g)(x)$, and $(f - g)(x)$.

b. **GRAPHICAL** Graph $f(x)$, $g(x)$, and $(f + g)(x)$ on the same coordinate grid.

c. **GRAPHICAL** Graph $f(x)$, $g(x)$, and $(f - g)(x)$ on the same coordinate grid.

d. **VERBAL** Describe the relationship among the graphs of $f(x)$, $g(x)$, $(f + g)(x)$, and $(f - g)(x)$.

Real-World Link

Recently, 21.7 million, or about 58% of the 16- to 24-year-olds, were employed in the United States.

Source: U.S. Bureau of Labor and Statistics

51. **EMPLOYMENT** The number of women and men age 16 and over employed each year in the United States can be modeled by the following equations, where x is the number of years since 1994 and y is the number of people in thousands.
women: $y = 1086.4x + 56{,}610$
men: $y = 999.2x + 66{,}450$

a. Write a function that models the total number of men and women employed in the United States during this time.

b. If f is the function for the number of men, and g is the function for the number of women, what does $(f - g)(x)$ represent?

If $f(x) = x + 2$, $g(x) = -4x + 3$, and $h(x) = x^2 - 2x + 1$, find each value.

52. $(f \cdot g \cdot h)(3)$ **53.** $[(f + g) \cdot h](1)$ **54.** $\left(\frac{h}{fg}\right)(-6)$

55. $[f \circ (g \circ h)](2)$ **56.** $[g \circ (h \circ f)](-4)$ **57.** $[h \circ (f \circ g)](5)$

58. **MULTIPLE REPRESENTATIONS** You will explore $(f \cdot g)$ and $\left(\frac{f}{g}\right)$ if $f(x) = x^2 + 1$ and $g(x) = x - 3$.

a. **TABULAR** Copy and complete the table.

b. **GRAPHICAL** Graph $(f \cdot g)$ and $\left(\frac{f}{g}\right)$ on the same coordinate grid.

c. **VERBAL** Explain the relationship between $(f \cdot g)$ and $\left(\frac{f}{g}\right)$.

x	$(f \cdot g)(x)$	$\left(\frac{f}{g}\right)(x)$
1		
2		
3		
4		
5		

H.O.T. Problems

Use **H**igher-**O**rder **T**hinking Skills

59. **OPEN ENDED** Write two functions $f(x)$ and $g(x)$ such that $(f \circ g)(4) = 0$.

60. **FIND THE ERROR** Chris and Tobias are finding the composition $(f \circ g)(x)$, where $f(x) = x^2 + 2x - 8$ and $g(x) = x^2 + 8$. Is either of them correct? Explain your reasoning.

Chris

$(f \circ g)(x) = f[g(x)]$
$= (x^2 + 8)^2 + 2x - 8$
$= x^4 + 16x^2 + 64 + 2x - 8$
$= x^4 + 16x^2 + 2x + 58$

Tobias

$(f \circ g)(x) = f[g(x)]$
$= (x^2 + 8)^2 + 2(x^2 + 8) - 8$
$= x^4 + 16x^2 + 64 + 2x^2 + 16 - 8$
$= x^4 + 18x^2 + 72$

61. **CHALLENGE** Given $f(x) = \sqrt{x^3}$ and $g(x) = \sqrt{x^6}$, determine the restrictions on the domain for the following.

a. $g(x) \cdot g(x)$ **b.** $f(x) \cdot f(x)$

62. **REASONING** State whether each statement is *sometimes*, *always*, or *never* true. Explain your reasoning.

a. The domain of two functions $f(x)$ and $g(x)$ that are composed $g[f(x)]$ is restricted by the domain of $f(x)$.

b. The domain of two functions $f(x)$ and $g(x)$ that are composed $g[f(x)]$ is restricted by the domain of $g(x)$.

63. **WRITING IN MATH** Explain why a person would perform a composition of functions. Include a real-world example that you could solve by using composition of functions.

NYSRE PRACTICE A2.A.42

64. What is the value of x in the equation $7(x-4) = 44 - 11x$?

A 1
B 2
C 3
D 4

65. If $g(x) = x^2 + 9x + 21$ and $h(x) = 2(x+5)^2$, which is an equivalent form of $h(x) - g(x)$?

F $k(x) = -x^2 - 11x - 29$
G $k(x) = x^2 + 11x + 29$
H $k(x) = x + 4$
J $k(x) = x^2 + 7x + 11$

66. SHORT RESPONSE In his first three years of coaching basketball at North High School, Coach Lucas' team won 8 games the first year, 17 games the second year, and 6 games the third year. How many games does the team need to win in the fourth year so the coach's average will be 10 wins per year?

67. ACT/SAT What is the value of $f[g(6)]$ if $f(x) = 2x + 4$ and $g(x) = x^2 + 5$?

A 38
B 43
C 86
D 261

Spiral Review

Find all rational zeros of each function. (Lesson 6-8)

68. $f(x) = 2x^3 - 13x^2 + 17x + 12$

69. $f(x) = x^3 - 3x^2 - 10x + 24$

70. $f(x) = x^4 - 4x^3 - 7x^2 + 34x - 24$

71. $f(x) = 2x^3 - 5x^2 - 28x + 15$

State the possible number of positive real zeros, negative real zeros, and imaginary zeros of each function. (Lesson 6-7)

72. $f(x) = 2x^4 - x^3 + 5x^2 + 3x - 9$

73. $f(x) = -4x^4 - x^2 - x + 1$

74. $f(x) = 3x^4 - x^3 + 8x^2 + x - 7$

75. $f(x) = 2x^4 - 3x^3 - 2x^2 + 3$

76. MANUFACTURING A box measures 12 inches by 16 inches by 18 inches. The manufacturer will increase each dimension of the box by the same number of inches and have a new volume of 5985 cubic inches. How much should be added to each dimension? (Lesson 6-7)

Solve each system of equations. (Lesson 3-5)

77. $x + 4y - z = 6$
$3x + 2y + 3z = 16$
$2x - y + z = 3$

78. $2a + b - c = 5$
$a - b + 3c = 9$
$3a - 6c = 6$

79. $y + z = 4$
$2x + 4y - z = -3$
$3y = -3$

80. INTERNET A webmaster estimates that the time, in seconds, to connect to the server when n people are connecting is given by $t(n) = 0.005n + 0.3$. Estimate the time to connect when 50 people are connecting. (Lesson 2-2)

Skills Review

Solve each equation or formula for the specified variable. (Lesson 1-3)

81. $5x - 7y = 12$, for x

82. $3x^2 - 6xy + 1 = 4$, for y

83. $4x + 8yz = 15$, for x

84. $D = mv$, for m

85. $A = k^2 + b$, for k

86. $(x+2)^2 - (y+5)^2 = 4$, for y

7-2 Inverse Functions and Relations

Then

You transformed and solved equations for a specific variable. (Lesson 1-3)

Now

- Find the inverse of a function or relation.
- Determine whether two functions or relations are inverses.

NYS Core Curriculum

A2.A.44 Define the inverse of a function
A2.A.45 Determine the inverse of a function and use composition to justify the result

New Vocabulary

inverse relation
inverse function

NY Math Online

glencoe.com
- Extra Examples
- Personal Tutor
- Self-Check Quiz
- Homework Help

Why?

The table shows the value of $1 (U.S.) compared to Canadian dollars and Mexican pesos.

The equation $p = 10.75d$ represents the number of pesos p you can receive for every U.S. dollar d. To determine how many U.S. dollars you can receive for one Mexican peso, solve the equation $p = 10.75d$ for d. The result, $d \approx 0.09p$, is the inverse function.

	U.S.	Canada	Mexico
U.S.		1.05	10.75
Canada	0.95		10.26
Mexico	0.09	0.10	

Find Inverses Recall that a relation is a set of ordered pairs. The **inverse relation** is the set of ordered pairs obtained by exchanging the coordinates of each ordered pair. The domain of a relation becomes the range of its inverse, and the range of the relation becomes the domain of its inverse.

Key Concept — Inverse Relations (For Your FOLDABLE)

Words Two relations are inverse relations if and only if whenever one relation contains the element (a, b), the other relation contains the element (b, a).

Example A and B are inverse relations.

$A = \{(1, 5), (2, 6), (3, 7)\}$ $B = \{(5, 1), (6, 2), (7, 3)\}$

EXAMPLE 1 Find an Inverse Relation

GEOMETRY The vertices of $\triangle ABC$ can be represented by the relation $\{(1, -2), (2, 5), (4, -1)\}$. Find the inverse of this relation. Describe the graph of the inverse.

Graph the relation. To find the inverse, exchange the coordinates of the ordered pairs. The inverse of the relation is $\{(-2, 1), (5, 2), (-1, 4)\}$.

Plotting these points shows that the ordered pairs describe the vertices of $\triangle A'B'C'$ after a reflection in the line $y = x$.

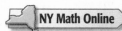

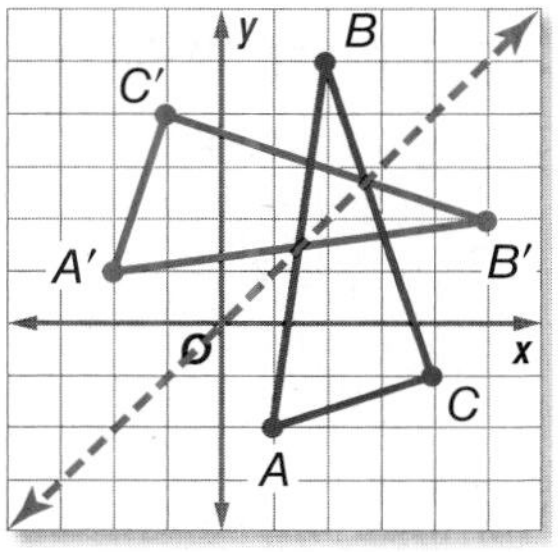

Check Your Progress

1. **GEOMETRY** The ordered pairs of the relation $\{(-8, -3), (-8, -6), (-3, -6)\}$ are the coordinates of the vertices of a right triangle. Find the inverse of this relation. Describe the graph of the inverse.

Personal Tutor glencoe.com

As with relations, the ordered pairs of **inverse functions** are also related. We can write the inverse of the function $f(x)$ as $f^{-1}(x)$.

Key Concept — Property of Inverses

For Your FOLDABLE

Words If f and f^{-1} are inverses, then $f(a) = b$ if and only if $f^{-1}(b) = a$.

Example Let $f(x) = x - 4$ and represent its inverse as $f^{-1}(x) = x + 4$.

Evaluate $f(6)$.
$f(x) = x - 4$
$f(6) = 6 - 4$ or 2

Evaluate $f^{-1}(2)$.
$f^{-1}(x) = x + 4$
$f^{-1}(2) = 2 + 4$ or 6

Because $f(x)$ and $f^{-1}(x)$ are inverses, $f(6) = 2$ and $f^{-1}(2) = 6$.

ReadingMath

Inverse Functions f^{-1} is read *f inverse* or *the inverse of f*. Note that -1 is *not* an exponent.

When the inverse of a function is a function, the original function is one-to-one. Recall that the vertical line test can be used to determine whether a relation is a function. Similarly, the *horizontal line test* can be used to determine whether the inverse of a function is also a function.

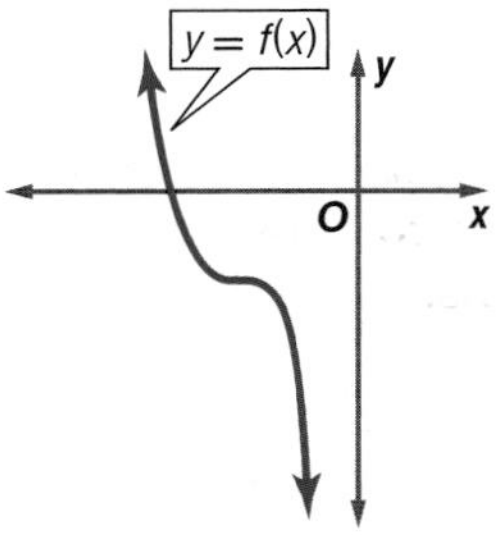

No horizontal line can be drawn so that it passes through more than one point. The inverse of $y = f(x)$ is a function.

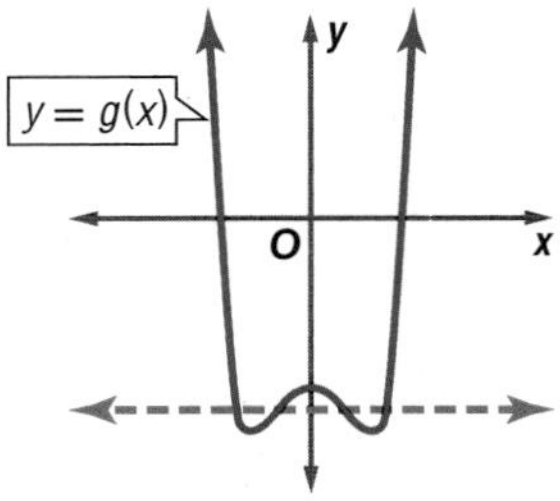

A horizontal line can be drawn that passes through more than one point. The inverse of $y = g(x)$ is not a function.

The inverse of a function can be found by exchanging the domain and the range.

EXAMPLE 2 Find and Graph an Inverse

Find the inverse of each function. Then graph the function and its inverse.

a. $f(x) = 2x - 5$

Step 1 Rewrite the function as an equation relating x and y.
$f(x) = 2x - 5 \rightarrow y = 2x - 5$

Step 2 Exchange x and y in the equation. $x = 2y - 5$

Step 3 Solve the equation for y.

$x = 2y - 5$	**Inverse of $y = 2x - 5$**
$x + 5 = 2y$	**Add 5 to each side.**
$\frac{x+5}{2} = y$	**Divide each side by 2.**

Step 4 Replace y with $f^{-1}(x)$.
$y = \frac{x+5}{2} \rightarrow f^{-1}(x) = \frac{x+5}{2}$

The inverse of $f(x) = 2x - 5$ is $f^{-1}(x) = \frac{x+5}{2}$.
The graph of $f^{-1}(x) = \frac{x+5}{2}$ is the reflection of the graph of $f(x) = 2x - 5$ in the line $y = x$.

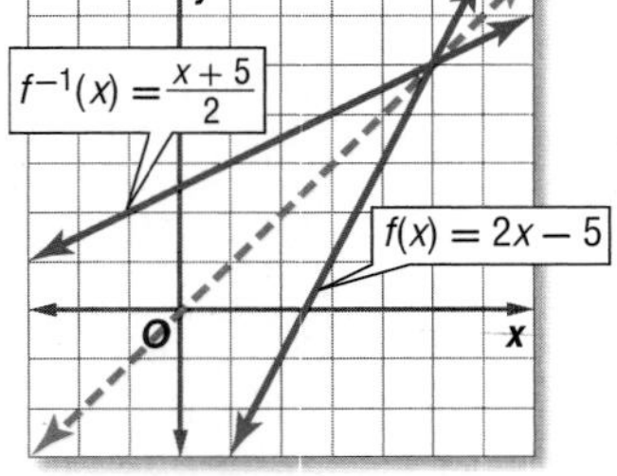

StudyTip

Functions The inverse of the function in part **b** is not a function since it does not pass the vertical line test.

b. $f(x) = x^2 + 1$

Step 1 $f(x) = x^2 + 1 \rightarrow y = x^2 + 1$

Step 2 $x = y^2 + 1$

Step 3
$$x = y^2 + 1$$
$$x - 1 = y^2$$
$$\pm\sqrt{x - 1} = y$$
Take the square root of each side.

Step 4 $y = \pm\sqrt{x - 1} \rightarrow f^{-1}(x) = \pm\sqrt{x - 1}$

Graph $f^{-1}(x) = \pm\sqrt{x - 1}$ by reflecting the graph of $f(x) = x^2 + 1$ in the line $y = x$.

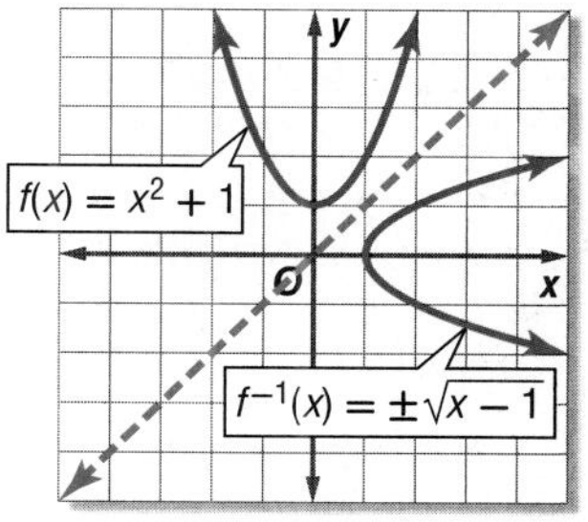

Check Your Progress

Find the inverse of each function. Then graph the function and its inverse.

2A. $f(x) = \frac{x - 3}{5}$ **2B.** $f(x) = 3x^2$

Personal Tutor glencoe.com

Review Vocabulary

identity function the function $f(x) = x$ (Lesson 2-7)

Verifying Inverses You can determine whether two functions are inverses by finding both of their compositions. If both compositions equal the identity function $I(x) = x$, then the functions are inverse functions.

Key Concept Inverse Functions

For Your FOLDABLE

Words Two functions f and g are inverse functions if and only if both of their compositions are the identity function.

Symbols $f(x)$ and $g(x)$ are inverses if and only if $[f \circ g](x) = x$ and $[g \circ f](x) = x$.

EXAMPLE 3 Verify that Two Functions are Inverses

Determine whether each pair of functions are inverse functions. Explain your reasoning.

Watch Out!

Inverse Functions Be sure to check both $[f \circ g](x)$ and $[g \circ f](x)$ to verify that functions are inverses. By definition, both compositions must be the identity function.

a. $f(x) = 3x + 9$ **and** $g(x) = \frac{1}{3}x - 3$

Verify that the compositions of $f(x)$ and $g(x)$ are identity functions.

$$[f \circ g](x) = f[g(x)] \quad\quad [g \circ f](x) = g[f(x)]$$
$$= f\left(\frac{1}{3}x - 3\right) \quad\quad = g(3x + 9)$$
$$= 3\left(\frac{1}{3}x - 3\right) + 9 \quad\quad = \frac{1}{3}(3x + 9) - 3$$
$$= x - 9 + 9 \text{ or } x \quad\quad = x + 3 - 3 \text{ or } x$$

The functions are inverses because $[f \circ g](x) = [g \circ f](x) = x$.

b. $f(x) = 4x^2$ **and** $g(x) = 2\sqrt{x}$

$$[f \circ g](x) = f\left(2\sqrt{x}\right)$$
$$= 4\left(2\sqrt{x}\right)^2$$
$$= 4(4x) \text{ or } 16x$$

Because $[f \circ g](x) \neq x$, $f(x)$ and $g(x)$ are not inverses.

Check Your Progress

3A. $f(x) = 3x - 3,\ g(x) = \frac{1}{3}x + 4$ **3B.** $f(x) = 2x^2 - 1,\ g(x) = \sqrt{\frac{x + 1}{2}}$

Personal Tutor glencoe.com

Check Your Understanding

Example 1 p. 417 — **Find the inverse of each relation.**

1. $\{(-9, 10), (1, -3), (8, -5)\}$
2. $\{(-2, 9), (4, -1), (-7, 9), (7, 0)\}$

Example 2 pp. 418–419 — **Find the inverse of each function. Then graph the function and its inverse.**

3. $f(x) = -3x$
4. $g(x) = 4x - 6$
5. $h(x) = x^2 - 3$

Example 3 p. 419 — **Determine whether each pair of functions are inverse functions. Write *yes* or *no*.**

6. $f(x) = x - 7$, $g(x) = x + 7$
7. $f(x) = \frac{1}{2}x + \frac{3}{4}$, $g(x) = 2x - \frac{4}{3}$
8. $f(x) = 2x^3$, $g(x) = \frac{1}{3}\sqrt{x}$

Practice and Problem Solving

● = **Step-by-Step Solutions** begin on page R20. **Extra Practice** begins on page 947.

Example 1 p. 417 — **Find the inverse of each relation.**

9. $\{(-8, 6), (6, -2), (7, -3)\}$
10. $\{(7, 7), (4, 9), (3, -7)\}$
11. $\{(8, -1), (-8, -1), (-2, -8), (2, 8)\}$
12. $\{(4, 3), (-4, -4), (-3, -5), (5, 2)\}$
13. $\{(1, -5), (2, 6), (3, -7), (4, 8), (5, -9)\}$
14. $\{(3, 0), (5, 4), (7, -8), (9, 12), (11, 16)\}$

Example 2 pp. 418–419 — **Find the inverse of each function. Then graph the function and its inverse.**

15. $f(x) = x + 2$
16. $g(x) = 5x$
17. $y = -2x + 1$
18. $h(x) = \frac{x - 4}{3}$
19. $y = -\frac{5}{3}x - 8$
20. $g(x) = x + 4$
21. $f(x) = 4x$
22. $y = -8x + 9$
23. $f(x) = 5x^2$
24. $h(x) = x^2 + 4$
25. $f(x) = \frac{1}{2}x^2 - 1$
26. $y = (x + 1)^2 + 3$

Example 3 p. 419 — **Determine whether each pair of functions are inverse functions. Write *yes* or *no*.**

27. $f(x) = 2x + 3$, $g(x) = 2x - 3$
28. $f(x) = 4x + 6$, $g(x) = \frac{x - 6}{4}$
29. $f(x) = -\frac{1}{3}x + 3$, $g(x) = -3x + 9$
30. $f(x) = -6x$, $g(x) = \frac{1}{6}x$
31. $f(x) = \frac{1}{2}x + 5$, $g(x) = 2x - 10$
32. $f(x) = \frac{x + 10}{8}$, $g(x) = 8x - 10$
33. $f(x) = 4x^2$, $g(x) = \frac{1}{2}\sqrt{x}$
34. $f(x) = \frac{1}{3}x^2 + 1$, $g(x) = \sqrt{3x - 3}$
35. ● $f(x) = x^2 - 9$, $g(x) = x + 3$
36. $f(x) = \frac{2}{3}x^3$, $g(x) = \sqrt{\frac{2}{3}x}$
37. $f(x) = (x + 6)^2$, $g(x) = \sqrt{x} - 6$
38. $f(x) = 2\sqrt{x - 5}$, $g(x) = \frac{1}{4}x^2 - 5$

39. **FUEL** The average miles traveled for every gallon g of gas consumed by Leroy's car is represented by the function $m(g) = 28g$.
    a. Find a function $c(g)$ to represent the cost per gallon of gasoline.
    b. Use inverses to determine the function used to represent the cost per mile traveled in Leroy's car.

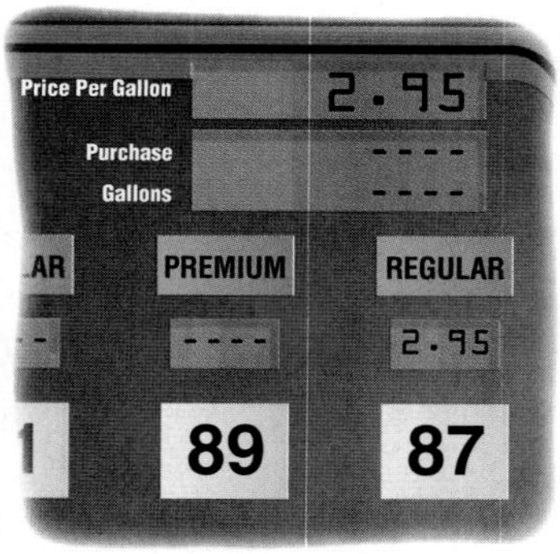

40. **SHOES** The shoe size for the average U.S. teen or adult male can be determined using the formula $M(x) = 3x - 22$, where x is length of a foot in measured inches. The shoe size for the average U.S. teen or adult female can be found by using the formula $F(x) = 3x - 21$.

a. Find the inverse of each function.

b. If Lucy wears a size $7\frac{1}{2}$ shoe, how long are her feet?

Real-World Link

Until the nineteenth century, there was no distinction between right and left shoes. The first right and left shoes were made in Philadelphia, Pennsylvania.

Source: Fact Monster

41. **GEOMETRY** The formula for the area of a circle is $A = \pi r^2$.

a. Find the inverse of the function.

b. Use the inverse to find the radius of a circle with an area of 36 square centimeters.

Use the horizontal line test to determine whether the inverse of each function is also a function.

42. $f(x) = 2x^2$
43. $f(x) = x^3 - 8$
44. $g(x) = x^4 - 6x^2 + 1$
45. $h(x) = -2x^4 - x - 2$
46. $g(x) = x^5 + x^2 - 4x$
47. $h(x) = x^3 + x^2 - 6x + 12$

48. **SHOPPING** Felipe bought a used car. The sales tax rate was 7.25% of the selling price, and he paid \$350 in processing and registration fees. Find the selling price if Felipe paid a total of \$8395.75.

49. **TEMPERATURE** A formula for converting degrees Celsius to Fahrenheit is $F(x) = \frac{9}{5}x + 32$.

a. Find the inverse $F^{-1}(x)$. Show that $F(x)$ and $F^{-1}(x)$ are inverses.

b. Explain what purpose $F^{-1}(x)$ serves.

50. **MEASUREMENT** There are approximately 1.852 kilometers in a mile.

a. Write a function that converts miles to kilometers.

b. Find the inverse of the function that converts kilometers back to miles.

c. Using composition of functions, verify that these two functions are inverses.

51. **MULTIPLE REPRESENTATIONS** Consider the functions $y = x^n$ for $n = 0, 1, 2, \ldots$.

a. **GRAPHING** Use a graphing calculator to graph $y = x^n$ for $n = 0, 1, 2, 3$, and 4.

b. **TABULAR** For which values of n is the inverse a function? Record your results in a table.

c. **ANALYTICAL** Make a conjecture about the values of n for which the inverse of $f(x) = x^n$ is a function. Assume that n is a whole number.

H.O.T. Problems

Use Higher-Order Thinking Skills

52. **REASONING** If a relation is *not* a function, then its inverse is *sometimes, always,* or *never* a function. Explain your reasoning.

53. **OPEN ENDED** Give an example of a function and its inverse. Verify that the two functions are inverses.

54. **CHALLENGE** Give an example of a function that is its own inverse.

55. **PROOF** Show that the inverse of a linear function $y = mx + b$, where $m \neq 0$ and $x \neq b$, is also a linear function.

56. **WRITING IN MATH** Suppose you have a composition of two functions that are inverses. When you put in a value of 5 for x, why is the result always 5?

NYSRE PRACTICE A2.A.44, A2.A.45

57. SHORT RESPONSE If the length of a rectangular television screen is 24 inches and its height is 18 inches, what is the length of its diagonal in inches?

58. GEOMETRY If the base of a triangle is represented by $2x + 5$ and the height is represented by $4x$, which expression represents the area of the triangle?

A $(2x + 5) + (4x)$

B $(2x + 5)(4x)$

C $\frac{1}{2}(2x + 5) + (4x)$

D $\frac{1}{2}(2x + 5)(4x)$

59. Which expression represents $f[g(x)]$ if $f(x) = x^2 + 3$ and $g(x) = -x + 1$?

F $x^2 - x + 2$

G $-x^2 - 2$

H $-x^3 + x^2 - 3x + 3$

J $x^2 - 2x + 4$

60. ACT/SAT Which of the following is the inverse of $f(x) = \frac{3x - 5}{2}$?

A $g(x) = \frac{2x + 5}{3}$

B $g(x) = \frac{3x + 5}{2}$

C $g(x) = 2x + 5$

D $g(x) = \frac{2x - 5}{3}$

Spiral Review

If $f(x) = 3x + 5$, $g(x) = x - 2$, and $h(x) = x^2 - 1$, find each value. (Lesson 7-1)

61. $g[f(3)]$

62. $f[h(-2)]$

63. $h[g(1)]$

64. CONSTRUCTION A picnic area has the shape of a trapezoid. The longer base is 8 more than 3 times the length of the shorter base, and the height is 1 more than 3 times the shorter base. What are the dimensions if the area is 4104 square feet? (Lesson 6-8)

Find the value of c that makes each trinomial a perfect square. Then write the trinomial as a perfect square. (Lesson 5-5)

65. $x^2 + 34x + c$

66. $x^2 - 11x + c$

Simplify. (Lesson 5-4)

67. $(3 + 4i)(5 - 2i)$

68. $(\sqrt{6} + i)(\sqrt{6} - i)$

69. $\frac{1 + i}{1 - i}$

70. $\frac{4 - 3i}{1 + 2i}$

Refer to quadrilateral $QRST$ shown at the right. (Lesson 4-4)

71. Write the vertex matrix. Multiply the vertex matrix by -1.

72. Graph the preimage and image.

73. What type of transformation does the graph represent?

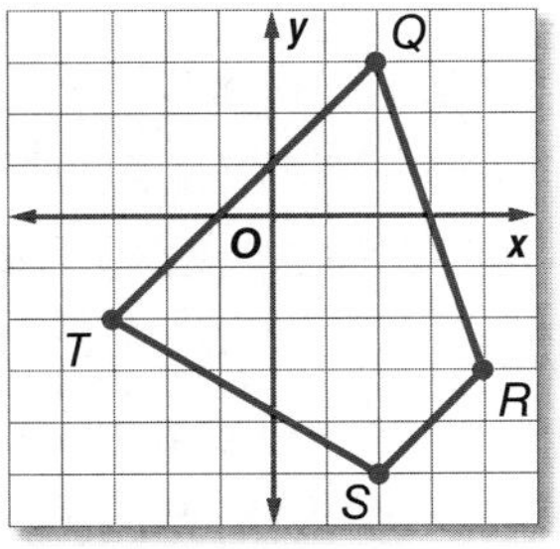

Skills Review

Graph each inequality. (Lesson 2-7)

74. $y > \frac{3}{4}x - 2$

75. $y \leq -3x + 2$

76. $y < -x - 4$

EXTEND
7-2 Graphing Technology Lab
Inverse Functions and Relations

NY Math Online glencoe.com
Other Calculator Keystrokes

NYS Core Curriculum **A2.A.45** Determine the inverse of a function and use composition to justify the result

You can use a TI-83/84 Plus graphing calculator to compare a function and its inverse using tables and graphs. Note that before you enter any values in the calculator, you should clear all lists.

ACTIVITY 1 Graph Inverses with Ordered Pairs

Graph $f(x) = \{(1, 2), (2, 4), (3, 6), (4, 8), (5, 10), (6, 12)\}$ and its inverse.

Step 1 Enter the x-values in L1 and the y-values in L2. Then graph the function.

KEYSTROKES: [STAT] [ENTER] 1 [ENTER] 2 [ENTER] 3 [ENTER] 4 [ENTER] 5 [ENTER] 6 [ENTER] [▶] 2 [ENTER] 4 [ENTER] 6 [ENTER] 8 [ENTER] 10 [ENTER] 12 [ENTER] [2nd] [STAT PLOT] [ENTER] [ENTER] [GRAPH]

Adjust the window to reflect the domain and range.

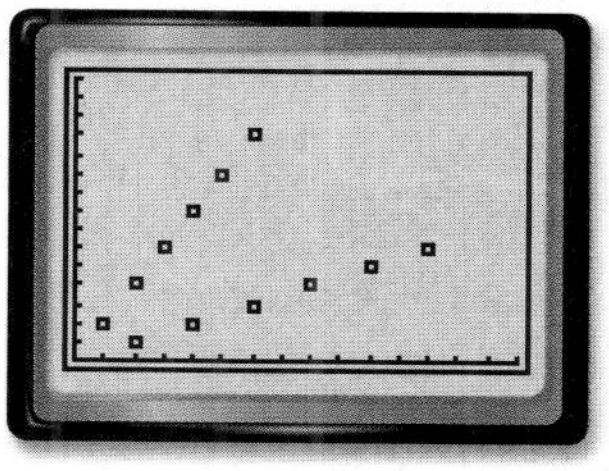

Step 2 Define the inverse function by setting Xlist to L2 and Ylist to L1. Then graph the inverse function.

KEYSTROKES: [2nd] [STAT PLOT] [▼] [ENTER] [ENTER] [▼] [▼] [2nd] [L2] [▼] [2nd] [L1] [GRAPH]

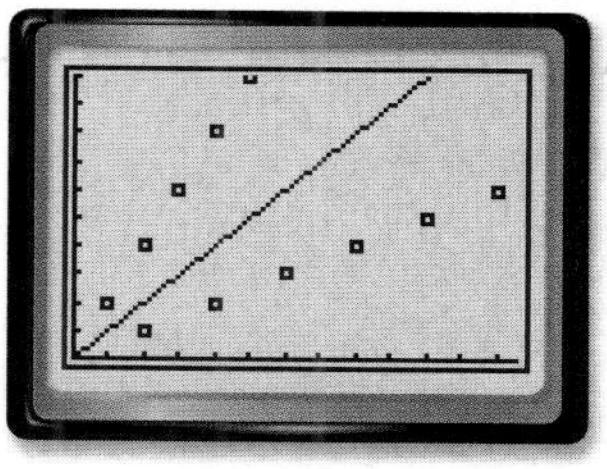

Step 3 Graph the line $y = x$.

KEYSTROKES: [Y=] [X,T,θ,n] [GRAPH]

ACTIVITY 2 Graph Inverses with Function Notation

Graph $f(x) = 3x$ and its inverse $g(x) = \frac{x}{3}$.

Step 1 Clear the data from Activity 1.

KEYSTROKES: [2nd] [STAT PLOT] [ENTER] [▶] [ENTER] [▲] [▶] [ENTER] [▶] [ENTER] [2nd] [QUIT]

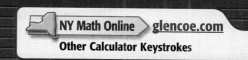

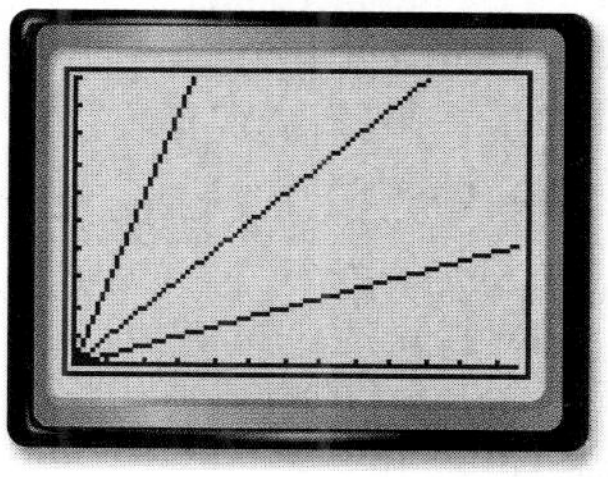

Step 2 Enter $f(x)$ as Y1, $g(x)$ as Y2, and $y = x$ as Y3. Then graph.

KEYSTROKES: [Y=] 3 [X,T,θ,n] [ENTER] [X,T,θ,n] [÷] 3 [ENTER] [X,T,θ,n] [ZOOM] 6

Exercises

Graph each function $f(x)$ and its inverse $g(x)$. Then graph $f \circ g(x)$.

1. $f(x) = 5x$

2. $f(x) = x - 3$

3. $f(x) = 2x + 1$

4. $f(x) = \frac{1}{2}x + 3$

5. $f(x) = x^2$

6. $f(x) = x^2 - 3$

7. What is the relationship between the graphs of a function and its inverse?

8. **MAKE A CONJECTURE** For any function $f(x)$ and its inverse $g(x)$, what is $(f \circ g)(x)$?

7-3 Square Root Functions and Inequalities

Then
You simplified expressions with square roots. (Lesson 5-4)

Now
- Graph and analyze square root functions.
- Graph square root inequalities.

NYS Core Curriculum

A2.A.41 Use functional notation to evaluate functions for given values in the domain

A2.A.46 Perform transformations with functions and relations: $f(x + a)$, $f(x) + a$, $f(-x)$, $-f(x)$, $af(x)$

New Vocabulary
square root function

radical function

square root inequality

NY Math Online

glencoe.com
- Extra Examples
- Personal Tutor
- Self-Check Quiz
- Homework Help

Why?

With guitars, pitch is dependent on string length and string tension. The longer the string, the higher the tension needed to produce a desired pitch. Likewise, the heavier the string, the higher the tension needed to reach a desired pitch.

This can be represented by the square root function $f = \frac{1}{2L}\sqrt{\frac{T}{P}}$, where T is the tension, P is the mass of the string, L is the length of the string, and f is the pitch.

Square Root Functions If a function contains the square root of a variable, it is called a **square root function**. The square root function is a type of **radical function**.

Key Concept — Parent Function of Square Root Functions
For Your FOLDABLE

Parent function: $f(x) = \sqrt{x}$

Domain: $\{x \mid x \geq 0\}$

Range: $\{y \mid y \geq 0\}$

Intercepts: $x = 0, y = 0$

Not defined: $x < 0$

End behavior: $x \to 0, y \to 0$

$x \to +\infty, y \to +\infty$

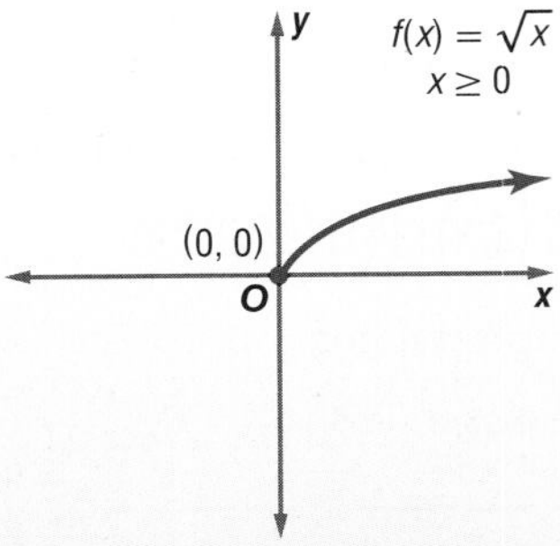

The domain of a square root function is limited to values for which the function is defined.

EXAMPLE 1 Identify Domain and Range

Identify the domain and range of $f(x) = \sqrt{x + 4}$.

The domain only includes values for which the radicand is nonnegative.

$x + 4 \geq 0$ **Write an inequality.**

$x \geq -4$ **Subtract 4 from each side.**

Thus, the domain is $\{x \mid x \geq -4\}$.

Find $f(-4)$ to determine the lower limit of the range.

$f(-4) = \sqrt{-4 + 4}$ or 0

So, the range is $\{y \mid y \geq 0\}$.

Check Your Progress

Identify the domain and range of each function.

1A. $f(x) = \sqrt{x - 3}$

1B. $f(x) = \sqrt{x + 6} + 2$

Personal Tutor glencoe.com

The same techniques used to transform the graph of other functions you have studied can be applied to the graphs of square root functions.

Key Concept — Transformations of Square Root Functions

For Your FOLDABLE

$$f(x) = a\sqrt{x - h} + k$$

***h*, Horizontal Translation**	***k*, Vertical Translation**
$\|h\|$ units right if h is positive $\|h\|$ units left if h is negative	$\|k\|$ units up if k is positive $\|k\|$ units down if k is negative
The domain is $\{x \mid x \geq h\}$.	The range is $\{y \mid y \geq k\}$.

***a*, Orientation and Shape**

- If $a < 0$, the graph is reflected across the x-axis.
- If $|a| > 1$, the graph is vertically expanded.
- If $0 < |a| < 1$, the graph is vertically compressed.

StudyTip

Domain and Range The limits on the domain and range also represent the initial point of the graph of a square root function.

EXAMPLE 2 Graph Square Root Functions

Graph each function. State the domain and range.

a. $y = \sqrt{x - 2} + 5$

The minimum point is at $(h, k) = (2, 5)$. Make a table of values for $x \geq 2$, and graph the function. The graph is the same shape as $f(x) = \sqrt{x}$, but is translated 2 units right and 5 units up. Notice the end behavior. As x increases, y increases.

The domain is $\{x \mid x \geq 2\}$ and the range is $\{y \mid y \geq 5\}$.

x	y
2	5
3	6
4	6.4
5	6.7
6	7
7	7.2
8	7.4

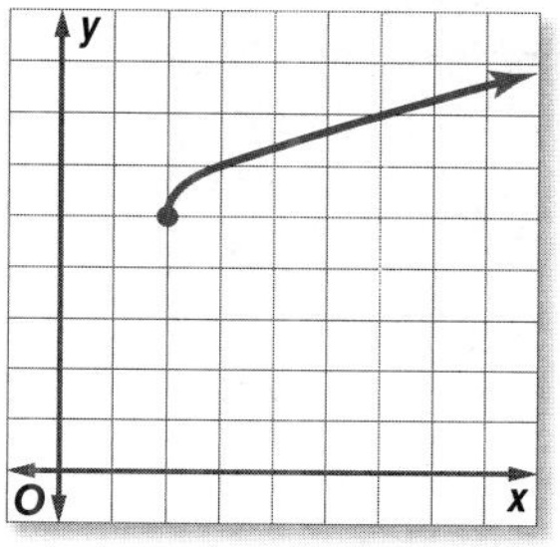

ReadingMath

Power Functions Square root functions are also *power functions* when the exponent is a fraction. In the case of square roots, the exponent is $\frac{1}{2}$.

b. $y = -2\sqrt{x + 3} - 1$

The minimum domain value is at h or -3. Make a table of values for $x \geq -3$, and graph the function. Because a is negative, the graph is similar to $f(x) = \sqrt{x}$, but is reflected in the line $f(x) = -1$. Because $|a| > 1$, the graph is vertically compressed. It is also translated 3 units left and 1 unit down.

x	y
−3	−1
−2	−3
−1	−3.8
0	−4.5
1	−5
2	−5.5
3	−5.9

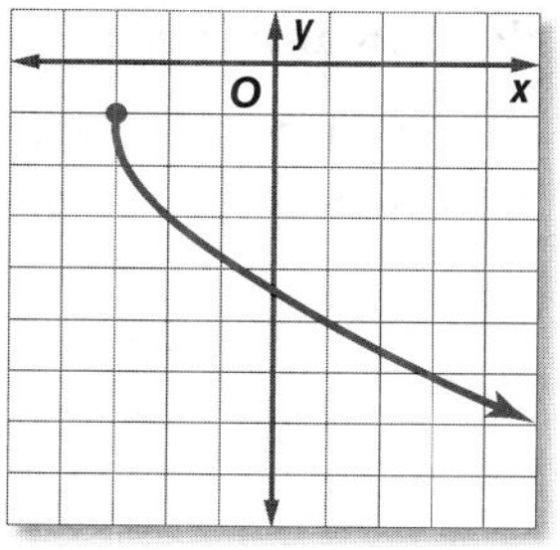

The domain is $\{x \mid x \geq -3\}$ and the range is $\{y \mid y \leq -1\}$.

Check Your Progress

2A. $f(x) = 2\sqrt{x + 4}$

2B. $f(x) = \frac{1}{4}\sqrt{x - 5} + 3$

Personal Tutor glencoe.com

Real-World EXAMPLE 3 Use Graphs to Analyze Square Root Functions

MUSIC **Refer to the application at the beginning of the lesson. The pitch, or frequency, measured in hertz (Hz) of a certain string can be determined by $f(T) = \frac{1}{1.28}\sqrt{\frac{T}{0.0000708}}$, where T is tension in kilograms.**

a. Graph the function for tension in the domain $\{T \mid 0 \leq T \leq 10\}$.

Make a table of values for $0 \leq T \leq 10$ and graph.

T	$f(T)$
0	0
1	92.8
2	131.3
3	160.8
4	185.7
5	207.6

T	$f(T)$
6	227.4
7	245.7
8	262.6
9	278.5
10	293.6

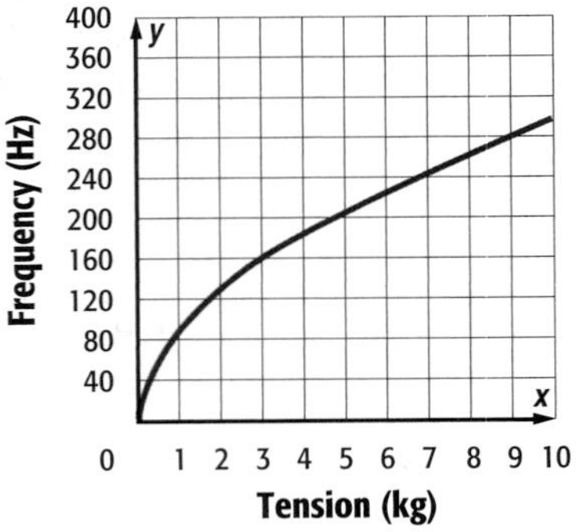

b. How much tension is needed for a pitch of over 200 Hz?

According to the graph and the table, more than 4.5 kilograms of tension is needed for a pitch of more than 200 hertz.

Real-World Link

On every string, the guitar player has an option of decreasing the length of the string in about 24 different ways. This will produce 24 different frequencies on each string.

Source: *Guitar World*

Check Your Progress

3. **MUSIC** The frequency of vibrations for a certain guitar string when it is plucked can be determined by $F = 200\sqrt{T}$, where F is the number of vibrations per second and T is the tension measured in pounds. Graph the function for $0 \leq T \leq 10$. Then determine the frequency for $T = 3$, 6, and 9 pounds.

Personal Tutor glencoe.com

Problem-Solving Tip

Make a Table Making a table is a good way to organize ordered pairs in order to see the general behavior of a graph.

Square Root Inequalities A **square root inequality** is an inequality involving square roots. They are graphed using the same method as other inequalities.

EXAMPLE 4 Graph a Square Root Inequality

Graph $y < \sqrt{x - 4} - 6$.

Graph the boundary $y = \sqrt{x - 4} - 6$.

The domain is $\{x \mid x \geq 4\}$. Because y is *less than,* the shaded region should be *below* the boundary and within the domain.

CHECK Select a point in the shaded region, and verify that it is a solution of the inequality.

Test $(7, -5)$: $-5 \stackrel{?}{<} \sqrt{7 - 4} - 6$

$-5 \stackrel{?}{<} \sqrt{3} - 6$

$-5 < -4.27$ ✓

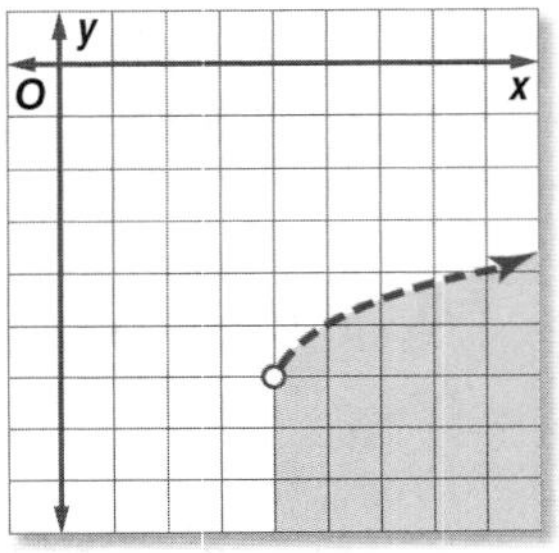

Check Your Progress

Graph each inequality.

4A. $f(x) \geq \sqrt{2x + 1}$

4B. $f(x) < -\sqrt{x + 2} - 4$

Personal Tutor glencoe.com

Check Your Understanding

Example 1 p. 424

Identify the domain and range of each function.

1. $f(x) = \sqrt{4x}$
2. $f(x) = \sqrt{x - 5}$
3. $f(x) = \sqrt{x + 8} - 2$

Example 2 p. 425

Graph each function. State the domain and range.

4. $f(x) = \sqrt{x} - 2$
5. $f(x) = 3\sqrt{x - 1}$
6. $f(x) = \frac{1}{2}\sqrt{x + 4} - 1$
7. $f(x) = -\sqrt{3x - 5} + 5$

Example 3 p. 426

8. **OCEAN** The speed that a tsunami, or tidal wave, can travel is modeled by the equation $v = 356\sqrt{d}$, where v is the speed in kilometers per hour and d is the average depth of the water in kilometers. A tsunami is found to be traveling at 145 kilometers per hour. What is the average depth of the water? Round to the nearest hundredth of a kilometer.

Example 4 p. 426

Graph each inequality.

9. $f(x) \geq \sqrt{x} + 4$
10. $f(x) \leq \sqrt{x - 6} + 2$
11. $f(x) < -2\sqrt{x + 3}$
12. $f(x) > \sqrt{2x - 1} - 3$

Practice and Problem Solving

● = **Step-by-Step Solutions** begin on page R20.
Extra Practice begins on page 947.

Example 1 p. 424

Identify the domain and range of each function.

13. $f(x) = -\sqrt{2x} + 2$
14. $f(x) = \sqrt{x} - 6$
15. $f(x) = 4\sqrt{x - 2} - 8$
16. $f(x) = \sqrt{x + 2} + 5$
17. $f(x) = \sqrt{x - 4} - 6$
18. $f(x) = -\sqrt{x - 6} + 5$

Example 2 p. 425

Graph each function. State the domain and range.

19. $f(x) = \sqrt{6x}$
20. $f(x) = -\sqrt{5x}$
21. $f(x) = \sqrt{x - 8}$
22. $f(x) = \sqrt{x + 1}$
23. $f(x) = \sqrt{x + 3} + 2$
24. $f(x) = \sqrt{x - 4} - 10$
25. $f(x) = 2\sqrt{x - 5} - 6$
26. $f(x) = \frac{3}{4}\sqrt{x + 12} + 3$
27. $f(x) = -\frac{1}{5}\sqrt{x - 1} - 4$
28. $f(x) = -3\sqrt{x + 7} + 9$

Example 3 p. 426

29. **SKYDIVING** The approximate time t in seconds that it takes an object to fall a distance of d feet is given by $t = \sqrt{\frac{d}{16}}$. Suppose a parachutist falls 11 seconds before the parachute opens. How far does the parachutist fall during this time?

30. **ROLLER COASTERS** The velocity of a roller coaster as it moves down a hill is $V = \sqrt{v^2 + 64h}$, where v is the initial velocity in feet per second and h is the vertical drop in feet. The designer wants the coaster to have a velocity of 90 feet per second when it reaches the bottom of the hill.
    a. If the initial velocity of the coaster at the top of the hill is 10 feet per second, write an equation that models the situation.
    b. How high should the designer make the hill?

Example 4 p. 426

Graph each inequality.

31. $y < \sqrt{x - 5}$

32. $y > \sqrt{x + 6}$

33 $y \geq -4\sqrt{x + 3}$

34. $y \leq -2\sqrt{x - 6}$

35. $y > 2\sqrt{x + 7} - 5$

36. $y \geq 4\sqrt{x - 2} - 12$

37. $y \leq 6 - 3\sqrt{x - 4}$

38. $y < \sqrt{4x - 12} + 8$

39. PHYSICS The kinetic energy of an object is the energy produced due to its motion and mass. The formula for kinetic energy, measured in joules j, is $E = 0.5mv^2$, where m is the mass in kilograms and v is the velocity of the object in meters per second.

a. Solve the above formula for v.

b. If a 1500-kilogram vehicle is generating 1 million joules of kinetic energy, how fast is it traveling?

c. *Escape velocity* is the minimum velocity at which an object must travel to escape the gravitational field of a planet or other object. Suppose a 100,000-kilogram ship must have a kinetic energy of 3.624×10^{14} joules to escape the gravitational field of Jupiter. Estimate the escape velocity of Jupiter.

Real-World Link

The leading cause of driving accidents among teenagers is distractions. Thirteen percent of teens admit to text messaging while driving.

Source: Allstate Foundation

40. DRIVING After an accident, police can determine how fast a car was traveling before the driver put on his or her brakes by using the equation $v = \sqrt{30fd}$. In this equation, v represents the speed in miles per hour, f represents the coefficient of friction, and d represents the length of the skid marks in feet. The coefficient of friction varies depending on road conditions. Assume that $f = 0.6$.

a. Find the speed of a car that skids 25 feet.

b. If your car is going 35 miles per hour, how many feet would it take you to stop?

c. If the speed of a car is doubled, will the skid be twice as long? Explain.

Write the square root function represented by each graph.

41.

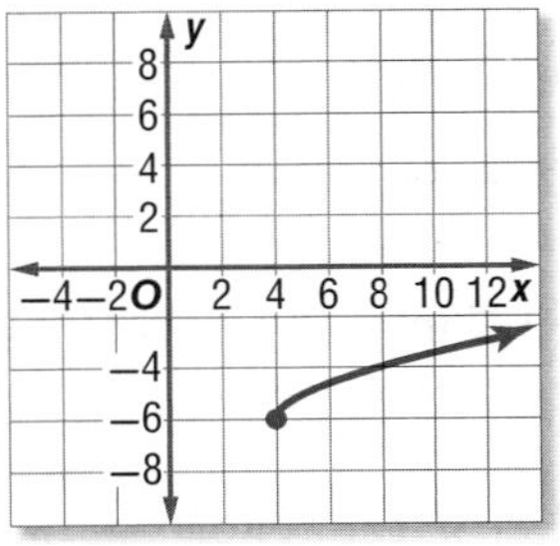

42.

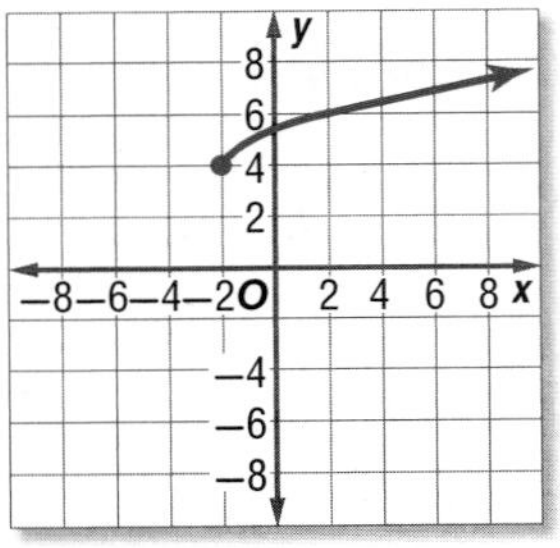

43.

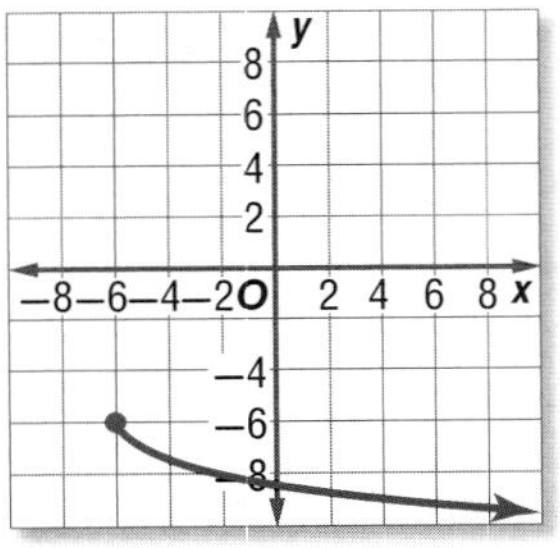

44. MULTIPLE REPRESENTATIONS In this problem, you will use the following functions to investigate transformations of square root functions.

$f(x) = 4\sqrt{x - 6} + 3$ $g(x) = \sqrt{16x + 1} - 6$ $h(x) = \sqrt{x + 3} + 2$

a. GRAPHICAL Graph each function on the same set of axes.

b. ANALYTICAL Identify the transformation on the graph of the parent function. What values caused each transformation?

c. ANALYTICAL Which functions appear to be stretched or compressed vertically? Explain your reasoning.

d. VERBAL The two functions that are stretched appear to be stretched by the same magnitude. How is this possible?

e. TABULAR Make a table of the rate of change for all three functions between 8 and 12 as compared to 12 and 16. What generalization about rate of change in square root functions can be made as a result of your findings?

Real-World Link

Pendulum clocks have been used to keep time since 1656, and they have not changed dramatically since then. These were the first clocks made that kept time accurately.

Source: HowStuffWorks, Inc.

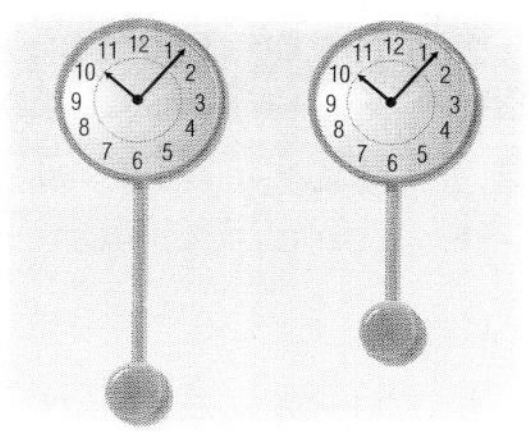

45. PENDULUMS The period of a pendulum can be represented by $T = 2\pi\sqrt{\frac{L}{g}}$, where T is the time in seconds, L is the length in feet, and g is gravity, 32 feet per second squared.

a. Graph the function for $0 \leq L \leq 10$.

b. What is the period for lengths of 2, 5, and 8 feet?

46. PHYSICS Using the function $m = \frac{m_0}{\sqrt{1 - \left(\frac{v^2}{c^2}\right)}}$, Einstein's theory of relativity states that the apparent mass m of a particle depends on its velocity v. An object that is traveling extremely fast, close to the speed of light c, will *appear* to have more mass compared to its mass at rest, m_0.

a. Use a graphing calculator to graph the function for a 10,000-kilogram ship for the domain $0 \leq v \leq 300{,}000{,}000$. Use 300 million meters per second for the speed of light.

b. What viewing window did you use to view the graph?

c. Determine the apparent mass m of the ship for speeds of 100 million, 200 million, and 299 million meters per second.

H.O.T. Problems

Use **H**igher-**O**rder **T**hinking Skills

47. CHALLENGE Write an equation for a square root function with a domain of $\{x \mid x \geq -4\}$, a range of $\{y \mid y \geq 6\}$, and that passes through (5, 3).

48. REASONING For what positive values of a are the domain and range of $f(x) = \sqrt[a]{x}$ the set of real numbers?

49. OPEN ENDED Write a square root function for which the domain is $\{x \mid x \geq 8\}$ and the range is $\{y \mid y \leq 14\}$.

50. WRITING IN MATH Explain why there are limitations on the domain and range of square root functions.

51. FIND THE ERROR Molly and Cleveland are graphing $y \leq \sqrt{5x + 15}$. Is either of them correct? Explain your reasoning.

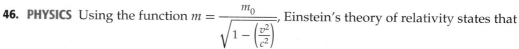

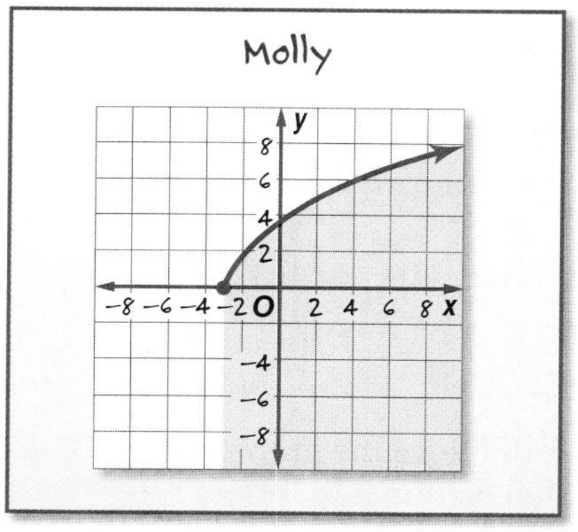

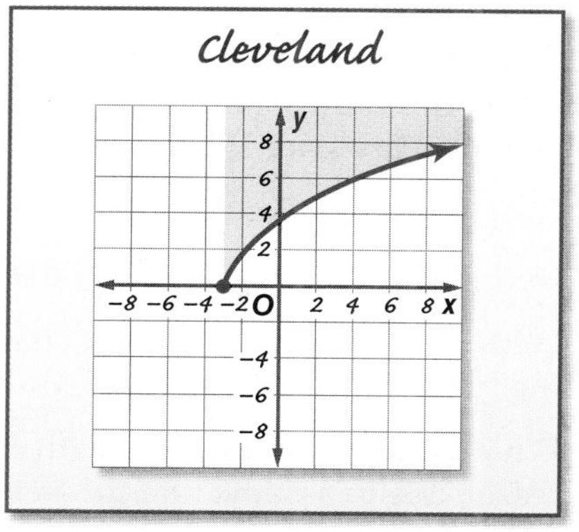

52. WRITING IN MATH Explain why $y = \pm\sqrt{x}$ is not a function.

53. OPEN ENDED Write an equation of a relation that contains a radical and its inverse such that:

a. the original relation is a function, and its inverse is not a function.

b. the original relation is not a function, and its inverse is a function.

NYSRE PRACTICE A2.A.41

54. The expression $\frac{-64x^6}{8x^3}$, $x \neq 0$, is equivalent to

A $8x^2$
B $8x^3$
C $-8x^2$
D $-8x^3$

55. PROBABILITY For a game, Patricia must roll a standard die and draw a card from a deck of 26 cards, each card having a letter of the alphabet on it. What is the probability that Patricia will roll an odd number and draw a letter in her name?

F $\frac{2}{3}$
G $\frac{3}{26}$
H $\frac{1}{13}$
J $\frac{1}{26}$

56. SHORT RESPONSE What is the product of $(d + 6)$ and $(d - 3)$?

57. ACT/SAT Given the graph of the square root function below, which must be true?

I. The domain is all real numbers.
II. The function is $y = \sqrt{x} + 3.5$.
III. The range is about $\{y \mid y \geq 3.5\}$.

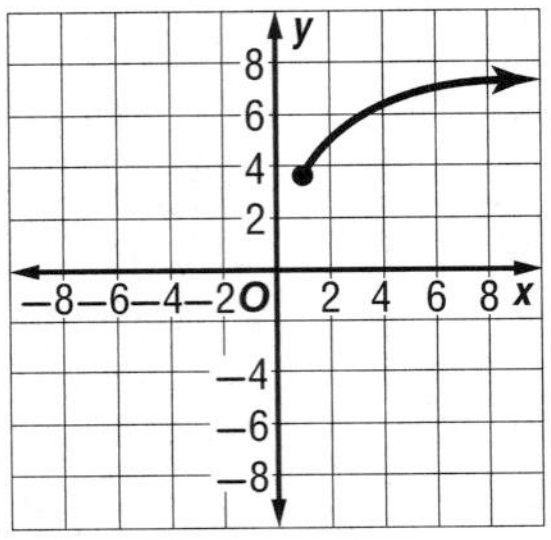

A I only
B I, II, and III
C II and III only
D III only

Spiral Review

Determine whether each pair of functions are inverse functions. Write *yes* or *no*. (Lesson 7-2)

58. $f(x) = 2x$
$g(x) = \frac{1}{2}x$

59. $f(x) = 3x - 7$
$g(x) = \frac{1}{3}x - \frac{7}{16}$

60. $f(x) = \frac{3x + 2}{5}$
$g(x) = \frac{5x - 2}{3}$

61. TIME The formula $h = \frac{m}{60}$ converts minutes m to hours h, and $d = \frac{h}{24}$ converts hours h to days d. Write a function that converts minutes to days. (Lesson 7-1)

62. CABLE TV The number of households in the United States with cable TV after 1985 can be modeled by the function $C(t) = -43.2t^2 + 1343t + 790$, where t represents the number of years since 1985. (Lesson 6-4)

a. Graph this equation for the years 1985 to 2005.

b. Describe the turning points of the graph and its end behavior.

c. What is the domain of the function? Use the graph to estimate the range for the function.

d. What trends in households with cable TV does the graph suggest? Is it reasonable to assume that the trend will continue indefinitely?

Skills Review

Determine whether each number is *rational* or *irrational*. (Lesson 1-2)

63. 6.34

64. 3.787887888

65. 5.333…

66. 1.25

7-4

nth Roots

Why?

According to a world-wide injury prevention study, the number of collisions between bicycles and automobiles increased as the number of bicycles per intersection increased. The relationship can be expressed using the equation $c = \sqrt[5]{b^2}$, where b is the number of bicycles and c is the number of collisions.

Then

You worked with square root functions. (Lesson 7-3)

Now

- Simplify radicals.
- Use a calculator to approximate radicals.

NYS Core Curriculum

A2.A.13 Simplify radical expressions

New Vocabulary

nth root
radical sign
index
radicand
principal root

NY Math Online

glencoe.com
- Extra Examples
- Personal Tutor
- Self-Check Quiz
- Homework Help

Simplify Radicals Finding the square root of a number and squaring a number are inverse operations. To find the square root of a number a, you must find a number with a square of a. Similarly, the inverse of raising a number to the nth power is finding the **nth root** of a number.

Powers	Factors	Words	Roots
$x^3 = 64$	$4 \cdot 4 \cdot 4 = 64$	4 is a cube root of 64.	$\sqrt[3]{64} = 4$
$x^4 = 625$	$5 \cdot 5 \cdot 5 \cdot 5 = 625$	5 is a fourth root of 625.	$\sqrt[4]{625} = 5$
$x^5 = 32$	$2 \cdot 2 \cdot 2 \cdot 2 \cdot 2 = 32$	2 is a fifth root of 32.	$\sqrt[5]{32} = 2$
$a^n = b$	$\underbrace{a \cdot a \cdot a \cdot \dots \cdot a}_{n \text{ factors of } a} = b$	a is an nth root of b.	$\sqrt[n]{b} = a$

This pattern suggests the following formal definition of an nth root.

Key Concept **Definition of nth Root** For Your FOLDABLE

Words For any real numbers a and b, and any positive integer n, if $a^n = b$, then a is an nth root of b.

Example Because $(-3)^4 = 81$, -3 is a fourth root of 81 and 3 is a principal root.

The symbol $\sqrt[n]{\ }$ indicates an nth root.

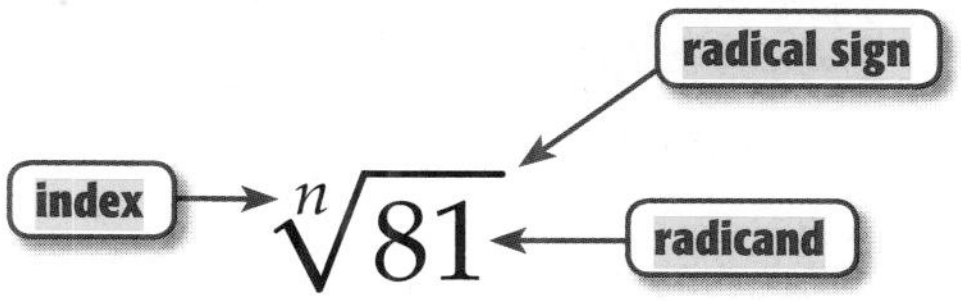

Some numbers have more than one real nth root. For example, 64 has two square roots, 8 and -8, since 8^2 and $(-8)^2$ both equal 64. When there is more than one real root and n is even, the nonnegative root is called the **principal root**.

Some examples of nth roots are listed below.

$\sqrt{25} = 5$ — $\sqrt{25}$ indicates the principal square root of 25.

$-\sqrt{25} = -5$ — $-\sqrt{25}$ indicates the opposite of the principal square root of 25.

$\pm\sqrt{25} = \pm 5$ — $\pm\sqrt{25}$ indicates both square roots of 25.

Key Concept Real *n*th Roots

For Your FOLDABLE

Suppose n is an integer greater than 1, and a is a real number.

a	n is even.	n is odd.
$a > 0$	1 unique positive and 1 unique negative real root: $\pm\sqrt[n]{a}$; positive root is principal root	1 unique positive and 0 negative real root: $\sqrt[n]{a}$
$a < 0$	0 real roots	0 positive and 1 negative real root: $\sqrt[n]{a}$
$a = 0$	1 real root: $\sqrt[n]{0} = 0$	1 real root: $\sqrt[n]{0} = 0$

EXAMPLE 1 Find Roots

Simplify.

a. $\pm\sqrt{16y^4}$

$\pm\sqrt{16y^4} = \pm\sqrt{(4y^2)^2}$

$= \pm 4y^2$

The square roots of $16y^4$ are $\pm 4y^2$.

b. $-\sqrt{(x^2-6)^8}$

$-\sqrt{(x^2-6)^8} = -\sqrt{[(x^2-6)^4]^2}$

$= -(x^2-6)^4$

The opposite of the principal square root of $(x^2-6)^8$ is $-(x^2-6)^4$.

c. $\sqrt[5]{243a^{20}b^{25}}$

$\sqrt[5]{243a^{20}b^{25}} = \sqrt[5]{(3a^4b^5)^5}$

$= 3a^4b^5$

The fifth root of $243a^{20}b^{25}$ is $3a^4b^5$.

d. $\sqrt{-16x^4y^8}$

$\sqrt[2]{-16x^4y^8}$ ← *b* is negative. ← *n* is even.

There are no real roots since $\sqrt{-16}$ is not a real number. However, there are two imaginary roots, $4ix^2y^4$ and $-4ix^2y^4$.

Review Vocabulary

pure imaginary numbers Square roots of negative real numbers; for any positive real number b, $\sqrt{-b^2} = \sqrt{b^2} \cdot \sqrt{-1}$, or bi, where i is the imaginary unit. (Lesson 5-4)

✓ Check Your Progress

1A. $\pm\sqrt{36x^{10}}$

1B. $-\sqrt{(y+7)^{16}}$

Personal Tutor glencoe.com

When you find an even root of an even power and the result is an odd power, you must use the absolute value of the result to ensure that the answer is nonnegative.

StudyTip

Odd Index If n is odd, there is only one real root. Therefore, there is no principal root when n is odd, and absolute value symbols are never needed.

EXAMPLE 2 Simplify Using Absolute Value

Simplify.

a. $\sqrt[4]{y^4}$

$\sqrt[4]{y^4} = |y|$

Since y could be negative, you must take the absolute value of y to identify the principal root.

b. $\sqrt[6]{64(x^2-3)^{18}}$

$\sqrt[6]{64(x^2-3)^{18}} = 2|(x^2-3)^3|$

Since the index 6 is even and the exponent 3 is odd, you must use absolute value.

✓ Check Your Progress

2A. $\sqrt{36y^6}$

2B. $\sqrt[4]{16(x-3)^{12}}$

Personal Tutor glencoe.com

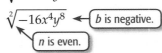

Approximate Radicals with a Calculator Recall that real numbers that cannot be expressed as terminating or repeating decimals are irrational numbers. Approximations for irrational numbers are often used in real-world problems.

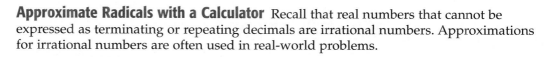

Real-World Link

77% of the employees in China commute by bicycle.

Source: International Bicycle Fund

Real-World EXAMPLE 3 Approximate Radicals

INJURY PREVENTION Refer to the beginning of the lesson.

a. If $c = \sqrt[5]{b^2}$ represents the number of collisions and b represents the number of bicycle riders per intersection, estimate the number of collisions at an intersection that has 1000 bicycle riders per week.

Understand You want to find out how many collisions there were.

Plan Let b be the number of bicycles. The number of collisions c is 1000.

Solve

$c = \sqrt[5]{b^2}$ **Original formula**

$= \sqrt[5]{1000^2}$ **$b = 1000$**

≈ 15.85 **Use a calculator.**

There are about 16 collisions per week at the intersection.

Check

$15.85 \stackrel{?}{=} \sqrt[5]{b^2}$ **$c = 15.85$**

$15.85^5 \stackrel{?}{=} b^2$ **Raise each side to the fifth power.**

$1{,}000{,}337 \stackrel{?}{=} b^2$ **Simplify.**

$1000 \approx b$ ✓ **Take the square root of each side.**

b. If the total number of collisions reported in one week is 21, estimate the number of bicycle riders that passed through that intersection.

$c = \sqrt[5]{b^2}$ **Original formula**

$21 = \sqrt[5]{b^2}$ **$c = 27$**

$21^5 = b^2$ **Raise each side to the fifth power.**

$4{,}084{,}101 = b^2$ **Simplify.**

$2021 \approx b$ **Take the square root of each side.**

Check Your Progress

3A. The surface area of a sphere can be determined from the volume of the sphere using the formula $S = \sqrt[3]{36\pi V^2}$, where V is the volume. Determine the surface area of a sphere with a volume of 200 cubic inches.

3B. If the surface area of a sphere is about 214.5 square inches, determine the volume.

Personal Tutor glencoe.com

Check Your Understanding

Examples 1 and 2 p. 432

Simplify.

1. $\pm\sqrt{100y^8}$

2. $-\sqrt{49u^8v^{12}}$

3. $\sqrt{(y-6)^8}$

4. $\sqrt[4]{16g^{16}h^{24}}$

5. $\sqrt{-16y^4}$

6. $\sqrt[6]{64(2y+1)^{18}}$

Example 3 p. 433

Use a calculator to approximate each value to three decimal places.

7. $\sqrt{58}$

8. $-\sqrt{76}$

9. $\sqrt[5]{-43}$

10. $\sqrt[4]{71}$

11. TELEVISION The radius r of the orbit of a television satellite is given by $\sqrt[3]{\frac{GMt^2}{4\pi^2}}$, where G is the universal gravitational constant, M is the mass of Earth, and t is the time it takes the satellite to complete one orbit. Find the radius of the satellite's orbit if G is $6.67 \times 10^{-11} N \cdot m^2/kg^2$, M is 5.98×10^{24} kg, and t is 2.6×10^6 seconds.

Practice and Problem Solving

● = **Step-by-Step Solutions** begin on page R20.
Extra Practice begins on page 947.

Examples 1 and 2 p. 432

Simplify.

12. $\pm\sqrt{121x^4y^{16}}$
13. $\pm\sqrt{225a^{16}b^{36}}$
14. $\pm\sqrt{49x^4}$
15. $-\sqrt{16c^4d^2}$
16. $-\sqrt{81a^{16}b^{20}c^{12}}$
17. $-\sqrt{400x^{32}y^{40}}$
18. $\sqrt{(x+15)^4}$
19. $\sqrt{(x^2+6)^{16}}$
20. $\sqrt{(a^2+4a)^{12}}$
21. $\sqrt[3]{8a^6b^{12}}$
22. $\sqrt[6]{d^{24}x^{36}}$
23. $\sqrt[3]{27b^{18}c^{12}}$
24. $-\sqrt{(2x+1)^6}$
25. $\sqrt{-(x+2)^8}$
26. $\sqrt[3]{-(y-9)^9}$
27. $\sqrt[6]{x^{18}}$
28. $\sqrt[4]{a^{12}}$
29. $\sqrt[3]{a^{12}}$
30. $\sqrt[4]{81(x+4)^4}$
31. $\sqrt[3]{(4x-7)^{24}}$
32. $\sqrt[3]{(y^3+5)^{18}}$
33. $\sqrt[4]{256(5x-2)^{12}}$
34. $\sqrt[8]{x^{16}y^8}$
35. $\sqrt[5]{32a^{15}b^{10}}$

Example 3 p. 433

36. SHIPPING An online book store wants to increase the size of the boxes it uses to ship orders. The new volume N is equal to the old volume V times the scale factor F cubed, or $N = V \cdot F^3$. What is the scale factor if the old volume was 0.8 cubic feet and the new volume is 21.6 cubic feet?

37. GEOMETRY The side length of a cube is determined by $r = \sqrt[3]{V}$, where V is the volume in cubic units. Determine the side length of a cube with a volume of 512 cm^3.

Use a calculator to approximate each value to three decimal places.

38. $\sqrt{92}$
39. $-\sqrt{150}$
40. $\sqrt{0.43}$
41. $\sqrt{0.62}$
42. $\sqrt[3]{168}$
43. $\sqrt[5]{-4382}$
44. $\sqrt[6]{(8912)^2}$
45. $\sqrt[5]{(4756)^2}$

46. GEOMETRY The radius r of a sphere with volume V can be found using the formula $r = \sqrt[3]{\frac{3V}{4\pi}}$.

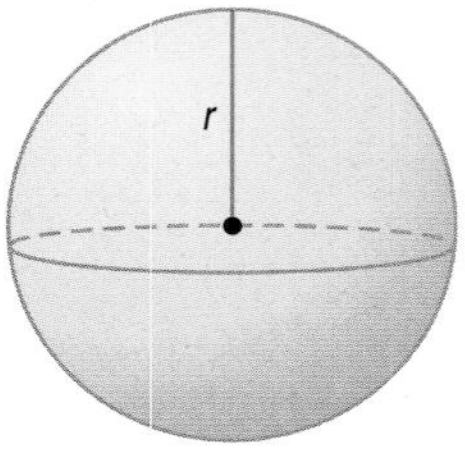

a. Determine the radius for volumes of 1000 cm^3, 8000 cm^3, and 64,000 cm^3.

b. How does the volume of the sphere change if the radius is doubled? Explain.

Simplify.

47. $\sqrt{196c^6d^4}$
48. $\sqrt{-64y^8z^6}$
49. $\sqrt[3]{-27a^{15}b^9}$
50. $\sqrt[4]{-16x^{16}y^8}$
51. $\sqrt{400x^{16}y^6}$
52. $\sqrt[3]{8c^3d^{12}}$
53. $\sqrt[3]{64(x+y)^6}$
54. $\sqrt[5]{-(y-z)^{15}}$

55. PHYSICS Johannes Kepler developed the formula $d = \sqrt[3]{6t^2}$, where d is the distance of a planet from the Sun in millions of miles and t is the number of Earth-days that it takes for the planet to orbit the Sun. If the length of a year on Mars is 687 Earth-days, how far from the Sun is Mars?

56. CHEMISTRY All matter is composed of atoms. The nucleus of an atom is the center portion of the atom that contains most of the mass of the atom. A theoretical formula for the radius r of the nucleus of an atom is $r = (1.3 \times 10^{-15})A^{\frac{1}{3}}$ meters, where A is the mass number of the nucleus. Find the radius of the nucleus for each atom in the table.

Atom	Mass Number
carbon	6
oxygen	8
sodium	11
aluminum	13
chlorine	17

Real-World Link

The komodo dragon is the largest lizard in the world, spanning up to over 10 feet in length. A large male can weigh over 500 pounds after a large meal.

Source: The Circle of the Dragon

57. **BIOLOGY** Kleiber's Law, $P = 73.3\sqrt[4]{m^3}$, shows the relationship between the mass m in kilograms of an organism and its metabolism P in Calories per day. Determine the metabolism for each of the animals listed at the right.

Animal	Mass (kg)
bald eagle	4.5
golden retriever	30
komodo dragon	72
bottlenose dolphin	156
Asian elephant	2300

58. **MULTIPLE REPRESENTATIONS** In this problem, you will use $f(x) = x^n$ and $g(x) = \sqrt[n]{x}$ to explore inverses.

a. **TABULAR** Make tables for $f(x)$ and $g(x)$ using $n = 3$, $n = 4$ and $-5 \leq x \leq 5$.

b. **GRAPHICAL** Graph the equations.

c. **ANALYTICAL** Which equations are functions? Which functions are one-to-one?

d. **ANALYTICAL** For what values of n are $g(x)$ and $f(x)$ inverses of each other?

e. **VERBAL** What conclusions can you make about $g(x) = \sqrt[n]{x}$ and $f(x) = x^n$ for all positive even values of n? for odd values of n?

H.O.T. Problems

Use Higher-Order Thinking Skills

59. **FIND THE ERROR** Ashley and Kimi are simplifying $\sqrt[4]{16x^4y^8}$. Is either of them correct? Explain your reasoning.

Ashley

$\sqrt[4]{16x^4y^8} = \sqrt[4]{(2xy^2)^4}$

$= 2|xy^2|$

Kimi

$\sqrt[4]{16x^4y^8} = \sqrt[4]{(2xy^2)^4}$

$= 2y^2|x|$

60. **CHALLENGE** Under what conditions is $\sqrt{x^2 + y^2} = x + y$ true?

61. **REASONING** Determine whether the statement $\sqrt[4]{(-x)^4} = x$ is *sometimes*, *always*, or *never* true.

62. **CHALLENGE** For what real values of x is $\sqrt[3]{x} > x$?

63. **OPEN ENDED** Write a number for which the principal square root and cube root are both integers.

64. **WRITING IN MATH** Explain when and why absolute value symbols are needed when taking an nth root.

65. **CHALLENGE** Write an equivalent expression in radical form for $16^{\frac{1}{3}}x^{\frac{1}{3}}y^{\frac{1}{3}}$. Simplify the radical.

CHALLENGE **Simplify each expression.**

66. $(0.0016)^{\frac{1}{4}}$

67. $(-0.0000001)^{\frac{1}{7}}$

68. $\dfrac{\sqrt[5]{-0.00032}}{\sqrt[3]{-0.027}}$

69. **CHALLENGE** Solve $-5a^{-\frac{1}{2}} = -125$ for a.

NYSRE PRACTICE A2.A.13

70. What is the value of w in the equation $\frac{1}{2}(4w + 36) = 3(4w - 3)$?

A 2
B 2.7
C 27
D 36

71. What is the product of the complex numbers $(5 + i)$ and $(5 - i)$?

F 24
G 26
H $25 - i$
J $26 - 10i$

72. EXTENDED RESPONSE A cylindrical cooler has a diameter of 9 inches and a height of 11 inches. Tate plans to use it for soda cans that have a diameter of 2.5 inches and a height of 4.75 inches.

a. Tate plans to place two layers consisting of 9 cans each into the cooler. What is the volume of the space that will not be filled with the cans?

b. Find the ratio of the volume of the cooler to the volume of the cans in part **a**.

73. ACT/SAT Which of the following is closest to $\sqrt[3]{7.32}$?

A 1.8
B 1.9
C 2.0
D 2.1

Spiral Review

Graph each function. (Lesson 7-3)

74. $y = \sqrt{x - 5}$

75. $y = \sqrt{x} - 2$

76. $y = 3\sqrt{x} + 4$

77. HEALTH The average weight of a baby born at a certain hospital is $7\frac{1}{2}$ pounds and the average length is 19.5 inches. One kilogram is about 2.2 pounds and 1 centimeter is about 0.3937 inches. Find the average weight in kilograms and the length in centimeters. (Lesson 7-2)

Simplify. (Lesson 6-1)

78. $(4c - 5) - (c + 11) + (-6c + 17)$

79. $(11x^2 + 13x - 15) - (7x^2 - 9x + 19)$

80. $(d - 5)(d + 3)$

81. $(2a^2 + 6)^2$

82. GAS MILEAGE The gas mileage y in miles per gallon for a certain vehicle is given by the equation $y = 10 + 0.9x - 0.01x^2$, where x is the speed of the vehicle between 10 and 75 miles per hour. Find the range of speeds that would give a gas mileage of at least 25 miles per gallon. (Lesson 5-8)

Write each equation in vertex form, if not already in that form. Identify the vertex, axis of symmetry, and direction of opening. Then graph the function. (Lesson 5-7)

83. $y = -6(x + 2)^2 + 3$

84. $y = -\frac{1}{3}x^2 + 8x$

85. $y = (x - 2)^2 - 2$

86. $y = 2x^2 + 8x + 10$

Skills Review

Find each product. (Lesson 6-1)

87. $(x + 4)(x + 5)$

88. $(y - 3)(y + 4)$

89. $(a + 2)(a - 9)$

90. $(a - b)(a - 3b)$

91. $(x + 2y)(x - y)$

92. $2(w + z)(w - 4z)$

EXTEND 7-4

Graphing Technology Lab
Graphing *n*th Root Functions

NY Math Online > glencoe.com
Other Calculator Keystrokes

You can use a graphing calculator to graph *n*th root functions.

EXAMPLE 1 Graph an *n*th Root Function

Graph $y = \sqrt[5]{x}$.

Enter the equation as **Y1** and graph.

KEYSTROKES: [Y=] 5 [MATH] 5 [X,T,θ,*n*] [GRAPH]

Another way to enter the equation is to use $y = x^{\frac{1}{5}}$.

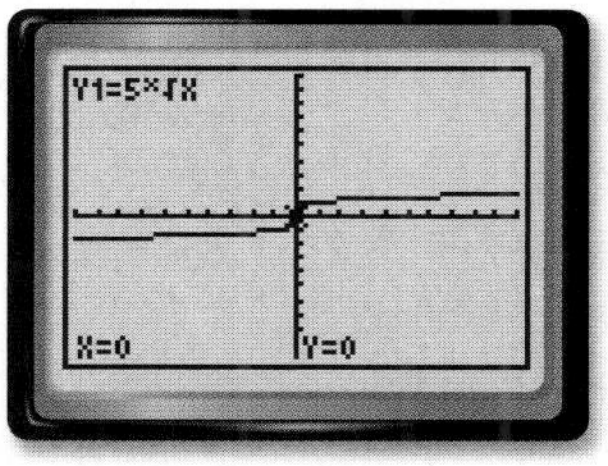

EXAMPLE 2 *n*th Root Functions with Different Roots

Graph and compare $y = \sqrt{x}$ and $y = \sqrt[4]{x}$.

Enter $y = \sqrt{x}$ as **Y1** and $y = \sqrt[4]{x}$ as **Y2**. Then graph.

KEYSTROKES: [Y=] [CLEAR] [2nd] [√] [X,T,θ,*n*] [ENTER] 4 [MATH] 5 [X,T,θ,*n*] [GRAPH]

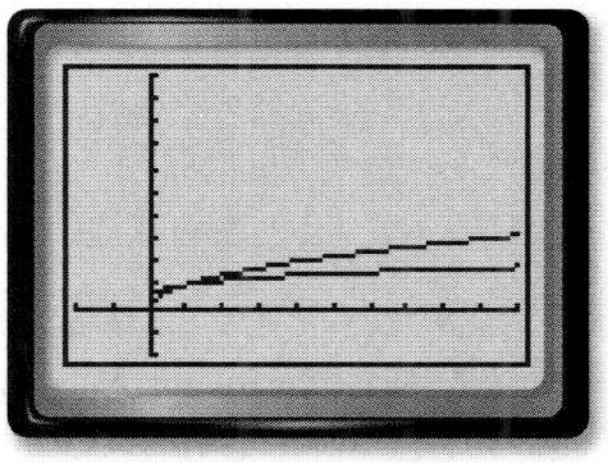

EXAMPLE 3 *n*th Root Functions with Different Radicands

Graph and compare $y = \sqrt[3]{x}$, $y = \sqrt[3]{x + 4}$, and $y = \sqrt[3]{x} + 4$.

Enter $y = \sqrt[3]{x}$ as **Y1**, $y = \sqrt[3]{x + 4}$ as **Y2**, and $y = \sqrt[3]{x} + 4$ as **Y3**. Then graph.

KEYSTROKES: [Y=] [CLEAR] 3 [MATH] 5 [X,T,θ,*n*] [ENTER] [CLEAR] 3 [MATH] 5 [(] [X,T,θ,*n*] [+] 4 [)] [ENTER] 3 [MATH] 5 [X,T,θ,*n*] [)] [+] 4 [ENTER] [ZOOM] 6

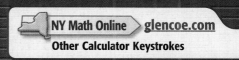

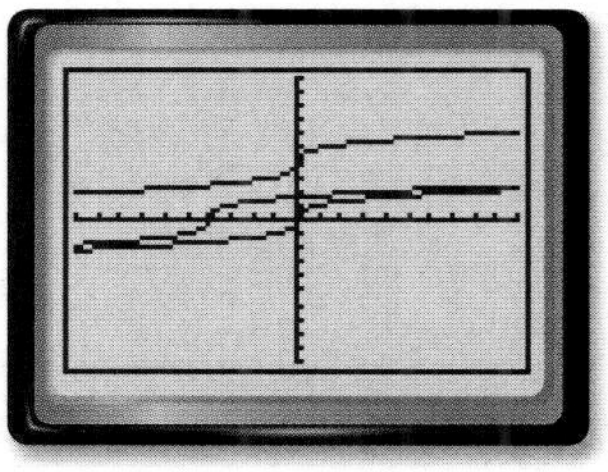

Exercises

Graph each function.

1. $y = \sqrt[4]{x}$
2. $y = \sqrt[4]{x + 2}$
3. $y = \sqrt[4]{x} + 2$
4. $y = \sqrt[5]{x}$
5. $y = \sqrt[5]{x - 5}$
6. $y = \sqrt[5]{x} - 5$
7. What is the effect of adding or subtracting a constant under the radical sign?
8. What is the effect of adding or subtracting a constant outside the radical sign?

Mid-Chapter Quiz

Lessons 7-1 through 7-4

NYS Core Curriculum
A2.A.13, A2.A.45

Given $f(x) = 2x^2 + 4x - 3$ and $g(x) = 5x - 2$, find each function.

1. $(f + g)(x)$
2. $(f - g)(x)$
3. $(f \cdot g)(x)$
4. $\left(\frac{f}{g}\right)(x)$
5. $[f \circ g](x)$
6. $[g \circ f](x)$

7. **SHOPPING** Mrs. Ross is shopping for her children's school clothes. She has a coupon for 25% off her total. The sales tax of 6% is added to the total after the coupon is applied.
   a. Express the total price after the discount and the total price after the tax using function notation. Let x represent the price of the clothing, $p(x)$ represent the price after the 25% discount, and $g(x)$ represent the price after the tax is added.
   b. Which composition of functions represents the final price, $p[g(x)]$ or $g[p(x)]$? Explain your reasoning.

Determine whether each pair of functions are inverse functions. Write *yes* or *no*.

8. $f(x) = 2x + 16$, $g(x) = \frac{1}{2}x - 8$
9. $g(x) = 4x + 15$, $h(x) = \frac{1}{4}x - 15$
10. $f(x) = x^2 - 5$, $g(x) = 5 + x^{-2}$
11. $g(x) = -6x + 8$, $h(x) = \frac{8 - x}{6}$

Find the inverse of each function, if it exists.

12. $h(x) = \frac{2}{5}x + 8$
13. $f(x) = \frac{4}{9}(x - 3)$
14. $h(x) = -\frac{10}{3}(x + 5)$
15. $f(x) = \frac{x + 12}{7}$

16. **JOBS** Louise runs a lawn care service. She charges \$25 for supplies plus \$15 per hour. The function $f(h) = 15h + 25$ gives the cost $f(h)$ for h hours of work.
   a. Find $f^{-1}(h)$. What is the significance of $f^{-1}(h)$?
   b. If Louise charges a customer \$85, how many hours did she work?

Graph each inequality.

17. $y < \sqrt{x - 5}$
18. $y \leq -2\sqrt{x}$
19. $y > \sqrt{x + 9} + 3$
20. $y \geq \sqrt{x + 4} - 5$

Graph each function. State the domain and range of each function.

21. $y = 2 + \sqrt{x}$
22. $y = \sqrt{x + 4} - 1$

23. **MULTIPLE CHOICE** What is the domain of $f(x) = \sqrt{2x + 5}$?
   A $\left\{x \middle| x > \frac{5}{2}\right\}$
   B $\left\{x \middle| x > -\frac{5}{2}\right\}$
   C $\left\{x \middle| x \geq \frac{5}{2}\right\}$
   D $\left\{x \middle| x \geq -\frac{5}{2}\right\}$

Simplify.

24. $\pm\sqrt{121a^4b^{18}}$
25. $\sqrt{(x^4 + 3)^{12}}$
26. $\sqrt[3]{27(2x - 5)^{15}}$
27. $\sqrt[5]{-(y - 6)^{20}}$
28. $\sqrt[3]{8(x + 4)^6}$
29. $\sqrt[4]{16(y + x)^8}$

30. **MULTIPLE CHOICE** The radius of the cylinder below is equal to the height of the cylinder. The radius r can be found using the formula $r = \sqrt[3]{\frac{V}{\pi}}$. Find the radius of the cylinder if the volume is 500 cubic inches.

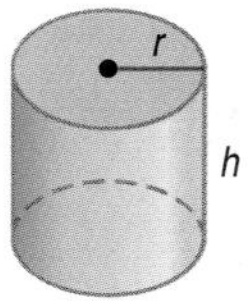

   F 2.53 inches
   G 5.42 inches
   H 7.94 inches
   J 24.92 inches

31. **PRODUCTION** The cost in dollars of producing x cell phones in a factory is represented by $C(p) = 5p + 60$. The number of cell phones produced in h hours is represented by $P(h) = 40h$.
   a. Find the composition function.
   b. Determine the cost of producing cell phones for 8 hours.

7-5 Operations with Radical Expressions

Then
You simplified expressions with nth roots. (Lesson 7-4)

Now
- Simplify radical expressions.
- Add, subtract, multiply, and divide radical expressions.

NYS Core Curriculum
A2.A.14 Perform addition, subtraction, multiplication, and division of radical expressions
A2.A.15 Rationalize denominators involving algebraic radical expressions
Also addresses A2.N.2, A2.N.4, and A2.N.5.

New Vocabulary
rationalizing the denominator
like radical expressions
conjugate

NY Math Online
glencoe.com
- Extra Examples
- Personal Tutor
- Self-Check Quiz
- Homework Help
- Math in Motion

Why?

Golden rectangles have been used by artists and architects to create beautiful designs. Many golden rectangles appear in the Parthenon in Athens, Greece. The ratio of the lengths of the sides of a golden rectangle is $\frac{2}{\sqrt{5}-1}$. In this lesson, you will learn to simplify radical expressions like $\frac{2}{\sqrt{5}-1}$.

Simplify Radicals The properties you have used to simplify radical expressions involving square roots also hold true for expressions involving nth roots.

Key Concept: Product Property of Radicals (For Your FOLDABLE)

Words For any real numbers a and b and any integer $n > 1$, $\sqrt[n]{ab} = \sqrt[n]{a} \cdot \sqrt[n]{b}$, if n is even and a and b are both nonnegative or if n is odd.

Examples $\sqrt{2} \cdot \sqrt{8} = \sqrt{16}$ or 4 and $\sqrt[3]{3} \cdot \sqrt[3]{9} = \sqrt[3]{27}$ or 3

In order for a radical to be in simplest form, the radicand must contain no factors that are nth powers of an integer or polynomial.

EXAMPLE 1 Simplify Expressions with the Product Property

Simplify.

a. $\sqrt{32x^8}$

$\sqrt{32x^8} = \sqrt{4^2 \cdot 2 \cdot (x^4)^2}$ **Factor into squares.**

$= \sqrt{4^2} \cdot \sqrt{(x^4)^2} \cdot \sqrt{2}$ **Product Property of Radicals**

$= 4x^4\sqrt{2}$ **Simplify.**

b. $\sqrt[4]{16a^{24}b^{13}}$

$\sqrt[4]{16a^{24}b^{13}} = \sqrt[4]{2^4 \cdot (a^6)^4(b^3)^4 \cdot b}$ **Factor into squares.**

$= \sqrt[4]{2^4} \cdot \sqrt[4]{(a^6)^4} \cdot \sqrt[4]{(b^3)^4} \cdot \sqrt[4]{b}$ **Product Property of Radicals**

$= 2a^6|b^3|\sqrt[4]{b}$ **Simplify.**

In this case, the absolute value symbols are not necessary because in order for $\sqrt[4]{16a^{24}b^{13}}$ to be defined, b must be nonnegative.

Thus, $\sqrt[4]{16a^{24}b^{13}} = 2a^6b^3\sqrt[4]{b}$.

Check Your Progress

1A. $\sqrt{12c^6d^3}$ **1B.** $\sqrt[3]{27y^{12}z^7}$

Personal Tutor glencoe.com

The Quotient Property of Radicals is another property used to simplify radicals.

Key Concept **Quotient Property of Radicals** — For Your FOLDABLE

Words For any real numbers a and $b \neq 0$ and any integer $n > 1$,

$\sqrt[n]{\frac{a}{b}} = \frac{\sqrt[n]{a}}{\sqrt[n]{b}}$, if all roots are defined.

Examples $\frac{\sqrt{27}}{\sqrt{3}} = \sqrt{9}$ or 3 $\quad \sqrt[3]{\frac{x^6}{8}} = \frac{\sqrt[3]{x^6}}{\sqrt[3]{8}} = \frac{x^2}{2}$ or $\frac{1}{2}x^2$

StudyTip

Exact Roots Exact roots occur when the powers of the constants and variables are all identical to or multiples of the index. For example, $\sqrt[3]{2} \cdot \sqrt[3]{2^2} = \sqrt[3]{2^3}$ or 2.

To eliminate radicals from a denominator or fractions from a radicand, you can use a process called **rationalizing the denominator**. To rationalize a denominator, multiply the numerator and denominator by a quantity so that the radicand has an exact root.

If the denominator is:	Multiply the numerator and denominator by:	Examples
$\sqrt{b}$	$\sqrt{b}$	$\frac{2}{\sqrt{3}} = \frac{2}{\sqrt{3}} \cdot \frac{\sqrt{3}}{\sqrt{3}}$ or $\frac{2\sqrt{3}}{3}$
$\sqrt[n]{b^x}$	$\sqrt[n]{b^{n-x}}$	$\frac{5}{\sqrt[3]{2}} = \frac{5}{\sqrt[3]{2}} \cdot \frac{\sqrt[3]{2^2}}{\sqrt[3]{2^2}}$ or $\frac{5\sqrt[3]{4}}{2}$

EXAMPLE 2 Simplify Expressions with the Quotient Property

Simplify.

a. $\sqrt{\frac{x^6}{y^7}}$

$\sqrt{\frac{x^6}{y^7}} = \frac{\sqrt{x^6}}{\sqrt{y^7}}$ **Quotient Property**

$= \frac{\sqrt{(x^3)^2}}{\sqrt{(y^3)^2 \cdot y}}$ **Factor into squares.**

$= \frac{\sqrt{(x^3)^2}}{\sqrt{(y^3)^2} \cdot \sqrt{y}}$ **Product Property**

$= \frac{x^3}{y^3\sqrt{y}}$ **Simplify.**

$= \frac{x^3}{y^3\sqrt{y}} \cdot \frac{\sqrt{y}}{\sqrt{y}}$ **Rationalize the denominator.**

$= \frac{x^3\sqrt{y}}{y^4}$ $\sqrt{y} \cdot \sqrt{y} = y$

b. $\sqrt[4]{\frac{6}{5x}}$

$\sqrt[4]{\frac{6}{5x}} = \frac{\sqrt[4]{6}}{\sqrt[4]{5x}}$ **Quotient Property**

$= \frac{\sqrt[4]{6}}{\sqrt[4]{5x}} \cdot \frac{\sqrt[4]{5^3x^3}}{\sqrt[4]{5^3x^3}}$ **Rationalize the denominator.**

$= \frac{\sqrt[4]{6 \cdot 5^3x^3}}{\sqrt[4]{5x \cdot 5^3x^3}}$ **Product Property**

$= \frac{\sqrt[4]{750x^3}}{\sqrt[4]{5^4x^4}}$ **Multiply.**

$= \frac{\sqrt[4]{750x^3}}{5x}$ $\sqrt[4]{5^4x^4} = 5x$

Check Your Progress

2A. $\frac{\sqrt{a^9}}{\sqrt{b^5}}$ **2B.** $\sqrt[5]{\frac{3}{4y}}$

Personal Tutor glencoe.com

Here is a summary of the rules used to simplify radicals.

Concept Summary — **Simplifying Radical Expressions** — For Your FOLDABLE

A radical expression is in simplified form when the following conditions are met.

- The index n is as small as possible.
- The radicand contains no factors (other than 1) that are nth powers of an integer or polynomial.
- The radicand contains no fractions.
- No radicals appear in a denominator.

Operations with Radicals You can use the Product and Quotient Properties to multiply and divide some radicals.

> Math *in Motion,* Animation glencoe.com

EXAMPLE 3 Multiply Radicals

Simplify $5\sqrt[3]{-12ab^4} \cdot 3\sqrt[3]{18a^2b^2}$.

$5\sqrt[3]{-12ab^4} \cdot 3\sqrt[3]{18a^2b^2} = 5 \cdot 3 \cdot \sqrt[3]{-12ab^4 \cdot 18a^2b^2}$	**Product Property of Radicals**
$= 15 \cdot \sqrt[3]{-2^2 \cdot 3 \cdot ab^4 \cdot 2 \cdot 3^2 \cdot a^2b^2}$	**Factor constants.**
$= 15 \cdot \sqrt[3]{-2^3 \cdot 3^3 \cdot a^3b^6}$	**Group into cubes if possible.**
$= 15 \cdot \sqrt[3]{-2^3} \cdot \sqrt[3]{3^3} \cdot \sqrt[3]{a^3} \cdot \sqrt[3]{b^6}$	**Product Property of Radicals**
$= 15 \cdot (-2) \cdot 3 \cdot a \cdot b^2$	**Simplify.**
$= -90ab^2$	**Multiply.**

Check Your Progress

Simplify.

3A. $6\sqrt{8c^3d^5} \cdot 4\sqrt{2cd^3}$

3B. $2\sqrt[4]{8x^3y^2} \cdot 3\sqrt[4]{2x^5y^2}$

Personal Tutor glencoe.com

Radicals can be added and subtracted in the same manner as monomials. In order to add or subtract, the radicals must be like terms. Radicals are **like radical expressions** if *both* the index and the radicand are identical.

Like: $\sqrt{3b}$ and $4\sqrt{3b}$

Unlike: $\sqrt{3b}$ and $\sqrt[3]{3b}$

Unlike: $\sqrt{2b}$ and $\sqrt{3b}$

> **StudyTip**
> **Adding and Subtracting Radicals** Simplify the individual radicals before attempting to combine like terms.

EXAMPLE 4 Add and Subtract Radicals

Simplify $\sqrt{98} - 2\sqrt{32}$.

$\sqrt{98} - 2\sqrt{32} = \sqrt{2 \cdot 7^2} - 2\sqrt{4^2 \cdot 2}$	**Factor using squares.**
$= \sqrt{7^2} \cdot \sqrt{2} - 2 \cdot \sqrt{4^2} \cdot \sqrt{2}$	**Product Property**
$= 7\sqrt{2} - 2 \cdot 4 \cdot \sqrt{2}$	**Simplify radicals.**
$= 7\sqrt{2} - 8\sqrt{2}$	**Multiply.**
$= -\sqrt{2}$	$(7 - 8)\sqrt{2} = (-1)(\sqrt{2})$

Check Your Progress

4A. $4\sqrt{8} + 3\sqrt{50}$

4B. $5\sqrt{12} + 2\sqrt{27} - \sqrt{128}$

Personal Tutor glencoe.com

Just as you can add and subtract radicals like monomials, you can multiply radicals using the FOIL method as you do when multiplying binomials.

EXAMPLE 5 Multiply Radicals

Simplify $(4\sqrt{3} + 5\sqrt{2})(3\sqrt{2} - 6)$.

$$\begin{aligned}(4\sqrt{3} + 5\sqrt{2})(3\sqrt{2} - 6) &= \overset{\textbf{F}}{4\sqrt{3} \cdot 3\sqrt{2}} + \overset{\textbf{O}}{4\sqrt{3} \cdot (-6)} + \overset{\textbf{I}}{5\sqrt{2} \cdot 3\sqrt{2}} + \overset{\textbf{L}}{5\sqrt{2} \cdot (-6)} \\ &= 12\sqrt{3 \cdot 2} - 24\sqrt{3} + 15\sqrt{2^2} - 30\sqrt{2} && \textbf{Product Property} \\ &= 12\sqrt{6} - 24\sqrt{3} + 30 - 30\sqrt{2} && \textbf{Simplify.}\end{aligned}$$

Check Your Progress

Simplify.

5A. $(6\sqrt{3} - 5)(2\sqrt{5} + 4\sqrt{2})$

5B. $(7\sqrt{2} - 3\sqrt{3})(7\sqrt{2} + 3\sqrt{3})$

Personal Tutor glencoe.com

StudyTip

Conjugates The product of conjugates is always a rational number.

Binomials of the form $a\sqrt{b} + c\sqrt{d}$ and $a\sqrt{b} - c\sqrt{d}$, where a, b, c, and d are rational numbers, are called **conjugates** of each other. You can use conjugates to rationalize denominators.

Real-World EXAMPLE 6 Use a Conjugate to Rationalize a Denominator

ARCHITECTURE Refer to the beginning of the lesson. Use a conjugate to rationalize the denominator and simplify $\frac{2}{\sqrt{5} - 1}$.

$$\begin{aligned}\frac{2}{\sqrt{5} - 1} &= \frac{2}{\sqrt{5} - 1} \cdot \frac{\sqrt{5} + 1}{\sqrt{5} + 1} && \sqrt{5} + 1 \text{ is the conjugate of } \sqrt{5} - 1. \\ &= \frac{2\sqrt{5} + 2(1)}{(\sqrt{5})^2 + 1(\sqrt{5}) - 1(\sqrt{5}) - 1(1)} && \textbf{Multiply.} \\ &= \frac{2\sqrt{5} + 2}{5 + \sqrt{5} - \sqrt{5} - 1} && \textbf{Simplify.} \\ &= \frac{2\sqrt{5} + 2}{4} && \textbf{Subtract.} \\ &= \frac{\sqrt{5} + 1}{2} && \textbf{Simplify.}\end{aligned}$$

Check Your Progress

6. GEOMETRY The area of the rectangle at the right is 900 ft^2. Write and simplify an equation for L in terms of x.

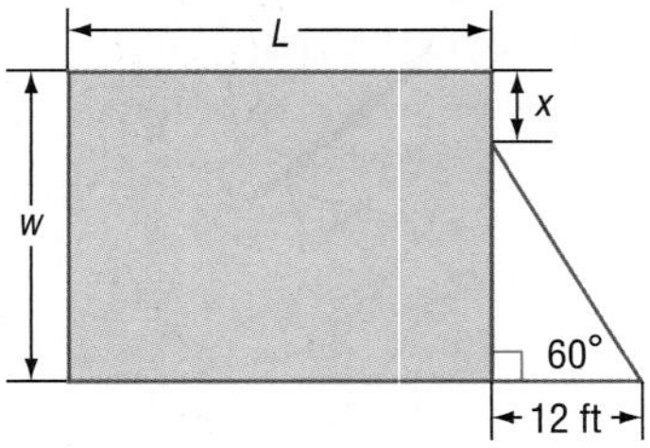

Personal Tutor glencoe.com

The Granger Collection, New York

Math History Link

Theano (c. 5th century B.C.) Theano is believed to have been the wife of Pythagoras. It is also believed that she directed a famous mathematics academy and carried on the work of Pythagoras after his death. Her most important work was on the idea of the golden mean, which is the irrational number $\frac{1 + \sqrt{5}}{2}$.

Check Your Understanding

Examples 1–5 pp. 439–442

Simplify.

1. $\sqrt{36ab^4c^5}$ **2.** $\sqrt{144x^7y^5}$ **3.** $\frac{\sqrt{c^5}}{\sqrt{d^9}}$ **4.** $\sqrt[4]{\frac{5x}{8y}}$

5. $5\sqrt{2x} \cdot 3\sqrt{8x}$ **6.** $4\sqrt{5a^5} \cdot \sqrt{125a^3}$

7. $3\sqrt[3]{36xy} \cdot 2\sqrt[3]{6x^2y^2}$ **8.** $\sqrt[4]{3x^3y^2} \cdot \sqrt[4]{27xy^2}$

9. $5\sqrt{32} + \sqrt{27} + 2\sqrt{75}$ **10.** $4\sqrt{40} + 3\sqrt{28} - \sqrt{200}$

11. $(4 + 2\sqrt{5})(3\sqrt{3} + 4\sqrt{5})$ **12.** $(8\sqrt{3} - 2\sqrt{2})(8\sqrt{3} + 2\sqrt{2})$

13. $\frac{5}{\sqrt{2} + 3}$ **14.** $\frac{8}{\sqrt{6} - 5}$

15. $\frac{4 + \sqrt{2}}{\sqrt{2} - 3}$ **16.** $\frac{6 - \sqrt{3}}{\sqrt{3} + 4}$

Example 6 p. 442

17. GEOMETRY Find the altitude of the triangle if the area is $189 + 4\sqrt{3}$ square centimeters.

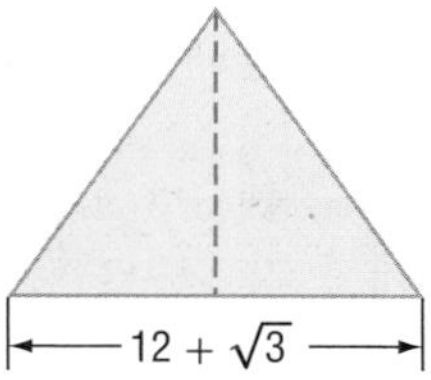

Practice and Problem Solving

● = **Step-by-Step Solutions** begin on page R20. **Extra Practice** begins on page 947.

Examples 1–4 pp. 439–441

Simplify.

18. $\sqrt{72a^8b^5}$ **19.** $\sqrt{9a^{15}b^3}$ **20.** $\sqrt{24a^{16}b^8c}$ **21.** $\sqrt{18a^6b^3c^5}$

22. $\frac{\sqrt{5a^5}}{\sqrt{b^{13}}}$ **23.** $\sqrt{\frac{7x}{10y^3}}$ **24.** $\frac{\sqrt[3]{6x^2}}{\sqrt[3]{5y}}$ **25.** $\sqrt[4]{\frac{7x^3}{4b^2}}$

26. $3\sqrt{5y} \cdot 8\sqrt{10yz}$ **27.** $2\sqrt{32a^3b^5} \cdot \sqrt{8a^7b^2}$ **28.** $6\sqrt{3ab} \cdot 4\sqrt{24ab^3}$

29. $5\sqrt{x^8y^3} \cdot 5\sqrt{2x^5y^4}$ **30.** $3\sqrt{90} + 4\sqrt{20} + \sqrt{162}$ **31.** $9\sqrt{12} + 5\sqrt{32} - \sqrt{72}$

32. $4\sqrt{28} - 8\sqrt{810} + \sqrt{44}$ **33.** $3\sqrt{54} + 6\sqrt{288} - \sqrt{147}$

34. GEOMETRY Find the perimeter of the rectangle.

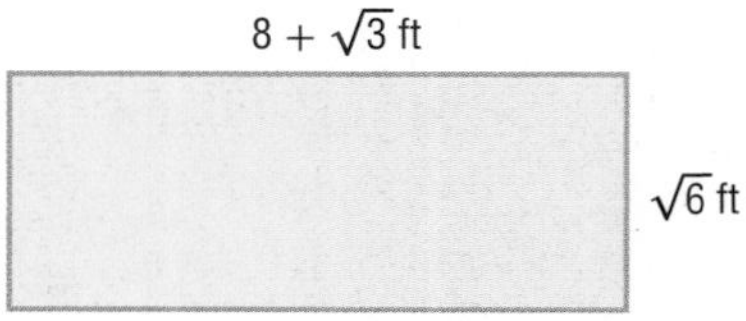

35 GEOMETRY Find the area of the rectangle.

36. GEOMETRY Find the exact surface area of a sphere with radius of $4 + \sqrt{5}$ inches.

Examples 5 and 6 p. 442

Simplify.

37. $(7\sqrt{2} - 3\sqrt{3})(4\sqrt{6} + 3\sqrt{12})$ **38.** $(8\sqrt{5} - 6\sqrt{3})(8\sqrt{5} + 6\sqrt{3})$

39. $(12\sqrt{10} - 6\sqrt{5})(12\sqrt{10} + 6\sqrt{5})$ **40.** $(6\sqrt{3} + 5\sqrt{2})(2\sqrt{6} + 3\sqrt{8})$

41. $\frac{6}{\sqrt{3} - \sqrt{2}}$ **42.** $\frac{\sqrt{2}}{\sqrt{5} - \sqrt{3}}$ **43.** $\frac{9 - 2\sqrt{3}}{\sqrt{3} + 6}$ **44.** $\frac{2\sqrt{2} + 2\sqrt{5}}{\sqrt{5} + \sqrt{2}}$

Simplify.

45. $\sqrt[3]{16y^4z^{12}}$

46. $\sqrt[3]{-54x^6y^{11}}$

47. $\sqrt[4]{162a^6b^{13}c}$

48. $\sqrt[4]{48a^9b^3c^{16}}$

49. $\sqrt[4]{\dfrac{12x^3y^2}{5a^2b}}$

50. $\dfrac{\sqrt[3]{36xy^2}}{\sqrt[3]{10xz}}$

51 $x + 1/\sqrt{x} - 1$

52. $\dfrac{x-2}{\sqrt{x^2-4}}$

53. $\dfrac{\sqrt{x}}{\sqrt{x^2-1}}$

Real-World Link

Fresh apples float because 25% of their volume is air.

54. APPLES The diameter of an apple is related to its weight and can be modeled by the formula $d = \sqrt[3]{3w}$, where d is the diameter in inches and w is the weight in ounces. Find the diameter of an apple that weighs 6.47 ounces.

Simplify each expression if *b* is an even number.

55. $\sqrt[b]{a^b}$

56. $\sqrt[b]{a^{4b}}$

57. $\sqrt[b]{a^{2b}}$

58. $\sqrt[b]{a^{3b}}$

59. MULTIPLE REPRESENTATIONS In this problem, you will explore operations with like radicals.

a. NUMERICAL Copy the diagram at the right on dot paper. Use the Pythagorean Theorem to prove that the length of the red segment is $\sqrt{2}$ units.

b. GRAPHICAL Extend the segment to represent $\sqrt{2} + \sqrt{2}$.

c. ANALYTICAL Use your drawing to show that $\sqrt{2} + \sqrt{2} \neq \sqrt{2+2}$ or 2.

d. GRAPHICAL Use the dot paper to draw a square with side lengths $\sqrt{2}$ units.

e. NUMERICAL Prove that the area of the square is $\sqrt{2} \cdot \sqrt{2} = 2$ square units.

H.O.T. Problems

Use Higher-Order Thinking Skills

60. FIND THE ERROR Twyla and Ben are simplifying $4\sqrt{32} + 6\sqrt{18}$. Is either of them correct? Explain your reasoning.

Twyla

$4\sqrt{32} + 6\sqrt{18}$

$= 4\sqrt{4^2 \cdot 2} + 6\sqrt{3^2 \cdot 2}$

$= 16\sqrt{2} + 18\sqrt{2}$

$= 34\sqrt{2}$

Ben

$4\sqrt{32} + 6\sqrt{18}$

$= 4\sqrt{16 \cdot 2} + 6\sqrt{9 \cdot 2}$

$= 64\sqrt{2} + 54\sqrt{2}$

$= 118\sqrt{2}$

61. CHALLENGE Show that $\dfrac{-1 - i\sqrt{3}}{2}$ is a cube root of 1.

62. REASONING For what values of a is $\sqrt{a} \cdot \sqrt{-a}$ a real number? Explain.

63. CHALLENGE Find four combinations of whole numbers that satisfy $\sqrt[a]{256} = b$.

64. OPEN ENDED Find a number other than 1 that has a positive whole number for a square root, cube root, and fourth root.

65. WRITING IN MATH Explain why absolute values may be unnecessary when an nth root of an even power results in an odd power.

NYSRE PRACTICE A2.A.14

66. PROBABILITY A six-sided number cube has faces with the numbers 1 though 6 marked on it. What is the probability that a number less than 4 will occur on one toss of the number cube?

A $\frac{1}{2}$ C $\frac{1}{4}$

B $\frac{1}{3}$ D $\frac{1}{5}$

67. When the number of a year is divisible by 4, the year is a leap year. However, when the year is divisible by 100, the year is not a leap year, unless the year is divisible by 400. Which is *not* a leap year?

F 1884 H 1904

G 1900 J 1940

68. SHORT RESPONSE Which property is illustrated by $4x + 0 = 4x$?

69. ACT/SAT The expression $\sqrt{180a^2b^8}$ is equivalent to which of the following?

A $5\sqrt{6}|a|b^4$

B $6\sqrt{5}|a|b^4$

C $3\sqrt{10}|a|b^4$

D $36\sqrt{5}|a|b^4$

Spiral Review

Simplify. (Lesson 7-4)

70. $\sqrt{81x^6}$ **71.** $\sqrt[3]{729a^3b^9}$ **72.** $\sqrt{(g+5)^2}$

73. Graph $y \leq \sqrt{x-2}$. (Lesson 7-3)

Solve each equation. (Lesson 6-5)

74. $x^4 - 34x^2 + 225 = 0$ **75.** $x^4 - 15x^2 - 16 = 0$ **76.** $x^4 + 6x^2 - 27 = 0$

77. $x^3 + 64 = 0$ **78.** $27x^3 + 1 = 0$ **79.** $8x^3 - 27 = 0$

80. MODELS A model car builder is building a display table for model cars. He wants the perimeter of the table to be 26 feet, but he wants the area of the table to be no more than 30 square feet. What could be the width of the table? (Lesson 5-8)

81. CONSTRUCTION Cho charges \$1500 to build a small deck and \$2500 to build a large deck. During the spring and summer, she built 5 more small decks than large decks. If she earned \$23,500 how many of each type of deck did she build? (Lesson 4-6)

82. FOOD The Hot Dog Grille offers the lunch combinations shown. Assume that the price of a combo meal is the same price as purchasing each item separately. Find the prices for a hot dog, a soda, and a bag of potato chips. (Lesson 3-5)

Lunch Combo Meals	
1. Two hot dogs, one soda	\$5.40
2. One hot dog, potato chips, one soda	\$4.35
3. Two hot dogs, two bags of chips	\$5.70

Skills Review

Evaluate each expression.

83. $2\left(\frac{1}{6}\right)$ **84.** $3\left(\frac{1}{8}\right)$ **85.** $\frac{1}{4} + \frac{1}{3}$

86. $\frac{1}{2} + \frac{3}{8}$ **87.** $\frac{2}{3} - \frac{1}{4}$ **88.** $\frac{5}{6} - \frac{2}{5}$

7-6 Rational Exponents

Then

You used properties of exponents. (Lesson 6-1)

Now

- Write expressions with rational exponents in radical form and vice versa.
- Simplify expressions in exponential or radical form.

NYS Core Curriculum

A2.A.10 Rewrite algebraic expressions with fractional exponents as radical expressions
A2.A.11 Rewrite algebraic expressions in radical form as expressions with fractional exponents
Also addresses A2.N.1, A2.N.3, and A2.A.8.

NY Math Online

glencoe.com
- Extra Examples
- Personal Tutor
- Self-Check Quiz
- Homework Help

Why?

The formula $C = c(1 + r)^n$ can be used to estimate the future cost of an item due to inflation. C represents the future cost, c represents the current cost, r is the rate of inflation, and n is the number of years for the projection.

For example, $C = c(1 + r)^{\frac{1}{2}}$ can be used to estimate the cost of a video game system in six months.

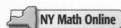

Rational Exponents and Radicals You know that squaring a number and taking the square root of a number are inverse operations. But how would you evaluate an expression that contains a fractional exponent such as the one above? You can investigate such an expression by assuming that fractional exponents behave as integral exponents.

$\left(b^{\frac{1}{2}}\right)^2 = b^{\frac{1}{2}} \cdot b^{\frac{1}{2}}$ **Write as a multiplication expression.**

$= b^{\frac{1}{2}+\frac{1}{2}}$ **Add the exponents.**

$= b^1$ or b **Simplify.**

Thus, $b^{\frac{1}{2}}$ is a number with a square equal to b. So $b^{\frac{1}{2}} = \sqrt{b}$.

Key Concept $b^{\frac{1}{n}}$ — For Your FOLDABLE

Words For any real number b and any positive integer n, $b^{\frac{1}{n}} = \sqrt[n]{b}$, except when $b < 0$ and n is even. When $b < 0$ and n is even, a complex root may exist.

Examples $27^{\frac{1}{3}} = \sqrt[3]{27}$ or 3 $\quad (-16)^{\frac{1}{2}} = \sqrt{-16}$ or $4i$

EXAMPLE 1 Radical and Exponential Forms

Simplify.

a. Write $x^{\frac{1}{6}}$ in radical form.

$x^{\frac{1}{6}} = \sqrt[6]{x}$ **Definition of $b^{\frac{1}{n}}$**

b. Write $\sqrt[4]{z}$ in exponential form.

$\sqrt[4]{z} = z^{\frac{1}{4}}$ **Definition of $b^{\frac{1}{n}}$**

Check Your Progress

1A. Write $a^{\frac{1}{5}}$ in radical form.

1B. Write $\sqrt[8]{c}$ in exponential form.

1C. Write $d^{\frac{7}{4}}$ in radical form.

1D. Write $\sqrt[3]{c^{-5}}$ in exponential form.

Personal Tutor glencoe.com

EXAMPLE 2 Evaluate Expressions with Rational Exponents

Evaluate each expression.

a. $81^{-\frac{1}{4}}$

$81^{-\frac{1}{4}} = \frac{1}{81^{\frac{1}{4}}}$ $b^{-n} = \frac{1}{b^n}$

$= \frac{1}{\sqrt[4]{81}}$ $81^{\frac{1}{4}} = \sqrt[4]{81}$

$= \frac{1}{\sqrt[4]{3^4}}$ $81 = 3^4$

$= \frac{1}{3}$ **Simplify.**

b. $216^{\frac{2}{3}}$

$216^{\frac{2}{3}} = (6^3)^{\frac{2}{3}}$ $216 = 6^3$

$= 6^{3 \cdot \frac{2}{3}}$ **Power of a Power**

$= 6^2$ **Multiply exponents.**

$= 36$ **Simplify.**

Check Your Progress

2A. $-3125^{-\frac{1}{5}}$

2B. $256^{\frac{3}{8}}$

Personal Tutor glencoe.com

Real-World Career

Economist Economists are employed in banking, finance, accounting, commerce, marketing, and business administration. Politicians often consult economists before enacting policy. Economists use shopping as one way of judging the economy. A bachelor's or master's degree in economics is required.

Examples 2a and 2b use the definition of $b^{\frac{1}{n}}$ and the properties of powers to evaluate an expression. Both methods suggest the following general definition of rational exponents.

Key Concept — Rational Exponents

For Your FOLDABLE

Words For any real nonzero number b, and any integers x and y, with $y > 1$, $b^{\frac{x}{y}} = \sqrt[y]{b^x} = \left(\sqrt[y]{b}\right)^x$, except when $b < 0$ and y is even. When $b < 0$ and y is even, a complex root may exist.

Examples $27^{\frac{2}{3}} = \left(\sqrt[3]{27}\right)^2 = 3^2$ or 9 $(-16)^{\frac{3}{2}} = \left(\sqrt{-16}\right)^3 = (4i)^3$ or $-64i$

Real-World EXAMPLE 3 Solve Equations with Rational Exponents

SHOPPING Refer to the beginning of the lesson. Suppose a video game system costs \$390 now. How much would the price increase in six months with an annual inflation rate of 5.3%?

$C = c(1 + r)^n$ **Original formula**

$= 390(1 + 0.053)^{\frac{1}{2}}$ $c = 390$, $r = 0.053$, and $n = \frac{6 \text{ months}}{12 \text{ months}}$ or $\frac{1}{2}$

≈ 400.20 **Use a calculator.**

In six months the price of the video game system will be \$400.20 − \$390.00 or \$10.20 more than its current price.

Check Your Progress

3. Suppose a gallon of milk costs \$2.99 now. How much would the price increase in 9 months with an inflation rate of 5.3%?

Personal Tutor glencoe.com

Simplify Expressions All of the properties of powers you learned in Lesson 6-1 apply to rational exponents. Write each expression with all positive exponents. Also, any exponents in the denominator of a fraction must be positive *integers*. So, it may be necessary to rationalize a denominator.

StudyTip

Simplifying Expressions When simplifying expressions containing rational exponents, leave the exponent in rational form rather than writing the expression as a radical.

EXAMPLE 4 Simplify Expressions with Rational Exponents

Evaluate each expression.

a. $a^{\frac{2}{7}} \cdot a^{\frac{4}{7}}$

$a^{\frac{2}{7}} \cdot a^{\frac{4}{7}} = a^{\frac{2}{7} + \frac{4}{7}}$ **Add powers.**

$= a^{\frac{6}{7}}$ **Add exponents.**

b. $b^{-\frac{5}{6}}$

$b^{-\frac{5}{6}} = \frac{1}{b^{\frac{5}{6}}}$ $b^{-n} = \frac{1}{b^n}$

$= \frac{1}{b^{\frac{5}{6}}} \cdot \frac{b^{\frac{1}{6}}}{b^{\frac{1}{6}}}$ **Why use** $\frac{b^{\frac{1}{6}}}{b^{\frac{1}{6}}}$**?**

$= \frac{b^{\frac{1}{6}}}{b^{\frac{6}{6}}}$ $b^{\frac{5}{6}} \cdot b^{\frac{1}{6}} = b^{\frac{5}{6} + \frac{1}{6}}$

$= \frac{b^{\frac{1}{6}}}{b}$ $b^{\frac{6}{6}} = b^1$ **or** b

Check Your Progress

4A. $p^{\frac{1}{4}} \cdot p^{\frac{9}{4}}$

4B. $r^{-\frac{4}{5}}$

Personal Tutor glencoe.com

When simplifying a radical expression, always use the least index possible. Using rational exponents makes this process easier, but the answer should be written in radical form.

EXAMPLE 5 Simplify Radical Expressions

Simplify each expression.

a. $\frac{\sqrt[4]{27}}{\sqrt{3}}$

$\frac{\sqrt[4]{27}}{\sqrt{3}} = \frac{27^{\frac{1}{4}}}{3^{\frac{1}{2}}}$ **Rational exponents**

$= \frac{(3^3)^{\frac{1}{4}}}{3^{\frac{1}{2}}}$ $27 = 3^3$

$= \frac{3^{\frac{3}{4}}}{3^{\frac{1}{2}}}$ **Power of a Power**

$= 3^{\frac{3}{4} - \frac{1}{2}}$ **Quotient of Powers**

$= 3^{\frac{1}{4}}$ **Simplify.**

$= \sqrt[4]{3}$ **Rewrite in radical form.**

b. $\sqrt[3]{64z^6}$

$\sqrt[3]{64z^6} = (64z^6)^{\frac{1}{3}}$ **Rational exponents**

$= (8^2 \cdot z^6)^{\frac{1}{3}}$ $64 = 8^2$

$= 8^{\frac{2}{3}} \cdot z^{\frac{6}{3}}$ **Power of a Power**

$= 4z^2$ $8^{\frac{2}{3}} = 4$

c. $\dfrac{x^{\frac{1}{2}} - 2}{3x^{\frac{1}{2}} + 2}$

$\dfrac{x^{\frac{1}{2}} - 2}{3x^{\frac{1}{2}} + 2} = \dfrac{x^{\frac{1}{2}} - 2}{3x^{\frac{1}{2}} + 2} \cdot \dfrac{3x^{\frac{1}{2}} - 2}{3x^{\frac{1}{2}} - 2}$ **$3x^{\frac{1}{2}} - 2$ is the conjugate of $3x^{\frac{1}{2}} + 2$.**

$= \dfrac{3x^{\frac{2}{2}} - 8x^{\frac{1}{2}} + 4}{9x^{\frac{2}{2}} - 4}$ **Multiply.**

$= \dfrac{3x - 8x^{\frac{1}{2}} + 4}{9x - 4}$ **Simplify.**

StudyTip

Radical Expressions Write the simplified expression in the same form as the beginning expression. When you start with a radical expression, end with a radical expression. When you start with an expression with rational exponents, end with an expression with rational exponents.

Check Your Progress

5A. $\dfrac{\sqrt[4]{32}}{\sqrt[3]{2}}$ **5B.** $\sqrt[3]{16x^4}$ **5C.** $\dfrac{y^{\frac{1}{2}} + 2}{y^{\frac{1}{2}} - 2}$

Personal Tutor glencoe.com

Concept Summary

For Your FOLDABLE

Expressions with Rational Exponents

An expression with rational exponents is simplified when all of the following conditions are met.

- It has no negative exponents.
- It has no fractional exponents in the denominator.
- It is not a complex fraction.
- The index of any remaining radical is the least number possible.

Check Your Understanding

Example 1 p. 446

Write each expression in radical form, or write each radical in exponential form.

1. $10^{\frac{1}{4}}$ **2.** $x^{\frac{3}{5}}$ **3.** $\sqrt[3]{15}$ **4.** $\sqrt[4]{7x^6y^9}$

Example 2 p. 447

Evaluate each expression.

5. $343^{\frac{1}{3}}$ **6.** $32^{-\frac{1}{5}}$ **7.** $125^{\frac{2}{3}}$ **8.** $\dfrac{24}{4^{\frac{3}{2}}}$

Example 3 p. 447

9. **GARDENING** If the area A of a square is known, then the lengths of its sides ℓ can be computed using $\ell = A^{\frac{1}{2}}$. You have purchased a 169 ft^2 share in a community garden for the season. What is the length of one side of your square garden?

Examples 4 and 5 p. 448

Simplify each expression.

10. $a^{\frac{3}{4}} \cdot a^{\frac{1}{2}}$ **11.** $\dfrac{x^{\frac{4}{5}}}{x^{\frac{1}{5}}}$ **12.** $\dfrac{b^3}{c^{\frac{1}{2}}} \cdot \dfrac{c}{b^{\frac{1}{3}}}$

13. $\sqrt[4]{9g^2}$ **14.** $\dfrac{\sqrt[5]{64}}{\sqrt[5]{4}}$ **15.** $\dfrac{g^{\frac{1}{2}} - 1}{g^{\frac{1}{2}} + 1}$

Practice and Problem Solving

= **Step-by-Step Solutions** begin on page R20.
Extra Practice begins on page 947.

Example 1 p. 446

Write each expression in radical form, or write each radical in exponential form.

16. $8^{\frac{1}{5}}$ **17.** $4^{\frac{2}{7}}$ **18.** $a^{\frac{3}{4}}$ **19.** $(x^3)^{\frac{3}{2}}$

20. $\sqrt{17}$ **21.** $\sqrt[4]{63}$ **22.** $\sqrt[3]{5xy^2}$ **23.** $\sqrt[4]{625x^2}$

Example 2 p. 447

Evaluate each expression.

24. $27^{\frac{1}{3}}$ **25.** $256^{\frac{1}{4}}$ **26.** $16^{-\frac{1}{2}}$ **27.** $81^{-\frac{1}{4}}$

Example 3 p. 447

28. **BASKETBALL** A women's regulation-sized basketball is slightly smaller than a men's basketball. The radius r of the ball that holds V cubic units of air is $\left(\frac{3V}{4\pi}\right)^{\frac{1}{3}}$.

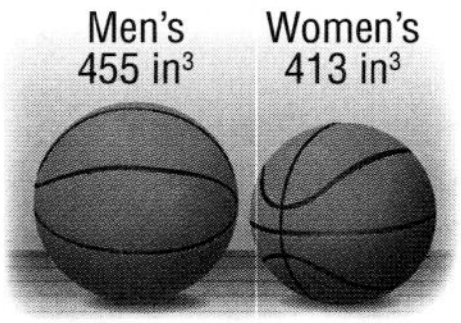

a. Find the radius of a women's basketball.

b. Find the radius of a men's basketball.

29. **GEOMETRY** The radius r of a sphere with volume V is given by $r = \left(\frac{3V}{4\pi}\right)^{\frac{1}{3}}$. Find the radius of a ball with a volume of 77 cm^3.

Examples 4 and 5 p. 448

Simplify each expression.

30. $x^{\frac{1}{3}} \cdot x^{\frac{2}{5}}$ **31.** $a^{\frac{4}{9}} \cdot a^{\frac{1}{4}}$ **32.** $b^{-\frac{3}{4}}$ **33.** $y^{-\frac{4}{5}}$ **34.** $\frac{\sqrt[8]{81}}{\sqrt[6]{3}}$

35 $\frac{\sqrt[4]{27}}{\sqrt[4]{3}}$ **36.** $\sqrt[4]{25x^2}$ **37.** $\sqrt[6]{81g^3}$ **38.** $\frac{h^{\frac{1}{2}} + 1}{h^{\frac{1}{2}} - 1}$ **39.** $\frac{x^{\frac{1}{4}} + 2}{x^{\frac{1}{4}} - 2}$

GEOMETRY **Find the area of each figure.**

40.

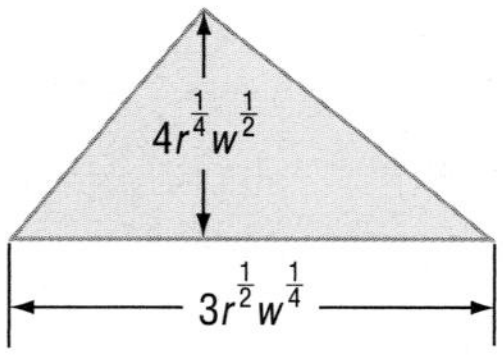

41.

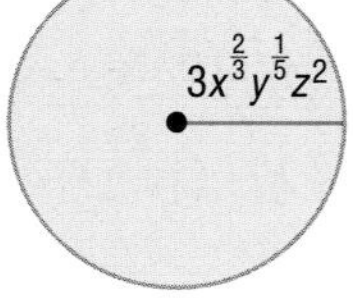

42. Find the simplified form of $18^{\frac{1}{2}} + 2^{\frac{1}{2}} - 32^{\frac{1}{2}}$.

43. What is the simplified form of $64^{\frac{1}{3}} - 32^{\frac{1}{3}} + 8^{\frac{1}{3}}$?

Simplify each expression.

44. $a^{\frac{7}{4}} \cdot a^{\frac{5}{4}}$ **45.** $x^{\frac{2}{3}} \cdot x^{\frac{8}{3}}$ **46.** $\left(b^{\frac{3}{4}}\right)^{\frac{1}{3}}$ **47.** $\left(y^{-\frac{3}{5}}\right)^{-\frac{1}{4}}$

48. $\sqrt[4]{64}$ **49.** $\sqrt[6]{216}$ **50.** $d^{-\frac{5}{6}}$ **51.** $w^{-\frac{7}{8}}$

Real-World Link

The most accurate method of determining the age of a deer is to look at a cross-section of a tooth. Each winter, deer develop a layer of cementum on the teeth, creating a ring, so each ring represents a year, just like the rings of a tree trunk.

Source: Whitetails

52. WILDLIFE A population of 100 deer is reintroduced to a wildlife preserve. Suppose the population does extremely well and the deer population doubles in two years. Then the number D of deer after t years is given by $D = 100 \cdot 2^{\frac{t}{2}}$.

a. How many deer will there be after $4\frac{1}{2}$ years?

b. Make a table that charts the population of deer every year for the next five years.

c. Make a graph using your table.

d. Using your table and graph, decide whether this is a reasonable trend over the long term. Explain.

Simplify each expression.

53. $\dfrac{f^{-\frac{1}{4}}}{4f^{\frac{1}{2}} \cdot f^{-\frac{1}{3}}}$ **54.** $\dfrac{g^{\frac{5}{2}}}{g^{\frac{1}{2}} + 2}$ **55.** $\dfrac{c^{\frac{2}{3}}}{c^{\frac{1}{6}}}$ **56.** $\dfrac{z^{\frac{4}{5}}}{z^{\frac{1}{2}}}$

57. $\sqrt{23} \cdot \sqrt[3]{23^2}$ **58.** $\sqrt[8]{36h^4j^4}$ **59.** $\sqrt{\sqrt{81}}$ **60.** $\sqrt[4]{\sqrt{256}}$

61. $\dfrac{ab}{\sqrt{c}}$ **62.** $\dfrac{xy}{\sqrt[3]{z}}$ **63.** $\dfrac{8^{\frac{1}{6}} - 9^{\frac{1}{4}}}{\sqrt{3} + \sqrt{2}}$ **64.** $\dfrac{x^{\frac{5}{3}} - x^{\frac{1}{3}}z^{\frac{4}{3}}}{x^{\frac{2}{3}} + z^{\frac{2}{3}}}$

65. MULTIPLE REPRESENTATIONS In this problem, you will explore the functions $f(x) = x^3$ and $g(x) = x^{\frac{1}{3}}$.

a. TABULAR Copy and complete the table to the right.

b. GRAPHICAL Graph $f(x)$ and $g(x)$.

c. VERBAL Explain the transformation between $f(x)$ and $g(x)$.

x	$f(x)$	$g(x)$
−2		
−1		
0		
1		
2		

H.O.T. Problems

Use Higher-Order Thinking Skills

66. REASONING Determine whether $-4^{-2} = (-4)^{-2}$ is *always*, *sometimes*, or *never* true. Explain your reasoning.

67. CHALLENGE Consider $\sqrt[4]{(-16)^3}$.

a. Explain why the expression is not a real number.

b. Find n such that $n\sqrt[4]{(-16)^3}$ is a real number.

68. OPEN ENDED Find two different expressions that equal 2 in the form $x^{\frac{1}{a}}$.

69. WRITING IN MATH Explain how it might be easier to simplify an expression using rational exponents rather than using radicals.

70. FIND THE ERROR Ayana and Kenji are simplifying $\dfrac{x^{\frac{3}{4}}}{x^{\frac{1}{2}}}$. Is either of them correct? Explain your reasoning.

Ayana

$\dfrac{x^{\frac{3}{4}}}{x^{\frac{1}{2}}} = x^{\frac{3}{4} + \frac{1}{2}}$

$= x^{\frac{3}{4} + \frac{2}{4}}$

$= x^{\frac{5}{4}}$

Kenji

$\dfrac{x^{\frac{3}{4}}}{x^{\frac{1}{2}}} = x^{\frac{3}{4} \div \frac{1}{2}}$

$= x^{\frac{3}{4} \cdot \frac{2}{1}}$

$= x^{\frac{3}{2}}$

NYSRE PRACTICE A2.A.10

71. The expression $\sqrt{56 - c}$ is equivalent to a positive integer when c is equal to

A 8
B −8
C 56
D 36

72. ACT/SAT Which of the following sentences is true about the graphs of $y = 2(x - 3)^2 + 1$ and $y = 2(x + 3)^2 + 1$?

F Their vertices are maximums.
G The graphs have the same shape with different vertices.
H The graphs have different shapes with different vertices.
J One graph has a vertex that is a maximum while the other graph has a vertex that is a minimum.

73. GEOMETRY What is the converse of the statement?
If it is summer, then it is hot outside.

A If it is not hot outside, then it is not summer.
B If it is not summer, then it is not hot outside.
C If it is hot outside, then it is summer.
D If it is hot outside, it is not summer.

74. SHORT RESPONSE If $3^5 \cdot p = 3^3$, then find p.

Spiral Review

Simplify. (Lesson 7-5)

75. $\sqrt{243}$

76. $\sqrt[3]{16y^3}$

77. $3\sqrt[3]{56y^6z^3}$

78. PHYSICS The speed of sound in a liquid is $s = \sqrt{\frac{B}{d}}$, where B is the bulk modules of the liquid and d is its density. For water, $B = 2.1 \times 10^9$ N/m^2 and $d = 10^3$ kg/m^3. Find the speed of sound in water to the nearest meter per second. (Lesson 7-4)

Find $p(-4)$ and $p(x + h)$ for each function. (Lesson 6-3)

79. $p(x) = x - 2$

80. $p(x) = -x + 4$

81. $p(x) = 6x + 3$

82. $p(x) = x^2 + 5$

83. $p(x) = x^2 - x$

84. $p(x) = 2x^3 - 1$

Solve each equation by factoring. (Lesson 5-3)

85. $x^2 - 11x = 0$

86. $x^2 + 6x - 16 = 0$

87. $4x^2 - 13x = 12$

88. $x^2 - 14x = -49$

89. $x^2 + 9 = 6x$

90. $x^2 - 3x = -\frac{9}{4}$

91. GEOMETRY A rectangle is inscribed in an isosceles triangle as shown. Find the dimensions of the inscribed rectangle with maximum area. (*Hint*: Use similar triangles.) (Lesson 5-1)

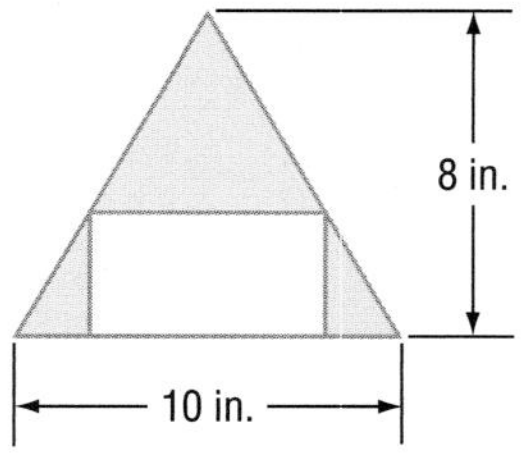

Skills Review

Find each power. (Lesson 7-5)

92. $(\sqrt{x - 3})^2$

93. $(\sqrt[3]{3x - 4})^3$

94. $(\sqrt[4]{7x - 1})^4$

95. $(\sqrt{x} - 4)^2$

96. $(2\sqrt{x} - 5)^2$

97. $(3\sqrt{x} + 1)^2$

7-7 Solving Radical Equations and Inequalities

Then

You solved polynomial equations. (Lesson 6-5)

Now

- Solve equations containing radicals.
- Solve inequalities containing radicals.

NYS Core Curriculum

A2.A.22 Solve radical equations

New Vocabulary

radical equation
extraneous solution
radical inequality

NY Math Online

glencoe.com

- Extra Examples
- Personal Tutor
- Self-Check Quiz
- Homework Help

Why?

When you jump, the time that you are in the air is your hang time. Hang time can be calculated in seconds t if you know the height h of the jump in feet. The formula for hang time is $t = 0.5\sqrt{h}$.

Michael Jordan had a hang time of about 0.98 second. How would you calculate the height of Jordan's jump?

Solve Radical Equations **Radical equations** include radical expressions. You can solve a radical equation by raising each side of the equation to a power.

Key Concept — Solving Radical Equations

For Your FOLDABLE

Step 1 Isolate the radical on one side of the equation.

Step 2 Raise each side of the equation to a power equal to the index of the radical to eliminate the radical.

Step 3 Solve the resulting polynomial equation.

When solving radical equations, the result may be a number that does not satisfy the original equation. Such a number is called an **extraneous solution**.

EXAMPLE 1 Solve Radical Equations

Solve each equation.

a. $\sqrt{x+2} + 4 = 7$

$\sqrt{x+2} + 4 = 7$	Original equation
$\sqrt{x+2} = 3$	Subtract 4 from each side to isolate the radical.
$(\sqrt{x+2})^2 = 3^2$	Square each side to eliminate the radical.
$x + 2 = 9$	Find the squares.
$x = 7$	Subtract 2 from each side.

CHECK

$\sqrt{x+2} + 4 = 7$	Original equation
$\sqrt{7+2} + 4 \stackrel{?}{=} 7$	Replace x with 7.
$7 = 7$ ✓	Simplify.

b. $\sqrt{x-12} = 2 - \sqrt{x}$

$\sqrt{x-12} = 2 - \sqrt{x}$	Original equation
$(\sqrt{x-12})^2 = (2 - \sqrt{x})^2$	Square each side.
$x - 12 = 4 - 4\sqrt{x} + x$	Find the squares.
$-16 = -4\sqrt{x}$	Isolate the radical.
$4 = \sqrt{x}$	Divide each side by -4.
$16 = x$	Evaluate the squares.

StudyTip

Solution Check You can use a graphing calculator to check solutions of equations. Graph each side of the original equation, and examine the intersection.

CHECK $\sqrt{x-12} = 2 - \sqrt{x}$

$\sqrt{16-12} \stackrel{?}{=} 2 - \sqrt{16}$

$\sqrt{4} \stackrel{?}{=} 2 - 4$

$2 \neq -2$ ✗

The solution does not check, so the equation has an extraneous solution. The graphs of $y = \sqrt{x-12}$ and $y = 2 - \sqrt{x}$ do not intersect, which confirms that there is no real solution.

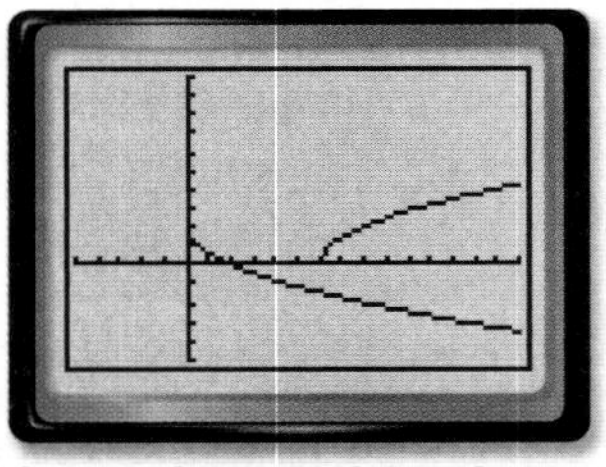

[−10, 30] scl: 2 by [−5, 10] scl: 1

Check Your Progress

1A. $5 = \sqrt{x-2} - 1$

1B. $\sqrt{x+15} = 5 + \sqrt{x}$

Personal Tutor glencoe.com

To undo a square root, you square the expression. To undo a cube root, you must raise the expression to the third power.

EXAMPLE 2 Solve a Cube Root Equation

Solve $2(6x-3)^{\frac{1}{3}} - 4 = 0$.

In order to remove the $\frac{1}{3}$ power, or cube root, you must first isolate it and then raise each side of the equation to the third power.

$2(6x-3)^{\frac{1}{3}} - 4 = 0$	**Original equation**
$2(6x-3)^{\frac{1}{3}} = 4$	**Add 4 to each side.**
$(6x-3)^{\frac{1}{3}} = 2$	**Divide each side by 2.**
$\left[(6x-3)^{\frac{1}{3}}\right]^3 = 2^3$	**Cube each side.**
$6x - 3 = 8$	**Evaluate the cubes.**
$6x = 11$	**Add 3 to each side.**
$x = \frac{11}{6}$	**Divide each side by 6.**

CHECK

$2(6x-3)^{\frac{1}{3}} - 4 = 0$	**Original equation**
$2\left(6 \cdot \frac{11}{6} - 3\right)^{\frac{1}{3}} - 4 \stackrel{?}{=} 0$	**Replace x with $\frac{11}{6}$.**
$2(8)^{\frac{1}{3}} - 4 \stackrel{?}{=} 0$	**Simplify.**
$2(2) - 4 \stackrel{?}{=} 0$	**The cube root of 8 is 2.**
$0 = 0$ ✓	**Subtract.**

Check Your Progress

Solve each equation.

2A. $(3n+2)^{\frac{1}{3}} + 1 = 0$

2B. $3(5y-1)^{\frac{1}{3}} - 2 = 0$

Personal Tutor glencoe.com

You can apply the methods used to solve square and cube root equations to solving equations with roots of any index. To undo an nth root, raise to the nth power.

Test-TakingTip

Substitute Values You could also solve the test question by substituting each answer for n in the equation to see if the solution is correct.

NYSRE EXAMPLE 3 A2.A.22

What is the solution of $3(\sqrt[4]{2n+6}) - 6 = 0$?

A -1 B 1 C 5 D 11

$3(\sqrt[4]{2n+6}) - 6 = 0$	**Original equation**
$3(\sqrt[4]{2n+6}) = 6$	**Add 6 to each side.**
$\sqrt[4]{2n+6} = 2$	**Divide each side by 3.**
$(\sqrt[4]{2n+6})^4 = 2^4$	**Raise each side to the fourth power.**
$2n + 6 = 16$	**Evaluate each side.**
$2n = 10$	**Subtract 6 from each side.**
$n = 5$	The answer is C.

Check Your Progress

3. What is the solution of $4(3x+6)^{\frac{1}{4}} - 12 = 0$?

F $x = 7$ G $x = 25$ H $x = 29$ J $x = 37$

Personal Tutor glencoe.com

Solve Radical Inequalities A **radical inequality** has a variable in the radicand. To solve radical inequalities, complete the following steps.

Key Concept **Solving Radical Inequalities** *For Your FOLDABLE*

Step 1 If the index of the root is even, identify the values of the variable for which the radicand is nonnegative.

Step 2 Solve the inequality algebraically.

Step 3 Test values to check your solution.

StudyTip

Radical Inequalities Since a principal square root is never negative, inequalities that simplify to the form $\sqrt{ax+b} \leq c$, where c is a negative number, have no solutions.

EXAMPLE 4 Solve a Radical Inequality

Solve $3 + \sqrt{5x - 10} \leq 8$.

Step 1 Since the radicand of a square root must be greater than or equal to zero, first solve $5x - 10 \geq 0$ to identify the values of x for which the left side of the inequality is defined.

$5x - 10 \geq 0$	**Set the radicand ≥ 0.**
$5x \geq 10$	**Add 10 to each side.**
$x \geq 2$	**Divide each side by 5.**

Step 2 Solve $3 + \sqrt{5x - 10} \leq 8$.

$3 + \sqrt{5x - 10} \leq 8$	**Original inequality**
$\sqrt{5x - 10} \leq 5$	**Isolate the radical.**
$5x - 10 \leq 25$	**Eliminate the radical.**
$5x \leq 35$	**Add 10 to each side.**
$x \leq 7$	**Divide each side by 5.**

Step 3 It appears that $2 \le x \le 7$. You can test some x-values to confirm the solution. Use three test values: one less than 2, one between 2 and 7, and one greater than 7. Organize the test values in a table.

$x = 0$	$x = 4$	$x = 9$
$3 + \sqrt{5(0) - 10} \overset{?}{\le} 8$ $3 + \sqrt{-5} \le 8$ ✗ Since $\sqrt{-5}$ is not a real number, the inequality is not satisfied.	$3 + \sqrt{5(4) - 10} \overset{?}{\le} 8$ $6.16 \le 8$ ✓ Since $6.16 \le 8$, the inequality is satisfied.	$3 + \sqrt{5(9) - 10} \overset{?}{\le} 8$ $8.92 \le 8$ ✗ Since $8.92 \not\le 8$, the inequality is not satisfied.

The solution checks. Only values in the interval $2 \le x \le 7$ satisfy the inequality. You can summarize the solution with a number line.

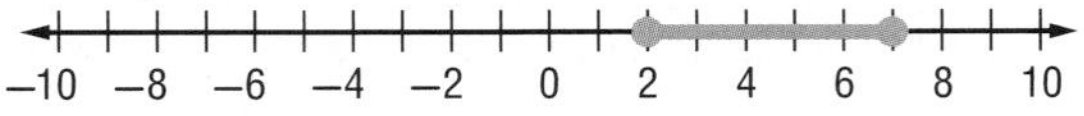

Check Your Progress

Solve each inequality.

4A. $\sqrt{2x + 2} + 1 \ge 5$

4B. $\sqrt{4x - 4} - 2 < 4$

Personal Tutor glencoe.com

Check Your Understanding

Examples 1 and 2 pp. 453–454

Solve each equation.

1. $\sqrt{x - 4} + 6 = 10$

2. $\sqrt{x + 13} - 8 = -2$

3. $8 - \sqrt{x + 12} = 3$

4. $\sqrt{x - 8} + 5 = 7$

5. $\sqrt[3]{x - 2} = 3$

6. $(x - 5)^{\frac{1}{3}} - 4 = -2$

7. $(4y)^{\frac{1}{3}} + 3 = 5$

8. $\sqrt[3]{n + 8} - 6 = -3$

9. $\sqrt{y} - 7 = 0$

10. $2 + 4z^{\frac{1}{2}} = 0$

11. $5 + \sqrt{4y - 5} = 12$

12. $\sqrt{2t - 7} = \sqrt{t + 2}$

13. PHYSICS The time T in seconds that it takes a pendulum to make a complete swing back and forth is given by the formula $T = 2\pi\sqrt{\frac{L}{g}}$, where L is the length of the pendulum in feet and g is the acceleration due to gravity, 32 feet per second squared.

a. In Tokyo, Japan, a huge pendulum in the Shinjuku building measures 73 feet 9.75 inches. How long does it take for the pendulum to make a complete swing?

b. A clockmaker wants to build a pendulum that takes 20 seconds to swing back and forth. How long should the pendulum be?

Example 3 p. 455

14. MULTIPLE CHOICE Solve $(2y + 6)^{\frac{1}{4}} - 2 = 0$.

A $y = 1$ **B** $y = 5$ **C** $y = 11$ **D** $y = 15$

Example 4 p. 455

Solve each inequality.

15. $\sqrt{3x + 4} - 5 \le 4$

16. $\sqrt{b - 7} + 6 \le 12$

17. $2 + \sqrt{4y - 4} \le 6$

18. $\sqrt{3a + 3} - 1 \le 2$

19. $1 + \sqrt{7x - 3} > 3$

20. $\sqrt{3x + 6} + 2 \le 5$

21. $-2 + \sqrt{9 - 5x} \ge 6$

22. $6 - \sqrt{2y + 1} < 3$

Practice and Problem Solving

= **Step-by-Step Solutions** begin on page R20.
Extra Practice begins on page 947.

Example 1 p. 453

Solve each equation.

23. $\sqrt{2x+5}-4=3$
24. $6+\sqrt{3x+1}=11$
25. $\sqrt{x+6}=5-\sqrt{x+1}$
26. $\sqrt{x-3}=\sqrt{x+4}-1$
27. $\sqrt{x-15}=3-\sqrt{x}$
28. $\sqrt{x-10}=1-\sqrt{x}$
29. $6+\sqrt{4x+8}=9$
30. $2+\sqrt{3y-5}=10$
31. $\sqrt{x-4}=\sqrt{2x-13}$
32. $\sqrt{7a-2}=\sqrt{a+3}$
33. $\sqrt{x-5}-\sqrt{x}=-2$
34. $\sqrt{b-6}+\sqrt{b}=3$

35. **GRAVITY** Isabel accidentally dropped her keys from the top of a Ferris wheel. The formula $t=\frac{1}{4}\sqrt{d-h}$ describes the time t in seconds at which the keys are h meters above the ground and Isabel is d meters above the ground. If Isabel was 65 meters high when she dropped the keys, how many meters above the ground will the keys be after 2 seconds?

Example 2 p. 454

Solve each equation.

36. $(5n-6)^{\frac{1}{3}}+3=4$
37. $(5p-7)^{\frac{1}{3}}+3=5$
38. $(6q+1)^{\frac{1}{4}}+2=5$
39. $(3x+7)^{\frac{1}{4}}-3=1$
40. $(3y-2)^{\frac{1}{5}}+5=6$
41. $(4z-1)^{\frac{1}{5}}-1=2$
42. $2(x-10)^{\frac{1}{3}}+4=0$
43. $3(x+5)^{\frac{1}{3}}-6=0$
44. $\sqrt[3]{5x+10}-5=0$
45. $\sqrt[3]{4n-8}-4=0$
46. $\frac{1}{7}(14a)^{\frac{1}{3}}=1$
47. $\frac{1}{4}(32b)^{\frac{1}{3}}=1$

Example 3 p. 455

48. **MULTIPLE CHOICE** Solve $\sqrt[4]{y+2}+9=14$.

A 23 **B** 53 **C** 123 **D** 623

49. **MULTIPLE CHOICE** Solve $(2x-1)^{\frac{1}{4}}-2=1$.

F 41 **G** 28 **H** 13 **J** 1

Example 4 p. 455

Solve each inequality.

50. $1+\sqrt{5x-2}>4$
51. $\sqrt{2x+14}-6\geq 4$
52. $10-\sqrt{2x+7}\leq 3$
53. $6+\sqrt{3y+4}<6$
54. $\sqrt{2x+5}-\sqrt{9+x}>0$
55. $\sqrt{d+3}+\sqrt{d+7}>4$
56. $\sqrt{3x+9}-2<7$
57. $\sqrt{2y+5}+3\leq 6$
58. $-2+\sqrt{8-4z}\geq 8$
59. $-3+\sqrt{6a+1}>4$
60. $\sqrt{2}-\sqrt{b+6}\leq -\sqrt{b}$
61. $\sqrt{c+9}-\sqrt{c}>\sqrt{3}$

62. **PENDULUMS** The formula $s=2\pi\sqrt{\frac{\ell}{32}}$ represents the swing of a pendulum, where s is the time in seconds to swing back and forth, and ℓ is the length of the pendulum in feet. Find the length of a pendulum that makes one swing in 1.5 seconds.

63. **FISH** The relationship between the length and mass of certain fish can be approximated by the equation $L=0.46\sqrt[3]{M}$, where L is the length in meters and M is the mass in kilograms. Solve this equation for M.

64. **HANG TIME** Refer to the information at the beginning of the lesson regarding hang time. Describe how the height of a jump is related to the amount of time in the air. Write a step-by-step explanation of how to determine the height of Jordan's 0.98-second jump.

65. **CONCERTS** The organizers of a concert are preparing for the arrival of 50,000 people in the open field where the concert will take place. Each person is allotted 5 square feet of space, so the organizers rope off a circular area of 250,000 square feet. Using the formula $A = \pi r^2$, where A represents the area of the circular region and r represents the radius of the region, find the radius of this region.

Real-World Link

The Florida High School Athletic Association sanctioned girls' weightlifting as a high school sport in 1997. To date, they are the only state to support the sport for girls.

Source: *The New York Tmes*

66. **WEIGHTLIFTING** The formula $M = 512 - 146{,}230B^{-\frac{8}{5}}$ can be used to estimate the maximum total mass that a weightlifter of mass B kilograms can lift using the snatch and the clean and jerk. According to the formula, how much does a person weigh who can lift at most 470 kilograms?

H.O.T. Problems

Use **H**igher-**O**rder **T**hinking Skills

67. **WHICH ONE DOESN'T BELONG?** Which equation does not have a solution?

$\sqrt{x-1} + 3 = 4$	$\sqrt{x+1} + 3 = 4$
$\sqrt{x-2} + 7 = 10$	$\sqrt{x+2} - 7 = -10$

68. **CHALLENGE** Lola is working to solve $(x+5)^{\frac{1}{4}} = -4$. She said that she could tell there was no real solution without even working the problem. Is Lola correct? Explain your reasoning.

69. **REASONING** Determine whether $\frac{\sqrt{(x^2)^2}}{-x} = x$ is *sometimes*, *always*, or *never* true when x is a real number. Explain your reasoning.

70. **OPEN ENDED** Select a whole number. Now work backward to write two radical equations that have that whole number as solutions. Write one square root equation and one cube root equation. You may need to experiment until you find a whole number you can easily use.

71. **WRITING IN MATH** Explain the relationship between the index of the root of a variable in an equation and the power to which you raise each side of the equation to solve the equation.

72. **OPEN ENDED** Write an equation that can be solved by raising each side of the equation to the given power.

a. $\frac{3}{2}$ power **b.** $\frac{5}{4}$ power **c.** $\frac{7}{8}$ power

73. **CHALLENGE** Solve $7^{3x-1} = 49^{x+1}$ for x. (*Hint:* $b^x = b^y$ *if and only if* $x = y$.)

REASONING Determine whether the following statements are *sometimes*, *always*, or *never* true. Explain your reasoning.

74. If the denominator of a rational root is odd, there will be extraneous roots.

75. If the denominator of a rational root is even, there will be extraneous roots.

NYSRE PRACTICE A2.A.22

76. What is an equivalent form of $\frac{4}{5+i}$?

A $\frac{10-2i}{13}$ C $\frac{6-i}{6}$

B $\frac{5-i}{6}$ D $\frac{6-i}{13}$

77. Which set of points describes a function?

F $\{(3, 0), (-2, 5), (2, -1), (2, 9)\}$

G $\{(-3, 5), (-2, 3), (-1, 5), (0, 7)\}$

H $\{(2, 5), (2, 4), (2, 3), (2, 2)\}$

J $\{(3, 1), (-3, 2), (3, 3), (-3, 4)\}$

78. **GRIDDED RESPONSE** The perimeter of an isosceles triangle is 56 inches. If one leg is 20 inches long, what is the measure of the base of the triangle?

79. **ACT/SAT** If $\sqrt{x+5}+1=4$, what is the value of x?

A 4

B 10

C 11

D 20

Spiral Review

Evaluate. (Lesson 7-6)

80. $27^{-\frac{2}{3}}$

81. $9^{\frac{1}{3}} \cdot 9^{\frac{5}{3}}$

82. $\left(\frac{8}{27}\right)^{-\frac{2}{3}}$

83. **GEOMETRY** The measures of the legs of a right triangle can be represented by the expressions $4x^2y^2$ and $8x^2y^2$. Use the Pythagorean Theorem to find a simplified expression for the measure of the hypotenuse. (Lesson 7-5)

Find the inverse of each function. (Lesson 7-2)

84. $f(x) = 3x - 4$

85. $f(x) = -2x - 3$

86. $f(x) = x^2$

87. $f(x) = (2x + 3)^2$

For each graph,

a. describe the end behavior,

b. determine whether it represents an odd-degree or an even-degree polynomial function, and

c. state the number of real zeros. (Lesson 6-3)

88.

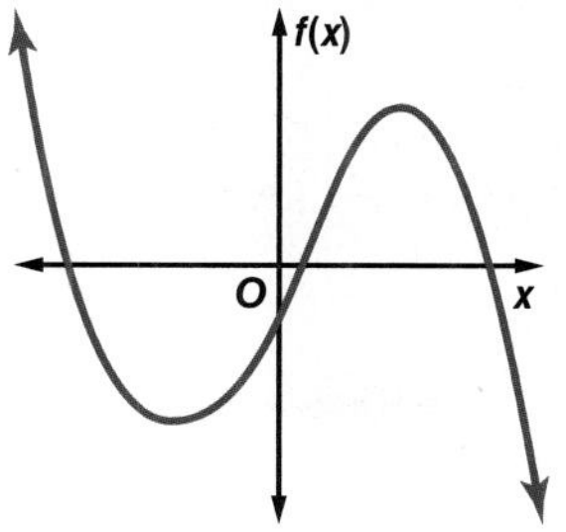

89.

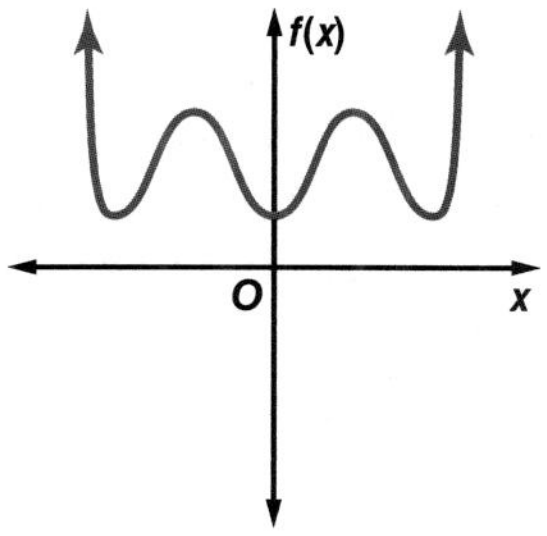

90.

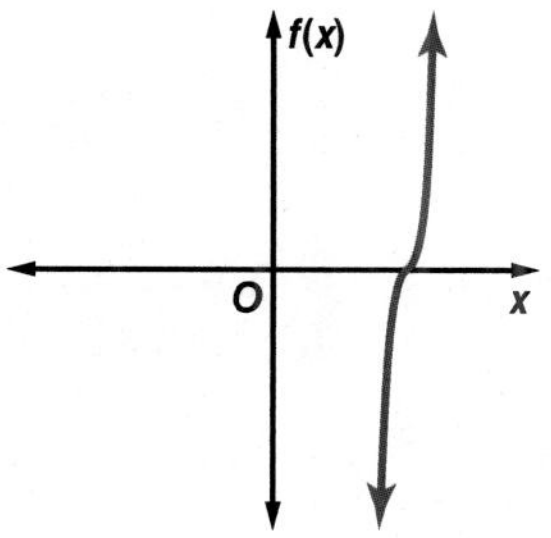

Skills Review

Solve each equation. Write in simplest form. (Lesson 1-3)

91. $\frac{8}{5}x = \frac{4}{15}$

92. $\frac{27}{14}y = \frac{6}{7}$

93. $\frac{3}{10} = \frac{12}{25}a$

94. $\frac{6}{7} = 9m$

95. $\frac{9}{8}b = 18$

96. $\frac{6}{7}n = \frac{3}{4}$

97. $\frac{1}{3}p = \frac{5}{6}$

98. $\frac{2}{3}q = 7$

EXTEND

7-7 Graphing Technology Lab
Solving Radical Equations and Inequalities

NY Math Online glencoe.com
Other Calculator Keystrokes

NYS Core Curriculum **A2.A.22 Solve radical equations**

You can use a TI-83/84 Plus graphing calculator to solve radical equations and inequalities. One way to do this is to rewrite the equation or inequality so that one side is 0. Then use the **zero** feature on the calculator.

EXAMPLE 1 Radical Equation

Solve $\sqrt{x} + \sqrt{x + 2} = 3$.

Step 1 Rewrite the equation.

- Subtract 3 from each side of the equation to get $\sqrt{x} + \sqrt{x + 2} - 3 = 0$.
- Enter the function $y = \sqrt{x} + \sqrt{x + 2} - 3$ in the Y= list.

KEYSTROKES: [Y=] [2nd] [x^2] [X,T,θ,n] [ )] [+] [2nd] [x^2] [X,T,θ,n] [+] 2 [ )] [−] 3 [ENTER]

Step 2 Use a table.

- You can use the **TABLE** function to locate intervals where the solution(s) lie. First, enter the starting value and the interval for the table.

KEYSTROKES: [2nd] [TBLSET] 0 [ENTER] 1 [ENTER]

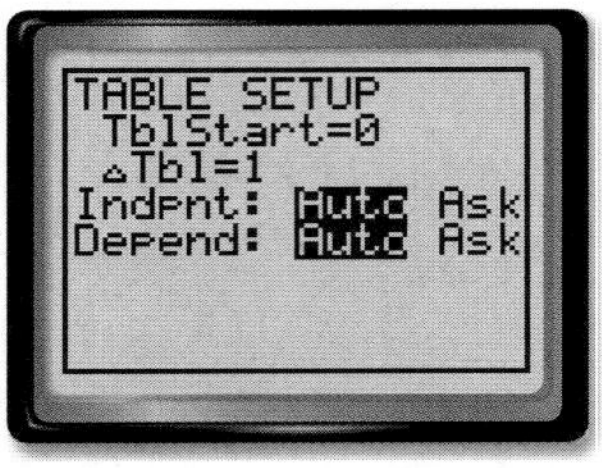

Step 3 Estimate the solution.

- Complete the table and estimate the solution(s).

KEYSTROKES: [2nd] [TABLE]

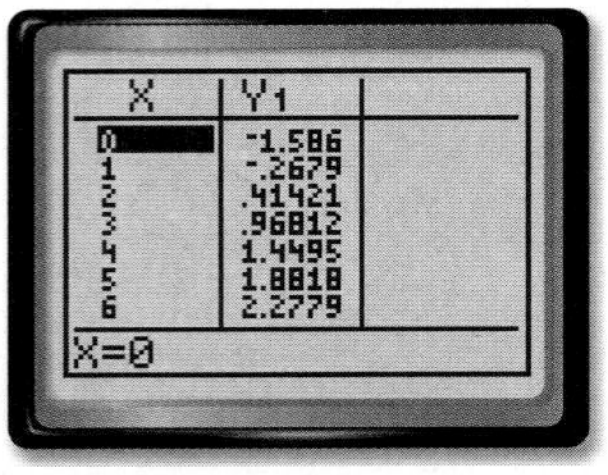

Since the function changes sign from negative to positive between $x = 1$ and $x = 2$, there is a solution between 1 and 2.

Step 4 Use the **zero** feature.

- Graph the function; then select **ZERO** from the **CALC** menu.

KEYSTROKES: [2nd] [CALC] 2

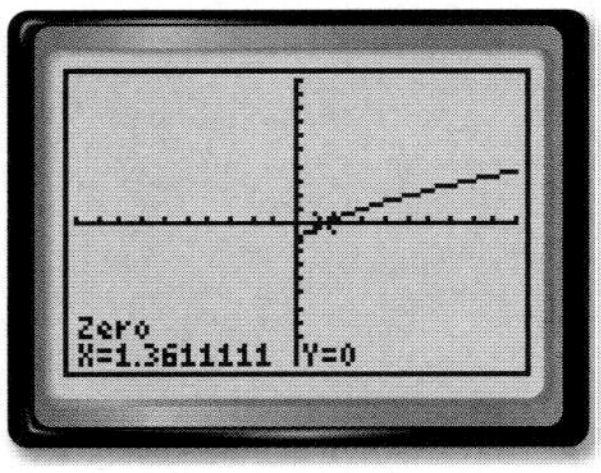

Place the cursor on a point at which $y < 0$ and press [ENTER] for the **LEFT BOUND**. Then place the cursor on a point at which $y > 0$ and press [ENTER] for the **RIGHT BOUND**. You can use the same point for the **GUESS** as for the **RIGHT BOUND**.

The solution is about 1.36. This is consistent with the estimate made by using the **TABLE**.

EXAMPLE 2 Radical Inequality

Solve $2\sqrt{x} > \sqrt{x+2} + 1$.

Step 1 Graph each side of the inequality and use the **trace** feature.

- In the Y= list, enter $y_1 = 2\sqrt{x}$ and $y_2 = \sqrt{x+2} + 1$. Then press [GRAPH].

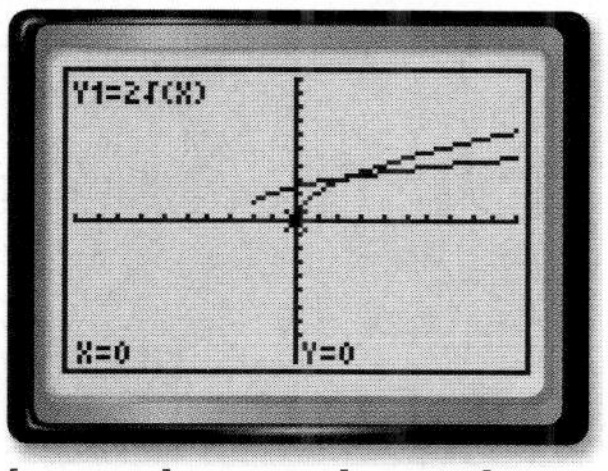

[−10, 10] scl: 1 by [−10, 10] scl: 1

- Press [TRACE]. You can use [▲] or [▼] to switch the cursor between the two curves.

The calculator screen above shows that, for points to the left of where the curves cross, Y1 < Y2 or $2\sqrt{x} < \sqrt{x+2} + 1$. To solve the original inequality, you must find points for which Y1 > Y2. These are the points to the right of where the curves cross.

Step 2 Use the **intersect** feature.

- You can use the **intersect** feature on the **CALC** menu to approximate the x-coordinate of the point at which the curves cross.

KEYSTROKES: [2nd] [CALC] 5

- Press [ENTER] for each of **FIRST CURVE?**, **SECOND CURVE?**, and **GUESS?**.

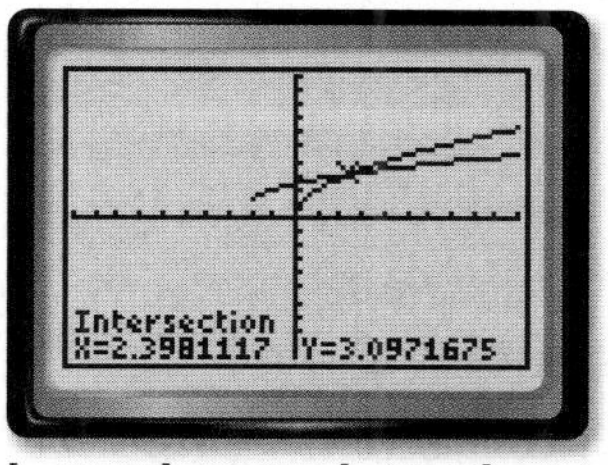

[−10, 10] scl: 1 by [−10, 10] scl: 1

The calculator screen shows that the x-coordinate of the point at which the curves cross is about 2.40. Therefore, the solution of the inequality is about $x > 2.40$. *Use the symbol > in the solution because the symbol in the original inequality is >.*

Step 3 Use the **table** feature to check your solution.

- Start the table at 2 and show x-values in increments of 0.1. Scroll through the table.

KEYSTROKES: [2nd] [TBLSET] 2 [ENTER] .1 [ENTER] [2nd] [TABLE]

Notice that when x is less than or equal to 2.4, Y1 < Y2. This verifies the solution $\{x \mid x > 2.40\}$.

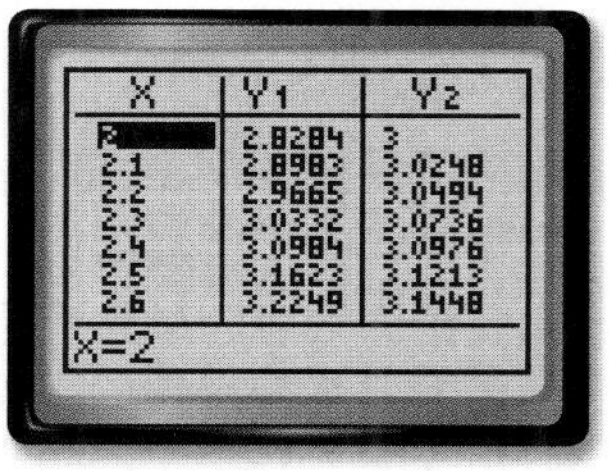

Exercises

Use a graphical method to solve each equation or inequality.

1. $\sqrt{x+4} = 3$
2. $\sqrt{3x-5} = 1$
3. $\sqrt{x+5} = \sqrt{3x+4}$
4. $\sqrt{x+3} + \sqrt{x-2} = 4$
5. $\sqrt{3x-7} = \sqrt{2x-2} - 1$
6. $\sqrt{x+8} - 1 = \sqrt{x+2}$
7. $\sqrt{x-3} \geq 2$
8. $\sqrt{x+3} > 2\sqrt{x}$
9. $\sqrt{x} + \sqrt{x-1} < 4$
10. **WRITING IN MATH** Explain how you could apply the technique in the first example to solving an inequality.

NY Math Online glencoe.com
• STUDY TO GO
• Vocabulary Review

Chapter Summary

Key Concepts

Operations on Functions (Lesson 7-1)

Operation	Definition
Sum	$(f + g)(x) = f(x) + g(x)$
Difference	$(f - g)(x) = f(x) - g(x)$
Product	$(f \cdot g)(x) = f(x) \cdot g(x)$
Quotient	$\left(\frac{f}{g}\right)(x) = \frac{f(x)}{g(x)}, g(x) \neq 0$
Composition	$[f \circ g](x) = f[g(x)]$

Inverse and Square Root Functions (Lessons 7-2 and 7-3)

- Reverse the coordinates of ordered pairs to find the inverse of a relation.
- Two functions are inverses if and only if both their compositions are the identity function.

Roots of Real Numbers (Lesson 7-4)

Real *n*th roots of b, $\sqrt[n]{b}$, or $-\sqrt[n]{b}$			
n	$\sqrt[n]{b}$ if $b > 0$	$\sqrt[n]{b}$ if $b < 0$	$\sqrt[n]{b}$ if $b = 0$
even	one positive root one negative root	no real roots	one real root, 0
odd	one positive root no negative roots	no positive roots one negative root	one real root, 0

Radicals (Lessons 7-5 through 7-7)

For any real numbers a and b and any integers n, x, and y, with $b \neq 0$, $n > 1$, and $y > 1$, the following are true.

- Product Property: $\sqrt[n]{ab} = \sqrt[n]{a} \cdot \sqrt[n]{b}$
- Quotient Property: $\sqrt[n]{\frac{a}{b}} = \frac{\sqrt[n]{a}}{\sqrt[n]{b}}$
- Rational Exponents: $b^{\frac{x}{y}} = \sqrt[y]{b^x} = \left(\sqrt[y]{b}\right)^x, b \geq 0$

FOLDABLES® Study Organizer

Be sure the Key Concepts are noted in your Foldable.

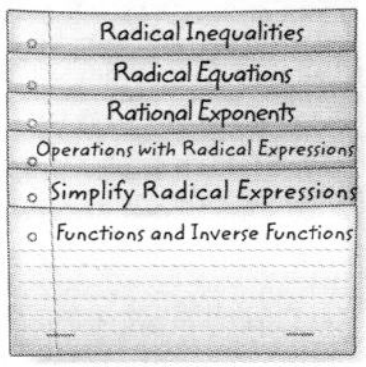

Key Vocabulary

composition of functions (p. 411)
conjugates (p. 442)
extraneous solution (p. 453)
index (p. 431)
inverse function (p. 417)
inverse relation (p. 417)
like radical expressions (p. 441)
***n*th root** (p. 431)
principal root (p. 431)
radical equation (p. 453)
radical function (p. 424)
radical inequality (p. 455)
radical sign (p. 431)
radicand (p. 431)
rationalizing the denominator (p. 440)
square root function (p. 424)
square root inequality (p. 426)

Vocabulary Check

Choose a word or term from the list above that best completes each statement.

1. If both compositions result in the ____________, then the functions are inverse functions.
2. Radicals are ____________ if *both* the index and the radicand are identical.
3. In a(n) ____________, the results of one function are used to evaluate a second function.
4. When there is more than one real root, the nonnegative root is called the ____________.
5. To eliminate radicals from a denominator or fractions from a radicand, you use a process called ____________.
6. Equations with radicals that have variables in the radicands are called ____________.
7. Two relations are ____________ if and only if one relation contains the element (b, a) when the other relation contains the element (a, b).
8. When solving a radical equation, sometimes you will obtain a number that does not satisfy the original equation. Such a number is called a(n) ____________.
9. The square root function is a type of ____________.

Lesson-by-Lesson Review

7-1 Operations on Functions (pp. 409–416)

A2.A.42

Find $[f \circ g](x)$ and $[g \circ f](x)$.

10. $f(x) = 2x + 1$
$g(x) = 4x - 5$

11. $f(x) = x^2 + 1$
$g(x) = x - 7$

12. $f(x) = x^2 + 4$
$g(x) = -2x + 1$

13. $f(x) = 4x$
$g(x) = 5x - 1$

14. $f(x) = x^3$
$g(x) = x - 1$

15. $f(x) = x^2 + 2x - 3$
$g(x) = x + 1$

16. MEASUREMENT The formula $f = \frac{y}{3}$ converts yards y to feet f and $n = \frac{f}{12}$ converts inches n to feet f. Write a composition of functions that converts yards to inches.

EXAMPLE 1

If $f(x) = x^2 + 3$ and $g(x) = 3x - 2$, find $g[f(x)]$ and $f[g(x)]$.

$g[f(x)] = 3(x^2 + 3) - 2$	Replace $f(x)$ with $x^2 + 3$.
$= 3x^2 + 9 - 2$	Multiply.
$= 3x^2 + 7$	Simplify.

$f[g(x)] = (3x - 2)^2 + 3$	Replace $g(x)$ with $3x - 2$.
$= 9x^2 - 12x + 4 + 3$	Multiply.
$= 9x^2 - 12x + 7$	Simplify.

7-2 Inverse Functions and Relations (pp. 417–422)

A2.A.44, A2.A.45

Find the inverse of each function. Then graph the function and its inverse.

17. $f(x) = 5x - 6$

18. $f(x) = -3x - 5$

19. $f(x) = \frac{1}{2}x + 3$

20. $f(x) = \frac{4x + 1}{5}$

21. $f(x) = x^2$

22. $f(x) = (2x + 1)^2$

23. SHOPPING Samuel bought a computer. The sales tax rate was 6% of the sale price, and he paid \$50 for shipping. Find the sale price if Samuel paid a total of \$1322.

Use the horizontal line test to determine whether the inverse of each function is also a function.

24. $f(x) = 3x^2$

25. $h(x) = x^3 - 3$

26. $g(x) = -3x^4 + 2x - 1$

27. $g(x) = 4x^3 - 5x$

28. $f(x) = -3x^5 + x^2 - 3$

29. $h(x) = 4x^4 + 7x$

30. BANKING During the last month, Jonathan has made two deposits of \$45, made a deposit of double his original balance, and has withdrawn \$35 five times. His balance is now \$189. Write an equation that models this problem. How much money did Jonathan have in his account at the beginning of the month?

EXAMPLE 2

Find the inverse of $f(x) = -2x + 7$.

Rewrite $f(x)$ as $y = -2x + 7$. Then interchange the variables and solve for y.

$x = -2y + 7$	Interchange the variables.
$2y = -x + 7$	Solve for y.
$y = \frac{-x + 7}{2}$	Divide each side by 2.
$f^{-1}(x) = \frac{-x + 7}{2}$	Rewrite using function notation.

EXAMPLE 3

Use the horizontal line test to determine whether the inverse of $f(x) = 2x^3 + 1$ is also a function.

Graph the function.

No horizontal line can be drawn so that it passes through more than one point. The inverse of this function is a function.

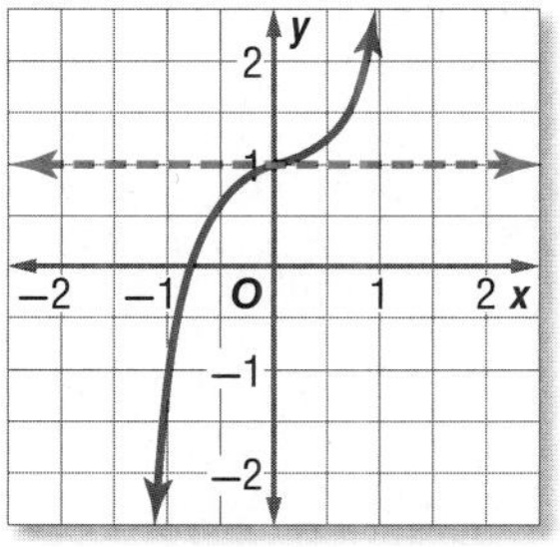

Study Guide and Review

7-3 Square Root Functions and Inequalities (pp. 424–430)

A2.A.41, A2.A.46

Graph each function. State the domain and range.

31. $f(x) = \sqrt{3x}$

32. $f(x) = -\sqrt{6x}$

33. $f(x) = \sqrt{x - 7}$

34. $f(x) = \sqrt{x + 5} - 3$

35. $f(x) = \frac{3}{4}\sqrt{x - 1} + 5$

36. $f(x) = -\frac{1}{3}\sqrt{x + 4} - 1$

37. GEOMETRY The area of a circle is given by the formula $A = \pi r^2$. What is the radius of a circle with an area of 300 square inches?

Graph each inequality.

38. $y \geq \sqrt{x} + 3$

39. $y < 2\sqrt{x - 5}$

40. $y > -\sqrt{x - 1} + 2$

EXAMPLE 4

Graph $f(x) = \sqrt{x + 1} - 2$. State the domain and range.

Identify the domain.

$x + 1 \geq 0$ **Write the radicand as greater than or equal to 0.**

$x \geq -1$ **Subtract 1 from each side.**

Make a table of values for $x \geq -1$ and graph the function.

x	y
−1	−2
0	−1
1	−0.59
2	−0.27
3	0
4	0.24
5	0.45

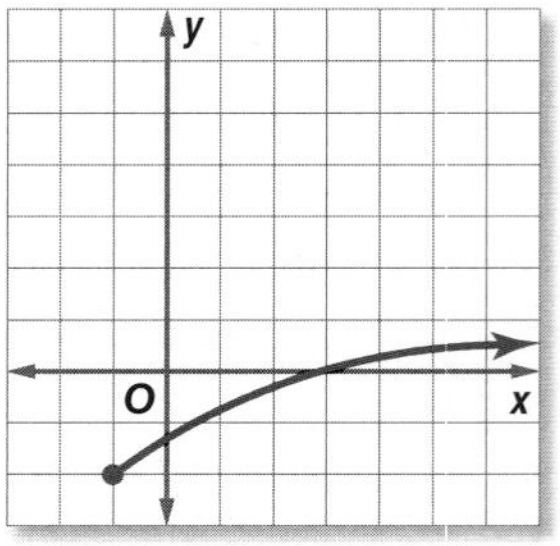

The domain is $\{x \mid x \geq -1\}$, and the range is $\{y \mid y \geq -2\}$.

7-4 nth Roots (pp. 431–436)

A2.A.13

Simplify.

41. $\pm\sqrt{121}$

42. $\sqrt[3]{-125}$

43. $\sqrt{(-6)^2}$

44. $\sqrt{-(x + 3)^4}$

45. $\sqrt[6]{(x^2 + 2)^{18}}$

46. $\sqrt[3]{27(x + 3)^3}$

47. $\sqrt[4]{a^8 b^{12}}$

48. $\sqrt[5]{243x^{10}y^{25}}$

49. PHYSICS The velocity v of an object can be defined as $v = \sqrt{\frac{2K}{m}}$, where m is the mass of an object and K is the kinetic energy in joules. Find the velocity in meters per second of an object with a mass of 17 grams and a kinetic energy of 850 joules.

EXAMPLE 5

Simplify $\sqrt{64x^6}$.

$\sqrt{64x^6} = \sqrt{(8x^3)^2}$ **$64x^6 = (8x^3)^2$**

$= 8|x^3|$ **Simplify.**

Use absolute value symbols because x could be negative.

EXAMPLE 6

Simplify $\sqrt[6]{4096x^{12}y^{24}}$.

$\sqrt[6]{4096x^{12}y^{24}} = \sqrt[6]{(4x^2y^4)^6}$ **$4096x^{12}y^{24} = (4x^2y^4)^6$**

$= 4x^2y^4$ **Simplify.**

MIXED PROBLEM SOLVING
For mixed problem-solving practice, see page 985.

7-5 Operations with Radical Expressions (pp. 439–445)

A2.A.14, A2.A.15

Simplify.

50. $\sqrt[3]{54}$

51. $\sqrt{144a^3b^5}$

52. $4\sqrt{6y} \cdot 3\sqrt{7x^2y}$

53. $6\sqrt{72} + 7\sqrt{98} - \sqrt{50}$

54. $\left(6\sqrt{5} - 2\sqrt{2}\right)\left(3\sqrt{5} + 4\sqrt{2}\right)$

55. $\dfrac{\sqrt{6m^5}}{\sqrt{p^{11}}}$

56. $\dfrac{3}{5 + \sqrt{2}}$

57. $\dfrac{\sqrt{3}}{\sqrt{5} - \sqrt{6}}$

58. **GEOMETRY** What are the perimeter and the area of the rectangle?

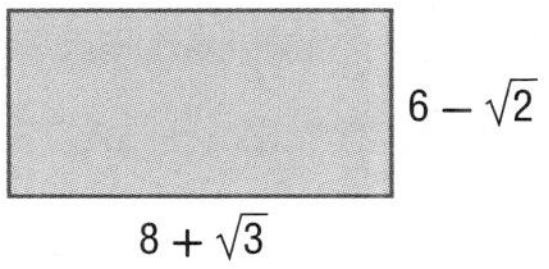

EXAMPLE 7

Simplify $2\sqrt[3]{18a^2b} \cdot 3\sqrt[3]{12ab^5}$.

$2\sqrt[3]{18a^2b} \cdot 3\sqrt[3]{12ab^5}$

$= (2 \cdot 3)\sqrt[3]{18a^2b \cdot 12ab^5}$	**Product Property**
$= 6\sqrt[3]{2^3 3^3 a^3 b^6}$	**Factor.**
$= 6 \cdot \sqrt[3]{2^3} \cdot \sqrt[3]{3^3} \cdot \sqrt[3]{a^3} \cdot \sqrt[3]{b^6}$	**Product Property**
$= 6 \cdot 2 \cdot 3 \cdot a \cdot b^2$	**Find cube roots.**
$= 36ab^2$	**Simplify.**

EXAMPLE 8

Simplify $\sqrt{\dfrac{x^4}{y^5}}$.

$\sqrt{\dfrac{x^4}{y^5}} = \dfrac{\sqrt{x^4}}{\sqrt{y^5}}$	**Quotient Property**
$= \dfrac{\sqrt{(x^2)^2}}{\sqrt{(y^2)^2} \cdot \sqrt{y}}$	**Factor into squares.**
$= \dfrac{x^2}{y^2\sqrt{y}} \cdot \dfrac{\sqrt{y}}{\sqrt{y}}$	**Rationalize the denominator.**
$= \dfrac{x^2\sqrt{y}}{y^3}$	$\sqrt{y} \cdot \sqrt{y} = y$

7-6 Rational Exponents (pp. 446–452)

A2.A.10, A2.A.11

Simplify each expression.

59. $x^{\frac{1}{2}} \cdot x^{\frac{2}{3}}$

60. $m^{-\frac{3}{4}}$

61. $\dfrac{d^{\frac{1}{6}}}{d^{\frac{3}{4}}}$

Simplify each expression.

62. $\dfrac{1}{y^{\frac{1}{4}}}$

63. $\sqrt[3]{\sqrt{729}}$

64. $\dfrac{x^{\frac{2}{3}} - x^{\frac{1}{3}}y^{\frac{2}{3}}}{x^{\frac{1}{3}}}$

65. **GEOMETRY** What is the area of the circle?

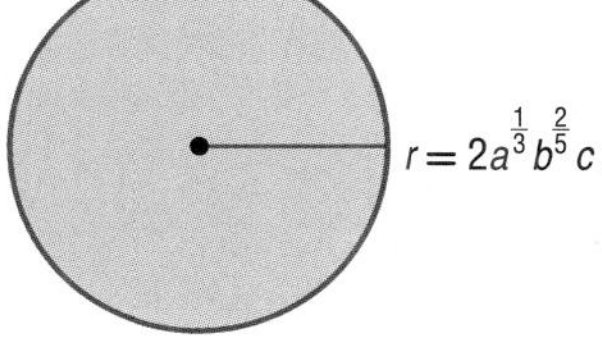

EXAMPLE 9

Simplify $a^{\frac{2}{3}} \cdot a^{\frac{1}{5}}$.

$a^{\frac{2}{3}} \cdot a^{\frac{1}{5}} = a^{\frac{2}{3} + \frac{1}{5}}$	**Product of Powers**
$= a^{\frac{13}{15}}$	**Add.**

EXAMPLE 10

Simplify $\dfrac{2a}{\sqrt[3]{b}}$.

$\dfrac{2a}{\sqrt[3]{b}} = \dfrac{2a}{b^{\frac{1}{3}}}$	**Rational exponents**
$= \dfrac{2a}{b^{\frac{1}{3}}} \cdot \dfrac{b^{\frac{2}{3}}}{b^{\frac{2}{3}}}$	**Rationalize the denominator.**
$= \dfrac{2ab^{\frac{2}{3}}}{b}$ or $\dfrac{2a\sqrt[3]{b^2}}{b}$	**Rewrite in radical form.**

7-7 Solving Radical Equations and Inequalities (pp. 453–459)

A2.A.22

Solve each equation.

66. $\sqrt{x-3}+5=15$

67. $-\sqrt{x-11}=3-\sqrt{x}$

68. $4+\sqrt{3x-1}=8$

69. $\sqrt{m+3}=\sqrt{2m+1}$

70. $\sqrt{2x+3}=3$

71. $(x+1)^{\frac{1}{4}}=-3$

72. $a^{\frac{1}{3}}-4=0$

73. $3(3x-1)^{\frac{1}{3}}-6=0$

74. PHYSICS The formula $t=2\pi\sqrt{\frac{\ell}{32}}$ represents the swing of a pendulum, where t is the time in seconds for the pendulum to swing back and forth and ℓ is the length of the pendulum in feet. Find the length of a pendulum that makes one swing in 2.75 seconds.

Solve each inequality.

75. $2+\sqrt{3x-1}<5$

76. $\sqrt{3x+13}-5\geq 5$

77. $6-\sqrt{3x+5}\leq 3$

78. $\sqrt{-3x+4}-5\geq 3$

79. $5+\sqrt{2y-7}<5$

80. $3+\sqrt{2x-3}\geq 3$

81. $\sqrt{3x+1}-\sqrt{6+x}>0$

EXAMPLE 11

Solve $\sqrt{2x+9}-2=5$.

$\sqrt{2x+9}-2=5$	**Original equation**
$\sqrt{2x+9}=7$	**Add 2 to each side.**
$(\sqrt{2x+9})^2=7^2$	**Square each side.**
$2x+9=49$	**Evaluate the squares.**
$2x=40$	**Subtract 9 from each side.**
$x=20$	**Divide each side by 2.**

EXAMPLE 12

Solve $\sqrt{2x-5}+2>5$.

$\sqrt{2x-5}\geq 0$	**Radicand must be ≥ 0.**
$2x-5\geq 0$	**Square each side.**
$2x\geq 5$	**Add 5 to each side.**
$x\geq 2.5$	**Divide each side by 2.**

The solution must be ≥ 2.5 to satisfy the domain restriction.

$\sqrt{2x-5}+2>5$	**Original inequality**
$\sqrt{2x-5}>3$	**Subtract 2 from each side.**
$(\sqrt{2x-5})^2>3^2$	**Square each side.**
$2x-5>9$	**Evaluate the squares.**
$2x>14$	**Add 5 to each side.**
$x>7$	**Divide each side by 2.**

Practice Test

NY Math Online glencoe.com
Chapter Test

Determine whether each pair of functions are inverse functions. Write *yes* or *no*. Explain your reasoning.

1. $f(x) = 3x + 8, g(x) = \frac{x - 8}{3}$
2. $f(x) = \frac{1}{3}x + 5, g(x) = 3x - 15$
3. $f(x) = x + 7, g(x) = x - 7$
4. $g(x) = 3x - 2, f(x) = \frac{x - 2}{3}$

5. **MULTIPLE CHOICE** Which inequality represents the graph below?

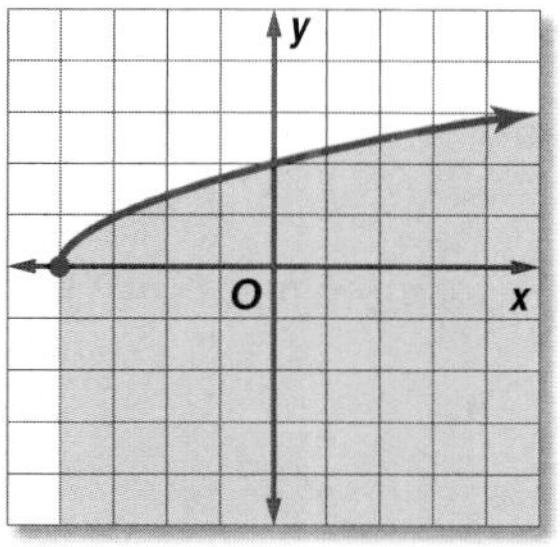

A $y \geq \sqrt{x + 4}$
B $y \leq \sqrt{x + 4}$
C $y \geq \sqrt{x - 4}$
D $y \leq \sqrt{x - 4}$

If $f(x) = 3x + 2$ and $g(x) = x^2 - 2x + 1$, find each value.

6. $(f + g)(x)$
7. $(f \cdot g)(x)$
8. $(f - g)(x)$
9. $\left(\frac{f}{g}\right)(x)$

Solve each equation.

10. $\sqrt{a + 12} = \sqrt{5a - 4}$
11. $\sqrt{3x} = \sqrt{x - 2}$
12. $4(\sqrt[4]{3x + 1}) - 8 = 0$
13. $\sqrt[3]{5m + 6} + 15 = 21$
14. $\sqrt{3x + 21} = \sqrt{5x + 27}$
15. $1 + \sqrt{x + 11} = \sqrt{2x + 15}$
16. $\sqrt{x - 5} = \sqrt{2x - 4}$
17. $\sqrt{x - 6} - \sqrt{x} = 3$

18. **MULTIPLE CHOICE** Which expression is equivalent to $125^{-\frac{1}{3}}$?

F -5
G $-\frac{1}{5}$
H $\frac{1}{5}$
J 5

Simplify.

19. $(2 + \sqrt{5})(6 - 3\sqrt{5})$
20. $(3 - 2\sqrt{2})(-7 + \sqrt{2})$
21. $\frac{12}{2 - \sqrt{3}}$
22. $\frac{m^{\frac{1}{2}} - 1}{2m^{\frac{1}{2}} + 1}$
23. $4\sqrt{3} - 8\sqrt{48}$
24. $5^{\frac{2}{3}} \cdot 5^{\frac{1}{2}} \cdot 5^{\frac{5}{6}}$
25. $\sqrt[6]{729a^9b^{24}}$
26. $\sqrt[5]{32x^{15}y^{10}}$
27. $w^{-\frac{4}{5}}$
28. $\frac{r^{\frac{2}{3}}}{r^{\frac{1}{6}}}$
29. $\frac{a^{-\frac{1}{2}}}{6a^{\frac{1}{3}} \cdot a^{-\frac{1}{4}}}$
30. $\frac{y^{\frac{3}{2}}}{y^{\frac{1}{2}} + 2}$

31. **MULTIPLE CHOICE** What is the area of the rectangle?

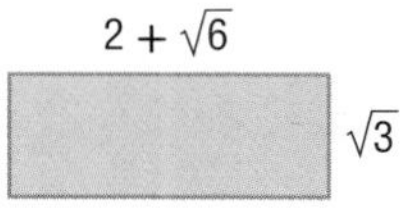

A $2\sqrt{3} + 3\sqrt{2}$ units^2
B $4 + 2\sqrt{6} + 2\sqrt{3}$ units^2
C $2\sqrt{3} + \sqrt{6}$ units^2
D $2\sqrt{3} + 3$ units^2

Solve each inequality.

32. $\sqrt{4x - 3} < 5$
33. $-2 + \sqrt{3m - 1} < 4$
34. $2 + \sqrt{4x - 4} \leq 6$
35. $\sqrt{2x + 3} - 4 \leq 5$
36. $\sqrt{b + 12} - \sqrt{b} > 2$
37. $\sqrt{y - 7} + 5 \geq 10$
38. $\sqrt{a - 5} - \sqrt{a + 7} \leq 4$
39. $\sqrt{c + 5} + \sqrt{c + 10} > 2$

40. **GEOMETRY** The area of a triangle with sides of length a, b, and c is given by $A = \sqrt{s(s - a)(s - b)(s - c)}$, where $s = \frac{1}{2}(a + b + c)$. What is the area of the triangle expressed in radical form?

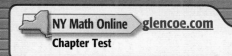

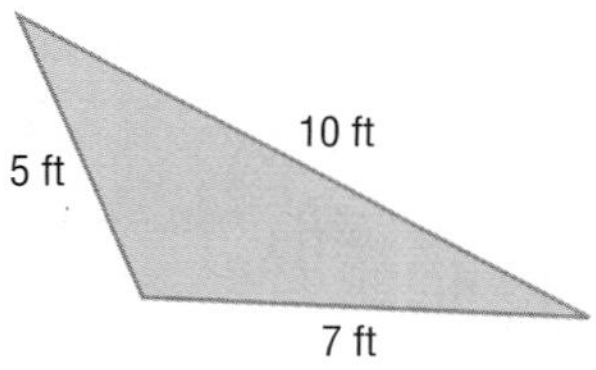

Preparing for Standardized Tests

Work Backward

In certain math problems, you are given information about an end result, but you need to find out something that happened earlier. You can work backward to solve problems like this.

Strategies for Working Backward

Step 1

Read the problem statement carefully.

Ask yourself:

- What information am I given?
- What am I being asked to solve?
- Does any of the information given relate to an end result?
- Am I being asked to solve for a quantity that occurred "earlier" in the problem statement?
- What operations are being used in the problem?

Step 2

Model the problem situation with an equation, an inequality, or a graph as appropriate. Then work backward to solve the problem.

- If needed, sketch a flow of events to show the sequence described in the problem statement.
- Use inverse operations to undo any operations while working backward until you arrive at your answer.

Step 3

Check by beginning with your answer and seeing if you arrive at the same result given in the problem statement.

EXAMPLE

Read the problem. Identify what you need to know. Then use the information in the problem to solve.

Maria bought a used car. The sales tax rate was 6.75% of the selling price, and she had to pay $450 in processing, title, and registration fees. If Maria paid a total of $15,768.63, what was the sale price of the car? Show your work.

Read the problem carefully. You know the total amount that Maria paid for the car after sales tax was applied and after she paid all of the other fees. You need to find the sale price of the car before taxes and fees.

Let x represent the sale price of the car and set up an equation. Use the work backward strategy to solve the problem.

Words	The sale price of the car plus the sales tax and other fees is equal to the final price.
Variable	Let x = sale price.
Equation	$1.0675x + 450 = 15{,}768.63$

Using the work backward strategy results in a simple equation. Use inverse operations to solve for x.

$$1.0675x + 450 = 15{,}768.63$$
$$1.0675x = 15{,}318.63$$
$$x \approx 14{,}350$$

Check your answer by working the problem forward. Begin with your answer and see if you get the same result as in the problem statement.

$14{,}350(1.0675) \approx 15{,}318.63$	**Compute the sales tax.**
$15{,}318.63 + 450 = 15{,}768.63$	**Add the other fees.**
$15{,}768.63 = 15{,}768.63$	**The result is the same.**

So, the sale price of the car was $14,350.

Exercises

Read the problem. Identify what you need to know. Then use the information in the problem to solve.

1. The equation $d = \frac{s^2}{30f}$ can be used to model the length of the skid marks left by a car when a driver applies the brakes to come to a sudden stop. In the equation, d is the length (in feet) of the skid marks left on the road, s is the speed of the car in miles per hour, and f is a coefficient of friction that describes the condition of the road. Suppose a car left skid marks that are 120 feet long.

a. Solve the equation for s, the speed of the car.

b. If the coefficient of friction for the road is 0.75, about how fast was the car traveling?

c. How fast was the car traveling if the coefficient of friction for the road is 1.1?

2. An object is shot straight upward into the air with an initial speed of 800 feet per second. The height h that the object will be after t seconds is given by the equation $h = -16t^2 + 800t$. When will the object reach a height of 10,000 feet?

A 10 seconds

B 25 seconds

C 100 seconds

D 625 seconds

3. Pedro is creating a scale drawing of a car. He finds that the height of the car in the drawing is $\frac{1}{32}$ of the actual height of the car x. Which equation best represents this relationship?

F $y = x - \frac{1}{32}$

G $y = -\frac{1}{32}x$

H $y = \frac{1}{32}x$

J $y = x + \frac{1}{32}$

CHAPTER 7

NYS Regents Exam Practice

Cumulative, Chapters 1 through 7

Multiple Choice

Answer all questions in this part. Select the answer that best completes the statement or answers the question.

1. Which of the following is *not* a zero of the polynomial function graphed below?

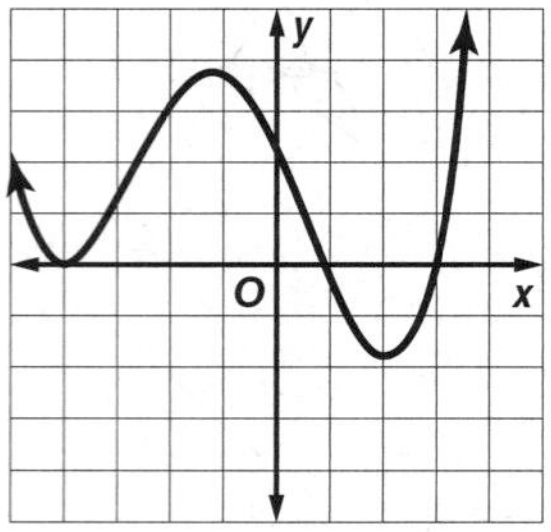

(1) -4

(2) -1

(3) 1

(4) 3

2. Simplify the following expression.

$$\frac{3+i}{1-i}$$

(1) $2-i$

(2) $\frac{1}{2}+\frac{2}{3}i$

(3) $1+2i$

(4) $\frac{3}{4}-\frac{3}{5}i$

3. The radius of a sphere with volume V can be found using the formula $r=\sqrt[3]{\frac{3V}{4\pi}}$. If the volume of a soccer ball is 5575 cubic centimeters, what is the radius of the ball? Round to the nearest centimeter.

(1) 10 cm

(2) 11 cm

(3) 12 cm

(4) 13 cm

Test-TakingTip

Question 3 Substitute 5575 for V in the formula and simplify to find the radius.

4. Solve the radical equation for x.

$$\sqrt{x+1}+1=\sqrt{x+6}$$

(1) 3

(2) 4

(3) 3, 6

(4) no solution

5. Which number line shows the solution to the inequality $|6n-3|<21$?

(1) 

(2) 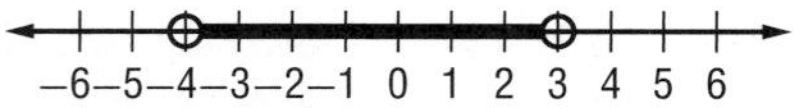

(3) 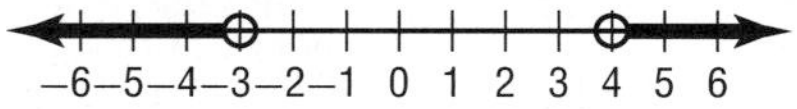

(4) 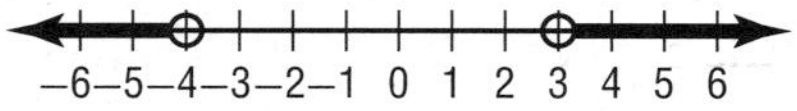

6. The equation $h=241m^{-\frac{1}{4}}$ predicts a mammal's heart rate h in beats per minute, based on its mass m in kilograms. The state mammal of New York is the North American beaver. These mammals can weigh up to 32 kilograms. What is the predicted heart rate, in beats per minute, of a 32-kilogram North American beaver? Round to the nearest whole number.

(1) 62 bpm

(2) 76 bpm

(3) 88 bpm

(4) 101 bpm

7. Find the inverse of the linear function $f(x)=x+3$.

(1) $f(x)=x-3$

(2) $f(x)=3x$

(3) $f(x)=\frac{x}{3}$

(4) $f(x)=3+x$

8. Between which two integers does $\sqrt{55}$ lie on a number line?

(1) between 6 and 7

(2) between 7 and 8

(3) between 8 and 9

(4) between 9 and 10

9. Last month a school bookstore sold 348 notebooks and folders for a total revenue of \$114. How many notebooks and folders were sold? Set up and solve a system of equations.

Bookstore Prices

Item	Cost
Notebook	\$0.50
Folder	\$0.25

(1) 82 notebooks, 266 folders

(2) 97 notebooks, 251 folders

(3) 108 notebooks, 240 folders

(4) 120 notebooks, 228 folders

10. A retail store discounts all remaining holiday items by 40% after the holiday passes. Irva also has a coupon good for \$10.00 off her next purchase from the store. If the coupon is applied *after* the store discount, which of the following functions can be used to find the final price of an item that originally cost d dollars?

(1) $P(d) = 0.6 \times (d + 10)$

(2) $P(d) = 0.6 \times (d - 10)$

(3) $P(d) = 0.4 \times (d - 10)$

(4) $P(d) = (0.6 \times d) - 10$

11. What is the degree of the polynomial below?

$$-x^5 - 4x^4 + 10x + 3$$

(1) 2 **(2)** 3 **(3)** 4 **(4)** 5

Open-Ended Response

Solve each problem. Clearly indicate the necessary steps, including appropriate formula substitutions, diagrams, graphs, charts, etc.

12. Find the altitude of the triangle below if the area is $50 + 18\sqrt{5}$ square centimeters.

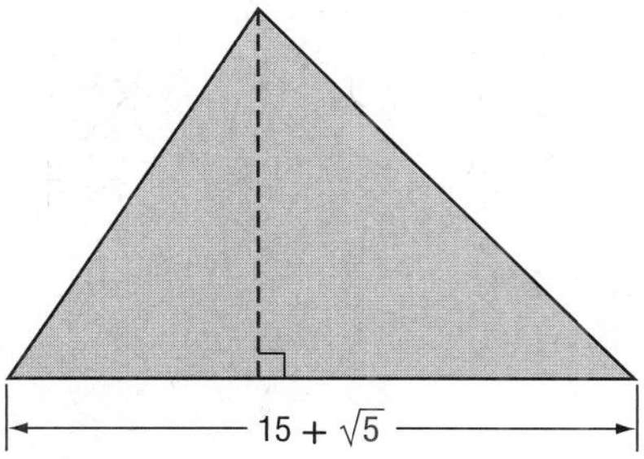

13. Use the triangle shown on the coordinate grid to answer each question.

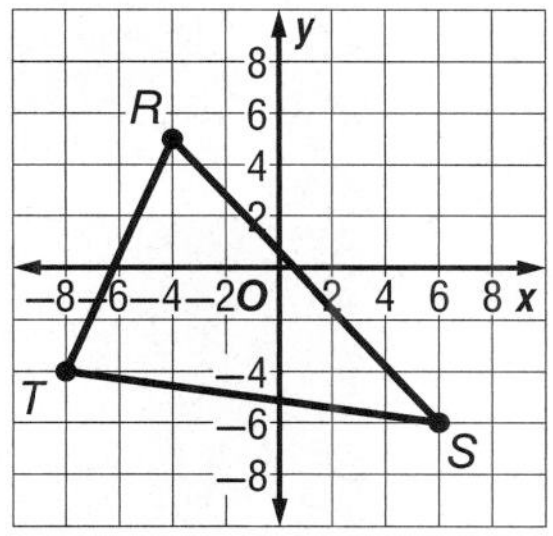

a. What are the vertices of triangle RST?

b. Write the coordinates in a vertex matrix.

c. Suppose the triangle is reflected across the line $y = x$. What matrix can you multiply by the vertex matrix to find the reflected vertices?

d. What are the vertices of the reflected triangle?

Need Extra Help?

If you missed Question...	1	2	3	4	5	6	7	8	9	10	11	12	13
Go to Lesson or Page...	7-1	1-1	6-1	7-2	7-4	7-3	5-7	7-2	3-1	4-4	7-7	7-4	6-1
NYS Core Curriculum	A2.A.50	A2.N.9	A2.A.13	A2.A.22	A2.A.1	A2.N.1	A2.A.45	A2.A.13	A.A.7	A2.A.42	A2.CM.3	A2.A.14	G.G.61

CHAPTER 8 Exponential and Logarithmic Functions and Relations

Then

In Chapter 2, you graphed functions and transformations of functions.

Now

In Chapter 8, you will:

- Graph exponential and logarithmic functions.
- Solve exponential and logarithmic equations and inequalities.
- Solve problems involving exponential growth and decay.

NYS Core Curriculum

A2.A.12, A2.A.18 Evaluate exponential and logarithmic expressions
A2.A.27 Solve exponential equations with and without common bases

Why?

SCIENCE Mathematics and science go hand in hand. Whether it is chemistry, biology, paleontology, zoology, or anthropology, you will need strong math skills. In this chapter, you will learn mathematical aspects of science such as computer viruses, populations of insects, bacteria growth, cell division, astronomy, tornados, and earthquakes.

Math *in Motion*, Animation glencoe.com

Get Ready for Chapter 8

Diagnose Readiness You have two options for checking Prerequisite Skills.

Text Option

Take the Quick Check below. Refer to the Quick Review for help.

QuickCheck

Simplify. Assume that no variable equals zero. (Lesson 6-1)

1. $a^4a^3a^5$
2. $(2xy^3z^2)^3$
3. $\dfrac{-24x^8y^5z}{16x^2y^8z^6}$
4. $\left(\dfrac{-8r^2n}{36n^3t}\right)^2$
5. **DENSITY** The density of an object is equal to the mass divided by the volume. An object has a mass of 7.5×10^3 grams and a volume of 1.5×10^3 cubic centimeters. What is the density of the object?

Find the inverse of each function. Then graph the function and its inverse. (Lesson 7-2)

6. $f(x) = 2x + 5$
7. $f(x) = x - 3$
8. $f(x) = -4x$
9. $f(x) = \frac{1}{4}x - 3$
10. $f(x) = \dfrac{x-1}{2}$
11. $y = \frac{1}{3}x + 4$

Determine whether each pair of functions are inverse functions.

12. $f(x) = x - 6$
    $g(x) = x + 6$
13. $f(x) = 2x + 5$
    $g(x) = 2x - 5$
14. **FOOD** A pizzeria charges \$12 for a medium cheese pizza and \$2 for each additional topping. If $f(x) = 2x + 12$ represents the cost of a medium pizza with x toppings, find $f^{-1}(x)$ and explain its meaning.

QuickReview

EXAMPLE 1

Simplify $\dfrac{(a^3bc^2)^2}{a^4a^2b^2bc^5c^3}$. Assume that no variable equals zero.

$\dfrac{(a^3bc^2)^2}{a^4a^2b^2bc^5c^3}$

$= \dfrac{a^6b^2c^4}{a^6b^3c^8}$ **Simplify the numerator by using the Power of a Power Rule and the denominator by using the Product of Powers Rule.**

$= \dfrac{1}{bc^4}$ or $b^{-1}c^{-4}$ **Simplify by using the Quotient of Powers Rule.**

EXAMPLE 2

Find the inverse of $f(x) = 3x - 1$.

Step 1 Replace $f(x)$ with y in the original equation: $f(x) = 3x - 1 \rightarrow y = 3x - 1$.

Step 2 Interchange x and y: $x = 3y - 1$.

Step 3 Solve for y.

$x = 3y - 1$	**Inverse**
$x + 1 = 3y$	**Add 1 to each side.**
$\dfrac{x+1}{3} = y$	**Divide each side by 3.**
$\frac{1}{3}x + \frac{1}{3} = y$	**Simplify.**

Step 4 Replace y with $f^{-1}(x)$.
$y = \frac{1}{3}x + \frac{1}{3} \rightarrow f^{-1}(x) = \frac{1}{3}x + \frac{1}{3}$

Online Option

NY Math Online Take a self-check Chapter Readiness Quiz at glencoe.com.

Get Started on Chapter 8

You will learn several new concepts, skills, and vocabulary terms as you study Chapter 8. To get ready, identify important terms and organize your resources. You may wish to refer to **Chapter 0** to review prerequisite skills.

FOLDABLES® Study Organizer

Exponential and Logarithmic Functions and Relations Make this Foldable to help you organize your Chapter 8 notes about exponential and logarithmic functions. Begin with two sheets of grid paper.

1. **Fold** in half along the width.

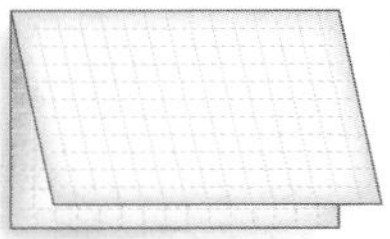

2. **On** the first sheet, cut 5 cm along the fold at the ends.

First Sheet

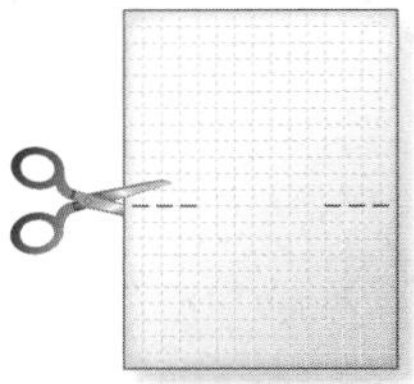

3. **On** the second sheet, cut in the center, stopping 5 cm from the ends.

Second Sheet

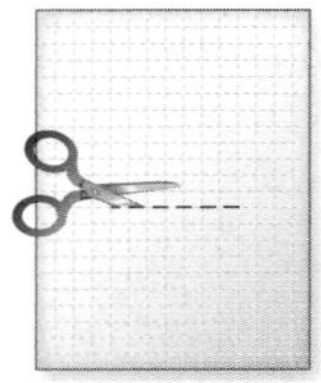

4. **Insert** the first sheet through the second sheet and align the folds. Label the pages with lesson numbers.

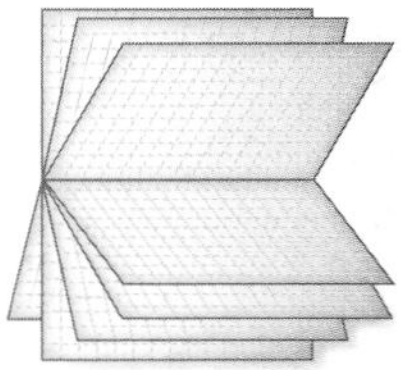

NY Math Online glencoe.com

- Study the chapter online
- Explore **Math in Motion**
- Get extra help from your own **Personal Tutor**
- Use **Extra Examples** for additional help
- Take a **Self-Check Quiz**
- **Review Vocabulary** in fun ways

New Vocabulary

English		Español
exponential function	• p. 475 •	función exponencial
exponential growth	• p. 475 •	crecimiento exponencial
asymptote	• p. 475 •	asíntota
growth factor	• p. 477 •	factor de crecimiento
exponential decay	• p. 477 •	desintegración exponencial
decay factor	• p. 478 •	factor de desintegración
exponential equation	• p. 485 •	ecuación exponencial
compound interest	• p. 486 •	interés compuesto
exponential inequality	• p. 487 •	desigualdad exponencial
logarithm	• p. 492 •	logaritmo
logarithmic function	• p. 493 •	función logarítmica
logarithmic equation	• p. 502 •	ecuación logarítmica
logarithmic inequality	• p. 503 •	desigualdad logarítmica
common logarithm	• p. 516 •	logaritmos communes
Change of Base Formula	• p. 518 •	fórmula del cambio de base
natural base, *e*	• p. 525 •	*e* base natural
natural base exponential function	• p. 525 •	base natural función exponencial
natural logarithm	• p. 525 •	logaritmo natural

Review Vocabulary

domain • p. P7 • dominio the set of all *x*-coordinates of the ordered pairs of a relation

function • p. P7 • función a relation in which each element of the domain is paired with exactly one element in the range

range • p. P7 • rango the set of all *y*-coordinates of the ordered pairs of a relation

$\{(-3, 1), (0, 2), (2, 4)\}$

Domain Range

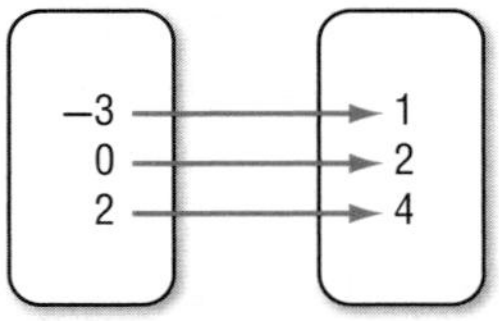

Multilingual eGlossary glencoe.com

8-1 Graphing Exponential Functions

Then

You graphed polynomial functions. (Lesson 6-4)

Now

- Graph exponential growth functions.
- Graph exponential decay functions.

NYS Core Curriculum

A2.A.6 Solve an application which results in an exponential function

A2.A.53 Graph exponential functions of the form $y = b^x$ for positive values of b, including $b = e$ *Also addresses A2.A.41.*

New Vocabulary

exponential function
exponential growth
asymptote
growth factor
exponential decay
decay factor

NY Math Online

glencoe.com

- Extra Examples
- Personal Tutor
- Self-Check Quiz
- Homework Help
- Math in Motion

Why?

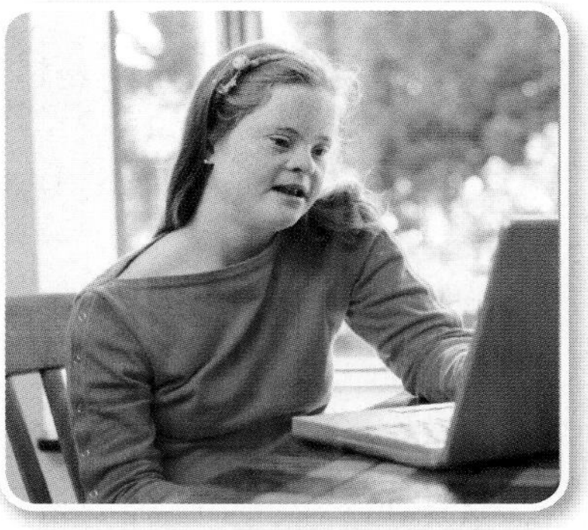

Have you ever received an e-mail that tells you to forward it to 5 friends? If each of those 5 friends then forwards it to 5 of their friends, who each forward it to 5 of their friends, the number of people receiving the e-mail is growing exponentially.

The equation $y = 5^x$ can be used to represent this situation, where x is the number of rounds that the e-mail has been forwarded.

Exponential Growth A function like $y = 5^x$, where the base is a constant and the exponent is the independent variable, is an **exponential function**. One type of exponential function is exponential growth. An **exponential growth** function is a function of the form $f(x) = b^x$, where $b > 1$. The graph of an exponential function has an **asymptote**, which is a line that the graph of the function approaches.

Key Concept — For Your FOLDABLE

Parent Function of Exponential Growth Functions

Parent function: $f(x) = b^x$, $b > 1$

Type of graph: continuous, one-to-one, and increasing

Domain: all real numbers

Range: all nonzero real numbers

Asymptote: x-axis

Intercept: (0, 1)

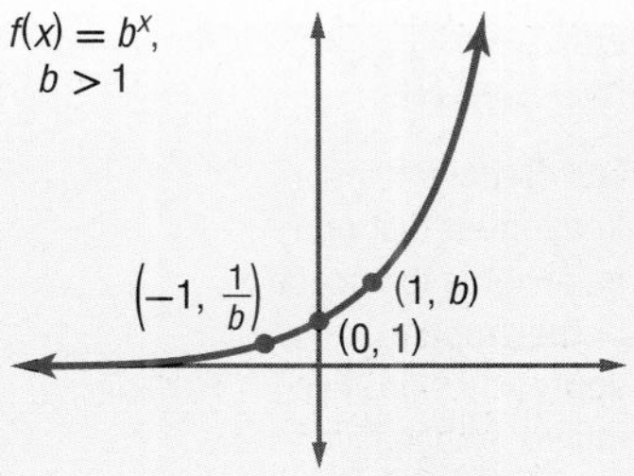

EXAMPLE 1 Graph Exponential Growth Functions

Graph $y = 3^x$. State the domain and range.

Make a table of values. Then plot the points and sketch the graph.

x	-3	-2	$-\frac{1}{2}$	0
$y = 3^x$	$3^{-3} = \frac{1}{27}$	$3^{-2} = \frac{1}{9}$	$3^{-\frac{1}{2}} = \frac{\sqrt{3}}{3}$	$3^0 = 1$

x	1	$\frac{3}{2}$	2
$y = 3^x$	$3^1 = 3$	$3^{\frac{3}{2}} = \sqrt{27}$	$3^2 = 9$

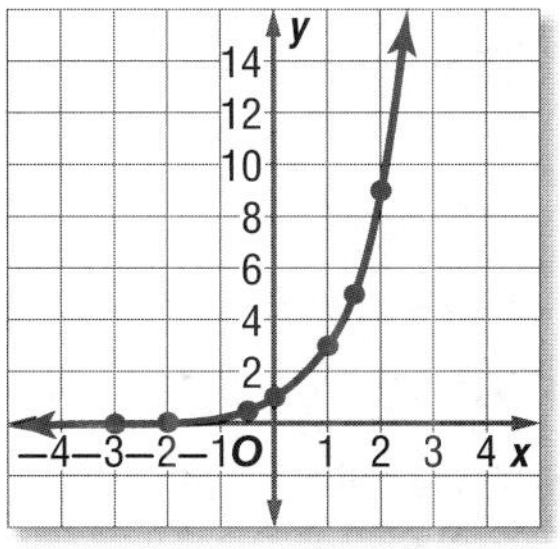

The domain is all real numbers, and the range is all positive real numbers.

Check Your Progress

1. Graph $y = 4^x$. State the domain and range.

Personal Tutor glencoe.com

The graph of $f(x) = b^x$ represents a parent graph of the exponential functions. The same techniques used to transform the graphs of other functions you have studied can be applied to the graphs of exponential functions.

StudyTip

Look Back To review **transformations of parent functions**, see Lesson 2-7.

Key Concept — Transformations of Exponential Functions

For Your FOLDABLE

$$f(x) = ab^{x-h} + k$$

h — Horizontal Translation	k — Vertical Translation
$\|h\|$ units right if h is positive $\|h\|$ units left if h is negative	$\|k\|$ units up if k is positive $\|k\|$ units down if k is negative

a — Orientation and Shape

If $a < 0$, the graph is reflected in the x-axis.

If $|a| > 1$, the graph is compressed.
If $0 < |a| < 1$, the graph is expanded.

EXAMPLE 2 Graph Transformations

Graph each function. State the domain and range.

a. $y = 2^x + 1$

The equation represents a translation of the graph of $y = 2^x$ one unit up.

x	$y = 2^x + 1$
−3	$2^{-3} + 1 = 1.125$
−2	$2^{-2} + 1 = 1.25$
−1	$2^{-1} + 1 = 1.5$
0	$2^0 + 1 = 2$
1	$2^1 + 1 = 3$
2	$2^2 + 1 = 5$
3	$2^3 + 1 = 9$

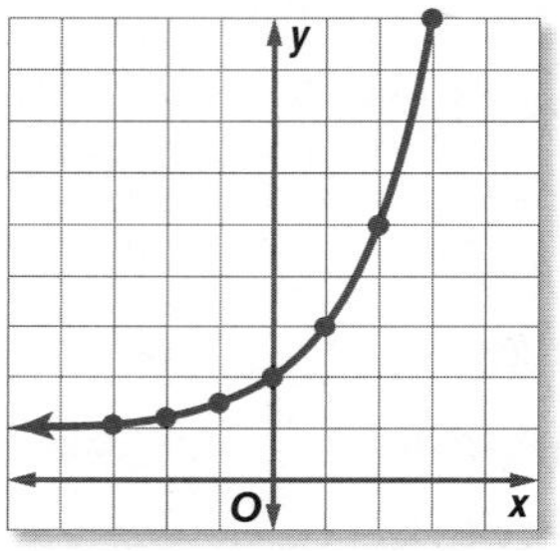

Domain = {all real numbers}

Range = $\{y \mid y > 1\}$

StudyTip

End Behavior Remember that end behavior is the action of the graph as x approaches positive infinity or negative infinity. In Example 2a, as x approaches infinity, y approaches infinity. In Example 2b, as x approaches infinity, y approaches negative infinity.

b. $y = -\frac{1}{2} \cdot 5^{x-2}$

The equation represents a transformation of the graph of $y = 5^x$. Graph $y = 5^x$ and transform the graph.

- $a = -\frac{1}{2}$: The graph is reflected in the x-axis and expanded.
- $h = 2$: The graph is translated 2 units right.
- $k = 0$: The graph is not translated vertically.

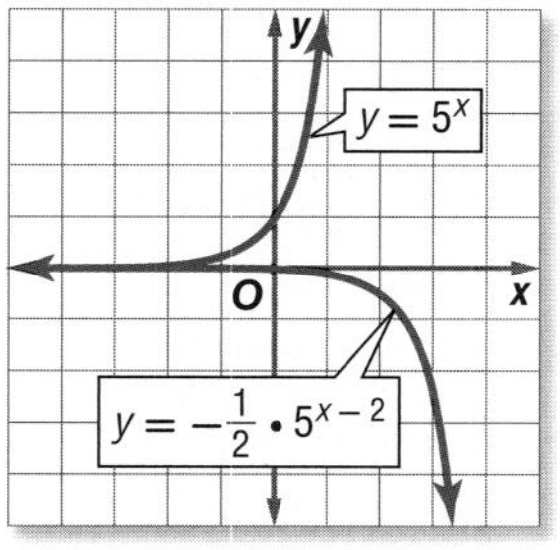

Domain = {all real numbers}

Range = $\{y \mid y < 0\}$

Check Your Progress

2A. $y = 2^{x+3} - 5$

2B. $y = 0.1(6)^x - 3$

Personal Tutor glencoe.com

You can model exponential growth with a constant percent increase over specific time periods using the following function.

$$A(t) = a(1 + r)^t$$

The function can be used to find the amount $A(t)$ after t time periods, where a is the initial amount and r is the percent of increase per time period. Note that the base of the exponential expression, $1 + r$, is called the **growth factor**.

The exponential growth function is often used to model population growth.

Real-World Link

The U.S. Census Bureau's American Community Survey is mailed to approximately 1 out of every 480 households.

Source: Census Bureau

Real-World EXAMPLE 3 Graph Exponential Growth Functions

CENSUS The first U.S. Census was conducted in 1790. At that time, the population was 3,929,214. Since then, the U.S. population has grown by approximately 2.03% annually. Draw a graph showing the population growth of the U.S. since 1790.

First, write an equation using $a = 3{,}929{,}214$, and $r = 0.0203$.

$y = 3{,}929{,}214(1.0203)^t$

Then graph the equation.

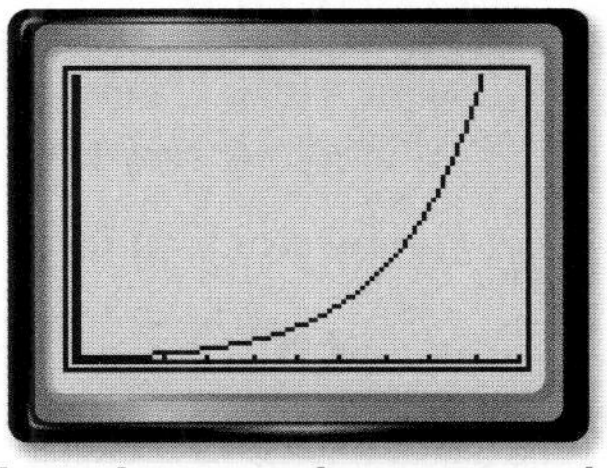

[0, 250] scl: 25 by [0, 400,000,000] scl: 40,000,000

Check Your Progress

3. **BIOLOGY** A lab technician starts a cell culture with 3500 cells. The number of cells doubles every 1.5 hours. Draw a graph to show the cell growth.

Personal Tutor glencoe.com

StudyTip

Rate of Growth When a population doubles during a time period, its rate of growth is 100%.

Exponential Decay The second type of exponential function is **exponential decay**.

Key Concept

For Your FOLDABLE

Parent Function of Exponential Decay Functions

Parent function: $f(x) = b^x$, $0 < b < 1$

Type of graph: continuous, one-to-one, and decreasing

Domain: all real numbers

Range: positive real numbers

Asymptote: x-axis

Intercept: (0, 1)

Model

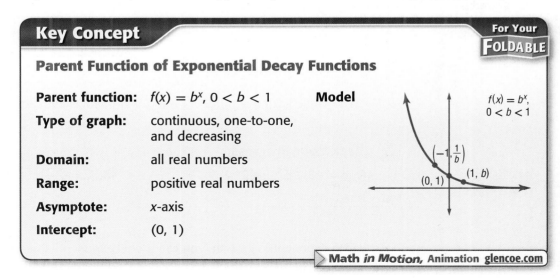

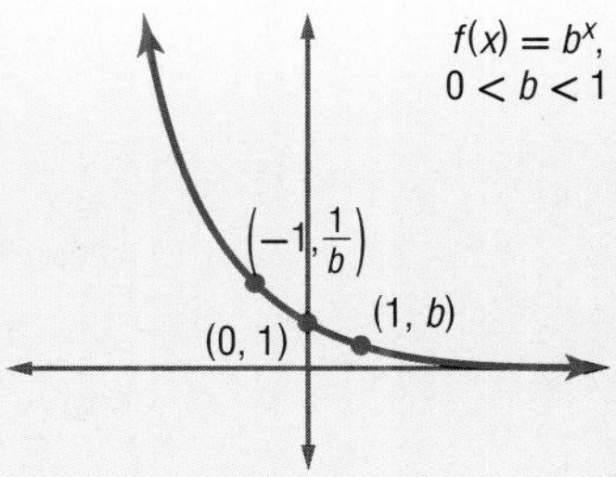

Math in Motion, Animation glencoe.com

The graphs of exponential decay functions can be transformed in the same manner as those of exponential growth.

EXAMPLE 4 Graph Exponential Decay Functions

Graph each function. State the domain and range.

a. $y = \left(\frac{1}{3}\right)^x$

x	$y = \left(\frac{1}{3}\right)^x$
-3	$\left(\frac{1}{3}\right)^{-3} = 27$
-2	$\left(\frac{1}{3}\right)^{-2} = 9$
$-\frac{1}{2}$	$\left(\frac{1}{3}\right)^{-\frac{1}{2}} = \sqrt{3}$
0	$\left(\frac{1}{3}\right)^{0} = 1$
1	$\left(\frac{1}{3}\right)^{1} = \frac{1}{3}$
$\frac{3}{2}$	$\left(\frac{1}{3}\right)^{\frac{3}{2}} = \sqrt{\frac{1}{27}}$
2	$\left(\frac{1}{3}\right)^{2} = \frac{1}{9}$

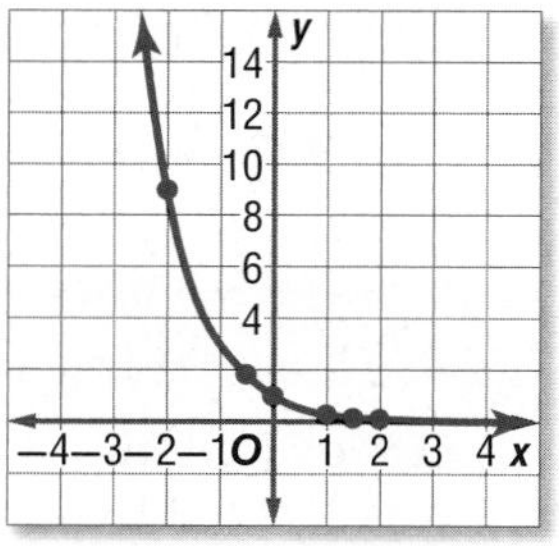

The domain is all real numbers, and the range is all positive real numbers.

b. $y = 2\left(\frac{1}{4}\right)^{x+2} - 3$

The equation represents a transformation of the graph of $y = \left(\frac{1}{4}\right)^x$.

Examine each parameter.

- $a = 2$: The graph is compressed.
- $h = -2$: The graph is translated 2 units left.
- $k = -3$: The graph is translated 3 units down.

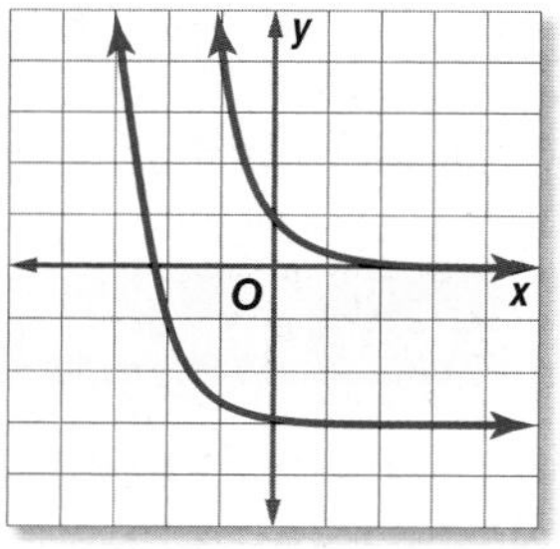

The range is all real numbers, and the domain is all real numbers greater than −3.

Check Your Progress

4A. $y = -3\left(\frac{2}{5}\right)^{x-4} + 2$

4B. $y = \frac{3}{8}\left(\frac{5}{6}\right)^{x-1} + 1$

Personal Tutor glencoe.com

StudyTip

Exponential Decay Be sure not to confuse a dilation in which $|a| < 1$ with exponential decay in which $0 < b < 1$.

Similar to exponential growth, you can model exponential decay with a constant percent increase over specific time periods using the following function.

$$A(t) = a(1 - r)^t$$

The base of the exponential expression, $1 - r$, is called the **decay factor**.

Real-World Link

After water, tea is the most consumed beverage in the U.S. It can be found in over 80% of American households. Just over half the American population drinks tea daily.

Source: The Tea Association of the USA

Real-World EXAMPLE 5 Graph Exponential Decay Functions

TEA A cup of green tea contains 35 milligrams of caffeine. The average teen can eliminate approximately 12.5% of the caffeine from their system per hour.

a. Draw a graph to represent the amount of caffeine remaining after drinking a cup of green tea.

$y = a(1 - r)^t$
$= 35(1 - 0.125)^t$
$= 35(0.875)^t$

Graph the equation.

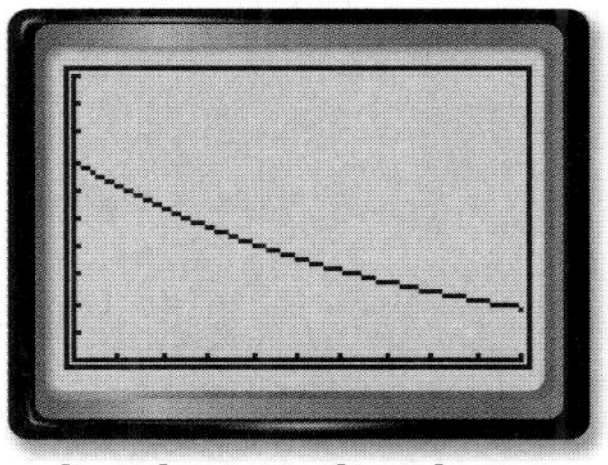

[0, 10] scl: 1 by [0, 50] scl: 5

b. Estimate the amount of caffeine in a teenager's body 3 hours after drinking a cup of green tea.

$y = 35(0.875)^t$	**Equation from part a**
$= 35(0.875)^3$	**Replace *t* with 3.**
≈ 23.45	**Use a calculator.**

The caffeine in a teenager will be about 23.45 milligrams after 3 hours.

Check Your Progress

5. A cup of black tea contains about 68 milligrams of caffeine. Draw a graph to represent the amount of caffeine remaining in the body after drinking a cup of black tea. Estimate the amount of caffeine in the body 2 hours after drinking a cup of black tea.

Personal Tutor glencoe.com

Check Your Understanding

Examples 1 and 2 pp. 475–476

Graph each function. State the domain and range.

1. $f(x) = 2^x$

2. $f(x) = 5^x$

3. $f(x) = 3^{x-2} + 4$

4. $f(x) = 2^{x+1} + 3$

5. $f(x) = 0.25(4)^x - 6$

6. $f(x) = 3(2)^x + 8$

Example 3 p. 477

7. SCIENCE A virus spreads through a network of computers such that each minute, 25% more computers are infected. If the virus began at only one computer, graph the function for the first hour of the spread of the virus.

Example 4 p. 478

Graph each function. State the domain and range.

8. $f(x) = 2\left(\frac{2}{3}\right)^{x-3} - 4$

9. $f(x) = -\frac{1}{2}\left(\frac{3}{4}\right)^{x+1} + 5$

10. $f(x) = -\frac{1}{3}\left(\frac{4}{5}\right)^{x-4} + 3$

11. $f(x) = \frac{1}{8}\left(\frac{1}{4}\right)^{x+6} + 7$

Example 5 p. 479

12. NEW CARS A new SUV depreciates in value each year by a factor of 15%. Draw a graph of the SUV's value for the first 20 years after the initial purchase.

Practice and Problem Solving

= **Step-by-Step Solutions** begin on page R20.
Extra Practice begins on page 947.

Examples 1 and 2
pp. 475–476

Graph each function. State the domain and range.

13. $f(x) = 2(3)^x$

14. $f(x) = -2(4)^x$

15. $f(x) = 4^{x+1} - 5$

16. $f(x) = 3^{2x} + 1$

17. $f(x) = -0.4(3)^{x+2} + 4$

18. $f(x) = 1.5(2)^x + 6$

Example 3
p. 477

19 **SCIENCE** The population of a colony of beetles grows 30% each week for 10 weeks. If the initial population is 65 beetles, graph the function that represents the situation.

Example 4
p. 478

Graph each function. State the domain and range.

20. $f(x) = -4\left(\frac{3}{5}\right)^{x+4} + 3$

21. $f(x) = 3\left(\frac{2}{5}\right)^{x-3} - 6$

22. $f(x) = \frac{1}{2}\left(\frac{1}{5}\right)^{x+5} + 8$

23. $f(x) = \frac{3}{4}\left(\frac{2}{3}\right)^{x+4} - 2$

24. $f(x) = -\frac{1}{2}\left(\frac{3}{8}\right)^{x+2} + 9$

25. $f(x) = -\frac{5}{4}\left(\frac{4}{5}\right)^{x+4} + 2$

Example 5
p. 479

26. **ATTENDANCE** The attendance for a basketball team declined at a rate of 5% per game throughout a losing season. Graph the function modeling the attendance if 15 home games were played and 23,500 people were at the first game.

27. **PHONES** The number of pay phones in use in the United States has been declining due, in large part, to increased usage of cell phones. The function $P(x) = 2.28(0.9^x)$ can be used to model the number of pay phones in millions x years since 1999.

a. Graph the function.

b. Explain what the y-intercept and the asymptote represent in this situation.

Real-World Link

Inventor William Gray installed the first coin-operated phone in 1889 at a bank in Hartford, Connecticut.

28. **HEALTH** A certain drug is taken once every ten days. Each day, 10% of the drug dissipates from the system.

a. Graph the function representing this situation.

b. At what point is 50% of the original amount still in the system?

c. How much of the original amount remains in the system after 9 days?

29. **NUMBER THEORY** A sequence of numbers follows a pattern in which the next number is 125% of the previous number. The first number in the pattern is 18.

a. Write the function that represents the situation.

b. Graph the function for the first 10 numbers.

c. What is the value of the tenth number? Round to the nearest whole number.

For each graph, $f(x)$ is the parent function and $g(x)$ is a transformation of $f(x)$. Use the graph to determine $g(x)$.

30. $f(x) = 3^x$

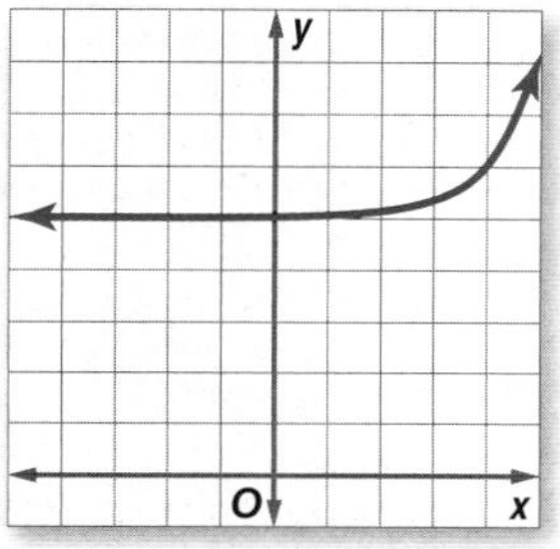

31. $f(x) = 2^x$

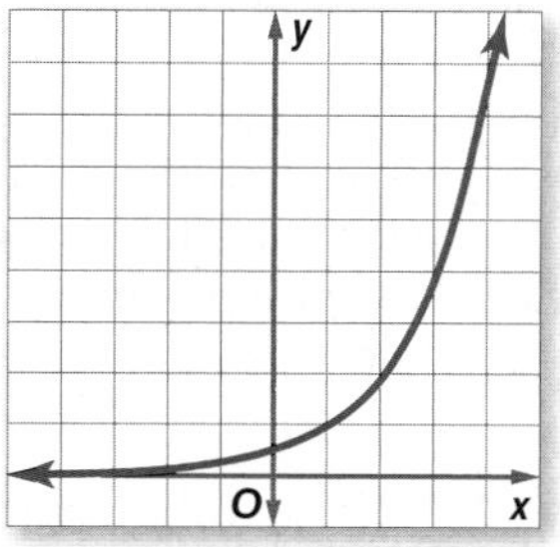

32. $f(x) = 4^x$

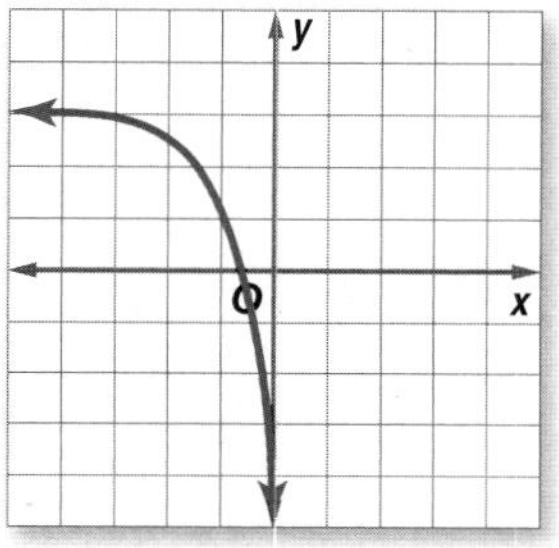

StudyTip

Parent Function For Exercise 33, it may be helpful to remember that the graph of any parent function $y = b^x$ passes through (0, 1).

33. **MULTIPLE REPRESENTATIONS** In this problem, you will use the tables below for exponential functions $f(x)$, $g(x)$, and $h(x)$.

x	−1	0	1	2	3	4	5
$f(x)$	2.5	2	1	−1	−5	−13	−29

x	−1	0	1	2	3	4	5
$g(x)$	5	11	23	47	95	191	383

x	−1	0	1	2	3	4	5
$h(x)$	3	2.5	2.25	2.125	2.0625	2.0313	2.0156

a. GRAPHICAL Graph the functions for $-1 \leq x \leq 5$ on separate graphs.

b. LOGICAL Which function(s) has a negative coefficient, a? Explain your reasoning.

c. LOGICAL Which function(s) is translated to the left?

d. ANALYTICAL Determine which functions are growth models and which are decay models.

H.O.T. Problems Use Higher-Order Thinking Skills

34. **REASONING** Determine whether each statement is *sometimes*, *always*, or *never* true. Explain your reasoning.

a. An exponential function of the form $y = ab^{x-h} + k$ has a y-intercept.

b. An exponential function of the form $y = ab^{x-h} + k$ has an x-intercept.

c. The function $f(x) = |b|^x$ is an exponential growth function if b is an integer.

35. **FIND THE ERROR** Vince and Grady were asked to graph $f(x) = -\frac{2}{3}\left(\frac{3}{4}\right)^{x-1}$. Is either of them correct? Explain your reasoning.

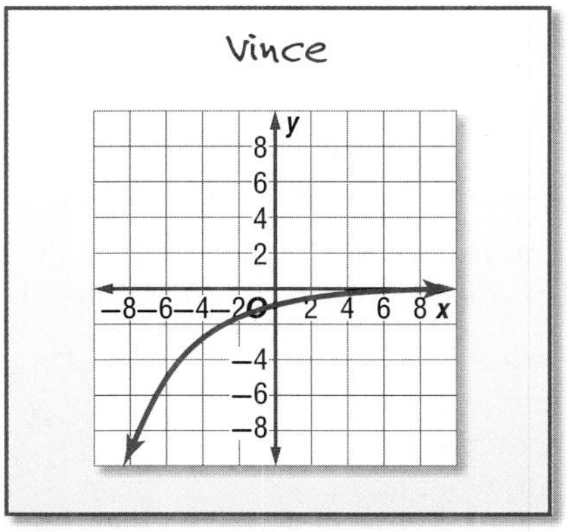

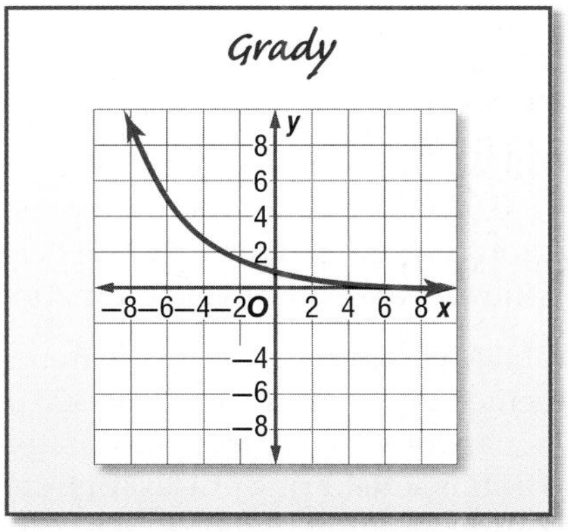

36. **CHALLENGE** A substance decays 35% each day. After 8 days, there are 8 milligrams of the substance remaining. How many milligrams were there initially?

37. **OPEN ENDED** Give an example of a value of b for which $f(x) = \left(\frac{8}{b}\right)^x$ represents exponential decay.

38. **WRITING IN MATH** Explain step by step how to translate $f(x) = ab^{x-h} + k$.

NYSRE PRACTICE Reinforcement of A.A.7

39. GRIDDED RESPONSE In the figure, $\overline{PO} \parallel \overline{RN}$, $ON = 12$, $MN = 6$, and $RN = 4$. What is the length of $\overline{PO}$?

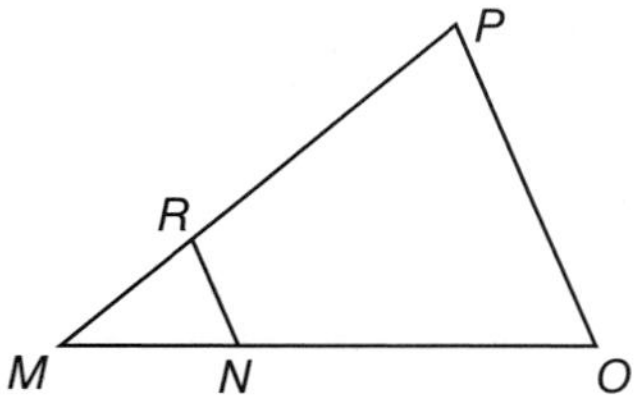

40. Ivan has enough money to buy 12 used CDs. If the cost of each CD was \$0.20 less, Ivan could buy 2 more CDs. How much money does Ivan have to spend on CDs?

A \$16.80 **C** \$15.80
B \$16.40 **D** \$15.40

41. One hundred students will attend the fall dance if tickets cost \$30 each. For each \$5 increase in price, 10 fewer students will attend. What price will deliver the maximum dollar sales?

F \$30
G \$35
H \$40
J \$45

42. Javier mows a lawn in 2 hours. Tonya mows the same lawn in 1.5 hours. About how many minutes will it take to mow the lawn if Javier and Tonya work together?

A 28 minutes
B 42 minutes
C 51 minutes
D 1.2 hours

Spiral Review

Solve each equation or inequality. (Lesson 7-7)

43. $\sqrt{y+5} = \sqrt{2y-3}$
44. $\sqrt{y+1} + \sqrt{y-4} = 5$
45. $10 - \sqrt{2x+7} \leq 3$
46. $6 + \sqrt{3y+4} < 6$
47. $\sqrt{d+3} + \sqrt{d+7} > 4$
48. $\sqrt{2y+5} - \sqrt{9+x} > 0$

Simplify. (Lesson 7-6)

49. $\dfrac{1}{y^{\frac{2}{5}}}$
50. $\dfrac{xy}{\sqrt[3]{z}}$
51. $\dfrac{3x+4x^2}{x^{-\frac{2}{3}}}$
52. $\sqrt[6]{27x^3}$
53. $\dfrac{\sqrt[4]{27}}{\sqrt[4]{3}}$
54. $\dfrac{a^{-\frac{1}{2}}}{6a^{\frac{1}{3}} \cdot a^{-\frac{1}{4}}}$

55. FOOTBALL The path of a football thrown across a field is given by the equation $y = -0.005x^2 + x + 5$, where x represents the distance, in feet, the ball has traveled horizontally and y represents the height, in feet, of the ball above ground level. About how far has the ball traveled horizontally when it returns to ground level? (Lesson 5-6)

56. COMMUNITY SERVICE A drug awareness program is being presented at a theater that seats 300 people. Proceeds will be donated to a local drug information center. If every two adults must bring at least one student, what is the maximum amount of money that can be raised? (Lesson 3-4)

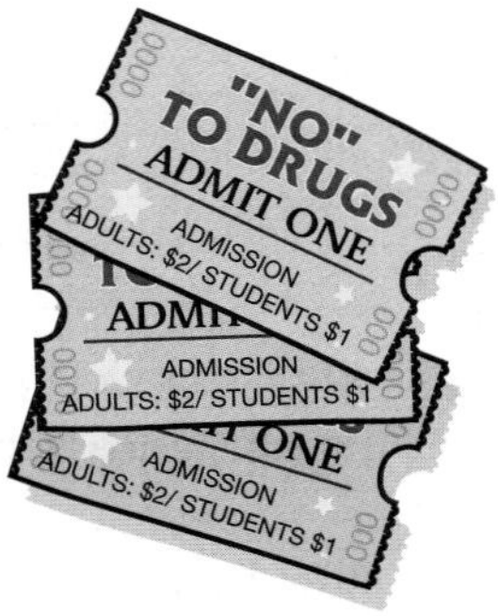

Skills Review

Simplify. Assume that no variable equals 0. (Lesson 6-1)

57. $f^{-7} \cdot f^4$
58. $(3x^2)^3$
59. $(2y)(4xy^3)$
60. $\left(\frac{3}{5}c^2f\right)\left(\frac{4}{3}cd\right)^2$

EXPLORE
8-2

Graphing Technology Lab
Solving Exponential Equations and Inequalities

NY Math Online glencoe.com
Other Calculator Keystrokes

NYS Core Curriculum **A2.A.27 Solve exponential equations with and without common bases**

You can use a TI-83/84 Plus graphing calculator to solve exponential equations by graphing or by using the table feature. To do this, you will write the equations as systems of equations.

ACTIVITY 1

Solve $3^{x-4} = \frac{1}{9}$.

Step 1 Graph each side of the equation as a separate function. Enter 3^{x-4} as Y1. Enter $\frac{1}{9}$ as Y2. Be sure to include parentheses around each exponent. Then graph the two equations.

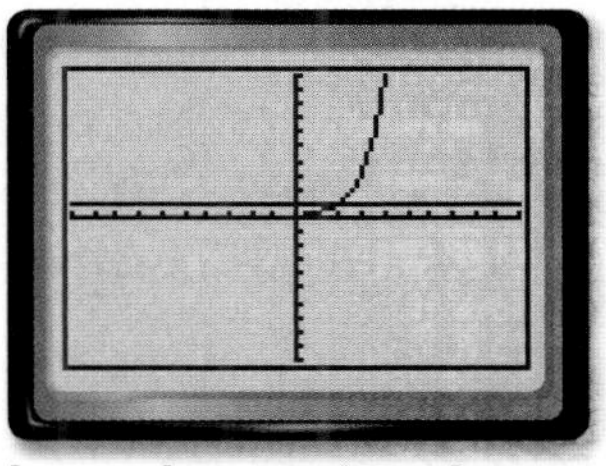

[−10, 10] scl: 1 by [−1, 1] scl: 0.1

Step 2 Use the intersect feature.

You can use the intersect feature on the CALC menu to approximate the ordered pair of the point at which the graphs cross.

The calculator screen shows that the x-coordinate of the point at which the curves cross is 2. Therefore, the solution of the equation is 2.

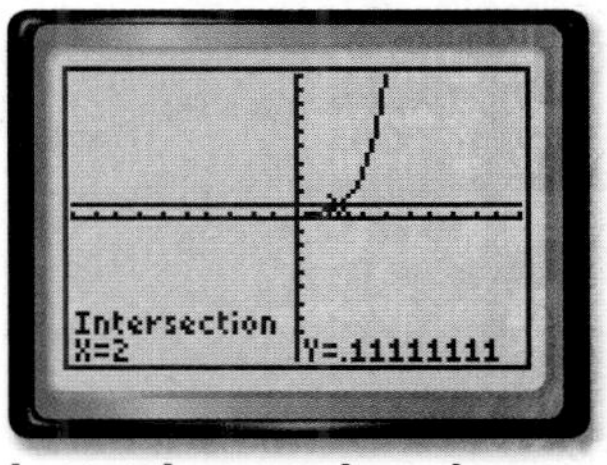

[−10, 10] scl: 1 by [−1, 1] scl: 0.1

Step 3 Use the TABLE feature.

You can also use the table feature to locate the point at which the curves intersect.

The table displays x-values and corresponding y-values for each graph. Examine the table to find the x-value for which the y-values of the graphs are equal.

At $x = 2$, both functions have a y-value of $0.\overline{1}$ or $\frac{1}{9}$. Thus, the solution of the equation is 2.

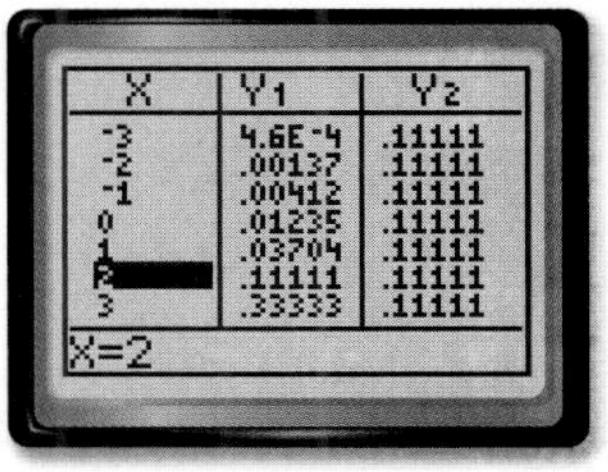

CHECK Substitute 2 for x in the original equation.

$3^{x-4} \stackrel{?}{=} \frac{1}{9}$	**Original equation**
$3^{2-4} \stackrel{?}{=} \frac{1}{9}$	**Substitute 2 for x.**
$3^{-2} \stackrel{?}{=} \frac{1}{9}$	**Simplify.**
$\frac{1}{9} = \frac{1}{9}$ ✓	**The solution checks.**

A similar procedure can be used to solve exponential inequalities using a graphing calculator.

ACTIVITY 2

Solve $2^{x-2} \geq 0.5^{x-3}$.

Step 1 Enter the related inequalities.

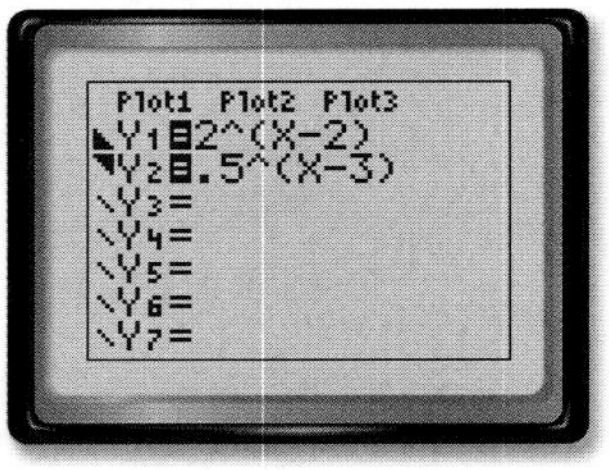

Rewrite the problem as a system of inequalities.

The first inequality is $2^{x-2} \geq y$ or $y \leq 2^{x-2}$. Since this inequality includes the *less than or equal to* symbol, shade below the curve.

First enter the boundary, and then use the arrow and ENTER keys to choose the shade below icon, ◣.

The second inequality is $y \geq 0.5^{x-3}$. Shade above the curve since this inequality contains *greater than or equal to.*

KEYSTROKES: Y= ◄ ◄ ENTER ENTER ENTER ► ► 2 ^ (X,T,θ,n − 2) ENTER ◄ ◄ ENTER ENTER ► ► .5 ^ (X,T,θ,n − 3)

Step 2 Graph the system.

KEYSTROKES: GRAPH

The x-values of the points in the region where the shadings overlap is the solution set of the original inequality. Using the **intersect** feature, you can conclude that the solution set is $\{x \mid x \geq 2.5\}$.

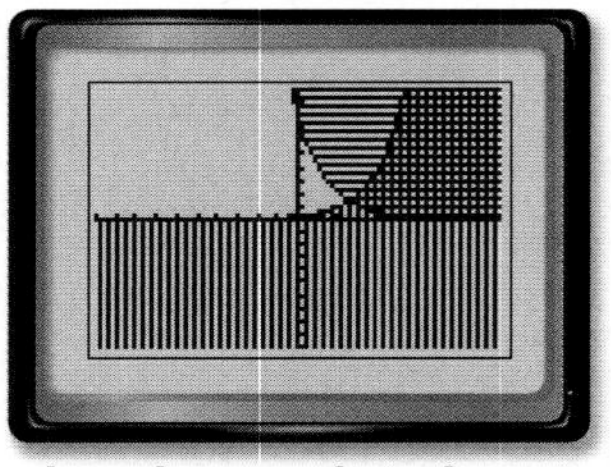

[−2, 8] scl: 1 by [−2, 8] scl: 1

Step 3 Use the **TABLE** feature.

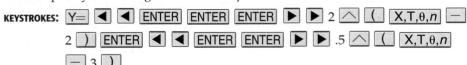

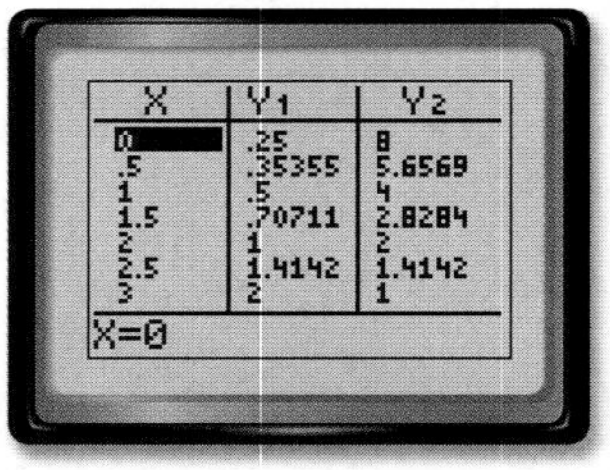

Verify using the **table** feature. Set up the table to show x-values in increments of 0.5.

KEYSTROKES: 2nd [TBLSET] 0 ENTER .5 ENTER 2nd [TABLE]

Notice that for x-values greater than $x = 2.5$, **Y1** > **Y2**. This confirms that the solution of the inequality is $\{x \mid x \geq 2.5\}$.

Exercises

Solve each equation or inequality.

1. $9^{x-1} = \frac{1}{81}$

2. $4^{x+3} = 2^{5x}$

3. $5^{x-1} = 2^{x}$

4. $3.5^{x+2} = 1.75^{x+3}$

5. $-3^{x+4} = -0.5^{2x+3}$

6. $6^{2-x} - 4 < -0.25^{x-2.5}$

7. $16^{x-1} > 2^{2x+2}$

8. $3^{x} - 4 \leq 5^{\frac{x}{2}}$

9. $5^{x+3} \leq 2^{x+4}$

10. **WRITING IN MATH** Explain why this technique of graphing a system of equations or inequalities works to solve exponential equations and inequalities.

8-2 Solving Exponential Equations and Inequalities

Then

You graphed exponential functions. (Lesson 8-1)

Now

- Solve exponential equations.
- Solve exponential inequalities.

NYS Core Curriculum

A2.A.12 Evaluate exponential expressions, including those with base *e*
A2.A.27 Solve exponential equations with and without common bases *Also addresses A2.A.6.*

New Vocabulary

exponential equation
compound interest
exponential inequality

NY Math Online

glencoe.com
- Extra Examples
- Personal Tutor
- Self-Check Quiz
- Homework Help

Why?

Membership on Internet social networking sites tends to increase exponentially. The membership growth of one Web site can be modeled by the equation $y = 2.2(1.37)^x$, where x is the number of years since 2004 and y is the number of members in millions.

You can use $y = 2.2(1.37)^x$ to determine how many members there will be in a given year, or to determine the year in which membership was at a certain level.

Solve Exponential Equations In an **exponential equation**, variables occur as exponents.

Key Concept — For Your FOLDABLE

Property of Equality for Exponential Functions

Words Let $b > 0$ and $b \neq 1$. Then $b^x = b^y$ if and only if $x = y$.

Example If $3^x = 3^5$, then $x = 5$. If $x = 5$, then $3^x = 3^5$.

The Equality Property can be used to solve exponential equations.

EXAMPLE 1 Solve Exponential Equations

Solve each equation.

a. $2^x = 8^3$

$2^x = 8^3$	**Original equation**
$2^x = (2^3)^3$	**Rewrite 8 as 2^3.**
$2^x = 2^9$	**Power of a Power**
$x = 9$	**Property of Equality for Exponential Functions**

b. $9^{2x-1} = 3^{6x}$

$9^{2x-1} = 3^{6x}$	**Original equation**
$(3^2)^{2x-1} = 3^{6x}$	**Rewrite 9 as 3^2.**
$3^{4x-2} = 3^{6x}$	**Power of a Power**
$4x - 2 = 6x$	**Property of Equality for Exponential Functions**
$-2 = 2x$	**Subtract $4x$ from each side.**
$-1 = x$	**Divide each side by 2.**

Check Your Progress

1A. $4^{2n-1} = 64$

1B. $5^{5x} = 125^{x+2}$

Personal Tutor glencoe.com

You can use information about growth or decay to write the equation of an exponential function.

Real-World EXAMPLE 2 Write an Exponential Function

SCIENCE Kristin starts an experiment with 7500 bacteria cells. After 4 hours, there are 23,000 cells.

a. Write an exponential function that could be used to model the number of bacteria after x hours if the number of bacteria changes at the same rate.

At the beginning of the experiment, the time is 0 hours and there are 7500 bacteria cells. Thus, the y-intercept, and the value of a, is 7500.

When $x = 4$, the number of bacteria cells is 23,000. Substitute these values into an exponential function to determine the value of b.

$y = ab^x$	**Exponential function**
$23{,}000 = 7500 \cdot b^4$	**Replace x with 4, y with 23,000, and a with 7500.**
$3.067 \approx b^4$	**Divide each side by 7500.**
$\sqrt[4]{3.067} \approx b$	**Take the 4th root of each side.**
$1.323 \approx b$	**Use a calculator.**

An equation that models the number of bacteria is $y = 7500(1.323)^x$.

b. How many bacteria cells can be expected in the sample after 12 hours?

$y = 7500(1.323)^x$	**Modeling equation**
$= 7500(1.323)^{12}$	**Replace x with 12.**
$\approx 215{,}665$	**Use a calculator.**

There will be approximately 215,665 bacteria cells after 12 hours.

Real-World Link

In 2005, the U.S. recycling rate of 32% percent prevented the release of approximately 49 million metric tons of carbon into the air—roughly the amount emitted annually by 39 million cars.

Source: Environmental Protection Agency

Check Your Progress

2. RECYCLING A manufacturer distributed 3.2 million aluminum cans in 2005.

A. In 2010, the manufacturer distributed 420,000 cans made from the recycled cans it had previously distributed. Assuming that the recycling rate continues, write an equation to model the distribution each year of cans that are made from recycled aluminum.

B. How many cans made from recycled aluminum can be expected in the year 2050?

Personal Tutor glencoe.com

Exponential functions are used in situations involving compound interest. **Compound interest** is interest paid on the principal of an investment and any previously earned interest.

Key Concept Compound Interest

For Your FOLDABLE

You can calculate compound interest using the following formula.

$$A = P\left(1 + \frac{r}{n}\right)^{nt},$$

where A is the amount in the account after t years, P is the principal amount invested, r is the annual interest rate, and n is the number of compounding periods each year.

EXAMPLE 3 Compound Interest

An investment account pays 4.2% annual interest compounded monthly. If $2500 is invested in this account, what will be the balance after 15 years?

Understand Find the total amount in the account after 15 years.

Plan Use the compound interest formula.

$P = 2500$, $r = 0.042$, $n = 12$, and $t = 15$

Solve

$A = P\left(1 + \frac{r}{n}\right)^{nt}$	**Compound Interest Formula**
$= 2500\left(1 + \frac{0.042}{12}\right)^{12 \cdot 15}$	**$P = 2500$, $r = 0.042$, $n = 12$, $t = 15$**
≈ 4688.87	**Use a calculator.**

Check Graph the corresponding equation $y = 2500(1.0035)^{12t}$. Use CALC: value to find y when $x = 15$.

The y-value 4688.8662 is very close to 4688.87, so the answer is reasonable.

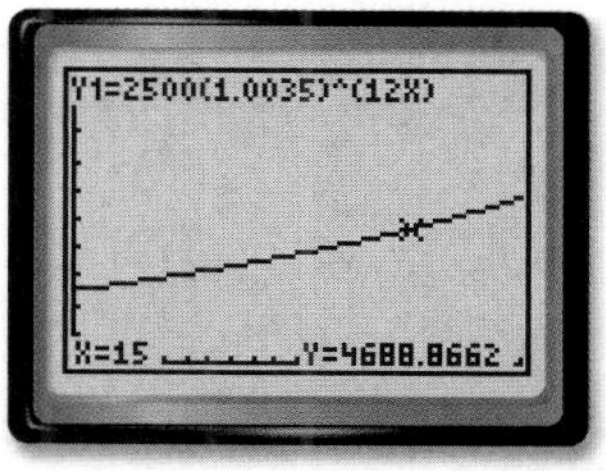

> **Watch Out!**
>
> **Percents** Remember to convert all percents to decimal form; 4.2% is 0.042.

Check Your Progress

3. Find the account balance after 20 years if $100 is placed in an account that pays 1.2% interest compounded twice a month.

Personal Tutor glencoe.com

Solve Exponential Inequalities An **exponential inequality** is an inequality involving exponential functions.

Key Concept

For Your FOLDABLE

Property of Inequality for Exponential Functions

Words Let $b > 1$. Then $b^x > b^y$ if and only if $x > y$, and $b^x < b^y$ if and only if $x < y$.

Example If $2^x \geq 2^6$, then $x > 6$. If $x > 6$, then $2^x \geq 2^6$.

This property also holds true for $\leq$ and $\geq$.

EXAMPLE 4 Solve Exponential Inequalities

Solve $16^{2x-3} < 8$.

$16^{2x-3} < 8$	**Original inequality**
$(2^4)^{2x-3} < 2^3$	**Rewrite 16 as 2^4 and 8 as 2^3.**
$2^{8x-12} < 2^3$	**Power of a Power**
$8x - 12 < 3$	**Property of Inequality for Exponential Functions**
$8x < 15$	**Add 12 to each side.**
$x < \frac{15}{8}$	**Divide each side by 8.**

Check Your Progress

Solve each inequality.

4A. $3^{2x-1} \geq \frac{1}{243}$

4B. $2^{x+2} > \frac{1}{32}$

Personal Tutor glencoe.com

Check Your Understanding

Example 1 p. 485

Solve each equation.

1. $3^{5x} = 27^{2x-4}$
2. $16^{2y-3} = 4^{y+1}$
3. $2^{6x} = 32^{x-2}$
4. $49^{x+5} = 7^{8x-6}$

Example 2 p. 486

5. **SCIENCE** Mitosis is a process in which one cell divides into two. The *Escherichia coli* is one of the fastest growing bacteria. It can reproduce itself in 15 minutes.
   a. Write an exponential function to represent the number of cells c after t minutes.
   b. If you begin with one *Escherichia coli* cell, how many cells will there be in one hour?

Example 3 p. 487

6. A certificate of deposit (CD) pays 2.25% annual interest compounded biweekly. If you deposit $500 into this CD, what will the balance be after 6 years?

Example 4 p. 487

Solve each inequality.

7. $4^{2x+6} \leq 64^{2x-4}$
8. $25^{y-3} \leq \left(\frac{1}{125}\right)^{y+2}$

Practice and Problem Solving

= **Step-by-Step Solutions** begin on page R20.
Extra Practice begins on page 947.

Example 1 p. 485

Solve each equation.

9. $8^{4x+2} = 64$
10. $5^{x-6} = 125$
11. $81^{a+2} = 3^{3a+1}$
12. $256^{b+2} = 4^{2-2b}$
13. $9^{3c+1} = 27^{3c-1}$
14. $8^{2y+4} = 16^{y+1}$

Example 2 p. 486

15. **MONEY** In 2003, My-Lien received $10,000 from her grandmother. Her parents invested all of the money, and by 2015, the amount will have grown to $16,780.
   a. Write an exponential function that could be used to model the money y. Write the function in terms of x, the number of years since 2003.
   b. Assume that the amount of money continues to grow at the same rate. What would be the balance in the account in 2025?

Write an exponential function for the graph that passes through the given points.

16. (0, 6.4) and (3, 100)
17. (0, 256) and (4, 81)
18. (0, 128) and (5, 371,293)
19. (0, 144), and (4, 21,609)

Example 3 p. 487

20. Find the balance of an account after 7 years if $700 is deposited into an account paying 4.3% interest compounded monthly.
21. Determine how much is in a retirement account after 20 years if $5000 was invested at 6.05% interest compounded weekly.
22. A savings account offers 0.7% interest compounded bimonthly. If $110 is deposited in this account, what will the balance be after 15 years?
23. A college savings account pays 13.2% annual interest compounded semiannually. What is the balance of an account after 12 years if $21,000 was initially deposited?

Example 4 p. 487

Solve each inequality.

24. $625 \geq 5^{a+8}$
25. $10^{5b+2} > 1000$
26. $\left(\frac{1}{64}\right)^{c-2} < 32^{2c}$
27. $\left(\frac{1}{27}\right)^{2d-2} \leq 81^{d+4}$
28. $\left(\frac{1}{9}\right)^{3t+5} \geq \left(\frac{1}{243}\right)^{t-6}$
29. $\left(\frac{1}{36}\right)^{w+2} < \left(\frac{1}{216}\right)^{4w}$

30. **SCIENCE** A mug of hot chocolate is 90°C at time $t = 0$. It is surrounded by air at a constant temperature of 20°C. If stirred steadily, its temperature in Celsius after t minutes will be $y(t) = 20 + 70(1.071)^{-t}$.

a. Find the temperature of the hot chocolate after 15 minutes.

b. Find the temperature of the hot chocolate after 30 minutes.

c. The optimum drinking temperature is 60°C. Will the mug of hot chocolate be at or below this temperature after 10 minutes?

31. **ANIMALS** Studies show that an animal will defend a territory, with area in square yards, that is directly proportional to the 1.31 power of the animal's weight in pounds.

a. If a 45-pound beaver will defend 170 square yards, write an equation for the area a defended by a beaver weighing w pounds.

b. Scientists believe that thousands of years ago, the beaver's ancestors were 11 feet long and weighed 430 pounds. Use your equation to determine the area defended by these animals.

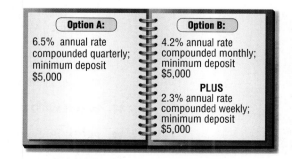

Real-World Link

One of the oldest living organisms on Earth, the General Sherman Tree in Sequoia National Park, California, is between 2100 and 2500 years old. It is the tallest of the giant sequoias, measuring approximately 84 meters.

Source: National Park Service

Solve each equation.

32. $\left(\frac{1}{2}\right)^{4x + 1} = 8^{2x + 1}$

33. $\left(\frac{1}{5}\right)^{x - 5} = 25^{3x + 2}$

34. $216 = \left(\frac{1}{6}\right)^{x + 3}$

35. $\left(\frac{1}{8}\right)^{3x + 4} = \left(\frac{1}{4}\right)^{-2x + 4}$

36. $\left(\frac{2}{3}\right)^{5x + 1} = \left(\frac{27}{8}\right)^{x - 4}$

37. $\left(\frac{25}{81}\right)^{2x + 1} = \left(\frac{729}{125}\right)^{-3x + 1}$

38. **POPULATION** In 1950, the world population was about 2.556 billion. By 1980, it had increased to about 4.458 billion.

a. Write an exponential function of the form $y = abx$ that could be used to model the world population y in billions for 1950 to 1980. Write the equation in terms of x, the number of years since 1950. (Round the value of b to the nearest ten-thousandth.)

b. Suppose the population continued to grow at that rate. Estimate the population in 2000.

c. In 2000, the population of the world was about 6.08 billion. Compare your estimate to the actual population.

d. Use the equation you wrote in Part a to estimate the world population in the year 2020. How accurate do you think the estimate is? Explain your reasoning.

39. **TREES** The diameter of the base of a tree trunk in centimeters varies directly with the $\frac{3}{2}$ power of its height in meters.

a. A young sequoia tree is 6 meters tall, and the diameter of its base is 19.1 centimeters. Use this information to write an equation for the diameter d of the base of a sequoia tree if its height is h meters high.

b. Refer to the information at the left. Find the diameter of the General Sherman Tree at its base.

40. **FINANCIAL DECISIONS** Mrs. Jackson has two different retirement investment plans from which to choose.

Option A:
6.5% annual rate compounded quarterly; minimum deposit $5,000

Option B:
4.2% annual rate compounded monthly; minimum deposit $5,000
PLUS
2.3% annual rate compounded weekly; minimum deposit $5,000

a. Write equations for Option A and Option B given the minimum deposits.

b. Draw a graph to show the balances for each investment option after t years.

c. Explain whether Option A or Option B is the better investment choice.

Real-World Link

In theory, if you repeated the process of cutting and stacking paper 42 times your stack of paper would reach the Moon.

41. **MULTIPLE REPRESENTATIONS** In this problem, you will explore the rapid increase of an exponential function. A large sheet of paper is cut in half, and one of the resulting pieces is placed on top of the other. Then the pieces in the stack are cut in half and placed on top of each other. Suppose this procedure is repeated several times.

a. **CONCRETE** Perform this activity and count the number of sheets in the stack after the first cut. How many pieces will there be after the second cut? How many pieces after the third cut? How many pieces after the fourth cut?

b. **TABULAR** Record your results in a table.

c. **SYMBOLIC** Use the pattern in the table to write an equation for the number of pieces in the stack after x cuts.

d. **ANALYTICAL** The thickness of ordinary paper is about 0.003 inch. Write an equation for the thickness of the stack of paper after x cuts.

e. **ANALYTICAL** How thick will the stack of paper be after 30 cuts?

H.O.T. Problems

Use **H**igher-**O**rder **T**hinking Skills

42. **WRITING IN MATH** In a problem about compound interest, describe what happens as the compounding period becomes more frequent while the principal and overall time remain the same.

43. **FIND THE ERROR** Beth and Liz are solving $6^{x-3} > 36^{-x-1}$. Is either of them correct? Explain your reasoning.

Beth

$6^{x-3} > 36^{-x-1}$

$6^{x-3} > (6^2)^{-x-1}$

$6^{x-3} > 6^{-2x-2}$

$x - 3 > -2x - 2$

$3x > 1$

$x > \frac{1}{3}$

Liz

$6^{x-3} > 36^{-x-1}$

$6^{x-3} > (6^2)^{-x-1}$

$6^{x-3} > 6^{-x+1}$

$x - 3 > -x + 1$

$2x > 4$

$x > 2$

44. **CHALLENGE** Solve for x: $16^{18} + 16^{18} + 16^{18} + 16^{18} + 16^{18} = 4^x$.

45. **OPEN ENDED** What would be a more beneficial change to a 5-year loan at 8% interest compounded monthly: reducing the term to 4 years or reducing the interest rate to 6.5%?

46. **REASONING** Determine whether the following statements are *sometimes*, *always*, or *never* true. Explain your reasoning.

a. $2^x > -8^{20x}$ for all values of x.

b. The graph of an exponential growth equation is increasing.

c. The graph of an exponential decay equation is increasing.

47. **OPEN ENDED** Write an exponential inequality with a solution of $x \leq 2$.

48. **PROOF** Show that $27^{2x} \cdot 81^{x+1} = 3^{2x+2} \cdot 9^{4x+1}$.

49. **WRITING IN MATH** If you were given the initial and final amounts of a radioactive substance and the amount of time that passes, how would you determine the rate at which the amount was increasing or decreasing in order to write an equation?

NYSRE PRACTICE Reinforcement of A.A.6, Reinforcement of A.A.24

50. $3 \times 10^{-4} =$

A $-30,000$ C -120
B 0.0003 D 0.00003

51. Which of the following could *not* be a solution to $5 - 3x < -3$?

F 2.5 H 3.5
G 3 J 4

52. GRIDDED RESPONSE The three angles of a triangle are $3x$, $x + 10$, and $2x - 40$. Find the measure of the smallest angle in the triangle.

53. ACT/SAT Which of the following is equivalent to $(x)(x)(x)(x)$ for all x?

A $4x$ C $x + 4$
B x^4 D $2x^2$

Spiral Review

Graph each function. (Lesson 8-1)

54. $y = 2(3)^x$

55. $y = 5(2)^x$

56. $4\left(\frac{1}{3}\right)^x$

Solve each equation. (Lesson 7-7)

57. $\sqrt{x + 5} - 3 = 0$

58. $\sqrt{3t - 5} - 3 = 4$

59. $\sqrt[4]{2x - 1} = 2$

60. $\sqrt{x - 6} - \sqrt{x} = 3$

61. $\sqrt[3]{5m + 2} = 3$

62. $(6n - 5)^{\frac{1}{3}} + 3 = -2$

63. $(5x + 7)^{\frac{1}{5}} + 3 = 5$

64. $(3x - 2)^{\frac{1}{5}} + 6 = 5$

65. $(7x - 1)^{\frac{1}{3}} + 4 = 2$

66. SALES A salesperson earns \$10 an hour plus a 10% commission on sales. Write a function to describe the salesperson's income. If the salesperson wants to earn \$1000 in a 40-hour week, what should his sales be? (Lesson 7-2)

67. STATE FAIR A dairy makes three types of cheese—cheddar, Monterey Jack, and Swiss—and sells the cheese in three booths at the state fair. At the beginning of one day, the first booth received x pounds of each type of cheese. The second booth received y pounds of each type of cheese, and the third booth received z pounds of each type of cheese. By the end of the day, the dairy had sold 131 pounds of cheddar, 291 pounds of Monterey Jack, and 232 pounds of Swiss. The table below shows the percent of the cheese delivered in the morning that was sold at each booth. How many pounds of cheddar cheese did each booth receive in the morning? (Lesson 3-5)

Type	Booth 1	Booth 2	Booth 3
Cheddar	40%	30%	10%
Monterey Jack	40%	90%	80%
Swiss	30%	70%	70%

Skills Review

Find $[g \circ h](x)$ and $[h \circ g](x)$. (Lesson 7-1)

68. $h(x) = 2x - 1$
$g(x) = 3x + 4$

69. $h(x) = x^2 + 2$
$g(x) = x - 3$

70. $h(x) = x^2 + 1$
$g(x) = -2x + 1$

71. $h(x) = -5x$
$g(x) = 3x - 5$

72. $h(x) = x^3$
$g(x) = x - 2$

73. $h(x) = x + 4$
$g(x) = |x|$

8-3 Logarithms and Logarithmic Functions

Then

You found the inverse of a function. (Lesson 7-2)

Now

- Evaluate logarithmic expressions.
- Graph logarithmic functions.

NYS Core Curriculum

A2.A.41 Use functional notation to evaluate functions for given values in the domain

A2.A.54 Graph logarithmic functions, using the inverse of the related exponential function

New Vocabulary

logarithm

logarithmic function

NY Math Online

glencoe.com

- Extra Examples
- Personal Tutor
- Self-Check Quiz
- Homework Help
- Math in Motion

Why?

Many scientists believe the extinction of the dinosaurs was caused by an asteroid striking Earth. Astronomers use the Palermo scale to classify objects near Earth based on the likelihood of impact. To make the comparing of several objects easier, the scale was developed using *logarithms*. The Palermo scale value of any object can be found using the equation $PS = \log_{10} R$, where R is the relative risk posed by the object.

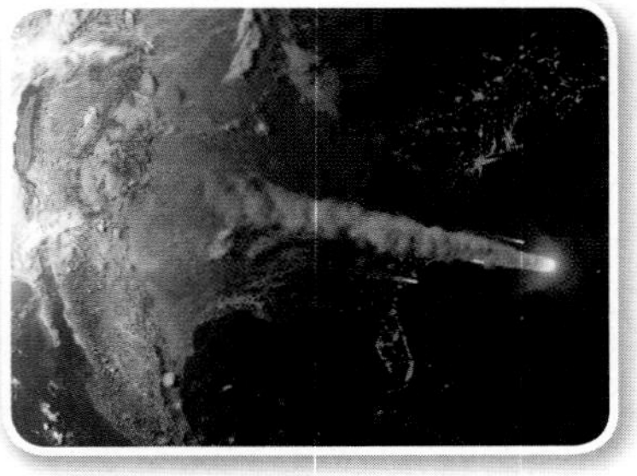

Logarithmic Functions and Expressions Consider the exponential function $f(x) = 2^x$ and its inverse. Recall that you can graph an inverse function by interchanging the x- and y-values in the ordered pairs of the function.

$y = 2^x$	
x	y
−3	$\frac{1}{8}$
−2	$\frac{1}{4}$
−1	$\frac{1}{2}$
0	1
1	2
2	4
3	8

$x = 2^y$	
x	y
$\frac{1}{8}$	−3
$\frac{1}{4}$	−2
$\frac{1}{2}$	−1
1	0
2	1
4	2
8	3

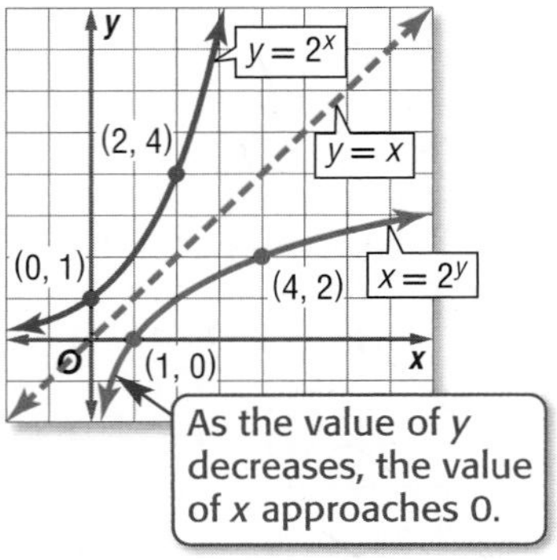

The inverse of $y = 2^x$ can be defined as $x = 2^y$. In general, the inverse of $y = b^x$ is $x = b^y$. In $x = b^y$, the variable y is called the **logarithm** of x. This is usually written as $y = \log_b x$, which is read *y equals log base b of x*.

Key Concept — Logarithm with Base b

For Your FOLDABLE

Words Let b and x be positive numbers, $b \neq 1$. The *logarithm of x with base b* is denoted $\log_b x$ and is defined as the exponent y that makes the equation $b^y = x$ true.

Symbols Suppose $b > 0$ and $b \neq 1$. For $x > 0$, there is a number y such that

$$\log_b x = y \text{ if and only if } b^y = x.$$

Example If $\log_3 27 = y$, then $3^y = 27$.

The definition of logarithms can be used to express logarithms in exponential form.

EXAMPLE 1 Logarithmic to Exponential Form

Write each equation in exponential form.

a. $\log_2 8 = 3$

$\log_2 8 = 3 \rightarrow 8 = 2^3$

b. $\log_4 \frac{1}{256} = -4$

$\log_4 \frac{1}{256} = -4 \rightarrow \frac{1}{256} = 4^{-4}$

Check Your Progress

1A. $\log_4 16 = 2$

1B. $\log_3 729 = 6$

Personal Tutor glencoe.com

The definition of logarithms can also be used to write exponential equations in logarithmic form.

EXAMPLE 2 Exponential to Logarithmic Form

Write each equation in logarithmic form.

a. $15^3 = 3375$

$15^3 = 3375 \rightarrow \log_{15} 3375 = 3$

b. $4^{\frac{1}{2}} = 2$

$4^{\frac{1}{2}} = 2 \rightarrow \log_4 2 = \frac{1}{2}$

Check Your Progress

2A. $4^3 = 64$

2B. $125^{\frac{1}{3}} = 5$

Personal Tutor glencoe.com

You can use the definition of a logarithm to evaluate a logarithmic expression.

EXAMPLE 3 Evaluate Logarithmic Expressions

Evaluate $\log_{16} 4$.

$\log_{16} 4 = y$	Let the logarithm equal y.
$4 = 16^y$	Definition of logarithm
$4^1 = 4^{2y}$	$16 = 4^2$
$1 = 2y$	Property of Equality for Exponential Functions
$\frac{1}{2} = y$	Divide each side by 2.

Thus, $\log_{16} 4 = \frac{1}{2}$.

Check Your Progress

Evaluate each expression.

3A. $\log_3 81$

3B. $\log_{\frac{1}{2}} 256$

Personal Tutor glencoe.com

> **Watch Out!**
>
> **Logarithmic Base** It is easy to get confused about which number is the base and which is the exponent in logarithmic equations. Consider highlighting each number as you solve to help organize your calculations.

Graphing Logarithmic Functions The function $y = \log_b x$, where $b \neq 1$, is called a **logarithmic function**. The graph of $f(x) = \log_b x$ represents a parent graph of the logarithmic functions.

Key Concept Parent Function of Logarithmic Functions

For Your FOLDABLE

Parent function:	$f(x) = \log_b x$	**Type of graph:**	continuous, one-to-one
Domain:	all positive real numbers	**Range:**	all real numbers
Asymptote:	y-axis	**Intercept:**	(1, 0)

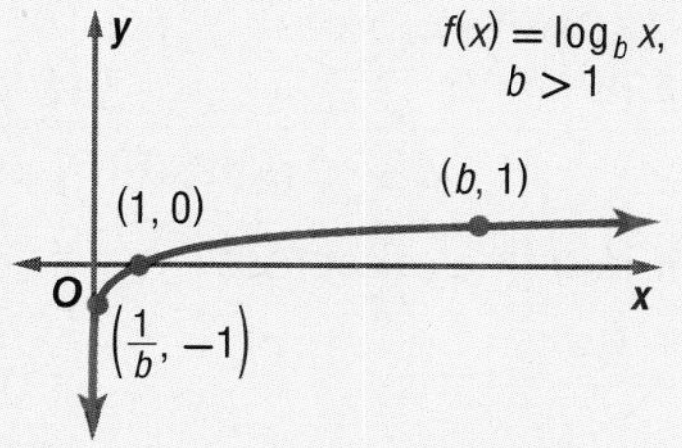

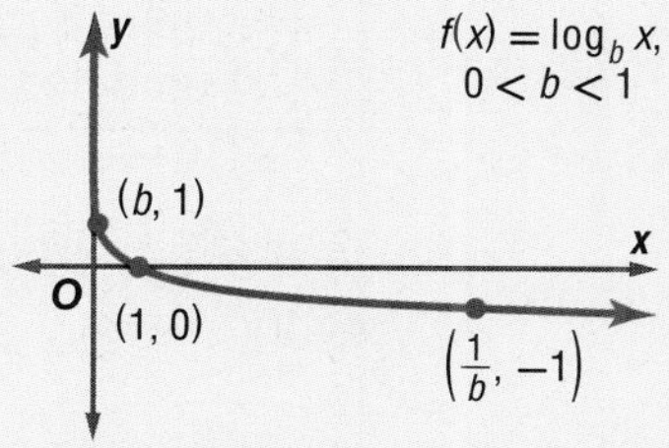

Math *in Motion*, Animation glencoe.com

StudyTip

Zero Exponent Recall that for any $b \neq 0$, $b^0 = 1$. Therefore, $\log_2 0$ is undefined because $2^x \neq 0$ for any x value.

StudyTip

Continuity Most exponential and logarithmic functions are continuous. In Example 2a, $f(x)$ is increasing from 0 to infinity.

EXAMPLE 4 Graph Logarithmic Functions

Graph each function.

a. $f(x) = \log_5 x$

Step 1 Identify the base.
$b = 5$

Step 2 Determine points on the graph.
Because $5 > 1$, use the points $\left(\frac{1}{b}, -1\right)$, $(1, 0)$, and $(b, 1)$.

Step 3 Plot the points and sketch the graph.
$\left(\frac{1}{b}, -1\right) \rightarrow \left(\frac{1}{5}, -1\right)$
$(1, 0)$
$(b, 1) \rightarrow (5, 1)$

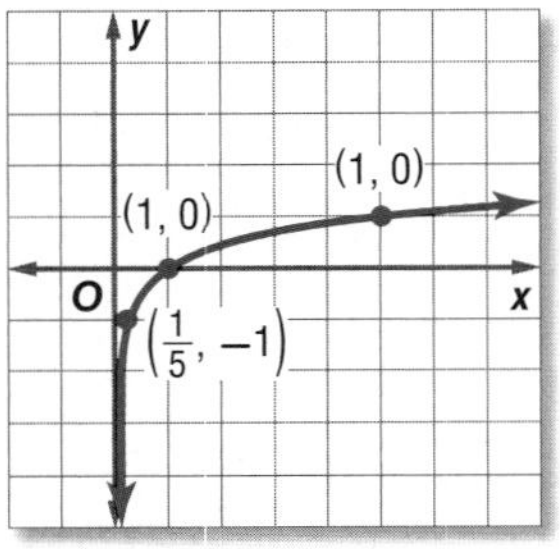

b. $f(x) = \log_{\frac{1}{3}} x$

Step 1 $b = \frac{1}{3}$

Step 2 $0 < \frac{1}{3} < 1$,

so use the points $\left(\frac{1}{3}, 1\right)$, $(1, 0)$ and $(3, -1)$.

Step 3 Sketch the graph.

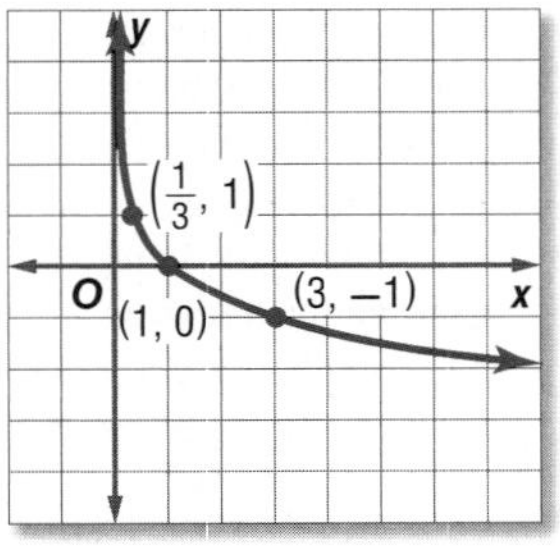

Check Your Progress

4A. $f(x) = \log_2 x$

4B. $f(x) = \log_{\frac{1}{8}} x$

Personal Tutor glencoe.com

The same techniques used to transform the graphs of other functions you have studied can be applied to the graphs of logarithmic functions.

Key Concept Transformations of Logarithmic Functions

For Your FOLDABLE

$f(x) = a \log_b (x - h) + k$

***h* – Horizontal Translation**	***k* – Vertical Translation**
$\lvert h \rvert$ units right if h is positive $\lvert h \rvert$ units left if h is negative	$\lvert k \rvert$ units up if k is positive $\lvert k \rvert$ units down if k is negative

***a* – Orientation and Shape**

If $a < 0$, the graph is reflected across the x-axis.

If $\lvert a \rvert > 1$, the graph is expanded vertically.
If $0 < \lvert a \rvert < 1$, the graph is compressed vertically.

StudyTip

End Behavior In Example 5a, as x approaches infinity, $f(x)$ approaches infinity.

EXAMPLE 5 Graph Logarithmic Functions

Graph each function.

a. $f(x) = 3 \log_{10} x + 1$

This represents a transformation of the graph of $f(x) = \log_{10} x$.

- $|a| = 3$: The graph expands vertically.
- $h = 0$: There is no horizontal shift.
- $k = 1$: The graph is translated 1 unit up.

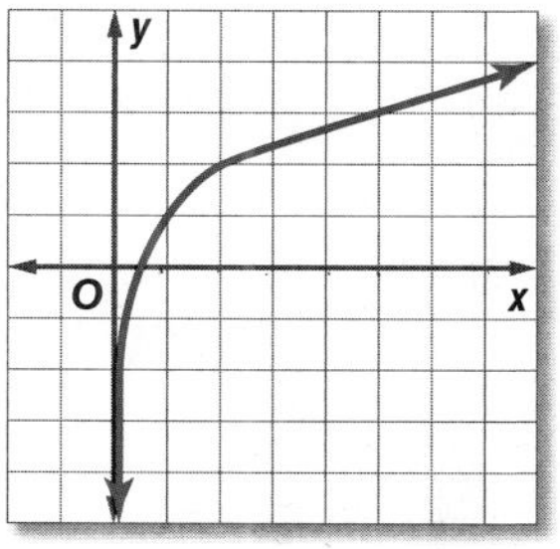

b. $f(x) = \frac{1}{2} \log_{\frac{1}{4}} (x - 3)$

This is a transformation of the graph of $f(x) = \log_{\frac{1}{4}} x$.

- $|a| = \frac{1}{2}$: The graph is compressed vertically.
- $h = 3$: The graph is translated 3 units to the right.
- $k = 0$: There is no vertical shift.

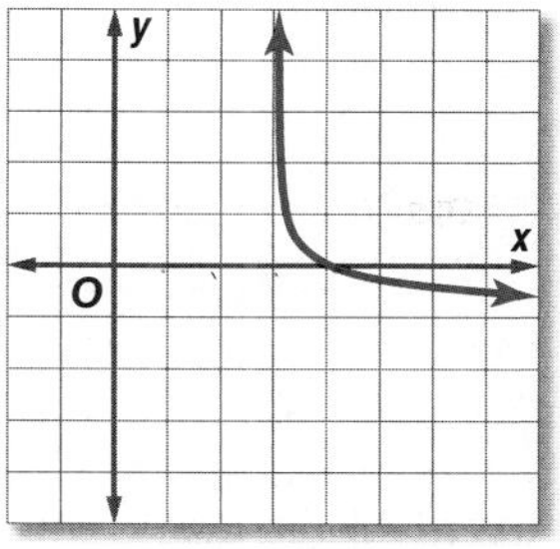

Check Your Progress

Graph each function.

5A. $f(x) = 2 \log_3 (x - 2)$

5B. $f(x) = \frac{1}{4} \log_{\frac{1}{2}} (x + 1) - 5$

Personal Tutor glencoe.com

Real-World EXAMPLE 6 Find Inverses of Exponential Functions

EARTHQUAKES The Richter scale measures earthquake intensity. The increase in intensity between each number is 10 times. For example, an earthquake with a rating of 7 is 10 times more intense than one measuring 6. The intensity of an earthquake can be modeled by $y = 10^{x-1}$, where x is the Richter scale rating.

a. Use the information at the left to find the intensity of the strongest recorded earthquake in the United States.

$y = 10^{x-1}$	Original equation
$= 10^{9.2-1}$	Substitute 9.2 for x.
$= 10^{8.2}$	Simplify.
$= 158{,}489{,}319.2$	Use a calculator.

b. Write an equation of the form $y = \log_{10} x + c$ for the inverse of the function.

$y = 10^{x-1}$	Original equation
$x = 10^{y-1}$	Replace x with y, replace y with x, and solve for y.
$y - 1 = \log_{10} x$	Definition of logarithm
$y = \log_{10} x + 1$	Add 1 to each side.

Check Your Progress

6. Write an equation for the inverse of the function $y = 0.5^x$.

Personal Tutor glencoe.com

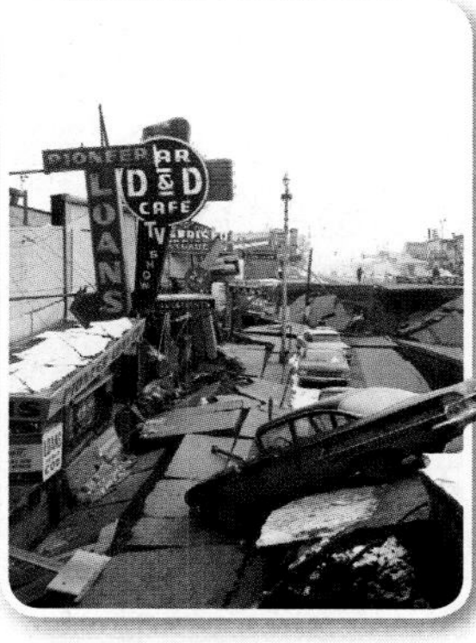

Real-World Link

The largest recorded earthquake in the United States was a magnitude 9.2 that struck Prince William Sound, Alaska, on Good Friday, March 28, 1964.

Source: United States Geological Survey

Check Your Understanding

Example 1 p. 492

Write each equation in exponential form.

1. $\log_8 512 = 3$
2. $\log_5 625 = 4$

Example 2 p. 493

Write each equation in logarithmic form.

3. $11^3 = 1331$
4. $16^{\frac{3}{4}} = 8$

Example 3 p. 493

Evaluate each expression.

5. $\log_{13} 169$
6. $\log_2 \frac{1}{128}$
7. $\log_6 1$

Examples 4 and 5 pp. 494–495

Graph each function.

8. $f(x) = \log_3 x$
9. $f(x) = \log_{\frac{1}{6}} x$
10. $f(x) = 4 \log_4 (x - 6)$
11. $f(x) = 2 \log_{\frac{1}{10}} x - 5$

Example 6 p. 495

12. **SCIENCE** Use the information at the beginning of the lesson. The Palermo scale value of any object can be found using the equation $PS = \log_{10} R$, where R is the relative risk posed by the object. Write an equation in exponential form for the inverse of the function.

Practice and Problem Solving

● = **Step-by-Step Solutions** begin on page R20. **Extra Practice** begins on page 947.

Example 1 p. 492

Write each equation in exponential form.

13. $\log_2 16 = 4$
14. $\log_7 343 = 3$
15. $\log_9 \frac{1}{81} = -2$
16. $\log_3 \frac{1}{27} = -3$
17. $\log_{12} 144 = 2$
18. $\log_9 1 = 0$

Example 2 p. 493

Write each equation in logarithmic form.

19. $9^{-1} = \frac{1}{9}$
20. $6^{-3} = \frac{1}{216}$
21. $2^8 = 256$
22. $4^6 = 4096$
23. $27^{\frac{2}{3}} = 9$
24. $25^{\frac{3}{2}} = 125$

Example 3 p. 493

Evaluate each expression.

25. $\log_3 \frac{1}{9}$
26. $\log_4 \frac{1}{64}$
27. $\log_8 512$
28. $\log_6 216$
29. $\log_{27} 3$
30. $\log_{32} 2$
31. $\log_9 3$
32. $\log_{121} 11$
33. $\log_{\frac{1}{5}} 3125$
34. $\log_{\frac{1}{8}} 512$
35. $\log_{\frac{1}{3}} \frac{1}{81}$
36. $\log_{\frac{1}{6}} \frac{1}{216}$

Examples 4 and 5 pp. 494–495

Graph each function.

37. $f(x) = \log_6 x$
38. $f(x) = \log_{\frac{1}{5}} x$
39. $f(x) = 4 \log_2 x + 6$
40. $f(x) = \log_{\frac{1}{9}} x$
41. $f(x) = \log_{10} x$
42. $f(x) = -3 \log_{\frac{1}{12}} x + 2$
43. $f(x) = 6 \log_{\frac{1}{8}} (x + 2)$
44. $f(x) = -8 \log_3 (x - 4)$
45. $f(x) = \log_{\frac{1}{4}} (x + 1) - 9$
46. $f(x) = \log_5 (x - 4) - 5$
47. $f(x) = -\frac{1}{6} \log_8 (x - 3) + 4$
48. $f(x) = -\frac{1}{3} \log_{\frac{1}{6}} (x + 2) - 5$

Example 6
p. 495

Real-World Link

The aperture setting, or f-stop, of a camera controls the amount of light exposure on film. Each step up in f-stop setting allows twice as much light exposure as the previous setting.

49. PHOTOGRAPHY The formula $n = \log_2 \frac{1}{p}$ represents the change in the f-stop setting n to use in less light where p is the fraction of sunlight.

a. Benito's camera is set up to take pictures in direct sunlight, but it is a cloudy day. If the amount of sunlight on a cloudy day is $\frac{1}{4}$ as bright as direct sunlight, how many f-stop settings should he move to accommodate less light?

b. Graph the function.

c. Use the graph in part b to predict what fraction of daylight Benito is accommodating if he moves down 3 f-stop settings. Is he allowing more or less light into the camera?

50. EDUCATION To measure a student's retention of knowledge, the student is tested after a given amount of time. A student's score on an Algebra 2 test t months after the school year is over can be approximated by $y(t) = 85 - 25 \log_{10} (t + 1)$, where $y(t)$ is the student's score as a percent.

a. What was the student's score at the time the school year ended ($t = 0$)?

b. What was the student's score after 2 months?

c. What was the student's score after 1 year?

Graph each function.

51. $f(x) = 4 \log_2 (2x - 4) + 6$

52. $f(x) = -3 \log_{12} (4x + 3) + 2$

53. $f(x) = 15 \log_{14} (x + 1) - 9$

54. $f(x) = 10 \log_5 (x - 4) - 5$

55. $f(x) = -\frac{1}{6} \log_8 (x - 3) + 4$

56. $f(x) = -\frac{1}{3} \log_6 (6x + 2) - 5$

57. ADVERTISING In general, the more money a company spends on advertising, the higher the sales. The amount of money in sales for a company, in thousands, can be modeled by the equation $S(a) = 10 + 20 \log_{10}(a + 1)$, where a is the amount of money spent on advertising in hundreds, when $a \geq 0$.

a. The value of $S(1) \approx 16$, which means that if \$100 is spent on advertising, \$16,000 is returned in sales. Find the values of $S(11)$, $S(21)$, and $S(31)$.

b. Interpret the meaning of each function value in the context of the problem.

c. Graph the function.

d. Use the graph in part c and your answers from part a to explain why the money spent in advertising becomes less "efficient" as it is used in larger amounts.

58. EDUCATION A teacher is conducting a study on different teaching methods. For Method 1, the students work individually to complete assignments, and Method 2 has students work cooperatively in groups of four to complete assignments. Knowledge retention of the students using Method 1 can be modeled by the equation $x(t) = 91 - 30 \log (t + 1)$. Knowledge retention of the students using Method 2 can be modeled by the equation $y(t) = 88 - 15 \log (t + 1)$.

a. Make a table that compares the average rate of retention for the two groups after 0, 1, 2, 6, 12, and 24 months.

b. Graph both of the graphs on your graphing calculator.

c. Based on the table of values and the graph, which teaching method helped students retain the material better for a long period of time?

59 CREDIT CARDS Jacy has spent \$2000 on a credit card. The credit card company charges 24% interest, compounded monthly. The credit card company uses $\log_{\left(1+\frac{0.24}{12}\right)} \frac{A}{2000} = 12t$ to determine how much time it will be until Jacy's debt reaches a certain amount, if A is the amount of debt after a period of time, and t is time in years.

a. Graph the function for Jacy's debt.

b. Approximately how long will it take Jacy's debt to double?

c. Approximately how long will it be until Jacy's debt triples?

Real-World Link

Approximately half of all credit card holders pay only their minimum monthly requirements. The average American household is solicited seven times a year by credit card companies.

Source: American Consumer Credit Counseling

H.O.T. Problems

Use **H**igher-**O**rder **T**hinking Skills

60. WHICH ONE DOESN'T BELONG? Find the expression that does not belong. Explain.

$\log_4 16$	$\log_2 16$	$\log_2 4$	$\log_3 9$

61. CHALLENGE Consider $y = \log_b x$ in which b, x, and y are real numbers. Zero can be in the domain *sometimes*, *always* or *never*. Justify your answer.

62. FIND THE ERROR Betsy says that all logarithmic graphs cross the y-axis at (0, 1) because any number to the zero power equals 1. Tyrone disagrees. Is either of them correct? Explain your reasoning.

63. REASONING Without using a calculator, compare $\log_7 51$, $\log_8 61$, and $\log_9 71$. Which of these is the greatest? Explain your reasoning.

64. OPEN ENDED Write a logarithmic expression of the form $y = \log_b x$ for each of the following conditions.

a. y is equal to 25.

b. y is negative.

c. y is between 0 and 1.

d. x is 1.

e. x is 0.

65. FIND THE ERROR Elisa and Matthew are evaluating $\log_{\frac{1}{7}} 49$. Is either of them correct? Explain your reasoning.

Elisa	Matthew
$\log_{\frac{1}{7}} 49 = y$	$\log_{\frac{1}{7}} 49 = y$
$\frac{1}{7}^{y} = 49$	$49^y = \frac{1}{7}$
$(7^{-1})^y = 7^2$	$(7^2)^y = (7)^{-1}$
$(7)^{-y} = 7^2$	$7^{2y} = (7)^{-1}$
$y = 2$	$2y = -1$
	$y = -\frac{1}{2}$

66. WRITING IN MATH A transformation of $\log_{10} x$ is $g(x) = a \log_{10}(x - h) + k$. Explain the process of graphing this transformation.

NYSRE PRACTICE Reinforcement of A.A.6

67. A rectangle is twice as long as it is wide. If the width of the rectangle is 3 inches, what is the area of the rectangle in square inches?

A 9
B 12
C 15
D 18

68. ACT/SAT Ichiro has some pizza. He sold 40% more slices than he ate. If he sold 70 slices of pizza, how many did he eat?

F 25
G 50
H 75
J 100

69. SHORT RESPONSE In the figure $AB = BC$, $CD = BD$, and angle $CAD = 70°$. What is the measure of angle ADC?

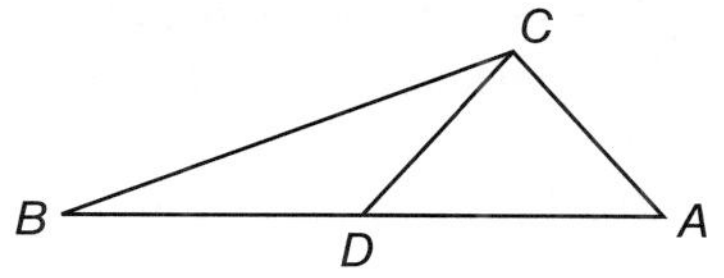

70. If $6x - 3y = 30$ and $4x = 2 - y$ then find $x + y$.

A -4
B -2
C 2
D 4

Spiral Review

Solve each inequality. Check your solution. (Lesson 8-2)

71. $3^{n-2} > 27$

72. $2^{2n} \leq \frac{1}{16}$

73. $16^n < 8^{n+1}$

74. $32^{5p+2} \geq 16^{5p}$

Graph each function. (Lesson 8-1)

75. $y = -\left(\frac{1}{5}\right)^x$

76. $y = -2.5(5)^x$

77. $y = 30^{-x}$

78. $y = 0.2(5)^{-x}$

79. GEOMETRY The area of a triangle with sides of length a, b, and c is given by $\sqrt{s(s-a)(s-b)(s-c)}$, where $s = \frac{1}{2}(a + b + c)$. If the lengths of the sides of a triangle are 6, 9, and 12 feet, what is the area of the triangle expressed in radical form? (Lesson 7-5)

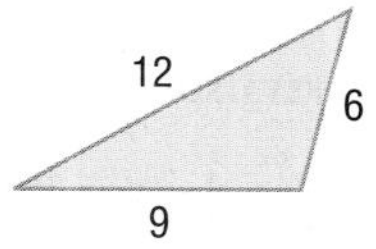

80. GEOMETRY The volume of a rectangular box can be written as $6x^3 + 31x^2 + 53x + 30$ when the height is $x + 2$. (Lesson 6-5)

a. What are the width and length of the box?

b. Will the ratio of the dimensions of the box always be the same regardless of the value of x? Explain.

81. AUTO MECHANICS Shandra is inventory manager for a local repair shop. She orders 6 batteries, 5 cases of spark plugs, and two dozen pairs of wiper blades and pays \$830. She orders 3 batteries, 7 cases of spark plugs, and four dozen pairs of wiper blades and pays \$820. The batteries are \$22 less than twice the price of a dozen wiper blades. Use augmented matrices to determine what the cost of each item on her order is. (Lesson 4-6)

Skills Review

Solve each equation or inequality. Check your solution. (Lesson 8-2)

82. $9^x = \frac{1}{81}$

83. $2^{6x} = 4^{5x+2}$

84. $49^{3p+1} = 7^{2p-5}$

85. $9^{x^2} \leq 27^{x^2-2}$

EXTEND 8-3

Graphing Technology Lab
Choosing the Best Model

NY Math Online glencoe.com
Other Calculator Keystrokes

NYS Core Curriculum **A2.S.6 Determine from a scatter plot whether a linear, logarithmic, exponential, or power regression model is most appropriate A2.S.7 Determine the function for the regression model, using appropriate technology, and use the regression function to interpolate and extrapolate from the data**

We can find exponential and logarithmic functions of best fit using a TI-83/84 Plus graphing calculator.

ACTIVITY

The population per square mile in the United States has changed dramatically over a period of years. The table shows the number of people per square mile for several years.

U.S. Population Density

Year	People per square mile	Year	People per square mile
1790	4.5	1900	21.5
1800	6.1	1910	26.0
1810	4.3	1920	29.9
1820	5.5	1930	34.7
1830	7.4	1940	37.2
1840	9.8	1950	42.6
1850	7.9	1960	50.6
1860	10.6	1970	57.5
1870	10.9	1980	64.0
1880	14.2	1990	70.3
1890	17.8	2000	80.0

Source: Northeast-Midwest Institute

a. Use a graphing calculator to enter the data. Then draw a scatter plot that shows how the number of people per square mile is related to the year.

Step 1 Enter the year into L1 and the people per square mile into L2.

KEYSTROKES: *See pages 94 and 95 to review how to enter lists.*

Be sure to clear the Y= list. Use the ▶ key to move the cursor from L1 to L2.

Step 2 Draw the scatter plot.

KEYSTROKES: *See pages 94 and 95 to review how to graph a scatter plot.*

Make sure that Plot 1 is on, the scatter plot is chosen, Xlist is L1, and Ylist is L2.

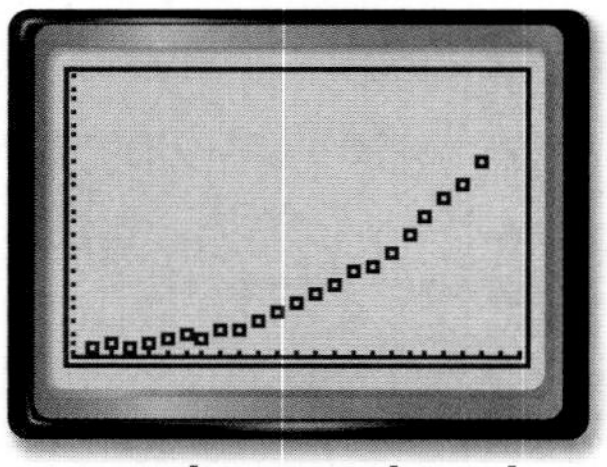

[1980, 2020] scl: 10 by [0, 115] scl: 5

Step 3 Find a regression equation.

To find an equation that best fits the data, use the regression feature of the calculator. Examine various regressions to determine the best model.

Recall that the calculator returns the correlation coefficient r, which is used to indicate how well the model fits the data. The closer r is to 1 or -1, the better the fit.

Linear regression

KEYSTROKES: STAT ▶ 4 ENTER

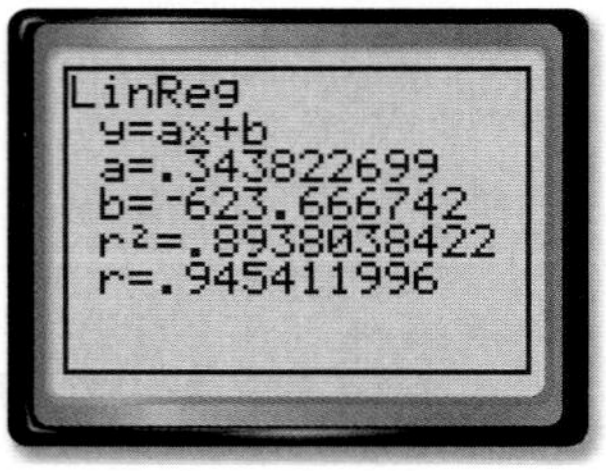

Quadratic regression

KEYSTROKES: STAT ▶ 5 ENTER

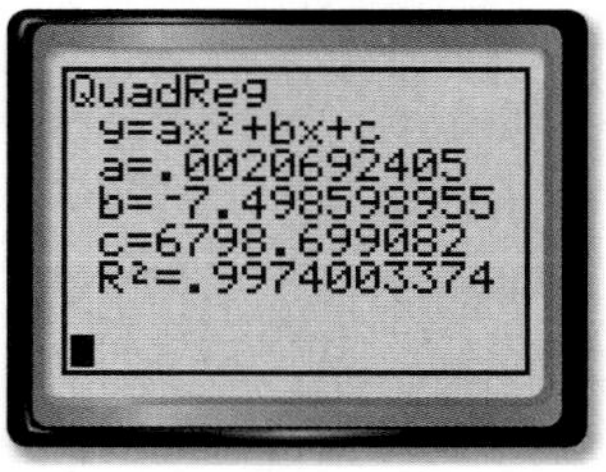

$r^2 = 0.9974003374$

$r = \sqrt{0.9974003374}$

$r \approx 0.9986993228$

Exponential regression

KEYSTROKES: STAT ▶ 0 ENTER

Power regression

KEYSTROKES: STAT ▶ ALPHA [A] ENTER

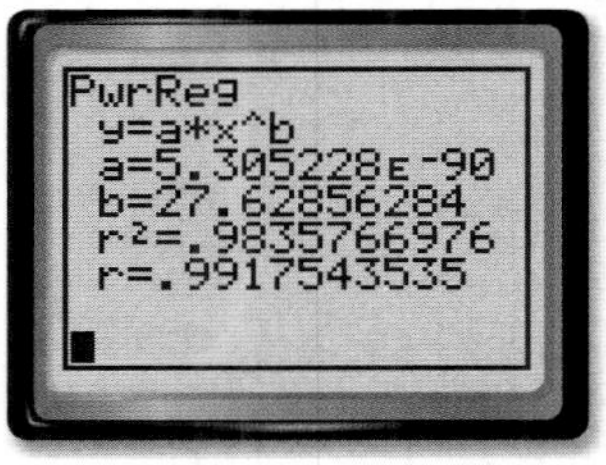

Compare the r-values.

Linear: 0.945411996

Exponential: 0.991887235

Quadratic: 0.9986993228

Power: 0.9917543535

The r-value of the quadratic regression is closest to 1, so it best models the data. You can examine the equation visually by graphing the regression equation with the scatter plot.

KEYSTROKES: STAT ▶ 5 ENTER Y= VARS 5 ▶ ▶ 1 GRAPH

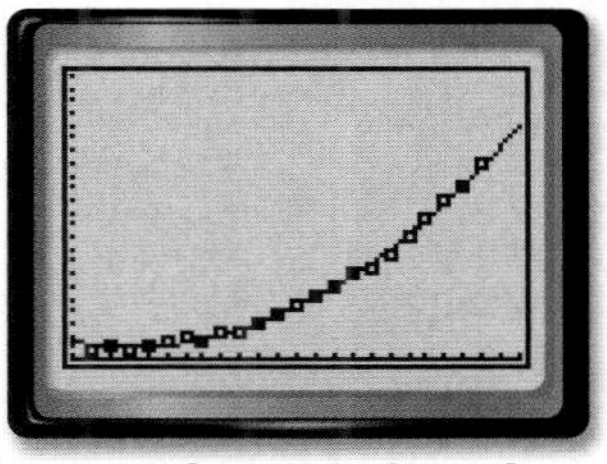

[1980, 2020] scl: 10 by [0, 115] scl: 5

b. If this trend continues, what will be the population per square mile in 2020?

To determine the population per square mile in 2020, find the value of y when $x = 2020$.

KEYSTROKES: 2nd [CALC] 2020 ENTER

If this trend continues, there will be approximately 94.9 people per square mile.

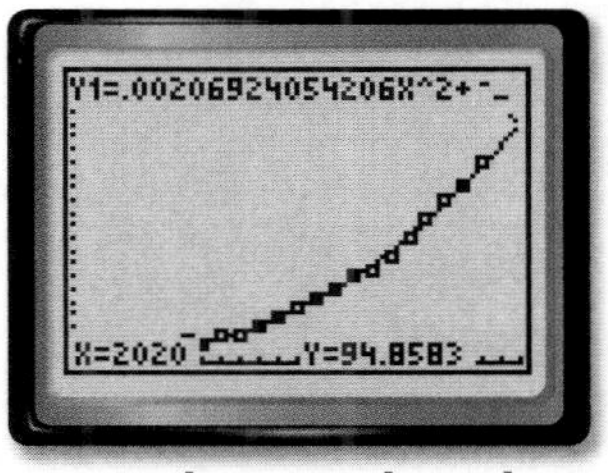

[1980, 2020] scl: 10 by [0, 115] scl: 5

Exercises

Jewel received $50 from her grandparents on her tenth birthday. Her mother deposited it into a new bank account for her. Both Jewel and her mother forgot about the account and made no further deposits or withdrawals. The table shows the account balance for several years.

Elapsed Time (years)	Balance
0	$50.00
2	$55.80
4	$64.80
6	$83.09
8	$101.40
10	$123.14
12	$162.67

1. Use a graphing calculator to draw a scatter plot of the data.
2. Calculate and graph a curve of fit for the data using an exponential regression.
3. Write the equation of best fit.
4. Based on the model, what will the account balance be after 25 years?
5. Is an exponential model the best fit for the data? Explain.

8-4 Solving Logarithmic Equations and Inequalities

Then

You evaluated logarithmic expressions. (Lesson 8-3)

Now

- Solve logarithmic equations.
- Solve logarithmic inequalities.

NYS Core Curriculum

A2.A.18 Evaluate logarithmic expressions in any base
A2.A.28 Solve a logarithmic equation by rewriting as an exponential equation
Also addresses A2.A.19.

New Vocabulary

logarithmic equation
logarithmic inequality

NY Math Online

glencoe.com

- Extra Examples
- Personal Tutor
- Self-Check Quiz
- Homework Help

Why?

Each year the National Weather Service documents about 1000 tornado touchdowns in the United States. The intensity of a tornado is measured on the Fujita scale. Tornados are divided into six categories according to their wind speed, path length, path width, and damage caused. Category F6 is the most destructive type of tornado.

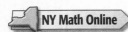

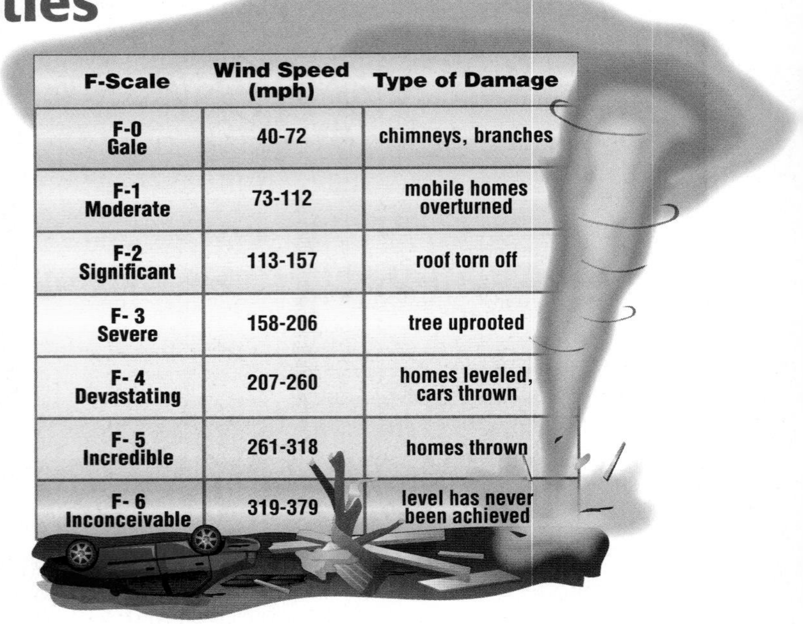

F-Scale	Wind Speed (mph)	Type of Damage
F-0 Gale	40-72	chimneys, branches
F-1 Moderate	73-112	mobile homes overturned
F-2 Significant	113-157	roof torn off
F- 3 Severe	158-206	tree uprooted
F- 4 Devastating	207-260	homes leveled, cars thrown
F- 5 Incredible	261-318	homes thrown
F- 6 Inconceivable	319-379	level has never been achieved

Solve Logarithmic Equations A **logarithmic equation** contains one or more logarithms. You can use the definition of a logarithm to help you solve logarithmic equations.

EXAMPLE 1 Solve a Logarithmic Equation

Solve $\log_{36} x = \frac{3}{2}$.

$\log_{36} x = \frac{3}{2}$ **Original equation**

$x = 36^{\frac{3}{2}}$ **Definition of logarithm**

$x = (6^2)^{\frac{3}{2}}$ **$36 = 6^2$**

$x = 6^3$ or 216 **Power of a Power**

Check Your Progress

Solve each equation.

1A. $\log_9 x = \frac{3}{2}$

1B. $\log_{16} x = \frac{5}{2}$

Personal Tutor glencoe.com

Use the following property to solve logarithmic equations that have logarithms with the same base on each side.

Key Concept

For Your FOLDABLE

Property of Equality for Logarithmic Functions

Symbols If b is a positive number other than 1, then $\log_b x = \log_b y$ if and only if $x = y$.

Example If $\log_5 x = \log_5 8$, then $x = 8$. If $x = 8$, then $\log_5 x = \log_5 8$.

NYSRE EXAMPLE 2 A2.A.18

Solve $\log_2 (x^2 - 4) = \log_2 3x$.

A -2 B -1 C 2 D 4

Read the Test Item

You need to find x for the logarithmic equation.

Solve the Test Item

$\log_2 (x^2 - 4) = \log_2 3x$	**Original equation**
$x^2 - 4 = 3x$	**Property of Equality for Logarithmic Functions**
$x^2 - 3x - 4 = 0$	**Subtract 3*x* from each side.**
$(x - 4)(x + 1) = 0$	**Factor.**
$x - 4 = 0$ or $x + 1 = 0$	**Zero Product Property**
$x = 4$ $x = -1$	**Solve each equation.**

CHECK Substitute each value into the original equation.

$x = 4$	$x = -1$
$\log_2 (4^2 - 4) \stackrel{?}{=} \log_2 3(4)$	$\log_2 [(-1)^2 - 4] \stackrel{?}{=} \log_2 3(-1)$
$\log_2 12 = \log_2 12$ ✓	$\log_2 (-3) \stackrel{?}{=} \log_2 (-3)$ ✗

The domain of a logarithmic function cannot be 0, so $\log_2 (-3)$ is undefined and -1 is an extraneous solution. The answer is D.

Test-TakingTip

Substitution To save time, you can substitute each answer choice in the original equation to find the one that results in a true statement.

Check Your Progress

2. Solve $\log_3 (x^2 - 15) = \log_3 2x$.

F -3 G -1 H 5 J 15

Personal Tutor glencoe.com

Solve Logarithmic Inequalities A **logarithmic inequality** is an inequality that involves logarithms. The following property can be used to solve logarithmic inequalities.

Key Concept

For Your FOLDABLE

Property of Inequality for Logarithmic Functions

If $b > 1$, $x > 0$, and $\log_b x > y$, then $x > b^y$.

If $b > 1$, $x > 0$, and $\log_b x < y$, then $0 < x < b^y$.

This property also holds true for $\leq$ and $\geq$.

EXAMPLE 3 Solve a Logarithmic Inequality

Solve $\log_3 x > 4$.

$\log_3 x > 4$	**Original inequality**
$x > 3^4$	**Property of Inequality for Logarithmic Functions**
$x > 81$	**Simplify.**

Check Your Progress

Solve each inequality.

3A. $\log_4 x \geq 3$

3B. $\log_2 x < 4$

Personal Tutor glencoe.com

The following property can be used to solve logarithmic inequalities that have logarithms with the same base on each side. Exclude from your solution set values that would result in taking the logarithm of a number less than or equal to zero in the original inequality.

Key Concept

For Your FOLDABLE

Property of Inequality for Logarithmic Functions

Symbols If $b > 1$, then $\log_b x > \log_b y$ if and only if $x > y$, and $\log_b x < \log_b y$ if and only if $x < y$.

Example If $\log_6 x > \log_6 35$, then $x > 35$.

This property also holds true for $\leq$ and $\geq$.

EXAMPLE 4 Solve Inequalities with Logarithms on Each Side

Solve $\log_4 (x + 3) > \log_4 (2x + 1)$.

$\log_4 (x + 3) > \log_4 (2x + 1)$	**Original inequality**
$x + 3 > 2x + 1$	**Property of Inequality for Logarithmic Functions**
$2 > x$	**Subtract $x + 1$ from each side.**

Exclude all values of x for which $x + 3 \leq 0$ or $2x + 1 \leq 0$. So, $x > -3$, $x > -\frac{1}{2}$, and $x < 2$. The solution set is $\left\{x \,\middle|\, -\frac{1}{2} < x < 2\right\}$.

Check Your Progress

4. Solve $\log_5 (2x + 1) \leq \log_5 (x + 4)$. Check your solution.

Personal Tutor glencoe.com

Check Your Understanding

Example 1 p. 502

Solve each equation.

1. $\log_8 x = \frac{4}{3}$

2. $\log_{16} x = \frac{3}{4}$

Example 2 p. 503

3. MULTIPLE CHOICE Solve $\log_5 (x^2 - 10) = \log_5 3x$.

A 10 **B** 2 **C** 5 **D** 2, 5

Examples 3 and 4 pp. 503–504

Solve each inequality.

4. $\log_5 x > 3$

5. $\log_8 x \leq -2$

6. $\log_4 (2x + 5) \leq \log_4 (4x - 3)$

7. $\log_8 (2x) > \log_8 (6x - 8)$

Practice and Problem Solving

= **Step-by-Step Solutions** begin on page R20. **Extra Practice** begins on page 947.

Examples 1 and 2 pp. 502–503

Solve each equation.

8. $\log_{81} x = \frac{3}{4}$

9. $\log_{25} x = \frac{5}{2}$

10. $\log_8 \frac{1}{2} = x$

11. $\log_6 \frac{1}{36} = x$

12. $\log_x 32 = \frac{5}{2}$

13. $\log_x 27 = \frac{3}{2}$

14. $\log_3 (3x + 8) = \log_3 (x^2 + x)$

15. $\log_{12} (x^2 - 7) = \log_{12} (x + 5)$

16. $\log_6 (x^2 - 6x) = \log_6 (-8)$

17. $\log_9 (x^2 - 4x) = \log_9 (3x - 10)$

18. $\log_4 (2x^2 + 1) = \log_4 (10x - 7)$

19. $\log_7 (x^2 - 4) = \log_7 (-x + 2)$

SCIENCE The equation for wind speed w, in miles per hour, near the center of a tornado is $w = 93 \log_{10} d + 65$, where d is the distance in miles that the tornado travels.

20. Write this equation in exponential form.

21. In May of 1999, a tornado devastated Oklahoma City with the fastest wind speed ever recorded. If the tornado traveled 525 miles, estimate the wind speed near the center of the tornado.

Examples 3 and 4
pp. 503–504

Solve each inequality.

22. $\log_6 x < -3$

23. $\log_4 x \geq 4$

24. $\log_3 x \geq -4$

25. $\log_2 x \leq -2$

26. $\log_5 x > 2$

27. $\log_7 x < -1$

28. $\log_2 (4x - 6) > \log_2 (2x + 8)$

29. $\log_7 (x + 2) \geq \log_7 (6x - 3)$

30. $\log_3 (7x - 6) < \log_3 (4x + 9)$

31. $\log_5 (12x + 5) \leq \log_5 (8x + 9)$

32. $\log_{11} (3x - 24) \geq \log_{11} (-5x - 8)$

33. $\log_9 (9x + 4) \leq \log_9 (11x - 12)$

34. SCIENCE The magnitude of an earthquake is measured on a logarithmic scale called the Richter scale. The magnitude M is given by $M = \log_{10} x$, where x represents the amplitude of the seismic wave causing ground motion.

a. How many times as great is the amplitude caused by an earthquake with a Richter scale rating of 8 as an aftershock with a Richter scale rating of 5?

b. In 1906, San Francisco was almost completely destroyed by a 7.8 magnitude earthquake. In 1911, an earthquake estimated at magnitude 8.1 occurred along the New Madrid fault in the Mississippi River Valley. How many times greater was the New Madrid earthquake than the San Francisco earthquake?

Math History Link

Zhang Heng (A.D. 78–139) The earliest known seismograph was invented by Zhang Heng in China in 132 B.C. It was a large brass vessel with a heavy pendulum and several arms that tripped when an earthquake tremor was felt. This helped determine the direction of the quake.

35. MUSIC The first key on a piano keyboard corresponds to a pitch with a frequency of 27.5 cycles per second. With every successive key, going up the black and white keys, the pitch multiplies by a constant. The formula for the frequency of the pitch sounded when the nth note up the keyboard is played is given by $n = 1 + 12 \log_2 \frac{f}{27.5}$.

a. A note has a frequency of 220 cycles per second. How many notes up the piano keyboard is this?

b. Another pitch on the keyboard has a frequency of 880 cycles per second. After how many notes up the keyboard will this be found?

36. MULTIPLE REPRESENTATIONS In this problem, you will explore the graphs shown: $y = \log_4 x$ and $y = \log_{\frac{1}{4}} x$.

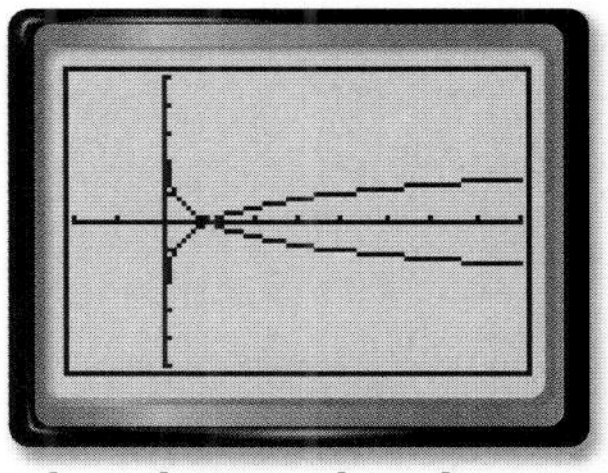

[−2, 8] scl: 1 by [−5, 5] scl: 1

a. ANALYTICAL How do the shapes of the graphs compare? How do the asymptotes and the x-intercepts of the graphs compare?

b. VERBAL Describe the relationship between the graphs.

c. GRAPHICAL Graph each function and $y = \log_4 x$ on the same graphing calculator screen. Then compare and contrast the graphs.

1. $y = \log_4 x + 2$ 2. $y = \log_4 (x + 2)$ 3. $y = 3 \log_4 x$

d. ANALYTICAL Describe the relationship between $y = \log_4 x$ and $y = -1(\log_4 x)$. What are a reasonable domain and range for each function?

e. LOGICAL What is a reasonable viewing window in order to see the trends of both functions?

Real-World Link

The maximum exposure time before your hearing is endangered at 85 decibels is 8 hours.

Source: National Institute for Occupational Safety and Health

37. **SOUND** The relationship between the intensity of sound I and the number of decibels β is $\beta = 10 \log_{10}\left(\frac{I}{10^{-12}}\right)$, where I is the intensity of sound in watts per square meter.

a. Find the number of decibels of a sound with an intensity of 1 watt per square meter.

b. Find the number of decibels of sound with an intensity of 10^{-2} watts per square meter.

c. The intensity of the sound of 1 watt per square meter is 100 times as much as the intensity of 10^{-2} watts per square meter. Why are the decibels of sound not 100 times as great?

Sound	Intensity	Decibels
pin drop	10^0	0
normal breathing	10^1	1
clothes dryer	10^6	6
subway train	10^{10}	10
firecracker	10^{12}	12

H.O.T. Problems

Use **H**igher-**O**rder **T**hinking Skills

38. **FIND THE ERROR** Ryan and Heather are solving $\log_3 x \geq -3$. Is either of them correct? Explain your reasoning.

Ryan

$\log_3 x \geq -3$

$x \geq 3^{-3}$

$x \geq \frac{1}{27}$

Heather

$\log_3 x \geq -3$

$\log_3 x \geq 3^{-3}$

$0 < x \leq \frac{1}{27}$

39. **CHALLENGE** Find $\log_3 27 + \log_9 27 + \log_{27} 27 + \log_{81} 27 + \log_{243} 27$.

40. **REASONING** The Property of Inequality for Logarithmic Functions states that when $b > 1$, $\log_b x > \log_b y$ if and only if $x > y$. What is the case for when $0 < b < 1$? Explain your reasoning.

41. **WRITING IN MATH** Explain how the domain and range of logarithmic functions are related to the domain and range of exponential functions.

42. **OPEN ENDED** Give an example of a logarithmic equation that has no solution.

43. **REASONING** Choose the appropriate term. Explain your reasoning. All logarithmic equations are of the form $y = \log_b x$.

a. If the base of a logarithmic equation is greater than 1 and the value of x is between 0 and 1, then the value for y is (*less than, greater than, equal to*) 0.

b. If the base of a logarithmic equation is between 0 and 1 and the value of x is greater than 1, then the value of y is (*less than, greater than, equal to*) 0.

c. There is (*no, one, infinitely many*) solution(s) for b in the equation $y = \log_b 0$.

d. There is (*no, one, infinitely many*) solution(s) for b in the equation $y = \log_b 1$.

44. **WRITING IN MATH** Explain why any logarithmic function of the form $y = \log_b x$ has an x-intercept of (1, 0) and no y-intercept.

NYSRE PRACTICE A2.S.12

45. Find x if $\frac{6.4}{x} = \frac{4}{7}$.

A 3.4
B 9.4
C 11.2
D 44.8

46. The monthly precipitation in Houston for part of a year is shown.

Month	Precipitation (in.)
April	3.60
May	5.15
June	5.35
July	3.18
August	3.83

Find the median precipitation.

F 3.60 in.
G 4.22 in.
H 3.83 in.
J 4.25 in.

47. Clara received a 10% raise each year for 3 consecutive years. What was her salary after the three raises if her starting salary was $12,000 per year?

A $14,520
B $15,972
C $16,248
D $16,410

48. ACT/SAT A vendor has 14 helium balloons for sale: 9 are yellow, 3 are red, and 2 are green. A balloon is selected at random and sold. If the balloon sold is yellow, what is the probability that the next balloon, selected at random, is also yellow?

Spiral Review

Evaluate each expression. (Lesson 8-3)

49. $\log_4 256$

50. $\log_2 \frac{1}{8}$

51. $\log_6 216$

52. $\log_3 27$

53. $\log_5 \frac{1}{125}$

54. $\log_7 2401$

Solve each equation or inequality. Check your solution. (Lesson 8-2)

55. $5^{2x+3} \leq 125$

56. $3^{3x-2} > 81$

57. $4^{4a+6} \leq 16^a$

58. $2.3^{x^2} = 66.6$

59. $3^{4x-7} = 4^{2x+3}$

60. $12^{x-4} \leq 2^{2-x}$

61. SHIPPING The height of a shipping cylinder is 4 feet more than the radius. If the volume of the cylinder is 5π cubic feet, how tall is it? Use the formula $V = \pi \cdot r^2 \cdot h$. (Lesson 6-8)

62. NUMBER THEORY Two complex conjugate numbers have a sum of 12 and a product of 40. Find the two numbers. (Lesson 5-4)

Skills Review

Simplify. Assume that no variable equals zero. (Lesson 6-1)

63. $x^5 \cdot x^3$

64. $a^2 \cdot a^6$

65. $(2p^2n)^3$

66. $(3b^3c^2)^2$

67. $\frac{x^4y^6}{xy^2}$

68. $\left(\frac{c^9}{d^7}\right)^0$

CHAPTER 8

Mid-Chapter Quiz

Lessons 8-1 through 8-4

NYS Core Curriculum
A2.A.53, A2.A.18

Graph each function. State the domain and range. (Lesson 8-1)

1. $f(x) = 3(4)^x$

2. $f(x) = -(2)^x + 5$

3. $f(x) = -0.5(3)^{x+2} + 4$

4. $f(x) = -3\left(\frac{2}{3}\right)^{x-1} + 8$

5. SCIENCE You are studying a bacteria population. The population originally started with 6000 bacteria cells. After 2 hours, there were 28,000 bacteria cells. (Lesson 8-1)

a. Write an exponential function that could be used to model the number of bacteria after x hours if the number of bacteria changes at the same rate.

b. How many bacteria cells can be expected after 4 hours?

6. MULTIPLE CHOICE Which exponential function has a graph that passes through the points at (0, 125) and (3, 1000)? (Lesson 8-1)

A $f(x) = 125(3)^x$

B $f(x) = 1000(3)^x$

C $f(x) = 125(1000)^x$

D $f(x) = 125(2)^x$

7. POPULATION In 1995, a certain city had a population of 45,000. It increased to 68,000 by 2007. (Lesson 8-2)

a. What is an exponential function that could be used to model the population of this city x years after 1995?

b. Use your model to estimate the population in 2015.

8. MULTIPLE CHOICE Find the value of x for $\log_3 (x^2 + 2x) = \log_3 (x + 2)$. (Lesson 8-3)

F $x = -2, 1$

G $x = -2$

H $x = 1$

J no solution

Graph each function. (Lesson 8-3)

9. $f(x) = 3 \log_2 (x - 1)$

10. $f(x) = -4 \log_3 (x - 2) + 5$

11. MULTIPLE CHOICE Which graph below is the graph of the function $f(x) = \log_3 (x + 5) + 3$? (Lesson 8-3)

A

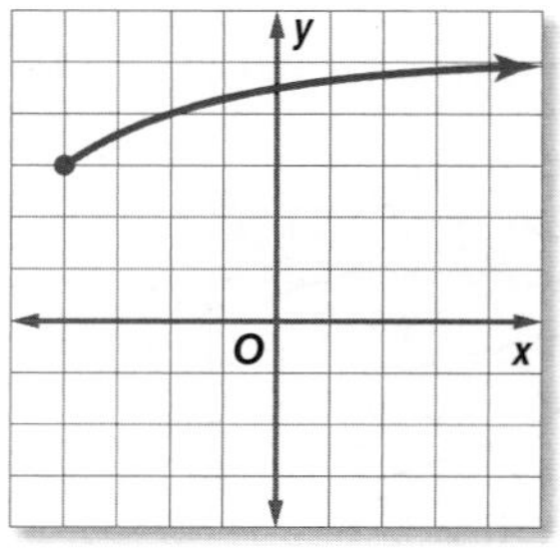

C

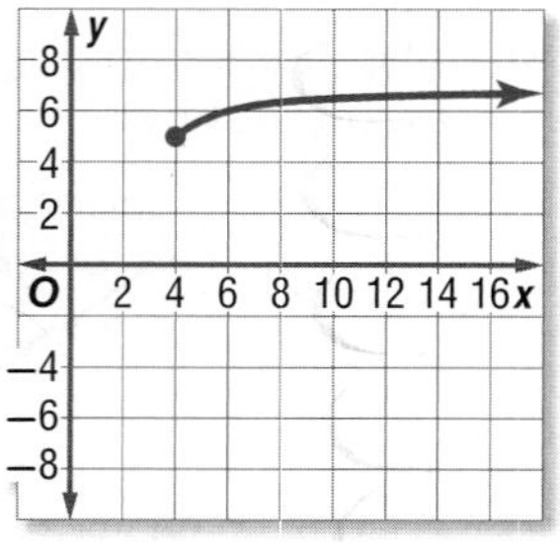

B

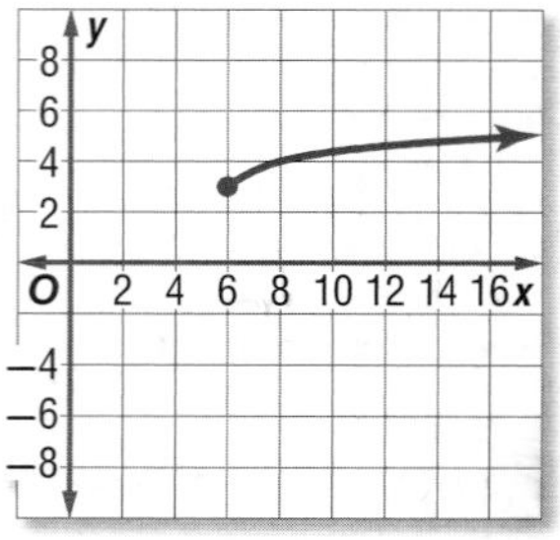

D

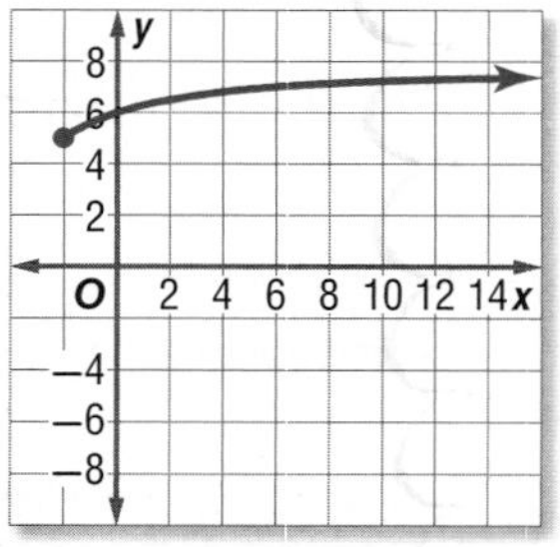

Evaluate each expression. (Lesson 8-3)

12. $\log_4 32$

13. $\log_5 5^{12}$

14. $\log_{16} 4$

15. Write $\log_9 729 = 3$ in exponential form. (Lesson 8-3)

Solve each equation or inequality. Check your solution. (Lessons 8-2 and 8-4)

16. $3^x = 27^2$

17. $4^{3x-1} = 16^x$

18. $\frac{1}{9} = 243^{2x+1}$

19. $16^{2x+3} < 64$

20. $\left(\frac{1}{32}\right)^{x+3} \geq 16^{3x}$

21. $\log_4 x = \frac{3}{2}$

22. $\log_7 (-x + 3) = \log_7 (6x + 5)$

23. $\log_2 x < -3$

24. $\log_8 (3x + 7) = \log_8 (2x - 5)$

8-5 Properties of Logarithms

Why?

The level of acidity in food is important to some consumers with sensitive stomachs. Most of the foods that we consume are more acidic than basic. The pH scale measures acidity; a low pH indicates an acidic solution, and a high pH indicates a basic solution. It is another example of a logarithmic scale based on powers of ten. Black coffee has a pH of 5, while neutral water has a pH of 7. Black coffee is one hundred times as acidic as neutral water, because $10^{7-5} = 10^2$ or 100.

Product	pH Level
Lemon Juice	2.1
Sauerkraut	3.5
Tomatoes	4.2
Black Coffee	5.0
Milk	6.4
Pure Water	7.0
Eggs	7.8
Milk of Magnesia	10.0

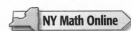

Then

You evaluated logarithmic expressions and solved logarithmic equations. (Lesson 8-4)

Now

- Simplify and evaluate expressions using the properties of logarithms.
- Solve logarithmic equations using the properties of logarithms.

NYS Core Curriculum

A2.A.18 Evaluate logarithmic expressions in any base
A2.A.19 Apply the properties of logarithms to rewrite logarithmic expressions in equivalent forms
Also addresses A2.A.28.

NY Math Online

glencoe.com

- Extra Examples
- Personal Tutor
- Self-Check Quiz
- Homework Help

Properties of Logarithms Since logarithms are exponents, the properties of logarithms can be derived from the properties of exponents. The Product Property of Logarithms can be derived from the Product of Powers Property of Exponents.

Key Concept — Product Property of Logarithms (For Your FOLDABLE)

Words The logarithm of a product is the sum of the logarithms of its factors.

Symbols For all positive numbers a, b, and x, where $x \neq 1$, $\log_x ab = \log_x a + \log_x b$.

Example $\log_2 (5)(6) = \log_2 5 + \log_2 6$

To show that this property is true, let $b^x = a$ and $b^y = c$. Then, using the definition of logarithm, $x = \log_b a$ and $y = \log_b c$.

$b^x b^y = ac$	**Substitution**
$b^{x+y} = ac$	**Product of Powers**
$\log_b b^{x+y} = \log_b ac$	**Property of Equality for Logarithmic Functions**
$x + y = \log_b ac$	**Inverse Property of Exponents and Logarithms**
$\log_b a + \log_b c = \log_b ac$	**Replace x with $\log_b a$ and y with $\log_b c$.**

You can use the Product Property of Logarithms to approximate logarithmic expressions.

EXAMPLE 1 Use the Product Property

Use $\log_4 3 \approx 0.7925$ to approximate the value of $\log_4 192$.

$\log_4 192 = \log_4 (4^3 \cdot 3)$	**Replace 192 with $64 \cdot 3$ or $4^3 \cdot 3$.**
$= \log_4 4^3 + \log_4 3$	**Product Property**
$= 3 + \log_4 3$	**Inverse Property of Exponents and Logarithms**
$= 3 + 0.7925$ or 3.7925	**Replace $\log_4 3$ with 0.7925.**

Check Your Progress

1. Use $\log_4 2 = 0.5$ to approximate the value of $\log_4 32$.

Personal Tutor glencoe.com

Recall that the quotient of powers is found by subtracting exponents. The property for the logarithm of a quotient is similar. Let $b^x = a$ and $b^y = c$. Then $\log_b a = x$ and $\log_b c = y$

$\frac{b^x}{b^y} = \frac{a}{c}$	
$b^{x-y} = \frac{a}{c}$	**Quotient Property**
$\log_b b^{x-y} = \log_b \frac{a}{c}$	**Property of Equality for Logarithmic Equations**
$x - y = \log_b \frac{a}{c}$	**Inverse Property of Exponents and Logarithms**
$\log_b a - \log_b c = \log_b \frac{a}{c}$	**Replace x with $\log_b a$ and y with $\log_b c$.**

Key Concept — Quotient Property of Logarithms

For Your FOLDABLE

Words The logarithm of a quotient is the difference of the logarithms of the numerator and the denominator.

Symbols For all positive numbers a, b, and x, where $x \neq 1$,
$\log_x \frac{a}{b} = \log_x a - \log_x b$.

Example $\log_2 \frac{5}{6} = \log_2 5 - \log_2 6$

Real-World EXAMPLE 2 Quotient Property

SCIENCE The pH of a substance is defined as the concentration of hydrogen ions $[H^+]$ in moles. It is given by the formula $\text{pH} = \log_{10} \frac{1}{H^+}$. Find the amount of hydrogen in a liter of acid rain that has a pH of 4.2.

Understand The formula for finding pH and the pH of the rain is given. You want to find the amount of hydrogen in a liter of this rain.

Plan Write the equation. Then, solve for $[H^+]$.

Solve

$\text{pH} = \log_{10} \frac{1}{H^+}$	**Original equation**
$4.2 = \log_{10} \frac{1}{H^+}$	**Substitute 4.2 for pH.**
$4.2 = \log_{10} 1 - \log_{10} H^+$	**Quotient Property**
$4.2 = 0 - \log_{10} H^+$	**$\log_{10} 1 = 0$**
$4.2 = -\log_{10} H^+$	**Simplify.**
$-4.2 = \log_{10} H^+$	**Multiply each side by −1.**
$10^{-4.2} = H^+$	**Definition of logarithm**

There are $10^{-4.2}$, or about 0.000063, mole of hydrogen in a liter of this rain.

Check

$4.2 = \log_{10} \frac{1}{H^+}$	**pH = 4.2**
$4.2 \stackrel{?}{=} \log_{10} \frac{1}{10^{-4.2}}$	**$H^+ = 10^{-4.2}$**
$4.2 \stackrel{?}{=} \log_{10} 1 - \log_{10} 10^{-4.2}$	**Quotient Property**
$4.2 \stackrel{?}{=} 0 - (-4.2)$	**Simplify.**
$4.2 = 4.2$ ✓	

Real-World Link

Acid rain is more acidic than normal rain. Smoke and fumes from burning fossil fuels rise into the atmosphere and combine with the moisture in the air to form acid rain. Acid rain can be responsible for the erosion of statues, as in the photo above.

Check Your Progress

2. SOUND The loudness L of a sound, measured in decibels, is given by $L = 10 \log_{10} R$, where R is the sound's relative intensity. Suppose one person talks with a relative intensity of 10^6 or 60 decibels. How much louder would 100 people be, talking at the same intensity?

Personal Tutor glencoe.com

Recall that the power of a power is found by multiplying exponents. The property for the logarithm of a power is similar.

Key Concept — **Power Property of Logarithms** — For Your FOLDABLE

Words The logarithm of a power is the product of the logarithm and the exponent.

Symbols For any real number p, and positive numbers m and b, where $b \neq 1$, $\log_b m^p = p \log_b m$.

Example $\log_2 6^5 = 5 \log_2 6$

StudyTip

Answer Check You can check this answer by evaluating $2^{4.6438}$ on a calculator. The calculator should give a result of about 25, since $\log_2 25 \approx 4.6438$ means $2^{4.6438} \approx 25$.

EXAMPLE 3 Power Property of Logarithms

Given $\log_2 5 \approx 2.3219$, approximate the value of $\log_2 25$.

$\log_2 25 = \log_2 5^2$	**Replace 25 with 5^2.**
$= 2 \log_2 5$	**Power Property**
$\approx 2(2.3219)$ or 4.6438	**Replace $\log_2 5$ with 2.3219.**

Check Your Progress

3. Given $\log_3 7 \approx 1.7712$, approximate the value of $\log_3 49$.

Personal Tutor glencoe.com

Solve Logarithmic Equations You can use the properties of logarithms to solve equations involving logarithms.

EXAMPLE 4 Solve Equations Using Properties of Logarithms

Solve $\log_6 x + \log_6 (x - 9) = 2$.

$\log_6 x + \log_6 (x - 9) = 2$	**Original equation**
$\log_6 x(x - 9) = 2$	**Product Property**
$x(x - 9) = 6^2$	**Definition of logarithm**
$x^2 - 9x - 36 = 0$	**Subtract 36 from each side.**
$(x - 12)(x + 3) = 0$	**Factor.**
$x - 12 = 0$ or $x + 3 = 0$	**Zero Product Property**
$x = 12$ $\quad$ $x = -3$	**Solve each equation.**

CHECK Substitute each value into the original equation.

$$\begin{aligned} \log_6 x + \log_6 (x - 9) &= 2 \\ \log_6 12 + \log_6 (12 - 9) &\stackrel{?}{=} 2 \\ \log_6 12 + \log_6 3 &\stackrel{?}{=} 2 \\ \log_6 (12 \cdot 3) &\stackrel{?}{=} 2 \\ \log_6 36 &\stackrel{?}{=} 2 \\ 2 &= 2 \checkmark \end{aligned}$$

$$\begin{aligned} \log_6 x + \log_6 (x - 9) &= 2 \\ \log_6 (-3) + \log_6 (-3 - 9) &\stackrel{?}{=} 2 \\ \log_6 (-3) + \log_6 (-12) &\stackrel{?}{=} 2 \end{aligned}$$

Because $\log_6 (-3)$ and $\log_6 (-12)$ are undefined, -3 is an extraneous solution.

Check Your Progress

4A. $2 \log_7 x = \log_7 27 + \log_7 3$

4B. $\log_6 x + \log_6 (x + 5) = 2$

Personal Tutor glencoe.com

Check Your Understanding

Examples 1 and 2 pp. 509–510

Use $\log_4 3 \approx 0.7925$ and $\log_4 5 \approx 1.1610$ to approximate the value of each expression.

1. $\log_4 18$

2. $\log_4 15$

3. $\log_4 \frac{5}{3}$

4. $\log_4 \frac{3}{4}$

Example 2 p. 510

5. MOUNTAIN CLIMBING As elevation increases, the atmospheric air pressure decreases. The formula for pressure based on elevation is $a = 15{,}500(5 - \log_{10} P)$, where a is the altitude in meters and P is the pressure in pascals (1 psi ≈ 6900 pascals). What is the air pressure at the summit in pascals for each mountain listed in the table at the right?

Mountain	Country	Height (m)
Everest	Nepal/Tibet	8850
Trisuli	India	7074
Bonete	Argentina/Chile	6872
McKinley	United States	6194
Logan	Canada	5959

Example 3 p. 511

Given $\log_3 5 \approx 1.465$ and $\log_5 7 \approx 1.2091$, approximate the value of each expression.

6. $\log_3 25$

7. $\log_5 49$

Example 4 p. 511

Solve each equation. Check your solutions.

8. $\log_4 48 - \log_4 n = \log_4 6$

9. $\log_3 2x + \log_3 7 = \log_3 28$

10. $3 \log_2 x = \log_2 8$

11. $\log_{10} a + \log_{10} (a - 6) = 2$

Practice and Problem Solving

= **Step-by-Step Solutions** begin on page R20.
Extra Practice begins on page 947.

Examples 1 and 2 pp. 509–510

Use $\log_4 2 = 0.5$ and $\log_4 3 \approx 0.7925$ to approximate the value of each expression.

12. $\log_4 30$

13. $\log_4 20$

14. $\log_4 \frac{2}{3}$

15. $\log_4 \frac{4}{3}$

16. $\log_4 9$

17. $\log_4 8$

Example 2 p. 510

18. SCIENCE In 1979, an earthquake near San Francisco registered approximately 5.9 on the Richter scale. The famous San Francisco earthquake of 1906 measured 8.3 in magnitude.

a. How much more intense was the 1906 earthquake than the 1979 earthquake?

b. Richter himself classified the 1906 earthquake as having a magnitude of 8.3. More recent research indicates it was most likely a 7.9. What is the difference in intensities?

Year	Location	Magnitude
1906	San Francisco	8.3
1923	Tokyo, Japan	8.3
1932	Gansu, China	7.6
1960	Chile	9.5
1964	Alaska	9.2
1979	San Francisco	5.9

Source: TLC

Example 3 p. 511

Given $\log_6 8 \approx 1.1606$ and $\log_7 9 \approx 1.1292$, approximate the value of each expression.

19. $\log_6 48$

20. $\log_7 81$

21. $\log_6 512$

22. $\log_7 729$

Example 4 p. 511

Solve each equation. Check your solutions.

23. $\log_3 56 - \log_3 n = \log_3 7$

24. $\log_2 (4x) + \log_2 5 = \log_2 40$

25. $5 \log_2 x = \log_2 32$

26. $\log_{10} a + \log_{10} (a + 21) = 2$

27 **PROBABILITY** In the 1930s, Dr. Frank Benford demonstrated a way to determine whether a set of numbers has been randomly chosen or manually chosen. If the sets of numbers were not randomly chosen, then the Benford formula, $P = \log_{10}\left(1 + \frac{1}{d}\right)$, predicts the probability of a digit d being the first digit of the set. For example, there is a 4.6% probability that the first digit is 9.

a. Rewrite the formula to solve for the digit if given the probability.

b. Find the digit that has a 9.7% probability of being selected.

c. Find the probability that the first digit is 1 ($\log_{10} 2 \approx 0.30103$).

Use $\log_5 3 \approx 0.6826$ and $\log_5 4 \approx 0.8614$ to approximate the value of each expression.

28. $\log_5 40$

29. $\log_5 30$

30. $\log_5 \frac{3}{4}$

31. $\log_5 \frac{4}{3}$

32. $\log_5 9$

33. $\log_5 16$

34. $\log_5 12$

35. $\log_5 27$

Solve each equation. Check your solutions.

36. $\log_3 6 + \log_3 x = \log_3 12$

37. $\log_4 a + \log_4 8 = \log_4 24$

38. $\log_{10} 18 - \log_{10} 3x = \log_{10} 2$

39. $\log_7 100 - \log_7 (y + 5) = \log_7 10$

40. $\log_2 n = \frac{1}{3}\log_2 27 + \log_2 36$

41. $3 \log_{10} 8 - \frac{1}{2}\log_{10} 36 = \log_{10} x$

Solve for n.

42. $\log_a 6n - 3 \log_a x = \log_a x$

43. $2 \log_b 16 + 6 \log_b n = \log_b (x - 2)$

Solve each equation. Check your solutions.

44. $\log_{10} z + \log_{10} (z + 9) = 1$

45. $\log_3 (a^2 + 3) + \log_3 3 = 3$

46. $\log_2 (15b - 15) - \log_2 (-b^2 + 1) = 1$

47. $\log_4 (2y + 2) - \log_4 (y - 2) = 1$

48. $\log_6 0.1 + 2 \log_6 x = \log_6 2 + \log_6 5$

49. $\log_7 64 - \log_7 \frac{8}{3} + \log_7 2 = \log_7 4p$

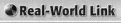

Real-World Link

Humpback whales are known for their songs. Scientists are studying these sounds to decipher their meaning, but they still do not know why whales sing. It is most likely that humpbacks sing to communicate with others.

50. ENVIRONMENT The humpback whale is an endangered species. Suppose there are 5000 humpback whales in existence today, and the population decreases at a rate of 4% per year.

a. Write a logarithmic function for the time in years based upon population.

b. After how long will the population drop below 1000? Round your answer to the nearest year.

State whether each equation is *true* or *false*.

51. $\log_8 (x - 3) = \log_8 x - \log_8 3$

52. $\log_5 22x = \log_5 22 + \log_5 x$

53. $\log_{10} 19k = 19 \log_{10} k$

54. $\log_2 y^5 = 5 \log_2 y$

55. $\log_7 \frac{x}{3} = \log_7 x - \log_7 3$

56. $\log_4 (z + 2) = \log_4 z + \log_4 2$

57. $\log_8 p^4 = (\log_8 p)^4$

58. $\log_9 \frac{x^2y^3}{z^4} = 2 \log_9 x + 3 \log_9 y - 4 \log_9 z$

59. PARADE An equation for loudness L, in decibels, is $L = 10 \log_{10} R$, where R is the relative intensity of the sound.

a. Solve $120 = 10 \log_{10} R$ to find the relative intensity of the Macy's Thanksgiving Day Parade with a loudness of 120 decibels depending on how close you are.

b. Some parents with young children want the decibel level lowered to 80. How many times less intense would this be? In other words, find the ratio of their intensities.

Real-World Link

According to the Jump Start Coalition, an advocate for financial literacy, one in every three high school seniors uses credit cards.

60. CREDIT CARDS The average American carries a credit card debt of approximately \$8600 with an annual percentage rate (APR) of 18.3%. The formula $m = \frac{b\left(\frac{r}{n}\right)}{1 - \left(1 + \frac{r}{n}\right)^{-nt}}$ can be used to compute the monthly payment m that is necessary to pay off a credit card balance b in a given number of years t, where r is the annual percentage rate and n is the number of payments per year.

a. What monthly payment should be made in order to pay off the debt in exactly three years? What is the total amount paid?

b. The equation $t = \frac{\log\left(1 - \frac{br}{mn}\right)}{-n \log\left(1 + \frac{r}{n}\right)}$ can be used to calculate the number of years necessary for a given payment schedule. Copy and complete the table.

Payment (m)	Years (t)
\$50	
\$100	
\$150	
\$200	
\$250	
\$300	

c. Graph the information in the table from part c.

d. If you could only afford to pay \$100 a month, will you be able to pay off the debt? If so, how long will it take? If not, why not?

e. What is the minimum monthly payment that will work toward paying off the debt?

H.O.T. Problems

Use Higher-Order Thinking Skills

61. OPEN ENDED Write a logarithmic expression for each condition. Then write the expanded expression.

a. a product and a quotient

b. a product and a power

c. a product, a quotient, and a power

62. PROOF Use the properties of exponents to prove the Quotient Property of Logarithms.

63. WRITING IN MATH Explain why the following are true.

a. $\log_b 1 = 0$ **b.** $\log_b b = 1$ **c.** $\log_b b^x = x$

64. CHALLENGE Simplify $\log_{\sqrt{a}} (a^2)$ to find an exact numerical value.

65. WHICH ONE DOESN'T BELONG? Find the expression that does not belong. Explain.

$\log_b 24 = \log_b 2 + \log_b 12$	$\log_b 24 = \log_b 20 + \log_b 4$
$\log_b 24 = \log_b 8 + \log_b 3$	$\log_b 24 = \log_b 4 + \log_b 6$

66. REASONING Use the properties of logarithms to prove that $\log_a \frac{1}{x} = -\log_a x$.

67. CHALLENGE Simplify $x^{3 \log_x 2 - \log_x 5}$ to find an exact numerical value.

68. WRITING IN MATH Explain how the properties of exponents and logarithms are related. Include examples like the one shown at the beginning of the lesson illustrating the Product Property, but with the Quotient Property and Power Property of Logarithms.

NYSRE PRACTICE A2.A.6

69. Find the mode of the data.

22, 11, 12, 23, 7, 6, 17, 15, 21, 19

A 11
B 15
C 16
D There is no mode.

70. **ACT/SAT** What is the effect on the graph of $y = 4x^2$ when the equation is changed to $y = 2x^2$?

F The graph is rotated 90 degrees about the origin.
G The graph is narrower.
H The graph is wider.
J The graph of $y = 2x^2$ is a reflection of the graph $y = 4x^2$ across the x-axis.

71. **SHORT RESPONSE** In $y = 6.5(1.07)^x$, x represents the number of years since 2000, and y represents the approximate number of millions of Americans 7 years of age and older who went camping two or more times that year. Describe how the number of millions of Americans who go camping is changing over time.

72. What are the x-intercepts of the graph of $y = 4x^2 - 3x - 1$?

A $-\frac{1}{4}$ and $\frac{1}{4}$
B -1 and $\frac{1}{4}$
C -1 and 1
D 1 and $-\frac{1}{4}$

Spiral Review

Solve each equation. Check your solutions. (Lesson 8-4)

73. $\log_5 (3x - 1) = \log_5 (2x^2)$
74. $\log_{10} (x^2 + 1) = 1$
75. $\log_{10} (x^2 - 10x) = \log_{10} (-21)$

Evaluate each expression. (Lesson 8-3)

76. $\log_{10} 0.001$
77. $\log_4 16^x$
78. $\log_3 27^x$

79. **ELECTRICITY** The amount of current in amperes I that an appliance uses can be calculated using the formula $I = \left(\frac{P}{R}\right)^{\frac{1}{2}}$, where P is the power in watts and R is the resistance in ohms. How much current does an appliance use if $P = 120$ watts and $R = 3$ ohms? Round to the nearest tenth. (Lesson 7-6)

Determine whether each pair of functions are inverse functions. Write *yes* or *no*. (Lesson 7-2)

80. $f(x) = x + 73$
$g(x) = x - 73$

81. $g(x) = 7x - 11$
$h(x) = \frac{1}{7}x + 11$

82. **SCULPTING** Antonio is preparing to make an ice sculpture. He has a block of ice that he wants to reduce in size by shaving off the same amount from the length, width, and height. He wants to reduce the volume of the ice block to 24 cubic feet. (Lesson 6-7)

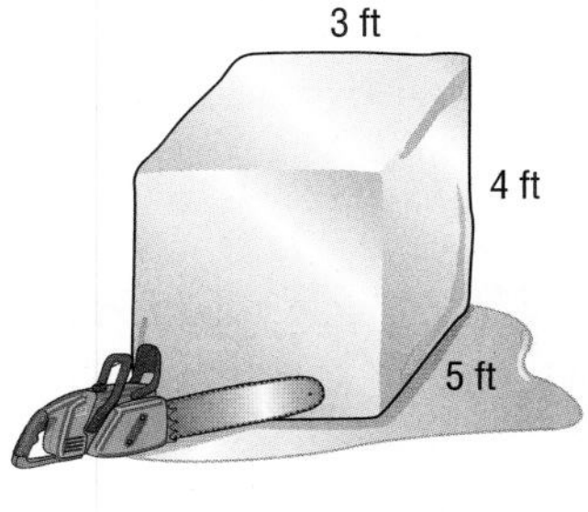

a. Write a polynomial equation to model this situation.

b. How much should he take from each dimension?

Skills Review

Solve each equation or inequality. Check your solution. (Lessons 9-1 through 9-4)

83. $3^{4x} = 3^{3 - x}$
84. $3^{2n} \leq \frac{1}{9}$
85. $3^{5x} \cdot 81^{1 - x} = 9^{x - 3}$
86. $49^x = 7^{x^2 - 15}$
87. $\log_2 (x + 6) > 5$
88. $\log_5 (4x - 1) = \log_5 (3x + 2)$

8-6 Common Logarithms

Then

You simplified expressions and solved equations using properties of logarithms. (Lesson 8-4)

Now

- Solve exponential equations and inequalities using common logarithms.
- Evaluate logarithmic expressions using the Change of Base Formula.

NYS Core Curriculum

A2.A.18 Evaluate logarithmic expressions in any base
A2.A.19 Apply the properties of logarithms to rewrite logarithmic expressions in equivalent forms
Also addresses A2.A.28.

New Vocabulary

common logarithm
Change of Base Formula

NY Math Online

glencoe.com

- Extra Examples
- Personal Tutor
- Self-Check Quiz
- Homework Help
- Math in Motion

Why?

Seismologists use the Richter scale to measure the strength or magnitude of earthquakes. The magnitude of an earthquake is determined using the logarithm of the amplitude of waves recorded by seismographs.

Richter Number	1	2	3	4	5	6	7	8
Intensity	10^1 Micro	10^2 Minor	10^3 Minor	10^4 Light	10^5 Moderate	10^6 Strong	10^7 Major	10^8 Great
Effect in Populated Areas	not felt, but recorded	generally not felt, loose hanging items sway	often felt, little to no damage	noticeable shaking, minor damage	slight damage to buildings over small regions	damage over regions up to 100 miles across	severe destruction over large areas	catastrophic destruction to areas several hundred miles across

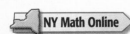

The logarithmic scale used by the Richter scale is based on the powers of 10. For example, a magnitude 6.4 earthquake can be represented by $6.4 = \log_{10} x$.

Common Logarithms You have seen that the base 10 logarithm function, $y = \log_{10} x$, is used in many applications. Base 10 logarithms are called **common logarithms**. Common logarithms are usually written without the subscript 10.

$$\log_{10} x = \log x, x > 0$$

Most scientific calculators have a LOG key for evaluating common logarithms.

EXAMPLE 1 Find Common Logarithms

Use a calculator to evaluate each expression to the nearest ten-thousandth.

a. log 5

KEYSTROKES: LOG 5 ENTER .6989700043

$\log 5 \approx 0.6990$

b. log 0.3

KEYSTROKES: LOG 0.3 ENTER −.5228787453

$\log 0.3 \approx -0.5229$

Check Your Progress

1A. $\log 7$

1B. $\log 0.5$

Personal Tutor glencoe.com

The common logarithms of numbers that differ by integral powers of ten are closely related. Remember that a logarithm is an exponent. For example, in the equation $y = \log x$, y is the power to which 10 is raised to obtain the value of x.

$$\begin{array}{rcl} \log x = y & \rightarrow \text{means} \rightarrow & 10^y = x \\ \log 1 = 0 & \text{since} & 10^0 = 1 \\ \log 10 = 1 & \text{since} & 10^1 = 10 \\ \log 10^m = m & \text{since} & 10^m = 10^m \end{array}$$

Real-World Career

Acoustical Engineer Acoustical engineers are concerned with reducing unwanted sounds, noise control, and making useful sounds. Examples of useful sounds are ultrasound, sonar, and sound reproduction. Employment in this field requires a minimum of a bachelor's degree.

Common logarithms are used in the measure of sound. Soft recorded music is about 36 decibels (dB).

Real-World EXAMPLE 2 Solve Logarithmic Equations

ROCK CONCERT The loudness L, in decibels, of a sound is $L = 10 \log \frac{I}{m}$, where I is the intensity of the sound and m is the minimum intensity of sound detectable by the human ear. Residents living several miles from a concert venue can hear the music at an intensity of 66.6 decibels. How many times the minimum intensity of sound detectable by the human ear was this sound, if m is defined to be 1?

$L = 10 \log \frac{I}{m}$ **Original equation**

$66.6 = 10 \log \frac{I}{1}$ **Replace L with 66.6 and m with 1.**

$6.66 = \log I$ **Divide each side by 10 and simplify.**

$I = 10^{6.66}$ **Exponential form**

$I = 4{,}570{,}882$ **Use a calculator.**

The sound heard by the residents was approximately 4,570,000 times the minimum intensity of sound detectable by the human ear.

Check Your Progress

2. EARTHQUAKES The amount of energy E in ergs that an earthquake releases is related to its Richter scale magnitude M by the equation $\log E = 11.8 + 1.5M$. Use the equation to find the amount of energy released by the 2004 Sumatran earthquake, which measured 9.0 on the Richter scale and led to a tsunami.

Personal Tutor glencoe.com

If both sides of an exponential equation cannot easily be written as powers of the same base, you can solve by taking the logarithm of each side.

EXAMPLE 3 Solve Exponential Equations Using Logarithms

Solve $4^x = 19$. Round to the nearest ten-thousandth.

$4^x = 19$ **Original equation**

$\log 4^x = \log 19$ **Property of Equality for Logarithmic Functions**

$x \log 4 = \log 19$ **Power Property of Logarithms**

$x = \frac{\log 19}{\log 4}$ **Divide each side by log 4.**

$x \approx \frac{1.2788}{0.6021}$ **Use a calculator.**

$x \approx 2.1239$ The solution is approximately 2.1239.

CHECK You can check this answer graphically by using a graphing calculator. Graph the line $y = 4^x$ and the line $y = 19$. Then use the calc menu to find the intersection of the two graphs. The intersection is very close to the answer that was obtained algebraically. ✓

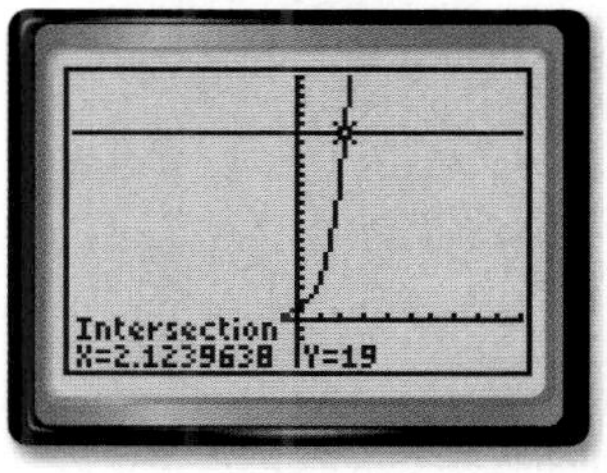

Check Your Progress

3A. $3^x = 15$

3B. $6^x = 42$

Personal Tutor glencoe.com

The same strategies that are used to solve exponential equations can be used to solve exponential inequalities.

StudyTip

Solving Inequalities Remember that the direction of an inequality must be switched if each side is multiplied or divided by a negative number. Since $5 \log 3 - \log 7 > 0$, the inequality does not change.

EXAMPLE 4 Solve Exponential Inequalities Using Logarithms

Solve $3^{5y} < 7^{y-2}$. Round to the nearest ten-thousandth.

$3^{5y} < 7^{y-2}$	Original inequality
$\log 3^{5y} < \log 7^{y-2}$	Property of Inequality for Logarithmic Functions
$5y \log 3 < (y - 2) \log 7$	Power Property of Logarithms
$5y \log 3 < y \log 7 - 2 \log 7$	Distributive Property
$5y \log 3 - y \log 7 < -2 \log 7$	Subtract $y \log 7$ from each side.
$y(5 \log 3 - \log 7) < -2 \log 7$	Distributive Property
$y < \dfrac{-2 \log 7}{5 \log 3 - \log 7}$	Divide each side by $5 \log 3 - \log 7$.
$\{y \mid y < -1.09717\}$	Use a calculator.

CHECK Test $y = -2$.

$3^{5y} < 7^{y-2}$	Original inequality
$3^{5(-2)} \overset{?}{<} 7^{(-2)-2}$	Replace y with -2.
$3^{-10} \overset{?}{<} 7^{-4}$	Simplify.
$\dfrac{1}{59049} < \dfrac{1}{2401}$ ✓	Negative Exponent Property

Check Your Progress

Solve each inequality. Round to the nearest ten-thousandth.

4A. $3^{2x} \geq 6^{x+1}$

4B. $4^{y} < 5^{2y+1}$

Personal Tutor glencoe.com

Change of Base Formula The **Change of Base Formula** allows you to write equivalent logarithmic expressions that have different bases.

Key Concept Change of Base Formula

For Your FOLDABLE

Symbols For all positive numbers a, b, and n, where $a \neq 1$ and $b \neq 1$,

$$\log_a n = \frac{\log_b n}{\log_b a}.$$

← log base b of original number (numerator)
← log base b of old base (denominator)

Example $\log_3 11 = \dfrac{\log_{10} 11}{\log_{10} 3}$

Math *in Motion,* Interactive Lab glencoe.com

To prove this formula, let $\log_a n = x$.

$a^x = n$	Definition of logarithm
$\log_b a^x = \log_b n$	Property of Equality for Logarithmic Functions
$x \log_b a = \log_b n$	Power Property of Logarithms
$x = \dfrac{\log_b n}{\log_b a}$	Divide each side by $\log_b a$.
$\log_a n = \dfrac{\log_b n}{\log_b a}$	Replace x with $\log_a n$.

The Granger Collection, New York

Math History Link

John Napier (1550–1617) John Napier was a Scottish mathematician and theologian who began the use of logarithms to aid in calculations. He is also known for popularizing the use of the decimal point.

The Change of Base Formula makes it possible to evaluate a logarithmic expression of any base by translating the expression into one that involves common logarithms.

EXAMPLE 5 **Change of Base Formula**

Express $\log_3 20$ in terms of common logarithms. Then round to the nearest ten-thousandth.

$\log_3 20 = \frac{\log_{10} 20}{\log_{10} 3}$ **Change of Base Formula**

≈ 2.7268 **Use a calculator.**

Check Your Progress

5. Express $\log_6 8$ in terms of common logarithms. Then round to the nearest ten-thousandth.

Personal Tutor glencoe.com

Check Your Understanding

Example 1 p. 516

Use a calculator to evaluate each expression to the nearest ten-thousandth.

1. log 5 **2.** log 21 **3.** log 0.4 **4.** log 0.7

Example 2 p. 517

5. **SCIENCE** The amount of energy E in ergs that an earthquake releases is related to its Richter scale magnitude M by the equation $\log E = 11.8 + 1.5M$. Use the equation to find the amount of energy released by the 1960 Chilean earthquake, which measured 8.5 on the Richter scale.

Example 3 p. 517

Solve each equation. Round to the nearest ten-thousandth.

6. $6^x = 40$ **7.** $2.1^{a+2} = 8.25$ **8.** $7^{x^2} = 20.42$ **9** $11^{b-3} = 5^b$

Example 4 p. 518

Solve each inequality. Round to the nearest ten-thousandth.

10. $5^{4n} > 33$ **11.** $6^{p-1} \leq 4^p$

Example 5 p. 519

Express each logarithm in terms of common logarithms. Then approximate its value to the nearest ten-thousandth.

12. $\log_3 7$ **13.** $\log_4 23$ **14.** $\log_9 13$ **15.** $\log_2 5$

Practice and Problem Solving

● = **Step-by-Step Solutions** begin on page R20.
Extra Practice begins on page 947.

Example 1 p. 516

Use a calculator to evaluate each expression to the nearest ten-thousandth.

16. log 3 **17.** log 11 **18.** log 3.2

19. log 8.2 **20.** log 0.9 **21.** log 0.04

Example 2 p. 517

22. **AUTO REPAIR** Loretta had a new muffler installed on her car. The noise level of the engine dropped from 85 decibels to 73 decibels.

a. How many times the minimum intensity of sound detectable by the human ear was the car with the old muffler, if m is defined to be 1?

b. How many times the minimum intensity of sound detectable by the human ear was the car with the new muffler? Find the percent of decrease of the intensity of the sound with the new muffler.

Example 3 p. 517

Solve each equation. Round to the nearest ten-thousandth.

23. $8^x = 40$

24. $5^x = 55$

25. $2.9^{a-4} = 8.1$

26. $9^{b-1} = 7^b$

27. $13^{x^2} = 33.3$

28. $15^{x^2} = 110$

Example 4 p. 518

Solve each inequality. Round to the nearest ten-thousandth.

29. $6^{3n} > 36$

30. $2^{4x} \leq 20$

31. $3^{y-1} \leq 4^y$

32. $5^{p-2} \geq 2^p$

Example 5 p. 519

Express each logarithm in terms of common logarithms. Then approximate its value to the nearest ten-thousandth.

33. $\log_7 18$

34. $\log_5 31$

35. $\log_2 16$

36. $\log_4 9$

37. $\log_3 11$

38. $\log_6 33$

39. **PETS** The number n of pet owners in thousands after t years can be modeled by $n = 35[\log_4 (t + 2)]$. Let $t = 0$ represent 2000. Use the Change of Base Formula to solve the following questions.

a. How many pet owners were there in 2010?

b. How long until there are 80,000 pet owners? When will this occur?

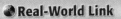

Real-World Link

The grizzly bear is a North American subspecies of the brown bear. Grizzlies are typically 800 pounds; however, they are quite fast and have been clocked at 30 miles an hour.

40. **GRIZZLY BEARS** Five years ago the grizzly bear population in a certain national park was 325. Today it is 450. Studies show that the park can support a population of 750.

a. What is the average annual rate of growth in the population if the grizzly bears reproduce once a year?

b. How many years will it take to reach the maximum population if the population growth continues at the same average rate?

Solve each equation or inequality. Round to the nearest ten-thousandth.

41. $3^x = 40$

42. $5^{3p} = 15$

43. $4^{n+2} = 14.5$

44. $8^{z-4} = 6.3$

45. $7.4^{n-3} = 32.5$

46. $3.1^{y-5} = 9.2$

47. $5^x \geq 42$

48. $9^{2a} < 120$

49. $3^{4x} \leq 72$

50. $7^{2n} > 52^{4n+3}$

51. $6^p \leq 13^{5-p}$

52. $2^{y+3} \geq 8^{3y}$

Express each logarithm in terms of common logarithms. Then approximate its value to the nearest ten-thousandth.

53. $\log_4 12$

54. $\log_3 21$

55. $\log_8 2$

56. $\log_6 7$

57. $\log_5 (2.7)^2$

58. $\log_7 \sqrt{5}$

59. **MUSIC** A musical cent is a unit in a logarithmic scale of relative pitch or intervals. One octave is equal to 1200 cents. The formula $n = 1200\left(\log_2 \frac{a}{b}\right)$ can be used to determine the difference in cents between two notes with frequencies a and b.

a. Find the interval in cents when the frequency changes from 443 Hertz (Hz) to 415 Hz.

b. If the interval is 55 cents and the beginning frequency is 225 Hz, find the final frequency.

Solve each equation. Round to the nearest ten-thousandth.

60. $10^{x^2} = 60$

61. $4^{x^2 - 3} = 16$

62. $9^{6y - 2} = 3^{3y + 1}$

63. $8^{2x - 4} = 4^{x + 1}$

64. $16^x = \sqrt{4^{x + 3}}$

65. $2^y = \sqrt{3^{y - 1}}$

Real-World Link

Worldwide, over 1.1 billion people lack access to safe drinking water because of industrial pollution, poor sanitation, and environmental factors.

Source: United Nations

66. ENVIRONMENTAL SCIENCE An environmental engineer is testing drinking water wells in coastal communities for pollution, specifically unsafe levels of arsenic. The safe standard for arsenic is 0.025 parts per million (ppm). Also, the pH of the arsenic level should be less than 9.5. The formula for hydrogen ion concentration is $\text{pH} = -\log H$. (*Hint*: 1 kilogram of water occupies approximately 1 liter. 1 ppm = 1 mg/kg.)

a. Suppose the hydrogen ion concentration of a well is 1.25×10^{-11}. Should the environmental engineer be worried about too high an arsenic content?

b. The environmental engineer finds 1 milligram of arsenic in a 3 liter sample, is the well safe?

c. What is the hydrogen ion concentration that meets the troublesome pH level of 9.5?

67. MULTIPLE REPRESENTATIONS In this problem, you will solve the exponential equation $4^x = 13$.

a. TABULAR Enter the function $y = 4^x$ into a graphing calculator, create a table of values for the function, and scroll through the table to find x when $y = 13$.

b. GRAPHICAL Graph $y = 4^x$ and $y = 13$ on the same screen. Use the **intersect** feature to find the point of intersection.

c. NUMERICAL Solve the equation algebraically. Do all of the methods produce the same result? Explain why or why not.

H.O.T. Problems

Use Higher-Order Thinking Skills

68. FIND THE ERROR Sam and Rosamaria are solving $4^{3p} = 10$. Is either of them correct? Explain your reasoning.

Sam

$$4^{3p} = 10$$
$$\log 4^{3p} = \log 10$$
$$p \log 4 = \log 10$$
$$p = \frac{\log 10}{\log 4}$$

Rosamaria

$$4^{3p} = 10$$
$$\log 4^{3p} = \log 10$$
$$3p \log 4 = \log 10$$
$$p = \frac{\log 10}{3 \log 4}$$

69. CHALLENGE Solve $\log_{\sqrt{a}} 3 = \log_a x$ for x and explain each step.

70. REASONING Write $\dfrac{\log_5 9}{\log_5 3}$ as a single logarithm.

71. PROOF Find the values of $\log_3 27$ and $\log_{27} 3$. Make and prove a conjecture about the relationship between $\log_a b$ and $\log_b a$.

72. WRITING IN MATH Explain how exponents and logarithms are related. Include examples like how to solve a logarithmic equation using exponents and how to solve an exponential equation using logarithms.

NYSRE PRACTICE Reinforcement of A.G.5

73. Which expression represents $f[g(x)]$ if $f(x) = x^2 + 4x + 3$ and $g(x) = x - 5$?

A $x^2 + 4x - 2$
B $x^2 - 6x + 8$
C $x^2 - 9x + 23$
D $x^2 - 14x + 6$

74. EXTENDED RESPONSE Colleen rented 3 documentaries, 2 video games, and 2 movies. The charge was \$16.29. The next week, she rented 1 documentary, 3 video games, and 4 movies for a total charge of \$19.84. The third week she rented 2 documentaries, 1 video game, and 1 movie for a total charge of \$9.14.

a. Write a system of equations to determine the cost to rent each item.

b. What is the cost to rent each item?

75. GEOMETRY If the surface area of a cube is increased by a factor of 9, what is the change in the length of the sides if the cube?

F The length is 2 times the original length.
G The length is 3 times the original length.
H The length is 6 times the original length.
J The length is 9 times the original length.

76. ACT/SAT Which of the following *most* accurately describes the translation of the graph $y = (x + 4)^2 - 3$ to the graph of $y = (x - 1)^2 + 3$?

A down 1 and to the right 3
B down 6 and to the left 5
C up 1 and to the left 3
D up 6 and to the right 5

Spiral Review

Solve each equation. Check your solutions. (Lesson 8-5)

77. $\log_5 7 + \frac{1}{2}\log_5 4 = \log_5 x$

78. $2 \log_2 x - \log_2 (x + 3) = 2$

79. $\log_6 48 - \log_6 \frac{16}{5} + \log_6 5 = \log_6 5x$

80. $\log_{10} a + \log_{10} (a + 21) = 2$

Solve each equation or inequality. (Lesson 8-4)

81. $\log_4 x = \frac{1}{2}$

82. $\log_{81} 729 = x$

83. $\log_8 (x^2 + x) = \log_8 12$

84. $\log_8 (3y - 1) < \log_8 (y + 5)$

85. SAILING The area of a triangular sail is $16x^4 - 60x^3 - 28x^2 + 56x - 32$ square meters. The base of the triangle is $x - 4$ meters. What is the height of the sail? (Lesson 6-2)

86. HOME REPAIR Mr. Turner is getting new locks installed. The locksmith charges \$85 for the service call, \$25 for each door, and each lock costs \$30. (Lesson 2-4)

a. Write an equation that represents the cost for x number of doors.

b. Mr. Turner wants the front, side, back, and garage door locks changed. How much will this cost?

Skills Review

Write an equivalent exponential equation. (Lesson 8-3)

87. $\log_2 5 = x$

88. $\log_4 x = 3$

89. $\log_5 25 = 2$

90. $\log_7 10 = x$

91. $\log_6 x = 4$

92. $\log_4 64 = 3$

Graphing Technology Lab
Solving Logarithmic Equations and Inequalities

NY Math Online glencoe.com
Other Calculator Keystrokes

NYS Core Curriculum **A2.A.27 Solve exponential equations with and without common bases**
Also addresses A2.A.28.

You have solved logarithmic equations algebraically. You can also solve logarithmic equations by graphing or by using a table. The TI-83/84 Plus has $y = \log_{10} x$ as a built-in function. Enter [Y=] [LOG] [X,T,θ,n] [GRAPH] to view this graph. To graph logarithmic functions with bases other than 10, you must use the Change of Base Formula, $\log_a n = \frac{\log_b n}{\log_b a}$.

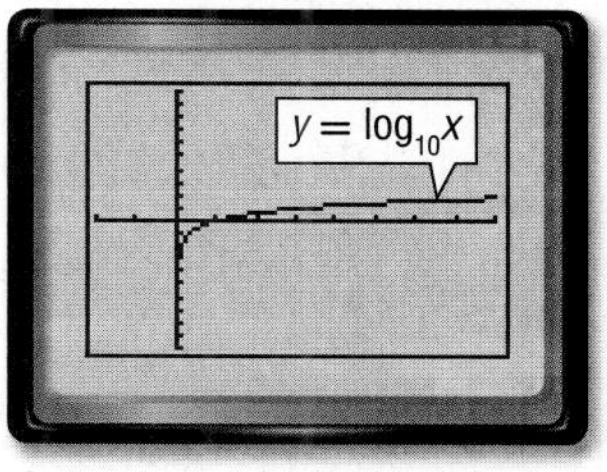

[−2, 8] scl: 1 by [−10, 10] scl: 1

ACTIVITY 1

Solve $\log_2 (6x - 8) = \log_3 (20x + 1)$.

Step 1 Graph each side of the equation.

Graph each side of the equation as a separate function. Enter $\log_2 (6x - 8)$ as **Y1** and $\log_3 (20x + 1)$ as **Y2**. Then graph the two equations.

KEYSTROKES: [Y=] [LOG] 6 [X,T,θ,n] [−] 8 [)] [÷] [LOG] 2 [)] [ENTER] [LOG] 20 [X,T,θ,n] [+] 1 [)] [÷] [LOG] 3 [)] [GRAPH]

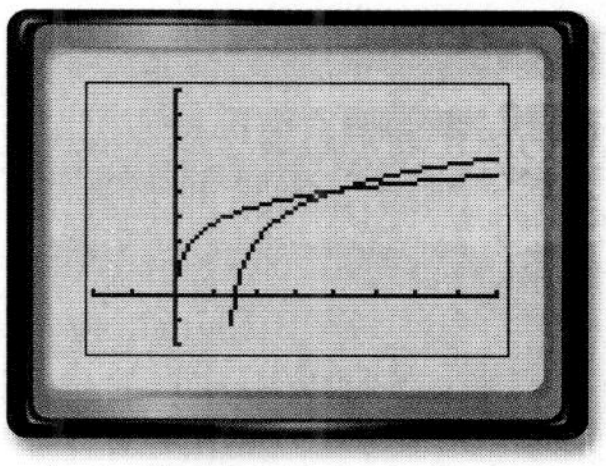

[−2, 8] scl: 1 by [−2, 8] scl: 1

Step 2 Use the **intersect** feature.

Use the **intersect** feature on the **CALC** menu to approximate the ordered pair of the point at which the curves intersect.

The calculator screen shows that the x-coordinate of the point at which the curves intersect is 4. Therefore, the solution of the equation is 4.

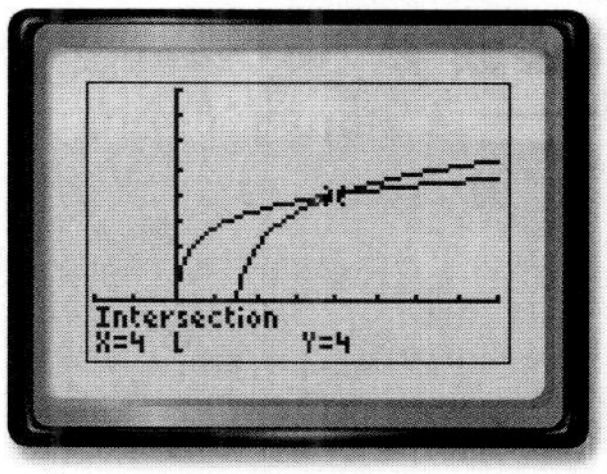

[−2, 8] scl: 1 by [−2, 8] scl: 1

Step 3 Use the **TABLE** feature.

Examine the table to find the x-value for which the y-values for the graphs are equal. At $x = 4$, both functions have a y-value of 4. Thus, the solution of the equation is 4.

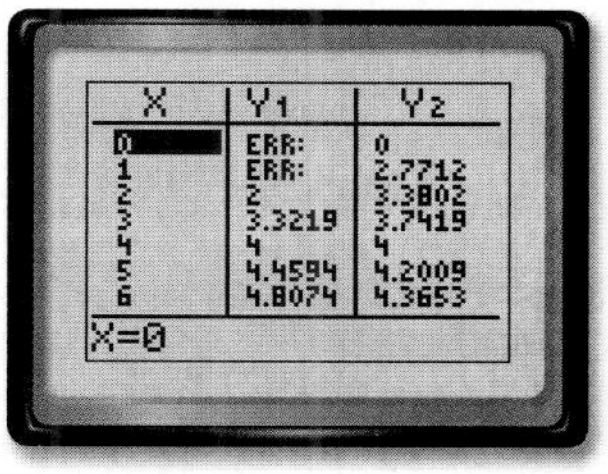

You can use a similar procedure to solve logarithmic inequalities using a graphing calculator.

ACTIVITY 2

Solve $\log_4 (10x + 1) < \log_5 (16 + 6x)$.

Step 1 Enter the inequalities.

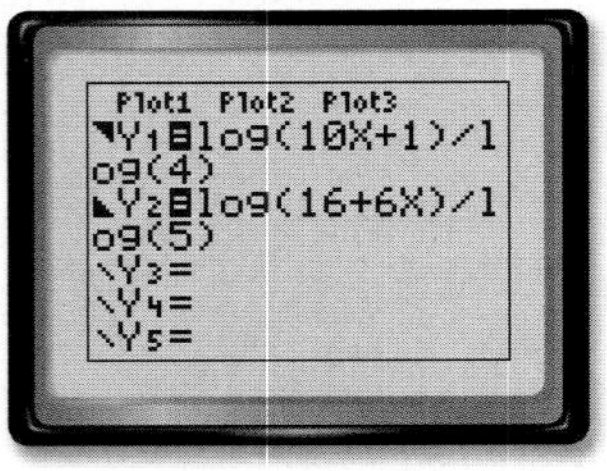

Rewrite the problem as a system of inequalities.

The first inequality is $\log_4 (10x + 1) < y$ or $y > \log_4 (10x + 1)$. Since this inequality includes the *greater than* symbol, shade above the curve.

First enter the boundary and then use the arrow and ENTER keys to choose the shade above icon, ◤.

The second inequality is $y < \log_5 (16 + 6x)$. Shade below the curve since this inequality contains *less than*.

KEYSTROKES: Y= ◀ ◀ ENTER ENTER ▶ ▶ LOG 10 X,T,θ,n + 1) ÷ LOG 4) ENTER ◀ ◀ ENTER ENTER ENTER ▶ ▶ LOG 16 + 6 X,T,θ,n) ÷ LOG 5)

Step 2 Graph the system.

KEYSTROKES: GRAPH

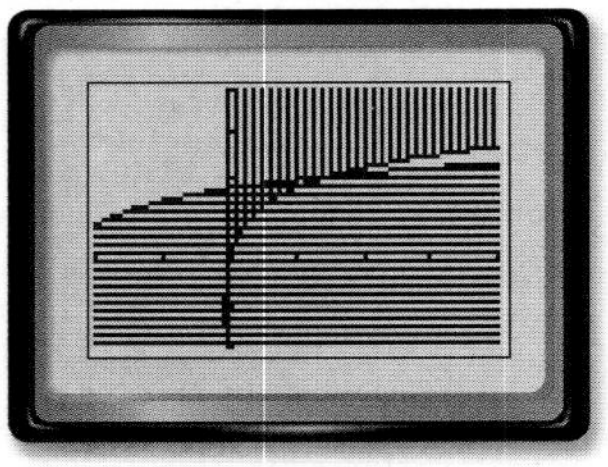

[−2, 4] scl: 1 by [−2, 4] scl: 1

The left boundary of the solution set is where the first inequality is undefined. It is undefined for $10x + 1 \leq 0$.

$$10x + 1 \leq 0$$
$$10x \leq -1$$
$$x \leq -\frac{1}{10}$$

Use the calculator's **intersect** feature to find the right boundary. You can conclude that the solution set is $\{x \mid -0.1 < x < 1.5\}$.

Step 3 Use the **TABLE** feature to check your solution.

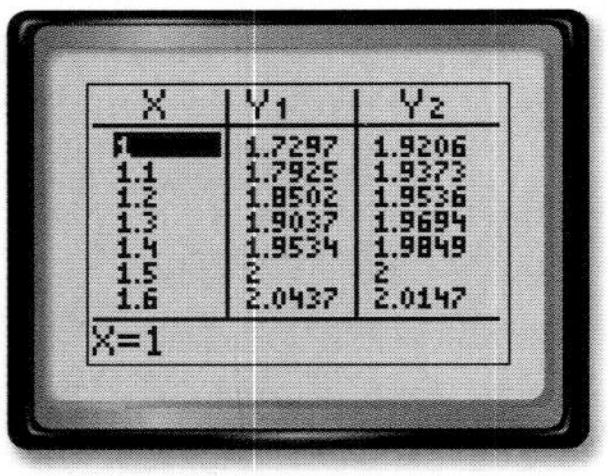

Start the table at −0.1 and show x-values in increments of 0.1. Scroll through the table.

KEYSTROKES: 2nd [TBLSET] −0.1 ENTER .5 ENTER 2nd [TBLSET]

The table confirms the solution of the inequality is $\{x \mid -0.1 < x < 1.5\}$.

Exercises

Solve each equation or inequality. Check your solution.

1. $\log_2 (3x + 2) = \log_3 (12x + 3)$

2. $\log_6 (7x + 1) = \log_4 (4x - 4)$

3. $\log_2 3x = \log_3 (2x + 2)$

4. $\log_{10} (1 - x) = \log_5 (2x + 5)$

5. $\log_4 (9x + 1) > \log_3 (18x - 1)$

6. $\log_3 (3x - 5) \geq \log_3 (x + 7)$

7. $\log_5 (2x + 1) < \log_4 (3x - 2)$

8. $\log_2 2x \leq \log_4 (x + 3)$

8-7 Base e and Natural Logarithms

Then

You worked with common logarithms. (Lesson 8-6)

Now

- Evaluate expressions involving the natural base and natural logarithm.
- Solve exponential equations and inequalities using natural logarithms.

NYS Core Curriculum

A2.A.12 Evaluate exponential expressions, including those with base e
A2.A.53 Graph exponential functions of the form $y = b$ for positive values of b, including $b = e$
Also addresses A2.A.18 and A2.A.19.

New Vocabulary

natural base, e
natural base exponential function
natural logarithm

NY Math Online

glencoe.com
- Extra Examples
- Personal Tutor
- Self-Check Quiz
- Homework Help

Why?

The St. Louis Gateway Arch in Missouri is in the form of an inverted catenary curve. A catenary curve directs the force of its weight along itself, so that:

- if a rope or chain is hanging, it is pulled into that shape, and,
- if a catenary is standing upright, it can support itself.

The equation for the catenary curve involves e, a special number that appears throughout mathematics and science.

Base e and Natural Logarithms Like π and $\sqrt{2}$, the number e is an irrational number. The value of e is 2.71828.... It is referred to as the **natural base, e**. An exponential function with base e is called a **natural base exponential function**.

Key Concept — Natural Base Functions (For Your FOLDABLE)

The function $f(x) = e^x$ is used to model continuous exponential growth.
The function $f(x) = e^{-x}$ is used to model continuous exponential decay.

The inverse of a natural base exponential function is called the **natural logarithm**. This logarithm can be written as $\log_e x$, but is more often abbreviated as $\ln x$.

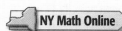

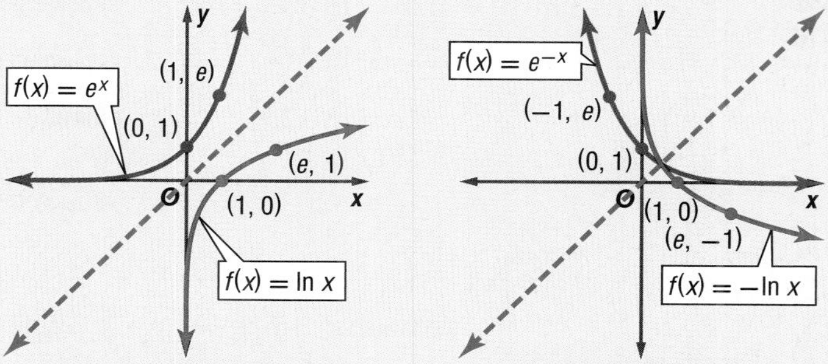

Exponential Growth — Exponential Decay

You can write an equivalent base e exponential equation for a natural logarithmic equation by using the fact that $\ln x = \log_e x$.

$$\ln 4 = x \quad \rightarrow \quad \log_e 4 = x \quad \rightarrow \quad e^x = 4$$

EXAMPLE 1 Write Equivalent Expressions

Write each exponential equation in logarithmic form.

a. $e^x = 8$

$e^x = 8 \quad \rightarrow \quad \log_e 8 = x$
$\ln 8 = x$

b. $e^5 = x$

$e^5 = x \quad \rightarrow \quad \log_e x = 5$
$\ln x = 5$

Check Your Progress

1A. $e^x = 9$

1B. $e^7 = x$

Personal Tutor glencoe.com

You can also write an equivalent natural logarithm equation for a natural base e exponential equation.

$$e^x = 12 \quad \rightarrow \quad \log_e 12 = x \quad \rightarrow \quad \ln 12 = x$$

EXAMPLE 2 Write Equivalent Expressions

Write each logarithmic equation in exponential form.

a. $\ln x \approx 0.7741$

$\ln x \approx 0.7741 \quad \rightarrow \quad \log_e x = 0.7741$

$x \approx e^{0.7741}$

b. $\ln 10 = x$

$\ln 10 = x \quad \rightarrow \quad \log_e 10 = x$

$10 = e^x$

Check Your Progress

2A. $\ln x \approx 2.1438$

2B. $\ln 18 = x$

Personal Tutor glencoe.com

The properties of logarithms you learned in Lesson 8-5 also apply to the natural logarithms. The logarithmic expressions below can be simplified into a single logarithmic term.

StudyTip

Simplifying When you simplify logarithmic expressions, verify that the logarithm contains no operations and no powers.

EXAMPLE 3 Simplify Expressions with *e* and the Natural Log

Write each expression as a single logarithm.

a. $3 \ln 10 - \ln 8$

$3 \ln 10 - \ln 8 = \ln 10^3 - \ln 8$	**Power Property of Logarithms**
$= \ln \frac{10^3}{8}$	**Quotient Property of Logarithms**
$= \ln 125$	**Simplify.**
$= \ln 5^3$	$5^3 = 125$
$= 3 \ln 5$	**Power Property of Logarithms**

CHECK Use a calculator to verify the solution.

KEYSTROKES: 3 [LN] 10 [)] [−] [LN] 8 [)] [ENTER]

KEYSTROKES: 3 [LN] 5 [)] [ENTER] 4.828313737 ✓

b. $\ln 40 + 2 \ln \frac{1}{2} + \ln x$

$\ln 40 + 2 \ln \frac{1}{2} + \ln x = \ln 40 + \ln \frac{1}{4} + \ln x$	**Power Property of Logarithms**
$= \ln \left(40 \cdot \frac{1}{4} \cdot x\right)$	**Product Property of Logarithms**
$= \ln 10x$	**Simplify.**

Check Your Progress

3A. $6 \ln 8 - 2 \ln 4$

3B. $2 \ln 5 + 4 \ln 2 + \ln 5y$

Personal Tutor glencoe.com

StudyTip

Look Back Refer to Lesson 7-2 to review **inverse functions**.

Because the natural base and natural log are inverse functions, they can be used to *undo* or eliminate each other.

$$e^{\ln x} = x \qquad \ln e^x = x$$

Equations and Inequalities with e and ln Equations and inequalities involving base e are easier to solve by using natural logarithms rather than by using common logarithms, because $\ln e = 1$.

EXAMPLE 4 Solve Base e Equations

Solve $4e^{-2x} - 5 = 3$. Round to the nearest ten-thousandth.

$4e^{-2x} - 5 = 3$	Original equation
$4e^{-2x} = 8$	Add 5 to each side.
$e^{-2x} = 2$	Divide each side by 4.
$\ln e^{-2x} = \ln 2$	Property of Equality for Logarithms
$-2x = \ln 2$	$\ln e^x = x$
$x = \frac{\ln 2}{-2}$	Divide each side by −2.
$x \approx -0.3466$	Use a calculator.

KEYSTROKES: [LN] 2 [ )] [÷] −2 [ENTER] −.34657359

StudyTip

Calculators Most calculators have an e^x and LN key for evaluating natural base and natural log expressions.

Check Your Progress

Solve each equation. Round to the nearest ten-thousandth.

4A. $3e^{4x} - 12 = 15$

4B. $4e^{-x} + 8 = 17$

Personal Tutor glencoe.com

Just like the natural logarithm can be used to eliminate e^x, the natural base exponential function can eliminate $\ln x$.

EXAMPLE 5 Solve Natural Log Equations and Inequalities

Solve each equation or inequality. Round to the nearest ten-thousandth.

a. $3 \ln 4x = 24$

$3 \ln 4x = 24$	Original equation
$\ln 4x = 8$	Divide each side by 3.
$e^{\ln 4x} = e^8$	Property of Equality for Exponential Functions
$4x = e^8$	$e^{\ln x} = x$
$x = \frac{e^8}{4}$	Divide each side by 4.
$x \approx 745.2395$	Use a calculator.

b. $\ln (x - 8)^4 < 4$

$\ln (x - 8)^4 < 4$	Original equation
$e^{\ln (x-8)^4} < e^4$	Write each side using exponents and base e.
$(x - 8)^4 < e^4$	$e^{\ln x} = x$
$x - 8 < e$	Property of Equality for Exponential Functions
$x < e + 8$	Add 8 to each side.
$x < 10.7183$	Use a calculator.

Check Your Progress

Solve each equation or inequality. Round to four decimal places.

5A. $5 \ln 6x = 8$

5B. $\ln (2x - 3)^3 > 6$

Personal Tutor glencoe.com

Real-World Link

The average cost of tuition at four-year public colleges is about $6000 per year.

Source: CNN

Interest compounded continuously can be found using e.

Key Concept — **Continuously Compounded Interest** — For Your FOLDABLE

Calculate continuously compounded interest using the following formula.

$$A = Pe^{rt},$$

where A is the amount in the account after t years, P is the principal amount invested, and r is the annual interest rate.

Real-World EXAMPLE 6 Solve Base e Inequalities

COLLEGE FUNDS When Angelina was born, her grandparents deposited $3000 into a college savings account paying 4% interest compounded continuously.

a. Assuming there are no deposits or withdrawals from the account, what will the balance be after 10 years?

$A = Pe^{rt}$	**Continuous Compounding Formula**
$= 3000e^{(0.04)(10)}$	**$P = 3000$, $r = 0.04$, and $t = 10$**
$= 3000e^{0.4}$	**Simplify.**
≈ 4475.47	**Use a calculator.**

The balance will be $4475.47.

b. How long will it take the balance to reach at least $10,000?

$A < Pe^{rt}$	**Continuous Compounding Formula**
$10{,}000 < 3000e^{(0.04)t}$	**$P = 3000$, $r = 0.04$, and $A = 10{,}000$**
$\frac{10}{3} < e^{0.04t}$	**Divide each side by 3000.**
$\ln \frac{10}{3} < \ln e^{0.04t}$	**Property of Equality of Logarithms**
$\ln \frac{10}{3} < 0.04t$	**$\ln e^x = x$**
$\frac{\ln \frac{10}{3}}{0.04} < t$	**Divide each side by 0.04.**
$30.099 < t$	**Use a calculator.**

It will take about 30 years to reach at least $10,000.

c. If her grandparents want Angelina to have $10,000 after 18 years, how much would they need to invest?

$10{,}000 = Pe^{(0.04)18}$	**$A = 10{,}000$, $r = 0.04$, and $t = 18$**
$\frac{10{,}000}{e^{0.72}} = P$	**Divide each side by $e^{0.72}$.**
$4867.52 \approx P$	**Use a calculator.**

They need to invest $4867.52.

StudyTip

Rounding Not rounding until the very end will provide a more accurate answer.

Check Your Progress

6. Use the information in Example 6 to answer the following.

A. If they invested $8000 at 3.75% interest compounded continuously, how much money would be in the account in 30 years?

B. If they could only deposit $10,000 in the account above, at what rate would the account need to grow in order for Angelina to have $30,000 in 18 years?

C. If Angelina's grandparents found an account that paid 5% compounded continuously and wanted her to have $30,000 after 18 years, how much would they need to deposit?

Personal Tutor glencoe.com

Check Your Understanding

Examples 1 and 2 pp. 525–526

Write an equivalent exponential or logarithmic function.

1. $e^x = 30$

2. $\ln x = 42$

3. $e^3 = x$

4. $\ln 18 = x$

Example 3 p. 526

Write each as a single logarithm.

5. $3 \ln 2 + 2 \ln 4$

6. $5 \ln 3 - 2 \ln 9$

7. $3 \ln 6 + 2 \ln 9$

8. $3 \ln 5 + 4 \ln x$

Example 4 p. 527

Solve each equation. Round to the nearest ten-thousandth.

9. $5e^x - 24 = 16$

10. $-3e^x + 9 = 4$

11. $3e^{-3x} + 4 = 6$

12. $2e^{-x} - 3 = 8$

Example 5 p. 527

Solve each equation or inequality. Round to the nearest ten-thousandth.

13. $\ln 3x = 8$

14. $-4 \ln 2x = -26$

15. $\ln (x + 5)^2 < 6$

16. $\ln (x - 2)^3 > 15$

17. $e^x > 29$

18. $5 + e^{-x} > 14$

Example 6 p. 528

19. **SCIENCE** A virus is spreading through a computer network according to the formula $v(t) = 30e^{0.1t}$, where v is the number of computers infected and t is the time in minutes. How long will it take the virus to infect 10,000 computers?

Practice and Problem Solving

● = **Step-by-Step Solutions** begin on page R20.
Extra Practice begins on page 947.

Examples 1 and 2 pp. 525–526

Write an equivalent exponential or logarithmic function.

20. $e^{-x} = 8$

21. $e^{-5x} = 0.1$

22. $\ln 0.25 = x$

23. $\ln 5.4 = x$

24. $e^{x-3} = 2$

25. $\ln (x + 4) = 36$

26. $e^{-2} = x^6$

27. $\ln e^x = 7$

Example 3 p. 526

Write each as a single logarithm.

28. $\ln 125 - 2 \ln 5$

29. $3 \ln 10 + 2 \ln 100$

30. $4 \ln \frac{1}{3} - 6 \ln \frac{1}{9}$

31 ● $7 \ln \frac{1}{2} + 5 \ln 2$

32. $8 \ln x - 4 \ln 5$

33. $3 \ln x^2 + 4 \ln 3$

Example 4 p. 527

Solve each equation. Round to the nearest ten-thousandth.

34. $6e^x - 3 = 35$

35. $4e^x + 2 = 180$

36. $3e^{2x} - 5 = -4$

37. $-2e^{3x} + 19 = 3$

38. $6e^{4x} + 7 = 4$

39. $-4e^{-x} + 9 = 2$

Examples 5 and 6 pp. 527–528

40. **DEPRECIATION** The value of a certain car depreciates according to $v(t) = 18500e^{-0.186t}$, where t is the number of years after the car is purchased new.

a. What will the car be worth in 18 months?

b. When will the car be worth half of its original value?

c. When will the car be worth less than \$1000?

Solve each inequality. Round to the nearest ten-thousandth.

41. $e^x \leq 8.7$

42. $e^x \geq 42.1$

43. $\ln (3x + 4)^3 > 10$

44. $4 \ln x^2 < 72$

45. $\ln (8x^4) > 24$

46. $-2 \ln (x - 6)^{-1} \leq 6$

47. **SAVINGS** Use the formula for continuously compounded interest.

a. If you deposited \$800 in an account paying 4.5% interest compounded continuously, how much money would be in the account in 5 years?

b. How long would it take you to double your money?

c. If you want to double your money in 9 years, what rate would you need?

d. If you want to open an account that pays 4.75% interest compounded continuously and have \$10,000 in the account 12 years after your deposit, how much would you need to deposit?

Real-World Link

Recently, the U.S. personal average saving rate was −1%. This means that Americans withdrew more money than they deposited.

Source: *Business Week*

Write the expression as a sum or difference of logarithms or multiples of logarithms.

48. $\ln 12x^2$ **49.** $\ln \frac{16}{125}$ **50.** $\ln \sqrt[5]{x^3}$ **51.** $\ln xy^4z^{-3}$

Use the natural logarithm to solve each equation.

52. $8^x = 24$ **53.** $3^x = 0.4$ **54.** $2^{3x} = 18$ **55.** $5^{2x} = 38$

56. **SCIENCE** Newton's Law of Cooling, which can be used to determine how fast an object will cool in given surroundings, is represented by $T(t) = T_s + T_0e^{-kt}$, where T_0 is the initial temperature of the object, T_s is the temperature of the surroundings, t is the time in minutes, and k is a constant value that depends on the type of object.

a. If a cup of coffee with an initial temperature of 180° is placed in a room with a temperature of 70°, then the coffee cools to 140° after 10 minutes, find the value of k.

b. Use this value of k to determine the temperature of the coffee after 20 minutes.

c. When will the temperature of the coffee reach 75°?

57. **MULTIPLE REPRESENTATIONS** In this problem, you will use $f(x) = e^x$ and $g(x) = \ln x$.

a. **GRAPHICAL** Graph both functions and their axis of symmetry, $y = x$, for $-5 \le x \le 5$. Then graph $a(x) = e^{-x}$ on the same graph.

b. **ANALYTICAL** The graphs of $a(x)$ and $f(x)$ are reflections along which axis? What function would be a reflection of $f(x)$ along the other axis?

c. **LOGICAL** Determine the two functions that are reflections of $g(x)$. Graph these new functions.

d. **VERBAL** We know that $f(x)$ and $g(x)$ are inverses. Are any of the other functions that we have graphed inverses as well? Explain your reasoning.

H.O.T. Problems

Use Higher-Order Thinking Skills

58. **CHALLENGE** Solve $4^x - 2^{x+1} = 15$ for x.

59. **PROOF** Prove $\ln ab = \ln a + \ln b$ for natural logarithms.

60. **REASONING** Determine whether $x > \ln x$ is *sometimes*, *always*, or *never* true. Explain your reasoning.

61. **OPEN ENDED** Express the value 3 using e^x and the natural log.

62. **WRITING IN MATH** Explain how the natural log can be used to solve a natural base exponential function.

NYSRE PRACTICE Reinforcement of A.G.5

63. Given the function $y = 2.34x + 11.33$, which statement best describes the effect of moving the graph down two units?

A The x-intercept decreases.
B The y-intercept decreases.
C The x-intercept remains the same.
D The y-intercept remains the same.

64. GRIDDED RESPONSE Aidan sells engraved necklaces over the Internet. He purchases 50 necklaces for \$400, and it costs him an additional \$3 for each personalized engraving. If he charges \$20 each, how many necklaces will he need to sell in order to make a profit of at least \$225?

65. Solve $|2x - 5| = 17$.

F $-6, -11$
G $-6, 11$
H $6, -11$
J $6, 11$

66. A local pet store sells rabbit food. The cost of two 5-pound bags is \$7.99. The total cost c of purchasing n bags can be found by—

A multiplying n by c.
B multiplying n by 5.
C multiplying n by the cost of 1 bag.
D dividing n by c.

Spiral Review

Solve each equation or inequality. Round to the nearest ten-thousandth. (Lesson 8-6)

67. $2^x = 53$

68. $2.3^{x^2} = 66.6$

69. $3^{4x - 7} < 4^{2x + 3}$

70. $6^{3y} = 8^{y - 1}$

71. $12^{x - 5} \geq 9.32$

72. $2.1^{x - 5} = 9.32$

73. SOUND Use the formula $L = 10 \log_{10} R$, where L is the loudness of a sound and R is the sound's relative intensity. Suppose the sound of one alarm clock is 80 decibels. Find out how much louder 10 alarm clocks would be than one alarm clock. (Lesson 8-5)

Given a polynomial and one of its factors, find the remaining factors of the polynomial. Some factors may not be binomials. (Lesson 6-6)

74. $x^3 + 5x^2 + 8x + 4;\ x + 1$

75. $x^3 + 4x^2 + 7x + 6;\ x + 2$

76. CRAFTS Mrs. Hall is selling crocheted items. She sells large afghans for \$60, baby blankets for \$40, doilies for \$25, and pot holders for \$5. She takes the following number of items to the fair: 12 afghans, 25 baby blankets, 45 doilies, and 50 pot holders. (Lesson 4-3)

a. Write an inventory matrix for the number of each item and a cost matrix for the price of each item.

b. Suppose Mrs. Hall sells all of the items. Find her total income as a matrix.

Skills Review

Solve each equation. (Lesson 9-2)

77. $2^{3x + 5} = 128$

78. $5^{n - 3} = \frac{1}{25}$

79. $\left(\frac{1}{9}\right)^m = 81^{m + 4}$

80. $\left(\frac{1}{7}\right)^{y - 3} = 343$

81. $10^{x - 1} = 100^{2x - 3}$

82. $36^{2p} = 216^{p - 1}$

EXPLORE **8-8**

Spreadsheet Lab
Compound Interest

NY Math Online glencoe.com
Other Calculator Keystrokes

You can use a spreadsheet to organize and display data. A spreadsheet is an easy way to track the amount of interest earned over a period of time.

Compound interest is earned not only on the original amount, but also on any interest that has been added to the principal.

ACTIVITY

Find the total amount of money after 5 years if you deposit $100 at 7% compounded annually.

Step 1 Label your columns as shown. The period is one year. Enter the starting values and the rate.

Step 2 Each row will be generated using formulas. Enter the formulas as shown.

Savings Account.xls

◇	A	B	C	D	E
1	End of Period	Principal	Interest	Balance	Rate per Period
2	0			$100.00	7%
3	=A2+1	=D2	=B3*E2	=C3+D2	

Sheet 1 | Sheet 2 | Sheet 3

Step 3 Use the FILL DOWN function to fill 4 additional rows.

Savings Account.xls

◇	A	B	C	D	E
1	End of Period	Principal	Interest	Balance	Rate per Period
2	0			$100.00	7%
3	1	$100.00	$7.00	$107.00	
4	2	$107.00	$7.49	$114.49	
5	3	$114.49	$8.01	$122.50	
6	4	$122.50	$8.58	$131.08	
7	5	$131.08	$9.18	$140.26	

Sheet 1 | Sheet 2 | Sheet 3

If you deposit $100 at 7% annual interest for 5 years, you will have $140.26 at the end of the 5 years.

Exercises

Find the total balance for each situation.

1. deposit $500 for 7 years at 5%

2. deposit $1000 for 5 years at 6%

3. deposit $200 for 2 years at 10%

4. deposit $800 for 3 years at 8%

5. borrow $10,000 for 5 years at 5.05%

6. borrow $25,000 for 30 years at 8%

8-8 Using Exponential and Logarithmic Functions

Then

You used exponential growth and decay formulas. (Lesson 8-1)

Now

- Use logarithms to solve problems involving exponential growth and decay.
- Use logarithms to solve problems involving logistic growth.

NYS Core Curriculum

A2.A.12 Evaluate exponential expressions, including those with base *e*

New Vocabulary

rate of continuous growth

rate of continuous decay

logistic growth model

NY Math Online

glencoe.com

- Extra Examples
- Personal Tutor
- Self-Check Quiz
- Homework Help

Why?

The ancient footprints of Acahualinca, discovered in Managua, Nicaragua, are believed to be the oldest human footprints in the world. Using carbon dating, scientists estimate that these footprints are 6000 years old.

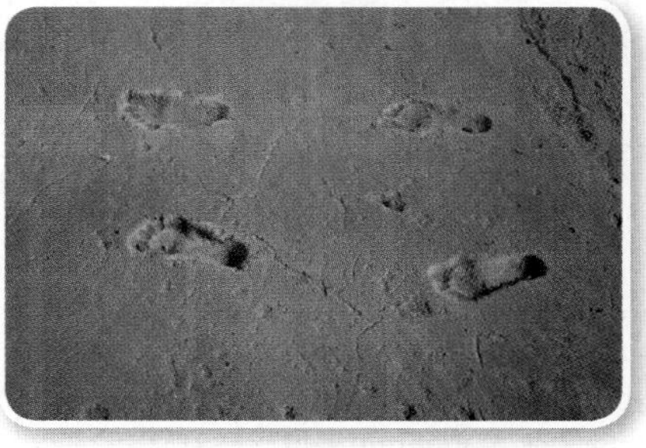

Exponential Growth and Decay Scientists and researchers frequently use alternate forms of the growth and decay formulas that you learned in Lesson 8-1.

Key Concept: Exponential Growth and Decay (For Your FOLDABLE)

Exponential Growth	Exponential Decay
Exponential growth can be modeled by the function $f(x) = ae^{kt}$, where a is the initial value, t is time in years, and k is a constant representing the **rate of continuous growth**.	Exponential decay can be modeled by the function $f(x) = ae^{-kt}$, where a is the initial value, t is time in years, and k is a constant representing the **rate of continuous decay**.

Real-World EXAMPLE 1 Exponential Decay

SCIENCE The half-life of a radioactive substance is the time it takes for half of the atoms of the substance to disintegrate. The half-life of Carbon-14 is 5730 years. Determine the value of k and the equation of decay for Carbon-14.

If a is the initial amount of the substance, then the amount y that remains after 5730 years can be represented by 12a or $0.5a$.

$y = ae^{-kt}$	**Exponential Decay Formula**
$0.5a = ae^{-k(5730)}$	**$y = 0.5a$ and $t = 5730$**
$0.5 = e^{-5730k}$	**Divide each side by a.**
$\ln 0.5 = \ln e^{-5730k}$	**Property of Equality for Logarithmic Functions**
$\ln 0.5 = -5730k$	**$\ln e^x = x$**
$\frac{\ln 0.5}{-5730} = k$	**Divide each side by −5730.**
$0.00012 \approx k$	**Use a calculator.**

Thus, the equation for the decay of Carbon-14 is $y = ae^{-0.00012t}$.

Check Your Progress

1. The half-life of Plutonium-239 is 24,000 years. Determine the value of k.

Personal Tutor glencoe.com

Now that the value of k for Carbon-14 is known, it can be used to date fossils.

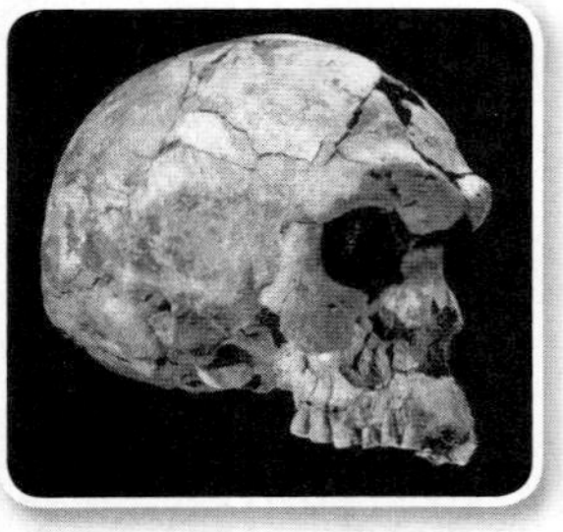

Real-World Link

The oldest modern human fossil, found in Ethiopia, is approximately 160,000 years old.

Source: National Public Radio

StudyTip

Carbon Dating When given a percent or fraction of decay, use an original amount of a or 1.

Real-World EXAMPLE 2 Carbon Dating

SCIENCE A paleontologist examining the bones of a prehistoric animal estimates that they contain 2% as much Carbon-14 as they would have contained when the animal was alive.

a. How long ago did the animal live?

Understand The formula for the decay of Carbon-14 is $y = ae^{kt}$. You want to find out how long ago the animal lived.

Plan Let a be the initial amount of Carbon-14 in the animal's body. The amount y that remains after t years is 2% of a or $0.02a$.

Solve

$y = ae^{-0.00012t}$	**Formula for the decay of Carbon-14**
$0.02a = ae^{-0.00012t}$	**$y = 0.02a$**
$0.02 = e^{-0.00012t}$	**Divide each side by a.**
$\ln 0.02 = \ln e^{-0.00012t}$	**Property of Equality for Logarithmic Functions**
$\ln 0.02 = -0.00012t$	**$\ln e^x = x$**
$\frac{\ln 0.02}{-0.00012} = t$	**Divide each side by -0.00012.**
$32{,}600 \approx t$	**Use a calculator.**

The animal lived about 32,600 years ago.

Check Use the formula to find the amount of a sample remaining after 32,600 years. Use an original amount of 1.

$y = ae^{-0.00012t}$	**Original equation**
$= 1e^{-0.00012(32{,}600)}$	**$a = 1$ and $t = 32{,}600$**
≈ 0.02 or 2% ✓	**Use a calculator.**

b. If prior research points to the animal being around 20,000 years old, how much Carbon-14 should be in the animal?

$y = ae^{-0.00012t}$	**Formula for the decay of Carbon-14**
$= 1e^{-0.00012(20{,}000)}$	**$a = 1$ and $t = 20{,}000$**
$= e^{-2.4}$	**Simplify.**
$= 0.09$ or 9%	**Use a calculator.**

Check Your Progress

2. Use the information in Example 2 to answer the following.

A. A specimen that originally contained 42 milligrams of Carbon-14 now contains 8 milligrams. How old is the fossil?

B. A wooly mammoth specimen was thought to be about 12,000 years old. How much Carbon-14 should be in the animal?

Personal Tutor glencoe.com

The exponential growth equation $y = ae^{kt}$ is identical to the continuously compounded interest formula you learned in Lesson 8-7.

Continuous Compounding	Population Growth
$A = Pe^{rt}$	$y = ae^{kt}$
P = initial amount	a = initial population
A = amount at time t	y = population at time t
r = interest rate	k = rate of continuous growth

Problem-SolvingTip

Use a Formula When dealing with population, it is almost always necessary to use an exponential growth or decay formula.

Real-World EXAMPLE 3 Continuous Exponential Growth

POPULATION In 2007, the population of Georgia was 9.36 million people. In 2000, it was 8.18 million.

a. Determine the value of k, Georgia's relative rate of growth.

$y = ae^{kt}$	Formula for continuous exponential growth
$9.36 = 8.18e^{k(7)}$	$y = 9.36$, $a = 8.18$, and $t = 2007 - 2000$ or 7
$\frac{9.36}{8.18} = e^{7k}$	Divide each side by 8.18.
$\ln \frac{9.36}{8.18} = \ln e^{7k}$	Property of Equality for Logarithmic Functions
$\ln \frac{9.36}{8.18} = 7k$	$\ln e^x = x$
$\frac{\ln \frac{9.36}{8.18}}{7} = k$	Divide each side by 7.
$0.01925 = k$	Use a calculator.

Georgia's relative rate of growth is about 0.01925 or about 2%.

b. When will Georgia's population reach 10 million people?

$y = ae^{kt}$	Formula for continuous exponential growth
$10 = 8.18e^{0.01925t}$	$y = 10$, $a = 8.18$, and $k = 0.01925$
$1.2225 = e^{0.01925t}$	Divide each side by 8.18.
$\ln 1.2225 = \ln e^{0.01925t}$	Property of Equality for Logarithmic Functions
$\ln 1.2225 = 0.01925t$	$\ln e^x = x$
$\frac{\ln 1.2225}{0.01925} = t$	Divide each side by 0.01925.
$10.436 \approx t$	Use a calculator.

Georgia's population will reach 10 million people by 2010.

c. Michigan's population in 2000 was 9.9 million and can be modeled by $y = 9.9e^{0.0028t}$. Determine when Georgia's population will surpass Michigan's.

$8.18e^{0.01925t} > 9.9e^{0.0028t}$	Formula for exponential growth
$\ln 8.18e^{0.01925t} > \ln 9.9e^{0.0028t}$	Property of Inequality for Logarithms
$\ln 8.18 + \ln e^{0.01925t} > \ln 9.9 + \ln e^{0.0028t}$	Product Property of Logarithms
$\ln 8.18 + 0.01925t > \ln 9.9 + 0.0028t$	$\ln e^x = x$
$0.01645t > \ln 9.9 - \ln 8.18$	Subtract $(0.0028t + \ln 8.18)$ from each side.
$t > \frac{\ln 9.9 - \ln 8.18}{0.01645}$	Divide each side by 0.01645.
$t > 11.6$	Use a calculator.

Georgia's population will surpass Michigan's by the year 2012.

StudyTip

Rounding Error In order to avoid any errors due to rounding, do not round until the very end of your calculations.

Check Your Progress

3. **BIOLOGY** A type of bacteria is growing exponentially according to the model $y = 1000e^{kt}$, where t is the time in minutes.

   A. If there are 1000 cells initially and 1650 cells after 40 minutes, find the value of k for the bacteria.

   B. Suppose a second type of bacteria is growing exponentially according to the model $y = 50e^{0.0432t}$. Determine how long it will be before the number of cells of this bacteria exceed the number of cells in the other bacteria.

Personal Tutor glencoe.com

Logistic Growth Refer to the equation representing Georgia's population in Example 3. According to the graph at the right, Georgia's population will be about one billion by the year 2130. Does this seem logical?

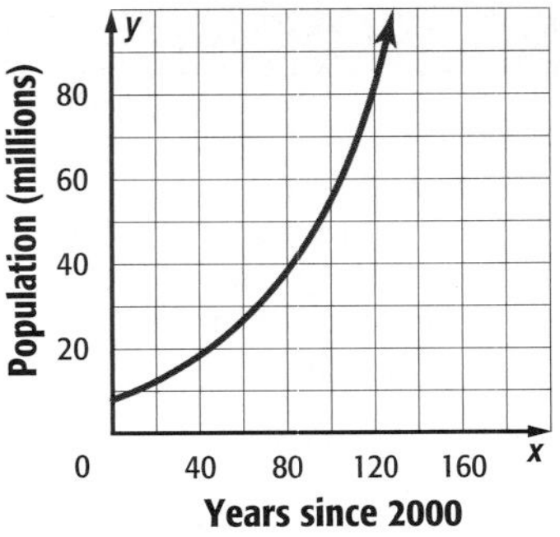

Populations cannot grow infinitely large. There are limitations, such as food supplies, war, living space, diseases, available resources, and so on.

Exponential growth is unrestricted, meaning it will increase without bound. A **logistic growth model**, however, represents growth that has a limiting factor. Logistic models are the most accurate models for representing population growth.

Key Concept — Logistic Growth Function

For Your FOLDABLE

Let a, b, and c be positive constants where $b < 1$. The logistic growth function is represented by $f(t) = \frac{c}{1 + ae^{-bt}}$, where t represents time.

Real-World Link

Phoenix is the fifth largest city in the country and has a population of 1.5 million.

EXAMPLE 4 Logistic Growth

The population of Phoenix, Arizona, in millions can be modeled by the logistic function $f(t) = \frac{2.0666}{1 + 1.66e^{-0.048t}}$, where t is the number of years after 1980.

a. Graph the function for $0 \leq t \leq 500$.

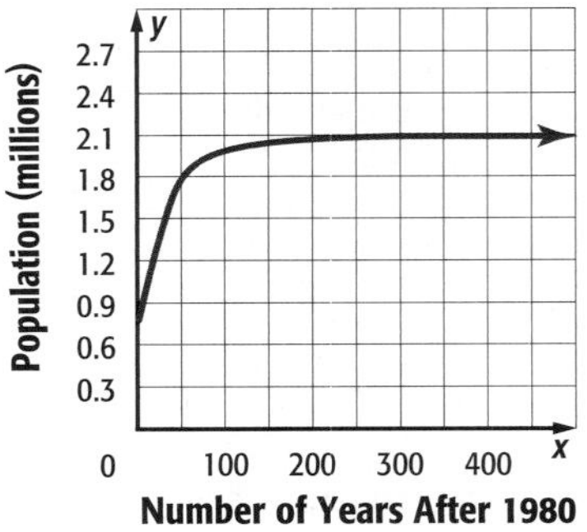

b. What is the horizontal asymptote?

The horizontal asymptote is at $y = 2.0666$.

c. Will the population of Phoenix increase indefinitely? If not, what will be their maximum population?

No. The population will reach a maximum of a little less than 2.0666 million people.

d. According to the function, when will the population of Phoenix reach 1.8 million people?

The graph indicates the population will reach 1.8 million people at $t \approx 50$. Replacing $f(t)$ with 1.8 and solving for t in the equation yields $t = 50.35$ years. So, the population of Phoenix will reach 1.8 million people by 2031.

StudyTip

Intersections To determine where the graph intersects 1.8 on the calculator, graph $y = 1.8$ on the same graph and select *intersection* in the CALC menu.

Check Your Progress

4. The population of a certain species of fish in a lake after t years can be modeled by the function $P(x) = \frac{1880}{1 + 1.42e^{-0.037t}}$, where $t \geq 0$.

A. Graph the function for $0 \leq t \leq 500$.

B. What is the horizontal asymptote?

C. What is the maximum population of the fish in the lake?

D. When will the population reach 1875?

Personal Tutor glencoe.com

Check Your Understanding

Examples 1 and 2 pp. 533–534

1. **PALEONTOLOGY** The half-life of Potassium-40 is about 1.25 billion years.
   a. Determine the value of k and the equation of decay for Potassium-40.
   b. A specimen currently contains 36 milligrams of Potassium-40. How long will it take the specimen to decay to only 15 milligrams of Potassium-40?
   c. How many milligrams of Potassium-40 will be left after 300 million years?
   d. How long will it take Potassium-40 to decay to one eighth of its original amount?

Example 3 p. 535

2. **SCIENCE** A certain food is dropped on the floor and is growing bacteria exponentially according to the model $y = 2e^{kt}$, where t is the time in seconds.
   a. If there are 2 cells initially and 8 cells after 20 seconds, find the value of k for the bacteria.
   b. The "5-second rule" says that if a person who drops food on the floor eats it within 5 seconds, there will be no harm. How much bacteria is on the food after 5 seconds?
   c. Would you eat food that had been on the floor for 5 seconds? Why or why not? Do you think that the information you obtained in this exercise is reasonable? Explain.

Example 4 p. 536

3. **ZOOLOGY** Suppose the red fox population in a restricted habitat follows the function $P(t) = \frac{16,500}{1 + 18e^{-0.085t}}$, where t represents the time in years.
   a. Graph the function for $0 \leq t \leq 200$.
   b. What is the horizontal asymptote?
   c. What is the maximum population?
   d. When does the population reach 16,450?

Practice and Problem Solving

● = **Step-by-Step Solutions** begin on page R20.
Extra Practice begins on page 947.

Examples 1 and 2 pp. 533–534

4. **SCIENCE** The half-life of Rubidium-87 is about 48.8 billion years.
   a. Determine the value of k and the equation of decay for Rubidium-87.
   b. A specimen currently contains 50 milligrams of Rubidium-87. How long will it take the specimen to decay to only 18 milligrams of Rubidium-87?
   c. How many milligrams of Rubidium-87 will be left after 800 million years?
   d. How long will it take Rubidium-87 to decay to one-sixteenth its original amount?

Example 3 p. 535

5. **BIOLOGY** A certain bacteria is growing exponentially according to the model $y = 80e^{kt}$, where t is the time in minutes.
   a. If there are 80 cells initially and 675 cells after 30 minutes, find the value of k for the bacteria.
   b. When will the bacteria reach a population of 6000 cells?
   c. If a second type of bacteria is growing exponentially according to the model $y = 35e^{0.0978t}$, determine how long it will be before the number of cells of this bacteria exceed the number of cells in the other bacteria.

Example 4 p. 536

6. **FORESTRY** The population of trees in a certain forest follows the function $f(t) = \frac{18000}{1 + 16e^{-0.084t}}$, where t is the time in years.
   a. Graph the function for $0 \leq t \leq 100$.
   b. When does the population reach 17500 trees?

7. **PALEONTOLOGY** A paleontologist finds a human bone and determines that the Carbon-14 found in the bone is 85% of that found in living bone tissue. How old is the bone?

8. **ANTHROPOLOGY** An anthropologist has determined that a newly discovered human bone is 8000 years old. How much of the original amount of Carbon-14 is in the bone?

9. **RADIOACTIVE DECAY** 100 milligrams of Uranium-238 are stored in a container. If Uranium-238 has a half-life of about 4.47 billion years, after how many years will only 10 milligrams be present?

Real-World Link

Oregon is known as "The Beaver State," and has a beaver pictured on the reverse side of its state flag. Oregon is the only state in the union that has a different pattern on the reverse side of its flag.

Source: *Oregon Blue Book*

10. **POPULATION GROWTH** The population of the state of Oregon has grown from 3.4 million in 2000 to 3.7 million in 2006.

a. Write an exponential growth equation of the form $y = ae^{kt}$ for Oregon, where t is the number of years after 2000.

b. Use your equation to predict the population of Oregon in 2020.

c. According to the equation, when will Oregon reach 6 million people?

11. **HALF-LIFE** A substance decays 99.9% of its total mass after 200 years. Determine the half-life of the substance.

12. **LOGISTIC GROWTH** The population in millions of the state of Ohio after 1900 can be modeled by $P(t) = \frac{12.95}{1 + 2.4e^{-kt}}$, where t is the number of years after 1900 and k is a constant.

a. If Ohio had a population of 10 million in 1970, find the value of k.

b. According to the equation, when will the population of Ohio reach 12 million?

13. **MULTIPLE REPRESENTATIONS** In this problem, you will explore population growth. The population growth of a country follows the exponential function $f(t) = 8e^{0.075t}$ or the logistic function $g(t) = \frac{400}{1 + 16e^{-0.025t}}$. The population is measured in millions and t is time in years.

a. **GRAPHICAL** Graph both functions for $0 \leq t \leq 100$.

b. **ANALYTICAL** Determine the intersection of the graphs. What is the significance of this intersection?

c. **ANALYTICAL** Which function is a more accurate estimate of the country's population 100 years from now? Explain your reasoning.

H.O.T. Problems

Use Higher-Order Thinking Skills

14. **OPEN ENDED** Give an example of a quantity that grows or decays at a fixed rate. Write a real-world problem involving the rate and solve by using logarithms.

15. **CHALLENGE** Solve $\frac{120{,}000}{1 + 48e^{-0.015t}} = 24e^{0.055t}$ for t.

16. **REASONING** Explain mathematically why $f(t) = \frac{c}{1 + 60e^{-0.5t}}$ approaches, but never reaches the value of c as $t \to +\infty$.

17. **OPEN ENDED** Give an example of a quantity that grows logistically and has limitations to growth. Explain why the quantity grows in this manner.

18. **WRITING IN MATH** Summarize the differences between exponential, continuous exponential, and logistic growth.

NYSRE PRACTICE A2.S.12

19. Kareem is making a circle graph showing the favorite ice cream flavors of customers at his store. The table summarizes the data. What central angle should Kareem use for the section representing chocolate?

Flavor	Customers
chocolate	35
vanilla	42
strawberry	7
mint chip	12
butter pecan	4

A 35 **C** 126
B 63 **D** 150

20. PROBABILITY Lydia has 6 books on her bookshelf. Two are literature books, one is a science book, two are math books, and one is a dictionary. What is the probability that she randomly chooses a science book and the dictionary?

F $\frac{1}{3}$ **H** $\frac{1}{12}$
G $\frac{1}{4}$ **J** $\frac{1}{15}$

21. ACT/SAT Peter has made a game for his daughter's birthday party. The playing board is a circle divided evenly into 8 sectors. If the circle has a radius of 18 inches, what is the approximate area of one of the sectors?

A 4 in^2 **C** 127 in^2
B 32 in^2 **D** 254 in^2

22. STATISTICS In a survey of 90 physical trainers, 15 said they went for a run at least 5 times per week. Of that group, 5 said they also swim during the week, and at least 25% run and swim every week. Which conclusion is valid based on the information given?

F The report is accurate because 15 out of 90 is 25%.
G The report is accurate because 5 out of 15 is 33%, which is at least 25%.
H The report is inaccurate because 5 out of 90 is only 5.6%.
J The report is inaccurate because no one knows if swimming is really exercising.

Spiral Review

Write an equivalent exponential or logarithmic equation. (Lesson 8-7)

23. $e^7 = y$ **24.** $e^{2n-4} = 36$ **25.** $\ln 5 + 4 \ln x = 9$

26. EARTHQUAKES The table shows the magnitude of some major earthquakes. (Lessons 8-5 and 8-6)

Year	Location	Magnitude
1939	Turkey	8.0
1963	Yugoslavia	6.0
1970	Peru	7.8
1988	Armenia	7.0
2004	Morocco	6.4

a. For which two earthquakes was the intensity of one 10 times that of the other? For which two was the intensity of one 100 times that of the other?

b. What would be the magnitude of an earthquake that is 1000 times as intense as the 1963 earthquake in Yugoslavia?

c. Suppose you know that $\log_7 2 \approx 0.3562$ and $\log_7 3 \approx 0.5646$. Describe two different methods that you could use to approximate $\log_7 2.5$. (You may use a calculator, of course.) Then describe how you can check your result.

Skills Review

Solve each equation. Write in simplest form. (Lesson 1-3)

27. $\frac{8}{5}x = \frac{4}{15}$ **28.** $\frac{27}{14}n = \frac{6}{7}$ **29.** $\frac{3}{10} = \frac{12}{25}a$ **30.** $\frac{6}{7} = 9p$

31. $\frac{9}{8}b = 18$ **32.** $\frac{6}{7}y = \frac{3}{4}$ **33.** $\frac{1}{3}z = \frac{5}{6}$ **34.** $\frac{2}{3}q = 7$

Graphing Technology Lab
Cooling

NY Math Online > glencoe.com
Other Calculator Keystrokes

In this lab, you will explore the type of equation that models the change in the temperature of water as it cools under various conditions.

Set Up the Lab

- Collect a variety of containers, such as a foam cup, a ceramic coffee mug, and an insulated cup.
- Boil water or collect hot water from a tap.
- Choose a container to test and fill with hot water. Place the temperature probe in the cup.
- Connect the temperature probe to your data collection device.

ACTIVITY Description

Step 1 Program the device to collect 20 or more samples in 1 minute intervals.

Step 2 Wait a few seconds for the probe to warm to the temperature of the water.

Step 3 Press the button to begin collecting data.

Analyze the Results

1. When the data collection is complete, graph the data in a scatter plot. Use time as the independent variable and temperature as the dependent variable. Write a sentence that describes the points on the graph.
2. Use the STAT menu to find an equation to model the data you collected. Try linear, quadratic, and exponential models. Which model appears to fit the data best? Explain.
3. Would you expect the temperature of the water to drop below the temperature of the room? Explain your reasoning.
4. Use the data collection device to find the temperature of the air in the room. Graph the function $y = t$, where t is the temperature of the room, along with the scatter plot and the model equation. Describe the relationship among the graphs. What is the meaning of the relationship in the context of the experiment?

Make a Conjecture

5. Do you think the results of the experiment would change if you used an insulated container for the water? Repeat the experiment to verify your conjecture.
6. How might the results of the experiment change if you added ice to the water? Repeat the experiment to verify your conjecture.

Chapter Summary

Key Concepts

Exponential Functions (Lessons 8-1 and 8-2)

- An exponential function is in the form $y = ab^x$, where $a \neq 0$, $b > 0$ and $b \neq 1$.
- Property of Equality for Exponential Functions: If b is a positive number other than 1, then $b^x = b^y$ if and only if $x = y$.
- Property of Inequality for Exponential Functions: If $b > 1$, then $b^x > b^y$ if and only if $x > y$, and $b^x < b^y$ if and only if $x < y$.

Logarithms and Logarithmic Functions (Lessons 8-3 through 8-6)

- Suppose $b > 0$ and $b \neq 1$. For $x > 0$, there is a number y such that $\log_b x = y$ if and only if $b^y = x$.
- The logarithm of a product is the sum of the logarithms of its factors.
- The logarithm of a quotient is the difference of the logarithms of the numerator and the denominator.
- The logarithm of a power is the product of the logarithm and the exponent.
- The Change of Base Formula: $\log_a n = \dfrac{\log_b n}{\log_b a}$

Natural Logarithms (Lesson 8-7)

- Since the natural base function and the natural logarithmic function are inverses, these two can be used to "undo" each other.

Using Exponential and Logarithmic Functions (Lesson 8-8)

- Exponential growth can be modeled by the function $f(x) = ae^{kt}$, where k is a constant representing the rate of continuous growth.
- Exponential decay can be modeled by the function $f(x) = ae^{-kt}$, where k is a constant representing the rate of continuous decay.

FOLDABLES® Study Organizer

Be sure the Key Concepts are noted in your Foldable.

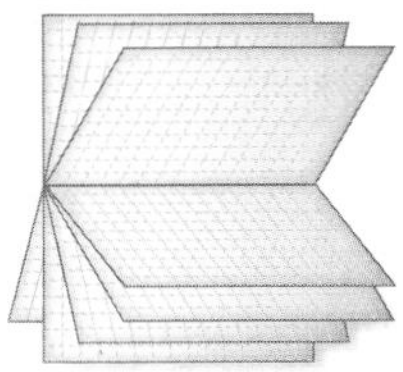

Key Vocabulary

asymptote (p. 475)
Change of Base Formula (p. 518)
common logarithm (p. 516)
compound interest (p. 486)
decay factor (p. 478)
exponential decay (p. 477)
exponential equation (p. 485)
exponential function (p. 475)
exponential growth (p. 475)
exponential inequality (p. 487)
growth factor (p. 477)
logarithmic equation (p. 502)
logarithmic function (p. 493)
logarithmic inequality (p. 503)
logarithm (p. 492)
logistic growth model (p. 536)
natural base, *e* (p. 525)
natural base exponential function (p. 525)
natural logarithm (p. 525)
rate of continuous decay (p. 533)
rate of continuous growth (p. 533)

Vocabulary Check

Choose a word or term from the list above that best completes each statement or phrase.

1. A function of the form $f(x) = b^x$ where $b > 1$ is a(n) ____________ function.
2. In $x = b^y$, the variable y is called the ____________ of x.
3. Base 10 logarithms are called ____________.
4. A(n) ____________ is an equation in which variables occur as exponents.
5. The ____________ allows you to write equivalent logarithmic expressions that have different bases.
6. The base of the exponential function, $A(t) = a(1 - r)^t$, $1 - r$ is called the ____________.
7. The function $y = \log_b x$, where $b > 0$ and $b \neq 1$, is called a(n) ____________.
8. An exponential function with base e is called the ____________.
9. The logarithm with base e is called the ____________.
10. The number e is referred to as the ____________.

Study Guide and Review

Lesson-by-Lesson Review

8-1 Graphing Exponential Functions (pp. 475–482)

A2.A.6, A2.A.53

Graph each function. State the domain and range.

11. $f(x) = 3^x$

12. $f(x) = -5(2)^x$

13. $f(x) = 3(4)^x - 6$

14. $f(x) = 3^{2x} + 5$

15. $f(x) = 3\left(\frac{1}{4}\right)^{x+3} - 1$

16. $f(x) = \frac{3}{5}\left(\frac{2}{3}\right)^{x-2} + 3$

17. POPULATION A city with a population of 120,000 decreases at a rate of 3% annually.

a. Write the function that represents this situation.

b. What will the population be in 10 years?

EXAMPLE 1

Graph $f(x) = -2(3)^x + 1$. State the domain and range.

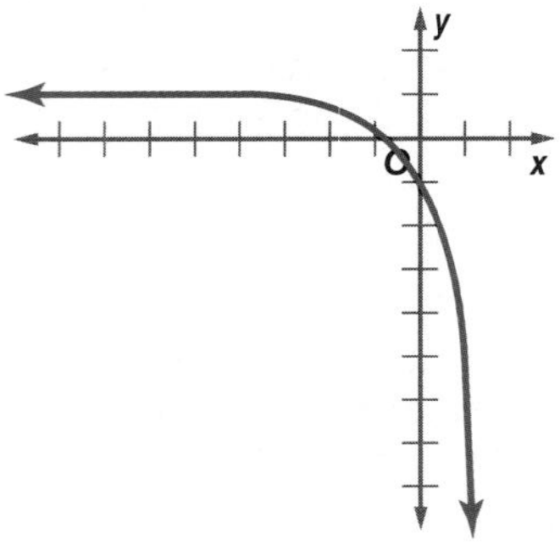

The domain is all real numbers, and the range is all real numbers less than 1.

8-2 Solving Exponential Equations and Inequalities (pp. 485–491)

A2.A.12, A2.A.27

Solve each equation or inequality.

18. $16^x = \frac{1}{64}$

19. $3^{4x} = 9^{3x+7}$

20. $64^{3n} = 8^{2n-3}$

21. $8^{3-3y} = 256^{4y}$

22. $9^{x-2} > \left(\frac{1}{81}\right)^{x+2}$

23. $27^{3x} \le 9^{2x-1}$

24. BACTERIA A bacteria population started with 5000 bacteria. After 8 hours there were 28,000 in the sample.

a. Write an exponential function that could be used to model the number of bacteria after x hours if the number of bacteria changes at the same rate.

b. How many bacteria can be expected in the sample after 32 hours?

EXAMPLE 2

Solve $4^{3x} = 32^{x-1}$ for x.

$4^{3x} = 32^{x-1}$	**Original equation**
$(2^2)^{3x} = (2^5)^{x-1}$	**Rewrite so each side has the same base.**
$2^{6x} = 2^{5x-5}$	**Simplify.**
$6x = 5x - 5$	**Property of Equality for Exponential Functions**
$x = -5$	

The solution is -5.

8-3 Logarithms and Logarithmic Functions (pp. 492–499)

A2.A.41, A2.A.54

25. Write $\log_2 \frac{1}{16} = -4$ in exponential form.

26. Write $10^2 = 100$ in logarithmic form.

Evaluate each expression.

27. $\log_4 256$

28. $\log_2 \frac{1}{8}$

Graph each function.

29. $f(x) = 2\log_{10} x + 4$

30. $f(x) = \frac{1}{6}\log_{\frac{1}{3}}(x - 2)$

EXAMPLE 3

Evaluate $\log_2 64$.

$\log_2 64 = y$	**Let the logarithm equal y.**
$64 = 2^y$	**Definition of logarithm**
$2^6 = 2^y$	**$64 = 2^6$**
$6 = y$	**Property of Equality for Exponential Functions**

MIXED PROBLEM SOLVING
For mixed problem-solving practice, see page 986.

8-4 Solving Logarithmic Equations and Inequalities (pp. 502–507)

A2.A.18, A2.A.28

Solve each equation or inequality.

31. $\log_{16} x = \frac{3}{2}$

32. $\log_2 \frac{1}{64} = x$

33. $\log_4 x < 3$

34. $\log_5 x < -3$

35. $\log_9 (3x - 1) = \log_9 (4x)$

36. $\log_2 (x^2 - 18) = \log_2 (-3x)$

37. $\log_3 (3x + 4) \leq \log_3 (x - 2)$

38. **EARTHQUAKE** The magnitude of an earthquake is measured on a logarithmic scale called the Richter scale. The magnitude M is given by $M = \log_{10} x$, where x represents the amplitude of the seismic wave causing ground motion. How many times as great is the amplitude caused by an earthquake with a Richter scale rating of 10 as an aftershock with a Richter scale rating of 7?

EXAMPLE 4

Solve $\log_{27} x < \frac{2}{3}$.

$\log_{27} x < \frac{2}{3}$	**Original inequality**
$x < 27^{\frac{2}{3}}$	**Logarithmic to Exponential Inequality**
$x < 9$	**Simplify.**

EXAMPLE 5

Solve $\log_5 (p^2 - 2) = \log_5 p$.

$\log_5 (p^2 - 2) = \log_5 p$	**Original equation**
$p^2 - 2 = p$	**Property of Equality**
$p^2 - p - 2 = 0$	**Subtract p from each side.**
$(p - 2)(p + 1) = 0$	**Factor.**
$p - 2 = 0$ or $p + 1 = 0$	**Zero Product Property**
$p = 2$ $\quad$ $p = -1$	**Solve each equation.**

8-5 Properties of Logarithms (pp. 509–515)

A2.A.18, A2.A.19

Use $\log_5 16 \approx 1.7227$ and $\log_5 2 \approx 0.4307$ to approximate the value of each expression.

39. $\log_5 8$

40. $\log_5 64$

41. $\log_5 4$

42. $\log_5 \frac{1}{8}$

43. $\log_5 \frac{1}{2}$

Solve each equation. Check your solution.

44. $\log_5 x - \log_5 2 = \log_5 15$

45. $3 \log_4 a = \log_4 27$

46. $2 \log_3 x + \log_3 3 = \log_3 36$

47. $\log_4 n + \log_4 (n - 4) = \log_4 5$

48. **SOUND** Use the formula $L = 10 \log_{10} R$, where L is the loudness of a sound and R is the sound's relative intensity, to find out how much louder 20 people talking would be than one person talking. Suppose the sound of one person talking has a relative intensity of 80 decibels.

EXAMPLE 6

Use $\log_5 16 \approx 1.7227$ and $\log_5 2 \approx 0.4307$ to approximate $\log_5 32$.

$\log_5 32 = \log_5 16 \cdot 2$	**Replace 32 with 16.**
$= \log_5 16 + \log_5 2$	**Product Property**
$\approx 1.7227 + 0.4307$	**Use a calculator.**
≈ 2.1534	

EXAMPLE 7

Solve $\log_3 3x + \log_3 4 = \log_3 36$.

$\log_3 3x + \log_3 4 = \log_3 36$	**Original equation**
$\log_3 3x(4) = \log_3 36$	**Product Property**
$3x(4) = 36$	**Definition of logarithm**
$12x = 36$	**Multiply.**
$x = 3$	**Divide each side by 12.**

8-6 Common Logarithms (pp. 516–522)

A2.A.18, A2.A.19

Solve each equation or inequality. Round to the nearest ten-thousandth.

49. $3^x = 15$

50. $6^{x^2} = 28$

51. $8^{m+1} = 30$

52. $12^{r-1} = 7^r$

53. $3^{5n} > 24$

54. $5^{x+2} \leq 3^x$

55. SAVINGS You deposited \$1000 into an account that pays an annual interest rate r of 5% compounded quarterly. Use $A = P\left(1 + \frac{r}{n}\right)^{nt}$.

a. How long will it take until you have \$1500 in your account?

b. How long it will take for your money to double?

EXAMPLE 8

Solve $5^{3x} > 7^{x+1}$.

$5^{3x} > 7^{x+1}$	**Original inequality**
$\log 5^{3x} > \log 7x + 1$	**Property of Inequality**
$3x \log 5 > (x + 1) \log 7$	**Power Property**
$3x \log 5 > x \log 7 + \log 7$	**Distributive Property**
$3x \log 5 - x \log 7 > \log 7$	**Subtract $x \log 7$.**
$x(3 \log 5 - \log 7) > \log 7$	**Distributive Property**
$x > \dfrac{\log 7}{3 \log 5 - \log 7}$	**Divide by $3 \log 5 - \log 7$.**
$x > 0.6751$	**Use a calculator.**

The solution set is $\{x \mid x > 0.6751\}$.

8-7 Base *e* and Natural Logarithms (pp. 525–531)

A2.A.12, A2.A.53

Solve each equation or inequality. Round to the nearest ten-thousandth.

56. $4e^x - 11 = 17$

57. $2e^{-x} + 1 = 15$

58. $\ln 2x = 6$

59. $2 + e^x > 9$

60. $\ln (x + 3)^5 < 5$

61. $e^{-x} > 18$

62. SAVINGS If you deposit \$2000 in an account paying 6.4% interest compounded continuously, how long will it take for your money to triple? Use $A = Pe^{rt}$.

EXAMPLE 9

Solve $3e^{5x} + 1 = 10$. Round to the nearest ten-thousandth.

$3e^{5x} + 1 = 10$	**Original equation**
$3e^{5x} = 9$	**Subtract 1 from each side.**
$e^{5x} = 3$	**Divide each side by 3.**
$\ln e^{5x} = \ln 3$	**Property of Equality**
$5x = \ln 3$	**$\ln e^x = x$**
$x = \dfrac{\ln 3}{5}$	**Divide each side by 5.**
$x \approx 0.2197$	**Use a calculator.**

8-8 Using Exponential and Logarithmic Functions (pp. 533–539)

A2.A.12

63. CARS Abe bought a used car for \$2500. It is expected to depreciate at a rate of 25% per year. What will be the value of the car in 3 years?

64. BIOLOGY For a certain strain of bacteria, k is 0.728 when t is measured in days. Using the formula $y = ae^{kt}$, how long will it take 10 bacteria to increase to 675 bacteria?

65. POPULATION The population of a city 20 years ago was 24,330. Since then, the population has increased at a steady rate each year. If the population is currently 55,250, find the annual rate of growth for this city.

EXAMPLE 10

A certain culture of bacteria will grow from 250 to 2000 bacteria in 1.5 hours. Find the constant k for the growth formula. Use $y = ae^{kt}$.

$y = ae^{kt}$	**Exponential Growth Formula**
$2000 = 250e^{k(1.5)}$	**Replace y with 2000, a with 250, and t with 1.5.**
$8 = e^{1.5k}$	**Divide each side by 250.**
$\ln 8 = \ln e^{1.5k}$	**Property of Equality**
$\ln 8 = 1.5k$	**Inverse Property**
$\dfrac{\ln 8}{1.5} = k$	**Divide each side by 1.5.**
$1.3863 \approx k$	**Use a calculator.**

Practice Test

NY Math Online glencoe.com
Chapter Test

Graph each function. State the domain and range.

1. $f(x) = 3^{x-3} + 2$
2. $f(x) = 2\left(\frac{3}{4}\right)^{x+1} - 3$

Solve each equation or inequality. Round to four decimal places if necessary.

3. $8^{c+1} = 16^{2c+3}$
4. $9^{x-2} > \left(\frac{1}{27}\right)^x$
5. $2^{a+3} = 3^{2a-1}$
6. $\log_2 (x^2 - 7) = \log_2 6x$
7. $\log_5 x > 2$
8. $\log_3 x + \log_3 (x - 3) = \log_3 4$
9. $6^{n-1} \leq 11^n$
10. $4e^{2x} - 1 = 5$
11. $\ln (x + 2)^2 > 2$

Use $\log_5 11 \approx 1.4899$ and $\log_5 2 \approx 0.4307$ to approximate the value of each expression.

12. $\log_5 44$
13. $\log_5 \frac{11}{2}$

14. **POPULATION** The population of a city 10 years ago was 150,000. Since then, the population has increased at a steady rate each year. The population is currently 185,000.
    a. Write an exponential function that could be used to model the population after x years if the population changes at the same rate.
    b. What will the population be in 25 years?

15. Write $\log_9 27 = \frac{3}{2}$ in exponential form.

16. **AGRICULTURE** An equation that models the decline in the number of U.S. farms is $y = 3{,}962{,}520(0.98)^x$, where x is the number of years since 1960 and y is the number of farms.
    a. How can you tell that the number is declining?
    b. By what annual rate is the number declining?
    c. Predict when the number of farms will be less than 1 million.

16. **MULTIPLE CHOICE** What is the value of $\log_4 \frac{1}{64}$?
    A -3
    B $-\frac{1}{3}$
    C $\frac{1}{3}$
    D 3

17. **SAVINGS** You put $7500 in a savings account paying 3% interest compounded continuously.
    a. Assuming there are no deposits or withdrawals from the account, what is the balance after 5 years?
    b. How long will it take your savings to double?
    c. In how many years will you have $10,000 in your account?

18. **MULTIPLE CHOICE** What is the solution of $\log_4 16 - \log_4 x = \log_4 8$?
    F $\frac{1}{2}$
    G 2
    H 4
    J 8

19. **MULTIPLE CHOICE** Which function is graphed below?

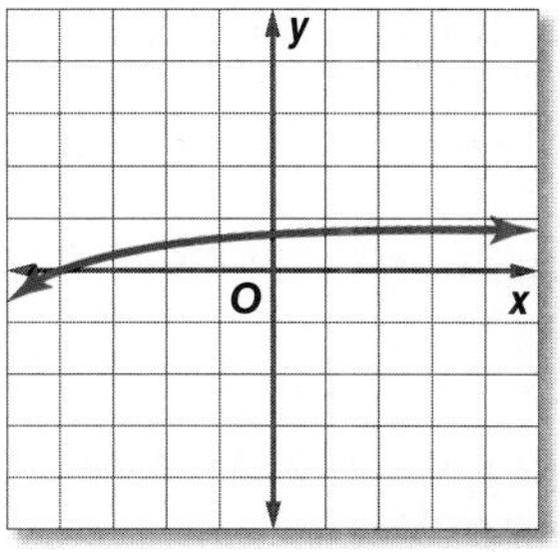

A $y = \log_{10} (x - 5)$
B $y = 5 \log_{10} x$
C $y = \log_{10} (x + 5)$
D $y = -5 \log_{10} x$

20. Write $2 \ln 6 + 3 \ln 4 - 5 \ln \left(\frac{1}{3}\right)$ as a single logarithm.

CHAPTER 8

Preparing for Standardized Tests

Using Technology

Your calculator can be a useful tool in taking standardized tests. Some problems that you encounter might have steps or computations that require the use of a calculator. A calculator may also help you solve a problem more quickly.

Strategies for Using Technology

Step 1

A calculator is a useful tool, but typically it should be used sparingly. Standardized tests are designed to measure your ability to reason and solve problems, not to measure your ability to punch keys on a calculator.

Before using a calculator, ask yourself:

- How would I normally solve this type of problem?
- Are there any steps that I cannot perform mentally or by using paper and pencil?
- Is a calculator absolutely necessary to solve this problem?
- Would a calculator help me solve this problem more quickly or efficiently?

Step 2

When might a calculator come in handy?

- solving problems that involve large, complex computations
- solving certain problems that involve graphing functions, evaluating functions, solving equations, and so on
- checking solutions of problems

EXAMPLE

Read the problem. Identify what you need to know. Then use the information in the problem to solve.

A certain can of soda contains 60 milligrams of caffeine. The caffeine is eliminated from the body at a rate of 15% per hour. What is the *half-life* of the caffeine? That is, how many hours does it take for half of the caffeine to be eliminated from the body?

A 4 hours

B 4.25 hours

C 4.5 hours

D 4.75 hours

Read the problem carefully. The problem can be solved using an exponential function. Use the exponential decay formula to model the problem and solve for the half-life of caffeine.

$$y = a(1 - r)^t$$

$$y = 60(1 - 1.015)^t$$

Half of 60 milligrams is 30. So, let $y = 30$ and solve for t.

$$30 = 60(1 - 0.15)^t$$

$$0.5 = (0.85)^t$$

Take the log of each side and use the power property.

$$\log 0.5 = \log (0.85)^t$$

$$\log 0.5 = t \log 0.85$$

$$\frac{\log 0.5}{\log 0.85} = t$$

At this point, it is necessary to use a calculator to evaluate the logarithms and solve the problem. Doing so shows that $t \approx 4.265$. So, the half-life of caffeine is about 4.25 hours. The correct answer is B.

Exercises

Read each problem. Identify what you need to know. Then use the information in the problem to solve.

1. Jason recently purchased a new truck for $34,750. The value of the truck decreases by 12% each year. What will the approximate value of the truck be 7 years after Jason purchased it?

A $13,775

B $13,890

C $14,125

D $14,200

2. A baseball is thrown upward at a velocity of 105 feet per second, releasing the baseball when it is 5 feet above the ground. The height of the baseball t seconds after being thrown is given by the formula $h(t) = -16t^2 + 105t + 5$. Find the time at which the baseball reaches its maximum height.

F 1.0 s

G 3.3 s

H 6.6 s

J 177.3 s

3. Lucinda deposited $2500 in a CD with the terms described below.

Super CD!

Earn 4.25% interest compounded daily!

(Minimum deposit of $1,000 over a period of at least 12 months.)

Use the formula below to solve for t, the number of years needed to earn $250 in interest with the CD.

$$2750 = 2500\left(1 + \frac{0.0425}{365}\right)^{365t}$$

A about 2.15 years

B about 2.24 years

C about 2.35 years

D about 2.46 years

CHAPTER 8 NYS Regents Exam Practice

Cumulative, Chapters 1 through 8

Multiple Choice

Answer all questions in this part. Select the answer that best completes the statement or answers the question.

1. Suppose there are only 2500 birds of a particular endangered species remaining in a region and the population decreases at a rate of about 3% each year. The logarithmic function $t = \log_{0.97}\left(\frac{p}{2500}\right)$ predicts how many years t it will be for the population to decease to a number p. About how long will it take for the population to reach 2000 birds? Round to the nearest whole number.

(1) 5 years

(2) 6 years

(3) 7 years

(4) 8 years

Test-Taking Tip

Question 1 Remember, $\log_b a$ can be evaluated as $(\log a)/(\log b)$. Use a scientific calculator to evaluate the expression.

2. The table shows the growth of a certain bacteria.

Time in Hours (x)	Number of Cells (N)
0	38
1	49
2	62
3	78
4	99
5	126

If N represents the number of cells at hour x, which equation *best* models this set of data?

(1) $N = 38.3(1.27)^x$

(2) $N = 1.27(38.3)^x$

(3) $N = 11.4x + 38.5$

(4) $N = 38.5x + 11.4$

3. In the system below, x represents New York's national rank according to its area, and y represents the area, in thousands of square kilometers.

$$\begin{bmatrix} 6 & -2 \\ -8 & 4 \end{bmatrix}\begin{bmatrix} x \\ y \end{bmatrix} = \begin{bmatrix} -120 \\ 348 \end{bmatrix}$$

Which of the following statements is true?

(1) New York is the 23rd largest state with an area of 97,000 square kilometers.

(2) New York is the 24th largest state with an area of 55,000 square kilometers.

(3) New York is the 26th largest state with an area of 102,000 square kilometers.

(4) New York is the 27th largest state with an area of 141,000 square kilometers.

4. Solve the equation for x to the nearest thousandth.

$$\ln 8x = 5$$

(1) 0.271

(2) 3.895

(3) 16.764

(4) 18.552

5. Which of the following is *not* a factor of the polynomial function graphed below?

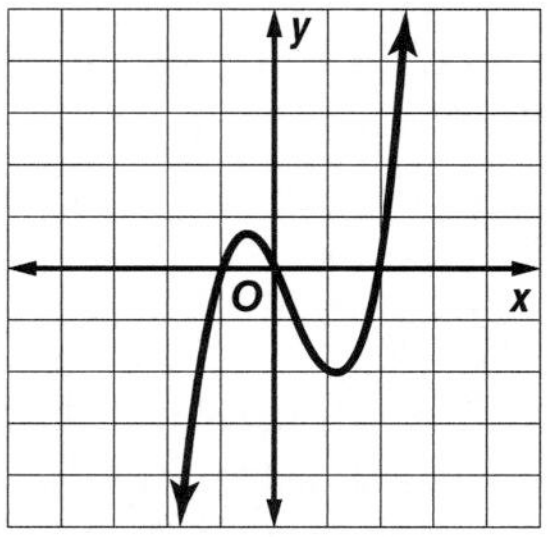

(1) $x - 1$

(2) $x - 2$

(3) $x + 1$

(4) x

6. The function $y = \left(\frac{1}{2}\right)^x$ is graphed below. What is the range of the function?

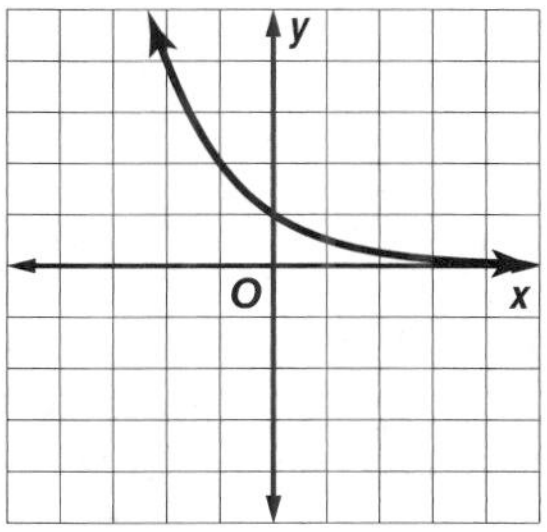

(1) all positive real numbers

(2) all real numbers

(3) all negative real numbers

(4) all real numbers greater than −3

7. What is the equation of the square root function graphed below?

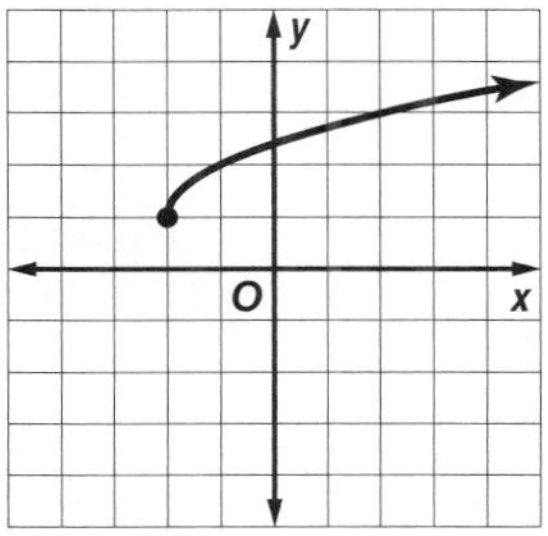

(1) $f(x) = \sqrt{x+1} + 2$

(2) $f(x) = \sqrt{x-1} + 2$

(3) $f(x) = \sqrt{x+2} + 1$

(4) $f(x) = \sqrt{x-2} + 1$

Open-Ended Response

Solve each problem. Clearly indicate the necessary steps, including appropriate formula substitutions, diagrams, graphs, charts, etc.

8. Suppose the number of whitetail deer in a particular region has increased at an annual rate of about 5% since 2000. There were 125,000 deer in 2000.

a. Write a function to model the number of whitetail deer t years after 2000.

b. About how many whitetail deer inhabited the region in 2005? Round your answer to the nearest hundred deer.

c. At this rate, about how many years will it take for the whitetail deer population to double? Round your answer to the nearest tenth.

9. The parent function $f(x) = 2^x$ is graphed on the coordinate grid below along with a transformation, $g(x)$. Write an equation to represent $g(x)$.

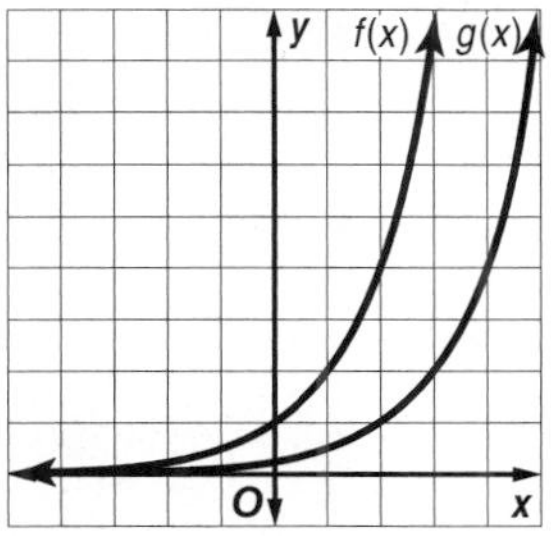

10. The radius of a sphere that has volume V is given by the expression $r = \left(\frac{3V}{4\pi}\right)^{\frac{1}{3}}$. What is the radius of a spherical globe that has a volume of 2150 cubic inches? Round your answer to the nearest inch.

Need Extra Help?

If you missed Question...	1	2	3	4	5	6	7	8	9	10
Go to Lesson or Page...	8-1	8-4	8-2	1-3	3-1	8-3	2-4	8-1	8-6	4-4
NYS Core Curriculum	A2.A.18	A2.S.7	A.A.10	A2.A.28	A2.A.50	A2.A.46	A2.A.46	A2.A.27	A2.A.53	A2.A.8

CHAPTER 9 Rational Functions and Relations

Then

In Chapter 5, you used factoring to solve quadratic equations and you graphed quadratic equations.

Now

In Chapter 9, you will:

- Simplify rational expressions.
- Graph rational functions.
- Solve direct, joint, and inverse variation problems.
- Solve rational equations and inequalities.

NYS Core Curriculum

A2.A.16 Perform arithmetic operations with rational expressions and rename to lowest terms
A2.A.23 Solve rational equations and inequalities

Why?

TRAVEL Whether you travel by boat, car, bicycle, or airplane, rational functions can be used to find distance traveled, time spent traveling, and speed. If you want to arrive at a destination on time, rational relations can tell you at what speed you need to travel to reach your goal. When graphing rational functions you see clearly how the speed at which you travel affects the time it takes to get there. Rational relations are used to read maps and estimate mileage while traveling.

Math *in Motion,* Animation glencoe.com

Get Ready for Chapter 9

Diagnose Readiness You have two options for checking Prerequisite Skills.

Text Option

Take the Quick Check below. Refer to the Quick Review for help.

*Quick*Check

Solve each equation. Write in simplest form. (Lesson 1-3)

1. $\frac{5}{14} = \frac{1}{3}x$

2. $\frac{1}{8}m = \frac{7}{3}$

3. $\frac{8}{5} = \frac{1}{4}k$

4. $\frac{10}{9}p = 7$

5. TRUCKS Martin used $\frac{1}{3}$ of his tank of gas in his truck to get to work. He began with a full tank of gas. If he had 18 gallons of gas left, how many gallons does his tank hold?

Simplify each expression.

6. $\frac{3}{4} - \frac{7}{8}$

7. $\frac{8}{9} - \frac{7}{6} + \frac{1}{3}$

8. $\frac{9}{10} - \frac{4}{15} + \frac{1}{3}$

9. $\frac{10}{3} + \frac{5}{6} + 3$

10. BAKING Annie baked cookies for a bake sale. She used $\frac{2}{3}$ cups of flour for one recipe and $4\frac{1}{2}$ cups of flour for the other recipe. How many cups total did she use?

Solve each proportion. (Concepts and Skills Bank 1)

11. $\frac{9}{12} = \frac{p}{36}$

12. $\frac{9}{18} = \frac{6}{m}$

13. $\frac{2}{7} = \frac{5}{k}$

14. SALES TAX Kirsten pays $4.40 tax on $55 worth of clothes. What amount of tax will she pay on $35 worth of clothes?

*Quick*Review

EXAMPLE 1

Solve $\frac{9}{11} = \frac{7}{8}r$. Write in simplest form.

$\frac{9}{11} = \frac{7}{8}r$

$\frac{72}{11} = 7r$ — **Multiply each side by 8.**

$\frac{72}{77} = r$ — **Divide each side by 7.**

Since the GCF of 72 and 77 is 1, the solution is in simplest form.

EXAMPLE 2

Simplify $\frac{1}{3} + \frac{3}{4} - \frac{5}{6}$.

$\frac{1}{3} + \frac{3}{4} - \frac{5}{6}$

$= \frac{1}{3}\left(\frac{4}{4}\right) + \frac{3}{4}\left(\frac{3}{3}\right) - \frac{5}{6}\left(\frac{2}{2}\right)$ — **The GCF of 3, 4, and 6 is 12.**

$= \frac{4}{12} + \frac{9}{12} - \frac{10}{12}$ — **Simplify.**

$= \frac{3}{12}$ — **Add and subtract.**

$= \frac{3 \div 3}{12 \div 3}$ or $\frac{1}{4}$ — **Simplify.**

EXAMPLE 3

Solve $\frac{5}{8} = \frac{u}{11}$.

$\frac{5}{8} = \frac{u}{11}$ — **Write the equation.**

$5(11) = 8u$ — **Find the cross products.**

$55 = 8u$ — **Simplify.**

$\frac{55}{8} = u$ — **Divide each side by 8.**

Since the GCF of 55 and 8 is 1, the answer is in simplified form. $u = \frac{55}{8}$ or $6\frac{7}{8}$.

Online Option

NY Math Online Take a self-check Chapter Readiness Quiz at glencoe.com.

Get Started on Chapter 9

You will learn several new concepts, skills, and vocabulary terms as you study Chapter 9. To get ready, identify important terms and organize your resources. You may wish to refer to **Chapter 0** to review prerequisite skills.

FOLDABLES® Study Organizer

Rational Functions and Relations Make this Foldable to help you organize your Chapter 9 notes about rational functions and relations. Begin with an $8\frac{1}{2}'' \times 11''$ sheet of grid paper.

1. **Fold** in thirds along the height.

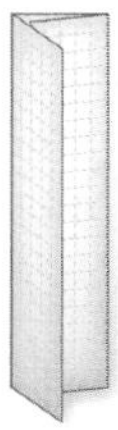

2. **Fold** the top edge down making a 2" tab at the top. Cut along the folds.

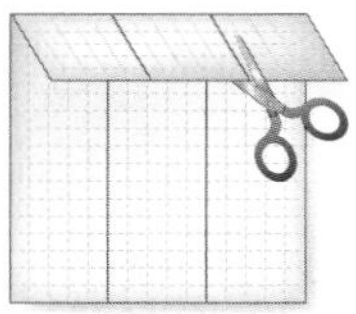

3. **Label** the outside tabs *Expressions*, *Functions*, and *Equations*. Use the inside tabs for definitions and notes.

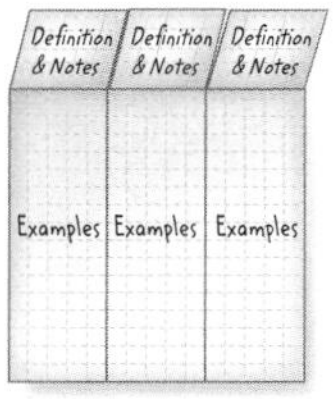

4. **Write** examples of each topic in the space below each tab.

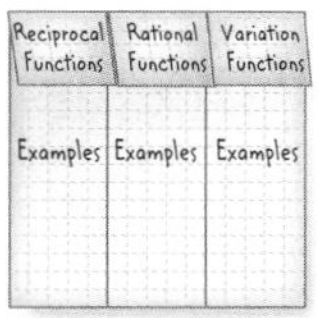

NY Math Online glencoe.com

- Study the chapter online
- Explore **Math in Motion**
- Get extra help from your own **Personal Tutor**
- Use **Extra Examples** for additional help
- Take a **Self-Check Quiz**
- **Review Vocabulary** in fun ways

New Vocabulary

English		Español
rational expression	• p. 553 •	expresión racional
complex fraction	• p. 556 •	fracción compleja
reciprocal function	• p. 569 •	función recíproco
hyperbola	• p. 569 •	hipérbola
asymptote	• p. 569 •	asíntota
rational function	• p. 577 •	función racional
vertical asymptote	• p. 577 •	asíntota vertical
horizontal asymptote	• p. 577 •	asíntota horizontal
oblique asymptote	• p. 579 •	asíntota oblicua
point discontinuity	• p. 580 •	discontinuidad evitable
direct variation	• p. 586 •	variación directa
constant of variation	• p. 586 •	constante de variación
joint variation	• p. 587 •	variación conjunta
inverse variation	• p. 588 •	variación inversa
combined variation	• p. 589 •	variación combinada
rational equation	• p. 594 •	ecuación racional
weighted average	• p. 596 •	media ponderada
rational inequality	• p. 599 •	desigualdad racional

Review Vocabulary

function • p. P4 • función a relation in which each element of the domain is paired with exactly one element of the range

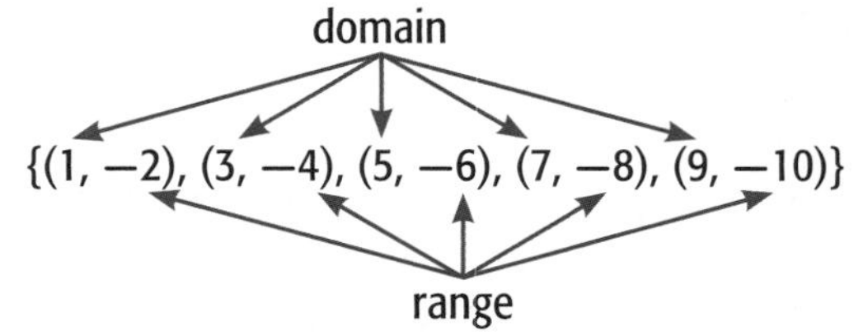

least common multiple • mínimo común múltiplo the least number that is a common multiple of two or more numbers

rational number • p. 11 • número racional a number expressed in the form $\frac{a}{b}$, where a and b are integers and $b \neq 0$

Multilingual eGlossary glencoe.com

9-1 Multiplying and Dividing Rational Expressions

Then
You factored polynomials. (Lesson 5-3)

Now
- Simplify rational expressions.
- Simplify complex fractions.

NYS Core Curriculum

A2.A.16 Perform arithmetic operations with rational expressions and rename to lowest terms
A2.A.17 Simplify complex fractional expressions

New Vocabulary
rational expression
complex fraction

NY Math Online

glencoe.com
- Extra Examples
- Personal Tutor
- Self-Check Quiz
- Homework Help
- Math in Motion

Why?

If a scuba diver goes to depths greater than 33 feet, the rational function $T(d) = \frac{1700}{d - 33}$ gives the maximum time a diver can remain at those depths and still surface at a steady rate with no stops. $T(d)$ represents the dive time in minutes, and d represents the depth in feet.

Simplify Rational Expressions A ratio of two polynomial expressions such as $\frac{1700}{d - 33}$ is called a **rational expression**.

Because variables in algebra often represent real numbers, operations with rational numbers and rational expressions are similar. Just as with reducing fractions, to simplify a rational expression, you divide the numerator and denominator by their greatest common factor (GCF).

$$\frac{8}{12} = \frac{2 \cdot \cancel{4}^{1}}{3 \cdot \cancel{4}_{1}} = \frac{2}{3} \qquad \text{GCF} = 4$$

$$\frac{x^2 - 4x + 3}{x^2 - 6x + 5} = \frac{(x - 3)\cancel{(x - 1)}^{1}}{(x - 5)\cancel{(x - 1)}_{1}} = \frac{(x - 3)}{(x - 5)} \qquad \text{GCF} = (x - 1)$$

EXAMPLE 1 Simplify a Rational Expression

a. Simplify $\frac{5x(x^2 + 4x + 3)}{(x - 6)(x^2 - 9)}$.

$$\frac{5x(x^2 + 4x + 3)}{(x - 6)(x^2 - 9)} = \frac{5x(x + 3)(x + 1)}{(x - 6)(x + 3)(x - 3)}$$ **Factor numerator and denominator.**

$$= \frac{5x(x + 1)}{(x - 6)\,(x - 3)} \cdot \frac{\cancel{(x + 3)}^{1}}{\cancel{(x + 3)}_{1}}$$ **Eliminate common factors.**

$$= \frac{5x(x + 1)}{(x - 6)(x - 3)}$$ **Simplify.**

b. Under what conditions is this expression undefined?

The original factored denominator is $(x - 6)(x + 3)(x - 3)$.
Determine the values that would make the denominator equal to 0.
These values are 6, −3, or 3, so the expression is undefined when $x = 6, 3$ or -3.

Check Your Progress

Simplify each expression. Under what conditions is the expression undefined?

1A. $\frac{4y(y - 3)(y + 4)}{y(y^2 - y - 6)}$

1B. $\frac{2z(z + 5)(z^2 + 2z - 8)}{(z - 1)(z + 5)(z - 2)}$

Personal Tutor glencoe.com

NYSRE EXAMPLE 2 A2.A.16, A2.A.17

For what value(s) is $\frac{x^2(x^2 - 5x - 14)}{4x(x^2 + 6x + 8)}$ undefined?

A $-2, -4$

B $-2, 7$

C $0, -2, -4$

D $0, -2, -4, 7$

Read the Test Item

You want to determine which values of x make the denominator equal to 0.

Solve the Test Item

With $4x$ in the denominator, x cannot equal 0. So, choices A and B can be eliminated. Next, factor the denominator.

$x^2 + 6x + 8 = (x + 2)(x + 4)$, so the denominator is $4x(x + 2)(x + 4)$.

Because the denominator equals 0 when $x = 0, -2$, and -4, the answer is C.

Test-Taking Tip

Eliminating Choices Sometimes you can save time by looking at the possible answers and eliminating choices.

Check Your Progress

2. For what value(s) of x is $\frac{x(x^2 + 8x + 12)}{-6(x^2 - 3x - 10)}$ undefined?

F $0, 5, -2$ **G** $5, -2$ **H** $0, -2, -6$ **J** $5, -2, -6$

Personal Tutor glencoe.com

Sometimes you can factor out -1 in the numerator or denominator to help simplify a rational expression.

EXAMPLE 3 Simplify Using −1

Simplify $\frac{(4w^2 - 3wy)(w + y)}{(3y - 4w)(5w + y)}$.

$$\frac{(4w^2 - 3wy)(w + y)}{(3y - 4w)(5w + y)} = \frac{w(4w - 3y)(w + y)}{(3y - 4w)(5w + y)}$$ **Factor.**

$$= \frac{w(-1)\overset{1}{\cancel{(3y - 4w)}}(w + y)}{\underset{1}{\cancel{(3y - 4w)}}(5w + y)}$$ **$4w - 3y = -1(3y - 4w)$**

$$= \frac{(-w)(w + y)}{5w + y}$$ **Simplify.**

Check Your Progress

Simplify each expression.

3A. $\frac{(xz - 4z)}{z^2(4 - x)}$

3B. $\frac{ab^2 - 5ab}{(5 + b)(5 - b)}$

Personal Tutor glencoe.com

The method for multiplying and dividing fractions also works with rational expressions. Remember that to multiply two fractions, you multiply the numerators and multiply the denominators. To divide two fractions, you multiply by the multiplicative inverse, or the reciprocal, of the divisor.

Multiplication

$$\frac{2}{9} \cdot \frac{15}{4} = \frac{\overset{1}{\cancel{2}} \cdot \overset{1}{\cancel{3}} \cdot 5}{\underset{1}{\cancel{3}} \cdot 3 \cdot \underset{1}{\cancel{2}} \cdot 2} = \frac{5}{3 \cdot 2} = \frac{5}{6}$$

Division

$$\frac{3}{5} \div \frac{6}{35} = \frac{3}{5} \cdot \frac{35}{6} = \frac{\overset{1}{\cancel{3}} \cdot \overset{1}{\cancel{5}} \cdot 7}{\underset{1}{\cancel{5}} \cdot 2 \cdot \underset{1}{\cancel{3}}} = \frac{7}{2}$$

The following table summarizes the rules for multiplying and dividing rational expressions.

Key Concept

For Your FOLDABLE

Multiplying Rational Expressions

Words To multiply rational expressions, multiply the numerators and multiply the denominators.

Symbols For all rational expressions $\frac{a}{b}$ and $\frac{c}{d}$ with $b \neq 0$ and $d \neq 0$, $\frac{a}{b} \cdot \frac{c}{d} = \frac{ac}{bd}$.

Dividing Rational Expressions

Words To divide rational expressions, multiply by the reciprocal of the divisor.

Symbols For all rational expressions $\frac{a}{b}$ and $\frac{c}{d}$ with $b \neq 0$, $c \neq 0$, and $d \neq 0$, $\frac{a}{b} \div \frac{c}{d} = \frac{a}{b} \cdot \frac{d}{c} = \frac{ad}{bc}$.

StudyTip

Eliminating Common Factors Be sure to eliminate factors from both the numerator and denominator.

EXAMPLE 4 Multiply and Divide Rational Expressions

Simplify each expression.

a. $\frac{6c}{5d} \cdot \frac{15cd^2}{8a}$

$\frac{6c}{5d} \cdot \frac{15cd^2}{8a} = \frac{2 \cdot 3 \cdot c \cdot 5 \cdot 3 \cdot c \cdot d \cdot d}{5 \cdot d \cdot 2 \cdot 2 \cdot 2 \cdot a}$ Factor.

$= \frac{\overset{1}{\cancel{2}} \cdot 3 \cdot c \cdot \overset{1}{\cancel{5}} \cdot 3 \cdot c \cdot \overset{1}{\cancel{d}} \cdot d}{\underset{1}{\cancel{5}} \cdot \underset{1}{\cancel{d}} \cdot \underset{1}{\cancel{2}} \cdot 2 \cdot 2 \cdot a}$ Eliminate common factors.

$= \frac{3 \cdot 3 \cdot c \cdot c \cdot d}{2 \cdot 2 \cdot a}$ Simplify.

$= \frac{9c^2d}{4a}$ Simplify.

b. $\frac{18xy^3}{7a^2b^2} \div \frac{12x^2y}{35a^2b}$

$\frac{18xy^3}{7a^2b^2} \div \frac{12x^2y}{35a^2b} = \frac{18xy^3}{7a^2b^2} \cdot \frac{35a^2b}{12x^2y}$ Multiply by reciprocal of the divisor.

$= \frac{2 \cdot 3 \cdot 3 \cdot x \cdot y \cdot y \cdot y \cdot 5 \cdot 7 \cdot a \cdot a \cdot b}{7 \cdot a \cdot a \cdot b \cdot b \cdot 2 \cdot 2 \cdot 3 \cdot x \cdot x \cdot y}$ Factor.

$= \frac{\overset{1}{\cancel{2}} \cdot \overset{1}{\cancel{3}} \cdot 3 \cdot \overset{1}{\cancel{x}} \cdot \overset{1}{\cancel{y}} \cdot y \cdot y \cdot 5 \cdot \overset{1}{\cancel{7}} \cdot \overset{1}{\cancel{a}} \cdot \overset{1}{\cancel{a}} \cdot \overset{1}{\cancel{b}}}{\underset{1}{\cancel{7}} \cdot \underset{1}{\cancel{a}} \cdot \underset{1}{\cancel{a}} \cdot \underset{1}{\cancel{b}} \cdot b \cdot \underset{1}{\cancel{2}} \cdot 2 \cdot \underset{1}{\cancel{3}} \cdot \underset{1}{\cancel{x}} \cdot x \cdot \underset{1}{\cancel{y}}}$ Eliminate common factors.

$= \frac{3 \cdot 5 \cdot y \cdot y}{2 \cdot b \cdot x}$ Simplify.

$= \frac{15y^2}{2bx}$ Simplify.

Check Your Progress

4A. $\frac{12c^3d^2}{21ab} \cdot \frac{14a^2b}{8c^2d}$

4B. $\frac{6xy}{15ab^2} \cdot \frac{21a^3}{18x^4y}$

4C. $\frac{16mt^2}{21a^4b^3} \div \frac{24m^3}{7a^2b^2}$

4D. $\frac{12x^4y^2}{40a^4b^4} \div \frac{6x^2y^4}{16a^2x}$

Personal Tutor glencoe.com

Sometimes you must factor the numerator and/or the denominator first before you can simplify a product or a quotient of rational expressions.

StudyTip

Factoring Polynomials When simplifying rational expressions, factors in one polynomial will often reappear in other polynomials. In Example 5a, $x - 8$ appears four times. Use this as a guide when factoring challenging polynomials.

EXAMPLE 5 Polynomials in the Numerator and Denominator

Simplify each expression.

a. $\dfrac{x^2 - 6x - 16}{x^2 - 16x + 64} \cdot \dfrac{x - 8}{x^2 + 5x + 6}$

$$\frac{x^2 - 6x - 16}{x^2 - 16x + 64} \cdot \frac{x - 8}{x^2 + 5x + 6} = \frac{(x - 8)(x + 2)}{(x - 8)(x - 8)} \cdot \frac{x - 8}{(x + 3)(x + 2)}$$ **Factor.**

$$= \frac{\overset{1}{\cancel{(x - 8)}}\overset{1}{\cancel{(x + 2)}}}{\underset{1}{\cancel{(x - 8)}}\underset{1}{\cancel{(x - 8)}}} \cdot \frac{\overset{1}{\cancel{x - 8}}}{(x + 3)\underset{1}{\cancel{(x + 2)}}}$$ **Eliminate common factors.**

$$= \frac{1}{x + 3}$$ **Simplify.**

b. $\dfrac{x^2 - 16}{12y + 36} \div \dfrac{x^2 - 12x + 32}{y^2 - 3y - 18}$

$$\frac{x^2 - 16}{12y + 36} \div \frac{x^2 - 12x + 32}{y^2 - 3y - 18} = \frac{x^2 - 16}{12y + 36} \cdot \frac{y^2 - 3y - 18}{x^2 - 12x + 32}$$ **Multiply by reciprocal.**

$$= \frac{(x + 4)(x - 4)}{12(y + 3)} \cdot \frac{(y - 6)(y + 3)}{(x - 4)(x - 8)}$$ **Factor.**

$$= \frac{(x + 4)\overset{1}{\cancel{(x - 4)}}}{12\underset{1}{\cancel{(y + 3)}}} \cdot \frac{(y - 6)\overset{1}{\cancel{(y + 3)}}}{\underset{1}{\cancel{(x - 4)}}(x - 8)}$$ **Eliminate common factors.**

$$= \frac{(x + 4)(y - 6)}{12(x - 8)}$$ **Simplify.**

Check Your Progress

5A. $\dfrac{8x - 20}{x^2 + 2x - 35} \cdot \dfrac{x^2 - 7x + 10}{4x^2 - 16}$

5B. $\dfrac{x^2 - 9x + 20}{x^2 + 10x + 21} \div \dfrac{x^2 - x - 12}{6x + 42}$

Personal Tutor glencoe.com

Simplify Complex Fractions A **complex fraction** is a rational expression with a numerator and/or denominator that is also a rational expression. The following expressions are complex fractions.

Math *in Motion*, Animation glencoe.com

$$\frac{\frac{c}{6}}{5d} \qquad \frac{\frac{8}{x}}{x - 2} \qquad \frac{\frac{x - 3}{8}}{\frac{x - 2}{x + 4}} \qquad \frac{\frac{4}{a} + 6}{\frac{12}{a} - 3}$$

To simplify a complex fraction, first rewrite it as a division expression.

EXAMPLE 6 Simplify Complex Fractions

Simplify each expression.

a. $\dfrac{\frac{a + b}{4}}{\frac{a^2 + b^2}{4}}$

$$\frac{\frac{a + b}{4}}{\frac{a^2 + b^2}{4}} = \frac{a + b}{4} \div \frac{a^2 + b^2}{4}$$ **Express as a division expression.**

$$= \frac{a + b}{4} \cdot \frac{4}{a^2 + b^2}$$ **Multiply by the reciprocal.**

$$= \frac{a + b}{\underset{1}{\cancel{4}}} \cdot \frac{\overset{1}{\cancel{4}}}{a^2 + b^2} \text{ or } \frac{a + b}{a^2 + b^2}$$ **Simplify.**

b. $\dfrac{\frac{x^2}{x^2 - y^2}}{\frac{4x}{y - x}}$

$\dfrac{\frac{x^2}{x^2 - y^2}}{\frac{4x}{y - x}} = \dfrac{x^2}{x^2 - y^2} \div \dfrac{4x}{y - x}$ **Express as a division expression.**

$= \dfrac{x^2}{x^2 - y^2} \cdot \dfrac{y - x}{4x}$ **Multiply by the reciprocal.**

$= \dfrac{x \cdot x}{(x + y)(x - y)} \cdot \dfrac{(-1)(x - y)}{4x}$ **Factor.**

$= \dfrac{x \cdot \cancel{x}^{1}}{(x + y)\cancel{(x - y)}_{1}} \cdot \dfrac{(-1)\cancel{(x - y)}^{1}}{4\cancel{x}_{1}}$ **Eliminate Factors.**

$= \dfrac{-x}{4(x + y)}$ **Simplify.**

Check Your Progress

Simplify each expression.

6A. $\dfrac{\frac{(x - 2)^2}{2(x^2 - 5x + 4)}}{\frac{x^2 - 4}{4x - 10}}$

6B. $\dfrac{\frac{x^2 - y^2}{y^2 - 49}}{\frac{y - x}{y + 7}}$

Personal Tutor glencoe.com

Check Your Understanding

Example 1 p. 553

Simplify each expression.

1. $\dfrac{x^2 - 5x - 24}{x^2 - 64}$

2. $\dfrac{c + d}{3c^2 - 3d^2}$

Example 2 p. 554

3. MULTIPLE CHOICE Identify all values of x for which $\dfrac{x + 7}{x^2 - 3x - 28}$ is undefined.

A $-7, 4$ B $7, 4$ C $4, -7, 7$ D $-4, 7$

Examples 3–6 pp. 554–557

Simplify each expression.

4. $\dfrac{y^2 + 3y - 40}{25 - y^2}$

5 $\dfrac{a^2x - b^2x}{by - ay}$

6. $\dfrac{27x^2y^4}{16yz^3} \cdot \dfrac{8z}{9xy^3}$

7. $\dfrac{12x^3y}{13ab^2} \div \dfrac{36xy^3}{26b}$

8. $\dfrac{x^2 - 4x - 21}{x^2 - 6x + 8} \cdot \dfrac{x - 4}{x^2 - 2x - 35}$

9. $\dfrac{a^2 - b^2}{3a^2 - 6a + 3} \div \dfrac{4a + 4b}{a^2 - 1}$

10. $\dfrac{\frac{a^3b^3}{xy^4}}{\frac{a^2b}{x^2y}}$

11. $\dfrac{\frac{4x}{x + 6}}{\frac{x^2 - 3x}{x^2 + 3x - 18}}$

12. MANUFACTURING The volume of a shipping container in the shape of a rectangular prism can be represented by the polnomial $6x^3 + 11x^2 + 4x$, where the height is x.

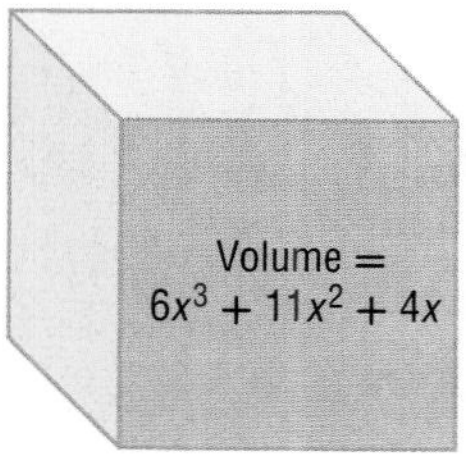

a. Find the length and width of the container.

b. Find the ratio of of the three dimensions of the container when $x = 2$.

c. Will the ratio of the three dimensions be the same for all values of x?

Practice and Problem Solving

● = **Step-by-Step Solutions** begin on page R20.
Extra Practice begins on page 947.

Example 1 p. 553

Simplify each expression.

13. $\frac{x(x-3)(x+6)}{x^2+x-12}$

14. $\frac{y^2(y^2+3y+2)}{2y(y-4)(y+2)}$

15. $\frac{(x^2-9)(x^2-z^2)}{4(x+z)(x-3)}$

16. $\frac{(x^2-16x+64)(x+2)}{(x^2-64)(x^2-6x-16)}$

17. $\frac{x^2(x+2)(x-4)}{6x(x^2+x-20)}$

18. $\frac{3y(y-8)(y^2+2y-24)}{15y^2(y^2-12y+32)}$

Example 2 p. 554

19. MULTIPLE CHOICE Identify all values of x for which $\frac{(x-3)(x+6)}{(x^2-7x+12)(x^2-36)}$ is undefined.

F 3, −6 G 4, 6 H −6, 6 J −6, 3, 4, 6

Example 3 p. 554

Simplify each expression.

20. $\frac{x^2-5x-14}{28+3x-x^2}$

21. $\frac{x^3-9x^2}{x^2-3x-54}$

22. $\frac{(x-4)(x^2+2x-48)}{(36-x^2)(x^2+4x-32)}$

23. $\frac{16-c^2}{c^2+c-20}$

24. GEOMETRY The cylinder at the right has a volume of $(x+3)(x^2-3x-18)\pi$ cubic centimeters. Find the height of the cylinder.

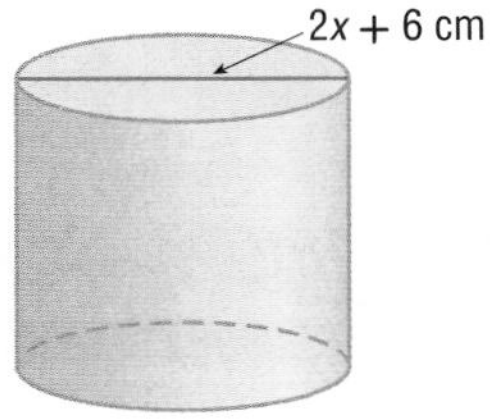

Examples 4–6 pp. 555–557

Simplify each expression.

25. $\frac{3ac^3f^3}{8a^2bcf^4} \cdot \frac{12ab^2c}{18ab^3c^2f}$

26. $\frac{14xy^2z^3}{21w^4x^2yz} \cdot \frac{7wxyz}{12w^2y^3z}$

27. $\frac{64a^2b^5}{35b^2c^3f^4} \div \frac{12a^4b^3c}{70abcf^2}$

28. $\frac{9x^2yz}{5z^4} \div \frac{12x^4y^2}{50xy^4z^2}$

29. $\frac{15a^2b^2}{21ac} \cdot \frac{14a^4c^2}{6ab^3}$

30. $\frac{14c^2f^5}{9a^2} \div \frac{35cf^4}{18ab^3}$

31 $\frac{y^2+8y+15}{y-6} \cdot \frac{y^2-9y+18}{y^2-9}$

32. $\frac{c^2-6c-16}{c^2-d^2} \div \frac{c^2-8c}{c+d}$

33. $\frac{x^2+9x+20}{8x+16} \cdot \frac{4x^2+16x+16}{x^2-25}$

34. $\frac{3a^2+6a+3}{a^2-3a-10} \div \frac{12a^2-12}{a^2-4}$

35. $\frac{\frac{x^2-9}{6x-12}}{\frac{x^2+10x+21}{x^2-x-2}}$

36. $\frac{\frac{y-x}{z^3}}{\frac{x-y}{6z^2}}$

37. $\frac{\frac{a^2-b^2}{b^3}}{\frac{b^2-ab}{a^2}}$

38. $\frac{\frac{x-y}{a+b}}{\frac{x^2-y^2}{b^2-a^2}}$

39. SOCCER At the end of her high school soccer career, Ashley had made 33 goals out of 121 attempts.

a. Write a ratio to represent the ratio of the number of goals made to goals attempted by Ashley at the end of her high school career.

b. Suppose Ashley attempted a goals and made m goals during her first year at college. Write a rational expression to represent the ratio of the number of career goals made to the number of career goals attempted at the end of her first year in college.

40. GEOMETRY Parallelogram F has an area of $8x^2 + 10x - 3$ square meters and a height of $2x + 3$ meters. Parallelogram G has an area of $6x^2 + 13x - 5$ square meters and a height of $3x - 1$ meters. Find the area of right triangle H.

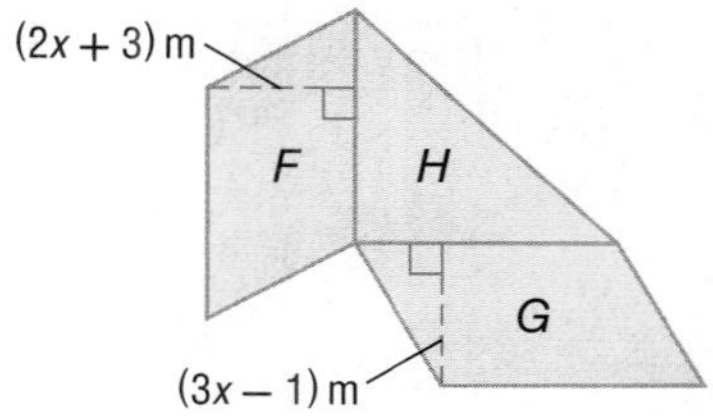

41. POLLUTION The thickness of an oil spill from a ruptured pipe on a rig is modeled by the function $T(x) = \dfrac{0.4(x^2 - 2x)}{x^3 + x^2 - 6x}$, where T is the thickness of the oil slick in meters and x is the distance from the rupture in meters.

a. Simplify the function.

b. How thick is the slick 100 meters from the rupture?

Real-World Link

In 1989, the Exxon Valdez spilled an estimated 11 million gallons of crude oil near the gulf coast of Alaska.

Source: Environmental Protection Agency

Simplify each expression.

42. $\dfrac{x^2 - 16}{3x^3 + 18x^2 + 24x} \cdot \dfrac{x^3 - 4x}{2x^2 - 7x - 4}$

43. $\dfrac{3x^2 - 17x - 6}{4x^2 - 20x - 24} \div \dfrac{6x^2 - 7x - 3}{2x^2 - x - 3}$

44. $\dfrac{9 - x^2}{x^2 - 4x - 21} \cdot \left(\dfrac{2x^2 + 7x + 3}{2x^2 - 15x + 7}\right)^{-1}$

45. $\left(\dfrac{2x^2 + 2x - 12}{x^2 + 4x - 5}\right)^{-1} \cdot \dfrac{2x^3 - 8x}{x^2 - 2x - 35}$

46. $\left(\dfrac{3xy^3z}{2a^2bc^2}\right)^3 \cdot \dfrac{16a^4b^3c^5}{15x^7yz^3}$

47. $\dfrac{20x^2y^6z^{-2}}{3a^3c^2} \cdot \left(\dfrac{16x^3y^3}{9acz}\right)^{-1}$

48. $\left(\dfrac{2xy^3}{3abc}\right)^{-2} \div \dfrac{6a^2b}{x^2y^4}$

49. $\dfrac{\dfrac{8x^2 - 10x - 3}{10x^2 + 35x - 20}}{\dfrac{2x^2 + x - 6}{4x^2 + 18x + 8}}$

50. $\dfrac{\dfrac{2x^2 + 7x - 30}{-6x^2 + 13x + 5}}{\dfrac{4x^2 + 12x - 72}{3x^2 - 11x - 4}}$

51. $\dfrac{\dfrac{4x^2 - 1}{3x^3 - 6x^2 - 24x}}{\dfrac{12x^2 + 12x - 9}{-2x^2 + 5x + 12}}$

52. GEOMETRY The area of the base of the rectangular prism at the right is 20 square centimeters.

a. Find the length of $\overline{BC}$ in terms of x.

b. If $DC = 3BC$, determine the area of the shaded region in terms of x.

c. Determine the volume of the prism in terms of x.

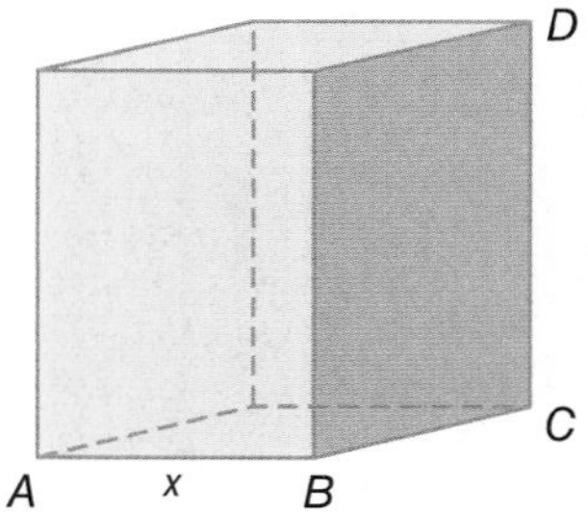

Simplify each expression.

53. $\dfrac{x^2 + 4x - 32}{2x^2 + 9x - 5} \cdot \dfrac{3x^2 - 75}{3x^2 - 11x - 4} \div \dfrac{6x^2 - 18x - 60}{x^3 - 4x}$

54. $\dfrac{8x^2 + 10x - 3}{3x^2 - 12x - 36} \div \dfrac{2x^2 - 5x - 12}{3x^2 - 17x - 6} \cdot \dfrac{4x^2 + 3x - 1}{4x^2 - 40x + 24}$

55. $\dfrac{4x^2 - 9x - 9}{3x^2 + 6x - 18} \div \dfrac{-2x^2 + 5x + 3}{x^2 - 4x - 32} \div \dfrac{8x^2 + 10x + 3}{6x^2 - 6x - 12}$

56. AIRPLANES Use the formula $d = rt$ and the following information. An airplane is traveling at a rate r of 500 miles per hour for a time t of $(6 + x)$ hours. A second airplane travels at the rate of $(540 + 90x)$ miles per hour for a time t of 6 hours.

a. Write a rational expression to represent the ratio of the distance d traveled by the first airplane to the distance d traveled by the second airplane.

b. Simplify the rational expression. What does this expression tell you about the distances traveled by the two airplanes?

c. Under what condition is the rational expression undefined? Describe what this condition would tell you about the two airplanes.

Real-World Link

The fastest train in the world is Japan's Shin Kansen line that travels at a top speed of 186 miles per hour.

Source: Yahoo

57. **TRAINS** Trying to get into a train yard one evening, all of the trains are backed up for 2 miles along a system of tracks. Assume that each car occupies an average of 75 feet of space on a track and that the train yard has 5 tracks.

a. Write an expression that could be used to determine the number of train cars involved in the backup.

b. How many train cars are involved in the backup?

c. Suppose that there are 8 attendants doing safety checks on each car, and it takes each vehicle an average of 45 seconds for each check. Approximately how many hours will it take for all the vehicles in the backup to exit?

58. **MULTIPLE REPRESENTATIONS** In this problem, you will investigate the graph of a rational function.

a. **ALGEBRAIC** Simplify $\frac{x^2 - 5x + 4}{x - 4}$.

b. **TABULAR** Let $f(x) = \frac{x^2 - 5x + 4}{x - 4}$. Use the expression you wrote in part **a** to write the related function $g(x)$. Use a graphing calculator to make a table for both functions for $0 \leq x \leq 10$.

c. **ANALYTICAL** What are $f(4)$ and $g(4)$? Explain the significance of these values.

d. **GRAPHICAL** Graph the functions on the graphing calculator. Use the TRACE function to investigate each graph, using the ▲ and ▼ keys to switch from one graph to the other. Compare and contrast the graphs.

e. **VERBAL** What conclusions can you draw about the expressions and the functions?

H.O.T. Problems Use Higher-Order Thinking Skills

59. **REASONING** Compare and contrast $\frac{(x - 6)(x + 2)(x + 3)}{x + 3}$ and $(x - 6)(x + 2)$.

60. **FIND THE ERROR** Troy and Beverly are simplifying $\frac{x + y}{x - y} \div \frac{4}{y - x}$. Is either of them correct? Explain your reasoning.

Troy

$$\frac{x+y}{x-y} \div \frac{4}{y-x} = \frac{x-y}{x+y} \cdot \frac{4}{y-x} = \frac{-4}{x+y}$$

Beverly

$$\frac{x+y}{x-y} \div \frac{4}{y-x} = \frac{x+y}{x-y} \cdot \frac{y-x}{4} = -\frac{x+y}{4}$$

61. **CHALLENGE** Find the value that makes the following statement true.

$$\frac{x - 6}{x + 3} \cdot \frac{?}{x - 6} = x - 2$$

62. **WHICH ONE DOESN'T BELONG?** Identify the expression that does not belong with the other three. Explain your reasoning.

$\frac{1}{x - 1}$	$\frac{x^2 + 3x + 2}{x - 5}$	$\frac{x + 1}{\sqrt{x + 3}}$	$\frac{x^2 + 1}{3}$

63. **REASONING** Determine whether the following statement is *sometimes, always,* or *never* true. Explain your reasoning.

A rational function that has a variable in the denominator is defined for all values of x.

64. **OPEN ENDED** Write a rational expression that simplifies to $\frac{x - 1}{x + 4}$.

65. **WRITING IN MATH** The rational expression $\frac{x^2 + 3x}{4x}$ is simplified to $\frac{x + 3}{4}$. Explain why this new expression is not defined for all values of x.

NYSRE PRACTICE A2.A.16, A2.A.17

66. SAT/ACT The Mason family wants to drive an average of 250 miles per day on their vacation. On the first five days, they travel 220 miles, 300 miles, 210 miles, 275 miles, and 240 miles. How many miles must they travel on the sixth day to meet their goal?

A 235 miles
B 255 miles
C 275 miles
D 315 miles

67. Which of the following equations gives the relationship between N and T in the table?

N	1	2	3	4	5	6
T	1	4	7	10	13	16

F $T = 2 - N$
G $T = 4 - 3N$
H $T = 3N + 1$
J $T = 3N - 2$

68. Cell phone calls cost 15 cents per minute for the first 12 minutes and 9 cents per minute thereafter. Which of the following represents the amount of money needed (in dollars) to talk for x minutes?

A $1.80 + 0.09(12 - x)$
B $1.80 + 0.09(x - 12)$
C $1.80 + 0.09x$
D $1.80 + 0.12x$

69. SHORT RESPONSE The area of a circle 6 meters in diameter exceeds the combined areas of a circle 4 meters in diameter and a circle 2 meters in diameter by how many square meters?

Spiral Review

70. ANTHROPOLOGY An anthropologist studying the bones of a prehistoric person finds there is so little remaining Carbon-14 in the bones that instruments cannot measure it. This means that there is less than 0.5% of the amount of Carbon-14 the bones would have contained when the person was alive. The half-life of Carbon-14 is 5760 years. How long ago did the person die? (Lesson 8-8)

Solve each equation. Round to the nearest ten thousandth. (Lesson 8-7)

71. $3e^x + 1 = 5$
72. $2e^x - 1 = 0$
73. $-3e^{4x} + 11 = 2$
74. $8 + 3e^{3x} = 26$

75. NOISE ORDINANCE A proposed city ordinance will make it illegal in a residential area to create sound that exceeds 72 decibels during the day and 55 decibels during the night. How many times as intense is the noise level allowed during the day as at night? (Lesson 8-3)

Simplify. (Lesson 7-5)

76. $\sqrt{50x^4}$
77. $\sqrt[3]{16y^3}$
78. $\sqrt{18x^2y^3}$
79. $\sqrt{40a^3b^4}$

80. AUTOMOBILES The length of the cargo space in a sport-utility vehicle is 4 inches greater than the height of the space. The width is 16 inches less than twice the height. The cargo space has a total volume of 55,296 cubic inches. (Lesson 6-8)

a. Write a polynomial function that represents the volume of the cargo space.

b. Will a package 34 inches long, 44 inches wide, and 34 inches tall fit in the cargo space? Explain.

Skills Review

Simplify. (Lesson 6-1)

81. $(2a + 3b) + (8a - 5b)$
82. $(x^2 - 4x + 3) - (4x^2 + 3x - 5)$
83. $(5y + 3y^2) + (-8y - 6y^2)$
84. $2x(3y + 9)$
85. $(x + 6)(x + 3)$
86. $(x + 1)(x^2 - 2x + 3)$

9-2 Adding and Subtracting Rational Expressions

Then

You added and subtracted polynomial expressions. (Lesson 6-2)

Now

- Determine the LCM of polynomials.
- Add and subtract rational expressions.

NYS Core Curriculum

A2.A.16 Perform arithmetic operations with rational expressions and rename to lowest terms
A2.A.17 Simplify complex fractional expressions

NY Math Online

glencoe.com
- Extra Examples
- Personal Tutor
- Self-Check Quiz
- Homework Help

Why?

As a fire engine moves toward a person, the pitch of the siren sounds higher to that person than it would if the fire engine were at rest. This is because the sound waves are compressed closer together, referred to as the *Doppler effect*. The Doppler effect can be represented by the rational expression $P_0\left(\frac{s_0}{s_0 - v}\right)$, where P_0 is the actual pitch of the siren, v is the speed of the fire truck, and s_0 is the speed of sound in air.

LCM of Polynomials Just as with rational numbers in fractional form, to add or subtract two rational expressions that have unlike denominators, you must first find the least common denominator (LCD). The LCD is the least common multiple (LCM) of the denominators.

To find the LCM of two or more numbers or polynomials, factor them. The LCM contains each factor the greatest number of times it appears as a factor.

Numbers

$$\frac{5}{6} + \frac{4}{9}$$

LCM of 6 and 9

$6 = 2 \cdot 3$
$9 = 3 \cdot 3$

LCM $= 2 \cdot 3 \cdot 3$ or 18

Polynomials

$$\frac{3}{x^2 - 3x + 2} + \frac{5}{2x^2 - 2}$$

LCM of $x^2 - 3x + 2$ and $2x^2 - 2$

$x^2 - 3x + 2 = (x - 1)(x - 2)$
$2x^2 - 2 = 2 \cdot (x - 1)(x + 1)$

LCM $= 2(x - 1)(x - 2)(x + 1)$

EXAMPLE 1 LCM of Monomials and Polynomials

Find the LCM of each set of polynomials.

a. $6xy$, $15x^2$, and $9xy^4$

$6xy = 2 \cdot 3 \cdot x \cdot y$ — Factor the first monomial.
$15x^2 = 3 \cdot 5 \cdot x^2$ — Factor the second monomial.
$9xy^4 = 3 \cdot 3 \cdot x \cdot y^4$ — Factor the third monomial.

LCM $= 2 \cdot 3 \cdot 3 \cdot 5 \cdot x^2 \cdot y^4$
$= 90x^2y^4$ — Use each factor the greatest number of times it appears. Then simplify.

b. $y^4 + 8y^3 + 15y^2$ and $y^2 - 3y - 40$

$y^4 + 8y^3 + 15y^2 = y^2(y + 5)(y + 3)$ — Factor the first polynomial.
$y^2 - 3y - 40 = (y + 5)(y - 8)$ — Factor the second polynomial.

LCM $= y^2(y + 5)(y + 3)(y - 8)$ — Use each factor the greatest number of times it appears as a factor.

Check Your Progress

1A. $12a^2b$, $15abc$, $8b^3c^4$

1B. $4a^2 - 12a - 16$ and $a^3 - 9a^2 + 20a$

Personal Tutor glencoe.com

Add and Subtract Rational Expressions As with fractions, rational expressions must have common denominators in order to be added or subtracted.

Key Concept

For Your FOLDABLE

Adding Rational Expressions

Words To add rational expressions, find the least common denominator (LCD). Rewrite each expression with the LCD. Then add.

Symbols For all $\frac{a}{b}$ and $\frac{c}{d}$, with $b \neq 0$ and $d \neq 0$, $\frac{a}{b} + \frac{c}{d} = \frac{ad}{bd} + \frac{bc}{bd} = \frac{ad + bc}{bd}$.

Subtracting Rational Expressions

Words To subtract rational expressions, find the least common denominator (LCD). Rewrite each expression with the LCD. Then subtract.

Symbols For all $\frac{a}{b}$ and $\frac{c}{d}$, with $b \neq 0$ and $d \neq 0$, $\frac{a}{b} - \frac{c}{d} = \frac{ad}{bd} - \frac{bc}{bd} = \frac{ad - bc}{bd}$.

EXAMPLE 2 Monomial Denominators

Simplify $\frac{3y}{2x^3} + \frac{5z}{8xy^2}$.

$\frac{3y}{2x^3} + \frac{5z}{8xy^2} = \frac{3y}{2x^3} \cdot \frac{4y^2}{4y^2} + \frac{5z}{8xy^2} \cdot \frac{x^2}{x^2}$ The LCD is $8x^3y^2$.

$= \frac{12y^3}{8x^3y^2} + \frac{5x^2z}{8x^3y^2}$ Multiply fractions.

$= \frac{12y^3 + 5x^2z}{8x^3y^2}$ Add the numerators.

StudyTip

Simplifying Rational Expressions After you add or subtract rational expressions, it is possible that the resulting expression can be further simplified.

Check Your Progress Simplify each expression.

2A. $\frac{4}{5a^3b^2} + \frac{9c}{10ab}$

2C. $\frac{3a^2}{16b^2} - \frac{8x}{5a^3b}$

Personal Tutor glencoe.com

The LCD is also used to combine rational expressions with polynomial denominators.

EXAMPLE 3 Polynomial Denominators

Simplify $\frac{5}{6x - 18} - \frac{x - 1}{4x^2 - 14x + 6}$.

$\frac{5}{6x - 18} - \frac{x - 1}{4x^2 - 14x + 6} = \frac{5}{6(x - 3)} - \frac{x - 1}{2(2x - 1)(x - 3)}$ Factor denominators.

$= \frac{5(2x - 1)}{6(x - 3)(2x - 1)} - \frac{(x - 1)(3)}{2(2x - 1)(x - 3)(3)}$ Multiply by missing factors.

$= \frac{10x - 5 - 3x + 3}{6(x - 3)(2x - 1)}$ Subtract numerators.

$= \frac{7x - 2}{6(x - 3)(2x - 1)}$ Simplify.

Check Your Progress Simplify each expression.

3A. $\frac{x - 1}{x^2 - x - 6} - \frac{4}{5x + 10}$

3B. $\frac{x - 8}{4x^2 + 21x + 5} + \frac{6}{12x + 3}$

Personal Tutor glencoe.com

One way to simplify a complex fraction is to simplify the numerator and the denominator separately, and then simplify the resulting expressions.

EXAMPLE 4 Complex Fractions with Different LCDs

Simplify $\dfrac{1+\frac{1}{x}}{1-\frac{x}{y}}$.

$$\frac{1+\frac{1}{x}}{1-\frac{x}{y}} = \frac{\frac{x}{x}+\frac{1}{x}}{\frac{y}{y}-\frac{x}{y}}$$ The LCD of the numerator is *x*. The LCD of the denominator is *y*.

$$= \frac{\frac{x+1}{x}}{\frac{y-x}{y}}$$ Simplify the numerator and denominator.

$$= \frac{x+1}{x} \div \frac{y-x}{y}$$ Write as a division expression.

$$= \frac{x+1}{x} \cdot \frac{y}{y-x}$$ Multiply by the reciprocal of the divisor.

$$= \frac{xy+y}{xy-x^2}$$ Simplify.

StudyTip

Undefined Terms Remember that there are restrictions on variables in the denominator.

Check Your Progress

Simplify each expression.

4A. $\dfrac{1-\frac{y}{x}}{\frac{1}{y}+\frac{1}{x}}$

4B. $\dfrac{\frac{c}{d}-\frac{d}{c}}{\frac{d}{c}+2}$

Personal Tutor glencoe.com

Another method of simplifying complex fractions is to find the LCD of all of the denominators. Then, the denominators are all eliminated by multiplying by the LCD.

EXAMPLE 5 Complex Fractions with Same LCD

Simplify $\dfrac{1+\frac{1}{x}}{1-\frac{x}{y}}$.

$$\frac{1+\frac{1}{x}}{1-\frac{x}{y}} = \frac{\left(1+\frac{1}{x}\right)}{\left(1-\frac{x}{y}\right)} \cdot \frac{xy}{xy}$$ The LCD of all of the denominators is *xy*. Multiply by $\frac{xy}{xy}$.

$$= \frac{xy+y}{xy-x^2}$$ Distribute *xy*.

Notice that the same problem is solved in Examples 4 and 5 using different methods, but both produce the same answer. So, how you solve problems similar to these is left up to your own discretion.

Check Your Progress

Simplify each expression.

5A. $\dfrac{1+\frac{2}{x}}{\frac{3}{y}-\frac{4}{x}}$

5B. $\dfrac{\frac{1}{d}-\frac{d}{c}}{\frac{1}{c}+6}$

5C. $\dfrac{\frac{1}{y}+\frac{1}{x}}{\frac{1}{y}-\frac{1}{x}}$

5D. $\dfrac{\frac{a}{b}+1}{1-\frac{b}{a}}$

Personal Tutor glencoe.com

Check Your Understanding

Example 1 p. 562

Find the LCM of each set of polynomials.

1. $16x, 8x^2y^3, 5x^3y$
2. $7a^2, 9ab^3, 21abc^4$
3. $3y^2 - 9y, y^2 - 8y + 15$
4. $x^3 - 6x^2 - 16x, x^2 - 4$

Examples 2 and 3 p. 563

Simplify each expression.

5. $\frac{12y}{5x} + \frac{5x}{4y^3}$
6. $\frac{5}{6ab} + \frac{3b^2}{14a^3}$
7. $\frac{7b}{12a} - \frac{1}{18ab^3}$
8. $\frac{y^2}{8c^2d^2} - \frac{3x}{14c^4d}$
9. $\frac{4x}{x^2 + 9x + 18} + \frac{5}{x + 6}$
10. $\frac{8}{y - 3} + \frac{2y - 5}{y^2 - 12y + 27}$
11. $\frac{4}{3x + 6} - \frac{x + 1}{x^2 - 4}$
12. $\frac{3a + 2}{a^2 - 16} - \frac{7}{6a + 24}$

13. **GEOMETRY** Find the perimeter of the rectangle.

Examples 4 and 5 p. 564

Simplify each expression.

14. $\frac{4 + \frac{2}{x}}{3 - \frac{2}{x}}$
15. $\frac{6 + \frac{4}{y}}{2 + \frac{6}{y}}$
16. $\frac{\frac{3}{x} + \frac{2}{y}}{1 + \frac{4}{y}}$
17. $\frac{\frac{2}{b} + \frac{5}{a}}{\frac{3}{a} - \frac{8}{b}}$

Practice and Problem Solving

● = **Step-by-Step Solutions** begin on page R20.
Extra Practice begins on page 947.

Example 1 p. 562

Find the LCM of each set of polynomials.

18. $24cd, 40a^2c^3d^4, 15abd^3$
19. $4x^2y^3, 18xy^4, 10xz^2$
20. $x^2 - 9x + 20, x^2 + x - 30$
21. $6x^2 + 21x - 12, 4x^2 + 22x + 24$

Examples 2 and 3 p. 563

Simplify each expression.

22. $\frac{5a}{24cf^4} + \frac{a}{36bc^4f^3}$
23. $\frac{4b}{15x^3y^2} - \frac{3b}{35x^2y^4z}$
24. $\frac{5b}{6a} + \frac{3b}{10a^2} + \frac{2}{ab^2}$
25. $\frac{4}{3x} + \frac{8}{x^3} + \frac{2}{5xy}$
26. $\frac{8}{3y} + \frac{2}{9} - \frac{3}{10y^2}$
27. $\frac{1}{16a} + \frac{5}{12b} - \frac{9}{10b^3}$
28. $\frac{8}{x^2 - 6x - 16} + \frac{9}{x^2 - 3x - 40}$
29. $\frac{6}{y^2 - 2y - 35} + \frac{4}{y^2 + 9y + 20}$
30. $\frac{12}{3y^2 - 10y - 8} - \frac{3}{y^2 - 6y + 8}$
31. $\frac{6}{2x^2 + 11x - 6} - \frac{8}{x^2 + 3x - 18}$
32. $\frac{2x}{4x^2 + 9x + 2} + \frac{3}{2x^2 - 8x - 24}$
33. $\frac{4x}{3x^2 + 3x - 18} - \frac{2x}{2x^2 + 11x + 15}$

34. **BIOLOGY** After a person eats something, the pH or acid level A of his or her mouth can be determined by the formula $A = \frac{20.4t}{t^2 + 36} + 6.5$, where t is the number of minutes that have elapsed since the food was eaten.
    a. Simplify the equation.
    b. What would the acid level be after 30 minutes?

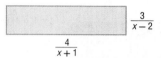

35. GEOMETRY Both triangles in the figure at the right are equilateral. If the area of the smaller triangle is 200 square centimeters and the area of the larger triangle is 300 square centimeters, find the minimum distance from A to B in terms of x and y and simplify.

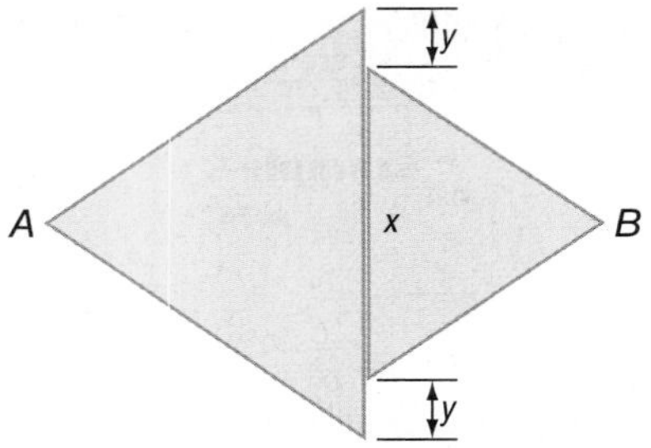

Examples 4 and 5
p. 564

Simplify each expression.

36. $\dfrac{\dfrac{2}{x-3}+\dfrac{3x}{x^2-9}}{\dfrac{3}{x+3}-\dfrac{4x}{x^2-9}}$

37. $\dfrac{\dfrac{4}{x+5}+\dfrac{9}{x-6}}{\dfrac{5}{x-6}-\dfrac{8}{x+5}}$

38. $\dfrac{\dfrac{5}{x+6}-\dfrac{2x}{2x-1}}{\dfrac{x}{2x-1}+\dfrac{4}{x+6}}$

39. $\dfrac{\dfrac{8}{x-9}-\dfrac{x}{3x+2}}{\dfrac{3}{3x+2}+\dfrac{4x}{x-9}}$

40. OIL PRODUCTION Managers of an oil company have estimated that oil will be pumped from a certain well at a rate based on the function $R(x)=\frac{20}{x}+\frac{200x}{3x^2+20}$, where $R(x)$ is the rate of production in thousands of barrels per year x years after pumping begins.

a. Simplify $R(x)$.

b. At what rate will oil be pumping from the well in 50 years?

Real-World Link

Prudhoe Bay, Alaska, is home to the largest oil field in the United States with cumulative production and reserves of over 13 billion barrels.

Source: Energy Information Administration

Find the LCM of each set of polynomials.

41. $12xy^4, 14x^4y^2, 5xyz^3, 15x^5y^3$

42. $-6abc^2, 18a^2b^2, 15a^4c, 8b^3$

43. $x^2-3x-28, 2x^2+9x+4, x^2-16$

44. $x^2-5x-24, x^2-9, 3x^2+8x-3$

Simplify each expression.

45. $\frac{1}{12a}+6-\frac{3}{5a^2}$

46. $\frac{5}{16y^2}-4-\frac{8}{3x^2y}$

47. $\frac{5}{6x^2+46x-16}+\frac{2}{6x^2+57x+72}$

48. $\frac{1}{8x^2-20x-12}+\frac{4}{6x^2+27x+12}$

49. $\frac{x^2+y^2}{x^2-y^2}+\frac{y}{x+y}-\frac{x}{x-y}$

50. $\frac{x^2+x}{x^2-9x+8}+\frac{4}{x-1}-\frac{3}{x-8}$

51. $\dfrac{\dfrac{2}{a-1}+\dfrac{3}{a-4}}{\dfrac{6}{a^2-5a+4}}$

52. $\dfrac{\dfrac{1}{x}+\dfrac{1}{y}}{\left(\dfrac{1}{x}-\dfrac{1}{y}\right)(x+y)}$

53. GEOMETRY An expression for the length of one rectangle is $\frac{x^2-9}{x-2}$. The length of a similar rectangle is expressed as $\frac{x+3}{x^2-4}$. What is the scale factor of the two rectangles? Write in simplest form.

54. KAYAKING Cameron is taking a 20-mile kayaking trip. He travels half the distance at one rate. The rest of the distance he travels 2 miles per hour slower.

a. If x represents the faster pace in miles per hour, write an expression that represents the time spent at that pace.

b. Write an expression for the amount of time spent at the slower pace.

c. Write an expression for the amount of time Cameron needed to complete the trip.

Find the slope of the line that passes through each pair of points.

55. $A\left(\frac{2}{p}, \frac{1}{2}\right)$ and $B\left(\frac{1}{3}, \frac{3}{p}\right)$

56. $C\left(\frac{1}{4}, \frac{4}{q}\right)$ and $D\left(\frac{5}{q}, \frac{1}{5}\right)$

57. $E\left(\frac{7}{w}, \frac{1}{7}\right)$ and $F\left(\frac{1}{7}, \frac{7}{w}\right)$

58. $G\left(\frac{6}{n}, \frac{1}{6}\right)$ and $H\left(\frac{1}{6}, \frac{6}{n}\right)$

Real-World Link

Macro photography is commonly described as being able to focus on a subject closely so that when a regular 4″ × 6″ print is made, the image is life-size or larger.

59. PHOTOGRAPHY The focal length of a lens establishes the field of view of the camera. The shorter the focal length is, the larger the field of view. For a camera with a fixed focal length of 70 mm to focus on an object x mm from the lens, the film must be placed a distance y from the lens. This is represented by $\frac{1}{x} + \frac{1}{y} = \frac{1}{70}$.

a. Express y as a function of x.

b. What happens to the focusing distance when the object is 70 mm away?

60. PHARMACOLOGY Two drugs are administered to a patient. The concentrations in the bloodstream of each are given by $f(t) = \frac{2t}{3t^2 + 9t + 6}$ and $g(t) = \frac{3t}{2t^2 + 6t + 4}$ where t is the time, in hours, after the drugs are administered.

a. Add the two functions together to determine a function for the total concentration of drugs in the patient's bloodstream.

b. What is the concentration of drugs after 8 hours?

61. DOPPLER EFFECT Refer to the application at the beginning of the lesson. George is equidistant from two fire engines traveling toward him from opposite directions.

a. Let x be the speed of the faster fire engine and y be the speed of the slower fire engine. Write and simplify a rational expression representing the difference in pitch between the two sirens according to George.

b. If one is traveling at 45 meters per second and the other is traveling at 70 meters per second, what is the difference in their pitches according to George? The speed of sound in air is 332 meters per second, and both engines have a siren with a pitch of 500 Hz.

62. RESEARCH A student studying learning behavior performed an experiment in which a rat was repeatedly sent through a maze. It was determined that the time it took the rat to complete the maze followed the rational function $T(x) = 4 + \frac{10}{x}$, where x represented the number of trials.

a. What is the domain of the function?

b. Graph the function for $0 \leq x \leq 10$.

c. Make a table of the function for $x = 20, 50, 100, 200$, and 400.

d. If it were possible to have an infinite number of trials, what do you think would be the rat's best time? Explain your reasoning.

H.O.T. Problems

Use Higher-Order Thinking Skills

63. CHALLENGE Simplify $\dfrac{5x^{-2} - \frac{x+1}{x}}{\frac{4}{3 - x^{-1}} + 6x^{-1}}$.

64. REASONING Determine whether the following statement is *true* or *false*. Explain your reasoning.

$$\frac{6}{x+2} + \frac{4}{x-3} = \frac{10x - 10}{(x+2)(x-3)} \textit{ for all values of } x.$$

65. OPEN ENDED Write three monomials with an LCM of $180a^4b^6c$.

66. WRITING IN MATH Explain how to add rational expressions that have unlike denominators.

NYSRE PRACTICE A2.A.16, A2.A.17

67. PROBABILITY A drawing is to be held to select the winner of a new bike. There are 100 seniors, 150 juniors, and 200 sophomores who had correct entries. The drawing will contain 3 tickets for each senior name, 2 for each junior, and 1 for each sophomore. What is the probability that a senior's ticket will be chosen?

A $\frac{1}{8}$ C $\frac{2}{7}$

B $\frac{2}{9}$ D $\frac{3}{8}$

68. SHORT RESPONSE Find the area of the figure.

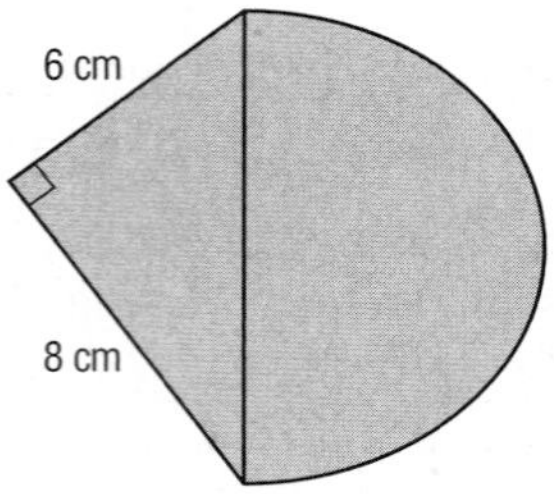

69. SAT/ACT If Mauricio receives b books in addition to the number of books he had, he will have t times as many as he had originally. In terms of b and t, how many books did Mauricio have at the beginning?

F $\frac{b}{t-1}$ H $\frac{t+1}{b}$

G $\frac{b}{t+1}$ J $\frac{b}{t}$

70. If $\frac{2a}{a} + \frac{1}{a} = 4$, then $a =$ ___.

A $-\frac{1}{8}$ C $\frac{1}{8}$

B $\frac{1}{2}$ D 2

Spiral Review

Simplify each expression. (Lesson 9-1)

71. $\frac{-4ab}{21c} \cdot \frac{14c^2}{22a^2}$ **72.** $\frac{x^2 - y^2}{6y} \div \frac{x+y}{36y^2}$ **73.** $\frac{n^2 - n - 12}{n+2} \div \frac{n-4}{n^2 - 4n - 12}$

74. BIOLOGY Bacteria usually reproduce by a process known as *binary fission*. In this type of reproduction, one bacterium divides, forming two bacteria. Under ideal conditions, some bacteria reproduce every 20 minutes. (Lesson 8-8)

a. Find the constant k for this type of bacteria under ideal conditions.

b. Write the equation for modeling the exponential growth of this bacterium.

Graph each function. State the domain and range of each function. (Lesson 7-3)

75. $y = -\sqrt{2x+1}$ **76.** $y = \sqrt{5x - 3}$ **77.** $y = \sqrt{x+6} - 3$

78. $y = 5 - \sqrt{x+4}$ **79.** $y = \sqrt{3x-6} + 4$ **80.** $y = 2\sqrt{3-4x} + 3$

Solve each equation. State the number and type of roots. (Lesson 6-7)

81. $3x + 8 = 0$ **82.** $2x^2 - 5x + 12 = 0$ **83.** $x^3 + 9x = 0$ **84.** $x^4 - 81 = 0$

Skills Review

Graph each function. (Lesson 5-7)

85. $y = 4(x+3)^2 + 1$ **86.** $y = -(x-5)^2 - 3$ **87.** $y = \frac{1}{4}(x-2)^2 + 4$

88. $y = \frac{1}{2}(x-3)^2 - 5$ **89.** $y = x^2 + 6x + 2$ **90.** $y = x^2 - 8x + 18$

9-3 Graphing Reciprocal Functions

Then

You graphed polynomial functions. (Lesson 6-5)

Now

- Determine properties of reciprocal functions.
- Graph transformations of reciprocal functions.

NYS Core Curriculum

A2.PS.6 Use a variety of strategies to extend solution methods to other problems

New Vocabulary

reciprocal function
hyperbola
asymptote

NY Math Online

glencoe.com

- Extra Examples
- Personal Tutor
- Self-Check Quiz
- Homework Help

Why?

The East High School Chorale wants to raise \$5000 to fund a trip to a national competition in Tampa. They have decided to sell candy bars. They will make a \$1 profit on each candy bar they sell, so they need to sell 5000 candy bars.

If c represents the number of candy bars each student has to sell and n represents the number of students, then $c = \frac{5000}{n}$.

Vertical and Horizontal Asymptotes The function $c = \frac{5000}{n}$ is a reciprocal function. A **reciprocal function** has an equation of the form $f(x) = \frac{1}{a(x)}$, where $a(x)$ is a linear function and $a(x) \neq 0$.

Key Concept — Parent Function of Reciprocal Functions

For Your FOLDABLE

Parent function: $f(x) = \frac{1}{x}$

Type of graph: **hyperbola**

Domain and range: all nonzero real numbers

Axes of symmetry: $x = 0$ and $f(x) = 0$

Intercepts: none

Not defined: $x = 0$ and $f(x) = 0$

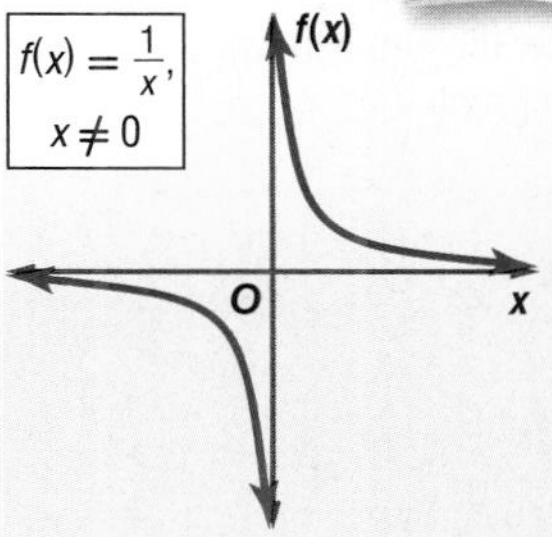

The domain of a reciprocal function is limited to values for which the function is defined.

Functions:	$f(x) = \frac{-3}{x+2}$	$g(x) = \frac{4}{x-5}$	$h(x) = \frac{3}{x}$
Not defined at:	$x = -2$	$x = 5$	$x = 0$

EXAMPLE 1 Limitations on Domain

Determine the values of x for which $f(x) = \frac{3}{x^2 - 6x + 5}$ is not defined.

Factor the denominator of the expression.

$\frac{3}{x^2 - 6x + 5} = \frac{3}{(x-1)(x-5)}$ The function is undefined for $x = 1$ and $x = 5$.

Check Your Progress

Determine the values of x for which each function is not defined.

1A. $\frac{2}{x^2 - 1}$

1B. $\frac{7}{x^2 + 6x + 8}$

Personal Tutor glencoe.com

The graphs of reciprocal functions may have breaks in continuity for excluded values. Some may have an **asymptote**, which is a line that the graph of the function approaches.

EXAMPLE 2 Determine Properties of Reciprocal Functions

Identify the asymptotes, domain, and range of each function.

a.

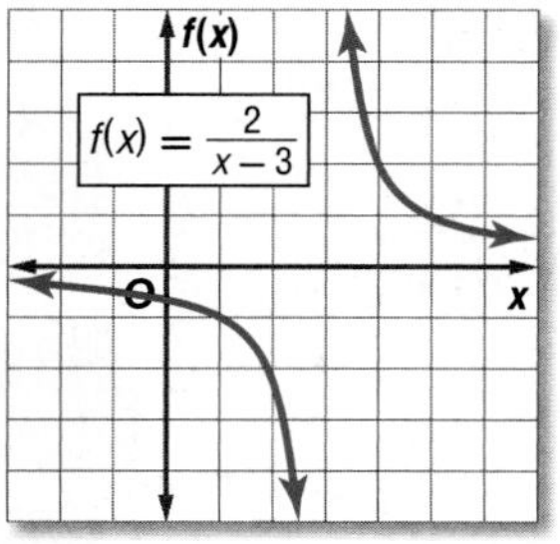

Identify x-values for which $f(x)$ is undefined.

$$x - 3 = 0$$
$$x = 3$$

$f(x)$ is not defined when $x = 3$. So there is an asymptote at $x = 3$.

From $x = 3$, as x-values decrease, $f(x)$-values approach 0, and as x-values increase, $f(x)$-values approach 0. So there is an asymptote at $f(x) = 0$.

The domain is all real numbers not equal to 3 and the range is all real numbers not equal to 0.

b.

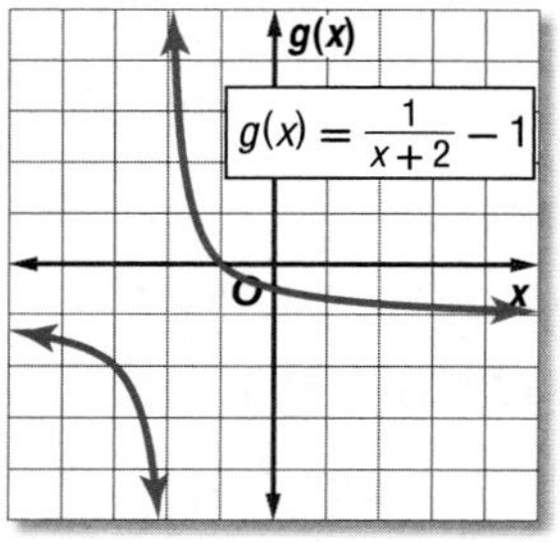

Identify x-values for which $g(x)$ is undefined.

$$x + 2 = 0$$
$$x = -2$$

$g(x)$ is not defined when $x = -2$. So there is an asymptote at $x = -2$.

From $x = -2$, as x-values decrease, $g(x)$-values approach -1, and as x-values increase, $g(x)$-values approach -1. So there is an asymptote at $g(x) = -1$.

The domain is all real numbers not equal to -2 and the range is all real numbers not equal to -1.

StudyTip

Asymptotes and Rational Functions Vertical asymptotes show where a function is undefined, while horizontal asymptotes show the end behavior of a graph.

Check Your Progress

2A.

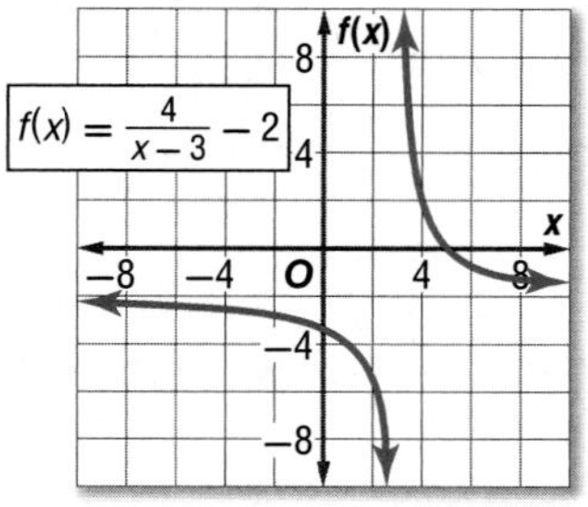

2B.

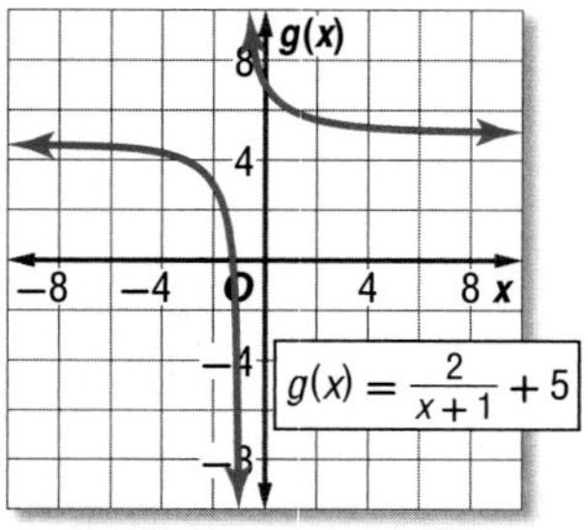

Personal Tutor glencoe.com

Transformations of Reciprocal Functions The same techniques used to transform the graphs of other functions you have studied can be applied to the graphs of reciprocal functions. In Example 2, note that the asymptotes have been moved along with the graphs of the functions.

Key Concept — Transformations of Reciprocal Functions

For Your FOLDABLE

$$f(x) = \frac{a}{x - h} + k$$

h **— Horizontal Translation**	***k*** **— Vertical Translation**
$\lvert h \rvert$ units right if h is positive $\lvert h \rvert$ units left if h is negative The *vertical* asymptote is at $x = h$.	$\lvert k \rvert$ units up if k is positive $\lvert k \rvert$ units down if k is negative The *horizontal* asymptote is at $f(x) = k$.

a **— Orientation and Shape**

If $a < 0$, the graph is reflected across the x-axis.

If $\lvert a \rvert > 1$, the graph is stretched vertically. If $0 < \lvert a \rvert < 1$, the graph is compressed vertically.

StudyTip

Asymptotes The asymptotes of a reciprocal function move with the graph of the function and intersect at (h, k).

EXAMPLE 3 Graph Transformations

Graph each function. State the domain and range.

a. $f(x) = \frac{2}{x - 4} + 2$

This represents a transformation of the graph of $f(x) = \frac{1}{x}$.

$a = 2$: The graph is expanded.

$h = 4$: The graph is translated 4 units right. There is an asymptote at $x = 4$.

$k = 2$: The graph is translated 2 units up. There is an asymptote at $f(x) = 2$.

Domain: $\{x \mid x \neq 4\}$ Range: $\{f(x) \mid f(x) \neq 2\}$

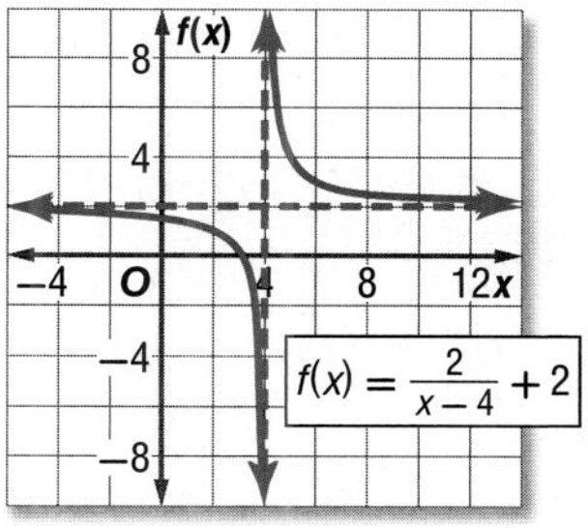

b. $f(x) = \frac{-3}{x + 1} - 4$

This represents a transformation of the graph of $f(x) = \frac{1}{x}$.

$a = -3$: The graph is expanded and reflected across the x-axis.

$h = -1$: The graph is translated 1 unit left. There is an asymptote at $x = -1$.

$k = -4$: The graph is translated 4 units down. There is an asymptote at $y = -4$.

Domain: $\{x \mid x \neq -1\}$ Range: $\{f(x) \mid f(x) \neq -4\}$

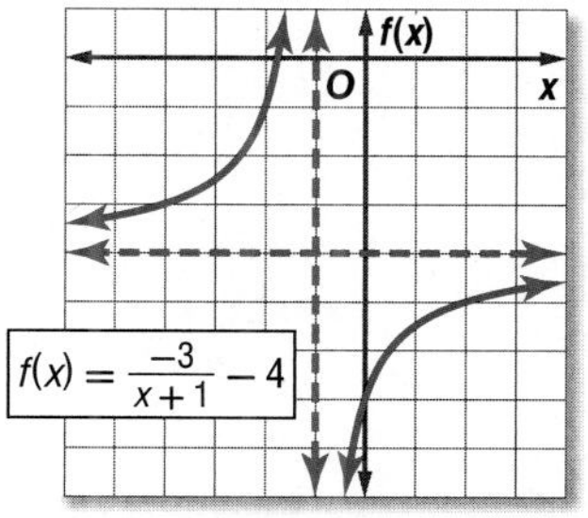

Check Your Progress

3A. $f(x) = \frac{-2}{x + 4} + 1$

3B. $g(x) = \frac{1}{3(x - 1)} - 2$

Personal Tutor glencoe.com

Reciprocal functions can be used to solve many real-world situations.

Real-World Career

Travel Agent Travel agents assess individual and business needs to help make the best possible travel arrangements. They may specialize by type of travel, such as leisure or business, or by destination, such as Europe or Africa. A high school diploma is required, and vocational training is preferred.

Real-World EXAMPLE 4 Write Equations

TRAVEL An airline has a daily nonstop flight between Los Angeles, California, and Sydney, Australia. A one-way trip is about 7500 miles.

a. Write an equation to represent the travel time from Los Angeles to Sydney as a function of flight speed. Then graph the equation.

Solve the formula $rt = d$ for t.

$rt = d$ **Original formula**

$t = \frac{d}{r}$ **Divide each side by r.**

$t = \frac{7500}{r}$ **$d = 7500$**

Graph the equation $t = \frac{7500}{r}$.

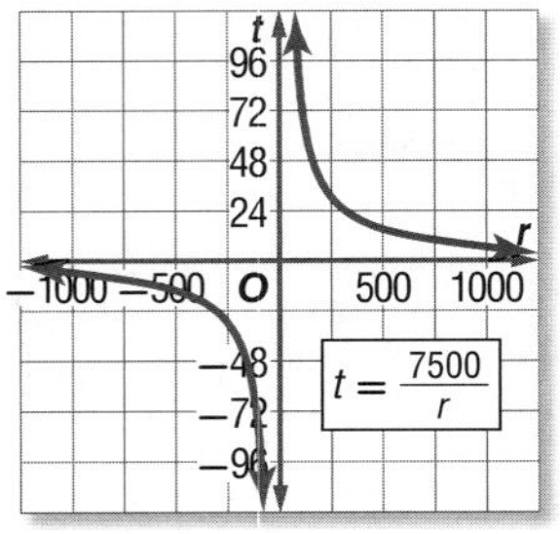

b. Explain any limitations to the range or domain in this situation.

In this situation, the range and domain are limited to all real numbers greater than zero because negative values do not make sense. There will be further restrictions to the domain because the aircraft has minimum and maximum speeds at which it can travel.

Check Your Progress

4. **HOMECOMING DANCE** The junior and senior class officers are sponsoring a homecoming dance. The total cost for the facilities and catering is \$45 per person plus a \$2500 deposit. Write and graph an equation to represent the average cost per person. Then explain any limitations to the domain and range.

Personal Tutor glencoe.com

Check Your Understanding

Examples 1 and 2 pp. 569–570

Identify the asymptotes, domain, and range of each function.

1.

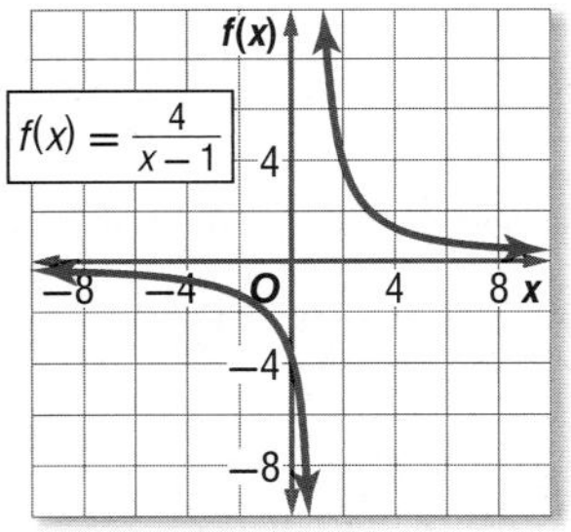

2.

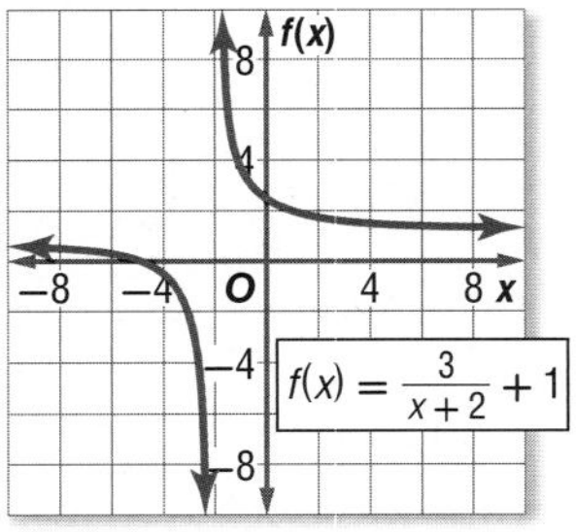

Example 3 p. 571

Graph each function. State the domain and range.

3. $f(x) = \frac{5}{x}$

4. $f(x) = \frac{2}{x + 3}$

5. $f(x) = \frac{-1}{x - 2} + 4$

Example 4 p. 572

6. **GROUP GIFT** A group of friends plans to get their youth group leader a gift certificate for a day at a spa. The certificate costs \$150.

   a. If c represents the cost for each friend and f represents the number of friends, write an equation to represent the cost to each friend as a function of how many friends give.

   b. Graph the function.

   c. Explain any limitations to the range or domain in this situation.

Practice and Problem Solving

= Step-by-Step Solutions begin on page R20.
Extra Practice begins on page 947.

Examples 1 and 2
pp. 569–570

Identify the asymptotes, domain, and range of each function.

7.

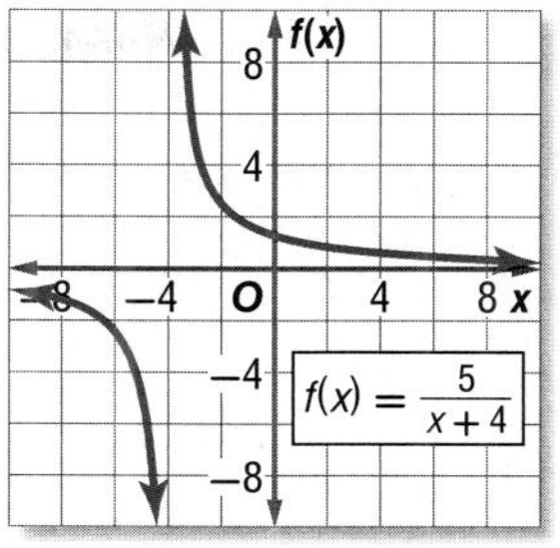

8.

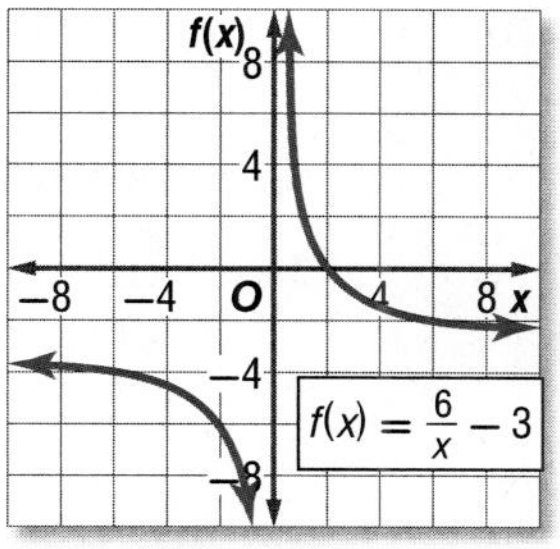

9.

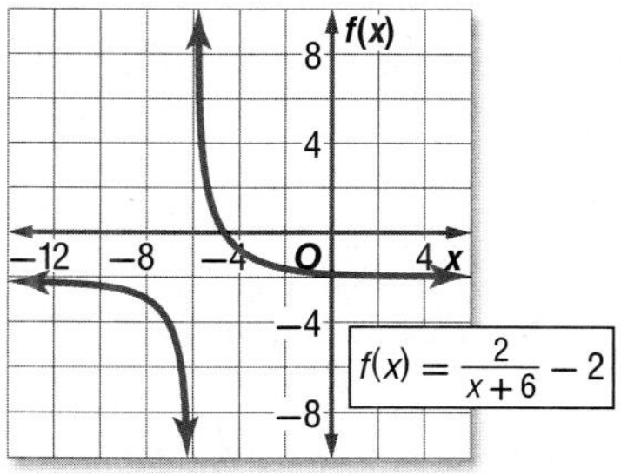

10.

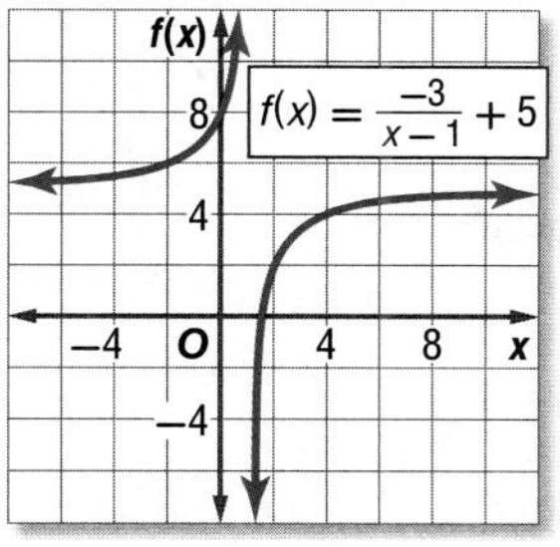

Example 3
p. 571

Graph each function. State the domain and range.

11. $f(x) = \frac{3}{x}$

12. $f(x) = \frac{-4}{x + 2}$

13. $f(x) = \frac{2}{x - 6}$

14. $f(x) = \frac{6}{x} - 5$

15 $f(x) = \frac{2}{x} + 3$

16. $f(x) = \frac{8}{x}$

17. $f(x) = \frac{-2}{x - 5}$

18. $f(x) = \frac{3}{x - 7} - 8$

19. $f(x) = \frac{9}{x + 3} + 6$

20. $f(x) = \frac{8}{x + 3}$

21. $f(x) = \frac{-6}{x + 4} - 2$

22. $f(x) = \frac{-5}{x - 2} + 2$

Example 4
p. 572

23. **CYCLING** Marina's New Year's resolution is to ride her bike 5000 miles.

a. If m represents the mileage Marina rides each day and d represents the number of days, write an equation to represent the mileage each day as a function of the number of days that she rides.

b. Graph the function.

c. If she rides her bike every day of the year, how many miles should she ride each day to meet her goal?

24. **CHEMISTRY** Parker has 200 grams of an unknown liquid. Knowing the density will help him discover what type of liquid this is.

a. Density of a liquid is found by dividing the mass by the volume. Write an equation to represent the density of this unknown as a function of volume.

b. Graph the function.

c. From the graph, identify the asymptotes, domain, and range of the function.

Graph each function. State the domain and range.

25. $f(x) = \frac{3}{2x - 4}$

26. $f(x) = \frac{5}{3x}$

27. $f(x) = \frac{2}{4x + 1}$

28. $f(x) = \frac{1}{2x + 3}$

29 **BASEBALL** The distance from the pitcher's mound to home plate is 60.5 feet.

a. If r represents the speed of the pitch and t represents the time it takes the ball to get to the plate, write an equation to represent the speed as a function of time.

b. Graph the function.

c. If a two-seam fastball reaches the plate in 0.48 second, what was its speed?

Real-World Link

Breaking the color barrier in 1947, Larry Doby became the first African-American player in the American League when he joined the Cleveland Indians. That same year, Doby also became the first African-American to play in the American Basketball League.

Graph each function. State the domain and range, and identify the asymptotes.

30. $f(x) = \frac{-3}{x + 7} - 1$ **31.** $f(x) = \frac{-4}{x + 2} - 5$ **32.** $f(x) = \frac{6}{x - 1} + 2$

33. $f(x) = \frac{2}{x - 4} + 3$ **34.** $f(x) = \frac{-7}{x - 8} - 9$ **35.** $f(x) = \frac{-6}{x - 7} - 8$

36. **AUTOMOBILES** Lawanda's car went 440 miles on one tank of gas.

a. If g represents the number of miles to the gallon that the car gets and t represents the size of the gas tank, write an equation to represent the miles to the gallon as a function of tank size.

b. Graph the function.

c. How many miles does the car get per gallon if it has a 15-gallon tank?

37. **MULTIPLE REPRESENTATIONS** Consider the functions $f(x) = \frac{1}{x}$ and $g(x) = \frac{1}{x^2}$.

a. **TABULAR** Make a table of values comparing the two functions.

b. **GRAPHICAL** Use the table of values to graph both functions.

c. **VERBAL** Compare and contrast the two graphs.

d. **ANALYTICAL** Make a conjecture about the difference between the graphs of reciprocal functions with an even exponent in the denominator and those with an odd exponent in the denominator.

H.O.T. Problems

Use Higher-Order Thinking Skills

38. **OPEN ENDED** Write a reciprocal function for which the graph has a vertical asymptote at $x = -4$ and a horizontal asymptote at $f(x) = 6$.

39. **REASONING** Compare and contrast the graphs of each pair of equations.

a. $y = \frac{1}{x}$ and $y - 7 = \frac{1}{x}$ **b.** $y = \frac{1}{x}$ and $y = 4\left(\frac{1}{x}\right)$ **c.** $y = \frac{1}{x}$ and $y = \frac{1}{x + 5}$

d. Without making a table of values, use what you observed in parts **a–c** to sketch a graph of $y - 7 = 4\left(\frac{1}{x + 5}\right)$.

40. **WHICH ONE DOESN'T BELONG?** Find the function that does not belong. Explain.

$f(x) = \frac{3}{x + 1}$	$g(x) = \frac{x + 2}{x^2 + 1}$	$h(x) = \frac{5}{x^2 + 2x + 1}$	$j(x) = \frac{20}{x - 7}$

41. **CHALLENGE** Write two different reciprocal functions with graphs having the same vertical and horizontal asymptotes. Then graph the functions.

42. **WRITING IN MATH** Refer to the beginning of the lesson. Explain how rational functions can be used in fundraising. Explain why only part of the graph is meaningful in the context of the problem.

NYSRE PRACTICE Reinforcement of A.A.15

43. SHORT RESPONSE What is the value of $(x + y)(x + y)$ if $xy = -3$ and $x^2 + y^2 = 10$?

44. GRIDDED RESPONSE If $x = 2y$, $y = 4z$, $2z = w$, and $w \neq 0$, then $\frac{x}{w} =$ ____.

45. If $c = 1 + \frac{1}{d}$ and $d > 1$, then c could equal ____.

A $\frac{5}{7}$

B $\frac{9}{7}$

C $\frac{15}{7}$

D $\frac{19}{7}$

46. SAT/ACT A car travels m miles at the rate of t miles per hour. How many hours does the trip take?

F $\frac{m}{t}$

G $m - t$

H $\frac{t}{m}$

J $t - m$

47. If $-1 < a < b < 0$, then which of the following has the greatest value?

A $a - b$

B $b - a$

C $a + b$

D $2b - a$

Spiral Review

48. BUSINESS A small corporation decides that 8% of its profits will be divided among its six managers. There are two sales managers and four nonsales managers. Fifty percent will be split equally among all six managers. The other 50% will be split among the four nonsales managers. Let p represent the profits. (Lesson 9-2)

a. Write an expression to represent the share of the profits each nonsales manager will receive.

b. Simplify this expression.

c. Write an expression in simplest form to represent the share of the profits each sales manager will receive.

Simplify each expression. (Lesson 9-1)

49. $\dfrac{\frac{p^3}{2n}}{-\frac{p^2}{4n}}$

50. $\dfrac{\frac{m+q}{5}}{\frac{m^2+q^2}{5}}$

51. $\dfrac{\frac{x+y}{2x-y}}{\frac{x+y}{2x+y}}$

Graph each function. State the domain and range. (Lesson 8-1)

52. $y = 2(3)^x$

53. $y = 5(2)^x$

54. $y = 0.5(4)^x$

55. $y = 4\left(\frac{1}{3}\right)^x$

Find $(f + g)(x)$, $(f - g)(x)$, $(f \cdot g)(x)$, and $\left(\frac{f}{g}\right)(x)$ for each $f(x)$ and $g(x)$. (Lesson 7-1)

56. $f(x) = x + 9$
$g(x) = x - 9$

57. $f(x) = 2x - 3$
$g(x) = 4x + 9$

58. $f(x) = 2x^2$
$g(x) = 8 - x$

59. GEOMETRY The width of a rectangular prism is w centimeters. The height is 2 centimeters less than the width. The length is 4 centimeters more than the width. If the volume of the prism is 8 times the measure of the length, find the dimensions of the prism. (Lesson 6-5)

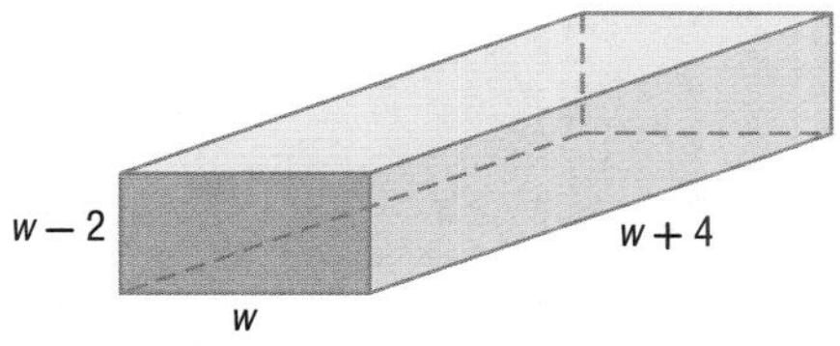

Skills Review

Graph each polynomial function. Estimate the x-coordinates at which the relative maxima and relative minima occur. State the domain and range for each function. (Lesson 6-4)

60. $f(x) = x^3 + 2x^2 - 3x - 5$

61. $f(x) = x^4 - 8x^2 + 10$

CHAPTER 9

Mid-Chapter Quiz

Lessons 9-1 through 9-3

NYS Core Curriculum
A2.A.16, A2.A.17

Simplify each expression. (Lesson 9-1)

1. $\frac{2x^2y^5}{7x^3yz} \cdot \frac{14xyz^2}{18x^4y}$

2. $\frac{24a^4b^6}{35ab^3} \div \frac{12abc}{7a^2c}$

3. $\frac{3x-3}{x^2+x-2} \cdot \frac{4x+8}{6x+18}$

4. $\frac{m^2+3m+2}{9} \div \frac{m+1}{3m+15}$

5. $\dfrac{\frac{r^2+3r}{r+1}}{\frac{3r}{3r+3}}$

6. $\dfrac{\frac{2y}{y^2-4}}{\frac{3}{y^2-4y+4}}$

7. **MULTIPLE CHOICE** For all $r \neq 0$, $\frac{r^2+6r+8}{r^2-4} =$ ___. (Lesson 9-1)

A $\frac{r-2}{r+4}$

B $\frac{r+4}{r-2}$

C $\frac{r+2}{r-4}$

D $\frac{r+4}{r+2}$

8. **MULTIPLE CHOICE** Identify all values of x for which $\frac{x^2-16}{(x^2-6x-27)(x+1)}$ is undefined. (Lesson 9-1)

F $-3, -1$

G $3, 1, -9$

H $-3, -1, 9$

J -1

9. What is the LCM of $x^2 - x$ and $3 - 3x$? (Lesson 9-2)

Simplify each expression. (Lesson 9-2)

10. $\frac{2x}{4x^2y} + \frac{x}{3xy^3}$

11. $\frac{3}{4m} + \frac{2}{3mn^2} - \frac{4}{n}$

12. $\frac{6}{r^2-3r-18} - \frac{1}{r^2+r-6}$

13. $\frac{3x+6}{x+y} + \frac{6}{-x-y}$

14. $\frac{x-4}{x^2-3x-4} + \frac{x+1}{2x-8}$

15. Determine the perimeter of the rectangle. (Lesson 9-2)

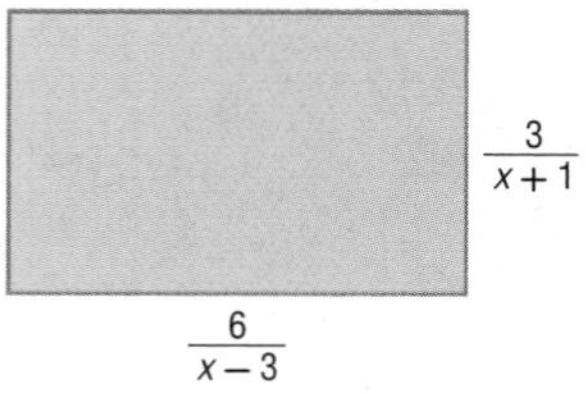

16. **TRAVEL** Lucita is going to a beach 100 miles away. She travels half the distance at one rate. The rest of the distance, she travels 15 miles per hour slower. (Lesson 9-2)

a. If x represents the faster pace in miles per hour, write an expression that represents the time spent at that pace.

b. Write an expression for the amount of time spent at the slower pace.

c. Write an expression for the amount of time Lucita needed to complete the trip.

Identify the asymptotes, domain, and range of each function. (Lesson 9-3)

17.

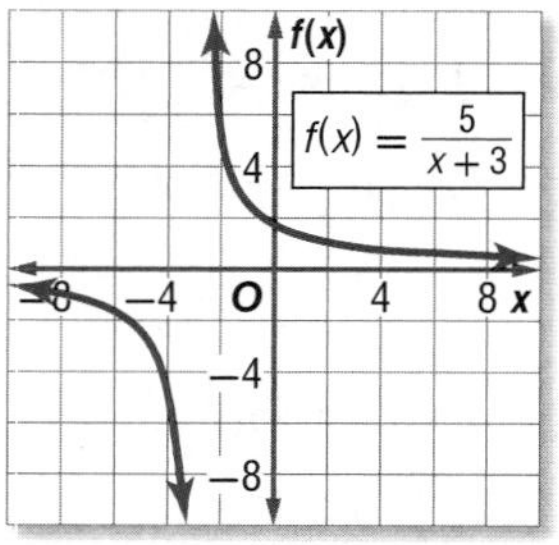

18.

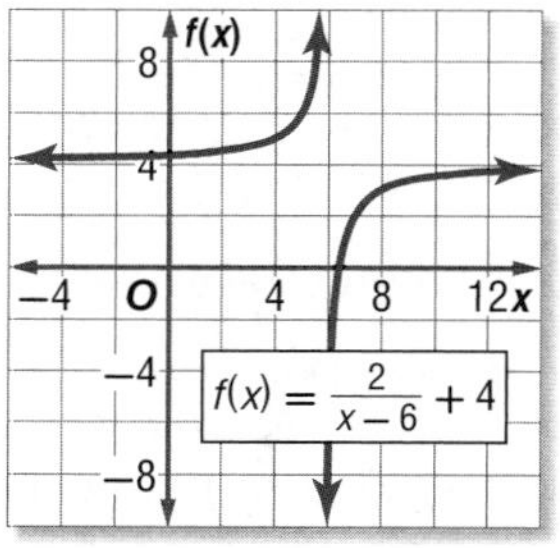

Graph each reciprocal function. State the domain and range. (Lesson 9-3)

19. $f(x) = \frac{4}{x}$

20. $f(x) = \frac{1}{3x}$

21. $f(x) = \frac{6}{x-1}$

22. $f(x) = \frac{-2}{x} + 4$

23. $f(x) = \frac{3}{x+2} - 5$

24. $f(x) = -\frac{1}{x-3} + 2$

25. **SANDWICHES** A group makes 45 sandwiches to take on a picnic. The number of sandwiches a person can eat depends on how many people go on the trip. (Lesson 9-3)

a. Write a function to represent this situation.

b. Graph the function.

9-4 Graphing Rational Functions

Then
You graphed reciprocal functions. (Lesson 9-3)

Now
- Graph rational functions with vertical and horizontal asymptotes.
- Graph rational functions with oblique asymptotes and point discontinuity.

NYS Core Curriculum

A2.A.41 Use functional notation to evaluate functions for given values in the domain

New Vocabulary
rational function
vertical asymptote
horizontal asymptote
oblique asymptote
point discontinuity

NY Math Online

glencoe.com
- Extra Examples
- Personal Tutor
- Self-Check Quiz
- Homework Help
- Math in Motion

Why?

Regina bought a digital camera and a photo printer for \$350. The manufacturer claims that ink and photo paper cost \$0.47 per photo. The rational function $C(p) = \frac{0.47p + 350}{p}$ can be used to determine the average cost $C(p)$ for printing p photos.

Vertical and Horizontal Asymptotes A **rational function** has an equation of the form $f(x) = \frac{a(x)}{b(x)}$, where $a(x)$ and $b(x)$ are polynomial functions and $b(x) \neq 0$.

In order to graph a rational function, it is helpful to locate the zeros and asymptotes. A zero of a rational function $f(x) = \frac{a(x)}{b(x)}$ occurs at every value of x for which $a(x) = 0$.

Key Concept: Vertical and Horizontal Asymptotes (For Your FOLDABLE)

Words If $f(x) = \frac{a(x)}{b(x)}$, $a(x)$ and $b(x)$ are polynomial functions with no common factors other than 1, and $b(x) \neq 0$, then:

- $f(x)$ has a **vertical asymptote** whenever $b(x) = 0$.
- $f(x)$ has at most one **horizontal asymptote**.
  - If the degree of $a(x)$ is greater than the degree of $b(x)$, there is no horizontal asymptote.
  - If the degree of $a(x)$ is less than the degree of $b(x)$, the horizontal asymptote is the line $y = 0$.
  - If the degree of $a(x)$ equals the degree of $b(x)$, the horizontal asymptote is the line $y = \frac{\text{leading coefficient of } a(x)}{\text{leading coefficient of } b(x)}$.

Examples

No horizontal asymptote

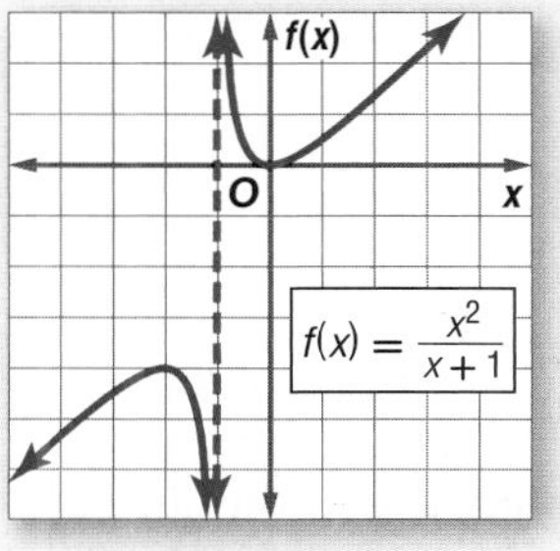

Vertical Asymptote:
$x = -1$

One horizontal asymptote

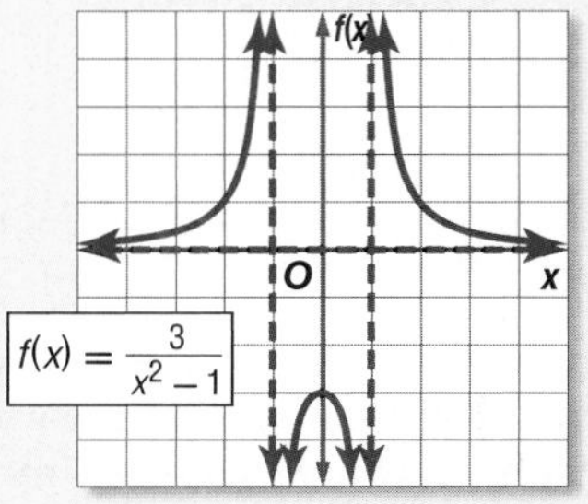

Vertical asymptotes:
$x = -1, x = 1$
Horizontal asymptote:
$f(x) = 0$

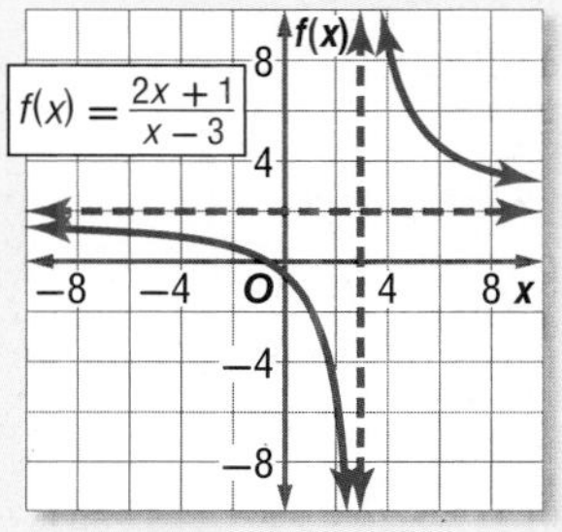

Vertical asymptote:
$x = 3$
Horizontal asymptote:
$f(x) = 2$

The asymptotes of a rational function can be used to draw the graph of the function. Additionally, the asymptotes can be used to divide a graph into regions to find ordered pairs on the graph.

EXAMPLE 1 Graph with no Horizontal Asymptote

Graph $f(x) = \frac{x^3}{x - 1}$.

Step 1 Find the zeros.

$x^3 = 0$ **Set $a(x) = 0$.**

$x = 0$ **Take the cube root of each side.**

There is a zero at $x = 0$.

Step 2 Draw the asymptotes.

Find the vertical asymptote.

$x - 1 = 0$ **Set $b(x) = 0$.**

$x = 1$ **Add 1 to each side.**

There is a vertical asymptote at $x = 1$.

The degree of the numerator is greater than the degree of the denominator. So, there is no horizontal asymptote.

StudyTip

Graphing Calculator The TABLE feature of a graphing calculator can be used to calculate decimal values for x and y.

Step 3 Draw the graph.

Use a table to find ordered pairs on the graph. Then connect the points.

x	$f(x)$
−3	6.75
−2	2.67
−1	0.5
0	0
0.5	−0.25
1.5	6.75
2	8
3	13.5

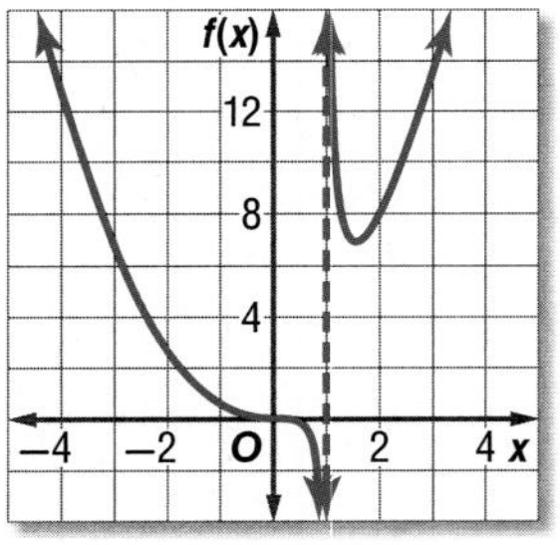

Check Your Progress

Graph each function.

1A. $f(x) = \frac{x^2 - x - 6}{x + 1}$

1B. $f(x) = \frac{(x + 1)^3}{(x + 2)^2}$

Personal Tutor glencoe.com

In the real world, sometimes values on the graph of a rational function are not meaningful. In the graph at the right, x-values such as time, distance, and number of people cannot be negative in the context of the problem. So, you do not even need to consider that portion of the graph.

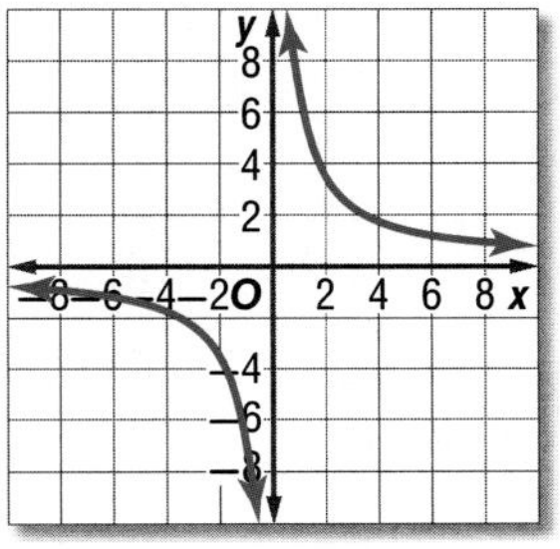

Real-World Career

U.S. Coast Guard Boatswain's Mate The most versatile member of the U.S. Coast Guard's operational team is the boatswain's mate. BMs are capable of performing almost any task. Training for BMs is accomplished through 12 weeks of intensive training.

Real-World EXAMPLE 2 Use Graphs of Rational Functions

AVERAGE SPEED A boat traveled upstream at r_1 miles per hour. During the return trip to its original starting point, the boat traveled at r_2 miles per hour. The average speed for the entire trip R is given by the formula $R = \frac{2r_1r_2}{r_1 + r_2}$.

a. Let r_1 be the independent variable, and let R be the dependent variable. Draw the graph if $r_2 = 10$ miles per hour.

The function is $R = \frac{2r_1(10)}{r_1 + (10)}$ or $R = \frac{20r_1}{r_1 + 10}$.

The vertical asymptote is $r_1 = -10$.

Graph the vertical asymptote and the function.

Notice that the horizontal asymptote is $R = 20$.

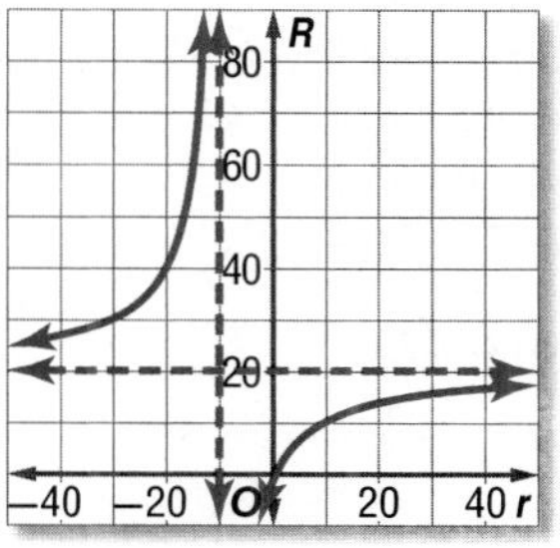

b. What is the R-intercept of the graph?

The R-intercept is 0.

c. What domain and range values are meaningful in the context of the problem?

In the problem context, speeds are nonnegative values. Therefore, only values of r_1 greater than or equal to 0 and values of R between 0 and 20 are meaningful.

Check Your Progress

2. SALARIES A company uses the formula $S(x) = \frac{45x + 25}{x + 1}$ to determine the salary in thousands of dollars of an employee during his xth year. Graph $S(x)$. What domain and range values are meaningful in the context of the problem? What is the meaning of the horizontal asymptote for the graph?

Personal Tutor glencoe.com

Oblique Asymptotes and Point Discontinuity An **oblique asymptote**, sometimes called a *slant asymptote,* is an asymptote that is neither horizontal nor vertical.

Key Concept Oblique Asymptotes

For Your FOLDABLE

Words If $f(x) = \frac{a(x)}{b(x)}$, $a(x)$ and $b(x)$ are polynomial functions with no common factors other than 1 and $b(x) \neq 0$, then $f(x)$ has an oblique asymptote if the degree of $a(x)$ minus the degree of $b(x)$ equals 1. The equation of the asymptote is $\frac{a(x)}{b(x)}$ with no remainder.

Example $f(x) = \frac{x^4 + 3x^3}{x^3 - 1}$

Vertical asymptote: $x = 1$

Oblique asymptote: $y = x + 3$

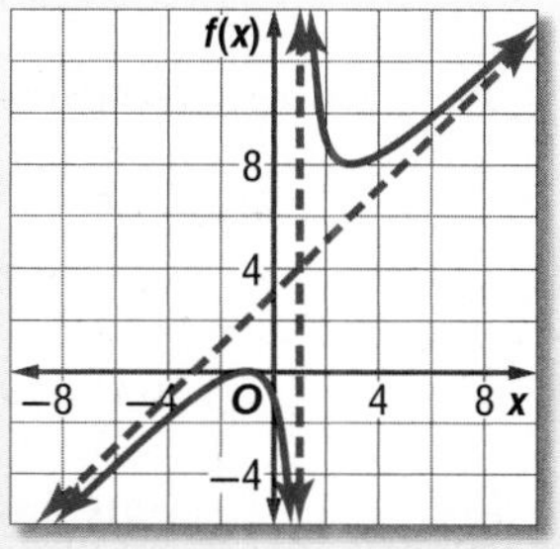

EXAMPLE 3 Determine Oblique Asymptotes

Graph $f(x) = \frac{x^2 + 4x + 4}{2x - 1}$.

StudyTip

Oblique Asymptotes Oblique asymptotes occur for rational functions that have a numerator polynomial that is one degree higher than the denominator polynomial.

Step 1 Find the zeros.

$x^2 + 4x + 4 = 0$	**Set $a(x) = 0$.**
$(x + 2)^2 = 0$	**Factor.**
$x + 2 = 0$	**Take the square root of each side.**
$x = -2$	**Subtract 2 from each side.**

There is a zero at $x = -2$.

Step 2 Find the asymptotes.

$2x - 1 = 0$	**Set $b(x) = 0$.**
$2x = 1$	**Add 1 to each side.**
$x = \frac{1}{2}$	**Divide each side by 2.**

There is a vertical asymptote at $x = \frac{1}{2}$.

The degree of the numerator is greater than the degree of the denominator, so there is no horizontal asymptote.

The difference between the degree of the numerator and the degree of the denominator is 1, so there is an oblique asymptote.

Divide the numerator by the denominator to determine the equation of the oblique asymptote.

The equation of the asymptote is the quotient excluding any remainder.

Thus, the oblique asymptote is the line $y = \frac{1}{2}x + \frac{9}{4}$.

$$\begin{array}{r} \frac{1}{2}x + \frac{9}{4} \\ 2x - 1 \overline{)\, x^2 + 4x + 4} \\ (-)x^2 - \frac{1}{2}x \quad \\ \hline \frac{9}{2}x + 4 \\ (-)\frac{9}{2}x - \frac{9}{4} \\ \hline \frac{25}{4} \end{array}$$

Step 3 Draw the asymptotes, and then use a table of values to graph the function.

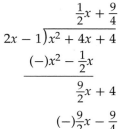

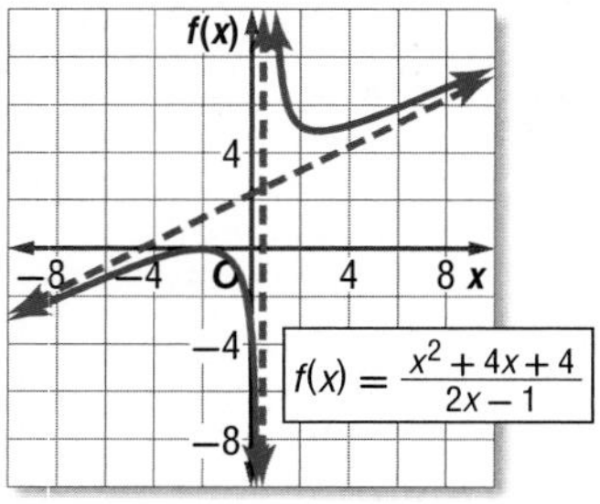

Check Your Progress

Graph each function.

3A. $f(x) = \frac{x^2}{x - 2}$

3B. $f(x) = \frac{x^3 - 1}{x^2 - 4}$

Personal Tutor glencoe.com

In some cases, graphs of rational functions may have **point discontinuity**, which looks like a hole in the graph. This is because the function is undefined at that point.

Key Concept — Point Discontinuity

For Your FOLDABLE

Words If $f(x) = \frac{a(x)}{b(x)}$, $b(x) \neq 0$, and $x - c$ is a factor of both $a(x)$ and $b(x)$, then there is a point discontinuity at $x = c$.

Example $f(x) = \frac{(x + 2)(x + 1)}{x + 1}$
$= x + 2; x \neq -1$

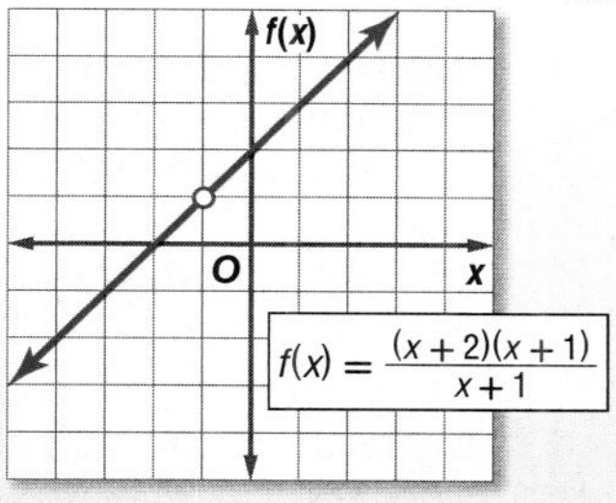

Math in Motion, Interactive Lab glencoe.com

EXAMPLE 4 Graph with Point Discontinuity

Graph $f(x) = \frac{x^2 - 16}{x - 4}$.

Notice that $\frac{x^2 - 16}{x - 4} = \frac{(x + 4)\overset{1}{\cancel{(x - 4)}}}{\underset{1}{\cancel{x - 4}}}$ or $x + 4$.

Therefore, the graph of $f(x) = \frac{x^2 - 16}{x - 4}$ is the graph of $f(x) = x + 4$ with a hole at $x = 4$.

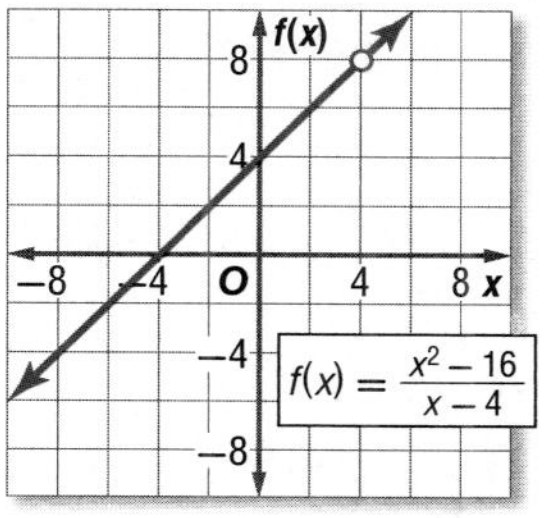

Watch Out!

Holes Remember that a common factor in the numerator and denominator can signal a hole.

Check Your Progress

Graph each function.

4A. $f(x) = \frac{x^2 + 4x - 5}{x + 5}$

4B. $f(x) = \frac{x^3 + 2x^2 - 9x - 18}{x^2 - 9}$

Personal Tutor glencoe.com

Check Your Understanding

Example 1 p. 578

Graph each function.

1. $f(x) = \frac{x^4 - 2}{x^2 - 1}$

2. $f(x) = \frac{x^3}{x + 2}$

Example 2 p. 579

3. FOOTBALL Eduardo plays football for his high school. So far this season, he has made 7 out of 11 field goals. He would like to improve his field goal percentage. If he can make x consecutive field goals, his field goal percentage can be determined using the function $P(x) = \frac{7 + x}{11 + x}$.

a. Graph the function.

b. What part of the graph is meaningful in the context of this problem?

c. Describe the meaning of the intercept of the vertical axis.

d. What is the equation of the horizontal asymptote? Explain its meaning with respect to Eduardo's field goal percentage.

Examples 3 and 4 pp. 580–581

Graph each function.

4. $f(x) = \frac{6x^2 - 3x + 2}{x}$

5 $f(x) = \frac{x^2 + 8x + 20}{x + 2}$

6. $f(x) = \frac{x^2 - 4x - 5}{x + 1}$

7. $f(x) = \frac{x^2 + x - 12}{x + 4}$

Practice and Problem Solving

= **Step-by-Step Solutions** begin on page R20. **Extra Practice** begins on page 947.

Example 1 p. 578

Graph each function.

8. $f(x) = \dfrac{x^4}{6x + 12}$

9. $f(x) = \dfrac{x^3}{8x - 4}$

10. $f(x) = \dfrac{x^4 - 16}{x^2 - 1}$

11. $f(x) = \dfrac{x^3 + 64}{16x - 24}$

Example 2 p. 579

12. **SCHOOL SPIRIT** As president of Student Council, Brandy is getting T-shirts made for a pep rally. Each T-shirt costs \$9.50, and there is a set-up fee of \$75. The student council plans to sell the shirts, but each of the 15 council members will get one for free.

a. Write a function for the average cost of a T-shirt to be sold. Graph the function.

b. What is the average cost if 200 shirts are ordered? if 500 shirts are ordered?

c. How many T-shirts must be ordered to bring the average cost under \$9.75?

Examples 2 and 3 pp. 579–580

Graph each function.

13. $f(x) = \dfrac{x}{x + 2}$

14. $f(x) = \dfrac{5}{(x - 1)(x + 4)}$

15. $f(x) = \dfrac{4}{(x - 2)^2}$

16. $f(x) = \dfrac{x - 3}{x + 1}$

17. $f(x) = \dfrac{1}{(x + 4)^2}$

18. $f(x) = \dfrac{2x}{(x + 2)(x - 5)}$

19. $f(x) = \dfrac{(x - 4)^2}{x + 2}$

20. $f(x) = \dfrac{(x + 3)^2}{x - 5}$

21. $f(x) = \dfrac{x^3 + 1}{x^2 - 4}$

22. $f(x) = \dfrac{4x^3}{2x^2 + x - 1}$

23. $f(x) = \dfrac{3x^2 + 8}{2x - 1}$

24. $f(x) = \dfrac{2x^2 + 5}{3x + 4}$

25. $f(x) = \dfrac{x^4 - 2x^2 + 1}{x^3 + 2}$

26. $f(x) = \dfrac{x^4 - x^2 - 12}{x^3 - 6}$

27. **ELECTRICITY** The current in amperes in an electrical circuit with three resistors in a series is given by the equation $I = \dfrac{V}{R_1 + R_2 + R_3}$, where V is the voltage in volts in a the circuit and R_1, R_2, and R_3 are the resistances in ohms of the three resistors.

a. Let R_1 be the independent variable, and let I be the dependent variable. Graph the function if $V = 120$ volts, $R_2 = 25$ ohms, and $R_3 = 75$ ohms.

b. Give the equation of the vertical asymptote and the R_1- and I-intercepts of the graph.

c. Find the value of I when the value of R_1 is 140 ohms.

d. What domain and range values are meaningful in the context of the problem?

Example 4 p. 581

Graph each function.

28. $f(x) = \dfrac{x^2 - 2x - 8}{x - 4}$

29. $f(x) = \dfrac{x^2 + 4x - 12}{x - 2}$

30. $f(x) = \dfrac{x^2 - 25}{x + 5}$

31. $f(x) = \dfrac{x^2 - 64}{x - 8}$

32. $f(x) = \dfrac{(x - 4)(x^2 - 4)}{x^2 - 6x + 8}$

33. $f(x) = \dfrac{(x + 5)(x^2 + 2x - 3)}{x^2 + 8x + 15}$

34. $f(x) = \dfrac{3x^4 + 6x^3 + 3x^2}{x^2 + 2x + 1}$

35. $f(x) = \dfrac{2x^4 + 10x^3 + 12x^2}{x^2 + 5x + 6}$

Real-World Link

The primary snow belt along the eastern Great Lakes gets anywhere from 60 to more than 110 inches of snow per year.

Source: Weathernet5

36. BUSINESS Liam purchased a snow plow for $4500 and plows the parking lots of local businesses. Each time he plows a parking lot, he incurs a cost of $50 for gas and maintenance.

a. Write and graph the rational function representing his average cost per customer as a function of the number of parking lots.

b. What are the asymptotes of the graph?

c. Why is the first quadrant in the graph the only relevant quadrant?

d. How many total parking lots does Liam need to plow for his average cost per parking lot to be less than $80?

37. INTERNET PHONES Kristina bought a new cell phone with Internet access. The phone cost $150, and her monthly usage charge is $30 plus $10 for the Internet access.

a. Write and graph the rational function representing her average monthly cost as a function of the number of months Kristina uses the phone.

b. What are the asymptotes of the graph?

c. Why is the first quadrant in the graph the only relevant quadrant?

d. After how many months will the average monthly charge be $45?

38. SOFTBALL Alana plays softball for Centerville High School. So far this season she has gotten a hit 4 out of 12 times at bat. She is determined to improve her batting average. If she can get x consecutive hits, her batting average can be determined using $B(x) = \frac{4 + x}{12 + x}$.

a. Graph the function.

b. What part of the graph is meaningful in the context of the problem?

c. Describe the meaning of the intercept of the vertical axis.

d. What is the equation of the horizontal asymptote? Explain its meaning with respect to Alana's batting average.

Graph each function.

39. $f(x) = \frac{x + 1}{x^2 + 6x + 5}$

40. $f(x) = \frac{x^2 - 10x - 24}{x + 2}$

41. $f(x) = \frac{6x^2 + 4x + 2}{x + 2}$

H.O.T. Problems

Use Higher-Order Thinking Skills

42. OPEN ENDED Sketch the graph of a rational function with a horizontal asymptote $y = 1$ and a vertical asymptote $x = -2$.

43. CHALLENGE Write a rational function for the graph at the right.

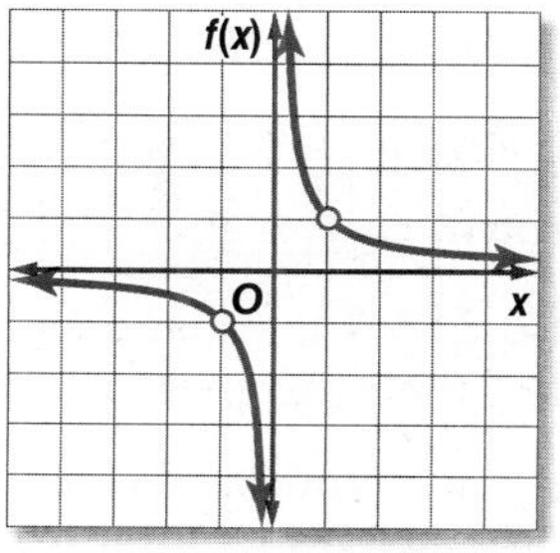

44. REASONING What is the difference between the graphs of $f(x) = x - 2$ and $g(x) = \frac{(x + 3)(x - 2)}{x + 3}$?

45. PROOF A rational function has an equation of the form $f(x) = \frac{a(x)}{b(x)}$, where $a(x)$ and $b(x)$ are polynomial functions and $b(x) \neq 0$. Show that $f(x) = \frac{x}{a - b} + c$ is a rational function.

46. WRITING IN MATH Explain how factoring can be used to determine the vertical asymptotes or point discontinuity of a rational function.

NYSRE PRACTICE A2.A.41

47. PROBABILITY Of the 6 courses offered by the music department at her school, Kaila must choose exactly 2 of them. How many different combinations of 2 courses are possible for Kaila if there are no restrictions on which 2 courses she can choose?

A 48 C 15
B 18 D 12

48. The projected sales of a game cartridge is given by the function $S(p) = \frac{3000}{2p + a}$, where $S(p)$ is the number of cartridges sold, in thousands, p is the price per cartridge, in dollars, and a is a constant.

If 100,000 cartridges are sold at $10 per cartridge, how many cartridges will be sold at $20 per cartridge?

F 20,000 H 60,000
G 50,000 J 150,000

49. GRIDDED RESPONSE Five distinct points lie in a plane such that 3 of the points are on line ℓ and 3 of the points are on a different line m. What is the total number of lines that can be drawn so that each line passes through exactly 2 of these 5 points?

50. GEOMETRY In the figure below, what is the value of $w + x + y + z$?

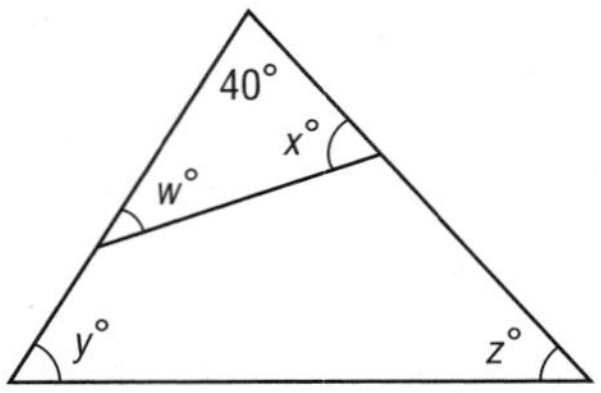

A 140 C 320
B 280 D 360

Spiral Review

Graph each function. State the domain and range. (Lesson 9-3)

51. $f(x) = \frac{-5}{x+2}$

52. $f(x) = \frac{4}{x-1} - 3$

53. $f(x) = \frac{1}{x+6} + 1$

Simplify each expression. (Lesson 9-2)

54. $\frac{m}{m^2 - 4} + \frac{2}{3m + 6}$

55. $\frac{y}{y+3} - \frac{6y}{y^2 - 9}$

56. $\frac{5}{x^2 - 3x - 28} + \frac{7}{2x - 14}$

57. $\frac{d-4}{d^2 + 2d - 8} - \frac{d+2}{d^2 - 16}$

Simplify each expression. (Lesson 7-6)

58. $y^{\frac{5}{3}} \cdot y^{\frac{7}{3}}$

59. $x^{\frac{3}{4}} \cdot x^{\frac{9}{4}}$

60. $\left(b^{\frac{1}{3}}\right)^{\frac{3}{5}}$

61. $\left(a^{-\frac{2}{3}}\right)^{-\frac{1}{6}}$

Skills Review

62. TRAVEL Mr. and Mrs. Wells are taking their daughter to college. The table shows their distances from home after various amounts of time. (Lesson 2-3)

a. Find the average rate of change in their distances from home between 1 and 3 hours after leaving home.

b. Find the average rate of change in their distances from home between 0 and 5 hours after leaving home.

Time (h)	Distance (mi)
0	0
1	55
2	110
3	165
4	165
5	225

EXTEND
9-4 Graphing Technology Lab
Graphing Rational Functions

NY Math Online glencoe.com
Other Calculator Keystrokes

A TI-83/84 Plus graphing calculator can be used to explore graphs of rational functions. These graphs have some features that never appear in the graphs of polynomial functions.

ACTIVITY 1 Graph with Asymptotes

Graph $y = \frac{8x - 5}{2x}$ in the standard viewing window. Find the equations of any asymptotes.

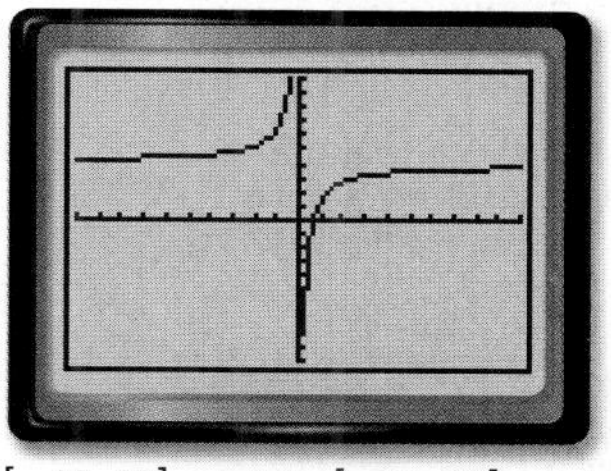

[−10, 10] scl: 1 by [−10, 10] scl: 1

Step 1 Enter the equation in the Y= list, and then graph.

KEYSTROKES: [Y=] [(] 8 [X,T,θ,n] [−] 5 [)] [÷] [(] 2 [X,T,θ,n] [)] [ZOOM] 6

Step 2 Examine the graph.

By looking at the equation, we can determine that if $x = 0$, the function is undefined. The equation of the vertical asymptote is $x = 0$. Notice what happens to the y-values as x grows larger and as x gets smaller. The y-values approach 4. So, the equation for the horizontal asymptote is $y = 4$.

ACTIVITY 2 Graph with Point Discontinuity

Graph $y = \frac{x^2 - 16}{x + 4}$ in the window [−5, 4.4] by [−10, 2] with scale factors of 1.

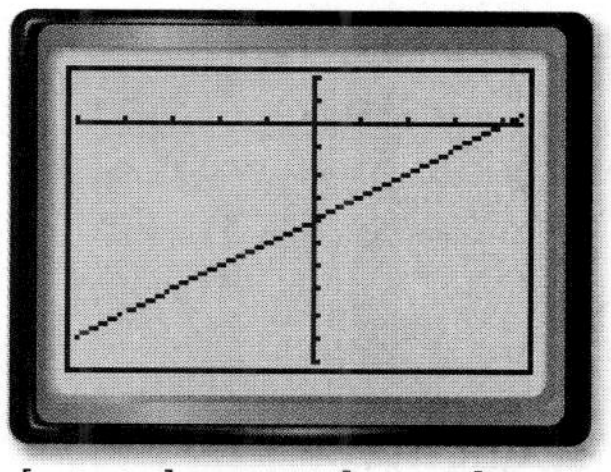

[−5, 4.4] scl: 1 by [−20, 2] scl: 1

Step 1 Because the function is not continuous, put the calculator in dot mode.

KEYSTROKES: [MODE] [▼] [▼] [▼] [▼] [▶] [ENTER]

Step 2 Examine the graph.

This graph looks like a line with a break in continuity at $x = -4$. This happens because the denominator is 0 when $x = -4$. Therefore, the function is undefined when $x = -4$.

If you **TRACE** along the graph, when you come to $x = -4$, you will see that there is no corresponding y-value.

Exercises

Use a graphing calculator to graph each function. Write the x-coordinates of any points of discontinuity and/or the equations of any asymptotes.

1. $f(x) = \frac{1}{x}$

2. $f(x) = \frac{x}{x + 2}$

3. $f(x) = \frac{2}{x - 4}$

4. $f(x) = \frac{2x}{3x - 6}$

5. $f(x) = \frac{4x + 2}{x - 1}$

6. $f(x) = \frac{x^2 - 9}{x + 3}$

9-5 Variation Functions

Then

You wrote and graphed linear equations.
(Lesson 2-4)

Now

- Recognize and solve direct and joint variation problems.
- Recognize and solve inverse and combined variation problems.

NYS Core Curriculum

A2.A.5 Use direct and inverse variation to solve for unknown values

New Vocabulary

direct variation
constant of variation
joint variation
inverse variation
combined variation

NY Math Online

glencoe.com

- Extra Examples
- Personal Tutor
- Self-Check Quiz
- Homework Help

Why?

While building skateboard ramps, Yu determined that the best ramps were the ones in which the length of the top of the ramp was 1.5 times as long as the height of the ramp.

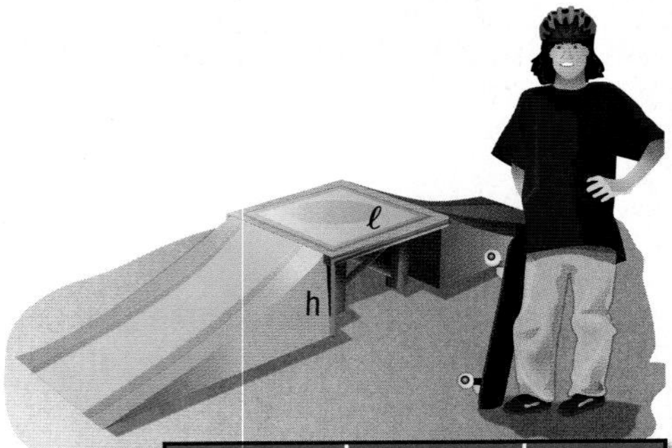

As shown in the table, the length of the top of the ramp depends on the height of a ramp. The length increases as the height increases, but the ratio remains the same, or is *constant*.

Length (ℓ)	Height (h)	Ratio $\frac{\ell}{h}$
3	2	1.5
6	4	1.5
9	6	1.5
12	8	1.5

The equation $\frac{\ell}{h} = 1.5$ can be written as $\ell = 1.5h$.

The length *varies directly* with the height of the ramp.

Direct Variation and Joint Variation The relationship given by $\ell = 1.5h$ is an example of direct variation. A **direct variation** can be expressed in the form $y = kx$. In this equation, k is called the **constant of variation**.

Notice that the graph of $\ell = 1.5h$ is a straight line through the origin. A direct variation is a special case of an equation written in slope-intercept form, $y = mx + b$. When $m = k$ and $b = 0$, $y = mx + b$ becomes $y = kx$. So the slope of a direct variation equation is its constant of variation.

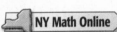

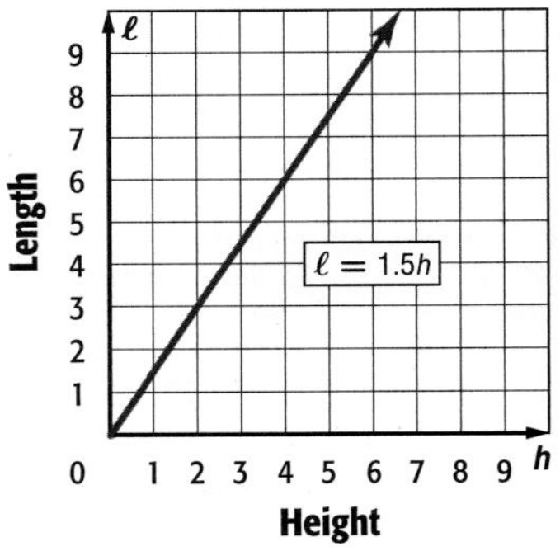

To express a direct variation, we say that y varies directly as x. In other words, as x increases, y increases or decreases at a constant rate.

Key Concept **Direct Variation** *For Your FOLDABLE*

Words y varies directly as x if there is some nonzero constant k such that $y = kx$. k is called the *constant of variation*.

Example If $y = 3x$ and $x = 7$, then $y = 3(7)$ or 21.

If you know that y varies directly as x and one set of values, you can use a proportion to find the other set of corresponding values.

$$y_1 = kx_1 \quad \text{and} \quad y_2 = kx_2$$

$$\frac{y_1}{x_1} = k \qquad \frac{y_2}{x_2} = k \qquad \text{Therefore, } \frac{y_1}{x_1} = \frac{y_2}{x_2}.$$

Using the properties of equality, you can find many other proportions that relate these same x- and y-values.

EXAMPLE 1 Direct Variation

If y varies directly as x and $y = 15$ when $x = -5$, find y when $x = 7$.

Use a proportion that relates the values.

$\frac{y_1}{x_1} = \frac{y_2}{x_2}$	**Direct variation**
$\frac{15}{-5} = \frac{y_2}{7}$	**$y_1 = 15$, $x_1 = -5$, and $x_2 = 7$**
$15(7) = -5(y_2)$	**Cross multiply.**
$105 = -5y_2$	**Simplify.**
$-21 = y_2$	**Divide each side by -5.**

Check Your Progress

1. If r varies directly as t and $r = -20$ when $t = 4$, find r when $t = -6$.

Personal Tutor glencoe.com

Another type of variation is joint variation. **Joint variation** occurs when one quantity varies directly as the product of two or more other quantities.

Key Concept Joint Variation

For Your FOLDABLE

Words y varies jointly as x and z if there is some nonzero constant k such that $y = kxz$.

Example If $y = 5xz$, $x = 6$ and $z = -2$, then $y = 5(6)(-2)$ or -60.

StudyTip

Joint Variation Some mathematicians consider joint variation a special type of combined variation.

If you know that y varies jointly as x and z and one set of values, you can use a proportion to find the other set of corresponding values.

$$y_1 = kx_1z_1 \quad \text{and} \quad y_2 = kx_2z_2$$

$$\frac{y_1}{x_1z_1} = k \qquad \frac{y_2}{x_2z_2} = k \qquad \text{Therefore, } \frac{y_1}{x_1z_1} = \frac{y_2}{x_2z_2}.$$

EXAMPLE 2 Joint Variation

Suppose y varies jointly as x and z. Find y when $x = 9$ and $z = 2$, if $y = 20$ when $z = 3$ and $x = 5$.

Use a proportion that relates the values.

$\frac{y_1}{x_1z_1} = \frac{y_2}{x_2z_2}$	**Joint variation**
$\frac{20}{5(3)} = \frac{y_2}{9(2)}$	**$y_1 = 20$, $x_1 = 5$, $z_1 = 3$, $x_2 = 9$, and $z_2 = 2$**
$20(9)(2) = 5(3)(y_2)$	**Cross multiply.**
$360 = 15y_2$	**Simplify.**
$24 = y_2$	**Divide each side by 15.**

Check Your Progress

2. Suppose r varies jointly as v and t. Find r when $v = 2$ and $t = 8$, if $r = 70$ when $v = 10$ and $t = 4$.

Personal Tutor glencoe.com

Inverse Variation and Combined Variation Another type of variation is inverse variation. If two quantities x and y show **inverse variation**, their product is equal to a constant k.

Inverse variation is often described as one quantity increasing while the other quantity is decreasing. For example, speed and time for a fixed distance vary inversely with each other; the faster you go, the less time it takes you to get there.

Key Concept **Inverse Variation** For Your FOLDABLE

Words y varies inversely as x if there is some nonzero constant k such that $xy = k$ or $y = \frac{k}{x}$, where $x \neq 0$ and $y \neq 0$.

Example If $xy = 2$, and $x = 6$, then $y = \frac{2}{6}$ or $\frac{1}{3}$.

StudyTip

Direct and Inverse Variation You can identify the type of variation by looking at a table of values for x and y. If the quotient $\frac{y}{x}$ has a constant value, y varies directly as x. If the product xy has a constant value, y varies inversely as x.

Suppose y varies inversely as x such that $xy = 6$ or $y = \frac{6}{x}$. The graph of this equation is shown at the right. Since k is a positive value, as the values of x increase, the values of y decrease.

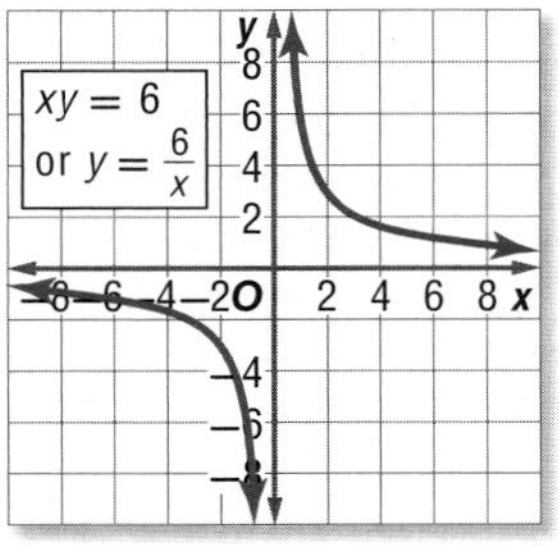

Notice that the graph of an inverse variation is a reciprocal function.

A proportion can be used with inverse variation to solve problems in which some quantities are known. The following proportion is only one of several that can be formed.

$x_1y_1 = k$ and $x_2y_2 = k$

$x_1y_1 = x_2y_2$ **Substitution Property of Equality**

$\frac{x_1}{y_2} = \frac{x_2}{y_1}$ **Divide each side by y_1y_2.**

EXAMPLE 3 **Inverse Variation**

If a varies inversely as b and $a = 28$ when $b = -2$, find a when $b = -10$.

Use a proportion that relates the values.

$\frac{a_1}{b_2} = \frac{a_2}{b_1}$ **Inverse Variation**

$\frac{28}{-10} = \frac{a_2}{-2}$ **$a_1 = 28$, $b_1 = -2$, and $b_2 = -10$**

$28(-2) = -10(a_2)$ **Cross multiply.**

$-56 = -10(a_2)$ **Simplify.**

$5\frac{3}{5} = a_2$ **Divide each side by -10.**

Check Your Progress

3. If x varies inversely as y and $x = 24$ when $y = 4$, find x when $y = 12$.

Personal Tutor glencoe.com

Inverse variation is often used in real-world situations.

Real-World EXAMPLE 4 Write and Solve an Inverse Variation

MUSIC The length of a violin string varies inversely as the frequency of its vibrations. A violin string 10 inches long vibrates at a frequency of 512 cycles per second. Find the frequency of an 8-inch violin string.

Let $v_1 = 10, f_1 = 512$, and $v_2 = 8$. Solve for f_2.

$v_1 f_1 = v_2 f_2$ **Original equation**

$10 \cdot 512 = 8 \cdot f_2$ **$v_1 = 10, f_1 = 512$, and $v_2 = 8$**

$\frac{5120}{8} = f_2$ **Divide each side by 8.**

$640 = f_2$ **Simplify.**

The 8-inch violin string vibrates at a frequency of 640 cycles per second.

Real-World Link

When you pluck a string, it vibrates back and forth. This causes mechanical energy to travel through the air in waves. The number of times per second these waves hit our ear is called the *frequency*. The more waves per second, the higher the pitch.

Check Your Progress

4. The apparent length of an object is inversely proportional to one's distance from the object. Earth is about 93 million miles from the Sun. Jupiter is about 483.6 million miles from the Sun. Find how many times as large the diameter of the Sun would appear on Earth as on Jupiter.

Personal Tutor glencoe.com

Another type of variation is combined variation. **Combined variation** occurs when one quantity varies directly and/or inversely as two or more other quantities.

If you know that y varies directly as x, y varies inversely as z and one set of values, you can use a proportion to find the other set of corresponding values.

$y_1 = \frac{kx_1}{z_1}$ and $y_2 = \frac{kx_2}{z_2}$

$\frac{y_1 z_1}{x_1} = k$ $\frac{y_2 z_2}{x_2} = k$ Therefore, $\frac{y_1 z_1}{x_1} = \frac{y_2 z_2}{x_2}$.

EXAMPLE 5 Combined Variation

Suppose f varies directly as g, and f varies inversely as h. Find g when $f = 18$ and $h = -3$, if $g = 24$ when $h = 2$ and $f = 6$.

First set up a correct proportion for the information given.

$f_1 = \frac{kg_1}{h_1}$ and $f_2 = \frac{kg_2}{h_2}$ ***g* varies directly as *f*, so *g* goes in the numerator. *h* varies inversely as *f*, so *h* goes in the denominator.**

$k = \frac{f_1 h_1}{g_1}$ and $k = \frac{f_2 h_2}{g_2}$ **Solve for *k*.**

$\frac{f_1 h_1}{g_1} = \frac{f_2 h_2}{g_2}$ **Set the two proportions equal to each other.**

$\frac{6(2)}{24} = \frac{18(-3)}{g_2}$ **$f_1 = 6, g_1 = 24, h_1 = 2, f_2 = 18$, and $h_2 = -3$**

$24(18)(-3) = 6(2)(g_2)$ **Cross multiply.**

$-1296 = 12g_2$ **Simplify.**

$-108 = g_2$ **Divide each side by 12.**

When $f = 18$ and $h = -3$, the value of g is -108.

StudyTip

Combined Variation Quantities that vary directly appear in the numerator. Quantities that vary inversely appear in the denominator.

Check Your Progress

5. Suppose p varies directly as r, and p varies inversely as t. Find t when $r = 10$ and $p = -5$, if $t = 20$ when $p = 4$ and $r = 2$.

Personal Tutor glencoe.com

Check Your Understanding

Examples 1–3 pp. 587–588

1. If y varies directly as x and $y = 12$ when $x = 8$, find y when $x = 14$.
2. Suppose y varies jointly as x and z. Find y when $x = 9$ and $z = -3$, if $y = -50$ when z is 5 and x is -10.
3. If y varies inversely as x and $y = -18$ when $x = 16$, find x when $y = 9$.

Example 4 p. 589

4. **TRAVEL** A map of Illinois is scaled so that 2 inches represents 15 miles. How far apart are Chicago and Rockford if they are 12 inches apart on the map?

Example 5 p. 589

5. Suppose a varies directly as b, and a varies inversely as c. Find b when $a = 8$ and $c = -3$, if $b = 16$ when $c = 2$ and $a = 4$.
6. Suppose d varies directly as f, and d varies inversely as g. Find g when $d = 6$ and $f = -7$, if $g = 12$ when $d = 9$ and $f = 3$.

Practice and Problem Solving

● = **Step-by-Step Solutions** begin on page R20.
Extra Practice begins on page 947.

Example 1 p. 587

If x varies directly as y, find x when $y = 8$.

7. $x = 6$ when $y = 32$
8. $x = 11$ when $y = -3$
9. $x = 14$ when $y = -2$
10. $x = -4$ when $y = 10$
11. **MOON** Astronaut Neil Armstrong, the first man on the Moon, weighed 360 pounds on Earth with all his equipment on, but weighed only 60 pounds on the Moon. Write an equation that relates weight on the Moon m with weight on Earth w.

Example 2 p. 587

If a varies jointly as b and c, find a when $b = 4$ and $c = -3$.

12. $a = -96$ when $b = 3$ and $c = -8$
13. $a = -60$ when $b = -5$ and $c = 4$
14. $a = -108$ when $b = 2$ and $c = 9$
15. $a = 24$ when $b = 8$ and $c = 12$
16. **TELEVISION** According to the A.C. Nielsen Company, the average American watches 4 hours of television a day.
    a. Write an equation to represent the average number of hours spent watching television by m household members during a period of d days.
    b. Assume that members of your household watch the same amount of television each day as the average American. How many hours of television would the members of your household watch in a week?

Example 3 p. 588

If f varies inversely as g, find f when $g = -6$.

17. $f = 15$ when $g = 9$
18. $f = 4$ when $g = 28$
19. $f = -12$ when $g = 19$
20. $f = 0.6$ when $g = -21$
21. **COMMUNITY SERVICE** Every year students at West High School collect canned goods for a local food pantry. They plan to distribute flyers to homes in the community asking for donations. Last year, 12 students were able to distribute 1000 flyers in four hours.
    a. Write an equation that relates the number of students s to the amount of time t it takes to distribute 1000 flyers.
    b. How long would it take 15 students to hand out the same number of flyers this year?

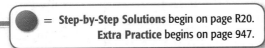

Example 4 p. 589

22. **BIRDS** When a group of snow geese migrate, the distance that they fly varies directly with the amount of time they are in the air.

a. A group of snow geese migrated 375 miles in 7.5 hours. Write a direct variation equation that represents this situation.

b. Every year, geese migrate 3000 miles from their winter home in the southwest United States to their summer home in the Canadian Arctic. Estimate the number of hours of flying time that it takes for the geese to migrate.

Example 5 p. 589

23. Suppose a varies directly as b, and a varies inversely as c. Find b when $a = 5$ and $c = -4$, if $b = 12$ when $c = 3$ and $a = 8$.

24. Suppose x varies directly as y, and x varies inversely as z. Find z when $x = 10$ and $y = -7$, if $z = 20$ when $x = 6$ and $y = 14$.

Determine whether each relation shows *direct* or *inverse* variation, or *neither*.

25.

x	y
4	12
8	24
16	48
32	96

26.

x	y
8	2
4	4
−2	−8
−8	−2

27.

x	y
2	4
3	9
4	16
5	25

28. If y varies inversely as x and $y = 6$ when $x = 19$, find y when $x = 2$.

29. If x varies inversely as y and $x = 16$ when $y = 5$, find x when $y = 20$.

30. Suppose a varies directly as b, and a varies inversely as c. Find b when $a = 7$ and $c = -8$, if $b = 15$ when $c = 2$ and $a = 4$.

31. Suppose x varies directly as y, and x varies inversely as z. Find z when $x = 8$ and $y = -6$, if $z = 26$ when $x = 8$ and $y = 13$.

State whether each equation represents a *direct, joint, inverse,* or *combined* variation. Then name the constant of variation.

32. $\frac{x}{y} = 2.75$ **33.** $fg = -2$ **34.** $a = 3bc$ **35.** $10 = \frac{xy^2}{z}$

36. $y = -11x$ **37.** $\frac{n}{p} = 4$ **38.** $9n = pr$ **39.** $-2y = z$

40. $a = 27b$ **41.** $c = \frac{7}{d}$ **42.** $-10 = gh$ **43.** $m = 20cd$

Real-World Link

Lake Tahoe is fed by 63 streams and 2 hot springs. According to research, with a volume of 39 trillion gallons of water, if the lake was ever drained it would take around 700 years to fill it again entirely.

Source: Lake Tahoe Visitor's Bureau

44. **CHEMISTRY** The volume of a gas v varies inversely as the pressure p and directly as the temperature t.

a. Write an equation to represent the volume of a gas in terms of pressure and temperature.

b. Is your equation a *direct, joint, inverse,* or *combined* variation?

c. A certain gas has a volume of 8 liters, a temperature of 275 Kelvin, and a pressure of 1.25 atmospheres. If the gas is compressed to a volume of 6 liters and is heated to 300 Kelvin, what will the new pressure be?

45. **VACATION** The time it takes the Levensteins to reach Lake Tahoe varies with their average rate of speed.

a. If they are 800 miles away, write and graph an equation relating their travel time to their average rate of speed.

b. Their goal is to arrive within 18 hours. What minimum average speed will accomplish this goal?

Real-World Link

As of 2007, digital audio players could hold up to 23,000 songs.

Source: *Consumer Reports*

46. MUSIC The maximum number of songs that a digital audio player can hold depends on the lengths and the quality of the songs that are recorded. A song will take up more space on the player if it is recorded at a higher quality, like from a CD, than at a lower quality, like from the Internet.

a. If a certain player has 5400 megabytes of storage space, write a function that represents the number of songs the player can hold as a function of the average size of the songs.

b. Is your function a *direct*, *joint*, *inverse*, or *combined* variation?

c. Suppose the average file size for a high-quality song is 8 megabytes and the average size for a low-quality song is 5 megabytes. Determine how many more songs the player can hold if they are low quality than if they are high quality.

47. GRAVITY According to the Law of Universal Gravitation, the attractive force F in newtons between any two bodies in the universe is directly proportional to the product of the masses m_1 and m_2 in kilograms of the two bodies and inversely proportional to the square of the distance d in meters between the bodies. That is, $F = \frac{Gm_1m_2}{d^2}$. G is the universal gravitational constant. Its value is 6.67×10^{-11} Nm^2/kg^2.

a. The distance between Earth and the Moon is about 3.84×10^8 meters. The mass of the Moon is 7.36×10^{22} kilograms. The mass of Earth is 5.97×10^{24} kilograms. What is the gravitational force that the Moon and Earth exert upon each other?

b. The distance between Earth and the Sun is about 1.5×10^{11} meters. The mass of the Sun is about 1.99×10^{30} kilograms. What is the gravitational force that the Sun and Earth exert upon each other?

c. Find the gravitational force exerted on each other by two 1000-kilogram iron balls at a distance of 0.1 meter apart.

H.O.T. Problems

Use Higher-Order Thinking Skills

48. FIND THE ERROR Jamil and Savannah are setting up a proportion to begin solving the combined variation in which z varies directly as x and z varies inversely as y. Who has set up the correct proportion? Explain your reasoning.

Jamil

$$z_1 = \frac{kx_1}{y_1} \text{ and } z_2 = \frac{kx_2}{y_2}$$

$$k = \frac{z_1y_1}{x_1} \text{ and } k = \frac{z_2y_2}{x_2}$$

$$\frac{z_1y_1}{x_1} = \frac{z_2y_2}{x_2}$$

Savannah

$$z_1 = \frac{kx_1}{y_1} \text{ and } z_2 = \frac{kx_2}{y_2}$$

$$k = \frac{z_1x_1}{y_1} \text{ and } k = \frac{z_2x_2}{y_2}$$

$$\frac{z_1x_1}{y_1} = \frac{z_2x_2}{y_2}$$

49. CHALLENGE If a varies inversely as b, c varies jointly as b and f, and f varies directly as g, how are a and g related?

50. REASONING Explain why some mathematicians consider every joint variation a combined variation, but not every combined variation a joint variation.

51. OPEN ENDED Describe three real-life quantities that vary jointly with each other.

52. WRITING IN MATH Determine the type(s) of variation(s) for which 0 cannot be one of the values. Explain your reasoning.

NYSRE PRACTICE A2.A.5

53. SAT/ACT Rafael left the dorm and drove toward the cabin at an average speed of 40 km/h. Monica left some time later driving in the same direction at an average speed of 48 km/h. After driving for five hours, Monica caught up with Rafael. How long did Rafael drive before Monica caught up?

A 2 hours
B 4 hours
C 6 hours
D 8 hours

54. 75% of 88 is the same as 60% of what number?

F 100
G 105
H 108
J 110

55. EXTENDED RESPONSE Audrey's hair is 7 inches long and is expected to grow at an average rate of 3 inches per year.

a. Make a table that shows the expected length of Audrey's hair after each of the first 4 years.

b. Write a function that can be used to determine the length of her hair after each year.

c. If she does not get a haircut, determine the length of her hair after 9 years.

56. Which of the following is equal to the sum of two consecutive even integers?

A 144
B 146
C 147
D 148

Spiral Review

Determine any vertical asymptotes and holes in the graph of each rational function. (Lesson 9-4)

57. $f(x) = \frac{1}{x^2 + 5x + 6}$

58. $f(x) = \frac{x + 2}{x^2 + 3x - 4}$

59. $f(x) = \frac{x^2 + 4x + 3}{x + 3}$

60. PHOTOGRAPHY The formula $\frac{1}{q} = \frac{1}{f} - \frac{1}{p}$ can be used to determine how far the film should be placed from the lens of a camera to create a perfect photograph. The variable q represents the distance from the lens to the film, f represents the focal length of the lens, and p represents the distance from the object to the lens. (Lesson 9-3)

a. Solve the formula for $\frac{1}{p}$.

b. Write the expression containing f and q as a single rational expression.

c. If a camera has a focal length of 8 centimeters and the lens is 10 centimeters from the film, how far should an object be from the lens so that the picture will be in focus?

Solve each equation. Check your solutions. (Lesson 8-5)

61. $\log_3 42 - \log_3 n = \log_3 7$

62. $\log_2(3x) + \log_2 5 = \log_2 30$

63. $2 \log_5 x = \log_5 9$

64. $\log_{10} a + \log_{10} (a + 21) = 2$

Given a polynomial and one of its factors, find the remaining factors of the polynomial. Some factors may not be binomials. (Lesson 6-6)

65. $2x^3 - 5x^2 - 28x + 15; x - 5$

66. $3x^3 + 10x^2 - x - 12; x + 3$

Skills Review

Find the LCM of each set of polynomials. (Lesson 5-3)

67. $a, 2a, a + 1$

68. $x, 4y, x - y$

69. $8, 24x, 12$

70. $x^4, 3x^2, 2xy$

71. $12a, 15, 4b^2$

72. $x + 2, x - 3, x^2 - x - 6$

9-6 Solving Rational Equations and Inequalities

Then

You simplified rational expressions. (Lesson 9-2)

Now

- Solve rational equations.
- Solve rational inequalities.

NYS Core Curriculum

A2.A.23 Solve rational equations and inequalities

New Vocabulary

rational equation
weighted average
rational inequality

NY Math Online

glencoe.com
- Extra Examples
- Personal Tutor
- Self-Check Quiz
- Homework Help
- Math in Motion

Why?

A gaming club charges \$20 per month for membership. Members also have to pay \$5 each time they visit the club. If a member visits the club x times in one month, then the charge for that month will be $20 + 5x$. The actual cost per visit will be $\frac{20 + 5x}{x}$.

To determine how many visits are needed for the cost per visit to be \$6, you would need to solve the equation $\frac{20 + 5x}{x} = 6$.

Solve Rational Equations Equations that contain one or more rational expressions are called **rational equations**. These equations are often easier to solve once the fractions are eliminated. You can eliminate the fractions by multiplying each side by the least common denominator (LCD).

EXAMPLE 1 Solve a Rational Equation

Solve $\frac{4}{x+3} + \frac{5}{6} = \frac{23}{18}$. Check your solution.

The LCD for the terms is $18(x + 3)$.

$$\frac{4}{x+3} + \frac{5}{6} = \frac{23}{18}$$ **Original equation**

$$18(x+3)\left(\frac{4}{x+3}\right) + 18(x+3)\left(\frac{5}{6}\right) = 18(x+3)\left(\frac{23}{18}\right)$$ **Multiply by LCD.**

$$18\cancel{(x+3)}^{1}\left(\frac{4}{\cancel{x+3}_{1}}\right) + \cancel{18}^{3}(x+3)\left(\frac{5}{\cancel{6}_{1}}\right) = \cancel{18}^{1}(x+3)\left(\frac{23}{\cancel{18}_{1}}\right)$$ **Divide common factors.**

$$72 + 15x + 45 = 23x + 69$$ **Multiply.**

$$15x + 117 = 23x + 69$$ **Simplify.**

$$48 = 8x$$ **Subtract 15*x* and 69.**

$$6 = x$$ **Divide.**

CHECK

$$\frac{4}{x+3} + \frac{5}{6} = \frac{23}{18}$$ **Original equation**

$$\frac{4}{6+3} + \frac{5}{6} \stackrel{?}{=} \frac{23}{18}$$ ***x* = 6**

$$\frac{4}{9} + \frac{5}{6} \stackrel{?}{=} \frac{23}{18}$$ **Simplify.**

$$\frac{8}{18} + \frac{15}{18} \stackrel{?}{=} \frac{23}{18}$$ **Simplify.**

$$\frac{23}{18} = \frac{23}{18} \checkmark$$ **Add.**

Check Your Progress

Solve each equation. Check your solution.

1A. $\frac{2}{x+3} + \frac{3}{2} = \frac{19}{10}$

1B. $\frac{7}{12} + \frac{9}{x-4} = \frac{55}{48}$

Personal Tutor glencoe.com

Multiplying each side of an equation by the LCD of rational expressions can yield results that are not solutions of the original equation. These are extraneous solutions.

EXAMPLE 2 Solve a Rational Equation

Solve $\frac{2x}{x+5} - \frac{x^2 - x - 10}{x^2 + 8x + 15} = \frac{3}{x+3}$. Check your solution.

The LCD for the terms is $(x + 3)(x + 5)$.

$$\frac{2x}{x+5} - \frac{x^2 - x - 10}{x^2 + 8x + 15} = \frac{3}{x+3}$$ **Original equation**

$$\frac{(x+3)(x+5)(2x)}{x+5} - \frac{(x+3)(x+5)(x^2 - x - 10)}{x^2 + 8x + 15} = \frac{(x+3)(x+5)3}{x+3}$$ **Multiply by LCD.**

$$\frac{(x+3)\overset{1}{\cancel{(x+5)}}(2x)}{\underset{1}{\cancel{x+5}}} - \frac{\overset{1}{\cancel{(x+3)}}\overset{1}{\cancel{(x+5)}}(x^2 - x - 10)}{\underset{1}{\cancel{x^2 + 8x + 15}}} = \frac{(x+5)\overset{1}{\cancel{(x+3)}}3}{\underset{1}{\cancel{x+3}}}$$ **Divide common factors.**

$$(x + 3)(2x) - (x^2 - x - 10) = 3(x + 5)$$ **Simplify.**

$$2x^2 + 6x - x^2 + x + 10 = 3x + 15$$ **Distribute.**

$$x^2 + 7x + 10 = 3x + 15$$ **Simplify.**

$$x^2 + 4x - 5 = 0$$ **Subtract $3x + 15$.**

$$(x + 5)(x - 1) = 0$$ **Factor.**

$$x + 5 = 0 \quad \text{or} \quad x - 1 = 0$$ **Zero Product Property**

$$x = -5 \qquad\qquad x = 1$$

CHECK Try $x = -5$.

$$\frac{2x}{x+5} - \frac{x^2 - x - 10}{x^2 + 8x + 15} = \frac{3}{x+3}$$

$$\frac{2(-5)}{-5+5} - \frac{(-5)^2 - (-5) - 10}{(-5)^2 + 8(-5) + 15} \stackrel{?}{=} \frac{3}{-5+3}$$

$$\frac{-10}{0} - \frac{25 + 5 - 10}{25 - 40 + 15} \neq -\frac{3}{2} \; ✗$$

Try $x = 1$.

$$\frac{2x}{x+5} - \frac{x^2 - x - 10}{x^2 + 8x + 15} = \frac{3}{x+3}$$

$$\frac{2(1)}{1+5} - \frac{1^2 - 1 - 10}{1^2 + 8(1) + 15} \stackrel{?}{=} \frac{3}{1+3}$$

$$\frac{2}{6} - \frac{-10}{24} \stackrel{?}{=} \frac{3}{4}$$

$$\frac{8}{24} + \frac{10}{24} \stackrel{?}{=} \frac{3}{4}$$

$$\frac{3}{4} = \frac{3}{4} \; ✓$$

When solving a rational equation, any possible solution that results in a zero in the denominator must be excluded from your list of solutions.

Since $x = -5$ results in a zero in the denominator, it is extraneous. Eliminate -5 from the list of solutions. The solution is 1.

Math History Link

Brook Taylor (1685–1731) English mathematician Taylor developed a theorem used in calculus known as Taylor's Theorem that relies on the remainders after computations with rational expressions.

Review Vocabulary

extraneous solutions solutions that do not satisfy the original equation (Lesson 7-7)

Check Your Progress

2A. $\frac{5}{y-2} + 2 = \frac{17}{6}$

2B. $\frac{2}{z+1} - \frac{1}{z-1} = \frac{-2}{z^2 - 1}$

2C. $\frac{7n}{3n+3} - \frac{5}{4n-4} = \frac{3n}{2n+2}$

2D. $\frac{1}{p-2} = \frac{2p+1}{p^2 + 2p - 8} + \frac{2}{p+4}$

Personal Tutor glencoe.com

The **weighted average** is a method for finding the mean of a set of numbers in which some elements of the set carry more importance, or weight, than others. Many real-world problems involving mixtures, work, distance, and interest can be solved by using rational equations.

StudyTip

Tables Tables like the one in Example 3 are useful in organizing and solving mixture, work, weighted average, and distance problems.

Real-World EXAMPLE 3 Mixture Problem

CHEMISTRY Mia adds a 70% acid solution to 12 milliliters of a solution that is 15% acid. How much of the 70% acid solution should be added to create a solution that is 60% acid?

Understand Mia needs to know how much of a solution needs to be added to an original solution to create a new solution.

Plan Each solution has a certain percentage that is acid. The percentage of acid in the final solution must equal the amount of acid divided by the total solution.

	Original	Added	New
Amount of Acid	0.15(12)	0.7(*x*)	0.15(12) + 0.7*x*
Total Solution	12	*x*	12 + *x*

$$\text{Percentage of acid in solution} = \frac{\text{amount of acid}}{\text{total solution}}$$

Solve

$$\frac{\text{percent}}{100} = \frac{\text{amount of acid}}{\text{total solution}} \quad \textbf{Write a proportion.}$$

$$\frac{60}{100} = \frac{0.15(12) + 0.7x}{12 + x} \quad \textbf{Substitute.}$$

$$\frac{60}{100} = \frac{1.8 + 0.7x}{12 + x} \quad \textbf{Simplify numerator.}$$

$$100(12 + x)\frac{60}{100} = 100(12 + x)\frac{1.8 + 0.7x}{12 + x} \quad \textbf{LCD is 100(12 + x). Multiply by LCD.}$$

$$\overset{}{\underset{1}{\cancel{100}}}(12 + x)\frac{60}{\underset{1}{\cancel{100}}} = 100\underset{1}{\cancel{(12 + x)}}\frac{1.8 + 0.7x}{\underset{1}{\cancel{12 + x}}} \quad \textbf{Divide common factors.}$$

$$(12 + x)60 = 100(1.8 + 0.7x) \quad \textbf{Simplify.}$$

$$720 + 60x = 180 + 70x \quad \textbf{Distribute.}$$

$$540 = 10x \quad \textbf{Subtract 60x and 180.}$$

$$54 = x \quad \textbf{Divide by 10.}$$

Check

$$\frac{60}{100} = \frac{0.15(12) + 0.7x}{12 + x} \quad \textbf{Original equation}$$

$$\frac{60}{100} \stackrel{?}{=} \frac{0.15(12) + 0.7(54)}{12 + 54} \quad \textbf{x = 54}$$

$$\frac{60}{100} \stackrel{?}{=} \frac{37.8}{66} \quad \textbf{Simplify.}$$

$$0.6 = 0.6 \checkmark \quad \textbf{Simplify.}$$

Mia needs to add 54 milliliters of the 70% acid solution.

Real-World Link

Training future politicians in chemistry and science will assure their ability to write science policy bills from an intelligent point of view, and will guarantee future funding for science.

Source: *Chemical Education Journal*

Check Your Progress

3. Jimmy adds a 65% fruit juice solution to 15 milliliters of a drink that is 10% fruit juice. How much of the 65% fruit juice solution must be added to create a fruit punch that is 35% fruit juice?

Personal Tutor glencoe.com

The formula relating distance, rate, and time can also be used to solve rational equations. The most common use is $d = rt$. However, it can also be represented by $r = \frac{d}{t}$ and $t = \frac{d}{r}$.

Real-World EXAMPLE 4 Distance Problem

ROWING Sandra is rowing a canoe on Stanhope Lake. Her rate in still water is 6 miles per hour. It takes Sandra 3 hours to travel 10 miles round trip. Assuming that Sandra rowed at a constant rate of speed, determine the rate of the current.

Understand We are given her speed in still water and the time it takes her to travel with the current and against it. We need to determine the speed of the current.

Plan She traveled 5 miles with the current and 5 miles against it. The formula that relates distance, rate, and time is $d = rt$, or $t = \frac{d}{r}$.

Time with the Current	Time Against the Current	Total Time
$\frac{5}{6+r}$	$\frac{5}{6-r}$	3 hours

Solve

$$\frac{5}{6+r} + \frac{5}{6-r} = 3$$ **Write the equation.**

$$(6+r)(6-r)\frac{5}{6+r} + (6+r)(6-r)\frac{5}{6-r} = (6+r)(6-r)3$$ **LCD = (6 + r)(6 − r) Multiply by LCD.**

$$\overset{1}{\cancel{(6+r)}}(6-r)\frac{5}{\underset{1}{\cancel{6+r}}} + (6+r)\overset{1}{\cancel{(6-r)}}\frac{5}{\underset{1}{\cancel{6-r}}} = (6+r)(6-r)3$$ **Divide common factors.**

$$(6-r)5 + (6+r)5 = (36-r^2)3$$ **Simplify.**

$$30 - 5r + 30 + 5r = 108 - 3r^2$$ **Distribute.**

$$60 = 108 - 3r^2$$ **Simplify.**

$$0 = -3r^2 + 48$$ **Subtract 10r.**

$$0 = -3(r+4)(r-4)$$ **Factor.**

$$0 = (r+4)(r-4)$$ **Divide each side by −3.**

$$r = 4 \text{ or } -4$$ **Zero Product Property**

Check

$$\frac{5}{6+r} + \frac{5}{6-r} = 3$$ **Original equation**

$$\frac{5}{6+4} + \frac{5}{6-4} \stackrel{?}{=} 3$$ **$r = 4$**

$$\frac{5}{10} + \frac{5}{2} \stackrel{?}{=} 3$$ **Simplify.**

$$\frac{1}{2} + \frac{5}{2} = \frac{6}{2} \checkmark$$ **Simplify.**

Since speed cannot be negative, the speed of the current is 4 miles per hour.

StudyTip

Distance Problems When distances involve round trips, the distance in one direction usually equals the distance in the other direction.

Check Your Progress

4. FLYING The speed of the wind is 20 miles per hour. If it takes a plane 7 hours to fly 2368 miles round trip, determine the plane's speed in still air.

Personal Tutor glencoe.com

Real-world problems that involve work can often be solved using rational equations.

Real-World Link

Since 1997, students from Rock Point School in Burlington, Vermont, spend a week servicing communities throughout the world. While working with Habitat for Humanity, the students spent the time in rural Tennessee, starting and completing the roof of a Habitat home in one week.

Source: Vermont Community Work

Real-World EXAMPLE 5 Work Problems

COMMUNITY SERVICE Every year, the junior and senior classes at Hillcrest High School build a house for the community. If it takes the senior class 24 days to complete a house and 18 days if they work with the junior class, how long would it take the junior class to complete a house if they worked alone?

Understand We are given how long it takes the senior class working alone and when the classes work together. We need to determine how long it would take the junior class by themselves.

Plan The senior class can complete 1 house in 24 days, so their rate is $\frac{1}{24}$ of a house per day.

The rate for the junior class is $\frac{1}{j}$.

The combined rate for both classes is $\frac{1}{18}$.

Senior Rate	Junior Rate	Combined Rate
$\frac{1}{24}$	$\frac{1}{j}$	$\frac{1}{18}$

Solve

$\frac{1}{24} + \frac{1}{j} = \frac{1}{18}$ **Write the equation.**

$72j\frac{1}{24} + 72j\frac{1}{j} = 72j\frac{1}{18}$ **LCD = 72*j*** **Multiply by LCD.**

$\overset{3}{\cancel{72}}j\frac{1}{\underset{1}{\cancel{24}}} + 72\overset{1}{\cancel{j}}\frac{1}{\underset{1}{\cancel{j}}} = \overset{4}{\cancel{72}}j\frac{1}{\underset{1}{\cancel{18}}}$ **Divide common factors.**

$3j + 72 = 4j$ **Distribute.**

$72 = j$ **Subtract 3*j*.**

Check Two methods are possible.

Method 1 Substitute values.

$\frac{1}{24} + \frac{1}{j} = \frac{1}{18}$ **Original equation**

$\frac{1}{24} + \frac{1}{72} \stackrel{?}{=} \frac{1}{18}$ ***j* = 72**

$\frac{3}{72} + \frac{1}{72} \stackrel{?}{=} \frac{4}{72}$ **LCD = 72**

$\frac{4}{72} = \frac{4}{72}$ ✓ **Simplify.**

Method 2 Use a calculator.

It would take the junior class 72 days to complete the house by themselves.

Math *in Motion*, Interactive Lab glencoe.com

Check Your Progress

5A. WORK It took Anthony and Travis 6 hours to rake the leaves together last year. The previous year it took Travis 10 hours to do it alone. How long will it take Anthony if he rakes them by himself this year?

5B. Noah and Owen paint houses together. If Noah can paint a particular house in 6 days and Owen can paint the same house in 5 days, how long would it take the two of them if they work together?

Personal Tutor glencoe.com

Solve Rational Inequalities To solve **rational inequalities**, which are inequalities that contain one or more rational expressions, follow these steps.

> **Key Concept** — **Solving Rational Inequalities** *For Your FOLDABLE*
>
> **Step 1** State the excluded values. These are the values for which the denominator is 0.
>
> **Step 2** Solve the related equation.
>
> **Step 3** Use the values determined from the previous steps to divide a number line into intervals.
>
> **Step 4** Test a value in each interval to determine which intervals contain values that satisfy the inequality.

EXAMPLE 6 Solve a Rational Inequality

Solve $\frac{x}{3} - \frac{1}{x-2} < \frac{x+1}{4}$.

Step 1 The excluded value for this inequality is 2.

Step 2 Solve the related equation.

$$\frac{x}{3} - \frac{1}{x-2} = \frac{x+1}{4} \quad \textbf{Related equation}$$

$$\overset{4}{\cancel{12}}(x-2)\frac{x}{\underset{1}{\cancel{3}}} - \overset{1}{12\cancel{(x-2)}}\frac{1}{\underset{1}{\cancel{x-2}}} = \overset{3}{\cancel{12}}(x-2)\frac{x+1}{\underset{1}{\cancel{4}}} \quad \textbf{LCD is } 12(x-2). \textbf{ Multiply by LCD.}$$

$$4x^2 - 8x - 12 = 3x^2 - 3x - 6 \quad \textbf{Distribute.}$$

$$x^2 - 5x - 6 = 0 \quad \textbf{Subtract } 3x^2 - 3x - 6.$$

$$(x-6)(x+1) = 0 \quad \textbf{Factor.}$$

$$x = 6 \text{ or } -1 \quad \textbf{Zero Product Property}$$

Step 3 Draw vertical lines at the excluded value and at the solutions to separate the number line into intervals.

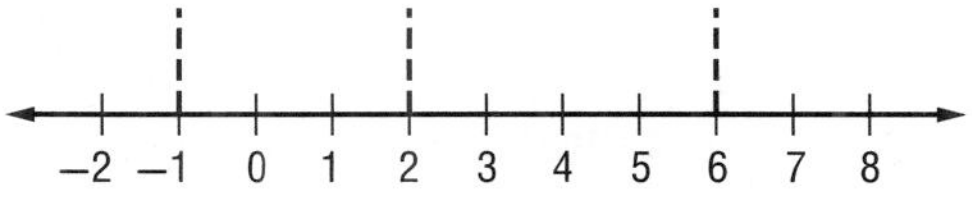

Step 4 Now test a sample value in each interval to determine whether the values in the interval satisfy the inequality.

Test $x = -3$.

$$\frac{-3}{3} - \frac{1}{-3-2} \overset{?}{<} \frac{-3+1}{4}$$
$$-1 + \frac{1}{5} \overset{?}{<} -\frac{2}{4}$$
$$-\frac{4}{5} < -\frac{1}{2} \checkmark$$

Test $x = 0$.

$$\frac{0}{3} - \frac{1}{0-2} \overset{?}{<} \frac{0+1}{4}$$
$$0 + \frac{1}{2} \overset{?}{<} \frac{1}{4}$$
$$\frac{1}{2} \not< \frac{1}{4}$$

Test $x = 4$.

$$\frac{4}{3} - \frac{1}{4-2} \overset{?}{<} \frac{4+1}{4}$$
$$\frac{4}{3} - \frac{1}{2} \overset{?}{<} \frac{5}{4}$$
$$\frac{5}{6} < \frac{5}{4}$$

Test $x = 8$.

$$\frac{8}{3} - \frac{1}{8-2} \overset{?}{<} \frac{8+1}{4}$$
$$\frac{32}{12} - \frac{2}{12} \overset{?}{<} \frac{27}{12}$$
$$\frac{30}{12} \not< \frac{27}{12}$$

The statement is true for $x = -3$ and $x = 4$. Therefore, the solution is $x < -1$ or $2 < x < 6$.

> **StudyTip**
>
> **Rational Inequalities** It is possible that none or all of the intervals will produce a true statement.

Check Your Progress **Solve each inequality.**

6A. $\frac{5}{x} + \frac{6}{5x} > \frac{2}{3}$

6B. $\frac{4}{3x} + \frac{7}{x} < \frac{5}{9}$

Personal Tutor glencoe.com

Check Your Understanding

Examples 1 and 2 pp. 594–595

Solve each equation. Check your solution.

1. $\frac{4}{7} + \frac{3}{x-3} = \frac{53}{56}$

2. $\frac{7}{3} - \frac{3}{x-5} = \frac{19}{12}$

3. $\frac{10}{2x+1} + \frac{4}{3} = 2$

4. $\frac{11}{4} - \frac{5}{y+3} = \frac{23}{12}$

5. $\frac{8}{x-5} - \frac{9}{x-4} = \frac{5}{x^2-9x+20}$

6. $\frac{14}{x+3} + \frac{10}{x-2} = \frac{122}{x^2+x-6}$

7. $\frac{14}{x-8} - \frac{5}{x-6} = \frac{82}{x^2-14x+48}$

8. $\frac{5}{x+2} - \frac{3}{x-2} = \frac{12}{x^2-4}$

Example 3 p. 596

9. MIXTURES Sara has 10 pounds of dried fruit selling for \$6.25 per pound. She wants to know how many pounds of mixed nuts selling for \$4.50 per pound she needs to make a trail mix selling for \$5 per pound.

a. Let m = the number of pounds of mixed nuts. Complete the following table.

	Pounds	Price per Pound	Total Price
Dried Fruit	10	\$6.25	6.25(10)
Mixed Nuts			
Trail Mix			

b. Write a rational equation using the last column of the table.

c. Solve the equation to determine how many pounds of mixed nuts are needed.

Example 4 p. 597

10. DISTANCE Alicia's average speed riding her bike is 11.5 miles per hour. She takes a round trip of 40 miles. It takes her 1 hour and 20 minutes with the wind and 2 hours and 30 minutes against the wind.

a. Write an expression for Alicia's time with the wind.

b. Write an expression for Alicia's time against the wind.

c. How long does it take to complete the trip?

d. Write and solve the rational equation to determine the speed of the wind.

Example 5 p. 598

11. WORK Kendal and Chandi wax cars. Kendal can wax a particular car in 60 minutes and Chandi can wax the same car in 80 minutes. They plan on waxing the same car together and want to know how long it will take.

a. How much will Kendal complete in 1 minute?

b. How much will Kendal complete in x minutes?

c. How much will Chandi complete in 1 minute?

d. How much will Chandi complete in x minutes?

e. Write a rational equation representing Kendal and Chandi working together on the car.

f. Solve the equation to determine how long it will take them to finish the car.

Example 6 p. 599

Solve each inequality. Check your solutions.

12. $\frac{3}{5x} + \frac{1}{6x} > \frac{2}{3}$

13. $\frac{1}{4c} + \frac{1}{9c} < \frac{1}{2}$

14. $\frac{4}{3y} + \frac{2}{5y} < \frac{3}{2}$

15. $\frac{1}{3b} + \frac{1}{4b} < \frac{1}{5}$

Practice and Problem Solving

● = Step-by-Step Solutions begin on page R20.
Extra Practice begins on page 947.

Examples 1 and 2
pp. 594–595

Solve each equation. Check your solutions.

16. $\frac{9}{x-7} - \frac{7}{x-6} = \frac{13}{x^2 - 13x + 42}$

17. $\frac{13}{y+3} - \frac{12}{y+4} = \frac{18}{y^2 + 7y + 12}$

18. $\frac{14}{x-2} - \frac{18}{x+1} = \frac{22}{x^2 - x - 2}$

19. $\frac{11}{a+2} - \frac{10}{a+5} = \frac{36}{a^2 + 7a + 10}$

20. $\frac{x}{2x-1} + \frac{3}{x+4} = \frac{-21}{2x^2 + 7x - 4}$

21. $\frac{2}{y-5} + \frac{y-1}{2y+1} = \frac{2}{2y^2 - 9y - 5}$

Examples 3–5
pp. 596–598

22. **CHEMISTRY** How many milliliters of a 20% acid solution must be added to 40 milliliters of a 75% acid solution to create a 30% acid solution?

23. **GROCERIES** Ellen bought 3 pounds of bananas for \$0.90 per pound. How many pounds of apples costing \$1.25 per pound must she purchase so that the total cost for fruit is \$1 per pound?

24. **BUILDING** Bryan's volunteer group can build a garage in 12 hours. Sequoia's group can build it in 16 hours. How long would it take them if they worked together?

Example 6
p. 599

Solve each inequality. Check your solutions.

25. $3 - \frac{4}{x} > \frac{5}{4x}$

26. $\frac{5}{3a} - \frac{3}{4a} > \frac{5}{6}$

27. $\frac{x-2}{x+2} + \frac{1}{x-2} > \frac{x-4}{x-2}$

28. $\frac{3}{4} - \frac{1}{x-3} > \frac{x}{x+4}$

29. $\frac{x}{5} + \frac{2}{3} < \frac{3}{x-4}$

30. $\frac{x}{x+2} + \frac{1}{x-1} < \frac{3}{2}$

31. **AIR TRAVEL** It takes a plane 20 hours to fly to its destination against the wind. The return trip takes 16 hours. If the plane's average speed in still air is 500 miles per hour, what is the average speed of the wind during the flight?

32. **INTEREST** Judie wants to invest \$10,000 in two different accounts. The risky account earns 9% interest, while the other account earns 5% interest. She wants to earn \$750 interest for the year. Of tables, graphs, or equations, choose the best representation needed and determine how much should be invested in each account.

33. **MULTIPLE REPRESENTATIONS** Consider $\frac{2}{x-3} + \frac{1}{x} = \frac{x-1}{x-3}$.

a. **ALGEBRAIC** Solve the equation for x. Were any values of x extraneous?

b. **GRAPHICAL** Graph $y_1 = \frac{2}{x-3} + \frac{1}{x}$ and $y_2 = \frac{x-1}{x-3}$ on the same graph for $0 < x < 5$.

c. **ANALYTICAL** For what value(s) of x do they intersect? Do they intersect where x is extraneous for the original equation?

d. **VERBAL** Use this knowledge to describe how you can use a graph to determine whether an apparent solution of a rational equation is extraneous.

Solve each equation. Check your solutions.

34. $\frac{2}{y+3} - \frac{3}{4-y} = \frac{2y-2}{y^2 - y - 12}$

35. $\frac{2}{y+2} - \frac{y}{2-y} = \frac{y^2+4}{y^2-4}$

H.O.T. Problems

Use Higher-Order Thinking Skills

36. **OPEN ENDED** Give an example of a rational equation that can be solved by multiplying each side of the equation by $4(x+3)(x-4)$.

37. **CHALLENGE** Solve $\frac{1 + \frac{9}{x} + \frac{20}{x^2}}{1 - \frac{25}{x^2}} = \frac{x+4}{x-5}$.

38. **WRITING IN MATH** While using the table feature on the graphing calculator to explore $f(x) = \frac{1}{x^2 - x - 6}$, the values -2 and 3 say "ERROR." Explain its meaning.

39. **REASONING** Explain why solutions of rational inequalities need to be checked.

NYSRE PRACTICE A2.A.23

40. Nine pounds of mixed nuts containing 55% peanuts were mixed with 6 pounds of another kind of mixed nuts that contain 40% peanuts. What percent of the new mixture is peanuts?

A 58% B 51% C 49% D 47%

41. Working alone, Dato can dig a 10-foot by 10-foot hole in five hours. Pedro can dig the same hole in six hours. How long would it take them if they worked together?

F 1.5 hours H 2.52 hours
G 2.34 hours J 2.73 hours

42. An aircraft carrier made a trip to Guam and back. The trip there took three hours and the trip back took four hours. It averaged 6 kilometers per hour on the return trip. Find the average speed of the trip to Guam.

A 6 km/h C 10 km/h
B 8 km/h D 12 km/h

43. SHORT RESPONSE If a line ℓ is perpendicular to a segment CD at point F and $CF = FD$, how many points on line ℓ are the same distance from point C as from point D?

Spiral Review

Determine whether each relation shows *direct* or *inverse* variation, or *neither.* (Lesson 9-5)

44.

x	y
14	3
28	1.5
56	0.75
112	0.375

45.

x	y
0.2	24
0.6	72
1.8	216
5.4	648

46.

x	y
12	18
24	36
36	18
72	9

Graph each function. (Lesson 9-4)

47. $f(x) = \frac{x + 4}{x^2 + 7x + 12}$

48. $f(x) = \frac{x^2 - 5x - 14}{x - 7}$

49. $f(x) = \frac{x^2 + 3x - 6}{x - 2}$

50. WEATHER The atmospheric pressure P, in bars, of a given height on Earth is given by using the formula $P = a \cdot e^{-\frac{k}{H}}$. In the formula, a is the surface pressure on Earth, which is approximately 1 bar, k is the altitude for which you want to find the pressure in kilometers, and H is always 7 kilometers. (Lesson 8-7)

a. Find the pressure for 2, 4, and 7 kilometers.

b. What do you notice about the pressure as altitude increases?

51. COMPUTERS Since computers have been invented, computational speed has multiplied by a factor of 4 about every three years. (Lesson 8-1)

a. If a typical computer operates with a computational speed s today, write an expression for the speed at which you can expect an equivalent computer to operate after x three-year periods.

b. Suppose your computer operates with a processor speed of 2.8 gigahertz and you want a computer that can operate at 5.6 gigahertz. If a computer with that speed is currently unavailable for home use, how long can you expect to wait until you can buy such a computer?

Skills Review

Determine whether the following are possible lengths of the sides of a right triangle. (Lesson 0-7)

52. 5, 12, 13

53. 60, 80, 100

54. 7, 24, 25

EXTEND

9-6 Graphing Technology Lab
Solving Rational Equations and Inequalities

NY Math Online glencoe.com
Other Calculator Keystrokes

NYS Core Curriculum **A2.A.23 Solve rational equations and inequalities**

You can use a TI-83/84 Plus graphing calculator to solve rational equations by graphing or by using the table feature. Graph both sides of the equation, and locate the point(s) of intersection.

ACTIVITY 1 Rational Equation

Solve $\frac{4}{x+1} = \frac{3}{2}$.

Step 1 Graph each side of the equation.

Graph each side of the equation as a separate function. Enter $\frac{4}{x+1}$ as Y1 and $\frac{3}{2}$ as Y2. Then graph the two equations.

KEYSTROKES: [Y=] 4 [÷] [(] [X,T,θ,n] [+] 1 [)] [ENTER] 3 [÷] 2 [ZOOM] 6

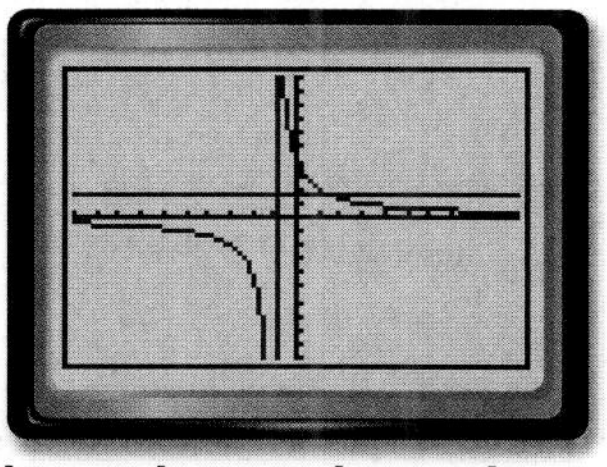

[−10, 10] scl: 1 by [−10, 10] scl: 1

Because the calculator is in connected mode, a vertical line may appear connecting the two branches of the hyperbola. This line is not part of the graph.

Step 2 Use the **intersect** feature.

The **intersect** feature on the **CALC** menu allows you to approximate the ordered pair of the point at which the graphs cross.

KEYSTROKES: [2nd] [CALC] 5

Select one graph and press [ENTER]. Select the other graph, press [ENTER], and press [ENTER] again.

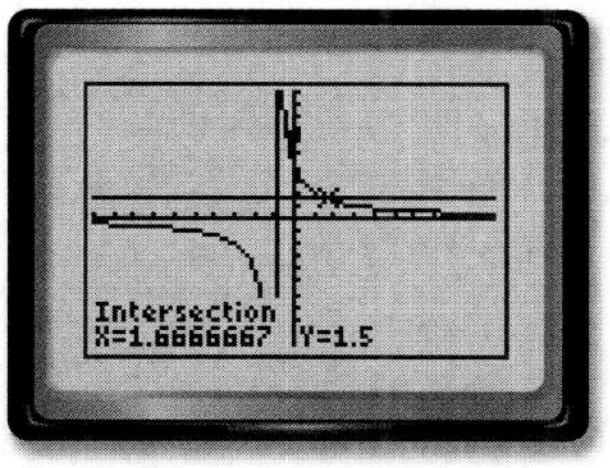

[−10, 10] scl: 1 by [−10, 10] scl: 1

The solution is $1\frac{2}{3}$.

Step 3 Use the **table** feature.

Verify the solution using the table feature. Set up the table to show x-values in increments of $\frac{1}{3}$.

KEYSTROKES: [2nd] [TblSet] 0 [ENTER] 1 [÷] 3 [ENTER] [2nd] [TABLE]

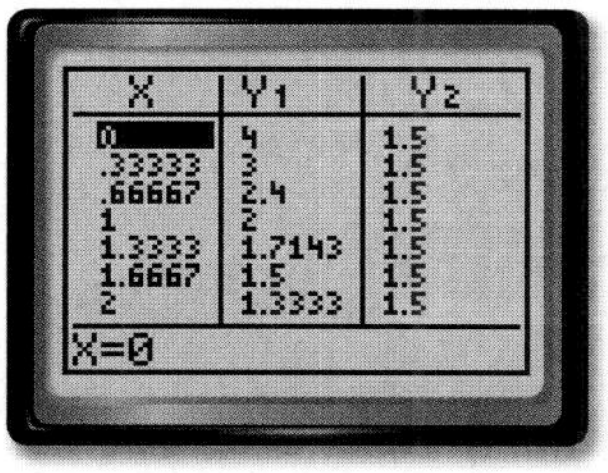

The table displays x-values and corresponding y-values for each graph. At $x = 1\frac{2}{3}$, both functions have a y-value of 1.5. Thus, the solution of the equation is $1\frac{2}{3}$.

You can use a similar procedure to solve rational inequalities using a graphing calculator.

ACTIVITY 2 Rational Inequality

Solve $\frac{3}{x} + \frac{7}{x} > 9$.

Step 1 Enter the inequalities.

Rewrite the problem as a system of inequalities.

The first inequality is $\frac{3}{x} + \frac{7}{x} > y$ or $y < \frac{3}{x} + \frac{7}{x}$. Since this inequality includes the *less than* symbol, shade below the curve. First enter the boundary and then use the arrow and [ENTER] keys to choose the shade below icon, ▙.

The second inequality is $y > 9$. Shade above the curve since this inequality contains *less than*.

KEYSTROKES: [◀] [◀] [ENTER] [ENTER] [ENTER] [▶] [▶] 3 [÷] [X,T,θ,n] [+] 7 [÷] [X,T,θ,n] [ENTER] [◀] [◀] [ENTER] [ENTER] [▶] [▶] 9 [GRAPH]

Step 2 Graph the system.

KEYSTROKES: [GRAPH]

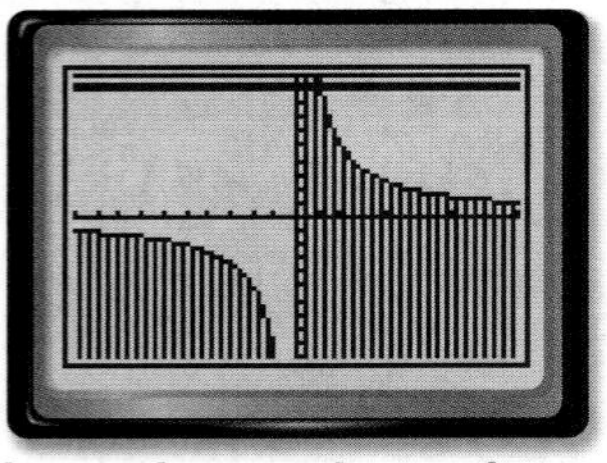

[−10, 10] scl: 1 by [−10, 10] scl: 1

The solution set of the original inequality is the set of x-values of the points in the region where the shadings overlap. Using the calculator's **intersect** feature, you can conclude that the solution set is $\left\{x \middle| 0 > x > 1\frac{1}{9}\right\}$.

Step 3 Use the TABLE feature.

Verify using the **table** feature. Set up the table to show x-values in increments of $\frac{1}{9}$.

KEYSTROKES: [2nd] [TblSet] 0 [ENTER] 1 [÷] 9 [ENTER] [2nd] [TABLE]

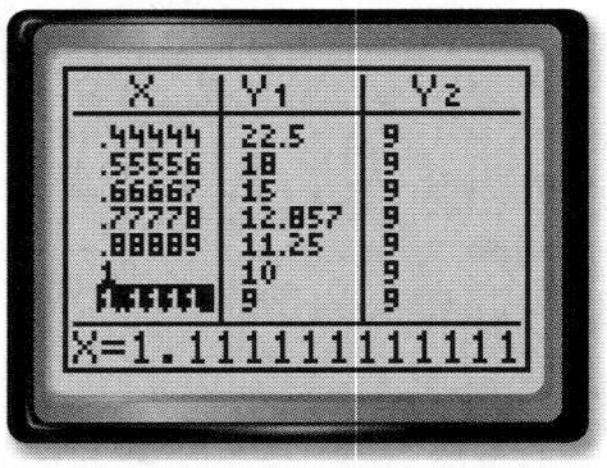

Scroll through the table. Notice that for x-values greater than 0 and less than $1\frac{1}{9}$, **Y1** > **Y2**. This confirms that the solution of the inequality is $\left\{x \middle| 0 > x > 1\frac{1}{9}\right\}$.

Exercises

Solve each equation or inequality.

1. $\frac{1}{x} + \frac{1}{2} = \frac{2}{x}$

2. $\frac{1}{x-4} = \frac{2}{x-2}$

3. $\frac{4}{x} = \frac{6}{x^2}$

4. $\frac{1}{1-x} = 1 - \frac{x}{x-1}$

5. $\frac{1}{x+4} = \frac{2}{x^2+3x-4} - \frac{1}{1-x}$

6. $\frac{1}{x} + \frac{1}{2x} > 5$

7. $\frac{1}{x-1} + \frac{2}{x} < 0$

8. $1 + \frac{5}{x-1} \leq 0$

9. $2 + \frac{1}{x-1} \geq 0$

NY Math Online glencoe.com
- STUDY TO GO
- Vocabulary Review

Chapter Summary

Key Concepts

Rational Expressions (Lessons 9-1 and 9-2)

- Multiplying and dividing rational expressions is similar to multiplying and dividing fractions.
- To simplify complex fractions, simplify the numerator and the denominator separately, and then simplify the resulting expression.

Reciprocal and Rational Functions (Lessons 9-3 and 9-4)

- A reciprocal function is of the form $f(x) = \frac{1}{a(x)}$, where $a(x)$ is a linear function and $a(x) \neq 0$.
- A rational function is of the form $\frac{a(x)}{b(x)}$, where $a(x)$ and $b(x)$ are polynomial functions and $b(x) \neq 0$.

Direct, Joint, and Inverse Variation (Lesson 9-5)

- Direct Variation: There is a nonzero constant k such that $y = kx$.
- Joint Variation: There is a number k such that $y = kxz$, where $x \neq 0$ and $z \neq 0$.
- Inverse Variation: There is a nonzero constant k such that $xy = k$ or $y = \frac{k}{x}$.

Rational Equations and Inequalities (Lesson 9-6)

- Eliminate fractions in rational equations by multiplying each side of the equation by the LCD.
- Possible solutions of a rational equation must exclude values that result in zero in the denominator.

FOLDABLES® Study Organizer

Be sure the Key Concepts are noted in your Foldable.

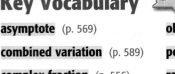

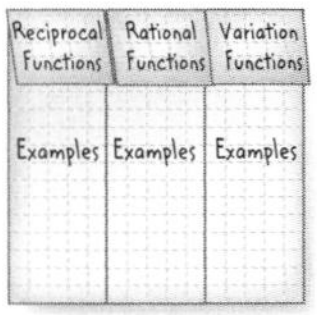

Key Vocabulary

asymptote (p. 569)
combined variation (p. 589)
complex fraction (p. 556)
constant of variation (p. 586)
direct variation (p. 586)
horizontal asymptote (p. 577)
inverse variation (p. 588)
joint variation (p. 587)
oblique asymptote (p. 579)
point discontinuity (p. 580)
rational equation (p. 594)
rational expression (p. 553)
rational function (p. 577)
rational inequality (p. 599)
reciprocal function (p. 569)
vertical asymptote (p. 577)
weighted average (p. 596)

Vocabulary Check

Choose a term from the list above that best completes each statement or phrase.

1. A(n) ___________ is a rational expression whose numerator and/or denominator contains a rational expression.
2. If two quantities show ___________, their product is equal to a constant k.
3. A(n) ___________ asymptote is a linear asymptote that is neither horizontal nor vertical.
4. A(n) ___________ can be expressed in the form $y = kx$.
5. Equations that contain one or more rational expressions are called ___________.
6. The graph of $y = \frac{x}{x+2}$ has a(n) ___________ at $x = -2$.
7. ___________ occurs when one quantity varies directly as the product of two or more other quantities.
8. A ratio of two polynomial expressions is called a(n) ___________.
9. ___________ looks like a hole in a graph because the graph is undefined at that point.
10. ___________ occurs when one quantity varies directly and/or inversely as two or more other quantities.

Study Guide and Review

Lesson-by-Lesson Review

9-1 Multiplying and Dividing Rational Expressions (pp. 553–561)

A2.A.16, A2.A.17

Simplify each expression.

11. $\frac{-16xy}{27z} \cdot \frac{15z^3}{8x^2}$

12. $\frac{x^2 - 2x - 8}{x^2 + x - 12} \cdot \frac{x^2 + 2x - 15}{x^2 + 7x + 10}$

13. $\frac{x^2 - 1}{x^2 - 4} \cdot \frac{x^2 - 5x - 14}{x^2 - 6x - 7}$

14. $\frac{x + y}{15x} \div \frac{x^2 - y^2}{3x^2}$

15. $\dfrac{\frac{x^2 + 3x - 18}{x + 4}}{\frac{x^2 + 7x + 6}{x + 4}}$

16. **GEOMETRY** A triangle has an area of $3x^2 + 9x - 54$ square centimeters. If the height of the triangle is $x + 6$ centimeters, find the length of the base.

EXAMPLE 1

Simplify $\frac{4a}{3b} \cdot \frac{9b^4}{2a^2}$.

$$\frac{4a}{3b} \cdot \frac{9b^4}{2a^2} = \frac{2 \cdot 2 \cdot a \cdot 3 \cdot 3 \cdot b \cdot b \cdot b \cdot b}{3 \cdot b \cdot 2 \cdot a \cdot a}$$

$$= \frac{6b^3}{a}$$

EXAMPLE 2

Simplify $\frac{r^2 + 5r}{2r} \div \frac{r^2 - 25}{6r - 12}$.

$$\frac{r^2 + 5r}{2r} \div \frac{r^2 - 25}{6r - 12} = \frac{r^2 + 5r}{2r} \cdot \frac{6r - 12}{r^2 - 25}$$

$$= \frac{r(r + 5)}{2r} \cdot \frac{6(r - 2)}{(r + 5)(r - 5)}$$

$$= \frac{3(r - 2)}{r - 5}$$

9-2 Adding and Subtracting Rational Expressions (pp. 562–568)

A2.A.16, A2.A.17

Simplify each expression.

17. $\frac{9}{4ab} + \frac{5a}{6b^2}$

18. $\frac{3}{4x - 8} - \frac{x - 1}{x^2 - 4}$

19. $\frac{y}{2x} + \frac{4y}{3x^2} - \frac{5}{6xy^2}$

20. $\frac{2}{x^2 - 3x - 10} - \frac{6}{x^2 - 8x + 15}$

21. $\frac{3}{3x^2 + 2x - 8} + \frac{4x}{2x^2 + 6x + 4}$

22. $\dfrac{\frac{3}{2x + 3} - \frac{x}{x + 1}}{\frac{2x}{x + 1} + \frac{5}{2x + 3}}$

23. **GEOMETRY** What is the perimeter of the rectangle?

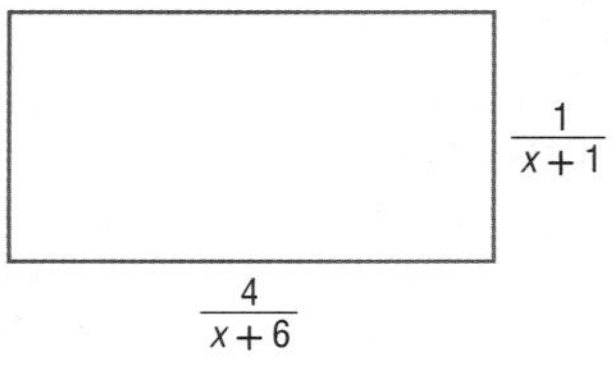

EXAMPLE 3

Simplify $\frac{3a}{a^2 - 4} - \frac{2}{a - 2}$.

$$\frac{3a}{a^2 - 4} - \frac{2}{a - 2} = \frac{3a}{(a - 2)(a + 2)} - \frac{2}{a - 2}$$

$$= \frac{3a}{(a - 2)(a + 2)} - \frac{2(a + 2)}{(a - 2)(a + 2)}$$

$$= \frac{3a - 2(a + 2)}{(a - 2)(a + 2)}$$ **Subtract numerators.**

$$= \frac{3a - 2a - 4}{(a - 2)(a + 2)}$$ **Distributive Property**

$$= \frac{a - 4}{(a - 2)(a + 2)}$$ **Simplify.**

MIXED PROBLEM SOLVING
For mixed problem-solving practice, see page 987.

9-3 Graphing Reciprocal Functions (pp. 569–575)

A.A.15

Graph each function. State the domain and range.

24. $f(x) = \frac{10}{x}$

25. $f(x) = -\frac{12}{x} + 2$

26. $f(x) = \frac{3}{x + 5}$

27. $f(x) = \frac{6}{x - 9}$

28. $f(x) = \frac{7}{x - 2} + 3$

29. $f(x) = -\frac{4}{x + 4} - 8$

30. CONSERVATION The student council is planting 28 trees for a service project. The number of trees each person plants depends on the number of student council members.

a. Write a function to represent this situation.

b. Graph the function.

EXAMPLE 4

Graph $f(x) = \frac{3}{x + 2} - 1$. State the domain and range.

$a = 3$: The graph is stretched vertically.
$h = -2$: The graph is translated 2 units left. There is an asymptote at $x = -2$.
$k: = -1$: The graph is translated 1 unit down. There is an asymptote is at $f(x) = -1$.

Domain: $\{x \mid x \neq -2\}$,
Range: $\{f(x) \mid f(x) \neq -1\}$

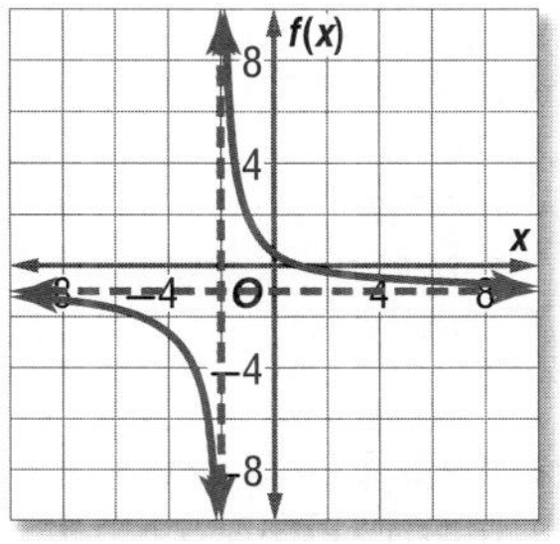

9-4 Graphing Rational Functions (pp. 577–584)

A2.A.41

Determine the equations of any vertical asymptotes and the values of x for any holes in the graph of each rational function.

31. $f(x) = \frac{3}{x^2 + 4x}$

32. $f(x) = \frac{x + 2}{x^2 + 6x + 8}$

33. $f(x) = \frac{x^2 - 9}{x^2 - 5x - 24}$

Graph each rational function.

34. $f(x) = \frac{x + 2}{(x + 5)^2}$

35. $f(x) = \frac{x}{x + 1}$

36. $f(x) = \frac{x^2 + 4x + 4}{x + 2}$

37. $f(x) = \frac{x - 1}{x^2 + 5x + 6}$

38. SALES Aliyah is selling magazine subscriptions. Out of the first 15 houses, she sold subscriptions to 10 of them. Suppose Aliyah goes to x more houses and sells subscriptions to all of them. The percentage of houses that she sold to out of the total houses can be determined using $P(x) = \frac{10 + x}{15 + x}$.

a. Graph the function.

b. What domain and range values are meaningful in the context of the problem?

EXAMPLE 5

Determine the equation of any vertical asymptotes and the values of x for any holes in the graph of $f(x) = \frac{x^2 - 1}{x^2 + 2x - 3}$.

$$\frac{x^2 - 1}{x^2 + 2x - 3} = \frac{(x - 1)(x + 1)}{(x - 1)(x + 3)}$$

The function is undefined for $x = 1$ and $x = -3$.
Since $\frac{(x - 1)(x + 1)}{(x - 1)(x + 3)} = \frac{x + 1}{x + 3}$, $x = -3$ is a vertical asymptote, and $x = 1$ represents a hole in the graph.

EXAMPLE 6

Graph $f(x) = \frac{3}{x(x - 1)}$.

The function is undefined for $x = 0$ and $x = 1$. Because $\frac{3}{x(x - 1)}$ is in simplest form, $x = 0$ and $x = 1$ are vertical asymptotes. Draw the two asymptotes and sketch the graph.

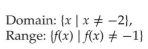

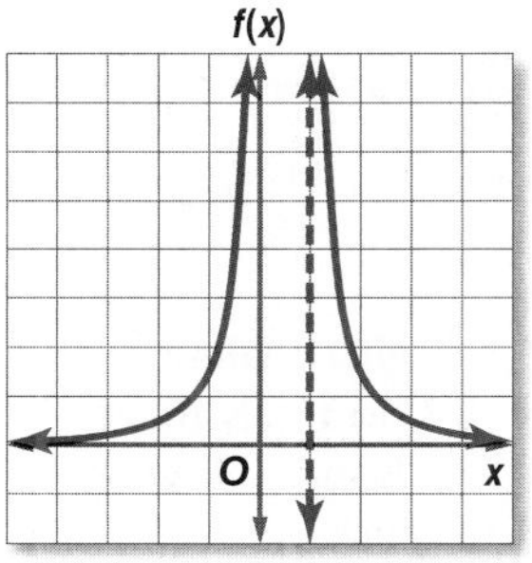

9-5 Variation Functions (pp. 586–593)

A2.A.5

39. If a varies directly as b and $b = 18$ when $a = 27$, find a when $b = 10$.

40. If y varies inversely as x and $y = 15$ when $x = 3.5$, find y when $x = -5$.

41. If y varies inversely as x and $y = -3$ when $x = 9$, find y when $x = 81$.

42. If y varies jointly as x and z, and $x = 8$ and $z = 3$ when $y = 72$, find y when $x = -2$ and $z = -5$.

43. If y varies jointly as x and z, and $y = 18$ when $x = 6$ and $z = 15$, find y when $x = 12$ and $z = 4$.

44. **JOBS** Lisa's earnings vary directly with how many hours she babysits. If she earns \$68 for 8 hours of babysitting, find her earnings after 5 hours of babysitting.

EXAMPLE 7

If y varies inversely as x and $x = 24$ when $y = -8$, find x when $y = 15$.

$\frac{x_1}{y_2} = \frac{x_2}{y_1}$ **Inverse variation**

$\frac{24}{15} = \frac{x_2}{-8}$ **$x_1 = 24$, $y_1 = -8$, $y_2 = 15$**

$24(-8) = 15(x_2)$ **Cross multiply.**

$-192 = 15x_2$ **Simplify.**

$-12\frac{4}{5} = x_2$ **Divide each side by 15.**

When $y = 15$, the value of x is $-12\frac{4}{5}$.

9-6 Solving Rational Equations and Inequalities (pp. 594–602)

A2.A.23

Solve each equation or inequality. Check your solutions.

45. $\frac{1}{3} + \frac{4}{x-2} = 6$

46. $\frac{6}{x+5} - \frac{3}{x-3} = \frac{6}{x^2+2x-15}$

47. $\frac{2}{x^2-9} = \frac{3}{x^2-2x-3}$

48. $\frac{4}{2x-3} + \frac{x}{x+1} = \frac{-8x}{2x^2-x-3}$

49. $\frac{x}{x+4} - \frac{28}{x^2+x-12} = \frac{1}{x-3}$

50. $\frac{x}{2} + \frac{1}{x-1} < \frac{x}{4}$

51. $\frac{1}{2x} - \frac{4}{5x} > \frac{1}{3}$

52. **YARD WORK** Lana can plant a garden in 3 hours. Milo can plant the same garden in 4 hours. How long will it take them if they work together?

EXAMPLE 8

Solve $\frac{3}{x+2} + \frac{1}{x} = 0$.

The LCD is $x(x + 2)$.

$$\frac{3}{x+2} + \frac{1}{x} = 0$$

$$x(x+2)\left(\frac{3}{x+2} + \frac{1}{x}\right) = x(x+2)(0)$$

$$x(x+2)\left(\frac{3}{x+2}\right) + x(x+2)\left(\frac{1}{x}\right) = 0$$

$$3(x) + 1(x+2) = 0$$

$$3x + x + 2 = 0$$

$$4x + 2 = 0$$

$$4x = -2$$

$$x = -\frac{1}{2}$$

Practice Test

NY Math Online glencoe.com
Chapter Test

Simplify each expression.

1. $\frac{r^2 + rs}{2r} \div \frac{r + s}{16r^2}$

2. $\frac{m^2 - 4}{3m^2} \cdot \frac{6m}{2 - m}$

3. $\frac{m^2 + m - 6}{n^2 - 9} \div \frac{m - 2}{n + 3}$

4. $\frac{\frac{x^2 + 4x + 3}{x^2 - 2x - 15}}{\frac{x^2 - 1}{x^2 - x - 20}}$

5. $\frac{x + 4}{6x + 3} + \frac{1}{2x + 1}$

6. $\frac{x}{x^2 - 1} - \frac{3}{2x + 2}$

7. $\frac{1}{y} + \frac{2}{7} - \frac{3}{2y^2}$

8. $\frac{2 + \frac{1}{x}}{5 - \frac{1}{x}}$

9. Identify the asymptotes, domain, and range of the function graphed.

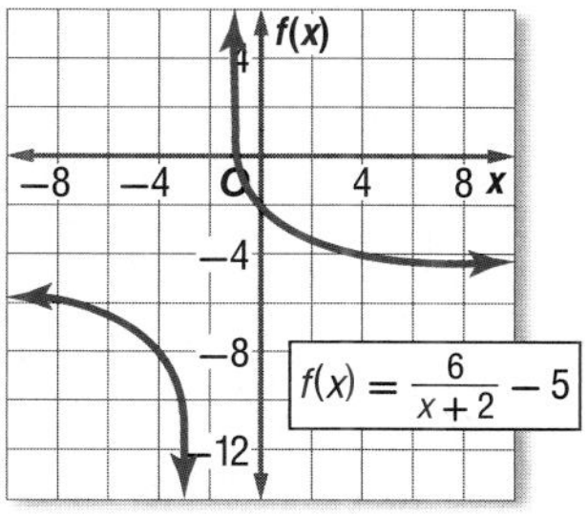

10. **MULTIPLE CHOICE** What is the equation for the vertical asymptote of the rational function $f(x) = \frac{x + 1}{x^2 + 3x + 2}$?

  A $x = -2$

  B $x = -1$

  C $x = 1$

  D $x = 2$

Graph each function.

11. $f(x) = -\frac{8}{x} - 9$

12. $f(x) = \frac{2}{x + 4}$

13. $f(x) = \frac{3}{x - 1} + 8$

14. $f(x) = \frac{5x}{x + 1}$

15. $f(x) = \frac{x}{x - 5}$

16. $f(x) = \frac{x^2 + 5x - 6}{x - 1}$

17. Determine the equations of any vertical asymptotes and the values of x for any holes in the graph of the function $f(x) = \frac{x + 5}{x^2 - 2x - 35}$.

18. Determine the equations of any oblique asymptotes in the graph of the function $f(x) = \frac{x^2 + x - 5}{x + 3}$.

Solve each equation or inequality.

19. $\frac{-1}{x + 4} = 6 - \frac{x}{x + 4}$

20. $\frac{1}{3} = \frac{5}{m + 3} + \frac{8}{21}$

21. $7 + \frac{2}{x} < -\frac{5}{x}$

22. $r + \frac{6}{r} - 5 = 0$

23. $\frac{6}{7} - \frac{3m}{2m - 1} = \frac{11}{7}$

24. $\frac{r + 2}{3r} = \frac{r + 4}{r - 2} - \frac{2}{3}$

25. If y varies inversely as x and $y = 18$ when $x = -\frac{1}{2}$, find x when $y = -10$.

26. If m varies directly as n and $m = 24$ when $n = -3$, find n when $m = 30$.

27. Suppose r varies jointly as s and t. If $s = 20$ when $r = 140$ and $t = -5$, find s when $r = 7$ and $t = 2.5$.

28. **BICYCLING** When Susan rides her bike, the distance that she travels varies directly with the amount of time she is biking. Suppose she bikes 50 miles in 2.5 hours. At this rate, how many hours would it take her to bike 80 miles?

29. **PAINTING** Peter can paint a house in 10 hours. Melanie can paint the same house in 9 hours. How long would it take if they worked together?

30. **MULTIPLE CHOICE** How many liters of a 25% acid solution must be added to 30 liters of an 80% acid solution to create a 50% acid solution?

  F 18

  G 30

  H 36

  J 66

31. What is the volume of the rectangular prism?

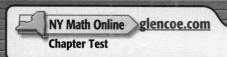

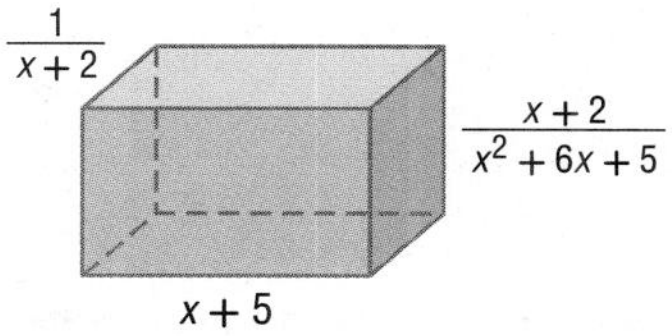

Preparing for Standardized Tests

Guess and Check

It is very important to pace yourself and keep track of how much time you have when taking a standardized test. If time is running short, or if you are unsure how to solve a problem, the guess-and-check strategy may help you determine the correct answer quickly.

Strategies for Guessing and Checking

Step 1

Carefully look over each possible answer choice and evaluate for reasonableness. Eliminate unreasonable answers.

Ask yourself:

- Are there any answer choices that are clearly incorrect?
- Are there any answer choices that are not in the proper format?
- Are there any answer choices that do not have the proper units for the correct answer?

Step 2

For the remaining answer choices, use the guess-and-check method.

- **Equations:** If you are solving an equation, substitute the answer choice for the variable and see if this results in a true number sentence.
- **System of Equations:** For a system of equations, substitute the answer choice for all variables and make sure all equations result in a true number sentence.

Step 3

Choose an answer choice and see if it satisfies the constraints of the problem statement. Identify the correct answer.

- If the answer choice you are testing does not satisfy the problem, move on to the next reasonable guess and check it.
- When you find the correct answer choice, stop.

EXAMPLE

Read the problem. Identify what you need to know. Then use the information in the problem to solve.

Solve: $\frac{2}{x-3} - \frac{4}{x+3} = \frac{8}{x^2-9}$.

A -1 | **C** 5
B 1 | **D** 7

The solution of the rational equation will be a real number. Since all four answer choices are real numbers, they are all possible correct answers and must be checked. Begin with the first answer choice and check it in the rational equation. Continue until you find the answer choice that results in a true number sentence.

	Check:
Guess: −1	$\frac{2}{(-1)-3} - \frac{4}{(-1)+3} = \frac{8}{(-1)^2-9}$ $-\frac{5}{2} \neq -1$ ✗

	Check:
Guess: 1	$\frac{2}{1-3} - \frac{4}{1+3} = \frac{8}{(1)^2-9}$ $-2 \neq -1$ ✗

	Check:
Guess: 5	$\frac{2}{5-3} - \frac{4}{5+3} = \frac{8}{(5)^2-9}$ $\frac{1}{2} = \frac{1}{2}$ ✓

If $x = 5$, the result is a true number sentence. So, the correct answer is C.

Exercises

Read each problem. Identify what you need to know. Then use the information in the problem to solve.

1. Solve: $\frac{2}{5x} - \frac{1}{2x} = -\frac{1}{2}$.

A $\frac{1}{10}$

B $\frac{1}{5}$

C $\frac{1}{4}$

D $\frac{1}{2}$

2. The sum of Kevin's, Anna's, and Tia's ages is 40. Anna is 1 year more than twice as old as Tia. Kevin is 3 years older than Anna. How old is Anna?

F 7   **H** 15

G 14   **J** 18

3. Determine the point(s) where the following rational function crosses the x-axis.

$$f(x) = \frac{2}{x-1} - \frac{x+4}{3}$$

A −5

B 4

C 2 or 3

D −5 or 2

4. Rafael's Theatre Company sells tickets for $10. At this price, they sell 400 tickets. Rafael estimates that they would sell 40 fewer tickets for each $2 price increase. What charge would give the most income?

F 10   **H** 15

G 13   **J** 20

NYS Regents Exam Practice

Cumulative, Chapters 1 through 9

Multiple Choice

Answer all questions in this part. Select the answer that best completes the statement or answers the question.

1. Which expression represents the perimeter of the rectangle below?

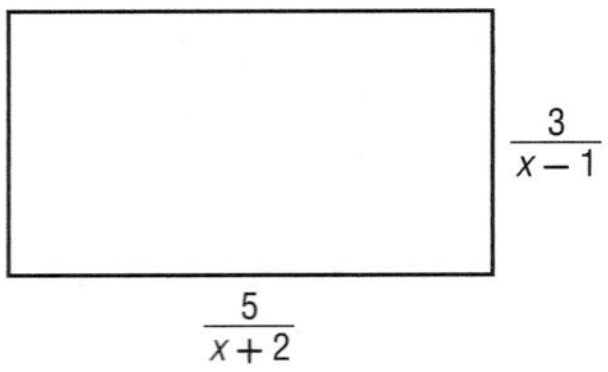

(1) $\frac{5x-1}{x^2+4x-6}$ (3) $\frac{6x+1}{x^2+2x-3}$

(2) $\frac{16x+2}{x^2+x-2}$ (4) $\frac{8x+1}{x^2+x-2}$

2. A polynomial function is graphed below. What is the degree of the polynomial?

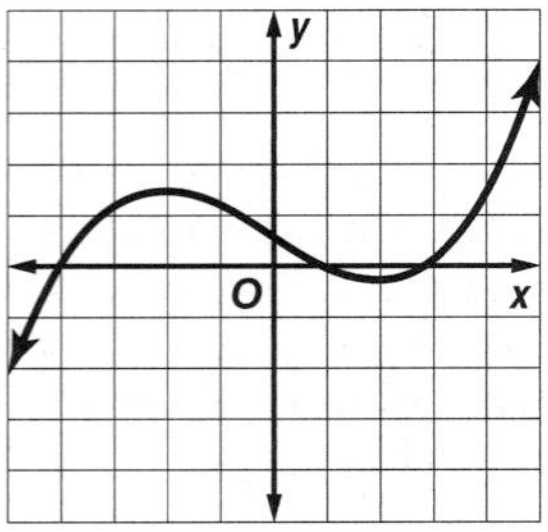

(1) 1

(2) 2

(3) 3

(4) 4

3. Simplify: $\dfrac{\frac{3}{x-4}}{1-\frac{2}{x-4}}$.

(1) $\frac{6}{x-4}$

(2) $\frac{4}{2x-1}$

(3) $\frac{3}{2}$

(4) $\frac{3}{x-6}$

4. Solve: $\frac{10}{x+2} = \frac{6}{x-2} = \frac{24}{x^2-4}$.

(1) 8 (3) 12

(2) 10 (4) 14

5. Reggie ordered a chicken sandwich, a bowl of soup, and a soft drink for lunch. If a 6% sales tax is added and Reggie leaves a 15% tip (after tax), what is the total cost of his lunch?

Dana's Diner Prices	
hamburger	$3.15
chicken sandwich	$4.55
french fries	$1.95
bowl of soup	$2.25
side salad	$3.95
soft drink	$1.50

(1) $9.76 (3) $10.23

(2) $10.12 (4) $11.17

6. What is the equation of the rational function graphed below?

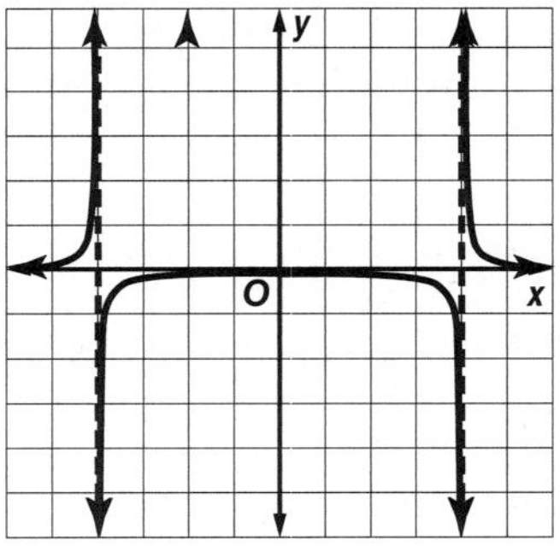

(1) $y = \frac{1}{x^2-16}$ (3) $y = \frac{1}{x^2-4}$

(2) $y = \frac{x^2}{16}$ (4) $y = \frac{x^2}{4}$

Test-Taking Tip

Question 6 Answer choices (2) and (4) are parabolas, so they can be eliminated as unreasonable.

7. Eduardo's father can mow the lawn on his riding mower in 40 minutes. It takes Eduardo 1 hour 30 minutes to mow the lawn with a push mower. Which of the following rational equations can be solved for the number of minutes t it would take them to mow the lawn working together?

(1) $\frac{t}{40} + \frac{t}{1.5} = 1$

(2) $\frac{t}{130} = 1$

(3) $\frac{t}{40} + \frac{t}{90} = 1$

(4) $\frac{t + 40}{t + 90} = 1$

8. Georgena is driving from Albany to Syracuse along I-90 to visit her brother at college. The total distance is about 144 miles, and the distance that Georgena has driven varies directly with the amount of time. If she has driven 93 miles in 1 hour 30 minutes, write a direct variation equation to model the situation.

(1) $d = \frac{t}{62}$

(2) $d = \frac{58}{t}$

(3) $d = 58t$

(4) $d = 62t$

9. What are the values of x and y in the equation below?

$$(8 - 4\mathbf{i}) - (x + y\mathbf{i}) = (5 - 2\mathbf{i})$$

(1) $x = 3, y = -2$

(2) $x = 3, y = -\mathbf{i}$

(3) $x = 3\mathbf{i}, y = -1$

(4) $x = -3, y = 2\mathbf{i}$

Open-Ended Response

Solve each problem. Clearly indicate the necessary steps, including appropriate formula substitutions, diagrams, graphs, charts, etc.

10. Simplify the complex fraction.

$$\frac{\frac{(x + 3)^2}{x^2 - 16}}{\frac{x + 3}{x + 4}}$$

11. Find the product, if possible.

$$\begin{bmatrix} 3 & 7 \\ 1 & 6 \end{bmatrix} \cdot \begin{bmatrix} 2 & 0 & -9 \\ -1 & 5 & 4 \end{bmatrix}$$

12. Heather is starting a small business giving tennis lessons on the weekends and evenings. Considering equipment expenses, travel, and court rental fees, she determines that the cost of running her business will be represented by the function $C(x) = 10x + 215$. The function C represents her cost, in dollars, when she has x clients taking lessons. Heather's income from giving lessons is given by the function $I(x) = 50x - 105$.

a. Solve the system of equations by elimination. How many clients will Heather need in order to break even?

b. Write a new profit function $P(x)$ by subtracting $C(x)$ from $I(x)$.

c. How much profit will Heather earn next week if she has 22 clients taking lessons?

Need Extra Help?

If you missed Question...	1	2	3	4	5	6	7	8	9	10	11	12
Go to Lesson or Page...	9-5	9-4	9-3	9-3	9-1	5-7	2-4	1-4	3-2	9-5	8-8	6-1
NYS Core Curriculum	A2.A.16	A2.CM.3	A2.A.17	A2.A.23	A2.PS.5	A2.A.46	A2.A.23	A2.A.5	A2.N.9	A2.A.17	A2.R.2	A.A.7

CHAPTER 10

Conic Sections

Then

In Chapter 3, you solved systems of linear equations algebraically and graphically.

Now

In Chapter 10, you will:

- Use the Midpoint and Distance Formulas.
- Write and graph equations of parabolas, circles, ellipses, and hyperbolas.
- Solve systems of quadratic equations and inequalities.

NYS Core Curriculum

A2.A.48, A2.A.49 Write equations of circles
A2.A.3 Solve systems of equations involving one linear equation and one quadratic equation algebraically

Why?

SPACE Conic sections are evident in many aspects of space. We can use the distance formula to measure the diameter of a crater on Mars. Equations of circles are used to pilot spacecraft and satellites in circular orbits around Earth and the Moon. Conic sections were used to discover that planets actually travel in elliptical paths, not circular as previously thought. Comets appear to travel along one branch of a hyperbola which can help us to predict when the comet will appear again.

Math *in Motion,* Animation glencoe.com

Get Ready for Chapter 10

Diagnose Readiness You have two options for checking Prerequisite Skills.

Text Option

Take the Quick Check below. Refer to the Quick Review for help.

QuickCheck

Solve each equation by completing the square. (Lesson 5-5)

1. $x^2 + 8x + 7 = 0$

2. $x^2 + 5x - 6 = 0$

3. $x^2 - 8x + 15 = 0$

4. $x^2 + 2x - 120 = 0$

5. $2x^2 + 7x - 15 = 0$

6. $2x^2 + 3x - 5 = 0$

7. $x^2 - \frac{3}{2}x - \frac{23}{16} = 0$

8. $3x^2 - 4x = 2$

Find the coordinates of the vertices of the image for each figure after the given translation. Then graph the preimage and image. (Lesson 4-4)

9. quadrilateral *ABCD* with vertices $A(-5, -1)$, $B(-4, 3)$, $C(2, 3)$, and $D(1, -1)$, translated 3 units right and 4 units down

10. triangle *EFG* with vertices $E(-2, 0)$, $F(5, 2)$, and $G(4, -3)$, translated 1 unit left and 2 units up

11. triangle *JKL* with vertices $J(1, 4)$, $K(2, -5)$, and $L(-6, -6)$, translated 4 units left and 2 units up

12. Triangle *XYZ* with vertices $X(-2, 2)$, $Y(3, 5)$, and $Z(5, -2)$ is translated so that X' is at $(1, -5)$. Find the coordinates of Y' and Z'.

13. LANDSCAPING Laura plots her shed plans on a grid with each unit equal to 1 foot. She places the corners at (100, 50), (110, 50), (100, 40), and (110, 40). She decides to move the shed up 10 feet and to the right 15 feet. What will be the new coordinates of the shed?

QuickReview

EXAMPLE 1

Solve $x^2 + 6x - 16 = 0$ by completing the square.

$$x^2 + 6x = 16$$
$$x^2 + 6x + 9 = 16 + 9$$
$$(x + 3)^2 = 25$$
$$x + 3 = \pm 5$$

$x + 3 = 5$ or $x + 3 = -5$

$x = 2$ $\quad$ $x = -8$

EXAMPLE 2

Find the coordinates of the vertices of the image of triangle *RST* with $R(1, 4)$, $S(4, 2)$, and $T(2, 0)$ if it is moved 2 units to the left and 1 unit up. Then graph *RST* and its image *R′S′T′*.

Write the vertex matrix for $\triangle RST$.

$$\begin{bmatrix} 1 & 4 & 2 \\ 4 & 2 & 0 \end{bmatrix}$$

Add the translation matrix $\begin{bmatrix} -2 & -2 & -2 \\ 1 & 1 & 1 \end{bmatrix}$ to the vertex matrix.

$$\begin{bmatrix} 1 & 4 & 2 \\ 4 & 2 & 0 \end{bmatrix} + \begin{bmatrix} -2 & -2 & -2 \\ 1 & 1 & 1 \end{bmatrix} =$$

$$\begin{bmatrix} -1 & 2 & 0 \\ 5 & 3 & 1 \end{bmatrix}$$

The vertices of $\triangle R'S'T'$ are $R'(-1, 5)$, $S'(2, 3)$, and $T'(0, 1)$.

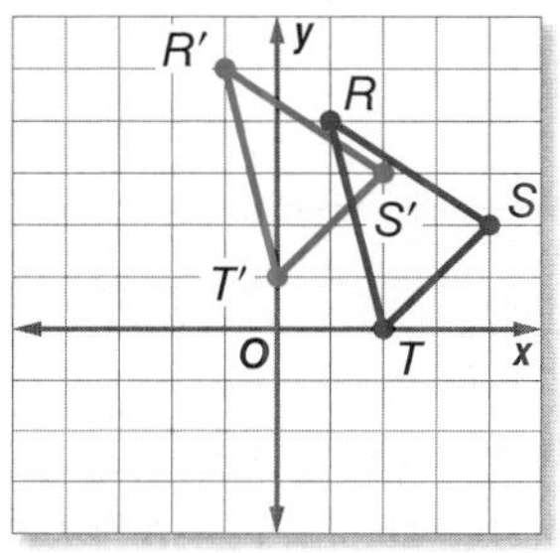

Online Option

NY Math Online Take a self-check Chapter Readiness Quiz at glencoe.com.

Get Started on Chapter 10

You will learn several new concepts, skills, and vocabulary terms as you study Chapter 10. To get ready, identify important terms and organize your resources. You may wish to refer to **Chapter 0** to review prerequisite skills.

FOLDABLES® Study Organizer

Conic Sections Make this Foldable to help you organize your Chapter 10 notes about conic sections. Begin with eight sheets of grid paper.

1. **Staple** the stack of grid paper along the top to form a booklet.

2. **Cut** seven lines from the bottom of the top sheet, six lines from the second sheet, and so on.

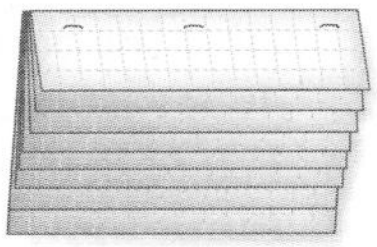

3. **Label** with lesson numbers as shown.

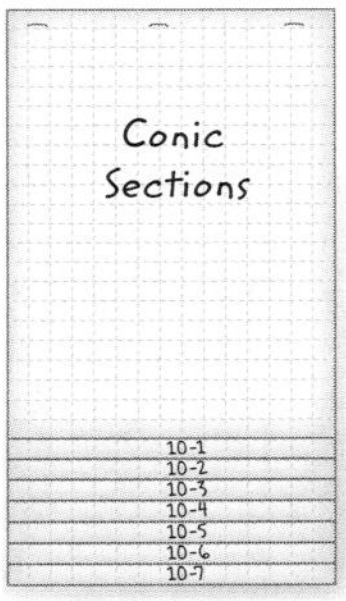

New Vocabulary

English		Español
parabola	• p. 623 •	parábola
focus	• p. 623 •	foco
directrix	• p. 623 •	directriz
latus rectum	• p. 623 •	latus rectum
circle	• p. 631 •	círculo
center of a circle	• p. 631 •	centro de un círculo
radius	• p. 631 •	radio
ellipse	• p. 639 •	elipse
foci	• p. 639 •	focos
major axis	• p. 639 •	eje mayor
minor axis	• p. 639 •	eje menor
center of an ellipse	• p. 639 •	centro de una elipse
vertices	• p. 639 •	vértices
co-vertices	• p. 639 •	co-vértices
constant sum	• p. 640 •	suma constante
hyperbola	• p. 648 •	hipérbola
transverse axis	• p. 648 •	eje transversal
conjugate axis	• p. 648 •	eje conjugado
constant difference	• p. 651 •	diferencia constante

Review Vocabulary

quadratic equation • p. 259 • ecuación cuadrática a quadratic function in the form $ax^2 + bx + c = 0$, where $a \neq 0$

system of equations • p. 135 • sistema de ecuaciones a set of equations with the same variables

x- and y-intercepts • p. 71 • intersecciónes x e y the x- or y-coordinate of the point at which a graph crosses the x- or y-axis

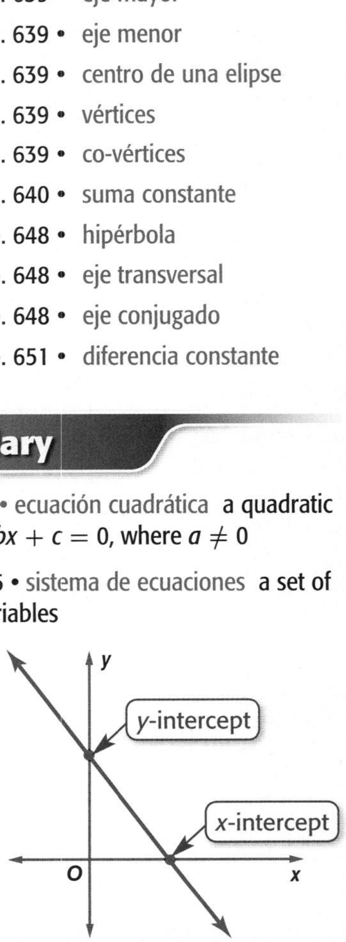

Multilingual eGlossary glencoe.com

NY Math Online glencoe.com

- Study the chapter online
- Explore **Math in Motion**
- Get extra help from your own **Personal Tutor**
- Use **Extra Examples** for additional help
- Take a **Self-Check Quiz**
- **Review Vocabulary** in fun ways

10-1 Midpoint and Distance Formulas

Then

You found the slope of a line passing through two points. (Lesson 2-3)

Now

- Find the midpoint of a segment on the coordinate plane.
- Find the distance between two points on the coordinate plane.

NYS Core Curriculum

Reinforcement of G.G.69 Investigate, justify, and apply the properties of triangles and quadrilaterals in the coordinate plane, using the distance, midpoint, and slope formulas

NY Math Online

glencoe.com

- Extra Examples
- Personal Tutor
- Self-Check Quiz
- Homework Help

Why?

The Zero Milestone in Washington, D.C., was established in 1919. It was intended to serve as the origin for all highway measures with highway markers across the United States displaying the distances from the Zero Milestone.

The Midpoint Formula Recall that point M is the midpoint of segment PQ if M is between P and Q and $PM = MQ$. There is a formula for the coordinates of the midpoint of a segment in terms of the coordinates of the endpoints.

Key Concept — **Midpoint Formula** (For Your FOLDABLE)

Words If a line segment has endpoints $P(x_1, y_1)$ and $Q(x_2, y_2)$, then the midpoint of the segment has coordinates

$$M\left(\frac{x_1 + x_2}{2}, \frac{y_1 + y_2}{2}\right).$$

Model

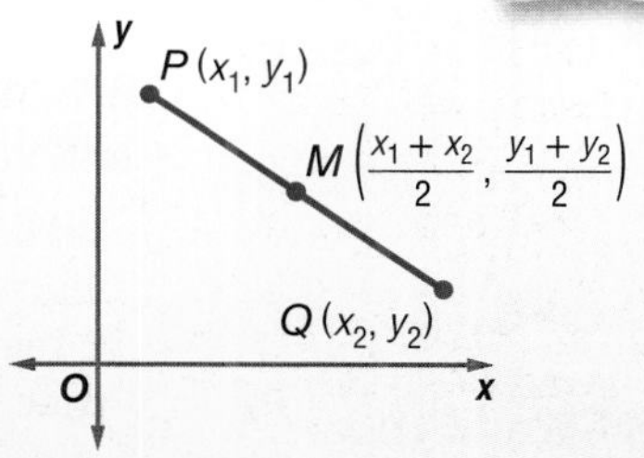

EXAMPLE 1 **Find a Midpoint**

Find the coordinates of M, the midpoint of $\overline{JK}$, for $J(-1, 2)$ and $K(6, 1)$.

Let J be (x_1, y_1) and K be (x_2, y_2).

$M\left(\frac{x_1 + x_2}{2}, \frac{y_1 + y_2}{2}\right)$ **Midpoint Formula**

$= M\left(\frac{-1 + 6}{2}, \frac{2 + 1}{2}\right)$ **$(x_1, y_1) = (-1, 2), (x_2, y_2) = (6, 1)$**

$= M\left(\frac{5}{2}, \frac{3}{2}\right)$ or $M\left(2\frac{1}{2}, 1\frac{1}{2}\right)$ **Simplify.**

Check Your Progress

1A. Find the coordinates of the midpoint of $\overline{AB}$ for $A(5, 12)$ and $B(-4, 8)$.

1B. Find the coordinates of the midpoint of $\overline{CD}$ for $C(4, 5)$ and $D(14, 13)$.

Personal Tutor glencoe.com

The Distance Formula The distance between two points, a and b, on a number line is $|a - b|$ or $|b - a|$. You can use this fact and the Pythagorean Theorem to derive a formula for the distance between two points on a coordinate plane.

StudyTip

Distance In mathematics, just as in real-world situations, distances are always nonnegative.

Let d represent the distance between (x_1, y_1) and (x_2, y_2).

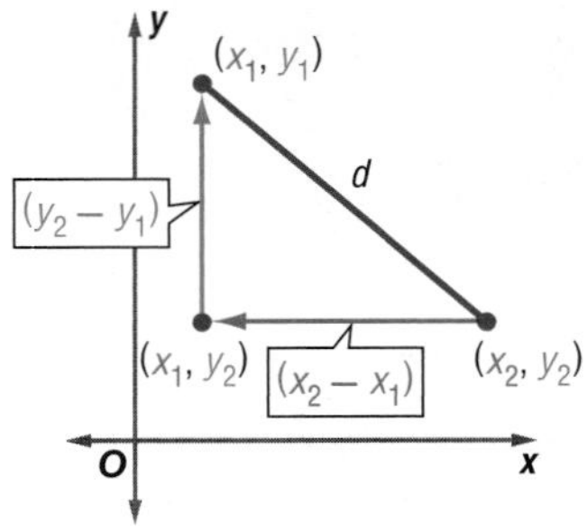

$c^2 = a^2 + b^2$	**Pythagorean Theorem**
$d^2 = \lvert x_2 - x_1\rvert^2 + \lvert y_2 - y_1\rvert^2$	**Substitute.**
$d^2 = (x_2 - x_1)^2 + (y_2 - y_1)^2$	$\lvert x_2 - x_1\rvert^2 = (x_2 - x_1)^2$, $\lvert y_2 - y_1\rvert^2 = (y_2 - y_1)^2$
$d = \sqrt{(x_2 - x_1)^2 + (y_2 - y_1)^2}$	**Find the nonnegative square root of each side.**

Key Concept — Distance Formula

For Your FOLDABLE

Words The distance between two points with coordinates (x_1, y_1) and (x_2, y_2) is given by $\sqrt{(x_2 - x_1)^2 + (y_2 - y_1)^2}$.

Model

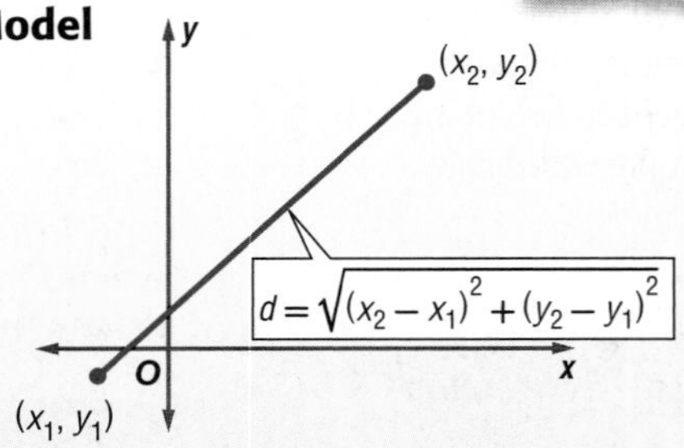

Real-World Link

There are more than 700 disc golf courses in the U.S. These courses are permanent installations, usually located in a public park, where players actually "drive" and "putt" with specially-styled discs.

Real-World EXAMPLE 2 — Find the Distance Between Two Points

DISC GOLF Troy's disc is 20 feet short and 8 feet to the right of the basket. On his first putt, the disc lands 2 feet to the left and 3 feet beyond the basket. If the disc went in a straight line, how far did it go?

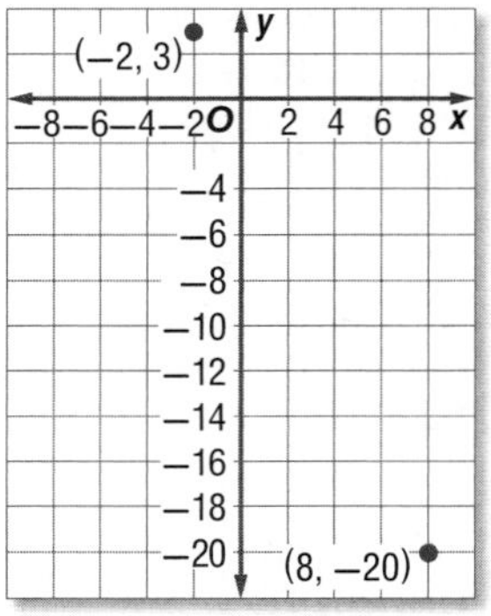

Model the situation. If the basket is at (0, 0), then the location of the disc is (8, −20). The location after the first putt is (−2, 3).

$d = \sqrt{(x_2 - x_1)^2 + (y_2 - y_1)^2}$	**Distance Formula**
$= \sqrt{(-2 - 8)^2 + [3 - (-20)]^2}$	$(x_1, y_1) = (8, -20)$ and $(x_2, y_2) = (-2, 3)$
$= \sqrt{(-10)^2 + 23^2}$	**Simplify.**
$= \sqrt{629}$ or about 25	

The disc traveled about 25 feet on his first putt.

Check Your Progress

2. Sharon hits a golf ball 12 feet past the hole and 3 feet to the left. Her first putt traveled to 2 feet beyond the cup and 1 foot to the right. How far did the ball travel on her first putt?

Personal Tutor glencoe.com

StudyTip

Midpoints The coordinates of the midpoint are the means of the coordinates of the endpoints.

There most likely will be problems involving the Midpoint and Distance Formulas on standardized tests you will have to take.

NYSRE EXAMPLE 3 Reinforcement of G.G.69

A coordinate grid is placed over a Florida map. St. Augustine is located at (3, 13), and Rockledge is located at (8, −1). If Port Orange is halfway between St. Augustine and Rockledge, which is closest to the distance in coordinate units from St. Augustine to Port Orange?

A 4.75 **B** 7.45 **C** 14.9 **D** 19

Read the Test Item

The question asks you to find the distance between one city and the midpoint. Find the midpoint, and then use the Distance Formula.

Solve the Test Item

Use the Midpoint Formula to find the coordinates of Port Orange.

$$\text{midpoint} = \left(\frac{3+8}{2}, \frac{13+(-1)}{2}\right) \quad \textbf{Midpoint Formula}$$

$$= (5.5, 6) \quad \textbf{Simplify.}$$

Use the Distance Formula to find the distance between St. Augustine (3, 13) and Port Orange (5.5, 6).

$$\text{distance} = \sqrt{(3-5.5)^2 + (13-6)^2} \quad \textbf{Distance Formula}$$

$$= \sqrt{(-2.5)^2 + 7^2} \quad \textbf{Evaluate exponents and add.}$$

$$= \sqrt{53.25} \text{ or about } 7.43 \quad \textbf{Simplify.}$$

The answer is B.

Test-TakingTip

Answer Check To check your answer, find the distance between Port Orange and Rockledge. Since Port Orange is at the midpoint, these distances should be equal.

Check Your Progress

3. The coordinates for points A and B are (−4, −5) and (10, −7), respectively. Find the distance between the midpoint of A and B and point B.

F $\sqrt{10}$ units **G** $5\sqrt{10}$ units **H** $\sqrt{50}$ units **J** $10\sqrt{5}$ units

Personal Tutor glencoe.com

Check Your Understanding

Example 1 p. 617

Find the midpoint of the line segment with endpoints at the given coordinates.

1. (−4, 7), (3, 9)

2. (8, 2), (−1, −5)

3. (11, 6), (18, 13.5)

4. (−12, −2), (−10.5, −6)

Example 2 p. 618

Find the distance between each pair of points with the given coordinates.

5. (3, −5), (13, −11)

6. (8, 1), (−2, 9)

7. (0.25, 1.75), (3.5, 2.5)

8. (−4.5, 10.75), (−6.25, −7)

Example 3 p. 619

9. MULTIPLE CHOICE The map of a mall is overlaid with a numeric grid. The kiosk for the cell phone store is halfway between The Ice Creamery and the See Clearly eyeglass store. If the ice cream store is at (2, 4) and the eyeglass store is at (78, 46), find the distance the kiosk is from the eyeglass store.

A 43.3 units **B** 47.2 units **C** 62.4 units **D** 94.3 units

Practice and Problem Solving

= Step-by-Step Solutions begin on page R20.
Extra Practice begins on page 947.

Example 1 p. 617

Find the midpoint of the line segment with endpoints at the given coordinates.

10. (20, 3), (15, 5)
11. (−27, 4), (19, −6)
12. (−0.4, 7), (11, −1.6)
13. (5.4, −8), (9.2, 10)
14. (−5.3, −8.6), (−18.7, 1)
15. (−6.4, −8.2), (−9.1, −0.8)

Example 2 p. 618

Find the distance between each pair of points with the given coordinates.

16. (1, 2), (6, 3)
17. (3, −4), (0, 12)
18. (−6, −7), (11, −12)
19. (−10, 8), (−8, −8)
20. (4, 0), (5, −6)
21. (7, 9), (−2, −10)
22. (−4, −5), (15, 17)
23. (14, −20), (−18, 25)

Example 3 p. 619

24. **TRACK AND FIELD** A shot put is thrown from the inside of a circle. A coordinate grid is placed over the shot put circle. The toe board is located at the front of the circle at (−4, 1), and the back of the circle is at (5, 2). If the center of the circle is halfway between these two points, what is the distance from the toe board to the center of the circle?

Find the midpoint of the line segment with endpoints at the given coordinates. Then find the distance between the points.

25. (−93, 15), (90, −15)
26. (−22, 42), (57, 2)
27. (−70, −87), (59, −14)
28. (−98, 5), (−77, 64)
29. (41, −45), (−25, 75)
30. (90, 60), (−3, −2)
31. (−1.2, 2.5), (0.34, −7)
32. (−7.54, 3.89), (4.04, −0.38)
33. $\left(-\frac{5}{12}, -\frac{1}{3}\right), \left(-\frac{17}{2}, -\frac{5}{3}\right)$
34. $\left(-\frac{5}{4}, -\frac{13}{2}\right), \left(-\frac{4}{3}, -\frac{5}{6}\right)$
35. $\left(-3\sqrt{2}, -4\sqrt{5}\right), \left(-3\sqrt{3}, 9\right)$
36. $\left(\frac{\sqrt{3}}{3}, \frac{\sqrt{2}}{4}\right), \left(\frac{-2\sqrt{3}}{3}, \frac{\sqrt{2}}{4}\right)$

37. **SPACE** Use the labeled points on the outline of the circular crater on Mars to estimate its diameter in kilometers.

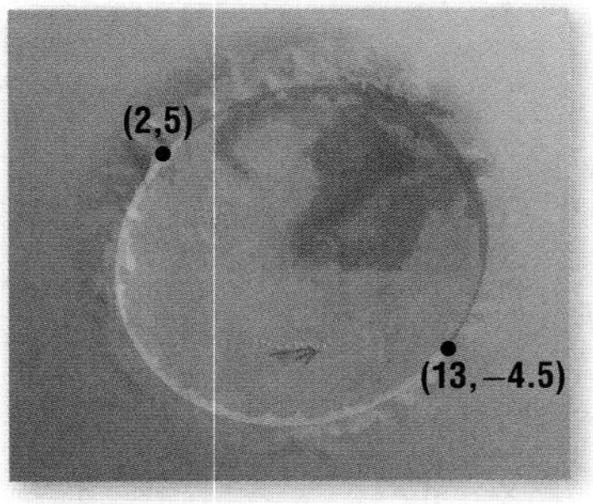

38. **GEOMETRY** Triangle *ABC* has vertices *A*(2, 1), *B*(−6, 5), and *C*(−2, −3).

a. An isosceles triangle has two sides with equal length. Is triangle *ABC* isosceles? Explain.

b An equilateral triangle has three sides of equal length. Is triangle *ABC* equilateral? Explain.

c. Triangle *EFG* is formed by joining the midpoints of the sides of triangle *ABC*. What type of triangle is *EFG*? Explain.

d. Describe any relationship between the lengths of the sides of the two triangles.

Real-World Link

U.S. Postal Service carriers travel over 1.2 billion miles every year delivering the mail.

Source: U.S. Postal Service

39. **PACKAGE DELIVERY** To determine the mileage between cities for their overnight delivery service, a package delivery service superimposes a coordinate grid over the United States. Each side of a grid unit is equal to 0.316 mile. Suppose the locations of two distribution centers are at (132, 428) and (254, 105). Find the actual distance between these locations to the nearest mile.

40. **HIKING** Orlando wants to hike from his camp to a waterfall. The waterfall is 5 miles south and 8 miles east of his campsite.

a. Use the Distance Formula to determine how far the waterfall is from the campsite.

b. Verify your answer in part a by using the Pythagorean Theorem to determine the distance between the campsite and the waterfall.

c. Orlando wants to stop for lunch halfway to the waterfall. If the camp is at the origin, where should he stop?

41. **MULTIPLE REPRESENTATIONS** Triangle XYZ has vertices $X(4, 9)$, $Y(8, -9)$, and $Z(-6, 5)$.

a. **CONCRETE** Draw $\triangle XYZ$ on a coordinate plane.

b. **NUMERICAL** Find the coordinates of the midpoint of each side of the triangle.

c. **GEOMETRIC** Find the perimeter of $\triangle XYZ$ and the perimeter of the triangle with vertices at the points found in part **b**.

d. **ANALYTICAL** How do the perimeters in part **c** compare?

H.O.T. Problems

Use **H**igher-**O**rder **T**hinking Skills

42. **CHALLENGE** Find the coordinates of the point that is three fourths of the way from $P(-1, 12)$ to $Q(5, -10)$.

43. **REASONING** Identify all the points that are equidistant from the point (5, 6). What figure does this make?

44. **REASONING** Triangle ABC is a right triangle.

a. Find the midpoint of the hypotenuse. Call it point Q.

b. Classify $\triangle BQC$ according to the lengths of its sides. Include sufficient evidence to support your conclusion.

c. Classify $\triangle BQA$ according to its angles.

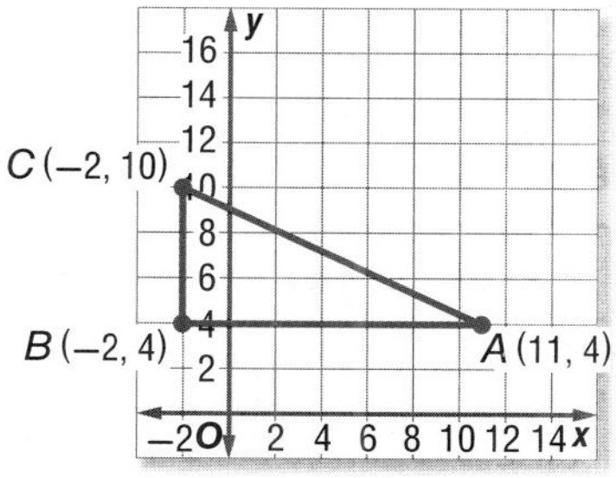

45. **OPEN ENDED** Plot two points, and find the distance between them. Does it matter which ordered pair is first when using the Distance Formula? Explain.

46. **WRITING IN MATH** Explain how the Midpoint Formula can be used to approximate the halfway point between two locations on a map.

NYSRE PRACTICE Reinforcement of G.G.69

47. **SHORT RESPONSE** You currently earn \$8.10 per hour and your boss gives you a 10% raise. What is your new hourly wage?

48. **ACT/SAT** A right circular cylinder has a radius of 3 and a height of 5. Which of the following dimensions of a rectangular solid will have a volume closest to that of the cylinder?

A 5, 5, 6
B 5, 6, 6
C 5, 5, 5
D 4, 5, 6

49. **GEOMETRY** If the sum of the lengths of the two legs of a right triangle is 49 inches and the hypotenuse is 41 inches, find the longer of the two legs.

F 9 in.
G 40 in.
H 42 in.
J 49 in.

50. Five more than 3 times a number is 17. Find the number.

A 3
B 4
C 5
D 6

Spiral Review

Solve each equation. Check your solutions. (Lesson 9-6)

51. $\frac{12}{v^2 - 16} - \frac{24}{v - 4} = 3$

52. $\frac{w}{w - 1} + w = \frac{4w - 3}{w - 1}$

53. $\frac{4n^2}{n^2 - 9} - \frac{2n}{n + 3} = \frac{3}{n - 3}$

54. **SWIMMING** When a person swims underwater, the pressure in his or her ears varies directly with the depth at which he or she is swimming. (Lesson 9-5)

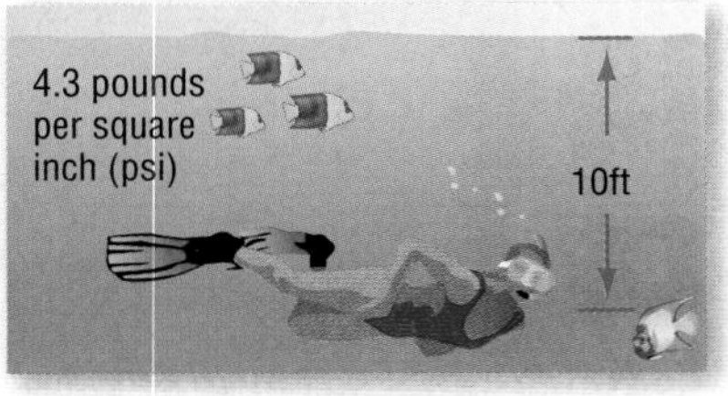

a. Write a direct variation equation that represents this situation.

b. Find the pressure at 60 feet.

c. It is unsafe for amateur divers to swim where the water pressure is more than 65 pounds per square inch. How deep can an amateur diver safely swim?

d. Make a table showing the number of pounds of pressure at various depths of water. Use the data to draw a graph of pressure versus depth.

Solve each equation or inequality. Round to the nearest ten-thousandth. (Lesson 8-6)

55. $9^{z - 4} = 6.28$

56. $8.2^{n - 3} = 42.5$

57. $2.1^{t - 5} = 9.32$

58. $8^{2n} > 52^{4n + 3}$

59. $7^{p + 2} \leq 13^{5 - p}$

60. $3^{y + 2} \geq 8^{3y}$

Solve each equation. (Lesson 7-7)

61. $(6n - 5)^{\frac{1}{3}} + 3 = -2$

62. $(5x + 7)^{\frac{1}{5}} + 3 = 5$

63. $(3x - 2)^{\frac{1}{5}} + 6 = 5$

Skills Review

Write each quadratic equation in vertex form. Then identify the vertex, axis of symmetry, and direction of opening. (Lesson 5-7)

64. $y = -x^2 - 4x + 8$

65. $y = x^2 - 6x + 1$

66. $y = -2x^2 + 20x - 35$

10-2 Parabolas

Then

You graphed quadratic functions. (Lessons 5-1 and 5-7)

Now

- Write equations of parabolas in standard form.
- Graph parabolas.

NYS Core Curriculum

A2.R.1 Use physical objects, diagrams, charts, tables, graphs, symbols, equations, or objects created using technology as representations of mathematical concepts

New Vocabulary

parabola
focus
directrix
latus rectum
standard form
general form

NY Math Online

glencoe.com

- Extra Examples
- Personal Tutor
- Self-Check Quiz
- Homework Help

Why?

Instead of expensive polished metal, glass, or an array of large mirrors to view images from space, frugal astronomers have begun using spinning basins of mercury.

While spinning, the surface of the mercury becomes *parabolic*, the perfect shape for a telescope's mirror.

Equations of Parabolas A **parabola** can be defined as the set of all points in a plane that are the same distance from a given point called the **focus** and a given line called the **directrix**.

The line segment through the focus of a parabola and perpendicular to the axis of symmetry is called the **latus rectum**. The endpoints of the latus rectum lie on the parabola.

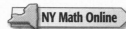

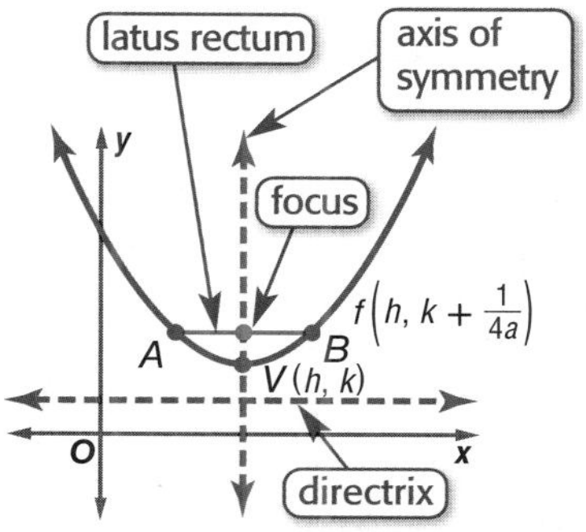

Key Concept Equations of Parabolas

For Your FOLDABLE

Form of Equation	$y = a(x - h)^2 + k$	$x = a(y - k)^2 + h$
Direction of Opening	upward if $a > 0$, downward if $a < 0$	right if $a > 0$, left if $a < 0$
Vertex	(h, k)	(h, k)
Axis of Symmetry	$x = h$	$y = k$
Focus	$\left(h, k + \frac{1}{4a}\right)$	$\left(h + \frac{1}{4a}, k\right)$
Directrix	$y = k - \frac{1}{4a}$	$x = h - \frac{1}{4a}$
Length of Latus Rectum	$\left\lvert\frac{1}{a}\right\rvert$ units	$\left\lvert\frac{1}{a}\right\rvert$ units

The **standard form** of the equation of a parabola with vertex (h, k) and axis of symmetry $x = h$ is $y = a(x - h)^2 + k$.

- If $a > 0$, k is the minimum value of the related function and the parabola opens upward.
- If $a < 0$, k is the maximum value of the related function and the parabola opens downward.

An equation of a parabola in the form $y = ax^2 + bx + c$ is the **general form**. Any equation in general form can be written in standard form. The shape of a parabola and the distance between the focus and directrix depend on the value of a in the equation.

Review Vocabulary

completing the square rewriting a quadratic expression as a perfect square trinomial (Lesson 5-5)

EXAMPLE 1 Analyze the Equation of a Parabola

Write $y = 2x^2 - 12x + 6$ in standard form. Identify the vertex, axis of symmetry, and direction of opening of the parabola.

$y = 2x^2 - 12x + 6$	**Original equation**
$= 2(x^2 - 6x) + 6$	**Factor 2 from the x- and x^2-terms.**
$= 2(x^2 - 6x + \blacksquare) + 6 - 2(\blacksquare)$	**Complete the square on the right side.**
$= 2(x^2 - 6x + 9) + 6 - 2(9)$	**The 9 added when you complete the square is multiplied by 2.**
$= 2(x - 3)^2 - 12$	**Factor.**

The vertex of this parabola is located at $(3, -12)$, and the equation of the axis of symmetry is $x = 3$. The parabola opens upward.

Check Your Progress

1. Write $y = 4x^2 + 16x + 34$ in standard form. Identify the vertex, axis of symmetry, and direction of opening of the parabola.

Personal Tutor glencoe.com

Graph Parabolas In Chapter 5, you learned that the graph of the quadratic equation $y = a(x - h)^2 + k$ is a transformation of the parent graph of $y = x^2$ translated h units horizontally and k units vertically, and reflected and/or dilated depending on the value of a.

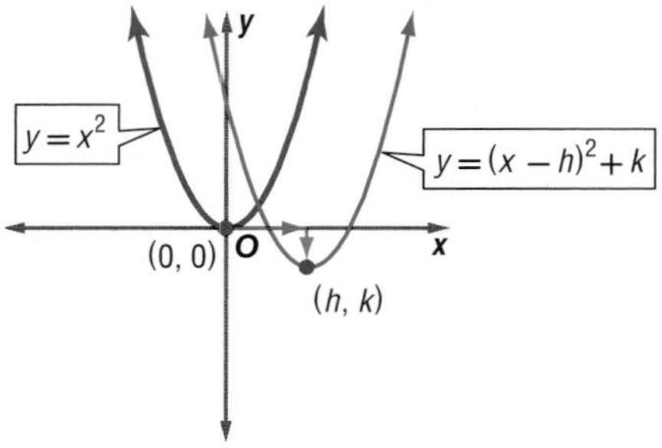

EXAMPLE 2 Graph Parabolas

Graph each equation.

a. $y = -3x^2$

For this equation, $h = 0$ and $k = 0$. The vertex is at the origin. Since the equation of the axis of symmetry is $x = 0$, substitute some small positive integers for x and find the corresponding y-values.

x	y
1	−3
2	−12
3	−27

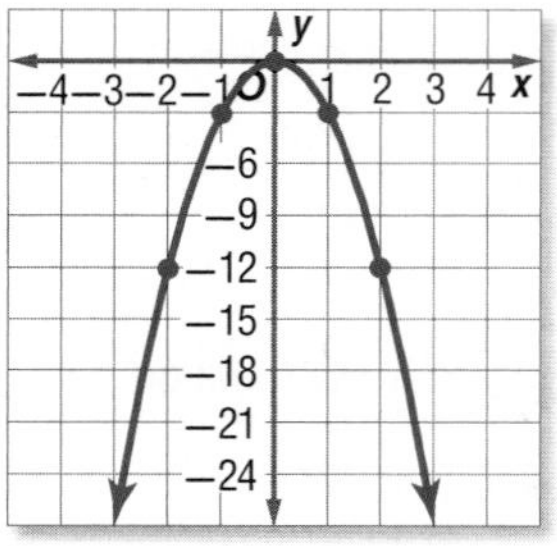

Since the graph is symmetric about the y-axis, the points at $(-1, -3)$, $(-2, -12)$, and $(-3, -27)$ are also on the parabola. Use all of these points to draw the graph.

b. $y = -3(x - 4)^2 + 5$

The equation is of the form $y = a(x - h)^2 + k$, where $h = 4$ and $k = 5$. The graph of this equation is the graph of $y = -3x^2$ in part a translated 4 units to the right and up 5 units. The vertex is now at $(4, 5)$.

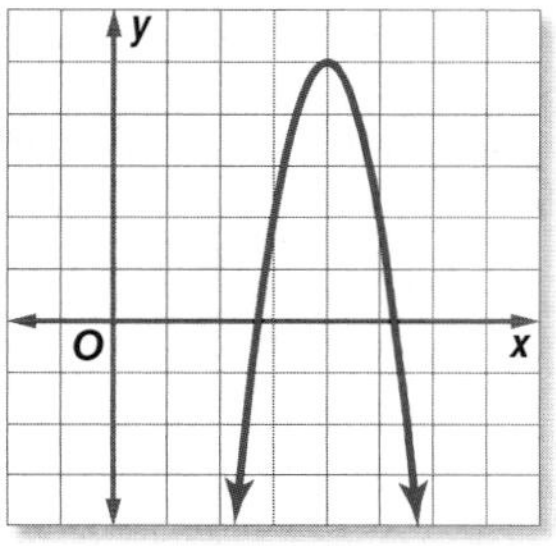

Check Your Progress

2A. $y = 2x^2$ **2B.** $y = 2(x - 1)^2 - 4$

Personal Tutor glencoe.com

Watch Out!

Translations It is easy to confuse the effect of a translation.

- If h is positive, translate the graph h units to the right.
- If h is negative, translate the graph h units to the left.
- If k is positive, translate the graph k units up.
- If k is negative, translate the graph k units down.

StudyTip

Graphing When graphing these functions, it may be helpful to sketch the graph of the parent function.

Equations of parabolas with vertical axes of symmetry have the parent function $y = x^2$ and are of the form $y = a(x - h)^2 + k$. These are functions. Equations of parabolas with horizontal axes of symmetry are of the form $x = a(y - k)^2 + h$ and are not functions. The parent graph for these equations is $x = y^2$.

EXAMPLE 3 Graph an Equation in General Form

Graph each equation.

a. $\mathbf{2x - y^2 = 4y + 10}$

Step 1 Write the equation in the form $x = a(y - k)^2 + h$.

$2x - y^2 = 4y + 10$	**Original equation**
$2x = y^2 + 4y + 10$	**Add y^2 to each side to isolate the x-term.**
$2x = (y^2 + 4y + \blacksquare) + 10 - \blacksquare$	**Complete the square.**
$2x = (y^2 + 4y + 4) + 10 - 4$	**Add and subtract 4, since $\left(\frac{4}{2}\right)^2 = 4$.**
$2x = (y + 2)^2 + 6$	**Factor and subtract.**
$x = \frac{1}{2}(y + 2)^2 + 3$	**$(h, k) = (3, -2)$**

Step 2 Use the equation to find information about the graph. Then draw the graph based on the parent graph, $x = x^2$.

vertex: $(3, -2)$

axis of symmetry: $y = -2$

focus: $\left(3 + \frac{1}{4\left(\frac{1}{2}\right)}, -2\right)$ or $(3.5, -2)$

directrix: $x = 3 - \frac{1}{4\left(\frac{1}{2}\right)}$ or 2.5

direction of opening: right, since $a > 0$

length of latus rectum: $\left|\frac{1}{\left(\frac{1}{2}\right)}\right|$ or 2 units

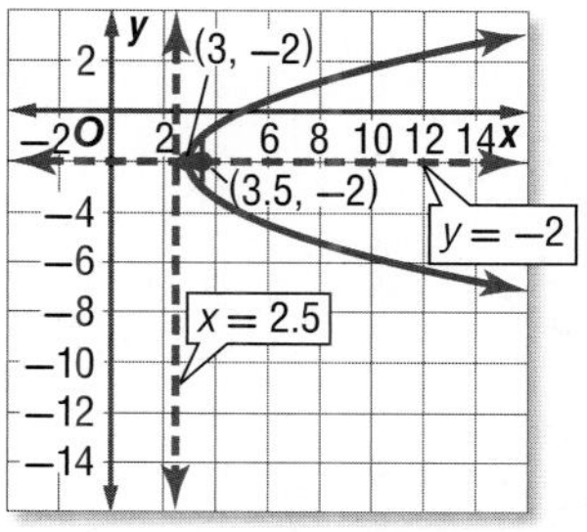

ReadingMath

latus rectum from the Latin *latus*, meaning side, and *rectum*, meaning straight

b. $\mathbf{y + 2x^2 + 32 = -16x - 1}$

Step 1

$y + 2x^2 + 32 = -16x - 1$	**Original equation**
$y = -2x - 16x - 33$	**Solve for y.**
$y = -2(x^2 + 8x + \blacksquare) - 33 - \blacksquare$	**Complete the square.**
$y = -2(x^2 + 8x + 16) - 33 - (-32)$	**Add and subtract −32.**
$y = -2(x + 4)^2 - 1$	**Factor and simplify.**

Step 2 vertex: $(-4, -1)$

axis of symmetry: $x = -4$

focus: $\left(-4, -\frac{9}{8}\right)$

directrix: $y = -\frac{7}{8}$

length of latus rectum: $\frac{1}{2}$ unit

opens down

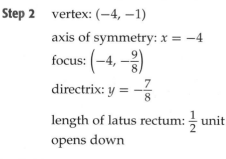

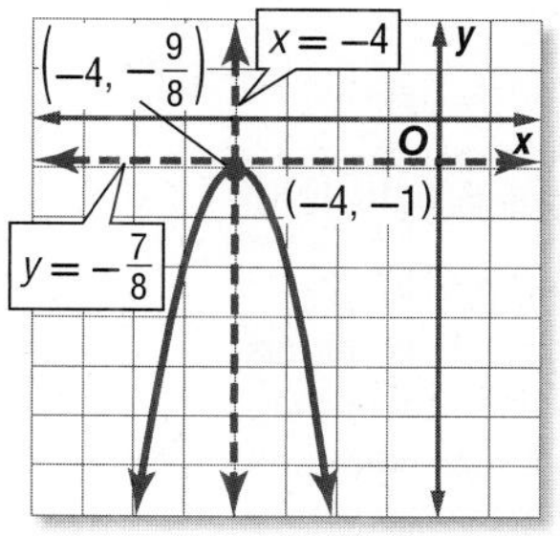

Check Your Progress

3A. $3x - y^2 = 4x + 25$

3B. $y = x^2 + 6x - 4$

Personal Tutor glencoe.com

You can use specific information about a parabola to write an equation and draw a graph.

EXAMPLE 4 Write an Equation of a Parabola

Write an equation for a parabola with a vertex (−2, −4) and directrix $y = 1$. Then graph the equation.

The directrix is a horizontal line, so the equation of the parabola is of the form $y = a(x - h)^2 + k$. Find a, h and k.

- The vertex is at (−2, −4), so $h = -2$ and $k = -4$.
- Use the equation of the directrix to find a.

$y = k - \frac{1}{4a}$	**Equation of directrix**
$1 = -4 - \frac{1}{4a}$	**Replace y with 1 and k with −4.**
$5 = -\frac{1}{4a}$	**Add 4 to each side.**
$20a = -1$	**Multiply each side by $4a$.**
$a = -\frac{1}{20}$	**Divide each side by 12.**

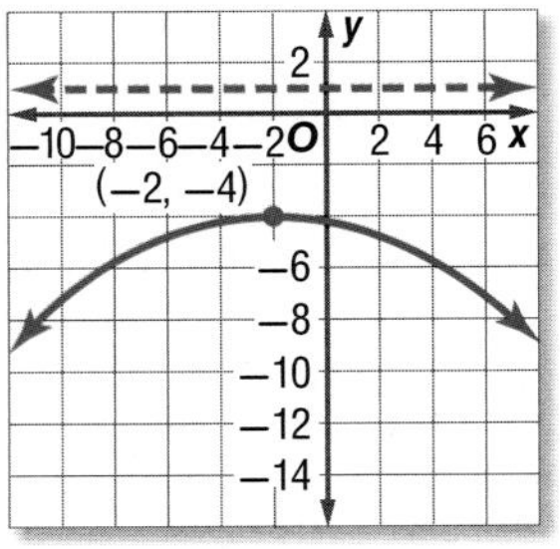

So, the equation of the parabola is $y = -\frac{1}{20}(x + 2)^2 - 4$.

Check Your Progress

Write an equation for each parabola described below. Then graph the equation.

4A. vertex (1, 3), focus (1, 5)

4B. focus (5, 6), directrix $x = -2$

Personal Tutor glencoe.com

Parabolas are often used in the real world.

Real-World EXAMPLE 5 Write an Equation for a Parabola

ENVIRONMENT **Solar energy may be harnessed by using parabolic mirrors. The mirrors reflect the rays from the Sun to the focus of the parabola. The focus of each parabolic mirror at the facility described at the left is 6.25 feet above the vertex. The latus rectum is 25 feet long.**

a. Assume that the focus is at the origin. Write an equation for the parabola formed by each mirror.

In order for the mirrors to collect the Sun's energy, the parabola must open upward. Therefore, the vertex must be below the focus.

focus: (0, 0) vertex: (0, −6.25)

The measure of the latus rectum is 25. So $25 = \left|\frac{1}{a}\right|$, and $a = \frac{1}{25}$.

Using the form $y = a(x - h)^2 + k$, an equation for the parabola formed by each mirror is $y = \frac{1}{25}x^2 - 6.25$.

b. Graph the equation.

Now use all of the information to draw a graph.

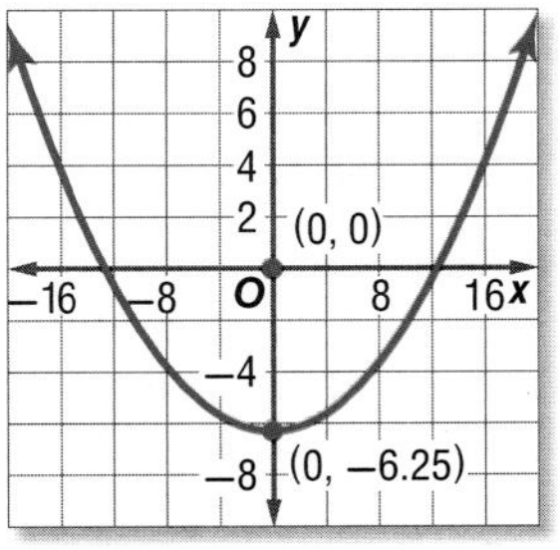

Check Your Progress

5. Write and graph an equation for a parabolic mirror that has a focus 4.5 feet above the vertex and a latus rectum that is 20 feet long, when the focus is at the origin.

Personal Tutor glencoe.com

Real-World Link

In California's Mojave Desert, parabolic mirrors are used to heat oil that flows through tubes placed at the focus. The heated oil is used to produce electricity.

Source: Solel

Check Your Understanding

Example 1 p. 624

Write each equation in standard form. Identify the vertex, axis of symmetry, and direction of opening of the parabola.

1. $y = 2x^2 - 24x + 40$
2. $y = 3x^2 - 6x - 4$
3. $x = y^2 - 8y - 11$
4. $x + 3y^2 + 12y = 18$

Examples 2 and 3 pp. 624–625

Graph each equation.

5. $y = (x - 4)^2 - 6$
6. $y = 4(x + 5)^2 + 3$
7. $y = -3x^2 - 4x - 8$
8. $x = 3y^2 - 6y + 9$

Example 4 p. 626

Write an equation for each parabola described below. Then graph the equation.

9. vertex (0, 2), focus (0, 4)
10. vertex (−2, 4), directrix $x = -1$
11. focus (3, 2), directrix $y = 8$
12. vertex (−1, −5), focus (−5, −5)

Example 5 p. 626

13. **ASTRONOMY** Consider a parabolic mercury mirror like the one described at the beginning of the lesson. The focus is 6 feet above the vertex and the latus rectum is 24 feet long.
    a. Assume that the focus is at the origin. Write an equation for the parabola formed by the parabolic microphone.
    b. Graph the equation.

Practice and Problem Solving

● = **Step-by-Step Solutions** begin on page R20. **Extra Practice** begins on page 947.

Example 1 p. 624

Write each equation in standard form. Identify the vertex, axis of symmetry, and direction of opening of the parabola.

14. $y = x^2 - 8x + 13$
15. $y = 3x^2 + 42x + 149$
16. $y = -6x^2 - 36x - 8$
17. $y = -3x^2 - 9x - 6$
18. $x = \frac{1}{3}y^2 - 3y + 4$
19. $x = \frac{2}{3}y^2 - 4y + 12$

Examples 2 and 3 pp. 624–625

Graph each equation.

20. $y = \frac{1}{3}x^2$
21. $y = -2x^2$
22. $y = -2(x - 2)^2 + 3$
23. $y = 3(x - 3)^2 - 5$
24. $x = \frac{1}{2}y^2$
25. $4x - y^2 = 2y + 13$

Example 4 p. 626

Write an equation for each parabola described below. Then graph the equation.

26. vertex (0, 1), focus (0, 4)
27. vertex (1, 8), directrix $y = 3$
28. focus (−2, −4), directrix $x = -6$
29. focus (2, 4), directrix $x = 10$
30. vertex (−6, 0), directrix $x = 2$
31. vertex (9, 6), focus (9, 5)

Example 5 p. 626

32. **BASEBALL** When a ball is thrown, the path it travels is a parabola. Suppose a baseball is thrown from ground level, reaches a maximum height of 50 feet, and hits the ground 200 feet from where it was thrown. Assuming this situation could be modeled on a coordinate plane with the focus of the parabola at the origin, find the equation of the parabolic path of the ball. Assume the focus is on ground level.
33. **SPACE** Ground antennas and satellites are used to relay signals between the NASA Mission Operations Center and the spacecraft it controls. One such dish is 146 feet in diameter. Its focus is 48 feet from the vertex.
    a. Sketch two options for the dish, one that opens up and one that opens left.
    b. Write two equations that model the sketches in part **a**.
    c. If you wanted to find the depth of the dish, does it matter which equation you use? Why or why not?

34. **UMBRELLAS** A beach umbrella has an arch in the shape of a parabola that opens downward. The umbrella spans 9 feet across and is $1\frac{1}{2}$ feet high. Write an equation of a parabola to model the arch, assuming that the origin is at the point where the pole and umbrella meet, beneath the vertex of the arch.

Real-World Link

Parabolic headlights use a special bulb with two filaments to produce high beams and low beams. The filament placed at the focus produces high beams, and the filament placed off the focus produces low beams.

Source: General Motors

35. **AUTOMOBILES** An automobile headlight contains a parabolic reflector. The light coming from the source bounces off the parabolic reflector and shines out the front of the headlight. The equation of the cross section of the reflector is $y = \frac{1}{12}x^2$. How far from the vertex should the filament for the high beams be placed?

36. **MULTIPLE REPRESENTATIONS** Start with a sheet of wax paper that is about 15 inches long and 12 inches wide.

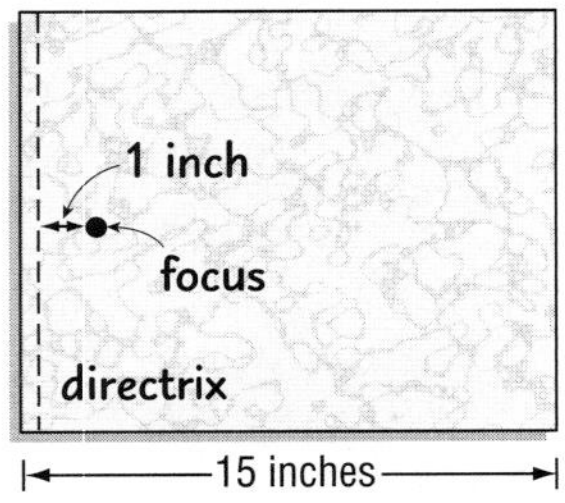

a. **CONCRETE** Make a line that is perpendicular to the sides of the sheet by folding the sheet near one end. Open up the paper again. This line is the directrix. Mark a point about midway between the sides of the sheet so that the distance from the directrix is about 1 inch. This is the focus.

b. **CONCRETE** Start with a new sheet of wax paper. Form another outline of a parabola with a focus that is about 3 inches from the directrix.

c. **CONCRETE** On a new sheet of a wax paper, form a third outline of a parabola with a focus that is about 5 inches from the directrix.

d. **VERBAL** Compare the shapes of the three parabolas. How does the distance between the focus and the directrix affect the shape of a parabola?

H.O.T. Problems Use Higher-Order Thinking Skills

37. **REASONING** How do you change the equation of the parent function $y = x^2$ to shift the graph to the right?

38. **OPEN ENDED** Two different parabolas have their vertex at $(-3, 1)$ and contain the point with coordinates $(-1, 0)$. Write two possible equations for these parabolas.

39. **FIND THE ERROR** Brianna and Russell are graphing $\frac{1}{4}y^2 + x = 0$. Is either of them correct? Explain your reasoning.

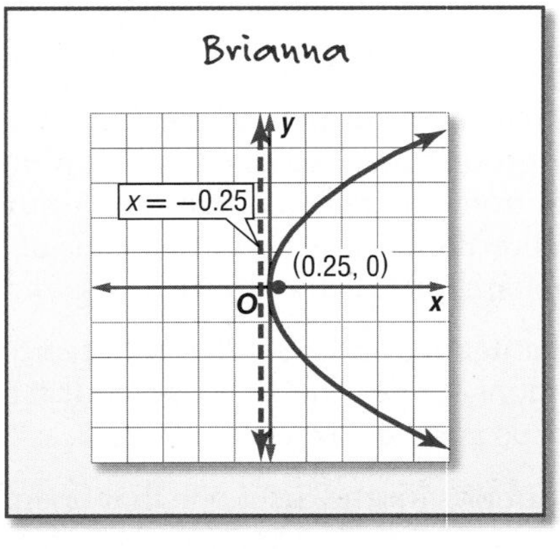

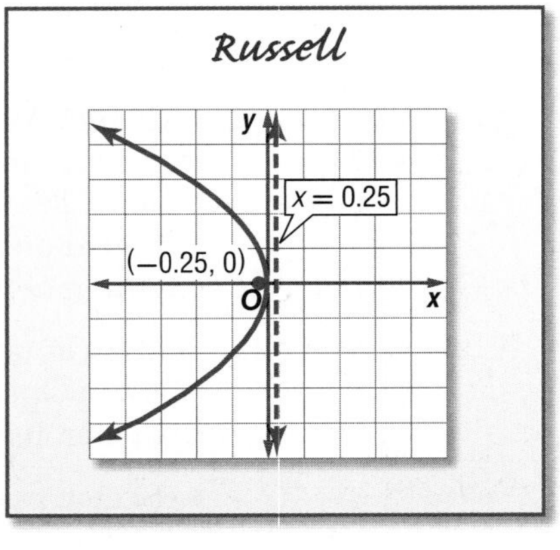

40. **WRITING IN MATH** How are parabolas used in televising sporting events? Explain why a televised sporting event filmed with a parabolic microphone is better than a televised sporting event filmed with a standard microphone.

NYSRE PRACTICE A2.R.1

41. A gardener is placing a fence around a 1320-square-foot garden. He ordered 148 feet of fencing. If he uses all the fencing, what is the length of the garden?

A 30 ft
B 34 ft
C 44 ft
D 46 ft

42. ACT/SAT When a number is divided by 5, the result is 7 more than the number. Find the number.

F $\frac{35}{4}$

G $-\frac{35}{4}$

H $\frac{28}{7}$

J $\frac{28}{4}$

43. GEOMETRY What is the area of the following square, if the length of $\overline{BD}$ is $2\sqrt{2}$?

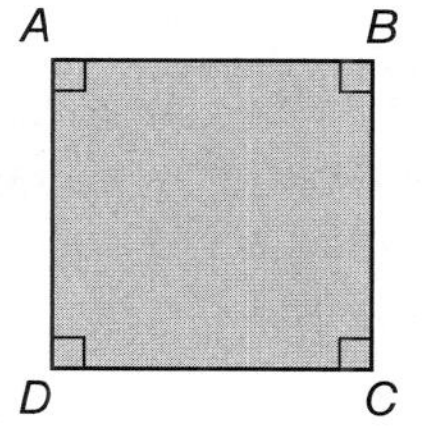

A 1
B 2
C 3
D 4

44. SHORT RESPONSE The measure of the smallest angle of a triangle is two thirds the measure of the middle angle. The measure of the middle angle is three sevenths of the measure of the largest angle. Find the largest angle's measure.

Spiral Review

45. GEOMETRY Find the perimeter of a triangle with vertices at (2, 4), (−1, 3), and (1, −3). (Lesson 10-1)

46. WORK A worker can powerwash a wall of a certain size in 5 hours. Another worker can do the same job in 4 hours. If the workers work together, how long would it take to do the job? Determine whether your answer is reasonable. (Lesson 9-6)

Solve each equation or inequality. Round to the nearest ten-thousandth. (Lesson 8-7)

47. $\ln(x + 1) = 1$

48. $\ln(x - 7) = 2$

49. $e^x > 1.6$

50. $e^{5x} \geq 25$

Simplify. (Lesson 7-4)

51. $\sqrt{0.25}$

52. $\sqrt[3]{-0.064}$

53. $\sqrt[4]{z^8}$

54. $-\sqrt[6]{x^6}$

List all of the possible rational zeros of each function. (Lesson 6-8)

55. $h(x) = x^3 + 8x + 6$

56. $p(x) = 3x^3 - 5x^2 - 11x + 3$

57. $h(x) = 9x^6 - 5x^3 + 27$

Skills Review

Simplify each expression. (Concepts and Skills Bank 2)

58. $\sqrt{24}$

59. $\sqrt{45}$

60. $\sqrt{252}$

61. $\sqrt{512}$

EXPLORE

10-3 Graphing Technology Lab: Equations of Circles

NYS Core Curriculum **A2.A.47** Determine the center-radius form for the equation of a circle in standard form

You can use a TI-*n*spire graphing calculator to examine characteristics of circles and the relationship with an equation of the circle.

ACTIVITY

Step 1 Draw a circle.

- From the Home screen, select **New Document**. Select **Graphs & Geometry**. Then press (menu) and select **Shapes**, and then select **Circle**. Place the pointer at the origin and press (enter) to set the center of the circle. Move the pointer out, creating a circle like the one shown.
- From (menu), select **Points & Lines**, and then **Point On** to place a point on the circle.
- Then, draw a radius by selecting (menu), **Points & Lines**, and then **Segment**.

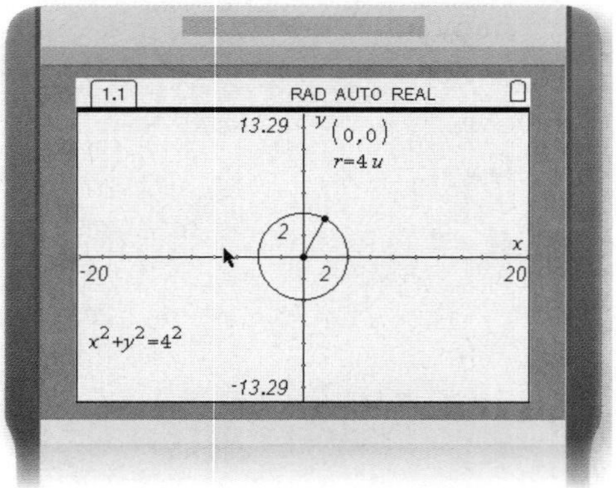

Step 2 Add labels.

- Under (menu), select **Actions**, then **Coordinates and Equations**. Use the pointer to select the center of the circle and display its coordinates. Move the coordinates out of the way.
- Display the length of the radius using (menu), then **Measurement**, and then **Length**.
- Use (menu), **Actions**, and **Coordinates and Equations** to display an equation of the circle.

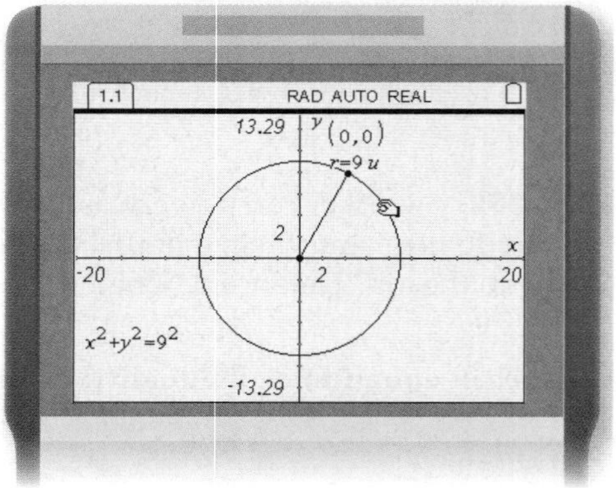

Step 3 Change the radius.

Move the pointer so that a point on the circle is highlighted, then press and hold until it is selected. Examine the equation of the circle. Then move the edge of the circle in. Make note of changes in the equation.

Step 4 Move the center of the circle.

Move the pointer so that the center of the circle is highlighted, then press and hold until it is selected. Move the center of the circle. Again, examine the equation of the circle.

Analyze the Results

1. How does moving the edge of the circle in or out affect the equation of the circle?
2. What effect does moving the center of the circle have on the equation?
3. Repeat the activity by placing the center of a circle in Quadrant I. Move the center to each of the other three quadrants. How does the equation change?
4. **MAKE A CONJECTURE** Without graphing, write an equation of each circle.

   a. center: $(4, 2)$, radius: 3 **b.** center: $(-1, 1)$, radius: 8

   c. center: $(-6, -5)$, radius: 2.5 **d.** center: (h, k), radius: r

10-3 Circles

Why?

When a rock is thrown into water, ripples move out from the center forming concentric circles. If the point where the rock entered the water is assigned coordinates, each ripple can be modeled by an equation of a circle.

Then
You graphed and wrote equations of parabolas. (Lesson 10-2)

Now
- Write equations of circles.
- Graph circles.

NYS Core Curriculum
A2.A.48 Write the equation of a circle, given its center and a point on the circle
A2.A.49 Write the equation of a circle from its graph *Also addresses A2.A.47.*

New Vocabulary
circle
center
radius

NY Math Online
glencoe.com
- Extra Examples
- Personal Tutor
- Self-Check Quiz
- Homework Help

Equations of Circles A **circle** is the set of all points in a plane that are equidistant from a given point in the plane, called the **center**. Any segment with endpoints at the center and a point on the circle is a **radius** of the circle.

Assume that (x, y) are the coordinates of a point on the circle at the right. The center is at (h, k), and the radius is r. You can find an equation of the circle by using the Distance Formula.

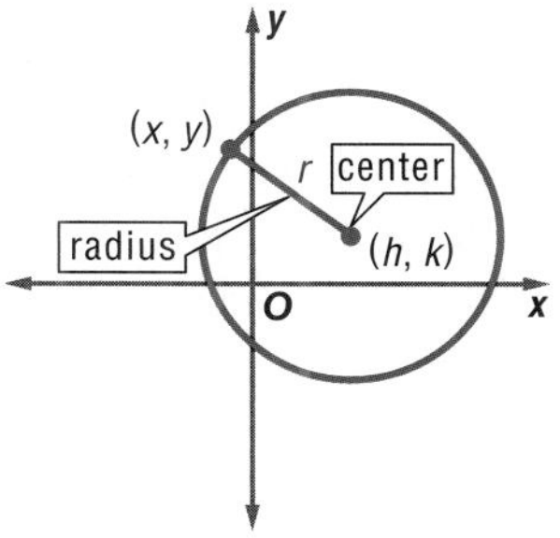

$\sqrt{(x_2 - x_1)^2 + (y_2 - y_1)^2} = d$	**Distance Formula**
$\sqrt{(x - h)^2 + (y - k)^2} = r$	**$(x_1, y_1) = (h, k)$, $(x_2, y_2) = (x, y)$, $d = r$**
$(x - h)^2 + (y - k)^2 = r^2$	**Square each side.**

Key Concept — Equations of Circles
For Your FOLDABLE

Standard Form of Equation	$x^2 + y^2 = r^2$	$(x - h)^2 + (y - k)^2 = r^2$
Center	$(0, 0)$	(h, k)
Radius	r	r

You can use the standard form of the equation of a circle to write an equation for a circle given the center and the radius or diameter.

Real-World EXAMPLE 1 — Write an Equation Given the Radius

DELIVERY Appliances + More offers free delivery within 35 miles of the store. The Jacksonville store is located 100 miles north and 45 miles east of the corporate office. Write an equation to represent the delivery boundary of the Jacksonville store if the origin of the coordinate system is the corporate office.

Since the corporate office is at (0, 0), the Jacksonville store is at (45, 100). The boundary of the delivery region is the circle centered at (45, 100) with radius 35 miles.

$(x - h)^2 + (y - k)^2 = r^2$	**Equation of a circle**
$(x - 45)^2 + (y - 100)^2 = 35^2$	**$(h, k) = (45, 100)$ and $r = 35$**
$(x - 45)^2 + (y - 100)^2 = 1225$	**Simplify.**

Check Your Progress

1. **WI-FI** A certain wireless transmitter has a range of 30 miles in any direction. If a Wi-Fi phone is 4 miles south and 3 miles west of headquarters, write an equation to represent the area within which the phone can operate via the Wi-Fi system.

Personal Tutor glencoe.com

You can write the equation of a circle when you know the location of the center and a point on the circle.

StudyTip

Center-Radius form Standard form is sometimes referred to as *center-radius form* because the center and radius of the circle are apparent in the equation.

EXAMPLE 2 Write an Equation from a Graph

Write an equation for the graph.

$(x - h)^2 + (y - k)^2 = r^2$	**Standard form**
$(2 + 3)^2 + (-1 - 1)^2 = r^2$	**$x = 2$, $y = -1$, $h = -3$, $k = 1$**
$(5)^2 + (-2)^2 = r^2$	**Simplify.**
$25 + 4 = r^2$	**Evaluate the exponents.**
$29 = r^2$	**Add.**

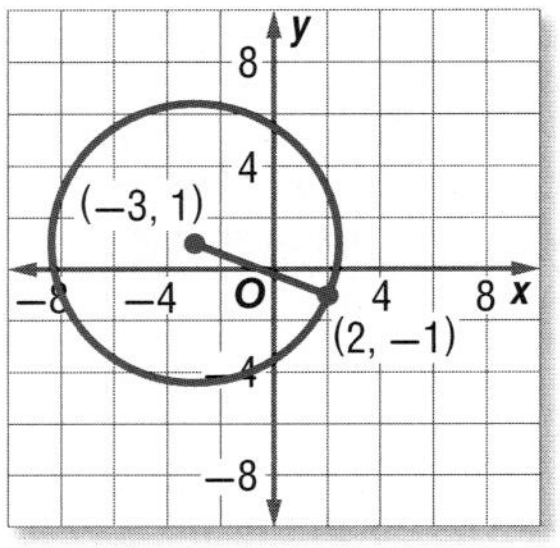

So, the equation of the circle is $(x + 3)^2 + (y - 1)^2 = 29$.

Check Your Progress

Write an equation for each graph.

2A.

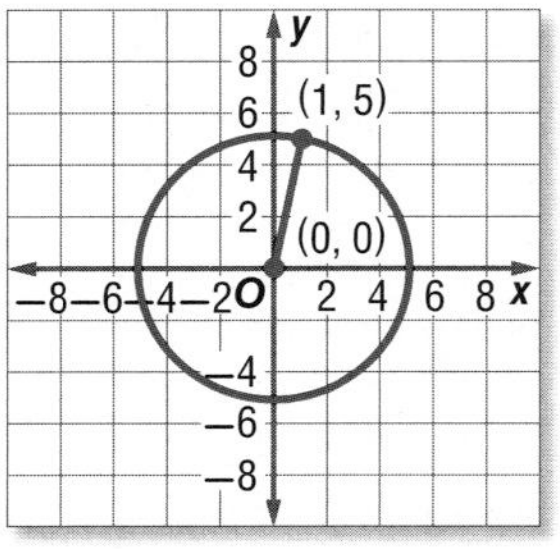

2B.

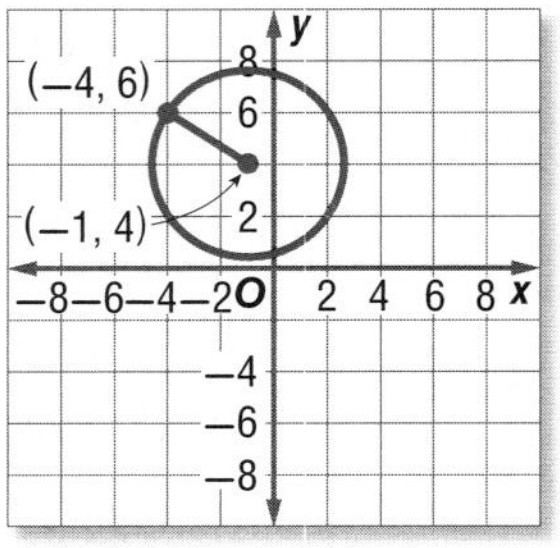

▷ Personal Tutor glencoe.com

You can use the Midpoint and Distance Formulas when you know the endpoints of the radius or diameter of a circle.

EXAMPLE 3 Write an Equation Given a Diameter

Write an equation for a circle if the endpoints of a diameter are at (7, 6) and (−1, −8).

Step 1 Find the center.

$(h, k) = \left(\frac{x_1 + x_2}{2}, \frac{y_1 + y_2}{2}\right)$	**Midpoint Formula**
$= \left(\frac{7 + (-1)}{2}, \frac{6 + (-8)}{2}\right)$	**$(x_1, y_1) = (7, 6)$, $(x_2, y_2) = (-1, -8)$**
$= \left(\frac{6}{2}, \frac{-2}{2}\right)$	**Add.**
$= (3, -1)$	**Simplify.**

Step 2 Find the radius.

$r = \sqrt{(x_2 - x_1)^2 + (y_2 - y_1)^2}$	**Distance Formula**
$= \sqrt{(3 - 7)^2 + (-1 - 6)^2}$	**$(x_1, y_1) = (7, 6)$, $(x_2, y_2) = (3, -1)$**
$= \sqrt{(-4)^2 + (-7)^2}$	**Subtract.**
$= \sqrt{65}$	**Simplify.**

The radius of the circle is $\sqrt{65}$ units, so $r^2 = 65$. Substitute h, k, and r^2 into the standard form of the equation of a circle. An equation of the circle is $(x - 3)^2 + (y + 1)^2 = 65$.

Check Your Progress

3. Write an equation for a circle if the endpoints of a diameter are at (3, −3) and (1, 5).

▷ Personal Tutor glencoe.com

StudyTip

Axis of Symmetry Every diameter in a circle is an axis of symmetry. There are infinitely many axes of symmetry in a circle.

Graph Circles You can use symmetry to help you graph circles.

EXAMPLE 4 Graph an Equation in Standard Form

Find the center and radius of the circle with equation $x^2 + y^2 = 100$. Then graph the circle.

- The center of the circle is at (0, 0), and the radius is 10.
- The table lists some integer values for x and y that satisfy the equation.
- Because the circle is centered at the origin, it is symmetric about the y-axis. Therefore, the points at (−6, 8), (−8, 6), and (−10, 0) lie on the graph.
- The circle is also symmetric about the x-axis, so the points (−6, −8), (−8, −6), (0, −10), (6, −8), and (8, −6) lie on the graph.
- Plot all of these points and draw the circle that passes through them.

x	y
0	10
6	8
8	6
10	0

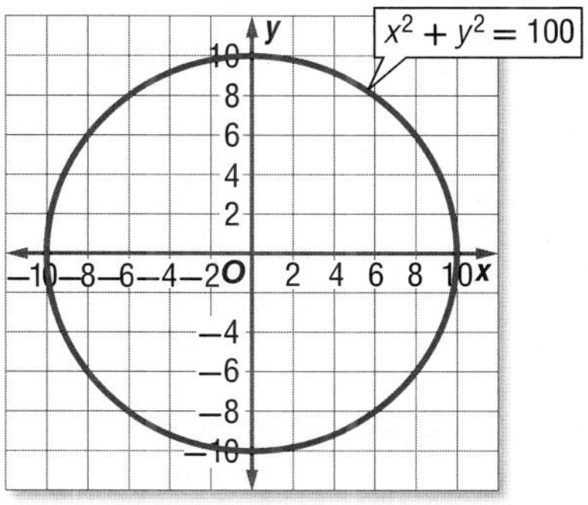

Check Your Progress

4. Find the center and radius of the circle with equation $x^2 + y^2 = 81$. Then graph the circle.

Personal Tutor glencoe.com

Circles with centers that are not (0, 0) can be graphed by using translations. The graph of $(x - h)^2 + (y - k)^2 = r^2$ is the graph of $x^2 + y^2 = r^2$ translated h units horizontally and k units vertically.

EXAMPLE 5 Graph an Equation Not in Standard Form

Find the center and radius of the circle with equation $x^2 + y^2 - 8x + 12y - 12 = 0$. Then graph the circle.

Complete the squares.

$$x^2 + y^2 - 8x + 12y - 12 = 0$$
$$x^2 - 8x + \blacksquare + y^2 + 12y + \blacksquare = 12 + \blacksquare + \blacksquare$$
$$x^2 - 8x + 16 + y^2 + 12y + 36 = 12 + 16 + 36$$
$$(x - 4)^2 + (y + 6)^2 = 64$$

The center of the circle is at (4, −6) and the radius is 8. The graph of $(x - 4)^2 + (y + 6)^2 = 64$ is the same as $x^2 + y^2 = 64$ translated 4 units to the right and down 6 units.

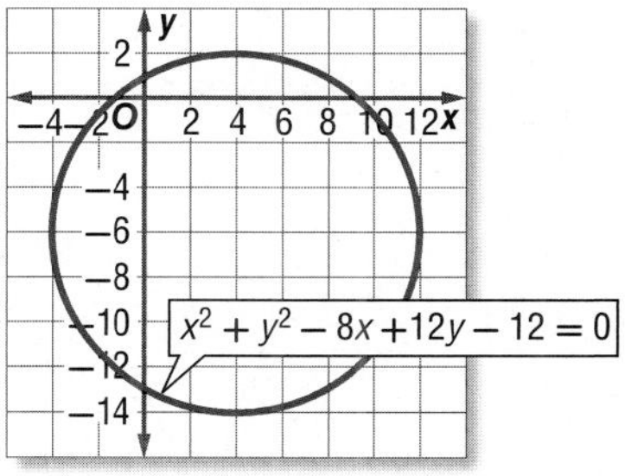

Check Your Progress

5. Find the center and radius of the circle with equation $x^2 + y^2 + 4x - 10y - 7 = 0$. Then graph the circle.

Personal Tutor glencoe.com

Check Your Understanding

Example 1 p. 631

1. **WEATHER** On average, the eye of a tornado is about 200 feet across. Suppose the center of the eye is at the point (72, 39). Write an equation to represent the boundary of the eye.

Write an equation for each circle given the center and radius.

2. center: $(-2, -6)$, $r = 4$ units
3. center: $(1, -5)$, $r = 3$ units

Example 2 p. 632

Write an equation for each graph.

4.

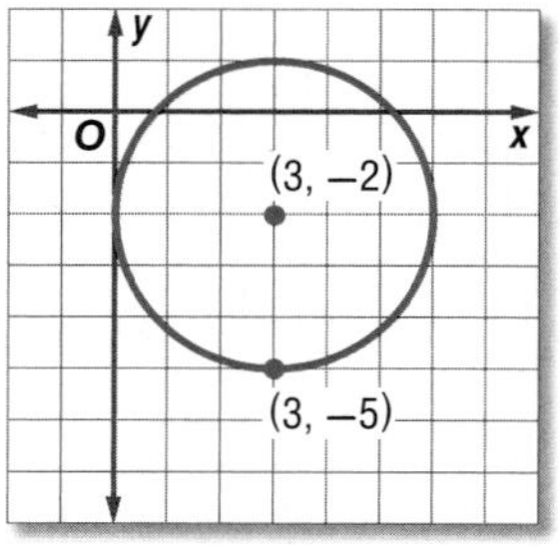

5.

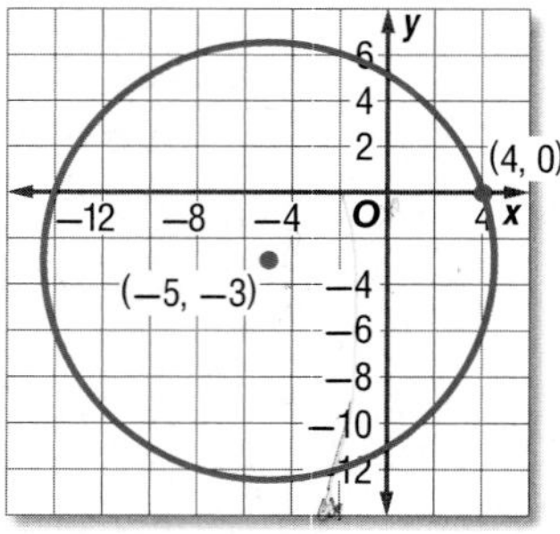

Example 3 p. 632

Write an equation for each circle given the endpoints of a diameter.

6. $(-1, -7)$ and $(0, 0)$
7. $(4, -2)$ and $(-4, -6)$

Examples 4 and 5 p. 633

Find the center and radius of each circle. Then graph the circle.

8. $x^2 + y^2 = 16$
9. $x^2 + (y - 7)^2 = 9$
10. $(x - 4)^2 + (y - 4)^2 = 25$
11. $x^2 + y^2 - 4x + 8y - 5 = 0$

Practice and Problem Solving

● = **Step-by-Step Solutions** begin on page R20.
Extra Practice begins on page 947.

Example 1 p. 631

Write an equation for each circle given the center and radius.

12. center: $(4, 9)$, $r = 6$
13. center: $(-3, 1)$, $r = 4$
14. center: $(-7, -3)$, $r = 13$
15. center: $(-2, -1)$, $r = 9$
16. center: $(1, 0)$, $r = \sqrt{15}$
17. center: $(0, -6)$, $r = \sqrt{35}$

18. **AIR TRAFFIC CONTROL** The radar for a county airport control tower is located at (5, 10) on a map. It can detect a plane up to 20 miles away. Write an equation for the outer limits of the detection area.

Example 2 p. 632

Write an equation for each graph.

19.

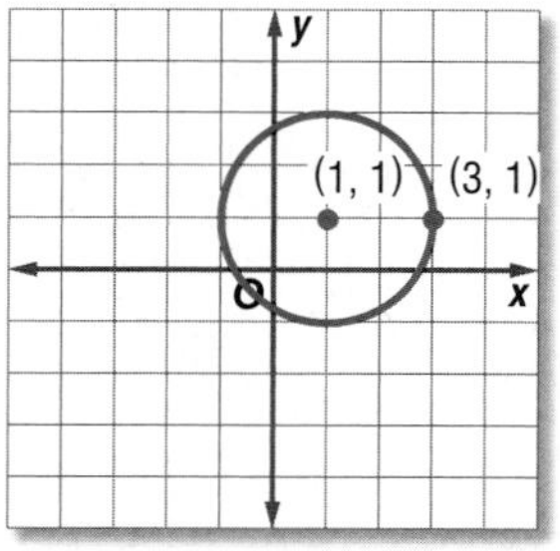

20.

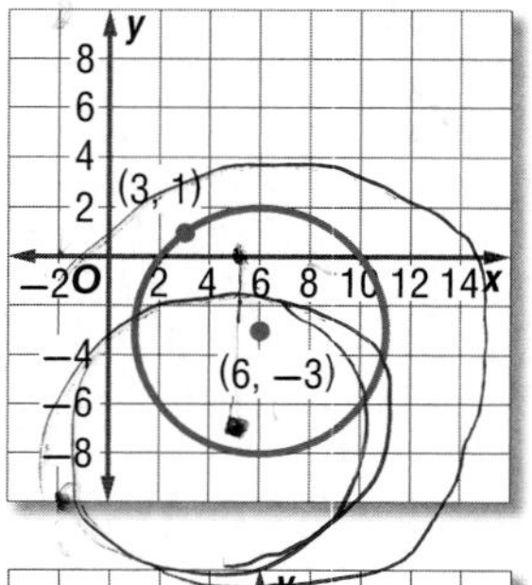

21.

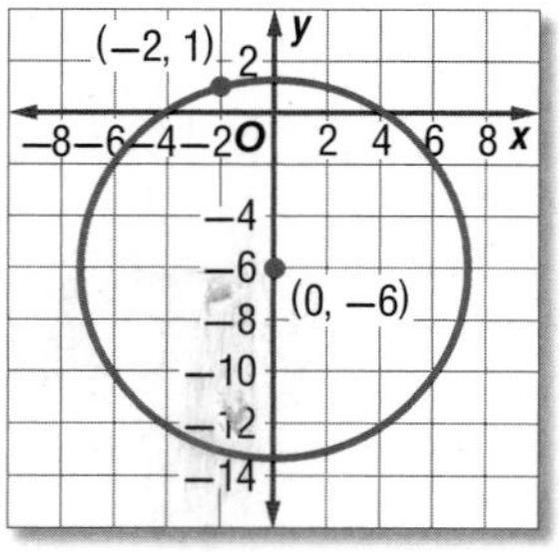

22.

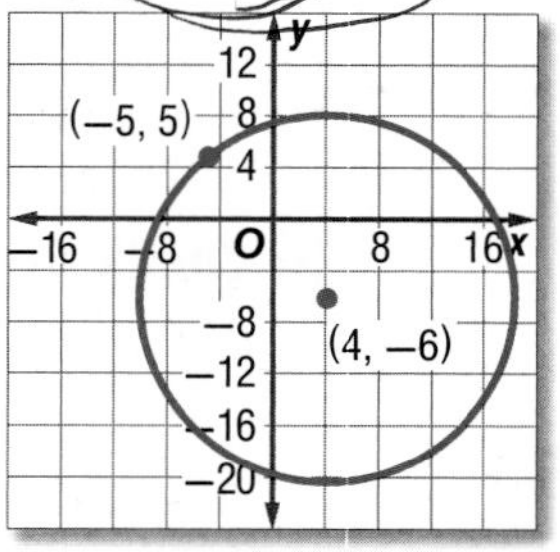

Example 3
p. 632

Write an equation for each circle given the endpoints of a diameter.

23. (2, 1) and (2, −4)

24. (−4, −10) and (4, −10)

25. (5, −7) and (−2, −9)

26. (−6, 4) and (4, 8)

27. (2, −5) and (6, 3)

28. (18, 11) and (−19, −13)

29. **LAWN CARE** A sprinkler waters a circular section of lawn.

a. Write an equation to represent the boundary of the sprinkler area if the endpoints of a diameter are at (−12, 16) and (12, −16).

b. What is the area of the lawn that the sprinkler waters?

30. **SPACE** Apollo 8 was the first manned spacecraft to orbit the Moon at an average altitude of 185 kilometers above the Moon's surface. Write an equation to model a single circular orbit of the command module if the endpoints of a diameter are at (1925, 0) and (−1925, 0). Let the center of the Moon be at the origin.

Examples 4 and 5
p. 633

Find the center and radius of each circle. Then graph the circle.

31. $x^2 + y^2 = 75$

32. $(x - 3)^2 + y^2 = 4$

33. $(x - 1)^2 + (y - 4)^2 = 34$

34. $x^2 + (y - 14)^2 = 144$

35. $(x - 5)^2 + (y + 2)^2 = 16$

36. $x^2 + y^2 = 256$

37. $(x - 4)^2 + y^2 = \frac{8}{9}$

38. $\left(x + \frac{2}{3}\right)^2 + \left(y - \frac{1}{2}\right)^2 = \frac{16}{25}$

39. $x^2 + y^2 + 4x = 9$

40. $x^2 + y^2 - 6y + 8x = 0$

41. $x^2 + y^2 + 2x + 4y = 9$

42. $x^2 + y^2 - 3x + 8y = 20$

43. $x^2 + y^2 + 6y = -50 - 14x$

44. $x^2 - 18x + 53 = 18y - y^2$

45. $2x^2 + 2y^2 - 4x + 8y = 32$

46. $3x^2 + 3y^2 - 6y + 12x = 24$

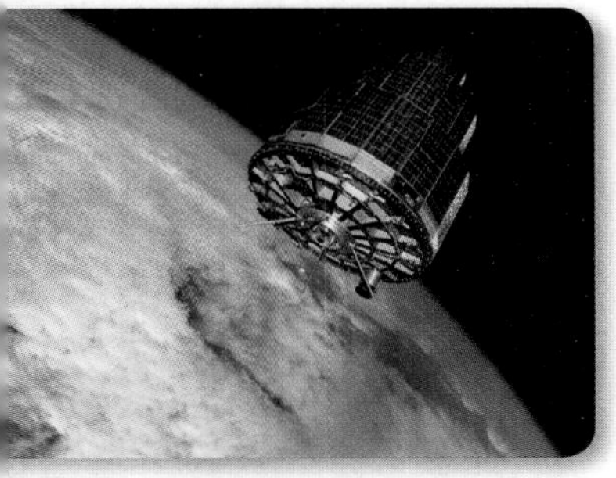

Real-World Link

There are over 2500 satellites in orbit around Earth. There are also over 10,000 human-made objects orbiting around Earth. These include a variety of pieces of satellite debris ranging from panels to old equipment.

47. **SPACE** A satellite is in a circular orbit 25,000 miles above Earth.

a. Write an equation for the orbit of this satellite if the origin is at the center of Earth. Use 8000 miles as the diameter of Earth.

b. Draw a sketch of Earth and the orbit to scale. Label your sketch.

48. **COMMUNICATIONS** Suppose an unobstructed radio station broadcast could travel 120 miles.

a. Write an equation to represent the boundary of the broadcast area.

b. If the transmission tower is relocated 40 miles east and 10 miles south of the current location, and an increased signal will transmit signals an additional 80 miles, what is an equation to represent the new broadcast area?

49. **GEOMETRY** Concentric circles are circles with the same center but different radii. Refer to the graph at the right where $\overline{AB}$ is a diameter of the circle.

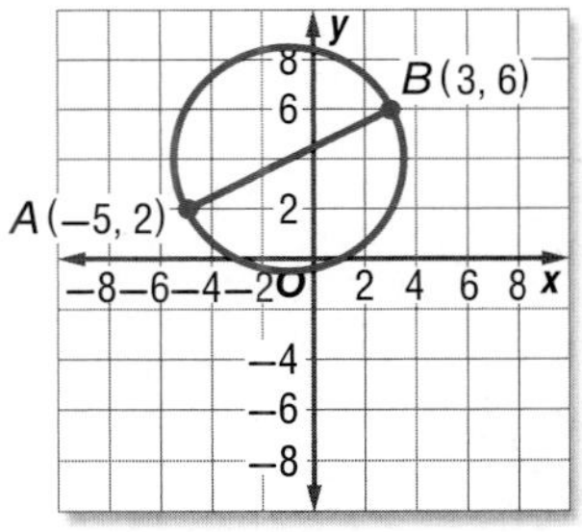

a. Write an equation of the circle concentric with the circle at the right, with radius 4 units greater.

b. Write an equation of the circle concentric with the circle at the right, with radius 2 units less.

c. Graph the circles from parts a and b on the same coordinate plane.

50. **EARTHQUAKES** The Rose Bowl is located about 35 miles west and about 40 miles north of downtown Los Angeles. Suppose an earthquake occurs with its epicenter about 55 miles from the stadium. Assume that the origin of a coordinate plane is located at the center of downtown Los Angeles. Write an equation for the set of points that could be the epicenter of the earthquake.

Write an equation for the circle that satisfies each set of conditions.

> **Review Vocabulary**
>
> **tangent** a line in the plane of a circle that intersects the circle in exactly one point

51. center $(9, -8)$, passes through $(19, 22)$

52. center $(-\sqrt{15}, 30)$, passes through the origin

53. center at $(8, -9)$, tangent to y-axis

54. center at $(2, 4)$, tangent to x-axis

55. center in the first quadrant; tangent to $x = 5$, the x-axis, and the y-axis

56. center in the second quadrant; tangent to $y = 1$, $y = 5$, and the y-axis

57. **MULTIPLE REPRESENTATIONS** Graph $y = \sqrt{9 - x^2}$ and $y = -\sqrt{9 - x^2}$ on the same graphing calculator screen.

- **a. VERBAL** Describe the graph formed by the union of these two graphs.
- **b. ALGEBRAIC** Write an equation for the union of the two graphs.
- **c. VERBAL** Most graphing calculators cannot graph the equation $x^2 + y^2 = 49$ directly. Describe a way to use a graphing calculator to graph the equation. Then graph the equation.
- **d. ANALYTICAL** Solve $(x - 2)^2 + (y + 1)^2 = 4$ for y. Why do you need two equations to graph a circle on a graphing calculator?
- **e. VERBAL** Do you think that it is easier to graph the equation in part d using graph paper and a pencil or using a graphing calculator? Explain.

Find the center and radius of each circle. Then graph the circle.

58. $x^2 - 12x + 84 = -y^2 + 16y$

59. $4x^2 + 4y^2 + 36y + 5 = 0$

60. $(x + \sqrt{5})^2 + y^2 - 8y = 9$

61. $x^2 + 2\sqrt{7}x + 7 + (y - \sqrt{11})^2 = 11$

H.O.T. Problems — Use Higher-Order Thinking Skills

62. FIND THE ERROR Heather says that $(x - 2)^2 + (y + 3)^2 = 36$ and $(x - 2) + (y + 3) = 6$ are equivalent equations. Carlota says that the equations are *not* equivalent. Is either of them correct? Explain your reasoning.

63. OPEN ENDED Consider graphs with equations of the form $(x - 3)^2 + (y - a)^2 = 64$. Assign three different values for a, and graph each equation. Describe all graphs with equations of this form.

64. REASONING Explain why the phrase "in a plane" is included in the definition of a circle. What would be defined if the phrase were *not* included?

65. OPEN ENDED Concentric circles have the same center, but most often, not the same radius. Write equations of two concentric circles. Then graph the circles.

66. REASONING Assume that (x, y) are the coordinates of a point on a circle. The center is at (h, k), and the radius is r. Find an equation of the circle by using the Distance Formula.

67. WRITING IN MATH The circle with equation $(x - a)^2 + (y - b)^2 = r^2$ lies in the first quadrant and is tangent to both the x-axis and the y-axis. Sketch the circle. Describe the possible values of a, b, and r. Do the same for a circle in Quadrants II, III, and IV. Discuss the similarities among the circles.

NYSRE PRACTICE A2.A.48, A2.A.49

68. **GRIDDED RESPONSE** Two circles, both with radii 6, have exactly one point in common. If A is a point on one circle and B is a point on the other circle, what is the maximum possible length for the line segment $\overline{AB}$?

69. In the senior class, there are 20% more girls than boys. If there are 180 girls, how many more girls than boys are there among the seniors?

A 30
B 36
C 90
D 144

70. A $1000 deposit is made at a bank that pays 2% compounded weekly. How much will you have in your account at the end of 10 years?

F $1200.00
G $1218.99
H $1221.36
J $1224.54

71. The mean of six numbers is 20. If one of the numbers is removed, the average of the remaining numbers is 15. What is the number that was removed?

A 42
B 43
C 45
D 48

Spiral Review

Graph each equation. (Lesson 10-2)

72. $y = -\frac{1}{2}(x - 1)^2 + 4$

73. $4(x - 2) = (y + 3)^2$

74. $(y - 8)^2 = -4(x - 4)$

Find the midpoint of the line segment with endpoints at the given coordinates. Then find the distance between the points. (Lesson 10-1)

75. $\left(-3, -\frac{2}{11}\right), \left(5, \frac{9}{11}\right)$

76. $\left(2\sqrt{3}, -5\right), \left(-3\sqrt{3}, 9\right)$

77. $(2.5, 4), (-2.5, 2)$

78. If y varies directly as x and $y = 8$ when $x = 6$, find y when $x = 15$. (Lesson 9-5)

79. If y varies jointly as x and z and $y = 80$ when $x = 5$ and $z = 8$, find y when $x = 16$ and $z = 2$. (Lesson 9-5)

80. If y varies inversely as x and $y = 16$ when $x = 5$, find y when $x = 20$. (Lesson 9-5)

Evaluate each expression. (Lesson 8-3)

81. $\log_9 243$

82. $\log_2 \frac{1}{32}$

83. $\log_3 \frac{1}{81}$

84. $\log_{10} 0.001$

85. **AMUSEMENT PARKS** The velocity v in feet per second of a roller coaster at the bottom of a hill is related to the vertical drop h in feet and the velocity v_0 in feet per second of the coaster at the top of the hill by the formula $v_0 = \sqrt{v^2 - 64h}$. (Lesson 7-5)

a. Explain why $v_0 = v - 8\sqrt{h}$ is not equivalent to the given formula.

b. What velocity must the coaster have at the top of the hill to achieve a velocity of 120 feet per second at the bottom?

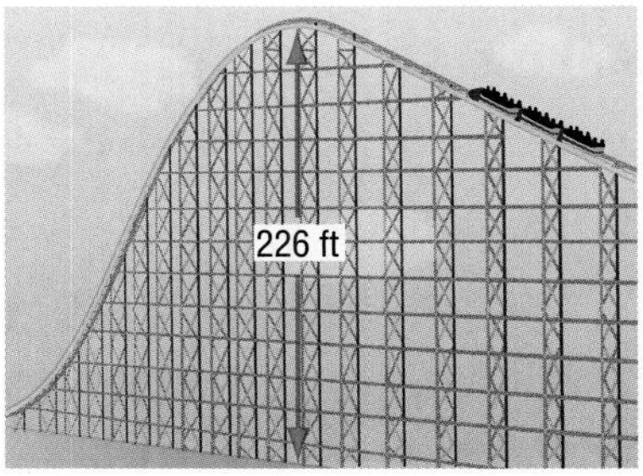

Skills Review

Solve each equation by completing the square. (Lesson 5-5)

86. $x^2 + 3x - 18 = 0$

87. $2x^2 - 3x - 3 = 0$

88. $x^2 + 2x + 6 = 0$

EXPLORE 10-4

Algebra Lab
Investigating Ellipses

NY Math Online glencoe.com
Math in Motion, Animation

Follow the steps below to construct another type of conic section.

ACTIVITY Make an Ellipse

Step 1 Place two thumbtacks in a piece of cardboard, about 1 foot apart.

Step 2 Tie a knot in a piece of string and loop it around the thumbtacks. Place your pencil in the string.

Step 3 Keep the string tight and draw a curve. Continue drawing until you return to your starting point.

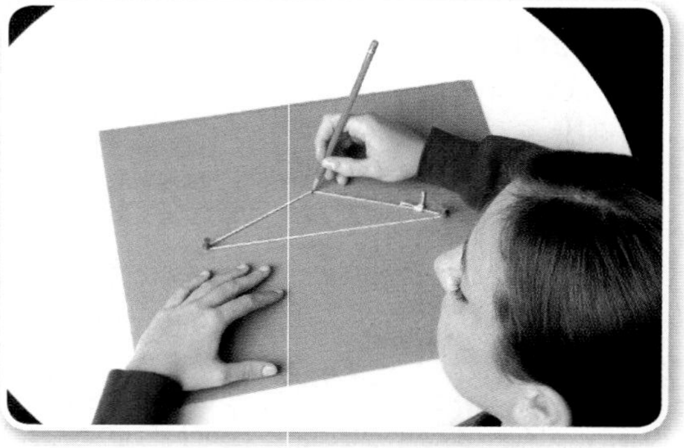

The curve you have drawn is called an **ellipse**. The points where the thumbtacks are located are called the **foci** of the ellipse. *Foci* is the plural of *focus*.

Model and Analyze

Place a large piece of grid paper on a piece of cardboard.

1. Place the thumbtacks at (7, 0) and (−7, 0). Choose a string long enough to loop around both thumbtacks. Draw an ellipse.
2. Repeat Exercise 1, but place the thumbtacks at (4, 0) and (−4, 0). Use the same loop of string and draw an ellipse. How does this ellipse compare to the one in Exercise 1?

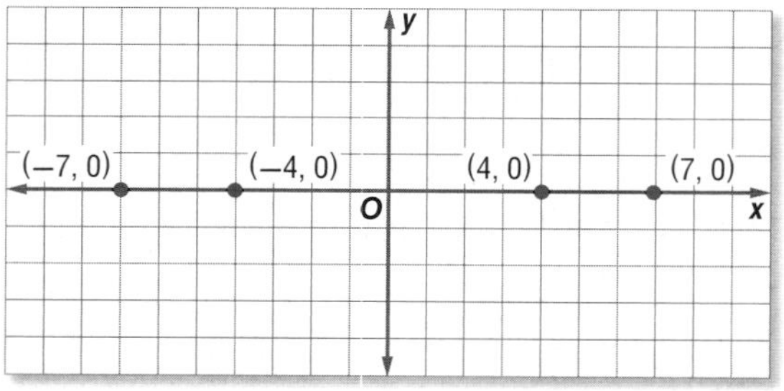

Place the thumbtacks at each set of points and draw an ellipse. You may change the length of the loop of string if you like.

3. (11, 0), (−11, 0)
4. (3, 0), (−3, 0)
5. (13, 3), (−9, 3)

Make a Conjecture

Describe what happens to the shape of an ellipse when each change is made.

6. The thumbtacks are moved closer together.
7. The thumbtacks are moved farther apart.
8. The length of the loop of string is increased.
9. The thumbtacks are arranged vertically.
10. One thumbtack is removed, and the string is looped around the remaining thumbtack.
11. Pick a point on one of the ellipses you have drawn. Use a ruler to measure the distances from that point to the points where the thumbtacks were located. Add the distances. Repeat for other points on the same ellipse. What relationship do you notice?
12. Could this activity be done with a rubber band instead of a piece of string? Explain.

10-4 Ellipses

Then

You graphed and wrote equations for circles. (Lesson 10-3)

Now

- Write equations of ellipses.
- Graph ellipses.

NYS Core Curriculum

A2.R.6 Use mathematics to show and understand physical phenomena (e.g., investigate sound waves using the sine and cosine functions)

New Vocabulary

ellipse
foci
major axis
minor axis
center
vertices
co-vertices
constant sum

NY Math Online

glencoe.com

- Extra Examples
- Personal Tutor
- Self-Check Quiz
- Homework Help

Why?

Mercury, like all of the planets of our solar system, does not orbit the Sun in a perfect circular path. At its farthest point, Mercury is about 43 million miles from the Sun. At its closest point, it is only about 28.5 million miles from the Sun. This orbit is in the shape of an ellipse with the Sun at a focus.

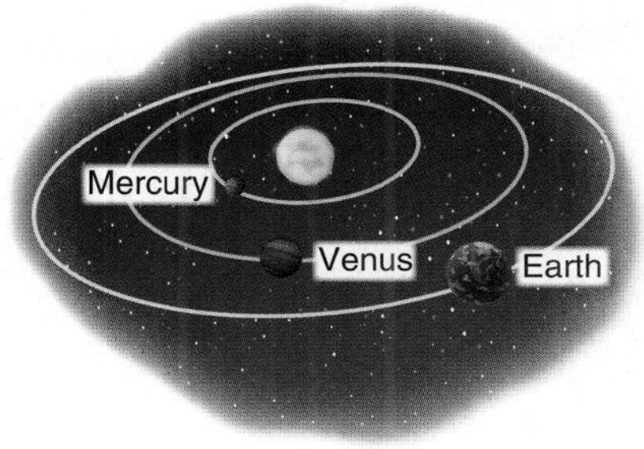

Equations of Ellipses An **ellipse** is the set of all points in a plane such that the sum of the distances from two fixed points is constant. These two points are called the **foci** of the ellipse.

Every ellipse has two axes of symmetry, the **major axis** and the **minor axis**. The axes are perpendicular at the **center** of the ellipse.

The foci of an ellipse always lie on the major axis. The endpoints of the major axis are the **vertices** of the ellipse and the endpoints of the minor axis are the **co-vertices** of the ellipse.

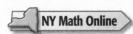

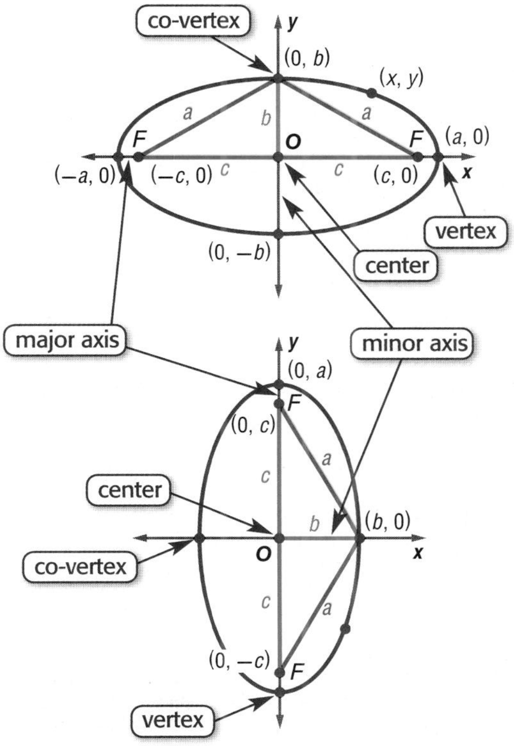

Key Concept

For Your FOLDABLE

Equations of Ellipses Centered at the Origin

Standard Form	$\frac{x^2}{a^2} + \frac{y^2}{b^2} = 1$	$\frac{y^2}{a^2} + \frac{x^2}{b^2} = 1$
Orientation	horizontal	vertical
Foci	$(c, 0), (-c, 0)$	$(0, c), (0, -c)$
Length of Major Axis	$2a$ units	$2a$ units
Length of Minor Axis	$2b$ units	$2b$ units

There are several important relationships among the many parts of an ellipse.

- The length of the major axis, $2a$ units, equals the sum of the distances from the foci to any point on the ellipse.
- The values of a, b, and c are related by the equation $c^2 = a^2 - b^2$.
- The distance from a focus to either co-vertex is a units.

The sum of the distances from the foci to any point on the ellipse, or the **constant sum**, must be greater than the distance between the foci.

StudyTip

Major Axis In standard form, if the x^2-term has the greater denominator, then the major axis is horizontal. If the y^2-term has the greater denominator, then it is vertical.

EXAMPLE 1 Write an Equation Given Vertices and Foci

Write an equation for the ellipse.

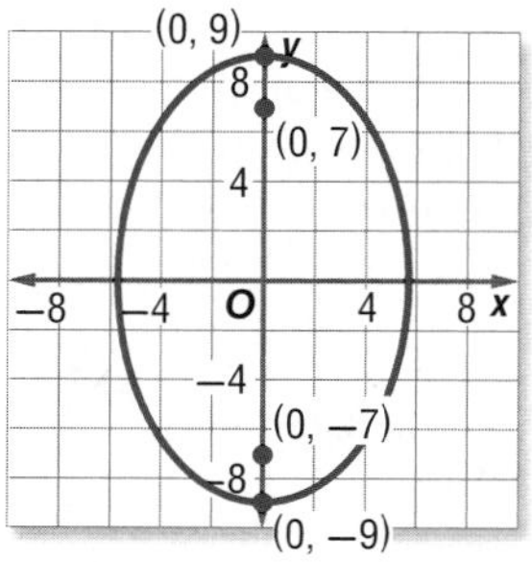

Step 1 Find the center.
The foci are equidistant from the center.
The center is at (0, 0).

Step 2 Find the value of a.
The vertices are (0, 9) and (0, −9), so the length of the major axis is 18.
The value of a is 18 ÷ 2 or 9, and $a^2 = 81$.

Step 3 Find the value of b.
We can use $c^2 = a^2 - b^2$ to find b.
The foci are 7 units from the center, so $c = 7$.

$c^2 = a^2 - b^2$ **Equation relating *a*, *b*, and *c***

$49 = 81 - b^2$ **$a = 9$ and $c = 7$**

$b^2 = 32$ **Solve for b^2.**

Step 4 Write the equation.
Because the major axis is vertical, a^2 goes with y and b^2 goes with x.
The equation for the ellipse is $\frac{y^2}{81} + \frac{x^2}{32} = 1$.

Check Your Progress

1. Write an equation for an ellipse with vertices at (−4, 0) and (4, 0) and foci at (2, 0) and (−2, 0).

Personal Tutor glencoe.com

Like other graphs, the graph of an ellipse can be translated. When the graph is translated h units right and k units up, the center of the translation is (h, k). This is equivalent to replacing x with $x - h$ and replacing y with $y - k$ in the parent function.

Key Concept Equations of Ellipses Centered at (*h*, *k*)

For Your FOLDABLE

Standard Form	$\frac{(x-h)^2}{a^2} + \frac{(y-k)^2}{b^2} = 1$	$\frac{(y-k)^2}{a^2} + \frac{(x-h)^2}{b^2} = 1$
Orientation	horizontal	vertical
Foci	$(h \pm c, k)$	$(h, k \pm c)$
Vertices	$(h \pm a, k)$	$(h, k \pm a)$
Co-vertices	$(h, k \pm b)$	$(h \pm b, k)$

We can use this information to determine the equations for ellipses. The original ellipse at the right is horizontal and has a major axis of 10 units, so $a = 5$.

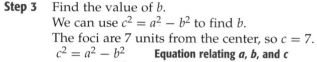

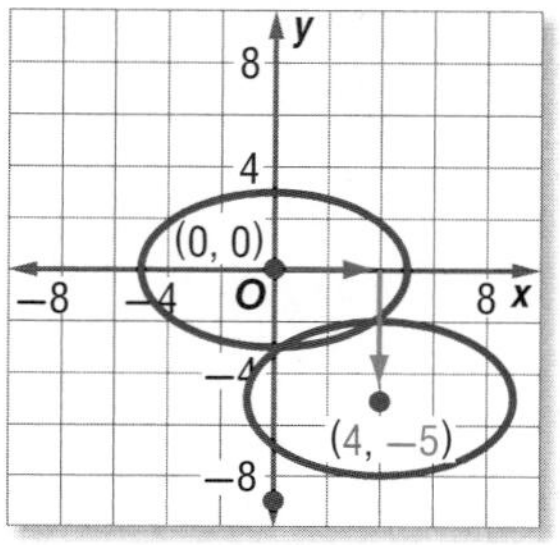

The length of the minor axis is 6 units, so $b = 3$.

The ellipse is translated 4 units right and 5 units down. So, the value of h is 4 and the value of k is −5.

The equation for the original ellipse is $\frac{x^2}{25} + \frac{y^2}{9} = 1$.

The equation for the translation is $\frac{(x-4)^2}{25} + \frac{(y+5)^2}{9} = 1$.

You can also determine the equation for an ellipse if you are given all four vertices.

EXAMPLE 2 Write an Equation Given the Lengths of the Axes

Write an equation for the ellipse with vertices at (6, −8) and (6, 4) and co-vertices at (3, −2) and (9, −2).

The x-coordinate is the same for both vertices, so the ellipse is vertical.

The center of the ellipse is at $\left(\frac{6+6}{2}, \frac{-8+4}{2}\right)$ or (6, −2).

The length of the major axis is $4 - (-8)$ or 12 units, so $a = 6$.

The length of the minor axis is $9 - 3$ or 6 units, so $b = 3$.

The equation for the ellipse is $\frac{(y+2)^2}{36} + \frac{(x-6)^2}{9} = 1$. $a^2 = 36, b^2 = 9$

Check Your Progress

2. Write an equation for the ellipse with vertices at (−3, 8) and (9, 8) and co-vertices at (3, 12) and (3, 4).

Personal Tutor glencoe.com

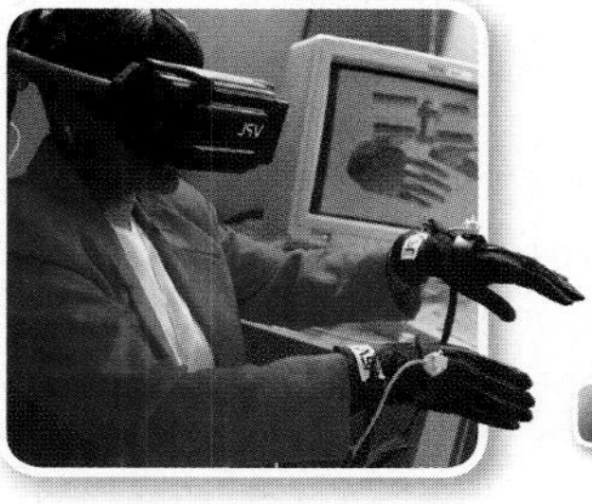

Real-World Career

Aerospace Technician
Aerospace technicians work for NASA, helping engineers research and develop virtual reality and verbal communication between humans and computer systems. Although a bachelor's degree is desired, on-the-job training is available.

Source: NASA

Many real-world phenomena can be represented by ellipses.

Real-World EXAMPLE 3 Write an Equation for an Ellipse

SPACE Refer to the application at the beginning of the lesson. Mercury's greatest distance from the Sun, or *aphelion*, is about 43 million miles. Mercury's closest distance, or *perihelion*, is about 28.5 million miles. The diameter of the Sun is about 870,000 miles. Use this information to determine an equation relating Mercury's elliptical orbit around the Sun in millions of miles.

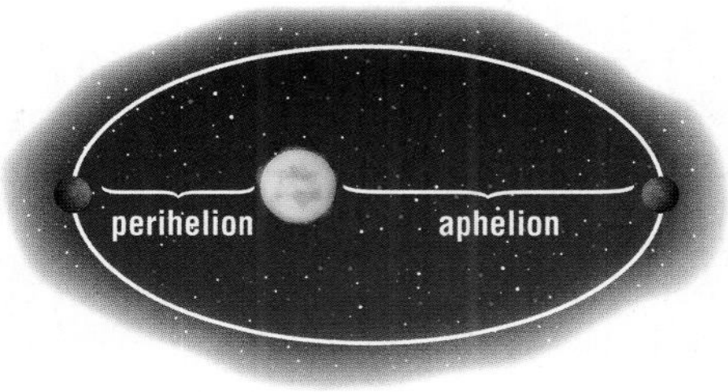

Understand We need to determine an equation representing Mercury's orbit around the Sun.

Plan Including the diameter of the Sun, the sum of the perihelion and aphelion equals the length on the major axis of the ellipse. We can use this information to determine the values of a, b, and c.

Solve Find the value of a.
The value of a is one half the length of the major axis.
$a = 0.5(43 + 28.5 + 0.87)$ or 36.185

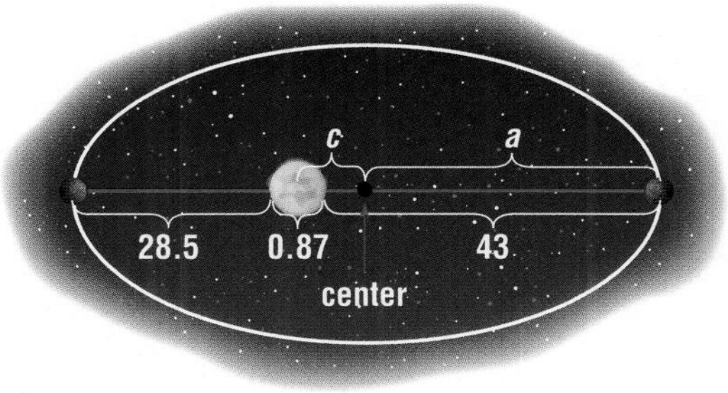

Find the value of c.
The value of c is the distance from the center of the ellipse to the focus.
This distance is equal to a minus the perihelion.
$c = 36.185 - 28.5$ or 7.685

Problem-Solving Tip

Draw a Diagram
Draw a diagram when the problem situation involves spatial reasoning or geometric figures.

(continued on the next page)

Find the value of b.

$c^2 = a^2 - b^2$	**Equation relating *a*, *b*, and *c***
$(7.685)^2 = (36.185)^2 - b^2$	**$c = 7.685$ and $a = 36.185$**
$59.0592 = 1309.3542 - b^2$	**Simplify.**
$b^2 = 1250.295$	**Solve for b^2.**
$b = 35.3595$	**Take the square root of each side.**

So, with the center of the orbit at the origin, the equation relating Mercury's orbit around the Sun can be modeled by $\frac{x^2}{1309.3542} + \frac{y^2}{1250.295} = 1$.

Check Use your answer to recalculate a, b, and c. Then determine the aphelion and perihelion based on your answer. Compare to the actual values.

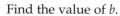

Real-World Link

Earth's orbit around the Sun is nearly circular, with only about a 3% difference between perihelion and aphelion.

Source: The Astronomer

Check Your Progress

3. SPACE Pluto's distance from the Sun is 2.757 billion miles at perihelion and about 4.583 billion miles at aphelion. Determine an equation relating Pluto's orbit around the Sun in billions of miles with the center of the horizontal ellipse at the origin.

Personal Tutor glencoe.com

Graph Ellipses When you are given an equation for an ellipse that is not in standard form, you can write it in standard form by completing the square for both x and y. Once the equation is in standard form, you can use it to graph the ellipse.

EXAMPLE 4 Graph an Ellipse

Find the coordinates of the center and foci, and the lengths of the major and minor axes of an ellipse with equation $25x^2 + 9y^2 + 250x - 36y + 436 = 0$. Then graph the ellipse.

Step 1 Write in standard form. Complete the square for each variable to write this equation in standard form.

$25x^2 + 9y^2 + 250x - 36y + 436 = 0$	**Original equation**
$25x^2 + 250x + 9y^2 - 36y = -436$	**Associative Property**
$25(x^2 + 10x) + 9(y^2 - 4y) = -436$	**Distributive Property**
$25(x^2 + 10x + \blacksquare) + 9(y^2 - 4y + \blacksquare) = -436 + 25(\blacksquare) + 9(\blacksquare)$	**Complete the squares.**
$25(x^2 + 10x + 25) + 9(y^2 - 4y + 4) = -436 + 25(25) + 9(4)$	**$5^2 = 25$ and $(-2)^2 = 4$**
$25(x + 5)^2 + 9(y - 2)^2 = 225$	**Write as perfect squares.**
$\frac{(x + 5)^2}{9} + \frac{(y - 2)^2}{25} = 1$	**Divide each side by 225.**

Step 2 Find the center.
$h = -5$ and $k = 2$, so the center of the ellipse is at $(-5, 2)$.

Step 3 Find the lengths of the axes and graph.
The ellipse is vertical.
$a^2 = 25$, so $a = 5$. $b^2 = 9$, so $b = 3$.
The length of the major axis is $2 \cdot 5$ or 10.
The length of the minor axis is $2 \cdot 3$ or 6.
The vertices are at $(-5, 7)$ and $(-5, -3)$.
The co-vertices are at $(-2, 2)$ and $(-8, 2)$.

Step 4 Find the foci.
$c^2 = 25 - 9$ or 16, so $c = 4$.
The foci are at $(-5, 6)$ and $(-5, -2)$.

Step 5 Graph the ellipse.
Draw the ellipse that passes through the vertices and co-vertices.

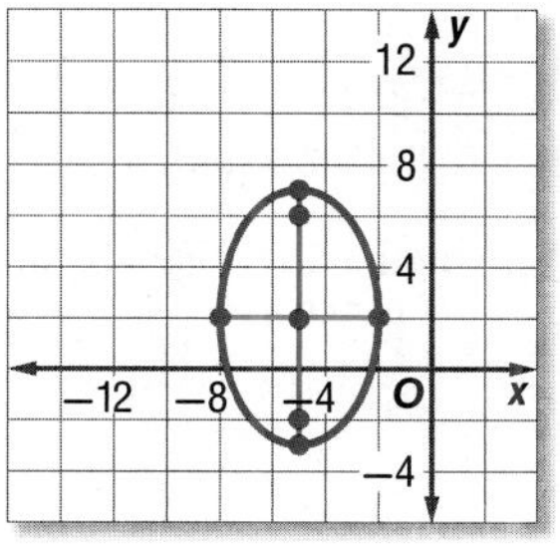

Check Your Progress

4. Find the coordinates of the center and foci and the lengths of the major and minor axes of the ellipse with equation $x^2 + 4y^2 - 2x + 24y + 21 = 0$. Then graph the ellipse.

Personal Tutor glencoe.com

Check Your Understanding

Example 1 **p. 640**

Write an equation for each ellipse.

1.

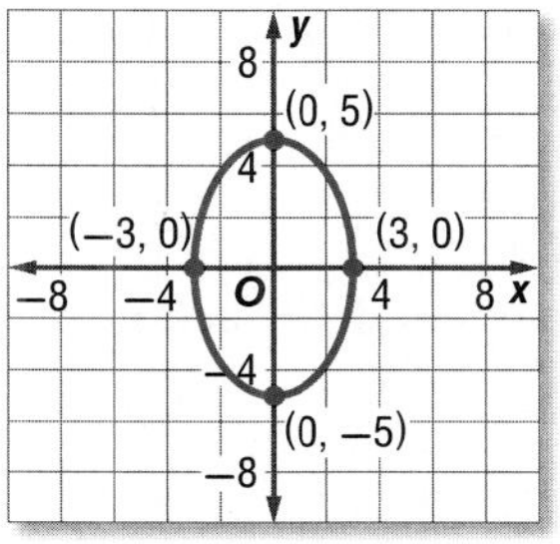

2.

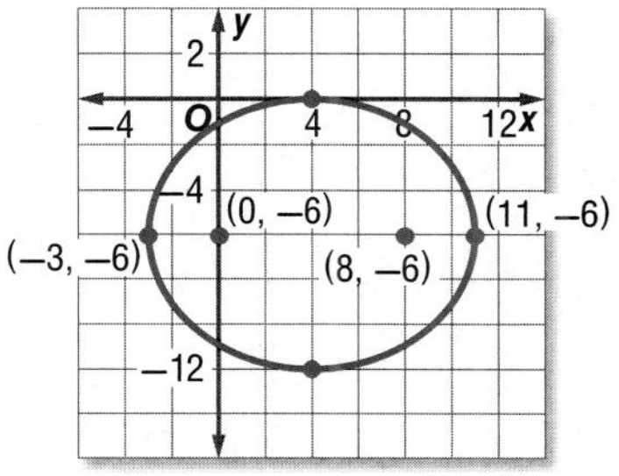

Example 2 **p. 641**

Write an equation for an ellipse that satisfies each set of conditions.

3. vertices at $(-2, -6)$ and $(-2, 4)$, co-vertices at $(-5, -1)$ and $(1, -1)$

4. vertices at $(-2, 5)$ and $(14, 5)$, co-vertices at $(6, 1)$ and $(6, 9)$

Example 3 **pp. 641–642**

5. **ARCHITECTURE** An architectural firm sent a proposal to a city for building a coliseum, shown at the right.

a. Determine the values of a and b.

b. Assuming that the center is at the origin, write an equation to represent the ellipse.

c. Determine the coordinates of the foci.

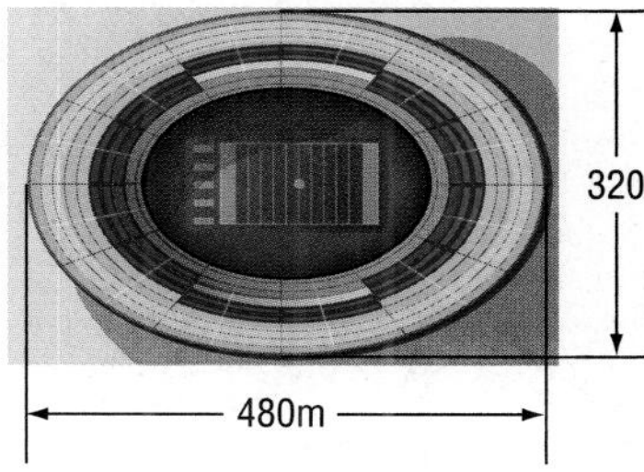

6. **SPACE** Earth's orbit is about 91.4 million miles at perihelion and about 94.5 million miles at aphelion. Determine an equation relating Earth's orbit around the Sun in millions of miles with the center of the horizontal ellipse at the origin.

Example 4 **pp. 642–643**

Find the coordinates of the center and foci and the lengths of the major and minor axes for the ellipse with the given equation. Then graph the ellipse.

7. $\frac{(y+1)^2}{64} + \frac{(x-5)^2}{28} = 1$

8. $\frac{(x+2)^2}{48} + \frac{(y-1)^2}{20} = 1$

9 $4x^2 + y^2 - 32x - 4y + 52 = 0$

10. $9x^2 + 25y^2 + 72x - 150y + 144 = 0$

Practice and Problem Solving

● = **Step-by-Step Solutions** begin on page R20.
Extra Practice begins on page 947.

Example 1 p. 640

Write an equation for each ellipse.

11.

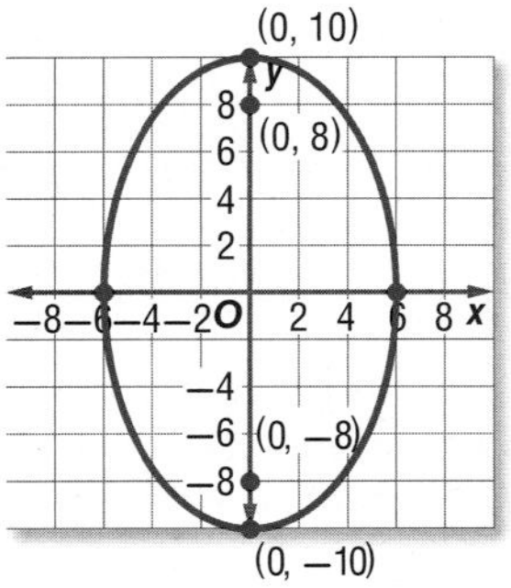

12.

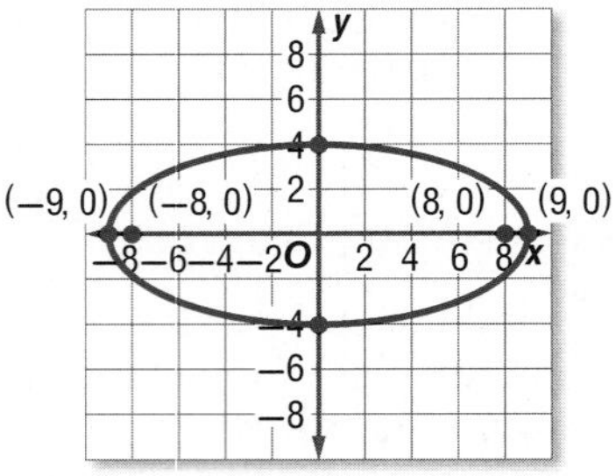

13.

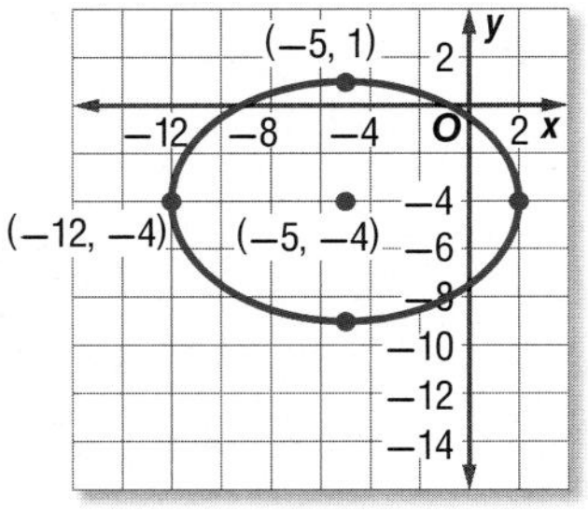

14.

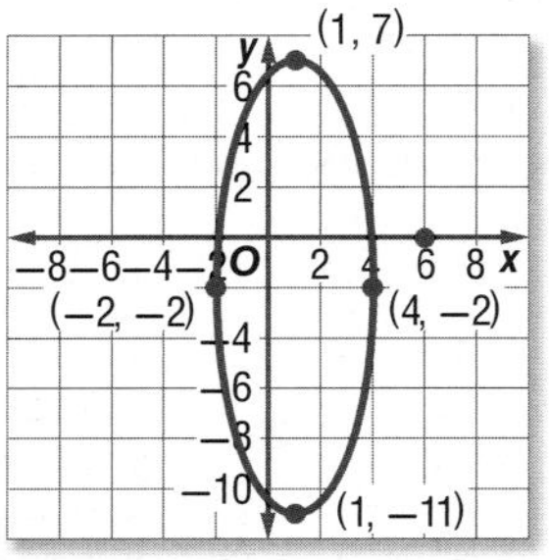

15.

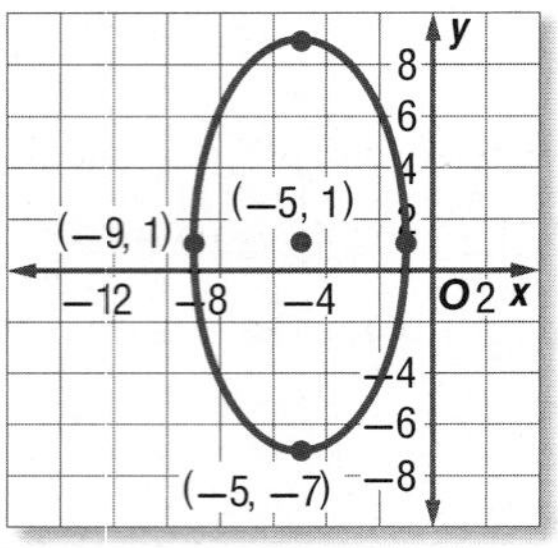

16.

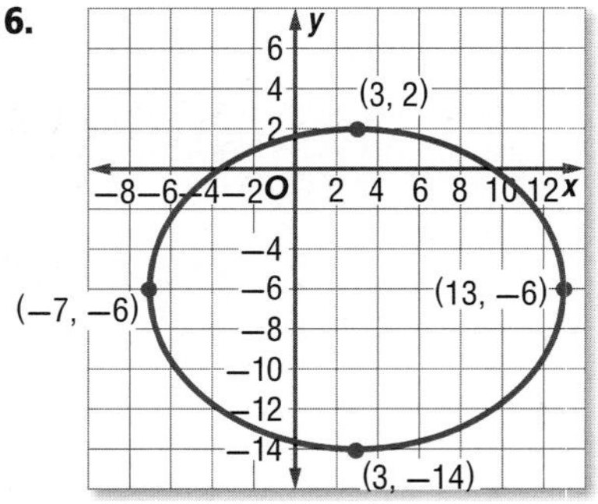

Example 2 p. 641

Write an equation for an ellipse that satisfies each set of conditions.

17. vertices at (−6, 4) and (12, 4), co-vertices at (3, 12) and (3, −4)

18. vertices at (−1, 11) and (−1, 1), co-vertices at (−4, 6) and (2, 6)

19. center at (−2, 6), vertex at (−2, 16), co-vertex at (1, 6)

20. center at (3, −4), vertex at (8, −4), co-vertex at (3, −2)

21. vertices at (4, 12) and (4, −4), co-vertices at (1, 4) and (7, 4)

22. vertices at (−11, 2) and (−1, 2), co-vertices at (−6, 0) and (−6, 4)

Example 3 pp. 641–642

23. **TUNNELS** The opening of a tunnel in the mountains can be modeled by semiellipses, or halves of ellipses. If the opening is 14.6 meters wide and 8.6 meters high, determine an equation to represent the opening with the center at the origin.

Example 4 pp. 642–643

Find the coordinates of the center and foci and the lengths of the major and minor axes for the ellipse with the given equation. Then graph the ellipse.

24. $\frac{(x-3)^2}{36} + \frac{(y-2)^2}{128} = 1$

25. $\frac{(x+6)^2}{50} + \frac{(y-3)^2}{72} = 1$

26. $\frac{x^2}{27} + \frac{(y-5)^2}{64} = 1$

27. $\frac{(x+4)^2}{16} + \frac{y^2}{75} = 1$

28. $3x^2 + y^2 - 6x - 8y - 5 = 0$

29. $3x^2 + 4y^2 - 18x + 24y + 3 = 0$

30. $7x^2 + y^2 - 56x + 6y + 93 = 0$

31. $3x^2 + 2y^2 + 12x - 20y + 14 = 0$

32. **SPACE** Like the planets, Halley's Comet travels around the Sun in an elliptical orbit. The aphelion is 3282.9 million miles and the perihelion is 54.87 million miles.

Write an equation for an ellipse that satisfies each set of conditions.

33. center at (−5, −2), focus at (−5, 2), co-vertex at (−8, −2)

34. center at (4, −3), focus at (9, −3), co-vertex at (4, −5)

35. foci at (−2, 8) and (6, 8), co-vertex at (2, 10)

36. foci at (4, 4) and (4, 14), co-vertex at (0, 9)

37 **GOVERNMENT** The Oval Office is located in the West Wing of the White House. It is an elliptical shaped room used as the main office by the President of the United States. The long axis is 10.9 meters long and the short axis is 8.8 meters long. Write an equation to represent the outer walls of the Oval Office. Assume that the center of the room is at the origin.

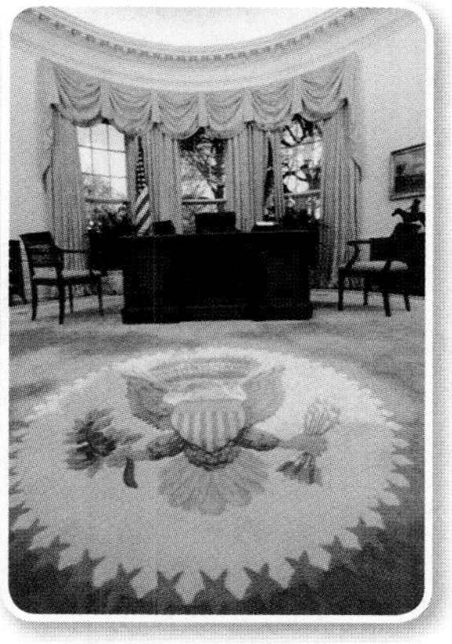

Real-World Link

President Taft established the first Oval Office in 1909. The room changed locations and was modified several times. Since 1934, though, the Oval Office has changed very little.

Source: The White House

38. **SOUND** A whispering gallery is an elliptical room in which a faint whisper at one focus that cannot be heard by other people in the room, can easily be heard by someone at the other focus. Suppose a gallery is 400 feet long and 120 feet wide. What is the distance between the foci?

39. **MULTIPLE REPRESENTATIONS** The *eccentricity* of an ellipse measures how circular the ellipse is.

a. **GRAPHICAL** Graph $\frac{x^2}{81} + \frac{y^2}{36} = 1$ and $\frac{x^2}{81} + \frac{y^2}{9} = 1$ on the same graph.

b. **VERBAL** Describe the difference between the two graphs.

c. **ALGEBRAIC** The eccentricity of an ellipse is $\frac{c}{a}$. Find the eccentricity for each.

d. **ANALYTICAL** Make a conjecture about the relationship between the value of an ellipse's eccentricity and the shape of the ellipse as compared to a circle.

H.O.T. Problems

Use Higher-Order Thinking Skills

40. **FIND THE ERROR** Serena and Karissa are determining the equation for an ellipse with foci at (−4, −11) and (−4, 5) and co-vertices at (2, −3) and (−10, −3). Is either of them correct? Explain your reasoning.

Serena	Karissa
$\frac{(x-4)^2}{64} + \frac{(y+3)^2}{36} = 1$	$\frac{(x+4)^2}{100} + \frac{(y+3)^2}{36} = 1$

41. **OPEN ENDED** Write an equation for an ellipse with a focus at the origin.

42. **CHALLENGE** When the values of a and b are equal, an ellipse is a circle. Use this information and your knowledge of ellipses to determine the formula for the area of an ellipse in terms of a and b.

43. **CHALLENGE** Determine an equation for an ellipse with foci at $(2, \sqrt{6})$ and $(2, -\sqrt{6})$ that passes through $(3, \sqrt{6})$.

44. **REASONING** What happens to the location of the foci as an ellipse becomes more circular? Explain your reasoning.

45. **REASONING** An ellipse has foci at (−7, 2) and (18, 2). If (2, 14) is a point on the ellipse, show that (2, −10) is also a point on the ellipse.

46. **WRITING IN MATH** Explain why the domain is $\{x \mid -a \leq x \leq a\}$ and the range is $\{y \mid -b \leq y \leq b\}$ for an ellipse with equation $\frac{x^2}{a^2} + \frac{y^2}{b^2} = 1$.

NYSRE PRACTICE A2.R.6

47. Multiply.

$$(2 + 3i)(4 + 7i)$$

A $8 + 21i$
B $-13 + 26i$
C $-6 + 10i$
D $13 + 12i$

48. The average lifespan of American women has been tracked, and the model for the data is $y = 0.2t + 73$, where $t = 0$ corresponds to 1960. What is the meaning of the y-intercept?

F In 2007, the average lifespan was 60.
G In 1960, the average lifespan was 58.
H In 1960, the average lifespan was 73.
J The lifespan is increasing 0.2 years every year.

49. GRIDDED RESPONSE If we decrease a number by 6 and then double the result, we get 5 less than the number. What is the number?

50. ACT/SAT The length of a rectangular prism is one inch greater than its width. The height is three times the length. Find the volume of the prism.

A $3x^3 + x^2 + 3x$
B $x^3 + x^2 + x$
C $3x^3 + 6x^2 + 3x$
D $3x^3 + 3x^2 + 3x$

Spiral Review

Write an equation for the circle that satisfies each set of conditions. (Lesson 10-3)

51. center $(8, -9)$, passes through $(21, 22)$

52. center at $(4, 2)$, tangent to x-axis

53. center in the second quadrant; tangent to $y = -1$, $y = 9$, and the y-axis

54. ENERGY A parabolic mirror is used to collect solar energy. The mirrors reflect the rays from the Sun to the focus of the parabola. The focus of a particular mirror is 9.75 feet above the vertex, and the latus rectum is 40 feet long. (Lesson 10-2)

a. Assume that the focus is at the origin. Write an equation for the parabola formed by the mirror.

b. One foot is exactly 0.3048 meter. Rewrite the equation for the mirror in meters.

c. Graph one of the equations for the mirror.

d. Which equation did you choose to graph? Explain why.

Simplify each expression. (Lesson 9-2)

55. $\frac{6}{d^2 + 4d + 4} + \frac{5}{d + 2}$

56. $\frac{a}{a^2 - a - 20} + \frac{2}{a + 4}$

57. $\frac{x}{x + 1} + \frac{3}{x^2 - 4x - 5}$

Solve each equation. (Lesson 8-4)

58. $\log_{10} (x^2 + 1) = 1$

59. $\log_b 64 = 3$

60. $\log_b 121 = 2$

Simplify. (Lesson 6-1)

61. $-5ab^2(-3a^2b + 6a^3b - 3a^4b^4)$

62. $2xy(3xy^3 - 4xy + 2y^4)$

63. $(4x^2 - 3y^2 + 5xy) - (8xy + 3y^2)$

64. $(10x^2 - 3xy + 4y^2) - (3x^2 + 5xy)$

Skills Review

Write an equation of the line passing through each pair of points. (Lesson 2-4)

65. $(-2, 5)$ and $(3, 1)$

66. $(7, 1)$ and $(7, 8)$

67. $(-3, 5)$ and $(2, 2)$

Mid-Chapter Quiz

Lessons 10-1 through 10-4

NYS Core Curriculum
A2.R.1, A2.A.48

Find the midpoint of the line segment with endpoints at the given coordinates. (Lesson 10-1)

1. $(7, 4), (-1, -5)$

2. $(-2, -9), (-6, 0)$

Find the distance between each pair of points with the given coordinates. (Lesson 10-1)

3. $(0, 6), (-2, 5)$

4. $(10, 1), (0, -4)$

5. HIKING Carla and Lance left their campsite and hiked 6 miles directly north and then turned and hiked 7 miles east to view a waterfall. (Lesson 10-1)

a. How far is the waterfall from their campsite?

b. Let the campsite be located at the origin on a coordinate grid. At the waterfall they decide to head directly back to the campsite. If they stop halfway between the waterfall and the campsite for lunch, at what coordinate will they stop for lunch?

Write each equation in standard form. Identify the vertex, axis of symmetry, and direction of opening of the parabola. (Lesson 10-2)

6. $y = 3x^2 - 12x + 21$

7. $x - 2y^2 = 4y + 6$

8. $y = \frac{1}{2}x^2 + 12x - 8$

9. $x = 3y^2 + 5y - 9$

10. BRIDGES Write an equation of a parabola to model the shape of the suspension cable of the bridge shown. Assume that the origin is at the lowest point of the cables. (Lesson 10-2)

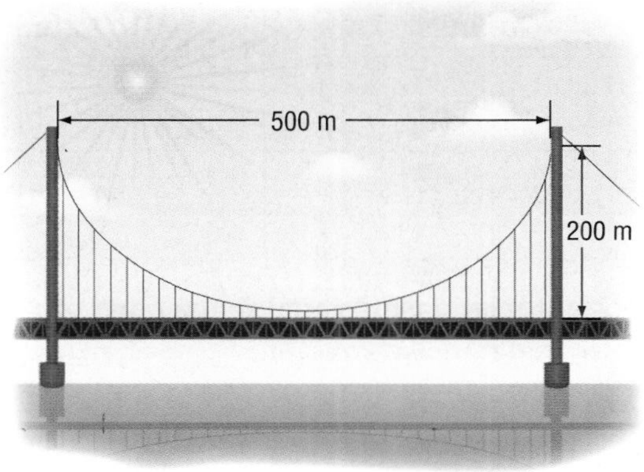

Identify the coordinates of the vertex and focus, the equation of the axis of symmetry and directrix, and the direction of opening of the parabola with the given equation. Then find the length of the latus rectum. (Lesson 10-2)

11. $y = x^2 + 6x + 5$

12. $x = -2y^2 + 4y + 1$

13. Find the center and radius of the circle with equation $(x - 1)^2 + y^2 = 9$. Then graph the circle. (Lesson 10-3)

14. Write an equation for a circle that has center at $(3, -2)$ and passes through $(3, 4)$. (Lesson 10-3)

15. Write an equation for a circle if the endpoints of a diameter are at $(8, 31)$ and $(32, 49)$. (Lesson 10-3)

16. MULTIPLE CHOICE What is the radius of the circle with equation $x^2 + 2x + y^2 + 14y + 35 = 0$? (Lesson 10-3)

A 2

B 4

C 8

D 16

Find the coordinates of the center and foci and the lengths of the major and minor axes of the ellipse with the given equation. Then graph the ellipse. (Lesson 10-4)

17. $\frac{(x + 4)^2}{16} + \frac{(y - 2)^2}{9} = 1$

18. $\frac{(x - 1)^2}{20} + \frac{(y + 2)^2}{4} = 1$

19. $4y^2 + 9x^2 + 16y - 90x + 205 = 0$

20. MULTIPLE CHOICE Which equation represents an ellipse with foci at $(-4, 12)$ and $(-4, -8)$ and endpoints of the major axis at $(-4, 6)$ and $(-4, -10)$? (Lesson 10-4)

F $\frac{(x - 2)^2}{36} + \frac{(y + 4)^2}{64} = 1$

G $\frac{(x + 4)^2}{64} + \frac{(y - 2)^2}{36} = 1$

H $\frac{(y - 2)^2}{64} + \frac{(x + 4)^2}{36} = 1$

J $\frac{(x - 2)^2}{64} + \frac{(y + 4)^2}{36} = 1$

10-5 Hyperbolas

Why?

Because Halley's Comet travels around the Sun in an elliptical path, it reappears in our sky. Other comets pass through our sky only once. Many of these comets travel in paths that resemble hyperbolas.

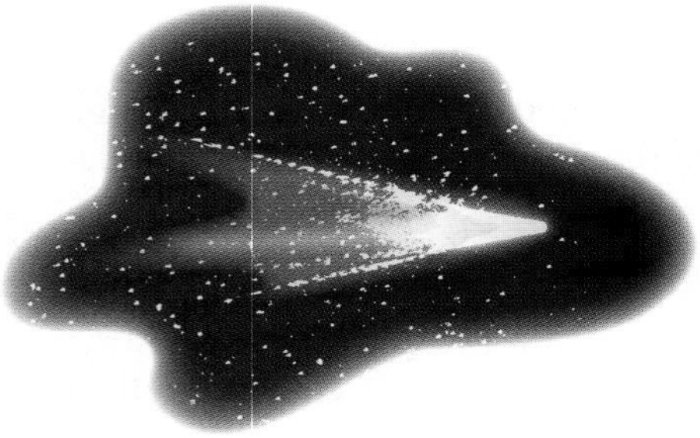

Then

You graphed and analyzed equations of ellipses. (Lesson 10-4)

Now

- Write equations of hyperbolas.
- Graph hyperbolas.

NYS Core Curriculum

A2.R.6 Use mathematics to show and understand physical phenomena (e.g., investigate sound waves using the sine and cosine functions)

New Vocabulary

hyperbola
transverse axis
conjugate axis
foci
vertices
co-vertices
constant difference

NY Math Online

glencoe.com

- Extra Examples
- Personal Tutor
- Self-Check Quiz
- Homework Help

Equations of Hyperbolas

Similar to an ellipse, a **hyperbola** is the set of all points in a plane such that the absolute value of the differences of the distances from the foci is constant.

Every hyperbola has two axes of symmetry, the **transverse axis** and the **conjugate axis**. The axes are perpendicular at the center of the hyperbola.

The **foci** of a hyperbola always lie on the transverse axis. The **vertices** are the endpoints of the transverse axis. The **co-vertices** are the endpoints of the conjugate axis.

As a hyperbola recedes from the center, both halves approach asymptotes.

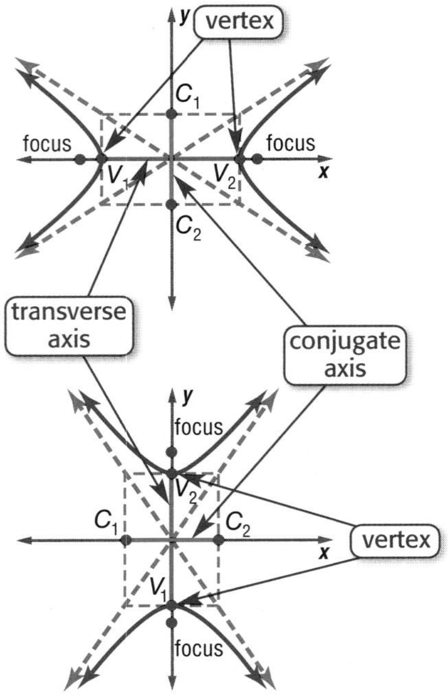

Key Concept

For Your FOLDABLE

Equations of Hyperbolas Centered at the Origin

Standard Form	$\frac{x^2}{a^2} - \frac{y^2}{b^2} = 1$	$\frac{y^2}{a^2} - \frac{x^2}{b^2} = 1$
Orientation	horizontal	vertical
Foci	$(\pm c, 0)$	$(1, \pm c)$
Length of Transverse Axis	$2a$ units	$2a$ units
Length of Conjugate Axis	$2b$ units	$2b$ units
Equations of Asymptotes	$y = \pm\frac{b}{a}x$	$y = \pm\frac{a}{b}x$

As with ellipses, there are several important relationships among the parts of hyperbolas.

- There are two axes of symmetry.
- The values of a, b, and c are related by the equation $c^2 = a^2 + b^2$.

EXAMPLE 1 Write an Equation Given Vertices and Foci

Write an equation for the hyperbola shown in the graph.

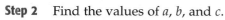

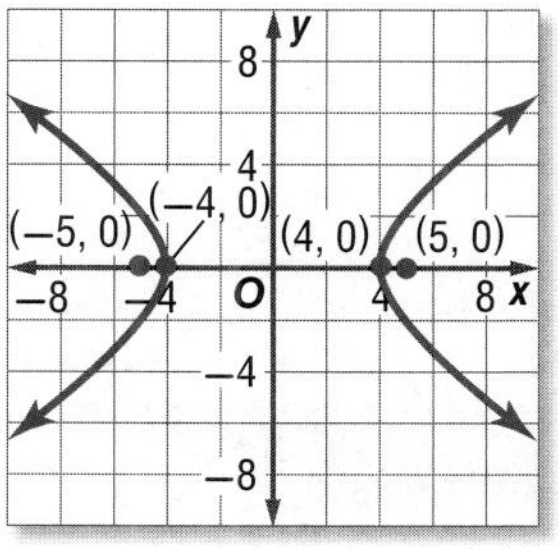

Step 1 Find the center.

The vertices are equidistant from the center.

The center is at (0, 0).

Step 2 Find the values of a, b, and c.

The value of a is the distance between a vertex and the center, or 4 units.

The value of c is the distance between a focus and the center, or 5 units.

$c^2 = a^2 + b^2$	**Equation relating a, b, and c for a hyperbola**
$5^2 = 4^2 + b^2$	**$c = 5$ and $a = 3$**
$9 = b^2$	**Subtract 4^2 from each side.**

Step 3 Write the equation.

The transverse axis is horizontal, so the equation is $\frac{x^2}{16} - \frac{y^2}{9} = 1$.

Check Your Progress

1. Write an equation for a hyperbola with vertices at (0, 6) and (0, −6) and foci at (0, 8) and (0, −8).

Personal Tutor glencoe.com

ReadingMath

Standard Form In the standard form of a hyperbola, the squared terms are subtracted. For an ellipse, they are added.

Hyperbolas can also be determined using the equations of their asymptotes.

EXAMPLE 2 Write an Equation Given Asymptotes

The asymptotes for a vertical hyperbola are $y = \frac{5}{3}x$ and $y = -\frac{5}{3}x$ and the vertices are at (0, 5) and (0, −5). Write the equation for the hyperbola.

Step 1 Find the center.

The vertices are equidistant from the center.

The center of the hyperbola is at (0, 0).

Step 2 Find the values of a and b.

The hyperbola is vertical, so $a = 5$.

From the asymptotes, $b = 3$.

The value of c is not needed.

Step 3 Write the equation.

The equation for the hyperbola is $\frac{y^2}{25} - \frac{x^2}{9} = 1$.

Check Your Progress

2. The asymptotes for a horizontal hyperbola are $y = \frac{7}{9}x$ and $y = -\frac{7}{9}x$. The vertices are (9, 0) and (−9, 0). Write an equation for the hyperbola.

Personal Tutor glencoe.com

Graphs of Hyperbolas Hyperbolas can be translated in the same manner as the other conic sections.

Key Concept — Equations of Hyperbolas Centered at (h, k)

For Your FOLDABLE

Standard Form	$\frac{(x-h)^2}{a^2} - \frac{(y-k)^2}{b^2} = 1$	$\frac{(y-k)^2}{a^2} - \frac{(x-h)^2}{b^2} = 1$
Orientation	horizontal	vertical
Foci	$(h \pm c, k)$	$(h, k \pm c)$
Vertices	$(h \pm a, k)$	$(h, k \pm a)$
Co-vertices	$(h, k \pm b)$	$(h \pm b, k)$
Equations of Asymptotes	$y - k = \pm\frac{b}{a}(x - h)$	$y - k = \pm\frac{a}{b}(x - h)$

StudyTip

Calculator You can graph a hyperbola on a graphing calculator by solving for y then graphing the two equations on the same screen.

EXAMPLE 3 Graph a Hyperbola

Graph $\frac{(x-3)^2}{4} - \frac{(y+2)^2}{16} = 1$. Identify the vertices, foci, and asymptotes.

Step 1 Find the center. The center is at (3, −2).

Step 2 Find a, b, and c. From the equation, $a^2 = 4$ and $b^2 = 16$, so $a = 2$ and $b = 4$.

$c^2 = a^2 + b^2$ — **Equation relating a, b, and c for a hyperbola**

$c^2 = 2^2 + 4^2$ — **$a = 2$, $b = 4$**

$c^2 = 20$ — **Simplify.**

$c = \sqrt{20}$ or about 4.47 — **Take the square root of each side.**

Step 3 Identify the vertices and foci. The hyperbola is horizontal and the vertices are 2 units from the center, so the vertices are at (1, −2) and (5, −2).
The foci are about 4.47 units from the center.
The foci are at (−1.47, −2) and (7.47, −2).

Step 4 Identify the asymptotes.

$y - k = \pm\frac{b}{a}(x - h)$ — **Equation for asymptotes of a horizontal hyperbola**

$y - (-2) = \pm\frac{4}{2}(x - 3)$ — **$a = 2$, $b = 4$, $h = 3$, and $k = -2$**

The equations for the asymptotes are $y = 2x - 8$ and $y = -2x + 4$.

Step 5 Graph the hyperbola. The hyperbola is symmetric about the transverse and conjugate axes. Use this symmetry to plot additional points for the hyperbola.

Use the asymptotes as a guide to draw the hyperbola that passes through the vertices and the other points.

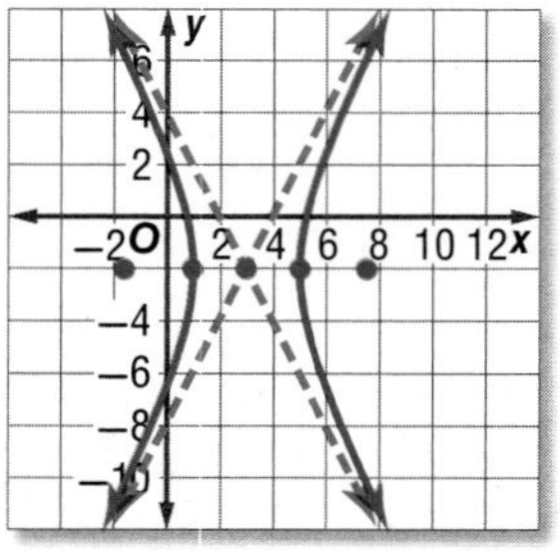

Check Your Progress

3. Graph $\frac{(y-4)^2}{9} - \frac{(x+3)^2}{25} = 1$. Identify the vertices, foci, and asymptotes.

Personal Tutor glencoe.com

In the equation for any hyperbola, the value of $2a$ represents the **constant difference**. This is the absolute value of the difference between the distances from any point on the hyperbola to the foci of the hyperbola.

Any point on the hyperbola at the right will have the same constant difference, $|y - x|$ or $2a$.

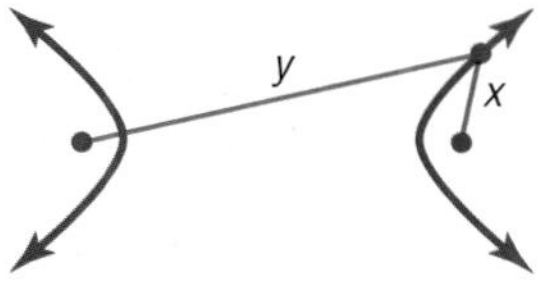

Real-World Link

Halley's Comet becomes visible to the unaided eye about every 76 years as it nears the Sun.

Source: NASA

Real-World EXAMPLE 4 Write an Equation of a Hyperbola

SPACE Earth and the Sun are 146 million kilometers apart. A comet follows a path that is one branch of a hyperbola. Suppose the comet is 30 million miles farther from the Sun than from Earth. Determine the equation of the hyperbola centered at the origin for the path of the comet.

Understand We need to determine the equation for the hyperbola.

Plan Find the center and the values of a and b. Once we have this information, we can determine the equation.

Solve The foci are Earth and the Sun, with the origin between them.

The value of c is $146 \div 2$ or 73.

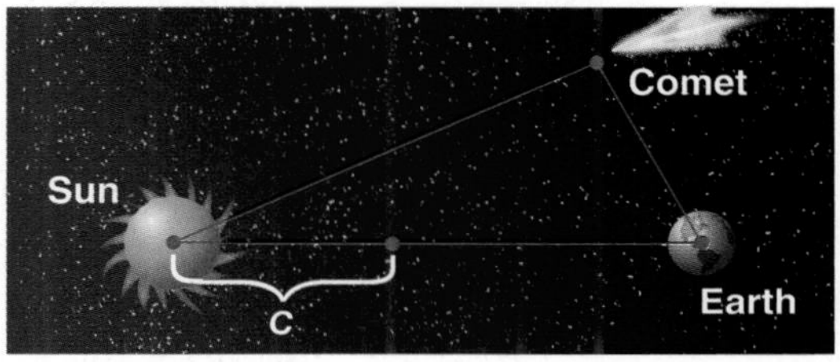

The difference of the distances from the comet to each body is 30. Therefore, a is $30 \div 2$ or 15 million miles.

$c^2 = a^2 + b^2$	**Equation relating a, b, and c for a hyperbola**
$73^2 = 15^2 + b^2$	**$a = 15$ and $c = 73$**
$5104 = b^2$	**Simplify.**

The equation of the hyperbola is $\frac{x^2}{225} - \frac{y^2}{5104} = 1$.

Since the comet is farther from the Sun, it is located on the branch of the hyperbola near Earth.

Check (21, 70) is a point that satisfies the equation.

The distance between this point and the Sun $(-73, 0)$ is $\sqrt{[21 - (-73)]^2 + (70 - 0)^2}$ or 117.2 million kilometers.

The distance between this point and Earth (73, 0) is $\sqrt{(21 - 73)^2 + (70 - 0)^2}$ or 87.2 million kilometers.

The difference between these distances is 30. ✓

StudyTip

Exact Locations A third receiving station is necessary to determine the plane's exact location.

Check Your Progress

4. **SEARCH AND RESCUE** Two receiving stations that are 150 miles apart receive a signal from a downed airplane. They determine that the airplane is 80 miles farther from station A than from station B. Determine the equation of the hyperbola centered at the origin on which the plane is located.

Personal Tutor glencoe.com

Check Your Understanding

Examples 1 and 2
p. 649

Write an equation for each hyperbola.

1.

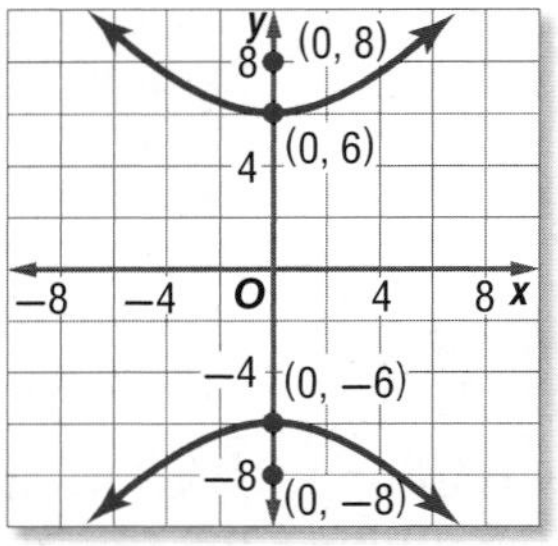

2.

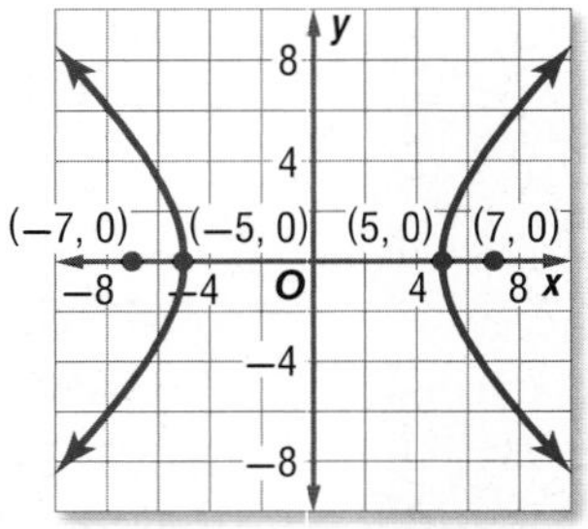

3.

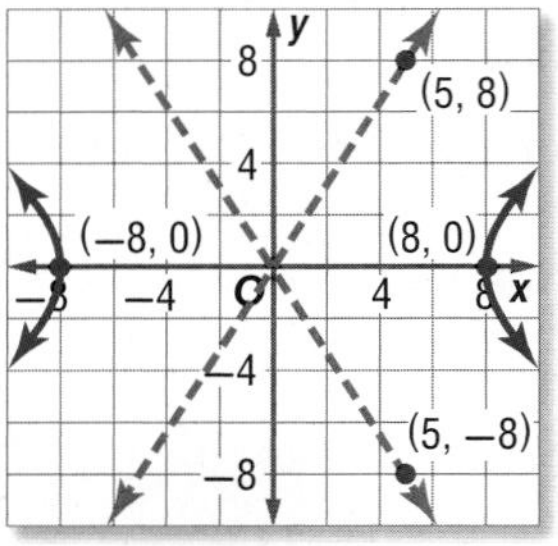

4.

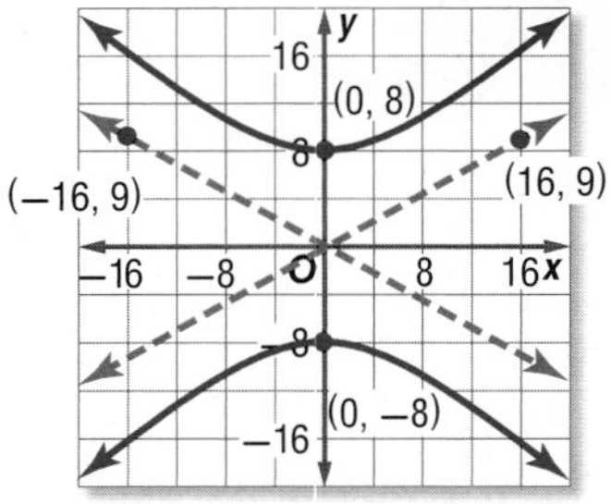

Example 3
p. 650

Graph each hyperbola. Identify the vertices, foci, and asymptotes.

5. $\frac{x^2}{64} - \frac{y^2}{49} = 1$

6. $\frac{y^2}{36} - \frac{x^2}{60} = 1$

7. $9y^2 + 18y - 16x^2 + 64x - 199 = 0$

8. $4x^2 + 24x - y^2 + 4y - 4 = 0$

Example 4
p. 651

9. **NAVIGATION** A ship determines that the difference of its distances from two stations is 60 nautical miles. Write an equation for a hyperbola on which the ship lies if the stations are at (−80, 0) and (80, 0).

Practice and Problem Solving

● = **Step-by-Step Solutions** begin on page R20.
Extra Practice begins on page 947.

Examples 1 and 2
p. 649

Write an equation for each hyperbola.

10.

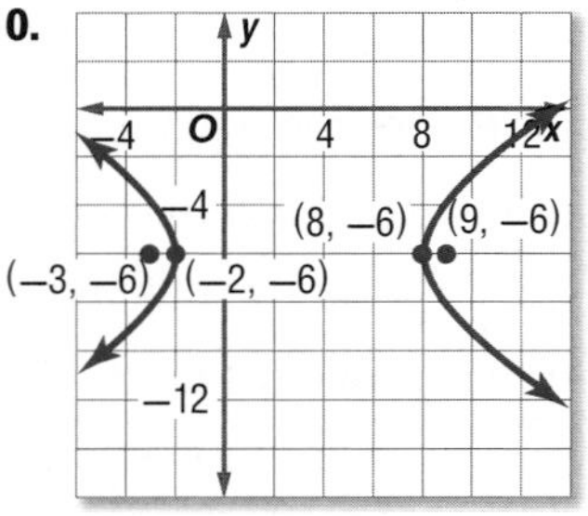

11

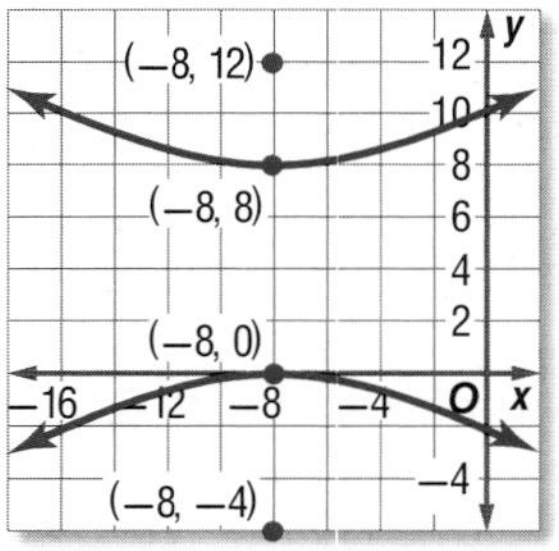

12.

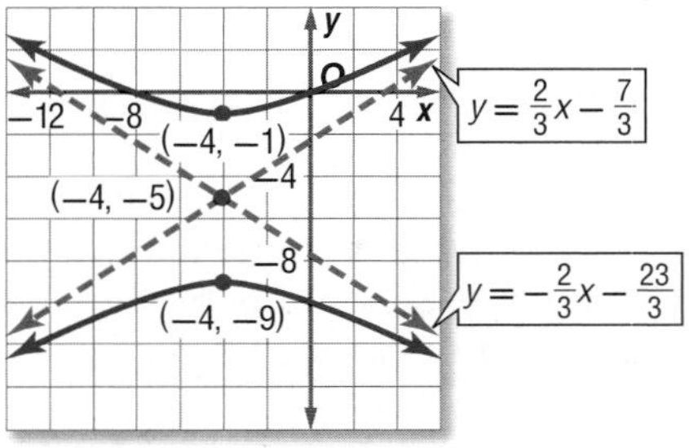

13.

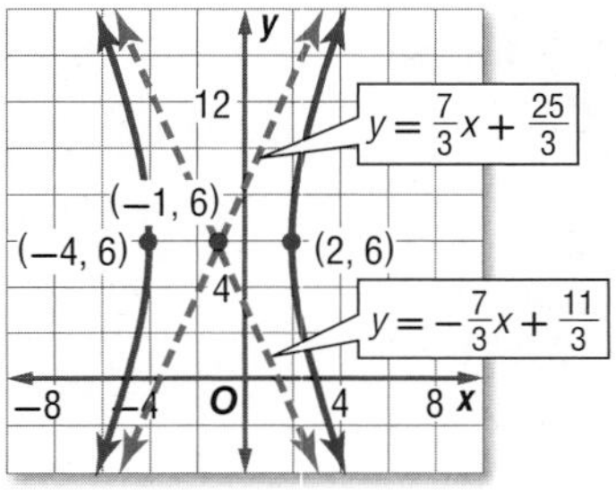

Example 3 p. 650

Graph each hyperbola. Identify the vertices, foci, and asymptotes.

14. $\frac{x^2}{36} - \frac{y^2}{4} = 1$

15. $\frac{y^2}{9} - \frac{x^2}{49} = 1$

16. $\frac{y^2}{36} - \frac{x^2}{25} = 1$

17. $\frac{x^2}{16} - \frac{y^2}{16} = 1$

18. $\frac{(x-3)^2}{16} - \frac{(y+1)^2}{4} = 1$

19. $\frac{(y+5)^2}{16} - \frac{(x+2)^2}{36} = 1$

20. $9y^2 - 4x^2 - 54y + 32x - 19 = 0$

21. $16x^2 - 9y^2 + 128x + 36y + 76 = 0$

22. $25x^2 - 4y^2 - 100x + 48y - 144 = 0$

23. $81y^2 - 16x^2 - 810y + 96x + 585 = 0$

Example 4 p. 651

24. NAVIGATION A ship determines that the difference of its distances from two stations is 80 nautical miles. Write an equation for a hyperbola on which the ship lies if the stations are at $(-100, 0)$ and $(100, 0)$.

Determine whether the following equations represent ellipses or hyperbolas.

25. $4x^2 = 5y^2 + 6$

26. $8x^2 - 2x = 8y - 3y^2$

27. $-5x^2 + 4x = 6y + 3y^2$

28. $7y - 2x^2 = 6x - 2y^2$

29. $6x - 7x^2 - 5y^2 = 2y$

30. $4x + 6y + 2x^2 = -3y^2$

31. SPACE Refer to the application at the beginning of the lesson. With the Sun as a focus and the center at the origin, a certain comet's path follows a branch of a hyperbola. If two of the coordinates of the path are (10, 0) and (30, 100) where the units are in millions of miles, determine the equation of the path.

32. COOLING Natural draft cooling towers are shaped like hyperbolas for more efficient cooling of power plants. The hyperbola in the tower at the right can be modeled by $\frac{x^2}{16} - \frac{y^2}{225} = 1$, where the units are in meters. Find the width of the tower at the top and at its narrowest point in the middle.

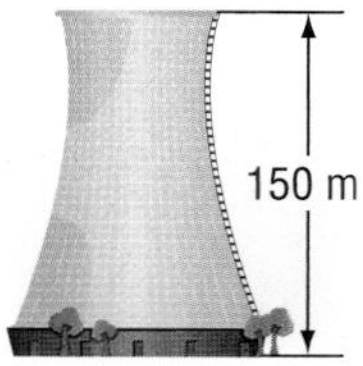

33. MULTIPLE REPRESENTATIONS Consider $xy = 16$.

a. **TABULAR** Make a table of values for the equation for $-12 \leq x \leq 12$.

b. **GRAPHICAL** Graph the hyperbola represented by the equation.

c. **LOGICAL** Determine and graph the asymptotes for the hyperbola.

d. **ANALYTICAL** What special property do you notice about the asymptotes? Hyperbolas that represent this property are called *rectangular hyperbolas*.

e. **ANALYTICAL** Without any calculations, what do you think will be the coordinates of the vertices for $xy = 25$? for $xy = 36$?

Real-World Link

Lightning strikes somewhere on the surface of Earth about 100 times every second.

Source: *National Geographic*

34. SEARCH AND RESCUE Two receiving stations that are 250 miles apart receive a signal from a downed airplane. They determine that the airplane is 70 miles farther from station *B* than from station *A*. Determine the equation of the horizontal hyperbola centered at the origin on which the plane is located.

35. WEATHER Luisa and Karl live exactly 3000 feet apart. While on the phone at their homes, Luisa hears thunder out of her window and Karl hears it 3 seconds later out of his. If sound travels 1100 feet per second, determine the equation for the horizontal hyperbola where the lightning is located.

Math History Link

Hypatia (415 B.C.–370 B.C.) Hypatia was a mathematician, scientist, and philosopher in Alexandria, Egypt. She is considered the first woman to write on mathematical topics. Hypatia edited the book *On the Conics of Apollonius*, adding her own problems and examples to clarify the topic for her students. This book developed the ideas of hyperbolas, parabolas, and ellipses.

36. ARCHITECTURE Large pillars with cross sections in the shape of hyperbolas were popular in ancient Greece. The curves can be modeled by the equation $\frac{x^2}{0.16} - \frac{y^2}{4} = 1$, where the units are in meters. If the pillars are 9 feet tall, find the width of the top of each pillar and the width of each pillar at the narrowest point in the middle. Round to the nearest hundredth of a foot.

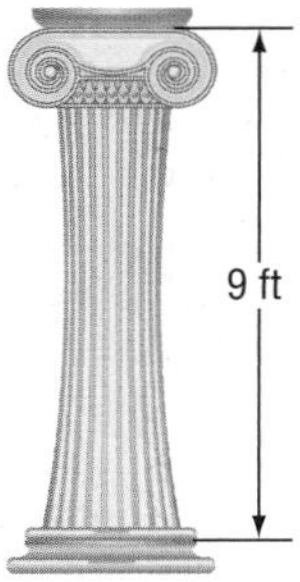

Write an equation for the hyperbola that satisfies each set of conditions.

37. vertices $(-8, 0)$ and $(8, 0)$, conjugate axis of length 20 units

38. vertices $(0, -6)$ and $(0, 6)$, conjugate axis of length 24 units

39. vertices $(6, -2)$ and $(-2, -2)$, foci $(10, -2)$ and $(-6, -2)$

40. vertices $(-3, 4)$ and $(-3, -8)$, foci $(-3, 9)$ and $(-3, -13)$

41. centered at the origin with a horizontal transverse axis of length 10 units and a conjugate axis of length 4 units

42. centered at the origin with a vertical transverse axis of length 16 units and a conjugate axis of length 12 units

43. TRIANGULATION While looking for their lost dog in the woods, Lae, Meg, and Cesar hear a bark. Meg hears it 2 seconds before Lae and Cesar hears it 3 seconds before Lae. With Lae at the origin, determine the exact location of their dog if sound travels 1100 feet per second.

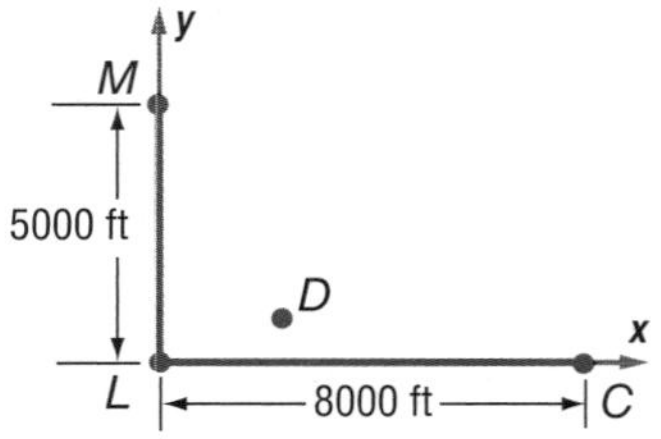

H.O.T. Problems

Use **H**igher-**O**rder **T**hinking Skills

44. FIND THE ERROR Simon and Gabriel are graphing $\frac{y^2}{25} - \frac{x^2}{4} = 1$. Is either of them correct? Explain your reasoning.

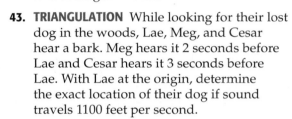

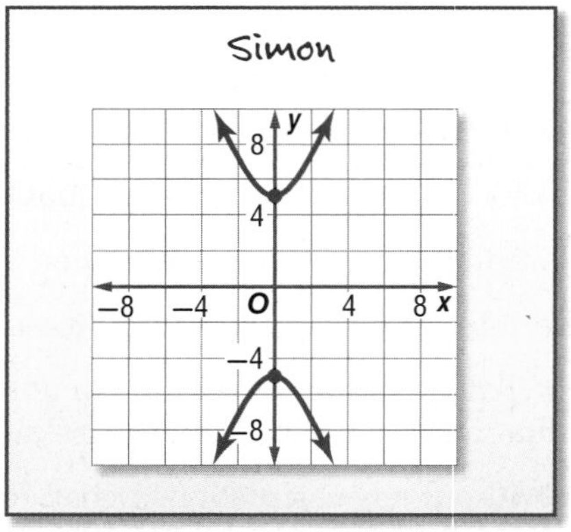

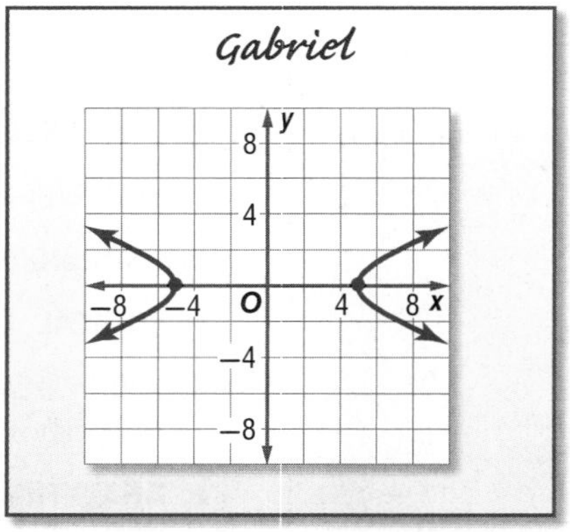

45. CHALLENGE The origin lies on a horizontal hyperbola. The asymptotes for the hyperbola are $y = -x + 1$ and $y = x - 5$. Find the equation for the hyperbola.

46. REASONING What happens to the location of the foci of a hyperbola as the value of a becomes increasingly smaller than the value of b? Explain your reasoning.

47. REASONING Consider $\frac{y^2}{36} - \frac{x^2}{16} = 1$. Describe the change in the shape of the hyperbola and the locations of the vertices and foci if 36 is changed to 9. Explain why this happens.

48. OPEN ENDED Write an equation for a hyperbola with a focus at the origin.

49. WRITING IN MATH Compare and contrast the characteristics of the equations and graphs of ellipses and hyperbolas.

NYSRE PRACTICE A2.R.6

50. You have 6 more dimes than quarters. You have a total of \$5.15. How many dimes do you have?

A 13
B 16
C 19
D 25

51. How tall is a tree that is 15 feet shorter than a pole three times as tall as the tree?

F 24.5 ft
G 22.5 ft
H 21.5 ft
J 7.5 ft

52. SHORT RESPONSE A rectangle is 8 feet long and 6 feet wide. If each dimension is increased by the same number of feet, the area of the new rectangle formed is 32 square feet more than the area of the original rectangle. By how many feet was each dimension increased?

53. ACT/SAT When the equation $y = 4x^2 - 5$ is graphed in the coordinate plane, the graph is which of the following?

A circle
B ellipse
C hyperbola
D parabola

Spiral Review

Write an equation for an ellipse that satisfies each set of conditions. (Lesson 10-4)

54. endpoints of major axis at (2, 2) and (2, −10), endpoints of minor axis at (0, −4) and (4, −4)

55. endpoints of major axis at (0, 10) and (0, −10), foci at (0, 8) and (0, −8)

Find the center and radius of the circle with the given equation. Then graph the circle. (Lesson 10-3)

56. $(x - 3)^2 + y^2 = 16$

57. $x^2 + y^2 - 6y - 16 = 0$

58. $x^2 + y^2 + 9x - 8y + 4 = 0$

59. BASKETBALL Zonta plays basketball for Centerville High School. So far this season, she has made 6 out of 10 free-throws. She is determined to improve her free-throw percentage. If she can make x consecutive free throws, her free-throw percentage can be determined using $P(x) = \frac{6 + x}{10 + x}$. (Lesson 9-4)

a. Graph the function.

b. What part of the graph is meaningful in the context of the problem?

c. Describe the meaning of the y-intercept.

d. What is the equation of the horizontal asymptote? Explain its meaning with respect to Zonta's shooting percentage.

Solve each equation. (Lesson 8-2)

60. $\left(\frac{1}{7}\right)^{y-3} = 343$

61. $10^{x-1} = 100^{2x-3}$

62. $36^{2p} = 216^{p-1}$

Graph each inequality. (Lesson 7-3)

63. $y \geq \sqrt{5x - 8}$

64. $y \geq \sqrt{x - 3} + 4$

65. $y < \sqrt{6x - 2} + 1$

Skills Review

66. Write an equation for a parabola with vertex at the origin that passes through (2, −8). (Lesson 5-7)

67. Write an equation for a parabola with vertex at (−3, −4) and y-intercept 8. (Lesson 5-7)

10-6 Identifying Conic Sections

Why?

Parabolas, circles, ellipses, and hyperbolas are called conic sections because they are the cross sections formed when a double cone is sliced by a plane.

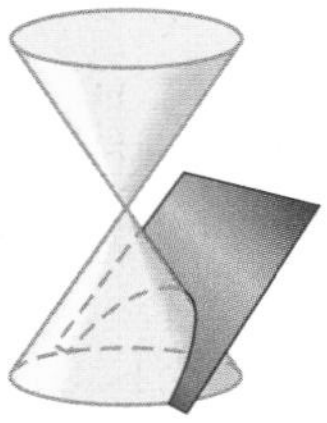

Parabola

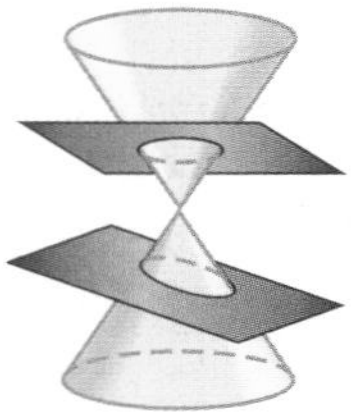

Circle and Ellipse

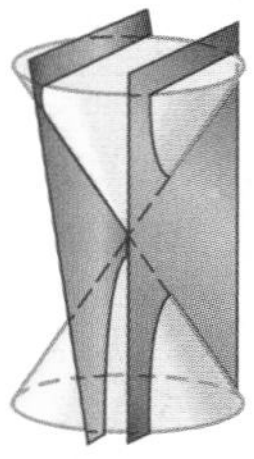

Hyperbola

Then

You analyzed different conic sections. (Lessons 10-2 through 10-5)

Now

- Write equations of conic sections in standard form.
- Identify conic sections from their equations.

NYS Core Curriculum

A2.CN.5 Understand how quantitative models connect to various physical models and representations

A2.R.1 Use physical objects, diagrams, charts, tables, graphs, symbols, equations, or objects created using technology as representations of mathematical concepts

NY Math Online

glencoe.com
- Extra Examples
- Personal Tutor
- Self-Check Quiz
- Homework Help

Conics in Standard Form The equation for any conic section can be written in the form $Ax^2 + Bxy + Cy^2 + Dx + Ey + F = 0$, where A, B, and C are not all zero. This general form can be converted to the standard forms below by completing the square.

Concept Summary — Standard Forms of Conic Sections (For Your FOLDABLE)

Conic Section	Standard Form of Equation	
Circle	$(x-h)^2 + (y-k) = r^2$	
	Horizontal Axis	**Vertical Axis**
Parabola	$y = a(x-h)^2 + k$	$x = a(y-k) + h$
Ellipse	$\frac{(x-h)^2}{a^2} + \frac{(y-k)^2}{b^2} = 1$	$\frac{(y-k)^2}{a^2} + \frac{(x-h)^2}{b^2} = 1$
Hyperbola	$\frac{(x-h)^2}{a^2} - \frac{(y-k)^2}{b^2} = 1$	$\frac{(y-k)^2}{a^2} - \frac{(x-h)^2}{b^2} = 1$

EXAMPLE 1 Rewrite an Equation of a Conic Section

Write $16x^2 - 25y^2 - 128x - 144 = 0$ in standard form. State whether the graph of the equation is a *parabola*, *circle*, *ellipse*, or *hyperbola*. Then graph the equation.

$16x^2 - 25y^2 - 128x - 144 = 0$ — **Original equation**

$16(x^2 - 8x + \blacksquare) - 25y^2 = 144 + 16(\blacksquare)$ — **Isolate terms.**

$16(x^2 - 8x + 16) - 25y^2 = 144 + 16(16)$ — **Complete the square.**

$16(x-4)^2 - 25y^2 = 400$ — **Perfect square**

$\frac{(x-4)^2}{25} - \frac{y^2}{16} = 1$ — **Divide each side by 400.**

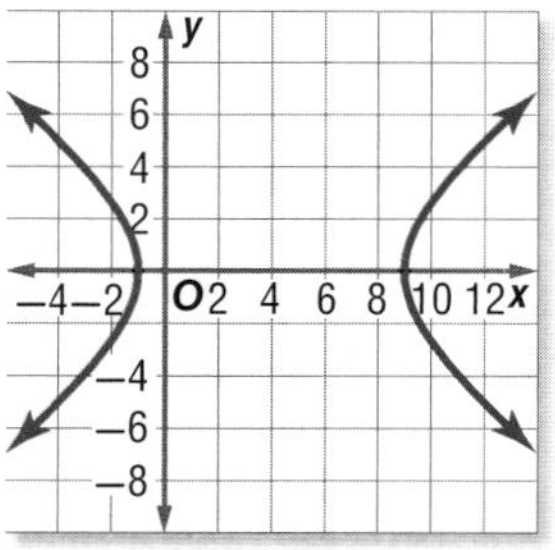

The graph is a hyperbola with its center at (4, 0).

Check Your Progress

1. Write $4x^2 + y^2 - 16x + 8y - 4 = 0$ in standard form. State whether the graph of the equation is a *parabola*, *circle*, *ellipse*, or *hyperbola*. Then graph the equation.

Personal Tutor glencoe.com

Review Vocabulary

discriminant the expression $b^2 - 4ac$ from the Quadratic Formula (Lesson 5-6)

Identify Conic Sections You can determine the type of conic without having to write $Ax^2 + Bxy + Cy^2 + Dx + Ey + F = 0$ in standard form. When there is an xy-term ($B \neq 0$), you can use the discriminant to identify the conic. $B^2 - 4AC$ is the discriminant of $Ax^2 + Bxy + Cy^2 + Dx + Ey + F = 0$.

Key Concept: Classify Conics with the Discriminant

For Your FOLDABLE

Discriminant	Conic Section
$B^2 - 4AC < 0$; $B = 0$ and $A = C$	circle
$B^2 - 4AC < 0$; either $B \neq 0$ or $A \neq C$	ellipse
$B^2 - 4AC = 0$	parabola
$B^2 - 4AC > 0$	hyperbola

StudyTip

Identifying Conics When there is no xy-term ($B = 0$), use A and C.
Parabola: A or $C = 0$ but not both.
Circle: $A = C$
Ellipse: A and C have the same sign but are not equal.
Hyperbola: A and C have opposite signs.

When $B = 0$, the conic will be either vertical or horizontal. When $B \neq 0$, the conic will be neither vertical nor horizontal.

Horizontal Ellipse: $B = 0$

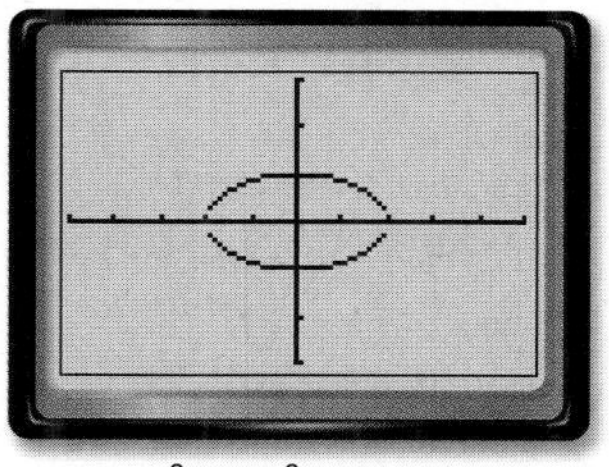

$x^2 + 4y^2 - 4 = 0$

Rotated Ellipse: $B \neq 0$

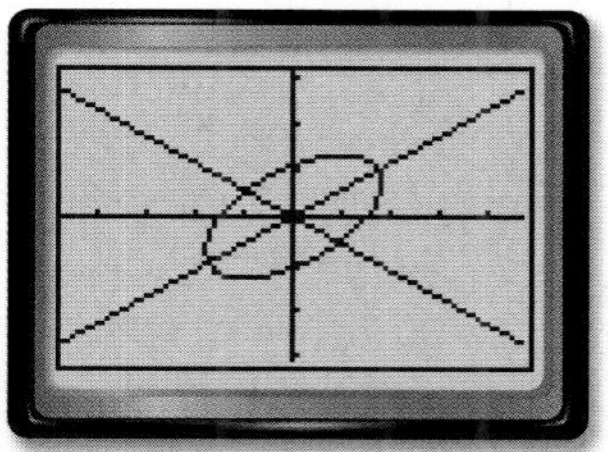

$7x^2 - 6\sqrt{3}xy + 13y^2 - 16 = 0$

EXAMPLE 2 Analyze an Equation of a Conic Section

Without writing in standard form, state whether the graph of each equation is a *parabola, circle, ellipse,* or *hyperbola.*

a. $y^2 + 4x^2 - 3xy + 4x - 5y - 8 = 0$

$A = 4$, $B = -3$, and $C = 1$
The discriminant is $(-3)^2 - 4(4)(1)$ or -7.
Because the discriminant is less than 0 and $B \neq 0$, the conic is an ellipse.

b. $3x^2 - 6x + 4y - 5y^2 + 2xy - 4 = 0$

$A = 3$, $B = 2$, and $C = -5$
The discriminant is $2^2 - 4(3)(-5)$ or 64.
Because the discriminant is greater than 0, the conic is a hyperbola.

c. $4y^2 - 8x + 6y - 14 = 0$

$A = 0$, $B = 0$, and $C = 4$
The discriminant is $0^2 - 4(0)(4)$ or 0.
Because the discriminant equals 0, the conic is a parabola.

Check Your Progress

2A. $8y^2 - 6x^2 + 4xy - 6x + 2y - 4 = 0$

2B. $3xy + 4x^2 - 2y + 9x - 3 = 0$

2C. $3x^2 + 16x - 12y + 2y^2 - 6 = 0$

Personal Tutor glencoe.com

Check Your Understanding

Example 1 p. 656

Write each equation in standard form. State whether the graph of the equation is a *parabola, circle, ellipse,* or *hyperbola*. Then graph the equation.

1. $x^2 + 4y^2 - 6x + 16y - 11 = 0$

2. $x^2 + y^2 + 12x - 8y + 36 = 0$

3. $9y^2 - 16x^2 - 18y - 64x - 199 = 0$

4. $6y^2 - 24y + 28 - x = 0$

Example 2 p. 657

Without writing in standard form, state whether the graph of each equation is a *parabola, circle, ellipse,* or *hyperbola*.

5. $4x^2 + 6y^2 - 3x - 2y = 12$

6. $5y^2 = 2x + 6y - 8 + 3x^2$

7. $8x^2 + 8y^2 + 16x + 24 = 0$

8. $4x^2 - 6y = 8x + 2$

9. $4x^2 - 3y^2 + 8xy - 12 = 2x + 4y$

10. $5xy - 3x^2 + 6y^2 + 12y = 18$

11. $8x^2 + 12xy + 16y^2 + 4y - 3x = 12$

12. $16xy + 8x^2 + 8y^2 - 18x + 8y = 13$

13. **AVIATION** A military jet performs for an air show. The path of the plane during one trick can be modeled by a conic section with equation $24x^2 + 1000y - 31{,}680x - 45{,}600 = 0$, where distances are represented in feet.

a. Identify the shape of the curved path of the jet. Write the equation in standard form.

b. If the jet begins its path upward or ascent at (0, 0), what is the horizontal distance traveled by the jet from the beginning of the ascent to the end of the descent?

c. What is the maximum height of the jet?

Practice and Problem Solving

● = **Step-by-Step Solutions** begin on page R20.
Extra Practice begins on page 947.

Example 1 p. 656

Write each equation in standard form. State whether the graph of the equation is a *parabola, circle, ellipse,* or *hyperbola*. Then graph the equation.

14. $3x^2 - 2y^2 + 18x + 8y - 35 = 0$

15. $3x^2 + 24x + 4y^2 - 40y + 52 = 0$

16. $x^2 + y^2 = 16 + 6y$

17. $32x + 28 = y - 8x^2$

18. $7x^2 - 8y = 84x - 2y^2 - 176$

19. $x^2 + 8y = 11 + 6x - y^2$

20. $4y^2 = 24y - x - 31$

21. $112y + 64x = 488 + 7y^2 - 8x^2$

22. $28x^2 + 9y^2 - 188 = 56x - 36y$

23. $25x^2 + 384y - 64y^2 + 200x = 1776$

Example 2 p. 657

Without writing in standard form, state whether the graph of each equation is a *parabola, circle, ellipse,* or *hyperbola*.

24. $4x^2 - 5y = 9x - 12$

25. $4x^2 - 12x = 18y - 4y^2$

26. $9x^2 + 12y = 9y^2 + 18y - 16$

27. $18x^2 - 16y = 12x - 4y^2 + 19$

28. $12y^2 - 4xy + 9x^2 = 18x - 124$

29. $5xy + 12x^2 - 16x = 5y + 3y^2 + 18$

30. $19x^2 + 14y = 6x - 19y^2 - 88$

31. $8x^2 + 20xy + 18 = 4y^2 - 12 + 9x$

32. $5x - 12xy + 6x^2 = 8y^2 - 24y - 9$

33. $18x - 24y + 324xy = 27x^2 + 3y^2 - 5$

34. **LIGHT** A lamp standing near a wall throws an arc of light in the shape of a conic section. Suppose the edge of the light can be represented by the equation $3y^2 - 2y - 4x^2 + 2x - 8 = 0$. Identify the shape of the edge of the light and graph the equation.

Match each graph with its corresponding equation.

35. 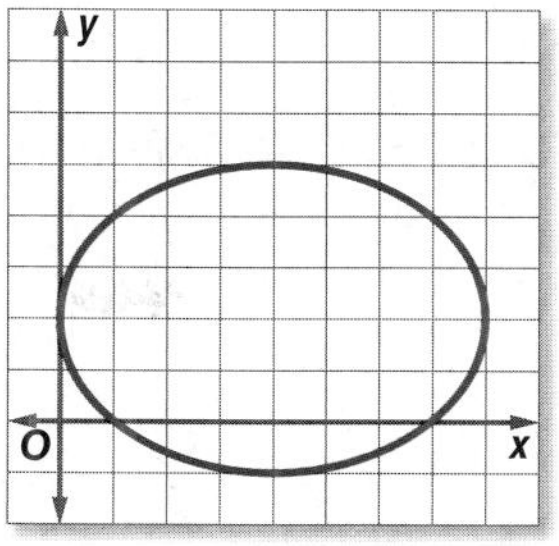

36. 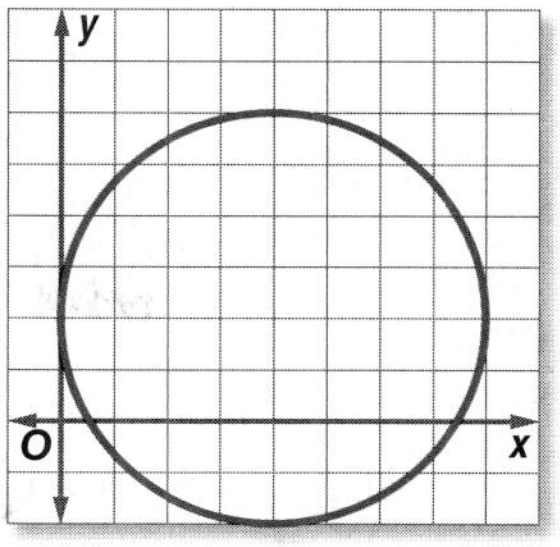

37. 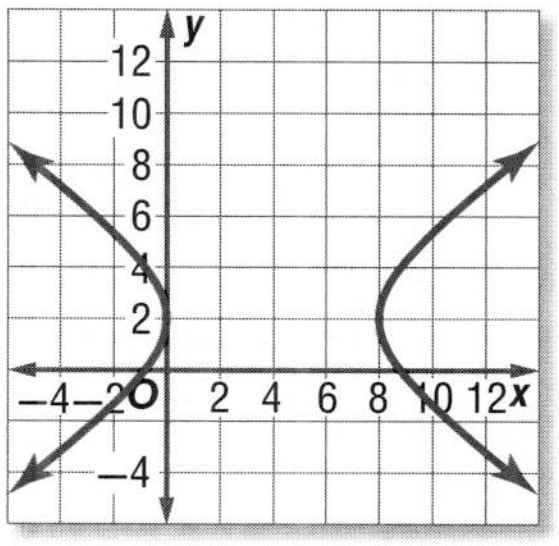

a. $x^2 + y^2 - 8x - 4y = -4$ **b.** $9x^2 - 16y^2 - 72x + 64y = 64$ **c.** $9x^2 + 16y^2 = 72x + 64y - 64$

For Exercises 38–41, match each situation with an equation that could be used to represent it.

a. $47.25x^2 - 9y^2 + 18y + 33.525 = 0$

b. $25x^2 + 100y^2 - 1900x - 2200y + 45{,}700 = 0$

c. $16x^2 - 90x + y - 0.25 = 0$

d. $x^2 + y^2 - 18x - 30y - 14{,}094 = 0$

38. COMPUTERS the boundary of a wireless network with a range of 120 feet

39. FITNESS the oval path of your foot on an exercise machine

40. COMMUNICATIONS the position of a cell phone between two cell towers

41. SPORTS the height of a football above the ground after being kicked

42. ENGINEERING The shape of the cables in a suspension bridge is approximately parabolic. If the towers for a planned bridge are 1000 meters apart and the lowest point of the suspension cables is 200 meters, write the equation in standard form with the origin at the vertex.

43. MULTIPLE REPRESENTATIONS Consider an ellipse with center $(3, -2)$, vertex $M(-1, -2)$, and co-vertex $N(3, -4)$.

a. ANALYTICAL Determine the standard form of the ellipse.

b. ALGEBRAIC Convert part **a** to $Ax^2 + Bxy + Cy^2 + Dx + Ey + F = 0$ form.

c. GRAPHICAL Graph the ellipse.

d. ANALYTICAL If the ellipse is rotated such that M is moved to $(3, -6)$, determine the location of N and the angle of rotation.

Real-World Link

The Golden Gate Bridge is an example of a suspension bridge. It took more than four years to complete and opened in 1937.

Source: Golden Gate Bridge Highway and Transportation District

H.O.T. Problems

Use **H**igher-**O**rder **T**hinking Skills

44. CHALLENGE When a plane passes through the vertex of a cone, a *degenerate* conic is formed.

a. Determine the type of conic represented by $4x^2 + 8y^2 = 0$.

b. Graph the conic.

c. Describe the difference between this degenerate conic and a standard conic.

45. REASONING Determine whether the following statement is *sometimes, always,* or *never* true. Explain your reasoning.

When a conic is vertical and $A = C$, it is a circle.

46. OPEN ENDED Write an equation of the form $Ax^2 + Bxy + Cy^2 + Dx + Ey + F = 0$, where $A = 9C$, that represents a parabola.

47. WRITING IN MATH Compare and contrast the graphs of the four types of conics and their corresponding equations.

NYSRE PRACTICE A2.CN.5, A2.R.1

48. ACT/SAT A class of 25 students took a science test. Ten students had a mean score of 80. The other students had an average score of 60. What is the average score of the whole class?

A 66
B 68
C 70
D 72

49. Six times a number minus 11 is 43. What is the number?

F 12
G 11
H 10
J 9

50. EXTENDED RESPONSE The amount of water remaining in a storage tank as it is drained can be represented by the equation $L = -4t^2 - 10t + 130$, where L represents the number of liters of water remaining and t represents the number of minutes since the drain was opened. How many liters of water were in the tank initially? Determine to the nearest tenth of a minute how long it will take for the tank to drain completely.

51. Ruben has a square piece of paper with sides 4 inches long. He rolled up the paper to form a cylinder. What is the volume of the cylinder?

A $\frac{4}{\pi}$
C 4π
B $\frac{16}{\pi}$
D 16π

Spiral Review

52. ASTRONOMY Suppose a comet's path can be modeled by a branch of the hyperbola with equation $\frac{y^2}{225} - \frac{x^2}{400} = 1$. Find the coordinates of the vertices and foci and the equations of the asymptotes for the hyperbola. Then graph the hyperbola. (Lesson 10-5)

Find the coordinates of the center and foci and the lengths of the major and minor axes for the ellipse with the given equation. Then graph the ellipse. (Lesson 10-4)

53. $\frac{y^2}{18} + \frac{x^2}{9} = 1$

54. $4x^2 + 8y^2 = 32$

55. $x^2 + 25y^2 - 8x + 100y + 91 = 0$

Graph each function. (Lesson 9-3)

56. $f(x) = \frac{3}{x}$

57. $f(x) = \frac{-2}{x + 5}$

58. $f(x) = \frac{6}{x - 2} - 4$

59. SPACE A radioisotope is used as a power source for a satellite. The power output P (in watts) is given by $P = 50e^{-\frac{t}{250}}$, where t is the time in days. (Lesson 8-8)

a. Is the formula for power output an example of exponential *growth* or *decay*? Explain your reasoning.

b. Find the power available after 100 days.

c. Ten watts of power are required to operate the equipment in the satellite. How long can the satellite continue to operate?

Skills Review

Solve each system of equations. (Lesson 3-2)

60. $6g - 8h = 50$
$6h = 22 - 4g$

61. $3u + 5v = 6$
$2u - 4v = -7$

62. $10m - 9n = 15$
$5m - 4n = 10$

EXTEND
10-6 Graphing Technology Lab
Identifying and Graphing Conic Sections

NY Math Online glencoe.com
Other Calculator Keystrokes

Objective
Graph conic sections using a graphing calculator.

You can use a TI-83/84 Plus application to make graphing conics on your graphing calculator without having to solve for y.

EXAMPLE Identify and Graph a Conic

Write $x^2 + y^2 + 4x - 6y = -4$ in standard form. State whether the graph of the equation is a *parabola, circle, ellipse,* or *hyperbola*. Then graph the equation.

Step 1 Write in standard form.

$x^2 + y^2 + 4x - 6y = -4$	**Original equation**
$(x^2 + 4x + \blacksquare) + (y^2 - 6y + \blacksquare) = -4$	**Isolate terms.**
$(x^2 + 4x + 4) + (y^2 - 6y + 9) = -4 + 4 + 9$	**Complete the square.**
$(x + 2)^2 + (y - 3)^2 = 9$	**Write as perfect squares.**

Step 2 Identify the conic.

The equation is in the form $(x - h)^2 + (y - k)^2 = r^2$, so the conic is a circle.

Step 3 Graph the equation.

Use the Conics application to graph the circle.

KEYSTROKES: Press [APPS]. Then use [▼] to select Conics and press [ENTER]. Press 1 to graph a circle, then press 1 to graph an equation of the form $(x - h)^2 + (y - k)^2 = r^2$.

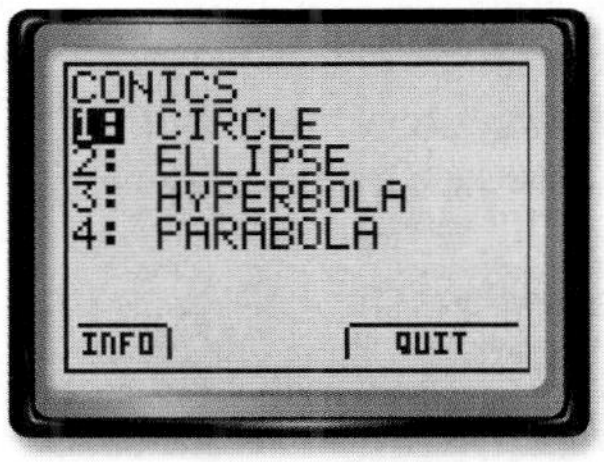

Enter the values for H, K, and R: press [(–)] 2 [ENTER] 3 [ENTER] 3 [ENTER] [GRAPH].

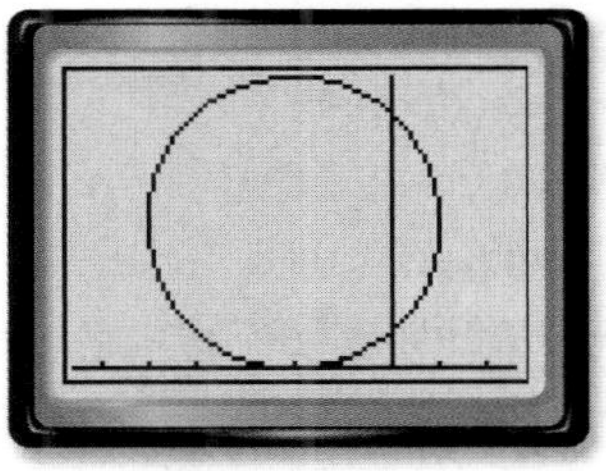

Exercises

Write each equation in standard form. State whether the graph of the equation is a *parabola, circle, ellipse,* or *hyperbola*. Then graph the equation.

1. $x^2 - y^2 + 8x = 16$
2. $y^2 - 2x^2 - 16 = 0$
3. $x^2 + 2y^2 = 2x + 8$
4. $x^2 - 8y + y^2 + 11 = 0$
5. $9y^2 + 18y = 25x^2 + 216$
6. $x^2 + 4y^2 + 2x - 24y + 33 = 0$
7. $3x^2 + 4y^2 + 8y = 8$
8. $x^2 + 4y^2 - 11 = 2(4y - x)$
9. $6x^2 - 24x - 5y^2 - 10y - 11 = 0$
10. $25y^2 + 9x^2 - 50y - 54x = 119$

10-7 Solving Linear-Nonlinear Systems

Then
You solved systems of linear equations. (Lessons 3-1 and 3-2)

Now
- Solve systems of linear and nonlinear equations algebraically and graphically.
- Solve systems of linear and nonlinear inequalities graphically.

NYS Core Curriculum

A2.A.3 Solve systems of equations involving one linear equation and one quadratic equation algebraically

NY Math Online

glencoe.com
- Extra Examples
- Personal Tutor
- Self-Check Quiz
- Homework Help
- Math in Motion

Why?

Ever wonder how law enforcement agencies can track a cell phone user's location? A person using a cell phone can be located in respect to three cellular towers. The respective coordinates and distances each tower is from the caller are used to pinpoint the caller's location. This is accomplished using a system of quadratic equations.

Systems of Equations When a system of equations consists of a linear and a nonlinear equation, the system may have zero, one, or two solutions. Some of the possible solutions are shown below.

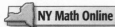

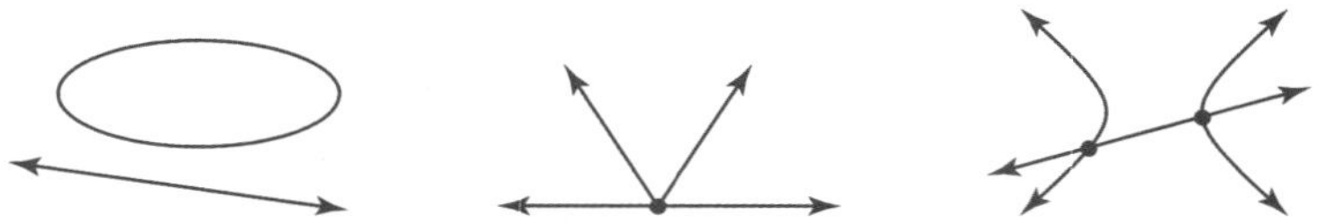

You can solve linear-quadratic systems by using graphical or algebraic methods. One way is to first solve for a variable in the linear equation and then substitute into the quadratic equation.

EXAMPLE 1 Linear-Quadratic System

Solve the system of equations.

$9x^2 + 25y^2 = 225$ **(1)**

$10y + 6x = 6$ **(2)**

Step 1 Solve the linear equation for y.

$10y + 6x = 6$	**Equation (2)**
$y = -0.6x + 0.6$	**Solve for *y*.**

Step 2 Substitute into the quadratic equation and solve for x.

$9x^2 + 25y^2 = 225$	**Quadratic equation**
$9x^2 + 25(-0.6x + 0.6)^2 = 225$	**Substitute $-0.6x + 0.6$ for *y*.**
$9x^2 + 25(0.36x^2 - 0.72x + 0.36) = 225$	**Simplify.**
$9x^2 + 9x^2 - 18x + 9 = 225$	**Distribute.**
$18x^2 - 18x - 216 = 0$	**Simplify.**
$x^2 - x - 12 = 0$	**Divide each side by 18.**
$(x - 4)(x + 3) = 0$	**Factor.**
$x = 4$ or -3	**Zero Product Property**

Step 3 Substitute x-values into the linear equation and solve for y.

$y = -0.6x + 0.6$	**Equation (2)**	$y = -0.6x + 0.6$
$= -0.6(4) + 0.6$	**Substitute the *x*-values.**	$= -0.6(-3) + 0.6$
$= -1.8$	**Simplify.**	$= 2.4$

The solutions of the system are $(4, -1.8)$ and $(-3, 2.4)$.

Check Your Progress

Solve each system of equations.

1A. $3y + x^2 - 4x - 17 = 0$
$3y - 10x + 38 = 0$

1B. $3(y - 4) - 2(x - 3) = -6$
$5x^2 + 2y^2 - 53 = 0$

Personal Tutor glencoe.com

Math *in Motion*, Interactive Lab glencoe.com

If a quadratic system contains two conic sections, the system may have anywhere from zero to four solutions. Some graphical representations are shown below.

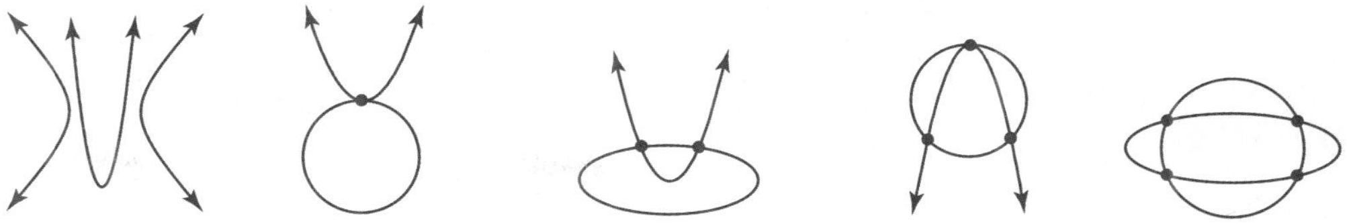

You can use elimination to solve quadratic-quadratic systems.

EXAMPLE 2 Quadratic-Quadratic System

Solve the system of equations.

$x^2 + y^2 = 45$ **(1)**

$y^2 - x^2 = 27$ **(2)**

$y^2 + x^2 = 45$	**Equation (1), Commutative Property**
$(+)\ y^2 - x^2 = 27$	**Equation (2)**
$2y^2 = 72$	**Add.**
$y^2 = 36$	**Divide each side by 2.**
$y = \pm 6$	**Take the square root of each side.**

Substitute 6 and −6 into one of the original equations and solve for x.

$x^2 + y^2 = 45$	**Equation (1)**	$x^2 + y^2 = 45$
$x^2 + 6^2 = 45$	**Substitute for *y*.**	$x^2 + (-6)^2 = 45$
$x^2 = 9$	**Subtract 36 from each side.**	$x^2 = 9$
$x = \pm 3$	**Take the square root of each side.**	$x = \pm 3$

The solutions are (−3, −6), (−3, 6), (3, −6), and (3, 6).

Check Your Progress **Solve each system of equations.**

2A. $x^2 + y^2 = 8$
$x^2 + 3y = 10$

2B. $3x^2 + 4y^2 = 48$
$2x^2 - y^2 = -1$

Personal Tutor glencoe.com

StudyTip

Graphing Calculators If you use ZSquare on the ZOOM menu, the graph of the first equation will look like a circle.

Systems of Inequalities Systems of quadratic inequalities can be solved by graphing.

EXAMPLE 3 Quadratic Inequalities

Solve the system of inequalities by graphing.

$\mathbf{x^2 + y^2 \le 49}$

$\mathbf{x^2 - 4y^2 > 16}$

The intersection of the graphs, shaded green, represents the solution of the system.

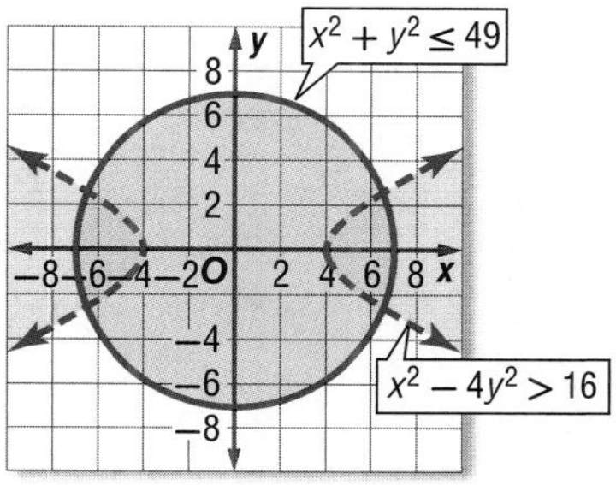

CHECK (6, 0) is in the shaded area. Use this point to check your solution.

$x^2 + y^2 \le 49$	$x^2 - 4y^2 > 16$
$6^2 + 0^2 \overset{?}{\le} 49$	$62 - 4(0)^2 \overset{?}{>} 16$
$36 \le 49$ ✓	$36 > 16$ ✓

Check Your Progress

Solve each system of inequalities by graphing.

3A. $5x^2 + 2y^2 \le 10$
$y \ge x^2 - 2x + 1$

3B. $x^2 - y^2 \le 8$
$x^2 + y^2 \ge 120$

Personal Tutor glencoe.com

Systems involving absolute value can also be solved by graphing.

EXAMPLE 4 Quadratics with Absolute Value

Solve the system of inequalities by graphing.

$y \geq |2x - 4|$

$y \leq -x^2 + 4x + 2$

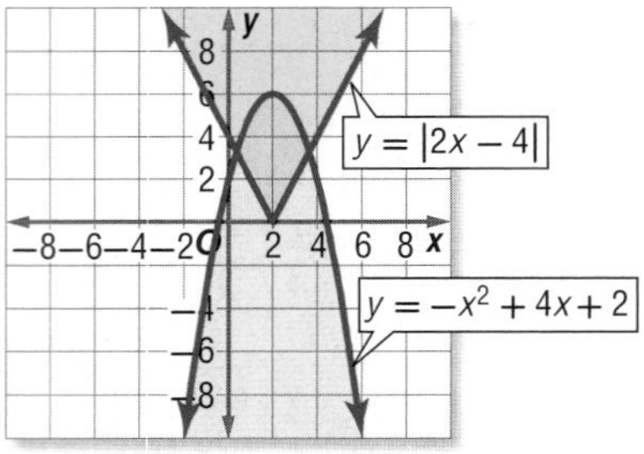

Graph the boundary equations. Then shade appropriately.

The intersection of the graphs, shaded green, represents the solution to the system.

CHECK (2, 4) is in the shaded area. Use the point to check your solution.

$y \geq |2x - 4|$	$y \leq -x^2 + 4x + 2$
$4 \overset{?}{\geq} |2(2) - 4|$	$4 \overset{?}{\leq} -(2)^2 + 4(2) + 2$
$4 \geq 0$ ✓	$4 \leq 6$ ✓

StudyTip

Graphing Calculator Like linear inequalities, systems of quadratic and absolute value inequalities can be checked with a graphing calculator.

Check Your Progress

Solve each system of inequalities by graphing.

4A. $y > |-0.5x + 2|$

$\frac{x^2}{16} + \frac{y^2}{36} \leq 1$

4B. $x^2 + y^2 \leq 49$

$y \geq |x^2 + 1|$

Personal Tutor glencoe.com

Check Your Understanding

Examples 1 and 2 pp. 662–663

Solve each system of equations.

1. $8y = -10x$
$y^2 = 2x^2 - 7$

2. $x^2 + y^2 = 68$
$5y = -3x + 34$

3. $y = 12x - 30$
$4x^2 - 3y = 18$

4. $6y^2 - 27 = 3x$
$6y - x = 13$

5. $x^2 + y^2 = 16$
$x^2 - y^2 = 20$

6. $y^2 - 2x^2 = 8$
$3y^2 + x^2 = 52$

7. $x^2 + 2y = 7$
$y^2 - x^2 = 8$

8. $4y^2 - 3x^2 = 11$
$3y^2 + 2x^2 = 21$

9. CELL PHONES Refer to the beginning of the lesson. A person using a cell phone can be located in respect to three cellular towers. In a coordinate system where one unit represents one mile, the location of the caller is determined to be 50 miles from the tower at the origin. The person is also 40 miles from a tower at (0, 30) and 13 miles from a tower at (35, 18). Where is the caller?

Examples 3 and 4 pp. 663–664

Solve each system of inequalities by graphing.

10. $6x^2 + 9(y - 2)^2 \leq 36$
$x^2 + (y + 3)^2 \leq 25$

11. $16x^2 + 4y^2 \leq 64$
$y \geq -x^2 + 2$

12. $4x^2 - 8y^2 \geq 32$
$y \geq |1.5x| - 8$

13. $x^2 + 8y^2 < 32$
$y < -|x - 2| + 2$

Practice and Problem Solving

● = **Step-by-Step Solutions** begin on page R20.
Extra Practice begins on page 947.

Examples 1 and 2
pp. 662–663

Solve each system of equations.

14. $3x^2 - 2y^2 = -24$
$2y = -3x$

15. $5x^2 + 4y^2 = 20$
$5y = 7x + 35$

16. $x^2 + 3x = -4y - 2$
$y = -2x + 1$

17. $y = 2x$
$4x^2 - 2y^2 = -36$

18. $2y = x + 10$
$y^2 - 4y = 5x + 10$

19. $9y = 8x - 19$
$8x + 11 = 2y^2 + 5y$

20. $2y^2 + 5x^2 = 26$
$2x^2 - y^2 = 5$

21. $x^2 + y^2 = 16$
$x^2 - 4x + y^2 = 12$

22. $x^2 + y^2 = 8$
$5y^2 = 3x^2$

23. $y^2 - x^2 + 3y = 26$
$x^2 + 2y^2 = 34$

24. $x^2 - y^2 = 25$
$x^2 + y^2 + 7 = 0$

25. $x^2 - 10x + 2y^2 = 47$
$y^2 - 2x^2 = -14$

26. FIREWORKS Two fireworks are set off simultaneously but from different altitudes. The height y in feet of one is represented by $y = -16t^2 + 120t + 10$, where t is the time in seconds. The height of the other is represented by $y = -16t^2 + 60t + 310$.

a. After how many seconds are the fireworks the same height?

b. What is that height?

Examples 3 and 4
pp. 663–664

Solve each system of inequalities by graphing.

27. $x^2 + y^2 \geq 36$
$x^2 + 9(y + 6)^2 \leq 36$

28. $-x > y^2$
$4x^2 + 14y^2 \leq 56$

29. $12x^2 - 4y^2 \geq 48$
$16(x - 4)^2 + 25y^2 < 400$

30. $8y^2 - 3x^2 \leq 24$
$2y > x^2 - 8x + 14$

31. $y > x^2 - 6x + 8$
$x \geq y^2 - 6y + 8$

32. $x^2 + y^2 \geq 9$
$25x^2 + 64y^2 \leq 1600$

33. $16(x - 3)^2 + 4y^2 \leq 64$
$y \leq -|x - 2| + 2$

34. $x^2 - 4x + y^2 + 6y = 23$
$y > |x - 2| - 6$

35. $2y - 4 \geq |x + 4|$
$12 - 2y > x^2 + 12x + 36$

36. $18y^2 - 3x^2 \leq 54$
$y \geq |2x| - 6$

37. $x^2 + y^2 < 16$
$y \geq |x - 2| + 6$

38. $x^2 > y - 2$
$y \leq |x + 8| - 4$

39. SPACE Two satellites are placed in orbit about Earth. The equations of the two orbits are $\frac{x^2}{(300)^2} + \frac{y^2}{(900)^2} = 1$ and $\frac{x^2}{(600)^2} + \frac{y^2}{(690)^2} = 1$, where distances are in kilometers and Earth is the center of each curve.

a. Solve each equation for y.

b. Use a graphing calculator to estimate the intersection points of the two orbits.

c. Compare the orbits of the two satellites.

Real-World Link

The first electronic monitoring device was introduced in 1983.

Source: Federal Probation

40. PETS Taci's dog was missing one day. Fortunately, he was wearing an electronic monitoring device. If the dog is 10 units from the tree, 13 units from the tower, and 20 units from the house, determine the coordinates of his location.

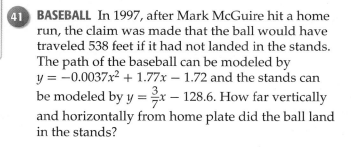

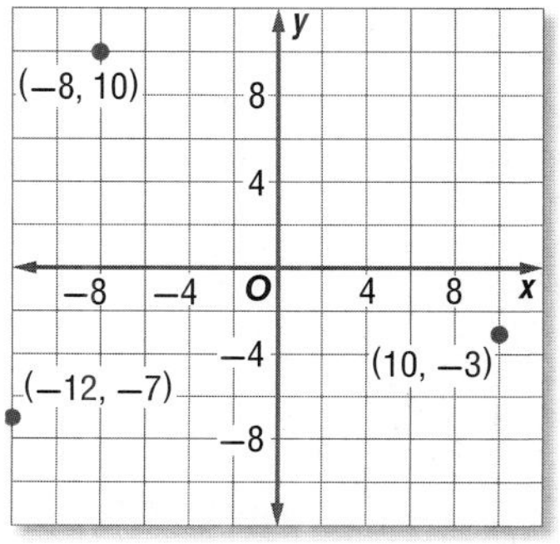

41. BASEBALL In 1997, after Mark McGuire hit a home run, the claim was made that the ball would have traveled 538 feet if it had not landed in the stands. The path of the baseball can be modeled by $y = -0.0037x^2 + 1.77x - 1.72$ and the stands can be modeled by $y = \frac{3}{7}x - 128.6$. How far vertically and horizontally from home plate did the ball land in the stands?

42. ADVERTISING The corporate logo for an automobile manufacturer is shown at the right. Write a system of three equations to model this logo.

Write a system of equations that satisfies each condition.

43. a circle and an ellipse that intersect at one point

44. a parabola and an ellipse that intersect at two points

45. a hyperbola and a circle that do not intersect

46. an ellipse and a parabola that intersect at three points

47. an ellipse and a hyperbola that intersect at four points

48. ECONOMICS Prices are often set on an equilibrium curve, where the supply of a certain product equals its corresponding demand by consumers. An economist represents the supply of a product with $y = p^2 + 10p$ and the corresponding demand with $-p^2 + 40p$, where p is the price. Determine the equilibrium price.

Real-World Link

With over 10.3 million participants, paintball is one of America's most popular extreme sports.

Source: *World and Regional Paintball Information Guide*

49. PAINTBALL The shape of a paintball field is modeled by $x^2 + 4y^2 = 10{,}000$ in yards where the center is at the origin. The teams are provided with short-range walkie-talkies with a maximum range of 80 yards. Are the teams capable of hearing each other anywhere on the field? Explain your reasoning graphically.

50. MULTIPLE REPRESENTATIONS

a. GRAPHICAL Sketch separate graphs of linear, quadratic, cubic, and quartic equations. Draw a straight line through each graph to maximize the number of intersections (solutions).

b. ANALYTICAL Use your graphs to fill in the first column of the table below.

Review Vocabulary

quartic equation a fourth-degree polynomial equation (Lesson 6-3)

Maximum Number of Solutions for each System

Type	linear	quadratic	cubic
linear			
quadratic			
cubic			
quartic			

c. GRAPHICAL Use a parabola and the original four graphs in part **a** to maximize the number of intersections with a quadratic.

d. ANALYTICAL Complete the table.

e. VERBAL What can you conclude about the relationship between a system of two equations and the greatest number of finite solutions for the system?

H.O.T. Problems

Use Higher-Order Thinking Skills

51. CHALLENGE Find all values of k for which the following system of equations has two solutions.

$$\frac{x^2}{a^2} + \frac{y^2}{b^2} = 1 \qquad x^2 + y^2 = k^2$$

52. REASONING When the vertex of a parabola lies on an ellipse, how many solutions can the quadratic system represented by the two graphs have? Explain your reasoning using graphs.

53. OPEN ENDED Write a system of equations, one a hyperbola and the other an ellipse, for which a solution is $(-4, 8)$.

54. WRITING IN MATH Explain how sketching the graph of a quadratic system can help you solve it.

NYSRE PRACTICE A2.A.3

55. SHORT RESPONSE Solve.

$$4y - 3y = 0$$
$$x^2 + y^2 = 25$$

56. You have 16 stamps. Some are postcard stamps that cost \$0.23, and the rest cost \$0.41. If you spent a total of \$5.30 on the stamps, how many postcard stamps do you have?

A 7

B 8

C 9

D 10

57. Ms. Talbot received a promotion and a 7.2% raise. Her new salary is \$53,600 a year. What was her salary before the raise?

F \$50,000

G \$53,600

H \$55,000

J \$57,500

58. ACT/SAT When a number is multiplied by $\frac{2}{3}$, the result is 188. Find the number.

A 292

B 282

C 272

D 262

Spiral Review

Match each equation with the situation that it could represent. (Lesson 10-6)

a. $9x^2 + 4y^2 - 36 = 0$

b. $0.004x^2 - x + y - 3 = 0$

c. $x^2 + y^2 - 20x + 30y - 75 = 0$

59. SPORTS the flight of a baseball

60. PHOTOGRAPHY the oval opening in a picture frame

61. GEOGRAPHY the set of all points 20 miles from a landmark

Find the coordinates of the vertices and foci and the equations of the asymptotes for the hyperbola with the given equation. Then graph the hyperbola. (Lesson 10-5)

62. $\frac{y^2}{16} - \frac{x^2}{25} = 1$

63. $\frac{(y-3)^2}{25} - \frac{(x-2)^2}{16} = 1$

64. $6y^2 = 2x^2 + 12$

Simplify each expression. (Lesson 9-1)

65. $\frac{12p^2 + 6p - 6}{4(p+1)^2} \div \frac{6p - 3}{2p + 10}$

66. $\frac{x^2 + 6x + 9}{x^2 + 7x + 6} \div \frac{4x + 12}{3x + 3}$

67. $\frac{r^2 + 2r - 8}{r^2 + 4r + 3} \div \frac{r - 2}{3r + 3}$

Graph each function. State the domain and range. (Lesson 8-1)

68. $f(x) = -\left(\frac{1}{5}\right)^x$

69. $y = -2.5(5)^x$

70. $f(x) = 2\left(\frac{1}{3}\right)^x$

Skills Review

Solve each equation or formula for the specified variable. (Lesson 1-3)

71. $d = rt$, for r

72. $x = \frac{-b}{2a}$, for a

73. $V = \frac{1}{3}\pi r^2 h$, for h

74. $A = \frac{1}{2}h(a + b)$, for b

CHAPTER 10 Study Guide and Review

NY Math Online glencoe.com
- STUDY TO GO
- Vocabulary Review

Chapter Summary

Key Concepts

Midpoint and Distance Formulas (Lesson 10-1)

- $M = \left(\frac{x_1 + x_2}{2}, \frac{y_1 + y_2}{2}\right)$
- $d = \sqrt{(x_2 - x_1)^2 + (y_2 - y_1)^2}$

Circles (Lesson 10-2)

- The equation of a circle with center (h, k) and radius r can be written in the form $(x - h)^2 + (y - k)^2 = r^2$.

Parabolas (Lesson 10-3)

- Standard Form: $y = a(x - h)^2 + k$
  $x = a(y - k)^2 + h$

Ellipses (Lesson 10-4)

- Standard Form: horizontal $\frac{(x - h)^2}{a^2} + \frac{(y - k)^2}{b^2} = 1$
  vertical $\frac{(y - k)^2}{a^2} + \frac{(x - h)^2}{b^2} = 1$

Hyperbolas (Lesson 10-5)

- Standard Form: horizontal $\frac{(x - h)^2}{a^2} - \frac{(y - k)^2}{b^2} = 1$
  vertical $\frac{(y - k)^2}{a^2} - \frac{(x - h)^2}{b^2} = 1$

Solving Quadratic Systems (Lesson 10-7)

- Systems of quadratic equations can be solved using substitution and elimination.
- A system of quadratic equations can have zero, one, two, three, or four solutions.

FOLDABLES® Study Organizer

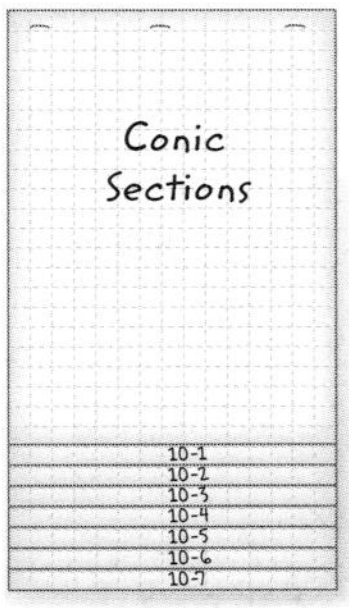

Be sure the Key Concepts are noted in your Foldable.

Key Vocabulary

center (of a circle) (p. 631)
center (of an ellipse) (p. 639)
circle (p. 631)
conjugate axis (p. 648)
constant difference (p. 651)
constant sum (p. 640)
co-vertices (of a hyperbola) (p. 648)
co-vertices (of an ellipse) (p. 639)
directrix (p. 623)
ellipse (p. 639)
foci (of a hyperbola) (p. 648)
foci (of an ellipse) (p. 639)
focus (p. 623)
hyperbola (p. 648)
latus rectum (p. 623)
major axis (p. 639)
minor axis (p. 639)
parabola (p. 623)
transverse axis (p. 648)
vertices (of a hyperbola) (p. 648)
vertices (of an ellipse) (p. 639)

Vocabulary Check

State whether each sentence is *true* or *false*. If *false*, replace the underlined term to make a true sentence.

1. The set of all points in a plane that are equidistant from a given point in a plane, called the focus, forms a circle.
2. A(n) ellipse is the set of all points in a plane such that the sum of the distances from the two fixed points is constant.
3. The endpoints of the major axis of an ellipse are the foci of the ellipse.
4. The radius is the distance from the center of a circle to any point on the circle.
5. The line segment through the focus of a parabola and perpendicular to the axis of symmetry is called the latus rectum.
6. Every hyperbola has two axes of symmetry, the transverse axis and the major axis.
7. A directrix is the set of all points in a plane that are equidistant from a given point in the plane, called the center.
8. A hyperbola is the set of all points in a plane such that the absolute value of the sum of the distances from any point on the hyperbola to two given points is constant.
9. A parabola can be defined as the set of all points in a plane that are the same distance from the focus and a given line called the directrix.
10. The major axis is the longer of the two axes of symmetry of an ellipse.

Lesson-by-Lesson Review

10-1 Midpoint and Distance Formulas (pp. 617–622)

G.G.69

Find the midpoint of the line segment with endpoints at the given coordinates.

11. $(-8, 6), (3, 4)$

12. $(-6, 0), (-1, 4)$

13. $\left(\frac{3}{4}, \frac{2}{3}\right), \left(-\frac{1}{3}, \frac{1}{4}\right)$

14. $(15, 20), (18, 21)$

Find the distance between each pair of points with the given coordinates.

15. $(10, -3), (1, -5)$

16. $(0, 6), (-9, 7)$

17. $\left(\frac{1}{4}, \frac{1}{2}\right), \left(\frac{3}{2}, \frac{5}{4}\right)$

18. $(5, -3), (7, -1)$

19. HIKING Marc wants to hike from his camp to a waterfall. The waterfall is 5 miles south and 8 miles east of his campsite.

a. How far away is the waterfall?

b. Marc wants to stop for lunch halfway to the waterfall. Where should he stop?

EXAMPLE 1

Find the midpoint of a line segment whose endpoints are at (−4, 8) and (10, −1).

Let $(x_2, y_2) = (-4, 8)$ and $(x_2, y_2) = (10, -1)$.

$$\left(\frac{x_1 + x_2}{2}, \frac{y_1 + y_2}{2}\right) = \left(\frac{-4 + 10}{2}, \frac{8 + (-1)}{2}\right)$$

$$= \left(\frac{6}{2}, \frac{7}{2}\right) \text{ or } \left(3, \frac{7}{2}\right)$$

EXAMPLE 2

Find the distance between $P(5, -3)$ and $Q(-1, 5)$.
Let $(x_1, y_1) = (5, -3)$ and $(x_2, y_2) = (-1, 5)$.

$d = \sqrt{(x_2 - x_1)^2 + (y_2 - y_1)^2}$ — **Distance Formula**

$= \sqrt{(-1 - 5)^2 + [5 - (-3)]^2}$ — **Substitute.**

$= \sqrt{36 + 64}$ — **Subtract.**

$= \sqrt{100}$ or 10 units — **Simplify.**

10-2 Parabolas (pp. 623–629)

A2.R.1

Graph each equation.

20. $y = 3x^2 + 24x - 10$

21. $3y - x^2 = 8x - 11$

22. $x = \frac{1}{2}y^2 - 4y + 3$

23. $x = y^2 - 14y + 25$

Write each equation in standard form. Identify the vertex, axis of symmetry, and direction of opening of the parabola.

24. $y = -\frac{1}{2}x^2$

25. $y = 4x^2 - 16x + 9$

26. $x - 6y = y^2 + 4$

27. $x = y^2 + 14y + 20$

28. SPORTS When a football is kicked, the path it travels is shaped like a parabola. Suppose a football is kicked from ground level, reaches a maximum height of 50 feet, and lands 200 feet away. Assuming the football was kicked at the origin, write an equation of the parabola that models the flight of the football.

EXAMPLE 3

Write each equation in standard form. Identify the vertex, axis of symmetry, and direction of opening of $3y - x^2 = 4x + 7$.

Write the equation in the form $y = a(x - h)^2 + k$ by completing the square.

$3y = x^2 + 4x + 7$ — **Isolate the terms with x.**

$3y = (x^2 + 4x + ■) + 7 - ■$ — **Complete the square.**

$3y = (x^2 + 4x + 4) + 7 - 4$ — $\left(\frac{4}{2}\right)^2 = 4$

$3y = (x + 2)^2 + 3$ — $(x^2 + 4x + 4) = (x + 2)$

$y = \frac{1}{3}(x + 2)^2 + 1$ — **Divide each side by 3.**

Vertex: $(-2, 1)$; axis of symmetry: $x = -2$; direction of opening: upward since $a > 0$.

Study Guide and Review

10-3 Circles (pp. 631–637)

A2.A.48, A2.A.49

Write an equation for the circle that satisfies each set of conditions.

29. center (−1, 6), radius 3 units

30. endpoints of a diameter (2, 5) and (0, 0)

31. endpoints of a diameter (4, −2) and (−2, −6)

Find the center and radius of each circle. Then graph the circle.

32. $(x + 5)^2 + y^2 = 9$

33. $(x - 3)^2 + (y + 1)^2 = 25$

34. $(x + 2)^2 + (y - 8)^2 = 1$

35. $x^2 + 4x + y^2 - 2y - 11 = 0$

36. SOUND A loudspeaker in a school is located at the point (65, 40). The speaker can be heard in a circle with a radius of 100 feet. Write an equation to represent the possible boundary of the loudspeaker sound.

EXAMPLE 4

Find the center and radius of the circle with equation $x^2 - 2x + y^2 + 6y + 6 = 0$. Then graph the circle.

Complete the squares.

$$x^2 - 2x + y^2 + 6y + 6 = 0$$
$$(x^2 - 2x + ■) + (y^2 + 6y + ■) = -6 + ■ + ■$$
$$(x^2 - 2x + 1) + (y^2 + 6y + 9) = -6 + 1 + 9$$
$$(x - 1)^2 + (y + 3)^2 = 4$$

The center of the circle is at (1, −3) and the radius is 2.

10-4 Ellipses (pp. 639–646)

A2.R.6

Find the coordinates of the center and foci and the lengths of the major and minor axes for the ellipse with the given equation. Then graph the ellipse.

37. $\frac{x^2}{9} + \frac{y^2}{36} = 1$

38. $\frac{y^2}{10} + \frac{x^2}{5} = 1$

39. $\frac{x^2}{36} + \frac{(y - 4)^2}{4} = 1$

40. $27x^2 + 9y^2 = 81$

41. $\frac{(x + 1)^2}{25} + \frac{(y - 2)^2}{16} = 1$

42. $9x^2 + 4y^2 + 54x - 8y + 49 = 0$

43. $9x^2 + 25y^2 - 18x + 50y - 191 = 0$

44. $7x^2 + 3y^2 - 28x - 12y = -19$

45. LANDSCAPING The Martins have a garden in their front yard that is shaped like an ellipse. The major axis is 16 feet and the minor axis is 10 feet. Write an equation to model the garden. Assume the origin is at the center of the garden.

EXAMPLE 5

Find the coordinates of the center and foci and the lengths of the major and minor axes for the ellipse with equation $9x^2 + 16y^2 - 54x + 32y - 47 = 0$. Then graph the ellipse.

First, convert to standard form.

$$9x^2 + 16y^2 - 54x + 32y - 47 = 0$$
$$9(x^2 - 6x + ■) + 16(y^2 + 2y + ■) = 47 + 9(■) + 16(■)$$
$$9(x^2 - 6x + 9) + 16(y^2 + 2y + 1) = 47 + 9(9) + 16(1)$$
$$9(x - 3)^2 + 16(y + 1)^2 = 144$$
$$\frac{(x - 3)^2}{16} + \frac{(y + 1)^2}{9} = 1$$

The center of the ellipse is (3, −1). The ellipse is horizontal. $a^2 = 16$, so $a = 4$. $b^2 = 9$, so $b = 3$. The length of the major axis is 2 • 4 or 8. The length of the minor axis is 2 • 3 or 6. To find the foci: $c^2 = 16 - 9$ or 7, so $c = \sqrt{7}$. The foci are $(3 + \sqrt{7}, -1)$ and $(3 - \sqrt{7}, -1)$.

MIXED PROBLEM SOLVING
For mixed problem-solving practice, see page 988.

10-5 Hyperbolas (pp. 648–655)

A2.R.6

Graph each hyperbola. Identify the vertices, foci, and asymptotes.

46. $\frac{y^2}{9} - \frac{x^2}{4} = 1$

47. $\frac{(x-3)^2}{1} - \frac{(y+2)^2}{4} = 1$

48. $\frac{(y+1)^2}{16} - \frac{(x-4)^2}{9} = 1$

49. $4x^2 - 9y^2 = 36$

50. $9y^2 - x^2 - 4x + 18y + 4 = 0$

51. **MIRRORS** A hyperbolic mirror is shaped like one branch of a hyperbola. It reflects light rays directed at one focus toward the other focus. Suppose a hyperbolic mirror is modeled by the upper branch of the hyperbola $\frac{y^2}{9} - \frac{x^2}{16} = 1$. A light source is located at $(-10, 0)$. Where should the light hit the mirror so that the light will be reflected to $(0, -5)$?

EXAMPLE 6

Graph $9x^2 - 4y^2 - 36x - 8y - 4 = 0$. Identify the vertices, foci, and asymptotes.

Complete the square.

$$9x^2 - 4y^2 - 36x - 8y - 4 = 0$$
$$9(x^2 - 4x + \blacksquare) - 4(y^2 + 2y + \blacksquare) = 4 + 9(\blacksquare) - 4(\blacksquare)$$
$$9(x^2 - 4x + 4) - 4(y^2 + 2y + 1) = 4 + 9(4) - 4(1)$$
$$9(x-2)^2 - 4(y+1)^2 = 36$$
$$\frac{(x-2)^2}{4} - \frac{(y+1)^2}{9} = 1$$

The center is at $(2, -1)$. The vertices are at $(0, -1)$ and $(4, -1)$. The foci are at $\left(2 + \sqrt{13}, -1\right)$ and $\left(2 - \sqrt{13}, -1\right)$. The equations of the asymptotes are $y + 1 = \pm\frac{3}{2}(x - 2)$.

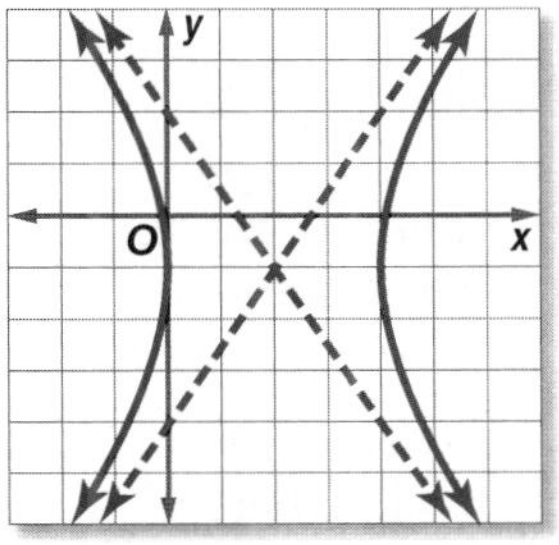

10-6 Identifying Conic Sections (pp. 656–660)

A2.CN.5, A2.R.1

Write each equation in standard form. State whether the graph of the equation is a *parabola, circle, ellipse,* or *hyperbola*. Then graph.

52. $3x^2 + 12x - y + 8 = 0$

53. $9x^2 + 16y^2 = 144$

54. $x^2 + y^2 - 8x - 2y + 8 = 0$

55. $-9x^2 + y^2 + 36x - 45 = 0$

Without writing the equation in standard form, state whether the graph of the equation is a *parabola, circle, ellipse,* or *hyperbola*.

56. $7x^2 + 9y^2 = 63$

57. $5y^2 + 2y + 4x - 13x^2 = 81$

58. $x^2 - 8x + 16 = 6y$

59. $x^2 + 4x + y^2 - 285 = 0$

60. **LIGHT** Suppose the edge of a shadow can be represented by the equation $16x^2 + 25y^2 - 32x - 100y - 284 = 0$.

a. What is the shape of the shadow?

b. Graph the equation.

EXAMPLE 7

Write $3x^2 + 3y^2 - 12x + 30y + 39 = 0$ in standard form. State whether the graph of the equation is a *parabola, circle, ellipse,* or *hyperbola*. Then graph the equation.

$$3x^2 + 3y^2 - 12x + 30y + 39 = 0$$
$$3(x^2 - 4x + \blacksquare) + 3(y^2 + 10y + \blacksquare) = -39 + 3(\blacksquare) + 3(\blacksquare)$$
$$3(x^2 - 4x + 4) + 3(y^2 + 10y + 25) = -39 + 3(4) + 3(25)$$
$$3(x-2)^2 + 3(y+5)^2 = 48$$
$$(x-2)^2 + (y+5)^2 = 16$$

In this equation $A = 3$ and $C = 3$. Since A and C are both positive and $A = C$, the graph is a circle. The center is at $(2, -5)$ and the radius is 4.

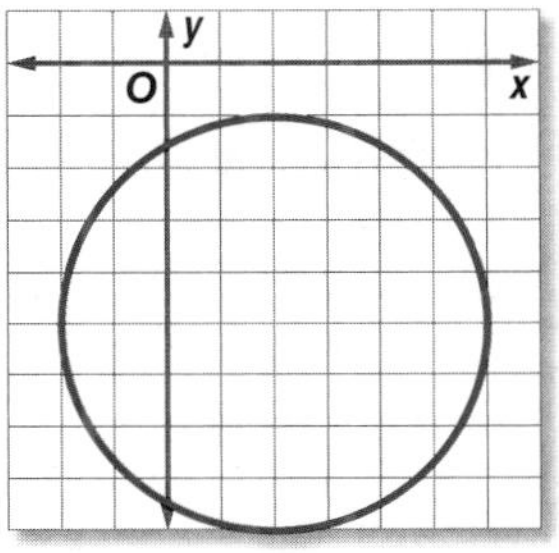

10-7 Solving Quadratics Systems (pp. 662–667)

A2.A.3

Solve each system of equations.

61. $x^2 + y^2 = 8$
$x + y = 0$

62. $x - 2y = 2$
$y^2 - x^2 = 2x + 4$

63. $y + x^2 = 4x$
$y + 4x = 16$

64. $3x^2 - y^2 = 11$
$x^2 + 4y^2 = 8$

65. $5x^2 + y^2 = 30$
$9x^2 - y^2 = -16$

66. $\frac{x^2}{30} + \frac{y^2}{6} = 1$
$x = y$

67. **PHYSICAL SCIENCE** Two balls are launched into the air at the same time. The heights they are launched from are different. The height y in feet of one is represented by $y = -16t^2 + 80t + 25$ where t is the time in seconds. The height of the other ball is represented by $y = -16t^2 + 30t + 100$.

a. After how many seconds are the balls at the same height?

b. What is this height?

68. **ARCHITECTURE** An architect is building the front entrance of a building in the shape of a parabola with the equation $y = -\frac{1}{10}(x - 10)^2 + 20$. While the entrance is being built, the construction team puts in two support beams with equations $y = -x + 10$ and $y = x - 10$. Where do the support beams meet the parabola?

Solve each system of inequalities by graphing.

69. $x^2 + y^2 < 64$
$x^2 + 16(y - 3)^2 < 16$

70. $x^2 + y^2 < 49$
$16x^2 - 9y^2 \geq 144$

71. $x + y < 4$
$9x^2 - 4y^2 \geq 36$

72. $x^2 + y^2 < 25$
$4x^2 - 9y^2 < 36$

73. $x^2 + y^2 < 36$
$4x^2 + 9y^2 > 36$

74. $y^2 < x$
$x^2 - 4y^2 < 16$

EXAMPLE 8

Solve the system of equations.
$x^2 + y^2 = 100$
$3x - y = 10$

Use substitution to solve the system.
First, rewrite $3x - y = 10$ as $y = 3x - 10$.

$$x^2 + y^2 = 100$$
$$x^2 + (3x - 10)^2 = 100$$
$$x^2 + 9x^2 - 60x + 100 = 100$$
$$10x^2 - 60x + 100 = 100$$
$$10x^2 - 60x = 0$$
$$10x(x - 6) = 0$$

$10x = 0$ or $x - 6 = 0$
$x = 0$ $\quad$ $x = 6$

Now solve for y.

$y = 3x - 10 = 3(0) - 10 = -10$

$y = 3x - 10 = 3(6) - 10 = 8$

The solutions of the system are $(0, -10)$ and $(6, 8)$.

EXAMPLE 9

Solve the system of inequalities by graphing.
$x^2 + y^2 \leq 9$
$2y \leq x^2 + 4$

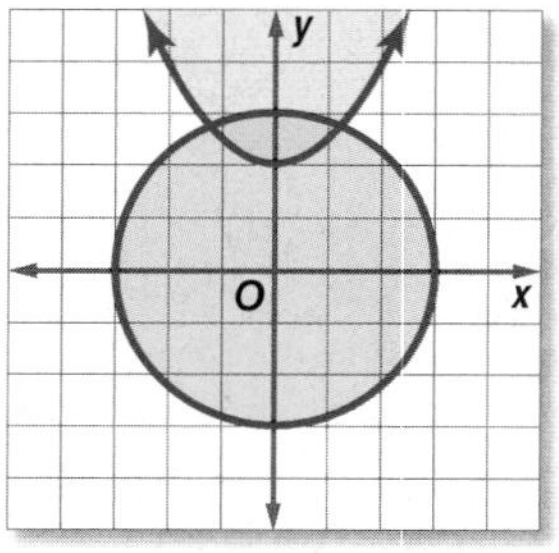

The solution is the green shaded region.

Practice Test

NY Math Online glencoe.com
Chapter Test

Find the midpoint of the line segment with endpoints at the given coordinates.

1. $(8, 3), (-4, 9)$
2. $\left(\frac{3}{4}, 0\right), \left(\frac{1}{2}, -1\right)$
3. $(-10, 0), (-2, 6)$

Find the distance between each pair of points with the given coordinates.

4. $(-5, 8), (4, 3)$
5. $\left(\frac{1}{3}, \frac{2}{3}\right), \left(-\frac{5}{6}, -\frac{11}{6}\right)$
6. $(4, -5), (4, 9)$

State whether the graph of each equation is a *parabola, circle, ellipse,* or *hyperbola*. Then graph the equation.

7. $y^2 = 64 - x^2$
8. $4x^2 + y^2 = 16$
9. $4x^2 - 9y^2 + 8x + 36y = 68$
10. $\frac{1}{2}x^2 - 3 = y$
11. $y = -2x^2 - 5$
12. $16x^2 + 25y^2 = 400$
13. $x^2 + 6x + y^2 = 16$
14. $\frac{y^2}{4} - \frac{x^2}{16} = 1$
15. $(x + 2)^2 = 3(y - 1)$
16. $4x^2 + 16y^2 + 32x + 63 = 0$

17. **MULTIPLE CHOICE** Which equation represents a hyperbola that has vertices at $(-3, -3)$ and $(5, -3)$ and a conjugate axis of length 6 units?

A $\frac{(y-1)^2}{16} - \frac{(x+3)^2}{9} = 1$

B $\frac{(x-1)^2}{16} - \frac{(y+3)^2}{9} = 1$

C $\frac{(y+1)^2}{16} - \frac{(x-3)^2}{9} = 1$

D $\frac{(x+1)^2}{16} - \frac{(y-3)^2}{9} = 1$

18. **CARPENTRY** Ellis built a window frame shaped like the top half of an ellipse. The window is 40 inches tall at its highest point and 160 inches wide at the bottom. What is the height of the window 20 inches from the center of the base?

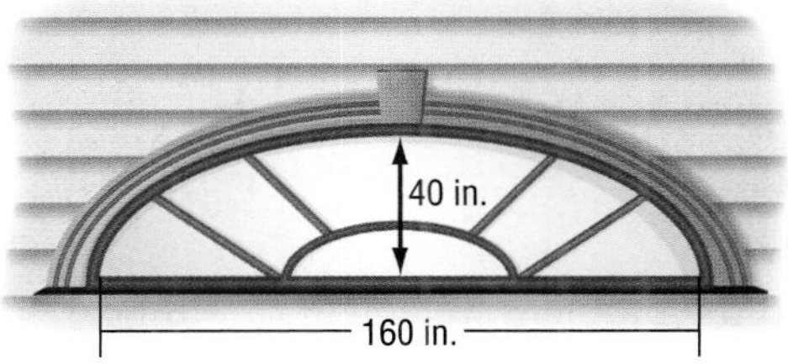

Solve each system of equations.

19. $x^2 + y^2 = 100$
$y = -x - 2$

20. $x^2 + 2y^2 = 11$
$x + y = 2$

21. $x^2 + y^2 = 34$
$y^2 - x^2 = 9$

Solve each system of inequalities.

22. $x^2 + y^2 \le 9$
$y > x^2 + 2$

23. $\frac{(x-2)^2}{4} - \frac{(y-4)^2}{9} \ge 1$
$x - 4y < 8$

24. **MULTIPLE CHOICE** Which is NOT the equation of a parabola?

F $y = 3x^2 + 5x - 3$

G $2y + 3x^2 + x - 9 = 0$

H $x = 3(y + 1)^2$

J $x^2 + 2y^2 + 6x = 10$

25. **FORESTRY** A forest ranger at an outpost in the Sam Houston National Forest and another ranger at the primary station both heard an explosion. The outpost and the primary station are 6 kilometers apart.

a. If one ranger heard the explosion 6 seconds before the other, write an equation that describes all the possible locations of the explosion. Place the two ranger stations on the x-axis with the midpoint between the stations at the origin. The transverse axis is horizontal. (*Hint*: The speed of sound is about 0.35 kilometer per second.)

b. Draw a sketch of the possible locations of the explosion. Include the ranger stations in the drawing.

CHAPTER 10 Preparing for Standardized Tests

Use a Formula

Sometimes it is necessary to use a formula to solve problems on standardized tests. In some cases you may even be given a sheet of formulas that you are permitted to reference while taking the test.

Strategies for Using a Formula

Step 1

Read the problem statement carefully.

Ask yourself:

- What am I being asked to solve?
- What information is given in the problem?
- Are there any formulas that I can use to help me solve the problem?

Step 2

Solve the problem and check your solution.

- Substitute the known quantities that are given in the problem statement into the formula.
- Simplify to solve for the unknown values in the formula.
- Check to make sure your answer makes sense. If time permits, check your answer.

EXAMPLE

Read the problem. Identify what you need to know. Then use the information in the problem to solve. Show your work.

What is the distance between points A and B on the coordinate plane? Round your answer to the nearest tenth if necessary.

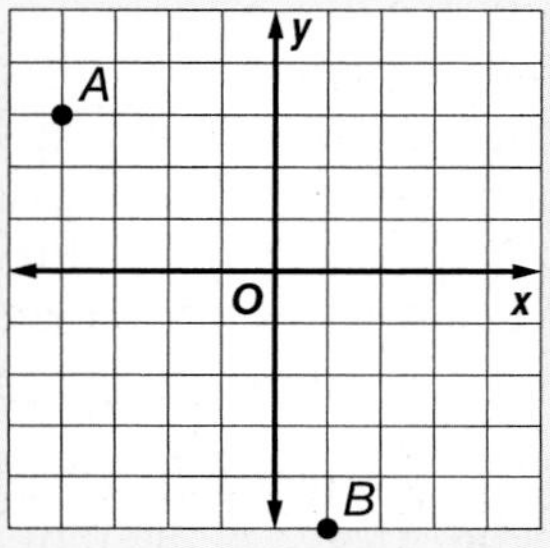

Scoring Rubric	
Criteria	**Score**
Full Credit: The answer is correct and a full explanation is provided that shows each step.	2
Partial Credit: • The answer is correct, but the explanation is incomplete. • The answer is incorrect, but the explanation is correct.	1
No Credit: Either an answer is not provided or the answer does not make sense.	0

Read the problem statement carefully. You are given the coordinates of two points on a coordinate plane and asked to find the distance between them. To solve this problem, you must use the **Distance Formula**.

Example of a 2-point response:

Use the Distance Formula to find the distance between points $A(-4, 3)$ and $B(1, -5)$.

$$d = \sqrt{(x_2 - x_1)^2 + (y_2 - y_1)^2}$$
$$= \sqrt{[1 - (-4)]^2 + [(-5) - 3]^2}$$
$$= \sqrt{5^2 + (-8)^2}$$
$$= \sqrt{25 + 64}$$
$$= \sqrt{89} \text{ or about } 9.4$$

The distance between points A and B is about 9.4 units.

The steps, calculations, and reasoning are clearly stated. The student also arrives at the correct answer. So, this response is worth the full 2 points.

Exercises

Read each problem. Identify what you need to know. Then use the information in the problem to solve. Show your work.

1. What is the midpoint of segment CD with endpoints $C(5, -12)$ and $D(-9, 4)$?

2. Katrina is making a map of her hometown on a coordinate plane. She plots the school at $S(7, 3)$ and the park at $P(-4, 12)$. If the scale of the map is 1 unit = 250 yards, what is the actual distance between the school and the park? Round to the nearest yard.

3. Mr. Washington is making a concrete table for his backyard. The tabletop will be circular with a diameter of 6 feet and a depth of 6 inches. How much concrete will Mr. Washington need to make the top of the table? Round to the nearest cubic foot.

4. What is the equation, in standard form, of the hyperbola graphed below?

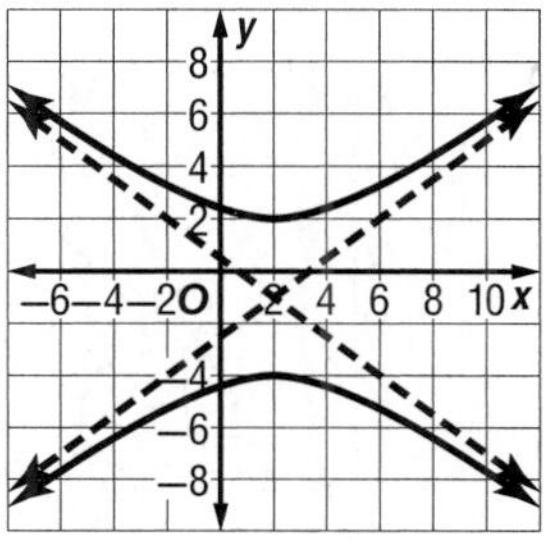

5. If the surface area of a cube is increased by a factor of 9, what is the change in the length of the sides of the cube?

   A The length is 2 times the original length.

   B The length is 3 times the original length.

   C The length is 6 times the original length.

   D The length is 9 times the original length.

CHAPTER 10

NYS Regents Exam Practice

Cumulative, Chapters 1 through 10

Multiple Choice

Answer all questions in this part. Select the answer that best completes the statement or answers the question.

1. Kimberly is making a map of New York on a coordinate grid. She plots point $B(3, 5)$ to represent Binghamton and point $S(20, 12)$ to represent Schenectady. If the scale of the map is 1 unit = 7 miles, what is the distance between Binghamton and Schenectady? Round to the nearest whole mile.

(1) 86 miles

(2) 102 miles

(3) 115 miles

(4) 129 miles

2. Solve $2x^2 - 5x + 3 = 0$ using the quadratic formula.

(1) 1, 1.5

(2) 1, 3

(3) −3, 1.5

(4) −1.5, 2

3. What is the equation of the hyperbola graphed below?

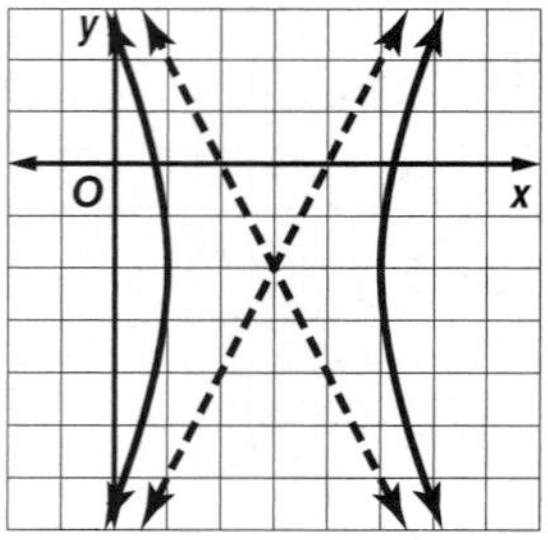

(1) $\frac{(x+3)^2}{4} - \frac{(y-2)^2}{16} = 1$

(2) $\frac{(x-3)^2}{4} - \frac{(y+2)^2}{16} = 1$

(3) $\frac{(y-3)^2}{4} - \frac{(x+2)^2}{16} = 1$

(4) $\frac{(y+3)^2}{4} - \frac{(x-2)^2}{16} = 1$

4. The total cost of reserving a campsite varies directly as the number of nights the site is rented as shown in the table.

Days	Total Cost
1	$24
2	$48
3	$72
4	$96

Which equation represents the direct variation?

(1) $y = x + 24$

(2) $y = 24x$

(3) $y = \frac{24}{x}$

(4) $y = 96x$

5. What are the vertices of the ellipse with equation $\frac{(x-3)^2}{36} + \frac{(y-2)^2}{144} = 1$?

(1) $(-3, 2)$ and $(9, 2)$

(2) $(-2, 3)$ and $(10, 3)$

(3) $(3, -10)$ and $(3, 14)$

(4) $(2, -11)$ and $(4, 13)$

6. Curtis bought a used car. The sales tax rate was 5.25% of the selling price, and he also had to pay $365 in registration fees. Find the selling price if Curtis spent a total of $14,700.05.

(1) $12,400

(2) $12,750

(3) $13,125

(4) $13,620

Test-TakingTip

Question 6 You know the final price but need to know the sales price. Work backward to find the solution.

7. Solve the system of equations below by graphing.

$$\begin{cases} 7x - 8y = 1 \\ 5x - 4y = 11 \end{cases}$$

(1) $(7, 6)$

(2) $(-4, 3)$

(3) $(2, 9)$

(4) $(5, -6)$

8. The quadratic system $\begin{cases} x^2 - 4y^2 = 9 \\ -x + 4y = 3 \end{cases}$ is graphed below. What are the solutions of the system?

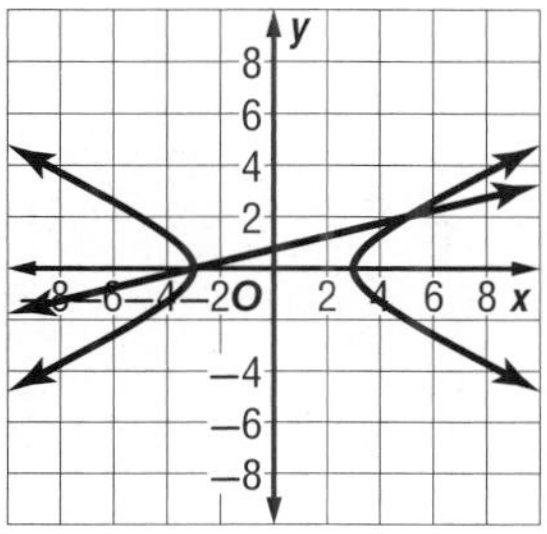

(1) (0, −3) and (5, 2)

(2) (0, −3) and (2, 5)

(3) (−3, 0) and (5, 2)

(4) (−3, 0) and (2, 5)

9. Which absolute value equation has the solutions shown on the number line below?

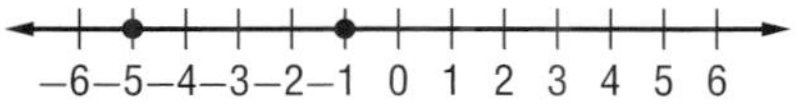

(1) $|x - 3| - 2 = 0$

(2) $|x + 3| - 2 = 0$

(3) $|x + 3| + 2 = 0$

(4) $|x - 3| + 2 = 0$

10. Simplify the expression below.

$$\frac{x^3 + 3x^2 - 5x - 4}{x + 4}$$

(1) $x^2 + 2x + 1$

(2) $x^2 - 3x - 4$

(3) $x^2 + x - 2$

(4) $x^2 - x - 1$

Open-Ended Response

Solve each problem. Clearly indicate the necessary steps, including appropriate formula substitutions, diagrams, graphs, charts, etc.

11. Use the graph of the circle below to answer each question.

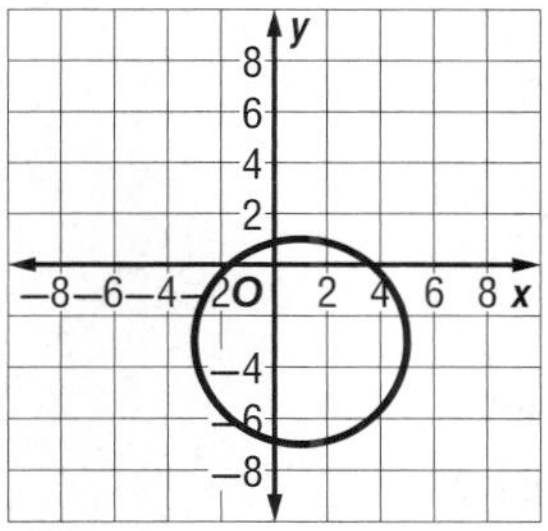

a. What is the center of the circle?

b. What is the radius of the circle?

c. Write an equation for the circle.

12. The soccer team is selling team merchandise this week to raise money for the program. For each T-shirt sold, the profit is \$7.50, and for each poster sold, the profit is \$3.75.

a. The team hopes to earn at least \$135 in profits from the sale. Let x represent the number of T-shirts sold and y the number of posters sold. Write an inequality to model the situation.

b. Graph the inequality you wrote in part **a** on a coordinate grid.

c. If the team sells 12 T-shirts and 21 posters this week, will they meet their goal? Explain.

Need Extra Help?

If you missed Question...	1	2	3	4	5	6	7	8	9	10	11	12
Go to Lesson or Page...	8-5	10-2	10-4	9-5	10-1	7-7	6-1	2-3	5-2	10-7	3-2	4-3
NYS Core Curriculum	G.G.69	A2.A.25	A2.A.46	A2.A.5	A2.R.6	A2.A.45	A.G.7	A2.A.3	A2.A.1	A.A.14	A2.A.49	A2.PS.8

CHAPTER 11

Sequences and Series

Then

In Chapter 1, you simplified and evaluated algebraic expressions.

Now

In Chapter 11, you will:

- Use arithmetic and geometric sequences and series.
- Use special sequences and iterate functions.
- Expand powers by using the Binomial Theorem.
- Prove statements by using mathematical induction.

NYS Core Curriculum

A2.A.29 Identify an arithmetic or geometric sequence and find the formula for its *n*th term
A2.A.33 Specify terms of a sequence, given its recursive definition
A2.A.35 Determine the sum of the first *n* terms of an arithmetic or geometric series

Why?

CONSERVATION AND NATURE Mathematics occurs in aspects of nature in astonishing ways. The Fibonacci sequence manifests itself in seeds, flowers, pine cones, fruits, and vegetables. Sequences and series can further help us conserve our natural resources by making water filtration systems more efficient.

Math *in Motion*, Animation glencoe.com

Get Ready for Chapter 11

Diagnose Readiness You have two options for checking Prerequisite Skills.

Text Option

Take the Quick Check below. Refer to the Quick Review for help.

*Quick*Check

Solve each equation. (Lesson 1-3)

1. $-6 = 7x + 78$
2. $768 = 3x^4$
3. $23 - 5x = 8$
4. $2x^3 + 4 = -50$
5. **PLANTS** Lauri has 48 plants for her two gardens. She plants 12 in the small garden. In the other garden she wants 4 plants in each row. How many rows will she have?

*Quick*Review

EXAMPLE 1

Solve $25 = 3x^3 + 400$.

$-375 = 3x^3$	**Subtract 400 from each side.**
$-125 = x^3$	**Divide each side by 3.**
$\sqrt[3]{-125} = \sqrt[3]{x^3}$	**Take the cube root of each side.**
$-5 = x$	**Simplify.**

Graph each function. (Lesson 0-1)

6. $\{(1, 3), (2, 5), (3, 7), (4, 9), (5, 11)\}$
7. $\{(1, -15), (2, -12), (3, -9), (4, -6), (5, -3)\}$
8. $\left\{(1, 27), (2, 9), (3, 3), (4, 1), \left(5, \frac{1}{3}\right)\right\}$
9. $\left\{(1, 1), (2, 2), \left(3, \frac{5}{2}\right), \left(4, \frac{11}{4}\right), \left(5, \frac{23}{8}\right)\right\}$
10. **DAYCARE** A child care center has expenses of \$125 per day. They charge \$50 per child per day. The function $P(c) = 50c - 125$ gives the amount of money the center makes when there are c children there. How much will they make if there are 8 children?

EXAMPLE 2

Graph the function $\{(1, 1), (2, 4), (3, 9), (4, 16), (5, 25)\}$. State the domain and range.

The domain of a function is the set of all possible x-values. So, the domain of the function is $\{1, 2, 3, 4, 5\}$. The range of a function is the set of all possible y-values. So, the range of this function is $\{1, 4, 9, 16, 25\}$.

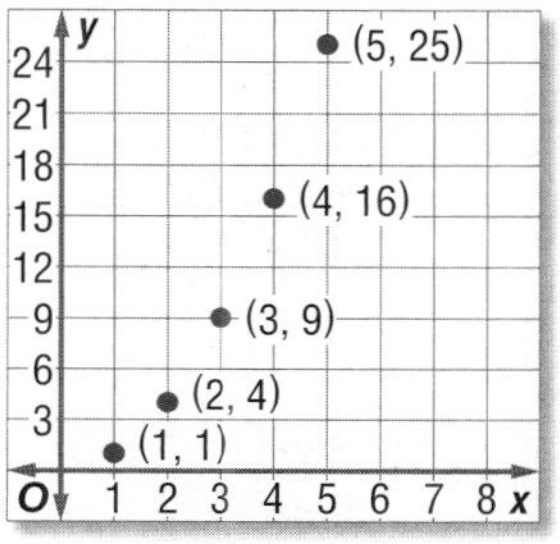

Evaluate each expression for the given value(s) of the variable(s). (Lesson 1-1)

11. $\frac{a}{3}(b + c)$ if $a = 9$, $b = -2$, and $c = -8$
12. $r + (n - 2)t$ if $r = 15$, $n = 5$, and $t = -1$
13. $x \cdot y^{z+1}$ if $x = -2$, $y = \frac{1}{3}$, and $z = 5$
14. $\frac{a(1 - bc)^2}{1 - b}$ if $a = -3$, $b = -4$, and $c = 1$

EXAMPLE 3

Evaluate $2 \cdot 3^{x+y}$ if $x = -2$ and $y = -3$.

$2 \cdot 3^{x+y} = 2 \cdot 3^{-2+-3}$	**Substitute.**
$= 2 \cdot 3^{-5}$	**Simplify.**
$= \frac{2}{3^5}$	**Rewrite with positive exponent.**
$= \frac{2}{243}$	**Evaluate the power.**

Online Option

NY Math Online Take a self-check Chapter Readiness Quiz at glencoe.com.

Get Started on Chapter 11

You will learn several new concepts, skills, and vocabulary terms as you study Chapter 11. To get ready, identify important terms and organize your resources. You may wish to refer to **Chapter 0** to review prerequisite skills.

FOLDABLES® Study Organizer

Sequences and Series Make this Foldable to help you organize your Chapter 11 notes about sequences and series. Begin with one $8\frac{1}{2}$" by 11" sheet of paper.

1. **Fold** in half, matching the short sides.

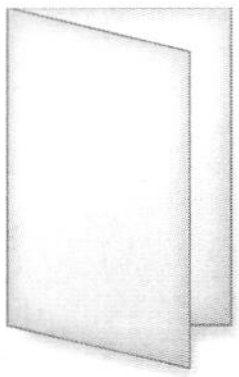

2. **Unfold** and fold the long side up 2 inches to form a pocket.

3. **Staple** or glue the outer edges to complete the pocket.

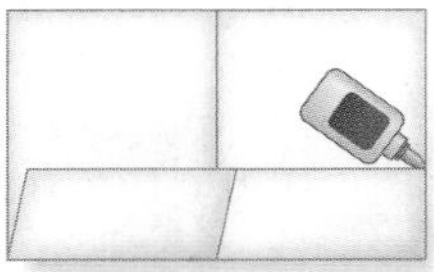

4. **Label** each side as shown. Use index cards to record notes and examples.

Sequences Series

NY Math Online glencoe.com

- Study the chapter online
- Explore **Math in Motion**
- Get extra help from your own **Personal Tutor**
- Use **Extra Examples** for additional help
- Take a **Self-Check Quiz**
- **Review Vocabulary** in fun ways

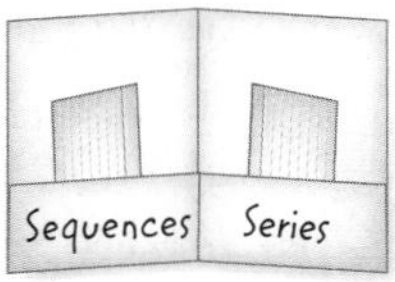

New Vocabulary

English		Español
sequence	• p. 681 •	sucesión
finite sequence	• p. 681 •	sucesión finita
infinite sequence	• p. 681 •	sucesión infinita
arithmetic sequence	• p. 681 •	sucesión arithmética
common difference	• p. 681 •	diferencia común
geometric sequence	• p. 683 •	sucesión geométrica
common ratio	• p. 683 •	razón común
arithmetic means	• p. 689 •	media arithmética
series	• p. 690 •	serie
arithmetic series	• p. 690 •	serie arithmética
partial sum	• p. 690 •	suma parcial
geometric means	• p. 697 •	media geométrica
geometric series	• p. 698 •	serie geométrica
convergent series	• p. 705 •	serie convergente
divergent series	• p. 705 •	serie divergente
recursive sequence	• p. 714 •	sucesión recursiva
iteration	• p. 716 •	iteración
mathematical induction	• p. 727 •	inducción mathemática
induction hypothesis	• p. 727 •	hipótesis inductiva

Review Vocabulary

coefficient • p. P7 • coeficiante the numerical factor of a monomial

coefficient

$$15x^3$$

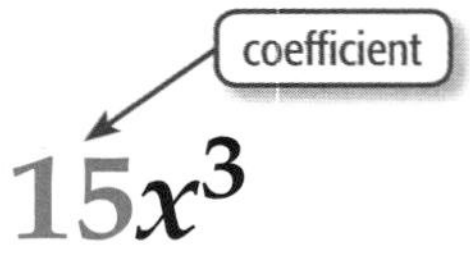

formula • p. 6 • fórmula a mathematical sentence that expresses the relationship between certain quantities

function • p. P4 • función a relation in which each element of the domain is paired with exactly one element in the range

Multilingual eGlossary glencoe.com

11-1

Sequences as Functions

Then
You analyzed linear and exponential functions. (Lessons 2-2 and 8-1)

Now
- Relate arithmetic sequences to linear functions.
- Relate geometric sequences to exponential functions.

NYS Core Curriculum

A2.A.30 Determine the common difference in an arithmetic sequence
A2.A.31 Determine the common ratio in a geometric sequence
Also addresses A2.A.32.

New Vocabulary
sequence
term
finite sequence
infinite sequence
arithmetic sequence
common difference
geometric sequence
common ratio

NY Math Online

glencoe.com
- Extra Examples
- Personal Tutor
- Self-Check Quiz
- Homework Help
- Math in Motion

Why?

During their routine, a high school marching band marches in rows. There is one performer in the first row, three performers in the next row, and five in the third row. This pattern continues for the rest of the rows.

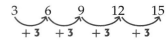

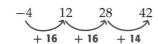

Arithmetic Sequences A **sequence** is a set of numbers in a particular order or pattern. Each number in a sequence is called a **term**. A sequence may be a **finite sequence** containing a limited number of terms, such as {−2, 0, 2, 4, 6}, or an **infinite sequence** that continues without end, such as {0, 1, 2, 3, …}. The first term of a sequence is denoted a_1, the second term is denoted a_2, and so on.

Key Concept — Sequences as Functions
For Your FOLDABLE

Words A sequence is a function in which the domain consists of natural numbers, and the range consists of real numbers.

Symbols

Domain:	$1 \quad 2 \quad 3 \quad \cdots \quad n$	the position of a term
Range:	$a_1 \quad a_2 \quad a_3 \quad \cdots \quad a_n$	the terms of the sequence

Examples

Finite Sequence	Infinite Sequence
{3, 6, 9, 12, 15}	{3, 6, 9, 12, 15, ...}
Domain: {1, 2, 3, 4, 5}	Domain: {all natural numbers}
Range: $\{y \mid 3 \leq y \leq 15\}$	Range: $\{y \mid y \geq 3\}$

Math *in Motion*, Interactive Labs glencoe.com

In an **arithmetic sequence**, each term is determined by adding a constant value to the previous term. This constant value is called the **common difference**.

Consider the sequence 3, 6, 9, 12, 15. This sequence is arithmetic because the terms share a common difference. Each term is 3 more than the previous term.

3 6 9 12 15
+3 +3 +3 +3

EXAMPLE 1 Identify Arithmetic Sequences

Determine whether each sequence is arithmetic.

a. 5, −6, −17, −28, …

5 −6 −17 −28
−11 −11 −11

The common difference is −11. The sequence is arithmetic.

b. −4, 12, 28, 42, …

−4 12 28 42
+16 +16 +14

There is no common difference. This is not an arithmetic sequence.

Check Your Progress

1A. 7, 12, 16, 20, …

1B. −6, 3, 12, 21, …

Personal Tutor glencoe.com

You can use the common difference to find terms of an arithmetic sequence.

EXAMPLE 2 Graph an Arithmetic Sequence

Consider the arithmetic sequence 18, 14, 10, … .

a. Find the next four terms of the sequence.

Step 1 To determine the common difference, subtract any term from the term directly after it. The common difference is $10 - 14$ or -4.

Step 2 To find the next term, add -4 to the last term. Continue to add -4 to find the following terms.

10 → 6 → 2 → −2 → −6

+(−4) +(−4) +(−4) +(−4)

The next four terms are 6, 2, −2, and −6.

b. Graph the first seven terms of the sequence.

The domain contains the terms {1, 2, 3, 4, 5, 6, 7} and the range contains the terms {18, 14, 10, 6, 2, −2, −6}. So, graph the corresponding ordered pairs.

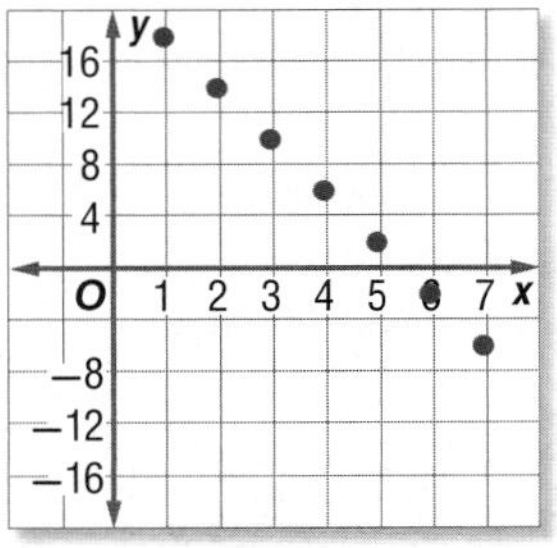

Check Your Progress

2. Find the next four terms of the arithmetic sequence 18, 11, 4, … . Then graph the first seven terms.

Personal Tutor glencoe.com

Notice that the graph of the terms of the arithmetic sequence lie on a line. An arithmetic sequence is a linear function in which the term number n is the independent variable, the term a_n is the dependent variable, and the common difference is the slope.

Real-World Link

Each year, about 100 bands compete in the Bands of America Grand National Championships.

Source: Bands of America

Real-World EXAMPLE 3 Find a Term

MARCHING BANDS Refer to the beginning of the lesson. Suppose the director wants to determine how many performers will be in the 14th row during the routine.

Understand Because the difference between any two consecutive rows is 2, the common difference for the sequence is 2.

Plan Use point-slope form to write an equation for the sequence. Let $m = 2$ and $(x_1, y_1) = (3, 5)$. Then solve for $x = 14$.

Solve

$(y - y_1) = m(x - x_1)$	**Point-slope form**
$(y - 5) = 2(x - 3)$	**$m = 2$ and $(x_1, y_1) = (3, 5)$**
$y - 5 = 2x - 6$	**Multiply.**
$y = 2x - 1$	**Add 5 to each side.**
$y = 2(14) - 1$	**Replace x with 14.**
$y = 28 - 1$ or 27	**Simplify.**

There will be 27 performers in the 14th row.

Check You can find the terms of the sequence by adding 2, starting with row 1, until you reach row 14.

Check Your Progress

3. MONEY Geraldo's employer offers him a pay rate of \$9 per hour with a \$0.15 raise every three months. How much will Geraldo earn per hour after 3 years?

Personal Tutor glencoe.com

Geometric Sequences Another type of sequence is a geometric sequence. In a **geometric sequence**, each term is determined by multiplying a nonzero constant by the previous term. This constant value is called the **common ratio**.

Consider the sequence $\frac{1}{16}, \frac{1}{4}, 1, 4, 16$. This sequence is geometric because the terms share a common ratio. Each term is 4 times as much as the previous term.

$$\frac{1}{16} \xrightarrow{\times 4} \frac{1}{4} \xrightarrow{\times 4} 1 \xrightarrow{\times 4} 4 \xrightarrow{\times 4} 16$$

EXAMPLE 4 Identify Geometric Sequences

Determine whether each sequence is geometric.

a. $-2, 6, -18, 54, \ldots$

Find the ratios of the consecutive terms.

$\frac{6}{-2} = -3 \qquad \frac{-18}{6} = -3 \qquad \frac{54}{-18} = -3$

The ratios are the same, so the sequence is geometric.

b. $8, 16, 24, 32, \ldots$

$\frac{16}{8} = 2 \qquad \frac{24}{16} = 1.5 \qquad \frac{32}{24} = 1.\overline{3}$

The ratios are not the same, so the sequence is not geometric.

Watch Out!

Ratios If you find the ratio of a term to the previous term, set up the remaining ratios the same way.

Check Your Progress

4A. $-8, 2, -0.5, 0.125, \ldots$

4B. $1, 3, 7, 15, \ldots$

Personal Tutor glencoe.com

You can use the common ratio to determine more terms of a geometric sequence.

EXAMPLE 5 Graph a Geometric Sequence

Consider the geometric sequence 32, 8, 2, ...

a. Find the next three terms of the sequence.

Step 1 Find the value of the common ratio: $\frac{2}{8}$ or $\frac{1}{4}$.

Step 2 To find the next term, multiply the previous term by $\frac{1}{4}$.

Continue multiplying by $\frac{1}{4}$ to find the following terms.

$$2 \xrightarrow{\times \frac{1}{4}} \frac{1}{2} \xrightarrow{\times \frac{1}{4}} \frac{1}{8} \xrightarrow{\times \frac{1}{4}} \frac{1}{32}$$

The next three terms are $\frac{1}{2}, \frac{1}{8}$, and $\frac{1}{32}$.

b. Graph the first six terms of the sequence.

Domain: {1, 2, 3, 4, 5, 6}

Range: $\left\{32, 8, 2, \frac{1}{2}, \frac{1}{8}, \frac{1}{32}\right\}$

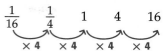

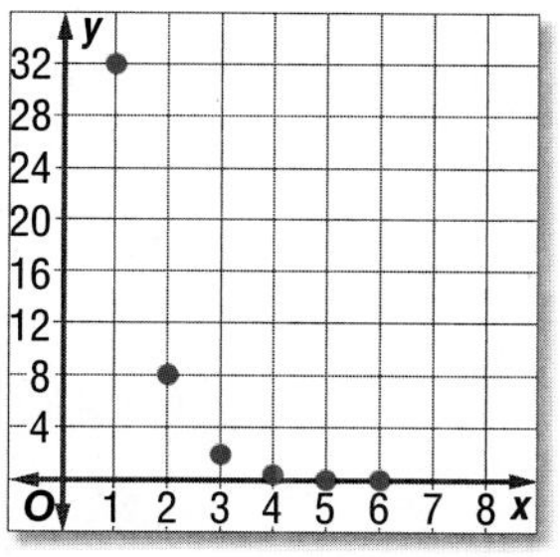

Check Your Progress

5. Find the next two terms of 7, 21, 63, ... Then graph the first six terms.

Personal Tutor glencoe.com

Review Vocabulary

exponential function a function of the form $f(x) = b^x$, where $b > 0$ and $b \neq 1$ (Lesson 8-1)

Examine the graph in Example 5. While the graph of an arithmetic sequence is linear, the graph of a geometric sequence is exponential and can be represented by $f(x) = r^x$, where r is the common ratio, $r > 0$, and $r \neq 1$.

Arithmetic

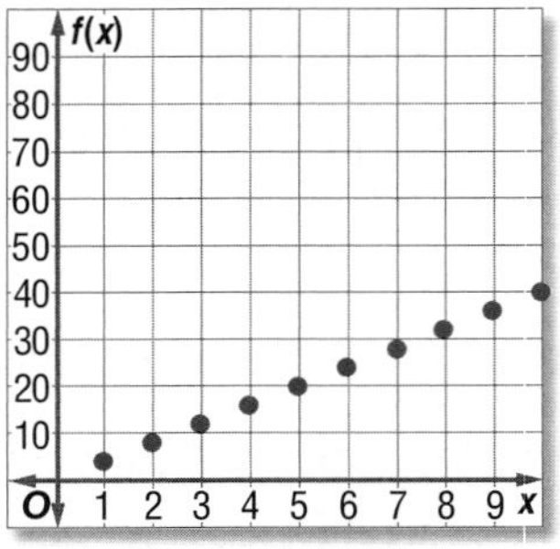

x	1	2	3	4	5	6	7	8	9	10
f(x)	4	8	12	16	20	24	28	32	36	40

Geometric

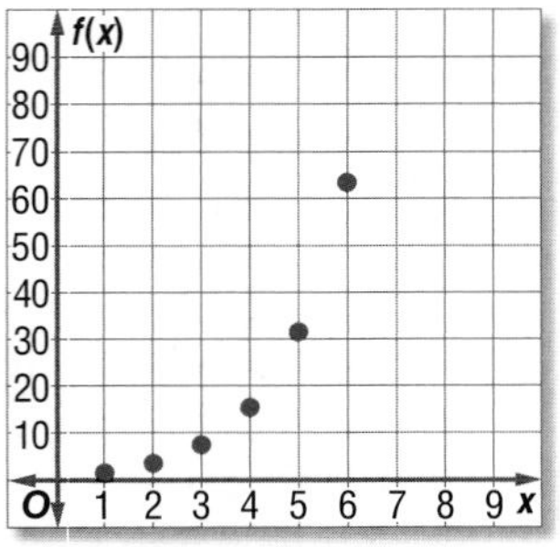

x	1	2	3	4	5	6
f(x)	2	4	8	16	32	64

The characteristics of arithmetic and geometric sequences can be used to classify sequences.

EXAMPLE 6 Classify Sequences

Determine whether each sequence is *arithmetic*, *geometric*, or *neither*. Explain your reasoning.

a. 16, 24, 36, 54, …

Check for a common difference.

$54 - 36 = 18$ $\quad$ $36 - 24 = 12$ ✗

Check for a common ratio.

$\frac{54}{36} = \frac{3}{2}$ $\quad$ $\frac{36}{24} = \frac{3}{2}$ $\quad$ $\frac{24}{16} = \frac{3}{2}$ ✓

Because there is a common ratio, the sequence is geometric.

b. 1, 4, 9, 16, …

Check for a common difference.

$16 - 9 = 7$ $\quad$ $9 - 4 = 5$ ✗

Check for a common ratio.

$\frac{16}{9} = 1.\overline{7}$ $\quad$ $\frac{9}{4} = 2.25$ ✗

Because there is no common difference or ratio, the sequence is neither arithmetic nor geometric.

c. 23, 17, 11, 5, …

Check for a common difference.

$5 - 11 = -6$ $\quad$ $11 - 17 = -6$ $\quad$ $17 - 23 = -6$ ✓

Because there is a common difference, the sequence is arithmetic.

Check Your Progress

6A. $\frac{5}{3}, 2, \frac{7}{3}, \frac{8}{3}, \ldots$

6B. $2, -\frac{3}{2}, \frac{9}{8}, -\frac{27}{32}, \ldots$

6C. $-4, 4, 5, -5, \ldots$

Personal Tutor glencoe.com

Check Your Understanding

Example 1 p. 681

Determine whether each sequence is arithmetic. Write *yes* or *no*.

1. $8, -2, -12, -22, \ldots$

2. $-19, -12, -5, 2, 9$

3. $1, 2, 4, 8, 16$

4. $0.6, 0.9, 1.2, 1.8, \ldots$

Example 2 p. 682

Find the next four terms of each arithmetic sequence. Then graph the sequence.

5. $6, 18, 30, \ldots$

6. $15, 6, -3, \ldots$

7. $-19, -11, -3, \ldots$

8. $-26, -33, -40, \ldots$

Example 3 p. 682

9. **SAVINGS** Kelly is saving her money to buy a car. She has \$250, and she plans to save \$75 per week from her job as a waitress.

a. How much will Kelly have saved after 8 weeks?

b. If the car costs \$2000, how long will it take her to save enough money at this rate?

Example 4 p. 683

Determine whether each sequence is geometric. Write *yes* or *no*.

10. $-8, -5, -1, 4, \ldots$

11. $4, 12, 36, 108, \ldots$

12. $27, 9, 3, 1, \ldots$

13. $7, 14, 21, 28, \ldots$

Example 5 p. 683

Find the next three terms of each geometric sequence. Then graph the sequence.

14. $8, 12, 18, 27, \ldots$

15. $8, 16, 32, 64, \ldots$

16. $250, 50, 10, 2, \ldots$

17. $9, -3, 1, -\frac{1}{3}, \ldots$

Example 6 p. 684

Determine whether each sequence is *arithmetic*, *geometric*, or *neither*. Explain your reasoning.

18. $5, 1, 7, 3, 9, \ldots$

19. $200, -100, 50, -25, \ldots$

20. $12, 16, 20, 24, \ldots$

Practice and Problem Solving

= **Step-by-Step Solutions** begin on page R20.
Extra Practice begins on page 947.

Example 1 p. 681

Determine whether each sequence is arithmetic. Write *yes* or *no*.

21. $\frac{1}{2}, \frac{1}{3}, \frac{1}{4}, \frac{1}{5}, \ldots$

22. $-9, -3, 0, 3, 9$

23. $14, -5, -19, \ldots$

24. $\frac{2}{9}, \frac{5}{9}, \frac{8}{9}, \frac{11}{9}, \ldots$

Example 2 p. 682

Find the next four terms of each arithmetic sequence. Then graph the sequence.

25. $-4, -1, 2, 5, \ldots$

26. $10, 2, -6, -14, \ldots$

27. $-5, -11, -17, -23, \ldots$

28. $-19, -2, 15, \ldots$

29. $\frac{1}{5}, \frac{4}{5}, \frac{7}{5}, \ldots$

30. $\frac{2}{3}, -\frac{1}{3}, -\frac{4}{3}$

Example 3 p. 682

31. **THEATER** There are 28 seats in the front row of a theater. Each successive row contains two more seats than the previous row. If there are 24 rows, how many seats are in the last row of the theater?

32. **EXERCISE** Mario began an exercise program to get back in shape. He plans to row 5 minutes on his rowing machine the first day and increase his rowing time by one minute and thirty seconds each day.

a. How long will he row on the 18th day?

b. On what day will Mario first row an hour or more?

c. Is it reasonable for this pattern to continue indefinitely? Explain.

Example 4 p. 683 **Determine whether each sequence is geometric. Write *yes* or *no*.**

33. 21, 14, 7, … **34.** 124, 186, 248, … **35** −27, 18, −12, …

36. 162, 108, 72, … **37.** $\frac{1}{2}, -\frac{1}{4}, 1, -\frac{1}{2}, \ldots$ **38.** −4, −2, 0, 2, …

Example 5 p. 683 **Find the next three terms of the sequence. Then graph the sequence.**

39. 0.125, −0.5, 2, … **40.** 18, 12, 8, … **41.** 64, 48, 36, …

42. 81, 108, 144, … **43.** $\frac{1}{3}, 1, 3, 9, \ldots$ **44.** 1, 0.1, 0.01, 0.001, …

Example 6 p. 684 **Determine whether each sequence is *arithmetic*, *geometric*, or *neither*. Explain your reasoning.**

45. 3, 12, 27, 48, … **46.** 1, −2, −5, −8, … **47.** 12, 36, 108, 324, …

48. $-\frac{2}{5}, -\frac{2}{25}, -\frac{2}{125}, -\frac{2}{625}, \ldots$ **49.** $\frac{5}{2}, 3, \frac{7}{2}, 4, \ldots$ **50.** 6, 9, 14, 21, …

51. READING Sareeta took an 800-page book on vacation. If she was already on page 112 and is going to be on vacation for 8 days, what is the minimum number of pages she needs to read per day to finish the book by the end of her vacation?

52. DEPRECIATION Tammy's car is expected to depreciate at a rate of 15% per year. If her car is currently valued at \$24,000, to the nearest dollar, how much will it be worth in 6 years?

53. PAPER FOLDING When a piece of paper is folded onto itself, it doubles in thickness. If a piece of paper that is 0.1 mm thick could be folded 37 times, how thick would it be?

Real-World Link

According to *Car and Driver*, a car loses 15% to 20% of its value each year.

H.O.T. Problems

Use Higher-Order Thinking Skills

54. REASONING Explain why the sequence 8, 10, 13, 17, 22 is not arithmetic.

55. OPEN ENDED Describe a real-life situation that can be represented by an arithmetic sequence with a common difference of 8.

56. CHALLENGE The sum of three consecutive terms of an arithmetic sequence is 6. The product of the terms is −42. Find the terms.

57. FIND THE ERROR Brody and Gen are determining whether the sequence 8, 8, 8, … is *arithmetic*, *geometric*, *neither*, or *both*. Is either of them correct? Explain your reasoning.

Brody	Gen
The sequence has a common difference of 0. The sequence is arithmetic.	The sequence has a common ratio of 1. The sequence is geometric.

58. OPEN ENDED Find a geometric sequence, an arithmetic sequence, and a sequence that is neither geometric nor arithmetic that begins 3, 9, … .

59. REASONING If a geometric sequence has a ratio r such that $|r| < 1$, what happens to the terms as n increases? What would happen to the terms if $|r| \geq 1$?

60. WRITING IN MATH Describe what happens to the terms of a geometric sequence when the common ratio is doubled. What happens when it is halved? Explain your reasoning.

NYSRE PRACTICE A2.A.30, A2.A.31

61. SHORT RESPONSE Mrs. Aguilar's rectangular bedroom measures 13 feet by 11 feet. She wants to purchase carpet for the bedroom that costs \$2.95 per square foot, including tax. How much will it cost to carpet her bedroom?

62. The pattern of filled circles and white circles below can be described by a relationship between two variables.

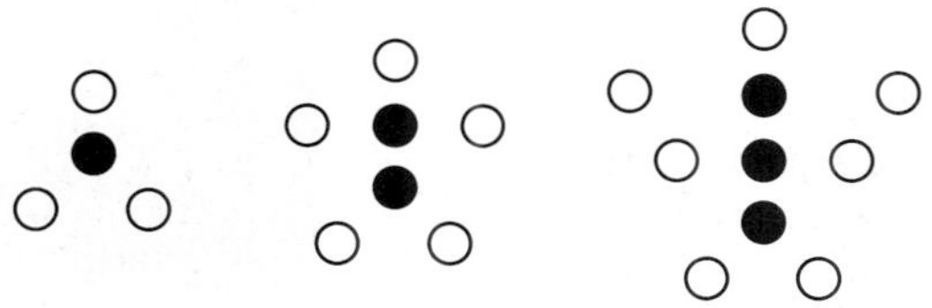

Which rule relates w, the number of white circles, to f, the number of dark circles?

A $w = 3f$
C $w = 2f + 1$
B $f = \frac{1}{2}w - 1$
D $f = \frac{1}{3}w$

63. ACT/SAT Donna wanted to determine the average of her six test scores. She added the scores correctly to get T, but divided by 7 instead of 6. Her average was 12 less than the actual average. Which equation could be used to determine the value of T?

F $6T + 12 = 7T$

G $\frac{T}{7} = \frac{T - 12}{6}$

H $\frac{T}{7} + 12 = \frac{T}{6}$

J $\frac{T}{6} = \frac{T - 12}{7}$

64. Find the next term in the geometric sequence $8, 6, \frac{9}{2}, \frac{27}{8}, \ldots$.

A $\frac{11}{8}$
C $\frac{9}{4}$
B $\frac{27}{16}$
D $\frac{81}{32}$

Spiral Review

Solve each system of equations. (Lesson 10-7)

65. $y = 5$
$y^2 = x^2 + 9$

66. $y - x = 1$
$x^2 + y^2 = 25$

67. $3x = 8y^2$
$8y^2 - 2x^2 = 16$

Write each equation in standard form. State whether the graph of the equation is a *parabola, circle, ellipse,* or *hyperbola.* Then graph the equation. (Lesson 10-6)

68. $6x^2 + 6y^2 = 162$

69. $4y^2 - x^2 + 4 = 0$

70. $x^2 + y^2 + 6y + 13 = 40$

Graph each function. (Lesson 9-4)

71. $f(x) = \frac{6}{(x - 2)(x + 3)}$

72. $f(x) = \frac{-3}{(x - 2)^2}$

73. $f(x) = \frac{x^2 - 36}{x + 6}$

74. HEALTH A certain medication is eliminated from the bloodstream at a steady rate. It decays according to the equation $y = ae^{-0.1625t}$, where t is in hours. Find the half-life of this substance. (Lesson 8-8)

Skills Review

Write an equation of each line. (Lesson 2-4)

75. passes through $(6, 4)$, $m = 0.5$

76. passes through $\left(2, \frac{1}{2}\right)$, $m = -\frac{3}{4}$

77. passes through $(0, -6)$, $m = 3$

78. passes through $(0, 4)$, $m = \frac{1}{4}$

79. passes through $(1, 3)$ and $\left(8, -\frac{1}{2}\right)$

80. passes through $(-5, 1)$ and $(5, 16)$

11-2 Arithmetic Sequences and Series

Then

You determined whether a sequence was arithmetic. (Lesson 11-1)

Now

- Use arithmetic sequences.
- Find sums of arithmetic series.

NYS Core Curriculum

A2.A.29 Identify an arithmetic or geometric sequence and find the formula for its *n*th term
A2.A.35 Determine the sum of the first *n* terms of an arithmetic or geometric series

New Vocabulary

arithmetic means
series
arithmetic series
partial sum
sigma notation

NY Math Online

glencoe.com
- Extra Examples
- Personal Tutor
- Self-Check Quiz
- Homework Help
- Math In Motion

Why?

In the 18th century, a teacher asked his class of elementary students to find the sum of the counting numbers 1 through 100. A pupil named Karl Gauss correctly answered within seconds, astonishing the teacher. Gauss went on to become a great mathematician.

He solved this problem by using an arithmetic series.

Arithmetic Sequences In Lesson 11-1, you used the point-slope form to find a specific term of an arithmetic sequence. It is possible to develop an equation for any term of an arithmetic sequence using the same process.

Consider the arithmetic sequence $a_1, a_2, a_3, \ldots, a_n$ in which the common difference is d.

$(y - y_1) = m(x - x_1)$	**Point-slope form**
$(a_n - a_1) = d(n - 1)$	**$(x, y) = (n, a_n)$, $(x_1, y_1) = (1, a_1)$, and $m = d$**
$a_n = a_1 + d(n - 1)$	**Add a_1 to each side.**

You can use this equation to find any term in an arithmetic sequence when you know the first term and the common difference.

Key Concept — *n*th Term of an Arithmetic Sequence (For Your FOLDABLE)

The *n*th term a_n of an arithmetic sequence in which the first term is a_1 and the common difference is d is given by the following formula, where n is any natural number.

$$a_n = a_1 + (n - 1)d$$

You will prove this formula in Exercise 80.

EXAMPLE 1 Find the *n*th Term

Find the 12th term of the arithmetic sequence 9, 16, 23, 30, … .

Step 1 Find the common difference.

$16 - 9 = 7 \qquad 23 - 16 = 7 \qquad 30 - 23 = 7$

So, $d = 7$.

Step 2 Find the 12th term.

$a_n = a_1 + (n - 1)d$	***n*th term of an arithmetic sequence**
$a_{12} = 9 + (12 - 1)(7)$	**$a_1 = 9$, $d = 7$, and $n = 12$**
$= 9 + 77$ or 86	**Simplify.**

Check Your Progress

Find the indicated term of each arithmetic sequence.

1A. $a_1 = -4, d = 6, n = 9$

1B. a_{20} for $a_1 = 15, d = -8$

Personal Tutor glencoe.com

If you are given some terms of an arithmetic sequence, you can write an equation for the nth term of the sequence.

EXAMPLE 2 Write Equations for the nth Term

Write an equation for the nth term of each arithmetic sequence.

a. 5, −13, −31, …

$d = -13 - 5$ or -18; 5 is the first term.

$a_n = a_1 + (n-1)d$	nth term of an arithmetic sequence
$a_n = 5 + (n-1)(-18)$	$a_1 = 5$ and $d = -18$
$a_n = 5 + (-18n + 18)$	Distributive Property
$a_n = -18n + 23$	Simplify.

b. $a_5 = 19$, $d = 6$

First, find a_1.

$a_n = a_1 + (n-1)d$	nth term of an arithmetic sequence
$19 = a_1 + (5-1)(6)$	$a_5 = 19$, $n = 5$, and $d = 6$
$19 = a_1 + 24$	Multiply.
$-5 = a_1$	Subtract 24 from each side.

Then write the equation.

$a_n = a_1 + (n-1)d$	nth term of an arithmetic sequence
$a_n = -5 + (n-1)(6)$	$a_1 = -5$ and $d = 6$
$a_n = -5 + (6n - 6)$	Distributive Property
$a_n = 6n - 11$	Simplify.

Check Your Progress

2A. 12, 3, −6, …

2B. $a_6 = 12$, $d = 8$

Personal Tutor glencoe.com

> **StudyTip**
>
> **Checking Solutions** Check your solution by using it to determine the first three terms of the sequence.

Sometimes you are given two terms of a sequence, but they are not consecutive terms of that sequence. The terms between any two nonconsecutive terms of an arithmetic sequence, called **arithmetic means**, can be used to find missing terms of a sequence.

> **ReadingMath**
>
> **arithmetic mean** the average of two or more numbers
>
> **arithmetic means** the terms between any two nonconsecutive terms of an arithmetic sequence

EXAMPLE 3 Find Arithmetic Means

Find the arithmetic means in the sequence −8, _?_, _?_, _?_, _?_, 22, … .

Step 1 Since there are four terms between the first and last terms given, there are $4 + 2$ or 6 total terms, so $n = 6$.

Step 2 Find d.

$a_n = a_1 + (n-1)d$	nth term of an arithmetic sequence
$22 = -8 + (6-1)d$	$a_1 = -8$, $a_6 = 22$, and $n = 6$
$30 = 5d$	Distributive Property
$6 = d$	Divide each side by 5.

Step 3 Use d to find the four arithmetic means.

$$-8 \xrightarrow{+6} -2 \xrightarrow{+6} 4 \xrightarrow{+6} 10 \xrightarrow{+6} 16 \xrightarrow{+6} 22$$

The arithmetic means are −2, 4, 10, and 16.

Check Your Progress

3. Find the five arithmetic means between −18 and 36.

Personal Tutor glencoe.com

Arithmetic Series A **series** is formed when the terms of a sequence are added. An **arithmetic series** is the sum of an arithmetic sequence. The sum of the first n terms is called the **partial sum** and is denoted S_n.

Key Concept **Partial Sum of an Arithmetic Series** For Your FOLDABLE

Given	The sum S_n of the first n terms is:
a_1 and a_n	$S_n = n\left(\frac{a_1 + a_n}{2}\right)$
a_1 and d	$S_n = \frac{n}{2}[2a_1 + (n - 1)d]$

Sometimes a_1, a_n, or n need to be determined before the sum of an arithmetic series can be found. When this occurs, use the formula for the nth term.

EXAMPLE 4 **Use the Sum Formulas**

Find the sum of 12 + 19 + 26 + ⋯ + 180.

Step 1 $a_1 = 12$, $a_n = 180$, and $d = 19 - 12$ or 7.
We need to find n before we can use one of the formulas.

$a_n = a_1 + (n - 1)d$ — **nth term of an arithmetic sequence**
$180 = 12 + (n - 1)(7)$ — **$a_n = 180$, $a_1 = 12$, and $d = 7$**
$168 = 7n - 7$ — **Simplify.**
$25 = n$ — **Solve for n.**

Step 2 Use either formula to find S_n.

$S_n = \frac{n}{2}[2a_1 + (n - 1)d]$ — **Sum formula**
$S_{25} = \frac{25}{2}[2(12) + (25 - 1)(7)]$ — **$n = 25$, $a_1 = 12$, and $d = 7$**
$S_{25} = 12.5(192)$ or 2400 — **Simplify.**

Check Your Progress

Find the sum of each arithmetic series.

4A. $2 + 4 + 6 \cdots + 100$

4B. $n = 16$, $a_n = 240$, and $d = 8$.

Personal Tutor glencoe.com

You can use a sum formula to find terms of a series.

Watch Out!

Common Difference Don't confuse the sign of the common difference in an arithmetic sequence. Check that the rule actually produces the terms of a sequence.

EXAMPLE 5 **Find the First Three Terms**

Find the first three terms of the arithmetic series in which $a_1 = 7$, $a_n = 79$, and $S_n = 430$.

Step 1 Find n.

$S_n = n\left(\frac{a_1 + a_n}{2}\right)$ — **Sum formula**
$430 = n\left(\frac{7 + 79}{2}\right)$ — **$S_n = 430$, $a_1 = 7$, and $a_n = 79$**
$430 = n(43)$ — **Simplify.**
$10 = n$ — **Divide each side by 43.**

Step 2 Find d.

$a_n = a_1 + (n - 1)d$	**nth term of an arithmetic sequence**
$79 = 7 + (10 - 1)d$	**$a_n = 79$, $a_1 = 7$, and $n = 10$**
$72 = 9d$	**Subtract 7 from each side.**
$8 = d$	**Divide each side by 9.**

Step 3 Use d to determine a_2 and a_3.

$a_2 = 7 + 8$ or 15 $\quad a_3 = 15 + 8$ or 23

The first three terms are 7, 15, and 23.

Check Your Progress

Find the first three terms of each arithmetic series.

5A. $S_n = 120, n = 8, a_n = 36$ **5B.** $a_1 = -24, a_n = 288, S_n = 5280$

Personal Tutor glencoe.com

ReadingMath

Sigma Notation The name comes from the Greek letter sigma, which is used in the notation.

The sum of a series can be written in shorthand by using **sigma notation**.

Key Concept Sigma Notation

For Your FOLDABLE

Symbols $\sum_{k=1}^{n} f(k)$ — last value of k: n; first value of k: $k = 1$; $f(k)$: formula for the terms of the series

Example $\sum_{k=1}^{12} (4k + 2) = [4(1) + 2] + [4(2) + 2] + [4(3) + 2] + \cdots + [4(12) + 2]$

$= 6 + 10 + 14 + \cdots + 50$

Math in Motion, Animation glencoe.com

NYSRE EXAMPLE 6 A2.A.29, A2.A.35

Find $\sum_{k=4}^{18} (6k - 1)$.

A 846 **B** 910 **C** 975 **D** 1008

Test-TakingTip

Solve a Simpler Problem Sometimes it is necessary to break a problem into parts, solve each part, then combine the solutions of the parts.

Read the Test Item

You need to find the sum of the series. Find a_1, a_n, and n.

Solve the Test Item

There are $18 - 4 + 1$ or 15 terms, so $n = 15$.

$a_1 = 6(4) - 1$ or 23 $\quad a_n = 6(18) - 1$ or 107

Find the sum.

$S_n = n\left(\frac{a_1 + a_n}{2}\right)$	**Sum formula**
$S_{15} = 15\left(\frac{23 + 107}{2}\right)$	**$n = 15$, $a_1 = 23$, and $a_n = 107$**
$S_{15} = 15(65)$ or 975	The correct answer is C.

Check Your Progress

6. **Find $\sum_{m=9}^{21} (5m + 6)$.**

F 972 **G** 1053 **H** 1281 **J** 1701

Personal Tutor glencoe.com

Check Your Understanding

Example 1 p. 688

Find the indicated term of each arithmetic sequence.

1. $a_1 = 14, d = 9, n = 11$
2. a_{18} for 12, 25, 38, ...

Example 2 p. 689

Write an equation for the nth term of each arithmetic sequence.

3. 13, 19, 25, ...
4. $a_5 = -12, d = -4$

Example 3 p. 689

Find the arithmetic means in each sequence.

5. 6, _?_, _?_, _?_, 42
6. −4, _?_, _?_, _?_, 8

Example 4 p. 690

Find the sum of each arithmetic series.

7. the first 50 natural numbers
8. $4 + 8 + 12 + \cdots + 200$
9. $a_1 = 12, a_n = 188, d = 4$
10. $a_n = 145, d = 5, n = 21$

Example 5 pp. 690, 691

Find the first three terms of each arithmetic series.

11. $a_1 = 8, a_n = 100, S_n = 1296$
12. $n = 18, a_n = 112, S_n = 1098$

Example 6 p. 691

13. **MULTIPLE CHOICE** Find $\sum_{k=1}^{12}(3k + 9)$.

A 45

B 78

C 342

D 410

Practice and Problem Solving

● = **Step-by-Step Solutions** begin on page R20. **Extra Practice** begins on page 947.

Example 1 p. 688

Find the indicated term of each arithmetic sequence.

14. $a_1 = -18, d = 12, n = 16$
15. $a_1 = -12, n = 66, d = 4$
16. $a_1 = 9, n = 24, d = -6$
17. a_{15} for −5, −12, −19, ...
18. a_{10} for −1, 1, 3, ...
19. a_{24} for 8.25, 8.5, 8.75, ...

Example 2 p. 689

Write an equation for the nth term of each arithmetic sequence.

20. 24, 35, 46, ...
21. 31, 17, 3, ...
22. $a_9 = 45, d = -3$
23. $a_7 = 21, d = 5$
24. $a_4 = 12, d = 0.25$
25. $a_5 = 1.5, d = 4.5$
26. 9, 2, −5, ...
27. $a_6 = 22, d = 9$
28. $a_8 = -8, d = -2$
29. $a_{15} = 7, d = \frac{2}{3}$
30. −12, −17, −22, ...
31. $a_3 = -\frac{4}{5}, d = \frac{1}{2}$

32. **SPORTS** José averaged 123 total pins per game in his bowing league this season. He is taking bowling lessons and hopes to bring his average up by 8 pins each new season.
    a. Write an equation to represent the nth term of the sequence.
    b. If the pattern continues, during what season will José average 187 per game?
    c. Is it reasonable for this pattern to continue indefinitely? Explain.

Example 3 p. 689

Find the arithmetic means in each sequence.

33. 24, _?_, _?_, _?_, _?_, −1
34. −6, _?_, _?_, _?_, _?_, 49
35. −28, _?_, _?_, _?_, _?_, 7
36. 84, _?_, _?_, _?_, _?_, 39
37. −12, _?_, _?_, _?_, _?_, _?_, −66
38. 182, _?_, _?_, _?_, _?_, _?_, 104

Example 4 p. 690

Find the sum of each arithmetic series.

39. the first 100 even natural numbers

40. the first 200 odd natural numbers

41. the first 100 odd natural numbers

42. the first 300 even natural numbers

43. $-18 + (-15) + (-12) + \cdots + 66$

44. $-24 + (-18) + (-12) + \cdots + 72$

45. $a_1 = -16, d = 6, n = 24$

46. $n = 19, a_n = 154, d = 8$

47. $n = 32, a_n = -86, S_n = 224$

48. $a_1 = 48, a_n = 180, S_n = 1368$

49. **SUMMER WORK** Wendy took a summer job making baskets at a factory. Because of the difficulty of work, her pay began at $150 per week and increased by $50 for each basket that she made. If she made one basket per week for eleven weeks, how much did she earn for the whole summer?

Example 5 pp. 690–691

Find the first three terms of each arithmetic series.

50. $a_1 = 3, a_n = 66, S_n = 759$

51. $n = 28, a_n = 228, S_n = 2982$

52. $a_1 = -72, a_n = 453, S_n = 6858$

53. $n = 30, a_n = 362, S_n = 4770$

54. $a_1 = 19, n = 44, S_n = 9350$

55. $a_1 = -33, n = 36, S_n = 6372$

56. **PRIZES** A radio station is offering a total of $8500 in prizes over ten hours. Each hour, the prize will increase by $100. Find the amounts of the first and last prize.

Example 6 p. 691

Find the sum of each arithmetic series.

57. $\sum_{k=1}^{16} (4k - 2)$

58. $\sum_{k=4}^{13} (4k + 1)$

59. $\sum_{k=5}^{16} (2k + 6)$

60. $\sum_{k=0}^{12} (-3k + 2)$

61. **LOANS** Daniela borrowed some money from her parents. She agreed to pay $50 at the end of the first month and $25 more each additional month for 12 months. How much does she pay in total after the 12 months?

62. **GRAVITY** When an object is in free fall and air resistance is ignored, it falls 16 feet in the first second, an additional 48 feet during the next second, and 80 feet during the third second. How many total feet will the object fall in 10 seconds?

Real-World Link

Some sources recommend that teens who borrow from their parents should sign a contract, deduct advances from future allowances, and pay interest on the loan.

Source: North Dakota State University

Use the given information to write an equation that represents the *n*th term in each arithmetic sequence.

63. The 100th term of the sequence is 245. The common difference is 13.

64. The eleventh term of the sequence is 78. The common difference is -9.

65. The sixth term of the sequence is -34. The 23rd term is 119.

66. The 25th term of the sequence is 121. The 80th term is 506.

67. **SEATING** The rectangular tables in a reception hall are often placed end-to-end to form one long table. The diagrams below show the number of people who can sit at each of the table arrangements.

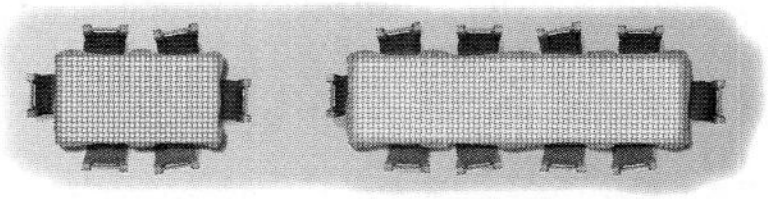

a. Make drawings to find the next three numbers as tables are added one at a time to the arrangement.

b. Write an equation representing the *n*th number in this pattern.

c. Is it possible to have seating for exactly 100 people with such an arrangement? Explain.

Real-World Link

An estimated 384,000 American high school students participate in cross country every year.

Source: National Federation of State High School Associations

68. **PERFORMANCE** A certain company pays its employees according to their performance. Belinda is paid a flat rate of \$200 per week plus \$24 for every unit she completes. If she earned \$512 in one week, how many units did she complete?

69 **SALARY** Terry currently earns \$28,000 per year. If Terry expects a \$4000 increase in salary every year, after how many years will he have a salary of \$100,000 per year?

70. **SPORTS** While training for cross country, Silvia plans to run 3 miles per day for the first week, and then increase the distance by a half mile each of the following weeks.

a. Write an equation to represent the *n*th term of the sequence.

b. If the pattern continues, during which week will she be running 10 miles per day?

c. Is it reasonable for this pattern to continue indefinitely? Explain.

71. **MULTIPLE REPRESENTATIONS** Consider $\sum_{k=1}^{x}(2k + 2)$.

a. **TABULAR** Make a table of the partial sums of the series for $1 \leq k \leq 10$.

b. **GRAPHICAL** Graph (*k*, partial sum).

c. **GRAPHICAL** Graph $f(x) = x^2 + 3x$ on the same grid.

d. **VERBAL** What do you notice about the two graphs?

e. **ANALYTICAL** What conclusions can you make about the relationship between quadratic functions and the sum of arithmetic series?

f. **ALGEBRAIC** Find the arithmetic series that relates to $g(x) = x^2 + 8x$.

Find the value of *x*.

72. $\sum_{k=3}^{x}(6k - 5) = 928$

73. $\sum_{k=5}^{x}(8k + 2) = 1032$

H.O.T. Problems

Use Higher-Order Thinking Skills

74. **FIND THE ERROR** Eric and Juana are determining the formula for the *n*th term for the sequence $-11, -2, 7, 16, \ldots$. Is either of them correct? Explain your reasoning.

Eric	Juana
$d = 16 - 7$ or 9, $a_1 = -11$ $a_n = -11 + (n - 1)9$ $= 9n - 20$	$d = 16 - 7$ or 9, $a_1 = -11$ $a_n = 9n - 11$

75. **REASONING** If *a* is the third term in an arithmetic sequence, *b* is the fifth term, and *c* is the eleventh term, express *c* in terms of *a* and *b*.

76. **CHALLENGE** There are three arithmetic means between *a* and *b* in an arithmetic sequence. The average of the arithmetic means is 16. What is the average of *a* and *b*?

77. **CHALLENGE** Find S_n for $(x + y) + (x + 2y) + (x + 3y) + \ldots$.

78. **OPEN ENDED** Write an arithmetic series with 8 terms and a sum of 324.

79. **WRITING IN MATH** Compare and contrast arithmetic sequences and series.

80. **PROOF** Prove the formula for the *n*th term of an arithmetic sequence.

81. **PROOF** Derive a sum formula that does not include a_1.

82. **PROOF** Derive the Alternate Sum Formula using the General Sum Formula.

NYSRE PRACTICE A2.A.29, A2.A.35

83. ACT/SAT The measures of the angles of a triangle form an arithmetic sequence. If the measure of the smallest angle is 36°, what is the measure of the largest angle?

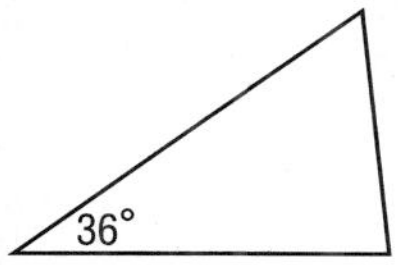

A 75°
B 84°
C 90°
D 97°

84. The area of a triangle is $\frac{1}{2}q^2 - 8$ and the height is $q + 4$. Which expression best describes the triangle's length?

F $(q + 1)$
G $(q + 2)$
H $(q - 3)$
J $(q - 4)$

85. The expression $1 + \sqrt{2} + \sqrt[3]{3}$ is equivalent to

A $\sum_{k=1}^{3} k^{\frac{1}{k}}$
B $\sum_{k=1}^{3} k^{k}$
C $\sum_{k=1}^{3} k^{-k}$
D $\sum_{k=1}^{3} \sqrt{k}$

86. SHORT RESPONSE Trevor can type a 200-word essay in 6 hours. Minya can type the same essay in $4\frac{1}{2}$ hours. If they work together, how many hours will it take them to type the essay?

Spiral Review

Determine whether each sequence is arithmetic. Write *yes* or *no*. (Lesson 11-1)

87. $-6, 4, 14, 24, \ldots$
88. $2, \frac{7}{5}, \frac{4}{5}, \frac{1}{5}, \ldots$
89. $10, 8, 5, 1, \ldots$

Solve each system of inequalities by graphing. (Lesson 10-7)

90. $x + 2y > 1$
$x^2 + y^2 \leq 25$

91. $x + y \leq 2$
$4x^2 - y^2 \geq 4$

92. $x^2 + y^2 \geq 4$
$4y^2 + 9x^2 \leq 36$

93. PHYSICS The distance a spring stretches is related to the mass attached to the spring. This is represented by $d = km$, where d is the distance, m is the mass, and k is the spring constant. When two springs with spring constants k_1 and k_2 are attached in a series, the resulting spring constant k is found by the equation $\frac{1}{k} = \frac{1}{k_1} + \frac{1}{k_2}$. (Lesson 9-6)

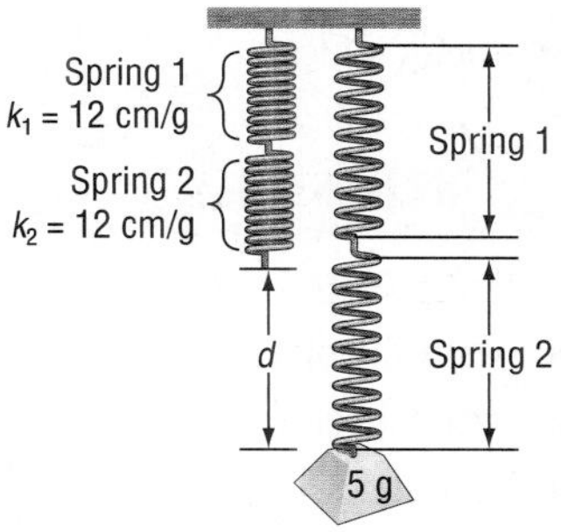

a. If one spring with constant of 12 centimeters per gram is attached in a series with another spring with constant of 8 centimeters per gram, find the resultant spring constant.

b. If a 5-gram object is hung from the series of springs, how far will the springs stretch? Is this answer reasonable in this context?

Graph each function. State the domain and range. (Lesson 8-1)

94. $f(x) = \frac{2}{3}(2^x)$
95. $f(x) = 4^x + 3$
96. $f(x) = 2\left(\frac{1}{3}\right)^x - 1$

Skills Review

Solve each equation. Round to the nearest ten-thousandth. (Lesson 8-6)

97. $5^x = 52$
98. $4^{3p} = 10$
99. $3^{n+2} = 14.5$
100. $16^{d-4} = 3^{3-d}$

11-3 Geometric Sequences and Series

Then
You determined whether a sequence was geometric. (Lesson 11-1)

Now
- Use geometric sequences.
- Find sums of geometric series.

NYS Core Curriculum

A2.A.29 Identify an arithmetic or geometric sequence and find the formula for its *n*th term
A2.A.35 Determine the sum of the first *n* terms of an arithmetic or geometric series

New Vocabulary
geometric means
geometric series

NY Math Online
glencoe.com
- Extra Examples
- Personal Tutor
- Self-Check Quiz
- Homework Help

Why?

Julian hears a song by a new band. He e-mails a link for the band's Web site to five of his friends. They each forward the link to five of their friends. The link is forwarded again following the same pattern. How many people will receive the link on the eighth round of E-mails?

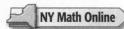

Geometric Sequences As with arithmetic sequences, there is a formula for the *n*th term of a geometric sequence. This formula can be used to determine any term of the sequence.

Key Concept ***n*th Term of a Geometric Sequence** For Your FOLDABLE

The *n*th term a_n of a geometric sequence in which the first term is a_1 and the common ratio is r is given by the following formula, where n is any natural number.

$$a_n = a_1 r^{n-1}$$

You will prove this formula in Exercise 68.

Real-World EXAMPLE 1 Find the *n*th Term

MUSIC If the pattern continues, how many E-mails will be sent in the eighth round?

Understand We need to determine the number of forwarded E-mails on the eighth round. Five E-mails were sent on the first round. Each of the five recipients sent five E-mails on the second round, and so on.

Plan This is a geometric sequence, and the common ratio is 5. Use the formula for the *n*th term of a geometric sequence.

Solve

$a_n = a_1 r^{n-1}$	*n*th term of a geometric sequence
$a_8 = 5(5)^{8-1}$	$a_1 = 5$, $r = 5$, and $n = 8$
$a_8 = 5(78{,}125)$ or $390{,}625$	$5^7 = 78{,}125$

Check Write out the first eight terms by multiplying by the common ratio.

5, 25, 125, 625, 3125, 15,625, 78,125, 390,625

There will be 390,625 E-mails sent on the 8th round.

Check Your Progress

1. **E-MAILS** Shira receives a joke in an E-mail that asks her to forward it to four of her friends. She forwards it, then each of her friends forwards it to four of their friends, and so on. If the pattern continues, how many people will receive the E-mail on the ninth round of forwarding?

Personal Tutor glencoe.com

Math History Link

Archytas (428–347 B.C.) Geometric sequences, or geometric progressions, were first studied by the Greek mathematician Archytas. His studies of these sequences came from his interest in music and octaves.

If you are given some of the terms of a geometric sequence, you can determine an equation for finding the *n*th term of the sequence.

EXAMPLE 2 Write an Equation for the *n*th Term

Write an equation for the *n*th term of each geometric sequence.

a. 0.5, 2, 8, 32, ...

$r = 8 \div 2$ or 4; 0.5 is the first term.

$a_n = a_1 r^{n-1}$ — *n*th term of a geometric sequence

$a_n = 0.5(4)^{n-1}$ — $a_1 = 0.5$ and $r = 4$

b. $a_4 = 5$ and $r = 6$

Step 1 Find a_1.

$a_n = a_1 r^{n-1}$ — *n*th term of a geometric sequence

$5 = a_1(6^{4-1})$ — $a_n = 5$, $r = 6$, and $n = 4$

$5 = a_1(216)$ — Evaluate the power.

$\frac{5}{216} = a_1$ — Divide each side by 216.

Step 2 Write the equation.

$a_n = a_1 r^{n-1}$ — *n*th term of a geometric sequence

$a_n = \frac{5}{216}(6)^{n-1}$ — $a_1 = \frac{5}{216}$ and $r = 6$

Check Your Progress

Write an equation for the *n*th term of each geometric sequence.

2A. $-0.25, 2, -16, 128, \ldots$

2B. $a_3 = 16$, $r = 4$

Personal Tutor glencoe.com

Like arithmetic means, **geometric means** are the terms between two nonconsecutive terms of a geometric sequence. The common ratio *r* can be used to find the geometric means.

ReadingMath

Geometric Means A geometric mean can also be represented geometrically. In the figure below, *h* is the geometric mean between *x* and *y*.

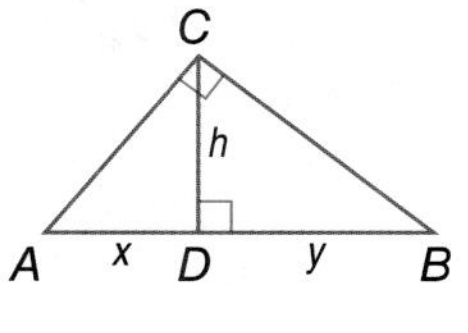

EXAMPLE 3 Find Geometric Means

Find three geometric means between 2 and 1250.

Step 1 Since there are three terms between the first and last term, there are 3 + 2 or 5 total terms, so $n = 5$.

Step 2 Find *r*.

$a_n = a_1 r^{n-1}$ — *n*th term of a geometric sequence

$1250 = 2r^{5-1}$ — $a_n = 1250$, $a_1 = 2$, and $n = 5$

$625 = r^4$ — Divide each side by 2.

$5 = r$ — Take the 4th root of each side.

Step 3 Use *r* to find the three geometric means.

2 → 10 → 50 → 250 → 1250 (each step $\times 5$)

The geometric means are 10, 50, and 250.

Check Your Progress

3. Find four geometric means between 0.5 and 512.

Personal Tutor glencoe.com

Geometric Series A **geometric series** is the sum of the terms of a geometric sequence. The sum of the first n terms of a series is denoted S_n. You can use either of the following formulas to find the partial sum S_n of the first n terms of a geometric series.

Key Concept: Partial Sum of a Geometric Series

For Your FOLDABLE

Given	The sum S_n of the first n terms is:
a_1 and n	$S_n = \frac{a_1 - a_1 r^n}{1 - r}, r \neq 1$
a_1 and a_n	$S_n = \frac{a_1 - a_n r}{1 - r}, r \neq 1$

Real-World EXAMPLE 4 Find the Sum of a Geometric Series

MUSIC Refer to the beginning of the lesson. If the pattern continues, what is the total number of E-mails sent in the eight rounds?

Five E-mails are sent in the 1st round and there are 8 rounds of E-mails. So, $a_1 = 5$, $r = 5$ and $n = 8$.

$S_n = \frac{a_1 - a_1 r^n}{1 - r}$ **Sum formula**

$S_8 = \frac{5 - 5 \cdot 5^8}{1 - 5}$ **$a_1 = 5$, $r = 5$, and $n = 8$**

$S_8 = \frac{-1{,}953{,}120}{-4}$ **Simplify the numerator and denominator.**

$S_8 = 488{,}280$ **Divide.**

There will be 488,280 E-mails sent after 8 rounds.

Check Your Progress **Find the sum of each geometric series.**

4A. $a_1 = 2, n = 10, r = 3$

4B. $a_1 = 2000, a_n = 125, r = \frac{1}{2}$

Personal Tutor glencoe.com

As with arithmetic series, sigma notation can also be used to represent geometric series.

> **Watch Out!**
>
> **Sigma Notation** Notice in Example 5 that you are being asked to evaluate the sum from the 3rd term to the 10th term.

EXAMPLE 5 Sum in Sigma Notation

Find $\sum_{k=3}^{10} 4(2)^{k-1}$.

Find a_1, r, and k. In the first term, $k = 3$ and $a_1 = 4 \cdot 2^{3-1}$ or 16. The base of the exponential function is r, so $r = 2$. There are $10 - 3 + 1$ or 8 terms, so $k = 8$.

$S_n = \frac{a_1 - a_1 r^k}{1 - r}$ **Sum formula**

$= \frac{16 - 16(2)^8}{1 - 2}$ **$a_1 = 16$, $r = 2$, and $k = 8$**

$= 4080$ **Use a calculator.**

Check Your Progress **Find each sum.**

5A. $\sum_{k=4}^{12} \frac{1}{4} \cdot 3^{k-1}$

5B. $\sum_{k=2}^{9} \frac{2}{3} \cdot 4^{k-1}$

Personal Tutor glencoe.com

You can use the formula for the sum of a geometric series to help find a particular term of the series.

EXAMPLE 6 **Find the First Term of a Series**

Find a_1 in a geometric series for which $S_n = 13{,}116$, $n = 7$, and $r = 3$.

$S_n = \frac{a_1 - a_1r^n}{1 - r}$ **Sum formula**

$13{,}116 = \frac{a_1 - a_1(3^7)}{1 - 3}$ **$S_n = 13{,}116$, $r = 3$, and $n = 7$**

$13{,}116 = \frac{a_1(1 - 3^7)}{1 - 3}$ **Distributive Property**

$13{,}116 = \frac{-2186a_1}{-2}$ **Subtract.**

$13{,}116 = 1093a_1$ **Simplify.**

$12 = a_1$ **Divide each side by 1093.**

Check Your Progress

6. Find a_1 in a geometric series for which $S_n = -26{,}240$, $n = 8$, and $r = -3$.

Personal Tutor glencoe.com

Check Your Understanding

Example 1 p. 696

1. INCOME Dean began working for his parents, earning $8 an hour. If his parents promised him a 5% raise every three months, how much will Dean be earning per hour after 5 years?

Example 2 p. 697

Write an equation for the *n*th term of each geometric sequence.

2. 2, 4, 8, …

3. 18, 6, 2, …

4. −4, 16, −64, …

5. $a_2 = 4$, $r = 3$

6. $a_6 = \frac{1}{8}$, $r = \frac{3}{4}$

7. $a_2 = -96$, $r = -8$

Example 3 p. 697

Find the geometric means of each sequence.

8. 0.25, _?_, _?_, _?_, 64

9. 0.20, _?_, _?_, _?_, 125

Example 4 p. 698

10. GAMES Miranda arranges some rows of dominoes so that after she knocks over the first one, each domino knocks over two more dominoes when it falls. If there are ten rows, how many dominoes does Miranda use?

Example 5 p. 698

Find the sum of each geometric series.

11. $\sum_{k=1}^{6} 3(4)^{k-1}$

12. $\sum_{k=1}^{8} 4\left(\frac{1}{2}\right)^{k-1}$

Example 6 p. 699

Find a_1 for each geometric series described.

13. $S_n = 85\frac{5}{16}$, $r = 4$, $n = 6$

14. $S_n = 91\frac{1}{12}$, $r = 3$, $n = 7$

15. $S_n = 1020$, $a_n = 4$, $r = \frac{1}{2}$

16. $S_n = 121\frac{1}{3}$, $a_n = \frac{1}{3}$, $r = \frac{1}{3}$

Practice and Problem Solving

= **Step-by-Step Solutions** begin on page R20. **Extra Practice** begins on page 947.

Example 1 p. 696

17. DEPRECIATION Brieanne's stock is expected to depreciate in value at a rate of 5% per year. If her stock is currently valued at \$24,000, how much will it be worth in 6 years?

Find a_n for each geometric sequence.

18. $a_1 = 2400, r = \frac{1}{4}, n = 7$

19. $a_1 = 800, r = \frac{1}{2}, n = 6$

20. $a_1 = \frac{2}{9}, r = 3, n = 7$

21. $a_1 = -4, r = -2, n = 8$

22. BIOLOGY A certain bacteria grows at a rate of 3 cells every 2 minutes. If there were 260 cells initially, how many are there after 21 minutes?

Example 2 p. 697

Write an equation for the *n*th term of each geometric sequence.

23. $-3, 6, -12, \ldots$

24. $288, -96, 32, \ldots$

25. $-1, 1, -1, \ldots$

26. $\frac{1}{3}, \frac{2}{9}, \frac{4}{27}, \ldots$

27. $8, 2, \frac{1}{2}, \ldots$

28. $12, -16, \frac{64}{3}, \ldots$

29. $a_3 = 28, r = 2$

30. $a_4 = -8, r = 0.5$

31. $a_6 = 0.5, r = 6$

32. $a_3 = 8, r = \frac{1}{2}$

33. $a_4 = 24, r = \frac{1}{3}$

34. $a_4 = 80, r = 4$

Example 3 p. 697

Find the geometric means of each sequence.

35. 810, ?, ?, ?, 10

36. 640, ?, ?, ?, 2.5

37. $\frac{7}{2}$, ?, ?, ?, $\frac{56}{81}$

38. $\frac{729}{64}$, ?, ?, ?, $\frac{324}{9}$

39. Find two geometric means between 3 and 375.

40. Find two geometric means between 16 and −2.

Example 4 p. 698

41. WATER TREATMENT A certain water filtration system can remove 70% of the contaminants each time a sample of water is passed through it. If the same water is passed through the system four times, what percent of the original contaminants will be removed from the water sample?

Find the sum of each geometric series.

42. $a_1 = 36, r = \frac{1}{3}, n = 8$

43. $a_1 = 16, r = \frac{1}{2}, n = 9$

44. $a_1 = 240, r = \frac{3}{4}, n = 7$

45. $a_1 = 360, r = \frac{4}{3}, n = 8$

46. VACUUMS A vacuum claims to pick up 80% of the dirt every time it is run over the carpet. Assuming this is true, what percent of the original amount of dirt is picked up after the seventh time the vacuum is run over the carpet?

Example 5 p. 698

Find the sum of each geometric series.

47. $\sum_{k=1}^{7} 4(-3)^{k-1}$

48. $\sum_{k=1}^{8} (-3)(-2)^{k-1}$

49. $\sum_{k=1}^{9} (-1)(4)^{k-1}$

50. $\sum_{k=1}^{10} 5(-1)^{k-1}$

Example 6 p. 699

Find a_1 for each geometric series described.

51. $S_n = -2912, r = 3, n = 6$

52. $S_n = -10{,}922, r = 4, n = 7$

53. $S_n = 1330, a_n = 486, r = \frac{3}{2}$

54. $S_n = 4118, a_n = 128, r = \frac{2}{3}$

55. $a_n = 1024, r = 8, n = 5$

56. $a_n = 1875, r = 5, n = 7$

Real-World Link

A metronome is a pendulum device that produces a pulse, usually to establish a steady beat or tempo for the performance of musical compositions.

Source: *Glide Magazine*

57. **REAL ESTATE** Colin's house has been appreciating in value at a rate of 8% per year. If this trend continues and his house is currently worth \$175,000, how much will it be worth in 15 years?

58. **CHEMISTRY** Radon has a half-life of about 4 days. This means that about every 4 days, half of the mass of radon decays into another element. How many grams of radon remain from an initial 60 grams after 4 weeks?

59. **COMPUTERS** A virus goes through a computer, infecting the files. If one file was infected initially and the total number of files infected doubles every minute, how many files will be infected in 20 minutes?

60. **GEOMETRY** In the figure, the sides of each equilateral triangle are twice the size of the sides of its inscribed triangle. If the pattern continues, find the sum of the perimeters of the first eight triangles.

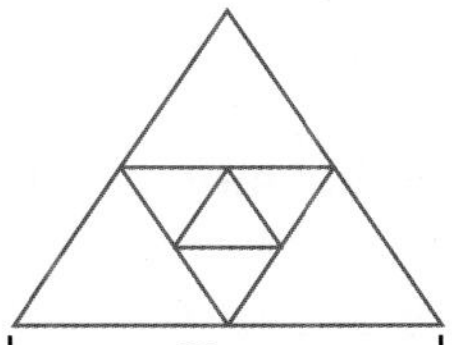

61. **PENDULUMS** The first swing of a pendulum travels 30 centimeters. If each subsequent swing travels 95% as far as the previous swing, find the total distance traveled by the pendulum after the 30th swing.

62. **PHONE CHAINS** A school established a phone chain in which every staff member calls two other staff members to notify them when the school closes due to weather. The first round of calls begins with the superintendent calling both principals. If there are 94 total staff members and employees at the school, how many rounds of calls are there?

63. **TELEVISIONS** High Tech Electronics advertises a weekly installment plan for the purchase of a popular brand of high definition television. The buyer pays \$5 at the end of the first week, \$5.50 at the end of the second week, \$6.05 at the end of the third week, and so on.

a. What will the payments be at the end of the 10th, 20th, and 40th weeks?

b. Find the total cost of the TV.

c. Why is the cost found in part **b** not entirely accurate?

H.O.T. Problems

Use Higher-Order Thinking Skills

64. **PROOF** Derive the Alternate Sum Formula using the General Sum Formula.

65. **PROOF** Derive a sum formula that does not include a_1.

66. **OPEN ENDED** Write a geometric series for which $r = \frac{3}{4}$ and $n = 6$.

67. **REASONING** Explain how $\sum_{k=1}^{10} 3(2)^{k-1}$ needs to be altered to refer to the same series if $k = 1$ changes to $n = 0$. Explain your reasoning.

68. **PROOF** Prove the formula for the nth term of a geometric sequence.

69. **CHALLENGE** The fifth term of a geometric sequence is $\frac{1}{27}$th of the eighth term. If the ninth term is 702, what is the eighth term?

70. **CHALLENGE** Use the fact that h is the geometric mean between x and y in the figure at the right to find h^4 in terms of x and y.

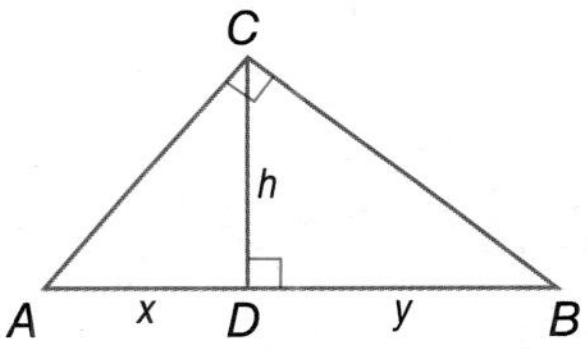

71. **OPEN ENDED** Write a geometric series with 6 terms and a sum of 252.

72. **WRITING IN MATH** Explain how you determine whether a series is *arithmetic*, *geometric*, *neither*, or *both*.

NYSRE PRACTICE A2.A.29, A2.A.35

73. Which of the following is closest to $\sqrt[3]{7.32}$?

A 1.8
B 1.9
C 2.0
D 2.1

74. The first term of a geometric series is -5, and the common ratio is -2. How many terms are in the series if its sum is $-6828\frac{1}{3}$?

F 5
G 9
H 10
J 12

75. SHORT RESPONSE Danette has a savings account. She withdraws half of the contents every year. After 4 years, she has \$2000 left. How much did she have in the savings account originally?

76. ACT/SAT The curve below could be part of the graph of which function?

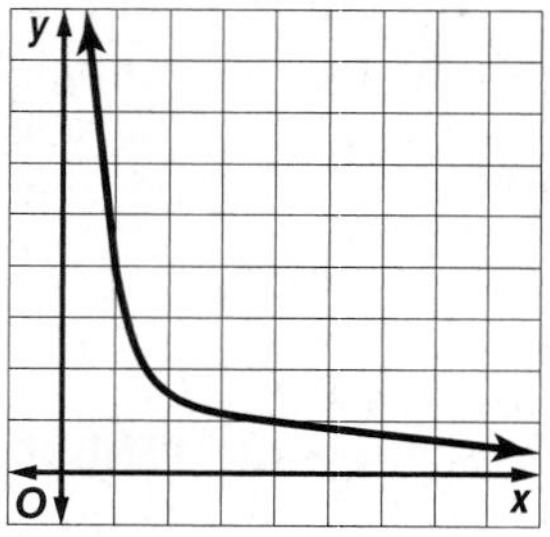

A $y = \sqrt{x}$
B $y = x^2 - 5x + 4$
C $xy = 4$
D $y = -x + 20$

Spiral Review

77. MONEY Elena bought a high-definition LCD television at the electronics store. She paid \$200 immediately and \$75 each month for a year and a half. How much did Elena pay in total for the TV? (Lesson 11-2)

Determine whether each sequence is *arithmetic, geometric,* or *neither*. Explain your reasoning. (Lesson 11-1)

78. $\frac{1}{10}, \frac{3}{5}, \frac{7}{20}, \frac{17}{20}, \ldots$

79. $-\frac{7}{25}, -\frac{13}{50}, -\frac{6}{25}, -\frac{11}{50}, \ldots$

80. $-\frac{22}{3}, -\frac{68}{9}, -\frac{208}{27}, -\frac{632}{81}, \ldots$

Find the center and radius of each circle. Then graph the circle. (Lesson 10-3)

81. $(x - 3)^2 + (y - 1)^2 = 25$

82. $(x + 3)^2 + (y + 7)^2 = 81$

83. $(x - 3)^2 + (y + 7)^2 = 50$

84. Suppose y varies jointly as x and z. Find y when $x = 9$ and $z = -5$, if $y = -90$ when $z = 15$ and $x = -6$. (Lesson 9-5)

85. SHOPPING A certain store found that the number of customers who will attend a sale can be modeled by $N = 125\sqrt[3]{100Pt}$, where N is the number of customers expected, P is the percent of the sale discount, and t is the number of hours the sale will last. Find the number of customers the store should expect for a sale that is 50% off and will last four hours. (Lesson 7-4)

Skills Review

Evaluate each expression if $a = -2$, $b = \frac{1}{3}$, and $c = -12$. (Lesson 1-1)

86. $\frac{3ab}{c}$

87. $\frac{a - c}{a + c}$

88. $\frac{a^3 - c}{b^2}$

89. $\frac{c + 3}{ab}$

Algebra Lab
Area Under a Curve

NY Math Online glencoe.com
Math in Motion, Animation

The Morgans are renovating the outside of their house so that there is an archway above the front entrance. Mr. Morgan made a scale drawing of the archway in which each line on the grid paper represents one foot of the actual archway. Mrs. Morgan modeled the shape of the top with the quadratic equation $y = -0.25x^2 + 3x$.

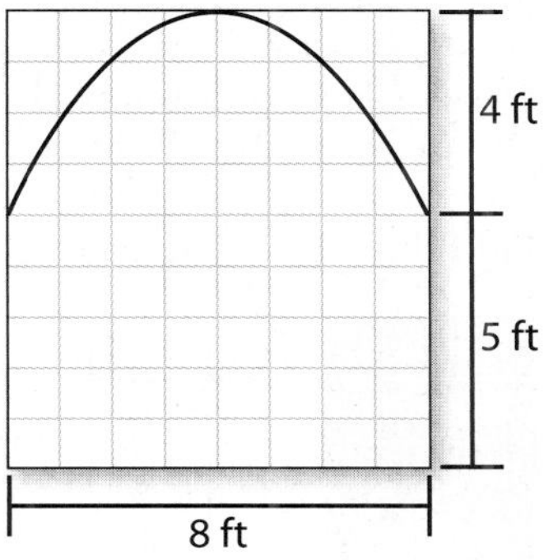

ACTIVITY

Find the area of the opening under the archway.

Method 1

Step 1 Make a table of values for $y = -0.25x^2 + 3x$. Then graph the equation.

x	0	1	2	3	4	5	6	7	8	9	10	11	12
y	0	2.75	5	6.75	8	8.75	9	8.75	8	6.75	5	2.75	0

Step 2 Divide the figure into regions.

To estimate the area inside the archway, you can divide the archway into rectangles as shown in red.

Because the left and right sides of the archway are 5 feet high and $y = 5$ when $x = 2$ and when $x = 10$, the opening of the entrance extends from $x = 2$ to $x = 10$.

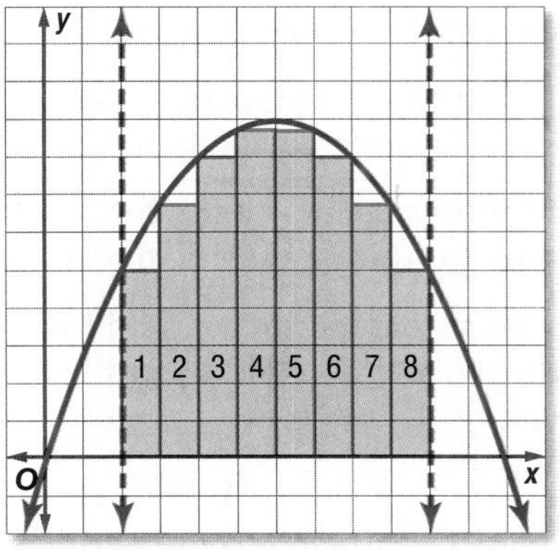

Step 3 Find the area of the regions.

Rectangle	1	2	3	4	5	6	7	8
Width (ft)	1	1	1	1	1	1	1	1
Height (ft)	5	6.75	8	8.75	8.75	8	6.75	5
Area (ft^2)	5	6.75	8	8.75	8.75	8	6.75	5

The approximate area of the archway is the sum of the areas of the rectangles.

$5 + 6.75 + 8 + 8.75 + 8.75 + 8 + 6.75 + 5 = 57 \text{ ft}^2$

Method 2

Step 1 Draw a second graph of the equation and divide into regions. Divide the archway into rectangles as shown in blue.

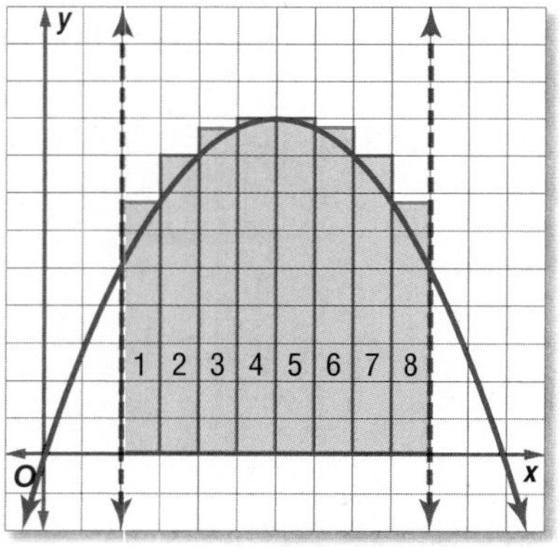

Step 2 Find the area of the regions.

Rectangle	1	2	3	4	5	6	7	8
Width (ft)	1	1	1	1	1	1	1	1
Height (ft)	6.75	8	8.75	9	9	8.75	8	6.75
Area (ft^2)	6.75	8	8.75	9	9	8.75	8	6.75

The approximate area of the archway is the sum of the areas of the rectangles.

$6.75 + 8 + 8.75 + 9 + 9 + 8.75 + 8 + 6.75 = 65 \text{ ft}^2$

Both Method 1 and Method 2 illustrate how to approximate the area under a curve within a specified interval.

Analyze the Results

1. Is the area of the regions calculated using Method 1 greater than or less than the actual area of the archway? Explain your reasoning.
2. Is the area of the regions calculated using Method 2 greater than or less than the actual area of the archway? Explain your reasoning.
3. Compare the area estimates for both methods. How could you find the best estimate for the area inside the archway? Explain your reasoning.
4. The diagram shows a third method for finding an estimate of the area of the archway. Is this estimate for the area greater than or less than the actual area? How does this estimate compare to the other two estimates of the area?

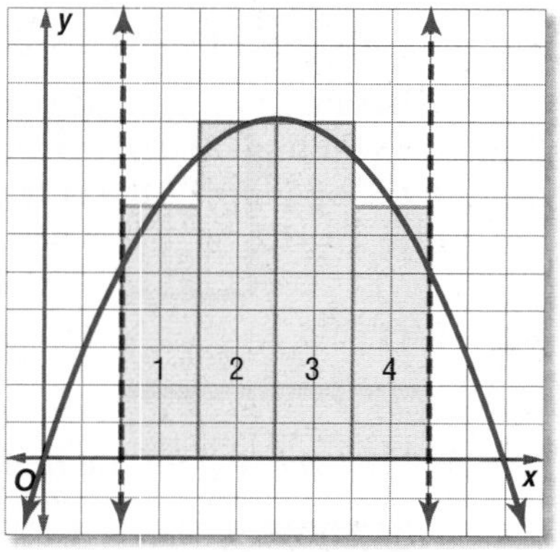

Exercises

Estimate the area described by any method. Make a table of values, draw graphs with rectangles, and make a table for the areas of the rectangles. Compare each estimate to the actual area.

5. the area under the curve for $y = -x^2 + 4$, from $x = -2$ to $x = 2$, and above the x–axis
6. the area under the curve for $y = x^3$, from $x = 0$ to $x = 4$, and above the x–axis
7. the area under the curve for $y = x^2$, from $x = -3$ to $x = 3$, and above the x–axis

11-4 Infinite Geometric Series

Why?

With their opponent on the 10-yard line, the defense is penalized half the distance to the goal, placing the ball on the 5-yard line. If they continue to be penalized in this way, where will the ball eventually be placed? Will they ever reach the goal line? How many total penalty yards will the defense have incurred? These questions can be answered by looking at infinite geometric series.

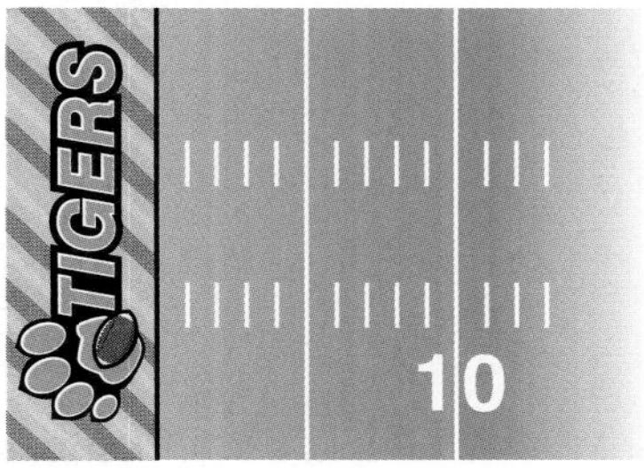

Then

You found sums of finite geometric series. (Lesson 11-4)

Now

- Find sums of infinite geometric series.
- Write repeating decimals as fractions.

NYS Core Curriculum

A2.A.34 Represent the sum of a series, using sigma notation
A2.A.35 Determine the sum of the first *n* terms of an arithmetic or geometric series
Also addresses A2.N.10.

New Vocabulary

convergent series
divergent series
infinity

NY Math Online

glencoe.com

- Extra Examples
- Personal Tutor
- Self-Check Quiz
- Homework Help

Infinite Geometric Series An infinite geometric series that has a sum is called a **convergent series**, because its sum converges to a specific value. An infinite geometric series that does *not* have a sum is called a **divergent series**.

When you evaluated the sum S_n of an infinite geometric series for the first n terms, you were finding the partial sum of the series. It is also possible to find the sum of an entire series. In the application above, it seems that the ball will eventually reach the goal line, and the defense will be penalized a total of 10 yards. This value is the actual sum of the infinite series $5 + 2.5 + 1.25 + \dots$. The graph of S_n for $1 \leq n \leq 10$ is shown on the left below. As n increases, S_n approaches 10.

Key Concept — Convergent and Divergent Series

For Your FOLDABLE

	Convergent Series	Divergent Series
Words	The sum approaches a finite value.	The sum does not approach a finite value.
Ratio	$\lvert r \rvert < 1$	$\lvert r \rvert \geq 1$
Example	$5 + 2.5 + 1.25 + \dots$	$\frac{1}{16} + \frac{1}{8} + \frac{1}{4} + \dots$

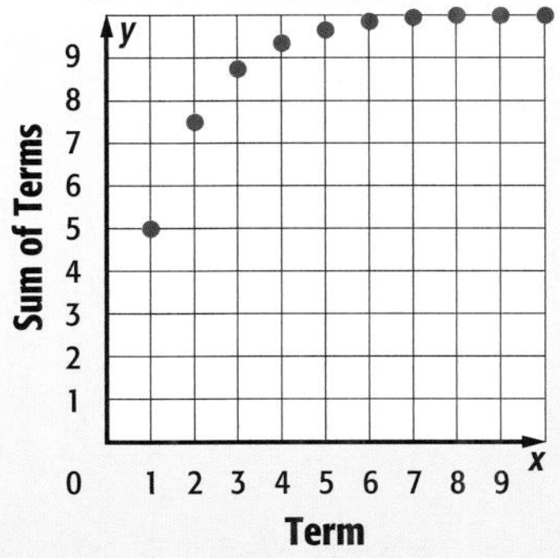

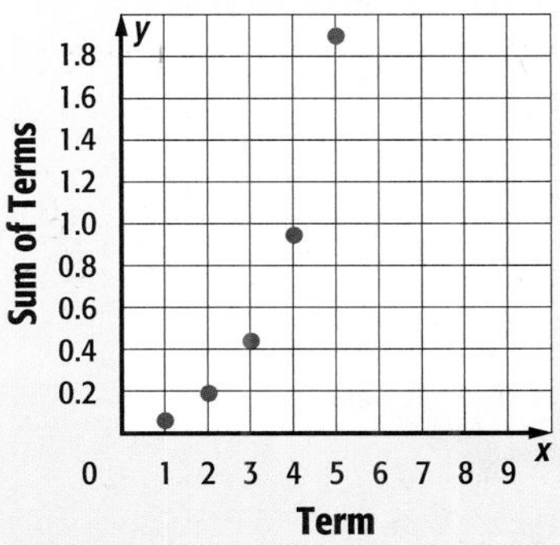

EXAMPLE 1 Convergent and Divergent Series

Determine whether each infinite geometric series is *convergent* or *divergent*.

a. $54 + 36 + 24 + \dots$

Find the value of r.

$r = \frac{36}{54}$ or $\frac{2}{3}$; since $-1 < \frac{2}{3} < 1$, the series is convergent.

StudyTip

Absolute Value Recall that $|r| < 1$ means $-1 < r < 1$.

b. $8 + 12 + 18 + \ldots$

$r = \frac{12}{8}$ or 1.5; since $1.5 > 1$, the series is divergent.

Check Your Progress

1A. $2 + 3 + 4.5 + \ldots$

1B. $100 + 50 + 25 + \ldots$

Personal Tutor glencoe.com

When $|r| < 1$, the value of r^n will approach 0 as n increases. Therefore, the partial sums of the infinite geometric series will approach $\frac{a_1 - a_1(0)}{1 - r}$ or $\frac{a_1}{1 - r}$.

Key Concept **Sum of an Infinite Geometric Series** For Your FOLDABLE

The sum S of an infinite geometric series with $|r| < 1$ is given by

$$S = \frac{a_1}{1 - r}.$$

If $|r| \geq 1$, the series has no sum.

When an infinite geometric series is divergent, $|r| \geq 1$ and the series has no sum because the value of r^n will increase infinitely as n increases.

The table at the right shows the partial sums for the divergent series $4 + 16 + 64 + \ldots$. As n increases, S_n increases rapidly without limit.

n	S_n
5	1364
10	87,380
15	1,431,655,764

EXAMPLE 2 Sum of an Infinite Series

Find the sum of each infinite series, if it exists.

a. $\frac{2}{3} + \frac{6}{15} + \frac{18}{75} + \ldots$

Step 1 Find the value of r to determine if the sum exists.

$r = \frac{6}{15} \div \frac{2}{3}$ or $\frac{3}{5}$ **Divide consecutive terms.**

Since $\left|\frac{3}{5}\right| < 1$, the sum exists.

Step 2 Use the formula to find the sum.

$S = \frac{a_1}{1 - r}$ **Sum formula**

$= \frac{\frac{2}{3}}{1 - \frac{3}{5}}$ $a_1 = \frac{2}{3}$ **and** $r = \frac{3}{5}$

$= \frac{2}{3} \div \frac{2}{5}$ or $\frac{5}{3}$ **Simplify.**

b. $6 + 9 + 13.5 + 20.25 + \ldots$

$r = \frac{9}{6}$ or 1.5; since $|1.5| \geq 1$, the series diverges and the sum does not exist.

StudyTip

Convergence and Divergence A sequence converges when the absolute value of a term is smaller than the absolute value of the previous term. An infinite arithmetic series will always be divergent.

Check Your Progress

2A. $4 - 2 + 1 - 0.5 + \ldots$

2B. $16 + 20 + 25 + \ldots$

Personal Tutor glencoe.com

Sigma notation can be used to represent infinite series. If a sequence goes to **infinity**, it continues without end. The infinity symbol ∞ is placed above the $\sum$ to indicate that a series is infinite.

EXAMPLE 3 Infinite Series in Sigma Notation

Find $\sum_{k=1}^{\infty} 18\left(\frac{4}{5}\right)^{k-1}$.

$S = \frac{a_1}{1-r}$ **Sum formula**

$= \frac{18}{1-\frac{4}{5}}$ $a_1 = 18$ **and** $r = \frac{4}{5}$

$= \frac{18}{\frac{1}{5}}$ or 90 **Simplify.**

Check Your Progress

3. Find $\sum_{k=1}^{\infty} 12\left(\frac{3}{4}\right)^{k-1}$.

Personal Tutor glencoe.com

Repeating Decimals A repeating decimal is the sum of an infinite geometric series. For instance, $0.\overline{45} = 0.454545\ldots$ or $0.45 + 0.0045 + 0.000045 + \ldots$. The formula for the sum of an infinite series can be used to convert the decimal to a fraction.

Problem-SolvingTip

Choose the Best Method of Computation In many cases, it is possible to solve a problem in more than one way. Use the method with which you are most comfortable.

EXAMPLE 4 Write a Repeating Decimal as a Fraction

Write $0.\overline{63}$ **as a fraction.**

Method 1 Use the sum of an infinite series.

$0.\overline{63} = 0.63 + 0.0063 + \ldots$

$= \frac{63}{100} + \frac{63}{10{,}000} + \ldots$

$S = \frac{a_1}{1-r}$ **Sum formula**

$= \frac{\frac{63}{100}}{1-\frac{1}{100}}$ $a_1 = \frac{63}{100}$ **and** $r = \frac{1}{100}$

$= \frac{63}{99}$ or $\frac{7}{11}$ **Simplify.**

Method 2 Use algebraic properties.

$x = 0.\overline{63}$	**Let** $x = 0.\overline{63}$.
$x = 0.636363\ldots$	**Write as a repeating decimal.**
$100x = 63.636363\ldots$	**Multiply each side by 100.**
$99x = 63$	**Subtract** x **from** $100x$ **and** $0.\overline{63}$ **from** $63.\overline{63}$.
$x = \frac{63}{99}$ or $\frac{7}{11}$	**Divide each side by 99.**

StudyTip

Repeating Decimals Every repeating decimal is a rational number and can be written as a fraction.

Check Your Progress

4. Write $0.\overline{21}$ as a fraction.

Personal Tutor glencoe.com

Check Your Understanding

Example 1 p. 705

Determine whether each infinite geometric series is *convergent* or *divergent*.

1. $16 - 8 + 4 - \ldots$
2. $32 - 48 + 72 - \ldots$
3. $0.5 + 0.7 + 0.98 + \ldots$
4. $1 + 1 + 1 + \ldots$

Example 2 p. 706

Find the sum of each infinite series, if it exists.

5. $440 + 220 + 110 + \ldots$
6. $520 + 130 + 32.5 + \ldots$
7. $\frac{1}{4} + \frac{3}{8} + \frac{9}{16} + \ldots$
8. $\frac{32}{9} + \frac{16}{3} + 8 + \ldots$

9. **MEDICINE** A certain drug has a half-life of 8 hours after it is administered to a patient. What percent of the drug is still in the patient's system after 24 hours?

Example 3 p. 707

Find the sum of each infinite series, if it exists.

10. $\sum_{k=1}^{\infty} 5 \cdot 4^{k-1}$
11. $\sum_{k=1}^{\infty} (-2) \cdot (0.5)^{k-1}$
12. $\sum_{k=1}^{\infty} 3 \cdot \left(\frac{4}{5}\right)^{k-1}$
13. $\sum_{k=1}^{\infty} \frac{1}{2} \cdot \left(\frac{3}{4}\right)^{k-1}$

Example 4 p. 707

Write each repeating decimal as a fraction.

14. $0.\overline{35}$
15. $0.\overline{642}$

Practice and Problem Solving

● = **Step-by-Step Solutions** begin on page R20.
Extra Practice begins on page 947.

Example 1 p. 705

Determine whether each infinite geometric series is *convergent* or *divergent*.

16. $21 + 63 + 189 + \ldots$
17. $480 + 360 + 270 + \ldots$
18. $\frac{3}{4} + \frac{9}{8} + \frac{27}{16} + \ldots$
19. $\frac{5}{6} + \frac{10}{9} + \frac{40}{27} + \ldots$
20. $0.1 + 0.01 + 0.001 + \ldots$
21. $0.008 + 0.08 + 0.8 + \ldots$

Example 2 p. 706

Find the sum of each infinite series, if it exists.

22. $18 + 21.6 + 25.92 + \ldots$
23. $-3 - 4.2 - 5.88 - \ldots$
24. $\frac{1}{2} + \frac{1}{6} + \frac{1}{18} + \ldots$
25. $\frac{12}{5} + \frac{6}{5} + \frac{3}{5} + \ldots$
26. $21 + 14 + \frac{28}{3} + \ldots$
27. $32 + 40 + 50 + \ldots$

28. **SWINGS** If Kerry does not push any harder after his initial swing, the distance traveled per swing will decrease by 10% with each swing. If his initial swing traveled 6 feet, find the total distance traveled when he comes to rest.

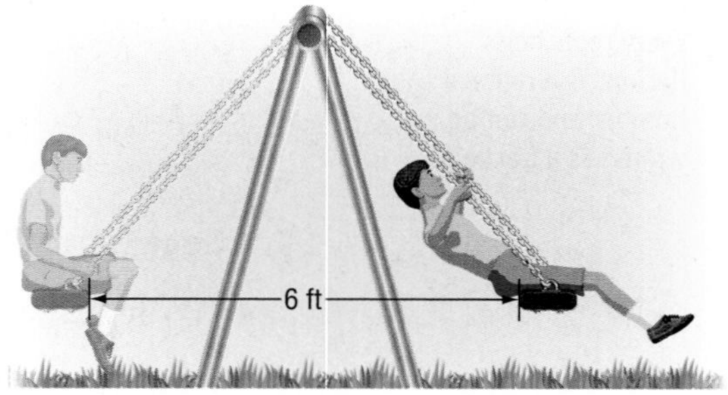

Example 3 p. 707

Find the sum of each infinite series, if it exists.

29. $\sum_{k=1}^{\infty} \frac{4}{3} \cdot \left(\frac{5}{4}\right)^{k-1}$

30. $\sum_{k=1}^{\infty} \frac{1}{4} \cdot 3^{k-1}$

31. $\sum_{k=1}^{\infty} \frac{5}{3} \cdot \left(\frac{3}{7}\right)^{k-1}$

32. $\sum_{k=1}^{\infty} \frac{2}{3} \cdot \left(\frac{4}{3}\right)^{k-1}$

33. $\sum_{k=1}^{\infty} \frac{8}{3} \cdot \left(\frac{5}{6}\right)^{k-1}$

34. $\sum_{k=1}^{\infty} \frac{1}{8} \cdot \left(\frac{1}{12}\right)^{k-1}$

Example 4 p. 707

Write each repeating decimal as a fraction.

35. $0.3\overline{21}$

36. $0.1\overline{45}$

37. $2.\overline{18}$

38. $4.\overline{96}$

39. $0.12\overline{14}$

40. $0.4\overline{336}$

41. **FANS** A fan is running at 10 revolutions per second. After it is turned off, its speed decreases at a rate of 75% per second. Determine the number of revolutions completed by the fan after it is turned off.

42. **INVESTMENT** Kamiko deposited $5000 into an account at the beginning of the year. The account earns 8% interest each year.

    a. How much money will be in the account after 20 years? (*Hint*: Let $5000(1 + 0.08)^1$ represent the end of the first year.)

    b. Is this series *convergent* or *divergent*? Explain.

Real-World Link

Batteries have been in use in the United States for over 100 years, and they are in demand like never before. As a result, about 3 billion single-use batteries are disposed of each year. One rechargeable battery can replace up to 100 single-use alkaline batteries.

43. **RECHARGEABLE BATTERIES** A certain rechargeable battery is advertised to recharge back to 99.9% of its previous capacity with every charge. If its initial capacity is 8 hours of life, how many total hours should the battery last?

Find the sum of each infinite series, if it exists.

44. $\frac{7}{5} + \frac{21}{20} + \frac{63}{80} + \ldots$

45. $\frac{15}{4} + \frac{5}{2} + \frac{5}{3} + \ldots$

46. $-\frac{16}{9} + \frac{4}{3} - 1 + \ldots$

47. $\frac{15}{8} + \frac{5}{2} + \frac{10}{3} + \ldots$

48. $\frac{21}{16} + \frac{7}{4} + \frac{7}{3} + \ldots$

49. $-\frac{18}{7} + \frac{12}{7} - \frac{8}{7} + \ldots$

50. **MULTIPLE REPRESENTATIONS** In this problem, you will use a square of paper that is at least 8 inches on a side.

    a. **CONCRETE** Let the square be one unit. Cut away one half of the square. Call this piece Term 1. Next, cut away one half of the remaining sheet of paper. Call this piece Term 2. Continue cutting the remaining paper in half and labeling the pieces with a term number as long as possible. List the fractions represented by the pieces.

    b. **NUMERICAL** If you could cut the squares indefinitely, you would have an infinite series. Find the sum of the series.

    c. **VERBAL** How does the sum of the series relate to the original square of paper?

51. **PHYSICS** In a physics experiment, a steel ball on a flat track is accelerated, and then allowed to roll freely. After the first minute, the ball has rolled 120 feet. Each minute the ball travels only 40% as far as it did during the preceding minute. How far does the ball travel?

52. **PENDULUMS** A pendulum travels 12 centimeters on its first swing and 95% of that distance on each swing thereafter. Find the total distance traveled by the pendulum when it comes to rest.

53. **TOYS** If a rubber ball can bounce back to 95% of its original height, what is the total vertical distance that it will travel if it is dropped from an elevation of 30 feet?

54. **CARS** During a maintenance inspection, a tire is removed from a car and spun on a diagnostic machine. When the machine is turned off, the spinning tire completes 20 revolutions the first second and 98% of the revolutions each additional second. How many revolutions does the tire complete before it stops spinning?

Real-World Link

In 2001, the federal government enacted a tax cut that sent checks ranging from $300 to $600 to every taxpayer in the United States.

Source: *The Economist*

55. **ECONOMICS** A state government decides to stimulate its economy by giving $500 to every adult. The government assumes that everyone who receives the money will spend 80% on consumer goods and that the producers of these goods will in turn spend 80% on consumer goods. How much money is generated for the economy for every $500 that the government provides?

56. **SCIENCE MUSEUM** An exhibit at a science museum offers visitors the opportunity to experiment with the motion of an object on a spring. One visitor pulls the object down and lets it go. The object travels 1.2 feet upward before heading back the other way. Each time the object changes direction, it decreases its distance by 20% when compared to the previous direction. Find the total distance traveled by the object.

Match each graph with its corresponding description.

57.

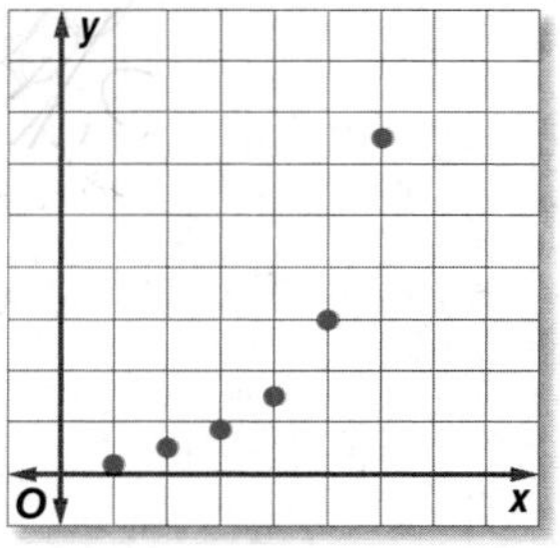

58.

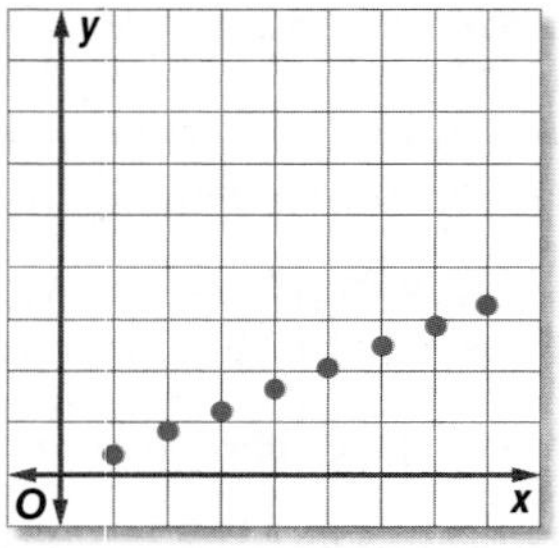

59.

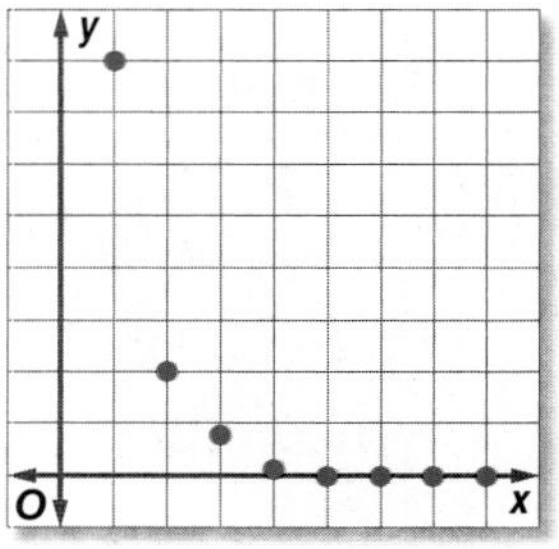

a. converging geometric series

b. diverging geometric series

c. converging arithmetic series

d. diverging arithmetic series

H.O.T. Problems

Use Higher-Order Thinking Skills

60. **FIND THE ERROR** Emmitt and Austin are asked to find the sum of $1 - 1 + 1 - \ldots$. Is either of them correct? Explain your reasoning.

Emmitt	Austin
The sum is 0 because the sum of each pair of terms in the sequence is 0.	*There is no sum because $\lvert r \rvert \geq 1$, and the series diverges.*

61. **PROOF** Derive the formula for the sum of an infinite geometric series.

62. **CHALLENGE** For what values of b does $3 + 9b + 27b^2 + 81b^3 + \ldots$ have a sum?

63. **REASONING** When does an infinite geometric series have a sum, and when does it not have a sum? Explain your reasoning.

64. **REASONING** Determine whether the following statement is *sometimes*, *always*, or *never* true. Explain your reasoning.

If the absolute value of a term of any geometric series is greater than the absolute value of the previous term, then the series is divergent.

65. **OPEN ENDED** Write an infinite series with a sum that converges to 9.

66. **OPEN ENDED** Write $3 - 6 + 12 - \ldots$ using sigma notation in two different ways.

67. **WRITING IN MATH** Explain why an arithmetic series is always divergent.

NYSRE PRACTICE A2.A.34, A2.A.35

68. ACT/SAT What is the sum of an infinite geometric series with a first term of 27 and a common ratio of $\frac{2}{3}$?

A 81 C 34

B 65 D 18

69. Adelina, Michelle, Masao, and Brandon each simplified the same expression at the board. Each student's work is shown below. The teacher said that while two of them had a correct answer, only one of them had arrived at the correct conclusion using correct steps.

Adelina's work

$$x^2x^{-5} = \frac{x^2}{x^{-5}} = x^7, x \neq 0$$

Masao's work

$$x^2x^{-5} = \frac{x^2}{x^5} = \frac{1}{x^3}, x \neq 0$$

Michelle's work

$$x^2x^{-5} = \frac{x^2}{x^{-5}} = x^{-3}, x \neq 0$$

Brandon's work

$$x^2x^{-5} = \frac{x^2}{x^5} = x^3, x \neq 0$$

Which is a completely accurate simplification?

F Adelina's work H Masao's work

G Michelle's work J Brandon's work

70. SHORT RESPONSE Evaluate $\log_8 60$ to the nearest hundredth.

71. GEOMETRY The radius of a large sphere was multiplied by a factor of $\frac{1}{3}$ to produce a smaller sphere.

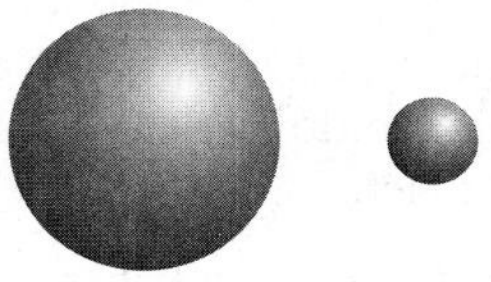

How does the volume of the smaller sphere compare to the volume of the larger sphere?

A The volume of the smaller sphere is $\frac{1}{9}$ as large.

B The volume of the smaller sphere is $\frac{1}{\pi^3}$ as large.

C The volume of the smaller sphere is $\frac{1}{27}$ as large.

D The volume of the smaller sphere is $\frac{1}{3}$ as large.

Spiral Review

72. GAMES An audition is held for a TV game show. At the end of each round, one half of the prospective contestants are eliminated from the competition. On a particular day, 524 contestants begin the audition. (Lesson 11-3)

a. Write an equation for finding the number of contestants who are left after n rounds.

b. Using this method, will the number of contestants who are to be eliminated always be a whole number? Explain.

73. CLUBS A quilting club consists of 9 members. Every week, each member must bring one completed quilt square. (Lesson 11-2)

a. Find the first eight terms of the sequence that describes the total number of squares that have been made after each meeting.

b. One particular quilt measures 72 inches by 84 inches and is being designed with 4-inch squares. After how many meetings will the quilt be complete?

Skills Review

Find each function value. (Lesson 2-1)

74. $f(x) = 5x - 9, f(6)$

75. $g(x) = x^2 - x, g(4)$

76. $h(x) = x^2 - 2x - 1, h(3)$

Graphing Technology Lab
Limits

NY Math Online glencoe.com
Other Calculator Keystrokes

You may have noticed that in some geometric sequences, the later the term in the sequence, the closer the value is to 0. Another way to describe this is that as n increases, a_n approaches 0. The value that the terms of a sequence approach, in this case 0, is called the **limit** of the sequence. Other types of infinite sequences may also have limits. But if the terms of a sequence do not approach a unique value, we say that the limit of the sequence does not exist.

You can use a TI-83/84 Plus graphing calculator to help find the limits of infinite sequences.

ACTIVITY

Find the limit of the geometric sequence $1, \frac{1}{4}, \frac{1}{16}, \ldots$.

Step 1 Enter the sequence.

The formula for this sequence is $a_n = \left(\frac{1}{4}\right)^{n-1}$.

- Position the cursor on **L1** in the **STAT EDIT 1: Edit...** screen and enter the formula **seq(N,N,1,10,1)**. This generates the values 1, 2, ..., 10 of the index **N**.
- Position the cursor on **L2** and enter the formula **seq((1/4)^(N-1),N,1,10,1)**. This generates the first ten terms of the sequence.

Notice that as n increases, the terms of the given sequence get closer and closer to 0. If you scroll down, you can see that for $n \geq 6$ the terms are so close to 0 that the calculator expresses them in scientific notation. This suggests that the limit of the sequence is 0.

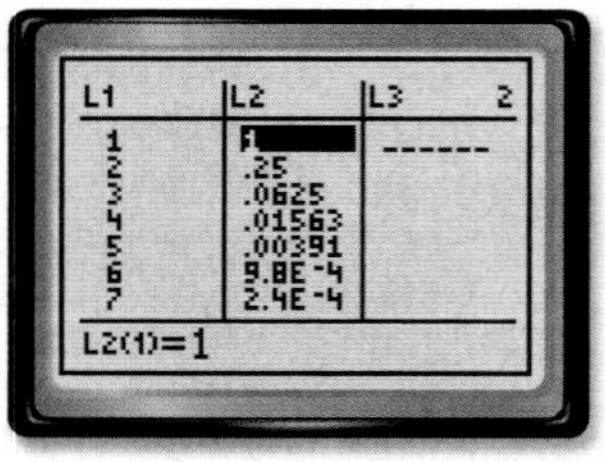

Step 2 Graph the sequence.

Use **STAT PLOT** to graph the sequence. Use **L1** as the **Xlist** and **L2** as the **Ylist**.

The graph also shows that, as n increases, the terms approach 0. In fact, for $n \geq 3$, the marks appear to lie on the horizontal axis. This strongly suggests that the limit of the sequence is 0.

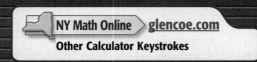

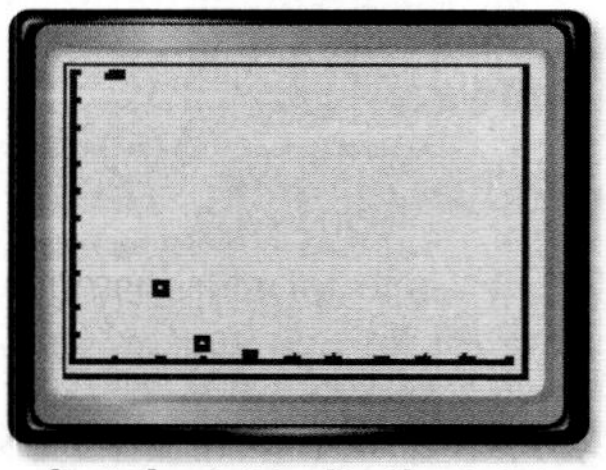

[0, 10] scl: 1 by [0, 1] scl: 0.1

Exercises

Find the limit of each sequence.

1. $a_n = \left(\frac{1}{3}\right)^n$

2. $a_n = \left(-\frac{1}{3}\right)^n$

3. $a_n = 5^n$

4. $a_n = \frac{1}{n^2}$

5. $a_n = \frac{3^n}{3^n + 1}$

6. $a_n = \frac{n^2}{n + 2}$

Mid-Chapter Quiz

Lessons 11-1 through 11-4

NYS Core Curriculum
A2.A.29, A2.A.35

Write an equation for the *n*th term of each arithmetic sequence. (Lesson 11-1)

1. $5, -3, -11, -19, -27 \ldots$

2. $\frac{1}{5}, \frac{7}{10}, \frac{6}{5}, \frac{17}{10}, \frac{11}{5} \ldots$

3. **HOUSING** Laura is a real estate agent. She needs to sell 15 houses in 6 months.
   a. By the end of the first 2 months she has sold 4 houses. If she sells 2 houses each month for the rest of the 6 months, will she meet her goal? Explain.
   b. If she has sold 5 houses by the end of the first month, how many will she have to sell on average each month in order to meet her goal?

4. **GEOMETRY** The figures below show a pattern of filled squares and white squares.

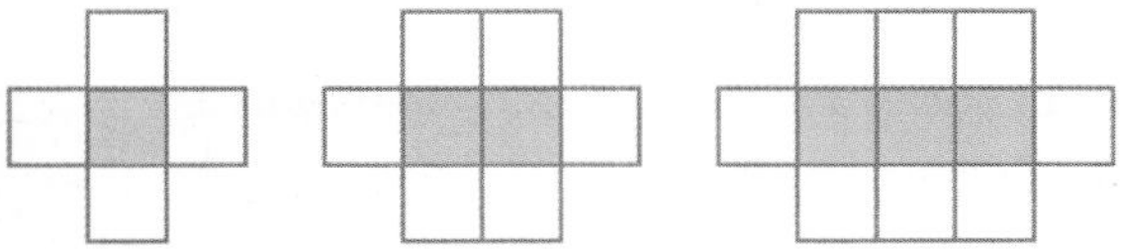

   a. Write an equation representing the *n*th number in this pattern where *n* is the number of white squares.
   b. Is it possible to have exactly 84 white squares in an arrangement? Explain.

Find the indicated term of each arithmetic sequence. (Lesson 11-1)

5. $a_1 = 10, d = -5, n = 9$

6. $a_1 = -8, d = 4, n = 99$

Find the sum of each arithmetic series. (Lesson 11-2)

7. $-15 + (-11) + (-7) + \cdots + 53$

8. $a_1 = -12, d = 8, n = 22$

9. What is the sum of the arithmetic series $\sum_{k=11}^{50} (-3k + 1)$? (Lesson 11-2)

10. **MULTIPLE CHOICE** What is the sum of the first 50 odd numbers? (Lesson 11-2)
    A 625
    B 2500
    C 2499
    D 2401

Find the indicated term for each geometric sequence. (Lesson 11-3)

11. $a_2 = 8, r = 2, a_8 = ?$

12. $a_3 = 0.5, r = 8, a_{10} = ?$

13. **MULTIPLE CHOICE** What are the geometric means of the sequence below?

    0.5, ______, ______, ______, 2048

    F 512.375, 1024.25, 1536.125
    G 683, 1365.5, 2048
    H 2, 8, 32
    J 4, 32, 256

14. **INCOME** Peter works for a house building company for 4 months per year. He starts out making \$3000 per month. At the end of each month, his salary increases by 5%. How much money will he make in those 4 months? (Lesson 11-3)

Evaluate the sum of each geometric series. (Lesson 11-4)

15. $\sum_{k=1}^{8} 3 \cdot 2^{k-1}$

16. $\sum_{k=1}^{9} 4 \cdot (-1)^{k-1}$

17. $\sum_{k=1}^{20} -2\left(\frac{2}{3}\right)^{k-1}$

Find S_n for each geometric series described. (Lesson 11-4)

18. $a_1 = 1296, a_n = 1, r = -\frac{1}{6}$

19. $a_1 = 343, a_n = -1, r = -\frac{1}{7}$

11-5 Recursion and Iteration

Then
You explored compositions of functions. (Lesson 7-1)

Now
- Recognize and use special sequences.
- Iterate functions.

NYS Core Curriculum

A2.A.33 Specify terms of a sequence, given its recursive definition
Also addresses A2.N.10.

New Vocabulary
Fibonacci sequence
recursive sequence
explicit formula
recursive formula
iteration

NY Math Online

glencoe.com
- Extra Examples
- Personal Tutor
- Self-Check Quiz
- Homework Help

Why?

The female honeybee is produced after the queen mates with a male, so the female has two parents, a male and a female. The male honeybee, however, is produced by the queen's unfertilized eggs and thus has only one parent, a female. The family tree for the honeybee follows a special sequence.

Generation	1	2	3	4	5	6
Ancestors	1	1	2	3	5	8

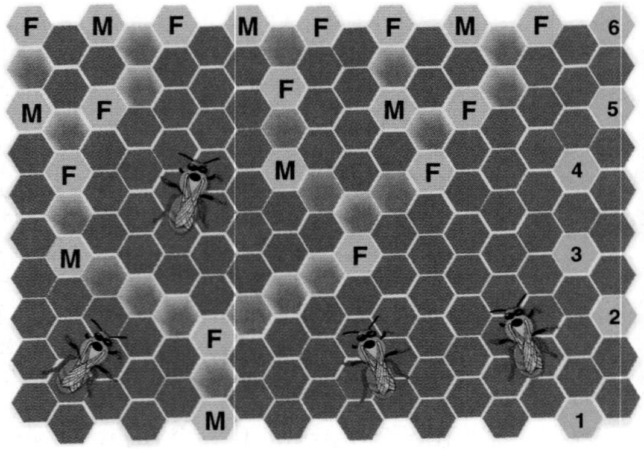

Special Sequences Notice that every term in the list of ancestors is the sum of the previous two terms. This special sequence is called the **Fibonacci sequence**, and it is found in many places in nature. The Fibonacci sequence is an example of a **recursive sequence**. In a recursive sequence, each term is determined by one or more of the previous terms.

The formulas you have used for sequences thus far have been explicit formulas. An **explicit formula** gives a_n as a function of n, such as $a_n = 3n + 1$. The formula that describes the Fibonacci sequence, $a_n = a_{n-2} + a_{n-1}$, is a **recursive formula**, which means that every term will be determined by one or more of the previous terms. An initial term must be given in a recursive formula.

Key Concept — Recursive Formulas for Sequences (For Your FOLDABLE)

Arithmetic Sequence $a_n = a_{n-1} + d$, where d is the common difference

Geometric Sequence $a_n = r \cdot a_{n-1}$, where r is the common ratio

EXAMPLE 1 Use a Recursive Formula

Find the first five terms of the sequence in which $a_1 = -3$ and $a_{n+1} = 4a_n - 2$, if $n \geq 1$.

$a_{n+1} = 4a_n - 2$	Recursive formula
$a_{1+1} = 4a_1 - 2$	$n = 1$
$a_2 = 4(-3) - 2$ or -14	$a_1 = -3$
$a_3 = 4(-14) - 2$ or -58	$a_2 = -14$
$a_4 = 4(-58) - 2$ or -234	$a_3 = -58$
$a_5 = 4(-234) - 2$ or -938	$a_4 = -234$

The first five terms of the sequence are −3, −14, −58, −234, and −938.

Check Your Progress

1. Find the first five terms of the sequence in which $a_1 = 8$ and $a_{n+1} = -3a_n + 6$, if $n \geq 1$.

Personal Tutor glencoe.com

In order to find a recursive formula, first determine the initial term. Then evaluate the pattern to generate the later terms.

StudyTip

Recursive Formula The recursive formula that generates a sequence does not include the value of the initial term.

EXAMPLE 2 Write Recursive Formulas

Write a recursive formula for each sequence.

a. **2, 10, 18, 26, 34, ...**

Step 1 Determine whether the sequence is arithmetic or geometric.
The sequence is arithmetic because each term after the first can be found by adding a common difference.

Step 2 Find the common difference.

$d = 10 - 2$ or 8

Step 3 Write the recursive formula.

$a_n = a_{n-1} + d$ **Recursive formula for arithmetic sequence**

$a_n = a_{n-1} + 8$ $\boldsymbol{d = 8}$

A recursive formula for the sequence is $a_n = a_{n-1} + 8$.

b. **16, 56, 196, 686, 2401, ...**

Step 1 Determine whether the sequence is arithmetic or geometric.
The sequence is geometric because each term after the first can be found after multiplying by a common ratio.

Step 2 Find the common ratio.

$r = \frac{56}{16}$ or 3.5

Step 3 Write the recursive formula.

$a_n = r \cdot a_{n-1}$ **Recursive formula for geometric sequence**

$a_n = 3.5a_{n-1}$ $\boldsymbol{r = 3.5}$

A recursive formula for the sequence is $a_n = 3.5a_{n-1}$.

c. $\boldsymbol{a_4 = 108}$ **and** $\boldsymbol{r = 3}$

Step 1 Determine whether the sequence is arithmetic or geometric.
Because r is given, the sequence is geometric.

Step 2 Write the recursive formula.

$a_n = r \cdot a_{n-1}$ **Recursive formula for geometric sequence**

$a_n = 3a_{n-1}$ $\boldsymbol{r = 3}$

A recursive formula for the sequence is $a_n = 3a_{n-1}$.

Check Your Progress

Write a recursive formula for each sequence.

2A. 8, 20, 50, 125, 312.5, ... **2B.** 8, 17, 26, 35, 44, ... **2C.** $a_3 = 16$ and $r = 4$

Personal Tutor glencoe.com

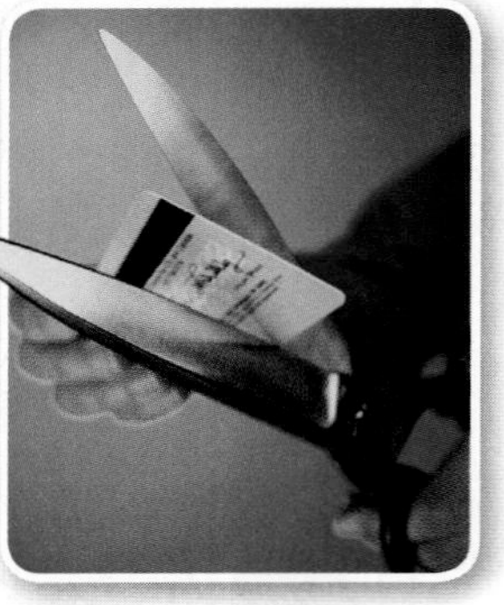

Real-World Link

In 2007, the average credit card debt for college students was about $2700.

Source: *Money Magazine*

Real-World EXAMPLE 3 Use a Recursive Formula

CREDIT CARDS Nate had $15,000 in credit card debt when he graduated from college. The balance increased by 2% each month due to interest, and Nate could only make payments of $400 per month. Write a recursive formula for the balance on his account each month. Then determine the balance after five months.

Step 1 Write the recursive formula.

Let a_n represent the balance on the account in the nth month. The initial balance a_1 is $15,000. After one month, interest is added and a payment is made.

initial balance	+	balance times 2%	−	monthly payment
$a_2 = a_1$	+	$(a_1 \times 0.02)$	−	400

$a_2 = 1.02a_1 - 400$

The formula is $a_n = 1.02a_{n-1} - 400$.

Step 2 Find the next five terms.

	Recursive formula
$a_n = 1.02a_{n-1} - 400$	
$a_2 = (15{,}000 \times 1.02) - 400$ or $14{,}900$	$a_1 = 15{,}000$
$a_3 = (14{,}900 \times 1.02) - 400$ or $14{,}798$	$a_2 = 14{,}900$
$a_4 = (14{,}798 \times 1.02) - 400$ or $14{,}694$	$a_3 = 14{,}798$
$a_5 = (14{,}694 \times 1.02) - 400$ or $14{,}588$	$a_4 = 14{,}694$
$a_6 = (14{,}588 \times 1.02) - 400$ or $14{,}480$	$a_5 = 14{,}588$

After the fifth month, the balance will be $14,480.

Check Your Progress

3. Write a recursive formula for a $10,000 debt, at 2.5% interest per month, with a $600 monthly payment. Then find the first five balances.

Personal Tutor glencoe.com

Iteration

Iteration is the process of repeatedly composing a function with itself. Consider x_0. The first iterate is $f(x_0)$, the second iterate is $f(f(x_0))$, the third iterate is $f(f(f(x_0)))$, and so on.

Iteration can be used to recursively generate a sequence. Start with the initial value x_0. Let $x_1 = f(x_0)$, $x_2 = f(f(x_0))$, and so on.

Review Vocabulary

composition of functions A function is performed, and then a second function is performed on the result of the first function. (Lesson 7-1)

EXAMPLE 4 Iterate a Function

Find the first three iterates x_1, x_2, and x_3 of $f(x) = 5x + 4$ for an initial value of $x_0 = 2$.

$x_1 = f(x_0)$	**Iterate the function.**
$= 5(2) + 4$ or 14	$x_0 = 2$
$x_2 = f(x_1)$	**Iterate the function.**
$= 5(14) + 4$ or 74	$x_1 = 14$
$x_3 = f(x_2)$	**Iterate the function.**
$= 5(74) + 4$ or 374	$x_2 = 74$

The first three iterates are 14, 74, and 374.

Check Your Progress

4. Find the first three iterates x_1, x_2, and x_3 of $f(x) = -3x + 8$ for an initial value of $x_0 = 6$.

Personal Tutor glencoe.com

Check Your Understanding

Example 1 p. 714

Find the first five terms of each sequence described.

1. $a_1 = 16, a_{n+1} = a_n + 4$
2. $a_1 = -3, a_{n+1} = a_n + 8$
3. $a_1 = 5, a_{n+1} = 3a_n + 2$
4. $a_1 = -4, a_{n+1} = 2a_n - 6$

Example 2 p. 715

Write a recursive formula for each sequence.

5. 3, 8, 18, 38, 78, …
6. 5, 14, 41, 122, 365, …

Example 3 p. 716

7. **FINANCING** Ben financed a $1500 rowing machine to help him train for the college rowing team. He could only make a $100 payment each month, and his bill increased by 1% due to interest at the end of each month.
   a. Write a recursive formula for the balance owed at the end of each month.
   b. Find the balance owed after the first four months.
   c. How much interest has accumulated after the first six months?

Example 4 p. 716

Find the first three iterates of each function for the given initial value.

8. $f(x) = 5x + 2, x_0 = 8$
9. $f(x) = -4x + 2, x_0 = 5$
10. $f(x) = 6x + 3, x_0 = -4$
11. $f(x) = 8x - 4, x_0 = -6$

Practice and Problem Solving

● = **Step-by-Step Solutions** begin on page R20.
Extra Practice begins on page 947.

Example 1 p. 714

Find the first five terms of each sequence described.

12. $a_1 = 10, a_{n+1} = 4a_n + 1$
13. $a_1 = -9, a_{n+1} = 2a_n + 8$
14. $a_1 = 12, a_{n+1} = a_n + n$
15. $a_1 = -4, a_{n+1} = 2a_n + n$
16. $a_1 = 6, a_{n+1} = 3a_n - n$
17. $a_1 = -2, a_{n+1} = 5a_n + 2n$
18. $a_1 = 7, a_2 = 10, a_{n+2} = 2a_n + a_{n+1}$
19. $a_1 = 4, a_2 = 5, a_{n+2} = 4a_n - 2a_{n+1}$
20. $a_1 = 4, a_2 = 3x, a_n = a_{n-1} + 4a_{n-2}$
21. $a_1 = 3, a_2 = 2x, a_n = 4a_{n-1} - 3a_{n-2}$
22. $a_1 = 2, a_2 = x + 3, a_n = a_{n-1} + 6a_{n-2}$
23. $a_1 = 1, a_2 = x, a_n = 3a_{n-1} + 6a_{n-2}$

Example 2 p. 715

Write a recursive formula for each sequence.

24. 16, 10, 7, 5.5, 4.75, …
25. 32, 12, 7, 5.75, …
26. 4, 15, 224, 50,175, …
27. 1, 2, 9, 730, …
28. 9, 33, 129, 513, …
29. 480, 128, 40, 18, …
30. 393, 132, 45, 16, …
31. 68, 104, 176, 320, …

Example 3 p. 716

32. **RETIREMENT** Mr. Edwards and his company deposit $20,000 into his retirement account at the end of each year. The account earns 8% interest before each deposit.
   a. Write a recursive formula for the balance in the account at the end of each year.
   b. Determine how much is in the account at the end of each of the first 8 years.

Example 4 p. 716

Find the first three iterates of each function for the given initial value.

33. $f(x) = 12x + 8, x_0 = 4$
34. $f(x) = -9x + 1, x_0 = -6$
35. $f(x) = -6x + 3, x_0 = 8$
36. $f(x) = 8x + 3, x_0 = -4$
37. $f(x) = -3x^2 + 9, x_0 = 2$
38. $f(x) = 4x^2 + 5, x_0 = -2$
39. $f(x) = 2x^2 - 5x + 1, x_0 = 6$
40. $f(x) = -0.25x^2 + x + 6, x_0 = 8$
41. $f(x) = x^2 + 2x + 3, x_0 = \frac{1}{2}$
42. $f(x) = 2x^2 + x + 1, x_0 = -\frac{1}{2}$

43 **FRACTALS** Consider the figures at the right. The number of blue triangles increases according to a specific pattern.

a. Write a recursive formula for the number of blue triangles in the sequence of figures.

b. How many blue triangles will be in the sixth figure?

44. **LOANS** Miguel's monthly car payment is \$234.85. The recursive formula $b_n = 1.005b_{n-1} - 234.85$ describes the balance left on the loan after n payments. Find the balance of the \$10,000 loan after each of the first eight payments.

45. **CONSERVATION** Suppose a lake is populated with 10,000 fish. A year later, 80% of the fish have died or been caught, and the lake is replenished with 10,000 new fish. If the pattern continues, will the lake eventually run out of fish? If not, will the population of the lake converge to any particular value? Explain.

Real-World Career

Conservation Scientist Conservation scientists work with private landowners and the government to manage, improve, and protect natural resources. A conservation scientist must have a bachelor's degree in forestry, biology, natural resource management, or the environmental sciences.

46. **GEOMETRY** Consider the pattern at the right.

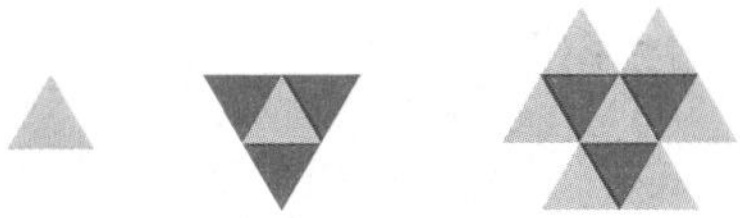

a. Write a sequence of the total number of triangles in the first six figures.

b. Write a recursive formula for the number of triangles.

c. How many triangles will be in the tenth figure?

47. **SPREADSHEETS** Consider the sequence with $x_0 = 20{,}000$ and $f(x) = 0.3x + 5000$.

a. Enter x_0 in cell A1 of your spreadsheet. Enter "= (0.3)*(A1) + 5000" in cell A2. What answer does it provide?

b. Copy cell A2, highlight cells A3 through A70, and paste. What do you notice about the sequence?

c. How do spreadsheets help analyze recursive sequences?

48. **VIDEO GAMES** The final monster in Helena's video game has 100 health points. During the final battle, the monster regains 10% of its health points after every 10 seconds. If Helena can inflict damage to the monster that takes away 10 health points every 10 seconds without getting hurt herself, will she ever kill the monster? If so, when?

H.O.T. Problems

Use Higher-Order Thinking Skills

49. **FIND THE ERROR** Marcus and Armando are finding the first three iterates of $f(x) = 5x - 3$ for an initial value of $x_0 = 4$. Is either of them correct? Explain.

Marcus	Armando
$f(4) = 5(4) - 3$ or 17	$f(4) = 5(4) - 3$ or 17
$f(17) = 5(17) - 3$ or 82	$f(17) = 5(17) - 3$ or 82
The first three iterates are 4, 17, and 82.	$f(82) = 5(82) - 3$ or 407
	The first three iterates are 17, 82, and 407.

50. **CHALLENGE** Find a recursive formula for 5, 23, 98, 401, ...

51. **REASONING** Is the statement *"If the first three terms of a sequence are identical, then the sequence is not recursive"* *sometimes*, *always*, or *never* true? Explain your reasoning.

52. **OPEN ENDED** Write a function for which the first three iterates are 9, 19, and 39.

53. **WRITING IN MATH** Explain the difference between a recursive sequence and a recursive formula.

NYSRE PRACTICE A2.A.33

54. GEOMETRY In the figure shown, $a + b + c = ?$

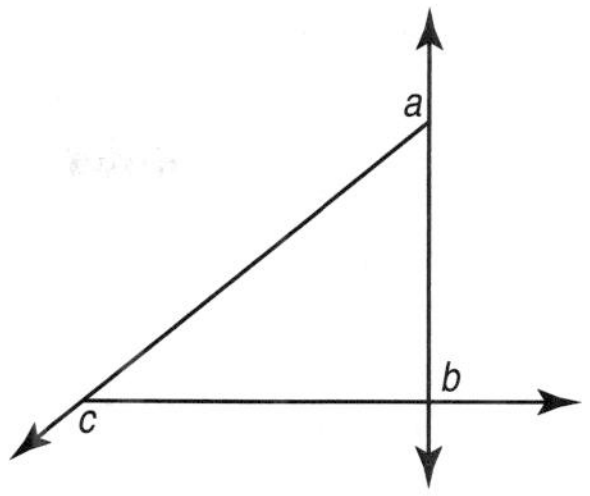

A 180°
B 270°
C 360°
D 450°

55. EXTENDED RESPONSE Bill launches a model rocket from ground level. The rocket's height h in meters is given by the equation $h = -4.9t^2 + 56t$, where t is the time in seconds after the launch.

a. What is the maximum height the rocket will reach? Round to the nearest tenth of a meter.

b. How long after it is launched will the rocket reach its maximum height? Round to the nearest tenth of a second.

c. How long after it is launched will the rocket land? Round to the nearest tenth of a second.

56. Which of the following is true about the graphs of $y = 3(x - 4)^2 + 5$ and $y = 3(x + 4)^2 + 5$?

F Their vertices are maximums.
G The graphs have the same shape with different vertices.
H The graphs have different shapes with different vertices.
J One graph has a vertex that is a maximum, while the other graph has a vertex that is a minimum.

57. Which factors could represent the length times the width?

$A = 16x^4 - 25y^2$

A $(4x - 5y)(4x - 5y)$
B $(4x + 5y)(4x - 5y)$
C $(4x^2 - 5y)(4x^2 + 5y)$
D $(4x^2 + 5y)(4x^2 + 5y)$

Spiral Review

Write each repeating decimal as a fraction. (Lesson 11-4)

58. $0.\overline{7}$ **59.** $5.\overline{126}$ **60.** $6.\overline{259}$

61. SPORTS Adrahan is training for a marathon, about 26 miles. He begins by running 2 miles. Then, when he runs every other day, he runs one and a half times the distance he ran the time before. (Lesson 11-3)

a. Write the first five terms of a sequence describing his training schedule.

b. When will he exceed 26 miles in one run?

c. When will he have run 100 total miles?

State whether the events are *independent* or *dependent*. (Lesson 0-4)

62. tossing a penny and rolling a number cube

63. choosing first and second place in an academic competition

Skills Review

Find each product. (Lesson 0-2)

64. $(y + 4)(y + 3)$ **65.** $(x - 2)(x + 6)$ **66.** $(a - 8)(a + 5)$

67. $(4h + 5)(h + 7)$ **68.** $(9p - 1)(3p - 2)$ **69.** $(2g + 7)(5g - 8)$

EXTEND 11-5

Spreadsheet Lab
Amortizing Loans

NY Math Online glencoe.com
Other Calculator Keystrokes

When a payment is made on a loan, part of the payment is used to cover the interest that has accumulated since the last payment. The rest is used to reduce the *principal*, or original amount of the loan. This process is called *amortization*. You can use a spreadsheet to analyze the payments, interest, and balance on a loan. A table that shows this kind of information is called an *amortization schedule*.

EXAMPLE

LOANS Gloria just bought a new computer for $695. The store is letting her make monthly payments of $60.78 at an interest rate of 9% for one year. How much will she still owe after six months?

Every month, the interest on the remaining balance will be $\frac{9\%}{12}$ or 0.75%. You can find the balance after a payment by multiplying the balance after the previous payment by 1 + 0.0075 or 1.0075 and then subtracting 60.78.

In a spreadsheet, the column of numbers represents the number of payments, and Column B shows the balance. Enter the interest rate and monthly payment in cells in Column A so that they can be easily updated if the information changes.

The spreadsheet at the right shows the formulas for the balances after each of the first six payments. After six months, Gloria still owes $355.28.

Excel sample.xls

◇	A	B	C
1	Interest Rate	=695*(1+A2)–A5	
2	0.0075	=B1*(1+A2)–A5	
3		=B2*(1+A2)–A5	
4	Monthly payment	=B3*(1+A2)–A5	
5	60.78	=B4*(1+A2)–A5	
6		=B5*(1+A2)–A5	
7			

Sheet 1 | Sheet 2 | Sheet 3

Model and Analyze

1. Let b_n be the balance left on Gloria's loan after n months. Write an equation relating b_n and b_{n+1}.
2. Payments at the beginning of a loan go more toward interest than payments at the end. What percent of Gloria's loan remains to be paid after half a year?
3. Extend the spreadsheet to the whole year. What is the balance after 12 payments? Why is it not 0?
4. Suppose Gloria decides to pay $70 every month. How long would it take her to pay off the loan?
5. Suppose that, based on how much she can afford, Gloria will pay a variable amount each month in addition to the $60.78. Explain how the flexibility of a spreadsheet can be used to adapt to this situation.
6. Ethan has a three-year, $12,000 motorcycle loan. The annual interest rate is 6%, and his monthly payment is $365.06. After fifteen months, he receives an inheritance which he wants to use to pay off the loan. How much does he owe at that point?

11-6 The Binomial Theorem

Then
You worked with combinations. (Concepts and Skills Bank, Lesson 0-5)

Now
- Use Pascal's triangle to expand powers of binomials.
- Use the Binomial Theorem to expand powers of binomials.

NYS Core Curriculum

A2.A.34 Represent the sum of a series, using sigma notation

New Vocabulary
Pascal's triangle

NY Math Online

glencoe.com
- Extra Examples
- Personal Tutor
- Self-Check Quiz
- Homework Help

Why?

A manager plans to hire 8 new employees. Not wanting to appear biased, the manager wants to hire a combination of males and females that has at least a 10% chance of occurring randomly. If there are an equal number of male and female applicants, is the probability of randomly hiring 6 men and 2 women less than 10%?

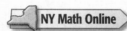

Pascal's Triangle In the 13th century, the Chinese discovered a pattern of numbers that would later be referred to as **Pascal's triangle**. This pattern can be used to determine the coefficients of an expanded binomial $(a + b)^n$.

$(a + b)^0$	1
$(a + b)^1$	1 1
$(a + b)^2$	1 2 1
$(a + b)^3$	1 3 3 1
$(a + b)^4$	1 4 6 4 1
$(a + b)^5$	1 5 10 10 5 1

For example, the expanded form of
$(a + b)^5 = 1a^5 + 5a^4b + 10a^3b^2 + 10a^2b^3 + 5ab^4 + 1b^5$.

Real-World EXAMPLE 1 Use Pascal's Triangle

Find the probability of hiring 6 men and 2 women by expanding $(m + f)^8$.

Write three more rows of Pascal's triangle and use the pattern to write the expansion.

5	1 5 10 10 5 1
6	1 6 15 20 15 6 1
7	1 7 21 35 35 21 7 1
8	1 8 28 56 70 56 28 8 1

$(m + f)^8 = m^8 + 8m^7f + 28m^6f^2 + 56m^5f^3 + 70m^4f^4 + 56m^3f^5 + 28m^2f^6 + 8mf^7 + f^8$

By adding the coefficients of the polynomial, we determine that there are 256 combinations of males and females that could be hired.

$28m^6f^2$ represents the number of combinations with 6 males and 2 females. Therefore, there is a $\frac{28}{256}$ or about an 11% chance of randomly hiring 6 males and 2 females.

Check Your Progress

1. Expand $(c + d)^9$.

Personal Tutor glencoe.com

The Binomial Theorem Instead of writing out row after row of Pascal's triangle, you can use the **Binomial Theorem** to expand a binomial. Recall that ${}_nC_r = \frac{n!}{r!(n - r)!}$.

Key Concept Binomial Theorem

For Your FOLDABLE

If n is a natural number, then $(a + b)^n =$

$${}_nC_0\, a^n b^0 + {}_nC_1\, a^{n-1} b^1 + {}_nC_2\, a^{n-2} b^2 + \cdots + {}_nC_n\, a^0 b^n = \sum_{k=0}^{n} \frac{n!}{k!(n-k)!} a^{n-k} b^k.$$

StudyTip

Combinations Recall that both ${}_nC_0$ and ${}_nC_n$ equal 1.

To use the theorem, replace n with the value of the exponent. Notice how the terms will follow the pattern of Pascal's triangle, and the coefficients will be symmetric.

EXAMPLE 2 Use the Binomial Theorem

Expand $(a + b)^7$.

Method 1 Use combinations.

Replace n with 7 in the Binomial Theorem.

$$(a + b)^7 = a^7 + {}_7C_1\, a^6 b + {}_7C_2\, a^5 b^2 + {}_7C_3\, a^4 b^3 + {}_7C_4\, a^3 b^4 + {}_7C_5\, a^2 b^5 + {}_7C_6\, a b^6 + b^7$$

$$= a^7 + \frac{7!}{6!} a^6 b + \frac{7!}{2!5!} a^5 b^2 + \frac{7!}{3!4!} a^4 b^3 + \frac{7!}{4!3!} a^3 b^4 + \frac{7!}{5!2!} a^2 b^5 + \frac{7!}{6!} a b^6 + b^7$$

$$= a^7 + 7a^6 b + 21a^5 b^2 + 35a^4 b^3 + 35a^3 b^4 + 21a^2 b^5 + 7ab^6 + b^7$$

Method 2 Use Pascal's triangle.

Use the Binomial Theorem to determine exponents, but instead of finding the coefficients by using combinations, look at the seventh row of Pascal's triangle.

6 1 6 15 20 15 6 1
7 1 7 21 35 35 21 7 1
8 1 8 28 56 70 56 28 8 1

$$(a + b)^7 = a^7 + 7a^6 b + 21a^5 b^2 + 35a^4 b^3 + 35a^3 b^4 + 21a^2 b^5 + 7ab^6 + b^7$$

Check Your Progress

2. Expand $(x + y)^{10}$.

Personal Tutor glencoe.com

When the binomial to be expanded has coefficients other than 1, the coefficients will no longer be symmetrical. In these cases, you may want to use the Binomial Theorem.

StudyTip

Graphing Calculator You can calculate ${}_nC_r$ by using a graphing calculator.

Press MATH and choose PRB 3.

EXAMPLE 3 Coefficients Other Than 1

Expand $(5a - 4b)^4$.

$(5a - 4b)^4$

$$= (5a)^4 + {}_4C_1\, (5a)^3(-4b) + {}_4C_2\, (5a)^2(-4b)^2 + {}_4C_3\, (5a)(-4b)^3 + {}_4C_4\, (-4b)^4$$

$$= 625a^4 + \frac{4!}{3!}(125a^3)(-4b) + \frac{4!}{2!2!}(25a^2)(16b^2) + \frac{4!}{3!}(5a)(-64b^3) + 256b^4$$

$$= 625a^4 - 2000a^3 b + 2400a^2 b^2 - 1280ab^3 + 256b^4$$

Check Your Progress

3. Expand $(3x + 2y)^5$.

Personal Tutor glencoe.com

Sometimes you may need to find only one term in a binomial expansion. To do this, you can use the summation formula for the Binomial Theorem, $\sum_{k=0}^{n} \frac{n!}{k!(n-k)!} a^{n-k} b^k$.

EXAMPLE 4 Determine a Single Term

Find the fifth term of $(y + z)^{11}$.

Step 1 Use the Binomial Theorem to write the expansion in sigma notation.

$$(y + z)^{11} = \sum_{k=0}^{11} \frac{11!}{k!(11 - k)!} y^{11 - k} z^k$$

Step 2 $\frac{11!}{k!(11 - k)!} y^{11 - k} z^k = \frac{11!}{4!(11 - 4)!} y^{11 - 4} z^4$ **For the fifth term, $k = 4$.**

$= 330y^7z^4$ **$C(11, 4) = 330$**

Check Your Progress

4. Find the sixth term of $(c + d)^{10}$.

Personal Tutor glencoe.com

Concept Summary — Binomial Expansion

For Your FOLDABLE

In a binomial expansion of $(a + b)^n$,

- there are $n + 1$ terms.
- n is the exponent of a in the first term and b in the last term.
- in successive terms, the exponent of a decreases by 1, and the exponent of b increases by 1.
- the sum of the exponents in each term is n.
- the coefficients are symmetric.

Check Your Understanding

Examples 1–3 pp. 721–722

Expand each binomial.

1. $(c + d)^5$

2. $(g + h)^7$

3. $(x - 4)^6$

4. $(2y - z)^5$

5. $(x + 3)^5$

6. $(y - 4z)^4$

7. **GENETICS** If a woman is equally as likely to have a baby boy or a baby girl, use binomial expansion to determine the probability that 5 of her 6 children are girls. Do not consider identical twins.

Example 4 p. 723

Find the indicated term of each expression.

8. fourth term of $(b + c)^9$

9. fifth term of $(x + 3y)^8$

10. third term of $(a - 4b)^6$

11. sixth term of $(2c - 3d)^8$

12. last term of $(5x + y)^5$

13. first term of $(3a + 8b)^5$

14. **FLOWERS** The color of a particular flower is determined by the combination of two genes, also called *alleles*. If the flower has two red alleles r, the flower is red. If the flower has two white alleles w, the flower is white. If the flower has one allele of each color, the flower will be pink. In a lab, two pink flowers are mated and eventually produce 1000 offspring. How many of the 1000 offspring will be pink?

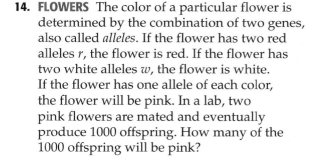

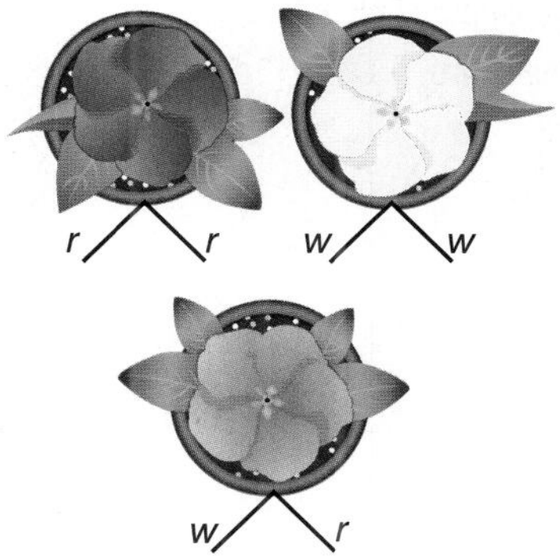

Practice and Problem Solving

= **Step-by-Step Solutions** begin on page R20.
Extra Practice begins on page 947.

Examples 1–3 pp. 721–722

Expand each binomial.

15. $(a - b)^6$
16. $(c - d)^7$
17. $(x + 6)^6$
18. $(y - 5)^7$
19. $(2a + 4b)^4$
20. $(3a - 4b)^5$

21. **COMMITTEES** If an equal number of men and women applied to be on a community planning committee and the committee needs a total of 10 people, find the probability that 7 of the members will be women. Assume that committee members will be chosen randomly.

22. **BASEBALL** If a pitcher is just as likely to throw a ball as a strike, find the probability that 11 of his first 12 pitches are balls.

Example 4 p. 723

Find the indicated term of each expression.

23. third term of $(x + 2z)^7$
24. fourth term of $(y - 3x)^6$
25. seventh term of $(2a - 2b)^8$
26. sixth term of $(4x + 5y)^6$
27. fifth term of $(x - 4)^9$
28. fourth term of $(c + 6)^8$

Expand each binomial.

29. $\left(x + \frac{1}{2}\right)^5$
30. $\left(x - \frac{1}{3}\right)^4$
31. $\left(2b + \frac{1}{4}\right)^5$
32. $\left(3c + \frac{1}{3}\right)^5$

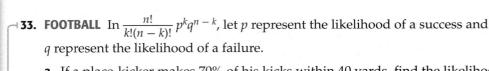

Real-World Link

As of 2007, Mike Vanderjagt of the Dallas Cowboys had the highest field goal percentage in the history of the NFL, at over 86%.

Source: NFL

33. **FOOTBALL** In $\frac{n!}{k!(n - k)!}p^k q^{n-k}$, let p represent the likelihood of a success and q represent the likelihood of a failure.

a. If a place-kicker makes 70% of his kicks within 40 yards, find the likelihood that he makes 9 of his next 10 attempts from within 40 yards.

b. If a quarterback completes 60% of his passes, find the likelihood that he completes 8 of his next 10 attempts.

c. If a team converts 30% of their two-point conversions, find the likelihood that they convert 2 of their next 5 conversions.

H.O.T. Problems

Use **H**igher-**O**rder **T**hinking Skills

34. **CHALLENGE** Find the sixth term of the expansion of $\left(\sqrt{a} + \sqrt{b}\right)^{12}$. Explain your reasoning.

35. **REASONING** Explain how the terms of $(x + y)^n$ and $(x - y)^n$ are the same and how they are different.

36. **REASONING** Determine whether the following statement is *true* or *false*. Explain your reasoning.

The eighth and twelfth terms of $(x + y)^{20}$ have the same coefficients.

37. **OPEN ENDED** Write a power of a binomial for which the second term of the expansion is $6x^4y$.

38. **WRITING IN MATH** Explain how to write out the terms of Pascal's triangle.

NYSRE PRACTICE A2.A.34

39. **PROBABILITY** A desk drawer contains 7 sharpened red pencils, 5 sharpened yellow pencils, 3 unsharpened red pencils, and 5 unsharpened yellow pencils. If a pencil is taken from the drawer at random, what is the probability that it is yellow, given that it is one of the sharpened pencils?

A $\frac{5}{12}$

B $\frac{7}{20}$

C $\frac{5}{8}$

D $\frac{1}{5}$

40. **SHORT RESPONSE** Two people are 17.5 miles apart. They begin to walk toward each other along a straight line at the same time. One walks at the rate of 4 miles per hour, and the other walks at the rate of 3 miles per hour. In how many hours will they meet?

41. **GEOMETRY** Kara has a cylindrical container that she needs to fill with dirt so she can plant some flowers.

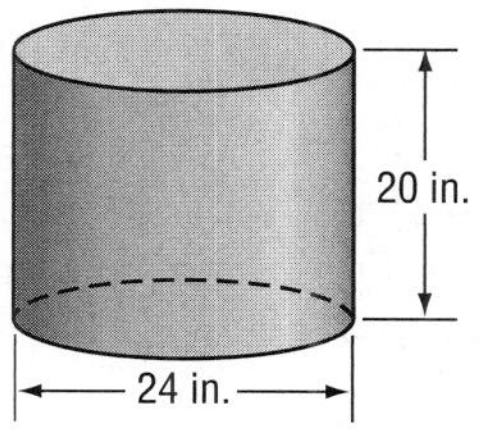

What is the volume of the cylinder in cubic inches rounded to the nearest cubic inch?

F 7540 H 30,159

G 9048 J 36,191

42. Which of the following is a linear function?

A $y = \frac{x+3}{x+2}$

B $y = (3x + 2)^2$

C $y = \frac{x+3}{2}$

D $y = |3x| + 2$

Spiral Review

Find the first five terms of each sequence. (Lesson 11-5)

43. $a_1 = -2, a_{n+1} = a_n + 5$

44. $a_1 = 3, a_{n+1} = 4a_n - 10$

45. $a_1 = 4, a_{n+1} = 3a_n - 6$

Find the sum of each infinite geometric series, if it exists. (Lesson 11-4)

46. $-6 + 3 - \frac{3}{2} + \ldots$

47. $\frac{3}{4} + \frac{1}{4} + \frac{1}{12} + \ldots$

48. $\sqrt{3} + 3 + \sqrt{27} + \ldots$

49. **TRAVEL** A trip between two towns takes 4 hours under ideal conditions. The first 150 miles of the trip is on an interstate, and the last 130 miles is on a highway with a speed limit that is 10 miles per hour less than on the interstate. (Lesson 9-6)

a. If x represents the speed limit on the interstate, write expressions for the time spent at that speed and for the time spent on the other highway.

b. Write and solve an equation to find the speed limits on the two highways.

Skills Review

State whether each statement is *true* or *false* when $n = 1$. Explain. (Lesson 1-1)

50. $\frac{(n+1)(n+1)}{2} = 2$

51. $3n + 5$ is even.

52. $n^2 - 1$ is odd.

EXTEND

11-6 Algebra Lab
Combinations and Pascal's Triangle

NY Math Online glencoe.com
Math *in Motion*, Animation

Recall that an arrangement or selection of objects in which order is not important is called a *combination*. For example, selecting 2 snacks from a choice of 6 is a combination of 6 objects taken 2 at a time and can be written ${}_6C_2$ or $C(6, 2)$.

ACTIVITY

A contestant on a game show has the opportunity to win up to five prizes, one for each of five rounds of the game. If the contestant wins a round, he or she may choose one prize. Determine the number of ways that prizes can be chosen.

Step 1 If a contestant does not win any rounds, he or she receives 0 prizes. This represents 5 items taken 0 at a time.

$${}_nC_r = \frac{n!}{(n-r)!\,r!} \qquad \textbf{Definition of combination}$$

$${}_5C_0 = \frac{5!}{(5-0)!\,0!} \qquad n = 5 \textbf{ and } r = 0$$

$$= \frac{120}{120(1)} \text{ or } 1 \qquad \textbf{5! = 120 and 0! = 1}$$

There is 1 way to receive 0 prizes.

If a contestant wins one round, any one of the prizes can be selected. If a contestant wins two rounds, two prizes can be chosen. If three rounds are won, three prizes can be chosen, and so on. In how many ways can 1 prize be chosen? 2 prizes? 3, 4, and 5 prizes? We can determine these answers by examining Pascal's triangle.

Step 2 Examine Pascal's triangle.

List Rows 0 through 5 of Pascal's triangle.

Row 0						1					
Row 1					1		1				
Row 2				1		2		1			
Row 3			1		3		3		1		
Row 4		1		4		6		4		1	
Row 5	1		5		10		10		5		1

The number of ways one prize can be chosen from 5 can be determined by looking at Row 5. The first number in Row 5 represents the number of ways to choose 0 prizes, the second number represents the number of ways to choose 1 prize, and so on.

Analyze the Results

1. Make a conjecture about how the numbers in one of the rows can be used to find the number of ways that 0, 1, 2, 3, 4, … n objects can be selected from n objects.
2. Suppose the rules of the game are changed so that there are 6 rounds and 6 prizes from which to choose. Find the number of ways that 0, 1, 2, 3, 4, 5, or 6 prizes can be chosen. Which row of Pascal's triangle can be used to find the answers?
3. Use Pascal's triangle to find ${}_8C_0$, ${}_8C_1$, ${}_8C_2$, ${}_8C_3$, ${}_8C_4$, ${}_8C_5$, ${}_8C_6$, ${}_8C_7$, and ${}_8C_8$. State the row number that you used to find the answers.

11-7 Proof by Mathematical Induction

Then

You have proved the sum of an arithmetic series. (Lesson 11-2)

Now

- Prove statements by using mathematical induction.
- Disprove statements by finding a counterexample.

NYS Core Curriculum

A2.RP.12 Apply inductive reasoning in making and supporting mathematical conjectures
A2.A.36 Apply the binomial theorem to expand a binomial and determine a specific term of a binomial expansion

New Vocabulary

mathematical induction
induction hypothesis

NY Math Online

glencoe.com
- Extra Examples
- Personal Tutor
- Self-Check Quiz
- Homework Help

Why?

When dominoes are set up closely and the first domino is knocked down, the rest of the dominoes come tumbling down. All that is needed with this setup is for the first domino to fall, and the rest will follow. The same is true with mathematical induction.

Mathematical Induction **Mathematical induction** is a method of proving statements involving natural numbers.

Key Concept — Mathematical Induction

For Your FOLDABLE

To prove that a statement is true for all natural numbers n,

Step 1 Show that the statement is true for $n = 1$.

Step 2 Assume that the statement is true for some natural number k. This assumption is called the **induction hypothesis**.

Step 3 Show that the statement is true for the next natural number $k + 1$.

EXAMPLE 1 Prove Summation

Prove that $1^3 + 2^3 + 3^3 + \cdots + n^3 = \dfrac{n^2(n+1)^2}{4}$.

Step 1 When $n = 1$, the left side of the equation is 1^3 or 1.

The right side is $\dfrac{1^2(1+1)^2}{4}$ or 1. Thus, the statement is true for $n = 1$.

Step 2 Assume that $1^3 + 2^3 + 3^3 + \cdots + k^3 = \dfrac{k^2(k+1)^2}{4}$ for a natural number k.

Step 3 Show that the given statement is true for $n = k + 1$.

$$1^3 + 2^3 + 3^3 + \cdots + k^3 = \frac{k^2(k+1)^2}{4} \quad \textbf{Inductive hypothesis}$$

$$1^3 + 2^3 + \cdots + k^3 + (k+1)^3 = \frac{k^2(k+1)^2}{4} + (k+1)^3 \quad \textbf{Add } (k+1)^3 \textbf{ to each side.}$$

$$= \frac{k^2(k+1)^2 + 4(k+1)^3}{4} \quad \textbf{The LCD is 4.}$$

$$= \frac{(k+1)^2[k^2 + 4(k+1)]}{4} \quad \textbf{Factor.}$$

$$= \frac{(k+1)^2(k^2 + 4k + 4)}{4} \quad \textbf{Simplify.}$$

$$= \frac{(k+1)^2(k+2)^2}{4} \quad \textbf{Factor.}$$

The last expression is the statement to be proved, where n has been replaced by $k + 1$. This proves the conjecture.

Check Your Progress

1. Prove that $1^2 + 2^2 + 3^2 + \cdots + n^2 = \dfrac{n(n+1)(2n+1)}{6}$.

Personal Tutor glencoe.com

Along with summation, mathematical induction can be used to prove divisibility.

EXAMPLE 2 Prove Divisibility

Prove that $8^n - 1$ is divisible by 7 for all natural numbers n.

Step 1 When $n = 1$, $8^n - 1 = 8^1 - 1$ or 7. Since 7 is divisible by 7, the statement is true for $n = 1$.

Step 2 Assume that $8^k - 1$ is divisible by 7 for some natural number k. This means that there is a natural number r such that $8^k - 1 = 7r$.

Step 3 Show that the statement is true for $n = k + 1$.

$8^k - 1 = 7r$ — **Inductive hypothesis**

$8^k = 7r + 1$ — **Add 1 to each side.**

$8(8^k) = 8(7r + 1)$ — **Multiply each side by 8.**

$8^{k+1} = 56r + 8$ — **Simplify.**

$8^{k+1} - 1 = 56r + 7$ — **Subtract 1 from each side.**

$8^{k+1} - 1 = 7(8r + 1)$ — **Factor.**

Since r is a natural number, $8r + 1$ is a natural number and $7(8r + 1)$ is divisible by 7. Therefore, $8^{k+1} - 1$ is divisible by 7.

This proves that 8^{n-1} is divisible by 7 for all natural numbers n.

StudyTip

Divisibility The r is a whole number used to show divisibility in proof. When a value equals $4r$, then it must be divisible by 4.

Check Your Progress

2. Prove that $7^n - 1$ is divisible by 6 for all natural numbers n.

Personal Tutor glencoe.com

Counterexamples Statements can be proved false by using mathematical induction. An easier method is by finding a counterexample, which is a specific case in which the statement is false.

Review Vocabulary

counterexample One of the synonyms of *counter* is to *contradict*, so a counterexample is an example that contradicts a hypothesis.

EXAMPLE 3 Use a Counterexample to Disprove

Find a counterexample to disprove the statement that $2^n + 2n^2$ is divisible by 4 for any natural number n.

Test different values of n.

n	$2^n + 2n^2$	Divisible by 4?
1	$2^1 + 2(1)^2 = 2 + 2$ or 4	yes
2	$2^2 + 2(2)^2 = 4 + 8$ or 12	yes
3	$2^3 + 2(3)^2 = 8 + 18$ or 26	no

The value $n = 3$ is a counterexample for the statement.

Check Your Progress

3. Find a counterexample to disprove $1^2 + 2^2 + 3^2 + \cdots + n^2 = \frac{n(3n - 1)}{2}$.

Personal Tutor glencoe.com

Check Your Understanding

Example 1 p. 727

Prove that each statement is true for all natural numbers.

1. $1 + 3 + 5 + \cdots + (2n - 1) = n^2$

2. $1 + 2 + 3 + \cdots + n = \dfrac{n(n+1)}{2}$

3. **NUMBER THEORY** A number is *triangular* if it can be represented visually by a triangular array.

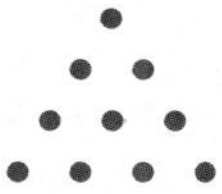

10

a. The first triangular number is 1. Find the next 5 triangular numbers.

b. Write a formula for the nth triangular number.

c. Prove that the sum of the first n triangular numbers equals $\dfrac{n(n+1)(n+2)}{6}$.

Example 2 p. 728

Prove that each statement is true for all natural numbers.

4. $10^n - 1$ is divisible by 9.

5. $4^n - 1$ is divisible by 3.

Example 3 p. 728

Find a counterexample to disprove each statement.

6. $3^n + 1$ is divisible by 4.

7. $2^n + 3^n$ is divisible by 4.

Practice and Problem Solving

● = **Step-by-Step Solutions** begins on page R20.
Extra Practice begins on page 947.

Example 1 p. 727

Prove that each statement is true for all natural numbers.

8. $\dfrac{1}{2} + \dfrac{1}{2^2} + \dfrac{1}{2^3} + \cdots + \dfrac{1}{2^n} = 1 - \dfrac{1}{2^n}$

9. $2 + 5 + 8 + \cdots + (3n - 1) = \dfrac{n(3n+1)}{2}$

10. $1 + 2 + 4 + \cdots + 2^{n-1} = 2^n - 1$

11. $1 + 5 + 9 + \cdots + (4n - 3) = n(2n - 1)$

12. $1 + 4 + 7 + \cdots + (3n - 2) = \dfrac{n(3n-1)}{2}$

13 $3 + 7 + 11 + \cdots + (4n - 1) = 2n^2 + n$

14. $\dfrac{1}{2} + \dfrac{1}{6} + \dfrac{1}{12} + \cdots + \dfrac{1}{n(n+1)} = \dfrac{n}{n+1}$

15. $1^2 + 3^2 + 5^2 + \cdots + (2n - 1)^2 = \dfrac{n(2n-1)(2n+1)}{3}$

16. **GEOMETRY** According to the Interior Angle Sum Formula, if a convex polygon has n sides, then the sum of the measures of the interior angles of a polygon equals $180(n - 2)$. Prove this formula for $n \geq 3$ using mathematical induction and geometry.

Example 2 p. 728

Prove that each statement is true for all natural numbers.

17. $5^n + 3$ is divisible by 4.

18. $9^n - 1$ is divisible by 8.

19. $12^n + 10$ is divisible by 11.

20. $13^n + 11$ is divisible by 12.

Example 3 p. 728

Find a counterexample to disprove each statement.

21. $1 + 2 + 3 + \cdots + n = n^2$

22. $1 + 8 + 27 + \cdots + n^3 = (2n + 2)^2$

23. $n^2 - n + 15$ is prime.

24. $n^2 + n + 23$ is prime.

Real-World Link

Fibonacci spirals are found everywhere in nature, because they provide optimal spacing and packaging.

Source: Mathematica

25 **NATURE** The terms of the Fibonacci sequence are found in many places in nature. The number of spirals of seeds in sunflowers is a Fibonacci number, as is the number of spirals of scales on a pinecone. The Fibonacci sequence begins 1, 1, 2, 3, 5, 8, ... Each element after the first two is found by adding the previous two terms. If f_n stands for the nth Fibonacci number, prove that $f_1 + f_2 + \cdots + f_n = f_{n+2} - 1$.

Prove that each statement is true for all natural numbers or find a counterexample.

26. $7^n + 5$ is divisible by 6.

27. $18^n - 1$ is divisible by 17.

28. $n^2 + 21n + 7$ is a prime number.

29. $n^2 + 3n + 3$ is a prime number.

30. $500 + 100 + 20 + \cdots + 4 \cdot 5^{4-n} = 625\left(1 - \frac{1}{5^n}\right)$

31. $\frac{1}{1 \cdot 2 \cdot 3} + \frac{1}{2 \cdot 3 \cdot 4} + \frac{1}{3 \cdot 4 \cdot 5} + \cdots + \frac{1}{n(n+1)(n+2)} = \frac{n(n+3)}{4(n+1)(n+2)}$

32. **CHECKERBOARDS** Refer to the figures below.

Figure 1

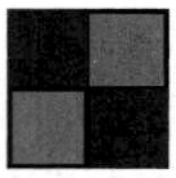

Figure 2

Figure 3

a. There is a total of 5 squares in the second figure. How many squares are there in the third figure?

b. Write a sequence for the first five figures.

c. How many squares are there in a standard 8 × 8 checkerboard?

d. Write a formula to represent the number of squares in an $n \times n$ grid.

H.O.T. Problems
Use **H**igher-**O**rder **T**hinking Skills

33. **CHALLENGE** Suggest a formula to represent $2 + 4 + 6 + \cdots + 2n$, and prove your hypothesis using mathematical induction.

REASONING **Determine whether the following statements are *true* or *false*. Explain.**

34. If you cannot find a counterexample to a statement, then it is true.

35. If a statement is true for $n = k$ and $n = k + 1$, then it is also true for $n = 1$.

36. **CHALLENGE** Prove $\sum_{k=1}^{n} k^3 = \left(\frac{n(n+1)}{2}\right)^2$.

37. **REASONING** Find a counterexample to $x^3 + 30 > x^2 + 20x$.

38. **OPEN ENDED** Write a sequence, the formula that produces it, and determine the formula for the sum of the terms of the sequence. Then prove the formula with mathematical induction.

39. **WRITING IN MATH** Explain how the concept of dominoes can help you understand the power of mathematical induction.

40. **WRITING IN MATH** Provide a real-world example other than dominoes that describes mathematical induction.

NYSRE PRACTICE A2.A.36

41. Which of the following is a counterexample to the statement below?

$$n^2 + n - 11 \text{ is prime.}$$

A $n = 2$ C $n = 5$
B $n = 4$ D $n = 6$

42. PROBABILITY Latisha wants to create a 7-character password. She wants to use an arrangement of the first 3 letters of her first name (lat), followed by an arrangement of the 4 digits in 1986, the year she was born. How many possible passwords can she create in this way?

F 72 H 288
G 144 J 576

43. GRIDDED RESPONSE A gear that is 8 inches in diameter turns a smaller gear that is 3 inches in diameter. If the larger gear makes 36 revolutions, how many revolutions does the smaller gear make in that time?

44. $3x + 8\overline{)3x^4 + 32x^3 + 46x^2 - 66x - 44} =$

A $2x^3 + 16x^2 - 12x - 12 - \frac{4}{3x + 8}$

B $x^3 + 8x^2 - 6x - 6 + \frac{4}{3x + 8}$

C $x^3 + 8x^2 - 6x - 6 - \frac{4}{3x + 8}$

D $x^3 - 8x^2 + 6x - 6 + \frac{4}{3x + 8}$

Spiral Review

Find the indicated term of each expansion. (Lesson 11-6)

45. fourth term of $(x + 2y)^6$
46. fifth term of $(a + b)^6$
47. fourth term of $(x - y)^9$

48. BIOLOGY In a particular forest, scientists are interested in how the population of wolves will change over the next two years. One model for animal population is the Verhulst population model, $p_{n+1} = p_n + rp_n(1 - p_n)$, where n represents the number of time periods that have passed, p_n represents the percent of the maximum sustainable population that exists at time n, and r is the growth factor. (Lesson 11-5)

a. To find the population of the wolves after one year, evaluate $p_1 = 0.45 + 1.5(0.45)(1 - 0.45)$.

b. Explain what each number in the expression in part **a** represents.

c. The current population of wolves is 165. Find the new population by multiplying 165 by the value in part **a**.

Find the exact solution(s) of each system of equations. (Lesson 10-7)

49. $x^2 + y^2 - 18x + 24y + 200 = 0$
$4x + 3y = 0$

50. $4x^2 + y^2 = 16$
$x^2 + 2y^2 = 4$

Skills Review

Evaluate each expression. (Lesson 0-5)

51. $P(8, 2)$
52. $P(9, 1)$
53. $P(12, 6)$
54. $C(5, 2)$
55. $C(8, 4)$
56. $C(20, 17)$
57. $P(12, 2)$
58. $P(7, 2)$
59. $C(8, 6)$
60. $C(9, 4) \cdot C(5, 3)$
61. $C(6, 1) \cdot C(4, 1)$
62. $C(10, 5) \cdot C(8, 4)$

CHAPTER 11

Study Guide and Review

NY Math Online glencoe.com
- STUDY TO GO
- Vocabulary Review

Chapter Summary

Key Concepts

Arithmetic Sequences and Series (Lessons 11-1 and 11-2)

- The *n*th term a_n of an arithmetic sequence with first term a_1 and common difference d is given by $a_n = a_1 + (n - 1)d$, where n is any positive integer.
- The sum S_n of the first n terms of an arithmetic series is given by $S_n = \frac{n}{2}[2a_1 + (n - 1)d]$ or $S_n = \frac{n}{2}(a_1 + a_n)$.

Geometric Sequences and Series (Lessons 11-3 and 11-4)

- The *n*th term a_n of a geometric sequence with first term a_1 and common ratio r is given by $a_n = a_1 \cdot r^{n-1}$, where n is any positive integer.
- The sum S_n of the first n terms of a geometric series is given by $S_n = \frac{a_1(1 - r^n)}{1 - r}$ or $S_n = \frac{a_1 - a_1 r^n}{1 - r}$, where $r \neq 1$.
- The sum S of an infinite geometric series with $-1 < r < 1$ is given by $S_n = \frac{a_1}{1 - r}$.

Recursion and Iteration (Lesson 11-5)

- In a recursive formula, each term is formulated from one or more previous terms.

The Binomial Theorem (Lesson 11-6)

- The Binomial Theorem:

$$(a + b)^n = \sum_{k=0}^{n} \frac{n!}{(n - k)!k!} a^{n-k} b^k$$

Mathematical Induction (Lesson 11-7)

- Mathematical induction is a method of proof used to prove statements about the positive integers.

FOLDABLES® Study Organizer

Be sure the Key Concepts are noted in your Foldable.

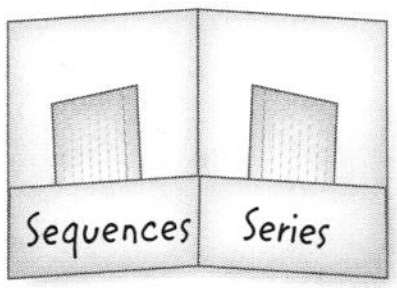

Key Vocabulary

arithmetic means (p. 689)
arithmetic sequence (p. 681)
arithmetic series (p. 690)
common difference (p. 681)
common ratio (p. 683)
convergent series (p. 705)
divergent series (p. 705)
explicit formula (p. 714)
Fibonacci sequence (p. 714)
finite sequence (p. 681)
geometric means (p. 697)
geometric sequence (p. 683)
geometric series (p. 698)
induction hypothesis (p. 727)
infinite sequence (p. 681)
infinity (p. 707)
iteration (p. 716)
mathematical induction (p. 727)
partial sum (p. 690)
Pascal's triangle (p. 721)
recursive formula (p. 714)
recursive sequence (p. 714)
series (p. 690)
sequence (p. 681)
sigma notation (p. 691)
term (p. 681)

Vocabulary Check

State whether each sentence is *true* or *false*. If *false*, replace the underlined term to make a true sentence.

1. An infinite geometric series that has a sum is called a convergent series.
2. Mathematical induction is the process of repeatedly composing a function with itself.
3. The arithmetic means of a sequence are the terms between any two non-successive terms of an arithmetic sequence.
4. A term is a list of numbers in a particular order.
5. The sum of the first n terms of a series is called the partial sum.
6. The formula $a_n = a_{n-2} + a_{n-1}$ is a recursive formula.
7. A geometric sequence is a sequence in which every term is determined by adding a constant value to the previous term.
8. An infinite geometric series that does not have a sum is called a partial sum.
9. Eleven and 17 are two geometric means between 5 and 23 in the sequence 5, 11, 17, 23.
10. Using the Binomial Theorem, $(x - 2)^4$ can be expanded to $x^4 - 8x^3 + 24x^2 - 32x + 16$.

Lesson-by-Lesson Review

11-1 Sequences as Functions (pp. 681–687)

A2.A.30, A2.A.31

Find the indicated term of each arithmetic sequence.

11. $a_1 = 9, d = 3, n = 14$

12. $a_1 = -3, d = 6, n = 22$

13. $a_1 = 10, d = -4, n = 9$

14. $a_1 = -1, d = -5, n = 18$

EXAMPLE 1

Find the 11th term of an arithmetic sequence if $a_1 = -15$ and $d = 6$.

$a_n = a_1 + (n - 1)d$	Formula for the *n*th term
$a_{11} = -15 + (11 - 1)6$	$n = 11, a_1 = -15, d = 6$
$a_{11} = 45$	Simplify.

11-2 Arithmetic Sequences and Series (pp. 688–695)

A2.A.29, A2.A.35

Find the arithmetic means in each sequence.

15. $-12, \_\_, \_\_, \_\_, 8$

16. $15, \_\_, \_\_, 29$

17. $12, \_\_, \_\_, \_\_, \_\_, -8$

18. $72, \_\_, \_\_, \_\_, 24$

19. BANKING Carson saves $40 every 2 months. If he saves at this rate for two years, how much will he have at the end of two years?

Find S_n for each arithmetic series.

20. $a_1 = 16, a_n = 48, n = 6$

21. $a_1 = 8, a_n = 96, n = 20$

22. $9 + 14 + 19 + \cdots + 74$

23. $16 + 7 + -2 + \cdots + -65$

24. DRAMA Laura has a drama performance in 12 days. She plans to practice her lines each night. On the first night she rehearses her lines 2 times. The next night she rehearses her lines 4 times. The third night she rehearses her lines 6 times. On the eleventh night, how many times has she rehearsed her lines?

Find the sum of each arithmetic series.

25. $\sum_{k=5}^{21}(3k - 2)$

26. $\sum_{k=0}^{10}(6k - 1)$

27. $\sum_{k=4}^{12}(-2k + 5)$

EXAMPLE 2

Find the two arithmetic means between 3 and 39.

$a_n = a_1 + (n - 1)d$	Formula for the *n*th term
$a_4 = 3 + (4 - 1)d$	$n = 4, a_1 = 3$
$39 = 3 + 3d$	$a_4 = 39$
$12 = d$	Simplify.

The arithmetic means are 3 + 12 or 15 and 15 + 12 or 27.

EXAMPLE 3

Find S_n for the arithmetic series with $a_1 = 18$, $a_n = 56$, and $n = 8$.

$S_n = \frac{n}{2}(a_1 + a_n)$	Sum formula
$S_8 = \frac{8}{2}(18 + 56)$	$n = 8, a_1 = 18, a_n = 56$
$= 296$	Simplify.

EXAMPLE 4

Evaluate $\sum_{k=3}^{15} 5k + 1$.

Use the formula $S_n = \frac{n}{2}(a_1 + a_n)$. There are 13 terms, $a_1 = 5(3) + 1$ or 16, and $a_{13} = 5(15) + 1$ or 76.

$$S_{13} = \frac{13}{2}(16 + 76)$$
$$= 598$$

Study Guide and Review

11-3 Geometric Sequences and Series (pp. 696–702)

A2.A.29, A2.A.35

Find the indicated term for each geometric sequence.

28. $a_1 = 5, r = 2, n = 7$

29. $a_1 = 11, r = 3, n = 3$

30. $a_1 = 128, r = -\frac{1}{2}, n = 5$

31. a_8 for $\frac{1}{8}, \frac{3}{8}, \frac{9}{8} \ldots$

Find the geometric means in each sequence.

32. 6, __, __, 162

33. 8, __, __, __, 648

34. −4, __, __, 108

35. SAVINGS Nolan has a savings account with a current balance of $1500. What would be Nolan's account balance after 4 years if he receives 5% interest annually?

Find S_n for each geometric series.

36. $a_1 = 15, r = 2, n = 4$

37. $a_1 = 9, r = 4, n = 6$

38. $5 - 10 + 20 - \ldots$ to 7 terms

39. $243 + 81 + 27 + \ldots$ to 5 terms

Evaluate the sum of each geometric series.

40. $\sum_{k=1}^{7} 3 \cdot (-2)^{n-1}$ **41.** $\sum_{k=1}^{8} -1\left(\frac{2}{3}\right)^{k-1}$

42. ADVERTISING Natalie is handing out fliers to advertise the next student council meeting. She hands out fliers to 4 people. Then, each of those 4 people hand out 4 fliers to 4 other people. Those 4 then hand out 4 fliers to 4 new people. If Natalie is considered the first round, how many people will have been given fliers after 4 rounds?

EXAMPLE 5

Find the sixth term of a geometric sequence for which $a_1 = 9$ and $r = 4$.

$a_n = a_1 \cdot r^{n-1}$	**Formula for the *n*th term**
$a_6 = 9 \cdot 4^{6-1}$	**$n = 6, a_1 = 9, r = 4$**
$a_6 = 9216$	

The sixth term is 9216.

EXAMPLE 6

Find two geometric means between 1 and 27.

$a_n = a_1 \cdot r^{n-1}$	**Formula for the *n*th term**
$a_4 = 1 \cdot r^{4-1}$	**$n = 4$ and $a_1 = 1$**
$27 = r^3$	**$a_4 = 27$**
$3 = r$	**Simplify.**

The geometric means are 1(3) or 3 and 3(3) or 9.

EXAMPLE 7

Find the sum of a geometric series for which $a_1 = 3$, $r = 5$, and $n = 11$.

$S_n = \frac{a_1 - a_1 r^n}{1 - r}$	**Sum formula**
$S_{11} = \frac{3 - 3 \cdot 5^{11}}{1 - 5}$	**$n = 11, a_1 = 3, r = 5$**
$S_{11} = 36{,}621{,}093$	**Use a calculator.**

EXAMPLE 8

Evaluate $\sum_{k=1}^{6} 2 \cdot (4)^{k-1}$.

$S_6 = \frac{2 - 2 \cdot 4^6}{1 - 4}$	**$n = 6, a_1 = 2, r = 4$**
$= \frac{-8190}{-3}$	**Simplify.**
$= 2730$	**Simplify.**

MIXED PROBLEM SOLVING
For mixed problem-solving practice, see page 989.

11-4 Infinite Geometric Series (pp. 705–711)

A2.A.34, A2.A.35

Find the sum of each infinite series, if it exists.

43. $a_1 = 8, r = \frac{3}{4}$

44. $\frac{5}{6} - \frac{20}{18} + \frac{80}{54} - \frac{320}{162} + \ldots$

45. $\sum_{k=1}^{\infty} 3\left(\frac{1}{2}\right)^{k-1}$

46. **PHYSICAL SCIENCE** Maddy drops a ball off of a building that is 60 feet high. Each time the ball bounces, it bounces back to $\frac{2}{3}$ its previous height. If the ball continues to follow this pattern, what will be the total distance that the ball travels?

EXAMPLE 9

Find the sum of the infinite geometric series for which $a_1 = 15$ and $r = \frac{1}{3}$.

$S = \frac{a_1}{1 - r}$ **Sum formula**

$= \frac{15}{1 - \frac{1}{3}}$ $a_1 = 15, r = \frac{1}{3}$

$= \frac{15}{\frac{2}{3}}$ or 22.5 **Simplify.**

11-5 Recursion and Special Sequences (pp. 714–719)

A2.A.33

Find the first five terms of each sequence.

47. $a_1 = -3, a_{n+1} = a_n + 4$

48. $a_1 = 5, a_{n+1} = 2a_n - 5$

49. $a_1 = 1, a_{n+1} = a_n + 5$

50. **SAVINGS** Sari has a savings account with a \$12,000 balance. She has a 5% interest rate that is compounded monthly. Every month Sari adds \$500 to the account. The recursive formula $b_n = 1.05b_{n-1} + 500$ describes the balance in Sari's savings account after n months. Find the balance of Sari's account after 3 months. Round your answer to the nearest penny.

Find the first three iterates of each function for the given initial value.

51. $f(x) = 2x + 1, x_0 = 3$

52. $f(x) = 5x - 4, x_0 = 1$

53. $f(x) = 6x - 1, x_0 = 2$

54. $f(x) = 3x + 1, x_0 = 4$

EXAMPLE 10

Find the first five terms of the sequence in which $a_1 = 1, a_{n+1} = 3a_n + 2$.

$a_{n+1} = 3a_n + 2$	**Recursive formula**
$a_{1+1} = 3a_1 + 2$	$n = 1$
$a_2 = 3(1) + 2$ or 5	$a_1 = 1$
$a_{2+1} = 3a_2 + 2$	$n = 2$
$a_3 = 3(5) + 2$ or 17	$a_2 = 5$
$a_{3+1} = 3a_3 + 2$	$n = 3$
$a_4 = 3(17) + 2$ or 53	$a_3 = 17$
$a_{4+1} = 3a_4 + 2$	$n = 4$
$a_5 = 3(53) + 2$ or 161	$a_4 = 53$

The first five terms of the sequence are 1, 5, 17, 53, and 161.

EXAMPLE 11

Find the first three iterates of the function $f(x) = 3x - 2$ for the initial value of $x_0 = 2$.

$x_1 = f(x_0)$	$x_2 = f(x_1)$	$x_3 = f(x_2)$
$= f(2)$	$= f(4)$	$= f(10)$
$= 3(2) - 2$	$= 3(4) - 2$	$= 3(10) - 2$
$= 4$	$= 10$	$= 28$

The first three iterates are 4, 10, and 28.

CHAPTER 11 Study Guide and Review

11-6 The Binomial Theorem (pp. 721–725)

A2.A.34

Expand each binomial.

55. $(a + b)^3$

56. $(y - 3)^7$

57. $(3 - 2z)^5$

58. $(4a - 3b)^4$

59. $\left(x - \frac{1}{4}\right)^5$

Find the indicated term of each expression.

60. third term of $(a + 2b)^8$

61. sixth term of $(3x + 4y)^7$

62. **EDUCATION** Mr. Collins is giving a 5-question multiple-choice quiz. Each question can be answered A, B, C, D, or E. How many ways could a student answer the questions using each answer A, B, C, D, or E once?

EXAMPLE 12

Expand $(x - 3y)^4$.

$$(x - 3y)^4$$
$$= x^4 + {}_4C_1x^3(-3y) + {}_4C_2x^2(-3y)^2 + {}_4C_3(-3y)^4 + {}_4C_4(-3y)^4$$
$$= x^4 + \frac{4!}{3!}x^3(-3y) + \frac{4!}{2!2!}x^2(9y^2) + \frac{4!}{3!}x(-27y^3) + 81y^4$$
$$= x^4 + -12x^3y + 54x^2y^2 + -108xy^3 + 81y^4$$

EXAMPLE 13

Find the fourth term of $(x + y)^8$.

Use the binomial Theorem to write the expansion in sigma notation.

$$(x + y)^8 = \sum_{k=0}^{8} \frac{8!}{k!(8-k)!}x^{8-k}y^k$$

For the fourth term, $k = 3$.

$$\frac{8!}{k!(8-k)!}x^{8-k}y^k = \frac{8!}{3!(8-3)!}x^{8-3}y^3$$
$$= 56x^5y^3$$

11-7 Proof and Mathematical Induction (pp. 727–731)

A2.A.36

Prove that each statement is true for all positive integers.

63. $2 + 6 + 12 + \cdots + n(n+1) = \frac{n(n+1)(n+2)}{3}$

64. $7^n - 1$ is divisible by 6.

65. $5^n - 1$ is divisible by 4.

Find a counterexample for each statement.

66. $8^n + 3$ is divisible by 11.

67. $6^{n+1} - 2$ is divisible by 17.

68. $n^2 + 2n + 4$ is prime.

69. $n + 19$ is prime.

EXAMPLE 14

Prove that $9^n + 3$ is divisible by 4.

Step 1 When $n = 1$, $9^n + 3 = 9^1 + 3$ or 12. Since 12 divided by 4 is 3, the statement is true for $n = 1$.

Step 2 Assume that $9^k + 3$ is divisible by 4 for some positive integer k. This means that $9^k + 3 = 4r$ for some whole number r.

Step 3
$$9^k + 3 = 4r$$
$$9^k = 4r - 3$$
$$9^{k+1} = 36r - 27$$
$$9^{k+1} + 3 = 36r - 27 + 3$$
$$9^{k+1} + 3 = 36r - 24$$
$$9^{k+1} + 3 = 4(9r - 6)$$

Since r is a whole number, $9r - 6$ is a whole number. Thus, $9^{k+1} + 3$ is divisible by 4, so the statement is true for $n = k + 1$.

Therefore, $9^n + 3$ is divisible by 4 for all positive integers n.

Practice Test

NY Math Online glencoe.com
Chapter Test

1. Find the next 4 terms of the arithmetic sequence 81, 72, 63, … .

2. Find the 25th term of an arithmetic sequence for which $a_1 = 9$ and $d = 5$.

3. **MULTIPLE CHOICE** What is the eighth term in the arithmetic sequence that begins 18, 20.2, 22.4, 24.6, …?

   A 26.8

   B 29

   C 31.2

   D 33.4

4. Find the four arithmetic means between −9 and 11.

5. Find the sum of the arithmetic series for which $a_1 = 11$, $n = 14$, and $a_n = 22$.

6. **MULTIPLE CHOICE** What is the next term in the geometric sequence below?

$$10, \frac{5}{2}, \frac{5}{8}, \frac{5}{32} \cdots$$

   F $\frac{13}{32}$

   G $\frac{5}{32}$

   H $\frac{5}{128}$

   J $\frac{5}{8}$

7. Find the three geometric means between 6 and 1536.

8. Find the sum of the geometric series for which $a_1 = 15$, $r = \frac{2}{3}$, and $n = 5$.

Find the sum of each series, if it exists.

9. $\sum_{k=2}^{12} (3k - 1)$

10. $\sum_{k=1}^{\infty} \frac{1}{2}(3^k)$

11. $45 + 37 + 29 + \cdots + -11$

12. $\frac{1}{8} + \frac{2}{24} + \frac{4}{72} + \ldots$

13. Write $0.\overline{65}$ as a fraction.

Find the first five terms of each sequence.

14. $a_1 = -1$, $a_{n+1} = 3a_n + 5$

15. $a_1 = 4$, $a_{n+1} = a_n + n$

16. **MULTIPLE CHOICE** What are the first 3 iterates of $f(x) = -5x + 4$ for an initial value of $x_0 = 3$?

   A 3, −11, 59

   B −11, 59, −291

   C −1, −6, −11

   D 59, −291, 1459

17. Expand $(2a - 3b)^4$.

18. What is the coefficient of the fifth term of $(m + 3n)^6$?

19. Find the fourth term of the expansion of $(c + d)^9$.

Prove that each statement is true for all positive integers.

20. $1 + 6 + 36 + \cdots + 6^{n-1} = \frac{1}{5}(6^n - 1)$.

21. $11^n - 1$ is divisible by 10.

22. Find a counterexample for the following statement.

$2^n + 4^n$ is divisible by 4.

23. **SCHOOL** There are an equal number of girls and boys in Mr. Marshall's science class. He needs to choose 8 students to represent his class at the science fair. What is the probability that 5 are boys?

24. **PENDULUM** Laurie swings a pendulum. The distance traveled per swing decreases by 15% with each swing. If the pendulum initially traveled 10 inches, find the total distance traveled when the pendulum comes to a rest.

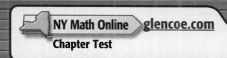

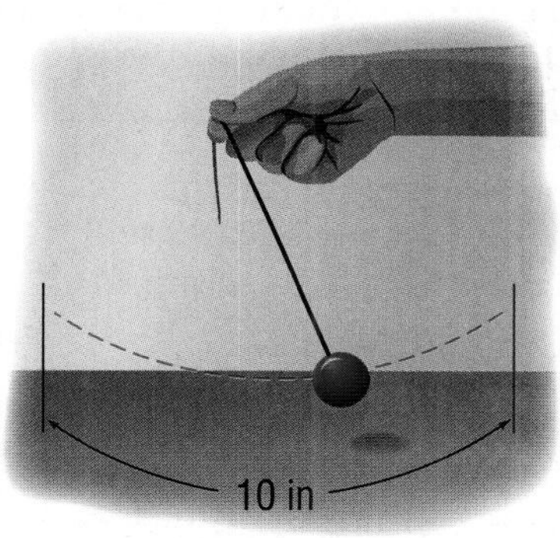

CHAPTER 11

Preparing for Standardized Tests

Look For a Pattern

One of the most common problem-solving strategies is to look for a pattern. The ability to recognize patterns, model them algebraically, and extend them is a valuable problem-solving tool.

Strategies for Looking For a Pattern

Step 1

Identify the pattern.

- Compare the numbers, shapes, or graphs in the pattern.
- **Ask yourself:** How are the terms of the pattern related?
- **Ask yourself:** Are there any common operations that lead from one term to the next?

Step 2

Generalize the pattern.

- Write a rule using words to describe how the terms of the pattern are generated.
- Assign variables and write an algebraic expression to model the pattern if appropriate.

Step 3

Find missing terms, extend the pattern, and solve the problem.

- Use your pattern or your rule to finding missing terms and/or extend the pattern to solve the problem.
- Check your answer to make sure it makes sense.

EXAMPLE

Read the problem. Identify what you need to know. Then use the information in the problem to solve.

Use the sequence of squares shown. How many squares will be needed to make the ninth figure of the sequence?

A 55 **C** 74

B 65 **D** 82

Figure 1 **Figure 2** **Figure 3**

Read the problem statement carefully. You are given three figures of a sequence and asked to find how many squares will be needed to make the ninth figure.

Look for a pattern in the figures of squares. Count the number of squares in each figure.

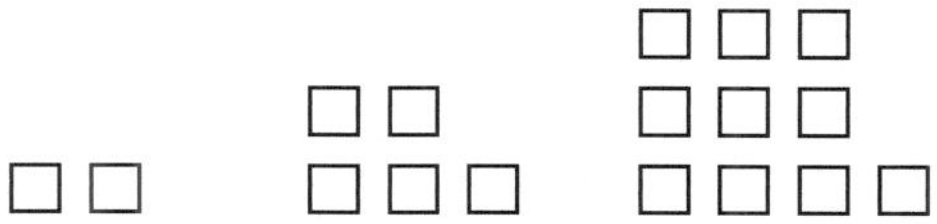

Write an expression to model this pattern.

Words	The number of squares is equal to the square of the figure number plus one.
Variable	Let n represent the figure number.
Equation	$a_n = n^2 + 1$

Use your expression to extend the pattern and find the number of squares in the ninth figure.

$a_9 = 9^2 + 1 = 82$

So, the ninth figure will have 82 squares. The correct answer is D.

Exercises

Read each problem. Use a pattern to solve the problem.

1. The numbers below form a famous mathematical sequence of numbers known as the Fibonacci sequence. What is the next Fibonacci number in the sequence?

1, 1, 2, 3, 5, 8, 13, 21, …

A 36

B 34

C 31

D 29

2. What is the missing number in the table?

n	a_n
1	0
2	2
3	6
4	12
5	??
6	30

F 17

G 18

H 20

J 21

CHAPTER 11 NYS Regents Exam Practice

Cumulative, Chapters 1 through 11

Multiple Choice

Answer all questions in this part. Select the answer that best completes the statement or answers the question.

1. Evaluate $\sum_{k=1}^{13}(12k - 3)$.

(1) 717

(2) 904

(3) 981

(4) 1053

2. The dairy industry is New York's leading agricultural sector. There are nearly 6000 dairy farms in the state producing 12 billion pounds of milk annually. Suppose a small dairy farmer has 68 cows and wants to increase this number by 6 cows per year. Which sequence shows how many cows he will own after n years?

(1) $a_n = 68 - 6n$

(2) $a_n = (68 + 6)n$

(3) $a_n = 68 + 6n$

(4) $a_n = 6 + 68n$

3. In which direction must the graph of $y = \frac{1}{x}$ be shifted to produce the graph of $y = \frac{1}{x+2}$?

(1) up

(2) down

(3) right

(4) left

4. When Jermaine drops a ball from a height of 12 feet, it rebounds to 75% of its previous height on each successive bounce. How high will the ball rebound after its third bounce? Round to the nearest tenth.

(1) 3.8 ft

(2) 4.5 ft

(3) 5.1 ft

(4) 6.8 ft

5. Without writing in standard form, state what type of graph is represented by the equation below.

$$8y^2 - 5y + 6x^2 = x - 1$$

(1) circle

(2) ellipse

(3) hyperbola

(4) parabola

6. Simplify the following expression.

$$\sqrt[3]{-27b^6c^{12}}$$

(1) $-3b^2c^4$

(2) $-3b^3c^6$

(3) $3b^2c^4$

(4) $3b^3c^6$

7. How many real roots does the quadratic function graphed below have?

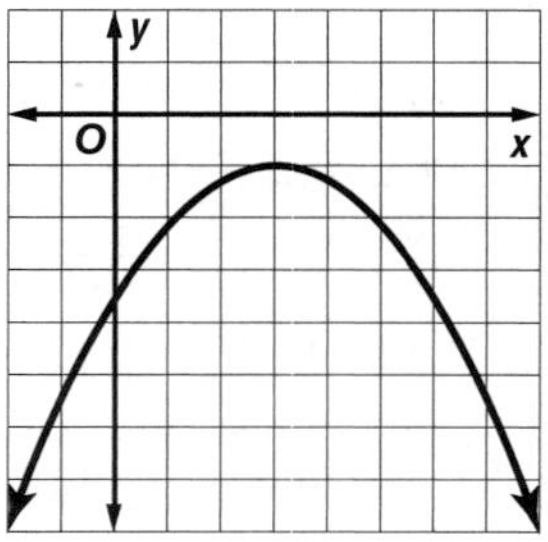

(1) 0

(2) 1

(3) 2

(4) 4

Test-Taking Tip

Question 7 The roots of an equation are the values where its graph crosses the x-axis.

8. The speed of light is approximately 3×10^8 meters per second. If the distance from the Sun to Saturn is about 1.43×10^{12} meters, how long does it take light from the Sun to reach Saturn? Round to the nearest whole number.

(1) 3856 seconds

(2) 4112 seconds

(3) 4389 seconds

(4) 4767 seconds

9. What is the equation of the line graphed below?

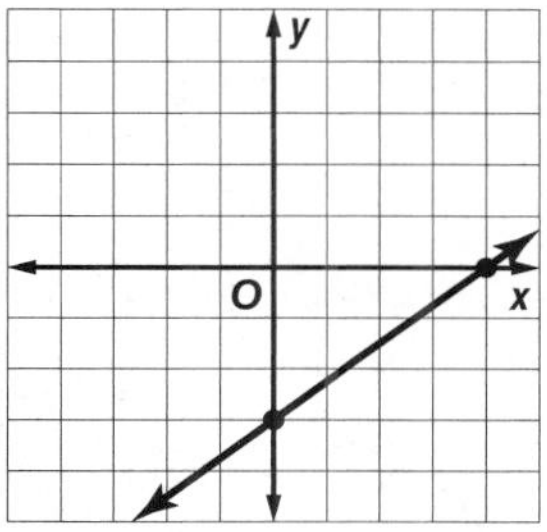

(1) $4x + 3y = 12$

(2) $4x - 3y = 12$

(3) $3x + 4y = 12$

(4) $3x - 4y = 12$

10. Matrix A is shown below. Which of the following shows the dimensions of matrix B for which the product AB exists?

$$A = \begin{bmatrix} -1 & 5 & 8 \\ 0 & 3 & -6 \end{bmatrix}$$

(1) 2×3

(2) 2×4

(3) 3×4

(4) 4×3

Open-Ended Response

Solve each problem. Clearly indicate the necessary steps, including appropriate formula substitutions, diagrams, graphs, charts, etc.

11. Consider the pattern below.

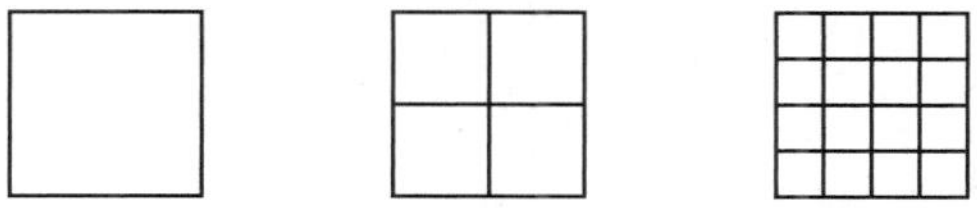

Figure 1 1 piece | Figure 2 4 pieces | Figure 3 16 pieces

a. Describe the pattern shown.

b. Write an expression for the number of pieces in Figure n.

c. Into how many pieces will the eighth figure of the pattern be divided?

12. Prove that the product of any two odd integers is odd.

13. Sandy inherited \$250,000 from her aunt in 1992. She invested the money and increased it as shown in the table below.

Year	Amount
1992	\$250,000
2000	\$329,202
2005	\$390,989

a. Write an exponential function that could be used to predict the amount of money A after investing for t years.

b. If the money continues to grow at the same rate, in what year will it be worth \$500,000?

Need Extra Help?

If you missed Question...	1	2	3	4	5	6	7	8	9	10	11	12	13
Go to Lesson or Page...	11-2	1-1	11-2	5-7	11-3	2-4	11-3	3-1	11-4	4-3	11-5	11-6	6-7
NYS Core Curriculum	A2.A.35	A2.A.30	A2.A.46	A2.A.35	A2.R.1	A2.A.13	A.G.8	A2.A.8	A.G.4	A2.R.2	A2.A.29	A2.RP.12	A2.A.6

CHAPTER 12 Probability and Statistics

Then

In Chapter 9, you calculated weighted averages.

Now

In Chapter 12, you will:

- Evaluate surveys, studies, and experiments.
- Create and use graphs of probability distributions.
- Use the Empirical Rule to find probabilities.
- Compare sample statistics and population statistics.

NYS Core Curriculum

A2.S.13, A2.S.14 Calculate theoretical and empirical probabilities
A2.S.3, A2.S.4 Calculate measures of central tendency and dispersion
A2.S.5 Know and apply the characteristics of the normal distribution

Why?

EDUCATION Probability and statistics are used in all facets of education. Surveys and experiments are done to find out which teaching methods promote the most learning. Statistics are used to determine grades when classes are curved, or when college professors weight their grades.

Math *in Motion,* Animation glencoe.com

Get Ready for Chapter 12

Diagnose Readiness You have two options for checking Prerequisite Skills.

Text Option

Take the Quick Check below. Refer to the Quick Review for help.

QuickCheck

State whether the events are *independent* or *dependent*. (Concepts and Skills Bank 4)

1. selecting a fiction book and a nonfiction book at the library

2. choosing a president, vice-president, secretary, and treasurer for Key Club, assuming that a person can hold only one office

3. choosing a model, color, and year of automobile

Determine whether each situation involves a *permutation* or a *combination*. (Lesson 0-5)

4. seven shoppers in line at a checkout counter

5. an arrangement of the letters in the word *intercept*

6. choosing 2 different pizza toppings from a list of 6

Expand each binomial. (Lesson 11-6)

7. $(a - 2)^4$

8. $(m - a)^5$

9. $(2b - x)^4$

10. $(2a + b)^6$

11. $(3x - 2y)^5$

12. $(3x + 2y)^4$

13. $\left(\frac{a}{2} + 2\right)^5$

14. $\left(3 + \frac{m}{3}\right)^5$

QuickReview

EXAMPLE 1

Mary wants to take 7 different classes next year. Assuming that each class is offered each period, how many different schedules could she have?

When Mary schedules a class for a given period, she cannot schedule that class for any other period. Therefore, the choices of which class to schedule each period are dependent events.

There are $7 \cdot 6 \cdot 5 \cdot 4 \cdot 3 \cdot 2 \cdot 1$ or 5040 different schedules that Mary could have.

EXAMPLE 2

Determine whether the situation involves a *permutation* or a *combination*.

choosing 6 students from a class of 25

Because the order of the students that are chosen does not matter, this is a combination.

EXAMPLE 3

Expand $(a + b)^4$.

Replace n with 4 in the Binomial Theorem.

$$(a + b)^4$$

$$= a^4 + {}_4C_1\, a^3b + {}_4C_2\, a^2b^2 + {}_4C_3\, ab^3 + {}_4C_4\, b^4$$

$$= a^4 + \frac{4!}{(4-1)! \cdot 1!}\, a^3b + \frac{4!}{(4-2)! \cdot 2!}\, a^2b^2 + \frac{4!}{(4-3)! \cdot 3!}\, ab^3 + \frac{4!}{(4-4)! \cdot 4!}\, b^4$$

$$= a^4 + \frac{4!}{3! \cdot 1!}\, a^3b + \frac{4!}{2! \cdot 2!}\, a^2b^2 + \frac{4!}{1! \cdot 3!}\, ab^3 + \frac{4!}{0! \cdot 4!}\, b^4$$

$$= a^4 + \frac{24}{6 \cdot 1}\, a^3b + \frac{24}{2 \cdot 2}\, a^2b^2 + \frac{24}{1 \cdot 6}\, ab^3 + \frac{24}{1 \cdot 4}\, b^4$$

$$= a^4 + 4a^3b + 6a^2b^2 + 4ab^3 + b^4$$

Online Option

NY Math Online Take a self-check Chapter Readiness Quiz at glencoe.com.

Get Started on Chapter 12

You will learn several new concepts, skills, and vocabulary terms as you study Chapter 12. To get ready, identify important terms and organize your resources. You may wish to refer to **Chapter 0** to review prerequisite skills.

FOLDABLES® Study Organizer

Probability and Statistics Make this Foldable to help you organize your Chapter 12 notes about probability and statistics. Begin with a sheet of $8\frac{1}{2}''$ by $11''$ paper.

1. **Fold** in half lengthwise.

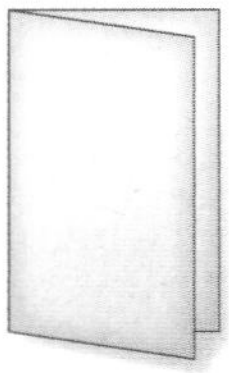

2. **Fold** the top to the bottom.

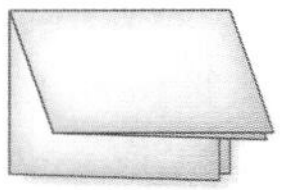

3. **Open.** Cut along the second fold to make two tabs.

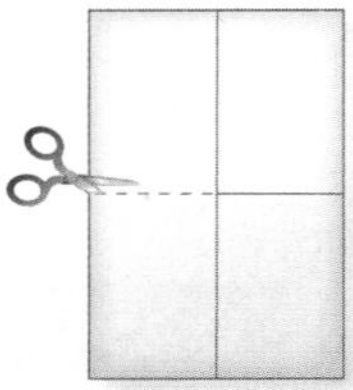

4. **Label** each tab as shown.

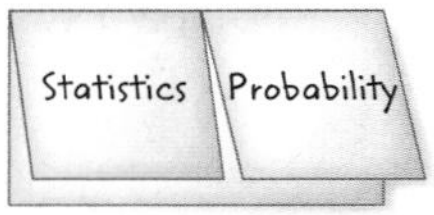

NY Math Online > glencoe.com

- Study the chapter online
- Explore **Math in Motion**
- Get extra help from your own **Personal Tutor**
- Use **Extra Examples** for additional help
- Take a **Self-Check Quiz**
- **Review Vocabulary** in fun ways

New Vocabulary

English		Español
survey	• p. 745 •	exámenes
population	• p. 745 •	población
sample	• p. 745 •	muestra
biased	• p. 745 •	en polarización negativa
unbiased	• p. 745 •	imparcial
observational study	• p. 746 •	estudio de observación
experiment	• p. 746 •	experimento
treatment group	• p. 746 •	grupo tratado
control group	• p. 746 •	grupo de control
parameter	• p. 752 •	parámetro
statistic	• p. 752 •	estadística
conditional probability	• p. 759 •	probabilidad condicional
relative frequency	• p. 760 •	frecuencia relativa
probability	• p. 764 •	probabilidad
random variable	• p. 766 •	variable aleatoria
expected value	• p. 767 •	valor previsto
normal distribution	• p. 773 •	distribución normal
skewed distribution	• p. 773 •	distribución asimétrica
inferential statistics	• p. 780 •	estadística deductiva
confidence interval	• p. 780 •	intervalo de la confianza
null hypothesis	• p. 781 •	hipótesis nula
alternative hypothesis	• p. 781 •	hipótesis alternativa
binomial distribution	• p. 786 •	distribución binomial
binomial experiment	• p. 786 •	experimento binomio

Review Vocabulary

combination • p. P12 • combinación an arrangement or selection of objects in which order is *not* important

permutation • p. P12 • permutación a group of objects or people arranged in a certain order

random • arbitrario Unpredictable, or not based on any predetermined characteristics of the population; When a die is tossed, a coin is flipped, or a spinner is spun, the outcome is a random event.

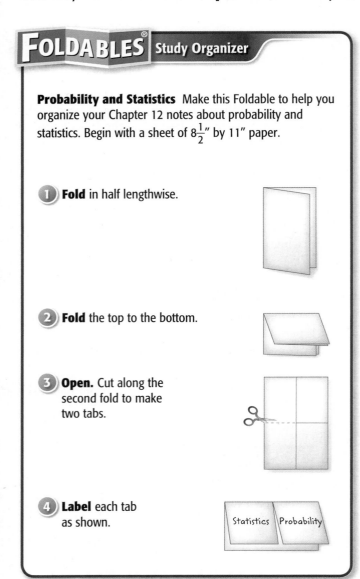

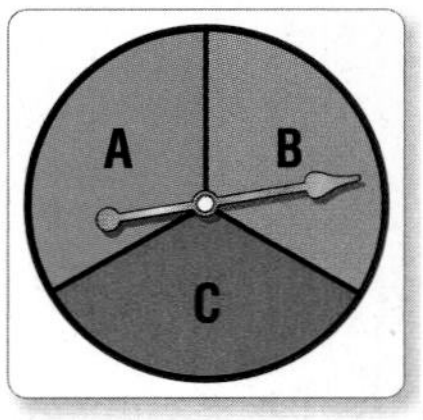

> Multilingual eGlossary glencoe.com

12-1

Experiments, Surveys, and Observational Studies

Then
You studied inductive and deductive reasoning. (Geometry)

Now
- Evaluate surveys, studies, and experiments.
- Distinguish between correlation and causation.

NYS Core Curriculum

A2.S.1 Understand the differences among various kinds of studies (e.g., survey, observation, controlled experiment)
A2.S.2 Determine factors which may affect the outcome of a survey

New Vocabulary
survey, population
census, sample
biased, unbiased
observational study
experiment
treatment group
control group
correlation, causation

NY Math Online
glencoe.com
- Extra Examples
- Personal Tutor
- Self-Check Quiz
- Homework Help

Why?

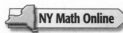

Students at a new school wanted to form a basketball team. In order to get funding for the program, they carried out a survey of students and their parents to determine who was in favor of starting the team.

Surveys, Studies, and Experiments **Surveys** are used to collect information. If everyone involved with the school were surveyed, then the survey would involve the entire **population**. A survey in which every member of the population is polled is called a **census**. If only 100 people selected at random from the school were surveyed, then the survey would involve a **sample**.

A survey is **biased** if its design favors certain outcomes. If the students above only surveyed basketball players and their parents, then the survey would be biased toward accepting the team. A sample is **unbiased** if it is *random*, or not based on any predetermined characteristics of the population. If they sent surveys to 100 students selected at random, then the survey would be unbiased.

Real-World EXAMPLE 1 Biased and Unbiased Samples

SURVEYS **State whether each survey would produce a random sample. Write *yes* or *no*. Explain.**

a. asking every tenth person coming out of a theater how many times a week they go to the theater to determine how often city residents support the performing arts

No; the people surveyed probably go to the theater more often than the average person.

b. surveying people going into a pet store to find out if the city's residents support the building and maintaining of a dog park

No; the people surveyed would probably be more likely than others to support pet activities.

c. A box contains the name of every student in the school. A hundred names are randomly pulled out of the box. Those students are asked their opinions on the new cafeteria rules.

Yes; everyone in the population has an equal chance to be part of the sample.

Check Your Progress

1A. asking every player at a golf course what sport they prefer to watch on TV

1B. calling 100 randomly selected numbers and asking for their opinions on a local tax

1C. going to a football game and asking 100 random fans about their favorite sport

Personal Tutor glencoe.com

To avoid bias in a survey, two things are needed: a strong random sample and unbiased survey techniques. A strong random sample is an unbiased sample with a large number of participants.

Real-World EXAMPLE 2 Survey Design

SCHOOL SURVEYS Christopher wants to determine the most desired location for the senior class trip. Which questions will get him the answer he is seeking?

a. Do you like Disneyland?
This question is biased in favor of Disneyland.

b. Which is better, King's Island or Cedar Point?
This question is biased because it only gives two options.

c. Where would you most like to go on the senior trip?
This is an unbiased question that will produce the answer he is seeking.

Check Your Progress

Which question will determine the most popular horror movie at school?

2A. Did you enjoy the last horror movie you saw?

2B. Which is better, romance or comedy?

2C. What is your favorite horror movie?

Personal Tutor glencoe.com

In an **observational study**, individuals are observed and no attempt is made to influence the results. In an **experiment**, something is intentionally done to people, animals, or objects, and then the response is observed.

Observational Study

- Find 100 people, 50 of whom have been taking a treatment.
- Collect the data.
- Analyze and interpret the data.

Experiment

- Find 100 people. Randomly select 50 people for treatment. Give the other 50 a placebo.
- Collect and analyze the data.

In an experiment, the people, animals, or objects given the treatment are called the **treatment group**. Those given the *placebo*, or false treatment, are the **control group**. The placebo is given so none of the participants will know which group he or she is in, and the experiment will be unbiased.

StudyTip

Experiments An experiment is biased when the participants know which group they are in.

Real-World EXAMPLE 3 Experiments and Observational Studies

EXPERIMENTS State whether each situation represents an *experiment* or an *observational study*. Identify the *control* group and the *treatment* group. If it is an experiment, determine whether there is bias.

a. Find 200 students, half of whom participated in extracurricular activities, and compare their grade-point averages.
This is an observational study. The students who participated in activities are the treated group and the other students are the control.

b. Find 200 people and randomly split them into two groups. One group jogs 2 miles per day and the other group does not jog at all.
This is an experiment because the people are put into groups at random. The treatment group is the joggers, and the control is the other group. This is a biased experiment because the participants all know which group they are in.

Check Your Progress

3. Find 80 college students, half of whom took a statistics course in high school, and compare their grades in a college statistics course.

Personal Tutor glencoe.com

How do you know when to use a survey, an observational study, or an experiment? A survey involves the random sampling of subjects from a population, while experiments involve the random assignment of treatments to subjects. In an experiment, you have control. In an observational study, you do not.

EXAMPLE 4 Experiments and Observational Studies

Determine whether each situation calls for a *survey*, an *observational study*, or an *experiment*. Explain the process.

a. You want to test a treatment for a disease.

This calls for an experiment. The test subjects are people with the disease. The treated group receives the treatment while the control group gets a placebo.

b. You want to find opinions on a presidential election.

This calls for a survey. It is best to call random numbers throughout the country in order to get an unbiased sample.

c. You want to find out if 10 years of smoking affects lung capacity.

This calls for an observational study. The lung capacity of people who have smoked for 10 years is compared to the lung capacity of an equal number of nonsmokers.

Check Your Progress

4. Two hundred randomly selected high school students rate their opinions regarding the new lunch rules from 1 (Totally Disagree) to 5 (Totally Agree).

Personal Tutor glencoe.com

Distinguish Between Correlation and Causation An observed association between the results of an experiment and the treatment does not necessarily imply that the treatment caused the results.

When there is a **correlation** between two events, the two events are related. When there is a **causation**, one event is shown to be the direct cause of another event. While a correlation between two events can be shown, causation is much more difficult to prove.

StudyTip

Causation If nothing else could have possibly caused the event, then you can assume causation.

EXAMPLE 5 Correlation Versus Causation

Determine whether the following statements show *correlation* or *causation*. Explain your reasoning.

a. Studies have shown that students are less energetic after they eat lunch.

Correlation; the statement ignores crucial factors that might have a causal influence on both.

b. If I lift weights, I can make the football team.

Correlation; there are more factors involved.

c. When the Sun is visible, we have daylight.

A good way to determine causation is to look for other alternatives that could cause daylight. Since there are none, it shows causation.

Check Your Progress

5. When I study, I will get an A.

Personal Tutor glencoe.com

Check Your Understanding

Example 1 p. 745 **State whether each survey would produce a random sample. Write *yes* or *no*. Explain.**

1. Survey every third person coming out of an ice cream shop to find people's favorite type of dessert.
2. A teacher sends every student whose last name ends with n to the blackboard.

Example 2 p. 746 **Determine the survey question that will best obtain the desired answer.**

3. Taylor wants to determine the most popular football team at the school.
   a. What is your favorite college football team?
   b. What is your favorite football team?
   c. Do you like the Dallas Cowboys or the Pittsburgh Steelers?

Example 3 p. 746 **State whether each situation represents an *experiment* or an *observational study*. Identify the *control* group and the *treatment* group. If it is an experiment, determine whether there is bias.**

4. A teacher has his first class complete review activities the day before the test. His second class does no review activities. He compares their test results.
5. Jaime finds 100 people, half of whom volunteer at a homeless shelter, and compares their average annual incomes.

Example 4 p. 747 **Determine whether the situation calls for a *survey*, an *observational study*, or an *experiment*. Explain the process.**

6. You want to test a drug that reverses male pattern baldness.
7. You want to find voters' opinions on recent legislation.

Example 5 p. 747 **Determine whether the following statements show *correlation* or *causation*. Explain.**

8. When I exercise, I am in a better mood.
9. If we have a Level 2 snow emergency, we do not have school.

Practice and Problem Solving

● = **Step-by-Step Solutions** begin on page R20.
Extra Practice begins on page 947.

Example 1 p. 745 **State whether each survey would produce a random sample. Write *yes* or *no*. Explain.**

10. A sporting goods store owner sends a survey to everyone whose address ends in a particular digit.
11. Students in an honors science class are asked what their favorite subject is.
12. Every other shopper coming out of a mall is surveyed to determine how much people spend during the holidays.
13. Every twentieth person coming out of your high school is asked for whom they will vote in the upcoming student council race.

Example 2 p. 746 **Determine the survey question that will best obtain the desired answer.**

14. Sabrina wants to determine interest in starting a chess club at her school.
    a. What day do you have free to stay after school?
    b. Do you like chess?
    c. Would you be willing to join a chess club at school?
15. Lauren wants to determine the most popular presidential candidate.
    a. For whom would you vote in the upcoming election?
    b. Do you prefer a particular political party?
    c. If you could vote, would you?

Example 3 p. 746

State whether each situation represents an *experiment* or an *observational study*. Identify the *control* group and the *treatment* group. If it is an experiment, determine whether there is bias.

16. Find 300 people and randomly split them into two groups. One group listens to Mozart for an hour every night before bed, and the other group does not listen to anything. Then compare how well they slept.

17. Find 250 students, half of whom are in the marching band, and compare the amounts of time spent on homework.

18. Find 100 students, half of whom are in the French Club, and compare their grades in French class.

Example 4 p. 747

Determine whether each situation calls for a *survey*, an *observational study*, or an *experiment*. Explain the process.

19. You want to find out if years of running affect knee movement.

20. You want to find out if drinking soda affects stomach linings.

21. You want to test a treatment that keeps deer out of your garden.

Example 5 p. 747

Determine whether the following statements show *correlation* or *causation*. Explain.

22. When it is very hot in the summer, there are ice cream vendors outside in New York.

23. Reading more will enable you to become more intelligent.

24. Researchers have concluded that Americans who speak more than one language are less likely to become ill.

25. Sleeping with your shoes on will cause you to have a headache.

26. **SELECTION BIAS** In a call-in poll, 81% of the more than 6000 respondents said that a certain businessman "symbolizes what makes the U.S.A. a great country." How is this an example of sampling bias?

27. **QUESTIONNAIRES** A company gives an exit questionnaire to employees who are leaving the company. One of the questions asks how the employee felt about his or her experience with the company. Is this survey biased? Explain why or why not.

H.O.T. Problems

Use **H**igher-**O**rder **T**hinking Skills

28. **FIND THE ERROR** Jordan and Kyle were asked to design an unbiased experiment. Is either of them correct? Explain your reasoning.

Jordan

- Get a group of 20 random people.
- Randomly put half of them on an all-fruit diet for 3 weeks.
- Compare their weight gain/loss at the end of the 3 weeks.

Kyle

- Get a group of 20 football players.
- Make half of them do 500 push-ups per day.
- Compare the number of push-ups each group can do after 3 weeks.

29. **CHALLENGE** How could a telephone survey introduce sampling bias into the results?

30. **WRITING IN MATH** Compare and contrast the random sampling of units from a population and the random assignment of treatments to experimental units.

31. **OPEN ENDED** Design one of each of the following.
    a. survey b. observational study c. experiment

32. **REASONING** How can bias occur in an experiment, and how does it affect the results? Provide an example to explain your reasoning.

Real-World Link

Phone surveys can cost up to five times as much as online surveys.

Source: Yahoo! Small Business

NYSRE PRACTICE A2.S.1, A2.S.2

33. GEOMETRY In $\triangle ABC$, $BC > AB$. Which of the following must be true?

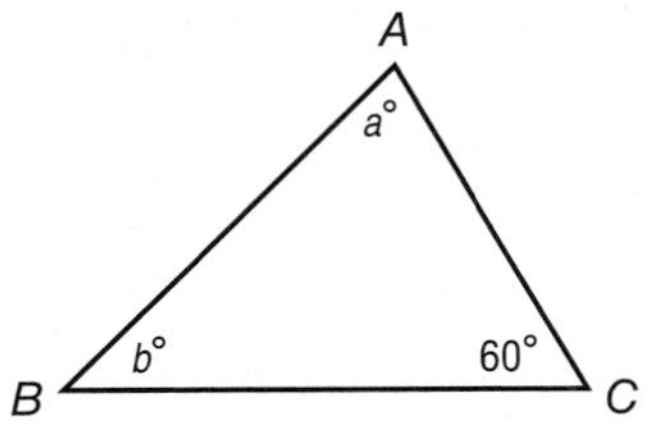

A $AB = BC$

B $AC < AB$

C $a > 60$

D $a = b$

34. SHORT RESPONSE What is the solution set of $4^{4x^2 - 2x - 4} = 4^{-2}$?

35. SAT/ACT A pie is divided evenly between 3 boys and a girl. If one boy gives one half of his share to the girl and a second boy keeps two thirds of his share and gives the rest to the girl, what portion will the girl have in all?

F $\frac{5}{24}$ G $\frac{11}{24}$ H $\frac{1}{2}$ J $\frac{13}{24}$

36. Which equation represents a hyperbola?

A $y^2 = 49 - x^2$

B $y = 49 - x^2$

C $y = 49x^2$

D $y = \frac{49}{x}$

Spiral Review

37. Prove that the statement $9^n - 1$ *is divisible by 8* is true for all natural numbers. (Lesson 11-7)

38. INTRAMURALS Ofelia is taking ten shots in the intramural free-throw shooting competition. How many sequences of hits and misses are there that result in her making eight shots and missing two? (Lesson 11-6)

Solve each system of equations. (Lesson 10-7)

39. $y = x + 3$; $y = 2x^2$

40. $x^2 + y^2 = 36$; $y = x + 2$

41. $y^2 + x^2 = 9$; $y = 7 - x$

42. $y + x^2 = 3$; $x^2 + 4y^2 = 36$

43. $x^2 + y^2 = 64$; $x^2 + 64y^2 = 64$

44. $y^2 = x^2 - 25$; $x^2 - y^2 = 7$

Find the distance between each pair of points with the given coordinates. (Lesson 10-1)

45. (9, −2), (12, −14)

46. (−4, −10), (−3, −11)

47. (1, −14), (−6, 10)

48. (−4, 9), (1, −3)

49. (2.3, −1.2), (−4.5, 3.7)

50. (0.23, 0.4), (0.68, −0.2)

Simplify. Assume that no variable equals 0. (Lesson 6-1)

51. $(5cd^2)(-c^4d)$

52. $(7x^3y^{-5})(4xy^3)$

53. $\frac{a^2n^6}{an^5}$

54. $(n^4)^4$

55. $\frac{-y^5z^7}{y^2z^5}$

56. $(-2r^2t)^3(3rt^2)$

Write a quadratic equation with the given root(s). Write the equation in the form $ax^2 + bx + c = 0$, where a, b, and c are integers. (Lesson 5-3)

57. −3, 9

58. $-\frac{1}{3}, -\frac{3}{4}$

59. 4, −5

Skills Review

60. TESTS Ms. Bonilla's class of 30 students took a biology test. If 20 of her students had an average of 83 on the test and the other students had an average score of 74, what was the average score of the whole class? (Lesson 9-6)

61. DRIVING During a 10-hour trip, Kwan drove 4 hours at 60 miles per hour and 6 hours at 65 miles per hour. What was her average rate, in miles per hour, for the entire trip? (Lesson 9-6)

EXTEND **12-1**

Graphing Technology Lab
Evaluating Published Data

NY Math Online glencoe.com
Other Calculator Keystrokes

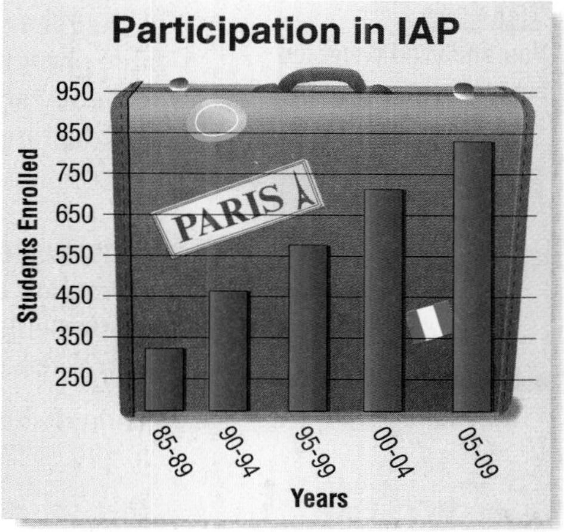

You can use a TI-83/84 Plus graphing calculator with the CelSheet application to evaluate data found in the media.

A newspaper ran a series of articles about high school students who study abroad for at least one semester. To support the claim that international study was gaining in popularity, the reporter presented the graph at the right. It includes information from a state university about the number of students earning credit through the university while studying abroad in the International Academic Programs.

Years	1985–1989	1990–1994	1995–1999	2000–2004	2005–2009
Students in IAP	316	451	561	704	823

ACTIVITY

Evaluate the graph of the data.

Step 1 Enter data in the CelSheet application.

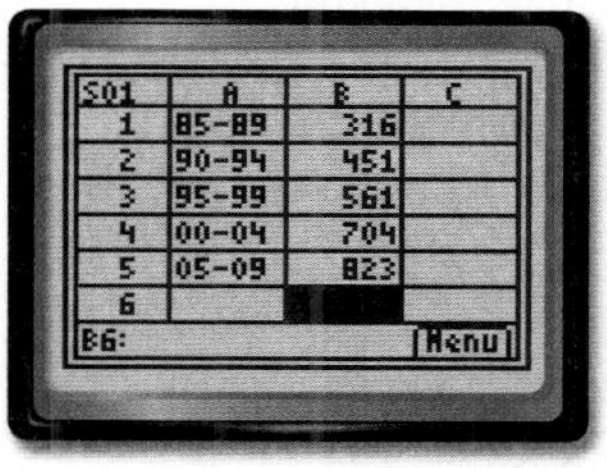

- Press [APPS] then press [▼] until **CelSheet** is highlighted.
  Press [ENTER]. Then press any key to exit the title and help pages.
- Press [ALPHA] ["] 85 [–] 89 [ENTER] to enter the first range of years into cell **A1**. Repeat for the remaining years.
- Use the arrow keys to highlight cell **B1**. Enter the data for each range of years.

Step 2 Make a bar graph of the data.

- Press [Graph] to access **Menu**. Press **4: Charts**. Then press **5: Bar**.
- Enter the **Category** range: press [ALPHA] **[A]** 1 [ALPHA] [:] [ALPHA] **[A]** 5 [ENTER].
- Enter **Series1**: press [ALPHA] **[B]** 1 [ALPHA] [:] [ALPHA] **[B]** 5 [ENTER].
- For **Ser1Name** enter **STDNTS**. At this prompt, the alpha is assumed.
- Use the arrow keys to scroll past **Series 2** and **Series 3** information. At **Title**, enter **IAP**. Press [ENTER] three times to display graph.
- Press [TRACE] then [▶] and [◀] to see information about each bar.

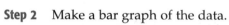

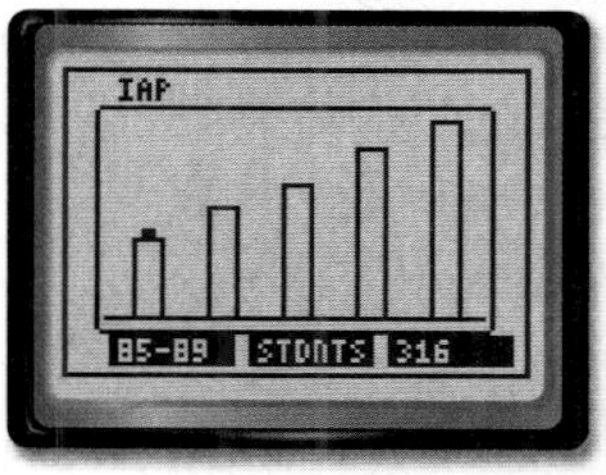

Analyze the Results

Compare your graph to the newspaper's graph.

1. Do the graphs display the same information?
2. Which graph seems to show a more dramatic increase? Why?
3. Why would the reporter choose to display the graph in this way? Is it acceptable? Why or why not?

12-2 Statistical Analysis

Then

You analyzed weighted averages. (Lesson 9-6)

Now

- Use measures of central tendency and variation to compare sets of data.
- Explore measures of variation.

NYS Core Curriculum

A2.S.4 Calculate measures of dispersion (range, quartiles, interquartile range, standard deviation, variance) for both samples and populations

New Vocabulary

univariate data
measure of central tendency
parameter
statistic
margin of sampling error
measure of variation
variance
standard deviation

NY Math Online

glencoe.com

- Extra Examples
- Personal Tutor
- Self-Check Quiz
- Homework Help

Why?

Denny has raced in 18 mountain bike races in the past year. His times are listed in the table shown. Which measure of central tendency should Denny use to describe the finishing times?

6:51	7:03	6:49	7:29	6:59	7:20
6:57	6:49	7:01	6:50	6:52	6:48
7:02	7:09	6:56	6:54	7:07	6:53

Measures of Central Tendency Data in one variable like the finishing times are called **univariate data**. These data can be described by a **measure of central tendency** because it represents the center or middle of the data. The most commonly used measures of central tendency are the *mean*, *median*, and *mode*.

When deciding which measure of central tendency to use to represent a set of data, look closely at the data values.

Key Concept: Measures of Central Tendency

For Your FOLDABLE

Use	Which Is...	When...
mean	the sum of the data divided by the number of items in the data set	The data set has no outliers.
median	the middle number of the ordered data, or the mean of the middle two numbers	The data set has outliers, but there are no big gaps in the middle of the data.
mode	the number or numbers that occur most often	The data set has many repeated numbers.

Real-World EXAMPLE 1 Measures of Central Tendency

a. RACING TIMES **Refer to the information above. Which measure of central tendency best represents the data, and why?**

Since the data are spread out and there do not appear to be any outliers, the mean best represents the data.

b. Which measure of central tendency best represents the data at the right, and why?

16	17	15	17
12	16	16	15
2	18	18	18
40	16	48	1

Since there are outliers and no big gaps in the middle, the median best represents the data.

Check Your Progress

1. RAFFLE A raffle is offering a grand prize worth $1000 and thirty other prizes worth $5 each. Which measure of central tendency best represents the data, and why?

Personal Tutor glencoe.com

Two types of measures can be applied to sets of data. A **parameter** is a measure that describes a characteristic of a *population*. One example of a parameter is the mean income of the United States. A **statistic** is a measure that describes a characteristic of a *sample*. An example of a statistic is the mean income of the people who live on your street.

Real-World Career

Teacher Teachers need a bachelor's degree and they need to be working towards their master's degree in education. High school teachers need to specialize in their chosen field. Many teachers also go through a rigorous testing process to be certified to teach in their states.

EXAMPLE 2 Samples Versus Populations

Determine whether each of the following represents a *population* or a *sample*.

a. The Nielsen Poll estimates the average number of hours of television watched per week for U.S. households.

This represents a sample because only a fraction of U.S. residents are polled.

b. A mathematics exam is given to every graduating senior in the country to analyze certain mathematics skills.

This represents a population because the exam tests *every* graduating senior.

Check Your Progress

2A. A teacher compares the scores on a test in her class.

2B. A teacher compares her class with the rest of the country on a national test.

Personal Tutor glencoe.com

When a single sample is drawn from a population, there is a risk of incurring a sampling error. As the size of the sample increases, the margin of error decreases. The **margin of sampling error** provides the interval that shows how much the responses from the sample would differ from the population.

Key Concept Margin of Sampling Error

For Your FOLDABLE

When a random sample n is taken from a population, the margin of sampling error can be approximated by $\pm\frac{1}{\sqrt{n}}$.

EXAMPLE 3 Margin of Sampling Error

In a random survey of 2148 people, 58% said that football is their favorite sport.

a. What is the margin of sampling error?

Margin of sampling error $= \pm\frac{1}{\sqrt{n}}$	**Margin of Sampling Error Formula**
$= \pm\frac{1}{\sqrt{2148}}$	$n = 2148$
$\approx \pm 0.0216$	**Simplify.**

The margin of sampling error is about $\pm 2.16\%$.

b. What is the likely interval that contains the percentage of the population that claims football is their favorite sport?

$0.58 + 0.0216 = 0.6016$ $\quad\quad$ $0.58 - 0.0216 = 0.5584$

The likely interval that contains the percentage of the population that claims football is their favorite sport is between 55.84% and 60.16%.

Check Your Progress

In a random survey of 3247 people, 41% said that they are satisfied with the government's performance.

3A. What is the margin of sampling error?

3B. What is the likely interval that contains the percentage of the population that is satisfied with the government?

Personal Tutor glencoe.com

Measures of Variation **Measures of variation** describe the *dispersion* or spread of a set of data. Two common measures of variation are the **variance** and **standard deviation**. These measures describe how closely a set of data clusters about the mean.

The sample mean $\overline{x}$, read *x bar*, and the population mean μ, or mu, are calculated the same way. The formulas for calculating the sample standard deviation s and the population standard deviation σ, or sigma, are given below.

Real-World Link

College professors often weigh their grades to put more emphasis on tests, essays, and research papers.

Source: The Ohio State University

Key Concept — Standard Deviation Formulas

For Your FOLDABLE

Sample

$$s = \sqrt{\frac{\sum_{k=1}^{n}(x_n - \overline{x})^2}{n-1}}$$

Population

$$\sigma = \sqrt{\frac{\sum_{k=1}^{n}(x_n - \mu)^2}{n}}$$

Real-World EXAMPLE 4 — Standard Deviation

TEST SCORES The Chapter 3 and Chapter 4 scores from Mr. Hoff's class both have a mean of 75. Find and compare their standard deviations.

Mr. Hoff's 2nd Period Chapter 3 Scores	Mr. Hoff's 2nd Period Chapter 4 Scores
85, 80, 75, 75, 70, 75, 75, 65, 75, 75, 75, 80, 75, 75, 70, 80, 70, 75, 75, 75, 75, 75, 75	100, 100, 90, 10, 100, 95, 10, 95, 100, 100, 85, 15, 95, 20, 95, 90, 100, 100, 90, 10, 100, 100, 25

a. Find the standard deviation for the Chapter 3 scores.

Step 1 This is a population. Since the mean of each set was 75, $\mu = 75$.

Step 2 Find the standard deviation.

$$\sigma = \sqrt{\frac{\sum_{k=1}^{n}(x_n - \mu)^2}{n}}$$ **Standard Deviation Formula**

$$= \sqrt{\frac{(85-75)^2 + (80-75)^2 + \ldots + (75-75)^2 + (75-75)^2}{23}} \approx 3.9$$

The class mean of the Chapter 3 test is 75 with a standard deviation of about 3.9.

StudyTip

Population Mean When the population mean μ is known, then the symbol can be interchanged with $\overline{x}$.

b. Use a calculator to find the standard deviation of the Chapter 4 scores.

Clear all lists. Then press STAT ENTER and enter each data value, pressing ENTER after each value. To view the statistics, press STAT ▶ 1 ENTER.

The class mean of the Chapter 4 test is 75 with a standard deviation of about 36.

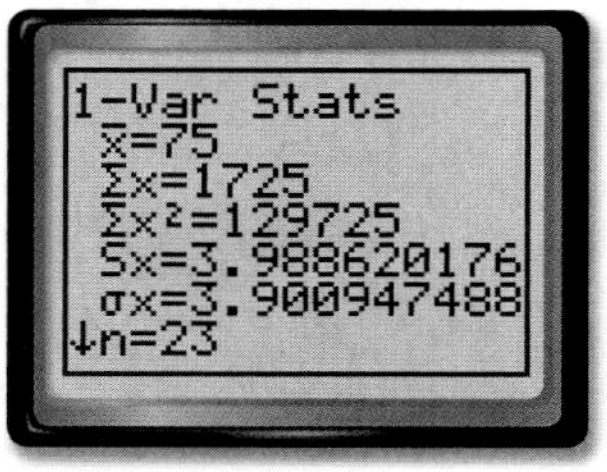

c. Compare the standard deviations of the two tests.

The standard deviation of the Chapter 4 test is far greater than for Chapter 3. Therefore, the scores are more dispersed in the Chapter 4 test, and they are much closer to the mean in the Chapter 3 test. Mr. Hoff can conclude that 75 is a stronger mean for Chapter 3, meaning that the majority of his students scored very close to 75.

Check Your Progress

4A. Calculate the mean and standard deviation of the population of data.

4B. Change 30 to 70. What should happen to the mean and standard deviation? Recalculate to confirm your results.

28	34	33	33	31
33	29	34	36	31
30	29	32	28	36
29	33	29	28	28
26	31	28	27	29

Personal Tutor glencoe.com

StudyTip

Standard Deviation The greater the standard deviation, the more the data deviate from the mean.

In a given set of data, the majority of the values fall within one standard deviation of the mean. Almost all of the data will fall within 2 standard deviations. Mr. Hoff's Chapter 3 scores had a mean of 75 and a standard deviation σ of 3.9. We can illustrate this graphically on a number line.

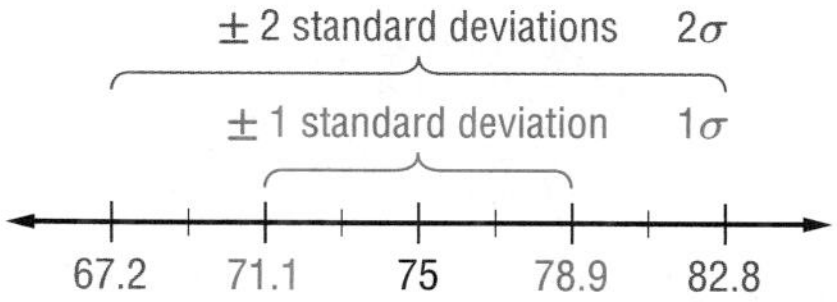

If Mr. Hoff were to compare his students' scores with other students throughout the country on a national test, the class would be considered a sample of all of the students who took the test. He would then need to calculate a sample mean $\bar{x}$ and a sample standard deviation σ.

Check Your Understanding

Example 1 p. 752

Which measure of central tendency best represents the data, and why?

1. {833, 796, 781, 776, 758}

2. {27.2, 36.8, 50.4, 71.6, 194.7}

3. {65, 21, 17, 52, 25, 17, 11, 22, 60, 44}

4. {53, 61, 46, 59, 61, 55, 49}

Example 2 p. 753

Determine whether each of the following represents a *population* or a *sample*.

5. Jerry's math club wants to compare their SAT scores.

6. The tennis team wants to compare their first-serve percentages with each other.

7. Jennifer conducts an online survey on political opinions.

8. Veronica compares the student-teacher ratios of all of the schools in her county.

Example 3 p. 753

9. **OLYMPICS** In a random survey of 5824 people, 29% said they will watch some of the Summer Olympics on television.

a. What is the margin of sampling error?

b. What is the likely interval that contains the percentage of the population that will watch the Summer Olympics on television?

Example 4 p. 754

10. **DRIVING** The maximum speed limits in miles per hour for interstate highways are given at the right.

a. Is this a sample or a population?

b. Find the standard deviation of the speeds.

Maximum Speed Limits Per State

70	65	75	70	70	75	65	65	70	70
55	75	65	65	65	70	65	70	65	65
65	70	70	70	70	65	75	75	65	65
75	65	70	70	65	75	65	65	75	65
70	75	65	65	70	70	65	75	65	65

Source: National Motorists Association

Practice and Problem Solving

● = **Step-by-Step Solutions** begin on page R20. **Extra Practice** begins on page 947.

Example 1 p. 752

Which measure of central tendency best represents the data, and why?

11. **NUTRITION** The table shows the number of Calories per serving of each vegetable.

Vegetable	Calories	Vegetable	Calories	Vegetable	Calories	Vegetable	Calories
asparagus	14	broccoli	25	cauliflower	10	lettuce	9
beans	30	cabbage	17	celery	17	spinach	9
bell pepper	20	carrots	28	corn	66	zucchini	17

12. **WEATHER** The table below shows daytime high temperatures for a week.

Day	Sun.	Mon.	Tues.	Wed.	Thurs.	Fri.	Sat.
Temperature	64°F	73°F	69°F	70°F	71°F	75°F	74°F

Example 2 p. 753

Determine whether each of the following represents a *population* or a *sample*.

13. Carissa calculates the average number of pineapples in 25 cans of pineapple.
14. The IRS calculates the mean income per household.
15. Middleburg Elementary School calculates the average height of all of its students.
16. Members of the football team want to compare their times in the 40 meter dash to those of the rest of the conference.
17. Jermaine asks 100 random people at the mall for their opinions on education.
18. The NFL compares the yards per game allowed by each team's defense.
19. Tomás compares the populations of every state.
20. Dona asks 400 random people what their favorite season is.

Example 3 p. 753

21. **MOVIES** A survey of 5669 random people found that 31% go to the movies at least once a month.
    a. What is the margin of sampling error?
    b. What is the likely interval that contains the percentage of the population that goes to the movies at least once a month?

22. **EXERCISE** A survey of 4213 people found that 78% exercise at least one hour each week.
    a. What is the margin of sampling error?
    b. What is the likely interval that contains the percentage of the population that do at least one hour of exercise each week?

Example 4 p. 754

23. **DOGSLED** The Iditarod is a 1150-mile dogsled race across Alaska. At the right are the winning times, in days, for recent years.

Iditarod Winning Times									
9	9	10	9	9	8	9	9	9	9
17	15	15	14	12	16	13	13	18	12
11	11	11	11	13	11	11	11		

    a. Is this a sample or a population?
    b. Find the standard deviation of the winning times.

24. **TRAINING** While training, Aiden recorded his times in the 40-meter dash. Find the standard deviation of the data.

40-Meter Dash Times									
4.8	4.9	4.8	4.7	5.0	4.9	4.8	4.9	4.8	5.0
5.0	5.1	4.8	4.9	4.6	4.8	4.7	4.9	4.8	4.8
5.0	4.9	4.9	5.0	4.9	5.0	4.8	4.8	4.7	4.6

Real-World Link

In 2007, the ACT was taken by more than 1.3 million students nationally with an average of 21.2.

Source: ACT, Inc

25. EDUCATION Below are ACT scores for a recent year.

Mean ACT Scores by State

20.2	21.3	21.5	20.4	21.6	20.3	22.5	21.5	17.8	20.5
20.0	21.7	21.3	20.2	21.6	22.0	21.6	20.3	19.8	22.6
20.8	22.4	21.4	22.2	18.8	21.5	21.7	21.7	21.2	22.5
21.2	20.1	22.3	20.3	21.2	21.4	20.6	22.5	21.8	21.9
19.3	21.5	20.5	20.3	21.5	22.7	20.9	22.5	22.2	21.4

Source: ACT, Inc

a. Compare the mean and median of the data.

b. Is this a sample or a population?

c. Find the standard deviation of the data. Round to the nearest hundredth.

d. Suppose the state with a mean score of 20.0 incorrectly reported the results. The score for the state is actually 22.5. How are the mean and median of the data affected by this change?

26. STUDENT-TEACHER RATIOS The table at the right shows the number of students in every math class at Principal Johnson's high school.

Students Per Math Class

25	27	26	26	19	27
24	23	19	28	25	24
20	22	22	24	26	18
28	29	29	26	24	24
23	23	25	25	29	28

a. Which measure of central tendency best represents the data? Why?

b. Is this a sample or a population?

c. Find the standard deviation of the data. Round to the nearest hundredth.

27. VACATIONS The table shows the number of annual vacation days for nine countries. Which measure of central tendency best represents the data? Justify your selection, and then find the measure of central tendency.

Annual Vacation Days

Country	Days	Country	Days
Brazil	34	Japan	25
Canada	26	Korea	25
France	37	U.K.	28
Germany	35	U.S.	13
Italy	42		

Source: *USA TODAY*

H.O.T. Problems

Use **H**igher-**O**rder **T**hinking Skills

28. OPEN ENDED Find and analyze a set of univariate real-world data of interest to you. Describe its measures of central tendency and variation.

29. CHALLENGE If 67% of the people surveyed responded positively and the likely interval that contains the percentage of the population is 64.8%–69.2%, how many people were surveyed?

30. REASONING A large outlier is eliminated from a set of data. How does this affect the mean and the standard deviation of the data? Explain.

31. REASONING With a linear transformation of data, all of the values are increased or decreased by the same value. If all of the values of the data are increased by 10, how does this affect the median, mean, and standard deviation? Explain.

32. WRITING IN MATH Compare and contrast the mean and median as measures of central tendency for a univariate data set.

33. REASONING The East basketball team has an average height of 6 feet with a standard deviation of 1.1 inches. The West basketball team has an average height of 6 feet with a standard deviation of 4.1 inches. Compare and contrast the heights of the players on the two teams.

NYSRE PRACTICE A2.S.4

34. STATISTICS In a set of nine different numbers, which of the following cannot affect the value of the median?

A doubling each number

B increasing each number by 10

C increasing the smallest number only

D increasing the largest number only

35. SHORT RESPONSE The average of the test scores of a class of c students is 80, and the average test scores of a class of d students is 85. When the scores of both classes are combined, the average score is 82. What is the value of $\frac{c}{d}$?

36. SAT/ACT What is the multiplicative inverse of $2i$?

F $-2i$ G $\frac{1}{2}$ H -2 J $\frac{-i}{2}$

37. Which equation best represents the graph?

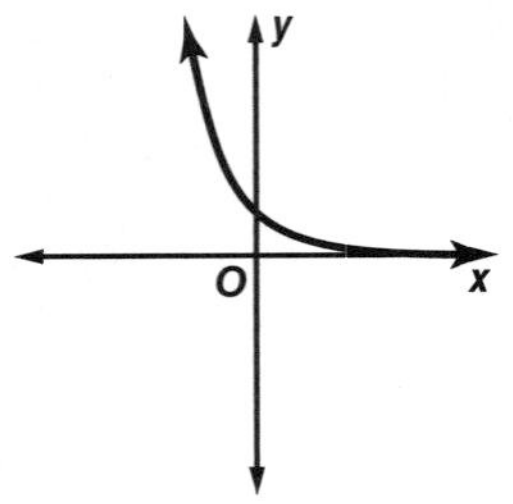

A $y = 4x$

B $y = x^2 + 4$

C $y = 4^{-x}$

D $y = -4^x$

Spiral Review

State whether each survey would produce a random sample. Explain. (Lesson 12-1)

38. the government sending a tax survey to everyone whose social security number ends in a particular digit

39. finding the heights of all the boys on the varsity basketball team to determine the average height of all the boys in your school

40. PARTIES Suppose each time a new guest arrives at a party, he or she shakes hands with each person already at the party. Prove that after n guests have arrived, a total of $\frac{n(n-1)}{2}$ handshakes have taken place. (Lesson 11-7)

41. ASTRONOMY The orbit of Pluto can be modeled by the equation $\frac{x^2}{39.5^2} + \frac{y^2}{38.3^2} = 1$, where the units are astronomical units. Suppose a comet is following a path modeled by the equation $x = y^2 + 20$. (Lesson 10-7)

a. Find the point(s) of intersection of the orbits of Pluto and the comet.

b. Will the comet necessarily hit Pluto? Explain.

c. Where do the graphs of $y = 2x + 1$ and $2x^2 + y^2 = 11$ intersect?

d. What are the coordinates of the points that lie on the graphs of both $x^2 + y^2 = 25$ and $2x^2 + 3y^2 = 66$?

Skills Review

Determine whether each situation involves a *permutation* or a *combination*. Then find the number of possibilities. (Lesson 0-5)

42. the winner of the first, second, and third runners-up in a contest with 8 finalists

43. selecting two of eight employees to attend a business seminar

44. an arrangement of the letters in the word *algebra*

45. placing an algebra book, a geometry book, a chemistry book, an English book, and a health book on a shelf

12-3 Conditional Probability

Then

You calculated probabilities. (Lesson PS 4)

Now

- Find probabilities of events given the occurrence of other events.
- Use contingency tables to find conditional probabilities.

NYS Core Curriculum

Reinforcement of A.S.18 Know the definition of conditional probability and use it to solve for probabilities in finite sample spaces

New Vocabulary

conditional probability
contingency table
relative frequency

NY Math Online

glencoe.com

- Extra Examples
- Personal Tutor
- Self-Check Quiz
- Homework Help

Why?

Alexis is testing a drug that protects people from getting sick. There are two groups; one group gets the experimental drug, while the other group receives a placebo.

After getting the results, Alexis needs to find the probability that a subject's staying healthy was a result of using the experimental drug.

This is an example of a conditional probability.

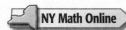

Conditional Probability The probability of an event given that another event has already occurred is called **conditional probability**. The conditional probability that event B occurs given that event A has already occurred can be represented by $P(B \mid A)$. This is read *the probability of B given A.*

Key Concept: Conditional Probability

For Your FOLDABLE

Given that A and B are dependent events, the conditional probability of an event B, given that an event A has already occurred, is defined as

$$P(B \mid A) = \frac{P(A \text{ and } B)}{P(A)}, \text{ where } P(A) \neq 0.$$

EXAMPLE 1 Conditional Probability

Carolina rolls a six-sided die. What is the probability that she has rolled a 3 given that she has rolled an odd number?

There are 6 possible results of rolling a six-sided die.

Let event A be that she rolled an odd number.
Let event B be that she rolled a 3.

$P(A) = \frac{1}{2}$ — **Three of the six outcomes are an odd number.**

$P(A \text{ and } B) = \frac{1}{6}$ — **One of the six outcomes is 3 and odd.**

$P(B \mid A) = \frac{P(A \text{ and } B)}{P(A)}$ — **Probability of B given A**

$= \frac{1}{6} \div \frac{1}{2} \text{ or } \frac{1}{3}$ — **$P(A) = \frac{1}{2}$ and $P(A \text{ and } B) = \frac{1}{6}$**

The probability of rolling a 3 given that the roll is odd is $\frac{1}{3}$.

Check Your Progress

1. Chen draws a card from a standard deck of 52 cards. Find the probability that he drew a king given that he drew a king, a queen, or a jack.

Personal Tutor glencoe.com

ReadingMath

Contingency Tables Contingency tables are also called *two-way frequency tables*.

Contingency Tables A **contingency table** records data in which different possible situations result in different possible outcomes. Each value represents the **relative frequency** of an outcome. These tables can be used to find conditional probabilities.

Real-World EXAMPLE 2 Contingency Tables

MEDICINE Find the probability that a test subject stayed healthy, given that he or she used an experimental drug.

Condition	Number of Subjects	
	Using Drug (D)	Using Placebo (P)
sick (S)	1600	1200
healthy (H)	800	400

There is a total of 1600 + 800 + 1200 + 400 or 4000 people in the study. We need to find the probability of H given that D occurs.

$P(H \mid D) = \dfrac{P(H \text{ and } D)}{P(D)}$ **Conditional Probability Formula**

$= \dfrac{800}{4000} \div \dfrac{2400}{4000}$ **$P(H$ and $D) = \dfrac{800}{4000}$ and $P(D) = \dfrac{1600 + 800}{4000}$**

$= \dfrac{800}{2400}$ or $\dfrac{1}{3}$ **Simplify.**

The probability that a subject stayed healthy given that he or she used the drug is $\frac{1}{3}$.

StudyTip

Independent Events If A and B are independent events, then $P(B \mid A) = P(B)$.

Check Your Progress

2. Find the probability that a test subject remained healthy, given that he or she used the placebo.

Personal Tutor glencoe.com

Contingency tables can be used to represent any number of possible situations.

NYSRE EXAMPLE 3 Reinforcement of A.S.18

The table below shows the number of students who are varsity athletes. Find the probability that a student is a varsity athlete given he or she is a junior.

A 19.8% **C** 11.5%

B 13.0% **D** 16.6%

Class	Freshman	Sophomore	Junior	Senior
varsity	7	22	36	51
nonvarsity	269	262	276	257

Read the Test Item

We need to find the probability that a student is a varsity athlete given that he or she is a junior. There is a total of 1180 students.

Solve the Test Item

$P(V \mid J) = \dfrac{P(V \text{ and } J)}{P(J)}$ **Conditional Probability Formula**

$= \dfrac{36}{1180} \div \dfrac{312}{1180}$ **$P(V$ and $J) = \dfrac{36}{1180}$ and $P(J) = \dfrac{36 + 276}{1180}$**

$\approx 11.5\%$ The correct answer is C.

Check Your Progress

3. Find the probability that a student plays varsity given that he or she is a freshman.

F 2.6% **G** 2.5% **H** 8.4% **J** 7.7%

Personal Tutor glencoe.com

Check Your Understanding

Example 1 p. 759

A bag contains 8 blue marbles, 6 red marbles, and 5 green marbles. The marbles are drawn one at a time. Find each probability.

1. The second marble is green, given that the first marble is blue and not replaced.
2. The second marble is red, given that the first marble is green and is replaced.
3. The third marble is red, given that the first two are red and blue and not replaced.
4. The third marble is green, given that the first two are red and are replaced.

Example 2 p. 760

5. **DRIVING TESTS** The table shows how students in Mr. Diaz's class fared on their first driving test. Some took a class to prepare, while others did not.

	Class	No Class
Passed	64	48
Failed	18	32

   a. Find the probability that Paige passed, given that she took the class.
   b. Find the probability that Elizabeth failed, given that she did not take the class.
   c. Find the probability that Terrence did not take the class, given that he passed.

Example 3 p. 760

6. **MULTIPLE CHOICE** The number of students who have attended a football game at North Coast High School is listed at the right. Find the probability that a student who has attended a game is a junior or a senior.

Class	Freshman	Sophomore	Junior	Senior
attended	48	90	224	254
not attended	182	141	36	8

A 90.8% B 91.6% C 86.2% D 96.9%

Practice and Problem Solving

● = **Step-by-Step Solutions** begin on page R20.
Extra Practice begins on page 947.

Example 1 p. 759

Three dice are rolled. Find each probability.

7. One of the rolls is a 6, given that all of the rolls are even.
8. One of the rolls is a 3, given that two of the rolls are odd.
9. Two of the rolls are 4s, given that all three rolls are the same.
10. At least two of the rolls are even, given that all three rolls are the same.

Example 2 p. 760

11. **SCHOOL CLUBS** King High School tallied the number of males and females that were members of at least one after school club. Find each probability.

	Clubs	No Clubs
Male	156	242
Female	312	108

   a. A student is a member of a club given that he is male.
   b. A student is not a member of a club given that she is female.
   c. A student is a male given that he is not a member of a club.

Example 3 p. 760

12. **MULTIPLE CHOICE** Naoko, Keisha, and Joshua compared the music on their MP3 players. Find the probability that a selected song is country given that it is not on Naoko's player.

Person	Rock	Country	R & B
Naoko	521	316	44
Keisha	119	145	302
Joshua	244	4	182

F 17.2% G 24.8% H 35.9% J 15.0%

Four coins are flipped. Find each probability.

13. Two are heads, given that at least one is tails.

14. Three are tails, given that at least one is heads.

15. None are heads, given that at least one is tails.

16. None are tails, given that at least three are heads.

17. **SOFTBALL** Paloma gets a hit 65% of the times she is at bat. What is the probability that she does not get a hit in five consecutive at-bats?

18. **WEATHER** According to the meteorologist, there will be a 20% chance of snow for each of the next five days. What is the probability that it snows for each of the next five days?

Real-World Link

Male teens spend about one fifth of their own money on video games.

Source: NOP World 2003

19. **COMPUTER GAMES** The table shows a distribution of computer games sold by a company.

a. Find the probability that a game is an action game, given that it is not a sports or role playing game.

b. Find the probability that a game is a family game, given that it is not a strategy or action game.

Type	*P*
strategy	0.19
children's	0.12
family	0.08
action	0.25
role playing	0.17
sports	0.16
other	0.03

20. **POP QUIZZES** His students have determined that Mr. Woodruff gives a pop quiz at the beginning of class 15% of the time. What is the probability that there will be no quizzes during a five-day week?

21. **FUNDRAISING** Mercedes and Victoria are trying to raise funds for their charity by calling numbers in the local phone book and asking for donations. They only reach 40% of the people they call. Of the people they reach, 20% promise to donate funds. Of the people who promise to donate, only 25% actually send money. What is the probability that a person who is called will actually contribute?

22. **HONORS CLASS** The probability that a student is in honors, given that he or she is in Mrs. Rollins' class, is $\frac{28}{51}$. The probability that a student is not in Mrs. Rollins' class, given that he or she is not in honors, is $\frac{33}{56}$. If there are 165 students that are neither in Mrs. Rollins' class nor in honors, how many students *are* in Mrs. Rollins' class and in honors?

H.O.T. Problems

Use Higher-Order Thinking Skills

23. **CHALLENGE** The probability that a student has a MyRoom page, given that he or she is a freshman, is $\frac{43}{55}$. The probability that a student does not have a MyRoom page, given that he or she is a sophomore, is $\frac{4}{27}$. Determine the probability that a student is a freshman or sophomore, given that he or she does have a MyRoom page.

24. **WRITING IN MATH** Explain the difference between conditional probability for dependent events and conditional probability for independent events. Provide examples of each type.

25. **REASONING** Which branches of a tree diagram represent conditional probability? Provide a sample tree diagram and explain your reasoning.

26. **REASONING** If a fair coin is flipped 20 times in a row and comes up heads every single time, what is the probability that it comes up heads on the 21st flip? Explain your reasoning.

27. **OPEN ENDED** Create a contingency table and calculate a conditional probability using the students in your class.

NYSRE PRACTICE Reinforcement of A.S.18

28. GEOMETRY If the perimeter of an equilateral triangle is 45, then what is the length of the altitude of the triangle?

A 9
B $9\sqrt{3}$
C 7.5
D $7.5\sqrt{3}$

29. Which expression is equivalent to $\dfrac{\frac{1}{4}+\frac{1}{4x}}{\frac{1}{x}+\frac{1}{4}}$?

F $\dfrac{x+1}{x+4}$
G 4
H $\dfrac{4x+4}{x+4}$
J $\dfrac{1}{4}$

30. SAT/ACT Aisha is late to practice 30% of the time each Wednesday because of French Club. What is the probability that she will be late on at least 3 of the next 5 Wednesdays?

A 3%
B 13%
C 16%
D 84%

31. SHORT RESPONSE If $\frac{12}{7}+\frac{15}{x}=1$, what is the value of x?

Spiral Review

32. MONEY The list shows the median income per capita in a recent year for 12 states in a region of the country. (Lesson 12-2)

a. Compare the mean and median for the region.

b. Find the standard deviation of the data. Round to the nearest hundredth.

c. Suppose the state's reported per capita income of \$22,861 is incorrect, and the actual value is \$24,861. How are the mean and median for the region affected?

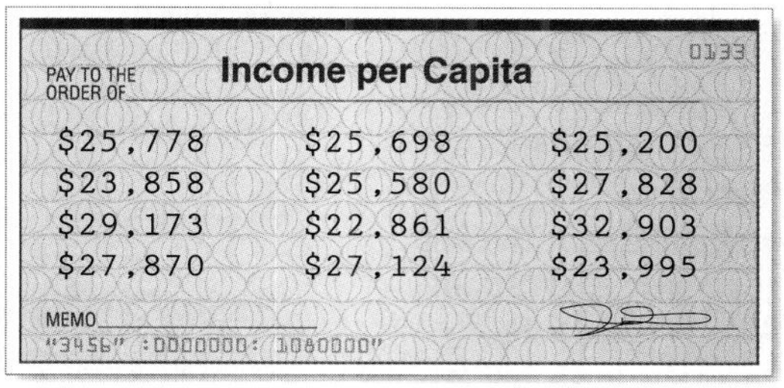

Income per Capita

\$25,778	\$25,698	\$25,200
\$23,858	\$25,580	\$27,828
\$29,173	\$22,861	\$32,903
\$27,870	\$27,124	\$23,995

Determine whether each situation would produce a random sample. Write *yes* or *no* and explain your answer. (Lesson 12-1)

33. surveying band members to find the most popular type of music at your school

34. surveying people coming into a post office to find out what color cars are most popular

35. ENTERTAINMENT A basketball team has a halftime promotion in which a fan gets to shoot a 3-pointer to try to win a jackpot. The jackpot starts at \$5000 for the first game and increases \$500 each time there is no winner. Ellis has tickets to the fifteenth game of the season. How much will the jackpot be for that game if no one wins by then? (Lesson 11-2)

Skills Review

Find x. Round to the nearest tenth if necessary. (Lesson 0-7)

36.
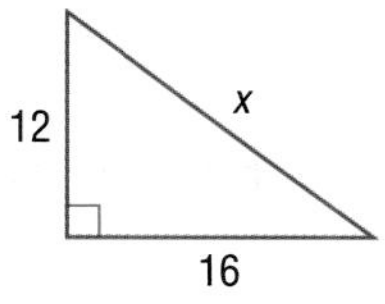

37.
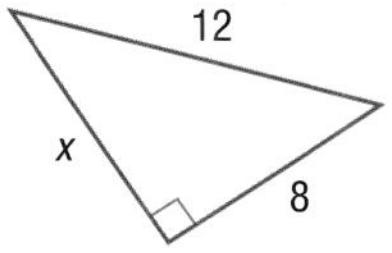

38.
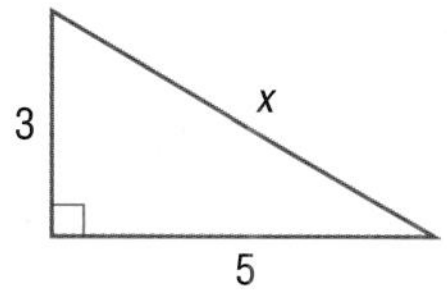

39.
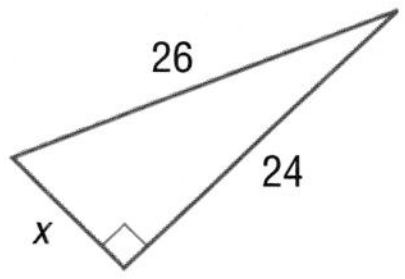

40.
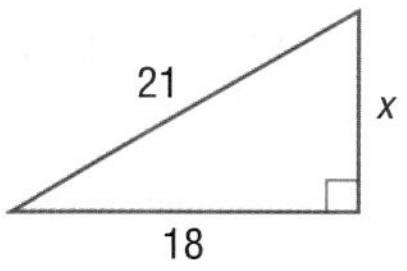

41.
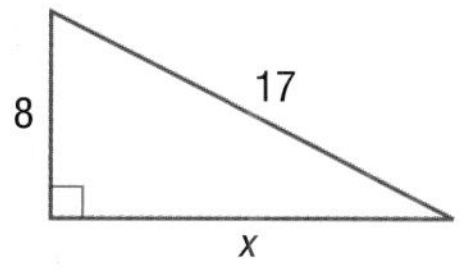

12-4 Probability and Probability Distributions

Then

You solved problems involving combinations and permutations. (Lesson 0-5)

Now

- Find probabilities by using combinations and permutations.
- Create and use graphs of probability distributions.

NYS Core Curriculum

A2.S.13 Calculate theoretical probabilities, including geometric applications
A2.S.14 Calculate empirical probabilities

New Vocabulary

probability, success
failure, sample space
random variable
probability distribution
uniform distribution
relative-frequency histogram
discrete probability distribution
theoretical probability
expected value

NY Math Online

glencoe.com
- Extra Examples
- Personal Tutor
- Self-Check Quiz
- Homework Help

Why?

Suppose the chances of passing a driving test the first time you take it are 5 in 6. The chances of passing the written part of the examination the first time are 9 in 10. What is the probability that you will pass both tests on the first try?

Probability The **probability** of an event is a ratio that measures the chances of the event occurring. A desired outcome is called a **success**. Any other outcome is called a **failure**. The set of all possible outcomes is called the **sample space**. The closer the probability of an event is to 1, the more likely the event is to occur.

Key Concept — Probability of Success and Failure (For Your FOLDABLE)

Words If an event can succeed in s ways and fail in f ways, then the probabilities of success $P(S)$ and of failure $P(F)$ are as follows.

Symbols $P(S) = \frac{s}{s+f}$ $P(F) = \frac{f}{s+f}$

EXAMPLE 1 Probability with Combinations

Twelve male and 16 female students have been selected as equal qualifiers for 6 college scholarships. If the qualifiers interviewed on the first day are to be chosen at random, what is the probability that 3 will be male and 3 will be female?

Step 1 Determine the number of successes.

$_{12}C_3$ **3 males chosen from 12 males**
$_{16}C_3$ **3 females chosen from 16 females**

Use combinations and the Fundamental Counting Principle to find s.

$_{12}C_3 \cdot {}_{16}C_3 = \frac{12!}{9!3!} \cdot \frac{16!}{13!3!}$ or 123,200 possible groups

Step 2 Determine the number of possibilities, $s + f$.

$_{28}C_6 = \frac{28!}{22!6!}$ or 376,740 total possible groups

Step 3 Find the probability.

$P(\text{3 males and 3 females}) = \frac{s}{s+f}$ **Probability of success**

$= \frac{123{,}200}{376{,}740}$ **$s = 123{,}200$ and $s + f = 376{,}740$**

≈ 0.327016 **Use a calculator.**

The probability of selecting 3 males and 3 females is about 0.327016 or 33%.

Check Your Progress

1. Three juniors and eleven seniors have been nominated for 4 spots to represent the school at a city-wide charity event. If the winners are drawn at random, what is the probability that 2 juniors and 2 seniors are selected?

Personal Tutor glencoe.com

Review Vocabulary

permutations and combinations When a group of objects or people is arranged in a certain order, the arrangement is called a *permutation*. An arrangement or selection of objects in which order is *not* important is called a *combination*. (Lesson 0-5)

Real-World EXAMPLE 2 Probability with Permutations

MUSIC Courtney has a playlist of 6 songs on her MP3 player. What is the probability that the player will randomly play her favorite song first, then her second favorite song, and the three least favorite songs last?

Step 1 Determine the number of successes.

$_1P_1$ **Play two favorite songs first and in order.**

$_3P_3$ **Play the least favorite songs last, but in any order.**

Use permutations and the Fundamental Counting Principle to find s.

$_1P_1 \cdot {}_3P_3 = 1! \cdot 3!$ or 6

Step 2 Determine the number of possibilities, $s + f$.

$_6P_6 = 6!$ or 720 possible orders of 6 songs

Step 3 Find the probability.

$P(\text{Courtney's desired order}) = \frac{s}{s+f}$ **Probability of success**

$= \frac{6}{720}$ **$s = 6$ and $s + f = 720$**

≈ 0.0083 **Use a calculator.**

The probability of the songs playing in Courtney's desired order is about 0.8%.

Check Your Progress

2. RACING Taryn, Stephanie, and Julie are in the 400-meter race with 5 other athletes. What is the probability that they all finish in the top three?

Personal Tutor glencoe.com

Sometimes, permutations and combinations are *both* used in determining a probability.

EXAMPLE 3 Probability with Combinations and Permutations

Suppose Hernanda pulls 5 marbles without replacement from a bag of 28 marbles in which 7 are red, 7 are black, 7 are blue, and 7 are white. What is the probability that 2 are of one color and 3 are of another color?

Step 1 Determine the number of successes.

$_4P_2$ **2 colors chosen from 4 if order matters**

$_7C_2$ **2 marbles of one color chosen from a group of 7**

$_7C_3$ **3 marbles of another color chosen from a group of 7**

Use permutations and combinations, along with the Fundamental Counting Principle, to find s.

$_4P_2 \cdot {}_7C_2 \cdot {}_7C_3 = 12 \cdot 21 \cdot 35$ or 8820

Step 2 Determine the number of possibilities, $s + f$.

$_{28}C_5 = 98{,}280$ ways to pull 5 marbles from a bag of 28

Step 3 Find the probability.

$P(\text{3 of one color, 2 of another}) = \frac{s}{s+f}$ **Probability of success**

$= \frac{8820}{98{,}280}$ **Substitute.**

≈ 0.0897 **Use a calculator.**

The probability of pulling 2 of one color and 3 of another is about 9%.

StudyTip

Replacement *With replacement* means that the object is *replaced*, or put back into the bag after it is drawn.

Check Your Progress

3. If 7 green marbles are added to the bag, what is the probability that Hernanda pulls out 4 of one color and 3 of another?

Personal Tutor glencoe.com

ReadingMath

Random Variables The notation $P(R = n)$ is used with random variables. $P(R = 4) = \frac{1}{6}$ is read *The probability that R equals 4 is one sixth.*

Probability Distributions The value of a **random variable** is the numerical outcome of a random event. A **probability distribution** for a particular random variable is a function that maps the sample space to its probabilities of the outcomes in the sample space.

A variable is said to be *random* if the sum of its probabilities is 1. The table below illustrates the probability distribution for rolling a die.

Sample space: $R = \{1, 2, 3, 4, 5, 6\}$

R = roll	1	2	3	4	5	6
Probability	$\frac{1}{6}$	$\frac{1}{6}$	$\frac{1}{6}$	$\frac{1}{6}$	$\frac{1}{6}$	$\frac{1}{6}$

$P(R = 4) = \frac{1}{6}$

A distribution in which all of the probabilities are equal is called a **uniform distribution**. For example, the distribution for rolling a die is uniform.

To help visualize a probability distribution, you can use a table of probabilities or a graph, called a **relative-frequency histogram**.

Real-World Link

There are currently 120 Crayola crayon colors. More than 400 have been introduced since 1903.

Source: Crayola

EXAMPLE 4 Probability Distribution

The spinner shows the probability distribution of the spinner landing on each color.

a. Create a relative-frequency histogram.

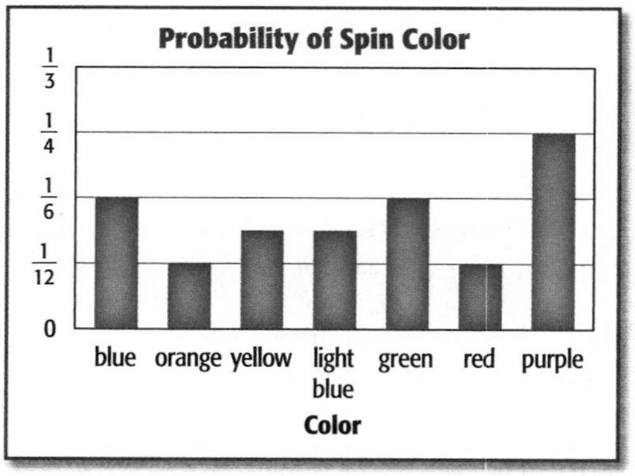

b. Use the graph to determine which outcome is most likely.

The most likely outcome is purple, and its probability is $\frac{1}{4}$.

c. Find P(blue or green).

The probability of spinning blue or green is $\frac{1}{6} + \frac{1}{6}$ or $\frac{1}{3}$.

Check Your Progress

Two six-sided dice are rolled, and their sum is recorded.

4A. Create a frequency table and a relative-frequency histogram of the data.

4B. Which outcomes are the least likely to occur? What is their probability?

4C. Find $P(5 \text{ or } 11)$.

Personal Tutor glencoe.com

Probability distributions like the one in Example 4 are called **discrete probability distributions** because there are only a finite number of possible outcomes. These distributions are commonly represented by histograms.

> **StudyTip**
> **Law of Large Numbers** The *Law of Large Numbers* states that as the number of trials increases, the experimental probability gets closer to the expected value.

The probabilities discussed here are **theoretical probabilities** because they are based on assumptions of what is expected to happen. The **expected value** $E(x)$ is the weighted average of the values in a probability distribution if the weight applied to each value is its theoretical probability. It tells you what you could expect in the "long run"—that is, after many trials.

EXAMPLE 5 Expected Value

Find the expected value of one roll of a die.

$E(x) = \left(1 \cdot \frac{1}{6}\right) + \left(2 \cdot \frac{1}{6}\right) + \left(3 \cdot \frac{1}{6}\right) + \left(4 \cdot \frac{1}{6}\right) + \left(5 \cdot \frac{1}{6}\right) + \left(6 \cdot \frac{1}{6}\right)$ **Weighted Average Formula**

$= \frac{1}{6} + \frac{2}{6} + \frac{3}{6} + \frac{4}{6} + \frac{5}{6} + \frac{6}{6}$ **Multiply.**

$= \frac{21}{6}$ or 3.5 **Add.**

Check Your Progress

5. Find the expected value of the sum of two dice.

Personal Tutor glencoe.com

Check Your Understanding

Example 1 p. 764

1. **ART** A museum curator at the Art Institute of Chicago is randomly selecting 4 paintings out of the 20 on display to showcase the work in a special exhibit. What is the probability that 3 of the 8 Paul Gauguin paintings are selected?

Example 2 p. 765

2. **TOURNAMENTS** Eight players entered a tournament. If the names are drawn randomly, what is the probability that the first four players selected are, in order, Alicia, Andrew, Marco, and Zack?

Example 3 p. 765

3. **CARDS** Suppose Justin draws 5 cards from a standard deck of 52 cards. What is the probability that those 5 cards contain 3 of one suit and two of another suit?

Example 4 p. 766

4. **FLOWERS** The table and relative-frequency histogram show the distribution of the number of red flowers if 4 seeds are planted.

R = Red	0	1	2	3	4
Probability	0.2	0.45	0.2	0.1	0.05

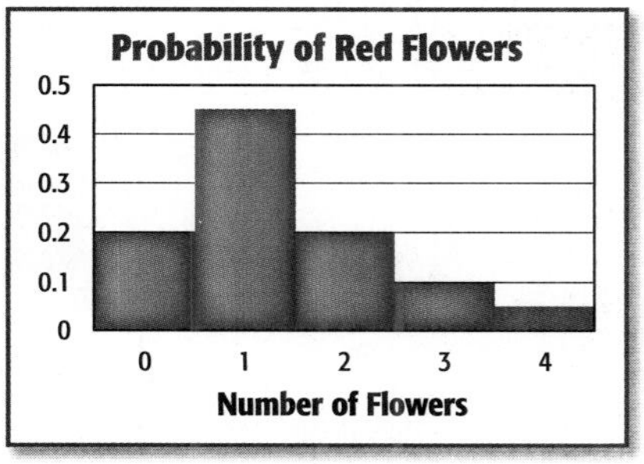

Find each of the following.

a. $P(R = 0)$

b. $P(R = 1)$

c. P(at least 2 are red)

d. P(only 1 flower is not red)

Example 5 p. 767

5. **PAPER CLIPS** In a box of paper clips, 25 are red, 15 are yellow, and 30 are green.

a. If 10 paper clips are selected, what is the expected number of yellow paper clips?

b. If 5 paper clips are selected, what is the expected number of green paper clips?

c. If 20 paper clips are selected, what is the expected number of red paper clips?

Practice and Problem Solving

● = **Step-by-Step Solutions** begin on page R20. **Extra Practice** begins on page 947.

Example 1 p. 764

6. **DRAWINGS** Twenty-four students entered a random drawing for 10 new calculators. What is the probability that 3 of the 5 students who entered from Mr. Kline's class won a calculator?

7. **RAFFLES** Fifty kids, including Lorena, Rebecca, and Melia, entered a raffle for 4 game consoles. What is the probability that two of these girls won?

8. **PERFORMANCES** During a magic show, the magician selects at random five members of the audience to assist in his performance. If there are 124 people in the audience, what is the probability that at least one of ten friends is selected?

Example 2 p. 765

9. **SEATING CHARTS** The new seating chart in Mr. Lian's class of 26 students was randomly generated. What is the probability that Jamila, Candace, and Haley are in the first, second, and third seats, respectively?

10. **LOTTERIES** In a lottery, 3 numbers from 1 through 10 are drawn without replacement, and the person who selects the correct numbers in the order in which they are drawn wins the prize. If Eva buys 5 different tickets, what is the probability that she will win?

Example 3 p. 765

11. **BALLOONS** A package of 48 balloons contains an equal number of red, white, blue, and purple balloons. If Shelby is given a handful of them to blow up, what is the probability that she gets 3 balloons of one color and 4 of another color?

12. **TRIVIA CONTESTS** Ten students from every grade level at West High were invited to a district-wide trivia contest. At the contest, 6 students are randomly selected to be alternates. What is the probability that 4 of these students are seniors and 2 are sophomores?

Example 4 p. 766

13. **RAFFLES** The table and relative-frequency histogram show the distribution of winning a raffle if 100 tickets are sold. There is 1 prize for first, 10 prizes for second, and 25 prizes for third. Find $P(Z > 0)$.

Z = Prize	no prize	1st	2nd	3rd
Probability	0.64	0.01	0.1	0.25

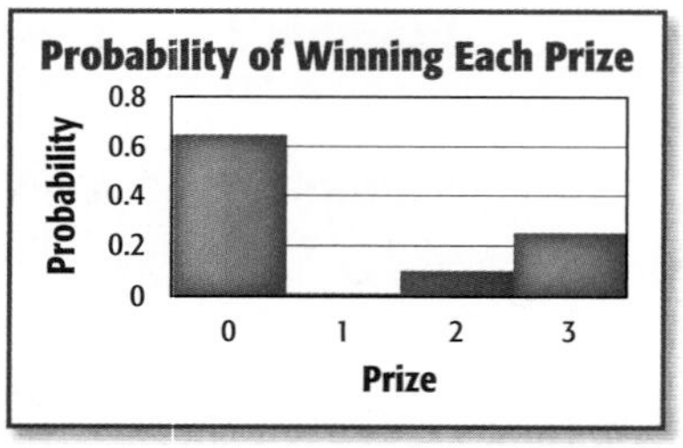

14. **MARBLES** Tai has a sack of 35 marbles. Eight are black, 12 are red, 9 are green, and the rest are white. Brianna pulls 2 marbles out of the bag.
    a. Create a frequency table and a relative-frequency histogram of the data.
    b. Which outcome is the most likely to occur?
    c. Find P(black and green).

Example 5 p. 767

15. **CARDS** In a standard deck of 52 cards, there are 4 different suits.
    a. If you are dealt 5 cards, what is the expected number of hearts?
    b. If you are dealt 7 cards, what is the expected number of spades?
    c. If you are dealt ten cards, what is the expected number of clubs?

16. **BILLIARD BALLS** In a rack of 16 billiard balls, there are 9 different colors, including the black eight ball and the white cue ball. Of the remaining 14 balls, 7 are striped and 7 are solid.

   a. If 5 balls are randomly selected, what is the expected number of stripes?

   b. If 4 balls are randomly selected, what is the expected number of white balls?

   c. If 6 balls are randomly selected, what is the expected number of balls that are neither white nor black?

17. **SNOW DAYS** The following probability distribution lists the probable number of snow days per school year at North High School. Use this information to determine the expected number of snow days per year.

Real-World Link

Before closing schools, superintendents use a system that could include field operatives traveling hazardous streets, Internet forecasts, and a school district's own Doppler radar.

Source: The Associated Press

Number of Snow Days Per Year

Days	0	1	2	3	4	5	6	7	8
Probability	0.1	0.1	0.15	0.15	0.25	0.1	0.08	0.05	0.02

18. **BASKETBALL** The distribution below lists the probability of the number of major upsets in the first round of a basketball tournament each year. Determine the expected number of upsets.

Number of Upsets Per Year

Upsets	0	1	2	3	4	5	6	7	8
Probability	$\frac{1}{32}$	$\frac{1}{16}$	$\frac{3}{32}$	$\frac{1}{8}$	$\frac{1}{8}$	$\frac{5}{16}$	$\frac{1}{8}$	$\frac{3}{32}$	$\frac{1}{32}$

19. **STUDENT GOVERNMENT** Based on previous data, the probability distribution of the number of students running for class president per year is listed at the right. Determine the expected number of students who will run.

Number of Students Running

Students	1	2	3	4	5	6
Probability	0.05	0.15	0.2	0.2	0.35	0.2

20. **MARBLES** In a bag of 25 marbles with an equal amount of red, blue, green, black, and clear marbles, what is the probability of pulling out 4 of one color and 2 of another?

21. **HISTOGRAMS** The histogram at the right shows the probability of each student winning a prize.

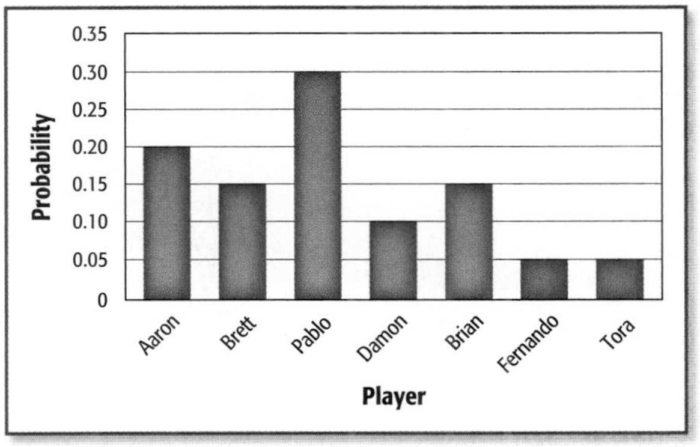

   a. Who has the best chance to win, and what is the probability?

   b. Which two players combined have the same chance of winning as Brian?

   c. Who has a better chance of winning, Damon or Brett?

   d. Find P(Aaron or Pablo).

   e. Find P(neither Damon nor Tora).

22. **CARDS** Three eights, 2 tens, 4 sixes, 3 fives, 2 twos, and a three are drawn from a deck of cards. If one card is drawn from these cards, what is its expected value?

Math History Link

Christian Huygens (1629–1695)
This Dutchman was the first to discuss games of chance. "Although in a pure game of chance the results are uncertain, the chance that one player has to win or to lose depends on a determined value." This became known as the *expected value*.

23. **VOLUNTEERING** Twenty girls and 25 boys sign up to volunteer at a shelter. If only eight students are allowed to go, what is the probability that 3 will be boys?

24. **HORSE RACING** In a race involving seven horses, Delsin randomly chose three horses to place first through third. What is the probability that he wins?

25. **MULTIPLE REPRESENTATIONS** In this problem, you will investigate geometric probability.

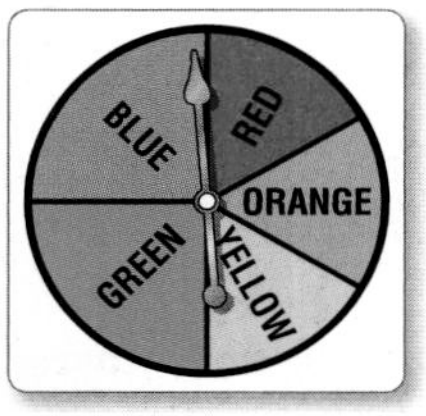

a. **TABULAR** The spinner shown has a radius of 2.5 inches. Copy and complete the table below.

Color	Probability	Sector Area	Total Area	Sector Area / Total Area
red				
orange				
yellow				
green				
blue				

b. **VERBAL** Make a conjecture about the relationship between the ratio of the area of the sector to the total area and the probability of the spinner landing on each color.

c. **ANALYTICAL** Consider the dartboard shown. Predict the probability of a dart landing in each area of the board. Assume that any dart thrown will land on the board and is equally likely to land at any point on the board.

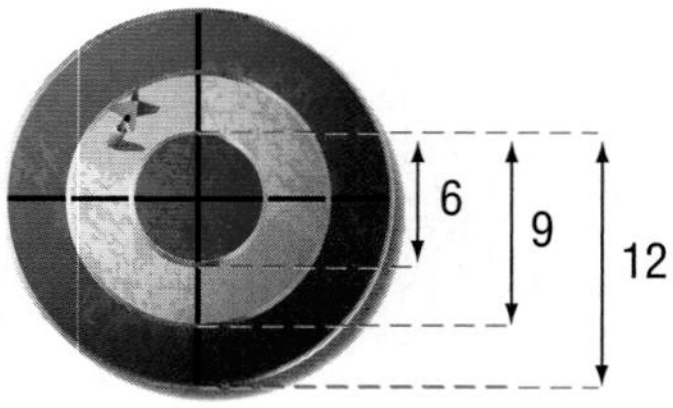

H.O.T. Problems

Use Higher-Order Thinking Skills

26. **FIND THE ERROR** Liana and Shannon created probability distributions for the sum of two spins on the spinner at the right. Is either of them correct? Explain your reasoning.

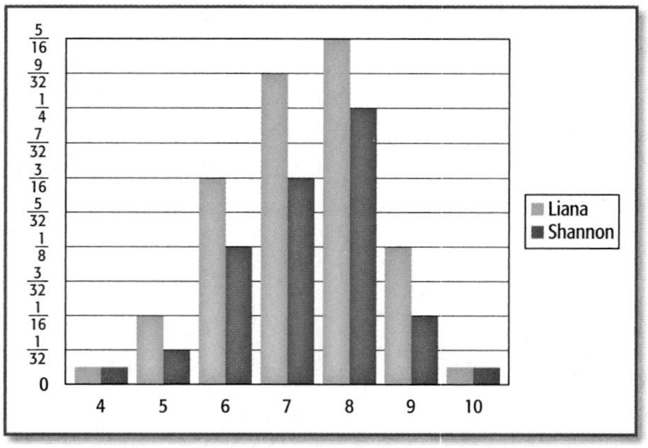

27. **REASONING** Determine whether the following statement is *true* or *false*. Explain your reasoning.

Theoretical probabilities are based on results of experiments.

28. **OPEN ENDED** Create a discrete probability distribution that shows five different outcomes and their associated probabilities.

NYSRE PRACTICE A2.S.13, A2.S.14

29. GRIDDED RESPONSE The height $f(x)$ of a bouncing ball after x bounces is represented by $f(x) = 140(0.8)^x$. How many times higher is the first bounce than the fifth bounce?

30. PROBABILITY Andres has a bag that contains 4 red, 6 yellow, 2 blue, and 4 green marbles. If he reaches into the bag and removes a marble without looking, what is the probability that it will not be yellow?

A $\frac{1}{8}$ C $\frac{3}{8}$

B $\frac{1}{4}$ D $\frac{5}{8}$

31. GEOMETRY Find the area of the shaded portion of the figure to the nearest square inch.

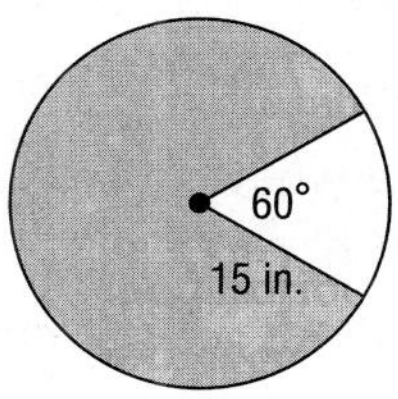

F 79

G 94

H 589

J 707

32. SAT/ACT If x and y are positive integers, which of the following expressions is equivalent to $\frac{(5^x)^y}{5^x}$?

A 1^y C 5^{xy-1}

B 5^y D 5^{xy-x}

Spiral Review

33. JOBS A computer company is interviewing 8 men and 7 women for 5 computer programming positions. If the applicants are chosen at random, what is the probability that the company will hire 3 men and 2 women? (Lesson 12-3)

Determine whether each of the following represents a *population* or a *sample*. (Lesson 12-2)

34. Shenae calculates the average number of people on 50 bus rides.

35. The U.S. Census Bureau conducts a demographics survey every 10 years.

36. East State College calculates the average cost of tuition of the entire student body.

37. Jared conducts a survey in his department to determine what time employees usually arrive in the morning.

Find the first five terms of each geometric sequence described. (Lesson 11-3)

38. $a_1 = 0.125, r = 1.5$

39. $a_1 = 0.5, r = 2.5$

40. $a_1 = 4, r = 0.5$

41. $a_1 = 12, r = \frac{1}{3}$

42. $a_1 = 21, r = \frac{2}{3}$

43. $a_1 = 80, r = \frac{5}{4}$

44. COMMUNICATION A microphone is placed at the focus of a parabolic reflector to collect sound for the television broadcast of a football game. Write an equation for the cross section, assuming that the focus is at the origin, the focus is 6 inches from the vertex, and the parabola opens to the right. (Lesson 10-2)

Solve each equation. Check your solutions. (Lesson 8-4)

45. $\log_9 x = \frac{3}{2}$

46. $\log_{\frac{1}{10}} x = -3$

47. $\log_b 9 = 2$

Skills Review

Find each percent. Round to the nearest tenth.

48. 65% of 27

49. 89% of 120

50. 11% of 30

51. 25% of 373

52. 77% of 200

53. 30% of 48

Mid-Chapter Quiz

Lessons 12-1 through 12-4

NYS Core Curriculum
A2.S.1, A2.S.13

State whether each survey would produce a random sample. Write *yes* or *no*. Explain. (Lesson 12-1)

1. Every other shopper coming out of a mall is surveyed to determine how many children they have.
2. Every tenth person in an office is surveyed to determine their feelings about their jobs.
3. Every other student in a high school is asked who their vote for Teacher of the Year is.
4. Henry surveys thirty random friends to determine who should be Homecoming Queen.

5. **MULTIPLE CHOICE** Determine which of the following statements show a *causation*. (Lesson 12-1)

 A If you practice every day, you can become a professional basketball player.

 B If you read your textbook, you will pass the test.

 C If you apply for ten different jobs, you will get an offer from at least one.

 D If you stand outside in the rain with no shelter, you will get wet.

State whether each situation represents an *experiment* or an *observational study*. If it is an experiment, identify the *control* group and the *treatment* group. Then determine whether there is bias. (Lesson 12-1)

6. Find 250 students, half of whom are on the honor roll, and compare their study habits.
7. Give a random half of the employees an extra hour lunch break every day and compare their attitudes towards work with their co-workers.

8. **MULTIPLE CHOICE** Determine which of the following represents a *population*. (Lesson 12-2)

 F Mr. Noble compares 100 random times in the 400 meter run in his gym classes.

 G Heather compares the ratings of every quarterback in the NFL.

 H Omar completes an online survey regarding the state of education in the United States.

 J A national newspaper sends out a survey with every paper asking for public opinion on the upcoming presidential election.

9. Which measure of central tendency best represents the data, and why? (Lesson 12-2)

Number of Years Playing an Instrument

2	2	3	2	4	1	2
2	3	1	3	4	2	1
3	2	3	2	3	1	4
2	3	4	1	1	1	0
1	2	1	2	2	2	3

10. **SCHOOL CLUBS** The table below shows the number of students who took algebra in eighth grade and the number who took calculus in high school. Use this information to determine the probability of each of the following. (Lesson 12-3)

	Did take calculus	Did not take calculus
Did take algebra	48	42
Did not take algebra	6	144

 a. Lori took calculus given that she took algebra in eighth grade.

 b. Kenny did not take algebra in eighth grade given that he did not take calculus.

Two dice are rolled. Determine each probability. (Lesson 12-3)

11. One of the rolls was odd, given that at least one was a 3.
12. One of the rolls was a 4, given one was even.
13. Both of the rolls were even, given at least one was a 2.
14. **FOOTBALL** The number of freshmen that make the varsity roster for Eddie's school each year is listed in the table below. Find the number of freshmen expected to make the next varsity roster. (Lesson 12-4)

Year	Freshmen	Year	Freshmen
2000	4	2004	4
2001	2	2005	2
2002	1	2006	3
2003	5	2007	2

15. **RAFFLES** Sixty students, including Michelle and her 8 friends, entered a raffle for 5 gift certificates. What is the probability that only Michelle or one of her friends wins a gift certificate? (Lesson 12-4)

12-5 The Normal Distribution

Why?

The graph lists the Scholastic Assessment Test (SAT) math scores for Ms. Fuentes's students. The data are clustered in the center, and the graph is shaped like a bell. This discrete probability distribution is close to being *normally* distributed.

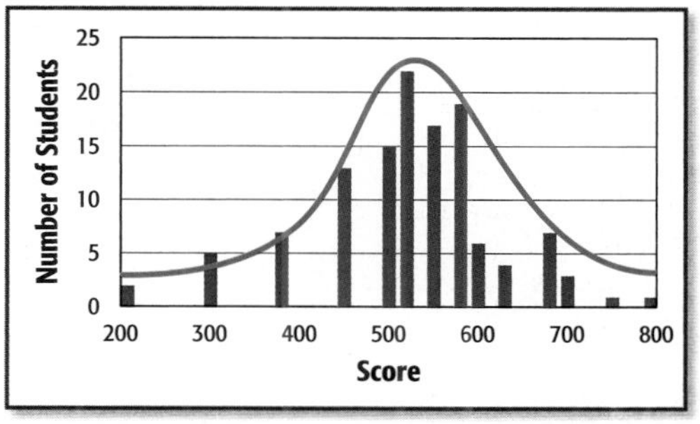

Then
You analyzed probability distributions. (Lesson 12-3)

Now
- Determine whether a set of data appears to be normally distributed or skewed.
- Use the Empirical Rule to find probabilities.

NYS Core Curriculum

A2.S.3 Calculate measures of central tendency with group frequency distributions
A2.S.5 Know and apply the characteristics of the normal distribution

New Vocabulary
continuous probability distribution
normal distribution
skewed distribution

NY Math Online
glencoe.com
- Extra Examples
- Personal Tutor
- Self-Check Quiz
- Homework Help
- Math in Motion

Normal and Skewed Distributions In a **continuous probability distribution**, the outcome can be any value in an interval of real numbers. A continuous probability distribution is best represented by a curve. The **normal distribution** is the most common example of a continuous probability distribution.

Key Concept — Characteristics of the Normal Distribution

For Your FOLDABLE

- The maximum occurs at the mean. The mean, median, and mode are equal.
- The distribution can extend from negative infinity to positive infinity, but never touches the *x*-axis.
- The population mean μ and standard deviation σ are used to determine probabilities. Probabilities are cumulative and are expressed as inequalities.
- Because the area under the normal curve represents probabilities, this area is 1.

While the normal distribution is continuous, discrete distributions like the one above can have a *normal* shape. Distributions with other shapes are called **skewed distributions**.

Normal Distribution

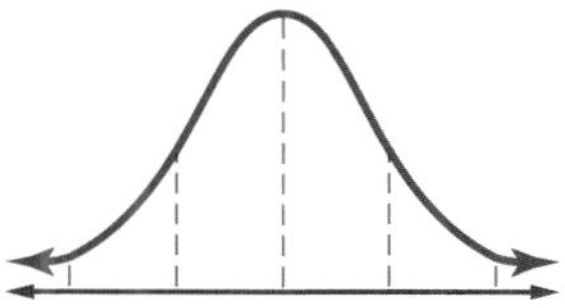

shaped like a bell and symmetric

Positively Skewed

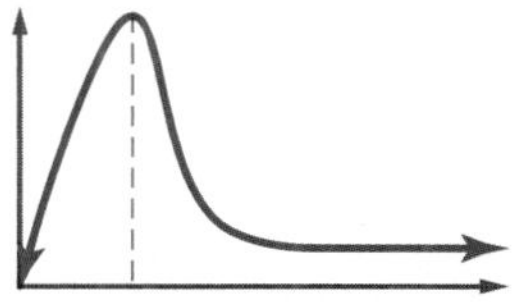

mass of distribution at the left and tail to the right

Negatively Skewed

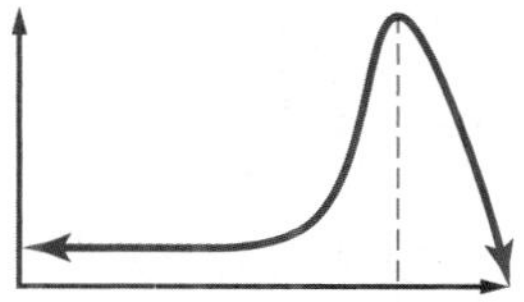

mass of distribution at the right and tail to the left

EXAMPLE 1 Classify a Data Distribution

Determine whether the following data appear to be *positively skewed, negatively skewed,* or *normally distributed.*

a.

10	12	13	15	13	15	15	14	14	16	15	18	16	18	19	16	14	13
16	16	15	14	18	17	11	19	17	18	13	15	14	21	14	15	15	17

Use the frequency table to make a histogram. Since the histogram is high in the middle and appears to be somewhat symmetric, the data are normally distributed.

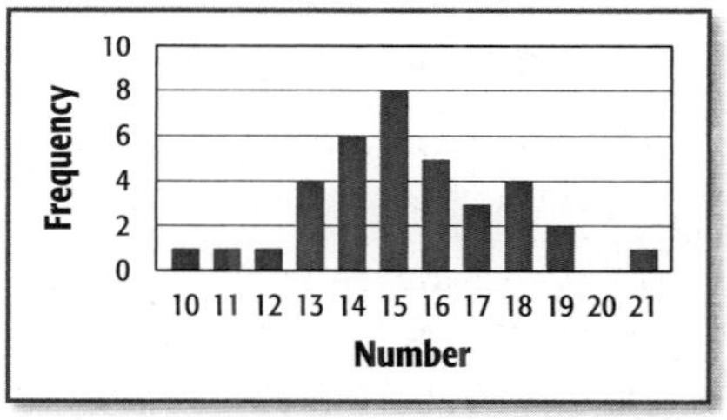

Math *in Motion*, Animation glencoe.com

b.

24	24	35	33	25	27	26	26	28	30	31	24	28	27	25	26	28	26
25	32	31	35	24	26	27	29	32	34	29	28	27	25	25	26	27	25

Use the frequency table to make a histogram. Since the histogram is high on the left and low in the middle and right, the data are positively skewed.

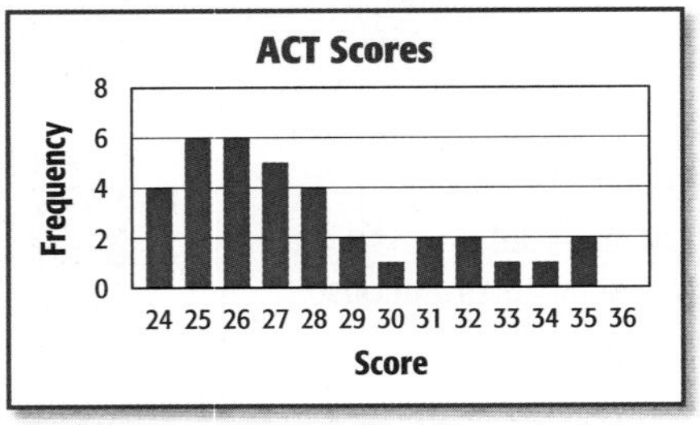

Check Your Progress

1. Determine whether the data at the right appear to be *positively skewed*, *negatively skewed*, or *normally distributed*.

Shoe Size	6	7	8	9	10	11	12
Frequency	4	8	9	7	4	2	3

Personal Tutor glencoe.com

The Empirical Rule The Empirical Rule describes other characteristics of normal distributions.

Key Concept The Empirical Rule

For Your FOLDABLE

A normal distribution with mean μ and standard deviation σ has the following properties.

- About 68% of the values are within 1σ of the mean.
- About 95% of the values are within 2σ of the mean.
- About 99% of the values are within 3σ of the mean.

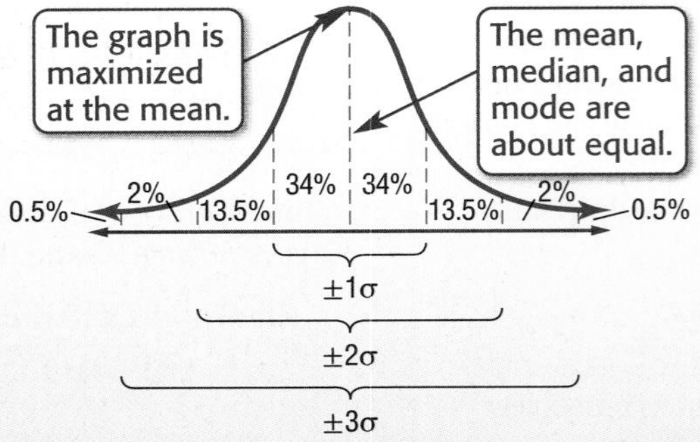

StudyTip

Normal Distributions In all of these cases, the number of data values must be large for the distribution to be approximately normal.

EXAMPLE 2 Normal Distribution

A normal distribution of data has a mean of 34 and standard deviation of 5. Find the probability that random value x is greater than 24, that is, $P(x > 24)$.

$\mu = 34$ and $\sigma = 5$

The probability that a randomly selected value in the distribution is greater than $\mu - 2\sigma$, that is, $34 - 2(5)$ or 24, is the shaded area under the normal curve.

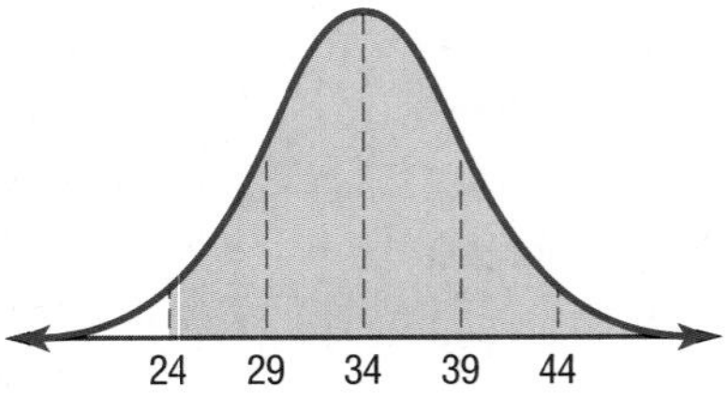

$$P(x > 24) = 13.5 + 34 + 34 + 13.5 + 2 + 0.5$$
$$= 97.5\%$$

Check Your Progress

2. Find the probability that a randomly selected value in the distribution above is less than 49.

Personal Tutor glencoe.com

Real-World Link

While the average adult American male is 5 feet 10 inches, the average height of adult males in the Netherlands is the highest worldwide, at almost 6 feet 1 inch.

Source: *Eurostats Statistical Yearbook*

A sample that is normally distributed can be represented by the normal curve as if it were a population.

Real-World EXAMPLE 3 Normally Distributed Sample

HEIGHTS The heights of 1800 teenagers are normally distributed with a mean of 66 inches and a standard deviation of 2 inches.

a. About how many teens are between 62 and 70 inches?

Draw a normal curve.

62 and 70 are 2σ away from the mean. Therefore, about 95% of the data are between 62 and 70.

Since $1800 \times 95\% = 1710$, we know that about 1710 of the teenagers are between 62 and 70 inches tall.

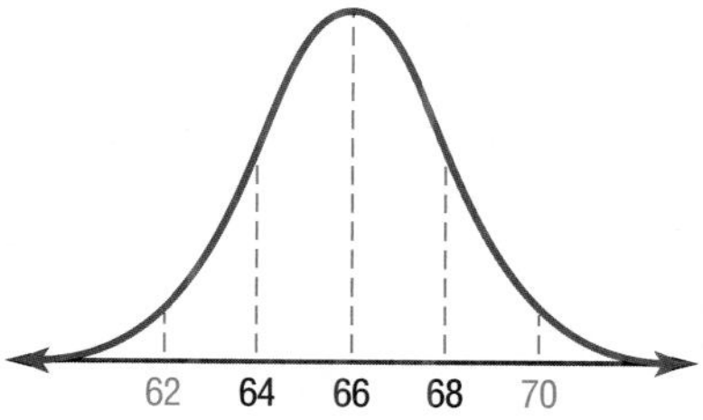

b. What is the probability that a teenager selected at random has a height greater than 68 inches?

From the curve, values greater than 68 are more than 1σ from the mean. 13.5% are between 1σ and 2σ, 2% are between 2σ and 3σ, and 0.5% are greater than 3σ.

So, the probability that a teenager selected at random has a height greater than 68 inches is $13.5 + 2 + 0.5$ or 16%.

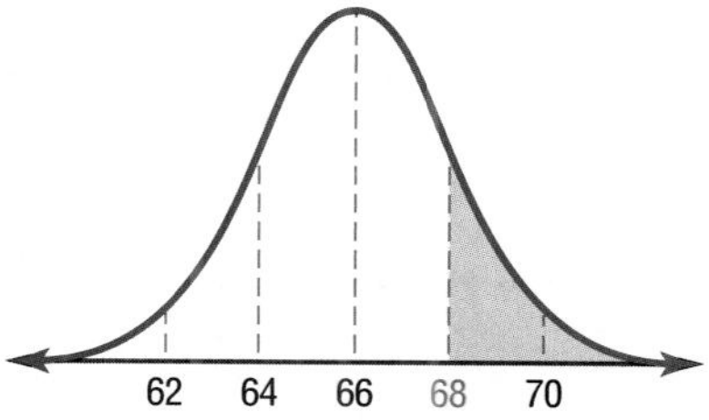

Check Your Progress

GRADES The grade-point averages of 1200 students at East High School are normally distributed with a mean of 2.6 and a standard deviation of 0.6.

3A. About how many students have a grade-point average between 2.0 and 3.2?

3B. What is the probability that a randomly selected student has an average less than 3.8?

Personal Tutor glencoe.com

Check Your Understanding

Example 1 pp. 773–774

1. ACT The table at the right shows recent composite ACT scores. Determine whether the data appear to be *positively skewed*, *negatively skewed*, or *normally distributed*.

Score	% of Students
33–36	1
28–32	9
24–27	19
20–23	29
16–19	27
13–15	12

Source: ACT, Inc.

Example 2 p. 774

2. A normal distribution of data has a mean of 161 and standard deviation of 12. Find the probability that random value x is less than 139, that is $P(x < 139)$.

Example 3 p. 775

3. SCHOOL Mr. Bash gave a quiz in his social studies class. The scores were normally distributed with a mean of 21 and a standard deviation of 2.

a. What percent would you expect to score between 19 and 23?

b. What percent would you expect to score between 23 and 25?

c. What is the probability that a student scored between 17 and 25?

Practice and Problem Solving

● = **Step-by-Step Solutions** begin on page R20.
Extra Practice begins on page 947.

Example 1 pp. 773–774

Determine whether the data appear to be *positively skewed, negatively skewed,* or *normally distributed.*

4.

20 Most Visited National Parks	
Visitors (millions)	**Number of Parks**
3–4	10
4–5	2
5–6	2
6–7	1
7–8	1
8+	4

5.

Tallest Buildings in the World	
Stories	**Number of Buildings**
0–39	1
40–59	11
60–79	35
80–99	9
100+	6

Example 2 p. 774

A normal distribution of data has each mean and standard deviation. Find each probability.

6. $\mu = 74, \sigma = 6, P(x > 86)$

7. $\mu = 13, \sigma = 0.4, P(x < 12.6)$

8. $\mu = 63, \sigma = 4, P(59 < x < 71)$

9. $\mu = 91, \sigma = 6, P(73 < x < 103)$

Example 3 p. 775

10. CAR BATTERIES The useful life of a certain car battery is normally distributed with a mean of 100,000 miles and a standard deviation of 10,000 miles. The company makes 20,000 batteries a month.

a. About how many batteries will last between 90,000 and 110,000 miles?

b. About how many batteries will last more than 120,000 miles?

c. About how many batteries will last less than 90,000 miles?

d. What is the probability that if you buy a car battery at random, it will last between 80,000 and 110,000 miles?

Real-World Link

Car batteries typically last between 5 and 7 years.

Source: National Tire and Battery

11. HEALTH The cholesterol level for adult males of a specific racial group is normally distributed with a mean of 158.3 and a standard deviation of 6.6.

a. About what percent of the males have cholesterol below 151.7?

b. How many of the 900 men in a study have cholesterol between 145.1 and 171.5?

12. FOOD The shelf life of a particular snack chip is normally distributed with a mean of 180 days and a standard deviation of 30 days.

a. About what percent of the product lasts between 150 and 210 days?

b. About what percent of the product lasts between 180 and 210 days?

c. About what percent of the product lasts less than 90 days?

d. About what percent of the product lasts more than 210 days?

13. VENDING A vending machine dispenses about 8 ounces of coffee. The amount varies and is normally distributed with a standard deviation of 0.3 ounce.

a. What percent of the time will you get more than 8 ounces of coffee?

b. What percent of the time will you get less than 8 ounces of coffee?

c. What percent of the time will you get between 7.4 and 8.6 ounces of coffee?

Real-World Link

Insurance rates for teenagers are affected by the type of car that is driven, the driving history, and the grades of the student.

Source: Esurance

Problem-Solving Tip

Use a Graph You can use a graph to visualize data, analyze trends, and make predictions.

14. **INSURANCE** The insurance industry uses various factors including age, type of car driven, and driving record to determine an individual's insurance rate. Suppose insurance rates for a sample population are normally distributed.
    a. If the mean annual cost per person is $829 and the standard deviation is $115, what is the range of rates you would expect 68% of the population to pay annually?
    b. If 900 people were sampled, how many would you expect to pay more than $1059 annually?
    c. Where on the distribution would you expect a person with several traffic citations to lie? Explain your reasoning.
    d. How do you think auto insurance companies use each factor to calculate risk?

15. **RAINFALL** Use the table at the right.
    a. Find the mean.
    b. Find the standard deviation.
    c. If the data are normally distributed, what percent of the time will annual precipitation in these cities be between 7.36 and 16.98 inches?

Average Annual Precipitation

City	Precipitation (in.)
Albuquerque	9
Boise	12
Phoenix	8
Reno	7
Salt Lake City	17
San Francisco	20

H.O.T. Problems

Use Higher-Order Thinking Skills

16. **FIND THE ERROR** A set of normally distributed tree diameters have mean 11.5 cm, standard deviation 2.5, and range 3.6 to 19.8. Monica and Hiroko are to find the range that represents the middle 68% of the data. Is either of them correct? Explain.

Monica

The data span 16.2 cm. 68% of 16.2 is about 11 cm. Center this 11-cm range around the mean of 11.5 cm. This 68% group will range from about 6 cm to about 17 cm.

Hiroko

The middle 68% span from $\mu + \sigma$ to $\mu - \sigma$. So we move 2.5 cm below 11.5 and then 2.5 cm above 11.5.

The 68% group will range from 9 cm to 14 cm.

17. **CHALLENGE** A case of digital audio players has an average battery life of 8.0 hours with a standard deviation of 0.7 hour. Eight of the players have a battery life greater than 10.1 hours. If the sample is normally distributed, how many players are in the case?

18. **WRITING IN MATH** Explain the difference between *positively skewed*, *negatively skewed*, and *normally distributed* sets of data and describe an example of each.

19. **REASONING** *True* or *false*: *According to the Empirical Rule, in a normal distribution, most of the data will fall within one standard deviation of the mean.* Explain.

20. **OPEN ENDED** Find a real-world data set that appears to represent a normal distribution. Describe the characteristics of the distribution, including its mean and standard deviation. Create a visual representation of the data.

21. **OPEN ENDED** Provide examples of a discrete probability distribution and a continuous probability distribution. Describe the differences between them.

22. **REASONING** The term *six sigma process* comes from the notion that if one has six standard deviations between the mean of a process and the nearest specification limit, there will be practically no items that fail to meet the specifications. Is this a true assumption? Explain.

NYSRE PRACTICE A2.S.3, A2.S.5

23. The lifetimes of 10,000 light bulbs are normally distributed. The mean lifetime is 300 days, and the standard deviation is 40 days. How many light bulbs will last between 260 and 340 days?

A 2500 **C** 5000
B 3400 **D** 6800

24. Which description best represents the graph?

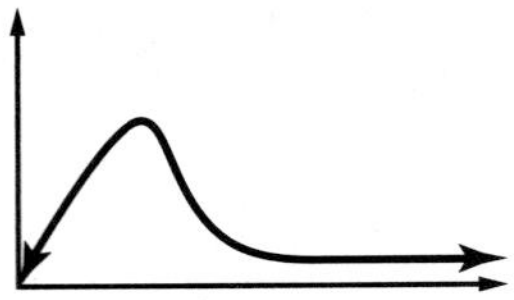

F negatively skewed **H** normal distribution
G no correlation **J** positively skewed

25. SHORT RESPONSE In the figure below, $RT = TS$ and $QR = QT$. What is the value of x?

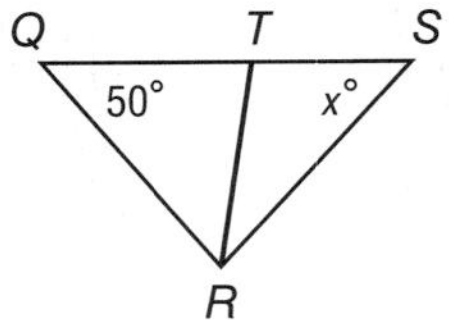

26. SAT/ACT The integer 99 can be expressed as a sum of n consecutive positive integers. The value of n could be which of the following?

I. 2
II. 3
III. 6

A I only **C** I and II only
B II only **D** I, II, and III

Spiral Review

27. SPEED A system collected and recorded the speed of drivers on a road near a school. The speeds were normally distributed with a mean of 37 miles per hour and a standard deviation of 4 miles per hour. Of the 425 cars sampled, how many would you expect were driving less than 33 miles per hour? (Lesson 12-4)

28. DOGS Three spaniels and eleven retrievers have been nominated for 4 spots to visit people at a hospital. If the winners are drawn at random, what is the probability that 2 spaniels and 2 retrievers are selected? (Lesson 12-3)

29. BRIDGES The Bayonne Bridge connects Staten Island, New York, to New Jersey. It has an arch in the shape of a parabola that opens downward. Write an equation of a parabola to model the arch, assuming that the origin is at the surface of the water, beneath the vertex of the arch. (Lesson 10-2)

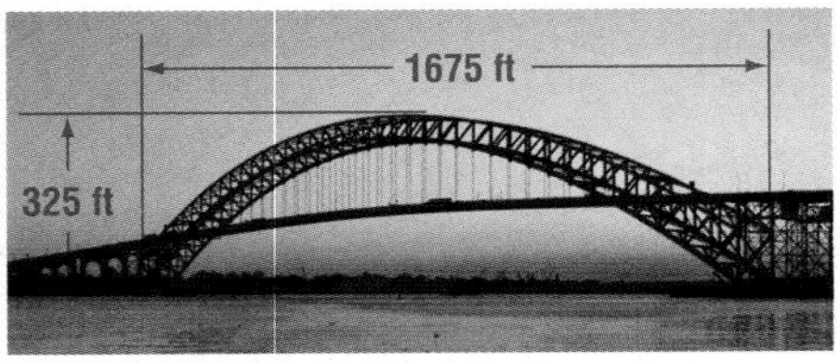

Identify the type of function represented by each graph. (Lesson 2-7)

30.

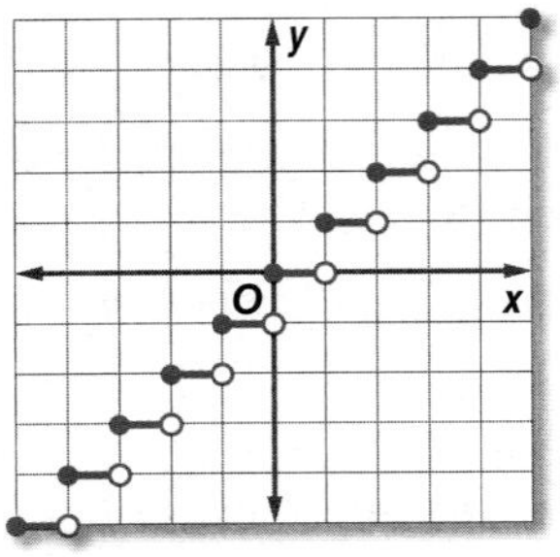

31.

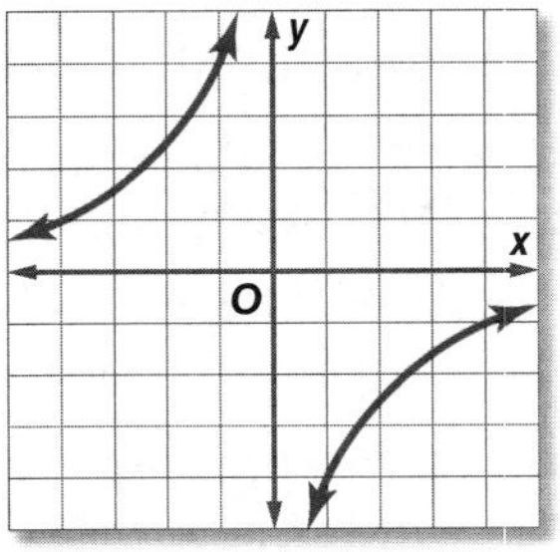

32.

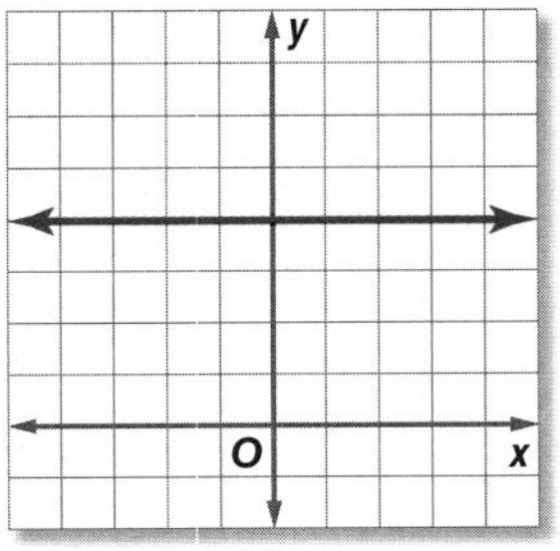

Skills Review

Find the standard deviation for each set of data. (Lesson 12-2)

33. {3, 11, 27, 14, 18, 19, 2, 33, 16, 12}

34. {45, 47, 49, 49, 51, 53, 46, 47, 50, 48}

35. {320, 400, 350, 410, 380, 390, 330, 400, 370, 360}

36. {505, 527, 512, 517, 509, 513, 522, 520, 516, 511}

EXTEND 12-5

Algebra Lab: The Empirical Rule and Percentiles

NY Math Online glencoe.com
Math *in Motion*, Animation

NYS Core Curriculum **A2.S.5 Know and apply the characteristics of the normal distribution**

If you know the mean and standard deviation of a normal distribution, you know that about 68%, 95%, and 99% of the data are within 1, 2, and 3 standard deviations of the mean, respectively. This is called the **Empirical Rule**. You can use the Empirical Rule to report percentiles. A **percentile** describes what percent of the data were at or below a given level.

Here is some additional information about percentiles.

- Percentiles measure rank from the bottom.
- There is no 0 percentile rank. The lowest score is at the 1st percentile.
- There is no 100th percentile rank. The highest score is at the 99th percentile.

ACTIVITY

A county-wide math contest was held for students in grades 9–12. Participants took at least three different tests during the competition. For the Problem-Solving Test, the scores were normally distributed with a mean of 30 and a standard deviation of 5.

Step 1 Draw a normal curve for the Problem-Solving Test scores, similar to the one shown at the right. Label the mean and the mean plus or minus multiples of the standard deviation. Label the percents as shown.

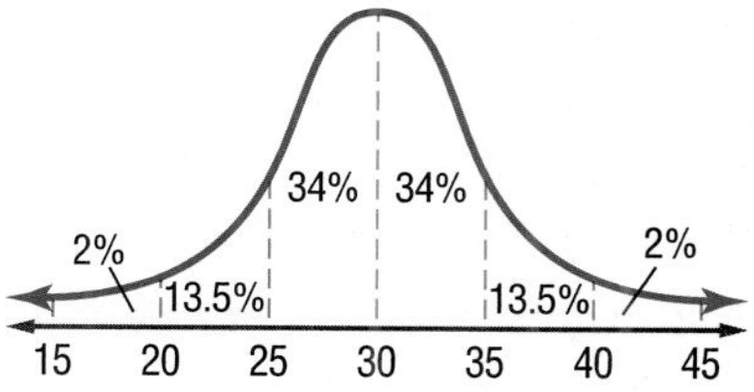

Step 2 The score of 30 is the mean. Looking at the diagram you can see that 50% of the scores are at or below the score of 30. You can say that a score of 30 is at the 50th percentile. What percent of the total scores was at or below a score of 25?

Step 3 What percent of the total scores was at or below a score of 40?

Step 4 What score is at the 99th percentile?
What score is at the 0th percentile?

Exercises

Make a drawing similar to the drawing in Step 1. Then find the percentiles or scores.

1. For the Geometry Test, the scores were normally distributed with a mean of 15 and a standard deviation of 2. Find the percentiles for the following scores: 21, 15, 13, 9.
2. For the Chemistry Test, the scores were normally distributed with a mean of 40 and a standard deviation of 4. Find the scores for the following percentiles: 99th, 2nd, 50th, 84th.

12-6

Hypothesis Testing

Why?

In a recent Gallup Poll published by *USA TODAY*, 1014 teens were surveyed. The end of the poll had the following disclaimer: "For results based on the total sample of national teens, one can say with 95% confidence that the margin of sampling error is +/–3 percentage points." To what are they referring?

Then

You found standard deviations. (Lesson 12-2)

Now

- Compare sample statistics and population statistics.
- Design experiments to test hypotheses.

NYS Core Curriculum

A2.CM.1 Communicate verbally and in writing a correct, complete, coherent, and clear design (outline) and explanation for the steps used in solving a problem

New Vocabulary

inferential statistics
statistical inference
confidence interval
hypothesis
null hypothesis
alternative hypothesis

NY Math Online

glencoe.com

- Extra Examples
- Personal Tutor
- Self-Check Quiz
- Homework Help

Confidence Interval While a distribution can provide general data about populations, it cannot give you any specifics. You can use **inferential statistics** to draw conclusions about a population by using a sample. When you use information from a sample to draw conclusions about the entire population, you are making a **statistical inference**.

To account for differences between sample statistics and population parameters, you can use an estimate. A **confidence interval** is an estimate of a population parameter stated as a range with a specific degree of certainty. Typically, statisticians use 90%, 95%, and 99% confidence intervals, but any other percentage can be considered. The most often used interval is 95%.

Key Concept — **95% Confidence Interval Formula** — For Your FOLDABLE

A 95% confidence interval estimate can be found by using the formula $CI = \bar{x} \pm 2 \cdot \frac{s}{\sqrt{n}}$, where $\bar{x}$ is the mean of the sample, s is the standard deviation of the sample, and n is the size of the sample.

These intervals are an estimation of the mean of the population, μ.

Real-World EXAMPLE 1 Find Confidence Intervals

SCHOOL WORK A sample of 200 students was asked for the average amount of time they spend on their homework during a week night. The mean time was 52.5 minutes with a standard deviation of 5.1 minutes. Determine a 95% confidence interval.

$CI = \bar{x} \pm 2 \cdot \frac{s}{\sqrt{n}}$ — **Confidence Interval Formula**

$= 52.5 \pm 2 \cdot \frac{5.1}{\sqrt{200}}$ — **$\bar{x} = 52.5$, $s = 5.1$, and $n = 200$**

$\approx 52.5 \pm 0.72$ — **Use a calculator.**

The 95% confidence interval is $51.78 \leq \mu \leq 53.22$.

Check Your Progress

1. Find a 95% confidence interval for $\bar{x} = 48.3$, $s = 6.4$, and $n = 80$.

Personal Tutor glencoe.com

Hypothesis Testing A hypothesis is an assumption that can be verified by testing. A specific hypothesis to be tested is called the **null hypothesis** H_0 (read *H null*). It is expressed as an equality and is considered true until evidence indicates otherwise.

If you conclude that the null hypothesis is false, then the alternative hypothesis must be true. The **alternative hypothesis** H_1 is mutually exclusive to the null hypothesis. It is stated as an inequality using $<$, $\leq$, $\neq$, $\geq$, or $>$. The alternative hypothesis represents the conclusion reached by rejecting the null hypothesis. Use these steps to test a hypothesis.

Key Concept **Hypothesis Testing** — For Your FOLDABLE

Step 1 State the null hypothesis H_0 and the alternative hypothesis H_1.

Step 2 Design the experiment.

Step 3 Conduct the experiment and collect the data.

Step 4 Find the confidence interval.

Step 5 Make the correct statistical inference. Accept the null hypothesis if the population parameter falls into the confidence interval.

StudyTip

Randomness and Inferential Statistics On any single observation, every outcome that is possible is equally likely to occur, but in the long run, predictable patterns of responses will emerge. This is the basis for inferential statistics.

Real-World EXAMPLE 2 Hypothesis Test

STUDENT COUNCIL Lindsey, the president of the junior class, has heard complaints that the cafeteria lunch line moves too slowly. The dining services coordinator assures her that the average wait time is 6 minutes. The students think it is much longer. Test the hypothesis that the average wait time is 6 minutes.

Step 1 State the hypotheses: H_0: $\mu = 6$ and H_1: $\mu > 6$.

Step 2 Design the experiment.
Lindsey will collect data and decide whether the data provide evidence for or against the null hypothesis. The results must differ enough from the null hypothesis to reject it, so Libby will use a 95% confidence level.

Step 3 Conduct the experiment and collect the data.
Lindsey selected a random sample of 40 students and measured their wait times. She found that $\bar{x} = 7.3$ and $s = 2.821$.

Step 4 Find the confidence interval.

$CI = \bar{x} \pm 2 \cdot \frac{s}{\sqrt{n}}$ — **Confidence Interval Formula**

$= 7.3 \pm 2 \cdot \frac{2.821}{\sqrt{40}}$ — **$n = 40$, $\bar{x} = 7.3$, and $s = 2.821$**

≈ 6.408 or 8.192 — **Use a calculator.**

This means that 95% of the time, the experiment will produce a mean between 6.408 and 8.192.

Step 5 Make the correct statistical inference.
The 95% confidence interval does not include H_0, so Lindsey can reject the null hypothesis. She can assume that the average wait time is longer than 6 minutes.

StudyTip

95% Confidence 95% confidence implies that for samples of the same size, approximately 95% of those samples will generate a confidence interval that captures the true mean.

Check Your Progress

2. FIRE DRILLS Ms. Guzman was told that it takes an average of 2 minutes to evacuate the building during a fire drill. She believes it takes longer. After collecting data for 20 fire drills, she arrives at a mean of 3.2 minutes and a standard deviation of 0.9 minute. Test the hypothesis that the average length of time is greater than 2 minutes.

Personal Tutor glencoe.com

Check Your Understanding

Example 1 p. 780

Find a 95% confidence interval for each of the following.

1. $\bar{x} = 90, s = 5.6$, and $n = 50$
2. $\bar{x} = 72, s = 4.6$, and $n = 100$
3. $\bar{x} = 84, s = 3.5$, and $n = 120$
4. $\bar{x} = 62.5, s = 2.3$, and $n = 150$
5. **VIDEO GAMES** A sample of 100 students was asked for the average amount of time they spend playing video games each day. The mean time was 75 minutes with a standard deviation of 5.1 minutes. Determine a 95% confidence interval.

Example 2 p. 781

Test each null hypothesis. Write *accept* or *reject*.

6. $H_0 = 10, H_1 < 10, n = 50, \bar{x} = 8.75$, and $\sigma = 0.9$
7. $H_0 = 48.8, H_1 > 48.8, n = 100, \bar{x} = 49$, and $\sigma = 1.5$
8. $H_0 = 75, H_1 > 75, n = 150, \bar{x} = 77$, and $\sigma = 2$
9. $H_0 = 90, H_1 > 90, n = 200, \bar{x} = 93$, and $\sigma = 3.5$
10. **SWIMMING** Yolanda's average time for the 400-meter butterfly was 8 minutes. She wants to test to see if that time is still accurate. After timing herself for 25 drills, she came to a mean of 8 minutes 10 seconds and a standard deviation of 30 seconds. Test the hypothesis that the average time is 8 minutes.

Practice and Problem Solving

● = Step-by-Step Solutions begin on page R20.
Extra Practice begins on page 947.

Example 1 p. 780

Find a 95% confidence interval for each of the following.

11. $\bar{x} = 26, s = 3.7$, and $n = 180$
12. $\bar{x} = 47, s = 5.9$, and $n = 200$
13. $\bar{x} = 58, s = 7.1$, and $n = 225$
14. $\bar{x} = 66, s = 6.3$, and $n = 250$
15. $\bar{x} = 92, s = 8.4$, and $n = 300$
16. $\bar{x} = 74, s = 6.8$, and $n = 350$
17. **MONEY** A sample of 500 students was asked for the average amount of money they spend a day. The mean amount was \$6.55 with a standard deviation of \$2.75.
    a. Determine a 95% confidence interval.
    b. Suppose the sample is expanded to 750 students, but the mean and standard deviation remain the same. Determine a new 95% confidence interval.
    c. How does a larger sample affect the confidence interval?

Example 2 p. 781

Test each null hypothesis. Write *accept* or *reject*.

18. $H_0 = 14, H_1 < 14, n = 80, \bar{x} = 12.75$, and $\sigma = 0.8$
19. $H_0 = 64.2, H_1 > 64.2, n = 200, \bar{x} = 64$, and $\sigma = 2.5$
20. $H_0 = 95, H_1 > 95, n = 150, \bar{x} = 97$, and $\sigma = 1.5$
21. $H_0 = 50, H_1 < 50, n = 400, \bar{x} = 49.5$, and $\sigma = 0.9$
22. $H_0 = 81, H_1 > 81, n = 300, \bar{x} = 81.5$, and $\sigma = 3.4$
23. $H_0 = 72, H_1 < 72, n = 350, \bar{x} = 71.7$, and $\sigma = 4.1$
24. **WALKING** Evan thought it took about 5 minutes to walk to school, while his sister Angela thought it took longer. They timed themselves for 40 days and calculated a mean of 5.8 minutes with a standard deviation of 0.6 minutes. Test the hypothesis.

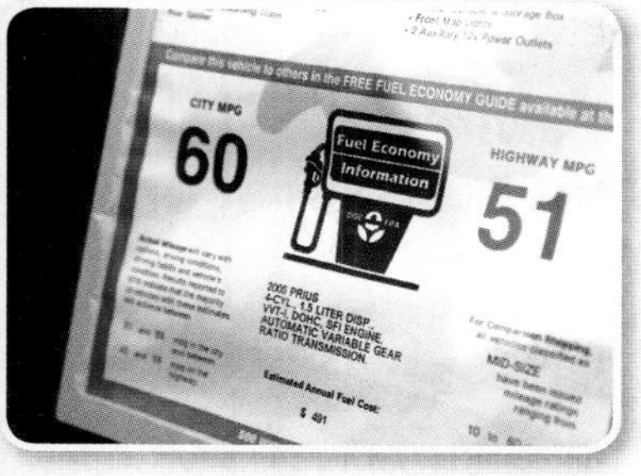

Real-World Link

Some hybrid vehicles can average over 50 miles per gallon in the city.

Source: EPA

25. **GAS MILEAGE** The manufacturer of Diana's car claimed that the car averages 28 miles per gallon in the city, but Diana believes it is less than that. The following data represent her calculations for the last 30 tanks of gas for her car. Conduct a hypothesis test to see if she is correct.

28.2	25.3	24.6	27.2	29.3	27.1	26.4	29.1	26.2	25.9
26.6	25.8	24.9	27.3	28.6	28.4	28.3	25.8	25.8	28.2
27.2	28.1	29.3	26.3	25.9	28.0	27.2	26.1	27.4	26.4

26. **QUALITY CONTROL** Grace is a quality tester for a manufacturing company. The company wants to claim that their new rechargeable battery lasts 8 hours. Grace tests 50 different batteries to see if they actually last fewer than 8 hours. Use the data below to conduct a hypothesis test.

8.1	7.9	7.8	8.0	8.2	7.8	7.7	8.1	7.8	7.7
8.2	8.4	8.2	7.8	7.7	7.7	7.9	8.3	8.1	8.0
8.0	7.9	7.9	8.4	8.1	8.2	8.0	7.6	7.7	7.9
7.8	7.9	8.0	8.0	8.1	8.1	8.2	7.6	7.8	7.8
8.0	8.1	8.1	7.9	7.9	7.9	8.1	7.8	7.8	8.0

StudyTip

Sample Size Larger samples will have a smaller error but will be more costly to obtain.

27. **COOKIES** A cookie manufacturer stated that there were 20 chocolate chips in every cookie. Lamar thought there were fewer than 20, so he tested 40 random cookies. Use the data below to conduct a hypothesis test.

21	19	20	20	19	19	18	21	19	17
19	18	18	20	20	19	18	20	18	19
21	21	19	17	17	18	19	19	19	21
22	21	21	20	20	19	17	17	17	20

28. **CANNED FOOD** The label on Leah's can of sliced peaches promises 12 slices in every can. Leah decides to test her hypothesis that there are more than 12 slices in every can by finding the number in 40 random cans. Test her hypothesis.

13	14	13	14	12	12	12	11	15	12
13	13	14	13	14	12	15	11	11	14
13	14	14	13	12	12	12	12	13	13
11	14	14	13	14	13	13	14	12	12

H.O.T. Problems

Use Higher-Order Thinking Skills

29. **CHALLENGE** A 95% confidence interval for the mean weight of a 20-ounce box of cereal was $19.932 \leq \overline{x} \leq 20.008$ with a sample standard deviation of 0.128 ounces. Determine the sample size that led to this interval.

30. **WRITING IN MATH** Describe how the size of the sample affects hypothesis testing.

31. **REASONING** Determine whether the following statement is *sometimes*, *always*, or *never* true. Explain your reasoning.

If a confidence interval contains H_0, then it is not rejected.

32. **OPEN ENDED** Conduct your own research study, and draw conclusions based on the results of a hypothesis test. Write a brief summary of your findings.

33. If $H_0 = 85$, $H_1 > 85$, $\overline{x} = 85.5$, and $n = 300$, what is the minimum sample standard deviation for which the null hypothesis will be accepted?

NYSRE PRACTICE Reinforcement of G.G.27

34. GEOMETRY In the graph below, line ℓ passes through the origin. What is the value of $\frac{a}{b}$?

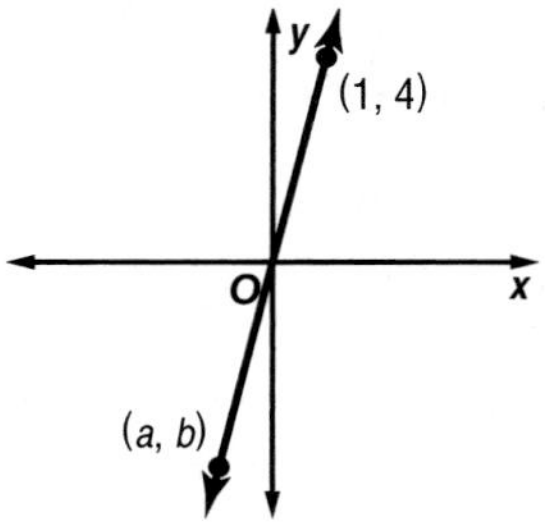

A 4 B $\frac{1}{4}$ C −4 D $-\frac{1}{4}$

35. SAT/ACT If $5 + i$ and $5 - i$ are the roots of $x^2 - 10x + c = 0$, what is the value of c?

F −25 H −26 G 25 J 26

36. The Service Club at Corey's school was founded 8 years ago. The number of members of the club by year is shown in the table. Which linear equation best models the data?

Year	Participation
0	11
2	13
4	15
6	19
8	22

A $y = 1.4x$ C $y = 1.6x$
B $y = 1.4x + 10.4$ D $y = 1.6x + 11.1$

37. SHORT RESPONSE Solve for x: $\log_2 (x - 6) = 3$.

Spiral Review

Find a 95% confidence interval for each of the following. (Lesson 12-5)

38. $\bar{x} = 56, s = 4.2$, and $n = 25$

39. $\bar{x} = 99, s = 3.8$, and $n = 50$

40. $\bar{x} = 75, s = 5.7$, and $n = 75$

41. $\bar{x} = 48.2, s = 6.1$, and $n = 150$

42. HEALTH The heights of students at Madison High School are normally distributed with a mean of 66 inches and a standard deviation of 2 inches. Of the 1080 students in the school, how many would you expect to be less than 62 inches tall? (Lesson 12-4)

Find a_n for each geometric sequence. (Lesson 11-3)

43. $a_1 = \frac{1}{3}, r = 3, n = 8$

44. $a_1 = \frac{1}{64}, r = 4, n = 9$

45. $a_4 = 16, r = 0.5, n = 8$

Write each equation in standard form. State whether the graph of the equation is a *parabola, circle, ellipse,* or *hyperbola*. Then graph the equation. (Lesson 10-6)

46. $4x^2 + 2y^2 = 8$

47. $x^2 = 8y$

48. $(x - 1)^2 - 9(y - 4)^2 = 36$

Write an equation in slope-intercept form for each graph. (Lesson 2-4)

49.

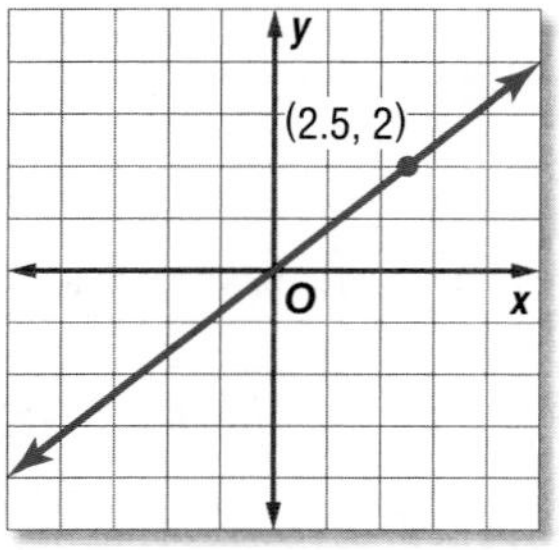

50.

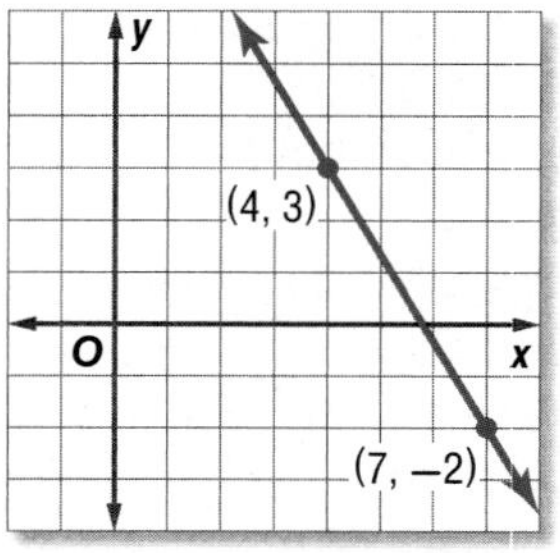

51.

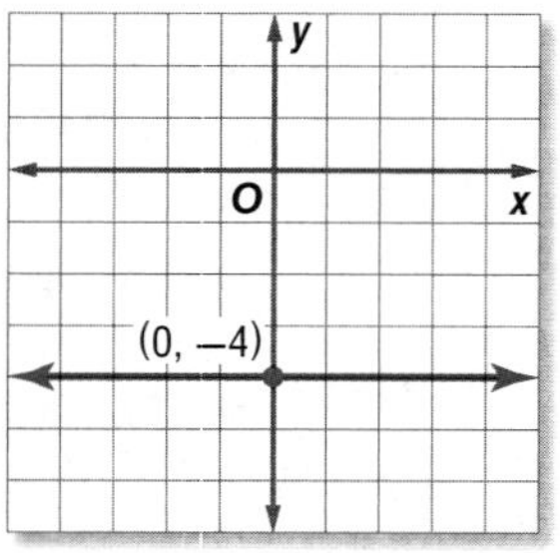

Skills Review

Expand each power. (Lesson 11-7)

52. $(a - b)^3$

53. $(m + n)^4$

54. $(r + n)^8$

EXPLORE

12-7 Algebra Lab: Simulations

NY Math Online glencoe.com
Math *in Motion*, Animation

NYS Core Curriculum **Reinforcement of A.S.21** Determine empirical probabilities based on specific sample data

A **simulation** uses a probability experiment to mimic a real-life situation. You can use a simulation to solve the following problem.

A fast food restaurant is offering one of six different food prize tickets on every soft drink cup. If the prizes are equally and randomly distributed, how many drinks, on average, would you have to buy in order to get at least one of each prize?

ACTIVITY

Work in pairs or small groups to complete Steps 1 through 4.

Step 1 Use the six numbers on a die to represent the six different food prizes.

Step 2 Roll the die and record which food prize was on the first soft drink cup. Use a tally sheet like the one shown at the right.

Step 3 Continue to roll the die and record the prize number until you have a complete set of food prizes. Stop as soon as you have a complete set. This is the end of one trial in your simulation. Record the number of drinks required for this trial.

Step 4 Repeat Steps 1, 2, and 3 until your group has carried out 25 trials. Use a new tally sheet for each trial.

Simulation Tally Sheet

Prize Number	Drinks Purchased
1	
2	
3	
4	
5	
6	
Total Needed	

Analyze the Data

1. Create two different statistical graphs of the data collected for 25 trials.

2. Determine the mean, median, maximum, minimum, and standard deviation of the total number of drinks needed in the 25 trials.

3. Combine the small-group results and determine the mean, median, maximum, minimum, and standard deviation of the number of drinks required for all the trials conducted by the class.

Make a Conjecture

4. If you carry out 25 additional trials, will your results be the same as in the first 25 trials? Explain.

5. Should the small-group results or the class results give a better idea of the average number of drinks required to get a complete set of food prizes? Explain.

6. If there were 8 prizes instead of 6, would you need to buy more drinks or fewer drinks on average?

7. **DESIGN A SIMULATION** What if one of the 6 prizes was more common than the other 5? For instance, suppose that one prize, a free ice cream sundae, appears on 25% of all the drinks and the other 5 prizes are equally and randomly distributed among the remaining 75% of the drinks. Design and carry out a new simulation to predict the average number of drinks you would need to buy to get a complete set. Include some measures of central tendency and dispersion with your data.

12-7 Binomial Distributions

Why?

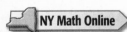

During a baseball game, a player who gets a hit 36% of the time is at the plate. However, he has been in a hitting slump recently, failing to get a hit in his last 20 at-bats. The commentators notice this and declare that he is "due" to get a hit this time. Is their assumption correct?

Then

You used the Binomial Theorem. (Lesson 11-6)

Now

- Find probabilities for binomial experiments.
- Find probabilities by using binomial distributions and expansions.

NYS Core Curriculum

A2.S.15 Know and apply the binomial probability formula to events involving the terms *exactly*, *at least*, and *at most*
A2.S.16 Use the normal distribution as an approximation for binomial probabilities
Also addresses A2.S.3.

New Vocabulary

binomial distribution
binomial experiment
experimental probability

NY Math Online

glencoe.com

- Extra Examples
- Personal Tutor
- Self-Check Quiz
- Homework Help

Binomial Experiments Another type of discrete probability distribution is the binomial distribution. A **binomial distribution** shows the probabilities of the outcomes of a binomial experiment. A **binomial experiment** is a random experiment with an outcome that is one of two simple events. In a binomial experiment, the following are true.

Key Concept — Binomial Experiments (For Your FOLDABLE)

- There are only two possible outcomes, success or failure.
- There is a fixed number of trials, n.
- The probability of success is the same in every trial.
- The trials are independent.
- The random variable is the number of successes in n trials.

The **experimental probability** is what is estimated from observed simulations or experiments. When a simulation is conducted, the observed data are analyzed and the experimental probability is determined from these results.

EXAMPLE 1 Design a Binomial Experiment

In a certain dice game, a player tries to roll a total of 7 or 11 with two dice. Design and conduct a binomial experiment for 10 rolls of the dice.

Step 1 Describe the trial for the situation.
Each roll is a trial. There will be 10 trials.

Step 2 Describe a success. What is the probability of a success?
A success is rolling 7 or 11. The probability of success is $\frac{6}{36} + \frac{2}{36}$ or $\frac{2}{9}$.

Step 3 Design and conduct a simulation to determine the experimental probability of rolling a 7 or 11 at least two out of ten times. Let s represent success, and let f represent failure.

Roll	1	2	3	4	5	6	7	8	9	10	Total Successes
Simulation 1	*f*	*f*	*f*	*f*	*f*	*s*	*f*	*f*	*s*	*f*	2
Simulation 2	*f*	*f*	*f*	*f*	*s*	*f*	*f*	*f*	*f*	*f*	1
Simulation 3	*s*	*s*	*f*	*f*	*f*	*f*	*f*	*f*	*f*	*f*	2
Simulation 4	*f*	*f*	*f*	*f*	*f*	*s*	*f*	*s*	*f*	*f*	2

Three of the four simulations produced at least 2 successes, so the experimental probability of rolling a 7 or 11 twice in 10 rolls is 75%.

Real-World Link

The Georgia State Fair, held in Macon since 1851, has raised over $4 million for education and charities.

Source: Georgia State Fair

Check Your Progress

1. Design and conduct an experiment, and then find the experimental probability of a fair coin landing on heads 6 out of 10 tosses.

Personal Tutor glencoe.com

Binomial distributions are often represented graphically, usually with a tree diagram.

Real-World EXAMPLE 2 Find a Probability

STATE FAIR Antonia earned 3 prize tokens for shooting baskets at the state fair. According to the advertisement, 30% of the tokens win prizes. Find the probability that *exactly* 2 of Antonia's tokens win a prize.

Each token has a probability of success of 0.3.
The probability of failure is 1 − 0.3 or 0.7.
The tree diagram shows all of the possibilities and the probability of each.

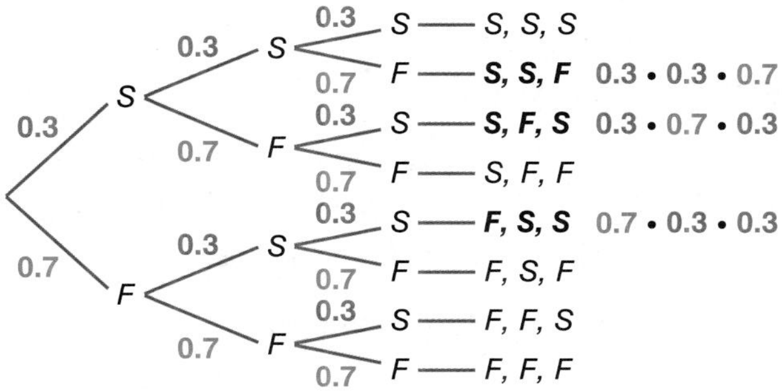

StudyTip

Tree Diagrams Multiply all of the probabilities along the branch to calculate its probability.

Three distinct branches of the tree indicate exactly two successes.
The sum of the probabilities of these branches will produce the overall probability.

$$0.3 \cdot 0.3 \cdot 0.7 + 0.3 \cdot 0.7 \cdot 0.3 + 0.7 \cdot 0.3 \cdot 0.3 = 0.189 \text{ or } 18.9\%$$

Check Your Progress

2. Becky bought 5 game cards at the store. Each game card has a 10% chance of winning. Find the probability that at least 2 of her game cards are winners.

Personal Tutor glencoe.com

Binomial Distribution The tree diagram in Example 2 is an example of a binomial distribution. This distribution can be simplified by the following formula.

Key Concept — Binomial Distribution Functions (For Your FOLDABLE)

The probability of x successes in n independent trials is

$$P(x) = C(n, x)\, s^x f^{n-x},$$

where s is the probability of success of an individual trial and f is the probability of failure on that same individual trial ($s + f = 1$).

The expected value of a binomial distribution can also be determined.

Key Concept — Expected Value of a Binomial Distribution (For Your FOLDABLE)

The expected value for a binomial distribution is $E(X) = ns$, where n is the total number of trials and s is the probability of success.

StudyTip

Selecting Probabilities Sometimes it is easier to find the probability of failure and subtract from 1 to get the probability of success. These probabilities are complements of each other.

StudyTip

Expected Values and Probability A probability is not a guarantee that something will occur. For example, if the probability shows that a ball team is expected to win 3 of the next 5 games, it does not mean they will.

EXAMPLE 3 Binomial Probability

A chocolate company makes boxes of assorted chocolates, 40% of which are dark chocolate on average. The production line mixes the chocolates randomly and packages 10 per box.

a. What is the probability that *at least* 3 chocolates in a box are dark?

A success is a dark chocolate, so $s = 0.4$ and $f = 1 - 0.4$ or 0.6.

Calculate the probability of the box having exactly 0, 1, or 2 dark chocolates, and then subtract that sum from 1.

$P(\geq 3 \text{ dark chocolates})$

$= 1 - P(< 3 \text{ dark chocolates})$ **$p + q = 1$**

$= 1 - [P(0) + P(1) + P(2)]$ **Mutually exclusive events**

$= 1 - [C(10, 0)(0.4)^0(0.6)^{10} + C(10, 1)(0.4)^1(0.6)^9 + C(10, 2)(0.4)^2(0.6)^8]$

$= 1 - 0.1673$ or 0.8327 **Simplify.**

The probability of at least 3 chocolates being dark is 0.8327 or 83.27%.

b. What is the expected number of dark chocolates in a box?

$E(X) = np$ **Expected Value of a Binomial Distribution**

$= 10(0.4)$ or 4   **$n = 10$ and $p = 0.4$**

The expected number of dark chocolates in a box is 4.

Check Your Progress

3. If 20% of the chocolates are white chocolates, what is the probability that at least one chocolate in a given box of 10 is a white chocolate?

Personal Tutor glencoe.com

You can find the full probability distribution for a binomial experiment by expanding the binomial.

EXAMPLE 4 Full Probability Distribution

Autumn ran out of time when she took her multiple-choice test so she randomly circled answers for the last 5 questions. Each question had 5 possible choices. Determine the probabilities associated with the number of answers she got correct on the last 5 questions.

We are asked to find the probability for each possible number of correct answers on the 5 she guessed on.

Expand the binomial $(s + f)^n$ with $n =$ the 5 questions.

There are five equal possibilities for each question, so $s = \frac{1}{5}$ or 0.2 and $f = \frac{4}{5}$ or 0.8.

$(s + f)^n$

$= 1s^5 + 5s^4f + 10s^3f^2 + 10s^2f^3 + 5sf^4 + 1f^5$

$= (0.2)^5 + 5(0.2)^4(0.8) + 10(0.2)^3(0.8)^2 + 10(0.2)^2(0.8)^3 + 5(0.2)(0.8)^4 + (0.8)^5$

$= 0.032\% + 0.64\% + 5.12\% + 20.48\% + 40.96\% + 32.768\%$

5 correct	4 correct	3 correct	2 correct	1 correct	0 correct
0.032%	0.64%	5.12%	20.48%	40.96%	32.768%

Check Your Progress

4. Ricky guessed on the last 6 questions of his test. Each question had 4 options. Determine the probabilities associated with the number of answers he got correct on the last 6 questions.

Personal Tutor glencoe.com

The graph of a binomial probability distribution can be drawn with the possible outcomes on the x-axis and their probabilities of success on the y-axis.

EXAMPLE 5 Graphing a Binomial Distribution

Graph the binomial probability distribution in Example 4. Describe the shape of the distribution.

List the number of correct answers along the x-axis.

The maximum probability is 40.96%, so the y-axis should range from 0 to 0.5.

The graph is positively skewed.

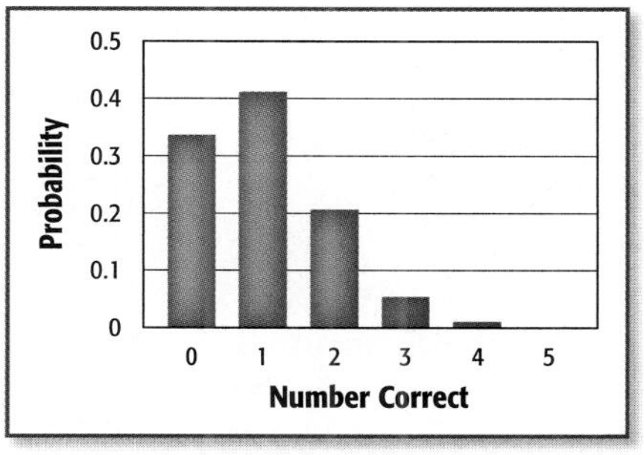

Check Your Progress

5. Graph the binomial probability distribution in Check Your Progress 4. Describe the shape of the distribution.

Personal Tutor glencoe.com

When the number of trials increases, a normal distribution can be used to approximate a binomial distribution.

Key Concept

For Your FOLDABLE

Normal Approximation of a Binomial Distribution

In a binomial distribution with n trials, a probability of success s, and a probability of failure f, such that $ns \geq 5$ and $nf \geq 5$, the binomial distribution can be approximated by a normal distribution with $\bar{x} = ns$ and $\sigma = \sqrt{nsf}$.

EXAMPLE 6 Normal Approximation of a Binomial Distribution

According to an online poll, 64% of middle-aged college graduates feel that their college years were the most exciting. Bernardo conducts a survey of 300 random middle-aged adults with college degrees. What is the probability that at least 200 of the responses will agree?

The number of people surveyed who say that their college years were the most exciting has a binomial distribution with $n = 300$, $s = 0.64$, and $f = 0.36$. Use a normal distribution to approximate the probability.

$\bar{x} = ns$ **Mean of a normal approximation**

$= 300(0.64)$ or 192 **$n = 300$ and $s = 0.64$**

$\sigma = \sqrt{nsf}$ **Standard deviation of a normal approximation**

$= \sqrt{300(0.64)(0.36)}$ **$n = 300$, $s = 0.64$, and $f = 0.36$**

≈ 8.31 **Use a calculator.**

200 is about 1 standard deviation greater than the mean, so the probability that at least 200 responses agree is 16%.

Check Your Progress

6. According to an online poll, 32% of adults feel that school should be in session year-round. Suki thinks the number should be lower, so she conducts a survey of 250 random adults. What is the probability that no more than 65 of the surveyed adults feel that school should be in session year-round?

Personal Tutor glencoe.com

StudyTip

Normal Approximations The normal approximation is used because as n increases, using the binomial distribution function to find the probability becomes a tedious process.

Check Your Understanding

Example 1 p. 786

1. CARDS Design and conduct an experiment, and use a table like the one at the right to find the experimental probability of drawing an ace or a king 1 out of 10 times when drawing from a deck of 52 cards with replacement.

Draw	Ace/King	Draw	Ace/King
1	Yes	6	No
2	No	7	Yes
3	No	8	No
4	No	9	No
5	No	10	No

Example 2 p. 787

2. PETS Chloe's cat is having kittens. The probability of a kitten being male is 0.5.

a. If Chloe's cat has 4 kittens, what is the probability that at least 3 will be male?

b. What is the expected number of males in a litter of 6?

Example 3 p. 788

3. BASEBALL What is the probability that at least 5 of the 25 students trying out for the baseball team are left-handed if 11% of the population is left-handed?

Example 4 p. 788

4. GUESSING Loranzo had to guess on the last 10 questions of his test. Luckily, they were true and false questions. Determine the probabilities associated with the number of answers he guessed correctly and make a table listing the probability for 0 correct, 1 correct, 2 correct, etc.

Example 5 p. 789

5. PARKING A poll at Steve's high school showed that 85% of the students were in favor of expanding the junior-senior parking lot. Steve asked 6 random students who participated in the poll if they were in favor of expanding the parking lot.

a. Graph the binomial probability distribution.

b. What is the probability that all 6 of the students were in favor of expansion?

Example 6 p. 789

6. SUMMER JOBS According to an online poll, 90% of high school upperclassmen have summer jobs. Tadeo thinks the number should be lower so he conducts a survey of 400 random upperclassmen. What is the probability that no more than 350 of the surveyed upperclassmen have summer jobs?

Practice and Problem Solving

● = **Step-by-Step Solutions** begin on page R20.
Extra Practice begins on page 947.

Example 1 p. 786

7. DICE Design and conduct an experiment, and then find the experimental probability of rolling a 7 with 2 six-sided dice 2 out of 10 times.

8. CARDS Design and conduct an experiment, and then find the experimental probability of drawing a face card out of a standard deck of cards 4 out of 10 times with replacement.

9. MARBLES Design and conduct an experiment using a bag of 4 blue, 3 green, and 5 red marbles. Then find the experimental probability of pulling out a red marble 6 out of 10 times with replacement.

Examples 2 and 3 pp. 787–788

10. MP3 PLAYERS According to a recent survey, 85% of high school students own an MP3 player. What is the probability that at least 6 of 10 random high school students own an MP3 player?

11. CARS According to a recent survey, 92% of high school seniors own their own car. What is the probability that fewer than 8 out of 10 random high school students own their own car?

12. **SENIOR PROM** According to a recent survey, 25% of high school upperclassmen think that the junior-senior prom is the most important event of the school year. What is the probability that no more than 3 out of 10 random high school upperclassmen think this way?

13. **FOOTBALL** A certain football team has won 75.7% of their games. Find the probability that they win at least 7 of their next 10 games.

14. **GARDENING** Peter is planting 24 irises in his front yard. The flowers he bought were a combination of two varieties, blue and white. The flowers are not blooming yet, but Peter knows that the probability of having a blue flower is 75%. What is the probability that at least 20 of the flowers will be blue?

15. **FOOTBALL** What is the probability that a field goal kicker makes at least 7 of his next 10 kicks from within 35 yards?

Range (yd)	Accuracy (%)
0–34	75
35–44	62
45+	20

16. **BABIES** Mr. and Mrs. Davis are planning to have 3 children and the probability of each child being a boy is 50%. What is the probability that they will have at least 2 boys?

Example 4 p. 788

17. **LAPTOPS** According to a recent survey, 95% of high school students own a laptop. Ten random students are chosen.
    - **a.** Determine the probabilities associated with the number of students who own a laptop.
    - **b.** What is the probability that at least 8 of the 10 students own a laptop?

18. **ATHLETICS** According to a recent survey, 80% of high school students have participated in at least one sport for their school. Six random students are chosen.
    - **a.** Determine the probabilities associated with the number of students playing in at least one sport.
    - **b.** What is the probability that no more than 2 of the students participated in a sport?

19. **CAR WASH** Some students are doing a car wash to raise money for the Spanish Club. They have determined that 65% of the time the customers donate more than the minimum amount for the car wash. What is the probability that at least 4 of the next 5 customers will donate more than the minimum?

20. **DRAWINGS** One in five students will win a prize in the class drawing. If there are 25 students in the drawing, including Jake, Leslie, Roberto, Ika, and Nicholas, what is the probability that at least one of them wins a prize?

Example 5 p. 789

21. **MUSIC** An online poll showed that 5% of adults still play vinyl records. Moe surveyed 8 random people from the population.
    - **a.** Graph the binomial probability distribution.
    - **b.** What is the probability that no more than 2 of the people surveyed still play vinyl records?

22. **PROMOTIONS** A beverage company has a promotion in which 30% of the bottles purchased during the promotion have bottle caps that win a free beverage. Melanie bought 10 bottles.
    - **a.** Graph the binomial probability distribution.
    - **b.** What is the probability that Melanie won at least 4 free beverages?

Example 6 p. 789

23. **REALITY SHOWS** According to an online poll, 70% of teens watch at least one reality show. Dillon surveyed 200 random teens. What is the probability that at least 150 of the teens surveyed watch at least one reality show?

24. **COLLEGE** A poll of students at Jacqui's school determined that 88% of the students wanted to go to college. Jacqui surveyed 150 random students from the school. What is the probability that at least 10 of the polled students did not want to go to college?

A binomial distribution has a 60% rate of success. There are 18 trials.

25. What is the probability that there will be at least 12 successes?

26. What is the probability that there will be 12 failures?

27. What is the expected number of successes?

28. **TENNIS** A player has won 85% of his matches over his career. Find each probability.
    a. He wins 3 of the next 5 matches.
    b. He wins at least 2 of his next 5 matches.
    c. He loses at least 1 of his next 5 matches.

Real-World Link

Sweepstakes hobbyists who enter by mail are often called "Sweepstakers" or "Sweepers." Many sweepstakers meet locally in "clubs" and nationally at a yearly convention, which 800 people attended in 2006.

Source: National Sweepstakes Convention

Each binomial distribution has n trials and p probability of success. Determine the most likely number of successes.

29. $n = 8, p = 0.6$
30. $n = 10, p = 0.4$
31. $n = 6, p = 0.8$
32. $n = 12, p = 0.55$
33. $n = 9, p = 0.75$
34. $n = 11, p = 0.35$

35. **SWEEPSTAKES** A beverage company is having a sweepstakes. The odds of winning selected prizes are shown at the right. If Ernesto purchases 8 beverages, what is the probability that he wins at least one prize?

Odds of Winning	
beverage	1 in 10
CD	1 in 200
hat	1 in 250
MP3 player	1 in 20,000
car	1 in 25,000,000

Each binomial distribution has n trials and p probability of success. Determine the probability of s successes.

36. $n = 8, p = 0.3, s \geq 2$
37. $n = 10, p = 0.2, s > 2$
38. $n = 6, p = 0.6, s \leq 4$
39. $n = 9, p = 0.25, s \leq 5$
40. $n = 10, p = 0.75, s < 8$
41. $n = 12, p = 0.1, s \geq 3$

H.O.T. Problems

Use Higher-Order Thinking Skills

42. **CHALLENGE** In a normal approximation of a binomial distribution, there is a 34% probability of there being between 60 and 66 successes. If the probability of success is 38%, how many trials are there?

43. **WRITING IN MATH** You poll a sample of your classmates to find out if they support using school funds for the science wing project. How could you use a binomial distribution to predict the number of people in the school who support the project?

44. **REASONING** Determine whether the following statement is *sometimes, always,* or *never* true. Explain your reasoning.

It is more beneficial to find the probability of failure and subtract it from 1 in order to determine the probability of success.

45. **OPEN ENDED** Describe a real-world setting within your school or community activities that seems to fit a binomial distribution. Identify the key components of your setting that connect to binomial distributions.

46. **WRITING IN MATH** Describe how binomial distributions are connected to Pascal's triangle.

47. **WRITING IN MATH** Explain the relationship between a binomial experiment and a binomial distribution.

NYSRE PRACTICE A2.S.15, A2.S.16

48. EXTENDED RESPONSE Carly is taking a 10-question multiple-choice test in which each question has four choices. If she guesses on each question, what is the probability that she will get

a. 7 questions correct?
b. 9 questions correct?
c. 0 questions correct?
d. 3 questions correct?

49. What is the maximum point of the graph of the equation $y = -2x^2 + 16x + 5$?

A $(-4, -59)$ C $(4, 37)$
B $(-4, -91)$ D $(4, 101)$

50. GEOMETRY On a number line, point X has coordinate -8 and point Y has coordinate 4. Point P is $\frac{2}{3}$ of the way from X to Y. What is the coordinate of P?

F -4 H 0
G -2 J 2

51. SAT/ACT The cost of 4 CDs is d dollars. At this rate, what is the cost, in dollars, of 36 CDs?

A $\frac{9d}{4}$ C $\frac{36}{d}$
B $\frac{d}{36}$ D $9d$

Spiral Review

52. BASKETBALL A basketball team wins 62% of its road games. Find the probability that they will win at least 6 of their next 11 road games. (Lesson 12-6)

Test each null hypothesis. Write *accept* or *reject*. (Lesson 12-5)

53. $H_0 = 33$, $H_1 > 33$, $n = 100$, $\bar{x} = 32.1$, and $\sigma = 1.2$

54. $H_0 = 5$, $H_1 < 5$, $n = 50$, $\bar{x} = 5.2$, and $\sigma = 0.8$

55. $H_0 = 0.04$, $H_1 > 0.04$, $n = 100$, $\bar{x} = 0.042$, and $\sigma = 0.1$

56. $H_0 = 300$, $H_1 < 300$, $n = 25$, $\bar{x} = 301$, and $\sigma = 1.5$

Find the missing value for each arithmetic sequence. (Lesson 11-1)

57. $a_5 = 12, a_{16} = 133, d = ?$

58. $a_9 = -34, a_{22} = 44, d = ?$

59. $a_4 = 18, a_n = 95, d = 7, n = ?$

60. $a_7 = -28, a_n = 76, d = 8, n = ?$

61. $a_6 = ?, a_{20} = 66, d = 7$

62. $a_8 = ?, a_{19} = 31, d = 8$

63. ASTROLOGY The table at the right shows the closest and farthest distances of Venus and Jupiter from the Sun in millions of miles. (Lesson 10-4)

Planet	Closest	Farthest
Venus	66.8	67.7
Jupiter	460.1	507.4

a. Write an equation for the orbit of each planet, assuming that the center of the orbit is the origin, the center of the Sun is a focus, and the Sun lies on the x-axis.

b. Which planet has an orbit that is closer to a circle?

Write an equivalent exponential or logarithmic function. (Lesson 8-7)

64. $e^{-x} = 5$

65. $e^2 = 6x$

66. $\ln e = 1$

67. $\ln 5.2 = x$

68. $e^{x+1} = 9$

69. $e^{-1} = x^2$

70. $\ln \frac{7}{3} = 2x$

71. $\ln e^x = 3$

Skills Review

72. MUSIC Tina owns 11 pop, 6 country, 16 rock, and 7 rap CDs. Find each probability if she randomly selects 4 CDs. (Lesson 12-3)

a. $P(2 \text{ rock})$

b. $P(1 \text{ rap})$

c. $P(1 \text{ rock and } 2 \text{ country})$

Study Guide and Review

NY Math Online glencoe.com
- STUDY TO GO
- Vocabulary Review

Chapter Summary

Key Concepts

Samples and Populations (Lessons 12-1 and 12-2)

- A sample is biased if its design favors certain outcomes.
- A sample is unbiased if it is random or unpredictable.

Standard Deviation	
Sample	Population
$\sqrt{\dfrac{\sum_{k=1}^{n}(x_n - \bar{x})^2}{n-1}}$	$\sqrt{\dfrac{\sum_{k=1}^{n}(x_n - \mu)^2}{n}}$

Conditional Probability (Lesson 12-3)

- The probability of an event given that another event has already occurred is the conditional probability.
- A contingency table records data in which different possible situations result in different possible outcomes.

Probability Distributions (Lessons 12-4, 12-5, and 12-7)

Sample	Population
uniform	All probabilities are equal.
discrete	finite number of possible outcomes
continuous	infinite number of possible outcomes
normal	symmetric curves
skewed	non-symmetric curves
binomial	Outcomes are one of two simple events.

Hypothesis Testing (Lesson 12-6)

- Inferential statistics draw conclusions about a population by using a sample.
- A hypothesis is an assumption that can be verified by testing.

FOLDABLES® Study Organizer

Be sure the Key Concepts are noted in your Foldable.

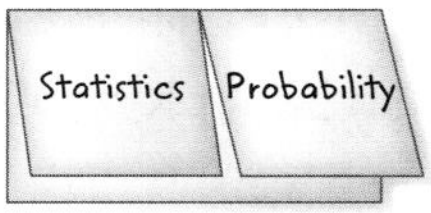

Key Vocabulary

alternative hypothesis (p. 781)
biased (p. 745)
binomial distribution (p. 786)
binomial experiment (p. 786)
causation (p. 747)
conditional probability (p. 759)
confidence interval (p. 780)
continuous probability distribution (p. 773)
control group (p. 746)
correlation (p. 747)
discrete probability distribution (p. 767)
expected value (p. 767)
experiment (p. 746)
inferential statistics (p. 780)
margin of sampling error (p. 753)
measure of central tendency (p. 752)
measure of variation (p. 754)
normal distribution (p. 773)
null hypothesis (p. 781)
observational study (p. 746)
parameter (p. 752)
population (p. 745)
probability (p. 764)
probability distribution (p. 766)
random variable (p. 766)
relative frequency (p. 760)
sample (p. 745)
skewed distribution (p. 773)
standard deviation (p. 754)
statistic (p. 752)
survey (p. 745)
theoretical probability (p. 767)
treatment group (p. 746)
unbiased (p. 745)

Vocabulary Check

Choose a word or term from the list above that best completes each statement.

1. A(n) ____________ for a particular random variable is a function that maps the sample space to the probabilities of the outcomes of the sample space.
2. When two events are related, there is a(n) ____________.
3. A survey is ____________ if its design favors certain outcomes.
4. The group given the placebo is the ____________.
5. The ____________ provides the interval that shows how much responses from the sample would differ from the population.
6. A probability distribution with only a finite number of possible outcomes is a(n) ____________.

Lesson-by-Lesson Review

12-1 Experiments, Surveys, and Observational Studies (pp. 745–750)

A2.S.1, A2.S.2

State whether each survey would produce a random sample. Write *yes* or *no*. Explain.

7. Every tenth shopper coming out of a hardware store is surveyed to determine his or her satisfaction with the store.
8. Every tenth person coming out of a high school is asked what their favorite class is.
9. A fast food restaurant asks their customers to complete a survey asking what their favorite fast food restaurant is.

Determine whether each situation calls for a *survey*, an *observational study*, or an *experiment*. Explain the process.

10. Find 100 students, half of which have part-time jobs, and compare their grade-point averages.
11. Find 100 people and randomly split them into two equal groups. One group eats a specific diet while the other group does not. Compare the results.

EXAMPLE 1

A car dealership selects 100 random customers who recently took their vehicles in for work and asks them how the service was. Would this produce a random sample? Explain.

Yes. Everyone in the population of customers has an equal chance to be part of the sample.

EXAMPLE 2

A teacher has his first class take their test while listening to headphones. His second class does not. He compares their test results. Is this a *survey*, an *observational study*, or an *experiment*? Explain the process.

Experiment. The treated group is the first class and the control is the second class. This is a biased experiment because the treated group knows who they are.

12-2 Statistical Analysis (pp. 752–758)

A2.S.4

Determine whether each of the following represents a *population* or a *sample*.

12. Jarred conducts an online survey on cancer.
13. The French club wants to compare their AP test scores with the national average.
14. The field hockey team wants to compare their scoring average with everyone else in the league.
15. **SEASONS** In a random survey of 3446 people, 34% said that spring is their favorite season. What is the margin of sampling error?
16. **SWIMMING** While practicing, Kelly kept track of her times in the 400-meter individual medley. Find the standard deviation of her practice times.

Times in Seconds					
301	311	320	308	312	307
313	315	309	308	304	302
311	313	313	316	314	306
329	326	319	310	306	309
320	318	315	318	314	309

EXAMPLE 3

A national poll estimates that the average number of hours spent per week sitting in traffic is four. Is this a *population* or a *sample*?

This represents a sample because only a fraction of the residents of the United States are polled.

EXAMPLE 4

In a random survey of 2645 people, 12% said that hockey is their favorite sport. What is the margin of sampling error?

$$\text{Margin of sampling error} = \pm\frac{1}{\sqrt{n}}$$
$$= \pm\frac{1}{\sqrt{2645}}$$
$$\approx \pm 0.0194$$

The margin of sampling error $\approx \pm 1.9\%$.

Study Guide and Review

12-3 Conditional Probability (pp. 759–763)

A.S.18

17. SOFTBALL Jillian gets a hit 65% of the times she is at bat. What is the probability that she does not get a hit in five consecutive at-bats?

18. BASEBALL The results of who made the varsity baseball team are listed in the table below. Find the probability of each.

	Yes	No
Left-Handed	6	5
Right-Handed	15	22

a. Peter made the team given that he is left-handed.

b. Paul is right-handed given he did not make the team.

EXAMPLE 5

MEDICINE Find the probability that Lori has a Health class, given that she is a freshman.

	Health Class	No Health Class
Freshmen	126	84
Sophomores	98	72

$P(H \mid F)$

$= \frac{P(H \text{ and } F)}{P(F)}$ **Conditional Probability**

$= \frac{126}{380} \div \frac{210}{380}$ $\mathbf{P(H \text{ and } F) = \frac{126}{380}, P(D) = \frac{210}{380}}$

$= \frac{126}{210}$ or $\frac{3}{5}$ **Simplify.**

12-4 Probability and Probability Distributions (pp. 764–771)

A2.S.13, A2.S.14

SPORTS CARDS Bob is moving and all of his sports cards are mixed up in a box. Twelve cards are baseball, eight are football, and five are basketball. If he reaches in the box and selects them at random, find each probability.

19. P(3 football)

20. P(3 baseball)

21. P(1 basketball, 2 football)

22. P(2 basketball, 1 baseball)

23. MARBLES Sammy has a sack of 25 marbles. Eight are black, 10 are red, 4 are green, and the rest are white. He pulls two marbles out of the bag.

a. Create a frequency table and a relative-frequency histogram of the data.

b. Which outcome is the most likely to occur?

c. Find P(black and green).

24. CARDS Three nines, 4 tens, 5 sixes, 4 fives, 2 twos, and a three are pulled from a deck of cards. If one card is drawn from the cards that were pulled out, what is its expected value?

EXAMPLE 6

Ramon has fie books on the floor, one for each of his classes: Algebra 2, chemistry, English, Spanish, and history. Ramon is going to put the books on a shelf. If he picks the books up at random and places them in a row on the same shelf, what is the probability that his English, Spanish, and Algebra 2 books will be the leftmost books on the shelf, but not necessarily in that order.

Step 1 Determine how many book arrangements meet the conditions.
$P(3, 3)$ **Place the 3 leftmost books.**
$P(2, 2)$ **Place the other 4 books.**

Step 2 Use the Fundamental Counting Principle to find the number of successes.
$P(3, 3) \cdot P(2, 2) = 3! \cdot 2!$ or 12

Step 3 Find the total number, $s + f$, of possible 5-book arrangements.
$P(5, 5) = 5!$ or 120 $\mathbf{s + f = 120}$

Step 4 Determine the probability.
$P = \frac{s}{s + f} = \frac{12}{120}$ or 0.1

The probability of placing English, Spanish, and Algebra 2 before the other four books is 0.1 or 10%.

MIXED PROBLEM SOLVING
For mixed problem-solving practice, see page 990.

12-5 The Normal Distribution (pp. 773–778)

A2.S.3, A2.S.5

A normal distribution of data has each mean and standard deviation. Find each probability.

25. $\mu = 121, \sigma = 9, P(x < 103)$

26. $\mu = 84, \sigma = 8, P(x > 108)$

27. $\mu = 181, \sigma = 12, P(x > 169)$

28. RUNNING TIMES The times in the 40-meter dash for a select group of professional football players is normally distributed with a mean of 4.7 and a standard deviation of 0.15.

a. About what percent of the players have times below 4.4?

b. About how many of the 800 players have times between 4.55 and 4.85?

EXAMPLE 7

A normal distribution of data has a mean of 78 and standard deviation of 5. Find the probability that random value x is greater than 83.

$\mu = 78$ and $\sigma = 5$

The probability that a randomly selected value in the distribution is greater than $\mu + \sigma$, that is, $78 + 5$ or 83, is $13.5\% + 2\% + 0.5\% = 16\%$

In the normal curve, this includes the area that is greater than $\mu + \sigma$.

12-6 Hypothesis Testing (pp. 780–784)

G.G.27

Find a 95% confidence interval for each of the following.

29. $\bar{x} = 23.3, s = 2.4$, and $n = 80$

30. $\bar{x} = 72.2, s = 5.8$, and $n = 120$

31. $\bar{x} = 81.4, s = 6.1$, and $n = 200$

32. INTERNET A sample of 300 students was asked for the average amount of time they spend online during a week night. The mean time was 64.3 minutes with a standard deviation of 7.3 minutes. Determine a 95% confidence interval.

Test each null hypothesis. Write *accept* or *reject*.

33. $H_0 = 60, H_1 < 60, n = 100, \bar{x} = 59.4$, and $s = 3.1$

34. $H_0 = 5.5, H_1 > 5.5, n = 80, \bar{x} = 5.8$, and $s = 0.7$

35. $H_0 = 32, H_1 < 32, n = 60, \bar{x} = 31.5$, and $s = 1.8$

36. INTERSECTIONS A light at an intersection is timed to let 10 cars turn left each rotation. Danny believes it is less than 10 and tests the light. After collecting data for 50 rotations, he arrives at a mean of 9.1 cars and a standard deviation of 0.8 cars. Test the hypothesis that the average number of cars is less than 10.

EXAMPLE 8

Find a 95% confidence interval for $\bar{x} = 65$, $s = 1.6$, and $n = 100$.

$CI = \bar{x} \pm 2 \cdot \frac{s}{\sqrt{n}}$ — **Confidence Interval Formula**

$= 65 \pm 2 \cdot \frac{1.6}{\sqrt{100}}$ — **$\bar{x} = 65, s = 1.6, n = 100$**

$= 65 \pm 0.32$ — **Simplify.**

The 95% confidence interval is $64.68 \leq \mu \leq 65.32$.

EXAMPLE 9

Test the null hypothesis. Write *accept* or *reject*. $H_0 = 8$, $H_1 < 8$, $n = 90$, $\bar{x} = 8.1$, and $s = 0.7$

$CI = \bar{x} \pm 2 \cdot \frac{s}{\sqrt{n}}$ — **Confidence Interval Formula**

$= 8.1 \pm 2 \cdot \frac{0.7}{\sqrt{90}}$ — **$\bar{x} = 8.1, s = 0.7, n = 90$**

$= 8.1 \pm 0.07$ — **Simplify.**

The 95% confidence interval is $8.03 \leq \mu \leq 8.17$. The confidence interval does not include H_0, so we reject the null hypothesis.

12-7 Binomial Distributions (pp. 786–793)

A2.S.15, A2.S.16

A binomial distribution has a 40% rate of success. There are 10 trials. Calculate the probability of each.

37. exactly 3 successes

38. less than 8 successes

39. no more than 3 successes

40. at least 4 successes

41. In a certain dice game, a player tries to roll a total of 3 or 10 with two dice. Kevin designed and conducted a binomial experiment for 7 rolls of the dice.

Simulation	1	2	3	4	5	6	7	Total Successes
1	f	f	f	f	f	f	f	0
2	f	s	f	f	s	f	f	2
3	s	f	f	f	f	f	f	1
4	f	f	f	s	f	s	f	2
5	f	s	f	f	f	f	s	2

From Kevin's experiment, what is the experimental probability of rolling a 3 or 10 twice in seven rolls?

42. **SENIOR TRIP** A poll of students at Ryan's school determined that 76% of the students wanted to go to a theme park for their senior trip. Ryan surveyed 180 random students from the school. What is the probability that at least 60 of the polled students did not want to go to a theme park?

43. **WORK** According to an online poll, 28% of adults feel that the standard 40-hour work-week should be increased. Sheila thinks the number should be lower, so she conducts a survey of 250 random adults. What is the probability that more than 55 of the surveyed adults feel that the standard 40-hour work-week should be increased?

44. **WATCHES** According to an online poll, 74% of adults wear watches. Timmy surveyed 200 random adults. What is the probability that at least 160 of the adults surveyed wear a watch?

EXAMPLE 10

A binomial distribution has a 55% rate of success. There are 8 trials. What is the probability that there will be at least 2 successes?

Calculate the probability of 0 and 1 successes.

$P(1) = C(8, 1)(0.55)^1(0.45)^7$
$= 0.01644$

$P(0) = C(8, 0)(0.55)^0(0.45)^8$
$= 0.00168$

$P(\geq 2 \text{ successes}) = 1 - P(< 2 \text{ successes})$
$= 1 - P(1) - P(0)$
$= 1 - 0.01644 - 0.00168$
$= 0.98188$

There is about a 98% probability that there will be at least 2 successes.

EXAMPLE 11

VACATIONS **According to an online poll, 70% of high school students take a vacation during the summer. Louie thinks the number should be lower so he conducts a survey of 650 random students. What is the probability that no more than 420 of the surveyed students go on a vacation in the summer?**

The number of people surveyed has a binomial distribution with $n = 650$, $s = 0.70$, and $f = 0.30$.

Use a normal distribution to approximate the probability.

Mean of a normal approximation

$\bar{x} = ns$
$= 650(0.70)$ or 455 — $n = 650, s = 0.7$

Standard deviation of a normal approximation

$\sigma = \sqrt{nsf}$
$= \sqrt{650(0.7)(0.3)}$ — $n = 300, s = 0.7, f = 0.3$
≈ 11.68 — Use a calculator.

420 is about 3 standard deviations less than the mean, so the probability that no more than 420 responses agree is 0.5%.

CHAPTER 12 Practice Test

NY Math Online glencoe.com
Chapter Test

Determine whether the following statements show *correlation* or *causation*. Explain.

1. When a baseball player hits the ball over the outfielder's head and into the bleachers, he has hit a home run.
2. When Jimmy is running in the hallways, he is late for class.

State whether each survey would produce a random sample. Write *yes* or *no*. Explain.

3. An online store surveys its customers asking how much money they spend online per month.
4. A teacher selects the names of 5 students from a hat to determine who gives their speeches in class that day.

Which measure of central tendency best represents the data, and why?

5.

AP Test Scores				
4	4	3	3	3
4	5	5	4	4
3	3	3	3	4
3	3	3	4	4
4	5	3	4	3

6.

Height in Inches				
61	64	62	61	64
63	65	61	66	73
74	63	62	65	61
61	62	66	63	61

Determine whether each of the following represents a *population* or a *sample*.

7. Olivia records the addresses of every student at her high school.
8. Bridgette compares her class's test results to the national average.
9. Joey separates the candy in his bag by color.
10. Paul asks 100 random people what their favorite movie is.

11. **MOVIES** A survey of 6225 random people found that 48% eat fast food at least once per week. What is the likely interval that contains the percentage of the population that eats fast food at least once per week?

    A 0.78% **B** 1.27% **C** ±1.27% **D** ±0.78%

A normal distribution of data has each mean and standard deviation. Find each probability.

12. $\mu = 54, \sigma = 5, P(x > 44)$
13. $\mu = 35, \sigma = 2.4, P(x < 37.4)$

14. **TESTS** Mr. Holt's class was given the opportunity to retake a test. He also held an optional review session at school the Sunday before. Some students improved and some did not.

	Improved	Did not Improve
Attended	12	3
Did not Attend	4	6

   a. Find the probability that Michael improved, given he attended the session.

   b. Find the probability that Melissa did not attend the session, given she did not improve.

A bag contains 10 blue marbles, 8 red marbles, and 12 green marbles. The marbles are drawn one at a time. Find each probability.

15. The second marble is red, given that the first marble is blue and not replaced.
16. The second marble is blue, given that the first marble is green and is replaced.

17. In a box of paper clips, 45 are red, 25 are yellow, and 30 are green. If 12 paperclips are drawn, what is the expected number of green paperclips?

    F 2.5 **G** 4.8 **H** 3.0 **J** 3.6

18. **DRAWINGS** Ten male and 12 female students have been selected for a drawing for 5 free mp3 players. If the five names will be drawn at random, what is the probability that 3 winners will be male and 2 will be female?

Test each null hypothesis. Write *accept* or *reject*.

19. $H_0 = 77, H_1 > 77, n = 150, \bar{x} = 78.1$, and $\sigma = 1.3$
20. $H_0 = 65, H_1 < 65, n = 120, \bar{x} = 64.8$, and $\sigma = 2.1$

21. A binomial distribution has a 65% rate of success. There are 15 trials. What is the probability that there will be at least 10 successes?

22. **WEATHER** The weatherman says that there is a 40% chance of snow for each of the next seven days. Find the probability that it snows at least 2 of those days.

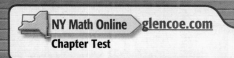

Solve Multi-Step Problems

Some problems that you will encounter on standardized tests require you to solve multiple parts in order to come up with the final solution. Use this lesson to practice these types of problems.

Strategies for Solving Multi-Step Problems

Step 1

Read the problem statement carefully.

Ask yourself:

- What am I being asked to solve? What information is given?
- Are there any intermediate steps that need to be completed before I can solve the problem?

Step 2

Organize your approach.

- List the steps you will need to complete in order to solve the problem.
- Remember that there may be more than one possible way to solve the problem.

Step 3

Solve and check.

- Work as efficiently as possible to complete each step and solve.
- If time permits, check your answer.

EXAMPLE

Read the problem. Identify what you need to know. Then use the information in the problem to solve.

There are 15 boys and 12 girls in Mrs. Lawrence's homeroom. Suppose a committee is to be made up of 6 randomly selected students. What is the probability that the committee will contain 3 boys and 3 girls? Round your answer to the nearest tenth of a percent.

A 27.2% **C** 31.5%

B 29.6% **D** 33.8%

Read the problem statement carefully. You are asked to find the probability that a committee will be made up of 3 boys and 3 girls. Finding this probability involves successfully completing several steps.

Step 1 Find the number of possible successes.

There are $C(15, 3)$ ways to choose 3 boys from 15, and there are $C(12, 3)$ to choose 3 girls from 12. Use the Fundamental Counting Principle to find s, the number of possible successes.

$s = C(15, 3) \times C(12, 3) = \frac{15!}{12!3!} \times \frac{12!}{9!3!}$ or 100,100

Step 2 Find the total number of possible outcomes.

Compute the number of ways 6 people can be chosen from a group of 27 students.

$C(27, 6) = 296{,}010$

Step 3 Compute the probability.

Find the probability by comparing the number of successes to the number of possible outcomes.

$P(3 \text{ boys, } 3 \text{ girls}) = \frac{100{,}100}{296{,}010} \approx 0.33816$

So, there is about a 33.8% chance of selecting 3 boys and 3 girls for the committee. The answer is D.

Exercises

Read the problem. Identify what you need to know. Then use the information in the problem to solve.

1. There are 52 cards in a standard deck. Of these, 4 of the cards are Aces. What is the probability of a randomly dealt 5-card hand containing a pair of Aces? Round your answer to the nearest whole percent.

   A 4%

   B 5%

   C 6%

   D 7%

2. According to the table, what is the probability that a randomly selected camper went on the horse ride, given that the camper is an 8th grader?

	Camp Activities		
Grade	Canoe Trip	Horse Ride	Nature Hike
6th	8	6	3
7th	5	4	7
8th	11	9	6

   F 0.361

   G 0.445

   H 0.423

   J 0.507

NYS Regents Exam Practice

Cumulative, Chapters 1 through 12

Multiple Choice

Answer all questions in this part. Select the answer that best completes the statement or answers the question.

1. Suppose the test scores on a final exam are normally distributed with a mean of 74 and a standard deviation of 3. What is the probability that a randomly selected test has a score higher than 80?

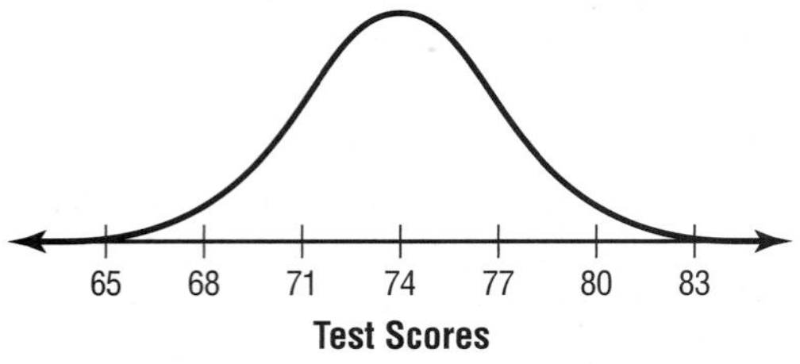

(1) 2.5%

(2) 13.5%

(3) 16%

(4) 34%

2. Segment AB has midpoint M with the coordinates shown below. What are the coordinates of B?

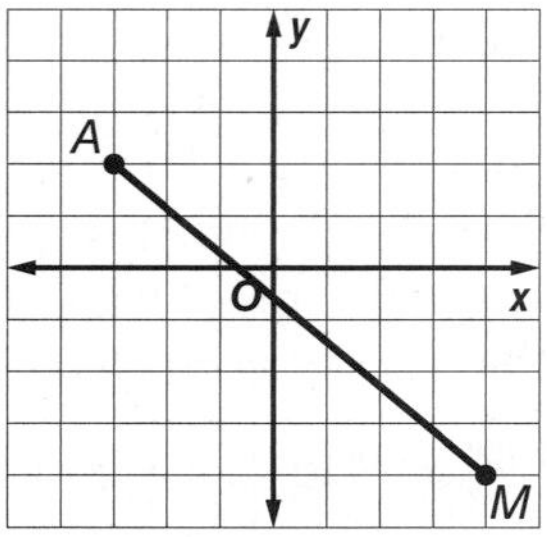

(1) $B\left(\frac{1}{2}, -1\right)$

(2) $B\left(-\frac{1}{2}, -1\right)$

(3) $B(10, -11)$

(4) $B(11, -10)$

3. Suppose a campaign wants to find opinions on favorite candidates in the upcoming city council elections. Which of the following best describes the method of data collection the campaign should employ?

(1) experiment

(2) observational study

(3) survey

(4) none of the above

4. Simplify: $\dfrac{\frac{2}{x-1}}{10 - \frac{5}{x-1}}$.

(1) $\dfrac{6}{5x - 10}$

(2) $\dfrac{2}{10x - 15}$

(3) $\dfrac{2}{3}$

(4) $\dfrac{4}{10x - 15}$

5. A company found that its daily profit P is given by the polynomial function $P = -2x^2 + 175x - 3000$, where x is the selling price for each unit of product. Which of the following is the *best* estimate of the price per unit the company should charge in order to maximize its daily profits?

(1) \$828.15

(2) \$756.15

(3) \$45.25

(4) \$43.75

6. Among New York residents age 25 and older, about 79.1% have completed their high school education. Suppose three New York residents age 25 and older are selected at random. What is the probability that all three have completed their high school education? Round to the nearest tenth.

(1) 44.3%

(2) 49.5%

(3) 56.8%

(4) 79.1%

7. The table shows the number of students at Bethany's high school who participate in after school activities.

Gender	After School Activities	No After School Activities
male	102	78
female	138	64

Suppose a student is selected at random. What is the probability that the student is involved in an after school activity given that the student is male? Express your answer as a percent. Round to the nearest tenth if necessary.

(1) 28.4% **(3)** 42.3%

(2) 39.5% **(4)** 56.7%

8. Which of the following geometric series does *not* converge to a sum?

(1) $\sum_{k=1}^{\infty} 100 \cdot \left(\frac{99}{100}\right)^{k-1}$ **(3)** $\sum_{k=1}^{\infty} \frac{5}{4} \cdot \left(\frac{1}{2}\right)^{k-1}$

(2) $\sum_{k=1}^{\infty} \frac{1}{12} \cdot \left(\frac{8}{7}\right)^{k-1}$ **(4)** $\sum_{k=1}^{\infty} (-4) \cdot \left(\frac{5}{6}\right)^{k-1}$

Test-TakingTip

Question 8 A geometric series converges to a sum if the common ratio *r* has an absolute value less than 1.

9. Andrea is using a coordinate grid to design a new deck for her backyard. The boundaries of the deck are given by the equations $y = 20$, $x = 16$, $y = 0$, $x = 0$, and $y = -x + 32$. If each unit of the coordinate grid represents 1 foot, what is the area of the deck?

(1) 276 ft^2 **(3)** 312 ft^2

(2) 290 ft^2 **(4)** 325 ft^2

Open-Ended Response

Solve each problem. Clearly indicate the necessary steps, including appropriate formula substitutions, diagrams, graphs, charts, etc.

10. Lawrence is taking a true-false test that has 10 questions. Each question has two possible answers: true or false. Lawrence forgot to study for the test, so he must guess at each answer.

a. What is the probability of guessing a correct answer on the test?

b. What is the expected number of correct answers if Lawrence guesses at each question?

c. What is the probability that Lawrence will get at least 7 questions correct? Round your answer to the nearest tenth of a percent.

11. A stone path that is *x* feet wide is built around a rectangular swimming pool. The pool is 12 feet wide and 25 feet long as shown below. If the combined area of the pool and the stone path is 558 square feet, what is the width of the walkway?

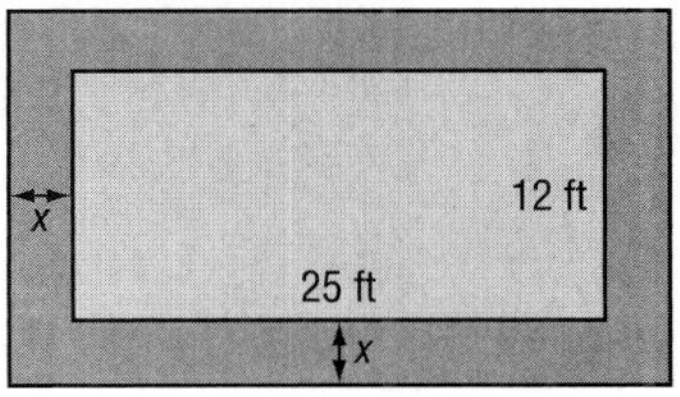

12. Give an example of two matrices, *A* and *B*, such that the product *AB* exists but the product *BA* does not exist.

Need Extra Help?												
If you missed Question...	1	2	3	4	5	6	7	8	9	10	11	12
Go to Lesson or Page...	12-5	10-1	2-4	5-7	12-2	6-1	7-6	3-1	8-1	11-2	12-1	1-4
NYS Core Curriculum	A2.A.16	G.G.69	A2.S.1	A2.A.17	A.A.8	A2.S.14	A.S.18	A2.A.35	A.A.7	A2.S.15	A2.A.26	A2.RP.10

CHAPTER 13 Trigonometric Functions

Then

Throughout this text, you have graphed and analyzed functions.

Now

In Chapter 13, you will:

- Find values of trigonometric functions.
- Solve problems by using right triangle trigonometry.
- Solve triangles by using the Law of Sines and Law of Cosines.
- Graph trigonometric functions.

NYS Core Curriculum

A2.A.55, A2.A.62 Find values of trigonometric functions
A2.A.69 Determine amplitude, period, frequency, and phase shift, given the graph or equation of a periodic function
A2.A.73 Solve for an unknown side or angle, using the Law of Sines or the Law of Cosines

Why?

WATER SPORTS Knowing trigonometric functions has practical applications in water sports. For instance, you can use right triangle trigonometry to find the height of a person parasailing. If you are familiar with angles and angle measures, then you have a better understanding of how impressive it is to be able to do a 540° rotation on a wakeboard.

Math *in Motion,* Animation glencoe.com

Get Ready for Chapter 13

Diagnose Readiness You have two options for checking Prerequisite Skills.

Text Option

Take the Quick Check below. Refer to the Quick Review for help.

*Quick*Check

Find the value of x to the nearest tenth. (Lesson 0-7)

1.

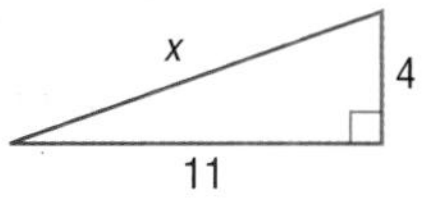

2.

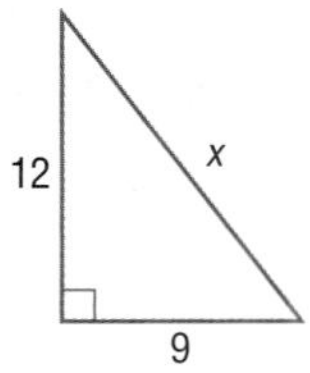

3.

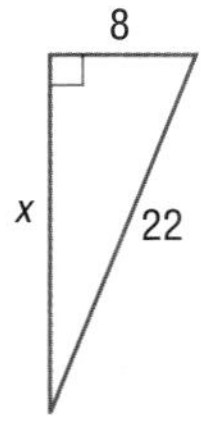

4. Laura has a rectangular garden in her backyard that measures 12 feet by 15 feet. She wants to put a rock walkway on the diagonal. How long will the walkway be? Round to the nearest tenth of a foot.

Find each missing measure. Write all radicals in simplest form. (Geometry)

5.

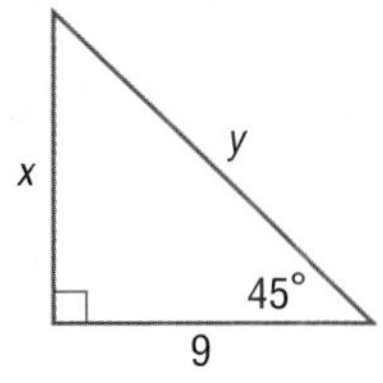

6.

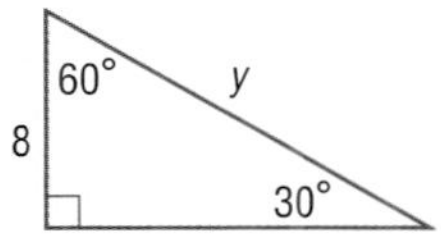

7.

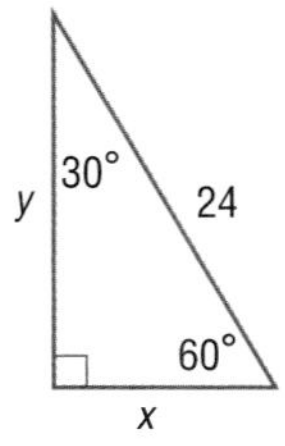

8. A ladder leans against a wall at a 45° angle. If the ladder is 12 feet long, how far up the wall does the ladder reach?

*Quick*Review

EXAMPLE 1

Find the missing measure of the right triangle.

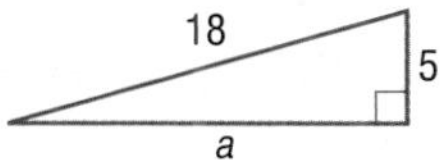

$c^2 = a^2 + b^2$	**Pythagorean Theorem**
$18^2 = a^2 + 5^2$	**Replace *c* with 18 and *b* with 5.**
$324 = a^2 + 25$	**Simplify.**
$299 = a^2$	**Subtract 25 from each side.**
$17.3 \approx a$	**Take the square root of each side.**

EXAMPLE 2

Find the missing measures. Write all radicals in simplest form.

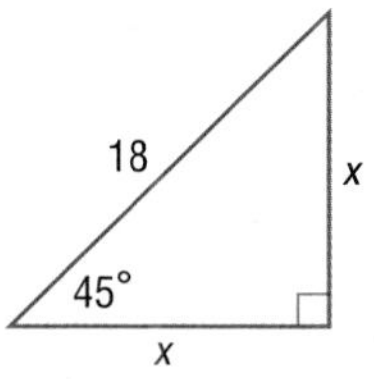

$x^2 + x^2 = 18^2$	**Pythagorean Theorem**
$2x^2 = 18^2$	**Combine like terms.**
$2x^2 = 324$	**Simplify.**
$x^2 = 162$	**Divide each side by 2.**
$x = \sqrt{162}$	**Take the square root of each side.**
$x = 9\sqrt{2}$	**Simplify.**

Online Option

NY Math Online Take a self-check Chapter Readiness Quiz at **glencoe.com**.

Get Started on Chapter 13

You will learn several new concepts, skills, and vocabulary terms as you study Chapter 13. To get ready, identify important terms and organize your resources. You may wish to refer to **Chapter 0** to review prerequisite skills.

FOLDABLES® Study Organizer

Trigonometric Functions Make this Foldable to help you organize your Chapter 13 notes about trigonometric functions. Begin with four pieces of grid paper.

1. **Stack** paper together and measure 2.5 inches from the bottom.

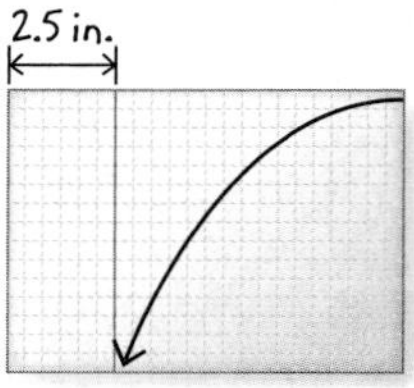

2. **Fold** on the diagonal.

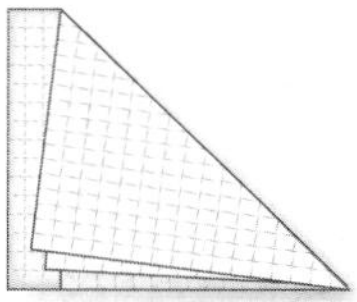

3. **Staple** along diagonal to form a book.

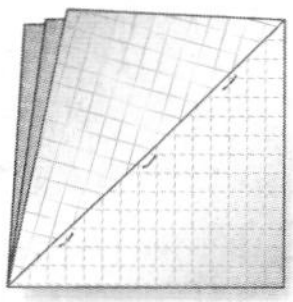

4. **Label** edge as Trigonometric Functions.

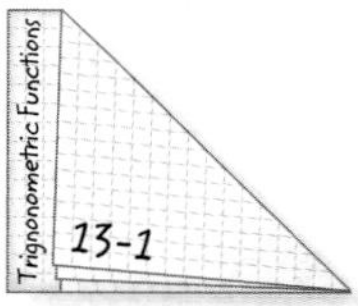

NY Math Online glencoe.com

- Study the chapter online
- Explore **Math in Motion**
- Get extra help from your own **Personal Tutor**
- Use **Extra Examples** for additional help
- Take a **Self-Check Quiz**
- **Review Vocabulary** in fun ways

New Vocabulary

English		Español
trigonometry	• p. 808 •	trigonometría
sine	• p. 808 •	seno
cosine	• p. 808 •	coseno
tangent	• p. 808 •	tangente
cosecant	• p. 808 •	cosecante
secant	• p. 808 •	secante
cotangent	• p. 808 •	cotangente
angle of elevation	• p. 812 •	ángulo de depresión
angle of depression	• p. 812 •	ángulo de elevación
standard position	• p. 817 •	posición estándar
radian	• p. 819 •	radián
Law of Sines	• p. 833 •	Ley de los senos
Law of Cosines	• p. 841 •	Ley de los cosenos
unit circle	• p. 848 •	círculo unitario
circular function	• p. 848 •	funciones circulares
periodic function	• p. 849 •	función periódica
cycle	• p. 849 •	ciclo
period	• p. 849 •	período
amplitude	• p. 855 •	amplitud
frequency	• p. 856 •	frecuencia

Review Vocabulary

acute angle • prior course • angulo agudo an angle with a measure between 0° and 90°

function • p. P4 • funcion a relation in which each element of the domain is paired with exactly one element in the range

inverse function • p. 417 • funcion inversa two functions *f* and *g* are inverse functions if and only if both of their compositions are the identity function

Pythagorean Theorem • p. P17 • Teorema de Pitágoras

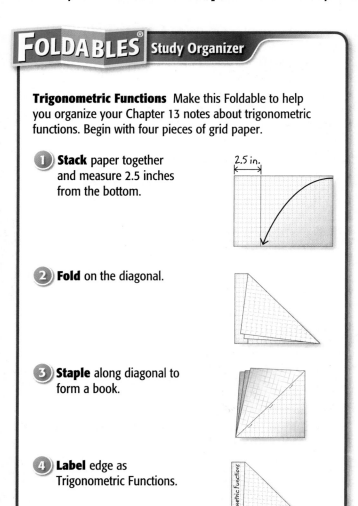

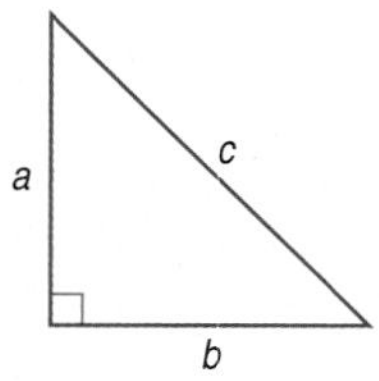

$a^2 + b^2 = c^2$

Multilingual eGlossary glencoe.com

EXPLORE 13-1

Spreadsheet Lab: Investigating Special Right Triangles

NY Math Online glencoe.com
Other Calculator Keystrokes

You can use a spreadsheet to investigate side measures of special right triangles.

ACTIVITY 45°-45°-90° Triangle

The legs of a 45°-45°-90° triangle, *a* and *b*, are equal in measure. What patterns do you observe in the ratios of the side measures of these triangles?

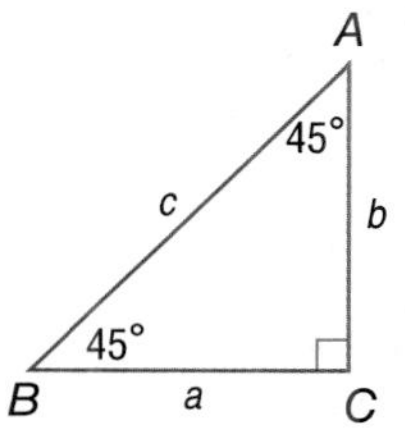

Step 1 Enter the indicated formulas in the spreadsheet. The formula uses the Pythagorean Theorem in the form $c = \sqrt{a^2 + b^2}$.

=SQRT(A2^2+B2^2) =B2/A2 =B2/C2 =A2/C2

Excel sample.xls

	A	B	C	D	E	F
1	a	b	c	b/a	b/c	a/c
2	1	1	1.414213562	1	0.707106781	0.707106781
3	2	2	2.828427125	1	0.707106781	0.707106781
4	3	3	4.242640687	1	0.707106781	0.707106781
5	4	4	5.656854249	1	0.707106781	0.707106781

Sheet 1 Sheet 2 Sheet 3

Step 2 Examine the results. Because 45°-45°-90° triangles share the same angle measures, these triangles are all similar. The ratios of the sides of these triangles are all the same. The ratios of side *b* to side *a* are 1. The ratios of side *b* to side *c* and of side *a* to side *c* are approximately 0.71.

Model and Analyze

Use the spreadsheet below for 30°-60°-90° triangles.

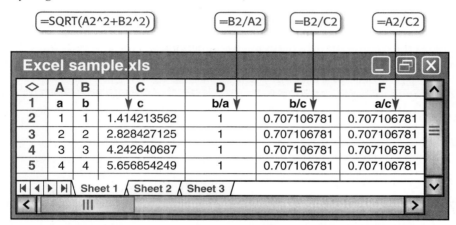

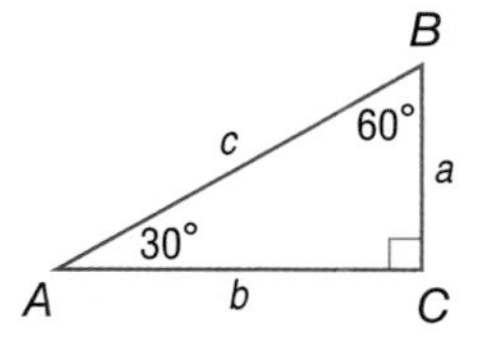

Excel sample.xls

	A	B	C	D	E	F
1	a	b	c	b/a	b/c	a/c
2	1		2			
3	2		4			
4	3		6			
5	4		8			

Sheet 1 Sheet 2 Sheet 3

1. Copy and complete the spreadsheet above.

2. Describe the relationship among the 30°-60°-90° triangles with the dimensions given.

3. What patterns do you observe in the ratios of the side measures of these triangles?

13-1 Trigonometric Functions in Right Triangles

Then

You used the Pythagorean Theorem to find side lengths of right triangles.

Now

- Find values of trigonometric functions.
- Use trigonometric functions to find side lengths and angle measures of right triangles.

NYS Core Curriculum

A2.A.55 Express and apply the six trigonometric functions as ratios of the sides of a right triangle
A2.A.58 Know and apply the co-function and reciprocal relationships between trigonometric ratios
Also addresses A2.A.66.

New Vocabulary

trigonometry
trigonometric ratio
trigonometric function
reciprocal functions

NY Math Online

glencoe.com

- Extra Examples
- Personal Tutor
- Self-Check Quiz
- Homework Help

Why?

The altitude of a person parasailing depends on the length of the tow rope ℓ and the angle the rope makes with the horizontal $x°$. If you know these two values, you can use a ratio to find the altitude of the person parasailing.

Trigonometric Functions for Acute Angles **Trigonometry** is the study of relationships among the angles and sides of a right triangle. A **trigonometric ratio** compares the side lengths of a right triangle. A **trigonometric function** has a rule given by a trigonometric ratio.

The Greek letter *theta* θ is often used to represent the measure of an acute angle in a right triangle. The *hypotenuse*, the *leg opposite* θ, and the *leg adjacent to* θ are used to define the six trigonometric functions.

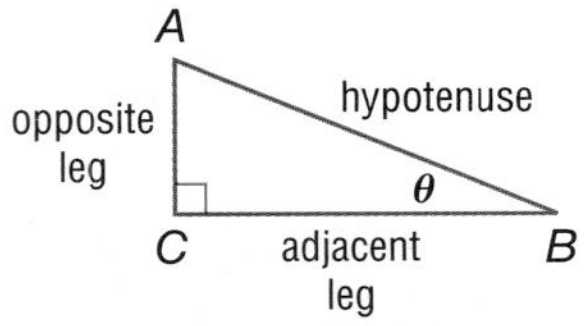

Key Concept: Trigonometric Functions in Right Triangles

For Your FOLDABLE

Words If θ is the measure of an acute angle of a right triangle, then the following trigonometric functions involving the opposite side *opp*, the adjacent side *adj*, and the hypotenuse *hyp* are true.

Symbols

sin (**sine**) $\theta = \frac{\text{opp}}{\text{hyp}}$ csc (**cosecant**) $\theta = \frac{\text{hyp}}{\text{opp}}$

cos (**cosine**) $\theta = \frac{\text{adj}}{\text{hyp}}$ sec (**secant**) $\theta = \frac{\text{hyp}}{\text{adj}}$

tan (**tangent**) $\theta = \frac{\text{opp}}{\text{adj}}$ cot (**cotangent**) $\theta = \frac{\text{adj}}{\text{opp}}$

Examples

$\sin\theta = \frac{4}{5}$ $\cos\theta = \frac{3}{5}$ $\tan\theta = \frac{4}{3}$

$\csc\theta = \frac{5}{4}$ $\sec\theta = \frac{5}{3}$ $\cot\theta = \frac{3}{4}$

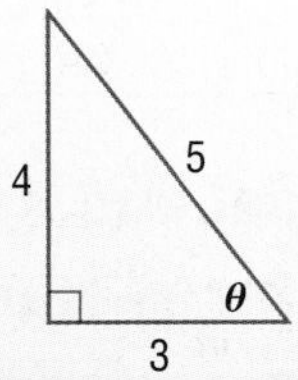

EXAMPLE 1 Evaluate Trigonometric Functions

Find the values of the six trigonometric functions for angle θ.

leg opposite θ: $BC = 8$ leg adjacent θ: $AC = 15$ hypotenuse: $AB = 17$

$\sin\theta = \frac{\text{opp}}{\text{hyp}} = \frac{8}{17}$ $\cos\theta = \frac{\text{adj}}{\text{hyp}} = \frac{15}{17}$ $\tan\theta = \frac{\text{opp}}{\text{adj}} = \frac{8}{15}$

$\csc\theta = \frac{\text{hyp}}{\text{opp}} = \frac{17}{8}$ $\sec\theta = \frac{\text{hyp}}{\text{adj}} = \frac{17}{15}$ $\cot\theta = \frac{\text{adj}}{\text{opp}} = \frac{15}{8}$

Check Your Progress

1. Find the values of the six trigonometric functions for angle B.

Personal Tutor glencoe.com

StudyTip

Memorize Trigonometric Ratios SOH-CAH-TOA is a mnemonic device for remembering the first letter of each word in the ratios for sine, cosine, and tangent.

$\sin\ \theta = \frac{\text{opp}}{\text{hyp}}$

$\cos\ \theta = \frac{\text{adj}}{\text{hyp}}$

$\tan\ \theta = \frac{\text{opp}}{\text{adj}}$

Notice that the cosecant, secant, and cotangent ratios are reciprocals of the sine, cosine, and tangent ratios, respectively. These are called the **reciprocal functions**. So, the following are also true.

$$\csc\theta = \frac{1}{\sin\theta} \qquad \sec\theta = \frac{1}{\cos\theta} \qquad \cot\theta = \frac{1}{\tan\theta}$$

The domain of any trigonometric function is the set of all acute angles θ of a right triangle. So, trigonometric functions depend only on the measures of the acute angles, not on the side lengths of a right triangle.

EXAMPLE 2 Find Trigonometric Ratios

In a right triangle, $\angle B$ is acute and $\sin B = \frac{5}{8}$. Find the value of $\tan B$.

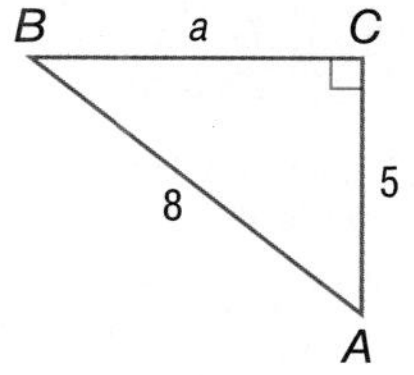

Step 1 Draw a right triangle and label one acute angle B. Since $\sin B = \frac{5}{8} = \frac{\text{opp}}{\text{hyp}}$, label the opposite side 5 and the hypotenuse 8.

Step 2 Use the Pythagorean Theorem to find a.

$a^2 + b^2 = c^2$	**Pythagorean Theorem**
$a^2 + 5^2 = 8^2$	**$b = 5$ and $c = 8$**
$a^2 + 25 = 64$	**Simplify.**
$a^2 = 39$	**Subtract 25 from each side.**
$a = \pm\sqrt{39}$	**Take the square root of each side.**
$a = \sqrt{39}$	**Length cannot be negative.**

Step 3 Find $\tan B$.

$\tan B = \frac{\text{opp}}{\text{adj}}$	**Tangent function**
$= \frac{5}{\sqrt{39}}$	**Replace *opp* with 5 and *adj* with $\sqrt{39}$.**
$= \frac{5\sqrt{39}}{39}$	**Rationalize the denominator.**

Check Your Progress

2. If $\tan B = \frac{3}{7}$, find the value of $\sin B$.

Personal Tutor glencoe.com

ReadingMath

Labeling Triangles Throughout this chapter, a capital letter is used to represent both a vertex of a triangle and the measure of the angle at that vertex. The same letter in lowercase is used to represent both the side opposite that angle and the length of the side.

Angles that measure 30°, 45°, and 60° occur frequently in trigonometry. The table below gives the values of three trigonometric functions for these angles.

Key Concept Trigonometric Values for Special Angles

For Your FOLDABLE

30°-60°-90°

$\sin 30° = \frac{1}{2}$ $\quad \cos 30° = \frac{\sqrt{3}}{2}$ $\quad \tan 30° = \frac{\sqrt{3}}{3}$

$\sin 60° = \frac{\sqrt{3}}{2}$ $\quad \cos 60° = \frac{1}{2}$ $\quad \tan 60° = \sqrt{3}$

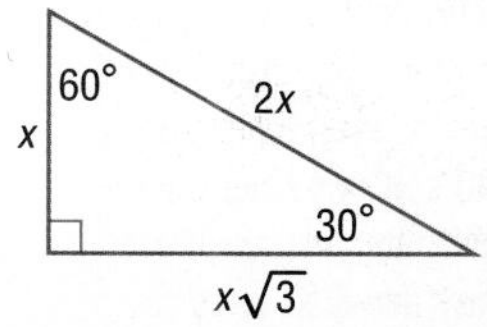

45°-45°-90°

$\sin 45° = \frac{\sqrt{2}}{2}$ $\quad \cos 45° = \frac{\sqrt{2}}{2}$ $\quad \tan 45° = 1$

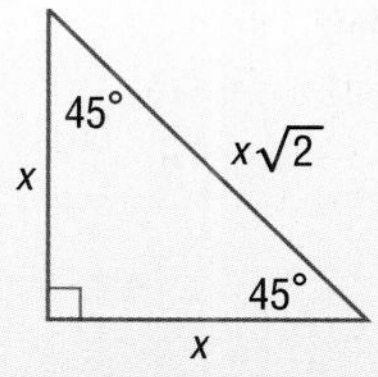

Use Trigonometric Functions You can use trigonometric functions to find missing side lengths and missing angle measures of right triangles.

EXAMPLE 3 Find a Missing Side Length

Use a trigonometric function to find the value of x. Round to the nearest tenth if necessary.

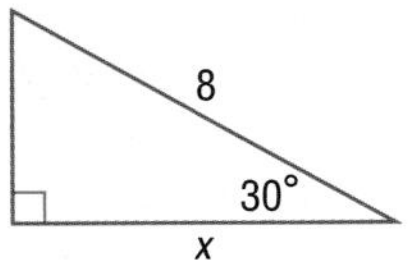

The length of the hypotenuse is 8. The missing measure is for the side adjacent to the 30° angle. Use the cosine function to find x.

$\cos \theta = \frac{\text{adj}}{\text{hyp}}$ **Cosine function**

$\cos 30° = \frac{x}{8}$ **Replace θ with 30°, *adj* with x, and *hyp* with 8.**

$\frac{\sqrt{3}}{2} = \frac{x}{8}$ **$\cos 30° = \frac{\sqrt{3}}{2}$**

$\frac{8\sqrt{3}}{2} = x$ **Multiply each side by 8.**

$6.9 \approx x$ **Use a calculator.**

StudyTip

Choose a Function If the length of the hypotenuse is unknown, then either the sine or cosine function must be used to find the missing measure.

Check Your Progress

3A.

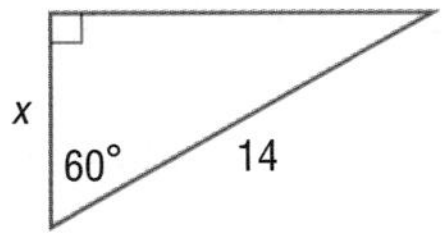

3B.

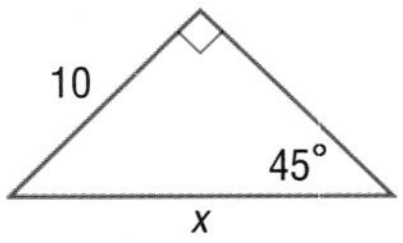

Personal Tutor glencoe.com

You can use a calculator to find the missing side lengths of triangles that do not have 30°, 45°, or 60° angles.

EXAMPLE 4 Find a Missing Side Length

BUILDINGS To calculate the height of a building, Joel walked 200 feet from the base of the building and used an inclinometer to measure the angle from his eye to the top of the building. If his eye level is at 6 feet, how tall is the building?

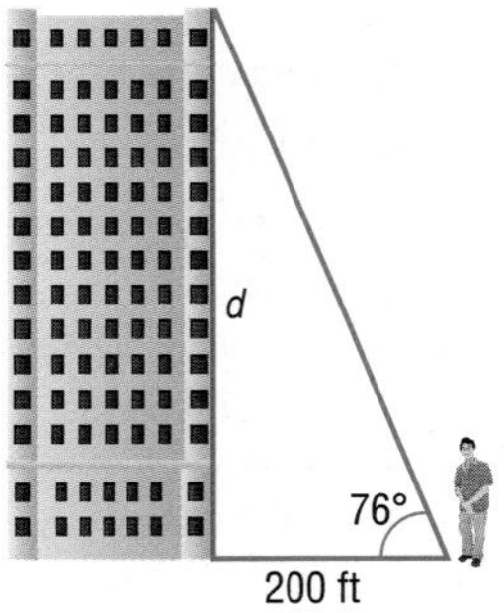

The measured angle is 76°. The side adjacent to the angle is 200 feet. The missing measure is the side opposite the angle. Use the tangent function to find d.

$\tan \theta = \frac{\text{opp}}{\text{adj}}$ **Tangent function**

$\tan 76° = \frac{d}{200}$ **Replace θ with 76°, *opp* with d, and *adj* with 200.**

$200 \tan 76° = d$ **Multiply each side by 200.**

$802 \approx d$ **Use a calculator to simplify: 200 [TAN] 76 [ENTER].**

Because the inclinometer was 6 feet above the ground, the height of the building is approximately 808 feet.

Real-World Link

Inclinometers measure the angle of Earth's magnetic field as well as the pitch and roll of vehicles, sailboats, and airplanes. They are also used for monitoring volcanoes and well drilling.

Source: *Science Magazine*

Check Your Progress

4. Use a trigonometric function to find the value of x. Round to the nearest tenth if necessary.

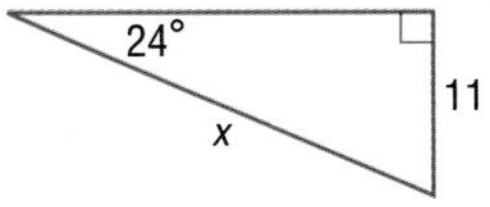

Personal Tutor glencoe.com

When solving equations like $3x = -27$, you use the inverse of multiplication to find x. You also can find angle measures by using the inverse of sine, cosine, or tangent.

Key Concept — Inverse Trigonometric Ratios

For Your FOLDABLE

Words If $\angle A$ is an acute angle and the sine of A is x, then the **inverse sine** of x is the measure of $\angle A$.

Symbols If $\sin A = x$, then $\sin^{-1} x = m\angle A$.

Example $\sin A = \frac{1}{2} \rightarrow \sin^{-1} \frac{1}{2} = m\angle A \rightarrow m\angle A = 30$

Words If $\angle A$ is an acute angle and the cosine of A is x, then the **inverse cosine** of x is the measure of $\angle A$.

Symbols If $\cos A = x$, then $\cos^{-1} x = m\angle A$.

Example $\cos A = \frac{\sqrt{2}}{2} \rightarrow \cos^{-1} \frac{\sqrt{2}}{2} = m\angle A \rightarrow m\angle A = 45$

Words If $\angle A$ is an acute angle and the tangent of A is x, then the **inverse tangent** of x is the measure of $\angle A$.

Symbols If $\tan A = x$, then $\tan^{-1} x = m\angle A$.

Example $\tan A = \sqrt{3} \rightarrow \tan^{-1} \sqrt{3} = m\angle A \rightarrow m\angle A = 60$

ReadingMath

Inverse Trigonometric Ratios The expression $\sin^{-1} x$ is read *the inverse sine of x* and is interpreted as the angle whose sine is x. Be careful not to confuse this notation with the notation for negative exponents; $\sin^{-1} x \neq \frac{1}{\sin x}$. Instead, this notation is similar to the notation for an inverse function, $f^{-1}(x)$.

If you know the sine, cosine, or tangent of an acute angle, you can use a calculator to find the measure of the angle, which is the inverse of the trigonometric ratio.

EXAMPLE 5 Find a Missing Angle Measure

Find the measure of each angle. Round to the nearest tenth if necessary.

a. $\angle N$

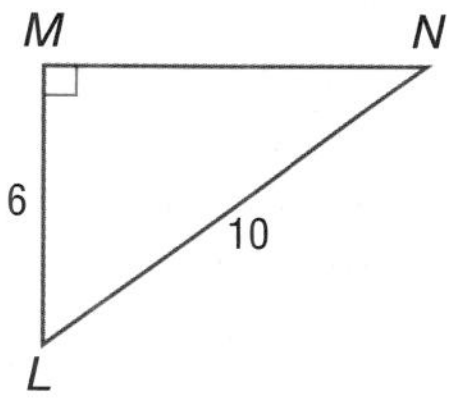

You know the measure of the side opposite $\angle N$ and the measure of the hypotenuse. Use the sine function.

$\sin N = \frac{6}{10}$	$\sin \theta = \frac{\text{opp}}{\text{hyp}}$
$\sin^{-1} \frac{6}{10} = m\angle N$	Inverse sine
$36.9 \approx m\angle N$	Use a calculator.

b. $\angle B$

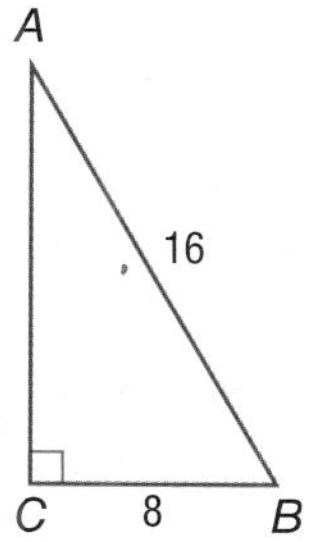

Use the cosine function.

$\cos B = \frac{8}{16}$	$\cos \theta = \frac{\text{adj}}{\text{hyp}}$
$\cos^{-1} \frac{8}{16} = m\angle B$	Inverse cosine
$60 = m\angle B$	Use a calculator.

Check Your Progress

Find x. Round to the nearest tenth if necessary.

5A.

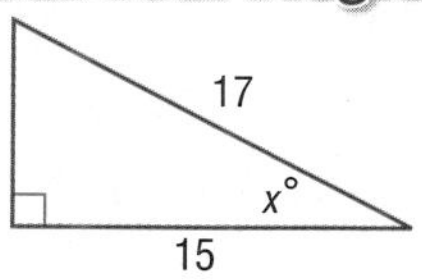

5B.

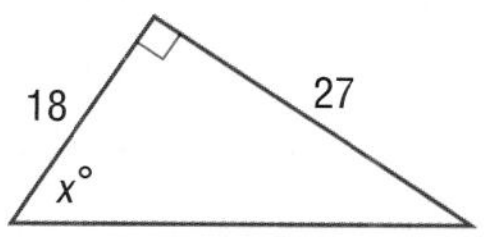

Personal Tutor glencoe.com

StudyTip

Angles of Elevation and Depression The angle of elevation and the angle of depression are congruent since they are alternate interior angles of parallel lines.

In the figure at the right, the angle formed by the line of sight from the swimmer and a line parallel to the horizon is called the **angle of elevation**. The angle formed by the line of sight from the lifeguard and a line parallel to the horizon is called the **angle of depression**.

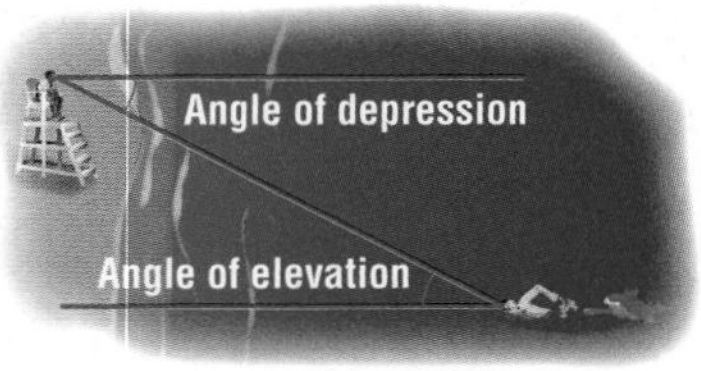

EXAMPLE 6 Use Angles of Elevation and Depression

a. GOLF A golfer is standing at the tee, looking up to the green on a hill. If the tee is 36 yards lower than the green and the angle of elevation from the tee to the hole is 12°, find the distance from the tee to the hole.

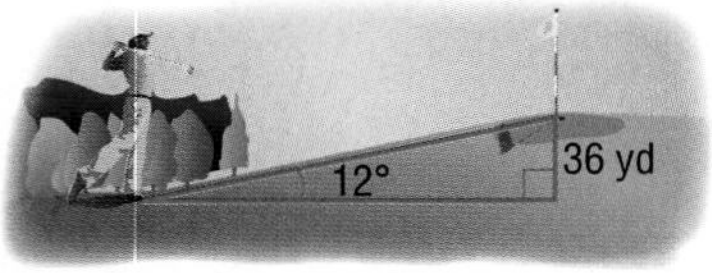

Write an equation using a trigonometric function that involves the ratio of the vertical rise (side opposite the 12° angle) and the distance from the tee to the hole (hypotenuse).

$\sin 12° = \frac{36}{x}$	$\sin \theta = \frac{\text{opp}}{\text{hyp}}$
$x \sin 12° = 36$	**Multiply each side by x.**
$x = \frac{36}{\sin 12°}$	**Divide each side by sin 12°.**
$x \approx 173.2$	**Use a calculator.**

So, the distance from the tee to the hole is about 173.2 yards.

Real-World Link

The steepest roller coasters in the world have angles of descent that are close to 90°.

Source: Ultimate Roller Coaster

b. ROLLER COASTER The hill of the roller coaster has an *angle of descent*, or an angle of depression, of 60°. Its vertical drop is 195 feet. Estimate the length of the hill.

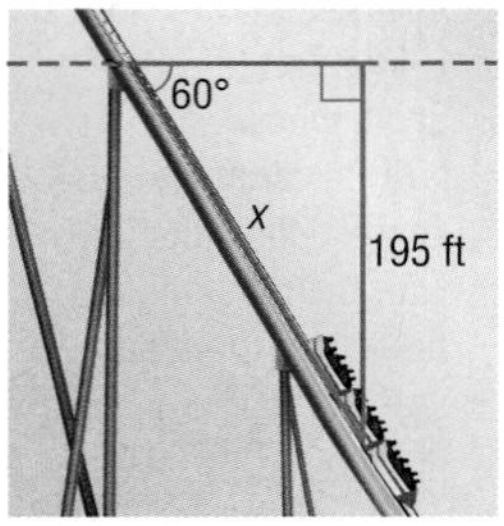

Write an equation using a trigonometric function that involves the ratio of the vertical drop (side opposite the 60° angle) and the length of the hill (hypotenuse).

$\sin 60° = \frac{195}{x}$	$\sin \theta = \frac{\text{opp}}{\text{hyp}}$
$x \sin 60° = 195$	**Multiply each side by x.**
$x = \frac{195}{\sin 60°}$	**Divide each side by sin 60°.**
$x \approx 225.2$	**Use a calculator.**

So, the length of the hill is about 225.2 feet.

Check Your Progress

6A. MOVING A ramp for unloading a moving truck has an angle of elevation of 32°. If the top of the ramp is 4 feet above the ground, estimate the length of the ramp.

6B. LADDERS A 14-ft long ladder is placed against a house at an angle of elevation of 72°. How high above the ground is the top of the ladder?

Personal Tutor glencoe.com

Check Your Understanding

Example 1 p. 808

Find the values of the six trigonometric functions for angle θ.

1. 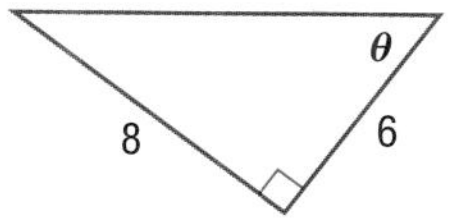

2. 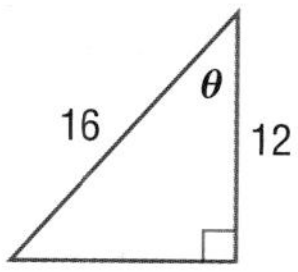

Example 2 p. 809

In a right triangle, $\angle A$ is acute.

3. If $\cos A = \frac{4}{7}$, what is $\sin A$?

4. If $\tan A = \frac{20}{21}$, what is $\cos A$?

Examples 3 and 4 p. 810

Use a trigonometric function to find the value of x. Round to the nearest tenth.

5. 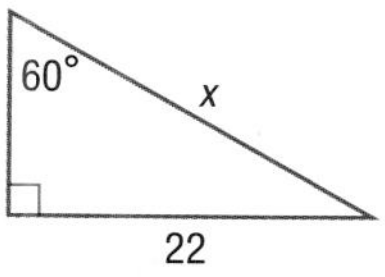

6. 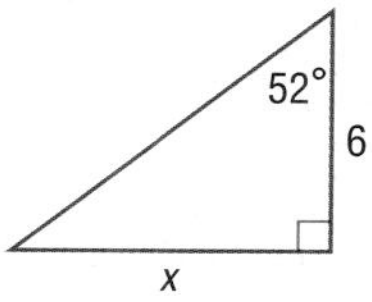

7. 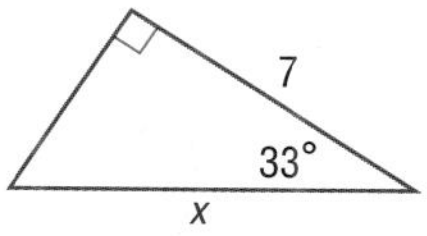

Example 5 p. 811

Find the value of x. Round to the nearest tenth.

8. 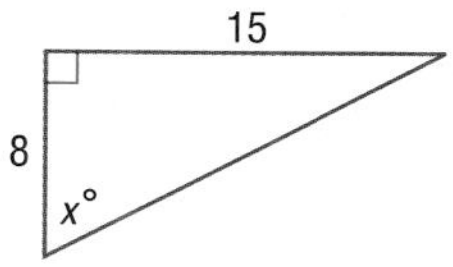

9. 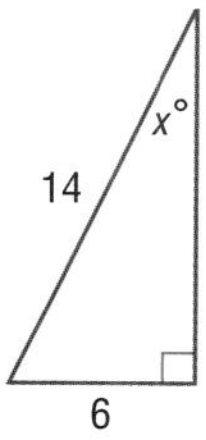

10. 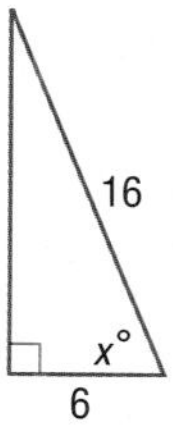

Example 6 p. 812

11. **GEOGRAPHY** Christian found two trees directly across from each other in a canyon. When he moved 100 feet from the tree on his side (parallel to the edge of the canyon), the angle formed by the tree on his side, Christian, and the tree on the other side was 70°. Find the distance across the canyon.

12. **LADDERS** The recommended angle of elevation for a ladder used in fire fighting is 75°. At what height on a building does a 21-foot ladder reach if the recommended angle of elevation is used? Round to the nearest tenth.

Practice and Problem Solving

● = **Step-by-Step Solutions** begin on page R20.
Extra Practice begins on page 947.

Example 1 p. 808

Find the values of the six trigonometric functions for angle θ.

13. 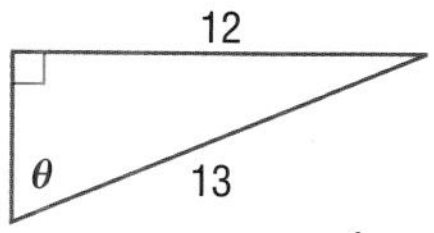

14. 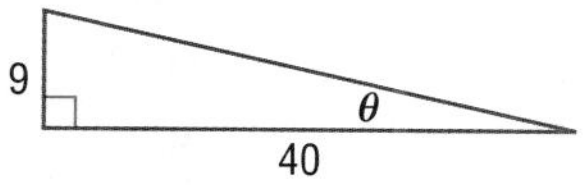

15. 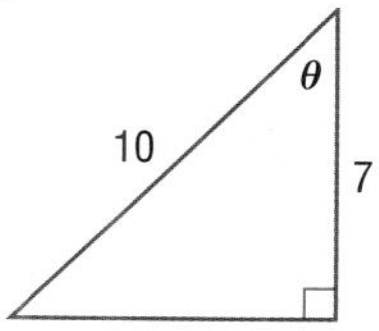

16. 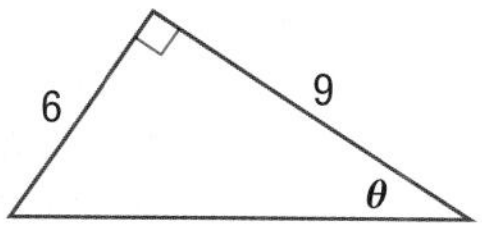

Example 2 p. 809

In a right triangle, $\angle A$ and $\angle B$ are acute.

17. If $\tan A = \frac{8}{15}$, what is $\cos A$?

18. If $\cos A = \frac{3}{10}$, what is $\tan A$?

19. If $\tan B = 3$, what is $\sin B$?

20. If $\sin B = \frac{4}{9}$, what is $\tan B$?

Examples 3 and 4
p. 810

Use a trigonometric function to find each value of x. Round to the nearest tenth.

21.

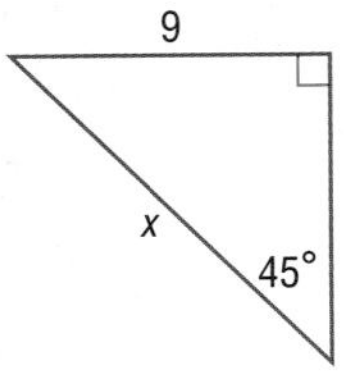

22.

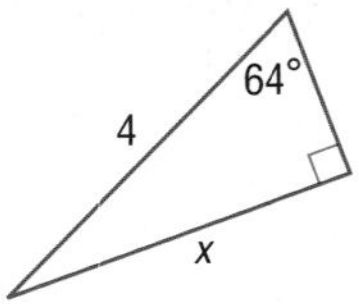

23

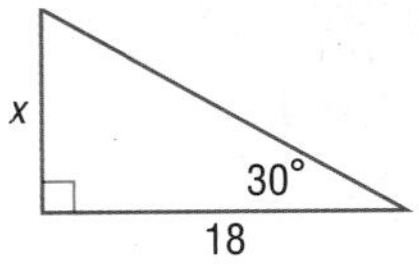

24.

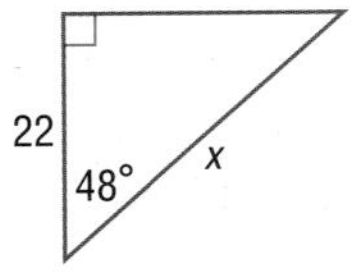

25.

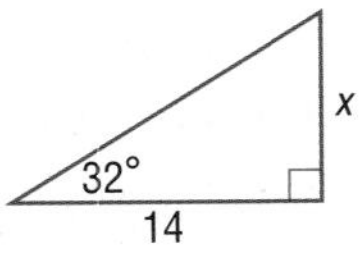

26.

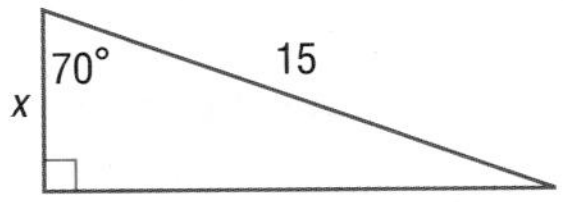

27. PARASAILING Refer to the beginning of the lesson and the figure at the right. Find a, the altitude of a person parasailing, if the tow rope is 250 feet long and the angle formed is 32°. Round to the nearest tenth.

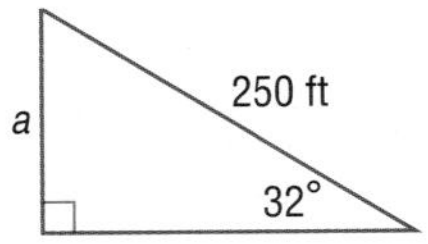

28. BRIDGES Devon wants to build a rope bridge between his treehouse and Cheng's treehouse. Suppose Devon's treehouse is directly behind Cheng's treehouse. At a distance of 20 meters to the left of Devon's treehouse, an angle of 52° is measured between the two treehouses. Find the length of the rope.

Example 5
p. 811

Find the value of x. Round to the nearest tenth.

29.

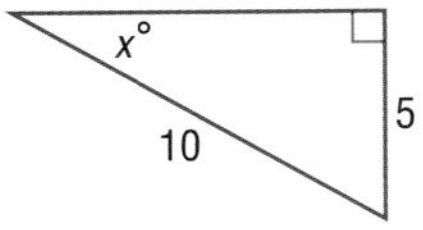

30.

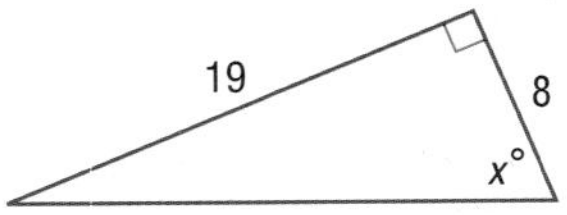

31.

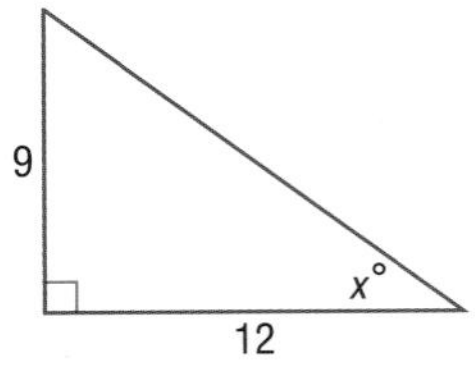

32.

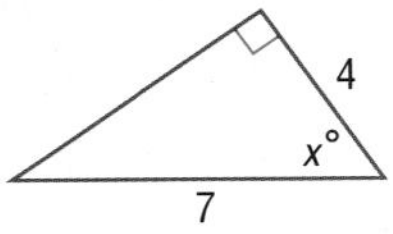

33.

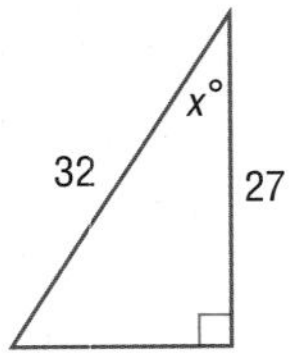

34.

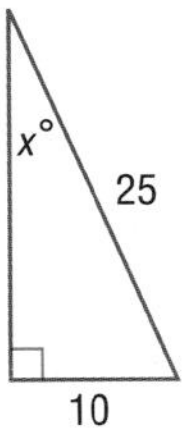

Example 6
p. 812

35. SQUIRRELS Adult flying squirrels can make glides of up to 160 feet. If a flying squirrel glides a horizontal distance of 160 feet and the angle of descent is 9°, find its change in height.

36. HANG GLIDING A hang glider climbs at a 20° angle of elevation. Find the change in altitude of the hang glider when it has flown a horizontal distance of 60 feet.

Use trigonometric functions to find the values of x and y. Round to the nearest tenth.

37.

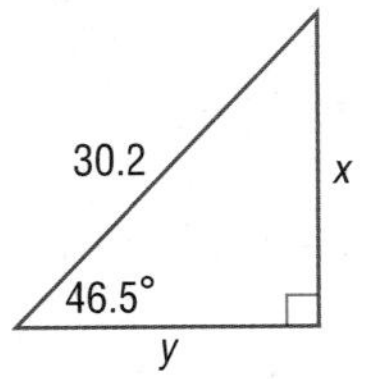

38.

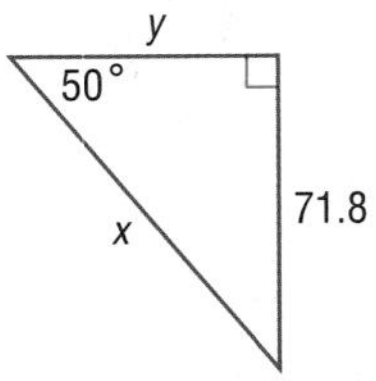

39.

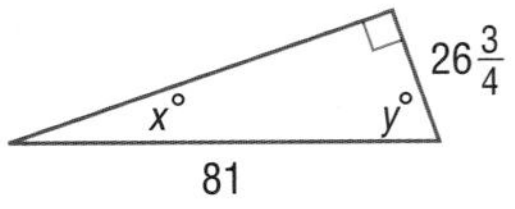

Solve each equation.

40. $\cos a = \frac{3}{19}$

41. $\sin n = \frac{9}{11}$

42. $\tan x = 15$

43. $\sin t = 0.35$

44. $\tan g = 0.125$

45. $\cos z = 0.98$

46. **MONUMENTS** A monument casts a shadow 24 feet long. The angle of elevation from the end of the shadow to the top of the monument is 50°.

a. Draw and label a right triangle to represent this situation.

b. Write a trigonometric function that can be used to find the height of the monument.

c. Find the value of the function to determine the height of the monument to the nearest tenth.

47. **NESTS** Tabitha's eyes are 5 feet above the ground as she looks up to a bird's nest in a tree. If the angle of elevation is 74.5° and she is standing 12 feet from the tree's base, what is the height of the bird's nest? Round to the nearest foot.

48. **RAMPS** Two bicycle ramps each cover a horizontal distance of 8 feet. One ramp has a 20° angle of elevation, and the other ramp has a 35° angle of elevation, as shown at the right.

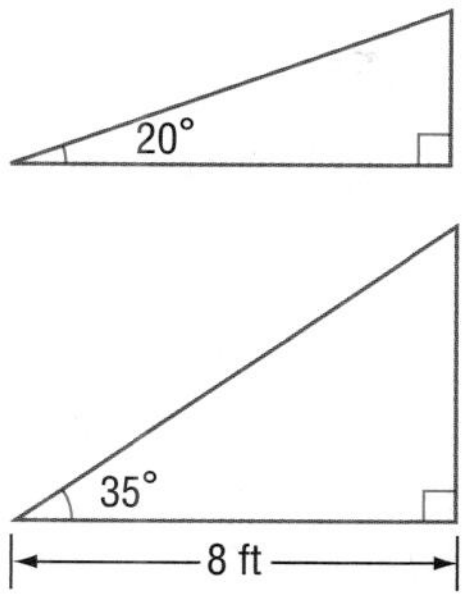

a. How much taller is the second ramp than the first? Round to the nearest tenth.

b. How much longer is the second ramp than the first? Round to the nearest tenth.

Real-World Link

A falcon can see an object 10 centimeters long from a distance of 1.5 kilometers away. A falcon can also see sharp images when diving at 100 miles per hour.

Source: Neuroscience for Kids

49. **FALCONS** A falcon at a height of 200 feet sees two mice A and B, as shown in the diagram.

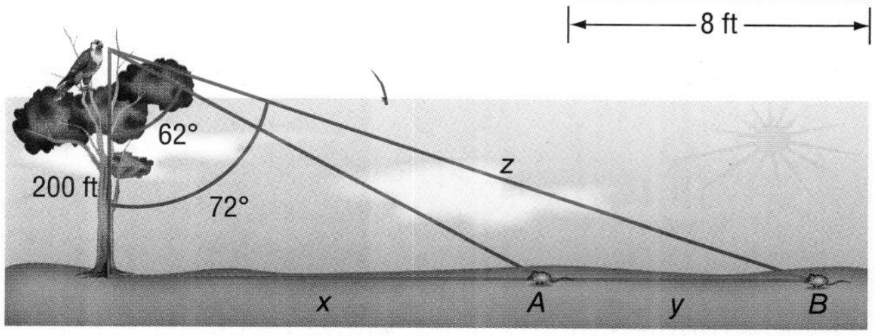

a. What is the approximate distance z between the falcon and mouse B?

b. How far apart are the two mice?

In $\triangle ABC$, $\angle C$ is a right angle. Use the given measurements to find the missing side lengths and missing angle measures of $\triangle ABC$. Round to the nearest tenth if necessary.

50. $m\angle A = 36°, a = 12$

51. $m\angle B = 31°, b = 19$

52. $a = 8, c = 17$

53. $\tan A = \frac{4}{5}, a = 6$

H.O.T. Problems

Use **H**igher-**O**rder **T**hinking Skills

54. **CHALLENGE** A line segment has endpoints $A(2, 0)$ and $B(6, 5)$, as shown in the figure at the right. What is the measure of the acute angle θ formed by the line segment and the x-axis? Explain how you found the measure.

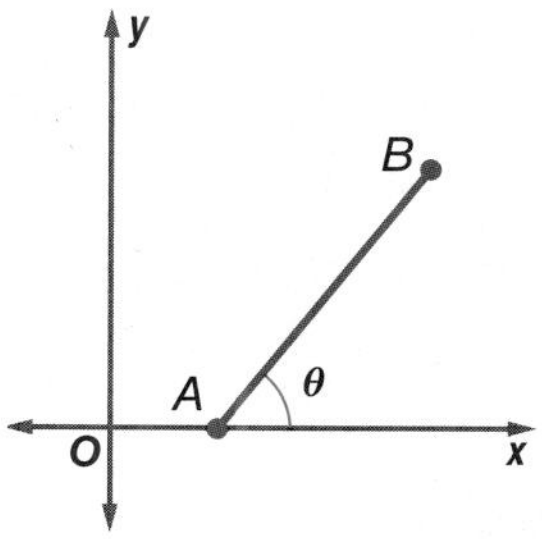

55. **REASONING** Determine whether the following statement is *true* or *false*. Explain your reasoning.

For any acute angle, the sine function will never have a negative value.

56. **OPEN ENDED** In right triangle ABC, $\sin A = \sin C$. What can you conclude about $\triangle ABC$? Justify your reasoning.

57. **WRITING IN MATH** A roof has a slope of $\frac{2}{3}$. Describe the connection between the slope and the angle of elevation θ that the roof makes with the horizontal. Then use an inverse trigonometric function to find θ.

NYSRE PRACTICE A2.A.55, A2.A.58

58. **EXTENDED RESPONSE** Your school needs 5 cases of yearbooks. Neighborhood Yearbooks lists a case of yearbooks at \$153.85 with a 10% discount on an order of 5 cases. Yearbooks R Us lists a case of yearbooks at \$157.36 with a 15% discount on 5 cases.

 a. Which company would you choose?

 b. What is the least amount that you would have to spend for the yearbooks?

59. **SHORT RESPONSE** As a fundraiser, the marching band sold T-shirts and hats. They sold a total of 105 items and raised \$1170. If the cost of a hat was \$10 and the cost of a T-shirt was \$15, how many T-shirts were sold?

60. A hot dog stand charges price x for a hot dog and price y for a drink. Two hot dogs and one drink cost \$4.50. Three hot dogs and two drinks cost \$7.25. Which matrix could be multiplied by $\begin{bmatrix} 4.50 \\ 7.25 \end{bmatrix}$ to find x and y?

 A $\begin{bmatrix} -1 & 1 \\ 2 & -1 \end{bmatrix}$ **C** $\begin{bmatrix} 1 & 2 \\ -1 & 3 \end{bmatrix}$

 B $\begin{bmatrix} 2 & -1 \\ -3 & 2 \end{bmatrix}$ **D** $\begin{bmatrix} 1 & -1 \\ -1 & 2 \end{bmatrix}$

61. **SAT/ACT** The length and width of a rectangle are in the ratio of 5:12. If the rectangle has an area of 240 square centimeters, what is the length, in centimeters, of its diagonal?

 F 26 **H** 30

 G 28 **J** 32

Spiral Review

62. **POLLS** A polling company wants to estimate how many people are in favor of a new environmental law. The polling company polls 20 people. The probability that a person is in favor of the law is 0.5. (Lesson 12-7)

 a. What is the probability that exactly 12 people are in favor of the new law?

 b. What is the expected number of people in favor of the law?

Text each null hypothesis. Write *accept* or *reject*. (Lesson 12-6)

63. $H_0 = 92$, $H_1 > 92$, $n = 80$, $\overline{x} = 92.75$, and $s = 2.8$

64. $H_0 = 48$, $H_1 > 48$, $n = 240$, $\overline{x} = 48.2$, and $s = 2.2$

65. $H_0 = 71$, $H_1 > 71$, $n = 180$, $\overline{x} = 72.4$, and $s = 3.5$

66. $H_0 = 55$, $H_1 < 55$, $n = 300$, $\overline{x} = 54.5$, and $s = 1.9$

Find each probability. (Lesson 12-4)

67. A city council consists of six Democrats, two of whom are women, and six Republicans, four of whom are men. A member is chosen at random. If the member chosen is a man, what is the probability that he is a Democrat?

68. Two boys and two girls are lined up at random. What is the probability that the girls are separated if a girl is on an end?

Skills Review

Find each product. Include the appropriate units with your answer. (Lesson 6-1B)

69. 4.3 miles$\left(\frac{5280\text{ feet}}{1\text{ mile}}\right)$

70. 8 gallons$\left(\frac{8\text{ pints}}{1\text{ gallon}}\right)$

71. $\left(\frac{5\text{ dollars}}{3\text{ meters}}\right)$21 meters

72. $\left(\frac{18\text{ cubic inches}}{5\text{ seconds}}\right)$24 seconds

73. 65 degrees$\left(\frac{10\text{ centimeters}}{3\text{ degrees}}\right)$

74. $\left(\frac{7\text{ liters}}{30\text{ minutes}}\right)$10 minutes

Angles and Angle Measure

Then

You used angles with degree measures. (Lesson 13-1)

Now

- Draw and find angles in standard position.
- Convert between degree measures and radian measures.

NYS Core Curriculum

A2.M.2 Convert between radian and degree measures
A2.A.61 Determine the length of an arc of a circle, given its radius and the measure of its central angle
Also addresses A2.M.1.

New Vocabulary

standard position
initial side
terminal side
coterminal angles
radian
central angle

NY Math Online

glencoe.com

- Extra Examples
- Personal Tutor
- Self-Check Quiz
- Homework Help
- Math in Motion

Why?

A sundial is an instrument that indicates the time of day by the shadow that it casts on a surface marked to show hours or fractions of hours. The shadow moves around the dial 15° every hour.

Angles in Standard Position An angle on the coordinate plane is in **standard position** if the vertex is at the origin and one ray is on the positive x-axis.

- The ray on the x-axis is called the **initial side** of the angle.
- The ray that rotates about the center is called the **terminal side**.

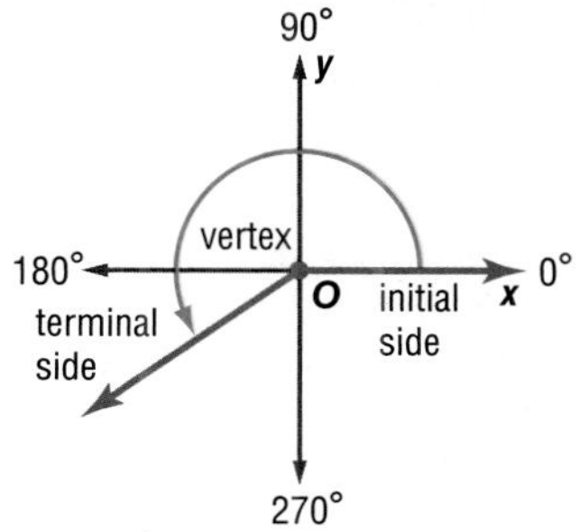

Key Concept — Angle Measures

For Your FOLDABLE

If the measure of an angle is positive, the terminal side is rotated counterclockwise.

If the measure of an angle is negative, the terminal side is rotated clockwise.

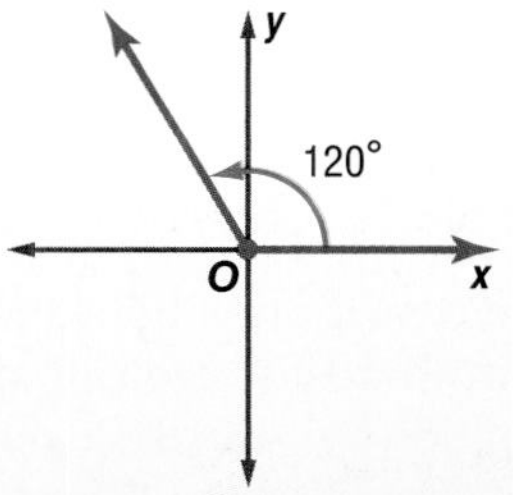

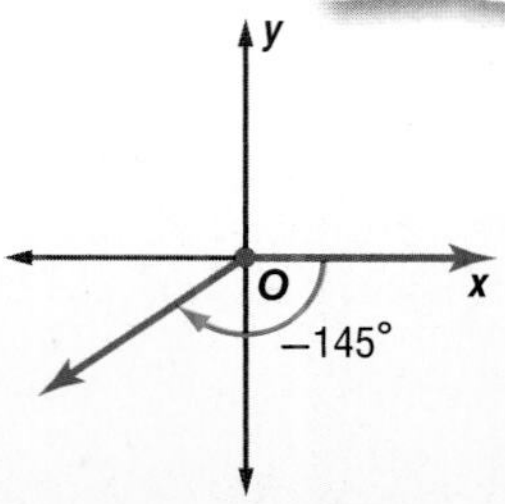

Math *in Motion*, Animation glencoe.com

EXAMPLE 1 Draw an Angle in Standard Position

Draw an angle with the given measure in standard position.

a. 215° $215° = 180° + 35°$

Draw the terminal side of the angle 35° counterclockwise past the negative x-axis.

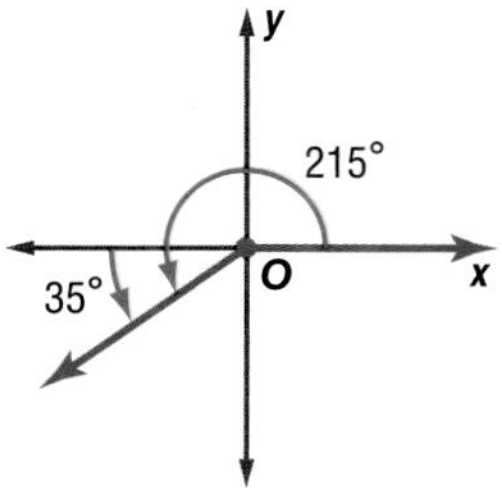

b. −40°

The angle is negative. Draw the terminal side of the angle 40° clockwise from the positive x-axis.

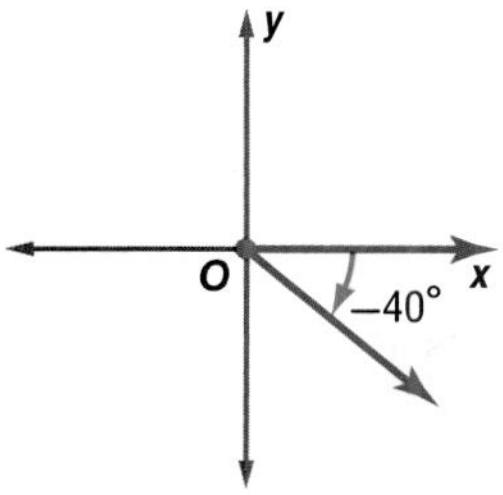

Check Your Progress

1A. 80°

1B. −105°

Personal Tutor glencoe.com

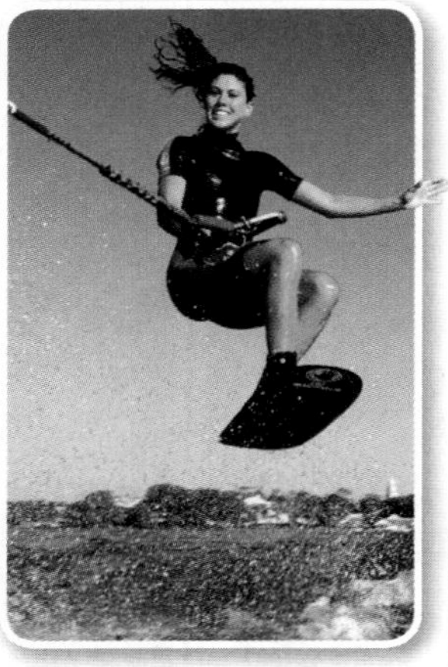

Real-World Link

Wakeboarding is one of the fastest-growing water sports in the United States. Participation increased more than 100% in recent years.

Source: King of Wake

The terminal side of an angle can make more than one complete rotation. For example, a complete rotation of 360° plus a rotation of 120° forms an angle that measures 360° + 120° or 480°.

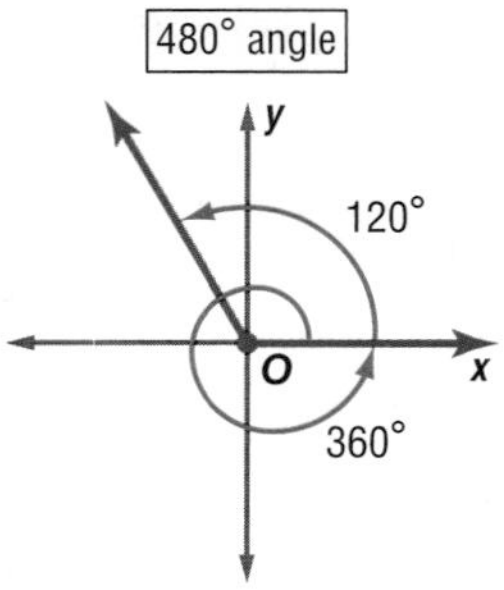

Real-World EXAMPLE 2 Draw an Angle in Standard Position

WAKEBOARDING ***Wakeboarding*** **is a combination of surfing, skateboarding, snowboarding, and water skiing. One maneuver involves a 540-degree rotation in the air. Draw an angle in standard position that measures 540°.**

540° = 360° + 180°

Draw the terminal side of the angle 180° past the positive x-axis.

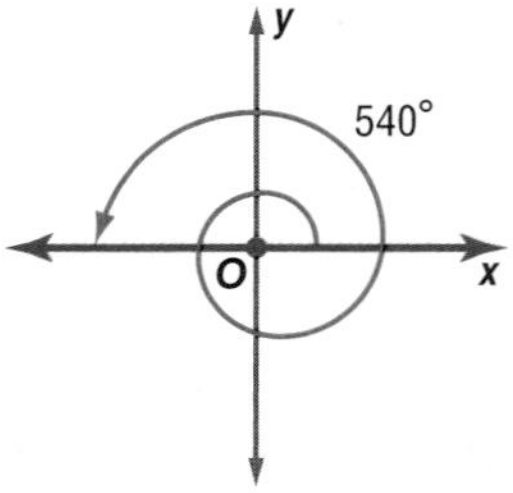

Check Your Progress

2. Draw an angle in standard position that measures 600°.

Personal Tutor glencoe.com

Two or more angles in standard position with the same terminal side are called **coterminal angles**. For example, angles that measure 60°, 420°, and −300° are coterminal, as shown in the figure at the right.

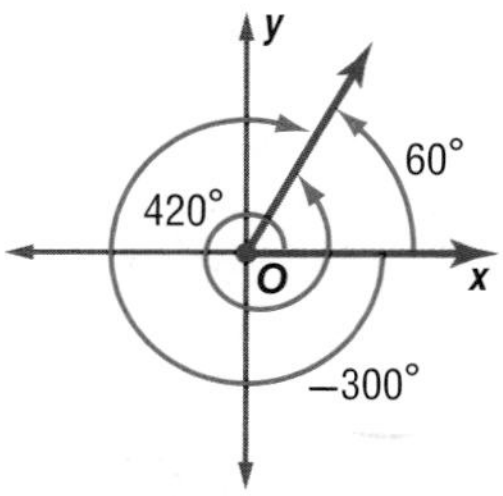

An angle that is coterminal with another angle can be found by adding or subtracting a multiple of 360°.

- 60° + 360° = 420°
- 60° − 360° = −300°

EXAMPLE 3 Find Coterminal Angles

Find an angle with a positive measure and an angle with a negative measure that are coterminal with each angle.

a. 130°

positive angle: 130° + 360° = 490° **Add 360°.**
negative angle: 130° − 360° = −230° **Subtract 360°.**

b. −200°

positive angle: −200° + 360° = 160° **Add 360°.**
negative angle: −200° − 360° = −560° **Subtract 360°.**

Check Your Progress

3A. 15° **3B.** −45°

Personal Tutor glencoe.com

ReadingMath

Angle of Rotation In trigonometry, an angle is sometimes referred to as an *angle of rotation.*

Convert Between Degrees and Radians Angles can also be measured in units that are based on arc length. One **radian** is the measure of an angle θ in standard position with a terminal side that intercepts an arc with the same length as the radius of the circle.

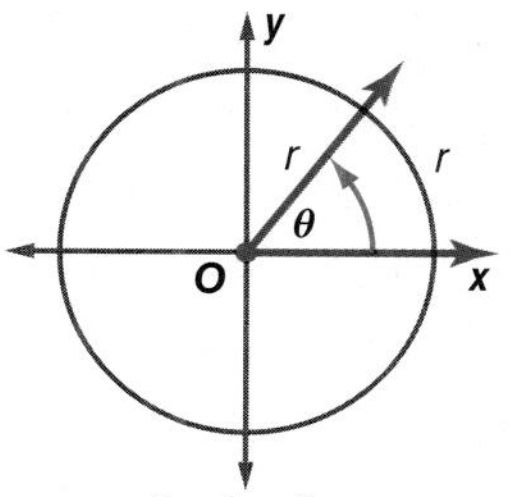

$\theta = 1$ radian

The circumference of a circle is $2\pi r$. So, one complete revolution around a circle equals 2π radians. Since 2π radians $= 360°$, degree measure and radian measure are related by the following equations.

$$2\pi \text{ radians} = 360° \qquad \pi \text{ radians} = 180°$$

StudyTip

Radians As with degrees, radians measure the amount of rotation from the initial side to the terminal side.

- The measure of an angle in radians is positive if its rotation is counterclockwise.
- The measure is negative if the rotation is clockwise.

Key Concept — Convert Between Degrees and Radians

For Your FOLDABLE

Degrees to Radians	Radians to Degrees
To convert from degrees to radians, multiply the number of degrees by $\frac{\pi \text{ radians}}{180°}$.	To convert from radians to degrees, multiply the number of radians by $\frac{180°}{\pi \text{ radians}}$.

Math *in Motion*, Interactive Lab glencoe.com

ReadingMath

Radian Measures The word *radian* is usually omitted when angles are expressed in radian measure. Thus, when no units are given for an angle measure, radian measure is implied.

EXAMPLE 4 Convert Between Degrees and Radians

Rewrite the degree measure in radians and the radian measure in degrees.

a. $-30°$

$$-30° = -30° \cdot \frac{\pi \text{ radians}}{180°}$$
$$= \frac{-30\pi}{180} \text{ or } -\frac{\pi}{6} \text{ radians}$$

b. $\frac{5\pi}{2}$

$$\frac{5\pi}{2} = \frac{5\pi}{2} \text{ radians} \cdot \frac{180°}{\pi \text{ radians}}$$
$$= \frac{900°}{2} \text{ or } 450°$$

Check Your Progress

4A. $120°$

4B. $-\frac{3\pi}{8}$

Personal Tutor glencoe.com

Concept Summary — Degrees and Radians

For Your FOLDABLE

The diagram shows equivalent degree and radian measures for special angles.

You may find it helpful to memorize the following equivalent degree and radian measures. The other special angles are multiples of these angles.

$30° = \frac{\pi}{6}$ $\qquad 45° = \frac{\pi}{4}$

$60° = \frac{\pi}{3}$ $\qquad 90° = \frac{\pi}{2}$

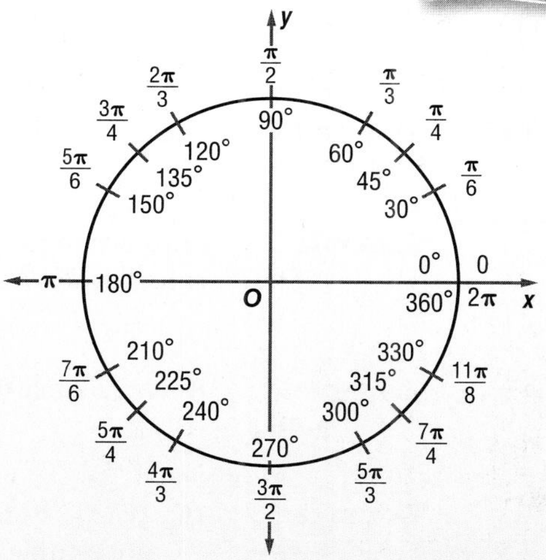

A **central angle** of a circle is an angle with a vertex at the center of the circle. If you know the measure of a central angle and the radius of the circle, you can find the length of the arc that is intercepted by the angle.

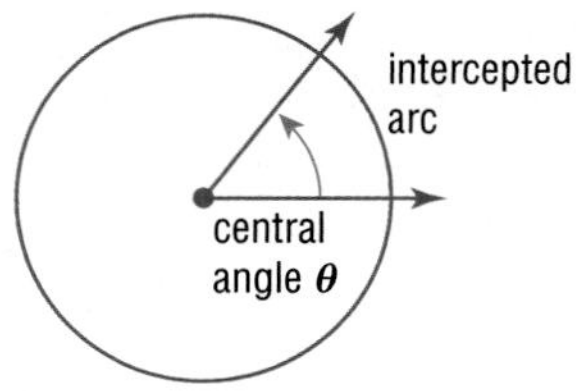

Key Concept — Arc Length

For Your FOLDABLE

Words For a circle with radius r and central angle θ (in radians), the arc length s equals the product of r and θ.

Symbols $s = r\theta$

Model

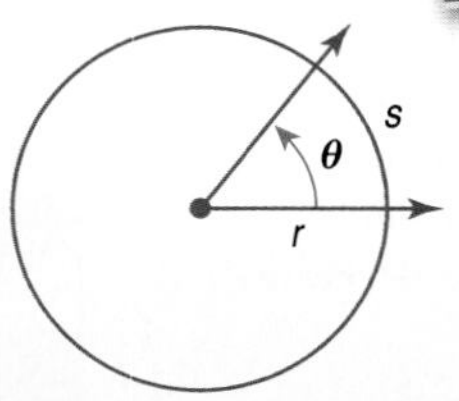

You will justify this formula in Exercise 52.

Real-World EXAMPLE 5 Find Arc Length

TRUCKS Monster truck tires have a radius of 33 inches. How far does a monster truck travel in feet after just three fourths of a tire rotation?

Step 1 Find the central angle in radians.

$\theta = \frac{3}{4} \cdot 2\pi$ or $\frac{3\pi}{2}$ — **The angle is $\frac{3}{4}$ of a complete rotation.**

Step 2 Use the radius and central angle to find the arc length.

$s = r\theta$	**Write the formula for arc length.**
$= 33 \cdot \frac{3\pi}{2}$	**Replace r with 33 and θ with $\frac{3\pi}{2}$.**
≈ 155.5 in.	**Use a calculator to simplify.**
≈ 13.0 ft	**Divide by 12 to convert to feet.**

So, the truck travels about 13 feet after three fourths of a tire rotation.

> **Watch Out!**
>
> **Arc Length** Remember to write the angle measure in radians, not degrees, when finding arc length. Also, recall that the number of radians in a complete rotation is 2π.

Check Your Progress

5. A circle has a diameter of 9 centimeters. Find the arc length if the central angle is 60°. Round to the nearest tenth.

Personal Tutor glencoe.com

Check Your Understanding

Examples 1 and 2 pp. 817–818

Draw an angle with the given measure in standard position.

1. 140° **2.** −60° **3.** 390°

Example 3 p. 818

Find an angle with a positive measure and an angle with a negative measure that are coterminal with each angle.

4. 25° **5.** 175° **6.** −100°

Example 4 p. 819

Rewrite each degree measure in radians and each radian measure in degrees.

7. $\frac{\pi}{4}$ **8.** 225° **9.** −40°

Example 5 p. 820

10. TENNIS A tennis player's swing moves along the path of an arc. If the radius of the arc's circle is 4 feet and the angle of rotation is 100°, what is the length of the arc? Round to the nearest tenth.

Practice and Problem Solving

● = **Step-by-Step Solutions** begin on page R20.
Extra Practice begins on page 947.

Examples 1 and 2 pp. 817–818

Draw an angle with the given measure in standard position.

11. 75° **12.** 160° **13.** −90°

14. −120° **15.** 295° **16.** 510°

17. GYMNASTICS A gymnast on the uneven bars swings to make a 240° angle of rotation.

18. FOOD The lid on a jar of pasta sauce is turned 420° before it comes off.

Example 3 p. 818

Find an angle with a positive measure and an angle with a negative measure that are coterminal with each angle.

19. 50° **20.** 95° **21.** 205°

22. 350° **23.** −80° **24.** −195°

Example 4 p. 819

Rewrite each degree measure in radians and each radian measure in degrees.

25 330° **26.** $\frac{5\pi}{6}$ **27.** $-\frac{\pi}{3}$

28. −50° **29.** 190° **30.** $-\frac{7\pi}{3}$

Example 5 p. 820

31. SKATEBOARDING The skateboard ramp at the right is called a *quarter pipe*. The curved surface is determined by the radius of a circle. Find the length of the curved part of the ramp.

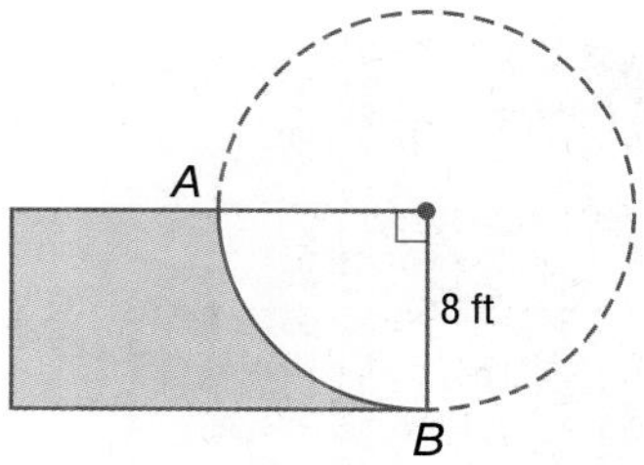

32. RIVERBOATS The paddlewheel of a riverboat has a diameter of 24 feet. Find the arc length of the circle made when the paddlewheel rotates 300°.

Find the length of each arc. Round to the nearest tenth.

33.

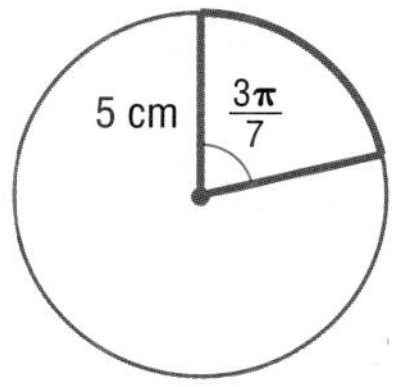

34.

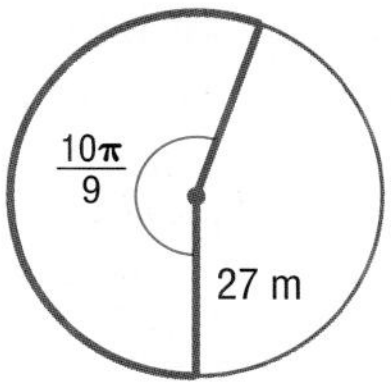

35. CLOCKS How long does it take for the minute hand on a clock to pass through 2.5π radians?

36. SUNDIALS Refer to the beginning of the lesson. A shadow moves around a sundial 15° every hour.

a. After how many hours is the angle of rotation of the shadow $\frac{8\pi}{5}$ radians?

b. What is the angle of rotation in radians after 5 hours?

c. A sundial has a radius of 8 inches. What is the arc formed by a shadow after 14 hours? Round to the nearest tenth.

Real-World Link

The Sundial Bridge at Turtle Bay in Redding, California, is the world's largest sundial. It is 700 feet long, 23 feet wide, and 217 feet high.

Source: Turtle Bay Exploration Park

Find an angle with a positive measure and an angle with a negative measure that are coterminal with each angle.

37. 620° **38.** −400° **39.** $-\frac{3\pi}{4}$ **40.** $\frac{19\pi}{6}$

41 **SWINGS** A swing has a 165° angle of rotation.

a. Draw the angle in standard position.

b. Write the angle measure in radians.

c. If the chains of the swing are $6\frac{1}{2}$ feet long, what is the length of the arc that the swing makes? Round to the nearest tenth.

d. Describe how the arc length would change if the lengths of the chains of the swing were doubled.

42. **MULTIPLE REPRESENTATIONS** Consider $A(-4, 0)$, $B(-4, 6)$, $C(6, 0)$, and $D(6, 8)$.

a. **GEOMETRIC** Draw $\triangle EAB$ and $\triangle ECD$ with E at the origin.

b. **ALGEBRAIC** Find the values of the tangent of $\angle BEA$ and the tangent of $\angle DEC$.

c. **ALGEBRAIC** Find the slope of $\overline{BE}$ and $\overline{ED}$.

d. **VERBAL** What conclusions can you make about the relationship between slope and tangent?

Rewrite each degree measure in radians and each radian measure in degrees.

43. $\frac{21\pi}{8}$ **44.** 124° **45.** −200° **46.** 5

47. **CAROUSELS** A carousel makes 5 revolutions per minute. The circle formed by riders sitting in the outside row has a radius of 17.2 feet. The circle formed by riders sitting in the inside row has a radius of 13.1 feet.

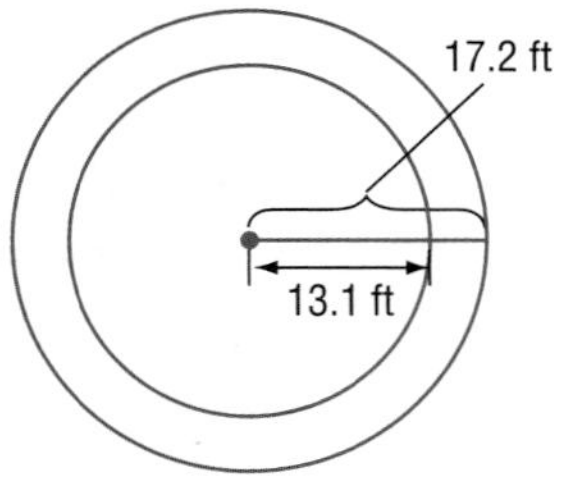

a. Find the angle θ in radians through which the carousel rotates in one second.

b. In one second, what is the difference in arc lengths between the riders sitting in the outside row and the riders sitting in the inside row?

Real-World Link

In the United States, carousels rotate counterclockwise, while carousels in Europe rotate clockwise.

Source: The Discovery Channel

H.O.T. Problems Use Higher-Order Thinking Skills

48. **FIND THE ERROR** Tarshia and Alan are writing an expression for the measure of an angle coterminal with the angle shown at the right. Is either of them correct? Explain your reasoning.

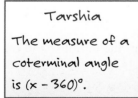

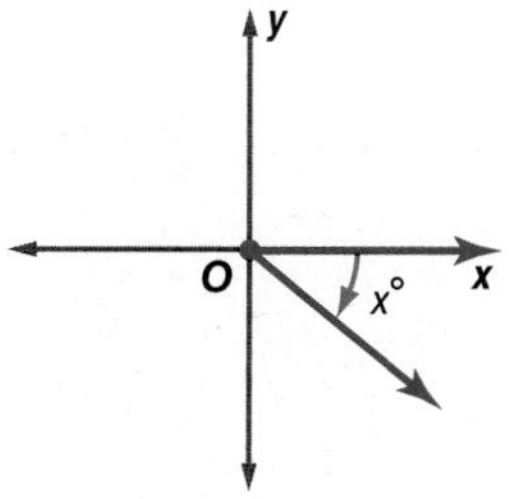

Tarshia	Alan
The measure of a coterminal angle is $(x - 360)°$.	The measure of a coterminal angle is $(360 - x)°$.

49. **CHALLENGE** A line makes an angle of $\frac{\pi}{2}$ radians with the positive x-axis at the point (2, 0). Find an equation for this line.

50. **REASONING** Express $\frac{1}{8}$ of a revolution in degrees and in radians. Explain your reasoning.

51. **OPEN ENDED** Draw and label an acute angle in standard position. Find two angles, one positive and one negative, that are coterminal with the angle.

52. **REASONING** Justify the formula for the length of an arc.

53. **WRITING IN MATH** Use a circle with radius r to describe what one degree and one radian represent. Then explain how to convert between the measures.

NYSRE PRACTICE A2.M.2, A2.A.61

54. SHORT RESPONSE If $(x + 6)(x + 8) - (x - 7)(x - 5) = 0$, find x.

55. Which of the following represents an inverse variation?

A

x	2	5	10	20	25	50
y	50	20	10	5	4	2

B

x	2	4	6	8	10	12
y	−4	−8	−12	−16	−20	−24

C

x	1	2	3	4	5	6
y	5	10	15	20	25	30

D

x	10	9	8	7	6	5
y	5	6	7	8	9	10

56. GEOMETRY If the area of the figure is 60 square units, what is the length of side $\overline{XZ}$?

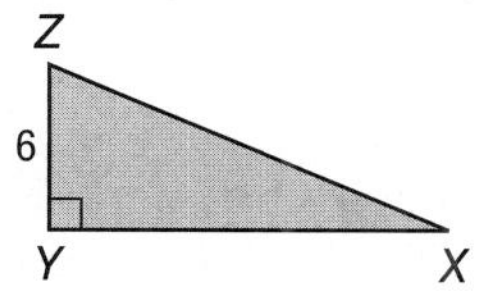

F $2\sqrt{34}$ H $4\sqrt{109}$
G $2\sqrt{109}$ J $4\sqrt{34}$

57. SAT/ACT The first term of a sequence is −6, and every term after the first is 8 more than the term immediately preceding it. What is the value of the 101st term?

A 788 C 802
B 794 D 808

Spiral Review

Find the values of the six trigonometric functions for angle θ. (Lesson 13-1)

58.

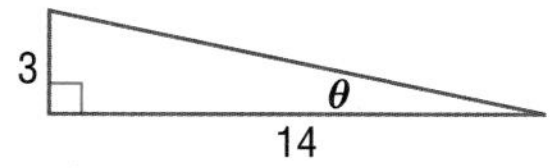

59.

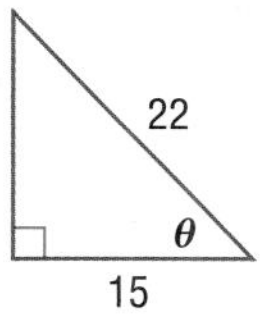

60.

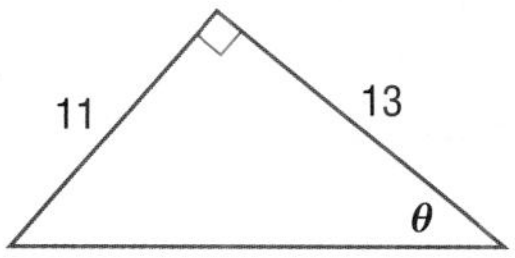

A binomial distribution has a 40% rate of success. There are 12 trials. (Lesson 12-7)

61. What is the probability that there will be at least 8 successes? exactly 5 failures?

62. What is the expected number of successes?

63. MANUFACTURING The sizes of CDs made by a company are normally distributed with a standard deviation of 1 millimeter. The CDs are supposed to be 120 millimeters in diameter, and they are made for drives that are 122 millimeters wide. (Lesson 12-5)

a. What percent of the CDs would you expect to be greater than 120 millimeters?

b. If the company manufactures 1000 CDs per hour, how many of the CDs made in one hour would you expect to be between 119 and 122 millimeters?

c. About how many CDs per hour will be too large to fit in the drives?

64. ECONOMICS If the rate of inflation is 2%, the cost of an item in future years can be found by iterating the function $c(x) = 1.02x$. Find the cost of a \$70 digital audio player in four years if the rate of inflation remains constant. (Lesson 11-5)

Skills Review

Use the Pythagorean Theorem to find the length of the hypotenuse for each right triangle with the given side lengths. (Lesson 0-7)

65. $a = 12, b = 15$ **66.** $a = 8, b = 17$ **67.** $a = 14, b = 11$

EXTEND 13-2

Geometry Lab
Areas of Parallelograms

NY Math Online glencoe.com
Math *in Motion*, Animation

NYS Core Curriculum **A2.A.74 Determine the area of a triangle or a parallelogram, given the measure of two sides and the included angle**

The area of any triangle can be found using the sine ratios in the triangle. A similar process can be used to find the area of a parallelogram.

ACTIVITY

Find the area of parallelogram *ABCD*.

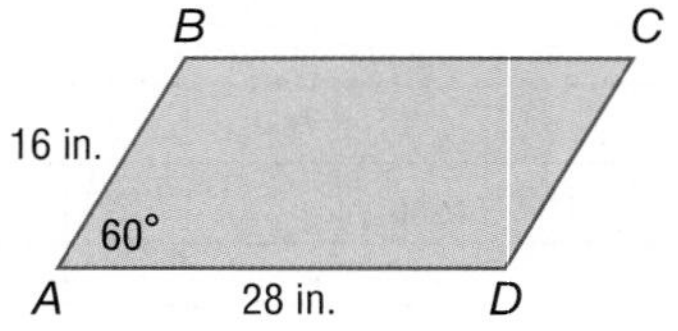

Step 1 Draw diagonal $\overline{BD}$.
$\overline{BD}$ divides the parallelogram into two congruent triangles, $\triangle ABD$ and $\triangle CDB$.

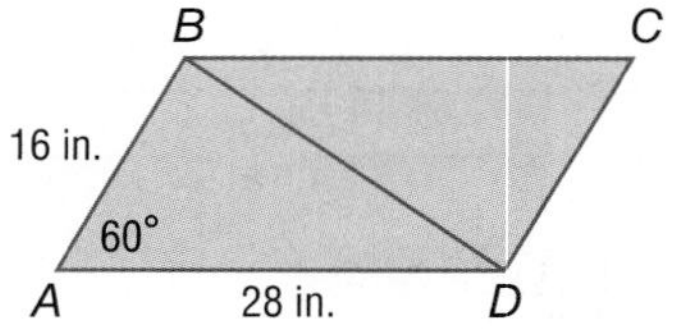

Step 2 Find the area of $\triangle ABD$.

$$\text{Area} = \frac{1}{2}(AB)(AD)\sin A \quad \textbf{Area of a triangle}$$

$$= \frac{1}{2}(16)(28)\sin 60° \quad AB = 16, AD = 28, \text{and } A = 60°$$

$$= 224\left(\frac{\sqrt{3}}{2}\right) \quad \textbf{Multiply and evaluate sin 60°.}$$

$$= 112\sqrt{3} \quad \textbf{Simplify.}$$

Step 3 Find the area of $\square ABCD$.

The area of $\square ABCD$ is equal to the sum of the areas of $\triangle ABD$ and $\triangle CDB$. Because $\triangle ABD \cong \triangle CDB$, the areas of $\triangle ABD$ and $\triangle CDB$ are equal. So, the area of $\square ABCD$ equals twice the area of $\triangle ABD$.

$2 \cdot 112\sqrt{3} = 224\sqrt{3}$ or about 387.98 square inches.

Exercises

For each of the following,

a. **find the area of each parallelogram.**

b. **find the area of each parallelogram when the included angle is half the given measure.**

c. **find the area of each parallelogram when the included angle is twice the given measure.**

1.

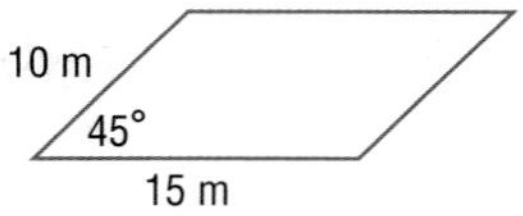

2.

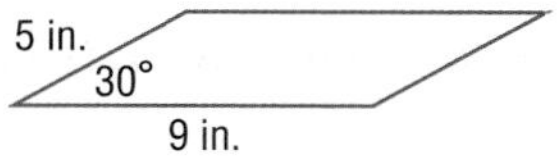

3.

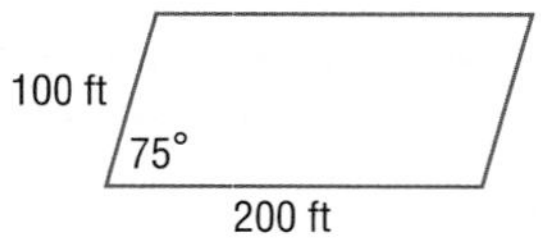

13-3 Trigonometric Functions of General Angles

Then

You found values of trigonometric functions for acute angles. (Lesson 13-1)

Now

- Find values of trigonometric functions for general angles.
- Find values of trigonometric functions by using reference angles.

NYS Core Curriculum

A2.A.57 Sketch and use the reference angle for angles in standard position A2.A.62 Find the value of trigonometric functions, if given a point on the terminal side of angle *Also addresses A2.A.56, A2.A.59, and A2.A.66.*

New Vocabulary

quadrantal angle

reference angle

NY Math Online

glencoe.com

- Extra Examples
- Personal Tutor
- Self-Check Quiz
- Homework Help
- Math in Motion

Why?

In the ride at the right, the cars rotate back and forth about a central point. The current positions of two of the arms can be described as 220° clockwise and 50° counterclockwise from the standard position.

Trigonometric Functions for General Angles You can find values of trigonometric functions for angles greater than 90° or less than 0°.

Key Concept — For Your FOLDABLE

Trigonometric Functions of General Angles

Let θ be an angle in standard position and let $P(x, y)$ be a point on its terminal side. Using the Pythagorean Theorem, $r = \sqrt{x^2 + y^2}$. The six trigonometric functions of θ are defined below.

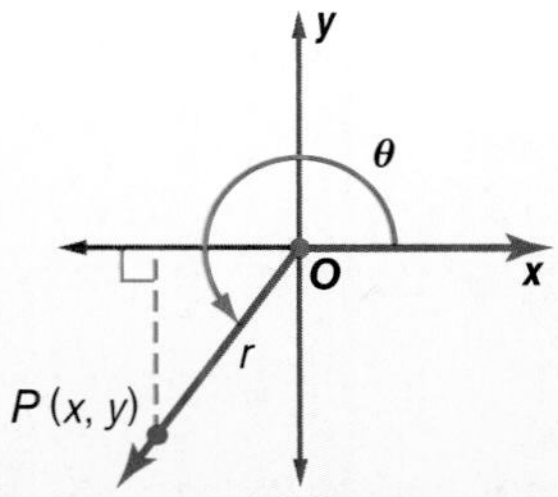

$\sin \theta = \frac{y}{r}$ $\quad$ $\cos \theta = \frac{x}{r}$ $\quad$ $\tan \theta = \frac{y}{x}, x \neq 0$

$\csc \theta = \frac{r}{y}, y \neq 0$ $\quad$ $\sec \theta = \frac{r}{x}, x \neq 0$ $\quad$ $\cot \theta = \frac{x}{y}, y \neq 0$

EXAMPLE 1 Evaluate Trigonometric Functions Given a Point

The terminal side of θ in standard position contains the point at (−3, −4). Find the exact values of the six trigonometric functions of θ.

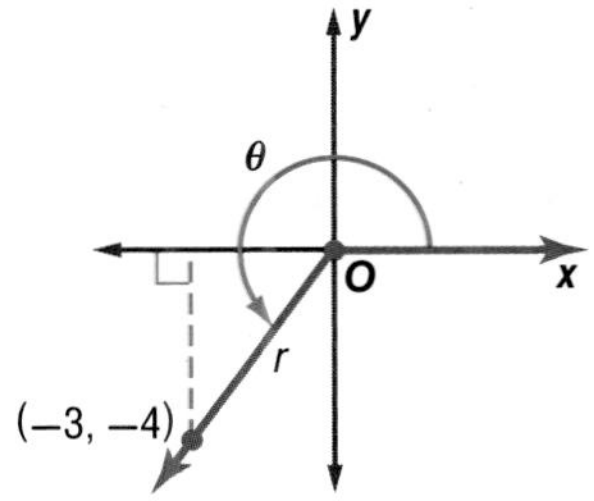

Step 1 Draw the angle, and find the value of r.

$$r = \sqrt{x^2 + y^2}$$
$$= \sqrt{(-3)^2 + (-4)^2}$$
$$= \sqrt{25} \text{ or } 5$$

Step 2 Use $x = -3$, $y = -4$, and $r = 5$ to write the six trigonometric ratios.

$\sin \theta = \frac{y}{r} = \frac{-4}{5}$ or $-\frac{4}{5}$ $\quad$ $\cos \theta = \frac{x}{r} = \frac{-3}{5}$ or $-\frac{3}{5}$ $\quad$ $\tan \theta = \frac{y}{x} = \frac{-4}{-3}$ or $\frac{4}{3}$

$\csc \theta = \frac{r}{y} = \frac{5}{-4}$ or $-\frac{5}{4}$ $\quad$ $\sec \theta = \frac{r}{x} = \frac{5}{-3}$ or $-\frac{5}{3}$ $\quad$ $\cot \theta = \frac{x}{y} = \frac{-3}{-4}$ or $\frac{3}{4}$

Check Your Progress

1. The terminal side of θ in standard position contains the point at (−6, 2). Find the exact values of the six trigonometric functions of θ.

Personal Tutor glencoe.com

If the terminal side of angle θ in standard position lies on the x- or y-axis, the angle is called a **quadrantal angle**.

StudyTip

Quadrantal Angles The measure of a quadrantal angle is a multiple of 90° or $\frac{\pi}{2}$.

Key Concept Quadrantal Angles

For Your FOLDABLE

$\theta = 0°$ or 0 radians	$\theta = 90°$ or $\frac{\pi}{2}$ radians	$\theta = 180°$ or π radians	$\theta = 270°$ or $\frac{3\pi}{2}$ radians
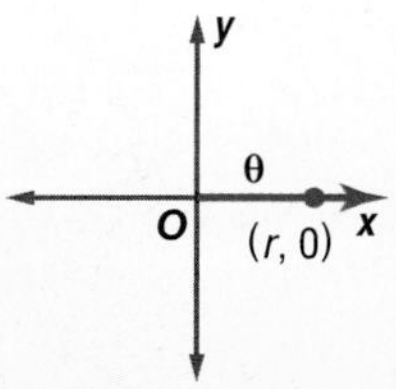	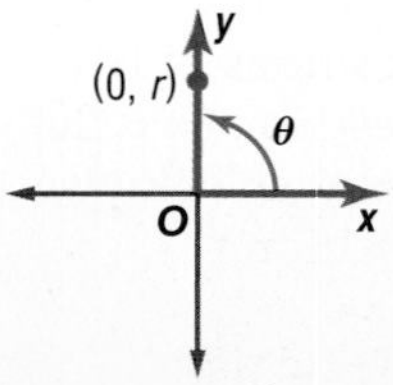	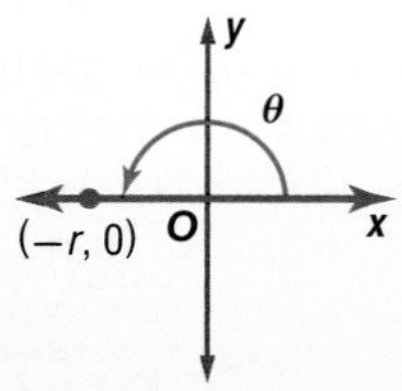	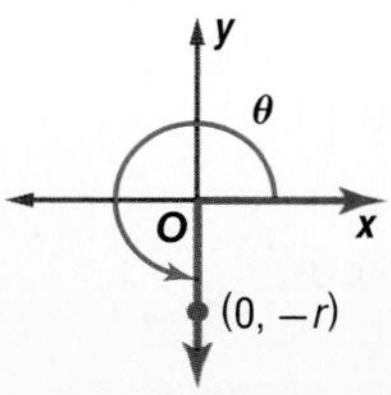

EXAMPLE 2 Quadrantal Angles

The terminal side of θ in standard position contains the point at (0, 6). Find the values of the six trigonometric functions of θ.

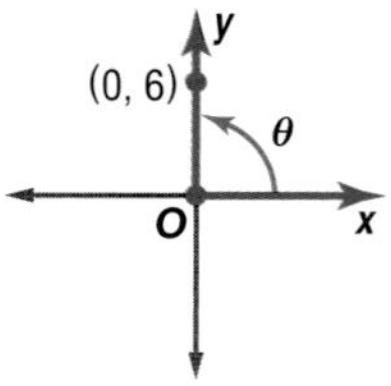

The point at (0, 6) lies on the positive y-axis, so the quadrantal angle θ is 90°. Use $x = 0$, $y = 6$, and $r = 6$ to write the trigonometric functions.

$\sin\theta = \frac{y}{r} = \frac{6}{6}$ or 1 $\quad \cos\theta = \frac{x}{r} = \frac{0}{6}$ or 0 $\quad \tan\theta = \frac{y}{x} = \frac{6}{0}$ undefined

$\csc\theta = \frac{r}{y} = \frac{6}{6}$ or 1 $\quad \sec\theta = \frac{r}{x} = \frac{6}{0}$ undefined $\quad \cot\theta = \frac{x}{y} = \frac{0}{6}$ or 0

Check Your Progress

2. The terminal side of θ in standard position contains the point at (−2, 0). Find the values of the six trigonometric functions of θ.

Personal Tutor glencoe.com

Trigonometric Functions with Reference Angles If θ is a nonquadrantal angle in standard position, its **reference angle** θ' is the acute angle formed by the terminal side of θ and the x-axis. The rules for finding the measures of reference angles for $0° < \theta < 360°$ or $0° < \theta < 2\pi$ are shown below.

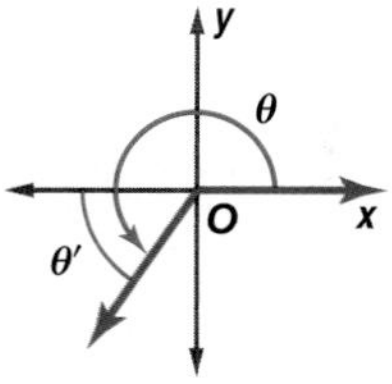

ReadingMath

Theta Prime θ' is read *theta prime*.

Math in Motion, Animation glencoe.com

Key Concept Reference Angles

For Your FOLDABLE

Quadrant I	Quadrant II	Quadrant III	Quadrant IV
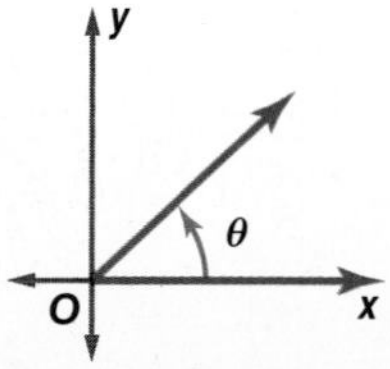	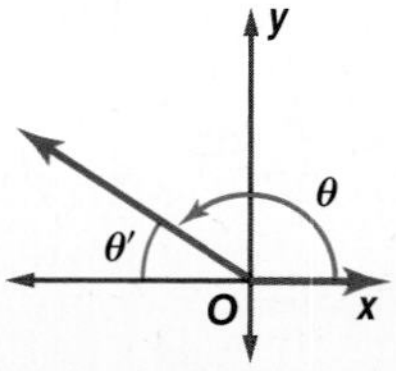	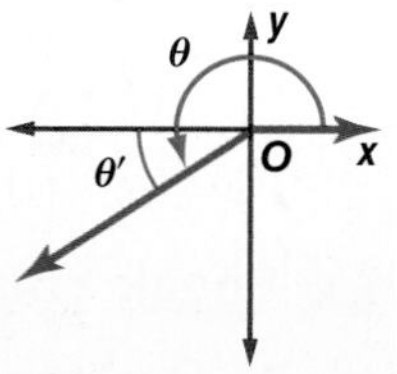	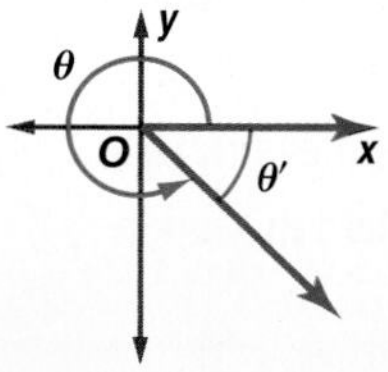
$\theta' = \theta$	$\theta' = 180° - \theta$ $\theta' = \pi - \theta$	$\theta' = \theta - 180°$ $\theta' = \theta - \pi$	$\theta' = 360° - \theta$ $\theta' = 2\pi - \theta$

If the measure of θ is greater than 360° or less than 0°, then use a coterminal angle with a positive measure between 0° and 360° to find the reference angle.

EXAMPLE 3 Find Reference Angles

Sketch each angle. Then find its reference angle.

a. 210°

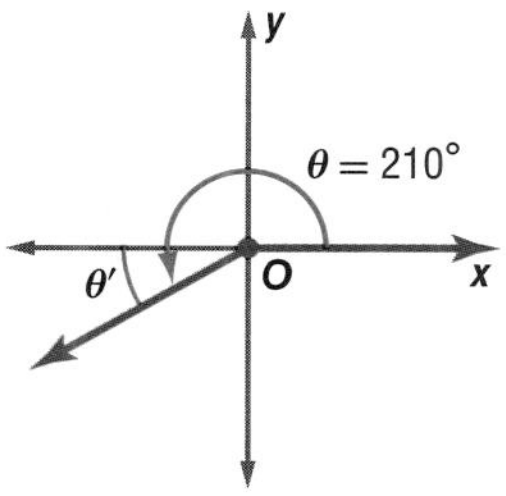

The terminal side of 210° lies in Quadrant III.

$\theta' = \theta - 180°$

$= 210° - 180°$ or $30°$

b. $-\frac{5\pi}{4}$

coterminal angle: $-\frac{5\pi}{4} + 2\pi = \frac{3\pi}{4}$

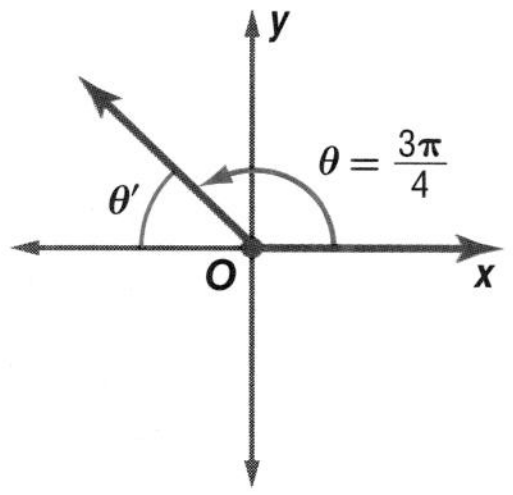

The terminal side of $\frac{3\pi}{4}$ lies in Quadrant III.

$\theta' = \pi - \theta$

$= \pi - \frac{3\pi}{4}$ or $\frac{\pi}{4}$

Check Your Progress

3A. −110°

3B. $\frac{2\pi}{3}$

Personal Tutor glencoe.com

> **StudyTip**
>
> **Graphing Angles** You can refer to the diagram in the Lesson 13-2 Concept Summary to help you sketch angles.

You can use reference angles to evaluate trigonometric functions for any angle θ. The sign of a function is determined by the quadrant in which the terminal side of θ lies. Use these steps to evaluate a trigonometric function for any angle θ.

Key Concept Evaluate Trigonometric Functions

For Your FOLDABLE

Step 1 Find the measure of the reference angle θ'.

Step 2 Evaluate the trigonometric function for θ'.

Step 3 Determine the sign of the trigonometric function value. Use the quadrant in which the terminal side of θ lies.

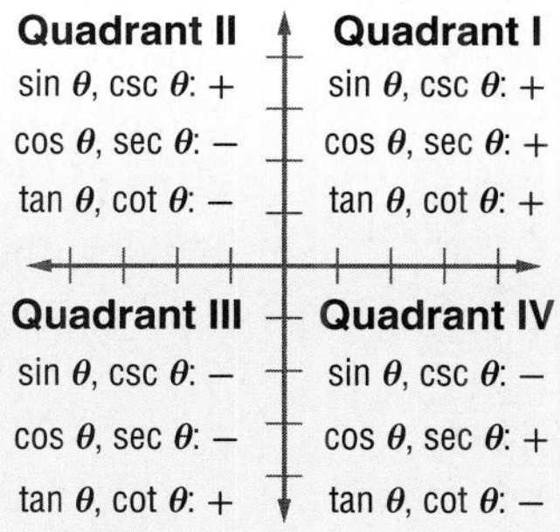

You can use the trigonometric values of angles measuring 30°, 45°, and 60° that you learned in Lesson 13-1.

Trigonometric Values for Special Angles

Sine	Cosine	Tangent	Cosecant	Secant	Cotangent
$\sin 30° = \frac{1}{2}$	$\cos 30° = \frac{\sqrt{3}}{2}$	$\tan 30° = \frac{\sqrt{3}}{3}$	$\csc 30° = 2$	$\sec 30° = \frac{2\sqrt{3}}{3}$	$\cot 30° = \sqrt{3}$
$\sin 45° = \frac{\sqrt{2}}{2}$	$\cos 45° = \frac{\sqrt{2}}{2}$	$\tan 45° = 1$	$\csc 45° = \sqrt{2}$	$\sec 45° = \sqrt{2}$	$\cot 45° = 1$
$\sin 60° = \frac{\sqrt{3}}{2}$	$\cos 60° = \frac{1}{2}$	$\tan 60° = \sqrt{3}$	$\csc 60° = \frac{2\sqrt{3}}{3}$	$\sec 60° = 2$	$\cot 60° = \frac{\sqrt{3}}{3}$

EXAMPLE 4 Use a Reference Angle to Find a Trigonometric Value

Find the exact value of each trigonometric function.

a. **cos 240°**

The terminal side of 240° lies in Quadrant III.

$\theta' = \theta - 180°$ — **Find the measure of the reference angle.**

$= 240° - 180°$ or $60°$ — $\theta = 240°$

$\cos 240° = -\cos 60°$ or $-\frac{1}{2}$ — **The cosine function is negative in Quadrant III.**

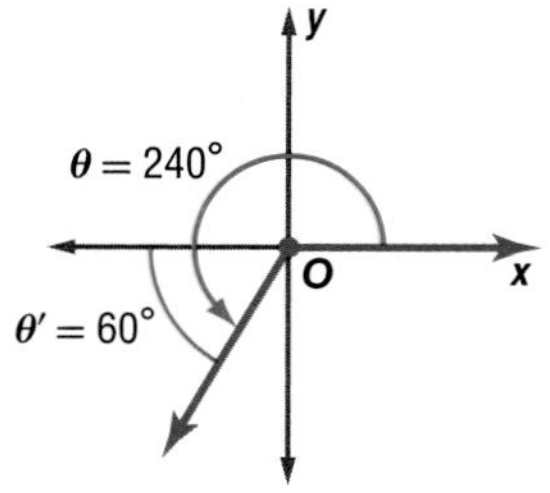

b. $\csc \frac{5\pi}{6}$

The terminal side of $\frac{5\pi}{6}$ lies in Quadrant II.

$\theta' = \pi - \theta$ — **Find the measure of the reference angle.**

$= \pi - \frac{5\pi}{6}$ or $\frac{\pi}{6}$ — $\theta = \frac{5\pi}{6}$

$\csc \frac{5\pi}{6} = \csc \frac{\pi}{6}$ — **The cosecant function is positive in Quadrant II.**

$= \csc 30°$ — $\frac{\pi}{6}$ **radians** $= 30°$

$= 2$ — $\csc 30° = \frac{1}{\sin 30}$

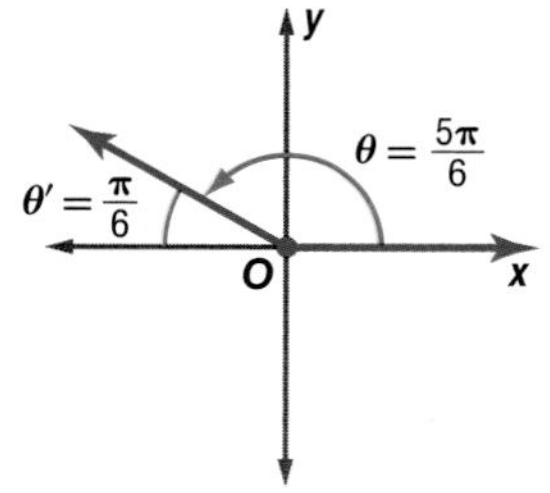

Check Your Progress

4A. cos 135°

4B. $\tan \frac{5\pi}{6}$

Personal Tutor glencoe.com

Real-World Link

On a swing ride, riders experience weightlessness just like the drop side of a roller coaster. The ride lasts one minute and reaches speeds of 60 miles per hour in both directions.

Source: Cedar Point

Real-World EXAMPLE 5 Use Trigonometric Functions

RIDES The swing arms of the ride at the right are 84 feet long and the height of the axis from which the arms swing is 97 feet. What is the total height of the ride at the peak of the arc?

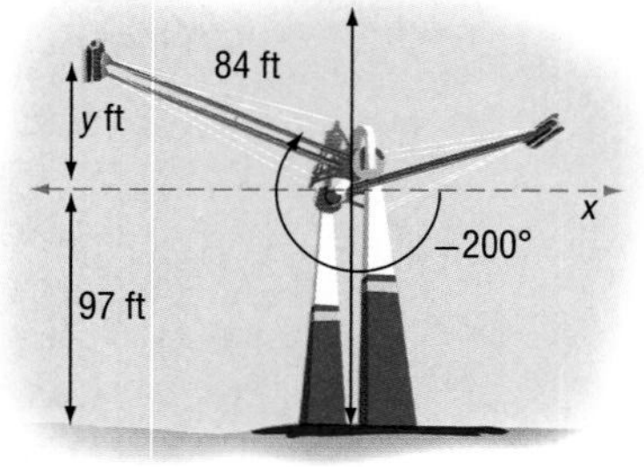

coterminal angle: $-200° + 360° = 160°$

reference angle: $180° - 160° = 20°$

$\sin \theta = \frac{y}{r}$ — **Sine function**

$\sin 20° = \frac{y}{84}$ — $\theta = 20°$ **and** $r = 84$

$84 \sin 20° = y$ — **Multiply each side by 84.**

$28.7 \approx y$ — **Use a calculator to solve for** y**.**

Since y is approximately 28.7 feet, the total height of the ride at its peak is $28.7 + 97$ or about 125.7 feet.

Check Your Progress

5. **RIDES** A similar ride that is smaller has swing arms that are 72 feet long. The height of the axis from which the arms swing is 88 feet, and the angle of rotation from the standard position is −195°. What is the total height of the ride at the peak of the arc?

Personal Tutor glencoe.com

Check Your Understanding

Examples 1 and 2 pp. 825–826

The terminal side of θ in standard position contains each point. Find the exact values of the six trigonometric functions of θ.

1. $(1, 2)$ **2.** $(-8, -15)$ **3.** $(0, -4)$

Example 3 p. 827

Sketch each angle. Then find its reference angle.

4. $300°$ **5.** $115°$ **6.** $-\frac{3\pi}{4}$

Example 4 p. 828

Find the exact value of each trigonometric function.

7. $\sin \frac{3\pi}{4}$ **8.** $\tan \frac{5\pi}{3}$ **9.** $\sec 120°$ **10.** $\sin 300°$

Example 5 p. 828

11. **ENTERTAINMENT** Alejandra opens her portable DVD player so that it forms a $125°$ angle. The screen is $5\frac{1}{2}$ inches long.

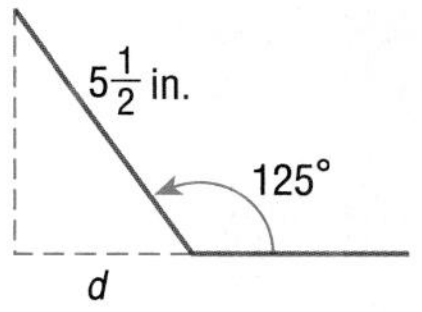

a. Redraw the diagram so that the angle is in standard position on the coordinate plane.

b. Find the reference angle. Then write a trigonometric function that can be used to find the distance to the wall d that she can place the DVD player.

c. Use the function to find the distance. Round to the nearest tenth.

Practice and Problem Solving

= **Step-by-Step Solutions** begin on page R20. **Extra Practice** begins on page 947.

Examples 1 and 2 pp. 825–826

The terminal side of θ in standard position contains each point. Find the exact values of the six trigonometric functions of θ.

12. $(5, 12)$ **13.** $(-6, 8)$ **14.** $(3, 0)$

15. $(0, -7)$ **16.** $(4, -2)$ **17.** $(-9, -3)$

Example 3 p. 827

Sketch each angle. Then find its reference angle.

18. $195°$ **19.** $285°$ **20.** $-250°$

21. $\frac{7\pi}{4}$ **22.** $-\frac{\pi}{4}$ **23.** $400°$

Example 4 p. 828

Find the exact value of each trigonometric function.

24. $\sin 210°$ **25.** $\tan 315°$ **26.** $\cos 150°$ **27.** $\csc 225°$

28. $\sin \frac{4\pi}{3}$ **29.** $\cos \frac{5\pi}{3}$ **30.** $\cot \frac{5\pi}{4}$ **31.** $\sec \frac{11\pi}{6}$

Example 5 p. 828

32. **SOCCER** A soccer player x feet from the goalie kicks the ball toward the goal, as shown in the figure. The goalie jumps up and catches the ball 7 feet in the air.

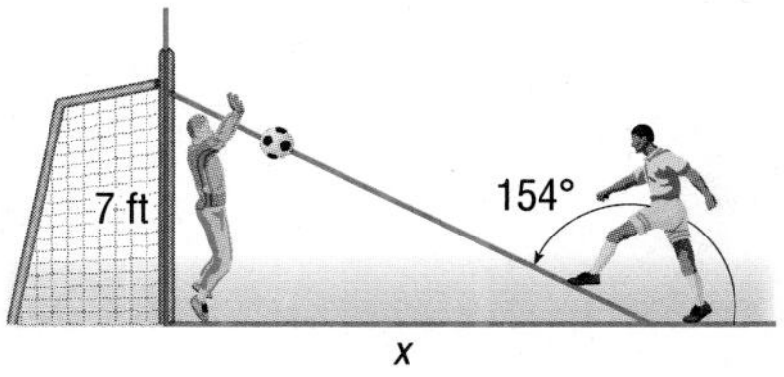

a. Find the reference angle. Then write a trigonometric function that can be used to find how far from the goalie the soccer player was when he kicked the ball.

b. About how far away from the goalie was the soccer player?

33. **SPRINKLER** A sprinkler rotating back and forth shoots water out a distance of 10 feet. From the horizontal position, it rotates 145° before reversing its direction. At a 145° angle, about how far from the sprinkler does the water reach?

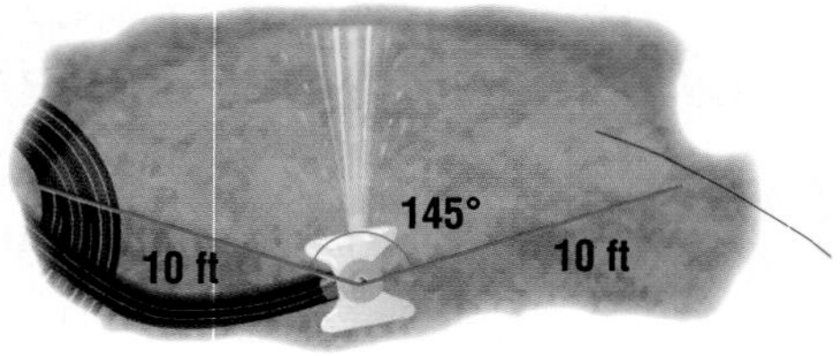

34. **BASKETBALL** The formula $R = \frac{V_0^2 \sin 2\theta}{32}$ gives the distance of a basketball shot with an initial velocity of V_0 feet per second at an angle θ with the ground.

   a. If the basketball was shot with an initial velocity of 24 feet per second at an angle of 75°, how far will the basketball travel?

   b. If the basketball was shot at an angle of 65° and traveled 10 feet, what was its initial velocity?

   c. If the basketball was shot with an initial velocity of 30 feet per second and traveled 12 feet, at what angle was it shot?

35. **PHYSICS** A rock is shot off the edge of a ravine with a slingshot at an angle of 65° and with an initial velocity of 6 meters per second. The equation that represents the horizontal distance of the rock x is $x = v_0 (\cos \theta)t$, where v_0 is the initial velocity, θ is the angle at which it is shot, and t is the time in seconds. About how far does the rock travel after 4 seconds?

36. **FERRIS WHEELS** The Wonder Wheel Ferris wheel at Coney Island has a radius of about 68 feet and is 15 feet off the ground. After a person gets on the bottom car, the Ferris wheel rotates 202.5° counterclockwise before stopping. How high above the ground is this car when it has stopped?

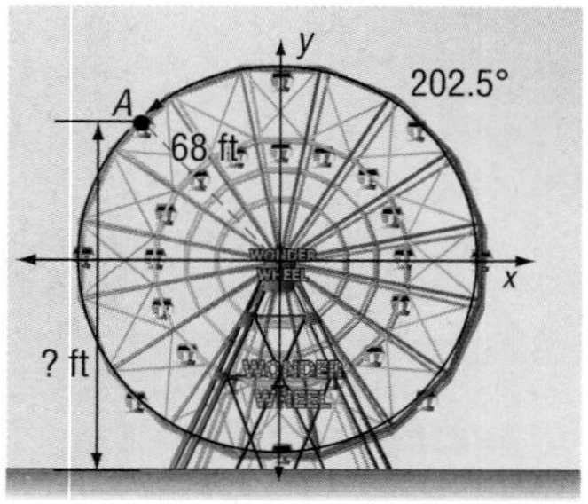

Real-World Link

The Wonder Wheel, which holds 144 people, is different from most Ferris wheels. It has eight stationary cars and sixteen cars that roll along tracks within the wheel's structure.

Source: Amusement Park History

Suppose θ is an angle in standard position whose terminal side is in the given quadrant. For each function, find the exact values of the remaining five trigonometric functions of θ.

37. $\sin \theta = \frac{4}{5}$, Quadrant II

38. $\tan \theta = -\frac{2}{3}$, Quadrant IV

39. $\cos \theta = -\frac{8}{17}$, Quadrant III

40. $\cot \theta = -\frac{12}{5}$, Quadrant IV

Find the exact value of each trigonometric function.

41. $\cot 270°$

42. $\csc 180°$

43. $\sin 570°$

44. $\tan \left(-\frac{7\pi}{6}\right)$

45. $\cos \left(-\frac{11\pi}{6}\right)$

46. $\cot \frac{9\pi}{4}$

H.O.T. Problems
Use Higher-Order Thinking Skills

47. **CHALLENGE** For an angle θ in standard position, $\sin \theta = \frac{\sqrt{2}}{2}$ and $\tan \theta = -1$. Can the value of θ be 225°? Justify your reasoning.

48. **REASONING** Determine whether $3 \sin 60° = \sin 180°$ is *true* or *false*. Explain your reasoning.

49. **REASONING** Use the sine and cosine functions to explain why cot 180° is undefined.

50. **OPEN ENDED** Give an example of a negative angle θ for which $\sin \theta > 0$ and $\cos \theta < 0$.

51. **WRITING IN MATH** Describe the steps for evaluating a trigonometric function for an angle θ that is greater than 90°. Include a description of a reference angle.

NYSRE PRACTICE A2.A.57, A2.A.62

52. SHORT RESPONSE If the sum of two numbers is 21 and their difference is 3, what is their product?

53. GEOMETRY D is the midpoint of $\overline{BC}$, and A and E are the midpoints of $\overline{BD}$ and $\overline{DC}$, respectively. If the length of $\overline{AE}$ is 12, what is the length of $\overline{BC}$?

A 6
B 12
C 24
D 48

54. The expression $(-6 + i)^2$ is equivalent to which of the following expressions?

F $-12i$
G $36 - i$
H $36 - 12i$
J $35 - 12i$

55. SAT/ACT Of the following, which is least?

A $1 + \frac{1}{4}$
B $1 - \frac{1}{4}$
C $\frac{1}{4} - 1$
D $1 \times \frac{1}{4}$

Spiral Review

Rewrite each radian measure in degrees. (Lesson 13-2)

56. $\frac{4}{3}\pi$ **57.** $\frac{11}{6}\pi$ **58.** $-\frac{17}{4}\pi$

Solve each equation. (Lesson 13-1)

59. $\cos a = \frac{13}{17}$ **60.** $\sin 30 = \frac{b}{6}$ **61.** $\tan c = \frac{9}{4}$

62. ARCHITECTURE A memorial being constructed in a city park will be a brick wall, with a top row of six gold-plated bricks engraved with the names of six local war veterans. Each row has two more bricks than the row above it. Prove that the number of bricks in the top n rows is $n^2 + 5n$. (Lesson 11-7)

63. LEGENDS There is a legend of a king who wanted to reward a boy for a good deed. The king gave the boy a choice. He could have \$1,000,000 at once, or he could be rewarded daily for a 30-day month, with one penny on the first day, two pennies on the second day, and so on, receiving twice as many pennies each day as the previous day. How much would the second option be worth? (Lesson 11-3)

Write an equation for each circle given the endpoints of a diameter. (Lesson 10-3)

64. (2, −4), (10, 2) **65.** (−1, −10), (−7, 6) **66.** (9, 0), (4, −7)

Simplify each expression. (Lesson 9-2)

67. $\frac{5}{x^2 + 6x + 8} + \frac{x}{x^2 - 3x - 28}$ **68.** $\frac{3x}{x^2 + 8x - 20} - \frac{6}{x^2 + 7x - 18}$ **69.** $\frac{4}{3x^2 + 12x} + \frac{2x}{x^2 - 2x - 24}$

Solve each equation or inequality. Round to the nearest ten-thousandth. (Lesson 8-6)

70. $8^x = 30$ **71.** $5^x = 64$ **72.** $3^{x+2} = 41$

Evaluate each expression. (Lesson 7-6)

73. $16^{-\frac{1}{4}}$ **74.** $27^{\frac{4}{3}}$ **75.** $25^{-\frac{5}{2}}$

Skills Review

Solve for x. (Concepts and Skills Bank 1)

76. $\frac{x + 2}{18} = \frac{x - 2}{9}$ **77.** $\frac{x + 5}{x - 1} = \frac{7}{4}$ **78.** $\frac{5}{x + 8} = \frac{15}{2x + 20}$

13-4 Law of Sines

Why?

Mars has hundreds of thousands of craters. These craters are named after famous scientists, science fiction authors, and towns on Earth. The craters named Wahoo, Wabash, and Naukan are shown in the figure. You can use trigonometry to find the distance between Wahoo and Naukan.

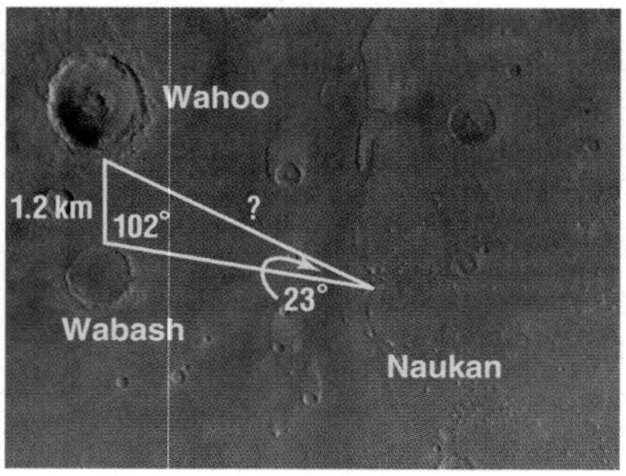

Then
You found side lengths and angle measures of right triangles. (Lesson 13-1)

Now
- Find the area of a triangle using two sides and an included angle.
- Use the Law of Sines to solve triangles.

NYS Core Curriculum

A2.A.73 Solve for an unknown side or angle, using the Law of Sines or the Law of Cosines A2.A.74 Determine the area of a triangle or a parallelogram, given the measure of two sides and the included angle *Also addresses A2.A.66 and A2.A.75.*

New Vocabulary
Law of Sines
solving a triangle

NY Math Online
glencoe.com
- Extra Examples
- Personal Tutor
- Self-Check Quiz
- Homework Help

Find the Area of a Triangle In the triangle at the right, $\sin A = \frac{h}{c}$, or $h = c \sin A$.

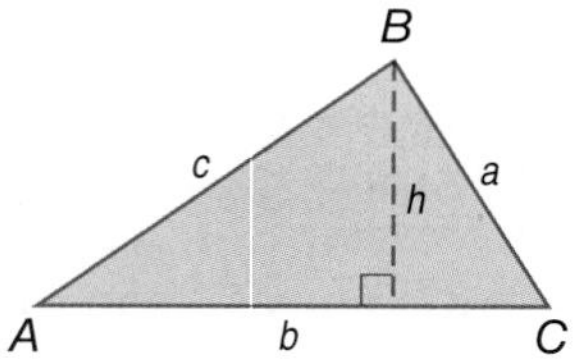

$\text{Area} = \frac{1}{2}bh$	**Formula for area of a triangle**
$\text{Area} = \frac{1}{2}b(c \sin A)$	**Replace *h* with *c* sin *A*.**
$\text{Area} = \frac{1}{2}bc \sin A$	**Simplify.**

You can use this formula or two other formulas to find the area of a triangle if you know the lengths of two sides and the measure of the included angle.

Key Concept: Area of a Triangle
For Your FOLDABLE

Words The area of a triangle is one half the product of the lengths of two sides and the sine of their included angle.

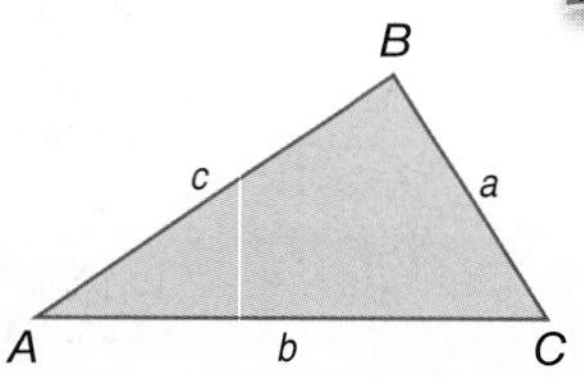

Symbols $\text{Area} = \frac{1}{2}bc \sin A$ $\text{Area} = \frac{1}{2}ac \sin B$ $\text{Area} = \frac{1}{2}ab \sin C$

EXAMPLE 1 Find the Area of a Triangle

Find the area of $\triangle ABC$ to the nearest tenth.

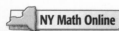

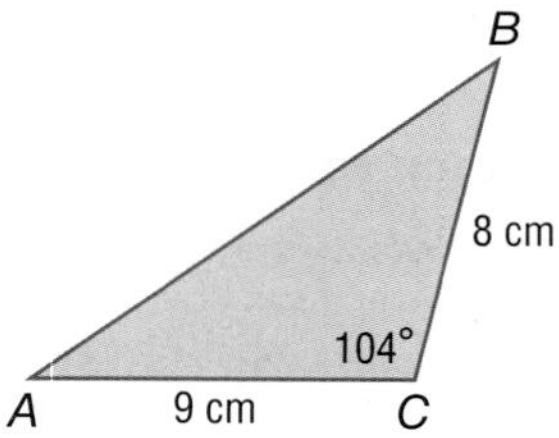

In $\triangle ABC$, $a = 8$, $b = 9$, and $C = 104°$.

$\text{Area} = \frac{1}{2}ab \sin C$	**Based on the known measures, use the third area formula.**
$= \frac{1}{2}(8)(9) \sin 104°$	**Substitution**
≈ 34.9	**Simplify.**

So, the area is about 34.9 square centimeters.

Check Your Progress

1. Find the area of $\triangle ABC$ to the nearest tenth if $A = 31°$, $b = 18$ meters, and $c = 22$ meters.

Personal Tutor glencoe.com

Use the Law of Sines to Solve Triangles You can use the area formulas to derive the **Law of Sines**, which shows the relationships between side lengths of a triangle and the sines of the angles opposite them.

$\frac{1}{2}bc \sin A = \frac{1}{2}ac \sin B = \frac{1}{2}ab \sin C$	**Set the area formulas equal to each other.**
$bc \sin A = ac \sin B = ab \sin C$	**Multiply each expression by 2.**
$\frac{bc \sin A}{abc} = \frac{ac \sin B}{abc} = \frac{ab \sin C}{abc}$	**Divide each expression by *abc*.**
$\frac{\sin A}{a} = \frac{\sin B}{b} = \frac{\sin C}{c}$	**Simplify.**

Key Concept Law of Sines

For Your FOLDABLE

In $\triangle ABC$, if sides with lengths a, b, and c are opposite angles with measures A, B, and C, respectively, then the following is true.

$$\frac{\sin A}{a} = \frac{\sin B}{b} = \frac{\sin C}{c}$$

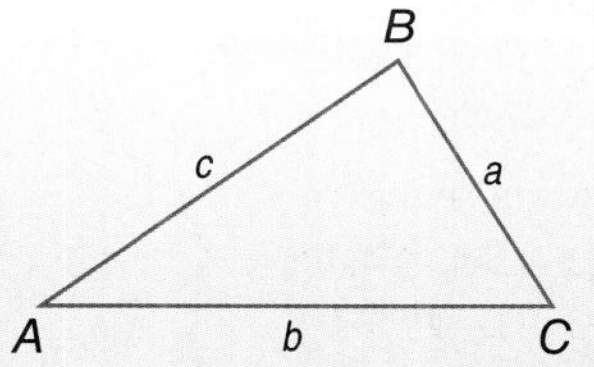

You can use the Law of Sines to solve a triangle if you know either one of the following.

- the measures of two angles and any side (angle-angle-side AAS or angle-side-angle ASA cases)

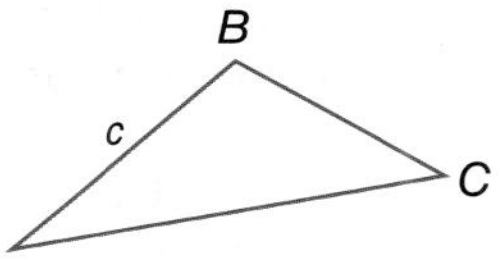

- the measures of two sides and the angle opposite one of the sides (side-side-angle SSA case)

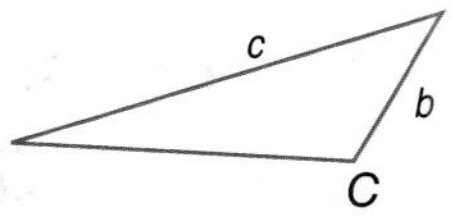

Using given measures to find all unknown side lengths and angle measures of a triangle is called **solving a triangle**.

StudyTip

Alternative Representations The Law of Sines may also be written as $\frac{a}{\sin A} = \frac{b}{\sin B} = \frac{c}{\sin C}$. So, the expressions below could also be used to solve the triangle in Example 2.

- $\frac{a}{\sin 55°} = \frac{3}{\sin 80°}$
- $\frac{b}{\sin 45°} = \frac{3}{\sin 80°}$

EXAMPLE 2 Solve a Triangle Given Two Angles and a Side

Solve $\triangle ABC$. Round to the nearest tenth if necessary.

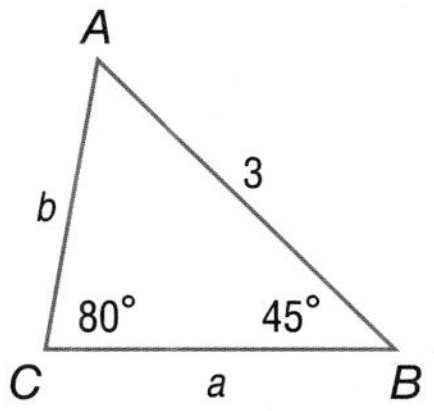

Step 1 Find the measure of the third angle.
$m\angle A = 180 - (80 + 45)$ or $55°$

Step 2 Use the Law of Sines to find side lengths a and b. Write an equation to find each variable.

$\frac{\sin A}{a} = \frac{\sin C}{c}$	**Law of Sines**	$\frac{\sin B}{b} = \frac{\sin C}{c}$
$\frac{\sin 55°}{a} = \frac{\sin 80°}{3}$	**Substitution**	$\frac{\sin 45°}{b} = \frac{\sin 80°}{3}$
$a = \frac{3 \sin 55°}{\sin 80°}$	**Solve for each variable.**	$b = \frac{3 \sin 45°}{\sin 80°}$
$a \approx 2.5$	**Use a calculator.**	$b \approx 2.2$

So, $A = 55°$, $a \approx 2.5$, and $b \approx 2.2$.

Check Your Progress

2. Solve $\triangle NPQ$ if $P = 42°$, $Q = 65°$, and $n = 5$.

Personal Tutor glencoe.com

StudyTip

Two Solutions A situation in which two solutions for a triangle exist is called the *ambiguous case*.

StudyTip

***A* is Acute** In the figures at the right, the altitude h is compared to a because h is the minimum distance from C to $\overline{AB}$ when A is acute.

$\sin A = \frac{\text{opp}}{\text{hyp}}$

$\sin A = \frac{h}{b}$

If you are given the measures of two angles and a side, exactly one triangle is possible. However, if you are given the measures of two sides and the angle opposite one of them, zero, one, or two triangles may be possible. So, when solving a triangle using the SSA case, zero, one, or two solutions are possible.

Key Concept Possible Triangles in SSA Case

For Your FOLDABLE

Consider a triangle in which a, b, and $m\angle A$ are given.

$\angle A$ is Acute.

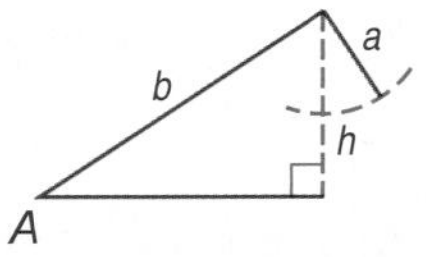

$a < h$
no solution

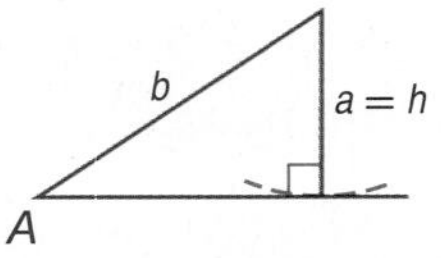

$a = h$
one solution

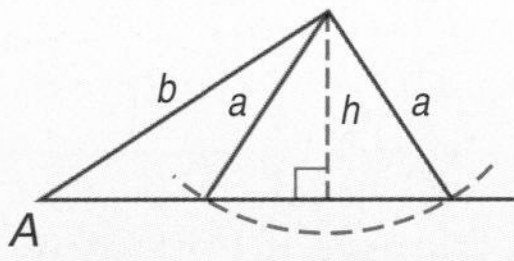

$h < a < b$
two solutions

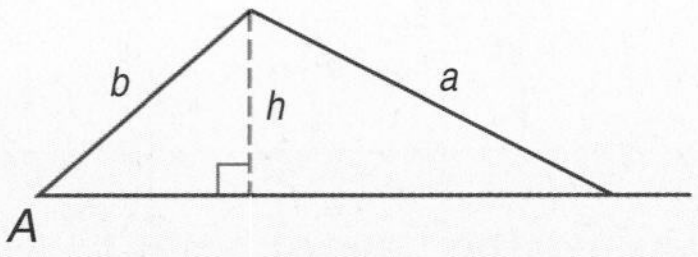

$a \geq b$
one solution

$\angle A$ is Right or Obtuse.

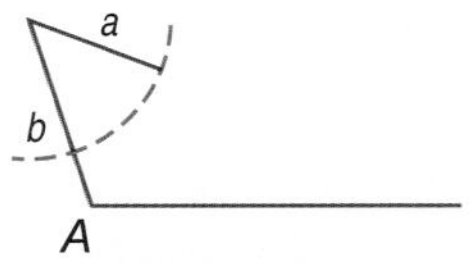

$a \leq b$
no solution

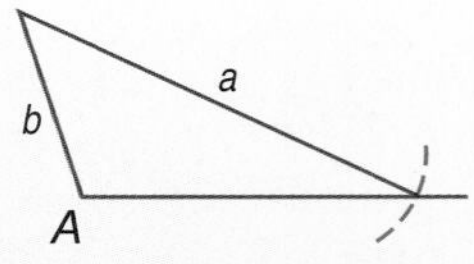

$a > b$
one solution

Since $\sin A = \frac{h}{b}$, you can use $h = b \sin A$ to find h in the acute triangles.

EXAMPLE 3 Solve a Triangle Given Two Sides and an Angle

Determine whether each triangle has *no* solution, *one* solution, or *two* solutions. Then solve the triangle. Round side lengths to the nearest tenth and angle measures to the nearest degree.

a. In $\triangle RST$, $R = 105°$, $r = 9$, and $s = 6$.

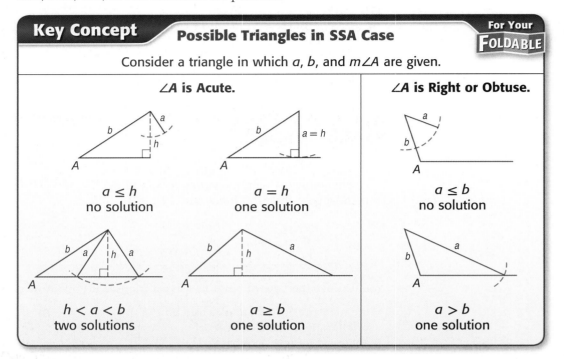

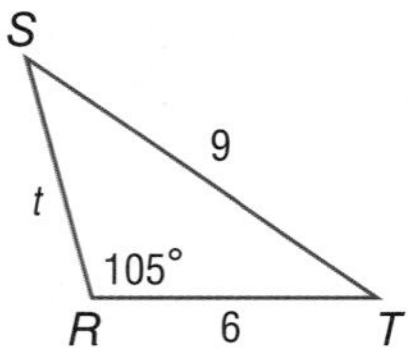

Because $\angle R$ is obtuse and $9 > 6$, you know that one solution exists.

Step 1 Use the Law of Sines to find $m\angle S$.

$\frac{\sin S}{6} = \frac{\sin 105°}{9}$ **Law of Sines**

$\sin S = \frac{6 \sin 105°}{9}$ **Multiply each side by 6.**

$\sin S \approx 0.6440$ **Use a calculator.**

$S \approx 40°$ **Use the $\sin^{-1}$ function.**

Step 2 Find $m\angle T$.

$m\angle T \approx 180 - (105 + 40)$ or $35°$

Step 3 Use the Law of Sines to find t.

$\frac{\sin 35°}{t} \approx \frac{\sin 105°}{9}$ **Law of Sines**

$t \approx \frac{9 \sin 35°}{\sin 105°}$ **Solve for t.**

$t \approx 5.3$ **Use a calculator.**

So, $S \approx 40°$, $T \approx 35°$, and $t \approx 5.3$.

b. In $\triangle ABC$, $A = 54°$, $a = 6$, and $b = 8$.

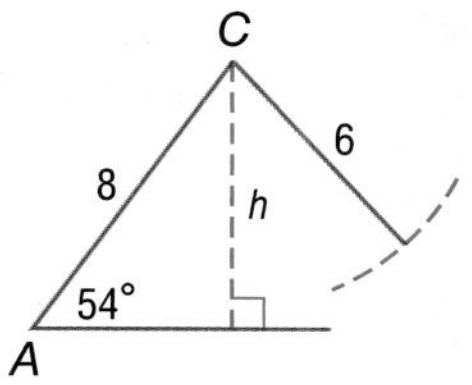

Since $\angle A$ is acute and $6 < 8$, find h and compare it to a.

$b \sin A = 8 \sin 54°$ — **$b = 8$ and $A = 54°$**

≈ 6.5 — **Use a calculator.**

Since $6 \leq 6.5$ or $a \leq h$, there is no solution.

c. In $\triangle ABC$, $A = 35°$, $a = 17$, and $b = 20$.

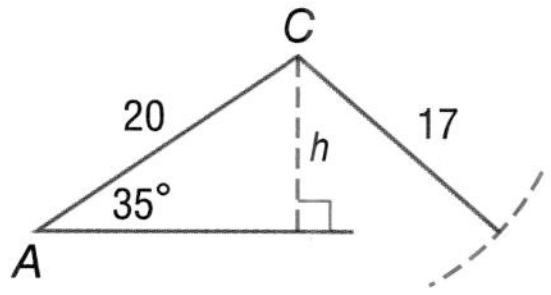

Since $\angle A$ is acute and $17 < 20$, find h and compare it to a.

$b \sin A = 20 \sin 35°$ — **$b = 20$ and $A = 35°$**

≈ 11.5 — **Use a calculator.**

Since $11.5 < 17 < 20$ or $h < a < b$, there are two solutions. So, there are two triangles to be solved.

Case 1 $\angle B$ is acute.

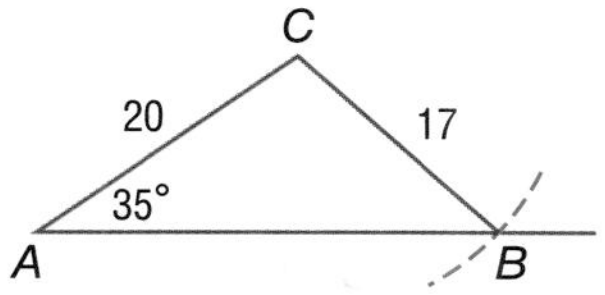

Step 1 Find $m\angle B$.

$\frac{\sin B}{20} = \frac{\sin 35°}{17}$ — **Law of Sines**

$\sin B = \frac{20 \sin 35°}{17}$ — **Solve for sin *B*.**

$\sin B \approx 0.6748$ — **Use a calculator.**

$B \approx 42°$ — **Find $\sin^{-1}$ 0.6748.**

Step 2 Find $m\angle C$.

$m\angle C \approx 180 - (35 + 42)$ or $103°$

Step 3 Find c.

$\frac{\sin 103°}{c} = \frac{\sin 35°}{17}$ — **Law of Sines**

$c = \frac{17 \sin 103°}{\sin 35°}$ — **Solve for *c*.**

$c \approx 28.9$ — **Simplify.**

Case 2 $\angle B$ is obtuse.

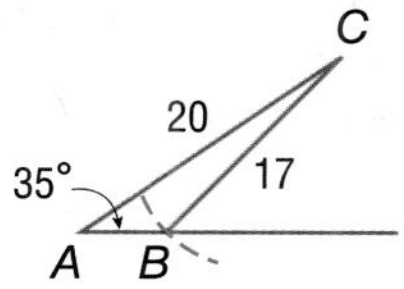

Step 1 Find $m\angle B$.

The sine function also has a positive value in Quadrant II. So, find an obtuse angle B for which $\sin B \approx 0.6748$.

$m\angle B \approx 180° - 42°$ or $138°$

Step 2 Find $m\angle C$.

$m\angle C \approx 180 - (35 + 138)$ or $7°$

Step 3 Find c.

$\frac{\sin 7°}{c} \approx \frac{\sin 35°}{17}$ — **Law of Sines**

$c \approx \frac{17 \sin 7°}{\sin 35°}$ — **Solve for *c*.**

$c \approx 3.6$ — **Simplify.**

So, one solution is $B \approx 42°$, $C \approx 103°$, and $c \approx 28.9$, and another solution is $B \approx 138°$, $C \approx 7°$, and $c \approx 3.6$.

StudyTip

Reference Angle In the triangle in Case 2, you are using the reference angle 42° to find the other value of *B*.

Check Your Progress

Determine whether each triangle has *no* solution, *one* solution, or *two* solutions. Then solve the triangle. Round side lengths to the nearest tenth and angle measures to the nearest degree.

3A. In $\triangle RST$, $R = 95°$, $r = 10$, and $s = 12$.

3B. In $\triangle MNP$, $N = 32°$, $n = 7$, and $p = 4$.

3C. In $\triangle ABC$, $A = 47°$, $a = 15$, and $b = 18$.

Personal Tutor glencoe.com

Real-World Link

High school and college baseball fields share the same infield dimensions as professional baseball fields. The outfield dimensions vary greatly.

Source: *Baseball Digest Magazine*

Real-World EXAMPLE 4 Use the Law of Sines to Solve a Problem

BASEBALL A baseball is hit between second and third bases and is caught at point B, as shown in the figure. How far away from second base was the ball caught?

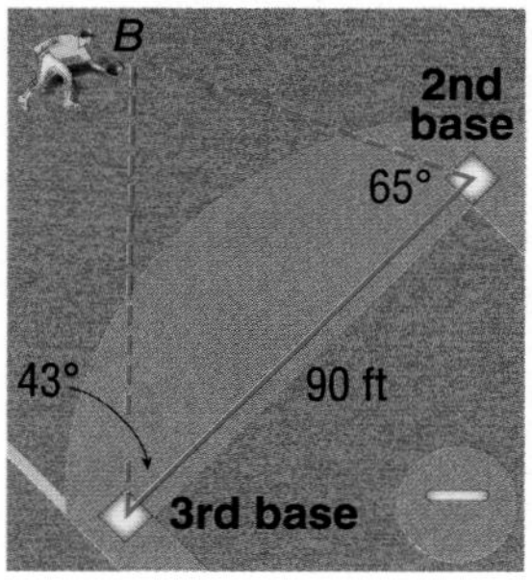

$\frac{\sin 72°}{90} = \frac{\sin 43°}{x}$ **Law of Sines**

$x \sin 72° = 90 \sin 43°$ **Cross products**

$x = \frac{90 \sin 43°}{\sin 72°}$ **Solve for x.**

$x \approx 64.5$ **Use a calculator.**

So, the distance is about 64.5 feet.

Check Your Progress

4. How far away from third base was the ball caught?

Personal Tutor glencoe.com

Check Your Understanding

Example 1 p. 832

Find the area of $\triangle ABC$ to the nearest tenth, if necessary.

1.

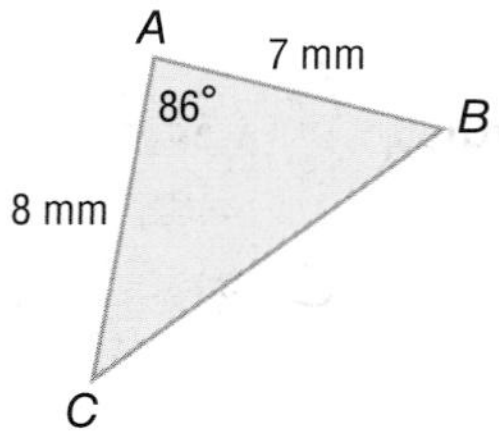

2.

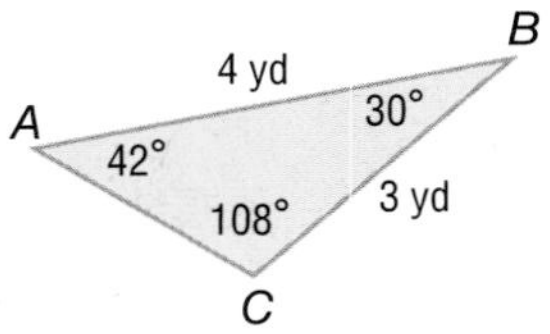

3. $A = 40°, b = 11$ cm, $c = 6$ cm

4. $B = 103°, a = 20$ in., $c = 18$ in.

Example 2 p. 833

Solve each triangle. Round side lengths to the nearest tenth and angle measures to the nearest degree.

5.

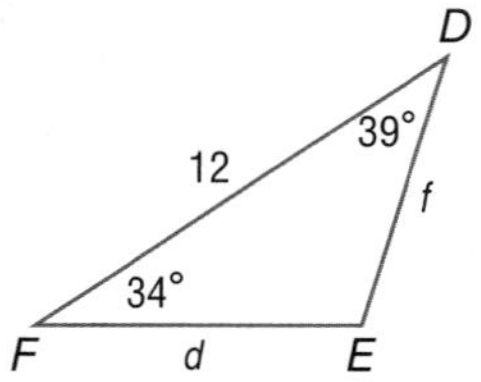

6.

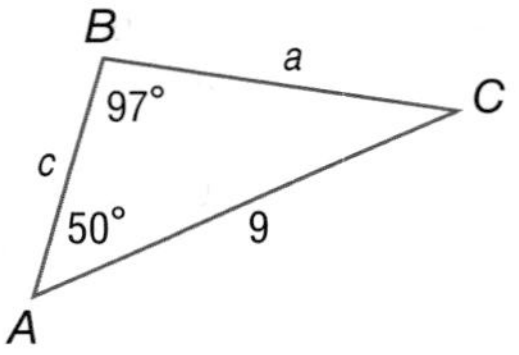

7. Solve $\triangle FGH$ if $G = 80°, H = 40°$, and $g = 14$.

Example 3 pp. 834–835

Determine whether each triangle has *no* solution, *one* solution, or *two* solutions. Then solve the triangle. Round side lengths to the nearest tenth and angle measures to the nearest degree.

8. $A = 95°, a = 19, b = 12$

9. $A = 60°, a = 15, b = 24$

10. $A = 34°, a = 8, b = 13$

11. $A = 30°, a = 3, b = 6$

Example 4 p. 836

12. SPACE Refer to the beginning of the lesson. Find the distance between the Wahoo Crater and the Naukan Crater on Mars.

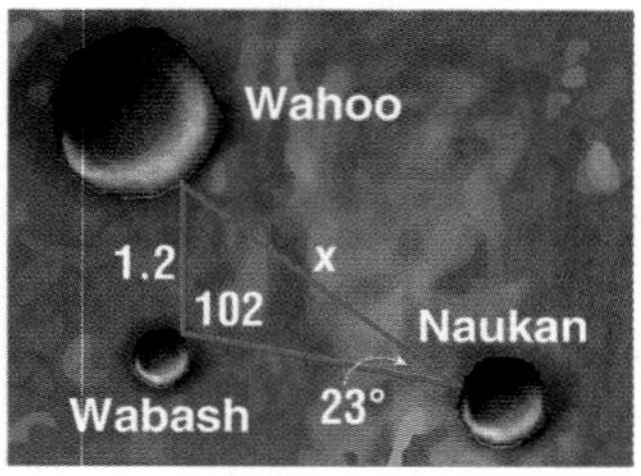

Practice and Problem Solving

● = **Step-by-Step Solutions** begin on page R20.
Extra Practice begins on page 947.

Example 1 p. 832

Find the area of $\triangle ABC$ to the nearest tenth.

13. 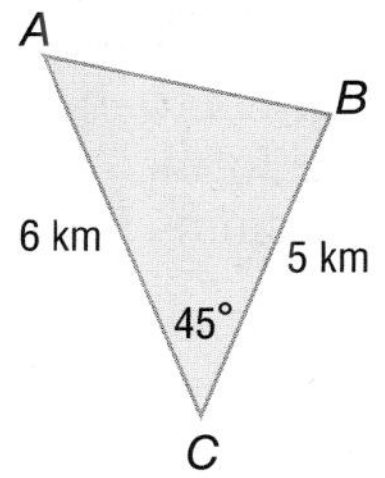

14. 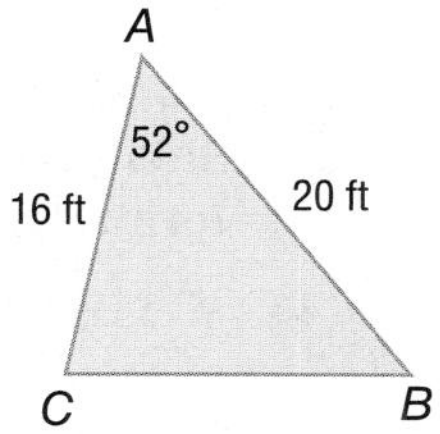

15. 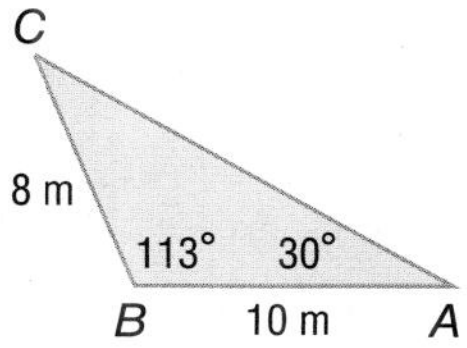

16. 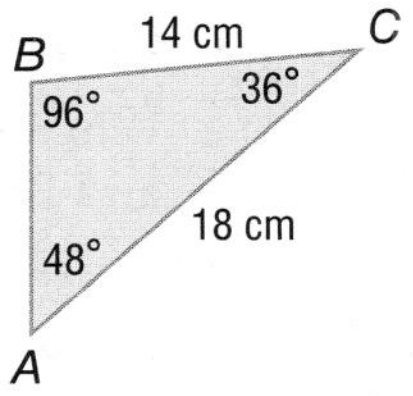

17. $C = 25°, a = 4$ ft, $b = 7$ ft

18. $A = 138°, b = 10$ in., $c = 20$ in.

19. $B = 92°, a = 14.5$ m, $c = 9$ m

20. $C = 116°, a = 2.7$ cm, $b = 4.6$ cm

Example 2 p. 833

Solve each triangle. Round side lengths to the nearest tenth and angle measures to the nearest degree.

21. 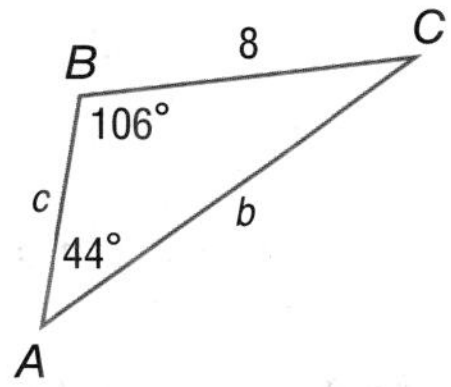

22. 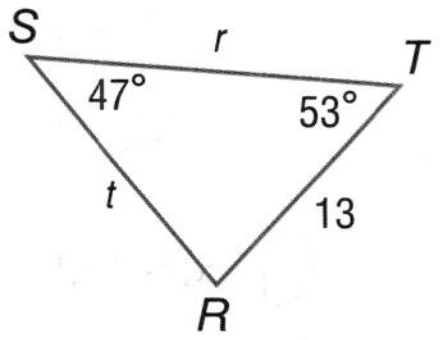

23. 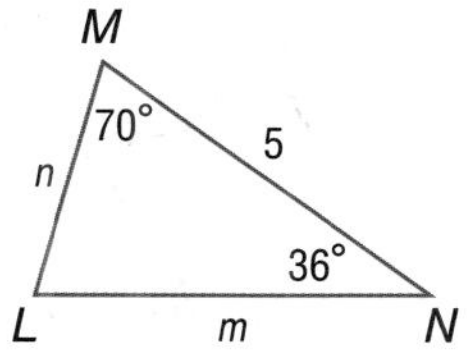

24. 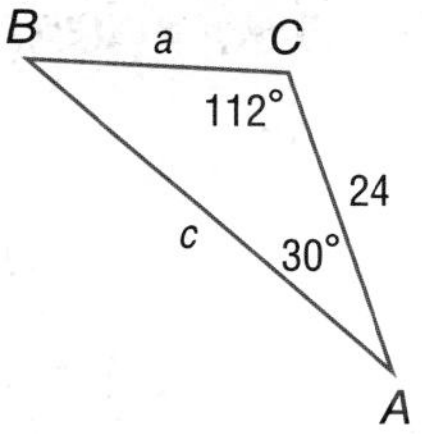

25. Solve $\triangle HJK$ if $H = 53°$, $J = 20°$, and $h = 31$.

26. Solve $\triangle NPQ$ if $P = 109°$, $Q = 57°$, and $n = 22$.

27. Solve $\triangle ABC$ if $A = 50°$, $a = 2.5$, and $C = 67°$.

28. Solve $\triangle ABC$ if $B = 18°$, $C = 142°$, and $b = 20$.

Example 3 pp. 834–835

Determine whether each triangle has *no* solution, *one* solution, or *two* solutions. Then solve the triangle. Round side lengths to the nearest tenth and angle measures to the nearest degree.

29. $A = 100°, a = 7, b = 3$

30. $A = 75°, a = 14, b = 11$

31. $A = 38°, a = 21, b = 18$

32. $A = 52°, a = 9, b = 20$

33. $A = 42°, a = 5, b = 6$

34. $A = 44°, a = 14, b = 19$

35. $A = 131°, a = 15, b = 32$

36. $A = 30°, a = 17, b = 34$

Example 4
p. 836

GEOGRAPHY In Hawaii, the distance from Hilo to Kailua is 57 miles, and the distance from Hilo to Captain Cook is 55 miles.

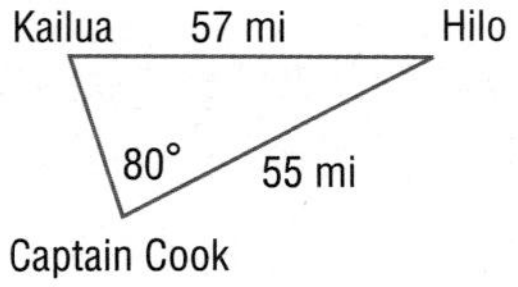

37. What is the measure of the angle formed at Hilo?

38. What is the distance between Kailua and Captain Cook?

39 **TORNADOES** Tornado sirens A, B, and C form a triangular region in one area of a city. Sirens A and B are 8 miles apart. The angle formed at siren A is 112°, and the angle formed at siren B is 40°. How far apart are sirens B and C?

40. **MYSTERIES** The Bermuda Triangle is a region of the Atlantic Ocean between Bermuda, Miami, Florida, and San Juan, Puerto Rico. It is an area where ships and airplanes have been rumored to mysteriously disappear.

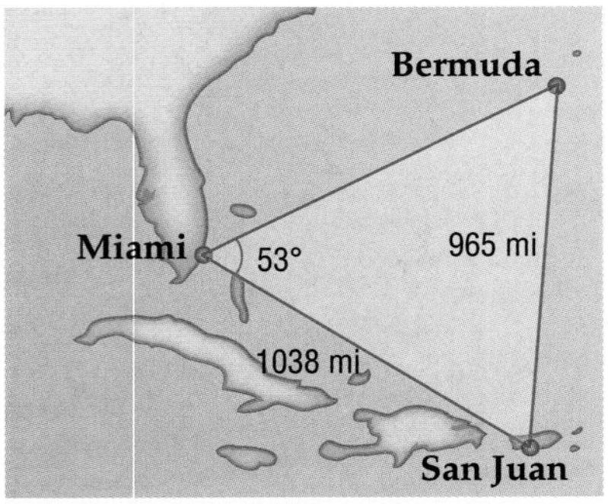

a. What is the distance between Miami and Bermuda?

b. What is the approximate area of the Bermuda Triangle?

Real-World Link

At 65 feet high, one of the tallest indoor rock climbing walls in the world is located in Bloomington, Illinois.

Source: HighBeam Research, Inc.

41. **BICYCLING** One side of a triangular cycling path is 4 miles long. The angle opposite this side is 64°. Another angle formed by the triangular path measures 66°.

a. Sketch a drawing of the situation. Label the missing sides a and b.

b. Write equations that could be used to find the lengths of the missing sides.

c. What is the perimeter of the path?

42. **ROCK CLIMBING** Savannah S and Leon L are standing 8 feet apart in front of a rock climbing wall, as shown at the right. What is the height of the wall? Round to the nearest tenth.

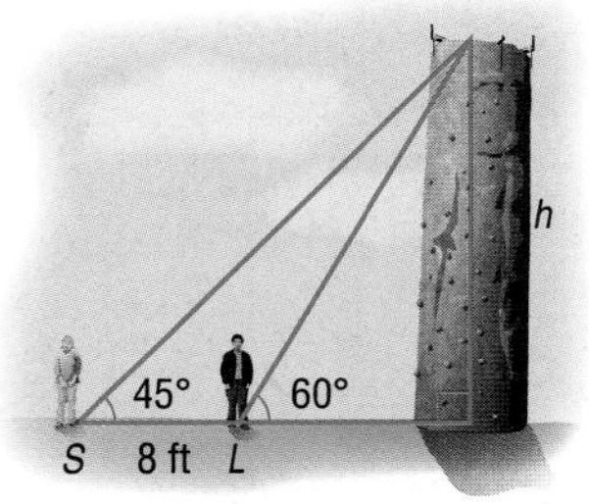

H.O.T. Problems

Use **H**igher-**O**rder **T**hinking Skills

43. **FIND THE ERROR** In $\triangle RST$, $R = 56°$, $r = 24$, and $t = 12$. Cameron and Gabriela are using the Law of Sines to find T. Is either of them correct? Explain your reasoning.

Cameron

$\frac{\sin T}{12} = \frac{\sin 56°}{24}$

$\sin T \approx 0.4145$

$T \approx 24.5°$

Gabriela

Since $r > t$, there is no solution.

44. **CHALLENGE** In $\triangle ABC$, $B = 30°$, and $a = 6$. How many triangles can be formed if $b = 5$? Explain your reasoning.

45. **CHALLENGE** Using the figure at the right, derive the formula Area $= \frac{1}{2}bc \sin A$.

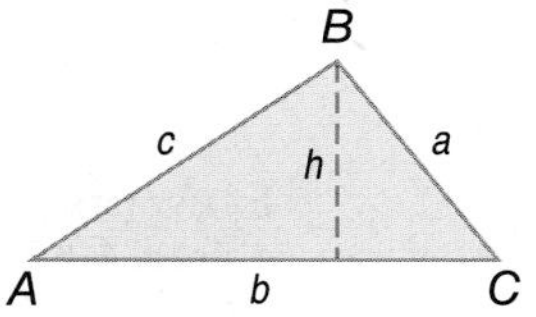

46. **REASONING** Find the side lengths of two different triangles ABC that can be formed if $A = 55°$ and $C = 20°$.

47. **WRITING IN MATH** Use the Law of Sines to explain why a and b do not have unique values in the figure shown.

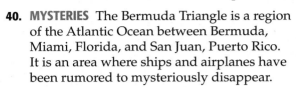

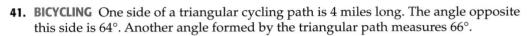

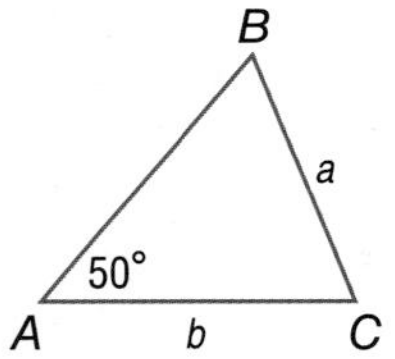

48. **OPEN ENDED** Given that $E = 62°$ and $d = 38$, find a value for e such that no triangle DEF can exist. Explain your reasoning.

NYSRE PRACTICE A2.A.73, A2.A.74

49. SHORT RESPONSE Given the graphs of $f(x)$ and $g(x)$, what is the value of $f(g(4))$?

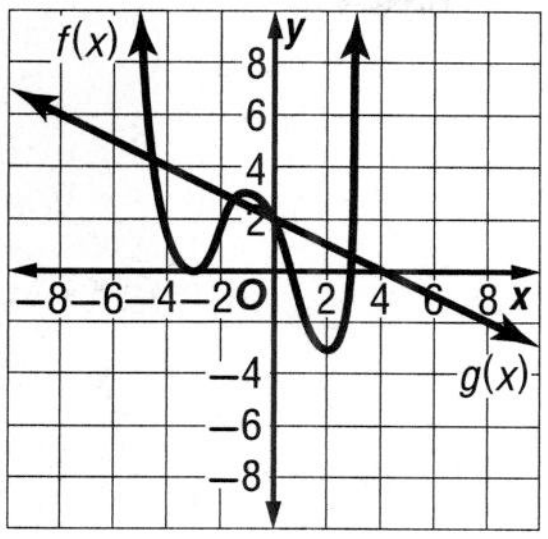

50. STATISTICS If the average of seven consecutive odd integers is n, what is the median of these seven integers?

A 0
B 7
C n
D $n - 2$

51. One zero of $f(x) = x^3 - 7x^2 - 6x + 72$ is 4. What is the factored form of the expression $x^3 - 7x^2 - 6x + 72$?

F $(x - 6)(x + 3)(x + 4)$
G $(x - 6)(x + 3)(x - 4)$
H $(x + 6)(x + 3)(x - 4)$
J $(x + 12)(x - 1)(x - 4)$

52. SAT/ACT Three people are splitting $48,000 using the ratio 5 : 4 : 3. What is the amount of the greatest share?

A $12,000
B $16,000
C $20,000
D $24,000

Spiral Review

Find the exact value of each trigonometric function. (Lesson 13-3)

53. $\sin 210°$
54. $\cos \frac{3}{4}\pi$
55. $\cot 60°$

Find an angle with a positive measure and an angle with a negative measure that are coterminal with each angle. (Lesson 13-2)

56. $125°$
57. $-32°$
58. $\frac{2}{3}\pi$

59. CLOCKS Jun's grandfather clock is broken. When she sets the pendulum in motion by holding it against the side of the clock and letting it go, it swings 24 centimeters to the other side, then 18 centimeters back, then 13.5 centimeters, and so on. What is the total distance that the pendulum swings before it stops? (Lesson 11-5)

Find the sum of each infinite series, if it exists. (Lesson 11-4)

60. $64 + 48 + 36 + \ldots$
61. $27 + 36 + 48 + \ldots$
62. $\sum_{n=1}^{\infty} 0.5(1.1)^n$

63. ASTRONOMY At its closest point, Earth is 91.8 million miles from the center of the Sun. At its farthest point, Earth is 94.9 million miles from the center of the Sun. Write an equation for the orbit of Earth, assuming that the center of the orbit is the origin and the Sun lies on the x-axis. (Lesson 10-4)

Simplify. (Lesson 7-4)

64. $\sqrt{(x - 4)^2}$
65. $\sqrt{(y + 2)^4}$
66. $\sqrt[3]{(a - b)^6}$

Skills Review

Evaluate each expression if $w = 6$, $x = -4$, $y = 1.5$, and $z = \frac{3}{4}$. (Lesson 1-1)

67. $w^2 + y^2 - 6xz$
68. $x^2 + z^2 + 5wy$
69. $wy + xz + w^2 - x^2$

EXTEND 13-4

Geometry Lab
Regular Polygons

NY Math Online glencoe.com
Math *in Motion*, Animation

You can use central angles of circles to investigate characteristics of regular polygons inscribed in a circle. Recall that a regular polygon is inscribed in a circle if each of its vertices lies on the circle.

ACTIVITY Collect the Data

Step 1 Use a compass to draw a circle with a radius of one inch.

Step 2 Inscribe an equilateral triangle inside the circle. To do this, use a protractor to measure three angles of 120° at the center of the circle, since $\frac{360°}{3} = 120°$. Then connect the points where the sides of the angles intersect the circle using a straightedge.

Step 3 The **apothem** of a regular polygon is a segment that is drawn from the center of the polygon perpendicular to a side of the polygon. Use the cosine of angle θ to find the length of an apothem, labeled a in the diagram.

Model and Analyze

1. Make a table like the one shown below and record the length of the apothem of the equilateral triangle. Inscribe each regular polygon named in the table in a circle with radius one inch. Copy and complete the table.

Number of Sides, n	θ	a	Number of Sides, n	θ	a
3	60		7		
4	45		8		
5			9		
6			10		

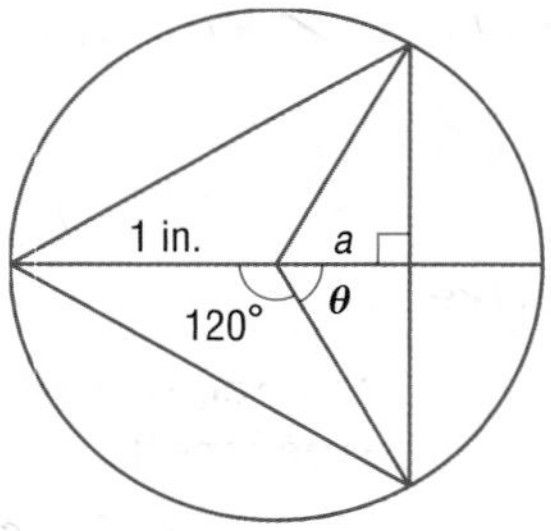

2. What do you notice about the measure of θ as the number of sides of the inscribed polygon increases?
3. What do you notice about the value of a?
4. **MAKE A CONJECTURE** Suppose you inscribe a 30-sided regular polygon inside a circle. Find the measure of angle θ.
5. Write a formula that gives the measure of angle θ for a polygon with n sides.
6. Write a formula that gives the length of the apothem of a regular polygon inscribed in a circle with radius one inch.
7. How would the formula you wrote in Exercise 5 change if the apothem of the circle was not one inch?

Extending the Investigation

Use the table to write each set of ordered pairs. Then use a graphing calculator to find a regression equation for the set of data.

8. (n, θ)

9. (θ, a)

13-5 Law of Cosines

Then

You solved triangles by using the Law of Sines. (Lesson 13-4)

Now

- Use the Law of Cosines to solve triangles.
- Choose methods to solve triangles.

NYS Core Curriculum

A2.A.66 Determine the trigonometric functions of any angle, using technology
A2.A.73 Solve for an unknown side or angle, using the Law of Sines or the Law of Cosines

New Vocabulary

Law of Cosines

NY Math Online

glencoe.com

- Extra Examples
- Personal Tutor
- Self-Check Quiz
- Homework Help

Why?

Submersibles, which are lowered into the ocean from ships, are used to take humans to depths they cannot reach by any other means. A submersible 520 meters from its ship shines a light on a shipwreck 338 meters away, as shown in the diagram. You can use trigonometry to find the distance from the ship to the shipwreck.

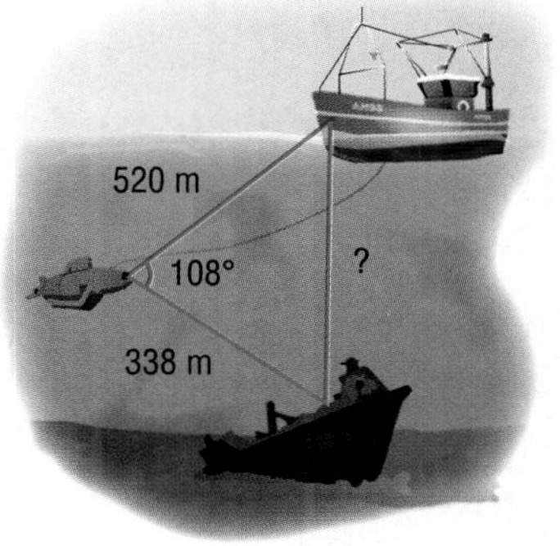

Use Law of Cosines to Solve Triangles You cannot use the Law of Sines to solve a triangle like the one shown above. You can use the **Law of Cosines** to solve a triangle if you know either one of the following.

- the measures of two sides and the included angle (side-angle-side SAS case)
- the measures of three sides (side-side-side SSS case)

Key Concept — Law of Cosines

For Your FOLDABLE

In $\triangle ABC$, if sides with lengths a, b, and c are opposite angles with measures A, B, and C, respectively, then the following are true.

$$a^2 = b^2 + c^2 - 2bc \cos A$$
$$b^2 = a^2 + c^2 - 2ac \cos B$$
$$c^2 = a^2 + b^2 - 2ab \cos C$$

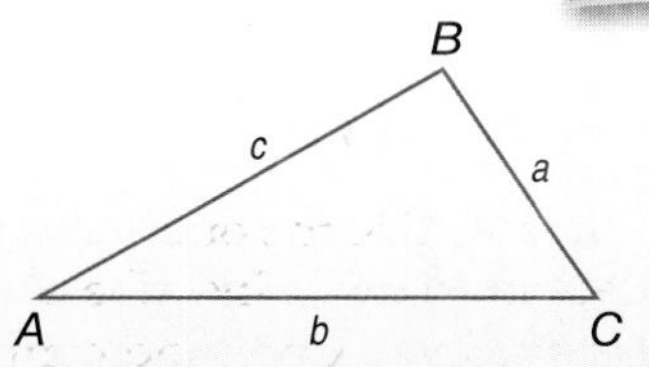

EXAMPLE 1 Solve a Triangle Given Two Sides and the Included Angle

Solve $\triangle ABC$.

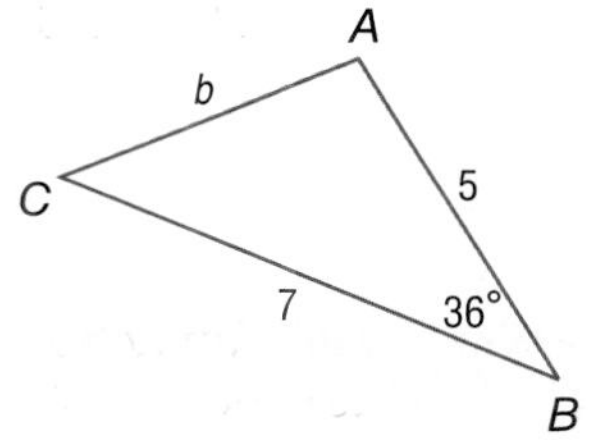

Step 1 Use the Law of Cosines to find the missing side length.

$b^2 = a^2 + c^2 - 2ac \cos B$ — Law of Cosines

$b^2 = 7^2 + 5^2 - 2(7)(5) \cos 36°$ — $a = 7, c = 5, B = 36°$

$b^2 \approx 17.4$ — Use a calculator to simplify.

$b \approx 4.2$ — Take the square root of each side.

Step 2 Use the Law of Sines to find a missing angle measure.

$\frac{\sin A}{7} \approx \frac{\sin 36°}{4.2}$ — $\frac{\sin A}{a} = \frac{\sin B}{b}$

$\sin A \approx \frac{7 \sin 36°}{4.2}$ — Multiply each side by 7.

$A \approx 78°$ — Use the $\sin^{-1}$ function.

Step 3 Find the measure of the other angle.

$m\angle C \approx 180° - (36° + 78°)$ or $66°$

So, $b \approx 4.2$, $A \approx 78°$, and $C \approx 66°$.

Check Your Progress

1. Solve $\triangle FGH$ if $G = 82°$, $f = 6$, and $h = 4$.

Personal Tutor glencoe.com

When you are only given the three side lengths of a triangle, you can solve it by using the Law of Cosines. The first step is to find the measure of the largest angle. This is done to ensure the other two angles are acute when using the Law of Sines.

EXAMPLE 2 Solve a Triangle Given Three Sides

Solve $\triangle ABC$.

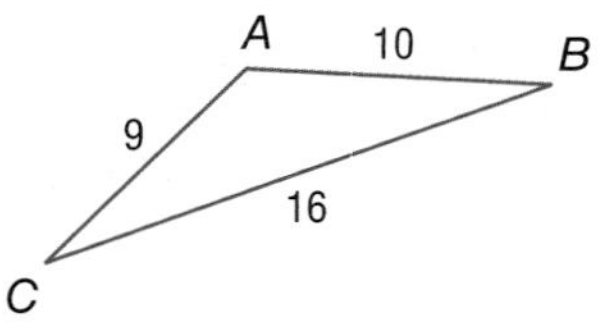

Step 1 Use the Law of Cosines to find the measure of the largest angle, $\angle A$.

$a^2 = b^2 + c^2 - 2bc \cos A$	**Law of Cosines**
$16^2 = 9^2 + 10^2 - 2(9)(10) \cos A$	**$a = 16$, $b = 9$, and $c = 10$**
$16^2 - 9^2 - 10^2 = -2(9)(10) \cos A$	**Subtract 9^2 and 10^2 from each side.**
$\frac{16^2 - 9^2 - 10^2}{-2(9)(10)} = \cos A$	**Divide each side by $-2(9)(10)$.**
$-0.4167 \approx \cos A$	**Use a calculator to simplify.**
$115° \approx A$	**Use the $\cos^{-1}$ function.**

Step 2 Use the Law of Sines to find the measure of $\angle B$.

$\frac{\sin B}{9} \approx \frac{\sin 115°}{16}$	$\frac{\sin B}{b} = \frac{\sin A}{a}$
$\sin B \approx \frac{9 \sin 115°}{16}$	**Multiply each side by 9.**
$\sin B \approx 0.5098$	**Use a calculator.**
$B \approx 31°$	**Use the $\sin^{-1}$ function.**

Step 3 Find the measure of $\angle C$.

$m\angle C \approx 180° - (115° + 31°)$ or about $34°$

So, $A \approx 115°$, $B \approx 31°$, and $C \approx 34°$.

Check Your Progress

1. Solve $\triangle ABC$ if $a = 5$, $b = 11$, and $c = 8$.

Personal Tutor glencoe.com

StudyTip

Alternative Method After finding *b* in Step 1, the Law of Cosines could be used again to find the measure of a second angle.

Review Vocabulary

oblique a triangle that has no right angle

Choose a Method to Solve Triangles You can use the Law of Sines and the Law of Cosines to solve problems involving oblique triangles. You need to know the measure of at least one side and any two other parts. If the triangle has a solution, you must decide whether to use the Law of Sines or the Law of Cosines to begin solving it.

Concept Summary: Solving Oblique Triangles

For Your FOLDABLE

Given	Begin by Using
two angles and any sides	Law of Sines
two sides and an angle opposite one of them	Law of Sines
two sides and their included angle	Law of Cosines
three sides	Law of Cosines

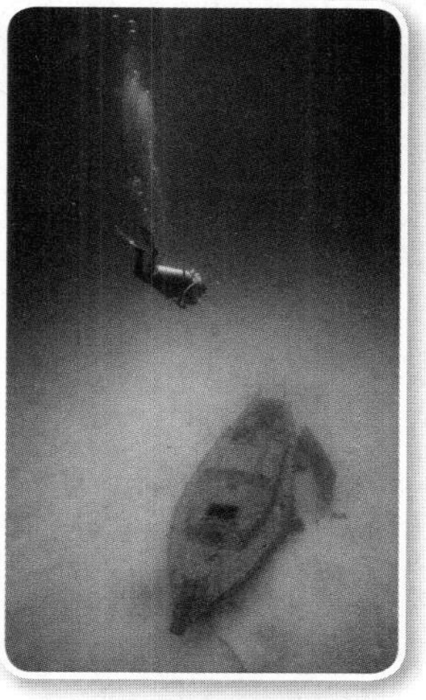

Real-World Link

The record for the deepest seawater scuba dive was 1044 feet, made by a scuba diver in the Red Sea.

Source: *Guinness Book of World Records*

Real-World EXAMPLE 3 Use the Law of Cosines

SCUBA DIVING A scuba diver looks up 20° and sees a turtle 9 feet away. She looks down 40° and sees a blue parrotfish 12 feet away. How far apart are the turtle and the blue parrotfish?

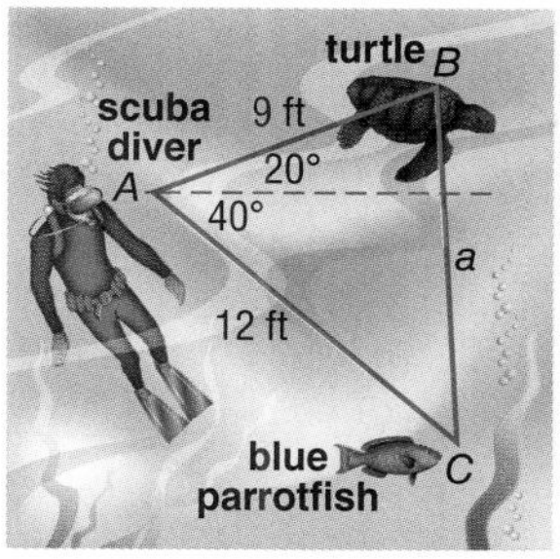

Understand You know the angles formed when the scuba diver looks up and when she looks down. You also know how far away the turtle and the blue parrotfish are from the scuba diver.

Plan Use the information to draw and label a diagram. Since two sides and the included angle of a triangle are given, you can use the Law of Cosines to solve the problem.

Solve

$a^2 = b^2 + c^2 - 2bc \cos A$	**Law of Cosines**
$a^2 = 12^2 + 9^2 - 2(12)(9) \cos 60$	**$b = 12$, $c = 9$, and $A = 60$**
$a^2 = 117$	**Use a calculator.**
$a \approx 10.8$	**Find the positive value of a.**

So, the turtle and the blue parrotfish are about 10.8 feet apart.

Check Using the Law of Sines, you can find that $B \approx 74°$ and $C \approx 46°$. Since $C < A < B$ and $c < a < b$, the solution is reasonable.

Check Your Progress

3. MARATHONS Amelia ran 6 miles in one direction. She then turned 79° and ran 7 miles. At the end of the run, how far was Amelia from her starting point?

Personal Tutor glencoe.com

Check Your Understanding

Examples 1 and 2 pp. 841–842

Solve each triangle. Round side lengths to the nearest tenth and angle measures to the nearest degree.

1.

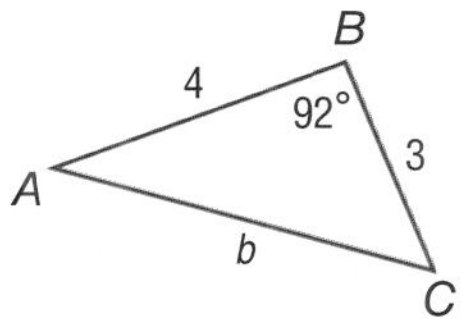

2.

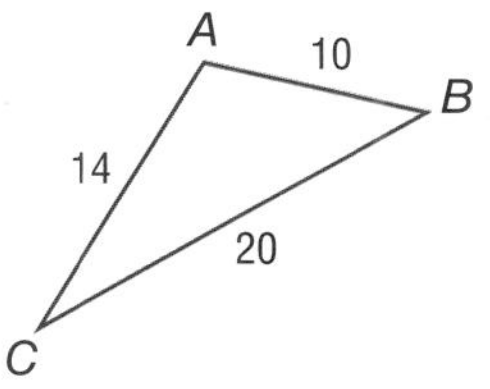

3. $a = 5, b = 8, c = 12$

4. $B = 110°, a = 6, c = 3$

Example 3 p. 843

Determine whether each triangle should be solved by beginning with the Law of *Sines* or the Law of *Cosines*. Then solve the triangle.

5.

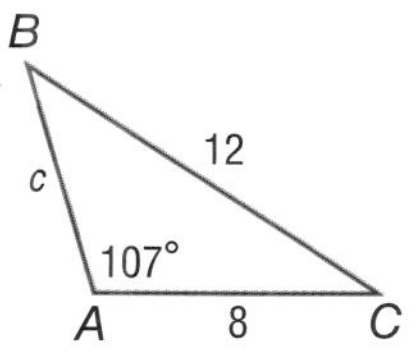

6.

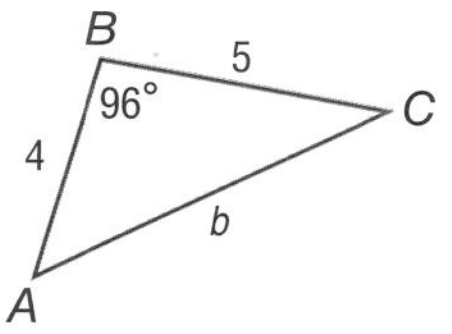

7. In $\triangle RST$, $R = 35°$, $s = 16$, and $t = 9$.

8. FOOTBALL In a football game, the quarterback is 20 yards from Receiver A. He turns 40° to see Receiver B, who is 16 yards away. How far apart are the two receivers?

Practice and Problem Solving

● = **Step-by-Step Solutions** begin on page R20.
Extra Practice begins on page 947.

Examples 1 and 2 pp. 841–842

Solve each triangle. Round side lengths to the nearest tenth and angle measures to the nearest degree.

9.

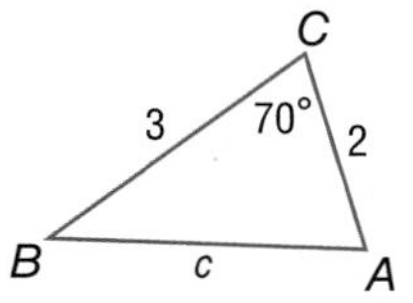

10.

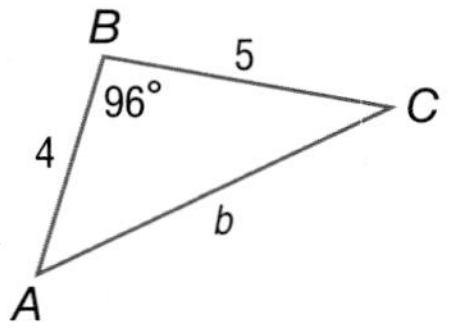

11.

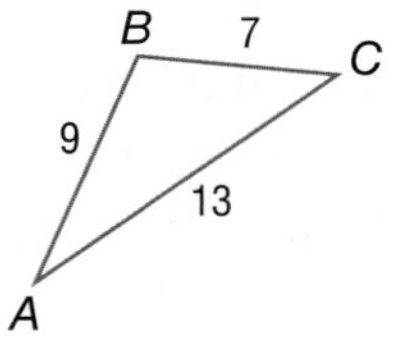

12.

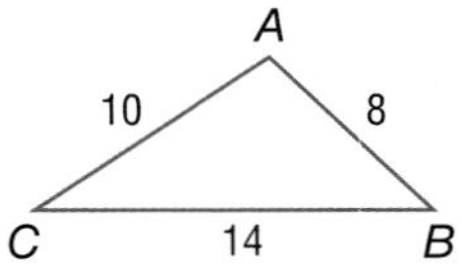

13. $A = 116°, b = 5, c = 3$

14. $C = 80°, a = 9, b = 2$

15. $f = 10, g = 11, h = 4$

16. $w = 20, x = 13, y = 12$

Example 3 p. 843

Determine whether each triangle should be solved by beginning with the Law of *Sines* or the Law of *Cosines*. Then solve the triangle.

17.

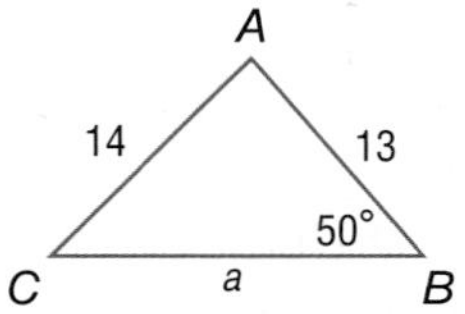

18.

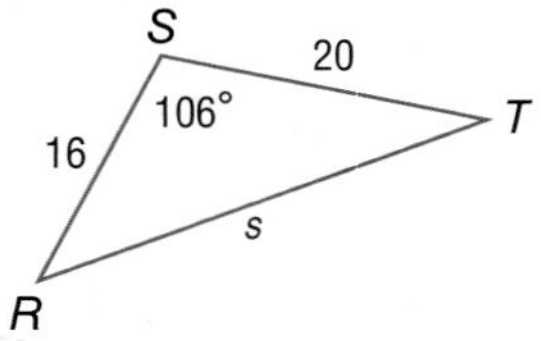

19.

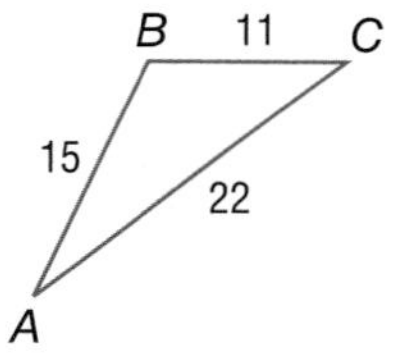

20.

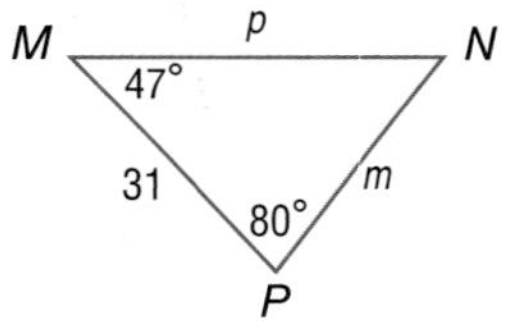

21. In $\triangle ABC$, $C = 84°$, $c = 7$, and $a = 2$.

22. In $\triangle HJK$, $h = 18$, $j = 10$, and $k = 23$.

23 **EXPLORATION** Refer to the beginning of the lesson. Find the distance between the ship and the shipwreck. Round to the nearest tenth.

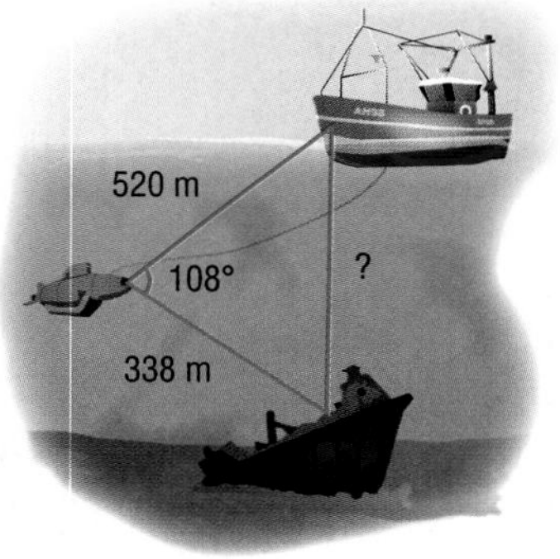

24. GEOMETRY A parallelogram has side lengths 8 centimeters and 12 centimeters. One angle between them measures 42°. To the nearest tenth, what is the length of the shorter diagonal?

25. RACING A triangular cross-country course has side lengths 1.8 kilometers, 2 kilometers, and 1.2 kilometers. What are the angles formed between each pair of sides?

Problem-Solving Tip

Work Backward
When the information needed to solve a problem is not given in an organized manner, begin with what you need to find and arrange the information from end to beginning.

26. SURVEYING A triangular plot of farm land measures 0.9 by 0.5 by 1.25 miles.

a. If the plot of land is fenced on the border, what will be the angles at which the fences of the three sides meet? Round to the nearest degree.

b. What is the area of the plot of land?

27. LAND Some land is in the shape of a triangle. The distances betwen each vertex of the triangle are 140 yd, 210 yd and 300 yd, respectively. Use the Law of Cosines to find the area of the land to the nearest square yard.

StudyTip

Rounding Sometimes rounding can lead to ambiguous answers, such as a triangle appearing to be 181 degrees.

28. **RIDES** Two bumper cars at an amusement park ride collide as shown below.

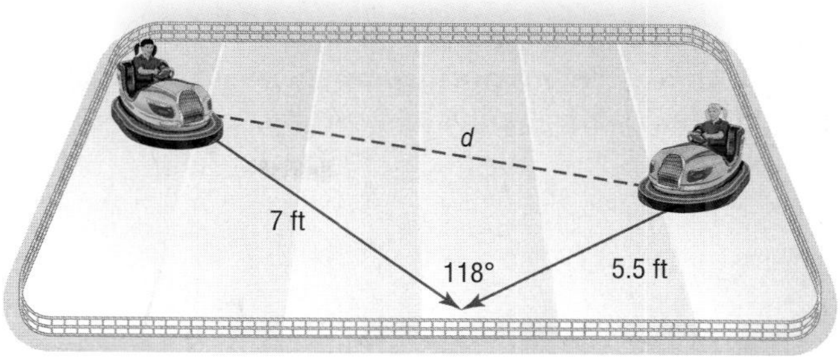

a. How far apart d were the two cars before they collided?

b. Before the collision, a third car was 10 feet from car 1 and 13 feet from car 2. Describe the angles formed by cars 1, 2, and 3 before the collision.

29. **PICNICS** A triangular picnic area is 11 yards by 14 yards by 10 yards.

a. Sketch and label a drawing to represent the picnic area.

b. Describe how you could find the area of the picnic area.

c. What is the area? Round to the nearest tenth.

30. **WATERSPORTS** A person on a personal watercraft makes a trip from point A to point B to point C traveling 28 miles per hour. She then returns from point C back to her starting point traveling 35 miles per hour. How many minutes did the entire trip take? Round to the nearest tenth.

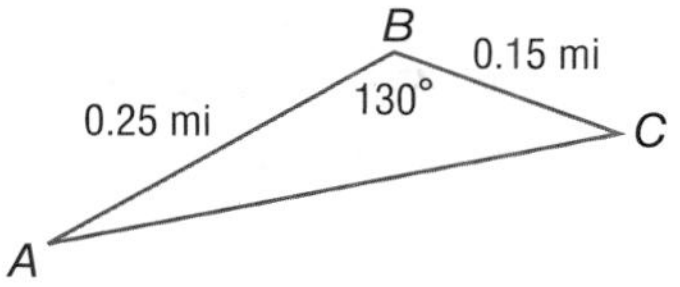

Real-World Link

Personal watercraft were first developed in the 1960s and became popular in the mid 1970s. Today, personal watercraft account for about $\frac{1}{3}$ of all powerboat sales.

Source: Personal Watercraft Industry Association

Solve each triangle. Round side lengths to the nearest tenth and angle measures to the nearest degree.

31.

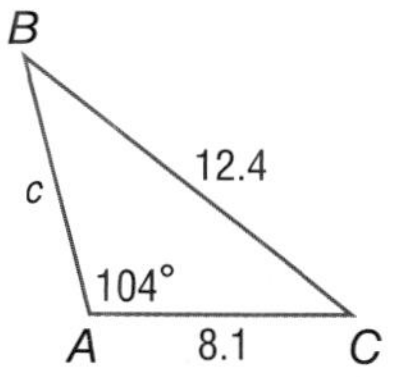

32.

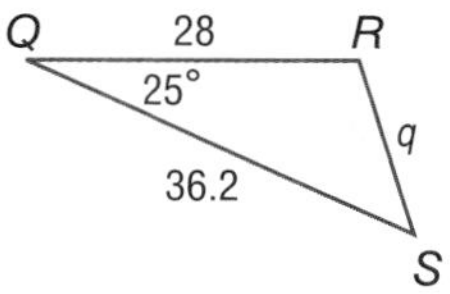

33.

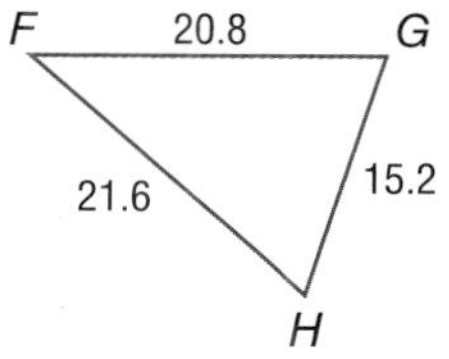

H.O.T. Problems Use Higher-Order Thinking Skills

34. **CHALLENGE** Use the figure and the Pythagorean Theorem to derive the Law of Cosines. Use the hints below.

- First, use the Pythagorean Theorem for $\triangle DBC$.
- In $\triangle ADB$, $c^2 = x^2 + h^2$.
- $\cos A = \frac{x}{c}$

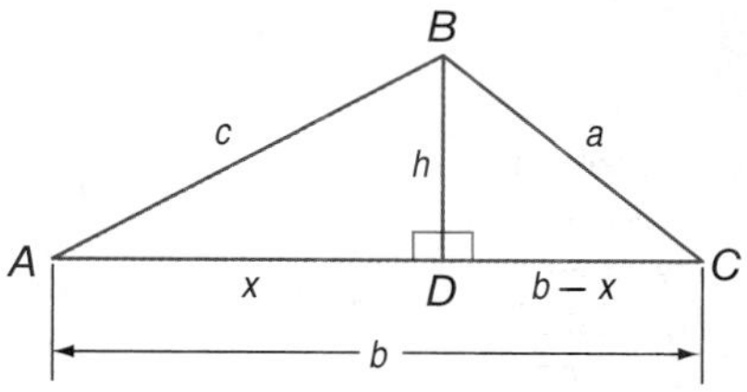

35. **REASONING** Three sides of a triangle measure 10.6 centimeters, 8 centimeters, and 14.5 centimeters. Explain how to find the measure of the largest angle. Then find the measure of the angle to the nearest degree.

36. **OPEN ENDED** Find three measures of a triangle so that the Law of Cosines can be used to solve the triangle. Then solve the triangle.

37. **WRITING IN MATH** Compare the circumstances in which you can use the Law of Sines and the Law of Cosines to solve a triangle.

NYSRE PRACTICE A2.A.66, A2.A.73

38. SAT/ACT If c and d are different positive integers and $4c + d = 26$, what is the sum of all possible values of c?

A 6
B 10
C 15
D 21

39. If $6^y = 21$, what is y?

F $\log 12 - \log 6$
G $\dfrac{\log 21}{\log 6}$
H $\dfrac{\log 6}{\log 21}$
J $\log\left(\dfrac{6}{21}\right)$

40. GEOMETRY Find the perimeter of the figure.

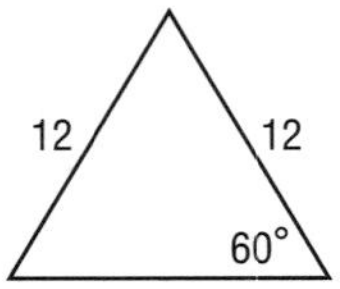

A 24 B 30 C 36 D 48

41. SHORT RESPONSE Solve the equation below for x.

$$\frac{1}{x-1} + \frac{5}{8} = \frac{23}{6x}$$

Spiral Review

Find the area of △ABC to the nearest tenth. (Lesson 13-4)

42.

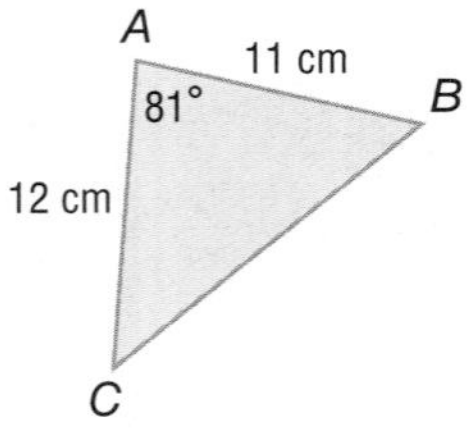

43.

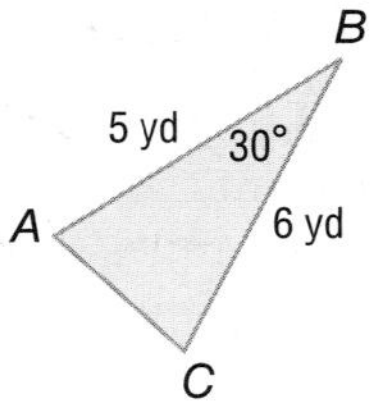

44.

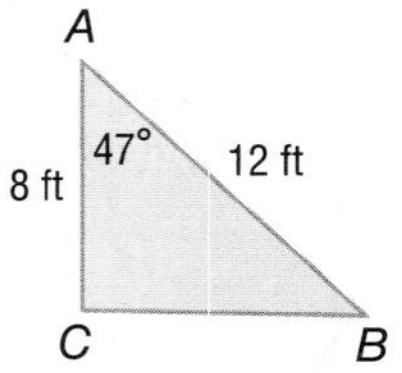

The terminal side of θ in standard position contains each point. Find the exact values of the six trigonometric functions of θ. (Lesson 13-3)

45. (8, 5) **46.** (−4, −2) **47.** (6, −9)

48. EDUCATION The Millersburg school board is negotiating a pay raise with the teachers' union. Three of the administrators have salaries of $90,000 each. However, a majority of the teachers have salaries of about $45,000 per year. (Lesson 12-2)

a. You are a member of the school board and would like to show that the current salaries are reasonable. Would you quote the mean, median, or mode as the "average" salary to justify your claim? Explain.

b. You are the head of the teachers' union and maintain that a pay raise is in order. Which of the mean, median, or mode would you quote to justify your claim? Explain your reasoning.

49. CELL PHONES A person using a cell phone can be located with respect to three cellular towers. In a coordinate system where a unit represents one mile, a caller is determined to be 50 miles from the tower at the origin. He is also 40 miles from a tower at (0, 30) and 13 miles from a tower at (35, 18). Where is the caller? (Lesson 10-7)

Without writing the equation in standard form, state whether the graph of each equation is a *parabola, circle, ellipse,* or *hyperbola.* (Lesson 10-6)

50. $x^2 + y^2 - 8x - 6y + 5 = 0$ **51.** $3x^2 - 2y^2 + 32y - 134 = 0$ **52.** $y^2 + 18y - 2x = -84$

Skills Review

Sketch each angle. Then find its reference angle. (Lesson 13-3)

53. 245° **54.** −15° **55.** $\frac{5}{4}\pi$

Mid-Chapter Quiz

Lessons 13-1 through 13-5

NYS Core Curriculum
A2.A.55, A2.A.73

Solve $\triangle XYZ$ by using the given measurements. Round measures of sides to the nearest tenth and measures of angles to the nearest degree. (Lesson 13-1)

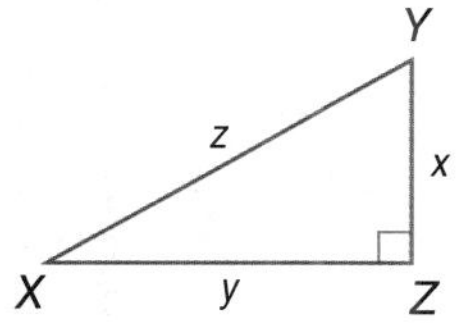

1. $Y = 65°, x = 16$

2. $X = 25°, x = 8$

3. Find the values of the six trigonometric functions for angle θ. (Lesson 13-1)

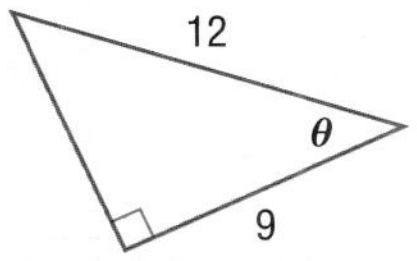

4. Draw an angle measuring $-80°$ in standard position. (Lesson 13-2)

Rewrite each degree measure in radians and each radian measure in degrees. (Lesson 13-2)

5. $215°$

6. $-350°$

7. $\frac{8\pi}{5}$

8. $\frac{9\pi}{2}$

9. **MULTIPLE CHOICE** What is the length of the arc below rounded to the nearest tenth? (Lesson 13-2)

A 4.2 cm

B 17.1 cm

C 53.9 cm

D 2638.9 cm

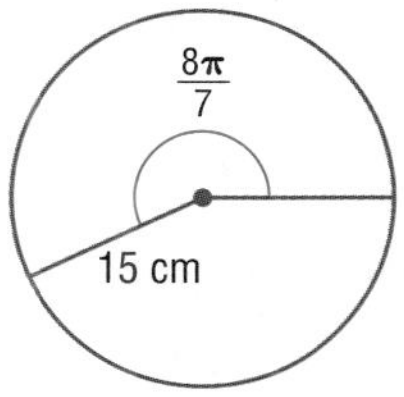

Find the exact value of each trigonometric function. (Lesson 13-3)

10. $\tan \pi$

11. $\cos \frac{3\pi}{4}$

The terminal side of θ in standard position contains each point. Find the exact values of the six trigonometric functions of θ. (Lesson 13-3)

12. $(0, -5)$

13. $(6, 8)$

14. **GARDEN** Lana has a garden in the shape of a triangle as pictured below. She wants to fill the garden with top soil. What is the area of the triangle? (Lesson 13-4)

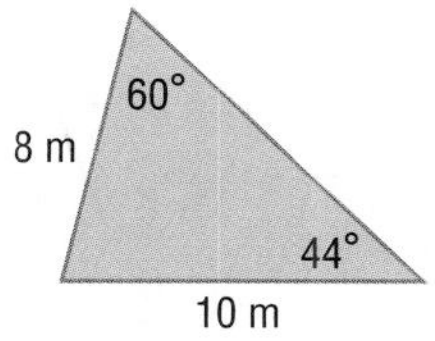

Determine whether each triangle has *no* solution, *one* solution, or *two* solutions. Then solve the triangle. Round side lengths to the nearest tenth and angle measures the nearest degree. (Lesson 13-4)

15. $A = 38°, a = 18, c = 25$

16. $A = 65°, a = 5, b = 7$

17. $A = 115°, a = 12, b = 8$

Solve each triangle. Round side lengths to the nearest tenth and angle measures to the nearest degree. (Lesson 13-5)

18.

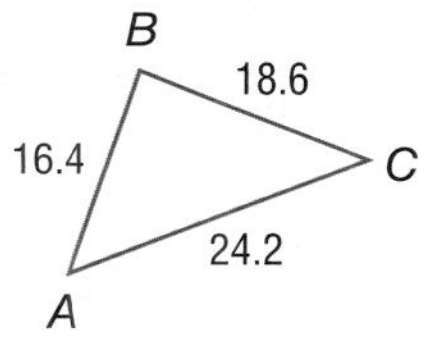

19.

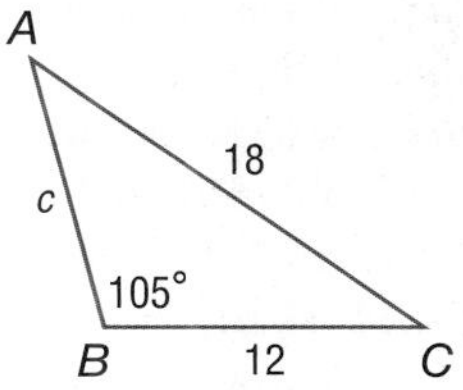

20. Eric and Zach are camping. Erik leaves Zach at the campsite and walks 4.5 miles. He then turns at a 120°angle and walks another 2.5 miles. If Eric were to walk directly back to Zach, how far would he walk? (Lesson 13-5)

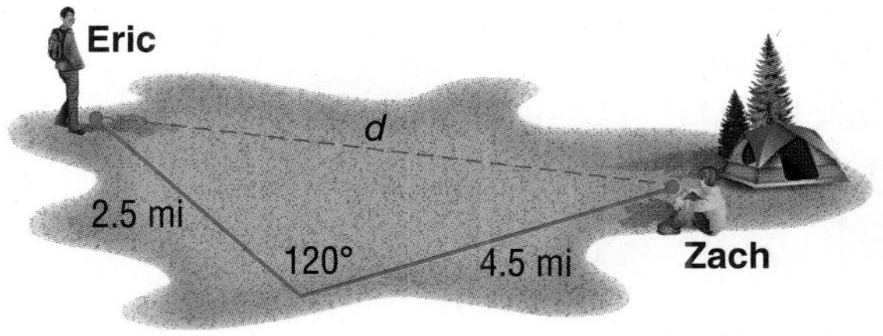

21. **MULTIPLE CHOICE** Suppose θ is an angle in standard position with $\cos \theta > 0$. In which quadrant(s) does the terminal side of θ lie? (Lesson 13-2)

F I

G II

H III

J I and IV

13-6 Circular Functions

Then

You evaluated trigonometric functions using reference angles. (Lesson 13-3)

Now

- Find values of trigonometric functions based on the unit circle.
- Use the properties of periodic functions to evaluate trigonometric functions.

NYS Core Curriculum

A2.A.56 Know the exact and approximate values of the sine, cosine, and tangent of 0°, 30°, 45°, 60°, 90°, 180°, and 270° angles
A2.A.60 Sketch the unit circle and represent angles in standard position

New Vocabulary

unit circle
circular function
periodic function
cycle
period

NY Math Online

glencoe.com

- Extra Examples
- Personal Tutor
- Self-Check Quiz
- Homework Help

Why?

The pedals on a bicycle rotate as the bike is being ridden. The height of a pedal is a function of time, as shown in the figure at the right.

Notice that the pedal makes one complete rotation every two seconds.

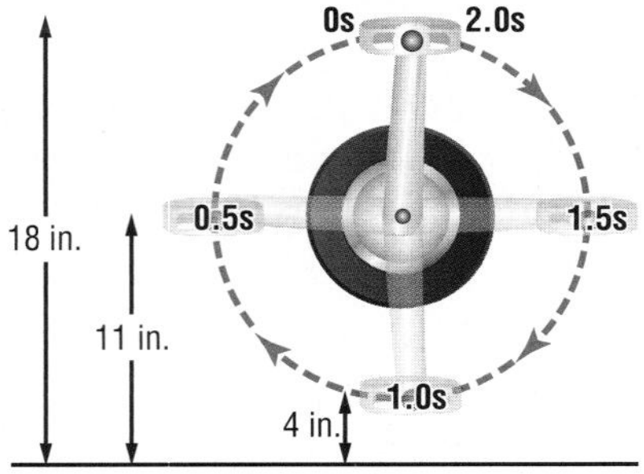

Circular Functions A **unit circle** is a circle with a radius of 1 unit centered at the origin on the coordinate plane. You can use a point P on the unit circle to generalize sine and cosine functions.

$$\sin \theta = \frac{y}{r} = \frac{y}{1} \text{ or } y \qquad \cos \theta = \frac{x}{r} = \frac{x}{1} \text{ or } x$$

So, the values of $\sin \theta$ and $\cos \theta$ are the y-coordinate and x-coordinate, respectively, of the point where the terminal side of θ intersects the unit circle.

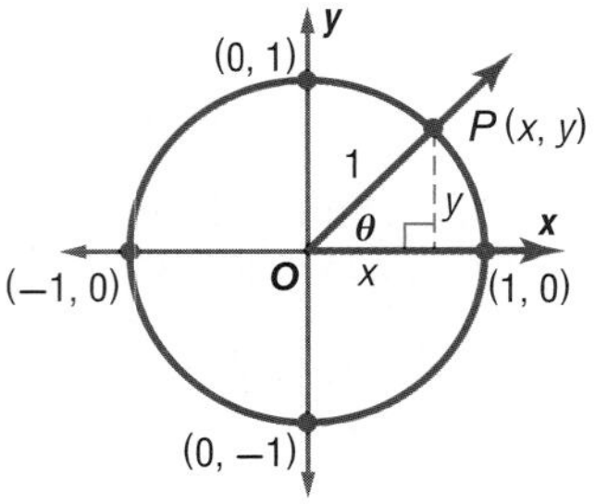

Key Concept — Functions on a Unit Circle

For Your FOLDABLE

Words If the terminal side of an angle θ in standard position intersects the unit circle at $P(x, y)$, then $\cos \theta = x$ and $\sin \theta = y$.

Symbols $P(x, y) = P(\cos \theta, \sin \theta)$

Example If $\theta = 120°$, $P(x, y) = P(\cos 120°, \sin 120°)$.

Model

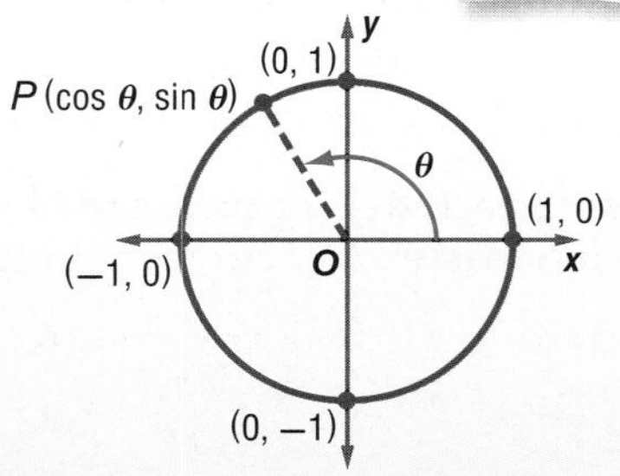

Both $\cos \theta = x$ and $\sin \theta = y$ are functions of θ. Because they are defined using a unit circle, they are called **circular functions**.

EXAMPLE 1 Find Sine and Cosine Given a Point on the Unit Circle

The terminal side of angle θ in standard position intersects the unit circle at $P\left(\frac{1}{2}, \frac{\sqrt{3}}{2}\right)$. Find $\cos \theta$ and $\sin \theta$.

$$P\left(\frac{1}{2}, \frac{\sqrt{3}}{2}\right) = P(\cos \theta, \sin \theta)$$

$$\cos \theta = \frac{1}{2} \qquad \sin \theta = \frac{\sqrt{3}}{2}$$

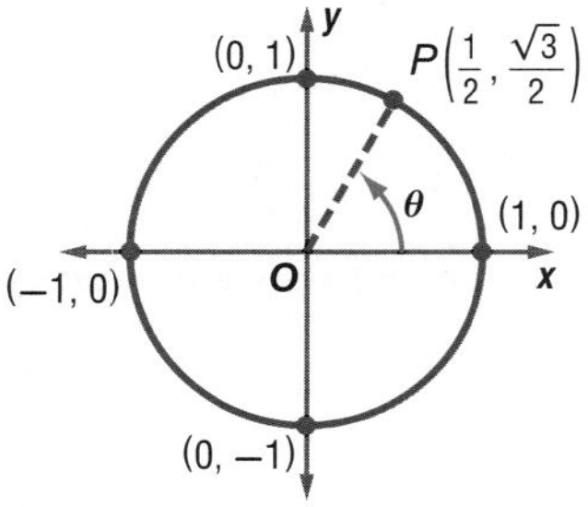

Check Your Progress

1. The terminal side of angle θ in standard position intersects the unit circle at $P\left(\frac{3}{5}, -\frac{4}{5}\right)$. Find $\cos \theta$ and $\sin \theta$.

Personal Tutor glencoe.com

Periodic Functions A **periodic function** has y-values that repeat at regular intervals. One complete pattern is a **cycle**, and the horizontal length of one cycle is a **period**.

StudyTip

Cycles A cycle can begin at any point on the graph of a periodic function. In Example 2, if the beginning of the cycle is at $\frac{\pi}{2}$, then the pattern repeats at $\frac{3\pi}{2}$. The period is $\frac{3\pi}{2} - \frac{\pi}{2}$ or π.

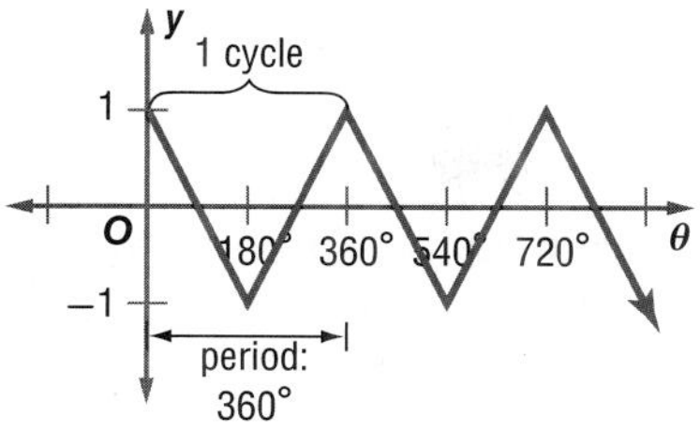

θ	y
0°	1
180°	−1
360°	1
540°	−1
720°	1

The cycle repeats every 360°.

EXAMPLE 2 Identify the Period

Determine the period of the function.

The pattern repeats at π, 2π, and so on. So, the period is π.

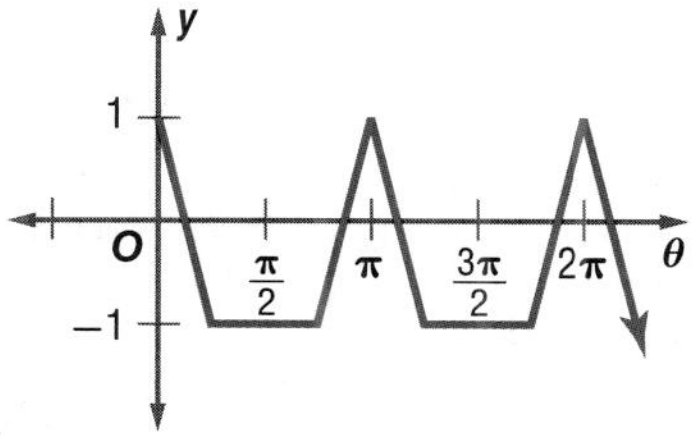

Check Your Progress

2. Graph a function with a period of 4.

Personal Tutor glencoe.com

Real-World Link

Most competitive cyclists pedal at rates of more than 200 rotations per minute. Most other people pedal at between 90 and 120 rotations per minute.

Source: SpringerLink

The rotations of wheels, pedals, carousels, and objects in space are all periodic.

Real-World EXAMPLE 3 Use Trigonometric Functions

CYCLING Refer to the beginning of the lesson. The height of a bicycle pedal varies periodically as a function of time, as shown in the figure.

a. Make a table showing the height of a bicycle pedal at 0, 0.5, 1.0, 1.5, 2.0, 2.5, and 3.0 seconds.

At 0 seconds, the pedal is 18 inches high. At 0.5 second, the pedal is 11 inches high. At 1.0 second, the pedal is 4 inches high, and so on.

Time (s)	Height (in.)
0	18
0.5	11
1.0	4
1.5	11
2.0	18
2.5	11
3.0	4

b. Identify the period of the function.

The period is the time it takes to complete one rotation. So, the period is 2 seconds.

c. Graph the function. Let the horizontal axis represent the time t and the vertical axis represent the height h in inches that the pedal is from the ground.

The maximum height of the pedal is 18 inches, and the minimum height is 4 inches. Because the period of the function is 2 seconds, the pattern of the graph repeats in intervals of 2 seconds.

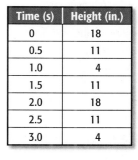

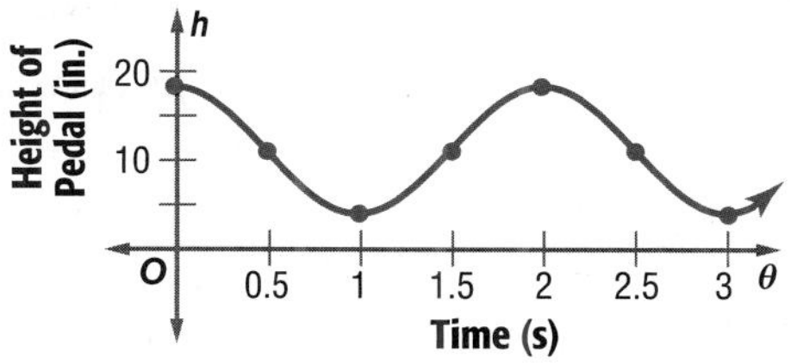

Check Your Progress

3. CYCLING Another cyclist pedals the same bike at a rate of 1 revolution per second.

A. Make a table showing the height of a bicycle pedal at times 0, 0.5, 1.0, 1.5, 2.0, 2.5, and 3.0 seconds.

B. Identify the period and graph the function.

Personal Tutor glencoe.com

StudyTip

Sine and Cosine To help you remember that for (*x*, *y*) on a unit circle, $x = \cos\theta$ and $y = \sin\theta$, notice that alphabetically *x* comes before *y* and *cosine* comes before *sine*.

The exact values of $\cos\theta$ and $\sin\theta$ for special angles are shown on the unit circle at the right. The cosine values are the *x*-coordinates of the points on the unit circle, and the sine values are the *y*-coordinates.

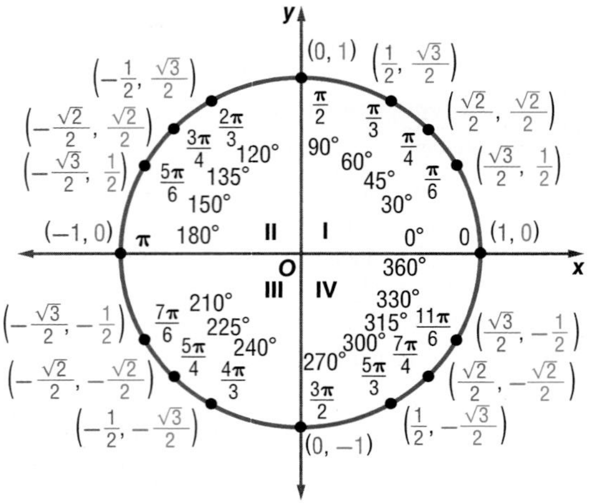

You can use this information to graph the sine and cosine functions. Let the horizontal axis represent the values of θ and the vertical axis represent the values of $\sin\theta$ or $\cos\theta$.

The cycles of the sine and cosine functions repeat every 360°. So, they are periodic functions. The period of each function is 360° or 2π.

Consider the points on the unit circle for $\theta = 45°$, $\theta = 150°$, and $\theta = 270°$.

$$(\cos 45°, \sin 45°) = \left(\frac{\sqrt{2}}{2}, \frac{\sqrt{2}}{2}\right)$$

$$(\cos 150°, \sin 150°) = \left(-\frac{\sqrt{3}}{2}, \frac{1}{2}\right)$$

$$(\cos 270°, \sin 270°) = (0, -1)$$

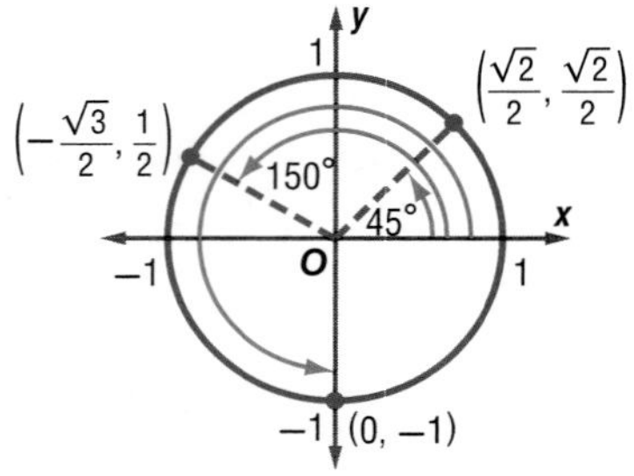

StudyTip

Radians The sine and cosine functions can also be graphed using radians as the units on the θ-axis.

These points can also be shown on the graphs of the sine and cosine functions.

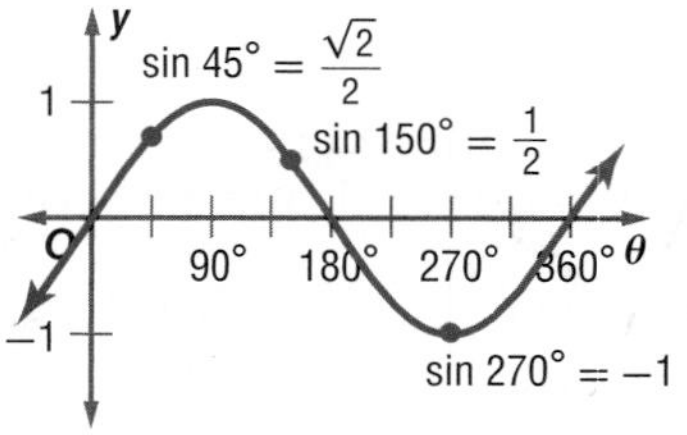

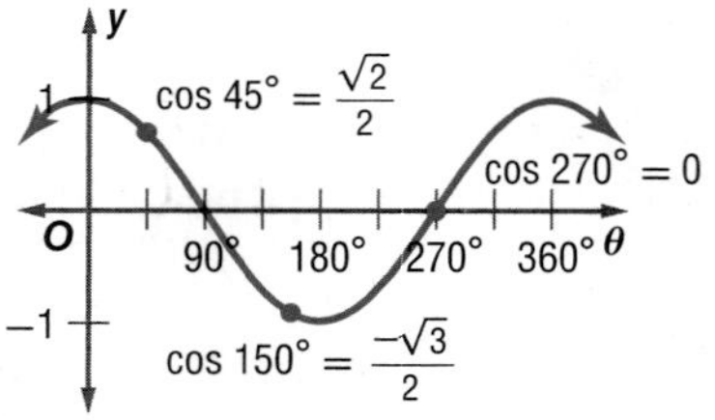

Since the period of the sine and cosine functions is 360°, the values repeat every 360°. So, $\sin(x + 360°) = \sin x$, and $\cos(x + 360°) = \cos x$.

EXAMPLE 4 Evaluate Trigonometric Functions

Find the exact value of each function.

a. cos 480°

$$\cos 480° = \cos(120° + 360°)$$
$$= \cos 120°$$
$$= -\frac{1}{2}$$

b. $\sin \frac{11\pi}{4}$

$$\sin\frac{11\pi}{4} = \sin\left(\frac{3\pi}{4} + \frac{8\pi}{4}\right)$$
$$= \sin\frac{3\pi}{4}$$
$$= \frac{\sqrt{2}}{2}$$

Check Your Progress

4A. $\cos\left(-\frac{3\pi}{4}\right)$

4B. $\sin 420°$

Personal Tutor glencoe.com

Check Your Understanding

Example 1 p. 848

The terminal side of angle θ in standard position intersects the unit circle at each point P. Find $\cos\theta$ and $\sin\theta$.

1. $P\left(\frac{15}{17}, \frac{8}{17}\right)$

2. $P\left(-\frac{\sqrt{2}}{2}, \frac{\sqrt{2}}{2}\right)$

Example 2 p. 849

Determine the period of each function.

3.

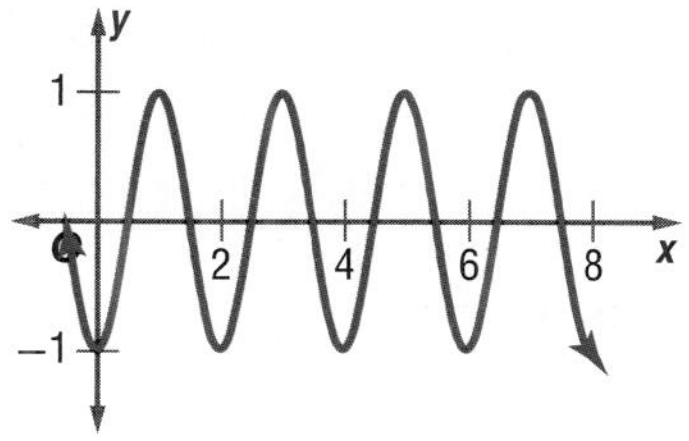

4.

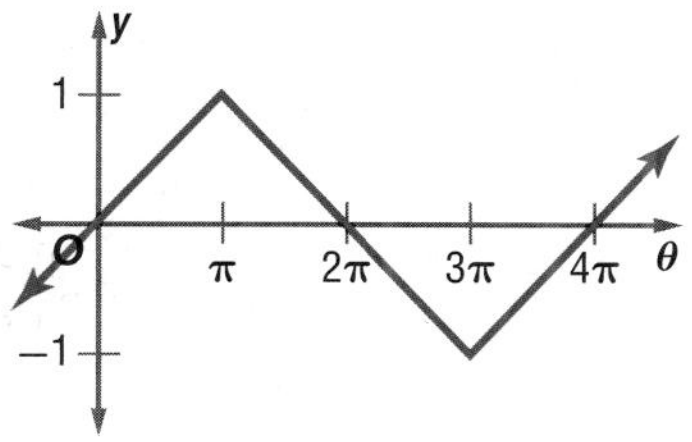

Example 3 p. 849

5. SWINGS The height of a swing varies periodically as the function of time. The swing goes forward and reaches its high point of 6 feet. It then goes backward and reaches 6 feet again. Its lowest point is 2 feet. The time it takes to swing from its high point to its low point is 1 second.

a. How long does it take for the swing to go forward and back one time?

b. Graph the height of the swing h as a function of time t.

Example 4 p. 850

Find the exact value of each function.

6. $\sin\frac{13\pi}{6}$

7. $\sin(-60°)$

8. $\cos 540°$

Practice and Problem Solving

= **Step-by-Step Solutions** begin on page R20.
Extra Practice begins on page 947.

Example 1 p. 848

The terminal side of angle θ in standard position intersects the unit circle at each point P. Find $\cos\theta$ and $\sin\theta$.

9. $P\left(\frac{6}{10}, -\frac{8}{10}\right)$

10. $P\left(-\frac{10}{26}, -\frac{24}{26}\right)$

11 $P\left(\frac{\sqrt{3}}{2}, \frac{1}{2}\right)$

12. $P\left(\frac{\sqrt{6}}{5}, \frac{\sqrt{19}}{5}\right)$

Example 2 p. 849

Determine the period of each function.

13.

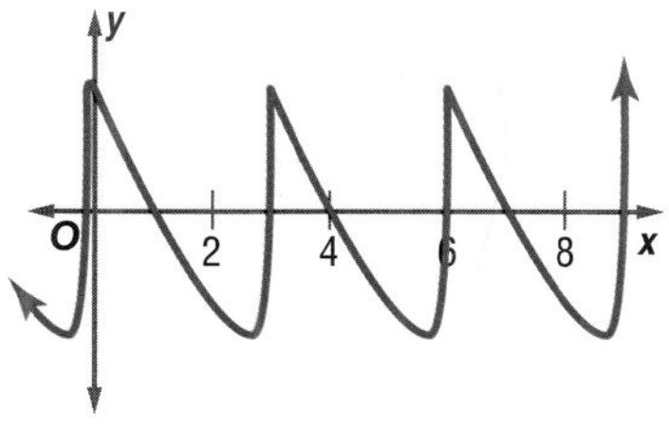

14.

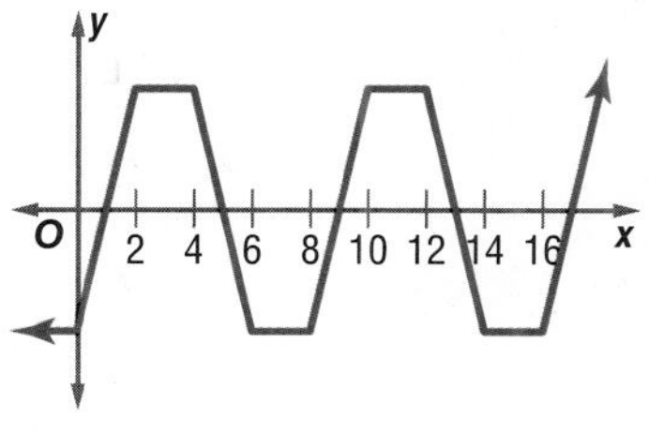

15.

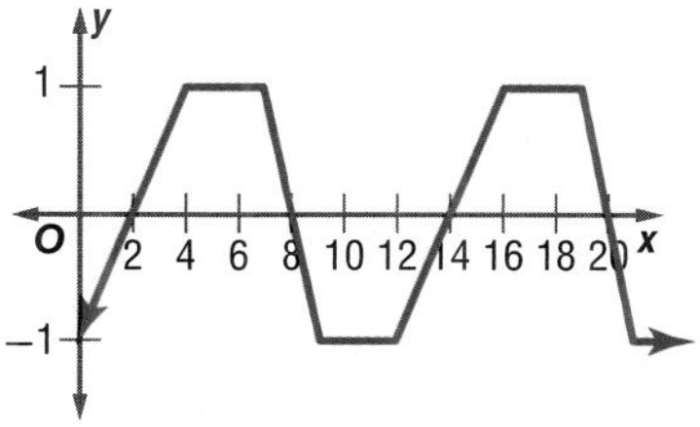

16.

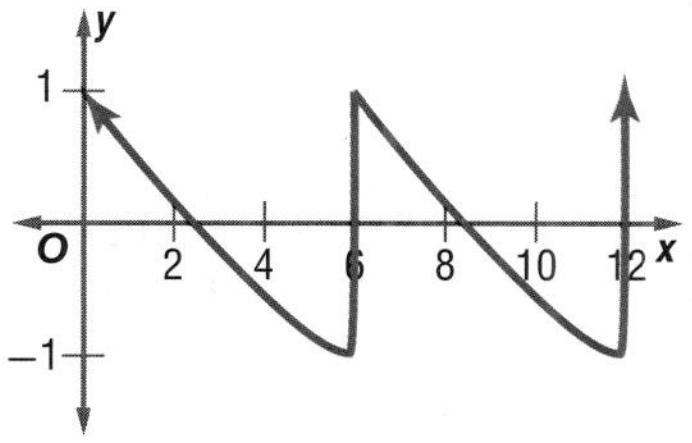

Determine the period of each function.

17.

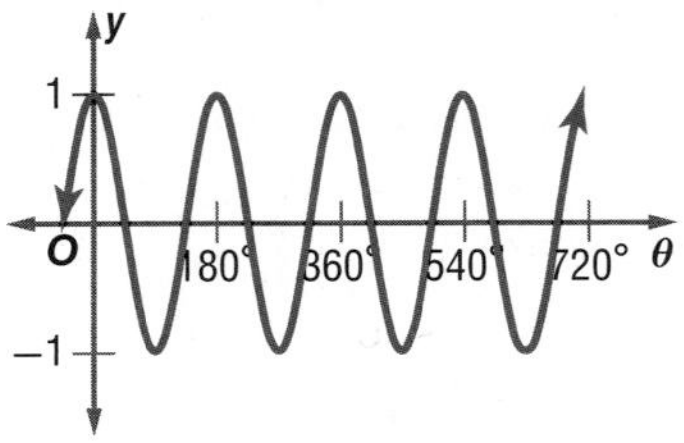

18.

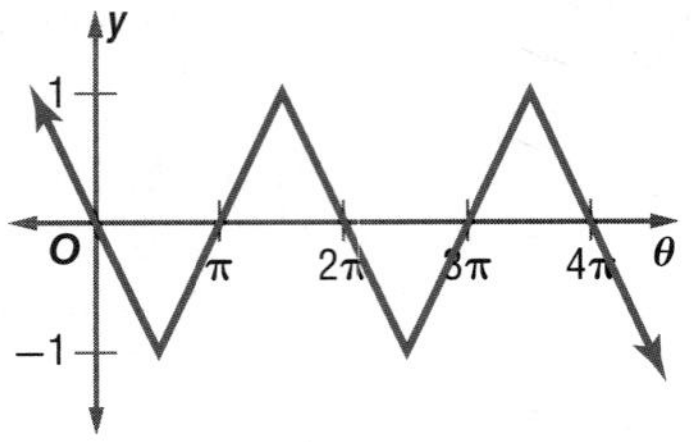

Example 3 p. 849

19. WEATHER In a city, the average high temperature for each month is shown in the table.

a. Sketch a graph of the function representing this situation.

b. Describe the period of the function.

Average High Temperatures

Month	Temperature (°F)	Month	Temperature (°F)
Jan	36	July	85
Feb.	41	Aug.	84
Mar.	52	Sept.	78
Apr.	64	Oct.	66
May	74	Nov.	52
Jun.	82	Dec.	41

Source: The Weather Channel

Example 4 p. 850

Find the exact value of each function.

20. $\sin \frac{7\pi}{3}$

21. $\cos(-60°)$

22. $\cos 450°$

23. $\sin \frac{11\pi}{4}$

24. $\sin(-45°)$

25. $\cos 570°$

26. ENGINES In the engine at the right, the distance d from the piston to the center of the circle, called the *crankshaft*, is a function of the speed of the piston rod. Point R on the piston rod rotates 150 times per second.

a. Identify the period of the function as a fraction of a second.

b. The shortest distance d is 0.5 inch, and the longest distance is 3.5 inches. Sketch a graph of the function. Let the horizontal axis represent the time t. Let the vertical axis represent the distance d.

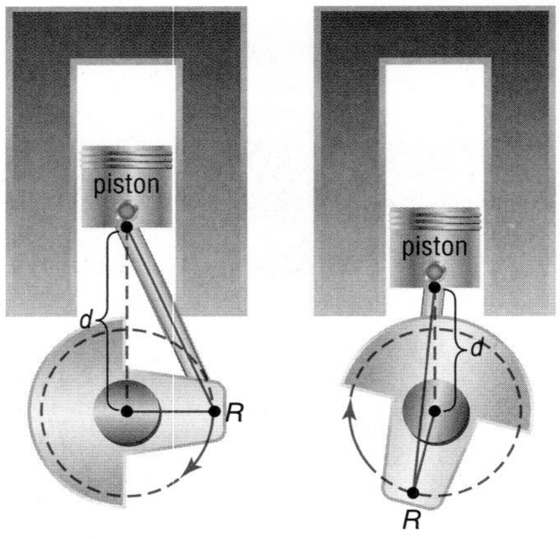

Real-World Link

Tornadoes can have whirling winds that reach 300 miles per hour. The forward speed of a tornado can reach up to 70 miles per hour.

Source: FEMA

27. TORNADOES A tornado siren makes 2.5 rotations per minute and the beam of sound has a radius of 1 mile. Ms. Miller's house is 1 mile from the siren. The distance of the sound beam from her house varies periodically as a function of time.

a. Identify the period of the function in seconds.

b. Sketch a graph of the function. Let the horizontal axis represent the time t from 0 seconds to 60 seconds. Let the vertical axis represent the distance d the sound beam is from Ms. Miller's house at time t.

28. FERRIS WHEEL A Ferris wheel in China has a diameter of approximately 520 feet. The height of a compartment h is a function of time t. It takes about 30 seconds to make one complete revolution. Let the height at the center of the wheel represent the height at time 0. Sketch a graph of the function.

Math History Link

Pauline Sperry (1885–1967) Pauline Sperry was born in Peabody, Massachusetts. During the 1920s, she wrote two textbooks, *Short Course in Spherical Trigonometry* and *Plane Trigonometry*. In 1923, she became the first woman to be promoted to assistant professor in the mathematics department at Berkeley.

29. **MULTIPLE REPRESENTATIONS** The terminal side of an angle in standard position intersects the unit circle at P, as shown in the figure.

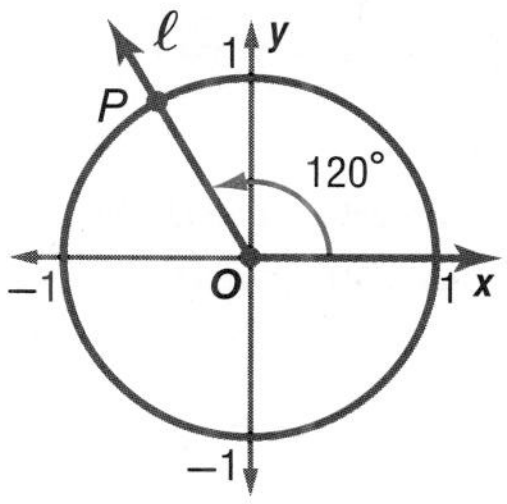

a. **GEOMETRIC** Copy the figure. Draw lines representing 30°, 60°, 150°, 210°, and 315°.

b. **TABULAR** Use a table of values to show the slope of each line to the nearest tenth.

c. **ANALYTICAL** What conclusions can you make about the relationship between the terminal side of the angle and the slope? Explain your reasoning.

30. **POGO STICK** A person is jumping up and down on a pogo stick at a constant rate. The difference between his highest and lowest points is 2 feet. He jumps 50 times per minute.

a. Describe the independent variable and dependent variable of the periodic function that represents this situation. Then state the period of the function in seconds.

b. Sketch a graph of the jumper's change in height in relation to his starting point. Assume that his starting point is halfway between his highest and lowest points. Let the horizontal axis represent the time t in seconds. Let the vertical axis represent the height h.

Find the exact value of each function.

31. $\cos 45° - \cos 30°$

32. $6(\sin 30°)(\sin 60°)$

33. $2 \sin \frac{4\pi}{3} - 3 \cos \frac{11\pi}{6}$

34. $\cos\left(-\frac{2\pi}{3}\right) + \frac{1}{3} \sin 3\pi$

35. $(\sin 45°)^2 + (\cos 45°)^2$

36. $\frac{(\cos 30°)(\cos 150°)}{\sin 315°}$

H.O.T. Problems

Use **H**igher-**O**rder **T**hinking Skills

37. **FIND THE ERROR** Francis and Benita are finding the exact value of $\cos \frac{-\pi}{3}$. Is either of them correct? Explain your reasoning.

Francis

$\cos \frac{-\pi}{3} = -\cos \frac{\pi}{3}$

$= -0.5$

Benita

$\cos \frac{-\pi}{3} = \cos\left(-\frac{\pi}{3} + 2\pi\right)$

$= \cos \frac{5\pi}{3}$

$= 0.5$

38. **CHALLENGE** A ray has its endpoint at the origin of the coordinate plane, and point $P\left(\frac{1}{2}, -\frac{\sqrt{3}}{2}\right)$ lies on the ray. Find the angle θ formed by the positive x-axis and the ray.

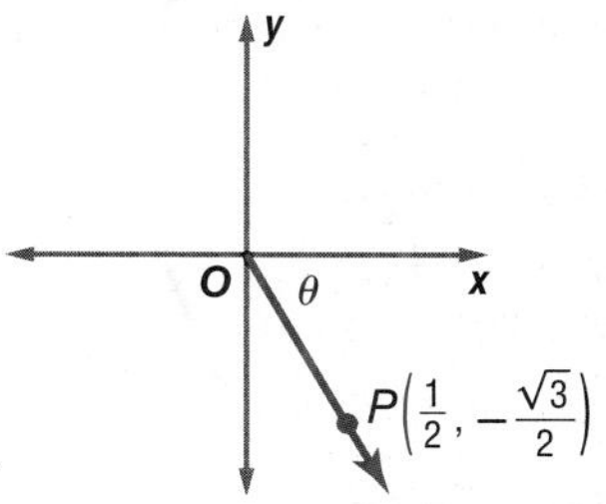

39. **REASONING** Is the period of a sine curve *sometimes*, *always*, or *never* a multiple of π? Justify your reasoning.

40. **OPEN ENDED** Draw the graph of a periodic function that has a maximum value of 10 and a minimum value of −10. Describe the period of the function.

41. **WRITING IN MATH** Explain how to determine the period of a periodic function from its graph. Include a description of a cycle.

NYSRE PRACTICE A2.A.56, A2.A.60

42. SHORT RESPONSE Describe the translation of the graph of $f(x) = x^2$ to the graph of $g(x) = (x + 4)^2 - 3$.

43. The rate of population decline of Hampton Cove is modeled by $P(t) = 24{,}000e^{-0.0064t}$, where t is time in years from this year and 24,000 is the current population. In how many years will the population be 10,000?

A 14 B 104 C 137 D 375

44. SAT/ACT If $d^2 + 8 = 21$, then $d^2 - 8 =$

F 5 G 13 H 31 J 161

45. STATISTICS If the average of three different positive integers is 65, what is the greatest possible value of one of the integers?

A 192 B 193 C 194 D 195

46. GRIDDED RESPONSE If $8xy + 3 = 3$, what is the value of xy?

Spiral Review

Solve each triangle. Round side lengths to the nearest tenth and angle measures to the nearest degree. (Lesson 13-5)

47.

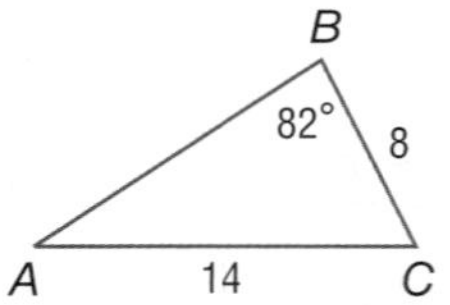

48.

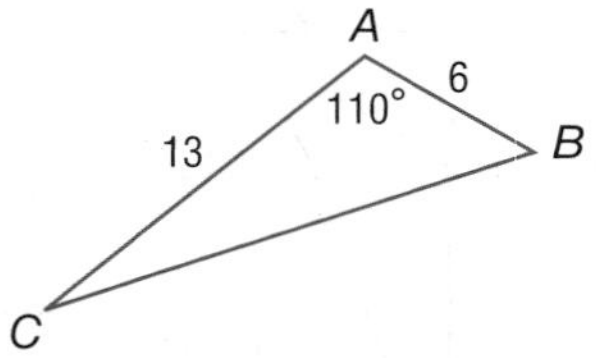

49.

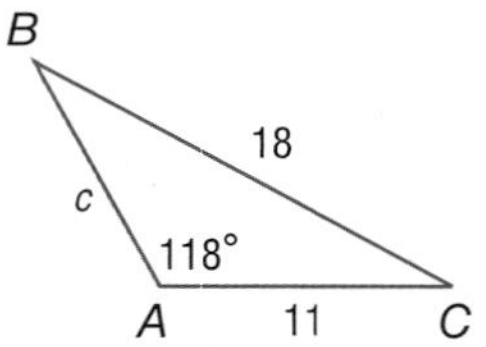

Determine whether each triangle has *no* solution, *one* solution, or *two* solutions. Then solve the triangle. Round side lengths to the nearest tenth and angle measures to the nearest degree. (Lesson 13-4)

50. $A = 72°, a = 6, b = 11$

51. $A = 46°, a = 10, b = 8$

52. $A = 110°, a = 9, b = 5$

A binomial distribution has a 70% rate of success. There are 10 trials. (Lesson 12-7)

53. What is the probability that there will be 3 failures?

54. What is the probability that there will be at least 7 successes?

55. What is the expected number of successes?

56. GAMES The diagram shows the board for a game in which spheres are dropped down a chute. A pattern of nails and dividers causes the disks to take various paths to the sections at the bottom. For each section, how many paths through the board lead to that section? (Lesson 11-6)

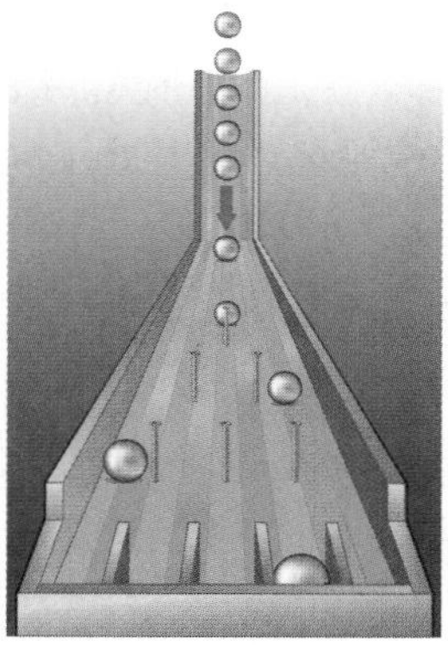

57. SALARIES Phillip's current salary is $40,000 per year. His annual pay raise is always a percent of his salary at the time. What would his salary be if he got four consecutive 4% increases? (Lesson 11-2)

Find the exact solution(s) of each system of equations. (Lesson 10-7)

58. $y = x + 2$
$y = x^2$

59. $4x + y^2 = 20$
$4x^2 + y^2 = 100$

Skills Review

Simplify each expression. (Lesson 1-4)

60. $\dfrac{240}{\left|1 - \frac{5}{4}\right|}$

61. $\dfrac{180}{\left|2 - \frac{1}{3}\right|}$

62. $\dfrac{90}{\left|2 - \frac{11}{4}\right|}$

13-7 Graphing Trigonometric Functions

Then

You examined periodic functions. (Lesson 13-6)

Now

- Describe and graph the sine, cosine, and tangent functions.
- Describe and graph other trigonometric functions.

NYS Core Curriculum

A2.A.70 Sketch and recognize one cycle of a function of the form $y = A \sin Bx$ or $y = A \cos Bx$
A2.A.71 Sketch and recognize the graphs of the functions $y = \sec(x)$, $y = \csc(x)$, $y = \tan(x)$, and $y = \cot(x)$ *Also addresses A2.R.6, A2.A.69, and A2.A.72.*

New Vocabulary

amplitude
frequency

NY Math Online

glencoe.com

- Extra Examples
- Personal Tutor
- Self-Check Quiz
- Homework Help
- Math in Motion

Why?

Visible light waves have different wavelengths or periods. Red has the longest wavelength and violet has the shortest wavelength.

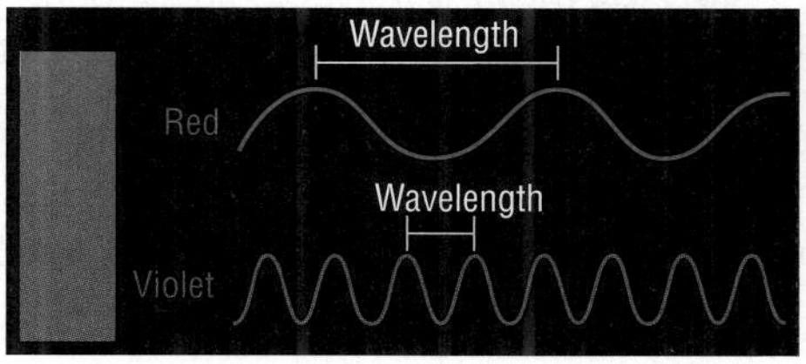

Sine, Cosine, and Tangent Functions Trigonometric functions can also be graphed on the coordinate plane. Recall that graphs of periodic functions have repeating patterns, or *cycles*. The horizontal length of each cycle is the *period*. The **amplitude** of the graph of a sine or cosine function equals half the difference between the maximum and minimum values of the function.

Key Concept: Sine and Cosine Functions

For Your FOLDABLE

Parent Function	$y = \sin \theta$	$y = \cos \theta$
Graph	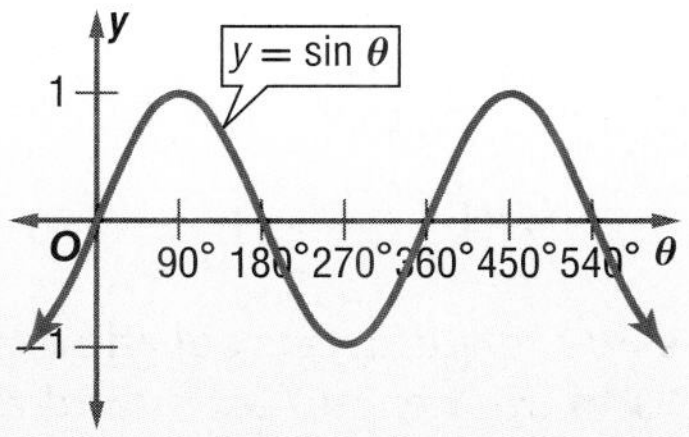	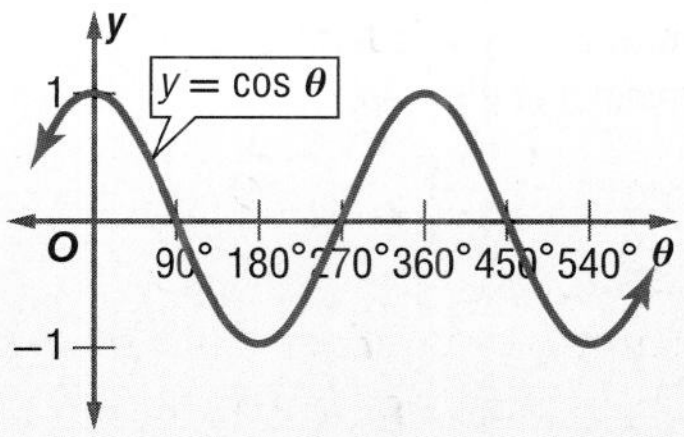
Domain	{all real numbers}	{all real numbers}
Range	$\{y \mid -1 \leq y \leq 1\}$	$\{y \mid -1 \leq y \leq 1\}$
Amplitude	1	1
Period	360°	360°

Math *in Motion*, Animation glencoe.com

As with other functions, trigonometric functions can be transformed. For the graphs of $y = a \sin b\theta$ and $y = a \cos b\theta$, the amplitude $= |a|$ and the period $= \frac{360°}{|b|}$.

EXAMPLE 1 Find Amplitude and Period

Find the amplitude and period of $y = 4 \cos 3\theta$.

amplitude: $|a| = |4|$ or 4

period: $\frac{360°}{|b|} = \frac{360°}{|3|}$ or 120°

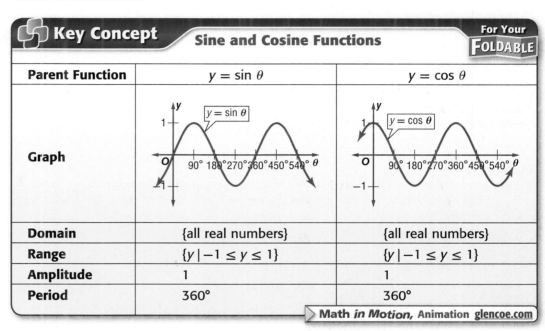

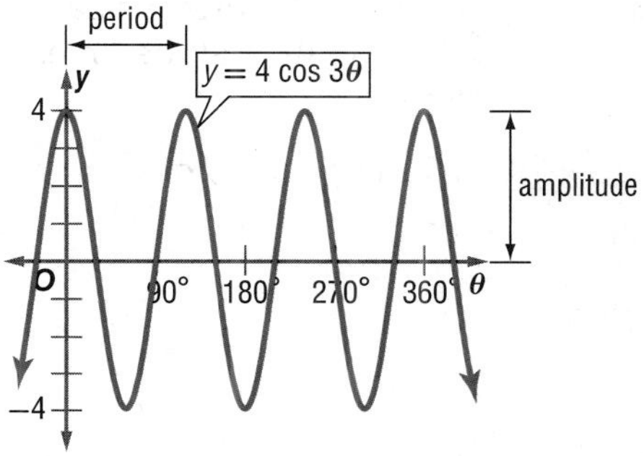

Check Your Progress

Find the amplitude and period of each function.

1A. $y = \cos \frac{1}{2}\theta$

1B. $y = 3 \sin 5\theta$

Personal Tutor glencoe.com

StudyTip

Periods In $y = a \sin b\theta$ and $y = a \cos b\theta$, b represents the number of cycles in 360°. In Example 1, the 3 in $y = 4 \cos 3\theta$ indicates that there are three cycles in 360°. So, there is one cycle in 120°.

StudyTip

Amplitude The graphs of $y = a \sin b\theta$ and $y = a \cos b\theta$ with amplitude of $|a|$ have maxima at $y = a$ and minima at $y = -a$.

Use the graphs of the parent functions to graph $y = a \sin b\theta$ and $y = a \cos b\theta$. Then use the amplitude and period to draw the appropriate sine and cosine curves. You can also use θ-intercepts to help you graph the functions.

The θ-intercepts of $y = a \sin b\theta$ and $y = a \cos b\theta$ in one cycle are as follows.

$y = a \sin b\theta$	$y = a \cos b\theta$
$(0, 0), \left(\frac{1}{2} \cdot \frac{360°}{b}, 0\right) \left(\frac{360°}{b}, 0\right)$	$\left(\frac{1}{4} \cdot \frac{360°}{b}, 0\right), \left(\frac{3}{4} \cdot \frac{360°}{b}, 0\right)$

EXAMPLE 2 Graph Sine and Cosine Functions

Graph each function.

a. $y = 2 \sin \theta$

Find the amplitude, the period, and the x-intercepts: $a = 2$ and $b = 1$.

amplitude: $|a| = |2|$ or 2 → The graph is stretched vertically so that the maximum value is 2 and the minimum value is −2.

period: $\frac{360°}{|b|} = \frac{360°}{|1|}$ or 360° → One cycle has a length of 360°.

x-intercepts: $(0, 0)$

$\left(\frac{1}{2} \cdot \frac{360°}{b}, 0\right) = (180°, 0)$

$\left(\frac{360°}{b}, 0\right) = (360°, 0)$

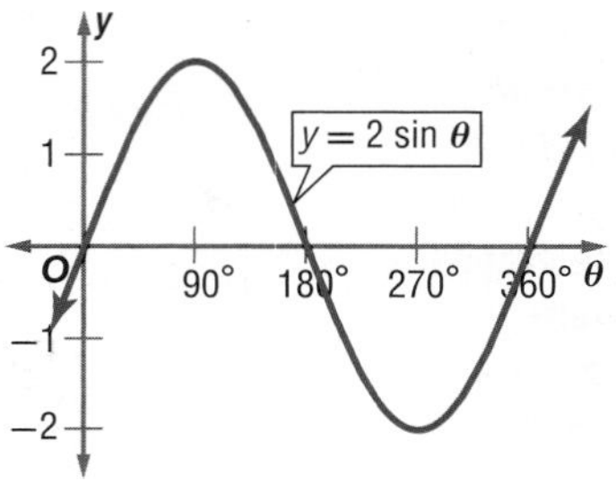

b. $y = \cos 4\theta$

amplitude: $|a| = |1|$ or 1

period: $\frac{360°}{|b|} = \frac{360°}{|4|}$ or 90°

x-intercepts: $\left(\frac{1}{4} \cdot \frac{360°}{b}, 0\right) = (22.5°, 0)$

$\left(\frac{3}{4} \cdot \frac{360°}{b}, 0\right) = (67.5°, 0)$

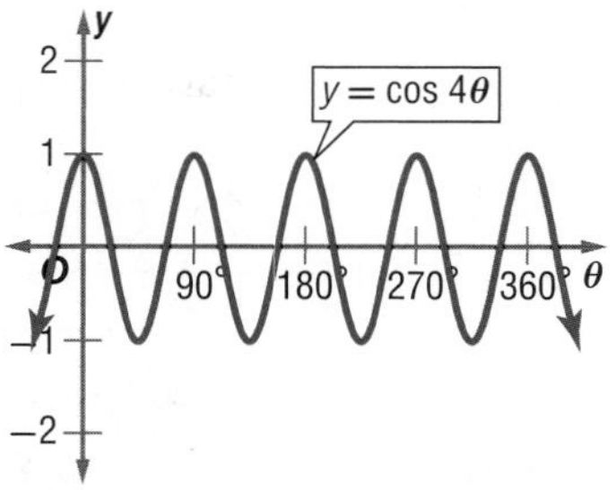

Check Your Progress

2A. $y = 3 \cos \theta$

2B. $y = \frac{1}{2} \sin 2\theta$

Personal Tutor glencoe.com

Trigonometric functions are useful for modeling real-world periodic motion such as electromagnetic waves or sound waves. Often these waves are described using *frequency*. **Frequency** is the number of cycles in a given unit of time.

The frequency of the graph of a function is the reciprocal of the period of the function. So, if the period of a function is $\frac{1}{100}$ second, then the frequency is 100 cycles per second.

Real-World Link

Elephants are able to hear sound coming from up to 5 miles away. Humans can hear sounds with frequencies between 20 Hz and 20,000 Hz.

Source: School for Champions

Real-World EXAMPLE 3 Model Periodic Situations

SOUND **Sound that has a frequency below the human range is known as *infrasound*. Elephants can hear sounds in the infrasound range, with frequencies as low as 5 hertz (Hz), or 5 cycles per second.**

a. Find the period of the function that models the sound waves.

There are 5 cycles per second, and the period is the time it takes for one cycle. So, the period is $\frac{1}{5}$ or 0.2 second.

b. Let the amplitude equal 1 unit. Write a sine equation to represent the sound wave y as a function of time t. Then graph the equation.

$\text{period} = \frac{2\pi}{|b|}$ **Write the relationship between the period and b.**

$0.2 = \frac{2\pi}{|b|}$ **Substitution**

$0.2|b| = 2\pi$ **Multiply each side by $|b|$.**

$b = 10\pi$ **Multiply each side by 5; b is positive.**

$y = a \sin b\theta$ **Write the general equation for the sine function.**

$y = 1 \sin 10\pi t$ **$a = 1$, $b = 10\pi$, and $\theta = t$**

$y = \sin 10\pi t$ **Simplify.**

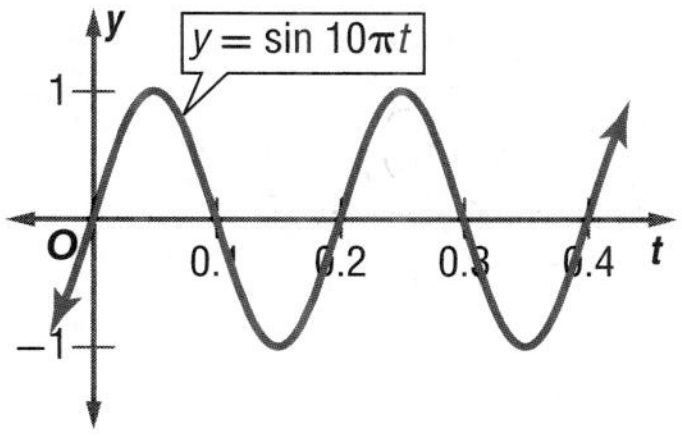

StudyTip

Amplitude and Period Note that the amplitude affects the graph along the vertical axis, and the period affects it along the horizontal axis.

Check Your Progress

3. SOUND Humans can hear sounds with frequencies as low as 20 hertz.

A. Find the period of the function.

B. Let the amplitude equal 1 unit. Write a cosine equation to model the sound waves. Then graph the equation.

Personal Tutor glencoe.com

Tangent is one of the trigonometric functions whose graphs have asymptotes.

Key Concept Tangent Function

For Your FOLDABLE

Parent Function	$y = \tan \theta$
Domain	$\{\theta \mid \theta \neq 90 + 180n$, n is an integer$\}$
Range	{all real numbers}
Amplitude	undefined
Period	180°
θ intercepts in one cycle	$(0, 0)$, $\left(\frac{1}{2} \cdot \frac{360°}{b}, 0\right)$, $\left(\frac{360°}{b}, 0\right)$

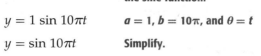

Graph

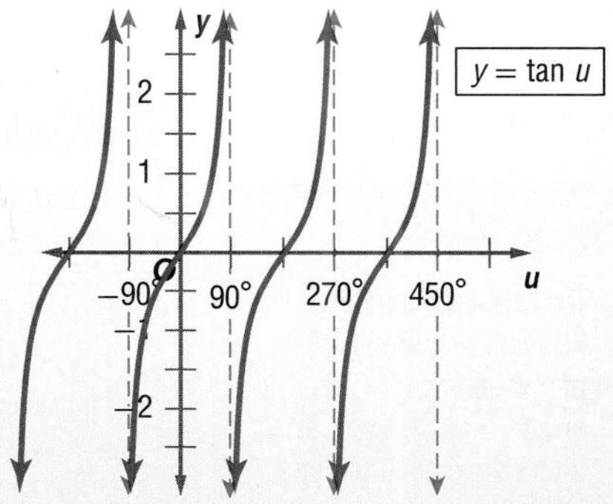

For the graph of $y = a \tan b\theta$, the period $= \frac{180°}{|b|}$, there is no amplitude, and the asymptotes are odd multiples of $\frac{180°}{2|b|}$.

StudyTip

Tangent The tangent function does not have an amplitude because it has no maximum or minimum values.

EXAMPLE 4 Graph Tangent Functions

Find the period of $y = \tan 2\theta$. Then graph the function.

period: $\frac{180°}{|b|} = \frac{180°}{|2|}$ or $90°$

asymptotes: $\frac{180°}{2|b|} = \frac{180°}{2|2|}$ or $45°$

Sketch asymptotes at $-1 \cdot 45°$ or $-45°$, $1 \cdot 45°$ or $45°$, $3 \cdot 45°$ or $135°$, and so on.

Use $y = \tan \theta$, but draw one cycle every $90°$.

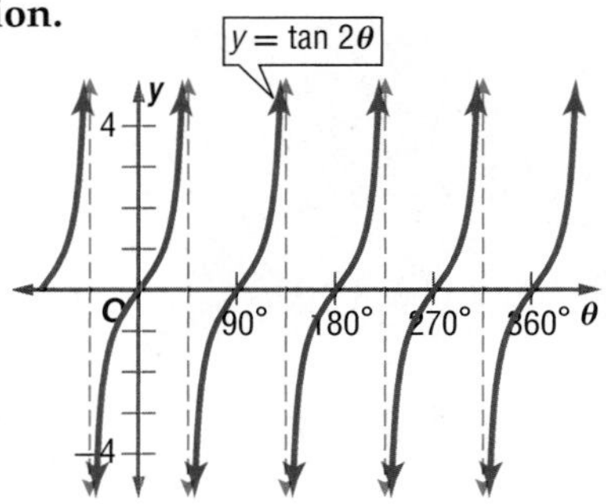

Check Your Progress

4. Find the period of $y = \frac{1}{2} \tan \theta$. Then graph the function.

Personal Tutor glencoe.com

Graphs of Other Trigonometric Functions The graphs of the cosecant, secant, and cotangent functions are related to the graphs of the sine, cosine, and tangent functions.

Key Concept Cosecant, Secant, and Cotangent Functions

For Your FOLDABLE

Parent Function	$y = \csc \theta$	$y = \sec \theta$	$y = \cot \theta$
Graph	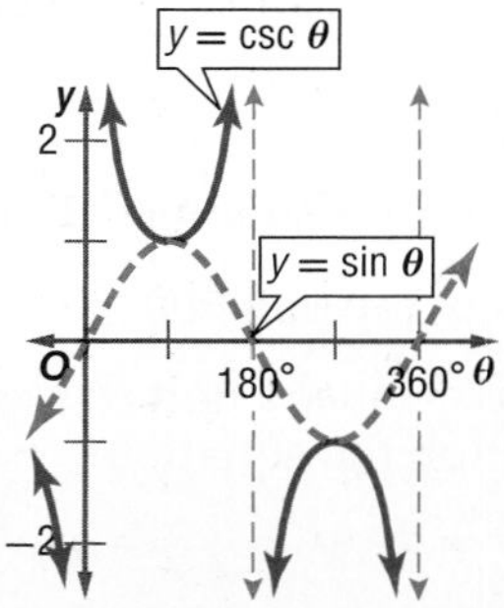	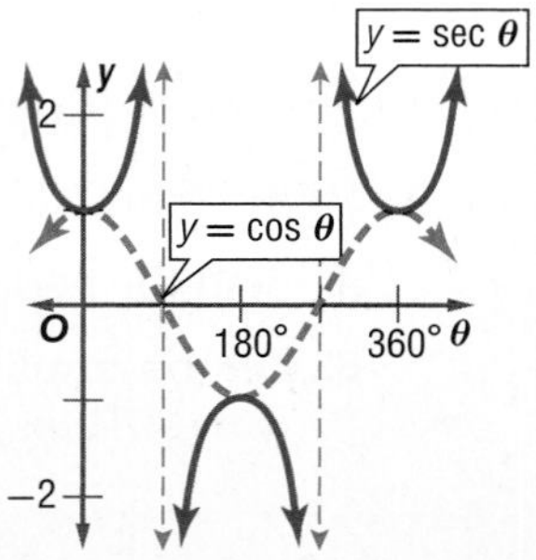	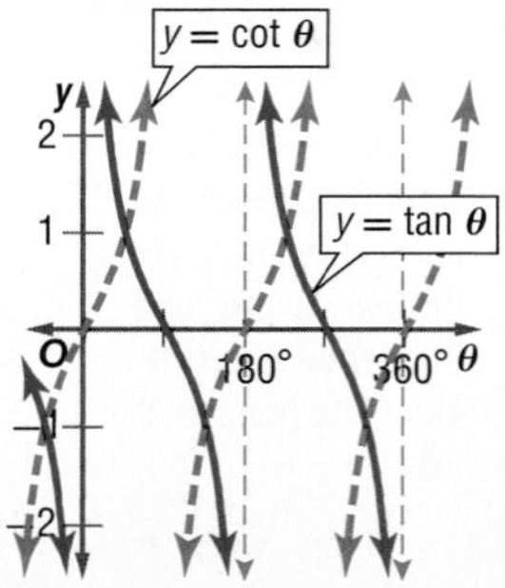
Domain	$\{\theta \mid \theta \neq 180n$, n is an integer$\}$	$\{\theta \mid \theta \neq 90 + 180n$, n is an integer$\}$	$\{\theta \mid \theta \neq 180n$, n is an integer$\}$
Range	$\{y \mid -1 > y \text{ or } y > 1\}$	$\{y \mid -1 > y \text{ or } y > 1\}$	{all real numbers}
Amplitude	undefined	undefined	undefined
Period	360°	360°	180°

StudyTip

Reciprocal Functions You can use the graphs of $y = \sin \theta$, $y = \cos \theta$, and $y = \tan \theta$ to graph the reciprocal functions, but these graphs are not part of the graphs of the cosecant, secant, and cotangent functions.

EXAMPLE 5 Graph Other Trigonometric Functions

Find the period of $y = 2 \sec \theta$. Then graph the function.

Since $2 \sec \theta$ is a reciprocal of $2 \cos \theta$, the graphs have the same period, $360°$. The vertical asymptotes occur at the points where $2 \cos \theta = 0$. So, the asymptotes are at $\theta = 90°$ and $\theta = 270°$.

Sketch $y = 2 \cos \theta$ and use it to graph $y = 2 \sec \theta$.

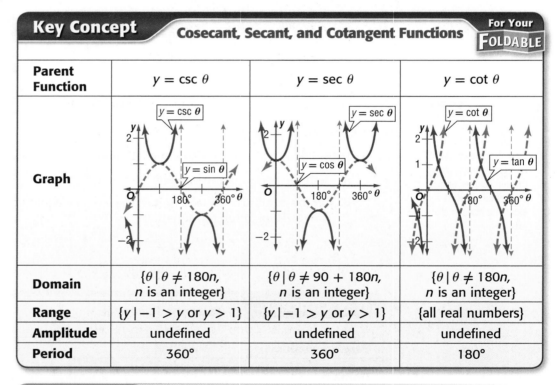

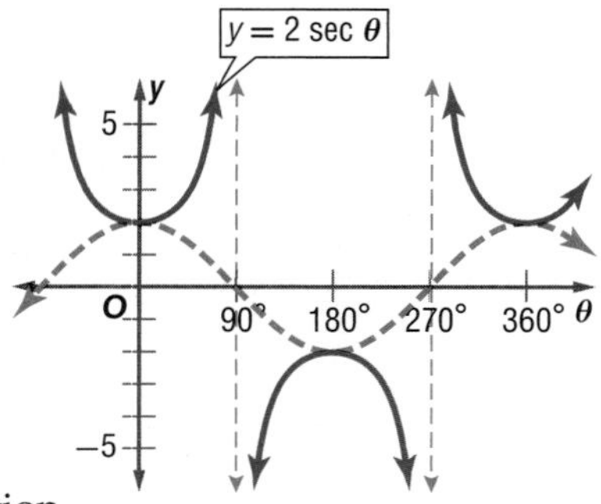

Check Your Progress

5. Find the period of $y = \csc 2\theta$. Then graph the function.

Personal Tutor glencoe.com

Check Your Understanding

Examples 1 and 2 pp. 855–856

Find the amplitude and period of each function. Then graph the function.

1. $y = 4 \sin \theta$
2. $y = \sin 3\theta$
3. $y = \cos 2\theta$
4. $y = \frac{1}{2} \cos 3\theta$

Example 3 p. 857

5. **SPIDERS** When an insect gets caught in a spider web, the web vibrates with a frequency of 14 hertz.
   a. Find the period of the function.
   b. Let the amplitude equal 1 unit. Write a sine equation to represent the vibration of the web y as a function of time t. Then graph the equation.

Examples 4 and 5 p. 858

Find the period of each function. Then graph the function.

6. $y = 3 \tan \theta$
7. $y = 2 \csc \theta$
8. $y = \cot 2\theta$

Practice and Problem Solving

● = **Step-by-Step Solutions** begin on page R20.
Extra Practice begins on page 947.

Examples 1 and 2 pp. 855–856

Find the amplitude and period of each function. Then graph the function.

9. $y = 2 \cos \theta$
10. $y = 3 \sin \theta$
11. $y = \sin 2\theta$
12. $y = \cos 3\theta$
13. $y = \cos \frac{1}{2}\theta$
14. $y = \sin 4\theta$
15. $y = \frac{3}{4} \cos \theta$
16. $y = \frac{3}{2} \sin \theta$
17. $y = \frac{1}{2} \sin 2\theta$
18. $y = 4 \cos 2\theta$
19. $y = 3 \cos 2\theta$
20. $y = 5 \sin \frac{2}{3}\theta$

Example 3 p. 857

21. **WAVES** A boat on a lake bobs up and down with the waves. The difference between the lowest and highest points of the boat is 8 inches. The boat is at *equilibrium* when it is halfway between the lowest and highest points. Each cycle of the periodic motion lasts 3 seconds.
    a. Write an equation for the motion of the boat. Let h represent the height in inches and let t represent the time in seconds. Assume that the boat is at equilibrium at $t = 0$ seconds.
    b. Draw a graph showing the height of the boat as a function of time.

22. **ELECTRICITY** The voltage supplied by an electrical outlet is a periodic function that *oscillates*, or goes up and down, between −165 volts and 165 volts with a frequency of 50 cycles per second.
    a. Write an equation for the voltage V as a function of time t. Assume that at $t = 0$ seconds, the current is 165 volts.
    b. Graph the function.

Examples 4 and 5 p. 858

Find the period of each function. Then graph the function.

23. $y = \tan \frac{1}{2}\theta$
24. $y = 3 \sec \theta$
25. $y = 2 \cot \theta$
26. $y = \csc \frac{1}{2}\theta$
27. $y = 2 \tan \theta$
28. $y = \sec \frac{1}{3}\theta$

Real-World Link

Of all of the music lessons that teens take, piano lessons are the most popular.

Source: *Piano Magazine*

29. **EARTHQUAKES** A seismic station detects an earthquake wave that has a frequency of 0.5 hertz and an amplitude of 1 meter.

- **a.** Write an equation involving sine to represent the height of the wave h as a function of time t. Assume that the equilibrium point of the wave, $h = 0$, is halfway between the lowest and highest points.
- **b.** Graph the function. Then determine the height of the wave after 20.5 seconds.

30. **PHYSICS** An object is attached to a spring as shown at the right. It oscillates according to the equation $y = 20 \cos \pi t$, where y is the distance in centimeters from its equilibrium position at time t.

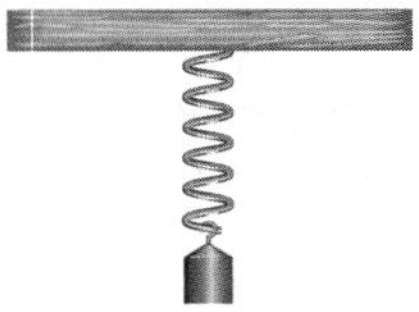

- **a.** Describe the motion of the object by finding the following: the amplitude in centimeters, the frequency in vibrations per second, and the period in seconds.
- **b.** Find the distance of the object from its equilibrium position at $t = \frac{1}{4}$ second.
- **c.** The equation $v = (-20 \text{ cm})(\pi \text{ rad/s}) \cdot \sin(\pi \text{ rad/s} \cdot t)$ represents the velocity v of the object at time t. Find the velocity at $t = \frac{1}{4}$ second.

31. **PIANOS** A piano string vibrates at a frequency of 130 hertz.

- **a.** Write and graph an equation using cosine to model the vibration of the string y as a function of time t. Let the amplitude equal 1 unit.
- **b.** Suppose the frequency of the vibration doubles. Do the amplitude and period increase, decrease, or remain the same? Explain.

Find the amplitude, if it exists, and period of each function. Then graph the function.

32. $y = 3 \sin \frac{2}{3}\theta$

33. $y = \frac{1}{2} \cos \frac{3}{4}\theta$

34. $y = 2 \tan \frac{1}{2}\theta$

35. $y = 2 \sec \frac{4}{5}\theta$

36. $y = 5 \csc 3\theta$

37. $y = 2 \cot 6\theta$

Identify the period of the graph and write an equation for each function.

38.

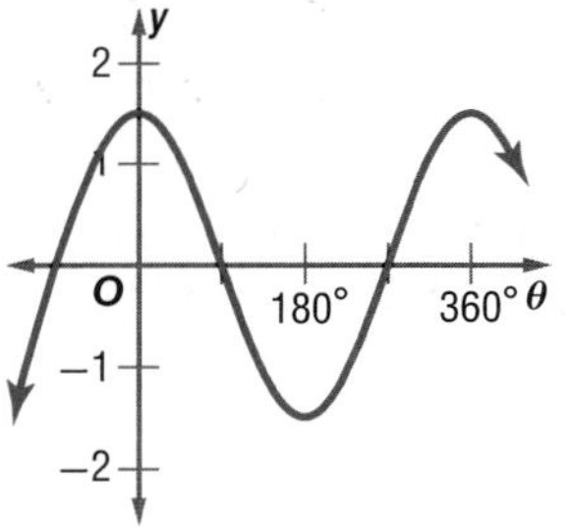

39.

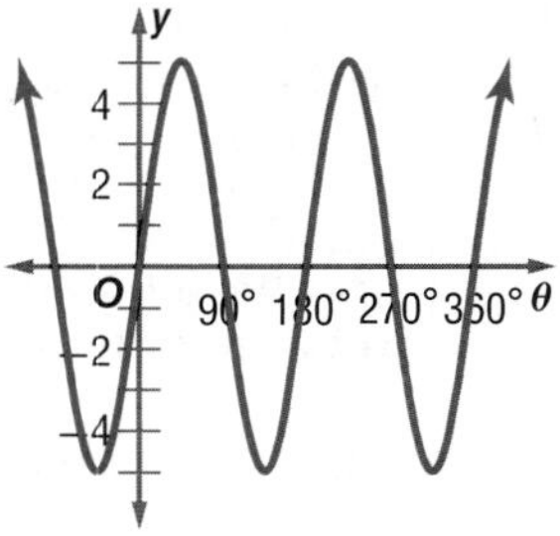

40.

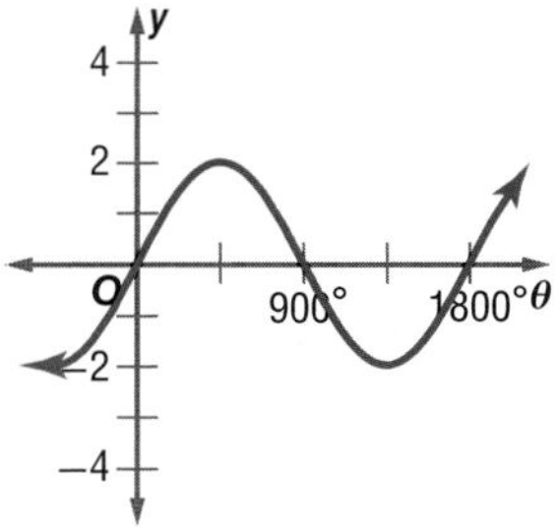

H.O.T. Problems

Use **H**igher-**O**rder **T**hinking Skills

41. **CHALLENGE** Describe the domain and range of $y = a \cos \theta$ and $y = a \sec \theta$, where a is any positive real number.

42. **REASONING** Compare and contrast the graphs of $y = \frac{1}{2} \sin \theta$ and $y = \sin \frac{1}{2}\theta$.

43. **OPEN ENDED** Write a trigonometric function that has an amplitude of 3 and a period of 180°. Then graph the function.

44. **WRITING IN MATH** Explain how to find the amplitude of $y = -2 \sin \theta$, and describe how the negative coefficient affects the graph.

NYSRE PRACTICE A2.A.70, A2.A.71

45. **SHORT RESPONSE** Find the 100,001st term of the sequence.

13, 20, 27, 34, 41, …

46. **STATISTICS** You bowled five games and had the following scores: 143, 171, 167, 133, and 156. What was your average?

A 147 B 153 C 154 D 156

47. Your city had a population of 312,430 ten years ago. If its current population is 418,270, by what percentage has it grown over the past 10 years?

F 25% G 34% H 66% J 75%

48. **SAT/ACT** If $h + 4 = b - 3$, then $(h - 2)^2 =$

A $h^2 + 4$
B $b^2 - 18b + 81$
C $b^2 - 14b + 49$
D $b^2 - 10b + 25$

Spiral Review

Find the exact value of each expression. (Lesson 13-6)

49. $\cos 120° - \sin 30°$

50. $3(\sin 45°)(\sin 60°)$

51. $4 \sin \frac{4\pi}{3} - 2 \cos \frac{\pi}{6}$

Solve each triangle. Round side lengths to the nearest tenth and angle measures to the nearest degree. (Lesson 13-5)

52.

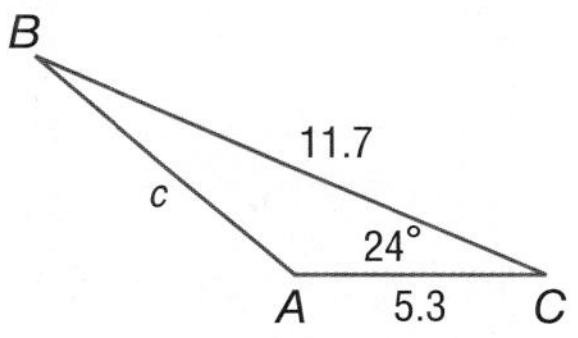

53.

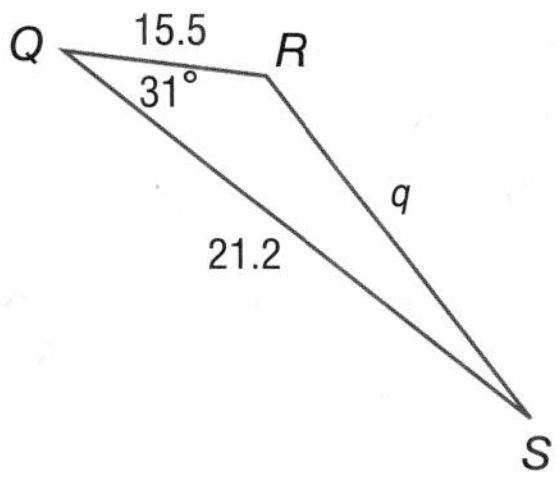

54.

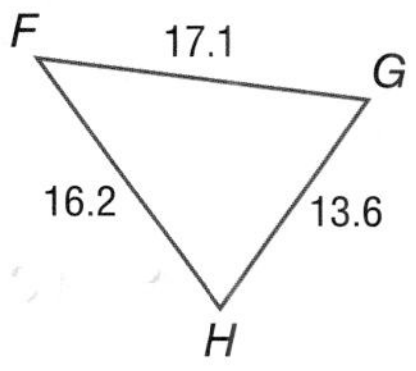

A bag contains 12 blue marbles, 9 red marbles, and 8 green marbles. The marbles are drawn one at a time. Find each probability. (Lesson 12-6)

55. The second marble is blue, given that the first marble is green and is replaced.

56. The third marble is green, given that the first two are red and blue and not replaced.

57. The third marble is red, given that the first two are red and not replaced.

58. **BANKING** Rita has deposited $1000 in a bank account. At the end of each year, the bank posts interest to her account in the amount of 3% of the balance, but then takes out a $10 annual fee. (Lesson 11-6)

a. Let b_0 be the amount Rita deposited. Write a recursive equation for the balance b_n in her account at the end of n years.

b. Find the balance in the account after four years.

Write an equation for an ellipse that satisfies each set of conditions. (Lesson 10-4)

59. center at (6, 3), focus at (2, 3), co-vertex at (6, 1)

60. foci at (2, 1) and (2, 13), co-vertex at (5, 7)

Skills Review

Graph each function. (Lesson 5-7)

61. $f(x) = 2(x - 3)^2 - 4$

62. $f(x) = \frac{1}{3}(x + 5)^2 + 2$

63. $f(x) = -3(x + 6)^2 + 7$

Graphing Technology Lab
Trigonometric Graphs

NY Math Online glencoe.com
Other Calculator Keystrokes

Objective
Use a graphing calculator to transform graphs of trigonometric functions.

NYS Core Curriculum

A2.A.66 Determine the trigonometric functions of any angle, using technology

You can use a TI-83/84 Plus graphing calculator to explore transformations of the graphs of trigonometric functions.

ACTIVITY

Step 1 Graph $y = \sin \theta$, $y = \sin \theta + 2$, and $y = \sin \theta - 3$ on the same coordinate plane. Let Y1 = $\sin \theta$, Y2 = $\sin \theta + 2$, and Y3 = $\sin \theta - 3$.

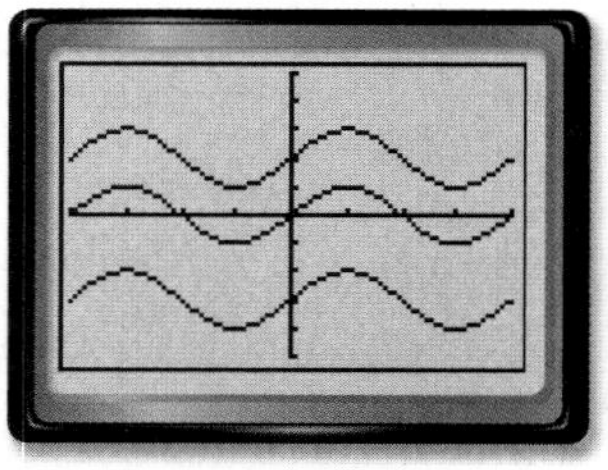

KEYSTROKES: [Y=] [SIN] [X,T,θ,n] [)] [ENTER]
[SIN] [X,T,θ,n] [)] [+] 2 [ENTER]
[SIN] [X,T,θ,n] [)] [−] 3 [GRAPH]

Describe the relationship among the graphs.

Step 2 Graph $y = \sin \theta$, $y = \sin (\theta + 45°)$, and $y = \sin (\theta - 90°)$ on the same coordinate plane. Let Y1 = $\sin \theta$, Y2 = $\sin (\theta + 45)$, and Y3 = $\sin (\theta - 90)$. **Be sure to clear the entries from Step 1.**

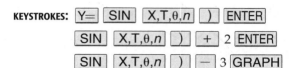

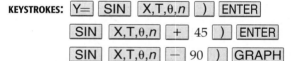

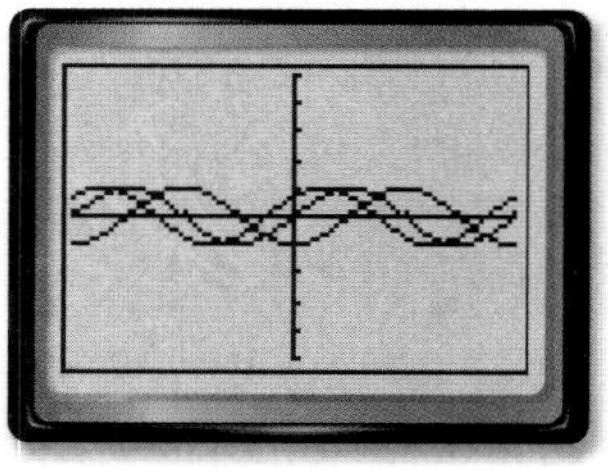

KEYSTROKES: [Y=] [SIN] [X,T,θ,n] [)] [ENTER]
[SIN] [X,T,θ,n] [+] 45 [)] [ENTER]
[SIN] [X,T,θ,n] [−] 90 [)] [GRAPH]

Describe the relationship among the graphs.

Model and Analyze

Repeat the activity for the cosine and tangent functions.

1. What is the effect of adding a constant to a trigonometric function?
2. What is the effect of adding a constant to θ in a trigonometric function?

Repeat the activity for each of the following. Describe the relationship between each pair of graphs.

3. $y = \sin \theta + 4$
   $y = \sin (2\theta) + 4$
4. $y = \cos \left(\frac{1}{2}\theta\right)$
   $y = \cos \frac{1}{2}(\theta + 45°)$
5. $y = 2 \sin \theta$
   $y = 2 \sin \theta - 1$
6. $y = \cos \theta - 3$
   $y = \cos (\theta - 90°) - 3$
7. Write a general equation for the sine, cosine, and tangent functions after changes in amplitude a, period b, horizontal position h, and vertical position k.

13-8 Translations of Trigonometric Graphs

Why?

The graphs at the right represent the waves in a bay during high and low tides. Notice that the shape of the waves does not change.

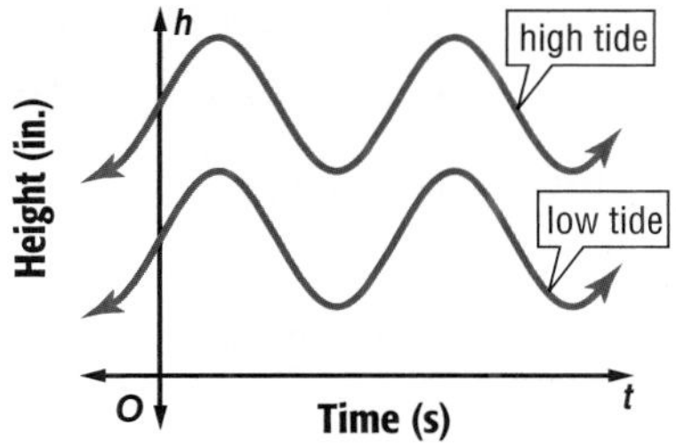

Then
You translated exponential functions. (Lesson 8-1)

Now
- Graph horizontal translations of trigonometric graphs and find phase shifts.
- Graph vertical translations of trigonometric graphs.

NYS Core Curriculum
A2.A.69 Determine amplitude, period, frequency, and phase shift, given the graph or equation of a periodic function
A2.A.72 Write the trigonometric function that is represented by a given periodic graph
Also addresses A2.A.70.

New Vocabulary
phase shift
vertical shift
midline

NY Math Online
glencoe.com
- Extra Examples
- Personal Tutor
- Self-Check Quiz
- Homework Help
- Math in Motion

Horizontal Translations Recall that a *translation* occurs when a figure is moved from one location to another on the coordinate plane without changing its orientation. A horizontal translation of a periodic function is called a **phase shift**.

Key Concept — Phase Shift
For Your FOLDABLE

Words The phase shift of the functions $y = a \sin b(\theta - h)$, $y = a \cos b(\theta - h)$, and $y = a \tan b(\theta - h)$ is h, where $b > 0$.

Models

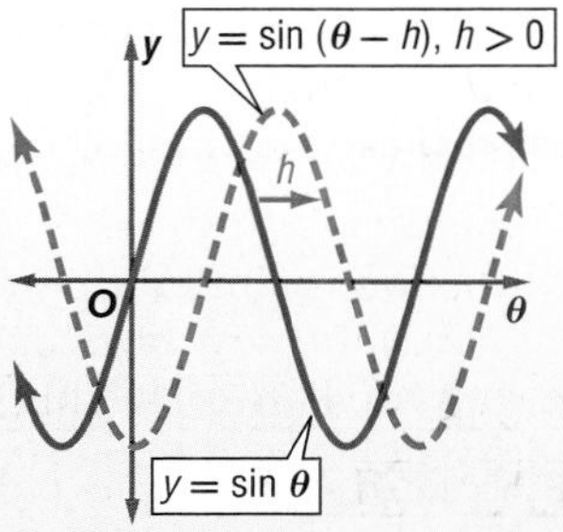

If $h > 0$, the shift is h units to the right.

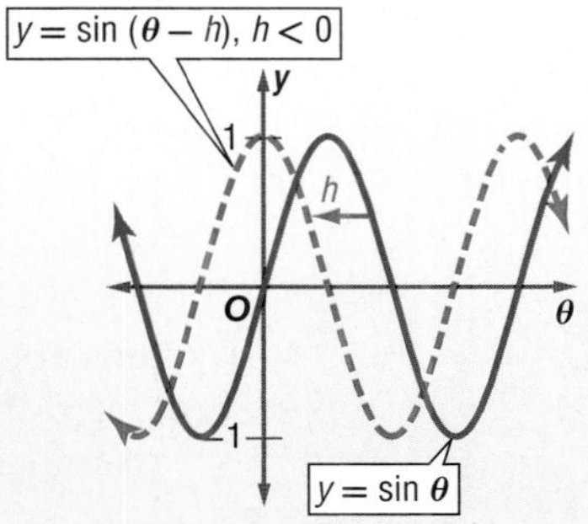

If $h < 0$, the shift is h units to the left.

Examples
$y = \cos(\theta - 90°)$ — **The phase shift is 90° to the right.**
$y = \tan(\theta + 30°)$ — **The phase shift is 30° to the left.**

Math *in Motion*, Animation glencoe.com

The secant, cosecant, and cotangent can be graphed using the same rules.

EXAMPLE 1 Graph Horizontal Translations

State the amplitude, period, and phase shift for $y = \sin(\theta - 90°)$. Then graph the function.

amplitude: $a = 1$

period: $\frac{360°}{|b|} = \frac{360°}{1}$ or $360°$

phase shift: $h = 90°$

Graph $y = \sin \theta$ shifted 90° to the right.

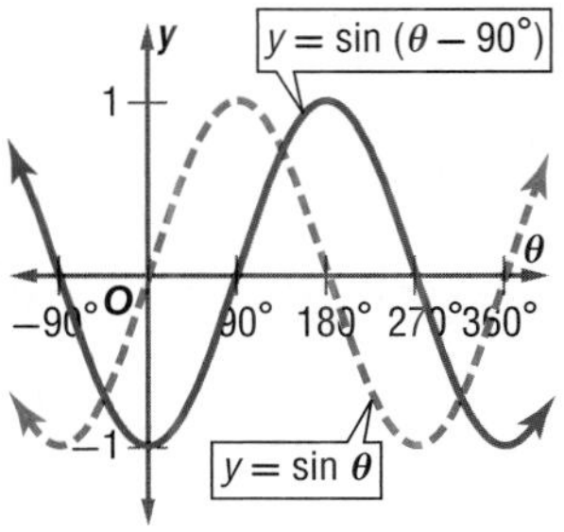

Check Your Progress

1. State the amplitude, period, and phase shift for $y = 2 \cos(\theta + 45°)$. Then graph the function.

Personal Tutor glencoe.com

Vertical Translations Recall that the graph of $y = x^2 + 5$ is the graph of the parent function $y = x^2$ shifted up 5 units. Similarly, graphs of trigonometric functions can be translated vertically through a **vertical shift**.

Key Concept — Vertical Shift

For Your FOLDABLE

Words The vertical shift of the functions $y = a \sin b\theta + k$, $y = a \cos b\theta + k$, and $y = a \tan b\theta + k$ is k.

Models

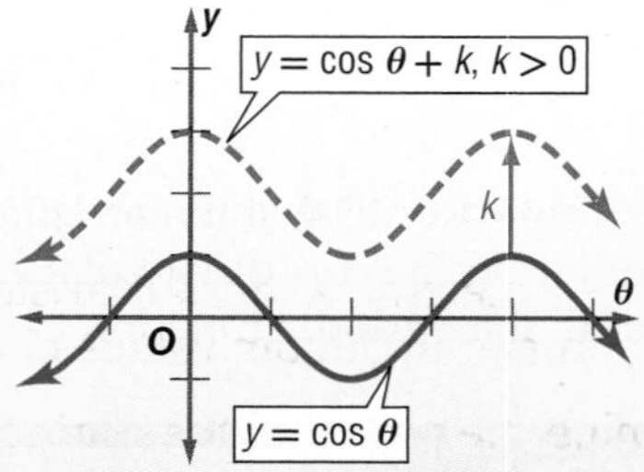

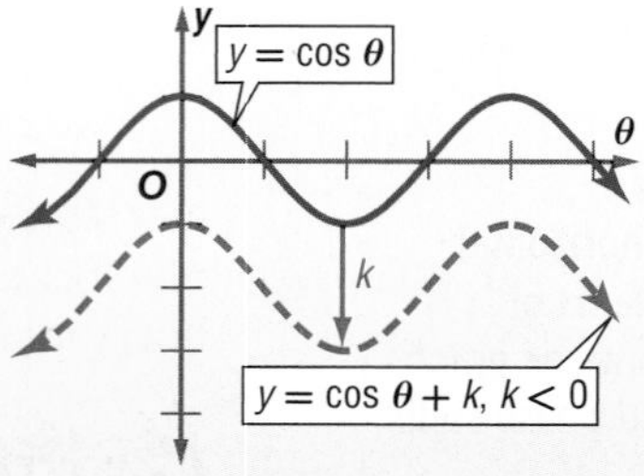

If $k > 0$, the shift is k units up. If $k < 0$, the shift is k units down.

Examples $y = \sin \theta + 4$ **The vertical shift is 4 units up.**
$y = \tan \theta - 3$ **The vertical shift is 3 units down.**

Math *in Motion*, Animation glencoe.com

The secant, cosecant, and cotangent can be graphed using the same rules.

StudyTip

Notation Note that $\sin(\theta + x) \neq \sin \theta + x$. The first expression indicates a phase shift. The second expression indicates a vertical shift.

When a trigonometric function is shifted up or down k units, the line $y = k$ is the new horizontal axis about which the graph oscillates. This line is called the **midline**, and it can be used to help draw vertical translations.

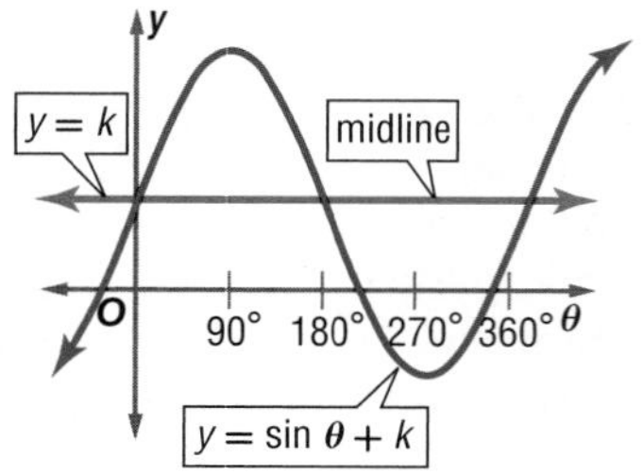

EXAMPLE 2 Graph Vertical Translations

State the amplitude, period, vertical shift, and equation of the midline for $y = \frac{1}{2} \cos \theta - 2$. Then graph the function.

amplitude: $|a| = \frac{1}{2}$

period: $\frac{2\pi}{|b|} = \frac{2\pi}{|1|}$ or 2π

vertical shift: $k = -2$

midline: $y = -2$

To graph $y = \frac{1}{2} \cos \theta - 2$, first draw the midline. Then use it to graph $y = \frac{1}{2} \cos \theta$ shifted 2 units down.

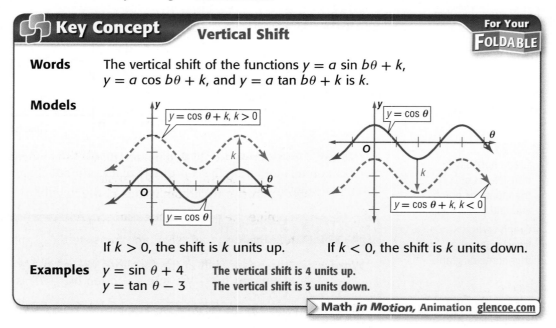

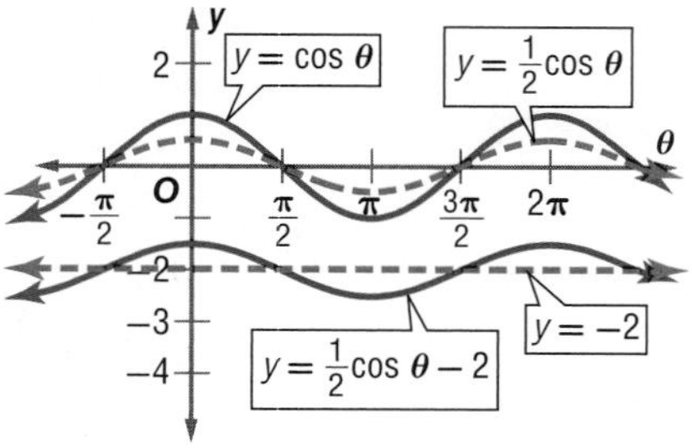

StudyTip

Using Color It may be helpful to first graph the parent function in one color. Next, apply the vertical shift and graph the function in another color. Then apply the change in amplitude and graph the function in the final color.

Check Your Progress

2. State the amplitude, period, vertical shift, and equation of the midline for $y = \tan \theta + 3$. Then graph the function.

Personal Tutor glencoe.com

You can use the following steps to graph trigonometric functions involving phase shifts and vertical shifts.

Concept Summary — Graph Trigonometric Functions

For Your FOLDABLE

$$y = a \sin b(\theta - h) + k$$

amplitude: a; period: b; phase shift: h; vertical shift: k

Step 1 Determine the vertical shift, and graph the midline.

Step 2 Determine the amplitude, if it exists. Use dashed lines to indicate the maximum and minimum values of the function.

Step 3 Determine the period of the function, and graph the appropriate function.

Step 4 Determine the phase shift, and translate the graph accordingly.

EXAMPLE 3 Graph Transformations

State the amplitude, period, phase shift, and vertical shift for $y = 3 \sin \frac{2}{3}(\theta - \pi) + 4$. Then graph the function.

amplitude: $|a| = 3$

period: $\frac{2\pi}{|b|} = \frac{2\pi}{\left|\frac{2}{3}\right|}$ or 3π — **The period indicates that the graph will be stretched.**

phase shift: $h = \pi$ — **The graph will shift π to the right.**

vertical shift: $k = 4$ — **The graph will shift 4 units up.**

midline: $y = 4$ — **The graph will oscillate around the line $y = 4$.**

Step 1 Graph the midline.

Step 2 Since the amplitude is 3, draw dashed lines 3 units above and 3 units below the midline.

Step 3 Graph $y = 3 \sin \frac{2}{3}\theta + 4$ using the midline as a reference.

Step 4 Shift the graph π to the right.

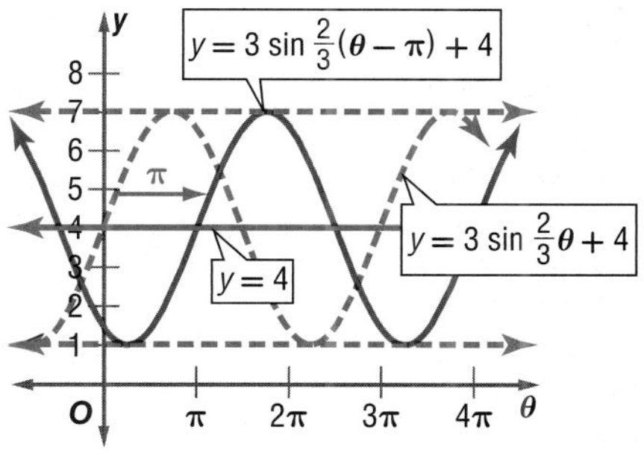

CHECK You can check the accuracy of your transformation by evaluating the function for various values of θ and confirming their location on the graph.

StudyTip

Verifying a Graph After drawing the graph of a trigonometric function, select values of θ and evaluate them in the equation to verify your graph.

Check Your Progress

3. State the amplitude, period, phase shift, and vertical shift for $y = 2 \cos \frac{1}{2}\left(\theta + \frac{\pi}{2}\right) - 2$. Then graph the function.

Personal Tutor glencoe.com

The sine wave occurs often in physics, signal processing, music, electrical engineering, and many other fields.

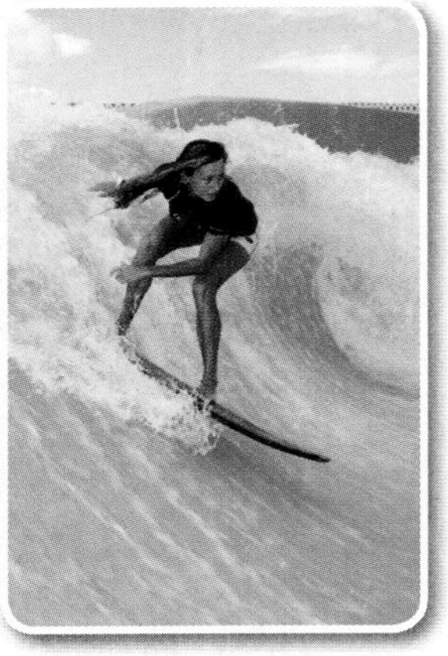

Real-World Link

In some wave pools, surfers can ride waves up to 70 meters.

Source: Orlando Wave Pool

Real-World EXAMPLE 4 Represent Periodic Functions

WAVE POOL The height of water in a wave pool oscillates between a maximum of 13 feet and a minimum of 5 feet. The wave generator pumps 6 waves per minute. Write a sine function that represents the height of the water at time t seconds. Then graph the function.

Step 1 Write the equation for the midline, and determine the vertical shift.

$y = \frac{13 + 5}{2}$ or 9 — **The midline lies halfway between the maximum and minimum values.**

Since the midline is $y = 9$, the vertical shift is $k = 9$.

Step 2 Find the amplitude.

$|a| = |13 - 9|$ or 4 — **Find the difference between the midline value and the maximum value.**

So, $a = 4$.

Step 3 Find the period.

Since there are 6 waves per minute, there is 1 wave every 10 seconds. So, the period is 10 seconds.

$10 = \frac{2\pi}{|b|}$ — **period $= \frac{2\pi}{|b|}$**

$|b| = \frac{2\pi}{10}$ — **Solve for $|b|$.**

$b = \pm\frac{\pi}{5}$ — **Simplify.**

Step 4 Write an equation for the function.

$h = a \sin b(t - h) + k$ — **Write the equation for sine relating height h and time t.**

$= 4 \sin \frac{\pi}{5}(t - 0) + 9$ — **Substitution: $a = 4, b = \frac{\pi}{5}, h = 0, k = 9$**

$= 4 \sin \frac{\pi}{5}t + 9$ — **Simplify.**

Then graph the function.

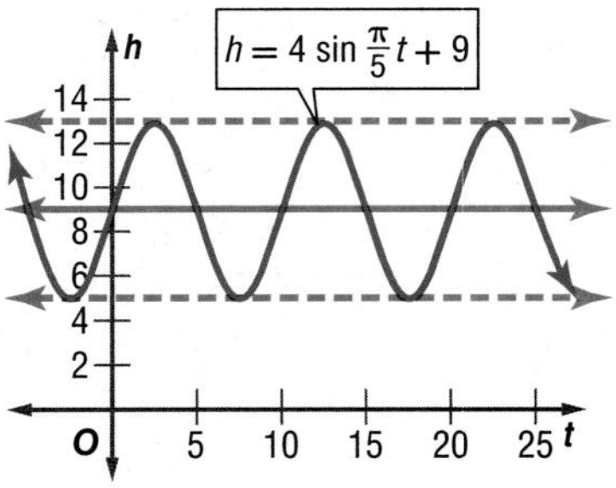

Check Your Progress

4. WAVE POOL The height of water in a wave pool oscillates between a maximum of 14 feet and a minimum of 6 feet. The wave generator pumps 5 waves per minute. Write a cosine function that represents the height of water at time t seconds. Then graph the function.

Personal Tutor glencoe.com

Check Your Understanding

Example 1 p. 863

State the amplitude, period, and phase shift for each function. Then graph the function.

1. $y = \sin(\theta - 180°)$
2. $y = \tan\left(\theta - \frac{\pi}{4}\right)$
3. $y = \sin\left(\theta - \frac{\pi}{2}\right)$
4. $y = \frac{1}{2}\cos(\theta + 90°)$

Example 2 p. 864

State the amplitude, period, vertical shift, and equation of the midline for each function. Then graph the function.

5. $y = \cos\theta + 4$
6. $y = \sin\theta - 2$
7. $y = \frac{1}{2}\tan\theta + 1$
8. $y = \sec\theta - 5$

Example 3 p. 865

State the amplitude, period, phase shift, and vertical shift for each function. Then graph the function.

9. $y = 2\sin(\theta + 45°) + 1$
10. $y = \cos 3(\theta - \pi) - 4$
11. $y = \frac{1}{4}\tan 2(\theta + 30°) + 3$
12. $y = 4\sin\frac{1}{2}\left(\theta - \frac{\pi}{2}\right) + 5$

Example 4 p. 866

13. **EXERCISE** While doing some moderate physical activity, a person's blood pressure oscillates between a maximum of 130 and a minimum of 90. The person's heart rate is 90 beats per minute. Write a sine function that represents the person's blood pressure P at time t seconds. Then graph the function.

Practice and Problem Solving

● = **Step-by-Step Solutions** begin on page R20.
Extra Practice begins on page 947.

Example 1 p. 863

State the amplitude, period, and phase shift for each function. Then graph the function.

14. $y = \cos(\theta + 180°)$
15. $y = \tan(\theta - 90°)$
16. $y = \sin(\theta + \pi)$
17. $y = 2\sin\left(\theta + \frac{\pi}{2}\right)$
18. $y = \tan\frac{1}{2}(\theta + 30°)$
19. $y = 3\cos\left(\theta - \frac{\pi}{3}\right)$

Example 2 p. 864

State the amplitude, period, vertical shift, and equation of the midline for each function. Then graph the function.

20. $y = \cos\theta + 3$
21. $y = \tan\theta - 1$
22. $y = \tan\theta + \frac{1}{2}$
23. $y = 2\cos\theta - 5$
24. $y = 2\sin\theta - 4$
25. $y = \frac{1}{3}\sin\theta + 7$

Example 3 p. 865

State the amplitude, period, phase shift, and vertical shift for each function. Then graph the function.

26. $y = 4\sin(\theta - 60°) - 1$
27. $y = \cos\frac{1}{2}(\theta - 90°) + 2$
28. $y = \tan(\theta + 30°) - 2$
29. $y = 2\tan 2\left(\theta + \frac{\pi}{4}\right) - 5$
30. $y = \frac{1}{2}\sin\left(\theta - \frac{\pi}{2}\right) + 4$
31. $y = \cos 3(\theta - 45°) + \frac{1}{2}$
32. $y = 3 + 5\sin 2(\theta - \pi)$
33. $y = -2 + 3\sin\frac{1}{3}\left(\theta - \frac{\pi}{2}\right)$

Example 4 p. 866

34. **EXERCISE** Suppose that while doing some moderate physical activity, a person's blood pressure is 130 over 90 and that the person has a heart rate of 90 beats per minute. Write a sine function that represents the person's blood pressure at time t seconds. Then graph the function.

35. LAKES A buoy marking the swimming area in a lake oscillates each time a speed boat goes by. Its distance d in feet from the bottom of the lake is given by $d = 1.8 \sin \frac{3\pi}{4}t + 12$, where t is the time in seconds. Graph the function. Describe the minimum and maximum distances of the buoy from the bottom of the lake when a boat passes by.

36. FERRIS WHEEL Suppose a Ferris wheel has a diameter of approximately 520 feet and makes one complete revolution in 30 minutes. Suppose the lowest car on the Ferris wheel is 5 feet from the ground. Let the height at the top of the wheel represent the height at time 0. Write an equation for the height of a car h as a function of time t. Then graph the function.

Write an equation for each translation.

37. $y = \sin x$, 4 units to the right and 3 units up

38. $y = \cos x$, 5 units to the left and 2 units down

39. $y = \tan x$, π units to the right and 2.5 units up

40. JUMP ROPE The graph at the right approximates the height of a jump rope h in inches as a function of time t in seconds. A maximum point on the graph is (1.25, 68), and a minimum point is (2.75, 2).

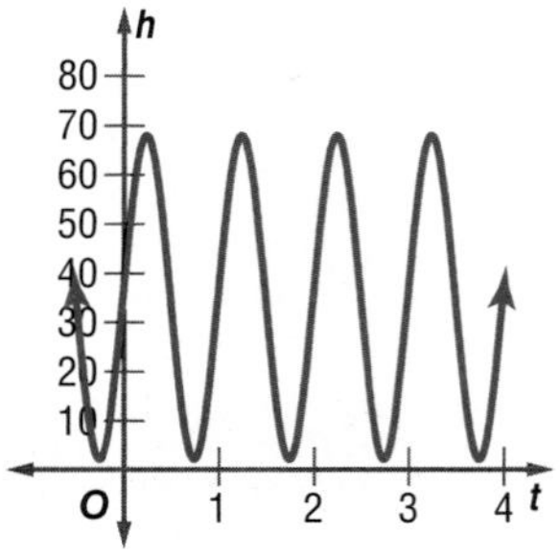

a. Describe what the maximum and minimum points mean in the context of the situation.

b. What are the equations for the midline, the amplitude, and the period of the function?

c. Write an equation for the function.

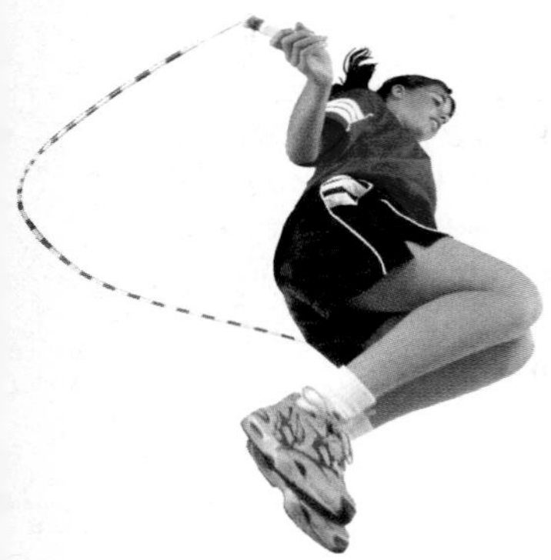

Real-World Link

The record for jump rope is 151,036 jumps of rope in 24 hours. That is an average of 104 skips every minute for the entire day.

Source: *Guinness Book of World Records*

41. CAROUSEL A horse on a carousel goes up and down 3 times as the carousel makes one complete rotation. The maximum height of the horse is 55 inches, and the minimum height is 37 inches. The carousel rotates once every 21 seconds. Assume that the horse starts and stops at its median height.

a. Write an equation to represent the height of the horse h as a function of time t.

b. Graph the function.

c. Use your graph to estimate the height of the horse after 8 seconds. Then use a calculator to find the height to the nearest tenth.

42. TEMPERATURES During one month, the outside temperature fluctuates between 40°F and 50°F. A cosine curve approximates the change in temperature, with a high of 50°F being reached every four days.

a. Describe the amplitude, period, and midline of the function that approximates the temperature y on day d.

b. Write a cosine function to estimate the temperature y on day d.

c. Sketch a graph of the function.

d. Estimate the temperature on the 7th day of the month.

Find a coordinate that represents a maximum for each graph.

43. $y = -2 \cos \left(x - \frac{\pi}{2}\right)$

44. $y = 4 \sin \left(x + \frac{\pi}{3}\right)$

45. $y = 3 \tan \left(x + \frac{\pi}{2}\right) + 2$

46. $y = -3 \sin \left(x - \frac{\pi}{4}\right) - 4$

Compare each pair of graphs.

47. $y = -\cos 3\theta$ and $y = \sin 3(\theta - 90°)$

48. $y = 2 + 0.5 \tan \theta$ and $y = 2 + 0.5 \tan(\theta + \pi)$

49. $y = 2 \sin\left(\theta - \frac{\pi}{6}\right)$ and $y = -2 \sin\left(\theta + \frac{5\pi}{6}\right)$

Identify the period of each function. Then write an equation for the graph using the given trigonometric function.

50. sine

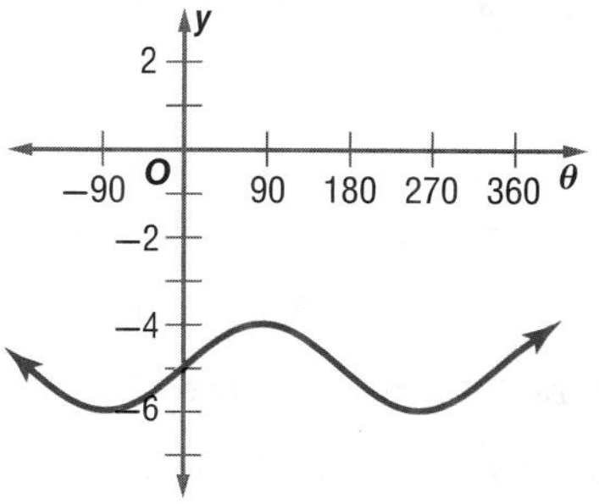

51. cosine

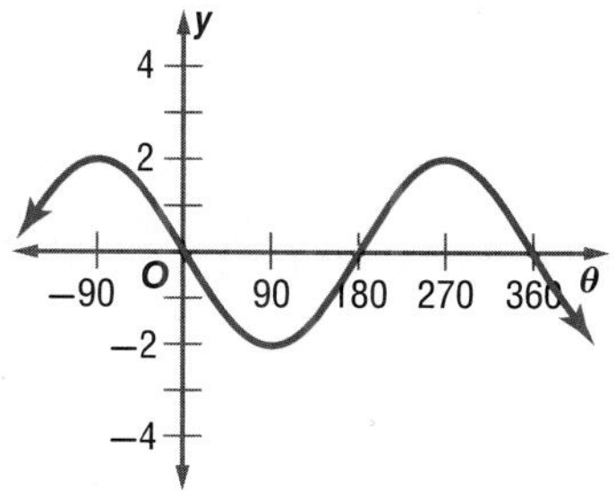

> **Watch Out!**
>
> **Parent Functions** Often the graph of a trigonometric function can be represented by more than one equation. For example, the graphs of $y = \cos \theta$ and $y = \sin(\theta + 90°)$ are the same.

52. cosine

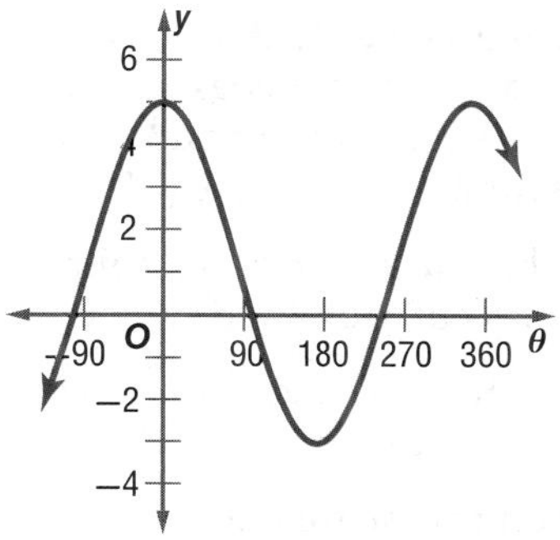

53 sine

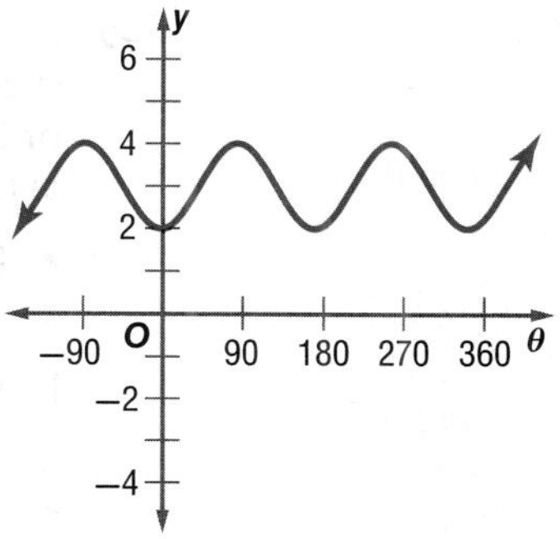

State the period, phase shift, and vertical shift. Then graph the function.

54. $y = \csc(\theta + \pi)$

55. $y = \cot \theta + 6$

56. $y = \cot\left(\theta - \frac{\pi}{6}\right) - 2$

57. $y = \frac{1}{2} \csc 3(\theta - 45°) + 1$

58. $y = 2 \sec \frac{1}{2}(\theta - 90°)$

59. $y = 4 \sec 2\left(\theta + \frac{\pi}{2}\right) - 3$

H.O.T. Problems Use Higher-Order Thinking Skills

60. CHALLENGE If you are given the amplitude and period of a cosine function, is it *sometimes, always,* or *never* possible to find the maximum and minimum values of the function? Explain your reasoning.

61. REASONING Describe how the graph of $y = 3 \sin 2\theta + 1$ is different from $y = \sin \theta$.

62. WRITING IN MATH Describe two different phase shifts that will translate the sine curve onto the cosine curve shown at the right. Then write an equation for the new sine curve using each phase shift.

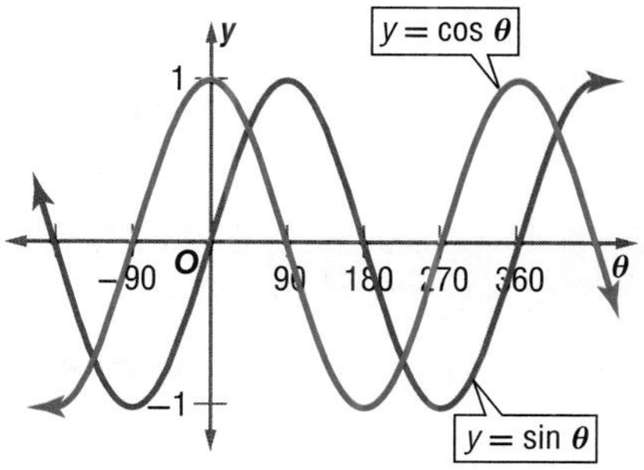

63. OPEN ENDED Write a periodic function that has an amplitude of 2 and midline at $y = -3$. Then graph the function.

64. REASONING How many different sine graphs pass through the origin $(n\pi, 0)$? Explain your reasoning.

NYSRE PRACTICE A2.A.69, A2.A.72

65. GRIDDED RESPONSE The expression $\frac{3x-1}{4}+\frac{x+6}{4}$ is how much greater than x?

66. Expand $(a-b)^4$.

A a^4-b^4

B $a^4-4ab+b^4$

C $a^4+4a^3b+6a^2b^2+4ab^3+b^4$

D $a^4-4a^3b+6a^2b^2-4ab^3+b^4$

67. Solve $\sqrt{x-3}+\sqrt{x+2}=5$.

F 7

G 0, 7

H 7, 13

J no solution

68. GEOMETRY Using the figures below, what is the average of a, b, c, d, and f?

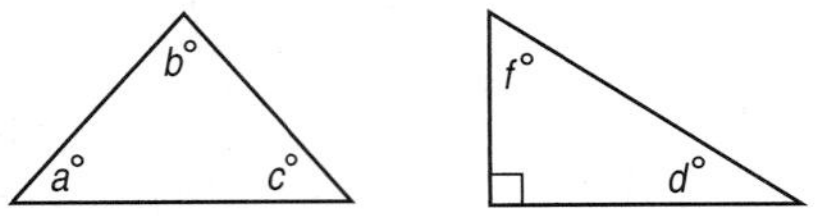

A 21   B 45   C 50   D 54

Spiral Review

Find the amplitude and period of each function. Then graph the function. (Lesson 13-7)

69. $y=2\cos\theta$

70. $y=3\sin\theta$

71. $y=\sin 2\theta$

Find the exact value of each expression. (Lesson 13-6)

72. $\sin\frac{4\pi}{3}$

73. $\sin(-30°)$

74. $\cos 405°$

State whether each situation represents an *experiment* or an *observational study*. Identify the *control* group and the *treated* group. Then determine whether there is bias. (Lesson 12-1)

75. Find 220 people and randomly split them into two groups. One group exercises for an hour a day and the other group does not. Then compare their body mass indexes.

76. Find 200 students, half of whom play soccer, and compare the amounts of time spent sleeping.

77. Find 100 students, half of whom have part-time jobs, and compare their grades.

78. GEOMETRY Equilateral triangle ABC has a perimeter of 39 centimeters. If the midpoints of the sides are connected, a smaller equilateral triangle results. Suppose the process of connecting midpoints of sides and drawing new triangles is continued indefinitely. (Lesson 11-4)

a. Write an infinite geometric series to represent the sum of the perimeters of all of the triangles.

b. Find the sum of the perimeters of all of the triangles.

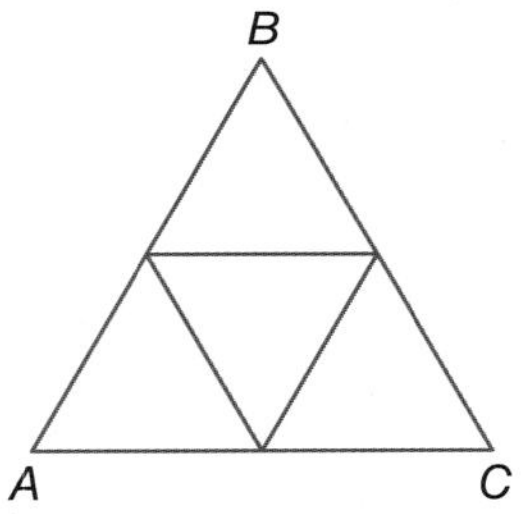

79. CONSTRUCTION A construction company will be fined for each day it is late completing a bridge. The daily fine will be \$4000 for the first day and will increase by \$1000 each day. Based on its budget, the company can only afford \$60,000 in total fines. What is the maximum number of days it can be late? (Lesson 11-3)

Skills Review

Find each value of θ. Round to the nearest degree. (Lesson 13-1)

80. $\sin\theta=\frac{7}{8}$

81. $\tan\theta=\frac{9}{10}$

82. $\cos\theta=\frac{1}{4}$

83. $\cos\theta=\frac{4}{5}$

84. $\sin\theta=\frac{5}{6}$

85. $\tan\theta=\frac{2}{7}$

13-9 Inverse Trigonometric Functions

Then

You graphed trigonometric functions. (Lesson 13-7)

Now

- Find values of inverse trigonometric functions.
- Solve equations by using inverse trigonometric functions.

NYS Core Curriculum

A2.A.64 Use inverse functions to find the measure of an angle, given its sine, cosine, or tangent
A2.A.65 Sketch the graph of the inverses of the sine, cosine, and tangent functions
Also addresses A2.A.63 and A2.A.68.

New Vocabulary

principal values
Arcsine function
Arccosine function
Arctangent function

NY Math Online

glencoe.com

- Extra Examples
- Personal Tutor
- Self-Check Quiz
- Homework Help
- Math in Motion

Why?

The leaning bookshelf at the right is 15 inches from the wall and reaches a height of 75 inches. In Lesson 13-1, you learned how to use the inverse of a trigonometric function to find the measure of acute angle θ.

$\tan \theta = \frac{15}{75}$ or 0.2 **Use the tangent function.**

Find an angle that has a tangent of 0.2.

2nd [TAN^{-1}] .2 ENTER 11.30993247

So, the measure of θ is about 11°.

Inverse Trigonometric Functions If you know the value of a trigonometric function for an angle, you can use the *inverse* to find the angle. Recall that an inverse function is the relation in which all values of x and y are reversed. The inverse of $y = \sin x$, $x = \sin y$, is graphed at the right.

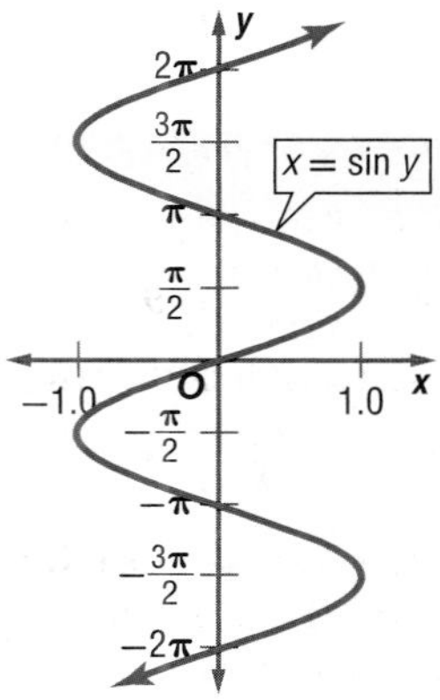

Notice that the inverse is not a function because there are many values of y for each value of x. If you restrict the domain of the sine function so that $-\frac{\pi}{2} \leq x \leq \frac{\pi}{2}$, then the inverse is a function.

The values in this restricted domain are called **principal values**. Trigonometric functions with restricted domains are indicated with capital letters.

- $y = \text{Sin } x$ if and only if $y = \sin x$ and $-\frac{\pi}{2} \leq x \leq \frac{\pi}{2}$.
- $y = \text{Cos } x$ if and only if $y = \cos x$ and $0 \leq x \leq \pi$.
- $y = \text{Tan } x$ if and only if $y = \tan x$ and $-\frac{\pi}{2} \leq x \leq \frac{\pi}{2}$.

You can use functions with restricted domains to define inverse trigonometric functions. The inverses of the sine, cosine, and tangent functions are the **Arcsine**, **Arccosine**, and **Arctangent** functions, respectively.

Key Concept Inverse Trigonometric Functions — For Your FOLDABLE

Inverse Function	Symbols	Domain	Range	Model
Arcsine	$y = \text{Arcsin } x$ $y = \text{Sin}^{-1} x$	$-1 \leq x \leq 1$	$-\frac{\pi}{2} \leq y \leq \frac{\pi}{2}$ $-90° \leq y \leq 90°$	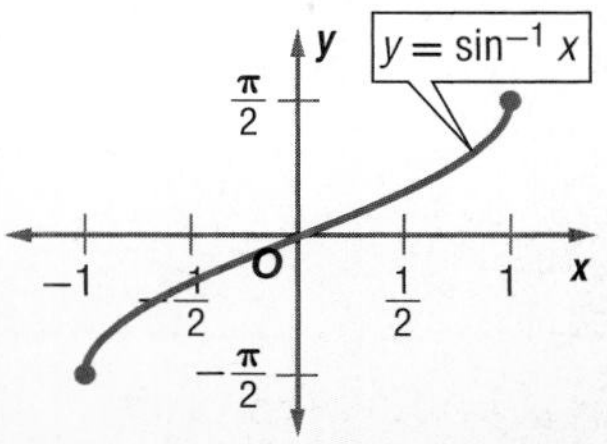
Arccosine	$y = \text{Arccos } x$ $y = \text{Cos}^{-1} x$	$-1 \leq x \leq 1$	$0 \leq y \leq \pi$ $0° \leq y \leq 180°$	
Arctangent	$y = \text{Arctan } x$ $y = \text{Tan}^{-1} x$	all real numbers	$-\frac{\pi}{2} \leq y \leq \frac{\pi}{2}$ $-90° \leq y \leq 90°$	

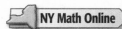

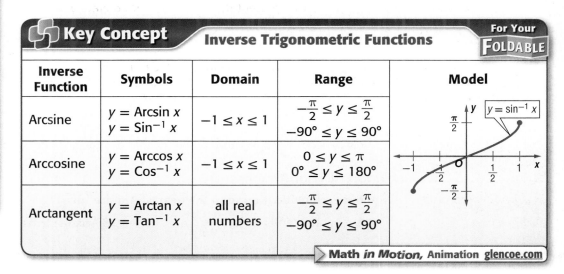

Math *in Motion,* Animation glencoe.com

Review Vocabulary

inverse functions If f and f^{-1} are inverse functions, then $f(a) = b$ if and only if $f^{-1}(b) = a$. (Lesson 7-2)

In the relation $y = \cos^{-1} x$, if $x = \frac{1}{2}$, $y = 60°$, $300°$, and all angles that are coterminal with those angles. In the function $y = \text{Cos}^{-1} x$, if $x = \frac{1}{2}$, $y = 60°$ only.

EXAMPLE 1 Evaluate Inverse Trigonometric Functions

Find each value. Write angle measures in degrees and radians.

a. $\text{Cos}^{-1}\left(-\frac{1}{2}\right)$

Find the angle θ for $0° \leq \theta \leq 180°$ that has a cosine value of $-\frac{1}{2}$.

Method 1 Use a unit circle.

Find a point on the unit circle that has an x-coordinate of $-\frac{1}{2}$.

When $\theta = 120°$, $\cos \theta = -\frac{1}{2}$.

So, $\text{Cos}^{-1}\left(-\frac{1}{2}\right) = 120°$ or $\frac{2\pi}{3}$.

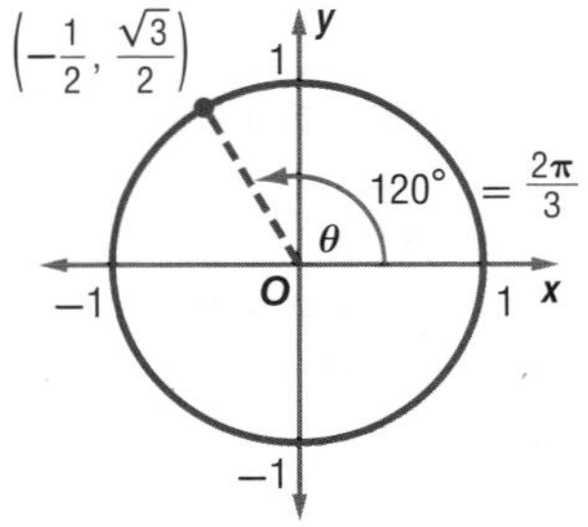

Method 2 Use a calculator.

KEYSTROKES: [2nd] [COS⁻¹] [(−)] 1 [÷] 2 [)] [ENTER] 120

Therefore, $\text{Cos}^{-1}\left(-\frac{1}{2}\right) = 120°$ or $\frac{2\pi}{3}$.

b. Arctan 1

Find the angle θ for $-90° \leq \theta \leq 90°$ that has a tangent value of 1.

KEYSTROKES: [2nd] [TAN⁻¹] 1 [ENTER] 45

Therefore, Arctan 1 = $45°$ or $\frac{\pi}{4}$.

Check Your Progress

1A. $\text{Cos}^{-1} 0$

1B. $\text{Arcsin}\left(-\frac{\sqrt{2}}{2}\right)$

Personal Tutor glencoe.com

When finding a value when there are multiple trigonometric functions involved, use the order of operations to solve.

EXAMPLE 2 Find a Trigonometric Value

Find $\tan\left(\text{Cos}^{-1} \frac{1}{2}\right)$. Round to the nearest hundredth.

Use a calculator.

KEYSTROKES: [TAN] [2nd] [COS⁻¹] 1 [÷] 2 [)] [ENTER] 1.732050808

So, $\tan\left(\text{Cos}^{-1} \frac{1}{2}\right) \approx 1.73$.

CHECK $\text{Cos}^{-1} \frac{1}{2} = 60°$ and $\tan 60° \approx 1.73$. So, the answer is correct.

Check Your Progress

Find each value. Round to the nearest hundredth.

2A. $\sin\left(\text{Tan}^{-1} \frac{3}{8}\right)$

2B. $\cos\left(\text{Arccos} -\frac{\sqrt{2}}{2}\right)$

Personal Tutor glencoe.com

StudyTip

Angle Measure Remember that when evaluating an inverse trigonometric function, the result is an angle measure.

Solve Equations by Using Inverses You can rewrite trigonometric equations to solve for the measure of an angle.

Test-Taking Tip

Eliminate Possibilities Because Sin θ is negative, look for angle measures in Quadrants III and IV.

NYSRE EXAMPLE 3 A2.A.64, A2.A.65

If Sin $\theta = -0.35$, find θ.

A $-20.5°$ B $-0.6°$ C $0.6°$ D $20.5°$

Read the Test Item

The sine of angle θ is -0.35. This can be written as Arcsin $(-0.35) = \theta$.

Solve the Test Item

Use a calculator.

KEYSTROKES: [2nd] [SIN^{-1}] [(–)] .35 [ENTER] −20.48731511

So, $\theta \approx -20.5°$. The answer is A.

Check Your Progress

3. If Tan $\theta = 1.8$, find θ.

F $0.03°$ G $29.1°$ H $60.9°$ J no solution

Personal Tutor glencoe.com

Inverse trigonometric functions can be used to determine angles of inclination, depression, and elevation.

Real-World EXAMPLE 4 Use Inverse Trigonometric Functions

WATER SKIING A water ski ramp is 6 feet tall and 9 feet long, as shown at the right. Write an inverse trigonometric function that can be used to find θ, the angle the ramp makes with the water. Then find the measure of the angle. Round to the nearest tenth.

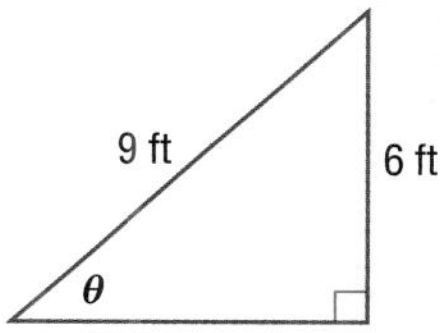

Because the measures of the opposite side and the hypotenuse are known, you can use the sine function.

$\sin \theta = \frac{6}{9}$ **Sine function**

$\theta = \text{Sin}^{-1} \frac{6}{9}$ **Inverse sine function**

$\theta \approx 41.8°$ **Use a calculator.**

So, the angle of the ramp is about $41.8°$.

CHECK Using your calculator, $\sin 41.8 \approx 0.66653 \approx \frac{6}{9}$. So, the answer is correct.

Check Your Progress

4. **SKIING** A ski trail is shown at the right. Write an inverse trigonometric function that can be used to find θ, the angle the trail makes with the ground in the valley. Then find the angle. Round to the nearest tenth.

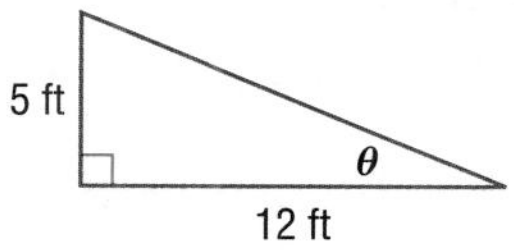

Personal Tutor glencoe.com

Check Your Understanding

Example 1 p. 872

Find each value. Write angle measures in degrees and radians.

1. $\text{Sin}^{-1}\frac{1}{2}$ **2.** $\text{Arctan}\,(-\sqrt{3})$ **3.** $\text{Arccos}\,(-1)$

Example 2 p. 872

Find each value. Round to the nearest hundredth if necessary.

4. $\cos\left(\text{Arcsin}\,\frac{4}{5}\right)$ **5.** $\tan\,(\text{Cos}^{-1}\,1)$ **6.** $\sin\left(\text{Sin}^{-1}\,\frac{\sqrt{3}}{2}\right)$

Example 3 p. 873

7. MULTIPLE CHOICE If $\sin\theta = 0.422$, find θ.

A 25° B 42° C 48° D 65°

Solve each equation. Round to the nearest tenth if necessary.

8. $\text{Cos}\,\theta = 0.9$ **9.** $\text{Sin}\,\theta = -0.46$ **10.** $\text{Tan}\,\theta = 2.1$

Example 4 p. 873

11. SNOWBOARDING A cross section of a superpipe for snowboarders is shown at the right. Write an inverse trigonometric function that can be used to find θ, the angle that describes the steepness of the superpipe. Then find the angle to the nearest degree.

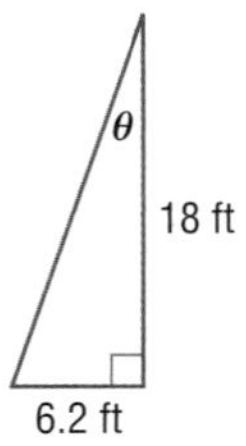

Practice and Problem Solving

● = **Step-by-Step Solutions** begin on page R20.
Extra Practice begins on page 947.

Example 1 p. 872

Find each value. Write angle measures in degrees and radians.

12. $\text{Arcsin}\left(\frac{\sqrt{3}}{2}\right)$ **13.** $\text{Arccos}\left(\frac{\sqrt{3}}{2}\right)$ **14.** $\text{Sin}^{-1}\,(-1)$

15. $\text{Tan}^{-1}\,\sqrt{3}$ **16.** $\text{Cos}^{-1}\left(-\frac{\sqrt{3}}{2}\right)$ **17.** $\text{Arctan}\left(-\frac{\sqrt{3}}{3}\right)$

Example 2 p. 872

Find each value. Round to the nearest hundredth if necessary.

18. $\tan\,(\text{Cos}^{-1}\,1)$ **19.** $\tan\left[\text{Arcsin}\left(-\frac{1}{2}\right)\right]$ **20.** $\cos\left(\text{Tan}^{-1}\,\frac{3}{5}\right)$

21. $\sin\,(\text{Arctan}\,\sqrt{3})$ **22.** $\cos\left(\text{Sin}^{-1}\,\frac{4}{9}\right)$ **23.** $\sin\left[\text{Cos}^{-1}\left(-\frac{\sqrt{2}}{2}\right)\right]$

Example 3 p. 873

Solve each equation. Round to the nearest tenth if necessary.

24. $\text{Tan}\,\theta = 3.8$ **25.** $\text{Sin}\,\theta = 0.9$ **26.** $\text{Sin}\,\theta = -2.5$

27. $\text{Cos}\,\theta = -0.25$ **28.** $\text{Cos}\,\theta = 0.56$ **29.** $\text{Tan}\,\theta = -0.2$

Example 4 p. 873

30. BOATS A boat is traveling west to cross a river that is 190 meters wide. Because of the current, the boat lands at point Q, which is 59 meters from its original destination point P. Write an inverse trigonometric function that can be used to find θ, the angle at which the boat veered south of the horizontal line. Then find the measure of the angle to the nearest tenth.

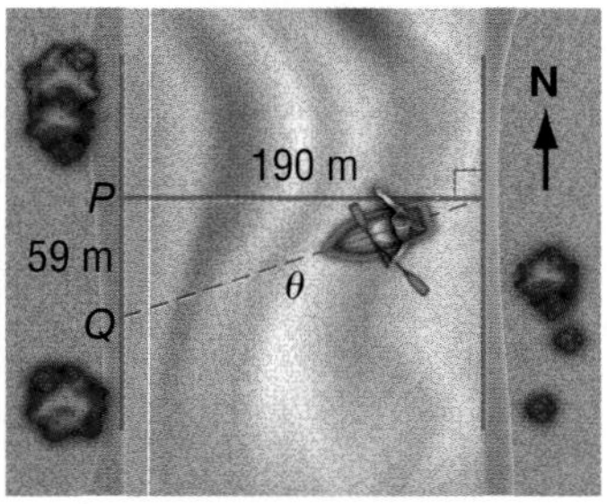

31. TREES A 24-foot tree is leaning 2.5 feet left of vertical, as shown in the figure. Write an inverse trigonometric function that can be used to find θ, the angle at which the tree is leaning. Then find the measure of the angle to the nearest degree.

32. DRIVING An expressway off-ramp curve has a radius of 52 meters and is designed for vehicles to safely travel at speeds up to 45 kilometers per hour (or 12.5 meters per second). The equation below represents the angle θ of the curve. What is the measure of the angle to the nearest degree?

$$\tan \theta = \frac{(12.5 \text{ m/s})^2}{(52 \text{ m})(9.8 \text{ m/s}^2)}$$

33. TRACK AND FIELD A shot-putter throws the shot with an initial speed of 15 meters per second. The expression $\frac{15 \text{ m/s} (\sin x)}{9.8 \text{ m/s}^2}$ represents the time in seconds at which the shot reached its maximum height. In the expression, x is the angle at which the shot was thrown. If the maximum height of the shot was reached in 1.0 second, at what angle was it thrown? Round to the nearest tenth.

Real-World Career

Sport Science Administrator A sport science administrator provides sport science information to players, coaches, and parents. He or she implements testing, training, and treatment programs for athletes. A master's degree in sport science or a related area is recommended.

Solve each equation for $0 \leq \theta \leq 2\pi$.

34. $\csc \theta = 1$

35. $\sec \theta = -1$

36. $\sec \theta = 1$

37. $\csc \theta = \frac{1}{2}$

38. $\cot \theta = 1$

39. $\sec \theta = 2$

40. MULTIPLE REPRESENTATIONS Consider $y = \text{Cos}^{-1} x$.

a. GRAPHICAL Sketch a graph of the function. Describe the domain and the range.

b. SYMBOLIC Write the function using different notation.

c. NUMERICAL Choose a value for x between -1 and 0. Then evaluate the inverse cosine function. Round to the nearest tenth.

d. ANALYTICAL Compare the graphs of $y = \cos x$ and $y = \text{Cos}^{-1} x$.

H.O.T. Problems

Use Higher-Order Thinking Skills

41. CHALLENGE Determine whether *cos (Arccos x) = x for all values of x* is *true* or *false*. If false, give a counterexample.

42. FIND THE ERROR Desiree and Oscar are solving $\cos \theta = 0.3$ where $90 < \theta < 180$. Is either of them correct? Explain your reasoning.

Desiree	Oscar
$\cos \theta = 0.3$	$\cos \theta = 0.3$
$\cos^{-1} 0.3 = 162.5°$	$\cos^{-1} 0.3 = 72.5°$

43. REASONING Explain how the domain of $y = \text{Sin}^{-1} x$ is related to the range of $y = \text{Sin } x$.

44. OPEN ENDED Write an equation with an Arcsine function and an equation with a Sine function that both involve the same angle measure.

45. WRITING IN MATH Compare and contrast the relations $y = \tan^{-1} x$ and $y = \text{Tan}^{-1} x$. Include information about the domains and ranges.

46. REASONING Explain how $\text{Sin}^{-1} 8$ and $\text{Cos}^{-1} 8$ are undefined while $\text{Tan}^{-1} 8$ is defined.

NYSRE PRACTICE A2.A.64, A2.A.65

47. Simplify $\dfrac{\frac{2}{x}+2}{\frac{2}{x}-2}$.

A $\dfrac{1+x}{1-x}$ C $\dfrac{1-x}{1+x}$

B $\dfrac{2}{x}$ D $-x$

48. SHORT RESPONSE What is the equation of the graph below?

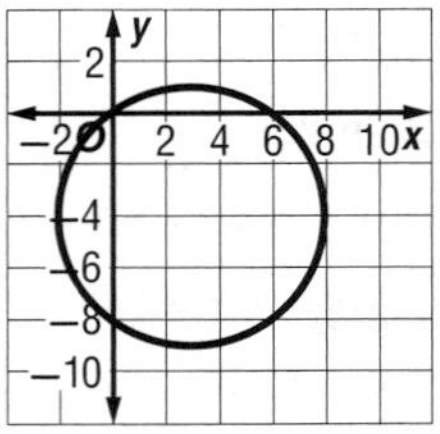

49. If $f(x) = 2x^2 - 3x$ and $g(x) = 4 - 2x$, what is $g[f(x)]$?

F $g[f(x)] = 4 + 6x - 8x^2$

G $g[f(x)] = 4 + 6x - 4x^2$

H $g[f(x)] = 20 - 26x + 8x^2$

J $g[f(x)] = 44 - 38x + 8x^2$

50. If g is a positive number, which of the following is equal to $12g$?

A $\sqrt{144g}$

B $\sqrt{12g^2}$

C $\sqrt{24g^2}$

D $6\sqrt{4g^2}$

Spiral Review

51. RIDES The Cosmoclock 21 is a huge Ferris wheel in Japan. The diameter is 328 feet. Suppose a rider enters the ride at 0 feet, and then rotates in 90° increments counterclockwise. The table shows the angle measures of rotation and the height above the ground of the rider. (Lesson 13-8)

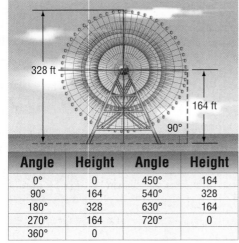

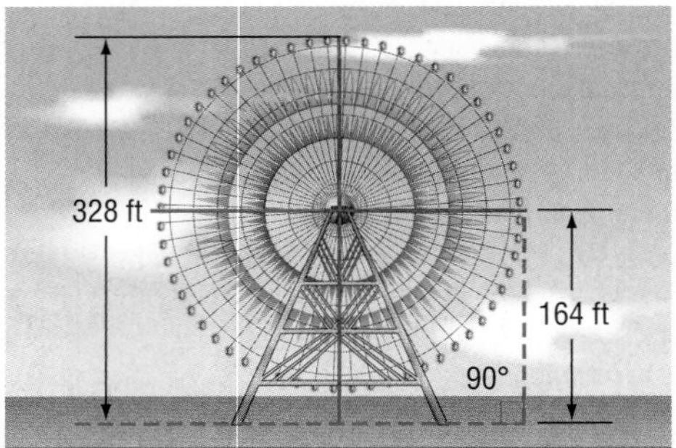

Angle	Height	Angle	Height
0°	0	450°	164
90°	164	540°	328
180°	328	630°	164
270°	164	720°	0
360°	0		

a. A function that models the data is $y = 164 \cdot [\sin (x - 90°)] + 164$. Identify the vertical shift, amplitude, period, and phase shift of the graph.

b. Write an equation using the sine that models the position of a rider on the Vienna Giant Ferris Wheel in Austria, with a diameter of 200 feet. Check your equation by plotting the points and the equation with a graphing calculator.

52. TIDES The world's record for the highest tide is held by the Minas Basin in Nova Scotia, Canada, with a tidal range of 54.6 feet. A tide is at equilibrium when it is at its normal level halfway between its highest and lowest points. Write an equation to represent the height h of the tide. Assume that the tide is at equilibrium at $t = 0$, that the high tide is beginning, and that the tide completes one cycle in 12 hours. (Lesson 13-7)

Solve each equation. (Lesson 8-4)

53. $\log_3 5 + \log_3 x = \log_3 10$

54. $\log_4 a + \log_4 9 = \log_4 27$

55. $\log_{10} 16 - \log_{10} 2t = \log_{10} 2$

56. $\log_7 24 - \log_7 (y + 5) = \log_3 8$

Skills Review

Find the exact value of each trigonometric function. (Lesson 13-3)

57. $\cos 3\pi$

58. $\tan 120°$

59. $\sin 300°$

60. $\sec \frac{7\pi}{6}$

CHAPTER 13 Study Guide and Review

NY Math Online glencoe.com
- STUDY TO GO
- Vocabulary Review

Chapter Summary

Key Concepts

Right Triangle Trigonometry (Lesson 13-1)

- $\sin \theta = \frac{\text{opp}}{\text{hyp}}$, $\cos \theta = \frac{\text{adj}}{\text{hyp}}$, $\tan \theta = \frac{\text{opp}}{\text{adj}}$,
  $\csc \theta = \frac{\text{hyp}}{\text{opp}}$, $\sec \theta = \frac{\text{hyp}}{\text{adj}}$, $\cot \theta = \frac{\text{adj}}{\text{opp}}$

Angle Measures and Trigonometric Functions of General Angles (Lessons 13-2 and 13-3)

- The measure of an angle is determined by the amount of rotation from the initial side to the terminal side.
- You can find the exact values of the six trigonometric functions of θ, given the coordinates of a point $P(x, y)$ on the terminal side of the angle.

Law of Sines and Law of Cosines (Lessons 13-4 and 13-5)

- $\frac{\sin A}{a} = \frac{\sin B}{b} = \frac{\sin C}{c}$
- $a^2 = b^2 + c^2 - 2bc \cos A$
  $b^2 = a^2 + c^2 - 2ac \cos B$
  $c^2 = a^2 + b^2 - 2ab \cos C$

Circular and Inverse Trigonometric Functions (Lessons 13-6 and 13-9)

- If the terminal side of an angle θ in standard position intersects the unit circle at $P(x, y)$, then $\cos \theta = x$ and $\sin \theta = y$.
- $y = \text{Sin } x$ if $y = \sin x$ and $-\frac{\pi}{2} \leq x \leq \frac{\pi}{2}$

Graphing Trigonometric Functions (Lesson 13-7)

- For trigonometric functions of the form $y = a \sin b\theta$ and $y = a \cos b\theta$, the amplitude is $|a|$, and the period is $\frac{360°}{|b|}$ or $\frac{2\pi}{|b|}$.
- The period of $y = a \tan b\theta$ is $\frac{180°}{|b|}$ or $\frac{\pi}{|b|}$.

FOLDABLES® Study Organizer

Be sure the Key Concepts are noted in your Foldable.

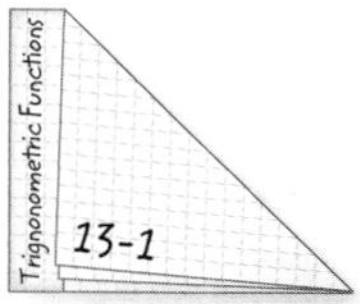

Key Vocabulary

amplitude (p. 855)
angle of depression (p. 812)
angle of elevation (p. 812)
Arccosine function (p. 871)
Arcsine function (p. 871)
Arctangent function (p. 871)
central angle (p. 820)
circular function (p. 848)
cosecant (p. 808)
cosine (p. 808)
cotangent (p. 808)
coterminal angles (p. 818)
cycle (p. 849)
frequency (p. 856)
initial side (p. 817)
Law of Cosines (p. 841)
Law of Sines (p. 833)
midline (p. 864)
period (p. 849)
periodic function (p. 849)
phase shift (p. 863)
principle values (p. 871)
quadrantal angle (p. 826)
radian (p. 819)
reference angle (p. 826)
secant (p. 808)
sine (p. 808)
solving a triangle (p. 833)
standard position (p. 817)
tangent (p. 808)
terminal side (p. 817)
trigonometric function (p. 808)
trigonometric ratio (p. 808)
trigonometry (p. 808)
unit circle (p. 848)
vertical shift (p. 864)

Vocabulary Check

State whether each sentence is *true* or *false*. If *false*, replace the underlined term to make a true sentence.

1. The <u>Law of Cosines</u> is used to solve a triangle when two angles and any sides are known.
2. An angle on the coordinate plane is in <u>standard position</u> if the vertex is at the origin and one ray is on the positive x-axis.
3. <u>Coterminal angles</u> are angles in standard position that have the same terminal side.
4. A horizontal translation of a periodic function is called a <u>phase shift</u>.
5. The inverse of the sine function is the <u>cosecant function</u>.
6. The <u>cycle</u> of the graph of a sine or cosine function equals half the difference between the maximum and minimum values of the function.

CHAPTER 13

Study Guide and Review

Lesson-by-Lesson Review

13-1 Right Triangle Trigonometry (pp. 808–816)

A2.A.55, A2.A.58

Solve $\triangle ABC$ by using the given measurements. Round measures of sides to the nearest tenth and measures of angles to the nearest degree.

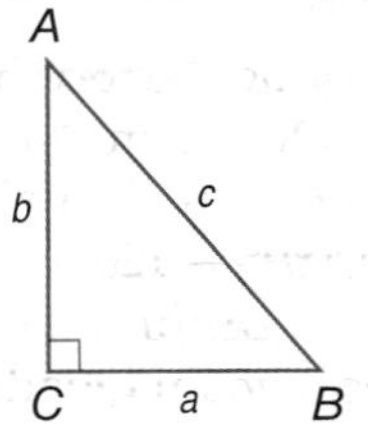

7. $c = 12, b = 5$

8. $a = 10, B = 55°$

9. $B = 75°, b = 15$

10. $B = 45°, c = 16$

11. $A = 35°, c = 22$

12. $\sin A = \frac{2}{3}, a = 6$

13. **TRUCK** The back of a moving truck is 3 feet off of the ground. What length does a ramp off the back of the truck need to be in order for the angle of elevation of the ramp to be 20°?

EXAMPLE 1

Solve $\triangle ABC$ by using the given measurements. Round measures of sides to the nearest tenth and measures of angles to the nearest degree.

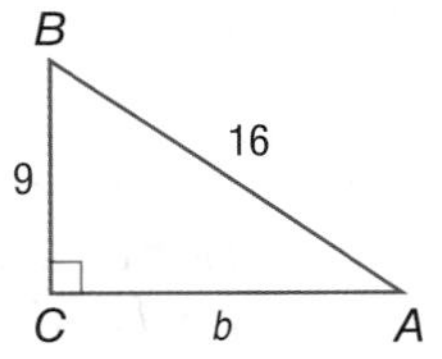

Find b.
$$a^2 + b^2 = c^2$$
$$9^2 + b^2 = 16^2$$
$$b = \sqrt{16^2 - 9^2}$$
$$b \approx 13.2$$

Find A. $\sin A = \frac{9}{16}$

Use a calculator.
To the nearest degree, $A = 34°$.

Find B.
$$34° + B \approx 90°$$
$$B \approx 56°$$

Therefore, $b \approx 13.2$, $A \approx 34°$, and $B \approx 56°$.

13-2 Angles and Angle Measures (pp. 817–823)

A2.M.2, A2.A.61

Rewrite each degree measure in radians and each radian measure in degrees.

14. 215°

15. $\frac{5\pi}{2}$

16. -3π

17. $-315°$

Find one angle with positive measure and one angle with negative measure coterminal with each angle.

18. 265°

19. $-65°$

20. $\frac{7\pi}{2}$

21. **BICYCLE** A bicycle tire makes 8 revolutions in one minute. The tire has a radius of 15 inches. Find the angle θ in radians through which the tire rotates in one second.

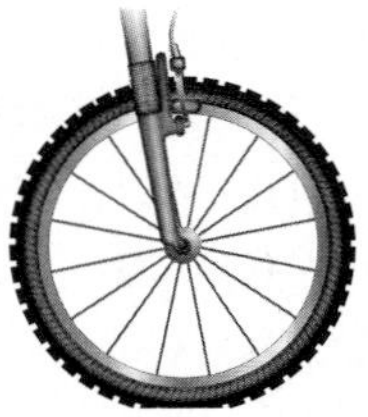

EXAMPLE 2

Rewrite 160° in radians.

$$160° = 160°\left(\frac{\pi \text{ radians}}{180°}\right)$$
$$= \frac{160\pi}{180} \text{ radians or } \frac{8\pi}{9}$$

EXAMPLE 3

Find one angle with positive measure and one angle with negative measure coterminal with 150°.

positive angle:

$150° + 360° = 510°$ **Add 360°.**

negative angle:

$150° - 360° = -210°$ **Subtract 360°.**

13-3 Trigonometric Functions of General Angles (pp. 825–831)

A2.A.57, A2.A.62

Find the exact value of each trigonometric function.

22. $\cos 135°$

23. $\tan 150°$

24. $\sin 2\pi$

25. $\cos \frac{3\pi}{2}$

The terminal side of θ in standard position contains each point. Find the exact values of the six trigonometric functions of θ.

26. $P(-4, 3)$

27. $P(5, 12)$

28. $P(16, -12)$

29. **BALL** A ball is thrown off the edge of a building at an angle of 70° and with an initial velocity of 5 meters per second. The equation that represents the horizontal distance of the ball x is $x = v_0(\cos \theta)t$, where v_0 is the initial velocity, θ is the angle at which it is thrown, and t is the time in seconds. About how far will the ball travel in 10 seconds?

EXAMPLE 4

Find the exact value of sin 120°.

Because the terminal side of 120° lies in Quadrant II, the reference angle θ' is $180° - 120°$ or 60°. The sine function is positive in Quadrant II, so $\sin 120° = \sin 60°$ or $\frac{\sqrt{3}}{2}$.

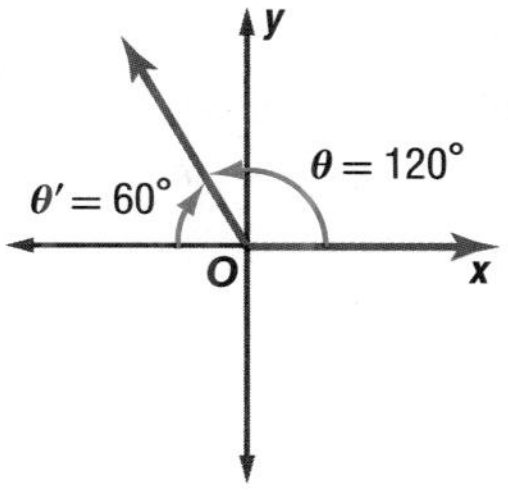

EXAMPLE 5

The terminal side of θ in standard position contains the point (6, 5). Find the exact values of the six trigonometric functions of θ.

$\sin \theta = \frac{y}{r}$ or $\frac{5\sqrt{61}}{61}$

$\cos \theta = \frac{x}{r}$ or $\frac{6\sqrt{61}}{61}$

$\tan \theta = \frac{y}{x}$ or $\frac{5}{6}$

$\csc \theta = \frac{r}{y}$ or $\frac{\sqrt{61}}{5}$

$\sec \theta = \frac{r}{x}$ or $\frac{\sqrt{61}}{6}$

$\cot \theta = \frac{x}{y}$ or $\frac{6}{5}$

13-4 Law of Sines (pp. 832–839)

A2.A.73, A2.A.74

Determine whether each triangle has *no* solution, *one* solution, or *two* solutions. Then solve each triangle. Round measures of sides to the nearest tenth and measures of angles to the nearest degree.

30. $C = 118°, c = 10, a = 4$

31. $A = 25°, a = 15, c = 18$

32. $A = 70°, a = 5, c = 16$

33. **BOAT** Kira and Mallory are standing on opposite sides of a river. How far is Kira from the boat? Round to the nearest tenth if necessary.

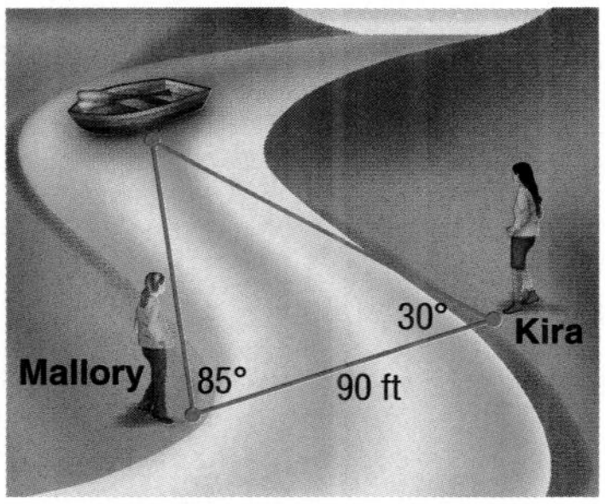

EXAMPLE 6

Solve $\triangle ABC$.

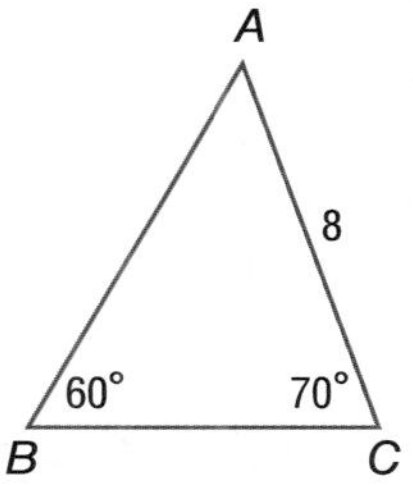

First, find the measure of the third angle.

$$60° + 70° + a = 180°$$
$$A = 50°$$

Now use the Law of Sines to find a and c. Write two equations, each with one variable.

$$\frac{\sin B}{b} = \frac{\sin C}{c} \qquad \frac{\sin B}{b} = \frac{\sin A}{a}$$

$$\frac{\sin 60°}{8} = \frac{\sin 70°}{c} \qquad \frac{\sin 60°}{8} = \frac{\sin 50°}{a}$$

$$c = \frac{8 \sin 70°}{\sin 60°} \qquad a = \frac{8 \sin 50°}{\sin 60°}$$

$$c \approx 8.9 \qquad a \approx 7.1$$

Therefore, $A = 50°$, $c \approx 8.9$, and $a \approx 7.1$.

Study Guide and Review

13-5 Law of Cosines (pp. 841–846)

A2.A.66, A2.A.73

Determine whether each triangle should be solved by beginning with the Law of *Sines* or Law of *Cosines*. Then solve each triangle. Round measures of sides to the nearest tenth and measures of angles to the nearest degree.

34.

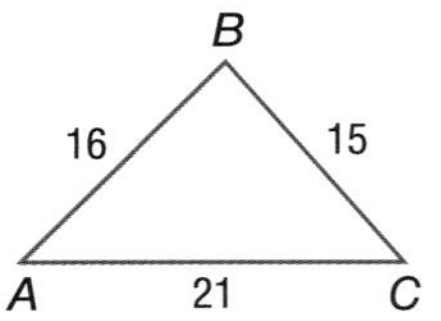

35.

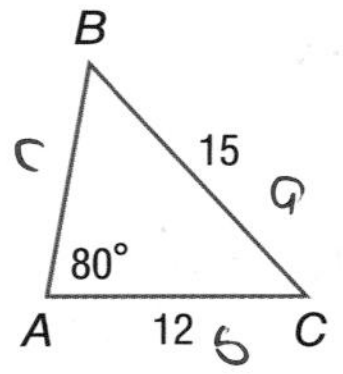

36. $C = 75°, a = 5, b = 7$

37. $A = 42°, a = 9, b = 13$

38. $b = 8.2, c = 15.4, A = 35°$

39. FARMING A farmer wants to fence a piece of his land. Two sides of the triangular field have lengths of 120 feet and 325 feet. The measure of the angle between those sides is 70°. How much fencing will the farmer need?

EXAMPLE 7

Solve $\triangle ABC$ for $C = 55°$, $b = 11$, and $a = 8$.

You are given the measure of two sides and the included angle. Begin by drawing a diagram and using the Law of Cosines to determine c.

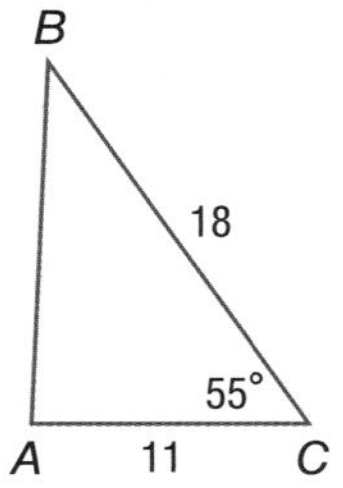

$c^2 = a^2 + b^2 - 2ab \cos C$

$c^2 = 8^2 + 11^2 - 2(8)(11) \cos 55°$

$c^2 \approx 84$

$c \approx 9.2$

Next, you can use the Law of Sines to find the measure of angle A.

$\frac{\sin A}{8} \approx \frac{\sin 55°}{9.2}$

$\sin A \approx \frac{8 \sin 55°}{9.2}$ or A is about 45.4°

The measure of the angle B is approximately $180 - (45.4 + 55)$ or 79.6°.

Therefore, $c \approx 9.2$, $A \approx 45.4°$, and $B \approx 79.6°$.

13-6 Circular Functions (pp. 848–854)

A2.A.56, A2.A.60

Find the exact value of each function.

40. $\cos(-210°)$

41. $(\cos 45°)(\cos 210°)$

42. $\sin -\frac{7\pi}{4}$

43. $\left(\cos \frac{\pi}{2}\right)\left(\sin \frac{\pi}{2}\right)$

44. Determine the period of the function.

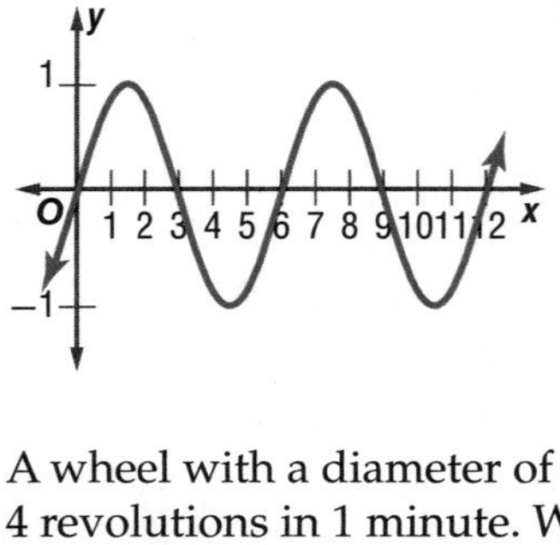

45. A wheel with a diameter of 18 inches completes 4 revolutions in 1 minute. What is the period of the function that describes the height of one spot on the outside edge of the wheel as a function of time?

EXAMPLE 8

Find the exact value of sin 510°.

$\sin 510° = \sin(310° + 150°)$

$= \sin 150°$

$= \frac{1}{2}$

EXAMPLE 9

Determine the period of the function below.

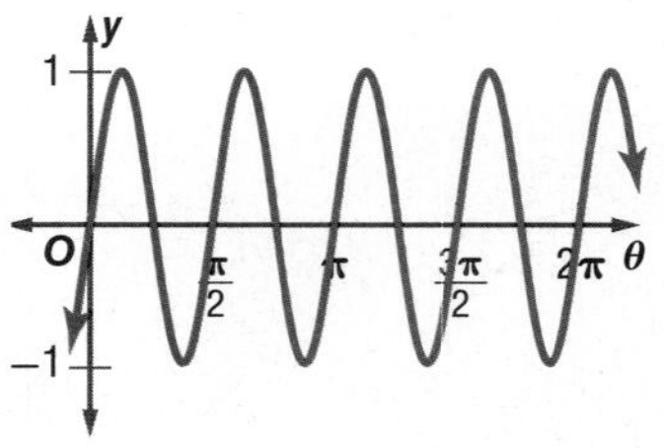

The pattern repeats itself at $\frac{\pi}{2}$, π, and so on. So, the period is $\frac{\pi}{2}$.

MIXED PROBLEM SOLVING
For mixed problem-solving practice, see page 991.

13-7 Graphing Trigonometric Functions (pp. 855–861)

A2.A.70, A2.A.71

Find the amplitude, if it exists, and period of each function. Then graph the function.

46. $y = 4 \sin 2\theta$

47. $y = \cos \frac{1}{2}\theta$

48. $y = 3 \csc \theta$

49. $y = 3 \sec \theta$

50. $y = \tan 2\theta$

51. $y = 2 \csc \frac{1}{2}\theta$

52. When Lauren jumps on a trampoline it vibrates with a frequency of 10 hertz. Let the amplitude equal 5 feet. Write a sine equation to represent the vibration of the trampoline y as a function of time t.

EXAMPLE 10

Find the amplitude and period of $y = 2 \cos 4\theta$. Then graph the function.

amplitude: $|a| = |2|$ or 2. The graph is stretched vertically so that the maximum value is 2 and the minimum value is −2.

period:
$\frac{360°}{|b|} = \frac{360°}{|4|}$ or 90°

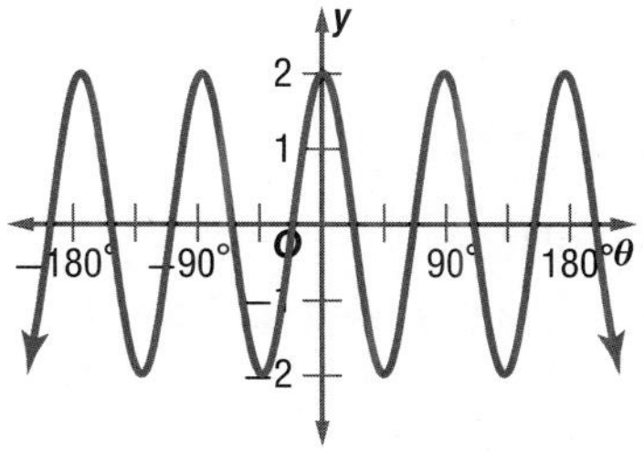

13-8 Translations of Trigonometric Graphs (pp. 863–870)

A2.A.69, A2.A.72

State the vertical shift, amplitude, period, and phase shift of each function. Then graph the function.

53. $y = 3 \sin [2(\theta - 90°)] + 1$

54. $y = \frac{1}{2} \tan [2(\theta - 30°)] - 3$

55. $y = 2 \sec \left[3\left(\theta - \frac{\pi}{2}\right)\right] + 2$

56. $y = \frac{1}{2} \cos \left[\frac{1}{4}\left(\theta + \frac{\pi}{4}\right)\right] - 1$

57. $y = \frac{1}{3} \sin \left[\frac{1}{3}(\theta - 90°)\right] + 2$

58. The graph below approximates the height y of a rope that two people are twirling as a function of time t in seconds. Write an equation for the function.

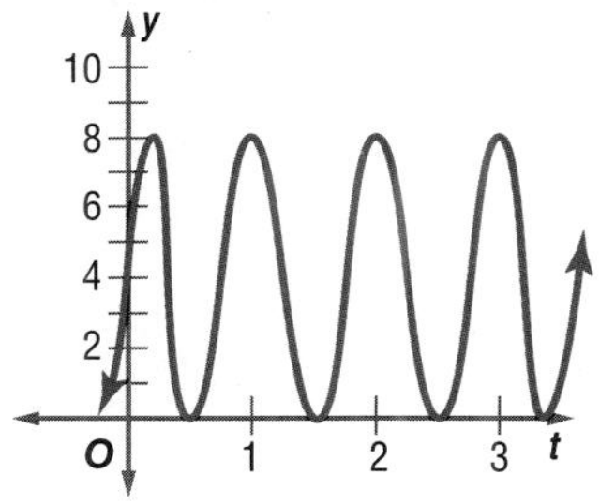

EXAMPLE 11

State the vertical shift, amplitude, period, and phase shift of $y = 2 \sin \left[3\left(\theta + \frac{\pi}{2}\right)\right] + 4$. Then graph the function.

Identify the values of k, a, b, and h.

$k = 4$, so the vertical shift is 4.

$a = 2$, so the amplitude is 2.

$b = 3$, so the period is $\frac{2\pi}{|3|}$ or $\frac{2\pi}{3}$.

$h = -\frac{\pi}{2}$, so the phase shift is $\frac{\pi}{2}$ to the left.

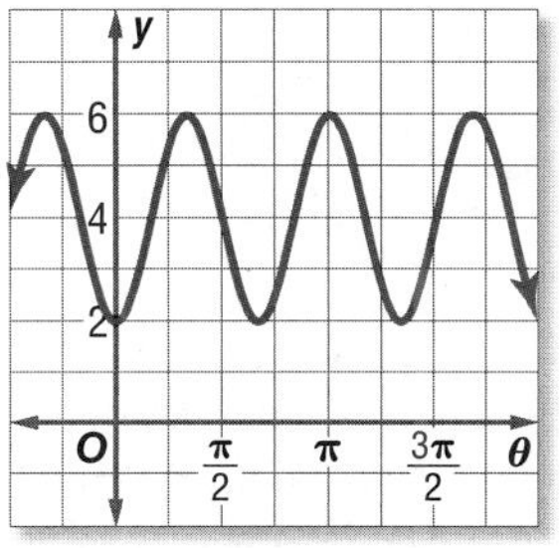

13-9 Inverse Trigonometric Functions (pp. 871–876)

A2.A.64, A2.A.65

Evaluate each inverse trigonometric function. Write angle measures in degrees and radians.

59. $\text{Sin}^{-1}(1)$

60. Arctan (0)

61. $\text{Arcsin}\ \frac{\sqrt{3}}{2}$

62. $\text{Cos}^{-1}\ \frac{\sqrt{2}}{2}$

63. $\text{Tan}^{-1}\ 1$

64. Arccos 0

65. A bicycle ramp is 5 feet tall and 10 feet long, as shown below. Write an inverse trigonometric function that can be used to find θ, the angle the ramp makes with the ground. Then find the angle. Round to the nearest hundredth.

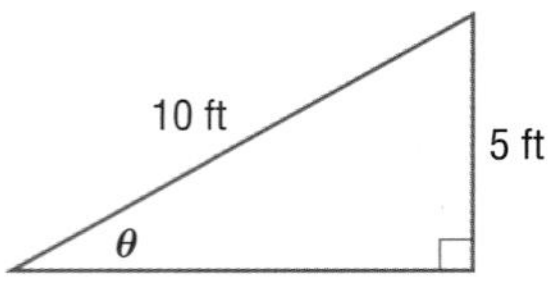

Evaluate each inverse trigonometric function. Round to the nearest hundredth if necessary.

66. $\tan\left(\text{Cos}^{-1}\frac{1}{3}\right)$

67. $\text{Sin}\left(\text{Arcsin} -\frac{\sqrt{2}}{2}\right)$

68. $\sin(\text{Tan}^{-1}\ 0)$

Solve each equation. Round to the nearest tenth if necessary.

69. $\text{Tan}\ \theta = -1.43$

70. $\text{Sin}\ \theta = 0.8$

71. $\text{Cos}\ \theta = 0.41$

EXAMPLE 12

Evaluate $\text{Cos}^{-1}\ \frac{1}{2}$. Write angle measures in degrees and radians.

Find the angle θ for $0° \le \theta \le 180°$ that has a cosine value of $\frac{1}{2}$.

Use a unit circle.

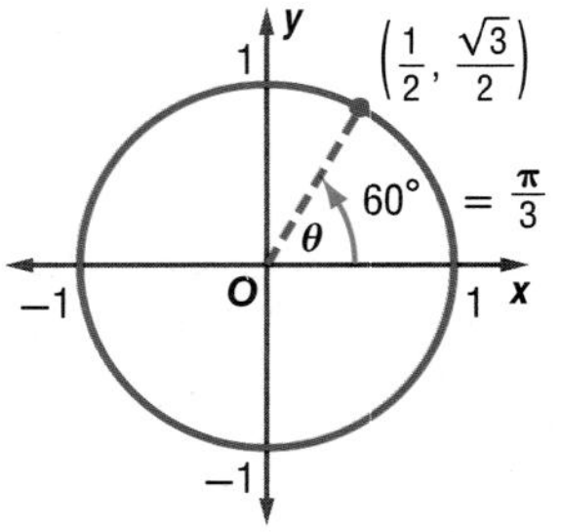

Find a point on the unit circle that has an x-coordinate of $\frac{1}{2}$. When $\theta = 60°$, $\text{Cos}\ \theta = \frac{1}{2}$.

So, $\text{Cos}^{-1} = 60°$ or $\frac{\pi}{3}$.

EXAMPLE 13

Evaluate $\sin\left(\text{Tan}^{-1}\ \frac{1}{2}\right)$. Round to the nearest hundredth.

Use a calculator.

KEYSTROKES: SIN 2nd [TAN^{-1}] 1 ÷ 2) ENTER

0.4472135955

So, $\sin\left(\text{Tan}^{-1}\ \frac{1}{2}\right) \approx 0.45$.

EXAMPLE 14

If $\text{Cos}\ \theta = 0.72$, find θ.

Use a calculator.

KEYSTROKES: 2nd [COS^{-1}] .72 ENTER 43.9455195623

So, $\theta \approx 43.9°$.

Practice Test

NY Math Online glencoe.com
Chapter Test

Solve $\triangle ABC$ by using the given measurements. Round measures of sides to the nearest tenth and measures of angles to the nearest degree.

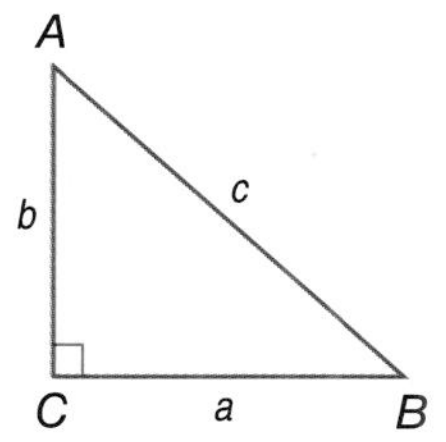

1. $A = 36°, c = 9$
2. $a = 12, A = 58°$
3. $B = 85°, b = 8$
4. $a = 9, c = 12$

Rewrite each degree measure in radians and each radian measure in degrees.

5. $325°$
6. $-175°$
7. $\frac{9\pi}{4}$
8. $-\frac{5\pi}{6}$

9. Determine whether $\triangle ABC$, with $A = 110°$, $a = 16$, and $b = 21$, has *no* solution, *one* solution, or *two* solutions. Then solve the triangle, if possible. Round measures of sides to the nearest tenth and measures of angles to the nearest degree.

Find the exact value of each function. Write angle measures in degrees.

10. $\cos(-90°)$
11. $\sin 585°$
12. $\cot \frac{4\pi}{3}$
13. $\sec\left(-\frac{9\pi}{4}\right)$
14. $\tan\left(\text{Cos}^{-1} \frac{4}{5}\right)$
15. $\text{Arccos } \frac{1}{2}$

16. The terminal side of angle θ in standard position intersects the unit circle at point $P\left(\frac{1}{2}, \frac{\sqrt{3}}{2}\right)$. Find $\cos \theta$ and $\sin \theta$.

17. **MULTIPLE CHOICE** What angle has a tangent and sine that are both negative?

 A 65°

 B 310°

 C 120°

 D 265°

18. **NAVIGATION** Airplanes and ships measure distance in nautical miles. The formula 1 nautical mile = $6077 - 31 \cos 2\theta$ feet, where θ is the latitude in degrees, can be used to find the approximate length of a nautical mile at a certain latitude. Find the length of a nautical mile when the latitude is 120°.

Find the amplitude and period of each function. Then graph the function.

19. $y = 2 \sin 3\theta$
20. $y = \frac{1}{2} \cos 2\theta$

21. **MULTIPLE CHOICE** What is the period of the function $y = 3 \cot \theta$?

 F 120°

 G 180°

 H 360°

 J 1080°

22. Determine whether $\triangle XYZ$, with $y = 15$, $z = 9$, and $X = 105°$, should be solved by beginning with the Law of Sines or Law of Cosines. Then solve the triangle. Round measures of sides to the nearest tenth and measures of angles to the nearest degree.

State the amplitude, period, and phase shift for each function. Then graph the function.

23. $y = \cos(\theta + 180)$
24. $y = \frac{1}{2} \tan\left(\theta - \frac{\pi}{2}\right)$

25. **WHEELS** A water wheel has a diameter of 20 feet. It makes one complete revolution in 45 seconds. Let the height at the top of the wheel represent the height at time 0. Write an equation for the height of point h in the diagram below as a function of time t.

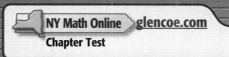

CHAPTER 13

Preparing for Standardized Tests

Using a Scientific Calculator

Scientific calculators and graphing calculators are powerful problem-solving tools. As you have likely seen, some test problems that you encounter have steps or computations that require the use of a scientific calculator.

Strategies for Using a Scientific Calculator

Step 1

Familiarize yourself with the various functions of a scientific calculator as well as when they should be used.

- **Scientific notation**—for calculating large numbers
- **Logarithmic and exponential functions**—growth and decay problems, compound interest
- **Trigonometric functions**—problems involving angles, triangle problems, indirect measurement problems
- **Square roots and *n*th roots**—distance on a coordinate plane, Pythagorean Theorem

Step 2

Use your scientific or graphing calculator to solve the problem.

- Remember to work as efficiently as possible. Some steps may be done mentally or by hand, while others must be done using your calculator.
- If time permits, check your answer.

EXAMPLE

Read the problem. Identify what you need to know. Then use the information in the problem to solve.

When Molly stands at a distance of 18 feet from the base of a tree, she forms an angle of 57° with the top of the tree. What is the height of the tree to the nearest tenth?

A 27.7 ft

B 28.5 ft

C 29.2 ft

D 30.1 ft

Read the problem carefully. You are given some measurements and asked to find the height of a tree. It may be helpful to first sketch a model of the problem.

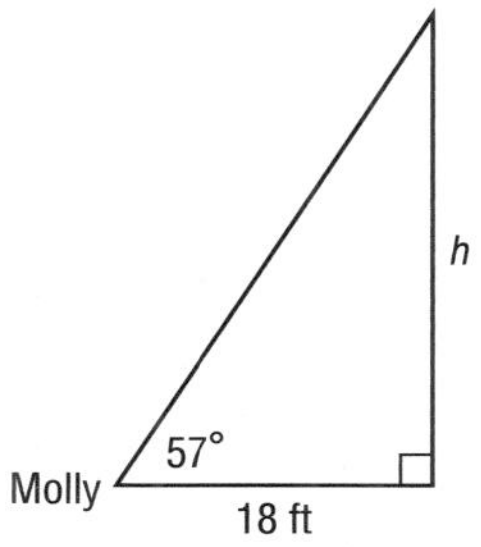

Use a trigonometric function to relate the lengths and the angle measure in the right triangle.

$\text{tangent } \theta = \frac{\text{opposite}}{\text{adjacent}}$ **Definition of tangent ratio**

$\tan 57° = \frac{h}{18}$ **Substitute.**

You need to evaluate tan 57° to solve for the height of a tree h. Use a scientific calculator.

$1.53986 \approx \frac{h}{18}$ **Use a calculator.**

$27.71748 \approx h$ **Multiply each side by 18.**

The height of the tree is about 27.7 feet. The correct answer is A.

Exercises

Read each problem. Identify what you need to know. Then use the information in the problem to solve.

1. An airplane takes off and climbs at a constant rate. After traveling 800 yards horizontally, the plane has climbed 285 yards vertically. What is the plane's angle of elevation during the takeoff and initial climb?

A 15.6°

B 18.4°

C 19.6°

D 22.3°

2. What is the angle of the bike ramp below?

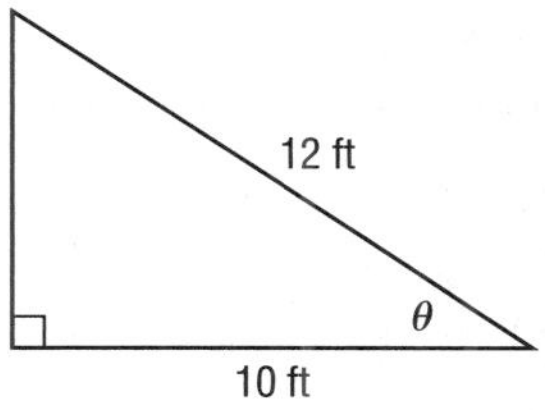

F 26.3°

G 28.5°

H 30.4°

J 33.6°

CHAPTER 13 NYS Regents Exam Practice

Cumulative, Chapters 1 through 13

Multiple Choice

Answer all questions in this part. Select the answer that best completes the statement or answers the question.

1. When a surveyor stands at a distance of 40 meters from the base of the Statue of Liberty, the angle of elevation to the torch is 66.7°. What is the height of the Statue of Liberty, including its base? Round to the nearest meter.

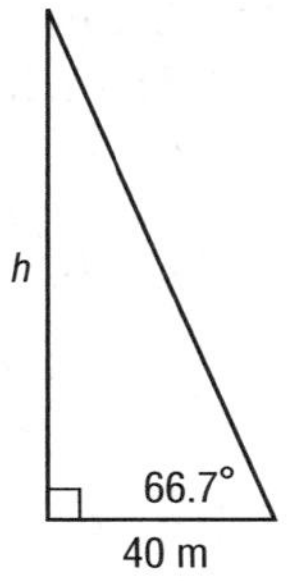

(1) 55 m

(2) 71 m

(3) 93 m

(4) 104 m

2. The table below shows the grades earned by students on a science test. Calculate the standard deviation of the test scores. Round to the nearest thousandth.

76	84	91	75	83
82	65	94	90	71
92	84	83	88	80
78	84	89	95	93

(1) 7.91 **(3)** 8.23

(2) 8.03 **(4)** 8.55

Test-TakingTip

Question 2 You can use a scientific calculator to find the standard deviation. Enter the data values as a list and calculate the 1-variable statistics.

3. Simplify: $\sqrt[5]{-64b^4c^{10}}$.

(1) $-2c^2\sqrt[5]{2b^4}$

(2) $-2c^5\sqrt[5]{2b^4}$

(3) $2c\sqrt[5]{b^4}$

(4) $2b^2c^5$

4. A leaf on the surface of a pond bobs up and down as the ripples of a wave pass it by. The position of the leaf is given by the function $h(t) = 0.75 \sin (2\pi t)$, where t is time in seconds and h is the height in inches. How long does it take to complete one full cycle from ripple to ripple?

(1) 0.75 second

(2) 1 second

(3) 2 seconds

(4) 4 seconds

5. What is the equation of the circle below?

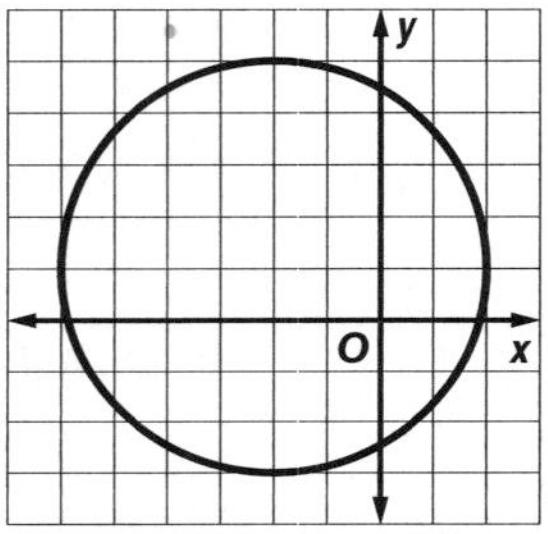

(1) $(x - 2)^2 + (y + 1)^2 = 3^2$

(2) $(x - 2)^2 + (y + 1)^2 = 4^2$

(3) $(x + 2)^2 + (y - 1)^2 = 3^2$

(4) $(x + 2)^2 + (y - 1)^2 = 4^2$

6. Use the Law of Cosines to solve for a. Round to the nearest tenth.

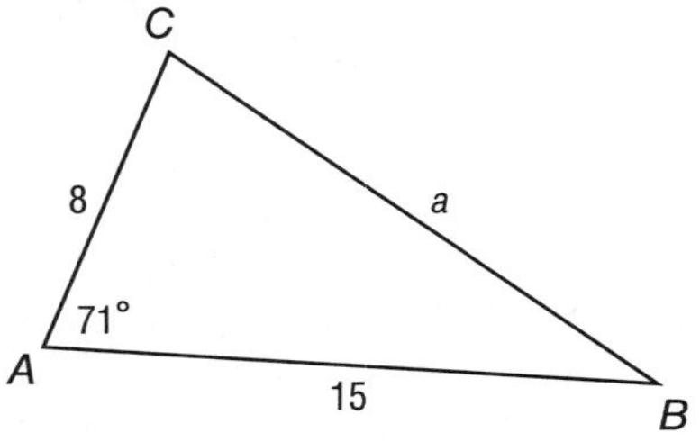

(1) 14.5

(2) 16.3

(3) 19.1

(4) 22.6

7. Josh can paint a small room in 1 hour 15 minutes. Craig can paint the same size room in 1 hour 30 minutes. How long would it take Josh and Craig to paint the room working together? Round to the nearest minute.

(1) 38 minutes
(2) 39 minutes
(3) 41 minutes
(4) 45 minutes

8. What is the period of the function graphed below?

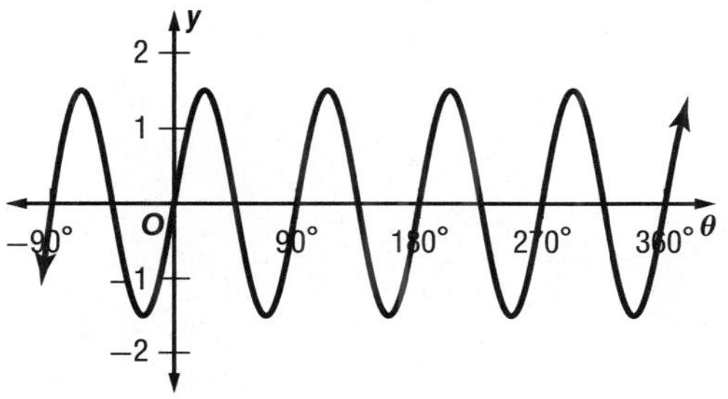

(1) 45°
(2) 90°
(3) 180°
(4) 360°

9. The quadratic equation $y = x^2 - 4x + 13$ has no real roots. Use the Quadratic Formula to find the complex roots of the equation.

(1) $1 \pm 4i$
(2) $2 \pm 5i$
(3) $3 \pm 2i$
(4) $2 \pm 3i$

10. Solve for y in the matrix equation below.

$$\begin{bmatrix} 42 & 5y+1 \\ 3x-7 & 50 \\ 12 & z+8 \end{bmatrix} = \begin{bmatrix} 42 & 26 \\ 35 & 50 \\ 12 & 17 \end{bmatrix}$$

(1) 5
(2) 7
(3) 10
(4) 12

Open-Ended Response

Solve each problem. Clearly indicate the necessary steps, including appropriate formula substitutions, diagrams, graphs, charts, etc.

11. Suppose a Ferris wheel has a diameter of 75 feet. The wheel rotates 12° each time a new passenger is picked up. Use this information to answer each question. Round your answers to the nearest tenth.

a. What is the circumference of the Ferris wheel?

b. How far would a passenger travel when the wheel rotates 12° to pick up the next passenger?

c. Suppose Mandy and Kaylee get in a car on the ride. Their friends Jason and Aaron get in 4 cars later. How far apart are they along the circumference of the wheel?

12. Katrina tosses ten coins. Use the Binomial Theorem to find the probability that she tosses exactly 7 heads and 3 tails. Express your answer as a fraction.

13. The volume of a rectangular prism is 350 cubic centimeters. The height of the prism is 2 centimeters less than the width. The length is 3 centimeters more than the width.

a. Sketch a model of the prism. Let w represent the width.

b. Use the volume of the prism to write an equation that can be solved for w.

c. Use the Rational Zero Theorem to list the possible zeroes of the equation.

d. Solve for w. What are the dimensions of the prism?

Need Extra Help?

If you missed Question...	1	2	3	4	5	6	7	8	9	10	11	12	13
Go to Lesson or Page...	13-1	1-3	13-2	2-4	13-3	3-5	13-4	4-1	5-1	6-2	7-3	8-2	9-5
NYS Core Curriculum	A2.A.55	A2.S.4	A2.A.13	A2.A.70	A2.A.49	A2.A.73	A2.A.23	A2.A.69	A2.A.25	A2.R.2	A2.A.61	A2.S.15	A2.A.26

CHAPTER 14

Trigonometric Identities and Equations

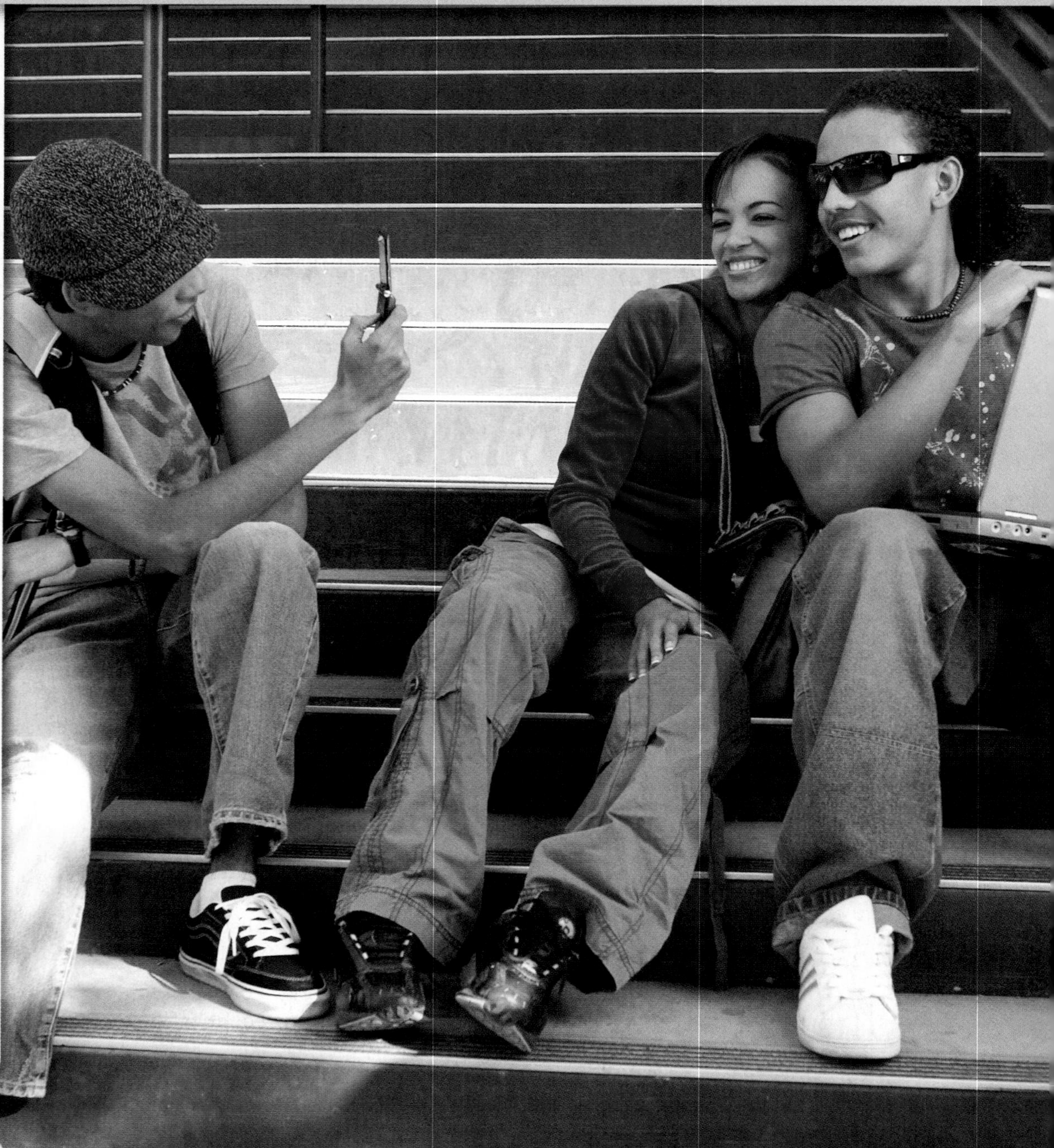

Then

In Chapter 13, you graphed trigonometric functions and determined the period, amplitude, phase shifts, and vertical shifts.

Now

In Chapter 14, you will:

- Use and verify trigonometric identities.
- Use the sum and difference of angles identities.
- Use the double- and half-angle identities.
- Solve trigonometric equations.

NYS Core Curriculum

A2.A.76, A2.A.77 Apply trigonometric formulas

A2.A.68 Solve trigonometric equations for all values of the variable from 0° to 360°

Why?

ELECTRONICS Many aspects of electronics can be modeled by trigonometric functions. Radio, television, cellular telephones, and wireless Internet all communicate through radio waves that are modeled by trigonometric functions. The amount of power in an electronic gadget can be found by using a trigonometric equation.

Math *in Motion,* Animation glencoe.com

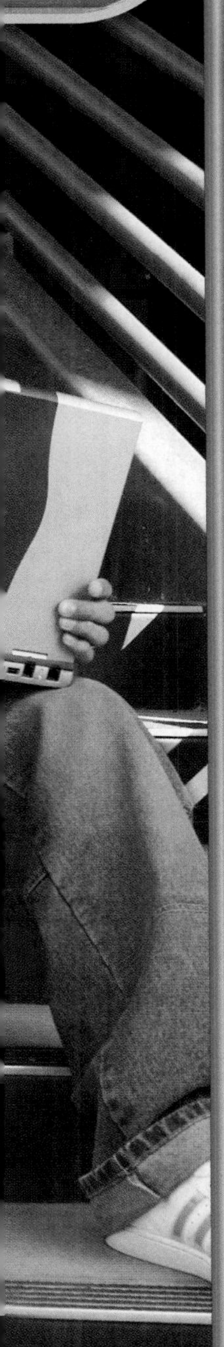

Get Ready for Chapter 14

Diagnose Readiness You have two options for checking Prerequisite Skills.

Text Option

Take the Quick Check below. Refer to the Quick Review for help.

*Quick*Check

Factor completely. If the polynomial is not factorable, write *prime*. (Lesson 6-5)

1. $-16a^2 + 4a$

2. $5x^2 - 20$

3. $x^3 + 9$

4. $2y^2 - y - 15$

5. GEOMETRY The area of a rectangular piece of cardboard is $x^2 + 6x + 8$ square inches. If the cardboard has a length of $(x + 4)$ inches, what is the width?

Solve each equation by factoring. (Lesson 5-3)

6. $x^2 + 6x = 0$

7. $x^2 + 2x - 35 = 0$

8. $x^2 - 9 = 0$

9. $x^2 - 7x + 12 = 0$

10. GARDENING Peyton is building a flower bed in her back yard. The area of the flower bed will be 42 square feet. Find the possible values for x.

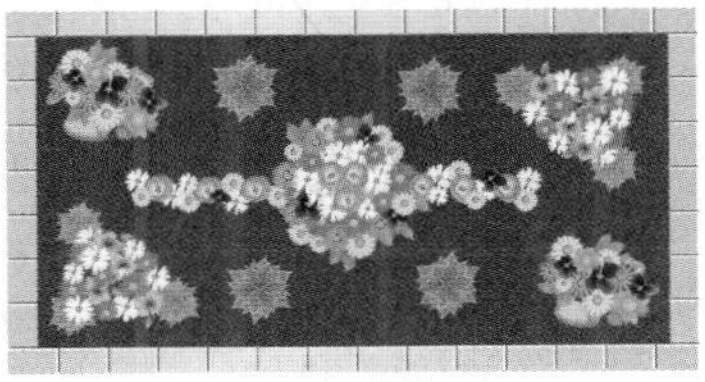

Find the exact value of each trigonometric function. (Lesson 13-3)

11. sin 45°

12. cos 225°

13. tan 150°

14. sin 120°

15. RIDES The distance from the highest point of a Ferris wheel to the ground can be found by multiplying 90 feet by sin 90°. What is the height of the Ferris wheel when it is halfway between the tallest point and the ground?

*Quick*Review

EXAMPLE 1

Factor $x^3 + 2x^2 - 24x$ completely.

$x^3 + 2x^2 - 24x = x(x^2 + 2x - 24)$

The product of the coefficients of the x terms must be -24, and their sum must be 2. The product of 6 and -4 is -24 and their sum is 2.

$x(x^2 + 2x - 24) = x(x + 6)(x - 4)$

EXAMPLE 2

Solve $x^2 + 6x + 5 = 0$ by factoring.

$x^2 + 6x + 5 = 0$ **Original equation**

$(x + 5)(x + 1) = 0$ **Factor.**

$x + 5 = 0$ or $x + 1 = 0$

$x = -5$ $\quad$ $x = -1$

The solution set is $\{-5, -1\}$.

EXAMPLE 3

Find the exact value of cos 135°.

The reference angle is 180° − 135° or 45°.

cos 45° is $\frac{\sqrt{2}}{2}$. Since 135° is in the second quadrant, cos 135° $= -\frac{\sqrt{2}}{2}$.

Online Option

NY Math Online Take a self-check Chapter Readiness Quiz at **glencoe.com**.

Get Started on Chapter 14

You will learn several new concepts, skills, and vocabulary terms as you study Chapter 14. To get ready, identify important terms and organize your resources. You may wish to refer to **Chapter 0** to review prerequisite skills.

FOLDABLES® Study Organizer

Trigonometric Identities and Equations Make this Foldable to help you organize your Chapter 14 notes about trigonometric identities and equations. Begin with one sheet of 11″ × 17″ paper and four sheets of grid paper.

1. **Fold** the short sides of the 11″ × 17″ paper to meet in the middle.

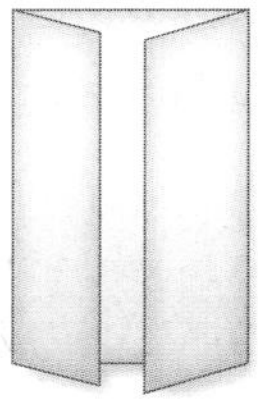

2. **Cut** each tab in half as shown.

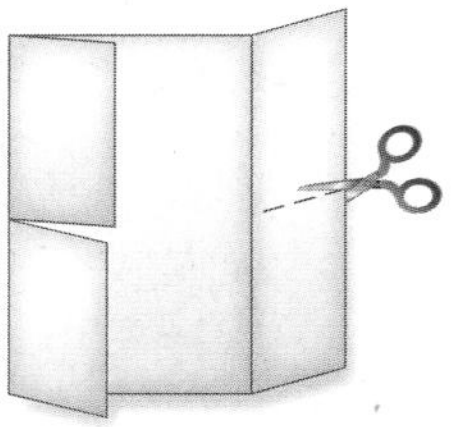

3. **Cut** four sheets of grid paper in half and fold the half-sheets in half.

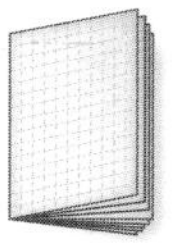

4. **Insert** two folded half-sheets under each of the four tabs and staple along the fold. Label each tab as shown.

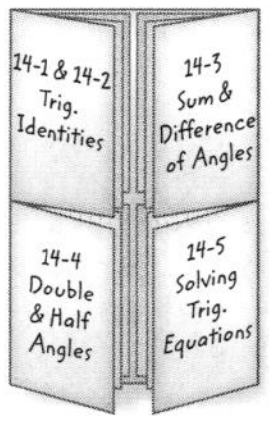

New Vocabulary

English		Español
trigonometric identity	p. 891	identidad trigométrica
quotient identity	p. 891	identidad de cociente
reciprocal identity	p. 891	identidad recíproca
Pythagorean identity	p. 891	identidad Pitagórica
cofunction identity	p. 891	identidad de función conjunta
negative angle identity	p. 891	identidad negativa de ángulo
trigonometric equation	p. 919	ecuación trigométrica

Review Vocabulary

formula • p. 6 • fórmula a mathematical sentence that expresses the relationship between certain quantities

identity • p. 229 • identidad an equality that remains true regardless of the values of any variables that are in it

trigonometric functions • p. 808 • funciones trigonométricas For any angle, with measure θ, a point $P(x, y)$ on its terminal side, $r = \sqrt{x^2 + y^2}$, the trigonometric functions of θ are as follows.

$\sin \theta = \frac{y}{r}$ $\cos \theta = \frac{x}{r}$ $\tan \theta = \frac{y}{x}$

$\csc \theta = \frac{r}{y}$ $\sec \theta = \frac{r}{x}$ $\cot \theta = \frac{x}{y}$

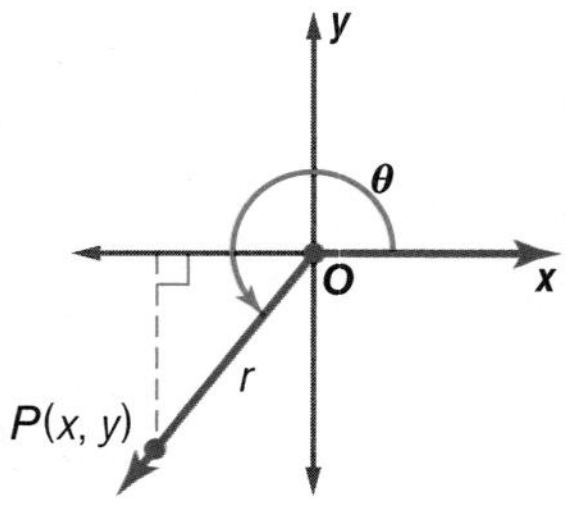

Multilingual eGlossary glencoe.com

NY Math Online glencoe.com

- Study the chapter online
- Explore **Math in Motion**
- Get extra help from your own **Personal Tutor**
- Use **Extra Examples** for additional help
- Take a **Self-Check Quiz**
- **Review Vocabulary** in fun ways

14-1 Trigonometric Identities

Why?

The amount of light that a source provides to a surface is called the *illuminance*. The illuminance E in foot candles on a surface is related to the distance R in feet from the light source.

The formula $\sec \theta = \frac{I}{ER^2}$, where I is the intensity of the light source measured in candles and θ is the angle between the light beam and a line perpendicular to the surface, can be used in situations in which lighting is important, as in photography.

Then

You evaluated trigonometric functions. (Lesson 13-7)

Now

- Use trigonometric identities to find trigonometric values.
- Use trigonometric identities to simplify expressions.

NYS Core Curriculum

A2.A.58 Know and apply the co-function and reciprocal relationships between trigonometric ratios

New Vocabulary

trigonometric identity

NY Math Online

glencoe.com

- Extra Examples
- Personal Tutor
- Self-Check Quiz
- Homework Help

Find Trigonometric Values The equation above can also be written as $E = \frac{I \cos \theta}{R^2}$. This is an example of a trigonometric identity. A **trigonometric identity** is an equation involving trigonometric functions that is true for all values for which every expression in the equation is defined.

If you can show that a specific value of the variable in an equation makes the equation false, then you have produced a *counterexample*. It only takes one counterexample to prove that an equation is not an identity.

Key Concept — Basic Trigonometric Identities

For Your FOLDABLE

Quotient Identities

$\tan \theta = \frac{\sin \theta}{\cos \theta}$, $\cos \theta \neq 0$ $\qquad$ $\cot \theta = \frac{\cos \theta}{\sin \theta}$, $\sin \theta \neq 0$

Reciprocal Identities

$\sin \theta = \frac{1}{\csc \theta}$, $\csc \theta \neq 0$ $\qquad$ $\csc \theta = \frac{1}{\sin \theta}$, $\sin \theta \neq 0$

$\cos \theta = \frac{1}{\sec \theta}$, $\sec \theta \neq 0$ $\qquad$ $\sec \theta = \frac{1}{\cos \theta}$, $\cos \theta \neq 0$

$\tan \theta = \frac{1}{\cot \theta}$, $\cot \theta \neq 0$ $\qquad$ $\cot \theta = \frac{1}{\tan \theta}$, $\tan \theta \neq 0$

Pythagorean Identities

$\cos^2 \theta + \sin^2 \theta = 1$ $\qquad$ $\tan^2 \theta + 1 = \sec^2 \theta$ $\qquad$ $\cot^2 \theta + 1 = \csc^2 \theta$

Cofunction Identities

$\sin\left(\frac{\pi}{2} - \theta\right) = \cos \theta$ $\qquad$ $\cos\left(\frac{\pi}{2} - \theta\right) = \sin \theta$ $\qquad$ $\tan\left(\frac{\pi}{2} - \theta\right) = \cot \theta$

Negative Angle Identities

$\sin(-\theta) = -\sin \theta$ $\qquad$ $\cos(-\theta) = \cos \theta$ $\qquad$ $\tan(-\theta) = -\tan \theta$

The identity $\tan \theta = \frac{\sin \theta}{\cos \theta}$ is true except for angle measures such as 90°, 270°, … , 90° + k180°, where k is an integer. The cosine of each of these angle measures is 0, so $\tan \theta$ is not defined when $\cos \theta = 0$. These identities are sometimes called *quotient identities*. An identity similar to this is $\cot \theta = \frac{\cos \theta}{\sin \theta}$.

You can use trigonometric identities to find exact values of trigonometric functions. You can find approximate values by using a graphing calculator.

EXAMPLE 1 Use Trigonometric Identities

a. Find the exact value of cos θ if sin $\theta = \frac{1}{4}$ and $90° < \theta < 180°$.

$\cos^2 \theta + \sin^2 \theta = 1$	**Pythagorean identity**
$\cos^2 \theta = 1 - \sin^2 \theta$	**Subtract $\sin^2 \theta$ from each side.**
$\cos^2 \theta = 1 - \left(\frac{1}{4}\right)^2$	**Substitute $\frac{1}{4}$ for sin θ.**
$\cos^2 \theta = 1 - \frac{1}{16}$	**Square $\frac{1}{4}$.**
$\cos^2 \theta = \frac{15}{16}$	**Subtract: $\frac{16}{16} - \frac{1}{16} = \frac{15}{16}$.**
$\cos \theta = \pm\frac{\sqrt{15}}{4}$	**Take the square root of each side.**

Since θ is in the second quadrant, cos θ is negative. Thus, $\cos \theta = -\frac{\sqrt{15}}{4}$.

CHECK Use a calculator to find an approximate answer.

Step 1 Find Arcsin $\frac{1}{4}$.

$\sin^{-1}\frac{1}{4} \approx 14.48°$ **Use a calculator.**

Because $90° < \theta < 180°$, $\theta \approx 180° - 14.48°$ or about 165.52°.

Step 2 Find cos θ.
Replace θ with 165.52°.
$\cos 165.52° \approx -0.97$

Step 3 Compare with the exact value.

$-\frac{\sqrt{15}}{4} \stackrel{?}{\approx} 0.97$

$-0.968 \approx 0.97$ ✓

b. Find the exact value of csc θ if cot $\theta = -\frac{3}{5}$ and $270° < \theta < 360°$.

$\cot^2 \theta + 1 = \csc^2 \theta$	**Pythagorean identity**
$\left(-\frac{3}{5}\right)^2 + 1 = \csc^2 \theta$	**Substitute $-\frac{3}{5}$ for cot θ.**
$\frac{9}{25} + 1 = \csc^2 \theta$	**Square $-\frac{3}{5}$.**
$\frac{34}{25} = \csc^2 \theta$	**Add: $\frac{9}{25} + \frac{25}{25} = \frac{34}{25}$.**
$\pm\frac{\sqrt{34}}{5} = \csc \theta$	**Take the square root of each side.**

Since θ is in the fourth quadrant, csc θ is negative. Thus, $\csc \theta = -\frac{\sqrt{34}}{5}$.

Check Your Progress

1A. Find sin θ if $\cos \theta = \frac{1}{3}$ and $270° < \theta < 360°$.

1B. Find sec θ if $\sin \theta = -\frac{2}{7}$ and $180° < \theta < 270°$.

Personal Tutor glencoe.com

StudyTip

Quadrants Here is a table to help you remember which ratios are positive and which are negative in each quadrant.

Function	+	−
sin θ	1, 2	3, 4
cos θ	1, 4	2, 3
tan θ	1, 3	2, 4
csc θ	1, 2	3, 4
sec θ	1, 4	2, 3
cot θ	1, 3	2, 4

Simplify Expressions Simplifying an expression that contains trigonometric functions means that the expression is written as a numerical value or in terms of a single trigonometric function, if possible.

StudyTip

Simplifying It is often easiest to write all expressions in terms of sine and/or cosine.

EXAMPLE 2 Simplify an Expression

Simplify $\frac{\sin\theta\csc\theta}{\cot\theta}$.

$$\frac{\sin\theta\csc\theta}{\cot\theta} = \frac{\sin\theta\cdot\frac{1}{\sin\theta}}{\frac{1}{\tan\theta}} \qquad \csc\theta = \frac{1}{\sin\theta} \text{ and } \cot\theta = \frac{1}{\tan\theta}$$

$$= \frac{1}{\frac{1}{\tan\theta}} \qquad \frac{\sin\theta}{\sin\theta} = 1$$

$$= \frac{1}{1}\cdot\frac{\tan\theta}{1} \text{ or } \tan\theta \qquad \frac{a}{b}\div\frac{c}{d} = \frac{a}{b}\cdot\frac{d}{c}$$

Check Your Progress

Simplify each expression.

2A. $\frac{\tan^2\theta\csc^2\theta - 1}{\sec^2\theta}$

2B. $\frac{\sec\theta}{\sin\theta}(1 - \cos^2\theta)$

Personal Tutor glencoe.com

Math History Link

Aryabhatta (476–550 A.D.) Among Indian mathematicians, Aryabhatta is probably the most famous. His name is closely associated with trigonometry. He was the first to introduce inverse trig functions and spherical trigonometry. Aryabhatta also calculated approximations for pi and trig functions.

Simplifying trigonometric expressions can be helpful when solving real-world problems.

Real-World EXAMPLE 3 Simplify and Use an Expression

LIGHTING Refer to the beginning of the lesson.

a. Solve the formula in terms of E.

$\sec\theta = \frac{I}{ER^2}$	Original equation
$ER^2\sec\theta = I$	Multiply each side by ER^2.
$ER^2\frac{1}{\cos\theta} = I$	$\frac{1}{\cos\theta} = \sec\theta$
$\frac{E}{\cos\theta} = \frac{I}{R^2}$	Divide each side by R^2.
$E = \frac{I\cos\theta}{R^2}$	Multiply each side by $\cos\theta$.

b. Is the equation in part a equivalent to $R^2 = \frac{I\tan\theta\cos\theta}{E}$? Explain.

$R^2 = \frac{I\tan\theta\cos\theta}{E}$	Original equation
$ER^2 = I\tan\theta\cos\theta$	Multiply each side by E.
$E = \frac{I\tan\theta\cos\theta}{R^2}$	Divide each side by R^2.
$E = \frac{I\frac{\sin\theta}{\cos\theta}\cos\theta}{R^2}$	$\tan\theta = \frac{\sin\theta}{\cos\theta}$
$E = \frac{I\sin\theta}{R^2}$	Simplify.

No; the equations are not equivalent. $R^2 = \frac{I\tan\theta\cos\theta}{E}$ simplifies to $E = \frac{I\sin\theta}{R^2}$.

Check Your Progress

3. Rewrite $\cot^2\theta - \tan^2\theta$ in terms of $\sin\theta$.

Personal Tutor glencoe.com

Check Your Understanding

Example 1 p. 892

Find the exact value of each expression if $0° < \theta < 90°$.

1. If $\cot \theta = 2$, find $\tan \theta$.
2. If $\sin \theta = \frac{4}{5}$, find $\cos \theta$.
3. If $\cos \theta = \frac{2}{3}$, find $\sin \theta$.
4. If $\cos \theta = \frac{2}{3}$, find $\csc \theta$.

Example 2 p. 893

Simplify each expression.

5. $\tan \theta \cos^2 \theta$
6. $\csc^2 \theta - \cot^2 \theta$
7. $\frac{\cos \theta \csc \theta}{\tan \theta}$

Example 3 p. 893

8. **OPTICS** When unpolarized light passes through polarized sunglass lenses, the intensity of the light is cut in half. If the light then passes through another polarized lens with its axis at an angle of θ to the first, the intensity of the light is again diminished. The intensity of the emerging light can be found by using the formula $I = I_0 - \frac{I_0}{\csc^2 \theta}$, where I_0 is the intensity of the light incoming to the second polarized lens, I is the intensity of the emerging light, and θ is the angle between the axes of polarization.

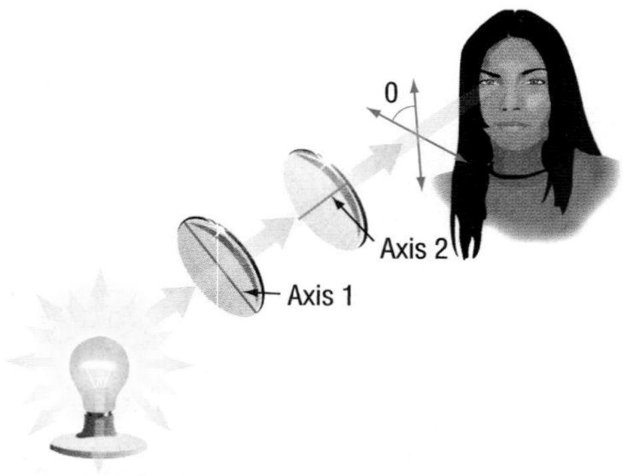

a. Simplify the formula in terms of $\cos \theta$.

b. Use the simplified formula to determine the intensity of light that passes through a second polarizing lens with axis at 30° to the original.

Practice and Problem Solving

● = **Step-by-Step Solutions** begin on page R20. **Extra Practice** begins on page 947.

Example 1 p. 892

Find the exact value of each expression if $0° < \theta < 90°$.

9. If $\cos \theta = \frac{3}{5}$, find $\csc \theta$.
10. If $\sin \theta = \frac{1}{2}$, find $\tan \theta$.
11. If $\sin \theta = \frac{3}{5}$, find $\cos \theta$.
12. If $\tan \theta = 2$, find $\sec \theta$.

Find the exact value of each expression if $180° < \theta < 270°$.

13. If $\cos \theta = -\frac{3}{5}$, find $\csc \theta$.
14. If $\sec \theta = -3$, find $\tan \theta$.
15. If $\cot \theta = \frac{1}{4}$, find $\csc \theta$.
16. If $\sin \theta = -\frac{1}{2}$, find $\cos \theta$.

Find the exact value of each expression if $270° < \theta < 360°$.

17. If $\cos \theta = \frac{5}{13}$, find $\sin \theta$.
18. If $\tan \theta = -1$, find $\sec \theta$.
19. If $\sec \theta = \frac{5}{3}$, find $\cos \theta$.
20. If $\csc \theta = -\frac{5}{3}$, find $\cos \theta$.

Example 2 p. 893

Simplify each expression.

21. $\sec \theta \tan^2 \theta + \sec \theta$
22. $\cos \left(\frac{\pi}{2} - \theta\right) \cot \theta$
23. $\cot \theta \sec \theta$
24. $\sin \theta (1 + \cot^2 \theta)$
25. $\sin \left(\frac{\pi}{2} - \theta\right) \sec \theta$
26. $\frac{\cos(-\theta)}{\sin(-\theta)}$

Example 3 p. 893

27. **ELECTRONICS** When there is a current in a wire in a magnetic field, such as in a hairdryer, a force acts on the wire. The strength of the magnetic field can be determined using the formula $B = \frac{F \csc \theta}{I\ell}$, where F is the force on the wire, I is the current in the wire, ℓ is the length of the wire, and θ is the angle the wire makes with the magnetic field. Rewrite the equation in terms of $\sin \theta$. (*Hint:* Solve for F.)

Simplify each expression.

28. $\frac{1 - \sin^2 \theta}{\sin^2 \theta}$

29. $\tan \theta \csc \theta$

30. $\frac{1}{\sin^2 \theta} - \frac{\cos^2 \theta}{\sin^2 \theta}$

31. $2(\csc^2 \theta - \cot^2 \theta)$

32. $(1 + \sin \theta)(1 - \sin \theta)$

33. $1 - 2 \sin^2 \theta$

34. **SUN** The ability of an object to absorb energy is related to a factor called the emissivity e of the object. The emissivity can be calculated by using the formula $e = \frac{W \sec \theta}{AS}$, where W is the rate at which a person's skin absorbs energy from the Sun, S is the energy from the Sun in watts per square meter, A is the surface area exposed to the Sun, and θ is the angle between the Sun's rays and a line perpendicular to the body.

a. Solve the equation for W. Write your answer using only $\sin \theta$ or $\cos \theta$.

b. Find W if $e = 0.80$, $\theta = 40°$, $A = 0.75$ m^2, and $S = 1000$ W/m^2. Round to the nearest hundredth.

Real-World Link

Research has shown that skin cancer is related to Sun exposure. If current trends continue, 1 in 5 Americans will develop skin cancer during their lifetimes.

Source: Skin Cancer Net

35. **MAPS** The map shows some of the buildings in Maria's neighborhood that she visits on a regular basis. The sine of the angle θ formed by the roads connecting the dance studio, the school, and Maria's house is $\frac{4}{9}$.

a. What is the cosine of the angle?

b. What is the tangent of the angle?

c. What are the sine, cosine, and tangent of the angle formed by the roads connecting the piano teacher's house, the school, and Maria's house?

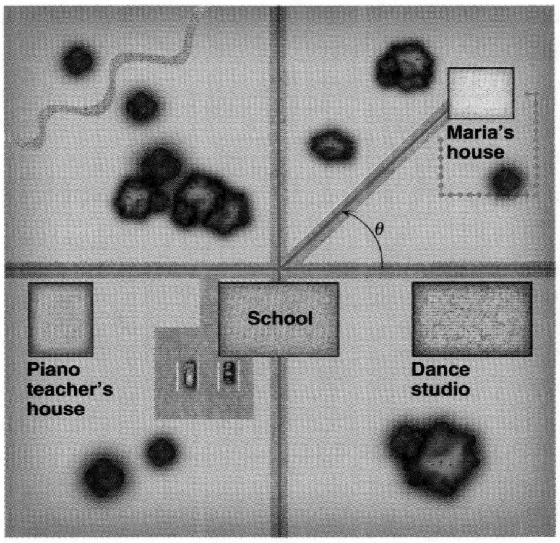

36. **MULTIPLE REPRESENTATIONS** In this problem, you will use a graphing calculator to determine whether an equation may be a trigonometric identity. Consider the trigonometric identity $\tan^2 \theta - \sin^2 \theta = \tan^2 \theta \sin^2 \theta$.

a. **TABULAR** Complete the table below.

θ	0°	30°	45°	60°
$\tan^2 \theta - \sin^2$				
$\tan^2 \theta \sin^2 \theta$				

b. **GRAPHICAL** Use a graphing calculator to graph $\tan^2 \theta - \sin^2 \theta = \tan^2 \theta \sin^2 \theta$ as two separate functions. Sketch the graph.

c. **ANALYTICAL** If the graphs of the two functions do not match, then the equation is not an identity. Do the graphs coincide?

d. **ANALYTICAL** Use a graphing calculator to determine whether the equation $\sec^2 x - 1 = \sin^2 x \sec^2 x$ may be an identity. (Be sure your calculator is in degree mode.)

37. SKIING A skier of mass m descends a θ-degree hill at a constant speed. When Newton's laws are applied to the situation, the following system of equations is produced: $F_n - mg \cos \theta = 0$ and $mg \sin \theta - \mu_k F_n = 0$, where g is the acceleration due to gravity, F_n is the normal force exerted on the skier, and μ_k is the coefficient of friction. Use the system to define μ_k as a function of θ.

Simplify each expression.

38. $\dfrac{\tan\left(\frac{\pi}{2} - \theta\right)\sec \theta}{1 - \csc^2 \theta}$

39. $\dfrac{\cos\left(\frac{\pi}{2} - \theta\right) - 1}{1 + \sin(-\theta)}$

40. $\dfrac{\sec \theta \sin \theta + \cos\left(\frac{\pi}{2} - \theta\right)}{1 + \sec \theta}$

41. $\dfrac{\cot \theta \cos \theta}{\tan(-\theta) \sin\left(\frac{\pi}{2} - \theta\right)}$

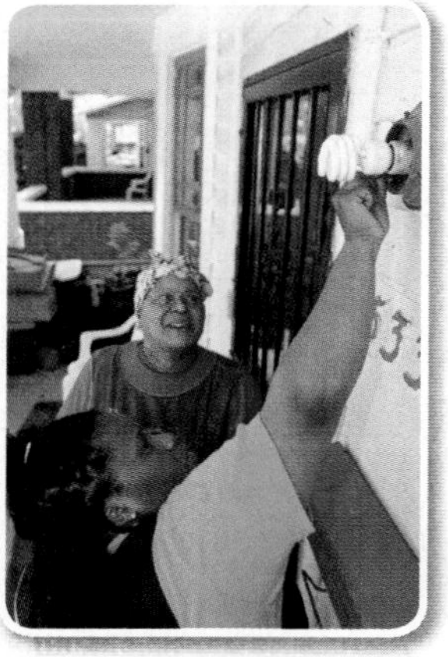

Real-World Link

Compact fluorescent lights, or CFLs, typically have a lifespan of between 6000 and 15,000 hours, whereas incandescent lights are usually manufactured to have a life span of 750 to 1000 hours.

Source: Osram

H.O.T. Problems

Use **H**igher-**O**rder **T**hinking Skills

42. FIND THE ERROR Clyde and Rosalina are debating whether an equation from their homework assignment is an identity. Clyde says that since he has tried ten specific values for the variable and all of them worked, it must be an identity. Rosalina argues that specific values could only be used as counterexamples to prove that an equation is not an identity. Is either of them correct? Explain your reasoning.

43. CHALLENGE Find a counterexample to show that $1 - \sin x = \cos x$ is *not* an identity.

44. REASONING Demonstrate how the formula about illuminance from the beginning of the lesson can be rewritten to show that $\cos \theta = \frac{ER^2}{I}$.

45. WRITING IN MATH Pythagoras is most famous for the Pythagorean Theorem. The identity $\cos^2 \theta + \sin^2 \theta = 1$ is an example of a Pythagorean identity. Why do you think that this identity is classified in this way?

46. PROOF Prove that $\tan(-a) = -\tan a$ by using the quotient and negative angle identities.

47. OPEN ENDED Write two expressions that are equivalent to $\tan \theta \sin \theta$.

48. REASONING Explain how you can use division to rewrite $\sin^2 \theta + \cos^2 \theta = 1$ as $1 + \cot^2 \theta = \csc^2 \theta$.

49. CHALLENGE Find $\cot \theta$ if $\sin \theta = \frac{3}{5}$ and $90° \leq \theta < 180°$.

50. FIND THE ERROR Jordan and Ebony are simplifying $\frac{\sin^2 \theta}{\cos^2 \theta + \sin^2 \theta}$. Is either of them correct? Explain your reasoning.

Jordan

$$\frac{\sin^2 \theta}{\cos^2 \theta + \sin^2 \theta} = \frac{\sin^2 \theta}{\cos^2 \theta} + \frac{\sin^2 \theta}{\sin^2 \theta}$$
$$= \tan^2 \theta + 1$$
$$= \sec^2 \theta$$

Ebony

$$\frac{\sin^2 \theta}{\cos^2 \theta + \sin^2 \theta} = \frac{\sin^2 \theta}{1}$$
$$= \sin^2 \theta$$

NYSRE PRACTICE A2.A.55, A2.A.12

51. Refer to the figure below. If $\cos D = 0.8$, what is the length of $\overline{DF}$?

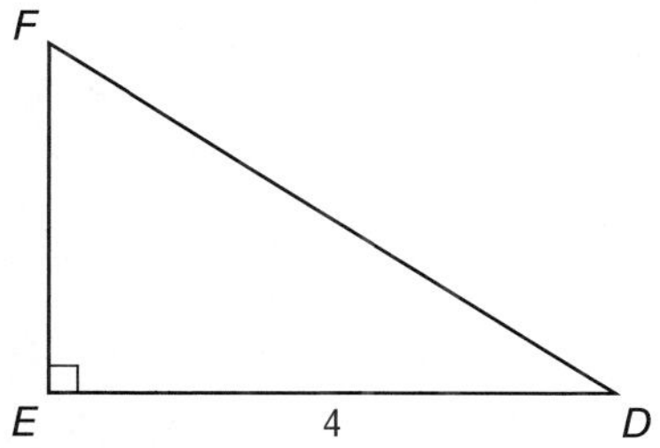

A 5
B 4
C 3.2
D $\frac{4}{5}$

52. PROBABILITY There are 16 green marbles, 2 red marbles, and 6 yellow marbles in a jar. How many yellow marbles need to be added to the jar in order to double the probability of selecting a yellow marble?

F 4
G 6
H 8
J 12

53. ACT/SAT Ella is 6 years younger than Amanda. Zoe is twice as old as Amanda. The total of their ages is 54. Which equation can be used to find Amanda's age?

A $x + (x - 6) + 2(x - 6) = 54$
B $x - 6x + (x + 2) = 54$
C $x - 6 + 2x = 54$
D $x + (x - 6) + 2x = 54$

54. Which of the following functions represents exponential decay?

F $y = 0.2(7)^x$
G $y = (0.5)^x$
H $y = 4(9)^x$
J $y = 5\left(\frac{4}{3}\right)^x$

Spiral Review

Find each value. Write angle measures in radians. Round to the nearest hundredth. (Lesson 13-9)

55. $\text{Cos}^{-1}\left(-\frac{1}{2}\right)$

56. $\text{Sin}^{-1}\frac{\pi}{2}$

57. $\text{Arctan}\frac{\sqrt{3}}{3}$

58. $\tan\left(\text{Cos}^{-1}\frac{6}{7}\right)$

59. $\sin\left(\text{Arctan}\frac{\sqrt{3}}{3}\right)$

60. $\cos\left(\text{Arcsin}\frac{3}{5}\right)$

61. TIDES The height of the water in a harbor rose to a maximum height of 15 feet at 6:00 P.M., and then dropped to a minimum level of 3 feet by 3:00 A.M. Assume that the water level can be modeled by the sine function. Write an equation that represents the height h of the water t hours after noon on the first day. (Lesson 13-8)

Evaluate the sum of each geometric series. (Lesson 11-3)

62. $\sum_{k=1}^{5} \frac{1}{4} \cdot 2^{k-1}$

63. $\sum_{k=1}^{7} 81\left(\frac{1}{3}\right)^{k-1}$

64. $\sum_{k=1}^{8} \frac{1}{3} \cdot 5^{k-1}$

Skills Review

Solve each equation. (Lesson 9-6)

65. $a + 1 = \frac{6}{a}$

66. $\frac{9}{t-3} = \frac{t-4}{t-3} + \frac{1}{4}$

67. $\frac{5}{x+1} - \frac{1}{3} = \frac{x+2}{x+1}$

14-2 Verifying Trigonometric Identities

Then

You used identities to find trigonometric values and simplify expressions. (Lesson 14-1)

Now

- Verify trigonometric identities by transforming one side of an equation into the form of the other side.
- Verify trigonometric identities by transforming each side of the equation into the same form.

NYS Core Curriculum

A2.A.67 Justify the Pythagorean identities

NY Math Online

glencoe.com

- Extra Examples
- Personal Tutor
- Self-Check Quiz
- Homework Help

Why?

While running on a circular track, Lamont notices that his body is not perpendicular to the ground. Instead, it leans away from a vertical position. The nonnegative acute angle θ that Lamont's body makes with the vertical is called the *angle of incline* and is described by the equation $\tan \theta = \frac{v^2}{gR}$.

This is not the only equation that describes the angle of incline in terms of trigonometric functions. Another such equation is $\sin \theta = \cos \frac{v^2}{gR}\theta$, where $0 \leq \theta \leq 90°$. Are these two equations completely independent of one another or are they merely different versions of the same relationship?

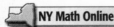

Transform One Side of an Equation You can use the basic trigonometric identities along with the definitions of the trigonometric functions to verify identities. If you wish to show an identity, you need to show that it is true for all values of θ.

Key Concept — For Your FOLDABLE

Verifying Identities by Transforming One Side

Step 1 Simplify one side of an equation until the two sides of the equation are the same. It is often easier to work with the more complicated side of the equation.

Step 2 Transform that expression into the form of the simpler side.

EXAMPLE 1 Transform One Side of an Equation

Verify that $\frac{\sin^2 \theta}{1 - \cos \theta} = 1 + \cos \theta$ is an identity.

$$\frac{\sin^2 \theta}{1 - \cos \theta} \stackrel{?}{=} 1 + \cos \theta$$ **Original equation**

$$\frac{1 + \cos \theta}{1 + \cos \theta} \cdot \frac{\sin^2 \theta}{1 - \cos \theta} \stackrel{?}{=} 1 + \cos \theta$$ **Multiply the numerator and denominator by $1 + \cos \theta$.**

$$\frac{\sin^2 \theta(1 + \cos \theta)}{1 - \cos^2 \theta} \stackrel{?}{=} 1 + \cos \theta$$ **$(1 + \cos \theta)(1 - \cos \theta) = 1 - \cos^2 \theta$**

$$\frac{\sin^2 \theta(1 + \cos \theta)}{\sin^2 \theta} \stackrel{?}{=} 1 + \cos \theta$$ **$\sin^2 \theta = 1 - \cos^2 \theta$**

$$1 + \cos \theta = 1 + \cos \theta \ \checkmark$$ **Divide the numerator and denominator by $\sin^2 \theta$.**

Check Your Progress

1. Verify that $\cot^2 \theta - \cos^2 \theta = \cot^2 \theta \cos^2 \theta$ is an identity.

Personal Tutor glencoe.com

When you verify a trigonometric identity, you are really working backward. In Example 1, consider the last step $1 + \cos\theta = 1 + \cos\theta$. Since that step is clearly true, you can conclude that the next-to-last step is also true, and so on, all the way back to the original equation.

Watch Out!

Simplify Separately Verifying an identity is like checking the solution of an equation. You must simplify one or both sides separately until they are the same.

NYSRE EXAMPLE 2 A2.A.58

$\dfrac{\cos\theta\csc\theta}{\tan\theta} =$

A $\cot\theta$ **B** $\csc\theta$ **C** $\cot^2\theta$ **D** $\csc^2\theta$

Read the Test Item

Find an expression that is always equal to the given expression. Notice that all of the answer choices involve either $\cot\theta$ or $\csc\theta$. So work toward eliminating the other trigonometric functions.

Solve the Test Item

Transform the given expression to match one of the choices.

$$\frac{\cos\theta\csc\theta}{\tan\theta} = \frac{\cos\theta\frac{1}{\sin\theta}}{\frac{\sin\theta}{\cos\theta}} \quad \textbf{csc } \theta = \frac{1}{\sin\theta} \textbf{ and tan } \theta = \frac{\sin\theta}{\cos\theta}$$

$$= \frac{\frac{\cos\theta}{\sin\theta}}{\frac{\sin\theta}{\cos\theta}} \quad \textbf{Multiply.}$$

$$= \frac{\cos\theta}{\sin\theta}\cdot\frac{\cos\theta}{\sin\theta} \quad \textbf{Invert the denominator and multiply.}$$

$$= \cot\theta\cdot\cot\theta \quad \textbf{cot } \theta = \frac{\cos\theta}{\sin\theta}$$

$$= \cot^2\theta \quad \textbf{Multiply.}$$

The answer is C.

Test-TakingTip

Checking Answers Verify your answer by choosing values for θ. Then evaluate the original expression and compare to your answer choice.

Check Your Progress

2. $\tan^2\theta(\cot^2\theta - \cos^2\theta) =$

F $\cot^2\theta$ **G** $\tan^2\theta$ **H** $\cos^2\theta$ **J** $\sin^2\theta$

Personal Tutor glencoe.com

Transform Each Side of an Equation Sometimes it is easier to transform each side of an equation separately into a common form. The following suggestions may be helpful as you verify trigonometric identities.

Key Concept **Suggestions for Verifying Identities** For Your FOLDABLE

- Substitute one or more basic trigonometric identities to simplify the expression.
- Factor or multiply as necessary. You may have to multiply both the numerator and denominator by the same trigonometric expression.
- Write each side of the identity in terms of sine and cosine only. Then simplify each side as much as possible.
- The properties of equality do not apply to identities as with equations. Do not perform operations to the quantities from each side of an unverified identity.

EXAMPLE 3 Verify by Transforming Each Side

Verify that $1 - \tan^4 \theta = 2\sec^2 \theta - \sec^4 \theta$ is an identity.

$1 - \tan^4 \theta \stackrel{?}{=} 2\sec^2 \theta - \sec^4 \theta$	**Original equation**
$(1 - \tan^2 \theta)(1 + \tan^2 \theta) \stackrel{?}{=} \sec^2 \theta\,(2 - \sec^2 \theta)$	**Factor each side.**
$[1 - (\sec^2 \theta - 1)]\sec^2 \theta \stackrel{?}{=} (2 - \sec^2 \theta)\sec^2 \theta$	**$1 + \tan^2 \theta = \sec^2 \theta$**
$(2 - \sec^2 \theta)\sec^2 \theta = (2 - \sec^2 \theta)\sec^2 \theta$ ✓	**Simplify.**

Check Your Progress

3. Verify that $\csc^2 \theta - \cot^2 \theta = \cot \theta \tan \theta$ is an identity.

Personal Tutor glencoe.com

Check Your Understanding

Examples 1 and 3 pp. 898, 900

Verify that each equation is an identity.

1. $\cot \theta + \tan \theta = \dfrac{\sec^2 \theta}{\tan \theta}$

2. $\cos^2 \theta = (1 + \sin \theta)(1 - \sin \theta)$

3. $\sin \theta = \dfrac{\sec \theta}{\tan \theta + \cot \theta}$

4. $\tan^2 \theta = \dfrac{1 - \cos^2 \theta}{\cos^2 \theta}$

5. $\tan^2 \theta \csc^2 \theta = 1 + \tan^2 \theta$

6. $\tan^2 \theta = (\sec \theta + 1)(\sec \theta - 1)$

Example 2 p. 899

7. **MULTIPLE CHOICE** Which expression can be used to form an identity with $\dfrac{\tan^2 \theta + 1}{\tan^2 \theta}$?

A $\sin^2 \theta$ **B** $\cos^2 \theta$ **C** $\tan^2 \theta$ **D** $\csc^2 \theta$

Practice and Problem Solving

● = **Step-by-Step Solutions** begin on page R20.
Extra Practice begins on page 947.

Example 1 p. 898

Verify that each equation is an identity.

8. $\cos^2 \theta + \tan^2 \theta \cos^2 \theta = 1$

9. $\cot \theta\,(\cot \theta + \tan \theta) = \csc^2 \theta$

10. $1 + \sec^2 \theta \sin^2 \theta = \sec^2 \theta$

11. $\sin \theta \sec \theta \cot \theta = 1$

12. $\dfrac{1 - \cos \theta}{1 + \cos \theta} = (\csc \theta - \cot \theta)^2$

13. $\dfrac{1 - 2\cos^2 \theta}{\sin \theta - \cos \theta} = \tan \theta - \cot \theta$

14. $\tan \theta = \dfrac{\sec \theta}{\csc \theta}$

15. $\cos \theta = \sin \theta \cot \theta$

16. $(\sin \theta - 1)(\tan \theta + \sec \theta) = -\cos \theta$

17. $\cos \theta \cos(-\theta) - \sin \theta \sin(-\theta) = 1$

Example 2 p. 899

18. **LADDER** Some students derived an expression for the length of a ladder that, when carried flat, could fit around a corner from a 5-foot-wide hallway into a 7-foot-wide hallway, as shown. They determined that the maximum length ℓ of a ladder that would fit was given by $\ell(\theta) = \dfrac{7 \sin \theta + 5 \cos \theta}{\sin \theta \cos \theta}$. When their teacher worked the problem, she concluded that $\ell(\theta) = 7 \sec \theta + 5 \csc \theta$. Are the two expressions equivalent?

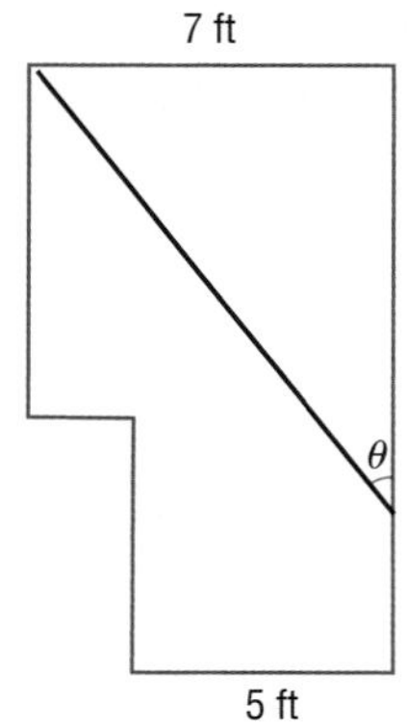

Example 3 p. 900

Verify that each equation is an identity.

19. $\sec\theta - \tan\theta = \dfrac{1-\sin\theta}{\cos\theta}$

20. $\dfrac{1+\tan\theta}{\sin\theta+\cos\theta} = \sec\theta$

21. $\sec\theta\csc\theta = \tan\theta + \cot\theta$

22. $\sin\theta + \cos\theta = \dfrac{2\sin^2\theta - 1}{\sin\theta - \cos\theta}$

23. $(\sin\theta + \cos\theta)^2 = \dfrac{2+\sec\theta\csc\theta}{\sec\theta\csc\theta}$

24. $\dfrac{\cos\theta}{1-\sin\theta} = \dfrac{1+\sin\theta}{\cos\theta}$

25. $\csc\theta - 1 = \dfrac{\cot^2\theta}{\csc\theta+1}$

26. $\cos\theta\cot\theta = \csc\theta - \sin\theta$

27. $\sin\theta\cos\theta\tan\theta + \cos^2\theta = 1$

28. $(\csc\theta - \cot\theta)^2 = \dfrac{1-\cos\theta}{1+\cos\theta}$

29. $\csc^2\theta = \cot^2\theta + \sin\theta\csc\theta$

30. $\dfrac{\sec\theta - \csc\theta}{\csc\theta\sec\theta} = \sin\theta - \cos\theta$

31. $\sin^2\theta + \cos^2\theta = \sec^2\theta - \tan^2\theta$

32. $\sec\theta - \cos\theta = \tan\theta\sin\theta$

33. TETHERBALL The diagram at the right represents a game of tetherball. As the ball rotates around the pole, a conical surface is swept out by line segment $\overline{SP}$. A formula for the relationship between the length L of the string and the angle θ that the string makes with the pole is given by the equation $L = \dfrac{g\sec\theta}{\omega^2}$. Is $L = \dfrac{g\tan\theta}{\omega^2\sin\theta}$ also an equation for the relationship between L and θ?

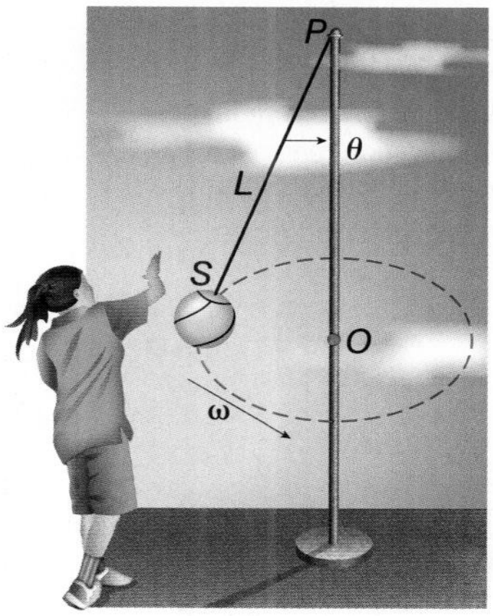

34. RUNNING A portion of a racetrack has the shape of a circular arc with a radius of 16.7 meters. As a runner races along the arc, the sine of her angle of incline θ is found to be $\frac{1}{4}$. Find the speed of the runner. Use the Angle of Incline Formula given at the beginning of the lesson, $\tan\theta = \frac{v^2}{gR}$, where $g = 9.8$ and R is the radius. (*Hint*: Find $\cos\theta$ first.)

Real-World Link

Running games were organized in ancient Egypt as early as 3800 B.C. The first marathon race, which was 24 miles long, was held during the 1896 Olympic Games in Athens.

When simplified, would the expression be equal to 1 or −1?

35. $\cot(-\theta)\tan(-\theta)$

36. $\sin\theta\csc(-\theta)$

37. $\sin^2(-\theta) + \cos^2(-\theta)$

38. $\sec(-\theta)\cos(-\theta)$

39 $\sec^2(-\theta) - \tan^2(-\theta)$

40. $\cot(-\theta)\cot\left(\frac{\pi}{2} - \theta\right)$

Simplify the expression to either a constant or a basic trigonometric function.

41. $\dfrac{\tan\left(\frac{\pi}{2}-\theta\right)\csc\theta}{\csc^2\theta}$

42. $\dfrac{1+\tan\theta}{1+\cot\theta}$

43. $(\sec^2\theta + \csc^2\theta) - (\tan^2\theta + \cot^2\theta)$

44. $\dfrac{\sec^2\theta - \tan^2\theta}{\cos^2 x + \sin^2 x}$

45. $\tan\theta\cos\theta$

46. $\cot\theta\tan\theta$

47. $\sec\theta\sin\left(\frac{\pi}{2} - \theta\right)$

48. $\dfrac{1+\tan^2\theta}{\csc^2\theta}$

49. PHYSICS When a firework is fired from the ground, its height y and horizontal displacement x are related by the equation $y = \dfrac{-gx^2}{2{v_0}^2\cos^2\theta} + \dfrac{x\sin\theta}{\cos\theta}$, where v_0 is the initial velocity of the projectile, θ is the angle at which it was fired, and g is the acceleration due to gravity. Rewrite this equation so that $\tan\theta$ is the only trigonometric function that appears in the equation.

Real-World Career

Electrician An electrician specializes in the wiring of electrical components. Electricians serve an apprenticeship lasting 3–5 years. Schooling in electrical theory and building codes is required. Certification requires work experience and a passing score on a written test.

50. ELECTRONICS When an alternating current of frequency f and peak current I_0 passes through a resistance R, then the power delivered to the resistance at time t seconds is $P = I_0^2 R \sin^2 2\pi ft$.

a. Write an expression for the power in terms of $\cos^2 2\pi$ ft.

b. Write an expression for the power in terms of $\csc^2 2\pi$ ft.

51. THROWING A BALL In this problem, you will investigate the path of a ball represented by the equation $h = \frac{v_0^2 \sin^2 \theta}{2g}$, where θ is the measure of the angle between the ground and the path of the ball, v_0 is its initial velocity in meters per second, and g is the acceleration due to gravity. The value of g is 9.8 m/s^2.

a. If the initial velocity of the ball is 47 meters per second, find the height of the ball at 30°, 45°, 60°, and 90°. Round to the nearest tenth.

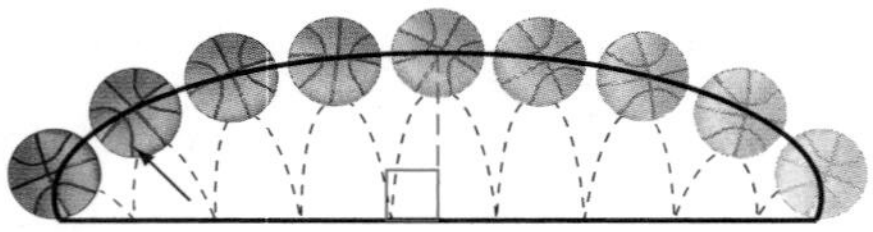

b. Graph the equation on a graphing calculator.

c. Show that the formula $h = \frac{v_0^2 \tan^2 \theta}{2g \sec^2 \theta}$ is equivalent to the one given above.

H.O.T. Problems Use Higher-Order Thinking Skills

52. WHICH ONE DOESN'T BELONG? Identify the equation that does not belong with the other three. Explain your reasoning.

$\sin^2 \theta + \cos^2 \theta = 1$	$1 + \cot^2 \theta = \csc^2 \theta$
$\sin^2 \theta - \cos^2 \theta = 2 \sin^2 \theta$	$\tan^2 \theta + 1 = \sec^2 \theta$

53. CHALLENGE Transform the right side of $\tan^2 \theta = \frac{\sin^2 \theta}{\cos^2 \theta}$ to show that $\tan^2 \theta = \sec^2 \theta - 1$.

54. WRITING IN MATH Explain why you cannot square each side of an equation when verifying a trigonometric identity.

55. REASONING Explain why $\sin^2 \theta + \cos^2 \theta = 1$ is an identity, but $\sin \theta = \sqrt{1 - \cos \theta}$ is not.

56. WRITE A QUESTION A classmate is having trouble trying to verify a trigonometric identity involving multiple trigonometric functions to multiple degrees. Write a question to help her work through the problem.

57. WRITING IN MATH Write about why you think terms of a trigonometric identity are often rewritten in terms of sine and cosine.

58. CHALLENGE Let $x = \frac{1}{2} \tan \theta$, where $-\frac{\pi}{2} < \theta < \frac{\pi}{2}$. Write $f(x) = \frac{x}{\sqrt{1 + 4x^2}}$ in terms of a single trigonometric function of θ.

59. REASONING Justify the three basic Pythagorean identities.

NYSRE PRACTICE A2.A.48, A2.S.4

60. ACT/SAT A small business owner must hire seasonal workers as the need arises. The following list shows the number of employees hired monthly for a 5-month period.

5, 14, 6, 8, 12

If the mean of these data is 9, what is the population standard deviation for these data? (Round your answer to the nearest tenth.)

A 2.6 **C** 8.6

B 5.7 **D** 12.3

61. Find the center and radius of the circle with equation $(x - 4)^2 + y^2 - 16 = 0$.

F $C(-4, 0)$; $r = 4$ units

G $C(-4, 0)$; $r = 16$ units

H $C(4, 0)$; $r = 4$ units

J $C(4, 0)$; $r = 16$ units

62. GEOMETRY The perimeter of a right triangle is 36 inches. Twice the length of the longer leg minus twice the length of the shorter leg is 6 inches. What are the lengths of all three sides?

A 3 in., 4 in., 5 in.

B 6 in., 8 in., 10 in.

C 9 in., 12 in., 15 in.

D 12 in., 16 in., 20 in.

63. Simplify $128^{\frac{1}{4}}$.

F $2\sqrt[4]{2}$ **H** 4

G $2\sqrt[4]{8}$ **J** $4\sqrt[4]{2}$

Spiral Review

Find the exact value of each expression. (Lesson 14-1)

64. $\tan \theta$, if $\cot \theta = 2$; $0° \leq \theta < 90°$

65. $\sin \theta$, if $\cos \theta = \frac{2}{3}$; $0° \leq \theta < 90°$

66. $\csc \theta$, if $\cos \theta = -\frac{3}{5}$; $90° < \theta < 180°$

67. $\cos \theta$, if $\sec \theta = \frac{5}{3}$; $270° < \theta < 360°$

68. ARCHITECTURE The support for a roof is shaped like two right triangles, as shown at the right. Find θ. (Lesson 13-9)

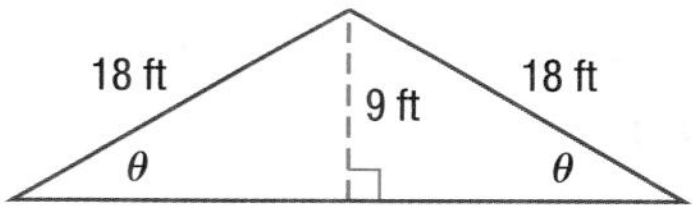

69. PROBABILITY An administrative assistant has 4 blue file folders, 3 red folders, and 3 yellow folders on her desk. Each folder contains different information, so two folders of the same color should be viewed as being different. She puts the file folders randomly in a box to take to a meeting. Find each probability. (Lesson 12-3)

a. P(4 blue, 3 red, 3 yellow, in that order)

b. P(first 2 blue, last 2 blue)

Find the coordinates of the vertices and foci and the equations of the asymptotes for the hyperbolas with the given equations. Then graph the hyperbola. (Lesson 10-5)

70. $\frac{y^2}{18} - \frac{x^2}{20} = 1$

71. $\frac{(y + 6)^2}{20} - \frac{(x - 1)^2}{25} = 1$

72. $x^2 - 36y^2 = 36$

Skills Review

Simplify. (Lesson 7-5)

73. $\frac{2 + \sqrt{2}}{5 - \sqrt{2}}$

74. $\frac{x + 1}{\sqrt{x^2 - 1}}$

75. $\frac{x - 1}{\sqrt{x} - 1}$

76. $\frac{-2 - \sqrt{3}}{1 + \sqrt{3}}$

14-3

Sum and Difference of Angles Identities

Then
You found values of trigonometric functions for general angles. (Lesson 13-3)

Now
- Find values of sine and cosine by using sum and difference identities.
- Verify trigonometric identities by using sum and difference identities.

NYS Core Curriculum

A2.A.76 Apply the angle sum and difference formulas for trigonometric functions

NY Math Online
glencoe.com
- Extra Examples
- Personal Tutor
- Self-Check Quiz
- Homework Help

Why?

Have you ever been using a wireless Internet provider and temporarily lost the signal? Waves that pass through the same place at the same time cause interference. Interference occurs when two waves combine to have a greater, or smaller, amplitude than either of the component waves.

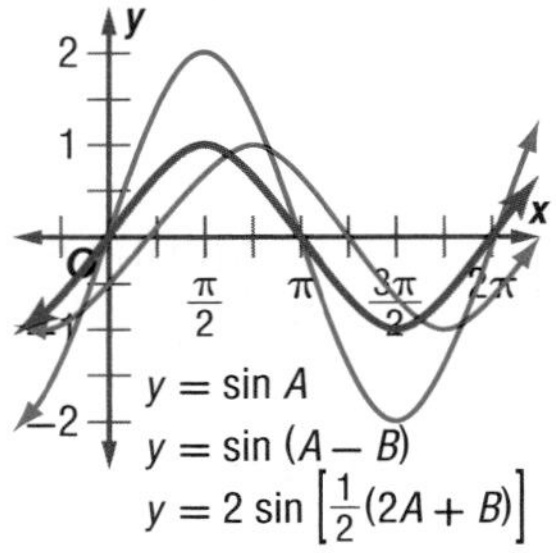

Sum and Difference Identities Notice that the third equation shown above involves the sum of A and B. It is often helpful to use formulas for the trigonometric values of the difference or sum of two angles. For example, you could find the exact value of sin 15° by evaluating sin (60° − 45°). Formulas exist that can be used to evaluate expressions like $\sin(A - B)$ or $\cos(A + B)$.

Key Concept — For Your FOLDABLE

Sum Identities
- $\sin(A + B) = \sin A \cos B + \cos A \sin B$
- $\cos(A + B) = \cos A \cos B - \sin A \sin B$
- $\tan(A + B) = \dfrac{\tan A + \tan B}{1 - \tan A \tan B}$

Difference Identities
- $\sin(A - B) = \sin A \cos B - \cos A \sin B$
- $\cos(A - B) = \cos A \cos B + \sin A \sin B$
- $\tan(A - B) = \dfrac{\tan A - \tan B}{1 + \tan A \tan B}$

EXAMPLE 1 Find Trigonometric Values

Find the exact value of each expression.

a. sin 105°

Use the identity $\sin(A + B) = \sin A \cos B + \cos A \sin B$.

$\sin 105° = \sin(60° + 45°)$ — **$A = 60°$ and $B = 45°$**

$= \sin 60° \cos 45° + \cos 60° \sin 45°$ — **Sum identity**

$= \left(\frac{\sqrt{3}}{2} \cdot \frac{\sqrt{2}}{2}\right) + \left(\frac{1}{2} \cdot \frac{\sqrt{2}}{2}\right)$ — **Evaluate each expression.**

$= \frac{\sqrt{6}}{4} + \frac{\sqrt{2}}{4}$ or $\frac{\sqrt{6} + \sqrt{2}}{4}$ — **Multiply.**

b. cos (−120°)

Use the identity $\cos(A - B) = \cos A \cos B + \sin A \sin B$.

$\cos(-120) = \cos(60° - 180°)$ — **$A = 60°$ and $B = 180°$**

$= \cos 60° \cos 180° + \sin 60° \sin 180°$ — **Difference identity**

$= \frac{1}{2} \cdot (-1) + \frac{\sqrt{3}}{2} \cdot 0$ — **Evaluate each expression.**

$= -\frac{1}{2}$ — **Multiply.**

Check Your Progress

1A. sin 15°

1B. cos (−15°)

Personal Tutor glencoe.com

You can use the sum and difference of angles identities to solve real-world applications.

Real-World EXAMPLE 2 Sum and Difference of Angles Identities

A geologist measures the angle between one side of a rectangular lot and the line from her position to the opposite corner of the lot as 30°. She then measures the angle between that line and the line to the point on the property where a river crosses as 45°. She stands 100 yards from the opposite corner of the property. How far is she from the point at which the river crosses the property line?

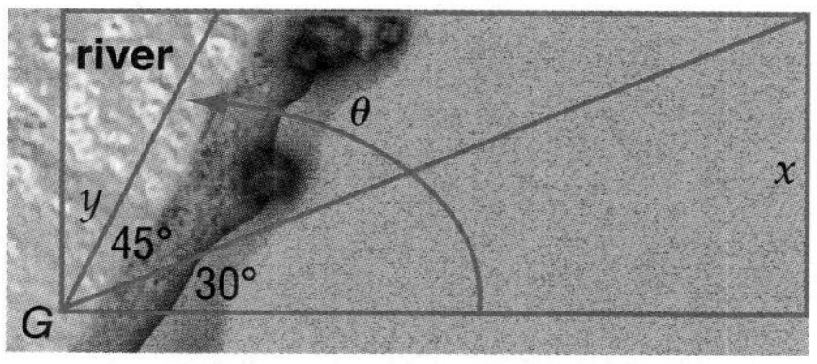

Understand The question asks for the distance between the geologist and the point where the river crosses the property line, or y.

Plan Draw a picture that labels all the things that you know from the information given.

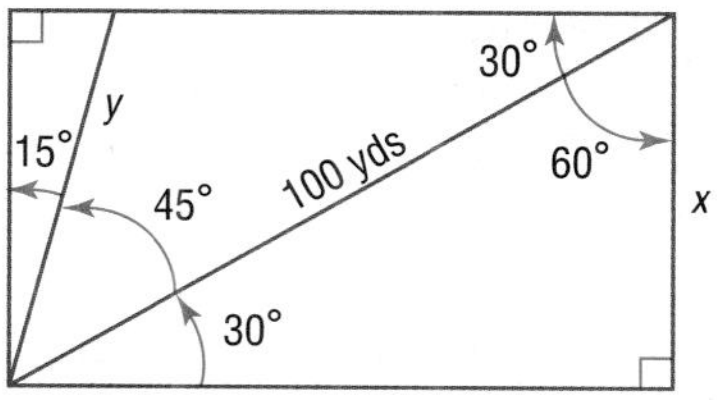

Solve Solve for x.

$\sin 30° = \frac{x}{100}$ **Definition of sine**

$x = 100 \sin 30°$

$x = 50$ **Since the lot is rectangular, opposite sides are equal.**

Now look at the triangle on the far left and solve for y.

$\cos 15° = \frac{50}{y}$ **Definition of cosine**

$\cos (45° - 30°) = \frac{50}{y}$ **15 = 45 − 30**

$\cos 45° \cos 30° + \sin 45° \sin 30° = \frac{50}{y}$ **Difference identity**

$\frac{\sqrt{2}}{2} \cdot \frac{\sqrt{3}}{2} + \frac{\sqrt{2}}{2} \cdot \frac{1}{2} = \frac{50}{y}$ **Evaluate.**

$\frac{\sqrt{6} + \sqrt{2}}{4} = \frac{50}{y}$ **Simplify.**

$(\sqrt{6} + \sqrt{2})y = 200$ **Cross products**

$y = \frac{200}{(\sqrt{6} + \sqrt{2})} \cdot \frac{(\sqrt{6} - \sqrt{2})}{(\sqrt{6} - \sqrt{2})}$

$y = 50(\sqrt{6} - \sqrt{2})$

$y = 50\sqrt{6} - 50\sqrt{2}$ or about 51.8

The geologist is about 51.8 yards from the point where the river crosses the property line.

Check Use a calculator to find the Arccos of $\frac{50}{51.8} \approx 15°$. ✓

Check Your Progress

2. Use the expression $E \sin (113.5° - \phi)$, where ϕ is the latitude of the location, and E is the amount of light energy. Determine the amount of light energy in West Hollywood, California, which is located at latitude of 34.1° N.

Personal Tutor glencoe.com

Problem-Solving Tip

Make a Model Make a model to visualize a problem situation. A model can be a drawing or a figure made of different objects, such as algebra tiles or folded paper.

Reading Math

Greek Letters The Greek letter phi, ϕ, is used in Check Your Progress 2 to represent latitude.

StudyTip

Make a List Make a list of the trigonometric values for the angles between 0° and 360° for which the sum and difference identities can be easily used. Use your list as a reference.

Verify Trigonometric Identities You can also use the sum and difference identities to verify identities.

EXAMPLE 3 Verify Trigonometric Identities

Verify that each equation is an identity.

a. $\cos (90° - \theta) = \sin \theta$

$\cos (90° - \theta) \stackrel{?}{=} \sin \theta$	**Original equation**
$\cos 90° \cos \theta + \sin 90° \sin \theta \stackrel{?}{=} \sin \theta$	**Sum identity**
$0 \cdot \cos \theta + 1 \cdot \sin \theta \stackrel{?}{=} \sin \theta$	**Evaluate each expression.**
$\sin \theta = \sin \theta$ ✓	**Simplify.**

b. $\sin \left(\theta + \frac{\pi}{2}\right) = \cos \theta$

$\sin \left(\theta + \frac{\pi}{2}\right) \stackrel{?}{=} \cos \theta$	**Original equation**
$\sin \theta \cos \frac{\pi}{2} + \cos \theta \sin \frac{\pi}{2} \stackrel{?}{=} \cos \theta$	**Sum identity**
$\sin \theta \cdot 0 + \cos \theta \cdot 1 \stackrel{?}{=} \cos \theta$	**Evaluate each expression.**
$\cos \theta = \cos \theta$ ✓	**Simplify.**

Check Your Progress

Verify that each equation is an identity.

3A. $\sin (90° - \theta) = \cos \theta$

3B. $\cos (90° + \theta) = -\sin \theta$

Personal Tutor glencoe.com

Check Your Understanding

Example 1 p. 904

Find the exact value of each expression.

1. cos 165°
2. cos 105°
3. cos 75°
4. sin (−30°)
5. sin 135°
6. sin (−210°)

Example 2 p. 905

7. **ELECTRONICS** Refer to the beginning of the lesson. *Constructive interference* occurs when two waves combine to have a greater amplitude than either of the component waves. *Destructive interference* occurs when the component waves combine to have a smaller amplitude. The first signal can be modeled by the equation $y = 20 \sin (3\theta + 45°)$. The second signal can be modeled by the equation $y = 20 \sin (3\theta + 225°)$.

a. Find the sum of the two functions.

b. What type of interference results when signals modeled by the two equations are combined?

Example 3 p. 906

Verify that each equation is an identity.

8. $\sin (90° + \theta) = \cos \theta$

9. $\cos \left(\frac{3\pi}{2} - \theta\right) = -\sin \theta$

10. $\tan \left(\theta + \frac{\pi}{2}\right) = -\cot \theta$

11. $\sin (\theta + \pi) = -\sin \theta$

Practice and Problem Solving

● = **Step-by-Step Solutions** begin on page R20.
Extra Practice begins on page 947.

Example 1 p. 904

Find the exact value of each expression.

12. sin 165° **13.** cos 135° **14.** $\cos \frac{7\pi}{12}$

15. $\sin \frac{\pi}{12}$ **16.** tan 195° **17.** $\cos\left(-\frac{\pi}{12}\right)$

Example 2 p. 905

18. ELECTRONICS In a certain circuit carrying alternating current, the formula $c = 2 \sin (120t)$ can be used to find the current c in amperes after t seconds.

a. Rewrite the formula using the sum of two angles.

b. Use the sum of angles formula to find the exact current at $t = 1$ second.

Example 3 p. 906

Verify that each equation is an identity.

19. $\cos\left(\frac{\pi}{2} + \theta\right) = -\sin \theta$ **20.** $\cos (60° + \theta) = \sin (30° - \theta)$

21. $\cos (180° + \theta) = -\cos \theta$ **22.** $\tan (\theta + 45°) = \frac{1 + \tan \theta}{1 - \tan \theta}$

Real-World Link

The most snow that Minneapolis, Minnesota, received in one year was 101.5 inches in 1983. On average, Minneapolis receives 40 inches of snow per year.

Source: Minnesota Climatology

23. WEATHER The monthly high temperatures for Minneapolis, Minnesota, can be modeled by the equation $y = 31.65 \sin\left(\frac{\pi}{6}x - 2.09\right) + 52.35$, where the months x are represented by January = 1, February = 2, and so on. The monthly low temperatures for Minneapolis can be modeled by the equation $y = 30.15 \sin\left(\frac{\pi}{6}x - 2.09\right) + 32.95$.

a. Write a new function by adding the expressions on the right side of each equation and dividing the result by 2.

b. What is the meaning of the function you wrote in part **a**?

Find the exact value of each expression.

24. tan 165° **25.** sec 1275° **26.** sin 735°

27. $\tan \frac{23\pi}{12}$ **28.** $\csc \frac{5\pi}{12}$ **29.** $\cot \frac{113\pi}{12}$

30. FORCE In the figure at the right, the effort F necessary to hold a safe in position on a ramp is given by $F = \frac{W(\sin A + \mu \cos A)}{\cos A - \mu \sin A}$, where W is the weight of the safe and $\mu = \tan \theta$. Show that $F = W \tan (A + \theta)$.

31 **QUILTING** As part of a quilt that is being made, the quilter places two right triangular swatches together to make a new triangular piece. One swatch has sides 6 inches, 8 inches, and 10 inches long. The other swatch has sides 8 inches, $8\sqrt{3}$ inches, and 16 inches long. The pieces are placed with the sides of eight inches against each other, as shown in the figure, to form triangle ABC.

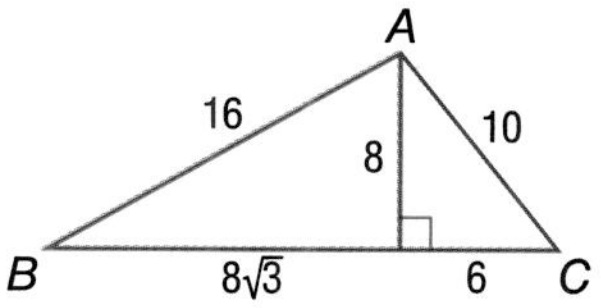

a. What is the exact value of the sine of angle BAC?

b. What is the exact value of the cosine of angle BAC?

c. What is the measure of angle BAC?

d. Is the new triangle formed from the two triangles also a right triangle?

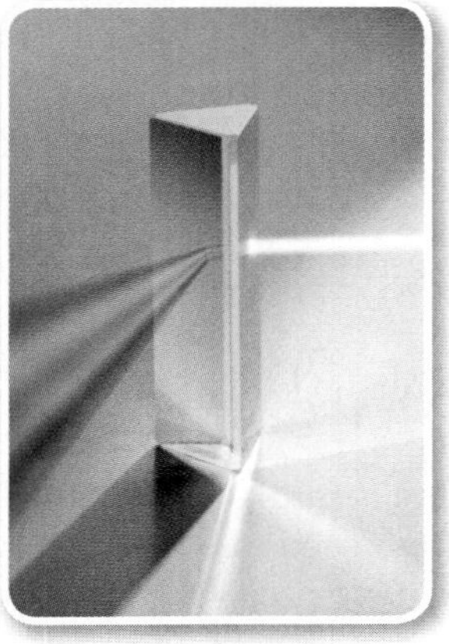

Real-World Link

Dispersive prisms break light into its constituent spectral colors. White light entering a prism is a mixture of different frequencies, each of which gets bent differently. Blue light is slowed down more than red light and will therefore be bent more than red light.

32. OPTICS When light passes symmetrically through a prism, the index of refraction n of the glass with respect to air is $n = \dfrac{\sin\left[\frac{1}{2}(a + b)\right]}{\sin \frac{b}{2}}$, where a is the measure of the deviation angle and b is the measure of the prism apex angle.

a. Show that for the prism shown, $n = \sqrt{3} \sin \frac{a}{2} + \cos \frac{a}{2}$.

b. Find n for the prism shown.

33. MULTIPLE REPRESENTATIONS In this problem, you will disprove the hypothesis that $\sin (A + B) = \sin A + \sin B$.

a. TABULAR Complete the table.

A	B	$\sin A$	$\sin B$	$\sin (A + B)$	$\sin A + \sin B$
30°	90°				
45°	60°				
60°	45°				
90°	30°				

b. GRAPHICAL Assume that B is always 15° less than A. Use a graphing calculator to graph $\sin (x + x - 15)$ and $\sin x + \sin (x - 15)$ on the same screen.

c. ANALYTICAL Determine whether $\cos (A + B) = \cos A + \cos B$ is an identity. Explain your reasoning.

Verify that each equation is an identity.

34. $\sin (A + B) = \dfrac{\tan A + \tan B}{\sec A \sec B}$

35. $\cos (A + B) = \dfrac{1 - \tan A \tan B}{\sec A \sec B}$

36. $\sec (A - B) = \dfrac{\sec A \sec B}{1 + \tan A \tan B}$

37 $\sin (A + B) \sin (A - B) = \sin^2 A - \sin^2 B$

H.O.T. Problems

Use Higher-Order Thinking Skills

38. REASONING Simplify the following expression without expanding any of the sums or differences.

$$\sin\left(\frac{\pi}{3} - \theta\right) \cos\left(\frac{\pi}{3} + \theta\right) - \cos\left(\frac{\pi}{3} - \theta\right) \sin\left(\frac{\pi}{3} + \theta\right)$$

39. WRITING IN MATH Use the information at the beginning of the lesson and in Exercise 7 to explain how the sum and difference identities are used to describe wireless Internet interference. Include an explanation of the difference between constructive and destructive interference.

40. CHALLENGE Derive an identity for $\cot (A + B)$ in terms of $\cot A$ and $\cot B$.

41. PROOF The figure at the right shows two angles A and B in standard position on the unit circle. Use the Distance Formula to find d, where $(x_1, y_1) = (\cos B, \sin B)$ and $(x_2, y_2) = (\cos A, \sin A)$.

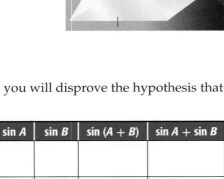

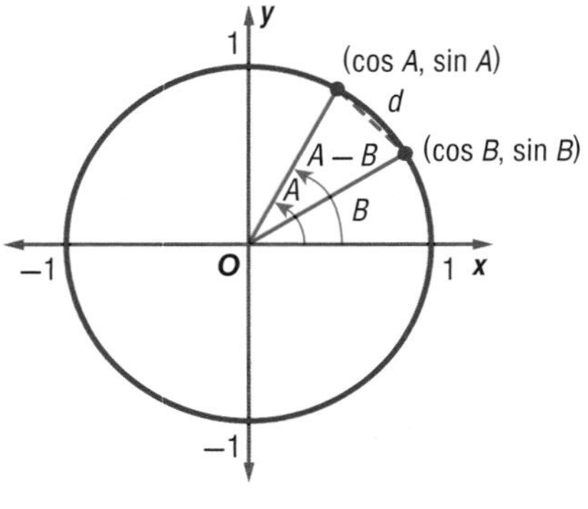

42. OPEN ENDED Consider the following theorem. *If A, B, and C are the angles of an oblique triangle, then tan A + tan B + tan C = tan A tan B tan C.* Choose values for A, B, and C. Verify that the conclusion is true for your specific values.

NYSRE PRACTICE A2.A.33, Reinforcement of A.S.18

43. GRIDDED RESPONSE The mean of seven numbers is 0. The sum of three of the numbers is -9. What is the sum of the remaining four numbers?

44. The variables a, b, c, d, and f are integers in a sequence, where $a = 2$ and $b = 12$. To find the next term, double the last term and add that result to one less than the next-to-last term. For example, $c = 25$, because $2(12) = 24$, $2 - 1 = 1$, and $24 + 1 = 25$. What is the value of f?

A 74

B 144

C 146

D 256

45. ACT/SAT Solve $x^2 - 5x < 14$.

F $\{x \mid -7 < x < 2\}$

G $\{x \mid -7 < x > 2\}$

H $\{x \mid -2 < x < 7\}$

J $\{x \mid -2 < x > 7\}$

46. PROBABILITY A math teacher is randomly distributing 15 yellow pencils and 10 green pencils. What is the probability that the first pencil she hands out will be yellow and the second pencil will be green?

A $\frac{1}{24}$

B $\frac{1}{4}$

C $\frac{2}{5}$

D $\frac{23}{25}$

Spiral Review

Verify that each equation is an identity. (Lesson 14-2)

47. $\frac{\sin \theta}{\tan \theta} + \frac{\cos \theta}{\cot \theta} = \cos \theta + \sin \theta$

48. $\sec \theta (\sec \theta - \cos \theta) = \tan^2 \theta$

Simplify each expression. (Lesson 14-1)

49. $\sin \theta \csc \theta - \cos^2 \theta$

50. $\cos^2 \theta \sec \theta \csc \theta$

51. $\cos \theta + \sin \theta \tan \theta$

52. GUITAR When a guitar string is plucked, it is displaced from a fixed point in the middle of the string and vibrates back and forth, producing a musical tone. The exact tone depends on the frequency, or number of cycles per second, that the string vibrates. To produce an A, the frequency is 440 cycles per second, or 440 hertz (Hz). (Lesson 13-6)

a. Find the period of this function.

b. Graph the height of the fixed point on the string from its resting position as a function of time. Let the maximum distance above the resting position have a value of 1 unit, and let the minimum distance below this position have a value of 1 unit.

Prove that each statement is true for all positive integers. (Lesson 11-7)

53. $4^n - 1$ is divisible by 3.

54. $5^n + 3$ is divisible by 4.

Skills Review

Solve each equation. (Lesson 7-7)

55. $7 + \sqrt{4x + 8} = 9$

56. $\sqrt{y + 21} - 1 = \sqrt{y + 12}$

57. $\sqrt{4z + 1} = 3 + \sqrt{4z - 2}$

Mid-Chapter Quiz

Lessons 14-1 through 14-3

NYS Core Curriculum
A2.A.58, A2.A.67, A2.A.76

Simplify each expression. (Lesson 14-1)

1. $\cot\theta\sec\theta$
2. $\frac{1-\cos^2\theta}{\sin^2\theta}$
3. $\frac{1}{\cos\theta}-\frac{\sin^2\theta}{\cos\theta}$
4. $\cos\left(\frac{\pi}{2}-\theta\right)\csc\theta$
5. **HISTORY** In 1861, the United States 34-star flag was adopted. For this flag, $\tan\theta=\frac{31.5}{51}$. Find $\sin\theta$.

Find the value of each expression. (Lesson 14-1)

6. $\sin\theta$, if $\cos\theta=\frac{3}{5}$; $0°<\theta<90°$
7. $\csc\theta$, if $\cot\theta=\frac{1}{2}$; $270°<\theta<360°$
8. $\tan\theta$, if $\sec\theta=\frac{4}{3}$; $0°<\theta<90°$
9. **MULTIPLE CHOICE** Which of the following is equivalent to $\frac{\cos\theta}{1-\sin^2\theta}$? (Lesson 14-1)

   A $\cos\theta$

   B $\csc\theta$

   C $\tan\theta$

   D $\sec\theta$

10. **AMUSEMENT PARKS** Suppose a child on a merry-go-round is seated on an outside horse. The diameter of the merry-go-round is 16 meters. The angle of inclination is represented by the equation $\tan\theta=\frac{v^2}{gR}$, where R is the radius of the circular path, v is the speed in meters per second, and g is 9.8 meters per second squared. (Lessoon 14-1)

    a. If the sine of the angle of inclination of the child is $\frac{1}{5}$, what is the angle of inclination made by the child?

    b. What is the velocity of the merry-go-round?

    c. If the speed of the merry-go-round is 3.6 meters per second, what is the value of the angle of inclination of a rider?

Verify that each of the following is an identity. (Lesson 14-2)

11. $\cot^2\theta+1=\frac{\cot\theta}{\cos\theta\cdot\sin\theta}$
12. $\frac{\cos\theta\csc\theta}{\cot\theta}=1$
13. $\frac{\sin\theta\tan\theta}{1-\cos\theta}=(1+\cos\theta)\sec\theta$
14. $\tan\theta(1-\sin\theta)=\frac{\cos\theta\sin\theta}{1+\sin\theta}$
15. **COMPUTER** The front of a computer monitor is usually measured along the diagonal of the screen as shown below. (Lesson 14-2)

    a. Find h.

    b. Using the diagram shown, show that $\cot\theta=\frac{\cos\theta}{\sin\theta}$.

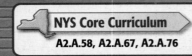

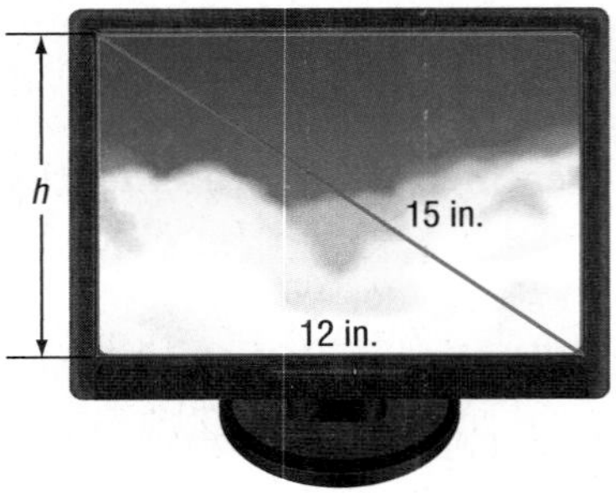

Verify that each of the following is an identity. (Lesson 14-2)

16. $\tan^2\theta+1=\frac{\tan\theta}{\cos\theta\cdot\sin\theta}$
17. $\frac{\sin\theta\cdot\sec\theta}{\sec\theta-1}=(\sec\theta+1)\cot\theta$
18. $\sin^2\theta\cdot\tan^2=\tan^2\theta-\sin^2\theta$
19. $\cot\theta(1-\cos\theta)=\frac{\cos\theta\cdot\sin\theta}{1+\cos\theta}$

Find the exact value of each expression. (Lesson 14-3)

20. $\cos 105°$
21. $\sin(-135°)$
22. $\tan 15°$
23. $\cot 75°$
24. **MULTIPLE CHOICE** What is the exact value of $\cos\frac{5\pi}{12}$? (Lesson 14-3)

    F $\sqrt{2}$

    G $\frac{\sqrt{6}+\sqrt{2}}{2}$

    H $\frac{\sqrt{6}-\sqrt{2}}{4}$

    J $\frac{\sqrt{6}+\sqrt{2}}{4}$

25. Verify that $\cos(30°-\theta)=\sin(60°+\theta)$ is an identity. (Lesson 14-3)

14-4 Double-Angle and Half-Angle Identities

Then

You found values of sine and cosine by using sum and difference identities. (Lesson 14-3)

Now

- Find values of sine and cosine by using double-angle identities.
- Find values of sine and cosine by using half-angle identities.

NYS Core Curriculum

A2.A.77 Apply the double-angle and half-angle formulas for trigonometric functions

NY Math Online

glencoe.com

- Extra Examples
- Personal Tutor
- Self-Check Quiz
- Homework Help

Why?

Chicago's Buckingham Fountain contains jets placed at specific angles that shoot water into the air to create arcs. When a stream of water shoots into the air with velocity v at an angle of θ with the horizontal, the model predicts that the water will travel a horizontal distance of $D = \frac{v^2}{g} \sin 2\theta$ and reach a maximum height of $H = \frac{v^2}{2g} \sin^2 \theta$. The ratio of H to D helps determine the total height and width of the fountain. Express $\frac{H}{D}$ as a function of θ.

Double-Angle Identities It is sometimes useful to have identities to find the value of a function of twice an angle or half an angle.

Key Concept — Double-Angle Identities

For Your FOLDABLE

The following identities hold true for all values of θ.

$\sin 2\theta = 2 \sin \theta \cos \theta$

$\cos 2\theta = \cos^2 \theta - \sin^2 \theta$
$\cos 2\theta = 2 \cos^2 \theta - 1$
$\cos 2\theta = 1 - 2 \sin^2 \theta$

$\tan 2\theta = \frac{2 \tan \theta}{1 - \tan^2 \theta}$

EXAMPLE 1 Double-Angle Identities

Find the exact value of $\sin 2\theta$ if $\sin \theta = \frac{2}{3}$ and θ is between 0° and 90°.

Step 1 Use the identity $\sin 2\theta = 2 \sin \theta \cos \theta$ to find the value of $\cos \theta$.

$\cos^2 \theta = 1 - \sin^2 \theta$ — **$\cos^2 \theta + \sin^2 \theta = 1$**

$\cos^2 \theta = 1 - \left(\frac{2}{3}\right)^2$ — **$\sin \theta = \frac{2}{3}$**

$\cos^2 \theta = \frac{5}{9}$ — **Subtract.**

$\cos \theta = \pm\frac{\sqrt{5}}{3}$ — **Take the square root of each side.**

Since θ is in the first quadrant, cosine is positive. Thus, $\cos \theta = \frac{\sqrt{5}}{3}$.

Step 2 Find $\sin 2\theta$.

$\sin 2\theta = 2 \sin \theta \cos \theta$ — **Double-angle identity**

$= 2\left(\frac{2}{3}\right)\left(\frac{\sqrt{5}}{3}\right)$ — **$\sin \theta = \frac{2}{3}$ and $\cos \theta = \frac{\sqrt{5}}{3}$**

$= \frac{4\sqrt{5}}{9}$ — **Multiply.**

Check Your Progress

1. Find the exact value of $\sin 2\theta$ if $\cos \theta = -\frac{1}{3}$ and $90° < \theta < 180°$.

Personal Tutor glencoe.com

EXAMPLE 2 Double-Angle Identities

Find the exact value of each expression if $\sin \theta = \frac{2}{3}$ and θ is between 0° and 90°.

a. $\cos 2\theta$

Since we know the values of $\cos \theta$ and $\sin \theta$, we can use any of the double-angle identities for cosine. We will use the identity $\cos 2\theta = 1 - 2 \sin^2 \theta$.

$$\cos 2\theta = 1 - 2\sin^2 \theta \quad \text{Double-angle identity}$$
$$= 1 - 2\left(\frac{2}{3}\right)^2 \text{ or } \frac{1}{9} \quad \sin \theta = \frac{2}{3}$$

b. $\tan 2\theta$

Step 1 Find $\tan \theta$ to use the double-angle identity for $\tan 2\theta$.

$$\tan \theta = \frac{\sin \theta}{\cos \theta} \quad \text{Definition of tangent}$$
$$= \frac{\frac{2}{3}}{\frac{\sqrt{5}}{3}} \quad \sin \theta = \frac{2}{3} \text{ and } \cos \theta = \frac{\sqrt{5}}{3}$$
$$= \frac{2}{\sqrt{5}} \text{ or } \frac{2\sqrt{5}}{5} \quad \text{Rationalize the denominator.}$$

Step 2 Find $\tan 2\theta$.

$$\tan 2\theta = \frac{2 \tan \theta}{1 - \tan^2 \theta} \quad \text{Double-angle identity}$$
$$= \frac{2\left(\frac{2\sqrt{5}}{5}\right)}{1 - \left(\frac{2\sqrt{5}}{5}\right)^2} \quad \tan \theta = \frac{2\sqrt{5}}{5}$$
$$= \frac{2\left(\frac{2\sqrt{5}}{5}\right)}{\frac{25}{25} - \frac{20}{25}} \quad \text{Square the denominator.}$$
$$= \frac{\frac{4\sqrt{5}}{5}}{\frac{1}{5}} \quad \text{Simplify.}$$
$$= \frac{4\sqrt{5}}{5} \cdot \frac{5}{1} \text{ or } 4\sqrt{5} \quad \frac{a}{b} \div \frac{c}{d} = \frac{a}{b} \cdot \frac{d}{c}$$

Check Your Progress

Find the exact value of each expression if $\cos \theta = -\frac{1}{3}$ and $90° < \theta < 180°$.

2A. $\cos 2\theta$

2B. $\tan 2\theta$

Personal Tutor glencoe.com

StudyTip

Deriving Formulas You can use the identity for $\sin (A + B)$ to find the sine of twice an angle θ, $\sin 2\theta$, and the identity for $\cos (A + B)$ to find the cosine of twice an angle θ, $\cos 2\theta$.

Half-Angle Identities It is sometimes useful to have identities to find the value of a function of half an angle.

Key Concept Half-Angle Identities

For Your FOLDABLE

The following identities hold true for all values of θ.

$$\sin \frac{\theta}{2} = \pm\sqrt{\frac{1 - \cos \theta}{2}} \qquad \cos \frac{\theta}{2} = \pm\sqrt{\frac{1 + \cos \theta}{2}} \qquad \tan \frac{\theta}{2} = \sqrt{\frac{1 - \cos \theta}{1 + \cos \theta}}, \cos \theta \neq -1$$

StudyTip

Choosing the Sign In the first step of the solution, you may want to determine the quadrant in which the terminal side of $\frac{\theta}{2}$ will lie. Then you can use the correct sign from that point on.

ReadingMath

Plus or Minus The first sign of the half-angle identity is read *plus or minus*. Unlike with the double-angle identities, you must determine the sign.

EXAMPLE 3 Half-Angle Identities

a. Find the exact value of $\cos \frac{\theta}{2}$ if $\sin \theta = -\frac{4}{5}$ and θ is in the third quadrant.

$\cos^2 \theta = 1 - \sin^2 \theta$	**Use a Pythagorean identity to find $\cos \theta$.**
$\cos^2 \theta = 1 - \left(-\frac{4}{5}\right)^2$	**$\sin \theta = -\frac{4}{5}$**
$\cos^2 \theta = 1 - \frac{16}{25}$	**Evaluate exponent.**
$\cos^2 \theta = \frac{9}{25}$	**Subtract.**
$\cos \theta = \pm\frac{3}{5}$	**Take the square root of each side.**

Since θ is in the third quadrant, $\cos \theta = -\frac{3}{5}$.

$\cos \frac{\theta}{2} = \pm\sqrt{\frac{1 + \cos \theta}{2}}$	**Half-angle identity**
$= \pm\sqrt{\frac{1 - \frac{3}{5}}{2}}$	**$\cos \theta = -\frac{3}{5}$**
$= \pm\sqrt{\frac{1}{5}}$	**Simplify.**
$= \pm\frac{1}{\sqrt{5}} \cdot \frac{\sqrt{5}}{\sqrt{5}}$ or $\pm\frac{\sqrt{5}}{5}$	**Rationalize the denominator.**

If θ is between 180° and 270°, $\frac{\theta}{2}$ is between 90° and 135°. So, $\cos \frac{\theta}{2}$ is $-\frac{\sqrt{5}}{5}$.

b. Find the exact value of cos 67.5°.

$\cos 67.5° = \cos \frac{135°}{2}$	**$67.5° = \frac{135°}{2}$**
$= \sqrt{\frac{1 + \cos 135°}{2}}$	**$\cos \frac{\theta}{2} = \pm\sqrt{\frac{1 + \cos \theta}{2}}$**
$= \sqrt{\frac{1 - \frac{\sqrt{2}}{2}}{2}}$	**67.5° is in Quadrant I; the value is positive.**
$= \sqrt{\frac{\frac{2}{2} - \frac{\sqrt{2}}{2}}{2}}$	**$1 = \frac{2}{2}$**
$= \sqrt{\frac{\frac{2 - \sqrt{2}}{2}}{2}}$	**Subtract fractions.**
$= \sqrt{\frac{2 - \sqrt{2}}{2} \cdot \frac{1}{2}}$	**$\frac{a}{b} \div \frac{c}{d} = \frac{a}{b} \cdot \frac{d}{c}$**
$= \sqrt{\frac{2 - \sqrt{2}}{4}}$	**Multiply.**
$= \frac{\sqrt{2 - \sqrt{2}}}{\sqrt{4}}$	**$\sqrt{\frac{a}{b}} = \frac{\sqrt{a}}{\sqrt{b}}$**
$= \frac{\sqrt{2 - \sqrt{2}}}{2}$	**Simplify.**

Check Your Progress

3. Find the exact value of $\sin \frac{\theta}{2}$ if $\sin \theta = \frac{2}{3}$ and θ is in the second quadrant.

Personal Tutor glencoe.com

Real-World Link

The City Hall Park Fountain in New York City is located in the heart of Manhattan in front of City Hall.

Source: *Fodor's*

Real-World EXAMPLE 4 Simplify Using Double-Angle Identities

FOUNTAIN **Refer to the beginning of the lesson. Find $\frac{H}{D}$.**

$\frac{H}{D} = \frac{\frac{v^2}{2g}\sin^2\theta}{\frac{v^2}{g}\sin 2\theta}$ — **Original equation**

$= \frac{\frac{v^2\sin^2\theta}{2g}}{\frac{v^2\sin 2\theta}{g}}$ — **Simplify the numerator and denominator.**

$= \frac{v^2\sin^2\theta}{2g} \cdot \frac{g}{v^2\sin 2\theta}$ — $\frac{a}{b} \div \frac{c}{d} = \frac{a}{b} \cdot \frac{d}{c}$

$= \frac{\sin^2\theta}{2\sin 2\theta}$ — **Simplify.**

$= \frac{\sin^2\theta}{4\sin\theta\cos\theta}$ — $\sin 2\theta = 2\sin\theta\cos\theta$

$= \frac{1}{4} \cdot \frac{\sin\theta}{\cos\theta}$ — **Simplify.**

$= \frac{1}{4}\tan\theta$ — $\frac{\sin\theta}{\cos\theta} = \tan\theta$

Check Your Progress

Find each value.

4A. $\sin 135°$

4B. $\cos \frac{7\pi}{8}$

Personal Tutor glencoe.com

Recall that you can use the sum and difference identities to verify identities. Double- and half-angle identities can also be used to verify identities.

EXAMPLE 5 Verify Identities

Verify that $\frac{\cos 2\theta}{1 + \sin 2\theta} = \frac{\cot\theta - 1}{\cot\theta + 1}$ is an identity.

$\frac{\cos 2\theta}{1 + \sin 2\theta} \stackrel{?}{=} \frac{\cot\theta - 1}{\cot\theta + 1}$ — **Original equation**

$\frac{\cos 2\theta}{1 + \sin 2\theta} \stackrel{?}{=} \frac{\frac{\cos\theta}{\sin\theta} - 1}{\frac{\cos\theta}{\sin\theta} + 1}$ — $\cot\theta = \frac{\cos\theta}{\sin\theta}$

$\frac{\cos 2\theta}{1 + \sin 2\theta} \stackrel{?}{=} \frac{\cos\theta - \sin\theta}{\cos\theta + \sin\theta}$ — **Multiply numerator and denominator by $\sin\theta$.**

$\frac{\cos 2\theta}{1 + \sin 2\theta} \stackrel{?}{=} \frac{\cos\theta - \sin\theta}{\cos\theta + \sin\theta} \cdot \frac{\cos\theta + \sin\theta}{\cos\theta + \sin\theta}$ — **Multiply the right side by 1.**

$\frac{\cos 2\theta}{1 + \sin 2\theta} \stackrel{?}{=} \frac{\cos^2\theta - \sin^2\theta}{\cos^2\theta + 2\cos\theta\sin\theta + \sin^2\theta}$ — **Multiply.**

$\frac{\cos 2\theta}{1 + \sin 2\theta} \stackrel{?}{=} \frac{\cos^2\theta - \sin^2\theta}{1 + 2\cos\theta\sin\theta}$ — **Simplify.**

$\frac{\cos 2\theta}{1 + \sin 2\theta} = \frac{\cos 2\theta}{1 + \sin 2\theta}$ ✓ — $\cos^2\theta - \sin^2\theta = \cos 2\theta$; $2\cos\theta\sin\theta = \sin 2\theta$

Check Your Progress

5. Verify that $4\cos^2 x - \sin^2 2x = 4\cos^4 x$.

Personal Tutor glencoe.com

Check Your Understanding

Examples 1–3 pp. 911–913

Find the exact values of $\sin 2\theta$, $\cos 2\theta$, $\sin \frac{\theta}{2}$, and $\cos \frac{\theta}{2}$.

1. $\sin \theta = \frac{1}{4}$; $0° < \theta < 90°$

2. $\sin \theta = \frac{4}{5}$; $90° < \theta < 180°$

3. $\cos \theta = -\frac{5}{13}$; $\frac{\pi}{2} < \theta < \pi$

4. $\cos \theta = \frac{3}{5}$; $270° < \theta < 360°$

5. $\tan \theta = -\frac{8}{15}$; $90° < \theta < 180°$

6. $\tan \theta = \frac{5}{12}$; $\pi < \theta < \frac{3\pi}{2}$

Find the exact value of each expression.

7. $\sin \frac{\pi}{8}$

8. $\cos 15°$

Example 4 p. 914

9. **SOCCER** A soccer player kicks a ball at an angle of 37° with the ground with an initial velocity of 52 feet per second. The distance d that the ball will go in the air if it is not blocked is given by $d = \frac{2v^2 \sin \theta \cos \theta}{g}$. In this formula, g is the acceleration due to gravity and is equal to 32 feet per second squared, and v is the initial velocity.

a. Simplify this formula by using a double-angle identity.

b. Using the simplified formula, how far will this ball go?

Example 5 p. 914

Verify that each equation is an identity.

10. $\tan \theta = \frac{1 - \cos 2\theta}{\sin 2\theta}$

11. $(\sin \theta + \cos \theta)^2 = 1 + 2 \sin \theta \cos \theta$

Practice and Problem Solving

● = **Step-by-Step Solutions** begin on page R20. **Extra Practice** begins on page 947.

Examples 1–3 pp. 911–913

Find the exact values of $\sin 2\theta$, $\cos 2\theta$, $\sin \frac{\theta}{2}$, and $\cos \frac{\theta}{2}$.

12. $\sin \theta = \frac{2}{3}$; $90° < \theta < 180°$

13. $\sin \theta = -\frac{15}{17}$; $\pi < \theta < \frac{3\pi}{2}$

14. $\cos \theta = \frac{3}{5}$; $\frac{3\pi}{2} < \theta < 2\pi$

15. $\cos \theta = \frac{1}{5}$; $270° < \theta < 360°$

16. $\tan \theta = \frac{4}{3}$; $180° < \theta < 270°$

17. $\tan \theta = -2$; $\frac{\pi}{2} < \theta < \pi$

Find the exact value of each expression.

18. $\sin 75°$

19. $\sin \frac{3\pi}{8}$

20. $\cos \frac{7\pi}{12}$

21. $\tan 165°$

22. $\tan \frac{5\pi}{12}$

23. $\tan 22.5°$

24. **GEOGRAPHY** The Mercator projection of the globe is a projection on which the distance between the lines of latitude increases with their distance from the equator. The calculation of the location of a point on this projection involves the expression $\tan\left(45° + \frac{L}{2}\right)$, where L is the latitude of the point.

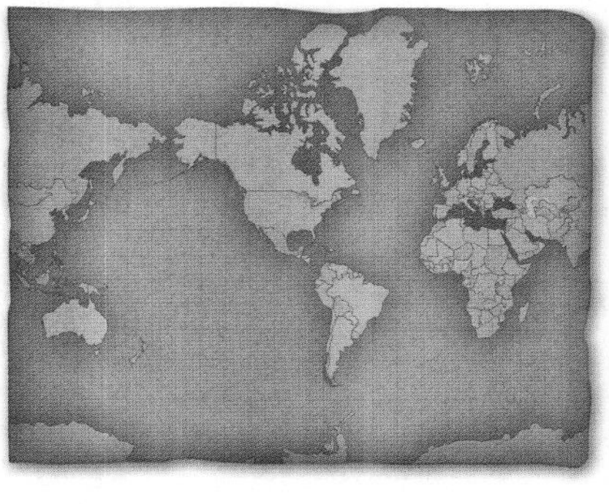

a. Write this expression in terms of a trigonometric function of L.

b. The latitude of Tallahassee, Florida, is 30° north. Find the value of the expression if $L = 30°$.

Example 4 p. 914

25 **ELECTRONICS** Consider an AC circuit consisting of a power supply and a resistor. If the current I_0 in the circuit at time t is $I_0 \sin t\theta$, then the power delivered to the resistor is $P = I_0^2 R \sin^2 t\theta$, where R is the resistance. Express the power in terms of $\cos 2t\theta$.

Example 5 p. 914

Verify that each equation is an identity.

26. $\tan 2\theta = \dfrac{2}{\cot \theta - \tan \theta}$

27. $1 + \dfrac{1}{2}\sin 2\theta = \dfrac{\sec \theta + \sin \theta}{\sec \theta}$

28. $\sin \dfrac{\theta}{2} \cos \dfrac{\theta}{2} = \dfrac{\sin \theta}{2}$

29. $\tan \dfrac{\theta}{2} = \dfrac{\sin \theta}{1 + \cos \theta}$

Real-World Link

On average, a place kick travels about 15 yards farther than a punt, assuming the kicker is at least 13 years old.

30. **FOOTBALL** Suppose a place kicker consistently kicks a football with an initial velocity of 95 feet per second. Prove that the horizontal distance the ball travels in the air will be the same for $\theta = 45° + A$ as for $\theta = 45° - A$. Use the formula given in Exercise 9.

Find the exact values of sin 2θ, cos 2θ, and tan 2θ.

31. $\cos \theta = \frac{4}{5}$; $0° < \theta < 90°$

32. $\sin \theta = \frac{1}{3}$; $0 < \theta < \frac{\pi}{2}$

33. $\tan \theta = -3$; $90° < \theta < 180°$

34. $\sec \theta = -\frac{4}{3}$; $90° < \theta < 180°$

35. $\csc \theta = -\frac{5}{2}$; $\frac{3\pi}{2} < \theta < 2\pi$

36. $\cot \theta = \frac{3}{2}$; $180° < \theta < 270°$

H.O.T. Problems

Use **H**igher-**O**rder **T**hinking Skills

37. **FIND THE ERROR** Teresa and Nathan are calculating the exact value of sin 15°. Is either of them correct? Explain your reasoning.

Teresa

$\sin (A - B) = \sin A \cos B - \cos A \sin B$

$\sin (45 - 30) = \sin 45 \cos 30 - \cos 45 \sin 30$

$= \dfrac{\sqrt{2}}{2} \cdot \dfrac{\sqrt{3}}{2} - \dfrac{\sqrt{2}}{2} \cdot \dfrac{1}{2}$

$= \dfrac{\sqrt{4}}{4}$

Nathan

$\sin \dfrac{A}{2} = \pm\sqrt{\dfrac{1 - \cos A}{2}}$

$\sin \dfrac{30}{2} = \pm\sqrt{\dfrac{1 - \frac{1}{2}}{2}}$

$= 0.5$

38. **CHALLENGE** Circle O is a unit circle. Use the figure to prove that $\tan \frac{1}{2}\theta = \dfrac{\sin \theta}{1 + \cos \theta}$.

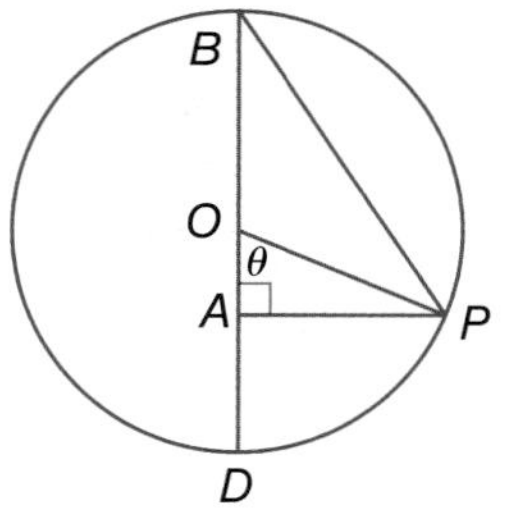

39. **WRITING IN MATH** Write a short paragraph about the conditions under which you would use each of the three identities for cos 2θ.

40. **PROOF** Use the formula for sin $(A + B)$ to derive the formula for sin 2θ, and use the formula for cos $(A + B)$ to derive the formula for cos 2θ.

41. **REASONING** Derive the half-angle identities from the double-angle identities.

42. **OPEN ENDED** Suppose a golfer consistently hits the ball so that it leaves the tee with an initial velocity of 115 feet per second and $d = \dfrac{2v^2 \sin \theta \cos \theta}{g}$. Explain why the maximum distance is attained when $\theta = 45°$.

NYSRE PRACTICE A2.A.1, A2.A.48

43. SHORT RESPONSE Perry drove to the gym at an average rate of 30 miles per hour. It took him 45 minutes. Going home, he took the same route, but drove at a rate of 45 miles per hour. How many miles is it to his house from the gym?

44. ACT/SAT Ms. Romero has a list of the yearly salaries of the staff members in her department. Which measure of data describes the middle income value of the salaries?

A mean

B median

C mode

D range

45. Identify the domain and range of the function $f(x) = |4x + 1| - 8$.

F $D = \{x \mid -3 \leq x \leq 1\}$, $R = \{y \mid y \geq -8\}$

G $D = \{\text{all real numbers}\}$, $R = \{y \mid y \geq -8\}$

H $D = \{x \mid -3 \leq x \leq 1\}$, $R = \{\text{all real numbers}\}$

J $D = \{\text{all real numbers}\}$, $R = \{\text{all real numbers}\}$

46. GEOMETRY Angel is putting a stone walkway around a circular pond. He has enough stones to make a walkway 144 feet long. If he uses all of the stones to surround the pond, what is the radius of the pond?

A $\frac{144}{\pi}$ ft

B $\frac{72}{\pi}$ ft

C 144π ft

D 72π ft

Spiral Review

Find the exact value of each expression. (Lesson 14-3)

47. $\sin 135°$

48. $\cos 105°$

49. $\sin 285°$

50. $\cos(-30°)$

51. $\sin(-240°)$

52. $\cos(-120°)$

Verify that each equation is an identity. (Lesson 14-2)

53. $\cot \theta + \sec \theta = \frac{\cos^2 \theta + \sin \theta}{\sin \theta \cos \theta}$

54. $\sin^2 \theta + \tan^2 \theta = (1 - \cos^2 \theta) + \frac{\sec^2 \theta}{\csc^2 \theta}$

Determine whether each triangle should be solved by beginning with the Law of *Sines* or Law of *Cosines*. Then solve each triangle. Round measures of sides to the nearest tenth and measures of angles to the nearest degree. (Lesson 13-5)

55.

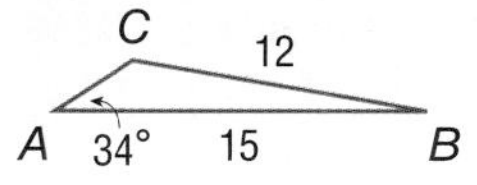

56.

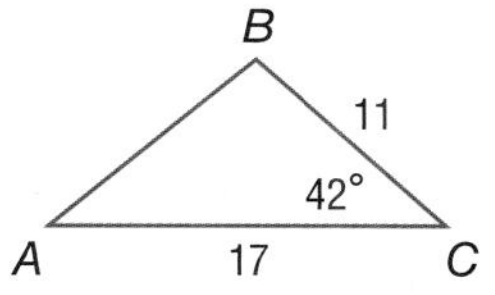

57.

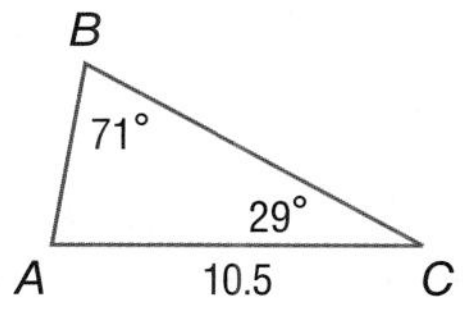

Skills Review

Solve each equation by factoring. (Lesson 5-3)

58. $x^2 + 5x - 24 = 0$

59. $x^2 - 3x - 28 = 0$

60. $x^2 - 4x = 21$

EXPLORE

14-5

Graphing Technology Lab
Solving Trigonometric Equations

NY Math Online glencoe.com
Other Calculator Keystrokes

The graph of a trigonometric function is made up of points that represent all values that satisfy the function. To solve a trigonometric equation, you need to find all values of the variable that satisfy the equation. You can use a TI-83/84 Plus graphing calculator to solve trigonometric equations by graphing each side of the equation as a function and then locating the points of intersection.

ACTIVITY 1 Real Solutions

Use a graphing calculator to solve $\sin x = 0.4$ if $0° \leq x < 360°$.

Step 1 Enter and graph related equations. Rewrite the equation as two equations, Y1 = $\sin x$ and Y2 = 0.4. Then graph the two equations. Because the interval is in degrees, set your calculator to degree mode.

KEYSTROKES: [MODE] [▼] [▼] [▶] [ENTER]
[Y=] [SIN] [X,T,θ,n] [ )]
[ENTER] 0.4 [ENTER] [GRAPH]

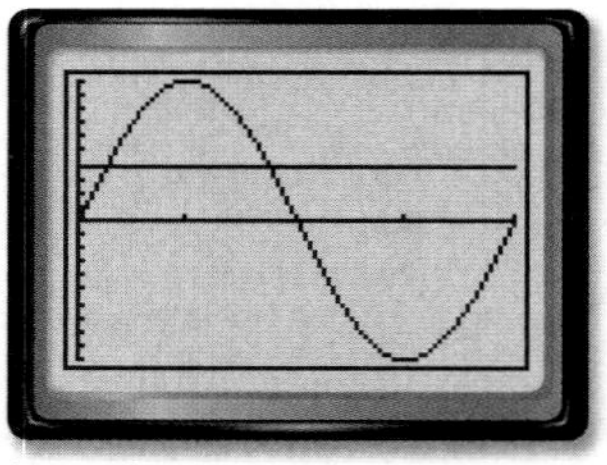

[0, 360] scl: 90 by [−15, 15] scl: 1

Step 2 Approximate the solutions. Based on the graph, you can see that there are two points of intersection in the interval $0° \leq x < 360°$. Use the CALC feature to determine the x-values at which the two graphs intersect.

The solutions are $x \approx 23.57°$ and $x \approx 156.4°$.

ACTIVITY 2 No Real Solutions

Use a graphing calculator to solve $\tan^2 x \cos x + 3 \cos x = 0$ if $0° \leq x < 360°$.

Step 1 Enter and graph related equations. The related equations to be graphed are $y_1 = \tan^2 x \cos x + 3 \cos x$ and $y_2 = 0$.

KEYSTROKES: [Y=] [TAN] [X,T,θ,n] [ )] [x^2]
[+] 3 [COS] [X,T,θ,n] [ )]
[ENTER] 0 [ENTER]

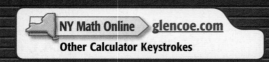

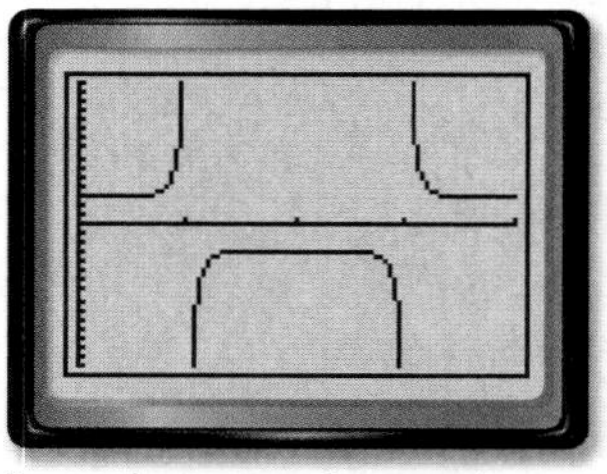

[0, 360] scl: 90 by [−15, 15] scl: 1

Step 2 These two functions do not intersect.

Therefore, the equation $\tan^2 x \cos x + 3 \cos x = 0$ has no real solutions.

Exercises

Use a graphing calculator to solve each equation for the values of x indicated.

1. $\sin x = 0.7$; $0° \leq x < 360°$
2. $\tan x = \cos x$; $0° \leq x < 360°$
3. $3 \cos x + 4 = 0.5$; $0° \leq x < 360°$
4. $0.25 \cos x = 3.4$; $-720° \leq x < 720°$
5. $\sin 2x = \sin x$; $0° \leq x < 360°$
6. $\sin 2x - 3 \sin x = 0$ if $-360° \leq x < 360°$

14-5 Solving Trigonometric Equations

Then

You verified trigonometric identities. (Lessons 14-2 through 14-4)

Now

- Solve trigonometric equations.
- Find extraneous solutions from trigonometric equations.

NYS Core Curriculum

A2.A.68 Solve trigonometric equations for all values of the variable from 0° to 360°

New Vocabulary

trigonometric equations

NY Math Online

glencoe.com

- Extra Examples
- Personal Tutor
- Self-Check Quiz
- Homework Help

Why?

When you ride a Ferris wheel that has a diameter of 40 meters and turns at a rate of 1.5 revolutions per minute, the height above the ground, in meters, of your seat after t minutes can be modeled by the equation

$$h = 21 - 20 \cos 3\pi t.$$

After the ride begins, how long is it before your seat is 31 meters above the ground for the first time?

Solve Trigonometric Equations So far in this chapter, we have studied a special type of trigonometric equation called an identity. Trigonometric identities are equations that are true for all values of the variable for which both sides are defined. In this lesson, we will examine **trigonometric equations** that are true for only certain values of the variable. Solving these equations resembles solving algebraic equations.

EXAMPLE 1 Solve Equations for a Given Interval

Solve $\sin \theta \cos \theta - \frac{1}{2} \cos \theta = 0$ if $0 \leq \theta \leq 180°$.

$\sin \theta \cos \theta - \frac{1}{2} \cos \theta = 0$ **Original equation**

$\cos \theta \left(\sin \theta - \frac{1}{2}\right) = 0$ **Factor.**

$\cos \theta = 0$ or $\sin \theta - \frac{1}{2} = 0$ **Zero Product Property**

$\theta = 90°$ or $270°$ $\quad\quad \sin \theta = \frac{1}{2}$

$\theta = 30°$ or $150°$

The solutions are 30°, 90°, and 150°.

CHECK You can check the answer by graphing $y = \sin \theta \cos \theta$ and $y = \frac{1}{2} \cos \theta$ in the same coordinate plane on a graphing calculator. Then find the points where the graphs intersect. You can see that there are infinitely many such points, but we are only interested in the points between 0° and 180°.

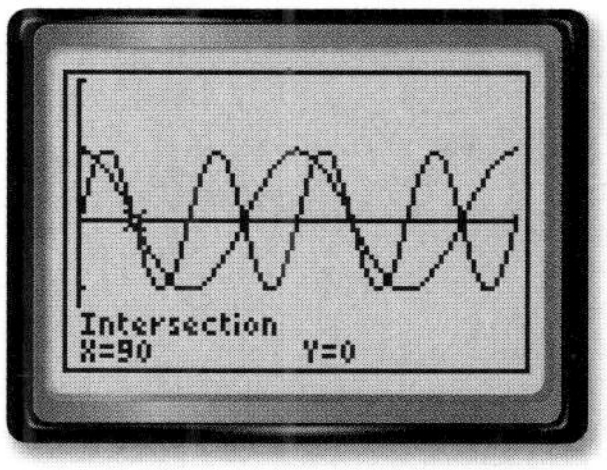

Check Your Progress

1. Find all solutions of $\sin 2\theta = \cos \theta$ if $0 \leq \theta \leq 2\pi$.

Personal Tutor glencoe.com

Trigonometric equations are usually solved for values of the variable between 0° and 360° or between 0 radians and 2π radians. There are solutions outside that interval. These other solutions differ by integral multiples of the period of the function.

EXAMPLE 2 Infinitely Many Solutions

Solve $\cos\theta + 1 = 0$ for all values of θ if θ is measured in radians.

$$\cos\theta + 1 = 0$$
$$\cos\theta = -1$$

Look at the graph of $y = \cos\theta$ to find solutions of $\cos\theta = -1$.

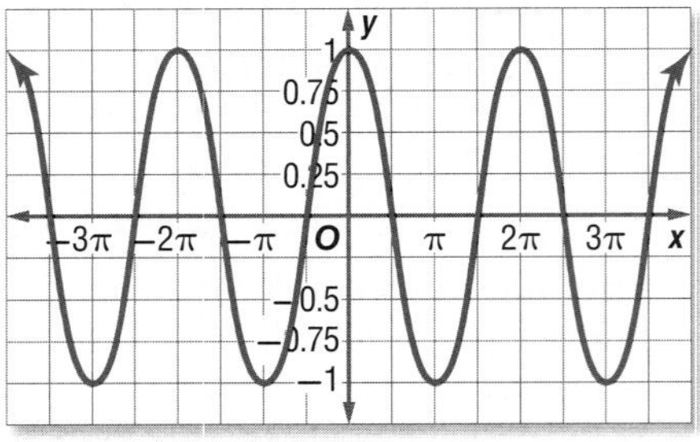

The solutions are π, 3π, 5π, and so on, and $-\pi$, -3π, -5π, and so on. The only solution in the interval 0 radians to 2π radians is π. The period of the cosine function is 2π radians. So the solutions can be written as $\pi + 2k\pi$, where k is any integer.

StudyTip

Expressing Solutions as Multiples The expression $\pi + 2k\pi$ includes 3π and its multiples, so it is not necessary to list them separately.

Check Your Progress

2A. Solve $\cos 2\theta + \cos\theta + 1 = 0$ for all values of θ if θ is measured in degrees.

2B. Solve $2\sin\theta = -1$ for all values of θ if θ is measured in radians.

Personal Tutor glencoe.com

Trigonometric equations are often used to solve real-world problems.

Real-World EXAMPLE 3 Solve Trigonometric Equations

AMUSEMENT PARKS Refer to the beginning of the lesson. How long after the Ferris wheel starts will your seat first be 31 meters above the ground?

$h = 21 - 20\cos 3\pi t$	**Original equation**
$31 = 21 - 20\cos 3\pi t$	**Replace h with 31.**
$10 = -20\cos 3\pi t$	**Subtract 21 from each side.**
$-\frac{1}{2} = \cos 3\pi t$	**Divide each side by −20.**
$\cos^{-1} -\frac{1}{2} = 3\pi t$	**Take the Arccosine.**
$\frac{2\pi}{3} = 3\pi t$ or $\frac{4\pi}{3} = 3\pi t$	**The Arccosine of $-\frac{1}{2}$ is $\frac{2\pi}{3}$ or $\frac{4\pi}{3}$.**
$\frac{2\pi}{3} + 2\pi k = 3\pi t$ or $\frac{4\pi}{3} + 2\pi k = 3\pi t$	**k is any integer.**
$\frac{2}{9} + \frac{2}{3}k = t$ $\frac{4}{9} + \frac{2}{3}k = t$	**Divide each term by 3π.**

The least positive value for t is obtained by letting $k = 0$ in the first expression. Therefore, $t = \frac{2}{9}$ of a minute or about 13 seconds.

Check Your Progress

3. How long after the Ferris wheel starts will your seat first be 41 meters above the ground?

Personal Tutor glencoe.com

Problem-SolvingTip

Look for a Pattern Look for patterns in your solutions. Look for pairs of solutions that differ by exactly π or 2π and write your solutions with the simplest possible pattern.

Extraneous Solutions Some trigonometric equations have no solution. For example, the equation $\cos \theta = 4$ has no solution because all values of $\cos \theta$ are between -1 and 1, inclusive. Thus, the solution set for $\cos \theta = 4$ is empty.

EXAMPLE 4 Determine Whether a Solution Exists

Solve each equation.

a. $\mathbf{2 \sin^2 \theta - 3 \sin \theta - 2 = 0}$ **if** $\mathbf{0 \le \theta \le 2\pi}$

$2 \sin^2 \theta - 3 \sin \theta - 2 = 0$ — **Original equation**

$(\sin \theta - 2)(2 \sin \theta + 1) = 0$ — **Factor.**

$\sin \theta - 2 = 0$ or $2 \sin \theta + 1 = 0$ — **Zero Product Property**

$\sin \theta = 2$ $\qquad$ $2 \sin \theta = -1$

This is not a solution since all values of sin θ are between −1 and 1, inclusive.

$\sin \theta = -\frac{1}{2}$

$\theta = \frac{7\pi}{6}$ or $\frac{11\pi}{6}$

The solutions are $\frac{7\pi}{6}$ or $\frac{11\pi}{6}$.

CHECK

$$2 \sin \theta - 3 \sin \theta - 2 = 0$$
$$2 \sin^2 \left(\frac{7\pi}{6}\right) - 3 \sin \left(\frac{7\pi}{6}\right) - 2 \stackrel{?}{=} 0$$
$$2\left(\frac{1}{4}\right) - 3\left(-\frac{1}{2}\right) - 2 \stackrel{?}{=} 0$$
$$\frac{1}{2} + \frac{3}{2} - 2 \stackrel{?}{=} 0$$
$$0 = 0 \checkmark$$

$$2 \sin^2 \theta - 3 \sin \theta - 2 = 0$$
$$2 \sin^2 \left(\frac{11\pi}{6}\right) - 3 \sin \left(\frac{11\pi}{6}\right) - 2 \stackrel{?}{=} 0$$
$$2\left(\frac{1}{4}\right) - 3\left(-\frac{1}{2}\right) - 2 \stackrel{?}{=} 0$$
$$\frac{1}{2} + \frac{3}{2} - 2 \stackrel{?}{=} 0$$
$$0 = 0 \checkmark$$

b. $\mathbf{\sin \theta = 1 + \cos \theta}$ **if** $\mathbf{0° \le \theta < 360°}$

$\sin \theta = 1 + \cos \theta$ — **Original equation**

$\sin^2 \theta = (1 + \cos \theta)^2$ — **Square each side.**

$1 - \cos^2 \theta = 1 + 2 \cos \theta + \cos^2 \theta$ — $\mathbf{\sin^2 \theta = 1 - \cos^2 \theta}$

$0 = 2 \cos \theta + 2 \cos^2 \theta$ — **Set the left side equal to 0.**

$0 = 2 \cos \theta\,(1 + \cos \theta)$ — **Factor.**

$1 + \cos \theta = 0$ or $2 \cos \theta = 0$ — **Zero Product Property**

$\cos \theta = -1$ $\qquad$ $\cos \theta = 0$

$\theta = 180$ $\qquad$ $\theta = 90°$ or $270°$

CHECK

$$\sin \theta = 1 + \cos \theta$$
$$\sin 90° \stackrel{?}{=} 1 + \cos 90°$$
$$1 \stackrel{?}{=} 1 + 0$$
$$1 = 1 \checkmark$$

$$\sin \theta = 1 + \cos \theta$$
$$\sin 180° \stackrel{?}{=} 1 + \cos 180°$$
$$0 \stackrel{?}{=} 1 + (-1)$$
$$0 = 0 \checkmark$$

$$\sin \theta = 1 + \cos \theta$$
$$\sin 270° \stackrel{?}{=} 1 + \cos 270°$$
$$-1 \stackrel{?}{=} 1 + 0$$
$$-1 \ne 1 \text{ ✗}$$

The solutions are 90° and 180°.

Check Your Progress

Solve each equation.

4A. $\sin^2 \theta + 2 \cos^2 \theta = 4$

4B. $\cos^2 \theta + 3 = 4 - \sin^2 \theta$

Personal Tutor glencoe.com

If an equation cannot be solved easily by factoring, try rewriting the expression using trigonometric identities. However, using identities and some algebraic operations, such as squaring, may result in extraneous solutions. So, it is necessary to check your solutions using the original equation.

StudyTip

Solving Trigonometric Equations Remember that *solving a trigonometric equation* means solving for all values of the variable.

EXAMPLE 5 Solve Trigonometric Equations by Using Identities

Solve $2 \sec^2 \theta - \tan^4 \theta = -1$ for all values of θ if θ is measured in degrees.

$2 \sec^2 \theta - \tan^4 \theta = -1$	**Original equation**
$2(1 + \tan^2 \theta) - \tan^4 \theta = -1$	**$\sec^2 \theta = 1 + \tan^2 \theta$**
$2 + 2 \tan^2 \theta - \tan^4 \theta = -1$	**Distributive Property**
$\tan^4 \theta - 2 \tan^2 \theta - 3 = 0$	**Set one side of the equation equal to 0.**
$(\tan^2 \theta - 3)(\tan^2 \theta + 1) = 0$	**Factor.**

$\tan^2 \theta - 3 = 0$ or $\tan^2 \theta + 1 = 0$ **Zero Product Property**

$\tan^2 \theta = 3$ $\qquad$ $\tan^2 \theta = -1$

$\tan \theta = \pm\sqrt{3}$ $\qquad$ This part gives no solutions since $\tan^2 \theta$ is never negative.

$\theta = 60° + 180°k$ and $\theta = -60° + 180°k$, where k is any integer. The solutions are $60° + 180°k$ and $-60° + 180°k$.

Check Your Progress

Solve each equation.

5A. $\sin \theta \cot \theta - \cos^2 \theta = 0$

5B. $\frac{\cos \theta}{\cot \theta} + 2 \sin^2 \theta = 0$

Personal Tutor glencoe.com

Check Your Understanding

Example 1 p. 919

Solve each equation if $0° \leq \theta \leq 360°$.

1. $2 \sin \theta + 1 = 0$

2. $\cos^2 \theta + 2 \cos \theta + 1 = 0$

3. $\cos 2\theta + \cos \theta = 0$

4. $2 \cos \theta = 1$

5. $\cos \theta = -\frac{\sqrt{3}}{2}$

6. $\sin 2\theta = -\frac{\sqrt{3}}{2}$

7. $\cos 2\theta = 8 - 15 \sin \theta$

8. $\sin \theta + \cos \theta = 1$

Example 2 p. 920

Solve each equation for all values of θ if θ is measured in radians.

9. $4 \sin^2 \theta - 1 = 0$

10. $2 \cos^2 \theta = 1$

11. $\cos 2\theta \sin \theta = 1$

12. $\sin \frac{\theta}{2} + \cos \frac{\theta}{2} = \sqrt{2}$

13. $\cos 2\theta + 4 \cos \theta = -3$

14. $\sin \frac{\theta}{2} + \cos \theta = 1$

Solve each equation for all values of θ if θ is measured in degrees.

15. $\cos 2\theta - \sin^2 \theta + 2 = 0$

16. $\sin^2 \theta - \sin \theta = 0$

17. $2 \sin^2 \theta - 1 = 0$

18. $\cos \theta - 2 \cos \theta \sin \theta = 0$

19. $\cos 2\theta \sin \theta = 1$

20. $\sin \theta \tan \theta - \tan \theta = 0$

Example 3 p. 920

21. **LIGHT** The number of hours of daylight d in Hartford, Connecticut, may be approximated by the equation $d = 3 \sin \frac{2\pi}{365}t + 12$, where t is the number of days after March 21.

a. On what days will Hartford have exactly $10\frac{1}{2}$ hours of daylight?

b. Using the results in part **a**, tell what days of the year have at least $10\frac{1}{2}$ hours of daylight. Explain how you know.

Examples 4 and 5
pp. 921, 922

Solve each equation.

22. $\sin^2 2\theta + \cos^2 \theta = 0$

23. $\tan^2 \theta + 2 \tan \theta + 1 = 0$

24. $\cos^2 \theta + 3 \cos \theta = -2$

25. $\sin 2\theta - \cos \theta = 0$

26. $\tan \theta = 1$

27. $\cos 8\theta = 1$

28. $\sin \theta + 1 = \cos 2\theta$

29. $2 \cos^2 \theta = \cos \theta$

Practice and Problem Solving

= **Step-by-Step Solutions** begin on page R20.
Extra Practice begins on page 947.

Example 1
p. 919

Solve each equation for the given interval.

30. $\cos^2 \theta = \frac{1}{4}; 0° \leq \theta \leq 360°$

31. $2 \sin^2 \theta = 1; 90° < \theta < 270°$

32. $\sin 2\theta - \cos \theta = 0; 0 \leq \theta \leq 2\pi$

33. $3 \sin^2 \theta = \cos^2 \theta; 0 \leq \theta \leq \frac{\pi}{2}$

34. $2 \sin \theta + \sqrt{3} = 0; 180° < \theta < 360°$

35. $4 \sin^2 \theta - 1 = 0; 180° < \theta < 360°$

Example 2
p. 920

Solve each equation for all values of θ if θ is measured in radians.

36. $\cos 2\theta + 3 \cos \theta = 1$

37 $2 \sin^2 \theta = \cos \theta + 1$

38. $\cos^2 \theta - \frac{3}{2} = \frac{5}{2} \cos \theta$

39. $3 \cos \theta - \cos \theta = 2$

Solve each equation for all values of θ if θ is measured in degrees.

40. $\sin \theta - \cos \theta = 0$

41. $\tan \theta - \sin \theta = 0$

42. $\sin^2 \theta = 2 \sin \theta + 3$

43. $4 \sin^2 \theta = 4 \sin \theta - 1$

Example 3
p. 920

44. **ELECTRONICS** One of the tallest structures in the world is a television transmitting tower located near Fargo, North Dakota, with a height of 2064 feet. What is the measure of θ if the length of the shadow is 1 mile?

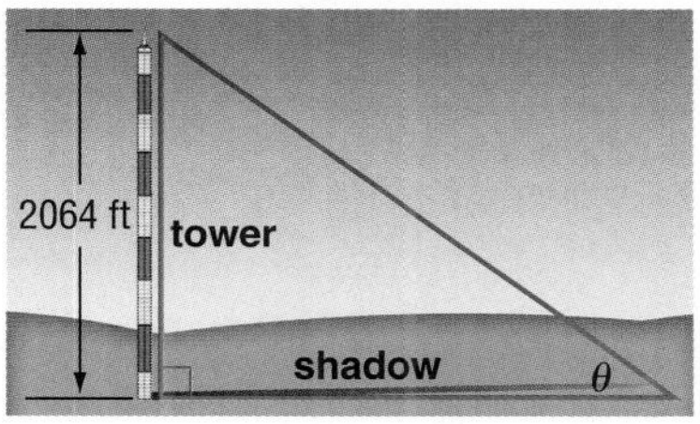

Examples 4 and 5
pp. 921, 922

Solve each equation.

45. $2 \sin^2 \theta = 3 \sin \theta + 2$

46. $2 \cos^2 \theta + 3 \sin \theta = 3$

47. $\sin^2 \theta + \cos 2\theta = \cos \theta$

48. $2 \cos^2 \theta = -\cos \theta$

49. **RIVERS** Due to ocean tides, the depth y in meters of the River Thames in London varies as a sine function of x, the hour of the day. On a certain day that function was $y = 3 \sin \left[\frac{\pi}{6}(x - 4)\right] + 8$, where $x = 0, 1, 2, \ldots, 24$ corresponds to 12:00 midnight, 1:00 A.M., 2:00 A.M., …, 12:00 midnight the next night.

a. What is the maximum depth of the River Thames on that day?

b. At what times does the maximum depth occur?

Solve each equation if θ is measured in radians.

50. $(\cos \theta)(\sin 2\theta) - 2 \sin \theta + 2 = 0$

51. $2 \sin^2 \theta + \left(\sqrt{2} - 1\right) \sin \theta = \frac{\sqrt{2}}{2}$

Solve each equation if θ is measured in degrees.

52. $\sin 2\theta + \frac{\sqrt{3}}{2} = \sqrt{3} \sin \theta + \cos \theta$

53. $1 - \sin^2 \theta - \cos \theta = \frac{3}{4}$

Solve each equation.

54. $2 \sin \theta = \sin 2\theta$

55. $\cos \theta \tan \theta - 2 \cos^2 \theta = -1$

Real-World Link

The sparkle of a diamond is created by refracted light. Light travels at different speeds in various media. When light rays pass from one medium to another in which they travel at a different velocity, the light is bent or refracted.

56. DIAMONDS According to Snell's Law, $n_1 \sin \boldsymbol{i} = n_2 \sin r$, where n_1 is the index of refraction of the medium the light is exiting, n_2 is the index of refraction of the medium the light is entering, $\boldsymbol{i}$ is the degree measure of the angle of incidence, and r is the degree measure of the angle of refraction.

a. The index of refraction of a diamond is 2.42, and the index of refraction of air is 1.00. If a beam of light strikes a diamond at an angle of 35°, what is the angle of refraction?

b. Explain how a gemologist might use Snell's Law to determine whether a diamond is genuine.

57. MUSIC A wave traveling in a guitar string can be modeled by the equation $D = 0.5 \sin (6.5x) \sin (2500t)$, where D is the displacement in millimeters at the position x meters from the left end of the string at time t seconds. Find the first positive time when the point 0.5 meter from the left end has a displacement of 0.01 millimeter.

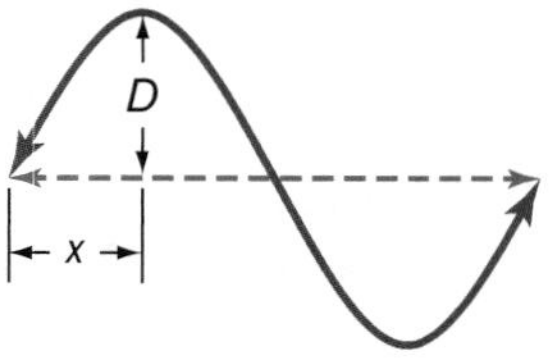

H.O.T. Problems

Use **H**igher-**O**rder **T**hinking Skills

58. FIND THE ERROR Jennifer and Tat are solving $2 \sin \theta \cos \theta = \sin \theta$ for $0° \leq \theta \leq 360°$. Is either of them correct? Explain your reasoning.

Jennifer

$2 \sin \theta \cos \theta = \sin \theta$

$\dfrac{2 \sin \theta \cos \theta}{\sin \theta} = \dfrac{\sin \theta}{\sin \theta}$

$2 \cos \theta = 1$

$\cos \theta = \dfrac{1}{2}$

$\theta = 0°, 60°, 180°, 300°$

Tat

$2 \sin \theta \cos \theta = \sin \theta$

$-\sin \theta = -\sin \theta$

$2 \cos \theta = 0$

$\cos \theta = 0$

$\theta = 90°, 270°$

59. CHALLENGE Solve $\sin 2x < \sin x$ for $0 \leq x \leq 2\pi$ without a calculator.

60. WRITING IN MATH Compare and contrast solving trigonometric equations with solving linear and quadratic equations. What techniques are the same? What techniques are different? How many solutions do you expect?

61. REASONING Explain why many trigonometric equations have infinitely many solutions.

62. OPEN ENDED Write an example of a trigonometric equation that has exactly two solutions if $0° \leq \theta \leq 360°$.

63. CHALLENGE How many solutions in the interval $0° \leq \theta \leq 360°$ should you expect for $a \sin (b\theta + c) + d = d\left(\frac{a}{2}\right)$, if $a \neq 0$ and b is a positive integer?

NYSRE PRACTICE A2.A.6, A2.S.12

64. EXTENDED RESPONSE Charles received \$2500 for a graduation gift. He put it into a savings account in which the interest rate was 5.5% per year.

a. How much did he have in his savings account after 5 years if he made no deposits or withdrawals?

b. After how many years will the amount in his savings account have doubled?

65. PROBABILITY Find the probability of rolling three 3s if a number cube is rolled three times.

A $\frac{1}{216}$ C $\frac{1}{6}$

B $\frac{1}{36}$ D $\frac{1}{4}$

66. Use synthetic substitution to find $f(-2)$ for the function below.

$$f(x) = x^4 + 10x^2 + x + 8$$

F 62 H 30

G 38 J 8

67. ACT/SAT The pattern of dots below continues infinitely, with more dots being added at each step.

Which expression can be used to determine the number of dots in the nth step?

A $2n$ C $n(n + 1)$

B $n(n + 2)$ D $2(n + 1)$

Spiral Review

Find the exact value of each expression. (Lesson 14-4)

68. $\cos 165°$

69. $\sin 22\frac{1}{2}°$

70. $\sin \frac{7\pi}{8}$

71. $\cos \frac{7\pi}{12}$

Verify that each equation is an identity. (Lesson 14-3)

72. $\sin (270° - \theta) = -\cos \theta$

73. $\cos (90° + \theta) = -\sin \theta$

74. $\cos (90° - \theta) = \sin \theta$

75. $\sin (90° - \theta) = \cos \theta$

76. WATER SAFETY A harbor buoy bobs up and down with the waves. The distance between the highest and lowest points is 4 feet. The buoy moves from its highest point to its lowest point and back to its highest point every 10 seconds. (Lesson 13-7)

a. Write an equation for the motion of the buoy. Assume that it is at equilibrium at $t = 0$ and that it is on the way up from the normal water level.

b. Draw a graph showing the height of the buoy as a function of time.

c. What is the height of the buoy after 12 seconds?

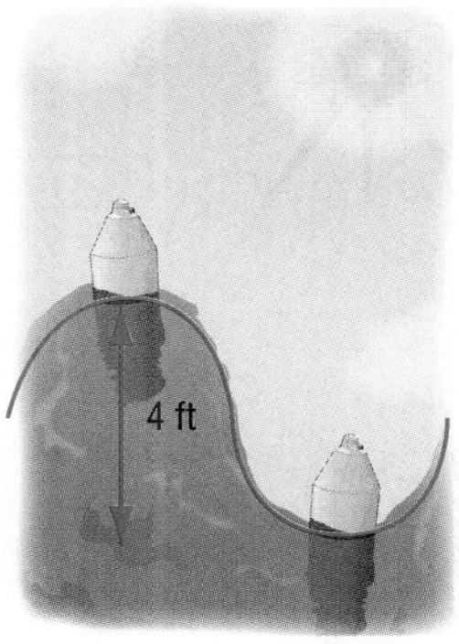

Find the first three terms of each arithmetic series described. (Lesson 11-2)

77. $a_1 = 17, a_n = 197, S_n = 2247$

78. $a_1 = -13, a_n = 427, S_n = 18{,}423$

79. $n = 31, a_n = 78, S_n = 1023$

80. $n = 19, a_n = 103, S_n = 1102$

Graph each rational function. (Lesson 9-4)

81. $f(x) = \frac{1}{(x + 3)^2}$

82. $f(x) = \frac{x + 4}{x - 1}$

83. $f(x) = \frac{x + 2}{x^2 - x - 6}$

CHAPTER 14 Study Guide and Review

NY Math Online glencoe.com
• STUDY TO GO
• Vocabulary Review

Chapter Summary

Key Concepts

Trigonometric Identities (Lessons 14-1, 14-2, and 14-5)

- Trigonometric identities describe the relationships between trigonometric functions.
- Trigonometric identities can be used to simplify, verify, and solve trigonometric equations and expressions.

Sum and Difference of Angles Identities
(Lesson 14-3)

- For all values of A and B:
  $\cos (A \pm B) = \cos A \cos B \pm \sin A \sin B$
  $\sin (A \pm B) = \sin A \sin B \pm \cos A \sin B$

Double-Angle and Half-Angle Identities
(Lesson 14-4)

- Double-angle identities:
  $\sin 2\theta = 2 \sin \theta \cos \theta$
  $\cos 2\theta = \cos^2 \theta - \sin^2 \theta$
  $\cos 2\theta = 1 - 2 \sin^2 \theta$
  $\cos 2\theta = 2 \sin^2 \theta - 1$
- Half-angle identities:
  $\sin \frac{\theta}{2} = \pm\sqrt{\frac{1 - \cos \theta}{2}}$
  $\cos \frac{\theta}{2} = \pm\sqrt{\frac{1 + \cos \theta}{2}}$
  $\tan \frac{\theta}{2} = \sqrt{\frac{1 - \cos \theta}{1 + \cos \theta}}, \cos \theta \neq -1$

FOLDABLES® Study Organizer

Be sure the Key Concepts are noted in your Foldable.

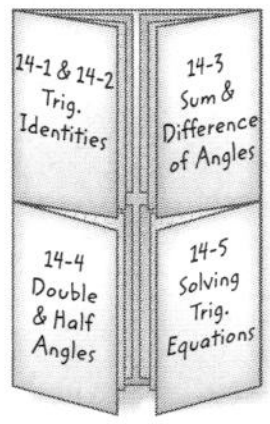

Key Vocabulary

cofunction identity (p. 891)
negative angle identity (p. 891)
Pythagorean identity (p. 891)
quotient identity (p. 891)
reciprocal identity (p. 891)
trigonometric equation (p. 919)
trigonometric identity (p. 891)

Vocabulary Check

Choose the correct term to complete each sentence.

1. The ___________ can be used to find the sine or cosine of 75° if the sine and cosine of 90° and 15° are known.
2. The identities $\tan \theta = \frac{\sin \theta}{\cos \theta}$ and $\cot \theta = \frac{\cos \theta}{\sin \theta}$ are examples of ___________.
3. A ___________ is an equation involving trigonometric functions that is true for all values for which every expression in the equation is defined.
4. The ___________ can be used to find sin 60° using 30° as a reference.
5. A ___________ is true for only certain values of the variable.
6. The ___________ formula can be used to find $\cos 33\frac{1}{2}°$.
7. The identities $\csc \theta = \frac{1}{\sin \theta}$ and $\sec \theta = \frac{1}{\cos \theta}$ are examples of ___________.
8. The ___________ can be used to find the sine or cosine of 120° if the sine and cosine of 90° and 30° are known.
9. $\cos^2 \theta + \sin^2 \theta = 1$ is an example of a ___________.

MIXED PROBLEM SOLVING
For mixed problem-solving practice, see page 992.

Lesson-by-Lesson Review

14-1 Trigonometric Identities (pp. 891–897)

A2.A.58

Find the value of each expression.

10. $\sin\theta$, if $\cos\theta = \frac{\sqrt{2}}{2}$ and $270° < \theta < 360°$

11. $\sec\theta$, if $\cot\theta = \frac{\sqrt{2}}{2}$ and $90° < \theta < 180°$

12. $\tan\theta$, if $\cot\theta = 2$ and $0° < \theta < 90°$

13. $\cos\theta$, if $\sin\theta = -\frac{3}{5}$ and $180° < \theta < 270°$

14. $\csc\theta$, if $\cot\theta = \frac{4}{5}$ and $270° < \theta < 360°$

15. SOCCER For international matches, the maximum dimensions of a soccer field are 110 meters by 75 meters. Find $\sin\theta$.

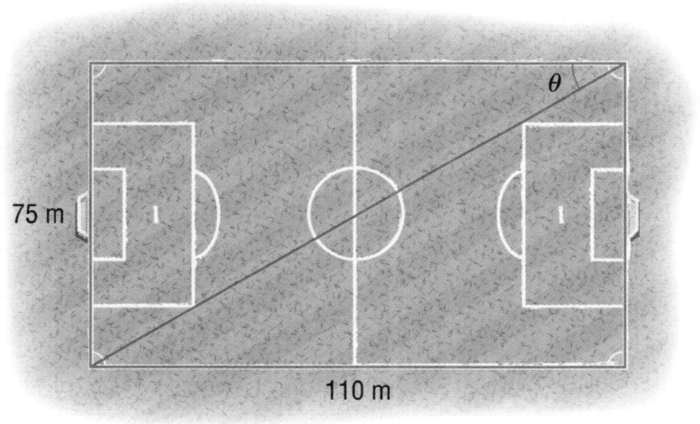

Simplify each expression.

16. $1 - \tan\theta\sin\theta\cos\theta$ **17.** $\tan\theta\csc\theta$

18. $\sin\theta + \cos\theta\cot\theta$ **19.** $\cos\theta\,(1 + \tan^2\theta)$

EXAMPLE 1

Find $\sin\theta$ if $\cos\theta = \frac{3}{4}$ and $0° < \theta < 90°$.

$\cos^2\theta + \sin^2\theta = 1$ — **Trigonometric identity**

$\sin^2\theta = 1 - \cos^2\theta$ — **Subtract $\cos^2\theta$ from each side.**

$\sin^2\theta = 1 - \left(\frac{3}{4}\right)^2$ — **Substitute $\frac{3}{4}$ for $\cos\theta$.**

$\sin^2\theta = 1 - \frac{9}{16}$ — **Square $\frac{3}{4}$.**

$\sin^2\theta = \frac{7}{16}$ — **Subtract.**

$\sin\theta = \pm\frac{\sqrt{7}}{4}$ — **Take the square root of each side.**

Because θ is in the first quadrant, $\sin\theta$ is positive. Thus, $\sin\theta = \frac{\sqrt{7}}{4}$.

EXAMPLE 2

Simplify $\cos\theta\sec\theta\cot\theta$.

$$\cos\theta\sec\theta\cot\theta = \cos\theta\left(\frac{1}{\cos\theta}\right)\left(\frac{\cos\theta}{\sin\theta}\right)$$
$$= \cot\theta$$

14-2 Verifying Trigonometric Identities (pp. 898–903)

A2.A.67

Verify that each of the following is an identity.

20. $\tan\theta\cos\theta + \cot\theta\sin\theta = \sin\theta + \cos\theta$

21. $\frac{\cos\theta}{\cot\theta} + \frac{\sin\theta}{\tan\theta} = \sin\theta + \cos\theta$

22. $\sec^2\theta - 1 = \frac{\sin^2\theta}{1 - \sin^2\theta}$

23. GEOMETRY The right triangle shown at the right is used in a special quilt. Use the measures of the sides of the triangle to show that $\tan^2\theta + 1 = \sec^2\theta$.

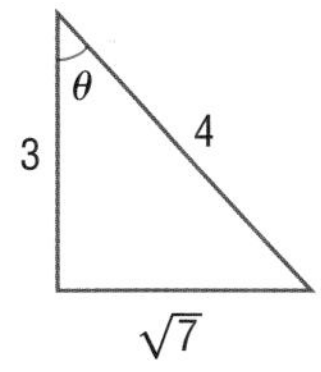

EXAMPLE 3

Verify that $\frac{\cos\theta + 1}{\sin\theta} = \cot\theta + \csc\theta$ is an identity.

$\frac{\cos\theta + 1}{\sin\theta} \stackrel{?}{=} \cot\theta + \csc\theta$ — **Original equation**

$\frac{\cos\theta}{\sin\theta} + \frac{1}{\sin\theta} \stackrel{?}{=} \cot\theta + \csc\theta$ — **Simplify.**

$\cot\theta + \csc\theta = \cot\theta + \csc\theta$ ✓ — **Simplify.**

CHAPTER 14 Study Guide and Review

14-3 Sum and Difference of Angles Identities (pp. 904–909)

A2.A.76

Find the exact value of each expression.

24. $\cos(-135°)$

25. $\cos 15°$

26. $\sin 210°$

27. $\sin 105°$

28. $\tan 75°$

29. $\cos 105°$

Verify that each of the following is an identity.

30. $\sin(\theta + 90) = \cos\theta$

31. $\sin\left(\frac{3\pi}{2} - \theta\right) = -\cos\theta$

32. $\tan(\theta - \pi) = \tan\theta$

EXAMPLE 4

Find the exact value of sin 75°.

Use $\sin(A + B) = \sin A \sin B + \cos A \sin B$.

$$\begin{aligned}\sin 75° &= \sin(30° + 45°)\\ &= \sin 30° \sin 45° + \cos 30° \sin 45°\\ &= \left(\frac{1}{2}\right)\left(\frac{\sqrt{2}}{2}\right) + \left(\frac{\sqrt{3}}{2}\right)\left(\frac{\sqrt{2}}{2}\right)\\ &= \frac{\sqrt{2}}{2} + \frac{\sqrt{6}}{4} \text{ or } \frac{\sqrt{2}+\sqrt{6}}{4}\end{aligned}$$

14-4 Double-Angle and Half-Angle Identities (pp. 911–917)

A2.A.77

Find the exact values of $\sin 2\theta$, $\cos 2\theta$, $\sin\frac{\theta}{2}$, and $\cos\frac{\theta}{2}$ for each of the following.

33. $\cos\theta = \frac{4}{5}$; $0° < \theta < 90°$

34. $\sin\theta = -\frac{1}{4}$; $180° < \theta < 270°$

35. $\cos\theta = \frac{2}{3}$; $\frac{\pi}{2} < \theta < \pi$

36. **BASEBALL** The infield of a baseball diamond is a square with side length 90 feet.

  a. Find the length of the diagonal.

  b. Write the ratio for sin 45° using the lengths of the baseball diamond.

  c. Use the formula $\sin\frac{\theta}{2} = \pm\sqrt{\frac{1-\cos\theta}{2}}$ to verify the ratio you wrote in part **b**.

EXAMPLE 5

Find the exact value of $\sin\frac{\theta}{2}$ if $\cos\theta = \frac{3}{5}$ and θ is in the second quadrant.

$$\sin\frac{\theta}{2} = \pm\sqrt{\frac{1-\cos\theta}{2}} \quad \textbf{Half-angle identity}$$

$$= \pm\sqrt{\frac{1-\frac{3}{5}}{2}} \quad \cos\theta = \frac{3}{5}$$

$$= \pm\sqrt{\frac{\frac{2}{5}}{2}} \quad \textbf{Subtract.}$$

$$= \pm\sqrt{\frac{1}{5}} \quad \textbf{Divide.}$$

$$= \pm\frac{\sqrt{5}}{5} \quad \textbf{Simplify.}$$

Since θ is in the second quadrant, $\sin\frac{\theta}{2} = \frac{\sqrt{5}}{5}$.

14-5 Solving Trigonometric Equations (pp. 919–925)

A2.A.68

Find all solutions of each equation for the given interval.

37. $2\cos\theta - 1 = 0$; $0° \le \theta < 360°$

38. $4\cos^2\theta - 1 = 0$; $0 \le \theta < 2\pi$

39. $\sin 2\theta + \cos\theta = 0$; $0° \le \theta < 360°$

40. $\sin^2\theta = 2\sin\theta + 3$; $0° \le \theta < 360°$

41. $4\cos^2\theta - 4\cos\theta + 1 = 0$; $0 \le \theta < 2\pi$

EXAMPLE 6

Find all solutions of $\sin 2\theta - \cos\theta = 0$ if $0 \le \theta < 2\pi$.

$$\sin 2\theta - \cos\theta = 0 \quad \textbf{Original equation}$$

$$2\sin\theta\cos\theta - \cos\theta = 0 \quad \textbf{Double-angle identity}$$

$$\cos\theta(2\sin\theta - 1) = 0 \quad \textbf{Factor.}$$

$\cos\theta = 0$ or $2\sin\theta - 1 = 0$

$\theta = \frac{\pi}{2}, \frac{3\pi}{2}$ $\sin\theta = \frac{1}{2}$; $\theta = \frac{\pi}{6}, \frac{5\pi}{6}$

Practice Test

NY Math Online > glencoe.com
Chapter Test

1. **MULTIPLE CHOICE** Which expression is equivalent to $\sin\theta + \cos\theta\cot\theta$?

   A $\cot\theta$ C $\sec\theta$

   B $\tan\theta$ D $\csc\theta$

2. Verify that $\cos(30° - \theta) = \sin(60° + \theta)$ is an identity.

3. Verify that $\cos(\theta - \pi) = -\cos\theta$.

4. **MULTIPLE CHOICE** What is the exact value of $\sin\theta$, if $\cos\theta = -\frac{3}{5}$ and $90° < \theta < 180°$?

   F $\frac{5}{3}$

   G $\frac{\sqrt{34}}{8}$

   H $-\frac{4}{5}$

   J $-\frac{24}{25}$

Find the value of each expression.

5. $\cot\theta$, if $\sec\theta = \frac{4}{3}$; $270° < \theta < 360°$

6. $\tan\theta$, if $\cos\theta = -\frac{1}{2}$; $90° < \theta < 180°$

7. $\sec\theta$, if $\csc\theta = 2$; $180° < \theta < 270°$

8. $\cot\theta$, if $\csc\theta = -\frac{5}{3}$; $270° < \theta < 360°$

9. $\sec\theta$, if $\sin\theta = \frac{1}{2}$; $0° \le \theta < 90°$

Verify that each of the following is an identity.

10. $\sin\theta(\cot\theta + \tan\theta) = \sec\theta$

11. $\frac{\cos^2\theta}{1 - \sin\theta} = \frac{\cos\theta}{\sec\theta - \tan\theta}$

12. $(\tan\theta + \cot\theta)^2 = \csc^2\theta\sec^2\theta$

13. $\frac{1 + \sec\theta}{\sec\theta} = \frac{\sin^2\theta}{1 - \cos\theta}$

14. $\frac{\sin\theta}{1 - \cos\theta} = \csc\theta + \cot\theta$

15. **MULTIPLE CHOICE** What is the exact value of $\tan\frac{\pi}{8}$?

    A $\frac{\sqrt{2 - \sqrt{3}}}{2}$

    B $\sqrt{2} - 1$

    C $1 - \sqrt{2}$

    D $-\frac{\sqrt{2 - \sqrt{3}}}{2}$

16. **HISTORY** Some researchers believe that the builders of ancient pyramids, such as the Great Pyramid of Khufu, may have tried to build the faces as equilateral triangles. Later they had to change to other types of triangles. Suppose a pyramid is built such that a face is an equilateral triangle of side length 18 feet.

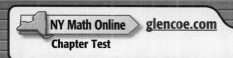

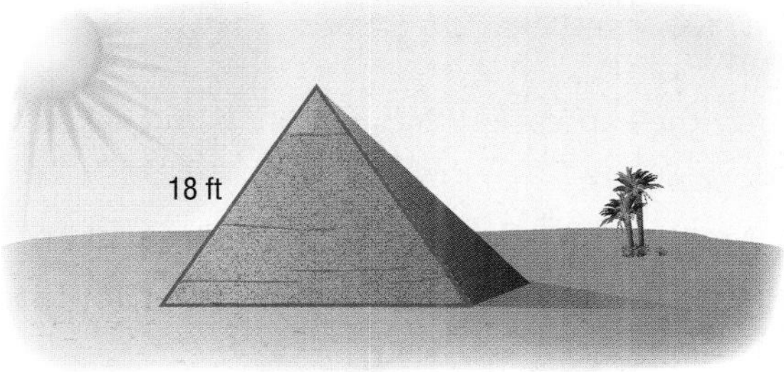

    a. Find the height of the equilateral triangle.

    b. Use the formula $\sin 2\theta = 2\sin\theta\cos\theta$ and the measures of the equilateral triangle and its height to show that $\sin 2(30°) = \sin 60°$. Find the exact values.

Find the exact value of each expression.

17. $\cos(-225°)$

18. $\sin 480°$

19. $\cos 75°$

20. $\sin 165°$

21. **ROCKETS** A model rocket is launched with an initial velocity of 20 meters per second. The range of a projectile is given by the formula $R = \frac{v^2}{g}\sin 2\theta$, where R is the range, v is the initial velocity, g is acceleration due to gravity or 9.8 meters per second squared, and θ is the launch angle. What angle is needed in order for the rocket to reach a range of 25 meters?

Solve each equation for all values of θ if θ is measured in radians.

22. $2\cos^2\theta - 3\cos\theta - 2 = 0$

23. $2\sin 3\theta - 1 = 0$

Solve each equation for $0° \le \theta \le 360°$ if θ is measured in degrees.

24. $\cos 2\theta + \cos\theta = 2$

25. $\sin\theta\cos\theta - \frac{1}{2}\sin\theta = 0$

CHAPTER 14

Preparing for Standardized Tests

Simplify Expressions

Some standardized test questions will require you to use the properties of algebra to simplify expressions. Follow the steps below to help prepare to solve these kinds of problems.

Strategies for Simplifying Expressions

Step 1

Study the expression that you are being asked to simplify.

Ask yourself:

- Are there any mathematical operations I can apply to help simplify the expression?
- Are there any laws or identities I can apply to help simplify the expression?

Step 2

Solve the problem and check your solution.

- Use the order of operations.
- Combine terms and factor as appropriate.
- Apply laws and identities.

Step 3

Check your solution if time permits.

- Retrace the steps in your work to make sure you answered the question thoroughly and accurately.
- If needed, sometimes you can use your scientific calculator to help you check your solution. Evaluate the original expression and your answer for some value and make sure they are the same.

EXAMPLE

Solve the problem below. Responses will be graded using the short-response scoring rubric shown.

Simplify the trigonometric expression shown below by writing it in terms of $\sin\theta$. Show your work to receive full credit.

$$\frac{\cos\theta}{\sec\theta + \tan\theta}$$

Scoring Rubric	
Criteria	**Score**
Full Credit: The answer is correct and a full explanation is provided that shows each step.	2
Partial Credit: • The answer is correct, but the explanation is incomplete. • The answer is incorrect, but the explanation is correct.	1
No Credit: Either an answer is not provided or the answer does not make sense.	0

Read the problem statement carefully. You are given a trigonometric expression and asked to simplify it by writing it in terms of sin θ. So, your final answer must contain only numbers and terms involving the sin θ. Show your work to receive full credit.

Example of a 2-point response:

Use trigonometric identities to simplify the expression.

$$\frac{\cos\theta}{\sec\theta + \tan\theta} = \frac{\cos\theta}{\frac{1}{\cos\theta} + \frac{\sin\theta}{\cos\theta}}$$ **Definition of sec θ and tan θ**

$$= \frac{\cos\theta}{\frac{1 + \sin\theta}{\cos\theta}}$$ **Simplify the denominator.**

$$= \frac{\cos^2\theta}{1 + \sin\theta}$$ **Simplify the complex fraction.**

$$= \frac{1 - \sin^2\theta}{1 + \sin\theta}$$ **Pythagorean identity**

$$= \frac{(1 + \sin\theta)(1 - \sin\theta)}{1 + \sin\theta}$$ **Factor.**

$$= 1 - \sin\theta$$ **Simplify.**

The simplified expression is $1 - \sin\theta$.

The steps, calculations, and reasoning are clearly stated. The student also arrives at the correct answer. So, this response is worth the full 2 points.

Exercises

Solve each problem. Show your work. Responses will be graded using the short-response scoring rubric given at the beginning of the lesson.

1. Simplify $\frac{\sec\theta}{\cot\theta + \tan\theta}$ by writing it in terms of $\sin\theta$.

2. What is $\frac{10a^{-3}}{29b^4} \div \frac{5a^{-5}}{16b^{-7}}$?

3. Write $\frac{y+1}{y-1} + \frac{y+2}{y-2} + \frac{y}{y^2 - 3y + 2}$ in simplest form.

4. Simplify $\frac{\cot^2\theta - \csc^2\theta}{\tan^2\theta - \sec^2\theta}$ by writing it as a constant.

5. Multiply $(-5 + 2i)(6 - i)(4 + 3i)$.

6. Simplify $(\cot\theta + 1)^2 - 2\cot\theta$ by writing it in terms of $\csc\theta$.

7. Express $\frac{4 - \sqrt{7}}{3 + \sqrt{7}}$ in simplest form.

CHAPTER 14

NYS Regents Exam Practice

Cumulative, Chapters 1 through 14

Multiple Choice

Answer all questions in this part. Select the answer that best completes the statement or answers the question.

1. Simplify the following expression by writing it in terms of cot θ.

$$\frac{\cos\theta\csc\theta}{\tan\theta}$$

(1) $\cot^2\theta$

(2) $2\cot\theta$

(3) $\frac{1}{\cot\theta}$

(4) $\frac{1}{\cot^2\theta}$

2. Suppose the P below is located on a unit circle. What is $\sin\theta$?

$$P\left(\frac{8}{17}, \frac{15}{17}\right)$$

(1) $\frac{8}{15}$

(2) $\frac{15}{17}$

(3) $\frac{8}{17}$

(4) $\frac{8}{17}$

3. A buoy bobs up and down with the waves of the ocean during a storm. The position of the buoy relative to sea level is given by the function $h(t) = 8.5\sin\frac{3}{4}\pi t$, where t is time in seconds and h is the height in feet. Which of the following is *not* a time at which the buoy is at the bottom of a wave (8.5 feet below sea level)?

(1) 2 seconds

(2) $4\frac{2}{3}$ seconds

(3) $7\frac{1}{3}$ seconds

(4) 9 seconds

4. The normal curve below shows the mean amount of snow, in inches, that New York City receives each winter. What is the probability that there will be more than 40 inches of snow in New York City next winter?

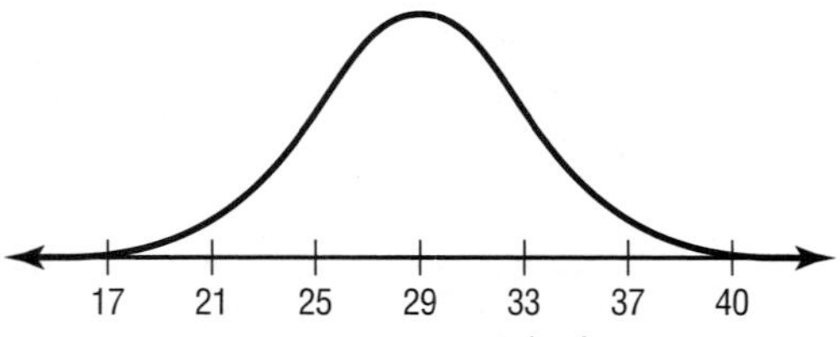

(1) 0.5%

(2) 2.5%

(3) 5%

(4) 16%

5. Use a sum or difference of angles identity to find the exact value of $\sin A$ in triangle ABC.

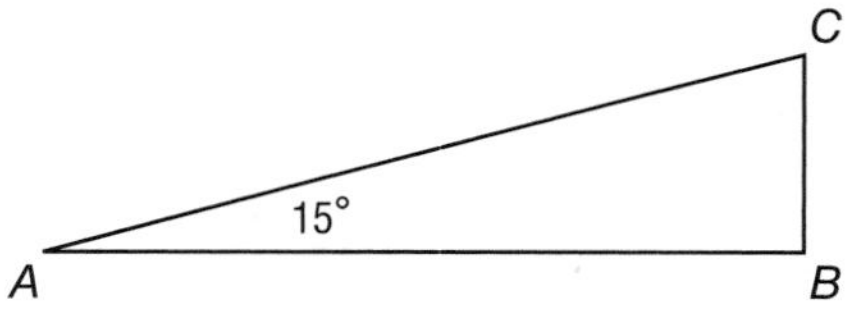

(1) $\frac{\sqrt{6}-\sqrt{2}}{2}$

(2) $\frac{\sqrt{6}+\sqrt{2}}{2}$

(3) $\frac{\sqrt{6}-\sqrt{2}}{4}$

(4) $\frac{\sqrt{6}+\sqrt{2}}{4}$

Test-Taking Tip

Question 5 You can check your answer using a scientific calculator. Find sin 15° and compare it to the value of your answer.

6. Evaluate $\sum_{k=1}^{50}(5k+9)$.

(1) 3149

(2) 6825

(3) 7872

(4) 10,145

7. What is the domain of the square root function graphed below?

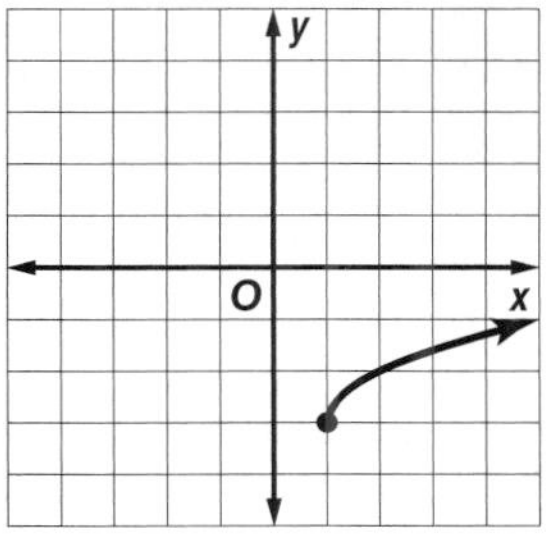

(1) $y \geq 1$
(2) $x \geq -3$
(3) $y \geq -3$
(4) $x \geq 1$

8. Matrix A contains the vertices of triangle LMN. Which matrix can you multiply by A in order to reflect the triangle over the x-axis?

$$A = \begin{bmatrix} 1 & 4 & -9 \\ 3 & 7 & -2 \end{bmatrix}$$

(1) $\begin{bmatrix} 1 & 0 \\ 0 & 1 \end{bmatrix}$
(2) $\begin{bmatrix} -1 & 0 \\ 0 & 1 \end{bmatrix}$
(3) $\begin{bmatrix} 1 & 0 \\ 0 & -1 \end{bmatrix}$
(4) $\begin{bmatrix} 0 & 1 \\ 1 & 0 \end{bmatrix}$

9. At 6:00 A.M. the outside temperature was 57°F. By 2:00 P.M. the temperature had climbed to 71°F. What was the hourly rate of change in the temperature?

(1) 1.5°F per hour
(2) 1.75°F per hour
(3) 2.15°F per hour
(4) 2.25°F per hour

Open-Ended Response

Solve each problem. Clearly indicate the necessary steps, including appropriate formula substitutions, diagrams, graphs, charts, etc.

10. Simplify $\cos \theta(1 + \tan^2 \theta)$. Show your work.

11. Use right triangle LMN below to show that $\sin 2N = \frac{2n\ell}{m^2}$.

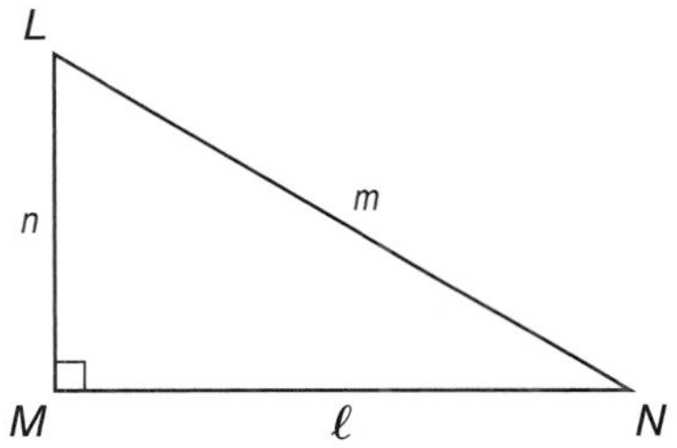

12. Use the clues below to answer each question.

Tim: "The sum of our ages is 52 years."

Lauren: "The combined ages of Tim and Amber is 12 more than my age."

Amber: "Tim is 2 years older than I am."

a. Write a system of equations that can be used to solve for the ages of each person.

b. Solve the system. What are the ages?

Need Extra Help?												
If you missed Question...	1	2	3	4	5	6	7	8	9	10	11	12
Go to Lesson or Page...	1-5	14-3	2-4	3-1	4-3	5-3	14-1	6-2	7-3	8-1	14-4	14-5
NYS Core Curriculum	A2.A.58	A2.A.55	A2.A.68	A2.S.5	A2.A.76	A2.A.35	A2.A.46	G.G.61	A.A.32	A2.A.58	A2.A.77	A.A.7

Student Handbook

Built-In Workbooks

Reference

How to Use the Student Handbook

The Student Handbook is the additional skill and reference material found at the end of the text. This Handbook can help you answer these questions.

What If I Need Problem Solving Practice?

You have probably used several different problem solving strategies in previous math courses. The **Problem-Solving Handbook** section provides examples and problems for refreshing your skills at using various strategies.

What If I Need More Practice?

You, or your teacher, may decide that working through some additional problems would be helpful. The **Extra Practice** section provides these problems for each lesson so you have ample opportunity to practice new skills.

What If I Have Trouble with Word Problems?

The **Mixed Problem Solving** portion of the book provides additional word problems that use the skills presented in each lesson. These problems give you real-world situations where math can be applied.

What if I Forget What I Learned Last Year?

Use the **Concepts and Skills Bank** section to refresh your memory about things you have learned in other math classes. Here's a list of the topics covered in your book.

1. Proportional Reasoning
2. Square Roots
3. Scientific Notation
4. Adding and Multiplying Probabilities
5. Bar and Line Graphs
6. Frequency Tables and Histograms
7. Stem-and-Leaf Plots
8. Box-and-Whisker Plots

What If I Need to Check a Homework Answer?

The answers to odd-numbered problems are included in **Selected Answers and Solutions**. Check your answers to make sure you understand how to solve all of the assigned problems.

What If I Forget a Vocabulary Word?

The **English-Spanish Glossary** provides a list of important or difficult words used throughout the textbook. It provides a definition in English and Spanish as well as the page number(s) where the word can be found.

What If I Need to Find Something Quickly?

The **Index** alphabetically lists the subjects covered throughout the entire textbook and the pages on which each subject can be found.

What if I Forget a Formula?

Inside the back cover of your math book is a list of **Formulas and Symbols** that are used in the book.

Problem-Solving Handbook

Problem-Solving Strategy: Look for a Pattern

There are many problem-solving strategies in mathematics. One of the most common is to **look for a pattern**. To use this strategy, analyze the first few numbers in a pattern and identify a rule that is used to go from the first number in the pattern to the second, and then to the third, and so on. Then use the rule to extend the pattern and find a solution.

Real-World EXAMPLE

A function passes through the points shown. List the coordinates of each point. Describe the pattern in the coordinates and predict the next point in the pattern in the positive direction.

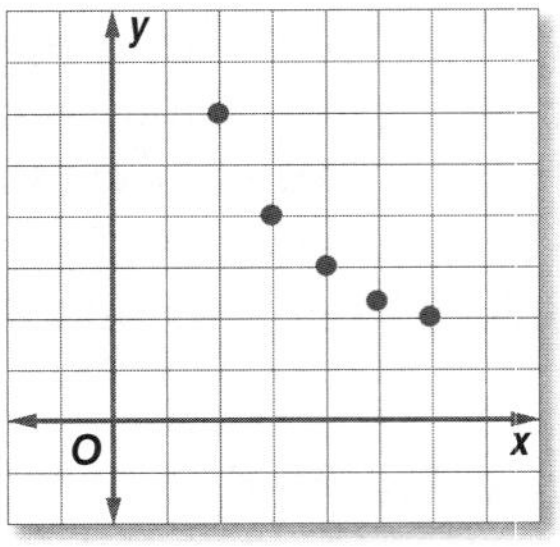

Step 1 List the coordinates of the points shown on the graph.

(2, 6), (3, 4), (4, 3), $\left(5, \frac{12}{5}\right)$, and (6, 2)

Step 2 Identify the pattern in the x-coordinates and the y-coordinates.

In each ordered pair, the product of the x- and y-coordinates is 12.

The next point is $\left(6 + 1, \frac{12}{6 + 1}\right)$ or $\left(7, \frac{12}{7}\right)$.

Practice

Solve each problem by looking for a pattern.

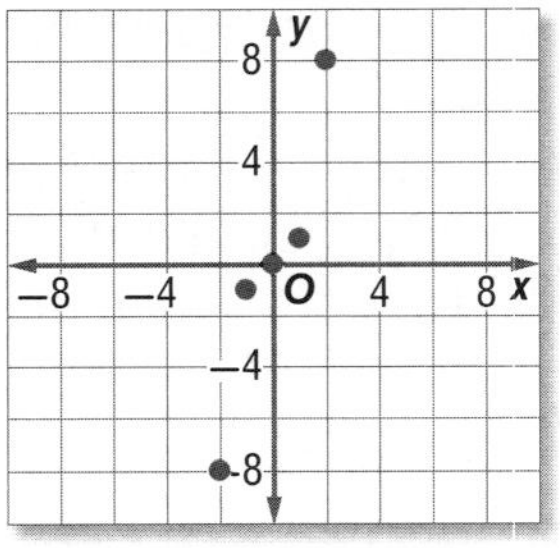

1. A function passes through the points shown.
   - **a.** Describe the pattern in the coordinates, and predict the next point in the pattern in the positive direction.
   - **b.** Predict the next point in the pattern in the negative direction.
2. Two workers can make two chairs in two days. How many chairs can 8 workers working at the same rate make in 20 days?
3. The sum of the measures of the angles of a triangle is 180°.
   - **a.** Use the sum of the angles of the triangles to determine the sum of the measures of the angles of each polygon by drawing all the diagonals from one vertex for a quadrilateral, a pentagon and a hexagon.
   - **b.** Write a rule to find the sum of the measures of the angles of an n-gon.

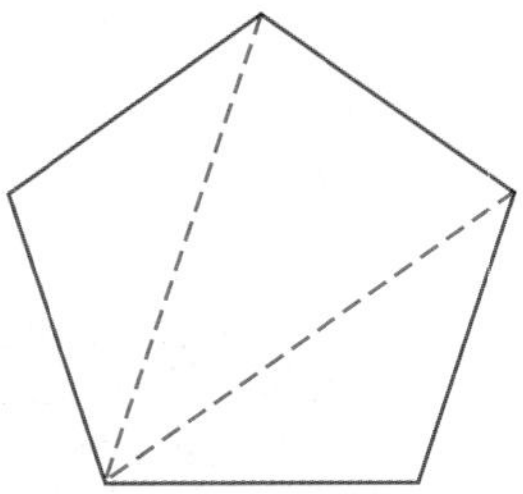

4. Courtney travels south on her bicycle riding 8 miles per hour. One hour later, her friend Horacio starts riding his bicycle from the same location. If he travels south at 10 miles per hour, how long will it take him to catch Courtney?
5. A ball bounces back 0.6 of its height on every bounce. If a ball is dropped from 200 feet, how high does it bounce on the fifth bounce? Round to the nearest tenth.

Problem-Solving Strategy: Create a Table

One strategy for solving problems is to **create a table**. A table allows you to organize information in an understandable way.

Real-World EXAMPLE

A fruit machine accepts dollars, and each piece of fruit costs 65 cents. If the machine gives only nickels, dimes, and quarters, what combinations of those coins are possible as change for a dollar?

The machine will give back $1.00 – $0.65 or 35 cents in change in a combination of nickels, dimes, and quarters.

Make a table showing different combinations of nickels, dimes, and quarters that total 35 cents. Organize the table by starting with the combinations that include the most quarters.

Quarters	Dimes	Nickels
1	1	0
1	0	2
0	3	1
0	2	3
0	1	5
0	0	7

The total for each combination of the coins is 35 cents. There are 6 combinations possible.

Practice

Solve each problem by creating a table.

1. How many ways can you make change for a half-dollar using only nickels, dimes, and quarters?

2. A penny, a nickel, a dime, and a quarter are in a purse. How many amounts of money are possible if you grab two coins at random?

3. Laura, Josie, and Marcus ate lunch together at the cafeteria. Each had a different item: a peanut butter sandwich, a hamburger, and a peanut butter and jelly sandwich. Josie ate the sandwich that had a bun. Laura does not like jelly. Marcus sat between the student eating the peanut butter sandwich and the student allergic to peanuts. Which student had which sandwich?

4. Johanna had a bag of four marbles. One marble is blue. Two marbles are green. One marble is orange. How many different ways are there to draw the marbles out of the bag one at a time?

5. At Midas High School, students are selling popcorn at a football game. Each small bag of popcorn is $1.25. Each large bag of popcorn is $2.25. Create a table to show the purchase prices of five bags of popcorn with every possible combination of large and/or small bags.

6. Aria asked her friends whether they used wrapping paper, gift bags, recycled paper, or no wrapping for birthday presents. Create a table to show how many students preferred each method of the 24 students she asked. Then predict how many would choose each method if 120 students were asked.

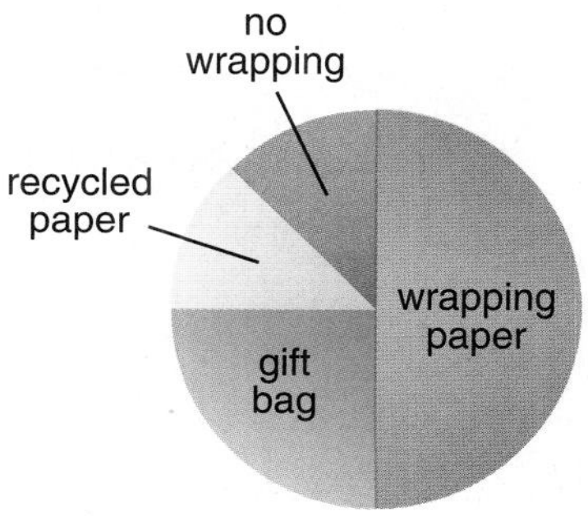

7. The equation for a semicircle is $y = \sqrt{16 - x^2}$. Create a table to show five ordered pairs with x-coordinates belonging to the set $\{-4, -2, 0, 2, 4\}$.

Problem-Solving Strategy: Make a Chart

Data presented in a problem can be organized by **making a chart**. This problem-solving strategy allows you to see patterns and relationships among data.

Real-World EXAMPLE

Find the time needed to stop each car using the data given in the diagram.

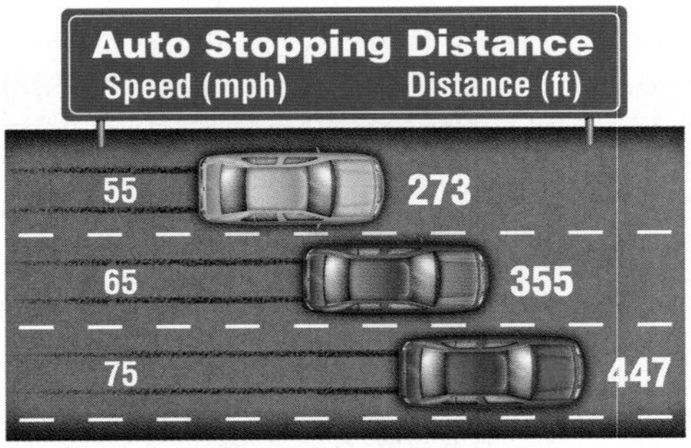

Source: AAA Foundation for Traffic Safety

Step 1 Make a chart that includes the information given—the speed and the distance—and a column to answer the question about time. Before setting up your chart, think about how speed, distance, and time are related. Include a column for Car 1, Car 2, and Car 3.

The distance formula is distance = rate of speed × time. So, use distance, rate, and time as three columns in your chart. Fill in the information given in the diagram.

Car	Distance (ft)	Rate of Speed (mph)	Time
1	273	55	
2	355	65	
3	447	75	

Step 2 Since the distance is listed in feet, find the rate of speed that it takes to stop in feet per second. To convert from miles per hour to feet per second, multiply by a conversion factor of 3600 s/5280 mi.

Car	Distance (ft)	Rate of Speed (ft/s)	Time
1	273	37.5	
2	355	44.3	
3	447	51.1	

Step 3 To find the time needed to stop, divide each distance by each rate of speed. Time will be in seconds.

Car	Distance (ft)	Rate of Speed (ft/s)	Time (s)
1	273	37.5	7.28
2	355	44.3	8.01
3	447	51.1	8.75

So, the time needed to stop would be about 7.28 seconds at 55 mph, 8.01 seconds at 65 mph, and 8.75 seconds at 75 mph.

Practice

Solve each problem by making a chart.

1. As the length of a square doubles, the area increases by a scale factor. Using the squares in the diagram, make a chart of each length and each area. Then find the scale factor.

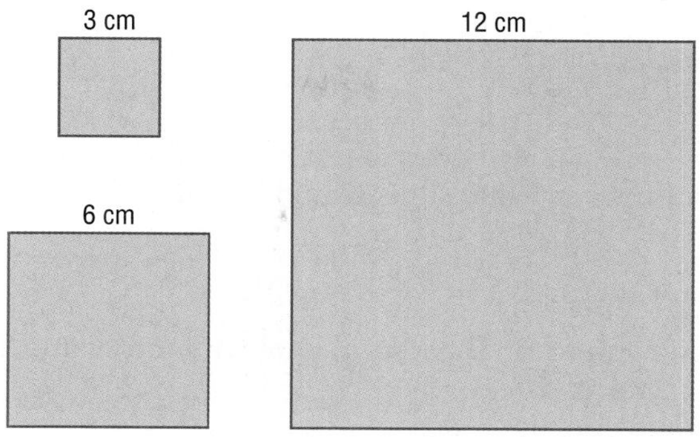

2. Given $f(x) = 2x + 3$ and $g(x) = -x - 3$, use a table to find $f(x) - g(x)$ for all positive integers less than or equal to 6.

3. **CONSTRUCTION** Nicole's new house has a small deck that measures 6 feet by 12 feet. She would like to build a larger deck. She can increase the dimensions in one-foot increments. Make a chart to show the possible dimensions if the area of the deck is doubled.

4. The diameters of the planets are given in the table. Make a box-and-whisker plot of the data. In which quartile are the data most closely clustered?

Planet	Mercury	Venus	Earth	Mars	Jupiter	Saturn	Uranus	Neptune
Diameter (mi)	3032	7521	7926	4194	88,736	74,978	32,193	30,775

5. The following table shows the official state reptile in each state. Make a tally chart that shows how many states have turtles (tortoise, terrapin), snakes, alligators, lizards, toads, or none. Find the ratio of states with alligators as their official state reptile compared to the states with no official state reptile.

AL	red-bellied turtle	LA	American alligator	OH	black racer
AK	none	ME	none	OK	collared lizard
AZ	ridge-nosed rattlesnake	MD	diamondback terrapin	OR	none
AR	none	MA	garter snake	PA	none
CA	desert tortoise	MI	painted turtle	RI	none
CO	none	MN	Blanding's turtle	SC	loggerhead turtle
CT	none	MS	American alligator	SD	none
DE	none	MO	three-toed box turtle	TN	eastern box turtle
FL	American alligator	MT	none	TX	horned lizard
GA	gopher tortoise	NE	none	UT	none
HI	none	NV	desert tortoise	VT	none
ID	none	NH	none	VA	none
IL	painted turtle	NJ	none	WA	none
IN	none	NM	whiptail lizard	WV	none
IA	none	NY	snapping turtle	WI	none
KS	ornate box turtle	NC	eastern box turtle	WY	horned toad
KY	none	ND	none		

Problem-Solving Strategy: Guess and Check

To solve some problems, you can make a reasonable guess and then check it in the problem. You can then use the results to improve your guess until you find the solution. This strategy is called **guess and check**.

Real-World EXAMPLE

Max needs to factor $2x^2 - x - 21$. He wants to use his graphing calculator to help him guess and check the answer. What steps would he use? What is the correct factorization?

Step 1 Use the calculator to graph $Y = 2x^2 - x - 21$.

Step 2 Find the root, or x-intercept, that has a whole number value. It is $x = -3$. Therefore, one factor is $(x + 3)$.

Step 3 By examining the first and last terms of the expression, guess the second factor. $(x + 3)(2x - 7)$

Step 4 Multiply the two factors to check for the desired product. $(x + 3)(2x - 7) = 2x^2 - 7x + 6x - 21 = 2x^2 - x - 21$.

Practice

Solve each problem by using the guess-and-check strategy.

Book	Price ($)
comic book	2.10
graphic novel	5.00
fiction book	5.90
atlas	22.40
calendar	16.00

1. Lacie bought four items at a school book fair. The prices of the books are shown in the table. Which four items did Lacie buy if she spent $25.20?

2. This year, Mr. Jefferson's age is a multiple of 11. Next year, his age will be a multiple of 9. How old is Mr. Jefferson this year?

3. Rafael is burning a CD for Selma. The CD will hold 35 minutes of music. Which songs should he select from the list to record the maximum time on the CD without going over?

Song	A	B	C	D	E	F	G	H	I	J
Time	5 min 4 s	9 min 10 s	4 min 12 s	3 min 9 s	3 min 44 s	4 min 30 s	5 min 0 s	7 min 21 s	4 min 33 s	5 min 58 s

4. Larry wrote down 4 different numbers whose sum was 12. Give two possible combinations for the numbers.

5. The Science Club sold candy bars and soft pretzels to raise money for an animal shelter. They raised a total of $62.75. They made $0.25 profit on each candy bar and $0.30 profit on each pretzel sold. How many of each did they sell?

6. The product of two consecutive even integers is 4224. Find the integers.

7. Jerrica has 8 coins, all dimes, quarters, and pennies. If she has $0.89, how many of each coin does she have?

8. One angle of a triangle is shown. Find the measures of the other two angles if their product is 1216.

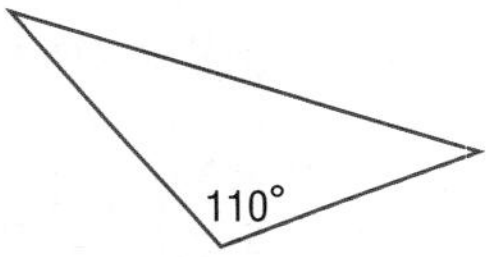

9. Anita sold tickets to the school musical. She had 2 bills worth $75 for the tickets she sold. If all the money was in $5 bills, $10 bills, and $20 bills, how many of each bill did she have?

Problem-Solving Strategy: Work Backward

On most problems, a set of conditions or facts is given and an end result must be found. However, some problems start with the result and ask for something that happened earlier. The strategy of **working backward** can be used to solve problems like this. To use this strategy, start with the end result and *undo* each step.

Real-World EXAMPLE

Kendrick spent half of the money he had this morning on lunch. After lunch, he loaned his friend a dollar. Now he has $1.50. How much money did Kendrick have this morning?

Start with the end result, $1.50, and work backward to find the amount Kendrick had this morning.

Kendrick now has $1.50. →	$1.50	
Undo the $1.00 he loaned to his friend. →	+ 1.00	Add $1.00 to undo giving his friend $1.00.
	$2.50	
Undo the half he spent for lunch. →	× 2	Multiply by 2 to undo spending half the original amount.
	$5.00	

Kendrick had $5.00 this morning.

CHECK Kendrick started with $5.00. If he spent half of that, or $2.50, on lunch, and loaned his friend $1.00, he would have $1.50 left. This matches the amount stated in the problem, so the solution is correct.

Practice

Solve each problem by working backward.

1. Tia used half of her allowance to buy a ticket to the class play. Then she spent $0.75 for an ice cream cone. Now she has $2.25 left. How much is her allowance?

2. Lawanda put $15 of her paycheck in savings. Then she spent one-half of what was left on clothes. She paid $24 for a concert ticket and later spent one-half of what was then left on a book. When she got home, she had $14 left. What was the amount of Lawanda's paycheck?

3. Mr. and Mrs. Delgado each own an equal number of shares of a stock. Mr. Delgado sells one-third of his shares for $2700. What was the total value of Mr. and Mrs. Delgado's stock before the sale?

4. A certain point was reflected across the x-axis, and then moved up three units and two units to the left. The final coordinates of the point are (3, 5). What were the original coordinates?

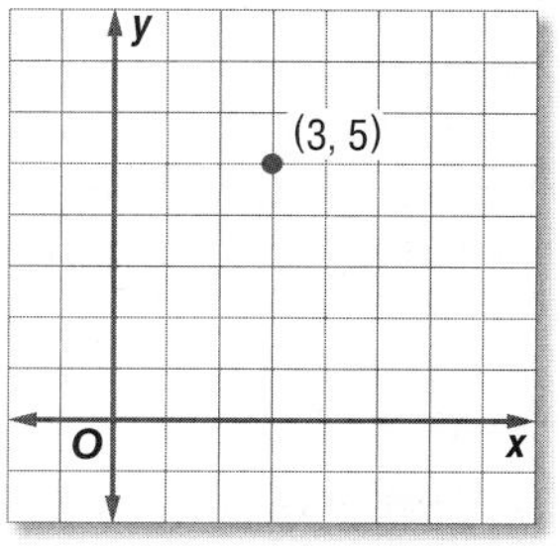

5. A certain bacteria doubles its population every 12 hours. After 3 full days, there are 1600 bacteria in a culture. How many bacteria were there at the beginning of the first day?

6. To catch a 7:30 A.M. bus, Don needs 30 minutes to get dressed, 30 minutes for breakfast, and 5 minutes to walk to the bus stop. What time should he wake up?

7. Troy lives $1\frac{1}{8}$ miles from his school. If he has walked $\frac{1}{4}$ mile to meet a friend and then they walked another $\frac{1}{2}$ mile to meet another friend, how far do they still need to walk to get to school?

Problem-Solving Strategy: Solve a Simpler Problem

One of the strategies you can use to solve a problem is to **solve a simpler problem**. To use this strategy, first solve a simpler or more familiar case of the problem. Then use the same concept and relationships to solve the original problem.

Real-World EXAMPLE

Find the sum of the numbers 1 through 500.

Consider a simpler problem. Find the sum of the numbers 1 through 10. Notice that you can group the addends into partial sums as shown below.

$1 + 2 + 3 + 4 + 5 + 6 + 7 + 8 + 9 + 10 = 55$

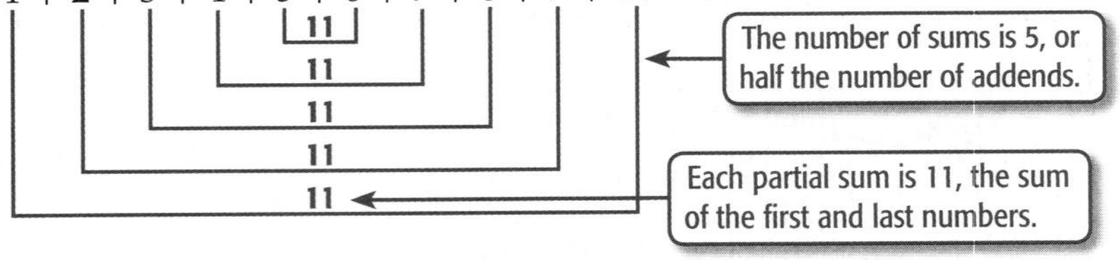
11
11
11
11
11
The number of sums is 5, or half the number of addends.
Each partial sum is 11, the sum of the first and last numbers.

The sum is 5×11 or 55.

Use the same concepts to find the sum of the numbers 1 through 500.

$1 + 2 + 3 + \cdots + 499 + 500 = 250 \times 501$

$= 125{,}250$

Multiply half the number of addends, 250, by the sum of the first and last numbers, 501.

Practice

Solve each problem by solving a simpler problem.

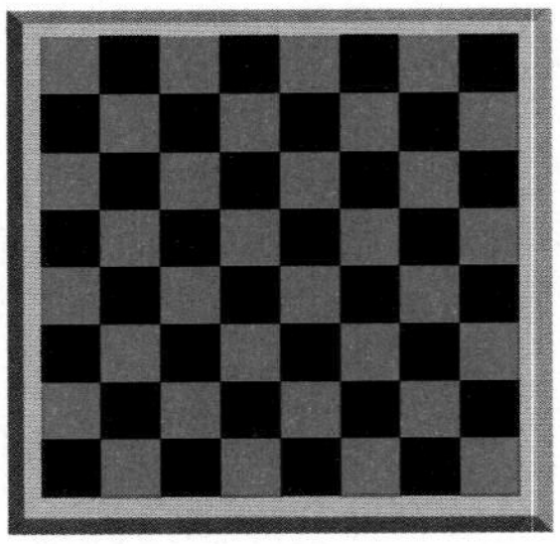

1. Find the number of squares of any size in the game board shown at the right.
2. Find the sum of the whole numbers through 1000.
3. How many links are needed to join 30 pieces of chain into one long chain?
4. Three people can pick six baskets of apples in one hour. How many baskets of apples can 2 people pick in one-half hour?

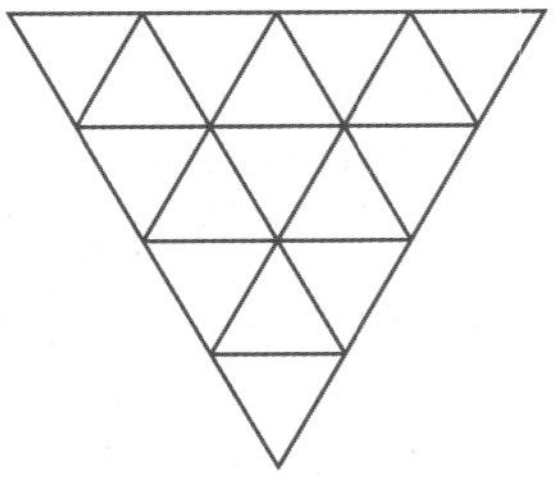

5. Find the number of triangles of any size in the figure at the right.
6. A shirt shop has 112 orders for T-shirt designs. Three designers can make 2 shirts in 2 hours. How many designers are needed to complete the orders in 8 hours?

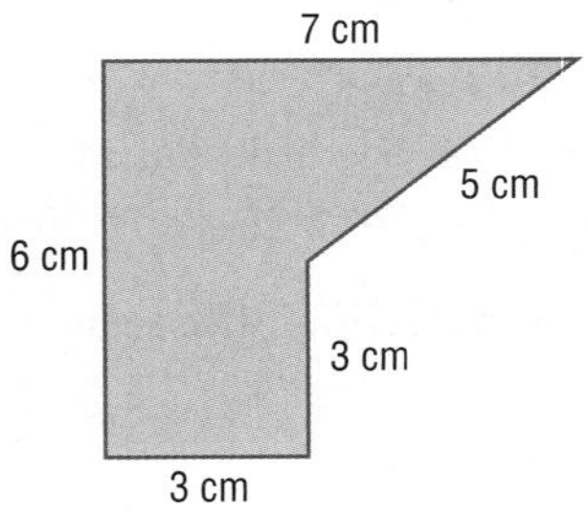
7 cm
5 cm
6 cm
3 cm
3 cm

7. Find the area of the composite figure at the right with the given measures.
8. Stamps for postcards cost \$0.24, and stamps for first-class letters cost \$0.41. Diego wants to send postcards and letters to 10 friends. If he has \$3.50 for stamps, how many postcards and how many letters can he send?

Problem-Solving Strategy: Draw a Diagram

Another strategy for solving problems is to **draw a diagram**. There will be times when a sketch or diagram will give you a better picture of how to tackle a mathematics problem. Adding details like units, labels, and numbers to the drawing or sketch can help you make decisions on how to solve the problem.

Real-World EXAMPLE

Cheng is designing a kite using similar triangles. One portion of the frame of the kite will look like the figure. If Cheng wants to place an upright strut 12 inches from the right angle, how long does it need to be?

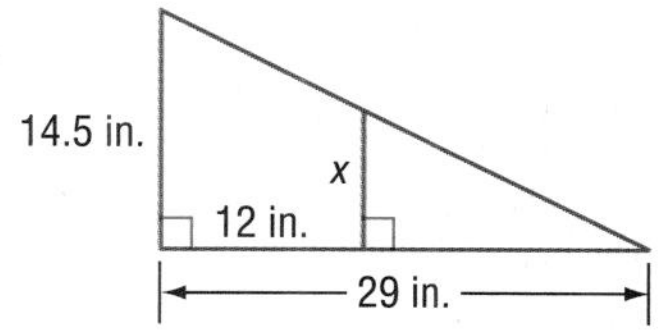

This problem can be solved using the concept of similar triangles. However, the two triangles overlap, making it difficult to compare the corresponding sides. It may be helpful to draw a diagram that shows the two triangles separately. Notice that the base of the triangle is 29 − 12 or 17 inches.

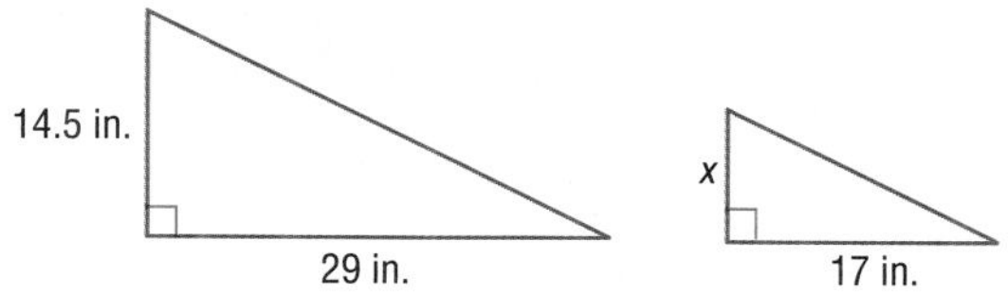

We can now set up the proportion $\frac{x}{14.5} = \frac{17}{29}$. Solving the proportion, we find that the strut should be 8.5 inches long.

Practice

Solve each problem by drawing a diagram.

1. Find the length of the horizontal side of the large triangle.

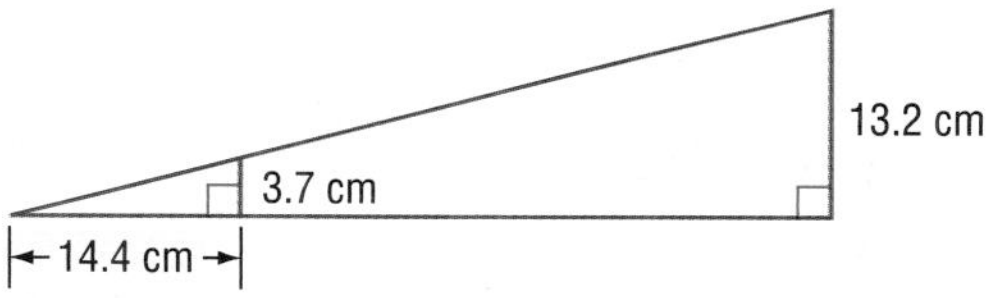

2. A 500-gallon water tank is being filled with water. Eighty gallons of water are in the tank after 4 minutes. How long will it take to fill the tank?
3. It takes 42 minutes to cut a 2-inch by 4-inch piece of wood into 7 equally sized pieces. How long will it take to cut a similar 2-inch by 4-inch piece into 4 equally sized pieces?
4. Find the number of line segments that can be drawn between any two vertices of an octagon.

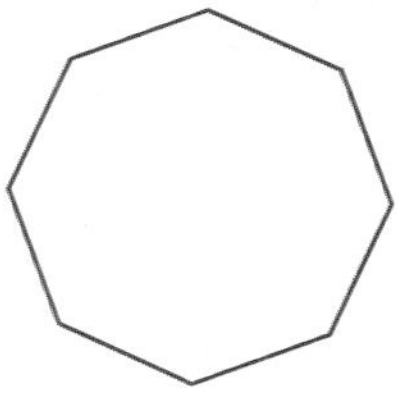

5. There are 7 people in a meeting. If each person shakes hands with all the other people, how many handshakes take place?
6. How many different teams of 3 players can be chosen from 8 players?
7. Nitarren is trying to decide how many 9-inch diameter pies will fit on her dessert table that measures 4 feet by 2 feet. How many pies could Nitarren fit on the table?
8. Two lines can share no points or one point. How many points can two circles share?
9. Which of these shapes can be made by slicing a cone with a plane—a circle, an oval, a rectangle, a square, a triangle?

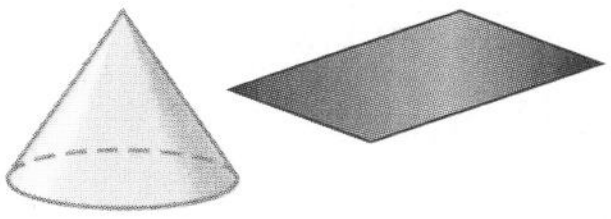

Problem-Solving Strategy: Use Estimation

When you need to make a decision on the basis of inexact information, a common strategy is to **use estimation**. Often estimation is used when an exact answer is not required or when mental math is used rather than a calculator or paper and pencil. You should use estimation to determine if your answer is reasonable.

Real-World EXAMPLE

Recently, 51 million international visitors came to the United States. Given the information in the table, estimate what percentage of the visitors were from Japan.

Source: Travel Industry Association

Step 1 Determine about how many international visitors came to the United States.

51 million or 51,000,000 would be an easier number to work with if it was rounded to only one digit in the ten millions place, so round 51,000,000 to 50,000,000.

Step 2 Determine from the chart the number of visitors who were from Japan.

According to the chart, 5,000,000 visitors to the United States were from Japan.

Step 3 To determine the percentage, divide.

$$\frac{\text{number of visitors from Japan}}{\text{total number of visitors}} = \frac{5{,}000{,}000}{50{,}000{,}000}$$
$$= 0.1 \text{ or } 10\%$$

About 10% of the visitors to the United States in 2000 were from Japan.

Practice

Solve each problem by using estimation.

1. If Café Mocha charges $0.98 for each cup of coffee, about how much money did Café Mocha earn in March?

Coffee of the Month Sales	
Month	**Number of Cups Sold**
January	850
February	765
March	587
April	500
May	387

2. The length of Fun Center's go-kart track is 843 feet. If Nadia circled the track 9 times, about how many feet did she travel?

3. In 2007, 82% of all Americans had at least one e-mail account. Of the 301,139,947 Americans, about how many checked their e-mail daily?

Source: UCLA Center for Communication Policy

4. Sarah solves the following problem. Her answer is 8 doses. Use estimation to determine whether or not Sarah's answer is correct. Explain your reasoning.

   How many $\frac{4}{5}$-ounce doses of medicine are in a 10-ounce jar?

Problem-Solving Strategy: Eliminate Unnecessary Information

A useful problem-solving strategy is to learn how to **eliminate unnecessary information**. If there is a diagram, it is important to determine if all or some of the information is necessary to find a solution.

Real-World EXAMPLE

Twila is making a quilt that shows a repeating design of a house on each block of the quilt. Which information is unnecessary to find the area of the white door of the house?

The dimensions needed to find the area of the door are the length and width of the door. So, the dimensions of $2\frac{1}{2}$ inches by $5\frac{1}{2}$ inches are needed. The other dimensions, such as the width of the window or the size of the square, are unnecessary.

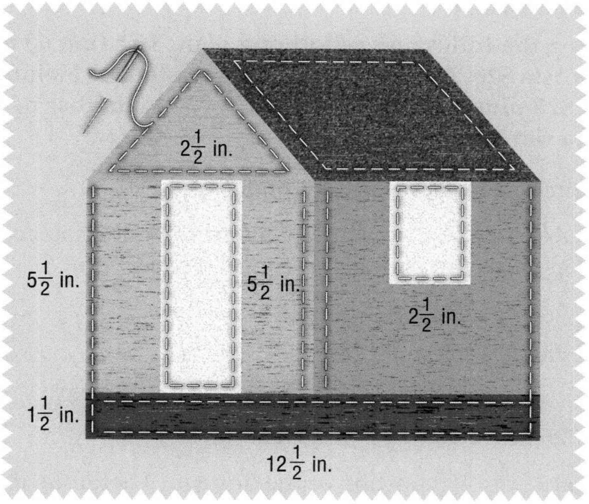

Practice

Solve each problem by eliminating unnecessary information.

1. Many school districts are banning electronic devices for students. The chart shows what percentage of schools have banned specific devices. If there are 723 districts in Ohio, what information is not necessary to determine about how many school districts have banned handheld games?

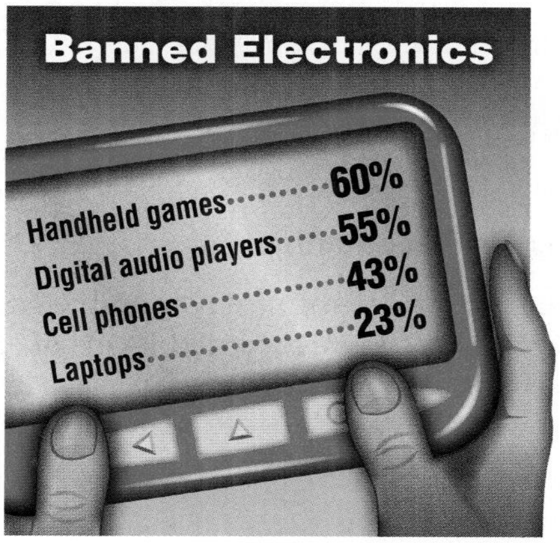

2. Miranda is making a mosaic with the tiles at the right. Which information is not needed to determine the area that each tile will cover?

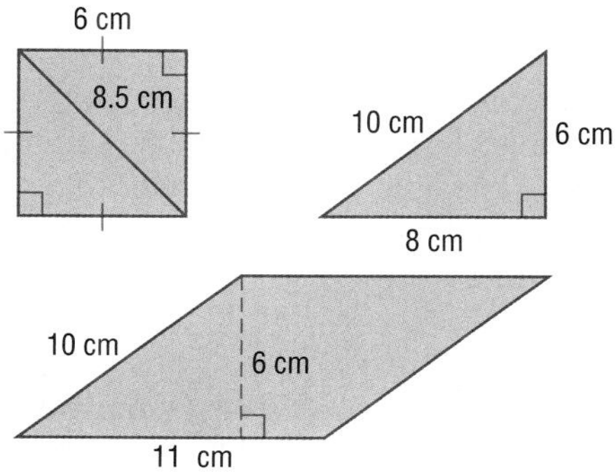

3. The Gemini North telescope was placed in Mauna Kea, Hawaii, in the year 2000. The Gemini South telescope was placed in Cerro Pachon, Chili, in the year 2001. Each of the twin telescopes are 8.1 meters in diameter. What information is not necessary to find the circumference of the base of the telescopes?

Problem-Solving Strategy: Write an Equation

A natural outcome of recognizing mathematical patterns and organizing data is to **write an equation**. Look at a set of data or read a word problem to determine which values are constants and which values vary. Figure out the dependent and independent variables in order to write an equation to reflect the given situation.

Real-World EXAMPLE

TelExcite Cell is offering the following cell phone plan. The first 60 minutes of phone use per month costs $20. Additional minutes, up to and including 400 minutes, the charge is $20 plus 10 cents a minute. For any number of minutes used past 400, the charge is a flat fee of $80.

a. Write a piecewise function to represent the plan.

Separate the domain as described. Pay careful attention to the category in which each boundary belongs. Write an expression for each condition.

$$f(x) = \begin{cases} 20 & \text{if } 0 \leq x \leq 60 \\ 20 + 0.10(x - 60) & \text{if } 60 < x \leq 400 \\ 80 & \text{if } x > 400 \end{cases}$$

b. Find the charge for 325 minutes of phone use.

325 minutes fits the $60 < x \leq 400$ portion of the domain. Therefore, the charge is $20 + 0.10(325 - 60)$ or $46.50.

c. When Anna uses more than 400 minutes of phone time in a month, she feels as if she does not get her money's worth if $80 is more than $20 plus 10 cents a minute she has used. For what amount of phone use does Anna feel the $80 charge is fair?

Solve $20 + 0.10(x - 60) = 80$ to find $x = 660$. Anna feels the $80 charge is fair for any number of minutes she uses over 660 minutes.

Practice

Solve each problem by writing an equation.

1. Isaac has been playing a massively multiplayer online game (MMOG) for the last two months. In the game, players are awarded levels of experience based on their actions in the game. For the first 10 hours of play, Isaac leveled 3 times each hour. For the next 5 hours of play, he only leveled once an hour. For the final nine hours of play, Isaac leveled once every three hours.
   a. Write a piecewise function L to show level as a function of time t.
   b. What level had Isaac reached after 18 hours of play?

2. The area of a rectangular parking lot is to be 20,000 square feet. The table shows some possible dimensions for the lot. Write an equation that can be used to find the width of the parking lot for any given length.

Length (ft)	Width (ft)
20	1000
50	400
100	200
200	100
400	50

3. Iceland spent approximately 8.8% of its Gross Domestic Product on public health expenditures in 2006, the highest percentage of all countries. If Iceland had a GDP of $11,380,000,000 in 2006, write an equation to show how much money was spent on public health expenditures.

Extra Practice

Lesson 1-1 Expressions and Formulas (pp. 5–10)

Evaluate each expression if $q = \frac{1}{2}$, $r = 1.2$, $v = -6$, and $t = 5$.

1. $qr - vt$
2. $qr \div vt$
3. $qrvt$
4. $qr + vt$
5. $\frac{3q}{4v}$
6. $\frac{5qr}{t}$
7. $\frac{2r(4v - 1)}{t}$
8. $\frac{4q^3v + 1}{t - 1}$

Evaluate each expression if $a = -0.5$, $b = 4$, $c = 5$, and $d = -3$.

9. $3b + 4d$
10. $ab^2 + c$
11. $bc + d \div a$
12. $7ab - 3d$
13. $ad + b^2 - c$
14. $\frac{4a + 3c}{3b}$
15. $\frac{3ab^2 - d^3}{a}$
16. $\frac{5a + ad}{bc}$

Lesson 1-2 Properties of Real Numbers (pp. 11–17)

Name the sets of numbers to which each number belongs. (Use N, W, Z, Q, I, and R.)

1. 8.2
2. -9
3. $\sqrt{36}$
4. $-\frac{1}{3}$
5. $\sqrt{2}$
6. $-0.\overline{24}$

Name the property illustrated by each equation.

7. $(4 + 9a)2b = 2b(4 + 9a)$
8. $3\left(\frac{1}{3}\right) = 1$
9. $a(3 - 2) = a \cdot 3 - a \cdot 2$
10. $(-3b) + 3b = 0$
11. $jk + 0 = jk$
12. $(2a)b = 2(ab)$

Simplify each expression.

13. $7r + 9t + 2r - 7t$
14. $6(2a + 3b) + 5(3a - 4b)$
15. $4(3x - 5y) - 8(2x + y)$
16. $0.2(5m - 8) + 0.3(6 - 2m)$
17. $\frac{1}{2}(7p + 3t) + \frac{3}{4}(6p - 4t)$
18. $\frac{4}{5}(3v - 2w) - \frac{1}{5}(7v - 2w)$

Lesson 1-3 Solving Equations (pp. 18–25)

Write an algebraic expression to represent each verbal expression.

1. twelve decreased by the square of a number
2. twice the sum of a number and negative nine
3. the product of the square of a number and 6
4. the square of the sum of a number and 11

Name the property illustrated by each statement.

5. If $a + 1 = 6$, then $3(a + 1) = 3(6)$.
6. If $x + (4 + 5) = 21$, then $x + 9 = 21$.
7. If $7x = 42$, then $7x - 5 = 42 - 5$.
8. If $3 + 5 = 8$ and $8 = 2 \cdot 4$, then $3 + 5 = 2 \cdot 4$.

Solve each equation. Check your solution.

9. $5t + 8 = 88$
10. $27 - x = -4$
11. $\frac{3}{4}y = \frac{2}{3}y + 5$
12. $8w - 3 = 5(2w + 1)$
13. $3(k - 2) = k + 4$
14. $0.5z + 10 = z + 4$
15. $8q - \frac{q}{3} = 46$
16. $-\frac{2}{7}r + \frac{3}{7} = 5$
17. $d - 1 = \frac{1}{2}(d - 2)$

Solve each equation or formula for the specified variable.

18. $C = \pi r$, for r
19. $I = Prt$, for t
20. $m = \frac{n - 2}{n}$, for n

Extra Practice

Lesson 1-4 Solving Absolute Value Equations (pp. 27–32)

Evaluate each expression if $x = -5$, $y = 3$, and $z = -2.5$.

1. $|2x|$
2. $|-3y|$
3. $|2x + y|$
4. $|y + 5z|$
5. $-|x + z|$
6. $8 - |5y - 3|$
7. $2|x| - 4|2 + y|$
8. $|x + y| - 6|z|$

Solve each equation. Check your solutions.

9. $|d + 1| = 7$
10. $|a - 6| = 10$
11. $2|x - 5| = 22$
12. $|t + 9| - 8 = 5$
13. $|p + 1| + 10 = 5$
14. $6|g - 3| = 42$
15. $2|y + 4| = 14$
16. $|3b - 10| = 2b$
17. $|3x + 7| + 4 = 0$
18. $|2c + 3| - 15 = 0$
19. $7 - |m - 1| = 3$
20. $3 + |z + 5| = 10$
21. $2|2d - 7| + 1 = 35$
22. $|3t + 6| + 9 = 30$
23. $|d - 3| = 2d + 9$
24. $|4y - 5| + 4 = 7y + 8$
25. $|2b + 4| - 3 = 6b + 1$
26. $|5t| + 2 = 3t + 18$

Lesson 1-5 Solving Inequalities (pp. 33–39)

Solve each inequality. Then graph the solution set on a number line.

1. $2z + 5 \leq 7$
2. $3r - 8 > 7$
3. $0.75b < 3$
4. $-3x > 6$
5. $2(3f + 5) \geq 28$
6. $-33 > 5g + 7$
7. $-3(y - 2) \geq -9$
8. $7a + 5 > 4a - 7$
9. $5(b - 3) \leq b - 7$
10. $3(2x - 5) < 5(x - 4)$
11. $8(2c - 1) > 11c + 22$
12. $2(d + 4) - 5 \geq 5(d + 3)$
13. $8 - 3t < 4(3 - t)$
14. $-x \geq \frac{x + 4}{7}$
15. $\frac{a + 8}{4} \leq \frac{7 + a}{3}$
16. $-y < \frac{y + 5}{2}$
17. $5(x - 1) - 4x \geq 3(3 - x)$
18. $6k - (4k + 7) > 5 - k$

Define a variable and write an inequality for each problem. Then solve.

19. The product of 7 and a number is greater than 42.
20. The difference of twice a number and 3 is at most 11.
21. The product of -10 and a number is greater than or equal to 20.
22. Thirty increased by a number is less than twice the number plus three.

Lesson 1-6 Solving Compound and Absolute Value Inequalities (pp. 41–48)

Write an absolute value inequality for each of the following. Then graph the solution set on a number line.

1. all numbers less than -9 and greater than 9
2. all numbers between -5.5 and 5.5
3. all numbers greater than or equal to -2 and less than or equal to 2

Solve each inequality. Graph the solution set on a number line.

4. $3m - 2 < 7$ or $2m + 1 > 13$
5. $2 < n + 4 < 7$
6. $-3 \leq y - 2 \leq 5$
7. $5t + 3 \leq -7$ or $5t - 2 \geq 8$
8. $7 \leq 4x + 3 \leq 19$
9. $4x + 7 < 5$ or $2x - 4 > 12$
10. $|7x| \geq 21$
11. $|8p| \leq 16$
12. $|7d| \geq -42$
13. $|a + 3| < 1$
14. $|t - 4| > 1$
15. $|2y - 5| < 3$
16. $|3d + 6| \geq 3$
17. $|4x - 1| < 5$
18. $|6v + 12| > 18$
19. $|2r + 4| < 6$
20. $|5w - 3| \geq 9$
21. $|z + 2| \geq 0$
22. $12 + |2q| < 0$
23. $|3h| + 15 < 0$
24. $|5n - 16| \geq 4$

Lesson 2-1 Relations and Functions (pp. 61–67)

State the domain and range of each relation. Then determine whether each relation is a *function*. If it is a function, determine if it is *one-to-one, onto, both,* or *neither.*

1.

Year	Population
1970	11,605
1980	13,468
1990	15,630
2000	18,140

2.

x	y
1	5
2	5
3	5
4	5

3.

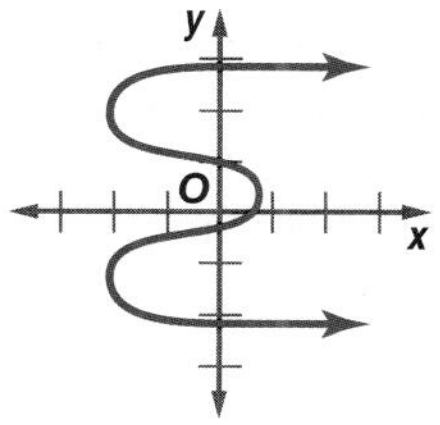

Graph each equation and determine the domain and range. Determine whether the equation is a *function,* is *one-to-one, onto, both,* or *neither.* Then state whether it is *discrete* or *continuous.*

4. $\{(1, 2), (2, 3), (3, 4), (4, 5)\}$

5. $\{(0, 3), (0, 2), (0, 1), (0, 0)\}$

6. $y = -x$

7. $y = 2x - 1$

8. $y = 2x^2$

9. $y = -x^2$

Find each value if $f(x) = x + 7$ and $g(x) = (x + 1)^2$.

10. $f(2)$

11. $g(-2)$

12. $f(a + 2)$

13. $g(b - 1)$

Lesson 2-2 Linear Relations and Functions (pp. 69–74)

State whether each equation or function is a linear function. Write *yes* or *no*. Explain your reasoning.

1. $\frac{x}{2} - y = 7$

2. $\sqrt{x} = y + 5$

3. $g(x) = \frac{2}{x - 3}$

4. $f(x) = 7$

Write each equation in standard form. Identify *A, B,* and *C.*

5. $x + 7 = y$

6. $x = -3y$

7. $5x = 7y + 3$

8. $-0.4x = 10$

Find the x-intercept and the y-intercept of the graph of each equation. Then graph the equation using intercepts.

9. $2x + y = 6$

10. $3x - 2y = -12$

11. $y = -x$

12. $y = -3$

Lesson 2-3 Rate of Change and Slope (pp. 76–82)

Find the slope of the line that passes through each pair of points.

1. $(0, 3), (5, 0)$

2. $(2, 8), (2, -8)$

3. $(1.5, -1), (3, 1.5)$

4. $(-3, c), (4, c)$

Determine the rate of change of each graph.

7.

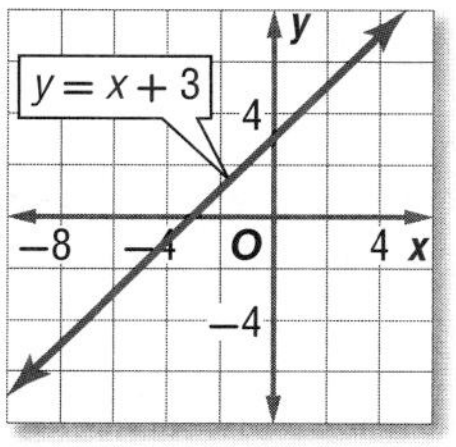

8.

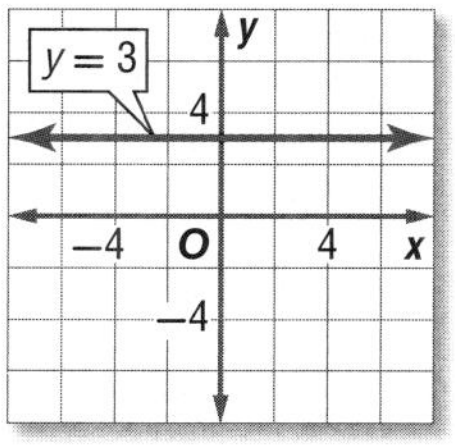

Find the rate of change for each set of data.

9.

Time (sec)	5	10	15	20
Distance (m)	15	30	45	60

10.

Cost ($)	10	20	30	40
Weight (lb)	2	4	6	8

Lesson 2-4 Writing Linear Equations (pp. 83–89)

Write an equation of each line.

1.

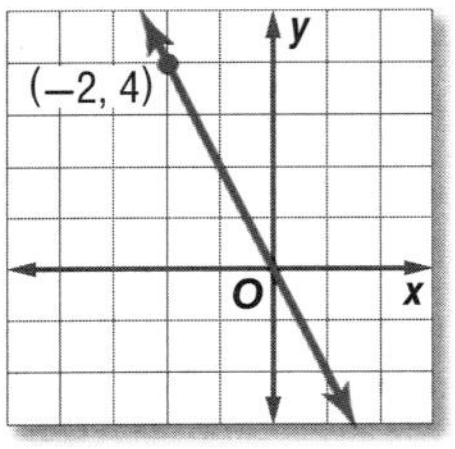

2.

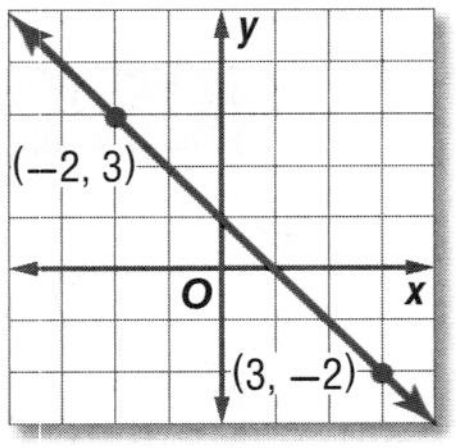

Write an equation in slope-intercept form for the line that satisfies each set of conditions.

3. slope -1, passes through $(7, 2)$

4. slope $\frac{3}{4}$, passes through the origin

5. passes through $(1, -3)$ and $(-1, 2)$

6. x-intercept -5, y-intercept 2

7. passes through $(1, 1)$, parallel to the graph of $2x + 3y = 5$

8. passes through $(0, 0)$, perpendicular to the graph of $2y + 3x = 4$

Lesson 2-5 Scatter Plots and Lines of Regression (pp. 92–98)

Complete parts a–c for each set of data in Exercises 1–3.

a. **Make a scatter plot and a line of fit, and describe the correlation.**

b. **Use two ordered pairs to write a prediction equation.**

c. **Use your prediction equation to predict the missing value.**

1. Telephone Costs

Minutes	Cost ($)
1	0.20
3	0.52
4	0.68
6	1.00
9	1.48
15	?

2. Washington

Year	Population
1960	2,853,214
1970	3,413,244
1980	4,132,353
1990	4,866,669
2000	5,894,121
2010	?

Source: *The World Almanac*

3. Federal Minimum Wage

Year	Wage
1981	$3.35
1990	$3.80
1991	$4.25
1996	$4.75
1997	$5.15
2015	?

Source: *The World Almanac*

Lesson 2-6 Special Functions (pp. 101–107)

Write the piecewise function shown in each graph.

1.

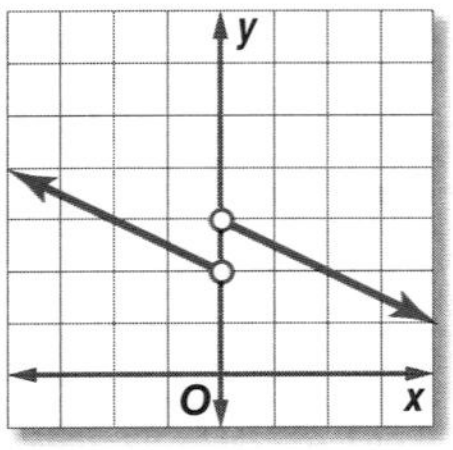

2.

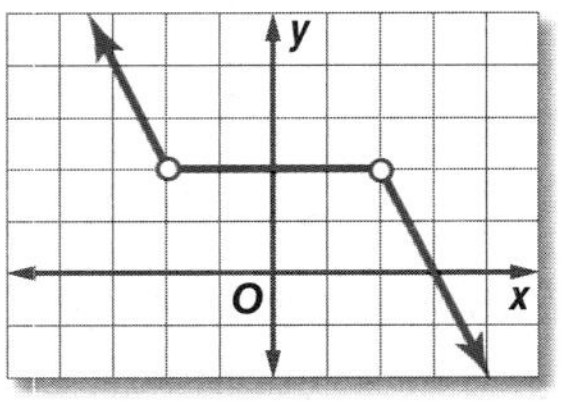

Graph each function. Identify the domain and range.

3. $f(x) = [\![x + 5]\!]$

4. $g(x) = [\![x]\!] - 2$

5. $f(x) = -2[\![x]\!]$

6. $h(x) = |x| - 3$

7. $h(x) = |x - 1|$

8. $g(x) = |2x| + 2$

9. $h(x) = \begin{cases} x \text{ if } x < -2 \\ 4 \text{ if } x \geq -2 \end{cases}$

10. $f(x) = \begin{cases} -3 \text{ if } x \leq 1 \\ -x \text{ if } x > 1 \end{cases}$

Lesson 2-7 Parent Functions and Transformations (pp. 109–116)

Identify the type of function represented by each graph.

1.

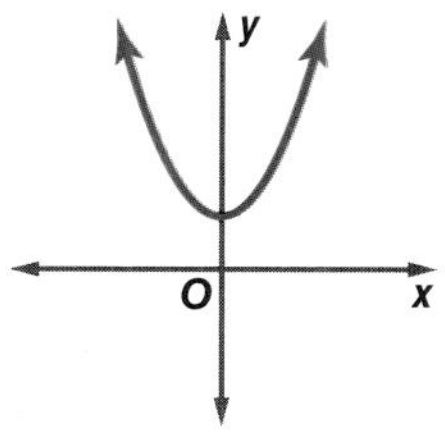

2.

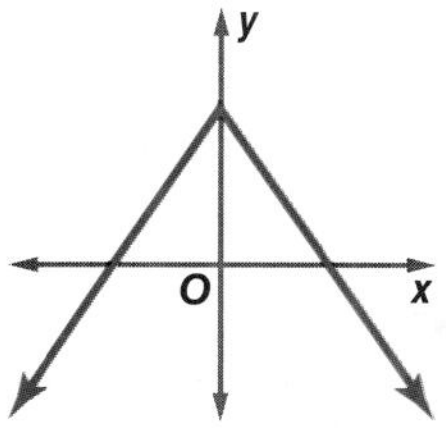

3.

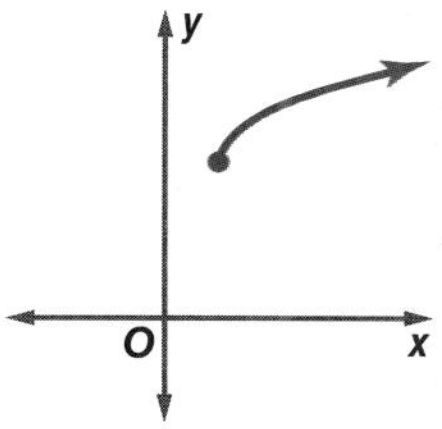

Describe the translation of each function. Then graph the function.

4. $y = |x| + 3$

5. $y = x^2 - 2$

Describe the reflection in each function. Then graph the function.

6. $y = -x^2$

7. $y = -x$

Describe the dilation in each function. Then graph the function.

8. $y = \frac{3}{4}x$

9. $y = \frac{2}{x}$

Lesson 2-8 Graphing Linear and Absolute Value Inequalities (pp. 117–121)

Graph each inequality.

1. $y \geq x - 2$

2. $y < -3x - 1$

3. $4y \leq -3x + 8$

4. $3x > y$

5. $x + 2 \geq y - 7$

6. $2x < 5 - y$

7. $y > \frac{1}{5}x - 8$

8. $2y - 5x \leq 8$

9. $-2x + 5 \leq \frac{2}{3}y$

10. $3x + 2y \geq 0$

11. $x \leq 2$

12. $\frac{y}{2} \leq x - 1$

13. $y - 3 < 5$

14. $y \geq -|x|$

15. $|x| \leq y + 3$

16. $y > |5x - 3|$

17. $y \leq |8 - x|$

18. $y < |x + 3| - 1$

19. $y + |2x| \geq 4$

20. $y \geq |2x - 1| + 5$

21. $y < \left|\frac{2x}{3}\right| - 1$

Lesson 3-1 Solving Systems of Equations by Graphing (pp. 135–141)

Solve each system of equations by graphing or by using a table.

1. $x + 3y = 18$
$-x + 2y = 7$

2. $x - y = 2$
$2x - 2y = 10$

3. $2x + 6y = 6$
$\frac{1}{3}x + y = 1$

4. $x + 3y = 0$
$2x + 6y = 5$

5. $2x - y = 7$
$\frac{2}{5}x - \frac{4}{3}y = -2$

6. $y = \frac{1}{3}x + 1$
$y = 4x + 1$

Graph each system of equations and describe it as *consistent and independent, consistent and dependent,* or *inconsistent.*

7. $2x + 3y = 5$
$-6x - 9y = -15$

8. $x - 2y = 4$
$y = x - 2$

9. $y = 0.5x$
$2y = x + 4$

10. $9x - 5 = 7y$
$4.5x - 3.5y = 2.5$

11. $\frac{3}{4}x - y = 0$
$\frac{1}{3}y + \frac{1}{2}x = 6$

12. $\frac{2}{3}x = \frac{5}{3}y$
$2x - 5y = 0$

Lesson 3-2 Solving Systems of Equations Algebraically (pp. 143–150)

Solve each system of equations by using substitution.

1. $2x + 3y = 10$
$x + 6y = 32$

2. $x = 4y - 10$
$5x + 3y = -4$

3. $3x - 4y = -27$
$2x + y = -7$

Solve each system of equations by using elimination.

4. $7x + y = 9$
$5x - y = 15$

5. $r + 5t = -17$
$2r - 6t = -2$

6. $6n + 8q = 20$
$5n - 4q = -26$

Solve each system of equations by using either substitution or elimination.

7. $2x - 3y = 7$
$3x + 6y = 42$

8. $2a + 5b = -13$
$3a - 4b = 38$

9. $3c + 4d = -1$
$6c - 2d = 3$

10. $7x - y = 35$
$y = 5x - 19$

11. $3m + 4n = 28$
$5m - 3n = -21$

12. $x = 2y - 1$
$4x - 3y = 21$

13. $2.5x + 1.5y = -2$
$3.5x - 0.5y = 18$

14. $\frac{5}{2}x + \frac{1}{3}y = 13$
$\frac{1}{2}x - y = -7$

15. $\frac{2}{7}c - \frac{4}{3}d = 16$
$\frac{4}{7}c + \frac{8}{3}d = -16$

Lesson 3-3 Solving Systems of Inequalities by Graphing (pp. 151–157)

Solve each system of inequalities by graphing.

1. $x \leq 5$
$y \geq -3$

2. $y < 3$
$y - x \geq -1$

3. $x + y < 5$
$x < 2$

4. $y + x < 2$
$y \geq x$

5. $x + y \leq 2$
$y - x \leq 4$

6. $y \leq x + 4$
$y - x \geq 1$

7. $y < \frac{1}{3}x + 5$
$y > 2x + 1$

8. $y + x \geq 1$
$y - x \geq -1$

9. $|x| > 2$
$|y| \leq 5$

10. $|x - 3| \leq 3$
$4y - 2x \leq 6$

11. $4x + 3y \geq 12$
$2y - x \geq -1$

12. $y \leq -1$
$3x - 2y \geq 6$

Find the coordinates of the vertices of the triangle formed by each system of inequalities.

13. $y \leq 3$
$x \leq 2$
$y \geq -\frac{3}{2}x + 3$

14. $y \geq -1$
$y \leq x$
$y \leq -x + 4$

15. $y \leq \frac{1}{3}x + \frac{7}{3}$
$4x - y \leq 5$
$y \geq -\frac{3}{2}x + \frac{1}{2}$

Lesson 3-4 Optimization with Linear Programming (pp. 160–166)

Graph each system of inequalities. Name the coordinates of the vertices of the feasible region. Find the maximum and minimum values of the given function for this region.

1. $4x - 5y \leq -10$
$y \leq 6$
$2x + y \geq 2$
$f(x, y) = x + y$

2. $x \leq 5$
$y \geq 2$
$2x - 5y \geq -10$
$f(x, y) = 3x + y$

3. $x - 2y \geq -7$
$x + y \leq 8$
$y \geq 5x + 8$
$f(x, y) = 3x - 4y$

4. $y \leq 4x + 6$
$x + 4y \geq 7$
$2x + y \leq 7$
$f(x, y) = 2x - y$

5. $y \geq 0$
$y \leq 5$
$y \leq -x + 7$
$5x + 3y \geq 20$
$f(x, y) = x + 2y$

6. $y \geq 0$
$3x - 2y \geq 0$
$x + 3y \leq 11$
$2x + 3y \leq 16$
$f(x, y) = 4x + y$

Lesson 3-5 Systems of Equations in Three Variables (pp. 167–173)

Solve each system of equations.

1. $4x + 2y - 6z = -38$
$5x - 4y + z = -18$
$x + 3y + 7z = 38$

2. $u + 3v + w = 14$
$2u - v + 3w = -9$
$4u - 5v - 2w = -2$

3. $x + y = -6$
$x + z = -2$
$y + z = 2$

4. $5a = 5$
$6b - 3c = 15$
$2a + 7c = -5$

5. $w + 2t = 5$
$7r - 3w + t = 20$
$2t = 8$

6. $2u - 3v = 13$
$3v + w = -3$
$4u - w = 2$

7. $4a + 2b - c = 5$
$2a + b - 5c = -11$
$a - 2b + 3c = 6$

8. $x + 2y - z = 1$
$x + 3y + 2z = 7$
$2x + 6y + z = 8$

9. $2x + y - z = 7$
$3x - y + 2z = 15$
$x - 4y + z = 2$

Lesson 4-1 Introduction to Matrices (pp. 185–191)

State the dimensions of each matrix.

1. $\begin{bmatrix} 0 & 3 & -3 & 1 \\ -1 & 4 & 5 & 0 \end{bmatrix}$

2. $\begin{bmatrix} 0 \\ -1 \\ 3 \\ -5 \end{bmatrix}$

3. $\begin{bmatrix} -3 & 2 \\ 1 & 7 \\ 19 & 11 \end{bmatrix}$

Identify each element of matrix $A = \begin{bmatrix} \mathbf{0} & \mathbf{-6} & \mathbf{2} & \mathbf{11} \\ \mathbf{2} & \mathbf{3} & \mathbf{0} & \mathbf{9} \\ \mathbf{7} & \mathbf{4} & \mathbf{3} & \mathbf{12} \end{bmatrix}$.

4. a_{32}

5. a_{11}

6. a_{33}

7. a_{24}

Lesson 4-2 Operations with Matrices (pp. 193–199)

Perform the indicated matrix operations. If the matrix does not exist, write *impossible*.

1. $\begin{bmatrix} 3 & 5 \\ -7 & 2 \end{bmatrix} + \begin{bmatrix} -2 & 6 \\ 8 & -1 \end{bmatrix}$

2. $\begin{bmatrix} 0 & -1 & 3 \end{bmatrix} + \begin{bmatrix} 5 \\ -2 \\ -3 \end{bmatrix}$

3. $\begin{bmatrix} 45 & 36 & 18 \\ 63 & 29 & 5 \end{bmatrix} - \begin{bmatrix} 45 & -2 & 36 \\ 18 & 9 & -10 \end{bmatrix}$

4. $4\begin{bmatrix} -8 & 2 & 9 \end{bmatrix} - 3\begin{bmatrix} 2 & -7 & 6 \end{bmatrix}$

5. $5\begin{bmatrix} 6 & -2 \\ 5 & 4 \end{bmatrix} - 2\begin{bmatrix} 6 & -2 \\ 5 & 4 \end{bmatrix} + 4\begin{bmatrix} 7 & -6 \\ -4 & 2 \end{bmatrix}$

6. $1.3\begin{bmatrix} 3.7 \\ -5.4 \end{bmatrix} + 4.1\begin{bmatrix} 6.4 \\ -3.7 \end{bmatrix} - 6.2\begin{bmatrix} -0.8 \\ 7.4 \end{bmatrix}$

Use matrices *A*, *B*, *C*, *D*, and *E* to find the following.

$A = \begin{bmatrix} 1 & 0 \\ 0 & 1 \end{bmatrix}, B = \begin{bmatrix} 1 & 0 \\ -0 & -1 \end{bmatrix}, C = \begin{bmatrix} 2 & -2 \\ -3 & 3 \end{bmatrix}, D = \begin{bmatrix} -2 & 2 \\ 3 & -3 \end{bmatrix}, E = \begin{bmatrix} 5 & -3 \\ -2 & 4 \end{bmatrix}$

7. $A + B$

8. $C + D$

9. $A - B$

10. $4B$

11. $D - C$

12. $E + 2A$

13. $D - 2B$

14. $2A + 3E - D$

Extra Practice

Lesson 4-3 Multiplying Matrices (pp. 200–207)

Find each product, if possible.

1. $\begin{bmatrix} -3 & 4 \end{bmatrix} \cdot \begin{bmatrix} -1 \\ 2 \end{bmatrix}$

2. $\begin{bmatrix} 2 & -4 \\ 0 & 5 \end{bmatrix} \cdot \begin{bmatrix} 1 & 3 \\ -2 & -1 \end{bmatrix}$

3. $\begin{bmatrix} 1 & 3 \\ -2 & -1 \end{bmatrix} \cdot \begin{bmatrix} 2 & -4 \\ 0 & 5 \end{bmatrix}$

4. $\begin{bmatrix} 3 & 2 \\ 5 & 2 \end{bmatrix} \cdot \begin{bmatrix} -8 \\ 15 \end{bmatrix}$

5. $\begin{bmatrix} -1 \\ 2 \\ 1 \end{bmatrix} \cdot \begin{bmatrix} 7 & 6 & 1 \\ 2 & -4 & 0 \end{bmatrix}$

6. $\begin{bmatrix} 0 & 1 & -2 \\ 5 & 3 & -4 \\ -1 & 0 & 0 \end{bmatrix} \cdot \begin{bmatrix} 1 & -3 & 0 \\ 2 & 0 & -1 \\ 0 & 1 & -2 \end{bmatrix}$

7. $\begin{bmatrix} 3 & -2 \\ 4 & 5 \end{bmatrix} \cdot \begin{bmatrix} 1 & 0 \\ 0 & 1 \end{bmatrix}$

8. $\begin{bmatrix} -1 & 0 & 2 \\ -6 & 5 & -3 \end{bmatrix} \cdot \begin{bmatrix} -2 \\ 1 \\ 7 \end{bmatrix}$

Lesson 4-4 Transformations with Matrices (pp. 209–217)

1. The vertices of quadrilateral *ABCD* are $A(1, 1)$, $B(-2, 3)$, $C(-4, -1)$, and $D(2, -3)$. The quadrilateral is dilated so that its perimeter is 2 times the original perimeter.

- **a.** Write the coordinates for *ABCD* in a vertex matrix.
- **b.** Find the coordinates of the image $A'B'C'D'$.
- **c.** Graph *ABCD* and $A'B'C'D'$.

2. The vertices of $\triangle MQN$ are $M(2, 4)$, $Q(3, -5)$, and $N(1, -1)$.

- **a.** Write the coordinates of $\triangle MQN$ in a vertex matrix.
- **b.** Write the reflection matrix for reflecting over the line $y = x$.
- **c.** Find the coordinates of $\triangle M'Q'N'$ after the reflection.
- **d.** Graph $\triangle MQN$ and $\triangle M'Q'N'$.
- **e.** Write a rotation matrix for rotating $\triangle MQN$ 90° counterclockwise about the origin.
- **f.** Find the coordinates of $\triangle M'Q'N'$ after the rotation.
- **g.** Graph $\triangle MQN$ and $\triangle M'Q'N'$.

Lesson 4-5 Determinants and Cramer's Rule (pp. 220–228)

Evaluate each determinant using diagonals.

1. $\begin{vmatrix} 2 & -3 & 5 \\ 1 & -2 & -7 \\ -1 & 4 & -3 \end{vmatrix}$

2. $\begin{vmatrix} 0 & -1 & 2 \\ -2 & 1 & 0 \\ 2 & 0 & -1 \end{vmatrix}$

3. $\begin{vmatrix} 4 & 3 & -2 \\ 2 & 5 & -8 \\ 6 & 4 & -1 \end{vmatrix}$

4. $\begin{vmatrix} -3 & 0 & 2 \\ 1 & -2 & -1 \\ 0 & 5 & 0 \end{vmatrix}$

5. $\begin{vmatrix} 3 & 2 & -1 \\ 2 & 3 & 0 \\ -1 & 0 & 3 \end{vmatrix}$

6. $\begin{vmatrix} 1 & 0 & 0 \\ 0 & 1 & 0 \\ 0 & 0 & 1 \end{vmatrix}$

7. $\begin{vmatrix} 6 & 4 & -1 \\ 2 & 5 & -8 \\ 4 & 3 & -2 \end{vmatrix}$

8. $\begin{vmatrix} 6 & 12 & 15 \\ 9 & 3 & 14 \\ 5 & 6 & 3 \end{vmatrix}$

Use Cramer's Rule to solve each system of equations.

9. $5x - y = 7$
$8x + 2y = 4$

10. $3m + t = 4$
$2m + 2t = 3$

11. $6c + 5d = 7$
$3c - 10d = -4$

12. $3a - 5b = 1$
$a + 3b = 5$

13. $2r - 7t = 24$
$-r + 8t = -21$

14. $x + y = -3$
$3x - 10y = 43$

15. $2m - 3t = 3$
$-4m + 9t = -8$

16. $x + y = 1$
$2x - 2y = -12$

Lesson 4-6 Inverse Matrices and Systems of Equations (pp. 229–235)

Determine whether each pair of matrices are inverses of each other.

1. $A = \begin{bmatrix} -7 & -6 \\ 8 & 7 \end{bmatrix}, B = \begin{bmatrix} -7 & -6 \\ 8 & 7 \end{bmatrix}$

2. $C = \begin{bmatrix} -3 & 4 \\ 2 & -2 \end{bmatrix}, D = \begin{bmatrix} -2 & -2 \\ -4 & -3 \end{bmatrix}$

Find the inverse of each matrix, if it exists.

3. $\begin{bmatrix} 2 & 4 \\ 2 & 3 \end{bmatrix}$

4. $\begin{bmatrix} 8 & -5 \\ -6 & 4 \end{bmatrix}$

5. $\begin{bmatrix} 10 & 3 \\ 5 & -2 \end{bmatrix}$

6. $\begin{bmatrix} -3 & 4 \\ -4 & 8 \end{bmatrix}$

Use a matrix equation to solve each system of equations.

7. $4c - 3d = -1$
$5c - 2d = 39$

8. $x + 2y - z = 6$
$-2x + 3y + z = 1$
$x + y + 3z = 8$

9. $2a - 3b - c = 4$
$4a + b + c = 15$
$a - b - c = -2$

Lesson 5-1 Graphing Quadratic Functions (pp. 249–257)

Complete parts a–c for each quadratic function.

a. Find the y-intercept, the equation of the axis of symmetry, and the x-coordinate of the vertex.

b. Make a table of values that includes the vertex.

c. Use this information to graph the function.

1. $f(x) = 6x^2$

2. $f(x) = -x^2$

3. $f(x) = x^2 + 5$

4. $f(x) = -x^2 - 2$

5. $f(x) = 2x^2 + 1$

6. $f(x) = -3x^2 + 6x$

7. $f(x) = x^2 + 6x - 3$

8. $f(x) = x^2 - 2x - 8$

9. $f(x) = -3x^2 - 6x + 12$

Determine whether each function has a *maximum* or a *minimum* value and find that value. Then state the domain and range of the function.

10. $f(x) = 9x^2$

11. $f(x) = 9 - x^2$

12. $f(x) = x^2 - 5x + 6$

13. $f(x) = 2 + 7x - 6x^2$

14. $f(x) = 4x^2 - 9$

15. $f(x) = x^2 + 2x + 1$

Lesson 5-2 Solving Quadratic Equations by Graphing (pp. 259–266)

Use the related graph of each equation to determine its solutions.

1. $x^2 + x - 6 = 0$

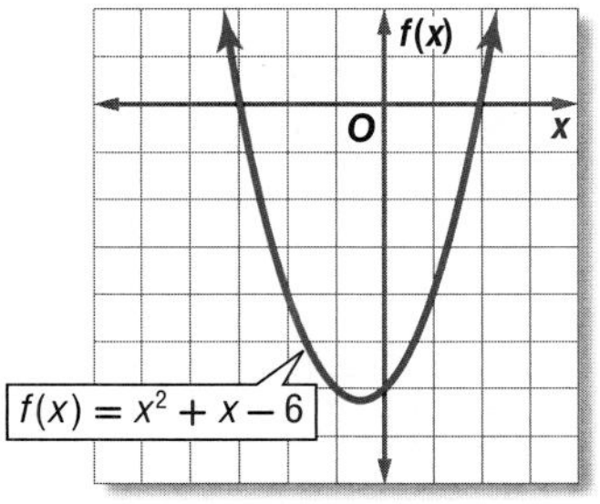

2. $-2x^2 = 0$

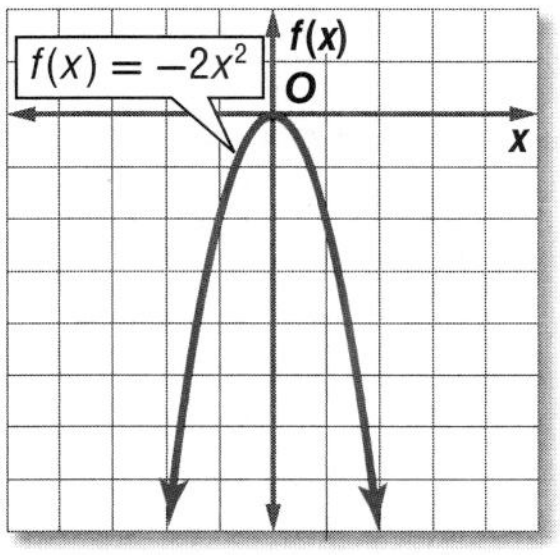

3. $x^2 - 4x - 5 = 0$

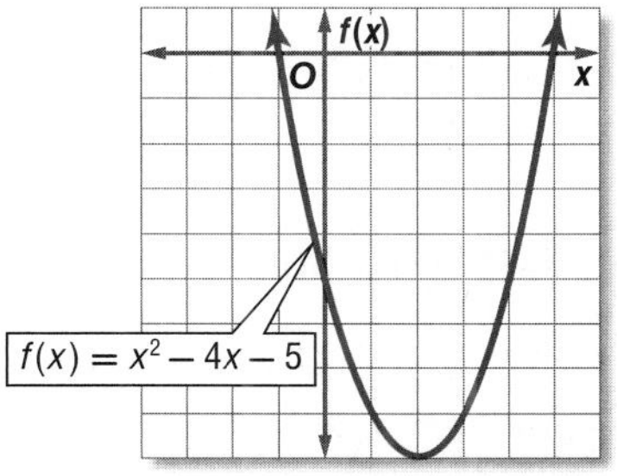

Solve each equation. If exact roots cannot be found, state the consecutive integers between which the roots are located.

4. $x^2 - 2x = 0$

5. $x^2 + 8x - 20 = 0$

6. $-2x^2 + 10x - 5 = 0$

7. $-5x + 2x^2 - 3 = 0$

8. $3x^2 - x + 8 = 0$

9. $-x^2 + 2 = 7x$

10. $4x^2 - 4x + 1 = 0$

11. $4x + 1 = 3x^2$

12. $x^2 = -9x$

Lesson 5-3 Solving Quadratic Equations by Factoring (pp. 268–275)

Solve each equation by factoring.

1. $x^2 + 7x + 10 = 0$
2. $3x^2 = 75x$
3. $2x^2 + 7x = 9$
4. $8x^2 = 48 - 40x$
5. $5x^2 = 20x$
6. $16x^2 - 64 = 0$
7. $24x^2 - 15 = 2x$
8. $x^2 = 72 - x$
9. $4x^2 + 9 = 12x$
10. $2x^2 - 8x = 0$
11. $8x^2 + 10x = 3$
12. $12x^2 - 5x = 3$
13. $x^2 + 9x + 14 = 0$
14. $9x^2 + 1 = 6x$
15. $6x^2 + 7x = 3$
16. $x^2 - 4x = 21$

Write a quadratic equation in standard form with the given roots.

17. $2, 1$
18. $-3, 4$
19. $-1, -7$
20. $-1, \frac{1}{2}$
21. $-5, \frac{1}{4}$
22. $-\frac{1}{3}, -\frac{1}{2}$

Lesson 5-4 Complex Numbers (pp. 276–282)

Simplify.

1. $\sqrt{-289}$
2. $\sqrt{-\frac{25}{121}}$
3. $\sqrt{-625b^8}$
4. $\sqrt{-\frac{28t^6}{27w^5}}$
5. $(7i)^2$
6. $(6i)(-2i)(11i)$
7. $(\sqrt{-8})(\sqrt{-12})$
8. $-i^{22}$
9. $i^{17} \cdot i^{12} \cdot i^{26}$
10. $(14 - 5i) + (-8 + 19i)$
11. $(7i) - (2 + 3i)$
12. $(2 + 2i) - (5 + i)$
13. $(7 + 3i)(7 - 3i)$
14. $(8 - 2i)(5 + i)$
15. $(6 + 8i)^2$
16. $\frac{3}{6 - 2i}$
17. $\frac{5i}{3 + 4i}$
18. $\frac{3 - 7i}{5 + 4i}$

Solve each equation.

19. $x^2 + 8 = 3$
20. $\frac{4x^2}{49} + 6 = 3$
21. $8x^2 + 5 = 1$
22. $12 - 9x^2 = 38$
23. $9x^2 + 7 = 4$
24. $\frac{1}{2}x^2 + 1 = 0$

Lesson 5-5 Completing the Square (pp. 284–290)

Find the value of *c* that makes each trinomial a perfect square. Then write the trinomial as a perfect square.

1. $x^2 - 4x + c$
2. $x^2 + 20x + c$
3. $x^2 - 11x + c$
4. $x^2 - \frac{2}{3}x + c$
5. $x^2 + 30x + c$
6. $x^2 + \frac{3}{8}x + c$
7. $x^2 - \frac{2}{5}x + c$
8. $x^2 - 3x + c$

Solve each equation by completing the square.

9. $x^2 + 3x - 4 = 0$
10. $x^2 + 5x = 0$
11. $x^2 + 2x - 63 = 0$
12. $3x^2 - 16x - 35 = 0$
13. $x^2 + 7x + 13 = 0$
14. $5x^2 - 8x + 2 = 0$
15. $x^2 - 6x + 11 = 0$
16. $x^2 - 12x + 36 = 0$
17. $8x^2 + 13x - 4 = 0$
18. $3x^2 + 5x + 6 = 0$
19. $x^2 + 14x - 1 = 0$
20. $4x^2 - 32x + 15 = 0$
21. $3x^2 - 11x - 4 = 0$
22. $x^2 + 8x - 84 = 0$
23. $x^2 - 7x + 5 = 0$
24. $x^2 + 3x - 8 = 0$
25. $x^2 - 5x - 10 = 0$
26. $3x^2 - 12x + 4 = 0$
27. $x^2 + 20x + 75 = 0$
28. $x^2 - 5x - 24 = 0$
29. $2x^2 + x - 21 = 0$

Lesson 5-6 The Quadratic Formula and the Discriminant (pp. 292–300)

Complete parts a–c for each quadratic equation.

a. **Find the value of the discriminant.**

b. **Describe the number and type of roots.**

c. **Find the exact solutions by using the Quadratic Formula.**

1. $x^2 + 7x + 13 = 0$
2. $6x^2 + 6x - 21 = 0$
3. $5x^2 - 5x + 4 = 0$
4. $9x^2 + 42x + 49 = 0$
5. $4x^2 - 16x + 3 = 0$
6. $2x^2 = 5x + 3$
7. $x^2 + 81 = 18x$
8. $3x^2 - 30x + 75 = 0$
9. $24x^2 + 10x = 43$
10. $9x^2 + 4 = 2x$
11. $7x = 8x^2$
12. $18x^2 = 9x + 45$
13. $x^2 - 4x + 4 = 0$
14. $4x^2 + 16x + 15 = 0$
15. $x^2 - 6x + 13 = 0$

Solve each equation by using the Quadratic Formula.

16. $x^2 + 4x + 29 = 0$
17. $4x^2 + 3x - 2 = 0$
18. $2x^2 + 5x = 9$
19. $x^2 = 8x - 16$
20. $7x^2 = 4x$
21. $2x^2 + 6x + 5 = 0$
22. $9x^2 - 30x + 25 = 0$
23. $3x^2 - 4x + 2 = 0$
24. $3x^2 = 108x$

Extra Practice

Lesson 5-7 Transformations with Quadratic Functions (pp. 305–310)

Write each quadratic function in vertex form. Then identify the vertex, axis of symmetry, and direction of opening.

1. $y = (x + 6)^2 - 1$
2. $y = 2(x - 8)^2 - 5$
3. $y = -(x + 1)^2 + 7$
4. $y = -9(x - 7)^2 + 3$
5. $y = -x^2 + 10x - 3$
6. $y = -2x^2 + 16x + 7$

Graph each function.

7. $y = x^2 - 2x + 4$
8. $y = -3x^2 + 18x$
9. $y = -2x^2 - 4x + 1$
10. $y = 2x^2 - 8x + 9$
11. $y = \frac{1}{3}x^2 + 2x + 7$
12. $y = x^2 + 6x + 9$
13. $y = x^2 + 3x + 6$
14. $y = -0.5x^2 + 4x - 3$
15. $y = -2x^2 - 8x - 1$

Lesson 5-8 Quadratic Inequalities (pp. 312–318)

Graph each inequality.

1. $y \leq 5x^2 + 3x - 2$
2. $y > -3x^2 + 2$
3. $y \geq x^2 - 8x$
4. $y \geq -x^2 - x + 3$
5. $y \leq 3x^2 + 4x - 8$
6. $y \leq -5x^2 + 2x - 3$
7. $y > 4x^2 + x$
8. $y \geq -x^2 - 3$

Solve each inequality by graphing.

9. $x^2 - 4 \leq 0$
10. $-x^2 + 6x - 9 \geq 0$
11. $x^2 + 4x - 5 < 0$

Solve each inequality algebraically.

12. $x^2 - 1 < 0$
13. $10x^2 - x - 2 \geq 0$
14. $-x^2 - 5x - 6 > 0$
15. $-3x^2 \geq 5$
16. $x^2 - 2x - 8 \leq 0$
17. $2x^2 \geq 5x + 12$
18. $x^2 + 3x - 4 > 0$
19. $2x - x^2 \leq -15$

Lesson 6-1 Operations with Polynomials (pp. 333–339)

Simplify. Assume that no variable equals 0.

1. $x^7 \cdot x^3 \cdot x$
2. $m^8 \cdot m \cdot m^{10}$
3. $7^5 \cdot 7^2$
4. $(-3)^4(-3)$
5. $\frac{t^{12}}{t}$
6. $-\frac{16x^8}{8x^2}$
7. $\frac{6^5}{6^3}$
8. $\frac{p^5t^7}{p^2t^5}$
9. $-(m^3)^8$
10. $(3^5)^7$
11. -3^4
12. $(abc)^3$
13. $(5x^4)^{-2}$
14. $(-3)^{-2}$
15. -3^{-2}
16. $\frac{x}{x^7}$
17. $-\left(\frac{x}{5}\right)^2$
18. $\left(\frac{5a^7}{2b^5c}\right)^3$
19. $\frac{1}{x^{-3}}$
20. $\frac{5^6a^{x+y}}{5^4a^{x-y}}$

Simplify.

21. $(4x^3 + 5x - 7x^2) + (-2x^3 + 5x^2 - 7y^2)$
22. $(2x^2 - 3x + 11) + (7x^2 + 2x - 8)$
23. $(-3x^2 + 7x + 23) + (-8x^2 - 5x + 13)$
24. $(-3x^2 + 7x + 23) - (-8x^2 - 5x + 13)$
25. $\frac{7}{uw}\left(4u^2w^3 - 5uw + \frac{w}{7u}\right)$
26. $-4x^5(-3x^4 - x^3 + x + 7)$
27. $(2x - 3)(4x + 7)$
28. $(3x - 5)(-2x - 1)$
29. $(3x - 5)(2x - 1)$
30. $(2x + 5)(2x - 5)$
31. $(-5x + 10)(-5x - 10)$
32. $(4x - 3)^2$
33. $(5x + 6)^2$
34. $(-x + 1)^2$
35. $\frac{3}{4}x(x^2 + 4x + 14)$
36. $-\frac{1}{2}a^2(a^3 - 6a^2 + 5a)$

Lesson 6-2 Dividing Polynomials (pp. 341–347)

Simplify.

1. $\frac{18r^3t^2 + 36r^2t^3}{9r^2t^2}$
2. $\frac{15v^3w^2 - 5v^4w^3}{-5v^4w^3}$
3. $\frac{x^2 - x + 1}{x}$
4. $(5bh + 5ch) \div (b + c)$
5. $(25c^4d + 10c^3d^2 - cd) \div 5cd$
6. $(16f^{18} + 20f^9 - 8f^6) \div 4f^3$
7. $(33m^5 + 55mp^5 - 11m^3)(11m)^{-1}$
8. $(8g^3 + 19g^2 - 12g + 9) \div (g + 3)$
9. $(p^{21} + 3p^{14} + p^7 - 2)(p^7 + 2)^{-1}$
10. $(q^4 + 8q^3 + 3q + 17) \div (q + 8)$
11. $(15v^3 + 8v^2 - 21v + 6) \div (5v - 4)$
12. $(-2x^3 + 15x^2 - 10x + 3) \div (x + 3)$
13. $(5k^3 + k^2 - 7) \div (k + 1)$
14. $(t^4 - 2t^3 + t^2 - 3t + 2) \div (t - 2)$
15. $(z^4 - 3z^3 - z^2 - 11z - 4) \div (z - 4)$
16. $(3r^4 - 6r^3 - 2r^2 + r - 6) \div (r + 1)$
17. $(2b^3 - 11b^2 + 12b + 9) \div (b - 3)$

Lesson 6-3 Polynomial Functions (pp. 348–355)

Find $p(5)$ and $p(-1)$ for each function.

1. $p(x) = 7x - 3$
2. $p(x) = -3x^2 + 5x - 4$
3. $p(x) = 5x^4 + 2x^2 - 2x$
4. $p(x) = -13x^3 + 5x^2$
5. $p(x) = x^6 - 2$
6. $p(x) = \frac{2}{3}x^2 + 5x$
7. $p(x) = x^3 + x^2 - x + 1$
8. $p(x) = x^4 - x^2 - 1$
9. $p(x) = 1 - x^3$

If $p(x) = -2x^2 + 5x + 1$ and $m(x) = x^3 - 1$, find each value.

10. $m(n)$
11. $p(2b)$
12. $m(z^3)$
13. $p(3m^2)$
14. $m(x + 1)$
15. $p(3 - x)$
16. $m(a^2 - 2)$
17. $3m(h - 3)$
18. $5[p(c - 4)]$
19. $m(n - 2) + m(n^2)$
20. $-3p(4a) - p(a)$
21. $2[m(d^2 + 1)] + 3m(d)$

Lesson 6-4 Analyzing Graphs of Polynomial Functions (pp. 357–364)

Complete each of the following.

a. Graph each function by making a table of values.

b. Determine the consecutive integer values of x between which each real zero is located.

c. Estimate the x-coordinates at which the relative maxima and relative minima occur.

1. $f(x) = x^3 + x^2 - 3x$

2. $f(x) = -x^4 + x^3 + 5$

3. $f(x) = x^3 - 3x^2 + 8x - 7$

4. $f(x) = 2x^5 + 3x^4 - 8x^2 + x + 4$

5. $f(x) = x^4 - 5x^3 + 6x^2 - x - 2$

6. $f(x) = 2x^6 + 5x^4 - 3x^2 - 5$

7. $f(x) = -x^3 - 8x^2 + 3x - 7$

8. $f(x) = -x^4 - 3x^3 + 5x$

9. $f(x) = x^5 - 7x^4 - 3x^3 + 2x^2 - 4x + 9$

10. $f(x) = x^4 - 5x^3 + x^2 - x - 3$

11. $f(x) = x^4 - 128x^2 + 960$

12. $f(x) = -x^5 + x^4 - 208x^2 + 145x + 9$

13. $f(x) = x^5 - x^3 - x + 1$

14. $f(x) = x^3 - 2x^2 - x + 5$

15. $f(x) = 2x^4 - x^3 + x^2 - x + 1$

16. $f(x) = -x^3 - x^2 - x - 1$

Lesson 6-5 Solving Polynomial Equations (pp. 368–375)

Factor completely. If the polynomial is not factorable, write *prime*.

1. $14a^3b^3c - 21a^2b^4c + 7a^2b^3c$

2. $10ax - 2xy - 15ab + 3by$

3. $x^2 + x - 42$

4. $2x^2 + 5x + 3$

5. $6x^2 + 71x - 12$

6. $6x^4 - 12x^3 + 3x^2$

7. $x^2 - 6x + 2$

8. $x^2 - 2x - 15$

9. $6x^2 + 23x + 20$

10. $24x^2 - 76x + 40$

11. $6p^2 - 13pt - 28t^2$

12. $2x^2 - 6x + 3$

13. $x^2 + 49 - 14x$

14. $9x^2 - 64$

15. $36 - t^{10}$

16. $x^2 + 16$

17. $a^4 - 81b^4$

18. $3a^3 + 12a^2 - 63a$

19. $x^3 - 8x^2 + 15x$

20. $x^2 + 6x + 9$

21. $18x^3 - 8x$

22. $3x^2 - 42x + 40$

23. $2x^2 + 4x - 1$

24. $2x^3 + 6x^2 + x + 3$

25. $35ac - 3bf - 7af + 15bc$

26. $5h^2 - 10hj + h - 2j$

Lesson 6-6 The Remainder and Factor Theorems (pp. 377–382)

Use synthetic substitution to find $f(3)$ and $f(-4)$ for each function.

1. $f(x) = x^2 - 6x + 2$

2. $f(x) = x^3 + 5x - 6$

3. $f(x) = x^3 - x^2 - 3x + 1$

4. $f(x) = -3x^3 + 5x^2 + 7x - 3$

5. $f(x) = 3x^5 - 5x^3 + 2x - 8$

6. $f(x) = 10x^3 + 2$

Given a polynomial and one of its factors, find the remaining factors of the polynomial.

7. $(x^3 - x^2 + x + 14); (x + 2)$

8. $(5x^3 - 17x^2 + 6x); (x - 3)$

9. $(2x^3 + x^2 - 41x + 20); (x - 4)$

10. $(x^3 - 8); (x - 2)$

11. $(x^2 + 6x + 5); (x + 1)$

12. $(x^4 + x^3 + x^2 + x); (x + 1)$

13. $(x^3 - 8x^2 + x + 42); (x - 7)$

14. $(x^4 + 5x^3 - 27x - 135); (x - 3)$

15. $(2x^3 - 15x^2 - 2x + 120); (2x + 5)$

16. $(6x^3 - 17x^2 + 6x + 8); (3x - 4)$

17. $(10x^3 + x^2 - 46x + 35); (5x - 7)$

18. $(x^3 + 9x^2 + 23x + 15); (x + 1)$

Lesson 6-7 Roots and Zeros (pp. 383–390)

Solve each equation. State the number and type of roots.

1. $-5x - 7 = 0$
2. $3x^2 + 10 = 0$
3. $x^4 - 2x^3 = 23x^2 - 60x$

State the number of positive real zeros, negative real zeros, and imaginary zeros of each function.

4. $f(x) = 5x^8 - x^6 + 7x^4 - 8x^2 - 3$
5. $f(x) = 6x^5 - 7x^2 + 5$
6. $f(x) = -2x^6 - 5x^5 + 8x^2 - 3x + 1$
7. $f(x) = 4x^3 + x^2 - 38x + 56$
8. $f(x) = 3x^4 - 5x^3 + 2x^2 - 7x + 5$
9. $f(x) = x^5 - x^4 + 7x^3 - 25x^2 + 8x - 13$

Find all zeros of each function.

10. $f(x) = x^3 - 7x^2 + 16x - 10$
11. $f(x) = 10x^3 + 7x^2 - 82x + 56$
12. $f(x) = x^3 - 16x^2 + 79x - 114$
13. $f(x) = -3x^3 + 6x^2 + 5x - 8$
14. $f(x) = 24x^3 + 64x^2 + 6x - 10$
15. $f(x) = 2x^3 + 2x^2 - 34x + 30$

Lesson 6-8 Rational Zero Theorem (pp. 391–396)

List all of the possible rational zeros of each function.

1. $f(x) = 3x^5 - 7x^3 - 8x + 6$
2. $f(x) = 4x^3 + 2x^2 - 5x + 8$
3. $f(x) = 6x^9 - 7$

Find all of the rational zeros of each function.

4. $f(x) = x^4 + 3x^3 - 7x^2 - 27x - 18$
5. $f(x) = 6x^4 - 31x^3 - 119x^2 + 214x + 560$
6. $f(x) = 20x^4 - 16x^3 + 11x^2 - 12x - 3$
7. $f(x) = 2x^4 - 30x^3 + 117x^2 - 75x + 280$
8. $f(x) = 3x^4 + 8x^3 + 9x^2 + 32x - 12$
9. $f(x) = x^5 - x^4 + x^3 + 3x^2 - x$

Find all of the zeros of each function.

10. $f(x) = x^4 + 8x^2 - 9$
11. $f(x) = 3x^4 - 9x^2 - 12$
12. $f(x) = 4x^4 + 19x^2 - 63$

Lesson 7-1 Operations on Functions (pp. 409–416)

Find $(f + g)(x)$, $(f - g)(x)$, $(f \cdot g)(x)$, and $\left(\frac{f}{g}\right)(x)$ for each $f(x)$ and $g(x)$.

1. $f(x) = 3x + 5$, $g(x) = x - 3$
2. $f(x) = \sqrt{x}$, $g(x) = x^2$
3. $f(x) = x^2 - 5$, $g(x) = x^2 + 5$
4. $f(x) = x^2 + 1$, $g(x) = x + 1$

For each pair of functions, find $f \circ g$ and $g \circ f$, if they exist.

5. $f = \{(-1, 1), (2, -1), (-3, 5)\}$, $g = \{(1, -1), (-1, 2), (5, -3)\}$
6. $f = \{(0, 6), (5, -8), (-9, 2)\}$, $g = \{(-8, 3), (6, 4), (2, 1)\}$
7. $f = \{(8, 2), (6, 5), (-3, 4), (1, 0)\}$, $g = \{(2, 8), (5, 6), (4, -3), (0, 1)\}$
8. $f = \{(10, 4), (-1, 2), (5, 6), (-1, 0)\}$, $g = \{(-4, 10), (2, -9), (-7, 5), (-2, -1)\}$

Find $[g \circ h](x)$ and $[h \circ g](x)$, if they exist.

9. $g(x) = 8 - 2x$, $h(x) = 3x$
10. $g(x) = x^2 - 7$, $h(x) = 3x + 2$
11. $g(x) = 2x + 7$, $h(x) = \frac{x - 7}{2}$
12. $g(x) = 3x + 2$, $h(x) = 5 - 3x$

Lesson 7-2 Inverse Functions and Relations (pp. 417–422)

Find the inverse of each relation.

1. $\{(-2, 7), (3, 0), (5, -8)\}$
2. $\{(-3, 9), (-2, 4), (3, 9), (-1, 1)\}$

Find the inverse of each function. Then graph the function and its inverse.

3. $f(x) = x - 7$
4. $y = 2x + 8$
5. $g(x) = 3x - 8$
6. $y = -5x - 6$
7. $y = -2$
8. $g(x) = 5 - 2x$
9. $h(x) = \frac{x}{5} + 1$
10. $h(x) = -\frac{2}{3}x$
11. $y = \frac{x - 5}{3}$
12. $y = \frac{1}{2}x - 1$
13. $f(x) = \frac{3x + 8}{4}$
14. $g(x) = \frac{2x - 1}{3}$

Determine whether each pair of functions are inverse functions. Write *yes* or *no*.

15. $f(x) = \frac{2x - 3}{5}$, $g(x) = \frac{3x - 5}{3}$
16. $f(x) = 5x - 6$, $g(x) = \frac{x + 6}{5}$
17. $f(x) = 6 - 3x$, $g(x) = 2 - \frac{1}{3}x$
18. $f(x) = 3x - 7$, $g(x) = \frac{1}{3}x + 7$

Lesson 7-3 Square Root Functions and Inequalities (pp. 424–430)

Graph each function. State the domain and range.

1. $y = \sqrt{x - 4}$
2. $y = \sqrt{x + 3} - 1$
3. $y = \frac{1}{3}\sqrt{x + 2}$
4. $y = \sqrt{2x + 5}$
5. $y = -\sqrt{4x}$
6. $y = 2\sqrt{x}$
7. $y = -3\sqrt{x}$
8. $y = \sqrt{x} + 5$
9. $y = \sqrt{2x} - 1$
10. $y = 5\sqrt{x} + 1$
11. $y = \sqrt{x + 1} - 2$
12. $y = 6 - \sqrt{x + 3}$

Graph each inequality.

13. $y > \sqrt{2x}$
14. $y \leq \sqrt{-5x}$
15. $y \geq \sqrt{x + 6} + 6$
16. $y < \sqrt{3x + 1} + 2$
17. $y \geq \sqrt{8x - 3} + 1$
18. $y < \sqrt{5x - 1} + 3$

Lesson 7-4 *n*th Roots (pp. 431–436)

Use a calculator to approximate each value to three decimal places.

1. $\sqrt{289}$
2. $\sqrt{7832}$
3. $\sqrt[4]{0.0625}$
4. $\sqrt[3]{-343}$
5. $\sqrt[10]{32^4}$
6. $\sqrt[3]{49}$
7. $\sqrt[5]{5}$
8. $-\sqrt[4]{25}$

Simplify.

9. $\sqrt{9h^{22}}$
10. $\sqrt[5]{0}$
11. $\sqrt{\frac{16}{9}}$
12. $\sqrt{\left(-\frac{2}{3}\right)^4}$
13. $\sqrt[5]{-32}$
14. $-\sqrt{-144}$
15. $\sqrt[4]{a^{16}b^8}$
16. $\pm\sqrt[4]{81x^4}$
17. $\sqrt[5]{\frac{1}{100{,}000}}$
18. $\sqrt[3]{-d^6}$
19. $\sqrt[5]{n^{25}q^{15}r^5s^{20}}$
20. $\sqrt[4]{(2x^2 - y^8)^8}$
21. $\pm\sqrt{16m^6p^2}$
22. $-\sqrt[3]{(2x - y)^3}$
23. $\sqrt[4]{(r + t)^4}$
24. $\sqrt{9a^2 + 6a + 1}$
25. $\sqrt{4y^2 + 12y + 9}$
26. $-\sqrt{x^2 - 2x + 1}$
27. $\pm\sqrt{x^2 + 2x + 1}$
28. $\sqrt[3]{a^3 + 6a^2 + 12a + 8}$

Lesson 7-5 Operations with Radical Expressions (pp. 439–445)

Simplify.

1. $\sqrt{75}$
2. $7\sqrt{12}$
3. $\sqrt[3]{81}$
4. $\sqrt{5r^5}$
5. $\sqrt[4]{7^8x^5y^6}$
6. $3\sqrt{5} + 6\sqrt{5}$
7. $\sqrt{18} - \sqrt{50}$
8. $4\sqrt[3]{32} + \sqrt[3]{500}$
9. $(\sqrt{12})(\sqrt{27})$
10. $3\sqrt{12} + 2\sqrt{300}$
11. $\sqrt[3]{54} - \sqrt[3]{24}$
12. $\sqrt{10}(2 - \sqrt{5})$
13. $-\sqrt{3}(2\sqrt{6} - \sqrt{63})$
14. $(5 + \sqrt{2})(3 + \sqrt{3})$
15. $(2 + \sqrt{5})(2 - \sqrt{5})$
16. $(8 + \sqrt{11})^2$
17. $(\sqrt{3} + \sqrt{6})(\sqrt{3} - \sqrt{6})$
18. $(\sqrt{8} + \sqrt{13})^2$
19. $(1 - \sqrt{7})(4 + \sqrt{7})$
20. $(5 - 2\sqrt{7})^2$
21. $\sqrt{\dfrac{3m^3}{24f^5}}$
22. $\dfrac{\sqrt{18}}{\sqrt{32}}$
23. $2\sqrt[3]{\dfrac{r^5}{2n^2t}}$
24. $\sqrt[3]{\dfrac{4}{7}}$

Lesson 7-6 Rational Exponents (pp. 446–452)

Write each expression in radical form or write each radical in exponential form.

1. $10^{\frac{1}{3}}$
2. $8^{\frac{1}{4}}$
3. $a^{\frac{2}{3}}$
4. $(b^2)^{\frac{3}{4}}$
5. $\sqrt{35}$
6. $\sqrt[4]{32}$
7. $3\sqrt{27a^2x}$
8. $\sqrt[5]{25ab^3c^4}$

Evaluate each expression.

9. $2401^{\frac{1}{4}}$
10. $27^{\frac{4}{3}}$
11. $(-32)^{\frac{2}{5}}$
12. $-81^{\frac{3}{4}}$
13. $(-125)^{-\frac{2}{3}}$
14. $16^{\frac{5}{2}} \cdot 16^{\frac{1}{2}}$
15. $8^{-\frac{2}{3}} \cdot 64^{\frac{1}{6}}$
16. $\left(\dfrac{48}{1875}\right)^{-\frac{5}{4}}$

Simplify each expression.

17. $7^{\frac{5}{9}} \cdot 7^{\frac{4}{9}}$
18. $32^{\frac{2}{3}} \cdot 32^{\frac{3}{5}}$
19. $\left(k^{\frac{8}{5}}\right)^5$
20. $x^{\frac{2}{5}} \cdot x^{\frac{8}{5}}$
21. $m^{\frac{2}{5}} \cdot m^{\frac{4}{5}}$
22. $\left(p^{\frac{5}{4}} \cdot r^{\frac{7}{2}}\right)^{\frac{8}{3}}$
23. $\left(4^{\frac{9}{2}}c^{\frac{3}{2}}\right)^2$
24. $\dfrac{7^{\frac{3}{4}}}{7^{\frac{5}{3}}}$
25. $\dfrac{1}{t^{\frac{9}{5}}}$
26. $a^{-\frac{8}{7}}$
27. $\dfrac{r}{r^{\frac{7}{5}}}$
28. $\sqrt[4]{36}$

Lesson 7-7 Solving Radical Equations and Inequalities (pp. 453–459)

Solve each equation.

1. $\sqrt{x} = 16$
2. $\sqrt{z + 3} = 7$
3. $\sqrt[3]{a + 5} = 1$
4. $g\sqrt{5} + 4 = g + 4$
5. $\sqrt{x - 8} = \sqrt{13 + x}$
6. $\sqrt{3z - 5} - 3 = 1$
7. $(5n - 1)^{\frac{1}{2}} = 0$
8. $(7x - 6)^{\frac{1}{3}} + 1 = 3$
9. $(6a - 8)^{\frac{1}{4}} + 9 = 10$

Solve each inequality.

10. $\sqrt{3x + 9} > 2$
11. $\sqrt{3n - 1} \leq 5$
12. $2 - 4\sqrt{21 - 6c} < -6$
13. $\sqrt{5y + 4} > 8$
14. $\sqrt{2w + 3} + 5 \geq 7$
15. $\sqrt{2c + 3} - 7 > 0$
16. $\sqrt{5y + 1} + 6 < 10$
17. $\sqrt{3n + 1} - 2 \leq 6$
18. $\sqrt{y - 5} - \sqrt{y} \geq 1$

Lesson 8-1 Graphing Exponential Functions (pp. 475–482)

Sketch the graph of each function. State the domain and range.

1. $y = 3(5)^x$ **2.** $y = 0.5(2)^x$ **3.** $y = 3\left(\frac{1}{4}\right)^x$ **4.** $y = 2(1.5)^x$

5. $y = 3(4)^x$ **6.** $y = (0.5)^x$ **7.** $y = 0.3(5)^x$ **8.** $y = \left(\frac{1}{5}\right)^x$

Lesson 8-2 Solving Exponential Equations and Inequalities (pp. 485–491)

Write an exponential function for the graph that passes through the given points.

1. (0, 6) and (2, 54) **2.** (0, −4) and (−4, −64) **3.** (0, 1.5) and (3, 40.5)

Solve each equation or inequality. Check your solution.

4. $27^{2x-1} = 3$ **5.** $8^{2+x} \geq 2$ **6.** $4^{2x+5} < 8^{x+1}$ **7.** $6^{x+1} = 36^{x-1}$

8. $10^{x-1} > 100^{4-x}$ **9.** $\left(\frac{1}{5}\right)^{x-3} = 125$ **10.** $2^{x^2+1} = 32$ **11.** $36^x = 6^{x^2-3}$

Lesson 8-3 Logarithms and Logarithmic Functions (pp. 492–499)

Write each equation in logarithmic form.

1. $3^5 = 243$ **2.** $10^3 = 1000$ **3.** $4^{-3} = \frac{1}{64}$

Write each equation in exponential form.

4. $\log_2 \frac{1}{8} = -3$ **5.** $\log_{25} 5 = \frac{1}{2}$ **6.** $\log_7 \frac{1}{7} = -1$

Evaluate each expression.

7. $\log_4 16$ **8.** $\log_{10} 10{,}000$ **9.** $\log_3 \frac{1}{9}$ **10.** $\log_2 1024$

11. $\log_6 6^5$ **12.** $\log_{\frac{1}{2}} 8$ **13.** $\log_{11} 121$ **14.** $5^{\log_5 10}$

Graph each function.

15. $f(x) = \log_4 x$ **16.** $f(x) = \log_{\frac{1}{5}} x$ **17.** $f(x) = 3 \log_3 (x - 2)$ **18.** $f(x) = 2 \log_{\frac{1}{2}} x - 5$

Lesson 8-4 Solving Logarithmic Equations and Inequalities (pp. 502–507)

Solve each equation or inequality. Check your solutions.

1. $\log_8 b = 2$ **2.** $\log_4 x < 3$ **3.** $\log_{\frac{1}{9}} n = -\frac{1}{2}$

4. $\log_x 7 = 1$ **5.** $\log_{\frac{2}{3}} a < 3$ **6.** $\log_2 (x^2 - 9) = 4$

7. $\log_9 x = 2$ **8.** $\log_{25} n = \frac{3}{2}$ **9.** $\log_{\frac{1}{7}} x = -1$

10. $\log_4 x < 2$ **11.** $\log_3 (2x - 1) \leq 2$ **12.** $\log_{16} x \geq \frac{1}{4}$

13. $\log_5 (3x - 1) = \log_5 (2x^2)$ **14.** $\log_{10} (x^2 - 10x) = \log_{10} (-21)$

15. $\log_2 (3x - 5) > \log_2 (x + 7)$ **16.** $\log_2 c > 8$

17. $\log_{64} y \leq \frac{1}{2}$ **18.** $\log_5 (5x - 7) \leq \log_5 (2x + 5)$

19. $\log_{\frac{1}{3}} p < 0$ **20.** $\log_2 (3x - 8) \geq 6$

21. $\log_6 (2x - 3) = \log_6 (x + 2)$ **22.** $\log_7 (x^2 + 36) = \log_7 100$

23. $\log_2 (4y - 10) \geq \log_2 (y - 1)$ **24.** $\log_{10} (a^2 - 6) > \log_{10} a$

Lesson 8-5 Properties of Logarithms (pp. 509–515)

Use $\log_3 5 \approx 1.4651$ and $\log_3 7 \approx 1.7712$ to approximate the value of each expression.

1. $\log_3 \frac{7}{5}$ **2.** $\log_3 245$ **3.** $\log_3 35$

Solve each equation. Check your solutions.

4. $\log_2 x + \log_2 (x - 2) = \log_2 3$
5. $\log_3 x = 2 \log_3 3 + \log_3 5$
6. $\log_5 (x^2 + 7) = \frac{2}{3} \log_5 64$
7. $\log_2 (x^2 - 9) = 4$
8. $\log_3 (x + 2) + \log_3 6 = 3$
9. $\log_6 x + \log_6 (x - 5) = 2$
10. $\log_5 (x + 3) = \log_5 8 - \log_5 2$
11. $2 \log_3 x - \log_3 (x - 2) = 2$
12. $\log_6 x = \frac{3}{2} \log_6 9 + \log_6 2$
13. $\log_8 (x + 6) + \log_8 (x - 6) = 2$
14. $\log_3 14 + \log_3 x = \log_3 42$
15. $\log_{10} x = \frac{1}{2} \log_{10} 81$

Lesson 8-6 Common Logarithms (pp. 516–522)

Use a calculator to evaluate each expression to four decimal places.

1. log 55 **2.** log 6.7 **3.** log 3.3
4. log 0.08 **5.** log 9.9 **6.** log 0.6

Solve each equation or inequality. Round to four decimal places.

7. $2^x = 15$ **8.** $4^{2a} > 45$ **9.** $7^{2x} = 35$
10. $11^{x + 4} > 57$ **11.** $1.5^{a - 7} = 9.6$ **12.** $3^{b^2} = 64$
13. $7^{3c} < 35^{2c - 1}$ **14.** $5^{m^2 + 1} = 30$ **15.** $7^{3y - 1} < 2^{2y + 4}$
16. $9^{n - 3} = 2^{n + 3}$ **17.** $11^{t + 1} \leq 22^{t + 3}$ **18.** $2^{3a - 1} = 3^{a + 2}$

Express each logarithm in terms of common logarithms. Then approximate its value to four decimal places.

19. $\log_3 21$ **20.** $\log_4 62$ **21.** $\log_5 28$ **22.** $\log_2 25$

Lesson 8-7 Base *e* and Natural Logarithms (pp. 525–531)

Use a calculator to evaluate each expression to four decimal places.

1. e^3 **2.** $e^{0.75}$ **3.** e^{-4} **4.** $e^{-2.5}$
5. ln 5 **6.** ln 8 **7.** ln 8.4 **8.** ln 0.6

Write an equivalent exponential or logarithmic equation.

9. $e^x = 10$ **10.** $\ln x \approx 2.3026$ **11.** $e^3 = 9x$ **12.** $\ln 0.2 = x$

Solve each equation or inequality. Round to the nearest ten-thousandth.

13. $25e^x = 1000$ **14.** $e^{0.075x} > 25$ **15.** $e^x < 3.8$
16. $-2e^x + 5 = 1$ **17.** $5 + 4e^{2x} = 17$ **18.** $e^{-3x} \leq 15$
19. $\ln 7x = 10$ **20.** $\ln 4x = 8$ **21.** $3 \ln 2x \geq 9$
22. $\ln (x + 2) = 4$ **23.** $\ln (2x + 3) > 0$ **24.** $\ln (3x - 1) = 5$

Lesson 8-8 Using Exponential and Logarithmic Functions (pp. 533–539)

1. **FARMING** Mr. Rogers purchased a combine for \$175,000 for his farming operation. It is expected to depreciate at a rate of 18% per year. What will be the value of the combine in 3 years?

2. **REAL ESTATE** The Jacksons bought a house for \$65,000 in 1992. Houses in the neighborhood have appreciated at the rate of 4.5% a year. How much is the house worth in 2003?

3. **POPULATION** In 1960, the population of a city was 50,000. Since then, the population has increased by 2.25% per year. If it continues to grow at this rate, what will the population be in 2015?

4. **BEARS** In a particular state, the population of black bears has been decreasing at the rate of 0.75% per year. In 1995, it was estimated that there were 400 black bears in the state. If the population continues to decline at the same rate, what will the population be in 2015?

Lesson 9-1 Multiplying and Dividing Rational Expressions (pp. 553–561)

Simplify each expression.

1. $\frac{25xy^2}{15y}$
2. $\frac{-4a^2b^3}{28ab^4}$
3. $\frac{(-2cd^3)^2}{8c^2d^5}$
4. $\frac{3x^3}{-2} \cdot \frac{-4}{9x}$
5. $\frac{21x^2}{-5} \cdot \frac{10}{7x^3}$
6. $\frac{2u^2}{3} \div \frac{6u^3}{5}$
7. $\frac{15x^3}{14} \div \frac{18x}{7}$
8. $\frac{xy^2}{2} \cdot \frac{x^2}{2y} \cdot \frac{2}{x^2y}$
9. $axy \div \frac{ax}{y}$
10. $\frac{9u^2}{28v} \div \frac{27u^2}{8v^2}$
11. $\frac{x^2-4}{4x^2-1} \cdot \frac{2x-1}{x+2}$
12. $\frac{x^2-1}{2x^2-x-1} \div \frac{x^2-4}{2x^2-3x-2}$
13. $\frac{2x^2+x-1}{2x^2+3x-2} \div \frac{x^2-2x+1}{x^2+x-2}$
14. $\frac{\frac{(ab)^2}{c}}{\frac{xa^3b}{cx^2}}$
15. $\frac{x^4-y^4}{x^3+y^3} \div \frac{x^3-y^3}{x+y}$

Lesson 9-2 Adding and Subtracting Rational Expressions (pp. 562–568)

Find the LCM of each set of polynomials.

1. $2a^2b, 4ab^2, 20a$
2. $x^2 - 4x - 12, x^2 + 7x + 10$

Simplify each expression.

3. $\frac{12}{7d} - \frac{3}{14d}$
4. $\frac{x+1}{x} - \frac{x-1}{x^2}$
5. $\frac{2x+1}{4x^2} - \frac{x+3}{6x}$
6. $\frac{7x}{13y^2} + \frac{4y}{6x^2}$
7. $\frac{x}{x-1} + \frac{1}{1-x}$
8. $\frac{1}{3v^2} + \frac{1}{uv} + \frac{3}{4u^2}$
9. $\frac{1}{x^2-x} + \frac{1}{x^2+x}$
10. $\frac{1}{x^2-1} - \frac{1}{(x-1)^2}$
11. $\frac{5}{x} - \frac{3}{x+5}$
12. $y - 1 + \frac{1}{y-1}$
13. $3m + 1 - \frac{2m}{3m+1}$
14. $\frac{3x}{x-y} + \frac{4x}{y-x}$
15. $\frac{4}{a^2-4} - \frac{3}{a^2+4a+4}$
16. $\frac{4}{3-3z^2} - \frac{2}{z^2+5z+4}$
17. $\frac{2c}{c^2-9} - \frac{1}{c^2+6c+9}$
18. $\frac{\frac{1}{x+y}}{\frac{1}{x}+\frac{1}{y}}$
19. $\frac{1-\frac{1}{x+1}}{1+\frac{1}{x-1}}$
20. $\frac{4+\frac{1}{x-2}}{3-\frac{1}{x-2}}$

Extra Practice

Lesson 9-3 Graphing Reciprocal Functions (pp. 569–575)

Graph each function. State the domain and range.

1. $f(x) = \frac{1}{x}$
2. $f(x) = \frac{3}{x}$
3. $f(x) = \frac{1}{x + 2}$
4. $f(x) = \frac{-5}{x + 1}$
5. $f(x) = \frac{-1}{x - 4}$
6. $f(x) = \frac{3}{x + 2}$
7. $f(x) = \frac{7}{x + 10}$
8. $f(x) = \frac{4}{7 - x}$

Lesson 9-4 Graphing Rational Functions (pp. 577–584)

Determine the equations of any vertical asymptotes and the values of x for any holes in the graph of each rational function.

1. $f(x) = \frac{1}{x + 4}$
2. $f(x) = \frac{x - 2}{x + 3}$
3. $f(x) = \frac{5}{(x + 1)(x - 8)}$
4. $f(x) = \frac{x}{x + 2}$
5. $f(x) = \frac{x^2 - 4}{x + 2}$
6. $f(x) = \frac{x^2 + x - 6}{x^2 + 8x + 15}$

Graph each rational function.

7. $f(x) = \frac{1}{x - 5}$
8. $f(x) = \frac{3x}{x + 1}$
9. $f(x) = \frac{x^2 - 16}{x - 4}$
10. $f(x) = \frac{x}{x - 6}$
11. $f(x) = \frac{1}{(x - 3)^2}$
12. $f(x) = \frac{2}{(x + 3)(x - 4)}$
13. $f(x) = \frac{x + 4}{x^2 - 1}$
14. $f(x) = \frac{x^2 + 5x - 14}{x^2 + 9x + 14}$

Lesson 9-5 Variation Functions (pp. 586–593)

State whether each equation represents a *direct, joint,* or *inverse* variation. Then name the constant of variation.

1. $xy = 10$
2. $\frac{x}{7} = y$
3. $\frac{x}{y} = -6$
4. $10x = y$
5. $x = \frac{2}{y}$
6. $A = \ell w$
7. $\frac{1}{4}b = -\frac{3}{5}c$
8. $D = rt$
9. If y varies directly as x and $y = 16$ when $x = 4$, find y when $x = 12$.
10. If x varies inversely as y and $x = 12$ when $y = -3$, find x when $y = -18$.
11. If m varies directly as w and $m = -15$ when $w = 2.5$, find m when $w = 12.5$.
12. If y varies jointly as x and z and $y = 10$ when $z = 4$ and $x = 5$, find y when $x = 4$ and $z = 2$.

Lesson 9-6 Solving Rational Equations and Inequalities (pp. 594–602)

Solve each equation or inequality. Check your solutions.

1. $\frac{x}{x - 3} = \frac{1}{4}$
2. $\frac{5}{x} + \frac{3}{5} = \frac{2}{x}$
3. $\frac{5}{b - 2} < 5$
4. $\frac{4}{a + 3} > 2$
5. $\frac{x - 2}{x} = \frac{x - 4}{x - 6}$
6. $-6 - \frac{8}{n} < n$
7. $\frac{2}{d} + \frac{1}{d - 2} = 1$
8. $\frac{1}{2 + 3x} + \frac{2}{2 - 3x} = 0$
9. $\frac{1}{n + 1} + \frac{1}{n - 1} = \frac{2}{n^2 - 1}$
10. $\frac{p}{p + 1} + \frac{3}{p - 3} + 1 = 0$
11. $\frac{5z + 2}{z^2 - 4} = \frac{-5z}{2 - z} + \frac{2}{z + 2}$
12. $\frac{1}{x - 3} + \frac{2}{x^2 - 9} = \frac{5}{x + 3}$
13. $\frac{1}{m^2 - 1} = \frac{2}{m^2 + m - 2}$
14. $\frac{12}{x^2 - 16} - \frac{24}{x - 4} = 3$
15. $n + \frac{1}{n + 3} = \frac{n^2}{n - 1}$

Lesson 10-1 Midpoint and Distance Formulas (pp. 617–622)

Find the midpoint of the line segment with endpoints at the given coordinates.

1. $(7, -3), (-11, 13)$
2. $(16, 29), (-7, 2)$
3. $(43, -18), (-78, -32)$
4. $(-7.54, 3.42), (4.89, -9.28)$
5. $\left(\frac{1}{2}, \frac{1}{4}\right), \left(\frac{2}{3}, \frac{3}{5}\right)$
6. $\left(-\frac{1}{4}, \frac{2}{3}\right), \left(-\frac{1}{2}, -\frac{1}{2}\right)$

Find the distance between each pair of points with the given coordinates.

7. $(5, 7), (3, 19)$
8. $(-2, -1), (5, 3)$
9. $(-3, 15), (7, -8)$
10. $(6, -3), (-4, -9)$
11. $(3.89, -0.38), (4.04, -0.18)$
12. $(5\sqrt{3}, 2\sqrt{2}), (-11\sqrt{3}, -4\sqrt{2})$
13. $\left(\frac{1}{4}, 0\right), \left(-\frac{2}{3}, \frac{1}{2}\right)$
14. $\left(4, -\frac{5}{6}\right), \left(-2, \frac{1}{6}\right)$
15. A circle has a radius with endpoints at $(-3, 1)$ and $(2, -5)$. Find the circumference and area of the circle. Write the answer in terms of π.
16. Triangle ABC has vertices $A(0, 0)$, $B(-3, 4)$, and $C(2, 6)$. Find the perimeter of the triangle.

Extra Practice

Lesson 10-2 Parabolas (pp. 623–629)

Write each equation in standard form. Identify the vertex, the axis of symmetry, and the direction of opening of the parabola.

1. $y + 4 = x^2$
2. $y = 5(x + 2)^2$
3. $4(y + 2) = 3(x - 1)^2$
4. $5x + 3y^2 = 15$
5. $y = 2x^2 - 8x + 7$
6. $x = 2y^2 - 8y + 7$
7. $3(x - 8)^2 = 5(y + 3)$
8. $x = 3(y + 4)^2 + 1$
9. $8y + 5x^2 + 30x + 101 = 0$
10. $x = -\frac{1}{5}y^2 + \frac{8}{5}y - 7$
11. $6x = y^2 - 6y + 39$
12. $-8y = x^2$
13. $y = 4x^2 + 24x + 38$
14. $y = x^2 - 6x + 3$
15. $y = x^2 + 4x + 1$

Write an equation for each parabola described below. Then graph the equation.

16. focus $(1, 1)$, directrix $y = -1$
17. vertex $(-1, 2)$, directrix $y = -4$

Lesson 10-3 Circles (pp. 631–637)

Write an equation for the circle that satisfies each set of conditions.

1. center $(3, 2)$, $r = 5$ units
2. center $(-5, 8)$, $r = 3$ units
3. center $(1, -6)$, $r = \frac{2}{3}$ units
4. center $(0, 7)$, tangent to x-axis
5. center $(-2, -4)$, tangent to y-axis
6. endpoints of a diameter at $(-9, 0)$ and $(2, -5)$
7. endpoints of a diameter at $(4, 1)$ and $(-3, 2)$
8. center $(6, -10)$, passes through origin
9. center $(0.8, 0.5)$, passes through $(2, 2)$

Find the center and radius of each circle. Then graph the circle.

10. $x^2 + y^2 = 36$
11. $(x - 5)^2 + (y + 4)^2 = 1$
12. $x^2 + 3x + y^2 - 5y = 0.5$
13. $x^2 + y^2 = 14x - 24$
14. $x^2 + y^2 = 2(y - x)$
15. $x^2 + 10x + (y - \sqrt{3})^2 = 11$
16. $x^2 + y^2 = 4x + 9$
17. $x^2 + y^2 - 6x + 4y = 156$
18. $x^2 + y^2 - 2x + 7y = 1$

Lesson 10-4 Ellipses (pp. 639–646)

Write an equation for the ellipse that satisfies each set of conditions.

1. endpoints of major axis at $(-2, 7)$ and $(4, 7)$, endpoints of minor axis at $(1, 5)$ and $(1, 9)$
2. endpoints of minor axis at $(1, -4)$ and $(1, 5)$, endpoints of major axis at $(-4, 0.5)$ and $(6, 0.5)$
3. major axis 24 units long and parallel to the y-axis, minor axis 4 units long, center at $(0, 3)$

Find the coordinates of the center and foci and the lengths of the major and minor axes for the ellipse with the given equation. Then graph the ellipse.

4. $\frac{x^2}{36} + \frac{y^2}{81} = 1$
5. $\frac{x^2}{121} + \frac{(y-5)^2}{16} = 1$
6. $\frac{(x+2)^2}{12} + \frac{(y+1)^2}{16} = 1$
7. $8x^2 + 2y^2 = 32$
8. $7x^2 + 3y^2 = 84$
9. $9x^2 + 16y^2 = 144$
10. $169x^2 - 338x + 169 + 25y^2 = 4225$
11. $x^2 + 4y^2 + 8x - 64y = -128$
12. $4x^2 + 5y^2 = 6(6x + 5y) + 658$
13. $9x^2 + 16y^2 - 54x + 64y + 1 = 0$

Extra Practice

Lesson 10-5 Hyperbolas (pp. 648–655)

Graph each hyperbola. Identify the vertices, foci, and asymptotes.

1. $\frac{y^2}{25} - \frac{x^2}{9} = 1$
2. $\frac{x^2}{4} - \frac{y^2}{9} = 1$
3. $\frac{x^2}{81} - \frac{y^2}{36} = 1$
4. $\frac{(x-4)^2}{64} - \frac{(y+1)^2}{16} = 1$
5. $\frac{(y-7)^2}{2.25} - \frac{(x-3)^2}{4} = 1$
6. $(x+5)^2 - \frac{(y+3)^2}{48} = 1$
7. $x^2 - 9y^2 = 36$
8. $4x^2 - 9y^2 = 72$
9. $49x^2 - 16y^2 = 784$
10. $576y^2 = 49x^2 + 490x + 29{,}449$
11. $25(y+5)^2 - 20(x-1)^2 = 500$

Write an equation for the hyperbola that satisfies each set of conditions.

12. vertices $(-3, 0)$ and $(3, 0)$; conjugate axis of length 8 units
13. vertices $(0, -7)$ and $(0, 7)$; conjugate axis of length 25 units
14. center $(0, 0)$; horizontal transverse axis of length 12 units and a conjugate axis of length 10 units

Lesson 10-6 Identifying Conic Sections (pp. 656–660)

Write each equation in standard form. State whether the graph of the equation is a *parabola, circle, ellipse,* or *hyperbola.* Then graph the equation.

1. $9x^2 - 36x + 36 = 4y^2 + 24y + 72$
2. $x^2 + 4x + 2y^2 + 16y + 32 = 0$
3. $x^2 + 6x + y^2 - 6y + 9 = 0$
4. $9y^2 = 25x^2 + 400x + 1825$
5. $2y^2 + 12y - x + 6 = 0$
6. $x^2 + y^2 = 10x + 2y + 23$
7. $3x^2 + y = 12x - 17$
8. $9x^2 - 18x + 16y^2 + 160y = -265$
9. $x^2 + 10x + 5 = 4y^2 + 16$
10. $\frac{(y-5)^2}{4} - (x+1)^2 = 4$
11. $9x^2 + 49y^2 = 441$
12. $4x^2 - y^2 = 4$

Without writing in standard form, state whether the graph of each equation is a *parabola, circle, ellipse,* or *hyperbola.*

13. $(x+3)^2 = 8(y+2)$
14. $x^2 + 4x + y^2 - 8y = 2$
15. $2x^2 - 13y^2 + 5 = 0$
16. $16(x-3)^2 + 81(y+4)^2 = 1296$

Lesson 10-7 Solving Linear-Nonlinear Systems (pp. 662–667)

Solve each system of inequalities by graphing.

1. $x^2 - 16y^2 \geq 16$
$x^2 + y^2 \leq 49$

2. $16x^2 + 25y^2 \leq 400$
$y \leq x - 2$

3. $y \geq x + 3$
$x^2 + y^2 < 25$

4. $4x^2 + (y - 3)^2 \leq 16$
$x + 2y \geq 4$

Solve each system of equations.

5. $x^2 + y^2 = 16$
$x - y = -3$

6. $x - y^2 = 0$
$(x + 3)^2 + y^2 = 53$

7. $4x^2 - 3(y + 2)^2 = 12$
$x^2 - y^2 = 11$

8. $2(x - 1)^2 + 5y^2 = 10$
$x - y = -1$

9. $x^2 + y^2 = 13$
$x^2 - y^2 = -5$

10. $x^2 - 5y^2 = 25$
$x - y = 4$

11. $x^2 + y = 0$
$x + y = -2$

12. $x^2 - 9y^2 = 36$
$x - y = 0$

13. $4x^2 + 6y^2 = 360$
$x - y = 0$

Lesson 11-1 Sequences as Functions (pp. 681–687)

Find the next four terms of each arithmetic sequence.

1. 9, 7, 5, ... **2.** 3, 4.5, 6, ... **3.** 40, 35, 30, ... **4.** 2, 5, 8, ...

Find the next two terms of each geometric sequence.

5. 5, 15, 45, ... **6.** 2, 10, 50, ... **7.** 64, 16, 4, ...

8. −9, 27, −81, ... **9.** 0.5, 0.75, 1.125, ... **10.** $\frac{1}{2}, -\frac{3}{8}, \frac{9}{32}, \ldots$

Lesson 11-2 Arithmetic Sequences and Series (pp. 688–695)

Find the sum of each arithmetic series.

1. $a_1 = 3, a_n = 20, n = 6$ **2.** $a_1 = 90, a_n = -4, n = 10$ **3.** $a_1 = 16, a_n = 14, n = 12$

4. $a_1 = -1, d = 10, n = 30$ **5.** $a_1 = 4, d = -5, n = 11$ **6.** $a_1 = 5, d = -\frac{1}{2}, n = 17$

Find the sum of each arithmetic series.

7. $\sum_{k=1}^{6}(k + 2)$ **8.** $\sum_{k=5}^{10}(2k - 5)$ **9.** $\sum_{k=1}^{5}(40 - 2k)$ **10.** $\sum_{k=8}^{12}(6 - 3k)$

Find the first three terms of each arithmetic series.

11. $a_1 = 11, a_n = 38, S_n = 245$ **12.** $n = 12, a_n = 13, S_n = -42$ **13.** $n = 11, a_n = 5, S_n = 0$

Find the first five terms of each arithmetic sequence.

14. $a_1 = 1, d = 7$ **15.** $a_1 = -5, d = 2$ **16.** $a_1 = 1.2, d = 3.7$ **17.** $a_1 = -\frac{5}{4}, d = -\frac{1}{2}$

Find the indicated term of each arithmetic sequence.

18. $a_1 = 4, d = 5, n = 10$ **19.** $a_1 = -30, d = -6, n = 5$ **20.** $a_1 = -3, d = 32, n = 8$

Write an equation for the *n*th term of each arithmetic sequence.

21. 3, 5, 7, 9, ... **22.** 2, −1, −4, −7, ... **23.** 20, 28, 36, 44, ...

Find the arithmetic means in each sequence.

24. 2, __?__, __?__, __?__, 34 **25.** 0, __?__, __?__, __?__, −28 **26.** −10, __?__, __?__, __?__, 14

Lesson 11-3 Geometric Sequences and Series (pp. 696–702)

Find the first five terms of each geometric sequence.

1. $a_1 = -2, r = 6$ **2.** $a_1 = 4, r = -5$ **3.** $a_1 = 0.8, r = 2.5$ **4.** $a_1 = -\frac{1}{3}, r = -\frac{3}{5}$

Find a_n for each geometric sequence.

5. $a_1 = 5, r = 7, n = 6$ **6.** $a_1 = 200, r = -\frac{1}{2}, n = 10$ **7.** $a_1 = 60, r = -2, n = 4$

Write an equation for the *n*th term of each geometric sequence.

8. 20, 40, 80, … **9.** $-\frac{1}{2}, -\frac{1}{8}, -\frac{1}{32}, \ldots$

Find the geometric means in each sequence.

10. 1, _?_, _?_, _?_, 81 **11.** 5, _?_, _?_, _?_, 6480

Find S_n for each geometric series described.

12. $a_1 = \frac{1}{81}, r = 3, n = 6$ **13.** $a_1 = 1, r = -2, n = 7$ **14.** $a_1 = 5, r = 4, n = 5$

15. $a_1 = -27, r = -\frac{1}{3}, n = 6$ **16.** $a_1 = 1000, r = \frac{1}{2}, n = 7$ **17.** $a_1 = 125, r = -\frac{2}{5}, n = 5$

18. $a_1 = 10, r = 3, n = 6$ **19.** $a_1 = 1250, r = -\frac{1}{5}, n = 5$ **20.** $a_1 = 1215, r = \frac{1}{3}, n = 5$

21. $a_1 = 16, r = \frac{3}{2}, n = 5$ **22.** $a_1 = 7, r = 2, n = 7$ **23.** $a_1 = -\frac{3}{2}, r = -\frac{1}{2}, n = 6$

Find the sum of each geometric series.

24. $\sum_{k=1}^{5} 2^k$ **25.** $\sum_{k=0}^{3} 3^{-k}$ **26.** $\sum_{k=0}^{3} 2(5^k)$ **27.** $\sum_{k=2}^{5} -(-3)^{k-1}$

Find the indicated term for each geometric series described.

28. $S_n = 300, a_n = 160, r = 2; a_1$ **29.** $S_n = -171, n = 9, r = -2; a_5$ **30.** $S_n = -4372, a_n = -2916, r = 3; a_4$

Lesson 11-4 Infinite Geometric Series (pp. 705–711)

Find the sum of each infinite geometric series, if it exists.

1. $a_1 = 54, r = \frac{1}{3}$ **2.** $a_1 = 2, r = -1$ **3.** $a_1 = 1000, r = -0.2$

4. $a_1 = 7, r = \frac{3}{7}$ **5.** $49 + 14 + 4 + \ldots$ **6.** $\frac{3}{4} + \frac{1}{2} + \frac{1}{3} + \ldots$

7. $12 - 4 + \frac{4}{3} - \ldots$ **8.** $3 - 9 + 27 - \ldots$ **9.** $3 - 2 + \frac{4}{3} - \ldots$

10. $\sum_{k=1}^{\infty} 3\left(\frac{1}{4}\right)^{k-1}$ **11.** $\sum_{k=1}^{\infty} 5\left(-\frac{1}{10}\right)^{k-1}$ **12.** $\sum_{k=1}^{\infty} -\frac{2}{3}\left(-\frac{3}{4}\right)^{k-1}$

Write each repeating decimal as a fraction.

13. $0.\overline{4}$ **14.** $0.\overline{27}$ **15.** $0.\overline{123}$

16. $0.\overline{645}$ **17.** $0.\overline{67}$ **18.** $0.8\overline{53}$

Extra Practice

Lesson 11-5 Recursion and Iteration (pp. 714–719)

Find the first five terms of each sequence described.

1. $a_1 = 4, a_{n+1} = 2a_{n+1}$
2. $a_1 = 6, a_{n+1} = a_n + 7$
3. $a_1 = 16, a_{n+1} = a_n + (n + 4)$
4. $a_1 = 1, a_{n+1} = \frac{n}{n+2} \cdot a_n$
5. $a_1 = -\frac{1}{2}, a_{n+1} = 2a_n + \frac{1}{4}$
6. $a_1 = \frac{1}{3}, a_2 = \frac{1}{4}, a_{n+1} = a_n + a_{n-1}$

Find the first three iterates of each function for the given initial value.

7. $f(x) = 3x - 1, x_0 = 3$
8. $f(x) = 2x^2 - 8, x_0 = -1$
9. $f(x) = 4x + 5, x_0 = 0$
10. $f(x) = 3x^2 + 1, x_0 = 1$
11. $f(x) = x^2 + 4x + 4, x_0 = 1$
12. $f(x) = x^2 + 9, x_0 = 2$
13. $f(x) = 2x^2 + x + 1, x_0 = -\frac{1}{2}$
14. $f(x) = 3x^2 + 2x - 1, x_0 = \frac{2}{3}$

Extra Practice

Lesson 11-6 The Binomial Theorem (pp. 721–725)

Expand each binomial.

1. $(z - 3)^5$
2. $(m + 1)^4$
3. $(x + 6)^4$
4. $(z - y)^2$
5. $(m + p)^5$
6. $(a - b)^4$
7. $(2n + 1)^4$
8. $(3n - 4)^3$
9. $(2n - m)^0$
10. $(4x - a)^4$
11. $(3r - 4t)^5$
12. $\left(\frac{b}{2} - 1\right)^4$

Find the indicated term of each expression.

13. sixth term of $(x + 3)^8$
14. fourth term of $(x - 2)^7$
15. fifth term of $(a + b)^6$
16. fourth term of $(x - y)^9$
17. sixth term of $(x + 4y)^7$
18. fifth term of $(3x + 5y)^{10}$

Lesson 11-7 Proof by Mathematical Induction (pp. 727–731)

Prove that each statement is true for all natural numbers.

1. $2 + 4 + 6 + \cdots + 2n = n^2 + n$
2. $1^3 + 3^3 + 5^3 + \cdots + (2n - 1)^3 = n^2(2n^2 - 1)$
3. $\frac{1}{1 \cdot 3} + \frac{1}{2 \cdot 4} + \frac{1}{3 \cdot 5} + \cdots + \frac{1}{n(n+2)} = \frac{n(3n+5)}{4(n+1)(n+2)}$
4. $1 \cdot 3 + 2 \cdot 4 + 3 \cdot 5 + \cdots + n(n + 2) = \frac{n(n+1)(2n+7)}{6}$
5. $\frac{5}{1 \cdot 2} \cdot \frac{1}{3} + \frac{7}{2 \cdot 3} \cdot \frac{1}{3^2} + \frac{9}{3 \cdot 4} \cdot \frac{1}{3^3} + \cdots + \frac{2n+3}{n(n+1)} \cdot \frac{1}{3^n} = 1 - \frac{1}{3^n(n+1)}$

Find a counterexample to disprove each statement.

6. $n^2 + 2n - 1$ is divisible by 2.
7. $2^n + 3^n$ is prime.
8. $2^{n-1} + n = 2^n + 2 - n$ for all integers $n \geq 2$
9. $3^n - 2n = 3^n - 2^n$ for all integers $n \geq 1$

Extra Practice

Lesson 12-1 Experiments, Surveys, and Observational Studies (pp. 745–750)

Determine whether each situation would produce a random sample. Write *yes* or *no* and explain your answer.

1. finding the most often prescribed pain reliever by asking all of the doctors at a hospital
2. taking a poll of the most popular baby girl names this year by studying birth announcements in newspapers from different cities across the country
3. polling people who are leaving a pizza parlor about their favorite restaurant in the city

State whether each situation represents an *experiment* or an *observational study*. Identify the control group and the treatment group. Then determine if there is bias.

4. A researcher stood at a busy intersection to see if the color of the automobile that a person drives is related to running a red light.
5. Subjects were randomly assigned to two groups, and one group was given an herb and the other group a placebo. After six months the number of respiratory tract infections each group had were compared.

Determine if the following statements show a *correlation* or *causation*.

6. The more hours I work, the more spending money I have.
7. When it snows heavily in the winter, there are snow plows outside.

Lesson 12-2 Statistical Analysis (pp. 752–758)

Which measure of central tendency best represents the data? Explain.

1. {86, 71, 74, 65, 45, 42, 76}
2. {20, 16, 15, 14, 24, 23, 25, 10, 19, 89}
3. {1, 2, 2, 2, 2, 4, 5}
4. {27, 33, 29, 37, 31, 28, 30, 34, 35}
5. **TEMPERATURES** The high temperatures at the right were recorded during a 38-day cold period in Cleveland, Ohio. Find the standard deviation.

29°	26°	17°	12°	5°	4°	25°	17°
23°	18°	13°	6°	25°	20°	27°	22°
26°	30°	31°	2°	12°	27°	16°	27°
16°	30°	6°	16°	5°	0°	5°	29°
18°	16°	22°	29°	8°	23°		

Lesson 12-3 Conditional Probability (pp. 759–763)

Use the table to answer Exercises 1 and 2.

Teacher Ratings				
	Explains well; Easy tests	Explains well; Hard tests	Explains poorly; Easy tests	Explains poorly; Hard tests
Junior	59	63	52	31
Senior	61	47	79	26

1. One of these students is randomly selected. If it is known that the student is a junior, find the probability that the student rated the teacher as "Explains well; Easy tests".
2. One of these students is randomly selected. If it is known that the student rated the teacher as "Explains poorly; Hard tests", find the probability that the student is a senior.
3. In a pizza restaurant, 95% of the customers order pizza. If 65% of the customers order pizza and a salad, find the probability that a customer will order a salad, given that he or she orders a pizza.

Lesson 12-4 Probability and Probability Distributions (pp. 764–771)

1. A small airline company employs 11 male flight attendants and 13 female flight attendants. As an economic move, the management decides to lay off two workers. Find the probability that both laid off workers will be male flight attendants.

2. Seven scientists are available to be the chairperson and assistant chairperson for a research project. If the names are drawn randomly, what is the probability that the two selected will be George and Sarah in that order?

3. The table shows the distribution of the number of heads tossed when tossing a coin three times.

X = Number of Heads	0	1	2	3
Probability	$\frac{1}{8}$	$\frac{3}{8}$	$\frac{3}{8}$	$\frac{1}{8}$

Find each of the following.

a. $P(X = 1)$ **b.** $P(X = 2)$ **c.** $P(X \neq 3)$

Lesson 12-5 The Normal Distribution (pp. 773–778)

1. Determine whether the data in the table appear to be *positively skewed, negatively skewed*, or *normally distributed*. The average size of a farm in each U.S. state was determined.

Acres	85–559	560–1034	1035–1509	1510–1984	1985–2459	2460–2934	2935–3409	3410–3884
States	37	4	3	1	2	1	0	2

Source: *The World Almanac*

2. The diameters of metal fittings made by a machine are normally distributed. The diameters have a mean of 7.5 centimeters and a standard deviation of 0.5 centimeters.
   a. What percent of the fittings have diameters between 7.0 and 8.0 centimeters?
   b. What percent of the fittings have diameters between 7.5 and 8.0 centimeters?
   c. What percent of the fittings have diameters greater than 6.5 centimeters?
   d. Of 100 fittings, how many will have a diameter between 6.0 and 8.5 centimeters?

3. A college extrance exam was administered at a state university. The scores were normally distributed with a mean of 510, and a standard deviation of 80.
   a. What percent would you expect to score above 510?
   b. What percent would you expect to score between 430 and 590?
   c. What is the probability that a student chosen at random scored between 350 and 670?

Lesson 12-6 Hypothesis Testing (pp. 780–784)

Find a 95% confidence interval for each of the following.

1. $\overline{X} = 93, s = 2.7$, and $n = 125$

2. $\overline{X} = 56, s = 1.9$, and $n = 75$

3. A sample of 200 students was asked for the average amount of time they spent studying or working on homework assignments per day. The mean time was 50 minutes with a standard deviation of 6.7 minutes. Determine a 95% confidence interval.

Test each of the following hypotheses.

4. $H_0 = 45, H_1 < 45, n = 75, \overline{X} = 39.5$, and $\sigma = 1.2$

5. $H_0 = 17.9, H_1 > 17.9, n = 125, \overline{X} = 18.2$, and $\sigma = 1.6$

Lesson 12-7 Binomial Distributions (pp. 786–793)

1. A fair coin is tossed 8 times. What is the probability of having at least 5 tails?
2. A test is made of 20 true or false questions. If a student guesses at random, what is the probability that the student will answer at least 12 questions correctly?
3. The Medicare program pays for 43% of all prescriptions filled at Bill's Pharmacy. If Bill's Pharmacy expects to fill 375 prescriptions this week, about how many can be expected to be paid by Medicare?
4. A research organization decides to mail 10 questionnaires to people selected at random. If the probability of any one person answering the questionnaire is $\frac{1}{7}$, find the probability that exactly 3 people will answer the questionnaire.

Lesson 13-1 Right Triangle Trigonometry (pp. 808–816)

Find the values of the six trigonometric functions for angle θ.

1.

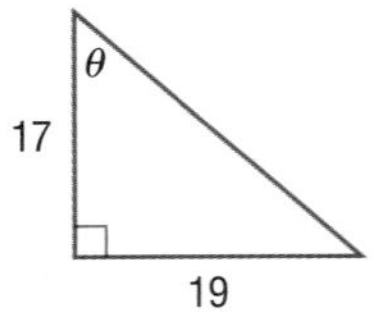

2.

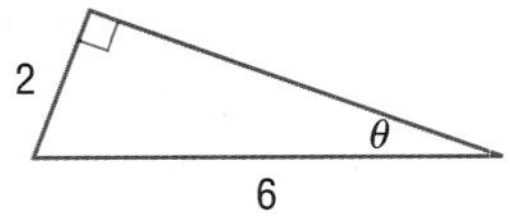

3.

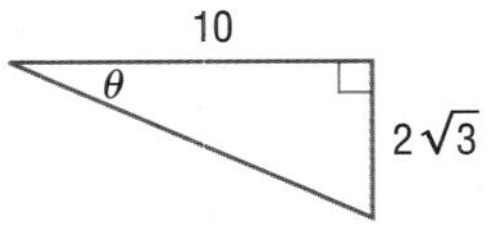

Solve $\triangle ABC$ using the diagram at the right and the given measurements. Round measures of sides to the nearest tenth and measures of angles to the nearest degree.

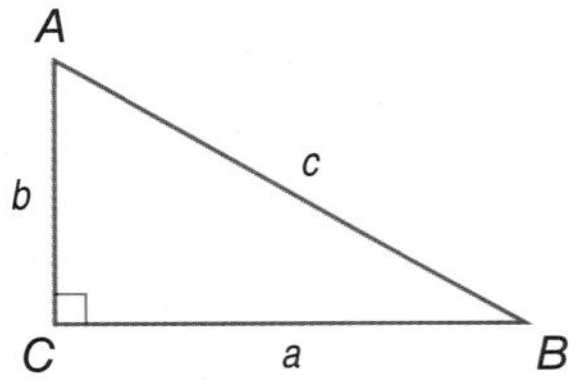

4. $B = 42°, c = 30$
5. $A = 84°, a = 4$
6. $B = 19°, b = 34$
7. $A = 75°, c = 55$
8. $b = 24, c = 36$
9. $a = 51, c = 115$
10. $\cos B = \frac{2}{5}, a = 12$
11. $\tan A = \frac{3}{2}, b = 22$

Lesson 13-2 Angles and Angle Measure (pp. 817–823)

Draw an angle with the given measure in standard position.

1. 60°
2. 250°
3. 315°
4. 150°

Rewrite each degree measure in radians and each radian measure in degrees.

5. −135°
6. −315°
7. 45°
8. 80°
9. 24°
10. −54°
11. $-\pi$
12. $\frac{9\pi}{4}$
13. $\frac{3\pi}{2}$
14. $-\frac{7\pi}{2}$
15. $\frac{9\pi}{10}$
16. $\frac{17\pi}{30}$
17. $\frac{7\pi}{12}$
18. 1
19. $-2\frac{1}{3}$

Find an angle with a positive measure and an angle with a negative measure that are coterminal with each angle.

20. 50°
21. −75°
22. 125°
23. −400°
24. 550°
25. 3π
26. -2π
27. $\frac{2\pi}{3}$
28. $\frac{12\pi}{5}$
29. 0

Lesson 13-3 Trigonometric Functions of General Angles (pp. 825–831)

The terminal side of θ in standard position contains each point. Find the exact values of the six trigonometric functions of θ.

1. $P(3, -4)$ **2.** $P(1, \sqrt{3})$ **3.** $P(0, 24)$ **4.** $P(-5, -5)$ **5.** $P(\sqrt{2}, -\sqrt{2})$

Find the exact value of each trigonometric function.

6. $\cos 225°$ **7.** $\sin\left(-\frac{5\pi}{3}\right)$ **8.** $\tan \frac{7\pi}{6}$ **9.** $\tan(-300°)$ **10.** $\cos \frac{7\pi}{4}$

Suppose θ is an angle in standard position whose terminal side is in the given quadrant. For each function, find the exact values of the remaining five trigonometric functions of θ.

11. $\cos\theta = -\frac{1}{3}$; Quadrant III **12.** $\sec\theta = 2$; Quadrant IV **13.** $\sin\theta = \frac{2}{3}$; Quadrant II

14. $\tan\theta = -4$; Quadrant IV **15.** $\csc\theta = -5$; Quadrant III **16.** $\cot\theta = -2$; Quadrant II

17. $\tan\theta = \frac{1}{3}$; Quadrant III **18.** $\cos\theta = \frac{1}{4}$; Quadrant I **19.** $\csc\theta = -\frac{5}{2}$; Quadrant IV

Lesson 13-4 Law of Sines (pp. 832–839)

Find the area of $\triangle ABC$ to the nearest tenth.

1. $a = 11$ m, $b = 13$ m, $C = 31°$ **2.** $a = 15$ ft, $b = 22$ ft, $C = 90°$ **3.** $a = 12$ cm, $b = 12$ cm, $C = 50°$

Solve each triangle. Round to the nearest tenth, if necessary.

4. $A = 18°, B = 37°, a = 15$ **5.** $A = 60°, C = 25°, c = 3$ **6.** $B = 40°, C = 32°, b = 10$

7. $B = 10°, C = 23°, c = 8$ **8.** $A = 12°, B = 60°, b = 5$ **9.** $A = 35°, C = 45°, a = 30$

Determine whether each triangle has *no* solution, *one* solution, or *two* solutions. Then solve the triangle. Round side lengths to the nearest tenth and angle measures to the nearest degree.

10. $A = 40°, B = 60°, c = 20$ **11.** $B = 70°, C = 58°, a = 84$ **12.** $A = 40°, a = 5, b = 12$

13. $A = 58°, a = 26, b = 29$ **14.** $A = 38°, B = 63°, c = 15$ **15.** $A = 150°, a = 6, b = 8$

16. $A = 57°, a = 12, b = 19$ **17.** $A = 25°, a = 125, b = 150$ **18.** $C = 98°, a = 64, c = 90$

19. $A = 40°, B = 60°, c = 20$ **20.** $A = 132°, a = 33, b = 50$ **21.** $A = 5\ 45°, a = 83, b = 79$

Lesson 13-5 Law of Cosines (pp. 841–846)

Determine whether each triangle should be solved by beginning with the Law of Sines or Law of Cosines. Then solve the triangle. Round side lengths to the nearest tenth and angle measures to the nearest degree.

1.

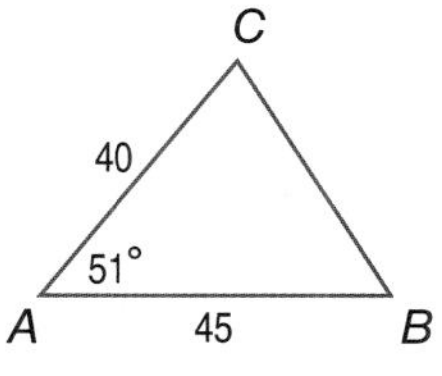

2.

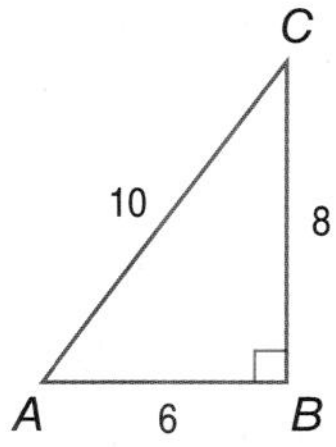

3.

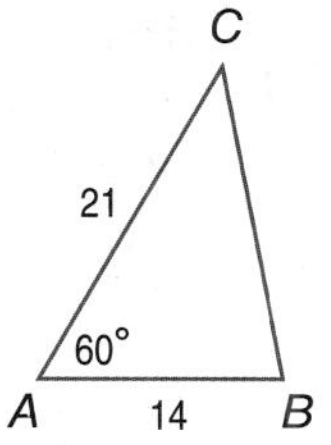

4. $a = 14, b = 15, c = 16$ **5.** $B = 41°, C = 52°, c = 27$ **6.** $a = 19, b = 24.3, c = 21.8$

7. $A = 112°, a = 32, c = 20$ **8.** $b = 8, c = 7, A = 28°$ **9.** $a = 5, b = 6, c = 7$

10. $C = 25°, a = 12, b = 9$ **11.** $a = 8, A = 49°, B = 58°$ **12.** $A = 42°, b = 120, c = 160$

13. $c = 10, A = 35°, C = 65°$ **14.** $a = 10, b = 16, c = 19$ **15.** $B = 45°, a = 40, c = 48$

16. $B = 100°, a = 10, c = 8$ **17.** $A = 40°, B = 45°, c = 4$ **18.** $A = 20°, b = 100, c = 84$

Lesson 13-6 Circular Functions (pp. 848–854)

The terminal side of angle θ in standard position intersects the unit circle at each point P. Find $\sin\theta$ and $\cos\theta$.

1. $P\left(\frac{4}{5}, \frac{3}{5}\right)$ **2.** $P\left(\frac{12}{3}, -\frac{5}{13}\right)$ **3.** $P\left(-\frac{8}{17}, -\frac{15}{17}\right)$ **4.** $P\left(\frac{3}{7}, \frac{2\sqrt{10}}{7}\right)$ **5.** $P\left(-\frac{2}{3}, \frac{\sqrt{5}}{3}\right)$

Find the exact value of each function.

6. $\sin 210°$ **7.** $\cos 150°$ **8.** $\cos(2135°)$ **9.** $\cos\frac{3\pi}{4}$

10. $\sin 570°$ **11.** $\sin 390°$ **12.** $\sin\frac{4\pi}{3}$ **13.** $\cos -\frac{7\pi}{3}$

14. $\cos 30° + \cos 60°$ **15.** $5(\sin 45°)(\cos 45°)$ **16.** $\frac{\sin 210° + \cos 240°}{3}$ **17.** $\frac{6\cos 120° + 4\sin 150°}{5}$

Determine the period of each function.

18.

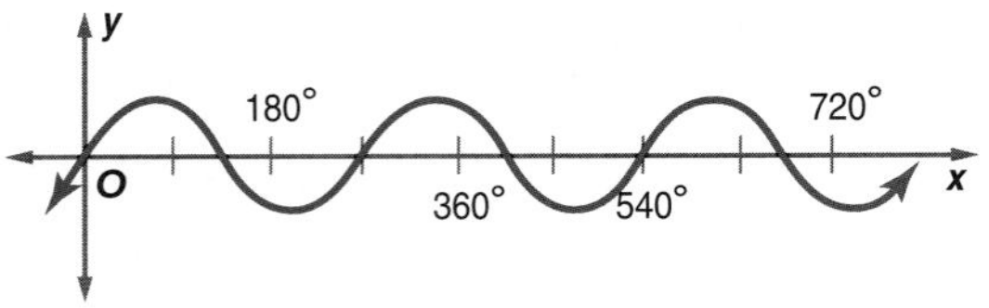

19.

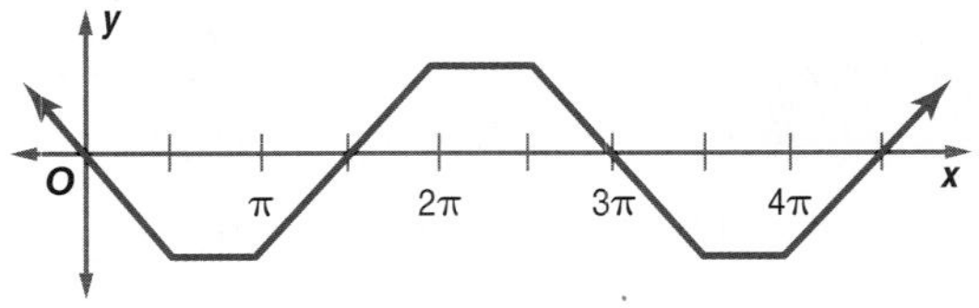

Lesson 13-7 Graphing Trigonometric Functions (pp. 855–861)

Find the amplitude and period of each function. Then graph the function.

1. $y = 2\cos\theta$ **2.** $y = \frac{1}{3}\sin\theta$ **3.** $y = \sin 3\theta$

4. $y = 3\sec\theta$ **5.** $y = \sec\frac{1}{3}\theta$ **6.** $y = 2\csc\theta$

7. $y = 3\tan\theta$ **8.** $y = 3\sin\frac{2}{3}\theta$ **9.** $y = 2\sin\frac{1}{5}\theta$

10. $y = 3\sin 2\theta$ **11.** $y = \frac{1}{2}\cos\frac{3}{4}\theta$ **12.** $y = 5\csc 3\theta$

13. $y = 2\cot 6\theta$ **14.** $y = 2\csc 6\theta$ **15.** $y = 3\tan\frac{1}{3}\theta$

Lesson 13-8 Translations of Trigonometric Graphs (pp. 863–870)

State the amplitude, period, and phase shift for each function. Then graph the function.

1. $y = \sin(\theta + 60°)$ **2.** $y = \cos(\theta - 90°)$

3. $y = \tan\left(\theta + \frac{\pi}{2}\right)$ **4.** $y = \sin\theta + \frac{\pi}{6}$

State the amplitude, period, vertical shift, and the equation of the midline for each function. Then graph the function.

5. $y = \cos\theta + 3$ **6.** $y = \sin\theta - 2$ **7.** $y = \sec\theta + 5$

8. $y = \csc\theta - 6$ **9.** $y = 2\sin\theta - 4$ **10.** $y = \frac{1}{3}\sin\theta + 7$

State the amplitude, period, vertical shift, and phase shift for each function. Then graph the function.

11. $y = 3\cos[2(\theta + 30°)] + 4$ **12.** $y = 2\tan[3(\theta - 60°)] - 2$ **13.** $y = \frac{1}{2}\sin[4(\theta + 45°)] + 1$

14. $y = \frac{2}{5}\cos[6(\theta + 45°)] - 5$ **15.** $y = 6 - 2\sin\left[3\left(\theta + \frac{\pi}{2}\right)\right]$ **16.** $y = 3 + 3\cos\left[2\left(\theta - \frac{\pi}{3}\right)\right]$

Lesson 13-9 Inverse Trigonometric Functions (pp. 871–876)

Solve each equation.

1. $y = \text{Sin}^{-1} -\frac{\sqrt{2}}{2}$
2. $\text{Tan}^{-1}(1) = x$
3. $a = \text{Arccos}\left(\frac{\sqrt{3}}{2}\right)$
4. $\text{Arcsin}(0) = x$
5. $y = \text{Cos}^{-1}\frac{1}{2}$
6. $y = \text{Sin}^{-1}(1)$

Find each value. Round to the nearest hundredth if necessary.

7. $\text{Arccos}\left(-\frac{\sqrt{2}}{2}\right)$
8. $\text{Sin}^{-1}(-1)$
9. $\cos\left[\text{Arcsin}\left(\frac{\sqrt{2}}{2}\right)\right]$
10. $\tan\left[\text{Sin}^{-1}\left(\frac{15}{13}\right)\right]$
11. $\sin\left[\text{Arccos}\frac{1}{2}\right]$
12. $\sin\left[\text{Arccos}\left(\frac{5}{17}\right)\right]$
13. $\sin\left[\text{Tan}^{-1}\left(\frac{5}{12}\right)\right]$
14. $\tan\left[\text{Arccos} -\left(\frac{\sqrt{3}}{2}\right)\right]$
15. $\sin^{-1}[\text{Cos}^{-1}(1) - 1]$
16. $\text{Cos}^{-1}\left[\tan\frac{\pi}{4}\right]$
17. $\cos\left[\text{Sin}^{-1}\frac{1}{2}\right]$
18. $\sin[\text{Cos}^{-1}(0)]$

Lesson 14-1 Trigonometric Identities (pp. 891–897)

Find the value of each expression.

1. $\sin\theta$, if $\cos\theta = \frac{4}{5}$; $0° \leq \theta \leq 90°$
2. $\tan\theta$, if $\sin\theta = \frac{1}{2}$; $0° \leq \theta \leq 90°$
3. $\csc\theta$, if $\sin\theta = \frac{3}{4}$; $90° \leq \theta \leq 180°$
4. $\cos\theta$, if $\tan\theta = 24$; $90° \leq \theta \leq 180°$
5. $\sec\theta$, if $\tan\theta = 24$; $90° \leq \theta \leq 180°$
6. $\sin\theta$, if $\cot\theta = -\frac{1}{4}$; $270° \leq \theta \leq 360°$
7. $\tan\theta$, if $\sec\theta = 23$; $90° \leq \theta \leq 180°$
8. $\sin\theta$, if $\cos\theta = \frac{3}{5}$; $270° \leq \theta \leq 360°$
9. $\cos\theta$, if $\sin\theta = -\frac{1}{2}$; $270° \leq \theta \leq 360°$
10. $\csc\theta$, if $\cot\theta = -\frac{1}{4}$; $90° \leq \theta \leq 180°$

Simplify each expression.

11. $\csc^2\theta - \cot^2\theta$
12. $\sin\theta\tan\theta\csc\theta$
13. $\tan\theta\csc\theta$
14. $\sec\theta\cot\theta\cos\theta$
15. $\cos\theta(1 - \cos^2\theta)$
16. $\frac{1 - \sin^2\theta}{\cos^2\theta}$
17. $\frac{\sin^2\theta + \cos^2\theta}{\cos^2}$
18. $\frac{1 + \tan^2\theta}{1 + \cot^2\theta}$
19. $\frac{1}{1 + \sin\theta} + \frac{1}{1 - \sin\theta}$

Lesson 14-2 Verifying Trigonometric Identities (pp. 898–903)

Verify that each equation is an identity.

1. $\sin^2\theta + \cos^2\theta + \tan^2\theta = \sec^2\theta$
2. $\frac{\tan\theta}{\sin\theta} = \sec\theta$
3. $\frac{\tan\theta}{\cot\theta} = \tan^2\theta$
4. $\csc^2\theta(1 - \cos^2\theta) = 1$
5. $1 - \cot^4\theta = 2\csc^2\theta - \csc^4\theta$
6. $\sin^4\theta - \cos^4\theta = \sin^2\theta - \cos^2\theta$
7. $\sin^2\theta + \cot^2\theta\sin^2\theta = 1$
8. $\frac{\cos\theta}{\csc\theta} - \frac{\csc\theta}{\sec\theta} = -\frac{\cos^3\theta}{\sin\theta}$
9. $\frac{\cos\theta}{\sec\theta} - \frac{1 + \cos\theta}{\sec\theta + 1} = 2\cot^2\theta$
10. $\frac{1 + \cos\theta}{\sin\theta} = \frac{\sin\theta}{1 - \cos\theta}$
11. $\sec\theta + \tan\theta = \frac{\cos\theta}{1 - \sin\theta}$
12. $\tan\theta + \cot\theta = \csc\theta\sec\theta$
13. $\frac{\cot^2\theta}{1 + \cot^2\theta} = 1 - \sin^2\theta$
14. $\frac{\tan\theta\, 2\sin\theta}{\sec\theta} = \frac{\sin^3\theta}{1 + \cos\theta}$
15. $\sin^2\theta(1 - \cos^2\theta) = \sin 4\theta$
16. $\sin^2\theta + \sin^2\theta\tan^2\theta = \tan^2\theta$
17. $\frac{\sec\theta - 1}{\sec\theta + 1} + \frac{\cos\theta - 1}{\cos\theta + 1} = 0$
18. $\tan^2\theta(1 - \sin^2\theta) = \sin^2\theta$

Lesson 14-3 Sum and Difference of Angles Identities (pp. 904–909)

Find the exact value of each expression.

1. $\sin 195°$
2. $\cos 285°$
3. $\sin 255°$
4. $\sin 105°$
5. $\cos 15°$
6. $\sin 15°$
7. $\cos 375°$
8. $\sin 165°$
9. $\sin(-225°)$
10. $\cos(-210°)$
11. $\cos(-225°)$
12. $\sin(-30°)$

Verify that each equation is an identity.

13. $\sin(90° + \theta) = \cos\theta$
14. $\cos(180° - \theta) = -\cos\theta$
15. $\sin(p + \theta) = -\sin\theta$
16. $\sin(\theta + 30°) + \sin(\theta + 60°) = \sqrt{3} + \frac{1}{2}(\sin\theta + \cos\theta)$
17. $\cos(30° - \theta) + \cos(30° + \theta) = \sqrt{3}\cos\theta$

Lesson 14-4 Double-Angle and Half-Angle Identities (pp. 911–917)

Find the exact value of $\sin 2\theta$, $\cos 2\theta$, $\sin\frac{\theta}{2}$, and $\cos\frac{\theta}{2}$ for each of the following.

1. $\cos\theta = \frac{7}{25}$; $0 < \theta < 90°$
2. $\sin\theta = \frac{2}{7}$; $0 < \theta < 90°$
3. $\cos\theta = -\frac{1}{8}$; $180° < \theta < 270°$
4. $\sin\theta = -\frac{5}{13}$; $270° < \theta < 360°$

Find the exact value of each expression by using the half-angle formulas.

5. $\sin 75°$
6. $\cos 75°$
7. $\sin\frac{\pi}{8}$
8. $\cos\frac{13\pi}{12}$
9. $\cos 22.5°$
10. $\cos\frac{\pi}{4}$

Verify that each equation is an identity.

11. $\frac{\sin 2\theta}{2\sin^2\theta} = \cot\theta$
12. $1 + \cos 2\theta = \frac{2}{1 + \tan^2\theta}$
13. $\csc\theta\sec\theta = 2\csc 2\theta$
14. $\sin 2\theta(\cot\theta + \tan\theta) = 2$
15. $\frac{1 - \tan^2\theta}{1 + \tan^2\theta} = \cos 2\theta$
16. $\frac{\csc\theta + \sin\theta}{\csc\theta - \sin\theta} = \frac{1 + \sin 2\theta}{\cos 2\theta}$

Lesson 14-5 Solving Trigonometric Equations (pp. 919–925)

Find all the solutions for each equation for $0° \leq \theta < 360°$.

1. $\cos\theta = -\frac{\sqrt{3}}{2}$
2. $\sin 2\theta = -\frac{\sqrt{3}}{2}$
3. $\cos 2\theta = 8 - 15\sin\theta$

Solve each equation for all values of θ if θ is measured in radians.

4. $\cos 2\theta\sin\theta = 1$
5. $\sin\frac{\theta}{2} + \cos\frac{\theta}{2} = \sqrt{2}$
6. $\cos 2\theta + 4\cos\theta = -3$

Solve each equation for all values of θ if θ is measured in degrees.

7. $2\sin^2\theta - 1 = 0$
8. $\cos\theta - 2\cos\theta\sin\theta = 0$
9. $\cos 2\theta\sin\theta = 1$

Solve each equation.

10. $\tan\theta = 1$
11. $\cos 8\theta = 1$
12. $\sin\theta + 1 = \cos 2\theta$
13. $8\sin\theta\cos\theta = 2\sqrt{3}$
14. $\cos\theta = 1 + \sin\theta$
15. $2\cos^2\theta = \cos\theta$

Mixed Problem Solving

Chapter 1 Equations and Inequalities (pp. 2–57)

1. **PACKAGING** A can is 5 inches tall and has a diameter of 4 inches. Its volume can be determined by the equation $V = \pi r^2 h$. What is the volume of the can? (Lesson 1-1)

2. **SIGN** A yield sign is 30 inches on a side and 26 inches tall. The formula for the area of a triangle is $A = \frac{1}{2}bh$. Find the area of a yield sign. (Lesson 1-1)

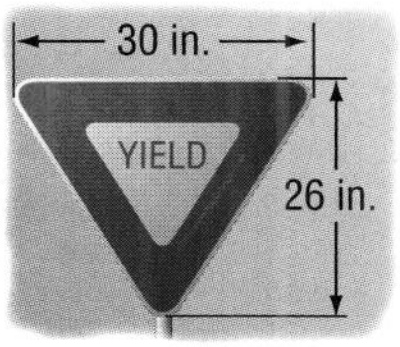

3. **BILLBOARDS** A standard roadside billboard is 14 feet by 48 feet. The area of a rectangle is $A = \ell w$. Find the area of a standard billboard. (Lesson 1-1)

4. **AREA** The area of a trapezoid is given by the formula $A = \frac{1}{2}(a + b)h$. Find A when $a = 2$ cm, $b = 5$ cm, and $h = 4$ cm. (Lesson 1-1)

5. **CLOTHING** A T-shirt shop has shirts on sale for \$9.99 each. Nina buys 2 of these shirts, Latisha buys 3, and Addie buys 1. (Lesson 1-2)

   a. Illustrate the Distributive Property by writing two expressions to represent the cost of these shirts.

   b. Use the Distributive Property to find how much money the store received from selling these shirts.

6. **TEMPERATURE** The formula for changing temperature in degrees Fahrenheit F to degrees Celsius C is given by the formula $C = \frac{5}{9}(F - 32)$. Find C if $F = 68°$. (Lesson 1-3)

7. **ALGEBRA** Write an algebraic equation to represent the sentence *The quotient of two numbers is equal to the sum of those numbers.* (Lesson 1-3)

8. **ICE HOCKEY** An ice hockey stick is 175 centimeters long, give or take 25 centimeters. Write and solve an absolute value equation to determine the least and greatest possible lengths of an ice hockey stick. (Lesson 1-4)

9. **SOCCER** A regulation soccer ball is an air-filled sphere with a circumference of 68 to 70 centimeters. Write an absolute value equation to show the range of sizes allowable for a regulation soccer ball. (Lesson 1-4)

10. **COFFEE** Coffee beans are typically roasted at 455°F, plus or minus 85°F. Write and solve an absolute value equation to determine the least and greatest temperatures at which coffee beans are typically roasted. (Lesson 1-4)

11. **PHONES** Frieda's cell phone plan costs \$40 per month plus \$0.50 for each minute she goes beyond her free minutes. How many minutes can she go beyond her free minutes and still pay less than \$65? (Lesson 1-5)

    a. Write an inequality to solve this problem.

    b. Solve the problem.

12. **DATA** Melissa's wireless handheld device service plan costs \$48 per month plus \$0.01 per kilobyte over her free usage. (Lesson 1-5)

    a. Write an inequality to show how much she can go over her free usage and still pay less than \$58.

    b. Solve the inequality.

13. **MONEY** Amy needs a new computer. They are on sale for \$350 but could vary in price by as much as \$125 from the sale price. (Lesson 1-6)

    a. Write an absolute value inequality to describe this situation.

    b. Solve the inequality to find the range of the prices of computers.

1. **POLITICS** The table below shows the population of several states and the number of U.S. representatives from those states. (Lesson 2-1)

State	Population (millions)	Number of Representatives
Alabama	4.45	7
Delaware	0.78	1
Indiana	6.08	9
Michigan	9.94	15
New York	18.98	29
Ohio	11.35	18

   a. Make a graph of the data with population on the horizontal axis and representatives on the vertical axis.

   b. Identify the domain and range.

   c. Is the relation *discrete* or *continuous*?

   d. Does the graph represent a function? Explain your reasoning.

2. **TUTORING** Katrina is starting a business tutoring students in math. She rents an office for $450 per month and charges $40 per hour per student. She has 12 students each week. (Lesson 2-2)

   a. Write an equation representing the situation if each student is tutored x hours per week.

   b. How much profit will Katrina make after 4 weeks?

3. **DRIVING** When driving up a certain hill, you rise 15 feet for every 1000 feet you drive forward. What is the slope of the road? (Lesson 2-3)

4. **CELL PHONE** Mario bought a cell phone for $125. Monthly expenses for the cell phone total $85 per month. Write an equation that represents the total cost of buying and owning the cell phone for x months. (Lesson 2-4)

5. **GAMES** In Scott's favorite online game, virtual cash is earned from playing games. Playing a game earns 3 virtual dollars. Each 250 points scored in the game earns 1 additional virtual dollar. (Lesson 2-4)

   a. Write an equation that models the virtual dollars earned d for a game score of g points.

   b. How many virtual dollars did Scott earn when he had a game score of 12,500?

6. **BASEBALL** The table below shows the attendance for the Los Angeles Angels' home games. (Lesson 2-5)

Game	Attendance	Game	Attendance
1	43,906	4	34,970
2	42,463	5	31,397
3	35,701	6	30,876

   a. Make a scatter plot of the data.

   b. Find a regression equation for the data.

7. **RECREATION** The charge for renting inline skates from a rental shop for different amounts of time is shown. (Lesson 2-6)

Inline Skate Rentals	
Time	Price ($)
1 hour	5
$\frac{1}{2}$ day	10
full day	17
full week	60

   a. Identify the type of function that models this situation.

   b. Write a function for the situation.

   c. Graph the function.

8. **HEALTH** Shooting baskets can burn up to 5.1 Calories per minute. The equation to represent how many Calories a person burns after m minutes of shooting baskets is $C(m) = 5.1m$. (Lesson 2-7)

   a. Identify the transformation in the function.

   b. Graph the function.

9. **CAR MAINTENANCE** Jerome needs to buy gas and oil for his car. Gas costs $4.03 a gallon and oil costs $2.99 a quart. He has $65 to spend. (Lesson 2-8)

   a. Write an inequality to represent the situation, where g is the number of gallons of gas he buys and q is the number of quarts of oil.

   b. Graph the inequality.

   c. Can Jerome buy 12 gallons of gasoline and 8 quarts of oil?

Chapter 3 Systems of Equations and Inequalities (pp. 132–181)

1. **RECREATION** The admission fees for a fair are as shown in the table. On a certain day, 3400 people enter the fair and $12,250 is collected. (Lesson 3-1)

Admission Fee	
Age	**Cost**
adult	$5.00
child	$2.50

   a. Write a system of equations to represent this situation.

   b. How many adults and children attended?

2. **LANDSCAPING** The school district placed two orders with a nursery. The first order was for 13 bushes and 4 trees and totaled $487. The second order was for 6 bushes and 2 trees and totaled $232. The bills do not list the per-item prices. (Lesson 3-1)

   a. Write a system of equations to represent this situation.

   b. What are the prices of one bush and of one tree?

3. **DRAMA** Tickets to the next drama performance are as shown in the table. For the two-day run of the performance, 718 tickets are sold. The Drama Club collected $4269. (Lesson 3-2)

Tickets	
Age	**Cost**
adult	$8
child	$3

   a. Write a system of equations to represent this situation.

   b. Find how many adults and children attended.

4. **NUMBERS** The sum of two numbers is 95. One number is 16 less than twice the other. (Lesson 3-2)

   a. Write a system of equations to represent this situation.

   b. Find the numbers.

5. **CANDY** The most Jay can spend on chocolates is $35. White chocolates sell for $9.00 per pound, and dark chocolates sell for $7.50 per pound. He needs to buy at least 25 pounds of chocolate. (Lesson 3-3)

   a. Graph the region that shows how many pounds of each type of chocolate he can purchase.

   b. Give an example of three different purchases he can make.

6. **FENCING** Jermaine's dog needs at least 20 square meters of space to run. Jermaine's dad only has 20 meters of fence to build the run. (Lesson 3-4)

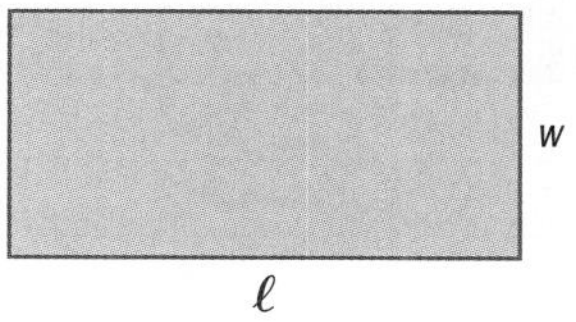

   a. Write a system of inequalities to represent the possible measurements of the dog run.

   b. Make a graph showing the feasible region and list the coordinates of the vertices of the feasible region.

   c. What is the largest rectangle that Jermaine can build for his dog?

7. **BUSINESS** Maura and Jeffrey use a shipping center for the pet accessories they sell. They want to track the sizes of the packages shipped.

Joshua's Shipping Center	
Weight of Package	**Cost**
≤ 5 lb	$2.90
$5 < \text{weight} < 10$ lb	$5.20
≥ 10 lb	$8.00

The number of smallest packages shipped is 50% more than the number of largest packages shipped. One-day shipping charges for 300 packages is $1508. (Lesson 3-5)

   a. Write a system of equations to represent this situation.

   b. Find the number of packages shipped in each weight category.

8. **PARKING** A parking lot has an area of 600 square meters. A car requires 6 square meters of space and a bus requires 30 square meters of space. The attendant can handle no more than 60 vehicles. If a car is charged $3 to park and a bus is charged $8, how many of each should the attendant accept to maximize income? (Lesson 3-5)

9. **EDUCATION** In 2000, Harvard, Yale, and Stanford had combined endowments of $38.1 billion. Harvard had $0.1 billion more than Yale and Stanford together. Stanford's endowments trailed Harvard's by $10.2 billion. What were the endowments of each university? (Lesson 3-5)

1. **CELL PHONE USE** Four students' weekly cell phone use is summarized in the table. (Lesson 4-1)

Name	Cell Phone Minutes	Text Messages	Picture Messages
Lila	131	212	85
José	95	189	25
Tony	147	208	78
Abril	185	247	93

   a. Organize the cell phone usage in a matrix.

   b. Which student uses their cell phone the most?

   c. Add the elements of each row and interpret the results.

2. **BUSINESS** An electronics store lists the prices for their most popular product in three categories at three stores. The manager has decided to put these products on sale for 25% off. (Lesson 4-2)

Store	Television	DVD Player	CD Player
1	582	132	85
2	621	129	89
3	594	131	95

   a. Write a matrix C to represent the current prices.

   b. What scalar can be used to determine a matrix N to represent the new prices?

   c. Find N.

   d. What is $N - C$? What does this represent in this situation?

3. **FUNDRAISING** Logan's band is selling boxes of fruit to raise money to go to Washington, D.C. The navel oranges are $17 a box; red grapefruit are $16 a box; and tangelos are $27 a box. (Lesson 4-3)

Name	Oranges	Grapefruit	Tangelos
Logan	22	15	8
Lizzie	20	18	12
Jordan	10	3	7
Katherine	19	17	10

   a. Write a matrix for the number of each type of fruit Logan and her friends sold.

   b. Use matrix multiplication to find the money raised by each person for each type of fruit.

   c. Find the total amount of money the friends raised for their trip.

4. **SPORTS** The Westfall Youth Baseball and Softball League charges the following registration fees: ages 7–8, $45; ages 9–10, $55; and ages 11–14, $65. (Lesson 4-3)

Team Members		
Age	Baseball	Softball
7–8	350	280
9–10	320	165
11–14	180	120

   a. Write a matrix for the registration fees and a matrix for the number of players.

   b. Find the total amount of money the league received from baseball and softball registrations.

5. **SIGNS** For the election for student council, Michael wants to enlarge his campaign logo. He graphs it with vertices at $A(6, 0)$, $B(5, 7)$, and $C(0, 0)$. He enlarges it so that the perimeter is three times the original perimeter. State the coordinates of the vertices of the dilated image. (Lesson 4-4)

6. **ADVERTISING** Julie and Tonya are advertising a friend's new club. They are hanging flyers on poles at (3, 1), (4, 15), and (7, 12) according to their map. Each unit represents 1 kilometer. If they hang flyers everywhere, what is the area of town in which they are advertising? (Lesson 4-5)

7. **GEOGRAPHY** Dominic is calculating the amount of forest that is part of a metro park. He lays a coordinate grid in which 1 unit = 10 miles over a map of the forest with the origin at the entrance to the park. The coordinates of the two other exits are (7, 5) and (2.5, 10). Estimate the area of the forest. (Lesson 4-5)

8. **PILOT TRAINING** Flight instruction costs $115 per hour, and the simulator costs $65 per hour. Chia spent 4 more hours in airplane training than in the simulator. If Chia spent $4780, how much time did she spend training in an airplane and in a simulator? (Lesson 4-6)

9. **NUMBERS** The sum of the digits in a two-digit number is 14. The number itself is 2 greater than 11 times the tens digit. Find the number. (*Hint:* The number can be written as $10a + b$, where a is the tens digit and b is the ones digit.) (Lesson 4-6)

1. **ECONOMICS** A souvenir shop sells about 300 coffee mugs per month for \$7.50 each. The shop owner estimates that for each \$0.75 increase in the price, he will sell about 15 fewer coffee mugs per month. (Lesson 5-1)

   a. How much should the owner charge for each mug in order to maximize the monthly income from their sales?

   b. What is the maximum monthly income the owner can expect to make from the mugs?

2. **PHYSICS** An object is fired straight up from the top of a 150-foot tower at a velocity of 75 feet per second. The height $h(t)$ of the object t seconds after firing is given by $h(t) = -16t^2 + 75t + 150$. (Lesson 5-2)

   a. What are the domain and range of the function?

   b. What domain and range values are reasonable in the given situation?

   c. Find the maximum height reached by the object and the time that the height is reached.

   d. Interpret the meaning of the y-intercept in the context of this problem.

3. **ARCHERY** An arrow is shot straight upward with a velocity of 85 feet per second. Use the formula $h(t) = v_0t - 16t^2$, where $h(t)$ is the height of an object in feet, v_0 is the object's initial velocity in feet per second, and t is the time in seconds. (Lesson 5-2)

   a. Assuming the archer is 6 feet tall, how long after the arrow is released does it hit the ground?

   b. How high does the arrow reach?

4. **NUMBERS** The sum of an integer and its square is 42. Find the integer. (Lesson 5-3)

5. **REMODELING** Sandy's closet was supposed to be 10 feet by 12 feet. The architect decided that this would not work and reduced the dimensions by the same amount x on each side. The area of the new closet is 63 square feet. (Lesson 5-3)

   a. Write a quadratic equation that represents the area of Sandy's closet now.

   b. Find the new dimensions of Sandy's closet.

6. **ELECTRICITY** In an AC circuit, voltage V, current C, and impedance I are related by the formula $V = CI$. (Lesson 5-4)

   a. Find the voltage in a circuit with current $3 + 5j$ amps and impedance $4 - 2j$ ohms.

   b. The voltage in a circuit is $76 - 10j$ volts, and the impedance is $8 - 7j$ ohms. What is the current?

7. **ARCHITECTURE** An architect's blueprints call for a room to be 12 feet by 12 feet. The customer would like the room to be square with an area of 275 square feet. How much will this add to each dimension? (Lesson 5-5)

8. **INSECTS** For a certain insect, the survival rate depends on temperature. A model of the number of larvae $N(t)$ that survive is given by $N(t) = -0.6t^2 + 32.1t - 350$, where t is the temperature in degrees Celsius. Find the range of temperatures where the insects can survive. (Lesson 5-6)

9. **ARCHITECTURE** An architect designed a pool that is fenced on three sides. If she uses 60 yards of fencing to enclose an area of 352 square yards, then the dimensions ℓ and w can be modeled by $w = \ell^2 - 30\ell + 176$. Graph this function. (Lesson 5-7)

10. **ROCKETS** The height $h(t)$ of a model rocket in feet t seconds after its launch can be represented by the function $h(t) = -16t^2 + 96t + 0.75$. During what interval is the rocket at least 100 feet above the ground? (Lesson 5-8)

11. **FUNDRAISING** The girls softball team is sponsoring a fundraising trip to see a baseball game. In order to earn a profit, they will charge \$15 per person if all seats on the bus are sold, but for each empty seat, they will increase the price by \$1.50 per person. (Lesson 5-8)

   a. Write a quadratic function giving the softball team's profit $P(n)$ from this fundraiser as a function of the number of passengers n.

   b. What is the minimum number of passengers needed for the team not to lose money?

   c. What is the maximum profit the team can earn, and how many passengers will it take to achieve this maximum?

1. **E-SALES** A small online retailer estimates that the cost, in dollars, associated with selling x units of a particular product is given by the expression $0.001x^2 + 5x + 500$. The revenue from selling x units is given by $10x$. (Lesson 6-1)

   a. Write a polynomial to represent the profit generated by the product.

   b. Find the profit from sales of 1850 units.

2. **INVESTMENT** Charlene made $1235 working a summer job. She wants to invest it during the school year in a savings account that has an annual interest rate of 1.9% and a money market account that pays 3.9% per year. Write a polynomial to represent the amount of interest she will earn in one year if she invests x dollars in the savings account. (Lesson 6-1)

3. **CARS** The number of cars produced in a plant each day can be estimated by $6x^2 - 3x - 1$, where x is the number of worker teams. Divide by x to find the average number of cars produced per team. (Lesson 6-2)

4. **WOODWORKING** Arthur is building a rectangular table with an area of $3x^2 - 17x - 28$ square feet. If the length of the table is $3x + 4$ feet, what should the width of the table be? (Lesson 6-2)

5. Consider the following graph. (Lesson 6-3)

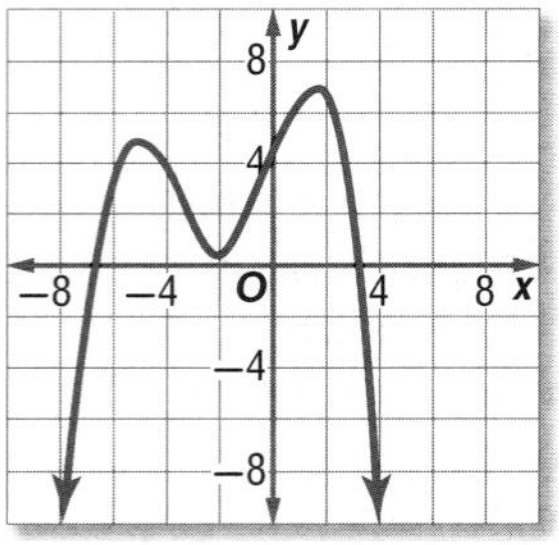

   a. Describe the end behavior.

   b. Determine whether it represents an odd-degree or an even-degree function.

   c. State the number of real zeros.

6. **HEALTH** The weight w, in pounds, of a patient during a three-week illness is modeled by the cubic equation $w(n) = 0.3n^3 - 0.4n^2 + 95$, where n is the number of weeks since the patient became ill. (Lesson 6-4)

   a. Graph the function.

   b. Describe the turning points of the graph and its end behavior.

   c. What trends in the patient's weight does the graph suggest?

7. **SAILING** The area of a right triangular sail is $x^2 + 7x + 10$. The length of one leg of the sail is $2x + 10$. Find the length of the other leg. (Lesson 6-5)

8. **LANDSCAPING** A brick border that is x feet wide is built around a rectangular patio. The patio is 6 feet wide and 8 feet long. The combined area of the patio and the brick border is 120 square feet. What is the width of the brick border? (Lesson 6-5)

9. **NUMBER SENSE** If the expression $ax^4 + bx^3 - x^2 + 2x + 3$ is divided by $x^2 + x - 2$, there is a remainder of $4x + 3$. Find the values of a and b. (Lesson 6-6)

10. **CELL PHONES** DeQuan found that the equation $f(x) = 4x^3 + 4x^2 - 8x$ could be used to find the production volume of a certain type of cell phone, where x is time in minutes. Find the roots and type of roots for this equation. (Lesson 6-7)

11. **NUMBER SENSE** Write a polynomial function of least degree with real coefficients having zeros -1 and $6 - 3i$. (Lesson 6-7)

12. **GEOMETRY** The volume of a rectangular solid is 160 cubic inches. The width is 3 inches more than the height, and the length is 1 inch less than the height. Find the dimensions of the solid. (Lesson 6-8)

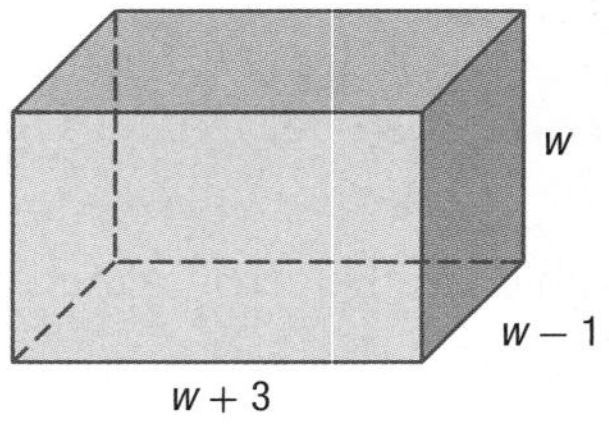

13. **FOOD** A restaurant orders spaghetti sauce in cylindrical metal cans. The volume of each can is about 160π cubic inches, and the height of the can is 6 inches more than the radius. (Lesson 6-8)

   a. Write a polynomial equation that represents the volume of a can. Use the formula for the volume of a cylinder, $V = \pi r^2 h$.

   b. What are the possible values of r? Which values are reasonable here?

   c. Find the dimensions of the can.

1. **TAXES** Claire has \$150 deducted from every paycheck for retirement. She can have this deduction taken before state taxes are applied, which reduces her taxable income. Her state income tax is 7%. If Claire earns \$1450 every pay period, find the difference in her net income if she has the retirement deduction taken before or after state taxes. (Lesson 7-1)

2. **SHOPPING** The Sound Loft is offering both an in-store \$25 rebate and a 25% discount on a stereo system that normally sells for \$450. Which provides the better price: taking the discount before or after the rebate? (Lesson 7-1)

3. **COMMISSION** Adelina works forty hours a week at a furniture store. She receives a \$220 weekly salary, plus a 3% commission on sales over \$5000. Assume that she sells enough this week to get the commission. (Lesson 7-1)

   a. Given the functions $f(x) = 0.03x$ and $g(x) = x - 5000$, which of $(f \circ g)(x)$ and $(g \circ f)(x)$ represents her commission?

   b. What is Adelina's commission on weekly sales of \$11,675?

   c. How much is Adelina paid for working a week in which her sales total \$17,381?

4. **GEOMETRY** The formula for the circumference of a circle is $C = 2\pi r$. (Lesson 7-2)

   a. Find the inverse of the function.

   b. Use the inverse to find the radius of a circle with a circumference of 150 inches.

5. **PHYSICS** The formula $F = ma$ represents the force in newtons on an object of mass m in kilograms accelerating at a in meters per second squared. (Lesson 7-2)

   a. Find the inverse of the function.

   b. Colby and his sled weigh 55 kilograms. They hit the bottom of a hill with a force of 275 newtons. How fast were they accelerating?

6. **PENDULUM** The length of time it takes for a pendulum on a clock to make a complete motion is found by $f(\ell) = 2\pi\sqrt{\left(\frac{\ell}{g}\right)}$, where ℓ is the length of the pendulum and g is the acceleration due to gravity. (Lesson 7-3)

   a. Graph the function.

   b. How long is the pendulum for a period of 1.9 seconds, if $g = 9.8$ meters per second squared?

7. **PACKAGING** Leroy needs a box to hold 3375 cubic centimeters. He is shipping cell phone parts and the packing material. If he makes the box a cube, how long is each side? (Lesson 7-4)

   a. Write an equation to represent the situation.

   b. Find the length of each side.

8. **GEOMETRY** Use the rectangle shown. (Lesson 7-5)

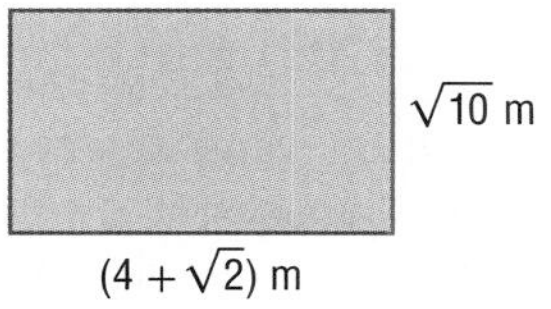

   a. Find the perimeter of the rectangle.

   b. Find the area of the rectangle.

9. **PRICES** The formula $C = c(1 + r)^n$ can be used to estimate the future cost of an item due to inflation. C represents the future cost, c represents the current cost, r is the rate of inflation, and n is the number of years for the projection. (Lesson 7-6)

   a. Suppose a gallon of gas costs \$3.98 now. How much would the price of gas be in 6 months with an inflation rate of 2.7%?

   b. Suppose a car costs \$14,500 now. How much would the price of gas be in 6 months with an inflation rate of 2.9%?

10. **BODIES** The formula $p = \sqrt[3]{\frac{m}{h}}$ measures the ponderal index p, which is a measure of a person's body based on height h in meters and mass m in kilograms. A person who is 1.8 meters tall has a ponderal index of about 2.29. How much does the person weigh in kilograms? (Lesson 7-7)

11. **GRAVITY** Hugo drops his keys from the top of a Ferris wheel. The formula $t = \frac{1}{4}\sqrt{65 - h}$ describes the time t in seconds when the keys are h feet above the boardwalk. If Hugo was 65 meters high when he dropped the keys, how many meters above the boardwalk will the keys be after 2 seconds? (Lesson 7-7)

Exponential and Logarithmic Functions and Relations (pp. 472–542)

1. **CAFFEINE** A cup of coffee contains 95 milligrams of caffeine. The average teen can eliminate approximately 12.5% of the caffeine from their system per hour. (Lesson 8-1)
   a. Write an equation to represent the amount of caffeine remaining after drinking a cup of coffee.
   b. Draw a graph for the equation.

2. **BIOLOGY** A bacteria culture used in biology class grows continuously at a rate of 4.6% per day. The teacher starts with 250 bacteria. (Lesson 8-1)
   a. Write an equation to represent the amount of bacteria as time passes.
   b. Draw a graph for the equation.

3. **POPULATION** Every ten years, the Bureau of the Census counts the number of people living in the U.S. In 1790, the population of the U.S. was 3.93 million. By 1800, this number had grown to 5.31 million. (Lesson 8-2)
   a. Write an exponential function that could be used to model the U.S. population y in millions for 1790 to 1800. Write the equation in terms of x, the number of decades since 1790.
   b. Assume that the U.S. population continued to grow at least that rapidly. Estimate the population for the years 1820, 1840, and 1860. Then compare your estimates with the actual population for those years, which were 9.64, 17.06, and 31.44 million, respectively.

4. **CHEMISTRY** The equation to determine the pH of a substance is $\text{pH} = -\log_{10}[\text{H}^+]$. (Lesson 8-3)
   a. What is the concentration of the hydrogen ion $[\text{H}^+]$ if the pH of an aqueous solution is -3.30?
   b. What is the concentration of the hydrogen ion in an aqueous solution with pH = 13.22?
   c. What is the concentration of the hydrogen ion in room temperature water with a pH of 7?
   d. The runoff from a highway is measured at a pH of -3.6. What is the concentration of the hydrogen ions in this solution?
   e. What is the pH of an aqueous solution in which $[\text{H}^+] = 2.7 \times 10^{-3}$ Moles?
   f. If the pH of an aqueous solution is 6.52, what is the concentration of hydrogen ions?

5. **EARTHQUAKES** The Richter scale is based on logarithms. The equation is $R = \log_{10} M$, where M is the amount of ground movement and R is the strength of the earthquake on the Richter scale. (Lesson 8-4)
   a. How much ground movement is there in an earthquake that registers 5?
   b. How many times as much ground movement is there in a level 5 as in a level 3?

6. **STAR LIGHT** The brightness, or apparent magnitude m of a star or planet is given by $m = 6 - 2.5 \log_{10} \frac{L}{L_0}$, where L is the amount of light coming to Earth from the star or planet and L_0 is the amount of light from a sixth magnitude star. Find the difference in the magnitudes of Sirius and the crescent moon. (Lesson 8-5)

Moon
Sirius
The cresent moon is about 100 times as bright as the brightest star, Sirius.

7. **FLIGHT** An airplane takes off from an airport at sea level and its altitude h, in feet, at time t, in minutes, is given by $h = 2000 \ln (t + 1)$. Find the altitude at time $t = 2$ minutes. (Lesson 8-6)

8. **MONEY** The amount A in an account after t years when interest is compounded continuously is found using the formula $A = Pe^{rt}$, where P is the amount of principal and r is the annual interest rate. When Saul was born, his parents deposited $4000 into an account paying 5% interest compounded continuously. (Lesson 8-7)
   a. If the account is not touched, what is the balance after 10 years?
   b. If they want to have $15,000 after 18 years, how much would they need to invest?

9. **OLYMPICS** In 1928, the winning women's high jump was 62.5 inches, while the winning men's jump was 76.5 inches. Since then, the winning jump for women has increased by about 0.38% per year, while the winning jump for men has increased at a slower rate, 0.3%. If these rates continue, when will the women's winning high jump be higher than the men's? (Lesson 8-8)

Chapter 9 Rational Functions and Relations (pp. 550–613)

1. **BASKETBALL** At the end of the 2009–2010 season, a professional basketball player had made 5422 field goals out of 12,138 attempts during his NBA career. (Lesson 9-1)
   a. Write a ratio to represent the number of career field goals made to career field goals attempted by the professional basketball player at the end of the 2009–2010 season.
   b. Suppose the professional basketball player attempted a field goals and made m field goals during the 2010–2011 season. Write a rational expression to represent the ratio of the number of career field goals made to the number of career field goals attempted at the end of the 2010–2011 season.

2. **GEOMETRY** In the figure, the area of the rectangle is 111 square centimeters and the triangle is equilateral. Find the perimeter of the figure in terms of x. (Lesson 9-2)

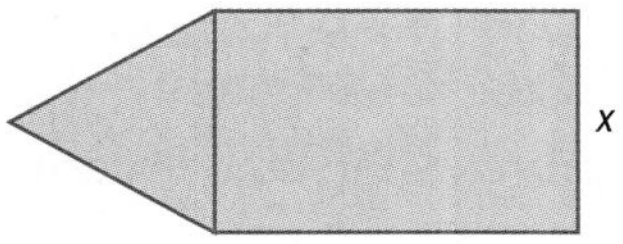

3. **GEOMETRY** Find the perimeter of the rectangle. (Lesson 9-2)

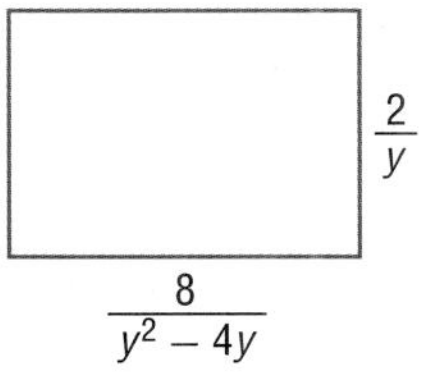

4. **AREA** A rectangle has an area equal to 100 cm^2 and a width x. (Lesson 9-3)
   a. Write an expression for the perimeter P in terms of x.
   b. Graph the perimeter as a function of x.
   c. What domain and range values are meaningful in the context of the problem?

5. **TRAVEL** A boat traveled upstream at r_1 miles per hour. During the return trip to its original starting point, the boat traveled at r_2 miles per hour. The average speed for the entire trip R is given by the formula $R = \frac{(2r_1r_2)}{(r_1 + r_2)}$. (Lesson 9-4)
   a. Draw the graph if $r_2 = 15$ miles per hour.
   b. What is the R-intercept of the graph?
   c. What domain and range values are meaningful in the context of the problem?

6. **NUTRITION** There are about 200 Calories in 50 grams of Swiss cheese. Susan ate 70 grams of this cheese. About how many Calories were in the cheese that she ate if the number of Calories varies directly as the weight of the cheese? (Lesson 9-5)

7. **PHYSICS** According to Hooke's law, the force needed to stretch a spring is proportional to the amount the spring is stretched. If 50 pounds stretches a spring five inches, how much will the spring be stretched by a force of 120 pounds? (Lesson 9-5)

8. **PLANETS** Kepler's Third Law of Planetary Motion states that the square of the time required for a planet to make one revolution about the Sun varies directly as the cube of the average distance of the planet from the Sun. If you assume that Mars is 1.5 times as far from the Sun as is Earth, find the approximate length of a Martian year. (Lesson 9-5)

9. **BIKING** Two cyclists start at the same time from opposite ends of a course that is 45 miles long. One cyclist is riding at 14 mph and the second cyclist is riding at 16 mph. How long after they begin will they meet? (Lesson 9-6)

10. **MONEY** An investment advisor deposited \$50,000 into two simple interest accounts. On the tax-free account the annual interest rate is 7%, and on the money market fund the annual simple interest rate is 13%. (Lesson 9-6)
    a. Let x equal the amount put into the tax-free account. Write an expression for when both accounts earn the same amount of interest.
    b. How much should be invested in the tax-free account so that both accounts earn the same interest?
    c. How much should be invested in the market fund so that both accounts earn the same interest?

11. **CYCLING** On a particular day, the wind added 3 kilometers per hour to Alfonso's rate when he was cycling with the wind and subtracted 3 kilometers per hour from his rate on his return trip. Alfonso found that in the same amount of time he could cycle 36 kilometers with the wind, he could go only 24 kilometers against the wind. What is his normal bicycling speed with no wind? Determine whether your answer is reasonable. (Lesson 9-6)

Chapter 10 Conic Sections (pp. 614–677)

1. **NEIGHBORHOOD** Katie and Angela agree to meet halfway between their houses. On a map, Katie lives at $(1, -7)$. Angela lives at $(-5, -3)$. (Lesson 10-1)
   a. How far apart do Katie and Angela live?
   b. Where is the halfway point?

2. **KEEP AWAY** Stephanie, standing at $(2, 3)$, threw a ball to Heather, standing at $(4, 1)$. (Lesson 10-1)
   a. How far did Stephanie throw the ball?
   b. If Stephanie and Heather are keeping Alex from catching the ball and he is standing exactly between them, where is Alex standing?

3. **FLASHLIGHT** The parabolic reflector in a flashlight reflects the light from the bulb and makes it more intense. If the vertex of the parabola is at the origin, the focus is at $(0, 2)$. (Lesson 10-2)
   a. Write an equation for the parabola formed by this flashlight reflector.
   b. Graph the equation.

4. **INTERNET** A bagel shop offers free WiFi. Their transmitter has a 60-meter range in any direction. (Lesson 10-3)
   a. Write an equation to represent the area in which people can access the free WiFi. Place the bagel shop at the origin.
   b. Graph the equation.

5. **SPACE** Saturn's distance from the Sun at the aphelion is about 930 million miles. Saturn's distance at the perihelion is 839 million miles. (Lesson 10-4)
   a. Determine an equation relating Saturn's orbit around the Sun in millions of miles with the center of the horizontal ellipse at the origin.
   b. Graph the equation.

6. **VENUS** At its closest point, Venus is 0.719 astronomical units from the Sun. At its farthest point, Venus is 0.728 astronomical units from the Sun. Write an equation for the orbit of Venus. Assume that the center of the orbit is the origin, the Sun lies on the x-axis, and the radius of the Sun is 400,000 miles. (Lesson 10-4)

7. **BATS** Two bats that are 300 feet apart send out signals looking for prey. The concentric sound waves meet in a hyperbolic shape. The bats determine that a swarm of gnats is 80 feet farther from the first bat than from the second bat. Determine the equation of the hyperbola centered at the origin on which the swarm is located. (Lesson 10-5)

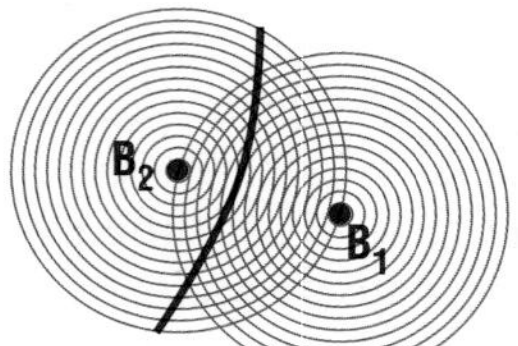

   a. Identify the vertices, foci, and asymptotes of the hyperbola.
   b. Write the equation.

8. **DISCS** The movement of a flying disc can be described by the equation: $y + \left(\frac{1}{80}\right)x^2 + x = 0$, where x represents the time in seconds. (Lesson 10-6)
   a. Write the equation in standard form.
   b. State whether the graph of the equation is a *parabola*, *circle*, *ellipse*, or *hyperbola*.
   c. Graph the equation.

9. **EARTHQUAKE** Seismographs are used to measure the intensity of earthquakes. Three seismographs can be used to find the center of the earthquake. Determine the location of the center of the earthquake on the grid. (Lesson 10-7)

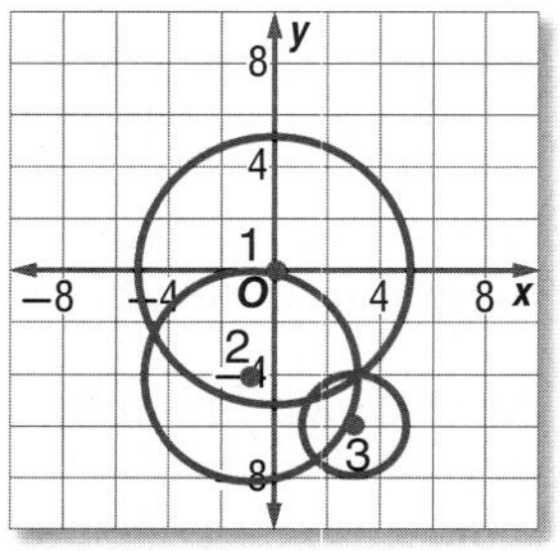

10. **ROCKETS** Two rockets are launched at the same time, but from different heights. The height y in feet of one rocket after t seconds is given by $y = -16t^2 + 150t + 5$. The height of the other rocket is given by $y = -16t^2 + 160t$. After how many seconds are the rockets at the same height? (Lesson 10-7)

1. **FIELD HOUSE** The field house has a section where the seating can be arranged so that the first row has 11 seats, the second row has 15 seats, the third row has 19 seats, and so on. There is sufficient space for 30 rows in the section. (Lesson 11-1)
   a. How many seats are in the last row?
   b. How many seats are in the whole section?

2. **SALARY** Mr. Patton starts his job at \$40,000 per year. He receives a \$1000 raise every year until he retires after 35 years. (Lesson 11-2)
   a. How much is Mr. Patton making the last year he works?
   b. How much does Mr. Patton make over the course of his career?

3. **ART** Alberta is making a beadwork design consisting of rows of colored beads. The first row consists of 10 beads, and each consecutive row will have 15 more beads than the previous row. (Lesson 11-2)
   a. Write an equation for the number of beads in the nth row.
   b. Find the number of beads in the design if it contains 25 rows.

4. **PHYSICAL SCIENCE** A ball bounced in place recovers a certain percentage of its original height. Suppose a ball that recovers 70% of its height is dropped from 200 feet. After its first bounce, it reaches a height of 140 feet. After the second bounce, it reaches a height of 98 feet. (Lesson 11-3)
   a. How high will the ball bounce on its sixth bounce, which is the seventh height?
   b. What is the sum of the heights the ball has bounced through on its sixth bounce?

5. **RUBBER BALL** A rubber ball is dropped from a height of 10 meters. Supposed it rebounds one half the distance after each fall. Find the total distance the ball travels. (Lesson 11-4)

6. **BUILDING** A 25-story building has a basement that is 4 feet below street level. Each floor is 13 feet high. (Lesson 11-5)
   a. Write a recursive formula for the height of each floor.
   b. Find the height of the first 5 floors.

Floor Number	Basement (0)	1	2	3	4	5
Height (ft)	−4					

7. **FISH** In a city fishing pond, the city starts with 4000 trout. 20% of the fish are caught each year, so the city adds 1000 fish every spring. (Lesson 11-5)
   a. Write a recursive formula for the number of fish in the pond.
   b. Determine the number of fish in the pond after 5 years.
   c. How many fish are needed to maintain a constant population?

8. **COMPETITION** Seventy-five teams take part in a competition organized so that teams meet one-on-one with the defeated team getting dropped out of the competition. How many games are needed before one team is declared a winner? (Lesson 11-6)

9. **SCHOOL** Mr. Hopkins is giving a five-question quiz. How many ways could a student answer the questions with three trues and two falses? (Lesson 11-6)

10. **GEOMETRY** Write an expanded expression for the volume of the cube. (Lesson 11-6)

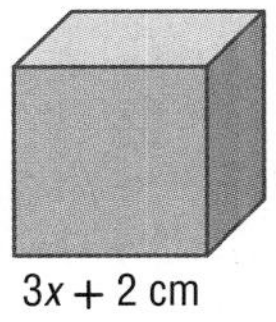

11. **PASCAL'S TRIANGLE** Study the first eight rows of Pascal's triangle. Write the sum of the terms in each row as a list. Make a conjecture about the sums of the rows of Pascal's triangle. (Lesson 11-7)

12. **NUMBER THEORY** Two statements that can be proved using mathematical induction are $\frac{1}{3} + \frac{1}{3^2} + \frac{1}{3^3} + \cdots + \frac{1}{3^n} = \frac{1}{2}\left(1 - \frac{1}{3^n}\right)$ and $\frac{1}{4} + \frac{1}{4^2} + \frac{1}{4^3} + \cdots + \frac{1}{4^n} = \frac{1}{3}\left(1 - \frac{1}{4^n}\right)$. Write and prove a conjecture involving $\frac{1}{5}$ that is similar to the statements. (Lesson 11-7)

Chapter 12 Probability and Statistics (pp. 742–803)

Determine if the statements show a *correlation* or a *causation*. (Lesson 12-1)

1. People who own red cars are twice as likely to have an accident as people who own blue cars.
2. An apple that is red has color.
3. Tom is in a room that is not empty.
4. Men who have beards are happier than men who do not.

5. **FOOTBALL** The tables show the Ohio State University football scores in the 2006 and 2007 seasons. (Lesson 12-2)

2006 Season

35	24	37	28
38	35	38	44
44	17	54	42

2007 Season

38	20	33	58
30	23	48	24
37	38	28	14

   a. Find the mean for the 2006 season.
   b. Find the mean for the 2007 season.
   c. Find the standard deviation of scores from 2006.
   d. Find the standard deviation of scores from 2007.
   e. How do the standard deviations compare?
   f. What conclusion can you make about the two seasons?

6. **DRIVING** The table shows the number of students who have a driver's license and the number who own a personal vehicle. (Lesson 12-3)

Own?	License	No License
Y	148	2
N	124	86

   a. Find the probability that Harry owns a vehicle, given that he has a license.
   b. Find the probability that Anton does not have a license, given that he does not own a vehicle.
   c. Find the probability that Morgan does not own a vehicle, given that she does not have a license.

7. **MARBLES** A jar contains 3 white and 6 red marbles, all of equal size. Three marbles are drawn at random without replacement. What is the probability that at least 2 marbles drawn are red? (Lesson 12-4)

8. **DRAWINGS** Twenty-four students entered the drawing below. What is the probability that 2 of 3 friends who entered won a ticket? (Lesson 12-4)

FREE
Concert Tickets!
Five tickets to tonight's show are being given away!
Enter to win!

9. **PINEAPPLES** The graph of the normal distribution of the number of pineapple rings in each can is shown. (Lesson 12-5)

Number of Pineapple Rings Per Can

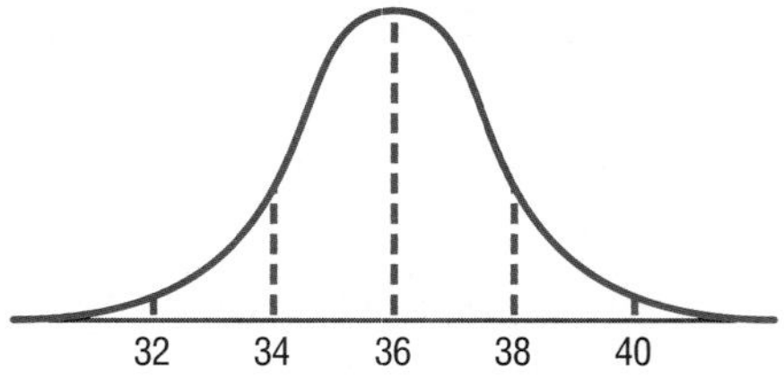

   a. Find the mean and standard deviation.
   b. What percent of the cans can be expected to contain at least 34 pineapple rings?
   c. In a random sample of 840 cans, how many can be expected to contain less than 40 pineapple rings?

10. **PHONES** A sample of 144 students was asked for the average amount of time they spent talking on the phone every day. The mean time was 85 minutes with a standard deviation of 3.5 minutes. Determine a 95% confidence interval. (Lesson 12-6)

11. **COINS** A fair coin is tossed 10 times. What is the probability that exactly 6 heads will occur? (Lesson 12-7)

12. **TELEPHONES** In the early days of telephone service, there was a probability of 0.8 of success in any attempt to make a telephone call. Calculate the probability of having 7 successes in 10 attempts. (Lesson 12-7)

1. **TREES** From a point on the ground 25 feet from the foot of a tree, the angle of elevation of the top of the tree is 32°. Find the height of the tree to the nearest foot. (Lesson 13-1)

2. **LADDER** A ladder 6 feet long leans against a wall and makes an angle of 71° with the ground. Find to the nearest tenth of a foot how high up the wall the ladder will reach. (Lesson 13-1)

3. **BICYCLES** A bicycle tire has a diameter of 28 inches. How far does a bicycle travel in feet after $1\frac{1}{4}$ tire rotations? (Lesson 13-2)

4. **MUSIC** A metronome is a metal bar that is 5 inches long that swings back and forth to keep track of beats for music.

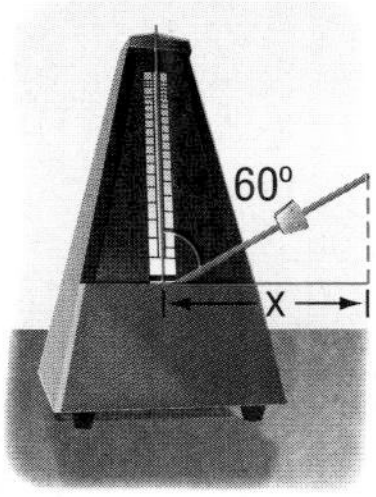

From the vertical position, it rotates 60° in each direction. How far outside the box, x, does the metronome arm reach? (Lesson 13-3)

5. **BOATS** Samuel sat on the deck of a river steamboat. As the paddle wheel turned, he noticed that a piece of seaweed was caught on one of the paddles. He started to keep track of the time and position above the water of the seaweed. When his stopwatch read 4 seconds, the seaweed was at its highest point, 16 feet above the surface of the water. The wheel's diameter was 18 ft and it completed its revolution every 10 seconds. (Lesson 13-3)
   - **a.** Make a table showing the height of the seaweed at 4, 6.5, 9, 11.5, 14, and 16.5 seconds.
   - **b.** Make a graph of the function. Let the horizontal axis represent the time t and the vertical axis represent the height h in inches that the paddle is from the water.

6. **TOWER** Miguel needs to measure the height of the Leaning Tower of Pisa. He walks exactly 200 feet from the base of the tower and looks up. The angle from the ground to the top of the tower is 44.3°. The Leaning Tower of Pisa leans about 4° towards Miguel. How tall is the Leaning Tower of Pisa? (Lesson 13-4)

7. **SURVEYING** To approximate the length of a quarry, a surveyor starts at one end of the lake and walks 245 yards. He then turns 110° and walks 270 yards until he arrives at the other end of the lake. Approximately how long is the lake? (Lesson 13-5)

8. **FISH** Goldfish can hear sound with a lower frequency than humans can hear. Goldfish can hear as low as 20 hertz, or 20 cycles per second. (Lesson 13-6)
   - **a.** Find the period of the function that models the sound waves.
   - **b.** Let the amplitude equal 1 unit. Write a sine function to represent the sound wave y as a function of time t.
   - **c.** Graph the function.

9. **BIOLOGY** In a certain wildlife refuge, the population of field mice can be modeled by $y = 3000 + 1250 \sin \frac{\pi}{6}t$, where y represents the number of mice and t represents the number of months past March 1 of a given year. (Lesson 13-7)
   - **a.** Determine the period of the function. What does this period represent?
   - **b.** What is the maximum number of mice, and when does this occur?

10. **ECONOMY** An economist indicates that the demand for temporary employment (measured in thousands of job applications per week) in Sid's county can be modeled by the function $d = 4.3 \sin (0.82t + 0.3) + 7.3$, where t is the time in years since January 1995. (Lesson 13-8)
    - **a.** State the amplitude, period, and vertical shift of the function.
    - **b.** Interpret the meaning of the function in the context of the situation.

11. **SQUIRRELS** A squirrel in a 10-foot tree was watching a cat that was sitting 9 feet from the base of the tree. At what angle was the squirrel looking? (Lesson 13-9)

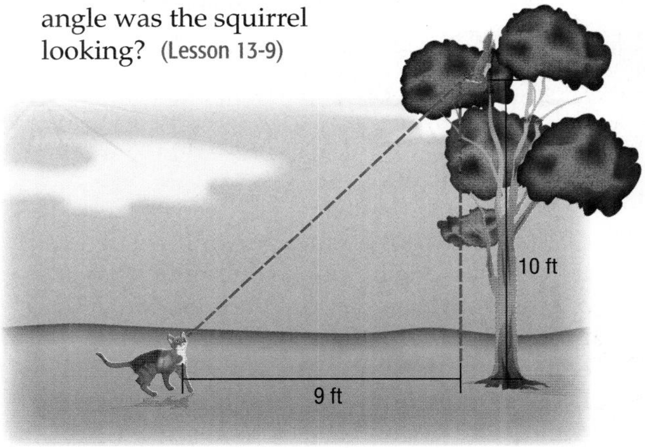

Mixed Problem Solving

1. GEOMETRY The area of a triangle can be determined by the formula $A = \frac{1}{2}ab \sin C$. (Lesson 14-1)

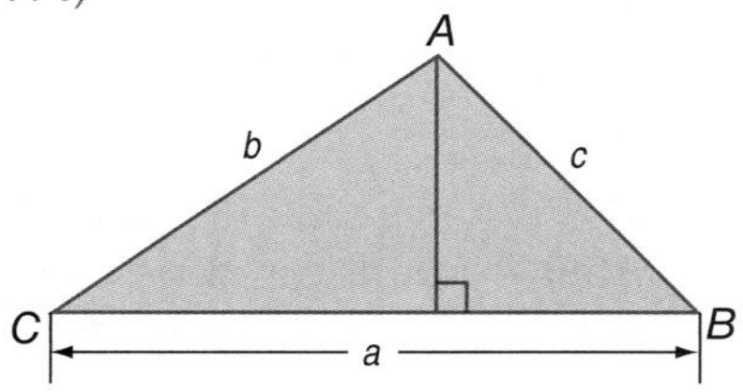

Rewrite the formula in terms of csc *C*.

2. SCIENCE Snell's Law describes the relationship between the angles of incidence and refraction passing through a boundary between two different media, such as air and glass.

$$\frac{\sin \theta_1}{\sin \theta_2} = \frac{v_1}{v_2}$$

Rewrite Snell's Law in terms of csc θ. (Lesson 14-1)

3. ROCKETS In the formula $h = \frac{v^2 \sin^2 \theta}{2g}$, h is the maximum height reached by a rocket, θ is the angle between the ground and the initial path of the object, v is the rocket's initial velocity, and g is the acceleration due to gravity. Verify the identity $\frac{v^2 \sin^2 \theta}{2g} = \frac{v^2 \cos^2 \theta}{2g \cot^2 \theta}$. (Lesson 14-2)

4. NEIGHBORHOOD The following grid is a map of Newtown. Francine lives at *O*. Norman lives at *N*. Mark lives at *M*. School is at *J*. $\angle a$ is 38° and $\angle b$ is 42°. Francine lives 1 mile from school. Round to the nearest hundredth, if necessary. (Lesson 14-3)

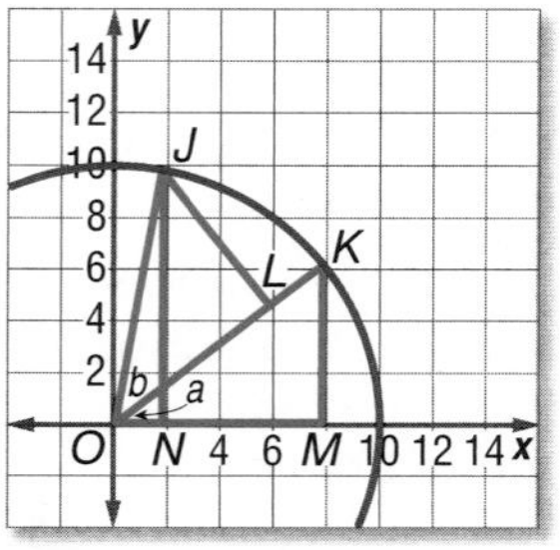

a. How far does Mark live from Francine?

b. How much closer does Norman live to school than Francine?

5. COMMUNICATION A radio transmitter sends out two signals, one for voice communication and another for data. Suppose the equation of the voice wave is $v = 10 \sin (2t - 30°)$, and the equation of the data wave is $d = 10 \cos (2t + 60°)$. Draw a graph of the waves when they are combined. (Lesson 14-3)

6. PHYSICAL SCIENCE Scott was solving a series of equations concerning maximizing the distance for throwing a rock off a bridge when he came across this equation.

$$x = \frac{1}{g(v^2 \sin \theta \cos \theta)}$$

Use the double-angle identity to simplify this equation. (Lesson 14-4)

7. AVIATION When a jet travels at speeds greater than the speed of sound, a sonic boom is created by the sound waves forming a cone behind the jet. If θ is the measure of the angle at the vertex of the cone, then the Mach number M can be determined using the formula $\sin \frac{\theta}{2} = \frac{1}{M}$. Find the Mach number of a jet if a sonic boom is created by a cone with a vertex angle of 75°. (Lesson 14-4)

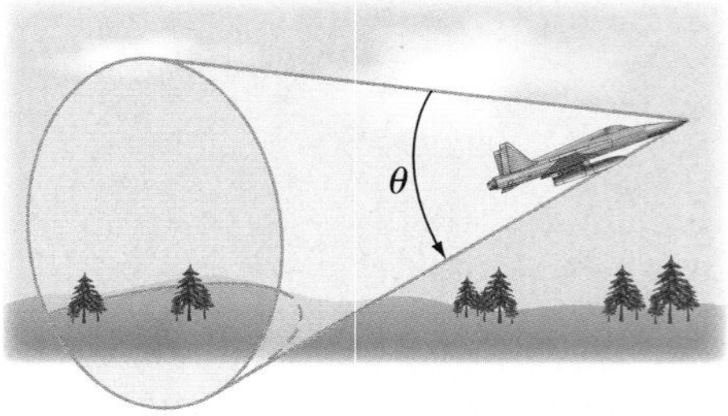

8. LADDER A ladder is leaning against the wall. What is the measure of the angle *A* that the ladder makes with the ground? (Lesson 14-5)

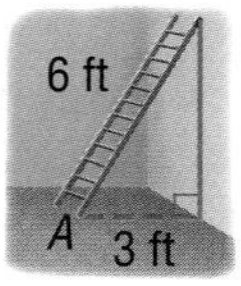

9. LAKES Determine the length ℓ of the lake. (Lesson 14-5)

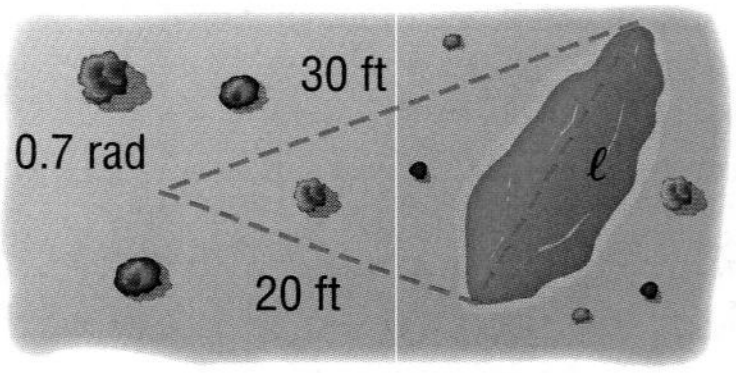

10. MOUNTAIN Determine the distance *d* to the top of the mountain. (Lesson 14-5)

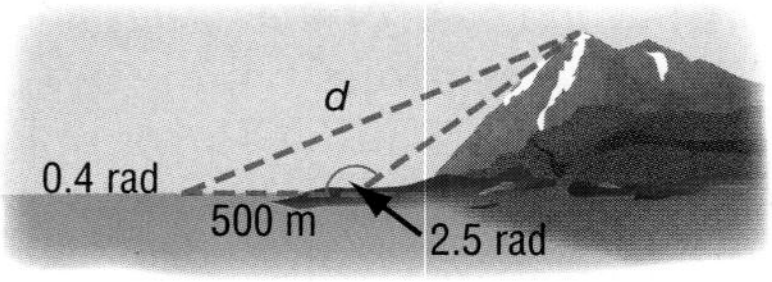

Concepts and Skills Bank

1 Proportional Reasoning

Two quantities are **proportional** if they have a constant ratio or rate. Relationships in which the ratios or rates are not constant are said to be **nonproportional**.

proportional

Distance (m)	3	9	12	21
Time (sec)	1	3	4	7

$$\frac{3}{1} = \frac{9}{3} = \frac{12}{4} = \frac{21}{7}$$

nonproportional

Distance (m)	1	4	7	9
Time (sec)	2	5	8	10

$$\frac{1}{2} \neq \frac{4}{5} \neq \frac{7}{8} \neq \frac{9}{10}$$

When two ratios are equal, they form a **proportion**. Consider the following proportion.

$$\frac{a}{b} = \frac{c}{d}$$

$$\frac{a}{\not{b}} \cdot \not{b}d = \frac{c}{\not{d}} \cdot b\not{d}$$ **Multiply each side by *bd*.**

$$ad = cb$$ **Simplify.**

The products *ad* and *cb* are called the **cross products** of this proportion. You can use cross products to determine whether two ratios form a proportion.

EXAMPLE 1

Use cross products to determine whether each pair of ratios forms a proportion.

a. $\frac{4.8}{6.4}, \frac{2.1}{2.8}$

$\frac{4.8}{6.4} \stackrel{?}{=} \frac{2.1}{2.8}$ **Write a proportion.**

$4.8 \cdot 2.8 \stackrel{?}{=} 2.1 \cdot 6.4$ **Find the cross products.**

$13.44 = 13.44$ ✓ **Simplify.**

The cross products are equal, so the ratios form a proportion.

b. $\frac{6.25}{7.5}, \frac{9.75}{12}$

$\frac{6.25}{7.5} \stackrel{?}{=} \frac{9.75}{12}$ **Write a proportion.**

$6.25 \cdot 12 \stackrel{?}{=} 9.75 \cdot 7.5$ **Find the cross products.**

$75 \neq 73.125$ ✗ **Simplify.**

The cross products are not equal, so the ratios do not form a proportion.

You can use cross products to solve a proportion in which one of the quantities is not known.

EXAMPLE 2

Solve each proportion.

a. $\frac{9}{15} = \frac{n}{25}$

$\frac{9}{15} = \frac{n}{25}$	**Original proportion**
$9 \cdot 25 = n \cdot 15$	**Cross products**
$225 = 15n$	**Multiply.**
$\frac{225}{15} = n$	**Divide each side by 15.**
$15 = n$	**Simplify.**

b. $\frac{10}{8.4} = \frac{5}{x}$

$\frac{10}{8.4} = \frac{5}{x}$	**Original proportion**
$10 \cdot x = 5 \cdot 8.4$	**Cross products**
$10x = 42$	**Multiply.**
$x = \frac{42}{10}$	**Divide each side by 10.**
$x = 4.2$	**Simplify.**

Exercises

Determine whether each pair of ratios forms a proportion. Write *yes* or *no*.

1. $\frac{3}{2}, \frac{21}{14}$
2. $\frac{8}{9}, \frac{12}{18}$
3. $\frac{2.3}{3.4}, \frac{3.0}{3.6}$
4. $\frac{5}{2}, \frac{4}{1.6}$
5. $\frac{21.1}{14.4}, \frac{1.1}{1.2}$
6. $\frac{4.2}{5.6}, \frac{1.68}{2.24}$
7. $\frac{4}{11}, \frac{12}{33}$
8. $\frac{16}{17}, \frac{8}{9}$

Solve each proportion.

9. $\frac{p}{6} = \frac{24}{36}$
10. $\frac{w}{11} = \frac{14}{22}$
11. $\frac{4}{10} = \frac{8}{a}$
12. $\frac{18}{12} = \frac{24}{q}$
13. $\frac{5}{h} = \frac{10}{30}$
14. $\frac{51}{z} = \frac{17}{7}$
15. $\frac{7}{45} = \frac{x}{9}$
16. $\frac{2}{15} = \frac{c}{72}$
17. $\frac{7}{5} = \frac{10.5}{b}$
18. $\frac{16}{7} = \frac{4.8}{h}$
19. $\frac{2}{9.4} = \frac{0.2}{v}$
20. $\frac{9}{7.2} = \frac{3.5}{k}$

21. DRIVING Marci drove 238 miles in 3.5 hours. At that rate, how long will it take her to drive an additional 87 miles?

22. TUTORING Amanda earns $28.50 tutoring for 3 hours. Write an equation relating her earnings m to the number of hours h she tutors. How much would Amanda earn tutoring for 2 hours? for 4.5 hours?

23. PHOTOGRAPHY A 3 inch-by-5 inch photo is enlarged so that the length of the new photo is 7 inches. Find the width of the new photo.

24. TRAVEL The Lehmans' minivan requires 5 gallons of gasoline to travel 120 miles. How much gasoline will they need for a 350-mile trip?

② Square Roots

You can estimate square roots by using perfect squares.

EXAMPLE 1

Estimate $\sqrt{45}$ to the nearest whole number.

- The first perfect square less than 45 is 36. $\sqrt{36} = 6$
- The first perfect square greater than 45 is 49. $\sqrt{49} = 7$
- Plot each square root on a number line.

The square root of 45 is between the whole numbers 6 and 7. Because 45 is closer to 49 than 36, you can expect that $\sqrt{45}$ is closer to 7 than 6.

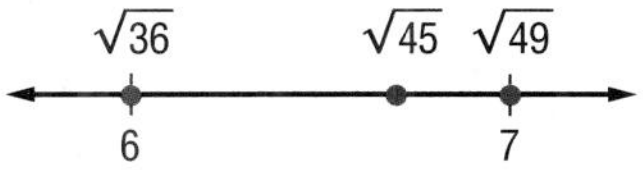

A radical expression is an expression that contains a square root. The expression is in simplest form when the following conditions have been met.

- No radicands have perfect square factors other than 1.
- No radicands contain fractions.
- No radicals appear in the denominator of a fraction.

The **Product Property of Square Roots** states that for two real numbers a and b, where $a \geq 0$ and $b \geq 0$, $\sqrt{ab} = \sqrt{a} \cdot \sqrt{b}$.

EXAMPLE 2

Simplify.

a. $\sqrt{96}$

$\sqrt{96}$

$= \sqrt{2 \cdot 2 \cdot 2 \cdot 2 \cdot 6}$ — **Prime factorization of 96**

$= \sqrt{2^2} \cdot \sqrt{2^2} \cdot \sqrt{6}$ — **Product Property of Square Roots**

$= 2 \cdot 2 \cdot \sqrt{6}$ — **$\sqrt{2^2} = 2$**

$= 4\sqrt{6}$ — **Simplify.**

b. $\sqrt{32} \cdot \sqrt{75}$

$\sqrt{32} \cdot \sqrt{75}$

$= \sqrt{32 \cdot 75}$ — **Product Property**

$= \sqrt{2 \cdot 2 \cdot 2 \cdot 2 \cdot 2 \cdot 3 \cdot 5 \cdot 5}$ — **Prime factorization**

$= \sqrt{2^2} \cdot \sqrt{2^2} \cdot \sqrt{2} \cdot \sqrt{3} \cdot \sqrt{5^2}$ — **Product Property**

$= 2 \cdot 2 \cdot 5 \cdot \sqrt{2} \cdot \sqrt{3}$ — **$\sqrt{2^2} = 2$ and $\sqrt{5^2} = 5$**

$= 20 \cdot \sqrt{2} \cdot \sqrt{3}$ — **Simplify.**

$= 20\sqrt{2 \cdot 3}$ or $20\sqrt{6}$ — **Product Property**

The **Quotient Property of Square Roots** states that for any real numbers a and b, where $a \geq 0$ and $b \geq 0$, $\sqrt{\frac{a}{b}} = \frac{\sqrt{a}}{\sqrt{b}}$.

EXAMPLE 3

Simplify $\sqrt{\frac{9}{64}}$.

$\sqrt{\frac{9}{64}} = \frac{\sqrt{9}}{\sqrt{64}}$ — **Quotient Property of Square Roots**

$= \frac{3}{8}$ — **Simplify.**

Concepts and Skills Bank

Rationalizing the denominator of a radical expression is a method used to eliminate radicals from the denominator of a fraction. To rationalize the denominator, multiply the expression by a fraction equivalent to 1 such that the resulting denominator is a perfect square.

EXAMPLE 4

Simplify $\frac{4}{\sqrt{5}}$.

$\frac{4}{\sqrt{5}} = \frac{4}{\sqrt{5}} \cdot \frac{\sqrt{5}}{\sqrt{5}}$ **Multiply by $\frac{\sqrt{5}}{\sqrt{5}}$.**

$= \frac{4\sqrt{5}}{5}$ **Simplify.**

Conjugates can be used to simplify radical expressions. Conjugates are binomials of the form $p\sqrt{a} + r\sqrt{b}$ and $p\sqrt{a} - r\sqrt{b}$.

EXAMPLE 5

Simplify $\frac{3}{1-\sqrt{2}}$.

$\frac{3}{1-\sqrt{2}} = \frac{3}{1-\sqrt{2}} \cdot \frac{1+\sqrt{2}}{1+\sqrt{2}}$ **$\frac{1+\sqrt{2}}{1+\sqrt{2}} = 1$**

$= \frac{3(1+\sqrt{2})}{1^2 - (\sqrt{2})^2}$ **$(a-b)(a+b) = a^2 - b^2$**

$= \frac{3+3\sqrt{2}}{1-2}$ **$(\sqrt{2})^2 = 2$**

$= \frac{3+3\sqrt{2}}{-1}$ or $-\frac{3+3\sqrt{2}}{1}$ **Simplify.**

Exercises

Estimate each square root to the nearest whole number.

1. $\sqrt{66}$ **2.** $\sqrt{103}$ **3.** $\sqrt{79}$ **4.** $\sqrt{95}$

5. $\sqrt{54}$ **6.** $\sqrt{125}$ **7.** $\sqrt{200}$ **8.** $\sqrt{396}$

Simplify.

9. $\sqrt{20}$ **10.** $\sqrt{52}$ **11.** $\sqrt{18}$ **12.** $\sqrt{24}$

13. $\sqrt{80}$ **14.** $\sqrt{75}$ **15.** $2\sqrt{32}$ **16.** $10\sqrt{90}$

17. $\sqrt{2} \cdot \sqrt{8}$ **18.** $\sqrt{3} \cdot \sqrt{18}$ **19.** $\sqrt{5} \cdot \sqrt{6}$ **20.** $\sqrt{3} \cdot \sqrt{8}$

21. $3\sqrt{10} \cdot 4\sqrt{10}$ **22.** $7\sqrt{30} \cdot 2\sqrt{6}$ **23.** $2\sqrt{3} \cdot 5\sqrt{27}$ **24.** $12\sqrt{5} \cdot 3\sqrt{45}$

25. $\sqrt{\frac{81}{49}}$ **26.** $\sqrt{\frac{25}{64}}$ **27.** $\sqrt{\frac{42}{121}}$ **28.** $\sqrt{\frac{3}{10}}$

29. $\frac{4}{\sqrt{6}}$ **30.** $\frac{\sqrt{14}}{\sqrt{5}}$ **31.** $\sqrt{\frac{2}{7}} \cdot \sqrt{\frac{7}{3}}$ **32.** $\sqrt{\frac{3}{5}} \cdot \sqrt{\frac{6}{4}}$

33. $\frac{3}{2+\sqrt{2}}$ **34.** $\frac{7}{3-\sqrt{7}}$ **35.** $\frac{18}{6-\sqrt{2}}$ **36.** $\frac{3\sqrt{3}}{\sqrt{6}-2}$

37. $\frac{10}{\sqrt{7}+\sqrt{2}}$ **38.** $\frac{2}{\sqrt{3}+\sqrt{6}}$ **39.** $\frac{4}{4-3\sqrt{3}}$ **40.** $\frac{3\sqrt{7}}{5\sqrt{3}+3\sqrt{5}}$

3 Scientific Notation

A number is expressed in **scientific notation** when it is written as the product of a factor and a power of 10. The factor must be greater than or equal to 1 and less than 10.

$$a \times 10^n, \text{ where } 1 \leq a < 10 \text{ and } n \text{ is an integer}$$

EXAMPLE 1

Express each number in scientific notation.

a. 32,500,000

$32{,}500{,}000 = 3.25 \times 10{,}000{,}000$ **The decimal point moves 7 places.**

$= 3.25 \times 10^7$ **$10{,}000{,}000 = 10^7$**

b. 0.00625

$0.00625 = 6.25 \times 0.001$ **The decimal point moves 3 places.**

$= 6.25 \times 10^{-3}$ **$0.001 = 10^{-3}$**

You can use a calculator to evaluate expressions involving scientific notation.

EXAMPLE 2

Evaluate.

a. $(6.58 \times 10^6)(3.97 \times 10^4)$

Enter the expression in a calculator. Use the [2nd] [EE] function to enter the base factor and the power of 10.

KEYSTROKES: 6.58 [2nd] [EE] 6 [×] 3.97 [2nd] [EE] 4 [ENTER] 2.61226E11

So, $(6.58 \times 10^6)(3.97 \times 10^4) = 2.61226 \times 10^{11}$.

b. $\dfrac{4.77 \times 10^{-8}}{1.02 \times 10^{-4}}$

KEYSTROKES: 4.77 [2nd] [EE] [(−)] 8 [÷] 1.02 [2nd] [EE] [(−)] 4 [ENTER] 4.676470588E−4

So, $\dfrac{4.77 \times 10^{-8}}{1.02 \times 10^{-4}} = 4.676470588 \times 10^{-4}$.

Exercises

Express each number in scientific notation.

1. 2,000,000
2. 499,000
3. 0.006
4. 0.0125
5. 50,000,000
6. 39,560
7. 0.000078
8. 0.000425

Evaluate.

9. $(4.24 \times 10^2)(5.72 \times 10^4)$
10. $(3.347 \times 10^{-1})(5.689 \times 10^{-3})$
11. $(1.399 \times 10^5)(1.5 \times 10^{-4})$
12. $\dfrac{9.01 \times 10^{-2}}{2.505 \times 10^3}$
13. $\dfrac{6.1 \times 10^4}{7.32 \times 10^7}$
14. $\dfrac{6.02 \times 10^{-12}}{9.931 \times 10^5}$

④ Adding and Multiplying Probabilities

A single event, like rolling a 5 on a die, is called a **simple event**. Rolling a 5 on a die and drawing a 2 from a standard deck of cards is an example of a **compound event**, which is made up of two or more simple events. The roll of a die does not affect the selection of a card. These two events are called **independent events**. To find the probability of independent events, multiply the individual probabilities.

EXAMPLE 1

In a board game, three dice are rolled to determine the number of moves for the players. What is the probability that the first die shows a 6, the second die shows a 6, and the third die does not?

Let A be the event that the first die shows a 6. $\rightarrow$ $P(A) = \frac{1}{6}$

Let B be the event that the second die shows a 6. $\rightarrow$ $P(B) = \frac{1}{6}$

Let C be the event that the third die does *not* show a 6. $\rightarrow$ $P(C) = \frac{5}{6}$

$P(A, B, \text{and } C) = P(A) \cdot P(B) \cdot P(C)$ **Probability of independent events**

$= \frac{1}{6} \cdot \frac{1}{6} \cdot \frac{5}{6}$ **Substitute.**

$= \frac{5}{216}$ **Multiply.**

The probability that the first and second dice show a 6 and the third die does not is $\frac{5}{216}$.

When the outcome of one event affects the outcome of another event, such as drawing two cards from a standard deck without replacement, the events are **dependent events**.

EXAMPLE 2

A bag contains 12 red marbles, 9 blue marbles, 11 yellow marbles, and 8 green marbles. Three marbles are randomly drawn from the bag one at a time and not replaced. Find the probability that red, blue, and green marbles are selected in order.

The selection of the first marble affects the selection of the next marble because there is one less marble from which to choose. So, the events are dependent.

First marble: $P(\text{red}) = \frac{12}{40}$ or $\frac{3}{10}$ $\leftarrow$ **number of red marbles** / $\leftarrow$ **total number of marbles**

Second marble: $P(\text{blue}) = \frac{9}{39}$ or $\frac{3}{13}$ $\leftarrow$ **number of blue marbles** / $\leftarrow$ **number of marbles remaining**

Third marble: $P(\text{green}) = \frac{8}{38}$ or $\frac{4}{19}$ $\leftarrow$ **number of green marbles** / $\leftarrow$ **number of marbles remaining**

$P(\text{red, blue, green}) = P(\text{red}) \cdot P(\text{blue}) \cdot P(\text{green})$ **Probability of dependent events**

$= \frac{3}{10} \cdot \frac{3}{13} \cdot \frac{4}{19}$ **Substitute.**

$= \frac{36}{2470}$ or $\frac{18}{1235}$ **Multiply and simplify.**

The probability of selecting red, blue, and green marbles in order is $\frac{18}{1235}$.

Events that cannot occur at the same time are called **mutually exclusive**. Suppose you want to find the probability of rolling a 2 *or* a 4 on a die. Because a die cannot show both a 2 and a 4 at the same time, the events are mutually exclusive. To determine the probability of mutually exclusive events, add the individual probabilities.

EXAMPLE 3

Keisha has a stack of 8 baseball cards, 5 basketball cards, and 6 hockey cards. If she selects a card at random from the stack, what is the probability that it is a baseball or a hockey card?

These are mutually exclusive events, because the card cannot be both a baseball card *and* a hockey card.

$P(\text{baseball or hockey}) = P(\text{baseball}) + P(\text{hockey})$ **Mutually exclusive events**

$= \frac{8}{19} + \frac{6}{19} \text{ or } \frac{14}{19}$ **Substitute and add.**

The probability that Keisha selects a baseball or a hockey card is $\frac{14}{19}$.

What is the probability of drawing a king or a spade from a standard deck of cards? Since it is possible to draw a card that is both a king and a spade, these events are not mutually exclusive. These are called **inclusive events**. To determine the probability of inclusive events, add the individual probabilities and subtract the probability of both conditions.

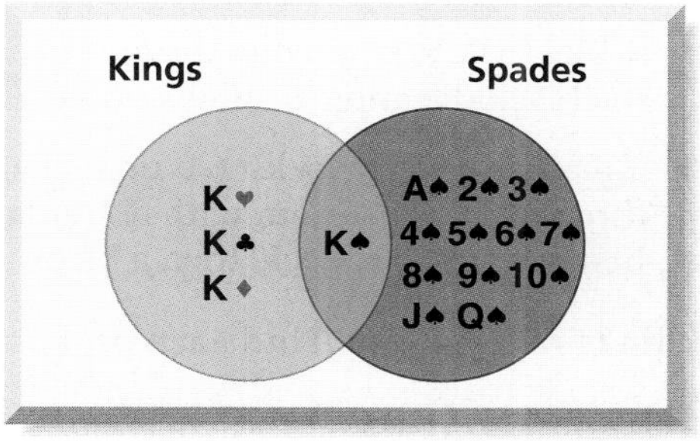

EXAMPLE 4

EDUCATION **Suppose that of 1400 students, 550 take Spanish, 700 take biology, and 400 take both Spanish and biology. What is the probability that a student selected at random takes Spanish or biology?**

$P(\text{Spanish}) = \frac{550}{1400}$ $P(\text{biology}) = \frac{700}{1400}$ $P(\text{Spanish and biology}) = \frac{400}{1400}$

$P(\text{Spanish or biology}) = P(\text{Spanish}) + P(\text{biology}) - P(\text{Spanish and biology})$

$= \frac{550}{1400} + \frac{700}{1400} - \frac{400}{1400}$ **Substitute.**

$= \frac{850}{1400}$ **Add and subtract.**

$= \frac{17}{28}$ **Simplify.**

The probability that a student selected at random takes Spanish or biology is $\frac{17}{28}$.

Probabilities based on known characteristics or facts are called **theoretical probabilities**. Probabilities that are based on the outcomes obtained by conducting an experiment are called **experimental probabilities**. Theoretical probabilities tell you what *should happen*, while experimental probabilities tell you what *actually happened*.

EXAMPLE 5

Find each probability.

a. What is the theoretical probability of rolling a double 6 using two dice?

$P(6 \text{ and } 6) = \frac{1}{6} \cdot \frac{1}{6}$ or $\frac{1}{36}$

b. The graph shows the results of an experiment in which two number cubes were rolled. Based on the experiment, what is the probability that a sum of 6 or 12 occurs?

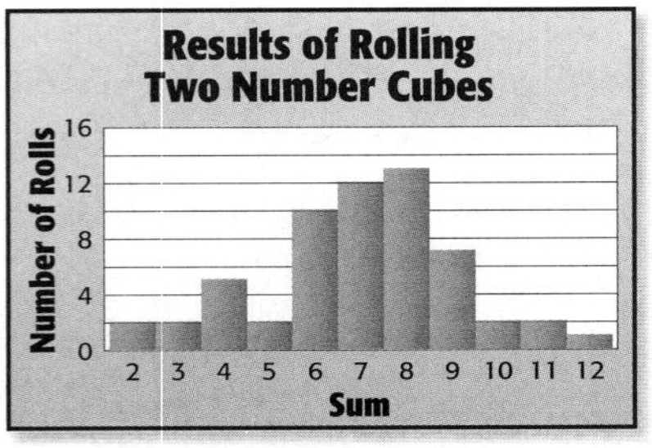

These are mutually exclusive events, because the sum cannot be both 6 and 12.

$P(\text{sum is 6 or 12}) = P(\text{sum is 6}) + P(\text{sum is 12})$ **Mutually exclusive events**

$= \frac{10}{58} + \frac{1}{58}$ or $\frac{11}{58}$ **Substitute and add.**

Exercises

Determine whether the events are *independent* or *dependent*. Then find the probability.

1. A black die and a white die are rolled. What is the probability that a 3 shows on the black die and a 5 shows on the white die?
2. Yana has 4 black socks, 6 blue socks, and 8 white socks in his drawer. If he selects three socks at random with no replacement, what is the probability that he will first select a blue sock, then a black sock, and then another blue sock?

A die is rolled twice. Find each probability.

3. P(2, then 3)
4. P(no 6s)
5. P(two of the same number)
6. P(two 4s)
7. P(1, then any number)
8. P(two different numbers)

There are 8 action, 3 comedy, and 5 children's DVDs on a shelf. Suppose two DVDs are selected at random from the shelf. Find each probability.

9. P(2 action DVDs), if replacement occurs
10. P(2 action DVDs), if no replacement occurs
11. P(a comedy DVD, then a children's DVD), if no replacement occurs

Six girls and eight boys walk into a video store at the same time. There are six salespeople available to help them. Find the probability that the salespeople will first help the given numbers of girls and boys.

12. P(4 girls, 2 boys or 4 boys, 2 girls)
13. P(5 girls, 1 boy or 5 boys, 1 girl)
14. P(all girls or all boys)
15. P(at least 4 boys)

Two cards are drawn from a standard deck of cards. Find each probability.

16. P(both queens or both red)
17. P(both jacks or both face cards)
18. P(both face cards or both black)
19. P(both either black or ace)

Determine whether each probability is *theoretical* or *experimental*. Then find the probability.

20. Two dice are rolled. What is the probability that the sum will be 10?
21. A baseball player has 126 hits in 410 at-bats this season. What is the probability that he gets a hit in his next at-bat?
22. A hand of 2 cards is dealt from a standard deck of cards. What is the probability that both cards are clubs?

⑤ Bar and Line Graphs

A **bar graph** compares different categories of data by showing each as a bar whose length is related to the frequency. A **double bar graph** compares two sets of data. Another way to represent data is by using a **line graph**. A line graph usually shows how data changes over a period of time.

EXAMPLE 1

The table shows the average age at which Americans marry for the first time. Make a double bar graph to display the data.

Step 1 Draw a horizontal and a vertical axis and label them as shown.

Step 2 Draw side-by-side bars to represent each category.

Average Age to Marry		
Year	1990	2007
men	26	27
women	22	25

The legend indicates that the blue bars refer to men and the red bars refer to women.

The side-by-side bars compare the ages of men and women for each year.

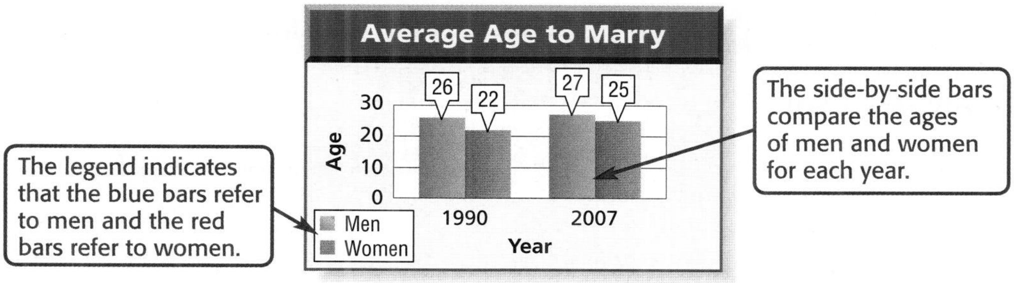

EXAMPLE 2

The table shows Mark's height at 2-year intervals. Make a line graph to display the data.

Age	2	4	6	8	10	12	14	16
Height (feet)	2.8	3.5	4.0	4.6	4.9	5.2	5.8	6

Step 1 Draw a horizontal and a vertical axis. Label them as shown.

Step 2 Plot the points.

Step 3 Draw a line connecting each pair of consecutive points.

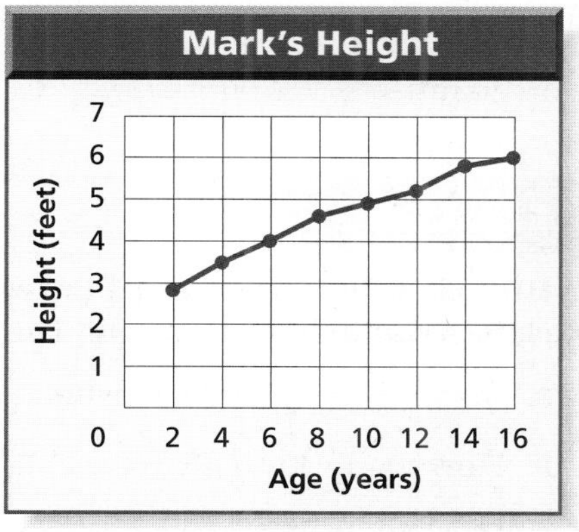

Exercises

1. The table below shows the life expectancy for Americans born in each year listed. Make a double bar graph to display the data.

Life Expectancy		
Year of Birth	Male	Female
1980	70.0	77.5
1985	71.2	78.2
1990	71.8	78.8
1995	72.5	78.9
2004	75.2	80.4

Source: *World Almanac*

2. The amount of money in Becky's savings account from August through March is shown in the table below. Make a line graph to display the data.

Month	Amount	Month	Amount
August	$300	December	$780
September	$400	January	$800
October	$700	February	$950
November	$780	March	$900

6 Frequency Tables and Histograms

A **frequency table** shows how often an item appears in a set of data. A tally mark is used to record each response. The total number of marks for a given response is the *frequency* of that response.

EXAMPLE 1

TELEVISION Use the frequency table.

a. How many more chose sports programs than news?

b. Which two programs together have the same frequency as adventures?

Favorite Television Shows		
Program	**Tally**	**Frequency**
Sports	𝍸 ||	7
Mysteries	𝍸	4
Soap operas	||||	5
News	𝍸	5
Quiz shows	𝍸 |	6
Music videos	||	2
Adventure	𝍸 ||||	9
Comedies	𝍸 ||	7

a. Seven people chose sports. Five people chose news. $7 - 5 = 2$, so 2 more people chose sports than news.

b. As many people chose adventures as the following pairs of programs.

sports and music videos
mysteries and news
mysteries and soap operas
comedies and music videos

Frequencies can be shown in a bar graph called a histogram. A **histogram** differs from other bar graphs in that no space is between the bars and the bars usually represent numbers grouped by intervals.

EXAMPLE 2

FITNESS A PE teacher tested the number of sit-ups students in two classes could do in 1 minute. The results are shown.

a. Make a histogram of the data. Title the histogram.

b. How many students were able to do 25–29 sit-ups in 1 minute?

c. How many students were unable to do 10 sit-ups in 1 minute?

d. Between which two consecutive intervals does the greatest increase in frequency occur? What is the increase?

Number of Sit-Ups	Frequency
0–4	8
5–9	12
10–14	15
15–19	6
20–24	18
25–29	10

a. Use the same intervals as those in the frequency table on the horizontal axis. Label the vertical axis with a scale that includes the frequency numbers from the table.

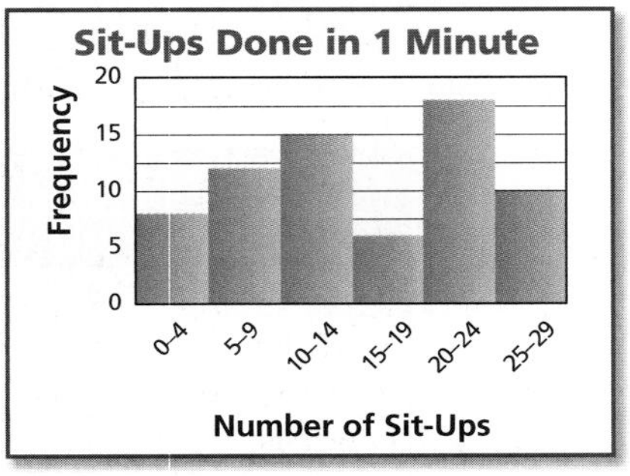

b. Ten students were able to do 25–29 sit-ups in 1 minute.

c. Add the students who did 0–4 sit-ups and 5–9 sit-ups. So $8 + 12$ or 20 students were unable to do 10 sit-ups in 1 min.

d. The greatest increase is between intervals 15–19 and 20–24. These frequencies are 6 and 18. So the increase is $18 - 6 = 12$.

Exercises

Display each set of data in a histogram.

1.

Weekly Study Time		
Time (hr)	**Tally**	**Frequency**
0–3	‖	2
4–6	‖‖	3
7–9	卌 ‖‖	8
10–12	卌 卌 ‖	12
13–15	卌 卌	10

2.

Weekly Allowance		
Amount	**Tally**	**Frequency**
\$0–\$5	卌 卌 |	11
\$6–\$11	卌 ||||	9
\$12–\$17	卌 |||	8
\$18–\$23	|||	3
\$24–\$29	卌	5

3. ART The prices, in dollars, of paintings sold at an art auction are shown.

1800	750	600	600	1800	1350	300	1200	750	600	750	2700
600	750	300	750	600	450	2700	1200	600	450	450	300

a. Make a frequency table of the data.

b. What price was paid most often for the artwork?

c. What is the average price paid for artwork at this auction?

d. How many paintings sold for at least \$600 and no more than \$1200?

4. PETS Refer to the table.

Number of Pets Per Family							
1	2	3	1	0	2	1	0
1	0	1	4	1	2	0	0
0	1	1	2	2	5	1	0

a. Use a frequency table to make a histogram of the data.

b. How many families own two to three pets?

c. How many families own more than three pets?

d. To the nearest percent, what percent of families own no pets?

e. Name the median, mode, and range of the data.

5. TREES Use the histogram shown.

a. Which interval contains the most evergreen seedlings?

b. Which intervals contain an equal number of trees?

c. Which intervals contain 95% of the data?

d. Between which two consecutive intervals does the greatest increase in frequency occur? What is the increase?

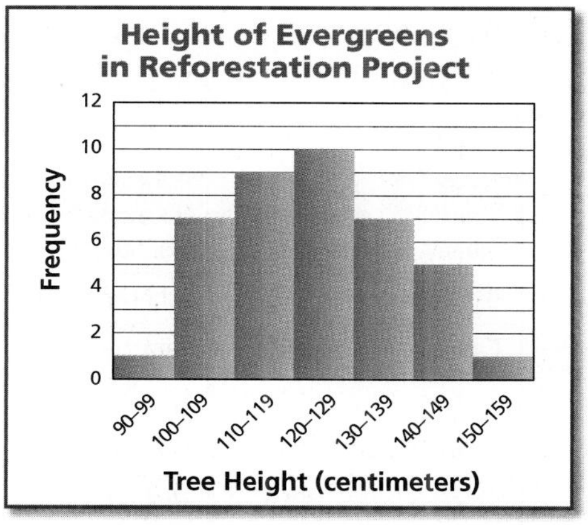

6. MARKET RESEARCH Johanna is a civil engineer studying traffic patterns. She counts the number of cars that make it through one green light cycle during rush hour. Organize her data into a frequency table, and then make a histogram.

15	16	10	8	8	14	9	7	6	9
10	11	14	10	7	8	9	11	14	10

Concepts and Skills Bank

7 Stem-and-Leaf Plots

In a **stem-and-leaf plot**, data are organized in two columns. The greatest place value of the data is used for the stems. The next greatest place value forms the leaves. Stem-and-leaf plots are useful for organizing long lists of numbers.

EXAMPLE 1

Isabella has collected data on the GPAs (grade point average) of the 16 students in the art club. Display the data in a stem-and-leaf plot.
{4.0, 3.9, 3.1, 3.9, 3.8, 3.7, 1.8, 2.6, 4.0, 3.9, 3.5, 3.3, 2.9, 2.5, 1.1, 3.5}

Step 1 Find the least and greatest numbers. Then identify the greatest place-value digit in each number, in this case, ones.

least data: 1.1 — The least number has 1 in the ones place.

greatest data: 4.0 — The greatest number has 4 in the ones place.

Step 2 Draw a vertical line and write the stems from 1 to 4 to the left of the line.

Step 3 Write the leaves to the right of the line with the corresponding stem. For example, write 0 to the right of 4 for 4.0. Arrange the leaves so they are ordered from least to greatest.

Step 4 Include a key or an explanation.

Stem	Leaf
1	1 8
2	5 6 9
3	1 3 5 5 7 8 9 9 9
4	0 0 *3|1 = 3.1*

Exercises

1. The stem-and-leaf plot at the right shows Charmaine's scores for her favorite computer game.
   a. What are Charmaine's highest and lowest scores?
   b. Which score(s) occurred most frequently?
   c. How many scores were above 115?

Stem	Leaf
9	0 0 0 1 3 4 5 5 7 8 8 8 9 9
10	0 3 4 4 5 6 9
11	0 3 9 9
12	1 2 6
13	0 *12|6 = 126*

2. The class scores on a 50-item test are shown in the table at the right. Make a stem-and-leaf plot of the data.

Test Scores

45	15	30	40	28	35
39	29	38	18	43	49
46	44	48	35	36	30

3. **GEOGRAPHY** The table shows the land area of each county in Wyoming. Round each area to the nearest hundred square miles and organize the data in a stem-and-leaf plot.

County	Area (mi^2)	County	Area (mi^2)	County	Area (mi^2)
Albany	4273	Hot Springs	2004	Sheridan	2523
Big Horn	3137	Johnson	4166	Sublette	4883
Campbell	4797	Laramie	2686	Sweetwater	10,425
Carbon	7896	Lincoln	4069	Teton	4008
Converse	4255	Natrona	5340	Unita	2082
Crook	2859	Niobrara	2626	Washakie	2240
Fremont	9182	Park	6942	Weston	2398
Goshen	2225	Platte	2085		

Source: *The World Almanac*

8 Box-and-Whisker Plots

In a set of data written in numerical order, **quartiles** are values that divide the data into four equal parts.

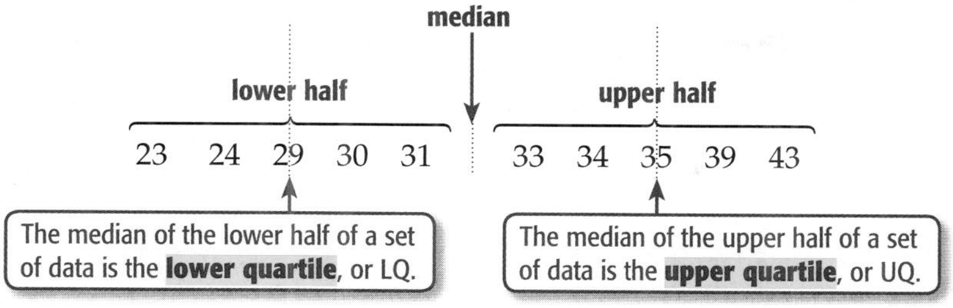

To make a **box-and-whisker plot**, draw a box around the quartile values, and draw lines or *whiskers* to represent the values in the lower fourth of the data and the upper fourth of the data.

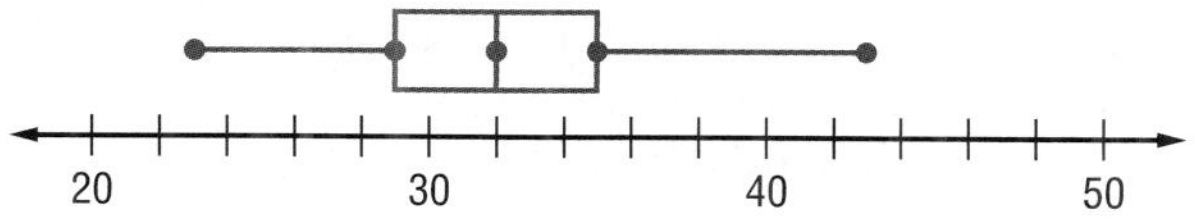

EXAMPLE 1

The amount spent in the cafeteria by 20 students is shown. Display the data in a box-and-whisker plot.

Amount Spent			
\$2.00	\$2.00	\$1.00	\$4.00
\$1.00	\$2.50	\$2.50	\$2.00
\$2.50	\$1.00	\$4.00	\$2.50
\$3.50	\$2.00	\$3.00	\$2.50
\$4.00	\$4.00	\$5.50	\$1.50

Step 1 Find the least and greatest number. Then draw a number line that covers the range of the data.

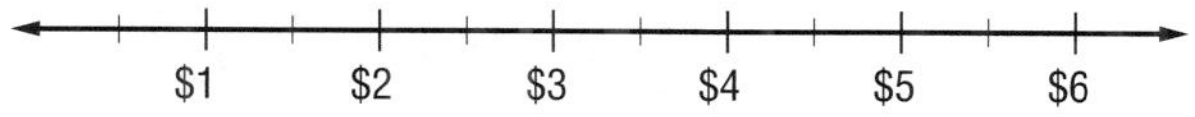

Step 2 Find the median, the extreme values, and the upper and lower quartiles. Mark these points above the number line.

1, 1, 1, 1.5, 2, 2, 2, 2, 2.5, 2.5, 2.5, 2.5, 2.5, 3, 3.5, 4, 4, 4, 4, 5.5

$LQ = \frac{2+2}{2}$ or 2 $M = \frac{2.5+2.5}{2}$ or 2.5 $UQ = \frac{3.5+4}{2}$ or 3.75

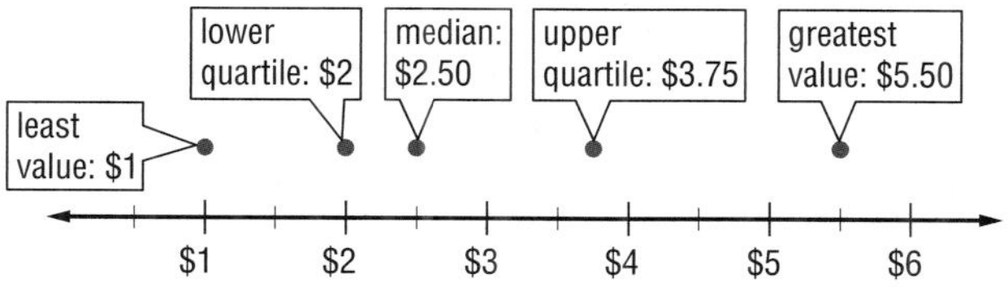

Step 3 Draw a box and the whiskers.

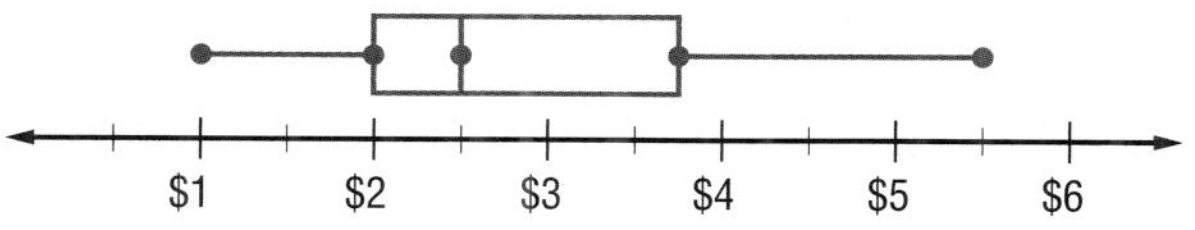

The **interquartile range (IQR)** is the range of the middle half of the data and contains 50% of the data in the set.

$$\text{interquartile range} = UQ - LQ$$

An **outlier** is any element of a set that is at least 1.5 interquartile ranges less than the lower quartile or greater than the upper quartile. The whisker representing the data is drawn from the box to the least or greatest value that is not an outlier. *Parallel box-and-whisker plots* can be used to compare two sets of data.

EXAMPLE 2

Two students are analyzing the number of hours they spent studying each day for the last month.

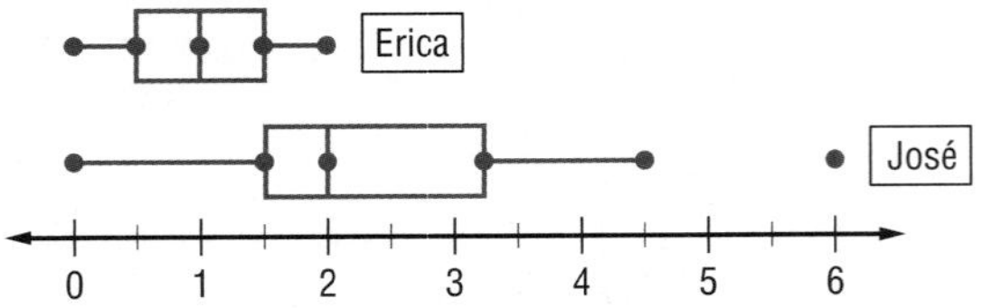

a. How does the time Erica spent studying compare to the time José spent studying?

Erica's daily study time was between a half hour and an hour and a half. José's daily study time was between an hour and a half and 3 hours 15 minutes. Thus, José's daily study time varies more than Erica's daily study time.

b. What was the greatest amount of time José studied in a day?

The greatest value in the plot is 6, so the greatest amount of time José studied in a day was 6 hours.

c. What is the interquartile range of this box-and-whisker plot?

The interquartile range is UQ − LQ. For this plot, the interquartile range is $3.25 - 1.5$ or 1.75 hours.

d. Identify any outliers in the data.

An outlier is at least 1.5(1.75) less than the lower quartile or more than the upper quartile. Since $3.25 + (1.5)(1.75) = 5.9$ and $6 > 5.9$, the value 6 is an outlier and was not included in the whisker.

Exercises

Tyler surveys 20 randomly chosen students at his school about how many miles they drive in an average day. The results are shown in the box-and-whisker plot.

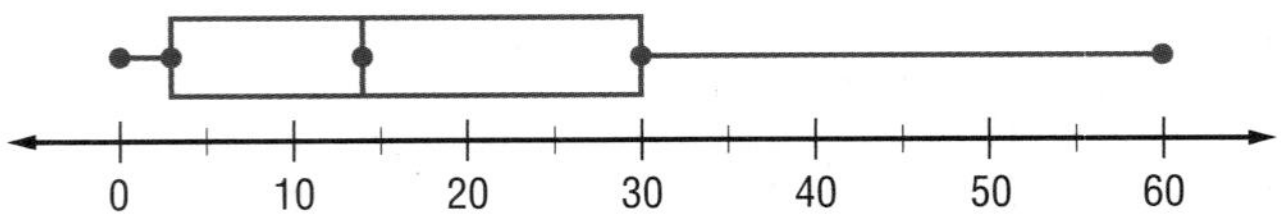

1. What percent of the students drive more than 30 miles in a day?
2. What is the interquartile range of the box-and-whisker plot?
3. Does a student at Tyler's school have a better chance to meet someone who drives the same mileage they do if they drive 50 miles in a day or 15 miles in a day? Why?
4. Carlos surveyed his friends to find the number of cans of soft drink they drink in an average week. Make a box-and-whisker plot of the data.
   {0, 0, 0, 1, 1, 1, 2, 2, 3, 4, 4, 5, 5, 7, 10, 10, 10, 11, 11}
5. The average life span of some animals commonly found in a zoo are given below. Make a box-and-whisker plot of the data.
   {1, 7, 7, 10, 12, 12, 15, 15, 18, 20, 20, 20, 25, 40, 100}

Key Concepts

Preparing for Advanced Algebra — Chapter 0

Independent Events (p. P9) If the outcome of an event does not affect the outcome of another event, the two events are independent.

Dependent Events (p. P9) If the outcome of an event does affect the outcome of another event, the two events are dependent.

Fundamental Counting Principle (p. P9) If event M can occur in m ways and is followed by event N that can occur in n ways, then the event M followed by event N can occur in $m \cdot n$ ways.

Permutations (p. P12) The number of permutations of n distinct objects taken r at a time is given by

$$P(n, r) = \frac{n!}{(n-r)!}.$$

Permutations with Repetition (p. P13) The number of permutations of n of which p are alike and q are alike is

$$\frac{n!}{p!\,q!}.$$

Combinations (p. P13) The number of combinations of n distinct objects taken r at a time is given by

$$C(n, r) = \frac{n!}{(n-r)!r!}.$$

Equations and Inequalities — Chapter 1

Order of Operations (p. 5)
- **Step 1** Evaluate expressions inside grouping symbols.
- **Step 2** Evaluate all powers.
- **Step 3** Multiply and/or divide from left to right.
- **Step 4** Add and/or subtract from left to right.

Properties of Equality (p. 19)
- **Reflexive** For any real number a, $a = a$.
- **Symmetric** For all real numbers a and b, if $a = b$, then $b = a$.
- **Transitive** For all real numbers a, b, and c, if $a = b$ and $b = c$, then $a = c$.
- **Substitution** If $a = b$, then a may be replaced by b and b may be replaced by a.

Absolute Value (p. 27) For any real number a, $|a| = a$ if $a \geq 0$, and $|a| = -a$ if $a < 0$.

Addition Property of Inequality (p. 34)
For any real numbers, a, b, and c:
If $a > b$, then $a + c > b + c$.
If $a < b$, then $a + c < b + c$.

Subtraction Property of Inequality (p. 34)
For any real numbers, a, b, and c:
If $a > b$, then $a - c > b - c$.
If $a < b$, then $a - c < b - c$.

'And' Compound Inequalities (p. 41) A compound inequality containing the word *and* is true if and only if *both* inequalities are true.

'Or' Compound Inequalities (p. 42) A compound inequality containing the word *or* is true if one or more of the inequalities is true.

Linear Relations and Functions — Chapter 2

One-to-one function (p. 61) Each element of the domain pairs to exactly one unique element of the range.

Onto function (p. 61) Each element of the range corresponds to an element of the domain.

Vertical Line Test (p. 62) If no vertical line intersects a graph in more than one point, the graph represents a function.

Standard Form of a Linear Equation (p. 70) The standard form of a linear equation is $Ax + By = C$, where A, B, and C are integers whose greatest common factor is 1, $A \geq 0$, and A and B are not both zero.

Slope of a Line (p. 78) The slope m of a line passing through (x_1, y_1) and (x_2, y_2) is given by $m = \frac{y_2 - y_1}{x_2 - x_1}$, where $x_1 \neq x_2$.

Slope-Intercept Form (p. 83) $y = mx + b$

slope (m) — y-intercept (b)

Point-Slope Form (p. 84)

$$y - y_1 = m(x - x_1)$$

slope (m); coordinates of point on line (x_1, y_1)

Parallel Lines (p. 85) Two nonvertical lines are parallel if and only if they have the same slope. All vertical lines are parallel.

Perpendicular Lines (p. 85) Two nonvertical lines are perpendicular if and only if the product of the slopes is -1. Vertical lines and horizontal lines are perpendicular.

Direct Variation (p. 90) y varies directly as x if there is some nonzero constant k such that $y = kx$. k is called the *constant of variation*.

Parent Function of Absolute Value (p. 103) $f(x) = |x|$, defined as

$$f(x) = \begin{cases} x \text{ if } x > 0 \\ 0 \text{ if } x = 0 \\ -x \text{ if } x < 0 \end{cases}$$

Parent Functions (p. 109)

Constant Function The general equation of a constant function is $f(x) = a$, where a is any number.

Identity Function The identity function $f(x) = x$ passes through all points with coordinates (a, a).

Absolute Value Function The parent function of absolute value functions is $f(x) = |x|$.

Quadratic Function The parent function of quadratic functions is $f(x) = x^2$.

Transformation of Functions (p. 112)

Translation	$f(x + h)$ $f(x - h)$	Translates graph h units left. Translates graph h units right.
	$f(x) + k$ $f(x) - k$	Translates graph k units up. Translates graph k units down.
Reflection	$-f(x)$ $f(-x)$	Reflects graph across the x-axis. Reflects graph across the y-axis.
Dilation	$a \cdot f(x), a > 1$ $a \cdot f(x), 0 < a < 1$	Expands graph vertically. Compresses graph vertically
	$f(ax), a > 1$ $f(ax), 0 < a < 1$	Compresses graph horizontally. Expands graph horizontally.

Systems of Equations and Inequalities

Chapter 3

Characteristics of Linear Systems (p. 138)

Consistent and independent intersecting lines; one solution
Consistent and dependent same line; infinitely many solutions
Inconsistent parallel lines; no solution

Substitution Method (p. 143)

Step 1 Solve one equation for one of the variables.
Step 2 Substitute the resulting expression into the other equation to replace the variable. Then solve the equation.
Step 3 Substitute to solve for the other variable.

Elimination Method (p. 144)

Step 1 Multiply one or both equations by a number to result in two equations that contain opposite terms.
Step 2 Add the equations, eliminating one variable. Then solve the equation.
Step 3 Substitute to solve for the other variable.

Solving Systems of Inequalities (p. 151)

Step 1 Graph each inequality, shading the correct area.
Step 2 Identify the region that is shaded for all of the inequalities. This is the solution of the system.

Feasible Regions (p. 160)

Bounded The feasible region is enclosed by the constraints. The maximum or minimum value of the related function *always* occurs at a vertex of the feasible region.
Unbounded The feasible region is open and can go on forever. Unbounded regions have either a maximum or a minimum.

Optimization with Linear Programming (p. 162)

Step 1 Define the variables.
Step 2 Write a system of inequalities.
Step 3 Graph the system of inequalities.
Step 4 Find the coordinates of the vertices of the feasible region.
Step 5 Write a linear function to be maximized or minimized.
Step 6 Substitute the coordinates of the vertices into the function.
Step 7 Select the greatest or least result. Answer the problem.

Adding and Subtracting Matrices (p. 193)

$$\underset{A}{\begin{bmatrix} a & b \\ c & d \end{bmatrix}} + \underset{B}{\begin{bmatrix} e & f \\ g & h \end{bmatrix}} = \underset{A+B}{\begin{bmatrix} a+e & b+f \\ c+g & d+h \end{bmatrix}}$$

$$\underset{A}{\begin{bmatrix} a & b \\ c & d \end{bmatrix}} - \underset{B}{\begin{bmatrix} e & f \\ g & h \end{bmatrix}} = \underset{A-B}{\begin{bmatrix} a-e & b-f \\ c-g & d-h \end{bmatrix}}$$

Multiplying by a Scalar (p. 194)

$$k \cdot A = kA$$

$$k\begin{bmatrix} a & b \\ c & d \end{bmatrix} = \begin{bmatrix} ka & kb \\ kc & kd \end{bmatrix}$$

Properties of Matrix Operations (p. 194) For any matrices A, B, and C for which the matrix sum is defined and any scalar k, the following properties are true.

Commutative Property of Addition	$A + B = B + A$
Associative Property of Addition	$(A + B) + C = A + (B + C)$
Left Scalar Distributive Property	$k(A + B) = kA + kB$
Right Scalar Distributive Property	$(A + B)k = kA + kB$

Multiplying Matrices (p. 201)

$$\underset{A}{\begin{bmatrix} a & b \\ c & d \end{bmatrix}} \cdot \underset{B}{\begin{bmatrix} e & f \\ g & h \end{bmatrix}} = \underset{AB}{\begin{bmatrix} ae+bg & af+bh \\ ce+dg & cf+dh \end{bmatrix}}$$

Properties of Matrix Multiplication (p. 204)

Associative Property of Matrix Multiplication	$(AB)C = A(BC)$
Associative Property of Scalar Multiplication	$k(AB) = (kA)B = A(kB)$
Left Distributive Property	$C(A + B) = CA + CB$
Right Distributive Property	$(A + B)C = AC + BC$

Reflection Matrices (p. 212)

For a reflection across the:	x-axis	y-axis	line $y = x$
Multiply the vertex matrix on the left by:	$\begin{bmatrix} 1 & 0 \\ 0 & -1 \end{bmatrix}$	$\begin{bmatrix} -1 & 0 \\ 0 & 1 \end{bmatrix}$	$\begin{bmatrix} 0 & 1 \\ 1 & 0 \end{bmatrix}$

Rotation Matrices (p. 212)

For a counterclockwise rotation about the origin of:	90°	180°	270°
Multiply the vertex matrix on the left by:	$\begin{bmatrix} 0 & -1 \\ 1 & 0 \end{bmatrix}$	$\begin{bmatrix} -1 & 0 \\ 0 & -1 \end{bmatrix}$	$\begin{bmatrix} 0 & 1 \\ -1 & 0 \end{bmatrix}$

Second-Order Determinant (p. 220)

$$\det\begin{bmatrix} a & b \\ c & d \end{bmatrix} = \begin{vmatrix} a & b \\ c & d \end{vmatrix} = ad - bc$$

Diagonal Rule (p. 221)

Step 1 Rewrite the first two columns to the right of the determinant.
Step 2 Draw diagonals, beginning with the upper left-hand element. Multiply the elements in each diagonal.
Step 3 Find the sum of the products of the elements in the diagonals.
Step 4 Repeat the process, beginning with the upper right-hand element.
Step 5 Subtract the second sum from the first sum.

Area of a Triangle (p. 222) The area of a triangle with vertices (a, b), (c, d), and (e, f) is $|A|$, where $A = \frac{1}{2}\begin{vmatrix} a & b & 1 \\ c & d & 1 \\ e & f & 1 \end{vmatrix}$.

Cramer's Rule (p. 223)
Let C be the coefficient matrix of the system. $\begin{matrix} ax + by = m \\ fx + gy = n \end{matrix} \rightarrow \begin{vmatrix} a & b \\ f & g \end{vmatrix}$

The solution of this system is $x = \dfrac{\begin{vmatrix} m & b \\ n & g \end{vmatrix}}{|C|}$, $y = \dfrac{\begin{vmatrix} a & m \\ f & n \end{vmatrix}}{|C|}$, if $C \neq 0$.

Cramer's Rule for a System of Three Equations (p. 224)
Let C be the coefficient matrix of the system. $\begin{matrix} ax + by + cz = m \\ fx + gy + hz = n \\ jx + ky + \ell z = p \end{matrix} \rightarrow \begin{vmatrix} a & b & c \\ f & g & h \\ j & k & \ell \end{vmatrix}$

The solution of this system is $x = \dfrac{\begin{vmatrix} m & b & c \\ n & g & h \\ p & k & \ell \end{vmatrix}}{|C|}$, $y = \dfrac{\begin{vmatrix} a & m & c \\ f & n & h \\ j & p & \ell \end{vmatrix}}{|C|}$, and

$$z = y = \dfrac{\begin{vmatrix} a & b & m \\ f & g & n \\ j & k & p \end{vmatrix}}{|C|}, \text{ if } C \neq 0.$$

Identity Matrix for Multiplication (p. 229)
If $A = \begin{bmatrix} a & b \\ c & d \end{bmatrix}$, then $I = \begin{bmatrix} 1 & 0 \\ 0 & 1 \end{bmatrix}$ such that

$$\begin{bmatrix} a & b \\ c & d \end{bmatrix} \cdot \begin{bmatrix} 1 & 0 \\ 0 & 1 \end{bmatrix} = \begin{bmatrix} 1 & 0 \\ 0 & 1 \end{bmatrix} \cdot \begin{bmatrix} a & b \\ c & d \end{bmatrix} = \begin{bmatrix} a & b \\ c & d \end{bmatrix}.$$

Inverse of a 2 × 2 Matrix (p. 230)
The inverse of matrix $A = \begin{bmatrix} a & b \\ c & d \end{bmatrix}$ is $A^{-1} = \dfrac{1}{ad - bc}\begin{bmatrix} d & -b \\ -c & a \end{bmatrix}$, where $ad - bc \neq 0$.

Quadratic Functions and Relations — Chapter 5

Graph of a Quadratic Function (p. 250) Consider the graph of $y = ax^2 + bx + c$, where $a \neq 0$.

- The y-intercept is $a(0)^2 + b(0) + c$ or c.
- The equation of the axis of symmetry is $x = -\frac{b}{2a}$.
- The x-coordinate of the vertex is $-\frac{b}{2a}$.

Maximum and Minimum Value (p. 252) The graph of $f(x) = ax^2 + bx + c$, where $a \neq 0$,

- opens up and has a minimum value when $a > 0$, and
- opens down and has a maximum value when $a < 0$.

Solutions of a Quadratic Equation (p. 260) A quadratic equation can have one real solution, two real solutions, or no real solutions.

FOIL Method for Multiplying Binomials (p. 268) To multiply two binomials, find the sum of the products of F the *First* terms, O the *Outer* terms, I the *Inner* terms, and L the *Last* terms.

Key Concepts

Zero Product Property (p. 271) For any real numbers a and b, if $ab = 0$, then either $a = 0$, $b = 0$, or both a and b equal zero.

Complex Numbers (p. 277) A complex number is any number that can be written in the form $a + bi$, where a and b are real numbers and i is the imaginary unit. A is called the real part, and b is called the imaginary part.

Completing the Square (p. 286) To complete the square for any quadratic expression of the form $x^2 + bx$, follow the steps below.

Step 1 Find one half of b, the coefficient of x.
Step 2 Square the result in Step 1.
Step 3 Add the result of Step 2 to $x^2 + bx$.

Quadratic Formula (p. 293) The solutions of a quadratic equation of the form $ax^2 + bx + c = 0$, where $a \neq 0$, are given by the following formula.

$$x = \frac{-b \pm \sqrt{b^2 - 4ac}}{2a}$$

Discriminant (p. 296) Consider $ax^2 + bx + c = 0$, where a, b, and c are rational numbers.

Value of Discriminant	$b^2 - 4ac > 0$; $b^2 - 4ac$ is a perfect square.	$b^2 - 4ac > 0$; $b^2 - 4ac$ is *not* a perfect square.	$b^2 - 4ac = 0$	$b^2 - 4ac < 0$
Type and Number of Roots	2 real, rational roots	2 real, irrational roots	1 real, rational root	2 complex roots

Sum and Product of Roots (p. 302) If the roots of $ax^2 + bx + c = 0$, with $a \neq 0$, are r_1 and r_2, then $r_1 + r_2 = -\frac{b}{a}$ and $r_1 \cdot r_2 = \frac{c}{a}$.

Transformations of Quadratic Functions (p. 307) $f(x) = a(x - h)^2 + k$

***h*, Horizontal Translation**	$\lvert h \rvert$ units to the right if h is positive $\lvert h \rvert$ units to the left if h is negative
***k*, Vertical Translation**	$\lvert k \rvert$ units up if k is positive $\lvert k \rvert$ units down if k is negative
***a*, Reflection**	If $a > 0$, the graph opens up. If $a < 0$, the graph opens down.
***a*, Dilation**	If $\lvert a \rvert > 1$, the graph is stretched vertically. If $0 < \lvert a \rvert < 1$, the graph is compressed vertically.

Chapter 6 Polynomials and Polynomial Functions

Properties of Exponents (p. 333)

Product of Powers $x^a \cdot x^b = x^{a+b}$

Quotient of Powers If $x \neq 0$, $\frac{x^a}{x^b} = x^{a-b}$.

Negative Exponent $x^{-a} = \frac{1}{x^a}$ and $\frac{1}{x^{-a}} = x^a$

Power of a Power $(x^a)^b = x^{ab}$

Power of a Product $(xy)^a = x^a y^a$

Power of a Quotient $\left(\frac{x}{y}\right)^a = \frac{x^a}{y^a}$, $y \neq 0$, and $\left(\frac{x}{y}\right)^{-a} = \left(\frac{y}{x}\right)^a$ or $\frac{y^a}{x^a}$, $x \neq 0$, $y \neq 0$

Zero Power $x^0 = 1$, $x \neq 0$

Simplifying Monomials (p. 334) A monomial expression is in simplified form when:
- there are no powers of powers,
- each base appears exactly once,
- all fractions are in simplest form, and
- there are no negative exponents.

Synthetic Division (p. 343)

Step 1 Write the coefficients of the dividend so that the degrees of the terms are in descending order. Write the constant r of the divisor $x - r$ in the box. Bring the first coefficient down.

Step 2 Multiply the first coefficient by r, and write the product under the second coefficient.

Step 3 Add the product and the second coefficient.

Step 4 Repeat Steps 2 and 3 until you reach a sum in the last column. The numbers along the bottom row are the coefficients of the quotient. The power of the first term is one less than the degree of the dividend. The final number is the remainder.

End Behavior of a Polynomial Function (p. 350)

Degree: even **Leading Coefficient:** positive **End Behavior:** $f(x) \to +\infty$ as $x \to -\infty$ $f(x) \to +\infty$ as $x \to +\infty$	**Degree:** odd **Leading Coefficient:** positive **End Behavior:** $f(x) \to -\infty$ as $x \to -\infty$ $f(x) \to +\infty$ as $x \to +\infty$
Degree: even **Leading Coefficient:** negative **End Behavior:** $f(x) \to -\infty$ as $x \to -\infty$ $f(x) \to -\infty$ as $x \to +\infty$	**Degree:** odd **Leading Coefficient:** negative **End Behavior:** $f(x) \to +\infty$ as $x \to -\infty$ $f(x) \to -\infty$ as $x \to +\infty$

Zeros of Even- and Odd-Degree Functions (p. 351) Odd-degree functions will always have an odd number of real zeros. Even-degree functions will always have an even number of real zeros or no real zeros at all.

Location Principle (p. 358) Suppose $y = f(x)$ represents a polynomial function and a and b are two real numbers such that $f(a) < 0$ and $f(b) > 0$. Then the function has at least one real zero between a and b.

Sum and Difference of Cubes (p. 368)

Sum of Two Cubes $a^3 + b^3 = (a + b)(a^2 - ab + b^2)$

Difference of Two Cubes $a^3 - b^3 = (a - b)(a^2 + ab + b^2)$

Factoring Techniques (p. 369)

Difference of Two Squares $a^2 - b^2 = (a + b)(a - b)$

Sum of Two Cubes $a^3 + b^3 = (a + b)(a^2 - ab + b^2)$

Difference of Two Cubes $a^3 - b^3 = (a - b)(a^2 + ab + b^2)$

Perfect Square Trinomials $a^2 + 2ab + b^2 = (a + b)^2$

$a^2 - 2ab + b^2 = (a - b)^2$

General Trinomials

$$acx^2 + (ad + bc)x + bd = (ax + b)(cx + d)$$

Grouping

$$ax + bx + ay + by = x(a + b) + y(a + b) = (a + b)(x + y)$$

Quadratic Form (p. 371) An expression that is in quadratic form can be written as $au^2 + bu + c$ for any numbers a, b, and c, $a \neq 0$, where u is some expression in x. The expression $au^2 + bu + c$ is called the quadratic form of the original expression.

Key Concepts

Remainder Theorem (p. 377) If a polynomial $P(x)$ is divided by $x - r$, the remainder is a constant $P(r)$, and

$$\begin{array}{ccccccc} \text{dividend} & \text{equals} & \text{quotient} & \text{times} & \text{divisor} & \text{plus} & \text{remainder} \\ P(x) & = & Q(x) & \cdot & (x - r) & + & P(r), \end{array}$$

where $Q(x)$ is a polynomial with degree one less than $P(x)$.

Factor Theorem (p. 379) The binomial $x - r$ is a factor of the polynomial $P(x)$ if and only if $P(r) = 0$.

Zeros, Factors, Roots, and Intercepts (p. 383) Let $P(x) = a_nx^n + \cdots + a_1x + a_0$ be a polynomial function. Then the following statements are equivalent.

- c is a zero of $P(x)$.
- c is a root or solution of $P(x) = 0$.
- $x - c$ is a factor of $a_nx^n + \cdots + a_1x + a_0$.
- If c is a real number, then $(c, 0)$ is an x-intercept of the graph of $P(x)$.

Fundamental Theorem of Algebra (p. 383) Every polynomial equation with degree greater than zero has at least one root in the set of complex numbers.

Corollary to the Fundamental Theorem of Algebra (p. 384) A polynomial equation of degree n has exactly n roots in the set of complex numbers, including repeated roots.

Descartes' Rule of Signs (p. 385) Let $P(x) = a_nx^n + \cdots + a_1x + a_0$ be a polynomial function with real coefficients. Then:

- the number of *positive* real zeros of $P(x)$ is the same as the number of changes in sign of the coefficients of the terms, or is less than this by an even number; and
- the number of *negative* real zeros of $P(x)$ is the same as the number of changes in sign of the coefficients of the terms of $P(-x)$, or is less than this by an even number.

Complex Conjugates Theorem (p. 387) Let a and b be real numbers, and $b \neq 0$. If $a + bi$ is a zero of a polynomial function with real coefficients, then $a - bi$ is also a zero of the function.

Rational Zero Theorem (p. 391) If $P(x)$ is a polynomial function with integral coefficients, then every rational zero of $P(x) = 0$ is of the form $\frac{p}{q}$, a rational number in simplest form, where p is a factor of the constant term and q is a factor of the leading coefficient.

Corollary to the Rational Zero Theorem (p. 391) If $P(x)$ is a polynomial function with integral coefficients, a leading coefficient of 1, and a nonzero constant term, then any rational zeros of $P(x)$ must be factors of the constant term.

Chapter 7

Inverses and Radical Functions and Relations

Operations on Functions (p. 409)

Addition	$(f + g)(x) = f(x) + g(x)$
Subtraction	$(f - g)(x) = f(x) - g(x)$
Multiplication	$(f \cdot g)(x) = f(x) \cdot g(x)$
Division	$\left(\frac{f}{g}\right)(x) = \frac{f(x)}{g(x)}, g(x) \neq 0$

Composition of Functions (p. 410) Suppose f and g are functions such that the range of g is a subset of the domain of f. Then the composition function $f \circ g$ can be described by

$$[f \circ g](x) = f[g(x)].$$

Inverse Relations (p. 417) Two relations are inverse relations if and only if whenever one relation contains the element (a, b), the other relation contains the element (b, a).

Key Concepts

Property of Inverse Functions (p. 418) If f and f^{-1} are inverse functions, then $f(a) = b$ if and only if $f^{-1}(b) = a$.

Inverse Functions (p. 419) Two functions f and g are inverse functions if and only if both of their compositions are the identity function.

Parent Function of Square Root Functions (p. 424)

Parent function:	$f(x) = \sqrt{x}$
Domain:	$\{x \mid x \geq 0\}$
Range:	$\{y \mid y \geq 0\}$
Intercepts:	$x = 0, y = 0$
Not defined:	$x < 0$
End behavior:	$x \to 0, y \to 0$ $x \to +\infty, y \to +\infty$

Transformations of Square Root Functions (p. 425) $f(x) = a\sqrt{x - h} + k$

h, Horizontal Translation	$\lvert h \rvert$ units to the right if h is positive $\lvert h \rvert$ units to the left if h is negative
k, Vertical Translation	$\lvert k \rvert$ units up if k is positive $\lvert k \rvert$ units down if k is negative
a, Orientation and Shape	If $a < 0$, the graph is reflected across the x-axis. If $\lvert a \rvert > 1$, the graph is stretched vertically. If $0 < \lvert a \rvert < 1$, the graph is compressed vertically.

Definition of nth Root (p. 431) For any real numbers a and b, and any positive integer n, if $a^n = b$, then a is an nth root of b.

Real nth Roots (p. 432)

a	n is even	n is odd
$a > 0$	1 positive and 1 negative real root: $\pm\sqrt[n]{a}$	1 positive and 0 negative real root: $\sqrt[n]{a}$
$a < 0$	0 real roots	0 positive and 1 negative real root: $\sqrt[n]{a}$
$a = 0$	1 real root: $\sqrt[n]{0} = 0$	1 real root: $\sqrt[n]{0} = 0$

Product Property of Radicals (p. 439) For any real numbers a and b and any integer $n > 1$,

1. if n is even and a and b are both nonnegative, then $\sqrt[n]{ab} = \sqrt[n]{a} \cdot \sqrt[n]{b}$, and
2. if n is odd, then $\sqrt[n]{ab} = \sqrt[n]{a} \cdot \sqrt[n]{b}$.

Quotient Property of Radicals (p. 440) For any real numbers a and $b \neq 0$ and any integer $n > 1$, $\sqrt[n]{\frac{a}{b}} = \frac{\sqrt[n]{a}}{\sqrt[n]{b}}$, if all roots are defined.

Simplifying Radical Expressions (p. 441) A radical expression is in simplified form when the following conditions are met.

- The index n is as small as possible.
- The radicand contains no factors (other than 1) that are nth powers of an integer or polynomial.
- The radicand contains no fractions.
- No radicals appear in a denominator.

$b^{\frac{1}{n}}$ (p. 446) For any real number b and any positive integer n, $b^{\frac{1}{n}} = \sqrt[n]{b}$, except when $b < 0$ and n is even. When $b < 0$ and n is even, a complex root may exist.

Rational Exponents (p. 447) For any real nonzero number b, and any integers x and y, with $y > 1$, $b^{\frac{x}{y}} = \sqrt[y]{b^x} = \left(\sqrt[y]{b}\right)^x$, except when $b < 0$ and y is even. When $b < 0$ and y is even, a complex root may exist.

Key Concepts

Expressions with Rational Exponents (p. 449) An expression with rational exponents is simplified when all of the following conditions are met.

- It has no negative exponents.
- It has no fractional exponents in the denominator.
- It is not a complex fraction.
- The index of any remaining radical is the least number possible.

Solving Radical Equations (p. 453)

Step 1 Isolate the radical on one side of the equation.
Step 2 Raise each side of the equation to a power equal to the index of the radical to eliminate the radical.
Step 3 Solve the resulting polynomial equation.

Solving Radical Inequalities (p. 455)

Step 1 If the index of the root is even, identify the values of the variable for which the radicand is nonnegative.
Step 2 Solve the inequality algebraically.
Step 3 Test values to check your solution.

Chapter 8

Exponential and Logarithmic Functions and Relations

Parent Function of Exponential Growth Functions (p. 475)

Parent function: $f(x) = b^x, b > 1$
Domain: all real numbers
Range: all nonzero real numbers
Asymptote: x-axis

Transformations of Exponential Functions (p. 476) $f(x) = ab^{x-h} + k$

***h*, Horizontal Translation**	$\lvert h \rvert$ units to the right if h is positive $\lvert h \rvert$ units to the left if h is negative
***k*, Vertical Translation**	$\lvert k \rvert$ units up if k is positive $\lvert k \rvert$ units down if k is negative
***a*, Orientation and Shape**	If $a < 0$, the graph is reflected across the x-axis. If $\lvert a \rvert > 1$, the graph is stretched vertically. If $0 < \lvert a \rvert < 1$, the graph is compressed vertically.

Parent Function of Exponential Decay Functions (p. 477)

Parent function: $f(x) = b^x, 0 < b < 1$
Domain: all real numbers
Range: positive real numbers
Asymptote: x-axis

Property of Equality for Exponential Functions (p. 485) Let $b > 0$ and $b \neq 1$. Then $b^x = b^y$ if and only if $x = y$.

Property of Inequality for Exponential Functions (p. 487) Let $b > 1$. Then $b^x > b^y$ if and only if $x > y$, and $b^x < b^y$ if and only if $x < y$.

Logarithm with Base *b* (p. 492)

$$\log_b x = y \text{ if and only if } b^y = x.$$

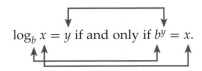

Parent Function of Logarithmic Functions (p. 493)
Parent function: $f(x) = \log_b x$
Range: all real numbers
Asymptote: y-axis

Transformations of Logarithmic Functions (p. 494) $f(x) = a \log_b (x - h) + k$

***h*, Horizontal Translation**	$\lvert h \rvert$ units to the right if h is positive $\lvert h \rvert$ units to the left if h is negative
***k*, Vertical Translation**	$\lvert k \rvert$ units up if k is positive $\lvert k \rvert$ units down if k is negative
***a*, Orientation and Shape**	If $a < 0$, the graph is reflected across the x-axis. If $\lvert a \rvert > 1$, the graph is stretched vertically. If $0 < \lvert a \rvert < 1$, the graph is compressed vertically.

Property of Equality for Logarithmic Functions (p. 502) If b is a positive number other than 1, then $\log_b x = \log_b y$ if and only if $x = y$.

Property of Inequality for Logarithmic Functions (p. 503)
If $b > 1$, $x > 0$, and $\log_b x > y$, then $x > b^y$.
If $b > 1$, $x > 0$, and $\log_b x < y$, then $0 < x < b^y$.

Property of Inequality for Logarithmic Functions (p. 504) If $b > 1$, then $\log_b x > \log_b y$ if and only if $x > y$, and $\log_b x < \log_b y$ if and only if $x < y$.

Product Property of Logarithms (p. 509) For all positive numbers a, b, and x, where $x \neq 1$, $\log_x ab = \log_x a + \log_x b$.

Quotient Property of Logarithms (p. 510) For all positive numbers a, b, and x, where $x \neq 1$, $\log_x \frac{a}{b} = \log_x a - \log_x b$.

Power Property of Logarithms (p. 511) For any real number p, and positive numbers m and b, where $b \neq 1$, $\log_b m^p = p \log_b m$.

Change of Base Formula (p. 518) For all positive numbers a, b, and n, where $a \neq 1$ and $b \neq 1$, $\log_a n = \frac{\log_b n}{\log_b a}$.

Natural Base Functions (p. 525) The function $f(x) = e^x$ is used to model continuous exponential growth. The function $f(x) = e^{-x}$ is used to model continuous exponential decay. The inverse of a natural base exponential function is called the natural logarithm. This logarithm can be written as $\log_e x$, but is more often abbreviated as $\ln x$.

Exponential Growth (p. 533) Exponential growth can be modeled by the function $f(x) = ae^{kt}$

Exponential Decay (p. 533) Exponential decay can be modeled by the function $f(x) = ae^{-kt}$

Logistic Growth Function (p. 536) Let a, b, and c be positive constants where $b < 1$. The logistic growth function is represented by $f(t) = \frac{c}{1 + ae^{-bt}}$, where t represents time.

Rational Functions and Relations — Chapter 9

Multiplying Rational Expressions (p. 553) For all rational expressions $\frac{a}{b}$ and $\frac{c}{d}$ with $b \neq 0$ and $d \neq 0$, $\frac{a}{b} \cdot \frac{c}{d} = \frac{ac}{bd}$.

Dividing Rational Expressions (p. 553) For all rational expressions $\frac{a}{b}$ and $\frac{c}{d}$ with $b \neq 0$, $c \neq 0$, and $d \neq 0$, $\frac{a}{b} \div \frac{c}{d} = \frac{a}{b} \cdot \frac{d}{c} = \frac{ad}{bc}$.

Adding Rational Expressions (p. 563) For all rational expressions $\frac{a}{b}$ and $\frac{c}{d}$ with $b \neq 0$, and $d \neq 0$, $\frac{a}{b} + \frac{c}{d} = \frac{ad}{bd} + \frac{bc}{bd} = \frac{ad + bc}{bd}$.

Subtracting Rational Expressions (p. 563) For all rational expressions $\frac{a}{b}$ and $\frac{c}{d}$ with $b \neq 0$, and $d \neq 0$, $\frac{a}{b} - \frac{c}{d} = \frac{ad}{bd} - \frac{bc}{bd} = \frac{ad - bc}{bd}$.

Parent Function of Reciprocal Functions (p. 567)

Parent function:	$f(x) = \frac{1}{x}$
Domain and range:	all nonzero real numbers
Axes of symmetry:	$x = 0$ and $f(x) = 0$
Not defined:	$x = 0$ and $f(x) = 0$

Transformations of Reciprocal Functions (p. 569) $f(x) = \frac{a}{x - h} + k$

***h*, Horizontal Translation**	$\lvert h \rvert$ units to the right if h is positive $\lvert h \rvert$ units to the left if h is negative The *vertical* asymptote is at $x = h$.
***k*, Vertical Translation**	$\lvert k \rvert$ units up if k is positive $\lvert k \rvert$ units down if k is negative The *horizontal* asymptote is at $f(x) = k$.
***a*, Orientation and Shape**	If $a < 0$, the graph is reflected across the x-axis. If $\lvert a \rvert > 1$, the graph is stretched vertically. If $0 < \lvert a \rvert < 1$, the graph is compressed vertically.

Vertical and Horizontal Asymptotes (p. 575) If $f(x) = \frac{a(x)}{b(x)}$, $a(x)$ and $b(x)$ are polynomial functions with no common factors other than 1, and $b(x) \neq 0$, then:

Vertical asymptote $f(x)$ has a vertical asymptote whenever $b(x) = 0$.

Horizontal asymptote $f(x)$ has at most one horizontal asymptote

- if the degree of $a(x)$ is greater than the degree of $b(x)$, there is no horizontal asymptote.
- if the degree of a(x) is less than the degree of $b(x)$, the horizontal asymptote is the line $y = 0$.
- if the degree of $a(x)$ equals the degree of $b(x)$, the horizontal asymptote is the line $y = \frac{\text{leading coefficient of } a(x)}{\text{leading coefficient of } b(x)}$.

Oblique Asymptotes (p. 577) If $f(x) = \frac{a(x)}{b(x)}$, $a(x)$ and $b(x)$ are polynomial functions with no common factors other than 1 and $b(x) \neq 0$, then $f(x)$ has an oblique asymptote if the degree of $a(x)$ minus the degree of $b(x)$ equals 1. The equation of the asymptote is $\frac{a(x)}{b(x)}$ with no remainder.

Point Discontinuity (p. 579) If $f(x) = \frac{a(x)}{b(x)}$, $b(x) \neq 0$, and $x - c$ is a factor of both $a(x)$ and $b(x)$, then there is a point discontinuity at $x = c$.

Direct Variation (p. 584) y varies directly as x if there is some nonzero constant k such that $y = kx$. k is called the constant of variation.

Joint Variation (p. 585) y varies jointly as x and z if there is some nonzero constant k such that $y = kxz$.

Inverse Variation (p. 586) y varies inversely as x if there is some nonzero constant k such that $xy = k$ or $y = \frac{k}{x}$, where $x \neq 0$ and $y \neq 0$.

Solving Rational Inequalities (p. 597)

Step 1 State the excluded values. These are the values for which the denominator is 0.

Step 2 Solve the related equation.

Step 3 Use the values determined from the previous steps to divide a number line into intervals.

Step 4 Test a value in each interval to determine which intervals contain values that satisfy the inequality.

Conic Sections — Chapter 10

Midpoint Formula (p. 615) If a line segment has endpoints $P(x_1, y_1)$ and $Q(x_2, y_2)$, then the midpoint of the segment has coordinates $M\left(\frac{x_1 + x_2}{2}, \frac{y_1 + y_2}{2}\right)$.

Distance Formula (p. 616) The distance between two points with coordinates (x_1, y_1) and (x_2, y_2) is given by $\sqrt{(x_2 - x_1)^2 + (y_2 - y_1)^2}$.

Equations of Parabolas (p. 621)

Form of Equation	$y = a(x - h)^2 + k$	$x = a(y - k)^2 + h$
Direction of Opening	upward if $a > 0$, downward if $a < 0$	right if $a > 0$, left if $a < 0$
Vertex	(h, k)	(h, k)
Axis of Symmetry	$x = h$	$y = k$
Focus	$\left(h, k + \frac{1}{4a}\right)$	$\left(h + \frac{1}{4a}, k\right)$
Directrix	$y = k - \frac{1}{4a}$	$x = h - \frac{1}{4a}$
Length of Latus Rectum	$\left\lvert\frac{1}{a}\right\rvert$ units	$\left\lvert\frac{1}{a}\right\rvert$ units

Equations of Circles (p. 629)

Standard Form of Equation	$x^2 + y^2 = r^2$	$(x - h)^2 + (y - k)^2 = r^2$
Center	$(0, 0)$	(h, k)
Radius	r	r

Equations of Ellipses Centered at the Origin (p. 637)

Standard Form	$\frac{x^2}{a^2} + \frac{y^2}{b^2} = 1$	$\frac{y^2}{a^2} + \frac{x^2}{b^2} = 1$
Orientation	horizontal	vertical
Foci	$(c, 0), (-c, 0)$	$(0, c), (0, -c)$
Length of Major Axis	$2a$ units	$2a$ units
Length of Minor Axis	$2b$ units	$2b$ units

Equations of Ellipses Centered at (h, k) (p. 638)

Standard Form	$\frac{(x - h)^2}{a^2} + \frac{(y - k)^2}{b^2} = 1$	$\frac{(y - k)^2}{a^2} + \frac{(x - h)^2}{b^2} = 1$
Orientation	horizontal	vertical
Foci	$(h \pm c, k)$	$(h, k \pm c)$
Vertices	$(h \pm a, k)$	$(h, k \pm a)$
Co-vertices	$(h, k \pm b)$	$(h \pm b, k)$

Equations of Hyperbolas Centered at the Origin (p. 646)

Standard Form	$\frac{x^2}{a^2} - \frac{y^2}{b^2} = 1$	$\frac{y^2}{a^2} - \frac{x^2}{b^2} = 1$
Orientation	horizontal	vertical
Foci	$(\pm c, 0)$	$(0, \pm c)$
Length of Transverse Axis	$2a$ units	$2a$ units
Length of Conjugate Axis	$2b$ units	$2b$ units
Equations of Asymptotes	$y = \pm \frac{b}{a}x$	$y = \pm \frac{a}{b}x$

Equations of Hyperbolas Centered at (h, k) (p. 648)

Standard Form	$\frac{(x-h)^2}{a^2} - \frac{(y-k)^2}{b^2} = 1$	$\frac{(y-k)^2}{a^2} - \frac{(x-h)^2}{b^2} = 1$
Orientation	horizontal	vertical
Foci	$(h \pm c, k)$	$(k, h \pm c)$
Vertices	$(h \pm a, k)$	$(k, h \pm a)$
Co-vertices	$(h, k \pm b)$	$(k \pm b, h)$
Equations of Asymptotes	$y - k = \pm \frac{b}{a}(x - h)$	$y - k = \pm \frac{a}{b}(x - h)$

Standard Forms of Conic Sections (p. 654)

Circle	$(x-h)^2 + (y-k)^2 = r^2$	
	Horizontal Axis	**Vertical Axis**
Parabola	$y = a(x-h)^2 + k$	$x = a(y-k) + h$
Ellipse	$\frac{(x-h)^2}{a^2} + \frac{(y-k)^2}{b^2} = 1$	$\frac{(y-k)^2}{a^2} + \frac{(x-h)^2}{b^2} = 1$
Hyperbola	$\frac{(x-h)^2}{a^2} - \frac{(y-k)^2}{b^2} = 1$	$\frac{(y-k)^2}{a^2} - \frac{(x-h)^2}{b^2} = 1$

Classify Conics with the Discriminant (p. 655)

Circle	$B^2 - 4AC < 0$; $B = 0$ and $A = C$
Ellipse	$B^2 - 4AC < 0$; either $B \neq 0$ or $A \neq C$
Parabola	$B^2 - 4AC = 0$
Hyperbola	$B^2 - 4AC > 0$

Sequences and Series — Chapter 11

Sequences as Functions (p. 681) A sequence is a function in which the domain consists of natural numbers and the range consists of real numbers.

$$\begin{array}{lccccc} \textbf{Domain:} & 1 & 2 & 3 & \ldots & n \\ & \downarrow & \downarrow & \downarrow & & \downarrow \\ \textbf{Range:} & a_1 & a_2 & a_3 & \ldots & a_n \end{array}$$

***n*th Term of an Arithmetic Sequence** (p. 688) The nth term a_n of an arithmetic sequence in which the first term is a_1 and the common difference is d is given by the following formula, where n is any natural number.

$$a_n = a_1 + (n - 1)d$$

Partial Sum of an Arithmetic Series (p. 690)

Given a_1 and a_n $\quad S_n = n\left(\frac{a_1 + a_n}{2}\right)$

Given a_1 and d $\quad S_n = \frac{n}{2}[2a_1 + (n - 1)d]$

Sigma Notation (p. 691)

$$\sum_{k=1}^{n} f(k)$$

last value of k → n

first value of k → $k = 1$

formula for the terms of the series → $f(k)$

***n*th Term of a Geometric Sequence** (p. 696) The nth term a_n of a geometric sequence in which the first term is a_1 and the common ratio is r is given by the following formula, where n is any natural number.

$$a_n = a_1 r^{n-1}$$

Partial Sum of a Geometric Series (p. 698)

Given a_1 and n $\quad S_n = \dfrac{a_1 - a_1 r^n}{1 - r}, r \neq 1$

Given a_1 and a_n $\quad S_n = \dfrac{a_1 - a_n r}{1 - r}, r \neq 1$

Convergent Series (p. 705) The sum approaches a finite value. $|r| < 1$

Divergent Series (p. 705) The sum does not approach a finite value. $|r| \geq 1$

Sum of an Infinite Geometric Series (p. 706) The sum S of an infinite geometric series with $|r| < 1$ is given by

$$S = \frac{a_1}{1 - r}.$$

If $|r| \geq 1$, the series has no sum.

Recursive Formulas for Sequences (p. 714)

Arithmetic Sequence $\quad a_n = a_{n-1} + d$, where d is the common difference

Geometric Sequence $\quad a_n = r \cdot a_{n-1}$, where r is the common ratio

Binomial Theorem (p. 722) If n is a natural number, then $(a + b)^n =$

$${}_nC_0\, a^n b^0 + {}_nC_1\, a^{n-1} b^1 + {}_nC_2\, a^{n-2} b^2 + \cdots + {}_nC_n\, a^0 b^n = \sum_{k=0}^{n} \frac{n!}{k!(n-k)!} a^{n-k} b^k.$$

Binomial Expansion (p. 723) In a binomial expansion of $(a + b)^n$,

- There are $n + 1$ terms.
- n is the exponent of a in the first term and b in the last term.
- In successive terms, the exponent of a decreases by 1, and the exponent of b increases by 1.
- The sum of the exponents in each term is n.
- The coefficients are symmetric.

Mathematical Induction (p. 727)

Step 1 Show that the statement is true for $n = 1$.

Step 2 Assume that the statement is true for some natural number k. This assumption is called the inductive hypothesis.

Step 3 Show that the statement is true for the next natural number $k + 1$.

Probability and Statistics — Chapter 12

Measures of Central Tendency (p. 752)

mean the sum of the data divided by the number of items in the data set

median the middle number of the ordered data, or the mean of the middle two numbers

mode the number or numbers that occur most often

Margin of Sampling Error (p. 753) When a random sample n is taken from a population, the margin of sampling error can be approximated by $\pm \frac{1}{\sqrt{n}}$.

Standard Deviation Formula (p. 754)

Sample $s = \sqrt{\frac{\sum_{k=1}^{n} (x_n - \overline{x})^2}{n-1}}$ **Population** $\sigma = \sqrt{\frac{\sum_{k=1}^{n} (x_n - \mu)^2}{n}}$

Conditional Probability (p. 759) Given that A and B are dependent events, the conditional probability of an event B, given that an event A has already occurred, is defined as

$$P(B \mid A) = \frac{P(A \text{ and } B)}{P(A)}, \text{ where } P(A) \neq 0.$$

Probability of Success and Failure (p. 764) If an event can succeed in s ways and fail in f ways, then the probabilities of success $P(S)$ and of failure $P(F)$ are as follows.

$$P(S) = \frac{s}{s+f} \qquad P(F) = \frac{f}{s+f}$$

Characteristics of the Normal Distribution (p. 773)

- The maximum occurs at the mean. The mean, median, and mode are equal.
- The distribution extends from negative infinity to positive infinity, but never touches the x-axis.
- The population mean μ and standard deviation σ are used to determine probabilities. Probabilities are cumulative and are expressed as inequalities.
- Because the area under the normal curve represents probabilities, this area is 1.

The Empirical Rule (p. 774) A normal distribution with mean μ and standard deviation σ has the following properties.

- About 68% of the values are within 1σ of the mean.
- About 95% of the values are within 2σ of the mean.
- About 99% of the values are within 3σ of the mean.

95% Confidence Interval Formula (p. 780) A 95% confidence interval estimate can be found by using the formula $CI = \overline{x} \pm 2 \cdot \frac{s}{\sqrt{n}}$, where $\overline{x}$ is the mean of the sample, s is the standard deviation of the sample, and n is the size of the sample.

Hypothesis Testing (p. 781)

Step 1 State the null hypothesis H_0 and the alternative hypothesis H_1.
Step 2 Design the experiment.
Step 3 Conduct the experiment and collect the data.
Step 4 Find the confidence interval.
Step 5 Make the correct statistical inference. Accept the null hypothesis if the population parameter falls into the confidence interval.

Bionomial Experiments (p. 786)

- There are only two possible outcomes, success or failure.
- There is a fixed number of trials, n.
- The probability of success is the same in every trial.
- The trials are independent.
- The random variable is the number of successes in n trials.

Key Concepts

Binomial Distribution Functions (p. 787) The probability of x successes in n independent trials is

$$P(x) = C(n, x)\, s^{x} f^{n-x},$$

where s is the probability of success of an individual trial and f is the probability of failure on that same individual trial ($s + f = 1$).

Expected Value of a Binomial Distribution (p. 787) The expected value for a binomial distribution is $E(X) = ns$, where n is the total number of trials and s is the probability of success.

Normal Approximation of a Binomial Distribution (p. 789) A binomial distribution with n trials and probability of success s and probability of failure f such that $ns \geq 5$ and $nf \geq 5$, then the binomial distribution can be approximated by a normal distribution with $\bar{x} = ns$ and $\sigma = \sqrt{nsf}$.

Trigonometric Functions — Chapter 13

Trigonometric Functions in Right Triangles (p. 810)

sine $\sin\theta = \dfrac{\text{opp}}{\text{hyp}}$ **cosine** $\cos\theta = \dfrac{\text{adj}}{\text{hyp}}$

tangent $\tan = \dfrac{\text{opp}}{\text{adj}}$ **cosecant** $\csc\theta = \dfrac{\text{hyp}}{\text{opp}}$

secant $\sec\theta = \dfrac{\text{hyp}}{\text{adj}}$ **cotangent** $\cot = \dfrac{\text{adj}}{\text{opp}}$

Inverse Trigonometric Ratios (p. 813)

inverse sine If $\sin A = x$, then $\sin^{-1} x = m\angle A$.
inverse cosine If $\cos A = x$, then $\cos^{-1} x = m\angle A$.
inverse tangent If $\tan A = x$, then $\tan^{-1} x = m\angle A$.

Angle Measures (p. 819) If the measure of an angle is positive, the terminal side is rotated counterclockwise. If the measure of an angle is negative, the terminal side is rotated clockwise.

Convert Degrees to Radians (p. 821) To convert from degrees to radians, multiply the number of degrees by $\dfrac{\pi \text{ radians}}{180°}$.

Convert Radians to Degrees (p. 821) To convert from radians to degrees, multiply the number of radians by $\dfrac{180°}{\pi \text{ radians}}$.

Arc Length (p. 822) For a circle with radius r and central angle θ (in radians), the arc length s equals the product of r and θ.

Trigonometric Functions of General Angles (p. 827) Let θ be an angle in standard position and let $P(x, y)$ be a point on its terminal side. Using the Pythagorean Theorem, $r = \sqrt{x^2 + y^2}$. The six trigonometric functions of θ are defined below.

$\sin\theta = \dfrac{y}{r}$ $\cos\theta = \dfrac{x}{r}$ $\tan\theta = \dfrac{y}{x}, x \neq 0$

$\csc\theta = \dfrac{r}{y}, y \neq 0$ $\sec\theta = \dfrac{r}{x}, x \neq 0$ $\cot\theta = \dfrac{x}{y}, y \neq 0$

Evaluate Trigonometric Functions (p. 829)

Step 1 Find the measure of the reference angle θ'.
Step 2 Evaluate the trigonometric function for θ'.
Step 3 Determine the sign of the trigonometric function value. Use the quadrant in which the terminal side of θ lies.

Area of a Triangle (p. 834)

$$\text{Area} = \frac{1}{2}bc \sin A$$

$$\text{Area} = \frac{1}{2}ac \sin B$$

$$\text{Area} = \frac{1}{2}ab \sin C$$

Law of Sines (p. 835) In $\triangle ABC$, if sides with lengths a, b, and c are opposite angles with measures A, B, and C, respectively, then the following is true.

$$\frac{\sin A}{a} = \frac{\sin B}{b} = \frac{\sin C}{c}$$

Law of Cosines (p. 843) In $\triangle ABC$, if sides with lengths a, b, and c are opposite angles with measures A, B, and C, respectively, then the following are true.

$$a^2 = b^2 + c^2 - 2bc \cos A$$
$$b^2 = a^2 + c^2 - 2ac \cos B$$
$$c^2 = a^2 + b^2 - 2ab \cos C$$

Sine and Cosine Functions on a Unit Circle (p. 850) If the terminal side of an angle θ in standard position intersects the unit circle at $P(x, y)$, then $\cos \theta = x$ and $\sin \theta = y$.
$P(x, y) = P(\cos \theta, \sin \theta)$

θ-Intercepts of the Sine and Cosine Functions (p. 858)

$y = a \sin b\theta$

$$(0, 0), \left(\frac{1}{2} \cdot \frac{360°}{b}, 0\right), \left(\frac{360°}{b}, 0\right)$$

$y = a \cos b\theta$

$$\left(\frac{1}{4} \cdot \frac{360°}{b}, 0\right), \left(\frac{3}{4} \cdot \frac{360°}{b}, 0\right)$$

Phase Shift (p. 865) The phase shift of the functions $y = a \sin b(\theta - h)$, $y = a \cos b(\theta - h)$, and $y = a \tan b(\theta - h)$ is h, where $b > 0$.
- If $h > 0$, the shift is h units to the right.
- If $h < 0$, the shift is h units to the left.

Vertical Shift (p. 866) The vertical shift of the functions $y = a \sin b\theta + k$, $y = a \cos b\theta + k$, and $y = a \tan b\theta + k$ is k.
- If $k > 0$, the shift is k units up.
- If $k < 0$, the shift is k units down.

Graph Trigonometric Functions (p. 867)

Step 1 Determine the vertical shift and graph the midline.
Step 2 Determine the amplitude, if it exists. Use dashed lines to indicate the maximum and minimum values of the function.
Step 3 Determine the period of the function and graph the appropriate function.
Step 4 Determine the phase shift and translate the graph accordingly.

Inverse Trigonometric Functions (p. 873)

Arcsine	$y = \text{Arcsin } x$
	$y = \text{Sin}^{-1} x$
Arccosine	$y = \text{Arccos } x$
	$y = \text{Cos}^{-1} x$
Arctangent	$y = \text{Arctan } x$
	$y = \text{Tan}^{-1} x$

Chapter 14

Trigonometric Identities and Equations

Basic Trigonometric Identities (p. 893)

Quotient Identities	$\tan\theta = \frac{\sin\theta}{\cos\theta}, \cos\theta \neq 0$		$\cot\theta = \frac{\cos\theta}{\sin\theta}, \sin\theta \neq 0$
Reciprocal Identities	$\csc\theta = \frac{1}{\sin\theta}, \sin\theta \neq 0$	$\sec\theta = \frac{1}{\cos\theta}, \cos\theta \neq 0$	$\cot\theta = \frac{1}{\tan\theta}, \tan\theta \neq 0$
Pythagorean Identities	$\cos^2\theta + \sin^2\theta = 1$	$\tan^2\theta + 1 = \sec^2\theta$	$\cot^2\theta + 1 = \csc^2\theta$
Cofunction Identities	$\sin\left(\frac{\pi}{2} - \theta\right) = \cos\theta$	$\cos\left(\frac{\pi}{2} - \theta\right) = \sin\theta$	$\tan\left(\frac{\pi}{2} - \theta\right) = \cot\theta$
Negative Angle Identities	$\sin(-\theta) = -\sin\theta$	$\cos(-\theta) = \cos\theta$	$\tan(-\theta) = -\tan\theta$

Verifying Identities by Transforming One Side (p. 900)

Step 1 Simplify one side of an equation until the two sides of the equation are the same. It is often easier to work with the more complicated side of the equation.

Step 2 Transform that expression into the form of the simpler side.

Suggestions for Verifying Identities (p. 901)

- Substitute one or more basic trigonometric identities to simplify the expression.
- Factor or multiply as necessary. You may have to multiply both the numerator and denominator by the same trigonometric expression.
- Write each side of the identity in terms of sine and cosine only. Then simplify each side as much as possible.
- The properties of equality do not apply to identities as they do with equations. Do not perform operations to the quantities from each side of an unverified identity.

Sum and Difference of Angles Identities (p. 906)

Sum Identities

- $\sin(A + B) = \sin A \cos B + \cos A \sin B$
- $\cos(A + B) = \cos A \cos B - \sin A \sin B$
- $\tan(A + B) = \frac{\tan A + \tan B}{1 - \tan A \tan B}$

Difference Identities

- $\sin(A - B) = \sin A \cos B - \cos A \sin B$
- $\cos(A - B) = \cos A \cos B + \sin A \sin B$
- $\tan(A + B) = \frac{\tan A - \tan B}{1 + \tan A \tan B}$

Double-Angle Identities (p. 913)

$\sin 2\theta = 2\sin\theta\cos\theta$

$\cos 2\theta = \cos^2\theta - \sin^2\theta$
$\cos 2\theta = 2\cos^2\theta - 1$
$\cos 2\theta = 1 - 2\sin^2\theta$

$\tan 2\theta = \frac{2\tan\theta}{1 - \tan^2\theta}$

Half-Angle Identities (p. 914)

$\sin\frac{\theta}{2} = \pm\sqrt{\frac{1 - \cos\theta}{2}}$

$\cos\frac{\theta}{2} = \pm\sqrt{\frac{1 + \cos\theta}{2}}$

$\tan\frac{\theta}{2} = \pm\sqrt{\frac{1 - \cos\theta}{1 + \cos\theta}}, \cos\theta \neq -1$

For Homework Help, go to Hotmath.com
Complete, step-by-step solutions of most odd-numbered exercises are provided free of charge.

Chapter 0 Preparing for Advanced Algebra

Page P5 Lesson 0-1

1. D = {1, 2, 3}, R = {6, 7, 10}; yes **3.** D = {1, 2}, R = {5, 7, 9}; no **5.** D = {−2, −1, 0, 3}, R = {−3, −2, 2}; yes **7.** D = {−1, 0, 1, 2, 3}, R = {−3, −2, −1, 2, 3, 4}; no **9.** I **11.** none

Page P6 Lesson 0-2

1. $a^2 + 6a + 8$ **3.** $h^2 - 16$ **5.** $b^2 + b - 12$ **7.** $r^2 - 5r - 24$ **9.** $p^2 + 16p + 64$ **11.** $2c^2 - 9c - 5$ **13.** $6m^2 - 7m - 20$ **15.** $2q^2 - 13q - 34$ **17a.** $n - 7$, $n + 2$ **17b.** $n^2 - 5n - 14$

Page P8 Lesson 0-3

1. $4x(3x + 1)$ **3.** $4ab(2b - 3)$ **5.** $(y + 3)(y + 9)$ **7.** $(3y + 1)(y + 4)$ **9.** $(3x + 4)(x + 8)$ **11.** $(y - 4)(y - 1)$ **13.** $2(3a - b)(a - 8b)$ **15.** $(2x - 3y)(9x - 2y)$ **17.** $(3x - 4)^2$ **19.** $(x + 12)(x - 12)$ **21.** $(4y + 1)(4y - 1)$ **23.** $4(3y + 2)(3y - 2)$

Pages P10 and P11 Lesson 0-4

1. independent **3.** independent **5.** 6 **7.** 12 **9.** 48 **11.** 60,480 **13.** 358,800 **15.** 60

Page P14 Lesson 0-5

1. 60 **3.** 2520 **5.** 6 **7.** 15,120 **9.** permutation; 5040 **11.** combination; 715 **13.** combination; 15 **15.** permutation; 3360 **17.** 840 ways **19.** 220 ways

Page P16 Lesson 0-6

1. similar **3.** neither **5.** similar **7.** 8; 21 **9.** 10.2; 13.6 **11.** $4\frac{1}{2}$ in.

Page P18 Lesson 0-7

1. 39 ft **3.** 8.3 cm **5.** 5 **7.** 9.2 **9.** 8.5 **11.** yes **13.** no **15.** yes **17.** about 2.66 m

Chapter 1 Equations and Inequalities

Page 3 Chapter 1 Get Ready

1. 12.25 **3.** −66.15 **5.** $1\frac{13}{15}$ **7.** $-1\frac{1}{3}$ **9.** $10\frac{1}{2}$ yd **11.** −64 **13.** 15.625 **15.** $\frac{2401}{81}$ **17.** $-\frac{3375}{256}$ **19.** true **21.** false **23.** yes

Pages 7–10 Lesson 1-1

1. 4.6 **3.** 18.4 **5.** 11.6 **7.** 0.96875 **9.** 0.6 **11.** $6\frac{4}{15}$ **13.** 28 **15.** −13.4 **17a.** 1524.6 mi **17b.** 720 mi **19.** 20

21 $\frac{b^2c^2}{ad} = \frac{(-0.8)^2(5)^2}{(-4)\left(\frac{1}{5}\right)}$ $a = -4, b = -0.8, c = 5, d = \frac{1}{5}$

$= \frac{(0.64)(25)}{(-0.8)}$ Evaluate the numerator and denominator separately.

$= \frac{16}{-0.8}$ Simplify the numerator.

$= -20$ Simplify the fraction.

23. 3.71 **25.** $\frac{1}{2}(x + 7)(2x)$ **27a.** 584,336,233.6 mi **27b.** 8761 h **27c.** yes; $\frac{8761}{24}$ = 365 days or 1 year **29.** 544 **31.** 13.8 **33.** 131.25 **35.** $6\pi x^3$

37 $t = 50 + \frac{n - 40}{4}$ Write the formula.

$= 50 + \frac{120 - 40}{4}$ $n = 120$

$= 50 + \frac{80}{4}$ Evaluate the numerator of the fraction.

$= 50 + 20$ Simplify the fraction.

$= 70$ Simplify.

If the number of chirps is 120, then the temperature is 70°F.

39a. \$3.91; \$5.36; \$7.31 **39b.** \$4.42; \$6.62; \$11.62; Sample answer: the average prices found in part **b** become increasingly higher with time.

41 $y = \sqrt{b^2\left(1 - \frac{x^2}{a^2}\right)}$ Write the equation.

$= \sqrt{8^2\left(1 - \frac{3^2}{6^2}\right)}$ $a = 6, b = 8, x = 3$

$= \sqrt{64\left(1 - \frac{9}{36}\right)}$ Evaluate the powers.

$= \sqrt{64\left(\frac{27}{36}\right)}$ Simplify inside the parentheses.

$= \sqrt{48}$ Simplify.

≈ 6.9 Use a calculator.

43. Lauren; −12 − 20 = −32. **45.** Subtract 8 from each side. Divide each side by 4. Add 12 to each side. Multiply each side by 3. Subtract 6 from each side. $k = -12$ **47.** Sample answer: $y\left(\frac{-4z}{x^2} - x\right) + z$ **49.** A table of on-base percentages is limited to those situations listed, while a formula can be used to find any on-base percentage. **51.** 9 mo **53.** G **55.** 10 cm **57.** $6x(x + 2)$ **59.** 3 and 11 **61.** 5 **63.** 11 **65.** −4 **67.** $\frac{5}{8}$

Pages 14–17 Lesson 1-2

1. N, W, Z, Q, R **3.** I, R **5.** Associative Property (×) **7.** Commutative Property (+) **9.** 7; $-\frac{1}{7}$ **11.** −3.8; $\frac{1}{3.8}$ **13a.** 22(2 + 4 + 3 + 1 + 5 + 6 + 7) or 22(2) + 22(4) + 22(3) + 22(1) + 22(5) + 22(6) + 22(7) **13b.** \$616 **13c.** If she continues to mow the same number of lawns, at the end of next week she will have the money. This may not be reasonable because not all the lawns she mowed this week may need to be mowed again next week.

15. $24a + 9b$ **17.** $-16x + 22y$ **19.** Q, R **21.** Q, R

23 $-\sqrt{144} = -12$ belongs to the set of integers (Z), the set of rationals (Q), and the set of reals (R).

25. I, R **27.** Distributive Property **29.** Inverse Property (×) **31.** $-12.1; \frac{1}{12.1}$ **33.** $-\frac{6}{13}; \frac{13}{6}$ **35.** $-\sqrt{15}; \frac{1}{\sqrt{15}}$ **37.** $12b + 6c$ **39.** $40x - 20y$ **41.** $28g - 48k$ **43.** $53(60 + 60)$; $53(60) + 53(60)$; $6360\ yd^2$

45 **a.** $(2 + 1)4.50 = (3)4.50$ Distributive Property
$= \$13.50$ Simplify.

b. The amount left over is \$20 − \$13.50 or \$6.50.
$\$6.50 \div 2 = 3.25$
Because Billie cannot buy part of a sandwich, she can buy 3 cold sandwiches.

c. In two weeks, or ten school days, Billie buys a hot lunch 3 times and buys a cold sandwich 3 times. She has to pack lunch $10 - (3 + 3) = 10 - 6$ or 4 times.

47. $\frac{27}{5}c - \frac{199}{20}d$ **49.** $-42x - 72y - 30z$

51 **a.** $-\sqrt{6}$ is an irrational number because the square root of 6 is not a perfect square.
3, or $\frac{3}{1}$, is a rational number, integer, whole number, and natural number.
$\frac{-15}{3}$, or -5, is a rational number and integer.
4.1, or $\frac{41}{10}$, is a rational number.
π is an irrational number.
0, or $\frac{0}{1}, \frac{0}{2}, \ldots$, is a rational number, integer, and whole number.
$\frac{3}{8}$ is a rational number.
$\sqrt{36}$, or 6, is a rational number, integer, whole number, and natural number.

irrational	rational	integer	whole	natural
$-\sqrt{6}, \pi$	$3, \frac{-15}{3}, 4.1, 0, \frac{3}{8}, \sqrt{36}$	$3, \frac{-15}{3}, 0, \sqrt{36}$	$3, 0, \sqrt{36}$	$3, \sqrt{36}$

b. Use a calculator to find the decimal form of $-\sqrt{6}$ and π; $-\sqrt{6} \approx -2.449$, $3 = 3.0$, $\frac{-15}{3} = -5$, $4.1 = 4.1$, $\pi \approx 3.14$, $0 = 0$, $\frac{3}{8} = 0.375$, $\sqrt{36} = 6$. Since $-5 < -2.449 < 0 < 0.375 < 3.14 < 4.1 < 6$, the numbers from least to greatest are $\frac{-15}{3}, -\sqrt{6}, 0, \frac{3}{8}, \pi, 4.1, \sqrt{36}$.

c. Draw a number line with tick marks at integers from −6 to 6. Then use the decimal forms in part d to graph each number.

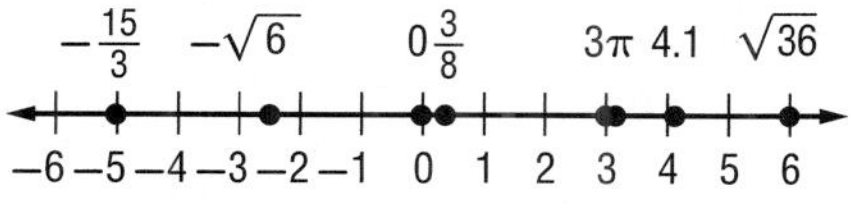

d. Sample answer: By converting the real numbers into decimal form, the decimal points can be easily lined up and the numbers compared.

53. $\sqrt{81}$; It is a rational number, while the other three are irrational numbers. **55.** No; Luna did not distribute the negative sign to the second term and Sophia switched the a and b terms because usually a comes first. The correct answer is $32a - 46b$.

57. Sample answer: $\sqrt{5} \cdot \sqrt{5} = 25$, which is not irrational **59.** Sample answer: (a) 3.2 and (b) $\sqrt{10}$

61. Sample answer: The Commutative Property does not hold for subtraction or division because order matters with these two operations. In addition or multiplication, the order does not matter. For example, $2 + 4 = 4 + 2$ and $2 \cdot 4 = 4 \cdot 2$. However, with subtraction, $2 - 4 \neq 4 - 2$, and with division, $2 \div 4 \neq 4 \div 2$. **63.** B **65.** B **67.** 24 **69.** about 2.66 m **71.** $3(3x^2 - x + 6)$ **73.** $10x(x - 2)$ **75.** $6(2x^2 - 3x - 4)$ **77.** $y^2 + y - 2$ **79.** $b^2 - 10b + 21$ **81.** $p^2 - 8p - 9$ **83.** $\frac{10}{9}$ **85.** 8 **87.** -1.176 **89.** -1.7

Pages 22–25 Lesson 1-3

1. $12[x + (-3)]$ **3.** The sum of five times a number and 7 equals 18. **5.** The difference between five times a number and the cube of that number is 12. **7.** Reflexive Property **9.** 53 **11.** −8 **13.** −6 **15.** 3 **17.** 4 **19.** $q = \frac{8r - 3}{5}$ **21.** B **23.** $8x^2$ **25.** $\frac{x}{4} + 5$ **27.** The quotient of the sum of 3 and a number and 4 is 5.

29 Let n = the number of home runs that Jacobs hit. Then $n + 6$ = the number of home runs that Cabrera hit.

$n + (n + 6) = 46$	Cabrera and Jacobs hit a combined total of 46 home runs.
$2n + 6 = 46$	Simplify.
$2n = 40$	Subtract 6 from each side.
$n = 20$	Divide each side by 2.

So, Jacobs hit 20 home runs and Cabrera hit $n + 6 = 20 + 6$ or 26 home runs.

31. Substitution **33.** Multiplication (=) **35.** 5 **37.** −3

39

$5(-2x - 4) - 3(4x + 5) = 97$	Original equation
$-10x - 20 - 12x - 15 = 97$	Apply the Distributive Property.
$-22x - 35 = 97$	Simplify the left side.
$-22x = 132$	Add 35 to each side.
$x = -6$	Divide each side by −22.

41. −3 **43.** s = length of a side; $5s = 100$; 20 in. **45.** $m = \frac{E}{c^2}$ **47.** $h = \frac{z}{\pi q^3}$ **49.** $a = \frac{y - bx - c}{x^2}$ **51a.** $V = \pi \cdot r \cdot r \cdot h$ **51b.** $h = \frac{V}{\pi r^2}$ **53.** −2 **55.** −4 **57.** $-\frac{117}{11}$ **59.** x = the cost of rent each month; $622 + 428 + 240 + 144 + 12x = 10{,}734$; \$775 per month

61 **a.** The integers from −5 to 5 are −5, −4, −3, −2, −1, 0, 1, 2, 3, 4, and 5. Draw a number line and plot a point at each integer.

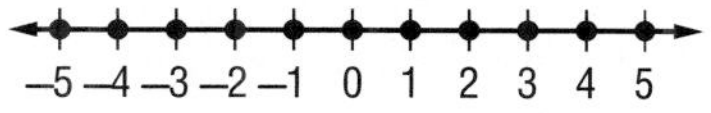

b. −5 and 5 are 5 units from zero, −4 and 4 are 4 units from zero, and so on.

Integer	Distance from Zero
−5	5
−4	4
−3	3
−2	2
−1	1
0	0
1	1
2	2
3	3
4	4
5	5

c. The points $(x, y) =$ (integer, distance from zero) are $(-5, 5)$, $(-4, 4)$, $(-3, 3)$, $(-2, 2)$, $(-1, 1)$, $(0, 0)$, $(1, 1)$, $(2, 2)$, $(3, 3)$, $(4, 4)$, and $(5, 5)$.

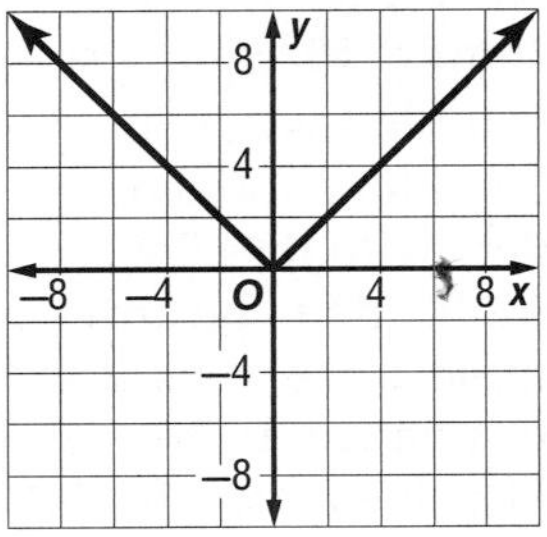

d. For positive integers, the distance from zero is the same as the integer. For negative integers, the distance is the integer with the opposite sign because distance is always positive.

63. $y_1 = y_2 - \sqrt{d^2 - (x_2 - x_1)^2}$ **65.** Sample answer: $3(x - 4) = 3x + 5$; $2(3x - 1) = 6x - 2$ **67.** D **69.** A **71.** $-3x + 6y + 6z$ **73.** 1 **75.** $4\frac{1}{5}$ **77.** $2x$ **79.** $-3\frac{2}{3}$ **81.** $-5x$

Pages 30–32 Lesson 1-4

1. 12 **3.** −108 **5a.** $|x - 78| = 2$ **5b.** least: 76°F; greatest: 80°F **5c.** 77°F; This would ensure a minimum temperature of 76°F. **7.** $\{15, -7\}$ **9.** $\varnothing$ **11.** $\left\{\frac{6}{5}, -\frac{4}{5}\right\}$ **13.** 2 **15.** 25 **17.** 9.2 **19.** 49.2 **21.** −63 **23.** $\{34, -8\}$ **25.** $\{4, -14\}$ **27.** $\{-2, -10\}$ **29.** 2

31

$2\|3x - 4\| + 8 = 6$	Original equation
$2\|3x - 4\| + 8 - 8 = 6 - 8$	Subtract 8 from each side.
$2\|3x - 4\| = -2$	Simplify.
$\|3x - 4\| = -1$	Divide each side by 2.

Because the absolute value of a number is always positive or zero, this sentence is never true. So, there is no solution.

33. $\varnothing$ **35.** $|x - 5.67| = 0.02$; heaviest: 5.69 g; lightest: 5.65 g **37.** 28 **39.** $\left\{1, \frac{1}{5}\right\}$ **41.** $-\frac{8}{3}$

43 The average altitude c is 100 ft. Since the altitude can be plus or minus 245 feet, the range r is 245.

$\|x - c\| = r$	Absolute value equation
$\|x - 100\| = 245$	$c = 100$ and $r = 245$

$|x - 100| = 245$ means $x - 100 = 245$ or $x - 100 = -245$.

Case 1	Case 2
$x - 100 = 245$	$x - 100 = -245$
$x - 100 + 100 = 245 + 100$	$x - 100 + 100 = -245 + 100$
$x = 345$	$x = -145$

The solutions are 345 and −145. This means the maximum is 345 ft above sea level; the minimum is −145 ft or 145 ft below sea level. The maximum is reasonable, but the minimum is not. Florida's lowest point should be at sea level where Florida meets the Atlantic Ocean and the Gulf of Mexico.

45. Ling; Ana included an extraneous solution. She would have caught this error if she had checked to see if her answers were correct by substituting the values into the original equation. **47.** Sometimes; this is only true for certain values of a. For example, it is true for $a = 8$; if $8 > 7$, then $11 > 10$. However, it is not true for $a = -8$; if $8 > 7$, then $5 \not> 10$. **49.** Always; starting with numbers between 1 and 5 and subtracting 3 will produce numbers between −2 and 2. These all have an absolute value less than or equal to 2. **51.** Sample answer: First, isolate the absolute value symbol by subtracting each side by c, and then dividing each side by a. You then have $|x - b|$ equals a mathematical expression. Take away the absolute value symbol, and form two new equations by setting $x - a$ equal to both the positive and negative values of the expression. Solve each equation for x. Then substitute each solution into the original equation, and confirm whether they are correct. **53.** $\frac{5}{8}$ **55.** C **57.** −2 **59a.** \$6800 **59b.** \$535.83 **59c.** 1 mo **61.** Distributive **63.** $10x + 2y$ **65.** $11m + 10a$ **67.** $32c - 46d$ **69.** 2 **71.** −8 **73.** $-\frac{4}{7}$

Pages 36–39 Lesson 1-5

1. $b < 8$

3. $x \le -6$

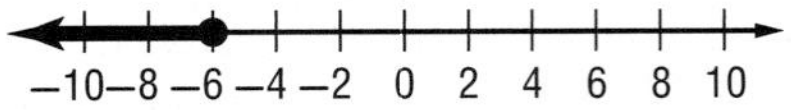

5. $w < 2$

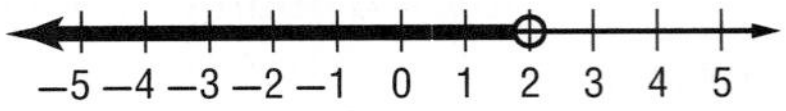

7

$s \ge \frac{s + 6}{5}$	Original inequality
$5s \ge s + 6$	Multiply each side by 5.
$4s \ge 6$	Subtract s from each side.
$s \ge 1.5$	Divide each side by 4.

The solution set is $\{s \mid s \ge 1.5\}$.

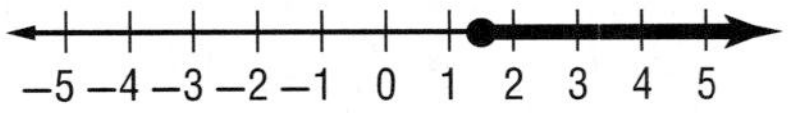

9. 40 bags

11. $n \le -3$

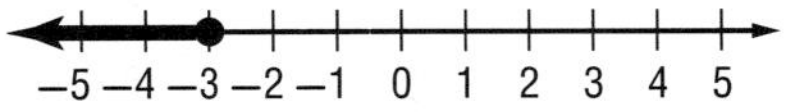

13. $t \le \frac{1}{2}$

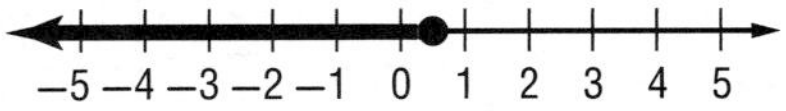

15. $k < 27$

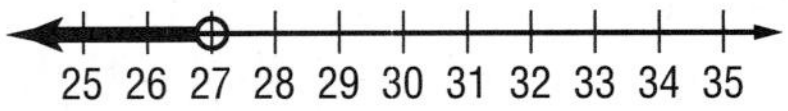

17. $z < 3$

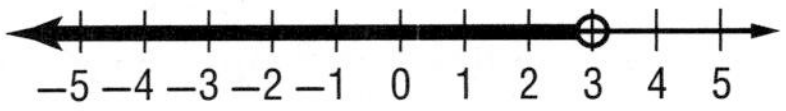

19

$12 < -4(3c - 6)$	Original inequality
$-3 > 3c - 6$	Divide each side by −4, reversing the inequality symbol.
$3 > 3c$	Add 6 to each side.
$1 > c$	Divide each side by 3.

The solution set is $\{c \mid c < 1\}$.

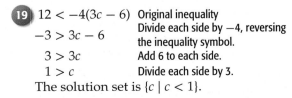

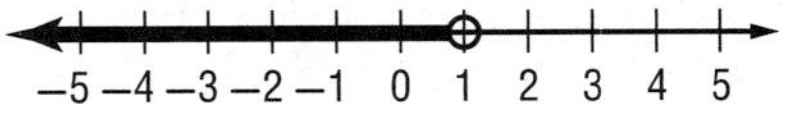

21. $z < 3$

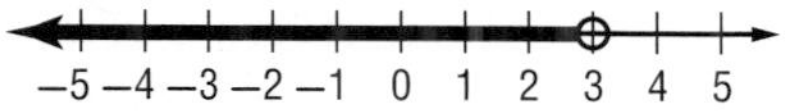

23. $3x - 12 < 21; x < 11$ **25.** $5x - 6 > x; x > 1.5$
27. 8 hours
29. $x > -\frac{3}{4}$

31. $y > 18.75$

33. $v > -4.5$

35. $r > -\frac{3}{4}$

37a. $250 + 0.03(500a) \geq 700$ **37b.** $a \geq 30$; He must sell at least 30 advertisements. **39.** $\frac{x}{3} + 4 \leq 2x + 12$; $x \geq -4.8$

41 a. Let d = the number of miles by which Jamie should increase her average daily run. Then $5 + d$ = her average daily distance after the increase.

3 times	average daily distance	is at least	length of a marathon
$3 \cdot$	$(5 + d)$	$\geq$	26.2

So, the inequality is $3(5 + d) \geq 26.2$.

b. $3(5 + d) \geq 26.2$ Original inequality
$5 + d \geq 8.73$ Divide each side by 3. Round to the nearest hundredth.
$d \geq 3.73$ Subtract 5 from each side.

In order to have enough endurance to run a marathon, Jamie should increase the distance of her average daily run by at least 3.73 miles.

43a. Sample answer:

Point	Resulting Statement	True or False
(0, 0)	$0 \geq 3$	False
(1, 1)	$1 \geq \frac{5}{2}$	False
(2, 2)	$2 \geq 2$	True
(3, 3)	$3 \geq \frac{3}{2}$	True
(4, 4)	$4 \geq 1$	True

43b. Sample answer:

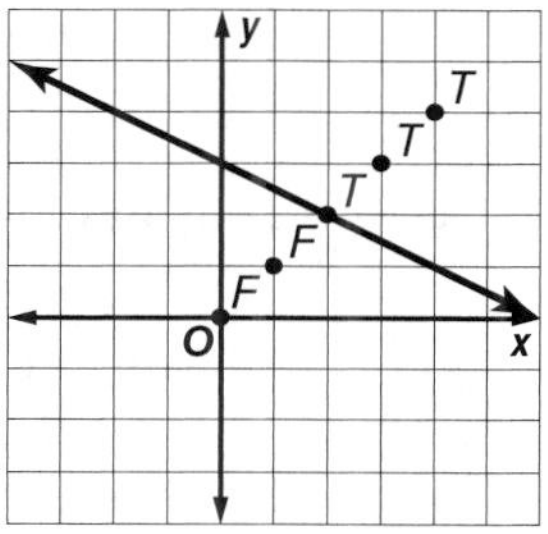

43c. Sample answer: The points on or above the line result in true statements, and the points below the line result in false statements. This is true for all points on the coordinate plane. **45.** No; Sample answer: Madlynn reversed the inequality sign when she added 1 to each side. Emilie did not reverse the inequality sign at all. **47.** Using the Triangle Inequality Theorem, we know that the sum of the lengths of any 2 sides of a triangle must be greater than the length of the remaining side. This generates 3 inequalities to examine.

$3x + 4 + 2x + 5 > 4x$
$x > -9$

$3x + 4 + 4x > 2x + 5$
$x > 0.2$

$2x + 5 + 4x > 3x + 4$
$x > -\frac{1}{3}$

In order for all 3 conditions to be true, x must be greater than 0.2. **49.** Sample answer: When one number is greater than another number, it is either more positive or less negative than that number. When these numbers are multiplied by a negative value, their roles are reversed. That is, the number that was more positive is now more negative than the other number. Thus, it is now *less than* that number and the inequality symbol needs to be reversed.
51. A **53.** D **55.** $\left\{-\frac{1}{3}, 3\right\}$ **57.** $|t - 3647.5| = 891.5$
59a. $SA = 2\pi r(r + h)$ **59b.** $78\pi\ cm^2$ **59c.** Sample answer: The formula in part b is quicker. **61.** $\{-9, 9\}$
63. $\left\{\frac{1}{2}, 7\right\}$ **65.** $\{-6, 2\}$

Pages 45–48 Lesson 1-6

1. $\{g \mid -12 < g < -2\}$

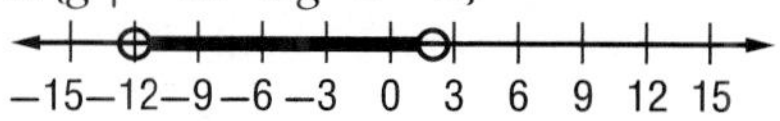

3. $\{z \mid z > -3 \text{ or } z < -6\}$

5. $\{c \mid c \geq 8 \text{ or } c \leq -8\}$

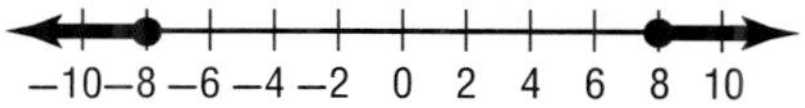

7. $\{z \mid -6 < z < 6\}$

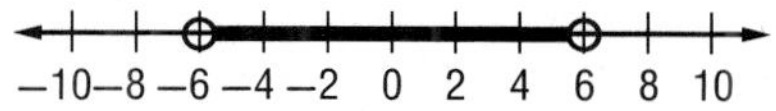

9. $\left\{v \,\middle|\, v > 3 \text{ or } v < -\frac{19}{3}\right\}$

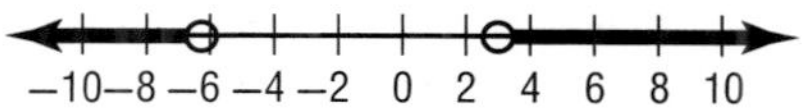

11. $43.96 \leq c \leq 77.94$; between \$43.96 and \$77.94

13. $\{d \mid -1 \leq d \leq 0.5\}$

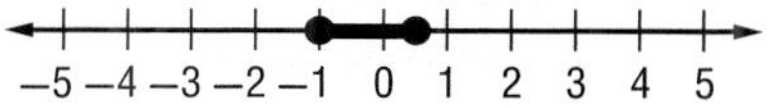

15. $\{y \mid y < -4 \text{ or } y > 7\}$

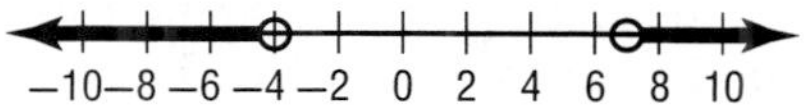

17. $\{k \mid -4 > k \text{ or } k > 4\}$

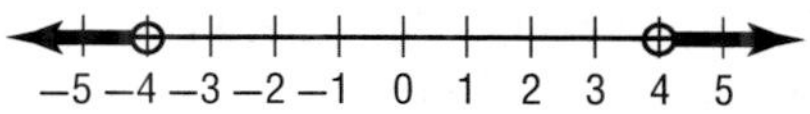

Selected Answers and Solutions

19 $|8t + 3| \leq 4$ is equivalent to $-4 \leq 8t + 3 \leq 4$.

$$-4 \leq 8t + 3 \leq 4$$
$$-4 - 3 \leq 8t + 3 - 3 \leq 4 - 3$$
$$-7 \leq 8t \leq 1$$
$$-\frac{7}{8} \leq \frac{8t}{8} \leq \frac{1}{8}$$
$$-\frac{7}{8} \leq t \leq \frac{1}{8}$$

The solution set is $\left\{t \,\middle|\, -\frac{7}{8} \leq t \leq \frac{1}{8}\right\}$.

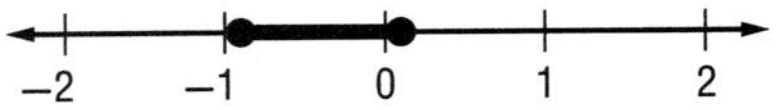

21. $\left\{j \,\middle|\, j \geq \frac{8}{5} \text{ or } j \leq -\frac{16}{5}\right\}$

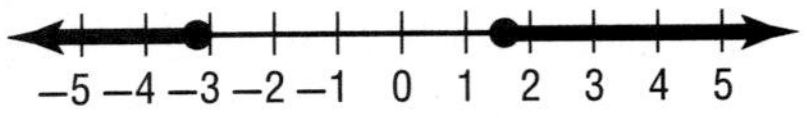

23. $|x - 1| \leq 5$ **25.** $|x + 9| \leq 3$ **27.** $|x - 2| \geq 10$
29. $|x + 3| > 1$

31 A healthy weight w for a fully grown female Labrador retriever is 55 pounds to 70 pounds. This can be represented by $55 \leq w \leq 70$.

33. $\{k \mid 2 < k < 4\}$

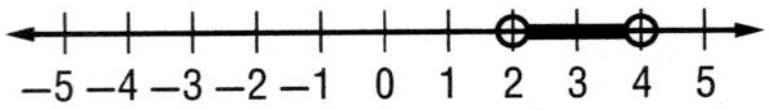

35. $\{h \mid h < -15 \text{ or } h > 15\}$

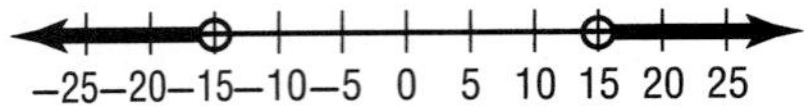

37. $\{z \mid z < -1 \text{ or } z > 5\}$

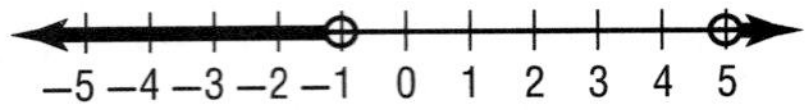

39. $\left\{f \,\middle|\, f > \frac{26}{5} \text{ or } f < -\frac{22}{5}\right\}$

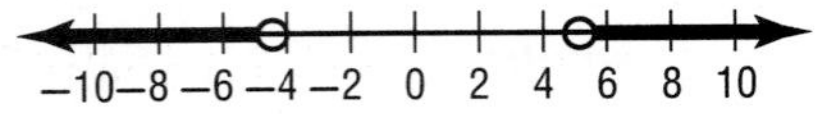

41. $|x + 5| \geq 4$ **43.** $6 \leq |x - 2| \leq 10$

45. $\{n \mid -7 < n < 1\}$

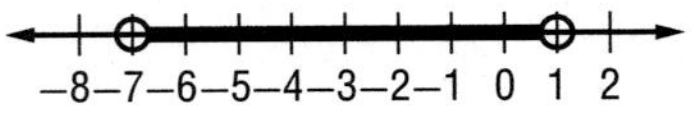

47. $\{x \mid \text{all real numbers}\}$

49. $\left\{g \,\middle|\, g \geq -\frac{2}{3}\right\}$

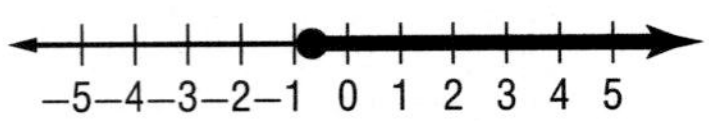

51. $|s - 88| > 38$; $\{s \mid s > 126 \text{ or } s < 50\}$ **53.** Sample answer: David; when Sarah converted the absolute value into two inequalities, she mistakenly switched the inequality symbols. **55.** False; sample answer: the graph of $x > 2$ and $x > 5$ is a ray bounded only on one end. **57.** true **59.** Sample answer: The graph on the left indicates a solution set from -3 to 5 for the inequality $|x - 1| \leq 4$. The graph on the right indicates a solution set of all numbers less than or equal to -3 or greater than or equal to 5 for the inequality $|x - 1| \geq 4$. **61.** Each of these has a non-empty solution set except for $x > 5$ and $x < 1$. There are no values of x that are simultaneously greater than 5 and less than 1. **63.** C **65.** 60 **67a.** $750 \leq x \leq 990$ **67b.** 110 g **69.** $\{0, 10\}$ **71.** $\varnothing$ **73.** Transitive (=)

Pages 49–52 Chapter 1 Study Guide and Review

1. false; nonnegative **3.** true **5.** true **7.** false; or **9.** true **11.** 3 **13.** 10 **15.** 21 **17.** 169.56 in^3 **19.** N, W, Z, Q, R **21.** $11x + 2y$ **23.** $5m + 41n$ **25.** -7 **27.** $\frac{3}{2}$ **29.** \$8 **31.** $m = \frac{r+5}{pn}$ **33.** 8 in. **35.** $\{2, 10\}$ **37.** $\left\{\frac{4}{3}, 14\right\}$

39. $a \geq -6$

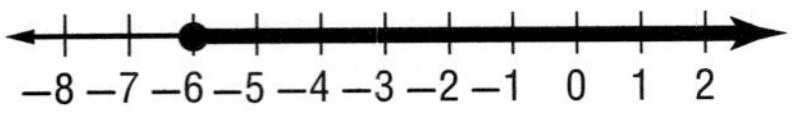

41. $x \leq -\frac{2}{9}$

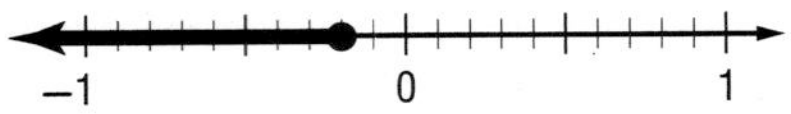

43. 3 or fewer slices each

45. $\{x \mid -2 < x < 4\}$

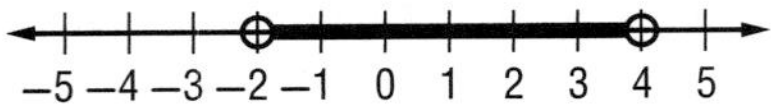

47. $\left\{m \,\middle|\, \frac{13}{5} \leq m < \frac{24}{5}\right\}$

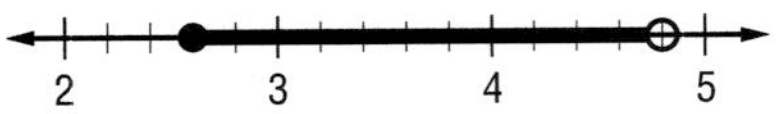

49. $\{p \mid -5 \leq p \leq 33\}$

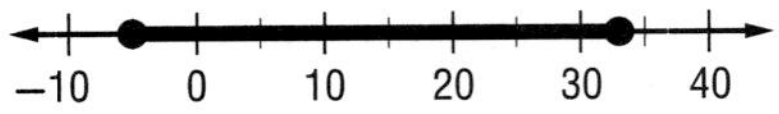

51. $\varnothing$ **53.** $20 \leq 2.50(3) + 1.25b \leq 30$; $10 \leq b \leq 18$

Chapter 2 Linear Relations and Functions

Page 59 Chapter 2 Get Ready

1. (4, 1) **3.** (0, 0) **5.** (−4, −4) **7.** −15 **9.** 10 **11.** \$40 **13.** $b = \frac{a}{3} - 3$ **15.** $x = \frac{8}{3} + \frac{4}{3}y$

Pages 64–67 Lesson 2-1

1. D = {5, 6, −2}, R = {3, −8, 1}; function; one-to-one

3 The domain is the set of x-values: $\{-2, 1, 4, 8\}$; the range is the set of y-values: $\{-4, -2, 6\}$. Since each element of the domain is paired with exactly one element of the range, the relation is a function. Since each element of the range corresponds to an element of the domain, the relation is an onto function.

5.

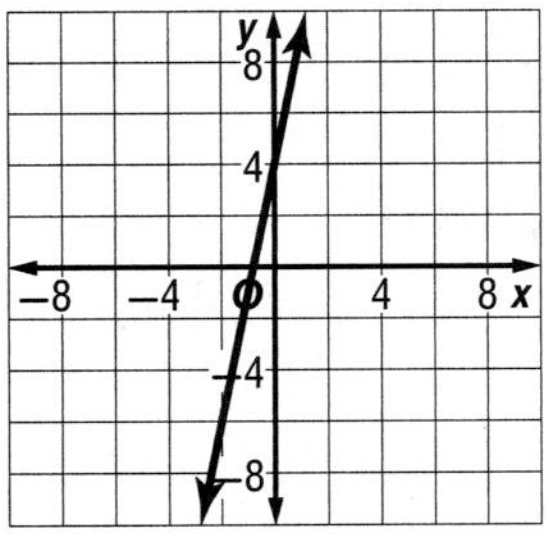

D = {all real numbers}, R = {all real numbers}; function; both; continuous

7. 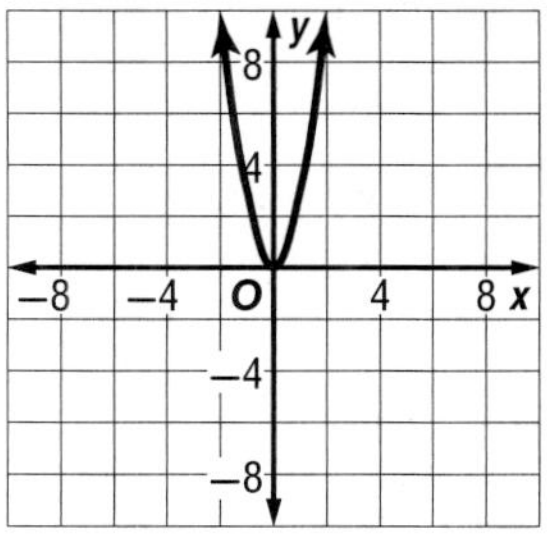

D = {all real numbers}, R = $\{y \mid y \geq 0\}$; function; neither; continuous **9.** 4 **11.** D = {−0.3, 0.4, 1.2}, R = {−6, −3, −1, 4}; not a function **13.** D = {−3, −1, 3, 5}, R = {−4, 0, 3}; function; one-to-one

15. 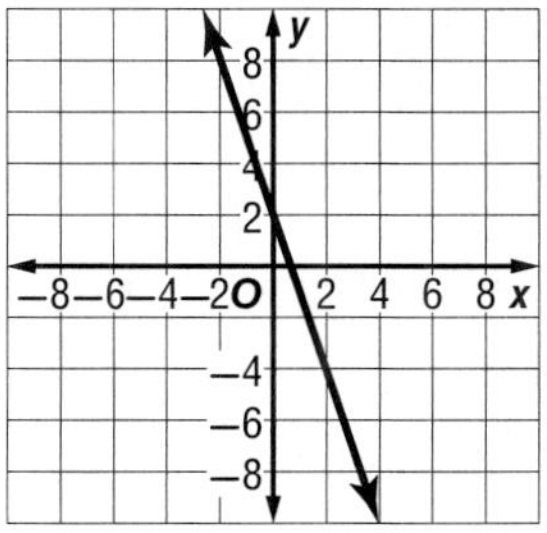

D = {all real numbers}, R = {all real numbers}; function; both; continuous

17. 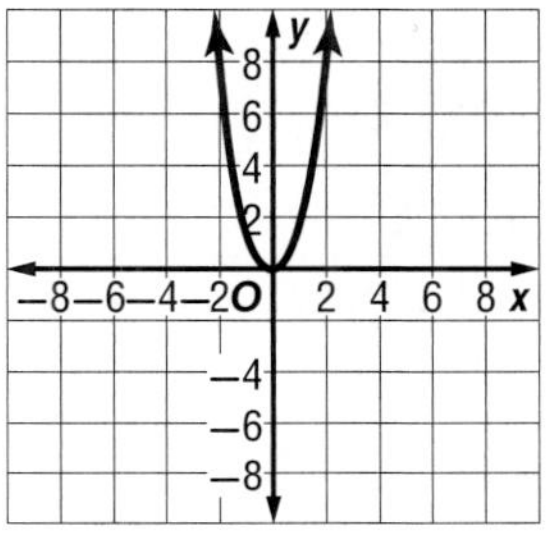

D = {all real numbers}, R = $\{y \mid y \geq 0\}$; function; both; continuous

19. 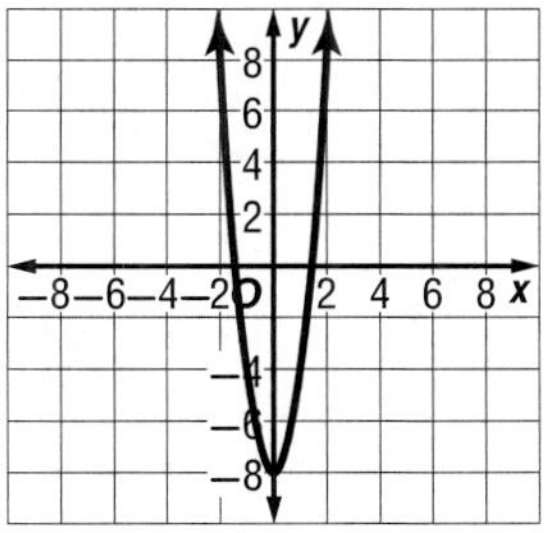

D = {all real numbers}, R = $\{y \mid y \geq -8\}$; function; both; continuous

21

$f(x) = 5x^3 + 1$	Original function
$f(-8) = 5(-8)^3 + 1$	Substitute −8 for each x.
$= 5(-512) + 1$	Evaluate $(-8)^3$.
$= -2560 + 1$	Multiply.
$= -2559$	Simplify.

23a. {(0, 1), (20, 1.6), (40, 2.2), (60, 2.8), (80, 3.4), (100, 4)}

23b. 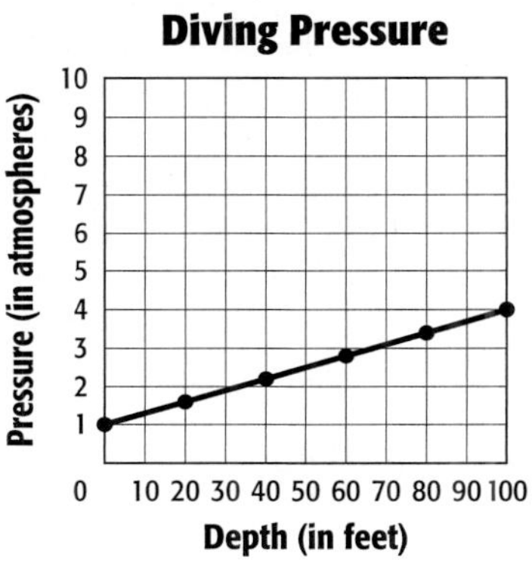

23c. D = $\{x \mid x \geq 0\}$, R = $\{y \mid y \geq 1\}$; continuous **23d.** Yes; each domain value is paired with only one range value so the relation is a function. **25.** 29 **27.** −72 **29.** −267 **31.** −4.5

33

$P(t) = 15 + 3t$	Original function
$P(8) = 15 + 3(8)$	Substitute 8 for each t.
$= 15 + 24$	Multiply.
$= 39$	Simplify.

After 8 months, Chaz will have 39 podcasts. **35.** Sample answer: Omar; Madison did not square the 3 before multiplying by −4. **37.** Never; if the graph crosses the y-axis twice, then there will be two separate y-values that correspond to $x = 0$, which violates the vertical line test. **39.** Sample answer: False; a function is onto and not one-to-one if all of the elements of the domain correspond to an element of the range, but more than one element of the domain corresponds to the same element of the range. **41.** A **43.** J **45.** $6 > y > 2$ **47.** $x > \frac{7}{4}$ or $x < -\frac{11}{4}$ **49.** $15x \leq 120$; She can buy up to 8 shirts. **51.** $\frac{3}{4}$ or $\frac{7}{4}$ **53.** 33a **55.** $10c + 36d$ **57.** 4 **59.** −4 **61.** −4 **63.** −6

Pages 71–74 Lesson 2-2

1. Yes; it can be written as $f(x) = \frac{x}{5} + \frac{12}{5}$. **3.** No; x has an exponent that is not 1.

5 **a.**

$m(x) = 0.75x$	Original function
$m(4) = 0.75(4)$	Substitute 4 for x.
$= 3$	Simplify.

If you have 4 CDs, you have 3 hours of music.

b.

$m(x) = 0.75x$	Original function
$6 = 0.75x$	Substitute 6 for $m(x)$.
$8 = x$	Divide each side by 0.75.

If the trip is 6 hours long, you should bring 8 CDs. **7.** $6x - y = -5$; $A = 6, B = -1, C = -5$ **9.** $8x + 9y = 6$; $A = 8, B = 9, C = 6$ **11.** $2x - 3y = 12$; $A = 2, B = -3, C = 12$

13. $\frac{5}{2}$; −10

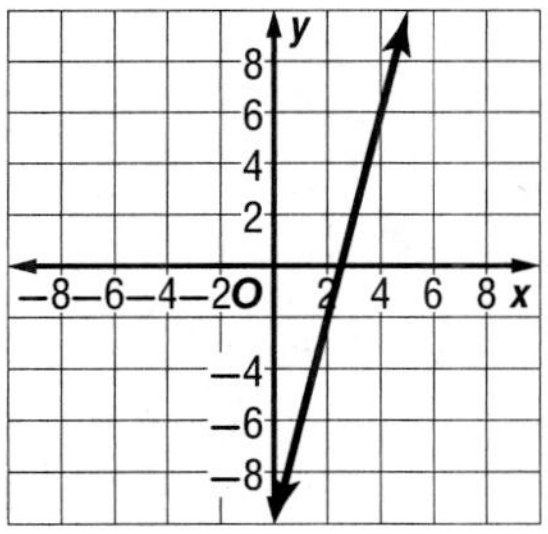

15. 7; $-\frac{21}{4}$

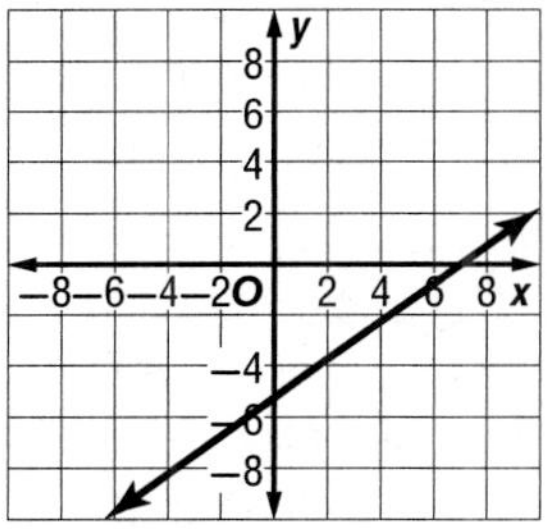

17. No; x has an exponent other than 1. **19.** No; x has an exponent other than 1. **21.** No; it cannot be written in $f(x) = mx + b$ form. **23.** No; it cannot be written in $f(x) = mx + b$ form; There is an xy term. **25a.** 260 m **25b.** Kingda Ka; Sample answer: The Kingda Ka travels 847.5 meters in 5 seconds, so it travels a greater distance in the same amount of time. **27.** $8x + 3y = -6$; $A = 8$, $B = 3$, $C = -6$ **29.** $2x + y = -11$; $A = 2$, $B = 1$, $C = -11$

31

$2.4y = -14.4x$	Original equation
$14.4x + 2.4y = 0$	Add $14.4x$ to each side.
$144x + 24y = 0$	Multiply each side by 10.
$6x + y = 0$	Divide each side by 24.

$A = 6, B = 1, C = 0$

33. $5x + 32y = 160$; $A = 5$, $B = 32$, $C = 160$

35. -0.5; -4

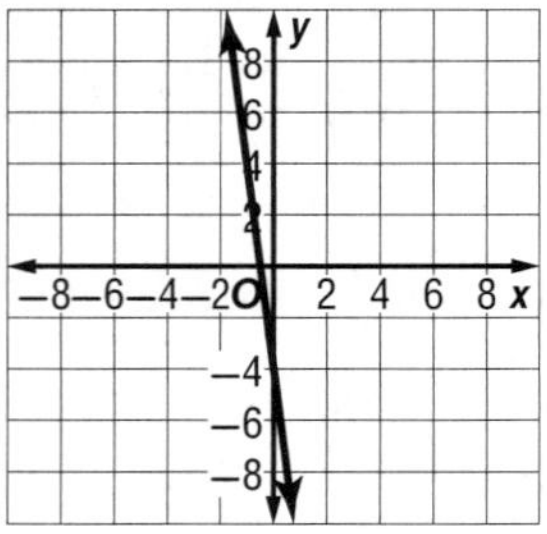

37. -7; 10.5

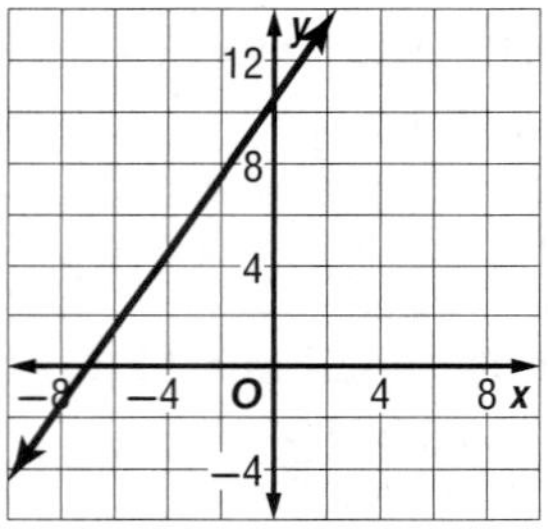

39. 12; -18

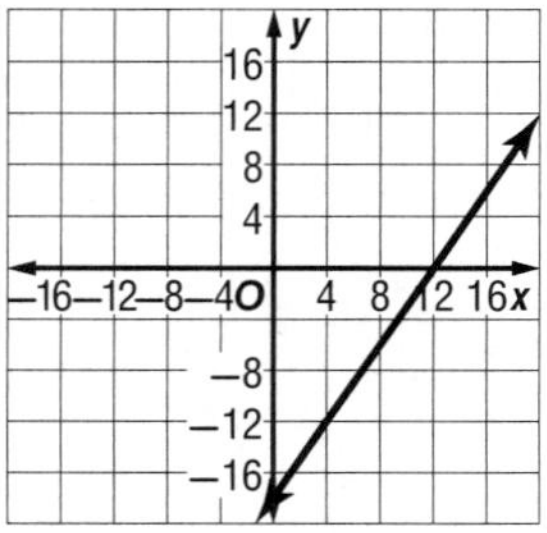

41a. $1.75m + 1.5n = 525$

41b.

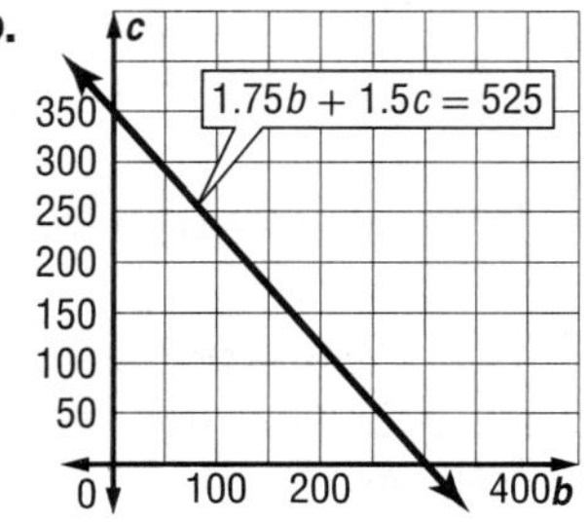

Yes; the graph passes the vertical line test. **41c.** No; the amount that Latonya will sell is 1.75 • 100 + 1.5 • 200, which is \$475.

43a. $y = 3x + 13$ **43b.** \$31 **45.** $4x - 40y = -59$; $A = 4$, $B = -40$, $C = -59$

47 The x-intercept is the value of x when $y = 0$.

$\frac{6x + 15}{4} = 3y - 12$	Original equation
$\frac{6x + 15}{4} = 3(0) - 12$	Substitute 0 for y.
$\frac{6x + 15}{4} = -12$	Simplify.
$6x + 15 = -48$	Multiply each side by 4.
$6x = -63$	Subtract 15 from each side.
$x = -\frac{63}{6}$	Divide each side by 6.
$x = -10.5$	Simplify.

The x-intercept is -10.5.

The y-intercept is the value of y when $x = 0$.

$\frac{6x + 15}{4} = 3y - 12$	Original equation
$\frac{6(0) + 15}{4} = 3y - 12$	Substitute 0 for x.
$\frac{15}{4} = 3y - 12$	Simplify.
$\frac{63}{4} = 3y$	Add 12 to each side.
$5.25 = y$	Divide each side by 3.

The y-intercept is 5.25.

49. $-1\frac{1}{75}$; $6\frac{1}{3}$ **51a.**

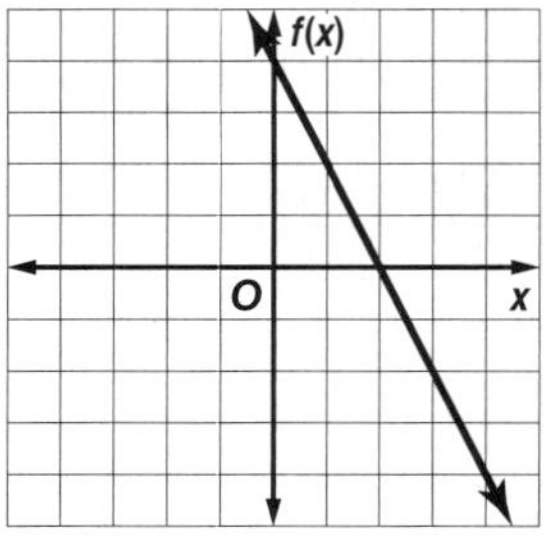

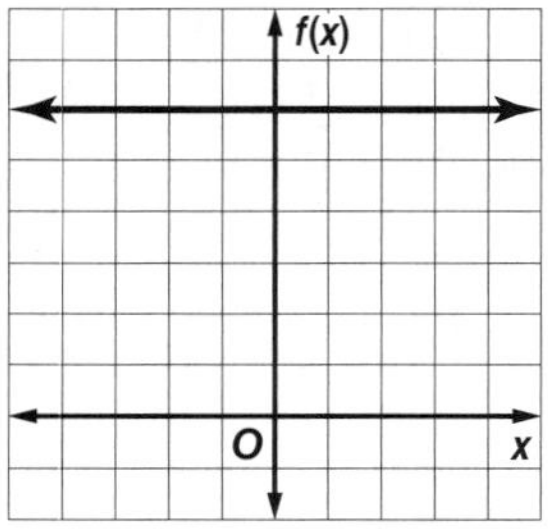

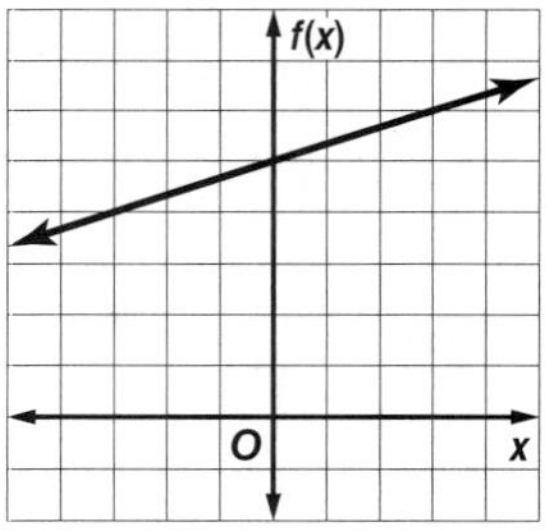

51b.

Function	One-to-One	Onto
$f(x) = -2x + 4$	yes	yes
$g(x) = 6$	no	no
$h(x) = \frac{1}{3}x + 5$	yes	yes

51c. No; horizontal lines are neither one-to-one nor onto because only one y-value is used and it is repeated for every x-value. Vertical lines are not one-to-one because only one x-value is used. Every other linear function is one-to-one and onto because every x-value has one unique y-value that is not used by any other x-element and every possible y-value is used. **53.** Sample answer: $f(x) = 2(x - 3)$ **55.** $y = 2xy$; Sample answer: $y = 2xy$ is not a linear function. **57.** C **59.** J **61.** D = {8, −4, −1}, R = {6, 3, 9}; not a function **63.** D = {−3, −4, 7}, R = {−1, −2, 9}; function; both **65.** 0.78 **67.** about −0.583 **69.** $\frac{2}{3}$ **71.** $-\frac{5}{4}$ **73.** $-\frac{1}{3}$ **75.** 9

Pages 79–82   Lesson 2-3

1. 6 feet/min **3a.** about 11,000 per year **3b.** about −5000 per year **3c.** The positive rate in part **a** represents an increase in the sales of digital cameras. The negative rate in part **b** represents a decrease in sales of film cameras. **5.** −3 **7.** $\frac{3}{5}$

9 Use the ordered pairs (3, 20) and (6, 40).

$$\text{rate of change} = \frac{\text{change in } y}{\text{change in } x}$$
$$= \frac{\text{change in height}}{\text{change in time}} \quad \begin{matrix}\leftarrow \text{mm} \\ \leftarrow \text{days}\end{matrix}$$
$$= \frac{40 - 20}{6 - 3}$$
$$= \frac{20}{3}$$

The rate of change is $\frac{20}{3}$ mm/day.

11a. 0.15°/h **11b.** −0.125°/h; Yes; the number should be negative because her temperature is dropping.
11c. Tuesday 8:00 A.M.–Tuesday 8:00 P.M. **13.** $\frac{14}{15}$
15. −2 **17.** $\frac{5}{3}$ **19.** 5

21 The line passes through (0, 20) and (5, 16).

$m = \frac{y_2 - y_1}{x_2 - x_1}$ Slope Formula
$= \frac{16 - 20}{5 - 0}$ $(x_1, y_1) = (0, 20), (x_2, y_2) = (5, 16)$
$= \frac{-4}{5}$ or −0.8 Simplify.

23. $\frac{4}{3}$ **25.** 3 **27.** $\frac{6}{5}$

29 $\text{slope} = \frac{\text{change in } y}{\text{change in } x}$
$= \frac{\text{change in vertical distance}}{\text{change in horizontal distance}}$
$= \frac{8.9}{2.8}$
≈ 3.2

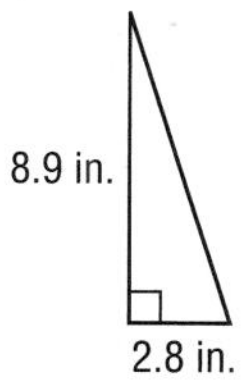

31. 9 **33.** 11 **35a.**

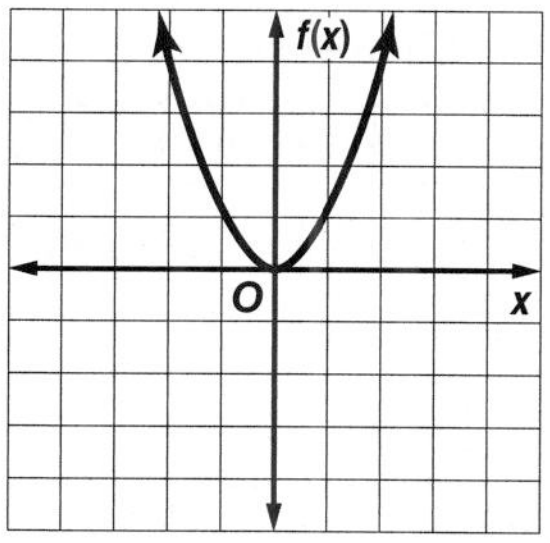

35b.

x	−4	−3	−2	−1	0	1	2	3	4
f(x)	16	9	4	1	0	1	4	9	16
slope		−7	−5	−3	−1	1	3	5	7

35c. Sample answer: The rate of change is not constant. The rate of change decreases as x approaches zero and then increases as x approaches infinity. **37.** Sample answer: Because the slope from (2, 3) to (5, 8) is the same as the slope from (5, 8) to (11, y), find the slope between each pair of points and set them equal to each other. Then solve for y.

$\frac{8 - 3}{5 - 2} = \frac{y - 8}{11 - 5}$
$\frac{5}{3} = \frac{y - 8}{6}$
$30 = 3(y - 8)$
$10 = y - 8$
$18 = y$

39. Sometimes; the slope of a vertical line is undefined.
41. $\frac{3}{2}$ **43.** G **45.** Yes; it can be written in $f(x) = mx + b$ form. **47.** No; it cannot be written in $f(x) = mx + b$ form. **49.** −46 **51.** 336 **53.** II
55. 3.5 **57.** $\frac{7}{3}$

Pages 86–89   Lesson 2-4

1. $y = 1.5x + 5$ **3.** $y = -2x + 11$ **5.** A

7 The slope of the given line is $\frac{7}{8}$. Lines that are parallel have the same slope, so the slope of the line parallel to the given line is $\frac{7}{8}$.

$y - y_1 = m(x - x_1)$ Point-slope form
$y - (-10) = \frac{7}{8}(x - 4)$ $(x_1, y_1) = (4, -10)$ and $m = \frac{7}{8}$
$y + 10 = \frac{7}{8}x - \frac{7}{2}$ Distributive Property
$y = \frac{7}{8}x - \frac{27}{2}$ Subtract 10 from each side.

9. $y = -\frac{1}{2}x + 5$ **11.** $y = 4.5x - 6.5$ **13.** $y = 4x - 15$
15. $y = -\frac{1}{4}x - 1$ **17.** $y = 2x - 2$ **19.** $y = -8x - 20$
21. $y = -0.5x + 3.4$ **23.** $y = \frac{1}{2}x$ **25.** $y = -\frac{1}{2}x + 6$
27. $y = 180x + 5900$ **29.** $y = -25x + 250$

31 First, find the slope. The line passes through (−6, 2) and (0, 6).

$m = \frac{y_2 - y_1}{x_2 - x_1}$ Slope Formula
$= \frac{6 - 2}{0 - (-6)}$ $(x_1, y_1) = (-6, 2), (x_2, y_2) = (0, 6)$
$= \frac{4}{6}$ or $\frac{2}{3}$ Simplify.

The graph intersects the y-axis at 6. So, $b = 6$. Substitute the values into the slope-intercept equation.

$y = mx + b$ Slope-intercept form
$y = \frac{2}{3}x + 6$ $m = \frac{2}{3}, b = 6$

33. 10 mi

35 **a.** Let x be the number of people Ms. Cooper recruits and let y be the amount of money she earns. Use the points (10, 100) and (14, 120) to represent this situation.

$m = \frac{y_2 - y_1}{x_2 - x_1}$ Slope Formula
$= \frac{120 - 100}{14 - 10}$ $(x_1, y_1) = (10, 100), (x_2, y_2) = (14, 120)$
$= 204$ or 5 Simplify.

Use the slope and either of the given points with the point-slope form to write the equation.

$y - y_1 = m(x - x_1)$ Point-slope form
$y - 100 = 5(x - 10)$ $(x_1, y_1) = (10, 100), m = 5$
$y - 100 = 5x - 50$ Distributive Property
$y = 5x + 50$ Add 100 to each side.

b. The y-intercept of the graph of $y = 5x + 50$ is 50. This represents the money Ms. Cooper would make if she had no recruits. So, \$50 is her daily salary.

c. Find the value of y when $x = 20$.

$y = 5x + 50$ Use the equation you found in part a.

$y = 5(20) + 50$ Replace x with 20.

$y = 150$ Simplify.

So, Ms. Cooper would earn \$150 in a day if she recruits 20 people.

37. Sample answer: Sometimes; while the two sets of parallel and perpendicular lines will always form a quadrilateral with four 90° angles, that figure will always be a rectangle, but not necessarily a square. **39.** Sample answer: $y - 0 = a\left(x + \frac{b}{a}\right)$

41. Sample answer: $y - d = -\frac{d}{c}(x - 0)$ **43.** A

45. G **47.** $-\frac{5}{3}$ **49.** $\frac{1}{5}$ **51.** $x \geq -4$ **53.** $x \geq -\frac{21}{13}$

55. yes **57.** $8c^2 + 8c - 30$ **59.** $-6a^2 + 7a + 20$

61. $\frac{3}{2}$ **63.** $\frac{1}{5}$ **65.** $-\frac{1}{9}$

Pages 95–98 Lesson 2-5

1 a. Graph the data as ordered pairs with the depth on the horizontal axis and the temperature on the vertical axis. Draw a line through two points that appear to represent the data well, such as (0, 22) and (2000, 6).

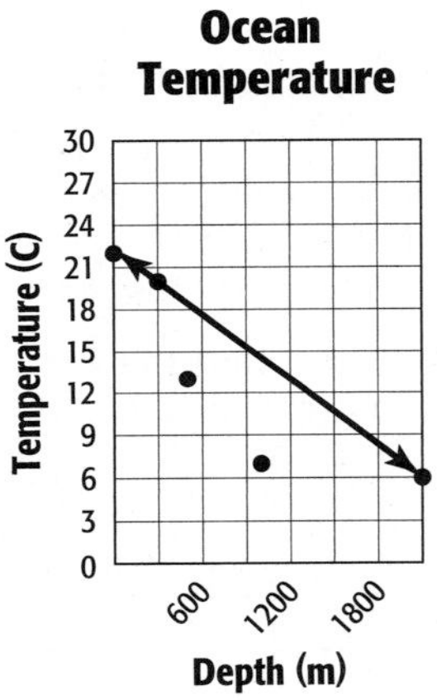

The data show a weak negative correlation.

b. Sample answer: Use (0, 22) and (2000, 6) to find an equation. First, find the slope of the line through (0, 22) and (2000, 6).

$m = \frac{y_2 - y_1}{x_2 - x_1}$ Slope Formula

$= \frac{6 - 22}{2000 - 0}$ $(x_1, y_1) = (0, 22), (x_2, y_2) = (2000, 6)$

≈ -0.008 Simplify.

Then write the equation.

$y - y_1 = m(x - x_1)$ Point-slope form

$y - 22 = -0.008(x - 0)$ $(x_1, y_1) = (0, 22), m = -0.008$

$y - 22 = -0.008x$ Simplify.

$y = -0.008x + 22$ Add 22 to each side.

c. Sample answer: Find y when $x = 2500$.

$y = -0.008x + 22$ Prediction equation

$= -0.008(2500) + 22$ $x = 2500$

$= -20 + 22$ or 2 Simplify.

So, at a depth of 2500 m, the temperature in the ocean is about 2°C.

3a.

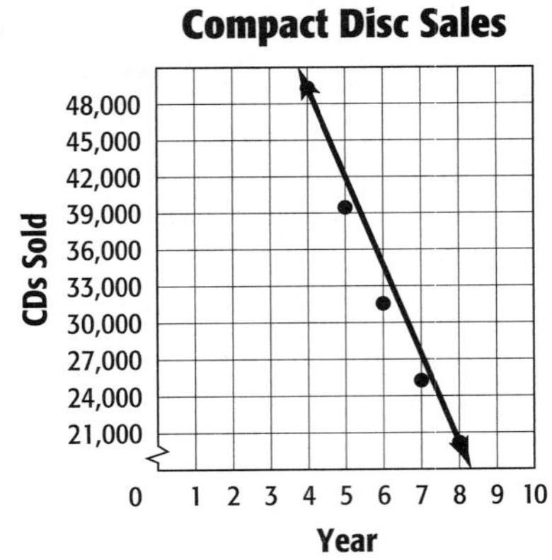

strong negative correlation

3b. Sample answer, using (4, 49,300) and (8, 20,193): $y = -7276.75x + 78{,}407$ **3c.** Sample answer: 12,916 CDs

5 a. Graph the data as ordered pairs with the month on the horizontal axis and the number of gallons sold on the vertical axis. Let 1 represent January, 2 represent February, and so on. Draw a line through two points that appear to represent the data well, such as (1, 37) and (8, 131).

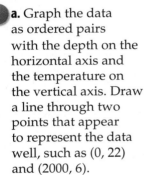

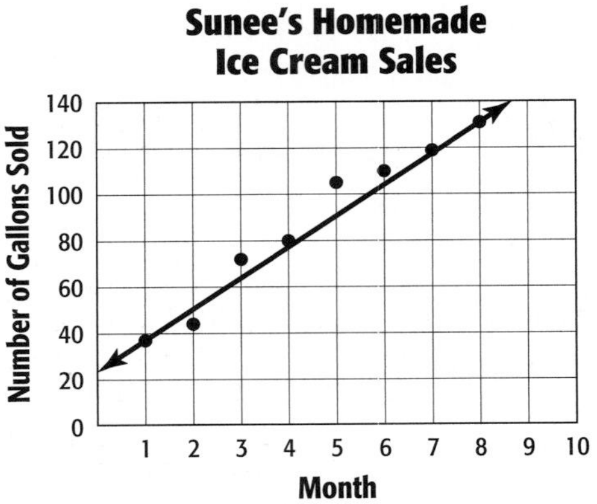

The data show a strong positive correlation.

b. Sample answer: Use (1, 37) and (8, 131) to find an equation. First, find the slope of the line through (1, 37) and (8, 131).

$m = \frac{y_2 - y_1}{x_2 - x_1}$ Slope Formula

$= \frac{131 - 37}{8 - 1}$ $(x_1, y_1) = (1, 37), (x_2, y_2) = (8, 131)$

$= \frac{94}{7}$ Simplify.

Then write the equation.

$y - y_1 = m(x - x_1)$ Point-slope form

$y - 37 = \frac{94}{7}(x - 1)$ $(x_1, y_1) = (1, 37), m = \frac{94}{7}$

$y - 37 = \frac{94}{7}x - \frac{94}{7}$ Distributive Property

$y = \frac{94}{7}x + \frac{165}{7}$ Add 37 to each side.

c. Sample answer: Find y when $x = 9$.

$y = \frac{94}{7}x + \frac{165}{7}$ Prediction equation

$= \frac{94}{7}(9) + \frac{165}{7}$ $x = 9$

$= \frac{1011}{7}$ Simplify.

≈ 144.4 Use a calculator.

So, about 144 gallons of ice cream are sold in September.

7. 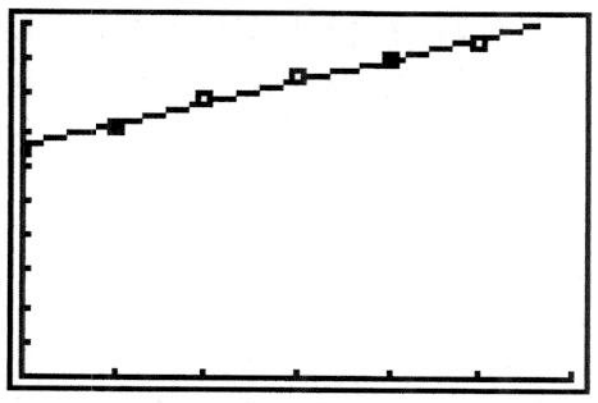

$y = 61.8x + 654$ (x is the number of years after 2002); \$1.148 million in sales

[2002, 2007] scl 1 by [0, 1000] scl 100

9 **a.** Graph the data as ordered pairs with the year on the horizontal axis and the attendance on the vertical axis.

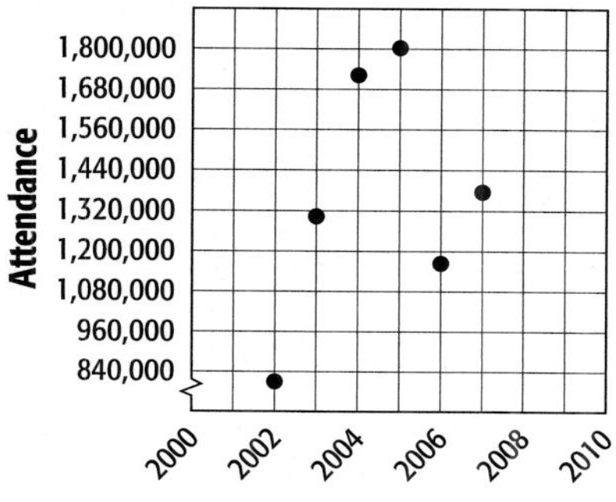

b. Use a graphing calculator to find a regression equation for the data. Enter the years in **L1** and the attendance in **L2**. Then select **LinReg(ax + b)** on the **STAT CALC** menu. The regression equation is approximately $y = 71{,}406.4x - 141{,}763{,}070.9$.
c. On a graphing calculator, copy the regression equation to the **Y=** list. Then use **VALUE** on the **CALC** menu to find y when $x = 2020$. Be sure to reset the window size to accommodate the x-value of 2020. When $x = 2020$, $y \approx 2{,}477{,}915$. So, in the year 2020, the attendance will be about 2,477,915.
d. Sample answer: The prediction is unreasonable. The attendance will not increase without bound because attendance is largely dependent on the team's winning status.

11a. $y = 3.1x - 6170$; $r = 0.63$ **11b.** about \$70.14 million
11c. $y = 2.6x - 5164$; $r = 0.986$ **11d.** about \$67.58 million
11e. Sample answer: The new equation has a correlation coefficient, 0.986, that is extremely close to 1, so this equation should accurately represent the data. **13.** Sample answer: If a and b have a positive correlation, then they are both increasing. If b and c have a negative correlation and b is increasing, then c must be decreasing. If c and d have a positive correlation and c is decreasing, then d must be decreasing. If a is increasing and d is decreasing, then they must have a negative correlation.

15. 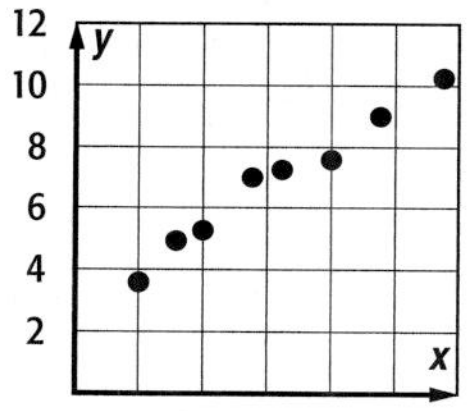

a; Sample answer: The data show a strong positive correlation which means that the correlation coefficient r should be close to 1.

17. 2 **19.** J **21.** $y = 2.5x - 6$ **23.** $y = -3x - 6$
25. 17.5 mi/hr **27.** 1.5 J/N **29.** 120 **31.** $\frac{17}{3}, -\frac{25}{3}$

Pages 104–107 Lesson 2-6

1. 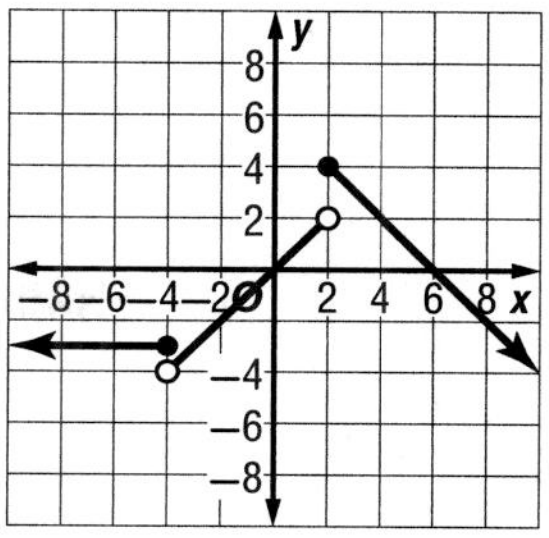

D = {all real numbers}; R = $\{y \mid y \leq 4\}$

3. $g(x) = \begin{cases} x + 4 \text{ if } x < -2 \\ -3 \text{ if } -2 \leq x \leq 3 \\ -2x + 12 \text{ if } x > 3 \end{cases}$

5 If the number of tickets sold is greater than 0 but less than or equal to 250, then the drama club must do 1 performance. If the number of tickets sold is greater than 250 but less than or equal to 500, then the drama club must do 2 performances, and so on. You can use the pattern to make a table, where x is the number of tickets sold and $P(x)$ is the number of performances. Then graph.

x	$P(x)$
$0 < x \leq 250$	1
$250 < x \leq 500$	2
$500 < x \leq 750$	3
$750 < x \leq 1000$	4
$1000 < x \leq 1250$	5

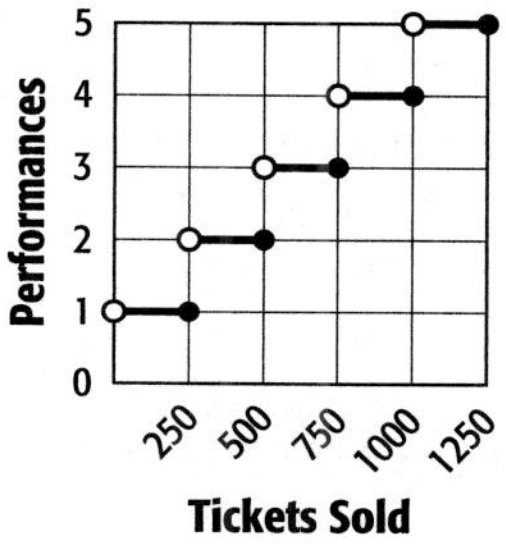

7. 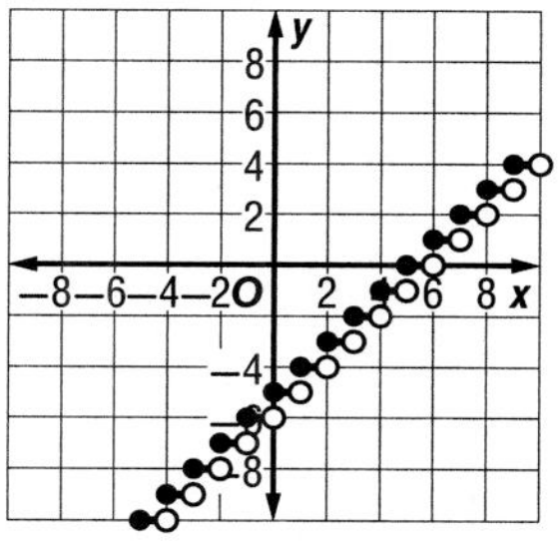

9. 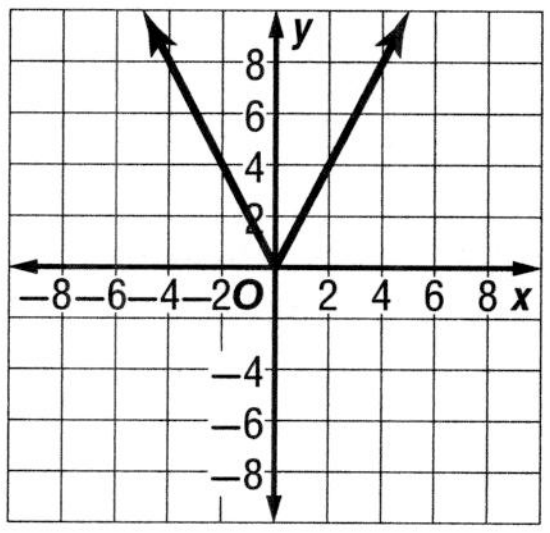

D = {all real numbers}; R = $\{f(x) \mid f(x) \geq 0\}$

11. 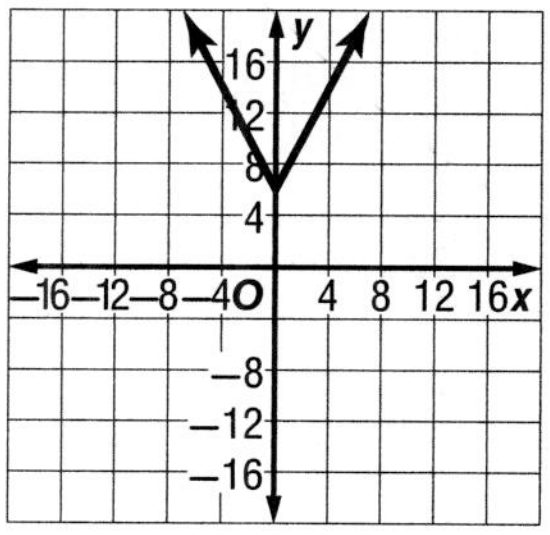

D = {all real numbers}; R = $\{s(x) \mid s(x) \geq 6\}$

13.

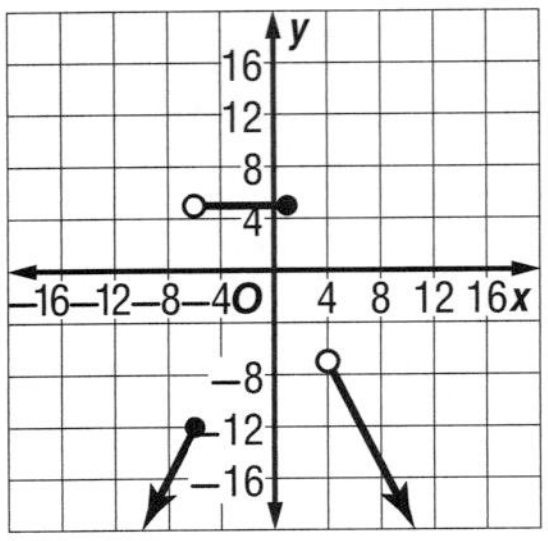

D = {x | $x \le 2$ or $x > 4$};
R = {$f(x)$ | $f(x) < -7$, or $f(x) = 5$}

15.

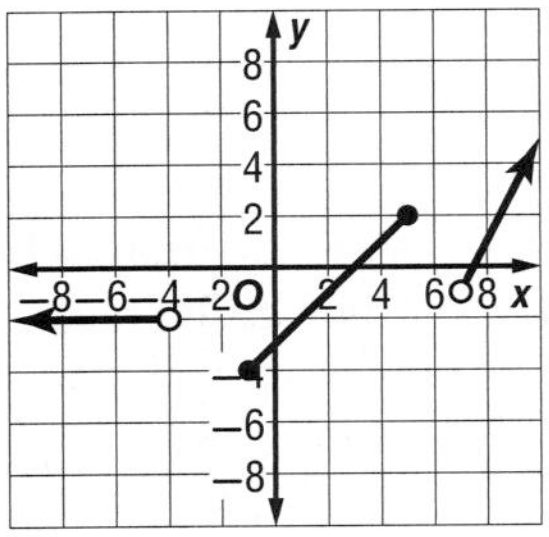

D = {x | $x < -4$, $-1 \le x \le 5$, or $x > 7$};
R = {$g(x)$ | $g(x) \ge -4$}

17 The left portion of the graph is the graph of $g(x) = -x - 4$. There is a circle at $(-3, -1)$, so the linear function is defined for {x | $x < -3$}. The middle portion of the graph is the graph of $g(x) = x + 1$. There are dots at $(-3, -2)$ and $(1, 2)$, so the linear function is defined for {x | $-3 \le x \le 1$}. The right portion of the graph is the graph of $g(x) = -6$. There is a circle at $(4, -6)$, so the linear function is defined for {x | $x > 4$}. Write the piecewise-defined function.

$$g(x) = \begin{cases} -x - 4 \text{ if } x < -3 \\ x + 1 \text{ if } -3 \le x \le 1 \\ -6 \text{ if } x > 4 \end{cases}$$

19. $g(x) = \begin{cases} 8 \text{ if } x \le -1 \\ 2x \text{ if } 4 \le x \le 6 \\ x - 8 \text{ if } x > 7 \end{cases}$

21.

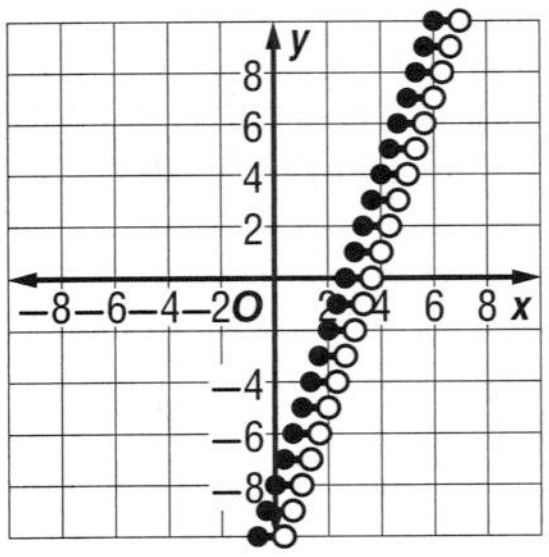

D = {all real numbers};
R = {all integers}

23.

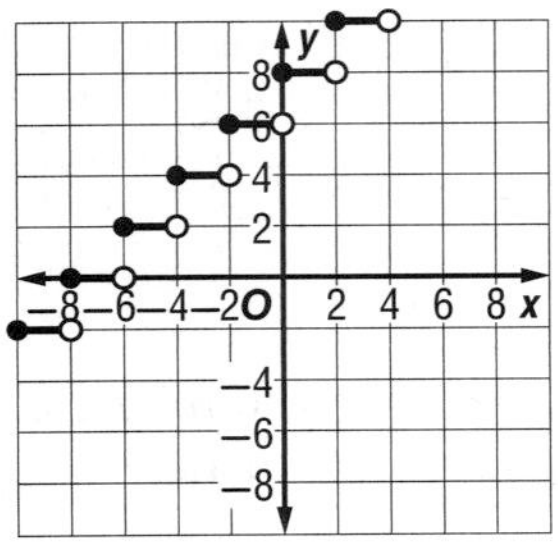

D = {all real numbers};
R = {0 and all even integers}

25.

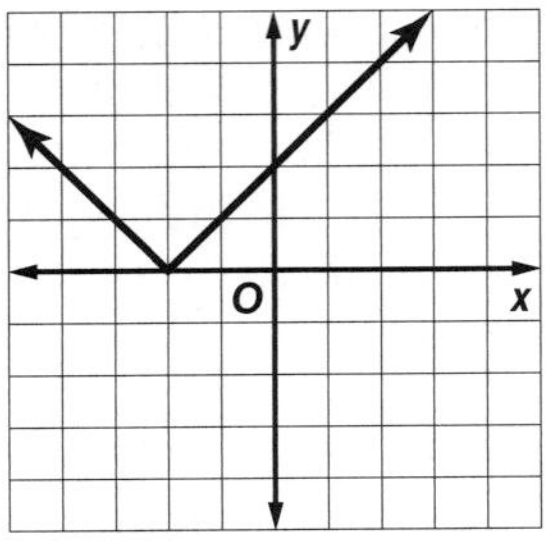

R = {all real numbers};
D = {$f(x)$ | $f(x) \ge 0$}

27.

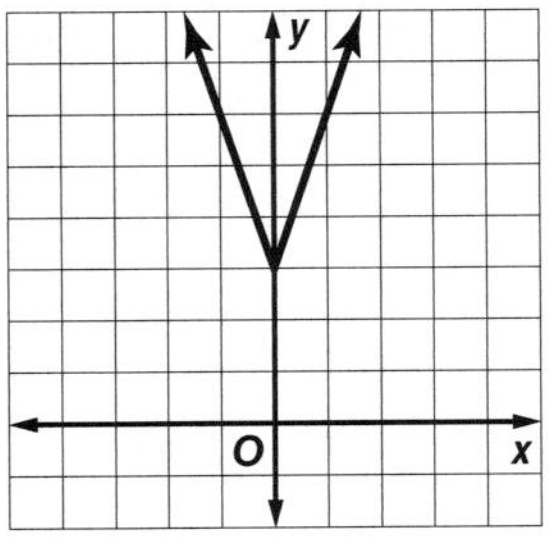

D = {all real numbers};
R = {$k(x)$ | $k(x) \ge 3$}

29.

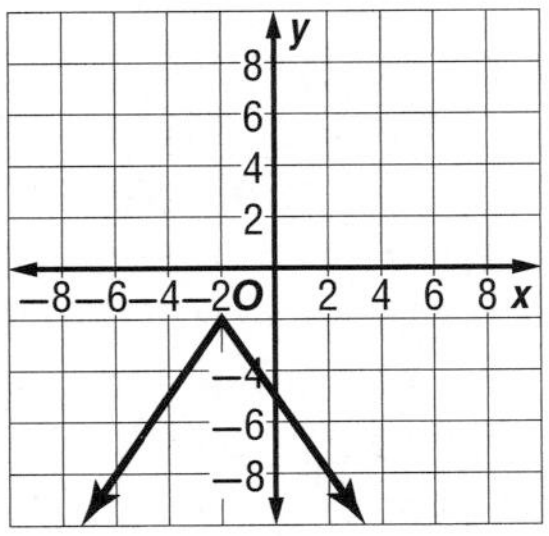

D = {all real numbers};
R = {$h(x)$ | $h(x) \le -2$}

31a. $f(a) = |a - 60|$ **31b.** {a | $a \ge 0$}

31c.

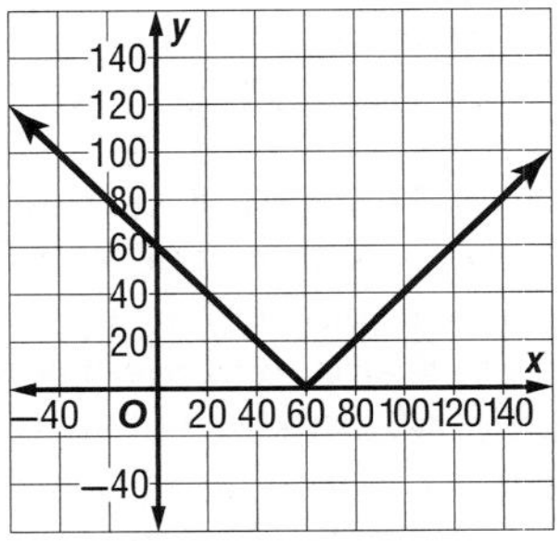

33 $f(x) = 0.5x$ if $x > 0$, $f(x) = 0$ if $x = 0$, and $f(x) = -0.5x$ if $x < 0$. So, according to the definition of absolute value, $f(x) = |0.5x|$.

35.

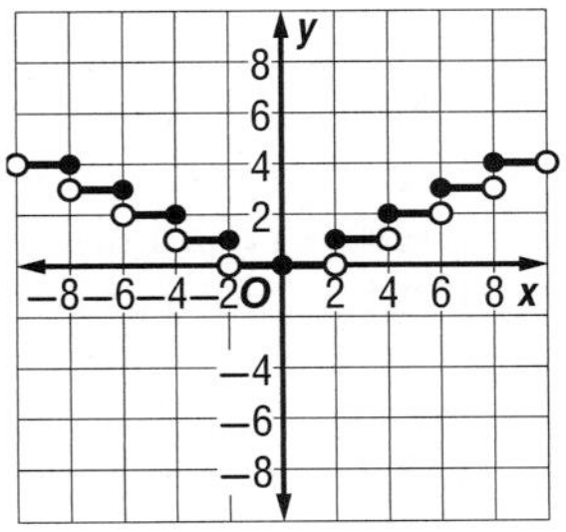

D = {all real numbers};
R = {$f(x)$ | $f(x) \ge 0$}

37.

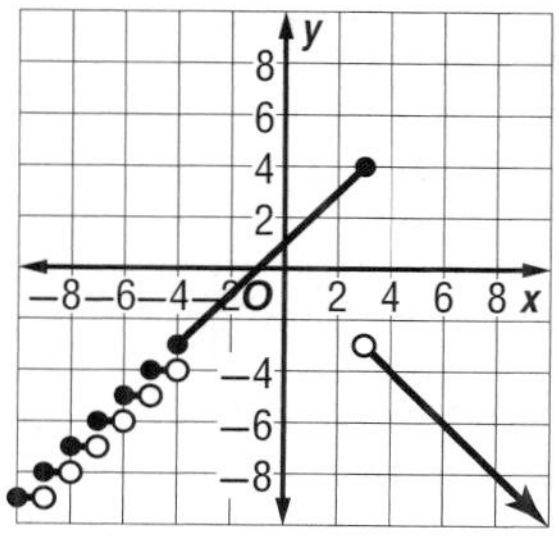

D = {all real numbers}; R = {$g(x) \mid g(x) \leq 4$}

39a.

x	−4	−3	−2	−1	0	1	2	3	4
f(x)	0	−1	−2	−3	−4	−3	−2	−1	0

x	−4	−3	−2	−1	0	1	2	3	4
g(x)	12	9	6	3	0	3	6	9	12

39b.

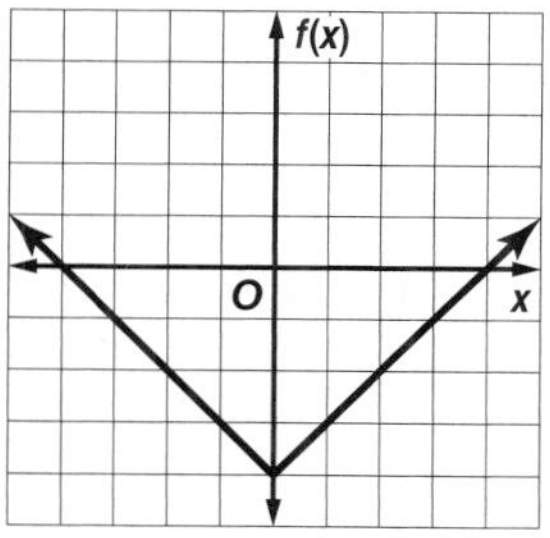

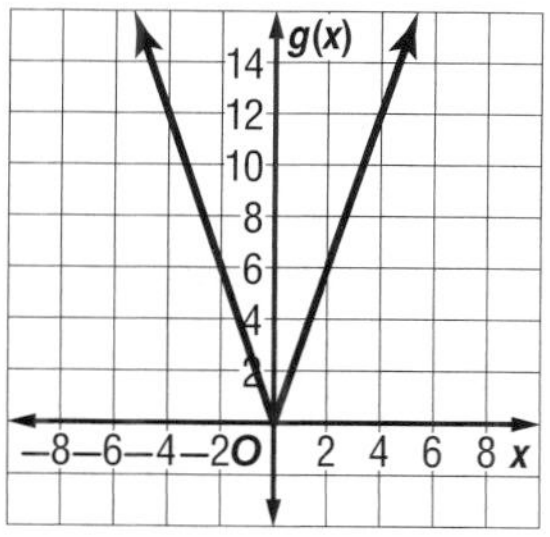

39c.

x	−4	−3	−2	−1	0	1	2	3	4
f(x)	0	−1	−2	−3	−4	−3	−2	−1	0
slope		−1	−1	−1	−1	1	1	1	1

x	−4	−3	−2	−1	0	1	2	3	4
f(x)	12	9	6	3	0	3	6	9	12
slope		−3	−3	−3	−3	3	3	3	3

39d. The two sections of an absolute value graph have opposite slopes. The slope is constant for each section of the graph.

41.

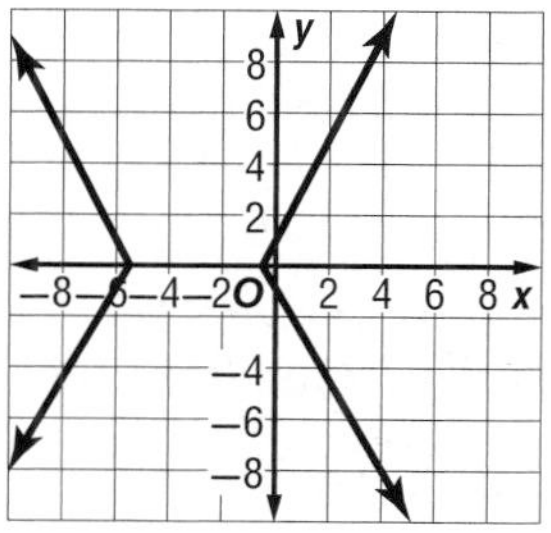

43. Sample answer: $f(x) = -|x - 2|$
45. $3n + 1$ **47.** J
49a. $y = 0.10x + 30.34$
49b. $r = 0.987$
49c. about 110
51. $y = -\frac{3}{2}x + 6$
53. $-8c + 6$ **55.** -99

57.

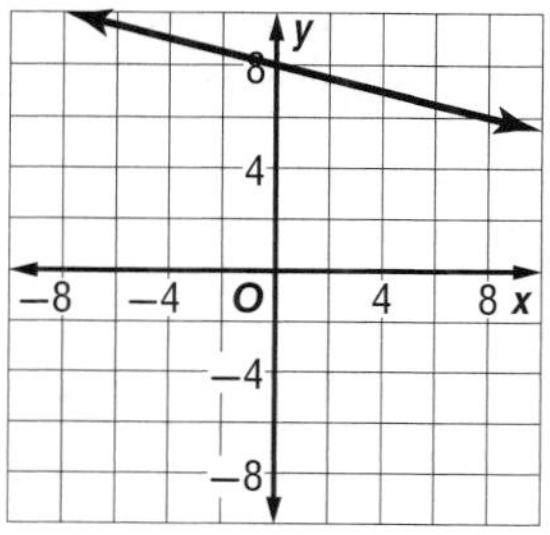

59.

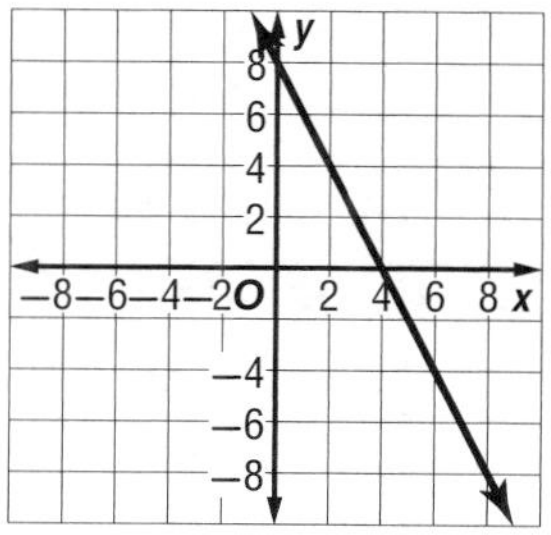

Pages 113–116 Lesson 2-7

1. linear

3. translation of the graph of $y = x^2$ down 4 units

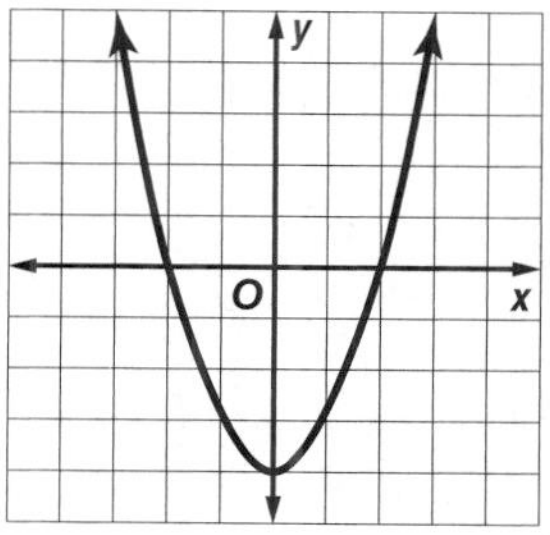

5. reflection of the graph of $y = |x|$ in the x-axis

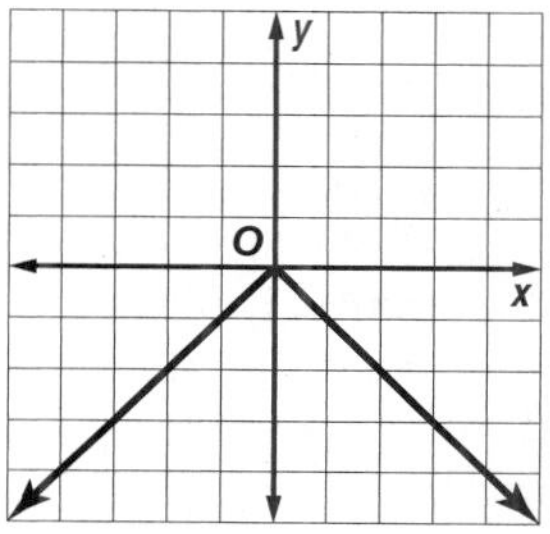

7. A dilation of the graph of $y = x$; the slope is not as steep as that of $y = x$.

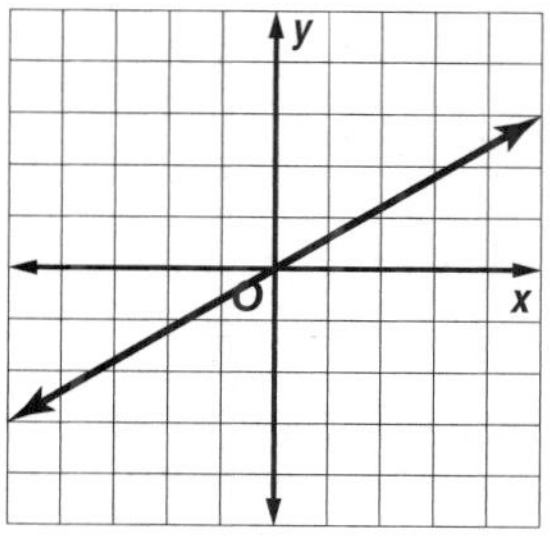

9. The function is a dilation and translation. The graph of $f(x) = \frac{1}{2}|x - 12|$ compresses the graph $f(x) = |x|$ vertically and translates it 12 units to the right.

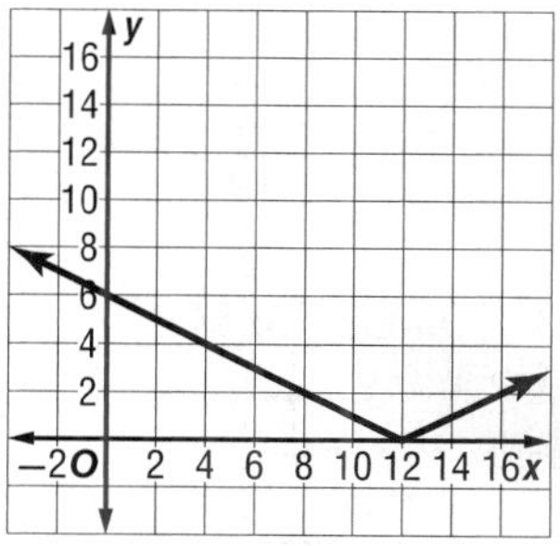

11 The graph is a curve that appears symmetrical. The graph represents a quadratic function.

13. linear

15. translation of the graph of $y = |x|$ down 3 units

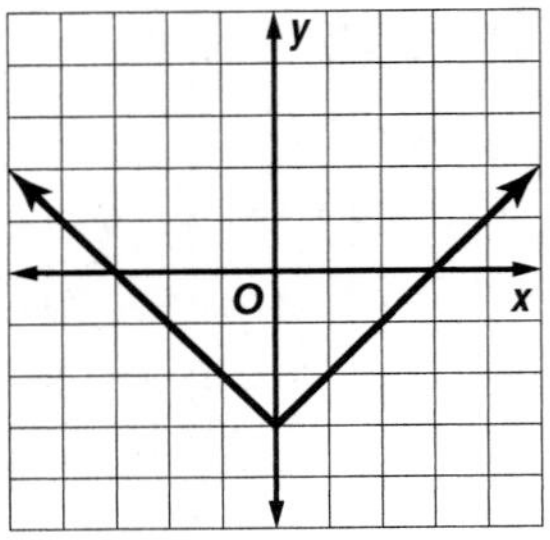

17. translation of the graph of $y = x$ up 2 units or left 2 units

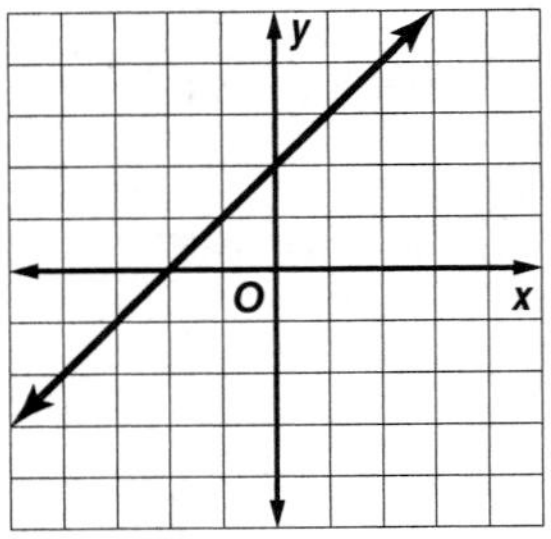

19. translation of the graph of $y = |x|$ left 6 units

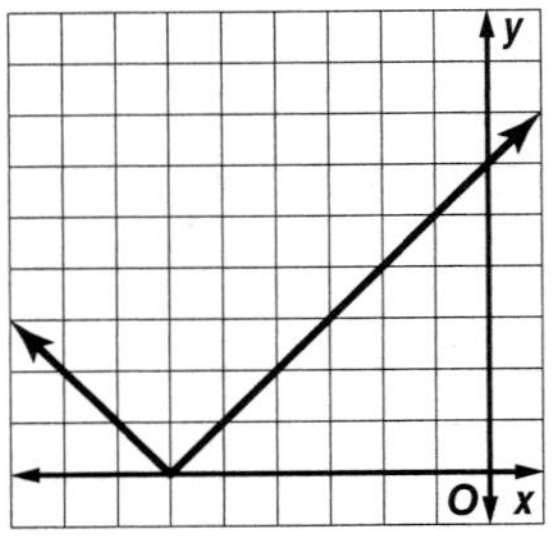

21. reflection of the graph of $y = x^2$ in the x-axis

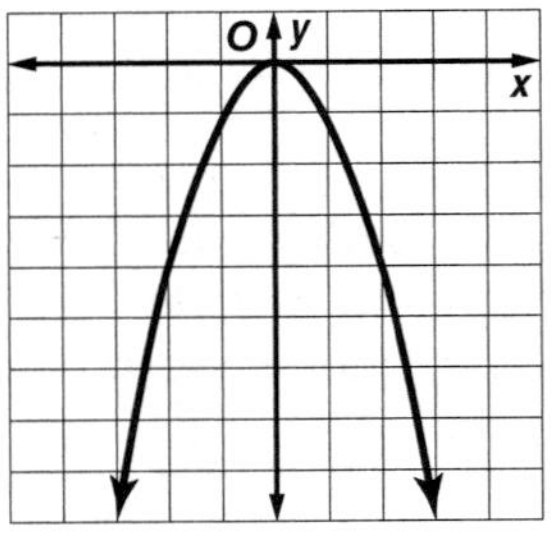

23. reflection of the graph of $y = |x|$ in the y-axis

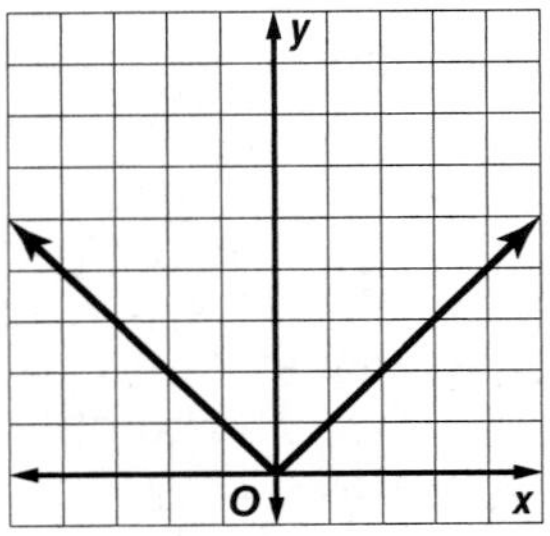

25. reflection of the graph of $y = x$ in the y-axis

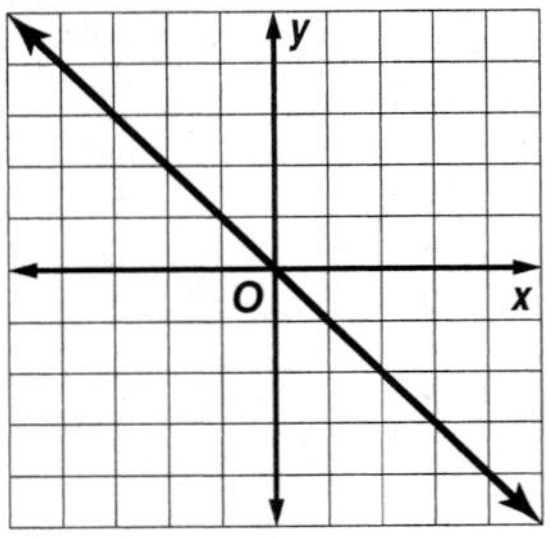

27. vertical stretch of the graph of $y = x$; The slope is steeper than that of $y = x$.

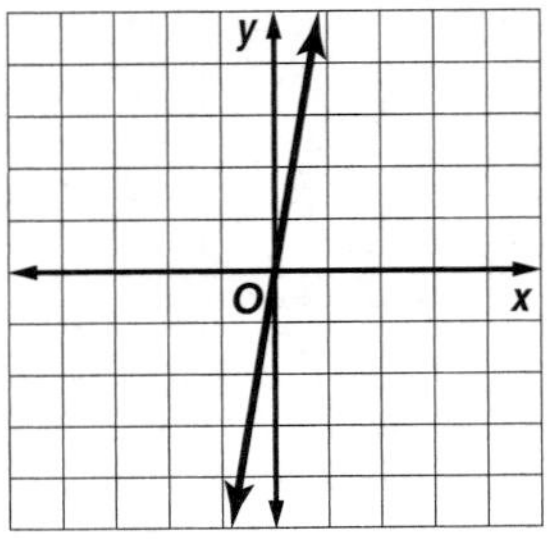

29. The dilation stretches the graph of $y = |x|$ vertically.

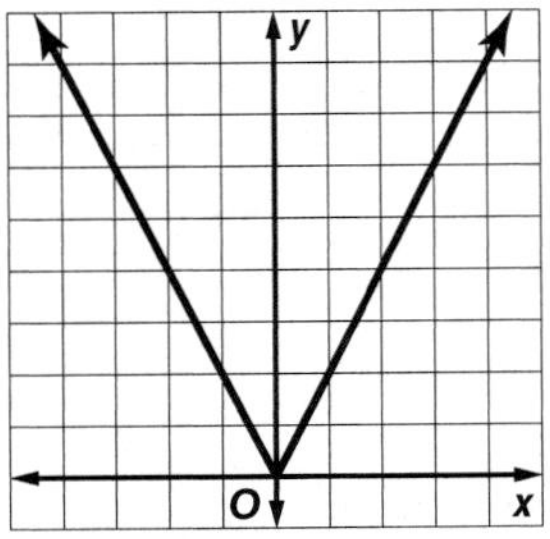

31. vertical compression of the graph of $y = x^2$

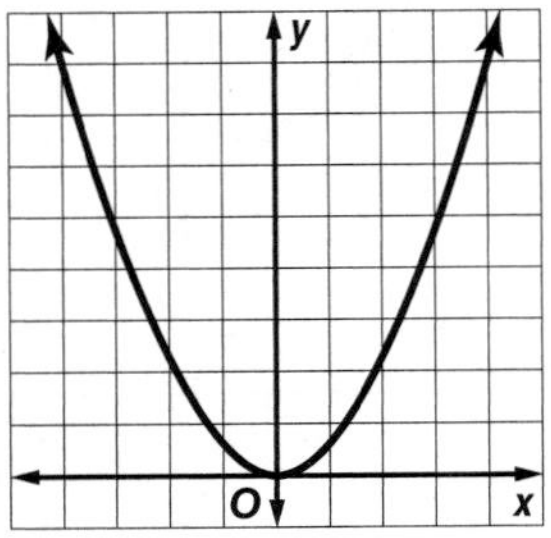

33. $y = x^2 + 1$ **35.** $y = x - 5$ **37.** $y = (x - 2)^2$

39 The blue line has a y-intercept of 4 and a slope of 1. So, an equation for the blue line is $y = x + 4$. The red line has a y-intercept of 2 and a slope of 1. So, an equation for the red line is $y = x + 2$. The red line is a translation of the blue line 2 units down.

41. $y = (x + 4)^2 - 6$ **43.** Sample answer: Since a vertical translation concerns only y-values and a horizontal translation concerns only x-values, order is irrelevant.

45. Sample graph:

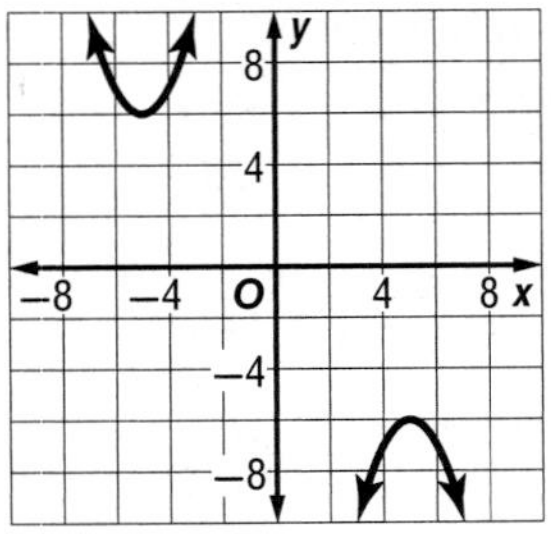

Sample answer: The figure in Quadrant II has been reflected and moved left 10 units.

47. Sample answer: It is not always true. When the axis of symmetry of the parabola is not along the y-axis, the graphs of the preimage and image will be different. **49.** G **51.** A

53.

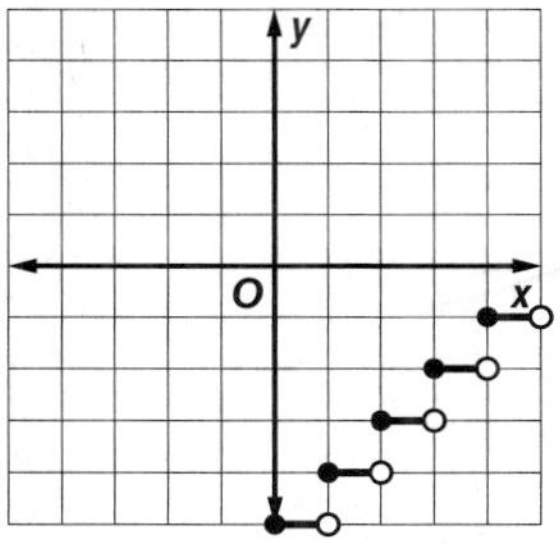

55a. $y = 7.83x - 15{,}605$ **55b.** $r = 0.953$ **55c.** about 118 people **57.** $2 > y > -3$ **59.** 20 **61.** yes **63.** -52 **65.** 84

Pages 119–121 Lesson 2-8

1.

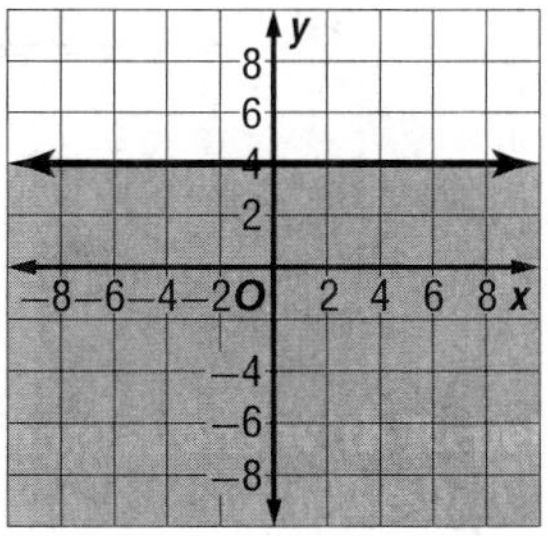

3.

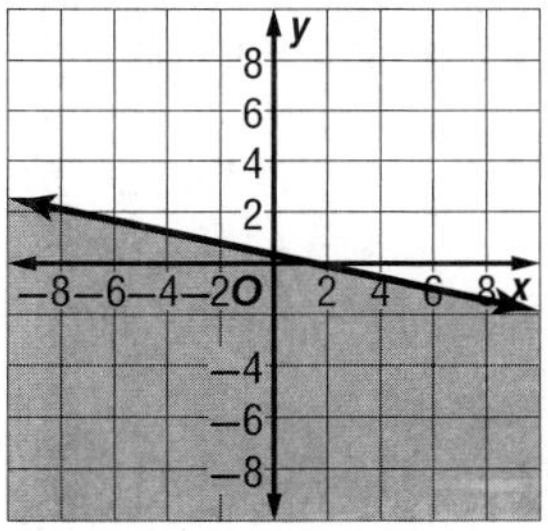

5a. $3.45g + 2.41q \leq 50$

5b.

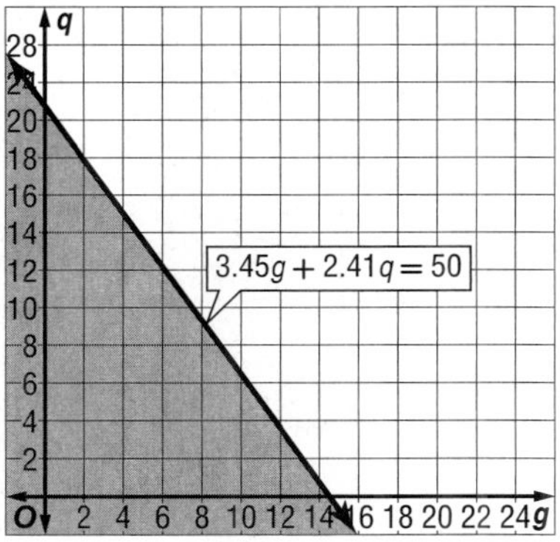

5c. No; (10, 8) is not in the shaded region.

7.

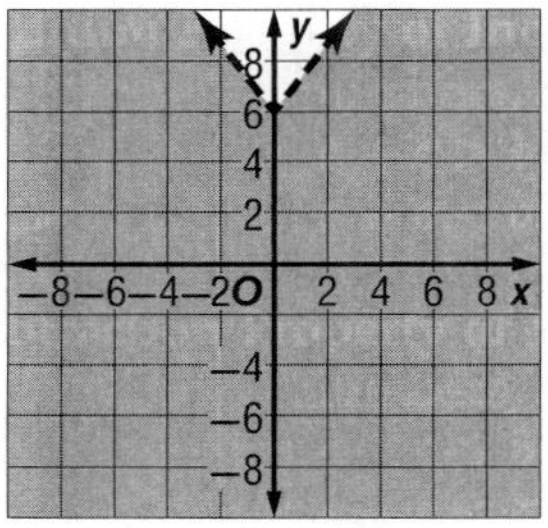

9.

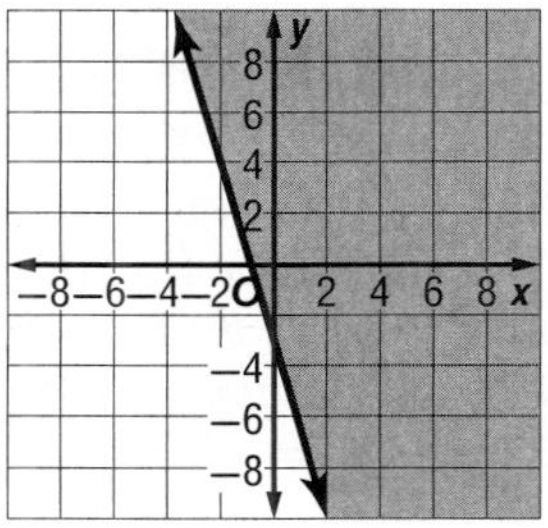

11.

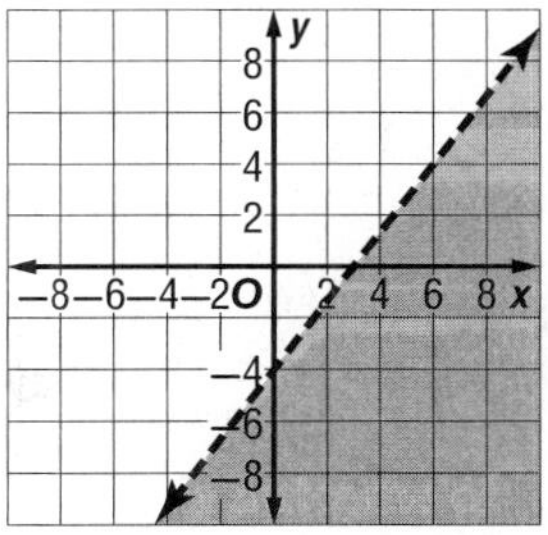

13.

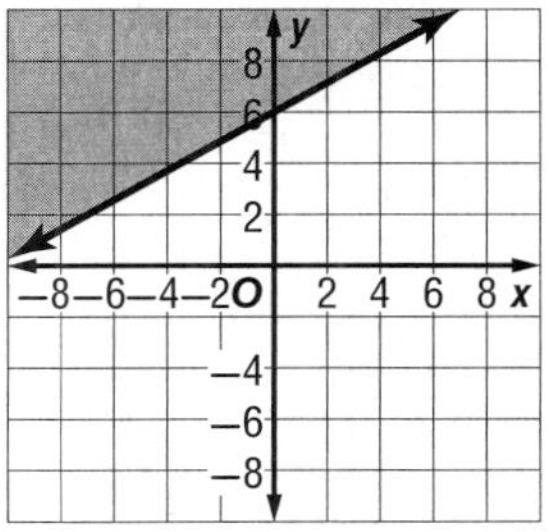

15.

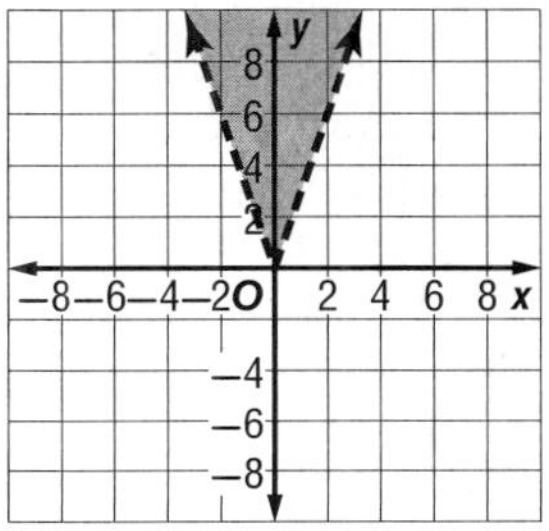

17 $y - 6 < |-2x|$ Original equation

$y < |-2x| + 6$ Add 6 to each side.

Since the inequality symbol is $<$, the boundary is dashed. Graph $y = |-2x| + 6$. Then test (0, 0).

$y < |-2x| + 6$ Write the inequality.

$0 \overset{?}{<} |-2(0)| + 6$ $(x, y) = (0, 0)$

$0 < 6$ ✓ True

The region that contains (0, 0) is shaded.

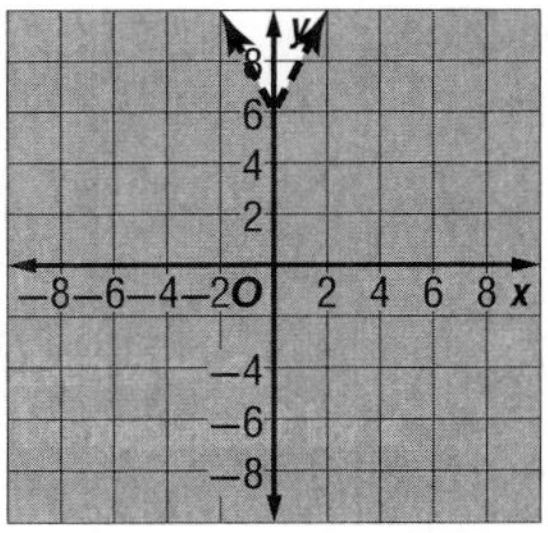

19.

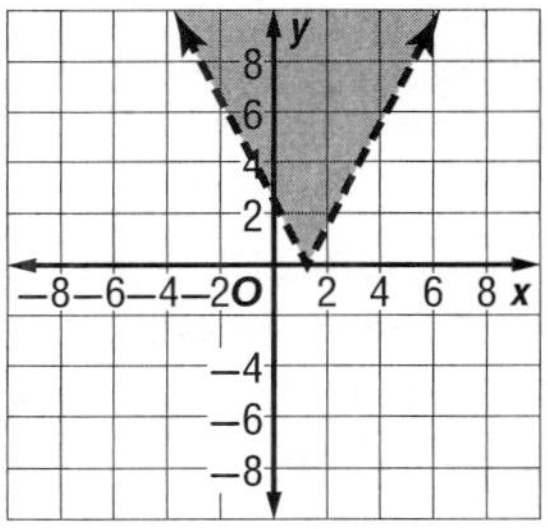

21a. $8a + 6b \geq 700$

21b.

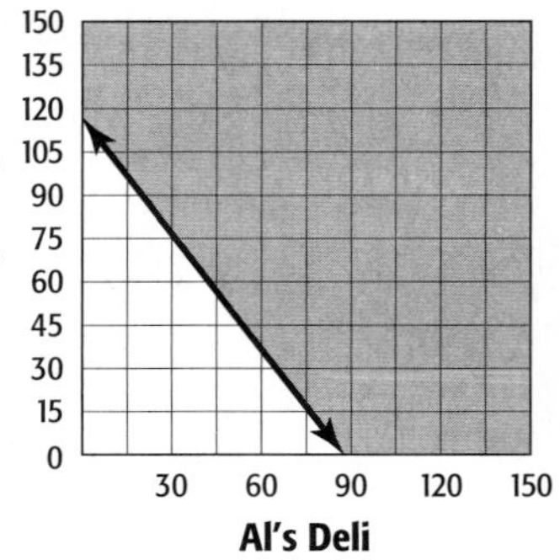

21c. yes

23.

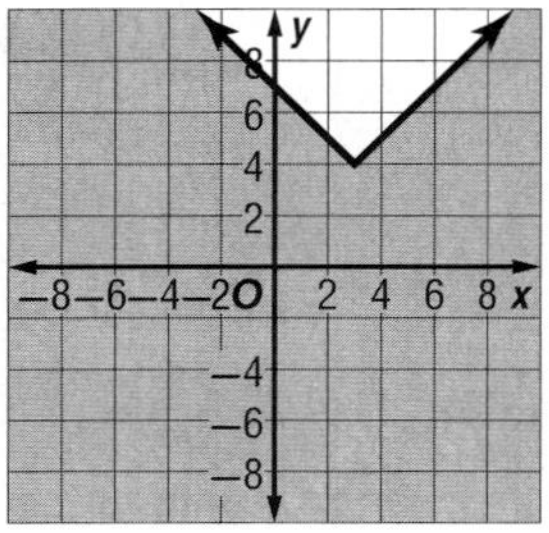

25.

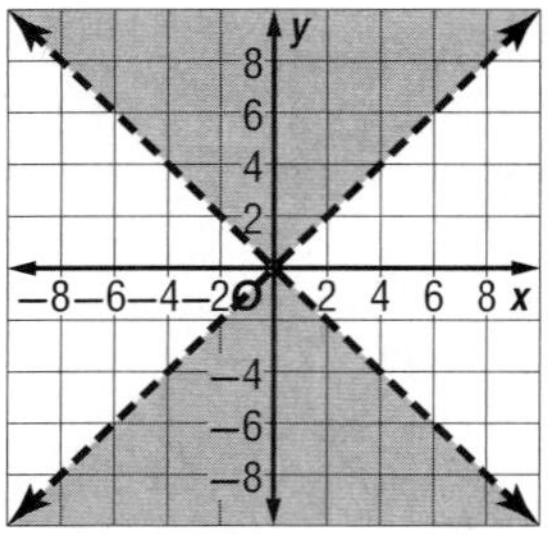

27. all real numbers (The graph would be shaded everywhere.)

29 **a.** $20d$ represents the cost of d DVDs and $15c$ represents the cost of c CDs. Because the sum can equal the maximum \$400, the inequality symbol is $\leq$. The inequality is $20d + 15c \leq 400$.
b. Graph the boundary $20d + 15c = 400$. Since the inequality symbol is $\leq$, the boundary is solid. Then test the point (0, 0).

$20d + 15c \leq 400$	Original inequality
$20(0) + 15(0) \overset{?}{\leq} 400$	$(c, d) = (0, 0)$
$0 \leq 400$ ✓	True

The region that contains (0, 0) is shaded.

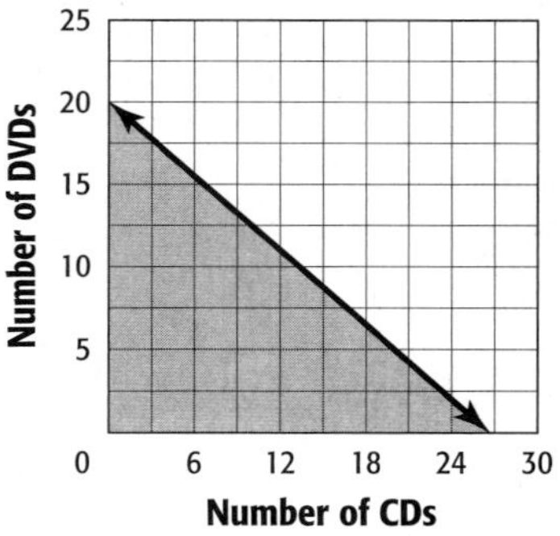

c. Sample answer: Since the points (18, 5), (12, 10), and (6, 15) lie inside the shaded region, they satisfy the inequality. This means that Susan can buy 18 CDs and 5 DVDs, 12 CDs and 10 DVDs, or 6 CDs and 15 DVDs.

31.

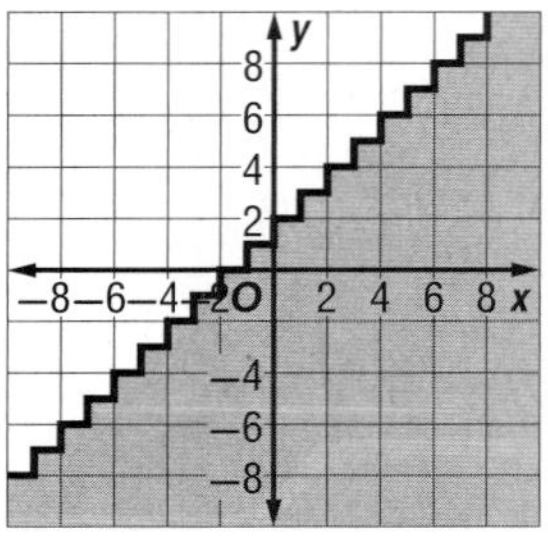

33. Sample answer: $|y| < x$ **35.** Paulo; $x - y \geq 2$ can be written as $y \leq x - 2$.

37. Sample answer: One possibility is when $|y| < 0$. In order for there to be a solution, the absolute value of y will need to be less than 0, and, by definition of absolute value, this is impossible. **39.** C **41.** J
43. $y = |x + 4| - 5$ **45.**

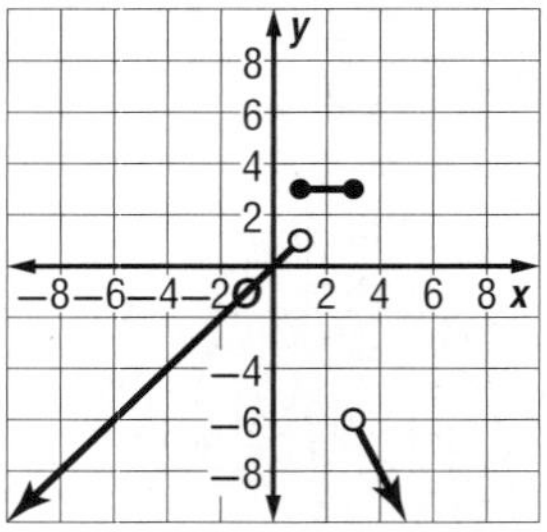

47.

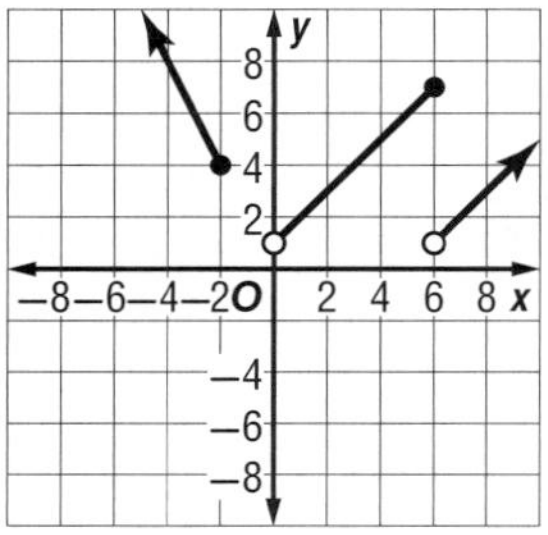

49. $4x - 15y = 6; A = 4, B = -15, C = 6$
51. 1000
53. $-6x^2 - 23x - 15$

55.

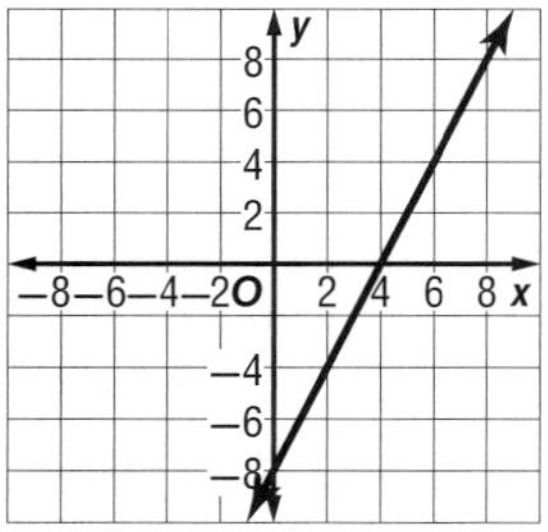

57.

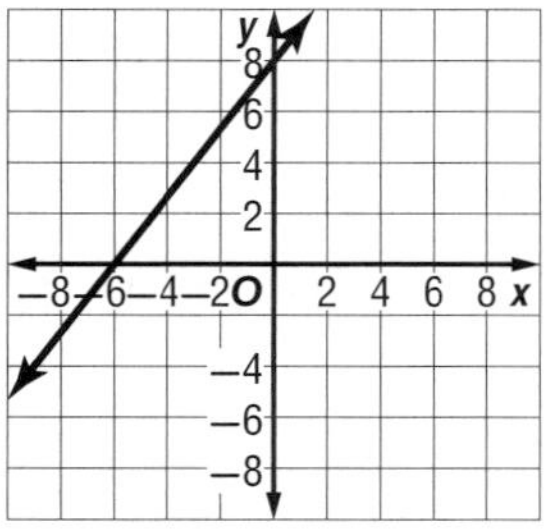

Pages 122–126 Chapter 2 Study Guide and Review

1. one-to-one **3.** identity **5.** scatter plot **7.** both **9.** not a function **11.** −10 **13.** 2 **15.** $3a + 2$ **17.** continuous function **19.** No; the variables have an exponent other than 1. **21.** yes **23.** No; x appears in a denominator. **25.** $12x - y = 0$; 12, −1, 0 **27.** $x - 2y = -3$; 1, −2, −3 **29.** 18 **31.** −1 **33.** $y = -2x - 11$ **35.** $y = 2x + 8$ **37.** $y = -4x + 25$

39. $y = 4x - 2$ **41.** $y = 32x + 250$ **43.** $y = -5x + 49$

45. 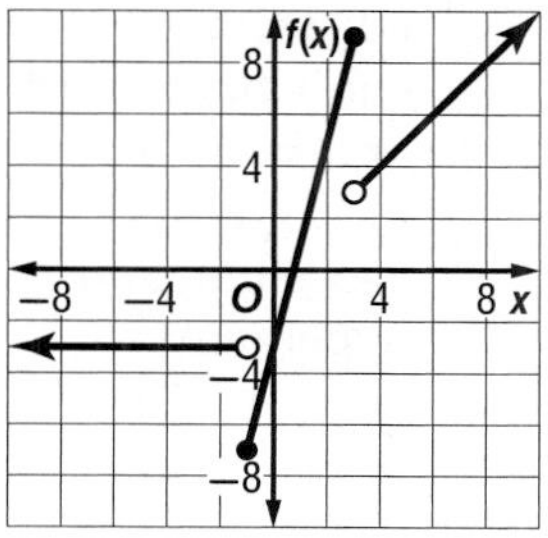

D = {all real numbers}, R = $\{f(x) \mid f(x) \geq -7\}$

47. 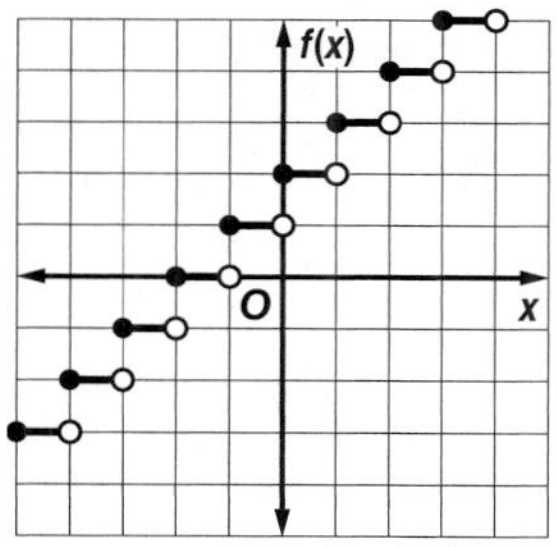

D = {all real numbers}, R = {all real numbers}

49. quadratic **51.** $y = x^2$ shifted down 3 units

53. parabola

55. 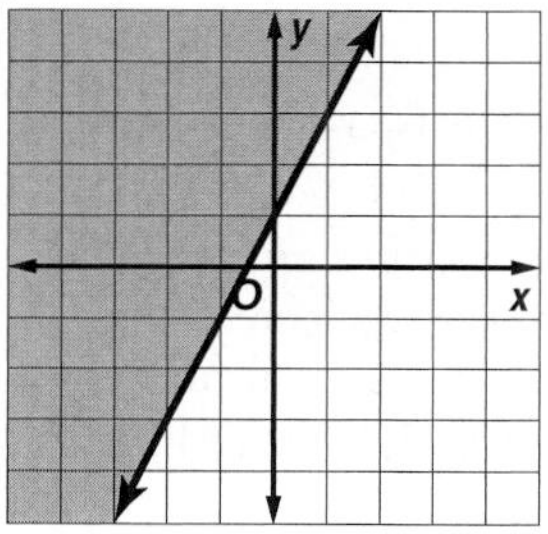

57. 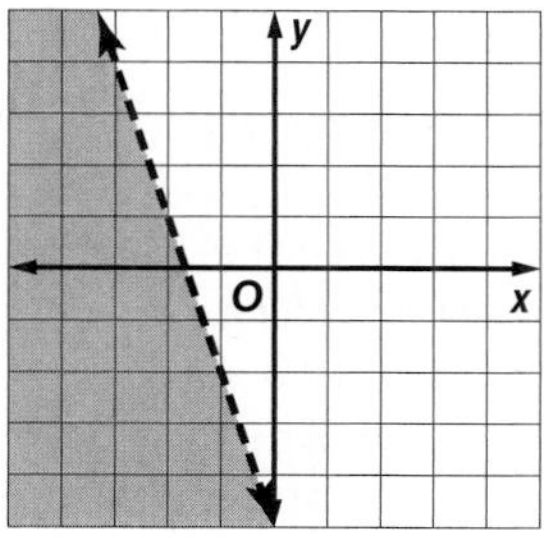

59. 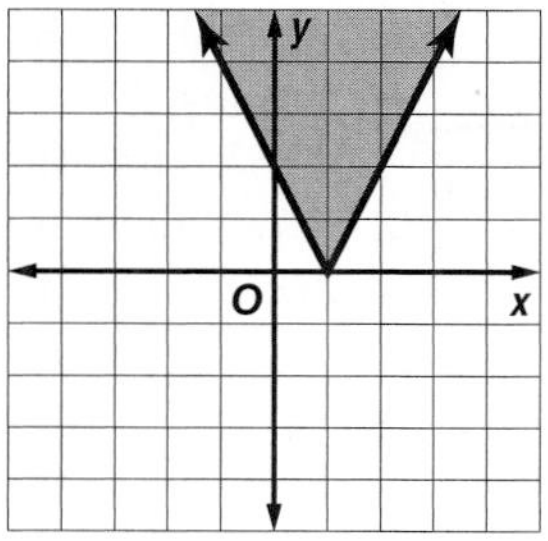

61. 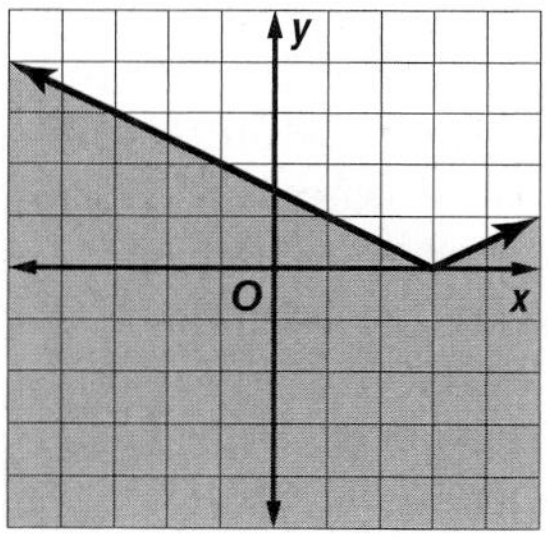

Chapter 3 Systems of Equations and Inequalities

Page 133 Chapter 3 Get Ready

1. 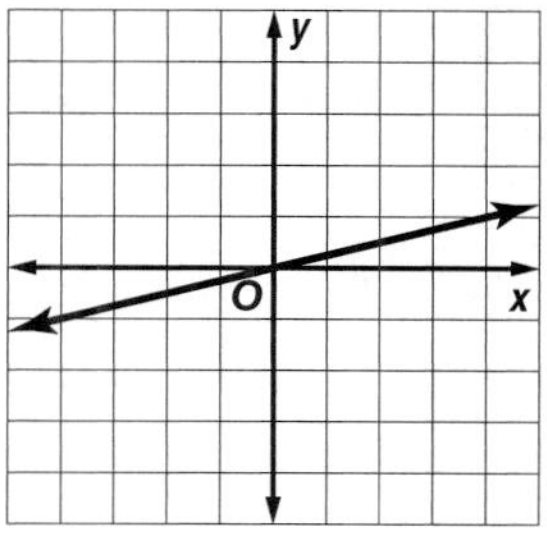

3. 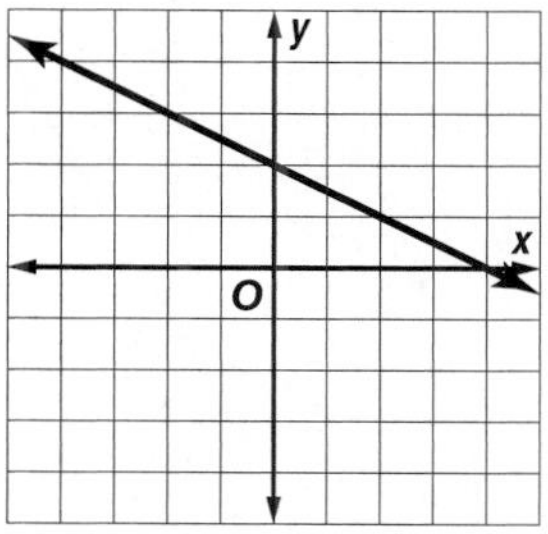

5. 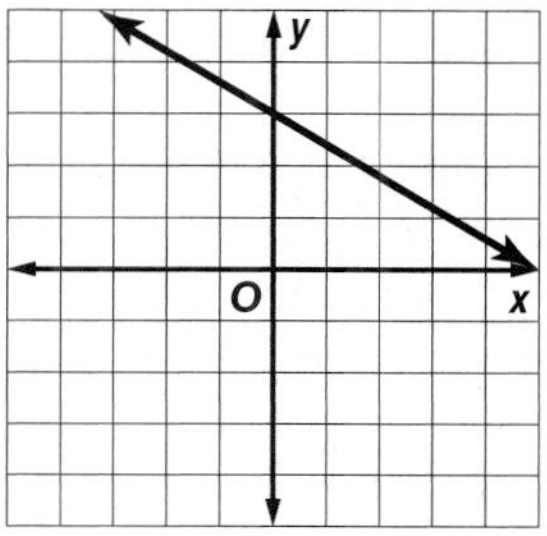

7a. $8.50a + 5.25c = 650$

7b. 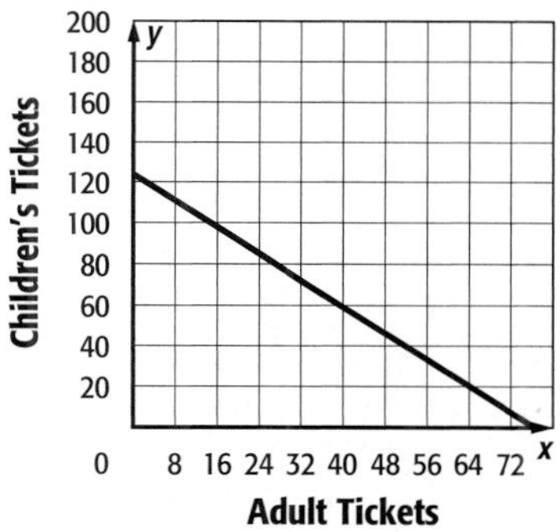

9. 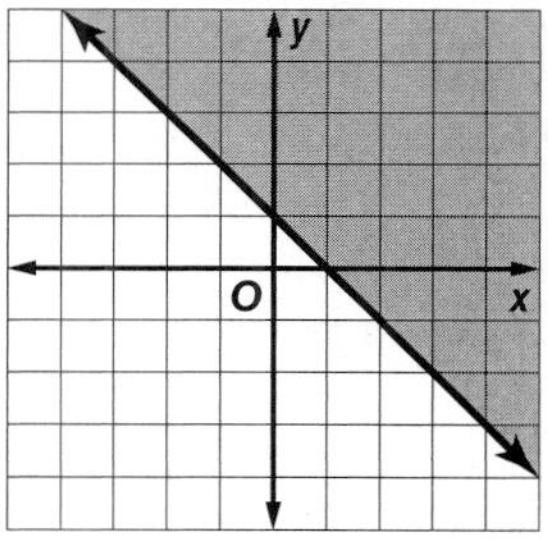

11. 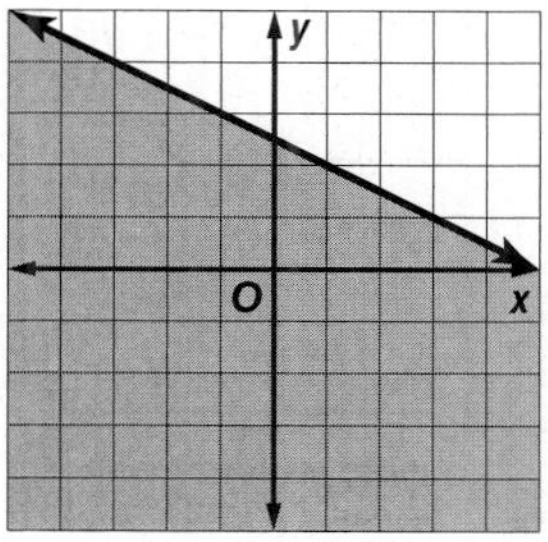

13. 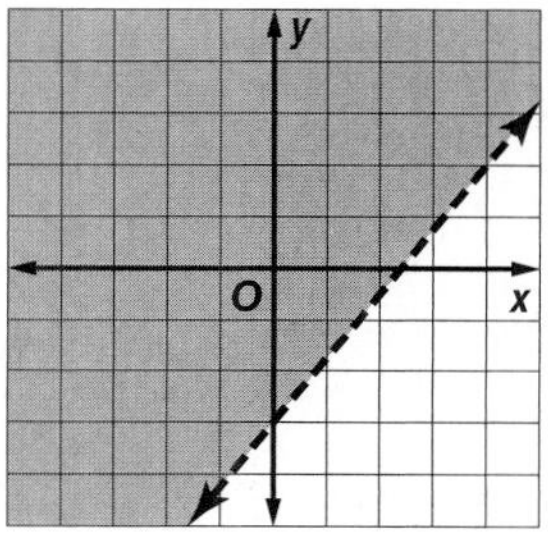

Pages 138–141 Lesson 3-1

1. (3, 5) **3.** (3, −3) **5.** (6, 7)

Selected Answers and Solutions

7 Write each equation in slope-intercept form. Then graph.

$4x + 5y = -41 \quad \rightarrow \quad y = -\frac{4}{5}x - \frac{41}{5}$

$3y - 5x = 5 \quad \rightarrow \quad y = \frac{5}{3}x + \frac{5}{3}$

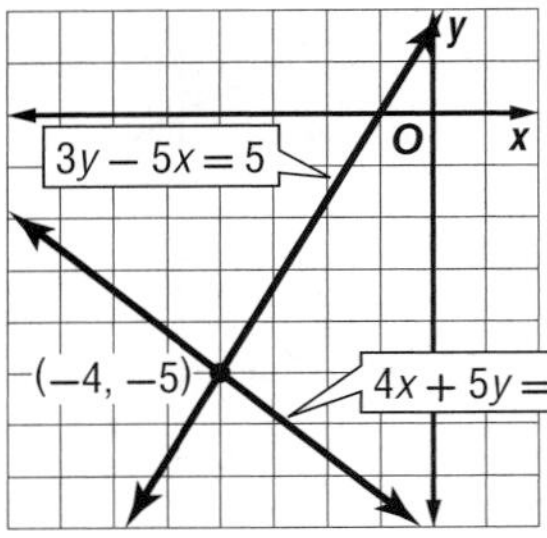

The graphs appear to intersect at (–4, –5). So, the solution is (–4, –5).

9a. $y = 0.15x + 2.70$, $y = 0.25x$ **9b.** $6.75 for 27 photos
9c. You should use EZ Online photos if you are printing more than 27 digital photos, and the local pharmacy if you are printing fewer than 27 photos.

11.

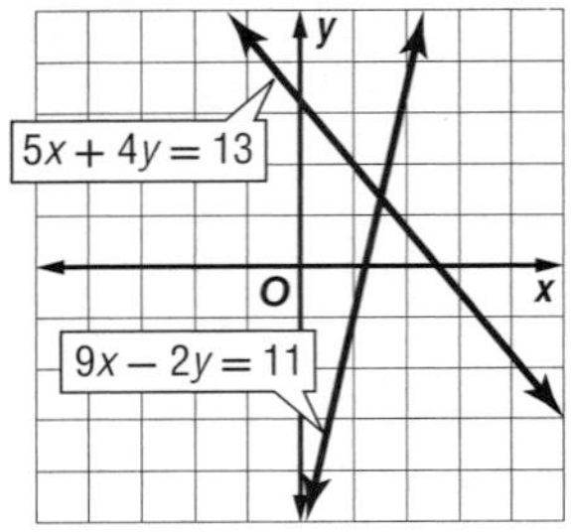

consistent and dependent

13. (–3, –12) **15.** (4, 3) **17.** (–3, –4) **19.** infinite solutions **21.** (–1.5, –2) **23.** $10 coupon for a purchase less than $66.67 and 15% discount coupon for a purchase over $66.67 **25.** (1, 3), (2, –1), (–2, –3)

27.

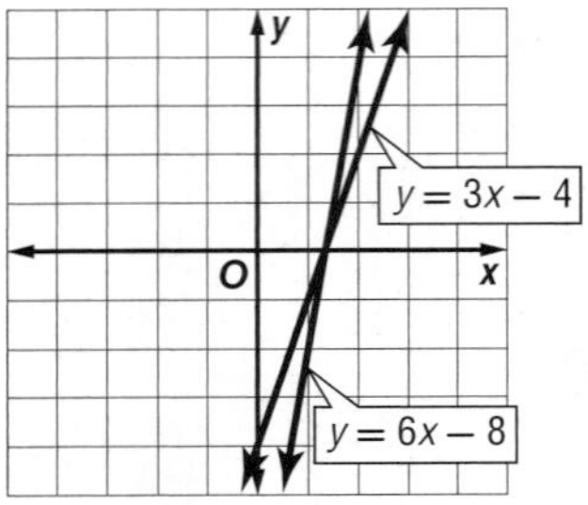

consistent and independent

29.

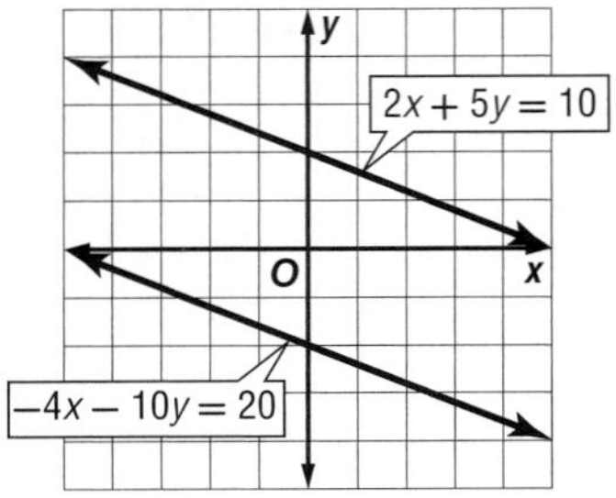

inconsistent

31 Write each equation in slope-intercept form. Then graph.

$-5x - 6y = 13 \quad \rightarrow \quad y = -\frac{5}{6}x - \frac{13}{6}$

$12y + 10x = -26 \quad \rightarrow \quad y = -\frac{5}{6}x - \frac{13}{6}$

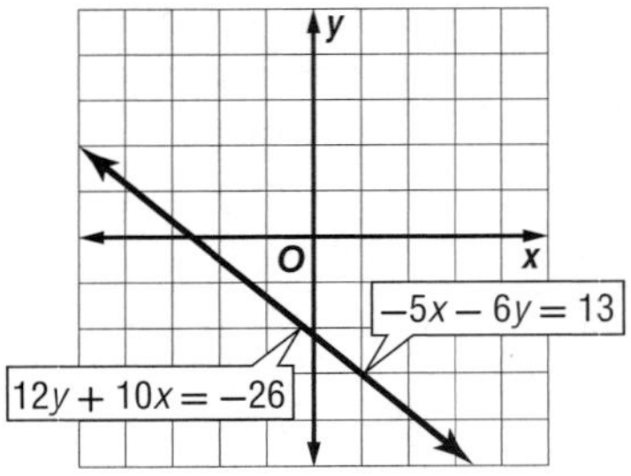

Because the equations are equivalent, their graphs are the same line. The system is consistent and dependent.

33.

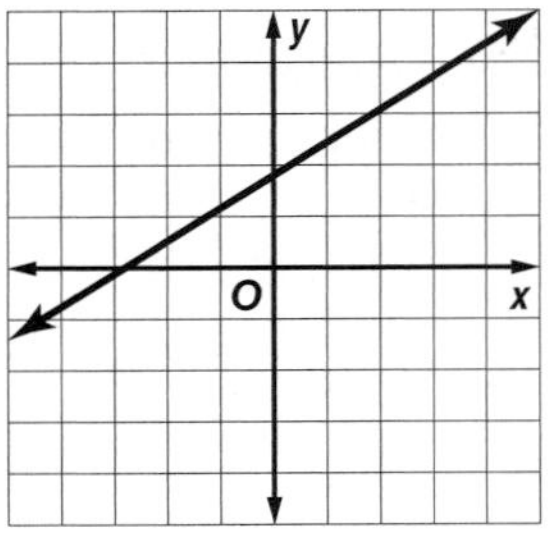

consistent and dependent
35. (4, 0.5) **37.** (–6, 3)
39. (15.03, 10.98)

41 **a.** Sample answer: For men, you can use (0, 10) and (40, 9.85) to write an equation.

$m = \frac{y_2 - y_1}{x_2 - x_1}$ Slope Formula

$= \frac{9.85 - 10}{40 - 0}$ $(x_1, y_1) = (0, 10), (x_2, y_2) = (40, 9.85)$

$= -0.00375$ Simplify.

At $x = 0$, $y = 10.0$. So, the y-intercept is 10.

$y = mx + b$ Slope-intercept form

$y = -0.00375x + 10$ $m = -0.00375, b = 10$

For women, you can use (0, 11.4) and (40, 10.93) to write an equation.

$m = \frac{y_2 - y_1}{x_2 - x_1}$ Slope Formula

$= \frac{10.93 - 11.4}{40 - 0}$ $(x_1, y_1) = (0, 11.4), (x_2, y_2) = (40, 10.93)$

$= -0.01175$ Simplify.

At $x = 0$, $y = 11.4$. So, the y-intercept is 11.4.

$y = mx + b$ Slope-intercept form

$y = -0.01175x + 11.4$ $m = -0.01175, b = 11.4$

b.

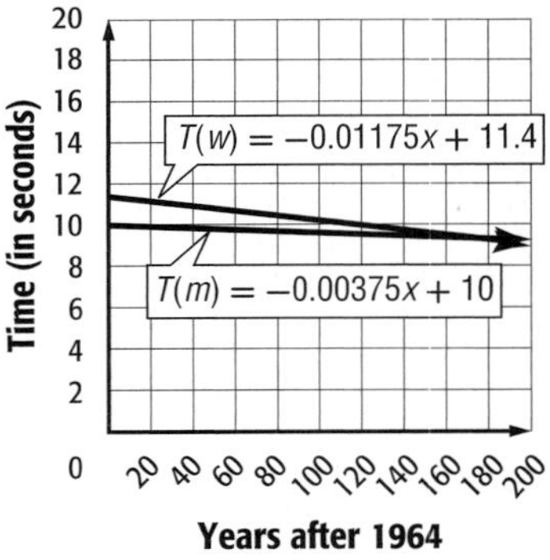

The graphs appear to intersect at (175, 9.3). This means that 175 years after 1964, the winning time for both men and women will be about 9.3 seconds. So, based on these data, the women's performance will catch up to the men's performance 175 years after 1964, or in the year 2139. The next Olympic year would be 2140.

c. Sample answer: No; it is unlikely that women's times will ever catch up to men's times because the times cannot continue to increase and decrease infinitely.

43. Alvin; Sample answer: Alvin used the **Intersect** command, while Victor used **Trace**.

45a. Sample answer: $y = 2x$; $y = 2x + 1$ **45b.** Sample answer: $y = x$; $y = x$ **45c.** Sample answer: $y = x$; $y = 3x - 1$ **47.** $12xy + 18y^2 - 15y$ **49.** J
51a. $10s + 15\ell \geq 350$

51b.

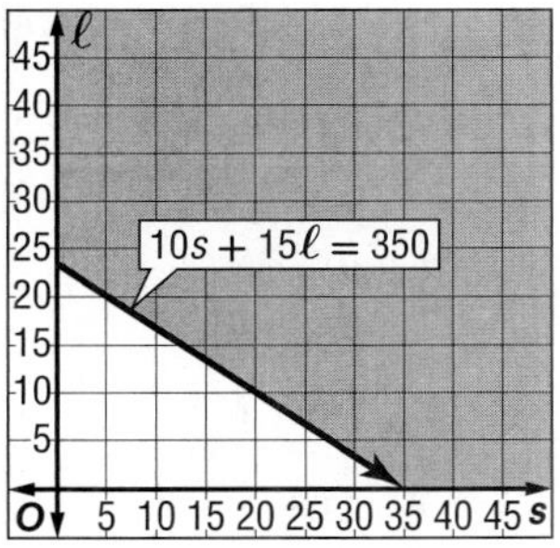

51c. no **53.** $-|x - 3|$ **55.** 7 **57.** 3.2 **59.** -8 **61.** $-x + 2$ **63.** $-2x + 3y$ **65.** $12x + 8y + 32$

Pages 146–150 Lesson 3-2

1. 250 T-shirts **3.** $(-2, 1)$

5 $2a + 8b = -8$ Multiply by 5. $\rightarrow$ $10a + 40b = -40$
$3a - 5b = 22$ Multiply by 8. $\rightarrow$ $24a - 40b = 176$

$10a + 40b = -40$ Equation 1 × 5
$(+)\ 24a - 40b = 176$ Equation 2 × 8
$34a = 136$ Add the equations.
$a = 4$ Divide each side by 34.

Substitute 4 for a into either original equation.

$2a + 8b = -8$ Equation 1
$2(4) + 8b = -8$ $a = 4$
$8 + 8b = -8$ Multiply.
$8b = -16$ Subtract 8 from each side.
$b = -2$ Divide each side by 8.

The solution is $(4, -2)$.

7. $(5, 1)$ **9.** $(-2, 7)$ **11.** $(-4, -3)$ **13.** no solution **15.** $(5, 2)$ **17.** $(6, -2)$ **19.** $(-4, 3)$ **21.** infinite solutions **23.** $(8, 4)$ **25.** $(-3, -1)$ **27a.** $x + y = 13$ and $4x + 2y = 38$ **27b.** 6 doubles games and 7 singles games **29.** $(0, 4)$ **31.** no solution **33.** $(8, -6)$ **35.** $(5, 4)$ **37.** $(-4, 5)$ **39.** $(6, -6)$ **41.** $(-10, 4)$

43 a. Let x represent the time in hours that Julian rides and let y represent the distance traveled.
Distance Julian travels after x hours: $y = 12x$
Distance Peter travels after x hours: $y = 16(x - 2)$

$y = 12x$ Equation representing Julian's distance
$16(x - 2) = 12x$ Substitute $16(x - 2)$ for y.
$16x - 32 = 12x$ Distributive Property
$4x - 32 = 0$ Subtract $12x$ from each side.
$4x = 32$ Add 32 to each side.
$x = 8$ Divide each side by 4.

So, after Julian rides 8 hours, or at 4:00 P.M., Peter catches up to him.

b. If Peter wants to catch up to Julian 1 hour sooner, or in 7 hours, then the total distance traveled y would be 12(7) or 84 miles.
Suppose Peter's speed remains the same but his starting time changes. Let t represent the number of hours after Julian that Peter starts.

$84 = 16(7 - t)$ Peter starts t hours after Julian.
$5.25 = 7 - t$ Divide each side by 16.
$-1.75 = -t$ Subtract 5.25 from each side.
$1.75 = t$ Divide each side by -1.

So, if Peter starts 1.75 hours after Julian and rides at a speed of 16 mph, then he will catch up to Julian in 7 hours. This answer is reasonable.
Suppose Peter's speed changes but his starting time remains the same. Let s represent Peter's new speed.

$84 = s(7 - 2)$ Peter cycles at s mph 2 hours after Julian.
$84 = s(5)$ Simplify.
$16.8 = s$ Divide each side by 5.

So, if Peter starts 2 hours after Julian and rides at a speed of 16.8 mph, then he will catch up to Julian in 7 hours. This answer is reasonable.

45. $(-5, 4)$ **47.** infinite solutions **49.** $(16, -8)$
51a. 7 16-ounce servings; If you drink 7 coffees, the price for each option is the same. **51b.** Sample answer: If you drink fewer than 9 coffees during that week, the disposable cup price is best. If you drink more than 9 coffees, the refillable mug price is best.
51c. Over a year's time, the refillable mug would be more economical because you would eventually have more than 7 coffees over the year. **53.** $m\angle A = 99$, $m\angle B = 81$

55 Let x represent the cost of an adult and y represent the cost of a student.
Van A: $2x + 5y = 77$
Van B: $2x + 7y = 95$

$2x + 5y = 77$ Multiply by -1. $\rightarrow$ $-2x - 5y = -77$

$-2x - 5y = -77$ Equation 1 × (-1)
$(+)\ 2x + 7y = 95$ Equation 2
$2y = 18$ Add the equations.
$y = 9$ Divide each side by 2.

$2x + 7y = 95$ Equation 2
$2x + 7(9) = 95$ $y = 9$
$2x + 63 = 95$ Multiply.
$2x = 32$ Subtract 63 from each side.
$x = 16$ Divide each side by 2.

The solution is $(16, 9)$. So, the cost of an adult is \$16 and the cost of a student is \$9.

57. $(-4.3, -6.8)$ **59.** $(3.3, -6.5)$

61 Find an equation for the diagonal that goes through (6, 3) and (2, 9).

Slope:
$$m = \frac{y_2 - y_1}{x_2 - x_1} = \frac{9 - 3}{2 - 6} = \frac{6}{-4} \text{ or } -\frac{3}{2}$$

Equation:
$$y - y_1 = m(x - x_1)$$
$$y - 3 = -\frac{3}{2}(x - 6)$$
$$y - 3 = -\frac{3}{2}x + 9$$
$$y = -\frac{3}{2}x + 12$$

Find an equation for the diagonal that goes through (3, 4) and (11, 18).

Slope:
$$m = \frac{y_2 - y_1}{x_2 - x_1} = \frac{18 - 4}{11 - 3} = \frac{14}{8} \text{ or } \frac{7}{4}$$

Equation:
$$y - y_1 = m(x - x_1)$$
$$y - 4 = \frac{7}{4}(x - 3)$$
$$y - 4 = \frac{7}{4}x - \frac{21}{4}$$
$$y = \frac{7}{4}x - \frac{5}{4}$$

Find the point of intersection of the diagonals.

$y = -\frac{3}{2}x + 12$ Equation 1

$\frac{7}{4}x - \frac{5}{4} = -\frac{3}{2}x + 12$ Substitute $\frac{7}{4}x - \frac{5}{4}$ for y.

$\frac{13}{4}x - \frac{5}{4} = 12$ Add $\frac{3}{2}x$ to each side.

$\frac{13}{4}x = \frac{53}{4}$ Add $\frac{5}{4}$ to each side.

$x = \frac{53}{13}$ Multiply each side by $\frac{4}{13}$.

$y = -\frac{3}{2}x + 12$ Equation 1

$y = -\frac{3}{2}\left(\frac{53}{13}\right) + 12$ $x = \frac{53}{13}$

$y = -\frac{159}{26} + 12$ Multiply.

$y = \frac{153}{26}$ Simplify.

The diagonals intersect at $\left(\frac{53}{13}, \frac{153}{26}\right)$.

63a.

Equation 1

x	y
0	$\frac{16}{3}$
1	5
2	$\frac{14}{3}$
3	$\frac{13}{3}$
4	4

Equation 2

x	y
0	−4
1	−2
2	0
3	2
4	4

Equation 3

x	y
0	10
1	5
2	0
3	−5
4	−10

63b. Equations 1 and 2 intersect at (4, 4), equations 2 and 3 intersect at (2, 0), and equations 1 and 3 intersect at (1, 5); there is no solution that satisfies all three equations.

63c.

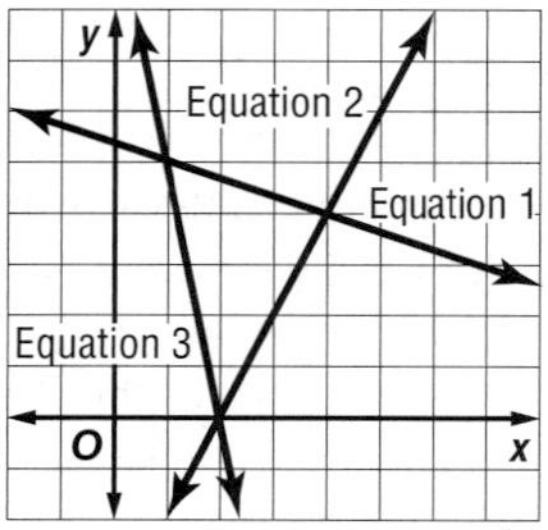

63d. If all three lines intersect at the same point, then the system has a solution. The system has no solution if the lines intersect at 3 different points, or if two or three lines are parallel.

65. $a \neq 0, b = 3$

67. Sample answer:

$4x + 5y = 21 \rightarrow 3(4x + 5y = 21)$

$3x - 2y = 10 \rightarrow 4(3x - 2y = 10)$

$12x + 15y = 63$

$(-)\ 12x - 8y = 40$

$23y = 23$

$y = 1$

$4x + 5(1) = 21$

$4x + 5 = 21$

$4x = 16$

$x = 4$

The solution is (4, 1).

69. 9 **71.** J **73a.** $y = 400$; $y = 150 + 5x$

73b.

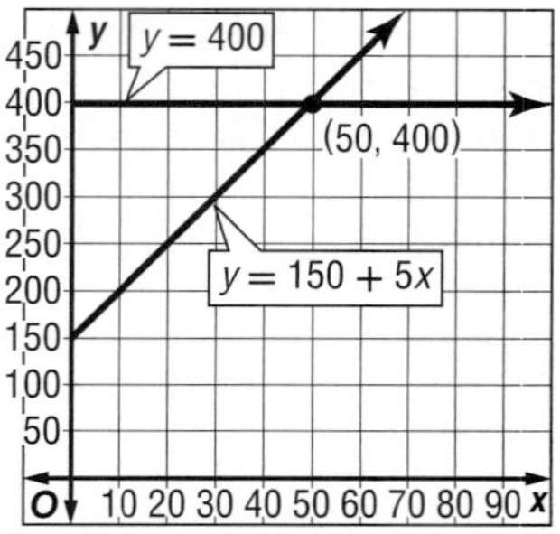

73c. It means that the options cost the same if you visit 50 times in a year.

73d. $400 per year

75.

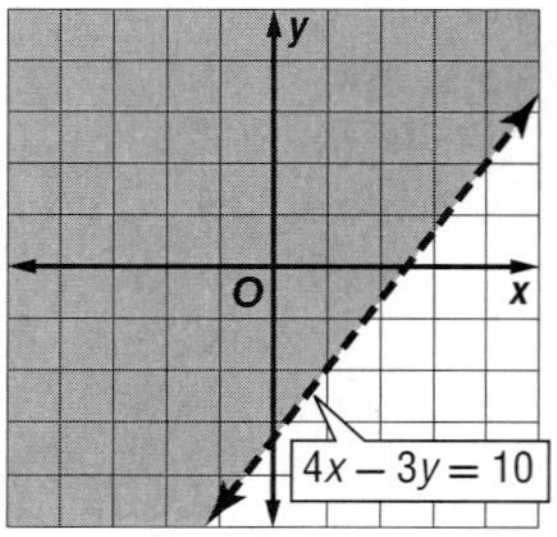

77. $y = -2x + 11$

79. $y = -4x - 25$

81. yes

83. no

Pages 154–157 Lesson 3-3

1.

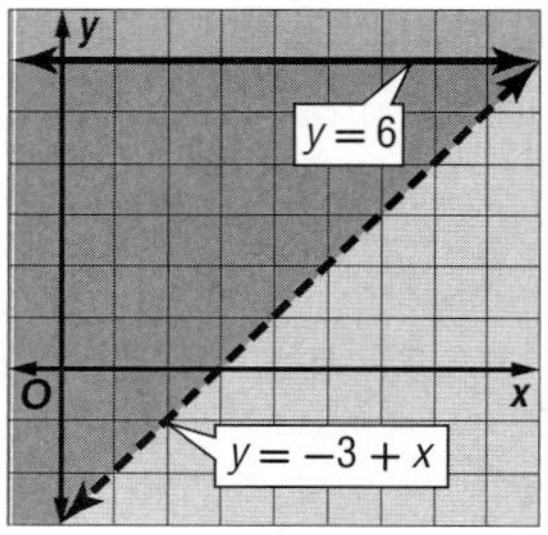

3.

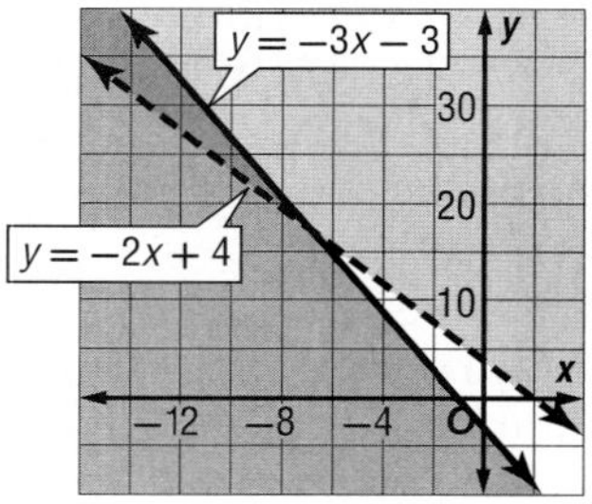

5.

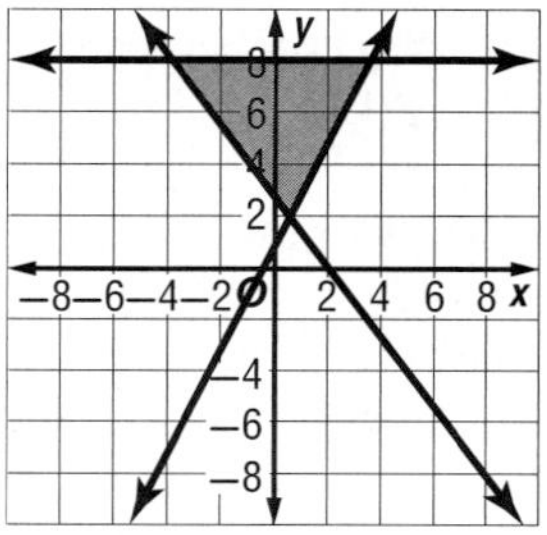

(3.5, 8), (−4, 8), (0.5, 2)

7.

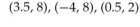

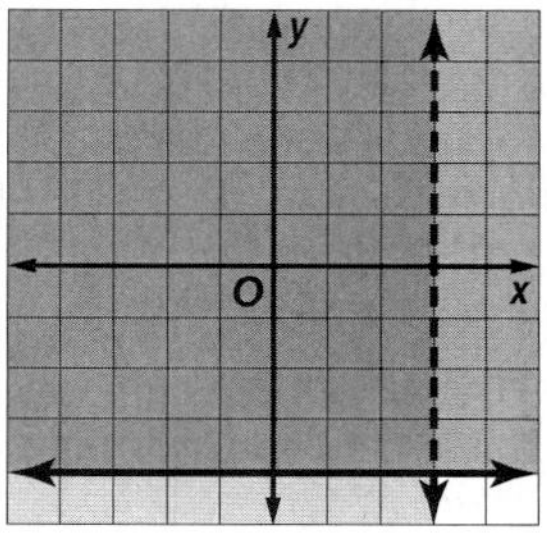

9 The solution of $y < -3x + 4$ is the region to the left of the boundary. The solution of $3y + x > -6$ is the region above the boundary. The intersection of the two regions is the solution of the system.

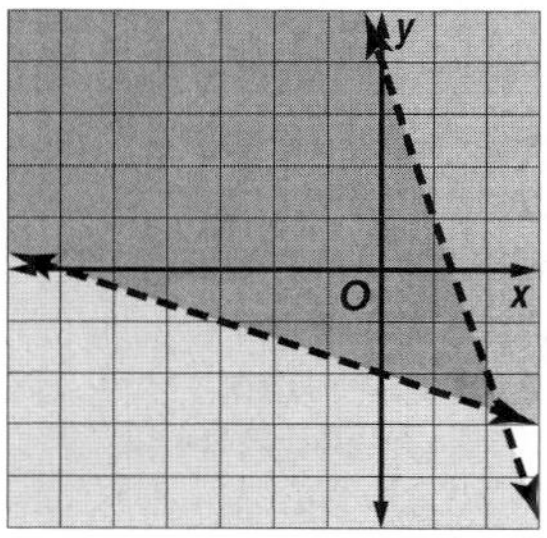

11.

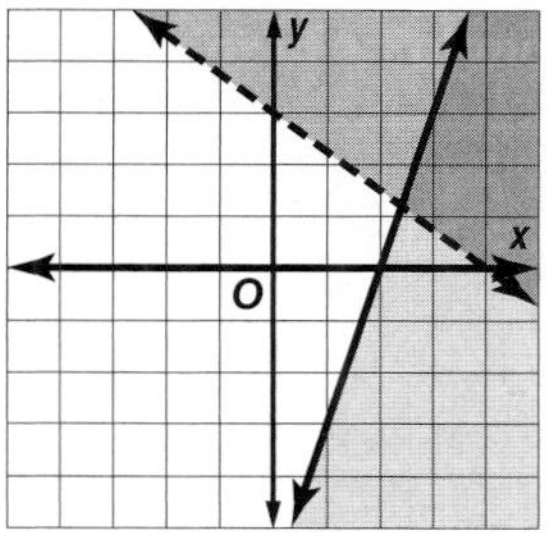

13.

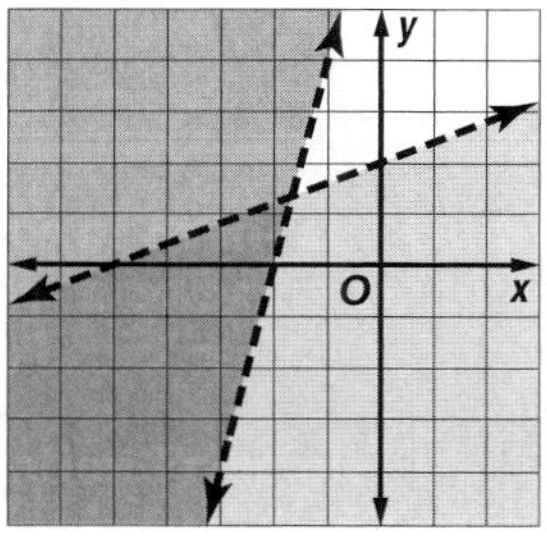

15.

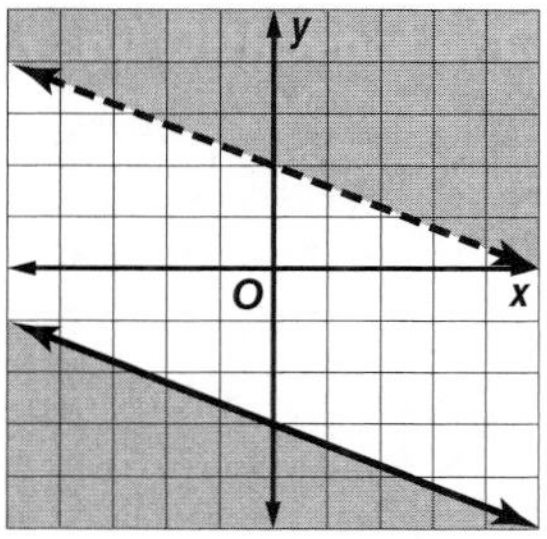

17.

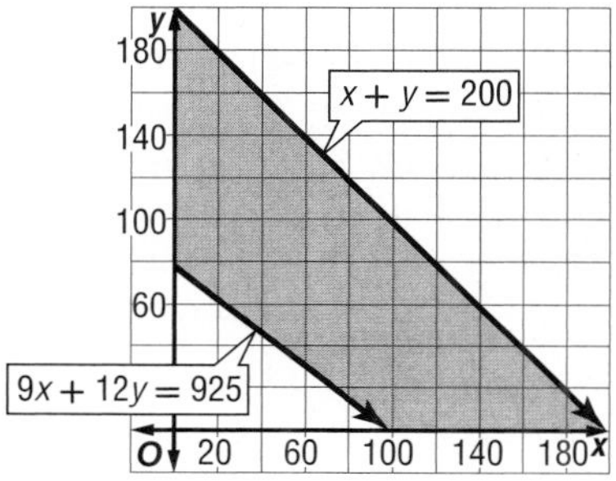

19.

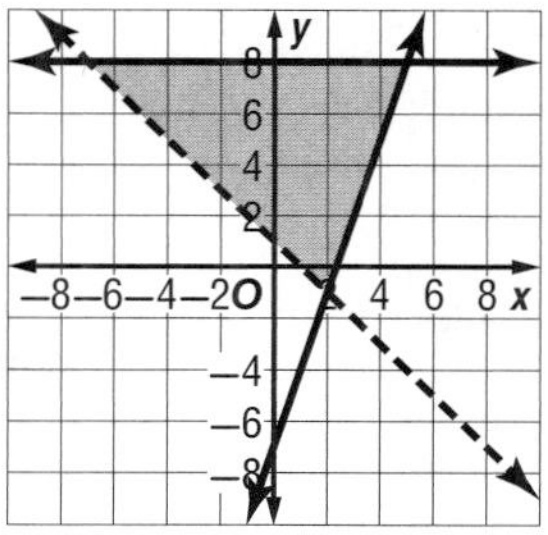

(2, −1), (5, 8), (−7, 8)

21.

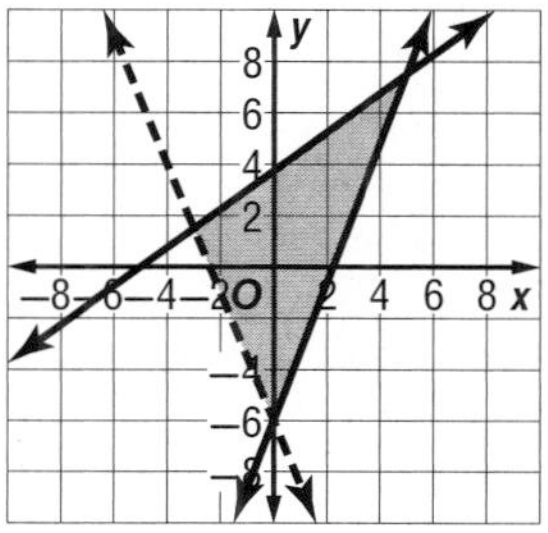

(−3, 1.5), (5, 7.5), (0, −6)

23.

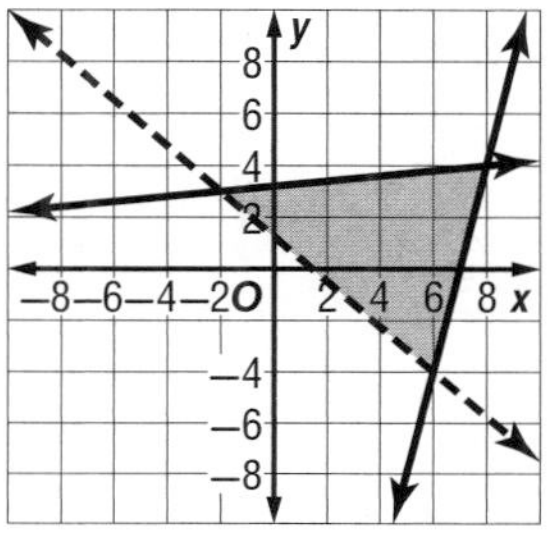

(8, 4), (6, −4), (−2, 3)

25 Let d represent the number of daytime minutes and n represent the number of nighttime minutes. Write a system of inequalities and then graph.

$d + n \leq 800$	Maximum number of minutes is 800.
$d \geq 2n$	At least twice as many daytime minutes as nighttime minutes
$n \geq 200$	At least 200 nighttime minutes

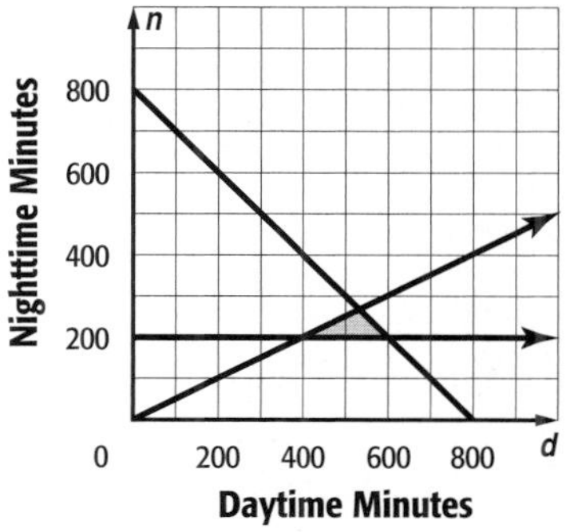

The intersection points are at (400, 200), (533.3, 266.7), and (600, 200).

$0.15d + 0.1n = 0.15(400) + 0.1(200)$	400 daytime min., 200 nighttime min.
$= 60 + 20$ or \$80	Simplify.
$0.15d + 0.1n = 0.15(533.3) + 0.1(266.7)$	533.3 daytime min., 266.7 nighttime min.
$\approx 80.0 + 26.7$ or \$106.70	Simplify.
$0.15d + 0.1n = 0.15(600) + 0.1(200)$	600 daytime min., 200 nighttime min.
$= 90 + 20$ or \$110	Simplify.

So, his maximum bill is \$110 and his minimum bill is \$80.

27a.

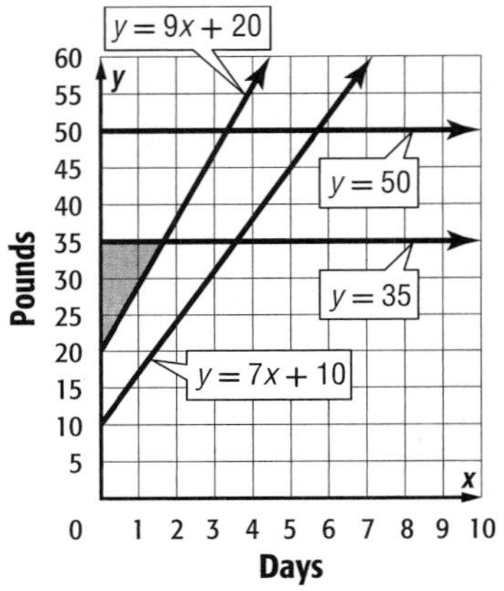

27b. $3\frac{1}{3}$ days **27c.** Marc; Jessica could last about a quarter of a day longer than Marc.

29.

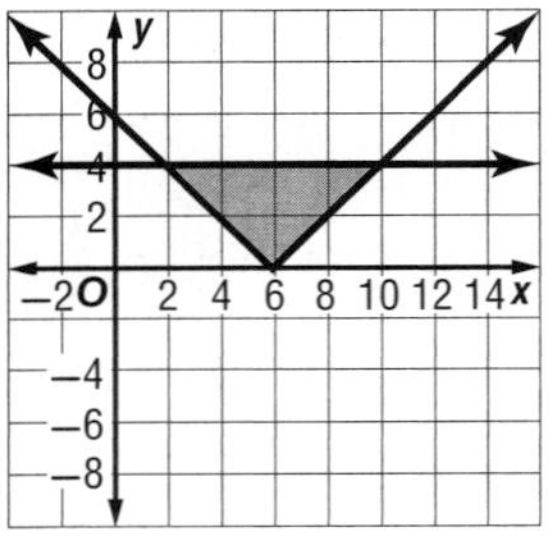

31.

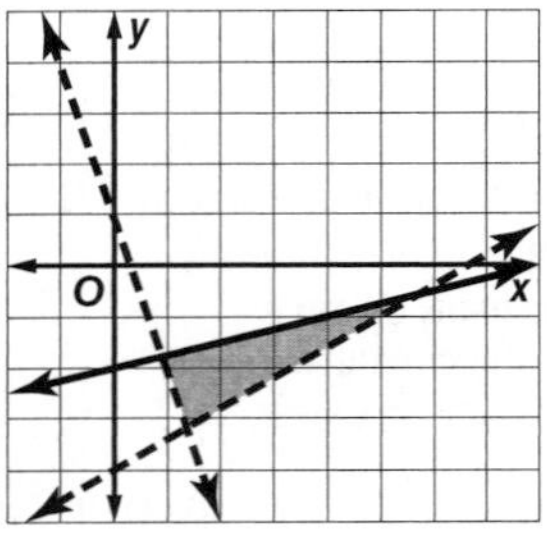

33.

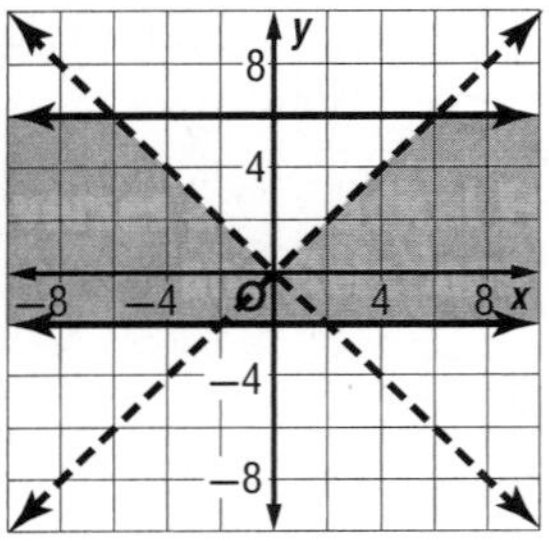

35.

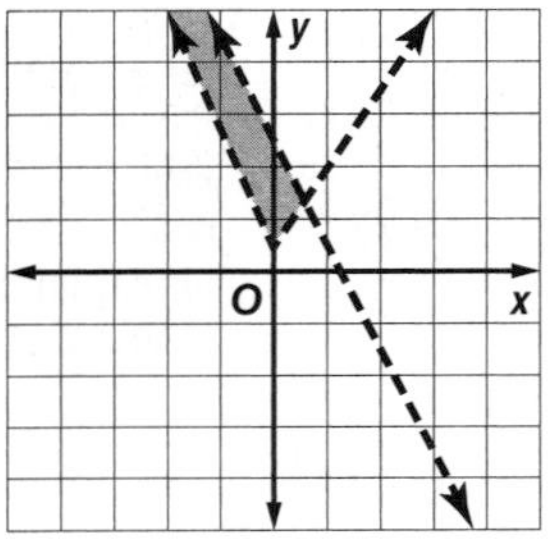

37.

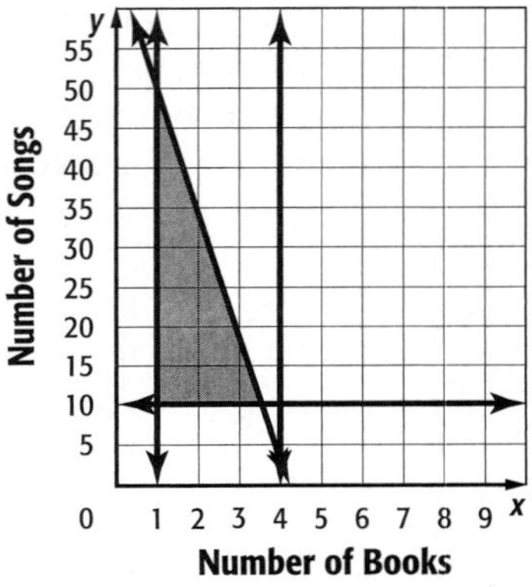

39. Let w = the number of hours writing, and let e = the number of hours exercising.
$w + e \leq 35$
$7 \leq e \leq 15$
$20 \leq w \leq 25$

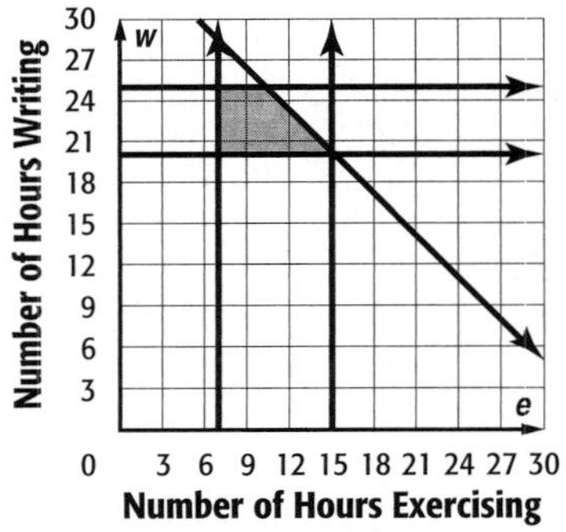

41. $(-6, -2), \left(-3\frac{13}{17}, 6\frac{16}{17}\right), \left(9\frac{1}{7}, 3\frac{5}{7}\right), (0.8, -8.8)$

43 Let x represent the amount in the fund that pays 6% interest and y represent the amount in the fund that pays 10% interest. Write a system of inequalities and then graph.

$x + y \leq 10{,}000$ Total amount invested is up to $10,000.
$0.06x + 0.10y \geq 740$ Total amount earned is at least $740.

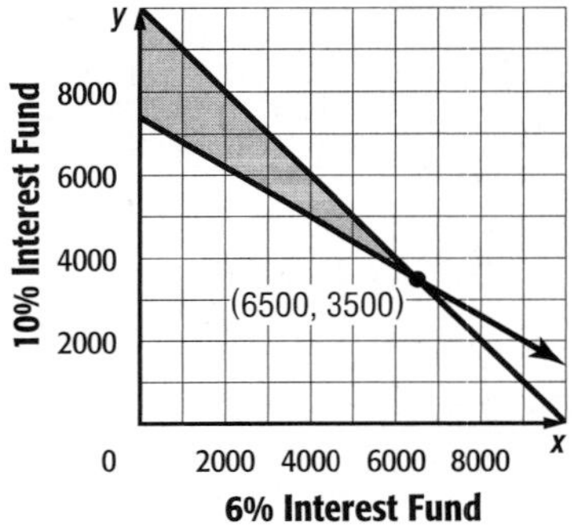

The least amount Mr. Hoffman can invest in the risky fund, or the 10% interest fund, is $3500.

45.

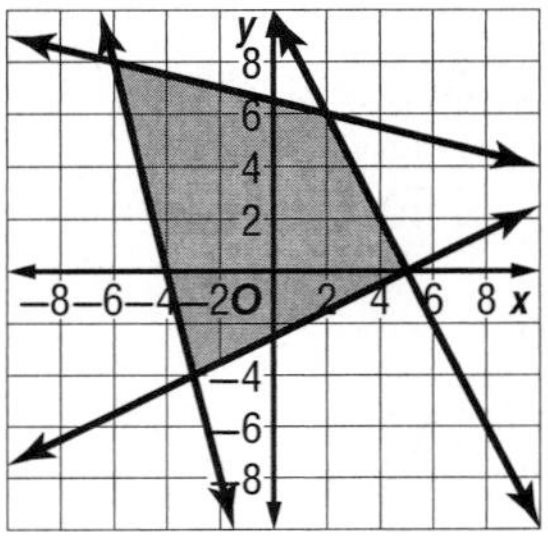

$(-3, -4), (-6, 8), (2, 6), (5, 0)$; 75 units^2

47. Sample answer:
$y \geq 2x - 6$;
$y \leq -0.5x + 4$;
$y \geq -3x - 6$

49. Sample answer: Shade each inequality in their standard way, by shading above the line if $y >$ and shading below the line if $y <$ (or you can use test points). Once you determine where to shade for each inequality, the area where *every* inequality needs to be shaded is the actual solution. This is only the shaded area. **51.** A **53.** $\frac{4}{5}z$ **55.** (1.5, 3), (3.5, 7), (8, 3), (10, 7) **57.** no solution

59.

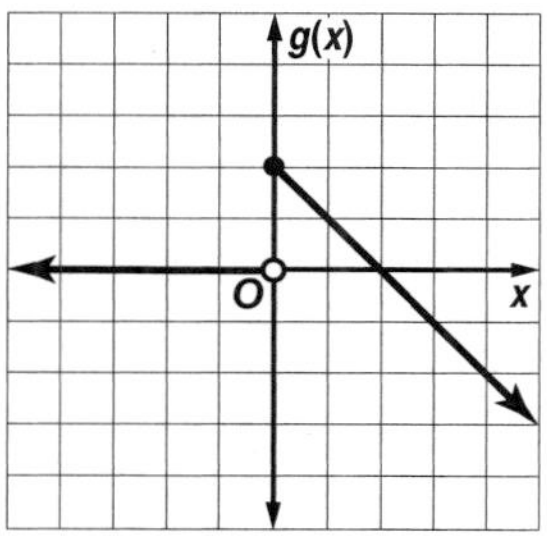

D = {all real numbers}, R = $\{g(x) \mid g(x) \leq 2\}$

61.

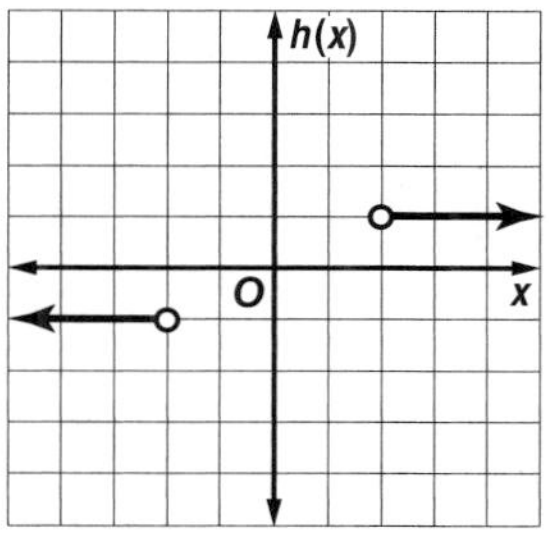

D = $\{x \mid x < -2 \text{ or } x > 2\}$, R = $\{-1, 1\}$
63. −1 **65.** 3 **67.** 4.5

Pages 163–166 ***Lesson 3-4***

1.

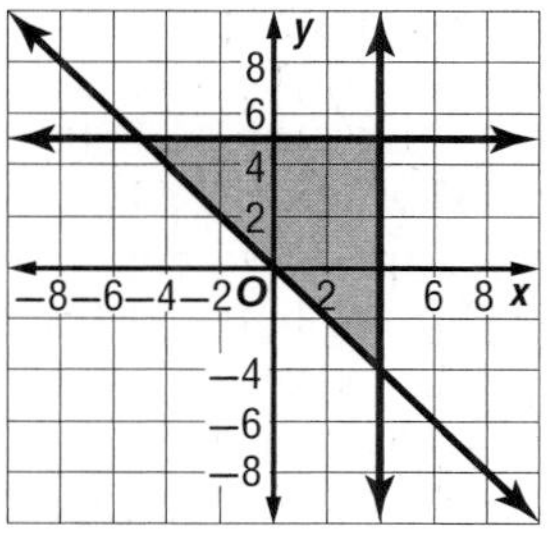

(4, 5), (4, −4), (−5, 5); max = 28, min = −35

3.

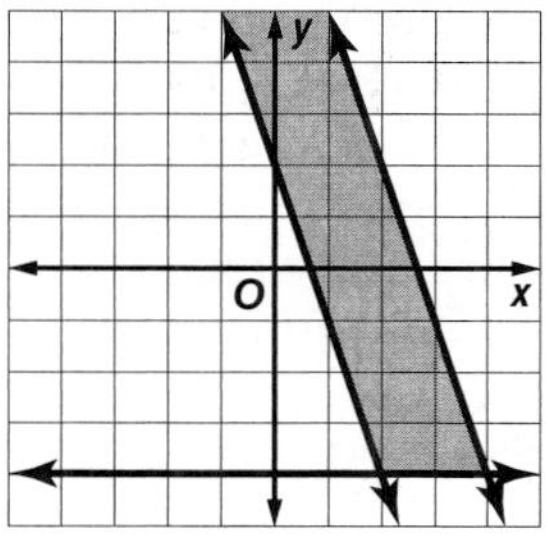

(2, −4), (4, −4); max does not exist, min = −52

5.

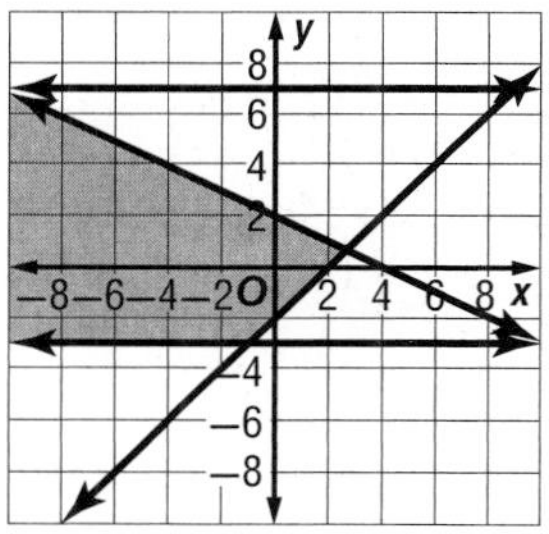

(4, 2), (−1, −3), (−6, 7); max does not exist; min = −30

7a. $g \geq 0, c \geq 0, 1.5g + c \leq 85, 2g + 0.5c \leq 40$

7b.

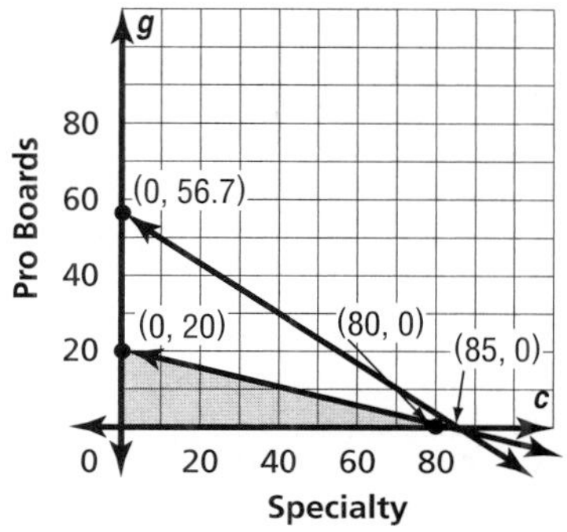

7c. (0, 0), (0, 20), (80, 0)

7d. $f(c, g) = 65c + 50g$

7e. 80 specialty boards, 0 pro boards; $5200

9 Graph the inequalities and locate the vertices.

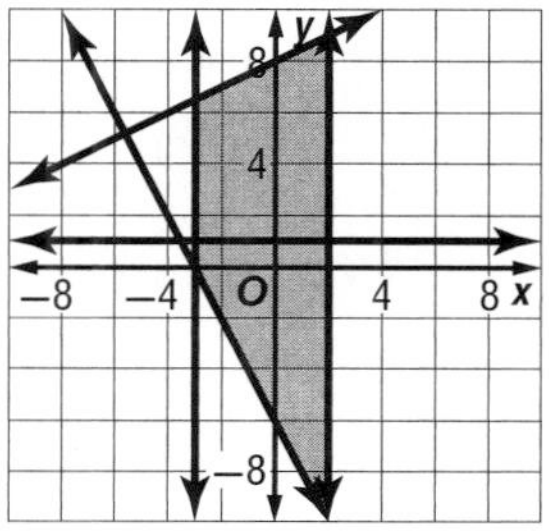

The vertices are at (2, −10), (−3, 0), (−3, 6.5), and (2, 9). Evaluate the function at each vertex.

(x, y)	−4x − 9y	f(x, y)
(2, −10)	−4(2) − 9(−10)	82
(−3, 0)	−4(−3) − 9(−0)	12
(−3, 6.5)	−4(−3) − 9(6.5)	70.5
(2, 9)	−4(2) − 9(9)	−89

The maximum value is 82 at (2, −10). The minimum value is −89 at (2, 9).

11.

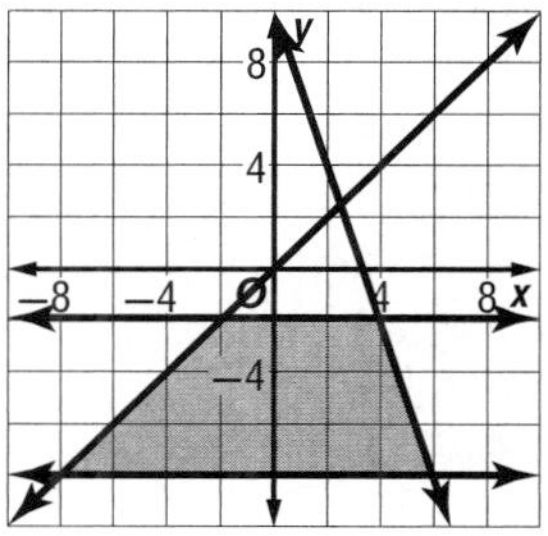

(6, −8), (4, −2), (−2, −2), (−8, −8); max = −8, min = −152

13.

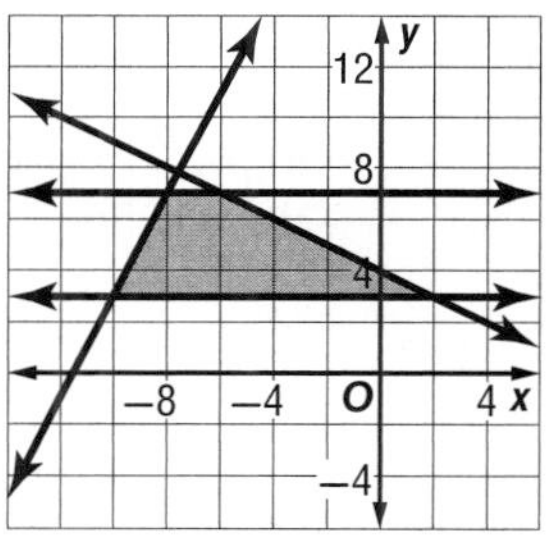

(−10, 3), (2, 3), (−6, 7), (−8, 7); max = 59, min = 9

15 Graph the inequalities and locate the vertices.

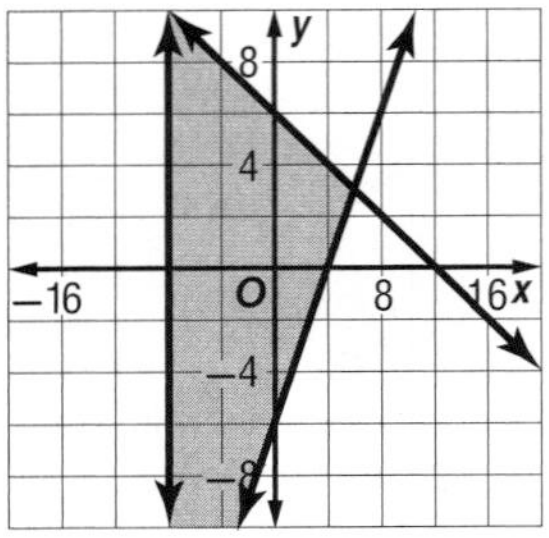

The vertices are at (6, 3), (−8, 10), and (−8, −18). Evaluate the function at each vertex.

(x, y)	10x − 6y	f(x, y)
(6, 3)	10(6) − 6(3)	42
(−8, 10)	10(−8) − 6(10)	−140
(−8, −18)	10(−8) − 6(−18)	28

The maximum value is 42 at (6, 3). The minimum value is −140 at (−8, 10).

17.

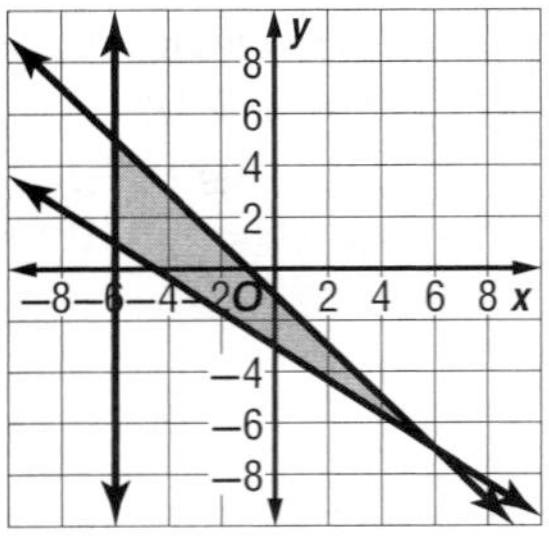

(−6, 1), (6, −7), (−6, 5); max = 48, min = 0

19.

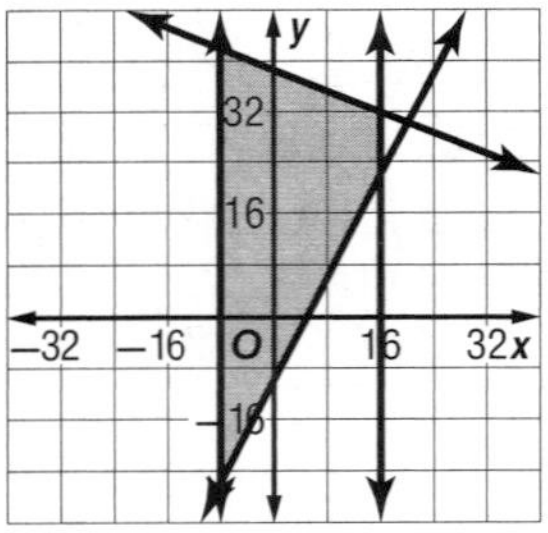

(−8, 44), (16, 32), (−8, −26), (16, 22); max = 672, min = −486

21.

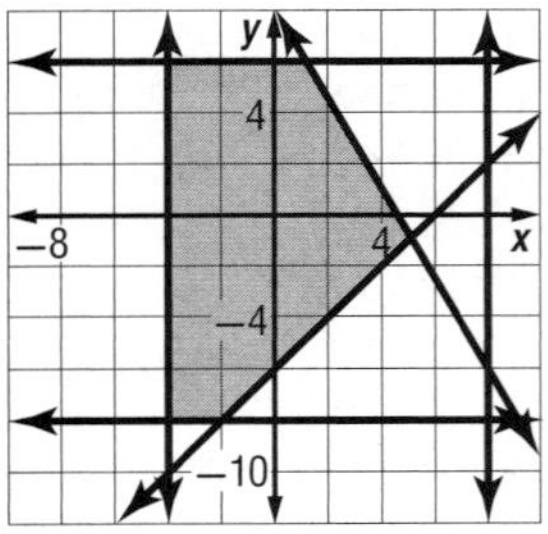

(5, −1), (1, 6), (−2, −8), (−4, −8), (−4, 6), max = 60, min = −112

23. 225 yellow cakes, 0 strawberry cakes

25a. $a \geq 0$, $b \geq 0$, $a + b \leq 45$, $4a + 5b \leq 200$

25b.

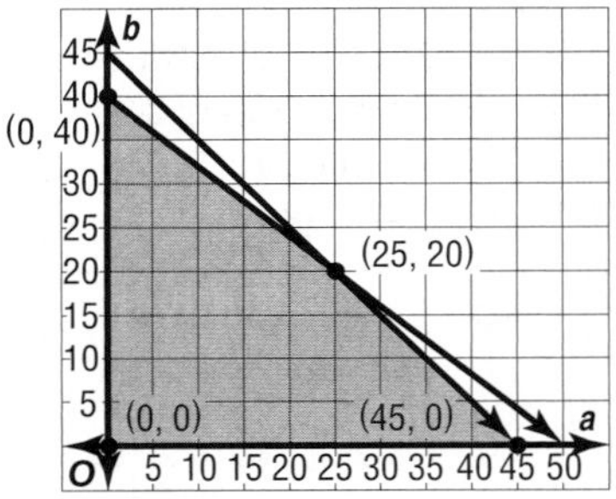

(0, 0), (0, 40), (25, 20), (45, 0)

25c. 25 sheds, 20 play houses

25d. $1250

27 **a.** Let x represent the number of small packages and y represent the number of large packages. Write a system of inequalities. Then graph the inequalities and locate the vertices.

$x \geq 0$	number of small packages ≥ 0
$y \geq 0$	number of large packages ≥ 0
$25x + 50y \leq 4200$	weight of packages ≤ 4200 lb
$3x + 5y \leq 480$	capacity of packages ≤ 480 cu ft

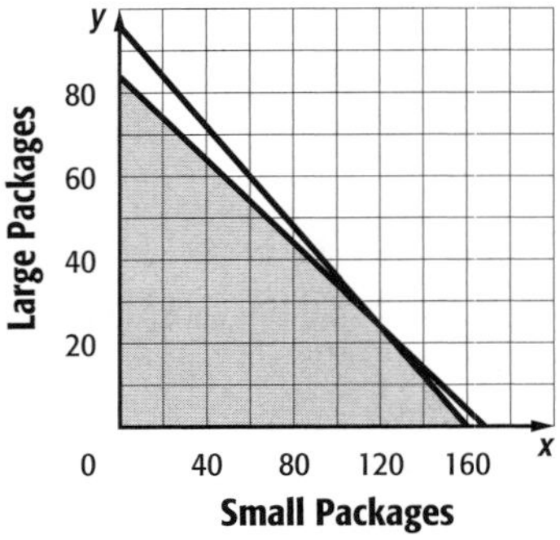

The vertices are at (0, 84), (120, 24), and (160, 0). Evaluate the function $f(x, y) = 5x + 8y$ at each vertex.

(x, y)	5x + 8y	f(x, y)
(0, 84)	5(0) + 8(84)	672
(120, 124)	5(120) + 8(24)	792
(160, 0)	5(160) + 8(0)	800

To maximize revenue, 160 small packages and 0 large packages should be placed on a train car.
b. The maximum revenue per train car is $800.
c. No; if revenue is maximized, the company will not deliver any large packages, and customers with large packages to ship will probably choose another carrier.

29. Sample answer: $-2 \geq y \geq -6$, $4 \leq x \leq 9$

31. b; The feasible region of Graph b is unbounded while the other three are bounded.

33. Sample answer: Even though the region is bounded, multiple maximums occur at *A* and *B* and all of the points on the boundary of the feasible region containing both *A* and *B*. This happened because that boundary of the region has the same slope as the function. **35.** $70.20 **37.** D

39.

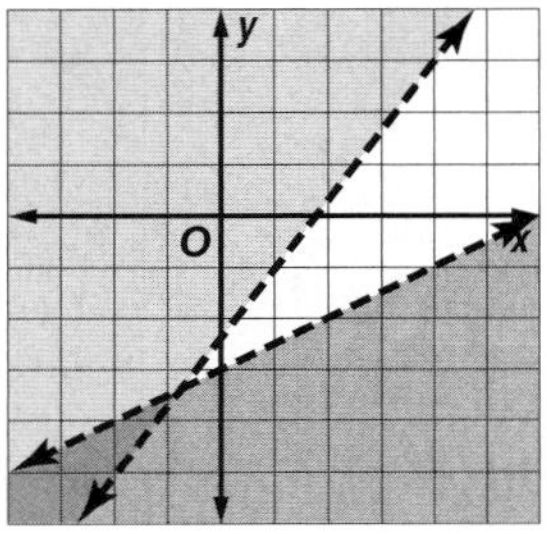

41. $7x + 15y = 330$, $8x + 16y = 360$; hats: 15, shirts: 15

43. $y = -\frac{1}{2}x + \frac{7}{2}$

45. 6; −2

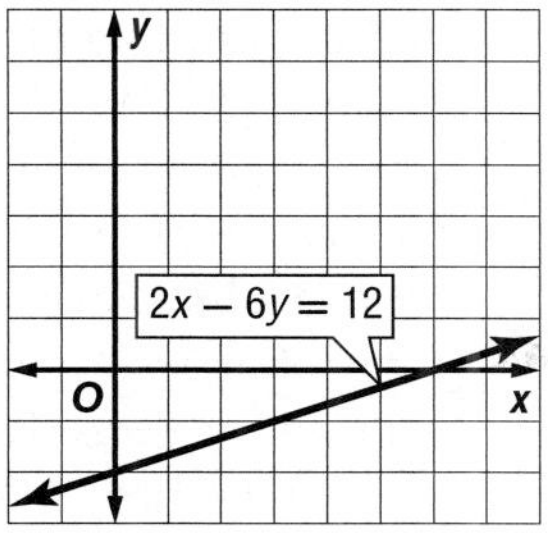

47. 5; 2

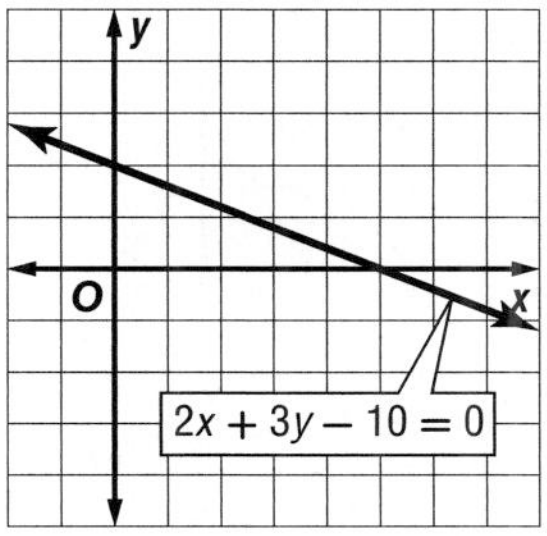

49. $\frac{1}{2}$; −2

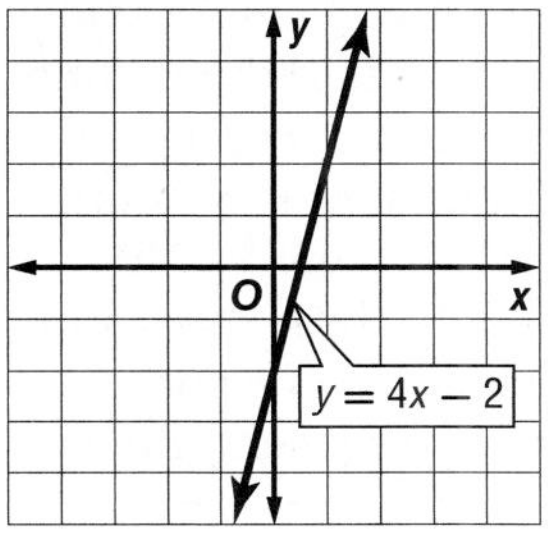

51. 9 **53.** −5 **55.** 15

Pages 171–173 Lesson 3-5

1. (−2, −3, 5) **3.** (−4, 3, 6) **5.** infinite solutions
7a. $s + d + t = 7$, $d = 2s$, $0.3s + 0.6d + 0.6t = 3.6$
7b. 2 sitcoms, 4 dramas, 1 talk show

9 $-a + 4b + 2c = -13$ Multiply by 4. → $-4a + 16b + 8c = -52$

$-4a + 16b + 8c = -52$ Equation 3 (× 4)
$(+)\ 4a + 5b - 6c = 2$ Equation 1
$21b + 2c = -50$

$-a + 4b + 2c = -13$ Multiply by −3. → $3a - 12b - 6c = 39$

$3a - 12b - 6c = 39$ Equation 3 × (−3)
$(+)\ -3a - 2b + 7c = -15$ Equation 2
$-14b + c = 24$

The resulting system of two equations and two variables is shown below.

$21b + 2c = -50$ New Equation 1
$-14b + c = 24$ New Equation 2

$-14b + c = 24$ Multiply by −2. → $28b - 2c = -48$

$21b + 2c = -50$ New Equation 1
$(+)\ 28b - 2c = -48$ New Equation 2 × (−2)
$49b = -98$ Add the equations.
$b = -2$ Divide each side by 49.

$-14b + c = 24$ New Equation 2
$-14(-2) + c = 24$ Replace b with −2.
$28 + c = 24$ Multiply.
$c = -4$ Subtract 28 from each side.

$-a + 4b + 2c = -13$ Original Equation 3
$-a + 4(-2) + 2(-4) = -13$ $b = -2$ and $c = -4$
$-a - 8 - 8 = -13$ Multiply.
$-a = 3$ Add 16 to each side.
$a = -3$ Multiply each side by −1.

The solution is (−3, −2, −4).

11. (−2, −1, 4) **13.** infinite solutions **15.** (−4, −1, 6) **17.** no solution **19.** infinite solutions **21.** roller coasters: 5; bumper cars: 1; water slides: 4

23 a = the amount invested in account A
b = the amount invested in account B
c = the amount invested in account C

$a + b + c = 100{,}000$ She invested a total of $100,000.
$a = c + 30{,}000$ She invested $30,000 more in account A than account C.
$0.04a + 0.08b + 0.1c = 6300$ The expected interest earned is $6300.

Substitute $a = c + 30{,}000$ in Equations 1 and 3.

$a + b + c = 100{,}000$ Equation 1
$c + 30{,}000 + b + c = 100{,}000$ $a = c + 30{,}000$
$30{,}000 + b + 2c = 100{,}000$ Add.
$b + 2c = 70{,}000$ Simplify.

$0.04a + 0.08b + 0.1c = 6300$ Equation 3
$0.04(c + 30{,}000) + 0.08b + 0.1c = 6300$ $a = c + 30{,}000$
$0.04c + 1200 + 0.08b + 0.1c = 6300$ Distribute.
$1200 + 0.08b + 0.14c = 6300$ Add.
$0.08b + 0.14c = 5100$ Simplify.

Solve the system of two equations in two variables.

$b + 2c = 70{,}000$ Multiply by −0.08. → $0.08b + 0.14c = 5100$

$-0.08b - 0.16c = -5600$
$(+)\ 0.08b + 0.14c = 5100$
$-0.02c = -500$
$c = 25{,}000$

Substitute to find b.

$b + 2c = 70{,}000$ Remaining equation in two variables.
$b + 2(25{,}000) = 70{,}000$ $c = 25{,}000$
$b + 50{,}000 = 70{,}000$ Distribute.
$b = 20{,}000$ Simplify.

Substitute to find a.

$a + b + c = 100{,}000$ Equation 1
$a + 20{,}000 + 25{,}000 = 100{,}000$ $b = 20{,}000$, $c = 25{,}000$
$a + 45{,}000 = 100{,}000$ Add.
$a = 55{,}000$ Simplify.

The solution is (55,000, 20,000, 25,000). She invested $55,000 in account A, $20,000 in account B, and $25,000 in account C.

25. $-3x^2 + 4x - 6$; $a = -3$, $b = 4$, $c = -6$

27. Sample answer:
$3x + 4y + z = -17$
$3(-5) + 4(-2) + 6 = -17$
$-15 + (-8) + 6 = -17$
$-23 + 6 = -17$

$$\begin{aligned} 2x - 5y - 3z &= -18 \\ 2(-5) - 5(-2) - 3(6) &= -18 \\ -10 + 10 - 18 &= -18 \\ -18 &= -18 \end{aligned}$$

$$\begin{aligned} -x + 3y + 8z &= 47 \\ -(-5) + 3(-2) + 8(6) &= 47 \\ 5 - 6 + 48 &= 47 \\ -1 + 48 &= 47 \end{aligned}$$

29. Sample answer: First, combine two of the original equations using elimination to form a new equation with three variables. Next, combine a different pair of the original equations using elimination to eliminate the same variable and form a second equation with three variables. Do the same thing with a third pair of the original equations. You now have a system of three equations with three variables. Follow the same procedure you learned in this section. Once you find the three variables, you need to use them to find the eliminated variable.
31. J **33.** A **35.** 16; −8 **37.** 9; −8 **39.** (6, 1)
41. (8, −5)

Pages 174–176 Chapter 3 Study Guide and Review

1. linear programming **3.** consistent **5.** unbounded
7. dependent **9.** inconsistent **11.** (0, 2) **13.** (−3, 4)
15. 4 h **17.** (−2, −7) **19.** (3, 5)

21.

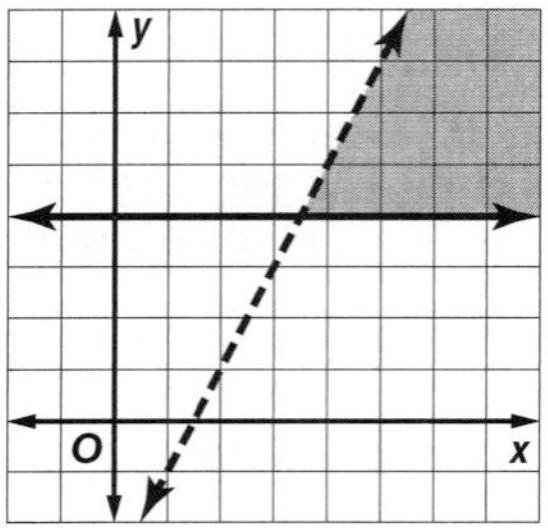

23.

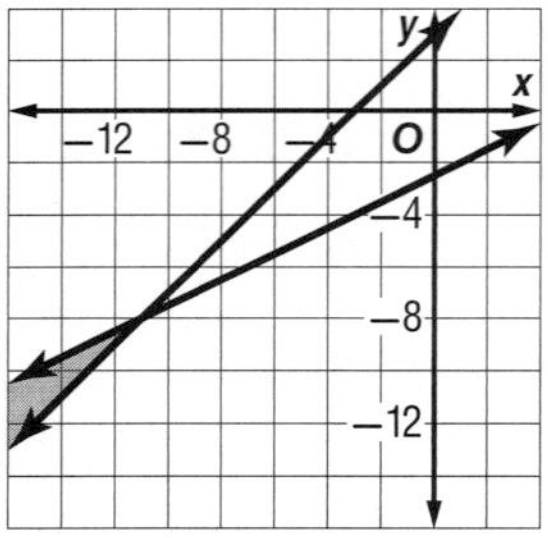

25.

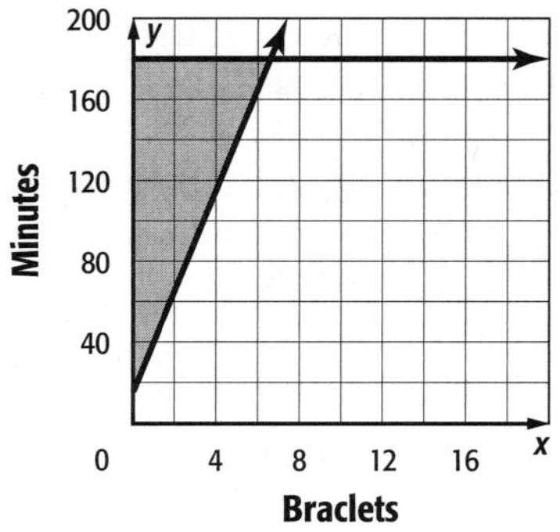

27. \$480; 12 outdoor, 16 indoor **29.** (5, −5, −4)

Chapter 4 Matrices

Page 183 Chapter 4 Get Ready

1. $-4, \frac{1}{4}$ **3.** −0.2, 5 **5.** $\frac{3}{4}, -\frac{4}{3}$ **7.** $6x + 12y$
9. $-19x + 6$ **11.** $17x - 3y - 9$ **13.** (1, 4) **15.** (−2, 6)

Pages 188–191 Lesson 4-1

1. 2 × 4 **3.** 3 × 2 **5.** 1 **7.** 9 **9.** 1 × 2 **11.** 2 × 4
13. 3 × 1 **15.** −9
17 Since b_{13} is the element in row 1, column 3 of matrix B, the value of b_{13} is $2x$.

19.

John	221	201	185	607
Hideo	168	233	159	560
Paulo	187	189	211	587

21a. $\begin{bmatrix} 3 & 2 & 2 & 1 \\ 4 & 3 & 2 & 3 \\ 5 & 5 & 4 & 4 \\ 1 & 5 & 5 & 2 \end{bmatrix}$

21b. Sample answer: Brand C; it was given the highest rating possible for cost and comfort, and a high rating for looks, and it will last a fairly long time.
21c. Sample answer: Yes; finding the sum of the rows and then calculating the average will provide an easy way to compare the data. **23.** $4x$ **25.** x
27 **a.** Write a matrix with three rows and two columns. Let the rows represent the seating areas and let the columns represent the days.

	Weekday	Weekend
Coach	249	259
Business	279	289
First Class	319	339

b. Write a matrix with two rows and three columns. Let the rows represent the days and let the columns represent the seating areas.

	Coach	Business	First Class
Weekday	249	279	319
Weekend	259	289	339

29. $x^2 + 4$ **31.** $-y$
33 **a.** Write a matrix with one row for each Planet and one column for each of the two types of distances.

	Sun	Earth
Mercury	36.00	57
Venus	67.24	26
Mars	141.71	35
Jupiter	483.88	370
Saturn	887.14	744
Uranus	1783.98	1607
Neptune	2796.46	2680

b. There are seven rows and two columns. So, the dimensions are 7 × 2.
c. The element a_{42} is in the fourth row and second column. So, the value is 370.

35. Sample answer: False; a square matrix with 4 columns has only 4 rows and cannot contain an element in a fifth row. **37.** No; element b_{32} is the second element in the third row, which is 2.

39. Sample answer:

	Hits	Walks	HR
Joe	95	12	8
John	102	16	5
Jim	109	13	12

41. D **43.** 780,000 ft^2 **45.** 45 first, 298 second, 147 third **47.** $-\frac{3}{2}$ **49.** $\frac{5}{3}$ **51.** $y^2 - 2y - 48$ **53.** −31 **55.** 52

Pages 196–199 Lesson 4-2

1. $\begin{bmatrix} 3 & -5 & 7 \end{bmatrix}$ **3.** $\begin{bmatrix} -2 & -18 \\ 11 & 13 \end{bmatrix}$

5. $\begin{bmatrix} 18 & 12 & 0 \\ -6 & 42 & -24 \\ -12 & -18 & 21 \end{bmatrix}$ **7.** $\begin{bmatrix} 20 & 4 \\ -14 & 38 \end{bmatrix}$

9. impossible **11a.** Test 1: $\begin{bmatrix} 85 \\ 75 \\ 96 \end{bmatrix}$ Test 2: $\begin{bmatrix} 72 \\ 74 \\ 83 \end{bmatrix}$

11b. $\begin{bmatrix} 157 \\ 149 \\ 179 \end{bmatrix}$ **11c.** $\begin{bmatrix} 13 \\ 1 \\ 13 \end{bmatrix}$

13 To find the sum of two matrices, they must have the same dimensions. Since the dimensions of the matrices are 2 × 2 and 3 × 2, it is impossible to find the sum.

15. $\begin{bmatrix} -24 \\ 10 \\ -3 \\ -7 \end{bmatrix}$ **17.** impossible **19.** $\begin{bmatrix} -7 \\ -32 \end{bmatrix}$

21a. Library A: $\begin{bmatrix} 10{,}000 \\ 5000 \\ 5000 \end{bmatrix}$; Library B: $\begin{bmatrix} 15{,}000 \\ 10{,}000 \\ 2500 \end{bmatrix}$;

Library C: $\begin{bmatrix} 4000 \\ 700 \\ 800 \end{bmatrix}$ **21b.** $\begin{bmatrix} 29{,}000 \\ 15{,}700 \\ 8300 \end{bmatrix}$ **21c.** $\begin{bmatrix} 6000 \\ 4300 \\ 4200 \end{bmatrix}$

21d. $\begin{bmatrix} 25{,}000 \\ 15{,}000 \\ 7500 \end{bmatrix}$; Sample answer: The sum represents the combined size of the two libraries.

23. $\begin{bmatrix} -8a & 32b & 8c - 8b \\ -104 & 80 & -40c \end{bmatrix}$

25 $-5\left(\begin{bmatrix} 4 & -8 \\ 8 & -9 \end{bmatrix} + \begin{bmatrix} 4 & -2 \\ -3 & -6 \end{bmatrix}\right)$

$= -5\begin{bmatrix} 4+4 & -8+(-2) \\ 8+(-3) & -9+(-6) \end{bmatrix}$ Add corresponding elements.

$= -5\begin{bmatrix} 8 & -10 \\ 5 & -15 \end{bmatrix}$ Simplify.

$= \begin{bmatrix} -5(8) & -5(-10) \\ -5(5) & -5(-15) \end{bmatrix}$ Distribute the scalar.

$= \begin{bmatrix} -40 & 50 \\ -25 & 75 \end{bmatrix}$ Multiply.

27. impossible **29.** $\begin{bmatrix} 68.6 & 19 \\ -19.99 & 18.3 \\ 11.83 & 38.7 \end{bmatrix}$

31. $\begin{bmatrix} 5 & 24 \\ -\frac{167}{12} & -10 \end{bmatrix}$

33 **a.** Write a matrix to show the American records and a matrix to show the world records. Then subtract.

$\begin{bmatrix} 24.63 \text{ s} \\ 53.99 \text{ s} \\ 1{:}57.41 \text{ min} \\ 8{:}16.22 \text{ min} \end{bmatrix} - \begin{bmatrix} 24.13 \text{ s} \\ 53.52 \text{ s} \\ 1{:}56.54 \text{ min} \\ 8{:}16.22 \text{ min} \end{bmatrix}$

$= \begin{bmatrix} 24.63 \text{ s} - 24.13 \text{ s} \\ 53.99 \text{ s} - 53.52 \text{ s} \\ 1{:}57.41 \text{ min} - 1{:}56.54 \text{ min} \\ 8{:}16.22 \text{ min} - 8{:}16.22 \text{ min} \end{bmatrix}$ Subtract corresponding elements.

$= \begin{bmatrix} 0.5 \text{ s} \\ 0.47 \text{ s} \\ 0.87 \text{ s} \\ 0 \text{ s} \end{bmatrix}$ Simplify.

b. In the 50-meter, the fastest American time is 0.5 second behind the world record. In the 100 m, the fastest American time is 0.47 second behind the world record. In the 200 m, the fastest American time is 0.87 second behind the world record. In the 800 m, the American and world records are the same. So, it was an American who set the world record.
c. In the 50-meter and 100-meter events, the fastest times were set at the Olympics. These times became world records.

35. To show that the Commutative Property of Matrix Addition is true for 2 × 2 matrices,

let $A = \begin{bmatrix} a & b \\ c & d \end{bmatrix}$ and $B = \begin{bmatrix} e & f \\ g & h \end{bmatrix}$. Show that $A + B = B + A$.

$A + B = \begin{bmatrix} a & b \\ c & d \end{bmatrix} + \begin{bmatrix} e & f \\ g & h \end{bmatrix}$ Substitution

$= \begin{bmatrix} a+e & b+f \\ c+g & d+h \end{bmatrix}$ Definition of matrix addition

$= \begin{bmatrix} e+a & f+b \\ g+c & h+d \end{bmatrix}$ Commutative Property of Addition for Real Numbers

$= \begin{bmatrix} e & f \\ g & h \end{bmatrix} + \begin{bmatrix} a & b \\ c & d \end{bmatrix}$ Definition of matrix addition

$= B + A$ Substitution

37. $\begin{bmatrix} 7 & 5 \\ -1 & -5 \end{bmatrix}$ **39.** Sample answer: $A = \begin{bmatrix} 6 & 1 \\ 6 & 3 \end{bmatrix}$ and $B = \begin{bmatrix} 3 & 2 \\ 4 & 2 \end{bmatrix}$ **41.** C **43.** F **45.** $4y$ **47.** does not exist **49.** (−2, 1, 6)

51.

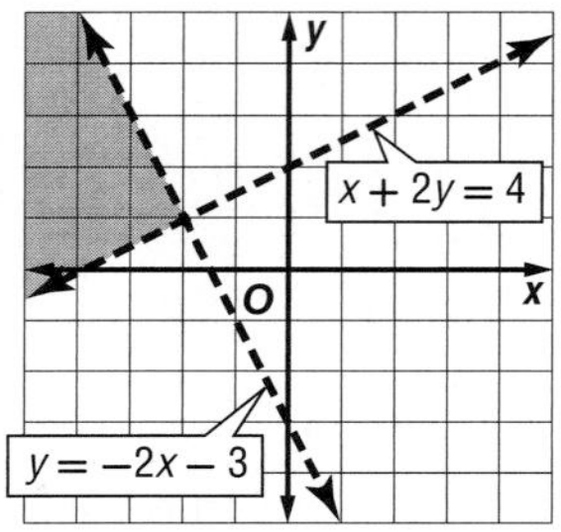

53.

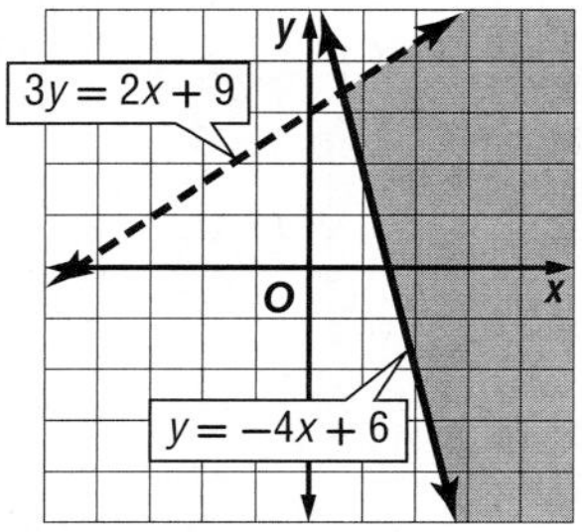

55. $350{,}349 - x = 15{,}991$; $334{,}358$ **57.** $-10a - b$

Pages 204–207 Lesson 4-3

1. 2×3 **3.** 8×10 **5.** $\begin{bmatrix} 0 & 44 \\ 8 & -34 \end{bmatrix}$

7 The product of a 2×1 matrix and a 1×3 matrix is a 2×3 matrix.

$$\begin{bmatrix} -9 \\ 6 \end{bmatrix} \cdot \begin{bmatrix} 1 & -10 & 1 \end{bmatrix}$$

$$= \begin{bmatrix} -9(-1) & -9(-10) & -9(1) \\ 6(-1) & 6(-10) & 6(1) \end{bmatrix}$$ Multiply the column by the row.

$$= \begin{bmatrix} 9 & 90 & -9 \\ -6 & -60 & 6 \end{bmatrix}$$ Simplify.

9. $\begin{bmatrix} -44 \\ 25 \end{bmatrix}$ **11.** $\begin{bmatrix} -16 & -1 \\ -6 & 10 \end{bmatrix}$

13. No; $\begin{bmatrix} 53 & -87 \\ -2 & -60 \end{bmatrix} \neq \begin{bmatrix} 62 & -33 \\ 28 & -69 \end{bmatrix}$.

15. 2×4 **17.** undefined **19.** undefined **21.** $[\ 26\ ]$

23
$$\begin{bmatrix} -3 & -7 \\ -2 & -1 \end{bmatrix} \cdot \begin{bmatrix} 4 & 4 \\ 9 & -3 \end{bmatrix}$$

$$= \begin{bmatrix} -3(4) + (-7)(9) & -3(4) + (-7)(-3) \\ & \end{bmatrix}$$ Multiply the 1st row in the 1st matrix by each column in the 2nd matrix.

$$\begin{bmatrix} -3 & -7 \\ -2 & -1 \end{bmatrix} \cdot \begin{bmatrix} 4 & 4 \\ 9 & -3 \end{bmatrix}$$

$$= \begin{bmatrix} -3(4) + (-7)(9) & -3(4) + (-7)(-3) \\ -2(4) + (-1)(9) & -2(4) + (-1)(-3) \end{bmatrix}$$ Multiply the 2nd row in the 1st matrix by each column in the 2nd matrix.

$$= \begin{bmatrix} -12 + (-63) & -12 + 21 \\ -8 + (-9) & -8 + 3 \end{bmatrix}$$ Multiply.

$$= \begin{bmatrix} -75 & 9 \\ -17 & -5 \end{bmatrix}$$ Add.

25. undefined **27.** $\begin{bmatrix} -40 & 64 \\ 22 & 1 \end{bmatrix}$

29a. $I = \begin{bmatrix} 3 & 2 & 2 \\ 2 & 3 & 1 \\ 4 & 3 & 0 \end{bmatrix}$, $C = \begin{bmatrix} 220 \\ 250 \\ 360 \end{bmatrix}$ **29b.** $\begin{bmatrix} \$1880 \\ \$1550 \\ \$1630 \end{bmatrix}$

29c. \$5060 **31.** $PQR = \begin{bmatrix} -22 & 240 \\ 44 & -12 \end{bmatrix}$ and

$RQP = \begin{bmatrix} 34 & -40 \\ -220 & -44 \end{bmatrix}$

33. No; $R(P + Q) = \begin{bmatrix} 34 & -6 \\ -64 & -30 \end{bmatrix}$ and

$PR + QR = \begin{bmatrix} 22 & 72 \\ 14 & -18 \end{bmatrix}$.

35 a. The bonuses can be found by multiplying the number of cars sold by the amount of bonus given for each new and used car sold.

Cars **Bonus**

$$C = \begin{bmatrix} 27 & 49 \\ 35 & 36 \\ 9 & 56 \\ 15 & 62 \end{bmatrix} \qquad B = \begin{bmatrix} 1000 \\ 500 \end{bmatrix}$$

$$CB = \begin{bmatrix} 27 & 49 \\ 35 & 36 \\ 9 & 56 \\ 15 & 62 \end{bmatrix} \cdot \begin{bmatrix} 1000 \\ 500 \end{bmatrix}$$ Write an equation.

$$= \begin{bmatrix} 27(1000) + 49(500) \\ 35(1000) + 36(500) \\ 9(1000) + 56(500) \\ 15(1000) + 62(500) \end{bmatrix}$$ Multiply columns by rows.

$$= \begin{bmatrix} 51{,}500 \\ 53{,}000 \\ 37{,}000 \\ 46{,}000 \end{bmatrix}$$ Simplify.

Westin earned the most, \$53,000.

b. total amount on bonuses = 51,500 + 53,000 + 37,000 + 46,000 or \$187,500

37. $\begin{bmatrix} -10 - 4.5y & 36.75 \\ 2x + 4 + 3y^2 & -6x - 4.5y - 12 \\ 3.6y + 26 & -83.4 \end{bmatrix}$

39. $\begin{bmatrix} -1.5x - 1.5y + 15 \\ y^2 + xy + 3x - 6 \\ 1.2x + 1.2y - 39 \end{bmatrix}$

41. undefined

43. $\begin{bmatrix} 15x + 69y - 12 \\ -18y^2 + 42.75y - 18xy + 20.25x \end{bmatrix}$

45a. A: \$421; B: \$274; C: \$150; D: \$68

45b. A: \$383.35; B: \$232.90; C: \$127.50; D: \$57.80

47a. $c(A + B)$

$= c\left(\begin{bmatrix} a & b \\ d & e \end{bmatrix} + \begin{bmatrix} w & x \\ y & z \end{bmatrix}\right)$ Substitution

$= c\begin{bmatrix} a + w & b + x \\ d + y & e + z \end{bmatrix}$ Definition of matrix addition

$= \begin{bmatrix} ca + cw & cb + cx \\ cd + cy & ce + cz \end{bmatrix}$ Definition of scalar multiplication

$= \begin{bmatrix} ca & cb \\ cd & ce \end{bmatrix} + \begin{bmatrix} cw & cx \\ cy & cz \end{bmatrix}$ Definition of matrix addition

$= cA + cB$ Substitution

47b. $$C(A+B) = \begin{bmatrix} a & b \\ c & d \end{bmatrix}\left(\begin{bmatrix} e & f \\ g & h \end{bmatrix} + \begin{bmatrix} j & k \\ m & n \end{bmatrix}\right)$$ Substitution

$$= \begin{bmatrix} a & b \\ c & d \end{bmatrix}\begin{bmatrix} e+j & f+k \\ g+m & h+n \end{bmatrix}$$ Definition of matrix addition

$$= \begin{bmatrix} a(e+j)+b(g+m) & a(f+k)+b(h+n) \\ c(e+j)+d(g+m) & c(f+k)+d(h+n) \end{bmatrix}$$ Definition of matrix multiplication

$$= \begin{bmatrix} ea+ja+gb+mb & fa+ka+hb+nb \\ ec+jc+gd+md & fc+kc+hd+nd \end{bmatrix}$$ Distributive Property

$$= \begin{bmatrix} ea+gb+ja+mb & fa+hb+ka+nb \\ ec+gd+jc+md & fc+hd+kc+nd \end{bmatrix}$$ Commutative Property of Addition

$$= \begin{bmatrix} ea+gb & fa+hb \\ ec+gd & fc+hd \end{bmatrix} + \begin{bmatrix} ja+mb & ka+nb \\ jc+md & kc+nd \end{bmatrix}$$ Definition of matrix addition

$$= CA + CB$$ Definition of matrix multiplication

$$(A+B)C = \left(\begin{bmatrix} a_{11} & a_{12} \\ a_{21} & a_{22} \end{bmatrix} + \begin{bmatrix} b_{11} & b_{12} \\ b_{21} & b_{22} \end{bmatrix}\right)\begin{bmatrix} c_{11} & c_{12} \\ c_{21} & c_{22} \end{bmatrix}$$ Substitution

$$= \begin{bmatrix} a_{11}+b_{11} & a_{12}+b_{12} \\ a_{21}+b_{21} & a_{22}+b_{22} \end{bmatrix}\begin{bmatrix} c_{11} & c_{12} \\ c_{21} & c_{22} \end{bmatrix}$$ Definition of matrix addition

$$= \begin{bmatrix} (a_{11}+b_{11})c_{11}+(a_{12}+b_{12})c_{21} & (a_{11}+b_{11})c_{12}+(a_{12}+b_{12})c_{22} \\ (a_{11}+b_{11})c_{11}+(a_{12}+b_{12})c_{21} & (a_{21}+b_{21})c_{12}+(a_{22}+b_{22})c_{22} \end{bmatrix}$$ Definition of matrix multiplication

$$= \begin{bmatrix} a_{11}c_{11}+b_{11}c_{11}+a_{12}c_{21}+b_{12}c_{21} & a_{11}c_{12}+b_{11}c_{12}+a_{12}c_{22}+b_{12}c_{22} \\ a_{21}c_{11}+b_{21}c_{11}+a_{22}c_{21}+b_{22}c_{21} & a_{21}c_{12}+b_{21}c_{12}+a_{22}c_{22}+b_{22}c_{22} \end{bmatrix}$$ Distributive Property

$$= \begin{bmatrix} a_{11}c_{11}+a_{12}c_{21}+b_{11}c_{11}+b_{12}c_{21} & a_{11}c_{12}+a_{12}c_{22}+b_{11}c_{12}+b_{12}c_{22} \\ a_{21}c_{11}+a_{22}c_{21}+b_{21}c_{11}+b_{22}c_{21} & a_{21}c_{12}+a_{22}c_{22}+b_{21}c_{12}+b_{22}c_{22} \end{bmatrix}$$ Commutative Property of Addition

$$= \begin{bmatrix} a_{11}c_{11}+a_{12}c_{21} & a_{11}c_{12}+a_{12}c_{22} \\ a_{21}c_{11}+a_{22}c_{21} & a_{21}c_{12}+a_{22}c_{22} \end{bmatrix} + \begin{bmatrix} b_{11}c_{11}+b_{12}c_{21} & b_{11}c_{12}+b_{12}c_{22} \\ b_{21}c_{11}+b_{22}c_{21} & b_{21}c_{12}+b_{22}c_{22} \end{bmatrix}$$ Definition of matrix addition

$$= AC + BC$$ Definition of matrix multiplication

47c. $$(AB)C = \left(\begin{bmatrix} a_{11} & a_{12} \\ a_{21} & a_{22} \end{bmatrix}\begin{bmatrix} b_{11} & b_{12} \\ b_{21} & b_{22} \end{bmatrix}\right)\begin{bmatrix} c_{11} & c_{12} \\ c_{21} & c_{22} \end{bmatrix}$$ Substitution

$$= \begin{bmatrix} a_{11}b_{11}+a_{12}b_{21} & a_{11}b_{12}+a_{12}b_{22} \\ a_{21}b_{11}+a_{22}b_{21} & a_{21}b_{12}+a_{22}b_{22} \end{bmatrix}\begin{bmatrix} c_{11} & c_{12} \\ c_{21} & c_{22} \end{bmatrix}$$ Definition of matrix multiplication

$$= \begin{bmatrix} (a_{11}b_{11}+a_{12}b_{21})c_{11}+(a_{11}b_{12}+a_{12}b_{22})c_{21} & (a_{11}b_{11}+a_{12}b_{21})c_{12}+(a_{11}b_{12}+a_{12}b_{22})c_{22} \\ (a_{21}b_{11}+a_{22}b_{21})c_{11}+(a_{21}b_{12}+a_{22}b_{22})c_{21} & (a_{21}b_{11}+a_{22}b_{21})c_{12}+(a_{21}b_{12}+a_{22}b_{22})c_{22} \end{bmatrix}$$ Definition of matrix multiplication

$$= \begin{bmatrix} a_{11}b_{11}c_{11}+a_{12}b_{21}c_{11}+a_{11}b_{12}c_{21}+a_{12}b_{22}c_{21} & a_{11}b_{11}c_{12}+a_{12}b_{21}c_{12}+a_{11}b_{12}c_{22}+a_{12}b_{22}c_{22} \\ a_{21}b_{11}c_{11}+a_{22}b_{21}c_{11}+a_{21}b_{12}c_{21}+a_{22}b_{22}c_{21} & a_{21}b_{11}c_{12}+a_{22}b_{21}c_{12}+a_{21}b_{12}c_{22}+a_{22}b_{22}c_{22} \end{bmatrix}$$ Distributive Property

$$= \begin{bmatrix} a_{11}b_{11}c_{11}+a_{11}b_{12}c_{21}+a_{12}b_{21}c_{11}+a_{12}b_{22}c_{21} & a_{11}b_{11}c_{12}+a_{11}b_{12}c_{22}+a_{12}b_{21}c_{12}+a_{12}b_{22}c_{22} \\ a_{21}b_{11}c_{11}+a_{21}b_{12}c_{21}+a_{22}b_{21}c_{11}+a_{22}b_{22}c_{21} & a_{21}b_{11}c_{12}+a_{21}b_{12}c_{22}+a_{22}b_{21}c_{12}+a_{22}b_{22}c_{22} \end{bmatrix}$$ Commutative Property of Addition

$$= \begin{bmatrix} a_{11}(b_{11}c_{11}+b_{12}c_{21})+a_{12}(b_{21}c_{11}+b_{22}c_{21}) & a_{11}(b_{11}c_{12}+b_{12}c_{22})+a_{12}(b_{21}c_{12}+b_{22}c_{22}) \\ a_{11}(b_{11}c_{11}+b_{12}c_{21})+a_{22}(b_{21}c_{11}+b_{22}b_{21}) & a_{21}(b_{11}c_{12}+b_{12}c_{22})+a_{22}(b_{21}c_{12}+b_{22}c_{22}) \end{bmatrix}$$ Definition of Distributive Property

$$= \begin{bmatrix} a_{11} & a_{12} \\ a_{21} & a_{22} \end{bmatrix}\begin{bmatrix} b_{11}c_{11}+b_{12}c_{21} & b_{11}c_{12}+b_{12}c_{22} \\ b_{21}c_{11}+b_{22}c_{21} & b_{21}c_{12}+b_{22}c_{22} \end{bmatrix}$$ Definition of matrix multiplication

$$= \begin{bmatrix} a_{11} & a_{12} \\ a_{21} & a_{22} \end{bmatrix}\left(\begin{bmatrix} b_{11} & b_{12} \\ b_{21} & b_{22} \end{bmatrix}\begin{bmatrix} c_{11} & c_{12} \\ c_{21} & c_{22} \end{bmatrix}\right)$$ Definition of matrix multiplication

$$= A(BC)$$ Substitution

47d.
$$c(AB) = c\left(\begin{bmatrix} a_{11} & a_{12} \\ a_{21} & a_{22} \end{bmatrix}\begin{bmatrix} b_{11} & b_{12} \\ b_{21} & b_{22} \end{bmatrix}\right) \quad \text{Substitution}$$
$$= c\begin{bmatrix} a_{11}b_{11} + a_{12}b_{21} & a_{11}b_{12} + a_{12}b_{22} \\ a_{21}b_{11} + a_{22}b_{21} & a_{21}b_{12} + a_{22}b_{22} \end{bmatrix} \quad \text{Definition of matrix addition}$$
$$= \begin{bmatrix} c(a_{11}b_{11} + a_{12}b_{21}) & c(a_{11}b_{12} + a_{12}b_{22}) \\ c(a_{21}b_{11} + a_{22}b_{21}) & c(a_{21}b_{12} + a_{22}b_{22}) \end{bmatrix} \quad \text{Definition of scalar multiplication}$$
$$= \begin{bmatrix} ca_{11}b_{11} + ca_{12}b_{21} & ca_{11}b_{12} + ca_{12}b_{22} \\ ca_{21}b_{11} + ca_{22}b_{21} & ca_{21}b_{12} + ca_{22}b_{22} \end{bmatrix} \quad \text{Distributive Property}$$
$$= \begin{bmatrix} ca_{11} & ca_{12} \\ ca_{21} & ca_{22} \end{bmatrix}\begin{bmatrix} b_{11} & b_{12} \\ b_{21} & b_{22} \end{bmatrix} \quad \text{Definition of matrix multiplication}$$
$$= c\begin{bmatrix} a_{11} & a_{12} \\ a_{21} & a_{22} \end{bmatrix}\begin{bmatrix} b_{11} & b_{12} \\ b_{21} & b_{22} \end{bmatrix} \quad \text{Definition of scalar multiplication}$$
$$= (cA)B \quad \text{Substitution}$$

$$c(AB) = c\left(\begin{bmatrix} a_{11} & a_{12} \\ a_{21} & a_{22} \end{bmatrix}\begin{bmatrix} b_{11} & b_{12} \\ b_{21} & b_{22} \end{bmatrix}\right) \quad \text{Substitution}$$
$$= c\begin{bmatrix} a_{11}b_{11} + a_{12}b_{21} & a_{11}b_{12} + a_{12}b_{22} \\ a_{21}b_{11} + a_{22}b_{21} & a_{21}b_{12} + a_{22}b_{22} \end{bmatrix} \quad \text{Definition of matrix multiplication}$$
$$= \begin{bmatrix} c(a_{11}b_{11} + a_{12}b_{21}) & c(a_{11}b_{12} + a_{12}b_{22}) \\ c(a_{21}b_{11} + a_{22}b_{21}) & c(a_{21}b_{12} + a_{22}b_{22}) \end{bmatrix} \quad \text{Definition of scalar multiplication}$$
$$= \begin{bmatrix} ca_{11}b_{11} + ca_{12}b_{21} & ca_{11}b_{12} + ca_{12}b_{22} \\ ca_{21}b_{11} + ca_{22}b_{21} & ca_{21}b_{12} + ca_{22}b_{22} \end{bmatrix} \quad \text{Distributive Property}$$
$$= \begin{bmatrix} a_{11}cb_{11} + a_{12}cb_{21} & a_{11}cb_{12} + a_{12}cb_{22} \\ a_{21}cb_{11} + a_{22}cb_{21} & a_{21}cb_{12} + a_{22}cb_{22} \end{bmatrix} \quad \text{Commutative Property}$$
$$= \begin{bmatrix} a_{11} & a_{12} \\ a_{21} & a_{22} \end{bmatrix}\begin{bmatrix} cb_{11} & cb_{12} \\ cb_{21} & cb_{22} \end{bmatrix} \quad \text{Definition of matrix multiplication}$$
$$= A(cB) \quad \text{Substitution}$$

49. $a = 2, b = 1, c = 3, d = 4$ **51.** 12 **53.** H

55. $\begin{bmatrix} 42 & -24 \\ -42 & -31 \end{bmatrix}$ **57.** $\begin{bmatrix} -80 & -44 \\ 68 & 4 \end{bmatrix}$ **59.** 2×2

61a. Sample answer: $y = 127.7x + 1348.3$

61b. Sample answer: $3391.5 **61c.** The value predicted by the equation is significantly lower than the one given in the graph.

63. translation 4 units right and 3 units up

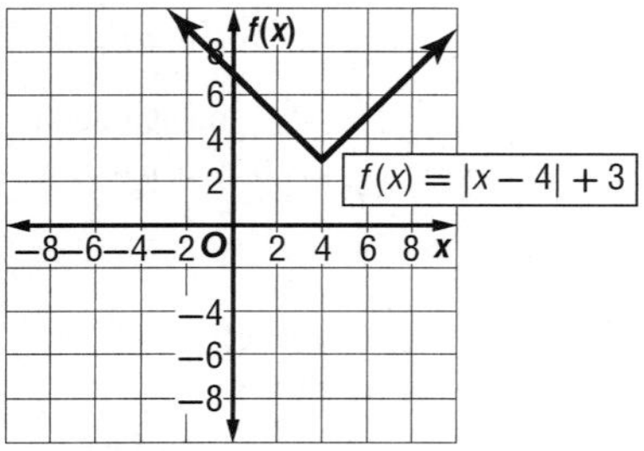

65. translation 2 units left and 6 units down

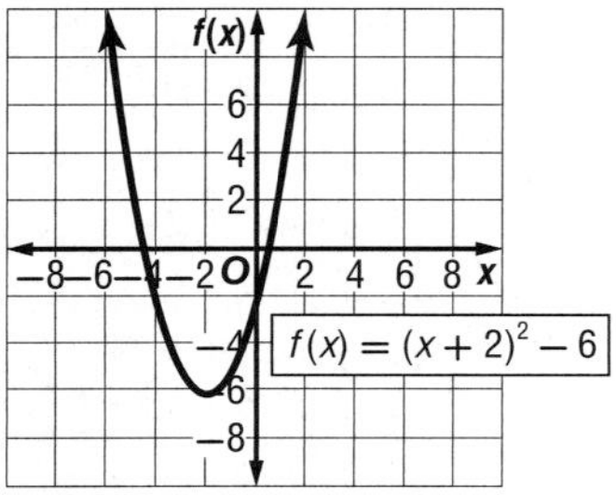

Pages 213–217 Lesson 4-4

1. $A'(2, -6)$, $B'(1, -1)$, $C'(6, -3)$

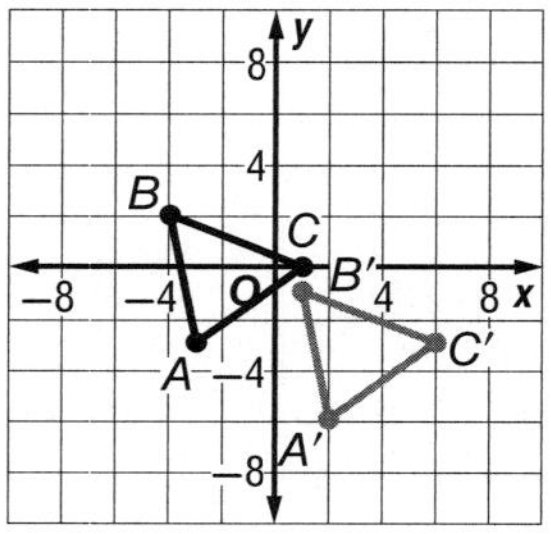

3 Find the translation matrix that moves $T(-3, -1)$ to $T'(-4, 1)$.

$$\overset{T}{\begin{bmatrix} -3 \\ -1 \end{bmatrix}} + \begin{bmatrix} x \\ y \end{bmatrix} = \overset{T'}{\begin{bmatrix} -4 \\ 1 \end{bmatrix}} \quad \text{Write an equation.}$$

$$\begin{bmatrix} -3 + x \\ -1 + y \end{bmatrix} = \begin{bmatrix} -4 \\ 1 \end{bmatrix} \quad \text{Add.}$$

The matrices are equal, so corresponding elements are equal.

$-3 + x = -4$ Solve for *x*. $\quad -1 + y = 1$ Solve for *y*.

$x = -1 \quad y = 2$

The translation matrix is $\begin{bmatrix} x \\ y \end{bmatrix} = \begin{bmatrix} -1 \\ 2 \end{bmatrix}$. So, rectangle *RSTU* is translated by adding -1 to the *x*-coordinate and adding 2 to the *y*-coordinate.

Vertex Matrix of *RSTU*		Translation Matrix	
$\begin{bmatrix} -3 & 1 & -3 & 1 \\ 2 & 2 & -1 & -1 \end{bmatrix}$	+	$\begin{bmatrix} -1 & -1 & -1 & -1 \\ 2 & 2 & 2 & 2 \end{bmatrix}$	=

Vertex Matrix of *R′S′T′U′*

$$\begin{bmatrix} -4 & 0 & -4 & 0 \\ 4 & 4 & 1 & 1 \end{bmatrix}$$

The vertices of *R′S′T′U′* are *R′*(−4, 4), *S′*(0, 4), *T′*(−4, 1), and *U′*(0, 1). So, the correct answer is A.

5. $S'\left(\frac{1}{2}, 0\right)$, $T'\left(2, -1\frac{1}{2}\right)$, $U'\left(\frac{1}{2}, -3\right)$, $V'\left(-1, -1\frac{1}{2}\right)$

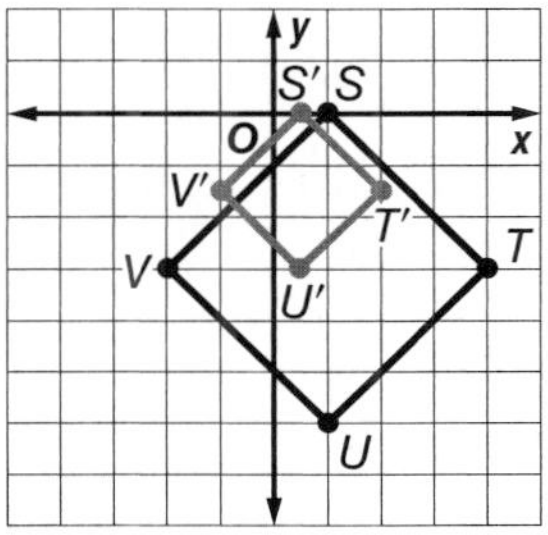

7. *P′*(2, −1), *Q′*(−4, 4), *R′*(−4, −1)

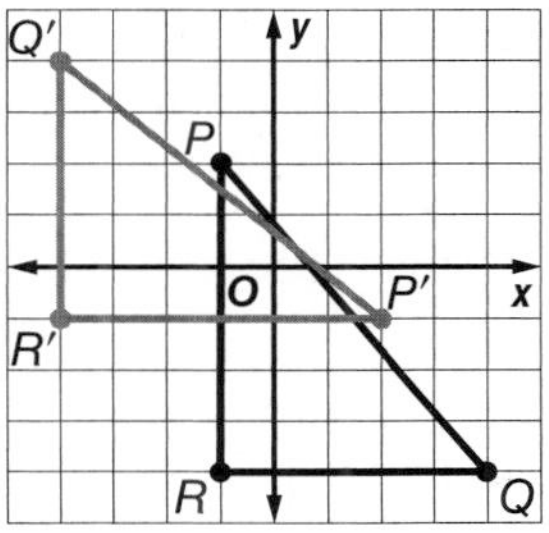

9. *A′*(−1, −4), *B′*(−4, −3), *C′*(−4, 0), *D′*(−1, 1)

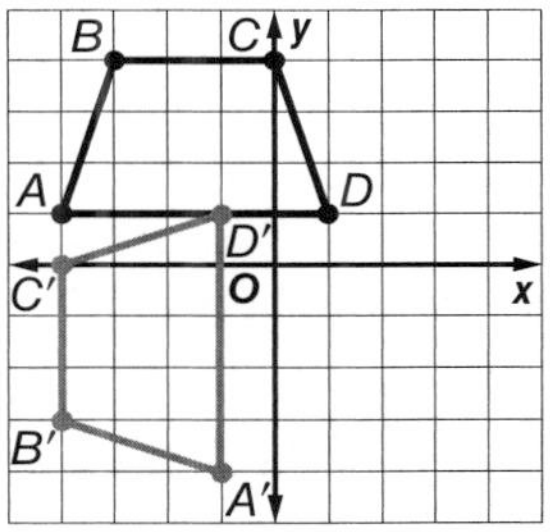

11. *E′*(−8, 4), *F′*(−2, −1), *G′*(−6, −3), *H′*(−7, −2)

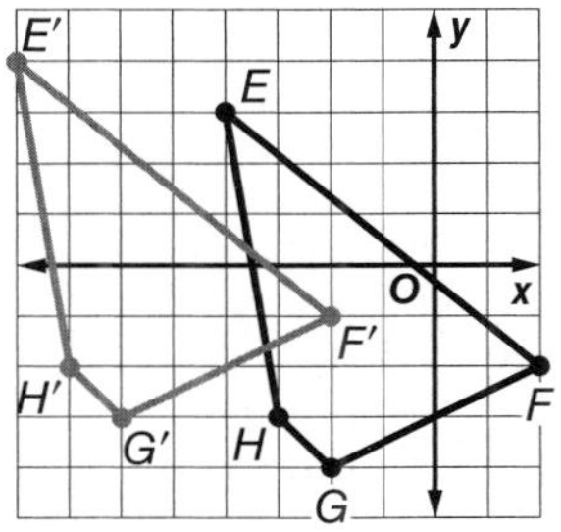

13. *J′*(4, 0), *K′*(5, −3), *L′*(2, −6)

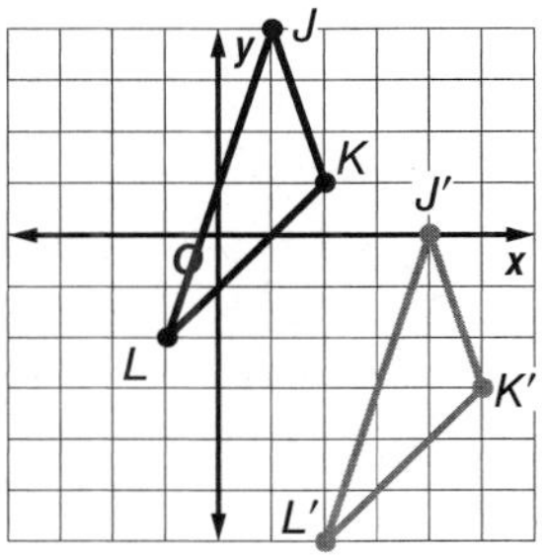

15. *W′*(1, 5), *X′*(1, 9), *Y′*(5, 9), *Z′*(5, 5)

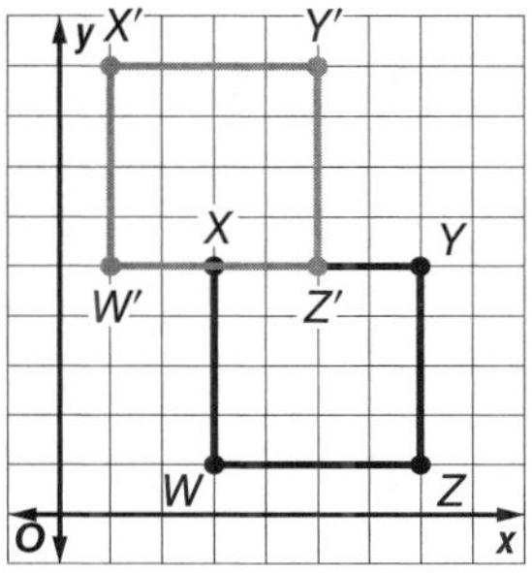

17 a. Three blocks east is 3 units to the right. Four blocks north is 4 units up. To translate 3 units to the right, add 3 to each *x*-coordinate. To translate 4 units up, add 4 to each *y*-coordinate. This can be represented by the translation matrix $\begin{bmatrix} 3 \\ 4 \end{bmatrix}$.

b. $\begin{bmatrix} 5.5 \\ 7 \end{bmatrix} + \begin{bmatrix} 3 \\ 4 \end{bmatrix} = \begin{bmatrix} 8.5 \\ 11 \end{bmatrix}$

The coordinates of the mall are (8.5, 11).

19. *D′*(0, 16), *E′*(8, 8), *F′*(0, 0), *G′*(−8, 8)

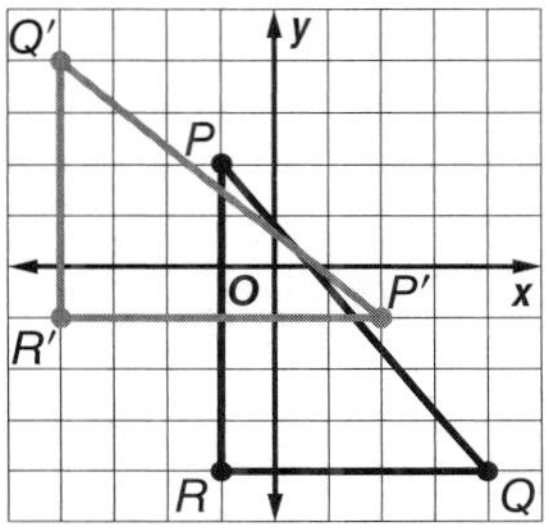

21. $Q'\left(1\frac{1}{2}, 1\right)$, *R′*(2, 0), $S'\left(0, \frac{1}{4}\right)$

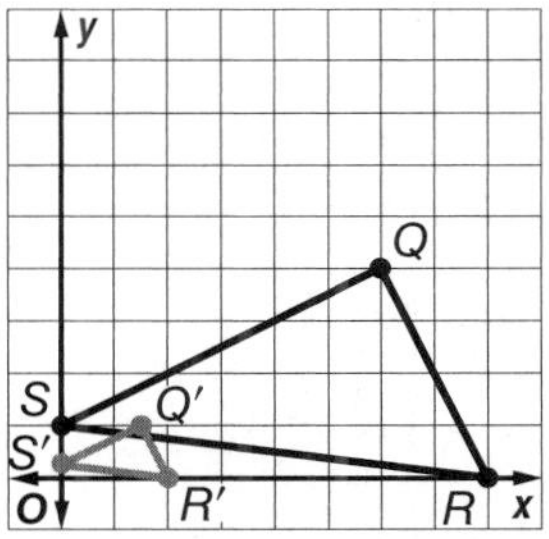

23

Reflection Matrix		Vertex Matrix of *DEF*		Vertex Matrix of *D′E′F′*
$\begin{bmatrix} 1 & 0 \\ 0 & -1 \end{bmatrix}$	·	$\begin{bmatrix} 7 & 4 & -3 \\ 4 & 0 & 2 \end{bmatrix}$	=	$\begin{bmatrix} 7 & 4 & -3 \\ -4 & 0 & -2 \end{bmatrix}$

The coordinates of the vertices of △*D′E′F′* are *D′*(7, −4), *E′*(4, 0), and *F′*(−3, −2).

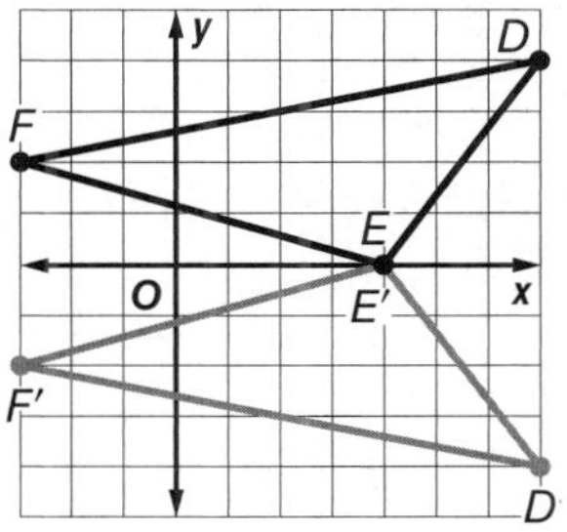

25. $L'(-1, -2), M'(5, -1), N'(4, 5), P'(-2, 4)$

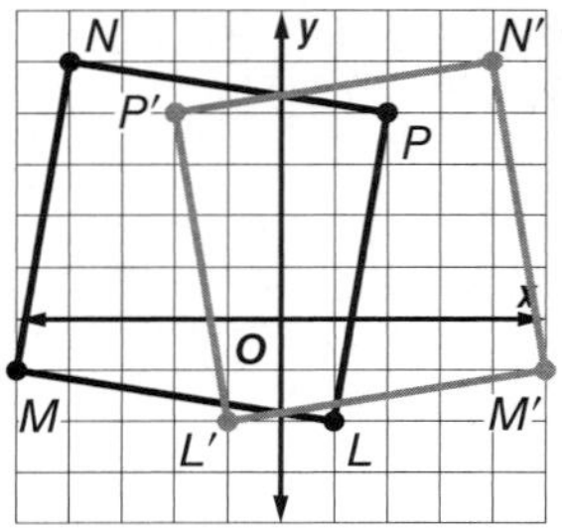

27. $T'(5, -4), U'(3, 1), V'(0, -2)$

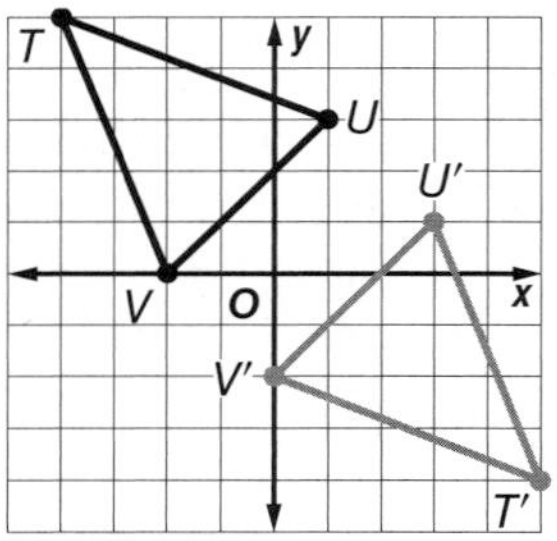

29. $X'(-1, -2), Y'(-1, -4), Z'(-5, -2)$

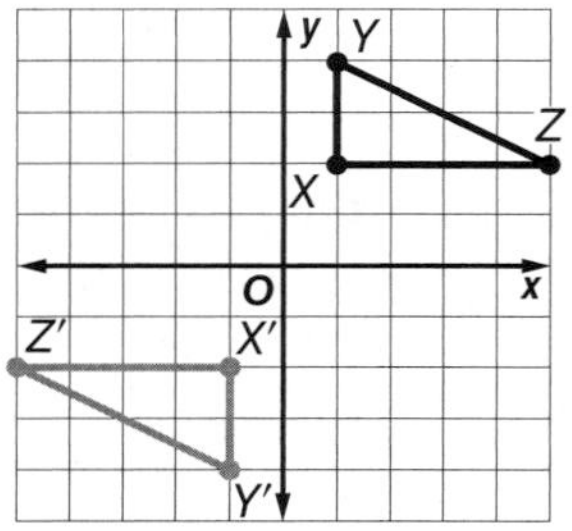

31. $E'(2, -3), F'(-4, -3), G'(-4, -2), H'(2, -2)$

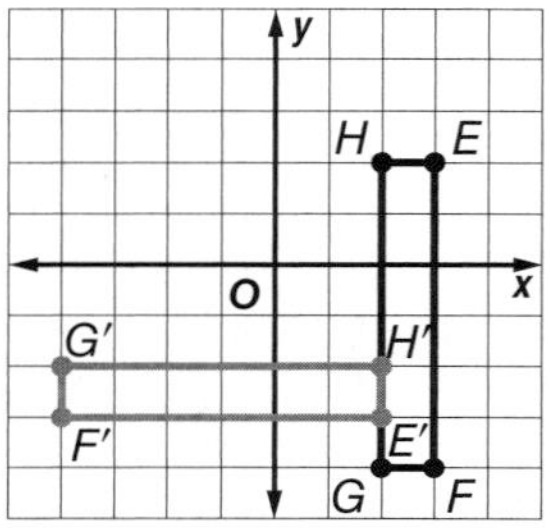

33. $N'(1, -2), P'(5, -3), Q'(5, -6), R'(1, -6)$
35. $(-165, 0)$
37. $X''(-8, -5), Y''(-6, -3), Z''(-5, -7)$
39. $A'(4, -4), B'(-4, -4), C'(-4, 4), D'(4, -4)$

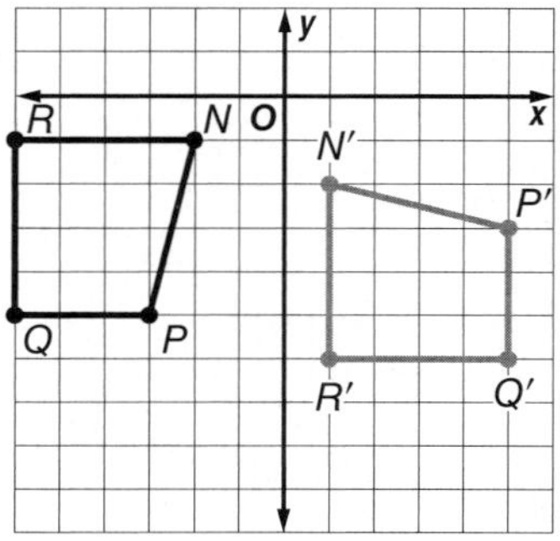

41 a. vertex matrix for $\triangle ABC$: $\begin{bmatrix} -2 & 4 & 2 \\ 3 & 0 & -5 \end{bmatrix}$

matrix for $\triangle A'B'C'$: $\begin{bmatrix} 1 & 0 \\ 0 & -1 \end{bmatrix} \cdot \begin{bmatrix} -2 & 4 & 2 \\ 3 & 0 & -5 \end{bmatrix} =$
$\begin{bmatrix} -2 & 4 & 2 \\ -3 & 0 & 5 \end{bmatrix}$

b.

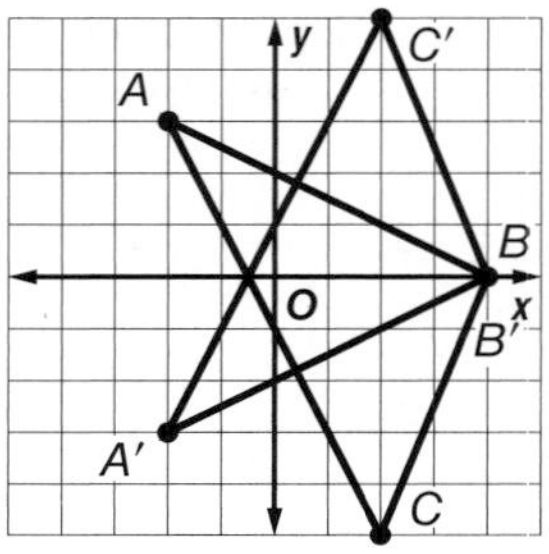

c. Conjecture: The reflection of $\triangle A'B'C'$ in the x-axis is $\triangle ABC$.

$$\begin{bmatrix} 1 & 0 \\ 0 & -1 \end{bmatrix} \cdot \begin{bmatrix} -2 & 4 & 2 \\ -3 & 0 & 5 \end{bmatrix} = \begin{bmatrix} -2 & 4 & 2 \\ 3 & 0 & -5 \end{bmatrix}$$

d. Reflection Matrix Reflection Matrix

$$\begin{bmatrix} 1 & 0 \\ 0 & -1 \end{bmatrix} \cdot \begin{bmatrix} 1 & 0 \\ 0 & -1 \end{bmatrix} = \begin{bmatrix} 1 & 0 \\ 0 & 1 \end{bmatrix}$$

The matrix is the identity matrix because reflecting a triangle in the same line twice produces the original figure.

43. Sometimes; the image and preimage are only congruent if the scale factor is 1 or −1. **45.** Sample answer: A reflection in the x-axis produces an image of $\begin{bmatrix} 3 \\ 2 \end{bmatrix}$. This is the same as applying the translation $\begin{bmatrix} 0 \\ 4 \end{bmatrix}$.

47. 180° counterclockwise rotation:

$$\begin{bmatrix} -1 & 0 \\ 0 & -1 \end{bmatrix} \cdot \begin{bmatrix} x_1 & x_2 & x_3 \\ y_1 & y_2 & y_3 \end{bmatrix}$$

$$= \begin{bmatrix} -1(x_1) + 0(y_1) & -1(x_2) + 0(y_2) & -1(x_3) + 0(y_3) \\ 0(x_1) - 1(y_1) & 0(x_2) - 1(y_2) & 0(x_3) - 1(y_3) \end{bmatrix}$$

$$= \begin{bmatrix} -x_1 & -x_2 & -x_3 \\ -y_1 & -y_2 & -y_3 \end{bmatrix}$$

Reflection in the x-axis:

$$\begin{bmatrix} 1 & 0 \\ 0 & -1 \end{bmatrix} \cdot \begin{bmatrix} x_1 & x_2 & x_3 \\ y_1 & y_2 & y_3 \end{bmatrix}$$

$$= \begin{bmatrix} 1(x_1) + 0(y_1) & 1(x_2) + 0(y_2) & 1(x_3) + 0(y_3) \\ 0(x_1) - 1(y_1) & 0(x_2) - 1(y_2) & 0(x_3) - 1(y_3) \end{bmatrix}$$

$$= \begin{bmatrix} x_1 & x_2 & x_3 \\ -y_1 & -y_2 & -y_3 \end{bmatrix}$$

followed by a reflection in the y-axis:

$$\begin{bmatrix} -1 & 0 \\ 0 & 1 \end{bmatrix} \cdot \begin{bmatrix} x_1 & x_2 & x_3 \\ -y_1 & -y_2 & -y_3 \end{bmatrix}$$

$$= \begin{bmatrix} -1(x_1) + 0(y_1) & -1(x_2) + 0(y_2) & -1(x_3) + 0(y_3) \\ 0(x_1) + 1(-y_1) & 0(x_2) + 1(-y_2) & 0(x_3) + 1(-y_3) \end{bmatrix}$$

$$= \begin{bmatrix} -x_1 & -x_2 & -x_3 \\ -y_1 & -y_2 & -y_3 \end{bmatrix}$$

49. B **51.** $\begin{bmatrix} \$29.99 & \$149.99 \\ \$39.99 & \$179.99 \\ \$49.99 & \$209.99 \\ \$69.99 & \$349.99 \end{bmatrix}, \begin{bmatrix} \$34.49 & \$172.49 \\ \$45.99 & \$206.99 \\ \$57.49 & \$241.49 \\ \$80.49 & \$402.49 \end{bmatrix}$

53. $\begin{bmatrix} 34 & 10 \\ -21 & -5 \end{bmatrix}$ **55.** impossible

57.

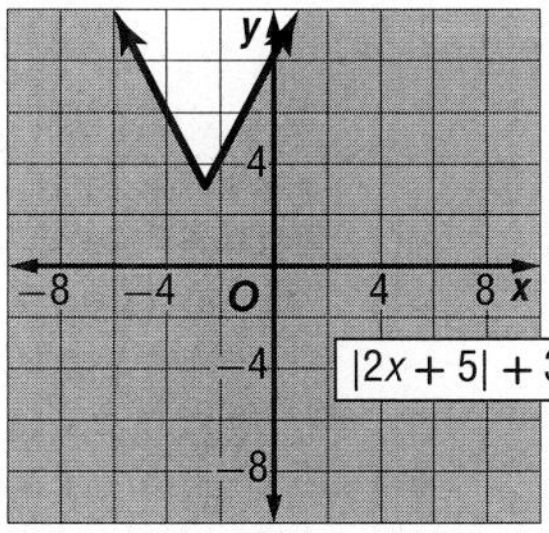

59.

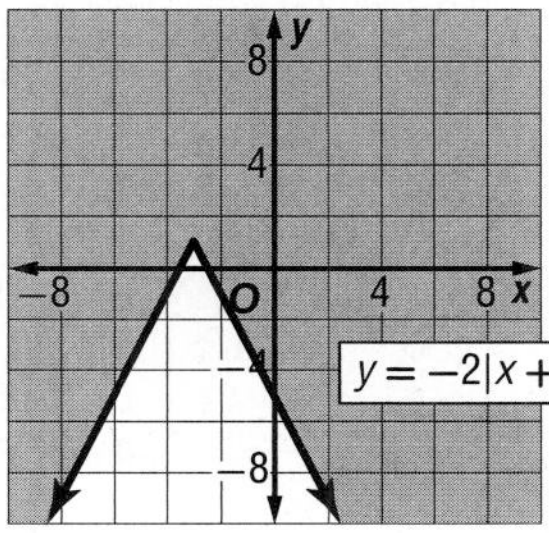

61. independent
63. (5, −1)

Pages 225–228 Lesson 4-5

1. 26 **3.** −128

5 Rewrite the first two columns to the right of the determinant. Then find the products of the elements of the diagonals.

$$\begin{vmatrix} 3 & -2 & 2 \\ -4 & 2 & -5 \\ -3 & 1 & 4 \end{vmatrix}\begin{matrix} 3 & -2 \\ -4 & 2 \\ -3 & 1 \end{matrix} \qquad \begin{matrix} 3(2)(4) = 24 \\ -2(-5)(-3) = -30 \\ 2(-4)(1) = -8 \end{matrix}$$

$$\begin{vmatrix} 3 & -2 & 2 \\ -4 & 2 & -5 \\ -3 & 1 & 4 \end{vmatrix}\begin{matrix} 3 & -2 \\ -4 & 2 \\ -3 & 1 \end{matrix} \qquad \begin{matrix} -3(2)(2) = -12 \\ 1(-5)(3) = -15 \\ 4(-4)(-2) = 32 \end{matrix}$$

The sum of the first group is 24 + (−30) + (−8) or −14. The sum of the second group is −12 + (−15) + 32 or 5. The first sum minus the second sum is −14 − 5 or −19.

7. −284 **9.** 72 **11.** 182 **13.** (6, −3) **15.** (4, −1)
17a. 17 units^2 **17b.** 520,625 mi^2 **19.** $\left(\frac{66}{7}, -\frac{116}{7}, -\frac{41}{7}\right)$
21. (−4, −2, 8) **23.** (−1, −3, 7) **25.** (4, 0, 8) **27.** 3
29. −135 **31.** −459 **33.** 0 **35.** 728 **37.** −952
39. (8, −5)

41
$$x = \frac{\begin{vmatrix} m & b \\ n & g \end{vmatrix}}{|C|} = \frac{\begin{vmatrix} -39 & -5 \\ 54 & 8 \end{vmatrix}}{\begin{vmatrix} -4 & -5 \\ 5 & 8 \end{vmatrix}}$$

$$x = \frac{-39(8) - 54(-5)}{-4(8) - 5(-5)} = \frac{-312 + 270}{-32 + 25} = \frac{-42}{-7} \text{ or } 6$$

$$y = \frac{\begin{vmatrix} a & m \\ f & n \end{vmatrix}}{|C|} = \frac{\begin{vmatrix} -4 & -39 \\ 5 & 54 \end{vmatrix}}{\begin{vmatrix} -4 & -5 \\ 5 & 8 \end{vmatrix}}$$

$$y = \frac{-4(54) - 5(-39)}{-4(8) - 5(-5)} = \frac{-216 + 195}{-32 + 25} = \frac{-21}{-7} \text{ or } 3$$

The solution of the system is (6, 3).

43. (−3, −7) **45.** (4, −2, 5) **47.** 6 **49.** 2 m^2
51. (4, 8, −5) **53.** $\left(-\frac{6187}{701}, -\frac{2904}{701}, -\frac{4212}{701}\right)$

55 **a.** Let x = the number of medium drinks. Then $2x$ = the number of small drinks. Let y = the number of large drinks.
number of drinks: $x + 2x + y = 1385 \rightarrow 3x + y = 1385$
total sales: $1.75x + 1.15(2x) + 2.25y = 2238.75$
$\rightarrow 4.05x + 2.25y = 2238.75$

$$x = \frac{\begin{vmatrix} m & b \\ n & g \end{vmatrix}}{|C|} = \frac{\begin{vmatrix} 1385 & 1 \\ 2238.75 & 2.25 \end{vmatrix}}{\begin{vmatrix} 3 & 1 \\ 4.05 & 2.25 \end{vmatrix}}$$

$$x = \frac{1385(2.25) - 2238.75(1)}{3(2.25) - 4.05(1)} = \frac{3116.25 - 2238.75}{6.75 - 4.05} = \frac{877.5}{2.7} \text{ or } 325$$

$$y = \frac{\begin{vmatrix} a & m \\ f & n \end{vmatrix}}{|C|} = \frac{\begin{vmatrix} 3 & 1385 \\ 4.05 & 2238.75 \end{vmatrix}}{\begin{vmatrix} 3 & 1 \\ 4.05 & 2.25 \end{vmatrix}}$$

$$y = \frac{3(2238.75) - 4.05(1385)}{3(2.25) - 4.05(1)} = \frac{6716.25 - 5609.25}{6.75 - 4.05} = \frac{1107}{2.7} \text{ or } 410$$

He sold 325 medium drinks, 2(325) or 650 small drinks, and 410 large drinks.

b. total sales = sales of small + sales of medium + sales of large
= 1.25(650 − 140) + 1.75(325 + 125) + 2.25(410 + 35)
= 637.50 + 787.50 + 1001.25
= $2426.25

c. It seems like it was a good move for the vendor. Although he sold 140 fewer small drinks, he sold 125 more medium drinks and 35 more large drinks. On the whole, he made $2,426.25 − $2,238.75 or $187.50 more this week than in the previous week.

57. Sample answer: There is no unique solution of the system. There are either infinite or no solutions.
59. 0 **61.** Sample answer: Given a 2 × 2 system of linear equations, if the determinant of the matrix of coefficients is 0, then the system does not have a unique solution. The system may have no solution and the graphical representation shows two parallel lines. The system may have infinitely many solutions in which the graphical representation will be the same line. **63.** H **65.** B
67. $E'(5, -2)$, $(-3, 5)$, $G'(-6, -1)$
69. no

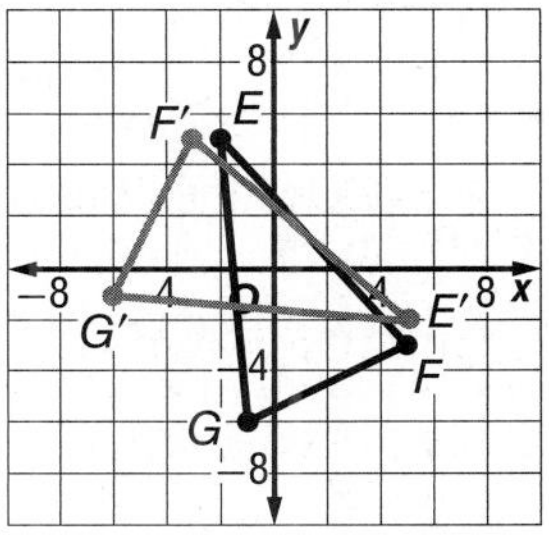

71. 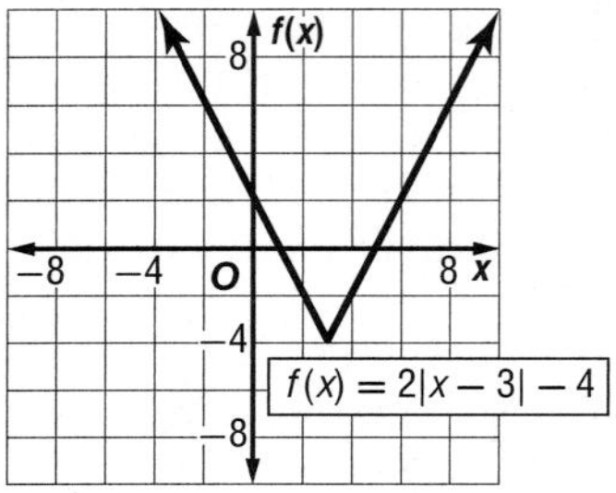

73. 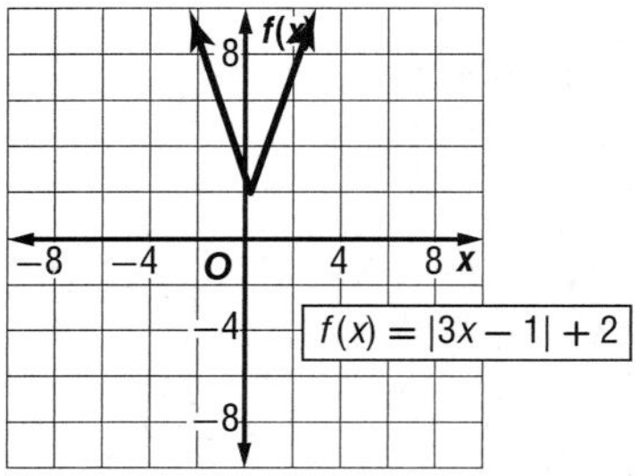

75. $\left(\frac{79}{25}, -\frac{56}{25}\right)$

Pages 233–235 Lesson 4-6

1. no **3.** yes **5.** $\begin{bmatrix} 0 & -1 \\ -\frac{1}{3} & -2 \end{bmatrix}$ **7.** $\begin{bmatrix} -\frac{1}{3} & 0 \\ \frac{5}{6} & \frac{1}{2} \end{bmatrix}$

9. $(-2, 5)$ **11.** $(1, -2)$ **13.** no **15.** no **17.** $\begin{bmatrix} \frac{1}{3} & 0 \\ 0 & \frac{1}{2} \end{bmatrix}$

19. $\begin{bmatrix} \frac{1}{3} & 0 \\ -\frac{5}{3} & 1 \end{bmatrix}$

21 $\begin{vmatrix} -5 & -4 \\ 4 & 2 \end{vmatrix} = -5(2) - 4(-4)$ or 6 Find the determinant.

Since the determinant $\neq 0$, the inverse exists.

$A^{-1} = \frac{1}{ad - bc}\begin{bmatrix} d & -b \\ -c & a \end{bmatrix}$ Definition of inverse

$= \frac{1}{-5(2) - (-4)(4)}\begin{bmatrix} 2 & 4 \\ -4 & -5 \end{bmatrix}$ $a = -5, b = -4, c = 4, d = 2$

$= \frac{1}{6}\begin{bmatrix} 2 & 4 \\ -4 & -5 \end{bmatrix}$ or $\begin{bmatrix} \frac{1}{3} & \frac{2}{3} \\ -\frac{2}{3} & -\frac{5}{6} \end{bmatrix}$ Simplify.

23. $\begin{bmatrix} \frac{9}{74} & \frac{5}{74} \\ -\frac{2}{37} & \frac{3}{37} \end{bmatrix}$ **25.** $\begin{bmatrix} \frac{7}{22} & \frac{4}{11} \\ \frac{4}{11} & \frac{3}{11} \end{bmatrix}$ **27.** no solution

29. $(-1, 5)$ **31.** no solution **33.** $(-5, 0)$ **35.** $\left(\frac{3}{4}, 3\right)$

37 a. Let x = the number of people who own CD players and let y = the number of people who own digital audio players. The following expressions represent the change in player ownership.

Keeping or switching to CD players: $0.35x + 0.12y$

Keeping or switching to digital audio players: $0.65x + 0.88y$

		From	
		CD	DAP
To	CD	0.35	0.12
	DAP	0.65	0.88

b. Let x = the number of people who will own CD players next year and let y = the number of people who will own digital audio players next year. Write a matrix equation.

$$\begin{bmatrix} 0.35 & 0.12 \\ 0.65 & 0.88 \end{bmatrix} \cdot \begin{bmatrix} 7748 \\ 17{,}252 \end{bmatrix} = \begin{bmatrix} x \\ y \end{bmatrix}$$

$$\begin{bmatrix} 4782 \\ 20{,}218 \end{bmatrix} = \begin{bmatrix} x \\ y \end{bmatrix}$$

So, about 20,218 people will own digital audio players next year.

c. Let x = the number of people who owned CD players last year and let y = the number of people who owned digital audio players last year. Write a matrix equation.

$$\begin{bmatrix} 0.35 & 0.12 \\ 0.65 & 0.88 \end{bmatrix} \cdot \begin{bmatrix} x \\ y \end{bmatrix} = \begin{bmatrix} 7748 \\ 17{,}252 \end{bmatrix}$$

Find the inverse of the coefficient matrix.

A^{-1}

$= \frac{1}{0.35(0.88) - 0.12(0.65)}\begin{bmatrix} 0.88 & -0.12 \\ -0.65 & 0.35 \end{bmatrix}$ $a = 0.35, b = 0.12, c = 0.65, d = 0.88$

$= \frac{1}{0.23}\begin{bmatrix} 0.88 & -0.12 \\ -0.65 & 0.35 \end{bmatrix}$ Simplify.

Multiply each side of the matrix equation by the inverse matrix.

$$\frac{1}{0.23}\begin{bmatrix} 0.88 & -0.12 \\ -0.65 & 0.35 \end{bmatrix} \cdot \begin{bmatrix} 0.35 & 0.12 \\ 0.65 & 0.88 \end{bmatrix} \cdot \begin{bmatrix} x \\ y \end{bmatrix} = \frac{1}{0.23}\begin{bmatrix} 0.88 & -0.12 \\ -0.65 & 0.35 \end{bmatrix} \cdot \begin{bmatrix} 7748 \\ 17{,}252 \end{bmatrix}$$

$$\begin{bmatrix} 1 & 0 \\ 0 & 1 \end{bmatrix} \cdot \begin{bmatrix} x \\ y \end{bmatrix} = \frac{1}{0.23}\begin{bmatrix} 4748 \\ 1002 \end{bmatrix}$$

$$\begin{bmatrix} x \\ y \end{bmatrix} = \begin{bmatrix} 20{,}643 \\ 4357 \end{bmatrix}$$

So, about 4357 people owned digital audio players last year.

39. The system would have to consist of two equations that are the same or one equation that is a multiple of the other.

41. Sample answer: $\begin{bmatrix} 2 & 3 \\ 4 & 6 \end{bmatrix} \cdot \begin{bmatrix} x \\ y \end{bmatrix} = \begin{bmatrix} 9 \\ 10 \end{bmatrix}$; any matrix that has a determinant equal to 0, such as $\begin{bmatrix} 1 & 0 \\ 0 & 1 \end{bmatrix}$ **43.** C **45.** $\left(\frac{1}{2}, \frac{1}{4}\right)$ **47.** −54 **49.** 551

51. 179 gal of skim and 21 gal of whole milk

53. absolute value

Pages 237–240 Chapter 4 Study Guide and Review

1. matrix **3.** translation **5.** dilation **7.** identity matrix **9.** determinant **11a.** $\begin{bmatrix} 64 & 108 & 31 \\ 42 & 9 & 68 \end{bmatrix}$

11b. 2×3 **11c.** 68 **11d.** 64 **11e.** The sum of column 1 is 106. This is the total number of customers for store A. The sum of column 2 is 117. This is the total number of customers for store B. **11f.** No, the stores are competing. **13.** $\begin{bmatrix} -3 & 27 \\ 9 & 12 \end{bmatrix}$ **15.** $[\ 62\ ]$ **17.** undefined

19. $A'(3, 7), B'(1, 1), C'(-3, 4)$ **21.** $A'(5, -4), B'(3, 2), C'(-1, -1)$ **23.** $A'(2, 12), B'(16, 12), C'(16, 4), D'(2, 4)$

25. −44 **27.** $(2, -3, 6)$

29. $\frac{1}{2}\begin{bmatrix} 2 & -4 \\ -3 & 7 \end{bmatrix}$ **31.** does not exist **33.** $(2, 1)$

Chapter 5 Quadratic Functions and Relations

Page 247 Chapter 5 Get Ready

1. 6 **3.** 4 **5.** 3 **7a.** $f(x) = 9x$ **7b.** 157,680 mi
9. $(x + 8)(x + 5)$ **11.** $(2x - 1)(x + 4)$ **13.** prime
15. $(x + 8)$ feet

Pages 254–257 Lesson 5-1

1a. y-int $= 0$; axis of symmetry: $x = 0$; x-coordinate $= 0$

1b.

x	f(x)
−2	12
−1	3
0	0
1	3
2	12

1c.

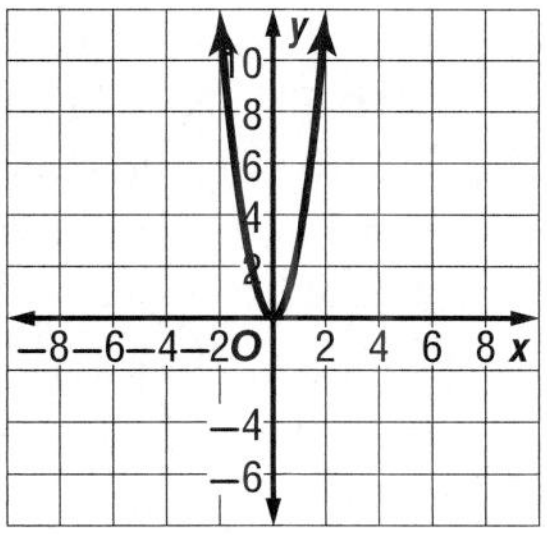

3a. y-int $= 0$; axis of symmetry: $x = 2$; x-coordinate $= 2$

3b.

x	f(x)
0	0
1	−3
2	−4
3	−3
4	0

3c.

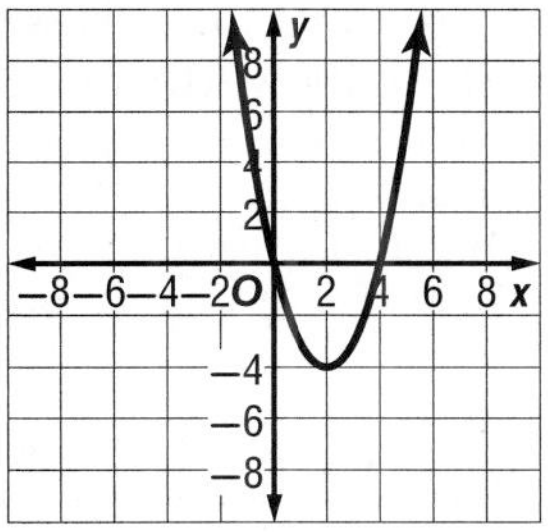

5a. y-int $= -3$; axis of symmetry: $x = 0.75$; x-coordinate $= 0.75$

5b.

x	f(x)
−1	7
0	−3
0.75	−5.25
1.5	−3
2.5	7

5c.

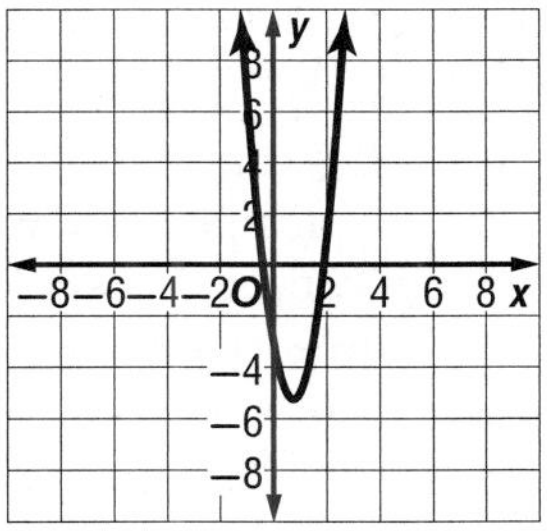

7. max $= 8$; D $= \{$all real numbers$\}$, R $= \{f(x) \mid f(x) \leq 8\}$
9. min $= -\frac{1}{3}$; D $= \{$all real numbers$\}$,
R $= \left\{f(x) \,\middle|\, f(x) \geq -\frac{1}{3}\right\}$ **11.** \$2.88 **13a.** y-int $= 0$;
axis of symmetry: $x = 0$; x-coordinate $= 0$

13b.

x	f(x)
−2	−8
−1	−2
0	0
1	−2
2	−8

13c.

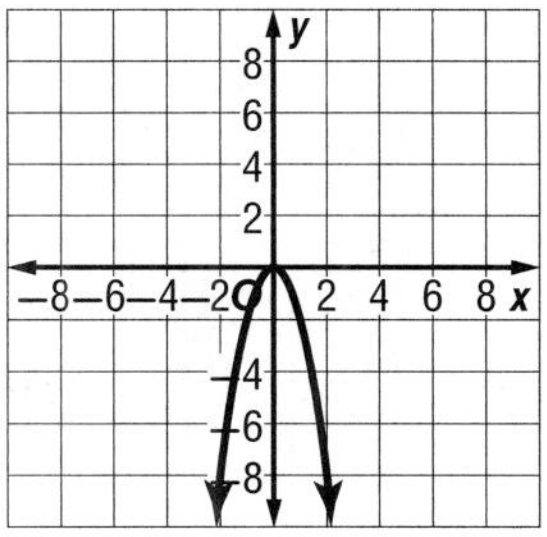

15a. y-int $= 3$; axis of symmetry: $x = 0$; x-coordinate $= 0$

15b.

x	f(x)
−2	7
−1	4
0	3
1	4
2	7

15c.

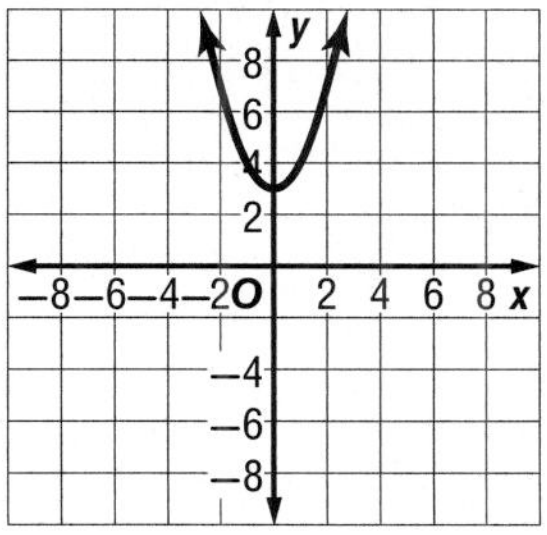

17a. y-int $= 5$; axis of symmetry: $x = 0$; x-coordinate $= 0$

17b.

x	f(x)
−2	−7
−1	2
0	5
1	2
2	−7

17c.

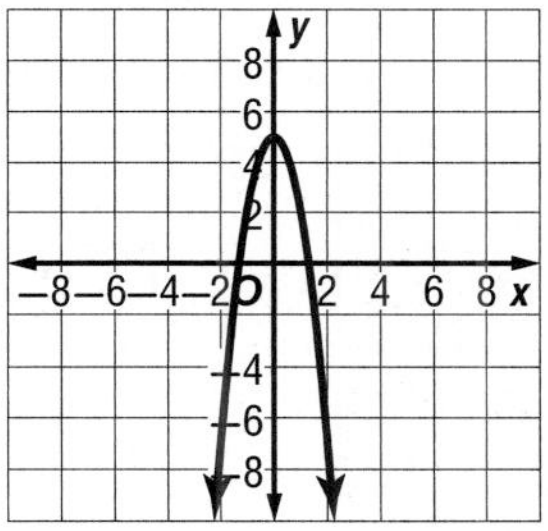

19 **a.** $f(x) = ax^2 + bx + c$
$\downarrow \quad \downarrow \quad \downarrow$
$f(x) = 1x^2 - 3x - 10$ $a = 1, b = -3, c = -10$
The y-intercept is $c = -10$.

$x = -\frac{b}{2a}$ Equation of the axis of symmetry

$= -\frac{(-3)}{2(1)}$ $a = 1$ and $b = -3$

$= \frac{3}{2}$ or 1.5 Simplify.

The equation of the axis of symmetry is $x = 1.5$. So, the x-coordinate of the vertex is 1.5.

b. Select five points, with the vertex in the middle and two points on either side of the vertex, including the y-intercept and its reflection.

x	f(x)	
0	−10	←reflection of y-intercept
1	−12	
1.5	−12.25	←vertex
2	−12	
3	−10	←y-intercept

c. Graph the five points from the table, connecting them with a smooth curve.

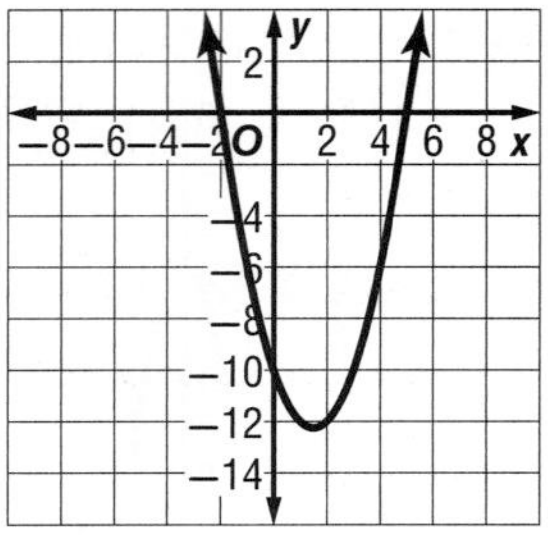

21a. y-int = 9; axis of symmetry: $x = 0.75$; x-coordinate = 0.75

21b.

x	$f(x)$
−1	4
0	9
0.75	10.125
1.5	9
2.5	4

21c.

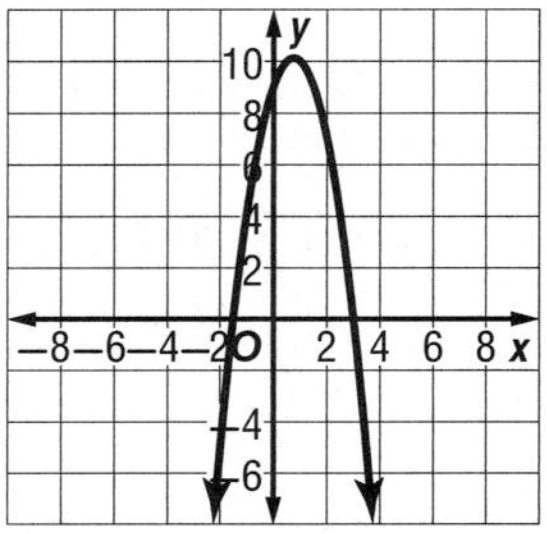

23. max = −12; D = {all real numbers}, R = $\{f(x) \mid f(x) \leq -12\}$ **25.** max = 13.25; D = {all real numbers}, R = $\{f(x) \mid f(x) \leq 13.25\}$ **27.** max = 7; D = {all real numbers}, R = $\{f(x) \mid f(x) \leq 7\}$
29. min = −9; D = {all real numbers}, R = $\{f(x) \mid f(x) \geq -9\}$ **31.** min = −74; D = {all real numbers}, R = $\{f(x) \mid f(x) \geq -74\}$ **33a.** y-int = −9; axis of symmetry: $x = 1.5$; x-coordinate of vertex = 1.5

33b.

x	$f(x)$
0	−9
1	−13
1.5	−13.5
2	−13
3	−9

33c.

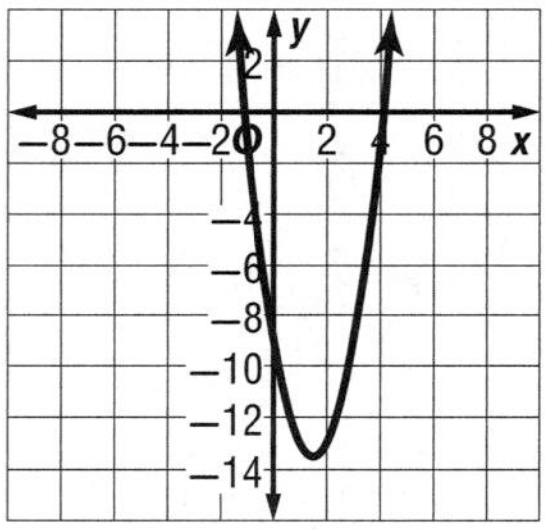

35a. y-int = 0; axis of symmetry: $x = \frac{5}{8}$; x-coordinate of vertex = $\frac{5}{8}$

35b.

x	$f(x)$
$-\frac{3}{4}$	−6
$\frac{1}{4}$	1
$\frac{5}{8}$	1.5625
1	1
2	−6

35c.

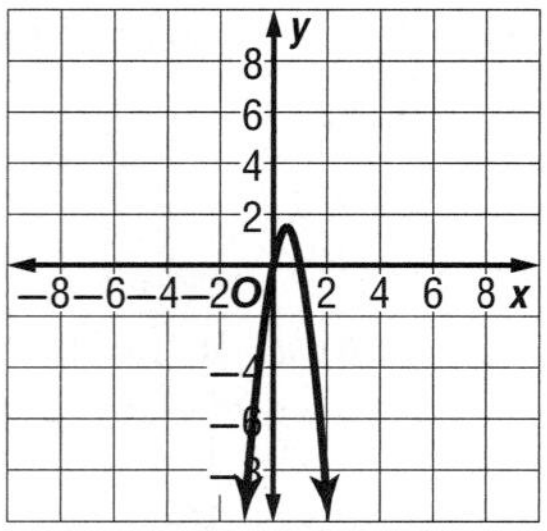

37a. y-int = 4; axis of symmetry: $x = -6$; x-coordinate of vertex = −6

37b.

x	$f(x)$
−10	−1
−8	−4
−6	−5
−4	−4
−2	−1

33c.

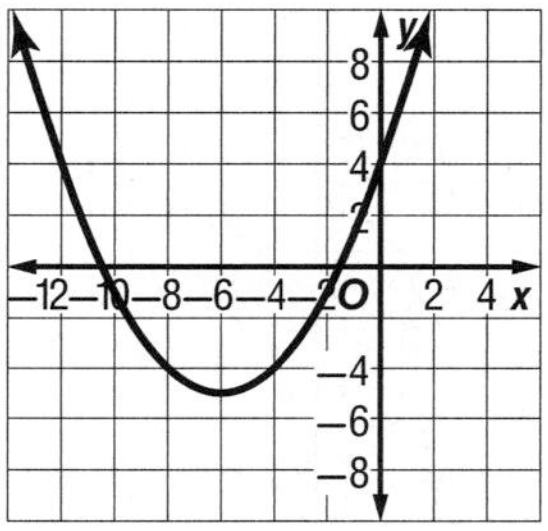

39a. y-int = −2.5; axis of symmetry: $x = -\frac{4}{3}$; x-coordinate of vertex = $-\frac{4}{3}$

39b.

x	$f(x)$
$-\frac{11}{3}$	3
$-\frac{8}{3}$	−2.5
$-\frac{4}{3}$	$-5\frac{1}{6}$
0	−2.5
1	3

39c.

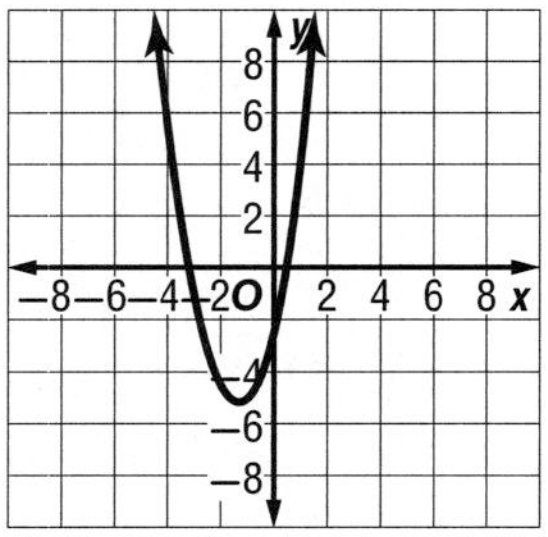

41a. $I(x) = -x^2 + 6x + 475$ **41b.** D = $\{x \mid x \geq 0\}$, R = $\{y \mid y \geq 0\}$ **41c.** \$11; Because the function has a maximum at $x = 3$, it is in the domain. Therefore, three \$0.50 increases is reasonable. **41d.** \$484
43. max = 23 **45.** max = −0.10 **47.** max = −4.11

49 $a = -5$, so the graph opens down and has a maximum value. The maximum value is the y-coordinate of the vertex.

$x = -\frac{b}{2a}$ Equation of the axis of symmetry

$= -\frac{4}{2(-5)}$ or 0.4 $a = -5$ and $b = 4$

The x-coordinate of the vertex is 0.4. Find the y-coordinate of the vertex by evaluating the function for $x = 0.4$.

$f(x) = -5x^2 + 4x - 8$ Original function

$= -5(0.4)^2 + 4(0.4) - 8$ $x = 0.4$

$= -7.2$ The maximum value of the function is −7.2.

The domain is all real numbers. The range is all real numbers less than or equal to the maximum value, or $\{f(x) \mid f(x) \leq -7.2\}$.

51. min = −9.375; D = {all real numbers}, R = $\{f(x) \mid f(x) \geq -9.375\}$ **53.** min = −23.5; D = {all real numbers}, R = $\{f(x) \mid f(x) \geq -23.5\}$
55. $f(x) = x^2 - 4x - 5$ **57.** $f(x) = x^2 - 6x + 8$

59 **a.** Let x = the number of 5-cent increases. Then $65 + 5x$ = price per can in cents and $600 - 100x$ = number of cans. Let $f(x)$ = income as a function of x.

Income = number of cans times the price per can

$f(x) \quad = \quad (600 - 100x) \quad \cdot \quad (65 + 5x)$

Solve for x in the equation.

$f(x) = (600 - 100x) \cdot (65 + 5x)$

$= 600(65) + 600(5x) + (-100x)(65) + (-100x)(5x)$ Distribute.

$= 39{,}000 + 3000x - 6500x - 500x^2$ Multiply.

$= 39{,}000 - 3500x - 500x^2$ Simplify.

So, the equation is $f(x) = 39{,}000 - 3500x - 500x^2$ or $f(x) = -500x^2 - 3500x + 39{,}000$.

$x = -\frac{b}{2a}$ Equation of the axis of symmetry

$x = -\frac{(-3500)}{2(-500)}$ or −3.5 $a = -500$ and $b = -3500$

$f(x) = -500x^2 - 3500x + 39{,}000$ Original function

$= -500(-3.5)^2 - 3500(-3.5) + 39{,}000$ $x = -3.5$

$= 45{,}125$ Simplify.

The coordinates of the vertex are (−3.5, 45,125). The y-intercept is 39,000.

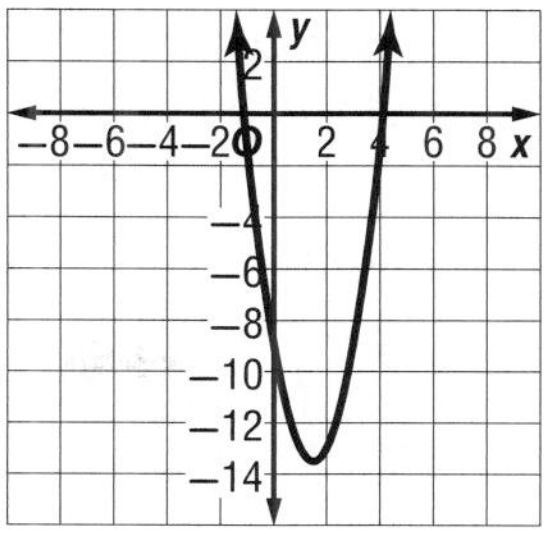

b. Let x = the number of 5-cent decreases. Then $65 - 5x$ = price per can and $600 + 100x$ = number of cans. Let $f(x)$ = income as a function of x.

Income = number of cans times the price per can

$f(x) \quad = \quad (600 + 100x) \quad \cdot \quad (65 - 5x)$

Solve for x in the equation.

$f(x) = (600 + 100x) \cdot (65 - 5x)$

$= 600(65) + 600(-5x) + (100x)(65) + (100x)(-5x)$ Distribute.

$= 39{,}000 - 3000x + 6500x - 500x^2$ Multiply.

$= 39{,}000 + 3500x - 500x^2$ Simplify.

So, the equation is $f(x) = 39{,}000 + 3500x - 500x^2$ or $f(x) = -500x^2 + 3500x + 39{,}000$.

$x = -\frac{b}{2a}$ Equation of the axis of symmetry

$x = -\frac{(3500)}{2(-500)}$ or 3.5 $a = -500$ and $b = 3500$

$f(x) = -500x^2 + 3500x + 39{,}000$ Original function

$= -500(-3.5)^2 + 3500(3.5) + 39{,}000$ $x = -3.5$

$= 45{,}125$ Simplify.

The coordinates of the vertex are (3.5, 45,125). So, Omar should have 3 price decreases and charge 65 − 3(5) or 50 cents. Or, he should have 4 price decreases and charge 65 − 4(5) or 45 cents.

c. $f(x) = -500x^2 + 3500x + 39{,}000$ Income function

$= -500(3)^2 + 3500(3) + 39{,}000$ Replace x with 3.

$= 45{,}000$ Simplify.

So, his income would be 45,000 cents or \$450 per week.

61. Sample answer: Madison is correct; when Trent found the x-coordinate of the vertex, he multiplied two negatives and mistakenly kept a negative.
63a. $a = 22$; $b = 26$; $c = -6$; $d = 2$ **63b.** 0
63c. maximum **65.** Sample answer: A function is quadratic if it has no other terms than a quadratic term, linear term, and constant term. The function has a maximum if the coefficient of the quadratic term is negative and has a minimum if the coefficient of the quadratic term is positive. **67.** J **69.** C

71. $\begin{bmatrix} -\frac{1}{4} & -\frac{1}{24} \\ 0 & \frac{1}{6} \end{bmatrix}$ **73.** 45 **75.** 0

77. No; it cannot be written as $y = mx + b$.
79. Yes; it is written in $y = mx + b$ form, $m = 0$.
81. −13

Pages 263–266 Lesson 5-2

1. no real solution
3. −4

5 Graph the related function $f(x) = x^2 - 3x - 18$. The equation of the axis of symmetry is $x = -\frac{(-3)}{2(1)}$ or 1.5. Make a table using x-values around 1.5. Then graph each point.

x	−3	0	1.5	3	6
$f(x)$	0	−18	−20.25	−18	0

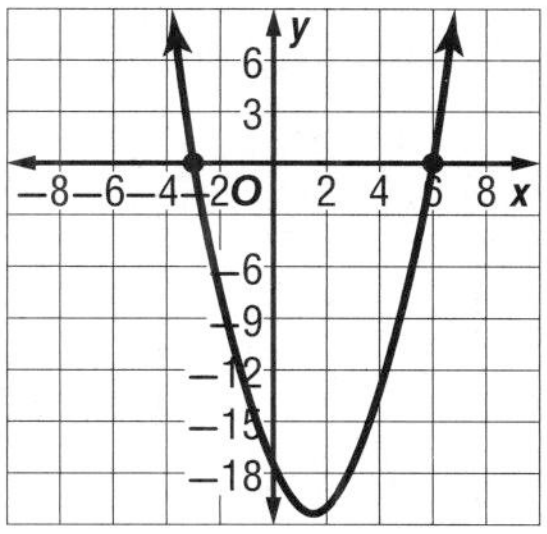

The zeros of the function are −3 and 6. So, the solutions of the equation are −3 and 6.

7.

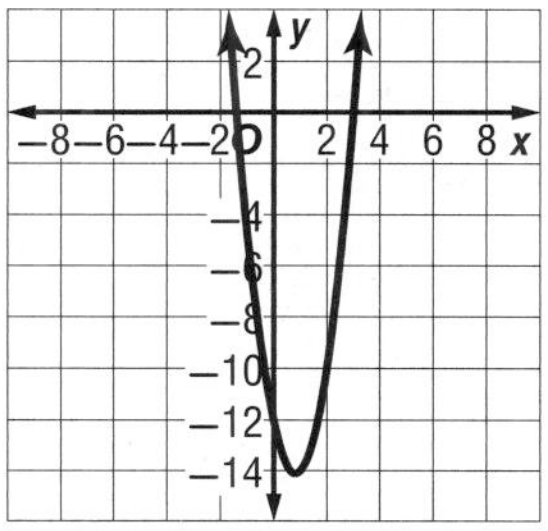

between −2 and −1, 3

9.

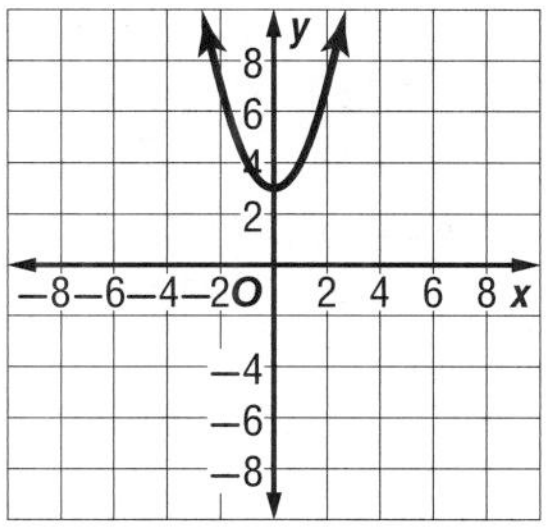

no real solution

11.

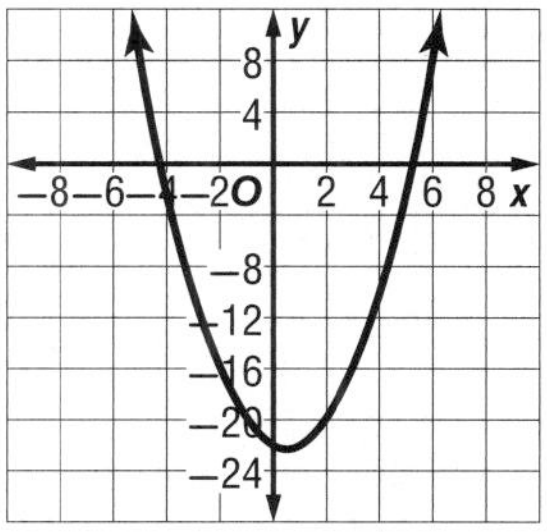

between −5 and −4, between 5 and 6

13. 5 seconds **15.** no real solution **17.** −2
19. −3, 4

21.

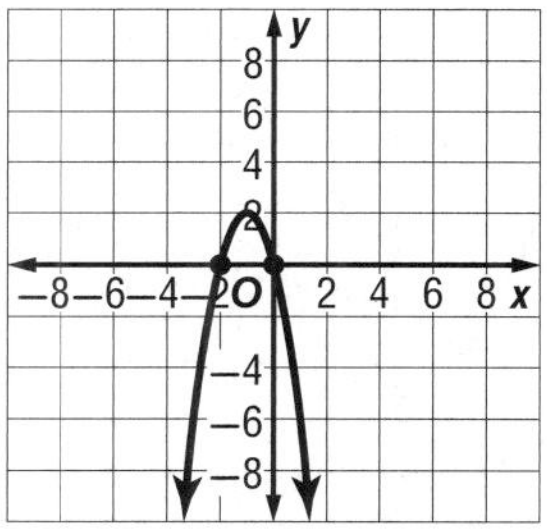

−2, 0

Selected Answers and Solutions

23. 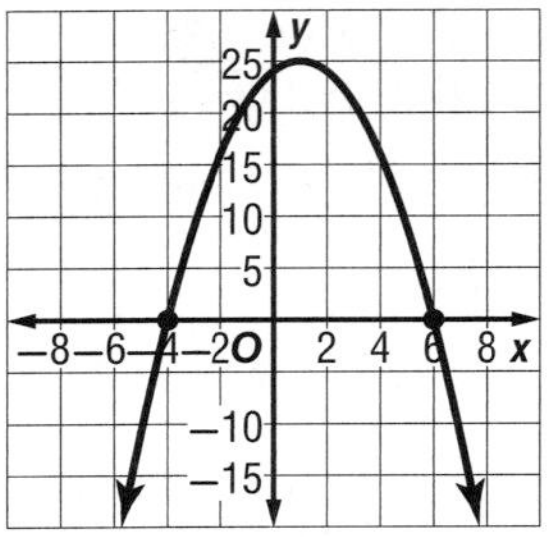

−4, 6

25. 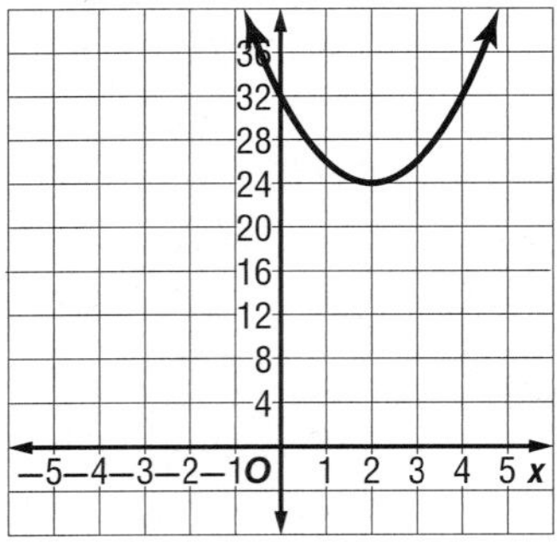

no real solution

27. 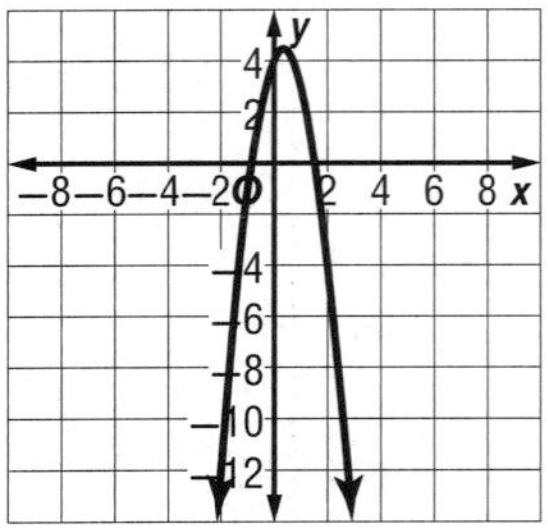

between −1 and 0, between 1 and 2

29. 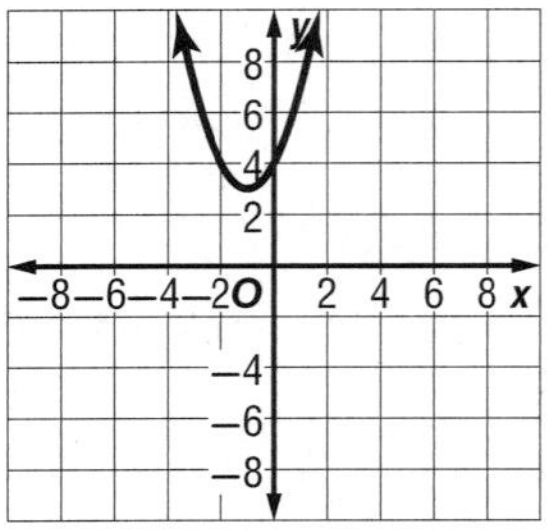

no real solution

31. between 0 and 1; between 2 and 3

33 Let x = one of the numbers. Then $-15 - x$ = the other number.

$x(-x - 15) = -54$ The product is −54.

$-x^2 - 15x = -54$ Distributive Property

$-x^2 - 15x + 54 = 0$ Add 54 to each side.

Graph the related function $f(x) = -x^2 - 15x + 54$. The equation of the axis of symmetry is $x = -\frac{(-15)}{2(-1)}$ or −7.5. Make a table using x-values around −7.5. Then graph each point.

x	−20	−15	−10	−7.5	−5	0	5
$f(x)$	−46	54	104	110.25	104	54	−46

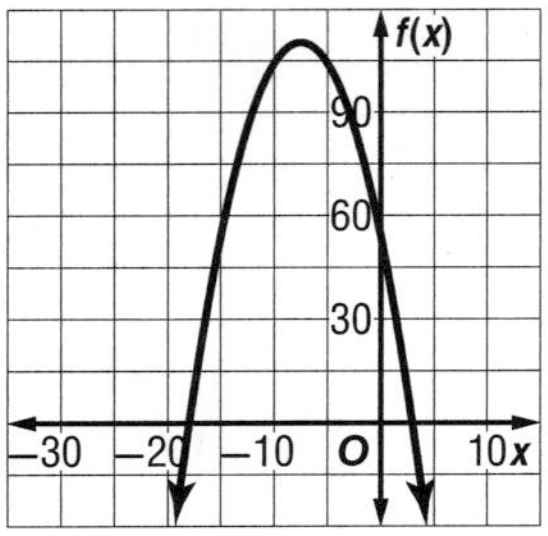

The zeros of the function are −18 and 3. So, the numbers are −18 and 3.

35. about −5.0 and 17.0 **37.** 11 and −19

39. about 3.4375 seconds

41. 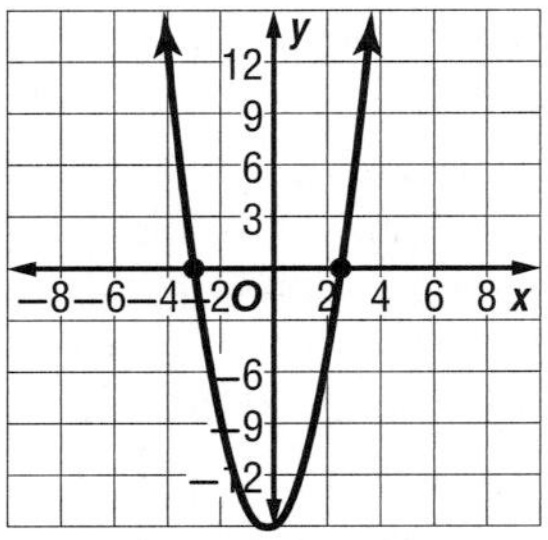

−3, between 2 and 3

43. 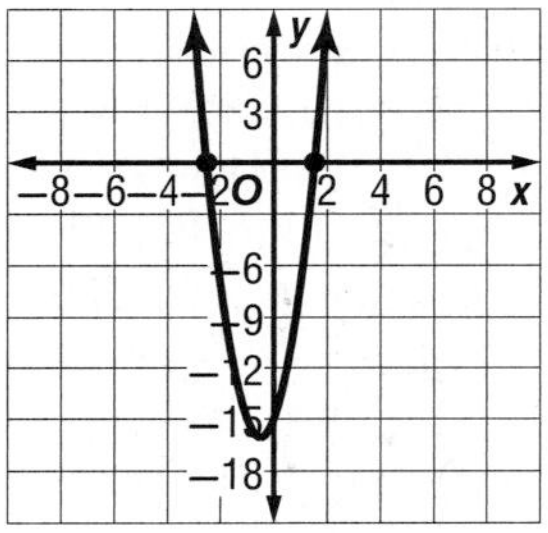

between −3 and −2, between 1 and 2

45. 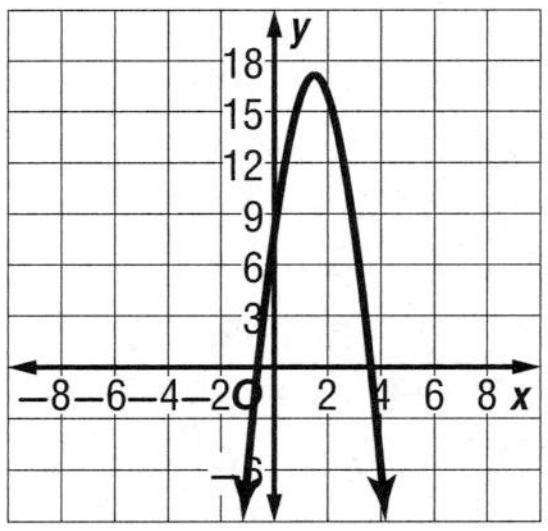

between −1 and 0, between 4 and 5

47. 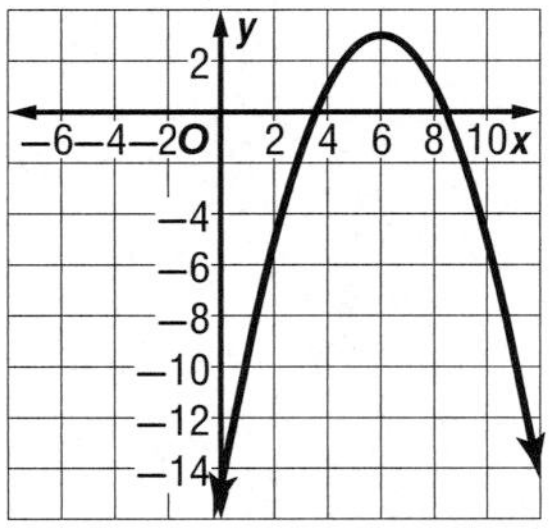

between 3 and 4, between 8 and 9

49 Find t when $h_0 = 60$ and $h(t) = 0$.

$h(t) = -16t^2 + h_0$ Original equation

$0 = -16t^2 + 60$ $h(t) = 0$ and $h_0 = 60$

Graph the related function $f(t) = -16t^2 + 60$ on a graphing calculator.

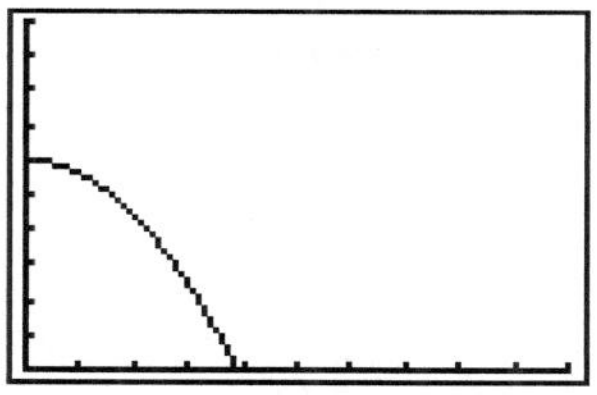

Use the **Zero** feature in the **CALC** menu to find the positive zero of the function, since time cannot be negative. $x \approx 1.94$, so it would take the balloon about 1.94 seconds to hit the ground. Tony's brother should be 4.4 ft/s • 1.94 s or about 8.5 feet from the target when Tony lets go of the balloon.

51. 25 seconds **53.** $k = 8$ **55.** $f(x) = -5x^2 + 30x + 80$ **57.** $\frac{1}{20}$ **59.** H **61.** maximum, -12; D = {all real numbers}, R = $\{f(x) \mid f(x) \leq -12\}$

63. maximum, 15; D = {all real numbers}, R = $\{f(x) \mid f(x) \leq 15\}$ **65.** no **67a.** The object is reflected over the x-axis, and then translated 6 units to the right. **67b.** Multiply the coordinates by $\begin{bmatrix} 1 & 0 \\ 0 & -1 \end{bmatrix}$, and then add the result to $\begin{bmatrix} 6 \\ 0 \end{bmatrix}$.

67c. Sample answer: No; since the translation does not change the y-coordinate, it does not matter whether or not you do the translation or reflection first. However, if the translation did change the y-coordinate, the order would be important.

67d. (17, −2), (23, 2) **69.** (−3, 4) **71.** $x \leq -\frac{8}{3}$

73. $x \leq \frac{19}{9}$ **75.** 1

Pages 272–275 Lesson 5-3

1. $x^2 + 3x - 40 = 0$ **3.** $6x^2 - 11x - 10 = 0$ **5.** $(6x - 1)(3x + 4)$ **7.** $(x - 7)(x + 3)$ **9.** $(4x - 3)(4x - 1)$ **11.** $6xy(2x - 3)$ **13.** 0, 9 **15.** 6 **17.** $x^2 - 14x + 49 = 0$ **19.** $5x^2 - 31x + 6 = 0$ **21.** $17c(3c^2 - 2)$ **23.** $3(x + 2)(x - 2)$

25 $48cg + 36cf - 4dg - 3df$ Original expression

$= (48cg + 36cf) + (-4dg - 3df)$ Group terms with common factors.

$= 12c(4g + 3f) + (-d)(4g + 3f)$ Factor the GCF from each group.

$= (12c - d)(4g + 3f)$ Distributive Property

27. $(x - 11)(x + 2)$ **29.** $(5x - 1)(3x + 2)$ **31.** $3(2x - 1)(3x + 4)$ **33.** $(3x + 5)(3x - 5)$ **35.** $3(5x + 2)(x - 6)$ **37.** $12x(y + 3)(y - 3)$ **39.** 8, −3 **41.** 2, −2 **43.** 5, $\frac{3}{4}$ **45.** 24 and 26 or −24 and −26 **47.** $x = 20$; 24 in. by 18 in. **49.** $-\frac{1}{2}, \frac{5}{6}$ **51.** $-\frac{3}{2}$ **53.** 6, −6

55 To find the number of movie screens that produces a profit, first find the number of movie screens in which the profit is zero. Solve $-x^2 + 48x - 512 = 0$.

$ac = -1(-512)$ or 512

$m = 16; p = 32$ $mp = 512$ and $ac = 512$; $m + p = 48$ and $b = 48$

$-x^2 + 16x + 32x - 512 = 0$ Write the pattern.

$(-x^2 + 16x) + (32x - 512) = 0$ Group terms with common factors.

$-x(x - 16) + 32(x - 16) = 0$ Group terms with common factors.

$(-x + 32)(x - 16) = 0$ Distributive Property

$-x + 32 = 0$ or $x - 16 = 0$ Zero Product Property

$x = 32$ $x = 16$ Solve each equation.

The solutions are 16 and 32. When there are 16 or 32 movie screens, the profit is zero. Since a is positive, the graph of the function opens down and has a maximum value. So, $P(x)$ is nonnegative for $16 \leq x \leq 32$. When there are 16 to 32 movie screens, the company will not lose money.

57. $25x^2 - 100x + 51 = 0$ **59.** $-3, \frac{1}{2}$ **61.** $1, -\frac{5}{4}$ **63.** $-\frac{3}{2}, \frac{5}{6}$ **65.** $x^2 - 6^2$; $(x + 6)(x - 6)$ **67.** 20 in. by 15 in. **69.** 13 cm **71.** $2(3 - 4y)(3a + 8b)$

73 $6a^2b^2 - 12ab^2 - 18b^3$ Original expression

$= 6b^2(a^2 - 2a - 3b)$ Factor the GCF, $6b^2$.

75. $2(2x - 3y)(8a + 3b)$ **77.** $(x + y)(x - y)(5a + 2b)$ **79.** Sample answer: Neither is correct; Gwen didn't have like terms in the parentheses in the third line; Morgan made a sign error in the fourth line.

81. $5x^2(2x - 3y)(4x^2 + 6xy + 9y^2)$

83. Sample answer:

$(x - p)(x - q) = 0$ Original equation

$x^2 - px - qx + pq = 0$ Multiply.

$x^2 - (p + q)x + pq = 0$ Simplify.

$x = -\frac{b}{2a}$ Formula for axis of symmetry

$x = -\frac{-(p + q)}{2(1)}$ $a = 1$ and $b = -(p + q)$

$x = \frac{p + q}{2}$ Simplify.

x is midway between p and q. Definition of midpoint

85. Sample answer: Always; in order to factor using perfect square trinomials, the coefficient of the linear term, bx, must be a multiple of 2, or even. **87.** 192 square units **89.** H **91.** −2, 4 **93.** −2

95.

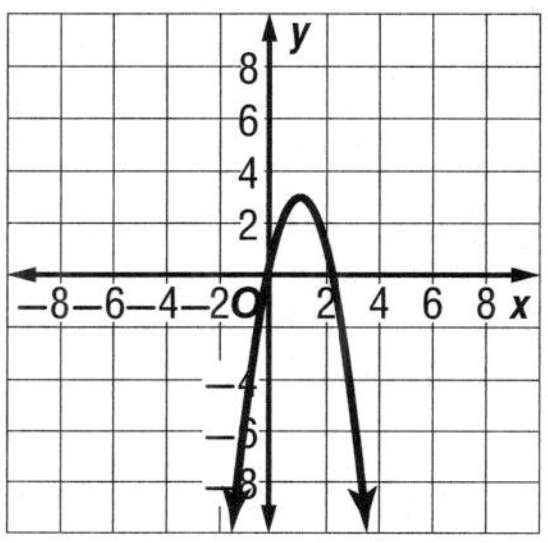

97a. $\begin{bmatrix} 72 & 49 \\ 68 & 63 \\ 90 & 56 \\ 86 & 62 \end{bmatrix}$

97b. $\begin{bmatrix} 96.50 \\ 99.50 \\ 118 \\ 117 \end{bmatrix}$

97c. juniors **97d.** \$431 **99.** 6

Pages 280–282 **Lesson 5-4**

1. $9i$ **3.** 12 **5.** 1 **7.** $\pm 2i\sqrt{2}$ **9.** 3, −2 **11.** $-3 + 2i$ **13.** $70 - 60i$ **15.** $\frac{1}{2} - \frac{1}{2}i$ **17.** $12 + 6j$ amps **19.** $13i$ **21.** $9i$ **23.** $-144i$ **25.** i **27.** −7 **29.** 9 **31.** $30 + 16i$ **33.** $1 + i$ **35.** $\frac{1}{3} - \frac{5}{3}i$

37

$3x^2 + 48 = 0$	Original equation
$3x^2 = -48$	Subtract 48 from each side.
$x^2 = -16$	Divide each side by 3.
$x = \pm\sqrt{-16}$	Square Root Property
$x = \pm 4i$	$\sqrt{-16} = \sqrt{16} \cdot \sqrt{-1}$ or $4i$

39. $\pm i\sqrt{5}$ **41.** $\pm 4i$ **43.** 2, −3 **45.** $\frac{4}{3}$, 4 **47.** 25, −2 **49.** $4i$ **51.** 8 **53.** $-21 + 15i$ **55.** $\frac{15}{13} + \frac{16}{13}i$ **57.** $11 + 23i$ **59.** $\frac{1}{7} - \frac{4\sqrt{3}}{7}i$

61

$V = C \cdot I$	Electricity formula
$= (3 + 6j) \cdot (5 - j)$	$C = 3 + 6j$ and $I = 5 - j$
$= 3(5) + 3(-j) + 6j(5) + 6j(-j)$	FOIL Method
$= 15 - 3j + 30j - 6j^2$	Multiply.
$= 15 - 3j + 30j - 6(-1)$	$j^2 = -1$
$= 21 + 27j$	Simplify.

The voltage is $21 + 27j$ volts.

63. $(3 + i)x^2 + (-2 + i)x - 8i + 7$

65a.

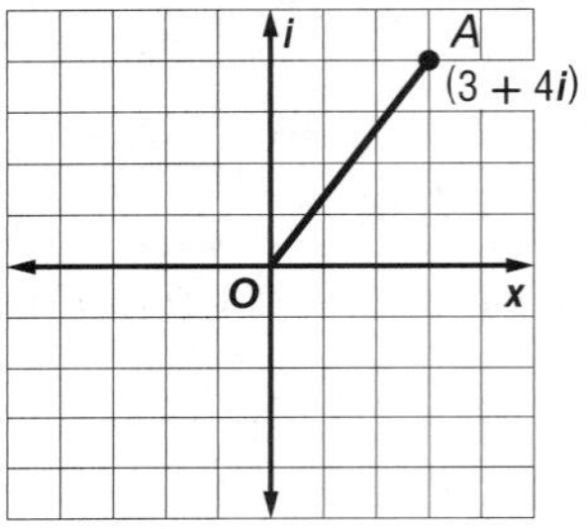

65b.

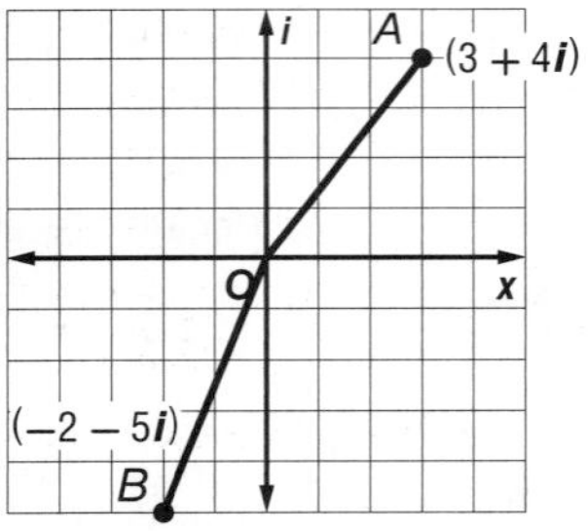

65c.

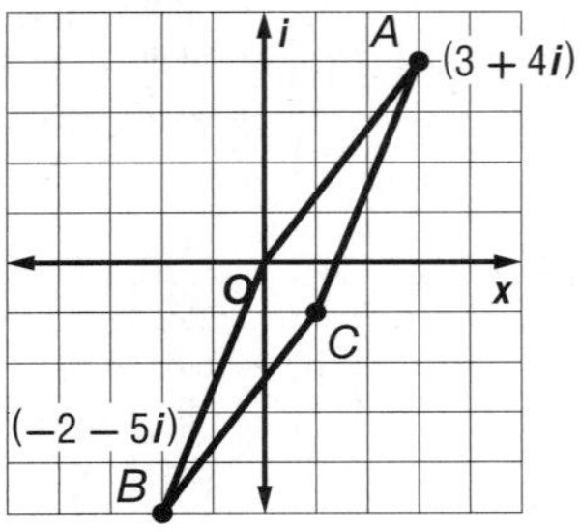

65d. $1 - i$; $A + B = C$ **67.** $-11 - 2i$ **69.** Sample answer: $(4 + 2i)(4 - 2i)$ **71a.** $\triangle CBE \cong \triangle ADE$

71b. $\angle AED \cong \angle CEB$ (Vertical angles)
$\overline{DE} \cong \overline{BE}$ (Both have length x.)
$\angle ADE \cong \angle CBE$ (Given)
Consecutive angles and the included side are all congruent, so the triangles are congruent by the ASA Property. **71c.** $\overline{EC} \cong \overline{EA}$ by CPCTC (corresponding parts of congruent triangles are congruent.) $EA = 7$, so $EC = 7$. **73.** H **75.** $-5, \frac{3}{2}$ **77.** $-\frac{1}{2}, \frac{4}{3}$ **79.** 3, 16 **81.** −9, −12

83.

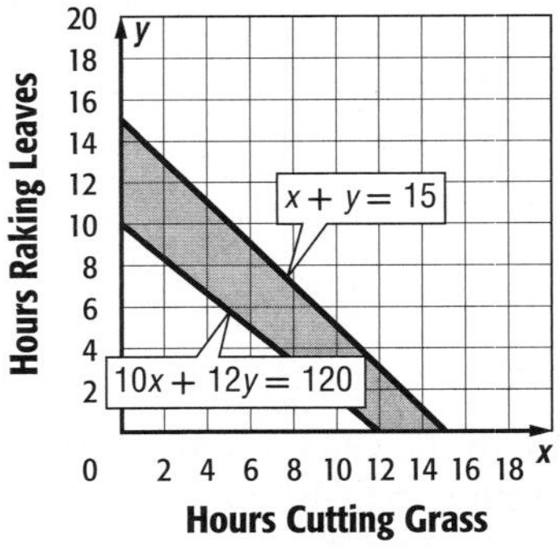

85. yes **87.** no **89.** yes

Pages 288–290 **Lesson 5-5**

1. {−8.45, −3.55} **3.** {−12.87, −5.13} **5.** 25 ft **7.** 6.25; $(x - 2.5)^2$ **9.** $\{2 - i\sqrt{5}, 2 + i\sqrt{5}\}$ **11.** {−4.37, 1.37} **13.** {−6.45, −1.55} **15.** {−1.47, 7.47} **17.** {−7.65, −2.35} **19.** {−1, 3} **21.** {4.67, 10.33} **23.** {−0.95, 3.95} **25.** {4, 5} **27.** 64; $(x + 8)^2$ **29.** 20.25; $(x + 4.5)^2$ **31.** {−4.61, 2.61} **33.** {1, 3} **35.** $\left\{\frac{3 - i\sqrt{31}}{4}, \frac{3 + i\sqrt{31}}{4}\right\}$

37

$3x^2 - 6x - 9 = 0$	Original equation
$x^2 - 2x - 3 = 0$	Divide by the coefficient of the quadratic term, 3.
$x^2 - 2x = 3$	Add 3 to each side.
$x^2 - 2x + 1 = 3 + 1$	Since $\left(\frac{-2}{2}\right)^2 = 1$, add 1 to each side.
$(x - 1)^2 = 4$	Write the left side as a perfect square.
$x - 1 = \pm 2$	Square Root Property
$x = \pm 2 + 1$	Add 1 to each side.
$x = 2 + 1$ or $x = -2 + 1$	Write as two equations.
$= 3$ $= -1$	Simplify.

The solution set is {−1, 3}.

39. $\{-2 - i\sqrt{7}, -2 + i\sqrt{7}\}$ **41.** $\{5 - 2i, 5 + 2i\}$ **43.** $\left\{\frac{7 - i\sqrt{47}}{4}, \frac{7 + i\sqrt{47}}{4}\right\}$ **45.** $\{2.65 - i\sqrt{1.5775}, 2.65 + i\sqrt{1.5775}\}$ **47.** {−0.89, 5.39} **49.** {2.38, 4.62} **51.** {−1.26, 0.26}

53 **a.** The time in which the firework explodes is the t-coordinate of the vertex.

$x = -\frac{b}{2a}$	Equation of the axis of symmetry
$x = -\frac{(25)}{2(-1.5)}$ or $8\frac{1}{3}$	$a = -1.5$ and $b = 25$

So, the firework explodes after $8\frac{1}{3}$ seconds.

b. The time in which the firework explodes is the d-coordinate of the vertex.

$d = -1.5t^2 + 25t$	Original function
$d = -1.5\left(8\frac{1}{3}\right)^2 + 25\left(8\frac{1}{3}\right)$	$t = 8\frac{1}{3}$
$d \approx 104.2$	Simplify.

So, the firework explodes at a height of about 104.2 feet.

55. 2.56; $(x - 1.6)^2$

57a.

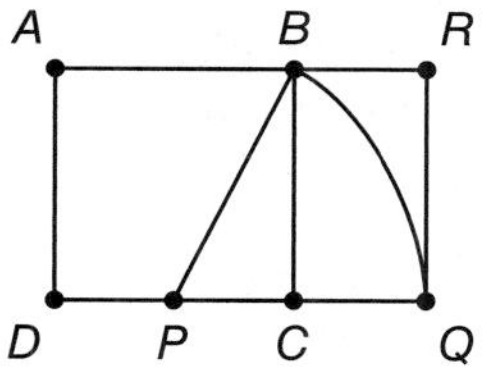

57b. $x = \frac{1 + \sqrt{5}}{2}$

57c.

CQ	*x*
2	$1 + \sqrt{5}$
3	$\frac{3 + 3\sqrt{5}}{2}$
4	$2 + 2\sqrt{5}$

57d. Sample answer: the x-values are multiples of $\frac{1 + \sqrt{5}}{2}$; $x = \frac{n(1 + \sqrt{5})}{2}$. **59.** $x = \frac{-b}{2} \pm \sqrt{\frac{b^2}{4} - c}$

61. Sample answer: $x^2 - \frac{2}{3}x + \frac{1}{9} = \frac{1}{4}$; $\left\{\frac{5}{6}, -\frac{1}{6}\right\}$

63. D **65.** 125 **67.** $39 + 80i$ **69.** $\frac{4}{39} - \frac{19}{39}i$

71. $5x^2 - 28x - 12 = 0$

73a.

	Evening	Matinee	Twilight
Adult	7.50	5.50	3.75
Child	4.50	4.50	3.75
Senior	5.50	5.50	3.75

73b. 3×3 **75.**

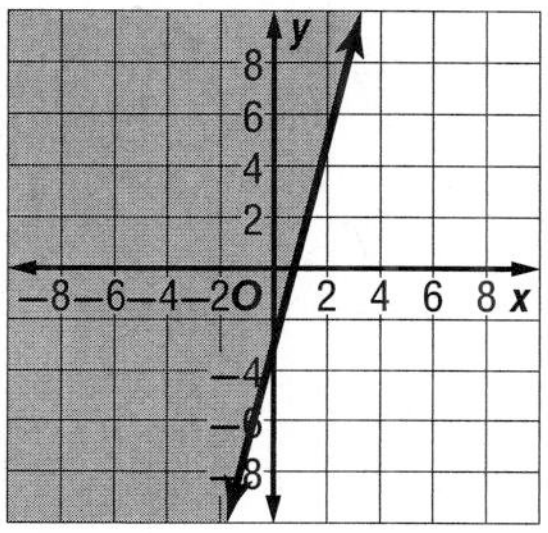

77.

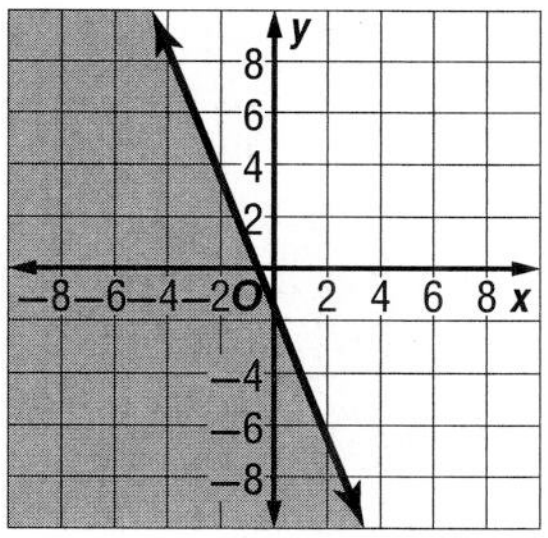

79. $f(x) = \begin{cases} -8 \text{ if } x \leq -2 \\ 2x - 6 \text{ if } -2 < x \leq 5 \\ x - 13 \text{ if } x > 5 \end{cases}$ **81.** -4 **83.** -136

Pages 297–300 Lesson 5-6

1

$x = \frac{-b \pm \sqrt{b^2 - 4ac}}{2a}$	Quadratic Formula
$= \frac{-12 \pm \sqrt{12^2 - 4(1)(-9)}}{2(1)}$	$a = 1, b = 12$, and $c = -9$
$= \frac{-12 \pm \sqrt{144 + 36}}{2}$	Multiply.
$= \frac{12 \pm \sqrt{180}}{2}$	Simplify.
$= \frac{-12 \pm 6\sqrt{5}}{2}$	$\sqrt{180} = 6\sqrt{5}$
$x = \frac{-12 + 6\sqrt{5}}{2}$ or $x = \frac{-12 - 6\sqrt{5}}{2}$	Write as two equations.
$= -6 + 3\sqrt{5}$ $\quad = -6 - 3\sqrt{5}$	Simplify.

The solutions are $-6 + 3\sqrt{5}$ and $-6 - 3\sqrt{5}$.

3. $\left(\frac{5 + \sqrt{57}}{8}, \frac{5 - \sqrt{57}}{8}\right)$ **5.** (1.5, −0.2)

7. $\left(\frac{2 + 2\sqrt{7}}{3}, \frac{2 - 2\sqrt{7}}{3}\right)$ **9.** about 2.5 seconds

11a. −36 **11b.** 2 complex roots **13a.** −76

13b. 2 complex roots **15.** $\frac{-3 \pm \sqrt{15}}{2}$ **17.** $\frac{-7 \pm \sqrt{129}}{8}$

19. $\frac{-3 \pm i\sqrt{71}}{8}$

21 **a.** $b^2 - 4ac = 3^2 - 4(2)(-3)$ $\quad a = 2, b = 3, c = -3$

$= 9 + 24$ or 33 $\quad$ Simplify.

b. The discriminate is positive and not a perfect square. So, there are 2 irrational roots.

c. $x = \frac{-b \pm \sqrt{b^2 - 4ac}}{2a}$	Quadratic Formula
$= \frac{-3 \pm \sqrt{3^2 - 4(2)(-3)}}{2(2)}$	$a = 2, b = 3$, and $c = -3$
$= \frac{-3 \pm \sqrt{9 + 24}}{4}$	Multiply.
$= \frac{-3 \pm \sqrt{33}}{4}$	Simplify.

23a. 49 **23b.** 2 rational **23c.** $\frac{1}{6}, -1$ **25a.** −87

25b. 2 complex **25c.** $\frac{3 \pm i\sqrt{87}}{6}$ **27a.** 36

27b. 2 rational **27c.** $1, -\frac{1}{5}$ **29a.** 1 **29b.** 2 rational

29c. $-1, -\frac{4}{3}$ **31a.** −16 **31b.** 2 complex **31c.** $-1 \pm 2i$

33a. 0 **33b.** about 2.3 seconds **35a.** 64

35b. 2 rational **35c.** $0, -\frac{8}{5}$ **37a.** 160 **37b.** 2 irrational

37c. $\frac{-1 \pm \sqrt{10}}{6}$ **39a.** 13.48 **39b.** 2 irrational

39c. $\frac{-0.7 \pm \sqrt{3.37}}{0.6}$

41

a. $y = -0.26x^2 - 0.55x + 91.81$	Original equation
$= -0.26(10)^2 - 0.55(10) + 91.81$	Replace x with 10.
$= 60.31$	Simplify.
$y = -0.26x^2 - 0.55x + 91.81$	Original equation
$= -0.26(15)^2 - 0.55(15) + 91.81$	Replace x with 15.
$= 25.06$	Simplify.

For 2010, the number of deaths per 100,000 is 60.31. For 2015, the number is 25.06.

b. $y = -0.26x^2 - 0.55x + 91.81$	Original equation
$50 = -0.26x^2 - 0.55x + 91.81$	Replace y with 50.
$0 = -0.26x^2 - 0.55x + 41.81$	Subtract 50 from each side.
$x = \frac{-b \pm \sqrt{b^2 - 4ac}}{2a}$	Quadratic Formula
$= \frac{-(-0.55) \pm \sqrt{(-0.55)^2 - 4(-0.26)(41.81)}}{2(-0.26)}$	$a = -0.26$, $b = -0.55$, and $c = 41.81$
$= \frac{0.55 \pm \sqrt{43.1799}}{-0.52}$	Simplify.

$x \approx -13.69$ or $x \approx 11.58$

Since the number of years after 2000 cannot be negative, the solution is 11.58. So, 11.58 years after 2000, or in 2011, the death rate will be 50 per 100,000.

c. $y = -0.26x^2 - 0.55x + 91.81$ Original equation

$0 = -0.26x^2 - 0.55x + 91.81$ Replace y with 0.

$x = \frac{-b \pm \sqrt{b^2 - 4ac}}{2a}$ Quadratic Formula

$= \frac{-(-0.55) \pm \sqrt{(-0.55)^2 - 4(-0.26)(91.81)}}{2(-0.26)}$ $a = -0.26$, $b = -0.55$, and $c = 91.81$

$= \frac{0.55 \pm \sqrt{95.7849}}{-0.52}$ Simplify.

$x \approx -19.78$ or $x \approx 17.76$

Since the number of years after 2000 cannot be negative, the solution is 17.76. So, 17.76 years after 2000, or in 2017, the death rate will be 0 per 100,000. Sample answer: This prediction is not reasonable because the death rate from cancer will never be 0 unless a cure is found. If and when a cure will be found cannot be predicted.

43. Jonathan is correct; you must first write the equation in the form $ax^2 + bx + c = 0$ to determine the values of a, b, and c. Therefore, the value of c is -7, not 7. **45a.** Sample answer: Always; when a and c are opposite signs, then ac will always be negative and $-4ac$ will always be positive. Since b^2 will also always be positive, then $b^2 - 4ac$ represents the addition of two positive values, which will never be negative. Hence, the discriminant can never be negative and the solutions can never be imaginary. **45b.** Sample answer: Sometimes; the roots will only be irrational if $b^2 - 4ac$ is not a perfect square. **47.** -0.75 **49.** B **51.** 112.5 in^2 **53.** 42.25; $(x + 6.5)^2$ **55.** $\frac{4}{25}$; $\left(x + \frac{2}{5}\right)^2$ **57.** $4i$ **59.** 27 hours of flight instruction and 23 hours in the simulator **61.** $y = x^2 + 1$ **63.** $y = |x + 3|$

Pages 308–310 *Lesson 5-7*

1. $y = (x + 3)^2 - 7$ **3.** $y = 4(x + 3)^2 - 12$

5.

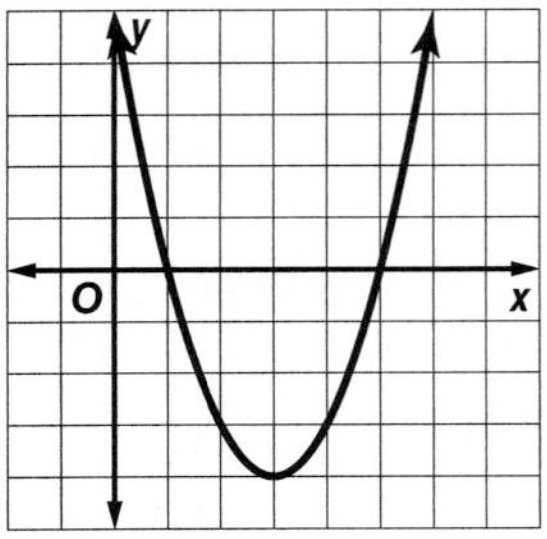

7.

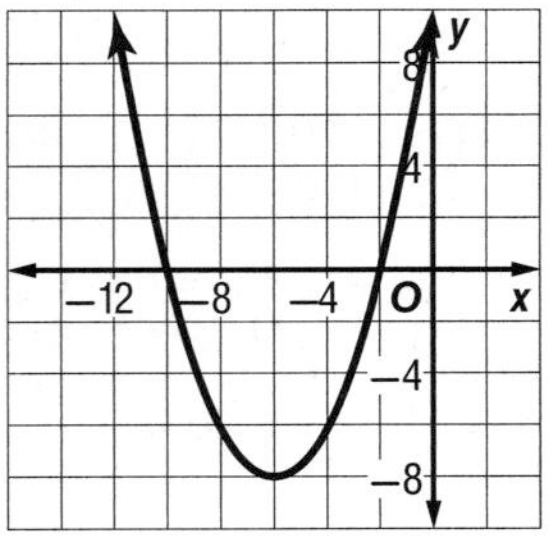

9. $y = (x - 3)^2 - 6$

11 $y = x^2 + 2x + 7$ Original equation.

$y = (x^2 + 2x + 1) + 7 - 1$ Complete the square by adding $\left(\frac{2}{2}\right)^2$ or 1. Balance the equation by subtracting 1.

$y = (x + 1)^2 + 6$ Write $x^2 + 2x + 1$ as a perfect square.

13. $y = (x + 4)^2$ **15.** $y = 3\left(x + \frac{5}{3}\right)^2 - \frac{25}{3}$

17. $y = -4(x + 3)^2 + 21$ **19.** $y = -(x + 2)^2 + 3$

21. $y = -15(x - 8.5)^2 + 4083.75$

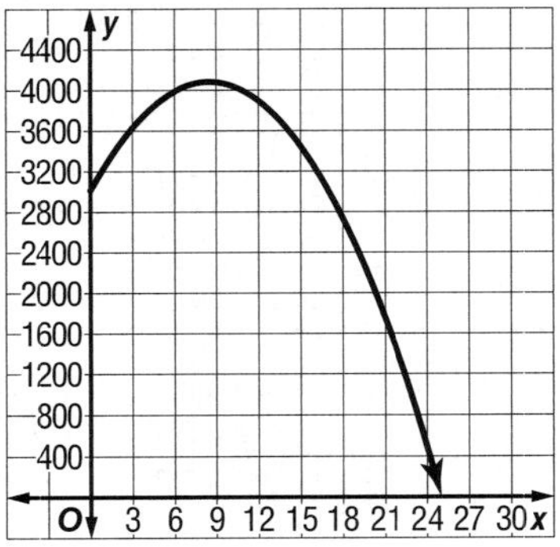

23.

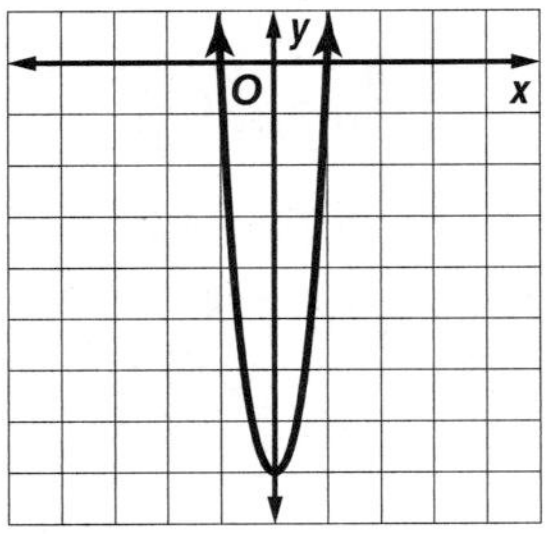

25.

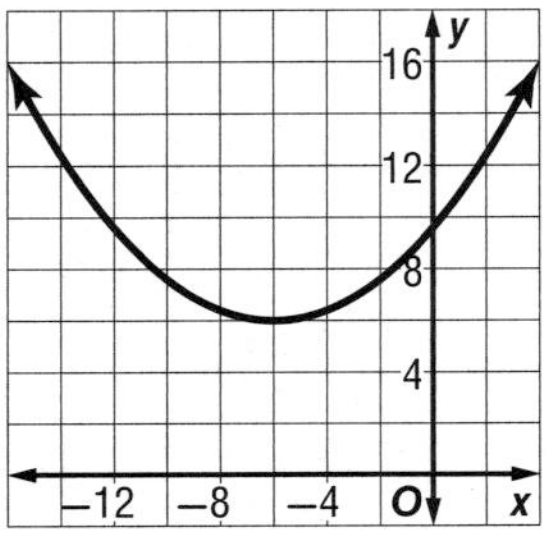

27.

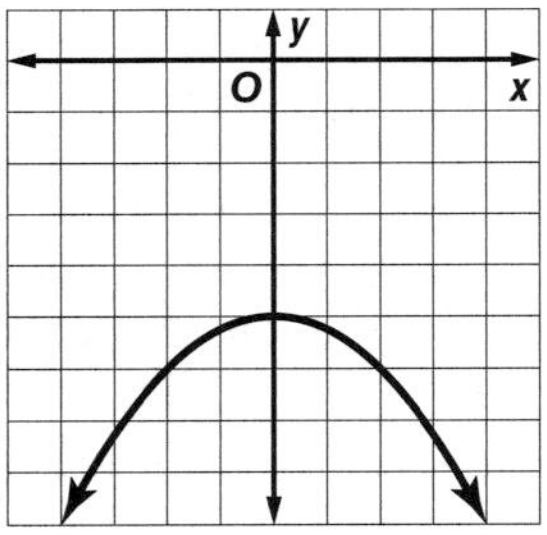

29.

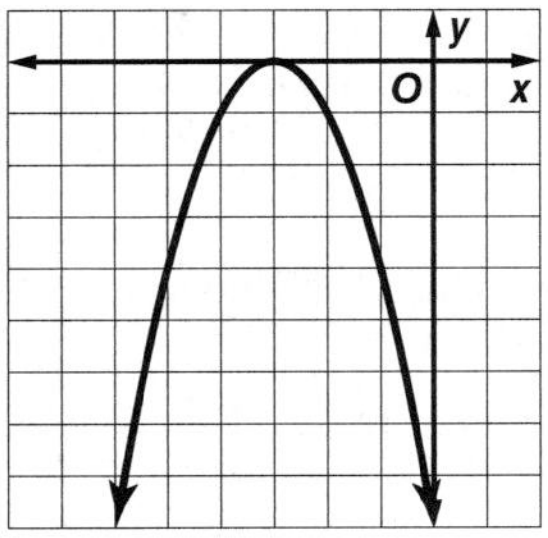

31.

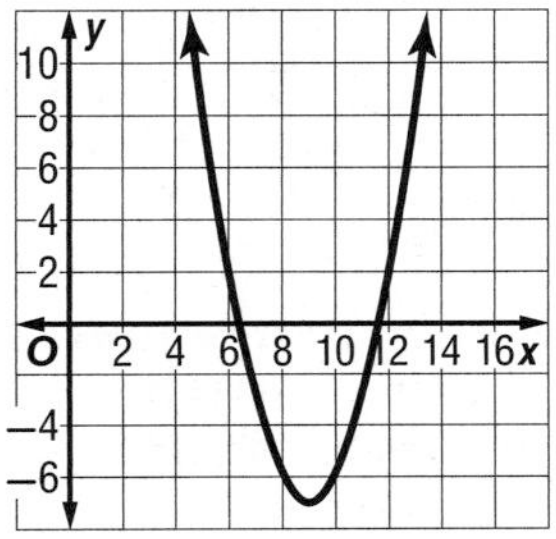

33. 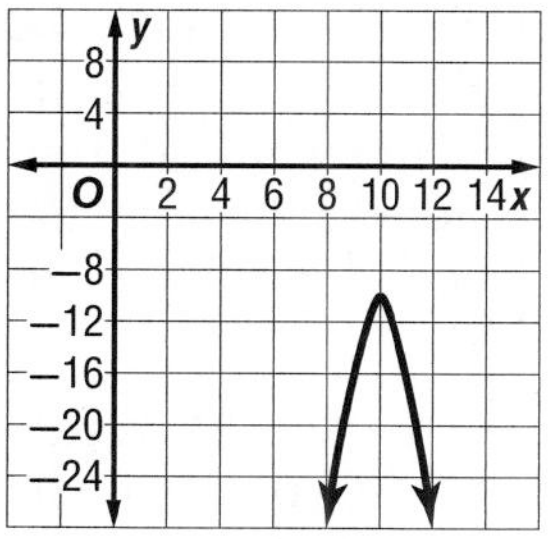

35. $y = 9(x - 6)^2 + 1$
37. $y = -\frac{2}{3}(x - 3)^2$
39. $y = \frac{1}{3}x^2 + 5$
41. $y = 3\left(x - \frac{2}{3}\right)^2 - \frac{10}{3}$; $\left(\frac{2}{3}, -\frac{10}{3}\right)$, $x = \frac{2}{3}$, opens up

43. $y = -(x + 2.35)^2 + 8.3225$; $(-2.35, 8.3225)$, $x = -2.35$, opens down **45.** $y = \left(x - \frac{1}{3}\right)^2 - 3$; $\left(\frac{1}{3}, -3\right)$, $x = \frac{1}{3}$, opens up

47 a.

$S(t) = \frac{1}{2}at^2 + v_0t$	Original equation
$S(t) = \frac{1}{2}(0.002)t^2 + 35t$	$a = 0.002$ and $v_0 = 35$
$S(t) = 0.001t^2 + 35t$	Simplify.
$S(t) = 0.001(t^2 + 35{,}000t)$	Group $ax^2 + bx$ and factor, dividing by a.
$S(t) = 0.001(t^2 + 35{,}000t + 17{,}500^2) - 306{,}250$	Complete the square by adding $17{,}500^2$ inside the parentheses. This is an overall addition of 306,250. Balance the equation by subtracting 306,250.
$S(t) = 0.001(t + 17{,}500)^2 - 306{,}250$	Write $t^2 + 35{,}000t + 17{,}500^2$ as a perfect square.

b.

$a \cdot t = v$	acceleration • time = velocity
$0.002\ \text{mi/s}^2 \cdot t = (68 - 35)\ \text{mi/h}$	Substitution
$0.002\ \text{mi/s}^2 \cdot t = 33\ \text{m/h}$	Simplify.
$0.002\ \text{mi/s}^2 \cdot t = \frac{33\ \text{mi}}{1\ \text{h}} \cdot \frac{1\ \text{h}}{3600\ \text{s}}$	Convert hours to seconds.
$0.002\ \text{mi/s}^2 \cdot t = 0.009\ \text{mi/s}$	Simplify.
$t = \frac{0.009\ \text{mi}}{\text{s}} \div \frac{0.002\ \text{mi}}{\text{s}^2}$	Divide each side by $0.002\ \text{mi/s}^2$
$t = \frac{0.009\ \text{mi}}{\text{s}} \cdot \frac{\text{s}^2}{0.002\ \text{mi}}$	Multiply by the reciprocal.
$t = 4.58\ \text{s}$	Simplify.

It will take about 4.58 seconds for Valerie to accelerate from 35 mi/h to 68 mi/h.

c. No; if we substitute $\frac{1}{8}$ for $S(t)$ and solve for t, we get 2.47 seconds. This is how long Valerie will be on the ramp. Since it will take her 4.58 seconds to accelerate to 68 mph, she will not be on the ramp long enough to accelerate to match the average expressway speed.

49. The equation of a parabola can be written in the form $y = ax^2 + bx + c$ with $a \neq 0$. For each of the three points, substitute the value of the x-coordinate for x in the equation and substitute the value of the y-coordinate for y in the equation. This will produce three equations in three variables a, b, and c. Solve the system of equations to find the values of a, b, and c. These values determine the quadratic equation.

51. Sample answer: The variable a represents different values for these functions, so making $a = 0$ will have a different effect on each function. For $f(x)$, when $a = 0$, the graph will be a horizontal line, $f(x) = k$. For $g(x)$, when $a = 0$, the graph will be linear, but not necessarily horizontal, $g(x) = bx + c$.

53. B **55.** D **57.** $\frac{-15 \pm \sqrt{561}}{8}$ **59.** $\frac{3 \pm \sqrt{39}}{5}$
61. 0.0025 **63.** minimum, $9\frac{1}{3}$
65. minimum, −12

67a.

	Weekday	Weekend
Single	60	79
Double	70	89
Suite	75	95

67b.

	Single	Double	Suite
Weekday	60	70	75
Weekend	79	89	95

69. 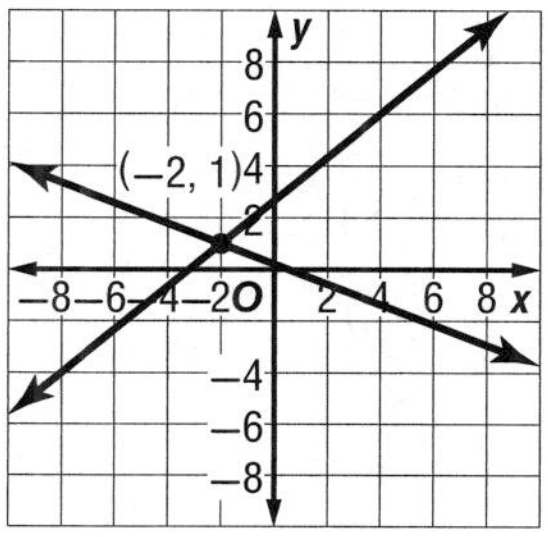

71. 9 **73.** 52 **75.** yes

Pages 315–318 Lesson 5-8

1. 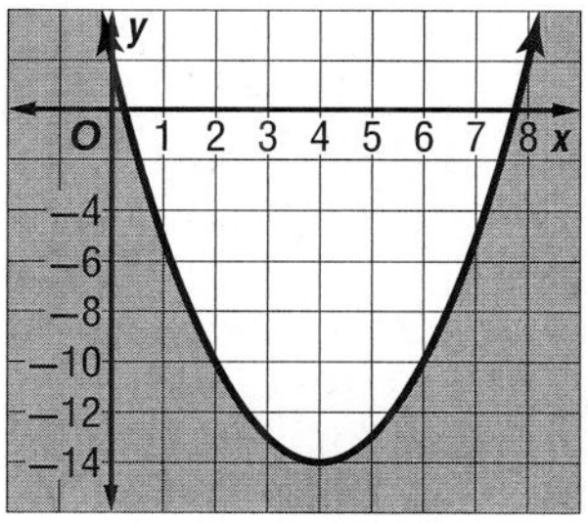

3. 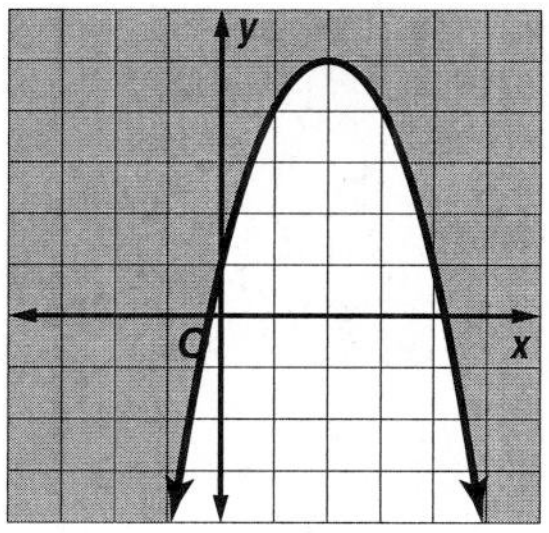

5. $\{x \mid -5 < x < -3\}$
7. $\{x \mid 0.29 \leq x \leq 1.71\}$
9. $\{x \mid -8 < x < 2\}$

11

$-x^2 + 12x = 28$	Related quadratic equation
$-x^2 + 12x - 28 = 0$	Subtract 28 from each side.
$x = \frac{-b \pm \sqrt{b^2 - 4ac}}{2a}$	Quadratic Formula
$= \frac{-12 \pm \sqrt{122 - 4(-1)(-28)}}{2(-1)}$	$a = -1$, $b = 12$, and $c = -28$
$x = \frac{-12 + \sqrt{32}}{-2}$ or $x = \frac{-12 - \sqrt{32}}{-2}$	Simplify and write as two equations.
≈ 3.17 ≈ 8.83	Simplify.

Plot 3.17 and 8.83 on a number line. Use dots since these values are solutions of the original inequality.

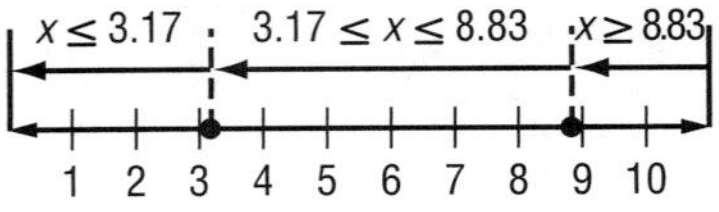

Test a value from each of the three intervals to see if it satisfies the original inequality.

$x \le 3.17$
Test $x = 0$.
$-x^2 + 12x \ge 28$
$-(0)^2 + 12(0) \ge 28$
$0 \not\ge 28$

$3.17 \le x \le 8.83$
Test $x = 5$.
$-x^2 + 12x \ge 28$
$-(5)^2 + 12(5) \ge 28$
$35 \ge 28$

$x \ge 8.83$
Test $x = 10$.
$-x^2 + 12x \ge 28$
$-(10)^2 + 12(10) \ge 28$
$20 \not\ge 28$

The solution set is $\{x \mid 3.17 \le x \le 8.83\}$.

13.

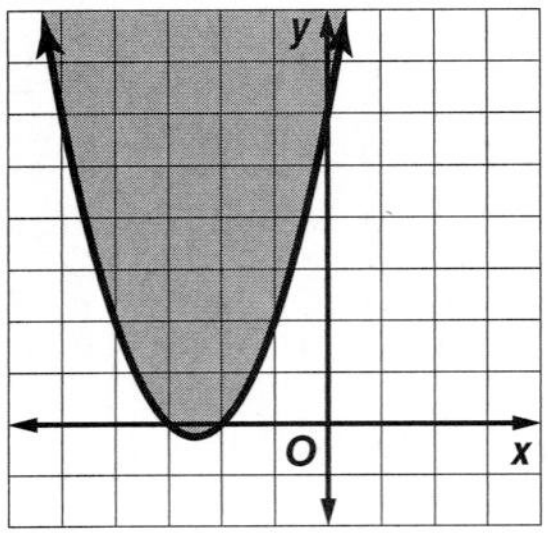

15.

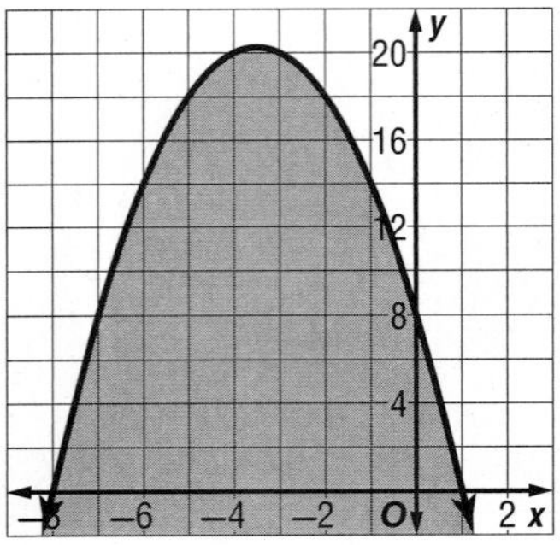

17.

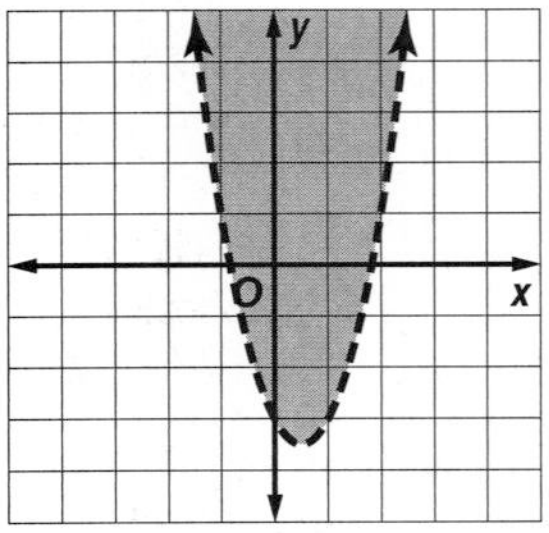

19. $\{x \mid 1.1 < x < 7.9\}$ **21.** $\{x \mid$ all real numbers$\}$
23. $\{x \mid x < -1.42$ or $x > 8.42\}$ **25.** $\varnothing$
27. $\{x \mid -0.73 < x < 2.73\}$ **29.** $\{x \mid -0.5 \le x \le 2.5\}$

31 The function describes the height of the arch. You want to find the values of x for which $f(x) \ge 7$.

$f(x) \ge 7$ Original inequality
$-x^2 + 6x + 1 \ge 7$ $f(x) = -x^2 + 6x + 1$
$-x^2 + 6x - 6 \ge 0$ Subtract 7 from each side.

Graph the related function $y = -x^2 + 6x - 6$ using a graphing calculator.

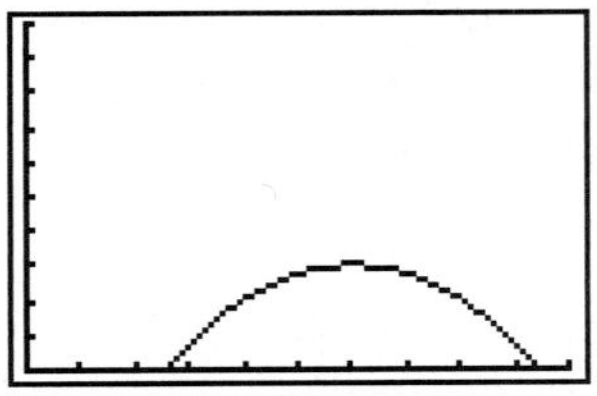

At $x \approx 1.26$ and $x \approx 4.73$, $f(x) \ge 7$. So, at about 1.26 ft to 4.73 ft from the sides of the arch, the height is at least 7 ft.

33. $\{x \mid 4 < x < 5\}$ **35.** $\{x \mid -1 < x < 2\}$ **37.** $\{x \mid x \le -2.32$ or $x \ge 4.32\}$ **39.** $\{x \mid x \le -1.58$ or $x \ge 1.58\}$
41. $\{x \mid$ all real numbers$\}$ **43.** $\{x \mid -2.84 < x < 0.84\}$

45a.

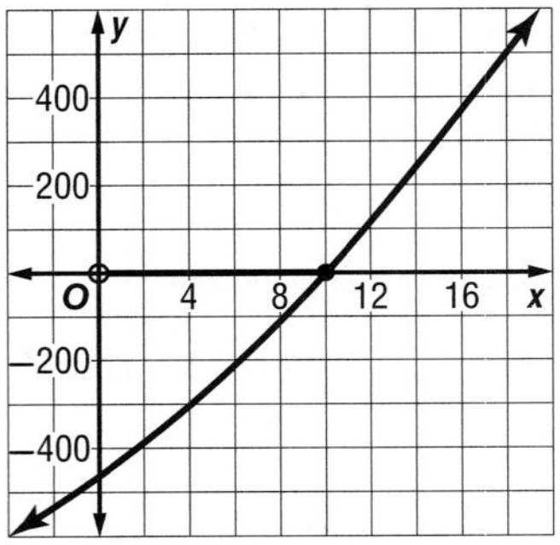

45b. greater than 0 ft but no more than 10.04 ft
47. $y \le -x^2 + 2x + 6$
49. $\{x \mid x < -1.06$ or $x > 7.06\}$

51

$11 = 4x^2 + 7x$ Related quadratic equation
$0 = 4x^2 + 7x - 11$ Subtract 11 from each side.
$x = \dfrac{-b \pm \sqrt{b^2 - 4ac}}{2a}$ Quadratic Formula
$x = \dfrac{-7 \pm \sqrt{7^2 - 4(4)(-11)}}{2(4)}$ $a = 4, b = 7,$ and $c = -11$
$x = \dfrac{-7 + \sqrt{255}}{8}$ or $x = \dfrac{-7 - \sqrt{255}}{8}$ Simplify and write as two equations.
$= 1$ $= -2.75$ Simplify.

Plot −2.75 and 1 on a number line. Use dots since these values are solutions of the original inequality.

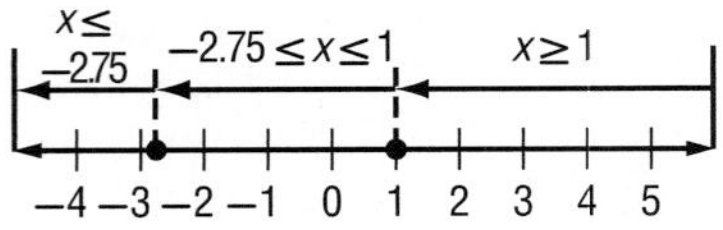

Test a value from each of the three intervals to see if it satisfies the original inequality.

$x \le -2.75$
Test $x = -3$.
$11 \le 4x^2 + 7x$
$11 \le 4(-3)^2 + 7(-3)$
$11 \le 15$

$-2.75 \le x \le 1$
Test $x = 0$.
$11 \le 4x^2 + 7x$
$11 \le 4(0)^2 + 7(0)$
$11 \not\le 0$

$x \ge 1$
Test $x = 2$.
$11 \le 4x^2 + 7x$
$11 \le 4(2)^2 + 7(2)$
$11 \le 30$

The solution set is $\{x \mid x \le -2.75$ or $x \ge 1\}$.

53. $\{x \mid x < 0.61$ or $x > 2.72\}$

55a.

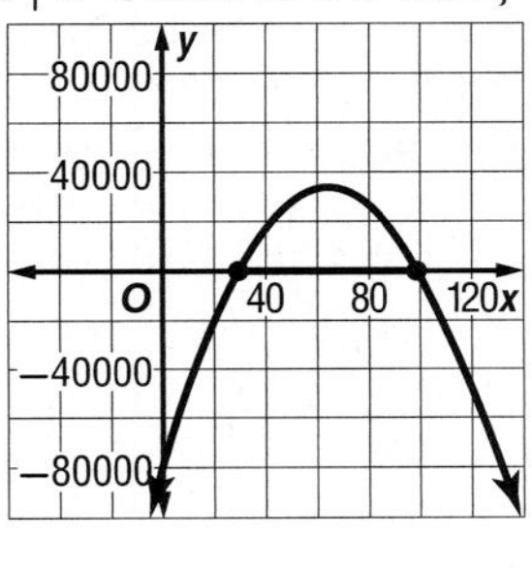

55b. from 30,000 to 98,000 digital audio players **55c.** The graph is shifted down 25,000 units. The manufacturer must sell from 48,000 to 81,000 digital audio players.

57a. Sample answer: $x^2 + 2x + 1 \geq 0$
57b. Sample answer: $x^2 - 4x + 6 < 0$ **59.** No; the graphs of the inequalities intersect the x-axis at the same points.

61.

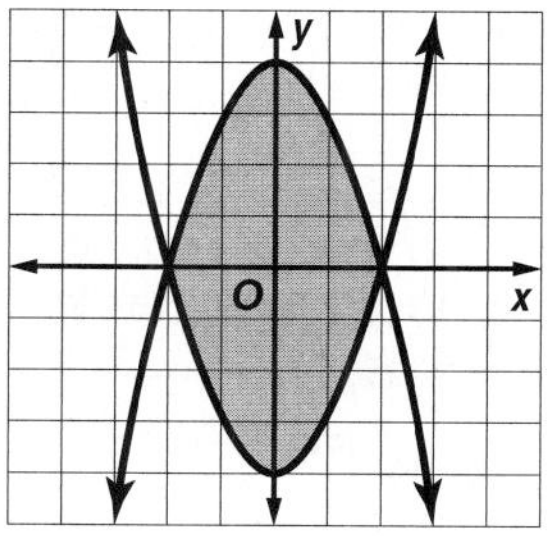

63. 15 **65.** G
67. $y = 2(x - 3)^2 - 4$
69. $y = 0.25(x + 4)^2 + 3$
71. -152; 2 complex roots **73.** $(-8, 7)$, $(-7, -8)$, and $(8, -7)$
75. $\begin{bmatrix} -28 & 60 \\ 20 & -70 \end{bmatrix}$
77. $-6x + 24$

79. $8y - 12z$ **81.** $2.5x + 3y$

Pages 320–324 Chapter 5 Study Guide and Review

1. false, standard form **3.** false, factored form **5.** false, completing the square **7.** true **9a.** y-int: 12; $x = -\frac{5}{2}$; $-\frac{5}{2}$

9b.

x	y
-3	6
-2	6
$-\frac{5}{2}$	$2\frac{3}{4}$
-1	8
0	12

9c.

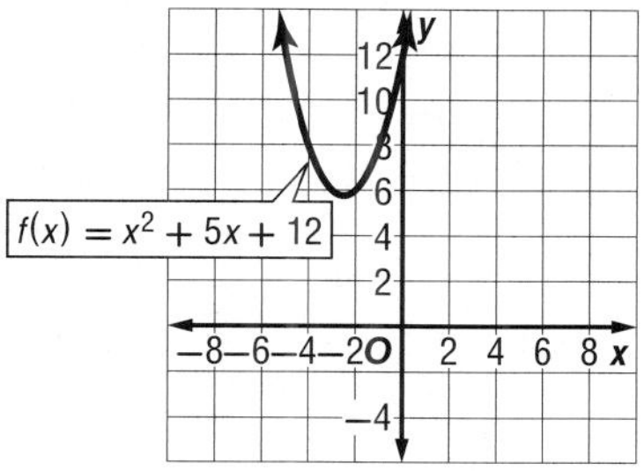

11a. y-int: -5; $x = \frac{9}{4}$; $\frac{9}{4}$

11b.

x	y
1	2
2	5
$\frac{9}{4}$	$4\frac{1}{8}$
3	4
4	-1

11c.

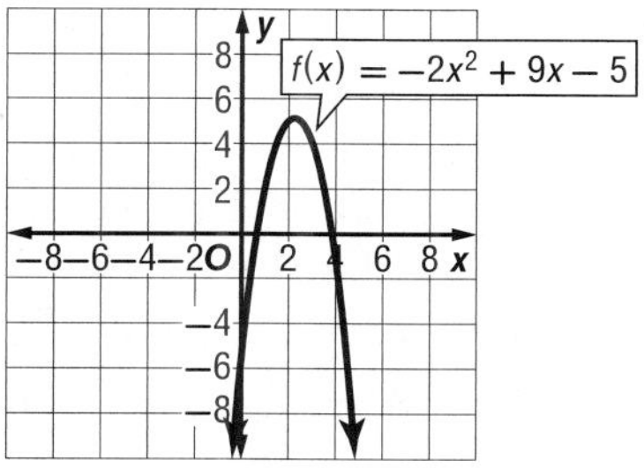

13. max; 1.25; D = {all real numbers}; R = $\{y \mid y \leq 1.25\}$ **15.** 75 T-shirts at \$15 each **17.** $\left(-1, \frac{3}{2}\right)$ **19.** 7.5 seconds **21.** $x^2 + 10x + 21 = 0$ **23.** $3x^2 - x - 2 = 0$ **25.** $4x^2 + 5x + 1 = 0$ **27.** $\left\{-\frac{1}{2}, 3\right\}$ **29.** $x = 12$; 9 ft by 14 ft **31.** $15 + 3i$ **33.** $28 + 3i$ **35.** $x = \pm 5i$ **37.** $x = \pm i\sqrt{5}$ **39.** $x = \pm \frac{1}{2}i$ **41.** 4; $(x - 2)^2$ **43.** 1.44; $(x + 1.2)^2$ **45.** $\frac{9}{25}$; $\left(x + \frac{3}{5}\right)^2$ **47.** $\{1 \pm i\sqrt{7}\}$ **49.** $\left\{1, -\frac{5}{2}\right\}$ **51a.** 0 **51b.** 1 real rational root **51c.** $\{5\}$ **53a.** 153 **53b.** 2 irrational real roots **53c.** $\left\{\frac{-3 \pm 3\sqrt{17}}{4}\right\}$ **55a.** -32 **55b.** 2 complex roots **55c.** $\{1 \pm 2i\sqrt{2}\}$ **57a.** -47 **57b.** 2 complex roots **57c.** $\left\{\frac{-5 \pm i\sqrt{47}}{4}\right\}$ **59.** $y = -3(x - 1)^2 + 5$; $(1, 5)$; $x = 1$; opens down **61.** $y = -\frac{1}{2}(x + 2)^2 + 14$; $(-2, 14)$; $x = -2$; opens down **63.** $f(x) = -x^2 + 10x$; 5 and 5

65.

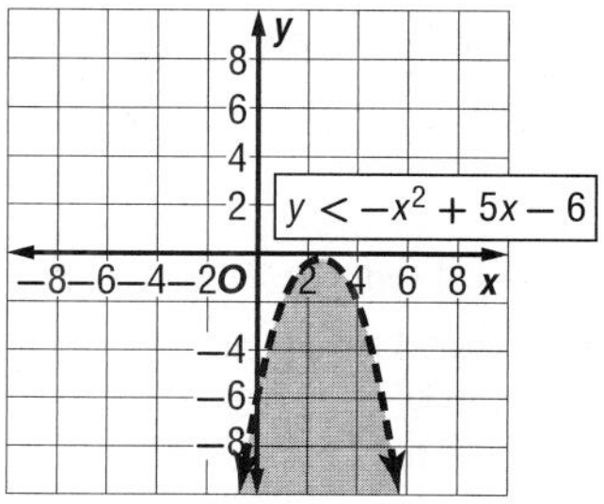

67.

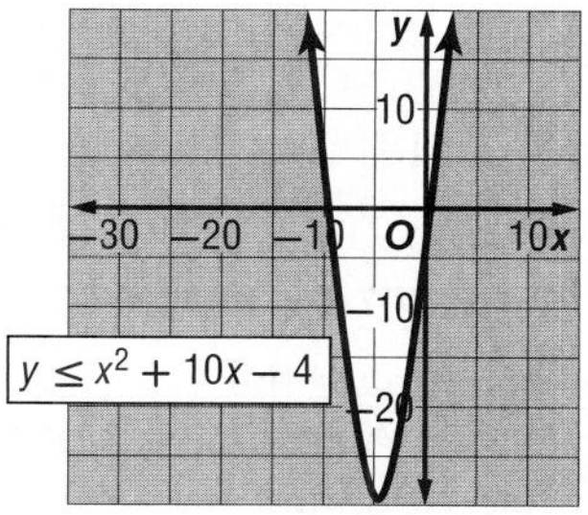

69. $\{x \mid x < -6 \text{ or } x > -2\}$ **71.** $\left\{x \mid x < -4 \text{ or } x > \frac{5}{2}\right\}$ **73.** $\left\{x \mid x < \frac{2}{3} \text{ or } x > 2\right\}$

Chapter 6 Polynomials and Polynomial Functions

Page 331 Chapter 6 Get Ready

1. $-5 + (-13)$ **3.** $5mr + (-7mp)$ **5.** $20 + (-2x)$ **7.** $-3b^2 - 2b + 1$ **9.** $-\frac{9}{4}z - \frac{15}{4}$ **11.** $-4, 2$ **13.** $-\frac{4}{3}, \frac{1}{2}$ **15.** about 1.77 seconds

Pages 337–339 Lesson 6-1

1. $-8a^5b^2$ **3.** $\frac{8a^6}{27b^3}$ **5.** yes, 1 **7.** no **9.** $-2x^2 - 6x + 3$ **11.** $8ab + 10a$ **13.** $n^2 - 2n - 63$ **15.** $750 - 2.5x$ **17.** $-8b^5c^3$ **19.** $-yz^2$ **21.** $\frac{a^2c^2}{2b^4}$ **23.** z^{18} **25.** yes; 3 **27.** no **29.** $3b^2 + 6b - 5$ **31.** $8x^3 + 4xy$

33 $(a + b)(a^3 - 3ab - b^2)$
$= a(a^3 - 3ab - b^2) + b(a^3 - 3ab - b^2)$ Distributive Property
$= a(a^3) - a(3ab) - a(b^2) + b(a^3) - b(3ab) - b(b^2)$ Distributive Property
$= a^4 - 3a^2b - ab^2 + a^3b - 3ab^2 - b^3$ Multiply.
$= a^4 + a^3b - 3a^2b - 4ab^2 - b^3$ Simplify.

35. $10c^3 - c^2 + 4c$ **37.** $12a^2b + 8a^2b^2 - 15ab^2 + 4b^2$ **39.** $4a^2x - 2a^2y + 10abx - 5aby + 6b^2x - 3b^2y$ **41.** $\frac{y^4}{81x^4}$ **43.** $\frac{x^6}{16y^{14}}$ **45.** b **47.** $\frac{2}{5}cd^4$ **49.** $\frac{1}{2}x^6y^3$

51 **a.** $d = rt$ distance = rate • time
$t = \frac{d}{r}$ Solve the formula for time.
$= \frac{2.367 \times 10^{21}\text{ m}}{3 \times 10^8\text{ m/s}}$ ← Distance from Andromeda to Earth ← Speed of light
$= \frac{2.367}{3} \cdot \frac{10^{21}}{10^8} \cdot \frac{\text{m}}{\text{m/s}}$ Separate to get powers of the same base.
$\approx 0.789 \cdot 10^{21-8}\text{ s}$ Subtract exponents.
$\approx 0.789 \times 10^{13}\text{ s}$ Simplify.
It takes about 0.789×10^{13} seconds or about 250,190.26 years.

b. $t = \frac{d}{r}$ Write the formula.

$= \frac{2.28 \times 10^{11} \text{ m}}{3 \times 10^{8} \text{ m/s}}$ ← Distance from the Sun to Mars ← Speed of light

$= \frac{2.28}{3} \cdot \frac{10^{11}}{10^{8}} \cdot \frac{\text{m}}{\text{m/s}}$ Separate to get powers of the same base.

$= 0.76 \times 10^{11-8} \text{ s}$ Subtract exponents.

$= 0.76 \times 10^{3} \text{ s}$ Simplify.

$= 760 \text{ s}$ Simplify.

It takes 760 seconds or about 12.67 minutes.

53. $2n^4 - 3n^3p + 6n^4p$ **55.** $b^3 + a^{-1}b + a^{-2}$ **57.** $2n^5 - 14n^3 + 4n^2 - 28$ **59.** $64n^3 - 240n^2 + 300n - 125$ **61a.** $0.1547x^2 + 8.8181x + 835.8$ **61b.** $0.0603x^2 - 10.5699x + 112.4$ **63.** 9 **65.** $\frac{1}{a^n} = \frac{a^0}{a^n} = a^{0-n} = a^{-n}$ **67.** Sample answer: We would have a 0 in the denominator, which makes the expression undefined. **69.** Sample answer: Astronomy deals with very large numbers that are sometimes difficult to work with because they contain so many digits. Properties of exponents make very large or very small numbers more manageable. As long as you know how far away a planet is from a light source, you can divide that distance by the speed of light to obtain how long it will take light to reach that planet. **71.** D **73.** C **75.** $x > 5$ or $x < -8$

77.

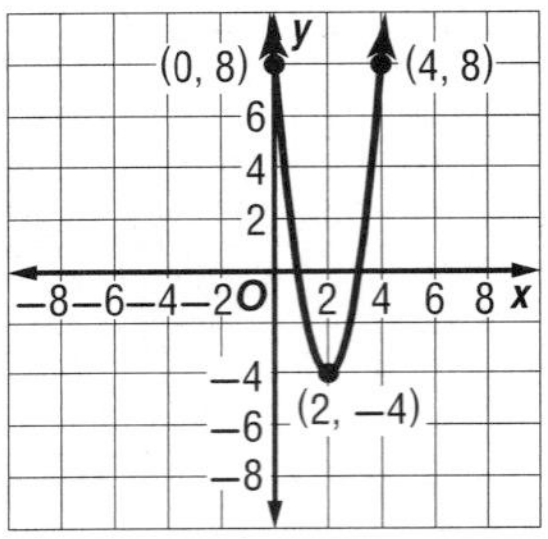

79.

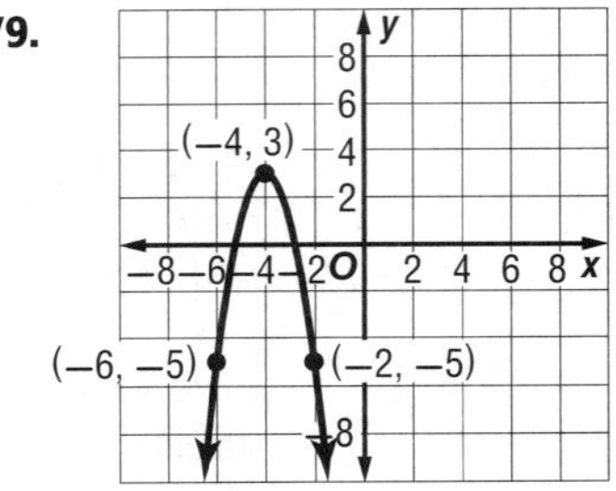

81. 42 **83.** 28 **85.** $\frac{7}{8}$ **87.** $\frac{1}{2}$ **89.** $\frac{5}{9}$ **91.** $4x(3ax^2 + 5bx + 8c)$ **93.** $(3y + 2)(4y + 3)$ **95.** $(2x - 3)(4a - 3)$

Pages 345–347 Lesson 6-2

1. $4y + 2x - 2$ **3.** $x - 8 - \frac{4}{x+2}$ **5.** $3z^3 - 15z^2 + 36z - 105 + \frac{309}{z+3}$ **7.** A **9.** $6a + 6 + \frac{21}{3a - 2}$ **11.** $3y + 5$ **13.** $x + 3y - 2$ **15.** $2a^2 + b - 3$ **17.** $3np - 6 + 7p$ **19.** $-w + 16 + \frac{1000}{w}$

21

$$\begin{array}{r} b^2 - 5b + 6 \\ b+1 \overline{)\, b^3 - 4b^2 + b - 2} \\ (-)\ b^3 + b^2 \qquad\qquad \\ \hline -5b^2 + b \qquad \\ (-)\ -5b^2 - 5b \qquad \\ \hline 6b - 2 \\ (-)\ 6b + 6 \\ \hline 8 \end{array}$$

The quotient is $b^2 - 5b + 6$, and the remainder is 8. So, the expression equals $b^2 - 5b + 6 - \frac{8}{b+1}$.

23. $x^4 + 4x^3 + 12x^2 + 52x + 208 + \frac{832}{x-4}$ **25.** $g^3 + 2g^2 + g + 2 - \frac{14}{g-2}$ **27.** $2x^4 + x^3 - x + \frac{2}{3} - \frac{2}{9x+3}$ **29.** $b^2 - 4b + 8 - \frac{8}{b+1}$ **31.** $2y^5 - y^4 + y^3 + y^2 - y - 3$ **33.** $V(t) = t^2 + 5t + 6$

35 a.

$$\begin{array}{r} 3500 \qquad \\ a^2 + 100 \overline{)\, 3500a^2 \qquad} \\ (-)\ 3500a^2 + 350{,}000 \\ \hline -350{,}000 \end{array}$$

The quotient is 3500, and the remainder is −350,000. So, the expression equals $3500 - \frac{350{,}000}{a^2 + 100}$.

b. $n = 3500 - \frac{350{,}000}{a^2 + 100}$ Write the equation.

$= 3500 - \frac{350{,}000}{15^2 + 100}$ $a = 15$

$= 3500 - \frac{350{,}000}{325}$ Simplify.

≈ 2423 subscriptions

37. $\frac{4c^2d - 3d}{2}$ **39.** $n^2 - n - 1$ **41.** $3z^4 - z^3 + 2z^2 - 4z + 9 - \frac{13}{z+2}$ **43.** Sample answer: Sharon; Jamal actually divided by $x + 3$. **45.** Sample answer: The degree of the quotient plus the degree of the divisor equals the degree of the dividend. **47.** $\frac{5}{x^2}$ does not belong with the other three. The other three expressions are polynomials. Since the denominator of $\frac{5}{x^2}$ contains a variable, it is not a polynomial. **49.** A **51.** 360 **53.** $3x^3 + 2x^2 + x + 4$ **55.** $23a^2 - 24a$ **57.** $8x^5y^8z^3$ **59.** 0 to 10 ft or 24 to 34 ft **61.** $4 \pm \sqrt{19}$ **63.** between −6 and −5; between −3 and −2 **65.** between −1 and 0; between 2 and 3 **67.** −21 **69.** −20 **71.** $-9d^2$

Pages 352–355 Lesson 6-3

1. degree = 6, leading coefficient = 11 **3.** not in one variable because there are two variables, x and y **5.** $w(5) = -247$; $w(-4) = 104$ **7.** $4y^9 - 5y^6 + 2$ **9.** $1536a^3 - 426a^2 - 144a + 82$ **11.** $f(x) \to -\infty$ as $x \to -\infty$. $f(x) \to +\infty$ as $x \to +\infty$. Since the end behavior is in opposite directions, it is an odd-degree function. The graph intersects the x-axis at three points, so there are three real zeros. **13.** not in one variable because there are two variables, x and y **15.** degree = 6, leading coefficient = −12 **17.** degree = 4, leading coefficient = −5

19. degree = 2, leading coefficient = 3 **21.** degree = 9, leading coefficient = 2 **23.** $p(-6) = 1227$; $p(3) = 66$
25. $p(-6) = -156$; $p(3) = 78$

27

$p(x) = -x^3 + 3x^2 - 5$	Original function
$p(-6) = -(-6)^3 + 3(-6)^2 - 5$	Replace x with -6.
$= 216 + 108 - 5$	Simplify.
$= 319$	Simplify.

$p(x) = -x^3 + 3x^2 - 5$	Original function
$p(-6) = -(3)^3 + 3(3)^2 - 5$	Replace x with 3.
$= -27 + 27 - 5$	Simplify.
$= -5$	Simplify.

29. $18a^2 - 12a + 3$ **31.** $2b^4 - 4b^2 + 3$ **33.** $-64y^3 + 144y^2 - 104y + 25$ **35.** $f(x) \to +\infty$ as $x \to -\infty$. $f(x) \to +\infty$ as $x \to +\infty$. Since the end behavior is in the same direction, it is an even-degree function. The graph intersects the x-axis at four points, so there are four real zeros. **37.** $f(x) \to -\infty$ as $x \to -\infty$. $f(x) \to +\infty$ as $x \to +\infty$. Since the end behavior is in opposite directions, it is an odd-degree function. The graph intersects the x-axis at one point, so there is one real zero. **39.** $f(x) \to -\infty$ as $x \to -\infty$. $f(x) \to -\infty$ as $x \to +\infty$. Since the end behavior is in the same direction, it is an even-degree function. The graph intersects the x-axis at two points, so there are two real zeros.

41

$KE(v) = 0.5mv^2$	Original function
$= 0.5(171)(11)^2$	Replace m with 171 and v with 11.
$= 10{,}345.5$	Simplify.

The kinetic energy is 10,345.5 kg-m/s or 10,345 joules.

43. $p(-2) = -16$; $p(8) = 1024$ **45.** $p(-2) = -0.5$; $p(8) = 3112$ **47.** D **49.** A **51.** $3a^3 - 24a^2 + 240a + 66$
53. $5a^6 - 298a^2 + 1008a - 928$

55a.

x	$p(x)$
−7	−585
−6	0
−4	240
−3	135
−2	0
0	−144
1	−105
2	0
4	240
6	0
7	−585

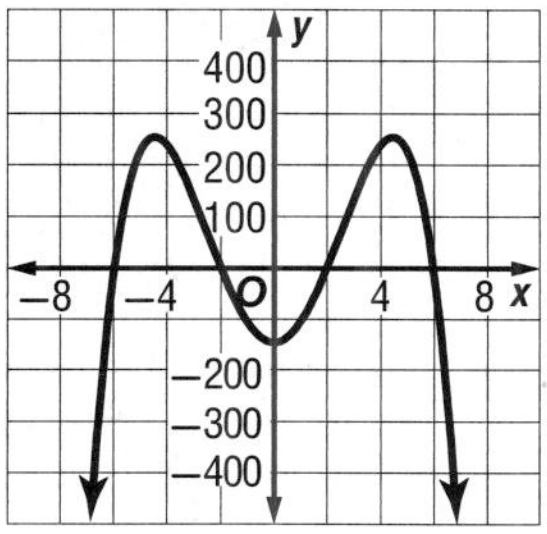

55b. −6, −2, 2, 6 **55c.** 2000 and 6000 items **55d.** Sample answer: The negative values should not be considered because the company will not produce negative items.

57 The degree, 4, is even and the leading coefficient, −5, is negative. So, $f(x) \to -\infty$ as $x \to -\infty$ and $f(x) \to -\infty$ as $x \to +\infty$.
59. $f(x) \to +\infty$ as $x \to -\infty$; $f(x) \to -\infty$ as $x \to +\infty$
61. $f(x) \to -\infty$ as $x \to -\infty$; $f(x) \to +\infty$ as $x \to +\infty$
63. Sample answer: Virginia is correct; an even function will have an even number of zeros and the double root represents 2 zeros. **65.** Sample answer: $f(x) \to +\infty$ as $x \to -\infty$; $f(x) \to +\infty$ as $x \to +\infty$; $\frac{f(x)}{g(x)}$ will become a 2-degree function with a positive leading coefficient. **67.** Sometimes; a polynomial function with four real roots may be a sixth-degree polynomial function with two imaginary roots. A polynomial function that has four real roots is at least a fourth-degree polynomial. **69.** Student A

71a.

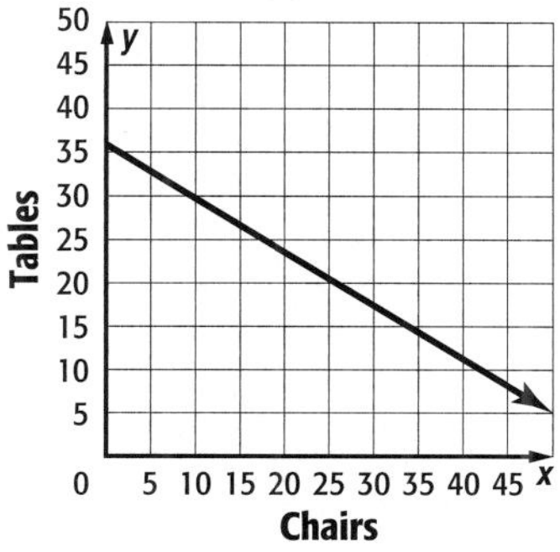

71b. $t = 0.5c$;

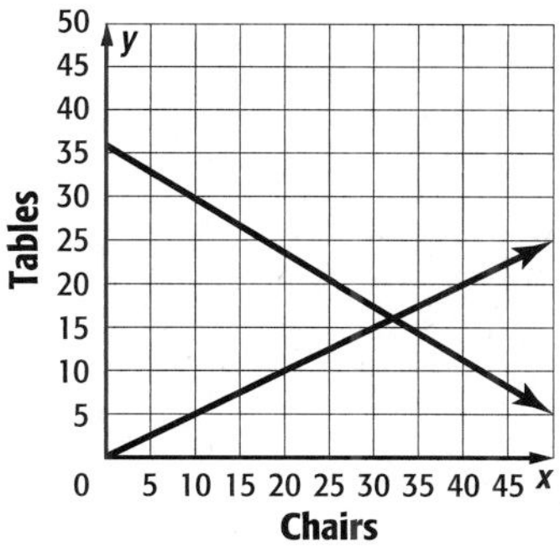

71c. 16 tables and 32 chairs **71d.** Sample answer: This can be determined by the intersection of the graphs. This point of intersection is the optimal amount of tables and chairs manufactured.
73. $2x^2y^2 + 4x^4y^4z^2$ **75.** $6c^3 - 1 + 4a^5cd^2$ **77.** yes; 6
79. $h(d) = -2d^2 + 4d + 6$; The graph opens downward and is narrower than the parent graph, and the vertex is at (1, 8). **81.** $x \le -\frac{2}{3}$ or $x \ge 2$ **83.** $\left(\frac{4}{3}, -\frac{4}{3}\right)$; minimum **85.** (8, 11); maximum

Pages 361–364 Lesson 6-4

1.

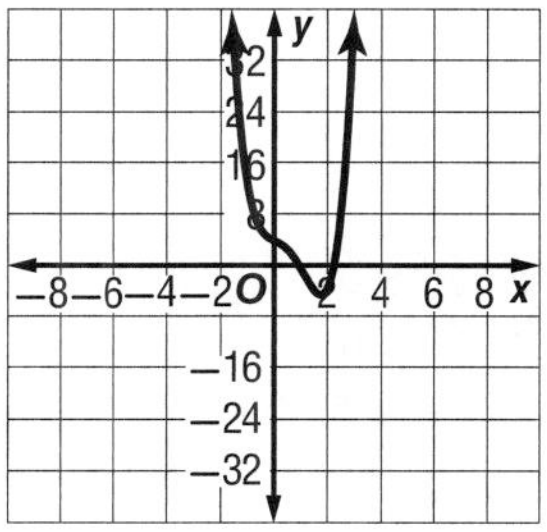

3.

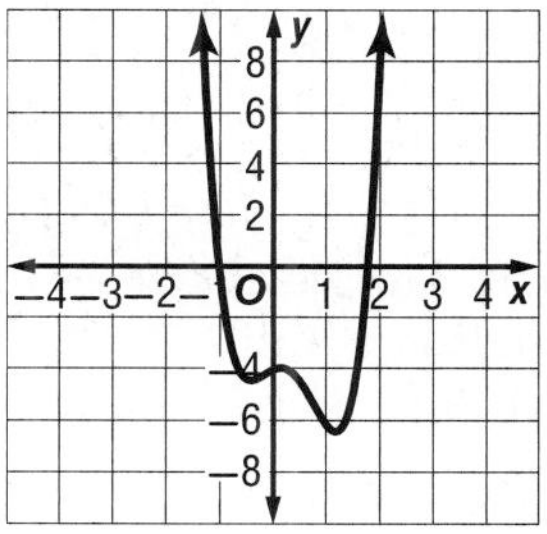

5. between −2 and −1

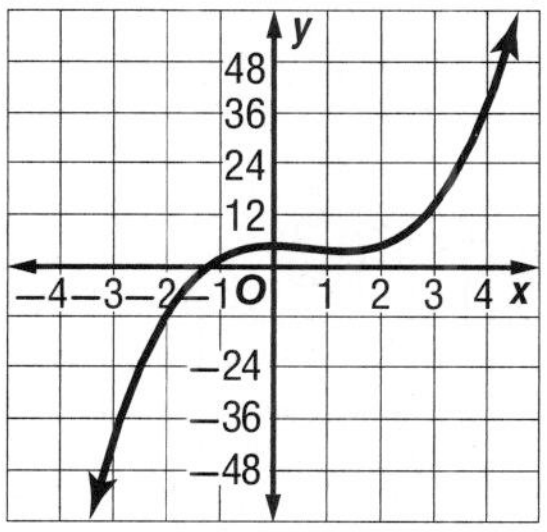

7. between 0 and 1 and between 2 and 3

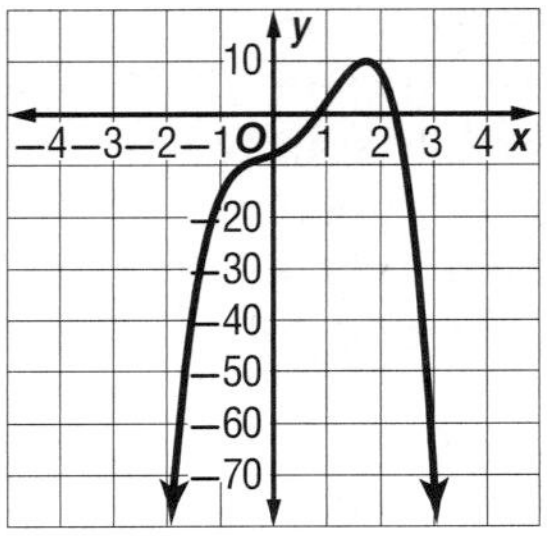

9. rel. max at $x \approx -1.8$; rel. min at $x \approx 1$; D = {all real numbers}, R = {all real numbers}

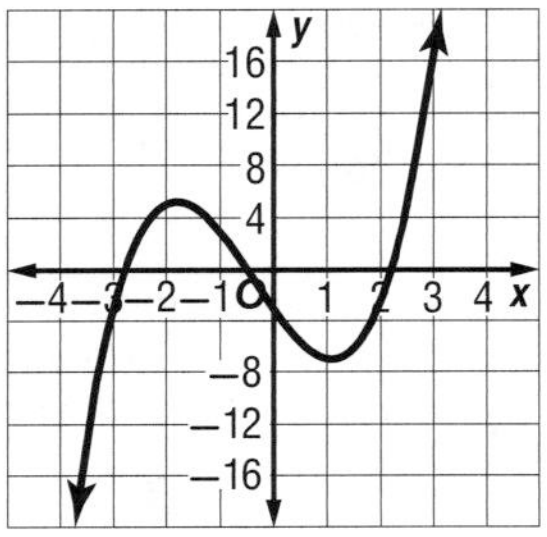

11. rel. max at $x \approx 2.5$; rel. min at $x \approx 0.5$; D = {all real numbers}, R = {all real numbers}

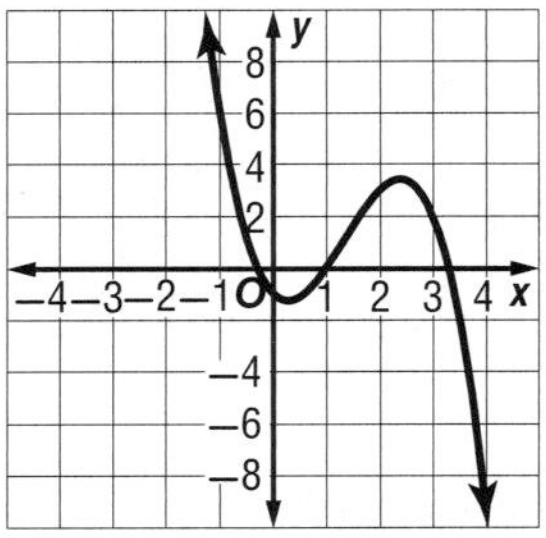

13a.

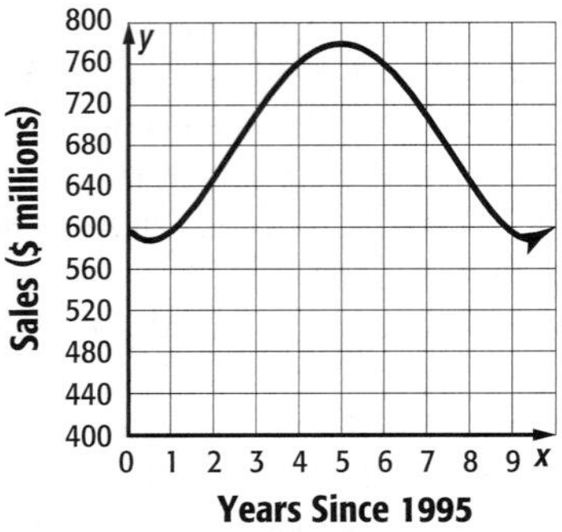

13b. Sample answer: The graph (music sales) increases until $x = 5$ (year 2000), then decreases until $x \approx 9.5$ (year 2004), and then increases indefinitely.

13c.

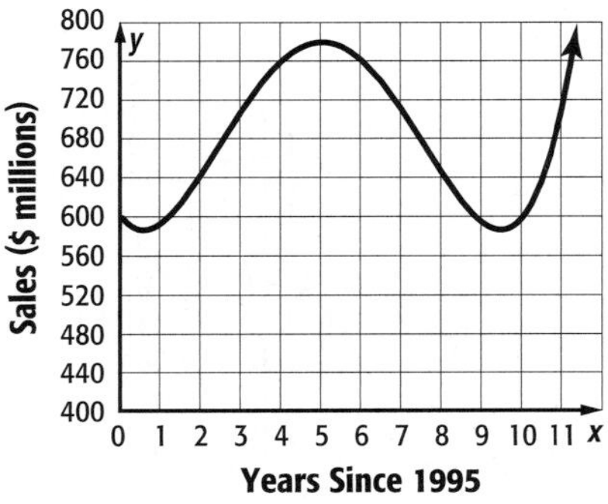

Sample answer: This suggests a dramatic increase in sales.

13d. Sample answer: No; with so many other forms of media on the market today, CD sales will not increase dramatically. In fact, the sales will probably decrease. The function appears to be accurate only until about 2005.

15 **a.** Since $f(x)$ is a third-degree polynomial function, it will have either 3 or 1 real zeros. Look at the values of $f(x)$ to locate the zeros. Then use the points to sketch the graph.

x	$f(x)$
−4	92
−3	41
−2	12
−1	−1
0	−4
1	−3
2	−4
3	−13
4	−36

← change in sign

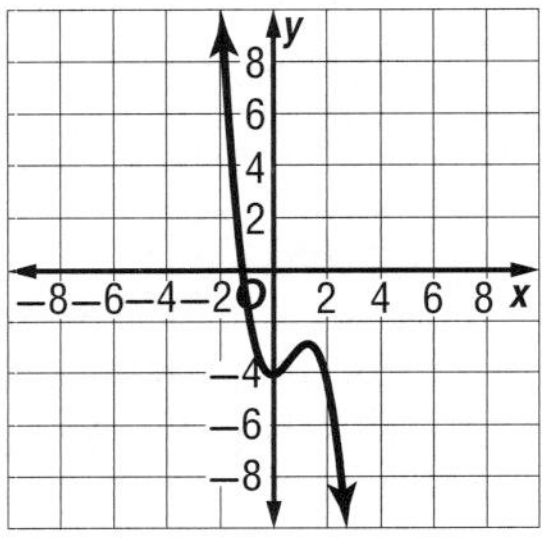

b. The value of $f(x)$ changes signs between $x = -2$ and $x = 1$. So, there is a zero between −2 and −1.
c. The value of $f(x)$ at $x = 0$ is less than the surrounding points, so there must be a relative minimum near $x = 0$. The value of $f(x)$ near $x = 1$ is greater than the surrounding points, so there must be a relative maximum near $x = 1$.

17a.

x	$f(x)$
−4	−155
−3	−80
−2	−33
−1	−8
0	1
1	0
2	−5
3	−8
4	−3
5	16

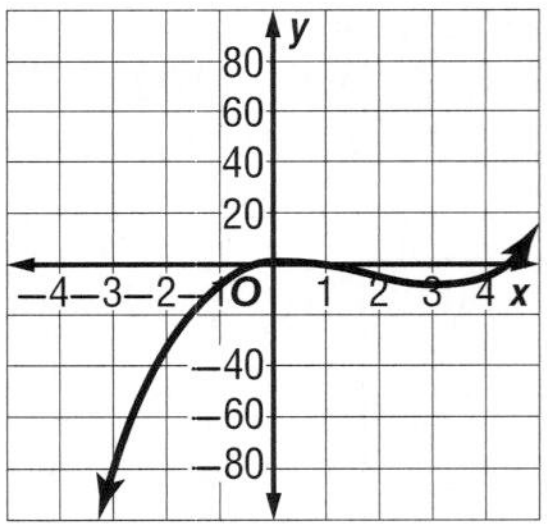

17b. at $x = 1$ and between $x = 4$ and $x = 5$ **17c.** rel. max: $x \approx \frac{1}{3}$, rel. min: $x = 3$

19a.

x	$f(x)$
−4	−176
−3	−77
−2	−22
−1	1
0	4
1	−1
2	−2
3	13
4	56

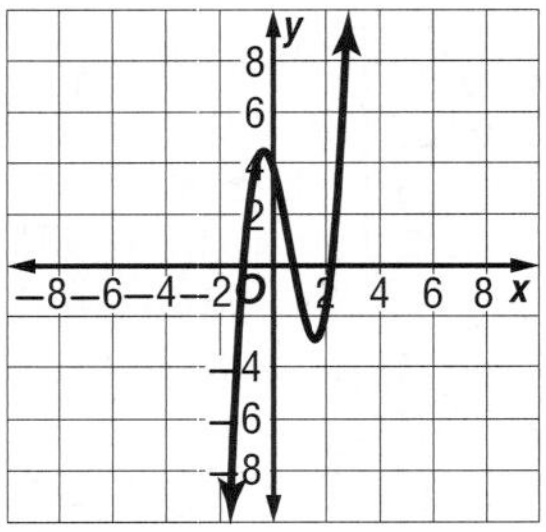

19b. between $x = -2$ and $x = -1$, between $x = 0$ and $x = 1$, and between $x = 2$ and $x = 3$
19c. rel. max: near $x = 0$; rel. min: near $x = 2$

21a.

x	$f(x)$
−4	372
−3	141
−2	36
−1	−3
0	−12
1	−3
2	36
3	141
4	372

21b. between $x = -2$ and $x = -1$ and between $x = 1$ and $x = 2$ **21c.** min: near $x = 0$ **23.** rel. max: $x = -2.73$; rel. min: $x = 0.73$ **25.** rel. max: $x = 1.34$, no rel. min

27. Sample answer:

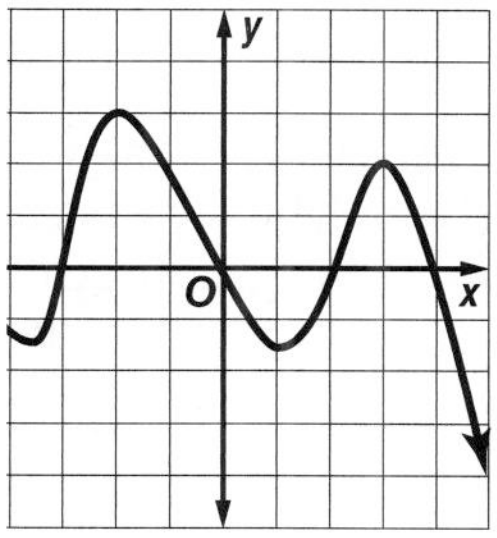

29. Sample answer:

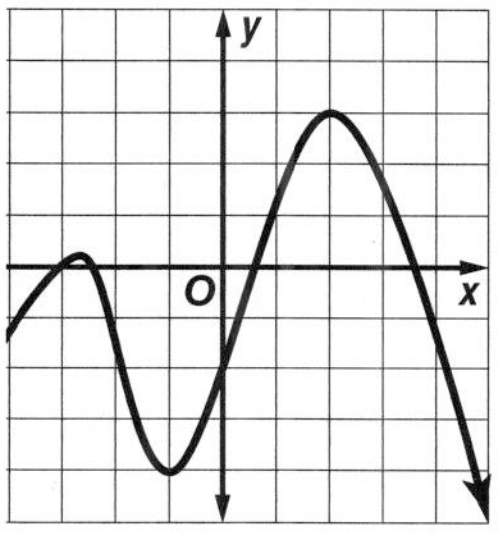

31. Sample answer:

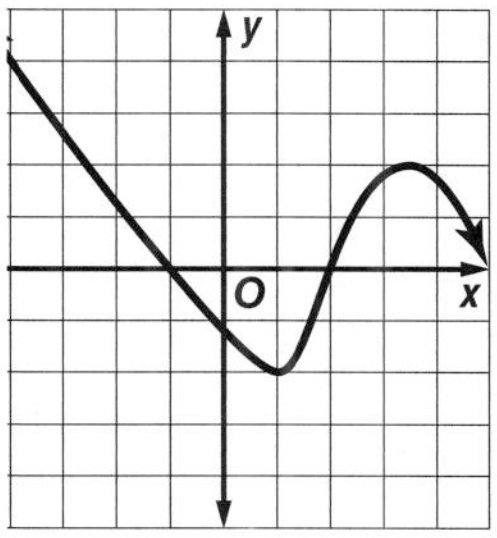

33 a.

x	$d(x)$
0	0
1	0.0145
2	0.056
3	0.1215
4	0.208
5	0.3125
6	0.432
7	0.5635
8	0.704
9	0.8505
10	1

b. Plot the points in the table and connect with a smooth curve.

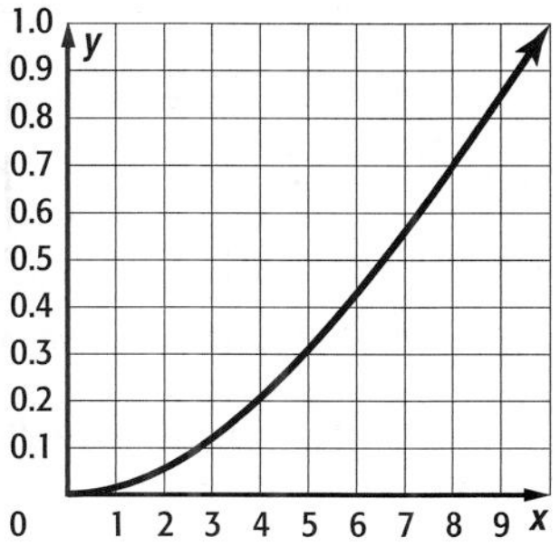

c. $d(x) \to +\infty$ as $x \to +\infty$; as x increases, $d(x)$ increases.
d. Sample answer: Since the diving board is only 10 feet long, x cannot be greater than 10. So, this trend cannot continue indefinitely.

35a. −2.5 (min), −0.5 (max), 1.5 (min) **35b.** −3.5, −1, 0, 3 **35c.** 4 **35d.** D = {all real numbers}; R = $\{y \mid y \geq -3.1\}$ **37a.** −3.5 (min), −2.5(max), −2 (min), −1 (max), 1 (min) **37b.** −3.75, −3.25, −2, −1.75, −0.25, 2.5 **37c.** 6 **37d.** D = {all real numbers}; R = $\{y \mid y \geq -4.9\}$ **39a.** −1.75 (max), 1 (min) **39b.** −3, −0.5, 2 **39c.** 3 **39d.** D = {all real numbers}; R = {all real numbers}

41.

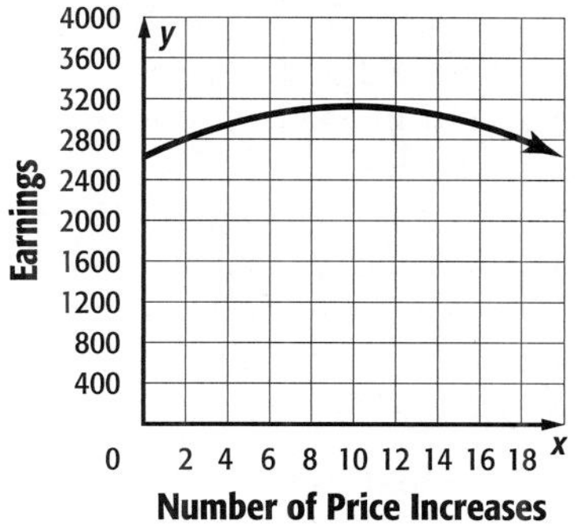

$1.25

43 a. Make a table of values and graph the function.

x	$f(x)$
−2	−16
−1.5	−4.5
−1	−4
−0.5	−4.6
0	−4
0.5	−3.3
1	−4
1.5	−3.5
2	8

← change in sign (between 1.5 and 2)

The value of $f(x)$ changes signs between $x = 1.5$ and $x = 2$. So, there is a zero between these values, at approximately 1.75.
The x-intercept ≈ 1.75.
The y-intercept is at −4.
The value of $f(x)$ at $x \approx -1.25$ and at $x \approx 0.5$ is greater than the surrounding points, so $x \approx -1.25$ and $x \approx 0.5$ are turning points. The value of $f(x)$ at $x \approx -0.5$ and at $x \approx 1.25$ is less than the surrounding points, so $x \approx -0.5$ and $x \approx 1.25$ are turning points.
b. The function does not have an axis of symmetry because the graph is not a parabola.
c. The function is increasing in the intervals $x \leq -1.25$, $-0.5 \leq x \leq 0.5$, and $x \geq 1.25$.
The function is decreasing in the intervals $-1.25 \leq x \leq -0.5$ and $0.5 \leq x \leq 1.25$.

45a. no zeros, no x-intercepts, y-intercept: 5; no turning points **45b.** no axis of symmetry **45c.** decreasing: $x \leq -4$; constant: $-4 < x \leq 0$; increasing: $x > 0$
47. As the x-values approach large positive or negative numbers, the term with the largest degree becomes more and more dominant in determining the value of $f(x)$.
49. Sample answer:

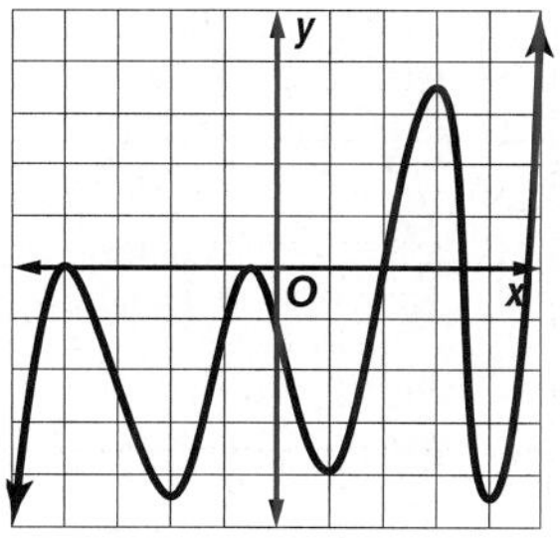

51. Sample answer: No; $f(x) = x^2 + x$ is an even degree, but $f(1) \neq f(-1)$. **53.** Sample answer: The degree will help determine whether the graph is even or odd and the maximum number of zeros and turning points for the graph. The leading coefficient determines the end behavior of the graph, and, along with the degree, builds the shape of the graph. The zeros and turning points allow for the plotting of specific points in the center of the graph. All of these things combine for an accurate sketch of the graph of a polynomial function. **55.** 95 **57.** C
59. $f(x) \to -\infty$ as $x \to -\infty$. $f(x) \to -\infty$ as $x \to +\infty$. Since the end behavior is in the same direction, it is an even-degree function. The graph intersects the x-axis at six points, so there are six real zeros.
61. $(x - 2)(x + 3)$ **63.** $2a^2 + a - 3 - \frac{3}{a - 1}$
65. $(x + 6)(x + 3)$ **67.** $(a + 8)(a - 2)$
69. $(3x - 4)(2x + 1)$

Pages 372–375 Lesson 6-5

1. $(a + b)(3x + 2y - z)$ **3.** prime
5. $12q(w - q)(w^2 + qw + q^2)$
7. $x^2(a + b)(a - b)(a^4 + a^2b^2 + b^4)$
9. $(2c - 5d)(4c^2 + 12cd + 25d^2)$
11. $4, -4, \pm\sqrt{3}$ **13.** -3 **15.** 5 ft **17.** not possible
19. $\sqrt{6}, -\sqrt{6}, 2\sqrt{3}, -2\sqrt{3}$
21. $x(4x + y)(16x^2 - 4xy + y^2)$
23. $y^3(x^2 + y^2)(x^4 - x^2y^2 + y^2)$ **25.** prime
27. $(6x^2 - 5y^2)(2a - 3b + 4c)$
29. $8x^5 - 25y^3 + 80x^4 - x^2y^3 + 200x^3 - 10xy^3$ Original expression
$= (8x^5 + 80x^4 + 200x^3) + (-25y^3 - x^2y^3 - 10xy^3)$
Group to find a GCF.
$= 8x^3(x^2 + 10x + 25) - y^3(25 + x^2 + 10x)$
Factor the GCF.
$= 8x^3(x + 5)^2 - y^3(x + 5)^2$ Perfect squares
$= (8x^3 - y^3)(x + 5)^2$ Distributive Property
$= (2x - y)(4x^2 + 2xy + y^2)(x + 5)^2$ Difference of cubes
31. $6, -6, \pm 2\sqrt{5}$ **33.** $\pm\sqrt{7}, \pm i\sqrt{13}$ **35.** $-\frac{1}{4}$
37. $-15(x^2)^2 + 18(x^2) - 4$ **39.** not possible
41. $4(2x^5)^2 + 1(2x^5) + 6$ **43.** $\pm\sqrt{5}, \pm i\sqrt{2}$
45. $\pm\frac{2\sqrt{3}}{3}, \pm\frac{\sqrt{15}}{3}$ **47.** $\pm\frac{\sqrt{6}}{6}, \pm i\frac{\sqrt{3}}{2}$
49. $(x^2 + 25)(x + 5)(x - 5)$ **51.** $x(x + 2)(x - 2)(x^2 + 4)$
53. $(5x + 4y + 5z)(3a - 2b + c)$
55. $x(x + 3)(x - 3)(3x + 2)(2x - 5)$ **57.** $x = 8$; 5, 8, 11
59. $\pm\frac{2\sqrt{3}}{3}, \pm i\frac{\sqrt{2}}{2}$ **61.** $\pm\frac{1}{3}, \pm i\frac{\sqrt{10}}{2}$ **63.** $3, -3, \pm i\frac{\sqrt{15}}{3}$
65
$x^6 - 26x^3 - 27 = 0$ Original equation
$(x^3)^2 - 26(x^3) - 27 = 0$ $(x^3)^2 = x^6$
$u^2 - 26u - 27 = 0$ Let $u = x^3$.
$(u - 27)(u + 1) = 0$ Factor.
$u = 27$ or $u = -1$ Zero Product Property
$x^3 = 27$ $x^3 = -1$ Replace u with x^3.
$x = 3$ $x = -1$ Take the cube root.
The solutions are -1 and 3.
67. $-1, 1, \pm\frac{1}{2}$ **69.** $\pm i\sqrt{5}, \pm i\sqrt{3}$ **71a.** 2 ft
71b. 176 ft^2 **71c.** 428 ft^2

73 a. $f(x) = (x + 6)[x + x + (x + 2) + (x + 2)] + x[x + (x + 2) + (x + 2)] + x(x + 2)$
$= (x + 6)(4x + 4) + x(3x + 4) + x(x + 2)$
$= 4x^2 + 24x + 4x + 24 + 3x^2 + 4x + x^2 + 2x$
$= 8x^2 + 34x + 24$
b. $f(x) = 8x^2 + 34x + 24$ Original function
$1366 = 8x^2 + 34x + 24$ Replace $f(x)$ with 1366.
$0 = 8x^2 + 34x - 1342$ Subtract 1366 from each side.
$0 = 2(4x^2 + 17x - 671)$ Factor.
$0 = 2(4x + 61)(x - 11)$ Perfect squares
$4x + 61 = 0$ or $x - 11 = 0$ Zero Product Property
$x = -15.25$ $x = 11$ Simplify.
Since distance cannot be negative, $x = 11$ ft.
75. $(x + 2)^3(x - 2)^3$ **77.** $(x + y)^3(x - y)^3$
79. $(6x^n + 1)^2$ **81.** Sample answer: $a = 1, b = -1$
83. Sample answer: The factors can be determined by the x-intercepts of the graph. An x-intercept of 5 represents a factor of $(x - 5)$. **85.** D **87.** D
89. rel max at $x \approx 1.75$, rel min at $x \approx 0$;

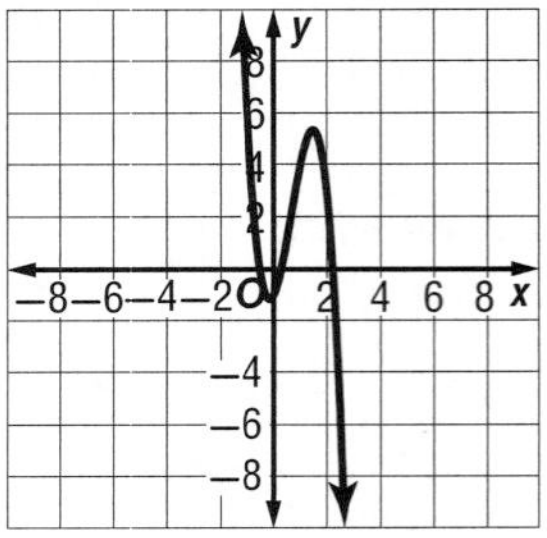

91. degree = 4; leading coefficient = 5
93. degree = 7; leading coefficient = −1 **95.** 18 skis and 10 snowboards **97.** $x + 2 - \frac{10}{x + 4}$
99. $8x^2 - 12x + 24 - \frac{42}{x + 2}$

Pages 380–382 Lesson 6-6

1. 58; −20 **3.** 12,526 **5.** $x + 4, x - 4$ **7.** $x - 5, 2x - 1$
9. 71; −6 **11.** −435; −15 **13.** −4150; 85 **15.** 647; −4
17. $(x - 1)^2$ **19.** $x - 4, x + 1$ **21.** $x + 6, 2x + 7$
23. $x + 1, x^2 + 2x + 3$ **25.** $x - 4, 3x - 2$
27 a.

1	−0.04	0.8	0.5	−1	0
		−0.04	0.76	1.26	0.26
	−0.04	0.76	1.26	0.26	0.26

$f(1) = 0.26$ ft/s; At 1 second, the speed of the boat is 0.26 ft/s.

2	−0.04	0.8	0.5	−1	0
		−0.08	1.44	3.88	5.76
	−0.04	0.72	1.94	2.88	5.76

$f(2) = 5.76$ ft/s; At 2 seconds, the speed of the boat is 5.76 ft/s.

3	−0.04	0.8	0.5	−1	0
		−0.12	2.04	7.62	19.86
	−0.04	0.68	2.54	6.62	19.86

$f(3) = 19.86$ ft/s; At 3 seconds, the speed of the boat is 19.86 ft/s.

b.

6	−0.04	0.8	0.5	−1	0
		−0.24	3.36	23.16	132.96
	−0.04	0.56	3.86	22.16	132.96

$f(6) = 132.96$ ft/s; This means the boat is traveling at 132.96 ft/s when it passes the second buoy.

29. $x + 2, x - 3, x^2 - x + 4$

31a. $9x^4 + 50x^3 + 51x^2 - 150x - 72$

31b.

x	f(x)
−5	−9922
−4	−4160
−3	−1242
−2	−112
−1	70
0	−72
1	−130
2	88
3	558
4	1040
5	1078

31c. There is a zero between $x = -2$ and $x = -1$ because $f(x)$ changes sign between the two values. There is another zero between $x = 1$ and $x = 2$ because $f(x)$ changes sign between the two values. Since we have located 2 zeros for the depressed polynomial and the degree is 4, the other 2 zeros are complex.

31d.

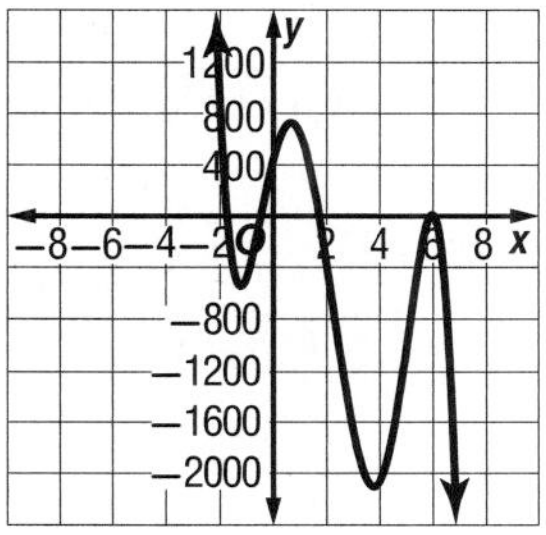

33

2	1	k	−17
		2	$2k + 4$
	1	$k + 2$	3

$-17 + 2k + 4 = 3$ Write an equation.
$-13 + 2k = 3$ Simplify.
$2k = 16$ Add 13 to each side.
$k = 8$ Divide each side by 2.

35. −3 **37.** $\pm\sqrt{6}, \pm\sqrt{3}$ **39a.** $x - c$ is a factor of $f(x)$. **39b.** $x - c$ is not a factor of $f(x)$. **39c.** $f(x) = x - c$ **41.** Sample answer: $f(x) = -x^3 + x^2 + x + 10$ **43.** Sample answer: A zero can be located using the Remainder Theorem and a table of values by determining when the output, or remainder, is equal to zero. For instance, if $f(6)$ leaves a remainder if 2 and $f(7)$ leaves a remainder of −1, then you know that there is a zero between $x = 6$ and $x = 7$. **45.** 4 **47.** B **49.** $\pm 3, \pm i\sqrt{3}$ **51a.** −1.5 (max), 0.5 (min), 2.5 (max) **51b.** −3.5, 3.75 **51c.** 4 **51d.** D = {all real numbers}; R = $\{y \mid y \leq 4.5\}$ **53a.** −3 (min), −1.5 (max), 1 (min) **53b.** −0.25, 3 **53c.** 4 **53d.** D = {all real numbers}; R = $\{y \mid y \geq 4.1\}$

55.

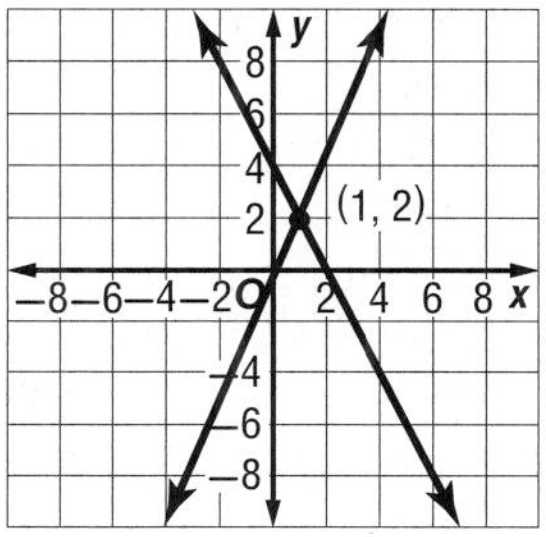

57.

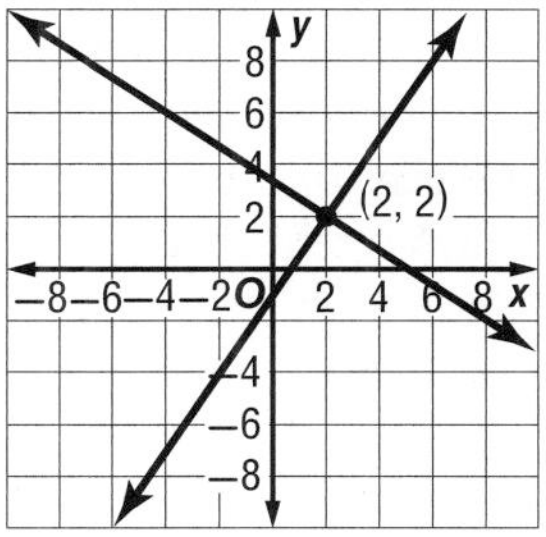

59. $4a^2 - 8a + 16$ **61.** $79a^2 - 58a + 12$ **63.** $-4a^4 - 24a^2 - 48a - 22$

Pages 388–390 Lesson 6-7

1. −2, 5; 2 real **3.** $-\frac{3}{2}, \frac{3}{2}, -\frac{3}{2}i, \frac{3}{2}i$; 2 real, 2 imaginary **5.** 3 or 1; 0; 0 or 2 **7.** 1 or 3; 0 or 2; 0, 2, or 4 **9.** −8, −2, 1 **11.** $-4, 6, -4i, 4i$ **13.** $x^3 - 9x^2 + 14x + 24$ **15.** $x^4 - 3x^3 - x^2 - 27x - 90$ **17.** $-2, \frac{3}{2}$; 2 real **19.** $-1, \frac{1 \pm i\sqrt{3}}{2}$; 1 real, 2 imaginary **21.** $-\frac{8}{3}$, 1; 2 real **23.** $-\frac{5}{2}, \frac{5}{2}, -\frac{5}{2}i, \frac{5}{2}i$; 2 real, 2 imaginary **25.** −2, −2, 0, 2, 2; 5 real

27 Find the number of sign changes for $f(x)$ and $f(-x)$.

$f(x) = x^4 - 5x^3 + 2x^2 + 5x + 7$

yes yes no no

Since there are 2 sign changes, the function has 0 or 2 positive real zeros.

$f(-x) = x^4 + 5x^3 + 2x^2 - 5x + 7$

no no yes yes

Since there are 2 sign changes, the function has 0 or 2 negative real zeros.

Since f(x) has degree 4, the function has 4 zeros. So, the function could have the following.
2 positive real zeros, 2 negative real zeros, 0 imaginary zeros
2 positive real zeros, 0 negative real zeros, 2 imaginary zeros
0 positive real zeros, 0 negative real zeros, 4 imaginary zeros

29. 0 or 2; 1, 2 or 4 **31.** 0 or 2; 0 or 2; 2, 4, or 6 **33.** −6, −2, 1 **35.** $-4, 7, -5i, 5i$ **37.** $4, 4, -2i, 2i$ **39.** $5, 4 + i, 4 - i$ **41.** $-\frac{1}{2}, \frac{1}{2}, -2i, 2i$ **43.** $x^3 - 2x^2 - 13x - 10$ **45.** $x^4 + 2x^3 + 5x^2 + 8x + 4$ **47.** $x^4 - x^3 - 20x^2 + 50x$

49 a. Find the number of sign changes for $P(x)$ and $P(-x)$.

$P(x) = -0.006x^4 + 0.15x^3 - 0.05x^2 - 1.8x$

yes yes no

Since there are 2 sign changes, the function has 0 or 2 positive real zeros.

$P(-x) = -0.006x^4 - 0.15x^3 - 0.05x^2 + 1.8x$

no no yes

Since there is 1 sign change, the function has 1 negative real zero.

Since $f(x)$ has degree 4, the function has 4 zeros. So, the function could have the following.
2 positive real zeros, 1 negative real zero, 1 imaginary zero
0 positive real zeros, 1 negative real zero, 3 imaginary zeros

b. Negative zeros do not make sense in the problem because the number of computers produced cannot be negative. Since the zeros are values of x for which $P(x) = 0$, nonnegative zeros represent numbers of computers produced per hour which lead to no profit for the manufacturer.

51. b **53a.** 3 or 1, 0, 2 or 0

53b.

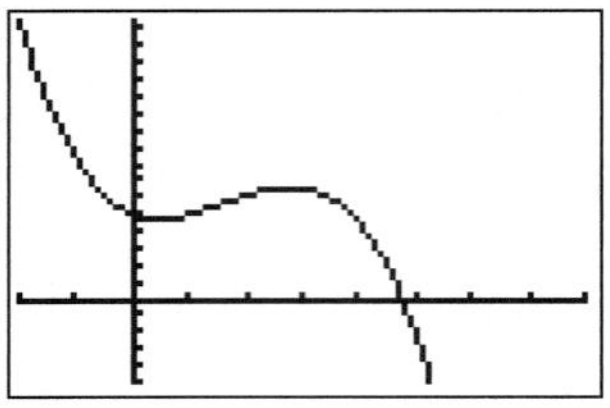

$[-10, 40]$ scl: 5 by $[-4000, 13{,}200]$ scl: 100

53c. 23.8; Sample answer: According to the model, the music hall will not earn any money after 2026.
55. 1 positive, 2 negative, 2 imaginary; Sample answer: The graph crosses the positive x-axis once, and crosses the negative x-axis twice. Because the degree of the polynomial is 5, there are 5 − 3 or 2 imaginary zeros. **57.** Sample answer: $f(x) = (x + 2i)(x - 2i)(3x + 5)(x + \sqrt{5})(x - \sqrt{5})$; Use conjugates for the imaginary and irrational values.
59a. $f(x) = x^4 + 4x^2 + 4$ **59b.** $f(x) = x^4 - 4x^2 + 4$
61. C **63.** H **65.** $f(-8) = -1638$; $f(4) = 342$
67. $f(-8) = -63{,}940$; $f(4) = 1868$
69. $(a^2 + b^2)(a^4 - a^2b^2 + b^4)$ **71.** $(a - 4)(a - 2)(5a + 2b)$
73. 0.25 s **75.** $\pm1, \pm2, \pm4, \pm8, \pm\frac{1}{5}, \pm\frac{2}{5}, \pm\frac{4}{5}, \pm\frac{8}{5}$

Pages 393–396 Lesson 6-8

1. $\pm1, \pm2, \pm3, \pm4, \pm6, \pm8, \pm12, \pm24$
3. 5 in. × 9 in. × 28 in.

5 If $\frac{p}{q}$ is a rational zero, then p is a factor of 12 and q is a factor of 2.
p: $\pm1, \pm2, \pm3, \pm4, \pm6, \pm12$ q: $\pm1, \pm2$
possible rational zeros: $\frac{p}{q} = \pm1, \pm2, \pm3, \pm4, \pm6, \pm12, \pm\frac{1}{2}, \pm\frac{3}{2}$

There are no changes of signs for $f(x)$, so there are no positive real zeros. There are 4 changes of signs for $f(-x)$, so there are 0, 2, or 4 negative real zeros. Make a table for synthetic division and test the possible negative values.

$\frac{p}{q}$	2	11	26	29	12
−1	2	9	17	12	0
−2	2	7	12	5	2
−3	2	5	11	−4	24
−4	2	3	14	−27	120
−6	2	−1	32	−163	990
−12	2	−13	182	−2155	25,872
$-\frac{1}{2}$	2	10	21	18.5	2.75
$-\frac{3}{2}$	2	8	14	8	0

Because $f(-1) = 0$ and $f\left(-\frac{3}{2}\right) = 0$, there are zeros at $-\frac{3}{2}$ and -1. **7.** $-\frac{1}{2}, \frac{-5 \pm i\sqrt{23}}{8}$ **9.** $-\frac{1}{2}, \frac{3}{2}, 1 + 2i, 1 - 2i$
11. $\pm1, \pm2, \pm4, \pm7, \pm8, \pm14, \pm28, \pm56$

13 If $\frac{p}{q}$ is a rational zero, then p is a factor of 35 and q is a factor of 3.
p: $\pm1, \pm5, \pm7, \pm35$ q: $\pm1, \pm3$
possible rational zeros: $\frac{p}{q} = \pm1, \pm5, \pm7, \pm35, \pm\frac{1}{3}, \pm\frac{5}{3}, \pm\frac{7}{3}, \pm\frac{35}{3}$

15. $\pm1, \pm2, \pm3, \pm6, \pm7, \pm14, \pm21, \pm42, \pm\frac{1}{2}, \pm\frac{3}{2}, \pm\frac{7}{2}, \pm\frac{21}{2}, \pm\frac{1}{4}, \pm\frac{3}{4}, \pm\frac{7}{4}, \pm\frac{21}{4}, \pm\frac{1}{8}, \pm\frac{3}{8}, \pm\frac{7}{8}, \pm\frac{21}{8}$
17. $\pm1, \pm2, \pm4, \pm8, \pm16, \pm32, \pm64, \pm128, \pm\frac{1}{2}, \pm\frac{1}{4}, \pm\frac{1}{8}, \pm\frac{1}{16}$ **19.** −5, −3, −2 **21.** $-5, \frac{3}{4}, 5$ **23.** −1, 2
25. $-\frac{1}{4}$ **27.** −7, 1, 3 **29.** $2, -1, i, -i$ **31.** $0, 3, -i, i$
33. $-2, \frac{4}{3}, \frac{-3 \pm i}{2}$ **35.** $3, \frac{2}{3}, -\frac{2}{3}, \frac{-3 \pm \sqrt{13}}{2}$
37. $-\frac{1}{2}, \frac{1}{3}, \frac{1}{2}, \frac{3}{4}$ **39a.** $V(x) = 324x^3 + 54x^2 - 19x - 2$
39b. $3, 1.05i, -4.22i$; 3 is the only reasonable value for x. The other two values are imaginary.

41 a.

$V = \pi r^2 h$	Volume of a cylinder
$V = \pi r^2(r + 6)$	$h = r + 6$
$V = \pi r^2(r) + \pi r^2(6)$	Distributive Property
$V = \pi r^3 + 6\pi r^2$	Simplify.

b.

$V = \pi r^3 + 6\pi r^2$	Volume of a cylinder
$160\pi = \pi r^3 + 6\pi r^2$	Substitute.
$0 = \pi r^3 + 6\pi r^2 - 160\pi$	Subtract 160π from each side.
$0 = r^3 + 6r^2 - 160$	Divide each side by π.

p: $\pm1, \pm2, \pm4, \pm5, \pm8, \pm20, \pm32, \pm40, \pm80, \pm160$
q: ±1
possible rational zeros: $\frac{p}{q} = \pm1, \pm2, \pm4, \pm5, \pm8, \pm20, \pm32, \pm40, \pm80, \pm160$

Make a table and test some possible rational zeros.

r	1	6	0	-160
1	1	7	7	-153
2	1	8	16	-128
3	1	9	27	-79
4	1	10	40	0

$V(r) = 0$, so there is a zero at $x = 4$. Use the Quadratic Formula to find zeros of the depressed polynomial $r^2 + 10r + 40$.

$$x = \frac{-b \pm \sqrt{b^2 - 4ac}}{2a} \quad \text{Quadratic Formula}$$
$$= \frac{-10 \pm \sqrt{10^2 - 4(1)(40)}}{2(1)} \quad a = 1, b = 10, \text{ and } c = 40$$
$$= \frac{-10 \pm \sqrt{100 - 160}}{2} \quad \text{Multiply.}$$
$$= \frac{-10 \pm \sqrt{-60}}{2} \quad \text{Simplify.}$$
$$= \frac{-10 \pm 2i\sqrt{15}}{2} \quad \sqrt{-60} = 2i\sqrt{15}$$
$$= -5 \pm 2i\sqrt{15}$$

Since the radius cannot be an imaginary number, the only reasonable value is $r = 4$ in.

c. Since $r = 4$ in., $h = r + 6$ or 10 in.

43a. $30x^3 - 478x^2 + 1758x - 7608 = 0$ **43b.** 1, 2, 3, 4, 6, 8, 12, 24, 317, 634, 951, 1268, 1902, 2536, 3804, 7608 **43c.** 2010 **43d.** No; Sample answer: Music sales decline from 1997 to 2005, then increase indefinitely. It is not reasonable to expect sales to increase forever. **45.** 2, 3, 3, -3, -4 **47.** Sample answer: $f(x) = x^4 - 12x^3 + 47x^2 - 38x - 58$ **49.** Sample answer: $f(x) = 4x^5 + 3x^3 + 8x + 18$ **51.** Sample answer: For any polynomial function, the constant term represents p and the leading coefficient represents q. The possible zeros of the function can be found with $\pm\frac{p}{q}$ where the fraction is every combination of factors of p and q. For example, if p is 4 and q is 3, then $\pm 4, \pm 2, \pm 1, \pm\frac{4}{3}, \pm\frac{2}{3}$, and $\pm\frac{1}{3}$ are all possible zeros. **53.** H **55.** 6 **57.** $x^4 + 4x^3 + 11x^2 - 64x - 80$ **59.** $(x - 1)(x + 2)(x + 1)$ **61.** $(x - 3)(x + 4)(x - i)$ **63a.** about 3.5 s **63b.** 430 ft **65.** $3x^3 + 12x$ **67.** 32 **69.** $18c + 2$

Pages 397–400 Chapter 6 Study Guide and Review

1. true **3.** false; depressed polynomial **5.** true **7.** true **9.** true **11.** $\frac{7x}{y^4}$ **13.** $r^2 + 8r - 5$ **15.** $m^3 - m^2p - mp^2 + p^3$ **17.** $3x^3 + 2x^2y^2 - 4xy$ **19.** $a^3 + 3a^2 - 4a + 2$ **21.** $x^2 + 3x - 40$ units^2 **23.** This is not a polynomial in one variable. It has two variables, x and y. **25.** $p(-2) = -3$; $p(x + h) = x^2 + 2xh + h^2 + 2x + 2h - 3$ **27.** $p(-2) = -25$; $p(x + h) = 3 - 5x^2 - 10xh - 5h^2 + x^3 + 3hx^2 + 3h^2x + h^3$

29a.

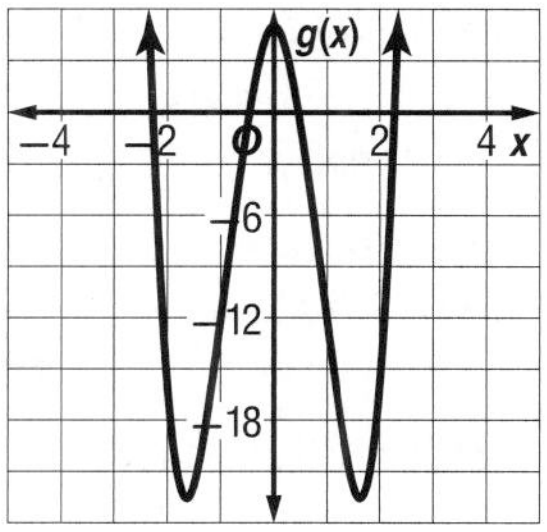

29b. between -3 and -2, between -1 and 0, between 0 and 1, between 2 and 3
29c. rel. max: $x \approx 0$; rel. min: $x \approx 1.62$ and $x \approx -1.62$
31a.

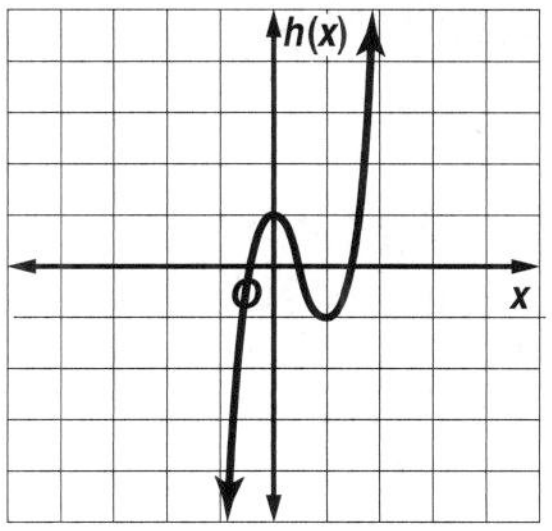

31b. between -1 and 0, between 0 and 1, and between 1 and 2 **31c.** rel. max: $x \approx 0$; rel. min: $x \approx 1$ **33.** 2 relative maxima and 1 relative minima **35.** prime

37. $(2y + z)(3a + 2b - c)$ **39.** $\pm\frac{\sqrt{3}}{2}, \pm\frac{\sqrt{2}}{2}$
41. $f(-2) = 1$; $f(4) = 13$ **43.** $f(-2) = 16$; $f(4) = 118$
45. $x + 2$ and $3x - 1$ **47.** $x + 3$, $x + 4$
49. positive real zeros: 0
negative real zeros: 4, 2, or 0
imaginary zeros: 4, 2, or 0
51. positive real zeros: 2 or 0
negative real zeros: 1
imaginary zeros: 4 or 2
53. $-2, -1 \pm \sqrt{2}$ **55.** $-2, \pm 2i$

Chapter 7 Inverses and Radical Functions and Relations

Page 407 Chapter 7 Get Ready

1. between 0 and 1, and between 3 and 4
3. between 1 and 2 seconds **5.** $3x + 2 - \frac{20}{x + 4}$
7. $3x^3 - 4x^2 + 5x - 3 + \frac{6}{x - 3}$

Pages 413–416 Lesson 7-1

1. $(f + g)(x) = 4x + 1$; $(f - g)(x) = -2x + 3$; $(f \cdot g)(x) = 3x^2 + 5x - 2$; $\left(\frac{f}{g}\right)(x) = \frac{x + 2}{3x - 1}, x \neq \frac{1}{3}$
3. $f \circ g$ is undefined; $g \circ f = \{(2, 8), (6, 13), (12, 11), (7, 15)\}$. **5.** $[f \circ g](x) = -15x + 18$; $[g \circ f](x) = -15x - 6$
7. Either way, she will have \$228.95 taken from her paycheck. If she takes the college savings plan deduction before taxes, \$76 will go to her college plan and \$152.95 will go to taxes. If she takes the college savings plan deduction after taxes, only \$62.70 will go to her college plan and \$166.25 will go to taxes.
9. $(f + g)(x) = 6x - 3$; $(f - g)(x) = -4x + 1$; $(f \cdot g)(x) = 5x^2 - 7x + 2$; $\left(\frac{f}{g}\right)(x) = \frac{x - 1}{5x - 2}, x \neq \frac{2}{5}$

11
$$(f + g)(x) = f(x) + g(x) \quad \text{Addition of functions}$$
$$= (3x) + (-2x + 6) \quad f(x) = 3x \text{ and } g(x) = -2x + 6$$
$$= x + 6 \quad \text{Simplify.}$$

$$(f - g)(x) = f(x) - g(x) \quad \text{Subtraction of functions}$$
$$= (3x) - (-2x + 6) \quad \text{Substitution}$$
$$= 5x - 6 \quad \text{Simplify.}$$

$$(f \cdot g)(x) = f(x) \cdot g(x) \quad \text{Multiplication of functions}$$
$$= (3x)(-2x + 6) \quad \text{Substitution}$$
$$= -6x^2 + 18x \quad \text{Simplify.}$$

$$\left(\frac{f}{g}\right)(x) = \frac{f(x)}{g(x)} \quad \text{Division of functions}$$
$$= \frac{3x}{-2x + 6}, x \neq 3 \quad \text{Substitution}$$

13. $(f + g)(x) = x^2 + x - 5$; $(f - g)(x) = x^2 - x + 5$; $(f \cdot g)(x) = x^3 - 5x^2$; $\left(\frac{f}{g}\right)(x) = \frac{x^2}{x - 5}, x \neq 5$
15. $(f + g)(x) = 4x^2 - 8x$; $(f - g)(x) = 2x^2 + 8x - 8$; $(f \cdot g)(x) = 3x^4 - 24x^3 + 8x^2 + 32x - 16$; $\left(\frac{f}{g}\right)(x) = \frac{3x^2 - 4}{x^2 - 8x + 4}, x \neq 4 \pm 2\sqrt{3}$ **17.** $f \circ g = \{(-4, 4)\}$; $g \circ f = \{(-8, 0), (0, -4), (2, -5), (-6, -1)\}$ **19.** $f \circ g$ is undefined; $g \circ f$ is undefined. **21.** $f \circ g$ is undefined; $g \circ f = \{(-4, 0), (1, 2)\}$. **23.** $f \circ g = \{(4, 6), (3, -8)\}$; $g \circ f$ is undefined. **25.** $f \circ g = \{(3, -1), (6, 11)\}$; $g \circ f = \{(-4, 5), (-2, 4), (-1, 8)\}$

27
$$[f \circ g](x) = f[g(x)] \quad \text{Composition of functions}$$
$$= f(x + 5) \quad \text{Replace } g(x) \text{ with } x + 5.$$
$$= 2(x + 5) \quad \text{Substitute } x + 5 \text{ for } x \text{ in } f(x).$$
$$= 2x + 10 \quad \text{Distributive Property}$$

$$[g \circ f](x) = g[f(x)] \quad \text{Composition of functions}$$
$$= g(2x) \quad \text{Replace } f(x) \text{ with } 2x.$$
$$= 2x + 5 \quad \text{Substitute } 2x \text{ for } x \text{ in } g(x).$$

29. $[f \circ g](x) = 3x - 2$; $[g \circ f](x) = 3x + 8$
31. $[f \circ g](x) = x^2 - 6x - 2$; $[g \circ f](x) = x^2 + 6x - 8$
33. $[f \circ g](x) = 4x^3 + 7$; $[g \circ f](x) = 64x^3 - 48x^2 + 12x + 1$
35. $[f \circ g](x) = 128x^4 + 96x^3 + 18x^2$; $[g \circ f](x) = 32x^4 + 6x^2$
37a. $p(x) = 0.65x$; $t(x) = 1.0625x$ **37b.** $[t \circ p](x)$; 35% would be deducted first, and then the sales tax would be applied to the price. **37c.** \$1587.75
39. $2(g \cdot f)(x) = 2x^3 - 4x^2 - 30x + 72$; D = {all real numbers} **41.** 25 **43.** 483 **45.** −5 **47.** $-30a + 5$
49. $-10a^2 + 10a + 1$

51 **a.** Let $w(x)$ represent the function for women and $m(x)$ represent the function for men.
$$(w + m)(x) = w(x) + m(x) \quad \text{Addition of functions}$$
$$= (1086.4x + 56{,}610) + (999.2x + 66{,}450) \quad \text{Substitution}$$
$$= 2085.6x + 123{,}060 \quad \text{Simplify.}$$
The equation $y = 2085.6x + 123{,}060$ models the total number.
b. $(f - g)(x)$ = the number of men employed in the U.S. − the number of women employed in the U.S. So, the function represents the difference in the number of men and women employed in the U.S.

53. 0 **55.** 1 **57.** 256 **59.** Sample answer: $f(x) = x - 9$, $g(x) = x + 5$ **61a.** D = {all real numbers} **61b.** D = $\{x \mid x \geq 0\}$ **63.** Compositions of functions are used when the value of a function is determined by another function. For example, the product of a manufacturing plant may have to go through several processes in a particular order, in which each process is described by a function.
65. G **67.** C **69.** −3, 2, 4 **71.** $-3, 5, \frac{1}{2}$ **73.** 1; 1; 2
75. 2 or 0; 2 or 0; 4, 2, or 0 **77.** (1, 2, 3) **79.** (3, −1, 5)
81. $x = \frac{12 + 7y}{5}$ **83.** $x = \frac{15 - 8yz}{4}$ **85.** $k = \pm\sqrt{A - b}$

Pages 420–422 ***Lesson 7-2***

1. {(10, −9), (−3, 1), (−5, 8)}
3. $f^{-1}(x) = -\frac{1}{3}x$

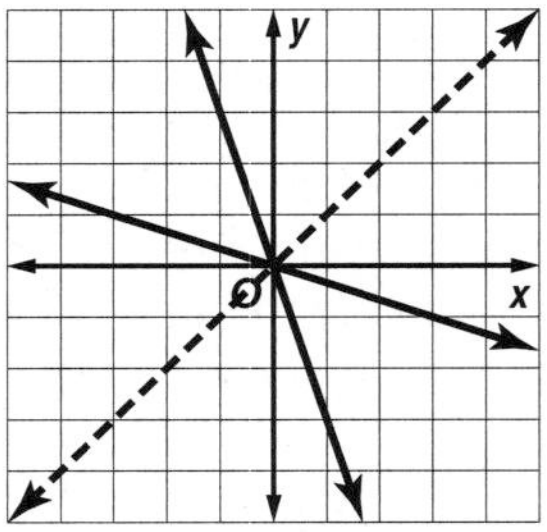

5. $h^{-1}(x) = \pm\sqrt{x + 3}$

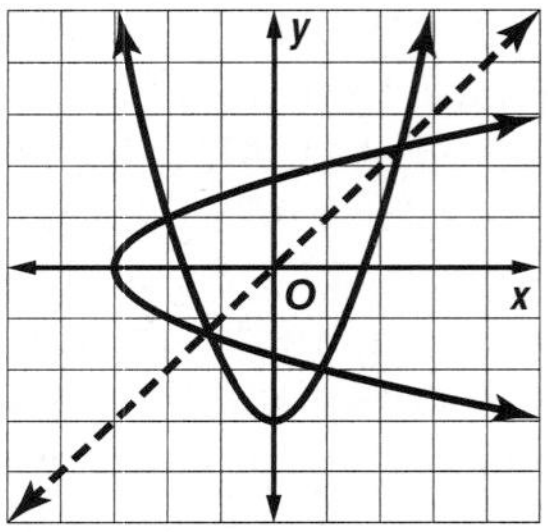

7. no **9.** {(6, −8), (−2, 6), (−3, 7)}
11. {(−1, 8), (−1, −8), (−8, −2), (8, 2)}
13. {(−5, 1), (6, 2), (−7, 3), (8, 4), (−9, 5)}
15. $f^{-1}(x) = x - 2$

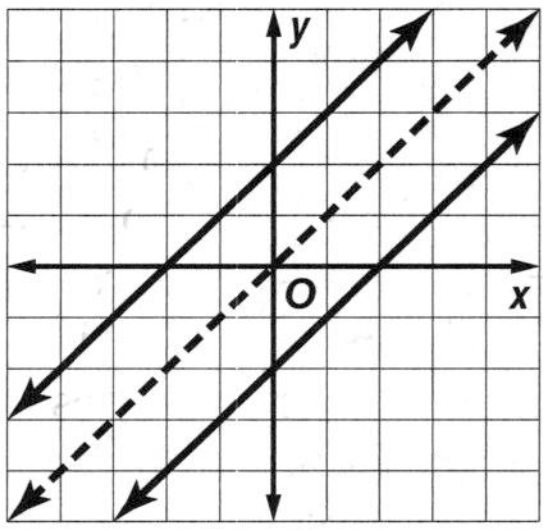

17. $y^{-1} = \frac{x - 1}{-2}$

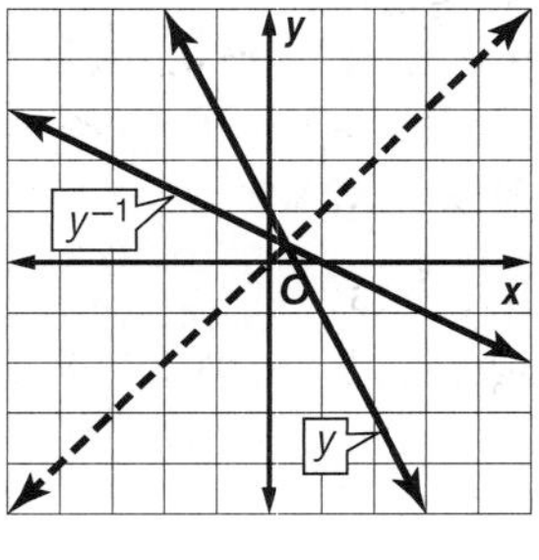

19. $y^{-1} = -\frac{3}{5}(x + 8)$

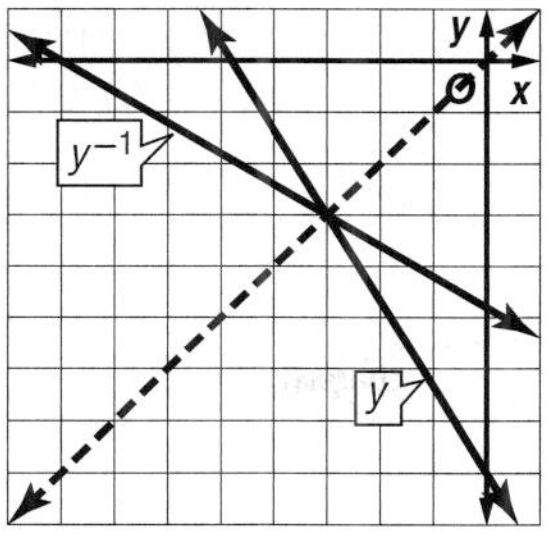

21. $f^{-1}(x) = \frac{1}{4}x$

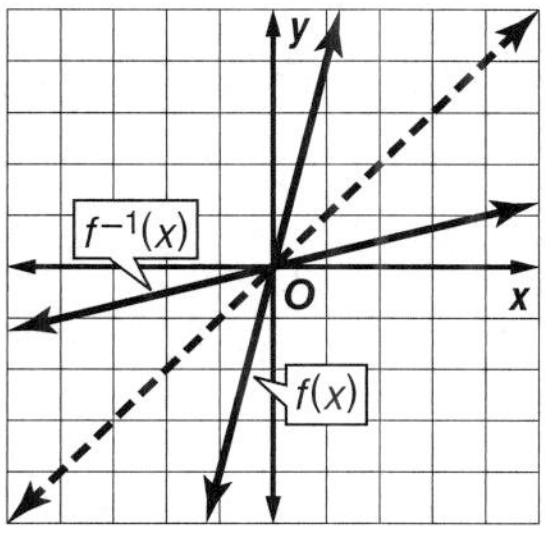

23. $f^{-1}(x) = \pm\sqrt{\frac{1}{5}x}$

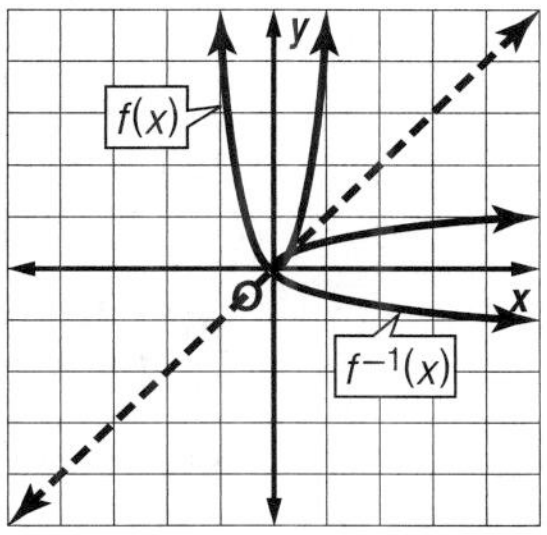

25. $f^{-1}(x) = \pm\sqrt{2x + 2}$

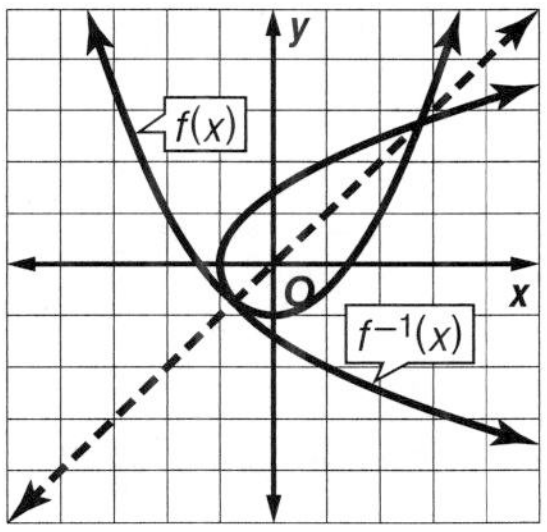

27. no **29.** yes **31.** yes **33.** yes

35 $[f \circ g](x) = f[g(x)]$ Composition of functions
$= f(x + 3)$ Replace $g(x)$ with $x + 3$.
$= (x + 3)^2 - 9$ Substitute $x + 3$ for x in $f(x)$.
$= x^2 + 6x + 9 - 9$ FOIL
$= x^2 + 6x$ Simplify.

$[g \circ f](x) = g[f(x)]$ Composition of functions
$= g(x^2 - 9)$ Replace $f(x)$ with $x^2 - 9$.
$= (x^2 - 9) + 3$ Substitute $x^2 - 9$ for x in $g(x)$.
$= x^2 - 6$ Simplify.

The functions are not inverses because $[f \circ g](x) \neq [g \circ f](x)$.

37. yes **39a.** $c(g) = 2.95g$ **39b.** $c(m) \approx 0.105m$

41 a. $A = \pi r^2$ Write the formula
$y = \pi x^2$ Write the formula using x and y.
$x = \pi y^2$ Exchange x and y in the equation.
$\frac{x}{\pi} = y^2$ Divide each side by π.
$\sqrt{\frac{x}{\pi}} = y$ Take the square root of each side.

So, replacing y with r and x with A, the inverse is $r = \sqrt{\frac{A}{\pi}}$.

b. $r = \sqrt{\frac{A}{\pi}}$ Write the inverse of the function.
$= \sqrt{\frac{36}{\pi}}$ $A = 36$
≈ 3.39 cm Simplify.

43. yes **45.** no **47.** no **49a.** $F^{-1}(x) = \frac{5}{9}(x - 32)$;
$F[F^{-1}(x)] = \frac{9}{5}\left[\frac{5}{9}(x - 32)\right] + 32 = x - 32 + 32 = x$;
$F^{-1}[F(x)] = \frac{5}{9}\left(\frac{9}{5}x + 32 - 32\right) = \frac{5}{9}\left(\frac{9}{5}x + 0\right) = x$.

49b. It can be used to convert Fahrenheit to Celsius.

51a.

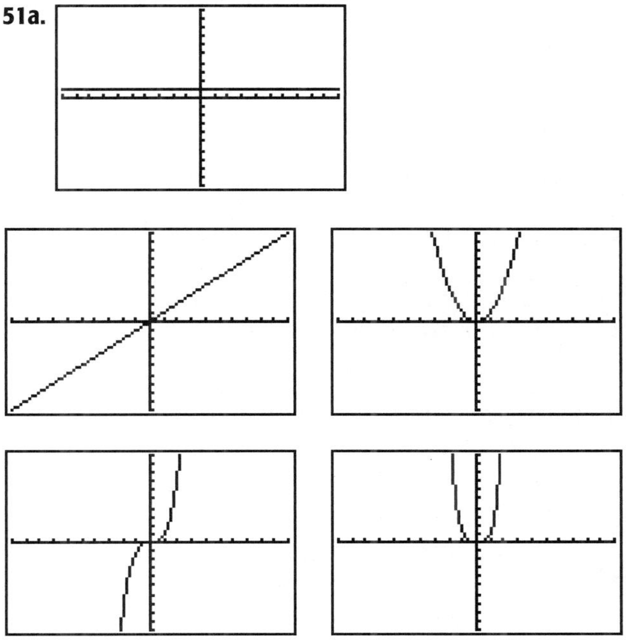

51b.

Function	Inverse a function?
$y = x^0$ or $y = 1$	no
$y = x^1$ or $y = x$	yes
$y = x^2$	no
$y = x^3$	yes
$y = x^4$	no

51c. n is odd. **53.** Sample answer: $f(x) = 2x$, $f^{-1}(x) = 0.5x$; $f[f^{-1}(x)] = f^{-1}[f(x)] = x$

55. $y^{-1} = \frac{x - b}{m}$ **57.** 30 in. **59.** J **61.** 12 **63.** 0
65. 289; $(x + 17)^2$ **67.** $23 + 14i$ **69.** i

71. $\begin{bmatrix} 2 & 4 & 2 & -3 \\ 3 & -3 & -5 & -2 \end{bmatrix}$; $\begin{bmatrix} -2 & -4 & -2 & 3 \\ -3 & 3 & 5 & 2 \end{bmatrix}$

73. 180° rotation

75.

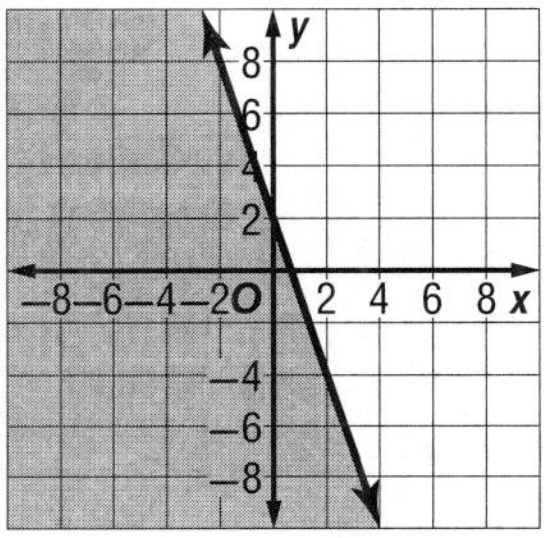

Pages 427–430 Lesson 7-3

1. $D = \{x \mid x \geq 0\}$; $R = \{y \mid y \geq 0\}$
3. $D = \{x \mid x \geq -8\}$; $R = \{y \mid y \geq -2\}$

5.

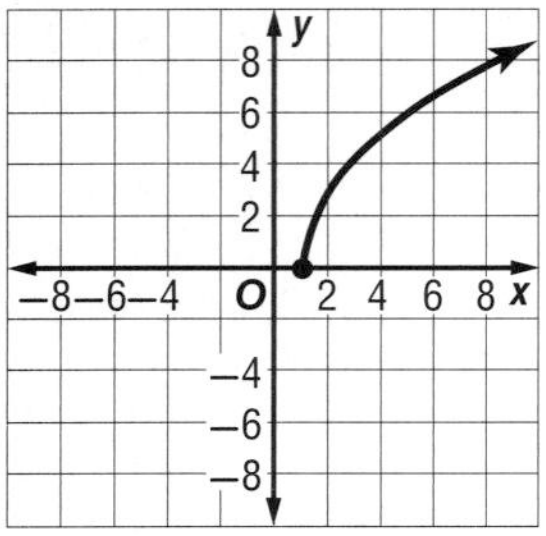

D = {x | $x \geq 1$};
R = {y | $y \geq 0$}

7.

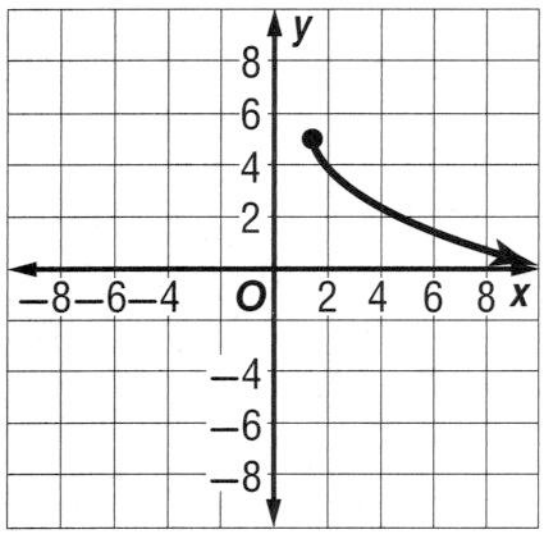

$D = \left\{x \,\middle|\, x \geq \frac{5}{3}\right\}$;
R = {y | $y \leq 5$}

9.

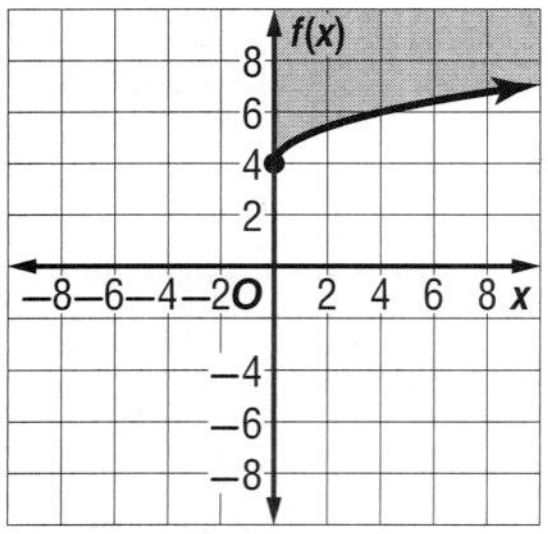

11.

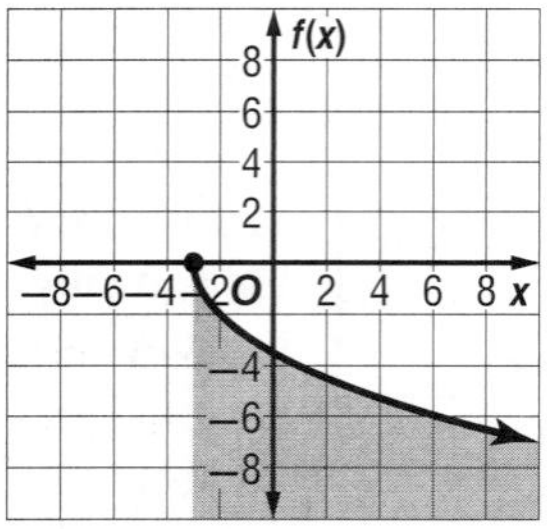

13. D = {x | $x \geq 0$}; R = {y | $y \leq 2$}

15 The domain only includes values for which the radicand is nonnegative.

$x - 2 \geq 0$ Write an inequality.

$x \geq 2$ Add 2 to each side.

The domain is {x | $x \geq 2$}. Find $f(2)$ to find the lower limit of the range.

$f(2) = 4\sqrt{2 - 2} - 8$

$= 0 - 8$ or -8

The range is {y | $y \geq 2$}.

17. D = {x | $x \geq 4$}; R = {y | $y \geq -6$}

19.

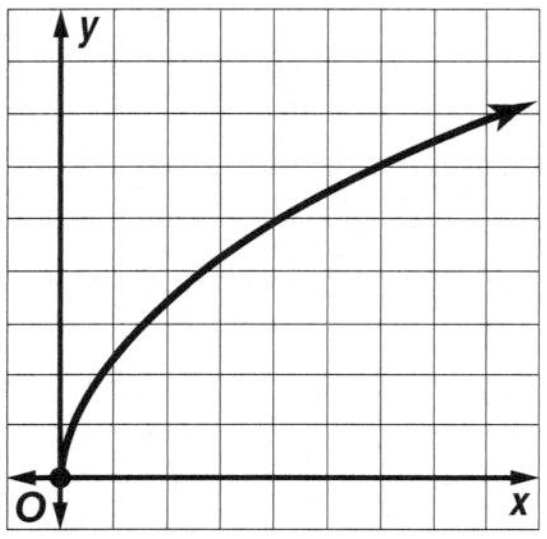

D = {x | $x \geq 0$};
R = {y | $y \geq 0$}

21.

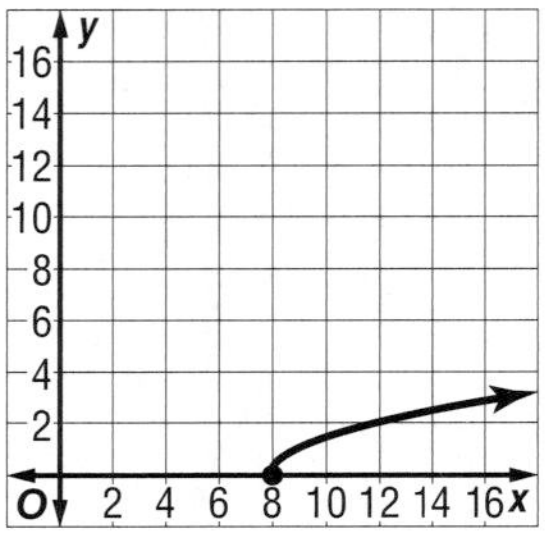

D = {x | $x \geq 8$};
R = {y | $y \geq 0$}

23.

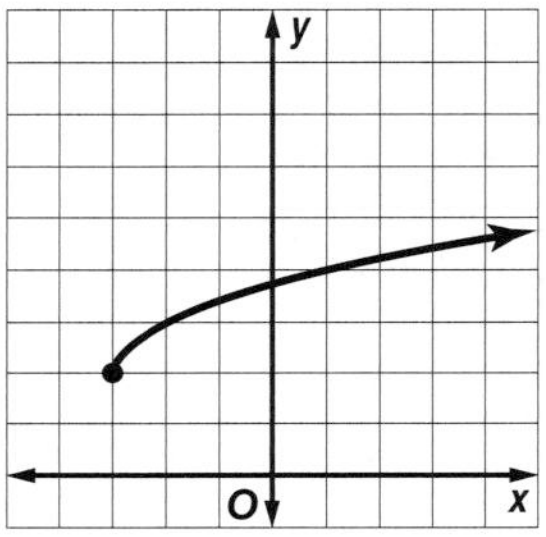

D = {x | $x \geq -3$};
R = {y | $y \geq 2$}

25.

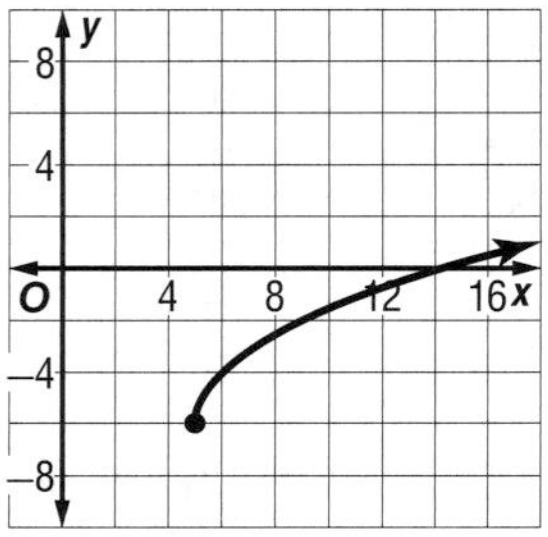

D = {x | $x \geq 5$};
R = {y | $y \geq -6$}

27.

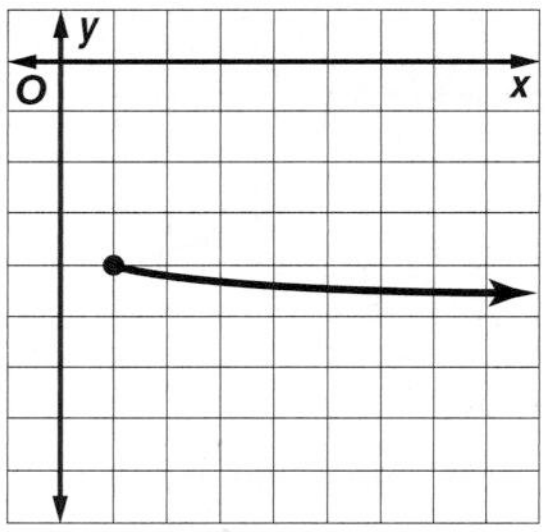

D = {x | $x \geq 1$};
R = {y | $y \leq -4$}

29. 1936 ft

31.

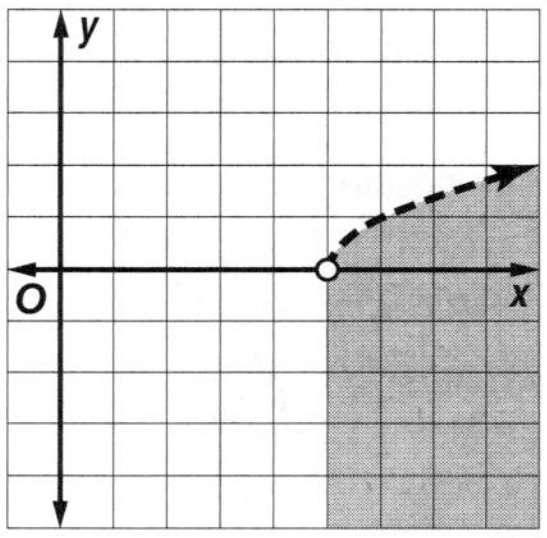

33 Graph the boundary $y = -4\sqrt{x + 3}$. The domain is {x | $x \geq -3$}. Because y is *greater than*, the shaded region should be *above* the boundary and within the domain.

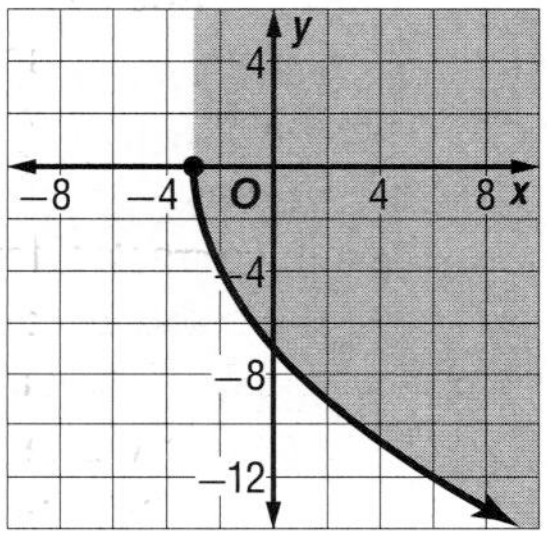

35. 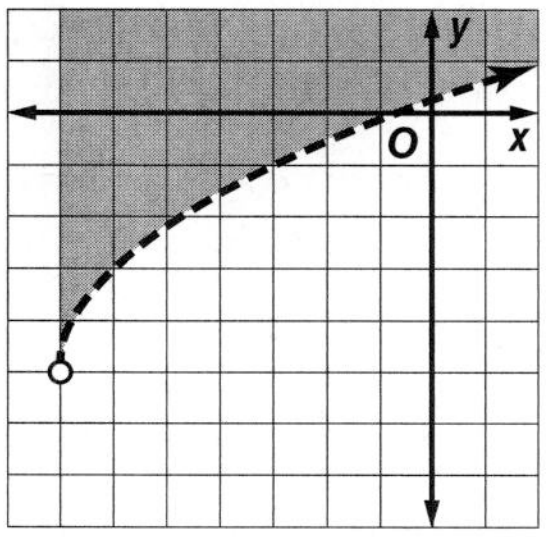

37. 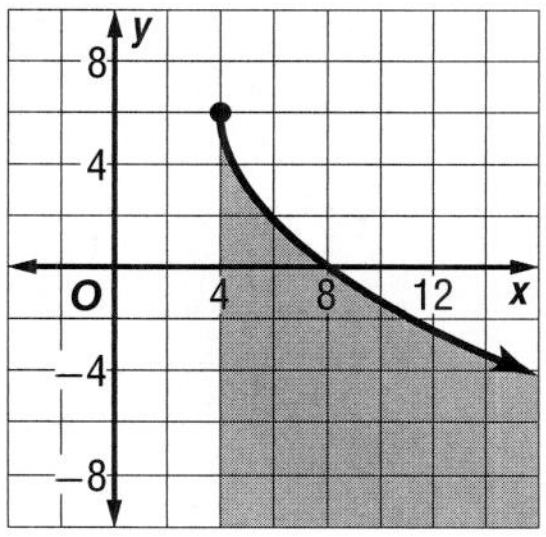

39a. $v = \sqrt{\frac{2E}{m}}$ **39b.** about 36.5 m/s **39c.** about 85,135 m/s **41.** $f(x) = \sqrt{x-4} - 6$

43. $f(x) = -\sqrt{x+6} - 6$

45 a. $T = 2\pi\sqrt{\frac{L}{g}}$ Original function

$T = 2\pi\sqrt{\frac{L}{32}}$ Replace g with 32.

Make a table of values for $0 \le L \le 10$. Graph the points and connect with a smooth curve.

L	*T*
0	0
1	1.11
2	1.57
3	1.92
4	2.22
5	2.48
6	2.72
7	2.94
8	3.14
9	3.33
10	3.51

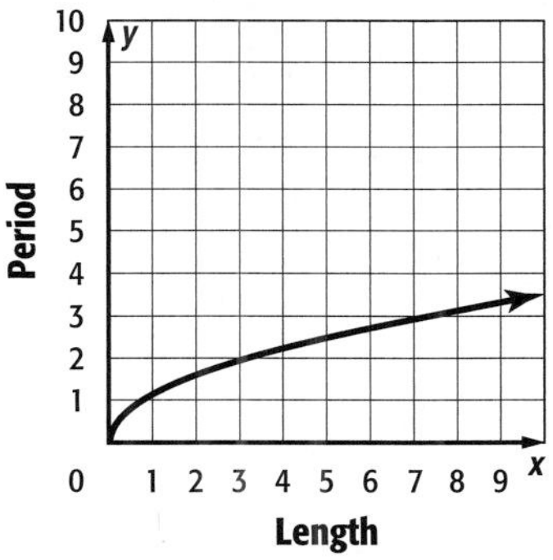

b. Use the table that you made in part a. (2, 1.57) means the period for a pendulum 2 feet long is about 1.57 seconds. (5, 2.48) means the period for a pendulum 5 feet long is about 2.48 seconds. (8, 3.14) means the period for a pendulum 8 feet long is about 3.14 seconds.

47. Sample answer: $f(x) = -\sqrt{x+4} + 6$

49. Sample answer: $f(x) = -\sqrt{x-8} + 14$

51. Molly is correct; Cleveland shaded incorrectly. He shaded above the graph when he should have shaded below. **53a.** Sample answer: The original is $y = x^2 + 2$ and inverse is $y = \pm\sqrt{x-2}$. **53b.** Sample answer: The original is $y = \sqrt{x} + 4$ and inverse is $y = (x-4)^2$. **55.** G **57.** D **59.** no **61.** $[d \circ h](m) = \frac{m}{1440}$ **63.** rational **65.** rational

Pages 433–436 Lesson 7-4

1. $\pm 10y^4$

3 $\sqrt{(y-6)^8} = \sqrt{[(y-6)^4]^2}$
$= (y-6)^4$

5. not a real number **7.** 7.616 **9.** −2.122

11. about 4.088×10^8 m

13 $\pm\sqrt{225a^{16}b^{36}} = \pm\sqrt{(15a^8b^{18})^2}$
$= \pm 15a^8b^{18}$

15. $-4c^2d$ **17.** $-20x^{16}y^{20}$ **19.** $(x^2+6)^8$ **21.** $2a^2b^4$ **23.** $3b^6c^4$ **25.** not a real number **27.** $|x^3|$ **29.** a^4 **31.** $(4x-7)^8$ **33.** $4|(5x-2)^3|$ **35.** $2a^3b^2$ **37.** 8 cm **39.** −12.247 **41.** 0.787 **43.** −5.350 **45.** 29.573 **47.** $14c^3d^2$ **49.** $-3a^5b^3$ **51.** $20x^8|y^3|$ **53.** $4(x+y)^2$ **55.** about 141 million mi

57 bald eagle:
$P = 73.3\sqrt[4]{m^3}$
$= 73.3\sqrt[4]{4.5^3}$
≈ 226.5 Cal/d

golden retriever:
$P = 73.3\sqrt[4]{m^3}$
$= 73.3\sqrt[4]{30^3}$
≈ 939.6 Cal/d

komodo dragon:
$P = 73.3\sqrt[4]{m^3}$
$= 73.3\sqrt[4]{72^3}$
≈ 1811.8 Cal/d

bottlenose dolphin:
$P = 73.3\sqrt[4]{m^3}$
$= 73.3\sqrt[4]{156^3}$
≈ 3235.5 Cal/d

Asian elephant:
$P = 73.3\sqrt[4]{m^3}$
$= 73.3\sqrt[4]{2300^3}$
$\approx 23{,}344.4$ Cal/d

59. Kimi is correct; Ashley's error was keeping the y^2 inside the absolute value symbol. **61.** Sample answer: Sometimes; when $x = -3$, $\sqrt[4]{(-x)^4} = |(-x)|$ or 3. When $x = 3$, $\sqrt[4]{(-x)^4} = |3|$ or 3. **63.** Sample answers: 1, 64 **65.** $2\sqrt[3]{2xy}$ **67a.** 0.2 **67b.** −0.1 **67c.** $\frac{2}{3}$ **69.** $\frac{1}{625}$ **71.** G **73.** B

75. 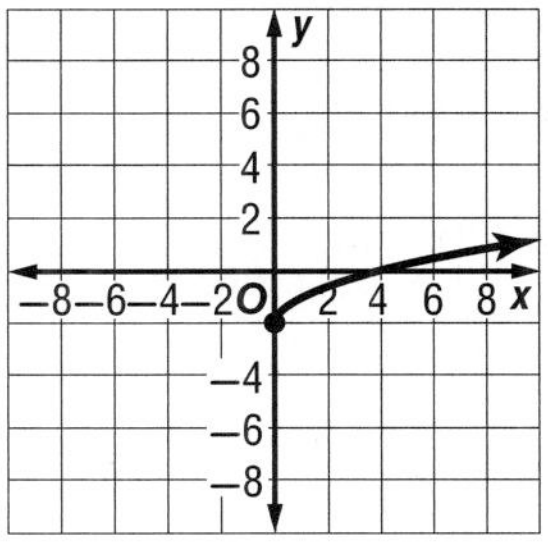

77. 3.41 kg and 49.53 cm
79. $4x^2 + 22x - 34$
81. $4a^4 + 24a^2 + 36$

83. (−2, 3); $x = -2$; down

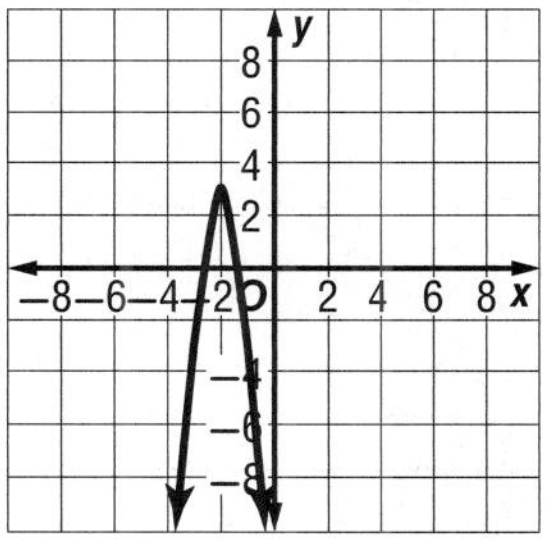

85. (2, −2); $x = 2$; up

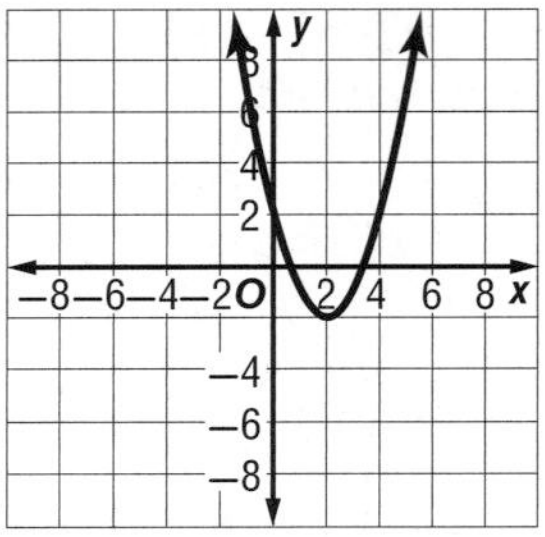

87. $x^2 + 9x + 20$ **89.** $a^2 - 7a - 18$ **91.** $x^2 + xy - 2y^2$

Pages 443–445 Lesson 7-5

1. $6b^2c^2\sqrt{ac}$ **3.** $\frac{c^2\sqrt{cd}}{d^5}$ **5.** $60x$ **7.** $36xy$
9. $20\sqrt{2} + 13\sqrt{3}$ **11.** $12\sqrt{3} + 16\sqrt{5} + 40 + 6\sqrt{15}$
13. $\frac{15 - 5\sqrt{2}}{7}$ **15.** $-2 - \sqrt{2}$ **17.** $32 - 2\sqrt{3}$ cm
19. $3a^7b\sqrt{ab}$ **21.** $3|a^3b|c^2\sqrt{2bc}$ **23.** $\frac{\sqrt{70xy}}{10y^2}$
25. $\frac{\sqrt[4]{28b^2x^3}}{2|b|}$ **27.** $32a^5b^3\sqrt{b}$ **29.** $25x^6y^3\sqrt{2xy}$
31. $18\sqrt{3} + 14\sqrt{2}$ **33.** $9\sqrt{6} + 72\sqrt{2} - 7\sqrt{3}$

35 $A = \ell w$ Area of a rectangle
$= (8 + \sqrt{3})(\sqrt{6})$ $\ell = 8 + \sqrt{3}$ and $w = \sqrt{6}$
$= 8 \cdot \sqrt{6} + \sqrt{3} \cdot \sqrt{6}$ Distributive Property
$= 8\sqrt{6} + \sqrt{18}$ Product Property
$= 8\sqrt{6} + 3\sqrt{2}$ ft^2 Simplify.

37. $56\sqrt{3} + 42\sqrt{6} - 36\sqrt{2} - 54$ **39.** 1260
41. $6\sqrt{3} + 6\sqrt{2}$ **43.** $\frac{20 - 7\sqrt{3}}{11}$ **45.** $2yz^4\sqrt[3]{2y}$
47. $3|a|b^3\sqrt[4]{2a^2bc}$ **49.** $\frac{\sqrt[4]{1500a^2b^3x^3y^2}}{5|a|b}$

51 $\frac{x+1}{\sqrt{x}-1} = \frac{x+1}{\sqrt{x}-1} \cdot \frac{\sqrt{x}+1}{\sqrt{x}+1}$ $\sqrt{x} + 1$ is the conjugate of $\sqrt{x} - 1$.
$= \frac{x(\sqrt{x}) + 1(x) + 1(\sqrt{x}) + 1(1)}{\sqrt{x}(\sqrt{x}) + 1(\sqrt{x}) + -1(\sqrt{x}) + (-1)(1)}$ Multiply.
$= \frac{x\sqrt{x} + x + \sqrt{x} + 1}{x + \sqrt{x} - \sqrt{x} - 1}$ Simplify.
$= \frac{x(\sqrt{x} + 1) + (\sqrt{x} + 1)}{x - 1}$ Simplify.
$= \frac{(x + 1)(\sqrt{x} + 1)}{x - 1}$ Simplify.

53. $\frac{\sqrt{x^3 - x}}{x^2 - 1}$ **55.** $|a|$ **57.** a^2

59a. $a^2 + b^2 = c^2$
$1^2 + 1^2 = c^2$
$2 = c^2$
$c = \sqrt{2}$

59b.

59c. $\sqrt{2} + \sqrt{2}$ units is the length of the hypotenuse of an isosceles right triangle with legs of length 2 units. Therefore, $\sqrt{2} + \sqrt{2} > 2$.

59d.

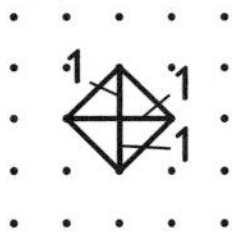

59e. The square creates 4 triangles with a base of 1 and a height of 1. Therefore the area of each triangle is $\frac{1}{2}bh = \frac{1}{2}(1)(1)$ or $\frac{1}{2}$. $4\left(\frac{1}{2}\right) = 2$. The area of the square is 2, so $\sqrt{2} \cdot \sqrt{2} = 2$.

61. $\left(\frac{-1 - i\sqrt{3}}{2}\right)^3 = \left(\frac{-1 - i\sqrt{3}}{2}\right) \cdot \left(\frac{-1 - i\sqrt{3}}{2}\right) \cdot \left(\frac{-1 - i\sqrt{3}}{2}\right)$

$= \frac{(-1 - i\sqrt{3})(-1 - i\sqrt{3})(-1 - i\sqrt{3})}{8}$

$= \frac{(1 + i\sqrt{3} + i\sqrt{3} + 3i^2)(-1 - i\sqrt{3})}{8}$

$= \frac{(2i\sqrt{3} - 2)(-1 - i\sqrt{3})}{8}$

$= \frac{-2i\sqrt{3} - 6i^2 + 2 + 2i\sqrt{3}}{8}$

$= \frac{-6i^2 + 2}{8} = \frac{8}{8}$ or 1

63. $\sqrt[1]{256} = 256$; $\sqrt[2]{256} = 16$; $\sqrt[4]{256} = 4$; $\sqrt[8]{256} = 2$
65. Sample answer: It is only necessary to use absolute values when it is possible that n could be odd or even and still be defined.
It is when the radicand must be nonnegative in order for the root to be defined that the absolute values are not necessary. **67.** G **69.** B **71.** $9ab^3$

73.

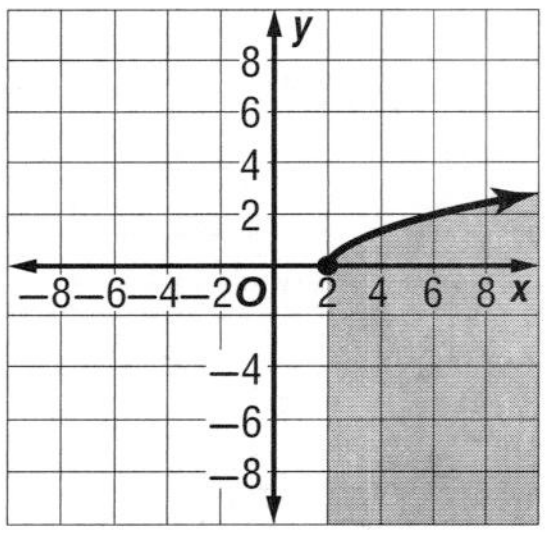

75. $-4, 4, -i, i$ **77.** $-4, 2 + 2i\sqrt{3}, 2 - 2i\sqrt{3}$
79. $\frac{3}{2}, \frac{-3 + 3i\sqrt{3}}{4}, \frac{-3 - 3i\sqrt{3}}{4}$ **81.** 9 small, 4 large **83.** $\frac{1}{3}$ **85.** $\frac{7}{12}$ **87.** $\frac{5}{12}$

Pages 449–452 Lesson 7-6

1. $\sqrt[4]{10}$ **3.** $15^{\frac{1}{3}}$ **5.** 7 **7.** 25

9 $\ell = A^{\frac{1}{2}}$ Write the formula.
$= 169^{\frac{1}{2}}$ $A = 169$
$= \sqrt{169}$ Write in radical form.
$= 13$ ft Simplify.

11. $x^{\frac{3}{5}}$ **13.** $\sqrt{3g}$ **15.** $\frac{g - 2g^{\frac{1}{2}} + 1}{g - 1}$ **17.** $\sqrt[7]{16}$ **19.** $\sqrt{x^9}$
21. $63^{\frac{1}{4}}$ **23.** $5x^{\frac{1}{2}}$ **25.** 4 **27.** $\frac{1}{3}$ **29.** about 2.64 cm
31. $a^{\frac{25}{36}}$ **33.** $\frac{y^{\frac{1}{5}}}{y}$

35 $\frac{\sqrt[4]{27}}{\sqrt[4]{3}} = \frac{27^{\frac{1}{4}}}{3^{\frac{1}{4}}}$ Rational exponents

$= \frac{(3^3)^{\frac{1}{4}}}{3^{\frac{1}{4}}}$ $27 = 3^3$

$= \frac{3^{\frac{3}{4}}}{3^{\frac{1}{4}}}$ Power of a power

$= 3^{\frac{3}{4} - \frac{1}{4}}$ Quotient of powers

$= 3^{\frac{2}{4}}$ Simplify.

$= 3^{\frac{1}{2}}$ Simplify.

$= \sqrt{3}$ Write in radical form.

37. $\sqrt[3]{9} \cdot \sqrt{g}$ **39.** $\frac{x + 4x^{\frac{3}{4}} + 8x^{\frac{1}{2}} + 16x^{\frac{1}{4}} + 16}{x - 16}$

41. $28.27x^{\frac{4}{3}}y^{\frac{2}{5}}z^4$ cm^2 **43.** $6 - 2 \cdot 4^{\frac{1}{3}}$ **45.** $x^{\frac{10}{3}}$ **47.** $y^{\frac{3}{20}}$

49. $\sqrt{6}$ **51.** $\frac{w^{\frac{1}{8}}}{w}$

53 $\frac{f^{-\frac{1}{4}}}{4f^{\frac{1}{2}} \cdot f^{-\frac{1}{3}}} = \frac{f^{\frac{1}{3}}}{4f^{\frac{1}{2}} \cdot f^{\frac{1}{4}}}$ $f^{-\frac{1}{4}} = \frac{1}{f^{\frac{1}{4}}} \cdot \frac{1}{f^{-\frac{1}{3}}} = f^{\frac{1}{3}}$

$= \frac{f^{-\frac{1}{3}}}{4f^{\frac{3}{4}}}$ Add powers.

$= \frac{f^{\frac{1}{3}}}{4f^{\frac{3}{4}}} \cdot \frac{f^{\frac{1}{4}}}{f^{\frac{1}{4}}}$ Multiply to rationalize the denominator.

$= \frac{f^{\frac{7}{12}}}{4f}$ $f^{\frac{1}{3}} \cdot f^{\frac{1}{4}} = f^{\frac{1}{3} + \frac{1}{4}}$

55. $c^{\frac{1}{2}}$ **57.** $23\sqrt[6]{23}$ **59.** 3 **61.** $\frac{ab\sqrt{c}}{c}$ **63.** $2\sqrt{6} - 5$

65a.

x	$f(x)$	$g(x)$
−2	−8	−1.26
−1	−1	−1
0	0	0
1	1	1
2	8	1.26

65b.

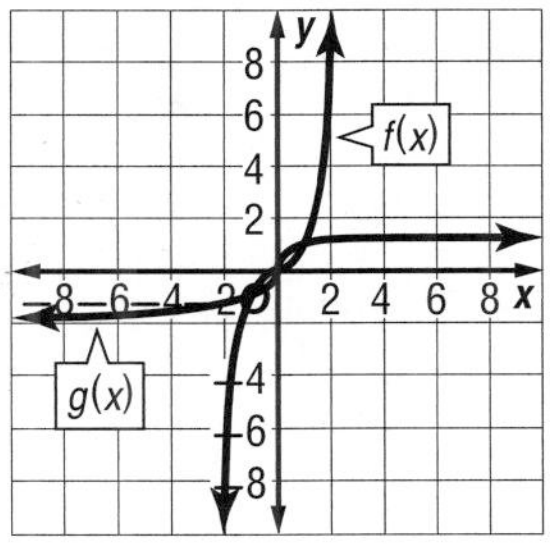

65c. It is a reflection of the line $y = x$. **67.** Never; the quantities are not the same. When the negative is enclosed inside of the parentheses and the base is raised to an even power, the answer is positive. When the negative is not enclosed inside of the parentheses and the base is raised to an even power, the answer is negative. **69.** Sample answer: $4^{\frac{1}{2}}$ and $16^{\frac{1}{4}}$. **71.** No; Ayana added the exponents and Kenji divided the exponents. The exponents should have been subtracted. **73.** G **75.** 3^{-2} **77.** $2y\sqrt[3]{2}$ **79.** 1449 m/s **81.** 8; $-x - h + 4$ **83.** 21; $x^2 + 2xh + h^2 + 5$ **85.** −129; $2x^3 + 6x^2h + 6xh^2 + 2h^3 - 1$ **87.** $\{-8, 2\}$ **89.** $\{7\}$ **91.** $\left\{\frac{3}{2}\right\}$ **93.** $x - 3$ **95.** $7x - 1$ **97.** $4x - 20\sqrt{x} + 25$

Pages 456–459 Lesson 7-7

1. 20 **3.** 13

5 $\sqrt[3]{x - 2} = 3$ Original equation

$(\sqrt[3]{x - 2})^3 = 3^3$ Raise each side to the third power.

$x - 2 = 27$ Evaluate each side.

$x = 29$ Add 2 to each side.

7. 2 **9.** 49 **11.** $\frac{27}{2}$ **13a.** about 9.5 s **13b.** about 324 feet **15.** $-\frac{4}{3} \le x \le \frac{77}{3}$ **17.** $1 \le y \le 5$ **19.** $x > 1$ **21.** $x \le -11$ **23.** 22 **25.** 3 **27.** no real solution **29.** $\frac{1}{4}$ **31.** 9 **33.** $\frac{81}{16}$ **35.** 1 m **37.** 3 **39.** 83 **41.** 61 **43.** 3 **45.** 18 **47.** 2 **49.** F **51.** $x \ge 43$ **53.** no real solution

55 $\sqrt{d + 3} + \sqrt{d + 7} > 4$ Original inequality

$\sqrt{d + 3} > 4 - \sqrt{d + 7}$ Subtract $\sqrt{d + 7}$ from each side.

$d + 3 > 16 - 8\sqrt{d + 7} + d + 7$ Square each side.

$\frac{5}{2} < \sqrt{d + 7}$ Simplify.

$\frac{25}{4} < d + 7$ Square each side.

$-\frac{3}{4} < d$ Subtract 7 from each side.

$d > -\frac{3}{4}$ Rewrite inequality.

57. $-\frac{5}{2} \le y \le 2$ **59.** $a > 8$ **61.** $0 \le c < 3$

63. $M = \left(\frac{L}{0.46}\right)^3$

65 $A = \pi r^2$ Original formula

$250{,}000 = \pi r^2$ Replace A with 250,000.

$79{,}577.5 \approx r^2$ Divide each side by π.

$\sqrt{79{,}577.5} \approx r$ Take the square root of each side.

$282.1 \approx r$ Use a calculator.

The radius is about 282 ft.

67. $\sqrt{x + 2} - 7 = -10$

69. never

$\frac{\sqrt{(x^2)^2}}{-x} = x$

$\frac{x^2}{-x} = x$

$x^2 = (x)(-x)$

$x^2 \ne -x^2$

71. They are reciprocals of each other. **73.** 3 **75.** Sometimes; Sample answer: When the radicand is negative, then there will be extraneous roots. **77.** G **79.** A **81.** 81 **83.** $4x^2y^2\sqrt{5}$ **85.** $f^{-1}(x) = \frac{-x - 3}{2}$ **87.** $f^{-1}(x) = \pm\frac{1}{2}\sqrt{x} - \frac{3}{2}$ **89a.** $f(x) \to +\infty$ as $x \to +\infty$, $f(x) \to +\infty$ as $x \to -\infty$ **89b.** even **89c.** 0 **91.** $\frac{1}{6}$ **93.** $\frac{5}{8}$ **95.** 16 **97.** $2\frac{1}{2}$

Pages 462–466 Chapter 7 Study Guide and Review

1. identity function **3.** composition of functions **5.** rationalizing the denominator **7.** inverse relations **9.** radical function **11.** $[f \circ g](x) = x^2 - 14x + 50$; $[g \circ f](x) = x^2 - 6$ **13.** $[f \circ g](x) = 20x - 4$; $[g \circ f](x) = 20x - 1$ **15.** $[f \circ g](x) = x^2 + 4x$; $[g \circ f](x) = x^2 + 2x - 2$

17. $f^{-1}(x) = \frac{x+6}{5}$

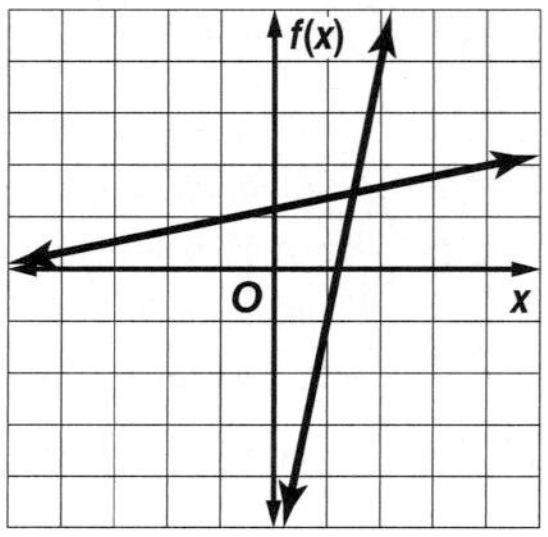

19. $f^{-1}(x) = 2x - 6$

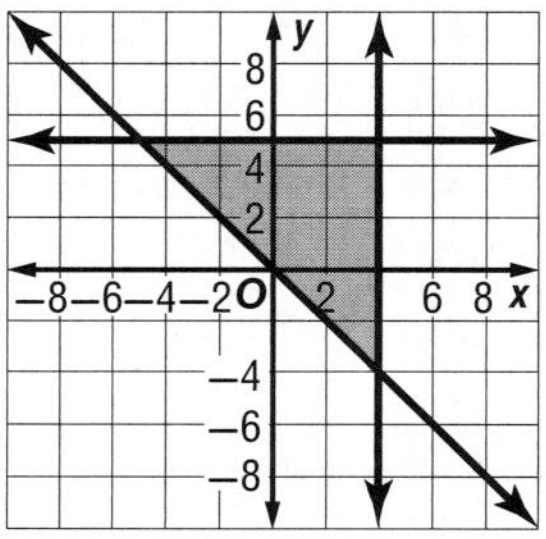

21. $f^{-1}(x) = \pm\sqrt{x}$

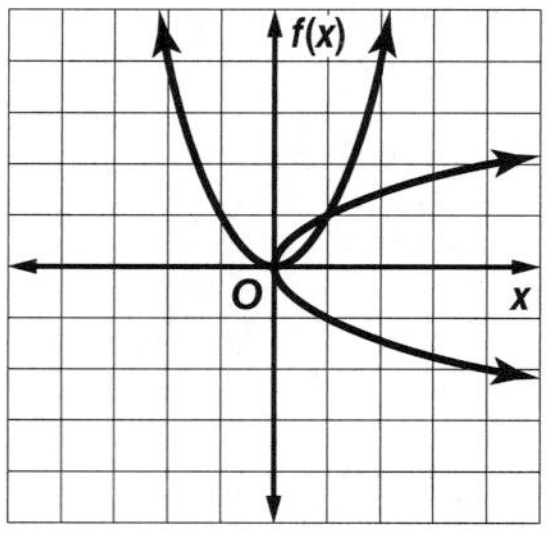

23. \$1200 **25.** yes **27.** no **29.** no

31.

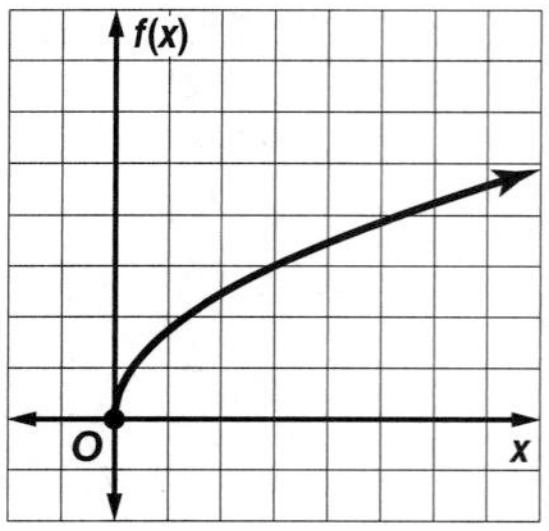

$D = \{x \mid x \geq 0\}$; $R = \{y \mid y \geq 0\}$

33.

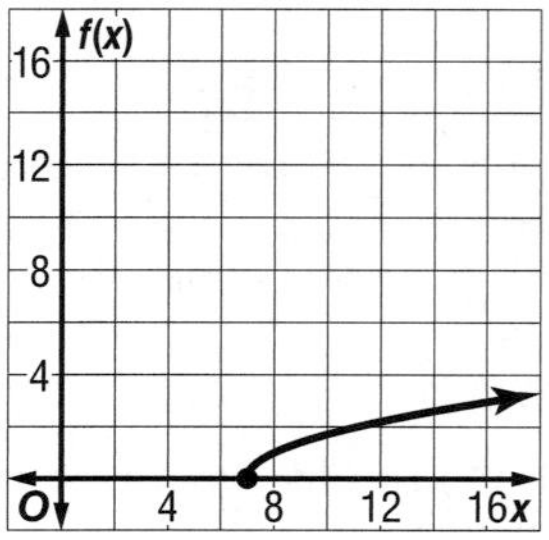

$D = \{x \mid x \geq 7\}$; $R = \{y \mid y \geq 0\}$

35.

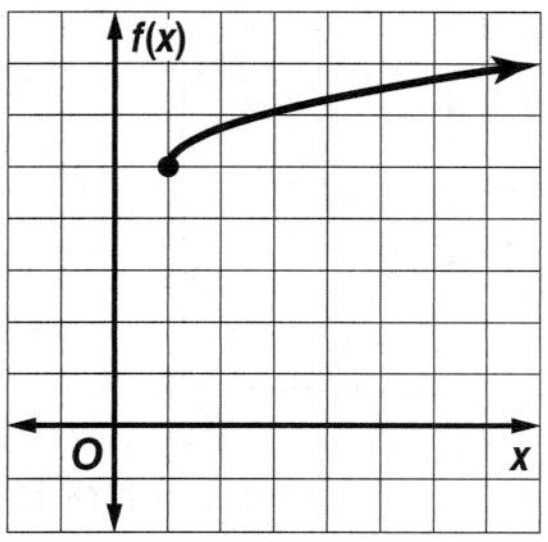

$D = \{x \mid x \geq 1\}$; $R = \{y \mid y \geq 5\}$

37. about 9.8 in.

39.

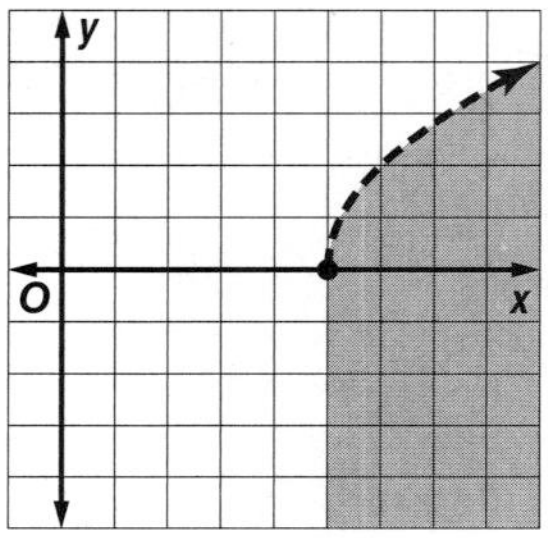

41. ± 11 **43.** 6 **45.** $(x^2 + 2)^3$ **47.** $a^2|b^3|$ **49.** 10 m/s **51.** $12ab^2\sqrt{ab}$ **53.** $80\sqrt{2}$ **55.** $\frac{m^2\sqrt{6mp}}{p^6}$ **57.** $-\sqrt{15} - 3\sqrt{2}$ units; $28 + 2\sqrt{3} - 2\sqrt{2}$ units^2 **59.** $x^{\frac{7}{6}}$ **61.** $\frac{d^{\frac{5}{12}}}{d}$ **63.** 3 **65.** $4a^{\frac{2}{3}}b^{\frac{4}{5}}c^2\pi$ units^2 **67.** $\frac{100}{9}$ **69.** 2 **71.** no solution **73.** 3 **75.** $\frac{1}{3} \leq x < \frac{10}{3}$ **77.** $x \geq \frac{4}{3}$ **79.** no solution **81.** $x > \frac{5}{2}$

Chapter 8 Exponential and Logarithmic Functions and Relations

Page 473 Chapter 8 Get Ready

1. a^{12} **3.** $\frac{-3x^6}{2y^3z^5}$ **5.** 5 g/cm^3

7. $f^{-1}(x) = x + 3$

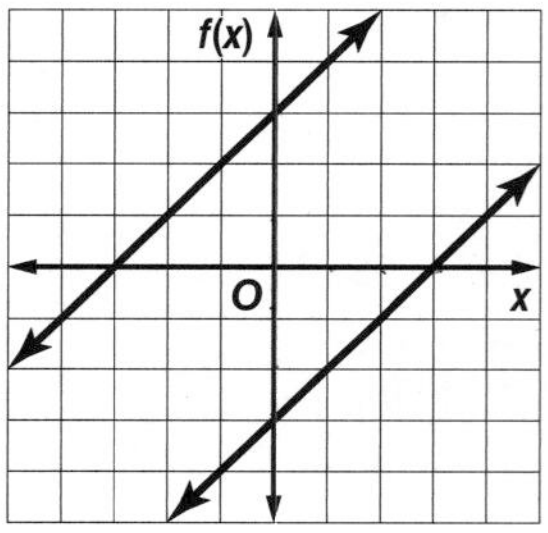

9. $f^{-1}(x) = 4x + 12$

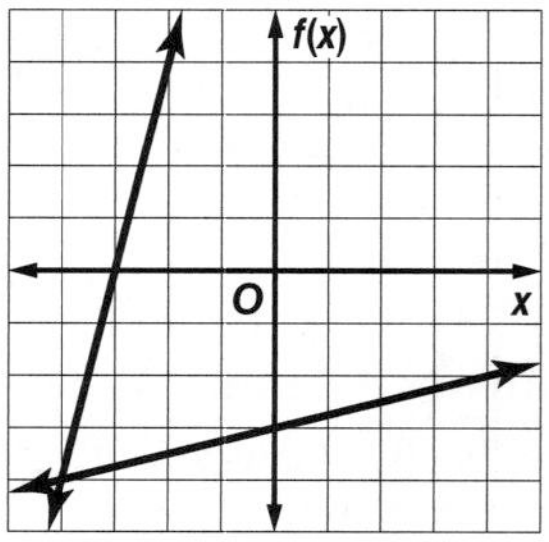

11. $f^{-1}(x) = 3x - 12$

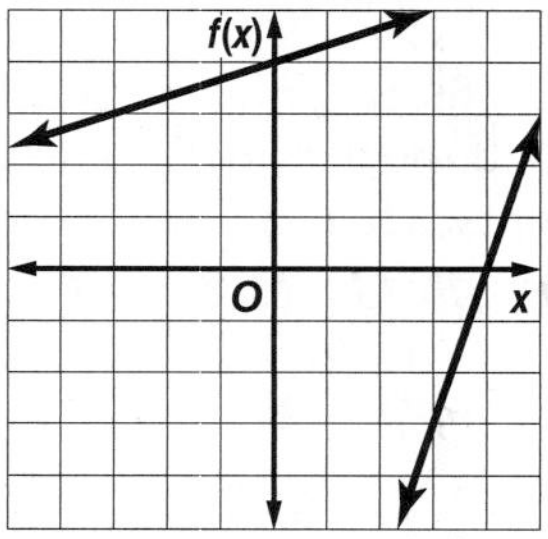

13. no

Pages 479–482   Lesson 8-1

1.

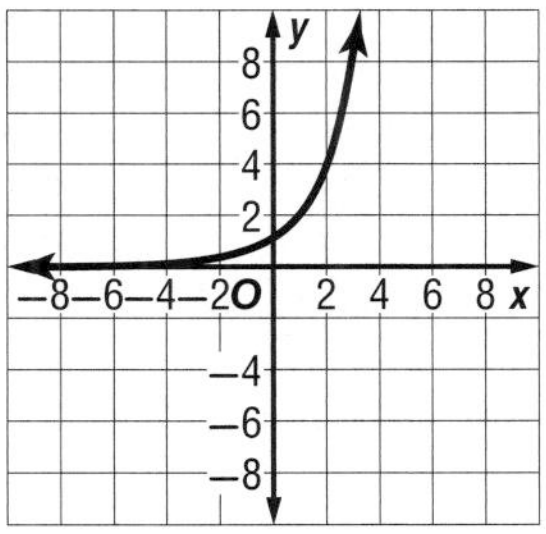

D = {all real numbers}; R = $\{y \mid y > 0\}$

3 Make a table of values. Then plot the points, and sketch the graph.

x	$f(x) = 3^{x-2} + 4$
−2	$3^{-2-2} + 4 = 4\frac{1}{81}$
−1	$3^{-1-2} + 4 = 4\frac{1}{27}$
0	$3^{0-2} + 4 = 4\frac{1}{9}$
1	$3^{1-2} + 4 = 4\frac{1}{3}$
2	$3^{2-2} + 4 = 5$
3	$3^{3-2} + 4 = 7$
4	$3^{-2-2} + 4 = 13$

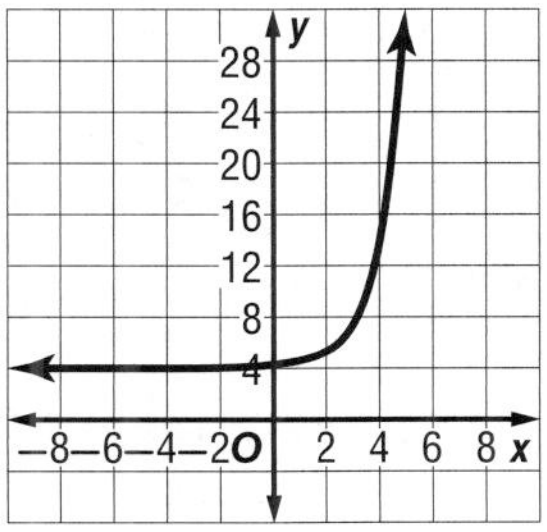

The domain is all real numbers and the range is all real numbers greater than 4.
D = {all real numbers}; R = $\{y \mid y > 4\}$

5.

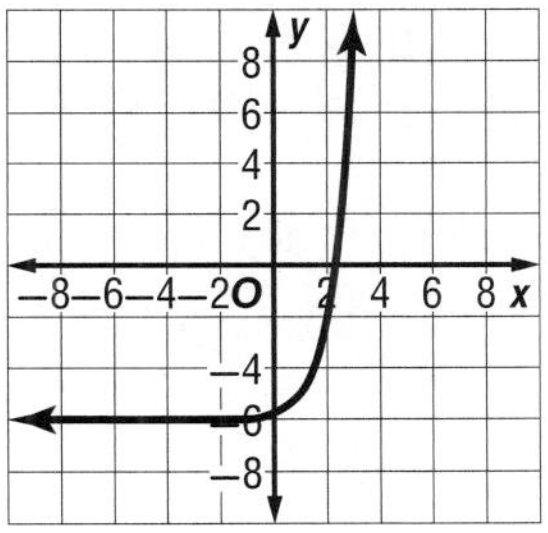

D = {all real numbers}; R = $\{y \mid y > -6\}$

7.

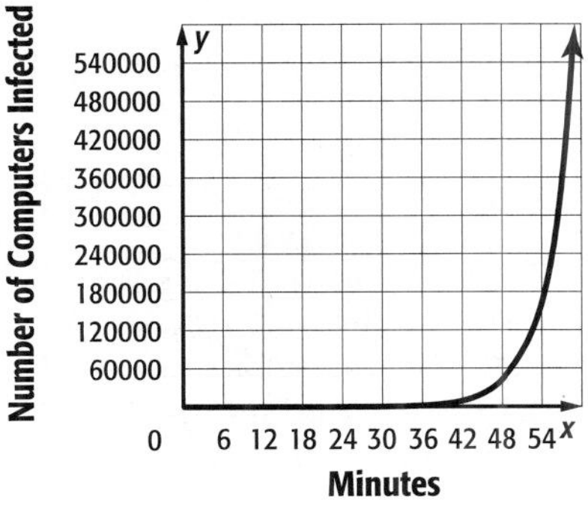

9.

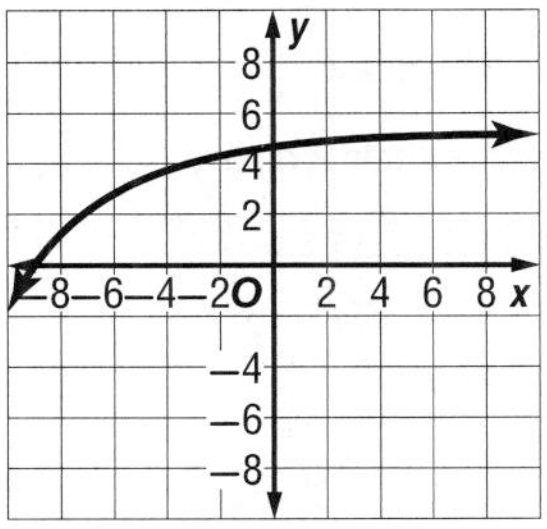

D = {all real numbers}; R = $\{y \mid y < 5\}$

11.

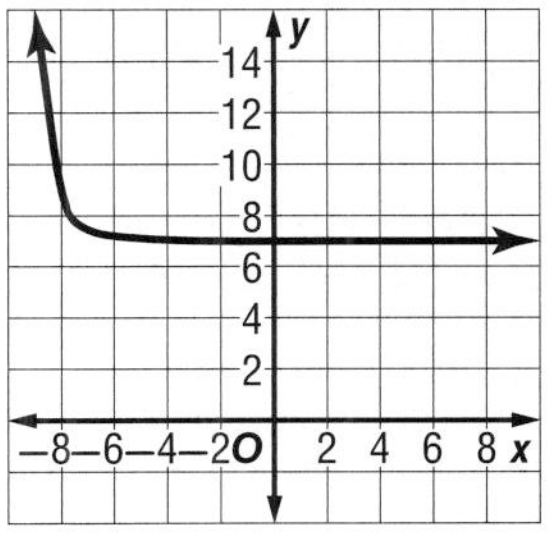

D = {all real numbers}; R = $\{y \mid y > 7\}$

13.

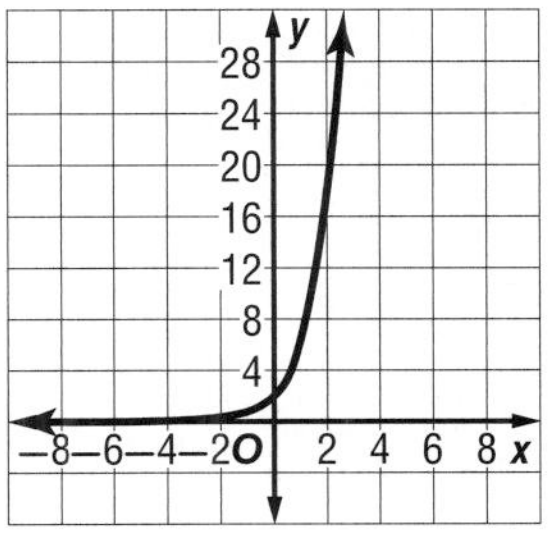

D = {all real numbers}; R = $\{y \mid y > 0\}$

15.

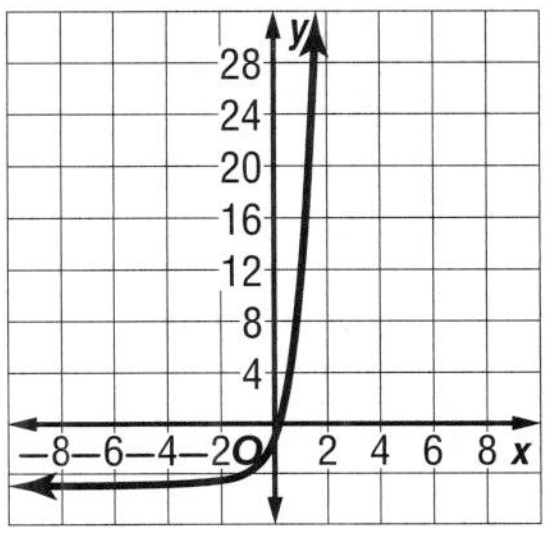

D = {all real numbers}; R = $\{y \mid y > -5\}$

17.

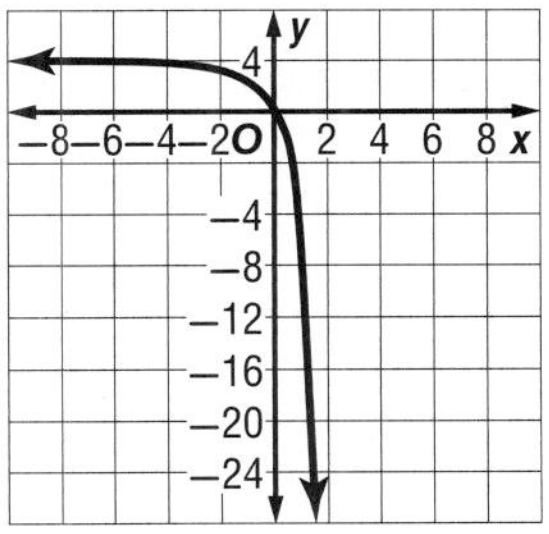

D = {all real numbers}; R = $\{y \mid y < 4\}$

19 $y = a(1 + r)^t$ Equation for exponential growth
$y = 65(1 + 0.3)^t$ $a = 65$ and $r = 0.3$
$y = 65(1.3)^t$ Simplify.

Make a table of values. Then plot the points, and sketch the graph.

x	$y = 65(0.3)^t$
0	$y = 65(1.3)^0 = 65$
2	$y = 65(1.3)^2 \approx 110$
4	$y = 65(1.3)^4 \approx 186$
6	$y = 65(1.3)^6 \approx 314$
8	$y = 65(1.3)^8 \approx 530$
10	$y = 65(1.3)^{10} \approx 896$

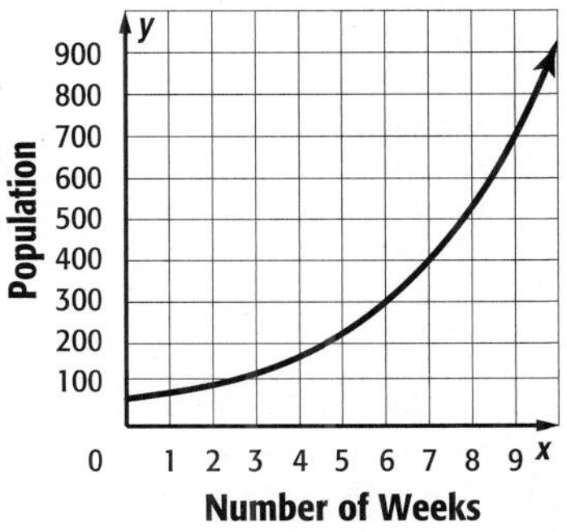

21.

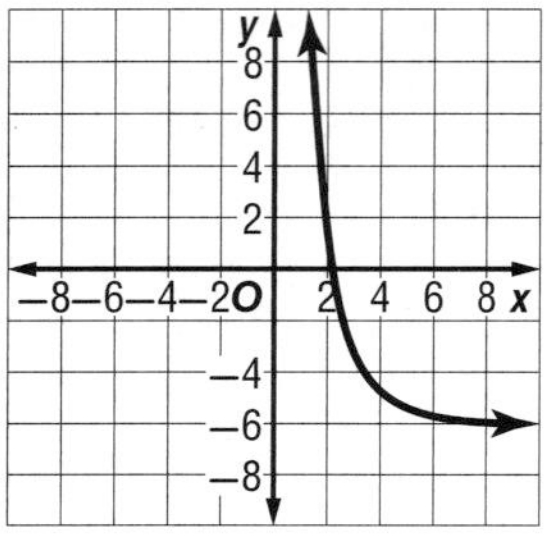

D = {all real numbers}; R = $\{y \mid y > -6\}$

23.

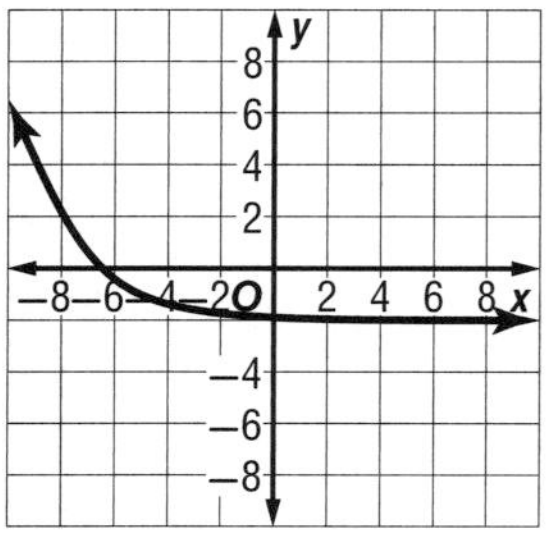

D = {all real numbers}; R = $\{y \mid y > -2\}$

25.

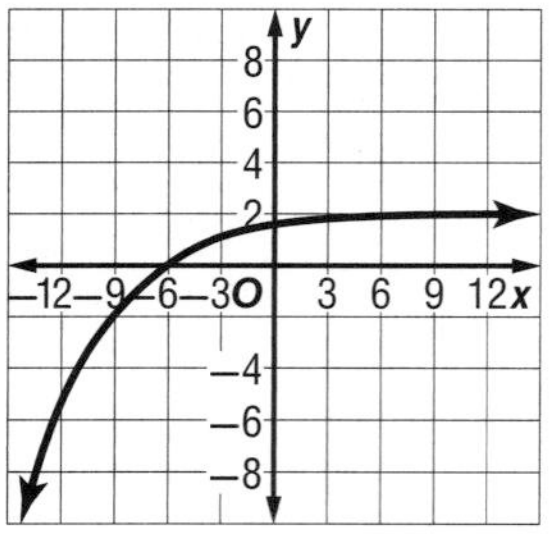

D = {all real numbers}; R = $\{y \mid y < 2\}$

27a.

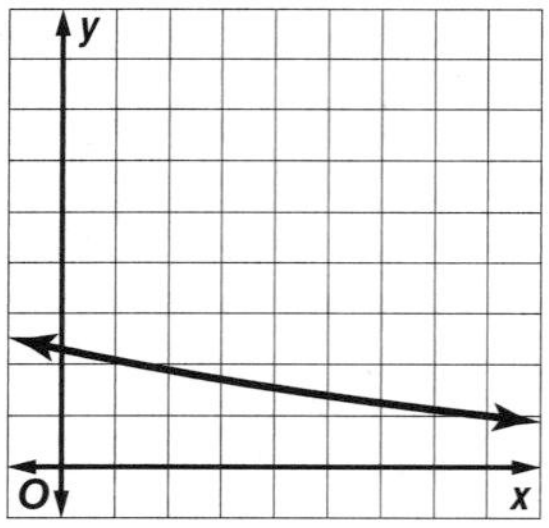

27b. The y-intercept represents the number of pay phones in 1999. The asymptote is the x-axis. The number of pay phones can approach 0, but will never equal 0. This makes sense as there will probably always be a need for some pay phones.

29a. $f(x) = 18(1.25)^{x-1}$

29b.

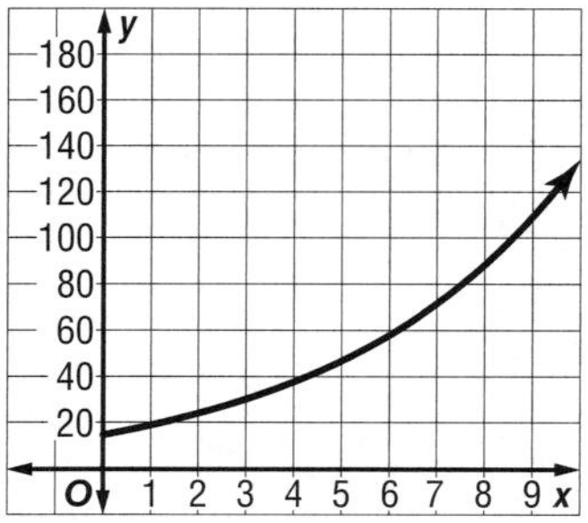

29c. 134 **31.** $g(x) = 4(2)^{x-3}$

33 a.

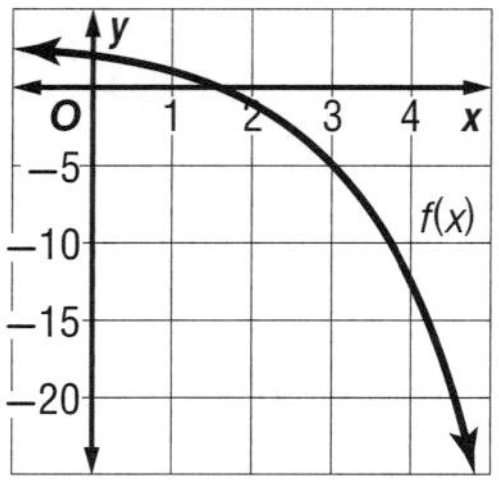

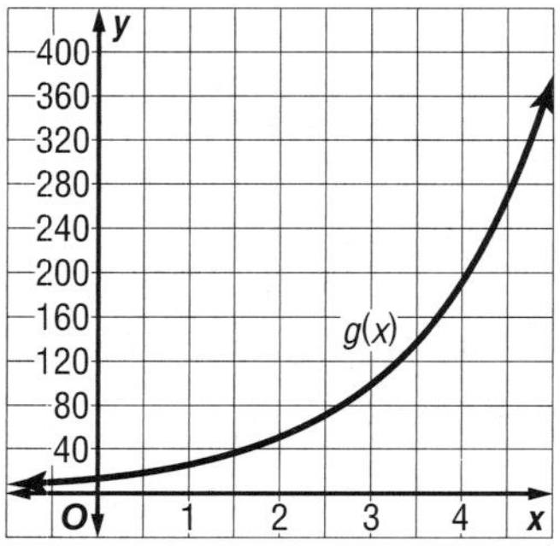

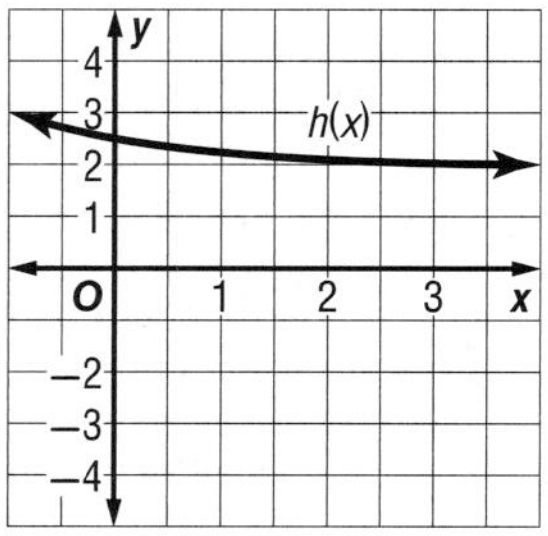

b. Sample answer: The graph of $f(x)$ appears to be the graph of $f(x) = b^x$ reflected across the x-axis. As the values of x increase, the output values decrease.
c. Sample answer: The graphs of $g(x)$ and $h(x)$ appear to be translated to the left.
d. Sample answer: $f(x)$ and $g(x)$ are growth and $h(x)$ is decay; the absolute value of the output is increasing for the growth functions and decreasing for the decay function.

35. Vince; Grady neglected to multiply by the negative sign. **37.** Sample answer: 10 **39.** 12 **41.** H

43. 8 **45.** $x \geq 21$ **47.** $d > -\frac{3}{4}$ **49.** $\frac{y^{\frac{3}{5}}}{y}$ **51.** $3x^{\frac{5}{3}} + 4x^{\frac{8}{3}}$

53. $\sqrt{3}$ **55.** about 204.88 ft **57.** $\frac{1}{f^3}$ **59.** $8xy^4$

Pages 488–491 Lesson 8-2

1. 12 **3.** −10 **5a.** $c = 2^{4t}$ **5b.** 16 cells **7.** $x \geq 4.5$ **9.** 0

11

$81^{a+2} = 3^{3a+1}$	Original equation
$(3^4)^{a+2} = 3^{3a+1}$	Rewrite 81 as 3^4.
$3^{4(a+2)} = 3^{3a+1}$	Power of a Power
$3^{4a+8} = 3^{3a+1}$	Distributive Property
$4a + 8 = 3a + 1$	Property of Equality for Exponential Functions
$a + 8 = 1$	Subtract $3a$ from each side.
$a = -7$	Subtract 8 from each side.

13. $\frac{5}{3}$ **15a.** $y = 10{,}000(1.045)^x$ **15b.** about \$26,336.52

17. $y = 256(0.75)^x$ **19.** $y = 144(3.5)^x$ **21.** \$16,755.63

23. \$97,362.61 **25.** $b > \frac{1}{5}$ **27.** $d \geq -1$ **29.** $w < \frac{2}{5}$

31 a.

$\frac{a}{w^{1.31}} = \frac{170}{45^{1.31}}$	Write a proportion
$a \cdot 45^{1.31} = w^{1.31} \cdot 170$	Cross Products Property
$a = \frac{w^{1.31} \cdot 170}{45^{1.31}}$	Divide each side by $45^{1.31}$.
$a = 1.16w^{1.31}$	Use a calculator.

b.

$a = 1.16w^{1.31}$	Write the equation.
$= 1.16(430)^{1.31}$	$w = 430$
≈ 3268	Use a calculator.

33. $\frac{1}{7}$ **35.** $-\frac{4}{13}$ **37.** 1 **39a.** $d = 1.30h^{\frac{3}{2}}$ **39b.** about 1001 cm **41a.** 2, 4, 8, 16

41b.

Cuts	Pieces
1	2
2	4
3	8
4	16

41c. $y = 2^x$ **41d.** $y = 0.003(2)^x$

41e. about 3,221,225.47 in. **43.** Sample answer: Beth; Liz added the exponents instead of multiplying them when taking the power of a power.
45. Reducing the term will be more beneficial. The multiplier is 1.3756 for the 4-year and 1.3828 for the 6.5%. **47.** Sample answer: $4^x \leq 4^2$
49. Sample answer: Divide the final amount by the initial amount. If n is the number of time intervals that pass, take the nth root of the answer. **51.** F **53.** B

55.
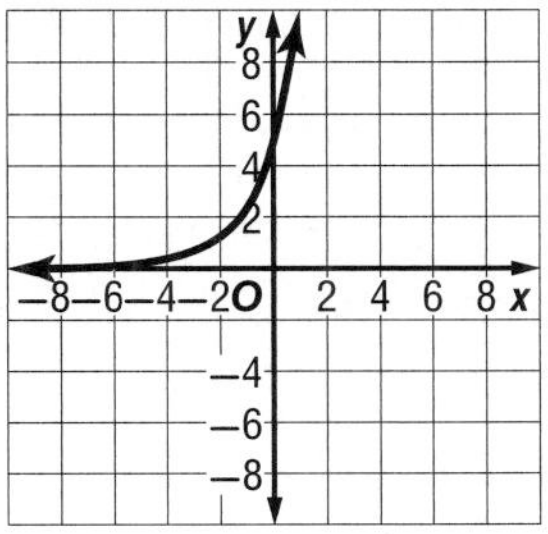

57. 4 **59.** 8.5 **61.** 5 **63.** 5 **65.** −1 **67.** booth 1, 190 lb; booth 2, 150 lb; booth 3, 100 lb **69.** $x^2 - 1$; $x^2 - 6x + 11$ **71.** $-15x - 5$; $-15x + 25$
73. $|x + 4|$; $|x| + 4$

Pages 496–499 Lesson 8-3

1. $8^3 = 512$ **3.** $\log_{11} 1331 = 3$ **5.** 2 **7.** 0

9.
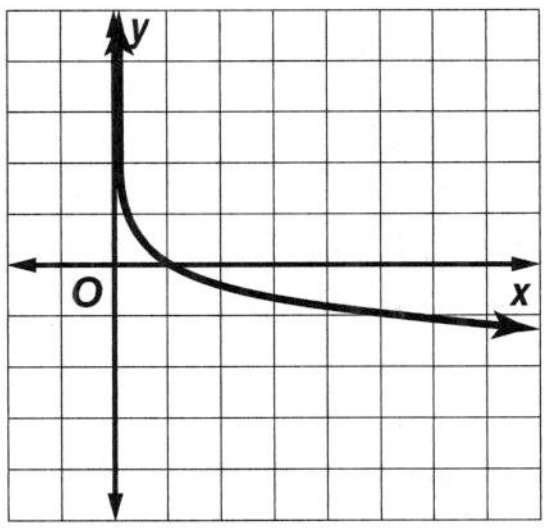

11.
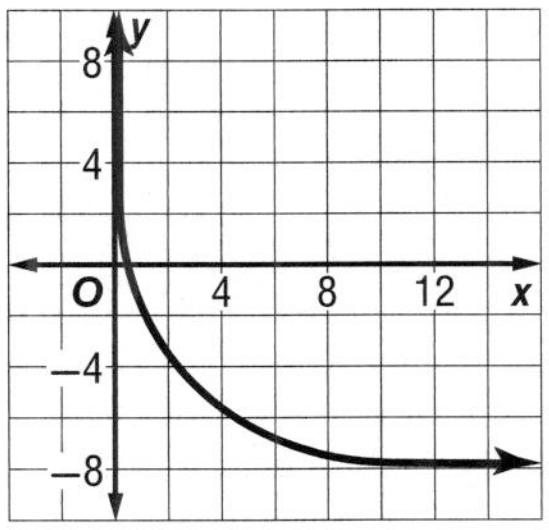

13. $2^4 = 16$ **15.** $9^{-2} = \frac{1}{81}$ **17.** $12^2 = 144$
19. $\log_9 \frac{1}{9} = -1$ **21.** $\log_2 256 = 8$ **23.** $\log_{27} 9 = \frac{2}{3}$
25. −2 **27.** 3 **29.** $\frac{1}{3}$ **31.** $\frac{1}{2}$

33 $\log_{\frac{1}{5}} 3125 = y$ — Let the logarithm equal y.
$3125 = \left(\frac{1}{5}\right)^y$ — Definition of logarithm
$5^5 = 5^{-1y}$ — $\frac{1}{5} = 5^{-1}$
$5 = -1y$ — Property of Equality for Exponential Functions
$-5 = y$ — Divide each side by −1.
So, $\log_{\frac{1}{5}} 3125 = -5$.

35. 4 **37.**
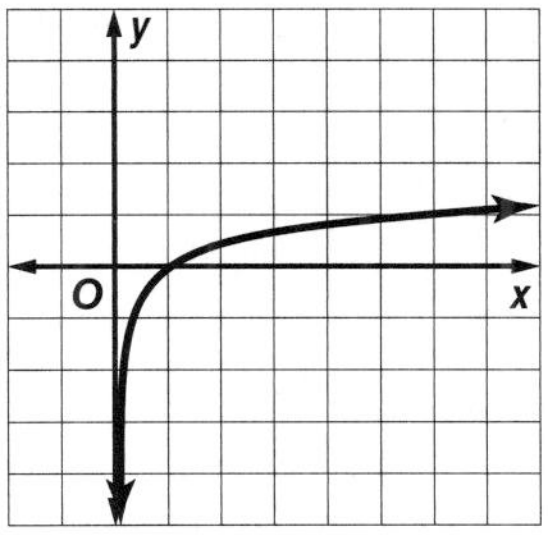

39.
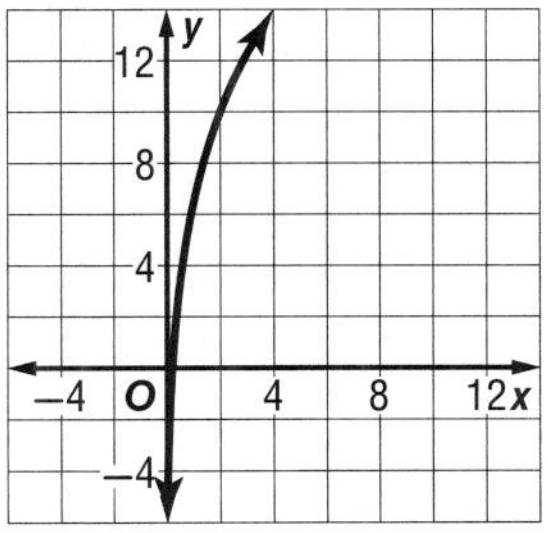

41.
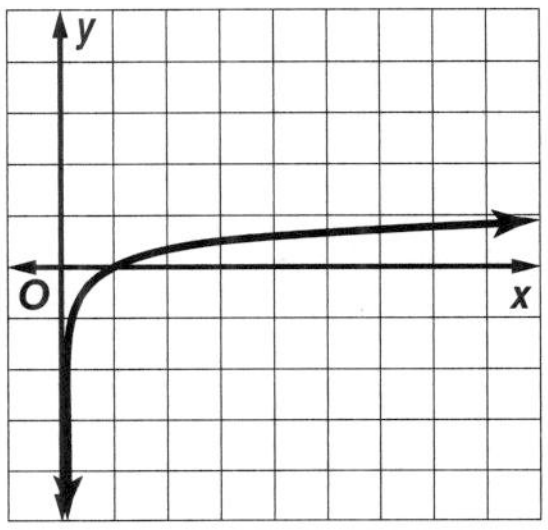

43.
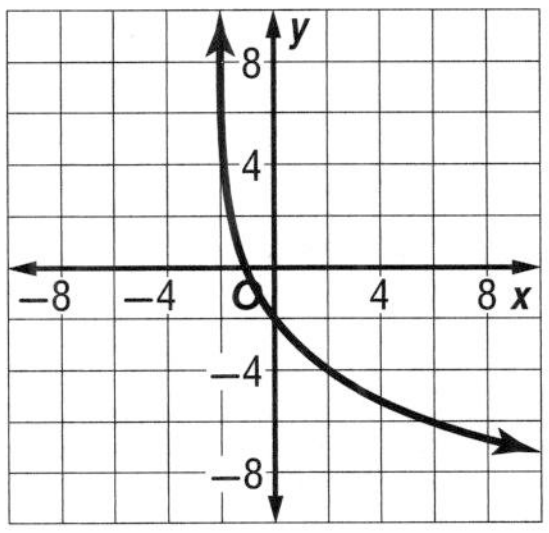

45.
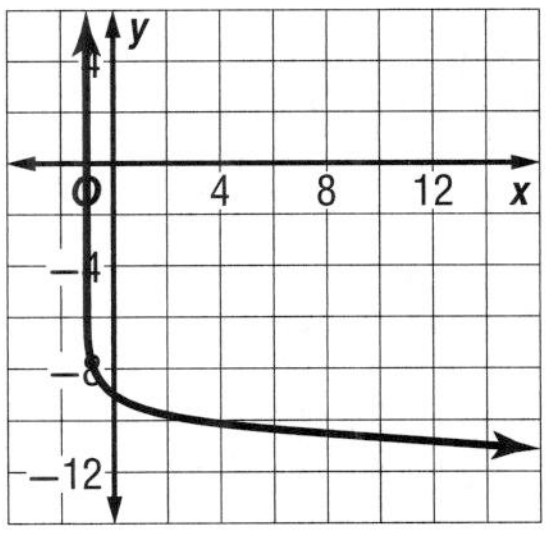

47.
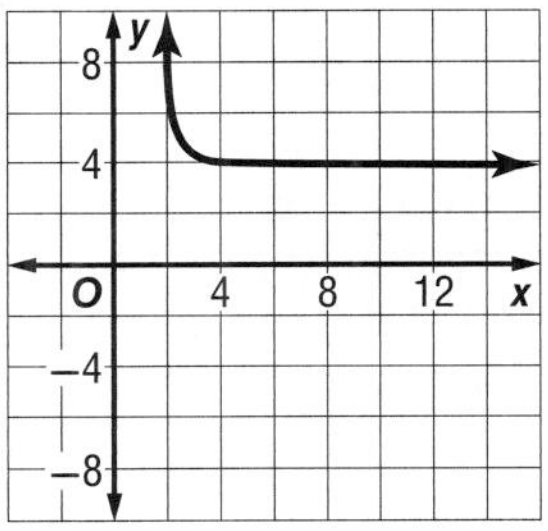

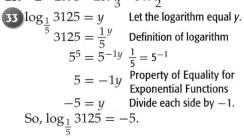

49a. 2

49b.

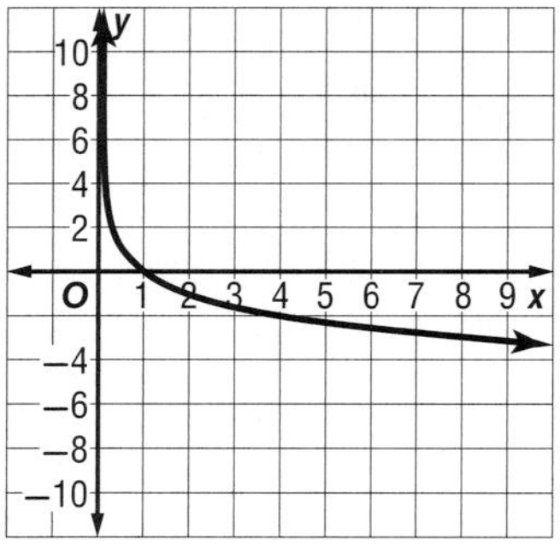

49c. less light; $\frac{1}{8}$

51 This represents a transformation of the graph of $f(x) = \log_2 x$.
$|a| = 4$: The graph expands vertically.
$h = 4$: The graph is translated 4 units to the right.
$k = 6$: The graph is translated 6 units up.

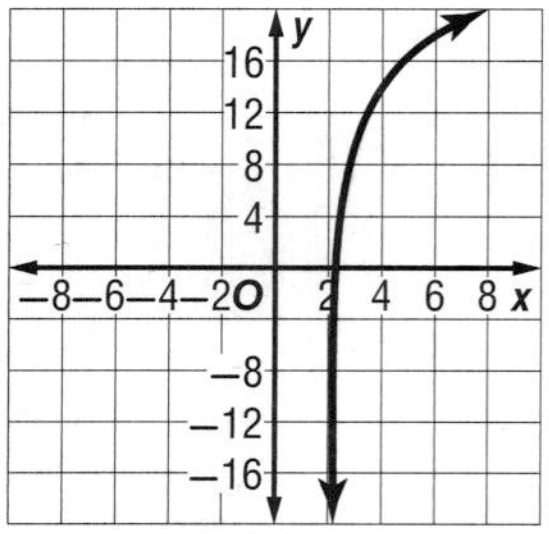

53.

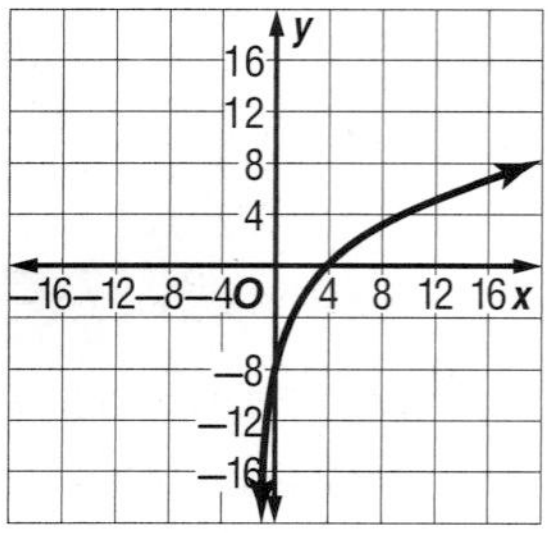

55.

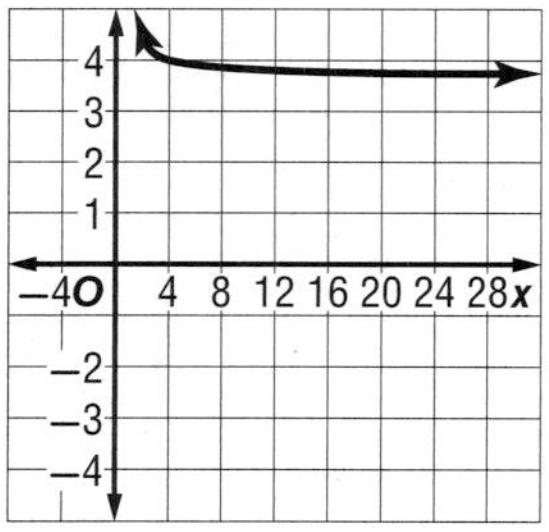

57a. $S(11) \approx 31.6$, $S(21) = 36.8$, $S(31) = 40.1$
57b. If \$1100 is spent on advertising, approximately \$31,600 is returned in sales. If \$2100 is spent on advertising, approximately \$36,800 is returned in sales. If \$3100 is spent on advertising, approximately \$40,100 is returned in sales.

57c.

Sales versus Money Spent on Advertising

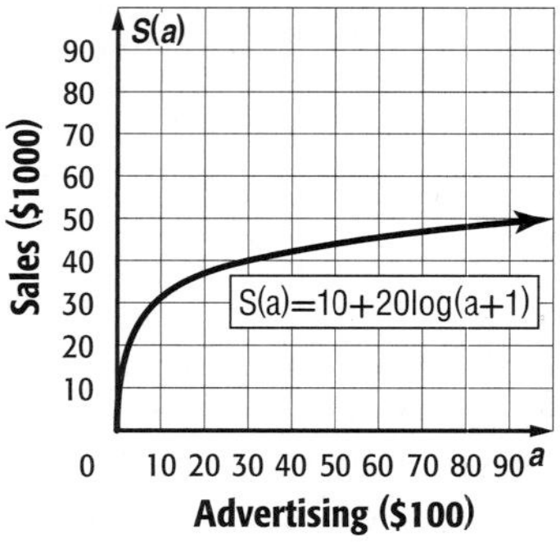

57d. Sample answer: Because eventually the graph plateaus, and no matter how much money you spend you are still returning about the same in sales.

59 **a.**
$\log_{\left(1+\frac{0.24}{12}\right)} \frac{A}{2000} = 12t$ Original formula
$\log_{1.02} \frac{A}{2000} = 12t$ Simplify.
$\frac{A}{2000} = 1.02^{12t}$ Definition of logarithm
$A = 2000 \cdot 1.02^{12t}$ Multiply each side by 2000.

Make a table of values. Then plot the points, and sketch the graph.

t	$A = 2000 \cdot 1.02^{12t}$
0	$A = 2000 \cdot 1.02^{12(0)} = 2000$
2	$A = 2000 \cdot 1.02^{12(2)} \approx 3217$
4	$A = 2000 \cdot 1.02^{12(4)} \approx 5174$
6	$A = 2000 \cdot 1.02^{12(6)} \approx 8322$
8	$A = 2000 \cdot 1.02^{12(8)} \approx 13{,}386$
10	$A = 2000 \cdot 1.02^{12(10)} \approx 21{,}530$

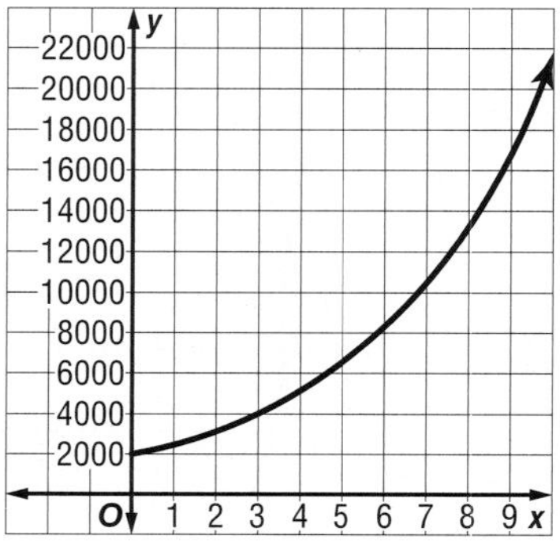

b. From the graph, $A = 4000$ at about $t = 3$. So, it will take approximately 3 years for the debt to double.
c. From the graph, $A = 6000$ at about $t = 4.5$. So, it will take approximately 4.5 years for the debt to triple.

61. Never; if zero were in the domain, the equation would be $y = \log_b 0$.
Then $b^y = 0$.
However, for any real number b, there is no real power that would let $b^y = 0$.

63. $\log_7 51$; Sample answer: $\log_7 51$ equals a little more than 2. $\log_8 61$ equals a little less than 2. $\log_9 71$ equals a little less than 2. Therefore, $\log_7 51$ is the greatest.

65. No; Elisa was closer. She should have $-y = 2$ or $y = -2$ instead of $y = 2$. Matthew used the definition of logarithms incorrectly. **67.** D
69. 80 **71.** $n > 5$ **73.** $n < 3$

75.

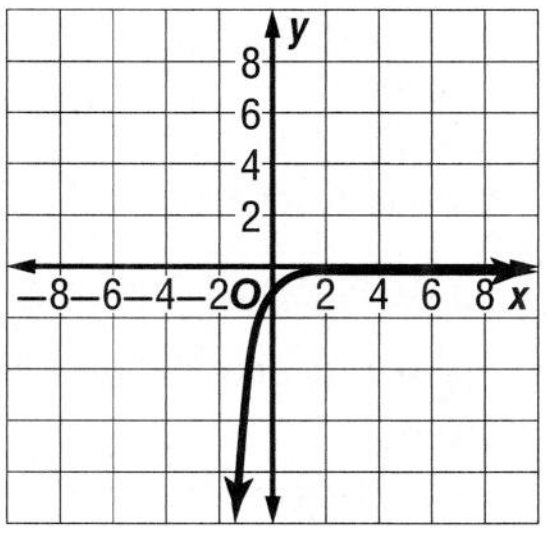

77.

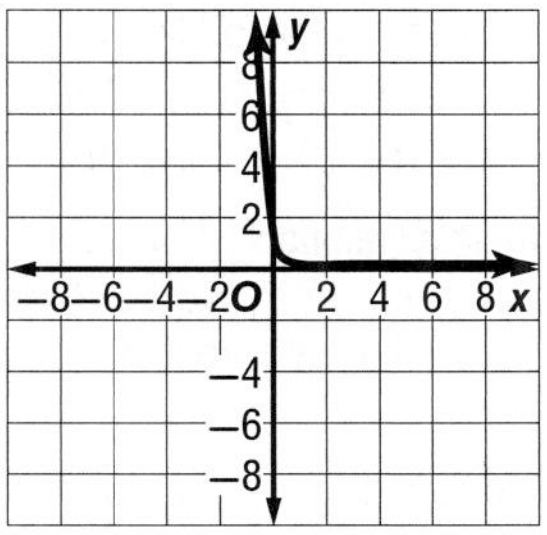

79. $\frac{27\sqrt{15}}{4}$ ft^2 **81.** batteries, \$74; spark plugs, \$58; wiper blades, \$48 **83.** −1 **85.** $x \leq -\sqrt{6}$ or $x \geq \sqrt{6}$

Pages 504–507 Lesson 8-4

1. 16 **3.** C **5.** $\left\{x \,\middle|\, 0 < x \leq \frac{1}{64}\right\}$ **7.** $\left\{x \,\middle|\, 2 > x > \frac{4}{3}\right\}$
9. 3125 **11.** −2 **13.** 9

15 $\log_{12}(x^2 - 7) = \log_{12}(x + 5)$ Original equation
$x^2 - 7 = x + 5$ Property of Equality for Exponential Functions
$x^2 - x - 7 = 5$ Subtract *x* from each side.
$x^2 - x - 12 = 0$ Subtract 5 from each side.
$(x - 4)(x + 3) = 0$ Factor.
$x - 4 = 0$ or $x + 3 = 0$ Zero Product Property
$x = 4$ $\quad x = -3$ Solve each equation.

17. 5 **19.** −2 or 3 **21.** 318 mph **23.** $\{x \mid x \geq 256\}$

25 $\log_2 x \leq -2$ Original inequality
$0 < x \leq 2^{-2}$ Property of Inequality for Exponential Functions
$0 < x \leq \frac{1}{4}$ Simplify.
The solution is $\left\{x \,\middle|\, 0 < x \leq \frac{1}{4}\right\}$.

27. $\left\{x \,\middle|\, 0 < x < \frac{1}{7}\right\}$ **29.** $\left\{x \,\middle|\, 1 \geq x > \frac{1}{2}\right\}$
31. $\left\{x \,\middle|\, -\frac{5}{12} < x \leq 1\right\}$ **33.** $\{x \mid x \geq 8\}$
35a. 37 **35b.** 61

37 **a.** $\beta = 10 \log_{10}\left(\frac{I}{10^{-12}}\right)$ Original equation
$= 10 \log_{10}\left(\frac{1}{10^{-12}}\right)$ $I = 10$
$= 10 \log_{10} 10^{12}$ Write $\frac{1}{10^{-12}}$ as 10^{12}.
$= 10(12)$ or 120 Definition of logarithm

b. $\beta = 10 \log_{10}\left(\frac{I}{10^{-12}}\right)$ Original equation
$= 10 \log_{10}\left(\frac{10^{-2}}{10^{-12}}\right)$ $I = 10^{-2}$
$= 10 \log_{10} 10^{10}$ Quotient of Powers Property
$= 10(10)$ or 100 Definition of logarithm

c. Sample answer: The power of the logarithm only changes by 2. The power is the answer to the logarithm. That 2 is multiplied by the 10 before the logarithm. So we expect the decibels to change by 20.

39. $6\frac{17}{20}$ **41.** The logarithmic function of the form $y = \log_b x$ is the inverse of the exponential function of the form $y = b^x$. The domain of one of the two inverse functions is the range of the other. The range of one of the two inverse functions is the domain of the other.
43a. less than **43b.** less than **43c.** no **43d.** infinitely many **45.** C **47.** B **49.** 4 **51.** 3 **53.** −3 **55.** $x \leq 0$
57. $a \leq -3$ **59.** $x \approx 7.3059$ **61.** 5 ft **63.** x^8 **65.** $8p^6n^3$
67. x^3y^4

Pages 512–515 Lesson 8-5

1. 2.085 **3.** 0.3685 **5.** Mt. Everest: 26,855.44 pascals; Mt. Trisuli: 34,963.34 pascals; Mt. Bonete: 36,028.42 pascals; Mt. McKinley: 39,846.22 pascals; Mt. Logan: 41,261.82 pascals **7.** 2.4181 **9.** 2
11. 13.4403 **13.** 2.1610

15 $\log_4 \frac{4}{3} = \log_4 4 - \log_4 3$ Quotient Property
$= 1 - \log_4 3$ Inverse Property of Exponents and Logarithms
$\approx 1 - 0.7925$ Replace $\log_4 3$ with 0.7925.
≈ 0.2075 Simplify.

17. 1.5 **19.** 2.1606 **21.** 3.4817 **23.** 8 **25.** 2

27 **a.** $P = \log_{10}\left(1 + \frac{1}{d}\right)$ Original equation
$10^P = 1 + \frac{1}{d}$ Definition of logarithm
$10^P - 1 = \frac{1}{d}$ Subtract 1 from each side.
$d(10^P - 1) = 1$ Multiply each side by *d*.
$d = \frac{1}{10^P - 1}$ Divide each side by $10^P - 1$.

b. $d = \frac{1}{10^P - 1}$ Write the formula.
$= \frac{1}{10^{0.097} - 1}$ $P = 0.097$
≈ 4 Use a calculator.

c. $P = \log_{10}\left(1 + \frac{1}{d}\right)$ Original equation
$= \log_{10}\left(1 + \frac{1}{1}\right)$ $d = 1$
$= \log_{10} 2$ Simplify.
≈ 0.30103 Replace $\log_{10} 2$ with 0.30103.
The probability is about 30.1%.

29. 2.1133 **31.** 0.1788 **33.** 1.7228 **35.** 2.0478 **37.** 3
39. 5 **41.** $85\frac{1}{3}$ **43.** $\left(\frac{x-2}{256}\right)^{\frac{1}{6}}$ **45.** $\sqrt{6}, -\sqrt{6}$ **47.** 5
49. 12 **51.** false **53.** false **55.** true **57.** false
59a. 10^{12} **59b.** 10^4 or about 10,000 times **61a.** Sample answer: $\log_b \frac{xz}{5} = \log_b x + \log_b z - \log_b 5$ **61b.** Sample answer: $\log_b m^4p^6 = 4 \log_b m + 6 \log_b p$

61c. Sample answer: $\log_b \frac{j^8k}{h^5} = 8\log_b j + \log_b k - 5\log_b h$

63a. $\log_b 1 = 0$, because $b^0 = 1$. **63b.** $\log_b b = 1$, because $b^1 = b$. **63c.** $\log_b b^x = x$, because $b^x = b^x$.
65. $\log_b 24 \neq \log_b 20 + \log_b 4$; all other choices are equal to $\log_b 24$.

67. $$\begin{aligned} x^{3\log_x 2 - \log_x 5} &= x^{3\log_x 2 - \log_x 5} \\ &= x^{\log_x 2^3 - \log_x 5} \\ &= x^{\log_x 8 - \log_x 5} \\ &= x^{\log_x \frac{8}{5}} \\ &= \frac{8}{5} \end{aligned}$$

69. D **71.** growing exponentially **73.** $\frac{1}{2}$, 1
75. no solution **77.** $2x$ **79.** 6.3 **81.** no **83.** $\frac{3}{5}$
85. 10 **87.** $x > 26$

Pages 519–522 Lesson 8-6

1. 0.6990 **3.** −0.3979 **5.** 3.55×10^{24} ergs **7.** 0.8442

9
$11^{b-3} = 5^b$	Original equation
$\log 11^{b-3} = \log 5^b$	Property of Equality for Logarithmic Functions
$(b - 3)\log 11 = b\log 5$	Power Property of Logarithms
$b\log 11 - 3\log 11 = b\log 5$	Distributive Property
$-3\log 11 = b\log 5 - b\log 11$	Subtract $b\log 11$ from each side.
$-3\log 11 = b(\log 5 - \log 11)$	Distributive Property
$\frac{-3\log 11}{\log 5 - \log 11} = b$	Divide each side by log 5 − log 11.
$9.1237 \approx b$	Use a calculator.

11. $\{p \mid p \leq 4.4189\}$ **13.** $\frac{\log 23}{\log 4} \approx 2.2618$
15. $\frac{\log 5}{\log 2} \approx 2.3219$ **17.** 1.0414 **19.** 0.9138 **21.** −1.3979
23. 1.7740 **25.** 5.9647 **27.** ±1.1691 **29.** $\{n \mid n > 0.6667\}$
31. $\{y \mid y \geq -3.8188\}$ **33.** $\frac{\log 18}{\log 7} \approx 1.4854$
35. $\frac{\log 16}{\log 2} = 4$ **37.** $\frac{\log 11}{\log 3} \approx 2.1827$

39 a.
$n = 35[\log_4(t + 2)]$	Original equation
$= 35[\log_4(10 + 2)]$	$t = 10$
$= 35[\log_4(12)]$	Simplify.
$= 35 \cdot \frac{\log_{10} 12}{\log_{10} 4}$	Change of Base Formula
≈ 62.737	Use a calculator.

In 2010, there are about 62.737 thousand, or 62,737 pet owners.

b.
$n = 35[\log_4(t + 2)]$	Original equation
$80 = 35[\log_4(t + 2)]$	$n = 80$
$2.2857 \approx \log_4(t + 2)$	Divide each side by 35.
$4^{2.2857} \approx t + 2$	Definition of logarithm
$22 \approx t$	

In 22 years after 2000, or in 2022, there will be 80,000 pet owners.
41. 3.3578 **43.** −0.0710 **45.** 4.7393 **47.** $\{x \mid x \geq 2.3223\}$
49. $\{x \mid x \leq 0.9732\}$ **51.** $\{p \mid p \leq 2.9437\}$

53. $\frac{\log 12}{\log 4} \approx 1.7925$ **55.** $\frac{\log 2}{\log 8} = 0.3333$
57. $\frac{\log 7.29}{\log 5} \approx 1.2343$ **59a.** 113.03 cents
59b. about 218 Hz

61
$4^{x^2-3} = 16$	Original equation
$4^{x^2-3} = 4^2$	Rewrite 16 as 4^2.
$x^2 - 3 = 2$	Property of Equality for Exponential Functions
$x^2 = 5$	Add 3 to each side.
$x = \sqrt{5}$	Take the square root of each side.
$\approx \pm 2.2361$	Use a calculator.

63. 3.5 **65.** −3.8188 **67a.** The solution is between 1.8 and 1.9. **67b.** (1.85, 13) **67c.** Yes; all methods produce the solution of 1.85. They all should produce the same result because you are starting with the same equation. If they do not, then an error was made.

69.
$\log_{\sqrt{a}} 3 = \log_a x$	Original equation
$\frac{\log_a 3}{\log_a \sqrt{a}} = \log_a x$	Change of Base Formula
$\log_a 3 - \log_a (a)^{\frac{1}{2}} = \log_a x$	Quotient Property of Logarithms
$\log_a \left(\frac{3}{\sqrt{a}}\right) = \log_a x$	Quotient Property of Logarithms
$x = \frac{3}{\sqrt{a}}$	Property of Equality for Logarithmic Functions
$x = \frac{3\sqrt{a}}{a}$	Rationalize the denominator.

71. $\log_3 27 = 3$ and $\log_{27} 3 = \frac{1}{3}$; Conjecture: $\log_a b = \frac{1}{\log_b a}$

Proof: $\log_a b \stackrel{?}{=} \frac{1}{\log_b a}$	Original statement
$\frac{\log_b b}{\log_b a} \stackrel{?}{=} \frac{1}{\log_b a}$	Change of Base Formula
$\frac{1}{\log_b a} = \frac{1}{\log_b a}$	Inverse Property of Exponents and Logarithms

73. B **75.** G **77.** 14 **79.** 15 **81.** 2 **83.** −4, 3
85. $32x^3 + 8x^2 - 24x + 16$ **87.** $2^x = 5$ **89.** $5^2 = 25$
91. $6^4 = x$

Pages 529–531 Lesson 8-7

1. $\ln 30 = x$ **3.** $\ln x = 3$ **5.** $7\ln 2$ **7.** $\ln 17496$
9. 2.0794 **11.** 0.1352 **13.** 993.6527 **15.** $\{x \mid x < 15.0855\}$
17. $\{x \mid x > 3.3673\}$ **19.** about 58 min
21. $\ln 0.1 = -5x$ **23.** $5.4 = e^x$ **25.** $e^{36} = x + 4$
27. $e^7 = e^x$ **29.** $7\ln 10$

31
$7\ln\frac{1}{2} + 5\ln 2 = 7\ln 2^{-1} + 5\ln 2$	Rewrite $\frac{1}{2}$ as 2^{-1}.
$= \ln 2^{-7} + \ln 2^5$	Power Property of Logarithms
$= \ln(2^{-7})(2^5)$	Product Property of Logarithms
$= \ln 2^{-2}$	Simplify.
$= -2\ln 2$	Power Property of Logarithms

33. $\ln 81x^6$ **35.** 3.7955 **37.** 0.6931 **39.** −0.5596
41. $\{x \mid x \leq 2.1633\}$ **43.** $\{x \mid x > 8.0105\}$
45. $\{x \mid x > 239.8802\}$

Selected Answers and Solutions

47 a. $A = Pe^{rt}$ Continuous Compounding Formula
$= 800e^{(0.045)(5)}$ $P = 800, r = 0.045, t = 5$
$= 800e^{0.225}$ Simplify.
≈ 1001.86 Use a calculator.

About \$1001.86 will be in the account.

b. $A = Pe^{rt}$ Continuous Compounding Formula
$1600 = 800e^{(0.045)t}$ $A = 2 \cdot 800$ or $1600, P = 800, r = 0.045$
$2 = e^{0.045t}$ Divide each side by 800.
$\ln 2 = \ln e^{0.045t}$ Property of Equality of Logarithms
$\ln 2 = 0.045t$ $\ln e^x = x$
$\frac{\ln 2}{0.045} = t$ Divide each side by 0.045.
$15.4 \approx t$ Use a calculator.

It would take about 15.4 years to double your money.

c. $A = Pe^{rt}$ Continuous Compounding Formula
$1600 = 800e^{r(9)}$ $A = 1600, P = 800, t = 9$
$2 = e^{9r}$ Divide each side by 800.
$\ln 2 = \ln e^{9r}$ Property of Equality of Logarithms
$\ln 2 = 9r$ $\ln e^x = x$
$\frac{\ln 2}{9} = r$ Divide each side by 9.
$0.077 \approx r$ Use a calculator.

You would need a rate of about 7.7%.

d. $A = Pe^{rt}$ Continuous Compounding Formula
$10{,}000 = Pe^{(0.0475)(12)}$ $A = 10{,}000, r = 0.0475, t = 12$
$10{,}000 = Pe^{0.57}$ Simplify.
$\frac{10{,}000}{e^{0.57}} = P$ Divide each side by $e^{0.57}$.
$5655.25 \approx P$ Use a calculator.

You would need to deposit about \$5655.25.

49. $4 \ln 2 - 3 \ln 5$ **51.** $\ln x + 4 \ln y - 3 \ln z$
53. -0.8340 **55.** 1.1301

57a.

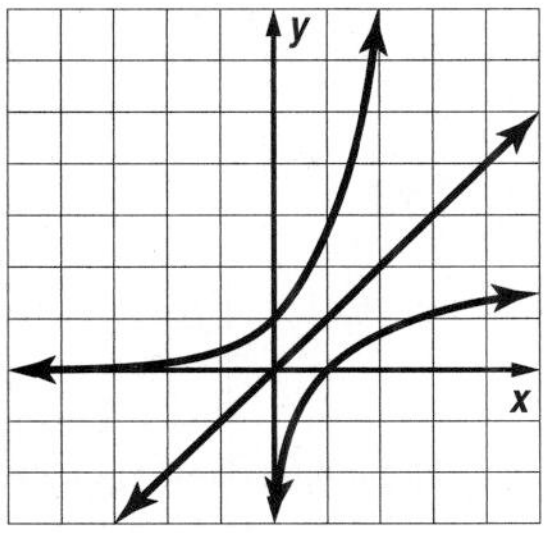

57b. y-axis; $a(x) = -e^x$

57c. $\ln(-x)$ is a reflection across the y-axis. $-\ln x$ is a reflection across the x-axis.

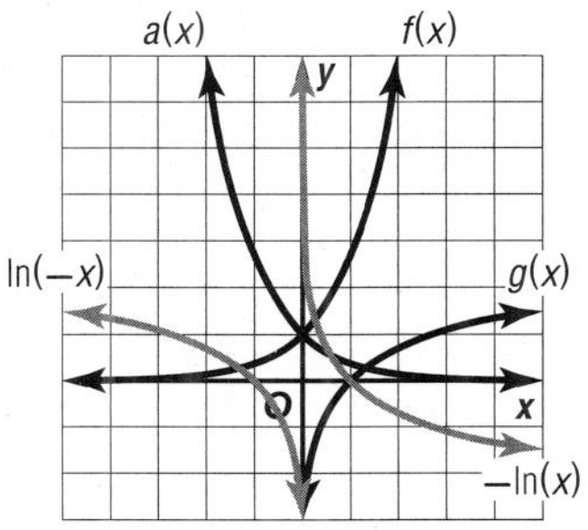

57d. Sample answer: No; these functions are reflections along $y = -x$, which indicates that they are not inverses.

59. Let $p = \ln a$ and $q = \ln b$. That means that $e^p = a$ and $e^q = b$.
$ab = e^p \times e^q$
$ab = e^{p+q}$ $\ln e^{(p+q)} = (p+q)\ln e = p + q$ $\ln e = 1$
$\ln(ab) = (p+q)$
$\ln(ab) = \ln a + \ln b$

61. Sample answer: $e^{\ln 3}$ **63.** B **65.** G **67.** 5.7279
69. $x < 7.3059$ **71.** $x \geq 5.8983$ **73.** 10 decibels
75. $x^2 + 2x + 3$ **77.** $\frac{2}{3}$ **79.** $-\frac{8}{3}$ **81.** $\frac{5}{3}$

Pages 537–539 Lesson 8-8

1a. 5.545×10^{-10} **1b.** 1,578,843,530 yr **1c.** about 30.48 mg **1d.** 3,750, 120,003 yr

3a.

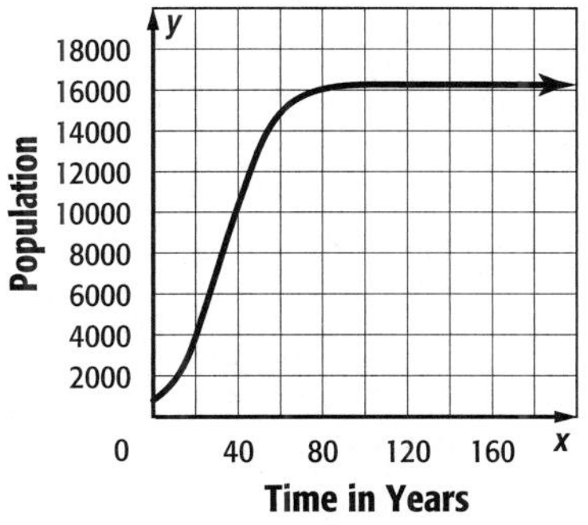

3b. $y = 16{,}500$
3c. 16,500
3d. about 102 years

5 a. $y = 80e^{kt}$ Original formula
$675 = 80e^{k(30)}$ $y = 675, t = 30$
$8.4375 = e^{30k}$ Divide each side by 80.
$\ln 8.4375 = \ln e^{30k}$ Property of Equality for Logarithmic Functions
$\ln 8.4375 = 30k$ $\ln e^x = x$
$\frac{\ln 8.4375}{30} = k$ Divide each side by 30.
$0.071 \approx k$ Use a calculator.

b. $y = 80e^{kt}$ Original formula
$6000 = 80e^{(0.071)t}$ $y = 6000, k \approx 0.071$
$75 = e^{0.071t}$ Divide each side by 80.
$\ln 75 = \ln e^{0.071t}$ Property of Equality for Logarithmic Functions
$\ln 75 = 0.071t$ $\ln e^x = x$
$\frac{\ln 75}{0.071} = t$ Divide each side by 0.071.
$60.8 \approx t$ Use a calculator.

The bacteria will reach a population of 6000 cells in about 60.8 minutes.

c. $35e^{0.0978t} > 80e^{0.071t}$ Formula for exponential growth
$\ln 35e^{0.0978t} > \ln 80e^{0.071t}$ Property of Inequality for Logarithms
$\ln 35 + \ln e^{0.0978t} > \ln 80 + \ln e^{0.071t}$ Product Property of Logarithms
$\ln 35 + 0.0978t > \ln 80 + 0.071t$ $\ln e^x = x$
$0.0268t > \ln 80 - \ln 35$ Subtract $(0.071t + \ln 35)$ from each side.
$t > \frac{\ln 80 - \ln 35}{0.0268}$ Divide each side by 0.0268.
$t > 30.85$ Use a calculator.

The number of cells of this bacteria exceed the number of cells in the other bacteria in about 30.85 minutes.

7

$y = ae^{-0.00012t}$	Equation for the decay of Carbon-14
$0.85a = ae^{-0.00012t}$	$y = 0.85a$
$0.85 = e^{-0.00012t}$	Divide each side by a.
$\ln 0.85 = \ln e^{-0.00012t}$	Property of Equality for Logarithmic Functions
$\ln 0.85 = -0.00012t$	$\ln e^x = x$
$\frac{\ln 0.85}{-0.00012} = t$	Divide each side by -0.00012.
$1354 \approx t$	Use a calculator.

The bone is about 1354 years old.

9. about 14.85 billion yr **11.** about 20.1 yr

13a.

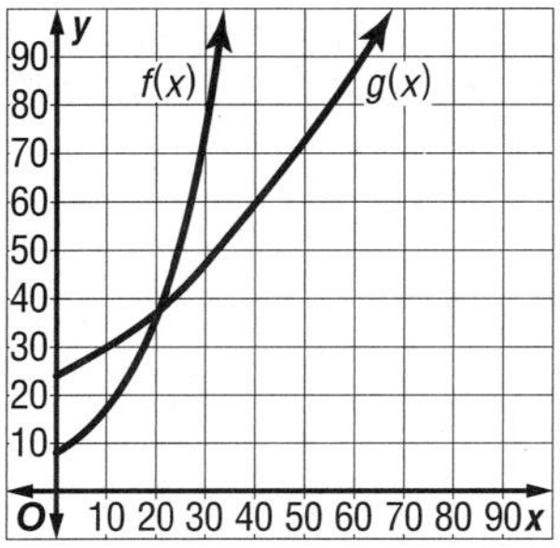

13b. The graphs intersect at $t = 20.79$. Sample answer: This intersection indicates the point at which both functions determine the same population at the same time.

13c. Sample answer: The logistic function $g(x)$ is a more accurate estimate of the country's population since $f(x)$ will continue to grow exponentially and $g(x)$ considers limitations on population growth such as food supply. **15.** $t \approx 113.45$ **17.** Sample answer: The spread of the flu throughout a small town. The growth of this is limited to the population of the town itself. **19.** C **21.** C **23.** $\ln y = 7$ **25.** $5x^4 = e^9$ **27.** $\frac{1}{6}$ **29.** $\frac{5}{8}$ **31.** 16 **33.** $2\frac{1}{2}$

Pages 541–546 Chapter 8 Study Guide and Review

1. exponential growth **3.** common logarithms **5.** change of base formula **7.** logarithmic function **9.** natural logarithm

11.

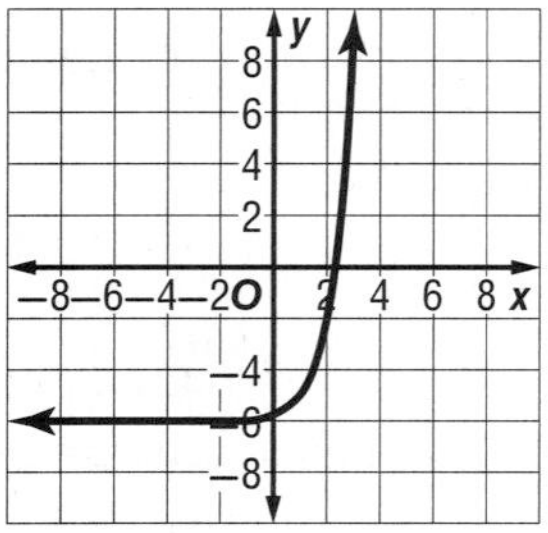

D = all real numbers
R = $\{y \mid y > 0\}$

13.

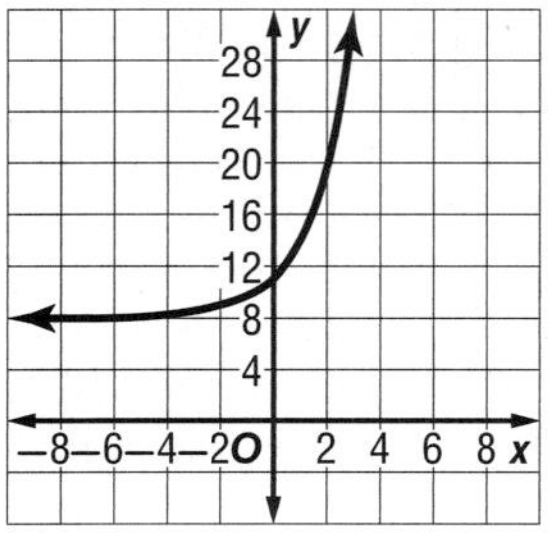

D = all real numbers
R = $\{y \mid y > -6\}$

15.

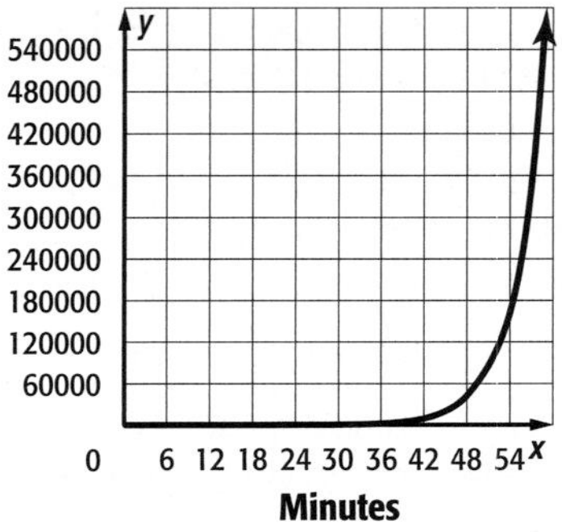

D = all real numbers
R = $\{y \mid y > -1\}$
17a. $f(x) = 120{,}000(0.97)^x$
17b. about 88,491
19. $x = -7$
21. $y = \frac{9}{41}$
23. $x \leq -\frac{2}{5}$
25. $6^0 = 1$

27. $\log_{10} 100 = 2$ **29.** 4

31.

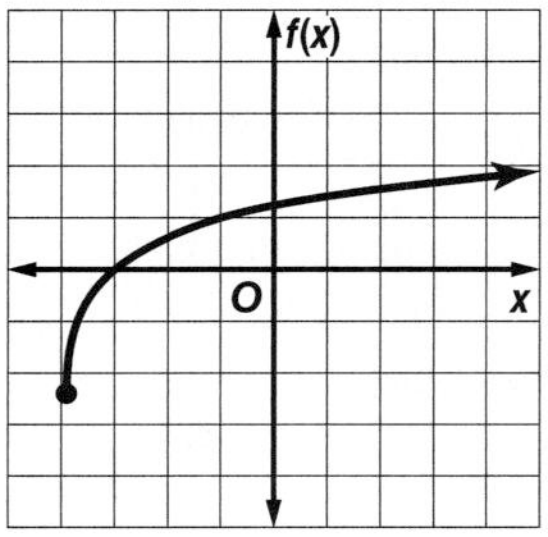

33. $x = 64$ **35.** $x < 64$ **37.** no solution **39.** $x \leq -3$ **41.** 1.2920 **43.** 0.8614 **45.** -0.4306 **47.** $a = 3$ **49.** $n = 5$ **51.** $x \approx 2.4650$ **53.** $m \approx 0.6356$ **55.** $n > 0.5786$ **57a.** about 8.2 years **57b.** about 13.9 years **59.** $x \approx -1.9459$ **61.** $x > 1.9459$ **63.** $x < -2.8904$ **65.** \$1054.69 **67.** about 4.1%

Chapter 9 Rational Functions and Relations

Page 551 Chapter 9 Get Ready

1. $x = \frac{15}{14}$ **3.** $k = \frac{32}{5}$ **5.** 27 gallons **7.** $\frac{1}{18}$ **9.** $\frac{43}{6}$ **11.** $p = 27$ **13.** $k = 17.5$

Pages 557–561 Lesson 9-1

1. $\frac{x+3}{x+8}$ **3.** D

5

$\frac{a^2x - b^2x}{by - ay} = \frac{x(a^2 - b^2)}{y(b - a)}$	Factor.
$= \frac{x(a - b)(a + b)}{y(b - a)}$	Factor.
$= \frac{-x(b - a)(a + b)}{y(b - a)}$	$a - b = -1(b - a)$
$= \frac{-x\cancel{(b - a)}^{1}(a + b)}{y\cancel{(b - a)}_{1}}$	Eliminate common factors.
$= \frac{-x(a + b)}{y}$	Simplify.

7. $\frac{2x^2}{3aby^2}$ **9.** $\frac{(a - b)(a + 1)}{12(a - 1)}$ **11.** 4 **13.** $\frac{x(x + 6)}{x + 4}$ **15.** $\frac{(x + 3)(x - z)}{4}$ **17.** $\frac{x(x + 2)}{6(x + 5)}$ **19.** J **21.** $\frac{x^2}{(x + 6)}$ **23.** $-\frac{c + 4}{c + 5}$ **25.** $\frac{c}{4ab^2f^2}$ **27.** $\frac{32b}{3ac^3f^2}$ **29.** $\frac{5a^4c}{3b}$

31

$\frac{y^2 + 8y + 15}{y - 6} \cdot \frac{y^2 - 9y + 18}{y^2 - 9}$	
$= \frac{(y + 3)(y + 5)}{y - 6} \cdot \frac{(y - 3)(y - 6)}{(y - 3)(y + 3)}$	Factor.
$= \frac{\cancel{(y + 3)}^{1}(y + 5)}{\cancel{y - 6}_{1}} \cdot \frac{\cancel{(y - 3)}^{1}\cancel{(y - 6)}^{1}}{\cancel{(y - 3)}_{1}\cancel{(y + 3)}_{1}}$	
$= y + 5$	Simplify.

33. $\dfrac{(x+4)(x+2)}{2(x-5)}$ **35.** $\dfrac{(x-3)(x+1)}{6(x+7)}$ **37.** $\dfrac{-a^2(a+b)}{b^4}$

39a. $\dfrac{33}{121}$ **39b.** $\dfrac{33+m}{121+a}$ **41a.** $T(x) = \dfrac{0.4}{x+3}$

41b. about 3.9 mm thick **43.** $\dfrac{1}{4}$ **45.** $\dfrac{x(x+2)(x-1)}{(x+3)(x-7)}$

47 $\dfrac{20x^2y^6z^{-2}}{3a^3c^2} \cdot \left(\dfrac{16x^3y^3}{9acz}\right)^{-1}$

$= \dfrac{20x^2y^6z^{-2}}{3a^3c^2} \cdot \dfrac{9acz}{16x^3y^3}$ $\quad \left(\dfrac{16x^3y^3}{9acz}\right)^{-1} = \dfrac{9acz}{16x^3y^3}$

$= \dfrac{20x^2y^6}{3a^3c^2z^2} \cdot \dfrac{9acz}{16x^3y^3}$ $\quad z^{-2} = \dfrac{1}{z^2}$

$= \dfrac{2 \cdot 2 \cdot 5 \cdot x \cdot x \cdot y \cdot y \cdot y \cdot y \cdot y \cdot y \cdot 3 \cdot 3 \cdot a \cdot c \cdot z}{3 \cdot a \cdot a \cdot a \cdot c \cdot c \cdot z \cdot z \cdot 2 \cdot 2 \cdot 2 \cdot 2 \cdot x \cdot x \cdot x \cdot y \cdot y \cdot y}$ Factor.

$= \dfrac{\cancel{2} \cdot \cancel{2} \cdot 5 \cdot \cancel{x} \cdot \cancel{x} \cdot \cancel{y} \cdot \cancel{y} \cdot \cancel{y} \cdot y \cdot y \cdot y \cdot \cancel{3} \cdot 3 \cdot \cancel{a} \cdot \cancel{c} \cdot \cancel{z}}{\cancel{3} \cdot \cancel{a} \cdot a \cdot a \cdot \cancel{c} \cdot c \cdot \cancel{z} \cdot z \cdot \cancel{2} \cdot \cancel{2} \cdot 2 \cdot 2 \cdot \cancel{x} \cdot \cancel{x} \cdot x \cdot \cancel{y} \cdot \cancel{y} \cdot \cancel{y}}$ Eliminate common factors.

$= \dfrac{5 \cdot y \cdot y \cdot y \cdot 3}{a \cdot a \cdot c \cdot z \cdot 2 \cdot 2 \cdot x}$ Simplify.

$= \dfrac{15y^3}{4a^2cxz}$ Simplify.

49. $\dfrac{2(4x+1)(2x+1)}{5(2x-1)(x+2)}$ **51.** $\dfrac{2x+1}{-9x(x+2)}$

53. $\dfrac{x(x-2)(x+8)}{2(2x-1)(3x+1)}$ **55.** $\dfrac{-2(x-8)(x+4)(x-2)(x+1)}{(2x+1)^2(x^2+2x-6)}$

57 a. $5 \text{ tracks} \cdot \dfrac{2 \text{ miles}}{1 \text{ track}} \cdot \dfrac{5280 \text{ feet}}{1 \text{ mile}} \cdot \dfrac{1 \text{ car}}{75 \text{ feet}}$

b. $5 \text{ tracks} \cdot \dfrac{2 \text{ miles}}{1 \text{ track}} \cdot \dfrac{5280 \text{ feet}}{1 \text{ mile}} \cdot \dfrac{1 \text{ car}}{75 \text{ feet}}$

$= \cancel{5} \,\cancel{\text{tracks}} \cdot \dfrac{2 \,\cancel{\text{miles}}}{1 \,\cancel{\text{track}}} \cdot \dfrac{\cancel{15} \cdot 352 \,\cancel{\text{feet}}}{1 \,\cancel{\text{mile}}} \cdot \dfrac{1 \text{ car}}{\cancel{5} \cdot \cancel{15} \,\cancel{\text{feet}}}$

$= \dfrac{1 \cdot 2 \cdot 352 \cdot 1 \text{ car}}{1 \cdot 1 \cdot 1 \cdot 1}$

$= 704$ cars

c. $704 \text{ cars} \cdot \dfrac{8 \text{ attendants}}{1 \text{ car}} \cdot \dfrac{45 \text{ s}}{1 \text{ attendant}} \cdot \dfrac{1 \text{ min}}{60 \text{ s}} \cdot \dfrac{60 \text{ min}}{1 \text{ h}}$

$= 704 \,\cancel{\text{cars}} \cdot \dfrac{8 \,\cancel{\text{attendants}}}{1 \,\cancel{\text{car}}} \cdot \dfrac{45 \,\cancel{\text{s}}}{1 \,\cancel{\text{attendant}}} \cdot \dfrac{1 \,\cancel{\text{min}}}{60 \,\cancel{\text{s}}} \cdot \dfrac{1 \text{ h}}{60 \,\cancel{\text{min}}}$

$= \dfrac{704 \cdot 8 \cdot 45 \cdot 1 \cdot 1 \text{ h}}{1 \cdot 1 \cdot 60 \cdot 60}$

$= 70.4$ hours

59. Sample answer: The two expressions are equivalent except that the rational expression is undefined at $x = 3$. **61.** $x^2 + x - 6$ **63.** Sample answer: Sometimes; with a denominator like $x^2 + 2$, in which the denominator cannot equal 0, the rational expression can be defined for all values of x. **65.** Sample answer: When the original expression was simplified, a factor of x was taken out of the denominator. If x were to equal 0, then this expression would be undefined. So, the simplified expression is also undefined for x.

67. J **69.** 4π **71.** 0.2877 **73.** 0.2747 **75.** $10^{1.7}$ or about 50 times **77.** $2y\sqrt[3]{2}$ **79.** $2ab^2\sqrt{10a}$

81. $10a - 2b$ **83.** $-3y - 3y^2$ **85.** $x^2 + 9x + 18$

Pages 565–568 Lesson 9-2

1. $80x^3y^3$ **3.** $3y(y-3)(y-5)$ **5.** $\dfrac{48y^4 + 25x^2}{20xy^3}$

7. $\dfrac{21b^4 - 2}{36ab^3}$ **9.** $\dfrac{9x+15}{(x+3)(x+6)}$ **11.** $\dfrac{x-11}{3(x+2)(x-2)}$

13. $\dfrac{14x - 10}{(x+1)(x-2)}$ **15.** $\dfrac{3y+2}{y+3}$ **17.** $\dfrac{2a+5b}{3b-8a}$

19. $180x^2y^4z^2$ **21.** $6(x+4)(2x-1)(2x+3)$

23. $\dfrac{28by^2z - 9bx}{105x^3y^4z}$ **25.** $\dfrac{20x^2y + 120y + 6x^2}{15x^3y}$

27. $\dfrac{15b^3 + 100ab^2 - 216a}{240ab^3}$

29 $\dfrac{6}{y^2 - 2y - 35} + \dfrac{4}{y^2 + 9y + 20}$

$= \dfrac{6}{(y-7)(y+5)} + \dfrac{4}{(y+4)(y+5)}$ Factor denominators.

$= \dfrac{6(y+4)}{(y-7)(y+5)(y+4)} + \dfrac{4(y-7)}{(y+4)(y+5)(y-7)}$ Multiply by missing factors.

$= \dfrac{6y + 24 + 4y - 28}{(y-7)(y+5)(y+4)}$ Add the numerators.

$= \dfrac{10y - 4}{(y-7)(y+5)(y+4)}$ Simplify.

31. $\dfrac{-10x - 10}{(2x-1)(x+6)(x-3)}$ **33.** $\dfrac{2x^2 + 32x}{3(x-2)(x+3)(2x+5)}$

35. $\dfrac{1000x + 800y}{x(x+2y)}$

37 $\dfrac{\dfrac{4}{x+5} + \dfrac{9}{x-6}}{\dfrac{5}{x-6} - \dfrac{8}{x+5}} = \dfrac{\dfrac{4(x-6)}{(x+5)(x-6)} + \dfrac{9(x+5)}{(x+5)(x-6)}}{\dfrac{5(x+5)}{(x+5)(x-6)} - \dfrac{8(x-6)}{(x+5)(x-6)}}$

$= \dfrac{\dfrac{4x - 24 + 9x + 45}{(x+5)(x-6)}}{\dfrac{5x + 25 - 8x + 48}{(x+5)(x-6)}}$ Simplify the numerator and denominator.

$= \dfrac{\dfrac{13x + 21}{(x+5)(x-6)}}{\dfrac{-3x + 73}{(x+5)(x-6)}}$ Combine like terms.

$= \dfrac{13x + 21}{(x+5)(x-6)} \div \dfrac{-3x + 73}{(x+5)(x-6)}$ Write as a division expression.

$= \dfrac{13x + 21}{(x+5)(x-6)} \cdot \dfrac{(x+5)(x-6)}{-3x + 73}$ Multiply by the reciprocal of the divisor.

$= \dfrac{13x + 21}{-3x + 73}$ Simplify.

39. $\dfrac{-x^2 + 33x + 16}{12x^2 + 11x - 27}$ **41.** $420x^5y^4z^3$

43. $(x+4)(x-4)(2x+1)(x-7)$ **45.** $\dfrac{360a^2 + 5a - 36}{60a^2}$

47. $\dfrac{42x + 41}{6(3x-1)(x+8)(2x+3)}$ **49.** 0 **51.** $\dfrac{5a - 11}{6}$

53. $(x-3)(x+2)$ to 1 **55.** $-\dfrac{3}{2}$ **57.** -1

59a. $y = \dfrac{70x}{x - 70}$ **59b.** Sample answer: When the object is 70 mm away, y needs to be 0, which is impossible.

61 a. $P_0\left(\dfrac{s_0}{s_0 - x}\right) - P_0\left(\dfrac{s_0}{s_0 - y}\right) = \dfrac{P_0s_0}{s_0 - x} - \dfrac{P_0s_0}{s_0 - y}$

$= \dfrac{P_0s_0(s_0 - y)}{(s_0 - x)(s_0 - y)} - \dfrac{P_0s_0(s_0 - x)}{(s_0 - x)(s_0 - y)}$

$= \dfrac{P_0s_0(s_0 - y) - P_0s_0(s_0 - x)}{(s_0 - x)(s_0 - y)}$

$= \dfrac{P_0s_0s_0 - P_0s_0y - P_0s_0s_0 - P_0s_0x}{(s_0 - x)(s_0 - y)}$

$= \dfrac{P_0s_0x - P_0s_0y}{(s_0 - x)(s_0 - y)}$

b. $\dfrac{P_0s_0x - P_0s_0y}{(s_0 - x)(s_0 - y)} = \dfrac{(500)(332)(70) - (500)(332)(45)}{(332 - 70)(332 - 45)}$ $P_0 = 500, s_0 = 332, x = 70, y = 45$

$= \dfrac{4{,}150{,}000}{75{,}194}$ Simplify.

≈ 55.2 Hz Simplify.

63. $\dfrac{-3x^3 - 2x^2 + 16x - 5}{4x^3 + 18x^2 - 6x}$ **65.** Sample answer: $20a^4b^2c$, $15ab^6$, $9abc$ **67.** D **69.** F **71.** $-\dfrac{4bc}{33a}$ **73.** $(n + 3)(n - 6)$

75. D = $\{x \mid x \geq -0.5\}$, R = $\{y \mid y \leq 0\}$

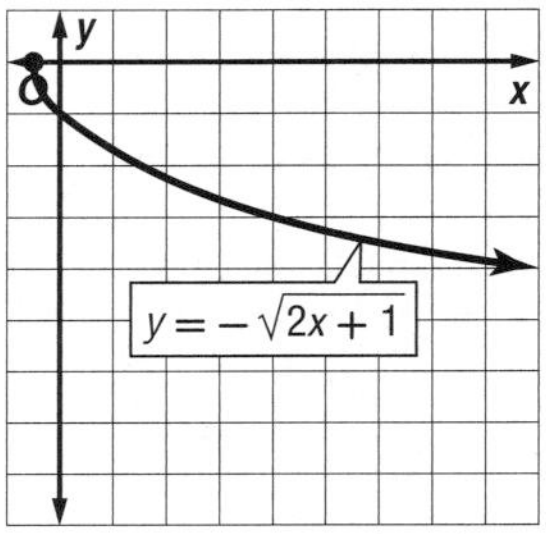

77. D = $\{x \mid x \geq -6\}$, R = $\{y \mid y \geq -3\}$

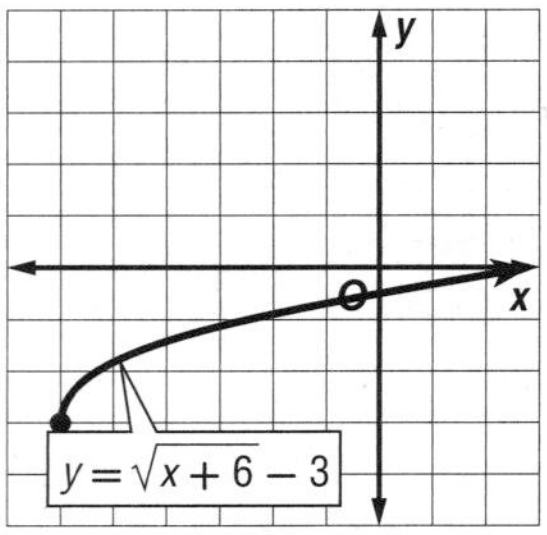

79. D = $\{x \mid x \geq 2\}$, R = $\{y \mid y \geq 4\}$

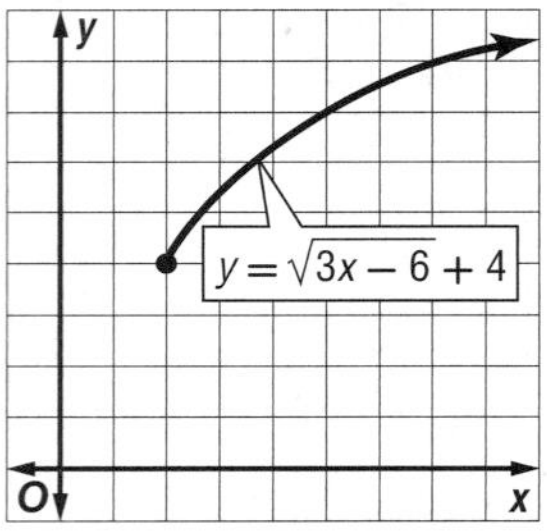

81. $-\frac{8}{3}$; 1 real **83.** 0, $3i$, $-3i$; 1 real, 2 imaginary

85.

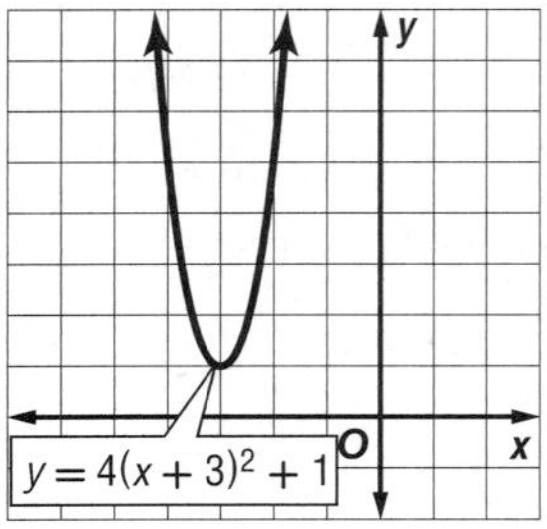

87.

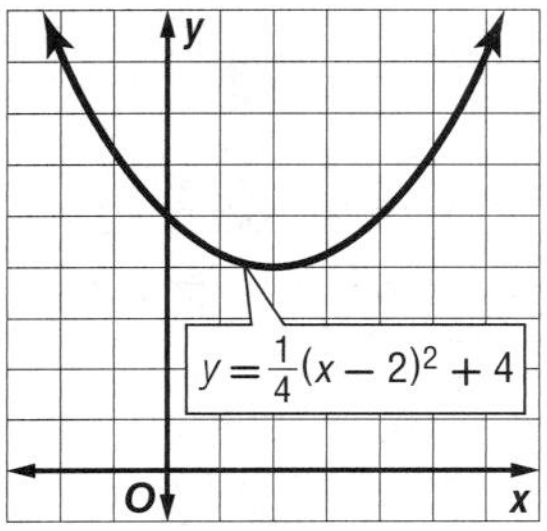

89.

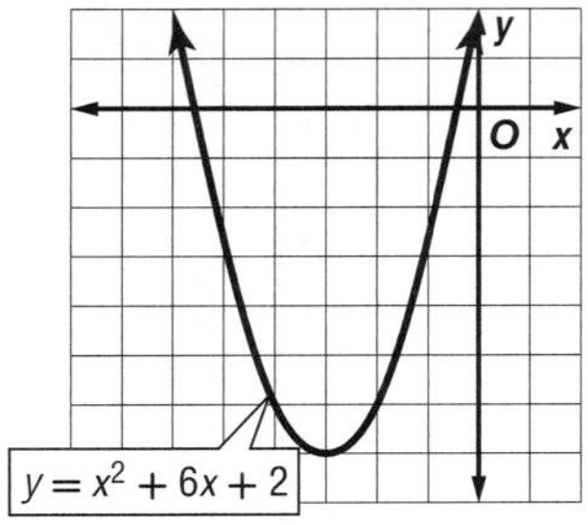

Pages 572–575 Lesson 9-3

1 $x - 1 = 0$

$x = 1$

$f(x)$ is not defined when $x = 1$. So, there is a vertical asymptote at $x = 1$.
From $x = 1$, as x-values decrease, $f(x)$ values approach 0, and as x-values increase, $f(x)$ values approach 0. So there is a horizontal asymptote at $f(x) = 0$. The domain is all real numbers not equal to 1 or D = $\{x \mid x \neq 1\}$. The range is all real numbers not equal to 0 or R = $\{f(x) \mid f(x) \neq 0\}$.

3.

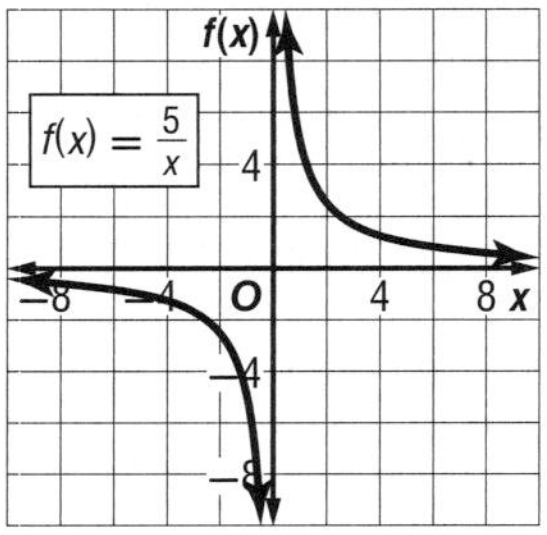

D = $\{x \mid x \neq 0\}$; R = $\{y \mid y \neq 0\}$

5.

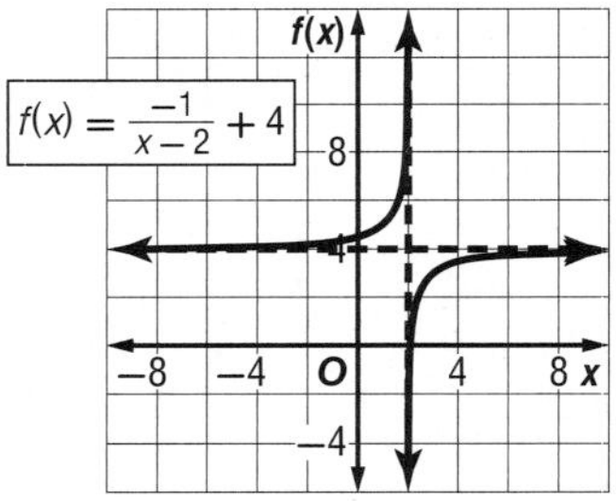

D = $\{x \mid x \neq 2\}$; R = $\{y \mid y \neq 4\}$

7. $x = -4$, $y = 0$; D = $\{x \mid x \neq -4\}$; R = $\{y \mid y \neq 0\}$
9. $x = -6$, $y = -2$; D = $\{x \mid x \neq -6\}$; R = $\{y \mid y \neq -2\}$

11.

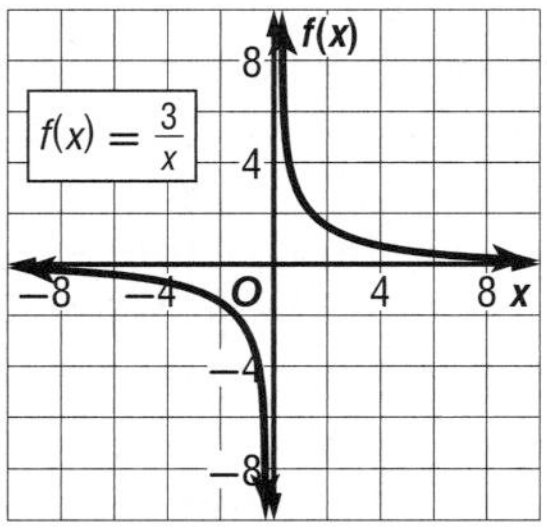

D = $\{x \mid x \neq 0\}$; R = $\{y \mid y \neq 0\}$

13.

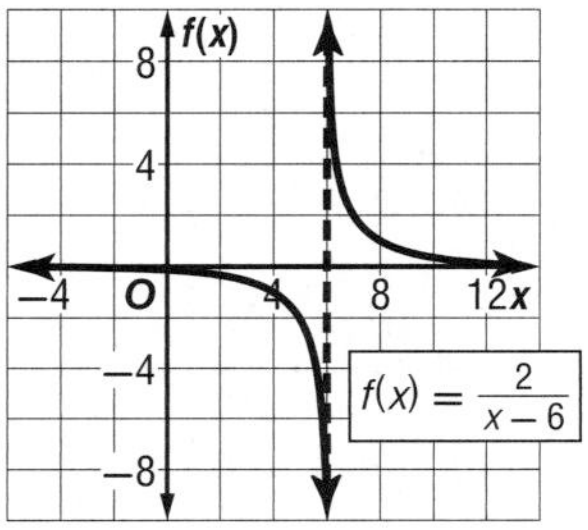

D = $\{x \mid x \neq 6\}$; R = $\{y \mid y \neq 0\}$

15 This represents a transformation of the graph of $f(x) = \frac{1}{x}$.
$a = 2$: The graph is expanded.
$k = 3$: The graph is translated 3 units up. There is a horizontal asymptote at $f(x) = 3$. Domain: $D = \{x \mid x \neq 0\}$. Range: $R = \{f(x) \mid f(x) \neq 3\}$.

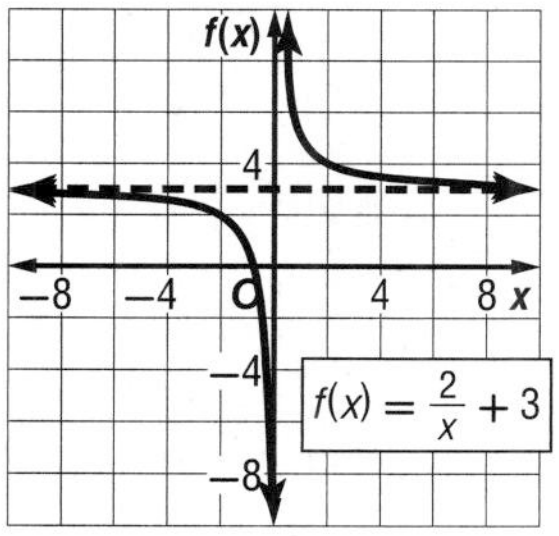

17.
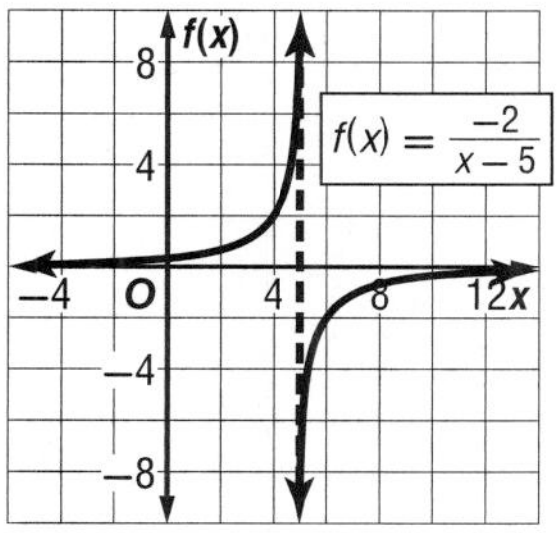

$D = \{x \mid x \neq 5\}$;
$R = \{y \mid y \neq 0\}$

19.
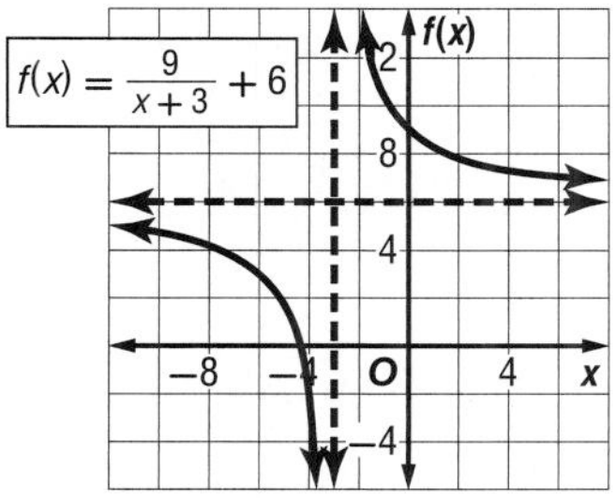

$D = \{x \mid x \neq -3\}$;
$R = \{y \mid y \neq 6\}$

21.
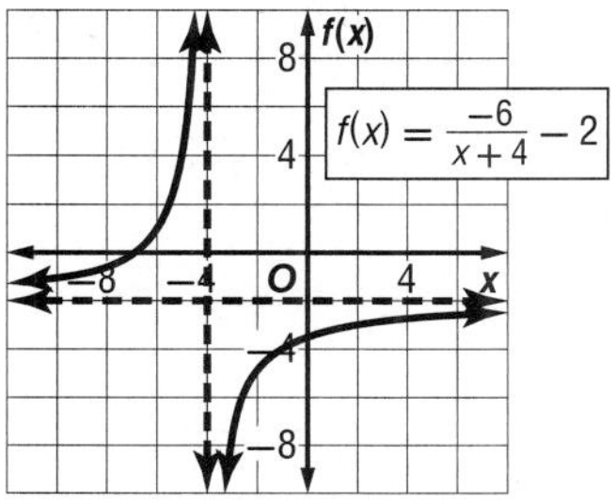

$D = \{x \mid x \neq -4\}$;
$R = \{y \mid y \neq -2\}$

23a. $m = \frac{5000}{d}$

23b.
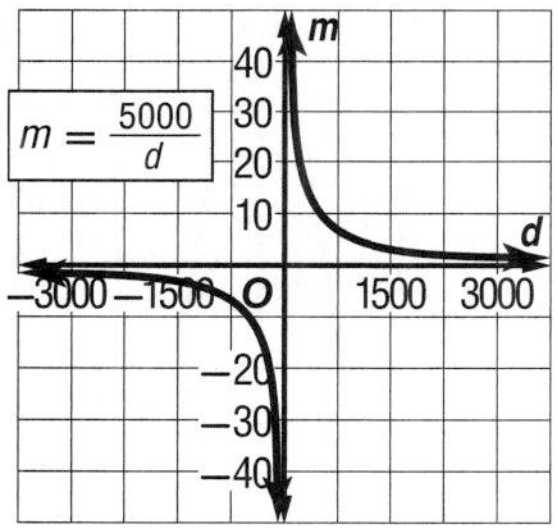

23c. 13.7 mi

25.
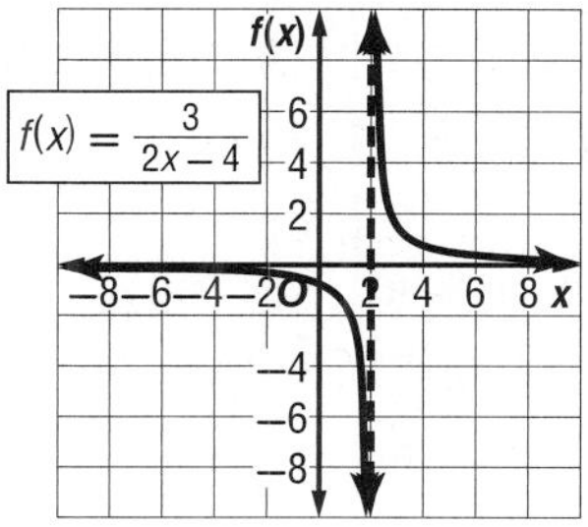

$D = \{x \mid x \neq 2\}$;
$R = \{y \mid y \neq 0\}$

27.
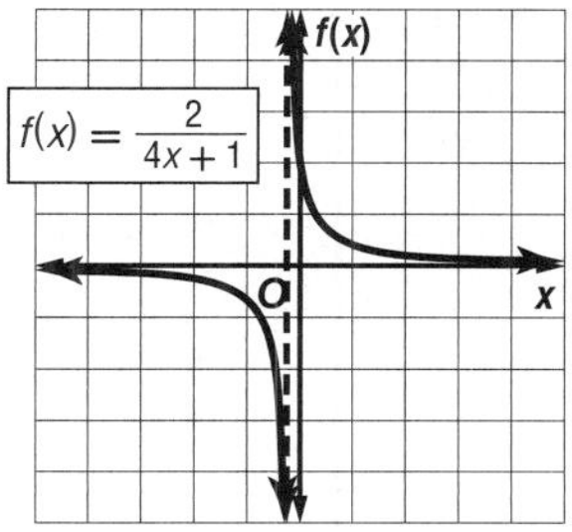

$D = \left\{x \,\middle|\, x \neq \frac{1}{4}\right\}$;
$R = \{y \mid y \neq 0\}$

29 **a.** $rt = d$ rate • time = distance
$rt = 60.5$ $d = 60.5$
$r = \frac{60.5}{t}$ Divide each side by t.

b. This represents a transformation of the graph of $f(x) = \frac{1}{x}$. There are asymptotes at $t = 0$ and $r = 0$. Since $a = 60.5$, the graph is expanded.

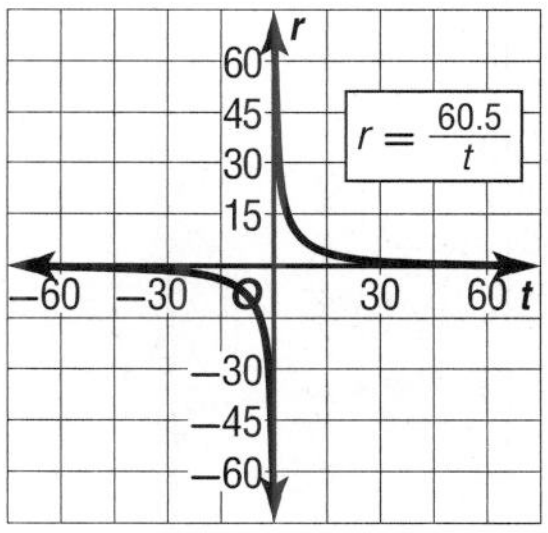

c. $r = \frac{60.5}{t}$ Write the equation.
$= \frac{60.5}{0.48}$ $t = 0.48$
≈ 126 ft/s Use a calculator.

31. $D = \{x \mid x \neq -2\}$;
$R = \{y \mid y \neq -5\}$;
$x = -2, y = -5$

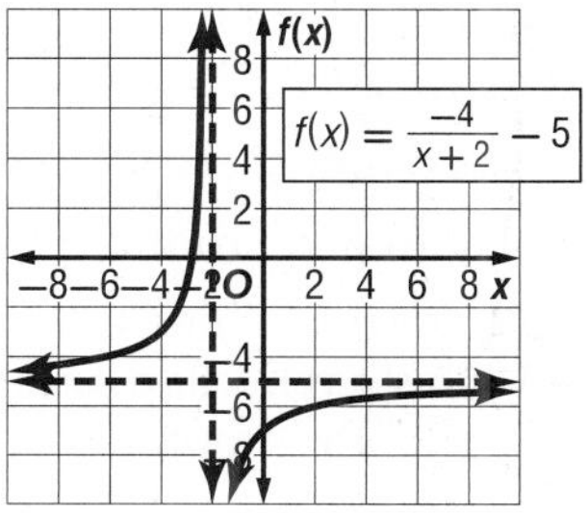

33. $D = \{x \mid x \neq 4\}$;
$R = \{y \mid y \neq 3\}$;
$x = 4, y = 3$

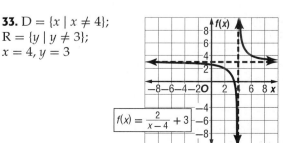

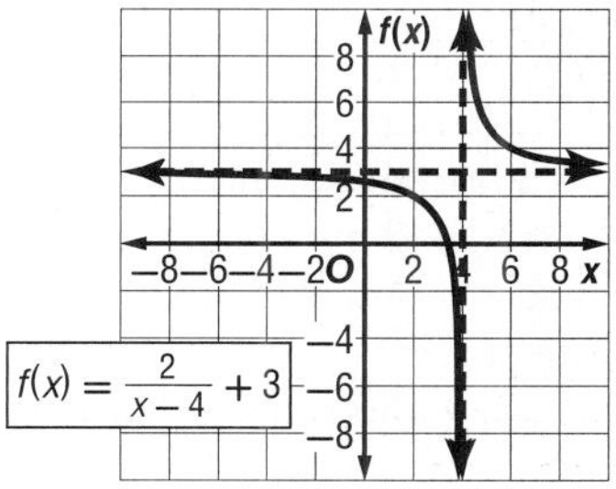

Selected Answers and Solutions

35. D = $\{x \mid x \neq 7\}$; R = $\{y \mid y \neq -8\}$; $x = 7, y = -8$

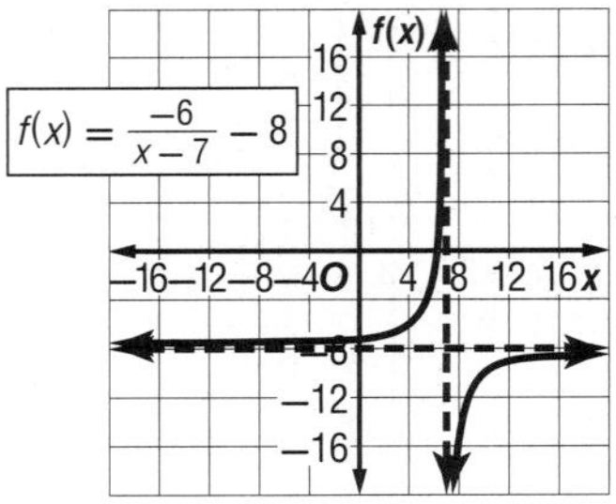

37a.

$f(x) = \frac{1}{x}$		$f(x) = \frac{1}{x^2}$	
x	**f(x)**	**x**	**f(x)**
−3	$-\frac{1}{3}$	−3	$\frac{1}{9}$
−2	$-\frac{1}{2}$	−2	$\frac{1}{4}$
−1	−1	−1	1
0	undefined	0	undefined
1	1	1	1
2	$\frac{1}{2}$	2	$\frac{1}{4}$
3	$\frac{1}{3}$	3	$\frac{1}{9}$

37b.

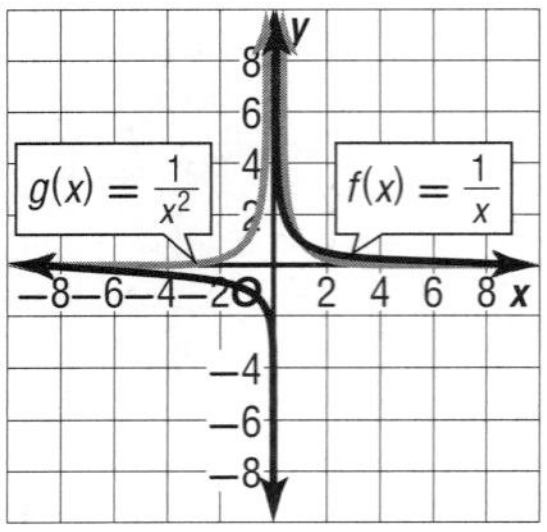

37c. The positive portion of $f(x) = \frac{1}{x^2}$ is similar to the graph of $f(x) = \frac{1}{x}$. Positive values of x produce positive values of $f(x)$. The negative portion of $f(x) = \frac{1}{x^2}$ appears to be a reflection of $f(x) = \frac{1}{x}$ over the x-axis. Negative values of x produce positive values of $f(x)$.

37d. Sample answer: When the exponent is even, the graph will show a reflection over the x-axis. When the exponent is odd, the graph will show a reflection over $y = x$. **39a.** The first graph has a vertical asymptote at $x = 0$ and a horizontal asymptote at $y = 0$. The second graph is translated 7 units up and has a vertical asymptote at $x = 0$ and a horizontal asymptote at $y = 7$. **39b.** Both graphs have a vertical asymptote at $x = 0$ and a horizontal asymptote at $y = 0$. The second graph is stretched by a factor of 4. **39c.** The first graph has a vertical asymptote at $x = 0$ and a horizontal asymptote at $y = 0$. The second graph is translated 5 units to the left and has a vertical asymptote at $x = -5$ and a horizontal asymptote at $y = 0$.

39d.

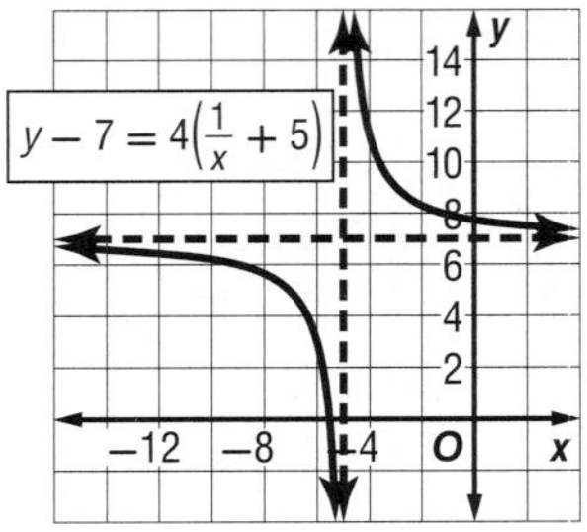

41. Sample answer: $f(x) = \frac{2}{x-3} + 4$ and $g(x) = \frac{5}{x-3} + 4$

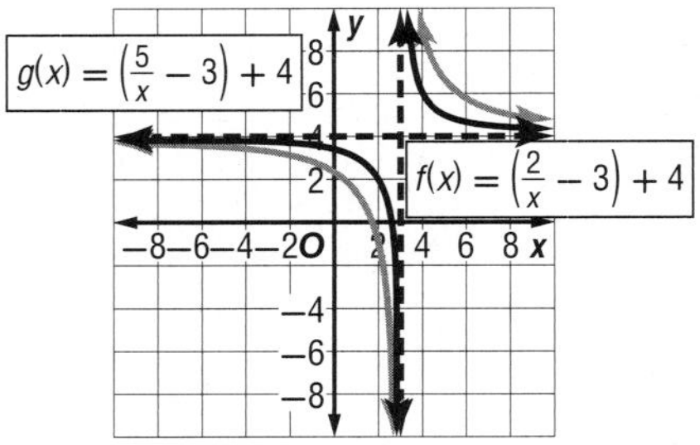

43. 4 **45.** B **47.** B **49.** $-2p$ **51.** $\frac{2x + y}{2x - y}$

53.

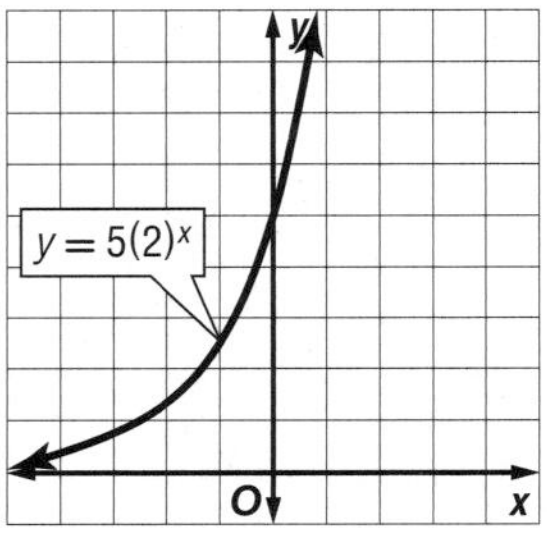

55.

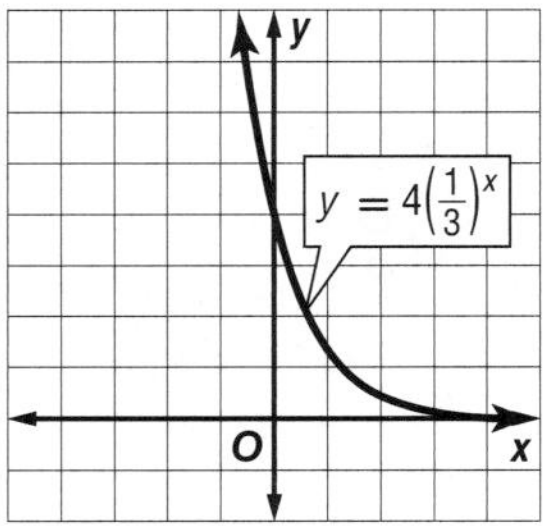

57. $6x + 6$; $-2x - 12$; $8x^2 + 6x - 27$; $\frac{2x-3}{4x+9}$, $x \neq -\frac{9}{4}$

59. $w = 4$ cm, $\ell = 8$ cm, $h = 2$ cm

61. Sample answer: rel. max at $x = 0$, rel. min at $x = -2$ and at $x = 2$; D = {all real numbers}, R = $\{f(x) \mid f(x) \geq -6\}$

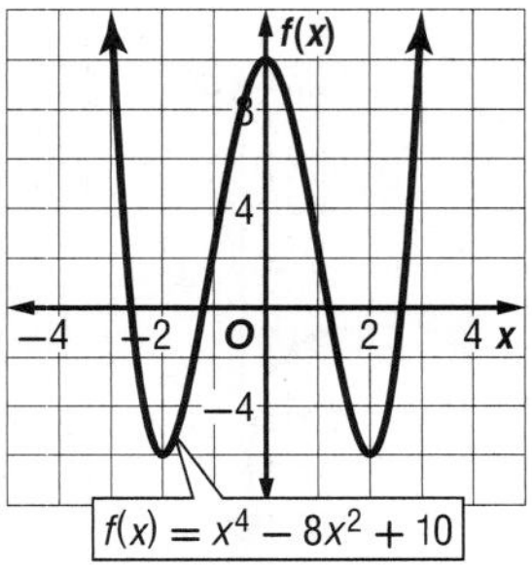

1.

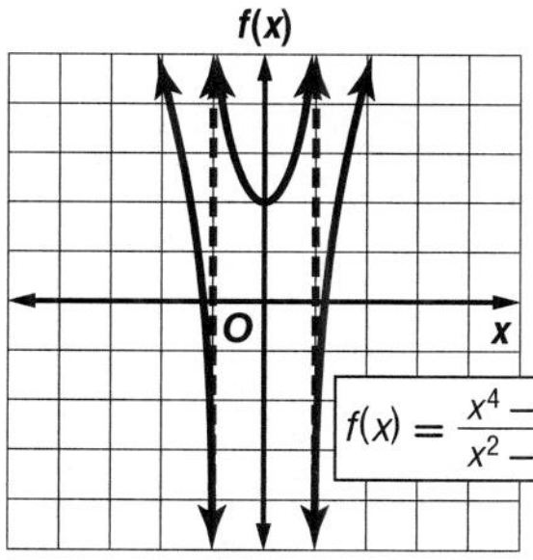

3a.

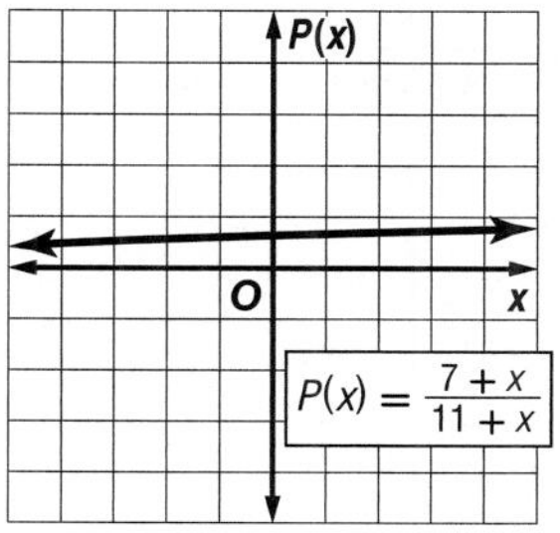

3b. The part in the first quadrant. **3c.** It represents his original field goal percentage of 63.6%. **3d.** $y = 1$; this represents 100% which he cannot achieve because he has already missed 4 field goals.

5 $x^2 + 8x + 20 = 0$ Set $a(x) = 0$.

Since $b^2 - 4ac = 8^2 - 4(1)(20)$ or -16, there are no real roots. So, there are no zeros.

$x + 2 = 0$ Set $b(x) = 0$.

$x = -2$ Subtract 2 from each side.

There is a vertical asymptote at $x = -2$. The degree of the numerator is greater than the degree of the denominator, so there is no horizontal asymptote. The difference between the degree of the numerator and the degree of the denominator is 1, so there is an oblique asymptote.

$$\begin{array}{r} x + 6 \\ x + 2\overline{)x^2 + 8x + 20} \\ \underline{(-)\ x^2 + 2x} \\ 6x + 10 \\ \underline{(-)\ 6x + 12} \\ -2 \end{array}$$

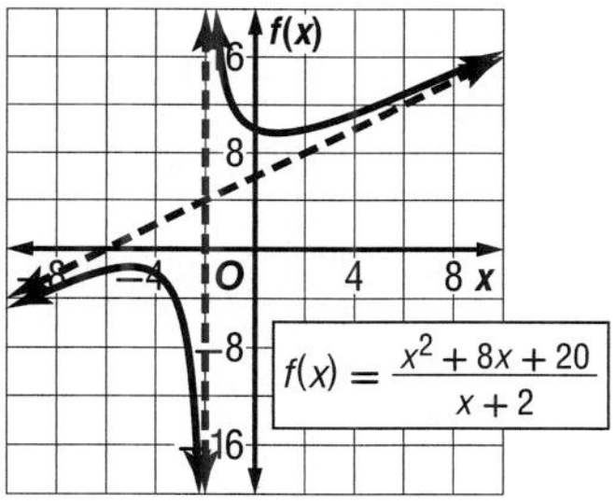

The oblique asymptote is $y = x + 6$.

7.

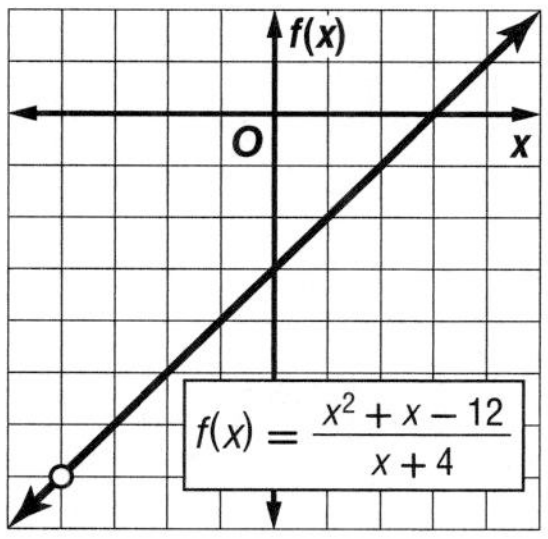

9.

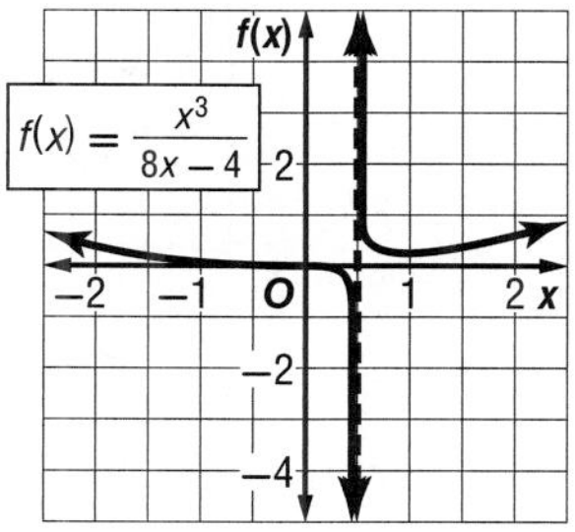

11.

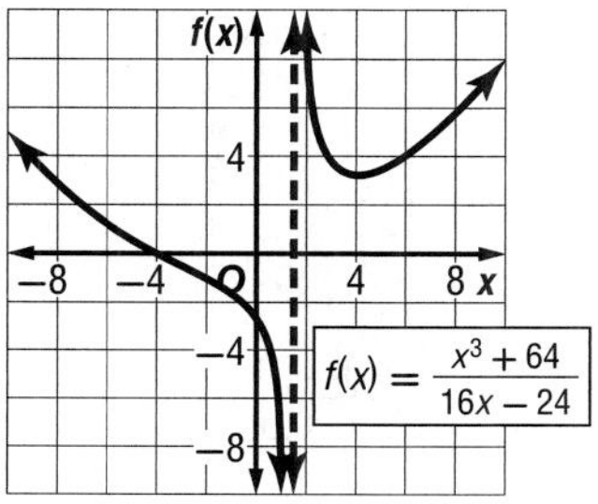

13.

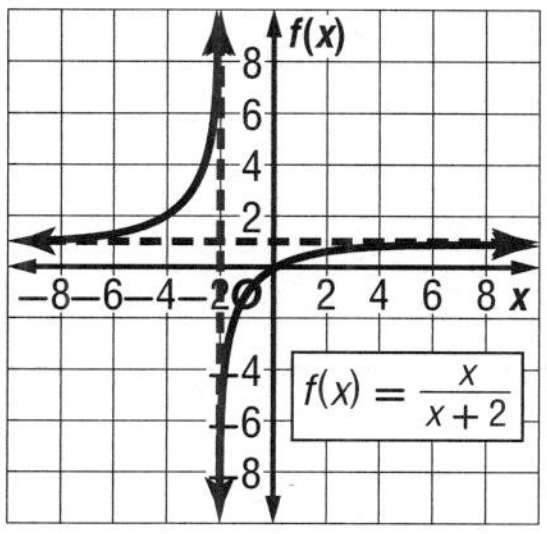

15 Since $a(x) = 4$, there are no zeros. The function is undefined for $x = 2$, so there is a vertical asymptote at $x = 2$. Since the degree of the numerator is less than the degree of the denominator, there is a horizontal asymptote at $f(x) = 0$. The difference between the degree of the numerator and the degree of the denominator is 2, so there is no oblique asymptote.

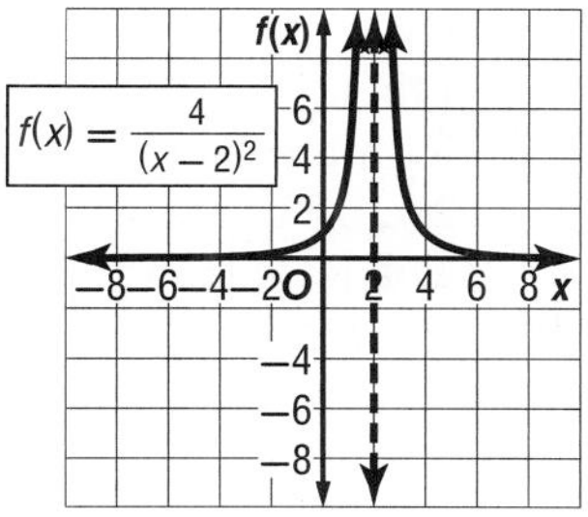

17.

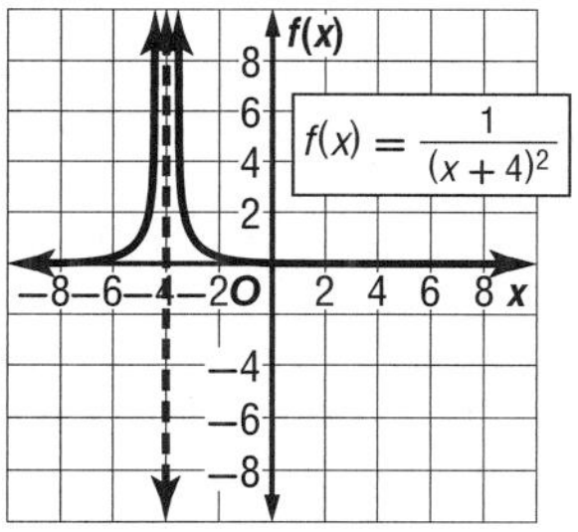

19.

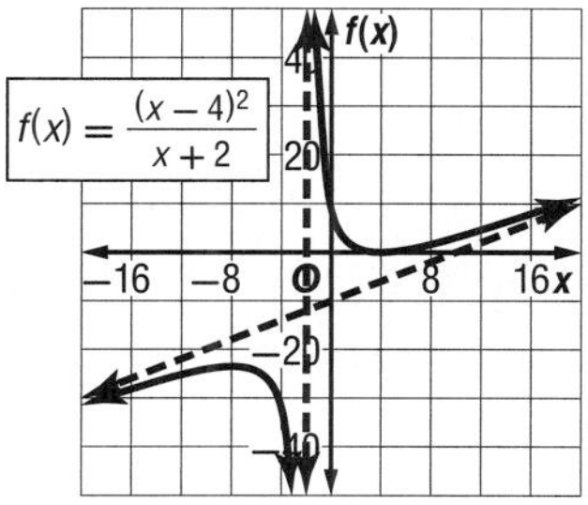

21.

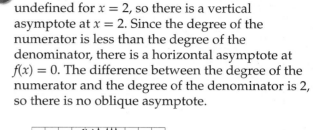

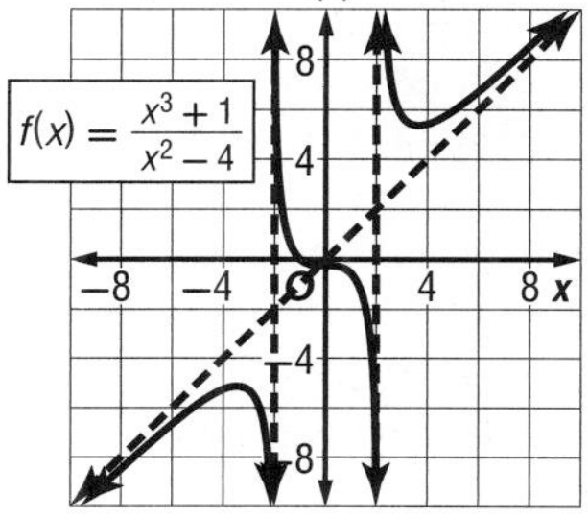

23.

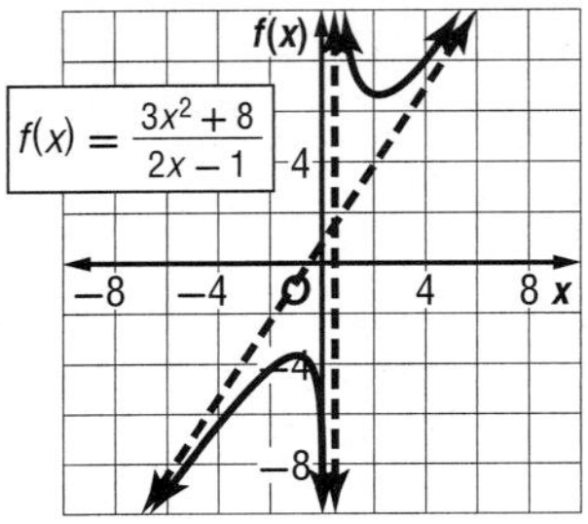

25.

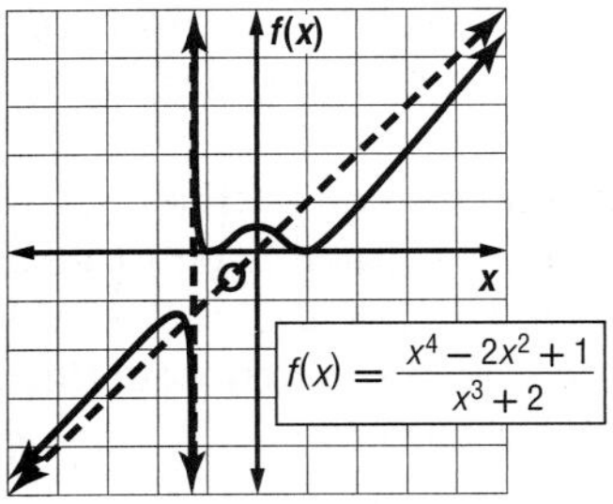

27a.

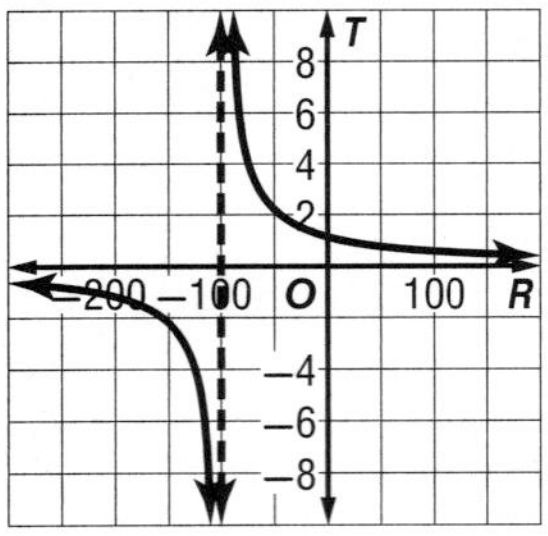

27b. $R_1 = -100$; no R_1-intercept; 1.2
27c. 0.5 amperes
27d. $R_1 \geq 0$ and $0 < I \leq 1.2$

29.

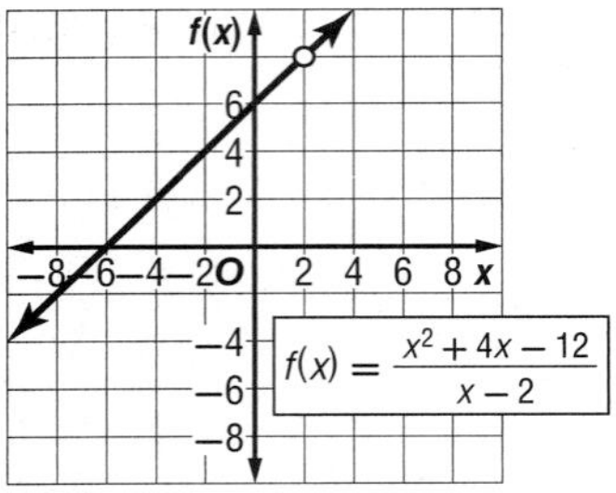

31.

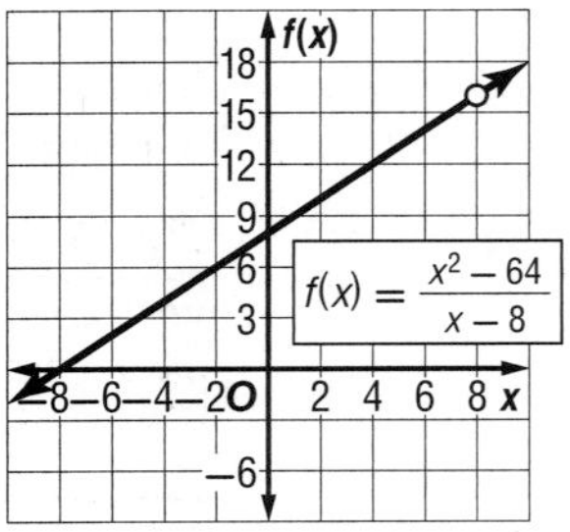

33.

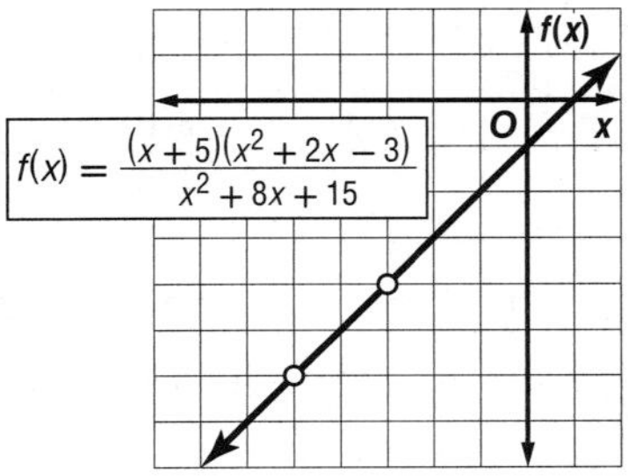

35.

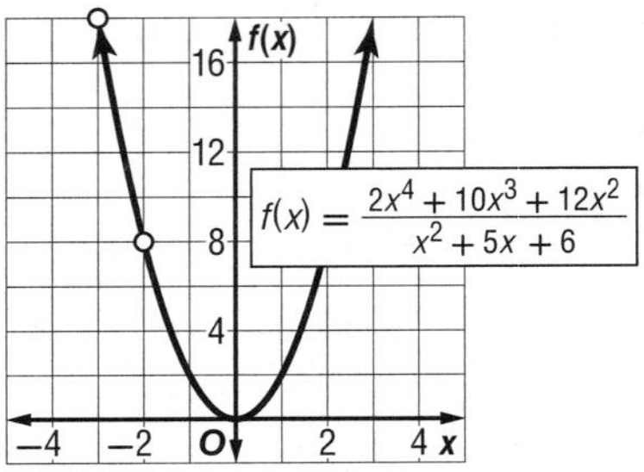

37 **a.** total cost = phone cost + monthly usage charge

$= 150 + 40x$

$$\text{average monthly cost} = \frac{\text{total cost}}{\text{number of months}}$$

$$f(x) = \frac{150 + 40x}{x}$$

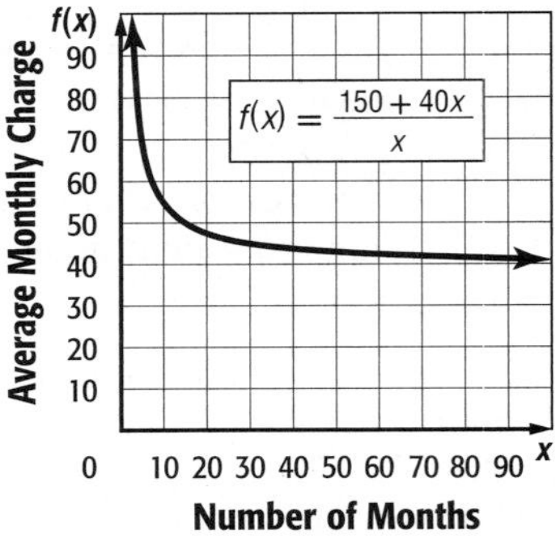

b. The vertical asymptote is $x = 0$. Since the degree of the numerator equals the degree of the denominator, the horizontal asymptote is at $f(x) = \frac{40}{1}$ or $f(x) = 40$.

c. Sample answer: The number of months and the average cost cannot have negative values.

d.

$f(x) = \frac{150 + 40x}{x}$	Write the equation.
$45 = \frac{150 + 40x}{x}$	$f(x) = 45$
$45x = 150 + 40x$	Multiply each side by x.
$5x = 150$	Subtract $40x$ from each side.
$x = 30$	Divide each side by 5.

After 30 months, the average monthly charge will be $45.

39.

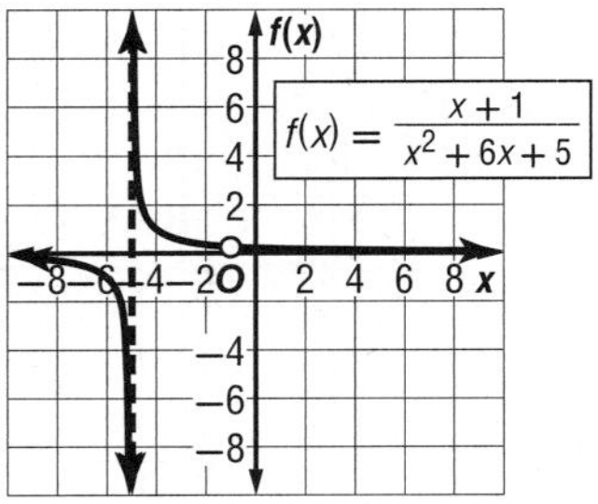

41.

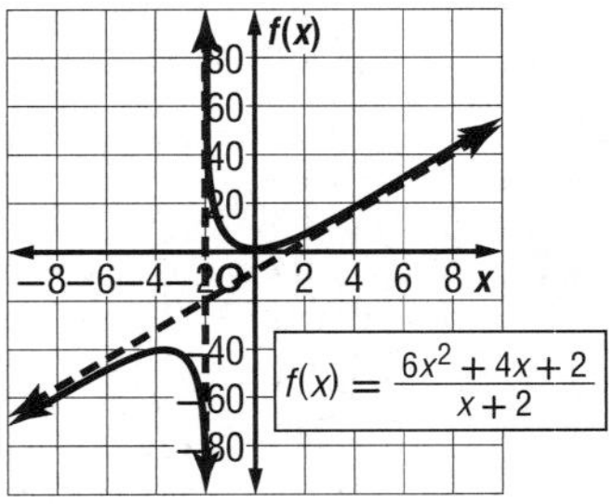

43. $f(x) = \frac{x^2 - 1}{x(x^2 - 1)}$

45. $f(x) = \frac{x}{a-b} + \frac{c(a-b)}{a-b}$
$= \frac{x + ca - cb}{a-b}$

47. C **49.** 4

51.

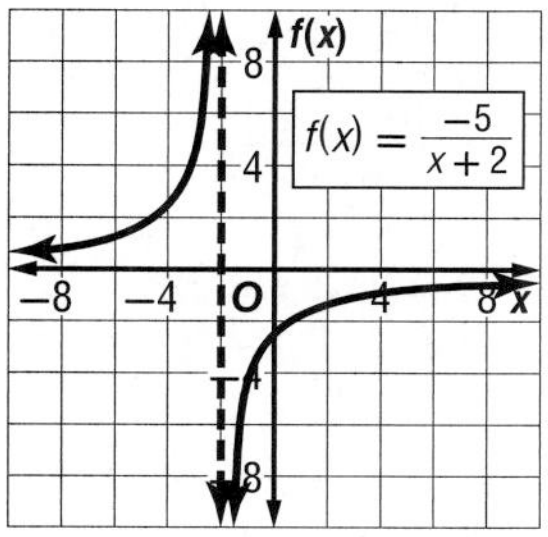

D = $\{x \mid x \neq -2\}$,
R = $\{y \mid y \neq 0\}$

53.

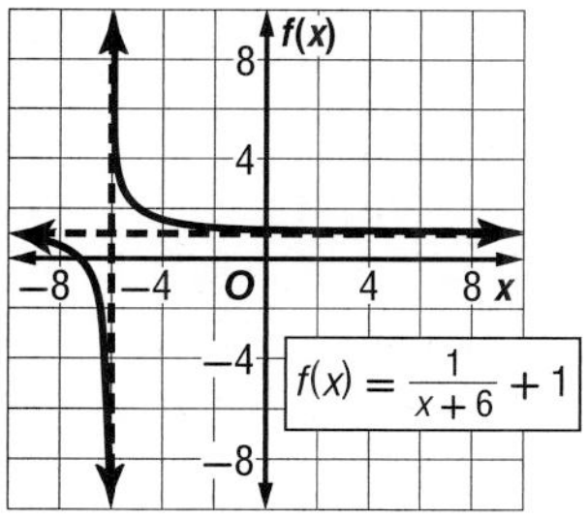

D = $\{x \mid x \neq -6\}$,
R = $\{y \mid y \neq 1\}$

55. $\frac{y(y-9)}{(y+3)(y-3)}$ **57.** $\frac{-8d+20}{(d-4)(d+4)(d-2)}$ **59.** x^3 **61.** $a^{\frac{1}{9}}$

Pages 590–593 Lesson 9-5

1. 21 **3.** −32 **5.** −48 **7.** 1.5 **9.** −56 **11.** $m = \frac{1}{6}w$

13

$\frac{a_1}{b_1c_1} = \frac{a_2}{b_2c_2}$	Joint variation
$\frac{-60}{-5(4)} = \frac{a_2}{4(-3)}$	$a_1 = -60, b_1 = -5, c_1 = 4, b_2 = 4, c_2 = -3$
$-60(4)(-3) = -5(4)(a_2)$	Cross multiply.
$720 = -20a_2$	Simplify.
$-36 = a_2$	Divide each side by −20.

15. −3 **17.** −22.5 **19.** 38 **21a.** $s = \frac{48}{t}$
21b. 3.2 hours **23.** −10 **25.** direct **27.** neither

29

$\frac{x_1}{y_2} = \frac{x_2}{y_1}$	Inverse variation
$\frac{16}{20} = \frac{x_2}{5}$	$x_1 = 16, y_1 = 5, y_2 = 20$
$16(5) = 20(x_2)$	Cross multiply.
$80 = 20x_2$	Simplify.
$4 = x_2$	Divide each side by 20.

31. −12 **33.** inverse; −2 **35.** combined; 10
37. direct; 4 **39.** direct; −2 **41.** inverse; 7 **43.** joint; 20

45a. $800 = rt$

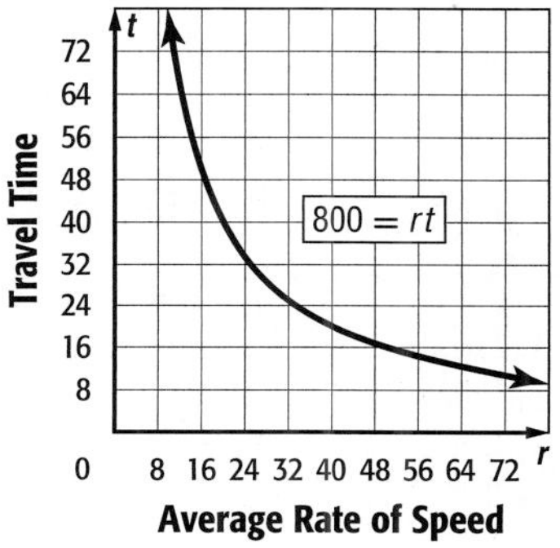

45b. $44.\overline{4}$ mph

47 **a.** $F = G\frac{m_1m_2}{d^2}$ Law of Universal Gravitation
$= (6.67 \times 10^{-11})\frac{(7.36 \times 10^{22})(5.97 \times 10^{24})}{(3.84 \times 10^8)^2}$
$\approx 2 \times 10^{20}$ newtons

b. $F = G\frac{m_1m_2}{d^2}$ Law of Universal Gravitation
$= (6.67 \times 10^{-11})\frac{(1.99 \times 10^{30})(5.97 \times 10^{24})}{(1.5 \times 10^{11})^2}$
$\approx 3.5 \times 10^{22}$ newtons

c. $F = G\frac{m_1m_2}{d^2}$ Law of Universal Gravitation
$= (6.67 \times 10^{-11})\frac{(1000)(1000)}{(0.1)^2}$
$= 6.67 \times 10^{-3}$ newtons

49. a and c are directly related. **51.** Sample answer: The force of an object varies jointly as its mass and acceleration. **53.** C

55a.

Month	Length
1	10
2	13
3	16
4	19

55b. $f(x) = 3x + 7$ **55c.** 34 in.
57. $x = -2, x = -3$
59. $x = -3$ **61.** 6 **63.** 3
65. $x + 3, x - \frac{1}{2}$ or $2x - 1$
67. $2a(a + 1)$ **69.** $24x$
71. $60ab^2$

Pages 600–602 Lesson 9-6

1. 11 **3.** 7

5 The LCD for the terms is $(x - 5)(x - 4)$.

$$\frac{8}{x-5} - \frac{9}{x-4} = \frac{5}{x^2 - 9x + 20}$$ Original equation

$$\frac{(x-5)(x-4)(8)}{x-5} - \frac{(x-5)(x-4)(9)}{x-4} = \frac{(x-5)(x-4)(5)}{x^2-9x+20}$$ Multiply by the LCD.

$$\frac{\cancel{(x-5)}^{1}(x-4)(8)}{\cancel{x-5}_{1}} - \frac{(x-5)\cancel{(x-4)}^{1}(9)}{\cancel{x-4}_{1}} = \frac{\cancel{(x-5)}^{1}\cancel{(x-4)}^{1}(5)}{\cancel{x^2-9x+20}_{1}}$$ Divide common factors.

$(x-4)(8) - (x-5)(9) = 5$	Simplify.
$8x - 32 - 9x + 45 = 5$	Distribute.
$-x + 13 = 5$	Simplify.
$-x = -8$	Subtract 13 from each side.
$x = 8$	Divide each side by −1.

7. 14

9a.

	pounds	price per pound	total price
dried fruit	10	\$6.25	6.25(10)
mixed nuts	m	\$4.50	4.5m
trail mix	10 + m	\$5.00	5(10 + m)

9b. $62.5 + 4.5m = 50 + 5m$ **9c.** 25 **11a.** $\frac{1}{60}$ **11b.** $\frac{x}{60}$
11c. $\frac{1}{80}$ **11d.** $\frac{x}{80}$ **11e.** $\frac{x}{60} + \frac{x}{80} = 1$ **11f.** about 34.3 min
13. $c < 0$, or $\frac{13}{18} < c$ **15.** $b < 0$, or $\frac{35}{12} < b$ **17.** 2 **19.** 4
21. ∅

23 cost of 3 pounds of bananas for \$0.90/pound = 0.9(3)
cost of x pounds of apples for \$1.25/pound = $1.25x$
total weight = $3 + x$

$\frac{\text{total cost}}{\text{total weight}} = 1$ Write an equation.

$\frac{0.9(3) + 1.25x}{3 + x} = 1$ Substitute.

$\frac{2.7 + 1.25x}{3 + x} = 1$ Simplify the numerator.

$\frac{(3 + x)(2.7 + 1.25x)}{3 + x} = (3 + x)(1)$ LCD is $(3 + x)$. Multiply by the LCD.

$\frac{\overset{1}{\cancel{(3 + x)}}(2.7 + 1.25x)}{\underset{1}{\cancel{3 + x}}} = (3 + x)(1)$ Divide out common factors.

$2.7 + 1.25x = 3 + x$ Simplify.

$2.7 + 0.25x = 3$ Subtract x from each side.

$0.25x = 0.3$ Subtract 2.7 from each side.

$x = 1.2$ Divide each side by 0.25.

She must purchase 1.2 pounds of apples.

25. $x < 0$ or $x > 1.75$ **27.** $x < -2$, or $2 < x < 14$
29. $x < -5$ or $4 < x < \frac{17}{3}$ **31.** 55.56 mph **33a.** 1; yes; 3

33b.
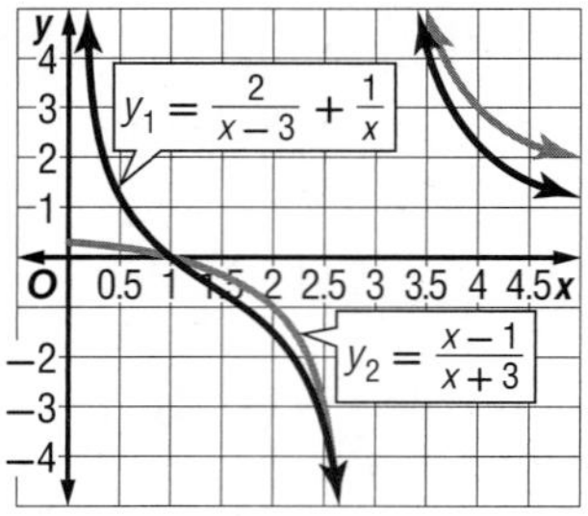

33c. 1; no
33d. Graph both sides of the equation. Where the graphs intersect, there is a solution. If they do not, then the possible solution is extraneous.
35. ∅ **37.** all real numbers except 5, −5, 0
39. Sample answer: Multiplying both sides of a rational inequality can produce extraneous solutions.
41. J **43.** all of the points **45.** direct

47.
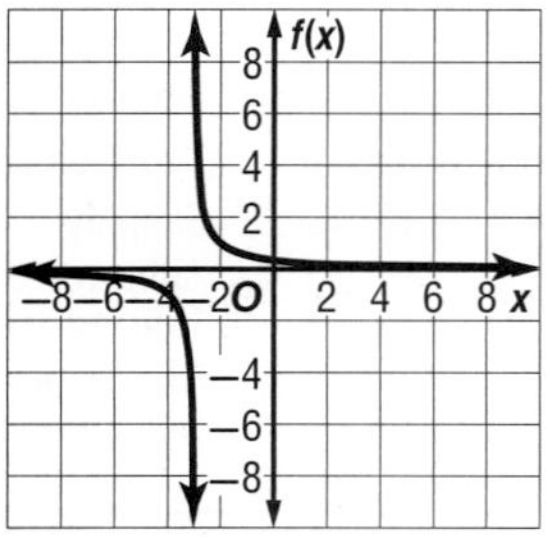

49.
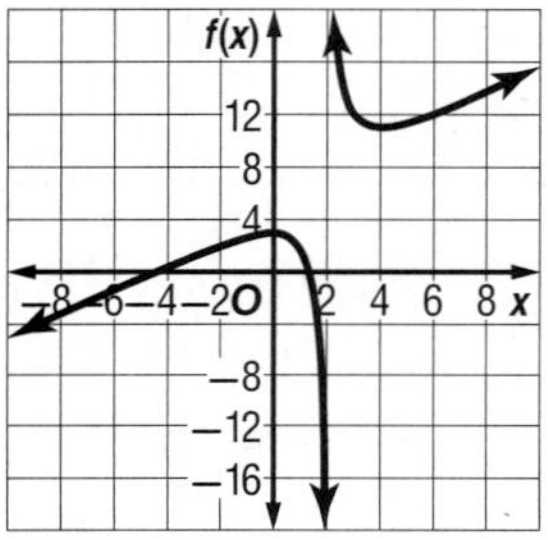

51a. $s \cdot 4^x$ **51b.** 0.5 three-yr periods or 1.5 yr **53.** yes

Pages 605–608 Chapter 9 Study Guide and Review

1. complex fraction **3.** oblique **5.** rational equations
7. Joint variation **9.** Point discontinuity **11.** $-\frac{10yz^2}{9x}$
13. $\frac{x-1}{x-2}$ **15.** $\frac{x-3}{x+1}$ **17.** $\frac{27b + 10a^2}{12ab^2}$ **19.** $\frac{3xy^3 + 8y^3 - 5x}{6x^2y^2}$
21. $\frac{12x^2 - 10x + 6}{2(x+2)(3x-4)(x+1)}$ **23.** $\frac{10x + 20}{(x+6)(x+1)}$

25.
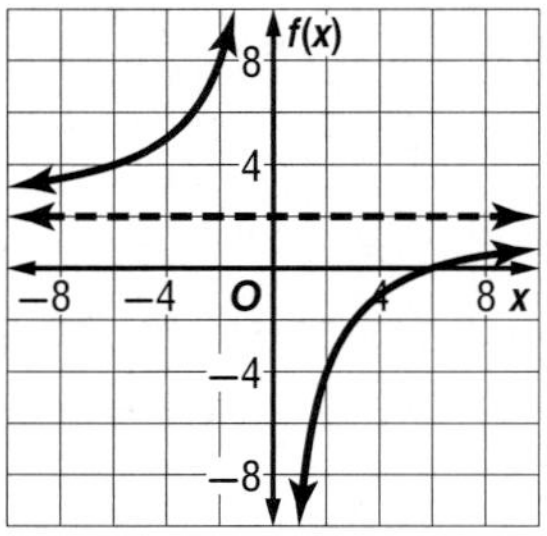

D = $\{x \mid x \neq 0\}$,
R = $\{y \mid y \neq 2\}$

27.
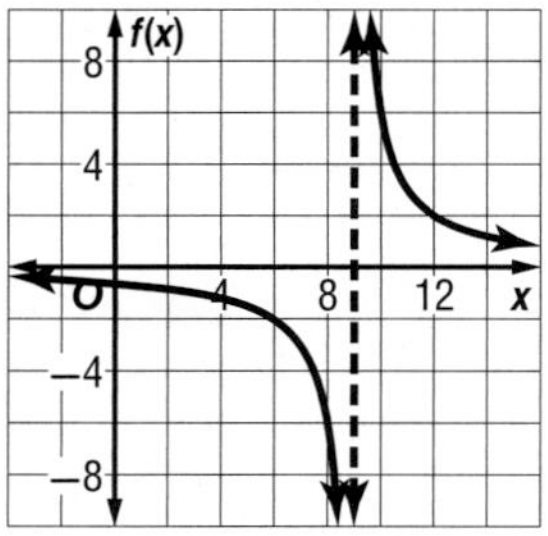

D = $\{x \mid x \neq 9\}$,
R = $\{y \mid y \neq 0\}$

29.
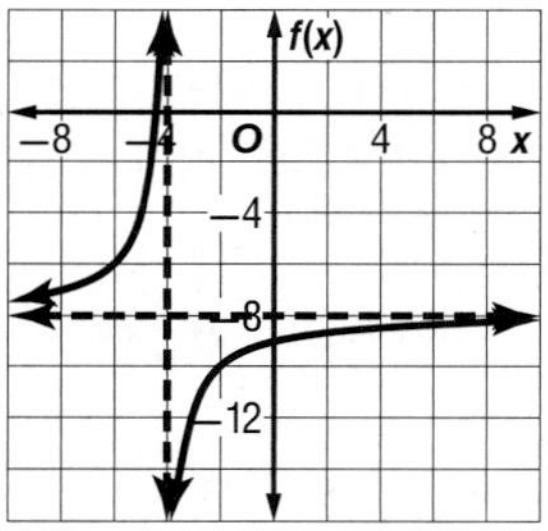

D = $\{x \mid x \neq -4\}$,
R = $\{y \mid y \neq -8\}$
31. $x = -4, x = 0$
33. $x = 8$; hole: $x = -3$

35.
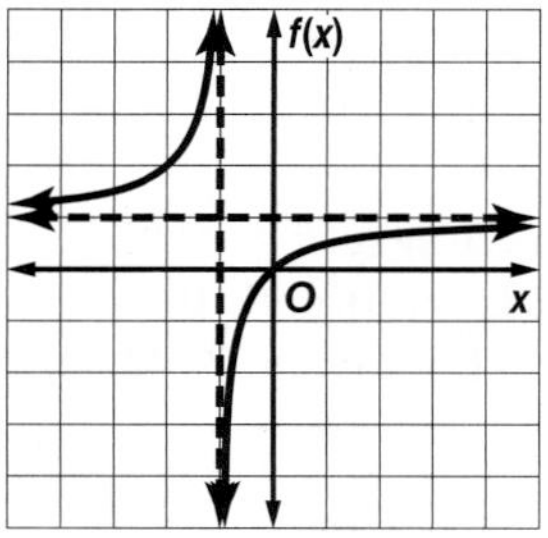

37.
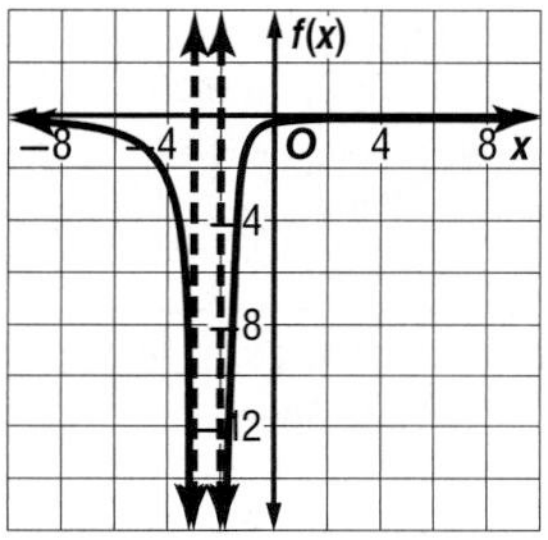

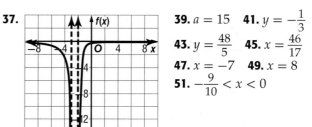

39. $a = 15$ **41.** $y = -\frac{1}{3}$
43. $y = \frac{48}{5}$ **45.** $x = \frac{46}{17}$
47. $x = -7$ **49.** $x = 8$
51. $-\frac{9}{10} < x < 0$

Chapter 10 Conic Sections

Page 615 Chapter 10 Get Ready

1. $\{-7, -1\}$ **3.** $\{3, 5\}$ **5.** $\left\{-5, \frac{3}{2}\right\}$ **7.** $\left\{\frac{3}{4} \pm \sqrt{2}\right\}$

9. $A'(-2, -5)$, $B'(-1, -1)$, $C'(5, -1)$, $D'(4, -5)$

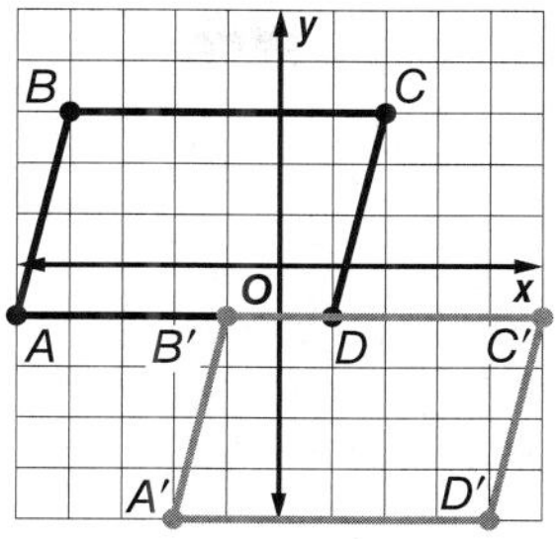

11. $J'(-3, 6)$, $K'(-2, -3)$, $L'(-10, -4)$

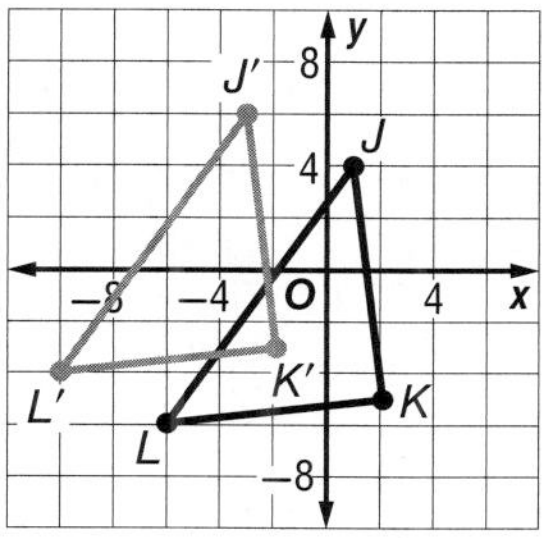

13. (110, 65), (120, 65), (110, 55), and (120, 55) **15.** (1, 3) **17.** (3, 14) **19.** (3, −1)

Pages 619–622 Lesson 10-1

1. $\left(-\frac{1}{2}, 8\right)$ **3.** (14.5, 9.75) **5.** 11.662 units

7 $d = \sqrt{(x_2 - x_1)^2 + (y_2 - y_1)^2}$ Distance Formula

$= \sqrt{(3.5 - 0.25)^2 + (2.5 - 1.75)^2}$ $(x_1, y_1) = (0.25, 1.75)$ and $(x_2, y_2) = (3.5, 2.5)$

$= \sqrt{3.25^2 + 0.75^2}$ Simplify.

$= \sqrt{11.125}$ or about 3.335 units Simplify.

9. A **11.** (−4, −1) **13.** (7.3, 1) **15.** (−7.75, −4.5) **17.** 16.279 units **19.** 16.125 units **21.** 21.024 units **23.** 38.833 units **25.** (−1.5, 0); 185.443 units **27.** (−5.5, −50.5); 148.223 units **29.** (8, 15); 136.953 units **31.** (−0.43, −2.25); 9.624 units

33 $\left(\frac{x_1 + x_2}{2}, \frac{y_1 + y_2}{2}\right)$ Midpoint Formula

$= \left(\frac{-\frac{5}{12} + \left(-\frac{17}{2}\right)}{2}, \frac{-\frac{1}{3} + \left(-\frac{5}{3}\right)}{2}\right)$ $(x_1, y_1) = \left(-\frac{5}{12}, -\frac{1}{3}\right)$ and $(x_2, y_2) = \left(-\frac{17}{2}, -\frac{5}{3}\right)$

$= \left(\frac{\frac{107}{12}}{2}, \frac{-2}{2}\right)$ Simplify.

$\approx (-4.458, -1)$ Simplify.

$d = \sqrt{(x_2 - x_1)^2 + (y_2 - y_1)^2}$ Distance Formula

$= \sqrt{\left(-\frac{17}{2} - \left(-\frac{5}{12}\right)\right)^2 + \left(-\frac{5}{3} - \left(-\frac{1}{3}\right)\right)^2}$ $(x_1, y_1) = \left(-\frac{5}{12}, -\frac{1}{3}\right)$ and $(x_2, y_2) = \left(-\frac{17}{2}, -\frac{5}{3}\right)$

$= \sqrt{\left(-\frac{97}{12}\right)^2 + \left(-\frac{4}{3}\right)^2}$ Simplify.

≈ 8.193 units Use a calculator.

35. (−4.719, 0.028); 17.97 units **37.** 14.53 km

39 $d = \sqrt{(x_2 - x_1)^2 + (y_2 - y_1)^2}$ Distance Formula

$= \sqrt{(254 - 132)^2 + (105 - 428)^2}$ $(x_1, y_1) = (132, 428)$ and $(x_2, y_2) = (254, 105)$

$= \sqrt{122^2 + (-323)^2}$ Simplify.

$= \sqrt{119{,}213}$ or about 345 units Simplify.

345 units • 0.316 mi/unit ≈ 109 mi

41a.

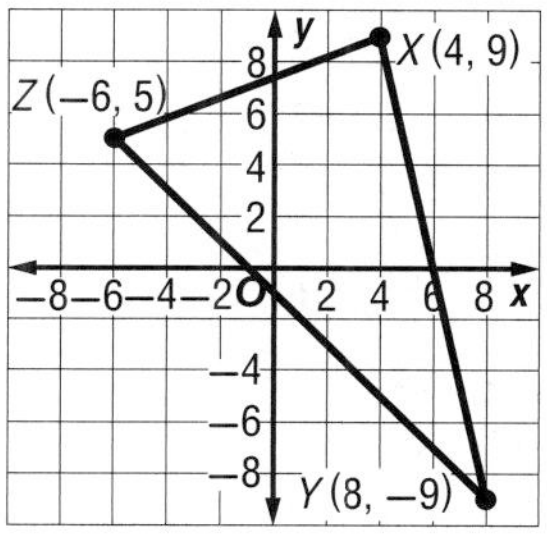

41b. midpoint of $\overline{XY} = (6, 0)$; midpoint of $\overline{YZ} = (1, -2)$; midpoint of $\overline{XZ} = (-1, 7)$ **41c.** The perimeter of $\triangle XYZ$ is $2\sqrt{29} + 14\sqrt{2} + 2\sqrt{85}$ units. perimeter $= \sqrt{29} + 7\sqrt{2} + \sqrt{85}$

41d. The perimeter of $\triangle XYZ$ is twice the perimeter of the smaller triangle. **43.** a shaded circle with center at (5, 6) **45.** See students' graphs; the distance from A to B equals the distance from B to A. Using the Distance Formula, the solution is the same no matter which ordered pair is used first. **47.** \$8.91 **49.** G **51.** −6, −2 **53.** $\frac{3}{2}$ **55.** 4.8362 **57.** 8.0086 **59.** $\{p \mid p \leq 1.9803\}$ **61.** −20 **63.** $\frac{1}{3}$ **65.** $y = (x - 3)^2 - 8$; $(3, -8)$; $x = 3$; up

Pages 627–629 Lesson 10-2

1. $y = 2(x - 6)^2 - 32$; vertex $(6, -32)$; axis of symmetry: $x = 6$; opens upward **3.** $x = (y - 4)^2 - 27$; vertex $(-27, 4)$; axis of symmetry: $y = 4$; opens right

5.

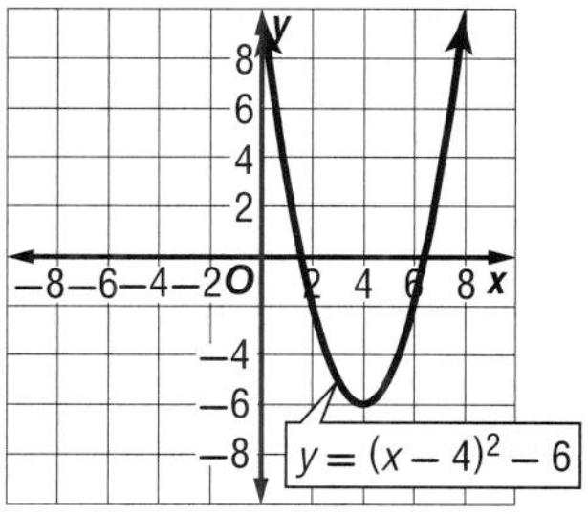

7.

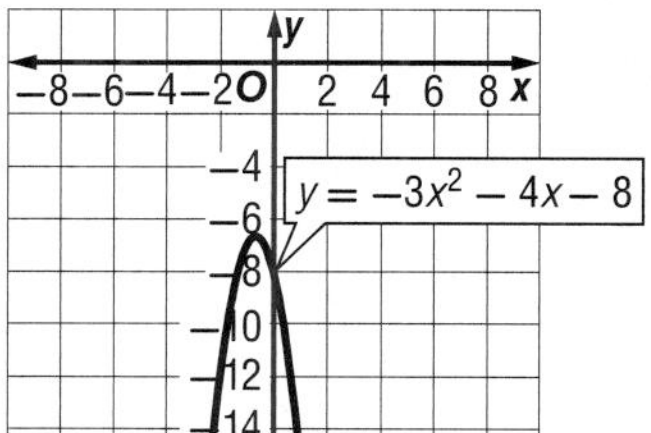

9. $y = \frac{1}{8}x^2 + 2$

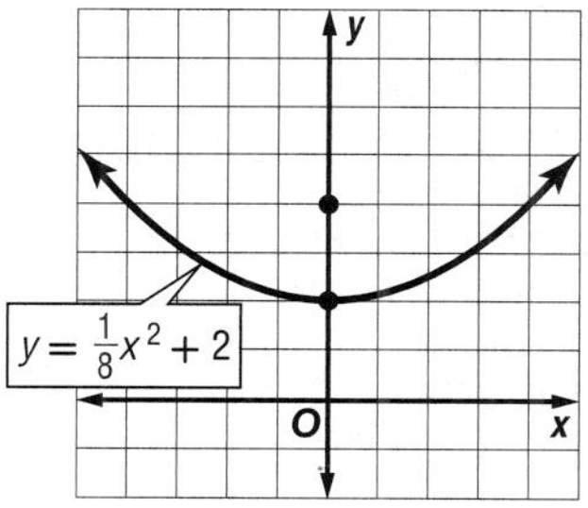

Selected Answers and Solutions

11. $y = -\frac{1}{12}(x - 3)^2 + 5$

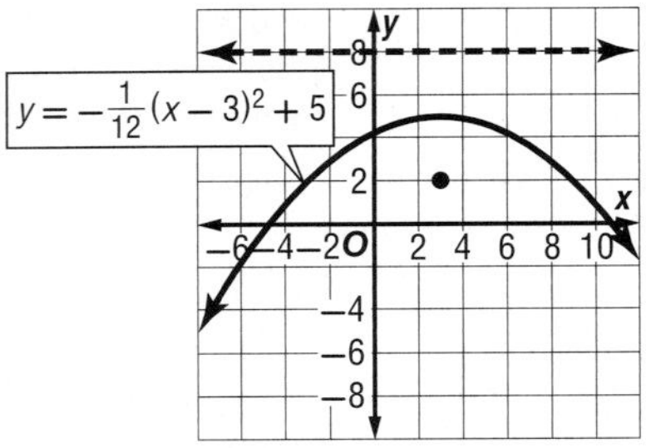

13a. $y = \frac{1}{24}x^2 - 6$

13b.

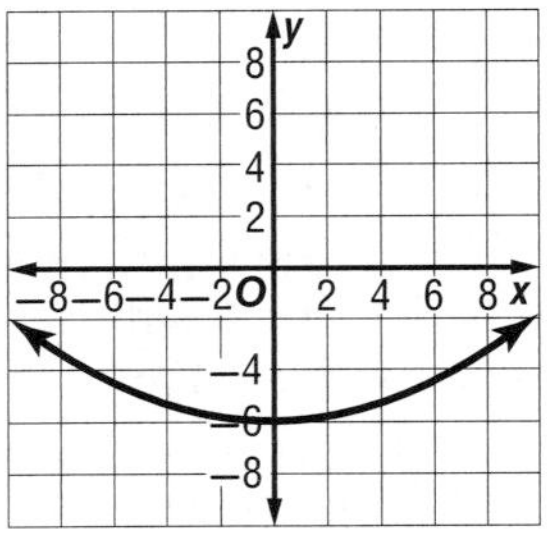

15. $y = 3(x + 7)^2 + 2$; vertex = $(-7, 2)$; axis of symmetry: $x = -7$; opens upward

17. $y = -3\left(x + \frac{3}{2}\right)^2 + \frac{3}{4}$; vertex = $\left(-\frac{3}{2}, \frac{3}{4}\right)$; axis of symmetry: $x = -\frac{3}{2}$; opens downward

19. $x = \frac{2}{3}(y - 3)^2 + 6$; vertex = $(6, 3)$; axis of symmetry: $y = 3$; opens right

21.

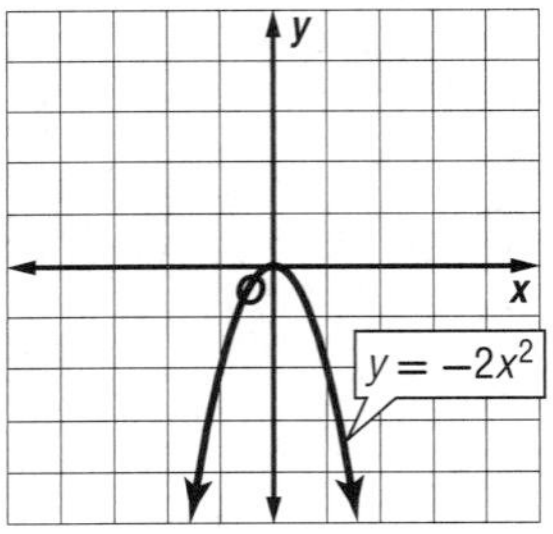

23.

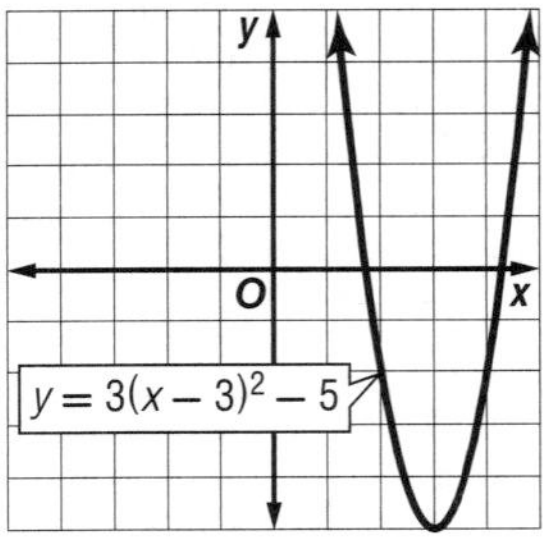

25.

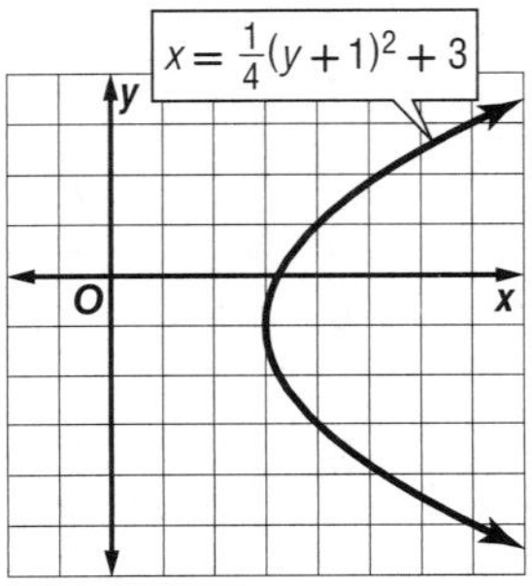

27. $y = \frac{1}{20}(x - 1)^2 + 8$

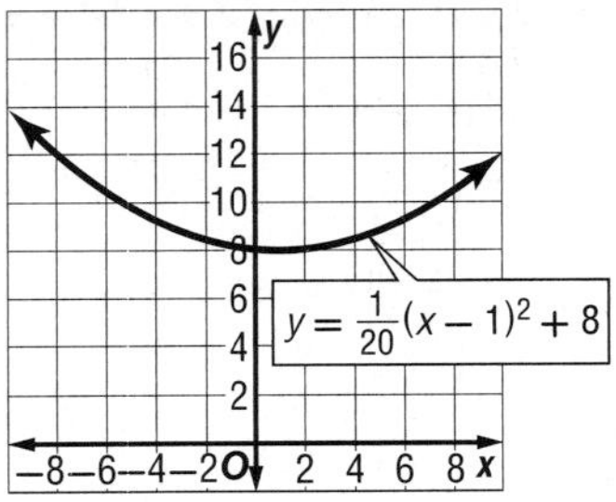

29 The directrix is a vertical line, so the equation of the parabola is of the form $x = a(y - k)^2 + h$. Since the vertex is equidistant from the focus and the directrix, the vertex is at (6, 4).

$x = h - \frac{1}{4a}$	Equation of directrix
$10 = 6 - \frac{1}{4a}$	$x = 10$ and $h = 6$
$4 = -\frac{1}{4a}$	Subtract 6 from each side.
$16a = -1$	Multiply each side by $4a$.
$a = -\frac{1}{16}$	Divide each side by 16.
$x = a(y - k)^2 + h$	Equation of a parabola
$x = -\frac{1}{16}(y - 4)^2 + 6$	$a = -\frac{1}{6}$, $(h, k) = (6, 4)$

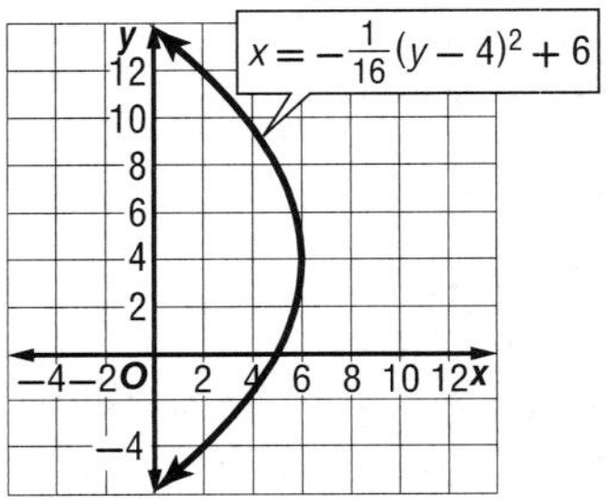

31. $y = -\frac{1}{4}(x - 9)^2 + 6$

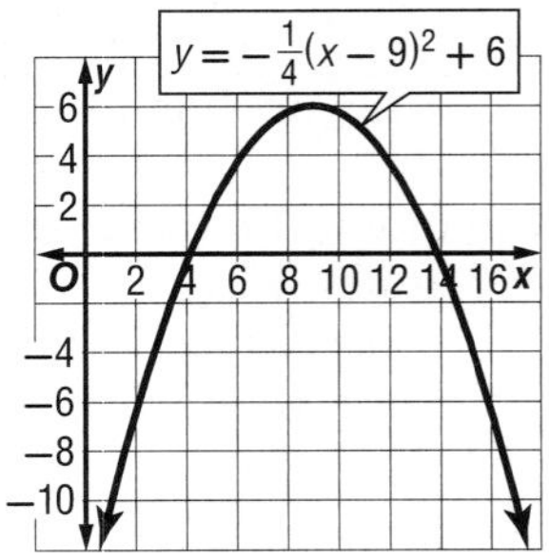

33a.

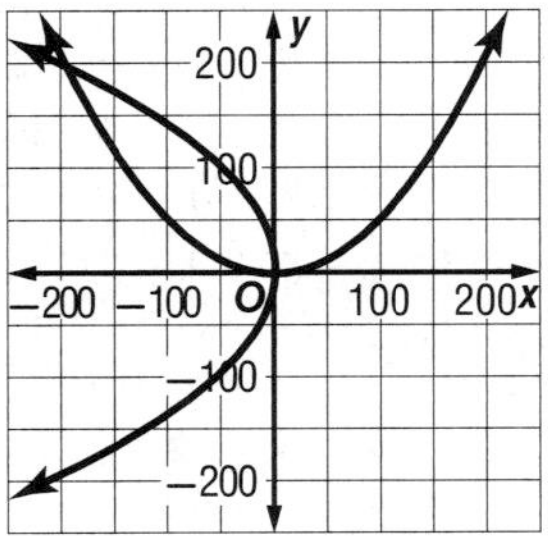

33b. $y = \frac{x^2}{192}$ and $x = \frac{y^2}{-192}$ **33c.** Sample answer: No; except for the direction in which they open, the graphs are identical.

35 The high beams should be placed at the focus.
The y-coordinate of the focus is $k + \frac{1}{4a}$.

$k + \frac{1}{4a} = 0 + \frac{1}{4\left(\frac{1}{12}\right)}$ $k = 0, a = \frac{1}{2}$

$= \frac{12}{4}$ or 3   Simplify.

The filament for the high beams should be placed 3 units above the vertex.

37. Rewrite it as $y = (x - h)^2$, where $h > 0$. **39.** Russell; the parabola should open to the left rather than to the right. **41.** C **43.** D **45.** $5\sqrt{2} + 3\sqrt{10}$ units **47.** 1.7183 **49.** $x > 0.4700$ **51.** 0.5 **53.** z^2 **55.** $\pm1, \pm2, \pm3, \pm6$ **57.** $\pm1, \pm\frac{1}{3}, \pm\frac{1}{9}, \pm3, \pm9, \pm27$ **59.** $3\sqrt{5}$ **61.** $16\sqrt{2}$

Pages 634–637 Lesson 10-3

1. $(x - 72)^2 + (y - 39)^2 = 10{,}000$
3. $(x - 1)^2 + (y + 5)^2 = 9$ **5.** $(x + 5)^2 + (y + 3)^2 = 90$
7. $x^2 + (y + 4)^2 = 20$
9. center: (0, 7); radius: 3

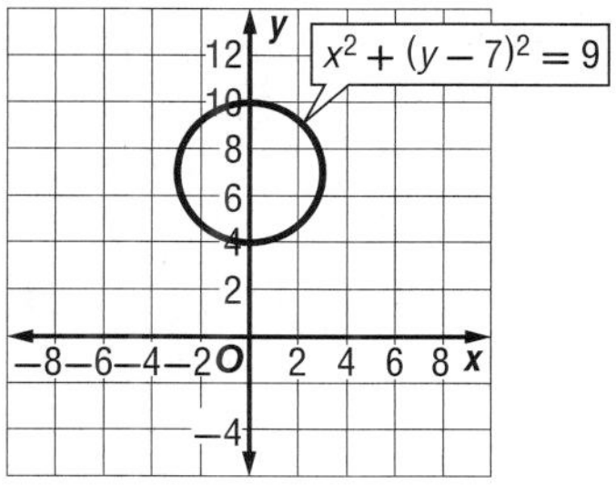

11. center: (2, −4); radius: 5

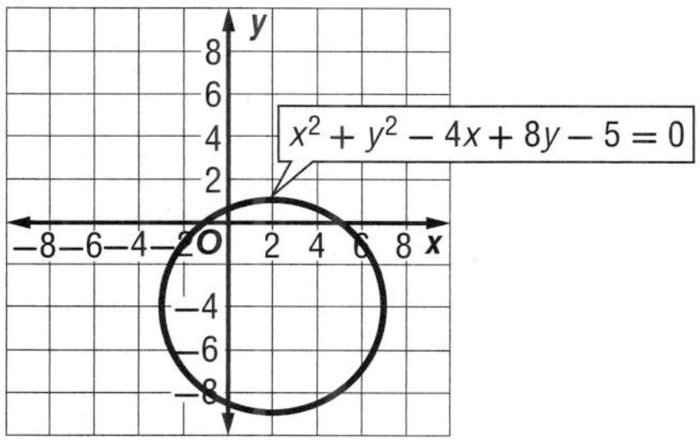

13. $(x + 3)^2 + (y - 1)^2 = 16$ **15.** $(x + 2)^2 + (y + 1)^2 = 81$

17

$(x - h)^2 + (y - k)^2 = r^2$	Equation of a circle
$(x - 0)^2 + [y - (-6)]^2 = (\sqrt{35})^2$	$(h, k) = (0, -6)$ and $r = \sqrt{35}$
$x^2 + (y + 6)^2 = 35$	Simplify.

19. $(x - 1)^2 + (y - 1)^2 = 4$ **21.** $x^2 + (y + 6) = 53$
23. $(x - 2)^2 + \left(y + \frac{3}{2}\right)^2 = \frac{25}{4}$ **25.** $\left(x - \frac{3}{2}\right)^2 + (y + 8)^2 = \frac{53}{4}$ **27.** $(x - 4)^2 + (y + 1)^2 = 20$

29 **a.**

$(h, k) = \left(\frac{x_1 + x_2}{2}, \frac{y_1 + y_2}{2}\right)$	Midpoint Formula
$= \left(\frac{-12 + 12}{2}, \frac{16 + (-16)}{2}\right)$	$(x_1, y_1) = (-12, 16)$ and $(x_2, y_2) = (12, -16)$
$= (0, 0)$	Simplify.
$r = \sqrt{(x_2 - x_1)^2 + (y_2 - y_1)^2}$	Distance Formula
$= \sqrt{[0 - (-12)]^2 + (0 - 16)^2}$	$(x_1, y_1) = (-12, 16)$ and $(x_2, y_2) = (0, 0)$
$= \sqrt{12^2 + 16^2}$	Subtract.
$\approx \sqrt{400}$	Simplify.

$(x - h)^2 + (y - k)^2 = r^2$	Equation of a circle
$(x - 0)^2 + (y - 0)^2 = (\sqrt{400})^2$	$(h, k) = (0, 0)$ and $r = \sqrt{400}$
$x^2 + y^2 = 400$	Simplify.

b.

$A = \pi r^2$	Area of a circle
$= \pi(\sqrt{400})^2$	$r = \sqrt{400}$
$= 400\pi$	Simplify.
≈ 1256.64 units^2	Use a calculator.

31. center: (0, 0); radius: $5\sqrt{3}$

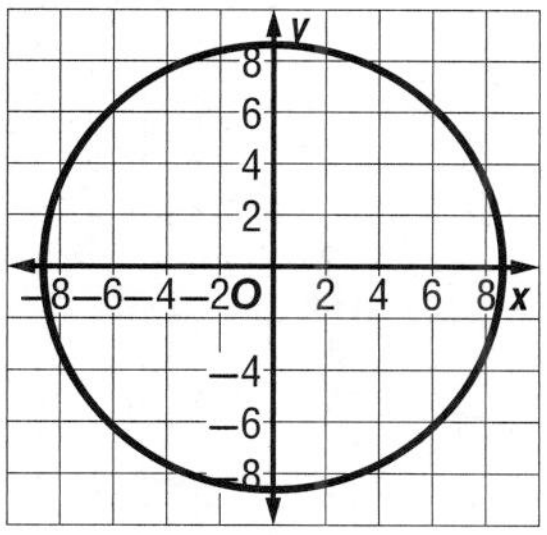

33. center: (1, 4); radius: $\sqrt{34}$

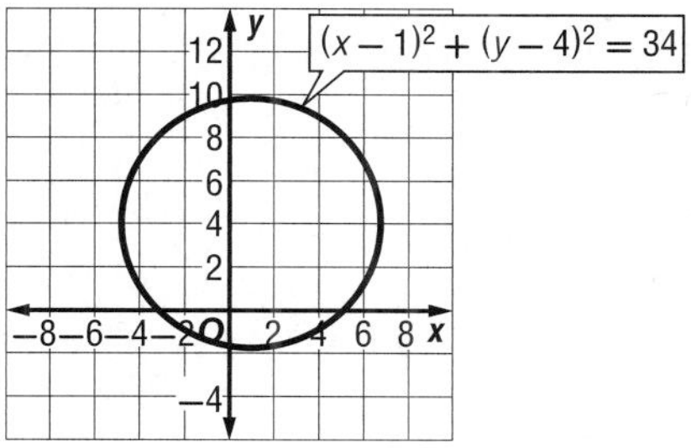

35. center: (5, −2); radius: 4

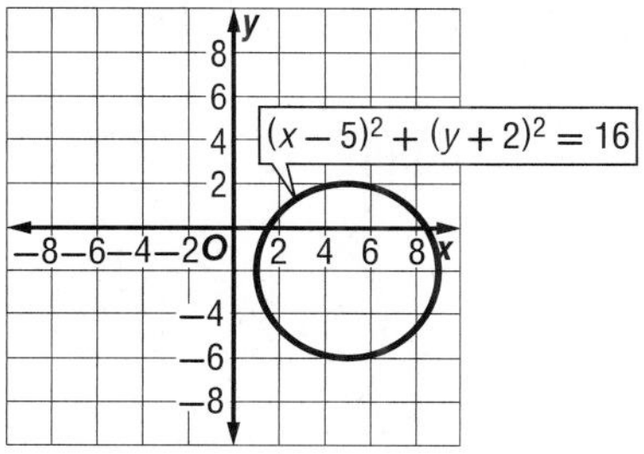

37. center: (4, 0); radius: $\frac{\sqrt{8}}{3}$

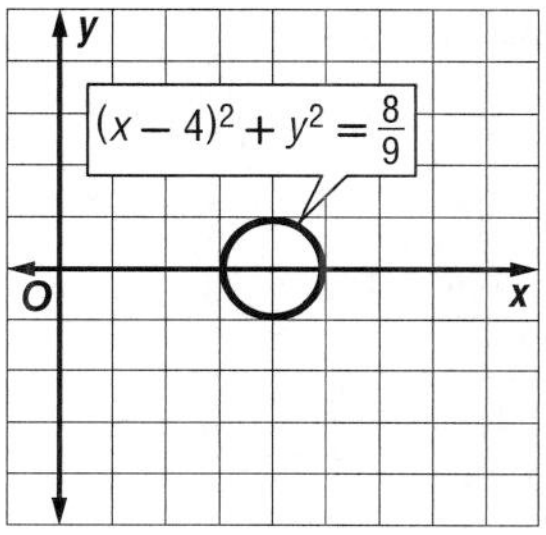

39. center: (−2, 0); radius: $\sqrt{13}$

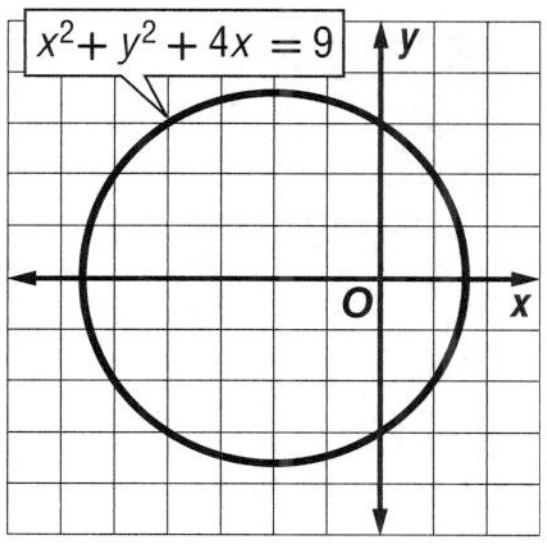

41. center: $(-1, -2)$; radius: $\sqrt{14}$

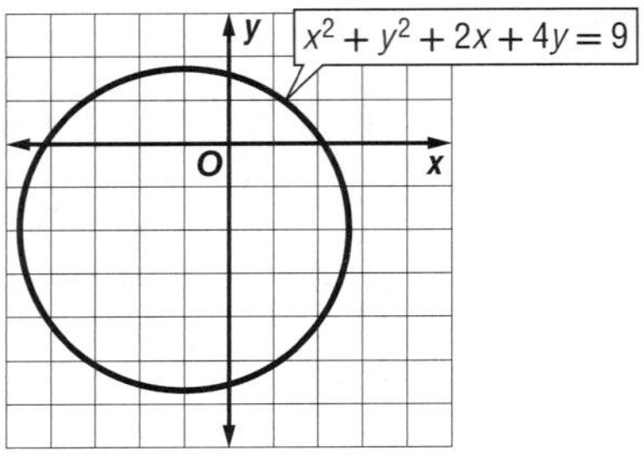

43. center: $(-7, -3)$; $2\sqrt{2}$ units

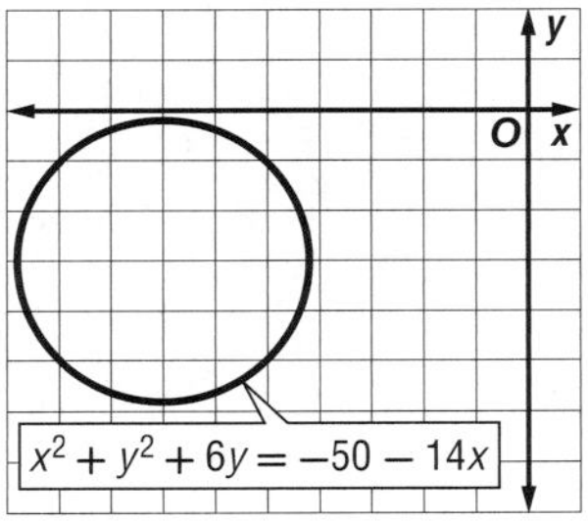

45. center: $(1, -2)$; radius: $\sqrt{21}$

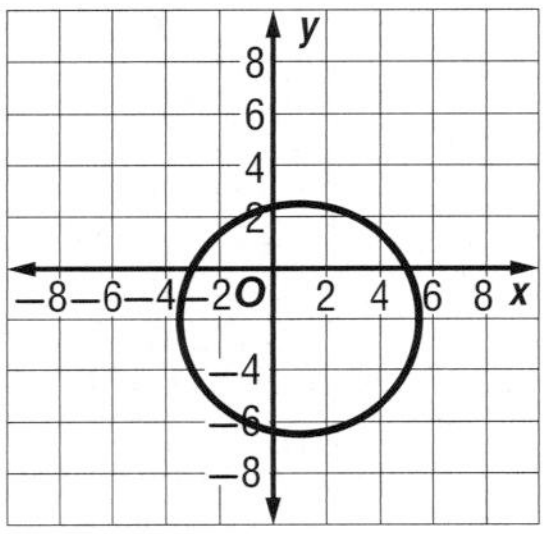

47a. $x^2 + y^2 = 841{,}000{,}000$

47b.

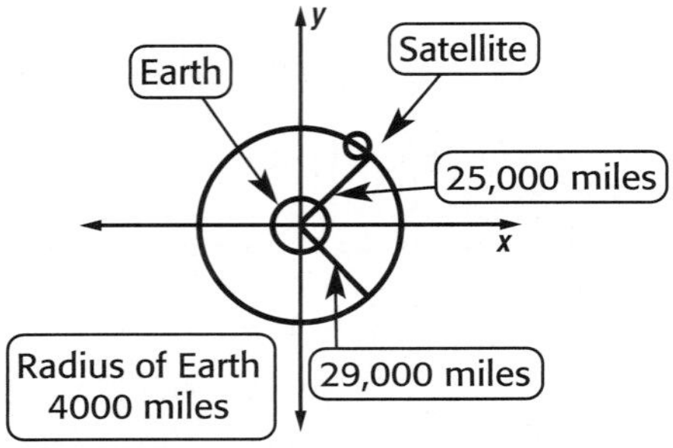

49a. $(x + 1)^2 + (y - 4)^2 = 36 + 16\sqrt{5}$

49b. $(x + 1)^2 + (y - 4)^2 = 24 - 8\sqrt{5}$

49c.

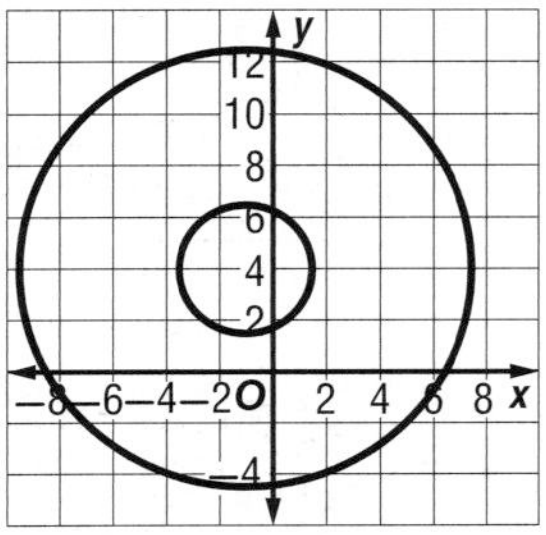

51 $r = \sqrt{(x_2 - x_1)^2 + (y_2 - y_1)^2}$ Distance Formula

$= \sqrt{(19 - 9)^2 + [22 - (-8)]^2}$ $(x_1, y_1) = (9, -8)$ and $(x_2, y_2) = (19, 22)$

$= \sqrt{10^2 + 30^2}$ Subtract.

$= \sqrt{1000}$ Simplify.

$(x - h)^2 + (y - k)^2 = r^2$ Equation of a circle

$(x - 9)^2 + [y - (-8)]^2 = (\sqrt{1000})^2$ $(h, k) = (9, -8)$ and $r = \sqrt{1000}$

$(x - 9)^2 + (y + 8)^2 = 1000$ Simplify.

53. $(x - 8)^2 + (y + 9)^2 = 64$ **55.** $(x - 2.5)^2 + (y - 2.5)^2 = 6.25$ **57a.** circle **57b.** $x^2 + y^2 = 9$

57c. Solve the equation for y: $y = \pm\sqrt{49 - x^2}$. Then graph the positive and negative answers.

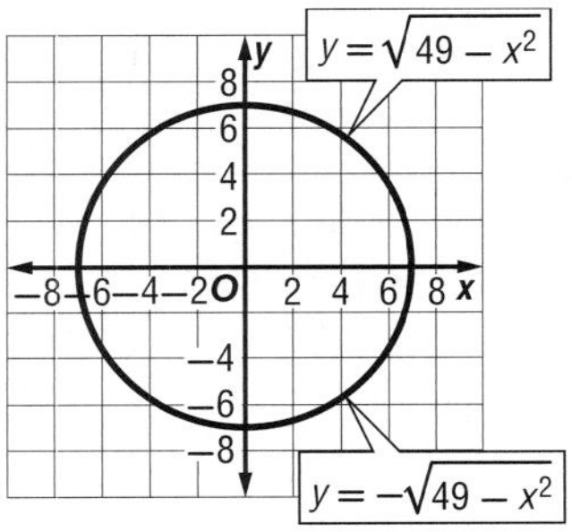

57d. $y = \pm\sqrt{4 - (x - 2)^2} - 1$; because when you solve for y you must take the square root resulting in both a positive and negative answer, so you have to enter the positive equation as **Y1** and the negative equation as **Y2**. **57e.** See students' work.

59. center: $\left(0, -\frac{9}{2}\right)$; radius: $2\sqrt{19}$

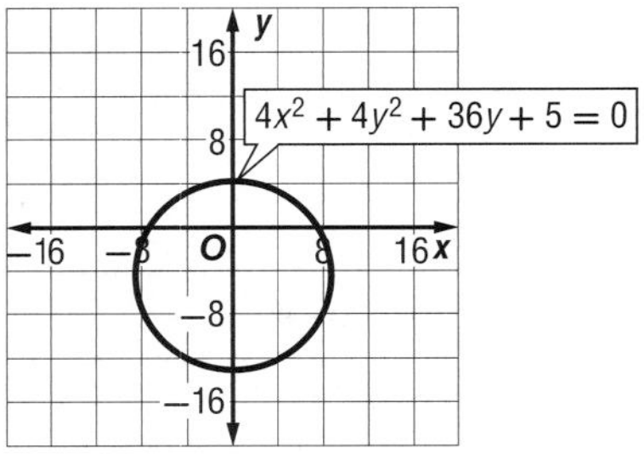

61. center: $(-\sqrt{7}, \sqrt{11})$; radius: $\sqrt{11}$

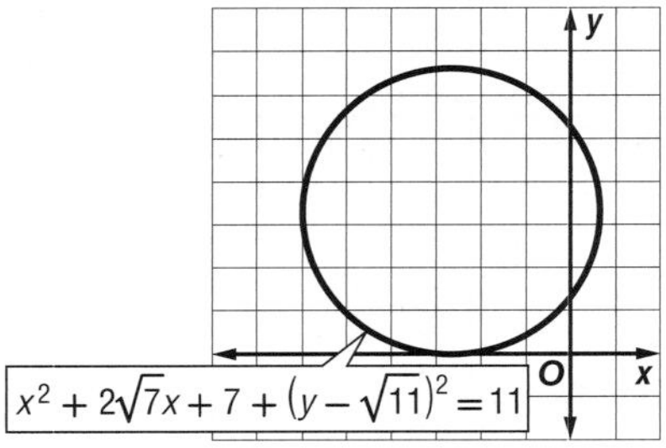

63. See students' work; circles with a radius of 8 and centers on the graph of $x = 3$.

65. Sample answer: $(x - 2)^2 + (y - 3)^2 = 25$ and $(x - 2)^2 + (y - 3)^2 = 36$

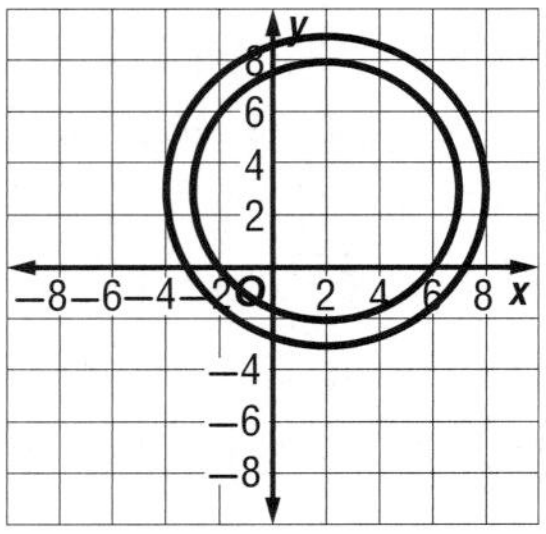

67. Quadrant I

$a > 0, b > 0, a = b, r > 0$

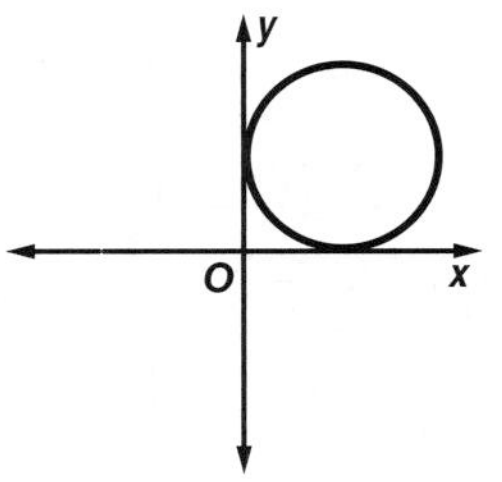

Quadrant II

$a < 0, b > 0, a = -b, r > 0$

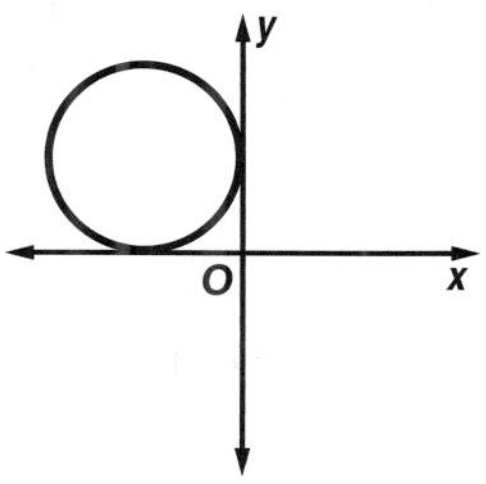

Quadrant III

$a < 0, b < 0, a = b, r > 0$

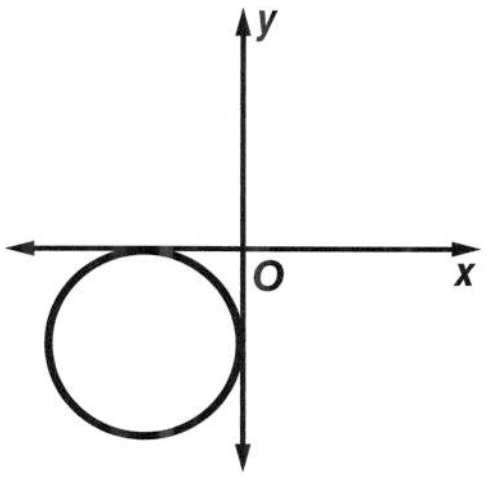

Quadrant IV

$a > 0, b < 0, a = -b, r > 0$

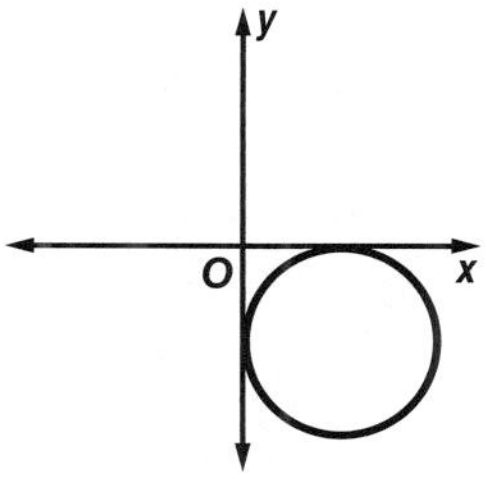

Sample answer: The circle is rotated 90° the origin from one quadrant to the next.

69. A **71.** C

73.

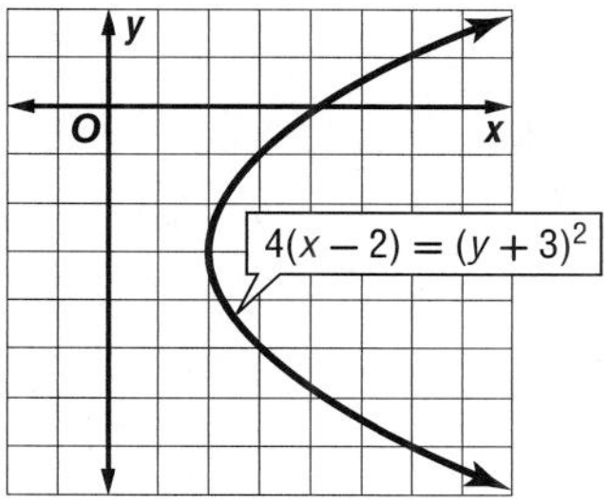

75. $\left(1, \frac{7}{22}\right)$; $\sqrt{65}$ units

77. (0, 3); $\sqrt{29}$ units

79. 64 **81.** $\frac{5}{2}$ **83.** −4

85a. The square root of a difference is not the difference of the square roots.

85b. 0 ft/s

87. $\left\{\frac{3 \pm \sqrt{33}}{4}\right\}$

Pages 643–646 Lesson 10-4

1. $\frac{y^2}{25} + \frac{x^2}{9} = 1$ **3.** $\frac{(y+1)^2}{25} + \frac{(x+2)^2}{9} = 1$

5a. $a = 240, b = 160$ **5b.** $\frac{x^2}{57{,}600} + \frac{y^2}{25{,}600} = 1$

5c. about (179, 0) and (−179, 0)

7. center (5, −1); foci (5, 5) and (5, −7); major axis: 16; minor axis: ≈ 10.58

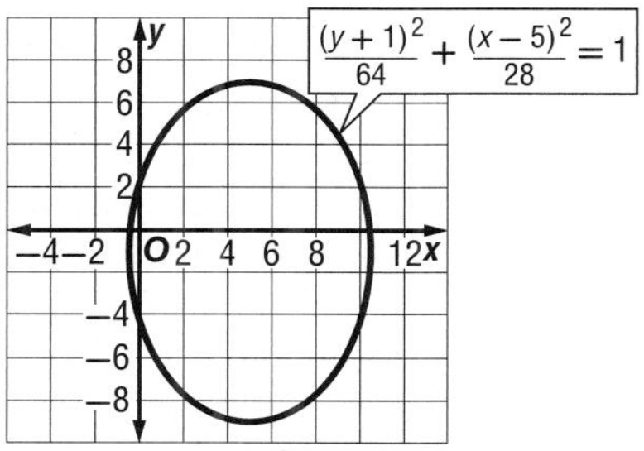

9 $4x^2 + y^2 - 32x - 4y + 52 = 0$ Original equation

$4x^2 - 32x + y^2 - 4y = -52$ Associative Property

$4(x^2 - 8x) + y^2 - 4y = -52$ Distributive Property

$4(x^2 - 8x + \blacksquare) + (y^2 - 4y + \blacksquare) = -52 + 4(\blacksquare) + (\blacksquare)$

Complete the squares.

$4(x^2 - 8x + 16) + (y^2 - 4y + 4) = -52 + 4(16) + (4)$

$(-4)^2 = 16$ and $(-2)^2 = 4$

$4(x - 4)^2 + (y - 2)^2 = 16$ Write as perfect squares.

$\frac{(x-4)^2}{4} + \frac{(y-2)^2}{16} = 1$ Divide each side by 16.

$h = 4$ and $k = 2$, so the center is at (4, 2).

The ellipse is vertical. $a^2 = 16$, so $a = 4$, and $b^2 = 4$, so $b = 2$.

$c^2 = 16 - 4$ or 12, so $c \approx 3.46$.

foci: (4, 2 + 3.46) or (4, 5.46); (4, 2 − 3.46) or (4, −1.46)

major axis: 2 • 4 or 8

minor axis: 2 • 2 or 4

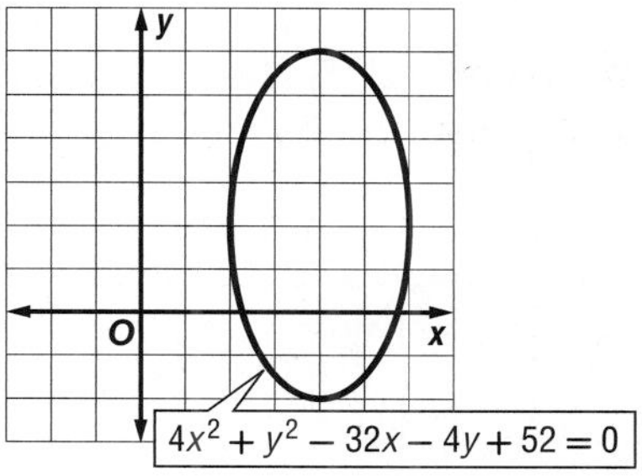

11. $\frac{y^2}{100} + \frac{x^2}{36} = 1$ **13.** $\frac{(x+5)^2}{49} + \frac{(y+4)^2}{25} = 1$

15. $\frac{(y-1)^2}{64} + \frac{(x+5)^2}{16} = 1$ **17.** $\frac{(x-3)^2}{81} + \frac{(y-4)^2}{64} = 1$

19 The x-coordinate is the same for both vertices, so the ellipse is vertical.

length of major axis: 16 − 6 or 10 units, so $a = 10$

length of minor axis: 1 − (−2) or 3 units, so $b = 3$

$\frac{(y-k)^2}{a^2} + \frac{(x-h)^2}{b^2} = 1$ Equation of a vertical ellipse

$\frac{(y-6)^2}{10^2} + \frac{[x-(-2)]^2}{3^2} = 1$ $(h, k) = (-2, 6), a = 10, b = 3$

$\frac{(y-6)^2}{100} + \frac{(x+2)^2}{9} = 1$ Simplify.

21. $\frac{(y-4)^2}{64} + \frac{(x-4)^2}{9} = 1$ **23.** $\frac{y^2}{73.96} + \frac{x^2}{53.29} = 1$

25. center (−6, 3); foci (−6, 7.69) and (−6, −1.69); major axis: ≈ 16.97; minor axis: ≈14.14

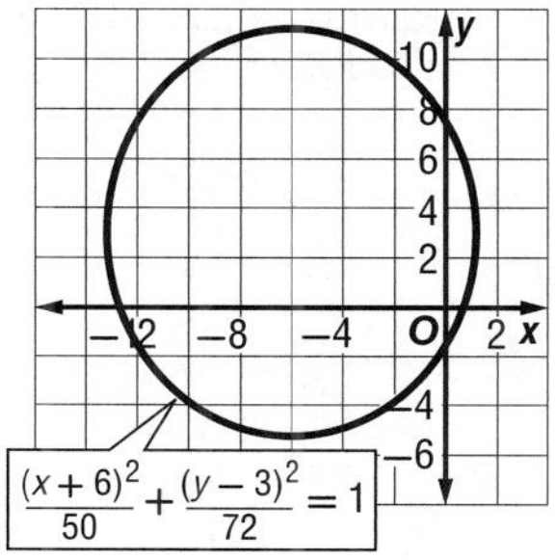

27. center (−4, 0); foci (−4, 7.68) and (−4, −7.68); major axis: ≈17.32; minor axis: 8

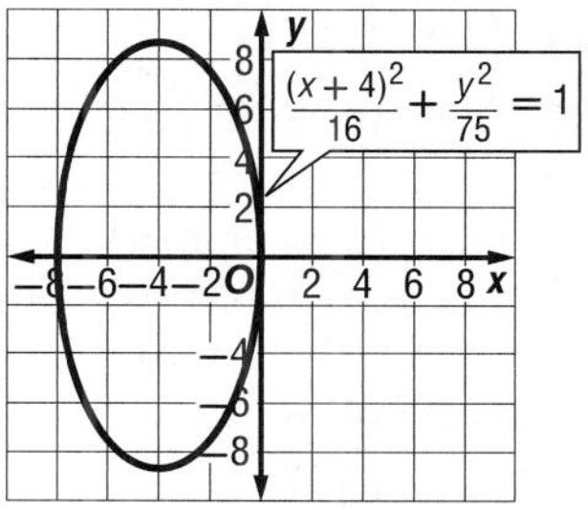

29. center (3, −3); foci (5.24, −3) and (0.76, −3); major axis: ≈8.94; minor axis: ≈7.75

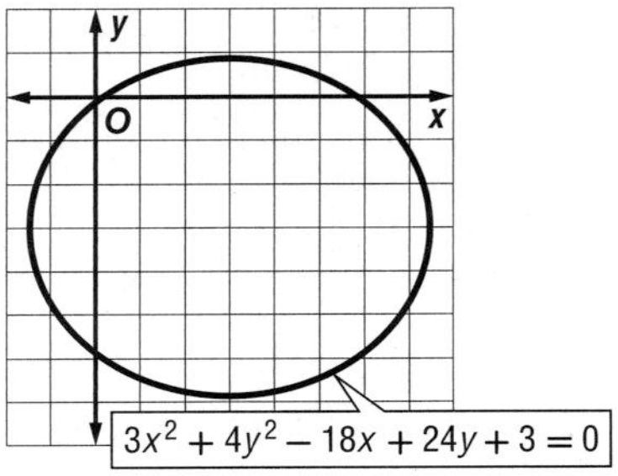

31. center (−2, 5); foci (−2, 7.83) and (−2, 2.17); major axis: ≈9.80; minor axis: 8

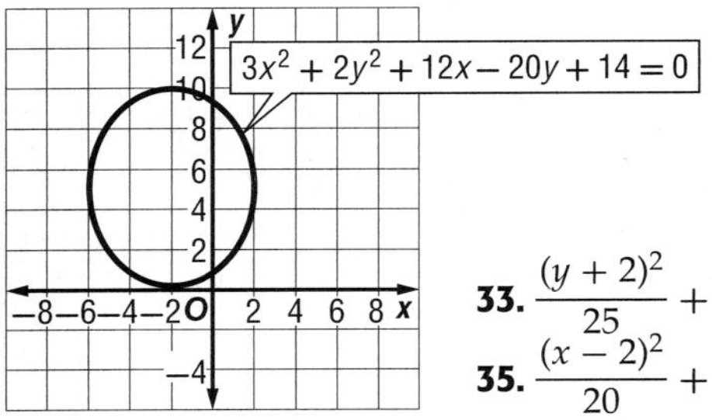

33. $\frac{(y+2)^2}{25} + \frac{(x+5)^2}{9} = 1$

35. $\frac{(x-2)^2}{20} + \frac{(y-8)^2}{4} = 1$

37 length of major axis $= 2a$

$$10.9 = 2a$$
$$5.45 = a$$

length of minor axis $= 2b$

$$8.8 = 2b$$
$$4.4 = b$$

$\frac{x^2}{a^2} + \frac{y^2}{b^2} = 1$ Equation of an ellipse

$\frac{x^2}{5.45^2} + \frac{y^2}{4.4^2} = 1$ Substitute.

$\frac{x^2}{29.7025} + \frac{y^2}{19.36} = 1$ Simplify.

39a.

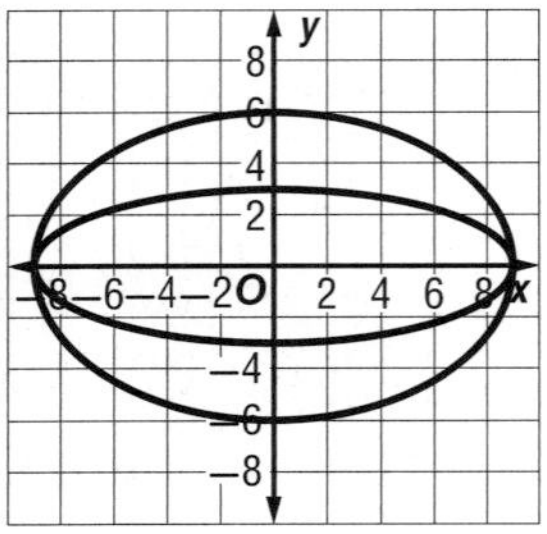

39b. Sample answer: The first graph is more circular than the second graph. **39c.** first graph: 0.745; second graph: 0.943 **39d.** Sample answer: The closer the eccentricity is to 0, the more circular the ellipse.

41. Sample answer: $\frac{(x+4)^2}{40} + \frac{y^2}{24} = 1$

43. $\frac{y^2}{9} + \frac{(x-2)^2}{3} = 1$

45. For any point on an ellipse, the sum of the distances from that point to the foci is constant by the definition of an ellipse. So, if (2, 14) is on the ellipse, then the sum of the distances from it to the foci will be a certain value consistent with every other point on the ellipse. The distance between (−7, 2) and (2, 14) is $\sqrt{(-7-2)^2 + (2-14)^2}$ or 15. The distance between (18, 2) and (2, 14) is $\sqrt{(18-2)^2 + (2-14)^2}$ or 20. The sum of these two distances is 35.
The distance between (−7, 2) and (2, −10) is $\sqrt{(-7-2)^2 + [2-(-10)]^2}$ or 15. The distance between (18, 2) and (2, −10) is $\sqrt{(18-2)^2 + [2-(-10)]^2}$ or 15. The sum of these distances is also 35. Thus, (2, −10) also lies on the ellipse.

47. B **49.** 7 **51.** $(x-8)^2 + (y+9)^2 = 1130$

53. $(x+5)^2 + (y-4)^2 = 25$ **55.** $\frac{5d+16}{(d+2)^2}$

57. $\frac{x^2 - 5x + 3}{(x-5)(x+1)}$ **59.** 4 **61.** $15a^3b^3 - 30a^4b^3 + 15a^5b^6$

63. $4x^2 - 3xy - 6y^2$ **65.** $y = -\frac{4}{5}x + \frac{17}{5}$

67. $y = -\frac{3}{5}x + \frac{16}{5}$

Pages 652–655 Lesson 10-5

1. $\frac{y^2}{36} - \frac{x^2}{28} = 1$ **3.** $\frac{x^2}{64} - \frac{y^2}{2y} = 1$

5.

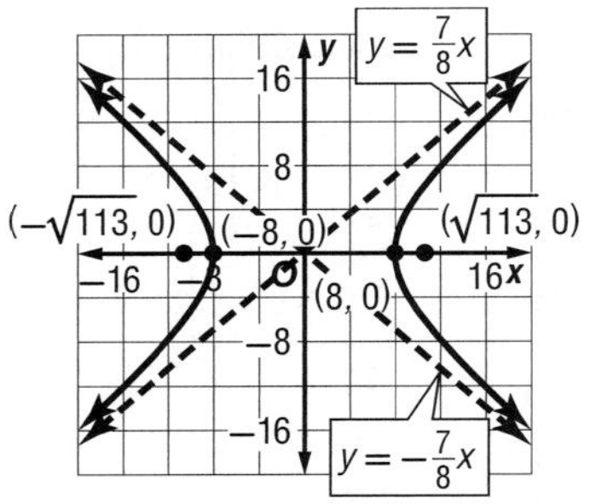

7.

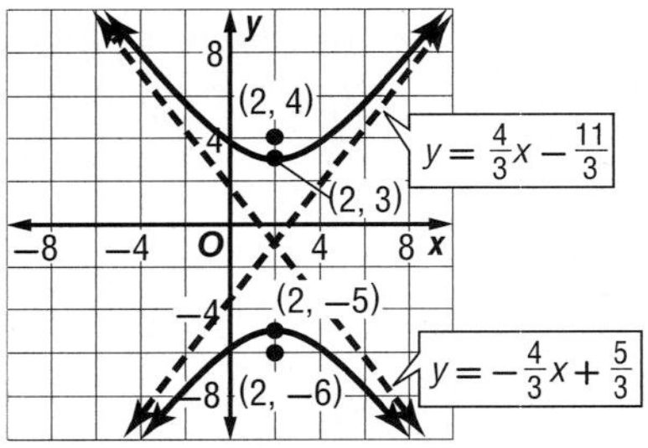

9. $\frac{x^2}{900} - \frac{y^2}{5500} = 1$

11 Since the vertices are equidistant from the center, the center is at (−8, 4). The value of a is the distance between a vertex and the center, or 4 units. The value of c is the distance between a focus and the center, or 8 units.

$c^2 = a^2 + b^2$ Equation relating a, b, and c for a hyperbola

$8^2 = 4^2 + b^2$ $c = 8$ and $a = 4$

$48 = b^2$ Subtract 4^2 from each side.

$\frac{(y-k)^2}{a^2} - \frac{(x-h)^2}{b^2} = 1$ Equation of a vertical hyperbola

$\frac{(y-4)^2}{4^2} - \frac{[x-(-8)]^2}{48} = 1$ $(h, k) = (-8, 4)$, $a = 4$, $b^2 = 48$

$\frac{(y-4)^2}{16} - \frac{(x+8)^2}{48} = 1$ Simplify.

13. $\frac{(x+1)^2}{9} - \frac{(y-6)^2}{49} = 1$

15.

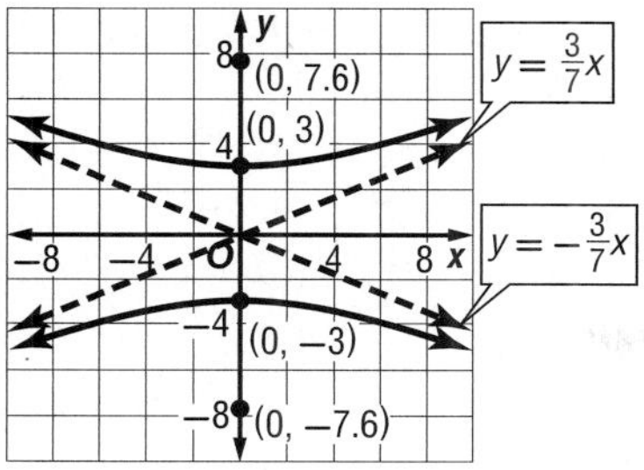

17.

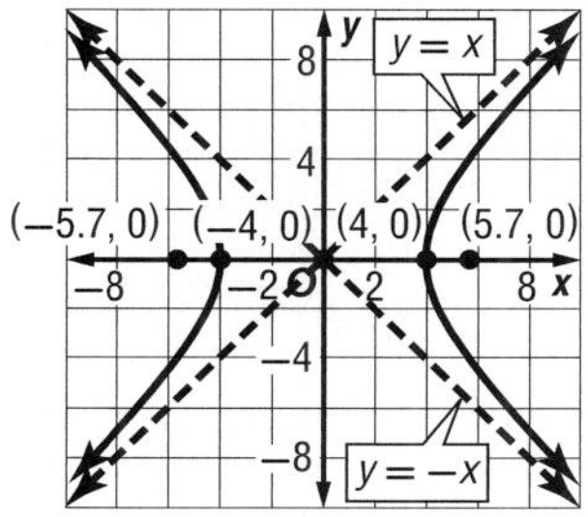

19.

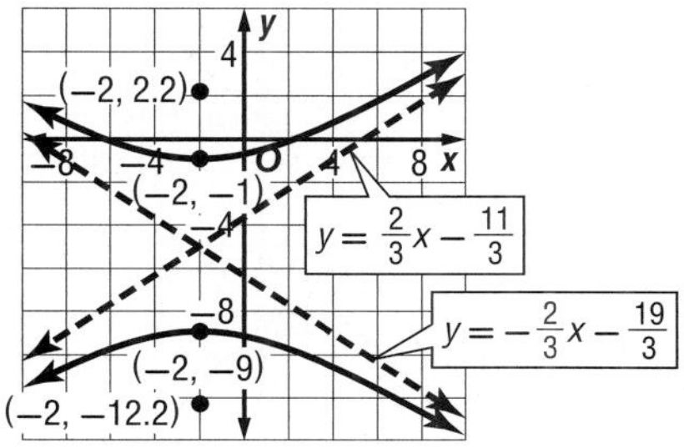

21.

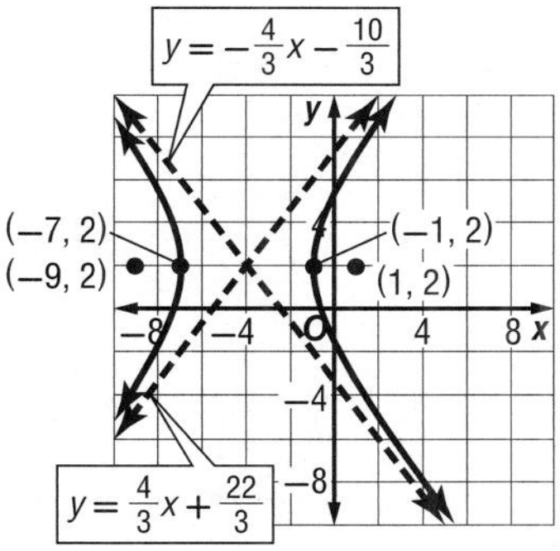

23.

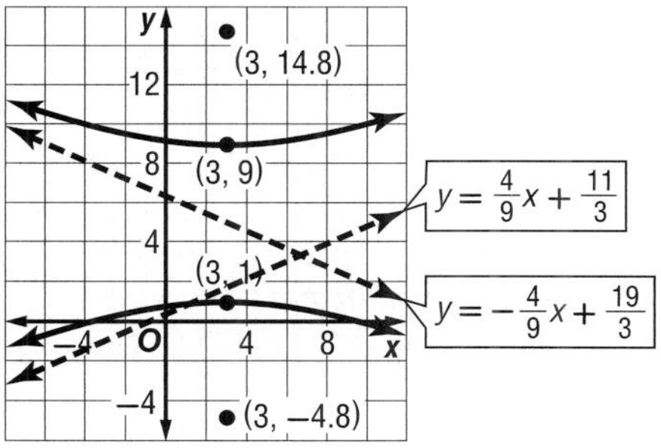

25. hyperbola **27.** ellipse **29.** ellipse

31 Because the center is at the origin, use the equation $\frac{x^2}{a^2} - \frac{y^2}{b^2} = 1$. The hyperbola intersects the x-axis at (10, 0), and one of the vertices is (10, 0). So, $a = 10$. The hyperbola also passes through (30, 100). Use $a = 10$, $x = 30$, and $y = 100$ to solve for b^2.

$\frac{x^2}{a^2} - \frac{y^2}{b^2} = 1$	Equation of an ellipse
$\frac{30^2}{10^2} - \frac{100^2}{b^2} = 1$	$a = 10, x = 30, y = 100$
$\frac{900}{100} - \frac{10{,}000}{b^2} = 1$	Evaluate exponents.
$9 - \frac{10{,}000}{b^2} = 1$	Simplify.
$-\frac{10{,}000}{b^2} = -8$	Subtract 9 from each side.
$-10{,}000 = -8b^2$	Multiply each side by b^2.
$\frac{-10{,}000}{-8} = b^2$	Divide each side by −8.
$1250 = b^2$	Simplify.

Substitute 1250 for b^2 in the equation $\frac{x^2}{a^2} - \frac{y^2}{b^2} = 1$. So, the equation of the path of the comet is $\frac{x^2}{100} - \frac{y^2}{1250} = 1$.

33a.

x	y
−12	−1.33
−10	−1.6
−8	−2
−6	−2.67
−4	−4
−2	−8
0	undef
2	8
4	4
6	2.67
8	2
10	1.6
12	1.33

33b.

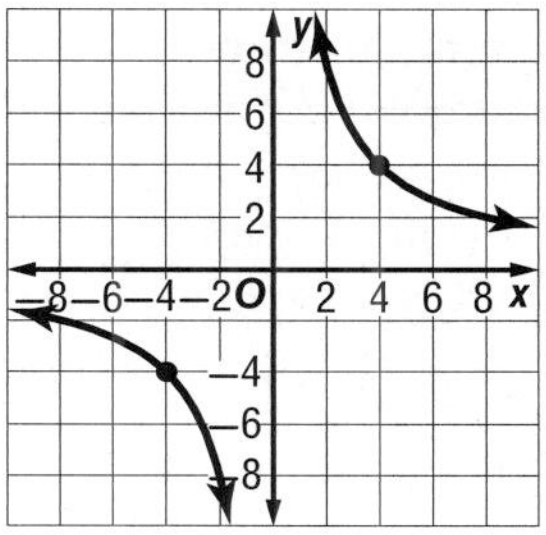

33c. The asymptotes are $y = 0$ and $x = 0$.

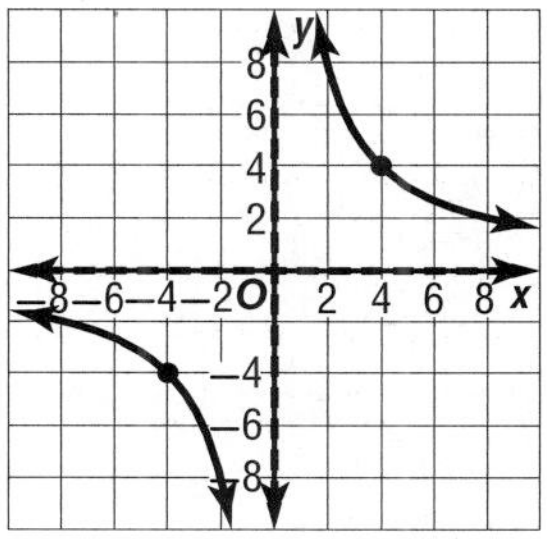

33d. They are perpendicular. **33e.** For $xy = 25$, the vertices will be at (5, 5) and (−5, −5), and for $xy = 36$, they will be at (−6, −6) and (6, 6).

35. $\frac{x^2}{1{,}322{,}500} - \frac{y^2}{927{,}500} = 1$ **37.** $\frac{x^2}{64} - \frac{y^2}{100} = 1$

39 The vertices are equidistant from the center. The center is at (2, −2). The value of a is the distance between a vertex and the center, or 4 units. The value of c is the distance between a focus and the center, or 8 units.

$c^2 = a^2 + b^2$	Equation relating a, b, and c for a hyperbola
$8^2 = 4^2 + b^2$	$c = 8$ and $a = 4$
$48 = b^2$	Subtract 4^2 from each side.

$$\frac{(x-h)^2}{a^2} - \frac{(y-k)^2}{b^2} = 1 \quad \text{Equation of a horizontal hyperbola}$$

$$\frac{(x-2)^2}{4^2} - \frac{[y-(-2)]^2}{48} = 1 \quad (h, k) = (2, -2), a = 4, b^2 = 48$$

$$\frac{(x-2)^2}{16} - \frac{(y+2)^2}{48} = 1 \quad \text{Simplify.}$$

41. $\frac{x^2}{25} - \frac{y^2}{4} = 1$ **43.** (2308, 826) **45.** $\frac{(x-3)^2}{5} - \frac{(y+2)^2}{5} = 1$ **47.** Sample answer: When 36 changes to 9, the vertical hyperbola widens (splits out from the y-axis faster). This is due to a smaller value of y being needed to produce the same value of x. The vertices are moved closer together due to the value of a decreasing from 6 to 3. The foci move farther from the vertices because the difference between c and a increased. **49.** Sample answer: The graphs of ellipses are closed in, while the branches of the hyperbolas extend without bound. There is always an upper and lower limit to the values of the coordinates of an ellipse, while maximum x- and y-values for the coordinates of hyperbolas are infinite. When both of the x^2 and y^2 terms are on the same side of the equation, the equation is for an ellipse if the signs of their coefficients are the same. Otherwise it is a hyperbola. (This is true only for conics that are not rotated.) **51.** J **53.** D

55. $\frac{y^2}{100} + \frac{x^2}{36} = 1$

57. (0, 3), 5 units

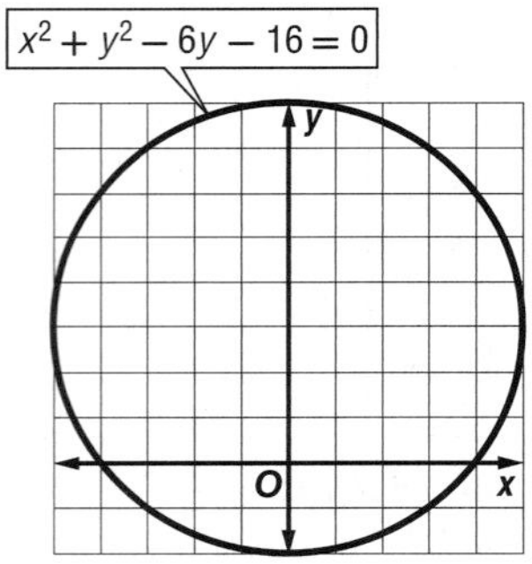

59a.

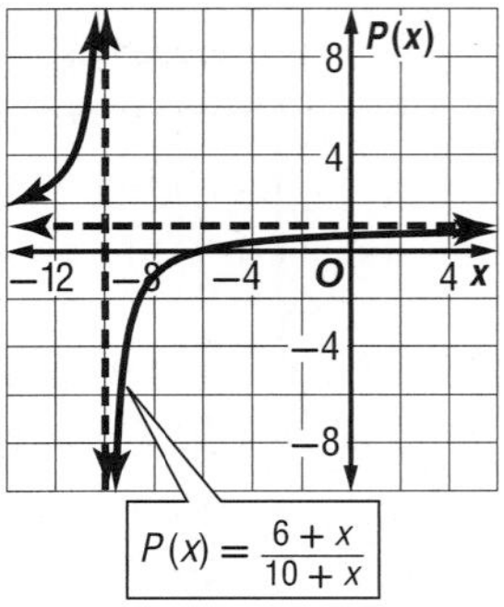

59b. the part in the first quadrant **59c.** It represents her original free-throw percentage of 60%. **59d.** $y = 1$; this represents 100%, which she cannot achieve because she has already missed 4 free throws. **61.** $\frac{5}{3}$

63.

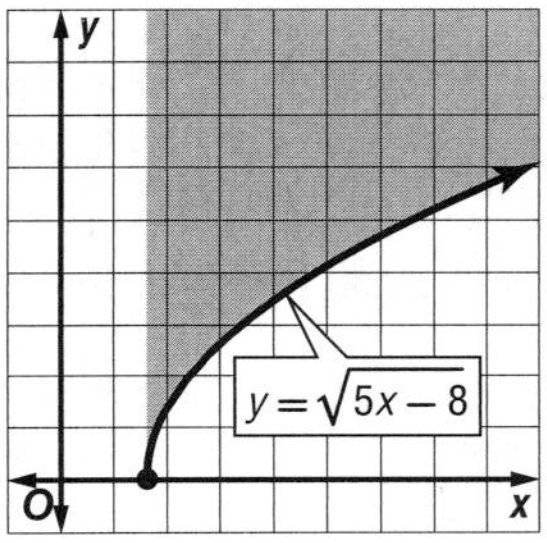

65.

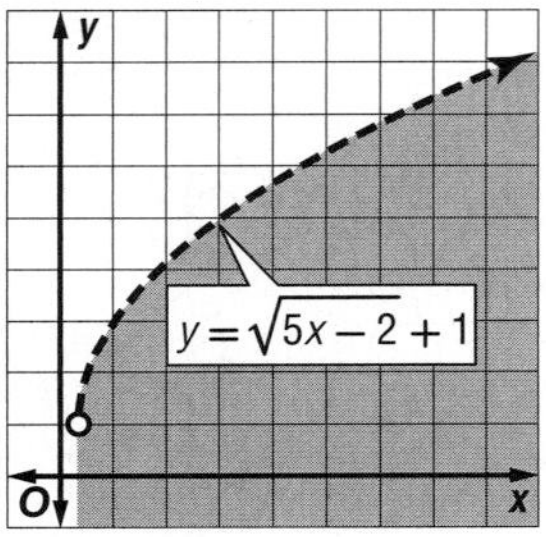

67. $y = 43(x + 3)^2 - 4$

Pages 658–660 Lesson 10-6

1. $\frac{(x-3)^2}{36} + \frac{(y+2)^2}{9} = 1$; ellipse

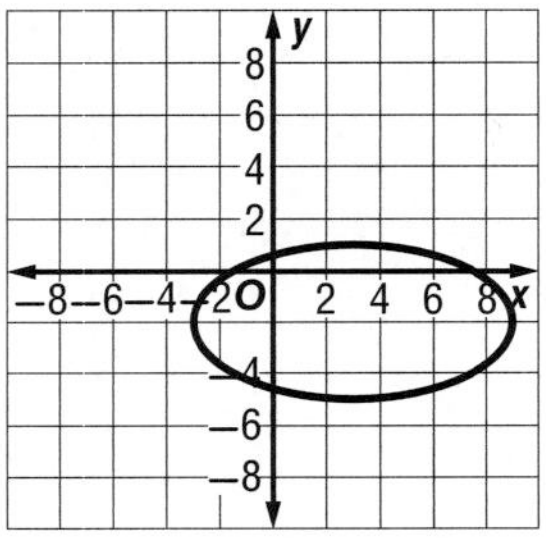

3. $\frac{(y-1)^2}{16} - \frac{(x+2)^2}{9} = 1$; hyperbola

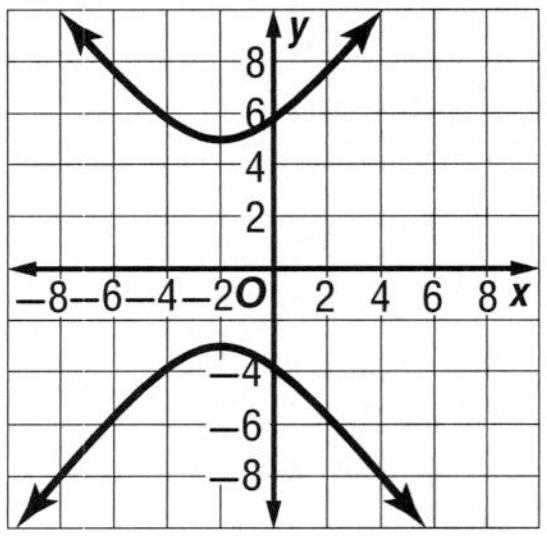

5. ellipse **7.** circle **9.** hyperbola **11.** ellipse
13a. parabola; $y = -0.024(x - 660)^2 + 10{,}500$
13b. about 1320 ft **13c.** 10,500 ft

15. $\frac{(x+4)^2}{32} + \frac{(y-5)^2}{24} = 1$; ellipse

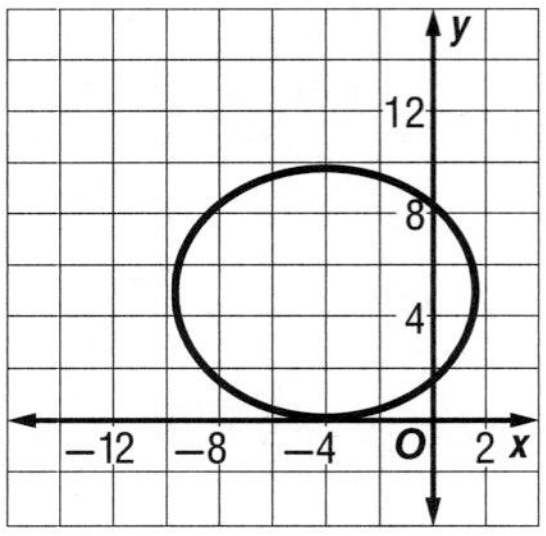

17. $y = 8(x + 2)^2 - 4$; parabola

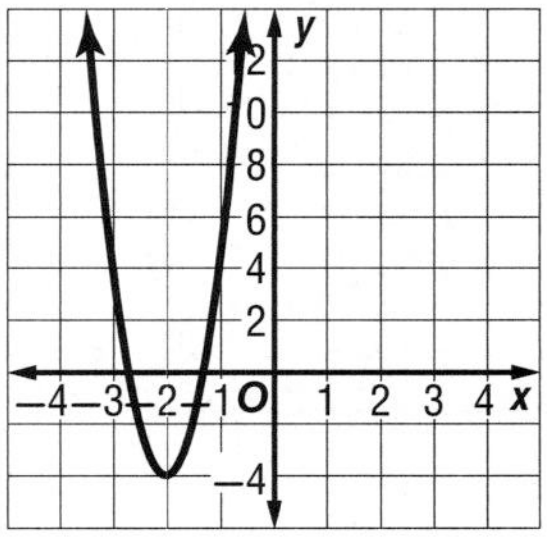

19. $(x - 3)^2 + (y + 4)^2 = 36$; circle

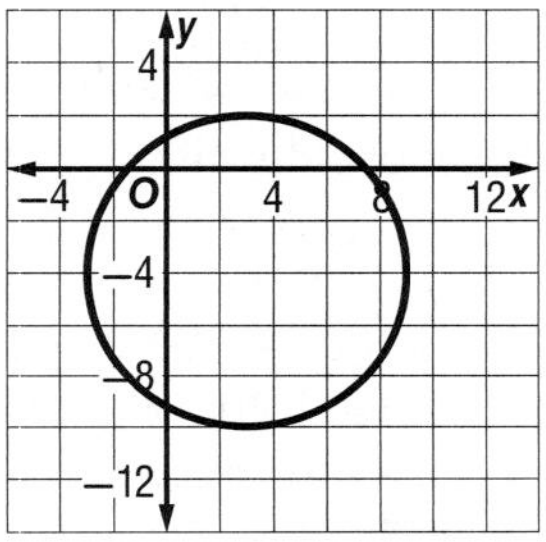

21. $\frac{(x+4)^2}{21} - \frac{(y-8)^2}{24} = 1$; hyperbola

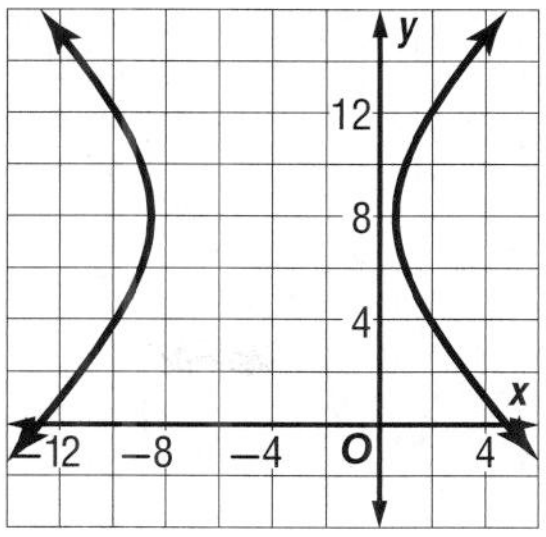

23. $\frac{(x+4)^2}{64} - \frac{(y-3)^2}{25} = 1$; hyperbola

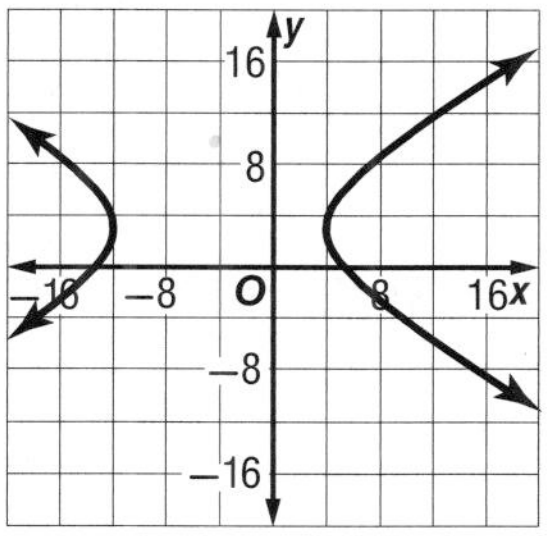

25. circle

27 $18x^2 - 16y = 12x - 4y^2 + 19$ Original equation
$18x^2 + 4y^2 - 12x - 16y - 19 = 0$ Standard form
$A = 18, B = 0$, and $C = 4$
$B^2 - 4AC = 0^2 - 4(18)(4)$ or -288
Since the discriminant is less than 0 and $A \neq C$, the conic is an ellipse.

29. hyperbola **31.** hyperbola **33.** hyperbola **35.** c **37.** b **39.** b

41 Equation c can be written as $y = -16x^2 + 90x + 0.25$. Since this is an equation of a parabola that opens downward, it could be used to represent the height of a football above the ground after being kicked.

43a. $\frac{(x-3)^2}{16} + \frac{(y+2)^2}{4} = 1$

43b. $x^2 + 4y^2 - 6x + 16y + 9 = 0$

43c.

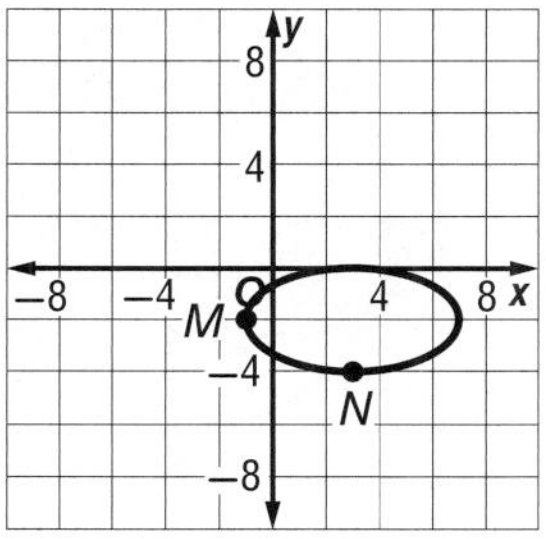

43d. $N(3, 0)$; 90° counterclockwise

45. Sample answer: Always; when a conic is vertical, $B = 0$. When this is true and $A = C$, the conic is a circle.

47. Sample answer: An ellipse is a flattened circle. Both circles and ellipses are enclosed regions, while hyperbolas and parabolas are not. A parabola has one branch, which is a smooth curve that never ends, and a hyperbola has two such branches that are reflections of each other. In standard form and when there is no xy-term: an equation for a parabola consists of only one squared term, an equation for a circle has values for A and C that are equal, an equation for an ellipse has values for A and C that are the same sign but not equal, and an equation for a hyperbola has values of A and C that have opposite signs. **49.** J **51.** B

53.

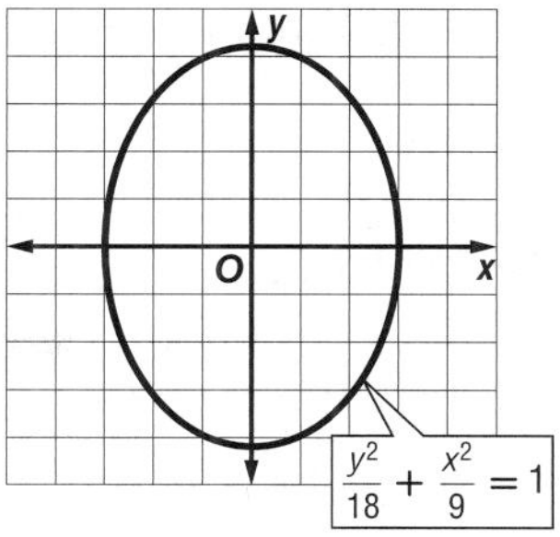

55.

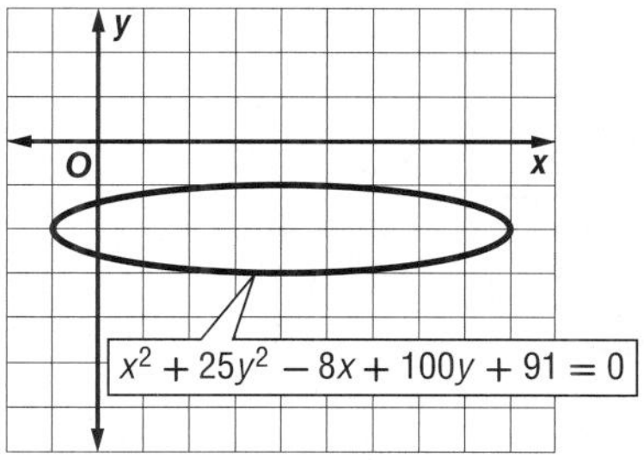

57.

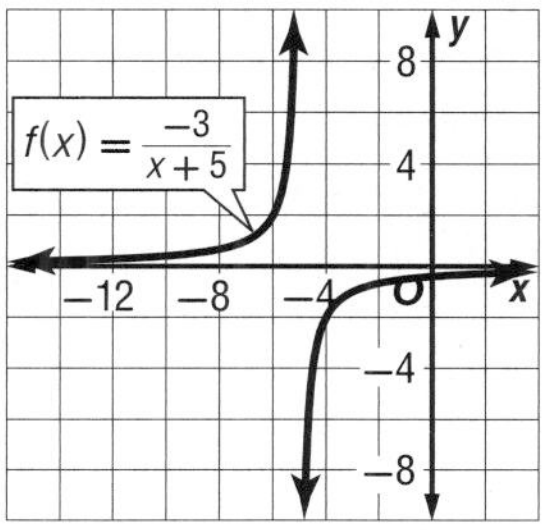

59a. Decay; the exponent is negative. **59b.** about 33.5 watts **59c.** about 402 days **61.** $\left(-\frac{1}{2}, \frac{3}{2}\right)$

Pages 664–667 Lesson 10-7

1. $(4, -5), (-4, 5)$ **3.** $(3, 6), (6, 42)$ **5.** no solution

7 Solve the second equation for x^2.
$y^2 - x^2 = 8 \rightarrow y^2 - 8 = x^2$
$x^2 + 2y = 7$ Write the first equation.
$y^2 - 8 + 2y = 7$ Substitute $y^2 - 8$ for x^2.
$y^2 + 2y - 15 = 0$ Subtract 7 from each side.
$(y - 3)(y + 5) = 0$ Factor.
$y - 3 = 0$ or $y + 5 = 0$ Zero Product Property
$y = 3$ $y = -5$ Solve each equation.
Substitute 3 and -5 into one of the original equations and solve for x.

$x^2 + 2y = 7$	$x^2 + 2y = 7$
$x^2 + 2(3) = 7$	$x^2 + 2(-5) = 7$
$x^2 + 6 = 7$	$x^2 - 10 = 7$
$x^2 = 1$	$x^2 = 17$
$x = \pm 1$	$x = \pm\sqrt{17}$

The solutions are $(-1, 3), (1, 3), (-\sqrt{17}, -5), (\sqrt{17}, -5)$.

9. $(40, 30)$ **11.**

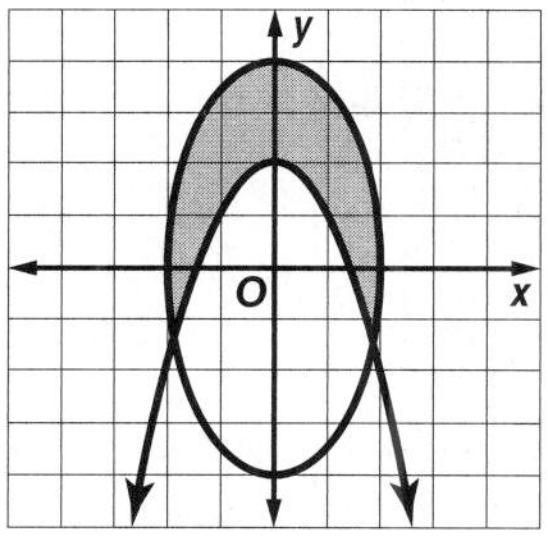

13.

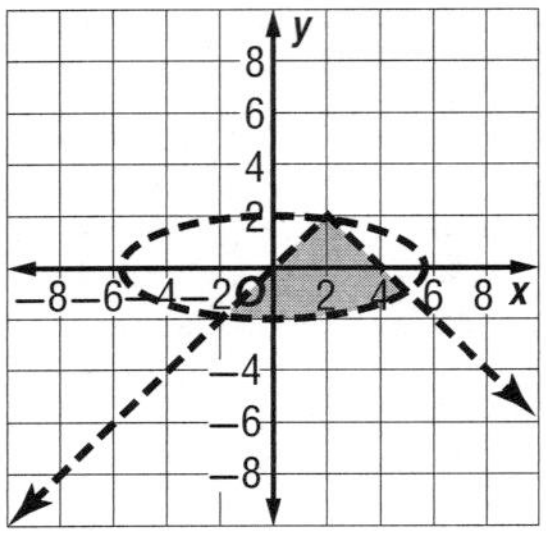

15. no solution
17. $(-3, -6), (3, 6)$
19. $(-1, -3), (8, 5)$
21. $(1, -\sqrt{15}), (1, \sqrt{15})$
23. $(-\sqrt{2}, 4), (\sqrt{2}, 4)$
25. $(5, -6), (5, 6), (-3, -2), (-3, 2)$

27.

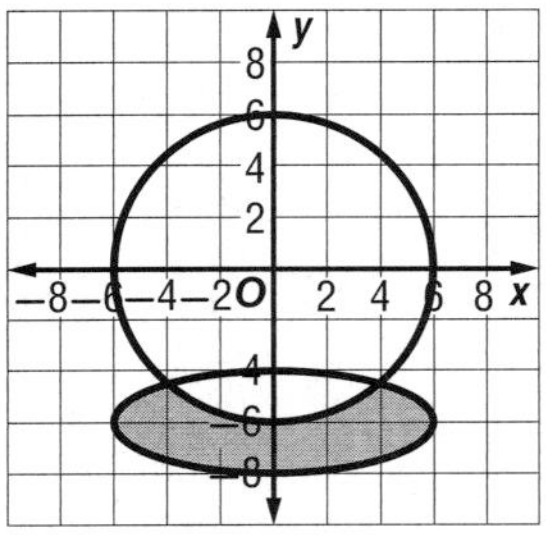

29.

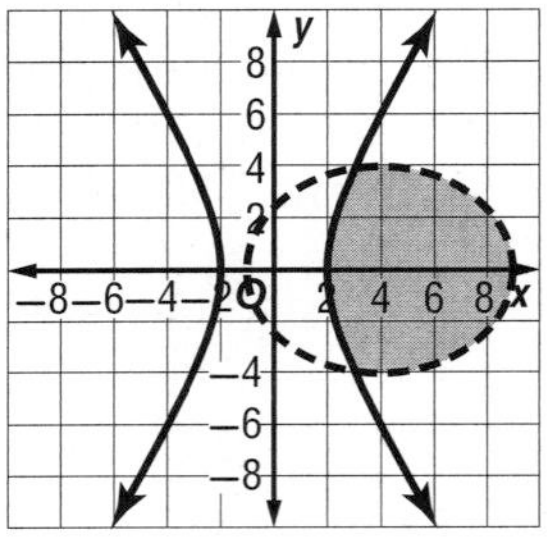

31.

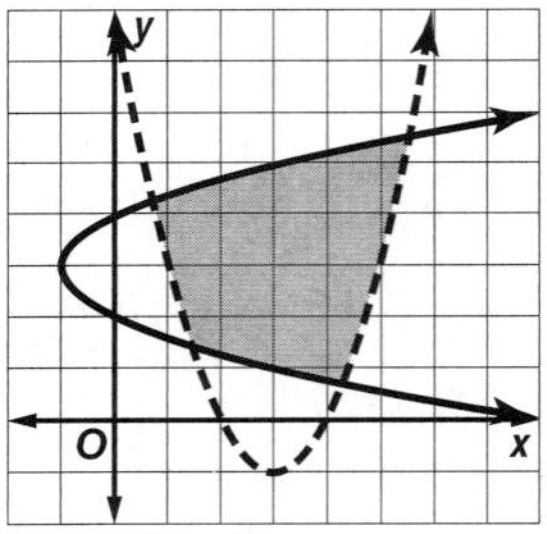

33.

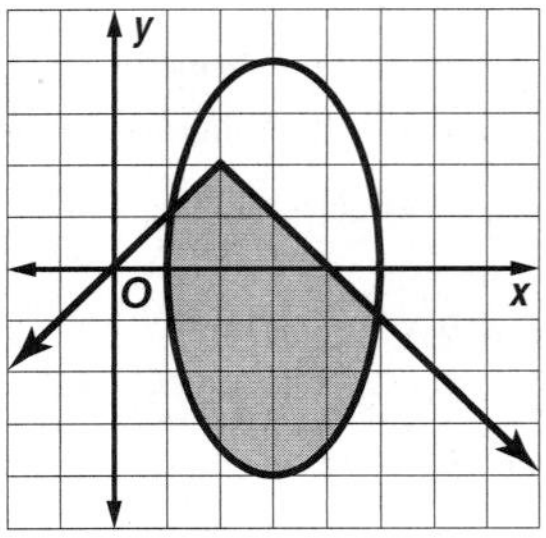

35.

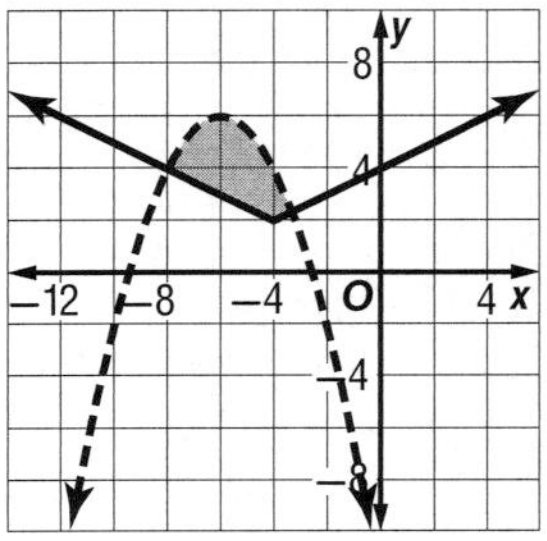

37. no solution

39a. $y = \pm 900\sqrt{1 - \dfrac{x^2}{(300)^2}}$; $y = \pm 690\sqrt{1 - \dfrac{x^2}{(600)^2}}$

39b. Sample answer: $(209, 646), (-209, 646), (-209, -646), (209, -646)$

39c. Sample answer: The orbit of the satellite modeled by the second equation is closer to a circle than the other orbit. The distance on the x-axis is twice as great for one satellite as for the other.

41 $y = -0.0037x^2 + 1.77x - 1.72$ — Path of baseball equation

$\frac{3}{7}x - 128.6 = -0.0037x^2 + 1.77x - 1.72$ — Substitute $\frac{3}{7}x - 128.6$ for y.

$-128.6 \approx -0.0037x^2 + 1.34x - 1.72$ — Substitute $\frac{3}{7}x$ from each side.

$0 \approx -0.0037x^2 + 1.34x + 126.88$ — Add 128.6 to each side.

$x = \dfrac{-b \pm \sqrt{b^2 - 4ac}}{2a}$ — Quadratic Formula

$x = \dfrac{-1.34 \pm \sqrt{1.34^2 - 4(-0.0037)(126.88)}}{2(-0.0037)}$ — $a = -0.0037$, $b = 1.34$, $c = 126.88$

$x \approx \dfrac{-1.34 \pm \sqrt{3.67}}{-0.0074}$ — Simplify.

$x \approx -78$ or $x \approx 440$ — Use a calculator.

Since the vertical distance cannot be negative, the ball landed about 440 ft from home plate.

$y = \frac{3}{7}x - 128.6$ — Original equation

$y = \frac{3}{7}(440) - 128.6$ — Replace x with 440.

$y \approx 60$ — Use a calculator.

The ball was 60 ft above the playing surface.

43 Sample answer: A circle with the equation $(x + 10)^2 + y^2 = 36$ has its center at $(-10, 0)$ and contains the point $(-4, 0)$. An ellipse with the equation $\frac{x^2}{16} + \frac{y^2}{36} = 1$ has its center at $(0, 0)$ and contains the point $(-4, 0)$. The circle and ellipse intersect only $(-4, 0)$.

45. Sample answer: $x^2 + y^2 = 1$ and $\frac{x^2}{16} - \frac{y^2}{36} = 1$

47. Sample answer: $\frac{x^2}{64} + \frac{y^2}{100} = 1$ and $x^2 - y^2 = 1$

49. Sample answer: No; if one player is in one of the shaded areas and the other player is in the other shaded area, they will not be able to hear each other.

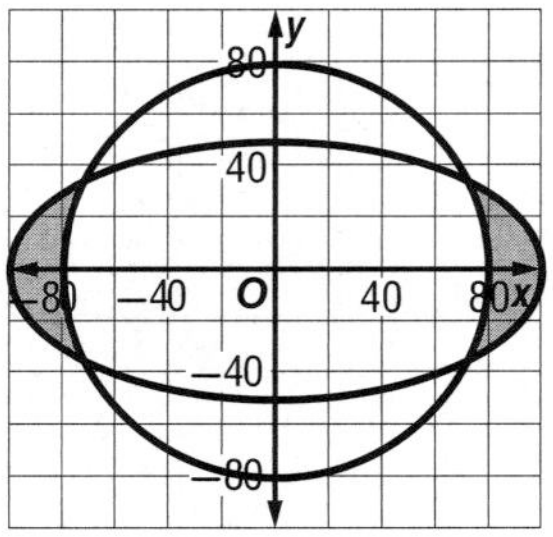

51. $k = a$ or $k = b$ **53.** Sample answer: $\frac{y^2}{128} + \frac{x^2}{32} = 1$ and $\frac{x^2}{8} + \frac{y^2}{64} = 1$ **55.** $(-3, -4), (3, 4)$ **57.** F **59.** b **61.** c

63. $(2, -2), (2, 8)$; $(2, 3 \pm \sqrt{41})$; $y - 3 = \pm\frac{5}{4}(x - 2)$

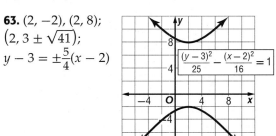

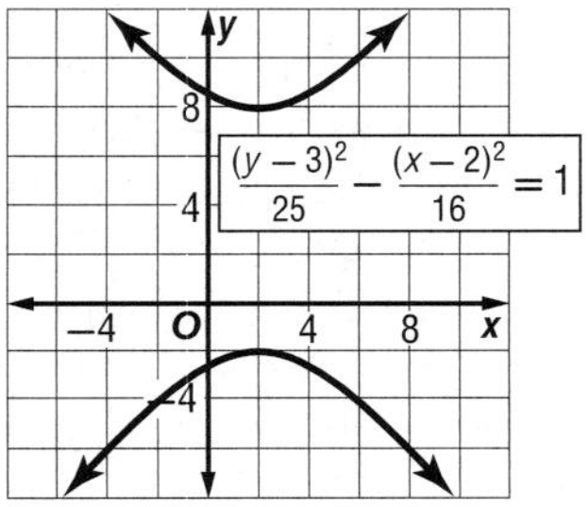

65. $\dfrac{p+5}{p+1}$ **67.** $\dfrac{3(r+4)}{r+3}$

69.

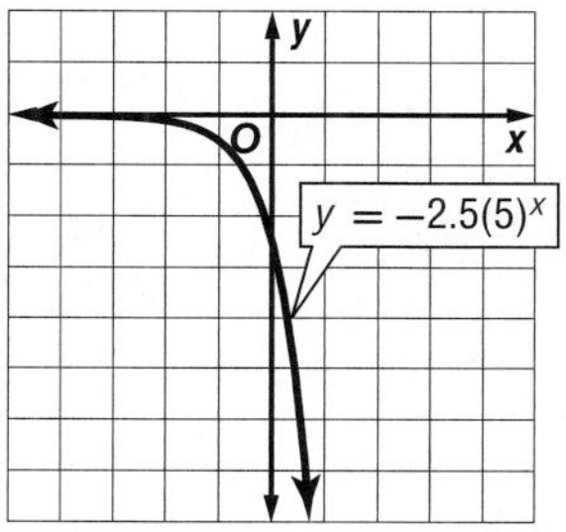

71. $\dfrac{d}{t} = r$ **73.** $\dfrac{3V}{\pi r^2} = h$

Pages 668–672 *Chapter 10* *Study Guide and Review*

1. false, center **3.** false, vertices **5.** true **7.** false, circle **9.** true **11.** $\left(-\frac{5}{2}, 5\right)$ **13.** $\left(\frac{5}{24}, \frac{5}{24}\right)$ **15.** $\sqrt{85}$ **17.** $\dfrac{\sqrt{34}}{4}$ **19a.** $\sqrt{89} \approx 9.4$ miles **19b.** $\left(4, -\frac{5}{2}\right)$

21.

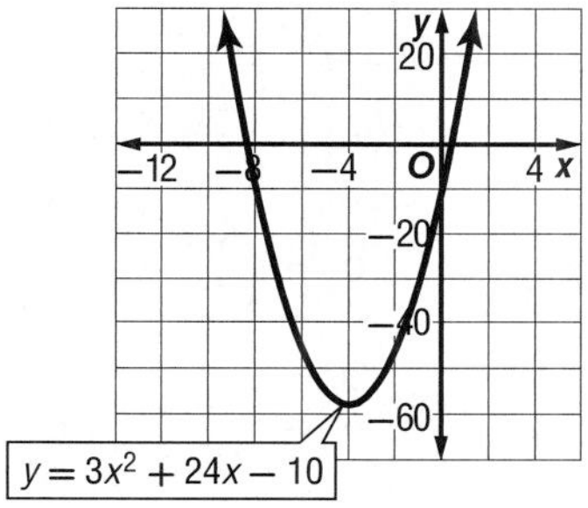

23.

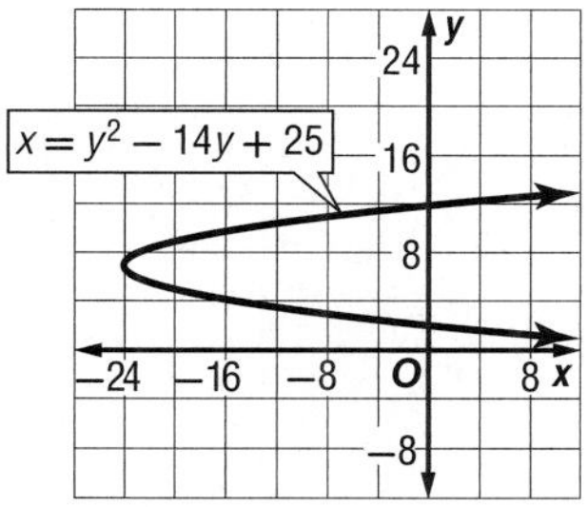

25. $y = 4(x-2)^2 - 7$; vertex: $(2, -7)$; axis of symmetry: $x = 2$; opens up **27.** $x = (y+7)^2 - 29$; vertex $= (-29, -7)$; axis of symmetry: $y = -7$; opens to the right **29.** $(x+1) + (y-6)^2 = 9$
31. $(x-3)^2 + (y+4)^2 = 5$

33. $(3, -1)$; $r = 5$

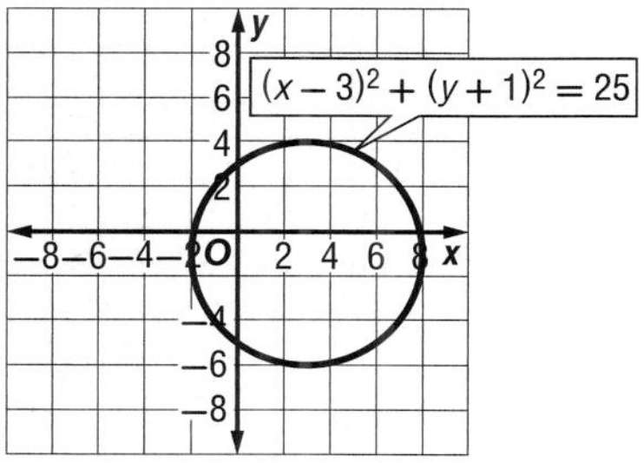

35. $(-2, 1)$; $r = 4$

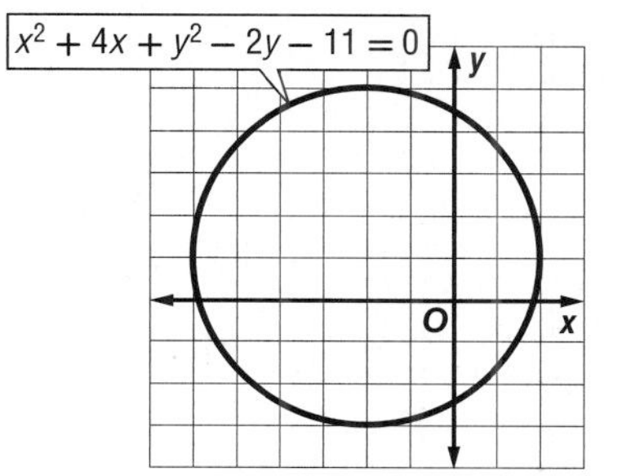

37. $(0, 0)$; $(0, \pm 3\sqrt{5})$; 12; 6

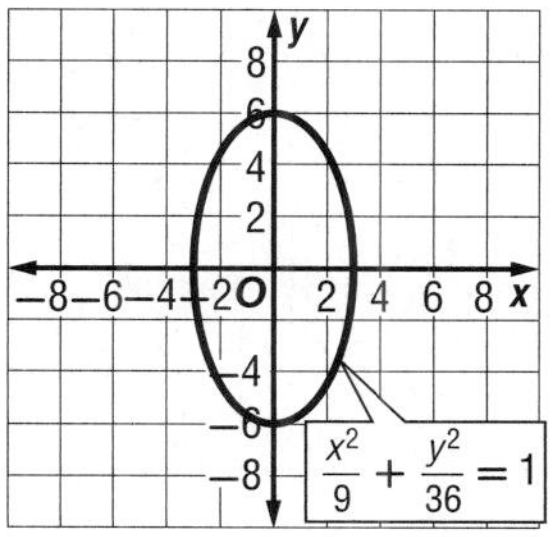

39. $(-1, 2)$; $(\pm\sqrt{41}, 0)$; 10; 8

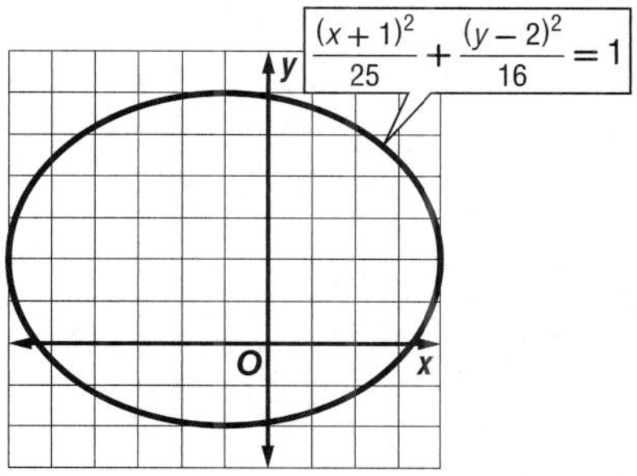

41. $(0, 0)$; $(0, \pm\sqrt{6})$; 6; $2\sqrt{3}$

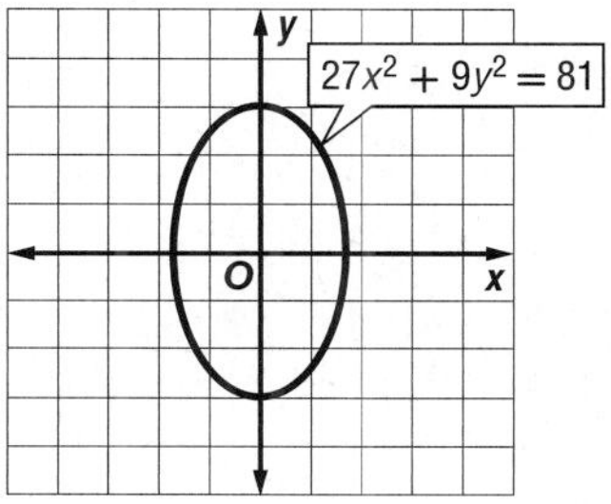

43. $(1, -1)$; $(\pm\sqrt{34}, 0)$; 10; 6

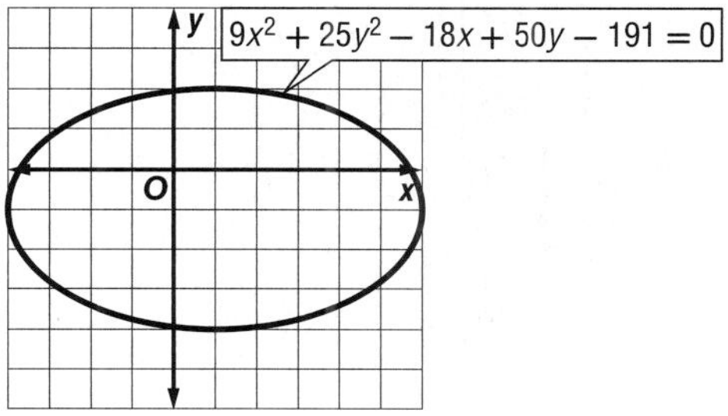

45. $\dfrac{x^2}{64} + \dfrac{y^2}{25} = 1$
47. $(2, -2)$, $(4, -2)$; $(3 \pm \sqrt{5}, -2)$; $y + 2 = \pm 2(x - 3)$

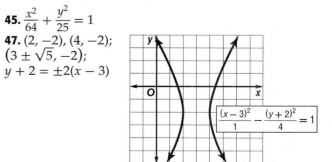

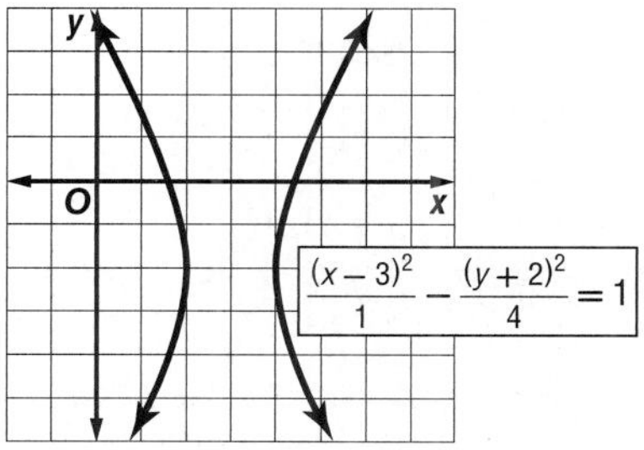

49. $(\pm 3, 0)$; $(\pm\sqrt{13}, 0)$; $y = \pm\frac{2}{3}x$

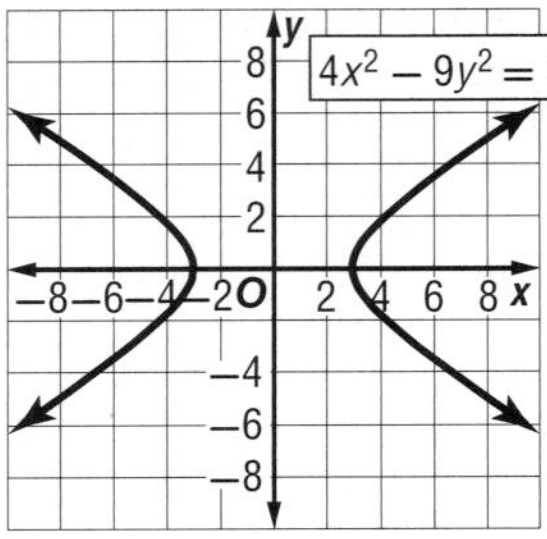

51. $\left(\frac{40 - 25\sqrt{5}}{5}, \frac{45 - 12\sqrt{5}}{5}\right)$

53. $\frac{x^2}{16} + \frac{y^2}{9} = 1$; ellipse

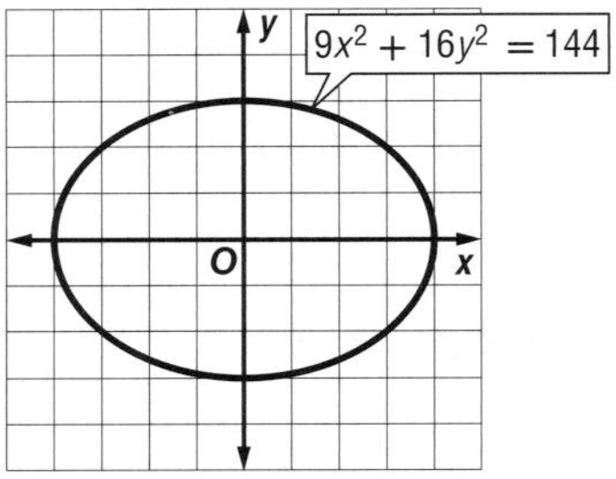

55. $\frac{y^2}{9} - \frac{(x-2)^2}{1} = 1$; hyperbola

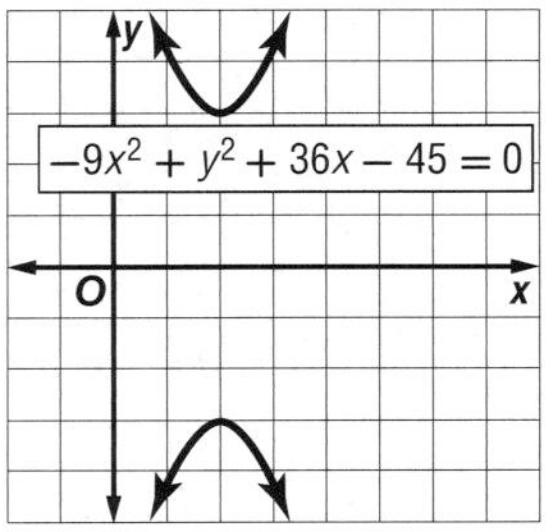

57. hyperbola **59.** circle **61.** $(2, -2), (-2, 2)$ **63.** $(4, 0)$ **65.** $(1, \pm 5), (-1, \pm 5)$ **67a.** 1.5 seconds **67b.** 109 feet

69.

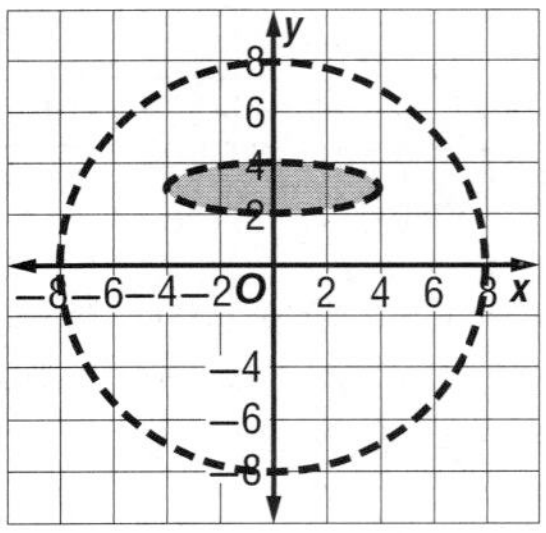

71.

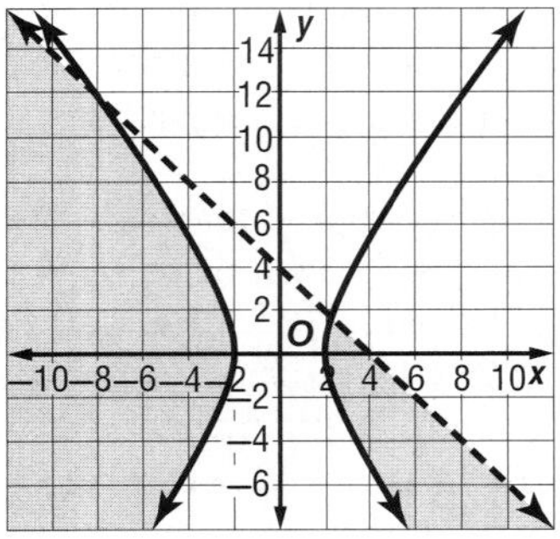

73.

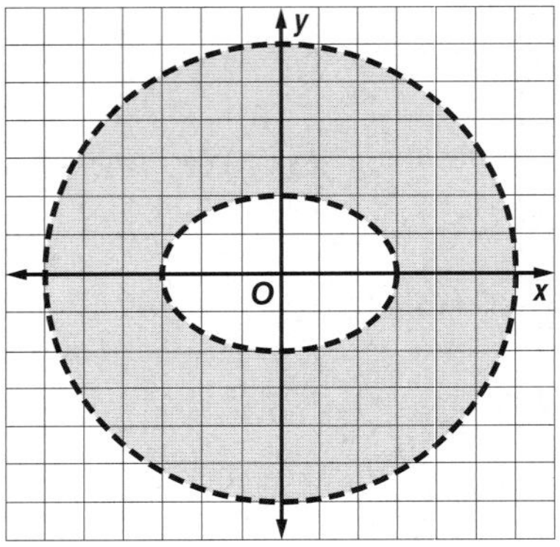

Chapter 11 Sequences and Series

Page 679 Chapter 11 Get Ready

1. $x = -12$ **3.** $x = 3$ **5.** 9 rows

7.

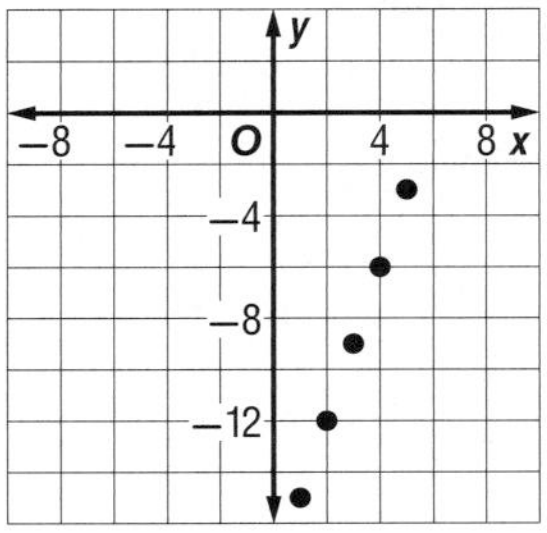

9.

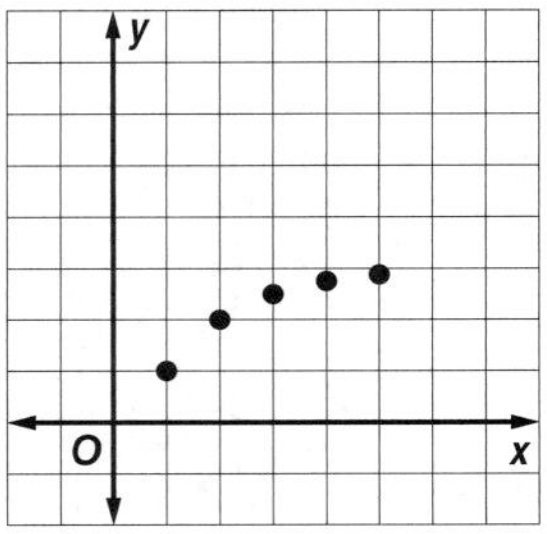

11. -30 **13.** $-\frac{2}{729}$

Pages 685–687 Lesson 11-1

1. yes **3.** no

5. 42, 54, 66, 78

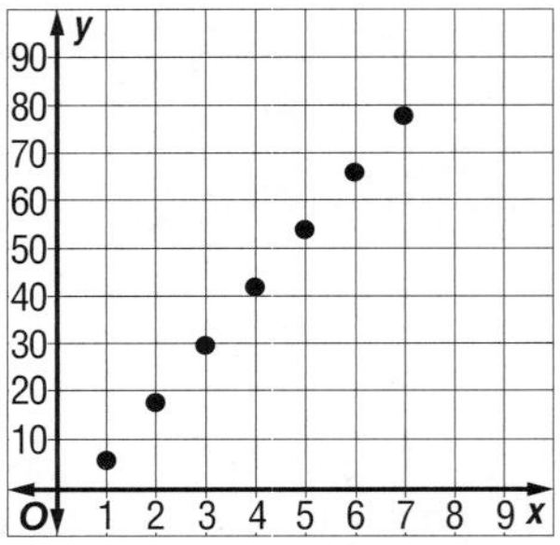

7. 5, 13, 21, 29

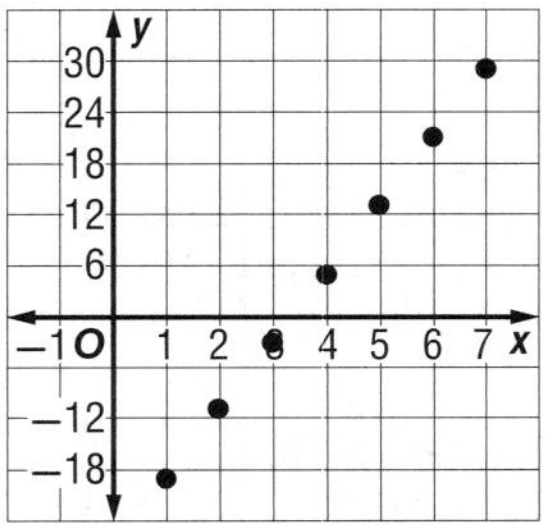

9a. \$850 **9b.** 24 wk **11.** yes **13.** no

15. 128, 256, 512

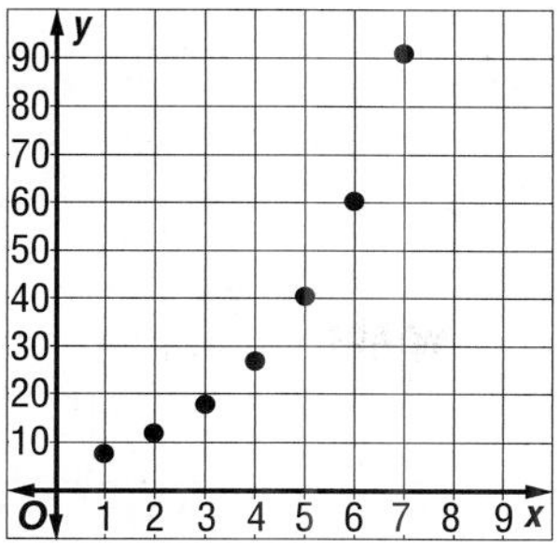

17. $\frac{1}{9}, -\frac{1}{27}, \frac{1}{81}$

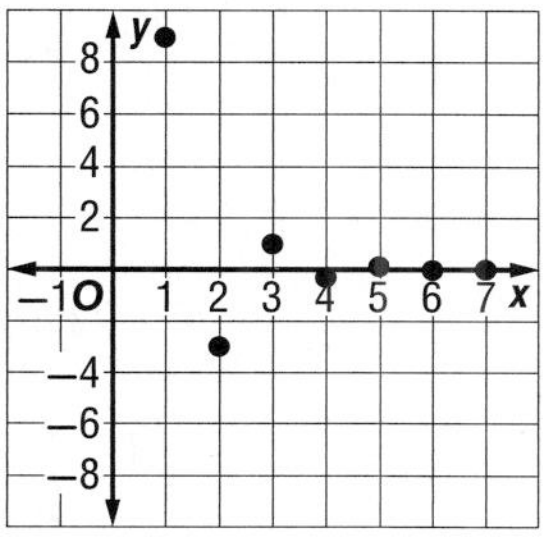

19. Geometric; the common ratio is $-\frac{1}{2}$. **21.** no
23. no

25. 8, 11, 14, 17

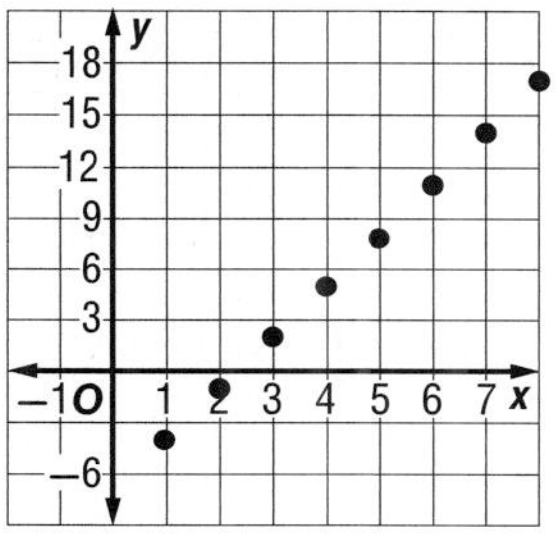

27. −29, −35, −41, −47

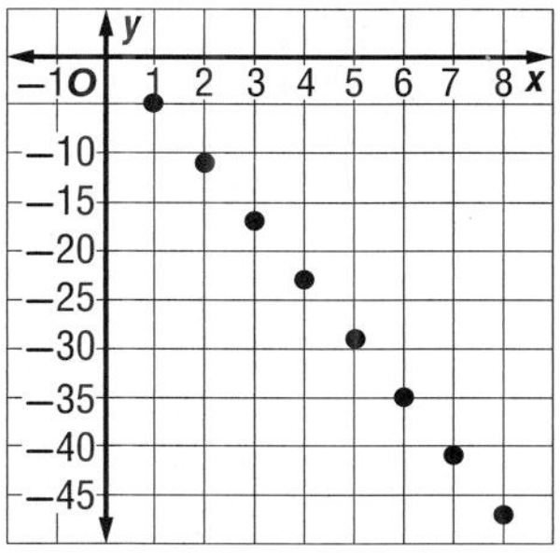

29. $2, \frac{13}{5}, \frac{16}{5}, \frac{19}{5}$

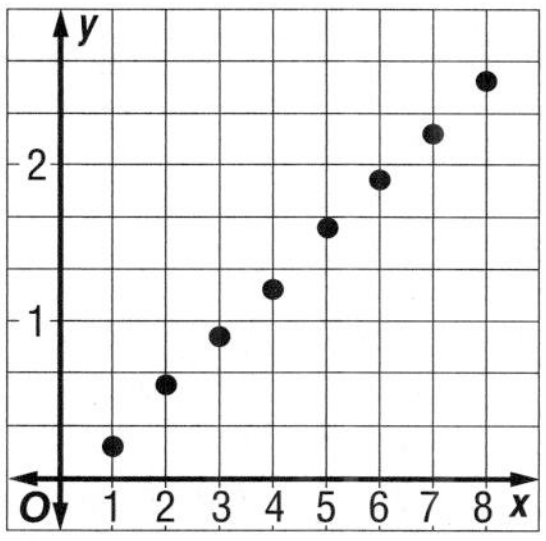

31 The difference between any two consecutive rows is 2, so the common difference is 2. Use point-slope form to write an equation for the sequence. Let $m = 2$ and $(x_1, y_1) = (1, 28)$. Solve for $x = 24$.

$(y - y_1) = m(x - x_1)$	Point-slope form
$(y - 28) = 2(x - 1)$	$m = 2$, $(x_1, y_1) = (1, 28)$
$y - 28 = 2x - 2$	Multiply.
$y = 2x + 26$	Add 28 to each side.
$y = 2(24) + 26$	Replace x with 24.
$y = 48 + 26$ or 74	Simplify.

There are 74 seats in the last row of the theater.

33. no

35 $\frac{18}{-27} = \frac{2}{3}$ and $\frac{-12}{18} = \frac{2}{3}$

Since the ratios of the consecutive terms are the same, the sequence is geometric.

37. no **39.** −8, 32, −128

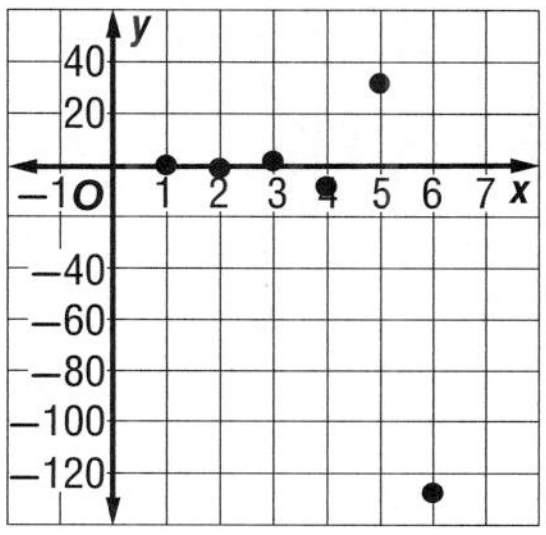

41. $27, \frac{81}{4}, \frac{243}{16}$

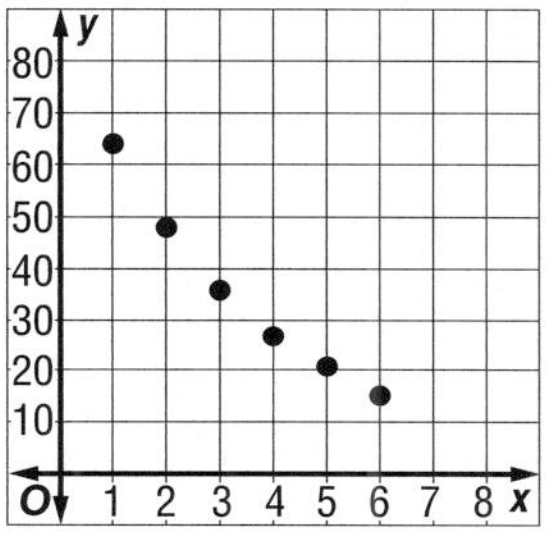

43. 27, 81, 243

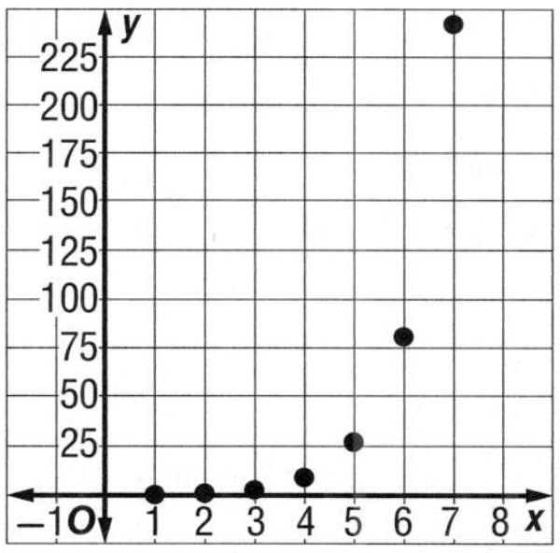

45. Neither; there is no common difference or ratio. **47.** Geometric; the common ratio is 3. **49.** Arithmetic; the common difference is $\frac{1}{2}$. **51.** 86 pg/day **53.** about 6872 km **55.** Sample answer: A babysitter earns \$20 for cleaning the house and \$8 extra for every hour she watches the children. **57.** Sample answer: Neither; the sequence is both arithmetic and geometric. **59.** Sample answer: If a geometric sequence has a ratio r such that $|r| < 1$, as n increases, the absolute value of the terms will decrease and approach zero because they are continuously being multiplied by a fraction. When $|r| \geq 1$, the absolute value of the terms will increase and approach infinity because they are continuously being multiplied by a value greater than 1. **61.** \$421.85 **63.** H **65.** (±4, 5) **67.** no solution

69. hyperbola

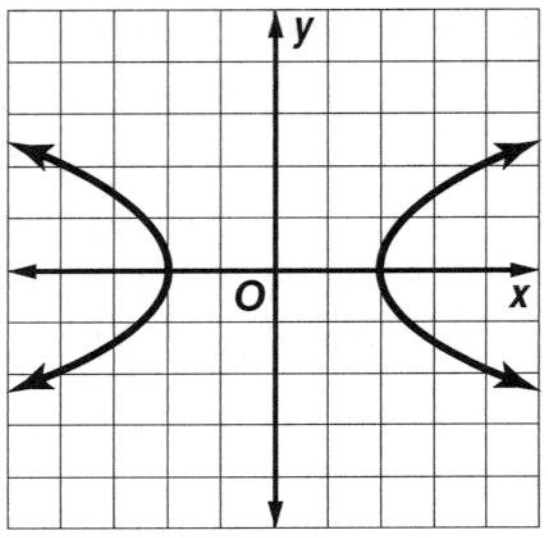

71.

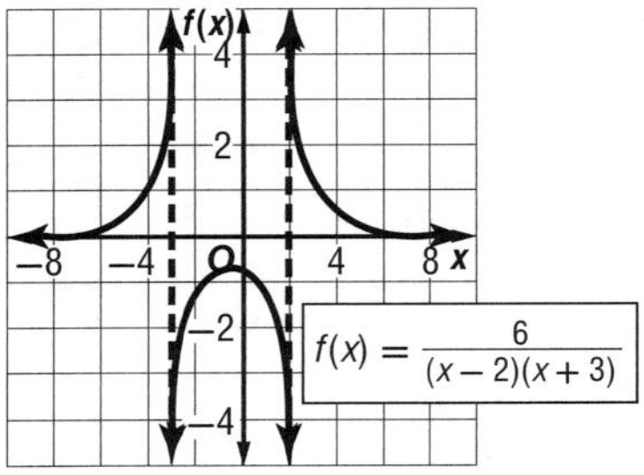

73.

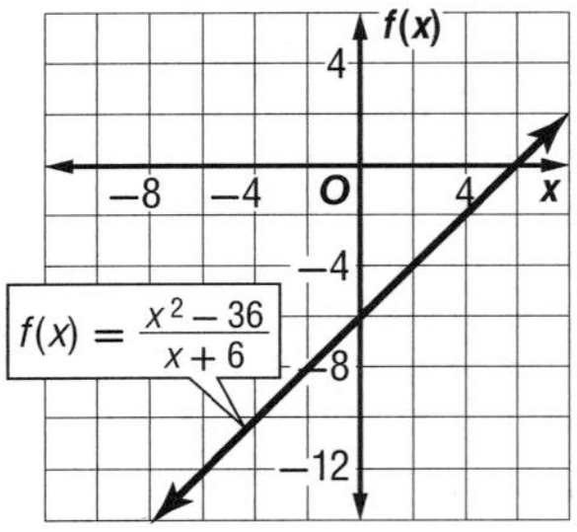

75. $y = 0.5x + 1$ **77.** $y = 3x - 6$ **79.** $y = -\frac{1}{2}x + \frac{7}{2}$

Pages 692–695 Lesson 11-2

1. 104 **3.** $a_n = 6n + 7$ **5.** 15, 24, 33 **7.** 1275 **9.** 4500
11. 8, 12, 16 **13.** C **15.** 248 **17.** −103 **19.** 14
21. $a_n = -14n + 45$

23

$a_n = a_1 + (n-1)d$	*n*th term of an arithmetic sequence
$21 = a_1 + (7-1)5$	$a_7 = 21$, $n = 7$, and $d = 5$
$21 = a_1 + 30$	Multiply.
$-9 = a_1$	Subtract 30 from each side.
$a_n = a_1 + (n-1)d$	*n*th term of an arithmetic sequence
$a_n = -9 + (n-1)5$	$a_1 = -9$ and $d = 5$
$a_n = -9 + (5n - 5)$	Distributive Property
$a_n = 5n - 14$	Simplify.

25. $a_n = 4.5n - 21$ **27.** $a_n = 9n - 32$ **29.** $a_n = \frac{2}{3}n - 3$
31. $a_n = \frac{1}{2}n - \frac{23}{10}$ **33.** 19, 14, 9, 4 **35.** −21, −14, −7, 0
37. −21, −30, −39, −48, −57 **39.** 10,100 **41.** 10,000
43. 696 **45.** 1272 **47.** 100, 94, 88 **49.** $4400

51

$S_n = n\left(\frac{a_1 + a_n}{2}\right)$	Sum Formula
$2982 = 28\left(\frac{a_1 + 228}{2}\right)$	$S_n = 2982$, $n = 28$, $a_n = 228$
$2982 = 14a_1 + 3192$	Simplify.
$-210 = 14a_1$	Subtract 3192 from each side.
$-15 = a_1$	Divide each side by 14.
$a_n = a_1 + (n-1)d$	*n*th term of an arithmetic sequence
$228 = -15 + (28-1)d$	$a_n = 228$, $a_1 = -15$, $n = 28$
$243 = 27d$	Add 15 to each side.
$9 = d$	Divide each side by 27.

$a_1 = -15$, $a_2 = -15 + 9$ or -6, $a_3 = -6 + 9$ or 3
The first three terms are −15, −6, 3.

53. −44, −30, −16 **55.** −33, −21, −9 **57.** 512
59. 324 **61.** $2250 **63.** $a_n = 13n - 1055$
65. $a_n = 9n - 88$
67a. 14, 18, 22

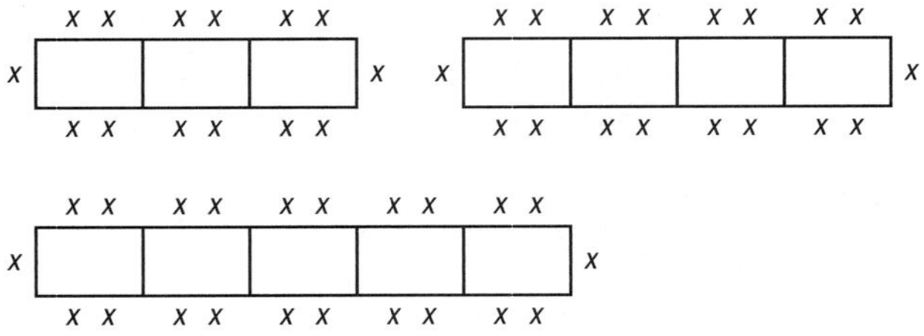

67b. $p_n = 4n + 2$ **67c.** No; there is no whole number n for which $4n + 2 = 100$.

69

$a_n = a_1 + (n-1)d$	*n*th term of an arithmetic sequence
$100{,}000 = 28{,}000 + (n-1)4000$	$a_n = 100{,}000$, $a_1 = 28{,}000$, $d = 4000$
$72{,}000 = (n-1)4000$	Subtract 28,000 from each side.
$18 = n - 1$	Divide each side by 4000.
$19 = n$	Add 1 to each side.

He will have a salary of $100,000 in the 19th year.

71a.

n	S_n
1	4
2	10
3	18
4	28
5	40
6	54
7	70
8	88
9	108
10	130

71b.

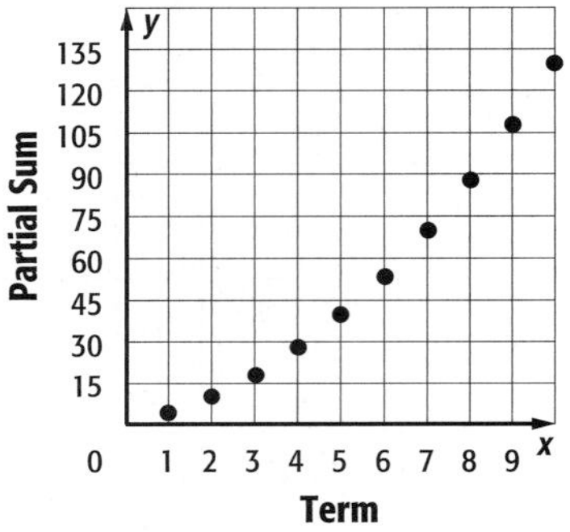

71c.

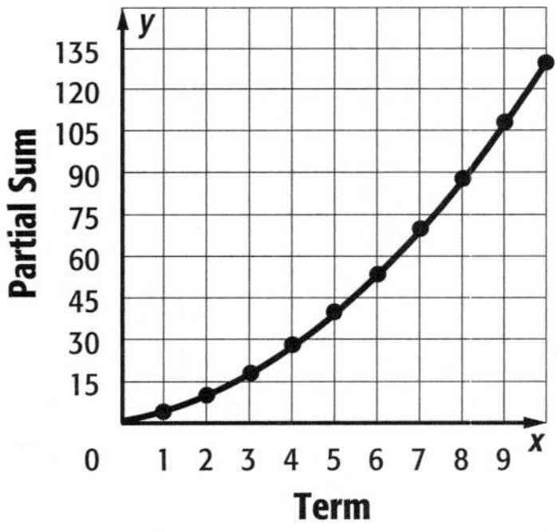

71d. Sample answer: The graphs cover the same range. The domain of the series is the natural numbers, while the domain of the quadratic function is all real numbers, $0 \leq x \leq 10$.

71e. Sample answer: For every partial sum of an arithmetic series, there is a corresponding quadratic function that shares the same range. **71f.** $\sum_{k=1}^{x} 2k + 7$

73. 16 **75.** $4b - 3a$ **77.** $S_n = nx + y\left(\frac{n^2 + n}{2}\right)$

79. Sample answer: An arithmetic sequence is a list of terms such that any pair of successive terms has a common difference. An arithmetic series is the sum of the terms of an arithmetic sequence.

81.
$$S_n = (a_1 + a_n) \cdot \left(\frac{n}{2}\right) \quad \text{General sum formula}$$
$$a_n = a_1 + (n-1)d \quad \text{Formula for } n\text{th term}$$
$$a_n - (n-1)d = a_1 \quad \text{Subtract } (n-1)d \text{ from both sides.}$$
$$S_n = [a_n - (n-1)d + a_n] \cdot \left(\frac{n}{2}\right) \quad \text{Substitution}$$
$$S_n = [2a_n - (n-1)d] \cdot \left(\frac{n}{2}\right) \quad \text{Simplify.}$$

83. B **85.** A **87.** yes **89.** no

91.

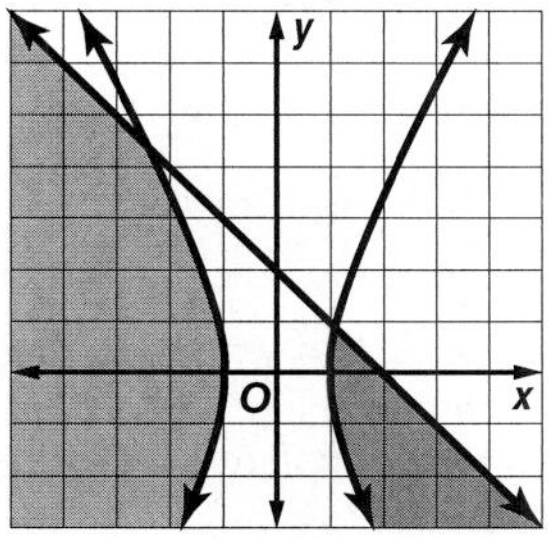

93a. 4.8 cm/g **93b.** 24 cm; The answer is reasonable. The object would stretch the first spring 60 cm and would stretch the second spring 40 cm. The object would have to stretch the combined springs less than it would stretch either of the springs individually.

95.

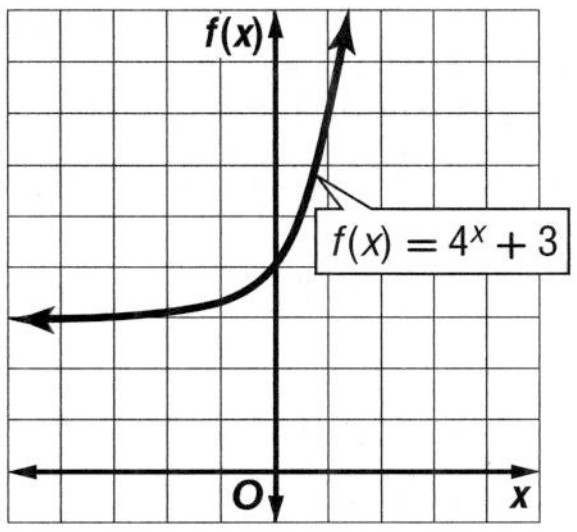

D = {all real numbers}, R = $\{y \mid y > 3\}$

97. 2.4550

99. 0.4341

Pages 699–702 Lesson 11-3

1. \$21.23 **3.** $a_n = 18 \cdot \left(\frac{1}{3}\right)^{n-1}$

5
$$a_n = a_1 r^{n-1} \quad n\text{th term of a geometric sequence}$$
$$4 = a_1(3^{2-1}) \quad a_n = 4, r = 3, \text{ and } n = 2$$
$$4 = a_1(3) \quad \text{Evaluate the power.}$$
$$\frac{4}{3} = a_1 \quad \text{Divide each side by 3.}$$
$$a_n = a_1 r^{n-1} \quad n\text{th term of a geometric sequence}$$
$$a_n = \frac{4}{3}(3)^{n-1} \quad a_1 = \frac{4}{3}, r = 3$$

7. $a_n = 12(-8)^{n-1}$ **9.** 1, 5, 25 **11.** 4095 **13.** $\frac{1}{16}$

15. 512 **17.** \$17,642.21 **19.** 12.5 **21.** 512

23. $a_n = (-3)(-2)^{n-1}$ **25.** $a_n = (-1)(-1)^{n-1}$

27. $a_n = 8 \cdot \left(\frac{1}{4}\right)^{n-1}$ **29.** $a_n = 7(2)^{n-1}$

31. $a_n = \frac{1}{15{,}552}(6)^{n-1}$ **33.** $a_n = 648\left(\frac{1}{3}\right)^{n-1}$

35. 270, 90, 30 **37.** $\frac{7}{3}, \frac{14}{9}, \frac{28}{27}$ **39.** 15 and 75 **41.** 97.3%

43. 31.9375 **45.** 9707.82 **47.** 2188 **49.** −87,381

51
$$S_n = \frac{a_1 - a_1 r^n}{1 - r} \quad \text{Sum formula}$$
$$-2912 = \frac{a_1 - a_1(3^6)}{1 - 3} \quad S_n = -2912, r = 3, \text{ and } n = 6$$
$$-2912 = \frac{a_1(1 - 3^6)}{1 - 3} \quad \text{Distributive Property}$$
$$-2912 = \frac{-728a_1}{-2} \quad \text{Subtract.}$$
$$-2912 = 364a_1 \quad \text{Simplify.}$$
$$-8 = a_1 \quad \text{Divide each side by 364.}$$

53. 64 **55.** 0.25

57
$$a_n = a_1 r^{n-1} \quad n\text{th term of a geometric sequence}$$
$$= 175{,}000(1.08^{15-1}) \quad a_1 = 175{,}000, r = 1.08, \text{ and } n = 15$$
$$\approx 514{,}009 \quad \text{Use a calculator.}$$

59. 524, 289 **61.** about 471 cm **63a.** \$11.79, \$30.58, \$205.72 **63b.** \$7052.15 **63c.** Each payment made is rounded to the nearest penny, so the sum of the payments will actually be more than the sum found in part b.

65.
$$S_n = \frac{a_1 - a_n r}{1 - r} \quad \text{Alternate sum formula}$$
$$a_n = a_1 \cdot r^{n-1} \quad \text{Formula for } n\text{th term}$$
$$\frac{a_n}{r^{n-1}} = a_1 \quad \text{Divide both sides by } r^{n-1}.$$
$$S_n = \frac{\frac{a_n}{r^{n-1}} - a_n r}{1 - r} \quad \text{Substitution.}$$
$$= \frac{\frac{a_n}{r^{n-1}} - \frac{a_n r \cdot r^{n-1}}{r^{n-1}}}{1 - r} \quad \text{Multiply by } \frac{\frac{r^{n-1}}{r^{n-1}}}{1}.$$
$$= \frac{\frac{a_n(1 - r^n)}{r^{n-1}}}{1 - r} \quad \text{Simplify.}$$
$$= \frac{a_n(1 - r^n)}{r^{n-1}(1 - r)} \quad \text{Divide by } 1 - r.$$
$$= \frac{a_n(1 - r^n)}{r^{n-1} - r^n} \quad \text{Simplify.}$$

67. Sample answer: $n - 1$ needs to change to n, and the 10 needs to change to a 9. When this happens, the terms for both series will be identical (a_1 in the first series will equal a_0 in the second series, and so on), and the series will be equal to each other. **69.** 234 **71.** Sample answer: 4 + 8 + 16 + 32 + 64 + 128 **73.** B **75.** \$32,000 **77.** \$1550 **79.** Arithmetic; the common difference is $\frac{1}{50}$.

81. (3, 1), 5 units

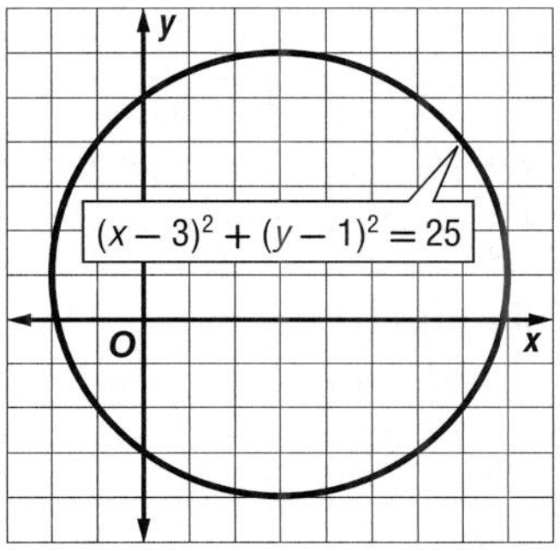

83. (3, −7), $5\sqrt{2}$ units

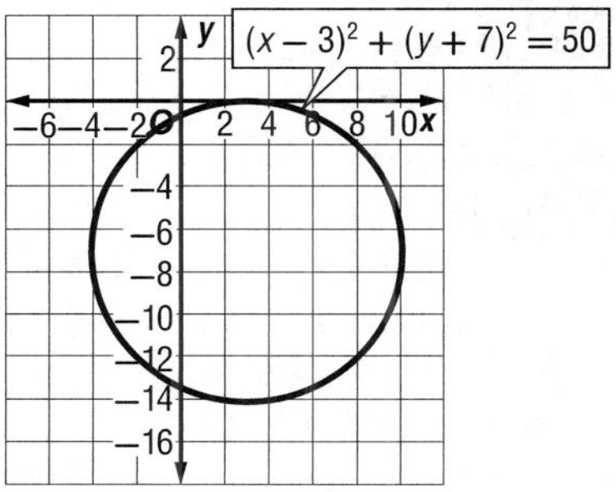

85. 731 customers **87.** $-\frac{5}{7}$ **89.** $\frac{27}{2}$

Pages 708–711 *Lesson 11-4*

1. convergent **3.** divergent **5.** 880 **7.** No sum exists. **9.** 12.5% **11.** −4 **13.** 2 **15.** $\frac{214}{333}$ **17.** convergent **19.** divergent **21.** divergent **23.** No sum exists.

25 $r = \frac{6}{5} \div \frac{12}{5}$ or $\frac{1}{2}$ Divide consecutive terms.

Since $\left|\frac{1}{2}\right| < 1$, the sum exists.

$S = \frac{a_1}{1 - r}$ Sum formula

$= \frac{\frac{12}{5}}{1 - \frac{1}{2}}$ $a_1 = \frac{12}{5}$ and $r = \frac{1}{2}$

$= \frac{12}{5} \div \frac{1}{2}$ or $\frac{24}{5}$ Simplify.

27. No sum exists. **29.** No sum exists. **31.** $\frac{35}{12}$ **33.** 16

35 $0.3\overline{21} = 0.3 + 0.021 + 0.00021 + \ldots$

$= \frac{3}{10} + \frac{21}{1000} + \frac{21}{100{,}000} + \ldots$

$S = \frac{a_1}{1 - r} + \frac{3}{10}$ Sum formula

$= \frac{\frac{21}{1000}}{1 - \frac{1}{100}} + \frac{3}{10}$ $a_1 = \frac{21}{1000}$ and $r = \frac{1}{100}$

$= \frac{2100}{99{,}000} + \frac{3}{10}$ Simplify.

$= \frac{107}{330} + \frac{3}{10}$ or $\frac{53}{165}$ Simplify.

37. $\frac{24}{11}$ **39.** $\frac{601}{4950}$ **41.** $\frac{40}{3}$ **43.** 8000 hrs **45.** $\frac{45}{4}$ **47.** No sum exists. **49.** $-\frac{54}{35}$ **51.** 200 ft **53.** 1170 ft

55 $S = \frac{a_1}{1 - r}$ Sum formula

$= \frac{500}{1 - 0.8}$ $a_1 = 500$ and $r = 0.8$

$= \frac{500}{0.2}$ or \$2500 Simplify.

57. B **59.** A **61.** Sample answer: The sum of a geometric series is $S_n = \frac{a_1 - a_1r^n}{1 - r}$. For an infinite series with $|r| < 1$, $r^n \to 0$ as $n \to \infty$. Thus, $S = \frac{a_1 - a_1(0)}{1 - r}$ or $\frac{a_1}{1 - r}$. **63.** Sample answer: An infinite geometric series has a sum when the common ratio is less than 1. When this occurs, the terms will approach 0 as n approaches infinity. With the future terms almost 0, the sum of the series will approach a limit. When the common ratio is 1 or greater, the terms will keep increasing and approach infinity as n approaches infinity and the sum of the series will have no limit. **65.** Sample answer: $3 + 2 + \frac{4}{3} + \ldots$ **67.** An arithmetic series has a common difference, so each term will eventually become more positive or more negative, but never approach 0. With the terms not approaching 0, the sum will never reach a limit and the series cannot converge. **69.** H **71.** C **73a.** 9, 18, 27, 36, 45, 54, 63, 72 **73b.** 42 meetings **75.** 12

Pages 717–719 *Lesson 11-5*

1. 16, 20, 24, 28, 32 **3.** 5, 17, 53, 161, 485 **5.** $a_{n+1} = 2a_n + 2; a_1 = 3$ **7a.** $a_n = 1.01a_{n-1} - 100$ **7b.** \$1415, \$1329.15, \$1242.44, \$1154.87 **7c.** \$77.08 **9.** −18, 74, −294 **11.** −52, −420, −3364 **13.** −9, −10, −12, −16, −24 **15.** −4, −7, −12, −21, −38

17 $a_{n+1} = 5a_n + 2n$ Recursive formula

$a_{1+1} = 5a_1 + 2(1)$ $n = 1$

$a_2 = 5(-2) + 2(1)$ or -8 $a_1 = -2$

$a_3 = 5(-8) + 2(2)$ or -36 $a_2 = -8, n = 2$

$a_4 = 5(-36) + 2(3)$ or -174 $a_3 = -36, n = 3$

$a_5 = 5(-174) + 2(4)$ or -862 $a_4 = -36, n = 4$

The first five terms are −2, −8, −36, −174, and −862.

19. 4, 5, 6, 8, 8 **21.** 3, $2x$, $8x - 9$, $26x - 36$, $80x - 117$ **23.** 1, x, $3x + 6$, $15x + 18$, $63x + 90$ **25.** $a_{n+1} = 0.25a_n + 4; a_1 = 32$ **27.** $a_{n+1} = (a_n)^3 + 1; a_1 = 1$ **29.** $a_{n+1} = 0.25a_n + 8; a_1 = 480$ **31.** $a_{n+1} = 2a_n - 32; a_1 = 84$ **33.** 56, 680, 8168 **35.** −45, 273, −1635 **37.** −3, −18, −963 **39.** 43, 3484, 24,259,093 **41.** 4.25, 29.5625, 936.0664

43 **a.** The number of blue triangles is 1, 3, and 9. The sequence is geometric because each term after the first can be found after multiplying by a common ratio, 3.

$a_n = r \cdot a_{n-1}$ Recursive formula for geometric sequence

$a_n = 3a_{n-1}$ $r = 3$

b. $a_1 = 1, a_2 = 3, a_3 = 9$

$a_n = 3a_{n-1}$ Recursive formula

$a_4 = 3a_3$ $n = 4$

$= 3(9)$ or 27 $a_3 = 9$

$a_5 = 3a_4$ $n = 5$

$= 3(27)$ or 81 $a_4 = 27$

$a_6 = 3a_5$ $n = 6$

$= 3(81)$ or 243 $a_5 = 27$

There will be 243 blue triangles in the sixth figure.

45. No; the population of fish will reach 12,500. Each year, 20% of 12,500 or 2500 fish plus 10,000 additional fish yields 12,500 fish.

47a. 11,000 **47b.** It converges to 7142.857.

47c. Sample answer: They make it easier to analyze recursive sequences because they can produce the first 100 terms instantaneously; it would take a long time to calculate the terms by hand. **49.** Armando; Marcus included x_0 with the iterates and only showed the first 2 iterates. **51.** Sample answer: Sometimes; the recursive formula could involve the first three terms. For example, 2, 2, 2, 8, 20, … is recursive with $a_{n+3} = a_n + a_{n+1} + 2a_{n+2}$.

53. Sample answer: In a recursive sequence, each term is determined by one or more of the previous terms. A recursive formula is used to produce the terms of the recursive sequence. **55a.** 160 m **55b.** 5.7 s **55c.** 11.4 s **57.** C **59.** $5\frac{14}{111}$

61a. 2, 3, 4.5, 6.75, 10.125 **61b.** the eighth session

61c. during the ninth session **63.** dependent
65. $x^2 + 4x - 12$ **67.** $4h^2 + 33h + 35$
69. $10g^2 + 19g - 56$

Pages 723–725 Lesson 11-6

1. $c^5 + 5c^4d + 10c^3d^2 + 10c^2d^3 + 5cd^4 + d^5$
3. $x^6 - 24x^5 + 240x^4 - 1280x^3 + 3840x^2 - 6144x + 4096$
5. $x^5 + 15x^4 + 90x^3 + 270x^2 + 405x + 243$
7. $\frac{3}{32}$ or 0.09375

9 $(x + 3y)^8 = \sum_{k=0}^{8} \frac{8!}{k!(8-k)!} x^{8-k}(3y)^k$

$\frac{8!}{k!(8-k)!} x^{8-k}(3y)^k = \frac{8!}{4!(8-4)!} x^{8-4}(3y)^4$ For the fifth term, $k = 4$.

$= 70x^4(81y^4)$ $C(8, 4) = 70$, $(3y)^4 = 81y^4$

$= 5670x^4y^4$ Simplify.

11. $-108{,}864c^3d^5$ **13.** $243a^5$ **15.** $a^6 - 6a^5b + 15a^4b^2 - 20a^3b^3 + 15a^2b^4 - 6ab^5 + b^6$ **17.** $x^6 + 36x^5 + 540x^4 + 4320x^3 + 19{,}440x^2 + 46{,}656x + 46{,}656$ **19.** $16a^4 + 128a^3b + 384a^2b^2 + 512ab^3 + 256b^4$

21 Let w represent the number of women and m represent the number of men.

$(w + m)^{10} = \sum_{k=0}^{10} \frac{10!}{k!(10-k)!} w^{10-k}m^k$

To find the probability that 7 members are women, find the term in which the exponent of w is 7. Since $10 - 3 = 7$, find the term in which $k = 3$, the fourth term.

$\frac{10!}{k!(10-k)!} w^{10-k}m^k = \frac{10!}{3!(10-3)!} w^{10-3}m^3$ For the fourth term, $k = 3$.

$= 120w^7m^3$ $C(10, 3) = 120$

The probability of choosing a woman is $\frac{1}{2}$ and the probability of choosing a man is $\frac{1}{2}$.

$120w^7m^3 = 120\left(\frac{1}{2}\right)^7\left(\frac{1}{2}\right)^3$ $w = \frac{1}{2}$ and $m = \frac{1}{2}$

$= 120\left(\frac{1}{2}\right)^{10}$ Product of Powers Property

$= \frac{120}{1024}$ $\left(\frac{1}{2}\right)^{10} = \frac{1}{1024}$

$= \frac{15}{128}$ Simplify.

The probability that 7 of the members will be women is $\frac{15}{128}$, or about 0.117.

23. $84x^5z^2$ **25.** $7168a^2b^6$ **27.** $32{,}256x^5$ **29.** $x^5 + \frac{5}{2}x^4 + \frac{5}{2}x^3 + \frac{5}{4}x^2 + \frac{5}{16}x + \frac{1}{32}$ **31.** $32b^5 + 20b^4 + 5b^3 + \frac{5}{8}b^2 + \frac{5}{128}b + \frac{1}{1024}$ **33a.** 0.121 **33b.** 0.121 **33c.** 0.309
35. Sample answer: While they have the same terms, the signs for $(x + y)^n$ will all be positive, while the signs for $(x - y)^n$ will alternate. **37.** Sample answer: $\left(x + \frac{6}{5}y\right)^5$ **39.** A **41.** G **43.** −2, 3, 8, 13, 18 **45.** 4, 6, 12, 30, 84 **47.** $1\frac{1}{8}$ **49a.** $\frac{150}{x}$; $\frac{130}{x - 10}$ **49b.** $\frac{150}{x} + \frac{130}{x - 10} = 4$; 75 mph, 65 mph **51.** true; $3(1) + 5 = 8$, which is even

Pages 729–731 Lesson 11-7

1. Step 1: When $n = 1$, the left side of the given equation is 1. The right side is 1^2 or 1, so the equation is true for $n = 1$.
Step 2: Assume that $1 + 3 + 5 + \cdots + (2k - 1) = k^2$ for some natural number k.
Step 3: $1 + 3 + 5 + \cdots + (2k - 1) + (2(k + 1) - 1)$
$= k^2 + (2(k + 1) - 1)$
$= k^2 + (2k + 2 - 1)$
$= k^2 + 2k + 1$
$= (k + 1)^2$
The last expression is the right side of the equation to be proved, where $n = k + 1$. Thus, the equation is true for $n = k + 1$. Therefore, $1 + 3 + 5 + \cdots + (2n - 1) = n^2$ for all natural numbers n.

3a. 3, 6, 10, 15, 21 **3b.** $a_n = \frac{n(n+1)}{2}$
3c. Step 1: When $n = 1$, the left side of the given equation is $\frac{1(1+1)}{2}$ or 1. The right side is $\frac{1(1+1)(1+2)}{6}$ or 1, so the equation is true for $n = 1$.
Step 2: Assume that $1 + 3 + 6 + \cdots + \frac{k(k+1)}{2} = \frac{k(k+1)(k+2)}{6}$ for some natural number k.
Step 3: $1 + 3 + 6 + \cdots + \frac{k(k+1)}{2} + \frac{(k+1)(k+1+1)}{2}$
$= \frac{k(k+1)(k+2)}{6} + \frac{(k+1)(k+1+1)}{2}$
$= \frac{k(k+1)(k+2)}{6} + \frac{3(k+1)(k+2)}{6}$
$= \frac{k^3 + 3k^2 + 2k}{6} + \frac{3k^2 + 9k + 6}{6}$
$= \frac{k^3 + 6k^2 + 11k + 6}{6}$
$= \frac{(k+1)(k+2)(k+3)}{6}$
The last expression is the right side of the equation to be proved, where $n = k + 1$. Thus, the equation is true for $n = k + 1$. Therefore, $1 + 3 + 6 + \cdots + \frac{n(n+1)}{2} = \frac{n(n+1)(n+2)}{6}$ for all natural numbers n.

5. Step 1: $4^1 - 1 = 3$, which is divisible by 3. The statement is true for $n = 1$.
Step 2: Assume that $4^k - 1$ is divisible by 3 for some natural number k. This means that $4^k - 1 = 3r$ for some whole number r.
Step 3: $4^k - 1 = 3r$
$4^k = 3r + 1$
$4^{k+1} = 12r + 4$
$4^{k+1} - 1 = 12r + 3$
$4^{k+1} - 1 = 3(4r + 1)$
Since r is a whole number, $4r + 1$ is a whole number. Thus, $4^{k+1} - 1$ is divisible by 3, so the statement is true for $n = k + 1$. Therefore, $4^n - 1$ is divisible by 3 for all natural numbers n. **7.** $n = 1$

9. Step 1: When $n = 1$, the left side of the given equation is 2. The right side is $\frac{1[3(1) + 1]}{2}$ or 2, so the equation is true for $n = 1$.
Step 2: Assume that $2 + 5 + 8 + \cdots + (3k - 1) = \frac{k(3k + 1)}{2}$ for some natural number k.

Step 3: $2 + 5 + 8 + \cdots + (3k - 1) + [3(k + 1) - 1]$

$$= \frac{k(3k + 1)}{2} + [3(k + 1) - 1]$$
$$= \frac{k(3k + 1) + 2[3(k + 1) - 1]}{2}$$
$$= \frac{3k^2 + k + 6k + 6 - 2}{2}$$
$$= \frac{3k^2 + 7k + 4}{2}$$
$$= \frac{(k + 1)(3k + 4)}{2}$$
$$= \frac{(k + 1)[3(k + 1) + 1]}{2}$$

The last expression is the right side of the equation to be proved, where $n = k + 1$. Thus, the equation is true for $n = k + 1$. Therefore, $2 + 5 + 8 + \cdots + (3n - 1) = \frac{n(3n + 1)}{2}$ for all natural numbers n.

11. Step 1: When $n = 1$, the left side of the given equation is 1. The right side is $1[2(1) - 1]$ or 1, so the equation is true for $n = 1$.
Step 2: Assume that $1 + 5 + 9 + \cdots + (4k - 3) = k(2k - 1)$ for some natural number k.
Step 3: $1 + 5 + 9 + \cdots + (4k - 3) + [4(k + 1) - 3]$

$$= k(2k - 1) + [4(k + 1) - 3]$$
$$= 2k^2 - k + 4k + 4 - 3$$
$$= 2k^2 + 3k + 1$$
$$= (k + 1)(2k + 1)$$
$$= (k + 1)[2(k + 1) - 1]$$

The last expression is the right side of the equation to be proved, where $n = k + 1$. Thus, the equation is true for $n = k + 1$. Therefore, $1 + 5 + 9 + \cdots + (4n - 3) = n(2n - 1)$ for all natural numbers n.

13 Test different values of n.

n	$1 + 2 + 3 + \cdots + n$	Equal n^2?
1	1	yes
2	$1 + 2 = 3$	no

The value $n = 2$ is a counterexample for the statement.

15. Step 1: When $n = 1$, the left side of the given equation is 1^2 or 1. The right side is $\frac{1[2(1) - 1][2(1) + 1]}{3}$ or 1, so the equation is true for $n = 1$.
Step 2: Assume that $1^2 + 3^2 + 5^2 + \cdots + (2k - 1)^2 = \frac{k(2k - 1)(2k + 1)}{3}$ for some natural number k.
Step 3: $1^2 + 3^2 + 5^2 + \cdots + (2k - 1)^2 + [2(k + 1) - 1]^2$

$$= \frac{k(2k - 1)(2k + 1)}{3} + [2(k + 1) - 1]^2$$
$$= \frac{k(2k - 1)(2k + 1) + 3(2k + 1)^2}{3}$$
$$= \frac{(2k + 1)[k(2k - 1) + 3(2k + 1)]}{3}$$
$$= \frac{(2k + 1)(2k^2 - k + 6k + 3)}{3}$$
$$= \frac{(2k + 1)(2k^2 + 5k + 3)}{3}$$
$$= \frac{(2k + 1)(k + 1)(2k + 3)}{3}$$
$$= \frac{(k + 1)[2(k + 1) - 1][2(k + 1) + 1]}{3}$$

The last expression is the right side of the equation to be proved, where $n = k + 1$. Thus, the equation is true for $n = k + 1$. Therefore, $1^2 + 3^2 + 5^2 + \cdots + (2n - 1)^2 = \frac{n(2n - 1)(2n + 1)}{3}$ for all natural numbers n.

17. Step 1: $5^1 + 3 = 8$, which is divisible by 4. The statement is true for $n = 1$.
Step 2: Assume $5^k + 3$ is divisible by 4 for some natural number k. This means that $5^k + 3 = 4r$ for some natural number r.
Step 3:

$$5^k + 3 = 4r$$
$$5^k = 4r - 3$$
$$5^{k+1} = 20r - 15$$
$$5^{k+1} + 3 = 20r - 12$$
$$5^{k+1} + 3 = 4(5r - 3)$$

Since r is a natural number, $5r - 3$ is a natural number. Thus, $5^{k+1} + 3$ is divisible by 4, so the statement is true for $n = k + 1$. Therefore, $5^n + 3$ is divisible by 4 for all natural numbers n.

19. Step 1: $12^1 + 10 = 22$, which is divisible by 11. The statement is true for $n = 1$.
Step 2: Assume that $12^k + 10$ is divisible by 11 for some natural number k. This means that $12^k + 10 = 11r$ for some natural number r.
Step 3:

$$12^k + 10 = 11r$$
$$12^k = 11r - 10$$
$$12^{k+1} = 132r - 120$$
$$12^{k+1} + 10 = 132r - 110$$
$$12^{k+1} + 10 = 11(12r - 10)$$

Since r is a natural number, $12r - 10$ is a natural number. Thus, $12^{k+1} + 10$ is divisible by 11, so the statement is true for $n = k + 1$. Therefore, $12^n + 10$ is divisible by 11 for all natural numbers n.

21. $n = 2$ **23.** $n = 1$

25 In the sequence 1, 1, 2, 3, 5, 8, …, $f_1 = 1, f_2 = 1, f_3 = 2, f_4 = 3, f_5 = 5, f_6 = 8, \ldots$.

$f_1 + f_2 + \cdots + f_n = f_{n+2} - 1$ Original equation
$f_1 = f_{1+2} - 1$ Let $n = 1$.
$f_1 = f_3 - 1$ Simplify.

Step 1: When $n = 1$, the left side of the given equation is f_1. The right side is $f_3 - 1$. Since $f_1 = 1$ and $f_3 = 2$, the equation becomes $1 = 2 - 1$ and is true for $n = 1$.
Step 2: Assume that $f_1 + f_2 + \cdots + f_k = f_{k+2} - 1$ for some natural number k.
Step 3: $f_1 + f_2 + \cdots + f_k + f_{k+1} = f_{k+2} - 1 + f_{k+1}$
$= f_{k+1} + f_{k+2} - 1$
$= f_{k+3} - 1$, since Fibonacci numbers are produced by adding the two previous Fibonacci numbers.
The last expression is the right side of the equation to be proved, where $n = k + 1$. Thus, the equation is true for $n = k + 1$. Therefore, $f_1 + f_2 + \cdots + f_n = f_{n+2} - 1$ for all natural numbers n.

27. Step 1: $18^1 - 1 = 17$, which is divisible by 17. The statement is true for $n = 1$.
Step 2: Assume that $18^k - 1$ is divisible by 17 for some natural number k. This means that $18^k - 1 = 17r$ for some natural number r.

Step 3: $18^k - 1 = 17r$

$$18^k = 17r + 1$$
$$18^{k+1} = 18(17r + 1)$$
$$18^{k+1} = 306r + 18$$
$$18^{k+1} - 1 = 306r + 17$$
$$18^{k+1} - 1 = 17(18r + 1)$$

Since r is a natural number, $18r + 1$ is a natural number. Thus, $18^{k+1} - 1$ is divisible by 17, so the statement is true for $n = k + 1$. Therefore, $18^n - 1$ is divisible by 17 for all natural numbers n. **29.** $n = 3$

31. Step 1: When $n = 1$, the left side of the given equation is $\frac{1}{1(1+1)(1+2)}$ or $\frac{1}{6}$. The right side is $\frac{1(1+3)}{4(1+1)(1+2)}$ or $\frac{1}{6}$, so the equation is true for $n = 1$.

Step 2: Assume that $\frac{1}{1 \cdot 2 \cdot 3} + \frac{1}{2 \cdot 3 \cdot 4} + \frac{1}{3 \cdot 4 \cdot 5} + \cdots + \frac{1}{k(k+1)(k+2)} = \frac{k(k+3)}{4(k+1)(k+2)}$ for some natural number k.

Step 3: $\frac{1}{1 \cdot 2 \cdot 3} + \frac{1}{2 \cdot 3 \cdot 4} + \cdots + \frac{1}{k(k+1)(k+2)} + \frac{1}{(k+1)(k+2)(k+3)}$

$$= \frac{k(k+3)}{4(k+1)(k+2)} + \frac{1}{(k+1)(k+2)(k+3)}$$
$$= \frac{k(k+3)(k+3)}{4(k+1)(k+2)(k+3)} + \frac{4}{4(k+1)(k+2)(k+3)}$$
$$= \frac{k^3 + 6k^2 + 9k + 4}{4(k+1)(k+2)(k+3)}$$
$$= \frac{(k+1)(k^2 + 5k + 4)}{4(k+1)(k+2)(k+3)}$$
$$= \frac{(k+1)(k+4)}{4(k+2)(k+3)}$$
$$= \frac{(k+1)[(k+1)+3]}{4[(k+1)+1][(k+1)+2]}$$

The last expression is the right side of the equation to be proved, where $n = k + 1$. Thus, the equation is true for $n = k + 1$. Therefore, $\frac{1}{1 \cdot 2 \cdot 3} + \frac{1}{2 \cdot 3 \cdot 4} + \frac{1}{3 \cdot 4 \cdot 5} + \cdots + \frac{1}{n(n+1)(n+2)} = \frac{n(n+3)}{4(n+1)(n+2)}$ for all natural numbers n.

33. $n(n + 1)$

Step 1: When $n = 1$, the left side of the given equation is 2(1) or 2. The right side is $1(1 + 1)$ or 2, so the equation is true for $n = 1$.

Step 2: Assume that $2 + 4 + 6 + \cdots + 2k = k(k + 1)$ for some natural number k.

Step 3: $2 + 4 + 6 + \cdots + 2k + 2(k + 1)$

$$= k(k+1) + 2(k+1)$$
$$= k^2 + k + 2k + 2$$
$$= k^2 + 3k + 2$$
$$= (k+1)(k+2)$$
$$= (k+1)[(k+1)+1]$$

The last expression is the right side of the equation to be proved, where $n = k + 1$. Thus, the equation is true for $n = k + 1$. Therefore, $2 + 4 + 6 + \cdots + n^2 = n(n + 1)$ for all natural numbers n. **35.** Sample answer: False; assume $k = 2$, just because a statement is true for $n = 2$ and $n = 3$ does not mean that it is true for $n = 1$.

37. $x = 3$ **39.** Sample answer: When dominoes are set up, after the first domino falls, the rest will fall as well. With induction, once it is proved that the statement is true for $n = 1$ (the first domino), $n = k$ (the second domino), and $n = k + 1$ (the next domino), it will be true for any integer value (any domino). **41.** B

43. 96 **45.** $160x^3y^3$ **47.** $-84x^6y^3$ **49.** (6, −8), (12, −16) **51.** 56 **53.** 665,280 **55.** 70 **57.** 132 **59.** 28 **61.** 24

Pages 732–736 Chapter 11 Study Guide and Review

1. true **3.** true **5.** true **7.** false, arithmetic sequence **9.** false, arithmetic means **11.** 48 **13.** −22 **15.** −7, −2, 3 **17.** 8, 4, 0, −4 **19.** \$480 **21.** 1040 **23.** −245 **25.** 629 **27.** −99 **29.** 99 **31.** $\frac{2187}{8}$ **33.** ±24, 72, ±216 **35.** \$1823.26 **37.** 12,285 **39.** 363 **41.** $-\frac{6305}{45{,}927}$ **43.** 32 **45.** 6 **47.** −3, 1, 5, 9, 13 **49.** 1, 6, 11, 16, 21 **51.** 7, 15, 31 **53.** 11, 65, 389 **55.** $a^3 + 3a^2b + 3ab^2 + b^3$

57. $-32z^5 + 240z^4 - 720z^3 + 1080z^2 - 810z + 243$

59. $x^5 - \frac{5}{4}x^4 + \frac{5}{8}x^3 - \frac{5}{32}x^2 + \frac{5}{256}x - \frac{1}{1024}$

61. $193{,}536x^2y^5$

63. Step 1: When $n = 1$, the left side of the equation is equal to 2. The right side of the equation is also equal to 2. So the equation is true for $n = 1$.

Step 2: Assume that $2 + 6 + 12 + \cdots + k(k + 1) = \frac{k(k+1)(k+2)}{3}$ for some positive integer k.

Step 3: $1*2 + 2*3 + \cdots + k(k + 1) + (k + 1)(k + 2) =$

$$\frac{k(k+1)(k+2)}{3} + (k+1)(k+2)$$
$$= \frac{k(k+1)(k+2)}{3} + \frac{3(k+1)(k+2)}{3}$$
$$= \frac{(k+1)[k(k+2) + 3(k+2)]}{3}$$
$$= \frac{(k+1)(k+2)(k+3)}{3}$$

The last expression is the right side of the equation to be proved, where $n = k + 1$. Thus, the equation is true for $n = k + 1$.

Therefore, $2 + 6 + 12 + \cdots + k(k + 1) = \frac{k(k+1)(k+2)}{3}$ for all positive integers n.

65. Step 1: When $n = 1$, $5^1 - 1 = 5$ or 4. Since 4 divided by 4 is 1, the statement is true for $n = 1$.

Step 2: Assume that $5^k - 1$ is divisible by 4 for some positive integer k. This means that $5^k - 1 = 4r$ for some whole number r.

Step 3: $5^k - 1 = 4r$

$$5^k = 4r + 1$$
$$5^{k+1} = 20r + 5$$
$$5^{k+1} - 1 = 20r + 5 - 1$$
$$5^{k+1} - 1 = 20r + 4$$
$$5^{k+1} - 1 = 4(5r + 1)$$

Since r is a whole number, $5r + 1$ is a whole number. Thus, $5^{k+1} - 1$ is divisible by 4, so the statement is true for $n = k + 1$.

Therefore, $5^n - 1$ is divisible by 4 for all positive integers n. **67.** $n = 2$ **69.** $n = 4$

Selected Answers and Solutions

Chapter 12 Probability and Statistics

Page 741 Chapter 12 Get Ready

1. independent **3.** independent **5.** permutation
7. $a^4 - 8a^3 + 24a^2 - 32a + 16$
9. $16b^4 - 32b^3 + 24b^2x^2 - 8bx^3 + x^4$ **11.** $243x^5 - 810x^4y + 1080x^3y^2 - 720x^2y^3 + 240xy^4 - 32y^5$
13. $\frac{a^5}{32} + \frac{5a^4}{8} + 5a^3 + 20a^2 + 40a + 32$

Pages 746–748 Lesson 12-1

1. No; the people surveyed would probably be more likely than others to love ice cream. **3.** b
5. Observational study; the people who volunteered at the shelter are the treatment group, and the other people are the control. **7.** Survey; it is best to call random numbers throughout the country in order to get an unbiased sample. **9.** Causation; the Level 2 emergency is a direct cause of the school closing.
11. No; the people surveyed would probably be more likely than others to like science. **13.** Yes; everyone in the population has an equal chance to be part of the sample.

15 Question **a** is an unbiased question that will elicit the desired response. Questions **b** and **c** are unbiased questions that will not get the desired response.

17 This is an observational study because no attempt is made to influence the results; the students who participated in marching band are the treatment group, and the other students are the control. This is unbiased because the students are not divided into groups. They do not know which group they are in.

19. Observational study; the people who run are the treatment group, and people who do not run are the control. **21.** Experiment; the test subjects are gardens with deer. The treatment group of gardens gets the treatment, while the control gets a placebo.
23. Correlation; while the two may be related, reading does not directly increase intelligence.
25. Correlation; while there may be a relationship between the two, one does not cause the other.
27. Sample answer: Yes; the majority of employees who leave are not happy about some facet of their employment. The majority of employees who are happy will not leave, and will thus not complete the questionnaire. Biased or not, the goal of these questionnaires is to determine why the employee left.
29. Sample answer: A telephone survey can introduce bias because unlisted phone numbers are not called and people without phones are not called.
31a. Sample answer: Survey 50 students at school on their opinions about changing to block scheduling. Sample: List all of the students at the school and randomly draw 50 names. Subject of survey: "Rate your opinion about block scheduling at school from 1 to 5, 1 being strongly against and 5 being strongly in favor."
31b. Sample answer: Observe 20 students, half of whom have study halls, and compare their grades at the end of the semester. Control: no study hall; treatment: study hall. **31c.** Sample answer: Select a sample of 20 random students with the common cold. Give half of them a pill and the other half a placebo, and compare the results after 3 weeks. Control: placebo; treatment: pill. **33.** C **35.** G
37. Step 1: $9^1 - 1 = 8$, which is divisible by 8. The statement is true for $n = 1$.
Step 2: Assume that $9^k - 1$ is divisible by 8 for some positive integer k. This means that $9^k - 1 = 8r$ for some whole number r.
Step 3:
$$9^k - 1 = 8r$$
$$9^k = 8r + 1$$
$$9^{k+1} = 72r + 9$$
$$9^{k+1} - 1 = 72r + 8$$
$$9^{k+1} - 1 = 8(9r + 1)$$
Since r is a whole number, $9r + 1$ is a whole number. Thus, $9^{k+1} - 1$ is divisible by 8, so the statement is true for $n = k + 1$.Therefore, $9^n - 1$ is divisible by 8 for all positive integers n. **39.** $\left(\frac{3}{2}, \frac{9}{2}\right)$, $(-1, 2)$
41. no solution **43.** $(\pm 8, 0)$ **45.** $3\sqrt{17}$ units
47. 25 units **49.** $\sqrt{70.25}$ units **51.** $-5c^5d^3$ **53.** an
55. $-y^3z^2$ **57.** $x^2 - 6x - 27 = 0$ **59.** $x^2 + x - 20 = 0$
61. 63

Pages 753–756 Lesson 12-2

1. Mean; there are no extreme values. **3.** Median; there is one value that is much less than the rest of the data. **5.** sample **7.** sample

9 **a.** Margin of sampling error

$= \pm\frac{1}{\sqrt{n}}$ Margin of Sampling Error Formula

$= \pm\frac{1}{\sqrt{5824}}$ $n = 5824$

$\approx \pm\ 0.0131$ Simplify.

b. $0.29 + 0.0131 = 0.3031 \approx 30.3\%$
$0.29 - 0.0131 = 0.2769 = 27.7\%$
The likely interval that contains the percentage of the population that will watch the Summer Olympics on television is between 27.7% and 30.3%.

11. Median; there is one value that is much greater than the rest of the data, 66. **13.** sample
15. population **17.** sample **19.** population

21 **a.** Margin of sampling error

$= \pm\frac{1}{\sqrt{n}}$ Margin of Sampling Error Formula

$= \pm\frac{1}{\sqrt{5669}}$ $n = 5669$

$\approx \pm\ 0.0133$ Simplify.

b. $0.31 + 0.0133 = 0.3233 \approx 32.3\%$
$0.31 - 0.0133 = 0.2967 \approx 29.7\%$
The likely interval that contains the percentage of the population that goes to the movies at least once a month is between 29.7% and 32.3%.

23a. sample **23b.** 2.7 **25a.** The mean is 21.182, and the median is 21.4. They are very close.
25b. population **25c.** 1.02 **25d.** The new mean is 21.232, and the median is 21.45. The mean and median are each slightly greater.

27 Since 13 is an outlier lowering the mean, the median best represents the data.
13, 25, 25, 26, (28), 34, 35, 37, 42
The median is 28.

29. 2066 **31.** Sample answer: The median will also increase by 10. For example, in a data set with a middle or median value of 18, if all of the data are increased by 10, the 18 increases to 28 and remains in the middle. Thus, the new median is 18 + 10 or 28. The mean will also increase by 10 because all of the data have increased by 10. For example, the data set of 2, 2, 2, 2, 2 has a mean of 2. If they are all increased by 10, then the new data set will be 12, 12, 12, 12, 12 and the mean will be 2 + 10 or 12. The standard deviation will be unaffected because even though the data all increase by 10, they are still the same distance from the mean, which also increased by 10.
33. Sample answer: While the average heights of both teams are the same, the West team will have more players that are much taller than 6 feet as well as more players that are shorter than 6 feet. For the East team, most of the players will be between 5 ft 11 and 6 ft 1, while for the West team, most of the players will be between 5 ft 8 and 6 ft 4. **35.** $\frac{3}{2}$ **37.** C
39. No; basketball players are more likely to be taller than the average high school student, so a sample of basketball players would not give representative heights for the whole school. **41a.** (39.2, ±4.4)
41b. No; the comet and Pluto may not be at either point of intersection at the same time.
41c. $\left(-\frac{5}{3}, -\frac{7}{3}\right)$, (1, 3) **41d.** (3, ±4), (−3, ±4)
43. combination; 28 **45.** permutation; 120

Pages 759–761 Lesson 12-3

1. $\frac{5}{18}$ **3.** $\frac{5}{17}$ **5a.** $\frac{32}{41}$ **5b.** $\frac{2}{5}$ **5c.** $\frac{3}{7}$ **7.** $\frac{12}{27}$ **9.** 0

11 a. There is a total of 156 + 312 + 242 + 108 or 818 people in the study. Find the probability that a student is a club member C given that the student is a male M.

$$P(C \mid M) = \frac{P(C \text{ and } M)}{P(M)} \quad \text{Conditional Probability Formula}$$
$$= \frac{156}{818} \div \frac{398}{818} \quad P(C \text{ and } M) = \frac{156}{818} \text{ and } P(M) = \frac{156 + 242}{818}$$
$$= \frac{156}{398} \text{ or } \frac{78}{199} \quad \text{Simplify.}$$

The probability that a student is a member of a club given that he is male is $\frac{78}{199}$ or about 39.2%.

b. There is a total of 156 + 312 + 242 + 108 or 818 people in the study. Find the probability that a student is not a club member N given that the student is a female F.

$$P(N \mid F) = \frac{P(N \text{ and } F)}{P(F)} \quad \text{Conditional Probability Formula}$$
$$= \frac{108}{818} \div \frac{420}{818} \quad P(N \text{ and } F) = \frac{156}{818} \text{ and } P(F) = \frac{312 + 108}{818}$$
$$= \frac{108}{420} \text{ or } \frac{9}{35} \quad \text{Simplify.}$$

The probability that a student is not a member of a club given that she is female is $\frac{9}{35}$ or about 25.7%.

c. There is a total of 156 + 312 + 242 + 108 or 818 people in the study. Find the probability that a student is a male M given that he is not a member of a club N.

$$P(M \mid N) = \frac{P(M \text{ and } N)}{P(N)} \quad \text{Conditional Probability Formula}$$
$$= \frac{242}{818} \div \frac{350}{818} \quad P(M \text{ and } N) = \frac{242}{818} \text{ and } P(N) = \frac{242 + 108}{818}$$
$$= \frac{242}{350} \text{ or } \frac{121}{175} \quad \text{Simplify.}$$

The probability that a student is male given that he is not a member of a club is $\frac{121}{175}$ or about 69.1%.

13. $\frac{2}{5}$ **15.** $\frac{1}{15}$

17 The probability that she does not get a hit is $P(N) = 0.35$. The probability that she does not get a hit in five consecutive at-bats is $P(N) \cdot P(N) \cdot P(N) \cdot P(N) \cdot P(N) = 0.35 \cdot 0.35 \cdot 0.35 \cdot 0.35 \cdot 0.35 \approx 0.0053$, or about 0.5%.

19a. 37.3% **19b.** 14.3% **21.** 2% **23.** $\frac{89}{109}$
25. Sample answer: The final branches represent conditional probability. For example, consider the probabilities of students being sophomores and licensed to drive. The 0.4 represents the probability that a student is licensed given he or she is a sophomore.

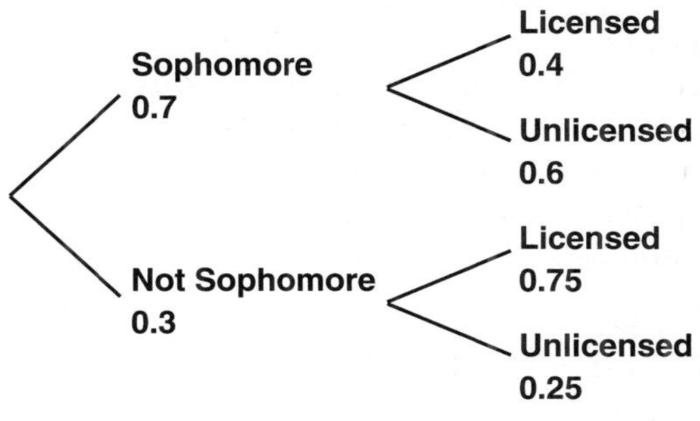

27. Sample answer:

Class	Male	Female
Freshman	6	9
Sophomore	8	5

the probability that a student is a female given she is a freshman: $\frac{3}{5}$

29. F **31.** −21 **33.** no **35.** $12,000 **37.** 8.9 **39.** 10
41. 15

Pages 765–769 Lesson 12-4

1. Determine the number of successes.
${}_8C_3$ 3 paintings chosen from 8 paintings
${}_{12}C_1$ 1 painting chosen from 20 − 8 or 12 paintings
${}_8C_3 \cdot {}_{12}C_1 = \frac{8!}{5!3!} \cdot \frac{12!}{11!1!}$ or 672 possible groups
Determine the total number of possibilities.
${}_{20}C_4 = \frac{20!}{16!4!}$ or 4845 ways to choose 4 paintings from 20 paintings
$P(\text{3 Paul Gauguin paintings, 1 other painting}) = \frac{s}{s+f}$ Probability of success
$= \frac{672}{4845}$ $s = 672$ and $s + f = 4845$
≈ 0.139 Use a calculator.
The probability that 3 of the 8 Paul Gauguin paintings are selected is about 13.9%.
3. about 10.3% **5a.** 2.14 **5b.** 2.14 **5c.** 7.14
7. $\frac{141}{230{,}300}$ or about 0.6% **9.** $\frac{1}{15{,}600}$

11 There are 12 red, 12 white, 12 blue, and 12 purple balloons.
Determine the number of successes.
${}_4P_2$ 2 colors chosen from 4 if order matters
${}_{12}C_3$ 3 red balloons chosen from a group of 12
${}_{12}C_4$ 4 blue balloons chosen from a group of 12
${}_4P_2 \cdot {}_{12}C_3 \cdot {}_{12}C_4 = 12 \cdot 220 \cdot 495$ or 1,306,800
Determine the total number of possibilities.
${}_{48}C_7 = 73{,}629{,}072$ ways to get 7 balloons from a package of 48
$P(\text{3 red, 4 blue}) = \frac{s}{s+f}$ Probability of success
$= \frac{1{,}306{,}800}{73{,}629{,}072}$ Substitute.
≈ 0.2 Use a calculator.
The probability is $\frac{1{,}306{,}800}{73{,}629{,}072}$ or about 2%.
13. 0.36 **15a.** 1.25 **15b.** 1.75 **15c.** 2.5

17 $E(x) = (0 \cdot 0.1) + (1 \cdot 0.1) + (2 \cdot 0.15) + (3 \cdot 0.15) + (4 \cdot 0.25) + (5 \cdot 0.1) + (6 \cdot 0.08) + (7 \cdot 0.05) + (8 \cdot 0.02)$
$= 0 + 0.1 + 0.3 + 0.45 + 1 + 0.5 + 0.48 + 0.35 + 0.16$
$= 3.34$ snow days
19. 4.7 **21a.** Pablo; 30% **21b.** Damon and Tora **21c.** Brett **21d.** 50% **21e.** 85%

23 ${}_{25}C_3$ 3 boys chosen from a group of 25
${}_{20}C_5$ 5 girls chosen from a group of 20
${}_{25}C_3 \cdot {}_{20}C_5 = 2300 \cdot 15{,}504 \cdot 495$ or 35,659,200
Determine the total number of possibilities.
${}_{45}C_8 = 215{,}553{,}195$ ways to choose 8 students from a group of 45
$P(\text{3 boys, 5 girls}) = \frac{s}{s+f}$ Probability of success
$= \frac{35{,}659{,}200}{215{,}553{,}195}$ Substitute.
≈ 0.165 Use a calculator.
The probability is about 16.5%.

25a. Sample answer:

Color	Probability	Sector Area	Total Area	Sector Area / Total Area
red	$\frac{1}{6}$	3.27 in^2	19.63 in^2	0.166
orange	$\frac{1}{6}$	3.27 in^2	19.63 in^2	0.166
yellow	$\frac{1}{6}$	3.27 in^2	19.63 in^2	0.166
green	$\frac{1}{4}$	4.91 in^2	19.63 in^2	0.25
blue	$\frac{1}{4}$	4.91 in^2	19.63 in^2	0.25

25b. Sample answer: The probability is equal to the ratio of the sector area to the total area. **25c.** red: $\frac{1}{9}$; yellow: $\frac{1}{3}$; blue: $\frac{15}{27}$ **27.** Sample answer: False; experimental probability is based on experiments, and theoretical probability is based on mathematical methods and assumptions.
29. 2.4 **31.** H **33.** 0.39 **35.** population **37.** sample
39. 0.5, 1.25, 3.125, 7.8125, 19.5313 **41.** 12, 4, $\frac{4}{3}$, $\frac{4}{9}$, $\frac{4}{27}$
43. 80, 100, 125, $\frac{625}{4}$, $\frac{3125}{16}$ **45.** 27 **47.** 3 **49.** 106.8
51. 93.3 **53.** 14.4

Pages 773–776 Lesson 12-5

1. normally distributed

3 **a.**

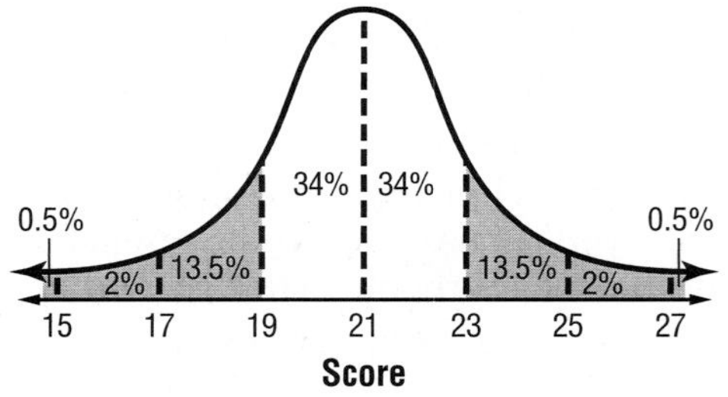

19 and 23 are 1σ away from the mean. Therefore, you would expect 34% + 34% or 68% to score between 19 and 23.
b. 23 and 25 are 1σ away from the mean and 2σ away from the mean, respectively. Therefore, you would expect 13.5% to score between 23 and 25.
c. 17 and 25 are 2σ away from the mean. Therefore, you would expect 13.5% + 34% + 34% + 13.5% or 95% to score between 17 and 25.
5. normally distributed

7

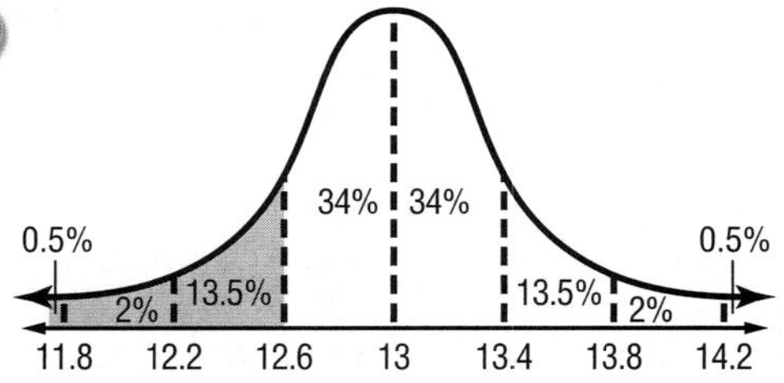

The probability that a randomly selected value in the distribution is less than $\mu - \sigma$, that is, 13 − 0.4 or 12.6, is the shaded area under the normal curve.
$P(x < 12.6) = 0.5\% + 2\% + 13.5\%$ or 16%

9. 97% **11a.** 16% **11b.** 855 **13a.** 50% **13b.** 50% **13c.** 95%

15 **a.** $\bar{x} = \dfrac{9 + 12 + 8 + 7 + 17 + 20}{6}$
≈ 12.17 in.

b. $\sigma = \sqrt{\dfrac{\sum_{k=1}^{n}(x_n - \mu)^2}{n}}$ Standard Deviation Formula

$$= \sqrt{\frac{(9 - 12.17)^2 + (12 - 12.17)^2 + \cdots + (17 - 12.17)^2 + (20 - 12.17)^2}{6}}$$

≈ 4.81

c. 7.36 and 16.98 are 1σ away from the mean. Therefore, you would expect the annual precipitation to be between 7.36 in. and 16.98 in. 34% + 34% or 68% of the time.

17. 1600 **19.** Sample answer: True; according to the Empirical Rule, 68% of the data lie within 1 standard deviation of the mean. **21.** Sample answer: A discrete probability distribution can be the uniform distribution of the roll of a die. In this type of distribution, there are only a finite number of possibilities. A continuous probability distribution can be the distribution of the lives of 400 batteries. In this distribution, there are an infinite number of possibilities. **23.** D **25.** 32.5 **27.** 68 **29.** about $y = -0.00046x^2 + 325$ **31.** inverse variation or rational **33.** 9.6 **35.** 30.7

Pages 780–782 Lesson 12-6

1. $88.42 \le \bar{x} \le 91.58$ **3.** $83.36 \le \bar{x} \le 84.64$
5. $73.98 \le \bar{x} \le 76.02$ **7.** accept **9.** reject
11. $25.45 \le \bar{x} \le 26.55$

13 $CI = \bar{x} \pm 2 \cdot \dfrac{s}{\sqrt{n}}$ Confidence Interval Formula
$= 58 \pm 2 \cdot \dfrac{7.1}{\sqrt{225}}$ $\bar{x} = 58$, $s = 7.1$, and $n = 225$
$\approx 58 \pm 0.95$ Use a calculator.

The 95% confidence interval is $57.05 \le \mu \le 58.95$.

15. $91.03 \le \bar{x} \le 92.97$ **17a.** $\$6.30 \le \bar{x} \le \$6.80$ **17b.** $\$6.35 \le \bar{x} \le \$6.75$ **17c.** Sample answer: A larger sample decreases the range of the confidence interval. **19.** accept **21.** reject **23.** accept

25 State the hypothesis: H_0: $\mu = 28$ and H_1: $\mu < 28$.
$\bar{x} = 27.03$
$s \approx 1.28$
$CI = \bar{x} \pm 2 \cdot \dfrac{s}{\sqrt{n}}$ Confidence Interval Formula
$= 27.03 \pm 2 \cdot \dfrac{1.28}{\sqrt{30}}$ $\bar{x} = 27.03$, $s \approx 1.28$, and $n = 30$
≈ 26.5626 or 27.4974 Use a calculator.

This means that 95% of the time, the experiment will produce a mean between 26.5626 and 27.4974. The 95% confidence interval does not include H_0, so Diana can reject the null hypothesis. She can assume that the car averages less than 28 miles per gallon in the city.

27. reject **29.** 45 **31.** Sample answer: Always; if the null hypothesis falls within the confidence interval, then it is accepted, not rejected. **33.** 4.33 **35.** J **37.** 14 **39.** $97.48 \le \bar{x} \le 100.07$ **41.** $47.2 \le \bar{x} \le 49.2$ **43.** 729 **45.** 1

47. parabola

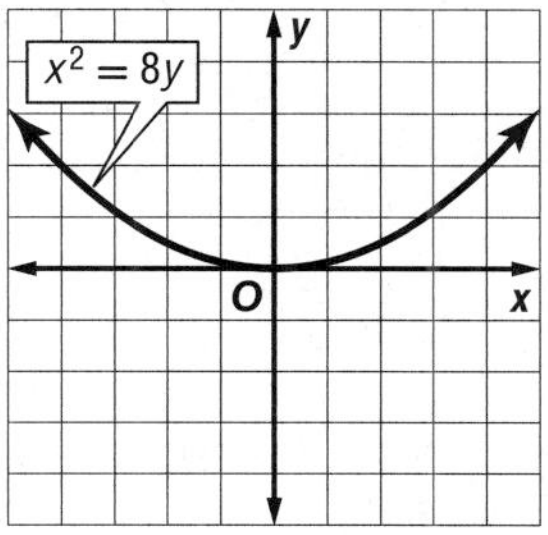

49. $y = 0.8x$
51. $y = -4$
53. $m^4 + 4m^3n + 6m^2n^2 + 4mn^3 + n^4$

Pages 788–791 Lesson 12-7

1.

K	-	-	-	-	-	A	-	-	-
-	-	-	A	K	-	K	-	-	-
-	-	K	A	-	-	A	-	-	-
-	A	-	-	-	-	-	-	-	K
-	K	-	-	-	-	-	-	-	K
-	-	K	-	-	A	-	-	-	K
-	-	-	-	-	-	-	-	-	-
-	-	-	-	-	-	-	-	-	-
-	-	-	-	-	-	-	-	-	-
-	-	-	-	-	A	-	-	-	-

An ace or a king was drawn at least once in 8 out of the 10 simulations, so the experimental probability of drawing at least one ace or king is 0.8. **3.** 13.3%

5a.

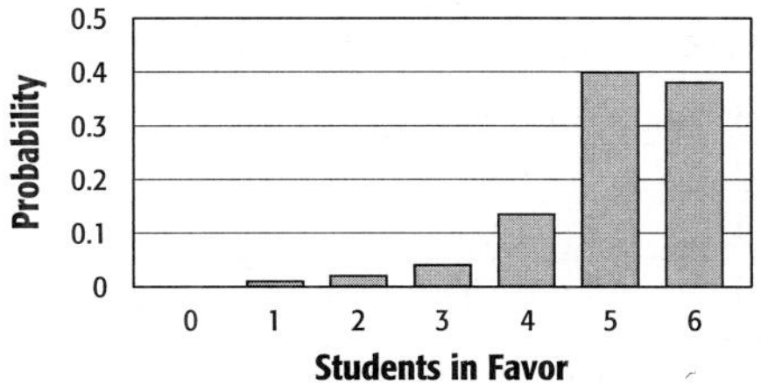

5b. 0.38

7.

5	11	5	3	8	6	7	11	6	8
8	3	6	5	6	5	2	8	7	8
6	7	7	9	11	8	8	4	7	6
7	8	9	5	8	7	8	5	5	8
10	6	9	6	7	6	5	8	4	9
7	4	4	8	10	7	9	7	4	8
5	3	11	12	7	11	7	7	3	7
7	12	6	2	6	8	9	9	6	9
8	6	10	6	8	7	11	6	8	9
6	9	7	6	3	6	6	11	7	6

A 7 was rolled at least 2 out of 10 times in 8 of 10 simulations, so the experimental probability is 0.8.

9.

-	R	R	R	-	-	-	-	R	-
-	-	-	-	-	-	-	R	R	-
-	R	R	-	R	-	-	R	-	-
R	-	-	R	R	R	-	R	-	-
-	R	-	-	-	-	-	-	-	R
-	R	-	R	R	-	-	R	R	R
R	-	R	-	R	R	-	R	-	-
-	R	-	-	R	-	R	-	R	-
R	-	-	R	-	-	R	R	-	R
-	-	-	R	-	-	-	-	-	-

A red marble was pulled six times in 1 out of the 10 simulations, so the experimental probability of pulling a red marble 6 out of 10 times is 0.1.

11 $s = 0.92$ and $f = 1 - 0.92$ or 0.08
Calculate the probability of 8, 9, or 10 random high school students owning their own car, and then subtract that sum from 1.

$$\begin{aligned} &P(< 8 \text{ students}) \\ &= 1 - P(\geq 8 \text{ students}) \\ &= 1 - [P(8) + P(9) + P(10)] \\ &= 1 - [C(10, 8)(0.92)^8(0.08)^2 + C(10, 9)(0.92)^9(0.08)^1 \\ &\quad + C(10, 10)(0.92)^{10}(0.08)^0] \\ &= 1 - 0.9599 \\ &= 0.0401 \text{ or } 4.01\% \end{aligned}$$

13. 0.792 or 79.2% **15.** about 77.6%

17a.

Number of Students	Probability
0	0.00000000001%
1	0.000000002%
2	0.000002%
3	0.000008%
4	0.0003%
5	0.006%
6	0.1%
7	1.0%
8	7.5%
9	31.5%
10	59.9%

17b. 98.9%

19 $s = 0.65$ and $f = 1 - 0.65$ or 0.35

$$\begin{aligned} (s + f)^5 &= 1s^5 + 5s^4f + 10s^3f^2 + 10s^2f^3 + 5sf^4 + 1f^5 \\ &= (0.65)^5 + 5(0.65)^4(0.35) + 10(0.65)^3(0.35)^2 + \\ &\quad 10(0.65)^2(0.35)^3 + 5(0.65)(0.35)^4 + (0.35)^5 \\ &\approx 11.6\% + 31.2\% + 33.6\% + 18.1\% + 4.9\% + 0.5\% \end{aligned}$$

5 customers, 4 customers, 3 customers, 2 customers, 1 customer, 0 customers

$$\begin{aligned} P(\geq 4 \text{ customers}) &= P(4 \text{ customers}) + P(5 \text{ customers}) \\ &\approx 31.2\% + 11.6\% \text{ or } 42.8\% \end{aligned}$$

The probability that at least 4 customers will donate more than the minimum is 42.8%.

21a.

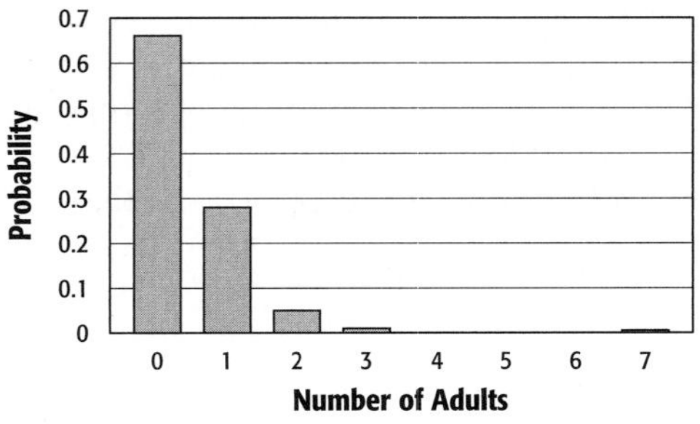

21b. 99.38% **23.** 16%

25 $s = 0.6$ and $f = 1 - 0.6$ or 0.4

$$\begin{aligned} P(\geq 12) &= P(12) + P(13) + P(14) + P(15) + P(16) + \\ &\quad P(17) + P(18) \\ &= C(18, 12)(0.6)^{12}(0.4)^6 + \\ &\quad C(18, 13)(0.6)^{13}(0.4)^5 + \\ &\quad \cdots + C(18, 18)(0.6)^{18}(0.4)^0 \\ &\approx 0.37 \text{ or } 37\% \end{aligned}$$

27. 10.8 **29.** 5 **31.** 5 **33.** 7 **35.** 60.3% **37.** 0.322 or 32.2% **39.** 0.99 or 99% **41.** 0.8891 or 88.91%
43. Sample answer: The poll will give you a percent of people supporting the addition. The percent of supporters represents the probability of success. You can use the formula for the expected number of successes in a binomial distribution with the total number of students in the school to predict the number that will support the science wing addition.
45. Sample answer: During May and June, lunches are held outside, weather permitting. Also during this time, there has historically been a 15% chance of rain. So, to determine the probability of not having rain for at least 24 of these 28 days, the binomial distribution would use $s = 0.85$, $f = 0.15$, and $n = 28$.
47. Sample answer: A binomial distribution shows the probabilities of the outcomes of a binomial experiment. **49.** C **51.** D **53.** reject **55.** accept
57. 11 **59.** 15 **61.** −34

63a. Venus: $\dfrac{x^2}{4522.5625} + \dfrac{y^2}{4522.765} = 1$;

Jupiter: $\dfrac{x^2}{234{,}014.06} + \dfrac{y^2}{233{,}454.74} = 1$

63b. Venus **65.** $2 = \ln 6x$ **67.** $e^x = 5.2$ **69.** $-1 = \ln x^2$
71. $e^3 = e^x$

Pages 792–796 Chapter 12 Study Guide and Review

1. probability distribution **3.** biased **5.** margin of sampling error **7.** Yes; everyone in the population of customers has an equal chance to be part of the sample. **9.** No; the sample is biased towards the restaurant. **11.** experiment **13.** population

15. ±1.7% **17.** 0.005 **19.** $\dfrac{14}{575}$ **21.** $\dfrac{7}{115}$

23a.

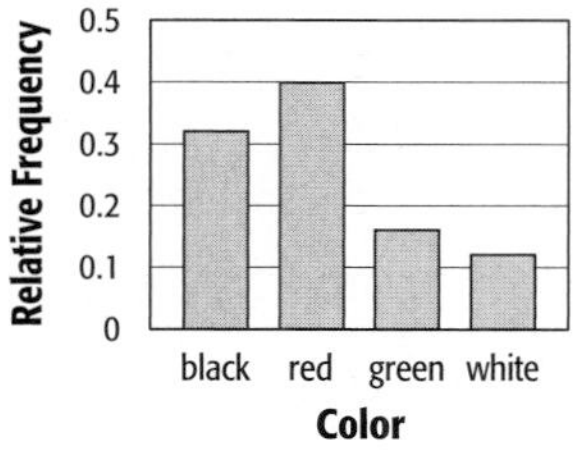

Color	black	red	green	white
Frequency	8	10	4	3

23b. red

23c. $\frac{12}{25}$ **25.** 97.5% **27.** 84% **29.** $22.8 \le \bar{x} \le 23.8$
31. $80.5 \le \bar{x} \le 82.3$ **33.** accept **35.** reject **37.** 21.5%
39. 40.4% **41.** 60% **43.** 97.5%

Chapter 13 Trigonometric Functions

Page 803 Chapter 13 Get Ready

1. 11.7 **3.** 20.5 **5.** $x = 9, y = 9\sqrt{2}$
7. $x = 12, y = 12\sqrt{3}$

Pages 811–814 Lesson 13-1

1. $\sin B = \frac{4}{5}$; $\cos B = \frac{3}{5}$; $\tan B = \frac{4}{3}$; $\csc B = \frac{5}{4}$; $\sec B = \frac{5}{3}$; $\cot B = \frac{3}{4}$ **3.** $\frac{\sqrt{33}}{7}$ **5.** 25.4 **7.** 8.3 **9.** 25.4
11. about 274.7 ft **13.** $\sin \theta = \frac{12}{13}$; $\cos \theta = \frac{5}{13}$; $\tan \theta = \frac{12}{5}$; $\csc \theta = \frac{13}{12}$; $\sec \theta = \frac{13}{5}$; $\cot \theta = \frac{5}{12}$
15. $\sin \theta = \frac{\sqrt{51}}{10}$; $\cos \theta = \frac{7}{10}$; $\tan \theta = \frac{\sqrt{51}}{7}$; $\csc \theta = \frac{10\sqrt{51}}{51}$; $\sec \theta = \frac{10}{7}$; $\cot \theta = \frac{7\sqrt{51}}{51}$

17 Since $\tan A = \frac{8}{15} = \frac{\text{opp}}{\text{adj}}$, label the opposite side 8 and the adjacent side 15.

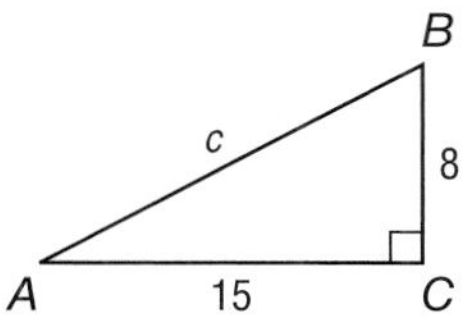

$a^2 + b^2 = c^2$ Pythagorean Theorem
$8^2 + 15^2 = c^2$ $a = 8$ and $b = 15$
$289 = c^2$ Simplify.
$17 = c$ Take the positive square root of each side.
$\cos A = \frac{\text{adj}}{\text{hyp}}$ Cosine function
$= \frac{15}{17}$ Replace adj with 15 and hyp with 17.

19. $\frac{3\sqrt{10}}{10}$ **21.** 12.7

23 $\tan \theta = \frac{\text{opp}}{\text{adj}}$ Tangent function
$\tan 30° = \frac{x}{18}$ Replace θ with 30°, opp with x, and adj with 18.
$\frac{\sqrt{3}}{3} = \frac{x}{18}$ $\tan 30° = \frac{\sqrt{3}}{3}$
$\frac{18\sqrt{3}}{3} = x$ Multiply each side by 18.
$10.4 \approx x$ Use a calculator.

25. 8.7 **27.** 132.5 ft **29.** 30 **31.** 36.9 **33.** 32.5 **35.** 25.3 ft higher **37.** $x = 21.9, y = 20.8$ **39.** $x = 19.3, y = 70.7$ **41.** 54.9 **43.** 20.5 **45.** 11.5

47

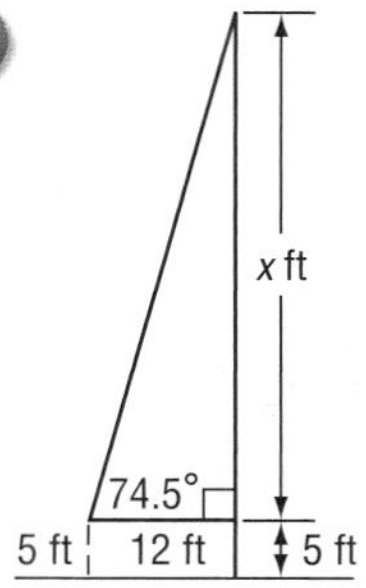

$\tan \theta = \frac{\text{opp}}{\text{adj}}$ Tangent function
$\tan 74.5° = \frac{x}{12}$ Replace θ with 74.5°, opp with x, and adj with 12.
$12 \cdot \tan 74.5° = x$ Multiply each side by 12.
$43 \approx x$ Use a calculator.

So, the height of the bird's nest is 43 + 5 or 48 feet.

49a. about 647.2 ft **49b.** about 239.4 ft
51. $m\angle A = 59°, a = 31.6, c = 36.9$ **53.** $m\angle A = 38.7°, m\angle B = 51.3°, b = 7.5, c = 9.6$ **55.** True; $\sin \theta = \frac{\text{opp}}{\text{hyp}}$ and the values of the opposite side and the hypotenuse of an acute triangle are positive, so the value of the sine function is positive. **57.** Sample answer: The slope describes the ratio of the vertical rise to the horizontal run of the roof. The vertical rise is opposite the angle that the roof makes with the horizontal. The horizontal run is the adjacent side. So, the tangent of the angle of elevation equals the ratio of the rise to the run, or the slope of the roof; $\theta = 33.7°$. **59.** 24 **61.** F **63.** reject **65.** reject **67.** $\frac{1}{2}$
69. 22,704 feet **71.** 35 dollars **73.** $216\frac{2}{3}$ centimeters

Pages 818–821 Lesson 13-2

1.

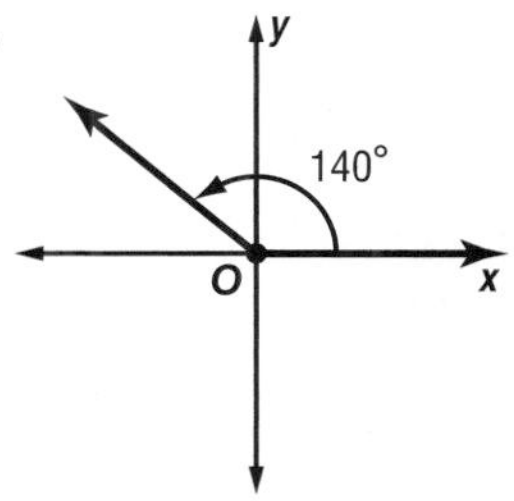

3.

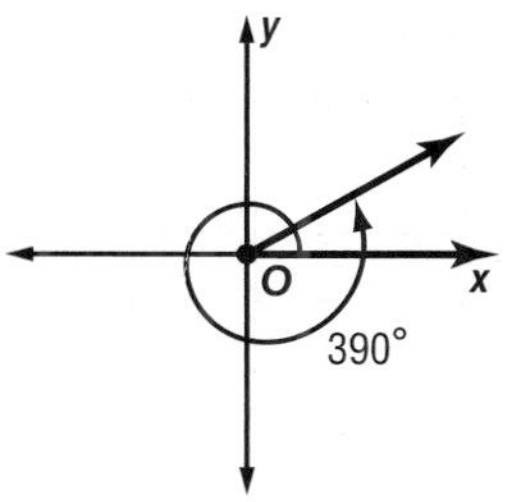

5 Sample answer:
positive angle: 175° + 360° = 535°
negative angle: 175° − 360° = −185°

7. 45° **9.** $-\frac{2\pi}{9}$

11.

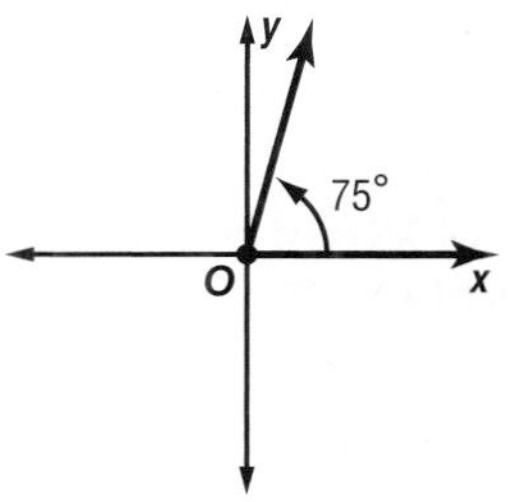

13.

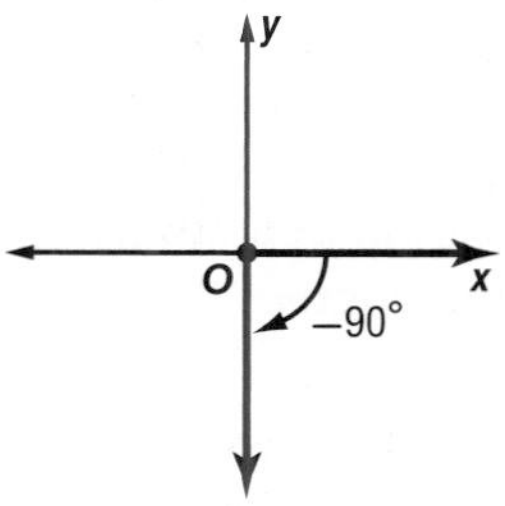

15.

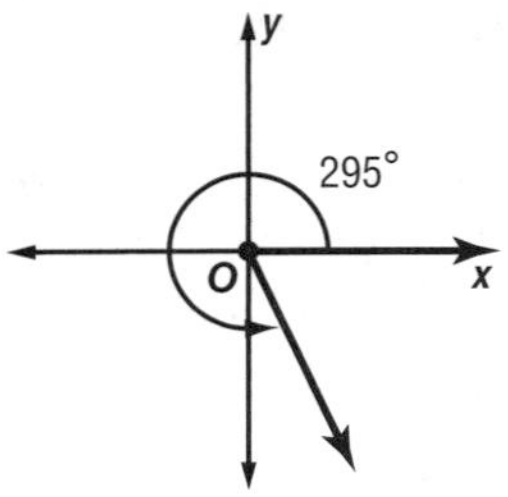

17.

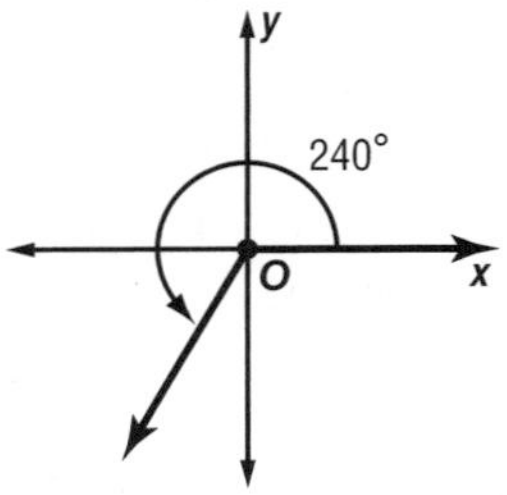

19. Sample answer: 410°, −310° **21.** Sample answer: 565°, −155° **23.** Sample answer: 280°, −440°

25 $330° = 330 \cdot \frac{\pi \text{ radians}}{180°}$

$= \frac{330\pi}{180}$ or $\frac{11\pi}{6}$ radians

27. −60° **29.** $\frac{19\pi}{18}$ **31.** about 12.6 ft **33.** 6.7 cm

35. 1 h 15 min **37.** Sample answer: 260°, −100°

39. Sample answer: $\frac{5\pi}{4}$, $-\frac{11\pi}{4}$

41 **a.**

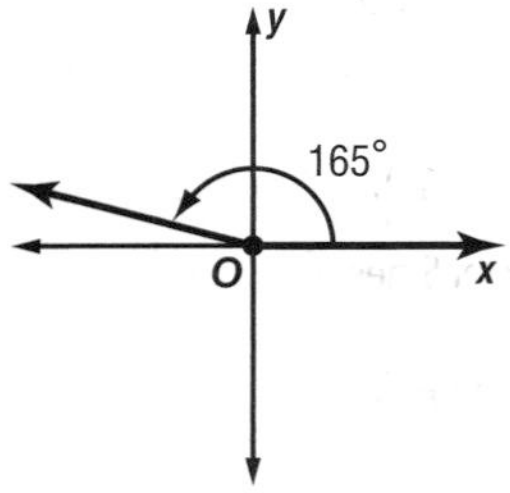

b. $165° = 165 \cdot \frac{\pi \text{ radians}}{180°}$

$= \frac{165\pi}{180}$ or $\frac{11\pi}{12}$ radians

c. $s = r\theta$ Formula for arc length

$= 6.5 \cdot \frac{11\pi}{12}$ $r = 6.5$ and $\theta = \frac{11\pi}{12}$

≈ 18.7 ft Use a calculator.

d. The arc length would double. Since $s = r\theta$, if r is doubled and θ remains unchanged, then the value of s is also doubled.

43. 472.5° **45.** $-\frac{10\pi}{9}$ **47a.** $\frac{\pi}{6}$ **47b.** 2.1 ft **49.** $x = 2$

51. Sample answer: 440° and −280°

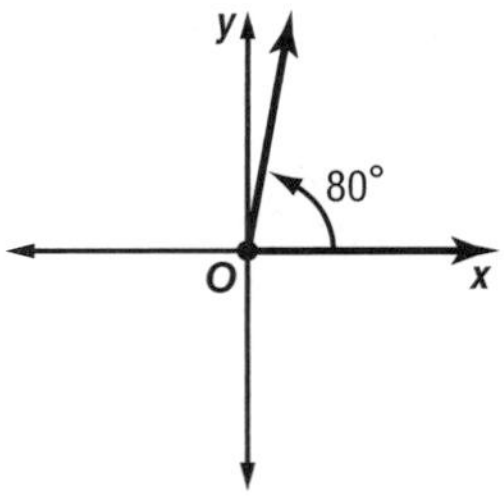

53. $\frac{\theta}{2\pi} = \frac{s}{2\pi r}$ Substitute.

$2\pi r\theta = 2\pi s$ Find the cross products.

$r\theta = s$ Divide each side by 2π.

One degree represents an angle measure that equals $\frac{1}{360}$ rotation around a circle. One radian represents the measure of an angle in standard position that intercepts an arc of length r. To change from degrees to radians, multiply the number of degrees by $\frac{\pi \text{ radians}}{180°}$. To change from radians to degrees, multiply the number of radians by $\frac{180°}{\pi \text{ radians}}$.

55. *A* **57.** *B* **59.** $\sin\theta = \frac{\sqrt{259}}{22}$, $\cos\theta = \frac{15}{22}$, $\tan\theta = \frac{\sqrt{259}}{15}$, $\csc\theta = \frac{22}{\sqrt{259}}$ or $\frac{22\sqrt{259}}{259}$, $\sec\theta = \frac{22}{15}$, $\cot\theta = \frac{15}{\sqrt{259}}$ or $\frac{15\sqrt{259}}{259}$ **61.** 0.37; 0.0145 **63a.** 50% **63b.** 815 **63c.** 25 **65.** $3\sqrt{41}$ **67.** $\sqrt{317}$

Pages 827–829 Lesson 13-3

1. $\sin\theta = \frac{2\sqrt{5}}{5}$, $\cos\theta = \frac{\sqrt{5}}{5}$, $\tan\theta = 2$, $\csc\theta = \frac{\sqrt{5}}{2}$, $\sec\theta = \sqrt{5}$, $\cot\theta = \frac{1}{2}$ **3.** $\sin\theta = -1$, $\cos\theta = 0$, $\tan\theta =$ undefined, $\csc\theta = -1$, $\sec\theta =$ undefined, $\cot\theta = 0$

5. 65°

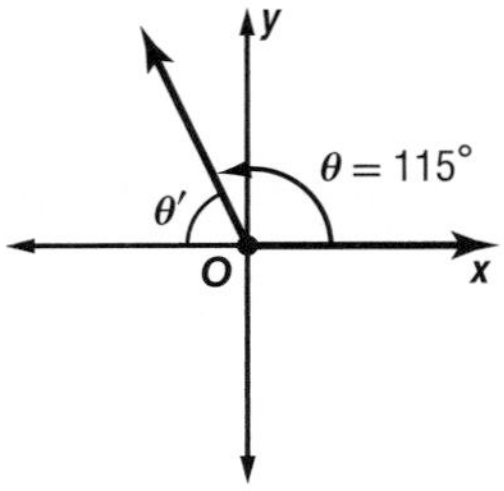

7. $\frac{\sqrt{2}}{2}$ **9.** −2

11a.

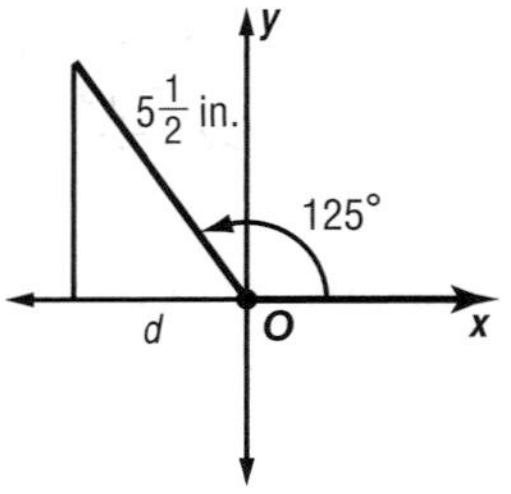

11b. 55°; $\cos 55° = \frac{d}{5\frac{1}{2}}$ **11c.** 3.2 in.

13

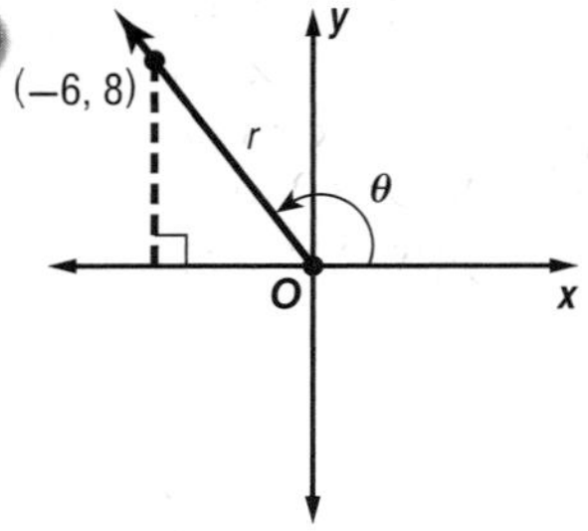

$r = \sqrt{x^2 + y^2}$

$= \sqrt{(-6)^2 + 8^2}$

$= \sqrt{100}$ or 10

Use $x = -6$, $y = 8$, and $r = 10$.

$\sin\theta = \frac{y}{r}$ $\cos\theta = \frac{x}{y}$

$= \frac{8}{10}$ or $\frac{4}{5}$ $= \frac{-6}{10}$ or $-\frac{3}{5}$

$\tan\theta = \frac{y}{x}$ $\csc\theta = \frac{r}{y}$

$= \frac{8}{-6}$ or $-\frac{4}{3}$ $= \frac{10}{8}$ or $\frac{5}{4}$

$\sec\theta = \frac{r}{x}$ $\cot\theta = \frac{x}{y}$

$= \frac{10}{-6}$ or $-\frac{5}{3}$ $= \frac{-6}{8}$ or $-\frac{3}{4}$

15. $\sin\theta = -1$, $\cos\theta = 0$, $\tan\theta =$ undefined, $\csc\theta = -1$, $\sec\theta =$ undefined, $\cot\theta = 0$

17. $\sin\theta = -\frac{\sqrt{10}}{10}$, $\cos\theta = -\frac{3\sqrt{10}}{10}$, $\tan\theta = \frac{1}{3}$, $\csc\theta = -\sqrt{10}$, $\sec\theta = -\frac{\sqrt{10}}{3}$, $\cot\theta = 3$

19. 75°

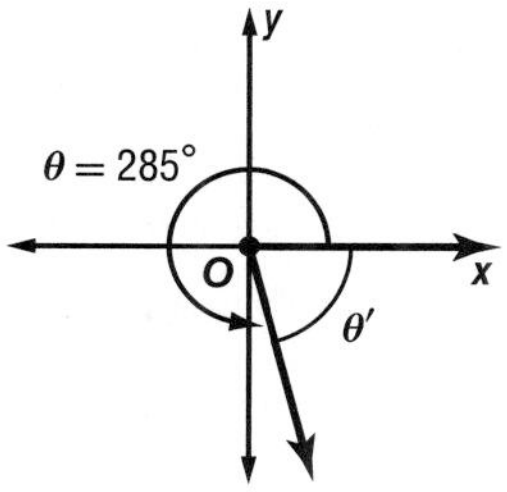

21. $\frac{\pi}{4}$

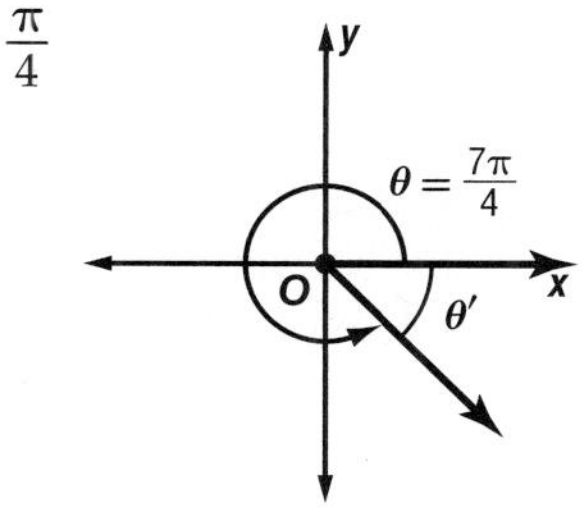

23. 40°

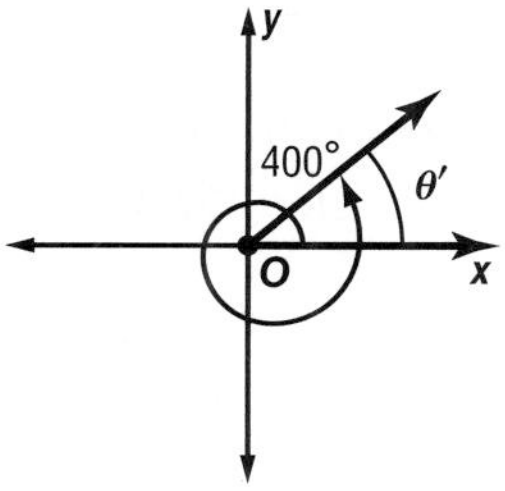

25. -1 **27.** $-\sqrt{2}$ **29.** $\frac{1}{2}$ **31.** $\frac{2\sqrt{3}}{3}$

33

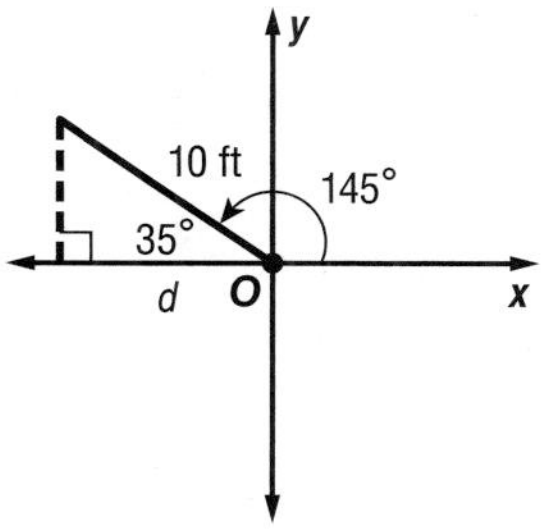

$\cos 35° = \frac{d}{10}$

$10 \cdot \cos 35° = d$

$8.2 \approx d$

The water reaches about 8.2 feet from the sprinkler.

35. about 10.1 m **37.** $\cos\theta = -\frac{3}{5}$, $\tan\theta = -\frac{4}{3}$, $\csc\theta = \frac{5}{4}$, $\sec\theta = -\frac{5}{3}$, $\cot\theta = -\frac{3}{4}$ **39.** $\sin\theta = -\frac{15}{17}$, $\tan\theta = \frac{15}{8}$, $\csc\theta = -\frac{17}{15}$, $\sec\theta = -\frac{17}{8}$, $\cot\theta = \frac{8}{15}$

41. 0 **43.** $-\frac{1}{2}$ **45.** $\frac{\sqrt{3}}{2}$ **47.** No; for $\sin\theta = \frac{\sqrt{2}}{2}$ and $\tan\theta = -1$, the reference angle is 45°. However, for $\sin\theta$ to be positive and $\tan\theta$ to be negative, the reference angle must be in the second quadrant. So, the value of θ must be 135° or an angle coterminal with 135°. **49.** Sample answer: $\cot 180 = \frac{1}{\tan 180} = \frac{1}{\frac{\sin 180}{\cos 180}} = \frac{\cos 180}{\sin 180}$. Since $\sin 180 = 0$, $\cot 180$ is undefined. **51.** Sample answer: First, sketch the angle and determine in which quadrant it is located. Then use the appropriate rule for finding its reference angle θ'. A reference angle is the acute angle formed by the terminal side of θ and the x-axis. Next, find the value of the trigonometric function for θ'. Finally, use the quadrant location to determine the sign of the trigonometric function value of θ. **53.** C **55.** C

57. 330° **59.** 40.1° **61.** 66.0° **63.** $10,737,418.23

65. $(x + 4)^2 + (y + 2)^2 = 73$ **67.** $\frac{x^2 + 7x - 35}{(x + 2)(x + 4)(x - 7)}$

69. $\frac{2(3x^2 + 2x - 12)}{3x(x + 4)(x - 6)}$ **71.** 2.5841 **73.** $\frac{1}{2}$ **75.** $\frac{1}{3125}$ **77.** 9

Pages 834–837 Lesson 13-4

1. 27.9 mm^2

3

$\text{Area} = \frac{1}{2}bc \sin A$	Area Formula
$= \frac{1}{2}(11)(6) \sin 40°$	Substitution
$\approx 21.2 \text{ cm}^2$	Simplify.

5. $E = 107°$, $d \approx 7.9$, $f \approx 7.0$ **7.** $F = 60°$, $f \approx 12.3$, $h \approx 9.1$ **9.** no solution **11.** one; $B = 90°$, $C = 60°$, $c \approx 5.2$ **13.** 10.6 km^2 **15.** 36.8 m^2 **17.** 5.9 ft^2 **19.** 65.2 m^2 **21.** $C = 30°$, $b \approx 11.1$, $c \approx 5.8$ **23.** $L = 74°$, $m \approx 4.9$, $n \approx 3.1$

25 $m\angle K = 180 - (53 + 20)$ or 107°

$\frac{\sin H}{h} = \frac{\sin J}{j}$	Law of Sines
$\frac{\sin 53°}{31} = \frac{\sin 20°}{j}$	Substitution
$j = \frac{31 \sin 20°}{\sin 53°}$	Solve for *j*.
$j \approx 13.3$	Use a calculator.
$\frac{\sin H}{h} = \frac{\sin K}{k}$	Law of Sines
$\frac{\sin 53°}{31} = \frac{\sin 107°}{k}$	Substitution
$k = \frac{31 \sin 107°}{\sin 53°}$	Solve for *k*.
$k \approx 37.1$	Use a calculator.

27. $B \approx 63°$, $b \approx 2.9$, $c \approx 3.0$ **29.** one; $B \approx 25°$, $C \approx 55°$, $c \approx 5.8$ **31.** one; $B \approx 32°$, $C \approx 110°$, $c \approx 32.1$ **33.** two; $B \approx 53°$, $C \approx 85°$, $c \approx 7.4$; $B \approx 127°$, $C \approx 11°$, $c \approx 1.4$ **35.** no solution **37.** about 28°

39

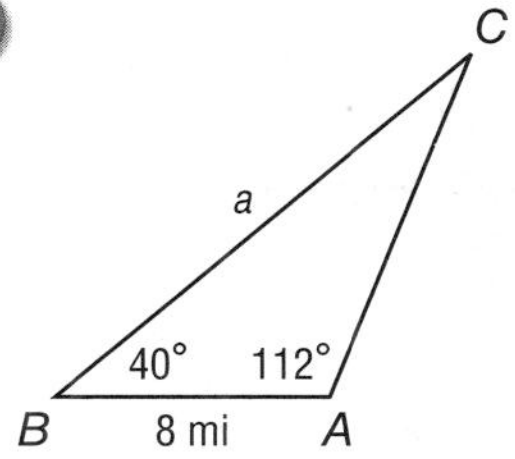

$m\angle C = 180 - (40 + 112)$ or 28°

$\frac{\sin A}{a} = \frac{\sin C}{c}$	Law of Sines
$\frac{\sin 112°}{a} = \frac{\sin 28°}{8}$	Substitution
$a = \frac{8 \sin 112°}{\sin 28°}$	Solve for *a*.
$a \approx 15.8$	Use a calculator.

Sirens *B* and *C* are about 15.8 miles apart.

41a. Sample answer:

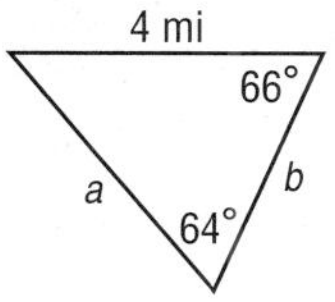

41b. Sample answer: $\frac{\sin 66°}{a} = \frac{\sin 64°}{4}$; $\frac{\sin 50°}{b} = \frac{\sin 64°}{4}$
41c. about 11.5 mi **43.** Cameron; R is acute and $r > t$, so there is one solution.
45. Sample answer:

$\text{Sin } A = \frac{\text{opposite}}{\text{hypotenuse}}$	Definition of sine
$\text{Sin } A = \frac{h}{c}$	h = opposite side, c = hypotenuse
$c \sin A = h$	Multiply both sides by c.
$\text{Area} = \frac{1}{2} \cdot \text{base} \cdot \text{height}$	Area of a triangle
$\text{Area} = \frac{1}{2}bh$	b = base, h = height
$\text{Area} = \frac{1}{2}bc \sin A$	Substitution

47. Sample answer: In the triangle, $B = 115°$. Using the Law of Sines, $\frac{\sin 50°}{a} = \frac{\sin 115°}{b}$. This equation cannot be solved because there are two unknown sides. To solve a triangle using the Law of Sines, two sides and an angle must be given or two angles and a side opposite one of the angles must be given.

49. 2 **51.** G **53.** $-\frac{1}{2}$ **55.** $\frac{\sqrt{3}}{3}$ **57.** 328°, −392° **59.** 96 cm **61.** No sum exists. **63.** $\frac{x^2}{8.7 \times 10^{15}} + \frac{y^2}{8.7 \times 10^{15}} = 1$ **65.** $(y + 2)^2$ **67.** 56.25 **69.** 26

Pages 841–844 Lesson 13-5

1. $A \approx 36°, C \approx 52°, b \approx 5.1$ **3.** $A \approx 18°, B \approx 29°, C \approx 133°$ **5.** Sines; $B \approx 40°, C \approx 33°, c \approx 6.8$

7 Since the lengths of two sides and the measure of the included angle are known, first use the Law of Cosines to find the missing side length.

$r^2 = s^2 + t^2 - 2st \cos R$	Law of Cosines
$r^2 = 16^2 + 9^2 - 2(16)(9) \cos 35°$	$s = 16, t = 9, R = 35°$
$r^2 \approx 101.1$	Use a calculator.
$r \approx 10.1$	Take the positive square root of each side.
$\frac{\sin R}{r} = \frac{\sin T}{t}$	Law of Sines
$\frac{\sin 35°}{10.1} = \frac{\sin T}{9}$	Substitution
$\frac{9 \sin 35°}{10.1} = \sin T$	Multiply each side by 9.
$31° \approx T$	Use the $\sin^{-1}$ function.

$m\angle S = 180 - (35° + 31°)$ or 114°

9. $A \approx 70°, B \approx 40°, c \approx 3.0$ **11.** $A \approx 31°, B \approx 108°, C \approx 41°$ **13.** $a \approx 6.9, B \approx 41°, C \approx 23°$ **15.** $F \approx 65°, G \approx 94°, H \approx 21°$ **17.** Sines; $C \approx 45°, A \approx 85°, a \approx 18.2$ **19.** Cosines; $s \approx 28.9, R \approx 42°, T \approx 32°$ **21.** Sines; $A \approx 17°, B \approx 79°, b \approx 6.9$

23 $d^2 = 338^2 + 520^2 - 2(338)(520) \cos 108°$
$d^2 \approx 493{,}269.7$
$d \approx 702.3$ m

25. 81°, 36°, 63° **27.** about 13,148 yd² **29a.** Sample answer:

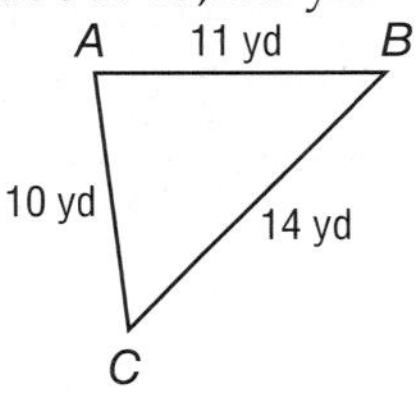

29b. Sample answer: Use the Law of Cosines to find the measure of $\angle A$. Then use the formula $\text{Area} = \frac{1}{2}bc \sin A$. **29c.** 54.6 yd²

31

$\frac{\sin A}{a} = \frac{\sin B}{b}$	Law of Sines
$\frac{\sin 104°}{12.4} = \frac{\sin B}{8.1}$	Substitution
$\frac{8.1 \sin 104°}{12.4} = \sin B$	Multiply each side by 8.1.
$39° \approx B$	Use the $\sin^{-1}$ function.

$m\angle C \approx 180 - (39° + 104°)$ or 37°

$\frac{\sin A}{a} = \frac{\sin C}{c}$	Law of Sines
$\frac{\sin 104°}{12.4} = \frac{\sin 37°}{c}$	Substitution
$c = \frac{12.4 \sin 37°}{\sin 104°}$	Solve for c.
$c \approx 7.7$	Use a calculator.

33. $F \approx 42°, G \approx 72°, H \approx 66°$ **35.** The longest side is 14.5 centimeters. Use the Law of Cosines to find the measure of the angle opposite the longest side; 102°. **37.** When two angles and a side are given or when two sides and an angle opposite one of the sides are given, you can use the Law of Sines to solve a triangle. When two sides and an included angle are given or when three sides are given, you can use the Law of Cosines to solve a triangle. **39.** G **41.** 4, $\frac{23}{15}$

43. 7.5 yd² **45.** $\sin \theta = \frac{5\sqrt{89}}{89}$, $\cos \theta = \frac{8\sqrt{89}}{89}$, $\tan \theta = \frac{5}{8}$, $\csc \theta = \frac{\sqrt{89}}{8}$, $\sec \theta = \frac{\sqrt{89}}{8}$, $\cot \theta = \frac{8}{5}$
47. $\sin \theta = -\frac{3\sqrt{13}}{13}$, $\cos \theta = \frac{2\sqrt{13}}{13}$, $\tan \theta = -1.5$, $\csc \theta = \frac{-\sqrt{13}}{3}$, $\sec \theta = \frac{\sqrt{13}}{2}$, $\cot \theta = -\frac{2}{3}$ **49.** (40, 30)
51. hyperbola

53.

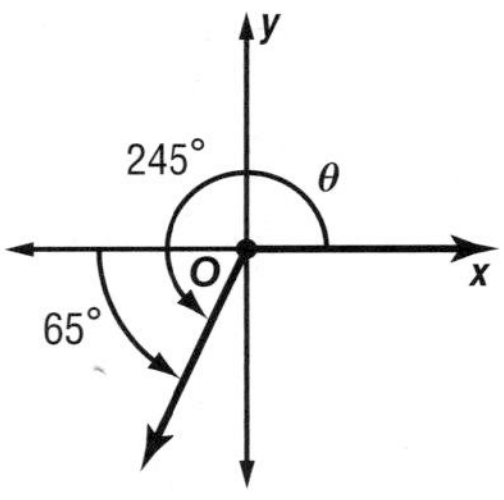

55.

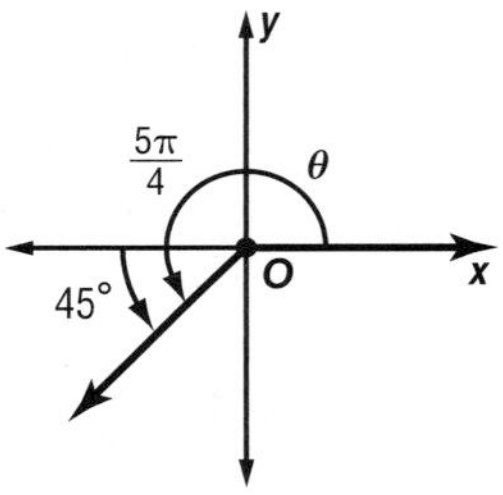

Pages 849–852 Lesson 13-6

1. $\cos \theta = \frac{15}{17}$, $\sin \theta = \frac{8}{17}$ **3.** 2 **5a.** 4 seconds
5b. Sample answer:

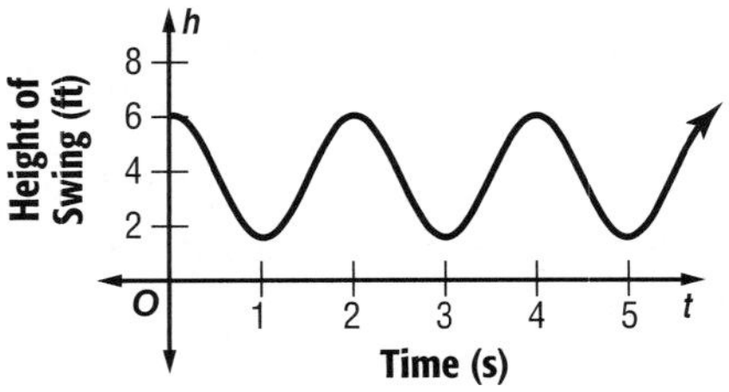

7. $-\frac{\sqrt{3}}{2}$ **9.** $\cos \theta = \frac{3}{5}$, $\sin \theta = -\frac{4}{5}$

11 $P\left(\frac{\sqrt{3}}{2}, \frac{1}{2}\right) = P(\cos\theta, \sin\theta)$

$\cos\theta = \frac{\sqrt{3}}{2}$ $\sin\theta = \frac{1}{2}$

13. 3 **15.** 12 **17.** 180°

19a. **Average High Temperatures**

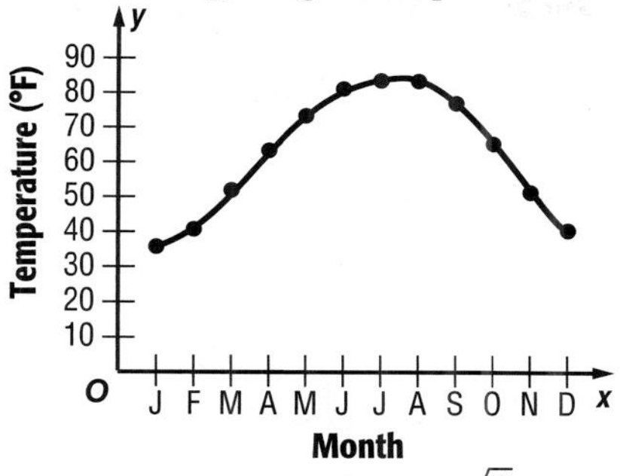

19b. 12 mo or 1 yr **21.** $\frac{1}{2}$ **23.** $\frac{\sqrt{2}}{2}$ **25.** $-\frac{\sqrt{3}}{2}$

27 a. The period is the time it takes to complete one rotation. So, the period is 60 seconds ÷ 2.5 or 24 seconds.

b. Since the siren is 1 mile from Ms. Miller's house and the beam of sound has a radius of 1 mile, then the minimum distance of the sound beam from the house is 0 miles and the maximum distance is 2 miles. Draw a sine curve with the pattern repeating every 24 seconds. Sample answer:

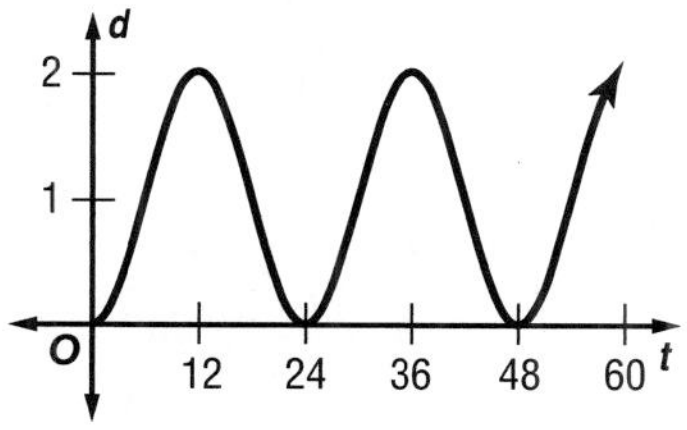

29a.

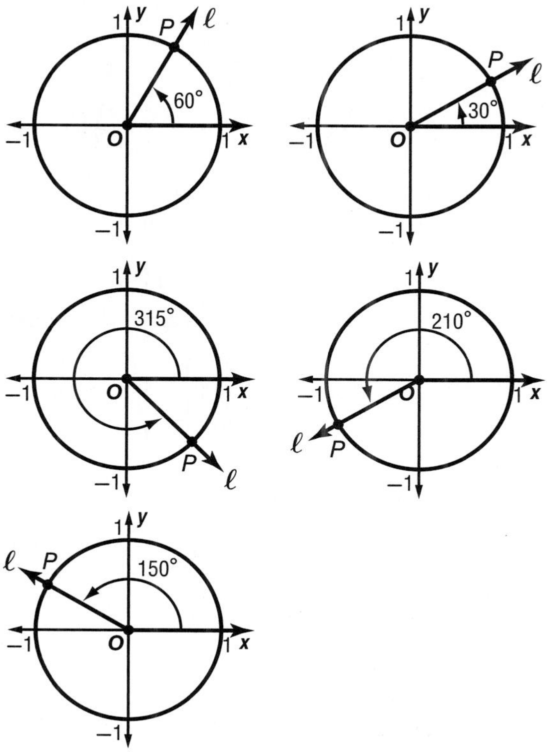

29b.

Angle	Slope
30	0.6
60	1.7
120	−1.7
150	−0.6
210	0.6
315	−1

29c. Sample answer: The slope corresponds to the tangent of the angle. For $\theta = 120°$, the x-coordinate of P is $-\frac{1}{2}$ and the y-coordinate is $\frac{\sqrt{3}}{2}$; slope = $\frac{\text{change in } y}{\text{change in } x}$. Since change in $x = -\frac{1}{2}$ and change in $y = \frac{\sqrt{3}}{2}$, slope $= \frac{\sqrt{3}}{2} \div \left(-\frac{1}{2}\right) = -\sqrt{3}$ or about -1.7.

31 $\cos 45° - \cos 30° = \frac{\sqrt{2}}{2} - \frac{\sqrt{3}}{2}$

$= \frac{\sqrt{2} - \sqrt{3}}{2}$

33. $-\frac{5\sqrt{3}}{2}$ **35.** 1 **37.** Benita; Francis incorrectly wrote $\cos\frac{-\pi}{3} = -\cos\frac{\pi}{3}$. **39.** Sometimes; the period of a sine curve could be $\frac{\pi}{2}$, which is not a multiple of π.

41. The period of a periodic function is the horizontal distance of the part of the graph that is nonrepeating. Each nonrepeating part of the graph is one cycle.

43. C **45.** A **47.** $A \approx 34°, C \approx 64°, c \approx 12.7$

49. $B \approx 33°, C \approx 29°, c \approx 9.9$ **51.** one solution; $B \approx 35°, C \approx 99°, c \approx 13.7$ **53.** 0.267 **55.** 7 **57.** $46,794.34

59. (5, 0), (−4, ±6) **61.** 108

Pages 857–859 Lesson 13-7

1. amplitude: 4; period: 360°

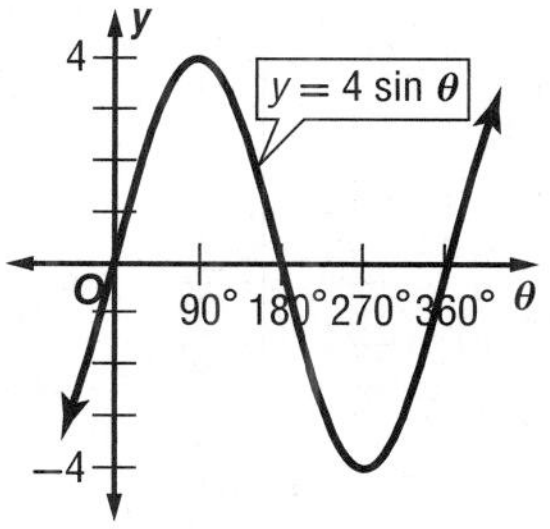

3. amplitude: 1; period: 180°

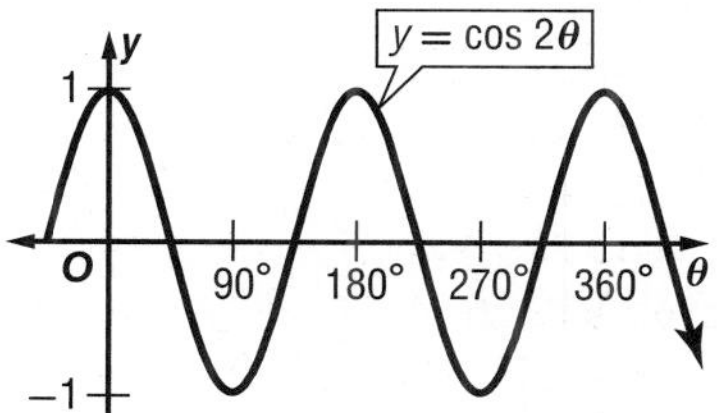

5a. $\frac{1}{14}$ or about 0.07 second

5b. $y = \sin 28\pi t$

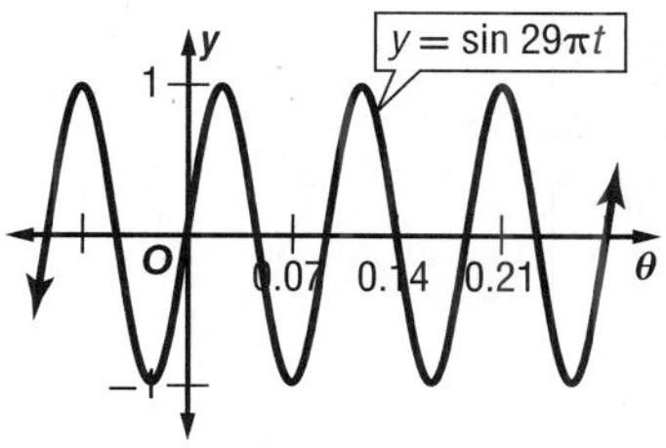

7. period: 360°

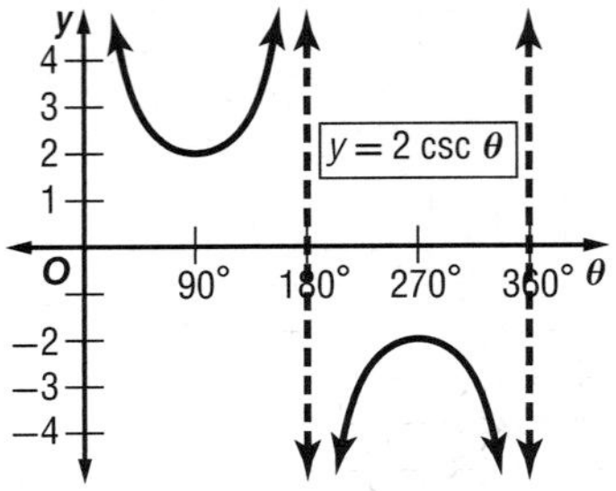

9. amplitude: 2; period: 360°

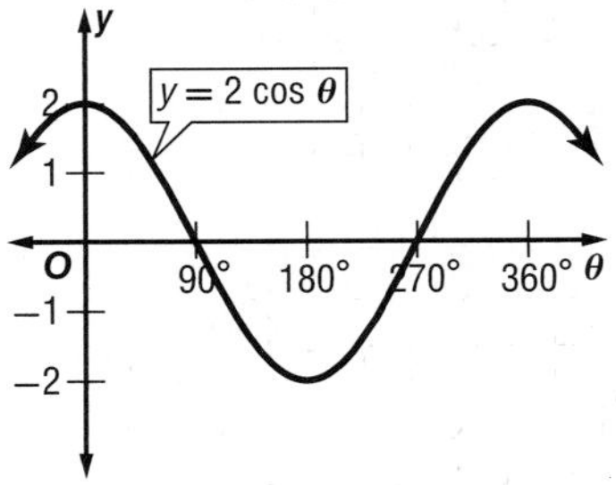

11. amplitude: 1; period: 180°

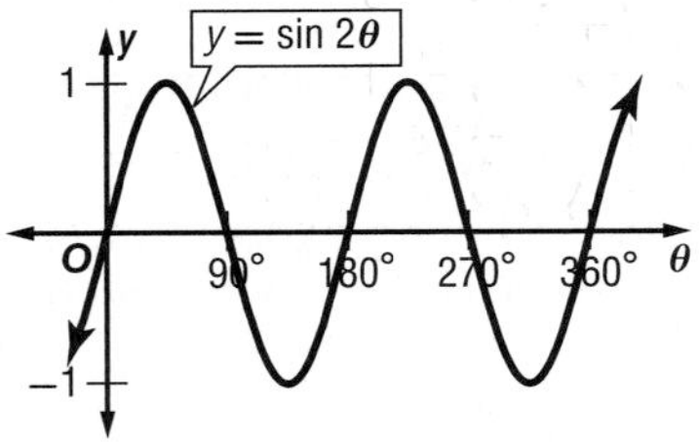

13. amplitude: 1; period: 720°

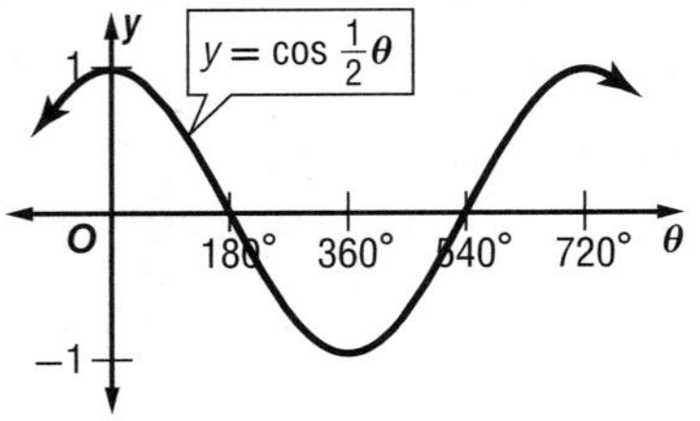

15. amplitude: $\frac{3}{4}$; period: 360°

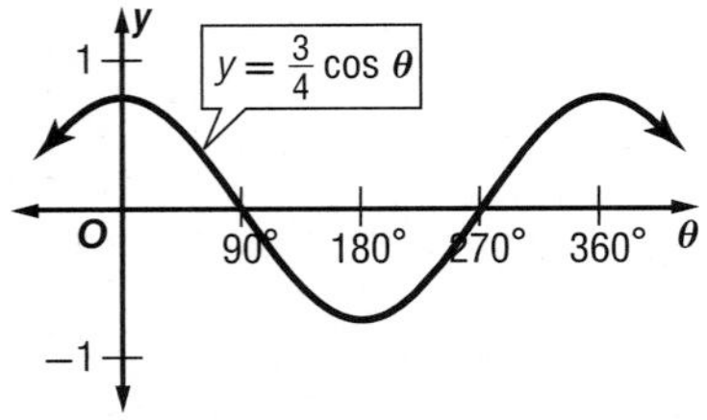

17 amplitude: $|a| = \left|\frac{1}{2}\right|$ or $\frac{1}{2}$

period: $\frac{360°}{|b|} = \frac{360°}{|2|}$
$= 180°$

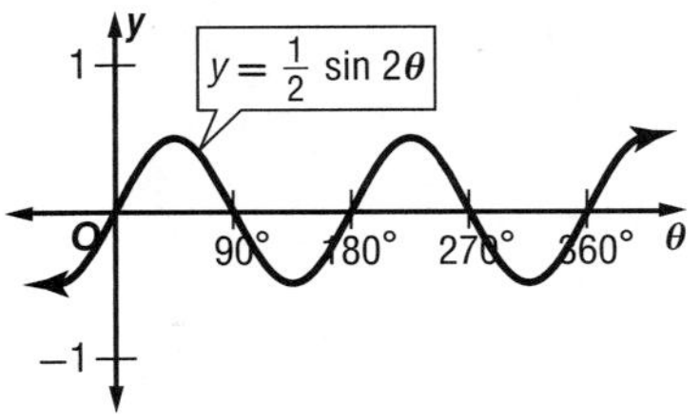

19. amplitude: 3; period: 180°

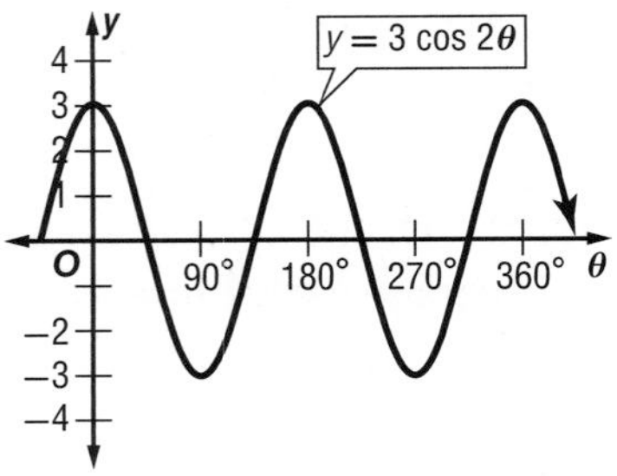

21a. $h = 4 \sin \frac{2}{3}\pi t$

21b.

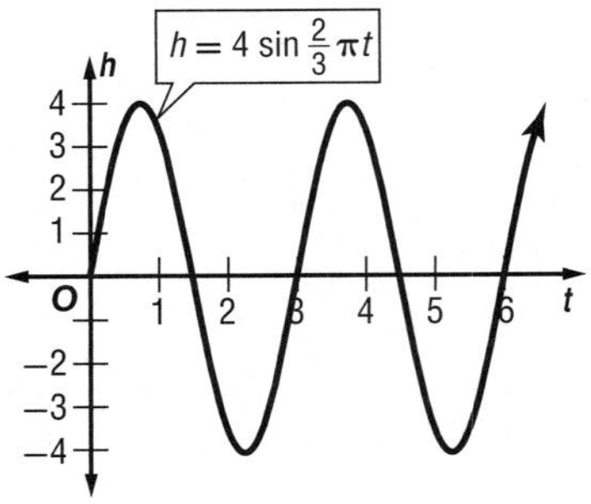

23. period: 360°

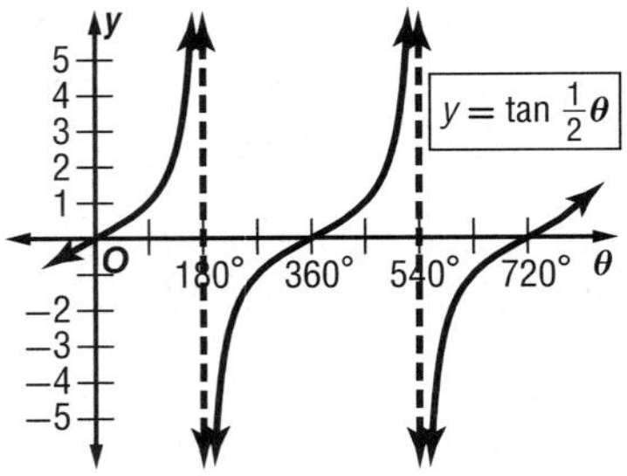

25. period: 180°

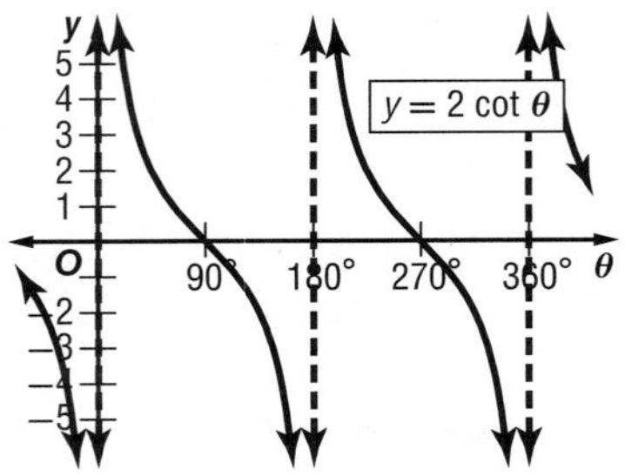

27. period: 180°

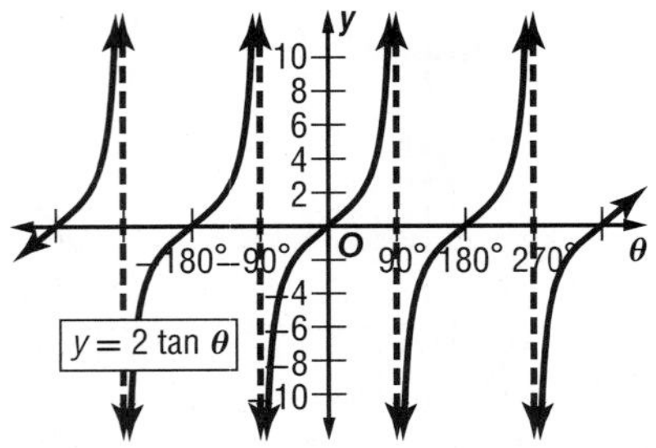

29 **a.** Since the frequency is 0.5, the period is $\frac{1}{0.5}$ or 2.

$\text{period} = \frac{2\pi}{|b|}$	Write the relationship between the period and b.
$2 = \frac{2\pi}{|b|}$	Substitution
$b = \pi$	Solve for b.
$y = a \sin b\theta$	General equation for the sine function
$h = 1 \sin \pi t$	Replace y with h, a with 1, b with π, and θ with t.
$h = \sin \pi t$	Simplify.

b.

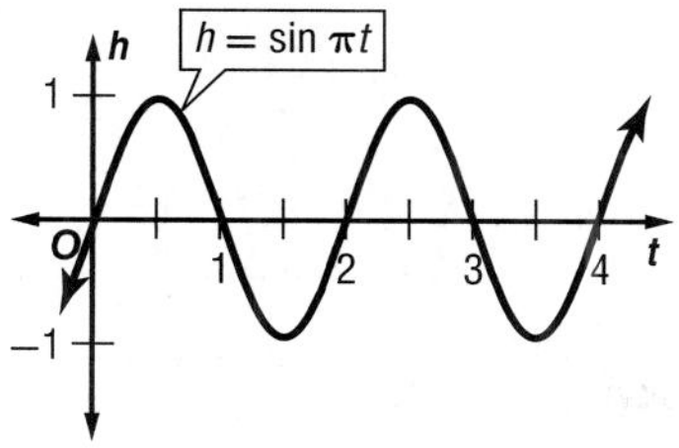

$h = \sin \pi t$
$= \sin (180 \cdot 20.5)$
$= 1$ m

31a. $y = \cos 260\pi t$ The amplitude remains the same. The period decreases because it is the reciprocal of the frequency.

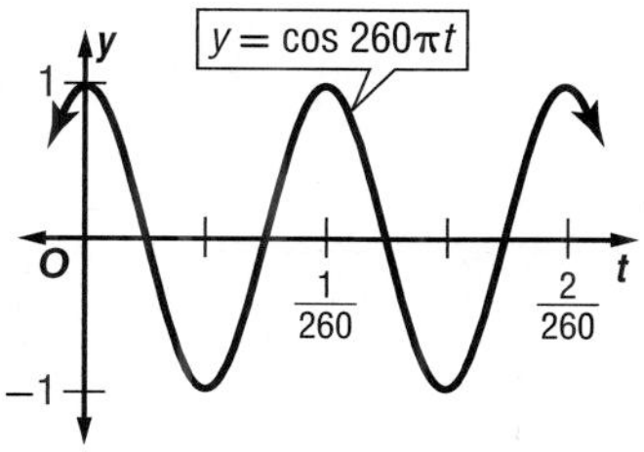

33. amplitude: $\frac{1}{2}$; period: 480°

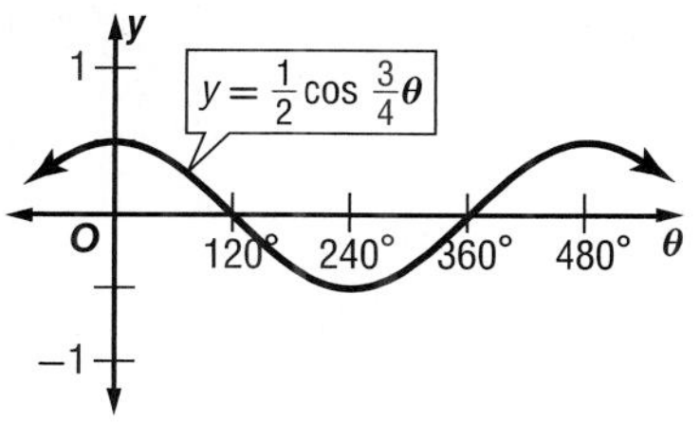

35. amplitude: does not exist; period: 450°

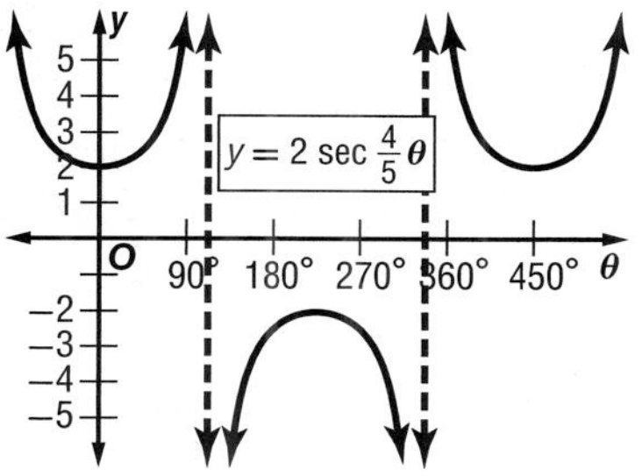

37. amplitude: does not exist; period: 30°

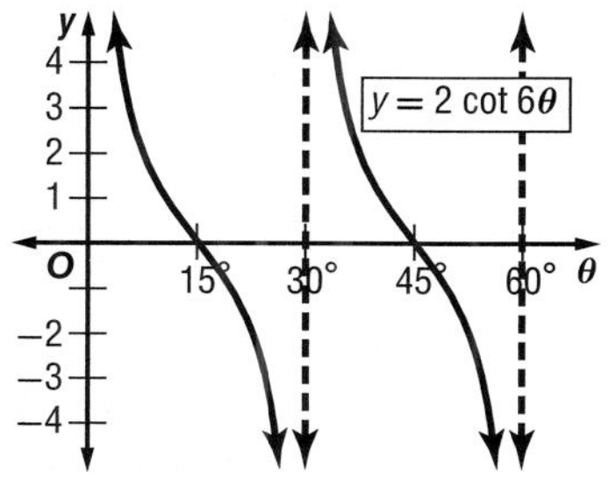

39. 180°; $y = 5 \sin 2\theta$ **41.** The domain of $y = a \cos \theta$ is the set of all real numbers. The domain of $y = a \sec \theta$ is the set of all real numbers except the values for which $\cos \theta = 0$. The range of $y = a \cos \theta$ is $-a \leq y \leq a$. The range of $y = a \sec \theta$ is $y \leq -a$ and $y \geq a$.

43. Sample answer: $y = 3 \sin 2\theta$

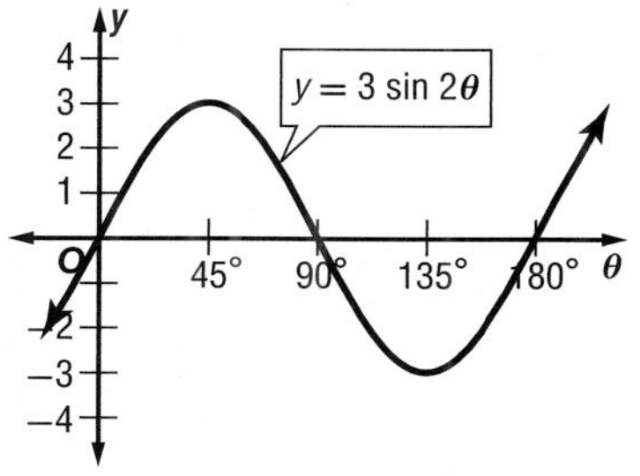

45. 700,013 **47.** G **49.** −1 **51.** $-3\sqrt{3}$ **53.** $R \approx 77°$, $S \approx 72°$, $q \approx 11.2$ **55.** $\frac{12}{29}$ **57.** $\frac{7}{27}$ **59.** $\frac{(x-6)^2}{12} + \frac{(y-3)^2}{4} = 1$

61.

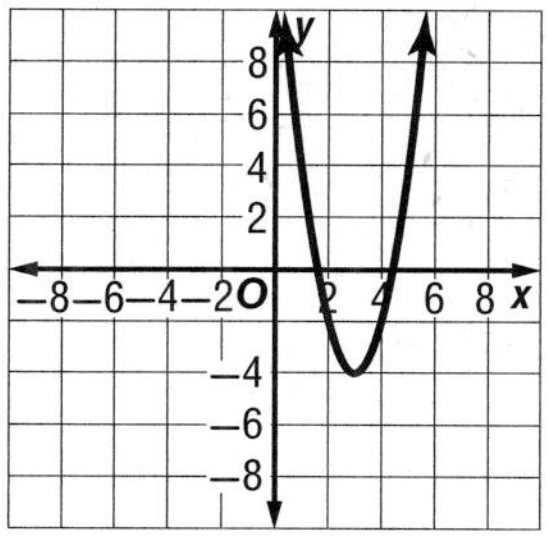

63.

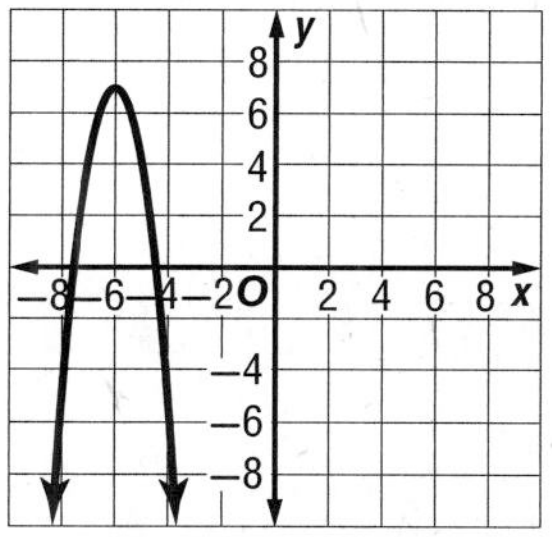

Pages 865–868 Lesson 13-8

1. 1; 360°; $h = 180°$

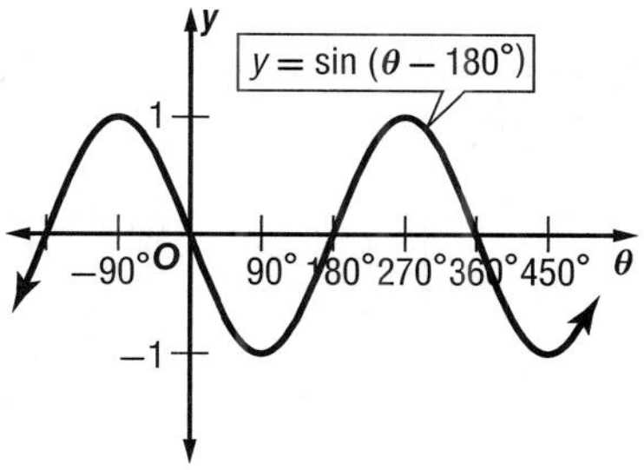

3. 1; 2π; $\frac{\pi}{2}$

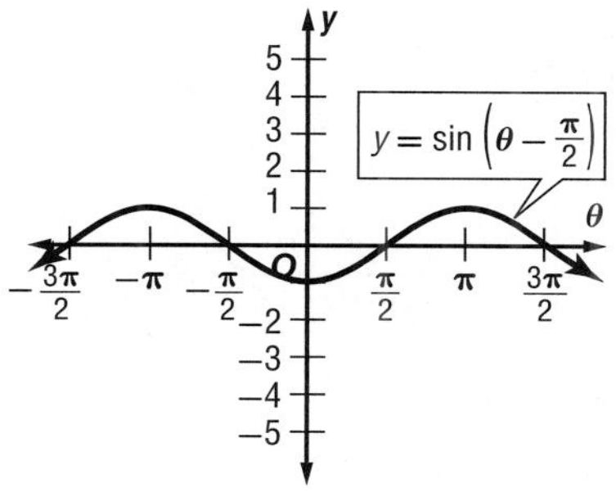

5. 1; 360°; $k = 4$; $y = 4$

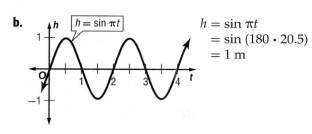

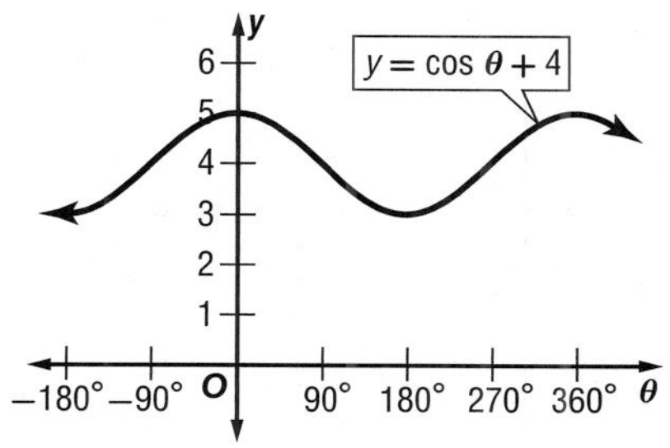

Selected Answers and Solutions

7. no amplitude; 180°; $k = 1$; $y = 1$

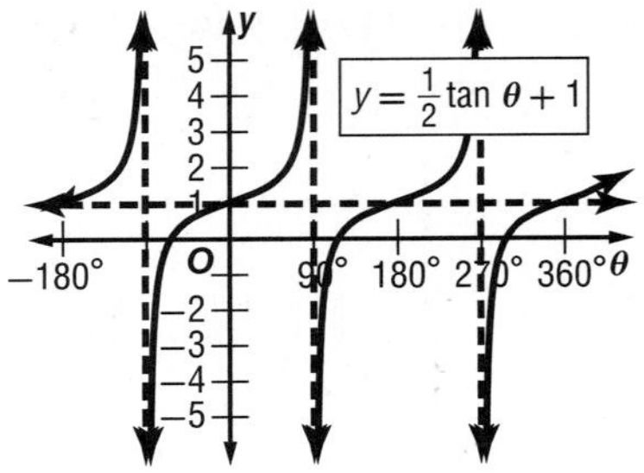

9. 2; 360°; $h = -45°$; $k = 1$

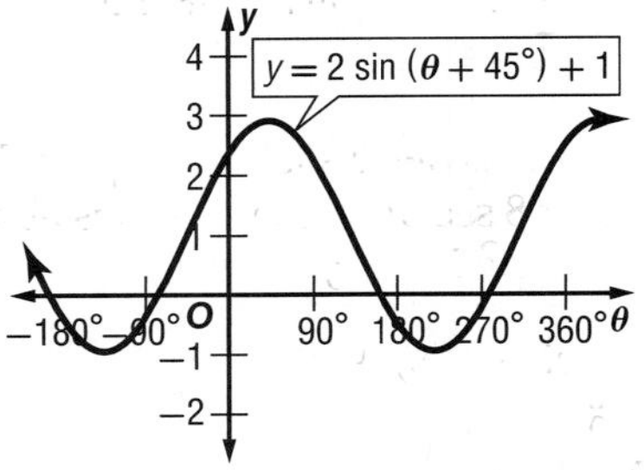

11. no amplitude; 90°; $h = -30°$; $k = 3$

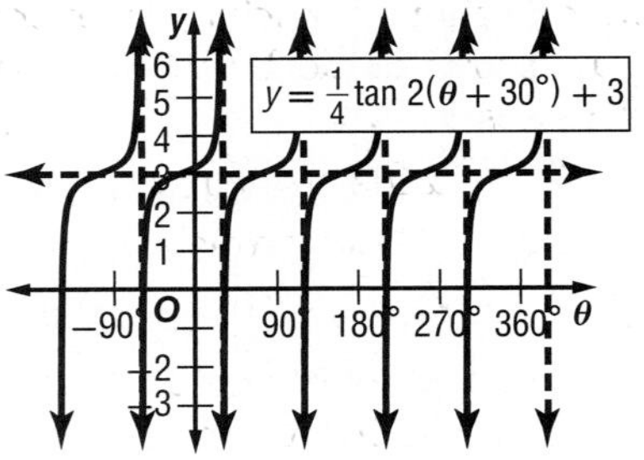

13. $P = 20\sin 3\pi t + 110$

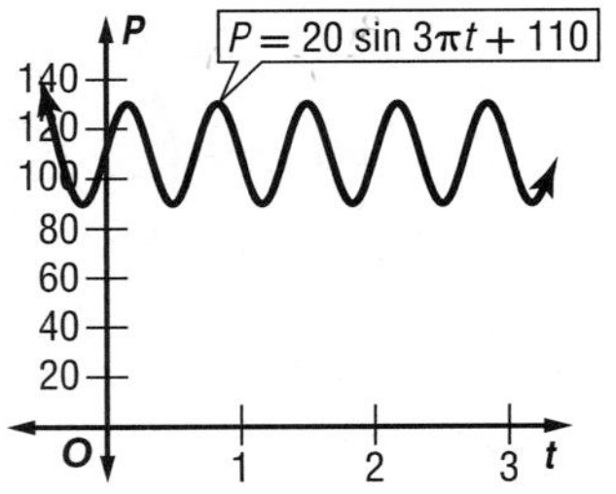

15. no amplitude; 180°; $h = 90°$

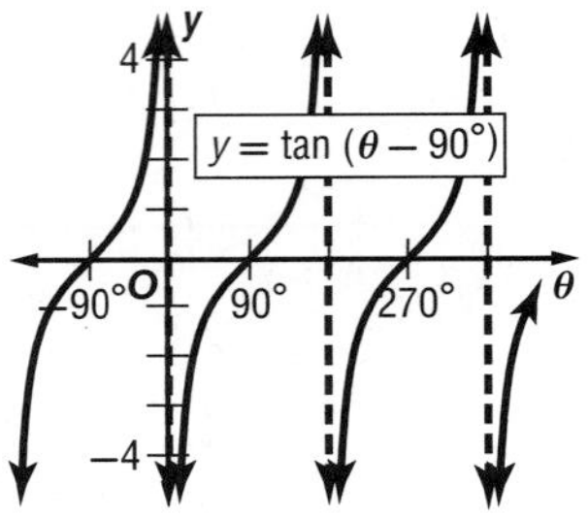

17. 2; 2π; $h = -\frac{\pi}{2}$

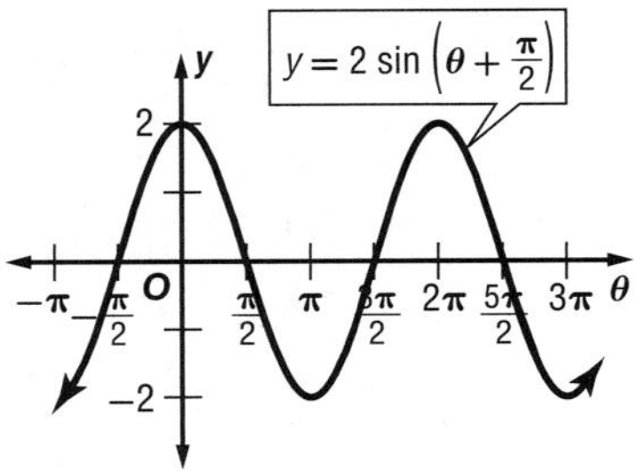

19. 3; 2π; $h = \frac{\pi}{3}$

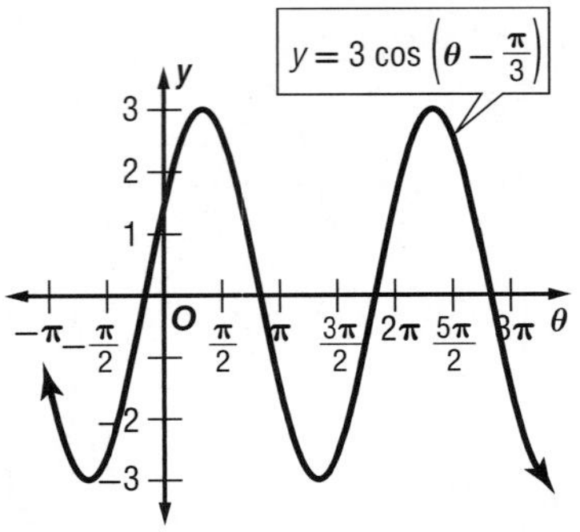

21. no amplitude; 180°; $k = -1$; $y = -1$

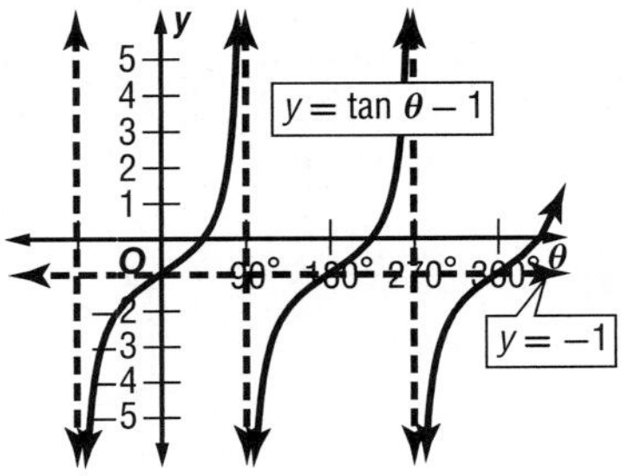

23 amplitude: $|a| = 2$

period: $\frac{360°}{|b|} = \frac{360°}{|1|}$ or 360°

vertical shift: $k = -5$

midline: $y = -5$

To graph $y = 2\cos\theta - 5$, first draw the midline. Then use it to graph $y = 2\cos\theta$ shifted 5 units down.

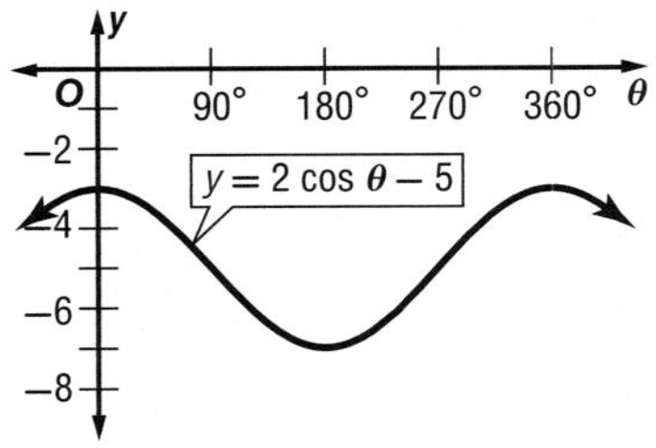

25. $\frac{1}{3}$; 360°; $k = 7$; $y = 7$

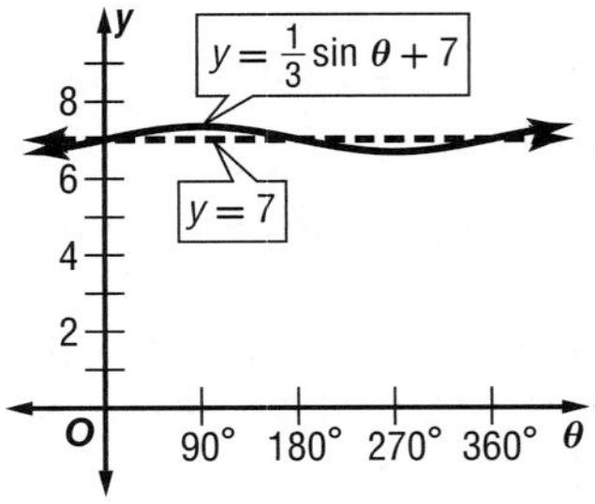

27. 1; 720°; $h = 90°$; $k = 2$

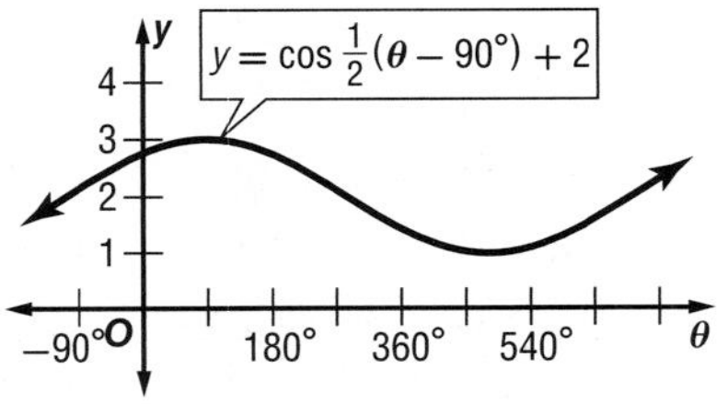

29. no amplitude; $\frac{\pi}{2}$; $h = -\frac{\pi}{4}$; $k = -5$

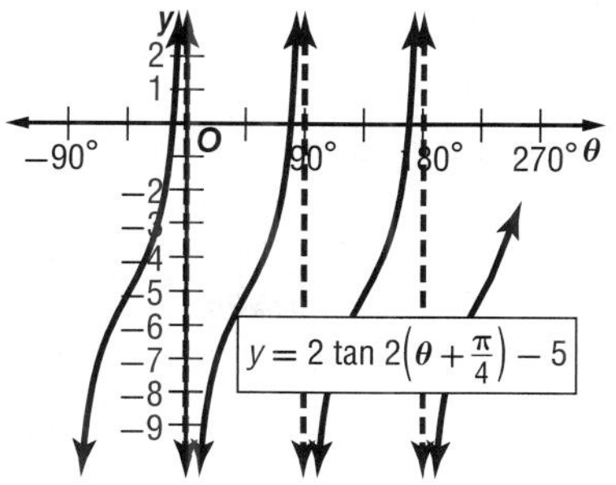

31. 1; 120°; $h = 45°$; $k = \frac{1}{2}$

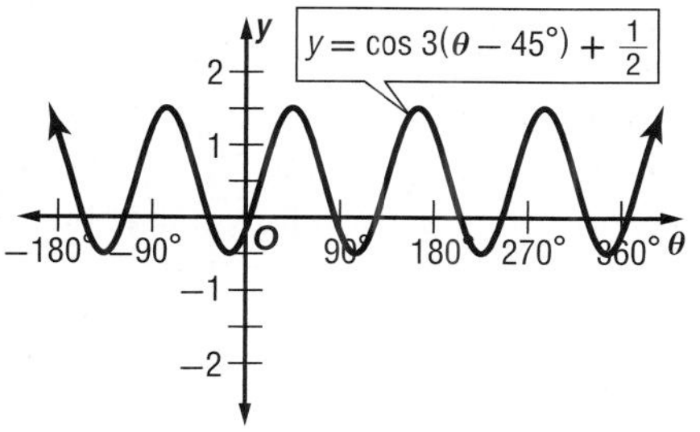

33. 3; 6π; $h = \frac{\pi}{2}$; $k = -2$

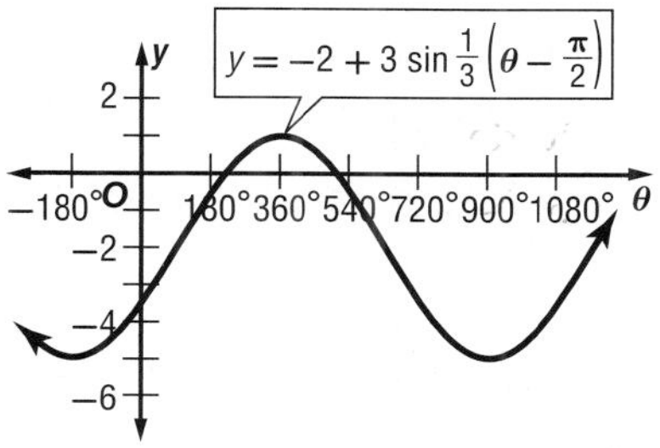

35.

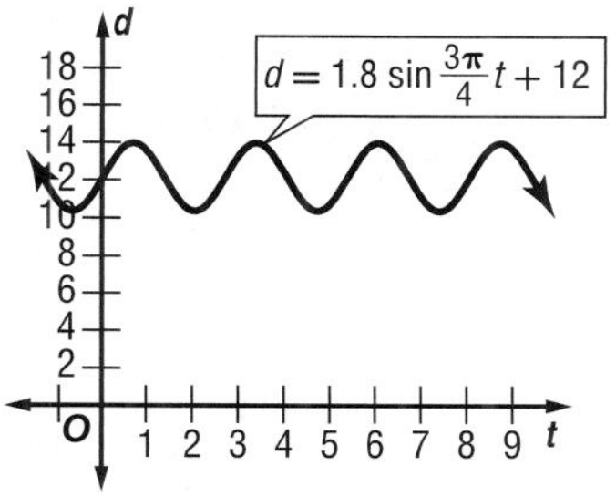

min: 10.2 ft; max: 13.8 ft **37.** $y = \sin(x - 4) + 3$ **39.** $y = \tan(x - \pi) + 2.5$

41 **a.** The midline lies halfway between the maximum and minimum values. So, $y = \frac{55 + 37}{2}$ or 46. Since the midline is $y = 46$, the vertical shift is $k = 46$.

The amplitude is the difference between the midline value and the maximum value. So, $|a| = |55 - 46|$ or 9.

Since the carousel rotates once every 21 seconds and a horse on the carousel goes up and down 3 times per rotation, it goes up and down every $21 \div 3$ or 7 seconds. So, the period is 7 seconds.

$\text{period} = \frac{2\pi}{|b|}$ Write the relationship between the period and b.

$7 = \frac{2\pi}{|b|}$ Substitution

$b = \frac{2\pi}{7}$ Solve for b.

$y = a\sin b\theta + k$ General equation for the sine function

$h = 9\sin\frac{2\pi}{7}t + 46$ Replace y with h, a with 9, b with $\frac{2\pi}{7}$, θ with t, and k with 46.

b.

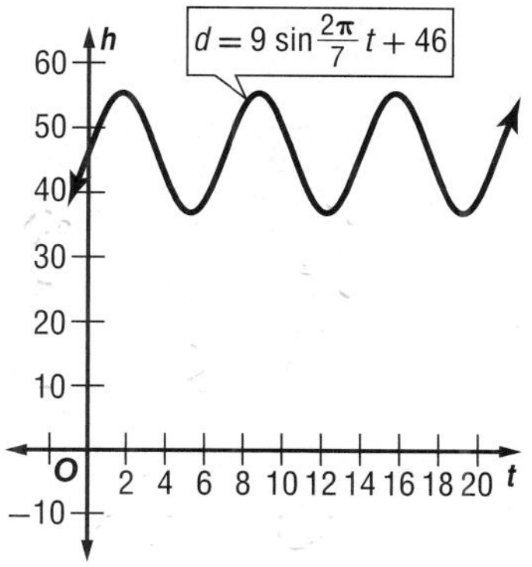

c. Sample answer: On the graph, when $t = 8$, $d \approx 53$. So after 8 seconds, the height is about 53 inches.

$h = 9\sin\frac{2\pi}{7}t + 46$

$= 9\sin\frac{2\pi}{7}(8) + 46$

≈ 53.0 in.

43. $\left(\frac{3\pi}{2}, 2\right)$ **45.** no maximum values **47.** The graphs are reflections of each other over the x-axis. **49.** The graphs are identical. **51.** 360°; Sample answer: $y = 4\cos\theta + 1$

53 The midline lies halfway between the maximum and minimum values. So $y = \frac{4 + 2}{2}$ or 3. Since the midline is $y = 3$, the vertical shift is $k = 3$.

The amplitude is the difference between the midline value and the maximum value. So $|a| = |4 - 3|$ or 1.

Since the cycle repeats every 180°, the period is 180°.

$\text{period} = \frac{360°}{|b|}$ Write the relationship between the period and b.

$180° = \frac{360°}{|b|}$ Substitution

$b = 2$ Solve for b.

The graph is the sine curve shifted 45° to the right. So the phase shift is $h = 45°$.

$y = a\sin b(\theta - h) + k$ General equation for the sine function

$y = 1\sin 2(\theta - 45°) + 3$ Replace a with 1, b with 2, h with 45°, and k with 3.

$y = \sin 2(\theta - 45°) + 3$ Simplify.

55. 180°; no phase shift; $k = 6$

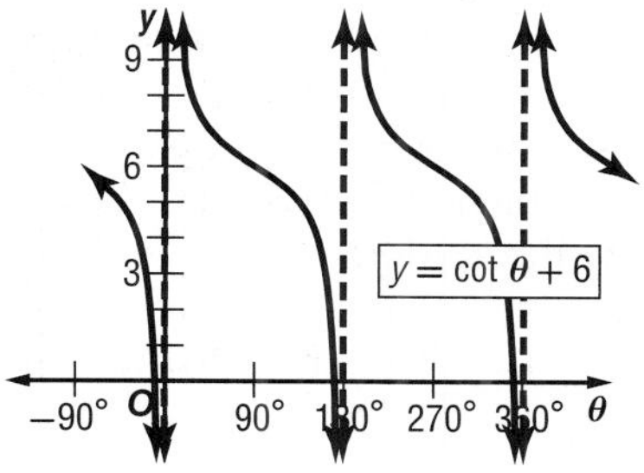

57. 120°; $h = 45°$; $k = 1$

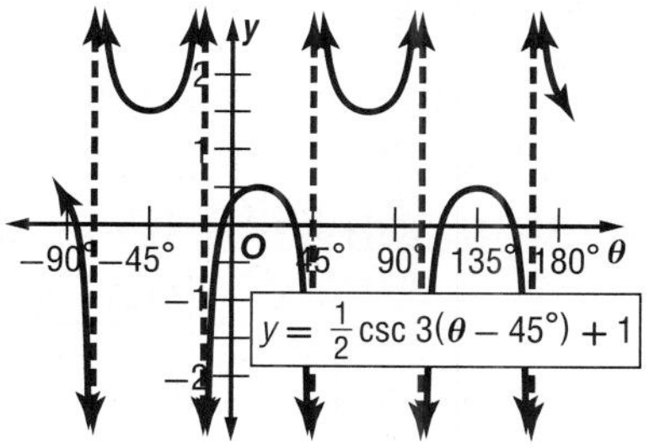

59. π; $h = -\frac{\pi}{2}$; $k = -3$

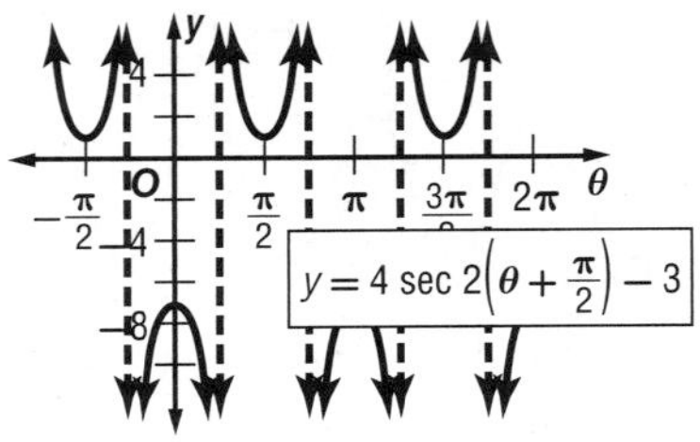

61. The graph of $y = 3 \sin 2\theta + 1$ has an amplitude of 3 rather than an amplitude of 1. It is shifted up 1 unit from the parent graph and is compressed so that it has a period of 180°.

63. Sample answer: $y = 2 \sin \theta - 3$

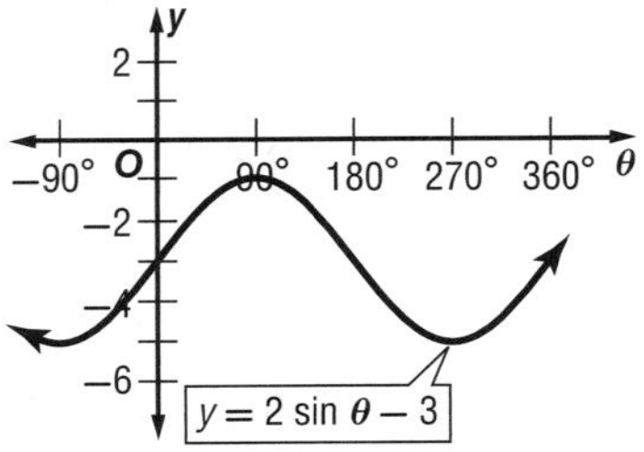

65. 1.25 **67.** F

69. amplitude: 2; period: 360°

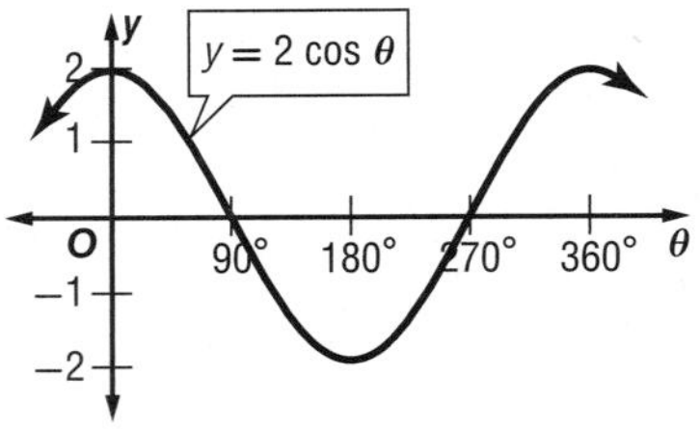

71. amplitude: 1; period: 180°

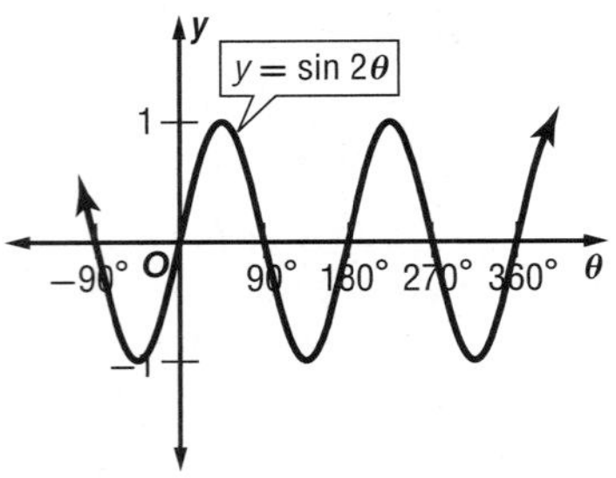

73. $-\frac{1}{2}$ **75.** Experiment; the people are put into groups at random. The treated group is the exercisers, and the control is the other group. This is a biased experiment because the participants all know which group they are in. **77.** Observational study; the students who have part-time jobs are the treated group, and the other students are the control; unbiased. **79.** 8 days **81.** 42° **83.** 37° **85.** 16°

Pages 872–874 *Lesson 13-9*

1. 30°; $\frac{\pi}{6}$ **3.** 180°; π **5.** 0 **7.** A **9.** −27.4°

11. Arctan $\frac{6.2}{18}$; 19°

13 KEYSTROKES: [2nd] [ENTER] [COS^{-1}] [2nd] [$\sqrt{\ }$] 3 [)] [÷] 2 [)] [ENTER] 30

So, $\text{Arccos}\left(\frac{\sqrt{3}}{2}\right) = 30°$ or $\frac{\pi}{6}$.

15. 60°; $\frac{\pi}{3}$ **17.** −30°; $-\frac{\pi}{6}$ **19.** −0.58 **21.** 0.87

23. 0.71 **25.** 64.2° **27.** 104.5° **29.** −11.3°

31. Arcsin $\frac{2.5}{24}$; 6°

33

$\frac{15 \text{ m/s} (\sin x)}{9.8 \text{ m/s}^2} = 1 \text{ s}$	Maximum height equals 1.
$\sin x = \frac{9.8}{15}$	Solve for sin *x*.
$x = \text{Sin}^{-1} \frac{9.8}{15}$	Inverse sine function
$x \approx 40.8°$	Use a calculator.

35. π **37.** no solution **39.** $\frac{\pi}{3}, \frac{5\pi}{3}$ **41.** false; $x = 2\pi$

43. The domain of $y = \text{Sin}^{-1} x$ is $-1 \le x \le 1$. This is the same as the range of $y = \text{Sin } x$. **45.** Sample answer: $y = \tan^{-1} x$ is a relation that has a domain of all real numbers and a range of all real numbers except odd multiples of $\frac{\pi}{2}$. The relation is not a function. $y = \text{Tan}^{-1} x$ is a function that has a domain of all real numbers and a range of $-\frac{\pi}{2} \le y \le \frac{\pi}{2}$.

47. A **49.** G **51a.** 164; 164; 360°, 90°

51b. $y = 100[\sin(x - 90°)] + 100$ **53.** 2 **55.** 4 **57.** −1

59. $-\frac{\sqrt{3}}{2}$

Pages 876–879 *Chapter 13* *Study Guide and Review*

1. false, Law of Sines **3.** true **5.** false, Arcsin function **7.** $a = 10.9$; $A = 65°$; $B = 25°$ **9.** $A = 15°$; $a = 4.0$; $c = 15.5$ **11.** $B = 55°$; $a = 12.6$; $b = 18.0$ **13.** about 8.8 feet **15.** 450° **17.** $-\frac{7\pi}{4}$ **19.** 295°, −425° **21.** $\frac{4\pi}{15}$ **23.** $-\frac{\sqrt{3}}{3}$ **25.** 0 **27.** $\sin \theta = \frac{12}{13}$, $\cos \theta = \frac{5}{13}$, $\tan \theta = \frac{12}{5}$, $\csc \theta = \frac{13}{12}$, $\sec \theta = \frac{13}{5}$, $\cot \theta = \frac{5}{12}$

29. about 17.1 meters **31.** two solutions; First solution: $C = 30°$, $B = 125°$, $b = 29.1$; second solution: $C = 150°$, $B = 5°$, $b = 3.1$ **33.** 105.5 ft **35.** Law of Sines; $B \approx 52°$, $C \approx 48°$, $c \approx 11.3$ **37.** Law of Sines; $B \approx 75°$, $C \approx 63°$, $c \approx 12.0$ **41.** $\frac{-\sqrt{6}}{4}$ **43.** 0 **45.** 20 seconds

47. amplitude: 1, period: 720°

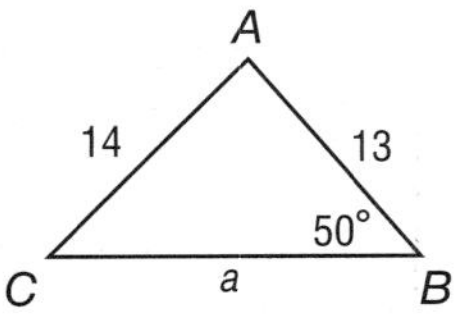

49. amplitude: not defined, period: 360°

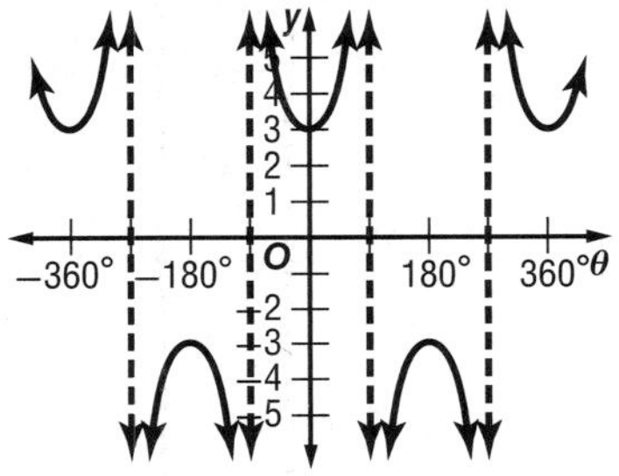

51. amplitude: not defined, period: 720°

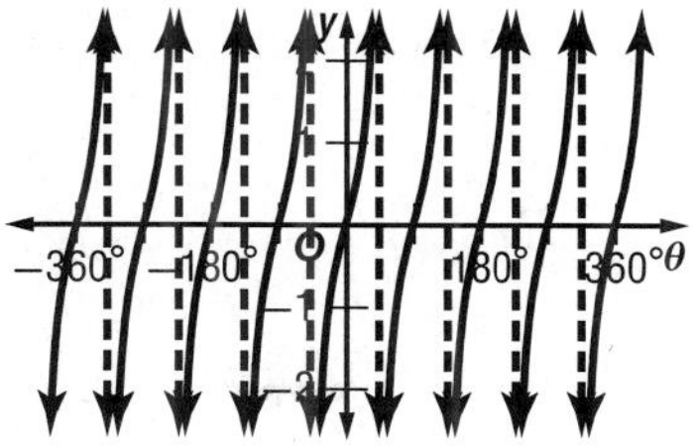

53. vertical shift: up 1; amplitude: 3; period 180°; phase shift: 90° right

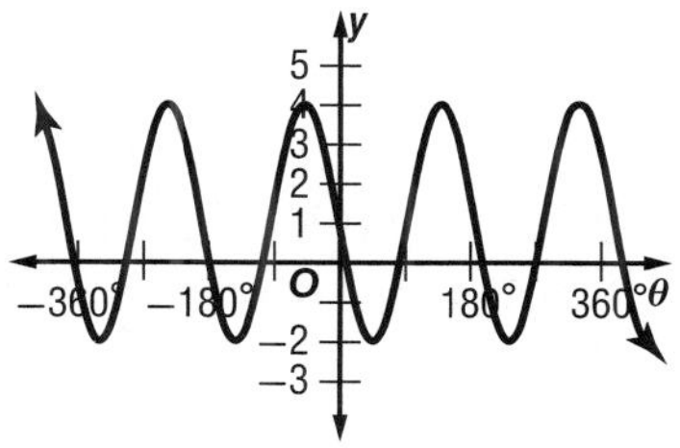

55. vertical shift: up 2 amplitude: not defined period: 120° phase shift: $\frac{\pi}{2}$° right

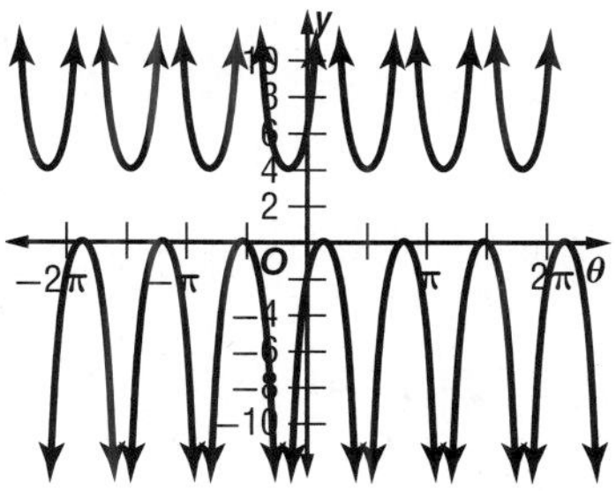

57. vertical shift: up 2 amplitude: $\frac{1}{3}$ period: 1080° phase shift: 90° right

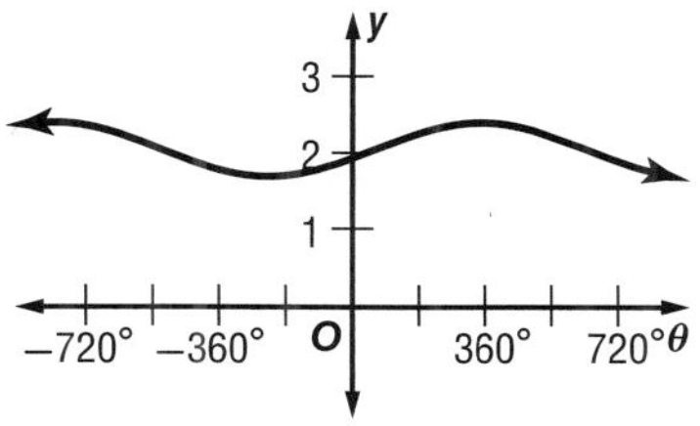

59. $90°, \frac{\pi}{2}$ **61.** $60°, \frac{\pi}{3}$ **63.** $45°, \frac{\pi}{4}$ **65.** $\sin\theta = \frac{5}{10}$; 30° **67.** −0.71 **69.** −55.0° **71.** 65.8°

Chapter 14 Trigonometric Identities and Equations

Page 889 Chapter 14 Get Ready

1. $-4a(a-1)$ **3.** prime **5.** $(x+2)$ in. **7.** $\{-7, 5\}$ **9.** $\{3, 4\}$ **11.** $\frac{\sqrt{2}}{2}$ **13.** $-\frac{\sqrt{3}}{3}$ **15.** 45 ft

Pages 894–897 Lesson 14-1

1. $\frac{1}{2}$ **3.** $\frac{\sqrt{5}}{3}$ **5.** $\sin\theta\cos\theta$ **7.** $\cot^2\theta$ **9.** $\frac{5}{4}$ **11.** $\frac{4}{5}$ **13.** $-\frac{5}{4}$

15 $\cot^2\theta + 1 = \csc^2\theta$ Pythagorean Identity

$\left(\frac{1}{4}\right)^2 + 1 = \csc^2\theta$ Substitute $\frac{1}{4}$ for $\cot\theta$.

$\frac{1}{16} + 1 = \csc^2\theta$ Square $\frac{1}{4}$.

$\frac{17}{16} = \csc^2\theta$ Add.

$\pm\frac{\sqrt{17}}{4} = \csc\theta$ Take the square root of each side.

Since θ is in the third quadrant, $\csc\theta$ is negative. So, $\csc\theta = -\frac{\sqrt{17}}{4}$.

17. $-\frac{12}{13}$ **19.** $\frac{3}{5}$ **21.** $\sec^3\theta$ **23.** $\csc\theta$ **25.** 1

27 $B = \frac{F\csc\theta}{I\ell}$ Original equation

$I\ell \cdot B = F\csc\theta$ Multiply each side by $I\ell$.

$I\ell B = F \cdot \frac{1}{\sin\theta}$ $\frac{1}{\sin\theta} = \csc\theta$

$I\ell B\sin\theta = F$ Multiply each side by $\sin\theta$.

The equation can be written as $F = I\ell B\sin\theta$.

29. $\sec\theta$ **31.** 2 **33.** $\cos 2\theta$ **35a.** $\frac{\sqrt{65}}{9}$ **35b.** $\frac{4\sqrt{65}}{65}$ **35c.** $\frac{4}{9}, \frac{\sqrt{65}}{9}, \frac{4\sqrt{65}}{65}$ **37.** $\mu_k = \tan\theta$

39 $\frac{\cos\left(\frac{\pi}{2}-\theta\right)-1}{1+\sin(-\theta)} = \frac{\sin\theta - 1}{1+\sin(-\theta)}$ $\cos\left(\frac{\pi}{2}-\theta\right) = \sin\theta$

$= \frac{\sin\theta - 1}{1-\sin\theta}$ $\sin(-\theta) = -\sin\theta$

$= \frac{\sin\theta - 1}{-1(\sin\theta - 1)}$ $1 - \sin\theta = -1(\sin\theta - 1)$

$= \frac{1}{-1}$ or -1 Simplify.

41. $-\cot^2\theta$ **43.** Sample answer: $x = 45°$ **45.** The functions $\cos\theta$ and $\sin\theta$ can be thought of as the lengths of the legs of a right triangle, and the number 1 can be thought of as the measure of the corresponding hypotenuse. **47.** Sample answer: $\frac{\sin\theta}{\cos\theta} \cdot \sin\theta$ and $\frac{\sin^2\theta}{\cos\theta}$ **49.** $-\frac{4}{3}$ **51.** A **53.** D **55.** 2.09 **57.** 0.52 **59.** 0.5 **61.** $h = 9 + 6\sin\left[\frac{\pi}{9}(t-1.5)\right]$ **63.** $\frac{1093}{9}$ **65.** −3, 2 **67.** 2

Pages 900–903 Lesson 14-2

1. $\cot\theta + \tan\theta \stackrel{?}{=} \frac{\sec^2\theta}{\tan\theta}$

$\tan\theta(\cot\theta + \tan\theta) \stackrel{?}{=} \sec^2\theta$

$1 + \tan^2\theta \stackrel{?}{=} \sec^2\theta$

$\sec^2\theta = \sec^2\theta$ ✓

3. $\sin\theta \stackrel{?}{=} \frac{\sec\theta}{\tan\theta + \cot\theta}$

$\frac{\sin\theta}{\sec\theta} \stackrel{?}{=} \frac{1}{\tan\theta + \cot\theta}$

$\frac{\sin\theta}{\sec\theta} \stackrel{?}{=} \frac{1}{\frac{\sin\theta}{\cos\theta} + \frac{\cos\theta}{\sin\theta}}$

$\frac{\sin\theta}{\sec\theta} \stackrel{?}{=} \frac{1}{\frac{\sin^2\theta + \cos^2\theta}{\sin\theta\cos\theta}}$

$\frac{\sin\theta}{\sec\theta} \stackrel{?}{=} \frac{\sin\theta\cos\theta}{\sin^2\theta\cos^2\theta}$

$\frac{\sin\theta}{\sec\theta} \stackrel{?}{=} \frac{\sin\theta\cos\theta}{1}$

$\frac{\sin\theta}{\sec\theta} = \frac{\sin\theta}{\sec\theta}$ ✓

5. $\tan^2\theta\csc^2\theta \stackrel{?}{=} 1 + \tan^2\theta$

$\frac{1 + \tan^2\theta}{\csc^2\theta} \stackrel{?}{=} \tan^2\theta$

$\frac{\sec^2\theta}{\csc^2\theta} \stackrel{?}{=} \tan^2\theta$

$\frac{\frac{1}{\cos^2\theta}}{\frac{1}{\sin^2\theta}} \stackrel{?}{=} \tan^2\theta$

$\frac{1}{\cos^2}\theta \cdot \sin^2\theta \stackrel{?}{=} \tan^2\theta$

$\tan^2\theta = \tan^2\theta$ ✓

7 $\dfrac{\tan^2\theta+1}{\tan^2\theta}=\dfrac{\sec^2\theta}{\tan^2\theta}$ $\quad\tan^2\theta+1=\sec^2\theta$

$=\dfrac{\frac{1}{\cos^2\theta}}{\frac{\sin^2\theta}{\cos^2\theta}}$ $\quad\sec^2\theta=\dfrac{1}{\cos^2\theta}$ and $\tan^2\theta=\dfrac{\sin^2\theta}{\cos^2\theta}$

$=\dfrac{1}{\cos^2\theta}\cdot\dfrac{\cos^2\theta}{\sin^2\theta}$ $\quad$ Invert the denominator and multiply.

$=\dfrac{1}{\sin^2\theta}$ $\quad$ Simplify.

$=\csc^2\theta$ $\quad\csc^2\theta=\dfrac{1}{\sin^2\theta}$

The answer is D.

9. $\cot\theta(\cot\theta+\tan\theta)\stackrel{?}{=}\csc^2\theta$

$\cot^2\theta+\cot\theta\tan\theta\stackrel{?}{=}\csc^2\theta$

$\cot^2\theta+\dfrac{\sin\theta}{\cos\theta}\cdot\dfrac{\cos\theta}{\sin\theta}\stackrel{?}{=}\csc^2\theta$

$\cot^2\theta+1\stackrel{?}{=}\csc^2\theta$

$\csc^2\theta=\csc^2\theta$ ✓

11. $\sin\theta\sec\theta\cot\theta\stackrel{?}{=}1$

$\sin\theta\cdot\dfrac{1}{\cos\theta}\cdot\dfrac{\cos\theta}{\sin\theta}\stackrel{?}{=}1$

$1\stackrel{?}{=}1$ ✓

13. $\dfrac{1-2\cos^2\theta}{\sin\theta-\cos\theta}\stackrel{?}{=}\tan\theta-\cot\theta$

$\dfrac{(1-\cos^2\theta)-\cos^2\theta}{\sin\theta\cos\theta}\stackrel{?}{=}\tan\theta-\cot\theta$

$\dfrac{\sin^2\theta-\cos^2\theta}{\sin\theta\cos\theta}\stackrel{?}{=}\tan\theta-\cot\theta$

$\dfrac{\sin^2\theta}{\sin\theta\cos\theta}-\dfrac{\cos^2\theta}{\sin\theta\cos\theta}\stackrel{?}{=}\tan\theta-\cot\theta$

$\dfrac{\sin\theta}{\cos\theta}-\dfrac{\cos\theta}{\sin\theta}\stackrel{?}{=}\tan\theta-\cot\theta$

$\tan\theta-\cot\theta=\tan\theta-\cot\theta$ ✓

15. $\cos\theta\stackrel{?}{=}\sin\theta\cot\theta$

$\cos\theta\stackrel{?}{=}\sin\theta\dfrac{\cos\theta}{\sin\theta}$

$\cos\theta=\cos\theta$ ✓

17. $\cos\theta\cos(-\theta)-\sin\theta\sin(-\theta)\stackrel{?}{=}1$

$\cos\theta\cos\theta-\sin\theta(-\sin\theta)\stackrel{?}{=}1$

$\cos^2\theta+\sin^2\theta=1$

$1=1$ ✓

19. $\sec\theta-\tan\theta\stackrel{?}{=}\dfrac{1-\sin\theta}{\cos\theta}$

$\dfrac{1}{\cos\theta}-\dfrac{\sin\theta}{\cos\theta}\stackrel{?}{=}\dfrac{1-\sin\theta}{\cos\theta}$

$\dfrac{1-\sin\theta}{\cos\theta}=\dfrac{1-\sin\theta}{\cos\theta}$ ✓

21. $\sec\theta\csc\theta\stackrel{?}{=}\tan\theta+\cot\theta$

$\dfrac{1}{\cos\theta}\cdot\dfrac{1}{\sin\theta}\stackrel{?}{=}\dfrac{\sin\theta}{\cos\theta}+\dfrac{\cos\theta}{\sin\theta}$

$\dfrac{1}{\cos\theta\sin\theta}\stackrel{?}{=}\dfrac{\sin^2\theta}{\sin\theta\cos\theta}+\dfrac{\cos^2\theta}{\sin\theta\cos\theta}$

$\dfrac{1}{\cos\theta\sin\theta}\stackrel{?}{=}\dfrac{\sin^2\theta+\cos^2\theta}{\sin\theta\cos\theta}$

$\dfrac{1}{\cos\theta\sin\theta}=\dfrac{1}{\cos\theta\sin\theta}$ ✓

23. $(\sin\theta+\cos\theta)^2\stackrel{?}{=}\dfrac{2+\sec\theta\csc\theta}{\sec\theta\csc\theta}$

$(\sin\theta+\cos\theta)^2(\sec\theta\csc\theta)\stackrel{?}{=}2+\sec\theta\csc\theta$

$(\sin^2\theta+2\sin\theta\cos\theta+\cos^2\theta)(\sec\theta\csc\theta)\stackrel{?}{=}2+\sec\theta\csc\theta$

$(1+2\sin\theta\cos\theta)(\sec\theta\csc\theta)\stackrel{?}{=}2+\sec\theta\csc\theta$

$\sec\theta\csc\theta+2\sin\theta\cos\theta\sec\theta\csc\theta\stackrel{?}{=}2+\sec\theta\csc\theta$

$2+\sec\theta\csc\theta=2+\sec\theta\csc\theta$ ✓

25. $\csc\theta-1\stackrel{?}{=}\dfrac{\cot^2\theta}{\csc\theta+1}$

$\csc\theta-1\stackrel{?}{=}\dfrac{\csc^2\theta-1}{\csc\theta+1}$

$(\csc\theta-1)(\csc\theta+1)\stackrel{?}{=}\csc^2\theta-1$

$\csc^2\theta-1=\csc^2\theta-1$ ✓

27. $\sin\theta\cos\theta\tan\theta+\cos^2\theta\stackrel{?}{=}1$

$\sin\theta\cos\theta\tan\theta\stackrel{?}{=}1-\cos^2\theta$

$\sin\theta\cos\theta\tan\theta\stackrel{?}{=}\sin^2\theta$

$\sin\theta\cos\theta\dfrac{\sin\theta}{\cos\theta}\stackrel{?}{=}\sin^2\theta$

$\sin^2\theta=\sin^2\theta$ ✓

29. $\csc^2\theta\stackrel{?}{=}\cot^2\theta+\sin\theta\csc\theta$

$\csc^2\theta\stackrel{?}{=}\cot^2\theta+\sin\theta\cdot\dfrac{1}{\sin\theta}$

$\csc^2\theta\stackrel{?}{=}\cot^2\theta+1$

$\csc^2\theta\stackrel{?}{=}\csc^2\theta$ ✓

31. $\sin^2\theta+\cos^2\theta\stackrel{?}{=}\sec^2\theta-\tan^2\theta$

$1\stackrel{?}{=}\tan^2\theta+1-\tan^2\theta$

$1=1$ ✓

33. yes

35 $\cot(-\theta)\tan(-\theta)=\dfrac{1}{\tan(-\theta)}\cdot\tan(-\theta)$ $\quad\dfrac{1}{\tan(-\theta)}=\cot(-\theta)$

$=1$ $\quad$ Simplify.

37. 1 **39.** 1 **41.** $\cos\theta$ **43.** 2 **45.** $\sin\theta$ **47.** 1

49. $y=-\dfrac{gx^2}{2v_0^{\,2}}(1+\tan^2\theta)+x\tan\theta$

51 **a.** $h=\dfrac{v_0^{\,2}\sin^2\theta}{2g}=\dfrac{47^2\sin^2\theta}{2(9.8)}$ $\quad$ Replace v_0 with 47 and g with 9.8.

$=\dfrac{2209\sin^2\theta}{19.6}$ $\quad$ Simplify.

$\dfrac{2209\sin^2 30°}{19.6}\approx 28.3$ m $\quad\theta=30°$

$\dfrac{2209\sin^2 45°}{19.6}\approx 56.4$ m $\quad\theta=45°$

$\dfrac{2209\sin^2 60°}{19.6}\approx 84.5$ m $\quad\theta=60°$

$\dfrac{2209\sin^2 90°}{19.6}\approx 112.7$ m $\quad\theta=90°$

b. Sample answer: Enter the equation $y=\dfrac{2209(\sin\theta)^2}{19.6}$.
Use the window Xmin = –600, Xmax = 600, Xscl = 60, Ymin = –10, Ymax = 10, Yscl = 1, Xres = 1.

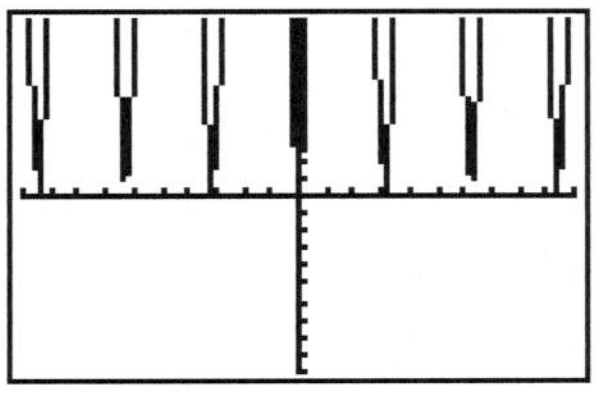

c. $\dfrac{v_0^{\,2}\tan^2\theta}{2g\sec^2\theta}\stackrel{?}{=}\dfrac{v_0^{\,2}\sin^2\theta}{2g}$

$\dfrac{v_0^{\,2}\left(\frac{\sin^2\theta}{\cos^2\theta}\right)}{2g\left(\frac{1}{\cos^2\theta}\right)}\stackrel{?}{=}\dfrac{v_0^{\,2}\sin^2\theta}{2g}$ $\quad\tan^2\theta=\dfrac{\sin^2\theta}{\cos^2\theta}$ and $\sec^2\theta=\dfrac{1}{\cos^2\theta}$

$\dfrac{v_0^{\,2}\sin^2\theta}{2g}=\dfrac{v_0^{\,2}\sin^2\theta}{2g}$ ✓ $\quad$ Simplify.

53. $\tan^2\theta = \frac{\sin^2\theta}{\cos^2\theta}$

$\tan^2\theta = \tan^2\theta$

$\tan^2\theta = \sec^2\theta - 1$

55. Sample answer: counterexample 45°, 30°

57. Sample answer: $\tan\theta\csc\theta = \sec\theta$ **59.** $\frac{1}{2}\sin\theta$

61. A **63.** C **65.** $\frac{1}{2}$ **67.** $\frac{5}{4}$ **69.** 30°

71. $(0, \pm 3\sqrt{2})$; $(0, \pm\sqrt{38})$; $y = \pm\frac{3\sqrt{10}}{10}$

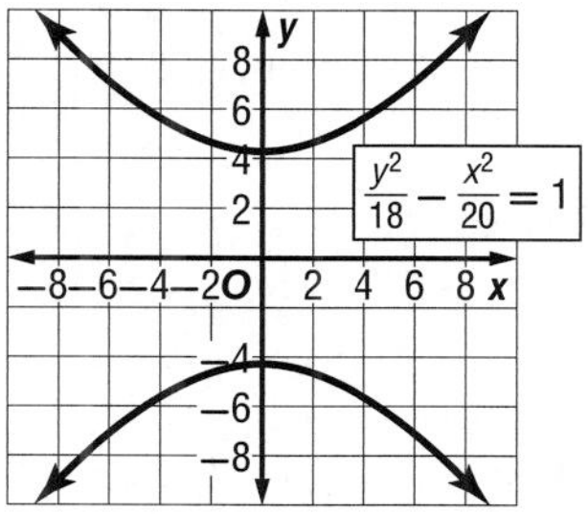

73. $(\pm 6, 0)$; $(\pm\sqrt{37}, 0)$; $y = \pm\frac{1}{6}x$

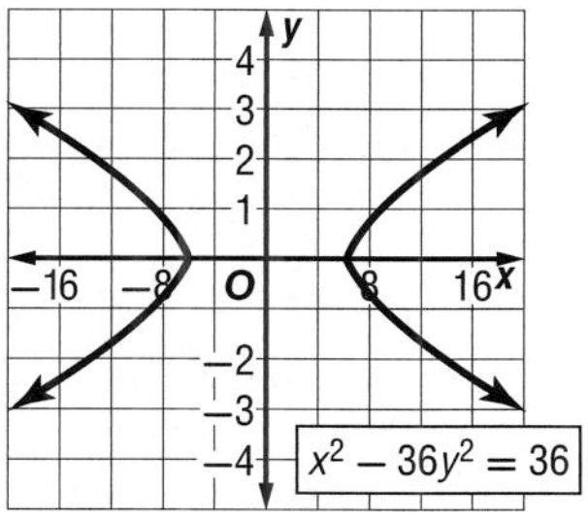

75. $\frac{\sqrt{x^2-1}}{x-1}$ **77.** $\frac{-1-\sqrt{3}}{2}$

Pages 906–909 Lesson 14-3

1 $\cos 165° = \cos(120° + 45°)$

$= \cos 120° \cos 45° - \sin 120° \sin 45°$

$= \left(-\frac{1}{2}\cdot\frac{\sqrt{2}}{2}\right) - \left(\frac{\sqrt{3}}{2}\cdot\frac{\sqrt{2}}{2}\right)$

$= -\frac{\sqrt{2}}{4} - \frac{\sqrt{6}}{4}$

$= -\frac{\sqrt{2}+\sqrt{6}}{4}$

3. $\frac{\sqrt{6}-\sqrt{2}}{4}$ **5.** $\frac{\sqrt{2}}{2}$ **7a.** 0 **7b.** The interference is destructive. The signals cancel each other completely.

9. $\cos\left(\frac{3\pi}{2} - \theta\right) \stackrel{?}{=} -\sin\theta$

$\cos\frac{3\pi}{2}\cos\theta + \sin\frac{3\pi}{2}\sin\theta \stackrel{?}{=} -\sin\theta$

$0\cdot\cos\theta - 1\cdot\sin\theta \stackrel{?}{=} -\sin\theta$

$-\sin\theta = -\sin\theta$ ✓

11. $\sin(\theta + \pi) \stackrel{?}{=} -\sin\theta$

$\sin\theta\cos\pi + \cos\theta\sin\pi \stackrel{?}{=} -\sin\theta$

$(\sin\theta)(-1) + (\cos\theta)(0) \stackrel{?}{=} -\sin\theta$

$-\sin\theta = -\sin\theta$ ✓

13. $-\frac{\sqrt{2}}{2}$ **15.** $\frac{\sqrt{6}-\sqrt{2}}{4}$ **17.** $\frac{\sqrt{2}+\sqrt{6}}{4}$

19. $\cos\left(\frac{\pi}{2} + \theta\right) \stackrel{?}{=} -\sin\theta$

$\cos\frac{\pi}{2}\cos\theta - \sin\frac{\pi}{2}\sin\theta \stackrel{?}{=} -\sin\theta$

$(0)(\cos\theta) - (1)(\sin\theta) \stackrel{?}{=} -\sin\theta$

$-\sin\theta = -\sin\theta$ ✓

21. $\cos(180° + \theta) \stackrel{?}{=} -\cos\theta$

$\cos 180°\cos\theta - \sin 180°\sin\theta \stackrel{?}{=} -\cos\theta$

$-1\cdot\cos\theta - 0\cdot\sin\theta \stackrel{?}{=} -\cos\theta$

$-\cos\theta = -\cos\theta$ ✓

23a. $y = 30.9\sin\left(\frac{\pi}{6}x - 2.09\right) + 42.65$ **23b.** The new function represents the average of the high and low temperatures for each month. **25.** $\sqrt{2} - \sqrt{6}$

27. $-2 + \sqrt{3}$ **29.** $2 - \sqrt{3}$

31 **a.** Let X be the endpoint of the segment that is 8 inches long.

$\sin(m\angle BAC)$

$= \sin(m\angle BAX + m\angle XAC)$

$= \sin(m\angle BAX)\cos(m\angle XAC) + \cos(m\angle BAX)\sin(m\angle XAC)$

$= \frac{8\sqrt{3}}{16}\cdot\frac{8}{10} + \frac{8}{16}\cdot\frac{6}{10}$ $\sin = \frac{\text{opp}}{\text{hyp}}$ and $\cos = \frac{\text{adj}}{\text{hyp}}$

$= \frac{4\sqrt{3}}{10} + \frac{3}{10}$ Multiply.

$= \frac{3+4\sqrt{3}}{10}$ Add.

b. Let X be the endpoint of the segment that is 8 inches long.

$\cos(m\angle BAC)$

$= \cos(m\angle BAX + m\angle XAC)$

$= \cos(m\angle BAX)\cos(m\angle XAC) - \sin(m\angle BAX)\sin(m\angle XAC)$

$= \frac{8}{16}\cdot\frac{8}{10} - \frac{8\sqrt{3}}{16}\cdot\frac{6}{10}$ $\sin = \frac{\text{opp}}{\text{hyp}}$ and $\cos = \frac{\text{adj}}{\text{hyp}}$

$= \frac{4}{10} - \frac{3\sqrt{3}}{10}$ Multiply.

$\approx \frac{4-3\sqrt{3}}{10}$ Add.

c. $\cos(m\angle BAC) = \frac{4-3\sqrt{3}}{10}$

$m\angle BAC = \text{Cos}^{-1}\left(\frac{3+4\sqrt{3}}{10}\right)$

$\approx 96.9°$

d. Since $m\angle BAC \neq 90$, the triangle formed is not a right triangle.

33a.

A	B	sin A	sin B	sin (A + B)	sin A + sin B
30°	90°	$\frac{1}{2}$	1	$\frac{\sqrt{3}}{2}$	$\frac{3}{2}$
45°	60°	$\frac{\sqrt{2}}{2}$	$\frac{\sqrt{3}}{2}$	$\frac{\sqrt{2}+\sqrt{6}}{4}$	$\frac{\sqrt{2}+\sqrt{3}}{2}$
60°	45°	$\frac{\sqrt{3}}{2}$	$\frac{\sqrt{2}}{2}$	$\frac{\sqrt{2}+\sqrt{6}}{4}$	$\frac{\sqrt{2}+\sqrt{3}}{2}$
90°	30°	1	$\frac{1}{2}$	$\frac{\sqrt{3}}{2}$	$\frac{3}{2}$

33b.

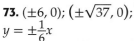

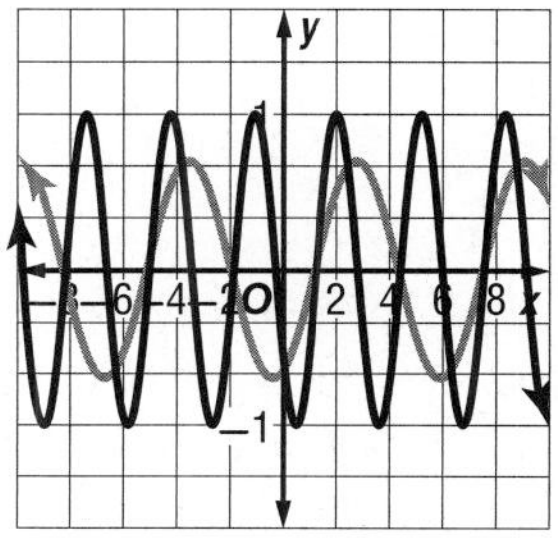

33c. No; a counterexample is: $\cos(30° + 45°) = \cos 30° + \cos 45°$, which equals $\frac{\sqrt{3}}{2} + \frac{\sqrt{2}}{2}$ or about 1.5731. Since a cosine value cannot be greater than 1, this statement must be false.

35 $\cos(A + B) \stackrel{?}{=} \dfrac{1 - \tan A \tan B}{\sec A \sec B}$

$\cos(A+B) \stackrel{?}{=} \dfrac{1 - \frac{\sin A}{\cos A} \cdot \frac{\sin B}{\cos B}}{\frac{1}{\cos A} \cdot \frac{1}{\cos B}}$ $\quad \tan A = \frac{\sin A}{\cos A}$ and $\tan B = \frac{\sin B}{\cos B}$; $\sec A = \frac{1}{\cos A}$ and $\sec B = \frac{1}{\cos B}$

$\cos(A+B) \stackrel{?}{=} \dfrac{1 - \frac{\sin A}{\cos A} \cdot \frac{\sin B}{\cos B}}{\frac{1}{\cos A} \cdot \frac{1}{\cos B}} \cdot \dfrac{\cos A \cos B}{\cos A \cos B}$ $\quad \frac{\cos A \cos B}{\cos A \cos B} = 1$

$\cos(A+B) \stackrel{?}{=} \left(1 - \frac{\sin A}{\cos A} \cdot \frac{\sin B}{\cos B}\right)\left(\frac{\cos A \cos B}{1}\right)\left(\frac{\cos A \cos B}{\cos A \cos B}\right)$

$\cos(A + B) \stackrel{?}{=} \dfrac{\cos A \cos B - \sin A \sin B}{1}$ $\quad$ Simplify.

$\cos(A + B) = \cos(A + B)$ ✓ $\quad$ Difference Identity

37. $\sin(A + B)\sin(A - B) \stackrel{?}{=} \sin^2 A - \sin^2 B$

$(\sin A \cos B + \cos A \sin B)(\sin A \cos B - \cos A \sin B) \stackrel{?}{=} \sin^2 A - \sin^2 B$

$(\sin A \cos B)^2 - (\cos A \sin B)^2 \stackrel{?}{=} \sin^2 A - \sin^2 B$

$\sin^2 B \cos^2 B - \cos^2 A \sin^2 B \stackrel{?}{=} \sin^2 A - \sin^2 B$

$\sin^2 A \cos^2 B + \sin^2 A \sin^2 B - \sin^2 A \sin^2 B - \cos^2 A \sin^2 B \stackrel{?}{=} \sin^2 A - \sin^2 B$

$\sin^2 A(\cos^2 B + \sin^2 B) - \sin^2 B(\sin^2 A + \cos^2 A) \stackrel{?}{=} \sin^2 A - \sin^2 B$

$(\sin^2 A)(1) - (\sin^2 B)(1) \stackrel{?}{=} \sin^2 A - \sin^2 B$

$\sin^2 A - \sin^2 B = \sin^2 A - \sin^2 B$ ✓

39. Sample answer: To determine wireless Internet interference, you need to determine the sine or cosine of the sum or difference of two angles. Interference occurs when waves pass through the same space at the same time. When the combined waves have a greater amplitude, constructive interference results. When the combined waves have a smaller amplitude, destructive interference results.

41. $d = \sqrt{(\cos\alpha - \cos\beta)^2 + (\sin\alpha - \sin\beta)^2}$

$d^2 = (\cos\alpha - \cos\beta)^2 + (\sin\alpha - \sin\beta)^2$

$d^2 = (\cos^2\alpha - 2\cos\alpha\cos\beta + \cos^2\beta) + (\sin^2\alpha - 2\sin\alpha\sin\beta + \sin^2\beta)$

$d^2 = \cos^2\alpha + \sin^2\alpha + \cos^2\beta + \sin^2\beta - 2\cos\alpha\cos\beta - 2\sin\alpha\sin\beta$

$d^2 = 1 + 1 - 2\cos\alpha\cos\beta - 2\sin\alpha\sin\beta$ $\quad \sin^2\alpha + \cos^2\alpha = 1$ and $\sin^2\beta + \cos^2\beta = 1$

$d^2 = 2 - 2\cos\alpha\cos\beta - 2\sin\alpha\sin\beta$

Now find the value of d^2 when the angle having measure $\alpha - \beta$ is in standard position on the unit circle, as shown in the figure below.

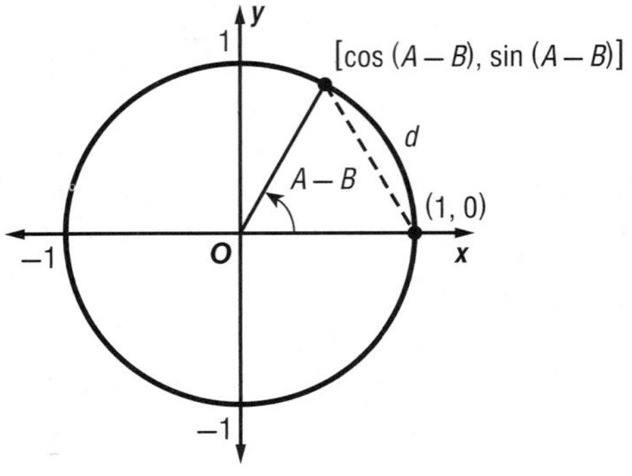

$d = \sqrt{[\cos(\alpha - \beta) - 1]^2 + [\sin(\alpha - \beta) - 0]^2}$

$d^2 = [\cos(\alpha - \beta) - 1]^2 + [\sin(\alpha - \beta) - 0]^2$

$= [\cos^2(\alpha - \beta) - 2\cos(\alpha - \beta) + 1] + \sin^2(\alpha - \beta)$

$= \cos^2(\alpha - \beta) + \sin^2(\alpha - \beta) - 2\cos(\alpha - \beta) + 1$

$= 1 - 2\cos(\alpha - \beta) + 1$

$= 2 - 2\cos(\alpha - \beta)$

43. 9 **45.** H

47. $\dfrac{\sin\theta}{\tan\theta} + \dfrac{\cos\theta}{\cot\theta} \stackrel{?}{=} \cos\theta + \sin\theta$

$\dfrac{\sin\theta}{1} + \dfrac{\cos\theta}{\frac{\cos\theta}{\sin\theta}} \stackrel{?}{=} \cos\theta + \sin\theta$

$\sin\theta \cdot \dfrac{\cos\theta}{\sin\theta} + \cos\theta \cdot \dfrac{\sin\theta}{\cos\theta} \stackrel{?}{=} \cos\theta + \sin\theta$

$\cos\theta + \sin\theta = \cos\theta + \sin\theta$ ✓

49. $\sin^2\theta$ **51.** $\sec\theta$

53. Step 1: $4^1 - 1 = 3$, which is divisible by 3. The statement is true for $n = 1$.
Step 2: Assume that $4^k - 1$ is divisible by 3 for some positive integer k. This means that $4^k - 1 = 3r$ for some whole number r.
Step 3: $4^k - 1 = 3r$

$4^k = 3r + 1$

$4^{k+1} = 12r + 4$

$4^{k+1} - 1 = 12r + 3$

$4^{k+1} - 1 = 3(4r + 1)$

Since r is a whole number, $4r + 1$ is a whole number. Thus, $4^{k+1} - 1$ is divisible by 3, so the statement is true for $n = k + 1$. Therefore, $4^n - 1$ is divisible by 3 for all positive integers n. **55.** −1 **57.** no solution

Pages 915–917 Lesson 14-4

1. $\frac{\sqrt{15}}{8}, \frac{7}{8}, \frac{\sqrt{8 - 2\sqrt{15}}}{4}, \frac{\sqrt{8 + 2\sqrt{15}}}{4}$ **3.** $-\frac{120}{169}, -\frac{119}{169}, \frac{3\sqrt{13}}{13}, \frac{2\sqrt{13}}{13}$ **5.** $-\frac{240}{289}, \frac{161}{289}, \frac{4\sqrt{17}}{17}, \frac{\sqrt{17}}{17}$ **7.** $\frac{\sqrt{2 - \sqrt{2}}}{2}$

9a. $d = \dfrac{v^2 \sin 2\theta}{g}$ **9b.** ≈ 81 ft

11. $(\sin\theta + \cos\theta)^2 \stackrel{?}{=} 1 + 2\sin\theta\cos\theta$

$(\sin\theta + \cos\theta)(\sin\theta + \cos\theta) \stackrel{?}{=} 1 + 2\sin\theta\cos\theta$

$\sin^2\theta + 2\sin\theta\cos\theta + \cos^2\theta \stackrel{?}{=} 1 + 2\sin\theta\cos\theta$

$1 + 2\sin\theta\cos\theta = 1 + 2\sin\theta\cos\theta$ ✓

13. $\frac{240}{289}, -\frac{161}{289}, \frac{5\sqrt{34}}{34}, -\frac{3\sqrt{34}}{34}$

15 $\sin^2\theta = 1 - \cos^2\theta$ $\quad \sin^2\theta + \cos^2\theta = 1$

$\sin^2\theta = 1 - \left(\frac{1}{5}\right)^2$ $\quad \cos\theta = \frac{1}{5}$

$\sin^2\theta = \frac{24}{25}$ $\quad$ Subtract.

$\sin\theta = \pm\frac{2\sqrt{6}}{5}$ $\quad$ Take the square root of each side.

Since θ is in the fourth quadrant, sine is negative. So, $\sin\theta = -\frac{2\sqrt{6}}{5}$.

$\sin 2\theta = 2\sin\theta\cos\theta$ $\quad$ Double-angle identity

$= 2\left(-\frac{2\sqrt{6}}{5}\right)\left(\frac{1}{5}\right)$ $\quad \sin\theta = -\frac{2\sqrt{6}}{5}$ and $\cos\theta = \frac{1}{5}$

$= -\frac{4\sqrt{6}}{25}$ $\quad$ Simplify.

$\cos 2\theta = 1 - 2\sin^2\theta$ $\quad$ Double-angle identity

$= 1 - 2\left(\frac{24}{25}\right)$ $\quad \sin^2\theta = \frac{24}{25}$

$= -\frac{23}{25}$ $\quad$ Simplify.

$$\sin\frac{\theta}{2} = \pm\sqrt{\frac{1-\cos\theta}{2}}$$ Half-angle identity

$$= \pm\sqrt{\frac{1-\frac{1}{5}}{2}}$$ $\cos\theta = \frac{1}{5}$

$$= \pm\sqrt{\frac{2}{5}}$$ Simplify.

$$= \pm\frac{\sqrt{2}}{\sqrt{5}} \cdot \frac{\sqrt{5}}{\sqrt{5}} \text{ or } \pm\frac{\sqrt{10}}{5}$$ Rationalize the denominator.

If θ is between 270° and 360°, $\frac{\theta}{2}$ is between 135° and 180°. So, $\sin\frac{\theta}{2}$ is $\frac{\sqrt{10}}{5}$.

$$\cos\frac{\theta}{2} = \pm\sqrt{\frac{1+\cos\theta}{2}}$$ Half-angle identity

$$= \pm\sqrt{\frac{1+\frac{1}{5}}{2}}$$ $\cos\theta = \frac{1}{5}$

$$= \pm\sqrt{\frac{3}{5}}$$ Simplify.

$$= \pm\sqrt{\frac{3}{5}} \cdot \frac{\sqrt{5}}{\sqrt{5}} \text{ or } \pm\frac{\sqrt{15}}{5}$$ Rationalize the denominator.

If θ is between 270° and 360°, $\frac{\theta}{2}$ is between 135° and 180°. So, $\cos\frac{\theta}{2}$ is $-\frac{\sqrt{15}}{5}$.

17. $-\frac{4}{5}, -\frac{3}{5}, \sqrt{\frac{\sqrt{5}+1}{2\sqrt{5}}}, \sqrt{\frac{\sqrt{5}-1}{2\sqrt{5}}}$ **19.** $\frac{\sqrt{2+\sqrt{2}}}{2}$

21. $\sqrt{3}-2$ **23.** $\sqrt{2}-1$

25 $P = I_0^2 R \sin^2 \theta t$ Original equation

$P = I_0^2 R(\cos^2 \theta t - \cos 2\theta t)$ $\sin^2 \theta t = \cos^2 \theta t - \cos 2\theta t$

$P = I_0^2 R\left(\frac{1}{2}\cos 2\theta t + \frac{1}{2} - \cos 2\theta t\right)$ $\cos^2 \theta t = \frac{1}{2}\cos 2\theta t + \frac{1}{2}$

$P = I_0^2 R\left(\frac{1}{2} - \frac{1}{2}\cos 2\theta t\right)$ Simplify.

$P = \frac{1}{2}I_0^2 R - \frac{1}{2}I_0^2 R\cos 2\theta t$ Distributive Property

27. $1 + \frac{1}{2}\sin 2\theta \stackrel{?}{=} \frac{\sec\theta + \sin\theta}{\sec\theta}$

$$\stackrel{?}{=} \frac{\frac{1}{\cos\theta} + \sin\theta}{\frac{1}{\cos\theta}}$$

$$\stackrel{?}{=} \frac{\frac{1}{\cos\theta} + \sin\theta}{\frac{1}{\cos\theta}} \cdot \frac{\cos\theta}{\cos\theta}$$

$$\stackrel{?}{=} 1 + \frac{1}{2} \cdot 2\sin\theta\cos\theta$$

$$\stackrel{?}{=} 1 + \frac{1}{2}\sin 2\theta \checkmark$$

29. $\tan\frac{\theta}{2} \stackrel{?}{=} \frac{\sin\theta}{1+\cos\theta}$

$$\tan\frac{\theta}{2} \stackrel{?}{=} \frac{\sin 2\left(\frac{\theta}{2}\right)}{1+\cos 2\left(\frac{\theta}{2}\right)}$$

$$\tan\frac{\theta}{2} \stackrel{?}{=} \frac{2\sin\frac{\theta}{2}\cos\frac{\theta}{2}}{1+2\cos^2\frac{\theta}{2}-1}$$

$$\tan\frac{\theta}{2} \stackrel{?}{=} \frac{2\sin\frac{\theta}{2}\cos\frac{\theta}{2}}{2\cos^2\frac{\theta}{2}}$$

$$\tan\frac{\theta}{2} \stackrel{?}{=} \frac{\sin\frac{\theta}{2}}{\cos\frac{\theta}{2}}$$

$$\tan\frac{\theta}{2} = \tan\frac{\theta}{2} \checkmark$$

31. $\frac{24}{25}, \frac{7}{25}, \frac{24}{7}$ **33.** $-\frac{3}{5}, -\frac{4}{5}, \frac{3}{4}$ **35.** $-\frac{4\sqrt{21}}{25}, \frac{17}{25}, -\frac{4\sqrt{21}}{17}$

37. No; Teresa incorrectly added the square roots, and Nathan used the half-angle identity incorrectly. He used sin 30° in the formula instead of first finding the cosine. **39.** If you are only given the value of $\cos\theta$, then $\cos 2\theta = 2\cos^2\theta - 1$ is the best identity to use. If you are only given the value of $\sin\theta$, then $\cos 2\theta = 1 - 2\sin^2\theta$ is the best identity to use. If you are given the values of both $\cos\theta$ and $\sin\theta$, then $\cos 2\theta = \cos^2\theta - \sin^2\theta$ works just as well as the other two.

41. $1 - 2\sin^2\theta = \cos 2\theta$ Double-angle identity

$1 - 2\sin^2\frac{A}{2} = \cos A$ Substitute $\frac{A}{2}$ for θ and A for 2θ.

$\sin^2\frac{A}{2} = \frac{1-\cos A}{2}$ Solve for $\sin^2\frac{A}{2}$.

$\sin\frac{A}{2} = \pm\sqrt{\frac{1-\cos A}{2}}$ Take the square root of each side.

Find $\cos\frac{A}{2}$.

$2\cos^2\theta - 1 = \cos 2\theta$ Double-angle identity

$2\cos^2\frac{A}{2} - 1 = \cos A$ Substitute $\frac{A}{2}$ for θ and A for 2θ.

$\cos^2\frac{A}{2} = \frac{1+\cos A}{2}$ Solve for $\cos^2\frac{A}{2}$.

$\cos\frac{A}{2} = \pm\sqrt{\frac{1+\cos A}{2}}$ Take the square root of each side.

43. 22.5 **45.** B **47.** $\frac{\sqrt{2}}{2}$ **49.** $\frac{-\sqrt{6}-\sqrt{2}}{4}$ **51.** $\frac{\sqrt{3}}{2}$

53. $\cot\theta + \sec\theta \stackrel{?}{=} \frac{\cos^2\theta + \sin\theta}{\sin\theta\cos\theta}$

$$\cot\theta + \sec\theta \stackrel{?}{=} \frac{\cos^2\theta}{\sin\theta\cos\theta} + \frac{\sin\theta}{\sin\theta\cos\theta}$$

$$\cot\theta + \sec\theta \stackrel{?}{=} \frac{\cos\theta}{\sin\theta} + \frac{1}{\cos\theta}$$

$$\cot\theta + \sec\theta = \cot\theta + \sec\theta$$

55. sines; $B \approx 102°$, $C \approx 44°$, $b \approx 21.0$ **57.** sines; $A = 80°$, $a \approx 10.9$, $c \approx 5.4$ **59.** $\{-4, 7\}$

Pages 922–925 ***Lesson 14-5***

1. 210°, 330° **3.** 60°, 180°, or 300° **5.** 150°, 210°

7 $\cos 2\theta = 8 - 15\sin\theta$ Original equation

$\cos 2\theta + 15\sin\theta - 8 = 0$ Add $15\sin\theta - 8$ to each side.

$1 - 2\sin^2\theta + 15\sin\theta - 8 = 0$ Double-angle identity

$-2\sin^2\theta + 15\sin\theta - 7 = 0$ Simplify.

$(-2\sin\theta + 1)(\sin\theta - 7) = 0$ Factor.

$-2\sin\theta + 1 = 0$ or $\sin\theta - 7 = 0$ Zero Product Property

$\sin\theta = \frac{1}{2}$ $\sin\theta = 7$

$\theta = 30°$ or $150°$ no solution since $0 \le \sin\theta \le 1$

9. $\pm\frac{\pi}{6} + 2k\pi$ or $\pm\frac{5\pi}{6} + 2k\pi$ **11.** $\frac{3\pi}{2} + 2k\pi$ **13.** $\pi + 2k\pi$ **15.** $90° + k \cdot 180°$ **17.** $45° + k \cdot 90°$ **19.** $270° + k \cdot 360°$

21a. There will be $10\frac{1}{2}$ hours of daylight 213 and 335 days after March 21; that is, on October 20 and February 19. **21b.** Every day from February 19 to October 20; sample explanation: Since the longest day of the year occurs around June 22, the days between February 19 and October 20 must increase in length until June 22 and then decrease in length until October 20. **23.** $\frac{3\pi}{4} + \pi k$ **25.** $\frac{\pi}{2} + \pi k, \frac{\pi}{6} + 2\pi k, \frac{5\pi}{6} + 2\pi k$ **27.** $0° + k \cdot 45°$ or $0 + k \cdot \frac{\pi}{4}$ **29.** $\pi + 2k\pi$ **31.** 135°, 225° **33.** $\frac{\pi}{6}$ **35.** 210°, 330°

37

$2\sin^2\theta = \cos\theta + 1$	Original equation
$2(1 - \cos^2\theta) = \cos\theta + 1$	$\sin^2\theta = 1 - \cos^2\theta$
$2 - 2\cos^2\theta = \cos\theta + 1$	Simplify.
$-2\cos^2\theta - \cos\theta + 1 = 0$	Subtract $\cos\theta + 1$ from each side and simplify.
$(-2\cos\theta - 1)(\cos\theta - 1) = 0$	Factor.
$-2\cos\theta - 1 = 0$ or $\cos\theta - 1 = 0$	Zero Product Property
$\cos\theta = \frac{1}{2}$ $\quad\quad$ $\cos\theta = 1$	

The solutions of $\cos\theta = \frac{1}{2}$ are $\frac{\pi}{3} + 2k\pi$ and $\frac{5\pi}{3} + 2k\pi$.

The solution of $\cos\theta = 1$ is $\pi + 2k\pi$.

39. $0 + 2k\pi$ **41.** $0° + k \cdot 180°$ **43.** $30° + k \cdot 360°, 150° + k \cdot 360°$ **45.** $\frac{7\pi}{6} + 2k\pi, \frac{11\pi}{6} + 2k\pi$ or $210° + k \cdot 360°, 330° + k \cdot 360°$ **47.** $0 + 2k\pi, \frac{\pi}{2} + 2k\pi, \frac{3\pi}{2} + 2k\pi$ or $0° + k \cdot 360°, 90° + k \cdot 360°, 270° + k \cdot 360°$

49a. 11 m **49b.** 7:00 A.M. and 7:00 P.M. **51.** $\frac{\pi}{6} + 2\pi k, \frac{5\pi}{6} + 2\pi k, \frac{5\pi}{4} + 2\pi k, \frac{7\pi}{4} + 2\pi k$ **53.** $120° + 360°k, 240° + 360°k$ **55.** $\frac{\pi}{6} + 2\pi k, \frac{5\pi}{6} + 2\pi k$

57

$D = 0.5 \sin(6.5x) \sin(2500t)$	Original equation
$0.01 = 0.5 \sin(6.5 \cdot 500) \sin(2500t)$	$D = 0.01$ mm and $x = 0.5 \cdot 1000$ or 500 mm
$0.01 = 0.5 \sin(3250) \sin(2500t)$	Simplify.
$0.1152 \approx \sin(2500t)$	Divide each side by 0.5 sin (3250).
$\text{Sin}^{-1}(0.1152) \approx 2500t$	Use the $\sin^{-1}$ function.
$6.6152 \approx 2500t$	Use a calculator.
$0.0026 \approx t$	Divide each side by 2500.

The time is about 0.0026 second.

59. $\frac{\pi}{3} < x < \pi$ or $\frac{5\pi}{3} < x < 2\pi$ **61.** All trigonometric functions are periodic. Adding the least common multiple of the periods of the functions that appear to any solution of the equation will always produce another solution. **63.** $2b$ **65.** A **67.** D

69. $\frac{\sqrt{2 - \sqrt{2}}}{2}$ **71.** $-\frac{\sqrt{2 - \sqrt{3}}}{2}$

73.
$$\cos(90° + \theta) \stackrel{?}{=} \cos 90° \cos\theta - \sin 90° \sin\theta$$
$$\stackrel{?}{=} 0 - 1\sin\theta$$
$$= -\sin\theta$$

75.
$$\sin(90° - \theta) \stackrel{?}{=} \cos\theta$$
$$\sin 90° \cos\theta - \cos 90° \sin\theta \stackrel{?}{=} \cos\theta$$
$$1 \cdot \cos\theta - 0 \cdot \sin\theta \stackrel{?}{=} \cos\theta$$
$$\cos\theta - 0 \stackrel{?}{=} \cos\theta$$
$$\cos\theta = \cos\theta \checkmark$$

77. 17, 26, 35 **79.** −12, −9, −6

81.

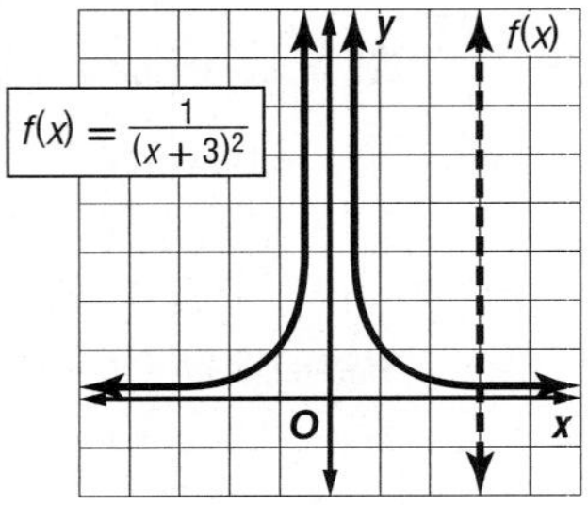

83.

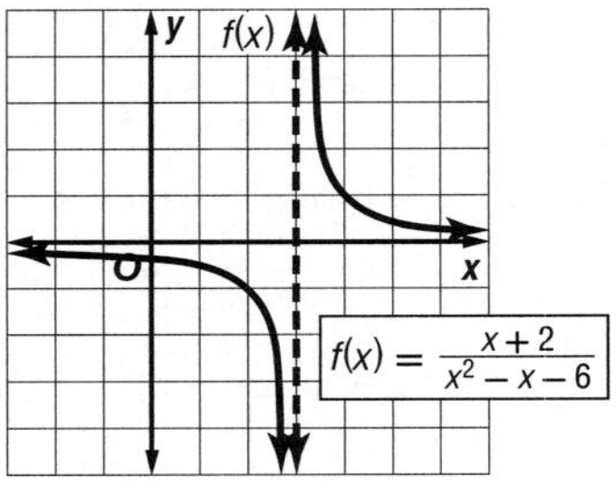

Pages 926–928 Chapter 14 Study Guide and Review

1. difference of angles identity **3.** trigonometric identity **5.** trigonometric equation **7.** reciprocal identities **9.** Pythagorean identity **11.** $-\sqrt{3}$ **13.** $-\frac{4}{5}$

15. First find the length of the diagonal: $75^2 + 110^2 = c^2$; $5625 + 12{,}100 = c^2$; $17{,}725 = c^2$; $c = 5\sqrt{709}$; $\sin\theta = \frac{110}{5\sqrt{709}} = \frac{15\sqrt{709}}{709}$ **17.** $\sec\theta$ **19.** $\cos\theta + \sin\theta$

21.
$$\frac{\cos\theta}{\cot\theta} + \frac{\sin\theta}{\tan\theta} = \sin\theta + \cos\theta$$
$$\cos\theta \div \frac{\cos\theta}{\sin\theta} + \sin\theta \div \frac{\sin\theta}{\cos\theta} = \sin\theta + \cos\theta$$
$$\cos\theta \cdot \frac{\sin\theta}{\cos\theta} + \sin\theta \cdot \frac{\cos\theta}{\sin\theta} = \sin\theta + \cos\theta$$
$$\sin\theta + \cos\theta = \sin\theta + \cos\theta$$

23. $\tan^2\theta + 1 = \left(\frac{\sqrt{7}}{3}\right)^2 + 1 = \frac{7}{9} + 1 = \frac{7}{9} + \frac{9}{9} = \frac{16}{9}$; $\sec^2\theta = \left(\frac{4}{3}\right)^2 = \frac{16}{9}$

25. $\frac{\sqrt{6} + \sqrt{2}}{4}$ **27.** $\frac{\sqrt{6} + \sqrt{2}}{4}$ **29.** $\frac{-\sqrt{6} + 2}{4}$

31.
$$\sin\left(\frac{3\pi}{2} - \theta\right) = -\cos\theta$$
$$\sin\frac{3\pi}{2}\cos\theta - \cos\frac{3\pi}{2}\sin\theta = -\cos\theta$$
$$(-1)\cos\theta - (0)\sin\theta = -\cos\theta$$
$$-\cos\theta = -\cos\theta \checkmark$$

33. $\sin 2\theta = \frac{24}{25}, \cos 2\theta = \frac{7}{25}, \sin\frac{\theta}{2} = \frac{\sqrt{10}}{10}$, and $\cos\frac{\theta}{2} = \frac{3\sqrt{10}}{10}$ **35.** $\sin 2\theta = -\frac{4\sqrt{5}}{9}, \cos 2\theta = -\frac{1}{9}, \sin\frac{\theta}{2} = \frac{\sqrt{30}}{6}$, and $\cos\frac{\theta}{2} = -\frac{\sqrt{6}}{6}$ **37.** 60°

39. 90°, 210°, 270°, 330° **41.** $\frac{\pi}{3}, \frac{5\pi}{3}$

Photo Credits

Cover Jason Reed/Getty Images, (bkgrnd)RubberBall Selects/Alamy; **iv** (tl tr bl br)The McGraw-Hill Companies; **v** (tl tr)The McGraw-Hill Companies; **xviii** Don Farrall/Getty Images; **xix** Thinkstock/CORBIS; **xx** Tom Stewart/CORBIS; **xxi** Kobal/The Picture Desk; **xxii** Andrew Weeks/SuperStock; **xxiii** Rodolfo Arpia/Alamy Images; **xxiv** Masterfile; **xxix** Digital Vision/Getty Images; **xxv** Burgess Blevins/Getty Images; **xxvi** Onne van der Wal/CORBIS; **xxvii** Photodisc/Alamy Images; **xxviii** Kim Karpeles/Alamy Images; **xxx** Charlie Munsey/CORBIS; **xxxi** Yellow Dog Productions/Getty Images; **P11** Photodisc/Getty Images; **P14** SW Productions/Getty Images; **2–3** Don Farrall/Getty Images; **5** moodboard/CORBIS; **8** Kevin Cooley/Getty Images; **13** Jon Riley/Getty Images; **15** CORBIS; **16** Glow Images/Alamy Images; **18** Bill Bachman/Alamy Images; **19** akg-images, London; **24** Mark Zylber/Alamy Images; **27** Onne van der Wal/CORBIS; **28** The Art Archive/Château de Beauregard Val de Loire/Gianni Dagli Orti/Picture Desk; **31** M. Timothy O'Keefe/Alamy Images; **36** Digital Vision/PunchStock; **37** Danita Delimont/Alamy Images; **38** Elise Amendola/AP Images; **41** Tom Stack/Tom Stack & Associates; **44** Jeff Greenberg/Index Stock Imagery; **46** Mark Scott/Alamy Images; **54** Comstock/PictureQuest; **58–59** Thinkstock/CORBIS; **63** Bernard Papon/AP Images; **65** image100/PunchStock; **66** Pixland/JupiterImages; **70** Akira Kaede/Getty Images; **73** David Young-Wolff / PhotoEdit; **77** Bonnie Kamin/PhotoEdit; **80** Phelan M. Ebenhack/AP Images; **81** (l)Brand X Pictures/PunchStock; **81** (r)Comstock Images/PictureQuest; **83** CORBIS; **87** Francois Gohier/Photo Researchers; **88** David Young-Wolff/PhotoEdit; **93** Jose Luis Pelaez Inc/Getty Images; **96** Michael Newman/PhotoEdit; **97** AP Images; **99** Jeff Cadge/Getty Images; **103** Jacques Boyer/Roger-Viollet/The Image Works; **106** Al Bello/Getty Images; **112** David Young-Wolff / PhotoEdit; **114** Mango Productions/CORBIS; **115** Bill Aron/PhotoEdit; **118** Thomas Barwick/Getty Images; **120** JupiterImages/Thinkstock/Alamy Images; **128** Rick Bowmer/AP Images; **132–133** Tom Stewart/CORBIS; **136** Jeff Greenberg/PhotoEdit; **140** George Silk/Time Life Pictures/Getty Images; **143** Adrian Sherratt/Alamy Images; **146** Brendan Hoffman/Alamy Images; **147** Alastair Grant/AP Images; **148** Visions of America, LLC/Alamy Images; **152** BananaStock/JupiterImages; **155** Jerry & Marcy Monkman/Earth Scenes; **156** Tracy Bennett/Bureau L.A. Collection/CORBIS; **162** Spencer Grant/PhotoEdit; **164** Brand X Pictures/PunchStock; **165** Thorsten Indra/Alamy Images; **170** Radius Images/Alamy Images; **172** Tony Freeman/PhotoEdit; **178** Christine Balderas/iStockphoto; **182–183** Kobal/The Picture Desk; **186** Stephen Dunn/Getty Images; **189** Scott T. Smith/CORBIS; **190** NASA, ESA, H. Weaver (JHU/APL), A. Stern (SwRI), and the HST Pluto Companion Search Team; **195** moodboard/CORBIS; **197** Kenneth Garrett/National Geographic Image Collection; **198** Darron Cummings/AP Images; **202** Stockbyte/Getty Images; **205** BananaStock/PunchStock; **206** Jose Fuste Raga/CORBIS; **209** CORBIS RF/Alamy Images; **215** Ian Timberlake/AFP/Getty Image; **216** Izzy Schwartz/Getty Images; **220** Moritsch, Marc (Via Hornocker, Maurice)/National Geographic Image Collection; **222** Digital Vision; **225** Photographers Choice RF/SuperStock; **226** Richard T. Nowitz/CORBIS; **227** BananaStock/PunchStock; **231** Courtesy Ichinoseki City Museum, Japan; **232** Ben Margo/AP Images; **234** (l)Ben Margot/AP Images; **234** (c)CORBIS/Punchstock; **234** (r)Andrew Hetherington/Getty Images; **246–247** Andrew Weeks/SuperStock; **253** Jennifer Page/AP Images; **255** Christopher Kerrigan/McGraw-Hill Companies; **256** Nice One Productions/CORBIS; **259** Bruce Laurance/Getty Images; **262** Emmanuel Duparcq/AFP/Getty Images; **265** StockShot/Alamy Images; **271** Stu Forster/Getty Images; **273** Randy Faris/CORBIS; **274** Eugene Hoshiko/AP Images; **279** Design Pics/Alamy Images; **281** Steve Allen/Getty Images; **284** Arthur S. Aubry/Getty Images; **289** Akira Kaede/Getty Images; **292** Amy Roh/AP Images; **294** British Library/akg-images, London; **299** Cris Haigh/Alamy Images; **309** Brand X Pictures; **312** Sharon Stabley/AP Images; **314** RubberBall/Alamy Images; **316** OJO Images Ltd/Alamy Images; **317** Keith Dannemiller/CORBIS; **319** Ed-Imaging; **326** Bobby Ristovski/iStockPhoto; **330–331** Rodolfo Arpia/Alamy Images; **333** Jerry Lodriguss/Photo Researchers; **336** ColorBlind Images/Getty Images; **338** Digital Vision/PunchStock; **346** Mark Kauffman/Sports Illustrated Set Number: X1388; **348** Tetra Images/Alamy Images; **349** Purestock/Alamy Images; **353** Bettmann/CORBIS; **354** David Young-Wolff/PhotoEdit; **357** Carnegie Library of Pittsburgh; **360** Randy Faris/CORBIS; **362** JupiterImages/Polka Dot/Alamy Images; **363** Rogelio V. Solis/AP Images; **369** Stock Montage/Getty Images; **373** Lawrence Migdale; **374** Jose Luis Pelaez, Inc./CORBIS; **377** blue jean images/Getty Images; **378** David Frazier/PhotoEdit; **380** Comstock/CORBIS; **383** Paul Sakuma/AP Images; **389** Ingram Publishing/SuperStock; **391** Blasius Erlinger/Getty Images; **394** Crystal Lagoons/www.crystal-lagoons.com; **395** J.Garcia/photocuisine/CORBIS; **402** KMITU/iStockphoto; **406–407** Masterfile; **412** Spencer Grant/PhotoEdit; **414** Tim Boyle/Getty Images; **415** BananaStock/PunchStock; **417** GeoStock/Getty Images; **421** Ryan McVay/Getty Images; **424** Ryan McVay/Getty Images; **426** Rob Wilkinson/Alamy Images; **428** Mikael Karlsson/Alamy Images; **429** Ken Reid/Getty Images; **431** JupiterImages; **433** Dennis Cox/Alamy Images; **435** Tom Uhlman/Visuals Unlimited; **439** akg-images, London; **442** The Granger Collection, NYC; **444** Paul Damien/Getty Images; **446** CORBIS; **447** Ariel Skelley/JupiterImages; **451** CORBIS; **453** John Swart/AP Images; **458** Duomo/CORBIS; **468** Superstudio/Getty Images; **472–473** Burgess Blevins/Getty Images; **475** Stockbyte/Alamy Images; **477** Creatas/SuperStock; **479** Core Imagery/Alamy Images; **480** Robert W. Ginn/PhotoEdit; **485** Bill Aron/PhotoEdit; **486** David Young-Wolff/PhotoEdit; **489** James Taylor/Alamy Images; **490** CORBIS; **492** Joe Tucciarone/Photo Researchers, Inc.; **495** Bettmann/CORBIS; **497** Andrew Paterson/Alamy Images; **498** David J. Green/Alamy Images; **505** Keren Su/CORBIS; **506** Ron Chapple/Thinkstock/Alamy Images; **510** Mark Leach/Alamy Images; **513** Digital Vision; **514** John Lund/Marc Romanelli/Getty Images; **517** Michael Newman/PhotoEdit; **518** The Granger Collection, New York; **520** Digital Vision; **521** Ingo Jezierski/CORBIS; **525** CORBIS; **528** Jose Luis Pelaez, Inc./CORBIS; **530** David Young-Wolff/PhotoEdit; **533** Nik Wheeler/CORBIS; **534** Tim White/AFP/Getty Images; **536** Lloyd Sutton/Masterfile; **538** (t)Oregon State Archive; **538** (b)Oregon State Archive; **540** Ed-Imaging; **546** ASP/Alamy Images; **550–551** Onne van der Wal/CORBIS; **553** Dave Fleetham/Tom Stack & Associates; **559** John Gaps III/AP Images; **560** Steve Allen Travel Photography/Alamy Images; **562** Philip Rostron/Masterfile; **566** Kenneth Garrett/National Geographic Image Collection; **567** Myrleen Ferguson Cate/PhotoEdit; **572** Bill Bachmann/PhotoEdit; **574** MLB Photos via Getty Images; **577** Myrleen Ferguson Cate/PhotoEdit; **579** PA1 Tom Sperduto, US Coast Guard; **583** Stefanie Weiss/AP Images; **589** RubberBall Productions; **591** Tim Fitzharris/Minden Pictures; **592** Tony West/Alamy Images; **594** Victor R. Caivano/AP Images; **595** Archive Photos/Getty Images; **596** Enigma/Alamy Images; **598** Paul Burns/Getty Images; **610** Stockbyte/Getty Images; **614–615** Photodisc/Alamy Images; **617** U.S. Department of Transportation; **618** Phil Noel/AP Images; **621** The McGraw-Hill Companies; **623** Chip Simons/Getty Images; **626** Brand X; **628** Derek Berwin/Getty Images; **631** Patrick Clark/Getty Images; **635** Brand X Pictures/PunchStock; **638** Ed-Imaging; **641** NASA-MSFC; **642** StockTrek/Getty Images; **644** James Houck/Alamy Images; **645** Wally McNamee/CORBIS SYGMA; **651** StockTrek/Getty Images; **653** R. Morley/PhotoLink/Getty Images; **654** Bettmann/CORBIS; **659** R. Morley/PhotoLink/Getty Images; **665** CLEO Photo/Alamy Images; **666** (l)Sean Murphy/Getty Images; **666** (r)Nick Ayliffe/Alamy Images; **674** Enrico Fianchini/iStockphoto; **678–679** Kim Karpeles/Alamy Images; **681** JupiterImages/Thinkstock / Alamy Images; **682** Thinkstock/JupiterImages; **686** Spencer Grant/PhotoEdit; **688** akg-images, London; **693** PhotoLink/Getty Images; **694** David Madison/Getty Images; **696** John Atashian/CORBIS; **697** Alinari Archives/The Image Works; **701** Nick Koudis/Getty Images; **709** Westend61/Alamy Images; **710** Blend Images/JupiterImages; **716** Thom Lang/CORBIS; **718** Stephen Alvarez/National Geographic Image Collection; **721** Somos/Veer/Getty Images; **724** Matt Slocum/AP Images; **727** Comstock Images/Alamy Images; **730** McGraw-Hill Companies; **738** Jose Luis Pelaez, Inc./CORBIS; **742–743** Digital Vision/Getty Images; **745** The McGraw-Hill Companies; **749** The McGraw-Hill Companies; **752** Brand X/JupiterImages; **753** image100/CORBIS; **754** Milk Photographie/CORBIS; **757** image100 Ltd; **759** Blend Images/SuperStock; **762** Kwame Zikomo/SuperStock; **764** Blend Images/SuperStock; **766** Dan Loh/AP Images; **769** Warren Westura/AP Images; **770** akg-images, London; **775** Ryan McVay/Getty Images; **776** Tony Freeman/PhotoEdit; **777** David Young-Wolff/PhotoEdit; **778** Noah Addis/Star Ledger/CORBIS; **780** Tony Anderson/Getty Images; **783** D. Hurst/Alamy Images; **785** Ed-Imaging; **786** Rim Light/PhotoLink/Getty Images; **787** Creative Expression/Alamy Images; **792** Skip Weisenburger/AP Images; **800** Charles Gupton/CORBIS; **804–805** Charlie Munsey/CORBIS; **808** isifa Image Service s.r.o./Alamy Images; **810** Edward Kinsman/Photo Researchers; **812** STAN HONDA/AFP/Getty Images; **815** Avico Ltd/Alamy Images; **817** Nick Koudis/Getty Images; **818** Grafton Marshall Smith/CORBIS; **821** Michael Burke/AP Images; **822** Digital Vision/Alamy Images; **825** Joseph Sohm/drr.net; **828** Jason Werling/AP Images; **830** age fotostock/SuperStock; **832** JPL/NASA; **836** Matthew Thayer/AP Images; **838** Image Source Black; **840** Ed-Imaging; **843** Federico Cabello/SuperStock; **845** CORBIS/SuperStock; **849** George S de Blonsky/Alamy Images; **852** Holger Otto/Reuters/CORBIS; **853** Kee Coleman/Courtesy of The Bancroft Library/University of California, Berkeley; **857** Digital Vision/PunchStock; **860** Brooke Fasani/CORBIS; **866** Stephen Frink Collection/Alamy Images; **868** Ingram Publishing/Alamy Images; **871** Pieter Estersohn/Beateworks/CORBIS; **875** Gene J. Puskar/PhotoEdit; **884** Cn Boon/Alamy Images; **888–889** Yellow Dog Productions/Getty Images; **891** Rubberball/JupiterImages; **893** Dinodia; **895** Thinkstock/CORBIS; **896** Jim West/Alamy Images; **898** S. Carmona/CORBIS; **901** George S de Blonsky/Alamy Images; **902** Marwood Jenkins/Alamy Images; **907** Balfour Studios/Alamy Images; **908** Thinkstock/Index Stock; **911** Steve Allen/Brand X/CORBIS; **914** Philip Scalia/Alamy Images; **916** Steve Kinderman/AP Images; **919** Wally McNamee/CORBIS; **924** JewelryStock/Alamy Images; **930** CORBIS/JupiterImages; **934** Eclipse Studios.

Math Online

A mathematics multilingual eGlossary is available at **glencoe.com**. The glossary includes the following languages:

Arabic	Haitian Creole	Russian	Vietnamese
Bengali	Hmong	Spanish	
Cantonese	Korean	Tagalog	
English	Portuguese	Urdu	

Cómo usar el glosario en español:

1. Busca el término en inglés que desees encontrar.
2. El término en español, junto con la definición, se encuentran en la columna de la derecha.

English / Español

A

absolute value (p. 27) A number's distance from zero on the number line, represented by x.

valor absoluto Distancia entre un número y cero en una recta numérica; se denota con $|x|$.

absolute value function (p. 103) A function written as $f(x) = x$, where $f(x) \geq 0$ for all values of $|x|$.

función del valor absoluto Una función que se escribe $f(x) = |x|$, donde $f(x) \geq 0$, para todos los valores de x.

algebraic expression (p. 5) An expression that contains at least one variable.

expresión algebraica Expresión que contiene al menos una variable.

alternative hypothesis (p. 781) Mutually exclusive to the null hypothesis. It is stated as an inequality using $<$, $\leq$, $>$, or $\geq$.

hipótesis alternativa Mutuamente exclusiva a la hipótesis nula. Se indica como usar de la desigualdad $<$, $\leq$, $>$, o $\geq$.

amplitude (p. 855) For functions in the form $y = a \sin b\,\theta$ or $y = a \cos b\,\theta$, the amplitude is $|a|$.

amplitud Para funciones de la forma $y = a$ sen $b\,\theta$ o $y = a \cos b\,\theta$, la amplitud es $|a|$.

angle of depression (p. 812) The angle between a horizontal line and the line of sight from the observer to an object at a lower level.

ángulo de depresión Ángulo entre una recta horizontal y la línea visual de un observador a una figura en un nivel inferior.

angle of elevation (p. 812) The angle between a horizontal line and the line of sight from the observer to an object at a higher level.

ángulo de elevación Ángulo entre una recta horizontal y la línea visual de un observador a una figura en un nivel superior.

Arccosine (p. 871) The inverse of $y = \cos x$, written as $x = \text{Arccos } y$.

arcocoseno La inversa de $y = \cos x$, que se escribe como $x = \arccos y$.

Arcsine (p. 871) The inverse of $y = \sin x$, written as $x = \text{Arcsin } y$.

arcoseno La inversa de $y = \text{sen } x$, que se escribe como $x = \text{arcsen } y$.

Arctangent (p. 871) The inverse of $y = \tan x$ written as $x = \text{Arctan } y$.

arcotangente La inversa de $y = \tan x$ que se escribe como $x = \arctan y$.

arithmetic mean (p. 689) The terms between any two nonconsecutive terms of an arithmetic sequence.

media aritmética Cualquier término entre dos términos no consecutivos de una sucesión aritmética.

arithmetic sequence (p. 681) A sequence in which each term after the first is found by adding a constant, the common difference d, to the previous term.

sucesión aritmética Sucesión en que cualquier término después del primero puede hallarse sumando una constante, la diferencia común d, al término anterior.

arithmetic series (p. 690) The indicated sum of the terms of an arithmetic sequence.

asymptote (pp. 475, 569) A line that a graph approaches but never crosses.

augmented matrix (p. 238) A coefficient matrix with an extra column containing the constant terms.

axis of symmetry (p. 250) A line about which a figure is symmetric.

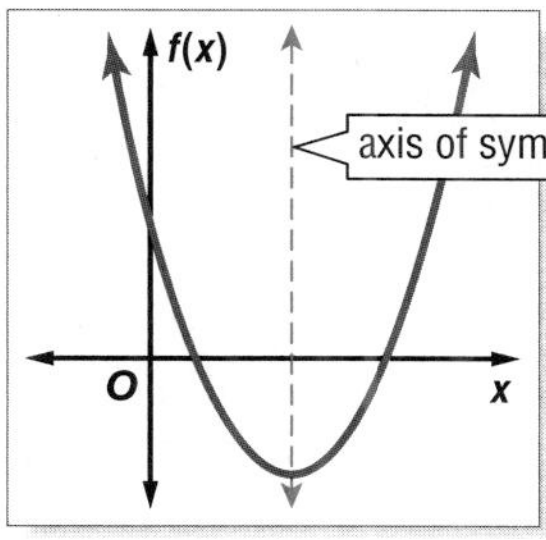

B

$b^{\frac{1}{n}}$ (p. 446) For any real number b and for any positive integer n, $b^{\frac{1}{n}} = \sqrt[n]{b}$, except when $b < 0$ and n is even .

bar graph (p. 1001) A graphic form using bars to make comparisons of statistics.

biased (p. 745) A sample or survey in which one or more parts of the population are favored over others.

binomial distribution (p. 786) A distribution that shows the probabilities of the outcomes of a binomial experiment.

binomial experiment (p. 786) An experiment in which there are exactly two possible outcomes for each trial, a fixed number of independent trials, and the probabilities for each trial are the same.

Binomial Theorem (p. 721) If n is a nonnegative integer, then $(a + b)^n = 1a^nb^0 + \frac{n}{1}a^{n-1}b^1 + \frac{n(n+1)}{1 \cdot 2}a^{n-2}b^2 + \cdots + 1a^0b^n$.

bivariate data (p. 92) Data with two variables.

boundary (p. 117) A line or curve that separates the coordinate plane into two regions.

bounded (p. 160) A region is bounded when the graph of a system of constraints is a polygonal region.

serie aritmética Suma específica de los términos de una sucesión aritmética.

asíntota Recta a la que se aproxima una gráfica, sin jamás cruzarla.

matriz ampliada Matriz coeficiente con una columna extra que contiene los términos constantes.

eje de simetría Recta respecto a la cual una figura es simétrica.

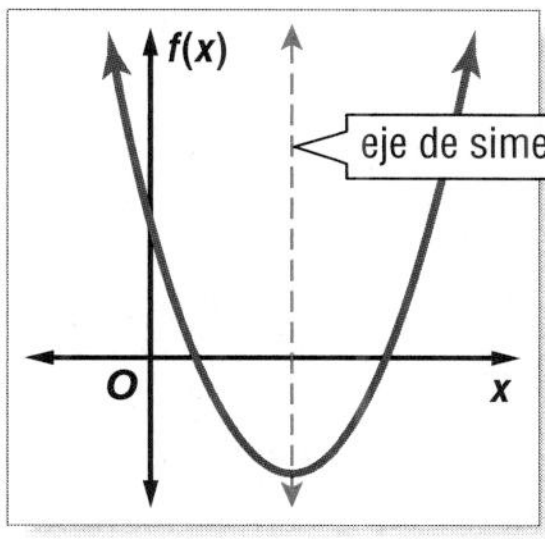

$b^{\frac{1}{n}}$ Para cualquier número real b y para cualquier entero positivo n, $b^{\frac{1}{n}} = \sqrt[n]{b}$, excepto cuando $b < 0$ y n es par.

gráfica de barras Tipo de gráfica que usa barras para comparar estadísticas.

sesgo Muestra o encuesta en la cual se prefiere una o más partes de la población sobre las otras.

distribución binómica Distribución que muestra las probabilidades de los resultados de un experimento binómico.

experimento binomial Experimento con exactamente dos resultados posibles para cada prueba, un número fijo de pruebas independientes y en el cual cada prueba tiene igual probabilidad.

teorema del binomio Si n es un entero no negativo, entonces $(a + b)^n = 1a^nb^0 + \frac{n}{1}a^{n-1}b^1 + \frac{n(n+1)}{1 \cdot 2}a^{n-2}b^2 + \cdots + 1a^0b^n$.

datos bivariados Datos con dos variables.

frontera Recta o curva que divide un plano de coordenadas en dos regiones.

acotada Una región está acotada cuando la gráfica de un sistema de restricciones es una región poligonal.

box-and-whisker plot (p. 1005) A diagram that divides a set of data into four parts using the median and quartiles. A box is drawn around the quartile values and whiskers extend from each quartile to the extreme data points.

diagrama de caja y patillas Diagrama que divide un conjunto de datos en cuatro partes usando la mediana y los cuartiles. Se dibuja una caja alrededor de los cuartiles y se extienden patillas de cada uno de ellos a los valores extremos.

break-even point (p. 136) The point at which the income equals the cost.

el punto de equilibrio Cuando el punto la renta iguala la causalidad del coste.

C

Cartesian coordinate plane (p. P4) A plane divided into four quadrants by the intersection of the x-axis and the y-axis at the origin.

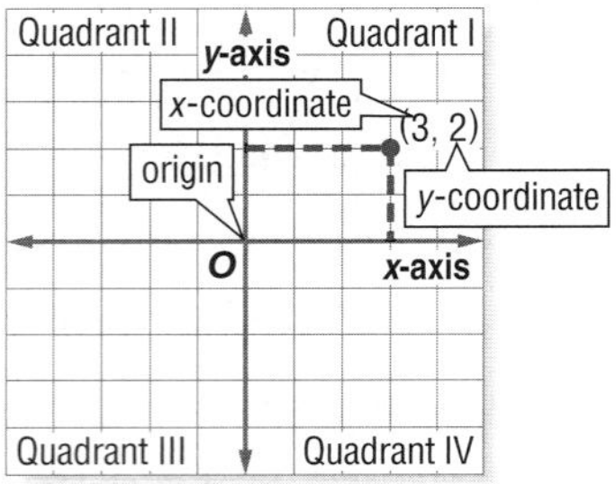

plano de coordenadas cartesiano Plano dividido en cuatro cuadrantes mediante la intersección en el origen de los ejes x y y.

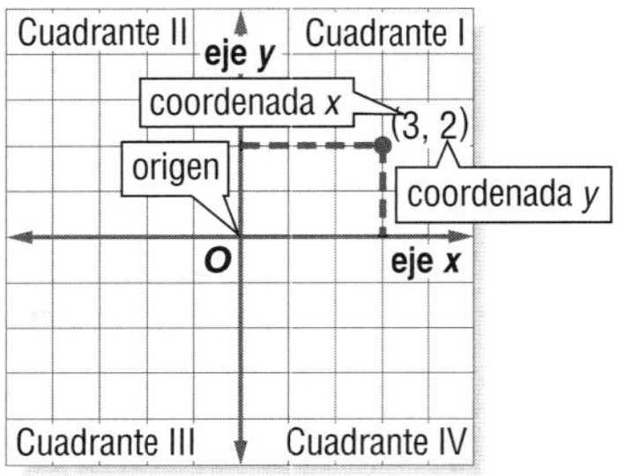

causation (p. 747) One event is shown to be the direct cause of another event.

acontecimiento Un acontecimiento se demuestra para ser la causa directa de otro.

census (p. 745) A survey in which every member of the population is polled.

censo Examen en el cual voten a cada miembro de la población.

center (pp. 631, 639)
1. circle—The given point on a plane from which the set of all points in the plane are equidistant.
2. ellipse—The point at which the axes are perpendicular.

centro
1. círculo—El conjunto de todos los puntos de un plano que son equidistantes.
2. elipse—El punto en el cual las ejes son perpendiculares.

center of a circle (p. 631) The point from which all points on a circle are equidistant.

centro de un círculo El punto desde el cual todos los puntos de un círculo están equidistantes.

center of a hyperbola (p. 648) The midpoint of the segment whose endpoints are the foci.

centro de una hipérbola Punto medio del segmento cuyos extremos son los focos.

center of an ellipse (p. 639) The point at which the major axis and minor axis of an ellipse intersect.

centro de una elipse Punto de intersección de los ejes mayor y menor de una elipse.

central angle (p. 820) An angle with a vertex at the center of the circle.

ángulo central Ángulo cuyo vértice es el centro del círculo.

Change of Base Formula (p. 518) For all positive numbers a, b, and n, where $a \neq 1$ and $b \neq 1$,

$$\log_a n = \frac{\log_b n}{\log_b a}.$$

fórmula del cambio de base Para todo número positivo a, b y n, donde $a \neq 1$ y $b \neq 1$,

$$\log_a n = \frac{\log_b n}{\log_b a}.$$

circle (p. 631) The set of all points in a plane that are equidistant from a given point in the plane, called the center.

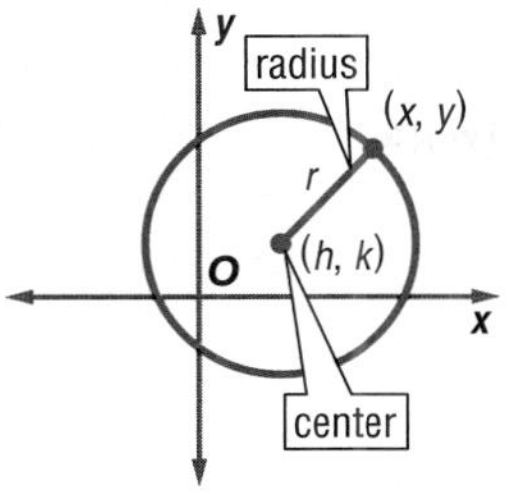

círculo Conjunto de todos los puntos en un plano que equidistan de un punto dado del plano llamado centro.

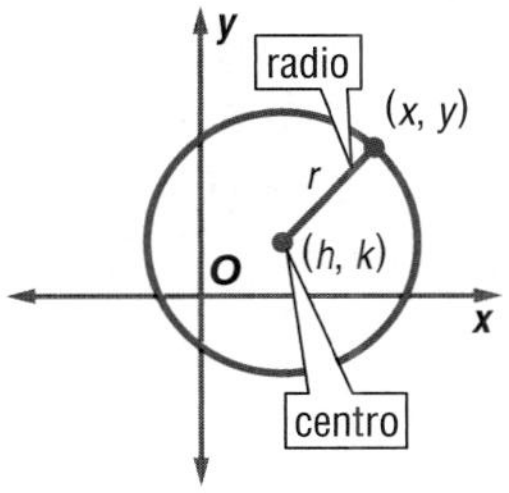

circular function (p. 848) A function defined using a unit circle.

funciones circulares Funciones definidas en un círculo unitario.

coefficient matrix (p. 223) A matrix that contains only the coefficients of a system of equations.

matriz coeficiente Una matriz que contiene solamente los coeficientes de un sistema de ecuaciones.

column matrix (p. 186) A matrix that has only one column.

matriz columna Matriz que sólo tiene una columna.

combination (p. P12) An arrangement of objects in which order is not important.

combinación Arreglo de elementos en que el orden no es importante.

combined variation (p. 589) When one quantity varies directly and/or inversely as two or more other quantities.

variación combinada Cuando una cantidad varía directamente e inverso como dos o más otras cantidades.

common difference (p. 681) The difference between the successive terms of an arithmetic sequence.

diferencia común Diferencia entre términos consecutivos de una sucesión aritmética.

common logarithms (p. 525) Logarithms that use 10 as the base.

logaritmos comunes El logaritmo de base 10.

common ratio (p. 683) The ratio of successive terms of a geometric sequence.

razón común Razón entre términos consecutivos de una sucesión geométrica.

completing the square (p. 285) A process used to make a quadratic expression into a perfect square trinomial.

completar el cuadrado Proceso mediante el cual una expresión cuadrática se transforma en un trinomio cuadrado perfecto.

complex conjugates (p. 279) Two complex numbers of the form $a + bi$ and $a - bi$.

conjugados complejos Dos números complejos de la forma $a + bi$ y $a - bi$.

complex fraction (p. 556) A rational expression whose numerator and/or denominator contains a rational expression.

fracción compleja Expresión racional cuyo numerador o denominador contiene una expresión racional.

complex number (p. 277) Any number that can be written in the form $a + bi$, where a and b are real numbers and i is the imaginary unit.

número complejo Cualquier número que puede escribirse de la forma $a + bi$, donde a y b son números reales e i es la unidad imaginaria.

composition of functions (p. 410) A function is performed, and then a second function is performed on the result of the first function. The composition of f and g is denoted by $f \cdot g$, and $[f \cdot g](x) = f[g(x)]$.

composición de funciones Se evalúa una función y luego se evalúa una segunda función en el resultado de la primera función. La composición de f y g se define con $f \cdot g$, y $[f \cdot g](x) = f[g(x)]$.

Glossary/Glosario

compound event (p. 998) Two or more simple events.

evento compuesto Dos o más eventos simples.

compound inequality (p. 41) Two inequalities joined by the word *and* or *or*.

desigualdad compuesta Dos desigualdades unidas por las palabras *y* u *o*.

compound interest (p. 486) Interest paid on the principal of an investment and any previously earned interest.

interés compuesto Interés obtenido tanto sobre la inversion inicial como sobre el interes conseguido.

conditional probability (p. 759) The probability of an event given that another event has already occurred.

probabilidad condicional La probabilidad de un acontecimiento condicion que ha occurido un cierto acontecimiento precedente.

confidence interval (p. 780) An estimate of a population parameter stated as a range with a specific degree of certainty.

intervalo de la confianza Una estimación de un parámetro de la población indicado como gama con un grado específico de la certeza.

conic section (p. 656) Any figure that can be obtained by slicing a double cone.

sección cónica Cualquier figura obtenida mediante el corte de un cono doble.

conjugates (p. 442) Binomials of the form $a\sqrt{b} + c\sqrt{d}$ and $a\sqrt{b} - c\sqrt{d}$, where a, b, c, and d are rational numbers.

conjugados Binomios de la forma $a\sqrt{b} + c\sqrt{d}$ y $a\sqrt{b} - c\sqrt{d}$, donde a, b, c, y d son números racionales.

conjugate axis (p. 648) The segment of length $2b$ units that is perpendicular to the transverse axis at the center.

eje conjugado El segmento de $2b$ unidades de longitud que es perpendicular al eje transversal en el centro.

consistent (p. 137) A system of equations that has at least one solution.

consistente Sistema de ecuaciones que posee por lo menos una solución.

constant difference (p. 651) The absolute value of the difference between the distances from any point on the hyperbola to the foci of the hyperbola.

diferencia constante El valor absoluto de la diferencia entre las distancias de cualquier punto en la hipérbola a los focos de la hipérbola.

constant function (p. 109) A linear function of the form $f(x) = b$.

función constante Función lineal de la forma $f(x) = b$.

constant matrix (p. 233) A matrix that contains the constants of a system.

matriz constante Una matriz que contiene las constantes de un sistema

constant of variation (pp. 90, 586) The constant k used with direct or inverse variation.

constante de variación La constante k que se usa en variación directa o inversa.

constant sum (p. 640) The sum of the distances from the foci to any point on the ellipse.

suma constante La suma de las distancias de los focos a cualquier punto en la elipse.

constant term (p. 249) In $f(x) = ax^2 + bx + c$, c is the constant term.

término constante En $f(x) = ax^2 + bx + c$, c es el término constante.

constraints (p. 160) Conditions given to variables, often expressed as linear inequalities.

restricciones Condiciones a que están sujetas las variables, a menudo escritas como desigualdades lineales.

contingency table (p. 760) Records data in which different possible situations result in different possible outcomes. Each value represents the relative frequency of an outcome.

table de contingencias Registra datos en que diferentes posibles situaciones resultan en distintos posibles resultados. Cada valor representa la frecuencia relative de un resultado.

continuous probability distribution (p. 773) The outcome can be any value in an interval of real numbers, represented by curves.

distribución de probabilidad continua El resultado puede ser cualquier valor de un intervalo de números reales, representados por curvas.

continuous relation (p. 62) A relation that can be graphed with a line or smooth curve.

relación continua Relación cuya gráfica puede ser una recta o una curva suave.

control goup (p. 746) In an experiment, those given the placebo, or false treatment.

grupo de control En un experimento, ésos dados el placebo, o el tratamiento falso.

convergent series (p. 705) An infinite series with a sum.

serie convergente Serie infinita con una suma.

correlation (p. 747) Two events ae related.

corelacion Dos acontecimientos son relacionados.

correlation coefficient (p. 94) A measure that shows how well data are modeled by a linear equation.

coeficiente de correlación Una medida que demuestra cómo los datos bien son modelados por una ecuación linear.

cosecant (p. 808) For any angle, with measure α, a point $P(x, y)$ on its terminal side, $r = \sqrt{x^2 + y^2}$, $\csc \alpha = \frac{r}{y}$.

cosecante Para cualquier ángulo de medida α, un punto $P(x, y)$ en su lado terminal, $r = \sqrt{x^2 + y^2}$, $\csc \alpha = \frac{r}{y}$.

cosine (p. 808) For any angle, with measure α, a point $P(x, y)$ on its terminal side, $r = \sqrt{x^2 + y^2}$, $\cos \alpha = \frac{x}{r}$.

coseno Para cualquier ángulo de medida α, un punto $P(x, y)$ en su lado terminal, $r = \sqrt{x^2 + y^2}$, $\cos \alpha = \frac{x}{r}$.

cotangent (p. 808) For any angle, with measure α, a point $P(x, y)$ on its terminal side, $r = \sqrt{x^2 + y^2}$, $\cot \alpha = \frac{x}{y}$.

cotangente Para cualquier ángulo de medida α, un punto $P(x, y)$ en su lado terminal, $r = \sqrt{x^2 + y^2}$, $\cot \alpha = \frac{x}{y}$.

coterminal angles (p. 818) Two angles in standard position that have the same terminal side.

ángulos coterminales Dos ángulos en posición estándar que tienen el mismo lado terminal.

co-vertices (pp. 639, 648)
ellipse—The endpoints of the minor axis.
hyperbola—The endpoints of the conjugate axis.

co-cimas
(elipse)—Puntos finales del eje de menor importancia.
(hipérbola)—Puntos finales del eje conyugal.

Cramer's Rule (p. 223) A method that uses determinants to solve a system of linear equations.

regla de Crámer Método que usa determinantes para resolver un sistema de ecuaciones lineales.

cross products (p. 993) If $\frac{a}{c} = \frac{b}{d}$, then $ad = bc$. If $ad = bc$, then $\frac{a}{c} = \frac{b}{d}$.

productos cruzados Si $\frac{a}{c} = \frac{b}{d}$, entonces $ad = bc$. Si $ad = bc$, entonces $\frac{a}{c} = \frac{b}{d}$.

cycle (p. 849) One complete pattern of a periodic function.

ciclo Un patrón completo de una función periódica.

D

decay factor (478) In exponential decay, the base of the exponential expression, $1 - r$.

factor de decaimiento En decaimiento exponencial, la base de la expresión exponencial, $1 - r$.

degree of a polynomial (p. 335) The greatest degree of any term in the polynomial.

grado de un polinomio Grado máximo de cualquier término del polinomio.

dependent (p. 137) When a system of linear equations has an infinite number of solutions.

dependiente Sistema de ecuaciones que posee un número infinito de soluciones.

dependent events (p. 998) The outcome of one event does affect the outcome of another event.

eventos dependientes El resultado de un evento afecta el resultado de otro evento.

dependent system (p. 137) A consistent system of equations that has an infinite number of solutions.

sistema dependiente Sistema de ecuaciones que posee un número infinito de soluciones.

dependent variable (p. 64) The other variable in a function, usually y, whose values depend on x.

variable dependiente La otra variable de una función, por lo general y, cuyo valor depende de x.

depressed polynomial (p. 379) The quotient when a polynomial is divided by one of its binomial factors.

polinomio reducido El cociente cuando se divide un polinomio entre uno de sus factores binomiales.

determinant (p. 220) A square array of numbers or variables enclosed between two parallel lines.

determinante Arreglo cuadrado de números o variábles encerrados entre dos rectas paralelas.

diagonal rule (p. 221) A method for finding the determinant of a third-order matrix.

regla diagonal Método para encontrar el determinante de una matriz third-order.

dilation (p. 111, 211) A transformation in which a geometric figure is enlarged or reduced.

homotecia Transformación en que se amplía o se reduce un figura geométrica.

dimensional analysis (p. 342) Performing operations with units.

anállisis dimensional Realizar operaciones con unidades.

dimension (p. 185) A description of the number of rows and columns of a matrix.

dimensión Una descripción del número de filas y de columnas de una matriz.

dimensions of a matrix (p. 185) The number of rows, m, and the number of columns, n, of the matrix written as $m \times n$.

tamaño de una matriz El número de filas, m, y columnas, n, de una matriz, lo que se escribe $m \times n$.

directrix (p. 623) See parabola.

directriz Véase parábola.

direct variation (pp. 90, 586) y varies directly as x if there is some nonzero constant k such that $y = kx$. k is called the constant of variation.

variación directa y varía directamente con x si hay una constante no nula k tal que $y = kx$. k se llama la constante de variación.

discrete probability distributions (p. 767) Probabilities that have a finite number of possible values.

distribución de probabilidad discreta Probabilidades que tienen un número finito de valores posibles.

discrete relation (p. 62) A relation in which the domain is a set of individual points.

relación discreta Relación en la cual el dominio es un conjunto de puntos individuales.

discriminant (p. 295) In the Quadratic Formula, the expression $b^2 - 4ac$.

discriminante En la fórmula cuadrática, la expresión $b^2 - 4ac$.

Distance Formula (p. 618) The distance between two points with coordinates (x_1, y_1) and (x_2, y_2) is given by $d = \sqrt{(x_2 - x_1)^2 + (y_2 - y_1)^2}$.

fórmula de la distancia La distancia entre dos puntos (x_1, y_1) and (x_2, y_2) viene dada por $d = \sqrt{(x_2 - x_1)^2 + (y_2 - y_1)^2}$.

divergent series (p. 705) An infinite geometric series that does *not* have sum.

serie divergente Serie geométrica infinita que no tiene suma.

domain (p. P4) The set of all x-coordinates of the ordered pairs of a relation.

dominio El conjunto de todas las coordenadas x de los pares ordenados de una relación.

dot plot (p. 92) Two sets of data plotted as ordered pairs in a coordinate plane.

diagrama del punto Dos conjuntos de datos gráficados como pares ordenados en un plano de coordenadas.

double bar graph (p. 1001) Compares two sets of data by showing each as a bar whose length is related to the frequency.

gráfica de barras dobles Compara dos conjuntos de datos al mostrar cada uno de ellos como una barra cuya longitud se relaciona con la frecuencia.

E

e (p. 527) The irrational number 2.71828…. *e* is the base of the natural logarithms.

e El número irracional 2.71828…. *e* es la base de los logaritmos naturales.

element (p. 185) Each value in a matrix.

elemento Cada valor de una matriz.

elimination method (p. 144) Eliminate one of the variables in a system of equations by adding or subtracting the equations.

método de eliminación Eliminar una de las variables de un sistema de ecuaciones sumando o restando las ecuaciones.

ellipse (p. 639) The set of all points in a plane such that the sum of the distances from two given points in the plane, called foci, is constant.

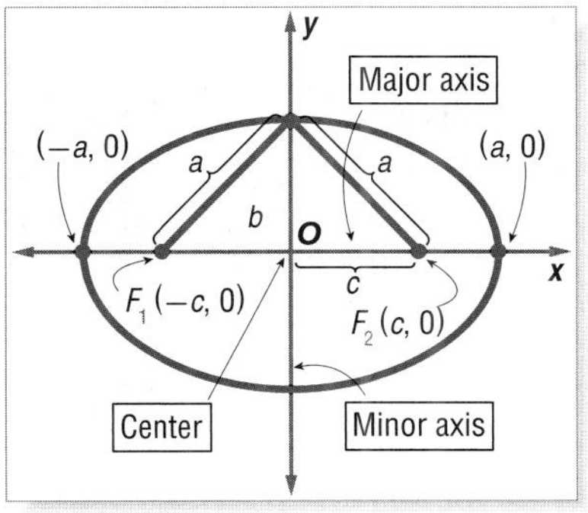

elipse Conjunto de todos los puntos de un plano en los que la suma de sus distancias a dos puntos dados del plano, llamados focos, es constante.

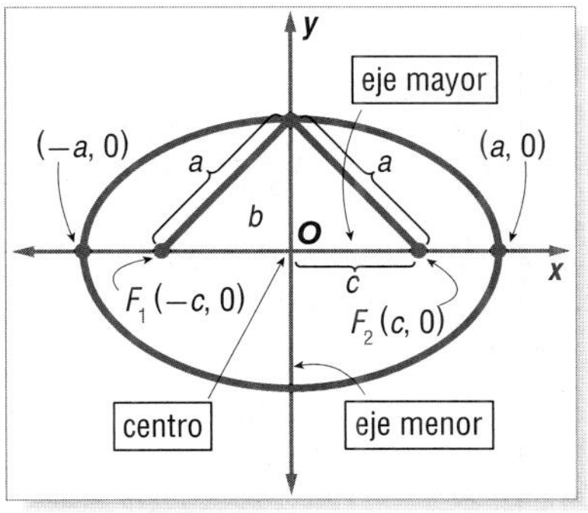

empty set (p. 28) The solution set for an equation that has no solution, symbolized by { } or ø.

conjunto vacío Conjunto solución de una ecuación que no tiene solución, denotado por { } o ø.

end behavior (p. 350) The behavior of the graph as x approaches positive infinity $(+\infty)$ or negative infinity $(-\infty)$.

comportamiento final El comportamiento de una gráfica a medida que x tiende a más infinito $(+\infty)$ o menos infinito $(-\infty)$.

equal matrices (p. 186) Two matrices that have the same dimensions and each element of one matrix is equal to the corresponding element of the other matrix.

matrices iguales Dos matrices que tienen las mismas dimensiones y en las que cada elemento de una de ellas es igual al elemento correspondiente en la otra matriz.

equation (p. 18) A mathematical sentence stating that two mathematical expressions are equal.

ecuación Enunciado matemático que afirma la igualdad de dos expresiones matemáticas.

event (p. 998) One or more outcomes of a trial.

evento Uno o más resultados de una prueba.

experiment (p. 746) Something that is intentionally done to people, animals, or objects, and then the response is observed.

experimento Algo se hace intencionalmente poblar, los animales, o los objetos, y entonces la respuesta se observa.

experimental probability (p. 786) What is estimated from observed simulations or experiments.

probabilidad experimental Qué se estima de simulaciones o de experimentos observados.

explicit formula (p. 714) Gives a_n as a function of n, such as $a_n = 3n + 1$.

exponential decay (p. 477) Exponential decay occurs when a quantity decreases exponentially over time.

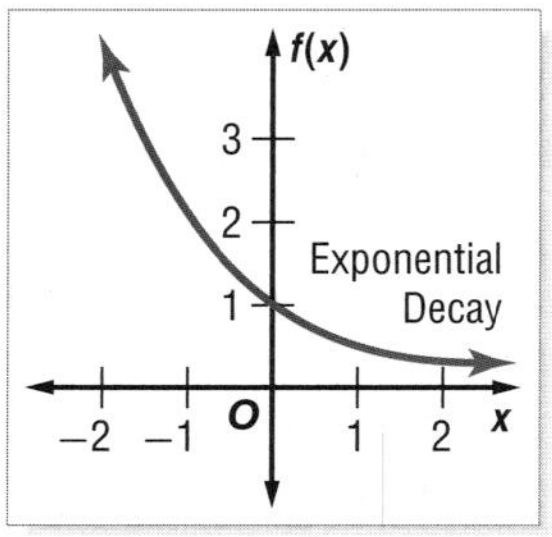

exponential equation (p. 485) An equation in which the variables occur as exponents.

exponential function (p. 475) A function of the form $y = ab^x$, where $a \neq 0$, $b > 0$, and $b \neq 1$.

exponential growth (p. 475) Exponential growth occurs when a quantity increases exponentially over time.

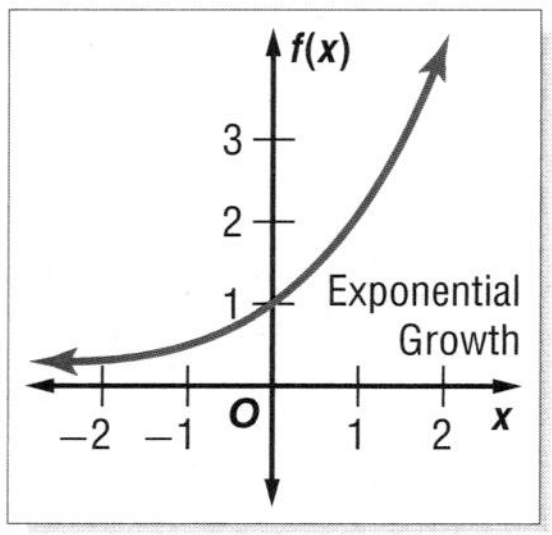

exponential inequality (p. 487) An inequality involving exponential functions.

extraneous solution (p. 453) A number that does not satisfy the original equation.

extrema (p. 358) The maximum and minimum values of a function.

un fórmula explícito Da en función de n, tal como $a_n = 3n + 1$.

desintegración exponencial Ocurre cuando una cantidad disminuye exponencialmente con el tiempo.

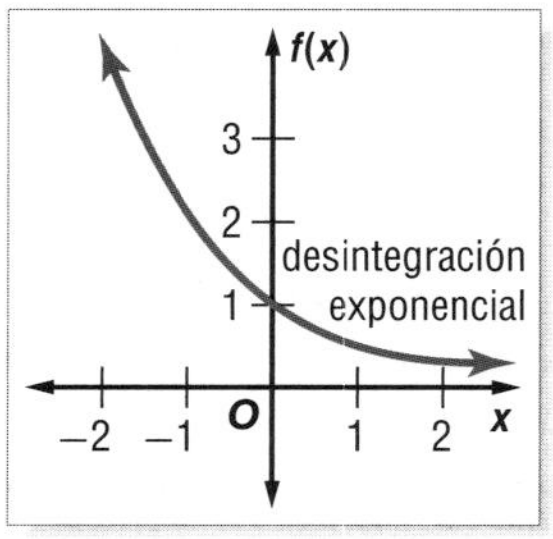

ecuación exponencial Ecuación en que las variables aparecen en los exponentes.

función exponencial Una función de la forma $y = ab^x$, donde $a \neq 0$, $b > 0$, y $b \neq 1$.

crecimiento exponencial El que ocurre cuando una cantidad aumenta exponencialmente con el tiempo.

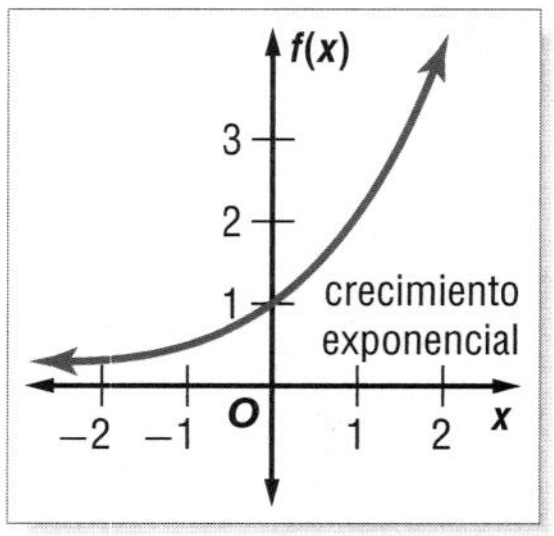

desigualdad exponencial Desigualdad que contiene funciones exponenciales.

solución extraña Número que no satisface la ecuación original.

extrema Son los valores máximos y mínimos de una función.

F

factored form (p. 268) The form of a polynomial showing all of its factors. $y = a(x - p)(x - q)$ is the factored form of a quadratic equation.

factorial (p. P10) If n is a positive integer, then $n! = n(n - 1)(n - 2) \ldots 2 \cdot 1$.

Factor Theorem (p. 379) If a polynomial $P(x)$ is divided by $x - r$, the remainder is a constant $P(r)$.

failure (p. 764) Any outcome other than the desired outcome.

family of graphs (p. 109) A group of graphs that displays one or more similar characteristics.

forma reducida La forma de un polinomio que demuestra todos sus factores. $y = a(x - p)(x - q)$ es la forma descompuesta en factores de una ecuación cuadrática.

factorial Si n es un entero positivo, entonces $n! = n(n - 1)(n - 2) \ldots 2 \cdot 1$.

teorema factor Si un polinomio $P(x)$ es dividido por $x - r$, el resto es un $P(r)$ constante.

fracaso Cualquier resultado distinto del deseado.

familia de gráficas Grupo de gráficas que presentan una o más características similares.

feasible region (p. 160) The intersection of the graphs in a system of constraints.

región viable Intersección de las gráficas de un sistema de restricciones.

Fibonacci sequence (p. 714) A sequence in which the first two terms are 1 and each of the additional terms is the sum of the two previous terms.

sucesión de Fibonacci Sucesión en que los dos primeros términos son iguales a 1 y cada término que sigue es igual a la suma de los dos anteriores.

finite sequence (p. 681) A sequence containing a limited number of terms.

secuencia finitas Una secuencia que contiene un número limitado de términos.

foci (pp. 639, 648)
Ellipse—The two fixed points from which the sum of the distances from a set of all points in a plane is constant.
Hyperbola—The two fixed points from which the difference of the distances from a set of all points in a plane is constant.

focos
(de elipse) Los dos puntos fijos de los cuales la suma de las distancias de un sistema de todos los puntos en un plano es constante.
(de hipérbola) Los dos puntos fijos de los cuales la diferencia de las distancias de un sistema de todos los puntos en un plano es constante.

focus (p. 623) See parabola, ellipse, hyperbola.

foco Véase parábola, elipse, hipérbola.

FOIL method (p. 268) The product of two binomials is the sum of the products of **F** the *first* terms, **O** the *outer* terms, **I** the *inner* terms, and **L** the *last* terms.

método FOIL El producto de dos binomios es la suma de los productos de los primeros (***F****irst*) términos, los términos exteriores (***O****uter*), los términos interiores (***I****nner*) y los últimos (***L****ast*) términos.

formula (p. 6) A mathematical sentence that expresses the relationship between certain quantities.

fórmula Enunciado matemático que describe la relación entre ciertas cantidades.

frequency (p. 856) The number of cycles in a given unit of time.

frecuencia El número de ciclos en una unidad del tiempo dada.

frequency table (p. 1002) A chart that indicates the number of values in each interval.

table de frecuencias Tabla que indica el número de valores en cada intervalo.

function (p. P4) A relation in which each element of the domain is paired with exactly one element in the range.

función Relación en que a cada elemento del dominio le corresponde un solo elemento del rango.

function notation (p. 64) An equation of y in terms of x can be rewritten so that $y = f(x)$. For example, $y = 2x + 1$ can be written as $f(x) = 2x + 1$.

notación funcional Una ecuación de y en términos de x puede escribirse en la forma $y = f(x)$. Por ejemplo, $y = 2x + 1$ puede escribirse como $f(x) = 2x + 1$.

Fundamental Counting Principle (p. P9) If event M can occur in m ways and is followed by event N that can occur in n ways, then event M followed by event N can occur in $m \cdot n$ ways.

principio fundamental de conteo Si el evento M puede ocurrir de m maneras y es seguido por el evento N que puede ocurrir de n maneras, entonces el evento M seguido por el evento N pueden ocurrir de $m \cdot n$ maneras.

G

general form (p. 623) An equation of a parabola in the form $y = ax^2 + bx + c$.

forma general Una ecuación de una parábola en la forma $y = ax^2 + bx + c$.

geometric mean (p. 697) The terms between any two nonsuccessive terms of a geometric sequence.

media geométrica Cualquier término entre dos términos no consecutivos de una sucesión geométrica.

geometric sequence (p. 683) A sequence in which each term after the first is found by multiplying the previous term by a constant r, called the common ratio.

sucesión geométrica Sucesión en que cualquier término después del primero puede hallarse multiplicando el término anterior por una constante r, llamada razón común.

geometric series (p. 698) The sum of the terms of a geometric sequence.

serie geométrica La suma de los términos de una sucesión geométrica.

greatest integer function (p. 102) A step function, written as $f(x) = [\![x]\!]$, where $f(x)$ is the greatest integer less than or equal to x.

función del máximo entero Una función etapa que se escribe $f(x) = [\![x]\!]$, donde $f(x)$ es el meaximo entero que es menor que o igual a x.

growth factor (p. 477) In exponential growth, the base of the exponential expression, $1 + r$.

factor del crecimiento En el crecimiento exponencial, la base de la expresión exponencial, $1 + r$.

H

histogram (p. 1002) A histogram uses bars to display numerical data that have been organized into equal intervals.

histograma Un histograma usa barras para exhibir datos numéricos que han sido organizados en intervalos iguales.

horizontal asymptote (p. 577) A horizontal line which a graph approaches.

asíntota horizontal Una linea horizontal a que un gráfico acerca.

hyperbola (p. 569, 648) The set of all points in the plane such that the absolute value of the difference of the distances from two given points in the plane, called foci, is constant.

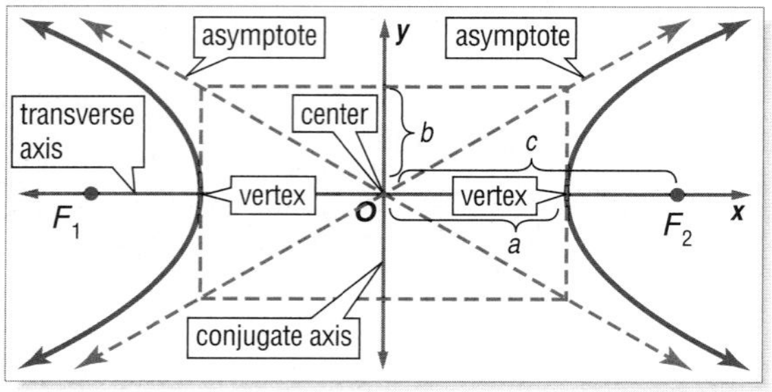

hipérbola Conjunto de todos los puntos de un plano en los que el valor absoluto de la diferencia de sus distancias a dos puntos dados del plano, llamados focos, es constante.

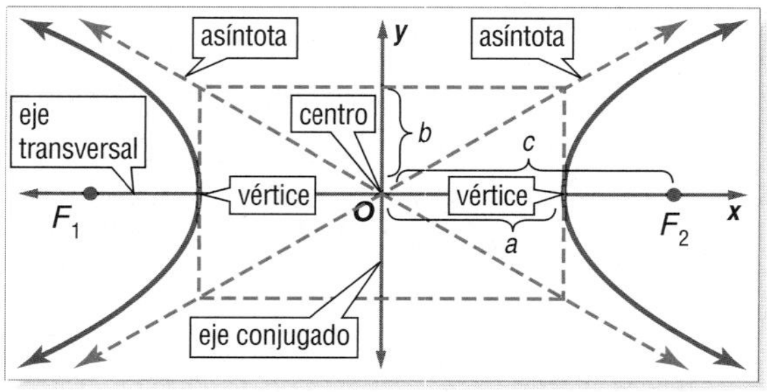

hypothesis (p. 781) An assumption about a population that can be verified by testing.

hipótesis Una asunción sobre una población que puede ser verificada probando.

I

identity function (p. 109) The function $I(x) = x$.

unción identidad La función $I(x) = x$.

identity matrix (p. 229) A square matrix that, when multiplied by another matrix, equals that same matrix. If A is any $n \times n$ matrix and I is the $n \times n$ identity matrix, then $A \cdot I = A$ and $I \cdot A = A$.

matriz identidad Matriz cuadrada que al multiplicarse por otra matriz, es igual a la misma matriz. Si A es una matriz de $n \times n$ e I es la matriz identidad de $n \times n$, entonces $A \cdot I = A$ y $I \cdot A = A$.

image (p. 209) The graph of an object after a transformation.

imagen Gráfica de una figura después de una transformación.

imaginary unit (p. 276) i, or the principal square root of -1.

unidad imaginaria i, o la raíz cuadrada principal de -1.

inclusive events (p. 999) Two events whose outcomes may be the same.

inconsistent (p. 137) A system of equations that has no solutions.

independent (p. 137) When a system of linear equations has exactly one solution.

independent events (p. 998) Events that do not affect each other.

independent system (p. 137) A system of equations that has exactly one solution.

independent variable (p. 64) In a function, the variable, usually x, whose values make up the domain.

index (p. 431) In nth roots, the value of n in the symbol $\sqrt[n]{}$. Indicates to what root the value under the radicand is being taken.

index of summation (p. 691) The variable used with the summation symbol. In the expression below, the index of summation is n.

$$\sum_{n=1}^{3} 4n$$

induction hypothesis (p. 727) The assumption that a statement is true for some positive integer k, where $k \geq n$.

inferential statistics (p. 780) Statistics like predictions and hypothesis testing are used to draw conclusions about a population by using a sample.

infinite geometric series (p. 705) A geometric series with an infinite number of terms.

infinite sequence (p. 681) A sequence that continues without end.

infinity (pp. 40, 707) Without bound, or continues without end.

initial side (p. 817) The fixed ray of an angle.

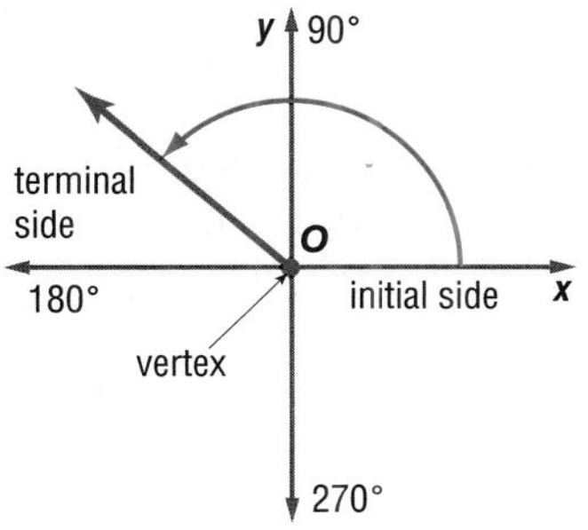

inclusivo Dos eventos que pueden tener los mismos resultados.

inconsistente Sistema de ecuaciones que no tiene solución alguna.

independiente Cuando un sistema de ecuaciones lineares tiene exactamente una solución.

eventos independientes Eventos que no se afectan mutuamente.

sistema independiente Sistema de ecuaciones que sólo tiene una solución.

variable independiente En una función, la variable, por lo general x, cuyos valores forman el dominio.

indice (de un radical) En las nth raíces, el valor de n en el símbolo $\sqrt[n]{}$. Indica qué raíz se está llevando el valor bajo radicand.

índice de suma Variable que se usa con el símbolo de suma. En la siguiente expresión, el índice de suma es n.

$$\sum_{n=1}^{3} 4n$$

hipótesis inductiva El suponer que un enunciado es verdadero para algún entero positivo k, donde $k \geq n$.

estadística deductiva La estadística como predicciones y la prueba de la hipótesis es utilizada para dibujar conclusiones sobre una población usando una muestra.

serie geométrica infinita Serie geométrica con un número infinito de términos.

secuencia infinita Una secuencia que continúa sin extremo.

infinito Sin límite, o continúa sin extremo.

lado inicial de un ángulo El rayo fijo de un ángulo.

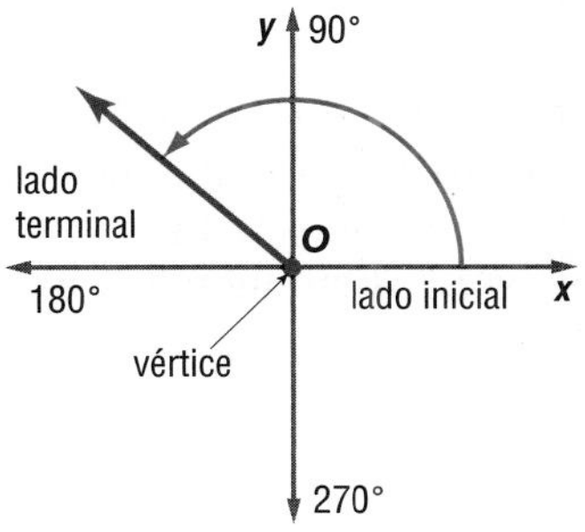

integer (p. 11) $\{\ldots, -3, -2, -1, 0, 1, 2, 3, \ldots\}$

número entero $\{\ldots, -3, -2, -1, 0, 1, 2, 3, \ldots\}$

interquartile range (IQR) (p. 1006) The range of the middle half of a set of data. It is the difference between the upper quartile and the lower quartile.

amplitud intercuartílica Amplitud de la mitad central de un conjunto de datos. Es la diferencia entre el cuartil superior y el inferior.

intersection (p. 41) The graph of a compound inequality containing *and*.

intersección Gráfica de una desigualdad compuesta que contiene la palabra *y*.

interval notation (p. 40) A way to describe the solution set of an inequality.

notación del intervalo Una manera de describir el sistema de la solución de una desigualdad.

inverse (p. 229) Two $n \times n$ matrices are inverses of each other if their product is the identity matrix.

inversa Dos matrices de $n \times n$ son inversas mutuas si su producto es la matriz identidad.

inverse function (p. 417) Two functions f and g are inverse functions if and only if both of their compositions are the identity function.

función inversa Dos funciones f y g son inversas mutuas si y sólo si las composiciones de ambas son la función identidad.

inverse of a trigonometric function (p. 871) The arccosine, arcsine, and arctangent relations.

inversa de una función trigonométrica Las relaciones arcocoseno, arcoseno y arcotangente.

inverse relation (p. 417) Two relations are inverse relations if and only if whenever one relation contains the element (a, b) the other relation contains the element (b, a).

relaciones inversas Dos relaciones son relaciones inversas mutuas si y sólo si cada vez que una de las relaciones contiene el elemento (a, b), la otra contiene el elemento (b, a).

inverse variation (p. 588) y varies inversely as x if there is some nonzero constant k such that $xy = k$ or $y = \frac{k}{x}$, where $x \neq 0$ and $y \neq 0$.

variación inversa y varía inversamente con x si hay una constante no nula k tal que $xy = k$ o $y = \frac{k}{x}$, donde $x \neq 0$ y $y \neq 0$.

irrational number (p. 11) A real number that is not rational. The decimal form neither terminates nor repeats.

número irracional Número que no es racional. Su expansión decimal no es ni terminal ni periódica.

iteration (p. 716) The process of composing a function with itself repeatedly.

iteración Proceso de componer una función consigo misma repetidamente.

J

joint variation (p. 587) y varies jointly as x and z if there is some nonzero constant k such that $y = kxz$.

variación conjunta y varía conjuntamente con x y z si hay una constante no nula k tal que $y = kxz$.

L

latus rectum (p. 623) The line segment through the focus of a parabola and perpendicular to the axis of symmetry.

latus rectum El segmento de recta que pasa por el foco de una parábola y que es perpendicular a su eje de simetría.

Law of Cosines (p. 841) Let $\triangle ABC$ be any triangle with a, b, and c representing the measures of sides, and opposite angles with measures A, B, and C, respectively. Then the following equations are true.

$a^2 = b^2 + c^2 - 2bc \cos A$
$b^2 = a^2 + c^2 - 2ac \cos B$
$c^2 = a^2 + b^2 - 2ab \cos C$

Ley de los cosenos Sea $\triangle ABC$ un triángulo cualquiera, con a, b, y c las longitudes de los lados y con ángulos opuestos de medidas A, B y C, respectivamente. Entonces se cumplen las siguientes ecuaciones.

$a^2 = b^2 + c^2 - 2bc \cos A$
$b^2 = a^2 + c^2 - 2ac \cos B$
$c^2 = a^2 + b^2 - 2ab \cos C$

Law of Sines (p. 833) Let $\triangle ABC$ be any triangle with a, b, and c representing the measures of sides opposite angles with measurements A, B, and C, respectively. Then $\frac{\sin A}{a} = \frac{\sin B}{b} = \frac{\sin C}{c}$.

Ley de los senos Sea $\triangle ABC$ cualquier triángulo con a, b y c las longitudes de los lados y con ángulos opuestos de medidas A, B y C, respectivamente. Entonces $\frac{\sin A}{a} = \frac{\sin B}{b} = \frac{\sin C}{c}$.

leading coefficient (p. 348) The coefficient of the term with the highest degree.

coeficiente líder Coeficiente del término de mayor grado.

like radical expressions (p. 441) Two radical expressions in which both the radicands and indices are alike.

expresiones radicales semejantes Dos expresiones radicales en que tanto los radicandos como los índices son semejantes.

limit (p. 712) The value that the terms of a sequence approach.

límite El valor al que tienden los términos de una sucesión.

linear correlation coefficient (p. 96) A value that shows how close data points are to a line.

coeficiente de correlación lineal Valor que muestra la cercanía de los datos a una recta.

linear equation (p. 69) An equation that has no operations other than addition, subtraction, and multiplication of a variable by a constant.

ecuación lineal Ecuación sin otras operaciones que las de adición, sustracción y multiplicación de una variable por una constante.

linear function (p. 69) A function whose ordered pairs satisfy a linear equation.

función lineal Función cuyos pares ordenados satisfacen una ecuación lineal.

linear inequality (p. 117) Resembles a linear equation, but with an inequality symbol instead of an equals symbol.

desigualdad linear Se asemeja a una ecuación linear, pero con un símbolo de la desigualdad en vez de una relación linear del símbolo de los iguales.

linear permutation (p. P12) The arrangement of objects or people in a line.

permutación lineal Arreglo de personas o figuras en una línea.

linear programming (p. 160) The process of finding the maximum or minimum values of a function for a region defined by inequalities.

programación lineal Proceso de hallar los valores máximo o mínimo de una función lineal en una región definida por las desigualdades.

linear relation (p. 69) A relation that has straight line graphs.

notación del intervalo Una manera de describir el sistema de la solución de una desigualdad.

linear term (p. 249) In the equation $f(x) = ax^2 + bx + c$, bx is the linear term.

término lineal En la ecuación $f(x) = ax^2 + bx + c$, el término lineal es bx.

line graph (p. 1001) A type of statistical graph used to show how values change over a period of time.

gráfica lineal Tipo de gráfica estadística que se usa para mostrar cómo cambian los valores durante un período de tiempo.

line of best fit (p. 94) A line that best matches a set of data.

recta de óptimo ajuste Recta que mejor encaja un conjunto de datos.

line of fit (p. 92) A line that closely approximates a set of data.

recta de ajuste Recta que se aproxima estrechamente a un conjunto de datos.

line of reflection (p. 111) The line over which a reflection flips a figure.

línea de la reflexión La línea excedente que una reflexión mueve de un tirón una figura.

Location Principle (p. 357) Suppose $y = f(x)$ represents a polynomial function and a and b are two numbers such that $f(a) < 0$ and $f(b) > 0$. Then the function has at least one real zero between a and b.

principio de ubicación Sea $y = f(x)$ una función polinómica con a y b dos números tales que $f(a) < 0$ y $f(b) > 0$. Entonces la función tiene por lo menos un resultado real entre a y b.

logarithm (p. 492) In the function $x = b^y$, y is called the logarithm, base b, of x. Usually written as $y = \log_b x$ and is read "y equals log base b of x."

logaritmo En la función $x = b^y$, y es el logaritmo en base b, de x. Generalmente escrito como $y = \log_b x$ y se lee "y es igual al logaritmo en base b de x."

logarithmic equation (p. 502) An equation that contains one or more logarithms.

ecuación logarítmica Ecuación que contiene uno o más logaritmos.

logarithmic function (p. 493) The function $y = \log_b x$, where $b > 0$ and $b \neq 1$, which is the inverse of the exponential function $y = bx$.

función logarítmica La función $y = \log_b x$, donde $b > 0$ y $b \neq 1$, inversa de la función exponencial $y = bx$.

logarithmic inequality (p. 503) An inequality that contains one or more logarithms.

desigualdad logarítmica Desigualdad que contiene uno o más logaritmos.

logistic growth model (p. 536) A growth model that represents growth that has a limiting factor. Logistic models are the most accurate models for representing population growth.

modelo logístico del crecimiento Un modelo del crecimiento que representa el crecimiento que tiene un factor limitador. Los modelos logísticos son los modelos más exactos para representar crecimiento de la población.

lower quartile (p. 1005) The median of the lower half of a set of data, indicated by LQ.

cuartil inferior Mediana de la mitad inferior de un conjunto de datos, se denota con CI.

M

major axis (p. 639) The longer of the two line segments that form the axes of symmetry of an ellipse.

eje mayor El más largo de dos segmentos de recta que forman los ejes de simetría de una elipse.

mapping (p. P4) How each member of the domain is paired with each member of the range.

transformaciones La correspondencia entre cada miembro del dominio con cada miembro del rango.

margin of sampling error (p. 753) The limit on the difference between how a sample responds and how the total population would respond.

margen de error muestral Límite en la diferencia entre las respuestas obtenidas con una muestra y cómo pudiera responder la población entera.

mathematical induction (p. 727) A method of proof used to prove statements about positive integers.

inducción matemática Método de demostrar enunciados sobre los enteros positivos.

matrix (p. 185) Any rectangular array of variables or constants in horizontal rows and vertical columns.

matriz Arreglo rectangular de variables o constantes en filas horizontales y columnas verticales.

matrix equation (p. 231) A matrix form used to represent a system of equations.

ecuación matriz Forma de matriz que se usa para representar un sistema de ecuaciones.

maximum value (p. 252) The y-coordinate of the vertex of the quadratic function $f(x) = ax^2 + bx + c$, where $a < 0$.

valor máximo La coordenada y del vértice de la función cuadrática $f(x) = ax^2 + bx + c$, where $a < 0$.

measure of central tendency (p. 752) A number that represents the center or middle of a set of data.

medida de tendencia central Número que representa el centro o medio de un conjunto de datos.

measure of variation (p. 754) A representation of how spread out or scattered a set of data is.

medida de variación Número que representa la dispersión de un conjunto de datos.

midline (p. 864) A horizontal axis used as the reference line about which the graph of a periodic function oscillates.

recta central Eje horizontal que se usa como recta de referencia alrededor de la cual oscila la gráfica de una función periódica.

minimum value (p. 252) The y-coordinate of the vertex of the quadratic function $f(x) = ax^2 + bx + c$, where $a > 0$.

minor axis (p. 639) The shorter of the two line segments that form the axes of symmetry of an ellipse.

mutually exclusive (p. 999) Two events that cannot occur at the same time.

valor mínimo La coordenada y del vértice de la función cuadrática $f(x) = ax^2 + bx + c$, donde $a > 0$.

eje menor El más corto de los dos segmentos de recta de los ejes de simetría de una elipse.

mutuamente exclusivos Dos eventos que no pueden ocurrir simultáneamente.

N

***n*th root** (p. 431) For any real numbers a and b, and any positive integer n, if $a^n = b$, then a is an nth root of b.

natural base (p. 525) The value of e is 2.71828. The base used for the LN logarithmic function.

natural base, *e* (p. 525) An irrational number approximately equal to 2.71828... .

natural base exponential function (p. 525) An exponential function with base e, $y = e^x$.

natural logarithm (p. 525) Logarithms with base e, written $\ln x$.

natural logarithmic function (p. 527) $y = \ln x$, the inverse of the natural base exponential function $y = e^x$.

natural number (p. 11) $\{1, 2, 3, 4, 5, \ldots\}$,

negative correlation (p. 92) When the values in a scatter plot are closely linked in a negative manner.

negative exponent (p. 312) For any real number $a \neq 0$ and any integer n, $a^{-n} = \frac{1}{a^n}$ and $\frac{1}{a^{-n}} = a^n$.

nonproportional relationship (p. 993) A relationship in which two ratios are not equal.

normal distribution (p. 773) A frequency distribution that often occurs when there is a large number of values in a set of data: about 68% of the values are within one standard deviation of the mean, 95% of the values are within two standard deviations from the mean, and 99% of the values are within three standard deviations.

Normal Distribution

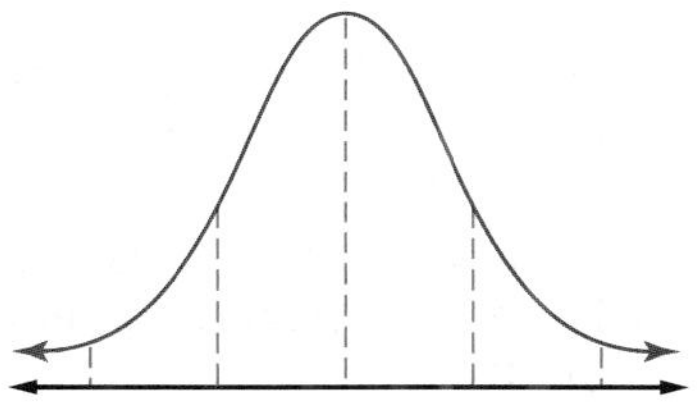

raíz *enésima* Para cualquier número real a y b y cualquier entero positivo n, si $a^n = b$, entonces a se llama una raíz *enésima* de b.

base natural El valor de e es 2.71828. La base usada para el número natural de la función logarítmica de LN.

base natural, *e* Número irracional aproximadamente igual a 2.71828... .

función exponencial natural La función exponencial de base e, $y = e^x$.

logaritmo natural Logaritmo de base e, el que se escribe $\ln x$.

función logarítmica natural $y = \ln x$, la inversa de la función exponencial natural $y = e^x$.

número natural $\{1, 2, 3, 4, 5, \ldots\}$,

corelación negativo Cuando los valores en un diagrama de dispersión se ligan de cerca de una manera negativa.

exponente negativo Para cualquier número real $a \neq 0$ cualquier entero positivo n, $a^{-n} = \frac{1}{a^n}$ y $\frac{1}{a^{-n}} = a^n$.

relación no proporcional Relación en la que dos razones no son iguales.

distribución normal Distribución de frecuencia que aparece a menudo cuando hay un número grande de datos: cerca del 68% de los datos están dentro de una desviación estándar de la media, 95% están dentro de dos desviaciones estándar de la media y 99% están dentro de tres desviaciones estándar de la media.

Distribución normal

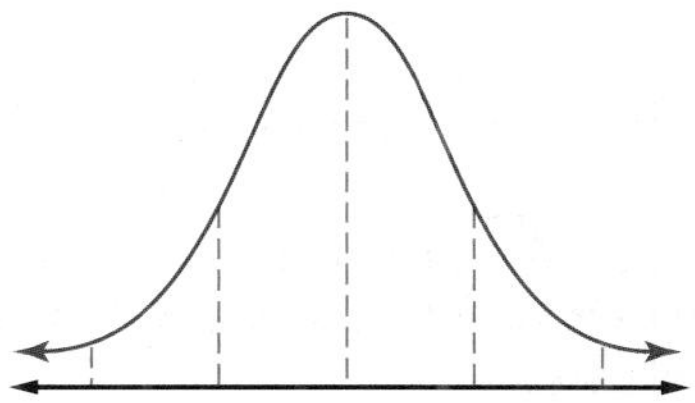

Glossary/Glosario

null hypothesis (p. 781) A specific hypothesis to be tested. It is expressed as an equality and is considered true until evidence indicates otherwise.

hipótesis nula Es una hipótesis específica que se probará. Se expresa como igualdad y se considera verdad hasta que la evidencia indica de otra manera.

O

oblique asymptote (p. 579) An asymptote that is neither horizontal nor vertical and is sometimes called a *slant asymptote*.

asíntota oblicuo Una asíntota que es ni horizontal ni la vertical y a veces se llama una *asíntota inclinada*.

observational study (p. 746) Individuals are observed and no attempt is made to influence the results.

estudio de observación Observan a los individuos y no se hace ninguna tentativa de influenciar los resultados.

one-to-one function (p. 61) **1.** A function where each element of the range is paired with exactly one element of the domain **2.** A function whose inverse is a function.

función biunívoca **1.** Función en la que a cada elemento del rango le corresponde sólo un elemento del dominio. **2.** Función cuya inversa es una función.

onto function (p. 61) Each element of the range corresponds to an element of the domain.

sobre la función Cada elemento de la gama corresponde a un elemento del dominio. Centro y un punto en el círculo. Índice del crecimiento continuo—la tarifa en la cual algo crece continuamente. El valor de k en la función exponencial del crecimiento, $f(x) = ae$.

open sentence (p. 18) A mathematical sentence containing one or more variables.

enunciado abierto Enunciado matemático que contiene una o más variables.

optimize (p. 162) To seek the optimal price or amount that is desired to minimize costs or maximize profits.

optimice Buscar el precio óptimo o ascender que se desea para reducir al mínimo costes o para maximizar de los beneficios.

ordered triple (p. 167) **1.** The coordinates of a point in space **2.** The solution of a system of equations in three variables x, y, and z.

triple ordenado **1.** Las coordenadas de un punto en el espacio **2.** Solución de un sistema de ecuaciones en tres variables x, y y z.

Order of Operations (p. 5)
Step 1 Evaluate expressions inside grouping symbols.
Step 2 Evaluate all powers.
Step 3 Do all multiplications and/or divisions from left to right.
Step 4 Do all additions and subtractions from left to right.

orden de las operaciones
Paso 1 Evalúa las expresiones dentro de símbolos de agrupamiento.
Paso 2 Evalúa todas las potencias.
Paso 3 Ejecuta todas las multiplicaciones y divisiones de izquierda a derecha.
Paso 4 Ejecuta todas las adiciones y sustracciones de izquierda a derecha.

outcomes (p. P9) The results of a probability experiment or an event.

resultados Lo que produce un experimento o evento probabilístico.

outlier (p. 95) A data point that does not appear to belong to the rest of the set.

valor atípico Dato que no parece pertenecer al resto el conjunto.

P

parabola (pp. 249, 623) The set of all points in a plane that are the same distance from a given point, called the focus, and a given line, called the directrix.

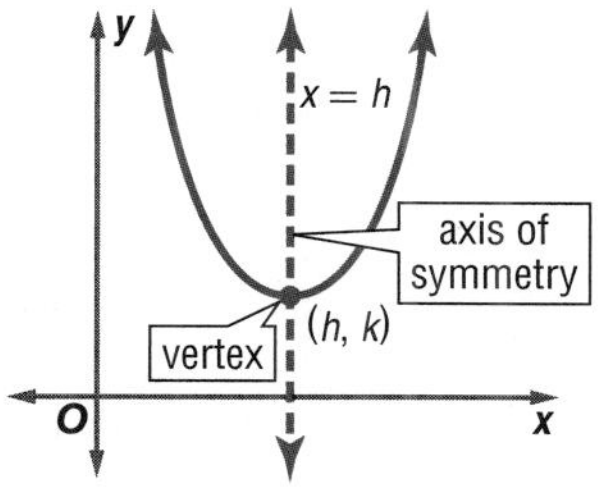

parallel lines (p. 85) Nonvertical coplanar lines with the same slope.

parameter (p. 752) A measure that describes a characteristic of a population.

parent function (p. 109) The simplest, most general function in a family of functions.

parent graph (p. 109) The simplest of graphs in a family.

partial sum (p. 690) The sum of the first n terms of a series.

Pascal's triangle (p. 721) A triangular array of numbers such that the $(n + 1)^{th}$ row is the coefficient of the terms of the expansion $(x + y)^n$ for $n = 0, 1, 2 \ldots$

period (p. 849) The least possible value of a for which $f(x) = f(x + a)$.

periodic function (p. 849) A function is called periodic if there is a number a such that $f(x) = f(x + a)$ for all x in the domain of the function.

permutation (p. P12) An arrangement of objects in which order is important.

perpendicular lines (p. 85) In a plane, any two oblique lines, the product of whose slopes is 21.

phase shift (p. 863) A horizontal translation of a trigonometric function.

piecewise-defined function (p. 101) A function that is written using two or more expressions.

piecewise-linear function (p. 102) Like a step function, it contains a single expression.

parábola Conjunto de todos los puntos de un plano que están a la misma distancia de un punto dado, llamado foco, y de una recta dada, llamada directriz.

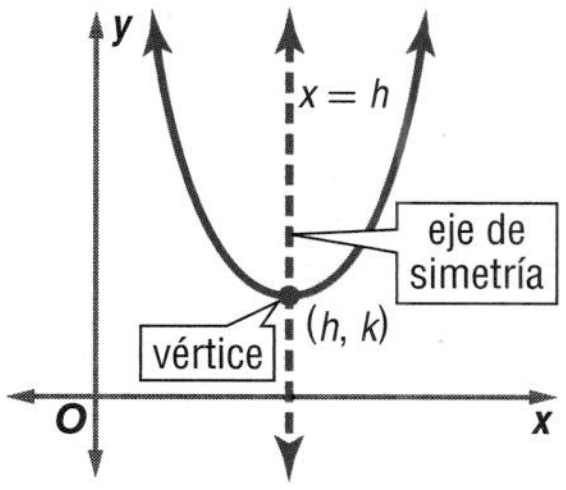

rectas paralelas Rectas coplanares no verticales con la misma pendiente.

parámetro Una medida que describe una característica de una población.

función del padre El más simple, la mayoría de la función general en una familia de funciones.

gráfica madre La gráfica más sencilla en una familia de gráficas.

suma parcial La suma de los primeros n términos de una serie.

triángulo de Pascal Arreglo triangular de números en el que la fila $(n + 1)^n$ proporciona los coeficientes de los términos de la expansión de $(x + y)^n$ para $n = 0, 1, 2 \ldots$

período El menor valor positivo posible para a, para el cual $f(x) = f(x + a)$.

función periódica Función para la cual hay un número a tal que $f(x) = f(x + a)$ para todo x en el dominio de la función .

permutación Arreglo de elementos en que el orden es importante.

rectas perpendiculares En un plano, dos rectas oblicuas cualesquiera cuyas pendientes tienen un producto igual a 21.

desvío de fase Traslación horizontal de una función trigonométrica.

función por trozos-definida Una función se escribe que usando dos o más expresiones.

función por partes lineal Como una función del paso, contiene una sola expression.

point discontinuity (p. 580) If the original function is undefined for $x = a$ but the related rational expression of the function in simplest form is defined for $x = a$, then there is a hole in the graph at $x = a$.

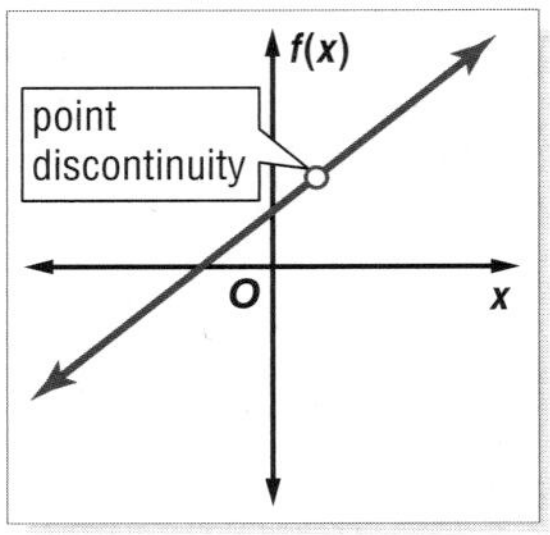

discontinuidad evitable Si la función original no está definida en $x = a$ pero la expresión racional reducida correspondiente de la función está definida en $x = a$, entonces la gráfica tiene una ruptura o corte en $x = a$.

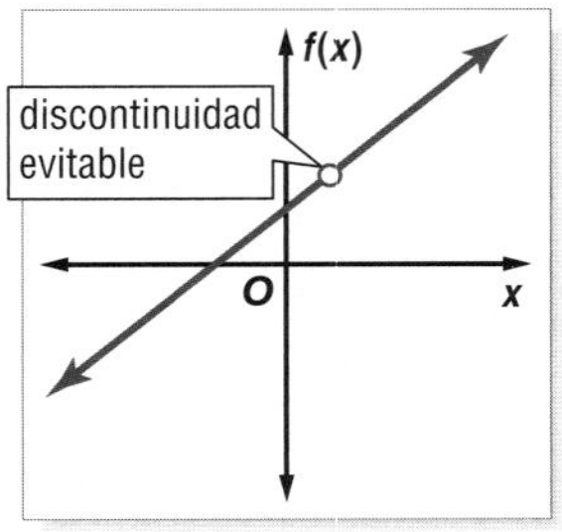

point-slope form (p. 84) An equation in the form $y - y_1 = m(x - x_1)$ where (x_1, y_1) are the coordinates of a point on the line and m is the slope of the line.

forma punto-pendiente Ecuación de la forma $y - y_1 = m(x - x_1)$ donde (x_1, y_1) es un punto en la recta y m es la pendiente de la recta.

polynomial function (p. 349) A function that is represented by a polynomial equation.

función polinomial Función representada por una ecuación polinomial.

polynomial in one variable (p. 348) $a_nx^n + a_{n-1}x^{n-1} + \cdots + a_2x^2 + a_1x + a_0$, where the coefficients $a_n, a_{n-1}, \ldots, a_0$ represent real numbers, and a_n is not zero and n is a nonnegative integer.

polinomio de una variable $a_nx^n + a_{n-1}x^{n-1} + \cdots + a_2x^2 + a_1x + a_0$, donde los coeficientes $a_n, a_{n-1}, \ldots, a_0$ son números reales, a_n no es nulo y n es un entero no negativo.

population (p. 745) An entire group of living things or objects.

población Un grupo entero de cosas o de objetos vivos.

positive correlation (p. 92) When the values in a scatter plot are closely linked in a positive manner.

correlación positivo Cuando los valores en un diagrama de la dispersión se ligan de cerca de una manera positiva.

power function (p. 349) An equation in the form $f(x) = ax^b$, where a and b are real numbers.

función potencia Ecuación de la forma $f(x) = ax^b$, donde a y b son números reales.

prediction equation (p. 92) An equation suggested by the points of a scatter plot that is used to predict other points.

ecuación de predicción Ecuación sugerida por los puntos de una gráfica de dispersión y que se usa para predecir otros puntos.

preimage (p. 209) The graph of an object before a transformation.

preimagen Gráfica de una figura antes de una transformación.

prime polynomial (p. 368) A polynomial that cannot be factored.

polinomio primero Un polinomio que no puede ser descompuesto en factores.

principal root (p. 431) The nonnegative root.

raíz principal La raíz no negativa.

principal values (p. 871) The values in the restricted domains of trigonometric functions.

valores principales Valores en los dominios restringidos de las funciones trigonométricas.

probability distribution (p. 766) A function that maps the sample space to the probabilities of the outcomes in the sample space for a particular random variable.

distribución de probabilidad Función que aplica el espacio muestral a las probabilidades de los resultados en el espacio muestral obtenidos para una variable aleatoria particular.

proportion (p. 993) A statement of equality of two or more ratios.

proporción Enunciado de la igualdad de dos o más razones.

proportional relationship (p. 993) The ratios of related terms are equal.

relación proporcional Relación en la que la razón entre los términos relacionados permanece igual.

pure imaginary number (p. 276) The square roots of negative real numbers. For any positive real number b, $\sqrt{-b^2} = \sqrt{b^2} \cdot \sqrt{-1}$, or bi.

número imaginario puro Raíz cuadrada de un número real negativo. Para cualquier número (real positivo b, $\sqrt{-b^2} = \sqrt{b^2} \cdot \sqrt{-1}$, ó bi.

Q

quadrantal angle (p. 826) An angle in standard position whose terminal side coincides with one of the axes.

ángulo de cuadrante Ángulo en posición estándar cuyo lado terminal coincide con uno de los ejes.

quadrants (p. P4) The four areas of a Cartesian coordinate plane.

cuadrantes Las cuatro regiones de un plano de coordenadas Cartesiano.

quadratic equation (p. 259) A quadratic function set equal to a value, in the form $ax^2 + bx + c$, where $a \neq 0$.

ecuación cuadrática Función cuadrática igual a un valor, de la forma $ax^2 + bx + c$, donde $a \neq 0$.

quadratic form (p. 371) For any numbers a, b, and c, except for $a = 0$, an equation that can be written in the form $a[f(x)^2] + b[f(x)] + c = 0$, where $f(x)$ is some expression in x.

forma de ecuación cuadrática Para cualquier número a, b, y c, excepto $a = 0$, una ecuación que puede escribirse de la forma $a[f(x)^2] + b[f(x)] + c = 0$, donde $f(x)$ es una expresión en x.

Quadratic Formula (p. 292) The solutions of a quadratic equation of the form $ax^2 + bx + c$, where $a \neq 0$, are given by the Quadratic Formula, which is $x = \frac{-b \pm \sqrt{b^2 - 4ac}}{2a}$.

fórmula cuadrática Las soluciones de una ecuación cuadrática de la forma a$ax^2 + bx + c$, donde $a \neq 0$, se dan por la fórmula cuadrática, que es $x = \frac{-b \pm \sqrt{b^2 - 4ac}}{2a}$.

quadratic function (pp. 109, 249) A function described by the equation $f(x) = ax^2 + bx + c$, where $a \neq 0$.

función cuadrática Función descrita por la ecuación $f(x) = ax^2 + bx + c$, donde $a \neq 0$.

quadratic inequality (p. 312) A quadratic equation in the form $y > ax^2 + bx + c$, $y \geq ax^2 + bx + c$, $y < ax^2 + bx + c$, or $y \leq ax^2 + bx + c$.

desigualdad cuadrática Ecuación cuadrática de la forma $y > ax^2 + bx + c$, $y \geq ax^2 + bx + c$, $y < ax^2 + bx + c$, $y \leq ax^2 + bx + c$.

quadratic term (p. 249) In the equation $f(x) = ax^2 + bx + c$, ax^2 is the quadratic term.

término cuadrático En la ecuación $f(x) = ax^2 + bx + c$, ax^2 el término cuadrático es ax^2.

quartic function (p. 350) A fourth-degree function.

función quartic Una función del cuarto-grado.

quartiles (p. 1005) The values that divide a set of data into four equal parts.

cuartiles Valores que dividen un conjunto de datos en cuatro partes iguales.

quintic function (p. 350) A fifth-degree function.

función quintic Una función del quinto-grado.

R

radian (p. 819) The measure of an angle θ in standard position whose rays intercept an arc of length 1 unit on the unit circle.

radián Medida de un ángulo θ en posición normal cuyos rayos intersecan un arco de 1 unidad de longitud en el círculo unitario.

radical equation (p. 453) An equation with radicals that have variables in the radicands.

ecuación radical Ecuación con radicales que tienen variables en el radicando.

radical function (p. 424) A function that contains the square root of a variable.

función radical Una función que contiene la raíz cuadrada de una variable.

radical inequality (p. 455) An inequality that has a variable in the radicand.

desigualdad radical Desigualdad que tiene una variable en el radicando.

radical sign (p. 431) In nth roots, the symbol $\sqrt[n]{}$.

signo radical El símbolo $\sqrt[n]{}$, que se usa par indicar la raíz cuadrada no negative o el símbolo por raíz enésima.

radicand (p. 431) In nth roots, the value inside in the symbol $\sqrt[n]{}$. Indicates the value that is being taken to the nth root.

radicando El número o la expressión que aparece debajo del signo radical.

radius (p. 631) Any segment whose endpoints are the center and a point on the circle.

radio Un segmento cuyos extremos son el centro y un punto del círculo.

random variable (p. 766) The outcome of a random process that has a numerical value.

variable aleatoria El resultado de un proceso aleatorio que tiene un valor numérico.

range (p. P4) The set of all y-coordinates of a relation.

rango Conjunto de todas las coordenadas y de una relación.

rate of change (p. 76) How much a quantity changes on average, relative to the change in another quantity, often time.

tasa de cambio Lo que cambia una cantidad en promedio, respecto al cambio en otra cantidad, por lo general el tiempo.

rate of continuous decay (p. 533) The rate at which something decays continuously. Represented by a constant k in the exponential decay function $f(x) = ae^{-kt}$, where a is the initial value, and t is time in years.

índice de desintegración continúa Ritmo al cual algo se desintegra continuamente. Representado por la constante k en la función de desintegración exponencial $f(x) = ae^{-kt}$, donde a es el valor inicial y t es el tiempo en años.

rate of continuous growth (p. 533) The rate at which something grows continuously. The value of k in the exponential growth function, $f(x) = ae^{kt}$.

el índice del crecimiento continuo Es la tasa en la cual algo crece continuamente. El valor de k en la función exponencial del crecimiento, $f(x) = ae^{kt}$. La tasa en la cual algo crece continuamente.

rational equation (p. 594) Any equation that contains one or more rational expressions.

ecuación racional Cualquier ecuación que contiene una o más expresiones racionales.

rational exponent (p. 446) For any nonzero real number b, and any integers m and n, with $n > 1$, $b^{\frac{m}{n}} = \sqrt[n]{b^m} = \left(\sqrt[n]{b}\right)^m$, except when $b < 0$ and n is even.

exponent racional Para cualquier número real no nulo b y cualquier entero m y n, con $n > 1$, $b^{\frac{m}{n}} = \sqrt[n]{b^m} = \left(\sqrt[n]{b}\right)^m$, excepto cuando $b < 0$ y n es par.

rational expression (p. 553) A ratio of two polynomial expressions.

expresión racional Razón de dos expresiones polinomiales.

rational function (p. 577) An equation of the form $f(x) = \frac{p(x)}{q(x)}$, where $p(x)$ and $q(x)$ are polynomial functions, and $q(x) \neq 0$.

función racional Ecuación de la forma $f(x) = \frac{p(x)}{q(x)}$, donde $p(x)$ y $q(x)$ son funciones polinomiales y $q(x) \neq 0$.

rational inequality (p. 599) Any inequality that contains one or more rational expressions.

desigualdad racional Cualquier desigualdad que contiene una o más expresiones racionales.

rationalizing the denominator (p. 440) To eliminate radicals from a denominator or fractions from a radicand.

racionalizar el denominador La eliminación de radicales de un denominador o de fracciones de un radicando.

rational number (p. 11) Any number $\frac{m}{n}$, where m and n are integers and n is not zero. The decimal form is either a terminating or repeating decimal.

número racional Cualquier número $\frac{m}{n}$, donde m y n son enteros y n no es cero. Su expansión decimal es o terminal o periódica.

Rational Zero Theorem (p. 391) Helps you choose some possible zeros of a polynomial function to test.

El teorema cero racional Ayudas usted elige algunos ceros posibles de una función polinómica para probar.

real numbers (p. 11) All numbers used in everyday life; the set of all rational and irrational numbers.

números reales Todos los números que se usan en la vida cotidiana; el conjunto de los todos los números racionales e irracionales.

reciprocal function (pp. 569, 809) Trigonometric functions that are reciprocals of each other.

funciones recíprocas Funciones trigonométricas de la función que son reciprocals de uno a.

recursive formula (p. 714) Each term is formulated from one or more previous terms.

fórmula recursiva Cada término proviene de uno o más términos anteriores.

recursive sequence (p. 714) A sequence in which each term is determined by one or more of the previous terms.

sucesión recursiva Una secuencia en la cual cada término es determinado por uno o más de los términos anteriores

reference angle (p. 826) The acute angle formed by the terminal side of an angle in standard position and the x-axis.

ángulo de referencia El ángulo agudo formado por el lado terminal de un ángulo en posición estándar y el eje x.

reflection (p. 111, 212) A transformation in which every point of a figure is mapped to a corresponding image across a line of symmetry.

reflexión Transformación en que cada punto de una figura se aplica a través de una recta de simetría a su imagen correspondiente.

reflection matrix (p. 214) A matrix used to reflect an object over a line or plane.

matriz de reflexión Matriz que se usa para reflejar una figura sobre una recta o plano.

regression line (p. 94) A line of best fit.

reca de regresión Una recta de óptimo ajuste.

relation (p. P4) A set of ordered pairs.

relación Conjunto de pares ordenados.

relative frequency (p. 760) In a contingency table, the frequency of occurrence for each data value.

histograma de frecuencia relativa Tabla de probabilidades o gráfica para asistir en la visualización de una distribución de probabilidad.

relative frequency histogram (p. 791) A table of probabilities or a graph to help visualize a probability distribution.

frecuencia relativa En una tabla de la contingencia, la frecuencia de la ocurrencia para cada valor de los datos.

relative maximum (p. 358) A point on the graph of a function where no other nearby points have a greater y-coordinate.

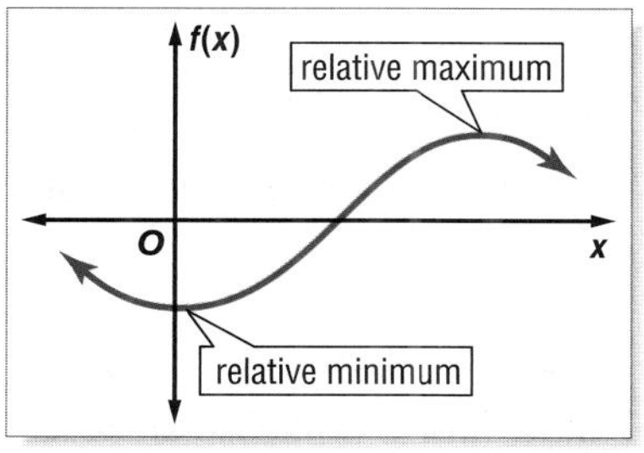

máximo relativo Punto en la gráfica de una función en donde ningún otro punto cercano tiene una coordenada y mayor.

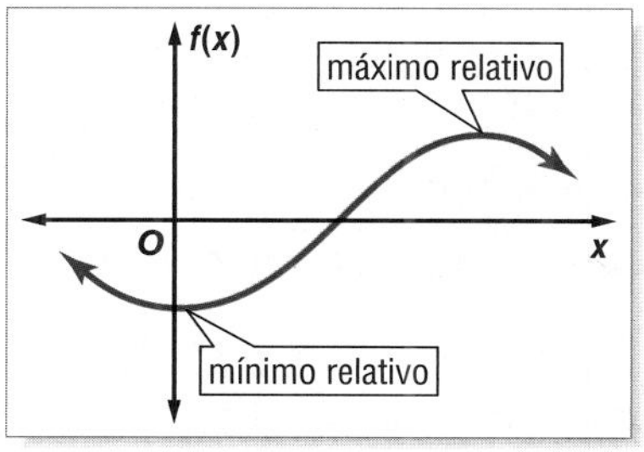

Glossary/Glosario

relative minimum (p. 358) A point on the graph of a function where no other nearby points have a lesser y-coordinate.

mínimo relativo Punto en la gráfica de una función en donde ningún otro punto cercano tiene una coordenada y menor.

root (p. 261) The solutions of a quadratic equation.

raíz Las soluciones de una ecuación cuadrática.

rotation (p. 212) A transformation in which an object is moved around a center point, usually the origin.

rotación Transformación en que una figura se hace girar alrededor de un punto central, generalmente el origen.

rotation matrix (p. 214) A matrix used to rotate an object.

matriz de rotación Matriz que se usa para hacer girar un objeto.

row matrix (p. 186) A matrix that has only one row.

matriz fila Matriz que sólo tiene una fila.

S

sample (p. 745) A part of a population.

muestra Parte de una población.

sample space (p. P9) The set of all possible outcomes of an experiment.

espacio muestral Conjunto de todos los resultados posibles de un experimento probabilístico.

scalar (p. 194) A constant.

escalar Una constante.

scalar multiplication (p. 194) Multiplying any matrix by a constant called a scalar; the product of a scalar k and an $m \times n$ matrix.

multiplicación por escalares Multiplicación de una matriz por una constante llamada escalar; producto de un escalar k y una matriz de $m \times n$.

scatter plot (p. 92) A set of data graphed as ordered pairs in a coordinate plane.

gráfica de dispersión Conjuntos de datos graficados como pares ordenados en un plano de coordenadas.

scientific notation (p. 997) The expression of a number in the form $a \times 10^n$, where $1 \leq a < 10$ and n is an integer.

notación científica Escritura de un número en la forma $a \times 10^n$, donde $1 \leq a < 10$ y n es un entero.

secant (p. 808) For any angle, with measure α, a point $P(x, y)$ on its terminal side, $r = \sqrt{x^2 + y^2}$, $\sec \alpha = \frac{r}{x}$.

secante Para cualquier ángulo de medida α, un punto $P(x, y)$ en su lado terminal, $r = \sqrt{x^2 + y^2}$, $\sec \alpha = \frac{r}{x}$.

second-order determinant (p. 220) The determinant of a 2×2 matrix.

determinante de segundo orden El determinante de una matriz de 2×2.

sequence (p. 681) A list of numbers in a particular order.

sucesión Lista de números en un orden particular.

series (p. 690) The sum of the terms of a sequence.

serie Suma específica de los términos de una sucesión.

set-builder notation (p. 35) The expression of the solution set of an inequality, for example $\{x \mid x > 9\}$.

notación de construcción de conjuntos Escritura del conjunto solucion de una desigualdad, por ejemplo, $\{x \mid x > 9\}$.

sigma notation (p. 691) For any sequence $a_1, a_2, a_3, \ldots$, the sum of the first k terms may be written $\sum_{n=1}^{k} a_n$, which is read "the summation from $n = 1$ to k of a_n." Thus, $\sum_{n=1}^{k} a_n = a_1 + a_2 + a_3 + \cdots + a_k$, where k is an integer value.

notación de suma Para cualquier sucesión $a_1, a_2, a_3, \ldots$, la suma de los k primeros términos puede escribirse $\sum_{n=1}^{k} a_n$, lo que se lee "la suma de $n = 1$ a k de los a_n." Así, $\sum_{n=1}^{k} a_n = a_1 + a_2 + a_3 + \cdots + a_k$, donde k es un valor entero.

simple event (p. 998) One event.

simplify (p. 333) To rewrite an expression without parentheses or negative exponents.

simulation (p. 785) The use of a probability experiment to mimic a real-life situation.

sine (p. 808) For any angle, with measure α, a point $P(x, y)$ on its terminal side, $r = \sqrt{x^2 + y^2}$, $\sin \alpha = \frac{y}{r}$.

skewed distribution (p. 773) A curve or histogram that is not symmetric.

Positively Skewed | Negatively Skewed

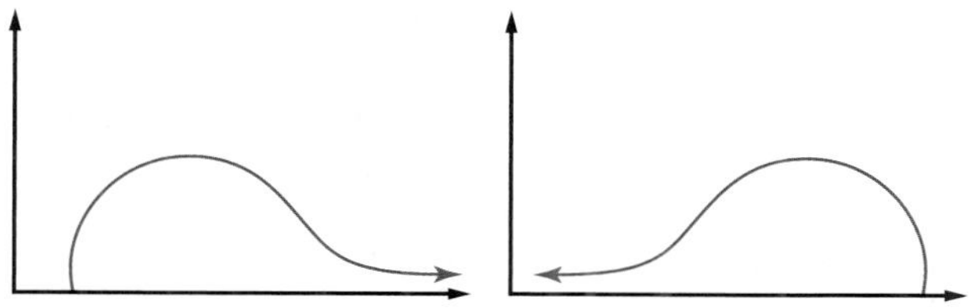

slope (p. 77) The ratio of the change in y-coordinates to the change in x-coordinates.

slope-intercept form (p. 83) The equation of a line in the form $y = mx + b$, where m is the slope and b is the y-intercept.

solution (p. 18) A replacement for the variable in an open sentence that results in a true sentence.

solving a right triangle (p. 833) The process of finding the measures of all of the sides and angles of a right triangle.

solving a triangle (p. 833) Using given measures to find all unknown side lengths and angle measures of a triangle.

square matrix (p. 186) A matrix with the same number of rows and columns.

square root (p. 995) For any real numbers a and b, if $a^2 = b$, then a is a square root of b.

square root function (p. 424) A function that contains a square root of a variable.

square root inequality (p. 426) An inequality involving square roots.

Square Root Property (p. 277) For any real number n, if $x^2 = n$, then $x = \pm\sqrt{n}$.

standard deviation (p. 754) The square root of the variance, represented by a.

evento simple Un solo evento.

reducir Escribir una expresión sin paréntesis o exponentes negativos.

simulación Uso de un experimento probabilístico para imitar una situación de la vida real.

seno Para cualquier ángulo de medida α, un punto $P(x, y)$ en su lado terminal, $r = \sqrt{x^2 + y^2}$, $\sin \alpha = \frac{y}{r}$.

distribución asimétrica Curva o histograma que no es simétrico.

Positivamente Alabeada | Negativamente Alabeada

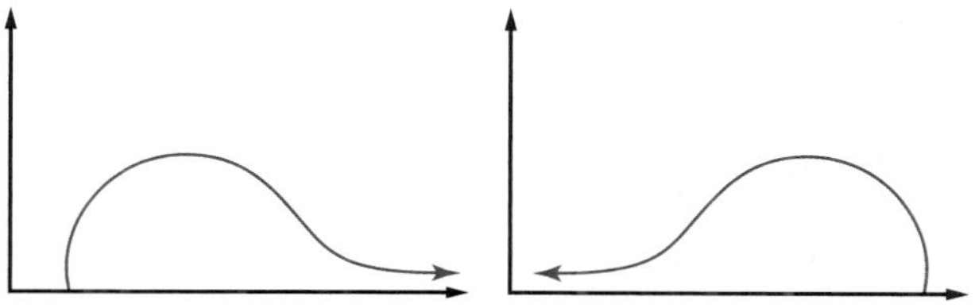

pendiente La razón del cambio en coordenadas y al cambio en coordenadas x.

forma pendiente-intersección Ecuación de una recta de la forma $y = mx + b$, donde m es la pendiente y b la intersección.

solución Sustitución de la variable de un enunciado abierto que resulta en un enunciado verdadero.

resolver un triángulo rectángulo Proceso de hallar las medidas de todos los lados y ángulos de un triángulo rectángulo.

resolver un triángulo Usar medidas dadas de hallar todas las medidas de las longitudes y laterales desconocidas del ángulo de un triángulo.

matriz cuadrada Matriz con el mismo número de filas y columnas.

raíz cuadrada Para cualquier número real a y b, si $a^2 = b$, entonces a es una raíz cuadrada de b.

función radical Función que contiene la raíz cuadrada de una variable.

desigualdad radical Desigualdad que presenta raíces cuadradas.

Propiedad de la raíz cuadrada Para cualquier número real n, si $x^2 = n$, entonces $x = \pm\sqrt{n}$.

desviación estándar La raíz cuadrada de la varianza, la que se escribe a.

standard form (pp. 70, 259, 623) **1.** A linear equation written in the form $Ax + By = C$, where A, B, and C are integers whose greatest common factor is 1, $A \geq 0$, and A and B are not both zero. **2.** A quadratic equation written in the form $ax^2 + bx + c = 0$, where a, b, and c are integers, and $a \neq 0$.

forma estándar **1.** Ecuación lineal escrita de la forma $Ax + By = C$, donde A, B, y C son enteros cuyo máximo común divisores 1, $A \geq 0$, y A y B no son cero simultáneamente. **2.** Una ecuación cuadrática escrita en la forma $ax^2 + bx + c = 0$, donde a, b, y c son números enteros, y $a \neq 0$.

standard position (p. 817) An angle positioned so that its vertex is at the origin and its initial side is along the positive x-axis.

posición estándar Ángulo en posición tal que su vértice está en el origen y su lado inicial está a lo largo del eje x positivo.

statistic (p. 752) A measure that describes a characteristic of a sample.

estadística Una medida que describe una característica de una muestra.

statistical inference (p. 780) Use inferential statistics like predictions and hypothesis testing. Use information from a sample to draw conclusions about a population.

inferencia estadística Utiliza la estadística deductiva como las predicciones y la hipótesis que prueban información de uso de una muestra para dibujar las conclusiones acerca de una población.

stem-and-leaf plot (p. 1004) A system used to condense a set of data where the greatest place value of the data forms the stem and the next greatest place value forms the leaves.

diagrama de tallo y hojas Sistema que se usa para condensar un conjunto de datos, en que el valor de posición máximo de los datos forma el tallo y el segundo valor de posición máximo forma las hojas.

step function (p. 102) A function whose graph is a series of line segments.

función etapa Función cuya gráfica es una serie de segmentos de recta.

substitution method (p. 143) A method of solving a system of equations in which one equation is solved for one variable in terms of the other.

método de sustitución Método para resolver un sistema de ecuaciones en que una de las ecuaciones se resuelve en una de las variables en términos de la otra.

success (p. 764) The desired outcome of an event.

éxito El resultado deseado de un evento.

survey (p. 745) Used to collect information about a population.

encuesta Reunía información acerca de una población.

synthetic division (p. 343) A method used to divide a polynomial by a binomial.

división sintética Método que se usa para dividir un polinomio entre un binomio.

synthetic substitution (p. 377) The use of synthetic division to evaluate a function.

sustitución sintética Uso de la división sintética para evaluar una función polinomial.

system of equations (p. 135) A set of equations with the same variables.

sistema de ecuaciones Conjunto de ecuaciones con las mismas variables.

system of inequalities (p. 151) A set of inequalities with the same variables.

sistema de desigualdades Conjunto de desigualdades con las mismas variables.

T

tangent (p. 808) **1.** A line that intersects a circle at exactly one point. **2.** For any angle, with measure α, a point $P(x, y)$ on its terminal side, $r = \sqrt{x^2 + y^2}$, $\tan \alpha = \frac{y}{x}$.

tangente **1.** Recta que interseca un círculo en un solo punto. **2.** Para cualquier ángulo, de medida α, un punto $P(x, y)$ en su lado terminal, $r = \sqrt{x^2 + y^2}$, $\tan \alpha = \frac{y}{x}$.

term (p. 681) **1.** The monomials that make up a polynomial. **2.** Each number in a sequence or series.

término **1.** Los monomios que constituyen un polinomio. **2.** Cada número de una sucesión o serie.

terminal side (p. 817) A ray of an angle that rotates about the center.

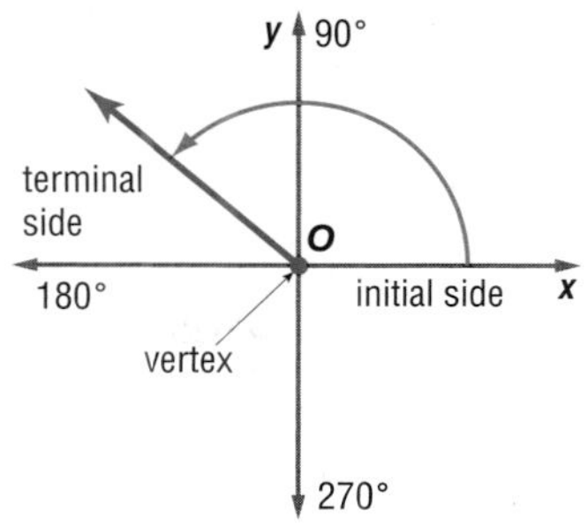

theoretical probability (p. 999) What should occur in a probability experiment.

third-order determinant (p. 221) Determinant of a 3×3 matrix.

transformation (p. 209) Functions that map points of a pre-image onto its image.

translation (p. 110, 209) A figure is moved from one location to another on the coordinate plane without changing its size, shape, or orientation.

translation matrix (p. 211) A matrix that represents a translated figure.

transverse axis (p. 648) The segment of length $2a$ whose endpoints are the vertices of a hyperbola.

treated group (p. 746) In an experiment, the people, animals, or objects given the treatment.

trigonometric equation (p. 921) An equation containing at least one trigonometric function that is true for some but not all values of the variable.

trigonometric functions (p. 808) For any angle, with measure α, a point $P(x, y)$ on its terminal side, $r = \sqrt{x^2 + y^2}$, the trigonometric functions of a are as follows.

$\sin \alpha = \frac{y}{r}$ $\cos \alpha = \frac{x}{r}$ $\tan \alpha = \frac{y}{x}$

$\csc \alpha = \frac{r}{y}$ $\sec \alpha = \frac{r}{x}$ $\cot \alpha = \frac{x}{y}$

trigonometric identity (p. 893) An equation involving a trigonometric function that is true for all values of the variable.

trigonometric ratio (p. 808) Compares the side lengths of a right triangle.

trigonometry (p. 808) The study of the relationships between the angles and sides of a right triangle.

turning point (p. 358) Point at which a graph turns. The location of relative maxima or minima.

lado terminal Rayo de un ángulo que gira alrededor de un centro.

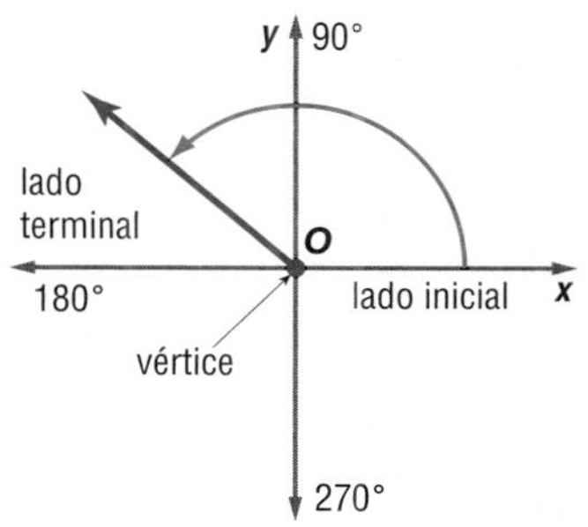

probabilidad teórica Lo que debería ocurrir en un experimento probabilístico.

determinante de tercer orden Determinante de una matriz de 3×3.

transformación Funciones que aplican puntos de una preimagen en su imagen.

traslación Se mueve una figura de un lugar a otro en un plano de coordenadas sin cambiar su tamaño, forma u orientación.

matriz de traslación Matriz que representa una figura trasladada.

eje transversal El segmento de longitud $2a$ cuyos extremos son los vértices de una hipérbola.

grupo tratado En un experimento, la gente, los animales, o los objetos dados el tratamiento.

ecuación trigonométrica Ecuación que contiene por lo menos una función trigonométrica y que sólo se cumple para algunos valores de la variable.

funciones trigonométricas Para cualquier ángulo, de medida α, un punto $P(x, y)$ en su lado terminal, $r = \sqrt{x^2 + y^2}$, las funciones trigonométricas de a son las siguientes.

$\sin \alpha = \frac{y}{r}$ $\cos \alpha = \frac{x}{r}$ $\tan \alpha = \frac{y}{x}$

$\csc \alpha = \frac{r}{y}$ $\sec \alpha = \frac{r}{x}$ $\cot \alpha = \frac{x}{y}$

identidad trigonométrica Ecuación que involucra una o más funciones trigonométricas y que se cumple para todos los valores de la variable.

razón trigonometric Compara las longitudes laterales de un triángulo derecho.

trigonometría Estudio de las relaciones entre los lados y ángulos de un triángulo rectángulo.

momento crucial Un punto en el cual un gráfico da vuelta. La localización de máximos o de mínimos relativos.

U

unbiased (p. 745) When a sample is random, or not based on any predetermined characteristics of the population.

No sesgada Muestra que se selecciona de modo que sea representative de la poblacion entera.

unbiased sample (p. 745) A sample in which every possible sample has an equal chance of being selected.

muestra no sesgada Muestra en que cualquier muestra posible tiene la misma posibilidad de seleccionarse.

unbounded (p. 160) A system of inequalities that forms a region that is open.

no acotado Sistema de desigualdades que forma una región abierta.

uniform distribution (p. 766) A distribution where all of the probabilities are the same.

distribución uniforme Distribución donde todas las probabilidades son equiprobables.

union (p. 42) The graph of a compound inequality containing *or*.

unión Gráfica de una desigualdad compuesta que contiene la palabra *o*.

unit analysis (p. 340) The process of including unit measurement when computing.

análisis de la unidad Proceso de incluir unidades de medida al computar.

unit circle (p. 848) A circle of radius 1 unit whose center is at the origin of a coordinate system.

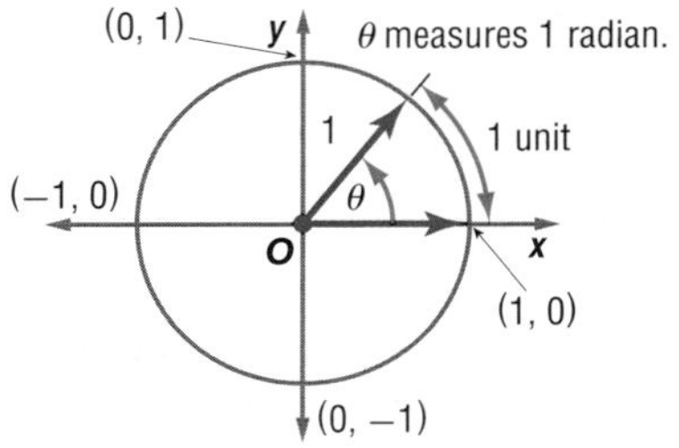

círculo unitario Círculo de radio 1 cuyo centro es el origen de un sistema de coordenadas.

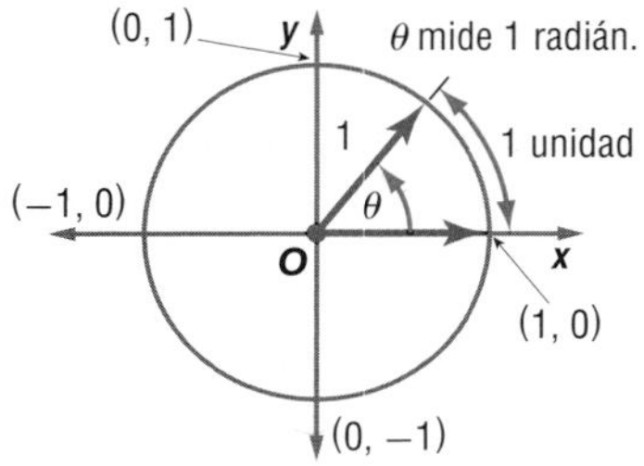

univariate data (p. 752) Data with one variable.

datos univariados Datos con una variable.

upper quartile (p. 1005) The median of the upper half of a set of data, indicated by UQ.

cuartil superior Mediana de la mitad superior de un conjunto de datos, denotada por CS.

V

variable (p. 5) Symbols, usually letters, used to represent unknown quantities.

variables Símbolos, por lo general letras, que se usan para representar cantidades desconocidas.

variable matrix (p. 231) A matrix that only contains the variables of a system of equations.

matriz variable Una matriz que contiene solamente las variables de un sistema de ecuaciones.

variance (p. 754) The mean of the squares of the deviations from the arithmetic mean.

varianza Media de los cuadrados de las desviaciones de la media aritmética.

vertex (pp. 160, 250) **1.** Any of the points of intersection of the graphs of the constraints that determine a feasible region. **2.** The point at which the axis of symmetry intersects a parabola. **3.** The point on each branch nearest the center of a hyperbola.

vértice **1.** Cualqeiera de los puntos de intersección de las gráficas que los contienen y que determinan una región viable. **2.** Punto en el que el eje de simetría interseca una parábola. **3.** El punto en cada rama más cercano al centro de una hipérbola.

vertex form (p. 305) A quadratic function in the form $y = a(x - h)^2 + k$, where (h, k) is the vertex of the parabola and $x = h$ is its axis of symmetry.

forma de vértice Función cuadrática de la forma $y = a(x - h)^2 + k$, donde (h, k) es el vértice de la parábola y $x = h$ es su eje de simetría.

vertex matrix (p. 209) A matrix used to represent the coordinates of the vertices of a polygon.

matriz de vértice Matriz que se usa para escribir las coordenadas de los vértices de un polígono.

vertical asymptote (p. 577) If the related rational expression of a function is written in simplest form and is undefined for $x = a$, then $x = a$ is a vertical asymptote.

asíntota vertical Si la expresión racional que corresponde a una función racional se reduce y está no definida en $x = a$, entonces $x = a$ es una asíntota vertical.

vertical line test (p. 62) If no vertical line intersects a graph in more than one point, then the graph represents a function.

prudba de la recta vertical Si ninguna recta vertical interseca una gráfica en más de un punto, entonces la gráfica representa una función.

vertical shift (p. 864) When graphs of trigonometric functions are translated vertically.

cambio vertical Cuando los gráficos de funciones trigonométricas son translado verticales.

vertices (pp. 639, 648) **Ellipse**—The endpoints of the major axis. **Hyperbola**—The endpoints of the transverse axis.

vértices **(de una elipse)** Son las puntos finales del eje principal. **(de una hipérbola)** Son las puntos finales del eje transversal.

W

weighted average (p. 596) A method for finding the mean of a set of numbers in which some elements of the set carry more importance, or weight, than others.

promedio ponderado Un método para encontrar el medio de un sistema de los números en los cuales algunos elementos del sistema llevan más importancia, o peso, que otros.

whole numbers (p. 11) $\{0, 1, 2, 3, 4, \ldots\}$

números naturales $\{0, 1, 2, 3, 4, \ldots\}$

X

x-intercept (p. 71) The x-coordinate of the point at which a graph crosses the x-axis.

intersección x La coordenada x del punto o puntos en que una gráfica interseca o cruza el eje x.

Y

y-intercept (p. 71) The y-coordinate of the point at which a graph crosses the y-axis.

intersección y La coordenada y del punto o puntos en que una gráfica interseca o cruza el eje y.

Z

zeros (p. 261) The x-intercepts of the graph of a quadratic equation; the points for which $f(x) = 0$.

ceros Las intersecciones x de la gráfica de una ecuación cuadrática; los puntos x para los que $f(x) = 0$.

zero matrix (p. 186) A matrix in which every element is zero.

matriz nula matriz cuyos elementos son todos igual a cero.

Index

A

B

C

D

Index

E

F

J

K

Index

L

M

Q

R

Index

S

Index

Index

T

U

V

W

X

Y

Z

Formulas

Coordinate Geometry

Midpoint	$M = \left(\frac{x_1 + x_2}{2}, \frac{y_1 + y_2}{2}\right)$	**Distance**	$d = \sqrt{(x_2 - x_1)^2 + (y_2 - y_1)^2}$
		Slope	$m = \frac{y_2 - y_1}{x_2 - x_1}, x_2 \neq x_1$

Matrices

Adding	$\begin{bmatrix} a & b \\ c & d \end{bmatrix} + \begin{bmatrix} e & f \\ g & h \end{bmatrix} = \begin{bmatrix} a+e & b+f \\ c+g & d+h \end{bmatrix}$	**Multiplying by a Scalar**	$k\begin{bmatrix} a & b \\ c & d \end{bmatrix} = \begin{bmatrix} ka & kb \\ kc & kd \end{bmatrix}$
Subtracting	$\begin{bmatrix} a & b \\ c & d \end{bmatrix} - \begin{bmatrix} e & f \\ g & h \end{bmatrix} = \begin{bmatrix} a-e & b-f \\ c-g & d-h \end{bmatrix}$	**Multiplying**	$\begin{bmatrix} a & b \\ c & d \end{bmatrix} \cdot \begin{bmatrix} e & f \\ g & h \end{bmatrix} = \begin{bmatrix} ab+bg & af+bh \\ ce+dg & cf+dh \end{bmatrix}$

Polynomials

Quadratic Formula	$x = \frac{-b \pm \sqrt{b^2 - 4ac}}{2a}, a \neq 0$	**Square of a Difference**	$(a - b)^2 = (a - b)(a - b) = a^2 - 2ab + b^2$
Square of a Sum	$(a + b)^2 = (a + b)(a + b) = a^2 + 2ab + b^2$	**Product of Sum and Difference**	$(a + b)(a - b) = (a - b)(a + b) = a^2 - b^2$

Logarithms

Product Property	$\log_x ab = \log_x a + \log_x b$	**Power Property**	$\log_b m^p = p \log_b m$
Quotient Property	$\log_x \frac{a}{b} = \log_x a - \log_x b, b \neq 0$	**Change of Base**	$\log_a n = \frac{\log_b n}{\log_b a}$

Conic Sections

Parabola	$y = a(x - h)^2 + k$ or $x = a(y - k)^2 + h$	**Ellipse**	$\frac{x^2}{a^2} + \frac{y^2}{b^2} = 1$ or $\frac{y^2}{a^2} + \frac{x^2}{b^2} = 1, a, b \neq 0$
Circle	$x^2 + y^2 = r^2$ or $(x - h)^2 + (y - k)^2 = r^2$	**Hyperbola**	$\frac{x^2}{a^2} - \frac{y^2}{b^2} = 1$ or $\frac{y^2}{a^2} - \frac{x^2}{b^2} = 1, a, b \neq 0$

Sequences and Series

***n*th term, Arithmetic**	$a_n = a_1 + (n - 1)d$	***n*th term, Geometric**	$a_n = a_1 r^{n-1}$
Sum of Arithmetic Series	$S_n = n\left(\frac{a_1 + a_n}{2}\right)$ or $S_n = \frac{n}{2}[2a_1 + (n - 1)d]$	**Sum of Geometric Series**	$S_n = \frac{a_1 - a_1 r^n}{1 - r}$ or $S_n = \frac{a_1 - a_n r}{1 - r}, r \neq 1$

Trigonometry

Law of Sines	$\frac{\sin A}{a} = \frac{\sin B}{b} = \frac{\sin C}{c}, a, b, c \neq 0$		
Law of Cosines	$a^2 = b^2 + c^2 - 2bc \cos A$	$b^2 = a^2 + c^2 - 2ac \cos B$	$c^2 = a^2 + b^2 - 2ab \cos C$
Trigonometric Functions	$\sin \theta = \frac{\text{opp}}{\text{hyp}}$	$\cos \theta = \frac{\text{adj}}{\text{hyp}}$	$\tan \theta = \frac{\text{opp}}{\text{adj}} = \frac{\sin \theta}{\cos \theta}$
	$\csc \theta = \frac{\text{hyp}}{\text{opp}} = \frac{1}{\sin \theta}$	$\sec \theta = \frac{\text{hyp}}{\text{adj}} = \frac{1}{\cos \theta}$	$\cot \theta = \frac{\text{adj}}{\text{opp}} = \frac{\cos \theta}{\sin \theta}$
Pythagorean Identities	$\cos^2 \theta + \sin^2 \theta = 1$	$\tan^2 \theta + 1 = \sec^2 \theta$	$\cot^2 \theta + 1 = \csc^2 \theta$

Symbols

$f(x) = \{$	piecewise-defined function
$f(x) = \lvert x \rvert$	absolute value function
$f(x) = [\![x]\!]$	function of greatest integer not greater than a
$f(x, y)$	f of x and y, a function with two variables, x and y
$\overrightarrow{AB}$	vector AB
i	the imaginary unit
$[f \circ g](x)$	f of g of x, the composition of functions f and g
$f^{-1}(x)$	inverse of $f(x)$
$b^{\frac{1}{n}} = \sqrt[n]{b}$	nth root of b
$\log_b x$	logarithm base b of x
$\log x$	common logarithm of x
$\ln x$	natural logarithm of x

$\sum$	sigma, summation
$\bar{x}$	mean of a sample
μ	mean of a population
s	standard deviation of a sample
σ	standard deviation of a population
$P(B \mid A)$	the probability of B given that A has already occurred
nPr	permutation of n objects taken r at a time
nCr	combination of n objects taken r at a time
$\text{Sin}^{-1}\, x$	Arcsin x
$\text{Cos}^{-1}\, x$	Arccos x
$\text{Tan}^{-1}\, x$	Arctan x

Parent Functions

Linear Functions

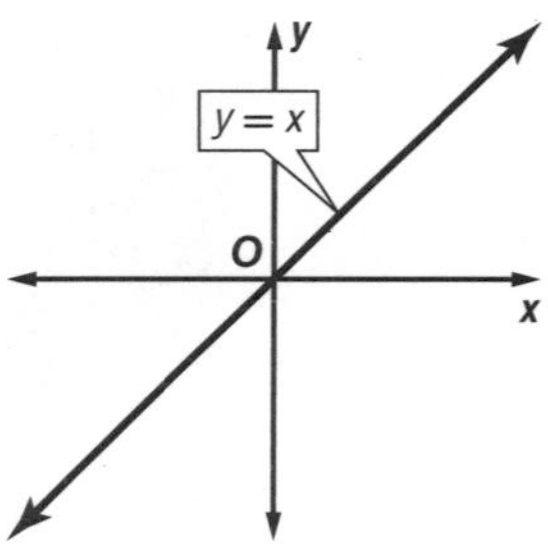

Absolute Value Functions

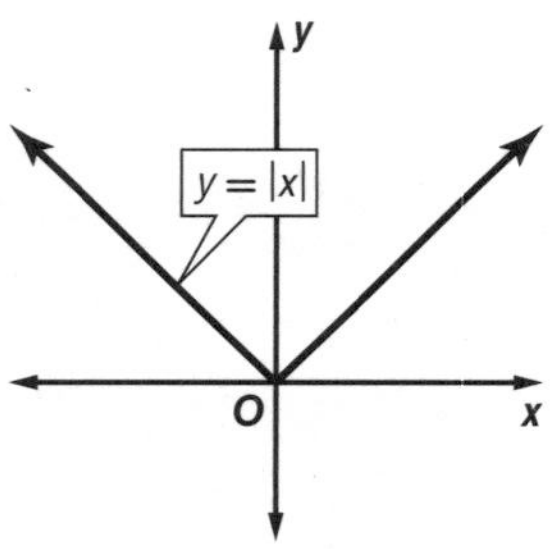

Quadratic Functions

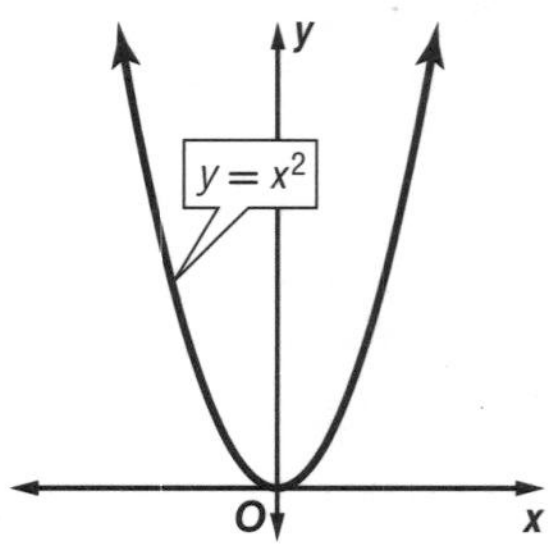

Exponential and Logarithmic Functions

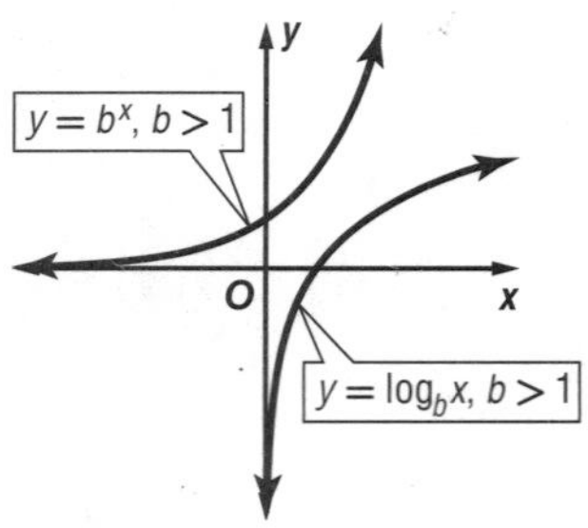

Square Root Functions

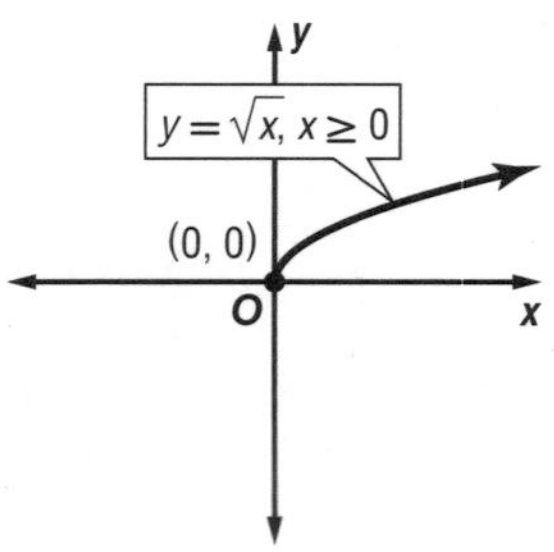

Reciprocal and Rational Functions

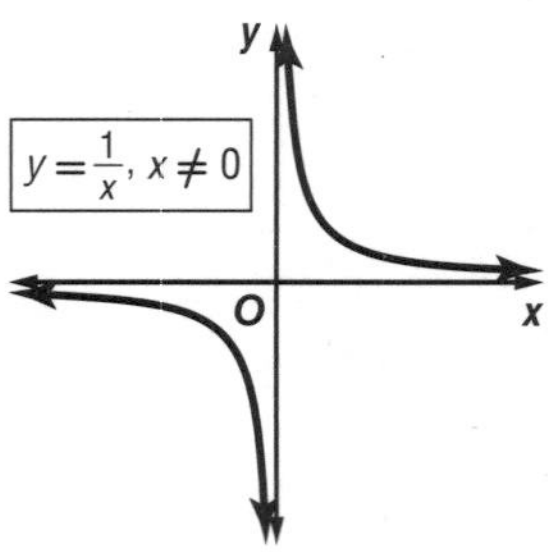